YAMAHA, MERCURY & MARINER

Outboards
1995-04 REPAIR MANUAL
ALL 4-STROKE ENGINES

SELOC

Managing Partners	Dean F. Morgantini, S.A.E.
	Barry L. Beck
Executive Editor	Kevin M. G. Maher, A.S.E.
Manager-Marine/Recreation	James R. Marotta, A.S.E.
Production Managers	Melinda Possinger
	Ronald Webb

Manufactured in USA
© 2005 Seloc Publications
104 Willowbrook Lane
West Chester, PA 19382
ISBN 0-89330-066-7
1234567890 4321098765

www.selocmarine.com
1-866-SELOC55

CONTENTS

1 — GENERAL INFORMATION, SAFETY AND TOOLS

HOW TO USE THIS MANUAL	1-2
BOATING SAFETY	1-4
BOATING EQUIPMENT (NOT REQUIRED BUT RECOMMENDED)	1-10
SAFETY IN SERVICE	1-13
TROUBLESHOOTING	1-13
SHOP EQUIPMENT	1-17
TOOLS	1-19
FASTENERS, MEASUREMENT, AND CONVERSIONS	1-27
SPECIFICATIONS	1-28

2 — MAINTENANCE AND TUNE-UP

GENERAL INFORMATION	2-2
LUBRICATION	2-6
ENGINE MAINTENANCE	2-12
BOAT MAINTENANCE	2-32
TUNE-UP	2-36
TIMING AND SYNCHRONIZATION	2-49
VALVE CLEARANCE	2-94
STORAGE	2-108
CLEARING A SUBMERGED MOTOR	2-112
SPECIFICATIONS	2-113

3 — FUEL SYSTEM

FUEL AND COMBUSTION BASICS	3-2
FUEL TANK AND LINES	3-6
CARBURETED FUEL SYSTEM	3-11
CARBURETOR SERVICE	3-17
ELECTRONIC FUEL INJECTION	3-50
FUEL PUMP (LOW PRESSURE) SERVICE	3-97
SPECIFICATIONS	3-105

4 — IGNITION AND ELECTRICAL

ELECTRICAL SYSTEMS	4-2
IGNITION SYSTEMS	4-8
CHARGING SYSTEM	4-35
CRANKING SYSTEM	4-39
ELECTRICAL SWITCH/SOLENOID SERVICE	4-51
SPECIFICATIONS	4-54
WIRING DIAGRAMS	4-62

5 — LUBRICATION AND COOLING

LUBRICATION SYSTEM	5-2
COOLING SYSTEM	5-10
SPECIFICATIONS	5-31

CONTENTS

6 POWERHEAD

POWERHEAD MECHANICAL	6-2
POWERHEAD REFINISHING	6-121
POWERHEAD BREAK-IN	6-136
SPECIFICATIONS	6-137

7 LOWER UNIT

LOWER UNIT	7-2
GEARCASE SERVICE - YAMAHA	7-5
GEARCASE SERVICE - MERCURY/MARINER	7-59
JET DRIVE	7-87

8 TRIM AND TILT

YAMAHA HYDRO TILT LOCK SYSTEM	8-2
YAMAHA TRIM & TILT SYSTEMS	8-2
YAMAHA SINGLE TILT RAM POWER TRIM/TILT SYSTEMS	8-3
YAMAHA LARGE MOTOR POWER TRIM/TILT SYSTEM	8-15
MERCURY/MARINER TRIM & TILT SYSTEMS	8-28
MERCURY/MARINER GAS ASSIST TILT SYSTEM	8-31
MERCURY/MARINER SINGLE RAM INTEGRAL POWER TILT/TRIM	8-35

9 REMOTE CONTROL

YAMAHA REMOTE CONTROLS	9-2
MERCURY/MARINER REMOTE CONTROLS	9-12
TILLER HANDLE	9-16
WIRING DIAGRAMS - YAMAHA	9-20
WIRING DIAGRAMS - MERCURY/MARINER	9-21

10 HAND REWIND STARTER

HAND REWIND STARTER	10-2

MASTER INDEX

MASTER INDEX	10-19

SAFETY NOTICE

Proper service and repair procedures are vital to the safe, reliable operation of all marine engines, as well as the personal safety of those performing repairs. This manual outlines procedures for servicing and repairing engines and drive systems using safe, effective methods. The procedures contain many NOTES, CAUTIONS and WARNINGS which should be followed, along with standard procedures, to minimize the possibility of personal injury or improper service which could damage the vehicle or compromise its safety.

It is important to note that repair procedures and techniques, tools and parts for servicing these engines, as well as the skill and experience of the individual performing the work, vary widely. It is not possible to anticipate all of the conceivable ways or conditions under which the engine may be serviced, or to provide cautions as to all possible hazards that may result. Standard and accepted safety precautions and equipment should be used during cutting, grinding, chiseling, prying, or any other process that can cause material removal or projectiles.

Some procedures require the use of tools specially designed for a specific task. Before substituting another tool or procedure, you must be completely satisfied that neither your personal safety, nor the performance of the vessel, will be endangered. All procedures covered in this manual requiring the use of special tools will be noted at the beginning of the procedure by means of an **OEM symbol**

Additionally, any procedure requiring the use of an electronic tester or scan tool will be noted at the beginning of the procedure by means of a **DVOM symbol**

Although information in this manual is based on industry sources and is complete as possible at the time of publication, the possibility exists that some manufacturers made later changes which could not be included here. While striving for total accuracy, Seloc Publishing cannot assume responsibility for any errors, changes or omissions that may occur in the compilation of this data. We must therefore warn you to follow instructions carefully, using common sense. If you are uncertain of a procedure, seek help by inquiring with someone in your area who is familiar with these motors before proceeding.

PART NUMBERS

Part numbers listed in this reference are not recommendations by Seloc Publishing for any particular product brand name, simply iterations of the manufacturer's suggestions. They are also references that can be used with interchange manuals and aftermarket supplier catalogs to locate each brand supplier's discrete part number.

SPECIAL TOOLS

Special tools are recommended by the manufacturers to perform a specific job. Use has been kept to a minimum, but, where absolutely necessary, they are referred to in the text by the part number of the manufacturer if at all possible; and also noted at the beginning of each procedure with one of the following symbols: **OEM** or **DVOM.**

The **OEM** symbol usually denotes the need for a unique tool purposely designed to accomplish a specific task, it will also be used, less frequently, to notify the reader of the need for a tool that is not commonly found in the average tool box.

The **DVOM** symbol is used to denote the need for an electronic test tool like an ohmmeter, multi-meter or, on cetain later engines, a scan tool.

These tools can be purchased, under the appropriate part number, from your local dealer or regional distributor, or an equivalent tool can be purchased locally from a tool supplier or parts outlet. Before substituting any tool for the one recommended, read the SAFETY NOTICE at the top of this page.
Providing the correct mix of service and repair procedures is an endless battle for any publisher of "How-To" information. Users range from first time do-it yourselfers to professionally trained marine technicians, and information important to one is frequently irrelevant to the other. The editors at Seloc Publishing strive to provide accurate and articulate information on all facets of marine engine repair, from the simplest procedure to the most complex. In doing this, we understand that certain procedures may be outside the capabilities of the average DIYer. Conversely we are aware that many procedures are unnecessary for a trained technician.

SKILL LEVELS

In order to provide all of our users, particularly the DIYers, with a feeling for the scope of a given procedure or task before tackling it we have included a rating system denoting the suggested skill level needed when performing a particular procedure. One of the following icons will be included at the beginning of most procedures:

EASY. These procedures are aimed primarily at the DIYer and can be classified, for the most part, as basic maintenance procedures; battery, fluids, filters, plugs, etc. Although certainly valuable to any experience level, they will generally be of little importance to a technician.

MODERATE. These procedures are suited for a DIYer with experience and a working knowledge of mechanical procedures. Even an advanced DIYer or professional technician will occasionally refer to these procedures. They will generally consist of component repair and service procedures, adjustments and minor rebuilds.

DIFFICULT. These procedures are aimed at the advanced DIYer and professional technician. They will deal with diagnostics, rebuilds and internal engine/drive components and will frequently require special tools.

SKILLED. These procedures are aimed at highly skilled technicians and should not be attempted without previous experience. They will usually consist of machine work, internal engine work and gear case rebuilds.

Please remember one thing when considering the above ratings—they are a guide for judging the complexity of a given procedure and are subjective in nature. Only you will know what your experience level is, and only you will know when a procedure may be outside the realm of your capability. First time DIYer, or life-long marine technician, we all approach repair and service differently so an easy procedure for one person may be a difficult procedure for another, regardless of experience level. All skill level ratings are meant to be used as a guide only! Use them to help make a judgement before undertaking a particular procedure, but by all means read through the procedure first and make your own decision—after all, our mission at Seloc is to make boat maintenance and repair easier for everyone whether you are changing the oil or rebuilding an engine. Enjoy boating!

ALL RIGHTS RESERVED

No part of this publication may be reproduced, transmitted or stored in any form or by any means, electronic or mechanical, including photocopy, recording, or by information storage or retrieval system, without prior written permission from the publisher.

ACKNOWLEDGMENTS

Seloc Publishing expresses appreciation to the following companies who supported the production of this book:
- Yamaha Motor Corporation—Cypress, CA
- Marine Mechanics Institute—Orlando, FL
- Belks Marine—Holmes, PA

Thanks to John Hartung and Judy Belk of Belk's Marine for for there assistance, guidance, patience and access to some of the motors photographed for this manual.

Seloc Publishing would like to express thanks to the fine companies who participate in the production of all our books:
- Hand tools supplied by Craftsman are used during all phases of our vehicle teardown and photography.
- Many of the fine specialty tools used in our procedures were provided courtesy of Lisle Corporation.
- Much of our shop's electronic testing equipment was supplied by Universal Enterprises Inc. (UEI).

1

GENERAL INFORMATION, SAFETY AND TOOLS

HOW TO USE THIS MANUAL	1-2
BOATING SAFETY	1-4
BOATING EQUIPMENT (NOT REQUIRED BUT RECOMMENDED)	1-10
SAFETY IN SERVICE	1-13
TROUBLESHOOTING	1-13
SHOP EQUIPMENT	1-17
TOOLS	1-19
FASTENERS, MEASUREMENTS AND CONVERSIONS	1-27
SPECIFICATIONS	1-28

BOATING EQUIPMENT (NOT REQUIRED BUT RECOMMENDED) ... 1-10
- ANCHORS ... 1-10
- BAILING DEVICES ... 1-10
- COMPASS ... 1-11
  - COMPASS PRECAUTIONS ... 1-12
  - INSTALLATION ... 1-11
  - SELECTION ... 1-11
- FIRST AID KIT ... 1-10
- OAR/PADDLE (SECOND MEANS OF PROPULSION) ... 1-10
- TOOLS AND SPARE PARTS 1-12
- VHF-FM RADIO ... 1-10

BOATING SAFETY ... 1-4
- COURTESY MARINE EXAMINATIONS ... 1-10
- REGULATIONS FOR YOUR BOAT ... 1-4
  - CAPACITY INFORMATION ... 1-5
  - CERTIFICATE OF COMPLIANCE ... 1-5
  - DOCUMENTING OF VESSELS ... 1-4
  - HULL IDENTIFICATION NUMBER ... 1-4
  - LENGTH OF BOATS ... 1-5
  - NUMBERING OF VESSELS ... 1-4
  - REGISTRATION OF BOATS ... 1-4
  - SALES AND TRANSFERS ... 1-4
  - VENTILATION ... 1-5
  - VENTILATION SYSTEMS ... 1-5
- REQUIRED SAFETY EQUIPMENT ... 1-6
  - FIRE EXTINGUISHERS ... 1-6
  - TYPES OF FIRES ... 1-6
  - PERSONAL FLOTATION DEVICES ... 1-6
  - SOUND PRODUCING DEVICES ... 1-9
  - VISUAL DISTRESS SIGNALS ... 1-9
  - WARNING SYSTEM ... 1-6

FASTENERS, MEASUREMENTS AND CONVERSIONS ... 1-27
- BOLTS, NUTS AND OTHER THREADED RETAINERS ... 1-27
- TORQUE ... 1-27
- STANDARD AND METRIC MEASUREMENTS ... 1-27

HOW TO USE THIS MANUAL ... 1-2
- AVOIDING THE MOST COMMON MISTAKES ... 1-4
- AVOIDING TROUBLE ... 1-2
- CAN YOU DO IT? ... 1-2
- DIRECTIONS AND LOCATIONS ... 1-3
- MAINTENANCE OR REPAIR? ... 1-2
- PROFESSIONAL HELP 1-3
- PURCHASING PARTS ... 1-3
- WHERE TO BEGIN ... 1-2
- WHY COMBINE YAMAHA AND MERCURY/MARINER? ... 1-2

SAFETY IN SERVICE ... 1-13
- DO'S ... 1-13
- DON'TS ... 1-13

SHOP EQUIPMENT ... 1-17
- CHEMICALS ... 1-17
  - CLEANERS ... 1-18
  - LUBRICANTS & PENETRANTS ... 1-18
  - SEALANTS ... 1-18
- SAFETY TOOLS ... 1-17
  - EYE AND EAR PROTECTION ... 1-17
  - WORK CLOTHES ... 1-17
  - WORK GLOVES ... 1-17

TOOLS ... 1-19
- HAND TOOLS ... 1-19
  - HAMMERS ... 1-22
  - PLIERS ... 1-22
  - SCREWDRIVERS ... 1-22
  - SOCKET SETS ... 1-19
  - WRENCHES ... 1-21
- OTHER COMMON TOOLS ... 1-22
- SPECIAL TOOLS ... 1-23
- ELECTRONIC TOOLS ... 1-23
  - BATTERY CHARGERS ... 1-23
  - BATTERY TESTERS ... 1-23
  - GAUGES ... 1-24
  - MULTI-METERS (DVOMS) ... 1-24
- MEASURING TOOLS ... 1-25
  - MICROMETERS & CALIPERS ... 1-25
  - DIAL INDICATORS ... 1-25
  - TELESCOPING GAUGES ... 1-26
  - DEPTH GAUGES ... 1-26

TROUBLESHOOTING ... 1-13
- BASIC OPERATING PRINCIPLES ... 1-13
  - 2-STROKE MOTORS ... 1-14
  - 4-STROKE MOTORS ... 1-16
  - COMBUSTION ... 1-16

SPECIFICATIONS ... 1-28
- BOLT TORQUE - TYPICAL METRIC BOLT TORQUE VALUES ... 1-28
- BOLT TORQUE - TYPICAL U.S. STANDARD BOLT TORQUE VALUES ... 1-28
- CONVERSION FACTORS ... 1-28

1-2 GENERAL INFORMATION, SAFETY AND TOOLS

HOW TO USE THIS MANUAL

This manual is designed to be a handy reference guide to maintaining and repairing your Yamaha or Mercury/Mariner Outboard. We strongly believe that regardless of how many or how few year's experience you may have, there is something new waiting here for you.

This manual covers the topics that a factory service manual (designed for factory trained mechanics) and a manufacturer owner's manual (designed more by lawyers than boat owners these days) covers. It will take you through the basics of maintaining and repairing your outboard, step-by-step, to help you understand what the factory trained mechanics already know by heart. By using the information in this manual, any boat owner should be able to make better informed decisions about what they need to do to maintain and enjoy their outboard.

Even if you never plan on touching a wrench (and if so, we hope that we can change your mind), this manual will still help you understand what a mechanic needs to do in order to maintain your engine.

Why Combine Yamaha and Mercury/Mariner?

To some people this combination of motors might seem strange, well at least at first glance. However, the truth of the matter is that just like Chevy sold Toyota's and Suzuki's badged as "GEOs" for many years (and later as Chevy's themselves), most Mercury/Mariner 4-strokes through 2004 have been based on powerheads produced by Yamaha and re-badged as a Mercury/Mariner. This actually applies to ALL Merc/Mariner 4-strokes EXCEPT for the smallest engine family, the 4/5/6 hp models and can be quickly seen by comparing the Specifications Charts on Mercury/Mariner models to the charts for Yamaha Models of the same size/HP.

This leads us to one interesting benefit of combining these manufacturers in our coverage. Because different engineers will approach the same problem (such as wear/failure analysis or component testing) differently, each company has provided slightly different technical information, specifications, and procedures for the same powerheads or components. Therefore, we have been able to compare and contrast the information provided by the technical departments of both companies, often giving you additional information on service, specifications and testing for each model.

Now keep in mind that when information from the 2 companies differs we've noted the actual difference. We cannot always be 100% certain that the technique, procedure or specification provided by the one company fully applies to the model sold by the other. However, having placed these outboards side-by-side during disassembly we've discovered that in many cases the motors are identical mechanically, often right down to the wiring harness.

So is it possible that Mercury/Mariner had the motors shipped with the flywheels not installed so they could install their OWN stators or do differences in test specifications relate simply to differences in the tolerance of the engineering departments? Though we cannot always be certain, we have our suspicion that more often than not the components are interchangeable. If so this gives you a potential alternate source of parts and special tools, which could become more advantageous as time goes on and used parts/tools become available.

Still, we have to remind you to use your best judgment when relying on a technique, specification or procedure that is specifically referenced as coming from the company engineering department for the sister model on which you are working.

Can You Do It?

If you are not the type who is prone to taking a wrench to something, NEVER FEAR. The procedures provided here cover topics at a level virtually anyone will be able to handle. And just the fact that you purchased this manual shows your interest in better understanding your outboard.

You may even find that maintaining your outboard yourself is preferable in most cases. From a monetary standpoint, it could also be beneficial. The money spent on hauling your boat to a marina and paying a tech to service the engine could buy you fuel for a whole weekend of boating. And, if you are really that unsure of your own mechanical abilities, at the very least you should fully understand what a marine mechanic does to your boat. You may decide that anything other than maintenance and adjustments should be performed by a mechanic (and that's your call), but if so you should know that every time you board your boat, you are placing faith in the mechanic's work and trusting him or her with your well-being, and maybe your life.

It should also be noted that in most areas a factory-trained mechanic will command a hefty hourly rate for off site service. If the tech comes to you this hourly rate is often charged from the time they leave their shop to the time that they return home. When service is performed at a boat yard, the clock usually starts when they go out to get the boat and bring it into the shop and doesn't end until it is tested and put back in the yard. The cost savings in doing the job yourself might be readily apparent at this point.

Of course, if even you're already a seasoned Do-It-Yourselfer or a Professional Technician, you'll find the procedures, specifications, special tips as well as the schematics and illustrations helpful when tackling a new job on a motor.

■ **To help you decide if a task is within your skill level, procedures will often be rated using a wrench symbol in the text. When present, the number of wrenches designates how difficult we feel the procedure to be on a 1-4 scale. For more details on the wrench icon rating system, please refer to the information under Skill Levels at the beginning of this manual.**

Where to Begin

Before spending any money on parts, and before removing any nuts or bolts, read through the entire procedure or topic. This will give you the overall view of what tools and supplies will be required to perform the procedure or what questions need to be answered before purchasing parts. So read ahead and plan ahead. Each operation should be approached logically and all procedures thoroughly understood before attempting any work.

Avoiding Trouble

Some procedures in this manual may require you to "label and disconnect . . . " a group of lines, hoses or wires. Don't be lulled into thinking you can remember where everything goes - you won't. If you reconnect or install a part incorrectly, the motor may operate poorly, if at all. If you hook up electrical wiring incorrectly, you may instantly learn a very expensive lesson.

A piece of masking tape, for example, placed on a hose and another on its fitting will allow you to assign your own label such as the letter "A", or a short name. As long as you remember your own code, you can reconnect the lines by matching letters or names. Do remember that tape will dissolve when saturated in some fluids (especially cleaning solvents). If a component is to be washed or cleaned, use another method of identification. A permanent felt-tipped marker can be very handy for marking metal parts; but remember that some solvents will remove permanent marker. A scribe can be used to carefully etch a small mark in some metal parts, but be sure NOT to do that on a gasket-making surface.

SAFETY is the most important thing to remember when performing maintenance or repairs. Be sure to read the information on safety in this manual.

Maintenance or Repair?

Proper maintenance is the key to long and trouble-free engine life, and the work can yield its own rewards. A properly maintained engine performs better than one that is neglected. As a conscientious boat owner, set aside a Saturday morning, at least once a month, to perform a thorough check of items that could cause problems. Keep your own personal log to jot down which services you performed, how much the parts cost you, the date, and the amount of hours on the engine at the time. Keep all receipts for parts purchased, so that they may be referred to in case of related problems or to determine operating expenses. As a do-it-yourselfer, these receipts are the only proof you have that the required maintenance was performed. In the event of a warranty problem (on new motors), these receipts can be invaluable.

It's necessary to mention the difference between maintenance and repair. Maintenance includes routine inspections, adjustments, and replacement of parts that show signs of normal wear. Maintenance compensates for wear or deterioration. Repair implies that something has broken or is not working. A need for repair is often caused by lack of maintenance.

For example: draining and refilling the gearcase oil is maintenance recommended by all manufacturers at specific intervals. Failure to do this

GENERAL INFORMATION, SAFETY AND TOOLS 1-3

can allow internal corrosion or damage and impair the operation of the motor, requiring expensive repairs. While no maintenance program can prevent items from breaking or wearing out, a general rule can be stated: MAINTENANCE IS CHEAPER THAN REPAIR.

Directions and Locations

◆ See Figure 1

Two basic rules should be mentioned here. First, whenever the Port side of the engine (or boat) is referred to, it is meant to specify the left side of the engine when you are sitting at the helm. Conversely, the Starboard means your right side. The Bow is the front of the boat and the Stern or Aft is the rear.

Most screws and bolts are removed by turning counterclockwise, and tightened by turning clockwise. An easy way to remember this is: righty-tighty; lefty-loosey. Corny, but effective. And if you are really dense (and we have all been so at one time or another), buy a ratchet that is marked ON and OFF (like Snap-on® ratchets), or mark your own. This can be especially helpful when you are bent over backwards, upside down or otherwise turned around when working on a boat-mounted component.

Professional Help

Occasionally, there are some things when working on an outboard that are beyond the capabilities or tools of the average Do-It-Yourselfer (DIYer). This shouldn't include most of the topics of this manual, but you will have to be the judge. Some engines require special tools or a selection of special parts, even for some basic maintenance tasks.

Talk to other boaters who use the same model of engine and speak with a trusted marina to find if there is a particular system or component on your engine that is difficult to maintain.

You will have to decide for yourself where basic maintenance ends and where professional service should begin. Take your time and do your research first (starting with the information contained within) and then make your own decision. If you really don't feel comfortable with attempting a procedure, DON'T DO IT. If you've gotten into something that may be over your head, don't panic. Tuck your tail between your legs and call a marine mechanic. Marinas and independent shops will be able to finish a job for you. Your ego may be damaged, but your boat will be properly restored to its full running order. So, as long as you approach jobs slowly and carefully, you really have nothing to lose and everything to gain by doing it yourself.

On the other hand, even the most complicated repair is within the ability of a person who takes their time and follows the steps of a procedure. A rock climber doesn't run up the side of a cliff, he/she takes it one step at a time and in the end, what looked difficult or impossible was conquerable. Worry about one step at a time.

Purchasing Parts

◆ See Figures 2 and 3

When purchasing parts there are two things to consider. The first is quality and the second is to be sure to get the correct part for your engine. To get quality parts, always deal directly with a reputable retailer. To get the proper parts always refer to the model number from the information tag on your engine prior to calling the parts counter. An incorrect part can adversely affect your engine performance and fuel economy, and will cost you more money and aggravation in the end.

Just remember a tow back to shore will cost plenty. That charge is per hour from the time the towboat leaves their home port, to the time they return to their home port. Get the picture. . .$$$?

So whom should you call for parts? Well, there are many sources for the parts you will need. Where you shop for parts will be determined by what kind of parts you need, how much you want to pay, and the types of stores in your neighborhood.

Your marina can supply you with many of the common parts you require. Using a marina as your parts supplier may be handy because of location (just walk right down the dock) or because the marina specializes in your particular brand of engine. In addition, it is always a good idea to get to know the marina staff (especially the marine mechanic).

The marine parts jobber, who is usually listed in the yellow pages or whose name can be obtained from the marina, is another excellent source for parts. In addition to supplying local marinas, they also do a sizeable business in over-the-counter parts sales for the do-it-yourselfer.

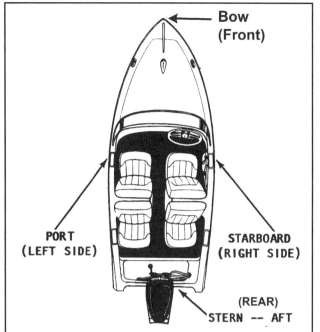

Fig. 1 Common terminology used for reference designation on boats of all size. These terms are used through out the text

Fig. 2 By far the most important asset in purchasing parts is a knowledgeable and enthusiastic parts person

Fig. 3 Parts catalogs, giving application and part number information, are provided by manufacturers for most replacement parts

1-4 GENERAL INFORMATION, SAFETY AND TOOLS

Almost every boating community has one or more convenient marine chain stores. These stores often offer the best retail prices and the convenience of one-stop shopping for all your needs. Since they cater to the do-it-yourselfer, these stores are almost always open weeknights, Saturdays, and Sundays, when the jobbers are usually closed.

The lowest prices for parts are most often found in discount stores or the auto department of mass merchandisers. Parts sold here are name and private brand parts bought in huge quantities, so they can offer a competitive price. Private brand parts are made by major manufacturers and sold to large chains under a store label. And, of course, more and more large automotive parts retailers are stocking basic marine supplies.

Avoiding the Most Common Mistakes

There are 3 common mistakes in mechanical work:

1. Following the incorrect order of assembly, disassembly or adjustment. When taking something apart or putting it together, performing steps in the wrong order usually just costs you extra time; however, it CAN break something. Read the entire procedure before beginning disassembly. Perform everything in the order in which the instructions say you should, even if you can't immediately see a reason for it. When you're taking apart something that is very intricate, you might want to draw a picture of how it looks when assembled at one point in order to make sure you get everything back in its proper position. When making adjustments, perform them in the proper order; often, one adjustment affects another, and you cannot expect satisfactory results unless each adjustment is made only when it cannot be changed by subsequent adjustments.

■ **Digital cameras are handy. If you've got access to one, take pictures of intricate assemblies during the disassembly process and refer to them during assembly for tips on part orientation.**

2. Over-torquing (or under-torquing). While it is more common for over-torquing to cause damage, under-torquing may allow a fastener to vibrate loose causing serious damage. Especially when dealing with plastic and aluminum parts, pay attention to torque specifications and utilize a torque wrench in assembly. If a torque figure is not available, remember that if you are using the right tool to perform the job, you will probably not have to strain yourself to get a fastener tight enough. The pitch of most threads is so slight that the tension you put on the wrench will be multiplied many times in actual force on what you are tightening.

3. Cross-threading. This occurs when a part such as a bolt is screwed into a nut or casting at the wrong angle and forced. Crossthreading is more likely to occur if access is difficult. It helps to clean and lubricate fasteners, then to start threading with the part to be installed positioned straight inward. Always start a fastener, etc. with your fingers. If you encounter resistance, unscrew the part and start over again at a different angle until it can be inserted and turned several times without much effort. Keep in mind that some parts may have tapered threads, so that gentle turning will automatically bring the part you're threading to the proper angle, but only if you don't force it or resist a change in angle. Don't put a wrench on the part until it has been tightened a couple of turns by hand. If you suddenly encounter resistance, and the part has not seated fully, don't force it. Pull it back out to make sure it's clean and threading properly.

BOATING SAFETY

In 1971 Congress ordered the U.S. Coast Guard to improve recreational boating safety. In response, the Coast Guard drew up a set of regulations.

Aside from these federal regulations, there are state and local laws you must follow. These sometimes exceed the Coast Guard requirements. This section discusses only the federal laws. State and local laws are available from your local Coast Guard. As with other laws, "Ignorance of the boating laws is no excuse." The rules fall into two groups: regulations for your boat and required safety equipment on your boat.

Regulations For Your Boat

Most boats on waters within Federal jurisdiction must be registered or documented. These waters are those that provide a means of transportation between two or more states or to the sea. They also include the territorial waters of the United States.

DOCUMENTING OF VESSELS

A vessel of five or more net tons may be documented as a yacht. In this process, papers are issued by the U.S. Coast Guard as they are for large ships. Documentation is a form of national registration. The boat must be used solely for pleasure. Its owner must be a citizen of the U.S., a partnership of U.S. citizens, or a corporation controlled by U.S. citizens. The captain and other officers must also be U.S. citizens. The crew need not be.

If you document your yacht, you have the legal authority to fly the yacht ensign. You also may record bills of sale, mortgages, and other papers of title with federal authorities. Doing so gives legal notice that such instruments exist. Documentation also permits preferred status for mortgages. This gives you additional security, and it aids in financing and transfer of title. You must carry the original documentation papers aboard your vessel. Copies will not suffice.

REGISTRATION OF BOATS

If your boat is not documented, registration in the state of its principal use is probably required. If you use it mainly on an ocean, a gulf, or other similar water, register it in the state where you moor it.

If you use your boat solely for racing, it may be exempt from the requirement in your state. Some states may also exclude dinghies, while others require registration of documented vessels and non-power driven boats.

All states, except Alaska, register boats. In Alaska, the U.S. Coast Guard issues the registration numbers. If you move your vessel to a new state of principal use, a valid registration certificate is good for 60 days. You must have the registration certificate (certificate of number) aboard your vessel when it is in use. A copy will not suffice. You may be cited if you do not have the original on board.

NUMBERING OF VESSELS

A registration number is on your registration certificate. You must paint or permanently attach this number to both sides of the forward half of your boat. Do not display any other number there.

The registration number must be clearly visible. It must not be placed on the obscured underside of a flared bow. If you can't place the number on the bow, place it on the forward half of the hull. If that doesn't work, put it on the superstructure. Put the number for an inflatable boat on a bracket or fixture. Then, firmly attach it to the forward half of the boat. The letters and numbers must be plain block characters and must read from left to right. Use a space or a hyphen to separate the prefix and suffix letters from the numerals. The color of the characters must contrast with that of the background, and they must be at least three inches high.

In some states your registration is good for only one year. In others, it is good for as long as three years. Renew your registration before it expires. At that time you will receive a new decal or decals. Place them as required by state law. You should remove old decals before putting on the new ones. Some states require that you show only the current decal or decals. If your vessel is moored, it must have a current decal even if it is not in use.

If your vessel is lost, destroyed, abandoned, stolen, or transferred, you must inform the issuing authority. If you lose your certificate of number or your address changes, notify the issuing authority as soon as possible.

SALES AND TRANSFERS

Your registration number is not transferable to another boat. The number stays with the boat unless its state of principal use is changed.

HULL IDENTIFICATION NUMBER

A Hull Identification Number (HIN) is like the Vehicle Identification Number (VIN) on your car. Boats built between November 1, 1972 and July 31, 1984 have old format HINs. Since August 1, 1984 a new format has been used.

Your boat's HIN must appear in two places. If it has a transom, the primary number is on its starboard side within two inches of its top. If it does not have a transom or if it was not practical to use the transom, the number is on the starboard side. In this case, it must be within one foot of the stern

GENERAL INFORMATION, SAFETY AND TOOLS 1-5

and within two inches of the top of the hull side. On pontoon boats, it is on the aft crossbeam within one foot of the starboard hull attachment. Your boat also has a duplicate number in an unexposed location. This is on the boat's interior or under a fitting or item of hardware.

LENGTH OF BOATS

For some purposes, boats are classed by length. Required equipment, for example, differs with boat size. Manufacturers may measure a boat's length in several ways. Officially, though, your boat is measured along a straight line from its bow to its stern. This line is parallel to its keel.

The length does not include bowsprits, boomkins, or pulpits. Nor does it include rudders, brackets, outboard motors, outdrives, diving platforms, or other attachments.

CAPACITY INFORMATION

◆ See Figure 4

Manufacturers must put capacity plates on most recreational boats less than 20 feet long. Sailboats, canoes, kayaks, and inflatable boats are usually exempt. Outboard boats must display the maximum permitted horsepower of their engines. The plates must also show the allowable maximum weights of the people on board. And they must show the allowable maximum combined weights of people, engine(s), and gear. Inboards and stern drives need not show the weight of their engines on their capacity plates. The capacity plate must appear where it is clearly visible to the operator when underway. This information serves to remind you of the capacity of your boat under normal circumstances. You should ask yourself, "Is my boat loaded above its recommended capacity" and, "Is my boat overloaded for the present sea and wind conditions?" If you are stopped by a legal authority, you may be cited if you are overloaded.

CERTIFICATE OF COMPLIANCE

◆ See Figure 4

Manufacturers are required to put compliance plates on motorboats greater than 20 feet in length. The plates must say, "This boat," or "This equipment complies with the U. S. Coast Guard Safety Standards in effect on the date of certification." Letters and numbers can be no less than one-eighth of an inch high. At the manufacturer's option, the capacity and compliance plates may be combined.

VENTILATION

A cup of gasoline spilled in the bilge has the potential explosive power of 15 sticks of dynamite. This statement, commonly quoted over 20 years ago, may be an exaggeration; however, it illustrates a fact. Gasoline fumes in the bilge of a boat are highly explosive and a serious danger. They are heavier than air and will stay in the bilge until they are vented out.

Because of this danger, Coast Guard regulations require ventilation on many powerboats. There are several ways to supply fresh air to engine and gasoline tank compartments and to remove dangerous vapors. Whatever the choice, it must meet Coast Guard standards.

■ **The following is not intended to be a complete discussion of the regulations. It is limited to the majority of recreational vessels. Contact your local Coast Guard office for further information.**

General Precautions

Ventilation systems will not remove raw gasoline that leaks from tanks or fuel lines. If you smell gasoline fumes, you need immediate repairs. The best device for sensing gasoline fumes is your nose. Use it! If you smell gasoline in a bilge, engine compartment, or elsewhere, don't start your engine. The smaller the compartment, the less gasoline it takes to make an explosive mixture.

Ventilation for Open Boats

In open boats, gasoline vapors are dispersed by the air that moves through them. So they are exempt from ventilation requirements.

To be "open," a boat must meet certain conditions. Engine and fuel tank compartments and long narrow compartments that join them must be open to the atmosphere." This means they must have at least 15 square inches of open area for each cubic foot of net compartment volume. The open area must be in direct contact with the atmosphere. There must also be no long, unventilated spaces open to engine and fuel tank compartments into which flames could extend.

Ventilation for All Other Boats

Powered and natural ventilation are required in an enclosed compartment with a permanently installed gasoline engine that has a cranking motor. A compartment is exempt if its engine is open to the atmosphere. Diesel powered boats are also exempt.

VENTILATION SYSTEMS

There are two types of ventilation systems. One is "natural ventilation." In it, air circulates through closed spaces due to the boat's motion. The other type is "powered ventilation." In it, air is circulated by a motor-driven fan or fans.

Natural Ventilation System Requirements

A natural ventilation system has an air supply from outside the boat. The air supply may also be from a ventilated compartment or a compartment open to the atmosphere. Intake openings are required. In addition, intake ducts may be required to direct the air to appropriate compartments.

The system must also have an exhaust duct that starts in the lower third of the compartment. The exhaust opening must be into another ventilated compartment or into the atmosphere. Each supply opening and supply duct, if there is one, must be above the usual level of water in the bilge. Exhaust openings and ducts must also be above the bilge water. Openings and ducts must be at least three square inches in area or two inches in diameter. Openings should be placed so exhaust gasses do not enter the fresh air intake. Exhaust fumes must not enter cabins or other enclosed, non-ventilated spaces. The carbon monoxide gas in them is deadly.

Intake and exhaust openings must be covered by cowls or similar devices. These registers keep out rain water and water from breaking seas. Most often, intake registers face forward and exhaust openings aft. This aids the flow of air when the boat is moving or at anchor since most boats face into the wind when properly anchored.

Power Ventilation System Requirements

◆ See Figure 5

Powered ventilation systems must meet the standards of a natural system, but in addition, they must also have one or more exhaust blowers. The blower duct can serve as the exhaust duct for natural ventilation if fan blades do not obstruct the air flow when not powered. Openings in engine compartment, for carburetion are in addition to ventilation system requirements.

Fig. 4 A U.S. Coast Guard certification plate indicates the amount of occupants and gear appropriate for safe operation of the vessel (and allowable engine size for outboard boats)

1-6 GENERAL INFORMATION, SAFETY AND TOOLS

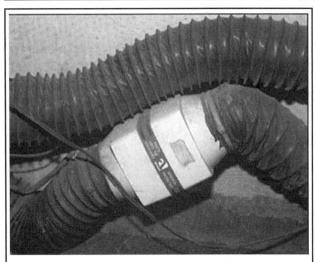

Fig. 5 Typical blower and duct system to vent fumes from the engine compartment

Required Safety Equipment

Coast Guard regulations require that your boat have certain equipment aboard. These requirements are minimums. Exceed them whenever you can.

TYPES OF FIRES

There are four common classes of fires:
- Class A - fires are of ordinary combustible materials such as paper or wood.
- Class B - fires involve gasoline, oil and grease.
- Class C - fires are electrical.
- Class D - fires involve ferrous metals

One of the greatest risks to boaters is fire. This is why it is so important to carry the correct number and type of extinguishers onboard.

The best fire extinguisher for most boats is a Class B extinguisher. Never use water on Class B or Class C fires, as water spreads these types of fires. Additionally, you should never use water on a Class C fire as it may cause you to be electrocuted.

FIRE EXTINGUISHERS

◆ See Figure 6

If your boat meets one or more of the following conditions, you must have at least one fire extinguisher aboard. The conditions are:
- Inboard or stern drive engines
- Closed compartments under seats where portable fuel tanks can be stored
- Double bottoms not sealed together or not completely filled with flotation materials
- Closed living spaces
- Closed stowage compartments in which combustible or flammable materials are stored
- Permanently installed fuel tanks
- Boat is 26 feet or more in length.

Contents of Extinguishers

Fire extinguishers use a variety of materials. Those used on boats usually contain dry chemicals, Halon, or Carbon Dioxide (CO2). Dry chemical extinguishers contain chemical powders such as Sodium Bicarbonate - baking soda.

Carbon dioxide is a colorless and odorless gas when released from an extinguisher. It is not poisonous but caution must be used in entering compartments filled with it. It will not support life and keeps oxygen from reaching your lungs. A fire-killing concentration of Carbon Dioxide can be lethal. If you are in a compartment with a high concentration of CO2, you will have no difficulty breathing. But the air does not contain enough oxygen to support life. Unconsciousness or death can result.

Halon Extinguishers

Some fire extinguishers and "built-in" or "fixed" automatic fire extinguishing systems contain a gas called Halon. Like carbon dioxide it is colorless and odorless and will not support life. Some Halons may be toxic if inhaled.

To be accepted by the Coast Guard, a fixed Halon system must have an indicator light at the vessel's helm. A green light shows the system is ready. Red means it is being discharged or has been discharged. Warning horns are available to let you know the system has been activated. If your fixed Halon system discharges, ventilate the space thoroughly before you enter it. There are no residues from Halon but it will not support life.

Although Halon has excellent fire fighting properties; it is thought to deplete the earth's ozone layer and has not been manufactured since January 1, 1994. Halon extinguishers can be refilled from existing stocks of the gas until they are used up, but high federal excise taxes are being charged for the service. If you discontinue using your Halon extinguisher, take it to a recovery station rather than releasing the gas into the atmosphere. Compounds such as FE 241, designed to replace Halon, are now available.

Fire Extinguisher Approval

Fire extinguishers must be Coast Guard approved. Look for the approval number on the nameplate. Approved extinguishers have the following on their labels: "Marine Type USCG Approved, Size..., Type..., 162.208/," etc. In addition, to be acceptable by the Coast Guard, an extinguisher must be in serviceable condition and mounted in its bracket. An extinguisher not properly mounted in its bracket will not be considered serviceable during a Coast Guard inspection.

Care and Treatment

Make certain your extinguishers are in their stowage brackets and are not damaged. Replace cracked or broken hoses. Nozzles should be free of obstructions. Sometimes, wasps and other insects nest inside nozzles and make them inoperable. Check your extinguishers frequently. If they have pressure gauges, is the pressure within acceptable limits? Do the locking pins and sealing wires show they have not been used since recharging?

Don't try an extinguisher to test it. Its valves will not reseat properly and the remaining gas will leak out. When this happens, the extinguisher is useless.

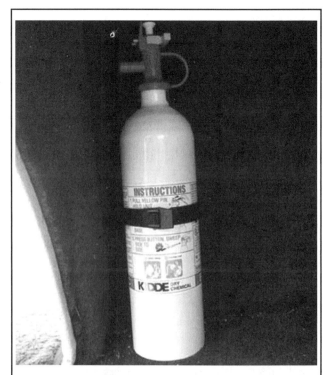

Fig. 6 An approved fire extinguisher should be mounted close to the operator for emergency use

GENERAL INFORMATION, SAFETY AND TOOLS 1-7

Weigh and tag carbon dioxide and Halon extinguishers twice a year. If their weight loss exceeds 10 percent of the weight of the charge, recharge them. Check to see that they have not been used. They should have been inspected by a qualified person within the past six months, and they should have tags showing all inspection and service dates. The problem is that they can be partially discharged while appearing to be fully charged.

Some Halon extinguishers have pressure gauges the same as dry chemical extinguishers. Don't rely too heavily on the gauge. The extinguisher can be partially discharged and still show a good gauge reading. Weighing a Halon extinguisher is the only accurate way to assess its contents.

If your dry chemical extinguisher has a pressure indicator, check it frequently. Check the nozzle to see if there is powder in it. If there is, recharge it. Occasionally invert your dry chemical extinguisher and hit the base with the palm of your hand. The chemical in these extinguishers packs and cakes due to the boat's vibration and pounding. There is a difference of opinion about whether hitting the base helps, but it can't hurt. It is known that caking of the chemical powder is a major cause of failure of dry chemical extinguishers. Carry spares in excess of the minimum requirement. If you have guests aboard, make certain they know where the extinguishers are and how to use them.

Using a Fire Extinguisher

A fire extinguisher usually has a device to keep it from being discharged accidentally. This is a metal or plastic pin or loop. If you need to use your extinguisher, take it from its bracket. Remove the pin or the loop and point the nozzle at the base of the flames. Now, squeeze the handle, and discharge the extinguisher's contents while sweeping from side to side. Recharge a used extinguisher as soon as possible.

If you are using a Halon or carbon dioxide extinguisher, keep your hands away from the discharge. The rapidly expanding gas will freeze them. If your fire extinguisher has a horn, hold it by its handle.

Legal Requirements for Extinguishers

You must carry fire extinguishers as defined by Coast Guard regulations. They must be firmly mounted in their brackets and immediately accessible.

A motorboat less than 26 feet long must have at least one approved hand-portable, Type B-1 extinguisher. If the boat has an approved fixed fire extinguishing system, you are not required to have the Type B-1 extinguisher. Also, if your boat is less than 26 feet long, is propelled by an outboard motor, or motors, and does not have any of the first six conditions described at the beginning of this section, it is not required to have an extinguisher. Even so, it's a good idea to have one, especially if a nearby boat catches fire, or if a fire occurs at a fuel dock.

A motorboat 26 feet to less than 40 feet long, must have at least two Type B-1 approved hand-portable extinguishers. It can, instead, have at least one Coast Guard approved Type B-2. If you have an approved fire extinguishing system, only one Type B-1 is required.

A motorboat 40 to 65 feet long must have at least three Type B-1 approved portable extinguishers. It may have, instead, at least one Type B-1 plus a Type B-2. If there is an approved fixed fire extinguishing system, two Type B-1 or one Type B-2 is required.

WARNING SYSTEM

Various devices are available to alert you to danger. These include fire, smoke, gasoline fumes, and carbon monoxide detectors. If your boat has a galley, it should have a smoke detector. Where possible, use wired detectors. Household batteries often corrode rapidly on a boat.

There are many ways in which carbon monoxide (a by-product of the combustion that occurs in an engine) can enter your boat. You can't see, smell, or taste carbon monoxide gas, but it is lethal. As little as one part in 10,000 parts of air can bring on a headache. The symptoms of carbon monoxide poisoning - headaches, dizziness, and nausea - are like seasickness. By the time you realize what is happening to you, it may be too late to take action. If you have enclosed living spaces on your boat, protect yourself with a detector.

PERSONAL FLOTATION DEVICES

Personal Flotation Devices (PFDs) are commonly called life preservers or life jackets. You can get them in a variety of types and sizes. They vary with their intended uses. To be acceptable, PFDs must be Coast Guard approved.

Type I PFDs

A Type I life jacket is also called an offshore life jacket. Type I life jackets will turn most unconscious people from facedown to a vertical or slightly backward position. The adult size gives a minimum of 22 pounds of buoyancy. The child size has at least 11 pounds. Type I jackets provide more protection to their wearers than any other type of life jacket. Type I life jackets are bulkier and less comfortable than other types. Furthermore, there are only two sizes, one for children and one for adults.

Type I life jackets will keep their wearers afloat for extended periods in rough water. They are recommended for offshore cruising where a delayed rescue is probable.

Type II PFDs

A Type II life jacket is also called a near-shore buoyant vest. It is an approved, wearable device. Type II life jackets will turn some unconscious people from facedown to vertical or slightly backward positions. The adult size gives at least 15.5 pounds of buoyancy. The medium child size has a minimum of 11 pounds. And the small child and infant sizes give seven pounds. A Type II life jacket is more comfortable than a Type I but it does not have as much buoyancy. It is not recommended for long hours in rough water. Because of this, Type IIs are recommended for inshore and inland cruising on calm water. Use them only where there is a good chance of fast rescue.

Type III PFDs

◆ See Figure 7

Type III life jackets or marine buoyant devices are also known as flotation aids. Like Type IIs, they are designed for calm inland or close offshore water where there is a good chance of fast rescue. Their minimum buoyancy is 15.5 pounds. They will **not** turn their wearers face up.

Type III devices are usually worn where freedom of movement is necessary. Thus, they are used for water skiing, small boat sailing, and fishing among other activities. They are available as vests and flotation coats. Flotation coats are useful in cold weather. Type IIIs come in many sizes from small child through large adult.

Life jackets come in a variety of colors and patterns - red, blue, green, camouflage, and cartoon characters. From purely a safety standpoint, the best color is bright orange. It is easier to see in the water, especially if the water is rough.

Type IV PFDs

◆ See Figures 8 and 9

Type IV ring life buoys, buoyant cushions and horseshoe buoys are Coast Guard approved devices called throwables. They are made to be thrown to people in the water, and should not be worn. Type IV cushions are often used as seat cushions. But, keep in mind that cushions are hard to hold onto in the water, thus, they do not afford as much protection as wearable life jackets.

The straps on buoyant cushions are for you to hold onto either in the water or when throwing them, they are **NOT** for your arms. A cushion should never be worn on your back, as it will turn you face down in the water.

Type IV throwables are not designed as personal flotation devices for unconscious people, non-swimmers, or children. Use them only in emergencies. They should not be used for, long periods in rough water.

Ring life buoys come in 18, 20, 24, and 30 in. diameter sizes. They usually have grab lines, but you will need to attach about 60 feet of polypropylene line to the grab rope to aid in retrieving someone in the water. If you throw a ring, be careful not to hit the person. Ring buoys can knock people unconscious

Type V PFDs

Type V PFDs are of two kinds, special use devices and hybrids. Special use devices include boardsailing vests, deck suits, work vests, and others. They are approved only for the special uses or conditions indicated on their labels. Each is designed and intended for the particular application shown on its label. They do not meet legal requirements for general use aboard recreational boats.

Hybrid life jackets are inflatable devices with some built-in buoyancy provided by plastic foam or kapok. They can be inflated orally or by cylinders

Fig. 7 Type III PFDs are recommended for inshore/inland use on calm water (where there is a good chance of fast rescue)

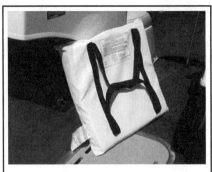

Fig. 8 Type IV buoyant cushions are thrown to people in the water. If you can squeeze air out of the cushion, it should be replaced

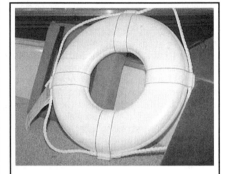

Fig. 9 Type IV throwables, such as this ring life buoy, are not designed for unconscious people, non-swimmers, or children

of compressed gas to give additional buoyancy. In some hybrids the gas is released manually. In others it is released automatically when the life jacket is immersed in water.

The inherent buoyancy of a hybrid may be insufficient to float a person unless it is inflated. The only way to find this out is for the user to try it in the water. Because of its limited buoyancy when deflated, a hybrid is recommended for use by a non-swimmer only if it is worn with enough inflation to float the wearer.

If they are to count against the legal requirement for the number of life jackets you must carry, hybrids manufactured before February 8, 1995 must be worn whenever a boat is underway and the wearer must not go below decks or in an enclosed space. To find out if your Type V hybrid must be worn to satisfy the legal requirement, read its label. If its use is restricted it will say, "REQUIRED TO BE WORN" in capital letters.

Hybrids cost more than other life jackets, but this factor must be weighed against the fact that they are more comfortable than Types I, II or III life jackets. Because of their greater comfort, their owners are more likely to wear them than are the owners of Type I, II or III life jackets.

The Coast Guard has determined that improved, less costly hybrids can save lives since they will be bought and used more frequently. For these reasons, a new federal regulation was adopted effective February 8, 1995. The regulation increases both the deflated and inflated buoyancys of hybrids, makes them available in a greater variety of sizes and types, and reduces their costs by reducing production costs.

Even though it may not be required, the wearing of a hybrid or a life jacket is encouraged whenever a vessel is underway. Like life jackets, hybrids are now available in three types. To meet legal requirements, a Type I hybrid can be substituted for a Type I life jacket. Similarly Type II and III hybrids can be substituted for Type II and Type III life jackets. A Type I hybrid, when inflated, will turn most unconscious people from facedown to vertical or slightly backward positions just like a Type I life jacket. Type II and III hybrids function like Type II and III life jackets. If you purchase a new hybrid, it should have an owner's manual attached that describes its life jacket type and its deflated and inflated buoyancys. It warns you that it may have to be inflated to float you. The manual also tells you how to don the life jacket and how to inflate it. It also tells you how to change its inflation mechanism, recommended testing exercises, and inspection or maintenance procedures. The manual also tells you why you need a life jacket and why you should wear it. A new hybrid must be packaged with at least three gas cartridges. One of these may already be loaded into the inflation mechanism. Likewise, if it has an automatic inflation mechanism, it must be packaged with at least three of these water sensitive elements. One of these elements may be installed.

Legal Requirements

A Coast Guard approved life jacket must show the manufacturer's name and approval number. Most are marked as Type I, II, III, IV or V. All of the newer hybrids are marked for type.

You are required to carry at least one wearable life jacket or hybrid for each person on board your recreational vessel. If your vessel is 16 feet or more in length and is not a canoe or a kayak, you must also have at least one Type IV on board. These requirements apply to all recreational vessels that are propelled or controlled by machinery, sails, oars, paddles, poles, or another vessel. Sailboards are not required to carry life jackets.

You can substitute an older Type V hybrid for any required Type I, II or III life jacket provided:

1. Its approval label shows it is approved for the activity the vessel is engaged in
2. It's approved as a substitute for a life jacket of the type required on the vessel
3. It's used as required on the labels
and
4. It's used in accordance with any requirements in its owner's manual (if the approval label makes reference to such a manual.)

A water skier being towed is considered to be on board the vessel when judging compliance with legal requirements.

You are required to keep your Type I, II or III life jackets or equivalent hybrids readily accessible, which means you must be able to reach out and get them when needed. All life jackets must be in good, serviceable condition.

General Considerations

The proper use of a life jacket requires the wearer to know how it will perform. You can gain this knowledge only through experience. Each person on your boat should be assigned a life jacket. Next, it should be fitted to the person who will wear it. Only then can you be sure that it will be ready for use in an emergency. This advice is good even if the water is calm, and you intend to boat near shore.

Boats can sink fast. There may be no time to look around for a life jacket. Fitting one on you in the water is almost impossible. Most drownings occur in inland waters within a few feet of safety. Most victims had life jackets, but they weren't wearing them.

Keeping life jackets in the plastic covers they came wrapped in, and in a cabin, assure that they will stay clean and unfaded. But this is no way to keep them when you are on the water. When you need a life jacket it must be readily accessible and adjusted to fit you. You can't spend time hunting for it or learning how to fit it.

There is no substitute for the experience of entering the water while wearing a life jacket. Children, especially, need practice. If possible, give your guests this experience. Tell them they should keep their arms to their sides when jumping in to keep the life jacket from riding up. Let them jump in and see how the life jacket responds. Is it adjusted so it does not ride up? Is it the proper size? Are all straps snug? Are children's life jackets the right sizes for them? Are they adjusted properly? If a child's life jacket fits correctly, you can lift the child by the jacket's shoulder straps and the child's chin and ears will not slip through. Non-swimmers, children, handicapped persons, elderly persons and even pets should always wear life jackets when they are aboard. Many states require that everyone aboard wear them in hazardous waters.

Inspect your lifesaving equipment from time to time. Leave any questionable or unsatisfactory equipment on shore. An emergency is no time for you to conduct an inspection.

Indelibly mark your life jackets with your vessel's name, number, and calling port. This can be important in a search and rescue effort. It could help concentrate effort where it will do the most good.

Care of Life Jackets

Given reasonable care, life jackets last many years. Thoroughly dry them before putting them away. Stow them in dry, well-ventilated places. Avoid the bottoms of lockers and deck storage boxes where moisture may collect. Air and dry them frequently.

GENERAL INFORMATION, SAFETY AND TOOLS

Life jackets should not be tossed about or used as fenders or cushions. Many contain kapok or fibrous glass material enclosed in plastic bags. The bags can rupture and are then unserviceable. Squeeze your life jacket gently. Does air leak out? If so, water can leak in and it will no longer be safe to use. Cut it up so no one will use it, and throw it away. The covers of some life jackets are made of nylon or polyester. These materials are plastics. Like many plastics, they break down after extended exposure to the ultraviolet light in sunlight. This process may be more rapid when the materials are dyed with bright dyes such as "neon" shades.

Ripped and badly faded fabrics are clues that the covering of your life jacket is deteriorating. A simple test is to pinch the fabric between your thumbs and forefingers. Now try to tear the fabric. If it can be torn, it should definitely be destroyed and discarded. Compare the colors in protected places to those exposed to the sun. If the colors have faded, the materials have been weakened. A life jacket covered in fabric should ordinarily last several boating seasons with normal use. A life jacket used every day in direct sunlight should probably be replaced more often.

SOUND PRODUCING DEVICES

All boats are required to carry some means of making an efficient sound signal. Devices for making the whistle or horn noises required by the Navigation Rules must be capable of a four-second blast. The blast should be audible for at least one-half mile. Athletic whistles are not acceptable on boats 12 meters or longer. Use caution with athletic whistles. When wet, some of them come apart and loose their "pea." When this happens, they are useless.

If your vessel is 12 meters long and less than 20 meters, you must have a power whistle (or power horn) and a bell on board. The bell must be in operating condition and have a minimum diameter of at least 200mm (7.9 in.) at its mouth.

VISUAL DISTRESS SIGNALS

◆ See Figure 10

Visual Distress Signals (VDS) attract attention to your vessel if you need help. They also help to guide searchers in search and rescue situations. Be sure you have the right types, and learn how to use them properly.

It is illegal to fire flares improperly. In addition, they cost the Coast Guard and its Auxiliary many wasted hours in fruitless searches. If you signal a distress with flares and then someone helps you, please let the Coast Guard or the appropriate Search And Rescue (SAR) Agency know so the distress report will be canceled.

Recreational boats less than 16 feet long must carry visual distress signals on coastal waters at night. Coastal waters are:
- The ocean (territorial sea)
- The Great Lakes
- Bays or sounds that empty into oceans
- Rivers over two miles across at their mouths upstream to where they narrow to two miles.

Recreational boats 16 feet or longer must carry VDS at all times on coastal waters. The same requirement applies to boats carrying six or fewer passengers for hire. Open sailboats less than 26 feet long without engines are exempt in the daytime as are manually propelled boats. Also exempt are boats in organized races, regattas, parades, etc. Boats owned in the United States and operating on the high seas must be equipped with VDS.

A wide variety of signaling devices meet Coast Guard regulations. For pyrotechnic devices, a minimum of three must be carried. Any combination can be carried as long as it adds up to at least three signals for day use and at least three signals for night use. Three day/night signals meet both requirements. If possible, carry more than the legal requirement.

■ The American flag flying upside down is a commonly recognized distress signal. It is not recognized in the Coast Guard regulations, though. In an emergency, your efforts would probably be better used in more effective signaling methods.

Types of VDS

VDS are divided into two groups; daytime and nighttime use. Each of these groups is subdivided into pyrotechnic and non-pyrotechnic devices.

Daytime Non-Pyrotechnic Signals

A bright orange flag with a black square over a black circle is the simplest VDS. It is usable, of course, only in daylight. It has the advantage of being a continuous signal. A mirror can be used to good advantage on sunny days. It can attract the attention of other boaters and of aircraft from great distances. Mirrors are available with holes in their centers to aid in "aiming." In the absence of a mirror, any shiny object can be used. When another boat is in sight, an effective VDS is to extend your arms from your sides and move them up and down. Do it slowly. If you do it too fast the other people may think you are just being friendly. This simple gesture is seldom misunderstood, and requires no equipment.

Daytime Pyrotechnic Devices

Orange smoke is a useful daytime signal. Hand-held or floating smoke flares are very effective in attracting attention from aircraft. Smoke flares don't last long, and are not very effective in high wind or poor visibility. As with other pyrotechnic devices, use them only when you know there is a possibility that someone will see the display.

To be usable, smoke flares must be kept dry. Keep them in airtight containers and store them in dry places. If the "striker" is damp, dry it out before trying to ignite the device. Some pyrotechnic devices require a forceful "strike" to ignite them.

All hand-held pyrotechnic devices may produce hot ashes or slag when burning. Hold them over the side of your boat in such a way that they do not burn your hand or drip into your boat.

Nighttime Non-Pyrotechnic Signals

An electric distress light is available. This light automatically flashes the international morse code SOS distress signal (••• — •••). Flashed four to six times a minute, it is an unmistakable distress signal. It must show that it is approved by the Coast Guard. Be sure the batteries are fresh. Dated batteries give assurance that they are current.

Under the Inland Navigation Rules, a high intensity white light flashing 50-70 times per minute is a distress signal. Therefore, use strobe lights on inland waters only for distress signals.

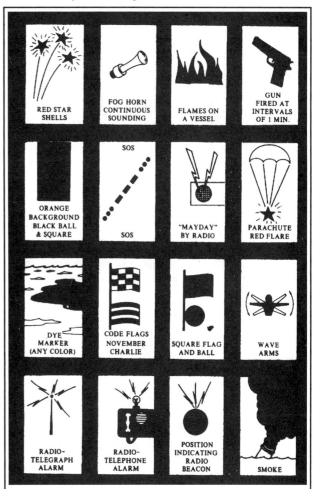

Fig. 10 Internationally accepted distress signals

1-10 GENERAL INFORMATION, SAFETY AND TOOLS

Nighttime Pyrotechnic Devices

◆ See Figure 11

Aerial and hand-held flares can be used at night or in the daytime. Obviously, they are more effective at night.

Currently, the serviceable life of a pyrotechnic device is rated at 42 months from its date of manufacture. Pyrotechnic devices are expensive. Look at their dates before you buy them. Buy them with as much time remaining as possible.

Like smoke flares, aerial and hand-held flares may fail to work if they have been damaged or abused. They will not function if they are or have been wet. Store them in dry, airtight containers in dry places. But store them where they are readily accessible.

Aerial VDSs, depending on their type and the conditions they are used in, may not go very high. Again, use them only when there is a good chance they will be seen.

A serious disadvantage of aerial flares is that they burn for only a short time; most burn for less than 10 seconds. Most parachute flares burn for less than 45 seconds. If you use a VDS in an emergency, do so carefully. Hold hand-held flares over the side of the boat when in use. Never use a road hazard flare on a boat; it can easily start a fire. Marine type flares are specifically designed to lessen risk, but they still must be used carefully.

Aerial flares should be given the same respect as firearms since they are firearms! Never point them at another person. Don't allow children to play with them or around them. When you fire one, face away from the wind. Aim it downwind and upward at an angle of about 60 degrees to the horizon. If there is a strong wind, aim it somewhat more vertically. Never fire it straight up. Before you discharge a flare pistol, check for overhead obstructions that might be damaged by the flare. An obstruction might deflect the flare to where it will cause injury or damage.

Disposal of VDS

Keep outdated flares when you get new ones. They do not meet legal requirements, but you might need them sometime, and they may work. It is illegal to fire a VDS on federal navigable waters unless an emergency exists. Many states have similar laws.

Emergency Position Indicating Radio Beacon (EPIRB)

There is no requirement for recreational boats to have EPIRBs. Some commercial and fishing vessels, though, must have them if they operate beyond the three-mile limit. Vessels carrying six or fewer passengers for hire must have EPIRBs under some circumstances when operating beyond the three-mile limit. If you boat in a remote area or offshore, you should have an EPIRB. An EPIRB is a small (about 6 to 20 in. high), battery-powered, radio transmitting buoy-like device. It is a radio transmitter and requires a license or an endorsement on your radio station license by the Federal Communications Commission (FCC). EPIRBs are either automatically activated by being immersed in water or manually by a switch.

Courtesy Marine Examinations

One of the roles of the Coast Guard Auxiliary is to promote recreational boating safety. This is why they conduct thousands of Courtesy Marine Examinations each year. The auxiliarists who do these examinations are well-trained and knowledgeable in the field.

These examinations are free and done only at the consent of boat owners. To pass the examination, a vessel must satisfy federal equipment requirements and certain additional requirements of the coast guard auxiliary. If your vessel does not pass the Courtesy Marine Examination, no report of the failure is made. Instead, you will be told what you need to correct the deficiencies. The examiner will return at your convenience to redo the examination.

If your vessel qualifies, you will be awarded a safety decal. The decal does not carry any special privileges, it simply attests to your interest in safe boating.

Fig. 11 Moisture-protected flares should be carried onboard any vessel for use as a distress signal

BOATING EQUIPMENT (NOT REQUIRED BUT RECOMMENDED)

Although not required by law, there are other pieces of equipment that are good to have onboard.

Oar/Paddle (Second Means of Propulsion)

All boats less than 16 feet long should carry a second means of propulsion. A paddle or oar can come in handy at times. For most small boats, a spare trolling or outboard motor is an excellent idea. If you carry a spare motor, it should have its own fuel tank and starting power. If you use an electric trolling motor, it should have its own battery.

Bailing Devices

All boats should carry at least one effective manual bailing device in addition to any installed electric bilge pump. This can be a bucket, can, scoop, hand-operated pump, etc. If your battery "goes dead" it will not operate your electric pump.

First Aid Kit

◆ See Figure 12

All boats should carry a first aid kit. It should contain adhesive bandages, gauze, adhesive tape, antiseptic, aspirin, etc. Check your first aid kit from time to time. Replace anything that is outdated. It is to your advantage to know how to use your first aid kit. Another good idea would be to take a Red Cross first aid course.

Anchors

◆ See Figure 13

All boats should have anchors. Choose one of suitable size for your boat. Better still, have two anchors of different sizes. Use the smaller one in calm water or when anchoring for a short time to fish or eat. Use the larger one when the water is rougher or for overnight anchoring.

Carry enough anchor line, of suitable size, for your boat and the waters in which you will operate. If your engine fails you, the first thing you usually should do is lower your anchor. This is good advice in shallow water where you may be driven aground by the wind or water. It is also good advice in windy weather or rough water, as the anchor, when properly affixed, will usually hold your bow into the waves.

VHF-FM Radio

Your best means of summoning help in an emergency or in case of a breakdown is a VHF-FM radio. You can use it to get advice or assistance from the Coast Guard. In the event of a serious illness or injury aboard your

GENERAL INFORMATION, SAFETY AND TOOLS 1-11

Fig. 12 Always carry an adequately stocked first aid kit on board for the safety of the crew and guests

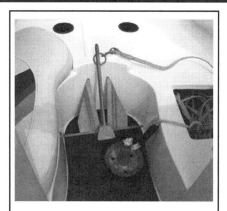

Fig. 13 Choose an anchor of sufficient weight to secure the boat without dragging

Fig. 14 Don't hesitate to spend a few extra dollars for a reliable compass

boat, the Coast Guard can have emergency medical equipment meet you ashore.

■ Although the VHF radio is the best way to get help, in this day and age, cell phones are a good backup source, especially for boaters on inland waters. You probably already know where you get a signal when boating, keep the phone charged, handy and off (so it doesn't bother you when boating right?). Keep phone numbers for a local dockmaster, coast guard, tow service or maritime police unit handy on board or stored in your phone directory.

Compass

SELECTION

◆ See Figure 14

The safety of the boat and her crew may depend on her compass. In many areas, weather conditions can change so rapidly that, within minutes, a skipper may find himself socked in by a fog bank, rain squall or just poor visibility. Under these conditions, he may have no other means of keeping to his desired course except with the compass. When crossing an open body of water, his compass may be the only means of making an accurate landfall.

During thick weather when you can neither see nor hear the expected aids to navigation, attempting to run out the time on a given course can disrupt the pleasure of the cruise. The skipper gains little comfort in a chain of soundings that does not match those given on the chart for the expected area. Any stranding, even for a short time, can be an unnerving experience.

A pilot will not knowingly accept a cheap parachute. By the same token, a good boater should not accept a bargain in lifejackets, fire extinguishers, or compass. Take the time and spend the few extra dollars to purchase a compass to fit your expected needs. Regardless of what the salesman may tell you, postpone buying until you have had the chance to check more than one make and model.

Lift each compass, tilt and turn it, simulating expected motions of the boat. The compass card should have a smooth and stable reaction.

The card of a good quality compass will come to rest without oscillations about the lubber's line. Reasonable movement in your hand, comparable to the rolling and pitching of the boat, should not materially affect the reading.

INSTALLATION

◆ See Figure 15

Proper installation of the compass does not happen by accident. Make a critical check of the proposed location to be sure compass placement will permit the helmsman to use it with comfort and accuracy. First, the compass should be placed directly in front of the helmsman, and in such a position that it can be viewed without body stress as he sits or stands in a posture of relaxed alertness. The compass should be in the helmsman's zone of comfort. If the compass is too far away, he may have to bend forward to watch it; too close and he must rear backward for relief.

Second, give some thought to comfort in heavy weather and poor visibility conditions during the day and night. In some cases, the compass position may be partially determined by the location of the wheel, shift lever and throttle handle.

Third, inspect the compass site to be sure the instrument will be at least two feet from any engine indicators, bilge vapor detectors, magnetic instruments, or any steel or iron objects. If the compass cannot be placed at least two feet (six feet would be better but on a small craft, let's get real two feet is usually pushing it) from one of these influences, then either the compass or the other object must be moved, if first order accuracy is to be expected.

Once the compass location appears to be satisfactory, give the compass a test before installation. Hidden influences may be concealed under the cabin top, forward of the cabin aft bulkhead, within the cockpit ceiling, or in a wood-covered stanchion.

Move the compass around in the area of the proposed location. Keep an eye on the card. A magnetic influence is the only thing that will make the card turn. You can quickly find any such influence with the compass. If the influence cannot be moved away or replaced by one of non-magnetic material, test to determine whether it is merely magnetic, a small piece of iron or steel, or some magnetized steel. Bring the north pole of the compass near the object, then shift and bring the south pole near it. Both the north and south poles will be attracted if the compass is demagnetized. If the object attracts one pole and repels the other, then the compass is magnetized. If your compass needs to be demagnetized, take it to a shop equipped to do the job PROPERLY.

After you have moved the compass around in the proposed mounting area, hold it down or tape it in position. Test everything you feel might affect the compass and cause a deviation from a true reading. Rotate the wheel from hard over-to-hard over. Switch on and off all the lights, radios, radio

Fig. 15 The compass is a delicate instrument which should be mounted securely in a position where it can be easily observed by the helmsman

1-12 GENERAL INFORMATION, SAFETY AND TOOLS

direction finder, radio telephone, depth finder and, if installed, the shipboard intercom. Sound the electric whistle, turn on the windshield wipers, start the engine (with water circulating through the engine), work the throttle, and move the gear shift lever. If the boat has an auxiliary generator, start it.

If the card moves during any one of these tests, the compass should be relocated. Naturally, if something like the windshield wipers causes a slight deviation, it may be necessary for you to make a different deviation table to use only when certain pieces of equipment are operating. Bear in mind, following a course that is off only a degree or two for several hours can make considerable difference at the end, putting you on a reef, rock or shoal.

Check to be sure the intended compass site is solid. Vibration will increase pivot wear.

Now, you are ready to mount the compass. To prevent an error on all courses, the line through the lubber line and the compass card pivot must be exactly parallel to the keel of the boat. You can establish the fore-and-aft line of the boat with a stout cord or string. Use care to transfer this line to the compass site. If necessary, shim the base of the compass until the stile-type lubber line (the one affixed to the case and not gimbaled) is vertical when the boat is on an even keel. Drill the holes and mount the compass.

COMPASS PRECAUTIONS

◆ See Figures 16, 17 and 18

Many times an owner will install an expensive stereo system in the cabin of his boat. It is not uncommon for the speakers to be mounted on the aft bulkhead up against the overhead (ceiling). In almost every case, this position places one of the speakers in very close proximity to the compass, mounted above the ceiling.

You probably already know that a magnet is used in the operation of the speaker. Therefore, it is very likely that the speaker, mounted almost under the compass in the cabin will have a very pronounced effect on the compass accuracy.

Consider the following test and the accompanying photographs as proof:
First, the compass was read as 190 degrees while the boat was secure in her slip.

Next, a full can of soda in an aluminum can was placed on one side and the compass read as 204 degrees, a good 14 degrees off.

Next, the full can was moved to the opposite side of the compass and again a reading was observed, this time as 189 degrees, 11 degrees off from the original reading.

Finally, the contents of the can were consumed, the can placed on both sides of the compass with NO effect on the compass reading.

Two very important conclusions can be drawn from these tests.
• Something must have been in the contents of the can to affect the compass so drastically.
• Keep even innocent things clear of the compass to avoid any possible error in the boat's heading.

■ Remember, a boat moving through the water at 10 knots on a compass error of just 5 degrees will be almost 1.5 miles off course in only ONE hour. At night, or in thick weather, this could very possibly put the boat on a reef, rock or shoal with disastrous results.

Tools and Spare Parts

◆ See Figures 19 and 20

Carry a few tools and some spare parts, and learn how to make minor repairs. Many search and rescue cases are caused by minor breakdowns that boat operators could have repaired. Carry spare parts such as propellers, fuses or basic ignition components (like spark plugs, wires or even ignition coils) and the tools necessary to install them.

Fig. 16 This compass is giving an accurate reading, right?

Fig. 17 ... well think again, as seemingly innocent objects may cause serious problems ...

Fig. 18 ... a compass reading off by just a few degrees could lead to disaster

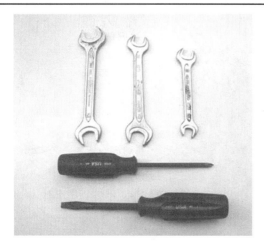

Fig. 19 A few wrenches, a screwdriver and maybe a pair of pliers can be very helpful to make emergency repairs

Fig. 20 A flashlight with a fresh set of batteries is handy when repairs are needed at night. It can also double as a signaling device

GENERAL INFORMATION, SAFETY AND TOOLS

SAFETY IN SERVICE

It is virtually impossible to anticipate all of the hazards involved with maintenance and service, but care and common sense will prevent most accidents.

The rules of safety for mechanics range from "don't smoke around gasoline," to "use the proper tool(s) for the job." The trick to avoiding injuries is to develop safe work habits and to take every possible precaution. Whenever you are working on your boat, pay attention to what you are doing. The more you pay attention to details and what is going on around you, the less likely you will be to hurt yourself or damage your boat.

Do's

- Do keep a fire extinguisher and first aid kit handy.
- Do wear safety glasses or goggles when cutting, drilling, grinding or prying, even if you have 20-20 vision. If you wear glasses for the sake of vision, wear safety goggles over your regular glasses.
- Do shield your eyes whenever you work around the battery. Batteries contain sulfuric acid. In case of contact with the eyes or skin, flush the area with water or a mixture of water and baking soda; then seek immediate medical attention.
- Do use adequate ventilation when working with any chemicals or hazardous materials.
- Do disconnect the negative battery cable when working on the electrical system. The secondary ignition system contains EXTREMELY HIGH VOLTAGE. In some cases it can even exceed 50,000 volts. Furthermore, an accidental attempt to start the engine could cause the propeller or other components to rotate suddenly causing a potentially dangerous situation.
- Do follow manufacturer's directions whenever working with potentially hazardous materials. Most chemicals and fluids are poisonous if taken internally.
- Do properly maintain your tools. Loose hammerheads, mushroomed punches and chisels, frayed or poorly grounded electrical cords, excessively worn screwdrivers, spread wrenches (open end), cracked sockets, or slipping ratchets can cause accidents.
- Likewise, keep your tools clean; a greasy wrench can slip off a bolt head, ruining the bolt and often harming your knuckles in the process.
- Do use the proper size and type of tool for the job at hand. Do select a wrench or socket that fits the nut or bolt. The wrench or socket should sit straight, not cocked.
- Do, when possible, pull on a wrench handle rather than push on it, and adjust your stance to prevent a fall.
- Do be sure that adjustable wrenches are tightly closed on the nut or bolt and pulled so that the force is on the side of the fixed jaw. Better yet, avoid the use of an adjustable if you have a fixed wrench that will fit.
- Do strike squarely with a hammer; avoid glancing blows.
- Do use common sense whenever you work on your boat or motor. If a situation arises that doesn't seem right, sit back and have a second look. It may save an embarrassing moment or potential damage to your beloved boat.

Dont's

- Don't run the engine in an enclosed area or anywhere else without proper ventilation - EVER! Carbon monoxide is poisonous; it takes a long time to leave the human body and you can build up a deadly supply of it in your system by simply breathing in a little every day. You may not realize you are slowly poisoning yourself.
- Don't work around moving parts while wearing loose clothing. Short sleeves are much safer than long, loose sleeves. Hard-toed shoes with neoprene soles protect your toes and give a better grip on slippery surfaces. Jewelry, watches, large belt buckles, or body adornment of any kind is not safe working around any craft or vehicle. Long hair should be tied back under a hat.
- Don't use pockets for toolboxes. A fall or bump can drive a screwdriver deep into your body. Even a rag hanging from your back pocket can wrap around a spinning shaft.
- Don't smoke when working around gasoline, cleaning solvent or other flammable material.
- Don't smoke when working around the battery. When the battery is being charged, it gives off explosive hydrogen gas. Actually, you shouldn't smoke anyway, it's bad for you. Instead, save the cigarette money and put it into your boat!
- Don't use gasoline to wash your hands; there are excellent soaps available. Gasoline contains dangerous additives that can enter the body through a cut or through your pores. Gasoline also removes all the natural oils from the skin so that bone dry hands will suck up oil and grease.
- Don't use screwdrivers for anything other than driving screws! A screwdriver used as a prying tool can snap when you least expect it, causing injuries. At the very least, you'll ruin a good screwdriver.

TROUBLESHOOTING

Troubleshooting can be defined as a methodical process during which one discovers what is causing a problem with engine operation. Although it is often a feared process to the uninitiated, there is no reason to believe that you cannot figure out what is wrong with a motor, as long as you follow a few basic rules.

To begin with, troubleshooting must be systematic. Haphazardly testing one component, then another, **might** uncover the problem, but it will more likely waste a lot of time. True troubleshooting starts by defining the problem and performing systematic tests to eliminate the largest and most likely causes first.

Start all troubleshooting by eliminating the most basic possible causes. Begin with a visual inspection of the boat and motor. If the engine won't crank, make sure that the kill switch or safety lanyard is in the proper position. Make sure there is fuel in the tank and the fuel system is primed before condemning the carburetor or fuel injection system. On electric start motors, make sure there are no blown fuses, the battery is fully charged, and the cable connections (at both ends) are clean and tight before suspecting a bad starter, solenoid or switch.

The majority of problems that occur suddenly can be fixed by simply identifying the one small item that brought them on. A loose wire, a clogged passage or a broken component can cause a lot of trouble and are often the cause of a sudden performance problem.

The next most basic step in troubleshooting is to test systems before components. For example, if the engine doesn't crank on an electric start motor, determine if the battery is in good condition (fully charged and properly connected) before testing the starting system. If the engine cranks, but doesn't start, you know already know the starting system and battery (if it cranks fast enough) are in good condition, now it is time to look at the ignition or fuel systems. Once you've isolated the problem to a particular system, follow the troubleshooting/testing procedures in the section for that system to test either subsystems (if applicable, for example: the starter circuit) or components (starter solenoid).

Basic Operating Principles

◆ See Figures 21 and 22

Before attempting to troubleshoot a problem with your motor, it is important that you understand how it operates. Once normal engine or system operation is understood, it will be easier to determine what might be causing the trouble or irregular operation in the first place. System descriptions are found throughout this manual, but the basic mechanical operating principles for both 2-stroke engines (like many outboards) and 4-stroke engines (like the outboards covered here) are given here. A basic understanding of both types of engines is useful not only in understanding and troubleshooting your outboard, but also for dealing with other motors in your life.

All motors covered by this manual (and probably MOST of the motors you own) operate according to the Otto cycle principle of engine operation. This means that all motors follow the stages of intake, compression, power and exhaust. But, the difference between a 2- and 4-stroke motor is in how many times the piston moves up and down within the cylinder to accomplish this. On 2-stroke motors (as the name suggests) the four cycles take place in 2 movements (one up and one down) of the piston. Again, as the name suggests, the cycles take place in 4 movements of the piston for 4-stroke motors.

1-14 GENERAL INFORMATION, SAFETY AND TOOLS

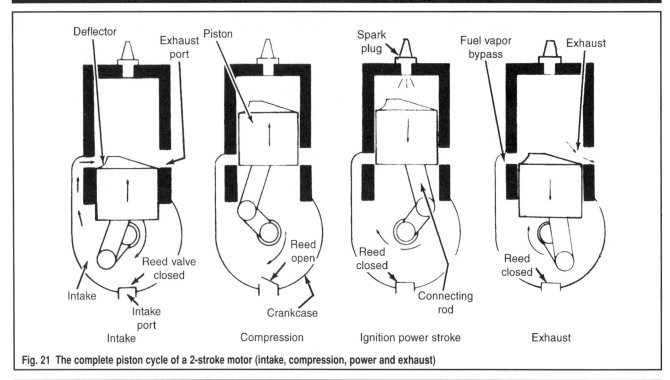

Fig. 21 The complete piston cycle of a 2-stroke motor (intake, compression, power and exhaust)

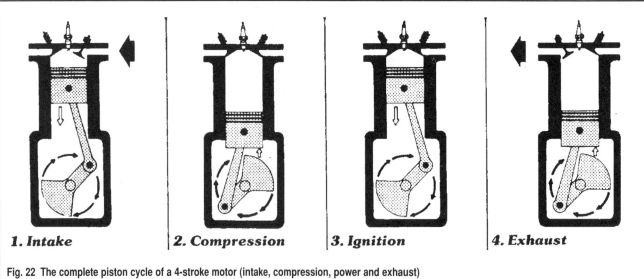

Fig. 22 The complete piston cycle of a 4-stroke motor (intake, compression, power and exhaust)

2-STROKE MOTORS

The 2-stroke engine differs in several ways from a conventional four-stroke (automobile or marine) engine.
1. The intake/exhaust method by which the fuel-air mixture is delivered to the combustion chamber.
2. The complete lubrication system.
3. The frequency of the power stroke.

Let's discuss these differences briefly (and compare 2-stroke engine operation with 4-stroke engine operation.)

Intake/Exhaust

◆ See Figures 23 thru 26

Two-stroke engines utilize an arrangement of port openings to admit fuel to the combustion chamber and to purge the exhaust gases after burning has been completed. The ports are located in a precise pattern in order for them to be open and closed off at an exact moment by the piston as it moves up and down in the cylinder. The exhaust port is located slightly higher than the fuel intake port. This arrangement opens the exhaust port first as the piston starts downward and therefore, the exhaust phase begins a fraction of a second before the intake phase.

Actually, the intake and exhaust ports are spaced so closely together that both open almost simultaneously. For this reason, some 2-stroke engines utilize deflector-type pistons. This design of the piston top serves two purposes very effectively.

First, it creates turbulence when the incoming charge of fuel enters the combustion chamber. This turbulence results in a more complete burning of the fuel than if the piston top were flat. The second effect of the deflector-type piston crown is to force the exhaust gases from the cylinder more rapidly. Although this configuration is used in many older outboards, it is generally found only on some of the smaller Yamaha motors. The majority of Yamaha motors are of the Loop (L) charged.

Loop charged motors, or as they are commonly called "loopers", differ in how the air/fuel charge is introduced to the combustion chamber. Instead of the charge flowing across the top of the piston from one side of the cylinder to the other (CV) the use a looping action on top of the piston as the charge is forced through irregular shaped openings cut in the piston's skirt. In a LV

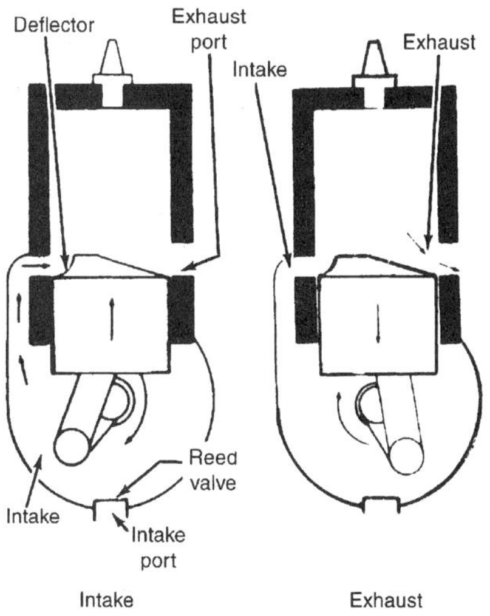

GENERAL INFORMATION, SAFETY AND TOOLS 1-15

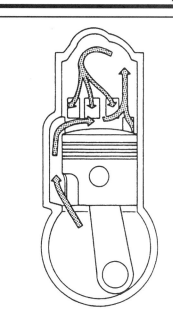

Fig. 23 The intake and exhaust cycles of a two-stroke engine - Cross flow (CV) design shown

Fig. 24 Cross-sectional view of a typical loop-charged cylinder, showing charge flow while piston is moving downward

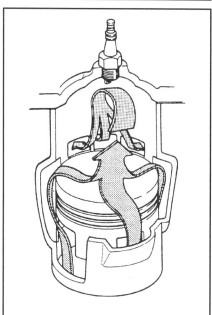

Fig. 25 Cutaway view of a typical loop-charged cylinder, depicting exhaust leaving the cylinder as the charge enters through 3 ports in the piston

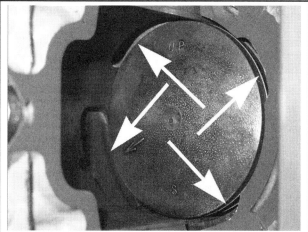

Fig. 26 The combustion chamber of a typical looper, notice the piston is far enough down the cylinder bore to reveal intake and exhaust ports

motor, the charge is forced out from the crankcase by the downward motion of the piston, through the irregular shaped openings and transferred upward by long, deep grooves in the cylinder wall. The charge completes its looping action by entering the combustion chamber, just above the piston, where the upward motion of the piston traps it in the chamber and compresses it for optimum ignition power.

Unlike the knife-edged deflector top pistons used in CV motors, the piston domes on Loop motors are relatively flat.

These systems of intake and exhaust are in marked contrast to individual intake and exhaust valve arrangement employed on four-stroke engines (and the mechanical methods of opening and closing these valves).

■ It should be noted here that there are some 2-stroke engines that utilize a mechanical valve train, though it is very different from the valve train employed by most 4-stroke motors. Rotary 2-stroke engines use a circular valve or rotating disc that contains a port opening around part of one edge of the disc. As the engine (and disc) turns, the opening aligns with the intake port at and for a predetermined amount of time, closing off the port again as the opening passes by and the solid portion of the disc covers the port.

Lubrication

A 2-stroke engine is lubricated by mixing oil with the fuel. Therefore, various parts are lubricated as the fuel mixture passes through the crankcase and the cylinder. In contrast, four-stroke engines have a crankcase containing oil. This oil is pumped through a circulating system and returned to the crankcase to begin the routing again.

Power Stroke

The combustion cycle of a 2-stroke engine has four distinct phases.
1. Intake
2. Compression
3. Power
4. Exhaust

The four phases of the cycle are accomplished with each up and down stroke of the piston, and the power stroke occurs with each complete revolution of the crankshaft. Compare this system with a four-stroke engine. A separate stroke of the piston is required to accomplish each phase of the cycle and the power stroke occurs only every other revolution of the crankshaft. Stated another way, two revolutions of the four-stroke engine crankshaft are required to complete one full cycle, the four phases.

Physical Laws

◆ See Figure 27

The 2-stroke engine is able to function because of two very simple physical laws.

One: Gases will flow from an area of high pressure to an area of lower pressure. A tire blowout is an example of this principle. The high-pressure air escapes rapidly if the tube is punctured.

Two: If a gas is compressed into a smaller area, the pressure increases, and if a gas expands into a larger area, the pressure is decreased.

If these two laws are kept in mind, the operation of the 2-stroke engine will be easier understood.

Actual Operation

◆ See Figure 21

■ The engine described here is of a carbureted type. EFI and HPDI motors operate similarly for intake of the air charge and for exhaust of the unburned gasses. Obviously though, the very nature of fuel injection changes the actual delivery of the fuel/oil charge.

1-16 GENERAL INFORMATION, SAFETY AND TOOLS

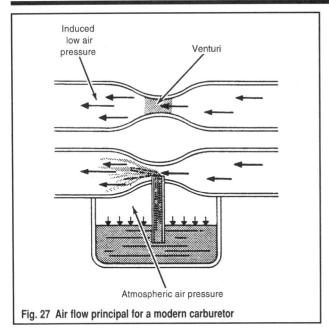

Fig. 27 Air flow principal for a modern carburetor

Beginning with the piston approaching top dead center on the compression stroke: the intake and exhaust ports are physically closed (blocked) by the piston. During this stroke, the reed valve is open (because as the piston moves upward, the crankcase volume increases, which reduces the crankcase pressure to less than the outside atmosphere (creates a vacuum under the piston). The spark plug fires; the compressed fuel-air mixture is ignited; and the power stroke begins.

As the piston moves downward on the power stroke, the combustion chamber is filled with burning gases. As the exhaust port is uncovered, the gases, which are under great pressure, escape rapidly through the exhaust ports. The piston continues its downward movement. Pressure within the crankcase (again, under the piston) increases, closing the reed valves against their seats. The crankcase then becomes a sealed chamber so the air-fuel mixture becomes compressed (pressurized) and ready for delivery to the combustion chamber. As the piston continues to move downward, the intake port is uncovered. The fresh fuel mixture rushes through the intake port into the combustion chamber striking the top of the piston where it is deflected along the cylinder wall. The reed valve remains closed until the piston moves upward again.

When the piston begins to move upward on the compression stroke, the reed valve opens because the crankcase volume has been increased, reducing crankcase pressure to less than the outside atmosphere. The intake and exhaust ports are closed and the fresh fuel charge is compressed inside the combustion chamber.

Pressure in the crankcase (beneath the piston) decreases as the piston moves upward and a fresh charge of air flows through the carburetor picking up fuel. As the piston approaches top dead center, the spark plug ignites the air-fuel mixture, the power stroke begins and one complete Otto cycle has been completed.

4-STROKE MOTORS

◆ See Figure 22

The 4-stroke motor may be easier to understand for some people either because of its prevalence in automobile and street motorcycle motors today or perhaps because each of the four strokes corresponds to one distinct phase of the Otto cycle. Essentially, a 4-stroke engine completes one Otto cycle of intake, compression, ignition/power and exhaust using two full revolutions of the crankshaft and four distinct movements of the piston (down, up, down and up).

Intake

The intake stroke begins with the piston near the top of its travel. As crankshaft rotation begins to pull the piston downward, the exhaust valve closes and the intake opens. As volume of the combustion chamber increases, a vacuum is created that draws in the air/fuel mixture from the intake manifold.

Compression

Once the piston reaches the bottom of its travel, crankshaft rotation will begin to force it upward. At this point the intake valve closes. As the piston rises in the bore, the volume of the sealed combustion chamber (both intake and exhaust valves are closed) decreases and the air/fuel mixture is compressed. This raises the temperature and pressure of the mixture and increases the amount of force generated by the expanding gases during the Ignition/Power stroke.

Ignition/Power

As the piston approaches top dead center (the highest point of travel in the bore), the spark plug will fire, igniting the air/fuel mixture. The resulting combustion of the air/fuel mixture forces the piston downward, rotating the crankshaft (causing other pistons to move in other phases/strokes of the Otto cycle on multi-cylinder motors).

Exhaust

As the piston approaches the bottom of the Ignition/Power stroke, the exhaust valve opens. When the piston begins its upward path of travel once again, any remaining unburned gasses are forced out through the exhaust valve. This completes one Otto cycle, which begins again as the piston passes top dead center, the intake valve opens and the Intake stroke starts.

COMBUSTION

Whether we are talking about a 2- or 4-stroke engine, all Otto cycle, internal combustion engines require three basic conditions to operate properly,
1. Compression
2. Ignition (Spark)
3. Fuel

A lack of any one of these conditions will prevent the engine from operating. A problem with any one of these will manifest itself in hard-starting or poor performance.

Compression

An engine that has insufficient compression will not draw an adequate supply of air/fuel mixture into the combustion chamber and, subsequently, will not make sufficient power on the power stroke. A lack of compression in just one cylinder of a multi-cylinder motor will cause the motor to stumble or run irregularly.

But, keep in mind that a sudden change in compression is unlikely in 2-stroke motors (unless something major breaks inside the crankcase, but that would usually be accompanied by other symptoms such as a loud noise when it occurred or noises during operation). On 4-stroke motors, a sudden change in compression is also unlikely, but could occur if the timing belt or chain was to suddenly break. Remember that the timing belt/chain is used to synchronize the valve train with the crankshaft. If the valve train suddenly ceases to turn, some intake and some exhaust valves will remain open, relieving compression in that cylinder.

Ignition (Spark)

Traditionally, the ignition system is the weakest link in the chain of conditions necessary for engine operation. Spark plugs may become worn or fouled, wires will deteriorate allowing arcing or misfiring, and poor connections can place an undue load on coils leading to weak spark or even a failed coil. The most common question asked by a technician under a no-start condition is: "do I have spark and fuel" (as they've already determined that they have compression).

A quick visual inspection of the spark plug(s) will answer the question as to whether or not the plug(s) is/are worn or fouled. While the engine is shut **OFF** a physical check of the connections could show a loose primary or secondary ignition circuit wire. An obviously physically damaged wire may also be an indication of system problems and certainly encourages one to inspect the related system more closely.

If nothing is turned up by the visual inspection, perform the Spark Test provided in the Ignition System section to determine if the problem is a lack of or a weak spark. If the problem is not compression or spark, it's time to look at the fuel system.

GENERAL INFORMATION, SAFETY AND TOOLS 1-17

Fuel

If compression and spark is present (and within spec), but the engine won't start or won't run properly, the only remaining condition to fulfill is fuel. As usual, start with the basics. Is the fuel tank full? Is the fuel stale? If the engine has not been run in some time (a matter of months, not weeks) there is a good chance that the fuel is stale and should be properly disposed of and replaced.

■ **Depending on how stale or contaminated (with moisture) the fuel is, it may be burned in an automobile or in yard equipment, though it would be wise to mix it well with a much larger supply of fresh gasoline to prevent moving your driveability problems to that motor. But it is better to get the lawn tractor stuck on stale gasoline than it would be to have your boat motor quit in the middle of the bay or lake.**

For hard starting motors, is the choke or primer system operating properly. Remember that the choke/prime should only be used for **cold** starts. A true cold start is really only the first start of the day, but it may be applicable to subsequent starts on cooler days, if the engine sat for more than a few hours and completely cooled off since the last use. Applying the primer to the motor for a hot start may flood the engine, preventing it from starting properly. One method to clear a flood is to crank the motor while the engine is at wide-open throttle (allowing the maximum amount of air into the motor to compensate for the excess fuel). But, keep in mind that the throttle should be returned to idle immediately upon engine start-up to prevent damage from over-revving.

Fuel delivery and pressure should be checked before delving into the carburetor(s) or fuel injection system. Make sure there are no clogs in the fuel line or vacuum leaks that would starve the motor of fuel.

Make sure that all other possible problems have been eliminated before touching the carburetor. It is rare that a carburetor will suddenly require an adjustment in order for the motor to run properly. It is much more likely that an improperly stored motor (one stored with untreated fuel in the carburetor) would suffer from one or more clogged carburetor passages sometime after shortly returning to service. Fuel will evaporate over time, leaving behind gummy deposits. If untreated fuel is left in the carburetor for some time (again typically months more than weeks), the varnish left behind by evaporating fuel will likely clog the small passages of the carburetor and cause problems with engine performance. If you suspect this, remove and disassemble the carburetor following procedures under Fuel System.

SHOP EQUIPMENT

Safety Tools

WORK GLOVES

◆ See Figure 28

Unless you think scars on your hands are cool, enjoy pain and like wearing bandages, get a good pair of work gloves. Canvas or leather gloves are the best. And yes, we realize that there are some jobs involving small parts that can't be done while wearing work gloves. These jobs are not the ones usually associated with hand injuries.

A good pair of rubber gloves (such as those usually associated with dish washing) or vinyl gloves is also a great idea. There are some liquids such as solvents and penetrants that don't belong on your skin. Avoid burns and rashes. Wear these gloves.

And lastly, an option. If you're tired of being greasy and dirty all the time, go to the drug store and buy a box of disposable latex gloves like medical professionals wear. You can handle greasy parts, perform small tasks, wash parts, etc. all without getting dirty! These gloves take a surprising amount of abuse without tearing and aren't expensive. Note however, that some people are allergic to the latex or the powder used inside some gloves, so pay attention to what you buy.

EYE AND EAR PROTECTION

◆ See Figures 29 and 30

Don't begin any job without a good pair of work goggles or impact resistant glasses! When doing any kind of work, it's all too easy to avoid eye injury through this simple precaution. And don't just buy eye protection and leave it on the shelf. Wear it all the time! Things have a habit of breaking, chipping, splashing, spraying, splintering and flying around. And, for some reason, your eye is always in the way!

If you wear vision-correcting glasses as a matter of routine, get a pair made with polycarbonate lenses. These lenses are impact resistant and are available at any optometrist.

Often overlooked is hearing protection. Engines and power tools are noisy! Loud noises damage your ears. It's as simple as that! The simplest and cheapest form of ear protection is a pair of noise-reducing ear plugs. Cheap insurance for your ears! And, they may even come with their own, cute little carrying case.

More substantial, more protection and more money is a good pair of noise reducing earmuffs. They protect from all but the loudest sounds. Hopefully those are sounds that you'll never encounter since they're usually associated with disasters.

WORK CLOTHES

Everyone has "work clothes." Usually these consist of old jeans and a shirt that has seen better days. That's fine. In addition, a denim work apron is a nice accessory. It's rugged, can hold some spare bolts, and you don't feel bad wiping your hands or tools on it. That's what it's for.

When working in cold weather, a one-piece, thermal work outfit is invaluable. Most are rated to below freezing temperatures and are ruggedly constructed. Just look at what local marine mechanics are wearing and that should give you a clue as to what type of clothing is good.

Chemicals

There is a whole range of chemicals that you'll find handy for maintenance and repair work. The most common types are: lubricants, penetrants and sealers. Keep these handy. There are also many chemicals that are used for detailing or cleaning.

When a particular chemical is not being used, keep it capped, upright and in a safe place. These substances may be flammable, may be irritants or

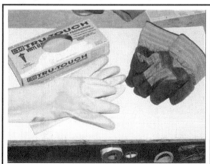

Fig. 28 Three different types of work gloves. The box contains latex gloves

Fig. 29 Don't begin major repairs without a pair of goggles for your eyes and earmuffs to protect your hearing

Fig. 30 Things have a habit of, splashing, spraying, splintering and flying around during repairs

1-18 GENERAL INFORMATION, SAFETY AND TOOLS

might even be caustic and should always be stored properly, used properly and handled with care. Always read and follow all label directions and be sure to wear hand and eye protection!

LUBRICANTS & PENETRANTS

◆ See Figure 31

Anti-seize is used to coat certain fasteners prior to installation. This can be especially helpful when two dissimilar metals are in contact (to help prevent corrosion that might lock the fastener in place). This is a good practice on a lot of different fasteners, BUT, NOT on any fastener that might vibrate loose causing a problem. If anti-seize is used on a fastener, it should be checked periodically for proper tightness.

Lithium grease, chassis lube, silicone grease or a synthetic brake caliper grease can all be used pretty much interchangeably. All can be used for coating rust-prone fasteners and for facilitating the assembly of parts that are a tight fit. Silicone and synthetic greases are the most versatile.

■ **Silicone dielectric grease is a non-conductor that is often used to coat the terminals of wiring connectors before fastening them. It may sound odd to coat metal portions of a terminal with something that won't conduct electricity, but here is it how it works. When the connector is fastened the metal-to-metal contact between the terminals will displace the grease (allowing the circuit to be completed). The grease that is displaced will then coat the non-contacted surface and the cavity around the terminals, SEALING them from atmospheric moisture that could cause corrosion.**

Silicone spray is a good lubricant for hard-to-reach places and parts that shouldn't be gooped up with grease.

Penetrating oil may turn out to be one of your best friends when taking something apart that has corroded fasteners. Not only can they make a job easier, they can really help to avoid broken and stripped fasteners. The most familiar penetrating oils are Liquid Wrench® and WD-40®. A newer penetrant, PB Blaster® works very well (and has become a mainstay in our shops). These products have hundreds of uses. For your purposes, they are vital!

Before disassembling any part, check the fasteners. If any appear rusted, soak them thoroughly with the penetrant and let them stand while you do something else (for particularly rusted or frozen parts you may need to soak them a few days in advance). This simple act can save you hours of tedious work trying to extract a broken bolt or stud.

SEALANTS

◆ See Figures 32 and 33

Sealants are an indispensable part for certain tasks, especially if you are trying to avoid leaks. The purpose of sealants is to establish a leak-proof bond between or around assembled parts. Most sealers are used in conjunction with gaskets, but some are used instead of conventional gasket material.

The most common sealers are the non-hardening types such as Permatex® No.2 or its equivalents. These sealers are applied to the mating surfaces of each part to be joined, then a gasket is put in place and the parts are assembled.

■ **A sometimes overlooked use for sealants like RTV is on the threads of vibration prone fasteners.**

One very helpful type of non-hardening sealer is the "high tack" type. This type is a very sticky material that holds the gasket in place while the parts are being assembled. This stuff is really a good idea when you don't have enough hands or fingers to keep everything where it should be.

The stand-alone sealers are the Room Temperature Vulcanizing (RTV) silicone gasket makers. On some engines, this material is used instead of a gasket. In those instances, a gasket may not be available or, because of the shape of the mating surfaces, a gasket shouldn't be used. This stuff, when used in conjunction with a conventional gasket, produces the surest bonds.

RTV does have its limitations though. When using this material, you will have a time limit. It starts to set-up within 15 minutes or so, so you have to assemble the parts without delay. In addition, when squeezing the material out of the tube, don't drop any glops into the engine. The stuff will form and set and travel around a cooling passage, possibly blocking it. Also, most types are not fuel-proof. Check the tube for all cautions.

CLEANERS

◆ See Figures 34 and 35

There are two basic types of cleaners on the market today: parts cleaners and hand cleaners. The parts cleaners are for the parts; the hand cleaners are for you. They are **not** interchangeable.

There are many good, non-flammable, biodegradable parts cleaners on the market. These cleaning agents are safe for you, the parts and the environment. Therefore, there is no reason to use flammable, caustic or toxic substances to clean your parts or tools.

As far as hand cleaners go; the waterless types are the best. They have always been efficient at cleaning, but they used to all leave a pretty smelly odor. Recently though, most of them have eliminated the odor and added stuff that actually smells good. Make sure that you pick one that contains lanolin or some other moisture-replenishing additive. Cleaners not only remove grease and oil but also skin oil.

■ **Most women already know to use a hand lotion when you're all cleaned up. It's okay. Real men DO use hand lotion too! Believe it or not, using hand lotion BEFORE your hands are dirty will actually make them easier to clean when you're finished with a dirty job. Lotion seals your hands, and keeps dirt and grease from sticking to your skin.**

Fig. 31 Keep a supply of anti-seize, penetrating oil, lithium grease, electronic cleaner and silicone spray

Fig. 32 Sealants are essential for preventing leaks

Fig. 33 On some engines, RTV is used instead of gasket material to seal components

GENERAL INFORMATION, SAFETY AND TOOLS 1-19

Fig. 34 Citrus hand cleaners not only work well, but they smell pretty good too. Choose one with pumice for added cleaning powe

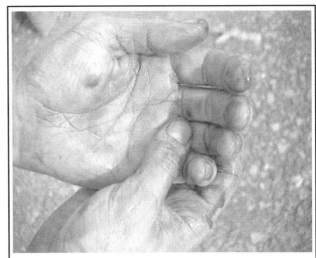

Fig. 35 The use of hand lotion seals your hands and keeps dirt and grease from sticking to your skin

TOOLS

◆ See Figure 36

Tools; this subject could fill a completely separate manual. The first thing you will need to ask yourself, is just how involved do you plan to get. If you are serious about maintenance and repair you will want to gather a quality set of tools to make the job easier, and more enjoyable. BESIDES, TOOLS ARE FUN!!!

Almost every do-it-yourselfer loves to accumulate tools. Though most find a way to perform jobs with only a few common tools, they tend to buy more over time, as money allows. So gathering the tools necessary for maintenance or repair does not have to be an expensive, overnight proposition.

When buying tools, the saying "You get what you pay for . . ." is absolutely true! Don't go cheap! Any hand tool that you buy should be drop forged and/or chrome vanadium. These two qualities tell you that the tool is strong enough for the job. With any tool, go with a name that you've heard of before, or, that is recommended buy your local professional retailer. Let's go over a list of tools that you'll need.

Most of the world uses the metric system. However, some American-built engines and aftermarket accessories use standard fasteners. So, accumulate your tools accordingly. Any good DIYer should have a decent set of both U.S. and metric measure tools.

■ Don't be confused by terminology. Most advertising refers to "SAE and metric", or "standard and metric." Both are misnomers. The Society of Automotive Engineers (SAE) did not invent the English system of measurement; the English did. The SAE likes metrics just fine. Both English (U.S.) and metric measurements are SAE approved. Also, the current "standard" measurement IS metric. So, if it's not metric, it's U.S. measurement.

Hand Tools

SOCKET SETS

◆ See Figures 37 thru 43

Socket sets are the most basic hand tools necessary for repair and maintenance work. For our purposes, socket sets come in three drive sizes: 1/4 inch, 3/8 inch and 1/2 inch. Drive size refers to the size of the drive lug on the ratchet, breaker bar or speed handle.

A 3/8 inch set is probably the most versatile set in any mechanic's toolbox. It allows you to get into tight places that the larger drive ratchets can't and gives you a range of larger sockets that are still strong enough for

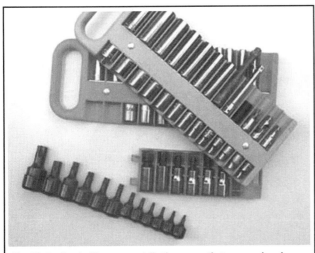

Fig. 36 Socket holders, especially the magnetic type, are handy items to keep tools in order

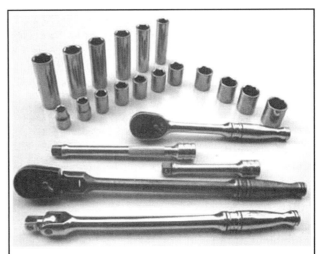

Fig. 37 A 3/8 in. socket set is probably the most versatile tool in any mechanic's tool box

1-20 GENERAL INFORMATION, SAFETY AND TOOLS

Fig. 38 A swivel (U-joint) adapter (left), a wobble-head adapter (center) and a 1/2 in.-to-3/8 in. adapter (right)

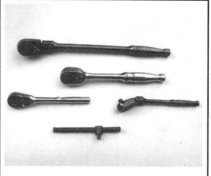

Fig. 39 Ratchets come in all sizes and configurations from rigid to swivel-headed

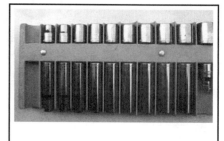

Fig. 40 Shallow sockets (top) are good for most jobs. But, some bolts require deep sockets (bottom)

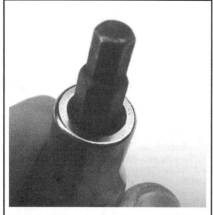

Fig. 41 Hex-head fasteners require a socket with a hex shaped driver

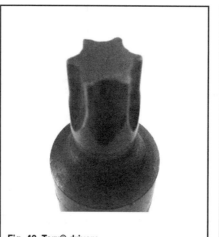

Fig. 42 Torx® drivers . . .

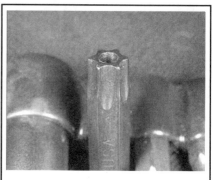

Fig. 43 . . . and tamper resistant drivers are required to remove special fasteners

heavy-duty work. The socket set that you'll need should range in sizes from 1/4 inch through 1 inch for standard fasteners, and a 6mm through 19mm for metric fasteners.

You'll need a good 1/2 inch set since this size drive lug assures that you won't break a ratchet or socket on large or heavy fasteners. Also, torque wrenches with a torque scale high enough for larger fasteners are usually 1/2 inch drive.

Plus, 1/4 inch drive sets can be very handy in tight places. Though they usually duplicate functions of the 3/8 in. set, 1/4 in. drive sets are easier to use for smaller bolts and nuts.

As for the sockets themselves, they come in shallow (standard) and deep lengths as well as 6 or 12 point. The 6 and 12 points designation refers to how many sides are in the socket itself. Each has advantages. The 6 point socket is stronger and less prone to slipping which would strip a bolt head or nut. 12 point sockets are more common, usually less expensive and can operate better in tight places where the ratchet handle can't swing far.

Standard length sockets are good for just about all jobs, however, some stud-head bolts, hard-to-reach bolts, nuts on long studs, etc., require the deep sockets.

Most marine manufacturers use recessed hex-head fasteners to retain many of the engine parts. These fasteners require a socket with a hex shaped driver or a large sturdy hex key. To help prevent torn knuckles, we would recommend that you stick to the sockets on any tight fastener and leave the hex keys for lighter applications. Hex driver sockets are available individually or in sets just like conventional sockets.

More and more, manufacturers are using Torx® head fasteners, which were once known as tamper resistant fasteners (because many people did not have tools with the necessary odd driver shape). Since Torx® fasteners have become commonplace in many DIYer tool boxes, manufacturers designed newer tamper resistant fasteners that are essentially Torx® head bolts that contain a small protrusion in the center (requiring the driver to contain a small hole to slide over the protrusion. Tamper resistant fasteners are often used where the manufacturer would prefer only knowledgeable mechanics or advanced Do-It-Yourselfers (DIYers) work.

Torque Wrenches

◆ See Figure 44

In most applications, a torque wrench can be used to ensure proper installation of a fastener. Torque wrenches come in various designs and most stores will carry a variety to suit your needs. A torque wrench should be used any time you have a specific torque value for a fastener. Keep in mind that because there is no worldwide standardization of fasteners, so charts or figure found in each repair section refer to the manufacturer's fasteners. Any general guideline charts that you might come across based on fastener size (they are sometimes included in a repair manual or with torque wrench packaging) should be used with caution. Just keep in mind that if you are using the right tool for the job, you should not have to strain to tighten a fastener.

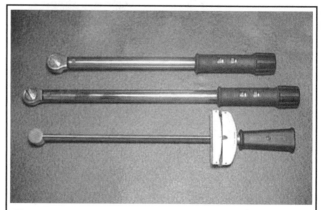

Fig. 44 Three types of torque wrenches. Top to bottom: a 3/8 in. drive beam type that reads in inch lbs., a 1/2 in. drive clicker type and a 1/2 in. drive beam type

GENERAL INFORMATION, SAFETY AND TOOLS

Beam Type
◆ See Figures 45 and 46

The beam type torque wrench is one of the most popular styles in use. If used properly, it can be the most accurate also. It consists of a pointer attached to the head that runs the length of the flexible beam (shaft) to a scale located near the handle. As the wrench is pulled, the beam bends and the pointer indicates the torque using the scale.

Click (Breakaway) Type
◆ See Figures 47 and 48

Another popular torque wrench design is the click type. The clicking mechanism makes achieving the proper torque easy and most use a ratcheting head for ease of bolt installation. To use the click type wrench you pre-adjust it to a torque setting. Once the torque is reached, the wrench has a reflex signaling feature that causes a momentary breakaway of the torque wrench body, sending an impulse to the operator's hand. But be careful, as continuing the turn the wrench after the momentary release will increase torque on the fastener beyond the specified setting.

Breaker Bars
◆ See Figure 49

Breaker bars are long handles with a drive lug. Their main purpose is to provide extra turning force when breaking loose tight bolts or nuts. They come in all drive sizes and lengths. Always take extra precautions and use the proper technique when using a breaker bar (pull on the bar, don't push, to prevent skinned knuckles).

WRENCHES
◆ See Figures 50 thru 54

Basically, there are 3 kinds of fixed wrenches: open end, box end, and combination.

Open-end wrenches have 2-jawed openings at each end of the wrench. These wrenches are able to fit onto just about any nut or bolt. They are extremely versatile but have one major drawback. They can slip on a worn or rounded bolt head or nut, causing bleeding knuckles and a useless fastener.

■ Line wrenches are a special type of open-end wrench designed to fit onto more of the fastener than standard open-end wrenches, thus reducing the chance of rounding the corners of the fastener.

Box-end wrenches have a 360° circular jaw at each end of the wrench. They come in both 6 and 12 point versions just like sockets and each type has some of the same advantages and disadvantages as sockets.

Combination wrenches have the best of both. They have a 2-jawed open end and a box end. These wrenches are probably the most versatile.

As for sizes, you'll probably need a range similar to that of the sockets, about 1/4 in. through 1 in. for standard fasteners, or 6mm through 19mm for metric fasteners. As for numbers, you'll need 2 of each size, since, in many instances, one wrench holds the nut while the other turns the bolt. On most fasteners, the nut and bolt are the same size so having two wrenches of the same size comes in handy.

■ Although you will typically just need the sizes we specified, there are some exceptions. Occasionally you will find a nut that is larger. For these, you will need to buy ONE expensive wrench or a very large adjustable. Or you can always just convince the spouse that we are talking about SAFETY here and buy a whole (read expensive) large wrench set.

One extremely valuable type of wrench is the adjustable wrench. An adjustable wrench has a fixed upper jaw and a moveable lower jaw. The lower jaw is moved by turning a threaded drum. The advantage of an adjustable wrench is its ability to be adjusted to just about any size fastener.

The main drawback of an adjustable wrench is the lower jaw's tendency to move slightly under heavy pressure. This can cause the wrench to slip if it is not facing the right way. Pulling on an adjustable wrench in the proper direction will cause the jaws to lock in place. Adjustable wrenches come in a large range of sizes, measured by the wrench length.

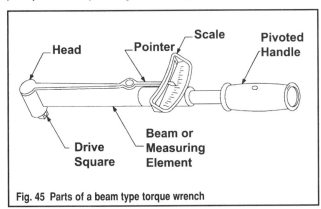

Fig. 45 Parts of a beam type torque wrench

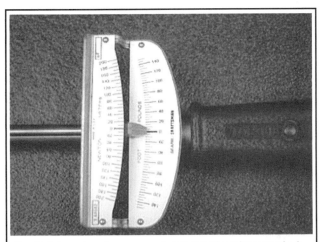

Fig. 46 A beam type torque wrench consists of a pointer attached to the head that runs the length of the flexible beam (shaft) to a scale located near the handle

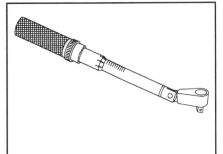

Fig. 47 A click type or breakaway torque wrench - note this one has a pivoting head

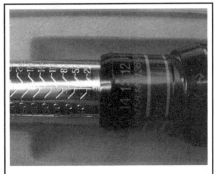

Fig. 48 Setting the torque on a click type wrench involves turning the handle until the specification appears on the dial

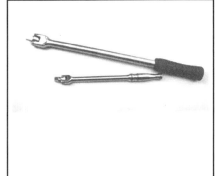

Fig. 49 Breaker bars are great for loosening large or stuck fasteners

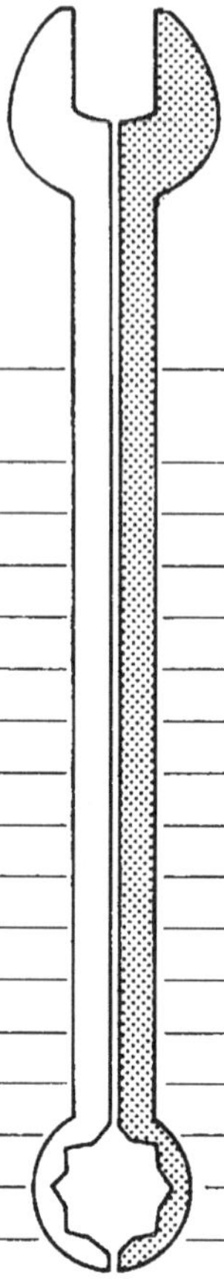

1-22 GENERAL INFORMATION, SAFETY AND TOOLS

INCHES	DECIMAL		DECIMAL	MILLIMETERS
1/8"	.125		.118	3mm
3/16"	.187		.157	4mm
1/4"	.250		.236	6mm
5/16"	.312		.354	9mm
3/8"	.375		.394	10mm
7/16"	.437		.472	12mm
1/2"	.500		.512	13mm
9/16"	.562		.590	15mm
5/8"	.625		.630	16mm
11/16"	.687		.709	18mm
3/4"	.750		.748	19mm
13/16"	.812		.787	20mm
7/8"	.875		.866	22mm
15/16"	.937		.945	24mm
1"	1.00		.984	25mm

Fig. 50 Comparison of U.S. measure and metric wrench sizes

PLIERS

♦ See Figure 55

Pliers are simply mechanical fingers. They are, more than anything, an extension of your hand. At least 3 pairs of pliers are an absolute necessity - standard, needle nose and slip joint.

In addition to standard pliers there are the slip-joint, multi-position pliers such as ChannelLock® pliers and locking pliers, such as Vise Grips®.

Slip joint pliers are extremely valuable in grasping oddly sized parts and fasteners. Just make sure that you don't use them instead of a wrench too often since they can easily round off a bolt head or nut.

Locking pliers are usually used for gripping bolts or studs that can't be removed conventionally. You can get locking pliers in square jawed, needle-nosed and pipe-jawed. Locking pliers can rank right up behind duct tape as the handy-man's best friend.

SCREWDRIVERS

You can't have too many screwdrivers. They come in 2 basic flavors, either standard or Phillips. Standard blades come in various sizes and thickness for all types of slotted fasteners. Phillips screwdrivers come in sizes with number designations from 1 on up, with the lower number designating the smaller size. Screwdrivers can be purchased separately or in sets.

HAMMERS

♦ See Figure 56

You need a hammer for just about any kind of work. You need a ball-peen hammer for most metal work when using drivers and other like tools. A plastic hammer comes in handy for hitting things safely. A soft-faced dead-blow hammer is used for hitting things safely and hard. Hammers are also VERY useful with non air-powered impact drivers.

Other Common Tools

There are a lot of other tools that every DIYer will eventually need (though not all for basic maintenance). They include:
- Funnels
- Chisels
- Punches
- Files
- Hacksaw
- Portable Bench Vise
- Tap and Die Set
- Flashlight
- Magnetic Bolt Retriever
- Gasket scraper
- Putty Knife
- Screw/Bolt Extractors
- Prybars

Hacksaws have just one use - cutting things off. You may wonder why you'd need one for something as simple as maintenance or repair, but you never know. Among other things, guide studs to ease parts installation can be made from old bolts with their heads cut off.

A tap and die set might be something you've never needed, but you will eventually. It's a good rule, when everything is apart, to clean-up all threads, on bolts, screws or threaded holes. Also, you'll likely run across a situation in which you will encounter stripped threads. The tap and die set will handle that for you.

Gasket scrapers are just what you'd think, tools made for scraping old gasket material off of parts. You don't absolutely need one. Old gasket material can be removed with a putty knife or single edge razor blade. However, putty knives may not be sharp enough for some really stubborn gaskets and razor blades have a knack of breaking just when you don't want them to, inevitably slicing the nearest body part! As the old saying goes, "always use the proper tool for the job". If you're going to use a razor to scrape a gasket, be sure to always use a blade holder.

GENERAL INFORMATION, SAFETY AND TOOLS 1-23

Fig. 51 Always use a backup wrench to prevent rounding flare nut fittings

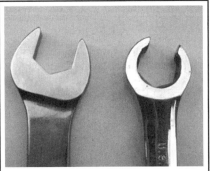

Fig. 52 Note how the flare wrench jaws are extended to grip the fitting tighter and prevent rounding

Fig. 53 Several types and sizes of adjustable wrenches

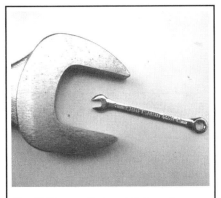

Fig. 54 You may find a nut that requires a particularly large or small wrench (that is usually available at your local tool store)

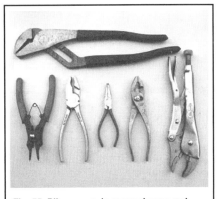

Fig. 55 Pliers come in many shapes and sizes. You should have an assortment on hand

Fig. 56 Three types of hammers. Top to bottom: ball peen, rubber dead-blow, and plastic

Putty knives really do have a use in a repair shop. Just because you remove all the bolts from a component sealed with a gasket doesn't mean it's going to come off. Most of the time, the gasket and sealer will hold it tightly. Lightly inserting a putty knife at various points between the two parts will break the seal without damage to the parts.

A small - 8-10 in. (20-25cm) long - prybar is extremely useful for removing stuck parts.

■ **Never use a screwdriver as a prybar! Screwdrivers are not meant for prying. Screwdrivers, used for prying, can break, sending the broken shaft flying!**

Screw/bolt extractors are used for removing broken bolts or studs that have broken off flush with the surface of the part.

Special Tools

◆ See Figure 57

Almost every marine engine around today requires at least one special tool to perform a certain task. In most cases, these tools are specially designed to overcome some unique problem or to fit on some oddly sized component.

When manufacturers go through the trouble of making a special tool, it is usually necessary to use it to ensure that the job will be done right. A special tool might be designed to make a job easier, or it might be used to keep you from damaging or breaking a part.

Don't worry, MOST maintenance procedures can either be performed without any special tools OR, because the tools must be used for such basic things, they are commonly available for a reasonable price. It is usually just the low production, highly specialized tools (like a super thin 7-point star-shaped socket capable of 150 ft. lbs. (203 Nm) of torque that is used only on the crankshaft nut of the limited production what-dya-callit engine) that tend to be outrageously expensive and hard to find. Hopefully, you will probably never need such a tool.

Special tools can be as inexpensive and simple as an adjustable strap wrench or as complicated as an ignition tester. A few common specialty tools are listed here, but check with your dealer or with other boaters for help in determining if there are any special tools for YOUR particular engine. There is an added advantage in seeking advice from others, chances are they may have already found the special tool you will need, and know how to get it cheaper (or even let you borrow it).

Electronic Tools

BATTERY TESTERS

The best way to test a non-sealed battery is using a hydrometer to check the specific gravity of the acid. Luckily, these are usually inexpensive and are available at most parts stores. Just be careful because the larger testers are usually designed for larger batteries and may require more acid than you will be able to draw from the battery cell. Smaller testers (usually a short, squeeze bulb type) will require less acid and should work on most batteries.

Electronic testers are available and are often necessary to tell if a sealed battery is usable. Luckily, many parts stores have them on hand and are willing to test your battery for you.

BATTERY CHARGERS

◆ See Figure 58

If you are a weekend boater and take your boat out every week, then you will most likely want to buy a battery charger to keep your battery fresh. There are many types available, from low amperage trickle chargers to electronically controlled battery maintenance tools that monitor the battery voltage to prevent over or undercharging. This last type is especially useful if you store your boat for any length of time (such as during the severe winter months found in many Northern climates).

Even if you use your boat on a regular basis, you will eventually need a battery charger. The charger should be used anytime the boat is going to be

1-24 GENERAL INFORMATION, SAFETY AND TOOLS

in storage for more than a few weeks or so. Never leave the dock or loading ramp without a battery that is fully charged.

Also, some smaller batteries are shipped dry and in a partial charged state. Before placing a new battery of this type into service it must be filled and properly charged. Failure to properly charge a battery (which was shipped dry) before it is put into service will prevent it from ever reaching a fully charged state.

MULTI-METERS (DVOMS)

◆ See Figure 59

Multi-meters or Digital Volt Ohmmeters (DVOMs) are an extremely useful tool for troubleshooting electrical problems. They can be purchased in either analog or digital form and have a price range to suit any budget. A multi-meter is a voltmeter, ammeter and ohmmeter (along with other features) combined into one instrument. It is often used when testing solid state circuits because of its high input impedance (usually 10 mega-ohms or more). A brief description of the multi-meter main test functions follows:

- Voltmeter - the voltmeter is used to measure voltage at any point in a circuit or to measure the voltage drop across any part of a circuit. Voltmeters usually have various scales and a selector switch to allow the reading of different voltage ranges. The voltmeter has a positive and a negative lead. To avoid the possibility of damage to the meter, whenever possible, connect the negative lead to the negative (-) side of the circuit (to ground or nearest the ground side of the circuit) and connect the positive lead to the positive (+) side of the circuit (to the power source or the nearest power source). Luckily, most quality DVOMs can adjust their own polarity internally and will indicate (without damage) if the leads are reversed. Note that the negative voltmeter lead will always be black and that the positive voltmeter will always be some color other than black (usually red).

- Ohmmeter - the ohmmeter is designed to read resistance (measured in ohms) in a circuit or component. Most ohmmeters will have a selector switch which permits the measurement of different ranges of resistance (usually the selector switch allows the multiplication of the meter reading by 10, 100, 1,000 and 10,000). Some ohmmeters are "auto-ranging" which means the meter itself will determine which scale to use. Since the meters are powered by an internal battery, the ohmmeter can be used like a self-powered test light. When the ohmmeter is connected, current from the ohmmeter flows through the circuit or component being tested. Since the ohmmeter's internal resistance and voltage are known values, the amount of current flow through the meter depends on the resistance of the circuit or component being tested. The ohmmeter can also be used to perform a continuity test for suspected open circuits. In using the meter for making continuity checks, do not be concerned with the actual resistance readings. Zero resistance, or any ohm reading, indicates continuity in the circuit. Infinite resistance indicates an opening in the circuit. A high resistance reading where there should be little or none indicates a problem in the circuit. Checks for short circuits are made in the same manner as checks for open circuits, except that the circuit must be isolated from both power and normal ground. Infinite resistance indicates no continuity, while zero resistance indicates a dead short.

※※ WARNING

Never use an ohmmeter to check the resistance of a component or wire while there is voltage applied to the circuit.

- Ammeter - an ammeter measures the amount of current flowing through a circuit in units called amperes or amps. At normal operating voltage, most circuits have a characteristic amount of amperes, called "current draw" which can be measured using an ammeter. By referring to a specified current draw rating, then measuring the amperes and comparing the two values; one can determine what is happening within the circuit to aid in diagnosis. An open circuit, for example, will not allow any current to flow, so the ammeter reading will be zero. A damaged component or circuit will have an increased current draw, so the reading will be high. The ammeter is always connected in series with the circuit being tested. All of the current that normally flows through the circuit must also flow through the ammeter; if there is any other path for the current to follow, the ammeter reading will not be accurate. The ammeter itself has very little resistance to current flow and, therefore, will not affect the circuit, but, it will measure current draw only when the circuit is closed and electricity is flowing. Excessive current draw can blow fuses and drain the battery, while a reduced current draw can cause motors to run slowly, lights to dim and other components to not operate properly.

GAUGES

Compression Gauge

◆ See Figure 60

An important element in checking the overall condition of your engine is to check compression. This becomes increasingly more important on outboards with high hours. Compression gauges are available as screw-in types and hold-in types. The screw-in type is slower to use, but eliminates the possibility of a faulty reading due to pressure escaping by the seal. A compression reading will uncover many problems that can cause rough running. Normally, these are not the sort of problems that can be cured by a tune-up.

Vacuum/Pressure Gauge/Pump

◆ See Figures 61, 62 and 63

Vacuum gauges are handy for discovering air leaks, late ignition or valve timing, and a number of other problems. A hand-held vacuum/pressure pump can be purchased at many automotive or marine parts stores and can be used for multiple purposes. The gauge can be used to measure vacuum or pressure in a line, while the pump can be used to manually apply vacuum or pressure to a solenoid or fitting. The hand-held pump can also be used to power-bleed trailer and tow vehicle brakes.

Fig. 57 Almost every marine engine around today requires at least one special tool to perform a certain task

Fig. 58 The Battery Tender® is more than just a battery charger, when left connected, it keeps your battery fully charged

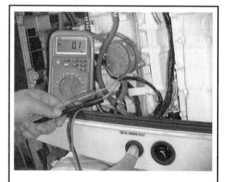

Fig. 59 Multi-meters, such as this one from UEI, are an extremely useful tool for troubleshooting electrical problems

GENERAL INFORMATION, SAFETY AND TOOLS

Fig. 60 Cylinder compression test results are extremely valuable indicators of internal engine condition

Fig. 61 Vacuum gauges are useful for troubleshooting including testing some fuel pumps

Fig. 62 You can also use the gauge on a hand-operated vacuum pump for tests

Measuring Tools

Eventually, you are going to have to measure something. To do this, you will need at least a few precision tools.

MICROMETERS & CALIPERS

Micrometers and calipers are devices used to make extremely precise measurements. The simple truth is that you really won't have the need for many of these items just for routine maintenance. But, measuring tools, such as an outside caliper can be handy during repairs. And, if you decide to tackle a major overhaul, a micrometer will absolutely be necessary.

Should you decide on becoming more involved in boat engine mechanics, such as repair or rebuilding, then these tools will become very important. The success of any rebuild is dependent, to a great extent on the ability to check the size and fit of components as specified by the manufacturer. These measurements are often made in thousandths and ten-thousandths of an inch.

Micrometers

◆ See Figures 64 and 65

A micrometer is an instrument made up of a precisely machined spindle that is rotated in a fixed nut, opening and closing the distance between the end of the spindle and a fixed anvil. When measuring using a micrometer, don't overtighten the tool on the part as either the component or tool may be damaged, and either way, an incorrect reading will result. Most micrometers are equipped with some form of thumbwheel on the spindle that is designed to freewheel over a certain light touch (automatically adjusting the spindle and preventing it from over-tightening).

Outside micrometers can be used to check the thickness of parts such shims or the outside diameter of components like the crankshaft journals. They are also used during many rebuild and repair procedures to measure the diameter of components such as the pistons. The most common type of micrometer reads in 1/1000 of an inch. Micrometers that use a vernier scale can estimate to 1/10 of an inch.

Inside micrometers are used to measure the distance between two parallel surfaces. For example, in powerhead rebuilding work, the "inside mike" measures cylinder bore wear and taper. Inside mikes are graduated the same way as outside mikes and are read the same way as well.

Remember that an inside mike must be absolutely perpendicular to the work being measured. When you measure with an inside mike, rock the mike gently from side to side and tip it back and forth slightly so that you span the widest part of the bore. Just to be on the safe side, take several readings. It takes a certain amount of experience to work any mike with confidence.

Metric micrometers are read in the same way as inch micrometers, except that the measurements are in millimeters. Each line on the main scale equals 1mm. Each fifth line is stamped 5, 10, 15 and so on. Each line on the thimble scale equals 0.01 mm. It will take a little practice, but if you can read an inch mike, you can read a metric mike.

Calipers

◆ See Figures 66, 67 and 68

Inside and outside calipers are useful devices to have if you need to measure something quickly and absolute precise measurement is not necessary. Simply take the reading and then hold the calipers on an accurate steel rule. Calipers, like micrometers, will often contain a thumbwheel to help ensure accurate measurement.

DIAL INDICATORS

◆ See Figure 69

A dial indicator is a gauge that utilizes a dial face and a needle to register measurements. There is a movable contact arm on the dial indicator. When the arm moves, the needle rotates on the dial. Dial indicators are calibrated to show readings in thousandths of an inch and typically, are used to measure end-play and run-out on various shafts and other components.

Dial indicators are quite easy to use, although they are relatively expensive. A variety of mounting devices are available so that the indicator can be used in a number of situations. Make certain that the contact arm is always parallel to the movement of the work being measured.

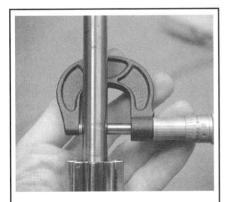

Fig. 64 Outside micrometers measure thickness, like shims or a shaft diameter

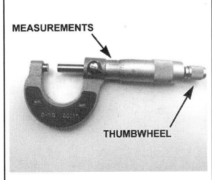

Fig. 65 Be careful not to over-tighten the micrometers always use the thumbwheel

Fig. 66 Calipers are the fast and easy way to make precise measurements

1-26 GENERAL INFORMATION, SAFETY AND TOOLS

Fig. 67 Calipers can also be used to measure depth . . .

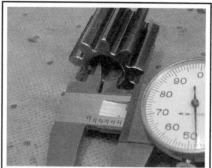

Fig. 68 . . . and inside diameter measurements, to 0.001 in. accuracy

Fig. 69 This dial indicator is measuring the end-play of a crankshaft during a powerhead rebuild

TELESCOPING GAUGES

◆ See Figure 70

A telescope gauge is really only used during rebuilding procedures (NOT during basic maintenance or routine repairs) to measure the inside of bores. It can take the place of an inside mike for some of these jobs. Simply insert the gauge in the hole to be measured and lock the plungers after they have contacted the walls. Remove the tool and measure across the plungers with an outside micrometer.

DEPTH GAUGES

◆ See Figure 71

A depth gauge can be inserted into a bore or other small hole to determine exactly how deep it is. One common use for a depth gauge is measuring the distance the piston sits below the deck of the block at top dead center. Some outside calipers contain a built-in depth gauge so you can save money and buy just one tool.

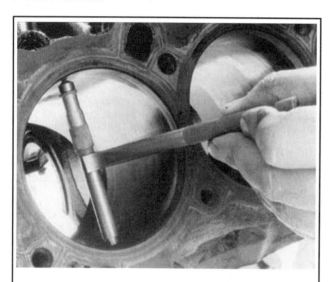

Fig. 70 Telescoping gauges are used during powerhead rebuilding procedures to measure the inside diameter of bores

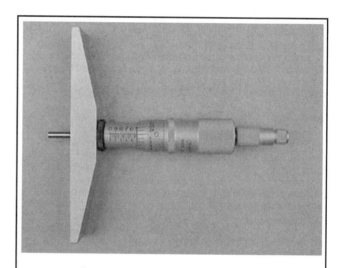

Fig. 71 Depth gauges are used to measure the depth of bore or other small holes

GENERAL INFORMATION, SAFETY AND TOOLS

FASTENERS, MEASUREMENTS AND CONVERSIONS

Bolts, Nuts and Other Threaded Retainers

◆ See Figures 72 and 73

Although there are a great variety of fasteners found in the modern boat engine, the most commonly used retainer is the threaded fastener (nuts, bolts, screws, studs, etc). Most threaded retainers may be reused, provided that they are not damaged in use or during the repair.

■ Some retainers (such as stretch bolts or torque prevailing nuts) are designed to deform when tightened or in use and should not be reused.

Whenever possible, we will note any special retainers which should be replaced during a procedure. But you should always inspect the condition of a retainer when it is removed and you should replace any that show signs of damage. Check all threads for rust or corrosion that can increase the torque necessary to achieve the desired clamp load for which that fastener was originally selected. Additionally, be sure that the driver surface itself (on the fastener) is not compromised from rounding or other damage. In some cases a driver surface may become only partially rounded, allowing the driver to catch in only one direction. In many of these occurrences, a fastener may be installed and tightened, but the driver would not be able to grip and loosen the fastener again. (This could lead to frustration down the line should that component ever need to be disassembled again).

If you must replace a fastener, whether due to design or damage, you must always be sure to use the proper replacement. In all cases, a retainer of the same design, material and strength should be used. Markings on the heads of most bolts will help determine the proper strength of the fastener. The same material, thread and pitch must be selected to assure proper installation and safe operation of the motor afterwards.

Thread gauges are available to help measure a bolt or stud's thread. Most part or hardware stores keep gauges available to help you select the proper size. In a pinch, you can use another nut or bolt for a thread gauge. If the bolt you are replacing is not too badly damaged, you can select a match by finding another bolt that will thread in its place. If you find a nut that will thread properly onto the damaged bolt, then use that nut as a gauge to help select the replacement bolt. If however, the bolt you are replacing is so badly damaged (broken or drilled out) that its threads cannot be used as a gauge, you might start by looking for another bolt (from the same assembly or a similar location) which will thread into the damaged bolt's mounting. If so, the other bolt can be used to select a nut; the nut can then be used to select the replacement bolt.

In all cases, be absolutely sure you have selected the proper replacement. Don't be shy, you can always ask the store clerk for help.

✷✷ WARNING

Be aware that when you find a bolt with damaged threads, you may also find the nut or tapped bore into which it was threaded has also been damaged. If this is the case, you may have to drill and tap the hole, replace the nut or otherwise repair the threads. Never try to force a replacement bolt to fit into the damaged threads.

Torque

Torque is defined as the measurement of resistance to turning or rotating. It tends to twist a body about an axis of rotation. A common example of this would be tightening a threaded retainer such as a nut, bolt or screw. Measuring torque is one of the most common ways to help assure that a threaded retainer has been properly fastened.

When tightening a threaded fastener, torque is applied in three distinct areas, the head, the bearing surface and the clamp load. About 50 percent of the measured torque is used in overcoming bearing friction. This is the friction between the bearing surface of the bolt head, screw head or nut face and the base material or washer (the surface on which the fastener is rotating). Approximately 40 percent of the applied torque is used in overcoming thread friction. This leaves only about 10 percent of the applied torque to develop a useful clamp load (the force that holds a joint together). This means that friction can account for as much as 90 percent of the applied torque on a fastener.

Standard and Metric Measurements

Specifications are often used to help you determine the condition of various components, or to assist you in their installation. Some of the most common measurements include length (in. or cm/mm), torque (ft. lbs., inch lbs. or Nm) and pressure (psi, in. Hg, kPa or mm Hg).

In some cases, that value may not be conveniently measured with what is available in your toolbox. Luckily, many of the measuring devices that are available today will have two scales so U.S. or Metric measurements may easily be taken. If any of the various measuring tools that are available to you do not contain the same scale as listed in your specifications, use the conversion factors that are provided in the Specifications section to determine the proper value.

The conversion factor chart is used by taking the given specification and multiplying it by the necessary conversion factor. For instance, looking at the first line, if you have a measurement in inches such as "free-play should be 2 in." but your ruler reads only in millimeters, multiply 2 in. by the conversion factor of 25.4 to get the metric equivalent of 50.8mm. Likewise, if a specification was given only in a Metric measurement, for example in Newton Meters (Nm), then look at the center column first. If the measurement is 100 Nm, multiply it by the conversion factor of 0.738 to get 73.8 ft. lb

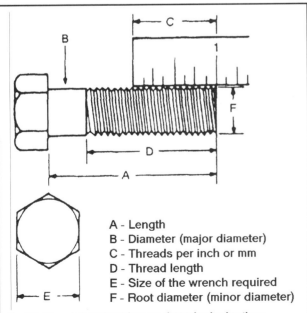

Fig. 72 Threaded retainer sizes are determined using these measurements

A - Length
B - Diameter (major diameter)
C - Threads per inch or mm
D - Thread length
E - Size of the wrench required
F - Root diameter (minor diameter)

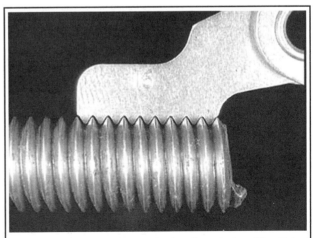

Fig. 73 Thread gauges measure the threads-per-inch and the pitch of a bolt or stud's threads

1-28 GENERAL INFORMATION, SAFETY AND TOOLS

SPECIFICATIONS

Metric Bolts

	4.6, 4.8						8.8			
Relative Strength Marking										
	Frequent Maximum Torque						Infrequent Maximum Torque			
Bolt Markings / Usage Bolt Size	Ft-Lb	Kgm	Nm				Ft-Lb	Kgm	Nm	
Thread Size x Pitch (mm)										
6 x 1.0	2–3	.2–.4	3–4				3–6	.4–.8	5–8	
8 x 1.25	6–8	.8–1	8–12				9–14	1.2–1.9	13–19	
10 x 1.25	12–17	1.5–2.3	16–23				20–29	2.7–4.0	27–39	
12 x 1.25	21–32	2.9–4.4	29–43				35–53	4.8–7.3	47–72	
14 x 1.5	35–52	4.8–7.1	48–70				57–85	7.8–11.7	77–110	
16 x 1.5	51–77	7.0–10.6	67–100				90–120	12.4–16.5	130–160	
18 x 1.5	74–110	10.2–15.1	100–150				130–170	17.9–23.4	180–230	
20 x 1.5	110–140	15.1–19.3	150–190				190–240	26.2–46.9	160–320	
22 x 1.5	150–190	22.0–26.2	200–260				250–320	34.5–44.1	340–430	
24 x 1.5	190–240	26.2–46.9	260–320				310–410	42.7–56.5	420–550	

SAE Bolts

SAE Grade Number	1 or 2			5			6 or 7		
Bolt Markings									
Manufacturers' marks may vary—number of lines always two less than the grade number.	Frequent Maximum Torque			Frequent Maximum Torque			Infrequent Maximum Torque		
Usage Bolt Size (Inches)—(Thread)	Ft-Lb	kgm	Nm	Ft-Lb	kgm	Nm	Ft-Lb	kgm	Nm
1/4 —20	5	0.7	6.8	8	1.1	10.8	10	1.4	13.5
—28	6	0.8	8.1	10	1.4	13.6			
5/16 —18	11	1.5	14.9	17	2.3	23.0	19	2.6	25.8
—24	13	1.8	17.6	19	2.6	25.7			
3/8 —16	18	2.5	24.4	31	4.3	42.0	34	4.7	46.0
—24	20	2.75	27.1	35	4.8	47.5			
7/16 —14	28	3.8	37.0	49	6.8	66.4	55	7.6	74.5
—20	30	4.2	40.7	55	7.6	74.5			
1/2 —13	39	5.4	52.8	75	10.4	101.7	85	11.75	115.2
—20	41	5.7	55.6	85	11.7	115.2			
9/16 —12	51	7.0	69.2	110	15.2	149.1	120	16.6	162.7
—18	55	7.6	74.5	120	16.6	162.7			
5/8 —11	83	11.5	112.5	150	20.7	203.3	167	23.0	226.5
—18	95	13.1	128.8	170	23.5	230.5			
3/4 —10	105	14.5	142.3	270	37.3	366.0	280	38.7	379.6
—16	115	15.9	155.9	295	40.8	400.0			
7/8 —9	160	22.1	216.9	395	54.6	535.5	440	60.9	596.5
—14	175	24.2	237.2	435	60.1	589.7			
1 —8	236	32.5	318.6	590	81.6	799.9	660	91.3	894.8
—14	250	34.6	338.9	660	91.3	849.8			

CONVERSION FACTORS

LENGTH-DISTANCE
- Inches (in.) x 25.4 = Millimeters (mm) x .0394 = Inches
- Feet (ft.) x .305 = Meters (m) x 3.281 = Feet
- Miles x 1.609 = Kilometers (km) x .0621 = Miles

VOLUME
- Cubic Inches (in3) x 16.387 = Cubic Centimeters x .061 = in3
- IMP Pints (IMP pt.) x .568 = Liters (L) x 1.76 = IMP pt.
- IMP Quarts (IMP qt.) x 1.137 = Liters (L) x .88 = IMP qt.
- IMP Gallons (IMP gal.) x 4.546 = Liters (L) x .22 = IMP gal.
- IMP Quarts (IMP qt.) x 1.201 = US Quarts (US qt.) x .833 = IMP qt.
- IMP Gallons (IMP gal.) x 1.201 = US Gallons (US gal.) x .833 = IMP gal.
- Fl. Ounces x 29.573 = Milliliters x .034 = Ounces
- US Pints (US pt.) x .473 = Liters (L) x 2.113 = Pints
- US Quarts (US qt.) x .946 = Liters (L) x 1.057 = Quarts
- US Gallons (US gal.) x 3.785 = Liters (L) x .264 = Gallons

MASS-WEIGHT
- Ounces (oz.) x 28.35 = Grams (g) x .035 = Ounces
- Pounds (lb.) x .454 = Kilograms (kg) x 2.205 = Pounds

PRESSURE
- Pounds Per Sq. In. (psi) x 6.895 = Kilopascals (kPa) x .145 = psi
- Inches of Mercury (Hg) x .4912 = psi x 2.036 = Hg
- Inches of Mercury (Hg) x 3.377 = Kilopascals (kPa) x .2961 = Hg
- Inches of Water (H₂O) x .07355 = Inches of Mercury x 13.783 = H₂O
- Inches of Water (H₂O) x .03613 = psi x 27.684 = H₂O
- Inches of Water (H₂O) x .248 = Kilopascals (kPa) x 4.026 = H₂O

TORQUE
- Pounds-Force Inches (in-lb) x .113 = Newton Meters (N·m) x 8.85 = in-lb
- Pounds-Force Feet (ft-lb) x 1.356 = Newton Meters (N·m) x .738 = ft-lb

VELOCITY
- Miles Per Hour (MPH) x 1.609 = Kilometers Per Hour (KPH) x .621 = MPH

POWER
- Horsepower (Hp) x .745 = Kilowatts x 1.34 = Horsepower

FUEL CONSUMPTION*
- Miles Per Gallon IMP (MPG) x .354 = Kilometers Per Liter (Km/L)
- Kilometers Per Liter (Km/L) x 2.352 = IMP MPG
- Miles Per Gallon US (MPG) x .425 = Kilometers Per Liter (Km/L)
- Kilometers Per Liter (Km/L) x 2.352 = US MPG

*It is common to covert from miles per gallon (mpg) to liters/100 kilometers (1/100 km), where mpg (IMP) x 1/100 km = 282 and mpg (US) x 1/100 km = 235.

TEMPERATURE
- Degree Fahrenheit (°F) = (°C x 1.8) + 32
- Degree Celsius (°C) = (°F − 32) x .56

2 MAINTENANCE & TUNE-UP

GENERAL INFORMATION	2-2
LUBRICATION	2-6
ENGINE MAINTENANCE	2-12
BOAT MAINTENANCE	2-32
TUNE-UP	2-36
TIMING AND SYNCHRONIZATION	2-49
VALVE CLEARANCE	2-94
STORAGE	2-108
CLEARING A SUBMERGED MOTOR	2-112
SPECIFICATIONS	2-113

BOAT MAINTENANCE 2-32
- BATTERIES 2-32
- FIBERGLASS HULL 2-35
- INTERIOR 2-35

CLEARING A SUBMERGED MOTOR 2-112

ENGINE MAINTENANCE 2-12
- ANODES (ZINCS) 2-29
- COOLING SYSTEM 2-13
- ENGINE COVERS 2-12
- ENGINE OIL AND FILTER 2-15
- FUEL FILTER 2-19
- JET DRIVE IMPELLER 2-28
- PROPELLER 2-21
  - REMOVAL & INSTALLATION 2-25
- TIMING BELT 2-31

GENERAL INFORMATION 2-2
- BEFORE/AFTER EACH USE 2-4
- ENGINE IDENTIFICATION 2-2
- MAINTENANCE COVERAGE IN THIS MANUAL 2-2
- MAINTENANCE EQUALS SAFETY 2-2
- OUTBOARDS ON SAIL BOATS 2-2

LUBRICATION 2-6
- GEARCASE (LOWER UNIT) OIL 2-9
- JET DRIVE BEARING 2-10
- LUBRICANTS 2-6
- LUBRICATING THE MOTOR 2-7
- LUBRICATION INSIDE THE BOAT 2-8
- POWER TRIM/TILT RESERVOIR 2-11

STORAGE 2-108
- RE-COMMISSIONING 2-111
- WINTERIZATION 2-108

TUNE-UP 2-36
- COMPRESSION TESTING 2-37
- ELECTRICAL SYSTEM CHECKS 2-45
- ELECTRONIC IGNITION SYSTEMS 2-45
- FUEL SYSTEM CHECKS 2-46
- IDLE SPEED ADJUSTMENT (MINOR ADJUSTMENTS) 2-48
- INTRODUCTION 2-36
- SPARK PLUG WIRES 2-43
- SPARK PLUGS 2-39
- TCI UNIT AIR GAP 2-44
- TUNE-UP SEQUENCE 2-36

TIMING AND SYNCHRONIZATION 2-49
- GENERAL INFORMATION 2-49
- PREPPING THE MOTOR 2-50
- SYNCHRONIZATION 2-50
- TIMING 2-49
- 2.5 HP YAMAHA 2-50
  - IDLE SPEED 2-50
  - IGNITION TIMING 2-50
- 4 HP YAMAHA 2-51
  - IDLE SPEED 2-51
  - IGNITION TIMING 2-52
  - STARTER LOCKOUT 2-51
  - THROTTLE CABLE 2-51
- 4/5/6 HP MERCURY/MARINER 2-52
  - IDLE MIXTURE 2-52
  - IDLE SPEED 2-52
  - IGNITION TIMING 2-53
  - TIMING POINTER 2-52
- 6/8 HP YAMAHA 2-54
  - IDLE SPEED 2-55
  - IGNITION TIMING 2-55
  - SHIFTER 2-54
  - STARTER LOCKOUT 2-55
  - THROTTLE AXLE LINK 2-54
  - THROTTLE CABLES 2-54
- 8/9.9 HP (232cc) YAMAHA 2-55
  - IDLE SPEED 2-56
  - IGNITION TIMING 2-59
  - NEUTRAL OPENING LIMIT 2-56
  - SHIFT ROD/CABLE 2-56
  - STARTER LOCKOUT 2-56
  - THROTTLE CABLE 2-57
  - THROTTLE LINK ADJUSTMENT 2-57
- 8/9.9 HP (232cc) MERCURY/MARINER ... 2-59
  - FULL THROTTLE LINK ROD 2-60
  - IDLE MIXTURE 2-60
  - IDLE SPEED 2-60
  - IGNITION TIMING 2-60
  - TILLER HANDLE 2-59
- 15 HP (323cc) YAMAHA 2-61
  - CONTROL LINKAGE 2-61
  - STARTER LOCKOUT 2-62
  - IDLE SPEED 2-62
  - IGNITION TIMING 2-62
- 9.9/15 HP (323cc) MERCURY/MARINER .. 2-63
  - IDLE SPEED 2-64
  - IGNITION TIMING 2-64
  - LINKAGE ADJUSTMENTS 2-63
- 25 HP 2-64
  - IDLE SPEED 2-65
  - IGNITION TIMING 2-66
  - LINKAGE 2-64
  - STARTER LOCKOUT 2-65
- 30/40 HP 3-CYL (CARB) 2-66
  - DASH POT 2-67
  - IDLE SPEED 2-66
  - IGNITION TIMING 2-68
  - SHIFT CABLE ADJUSTMENT 2-67
  - STARTER LOCKOUT 2-67
  - SYNCHRONIZATION 2-66
  - THROTTLE CABLE ADJUSTMENT ... 2-67
- 30/40 HP 3-CYL (EFI) & 40/50/60 HP 4-CYL (EFI) 2-68
  - IGNITION TIMING 2-69
  - THROTTLE & SHIFT CABLE 2-68
  - THROTTLE LINK IDLE SETTING .. 2-69
  - THROTTLE LINK WOT SETTING ... 2-69
- 40/45/50 HP (935cc) & 40/50/60 HP (996cc) 2-70
  - CARBURETOR 2-70
  - IGNITION TIMING 2-77
  - THROTTLE & SHIFT CABLES 2-70
- 75/80/90/100 HP 2-78
  - CARBURETOR 2-78
  - IGNITION TIMING 2-81
  - PICKUP TIMING 2-79
  - SHIFT CABLE 2-80
  - THROTTLE CABLE 2-81
- 115 HP MODELS 2-82
  - IGNITION TIMING 2-86
  - SHIFT CABLE 2-83
  - THROTTLE BODY PICKUP TIMING 2-82
  - THROTTLE CABLE 2-83
  - THROTTLE POSITION SENSOR 2-84
  - THROTTLE VALVE 2-84
- 150 HP YAMAHA 2-86
  - IGNITION TIMING 2-89
  - SHIFT CABLE 2-89
  - THROTTLE CABLE/LINK 2-87
  - THROTTLE POSITION SENSOR 2-86
  - THROTTLE VALVE 2-86
- 200/225 HP MODELS 2-90
  - IDLE SPEED 2-94
  - IGNITION TIMING 2-94
  - SHIFT CABLE 2-92
  - THROTTLE LINK & CABLE 2-91
  - THROTTLE VALVE & TPS 2-90

VALVE CLEARANCE 2-94
- VALVE LASH 2-94
- ADJUSTMENT 2-95

SPECIFICATIONS 2-113
- CAPACITIES 2-118
  - MERCURY/MARINER 2-118
  - YAMAHA 2-118
- GENERAL ENGINE 2-113
  - MERCURY/MARINER 2-113
  - YAMAHA 2-113
- LUBRICATION 2-117
  - MERCURY/MARINER 2-117
  - YAMAHA 2-117
- MAINTENANCE 2-116
- TUNE-UP 2-119
  - SPARK PLUG DIAGNOSIS 2-122
  - TUNE-UP - MERCURY/MARINER ... 2-120
  - TUNE-UP - YAMAHA 2-119
  - VALVE CLEARANCE - YAMAHA 2-121
  - VALVE CLEARANCE - MERCURY/MARINER 2-121

2-2 MAINTENANCE & TUNE-UP

GENERAL INFORMATION (WHAT EVERYONE SHOULD KNOW ABOUT MAINTENANCE)

At Seloc, we estimate that 75% of engine repair work can be directly or indirectly attributed to lack of proper care for the engine. This is especially true of care during the off-season period. There is no way on this green earth for a mechanical engine, particularly an outboard motor, to be left sitting idle for an extended period of time, say for six months, and then be ready for instant satisfactory service.

Imagine, if you will, leaving your car or truck for six months, and then expecting to turn the key, having it roar to life, and being able to drive off in the same manner as a daily occurrence.

Therefore it is critical for an outboard engine to either be run (at least once a month), preferably, in the water and properly maintained between uses or for it to be specifically prepared for storage and serviced again immediately before the start of the season.

Only through a regular maintenance program can the owner expect to receive long life and satisfactory performance at minimum cost.

Many times, if an outboard is not performing properly, the owner will "nurse" it through the season with good intentions of working on the unit once it is no longer being used. As with many New Year's resolutions, the good intentions are not completed and the outboard may lie for many months before the work is begun or the unit is taken to the marine shop for repair.

Imagine, if you will, the cause of the problem being a blown head gasket. And let us assume water has found its way into a cylinder. This water, allowed to remain over a long period of time, will do considerably more damage than it would have if the unit had been disassembled and the repair work performed immediately. Therefore, if an outboard is not functioning properly, do not stow it away with promises to get at it when you get time, because the work and expense will only get worse the longer corrective action is postponed. In the example of the blown head gasket, a relatively simple and inexpensive repair job could very well develop into major overhaul and rebuild work.

Maintenance Equals Safety

OK, perhaps no one thing that we do as boaters will protect us from risks involved with enjoying the wind and the water on a powerboat. But, each time we perform maintenance on our boat or motor, we increase the likelihood that we will find a potential hazard before it becomes a problem. Each time we inspect our boat and motor, we decrease the possibility that it could leave us stranded on the water.

In this way, performing boat and engine service is one of the most important ways that we, as boaters, can help protect ourselves, our boats, and the friends and family that we bring aboard.

Outboards On Sail Boats

Owners of sailboats pride themselves in their ability to use the wind to clear a harbor or for movement from Port A to Port B, or maybe just for a day sail on a lake. For some, the outboard is carried only as a last resort - in case the wind fails completely, or in an emergency situation or for ease of docking.

Therefore, in some cases, the outboard is stowed below, usually in a very poorly ventilated area, and subjected to moisture and stale air - in short, an excellent environment for "sweating" and corrosion.

If the owner could just take the time at least once every month, to pull out the outboard, clean it up, and give it a short run, not only would he/she have "peace of mind" knowing it will start in an emergency, but also maintenance costs will be drastically reduced.

Maintenance Coverage In This Manual

At Seloc, we strongly feel that every boat owner should pay close attention to this section. We also know that it is one of the most frequently used portions of our manuals. The material in this section is divided into sections to help simplify the process of maintenance. Be sure to read and thoroughly understand the various tasks that are necessary to keep your outboard in tip-top shape.

Topics covered in this section include:
1. General Information (What Everyone Should Know About Maintenance) - an introduction to the benefits and need for proper maintenance. A guide to tasks that should be performed before and after each use.

2. Lubrication Service - after the basic inspections that you should perform each time the motor is used, the most frequent form of periodic maintenance you will conduct will be the Lubrication Service. This section takes you through each of the various steps you must take to keep corrosion from slowly destroying your motor before your very eyes.

3. Engine Maintenance - the various procedures that must be performed on a regular basis in order to keep the motor and all of its various systems operating properly.

4. Boat Maintenance - the various procedures that must be performed on a regular basis in order to keep the boat hull and its accessories looking and working like new.

5. Tune-Up - also known as the pre-season tune-up, but don't let the name fool you. A complete tune-up is the best way to determine the condition of your outboard while also preparing it for hours and hours of hopefully trouble-free enjoyment. And if you use your boat enough during a single season, a second or even third tune-up could be required.

6. Winter Storage and Spring Commissioning Checklists - use these sections to guide you through the various parts of boat and motor maintenance that protect your valued boat through periods of storage and return it to operating condition when it is time to use it again.

7. Specification Charts - located at the end of the section are quick-reference, easy to read charts that provide you with critical information such as General Engine Specifications, Maintenance Intervals, Lubrication Service (intervals and lubricant types) and Capacities.

Engine Identification

ENGINE IDS THROUGHOUT THIS GUIDE

From 1995 to 2004 Yamaha produced more than 30 different 4-stroke models and Mercury/Mariner more than 20 different 4-stroke models just when looking at Horsepower, Gearcase Type and Fuel System. Other variances such as Manual vs. Electric, Tiller vs. Remote, or Manual Tilt vs. PTT might yield a dizzying array of models.

The GOOD news is that, when it all comes down to it you can identify motors by engine families. That is, you can usually identify a motor by the number of cylinders and size (in Cu. In or CCs) of the powerhead since there are so many similarities between models of a given family. And, when looked at it this way, there are only 13 Yamaha engine families and 10 Mercury/Mariner engine families. More importantly, of those 10 Merc engine families, NINE of them are shared with Yamaha!

Each manufacturer has unique single cylinder models, but shares most of the 2-4 cylinder inline engines and the V6 models.

In this service guide, we've included all of the 4-stroke models produced and sold worldwide by these companies between 1995 and 2004. We chose to do this because of the many similarities these motors have to each other. But, enough differences exist that many procedures will apply only to a sub-set of these motors. When this occurs, we'll either refer to the differences within a procedure or, if the differences are more significant, we'll break the motors out and give separate procedures. In order to prevent confusion, we try to sort and name the models in a way that is most easily understood.

■ **As we stated in the General Information, Safety & Tools section, under Why Combine Yamaha and Mercury/Mariner?, there are MANY similarities in parts, components, specifications and procedures on the common engine families sold by both manufacturers. And, although you'll have to determine for yourself when to use information tagged as being specific only to one or the other, we feel that most of the information is applicable to models from both manufacturers.**

Throughout this manual we will make reference to motors the easiest way possible. In some cases procedures may apply to Carbureted motors or all EFI motors. In other cases, they may apply to all 1-cylinder or all 2-cylinder motors (or all 3 or 4-cylinder motors, or perhaps all V6 motors, as applicable).

Like so many outboard manufacturers, we cannot sort engines by HP alone, since Yamaha and Merc both sold engines from multiple families under the same HP rating (such as the 40 hp 3-cyl and 40 hp 4-cyl models sold by both companies).

When it is necessary to distinguish between different types of motors that have the same number of cylinders, we'll differentiate using something else, like the HP rating or, since different motors may have the same rating, we'll use the HP rating plus the size (and in the case of the 747cc vs. 995cc 40 hp motor). In other instances, the HP rating and the number of cylinders (again that would be the case of the 3-cylinder vs. the 4-cylinder 40 hp motor) would be sufficient.

In most cases, the mechanical procedures will be similar or the same across different Hp ratings of the same engine family (of the same size). So it won't be uncommon to see a title or a procedure refer to 9.9/15 hp (323cc) motors or 75-100 hp (1596cc) motors. In both cases, we would be referring to all the motors of a particular family, including all models such as High Thrust (Yamaha) or Bigfoot (Merc), manual or electric start, remote or tiller control etc.

Starting in 2000 Yamaha began using fuel injection systems on some of their larger 4-stroke outboards, Mercury/Mariner followed suit by 2001. Currently, although each of these manufacturers has multiple fuel systems available on their 2-strokes, they both use a version one multi-port manifold Electronic Fuel Injection (EFI) system on their 4-stroke motors which we will refer to as EFI throughout this guide.

To help with proper engine identification, all of the engines covered here are listed in the General Engine and General Engine System Specifications charts for each of the manufacturers (one set for Yamaha and one for Mercury/Mariner) at the end of this section. In these charts, the engines are listed with their respective engine families, by horsepower rating, number of cylinders, engine type (inline or V), years of production and displacement (cubic inches and cubic centimeters or CCs).

■ **Most, but not all, charts throughout this guide come in two versions, one for Yamaha and another for Mercury/Mariner. You may wish to reference both to check for alternate specs.**

ENGINE MODEL & SERIAL NUMBERS

Yamaha

◆ See Figures 1 through 6

No matter whether you are trying to tell which version of a particular horsepower rated motor you have in order to follow the correct procedure or are trying to order replacement parts, the absolute best method in determining the year and model of the motor on which you are working is to start by referring to the engine serial number tag.

For all models covered here this ID tag (shown in the accompanying figures) is located on the side of the engine clamp or swivel/tilt brackets (port or starboard side depending upon the year and model). Most models are also equipped with a date of manufacture tag (located on the opposite side of the clamp or swivel/tilt bracket). Lastly, most models are also equipped with an Emissions Control Information label as well.

The engine model numbers are the manufacturer's key to engine changes. These alpha-numeric codes identify the year of manufacture, the horsepower rating, gearcase shaft length and various model/option differences (such as 4-stroke or High Thrust and starting/trim tilt options such as manual start/manual tilt or electric start power trim/tilt). If any correspondence or parts are required, the engine model number must be used for proper identification.

Remember that the model number establishes the model year for which the engine was produced, which is often not the year in which the motor was first installed on a boat. Also, keep in mind that a date of manufacture may be the year prior to the designated model year (as engines are often produced months before they are intended to be sold).

The engine model number tag also contains information used by the manufacturer internally as an engine family designation and a serial number (a unique sequential identifier given ONLY to that one motor).

When present, the emissions control information label states that the motor is in compliance with EPA emissions regulations for the model year of that engine. And, more importantly, it gives tune-up specifications that are vital to proper engine performance (that minimize harmful emissions). The specifications on this label may reflect changes that are made during production runs and are often not later reflected in a company's service literature. For this reason, specifications on the label always supersede those of a print or electronic manual. Typical specifications that are found on this label may include:

- Spark plug type and gap.
- Fuel recommendations.
- Idle speed settings

Fig. 1 A model ID tag, and often a date of manufacture tag, is found on the port . . .

Fig. 2 . . . and/or starboard side of most engine clamp or swivel/tilt brackets

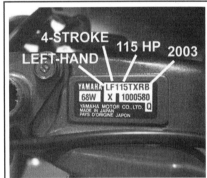

Fig. 3 The model ID tag provides critical information to identify and service the engine

Fig. 4 Keep in mind that the date of manufacture is often the year BEFORE the model year

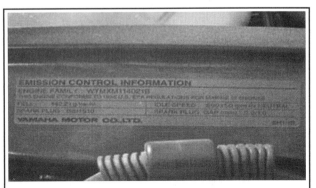

Fig. 5 When present, the emission control information label supersedes specifications listed elsewhere

2-4 MAINTENANCE & TUNE-UP

YAMAHA MODEL IDENTIFICATION DECODER

Model Description (1 or 2 digits / Alpha)	Prop Shaft HP (1-3 digits / Numerals)	Trim/Tilt Starting Method (1 digit - Alpha)	Shaft Length (1 digit - Alpha)	Method of Control (1 digit - Alpha)	Model Year (1 digit - Alpha)	Code Variant ① (1 digit / Numeral)
B = Inshore Series						
C = C Series	A number between	1986-90 ②	S = 15" (38cm)	1991-04 only	N = 1984	2 = 3 cyl (1997)
D = Twin Prop	2 = 2 hp	E = Electric Start		H = Tiller	K = 1985	
DX = Twin Prop / EFI	and		L = 20" (51cm)		J = 1986	3 = 3 cyl
E = Enduro Series	250 = 250 hp	T = Power Trim or		R = Remote	H = 1987	
F = 4-Stroke		Power Trim/Tilt	X = 25" (64cm)		G = 1988	1984-90
FT = 4-Stroke / High Thrust					F = 1989	JD = Jet Drive
L = Left-Hand Rotation		M = Tiller	U = 30" (76cm)		D = 1990	
LF = Left-Hand / 4-Stroke					P = 1991	
LX = Left-Hand / EFI		1991-04 ②	1991-04		Q = 1992	
LZ = Left-Hand / HPDI		E = Manual Tilt	J = Jet Drive		R = 1993	
P = Pro Series Model		w/ Electric Start			S = 1994	
PX = Pro Series / EFI					T = 1995	
S = Saltwater Series		M = Manual Tilt			U = 1996	
SX = Saltwater Series / EFI		w/ Manual Start			V = 1997	
T = 4-Stroke / High Thrust					W = 1998	
V = V-Max Series		T = Power Trim/Tilt			X = 1999	
VX = V-Max / EFI		w/ Electric Start			Y = 2000	
VZ = V-Max / HPDI					Z = 2001	
Z = HPDI		P = Power Tilt			A = 2002	
		w/ Electric Start			B = 2003	
					C = 2004	

① A special model variation code is used on 25 hp motors to distinquish between 2 cyl models (no code) and 3 cyl models (code present)
② For models through 1990 this portion of the code may be 1, 2 or 3 Alpha designators, all 1991 or later use only 1 digit at this part of the code

Fig. 6 Yamaha Model codes - 1986-04 models

- Possibly engine ignition timing (such as wide-open throttle and/or idle timing) specifications

Engines built since the 1991 model year contain a model code similar to that of earlier year production models, however there are more variations and therefore more possible codes. Most 1991 and later models will contain a 6-9 digit code. The code may or may not begin with a one or two digit alpha model description. This tells you what series (4-stroke, High Thrust, etc) to which the engine belongs. The next one, two or three digits will be numbers, representing the horsepower rating. The digit following the horsepower rating will be a single digit alpha code (E, M, T or P) identifying the starting and trim/tilt system on the motor. Following the starting and trim/tilt system identifier will be a single alpha identifier (S, L, X, U or J) representing gearcase shaft length (or type in the case of J for Jet Drive). Next, a single-digit, alpha identifier is used for the year. Finally, in some cases, a single check digit is used by the manufacturer to designate the 3-cylinder version of the 2-stroke 25 hp motor.

Refer to the accompanying illustration to interpret the various alpha digits found throughout the model code.

Mercury/Mariner

◆ See Figures 7 and 8

For many years Mercury/Mariner was a real headache for both us and the aftermarket. They paid very little attention to model year breaks, making changes or improvements to motors as they saw fit and simply noting in their literature that all models from a given serial number forward were now different.

We'd hate to think that the EPA has actually done us a favor, but in the late 90s, they required marine manufacturers to start affixing Emission Control Information Labels which INCLUDE the year of production for which the motor is certified. This helped our problems immensely (though Merc still does make rolling changes which require serial number identification, it just seems a little less prevalent these days).

No matter, in addition to the EPA label which can be found on most powerheads, ALL Mercury/Mariner 4-stroke motors also contain a Serial Number/Model Year ID tag on the starboard side of the transom mounting bracket.

This label first and foremost contains that all important Serial Number (which may be required for ID or for parts purchases at some point, so copy it down now).

Right under the serial number the label displays the model year. Directly below the model year is the Model Code (description).

Toward the bottom of the label includes information such as the year of manufacture (as it could be different than the model year) and the Certified Europe (CE) insignia.

In between, the label may contain other useful information such as engine weight, rated power output and operating parameters like Max engine RPM.

Before/After Each Use

As stated earlier, the best means of extending engine life and helping to protect yourself while on the water is to pay close attention to boat/engine maintenance. This starts with an inspection of systems and components before and after each time you use your boat.

A list of checks, inspections or required maintenance can be found in the Maintenance Intervals Chart at the end of this section. Some of these inspections or tasks are performed before the boat is launched, some only after it is retrieved and the rest, both times.

VISUALLY INSPECTING THE BOAT AND MOTOR

◆ See Figures 9 and 10

Both before each launch and immediately after each retrieval, visually inspect the boat and motor as follows:

1. **Check the fuel and oil levels** according to the procedures in this manual. Do NOT launch a boat without properly topped off fuel and crankcase oil. It is not worth the risk of getting stranded or of damage to the

MAINTENANCE & TUNE-UP 2-5

Fig. 7 A model ID tag is found on the starboard side of Mercury/Mariner motors

Fig. 8 The tag contains the Serial Number, Model Year, Model Code and other useful information

motor. Likewise, upon retrieval, check the oil and fuel levels while it is still fresh in your mind. This is a good way to track fuel consumption (one indication of engine performance). Oil consumption should be minimal (though usually notable during break-in), but all 4-stroke engines allow a small portion of oil to burn. Watch for sudden increases in the amount of oil burned and investigate further if found.

2. **Check for signs of fuel or oil leakage.** Probably as important as making sure enough fuel and oil is onboard, is the need to make sure that no dangerous conditions might arise due to leaks. Thoroughly check all hoses, fittings and tanks for signs of leakage. Oil leaks may cause the boat to become stranded, or worse, could destroy the motor if undetected for a significant amount of time. Fuel leaks can cause a fire hazard, or worse, an explosive condition. This check is not only about properly maintaining your boat and motor, but about helping to protect your life.

✱✱ CAUTION

On fuel injected motors fuel is pumped at high pressure through various lines under the motor cowl. The smallest leak will allow for fuel to spray in a fine, atomized and highly combustible stream from the damaged hose/fitting. It is critical that you remove the cowling and turn the key to the ON position (to energize the fuel pump and begin building system pressure) for a quick check before starting the motor (to ensure that no leaks are present). Even so, leaks may not show until the motor is operating so it is a good idea to either leave the cover off until the motor is running or to remove it again later in the day to double-check that you are leak free. Of course, if you DO remove the cover with the engine running, take GREAT care to prevent contact with any moving parts.

■ On V6 4-stroke motors, make a visual check for water in the fuel filter/strainer. There is a red ring in the filter unit on these models which will float if water is present. If so, remove the filter cup and drain the water. For details, please refer to Fuel Filter in this section.

3. **Inspect the boat hull and engine cases** for signs of corrosion or damage. Don't launch a damaged boat or motor. And don't surprise yourself dockside or at the launch ramp by discovering damage that went unnoticed last time the boat was retrieved. Repair any hull or case damage now.

4. **Check the battery** connections to make sure they are clean and tight. A loose or corroded connection will cause charging problems (damaging the system or preventing charging). There's only one thing worse than a dead battery dockside/launch ramp and that's a dead battery in the middle of a bay, river or worse, the ocean. Whenever possible, make a quick visual check of battery electrolyte levels (keeping an eye on the level will give some warning of overcharging problems). This is especially true if the

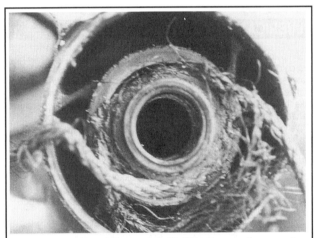

Fig. 9 Rope or fishing line entangled behind the propeller can cut through the seal, allowing water in or lubricant out

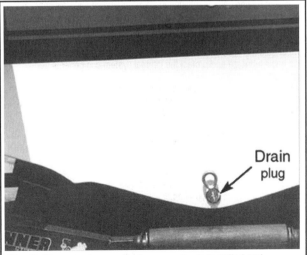

Fig. 10 Always make sure the transom plug is installed and tightened securely before a launch

2-6 MAINTENANCE & TUNE-UP

engine is operated at high speeds for extended periods of time.

5. **Check the propeller (impeller on jet drives) and gearcase.** Make sure the propeller shows no signs of damage. A broken or bent propeller may allow the engine to over-rev and it will certainly waste fuel. The gearcase should be checked before and after each use for signs of leakage. Check the gearcase oil for signs of contamination if any leakage is noted. Also, visually check behind the propeller for signs of entangled rope or fishing lines that could cut through the lower gearcase propeller shaft seal. This is a common cause of gearcase lubricant leakage, and eventually, water contamination that can lead to gearcase failure. Even if no gearcase leakage is noted when the boat is first retrieved, check again next time before launching. A nicked seal might not seep fluid right away when still swollen from heat immediately after use, but might begin seeping over the next day, week or month as it sat, cooled and dried out.

6. **Check all accessible fasteners for tightness.** Make sure all easily accessible fasteners appear to be tight. This is especially true for the propeller nut, any anode retaining bolts, all steering or throttle linkage fasteners and the engine clamps or mounting bolts. Don't risk loosing control or becoming stranded due to loose fasteners. Perform these checks before heading out, and immediately after you return (so you'll know if anything needs to be serviced before you want to launch again.)

7. **Check operation of all controls including the throttle/shifter, steering and emergency stop/start switch and/or safety lanyard.** Before launching, make sure that all linkage and steering components operate properly and move smoothly through their range of motion. All electrical switches (such as power trim/tilt) and especially the emergency stop system(s) must be in proper working order. While underway, watch for signs that a system is not working or has become damaged. With the steering, shifter or throttle, keep a watchful eye out for a change in resistance or the start of jerky/notchy movement.

8. **Check the water pump intake grate and water indicator.** The water pump intake grate should be clean and undamaged before setting out. Remember that a damaged grate could allow debris into the system that could destroy the impeller or clog cooling passages. Once underway, make sure the cooling indicator stream is visible at all times. Make periodic checks, including one final check before the motor is shut down each time. If a cooling indicator stream is not present at any point, troubleshoot the problem before further engine operation.

9. **If used in salt, brackish or polluted waters thoroughly rinse the engine (and hull), then flush the cooling system** according to the procedure in this section.

■ **Keep in mind that the cooling system can use attention, even if used in fresh waters. Sand, silt or other deposits can help clog passages, chemicals or pollutants can speed corrosion. It's a good idea to flush your motor after every use, regardless of where you use it.**

10. **Visually inspect all anodes** after each use for signs of wear, damage or to make sure they just plain didn't fall off (especially if you weren't careful about checking all the accessible fasteners the last time you launched).

11. **For Pete's sake, make sure the plug is in!** We shouldn't have to say it, but unfortunately we do. If you've been boating for any length of time, you've seen or heard of someone whose backed a trailer down a launch ramp, forgetting to check the transom drain plug before submerging (literally) the boat. Always make sure the transom plug is installed and tight before a launch.

LUBRICATION

An outboard motor's greatest enemy is corrosion. Face it, oil and water just don't mix and, as anyone who has visited a junkyard knows, metal and water aren't the greatest of friends either. To expose an engine to a harsh marine environment of water and wind is to expect that these elements will take their toll over time. But, there is a way to fight back and help prevent the natural process of corrosion that will destroy your beloved boat motor.

Various marine grade lubricants are available that serve two important functions in preserving your motor. Lubricants reduce friction on metal-to-metal contact surfaces and, they also displace air and moisture, therefore slowing or preventing corrosion damage. Periodic lubrication services are your best method of preserving an outboard motor. Marine lubricants are designed for the harsh environment to which outboards are exposed and are designed to stay in place, even when submerged in water (and we've got some bathing suits with marine grease stains to attest to this).

Lubrication takes place through various forms. For all engines, internal moving parts are lubricated by the engine oil which is held in an oil sump and pumped through oil passages. The gear oil for all motors and the engine crankcase oil should be periodically checked and replaced following the appropriate Engine Maintenance procedures. Perform these services based on time or engine use, as outlined in the Maintenance Intervals chart at the end of this section.

For motors equipped with power trim/tilt, the fluid level and condition in the reservoir should be checked periodically to ensure proper operation. Proper fluid level not only ensures that the system will function properly, but also helps lubricate and protect the internal system components from corrosion.

Most other forms of lubrication occur through the application of grease (Yamaha all-purpose Marine grease or one of Yamaha's Marine Greases such as Mercury Quicksilver Anti-Corrosion Grease, Mercury 2-4-C Lubricant with Teflon or Mercury Special Lubricant 101), either applied by hand (an old toothbrush can be helpful in preventing a mess) or using a grease gun to pump the lubricant into grease fittings (also known as zerk fittings). When using a grease gun, do not pump excessive amounts of grease into the fitting. Unless otherwise directed, pump until either the rubber seal (if used) begins to expand or until the grease just begins to seep from the joints of the component being lubricated (if no seal is used).

To ensure your motor is getting the protection it needs, perform a visual inspection of the various lubrication points at least once a week during regular seasonal operation (this assumes that the motor is being used at least once a week). Follow the recommendations given in the Lubrication Chart at the end of this section and perform the various lubricating services at least every 60 days when the boat is operated in fresh water or every 30 days when the boat is operated in salt, brackish or polluted waters. We said **at least** meaning you should perform these services more often, if a need is discovered by your weekly inspections.

■ **Jet drive models require one form of lubrication EVERY time that they are used. The jet drive bearing should be greased, following the procedure given in this section, after every day of boating. But don't worry, it only takes a minute once you've done it before.**

Lubricants

◆ See Figures 11 and 12

✱✱ WARNING

Yamaha lubricants recommended in the lubrication procedures are NOT interchangeable as each is designed to perform under different conditions. Mercury however seems to have 3 types of grease which MAY be interchangeable for different things. The Anti-Corrosion Grease, 2-4-C with Teflon and the Special 101 are often recommended interchangeably on one motor or with different lubricants serving the same purpose on different motors. Refer to the Mercury Lubrication Chart for more details.

All Purpose Marine Grease

Yamaha All-purpose marine grease and the 3 Mercury greases listed earlier are general outboard lubricants/greases, chemically formulated to resist salt water. They are recommended for application to bearings, bushings, and oil seals.

Gearcase Lubricant

Yamaha or Yamalube Gearcase Lubricant and Mercury Premium Blend Gear Lube contain high viscosity additives to protect the lower unit gears at high speed operation. The lubricant will extend gear life, reduce gear noise, minimize friction, and has a cooling affect on the lower unit moving parts.

Power Trim and Tilt Fluid

Yamaha or Yamalube power trim and tilt fluid and Quicksilver Power Trim and Steering Fluid are both highly refined hydraulic fluids. These products

MAINTENANCE & TUNE-UP 2-7

Fig. 11 Yamaha recommended lubricants and additives will not only keep the unit within the limits of the warranty but will also contribute to dependable performance and reduced maintenance costs

have a high detergent content and additives to keep seals pliable. A high grade automatic transmission fluid, Dexron® II, may also be used if the recommended fluid is not available.

Lubricating the Motor

◆ See Figures 13 thru 19

The first thing you should do upon purchasing a new or "new to you" motor is to remove the engine top cover and look for signs of grease. Note all components that have been freshly greased (or if the motor has been neglected that shows signs of wear or dirt/contamination that has collected on the remnants of old applied grease). If the motor shows signs of dirt, corrosion or wear, clean those components thoroughly and apply a fresh coat of grease.

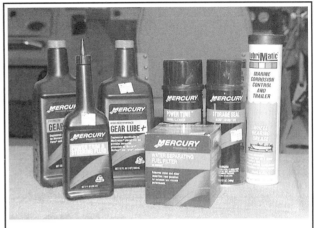

Fig. 12 Likewise, Mercury products are available for the proper care of your outboard

Thereafter, follow the recommendations in the accompanying lubrication charts (as they apply to the motor on which you are working) and grease all necessary surfaces regularly to keep them clean and well lubricated. As a general rule of thumb any point where two metallic mechanical parts connect and push, pull, turn, slide, pivot on each other should be greased. For most motors this will include shift and throttle cables and/or linkage, steering and swiveling points and items such as the cowl clamp bolts (on smaller motors) and top cover or cowling clamp levers.

■ For more information on greasing and lubrication points, check your owner's manual. Most owner's manuals will provide one or more illustrations to help you properly identify all necessary greasing points.

Points such as the swivel bracket and/or the tilt tube will normally be equipped with grease (zerk) fittings. For these, use a grease gun to carefully

Fig. 13 Lubrication points include all rotating...

Fig. 14 ... or sliding linkage points and cable ends

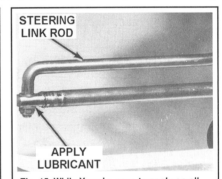

Fig. 15 While Yamaha says to apply an all purpose Marine Grease to the steering link rod joint ends Mercury recommends a light weight oil (both say to grease the link rod shaft itself however)

Fig. 16 The swivel bracket usually contains one or two grease fittings

Fig. 17 The propeller should be removed periodically to clean and re-grease the prop shaft splines

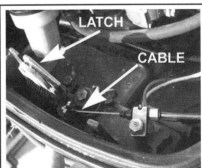

Fig.18 Any cowling clamp latches and/or cables should also be lubricated with an all purpose Marine Grease

2-8 MAINTENANCE & TUNE-UP

pump small amounts of grease into the fittings, displacing some of the older grease and lubricating the internal surfaces of the swivel and tilt tubes.

■ **Some models, mostly Mercs, don't contain grease fittings on the tilt tube, and for these motors you should apply a small amount of light weight engine oil, as it will more easily run down into the tube and lubricate the surface than would a grease.**

Some engine cowl levers require a dab of lubricant be applied manually over sliding surfaces, but many are also equipped with grease fittings for lubrication using a grease gun. A few of the larger outboards actually use a cable release system for the cowling, the cable ends and latches should all be greased periodically to prevent binding and wear.

Items without a grease fitting, such as the steering ram, cable ends, shifter and carburetor/throttle body linkage all must normally be greased by hand using a small dab of lubricant. Be sure not to over apply grease as it is just going to get over everything and exposed grease will tend to attract and hold dirt or other particles of general crud. For this reason it is always a good idea to wipe away the old grease before applying fresh lubricant to these surfaces.

Lubrication Inside the Boat

The following points inside the boat will also usually benefit from lubrication with a suitable marine grade grease:
- Remote control cable ends next to the hand nut. DO NOT over-lubricate the cable.
- Steering arm pivot socket.
- Exposed shaft of the cable passing through the cable guide tube.
- Steering link rod to steering cable.

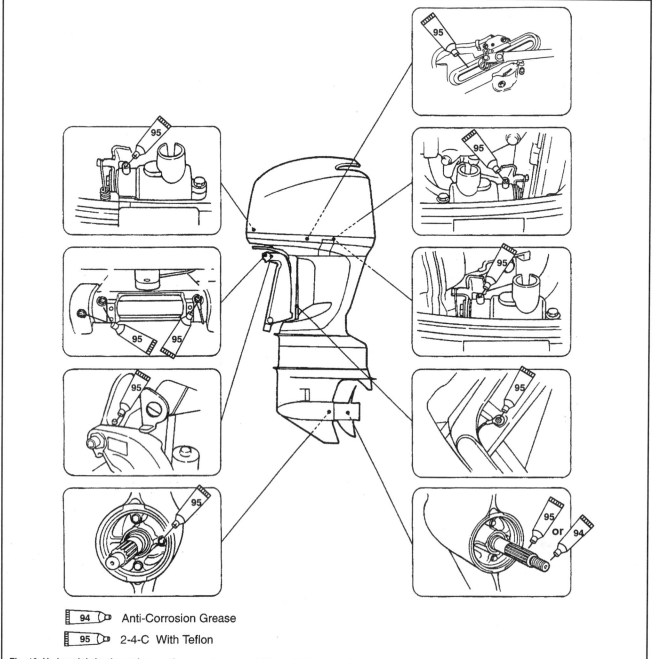

94 — Anti-Corrosion Grease
95 — 2-4-C With Teflon

Fig. 19 Various lubrication points on the powerhead should be maintained regularly to ensure a long service life (200/225 V6 shown, others similar)

MAINTENANCE & TUNE-UP

Gearcase (Lower Unit) Oil

◆ See Figures 20 and 21

Regular maintenance and inspection of the lower unit is critical for proper operation and reliability. A lower unit can quickly fail if it becomes heavily contaminated with water or excessively low on oil. The most common cause of a lower unit failure is water contamination.

Water in the lower unit is usually caused by fishing line or other foreign material, becoming entangled around the propeller shaft and damaging the seal. If the line is not removed, it will eventually cut the propeller shaft seal and allow water to enter the lower unit. Fishing line has also been known to cut a groove in the propeller shaft if left neglected over time. This area should be checked frequently.

OIL RECOMMENDATIONS

Use only Yamaha Gear Case Lube or an equivalent SAE-90W gearcase oil in Yamahas or Quicksilver Gear Lube - Premium Blend in Mercury/Mariner motors.

■ **Remember, it is this lower unit lubricant that prevents corrosion and lubricates the internal parts of the drive gears. Lack of lubrication due to water contamination or the improper type of oil can cause catastrophic lower unit failure.**

CHECKING GEARCASE OIL LEVEL & CONDITION

◆ See Figures 22 and 23

Visually inspect the gearcase before and after each use for signs of leakage. At least monthly, or as needed, remove the gearcase level plug in order to check the lubricant level and condition as follows:

1. Position the engine in the upright position with the motor shut off for at least 1 hour. Whenever possible, checking the level overnight cold will give a true indication of the level without having to account for heat expansion.
2. Disconnect the negative battery cable or remove the propeller for safety.

✳✳ CAUTION
Always observe extreme care when working anywhere near the propeller. Take steps to ensure that no accidental attempt to start the engine occurs while work is being performed or remove the propeller completely to be safe.

3. Position a small drain pan under the gearcase, then unthread the drain/filler plug at the bottom of the housing and allow a small sample (a teaspoon or less) to drain from the gearcase. Quickly install the drain/filler plug and tighten securely.
4. Examine the gear oil as follows:
    a. Visually check the oil for obvious signs of water. A small amount of moisture may be present from condensation, especially if a motor has been stored for some time, but a milky appearance indicates that either the fluid has not been changed in ages or the gearcase allowing some water to intrude. If significant water contamination is present, the first suspect is the propeller shaft seal.
    b. Dip an otherwise clean finger into the oil, then rub a small amount of the fluid between your finger and your thumb to check for the presence of debris. The lubricant should feel smooth. A **very** small amount of metallic shavings may be present, but should not really be felt. Large amounts of grit or metallic particles indicate the need to overhaul the gearcase looking for damaged/worn gears, shafts, bearings or thrust surfaces.

■ **If a large amount of lubricant escapes when the level/vent plug is removed in the next step, either the gearcase was seriously overfilled on the last service, the crankcase is still too hot from running the motor in gear (and the fluid is expanded) or a large amount of water has entered the gearcase. If the later is true, some water should escape before the oil and/or the oil will be a milky white in appearance (showing the moisture contamination).**

5. Next, remove the combination level/vent plug OR oil level plug (if separate) from the top of the gearcase and ensure the lubricant level is up to the bottom of the level/vent plug opening. A very small amount of fluid may be added through the level plug, but larger amounts of fluid should be added through the drain/filler plug opening to make certain that the case is properly filled. If necessary, add gear oil until fluid flows from the level/vent opening. If much more than 1 oz. (29 ml) is required to fill the gearcase, check the case carefully for leaks. Install any drain/filler plugs and/or the level/vent plug which were removed and tighten them securely.

Fig. 20 This lower unit was destroyed because the bearing carrier froze due to lack of lubrication

Fig. 21 Fishing line entangled behind the prop can actually cut through the seal

Fig. 22 Yamaha and Merc often label the vent/level plugs on their gearcases

2-10 MAINTENANCE & TUNE-UP

Fig. 23 Many models, like this Mercury have a separate vent and level plug

3. Remove the drain/filler plug from the lower end of the gear housing followed by the oil level/vent plug.
4. Allow the lubricant to completely drain from the lower unit.
5. If applicable, check the magnet end of the drain screw for metal particles. Some amount of metal is considered normal wear is to be expected but if there are signs of metal chips or excessive metal particles, the gearcase needs to be disassembled and inspected.
6. Inspect the lubricant for the presence of a milky white substance, water or metallic particles. If any of these conditions are present, the lower unit should be serviced immediately.
7. Place the outboard in the proper position for filling the lower unit (straight up and down). The lower unit should not list to either port or starboard and should be completely vertical.
8. Insert the lubricant tube into the oil drain hole at the bottom of the lower unit and inject lubricant until the excess begins to come out the oil level hole.

■ The lubricant must be filled from the bottom to prevent air from being trapped in the lower unit. Air displaces lubricant and can cause a lack of lubrication or a false lubricant level in the lower unit.

9. Oil should be squeezed in using a tube or with the larger quantities, by using a pump kit to fill the gearcase through the drain plug.

■ One trick that makes adding gearcase oil less messy is to install the level/vent plug BEFORE removing the pump from the drain/filler opening and threading the drain/filler plug back into position.

10. Using new gaskets/washers (if equipped) install the oil level/vent plug(s) first, then install the oil fill plug.
11. Wipe the excess oil from the lower unit and inspect the unit for leaks.
12. Place the used lubricant in a suitable container for transportation to an authorized recycling facility.

Jet Drive Bearing

◆ See Figure 27

Jet drive models covered by this manual require special attention to ensure that the driveshaft bearing remains properly lubricated.
The manufacturer recommends that you lubricate the jet drive bearing using a grease gun after EACH days use. However, at an absolute minimum, use the grease fitting every 10 hours (in fresh water) or 5 hours (in salt water). Also, after every 50 hours of fresh water operation or every 25 hours of salt/brackish/polluted water operation, the drive bearing grease must be replaced. Follow the appropriate procedure:

RECOMMENDED LUBRICANT

Use a suitable water-resistant NLGI No. 1 lubricant (Marine Grade grease).

■ One trick that makes adding gearcase oil less messy is to install the level/vent plug(s) BEFORE removing the pump from the drain/filler opening and threading the drain/filler plug back into position.

6. Once fluid is pumped into the gearcase, let the unit sit in a shaded area for at least 1 hour for the fluid to settle. Recheck the fluid level and, if necessary, add more lubricant.
7. Install the propeller and/or connect the negative battery cable, as applicable.

DRAINING AND FILLING

◆ See Figures 24, 25 and 26

1. Place a suitable container under the lower unit. It is usually a good idea to place the outboard in the tilted position so the drain plug is at the lowest position on the gearcase, this will help ensure the oil drains fully.
2. Loosen the oil level/vent plug(s) on the lower unit. This step is important! If the oil level/vent plug(s) cannot be loosened or removed, you cannot refill the gearcase with fluid and purge it of air.

■ Never remove the vent or filler plugs when the lower unit is hot. Expanded lubricant will be released through the hole.

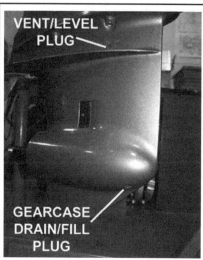

Fig. 24 The gearcase oil vent/level plug(s) is(are) on top, while the drain/fill plug is on the bottom

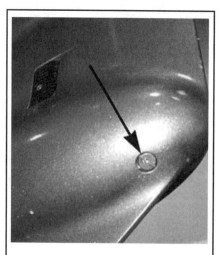

Fig. 25 Better view of a typical gearcase drain/fill plug (designed for a large flat-head screwdriver)

Fig. 26 When draining the lower unit, ensure it is fully tilted and a drain pan of adequate capacity is place to catch the lubricant

MAINTENANCE & TUNE-UP 2-11

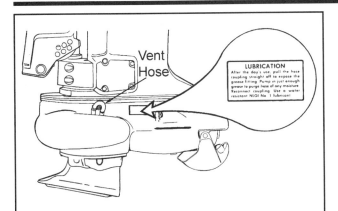

Fig. 27 Jet drive models require lubrication of the bearing after each day of use, sometimes they are equipped with a label on the housing to remind the owner

DAILY BEARING LUBRICATION

◆ See Figures 28 and 29

A grease fitting is located under a vent hose on the lower port side of the jet drive. Disconnect the hose from the fitting, then use a grease gun to apply enough grease to the fitting to **just** fill the vent hose. Basically, grease is pumped into the fitting until the old grease just starts to come out from the passages through the hose coupling, then reconnect the hose to the fitting.

■ Do not attempt to just grasp the vent hose and pull, as it is a tight fit and when it does come off, you'll probably go flying if you didn't prepare for it. The easier method of removing the vent hose from the fitting is to deflect the hose to one side and snap it free from the fitting.

GREASE REPLACEMENT

◆ See Figures 28, 29 and 30

A grease fitting is located under a vent hose on the lower port side of the jet drive. This grease fitting is utilized at the end of each day's use to add fresh grease to the jet drive bearing. But, every 50 or 25 hours and/or 30 or 15 days (depending if use is in fresh or salt/brackish/polluted waters), the grease should be completely replaced. This is very similar to the daily greasing, except that a lot more grease it used. Disconnect the hose from the fitting (by deflecting it to the side until it snaps free from the fitting), then use a grease gun to apply enough grease to the fitting until grease exiting the assembly fills the vent hose. Then, continue to pump grease into the fitting to force out all of the old grease (you can tell this has been accomplished when fresh grease starts to come out of the vent instead of old grease, which will be slightly darker due to minor contamination from normal use). When nothing but fresh grease comes out of the vent the fresh grease has completely displaced the old grease and you are finished. Be sure to securely connect the vent hose to the fitting.

Each time this is performed, inspect the grease for signs of moisture contamination or discoloration. A gradual increase in moisture content over a few services is a sign of seal wear that is beginning to allow some seepage. Very dark or dirty grease may indicate a worn seal (inspect and/or replace the seal, as necessary to prevent severe engine damage should the seal fail completely).

■ Keep in mind that some discoloration of the grease is expected when a new seal is broken-in. The discoloration should go away gradually after one or two additional grease replacement services.

Whenever the jet drive bearing grease is replaced, take a few minutes to apply some of that same water-resistant marine grease to the pivot points of the jet linkage.

Power Trim/Tilt Reservoir

RECOMMENDED LUBRICANT

Yamaha or Yamalube power trim and tilt fluid and Quicksilver Power Trim and Steering Fluid are both highly refined hydraulic fluids. These products have a high detergent content and additives to keep seals pliable. A high grade automatic transmission fluid, Dexron® II, may also be used if the recommended fluid is not available.

CHECKING FLUID LEVEL/CONDITION

◆ See Figures 31, 32 and 33

The fluid in the power trim/tilt reservoir should be checked periodically to ensure it is full and is not contaminated. To check the fluid, tilt the motor upward to the full tilt position, then manually engage the tilt support for safety and to prevent damage. Loosen and remove the filler plug using a suitable socket or wrench and make a visual inspection of the fluid. It should seem clear and not milky. The level is proper if, with the motor at full tilt, the level is even with the bottom of the filler plug hole.

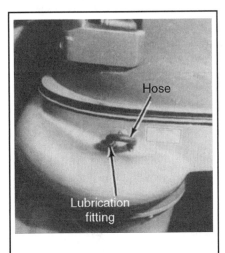

Fig. 28 The jet drive lubrication fitting is found under the vent hose

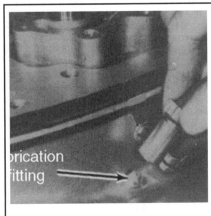

Fig. 29 Attach a grease gun to the fitting for lubrication

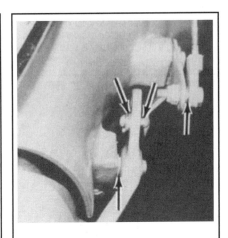

Fig. 30 Also, coat the pivot points of the jet linkage with grease periodically

2-12 MAINTENANCE & TUNE-UP

Fig. 31 On most medium sized motors, the PTT systems use only a single trim/tilt rod

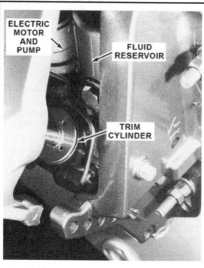

Fig. 32 More common, is the PTT system which uses one tilt and two trim rods. .

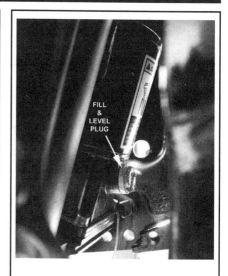

Fig. 33 . . . the fill plug is on the front of the reservoir

ENGINE MAINTENANCE

Engine Covers (Top Cover and Cowling)

REMOVAL & INSTALLATION

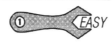

◆ See Figures 34 thru 39

Removal of the top cover is necessary for the most basic of maintenance and inspection procedures. The cover should come off before and after each use in order to perform these basic safety checks. Unlike some outboard manufacturers that use large and intricate multi-piece lower covers/cowlings, most Yamaha and Mercury/Mariner 4-stroke outboards simple use a low profile 1-piece cowling that does not require removal except during powerhead R&R. However there are a few exceptions on larger motors that are equipped with fore or aft or port and starboard lower covers which cover part of the intermediate housing or steering bracket. Removal of those lower covers is normally pretty straight forward in that most fasteners are exposed and obvious.

Although top cover removal will vary slightly from motor to motor, most have one, two or three latches located at the front and rear of the cowling. The latches must be released in order to free the cover from the cowling. On smaller motors, if only one lever is used the other end of the cover is usually secured by a tab that is held in place by friction once the lever is secured. On larger motors (mostly V6 models) one lever may be connected by cable to a release mechanism at both ends of the top cover. On most larger motors there are two or three separate latches.

Probably the most common form of cover retention on these outboards is the use of 2 separate levers, one at the front and one at the rear of the outboard. On some motors these levers are horizontal when latched and rotated 1/4 turn downward (or upward in some cases) to release. On most 4-stroke motors these latches are lifted up and outward to release or pushed down and inward to lock.

As mentioned in the lubrication section all cover latch mechanisms should be wiped clean and re-greased periodically to ensure proper operation and to prevent un-necessary wear or damage.

Whenever the top cover is removed, check the cover seal (a rubber insulator normally used between the cover and cowling) for wear or damage. Take the opportunity to perform a quick visual inspection for leaking hoses, chaffing wires etc. Get to know where things are placed under the cover so you'll recognize instantly if something is amiss. When installing the cover, care must be taken to ensure wiring and hoses are in their original positions to prevent damage. Once the cover is installed, make sure the latches grab it securely to prevent it from coming loose and flying off while underway.

■ On a few of the smaller outboards the cover is equipped with one or more drain holes. If equipped, check them periodically to ensure they are not blocked by debris.

Fig. 34 Most covers are secured by 1 or 2 latches. . .

Fig. 35 some are rotated 1/4 turn upward. . .

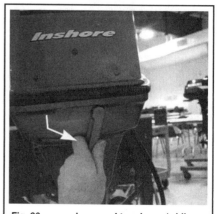

Fig. 36 . . . or downward to release (while others are pulled straight upward)

MAINTENANCE & TUNE-UP 2-13

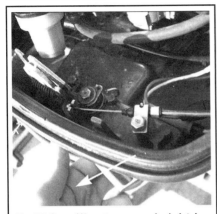

Fig. 37 Some V6 motors use a single latch, which when pulled outward...

Fig. 38 ... pulls on a cable that...

Fig. 39 ... actuates the latch on the other end of the cowling

One more point. The latches are adjustable one many of the larger motors. Make sure the latch engages securely so the top cover will not be potentially lost in use. If necessary, loosen the mounting bolts and reposition the latch components to ensure proper engagement and tighten the bolts securely.

Cooling System

FLUSHING THE COOLING SYSTEM

◆ See Figures 40 thru 44

The most important service that you can perform on your motor's cooling system is to flush it periodically using fresh, clean water. This should be done immediately following any use in salt, brackish or polluted waters in order to prevent mineral deposits or corrosion from clogging cooling passages. Even if you do not always boat in salt or polluted waters, get used to the flushing procedure and perform it often to ensure no silt or debris clogs your cooling system over time.

■ **Flush the cooling system after any use in which the motor was operated through suspended/churned-up silt, debris or sand.**

Although the flushing procedure should take place right away (dockside or on the trailer), be sure to protect the motor from damage due to possible thermal shock. If the engine has just been run under high load or at continued high speeds, allow time for it to cool to the point where the powerhead can be touched. Do not pump very cold water through a very hot engine, or you are just asking for trouble. If you trailer your boat short distances, the flushing procedure can probably wait until you arrive home or wherever the boat is stored, but ideally it should occur within an hour of use in salt water. Remember that the corrosion process begins as soon as the motor is removed from the water and exposed to air.

The flushing procedure is not used only for cooling system maintenance, but it is also a tool with which a technician can provide a source of cooling water to protect the engine (and water pump impeller) from damage anytime the motor needs to be run out of the water. **Never** start or run the engine out of the water, even for a few seconds, for any reason. Water pump impeller damage can occur instantly and damage to the engine from overheating can follow shortly thereafter. If the engine must be run out of the water for tuning or testing, always connect an appropriate flushing device **before** the engine is started and leave it turned on until **after** the engine is shut off.

✱✱ WARNING

ANYTIME the engine is run, the first thing you should do is check the cooling stream or water indicator. All Yamahas, Mercurys and Mariners are equipped with some form of a cooling stream indicator towards the aft portion of the lower engine cover. Anytime the engine is operating, a steady stream of water should come from the indicator, showing that the pump is supplying water to the engine for cooling. If the stream is ever absent, stop the motor and determine the cause before restarting.

As we stated earlier, flushing the cooling system consists of supplying fresh, clean water to the system in order to clean deposits from the internal passages. If the engine is running, the water does not normally have to be pressurized, as it is delivered through the normal water intake passages and the water pump (the system can self flush if supplied with clean water). Smaller, portable engines can be flushed by mounted them in a test tank (a sturdy, metallic 30 gallon drum or garbage pail filled with clean water).

Many of these models are equipped with a built-in flushing adapter. Normally flushing adapters are found on almost all 4-strokes. The flushing adapter is usually found just under the engine cowl, at the front (usually port, but sometimes starboard side) of the motor OR at the center of the cowling on the rear of the motor. However, keep in mind that, when present, the flushing adapter (at least for most Yamahas) is usually designed for cleaning the cooling system and should not be used when running the engine on a hose, as it may not allow sufficient water to reach the impeller (and could therefore allow water pump damage). In contrast, Mercury/Mariner recommends running some of their motors when flushing even through the adapter (this is to make sure the thermostat opens to fully flush internal passages). For more details, refer to your owner's manual.

When present the Yamaha flush adapters usually consist of a short length of hose which runs from a fitting on the powerhead to a female hose connection which is threaded onto a fitting on the outboard cowling. The adapter is used by unthreaded the female connection and attaching the male end of a garden hose to it. This allows you to supply fresh water from the garden hose directly into the powerhead cooling passages, bypassing the water pump. It is critical that you thread the hose back onto the cowling fitting before returning the motor to service. If the adapter hose is left unblocked, a certain amount of water from the water pump may bypass the powerhead during engine operation which could lead to overheating.

For almost all Yamaha motors and some Merc/Mariner models, it is a

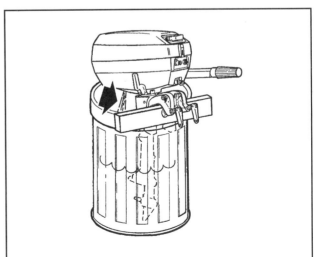

Fig. 40 All models may be flushed in a test tank. Smaller ones, in a garbage pail

2-14 MAINTENANCE & TUNE-UP

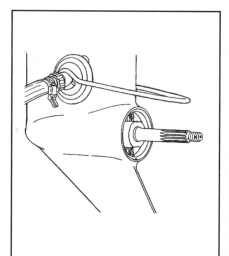

Fig. 41 The easiest way to flush most models is using a clamp-type adapter

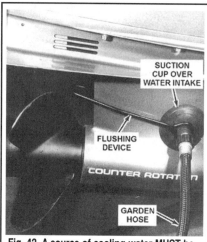

Fig. 42 A source of cooling water MUST be used anything the engine is run out of the water

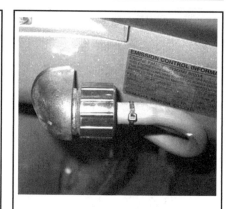

Fig. 43 Most 4-strokes (shown) are equipped with a flush adapter. . .

good idea to also have a generic (ear-muff type) flush adapter handy for service. The generic adapters fit over the engine water intakes on the gearcase (and resemble a pair of strange earmuffs with a hose fitting on one side). This will work on all motors whose water intake is on the sides of the gearcase. It may not work on the smallest motors if the intake is in the anti-ventilation plate, however those models are small enough to run in a sturdy metal trash can.

■ Jet drive models are equipped with a flushing port mounted under a flat head screw directly above the jet drive bearing grease fitting. For more information, please refer to FLUSHING JET DRIVES, later in this section.

■ When running the engine on a flushing adapter using a garden hose, make sure the hose delivers about 20-40 psi (140-300 kPa) of pressure.

Some of the smaller, portable motors covered by this manual utilize a water intake that is directly above the propeller. On these models the propeller must usually be removed before a clamp style flush adapter can be connected to the motor (unless the adapter is very thin and mounted so close to the anti-ventilation plate that it will not be hit by the propeller).

✱✱ CAUTION

For safety, the propeller should be removed ANYTIME the motor is run on the trailer or on an engine stand. We realize that this is not always practical when flushing the engine on the trailer, but cannot emphasize enough how much caution must be exercised to prevent injury to you or someone else. Either take the time to remove the propeller or take the time to make sure no-one or nothing comes close enough to it to become injured. Serious personal injury or death could result from contact with the spinning propeller.

1. Check the engine top case and, if necessary remove it to check the powerhead and ensure it is cooled enough to flush without causing thermal shock.
2. Prepare the engine for flushing depending on the method you are using as follows:
   a. If using a test tank, make sure the tank is made of sturdy material, then securely mount the motor to the tank. If necessary, position a wooden plank between the tank and engine clamp bracket for thickness. Fill the tank so the water level is at least 4 in. (10cm) above the anti-ventilation plate (above the water inlet).
   b. If using a flushing adapter of either the generic clamp-type or specific port-type for your model attach the water hose to the flush test adapter and connect the adapter to the motor following the instructions that came with the adapter. Unless your owner's manual says differently, if the motor is to be run (for flushing or testing), be sure to use the ear-muff style adapter, position the outboard vertically and remove the propeller, for safety. Also, be sure to position the water hose so it will not contact with moving parts (tie the hose out of the way with mechanic's wire or wire ties, as necessary).

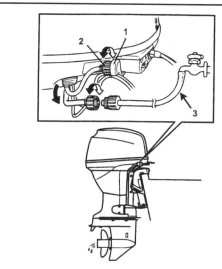

1 - Cowling Fitting
2 - Flushing Adapter
3 - Garden Hose

Fig. 44 . . . found somewhere on the underside or at the rear of the cowling

■ When using a clamp-type adapter, position the suction cup(s) over water intake grate(s) in such a way that they form tight seals. A little pressure seepage should not be a problem, but look to the water stream indicator once the motor is running to be sure that sufficient water is reaching the powerhead.

3. Unless using a test tank, turn the water on, making sure that pressure does not exceed 45 psi (300 kPa).
4. If using a test tank or if the motor must be run for testing/tuning procedures, start the engine and run in neutral until the motor reaches operating temperature. The motor will continue to run at fast idle until warmed.

✱✱ WARNING

As soon as the engine starts, check the cooling system indicator stream. It must be present and strong as long as the motor is operated. If not, stop the motor and rectify the problem before proceeding. Common problems could include insufficient water pressure or incorrect flush adapter installation.

MAINTENANCE & TUNE-UP 2-15

5. Flush the motor for at least 5-10 minutes or until the water exiting the engine is clear. When flushing while running the motor, check the engine temperature (using a gauge or carefully by touch) and stop the engine immediately if steam or overheating starts to occur. Make sure that carbureted motors slow to low idle for the last few minutes of the flushing procedure.

6. Stop the engine (if running), **then** shut the water off.

■ **If the motor is to be stored for any length of time more than a week or two, consider fogging the motor before it is shut off. Flushing and fogging a motor is one of the most important things you can do to ensure long life of the pistons, rods, crankshaft and bearings.**

7. Remove the adapter from the engine or the engine from the test tank, as applicable.

8. If flushing did not occur with the motor running (so the motor would already by vertical), be sure to place it in the full vertical position allowing the cooling system to drain. This is especially important if the engine is going to be placed into storage and could be exposed to freezing temperatures. Water left in the motor could freeze and crack the powerhead or gearcase.

FLUSHING JET DRIVES

◆ See Figure 45

Regular flushing of the jet drive will prolong the life of the powerhead, by clearing the cooling system of possible obstructions.

Jet drives are equipped with a plug on the port side just above the lubrication hose.

1. Remove the plug and gasket, install the flush adapter, connect the garden hose and turn on the water supply.

2. Start and operate the powerhead at a fast idle for about 15 minutes. Disconnect the flushing adapter and replace the plug.

■ **The procedure just described will only flush the powerhead cooling system, not the jet drive and impeller. To flush the jet drive unit, direct a stream of high pressure water through the intake grille.**

Engine Oil and Filter (4-Stroke)

OIL RECOMMENDATIONS

Yamaha recommends the use of a high quality 4-stroke motor oil of SAE 10W-30 or 10W-40 viscosity with an API rating of SE or higher (SF, SG, SH or SJ is acceptable) for all of their 4-stroke motors. However, if the engine is to be operated only in conditions of ambient temperatures above 68°F (20°C), using SAE 20W-40 is also acceptable on the following models: 9.9, 25, 30/40J/40, 40/45/50J/50, and the 60J/60 hp motors.

Mercury/Mariner recommends the use of a high quality 4-stroke motor oil of SAE 10W-30 viscosity with an API rating of SF or higher (SG, SH or CF-4, CE, CD, CDII is also acceptable) for all of their 4-stroke motors. However, if the engine is to be operated only in conditions of ambient temperatures above 40°F (4°C), using SAE 25W-40 is also acceptable on all models. Also, please note that for the 225 hp V6, Merc also notes that 10W-40 is acceptable for all temperature ranges, just like 10W-30.

The Society of Automotive Engineers (SAE) grade number indicates the viscosity of the engine oil; its resistance to flow at a given temperature. The lower the SAE grade number, the lighter the oil. For example, the monograde oils begin with SAE 5 weight, which is a thin, light oil, and continue in viscosity up to SAE 80 or 90 weight, which are heavy gear lubricants. These oils are also known as "straight weight", meaning they are of a single viscosity, and do not vary with engine temperature.

Multi-viscosity oils offer the important advantage of being adaptable to temperature extremes. These oils have designations such as 10W-40, 20W-50, etc. The 10W-40 means that in winter (the "W" in the designation) the oil acts like a thin 10 weight oil, allowing the engine to spin easily when cold and offering rapid lubrication. Once the engine has warmed up, however, the oil acts like a straight 40 weight, maintaining good lubrication and protection for the engine's internal components. A 20W-50 oil would therefore be slightly heavier than and not as ideal in cold weather as the 10W-40, but would offer better protection at higher rpm and temperatures because when warm it acts like a 50 weight oil. Whichever oil viscosity you choose when

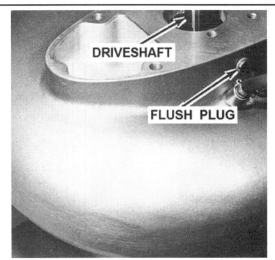

Fig. 45 Jet drives are equipped with a flushing plug on the side of the unit

changing the oil, make sure you are anticipating the temperatures your engine will be operating in until the oil is changed again.

The American Petroleum Institute (API) designation indicates the classification of engine oil used under certain given operating conditions. Only oils designated for use "Service SG, SH" or greater should be used. Oils of the SG, SH, SJ or its latest superseding oil type perform a variety of functions inside the engine in addition to the basic function as a lubricant. Through a balanced system of metallic detergents and polymeric dispersants, the oil prevents the formation of high and low temperature deposits and also keeps sludge and particles of dirt in suspension. Acids, particularly sulfuric acid, as well as other by-products of combustion, are neutralized. Both the SAE grade number and the API designation can be found on top of the oil bottle.

CHECKING OIL LEVEL

◆ See Figures 46 thru 49

One of the most important service items for a 4-stroke engine is maintaining the proper level of fresh, clean engine oil in the crankcase. Be certain to check the oil level both before and after each time the boat is used. In order to check the oil level the motor must be placed in the full vertical position. Because it takes some time for the oil to settle (and at least partially cool), the engine must be shut off for at least 30 minutes before truly accurate reading can be attained. If the boat is trailered, use the time for loading the boat onto the trailer and prepping the trailer for towing to allow the motor to cool. If the boat is kept in the water, take some time around the dock to secure lines, stow away items kept onboard and clean up the deck while waiting for the oil to settle/cool.

✱✱ WARNING

Running an engine with an improper oil level can cause significant engine damage. Although it is typically worse to run an engine with abnormally low oil, it can be just as harmful to run an engine that is overfilled. Don't take that risk, make checking the engine oil a regular part of your launch and recovery/docking routine.

All Mercury/Mariner and all Yamaha 4-stroke motors, with the exception of the Yamaha 2.5 hp model, are equipped with an automotive-style dipstick. The Yamaha 2.5 hp motor uses an oil level sight glass found on the port side of the motor, near where the tiller handle connects. All motors are equipped with some form of an oil filler cap located somewhere on the powerhead. For models equipped with a dipstick, the engine cover must be removed for access, but once removed it should be easy to locate the dipstick and filler cap. On most models the dipstick is found along the side of the motor. On larger motors it is normally found almost directly beneath the spin-on oil filter. The filler cap is usually found at the rear of the motor on or near the valve cover.

2-16 MAINTENANCE & TUNE-UP

Fig. 46 Locate the dipstick on the side of the powerhead...

Fig. 47 ... and remove it, holding it vertically as shown

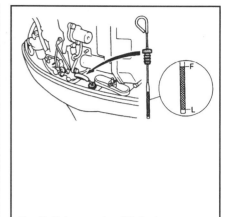

Fig. 48 Make sure the oil is in the acceptable range...

Fig. 49 ... if not, add it through the oil fill cap on the powerhead

1. Make sure the engine is in the full vertical position and has been shut off for at least 30 minutes. If possible, get in the habit of checking the oil with the engine cold from sitting overnight
2. Remove the engine cover.
3. For all models except the 2.5 hp, carefully pull the engine crankcase oil dipstick from the side (port or starboard, depending upon the model) of the engine. Wipe all traces of oil off the dipstick using a clean, lint free rag or cloth, then re-insert dipstick back into its opening until it is fully seated. Then, pull the dipstick out from the crankcase again and hold it vertically with the bottom end facing down in order to prevent a false oil reading.

■ Forget how your dad or buddy first taught you to read the level on a dipstick. It may be more convenient to hold it horizontally, but laying it down like that could allow oil to flow UPWARD giving a false high, or worse, false acceptable reading when in fact your engine needs oil. Last time we checked, oil won't flow UP a dipstick held vertically (but the high point of the oil will remain wet in contrast to the dry portion of the stick immediately above the wet line). So hold the dipstick vertically and you'll never run your engine with insufficient oil when you thought it was full.

4. On Yamaha 2.5 hp motors look at the sight glass, oil should be visible JUST about to the top of the glass. When adding or refilling the crankcase, always leave a very small air gap at the top of the sight glass so you can be sure it is not overfilled.
5. If the oil level is at or slightly below the top of the hatch-marks/full level mark on the dipstick (or sight glass), the oil level is fine. If not, add small amounts of oil through the filler cap until the level is correct. Add oil slowly, giving it time to settle into the crankcase before rechecking and again, don't overfill it either.

■ Dipstick markings on these outboards are normally two lines with a crosshatched area between them. The crosshatched area is the "acceptable" operating range, but you should try to maintain the level towards the top of the markings.

6. Visually check the oil on the dipstick (or in the sight glass) for water (a milky appearance will result from contamination with moisture) or a significant fuel odor. Both are signs that the powerhead likely needs overhaul to prevent damage.
7. On all but the Yamaha 2.5 hp motor, insert and properly seat the oil dipstick into the powerhead when you are finished.
8. If removed, install the oil fill cap and rotate it until it gently locks into position.

OIL CHANGE & FILTER SERVICE

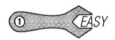

◆ See Figures 50 thru 57

Next to regular fluid level checks, the most important way to maintain a 4-stroke outboard motor is to change the engine crankcase oil (and change/clean the filter, if applicable) on a regular basis. The manufacturer recommends performing this service at every 100 hours or annually, right before storage, whichever comes first. For more information regarding engine oil, refer to OIL RECOMMENDATIONS earlier in this section.

All 4-stroke motors 8/9.9 hp (232cc) or larger utilize an oil filter which should be changed/cleaned whenever the engine oil is drained. Only the 8/9.9 hp (232cc) motor uses a serviceable filter; a reusable element mounted in a housing on the starboard side of the powerhead just behind the oil level dipstick. Although this element is reusable, Yamaha recommends replacing it annually to ensure yourself of proper protection (Mercury does not share this recommendation, but it can't hurt). All 9.9/15 hp (323cc) and larger motors utilize a disposable, automotive style, spin-on filter mounted to the side of the powerhead. The best method to remove the spin-on filter (resulting in fewest skinned knuckles) is a filter wrench, and our preference is the cap style that fits over the end of the filter. When purchasing a replacement oil filter check your local marine dealer or automotive parts dealer for a cap wrench that fits the filter.

Most people who have worked on their own machines, whether that is tractors, motorcycles, cars/trucks or boat motors, will tell you that oil should be changed hot. This seems to have always been the popular method, and it works well since hot oil flows better/faster and may remove more deposits that are still held in suspension. Of course, hot oil can be messy or even a bit dangerous to work with. Coupled with the sometimes difficult method of draining oil from an outboard, this might make it better in some instances to drain the oil cold. Of course, if this is desired, you'll have to leave more time for the oil to drain completely, thereby removing as much contaminants as possible from the crankcase. The choice is really yours, but be sure to take the appropriate steps to protect yourself either way.

■ If the engine is not being placed in storage after the oil change, it should be run to normal operating temperature (in a test tank or with a suitable flushing device) and inspected for leaks before returning it to

MAINTENANCE & TUNE-UP 2-17

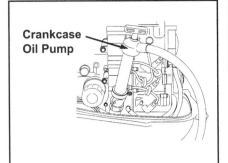

Fig. 50 Mercury allows two methods for oil changes, the first uses this Quicksilver crankcase oil pump...

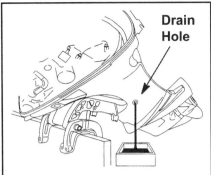

Fig. 51 ...the second involves removing the drain plug (normally found on the Port side for Merc/Mariner motors)

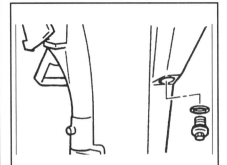

Fig. 52 Although the Merc pump might work on some Yamahas, the drain plug is usually conveniently located at the rear of the powerhead just underneath the cowling

service. If the engine is being placed into storage it should also be run using a flush device, but be sure not to run it too long. Just start and run the engine for a few minutes to thoroughly circulate the fresh oil, then prepare it for storage by fogging the motor.

If you decide to change oil with the engine hot, a source of cooling such as a test tank or flushing hose must be attached to the engine to prevent impeller or powerhead damage when running the engine to normal operating temperature. If you are lucky enough to store the boat (or live) close to waters in which to use the boat, you can simply enjoy a morning, evening or whole day on the water before changing the oil. The amount of time necessary to haul the boat and tow it to your work area should allow the oil to cool enough so that it won't be scalding hot, but still warm enough to flow well.

■ Although it is not recommended for normal service, the oil CAN be drained without removing the engine cases or servicing the filter (of course, since there is no filter on the smallest Yamaha motors, this doesn't apply to them). This might be desired if too much oil was added during a routine level check or if a small sample of oil is to be removed for inspection.

For all Mercury/Mariner motors 25 hp and larger, Mercury also sells the Quicksilver Crankcase Oil Pump which can be used to manually draw (pump) the oil from the crankcase and out the dipstick tube. Since these motors are mechanically identical to their Yamaha counterparts, you should be able to use this pump on Yamaha motors as well, if desired (and as a matter of fact, in the manual for the 2004 Yamaha 150 hp motors, Yamaha finally did show oil changing using either method). However, for the record, we prefer the drain plug changing method most of the time, to make sure you've really purged as much of the old oil from the crankcase as possible. Of course, if you use the motor enough in a single year/season, a mid-season oil change using the pump might be convenient and the best method to ensure proper care of a boat, especially one that is moored or slipped and not hauled for a service.

1. Prepare the engine and work area for the oil change by placing the motor in a fully vertical position over a large, flattened cardboard box (which can be used to catch any dripping oil missed by the drain pan). Have a drain pan, a few quarts larger than the oil capacity of the motor (refer to the Capacities Chart in this section) and a lot of clean rags or disposable shop towels handy.

2. Remove the upper engine cover and loosen the oil filler cap. If applicable, locate the oil filter.

 a. On 8/9.9 hp (232cc) motors the filter is found under a hex head cover threaded into a bore under the engine top cover.

 b. On 9.9/15 hp (323cc) and larger motors, the disposable spin-on oil filter is found at the center, side of the powerhead (it is often, though not always, found on the same side and directly above the dipstick).

 c. On V6 motors, the filter is located under a filter cover on the front, starboard side of the powerhead.

3. On V6 motors, remove the bolts (usually 4 bolts for Yamaha or 7 bolts for Merc models) securing the lower engine cover to the starboard side, then remove the cover for access to the drain plug. Next remove the bolt(s) securing the oil filter cover to the powerhead and remove the cover for access to the oil filter.

Fig. 53 For all 323cc (9.9/15 hp) or larger motors, the oil change also involves changing the automotive style spin-on filter

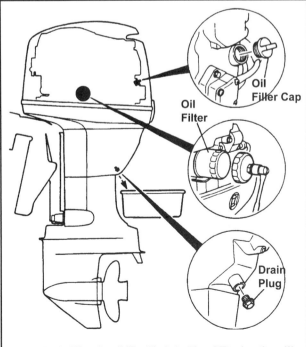

Fig. 54 Typical Yamaha oil filler (1), drain (2) and filter locations (3)

2-18 MAINTENANCE & TUNE-UP

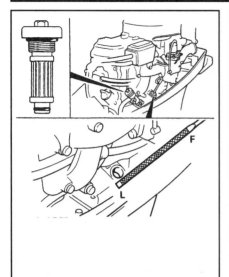

Fig. 55 A serviceable filter is threaded into a bore just behind the dipstick on 232cc (8/9.9 hp) motors...

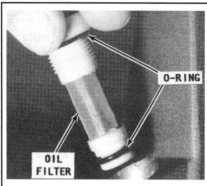

Fig. 56 ... the serviceable filter seals with O-rings on either end

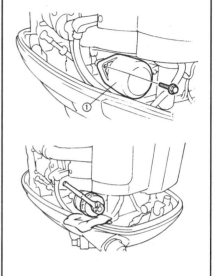

Fig. 57 The oil filter is located underneath a cover on V6 motors

4. Locate and remove the oil drain plug and gasket, as follows:
   a. For all Yamaha models except the V6, engine oil is drained by removing the plug from the oil pan at the lower rear of the powerhead, just below the cowling.

■ In order to improve oil flow, remove the oil fill cap.

   b. On the Merc/Mariner 4/5/6 hp motor the drain plug is located at the FRONT center of the motor, inboard of the tiller handle, just below the control panel.
   c. On all 8/9.9-115 hp Merc/Mariner motors the drain plug is located part way down the intermediate housing usually on the Port side of the motor, however there are some exceptions such as the 9.9/15 hp (323cc) motors for which the plug is on the Starboard side. To facilitate draining with less of a mess, turn the outboard so the drain plug faces as downward as possible.
   d. For all V6 motors, as noted in the previous step, the oil pan drain plug is located under the Starboard side cowling/apron
5. Either hold the drain pan under the powerhead and just above the ventilation plate or, in some cases, it may be possible to turn or trim the motor far to place the pan on the ground. It may also be possible to secure the pan carefully above the ventilation plate using some bungee cords. But honestly, at least on the Yamahas and the smallest of Mercs the least chance of a mess it just to be patient and hold it.
6. Inspect the drain plug and gasket for signs of damage. Replace the plug or gasket if any damage is found. Also, watch the draining oil for signs of contamination by moisture (a milky appearance will result), by fuel (a strong odor and thinner running oil would be present) or signs of metallic flakes/particles. A small amount of tiny metallic particles is a sign of normal wear, but large amounts or large pieces indicate internal engine damage and the need for an overhaul to determine and rectify the cause.
7. When it appears that the oil has drained, tilt the engine slightly and pivot it to one side or the other in order to ensure complete oil drainage.
8. Clean the drain plug, the engine and the gearcase. Place a new gasket onto the drain plug then carefully thread the plug into the opening. Tighten the plug securely.

■ Although it is not absolutely necessary to replace the gasket each time, it is a cheap way to help protect against possible leaks. We think it is a good idea.

9. For 8/9.9 hp (232cc) models remove and service the filter as follows:
   a. The filter is threaded into a boss on the powerhead, loosen it using a wrench or socket and carefully unthread it from the boss. Hold a small rag under the boss to catch oil as it drips from the element when you pull it out.
   b. Rinse the element using solvent and dry it using low pressure compressed air. If compressed air is not available, allow it to air dry for at least 15 minutes. Inspect the element for signs of clogging or damage and replace, if found.
   c. Apply a light coating of fresh 4-stroke engine oil to the filter element and O-rings, then install the element into the bore and thread it tightly into place.

■ Although the procedure for spin-on filters talks about placing a shop rag under the filter while it is removed, there is an alternate method to prevent a mess. If desired, loosen the filter slightly with a cap wrench, then slide a disposable Zip-Lock® or similar food storage bag completely over the filter and unthread it into the bag. Position a shop rag anyway, just to be sure to catch any stray oil that escapes. With a little practice, you'll find this method can be the best way to remove oil filters.

10. For models equipped with a disposable, spin-on filter element remove and service the filter as follows:
    a. Position a shop rag underneath, then place the oil filter wrench onto filter element.
    b. Loosen the spin-on element by turning the filter wrench counterclockwise, then remove the wrench and finish unthreading the element by hand. Remove the filter from the powerhead and clean up any spilled oil.
    d. Make sure the rubber gasket is not stuck to the oil filter mounting surface, then use a lint free shop rag to clean all dirt and oil from mounting surface.
    e. Apply a thin coating of engine oil to the sealing ring of the new oil filter, then thread the filter onto the adapter until the sealing washer touches the mounting surface. If you've got a torque wrench and a cap-type oil filter wrench, tighten the filter to 13 ft. lbs. 17.5 Nm. If not, tighten the filter by hand an additional 1/4-2/3 turn.
11. Clean up any spilled oil.
12. Refill the engine through the oil filler cap as described under Checking Engine Oil in this section. Add the oil gradually, checking the oil level frequently. Add oil until the level reaches the upper level of the hatched area (full mark) on the dipstick (or the upper portion of the sight glass on the Yamaha 2.5 hp motor).
13. Provide a temporary cooling system to the engine as detailed under Flushing The Cooling System, then start the engine and run it to normal operating temperature while visually checking for leakage.

■ If the engine is being placed into storage, don't run the motor too long, just long enough to use a can of fogging spray. Between the fresh oil circulated through the motor and the fogging spray coating the inside of the intake and combustion chambers you motor should sleep like a baby until next season.

14. Stop the motor and allow it to cool, then properly re-check the oil level after it has settled again into the crankcase.
15. For V6 motors, reinstall the starboard lower engine cover and the oil filter cover.

MAINTENANCE & TUNE-UP

Fuel Filter

◆ See Figure 58

A fuel filter is designed to keep particles of dirt and debris from entering the carburetor(s) or fuel injectors clogging the tiny internal passages. A small speck of dirt or sand can drastically affect the ability of the fuel system to deliver the proper amount of air and fuel to the engine. If a filter becomes clogged, it will quickly impede the flow of gasoline. This could cause lean fuel mixtures, hesitation and stumbling and idle problems in carburetors. Although a clogged fuel passage in a fuel injected engine could also cause lean symptoms and idle problems, dirt can also prevent a fuel injector from closing properly. A fuel injector that is stuck partially open by debris will cause the engine to run rich due to the unregulated fuel constantly spraying from the pressurized injector.

Regular cleaning or replacement of the fuel filter (depending on the type or types used) will decrease the risk of blocking the flow of fuel to the engine, which could leave you stranded on the water. It will also decrease the risk of damage to the small passages of a carburetor or fuel injector that could require more extensive and expensive replacement. Keep in mind that fuel filters are usually inexpensive and replacement is a simple task. Service your fuel filter on a regular basis to avoid fuel delivery problems.

We tried to find a pattern to tell us what type of filter will be installed on a powerhead, based on the year or model of the engine. However, Yamaha did not seem to have a whole lot of rhyme or reason when it comes to choosing a fuel filter for their outboards (or if they did, we couldn't figure it out). Loosely, we'll say that most smaller Yamaha outboards use disposable, inline filters (or fuel tank strainers on the smallest of outboards), while larger motors tend to use a serviceable filter element that can be removed from a housing/cup assembly for cleaning or replacement. However there are exceptions to this and some larger motors, including some 4-cylinder 4-stroke motors may be equipped with a non-serviceable inline filter.

Mercury/Mariner motors followed the same loose pattern as Yamaha. The smaller motors, up to about 60 hp TENDED to have a disposable inline filter, UNLESS they were fuel injected. Smaller fuel injected motors and pretty much ALL 75 hp and larger models used a serviceable cartridge filter assembly whether they were fuel injected or not. And, lastly, the smallest of the Merc motors, the 4/5/6 hp singles, have an additional fuel filter screen inside the fuel pump assembly.

Also, keep in mind that the type of fuel filter used on your boat/engine will vary not only with the year and model, but also with the accessories and rigging. Because of the number of possible variations it is impossible to accurately give instructions based on model. Instead, we will provide instructions for the different types of filters the manufacturer used on various families of motors or systems with which they are equipped. To determine what filter(s) are utilized by your boat and motor rigging, trace the fuel line from the tank to the fuel pump and then from the pump to the carburetor(s) or EFI fuel vapor separator tank. Most outboards place a filter (disposable or serviceable) found inline just before the fuel pump. Non-serviceable inline filters are replaced by simply removing the clamps, disconnecting the hoses and installing a new filter. When installing a new disposable inline filter, make sure the arrow on the filter points in the direction of fuel flow.

Some motors have a fuel filter mounted in the fuel tank itself. On small Yamaha motors with integral fuel tanks (specifically the 2.5 hp) the only fuel filter element is normally mounted on the fuel petcock or the fuel outlet fitting which must be removed from the tank in order to service it. For larger motors, in addition to the fuel filter mounted on the engine, a filter is usually found inside or near the fuel tank. Because of the large variety of differences in both portable and fixed fuel tanks, it is impossible to give a detailed procedure for removal and installation. However, keep in mind that most in-tank filters are simply a screen on the pickup line inside the fuel tank. Filters of this type rarely require service or attention, but if the tank is removed for cleaning the filter will usually only need to be cleaned and returned to service (assuming they are not torn or otherwise damaged).

■ Most EFI motors are equipped with additional fuel filter elements either on the high-pressure electric fuel pump or the pressure regulator. However these elements are not maintenance items and should only need attention during repair or replacement of these components.

Depending upon the boat rigging a fuel filter/water separator may be found inline between a boat mounted fuel tank and the motor. Boat mounted fuel filter/water separators are normally of the spin-on filter type (resembling automotive oil filters) but will vary greatly with boat rigging. Replacement of filters/separators is the same as a typical automotive oil filter replacement, just make sure to have a small drain basin handy to catch escaping fuel and make sure to coat the rubber gasket with a small dab of engine oil during installation.

Fig. 58 Boats with integrated fuel tanks will usually be rigged with an automotive style spin-on fuel filter/water separator

FUEL FILTER SERVICE

> ✱✱ **CAUTION**
>
> Observe all applicable safety precautions when working around fuel. Whenever servicing the fuel system, always work in a well-ventilated area. Do not allow fuel spray or vapors to come in contact with a spark or open flame. Do not smoke while working around gasoline. Keep a dry chemical fire extinguisher near the work area. Always keep fuel in a container specifically designed for fuel storage; also, always properly seal fuel containers to avoid the possibility of fire or explosion.

Yamaha Integral Fuel Tank Models

A small serviceable filter element is usually found on the fuel petcock or fuel tank outlet fitting on Yamaha integral tank models. This should include the 2.5 hp 4-stroke motor. To inspect, clean and/or replace this filter element, proceed as follows:

1. Drain the fuel in the tank into a suitable container.
2. It is usually necessary or just plain easier to work on with the fuel tank removed from the powerhead. Remove the mounting bolts and reposition the tank as necessary to disconnect a fuel line. In some cases it is easier to leave the fuel line connected to the tank and instead disconnect it from the other (carburetor) end.
3. Loosen the fuel petcock or fuel outlet fitting clamp screw or nut.
4. Remove the petcock from the clamp and tank.
5. Clean the filter assembly in solvent and blow it dry with compressed air. If excessively dirty or contaminated with water, replace the filter.

2-20 MAINTENANCE & TUNE-UP

6. Install the petcock on the clamp and tank, then connect the hose.
7. Tighten the fuel petcock clamp screw or nut.
8. Reposition the fuel tank and secure using the mounting bolts.
9. Check the fuel filter installation for leakage.

Disposable Inline Filters

◆ See Figure 59

As noted earlier, a number of outboards are equipped with a disposable inline filter. Generally speaking most of the smaller motors (Yamaha 4-40 hp and carbureted Mercury/Mariner 4-60 hp motors), as well as a few of the larger Yamaha motors (75-100 hp models), are equipped with a disposable, inline filter. Generally speaking, the filter is found inline just before the mechanical fuel pump.

This type of filter is a sealed canister type (usually plastic) and cannot be cleaned, so service is normally limited to replacement. Because of the relative ease and relatively low expense of a filter (when compared with the time and hassle of a carburetor overhaul) we encourage you to replace the filter at least annually.

When replacing the filter, release the hose clamps (they are usually equipped with spring-type clamps that are released by squeezing the tabs using a pair of pliers) and slide them back on the hose, past the raised portion of the filter inlet/outlet nipples.

Once a clamp is released, position a small drain pan or a shop towel under the filter and carefully pull the hose from the nipple. Allow any fuel remaining in the filter and fuel line to drain into the drain pan or catch fuel with the shop towel. Repeat on the other side, noting which fuel line connects to which portion of the filter (for assembly purposes). Inline filters are usually marked with an arrow indicating fuel flow. The arrow should point towards the fuel line that runs to the motor (not the fuel tank).

Before installation of the new filter, make sure the hoses are in good condition and not brittle, cracking and otherwise in need of replacement. During installation, be sure to fully seat the hoses, then place the clamps over the raised portions of the nipples to secure them. Spring clamps will weaken over time, so replace them if they've lost their tension. If wire ties or adjustable clamps were used, be careful not to over-tighten the clamp. If the clamp cuts into the hose, it's too tight; loosen the clamp or cut the wire tie (as applicable) and start again.

✳✳ CAUTION

Before returning the outboard to service, use the primer bulb to pressurize the system and check the filter/fittings for leaks.

Serviceable Canister-Type Inline Filters

◆ See Figure 60

It's probably safe to say that about half of these motors (including all EFI Mercury/Mariners and all models 75 hp and larger, except a few quirky Yamahas) are equipped with a serviceable canister-type inline filter. These can be identified by their design and shape, which varies from the typical inline filter. A typical, disposable inline filter will have a simple round canister to which the fuel lines attach at either end. The serviceable inline filters used by these motors usually have 2 fuel inlets at the top of a filter housing and the serviceable element is located in a bowl or cap that is threaded up from underneath the housing. When servicing these filters it may be possible to access and unthread the bowl without disconnecting any hoses, but if so, you'll have to hold the housing steady to prevent stressing and damaging the fuel lines. However, on some motors access to the bowl is impossible without first disconnecting one or more fuel hoses and repositioning the assembly.

■ On some models the bowl/canister itself threads into the housing, but on others a separate knurled locking is threaded over the bowl/canister and onto the housing to secure the assembly. If present, turn the LOCKRING to loosen the housing, not the bowl itself.

Serviceable filter elements should be cleaned carefully using solvent and inspected for clogging, tears or damage. If problems are found, normally just the element itself will require replacement. You really should always replace the O-ring, it's cheap insurance, but you CAN reuse it in a pinch, as long as it is not torn, cut or deformed. While you're servicing the filter always inspect and replace any damaged hose or clamp as you would with any other fuel filter.

Inline Engines

◆ See Figures 60, 61 and 62

■ It is usually possible to service the filter element without removing the canister from the motor, however if access is tight, follow the steps for removing the hose(s) from the assembly so it can be repositioned.

Although most filter assemblies are pretty straight forward, containing a filter element and a single O-ring seal, some models may differ containing additional components such as a water indicator float, a spring, or additional

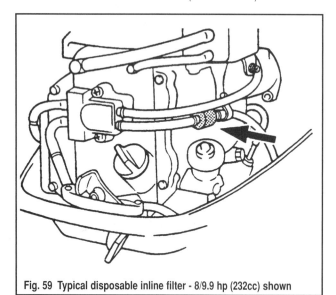

Fig. 59 Typical disposable inline filter - 8/9.9 hp (232cc) shown

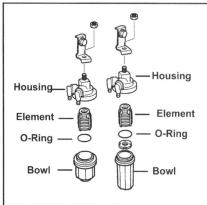

Fig. 60 Exploded view of some common serviceable canister-type fuel filters

Fig. 61 Typical serviceable canister-type filter assembly

Fig. 62 Access the element by unthreading the bowl from the housing

MAINTENANCE & TUNE-UP 2-21

seals. Keep track of the components as they are removed to ensure you will know how to properly assemble them.

1. If necessary for better access, remove the assembly from the powerhead as follows:

   a. On models with metal clips, slide each hose retaining clip off the filter assembly cover nipples with a pair of pliers. Disconnect the hoses from the cover and plug the hoses to prevent fuel leakage.

   b. On models with plastic clips, unsnap the plastic clips holding the hoses to the filter assembly cover. Disconnect the hoses from the cover and plug the hoses with golf tees to prevent fuel leakage.

   c. Remove the nut securing the filter cover to its mounting bracket. Remove the cover and canister assembly.

2. Unscrew the canister from the filter assembly cover. Remove the filter element from the canister.

3. Drain the canister and wipe the inside dry with a clean lint-free cloth or paper towel.

4. Inspect the cover O-ring. If necessary, remove and discard the old O-ring or gasket, as required.

To Install:

5. Clean the filter screen with solvent to remove any particles. If the filter screen is clogged or damaged, replace it.

6. Install the undamaged O-ring, or a new O-ring seal/gasket to the filter canister, as required.

7. Install the filter and the filter canister. Tighten the canister securely.

8. If the filter housing assembly was repositioned for access, install it as follows:

   a. Connect the inlet and outlet hoses to the canister cover.

   b. Install the nut securing the filter cover to its mounting bracket and tighten securely.

   c. On models with metal clips, slide each hose retaining clip onto the filter assembly cover nipples with a pair of pliers.

   d. On models with plastic clips, snap the plastic clips holding the hoses to the filter assembly cover.

9. Check the fuel filter installation for leakage by priming the fuel system with the fuel line primer bulb.

V6 Engines
◆ See Figures 63 and 64

Although most filter assemblies are pretty straight forward, containing a filter element and a single O-ring seal, some models may differ containing additional components such as a water indicator float, a spring, or additional seals. Keep track of the components as they are removed to ensure you will know how to properly assemble them.

The filter assembly found V6 motors is very similar to the serviceable filter used on many inline engines, however the major difference comes in the use of a lock-ring to secure the filter bowl. In addition most utilize a lock-ring retaining clip which bolts to the top of the housing to prevent the lock-ring from possibly loosening in service.

1. If equipped with a lock-ring retaining clip, loosen the bolt and remove the retaining clip so the lock-ring can be loosened.

2. Unscrew the ring nut securing the canister to the filter assembly cover.

3. Unscrew the canister from the filter assembly cover.

4. Remove the filter element and spring from the canister (along with any other components, such as a water indicator float).

5. Drain the canister and wipe the inside dry with a clean lint-free cloth or paper towel.

6. Inspect the cover O-ring. If necessary, remove and discard the old O-ring or gasket, as required.

To Install:

7. Clean the filter screen with solvent to remove any particles. If the filter screen is clogged or damaged, replace it.

8. Install the undamaged O-ring, or a new O-ring seal/gasket to the filter canister, as required.

9. Install the spring and filter into the filter canister (along with any other components as noted during removal).

10. Install the canister assembly onto the filter assembly cover and screw on the ring nut. Tighten the ring nut securely.

11. If equipped, install the lock-ring retaining clip and secure using the fastener.

12. Check the fuel filter installation for leakage by priming the fuel system with the fuel line primer bulb.

Propeller
◆ See Figures 65, 66 and 67

As you know, the propeller is actually what moves the boat through the water. The propeller operates in water in much the same manner as a wood screw or auger passing through wood. The propeller "bites" into the water as it rotates. Water passes between the blades and out to the rear in the shape of a cone. This "biting" through the water is what propels the boat.

All Yamaha and Mercury/Mariner outboards are equipped, from the factory, with a through-the-propeller exhaust, meaning that exhaust gas is routed out through the propeller.

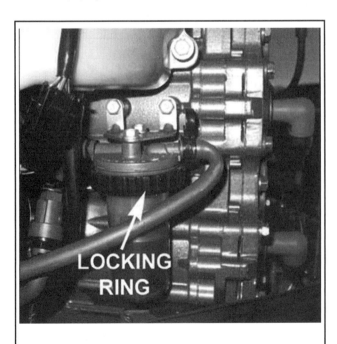

Fig. 63 Typical canister-type filter used on V6 motors

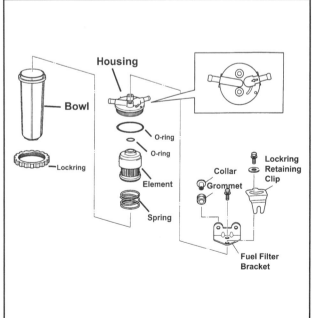

Fig. 64 Exploded view of a canister-type filter equipped with lockring and lockring retaining clip

2-22 MAINTENANCE & TUNE-UP

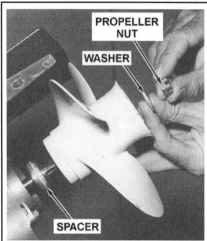

Fig. 65 Most Yamaha and some Merc/Mariner propellers are secured using a washer, castle nut and cotter pin

Fig. 66 An installed view of a nut retained propeller

Fig. 67 A severely damaged propeller (it's dead Jim)

GENERAL INFORMATION

Diameter and Pitch

◆ See Figures 68 and 69

Only two dimensions of the propeller are of real interest to the boat owner: diameter and pitch. These two dimensions are stamped on the propeller hub and always appear in the same order, the diameter first and then the pitch. Propellers furnished with the outboard by Yamaha also have a letter designation following the pitch size. This letter indicates the propeller type. For instance, the numbers and letter 9 7/8 x 10 1/2 - F stamped on the back of one blade indicates the propeller diameter to be 9 7/8 in., with a pitch of 10 1/2 in., and it is a Type F.

The diameter is the measured distance from the tip of one blade to the tip of the other.

The pitch of a propeller is the angle at which the blades are attached to the hub. This figure is expressed in inches of water travel for each revolution of the propeller. In our example of a 9 7/8 in. x 10 1/2 in., the propeller should travel 10 1/2 inches through the water each time it revolves. If the propeller action was perfect and there was no slippage, then the pitch multiplied by the propeller rpm would be the boat speed.

Most outboard manufacturers equip their units with a standard propeller, having a diameter and pitch they consider to be best suited to the engine and boat. Such a propeller allows the engine to run as near to the rated rpm and horsepower (at full throttle) as possible for the boat design.

The blade area of the propeller determines its load-carrying capacity. A two-blade propeller is used for high-speed running under very light loads.

A four-blade propeller is installed in boats intended to operate at low speeds under very heavy loads such as tugs, barges or large houseboats. The three-blade propeller is the happy medium covering the wide range between high performance units and load carrying workhorses.

Propeller Selection

There is no one propeller that will do the proper job in all cases. The list of sizes and weights of boats is almost endless. This fact, coupled with the many boat-engine combinations, makes the propeller selection for a specific purpose a difficult task. Actually, in many cases the propeller may be changed after a few test runs. Proper selection is aided through the use of charts set up for various engines and boats. These charts should be studied and understood when buying a propeller. However, bear in mind that the charts are based on average boats with average loads; therefore, it may be necessary to make a change in size or pitch, in order to obtain the desired results for the hull design or load condition.

Propellers are available with a wide range of pitch. Remember, a low pitch design takes a smaller bite of water than a high pitch propeller. This means the low pitch propeller will travel less distance through the water per revolution. However, the low pitch will require less horsepower and will allow the engine to run faster.

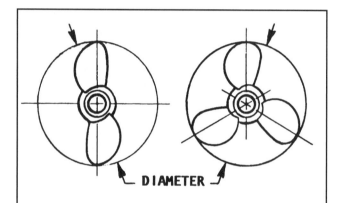

Fig. 68 Diameter and pitch are the two basic dimensions of a propeller. Diameter is measured across the circumference of a circle scribed by the propeller blades.

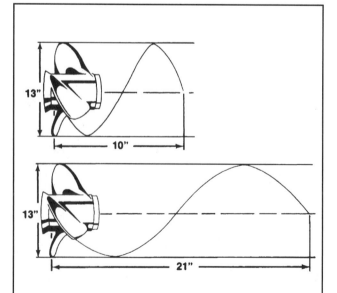

Fig. 69 This diagram illustrates the pitch dimension of a propeller. The pitch is the theoretical distance a propeller would travel through water if there were no friction

MAINTENANCE & TUNE-UP 2-23

All engine manufacturers design their units to operate with full throttle at, or in the upper end of the specified operating rpm. If the powerhead is operated at the rated rpm, several positive advantages will be gained.
- Spark plug life will be increased.
- Better fuel economy will be realized.
- Steering effort is often reduced.
- The boat and power unit will provide best performance.

Therefore, take time to make the proper propeller selection for the rated rpm of the engine at full throttle with what might be considered an average load. The boat will then be correctly balanced between engine and propeller throughout the entire speed range.

A reliable tachometer must be used to measure powerhead speed at full throttle, to ensure that the engine achieves full horsepower and operates efficiently and safely. To test for the correct propeller, make a test run in a body of smooth water with the lower unit in forward gear at full throttle. If the reading is above the manufacturer's recommended operating range, try propellers of greater pitch, until one is found allowing the powerhead to operate continually within the recommended full throttle range.

If the engine is unable to deliver top performance and the powerhead is properly tuned, then the propeller may not be to blame. Operating conditions have a marked effect on performance. For instance, an engine will lose rpm when run in very cold water. It will also lose rpm when run in salt water, as compared with fresh water. A hot, low-barometer day will also cause the engine to lose power.

Cavitation

◆ See Figure 70

Cavitation is the forming of voids in the water just ahead of the propeller blades. Marine propulsion designers are constantly fighting the battle against the formation of these voids, due to excessive blade tip speed and engine wear. The voids may be filled with air or water vapor, or they may actually be a partial vacuum. Cavitation may be caused by installing a piece of equipment too close to the lower unit, such as the knot indicator pickup, depth sounder or bait tank pickup.

Vibration

The propeller should be checked regularly to ensure that all blades are in good condition. If any of the blades become bent or nicked, this condition will set up vibrations in the drive unit and motor. If the vibration becomes very serious, it will cause a loss of power, efficiency, and boat performance. If the vibration is allowed to continue over a period of time, it can have a damaging effect on many of the operating parts.

Vibration in boats can never be completely eliminated, but it can be reduced by keeping all parts in good working condition and through proper maintenance and lubrication. Vibration can also be reduced in some cases by increasing the number of blades. For this reason, many racers use two-blade propellers, while luxury cruisers have four- and five-blade propellers installed.

Shock Absorbers

◆ See Figure 71

The shock absorber in the propeller plays a very important role in protecting the shafting, gears and engine against the shock of a blow, should the propeller strike an underwater object. The shock absorber allows the propeller to stop rotating at the instant of impact, while the power train continues turning.

How much impact the propeller is able to withstand, before causing the shock absorber to slip, is calculated to be more than the force needed to propel the boat, but less than the amount that could damage any part of the power train. Under normal propulsion loads of moving the boat through the water, the hub will not slip. However, it will slip if the propeller strikes an object with a force that would be great enough to stop any part of the power train.

If the power train was to absorb an impact great enough to stop rotation, even for an instant, something would have to give, resulting in severe damage. If a propeller is subjected to repeated striking of underwater objects, it will eventually slip on its clutch hub under normal loads. If the propeller should start to slip, a new shock absorber/cushion hub will have to be installed by a propeller repair shop.

Propeller Rake

◆ See Figure 72

If a propeller blade is examined on a cut extending directly through the center of the hub, and if the blade is set vertical to the propeller hub, the propeller is said to have a zero degree (0°.) rake. As the blade slants back, the rake increases. Standard propellers have a rake angle from 0° to 15°.

A higher rake angle generally improves propeller performance in a cavitating or ventilating situation. On lighter, faster boats, a higher rake often will increase performance by holding the bow of the boat higher.

Progressive Pitch

◆ See Figure 73

Progressive pitch is a blade design innovation that improves performance when forward and rotational speed is high and/or the propeller breaks the surface of the water.

Progressive pitch starts low at the leading edge and progressively increases to the trailing edge. The average pitch over the entire blade is the number assigned to that propeller. In the illustration of the progressive pitch, the average pitch assigned to the propeller would be 21.

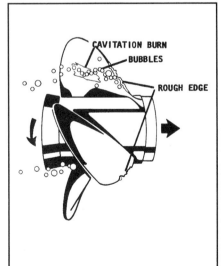

Fig. 70 Cavitation (air bubbles) can damage a prop

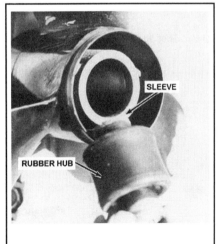

Fig. 71 A damaged rubber hub could cause the propeller to slip

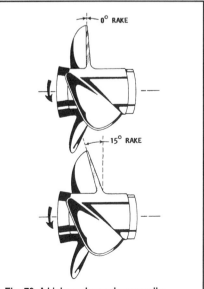

Fig. 72 A higher rake angle generally improves propeller performance

2-24 MAINTENANCE & TUNE-UP

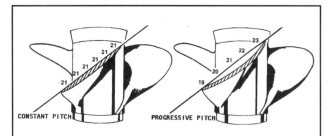

Fig. 73 Comparison of a constant and progressive pitch propeller. Notice how the pitch of the progressive propeller (right) changes to give the blade more thrust

Cupping

◆ See Figure 74

If the propeller is cast with an edge curl inward on the trailing edge, the blade is said to have a cup. In most cases, cupped blades improve performance. The cup helps the blades to HOLD and not break loose, when operating in a cavitating or ventilating situation.

A cup has the effect of adding to the propeller pitch. Cupping usually will reduce full-throttle engine speed about 150-300 rpm below that of the engine equipped with the same pitch propeller without a cup to the blade. A propeller repair shop is able to increase or decrease the cup on the blades. This change, as explained, will alter powerhead rpm to meet specific operating demands. Cups are rapidly becoming standard on propellers.

In order for a cup to be the most effective, the cup should be completely concave (hollowed) and finished with a sharp corner. If the cup has any convex rounding, the effectiveness of the cup will be reduced.

Rotation

◆ See Figure 75

Propellers are manufactured as right-hand (RH) rotation or left-hand (LH) rotation. The standard propeller for outboard units is RH rotation.

A right-hand propeller can easily be identified by observing it. Observe how the blade of the right-hand propeller slants from the lower left to upper right. The left-hand propeller slants in the opposite direction, from lower right to upper left.

When the RH propeller is observed rotating from astern the boat, it will be rotating clockwise when the outboard unit is in forward gear. The left-hand propeller will rotate counterclockwise.

High Performance Propellers

◆ See Figure 76 and 77

The term high performance is usually associated with, or has the connotation of, something used only for racing. The Yamaha high performance propeller does not fit this category and is not considered an aftermarket item.

The Yamaha high performance propeller is made of stainless steel with sophisticated designed blades, and carries an embossed **P** for positive identification.

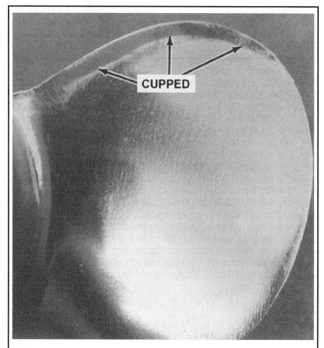

Fig. 74 Propeller with a cupped leading edge. Cupping gives the propeller a better hold in the water

The accompanying illustration of a high performance propeller clearly shows the unique design of the long blades and other features. This propeller is the weed less type, having extra sharp blades.

Installation of a high performance propeller requires raising the transom height, trim out after planing, and installation of a special design trim tab. Installation of a water pressure gauge is highly recommended, because raising the transom height will affect the amount of water entering the lower unit through the intake holes. An inadequate amount of water taken in will certainly cause powerhead cooling problems.

The high performance propeller has standard attaching hardware with the exception of the inner spacer, which is a special three-pronged design.

INSPECTION

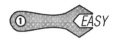

◆ See Figures 78, 79 and 80

The propeller should be inspected before and after each use to be sure the blades are in good condition. If any of the blades become bent or nicked, this condition will set up vibrations in the motor. Remove and inspect the propeller. Use a file to trim nicks and burrs. Take care not to remove any more material than is absolutely necessary.

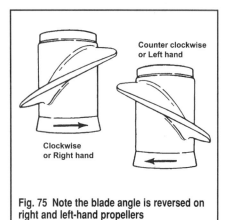

Fig. 75 Note the blade angle is reversed on right and left-hand propellers

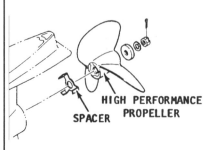

Fig. 76 Exploded view of a Yamaha high performance prop mounting (note 3-prong spacer)

Fig. 77 The blades on this high performance prop are narrow and have more pitch than a standard prop

MAINTENANCE & TUNE-UP 2-25

Fig. 78 This propeller is long overdue for repair or replacement

Fig. 79 Although minor damage can be dressed with a file...

Fig. 80 ...a propeller specialist should repair large nicks or damage

✱✱ CAUTION

Never run the engine with serious propeller damage, as it can allow for excessive engine speed and/or vibration that can damage the motor. Also, a damaged propeller will cause a reduction in boat performance and handling.

Also, check the rubber and splines inside the propeller hub for damage. If there is damage to either of these, take the propeller to your local marine dealer or a "prop shop". They can evaluate the damaged propeller and determine if it can be saved by rehubbing.

Additionally, the propeller should be removed AT LEAST every 100 hours of operation or at the end of each season, whichever comes first for cleaning, greasing and inspection. Whenever the propeller is removed, apply a fresh coating of a suitable water-resistant grease to the propeller shaft and the inner diameter of the propeller hub. This is necessary to prevent possible propeller seizure onto the shaft that could lead to costly or troublesome repairs. Also, whenever the propeller is removed, any material entangled behind the propeller should be removed before any damage to the shaft and seals can occur. This may seem like a waste of time at first, but the small amount of time involved in removing the propeller is returned many times by reduced maintenance and repair, including the replacement of expensive parts.

■ Propeller shaft greasing and debris inspection should occur more often depending upon motor usage. Frequent use in salt, brackish or polluted waters would make it advisable to perform greasing more often. Similarly, frequent use in areas with heavy marine vegetation, debris or potential fishing line would necessitate more frequent removal of the propeller to ensure the gearcase seals are not in danger of becoming cut.

REMOVAL & INSTALLATION

◆ See Figures 81 thru 91

The propeller is secured to the gearcase propshaft either by a nut and lockwasher or cotter pin. The propeller itself is driven by a splined connection to the shaft and the rubber drive hub found inside the propeller. The rubber hub provides a cushioning that allows softer shifts, but more importantly, it provides some measure of protection for the gearcase components in the event of an impact. In this way, the hope is that the propeller and hub will be sacrificed in the event of a collision, but the more expensive gearcase components will survive unharmed. Although these systems do supply a measure of protection, this, unfortunately, is not always the case and gearcase component damage will still occur with the right impact or with a sufficient amount of force.

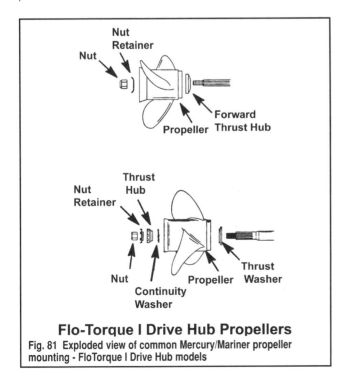

Fig. 81 Exploded view of common Mercury/Mariner propeller mounting - FloTorque I Drive Hub models

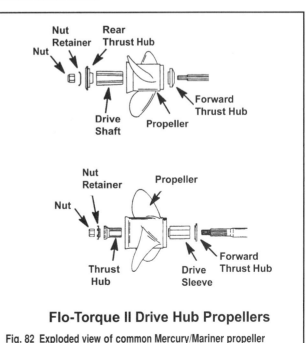

Fig. 82 Exploded view of common Mercury/Mariner propeller mounting - FloTorque II Drive Hub models

2-26 MAINTENANCE & TUNE-UP

Fig. 83 Most Yamaha props (and some Mercs) are secured by a nut and cotter pin

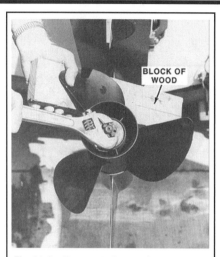

Fig. 84 In all cases, to loosen the nut use a block of wood to keep the prop from turning

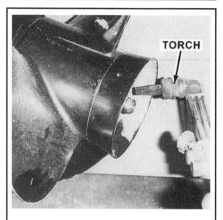

Fig. 85 A stuck propeller can be freed with heat. . .

✻✻ WARNING

Do not use excessive force when removing the propeller from the hub as excessive force can result in damage to the propeller, shaft and, even other gearcase components. If the propeller cannot be removed by normal means, consider having a reputable marine shop remove it. The use of heat or impacts to free the propeller will likely lead to damage.

■ Clean and lubricate the propeller and shaft splines using a high-quality, water-resistant, marine grease every time the propeller is removed from the shaft. This will help keep the hub from seizing to the shaft due to corrosion (which would require special tools to remove without damage to the shaft or gearcase.)

Many outboards are equipped with aftermarket propellers. Because of this, the attaching hardware may differ slightly from what is shown. Contact a reputable propeller shop or marine dealership for parts and information on other brands of propellers.

On almost all Yamaha outboards the propeller is held in place over the shaft splines by a large castellated nut. The nut is so named because, when viewed from the side, it appears similar to the upper walls or tower of a castle.

On most Mercury/Mariner outboards, the propeller is held in place over the shaft splines by a nut and some form of tabbed nut retainer/lock-washer. The most popular versions of these propellers are the FloTorque I and FloTorque II drive hubs, which come in various sizes and contain slightly different combinations of thrust hubs, retainers, drive sleeves etc.

■ When servicing all propellers, ESPECIALLY Mercury/Mariners, take note of the order of assembly and the direction each propeller retaining component faces as they are removed. Be sure to assemble each component in the same order and facing the same direction as noted during removal.

For safety, the nut is locked in place by a cotter pin or by tabs of a retainer that keep it from loosening while the motor is running. When equipped with a cotter pin, the pin passes through a hole in the propeller shaft, as well as through the notches in the sides of the castellated nut. When equipped with a tabbed retainer, the tabs (usually at least 3) are bend into position against the nut, keeping it from turning.

■ Be sure to install a new cotter pin anytime the propeller is removed and, perhaps more importantly, make sure the cotter pin is of the correct size and is made of materials designed for marine use. The tabbed retainer used by most Mercs can usually be reused, but inspect the tabs for weakening and replace it when one or more of the tabs become questionable.

Whenever working around the propeller, check for the presence of black rubber material in the drive hub (don't confuse bits of black soot/carbon deposits from the exhaust as rubber hub material) and spline grease. Presence of this material normally indicates that the hub has turned inside the propeller bore (have the propeller checked by a propeller repair shop). Keep in mind that a spun hub will not allow proper torque transfer from the motor to the propeller and will allow the engine to over-rev in attempting to produce thrust. If the propeller has spun on the hub it has been weakened and is more likely to fail completely in use.

1. For safety, disconnect the negative cable (if so equipped) and/or disconnect the spark plug leads from the plugs (ground the leads to prevent possible ignition damage should the motor be cranked at some point before the leads are reconnected to the spark plugs).

✻✻ CAUTION

Don't ever take the risk of working around the propeller if the engine could accidentally be started. Always take precautions such as disconnecting the spark plug leads and, if equipped, the negative battery cable.

2. If equipped with a cotter pin, cut the ends off the pin using wire cutters (as that is usually easier than trying to straighten them in most cases) or straighten the ends of the pin using a pair of pliers, whichever you prefer. Next, free the pin by grabbing the head with a pair of needle-nose pliers. Either tap on the pliers gently with a hammer to help free the pin from the nut or carefully use the pliers as a lever by prying back against the castellated nut. Discard the cotter pin once it is removed.

3. If equipped with a tabbed lock-washer/retainer, use a small punch to carefully bend the tabs away from the nut.

4. Place a block of wood between the propeller and the anti-ventilation housing to lock the propeller and shaft from turning, then loosen and remove the castellated nut. Note the orientation, then remove the washer and/or splined spacer (on most Mercs this would mean a nut retainer, followed by a drive sleeve on FloTorque II models), from the propeller shaft.

5. Slide the propeller from the shaft. If the prop is stuck, use a block of wood to prevent damage and carefully drive the propeller from the shaft.

■ If the propeller is completely seized on the shaft, have a reputable marine or propeller shop free it. Don't risk damage to the propeller or gearcase by applying excessive force.

6. Note the direction in which the inner or forward thrust washer is facing (since some motors or aftermarket props may use a thrust washer equipped with a fishing line trap that must face the proper direction if it is to protect the gearcase seal). Remove the thrust washer from the propshaft (if the washer appears stuck, tap lightly to free it from the propeller shaft).

7. Clean the thrust washer(s), propeller, drive sleeve and shaft splines (as equipped) of any old grease. Small amounts of corrosion can be removed carefully using steel wool or fine grit sandpaper.

8. Inspect the shaft for signs of damage including twisted splines or excessively worn surfaces. Rotate the shaft while looking for any deflection. Replace the propeller shaft if these conditions are found. Inspect the thrust washer for signs of excessive wear or cracks and replace, if found.

MAINTENANCE & TUNE-UP 2-27

Fig. 86 ... but you'll likely destroy the rubber hub

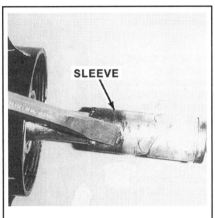

Fig. 87 If so, use a chisel to careful remove the frozen sleeve

Fig. 88 For installation position the inner spacer or thrust hub (as equipped) first...

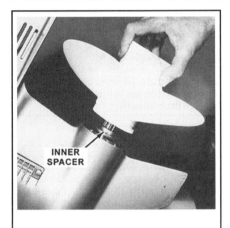

Fig. 89 ... on most models with the shoulder facing the prop

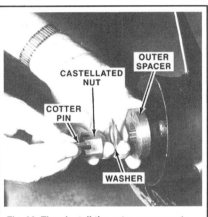

Fig. 90 Then install the outer components such as outer spacer, washer (if used), nut and cotter pin (as applicable)

Fig. 91 Assembled and ready to go!

To Install:

9. Apply a fresh coating of a water-resistant grease to all surfaces of the propeller shaft and to the splines inside the propeller hub.

■ **Some people prefer to use Anti-Seize on the hub splines, which is acceptable in most applications, but if used you should double-check after the first 10-20 hours of service to make sure the anti-seize is holding up to operating conditions.**

10. Position the inner (forward) thrust washer over the propshaft in the direction noted during removal. (Generally speaking, the flat shoulder should face rearward toward the propeller while the bevel faces forward toward the gearcase).
11. On Mercury/Mariner motors which contain the larger Flo-Torque II Drive Hub Propeller there is normally a drive sleeve which is installed at this point. On the smaller models the sleeve usually installed from the other side of the propeller.
12. Carefully slide the propeller onto the propshaft, rotating the propeller to align the splines. Push the propeller forward until it seats against the thrust washer.
13. Install the splined and/or plain spacer (or drive sleeve and nut retainer, as applicable) onto the propeller shaft, as equipped.
14. Place a block of wood between the propeller and housing to hold the prop from turning, then thread the castellated nut onto the shaft with the cotter pin grooves facing outward.
15. Tighten the nut to specification using a suitable torque wrench.

16. Yamaha propeller nut torque specifications are as follows:
- 2.5 and 4 hp motors: there is no specification, snug the nut making sure the cotter pin holes are exposed
- 6-15 hp motors (except the 9.9 hp model): 12.5 ft. lbs. (17 Nm)
- 9.9 hp motors: 9.4 ft. lbs. (13 Nm) for standard models or 15 ft. lbs. (21 Nm) for high-thrust models
- 25 hp motors: 25 ft. lbs. (35 Nm) for standard models or 28 ft. lbs. (39 Nm) for high-thrust models
- 30/40 hp (747cc) 3-cylinder motors: 29 ft. lbs. (39 Nm)
- 40-115 hp 4-cylinder motors: 22-25 ft. lbs. (30-35 Nm)
- 150 hp 4-cylinder motors: 38.4 ft. lbs. (52 Nm)
- V6 motors: 40 ft. lbs. (55 Nm)

17. Mercury/Mariner propeller nut torque specifications are as follows:
- 4/5/6 hp motors: 12.5 ft. lbs./150 inch lbs. (17 Nm)
- 8/9.9 hp (232cc) motors: 70 inch lbs. (7.9 Nm)
- 9.9/15 hp (323cc) motors: 100 inch lbs. (11.3 Nm) for standard models or 17 ft. lbs. (22.6 Nm) for Bigfoot models
- 25 hp and larger motors: 55 ft. lbs. (75 Nm)

18. For tabbed retainer models, bend the tabs up against the flats on the propeller. For most of the larger models you should be able to bend three tabs into position.
19. For cotter pin retained models, install a new cotter pin through the grooves in the nut that align with the hole in the propshaft. If the cotter pin hole and the grooves do not align, tighten or loosen the nut very slightly, just enough to align them. Once the cotter pin is inserted, spread the ends sufficiently to lock the pin in place.
20. Connect the spark plug leads and/or the negative battery cable, as applicable.

2-28 MAINTENANCE & TUNE-UP

Jet Drive Impeller

◆ See Figure 92

A jet drive motor uses an impeller enclosed in a jet drive housing instead of the propeller used by traditional gearcases. Outboard jet drives are designed to permit boating in areas prohibited to a boat equipped with a conventional propeller outboard drive system. The housing of the jet drive barely extends below the hull of the boat allowing passage in ankle deep water, white water rapids, and over sand bars or in shoal water which would foul a propeller drive.

The outboard jet drive provides reliable propulsion with a minimum of moving parts. It operates, simply stated, as water is drawn into the unit through an intake grille by an impeller. The impeller is driven by the driveshaft off the powerhead's crankshaft. Thrust is produced by the water that is expelled under pressure through an outlet nozzle that is directed away from the stern of the boat.

As the speed of the boat increases and reaches planing speed, only the very bottom of the jet drive where the intake grille is mounted facing downward remains in contact with the water.

The jet drive is provided with a reverse-gate arrangement and linkage to permit the boat to be operated in reverse. When the gate is moved downward over the exhaust nozzle, the pressure stream is deflected (reversed) by the gate and the boat moves sternward.

Conventional controls are used for powerhead speed, movement of the boat, shifting and power trim and tilt.

INSPECTION

Fig. 92 Instead of a traditional gearcase, jet drives use an impeller (it's sort of an enclosed propeller) mounted in a jet drive housing that just barely extends below the boat's hull

◆ See Figure 93

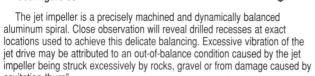

The jet impeller is a precisely machined and dynamically balanced aluminum spiral. Close observation will reveal drilled recesses at exact locations used to achieve this delicate balancing. Excessive vibration of the jet drive may be attributed to an out-of-balance condition caused by the jet impeller being struck excessively by rocks, gravel or from damage caused by cavitation "burn".

The term cavitation "burn" is a common expression used throughout the world among people working with pumps, impeller blades, and forceful water movement. These "burns" occur on the jet impeller blades from cavitation air bubbles exploding with considerable force against the impeller blades. The edges of the blades may develop small dime-size areas resembling a porous sponge, as the aluminum is actually "eaten" by the condition just described.

Excessive rounding of the jet impeller edges will reduce efficiency and performance. Therefore, the impeller and intake grate (that protects it from debris) should be inspected at regular intervals.

Before and after each use, make a quick visual inspection of the intake grate and impeller, looking for obvious signs of damage. Always clear any debris such as plastic bags, vegetation or other items that sometimes become entangled in the water intake grate before starting the motor. If the intake grate is damaged, do not operate the motor, or you will risk destroying the impeller if rocks or other debris are drawn upward by the jet drive. If possible, replace a damaged grate before the next launch. This makes inspection after use all that much more important. Imagine the disappointment if you only learn of a damaged grate while inspecting the motor immediately prior to the next launch.

An obviously damaged impeller should be removed and either repaired or replaced depending on the extent of the damage. If rounding is detected, the impeller can be placed on a work bench and the edges restored to as sharp a condition as possible, using a file. Draw the file in only one direction. A back-and-forth motion will not produce a smooth edge. Take care not to nick the smooth surface of the jet impeller. Excessive nicking or pitting will create water turbulence and slow the flow of water through the pump. For more details on impeller replacement or service, please refer to the information on Jet Drives in the Gearcase section of this manual.

CHECKING IMPELLER CLEARANCE

◆ See Figures 94 and 95

Proper operation of the jet drive depends upon the ability to create maximum thrust. In order for this to occur the clearance between the outer edge of the jet drive impeller and the water intake housing cone wall should

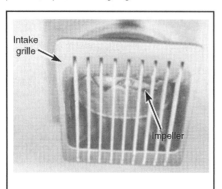

Fig. 93 Visually inspect the intake grate and impeller with each use.

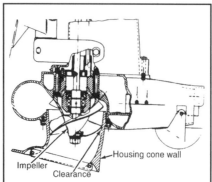

Fig. 94 Jet drive impeller clearance is the gap between the edges of the impeller and its housing

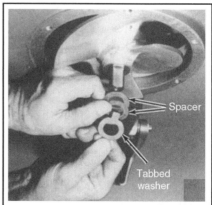

Fig. 95 Impeller clearance is adjusted using shims below and above the impeller

MAINTENANCE & TUNE-UP 2-29

be maintained at approximately 1/32 in. (0.79mm). This distance can be checked visually by shining a flashlight up through the intake grille and estimating the distance between the impeller and the casing cone, as indicated in the accompanying illustrations. But, it is not humanly possible to accurately measure this clearance by eye. Close observation between outings is fine to maintain a general idea of impeller condition, but, at least annually, the clearance must be measured using a set of long feeler gauges.

** CAUTION

Whenever working around the impeller, ALWAYS disconnect the negative battery cable and/or disconnect the spark plug leads to make sure the engine cannot be accidentally started during service. Failure to heed this caution could result in serious personal injury or death in the event that the engine is started.

When checking clearance, a feeler gauge larger than the clearance specification should not fit between the tips of the impeller and the housing. A gauge within specification should fit, but with a slight drag. A smaller gauge should fit without any interference whatsoever. Check using the feeler gauge at various points around the housing, while slowly rotating the impeller by hand.

After continued normal use, the clearance will eventually increase. In anticipation of this the manufacturer mounts the impeller deep in a tapered housing, and positions spacers beneath the impeller to hold it in position. The spacers are used to position the impeller along the driveshaft with the desired clearance between the jet impeller and the housing wall. When clearance has increased, spacers are removed from underneath the impeller and repositioned behind it, dropping the impeller slightly in the housing and thereby decreasing the clearance again.

If adjustment is necessary, refer to the Jet Drive procedures under Gearcase in this manual for impeller removal, shimming and installation procedures. Follow the appropriate parts of the Removal & Disassembly, as well as Assembly procedures for impeller service.

Anodes (Zincs)

◆ See Figures 96, 97 and 98

The idea behind anodes (also known as sacrificial anodes) is simple: When dissimilar metals are dunked in water and a small electrical current is leaked between or amongst them, the less-noble metal (galvanically speaking) is sacrificed (corrodes).

The zinc alloy of which most anodes are made is designed to be less noble than the aluminum alloy of which your outboard is constructed. If there's any electrolysis, and there almost always is, the inexpensive zinc anodes are consumed in lieu of the expensive outboard motor.

■ We say the zinc alloy of which MOST anodes are made because the anodes recommended for fresh water applications may also be made of a different material. Some manufacturers recommend that you use magnesium anodes for motors used solely in freshwater applications. Magnesium offers a better protection against corrosion for aluminum motors, however the ability to offer this protection generally makes it too active a material for use in salt water applications. Because zinc is the more common material many people (including us) will use the word zinc interchangeably with the word anode, don't let this confuse you as we are referring equally to all anodes, whether they are made of zinc or magnesium.

These zincs require a little attention in order to make sure they are capable of performing their function. Anodes must be solidly attached to a clean mounting site. Also, they must not be covered with any kind of paint, wax or marine growth.

INSPECTION

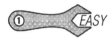

◆ See Figures 96, 97 and 98

Visually inspect all accessible anodes, especially gearcase mounted ones, before and after each use. You'll want to know right away if it has become loose or fallen off in service. Periodically inspect them closely to make sure they haven't eroded too much. At a certain point in the erosion process, the mounting holes start to enlarge, which is when the zinc might fall off. Obviously, once that happens your engine no longer has any protection. Generally, a zinc anode is considered worn if it has shrunken to 2/3 or less than the original size. To help judge this, buy a spare and keep it handy (in the boat or tow vehicle for comparison).

If you use your outboard in salt water or brackish water, and your zincs never seem to wear, inspect them carefully. Paint, wax or marine growth on zincs will insulate them and prevent them from performing their function properly. They must be left bare and must be installed onto bare metal of the motor. If the zincs are installed properly and not painted or waxed, inspect around them for sings of corrosion. If corrosion is found, strip it off immediately and repaint with a rust inhibiting paint. If in doubt, replace the zincs.

On the other hand, if your zinc seems to erode in no time at all, this may be a symptom of the zincs themselves. Each manufacturer uses a specific blend of metals in their zincs. If you are using zincs with the wrong blend of metals, they may erode more quickly or leave you with diminished protection.

At least annually or whenever an anode has been removed or replaced, check the mounting for proper electrical contact using a multi-meter. Set the multi-meter to check resistance (ohms), then connect one meter lead to the anode and the other to a good, unpainted or un-corroded ground on the motor. Resistance should be very low or zero. If resistance is high or infinite, the anode is insulated and cannot perform its function properly.

Fig. 96 Extensive corrosion of an anode suggests a problem or a complete disregard for maintenance

Fig. 97 Although most Yamaha's use a trim tab anode, others are normally found on the gearcase and powerhead

Fig. 98 Lead wires are used to protect bracketed components

2-30 MAINTENANCE & TUNE-UP

SERVICING

◆ See Figures 99 thru 107

Depending on your boat, motor and rigging, you may have anywhere from one to four (or even more) anodes. Regardless of the number, there are some fundamental rules to follow that will give your boat and motor's sacrificial anodes the ability to do the best job protecting your boat's underwater hardware that they can.

■ On most models (25 hp and larger) the trim tab serves as the gearcase anode. If the anode must be replaced, make an alignment mark between the anode and gearcase before removal, then transfer the mark to the replacement anode to preserve trim tab adjustment. If the boat pulls to one side after replacement and did not previously, refer to the Trim Tab Adjustment procedure in the Gearcase section to correct this condition.

All motors covered by this manual are equipped with at least one gearcase anode, normally mounted in, on, or near the anti-ventilation plate. Many of the motors covered by this manual (again generally 25 hp and

Fig. 99 Some trim tab anodes are bolted from below...

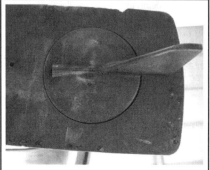

Fig. 100 ...while others are fastened from the top

Fig. 101 Remove the rubber cover...

Fig. 102 ...to access the anode retaining bolt

Fig. 103 Many motors have one...

Fig. 104 ..... or more anodes on the transom bracket

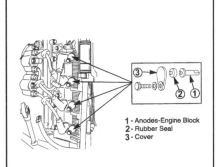

Fig. 105 Most of these models (25 hp and larger) use from 2 to 4 anodes in small bores on the engine block

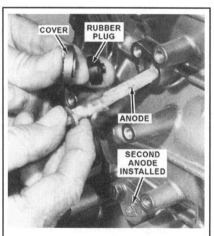

Fig. 106 To access block mounted anodes, remove the bolt, cover and rubber seal

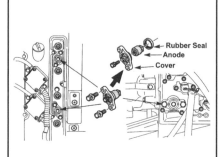

Fig. 107 The block mounted anodes mount a little differently on 75 hp and larger models, but it is the same idea

MAINTENANCE & TUNE-UP 2-31

larger) also have a powerhead mounted anode and/or an engine clamp bracket anode. Location of the powerhead zincs will vary slightly from motor-to-motor including mounting bosses specifically cast in the motor. Most multi-cylinder motors are equipped with one or more anodes on the engine mount clamp bracket.

Some people replace zincs annually. This may or may not be necessary, depending on the type of waters in which you boat and depending on whether or not the boat is hauled with each use or left in for the season. Either way, it is a good idea to remove zincs at least annually in order to make sure the mounting surfaces are still clean and free of corrosion.

■ **The 25 hp and larger motors are normally equipped with from 2 to 4 anodes mounted to the engine block under covers. Because they are not visible for daily inspection before and after each use, you should remove the covers and check them at least annually. During installation, tighten the cover retaining bolts to 70 inch lbs. (8 Nm).**

The first thing to remember is that zincs are electrical components and like all electrical components, they require good clean connections. So after you've undone the mounting hardware you want to get the zinc mounting sites clean and shiny.

Get a piece of coarse emery cloth or some 80-grit sandpaper. Thoroughly rough up the areas where the zincs attach (there's often a bit of corrosion residue in these spots). Make sure to remove every trace of corrosion as it could insulate the zinc from the motor.

Zincs are attached with stainless steel machine screws that thread into the mounting for the zincs. Over the course of a season, this mounting hardware is inclined to loosen. Mount the zincs and tighten the mounting hardware securely. Tap the zincs with a hammer hitting the mounting screws squarely. This process tightens the zincs and allows the mounting hardware to become a bit loose in the process. Now, do the final tightening. This will insure your zincs stay put for the entire season.

Timing Belt

INSPECTON

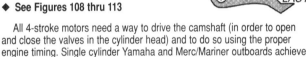

◆ See Figures 108 thru 113

All 4-stroke motors need a way to drive the camshaft (in order to open and close the valves in the cylinder head) and to do so using the proper engine timing. Single cylinder Yamaha and Merc/Mariner outboards achieve this using a gear attached to the camshaft that is timed to and driven off a gear on the crankshaft. However, more similar to automotive engines, all multi-cylinder Yamaha and Merc/Mariner outboards use a timing belt to synchronize the camshaft and crankshaft (for correct valve timing).

When equipped, the timing belt is a long life component that does not require much in the way or service, but we would recommend that you inspect it at least once every year. Although the manufacturer does not provide a specific replacement interval for all models, experience shows that it is wise to replace it at least every 5 years or anytime after about 1000 hours of operation, whichever comes first. Keep in mind, a timing belt that breaks or even slips a tooth will likely disable the motor, possibly stranding the boat.

✱✱ WARNING

Warnings that the manufacturer give against rotating the crankshaft with the timing belt removed makes us believe that some of these are interference motors. This means that if the belt were to snap suddenly during service with the engine running, there is a good chance that one or more valves would be left in the open position and would hit by a piston as it traveled upwards due to the remaining force of the spinning flywheel. This could destroy the piston, valve and/or cylinder head. Periodic inspection and replacement of the timing belt on interference motors is even more important. To help determine if your motor is an interference model, please refer to the Cylinder Head and Camshaft procedures found in the Powerhead section.

■ **We did not find any references leading us to believe that some of the other 4-stroke motors were interference, but that doesn't mean they are definitely not (except for the V6 which was specifically designed with pistons that are carved out of forged stock to prevent valve strikes), so use caution (and inspect the timing belt frequently).**

On some motors the timing belt may be partially visible at one point in the flywheel cover, but a thorough inspection is really only possible (and much easier) once the flywheel cover assembly is removed.

1. For safety when working around the flywheel, disconnect the negative battery cable and/or disconnect the leads from the spark plugs, then ground the leads on the powerhead.

■ **Although not absolutely necessary for this procedure, it is a good idea to remove the spark plugs at this time. Removing the spark plugs will relieve engine compression, making it easier to manually rotate the motor. Also, it presents a good opportunity to inspect, clean and/or replace the plugs.**

2. Remove the manual starter assembly or the flywheel cover, as applicable, for better access to the timing belt.
3. Use low-pressure compressed air to blow debris out from under the camshaft pulley, flywheel and timing belt.
4. Visually check the belt for worn, cracked or oil soaked surfaces. Slowly rotate the flywheel (by hand) while inspecting all of the timing belt cogs.
5. Visually check the camshaft pulley and flywheel teeth for worn, cracked, chipped or otherwise damaged surfaces.
6. On 9.9 and 15 hp motors use light thumb pressure at the center of the longest span on the timing belt. It should deflect no more than 0.40 in. (10mm) or the belt requires attention (adjustment or replacement if it cannot be adjusted, for more details please refer to the Powerhead section).

■ **Most larger models utilize a spring-loaded belt tensioner that automatically applies the correct amount of tension to the belt. If the belt seems loose, refer to the timing belt replacement information under Powerhead. The tensioner mounting bolts may be loosened (allowing the spring to readjust tension) and then retightened to see if it takes care of the excess slack. If this doesn't do the trick the belt should usually be replaced.**

Fig. 108 The flywheel cover should be removed...

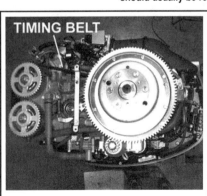

Fig. 109 ...in order to thoroughly inspect the timing belt

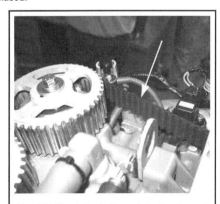

Fig. 110 Check the belt carefully for signs of wear, age or damage

2-32 MAINTENANCE & TUNE-UP

7. If the belt, pulleys or tensioner (if applicable) are damaged, replace them as described under Powerhead.
8. If removed, install the manual starter assembly or flywheel cover to the powerhead.
9. Install the spark plugs, then connect the leads followed by the negative battery cable and the engine cover.

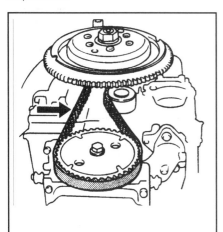

Fig. 111 Timing belt inspection on 8/9.9 and 9/15 hp motors includes measuring deflection

Fig. 112 Timing belt tension is set by the tensioner on most larger motors

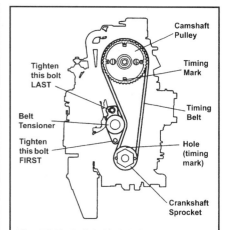

Fig. 113 Typical timing belt installation and tensioner components (40-50 hp/935cc models shown)

BOAT MAINTENANCE

Batteries

◆ See Figures 114 and 115

Batteries require periodic servicing, so a definite maintenance program will help ensure extended life. A failure to maintain the battery in good order can prevent it from properly charging or properly performing its job even when fully charged. Low levels of electrolyte in the cells, loose or dirty cable connections at the battery terminals or possibly an excessively dirty battery top can all contribute to an improperly functioning battery. So battery maintenance, first and foremost, involves keeping the battery full of electrolyte, properly charged and keeping the casing/connections clean of corrosion or debris.

If a battery charges and tests satisfactorily but still fails to perform properly in service, one of three problems could be the cause.

1. An accessory left on overnight or for a long period of time can discharge a battery.
2. Using more electrical power than the stator assembly or lighting coil can replace would slowly drain the battery during motor operation, resulting in an undercharged condition.
3. A defect in the charging system. A faulty stator assembly or lighting coil, defective regulator or rectifier or high resistance somewhere in the system could cause the battery to become undercharged.

■ For more information on marine batteries, please refer to BATTERY in the Ignition and Electrical Systems section.

MAINTENANCE

◆ See Figures 115 thru 118

Electrolyte Level

The most common and important procedure in battery maintenance is checking the electrolyte level. On most batteries, this is accomplished by removing the cell caps and visually observing the level in the cells. The bottom of each cell normally is equipped with a split vent which will cause the surface of the electrolyte to appear distorted when it makes contact.

Fig. 114 Explosive hydrogen gas is released from the batteries in a discharged state. This one exploded when something ignited the gas. Explosions can be caused by a spark from the battery terminals or jumper cables

Fig. 115 Ignoring a battery (and corrosion) to this extent is asking for it to fail

MAINTENANCE & TUNE-UP 2-33

When the distortion first appears at the bottom of the split vent, the electrolyte level is correct. Smaller marine batteries are sometimes equipped with translucent cases that are printed or embossed with high and low level markings on the side. On some of these, shining a flashlight through the battery case will help make it easier to determine the electrolyte level.

During hot weather and periods of heavy use, the electrolyte level should be checked more often than during normal operation. Add distilled water to bring the level of electrolyte in each cell to the proper level. Take care not to overfill, because adding an excessive amount of water will cause loss of electrolyte and any loss will result in poor performance, short battery life and will contribute quickly to corrosion.

■ **Never add electrolyte from another battery. Use only distilled water. Even tap water may contain minerals or additives that will promote corrosion on the battery plates, so distilled water is always the best solution.**

Although less common in marine applications than other uses today, sealed maintenance-free batteries also require electrolyte level checks, through the window built into the tops of the cases. The problem for marine applications is the tendency for deep cycle use to cause electrolyte evaporation and electrolyte cannot be replenished in a sealed battery. Although, more and more companies are producing a maintenance-free batteries for marine applications and their success should be noted.

The second most important procedure in battery maintenance is periodically cleaning the battery terminals and case.

Cleaning

Dirt and corrosion should be cleaned from the battery as soon as it is discovered. Any accumulation of acid film or dirt will permit a small amount of current to flow between the terminals. Such a current flow will drain the battery over a period of time.

Clean the exterior of the battery with a solution of diluted ammonia or a paste made from baking soda and water. This is a base solution that will neutralize any acid that may be present. Flush the cleaning solution off with plenty of clean water.

✱✱ WARNING

Take care to prevent any of the neutralizing solution from entering the cells as it will quickly neutralize the electrolyte (ruining the battery).

Poor contact at the terminals will add resistance to the charging circuit. This resistance may cause the voltage regulator to register a fully charged battery and thus cut down on the stator assembly or lighting coil output adding to the low battery charge problem.

At least once a season, the battery terminals and cable clamps should be cleaned. Loosen the clamps and remove the cables, negative cable first. On batteries with top mounted posts, if the terminals appear stuck, use a puller specially made for this purpose to ensure the battery casing is not damaged. NEVER pry a terminal off a battery post. Battery terminal pullers are inexpensive and available in most parts stores.

Clean the cable clamps and the battery terminal with a wire brush until all corrosion, grease, etc., is removed and the metal is shiny. It is especially important to clean the inside of the clamp thoroughly (a wire brush or brush part of a battery post cleaning tool is useful here), since a small deposit of foreign material or oxidation there will prevent a sound electrical connection and inhibit either starting or charging. It is also a good idea to apply some dielectric grease to the terminal, as this will aid in the prevention of corrosion.

After the clamps and terminals are clean, reinstall the cables, negative cable last, do not hammer the clamps onto battery posts. Tighten the clamps securely but do not distort them. To help slow or prevent corrosion, give the clamps and terminals a thin external coating of grease after installation.

Check the cables at the same time that the terminals are cleaned. If the insulation is cracked or broken or if its end is frayed, that cable should be replaced with a new one of the same length and gauge.

TESTING

◆ See Figure 119

A quick check of the battery is to place a voltmeter across the terminals. Although this is by no means a clear indication, it gives you a starting point when trying to troubleshoot an electrical problem that could be battery related. Most marine batteries will be of the 12 volt DC variety. They are constructed of 6 cells, each of which is capable of producing slightly more than two volts, wired in series so that total voltage is 12 and a fraction. A fully charged battery will normally show more than 12 and slightly less than 13 volts across its terminals. But keep in mind that just because a battery reads 12.6 or 12.7 volts does NOT mean it is fully charged. It is possible for it to have only a surface charge with very little amperage behind it to maintain that voltage rating for long under load. A discharged battery will read some value less than 12 volts, but can normally be brought back to 12 volts through recharging. Of course a battery with one or more shorted or un-chargeable cells will also read less than 12, but it cannot be brought back to 12+ volts after charging. For this reason, the best method to check battery condition on most marine batteries is through a specific gravity check or a load test.

Fig. 116 Place a battery terminal tool over posts, then rotate back and forth . . .

Fig. 117 . . . until the internal brushes expose a fresh, clean surface on the post

Fig. 118 Clean the insides of cable ring terminals using the tool's wire brush

2-34 MAINTENANCE & TUNE-UP

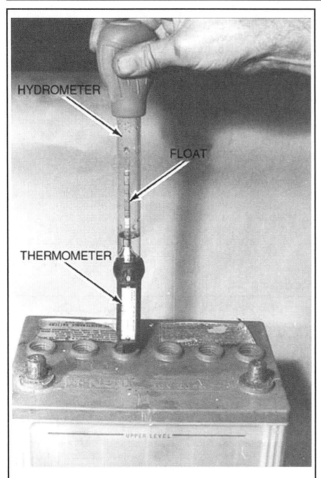

Fig. 119 A hydrometer is the best method for checking battery condition

A hydrometer is a device that measures the density of a liquid when compared to water (specific gravity). Hydrometers are used to test batteries by measuring the percentage of sulfuric acid in the battery electrolyte in terms of specific gravity. When the condition of the battery drops from fully charged to discharged, the acid is converted to water as electrons leave the solution and enter the plates, causing the specific gravity of the electrolyte to drop.

It may not be common knowledge but hydrometer floats are calibrated for use at 80°F (27°C). If the hydrometer is used at any other temperature, hotter or colder, a correction factor must be applied.

■ **Remember, a liquid will expand if it is heated and will contract if cooled. Such expansion and contraction will cause a definite change in the specific gravity of the liquid, in this case the electrolyte.**

A quality hydrometer will have a thermometer/temperature correction table in the lower portion, as illustrated in the accompanying illustration. By measuring the air temperature around the battery and from the table, a correction factor may be applied to the specific gravity reading of the hydrometer float. In this manner, an accurate determination may be made as to the condition of the battery.

When using a hydrometer, pay careful attention to the following points:
1. Never attempt to take a reading immediately after adding water to the battery. Allow at least 1/4 hour of charging at a high rate to thoroughly mix the electrolyte with the new water. This time will also allow for the necessary gases to be created.
2. Always be sure the hydrometer is clean inside and out as a precaution against contaminating the electrolyte.
3. If a thermometer is an integral part of the hydrometer, draw liquid into it several times to ensure the correct temperature before taking a reading.
4. Be sure to hold the hydrometer vertically and suck up liquid only until the float is free and floating.
5. Always hold the hydrometer at eye level and take the reading at the surface of the liquid with the float free and floating.
6. Disregard the slight curvature appearing where the liquid rises against the float stem. This phenomenon is due to surface tension.
7. Do not drop any of the battery fluid on the boat or on your clothing, because it is extremely caustic. Use water and baking soda to neutralize any battery liquid that does accidentally drop.
8. After drawing electrolyte from the battery cell until the float is barely free, note the level of the liquid inside the hydrometer. If the level is within the charged (usually green) band range for all cells, the condition of the battery is satisfactory. If the level is within the discharged (usually white) band for all cells, the battery is in fair condition.
9. If the level is within the green or white band for all cells except one, which registers in the red, the cell is shorted internally. No amount of charging will bring the battery back to satisfactory condition.
10. If the level in all cells is about the same, even if it falls in the red band, the battery may be recharged and returned to service. If the level fails to rise above the red band after charging, the only solution is to replace the battery.

■ **An alternate way of testing a battery is to perform a load test using a special Carbon-Pile Load Tester. These days most automotive and many marine parts stores contain a tester and will perform the check for free (hoping that your battery will fail and they can sell you another). Essentially a load test involves placing a specified load (current drain/draw) on a fully-charged battery and checking to see how it performs/recovers. This is the only way to test the condition of a sealed maintenance-free battery.**

STORAGE

If the boat is to be laid up (placed into storage) for the winter or anytime it is not going to be used for more than a few weeks, special attention must be given to the battery. This is necessary to prevent complete discharge and/or possible damage to the terminals and wiring. Before putting the boat in storage, disconnect and remove the batteries. Clean them thoroughly of any dirt or corrosion and then charge them to full specific gravity readings. After they are fully charged, store them in a clean cool dry place where they will not be damaged or knocked over, preferably on a couple blocks of wood. Storing the battery up off the deck, will permit air to circulate freely around and under the battery and will help to prevent condensation.

Never store the battery with anything on top of it or cover the battery in such a manner as to prevent air from circulating around the filler caps. All batteries, both new and old, will discharge during periods of storage, more so if they are hot than if they remain cool. Therefore, the electrolyte level and the specific gravity should be checked at regular intervals. A drop in the specific gravity reading is cause to charge them back to a full reading.

In cold climates, care should be exercised in selecting the battery storage area. A fully-charged battery will freeze at about 60°F below zero. The electrolyte of a discharged battery, almost dead, will begin forming ice at about 19°F above zero.

■ **For more information on batteries and the engine electrical systems, please refer to the Ignition and Electrical section of this manual.**

MAINTENANCE & TUNE-UP

Fiberglass Hull

INSPECTION AND CARE

◆ See Figures 120, 121 and 122

Fiberglass reinforced plastic hulls are tough, durable and highly resistant to impact. However, like any other material they can be damaged. One of the advantages of this type of construction is the relative ease with which it may be repaired.

A fiberglass hull has almost no internal stresses. Therefore, when the hull is broken or stove-in, it retains its true form. It will not dent to take an out-of-shape set. When the hull sustains a severe blow, the impact will be either absorbed by deflection of the laminated panel or the blow will result in a definite, localized break. In addition to hull damage, bulkheads, stringers and other stiffening structures attached to the hull may also be affected and therefore, should be checked. Repairs are usually confined to the general area of the rupture.

■ **The best way to care for a fiberglass hull is to wash it thoroughly, immediately after hauling the boat while the hull is still wet. The next best way to care for your hull is to give it a waxing a couple of times per season. Your local marina or boat supply store should be able to help you find some high quality boat soaps and waxes.**

A foul bottom can seriously affect boat performance. This is one reason why racers, large and small, both powerboat and sail, are constantly giving attention to the condition of the hull below the waterline.

In areas where marine growth is prevalent, a coating of vinyl, anti-fouling bottom paint should be applied if the boat is going to be left in the water for extended periods of time such as all or a large part of the season. If growth has developed on the bottom, it can be removed with a diluted solution of muriatic acid applied with a brush or swab and then rinsed with clear water. Always use rubber gloves when working with Muriatic acid and take extra care to keep it away from your face and hands. The fumes are toxic. Therefore, work in a well-ventilated area or if outside, keep your face on the windward side of the work.

■ **If marine growth is not too severe you may avoid the unpleasantness of working with muriatic acid by trying a power washer instead. Most marine vegetation can be removed by pressurized water and a little bit of scrubbing using a rough sponge (don't use anything that will scratch or damage the surface).**

Barnacles have a nasty habit of making their home on the bottom of boats that have not been treated with anti-fouling paint. Actually they will not harm the fiberglass hull but can develop into a major nuisance.

If barnacles or other crustaceans have attached themselves to the hull, extra work will be required to bring the bottom back to a satisfactory condition. First, if practical, put the boat into a body of fresh water and allow it to remain for a few days. A large percentage of the growth can be removed in this manner. If this remedy is not possible, wash the bottom thoroughly with a high-pressure fresh water source and use a scraper. Small particles of hard shell may still hold fast. These can be removed with sandpaper.

Interior

INSPECTION AND CARE

No one wants to walk around in bare feet on a boat whose deck or carpet is covered in fish guts right? It's not just a safety hazard, it's kind of nasty. Taking time to wash down and clean your boat's interior is just as important to the long term value of your boat as it is to your enjoyment. So take time, after every outing to make sure your baby is clean on the inside too.

Always try to find gentle cleaners for your vinyl and plastic seats. Harsh chemicals and abrasives will do more harm then good. Take care with guests aboard, as more than one brand of sun-tan lotion has been know to cause stains. Some people get carried away, forbidding things like cheesy coated chips/snacks or mustards on board. Don't let keeping your boat clean so much of an obsession that you forget to enjoy it, just keep a bottle of cleaner handy for quick spill clean-ups. And keeping it handy will ensure you'll be more likely to wipe things down after a fun-filled outing.

Be sure to always test a cleaner on a hidden or unexposed area of your carpet or vinyl before soaking things down with it. If it does not harm or damage the color of your finish, you're good to go.

When we trailer our boats, we sometimes find it more convenient to hit a spray-it-yourself car wash on the way home. This gives us a chance to spray down the boat hull, trailer and tow vehicle before we get home and turn our attention to engine flushing and wiping down/cleaning the interior.

If you're lucky enough to have snap out marine carpet, remove it and give it a good wash down once in a while. This allows you to spray down the deck as well. Hang the carpet to dry and reinstall once it is ready. If you've got permanently installed marine carpet, you can spray it down too, just make sure you can give it a chance to dry before putting the cover back on.

■ **For permanently installed marine carpet, try renting a rug steam-cleaner at least once a season and give it a good deep cleaning. We like to do it at the beginning and the end of each season!**

Fig. 120 The best way to care for a fiberglass hull is to wash it thoroughly

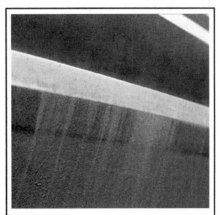

Fig. 121 If marine growth is a problem, apply a coating of anti-foul bottom paint

Fig. 122 Fiberglass, vinyl and rubber care products, like those from Meguiar's protect your boat

2-36 MAINTENANCE & TUNE-UP

TUNE-UP

Introduction

A proper tune-up is the key to long and trouble-free outboard life and the work can yield its own rewards. Studies have shown that a properly tuned and maintained outboard can achieve better fuel economy than an out-of-tune engine. As a conscientious boater, set aside a Saturday morning, say once a month, to check or replace items which could cause major problems later. Keep your own personal log to jot down which services you performed, how much the parts cost you, the date and the number of hours on the engine at the time. Keep all receipts for such items as oil and filters, so that they may be referred to in case of related problems or to determine operating expenses. These receipts are the only proof you have that the required maintenance was performed. In the event of a warranty problem on newer engines, these receipts will be invaluable.

The efficiency, reliability, fuel economy and enjoyment available from boating are all directly dependent on having your outboard tuned properly. The importance of performing service work in the proper sequence cannot be over emphasized. Before making any adjustments, check the specifications. Never rely on memory when making critical adjustments.

Before tuning any outboard, insure it has satisfactory compression. An outboard with worn or broken piston rings, burned pistons or scored cylinder walls, will not perform properly no matter how much time and expense is spent on the tune-up. Poor compression must be corrected or the tune-up will not give the desired results.

The extent of the engine tune-up is usually dependent on the time lapse since the last service. In this section, a logical sequence of tune-up steps will be presented in general terms. If additional information or detailed service work is required, refer to the section of this manual containing the appropriate instructions.

Tune-Up Sequence

A tune-up can be defined as pre-determined series of procedures (adjustments, tests and worn component replacements) that are performed to bring the engine operating parameters back to original condition. The series of steps are important, as the later procedures (especially adjustments) are dependant upon the earlier procedures. In other words, a procedure is performed only when subsequent steps would not change the result of that procedure (this is mostly for adjustments or settings that would be incorrect after changing another part or setting). For instance, fouled or excessively worn spark plugs may affect engine idle. If adjustments were made to the idle speed or mixture of a carbureted engine **before** these plugs were cleaned or replaced, the idle speed or mixture might be wrong after replacing the plugs. The possibilities of such an effect become much greater when dealing with multiple adjustments such as timing, idle speed and/or idle mixture. Therefore, be sure to follow each of the steps given here. Since many of the steps listed here are full procedures in themselves, refer to the procedures of the same name in this section for details.

■ Computer controlled ignition and fuel components on more and more modern outboards have lessoned the amount of steps necessary for a pre-season tune-up, but not completely eliminated the need for replacing worn components. EFI motors may not allow for many (or any) timing or mixture adjustments, however they still have mechanical and electrical components that wear making compression checks, spark plug/wire replacement, and component inspection steps all that much more important.

A complete pre-season tune-up should be performed at the beginning of each season or when the motor is removed from storage. Operating conditions, amount of use and the frequency of maintenance required by your motor may make one or more additional tune-ups necessary during the season. Perform additional tune-ups as use dictates.

1. Before starting, inspect the motor thoroughly for signs of obvious leaks, damage and loose or missing components. Make repairs, as necessary.

✱✱ CAUTION

We can't emphasize enough how important is this first step, ESPECIALLY on EFI motors. The high-pressure fuel system of these motors makes fuel system/line integrity an important safety issue.

2. Check all accessible bolts and fasteners and tighten any that are loose.

3. For many decades, a standard part of the outboard tune-up procedure was to re-torque the cylinder head bolts. Typically manufacturers advised you to follow the applicable steps of the cylinder cylinder head/cover removal and installation procedures. The cylinder head cover or rocker/valve cover must be removed for access to the head bolts. Each of the bolts would be loosened slightly using the reverse of the tightening sequence, then re-torqued using one or more passes of the tightening sequence, as directed. Refer to the procedures under Powerhead for details. However, Yamaha and Mercury are somewhat obscure about the need to do this on their modern outboards. They specifically mention that you should NOT do this on V6 4-stroke motors. But they do specifically mention that you SHOULD do this to the 30/40J/40 hp 4-strokes. However Yamaha does not specifically include or exclude the cylinder head bolts from the maintenance requirement of checking/tightening ALL BOLTS AND NUTS listed in most of the service charts for all of their other motors. So we'll leave it to your judgment whether or not you want to include the cylinder head/cover bolts on other motors, but it is probably not a bad idea to include it annually anyway (if in doubt, check with your local Yamaha or Merc/Mariner marine dealer).

■ Before installing the cylinder head (valve cover) on 4-stroke engines, check and adjust the valve lash, as necessary.

4. On all 4-stroke models, check and adjust the valve lash clearance. Since valve lash must be adjusted cold, it is best to perform this before starting and running the motor to warm it for a compression test.

5. Perform a compression check to make sure the motor is mechanically ready for a tune-up. An engine with low compression on one or more cylinder should be overhauled, not tuned. A tune-up will not be successful without sufficient engine compression. Refer to the Compression Testing in this section.

■ If this tune-up is occurring immediately after removing an engine from storage, be sure to start and run the motor using the old plugs first (if possible) while burning off the fogging oil. Then install the new spark plugs once the compression test is completed!

6. Since the spark plugs must be removed for the compression check, take the opportunity to inspect them thoroughly for signs of oil fouling, carbon fouling, damage due to detonation, etc. Clean and re-gap the plugs or, better yet, install new plugs as no amount of cleaning will precisely match the performance and life of new plugs. Refer to Spark Plugs, in this section. Also, this is a good time to check the spark plug wires as well. Please refer to Spark Plug Wires, in this section.

■ We don't care how little you use your motor, there is usually no excuse for not installing new plugs at the beginning of each season. If only because when storing a motor the fogging oil will go a long way to fouling even a decent set of spark plugs. Remember, the secondary ignition circuit is the most likely performance problem that occurs on an outboard.

7. Visually inspect all ignition system components for signs of obvious defects. Look for signs of burnt, cracked or broken insulation. Replace wires or components with obvious defects. If spark plug condition suggests weak or no spark on one or more cylinders, perform ignition system testing to eliminate possible worn or defective components. Refer to the Ignition System Inspection procedures in this section and the Ignition and Electrical System section.

8. Visually inspect all engine wiring and, if equipped, the battery and starter motor. A quick starter motor draw test can tell you a lot about the condition of your electrical starting system.

9. Remove and clean (on serviceable filters) or replace the inline filter and/or fuel pump filter, as equipped. Refer to the Fuel Filter procedures in this section. Perform a thorough inspection of the fuel system, hoses and components. Replace any cracked or deteriorating hoses. If carburetor adjustment or overhaul is necessary, perform these procedures before proceeding.

10. Pressurize the fuel system according to the procedures found in the Fuel System section, then check carefully for leaks. Again, this is important on all motors, but even more so on EFI motors.

MAINTENANCE & TUNE-UP 2-37

11. Perform engine Timing and Synchronization adjustments as described in this section.

■ **Although many of the motors covered here allow for certain ignition timing and, if applicable, carburetor adjustment procedures, none of them require the level of tuning attention that was once the norm. Many of the motors are equipped with electronic ignition systems that limit or eliminate timing adjustments. Carburetors used on most of these motors in the U.S. are EPA regulated and contain few mixture adjustments. The air/fuel mixture is completely computer controlled on fuel injected motors and allows for no adjustment.**

12. Except for jet drive models, remove the propeller in order to thoroughly check for leaks at the shaft seal. Inspect the propeller or impeller condition, look for nicks, cracks or other signs of damage and repair or replace, as necessary. If available, install a test wheel to run the motor in a test tank after completion of the tune-up. If no test wheel is available, lubricate the shaft/splines, then install the propeller or rotor. Refer to the procedures for Propeller or Jet Drive Impeller in this section, as applicable.

13. Change the gearcase oil as directed under the Gearcase Oil procedures in this section. If you are conducting a pre-season tune-up and the oil was changed immediately prior to storage this is not necessary. But, be sure to check the oil level and condition. Drain the oil anyway if significant contamination is present.

■ **Anytime large amounts of water or debris is present in the gearcase oil, be sure to troubleshoot and repair the problem before returning the gearcase to service. The presence of water may indicate problems with the seals, while debris could a sign that overhaul is required.**

14. Perform a test run of the engine to verify proper operation of the starting, fuel, oil and cooling systems. Although this can be performed using a flush/test adapter or even on the boat itself (if operating with a normal load/passengers), the preferred method is the use of a test tank. Keep in mind that proper operation without load and at low speed (on a flush/test adapter) doesn't really tell you how well the motor will run under load. If possible, run the engine, in a test tank using the appropriate test wheel. Monitor the cooling system indicator stream to ensure the water pump is working properly. Once the engine is fully warmed, slowly advance the engine to wide-open throttle, then note and record the maximum engine speed. Refer to the Tune-Up Specifications chart to compare engine speeds with the test propeller minimum rpm specifications. If engine speeds are below specifications, yet engine compression was sufficient at the beginning of this procedure, recheck the fuel and ignition system adjustments (as well as the condition of the prop if you're not using a test wheel).

Compression Testing

The quickest way to gauge the condition of an internal combustion engine is through a compression check. In order for an internal combustion engine to work properly, it must be able to generate sufficient compression in the combustion chamber to take advantage of the explosive force generated by the expanding gases after ignition. This is true on motors whether they are of the 2- or 4-stroke design.

If the combustion chambers or cylinder head valves are worn or damaged in some fashion as to allow pressure to escape, the engine cannot develop sufficient horsepower. Under these circumstances, combustion will not occur properly meaning that the air/fuel mixture cannot be set to maximize power and minimize emissions. Obviously, it is useless to try to tune an engine with extremely low or erratic compression readings, since a simple tune-up will not cure the problem. An engine with poor compression on one or more cylinders should be overhauled.

The pressure created in the combustion chamber may be measured with a gauge that remains at the highest reading it measures during the action of a one-way valve. This gauge is inserted into the spark plug hole and held or threaded in position while the motor is cranked. A compression test will uncover many mechanical problems that can cause rough running or poor performance.

If the powerhead shows any indication of overheating, such as discolored or scorched paint, inspect the cylinders visually through the spark plug hole for possible scoring. It is possible for a cylinder with satisfactory compression to be scored slightly. Also, check the water pump, since a faulty water pump can cause an overheating condition.

An even BETTER method of determining engine condition is to perform a leak down test with a set of gauges and an air compressor. In a leak down test each cylinder of a warmed motor is set to TDC of the compression stroke one at a time, while an adapter is used to pressurize the cylinder with air. The gauges are used to monitor the rate at which the air leaks out ("leaks down") to determine a percentage of pressure lost over a certain amount of time. Instructions come with most testers how to go about performing this test and what are the normally acceptable leak down rates. However, like a compression test a balanced leak percentage is desired between cylinders, and differences of more than 15-30% indicate a problem. No cylinder will leak 0% however and larger motors tend to leak more than smaller.

What is nice about this test is that you can perform further diagnosis by simply listening from wear the air is leaking. Air coming out the exhaust tells you the exhaust valve is the issue, air from the crankcase tells you that the piston and/or rings are the issue and air from the intake manifold (carb or throttle body) tells you that the intake valve is at fault. If air is coming from adjacent cylinders, suspect the head gasket.

TUNE-UP COMPRESSION CHECK

◆ See Figures 123, 124 and 125

■ **A compression check requires a compression gauge and a spark plug port adapter that matches the plug threads of your motor.**

When analyzing the results of a compression check, generally the actual amount of pressure measured during a compression check is not quite as important as the variation from cylinder-to-cylinder on the same motor. For multi-cylinder powerheads, Yamaha states that variations of up to 30 psi (207 kPa) may be considered normal. However, on single cylinder powerheads, a drop of about 15 psi (103 kPa) from the normal compression pressure you established when it was new is cause for concern (you did do a compression test on it when it was new, didn't you?). Mercury/Mariner states that differences between cylinders of more than 15% from highest to lowest reading are unacceptable. For single cylinder Mercs, you'll have to use the compression figures provided in the Powerhead Specifications Charts (and of course again, the test that you performed on the motor when it was new).

Ok, for the point of arguments sake let's say you bought the engine used or never checked compression the first season or so, assuming it wasn't something you needed to worry about. You're not alone, many of us have done that. But now that you're reading this it is your chance to take the data and note it for future.

For many years Yamaha did not publish much in the way of specific compression specifications (and Mercury was not always that much better) for the amount of compression each of their engines should generate, a general rule of thumb that can be applied is that generally most 4-stroke internal combustion engines should generate at least 100 psi (690 kPa). However, a quick check of the specifications that Yamaha HAS published to date shows a range of about 94 psi (660 kPa) to 181 psi (1250 kPa) on 4-stroke motors. Some manual start Mercury/Mariner models utilize a cylinder decompression device to help make starting physically easier, on these models, if the devise is left operational, readings will be well below 100 psi (690 kPa). When equipped, decompression devises will be mentioned for the Compression specifications found in the Powerhead Specifications charts.

■ **To see if Yamaha or Mercury/Mariner has published a specification for your particular engine, please check the Engine Specifications charts in the Powerhead section. If a specification is available it will be listed under Compression. For some Mercs, the specification is not a Minimum, but a Maximum and is not as useful as comparing the difference between cylinders.**

If a minimum compression specification is not available for your motor, put the most weight on a comparison of the readings from the other cylinders on the same motor (or readings when the motor was new).

When taking readings during the compression check, repeat the procedure a few times for each cylinder, recording the highest reading for that cylinder.

Then, for all multi-cylinder Yamaha motors, the compression reading on the lowest cylinder should within 30 psi (207 kPa) of the highest reading (and still within specification if one is given in the Engine Specifications chart for that motor). If not, consider 15-30% to be the absolute limit. If the reading in the lowest cylinder is less than 70% of the reading in the highest cylinder, it's time to find and remedy the cause.

2-38 MAINTENANCE & TUNE-UP

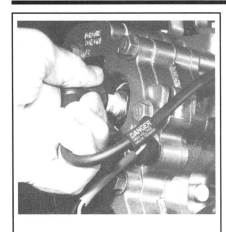

Fig. 123 Removing the high tension lead. Always use a twist and pull motion on the boot to prevent damage to the wire

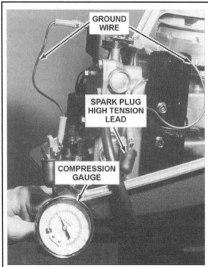

Fig. 124 The ignition must be disabled or all spark plugs must be grounded while making compression tests

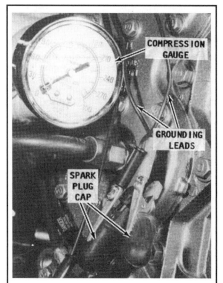

Fig. 125 The compression tester is threaded into an open spark plug port

For all multi-cylinder Mercury/Mariner motors, the compression reading on the lowest cylinder should within 15% of the highest. If the reading in the lowest cylinder is less than 85% of the reading in the highest cylinder, it's time to find and remedy the cause.

■ If the powerhead has been in storage for an extended period, the piston rings may have relaxed. This will often lead to initially low and misleading readings. Always run an engine to normal operating temperature to ensure that the readings are accurate.

■ If you've never removed the spark plugs from this cylinder head before, break each one loose and retighten them before starting the motor in order to make sure they will not seize in the head once it is warmed. Better yet, remove each one and coat the threads very lightly with some fresh anti-seize compound.

Prepare the engine for a compression test as follows:
1. Run the engine until it reaches operating temperature. The engine is at operating temperature a few minutes after the powerhead becomes warm to the touch and the stream of water exiting the cooling indicator becomes warm. If the test is performed on a cold engine, the readings will be considerably lower than normal, even if the engine is in perfect mechanical condition.
2. Label and disconnect the spark plug wires. Always grasp the molded cap and pull it loose with a twisting motion to prevent damage to the connection.
3. Clean all dirt and foreign material from around the spark plugs, and then remove all the plugs. Keep them in order by cylinder for later evaluation.

■ On some Yamaha and Mercury/Mariner motors you can disable the ignition system by leaving the safety lanyard disconnected but still use the starter motor to turn the motor. This is handy for things like compression tests or distributing fogging oil. To be certain use a spark plug gap tester on one lead and crank the motor using the keyswitch. If no spark is present, you're good to go, if not, you'll have to ground the spark plug leads to the cylinder head.

4. Ground the spark plug leads to the engine to render the ignition system inoperative while performing the compression check.

✳✳ CAUTION

Grounding the spark plug leads not only protects the ignition system from potential damage that may be caused by the excessive load placed on operating the system with the wires disconnected, but more importantly, protects you from the dangers of arcing. The ignition system operates at extremely high voltage and could cause serious shocks. Also, keep in mind that you're cranking an engine with open spark plug ports which could allow any remaining fuel vapors to escape become ignited by arcing current.

5. Insert a compression gauge into the No. 1, top, spark plug opening.
6. Move the throttle to the wide open position in order to make sure the throttle plates are not restricting air flow. If necessary you may have to spin the propeller shaft slowly by hand while advancing the throttle in order to get the shifter into gear.
7. Crank the engine with the starter through at least 4-5 complete strokes with the throttle at the wide-open position, to obtain the highest possible reading. Record the reading.

■ On electric start motors, it is very important to use a freshly charged cranking battery as a weakened battery will cause a slower than normal cranking speed, reducing the compression reading.

8. Repeat the test and record the compression for each cylinder.
9. A variation between cylinders is far more important than the actual readings. A variation of more than 30 psi (207 kPa) for Yamaha motors or 15% for Mercury/Mariner motors, between cylinders indicates the lower compression cylinder is defective (according to the manufacturer's recommended tolerances). Not all engines will exhibit the same compression readings. In fact, two identical engines may not have the same compression. Generally, the rule of thumb is that the lowest cylinder should be within about 15% of the highest (difference between the two readings).
10. If compression is low in one or more cylinders, the problem may be worn, broken, or sticking piston rings, scored pistons or worn cylinders.

LOW COMPRESSION

Compression readings that are generally low indicate worn, broken, or sticking piston rings, scored pistons or worn cylinders, and usually indicate an engine that has a lot of hours on it. Low compression in two adjacent cylinders (with normal compression in the other cylinders) indicates a blown head gasket between the low-reading cylinders. Other problems are possible (broken ring, hole burned in a piston), but a blown head gasket is most likely.

■ Use of an engine cleaner, available at any automotive parts house, will help to free stuck rings and to dissolve accumulated carbon. Follow the directions on the container.

To test a cylinder or motor further, add a few drops of engine oil to the cylinder and recheck.
• If compression is higher with oil added to the cylinder, suspect a worn or damaged piston, the cylinder and/or rings.
• If compression is the same as without oil, suspect one or more defective rings or a defective piston. It could also be a problem with the valve train.
• If compression is well above specification, suspect the carbon deposits are on the cylinder head and/or piston crown. Significant carbon deposits will lead to pre-ignition and other performance problems and should be cleaned (either using an additive or by removing the cylinder head for manual cleaning).

MAINTENANCE & TUNE-UP

Spark Plugs

◆ See Figure 126

The spark plug performs four main functions:
• First and foremost, it provides spark for the combustion process to occur.
• It also removes heat from the combustion chamber.
• Its removal provides access to the combustion chamber (for inspection or testing) through a hole in the cylinder head.
• It acts as a dielectric insulator for the ignition system.

It is important to remember that spark plugs do not create heat, they help remove it. Anything that prevents a spark plug from removing the proper amount of heat can lead to pre-ignition, detonation, premature spark plug failure and even internal engine damage.

In the simplest of terms, the spark plug acts as the thermometer of the engine. Much like a doctor examining a patient, this "thermometer" can be used to effectively diagnose the amount of heat present in each combustion chamber.

Spark plugs are valuable tuning tools, when interpreted correctly. They will show symptoms of other problems and can reveal a great deal about the engine's overall condition. Evaluating the appearance of the spark plug's firing tip, gives visual cues to determine the engine's overall operating condition, get a feel for air/fuel ratios and even diagnose driveability problems.

As spark plugs grow older, they lose their sharp edges as material from the center and ground electrodes slowly erodes away. As the gap between these two points grows, the voltage required to bridge this gap increases proportionately. The ignition system must work harder to compensate for this higher voltage requirement and hence there is a greater rate of misfires or incomplete combustion cycles. Each misfire means lost horsepower, reduced fuel economy and higher emissions. Replacing worn out spark plugs with new ones (with sharp new edges) effectively restores the ignition system's efficiency and reduces the percentage of misfires, restoring power, economy and reducing emissions.

■ Although spark plugs can typically be cleaned and regapped if they are not excessively worn, no amount of cleaning or regapping will return most spark plugs to original condition and it is usually best to just go ahead and replace them.

How long spark plugs last will depend on a variety of factors, including engine compression, fuel used, gap, center/ground electrode material and the conditions in which the outboard is operated.

SPARK PLUG HEAT RANGE

◆ See Figures 127, 128 and 129

Spark plug heat range is the ability of the plug to dissipate heat from the combustion chamber. The longer the insulator (or the farther it extends into the engine), the hotter the plug will operate; the shorter the insulator (the closer the electrode is to the engine's cooling passages) the cooler it will operate.

Selecting a spark plug with the proper heat range will ensure that the tip maintains a temperature high enough to prevent fouling, yet cool enough to prevent pre-ignition. A plug that absorbs little heat and remains too cool will quickly accumulate deposits of carbon since it won't be able to burn them off. This leads to plug fouling and consequently to misfiring. A plug that absorbs too much heat will have no deposits but, due to the excessive heat, the electrodes will burn away quickly and might also lead to pre-ignition or other ignition problems.

Pre-ignition takes place when plug tips get so hot that they glow sufficiently to ignite the air/fuel mixture before the actual spark occurs. This early ignition will usually cause a pinging during heavy loads and if not corrected, will result in severe engine damage. While there are many other things that can cause pre-ignition, selecting the proper heat range spark plug will ensure that the spark plug itself is not a hot-spot source.

■ The manufacturer recommended spark plugs are listed in the Tune-Up Specifications chart. Recommendations may also be present on one or more labels affixed to your motor. When the label disagrees with the chart, we'd normally defer to the label as it may reflect a change that was made mid-production and not reflected in the service literature.

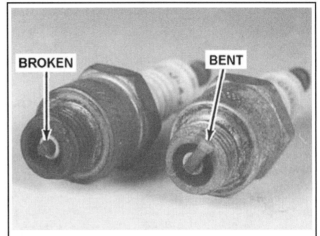

Fig. 126 Damaged spark plugs. Notice the broken electrode on the left plug. The electrode must be found and retrieved prior to returning the powerhead to service

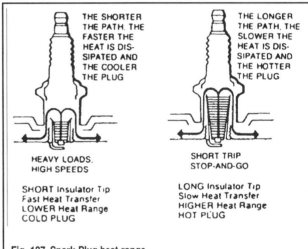

Fig. 127 Spark Plug heat range

Fig. 128 Many motors have a label which lists spark plug type and gap

2-40 MAINTENANCE & TUNE-UP

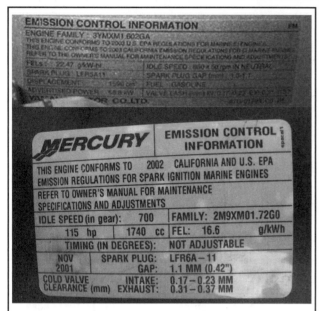

Fig. 129 Though these days, at least in the US, most motors have an Emission Control label which contains that information (Yamaha Top and Mercury Bottom)

REMOVAL & INSTALLATION

◆ See Figures 130 thru 135

■ New technologies in spark plug and ignition system design have greatly extended spark plug life over the years. But, spark plug life will still vary greatly with engine tuning, condition and usage. Mariner motors in general are a little tougher on plugs however, it is not uncommon for plugs to last up to 100 hours of operation or more on 4-strokes.

Typically spark plugs will require replacement once a season. The electrode on a new spark plug has a sharp edge but with use, this edge becomes rounded by wear, causing the plug gap to increase. As the gap increases, the plug's voltage requirement also increases. It requires a greater voltage to jump the wider gap and about two to three times as much voltage to fire a plug at high speeds than at idle.

■ Fouled plugs can cause hard-starting, engine mis-firing or other problems. You don't want that happening on the water. Take time, at least once a month to remove and inspect the spark plugs. Early signs of other tuning or mechanical problems may be found on the plugs that could save you from becoming stranded or even allow you to address a problem before it ruins the motor.

Tools needed for spark plug replacement include: a ratchet, short extension, spark plug socket (there are two types; either 13/16 inch or 5/8 inch, depending upon the type of plug), a combination spark plug gauge and gapping tool and a can of anti-seize type compound.

1. When removing spark plugs from multi-cylinder motors, work on one at a time. Don't start by removing the plug wires all at once, because unless you number them, they may become mixed up. Take a minute before you begin and number the wires with tape.
2. For safety, disconnect the negative battery cable or turn the battery switch **OFF**.
3. If the engine has been run recently, allow the engine to thoroughly cool (unless performing a compression check and then you should have already broken them loose once when cold and retightened them before warming the motor, so they should have less of a tendency to stick). Attempting to remove plugs from a hot cylinder head could cause the plugs to seize and damage the threads in the cylinder head, especially on aluminum heads!

■ To ensure an accurate reading during a compression check, the spark plugs must be removed from a hot engine. But, DO NOT force a plug if it feels like it is seized. Instead, wait until the engine has cooled, remove the plug and coat the threads lightly with anti-seize then reinstall and tighten the plug, then back off the tightened position a little less than 1/4 turn. With the plug(s) installed in this manner, re-warm the engine and conduct the compression check.

4. On some of the larger motors a plastic or composite spark plug cover is mounted over the cylinder head. If equipped, loosen the fasteners and then carefully remove the cover from the head to expose the spark plugs and wires.
5. Carefully twist the spark plug wire boot to loosen it, then pull the boot using a twisting motion to remove it from the plug. Be sure to pull on the boot and not on the wire, otherwise the connector located inside the boot may become separated from the high tension wire.

■ If removal is difficult (or on motors where the spark plug boot is hard to grip because of access) a spark plug wire removal tool is recommended as it will make removal easier and help prevent damage to the boot and wire assembly. Most tools have a wire loom that fits under the plug boot so the force of pulling upward is transmitted directly to the bottom of the boot.

6. Using compressed air (and safety glasses), blow debris from the spark plug area to assure that no harmful contaminants are allowed to enter the combustion chamber when the spark plug is removed. If compressed air is not available, use a rag or a brush to clean the area. Compressed air is available from both an air compressor or from compressed air in cans available at photography stores. In a pinch, blow up a balloon and use the escaping air to blow debris from the spark plug port(s).

■ Remove the spark plugs when the engine is cold, if possible, to prevent damage to the threads. If plug removal is difficult, apply a few drops of penetrating oil to the area around the base of the plug and allow it a few minutes to work.

Fig. 130 Spark plugs are threaded into ports on the cylinder heads

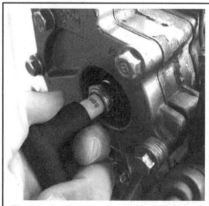

Fig. 131 Gently grasp the boot and pull the wire from the plug

Fig. 132 Then unthread the plug using a ratchet and socket

MAINTENANCE & TUNE-UP 2-41

Fig. 133 ALWAYS thread plugs by hand to prevent crossthreading...

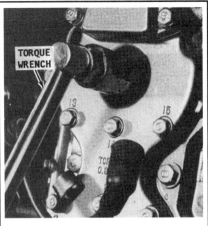

Fig. 134 ... then use a torque wrench to tighten the plug to spec

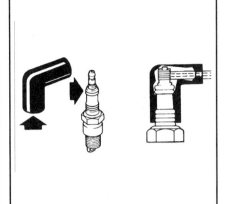

Fig. 135 To prevent corrosion, apply a small amount of grease to the plug and boot during installation

7. Using a spark plug socket that is equipped with a rubber insert to properly hold the plug, turn the spark plug counterclockwise to loosen and remove the spark plug from the bore.

※※ WARNING

Avoid the use of a flexible extension on the socket. Use of a flexible extension may allow a shear force to be applied to the plug. A shear force could break the plug off in the cylinder head, leading to costly and/or frustrating repairs. In addition, be sure to support the ratchet with your other hand - this will also help prevent the socket from damaging the plug.

8. Evaluate each cylinder's performance by comparing the spark condition. Check each spark plug to be sure they are from the same plug manufacturer and have the same heat range rating. Inspect the threads in the spark plug opening of the block and clean the threads before installing the plug.
9. When purchasing new spark plugs, always ask the dealer if there has been a spark plug change for the engine being serviced. Sometimes manufacturers will update the type of spark plug used in an engine to offer better efficiency or performance.
10. Always use a new gasket (if applicable). The gasket must be fully compressed on clean seats to complete the heat transfer process and to provide a gas tight seal in the cylinder.
11. Inspect the spark plug boot for tears or damage. If a damaged boot is found, the spark plug boot and possibly the entire wire will need replacement.
12. Check the spark plug gap prior to installing the plug. Most spark plugs do not come gapped to the proper specification.
13. Apply a thin coating of anti-seize on the thread of the plug. This is extremely important on aluminum head engines to prevent corrosion and heat from seizing the plug in the threads (which could lead to a damaged cylinder head upon removal).
14. Carefully thread the plug into the bore by hand. If resistance is felt before the plug completely bottoms, back the plug out and begin threading again.

※※ WARNING

Do not use the spark plug socket to thread the plugs. Always carefully thread the plug by hand or using an old plug wire/boot to prevent the possibility of crossthreading and damaging the cylinder head bore. An old plug wire/boot can be used to thread the plug if you turn the wire by hand. Should the plug begin to crossthread the wire will twist before the cylinder head would be damaged. This trick is useful when accessories or a deep cylinder head design prevents you from easily keeping fingers on the plug while it is threaded by hand.

15. For Yamaha motors carefully tighten the spark plug to specification using a torque wrench, as follows:
  • 9.9 hp (232cc) engines: 9-10 ft. lbs. (12-13 Nm)
  • 15-60 hp (323-996cc) engines: 13 ft. lbs./159 inch lbs. (18 Nm)
  • All other Yamaha engines: 18 ft. lbs. (25 Nm)

■ Whenever possible, spark plugs should be tightened to the factory torque specification. If a torque wrench is not available, and the plug you are installing is equipped with a crush washer, tighten the plug until the washer seats, then tighten it an additional 1/4 turn to crush the washer.

16. For Mercury/Mariner motors carefully tighten the spark plug to specification using a torque wrench, as follows:
  • 8/9.9 hp (232cc) engines: 9-10 ft. lbs. (12-13 Nm)
  • 9.9/15 (323cc) engines: 13 ft. lbs./159 inch lbs. (18 Nm)
  • 25 hp engines: 12.5 ft. lbs./150 inch lbs. (17 Nm)
  • Carbureted 30/40 hp 3-cyl motors and 40/45/50 hp (935cc) 4-cyl motors: 20 ft. lbs. (27 Nm)
  • EFI 30/40 3-cylinder motors, and all 40/50/60 hp (996cc) 4-cyl motors: 12.5 ft. lbs./150 inch lbs. (27 Nm)
  • 75-115 hp motors: 20 ft. lbs. (27 Nm)
  • 225 hp motors: 18 ft. lbs. (25 Nm)

17. Apply a small amount of a silicone dielectric grease or a suitable marine grease to the ribbed, ceramic portion of the spark plug lead and inside the spark plug boot to prevent sticking, then install the boot to the spark plug and push until it clicks into place. The click may be felt or heard. Gently pull back on the boot to assure proper contact.
18. If applicable, connect the negative battery cable or turn the battery switch **ON**.
19. Test run the outboard (using a test tank or flush fitting) and insure proper operation.

READING SPARK PLUGS

◆ See Figures 136 thru 141

Reading spark plugs can be a valuable tuning aid. By examining the insulator firing nose color, you can determine much about the engine's overall operating condition.

In general, a light tan/gray color tells you that the spark plug is at the optimum temperature and that the engine is in good operating condition.

Dark coloring, such as heavy black wet or dry deposits usually indicate a fouling problem. Heavy, dry deposits can indicate an overly rich condition, too cold a heat range spark plug, possible vacuum leak, low compression, overly retarded timing or too large a plug gap.

If the deposits are wet, it can be an indication of a breached head gasket, oil control from ring problems or an extremely rich condition, depending on what liquid is present at the firing tip.

Also look for signs of detonation, such as silver specs, black specs or melting or breakage at the firing tip.

Compare your plugs to the illustrations shown to identify the most common plug conditions.

2-42 MAINTENANCE & TUNE-UP

Fig. 136 A normally worn spark plug should have light tan or gray deposits on the firing tip (electrode)

Fig. 137 A carbon-fouled plug, identified by soft, sooty black deposits, may indicate an improperly tuned powerhead

Fig. 138 This spark plug has been left in the powerhead too long, as evidenced by the extreme gap. Plugs with such an extreme gap can cause misfiring and stumbling accompanied by a noticeable lack of power

Fouled Spark Plugs

A spark plug is "fouled" when the insulator nose at the firing tip becomes coated with a foreign substance, such as fuel, oil or carbon. This coating makes it easier for the voltage to follow along the insulator nose and leach back down into the metal shell, grounding out, rather than bridging the gap normally.

Fuel, oil and carbon fouling can all be caused by different things but in any case, once a spark plug is fouled, it will not provide voltage to the firing tip and that cylinder will not fire properly. In many cases, the spark plug cannot be cleaned sufficiently to restore normal operation. It is therefore recommended that fouled plugs be replaced.

Signs of fouling or excessive heat must be traced quickly to prevent further deterioration of performance and to prevent possible engine damage.

Overheated Spark Plugs

When a spark plug tip shows signs of melting or is broken, it usually means that excessive heat and/or detonation was present in that particular combustion chamber or that the spark plug was suffering from thermal shock.

Since spark plugs do not create heat by themselves, one must use this visual clue to track down the root cause of the problem. In any case, damaged firing tips most often indicate that cylinder pressures or temperatures were too high. Left unresolved, this condition usually results in more serious engine damage.

Detonation refers to a type of abnormal combustion that is usually preceded by pre-ignition. It is most often caused by a hot spot formed in the combustion chamber.

As air and fuel is drawn into the combustion chamber during the intake stroke, this hot spot will "pre-ignite" the air fuel mixture without any spark from the spark plugs.

Detonation

Detonation exerts a great deal of downward force on the pistons as they are being forced upward by the mechanical action of the connecting rods. When this occurs, the resulting concussion, shock waves and heat can be severe. Spark plug tips can be broken or melted and other internal engine components such as the pistons or connecting rods themselves can be damaged.

Left unresolved, engine damage is almost certain to occur, with the spark plug usually suffering the first signs of damage.

■ When signs of detonation or pre-ignition are observed, they are symptom of another problem. You must determine and correct the situation that caused the hot spot to form in the first place.

INSPECTION & GAPPING

◆ See Figures 142 and 143

A particular spark plug might fit hundreds of powerheads and although the factory will typically set the gap to a pre-selected setting, this gap may not be the right one for your particular powerhead.

Insufficient spark plug gap can cause pre-ignition, detonation, even engine damage. Too much gap can result in a higher rate of misfires, noticeable loss of power, plug fouling and poor economy.

■ Refer to the Tune-Up Specifications chart for spark plug gaps.

Check spark plug gap before installation. The ground electrode (the L-shaped one connected to the body of the plug) must be parallel to the center electrode and the specified size wire gauge must pass between the electrodes with a slight drag.

Do not use a flat feeler gauge when measuring the gap on a used plug, because the reading may be inaccurate. A round-wire type gapping tool is the best way to check the gap. The correct gauge should pass through the electrode gap with a slight drag. If you're in doubt, try a wire that is one size smaller and one larger. The smaller gauge should go through easily, while the larger one shouldn't go through at all.

Wire gapping tools usually have a bending tool attached. **USE IT!** This tool greatly reduces the chance of breaking off the electrode and is much more accurate. Never attempt to bend or move the center electrode. Also, be careful not to bend the side electrode too far or too often as it may weaken and break off within the engine, requiring removal of the cylinder head to retrieve it.

MAINTENANCE & TUNE-UP 2-43

Fig. 139 An oil-fouled spark plug indicates a powerhead with worn piston rings or a malfunctioning oil injection system that allows excessive oil to enter the combustion chamber

Fig. 140 A physically damaged spark plug may be evidence of severe detonation in that cylinder. Watch the cylinder carefully between services, as a continued detonation will not only damage the plug but will most likely damage the powerhead

Fig. 141 A bridged or almost bridged spark plug, identified by the build-up between the electrodes caused by excessive carbon or oil build up on the plug

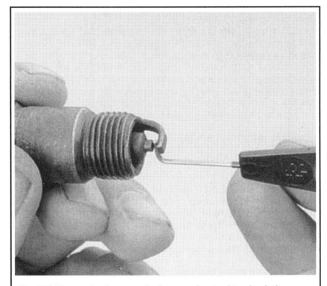

Fig. 142 Use a wire-type spark plug gapping tool to check the distance between center and ground electrodes

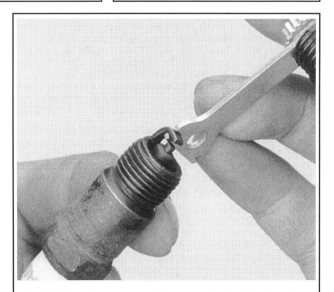

Fig. 143 Most plug gapping tools have an adjusting fitting used to bend the ground electrode

Spark Plug Wires

TESTING

Each time you remove the engine cover, visually inspect the spark plug wires for burns, cuts or breaks in the insulation. Check the boots on the coil and at the spark plug end. Replace any wire that is damaged.

Once a year, usually when you change your spark plugs, check the resistance of the spark plug wires with an ohmmeter. Wires with excessive resistance will cause misfiring and may make the engine difficult to start. In addition worn wires will allow arcing and misfiring in humid conditions.

■ Yamaha does not publish specifications for wire or resistor cap resistance on all models. Mercury/Mariner publishes specifications for wire resistance and cap resistance for all models where applicable, HOWEVER, on some of the models they only publish a combined spec of ignition coil secondary winding and wire, or secondary winding, wire AND cap. For more details, refer to the specification lists below.

MAINTENANCE & TUNE-UP

Remove the spark plug wire from the engine. Determine if the test requires you to check the resistor cap and/or the wire separately, or if you need to check the entire secondary circuit including the cap and/or wire by referring to the specifications following these paragraphs.

■ On models equipped with removable resistor caps, you can remove the cap by turning it counterclockwise while applying a light outward pressure on the cap itself. DO NOT pull hard or the lead may be damaged. Likewise, installation is accomplished by carefully threading the cap clockwise onto the high tension lead.

Test the components by connecting one lead of the ohmmeter to each end of the wire, cap or circuit. For models which are NOT listed below, use this general specification that resistance should NOT exceed 7000 ohms per foot of wire.

■ Keep in mind that just because a spark plug wire passes a resistance test doesn't mean that it is in good shape. Cracked or deteriorated insulation will allow the circuit to misfire under load, especially when wet. Always visually check wires to cuts, cracks or breaks in the insulation. If found, run the engine in a test tank or on a flush device either at night (looking for a bluish glow from the wires that would indicate arcing) or while spraying water on them while listening for an engine stumble.

The specifications that Yamaha publishes for resistance of the plugs wires and/or the spark plug caps are as follows:
• 2.5 hp and 4 hp motors use resistor caps on the ends of the spark plug leads, the caps themselves should have 4000-6000 ohms resistance.
• 6/8 hp motors use resistor caps on the ends of the spark plug leads, the caps themselves should have about 4300 ohms resistance.
• 40/45/50 hp motors use resistor caps on the ends of the spark plug leads for 1999-00 models, the caps should have 3750-6250 ohms resistance.
• For 75-100 hp (1596cc) motors the spark plug wire resistance varies by cylinder (which makes sense since the lengths vary). No. 1 lead should be 4500-10,700 ohms, the No. 2 lead should be 3300-8000 ohms, the No. 3 lead should be 3700-8900 ohms and the No. 4 lead should be 4300-10,200 ohms.
• For 150 hp motors the spark plug wire resistance varies by cylinder (which makes sense since the lengths vary). No. 1 lead should be 4600-10,900 ohms, the No. 2 lead should be 3300-8000 ohms, the No. 3 lead should be 3800-9300 ohms and the No. 4 lead should be 4200-10,000 ohms.
• 20/225 motors may or may not utilize removable caps. The factory gives no specs for the cap or for the wires themselves, all the factory says is to check secondary circuit resistance from one spark plug cap to the other. Specifications are the same for all cylinder pairs at 19,600-35,400 ohms.

■ Yamaha technical literature sometimes mentions that other models use spark plug resistor caps, however we could not locate any testing specifications other than those listed above. Because of the similarity between models, you MAY want to consider trying the specifications from Merc/Mariner in the case where Yamaha has not provided one, but use with caution.

The specifications that Mercury/Mariner publishes for resistance of the plugs wires and/or the spark plug caps are as follows:
• 4/5/6 hp motors use resistor caps on the ends of the spark plug leads, the caps themselves should have 3500-5200 ohms resistance. The secondary ignition circuit is checked w/out the wires attached. No spec is available for the wires themselves, but resistance should be minimal.

■ For more details on secondary ignition circuit testing please refer to the Ignition and Electrical System section and to the various Ignition Component Testing specifications charts.

• 8/9.9 hp (232cc) motors use resistor caps on the ends of the spark plug leads, the caps themselves should have about 4900-5100 ohms resistance. The secondary ignition circuit is checked WITH the wires attached (but not with caps) and should be 13,000-16,000 ohms.
• 9.9/15 hp (323cc) motors use resistor caps on the ends of the spark plug leads, the caps themselves should have about 3500-5200 ohms resistance. The secondary ignition circuit is checked WITH the wires attached (but not with caps) and should be 3900-5900 ohms.
• 25 hp motors use resistor caps on the ends of the spark plug leads, the caps themselves should have about 3500-5200 ohms. The secondary ignition circuit is checked WITH the wires attached (but not with caps) and should be 3500-4700 ohms.

• Carbureted 30/40 hp 3-cylinder motors MAY or MAY NOT use resistor caps on the ends of the spark plug leads. Check the side of the cap to be sure. If it is labeled VDFP, resistance of the cap itself should be 0. However if the cap is labeled VD05FP, then the caps themselves should have about 3500-5200 ohms resistance. The secondary ignition circuit is checked WITH the wires attached (but not with caps) and should be 2720-3680 ohms.
• EFI 30/40 hp 3-cylinder motors may or may not utilize removable caps. All the factory says is to check the wire itself (including the caps) from one end to the other and resistance should be 600-1100 ohms. Secondary resistance is checked at the coil itself WITHOUT any leads attached, and should be 3000-7000 ohms.
• 40/45/50 hp (935cc) motors use resistor caps on the ends of the spark plug leads, the caps themselves should have about 3500-5200 ohms resistance. The secondary ignition circuit is checked WITH the wires attached (but not with caps) and should be 3500-4700 ohms.
• Carbureted 40/50/60 hp (996cc) motors are normally equipped with resistor caps on the ends of the spark plug leads. Check the side of the cap to be sure, if the cap is labeled VD05FP, then the caps themselves should have about 4000-6000 ohms resistance. The secondary ignition circuit is checked WITH the wires attached (but not with caps) and should be 3500-4700 ohms.
• EFI 40/50/60 hp (996cc) motors may or may not utilize removable caps. All the factory says is to check the wire itself (including the caps) from one end to the other and resistance should be 600-1100 ohms. Secondary resistance is checked at the coil itself WITHOUT any leads attached, and should be 3000-7000 ohms.
• For 75/90 hp (1596cc) motors the spark plug wire resistance varies by cylinder (which makes sense since the lengths vary). No. 1 lead should be 4500-10,700 ohms, the No. 2 lead should be 3300-8000 ohms, the No. 3 lead should be 3700-8900 ohms and the No. 4 lead should be 4300-10,200 ohms.
• 115 hp motors may or may not utilize removable caps. The factory gives no specs for the cap or for the wires themselves, all the factory says is to check secondary circuit resistance from one spark plug cap to the other. Specifications however vary slightly between the assemblies for each pair of cylinders. The secondary circuit for the No. 1 and 4 should have 18,970-35,230 ohms, while the secondary circuit for the No. 2 and No. 3 cylinders should be 18,550-34,450 ohms.
• 225 hp motors may or may not utilize removable caps. The factory gives no specs for the cap or for the wires themselves, all the factory says is to check secondary circuit resistance from one spark plug cap to the other. Specifications are the same for all cylinder pairs at 19,600-35,400 ohms.

Regardless of resistance tests and visual checks, it is never a bad idea to replace spark plug leads at least every couple of years, and to keep the old ones around for spares. Think of spark plug wires as a relatively low cost item that whose replacement can also be considered maintenance.

REMOVAL & INSTALLATION

When installing a new set of spark plug wires, replace the wires one at a time so there will be no confusion. Coat the inside of the boots with a dielectric grease or a suitable marine grease to prevent sticking and corrosion. Install the boot firmly over the spark plug until it clicks into place. The click may be felt or heard. Gently pull back on the boot to assure proper contact. Repeat the process for each wire.

■ It is vitally important to route the new spark plug wire the same as the original and install it in a similar manner on the powerhead. Improper routing of spark plug wires may cause powerhead performance problems.

TCI Unit Air Gap

The TCI ignition system used on 2.5 and 4 hp Yamaha motors is contains one major component, the TCI unit. This component contains the primary ignition coil that is used to generate power to spark the ignition system. To eliminate the need for extra components, the small size and relative simplicity of these motors allows the TCI unit to not only function as the ignition coil, but as the power/charge coil itself. The unit is positioned under the flywheel so the primary coil can generate it's own ignition voltage/electric current in conjunction with the fixed magnet on attached to the rotating flywheel.

MAINTENANCE & TUNE-UP 2-45

In order for this ignition system to function properly there is one critical adjustment which should be checked at LEAST annually (or at each tune-up if the Ignition Timing check shows the motor is not properly timed). The TCI unit must be positioned with a precise gap between the unit and the flywheel itself. This gap is easily checked or adjusted using a set of feeler gauges.

In addition, the same basic checks that are made for all other Electronic Ignition systems hold true for the TCI ignition. Be sure to visually inspect the system, as detailed in this section.

ADJUSTMENT

◆ See Figures 144, 145 and 146  MODERATE

1. For 2.5 hp motors, remove the fuel tank from the top of the powerhead.
2. Remove the hand-rewind starter assembly from the top of the powerhead.
3. For 2.5 hp motors, remove the Flywheel, as detailed in the Ignition and Electrical System section.
4. There are 2 projections coming off the TCI unit, facing the flywheel (the flywheel ends of the projections are curved slightly to match the curve of the flywheel). The TCI unit air gap is the distance between these projections and the flywheel itself. Use an appropriate sized feeler gauge to check the gap as follows:
   - For 2.5 hp motors the gap should be 0.016-0.024 in. (0.4-0.6mm).
   - For 4 hp motors the gap should be 0.012-0.024 in. (0.3-0.6mm).

■ When checking the gap, a feeler gauge larger than the specification should not fit between the tips of the TCI unit and the flywheel. A gauge within specification should fit, but with a slight drag. A smaller gauge should fit without any interference whatsoever.

5. If adjustment is necessary, loosen the 2 mounting bolts (one is located over each of the TCI unit projections) and gently reposition by sliding each projection toward or away from the flywheel while holding the feeler gauge in position. Tighten the mounting bolts and recheck the gap to ensure the projections did not move while tightening the fasteners.
6. Install the flywheel, hand-rewind starter and fuel tank, as applicable.

Electronic Ignition Systems

INSPECTION

◆ See Figure 147 MODERATE

The ignition systems used on these Yamaha and Mercury/Mariner motors are known by various titles such as CDI, CDI Micro, TCI, TCI Micro, Digital Inductive (ECM555) and Inductive Electronic. And although there are minor differences between the components for each one, they are all forms of a modern pointless electronic ignition system and operating in a similar manner.

Modern electronic ignition systems have become one of the most reliable components on an outboard. There is very little maintenance involved in the operation of these ignition systems and even less to repair if they fail. Most systems are sealed and there is no option other than to replace failed components.

Just as a tune-up is pointless on an engine with no compression, installing new spark plugs will not do much for an engine with a damaged ignition system. At each tune-up, visually inspect all ignition system components for signs of obvious defects. Look for signs of burnt, cracked or broken insulation. Replace wires or components with obvious defects. If spark plug condition suggests weak or no spark on one or more cylinders, perform ignition system testing to eliminate possible worn or defective components.

If trouble is suspected, it is very important to narrow down the problem to the ignition system and replace the correct components rather than just replace parts hoping to solve the problem. Electronic components can be very expensive and are usually not returnable.

Refer to the "Ignition and Electrical Systems" section for more information on troubleshooting and repairing ignition systems.

These days, the only possible adjustment on most electronic ignition systems (not including the TCI air gap on the smallest) would be timing (Idle, Carb Pickup and/or WOT) and that varies by engine/model. For more details, please refer to the Timing and Synchronization procedure for your motor.

Various engines are equipped with either the Yamaha Microcomputer Ignition System (YMIS), the TCI Micro system or the Mercury/Mariner versions of Digital Inductive (ECM555) or Inductive Electronic ignitions. All but the TCI Micro are essentially a standard CDI type ignition with computer controls. The TCI Micro system is similar to the CDI Micro type system. In all cases, there are normally no few or no adjustable components in this system, but for more details on a particular engine, please refer to Timing and Synchronization in this section.

Electrical System Checks

CHECKING THE BATTERY

Difficulty in starting accounts for almost half of the service required on boats each year. Some years ago, a survey by Champion Spark Plug Company indicated that roughly one third of all boat owners experienced a "won't start" condition in a given year. When an engine won't start, most people blame the battery when, in fact, it may be that the battery has run down in a futile attempt to start an engine with other problems.

Maintaining your battery in peak condition may be though of as either tune-up or maintenance material. Most wise boaters will consider it to be both. A complete check up of the electrical system in your boat at the beginning of the boating season is a wise move. Continued regular maintenance of the battery will ensure trouble free starting on the water.

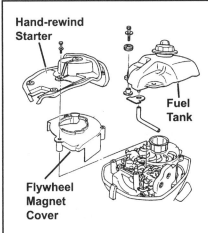

Fig. 144 To check the air gap, first you'll need access to the TCI unit and flywheel (2.5 hp motor shown)

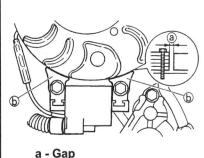

a - Gap
b - TCI Unit Projections

Fig. 145 The gap is the distance between the 2 projections on the TCI unit and the flywheel

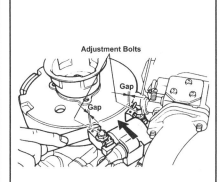

Fig. 146 If necessary, loosen the adjustment bolts and slide the projections gently against a feeler gauge held against the flywheel

2-46 MAINTENANCE & TUNE-UP

Details on battery service procedures are included under Batteries in Boat Maintenance. The following is a list of basic electrical system service procedures that should be performed as part of any tune-up.
- Check the battery for solid cable connections
- Check the battery and cables for signs of corrosion damage
- Check the battery case for damage or electrolyte leakage
- Check the electrolyte level in each cell
- Check to be sure the battery is fastened securely in position
- Check the battery's state of charge and charge as necessary
- Check battery voltage while cranking the starter. Voltage should remain above 9.5 volts
- Clean the battery, terminals and cables
- Coat the battery terminals with dielectric grease or terminal protector

CHECKING THE STARTER MOTOR

◆ See Figure 148

The starter motor system generally includes the battery, starter motor, solenoid, ignition switch and in most cases, a relay.

The frequency of starts governs how often the motor should be removed and reconditioned. Yamaha typically recommends removal and overhaul every 1000 hours. Mercury/Mariner doesn't mention an overhaul period, but obviously, when it breaks, it's time. Most starters do tend to give some warning signs, drawing a little more voltage, cranking a little slower, solenoid sticking or some other indication that it is nearing time for overhaul or replacement.

When checking the starter motor circuit during a tune-up, ensure the battery has the proper rating and is fully charged. Many starter motors are needlessly overhauled or replaced, when the battery is actually the culprit.

Connect one lead of a voltmeter to the positive terminal of the starter motor. Connect the other meter lead to a good ground on the engine. Check the battery voltage under load by turning the ignition switch to the START position and observing the voltmeter reading. If the reading is 9.5 volts or greater, and the starter motor fails to operate, repair or replace the starter motor.

CHECKING THE INTERNAL WIRING HARNESS

◆ See Figure 149

Corrosion is probably a boater's worst enemy. It is especially harmful to wiring harnesses and connectors. Small amounts of corrosion can cause havoc in an electrical system and make it appear as if major problems are present.

The following are a list of checks that should be performed as part of any tune-up.
- Perform a through visual check of all wiring harnesses and connectors on the vessel
- Check for frayed or chafed insulation, loose or corroded connections between wires and terminals
- Unplug all suspect connectors and check terminal pins to be sure they are not bent or broken, then lubricate and protect all terminal pins with dielectric grease to provide a water tight seal
- Check any suspect harness for continuity between the harness connection and terminal end. Repair any wire that shows no continuity (infinite resistance)

Fuel System Checks

FUEL INSPECTION

Most often, tune-ups are performed at the beginning of the boating season. If the fuel system in your boat was properly winterized, there should be no problem with starting the outboard for the first time in the spring. If problems exist, perform the following checks.

1. If the condition of the fuel is in doubt, drain, clean, and fill the tank with fresh fuel.
2. Visually check all fuel lines for kinks, leaks, deterioration or other damage.

> ✱✱ **CAUTION**
>
> We cannot over-emphasize how important it is to visually check the fuel lines and fittings for any sign of leakage. This is even more important on the high-pressure fuel circuits of the EFI. It is a good idea to prime all fuel systems using the primer bulb and, on EFI systems, check the lines/fittings of the high-pressure fuel circuit after the motor has been operated (meaning the system has been under normal operating pressures).

3. Disconnect the fuel lines and blow them out with compressed air to dislodge any contamination or other foreign material.
4. Check the line between the fuel pump and the carburetor (or the vapor separator tank on fuel injected motors) while the powerhead is operating and the line between the fuel tank and the pump when the powerhead is not operating. A leak between the tank and the pump many times will not appear when the powerhead is operating, because the suction created by the pump drawing fuel will not allow the fuel to leak. Once the powerhead is shut down and the suction no longer exists, fuel may begin to leak.

> ✱✱ **CAUTION**
>
> DO NOT do anything around the fuel system without first reviewing the warnings and safety precautions in the Fuel System section. Remember that fuel is highly combustible and all potential sources of ignition from cigarettes and open flame to sparks must be kept far away from the work area.

LOW-PRESSURE FUEL PUMP INSPECTION

◆ See Figure 150

ALL of these motors use a low-pressure fuel pump to draw fuel from a boat mounted or portable fuel tank and feed either the carburetor float bowls or fuel vapor separator tank of fuel injected motors. The low-pressure pumps operate in virtually the same manner for all Yamaha and Merc/Mariner motors (EXCEPT the V6 4-stroke), using a displacement-diaphragm design

Fig. 147 Typical ignition control unit, used on Yamaha models equipped with the CDI Micro (YMIS) system

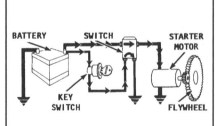

Fig. 148 Functional diagram of a typical cranking circuit

Fig. 149 Any time electrical gremlins are present, always check the harness connectors for pins which are bent, broken or corroded

MAINTENANCE & TUNE-UP 2-47

to alternately create a vacuum in the line from the fuel tank and generate pressure in the line going to the float bowl or vapor separator.

■ Unlike the diaphragm-displacement pump used on all other Yamaha motors, the V6 4-stroke motor uses an electric low-pressure pump. For more details on this pump, refer to the information on Electronic Fuel Injection in the Fuel System section.

On carbureted motors, if the powerhead operates as if the load on the boat is being constantly increased and decreased, even though an attempt is being made to hold a constant powerhead speed, the problem can most likely be attributed to the fuel pump.

Many times, a defective fuel pump diaphragm is mistakenly diagnosed as a problem in the ignition system. The most common problem is a tiny pin-hole in the diaphragm or a bent check valve inside the fuel pump.

If the fuel pump fails to perform properly, an insufficient fuel supply will be delivered to the carburetor or the vapor separator tank. This lack of fuel will cause the engine to run lean and loose rpm.

PRESSURE CHECK

◆ See Figure 151

■ If an integral fuel pump carburetor is installed, the fuel pressure cannot be checked.

1. Mount the outboard unit in a test tank, or on the boat in a body of water.

※※ CAUTION

Never operate the engine at high speed with a flush device attached. The engine, operating at high speed with such a device attached, would runaway from lack of a load on the propeller, causing extensive damage.

2. Install the fuel pressure gauge in the fuel line between the fuel pump and the carburetor or the vapor separator tank.
3. Start the engine and check the fuel pressure.

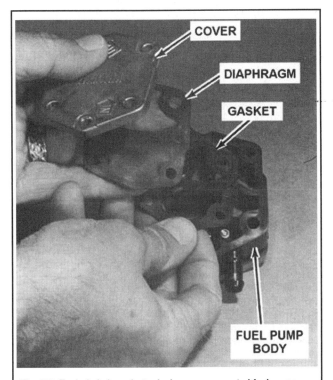

Fig. 150 Exploded view of a typical vacuum operated fuel pump - Even a tiny hole in the diaphragm can affect fuel pump performance

■ Remember, the powerhead will normally not start without the emergency tether in place behind the kill switch knob.

※※ CAUTION

Water must circulate through the lower unit to the engine any time the engine is run to prevent damage to the water pump in the lower unit. Just five seconds without water will damage the water pump.

4. Operate the powerhead at full throttle and check the pressure reading. The gauge should indicate at least 2 psi (14 kPa).

HIGH-PRESSURE FUEL PUMP INSPECTION

In addition to the all important visual inspection of the fuel system lines and fittings on outboards with fuel injection a quick check of the high-pressure fuel circuits will verify the ability of the system to operate properly. This check should be performed with each tune-up or at least annually at the beginning of the season.

PRESSURE CHECK

EFI engines utilize a 2-stage fuel system, the low-pressure circuit (fuel tank-to-pump and pump-to-vapor separator), as well as a high-pressure circuit (vapor separator-to-fuel injectors).

The low-pressure circuit of all fuel-injected motors works in the same manner as the low-pressure circuit used on carbureted models.

The high-pressure circuit of all fuel injected Yamaha motors works in a similar fashion using a submerged electric fuel pump located within the vapor separator tank to achieve fuel pressure roughly somewhere in the 30-45 psi (207-310 kPa) range depending upon the year and model. Most models are equipped with a fuel test port on the top of the separator tank, but there are exceptions where the port is remote mounted.

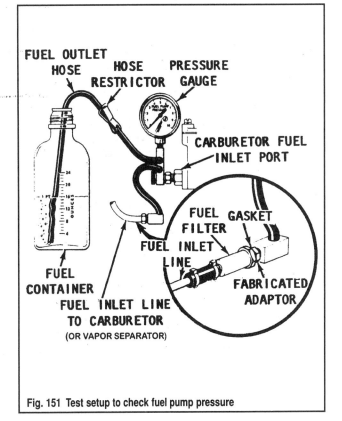

Fig. 151 Test setup to check fuel pump pressure

2-48 MAINTENANCE & TUNE-UP

EFI MOTORS

◆ See Figures 152 and 153

1. Locate the fuel pressure check valve/test fitting on the motor. On all models it will be a Schrader type valve fitting either located on the top of the vapor separator tank (60 hp and smaller motors) or on the fuel rail for 115 hp and larger motors. On the larger motors the check valve/test fitting is usually on top of a fuel rail, adjacent to the pressure regulator (115 hp and V6 motors), but on the 150 hp motors it is found a little more than halfway down the fuel rail, below the pressure regulator.

2. Remove the protective cap from the pressure test fitting, then cover the fitting with a clean, dry shop rag (to protect against possible fuel spray) and connect a fuel pressure gauge to the test fitting/check valve.

3. Provide a source of cooling water (using a test tank or flush fitting) and start the engine, allowing it to run at idle for about a minute. Observe and record the fuel pressure indicated on the gauge.

4. The fuel pressure should be as follows:
- For 60 hp and smaller motors: 42-44 psi (290-303 kPa)
- For 115 hp motors: 41-44 psi (283-304 kPa)
- For 150 hp motors: 37.7 psi (260 kPa)
- For 200/225 hp motors: 38-44 psi (262-304 kPa)

■ Be aware that Yamaha states that system pressure should be as high as 45 psi (310 kPa) when the pump runs before the engine is started and may, due to actuation of the pressure regulator drop as low as 35.6 psi (250 kPa) at idle. Also, note that Mercury/Mariner states the 44 psi (300 kPa) specification for 200/225 hp models is for a cold motor and that the pressure should drop to about 38 psi (262 kPa) on a warm motor.

5. Shut the engine off and, making sure a shop rag is in place to catch any escaping fuel spray, carefully disconnect the gauge.

6. If fuel pressure is not in specification, refer to the Fuel System section for further diagnosis.

■ Additional pump and pressure checks are available for the 60 hp and lower Mercury/Mariner EFI motors. For more details, please refer to Mercury/Mariner EFI Troubleshooting Charts in the Fuel System section.

CHECKING/CLEANING THE CARBURETORS OR THROTTLE BODIES

Visually inspect the carburetors or throttle bodies at each tune-up. Make sure there are no signs of gunk in the throttle bores. The throttle valves should move smoothly without sticking or binding. Linkage points should be lubricated as necessary.

Periodic carburetor adjustments are usually not necessary, as long as the outboard is running correctly. As a matter of fact, in recent service literature Yamaha finally came out and stated that "carburetor adjustments are not necessary on a properly operating motor" so they've finally admitted what we knew all along, "if it ain't broke, don't fix it." When it does come time for carburetor adjustments the mixture screw (when equipped) does not usually require adjustment unless the motor has been moved to/from altitude.

The timing and synchronization adjustments are covered elsewhere in this section and should at least be checked annually, even if not actual adjustment is made to the linkages.

■ Many carburetor adjustments require that the outboard unit is running in a test tank or with the boat in a body of water (and a helper to navigate the craft). For maximum performance, the idle rpm and other carburetor adjustments should be made under actual operating (load) conditions.

Idle Speed Adjustment (Minor Adjustments)

■ The engine idle speed is normally computer controlled on EFI motors, no adjustments are necessary or possible. For more details, please refer to Timing and Synchronization in this section or the information on fuel injection found in the Fuel System section.

Although the full timing and synchronization adjustments should be checked at each tune-up, idle speed can be adjusted independently on carbureted models using the carburetor throttle stop screw which physically determines the positioning of the throttle plate. When the screw is turned one direction (usually clockwise) the throttle plate is held open more, increasing

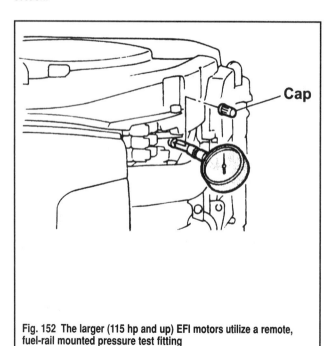

Fig. 152 The larger (115 hp and up) EFI motors utilize a remote, fuel-rail mounted pressure test fitting

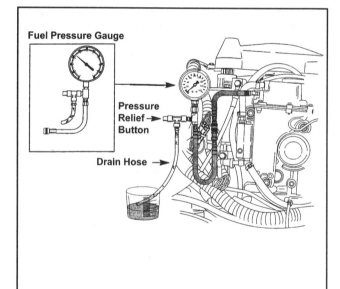

Fig. 153 For the smaller Mercury/Mariner EFI motors (60 hp and down) the high-pressure fuel circuit is checked using a test fitting on top of the vapor separator tank

idle speed. When the screw is turned in the opposite direction (normally counterclockwise) the plate is allowed to close further, lowering idle speed.

■ Don't rely on the boat mounted tachometer when setting idle speed, use a high quality shop tachometer to ensure accurate readings. For more details on tachometers refer to the information under Timing and Synchronization.

IDLE SPEED ADJUSTMENT

◆ See Figure 154

■ Remember, the powerhead should not start without the emergency tether in place behind the kill switch knob.

※※ CAUTION

Water must circulate through the lower unit to the engine any time the engine is run to prevent damage to the water pump in the lower unit. Just five seconds without water will damage the water pump. Never operate the engine at high speed with a flush device attached. The engine, operating at high speed with such a device attached, would runaway from lack of a load on the propeller, causing extensive damage.

The idle speed is regulated by the throttle stop screw. The screw sets the position of the throttle plate inside the carburetor throat. Typically on these models the throttle stop screw is a spring loaded screw threaded through a fixed braked, whose tip pushes against the carburetor (or usually lower carburetor when there are more than one) throttle arm.

※※ CAUTION

On multiple carburetor motors, MAKE SURE you've got the right screw. The other spring loaded screws that contact parts of the throttle linkage are SYNCHRONIZATION SCREWS and touching them may allow the carburetor linkage to come out of synchronization.

Fig. 154 A typical throttle stop screw - 75-100 hp motors shown

■ For additional information on your engine/carburetor, please refer to the Timing and Synchronization section and the Fuel System section.

1. Remove the cowling and attach a tachometer.
2. Start the engine and allow it to warm to operating temperature.
3. Note the idle speed on the tachometer. If the idle speed is not within specification, rotate the idle speed screw until the idle speed falls within specification. The idle speed specification is noted in the Tune-Up Specifications chart.
4. Rotating the idle speed screw clockwise increases powerhead speed, and rotating the screw counterclockwise decreases powerhead speed.

TIMING AND SYNCHRONIZATION

General Information

Any time the fuel system or the ignition system on a powerhead is serviced to replace a faulty part or any adjustments are made for any reason, powerhead timing and synchronization must be carefully checked and verified.

Depending on the engine (predominantly on carbureted motors without microcomputer controls), adjustment of the timing and synchronization can be extremely important to obtain maximum efficiency. The powerhead cannot perform properly and produce its designed horsepower output if the fuel and ignition systems have not been precisely adjusted. We say, depending on the engine because some of the models covered by this manual are equipped with a single carburetor or a computer controlled system (including the EFI motors) which require few, if any periodic adjustments once installed and properly set-up.

As a matter of fact, because of the EPA regulated carburetors used on most of the later models, very few adjustments are possible on most carburetors. Periodic mixture adjustments should not be necessary. However, any carburetor will require initial set-up and adjustment after disassembly or rebuilding. Also, any motor equipped with multiple carburetors will require synchronization with each other after the carburetors have been removed or separated. The multiple throttle valves used on EFI motors normally require some form of synchronization as well if they are removed or the linkage is disconnected for some other reason.

Although some of the motors covered by this manual utilize fully electronically controlled ignition and timing systems, most of the smaller 4-strokes allow for SOME form of timing adjustment. Care should be taken to ensure settings are correct during each tune-up.

■ Before making any adjustments to the ignition timing or synchronizing the ignition to the fuel system, both systems should be verified to be in good working order.

Timing

All outboard powerheads have some type of synchronization between the fuel and ignition systems. Most of the models covered here are equipped with a some form of an electronic advance ignition system. However, models where there is a mechanical advance type Capacitor Discharge Ignition (CDI) system use a series of link rods between the carburetor and the ignition base plate assembly. At the time the throttle is opened, the ignition base plate assembly is rotated by means of the link rod, thus advancing the timing.

On models equipped with some form of micro-computer or electronic advance, the control module decides when to advance or retard the timing, based on input from various sensors (often a crankshaft position sensor). Therefore, there is no link rod between the magneto control lever and the stator assembly.

2-50 MAINTENANCE & TUNE-UP

Many models have timing marks on the flywheel and CDI base. A timing light is normally used to check the ignition timing dynamically - with the powerhead operating. An alternate method is to check the static timing - with the powerhead not operating. This second method requires the use of a dial indicator gauge.

Various models have unique methods of checking ignition timing. These differences are explained in detail later in this section.

Synchronization

In simple terms, synchronization is timing the carburetion or throttle valves to the ignition (and to each other). As the throttle is advanced to increase powerhead rpm, the carburetor/throttle valve and the ignition systems are both advanced equally and at the same rate.

Any time the fuel system or the ignition system on a powerhead is serviced to replace a faulty part or any adjustments are made for any reason, powerhead timing and synchronization must be carefully checked and verified. For this reason the timing and synchronizing procedures have been separated from all others and presented alone in this section.

Before making adjustments with the timing or synchronizing, the ignition system should be thoroughly checked and the fuel system verified to be in good working order.

Prepping the Motor for Timing and Synchronization

Timing and synchronizing the ignition and fuel systems on an outboard motor are critical adjustments. The following equipment is essential and is called out repeatedly in this section. This equipment must be used as described, unless otherwise instructed by the equipment manufacturer. Naturally, the equipment is removed following completion of the adjustments.

For many of the adjustments on Yamaha motors the manufacturer recommends the use of a test propeller instead of the normal propeller in order to put a specific load on the engine and propeller shaft. The use of the test propeller prevents the engine from excessive rpm while applying a load of pre-set value.

- Dial Indicator - Top dead center (TDC) of the No. 1 (top) piston must be precisely known before the timing adjustment can be made on many models. TDC can only be determined through installation of a dial indicator into the No. 1 spark plug opening.
- Timing Light - During many procedures in this section, the timing mark on the flywheel must be aligned with a stationary timing mark on the engine while the powerhead is being cranked or is running. Only through use of a timing light connected to the No. 1 spark plug lead, can the timing mark on the flywheel be observed while the engine is operating.
- Tachometer - A tachometer connected to the powerhead must be used to accurately determine engine speed during idle and high-speed adjustment. Engine speed readings range from about 0-6,000 rpm in increments of 100 rpm. Choose a tachometer with solid state electronic circuits which eliminates the need for relays or batteries and contribute to their accuracy. For maximum performance, the idle rpm should be adjusted under actual operating conditions. Under such conditions it might be necessary to attach a tachometer closer to the powerhead than the one installed on the control panel.

■ An auxiliary tachometer can be connected by attaching it to the tachometer leads in the control panel. These leads are usually Black and Green. Connect the Black lead to the ground terminal of the auxiliary tachometer and the Green lead to the input or hot terminal of the auxiliary tachometer.

- Flywheel Rotation - The instructions may call for rotating the flywheel until certain marks are aligned with the timing pointer. When the flywheel must be rotated, always move the flywheel in the indicated direction (the normal direction of rotation, normally clockwise, but if in doubt you can always double-check by bumping the motor gently using the starter). If the flywheel should be rotated in the opposite direction, the water pump impeller vanes would be twisted. Should the powerhead be started with the pump tangs bent back in the wrong direction, the tangs may not have time to bend in the correct direction before they are damaged. The least amount of damage to the water pump will affect cooling of the powerhead
- Test Tank - Since the engine must be operated at various times and engine speeds during some procedures, a test tank or moving the boat into a body of water, is necessary. If installing the engine in a test tank, outfit the engine with an appropriate test propeller

✱✱ CAUTION

Water must circulate through the lower unit to the powerhead anytime the powerhead is operating to prevent damage to the water pump in the lower unit. Just a few seconds without water will damage the water pump impeller.

■ Remember the powerhead should not start without the emergency tether in place behind the kill switch knob.

✱✱ CAUTION

Never operate the powerhead above a fast idle with a flush attachment connected to the lower unit. Operating the powerhead at a high rpm with no load on the propeller shaft could cause the powerhead to runaway causing extensive damage to the unit.

■ Many adjustment procedures involve checking for proper operation BEFORE touching the current settings. In these cases, remember that no adjustment is necessary if the motor passes the checking portion of the procedure.

2.5 Hp Yamaha Models

IDLE SPEED

◆ See Figure 155

1. Mount the engine in a test tank or move the boat to a body of water.
2. Remove the cowling and connect a tachometer to the powerhead.
3. Start the engine and allow it to warm to operating temperature.
4. Check engine speed at idle. The powerhead should idle at the rpm specified in the Tune-up Specifications chart.

■ Each time the idle speed is adjusted, rev the engine a few times and let it fall back to idle speed for about 15 seconds to stabilize.

5. If adjustment is necessary, rotate the idle speed adjustment screw in small increments (no more than 1/2-1 turn at a time) until the powerhead idles at the required rpm. Turning the screw **IN** will raise the idle speed, while turning the screw **OUT** will lower the idle speed.

■ When turning the idle speed screw outward, make sure that the throttle link maintains contact with the screw. If the link looses contact with the screw at any point you will have to adjust the cable.

6. To adjust the throttle cable, make sure the throttle grip is fully closed, then loosen the cable stop/retaining screw on the end of the carburetor throttle lever. Adjust the engine idle speed, then tighten the throttle cable screw to secure the cable to the lever.
7. Check and adjust the Ignition Timing, as detailed in this section.

IGNITION TIMING

◆ See Figure 156 and 157

The ignition system on these models provides automatic ignition advance. Ignition timing is not adjustable, however it should still be checked periodically to ensure the system is operating properly.

1. Mount the engine in a test tank or move the boat to a body of water.
2. Remove the cowling, then connect a tachometer and a timing light to the powerhead.
3. Start the engine and allow it to warm to operating temperature. Place the engine in gear.
4. Check the idle speed and adjust as necessary.
5. Aim the timing light at the timing window (located on the top of the flywheel cover, at the front of the powerhead). If the timing is correct the magnet indent will be visible through the timing window.

MAINTENANCE & TUNE-UP 2-51

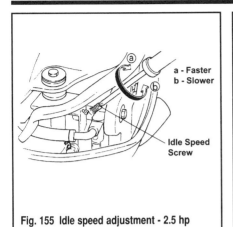

Fig. 155 Idle speed adjustment - 2.5 hp Yamaha

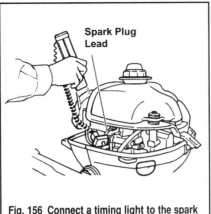

Fig. 156 Connect a timing light to the spark plug lead...

Fig. 157 ...then check that the magnet indent is visible through the timing window

6. The timing is automatically controlled by the TCI unit, so if incorrect either the TCI Unit Air Gap must be adjusted (according to the procedure in this section) or there is a problem with the ignition system.
7. Stop the engine, then remove the tachometer and timing light.

4 Hp Yamaha Models

CHECKING/ADJUSTING THE THROTTLE CABLE

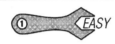

◆ See Figure 158

Before checking/adjusting the idle speed or ignition timing, perform a visual inspection of the throttle cable. Adjust a slack cable and/or replace a damaged cable before proceeding. If adjustment is necessary proceed as follows:
 1. Turn the throttle grip on the tiller arm to the fully closed position, making sure the throttle stop screw (threaded vertically downward into the carburetor boss, right next to the carburetor throttle arm) can be loosened.
 2. Loosen the throttle cable locking screw (mounted to the boss on the throttle lever arm through which the cable end is inserted).
 3. Put thumb pressure on the carburetor throttle lever to hold it in the fully closed position, then check the amount of throttle cable slack, there should be no noticeable slack. If necessary, gently pull the cable through the boss in the throttle lever and, still holding the throttle valve closed, gently tighten the throttle cable locking screw.
 4. Check the adjustment by opening the throttle using the tiller throttle grip. Make sure the throttle opens fully. Next, close the throttle using the tiller grip and verify that the carburetor throttle valve closes fully.
 5. After the cable has been adjusted, you must check/adjust the Idle Speed, as detailed in this section.

CHECKING/ADJUSTING THE STARTER LOCKOUT

◆ See Figure 159

This model contains a starter lockout safety feature to prevent the motor from being started while in gear. The system should be checked at each tune-up to ensure that it is adjusted and functioning properly. Checking is a simple matter of placing the motor in gear and gently attempting to start the motor. If the starter will not rotate, the system is functioning and no further attention is required. However, if the motor rotates in gear, then adjust the cable as follows:
 1. Begin with the engine not running and shifter positioned in **Neutral**.
 2. Loosen the locknuts slightly on either side of the lockout cable (at the adjustment bracket)
 3. Turn the adjuster on the end of the cable housing until the mark on the end of the lever aligns with the mark on the starter housing.
 4. Once positioned properly, retighten the locknuts on cable at either side of the adjustment bracket.
 5. Verify that the lockout is now working properly.

IDLE SPEED

◆ See Figures 160 and 161

 1. Make sure the throttle cable is not damaged and does not contain excessive slack. If the cable is replaced and/or contains excessive slack, adjust is according to the procedure earlier in this section.
 2. Mount the engine in a test tank or move the boat to a body of water.
 3. Remove the cowling for access, then connect a tachometer.

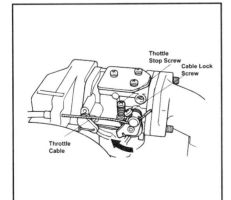

Fig. 158 Make sure there is no slack in the throttle cable - 4 hp Yamaha motors

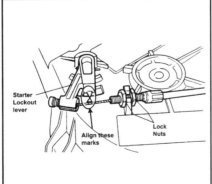

Fig. 159 Starter lockout cable adjustment - 4 hp Yamaha motors

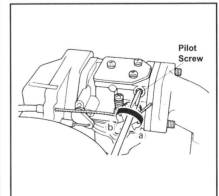

Fig. 160 Most non-US models are equipped with an adjustable pilot screw

MAINTENANCE & TUNE-UP

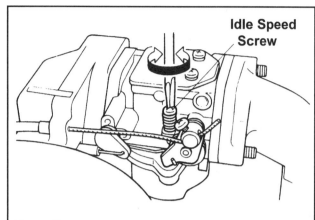

Fig. 161 Turn the idle speed adjustment screw to set the proper engine idle

4. For non-US models or any model whose carburetor contains an adjustable low-speed mixture (pilot) screw, adjust the pilot screw before starting the engine. Locate the pilot screw in the bore on the side, top of the carburetor and turn it inward until is just VERY lightly seats. Next, back out the screw the number of turns indicated in the Carburetor Set-Up Specifications chart, found in the Fuel System section.
5. Start the engine and allow it to warm to operating temperature.
6. Check engine speed at idle. The powerhead should idle at the rpm specified in the Tune-up Specifications chart.
7. Place the engine in gear and check engine trolling speed in the same manner.
8. If adjustment is necessary, rotate the idle speed adjustment screw (the vertical screw mounted in the boss next to the throttle lever, NOT the pilot screw) until the powerhead idles at the required rpm. Normally, turning the screw **IN** will raise the idle speed, while turning the screw **OUT** will lower the idle speed.
9. Check and adjust the Ignition Timing, as detailed in this section.

IGNITION TIMING

◆ See Figure 162

The ignition system on these models provides automatic ignition advance. Ignition timing is not adjustable, however it should still be checked periodically to ensure the system is operating properly.
1. Check the idle speed and adjust as necessary.
2. In addition to the tachometer already installed for idle speed adjustment, connect a timing light to the spark plug lead.
3. Start and run the engine, making sure it is fully warmed.
4. Aim the timing light at the timing window (on the side of the flywheel housing). The stationary pointer should be within the indicated range on the flywheel rotor assembly, if so the timing is correct.

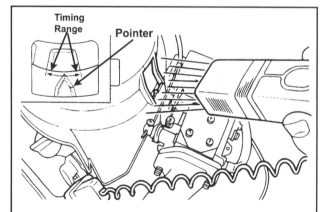

Fig. 162 The fixed pointer must align with the proper timing range - 4 hp Yamaha motors

5. Timing cannot be adjusted. If timing is incorrect, either the TCI Unit Air Gap requires adjustment (as detailed earlier in this section) or a fault has occurred in the TCI system.
6. Stop the engine, then remove the tachometer and timing light.

4/5/6 Hp Mercury/Mariner Models

IDLE MIXTURE

◆ See Figure 163

THE IDLE MIXTURE ADJUSTMENT IS NOT A PERIODIC MAINTENANCE SETTING. The screw should not require adjustment unless the carburetor has been rebuilt or replaced OR unless other mechanical operating parameters have changed drastically due to a significant amount of operating hours resulting in major engine wear OR a major engine repair. In other words, "if it ain't broke, don't fix it"
However, should an idle mixture adjustment become necessary, proceed as follows:

■ Most U.S. models after 2000 may not even be equipped with a pilot screw.

1. If not done during overhaul, turn the pilot mixture screw (located in a small horizontal bore on the inboard side of the carburetor, facing the throttle cables) to the initial setting. This means you need to LIGHTLY seat the screw (do NOT tighten it in the bore as you will crush distort the end of the needle rendering it useless for adjustment), then back it out 2 1/2-3 1/2 turns for models through 2000. For later models, refer to the Carburetor Set-Up Specifications Charts in the Fuel System section, as specs vary by model and market.
2. Mount the engine in a test tank or provide a suitable source of cooling water.
3. Start the engine and allow it to run at idle in forward gear until it reaches normal operating temperature.
4. With the engine still running in gear turn the pilot screw inward (CLOCKWISE) VERY slowly until the engine JUST STARTS to loose rpm, then back it out 1/4 turn.
5. Adjust the screw further as necessary for best performance, HOWEVER do NOT adjust it any leaner (do not turn it inward) than necessary to maintain a smooth idle. If the engine hesitates upon acceleration, readjust the pilot screw.
6. Proceed with Idle Speed adjustment.

IDLE SPEED

◆ See Figure 164

1. If not done already to adjust the Idle Mixture, mount the engine in a test tank or provide a suitable source of cooling water.
2. Connect a suitable tachometer to determine engine speed.
3. Start the engine and allow it to run at idle in forward gear until it reaches normal operating temperature.
4. The idle speed screw is the spring loaded screw lying flat on top of the carburetor, facing out, away from the powerhead. Tightening the screw pushes on the throttle lever holding it open further, thereby increasing idle speed. Loosening the screw allows the throttle lever to close further, lowering idle speed. Adjust the idle speed to the specification of 1300 rpm out of gear or 1100 rpm in gear.
5. If necessary check the Ignition Timing. If not, shut down the powerhead.

CHECKING THE TIMING POINTER

◆ See Figures 165 and 166

There is really only 1 reason why the timing pointer may be off, so it is not necessary to check it EVERYTIME you want to perform a quick timing check. The only reason the timing pointer really could be misaligned is if the flywheel is physically out of alignment with the crankshaft. This could occur

MAINTENANCE & TUNE-UP

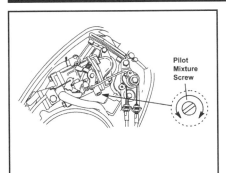

Fig. 163 The pilot screw is used to adjust idle mixture - 4/5/6 Hp Mercury/Mariner Models

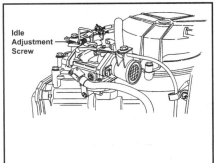

Fig. 164 The idle speed screw is located on top of the carburetor - 4/5/6 Hp Mercury/Mariner Models

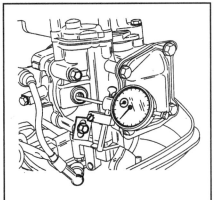

Fig. 165 Use a dial gauge to measure the precise point when the piston reaches TDC

for one of two reasons. One, it could have been installed incorrectly. Two, the flywheel key has sheared and the flywheel has turned independently of the crankshaft.

If you have no reason to suspect flywheel misalignment, proceed directly to Checking Ignition Timing. If the timing is way out, come back to this procedure and verify the pointer alignment before troubleshooting the ignition system.

1. Remove the spark plug.
2. Mount a dial indicator through the spark plug port so that it can contact the piston as it comes up in the bore.
3. Slowly rotate the flywheel in the normal direction of rotation (using the hand rewind starter) to bring the piston up to TDC. Watch the dial gauge to show you the highest point of travel, as that will be TDC.
4. Once the piston is at TDC, check the crankcase cover/cylinder block TDC mark (the longer mark to the left of the cover when looking at the motor) and make sure it is aligned with the timing mark on the flywheel.
5. If the marks are not properly aligned, double-check the setting on the dial indicator by rotating the flywheel again one additional turn, again watching the dial indicator to be sure you stop JUST at the piston reaches the top of its travel. If the marks are still misaligned, remove the flywheel to check for improper installation or a sheared flywheel drive key.
6. Remove the dial indicator and install the spark plug.

CHECKING IGNITION TIMING

◆ See Figure 167

■ The ignition timing on these models is set at assembly and cannot be changed. However, a periodic timing check is still recommended to make sure all ignition components are functioning properly.

✳✳ CAUTION

Because the timing check includes running the motor at Wide Open Throttle (WOT) it CANNOT be done on a flush fitting or the motor will likely be badly damaged by running overspeed without a load. You'll have to either mount it in a test tank or launch the craft to which it is attached.

1. If not done already to adjust the Idle Speed, mount the engine in a test tank so that it can be run at WOT.
2. Connect a suitable tachometer to determine engine speed and a timing light to help observe the timing marks.
3. Start the engine and allow it to run at idle in forward gear until it reaches normal operating temperature.
4. With the engine idling in Forward gear direct the timing light towards the timing marks to verify that idle timing is about 25 degrees Before Top Dead Center (BTDC). The timing mark must now align with the smaller mark to the far right of the crankcase cover/cylinder block (when looking at the motor).
5. With the engine still in Forward gear, advance the throttle until it is in the fully opened (WOT) position. Again, use the timing light to check that the motor is still at about 25 degrees BTDC.

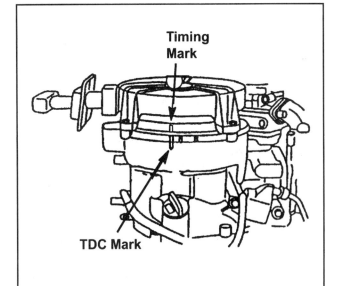

Fig. 166 With the piston at TDC the mark on the flywheel must align with the TDC mark on the crankcase cover

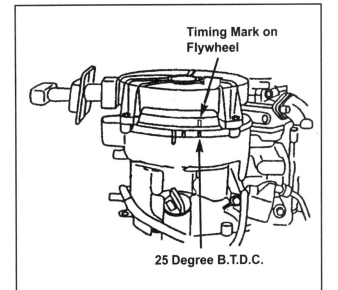

Fig. 167 At both Idle and WOT the flywheel timing mark should align with the 25 degree BTDC mark on the crankcase cover

2-54 MAINTENANCE & TUNE-UP

6. If the timing is off, first proceed with the Checking The Timing Pointer procedure earlier in this section to make sure the flywheel is properly installed/indexed. If the flywheel is not at fault, proceed with Ignition System diagnosis to determine the cause, but it is likely the ignition CDI Unit (also know as an Ignition Module or Power Pack).

7. Shut the powerhead down, then remove the tachometer and timing light.

6/8 Hp Yamaha Models

■ Many adjustment procedures involve checking for proper operation BEFORE touching the current settings. In these cases, remember that no adjustment is necessary if the motor passes the checking portion of the procedure.

CHECKING/ADJUSTING THE SHIFTER

◆ See Figure 168

Operate the shifter, checking for smooth, correct operation. Visually inspect the shift control cable for signs of damage or excessive wear and replace, if necessary. If the shifter handle exhibits incorrect positioning or if the shifter operates poorly, adjust the shifter cable, as follows:

■ Refer to the numbers in the accompanying illustration for component identification.

1. Move the shift handle so the shift rod lever (1) at the base of the powerhead is in the **Neutral** position, then remove the locknut (2)
2. Position the shift handle in **Neutral**.
3. While holding the shift handle to make sure it remains in **Neutral**, carefully turn the adjusting nut (4) in order to adjust the shift control cable (3) to the appropriate length.
4. Install and tighten the locknut to hold this position.
5. Slowly move the shift handle back-and-forth from **Forward** to **Reverse** making sure it shifts smoothly. If necessary, fine tune the adjustment.

■ Without the motor operating the shift handle may stick as you try to shift into Reverse. If this occurs, have an assistant SLOWLY rotate the propeller shaft as you shift.

CHECKING/ADJUSTING THE THROTTLE CABLES

◆ See Figure 169

Operate the throttle, checking for smooth, correct operation. Visually inspect the throttle control cables for signs of damage or excessive wear and replace, if necessary. If the throttle cables show excessive slack and/or the stoppers (a) and (b) at WOT, as well as (c) and (d) at idle, do not contact each other, adjust the throttle cables, as follows:

■ Refer to the numbers in the accompanying illustration for component identification.

1. Check and adjust the shifter, as necessary.
2. Turn the throttle grip to the fully-opened (WOT) position, then loosen the locknut (1) on the pull or accelerator throttle cable.
3. Adjust the length of the pull cable (2) using the adjusting nut (3) until the stopper (a) of the pulley (4) contacts the stopper (b) on the throttle cable retainer (5), then tighten the locknut.
4. Next, loosen the locknut (6) on the push or decelerator cable. Now adjust the length of the push cable (7) using the adjusting nut (8) until there is no clearance between the drum hole on the pulley (4) and the drum at the top of the push cable (7). Then, tighten the locknut.
5. Slowly open and close the throttle grip from the WOT to the idle positions. Make sure the stopper (a) contacts the stopper (b) at WOT and the stopper (c) contacts the stopper (d) at idle (fully closed). If not, repeat the adjustment procedure until the system works as noted.

CHECKING/ADJUSTING THE THROTTLE AXLE LINK

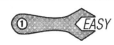

◆ See Figure 170

Visually inspect the throttle axle link (the shaft which runs between the throttle cable pulley and the carburetor throttle lever) for signs of bending or damage. Keep in mind that the stock part does contain a slight bend, so you're looking for signs of additional bending/kinking. Replace, if damaged.

Visually inspect the axle link to make sure it is properly operating the throttle. Twist the throttle grip to WOT and verify that the carburetor throttle lever moves to the fully opened position. Next twist the throttle grip to the idle or slowest position and make sure the carburetor throttle valve closes fully. If the throttle valve on the carburetor does not function as noted, adjust the throttle axle link, as follows:

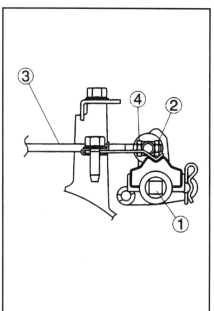

Fig. 168 Adjusting the shifter - 6/8 hp Yamaha models

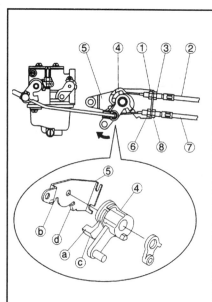

Fig. 169 Adjusting the throttle cables - 6/8 hp Yamaha models

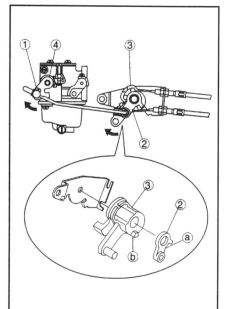

Fig. 170 Adjusting the throttle axle link - 6/8 hp Yamaha models

MAINTENANCE & TUNE-UP

■ Refer to the numbers in the accompanying illustration for component identification.

1. Check and adjust the throttle cables, as necessary.
2. Loosen the lockscrew (1) securing the axle link in position on the carburetor throttle lever.
3. Turn the throttle grip to the WOT position.
4. Hold the free axle lever (2) so the rear side face (a) is in contact with the actuating portion (b) of the pulley (3). Then, turn the carburetor throttle lever (4) fully clockwise until it stops and tighten the lockscrew (1) to hold the link in this position.
5. Operate the throttle grip from idle to the WOT position and back again. Visually check to make sure the carburetor throttle lever moves from the fully closed to fully open positions and back again. If necessary, repeat the adjustment to obtain the desired result.

IDLE SPEED

◆ See Figure 171

1. Check and adjust the throttle axle link, as necessary.
2. Mount the engine in a test tank or move the boat to a body of water.
3. Remove the cowling and connect a tachometer to the No. 1 cylinder spark plug lead.
4. Start the engine and allow it to warm to operating temperature.
5. Check engine speed at idle. The powerhead should idle at the rpm specified in the Tune-up Specifications chart.
6. If adjustment is necessary, rotate the idle adjustment screw (the vertical screw that contacts the carburetor throttle lever) until the powerhead idles at the required rpm. Turn the screw **inward** to increase idle speed or **outward** to decrease idle speed.

CHECKING/ADJUSTING THE STARTER LOCKOUT

◆ See Figure 172

This model contains a starter lockout safety feature to prevent the motor from being started while in gear. The system should be checked at each tune-up to ensure that it is adjusted and functioning properly. Checking is a simple matter of placing the motor in gear and gently attempting to start the motor. If the starter will not rotate, the system is functioning and no further attention is required. However, if the motor rotates in gear, then adjust the cable as follows:

■ Refer to the numbers in the accompanying illustration for component identification.

1. Begin with the engine not running and shifter positioned in **Neutral**.
2. Follow the cable from the top of the manual starter assembly, down to the point where it travels vertically through a horizontal bracket protruding from the powerhead. The nut on top of the bracket is the locknut, the nut on the bottom is the adjustment nut.
3. Loosen the locknut (top), then turn the adjusting nut (bottom) until the point (a) on the cable connector (c) aligns with the mark (b) on the manual starter housing.
4. Once positioned properly, retighten the locknut (top) to hold the cable in this position.
5. Verify that the lockout is now working properly.

IGNITION TIMING

◆ See Figure 173

The ignition system on these models provides automatic ignition advance. Ignition timing is not adjustable. However, ignition timing should be checked periodically to ensure proper powerhead operation.

1. Check the idle speed and adjust as necessary.
2. In addition to the tachometer already installed for idle speed adjustment, connect a timing light to the spark plug lead.
3. Start and run the engine, making sure it is fully warmed.
4. Place the engine in gear.

✱✱ WARNING

Remember, the motor MUST be operated in a test tank or with the boat launched in a body of water allowing it to run under load when running above idle. Failure to heed this warning will likely result in damage to the motor from overspeeding which will occur when the motor is raised toward WOT without a proper load on the propshaft.

5. Aim the timing light at the timing window, then run the motor at both idle and WOT to check the engine timing. The pointer should align with the appropriate markings on the flywheel for the noted engine operating speed. For timing specifications, please refer to the Tune-Up Specifications chart in this section.
6. Timing cannot be adjusted. If timing is incorrect, a fault has occurred in the CDI system.

8/9.9 Hp (232cc) Yamaha Models

Before making these adjustments make sure the valve clearances have been properly checked and adjusted. For details, please refer to Valve Clearance (4-Stroke Models), in this section.

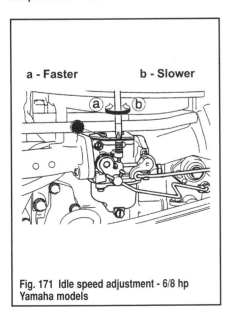

Fig. 171 Idle speed adjustment - 6/8 hp Yamaha models

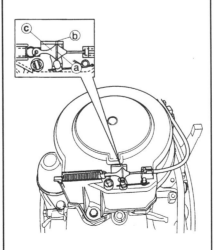

Fig. 172 Starter lockout cable adjustment - 6/8 hp Yamaha models

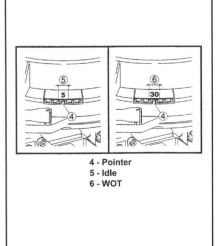

4 - Pointer
5 - Idle
6 - WOT

Fig. 173 Idle and WOT timing marks - 6/8 hp Yamaha models

2-56 MAINTENANCE & TUNE-UP

CHECKING/ADJUSTING THE SHIFT ROD/CABLE

◆ See Figure 174

Operate the shifter, checking for smooth, correct operation. Visually inspect the shift control rod/cable for signs of damage or excessive wear and replace, if necessary. If the shifter handle exhibits incorrect positioning or if the shifter operates poorly, adjust the shifter cable, as follows:

1. The shifter connects to the powerhead with a cable/rod linkage piece that is threaded onto the end of a rod and secured using a locknut. Start the adjustment by loosening the locknut and disconnecting the linkage from the powerhead.
2. Place the shift lever in **Neutral**, then also place the shift arm in Neutral.
3. Slowly turn the cable/rod end inward or outward until it aligns with the shift arm (with the arm and lever both in their current **Neutral** positions).
4. Fit the cable/rod end over the pin on the shift arm which is marked with an **M**, then push the end stopper forward to secure the pin.
5. Tighten the locknut and verify proper shifter operation.

CHECKING/ADJUSTING THE STARTER LOCKOUT

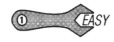

◆ See Figure 175

This model contains a starter lockout safety feature to prevent the motor from being started while in gear. The system should be checked at each tune-up to ensure that it is adjusted and functioning properly. Checking is a simple matter of placing the motor in gear and gently attempting to start the motor. If the starter will not rotate, the system is functioning and no further attention is required. However, if the motor rotates in gear, then adjust the cable as follows:

1. Begin with the engine not running and shifter positioned in **Neutral**.
2. Remove the hinge pin and pivot the front panel downward out of the way for access to the cable adjuster (a vertical bracket on the powerhead just in front of the panel through which the cable is run).
3. Loosen the locknut which contacts the vertical bracket (on the front panel side of the bracket).
4. Next, turn the adjuster nut (located just behind the locknut, a smidge closer to the front panel, it's part of the cable casing) until the arrow marks on the lock arms align with each other (see the accompanying illustration). To locate the lock arms, follow the cable from the adjuster bracket to the cable end which is inserted into a pulley that contains one of the lock arms. Got it?
5. Once positioned properly, retighten the locknut to hold the cable in this position.
6. Verify that the lockout is now working properly.

CHECKING/ADJUSTING THE NEUTRAL OPENING LIMIT

◆ See Figure 176

These motors use a type of mechanical limiter to prevent engine overspeed in **Neutral** by limiting the throttle position. The checking and adjustment procedures are pretty much the same, involving operating the motor in Neutral (in a test tank, using a flushing device or on a body of water) and making sure the engine will not go beyond about 3150-3250 rpm. To check and/or adjust the neutral opening limit, proceed as follows:

■ Refer to the numbers in the accompanying illustration for component identification.

1. Mount the engine in a test tank or launch the boat on a body of water.
2. Remove the cowling and attach a tachometer.
3. Set the shift lever to **Neutral** and start the engine.
4. Allow the engine to operate and come up to normal operating temperature.
5. Turn the throttle grip so the engine speed is 3150-3250 rpm. It should not go any further in **Neutral**.
6. If adjustment is necessary, maintain engine speed in the specified range and loosen the adjustment bolt (1) on the linkage.
7. Make sure the accelerator arm (2) and the shift rod link (3) are in contact. Reposition them slightly, as necessary, then tighten the adjustment bolt (1) to hold this position.
8. Twist the grip down to idle, then with the shifter still in **Neutral** rotate the throttle back up to speed and recheck operation. Readjust, as necessary.

IDLE SPEED

◆ See Figures 177 and 187

1. Check/set the Neutral Opening Limit, as necessary. Keep the motor in the test tank (or launched) and keep the tachometer attached.
2. For non-US/Swiss models or any model whose carburetor contains an adjustable low-speed mixture (pilot) screw, adjust the pilot screw before starting the engine. Locate the pilot screw mounted vertically on the top of the carburetor and turn it inward until is just VERY lightly seats. Next, back out the screw the number of turns indicated in the Carburetor Set-Up Specifications chart, found in the Fuel System section.
3. Start the engine and allow it to warm to operating temperature.
4. Check engine speed at idle. The powerhead should idle at the rpm specified in the Tune-up Specifications chart.
5. Place the engine in gear and check engine trolling speed in the same manner.

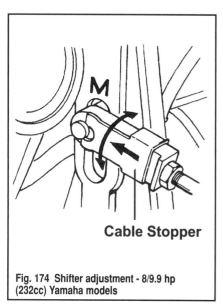

Fig. 174 Shifter adjustment - 8/9.9 hp (232cc) Yamaha models

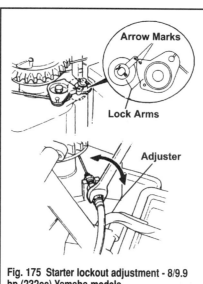

Fig. 175 Starter lockout adjustment - 8/9.9 hp (232cc) Yamaha models

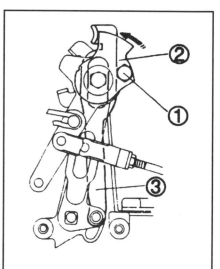

Fig. 176 Neutral opening limit adjustment - 8/9.9 hp (232cc) Yamaha models

MAINTENANCE & TUNE-UP 2-57

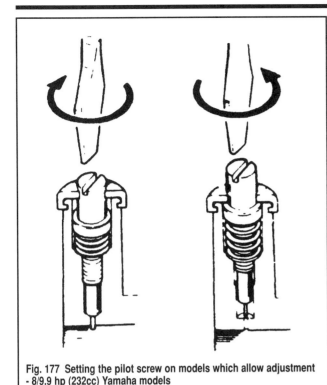

Fig. 177 Setting the pilot screw on models which allow adjustment - 8/9.9 hp (232cc) Yamaha models

6. If adjustment is necessary, rotate the idle speed adjustment screw (the horizontal screw mounted in the bracket next to the throttle lever, NOT the pilot screw) until the powerhead idles at the required rpm. Normally, turning the screw **IN** will raise the idle speed, while turning the screw **OUT** will lower the idle speed.

7. Check and adjust the Throttle Link, followed by the Ignition Timing, as detailed in this section.

THROTTLE LINK ADJUSTMENT

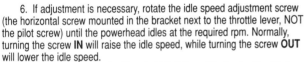

◆ See Figures 179 thru 182

With the engine idle speed properly adjusted, check to make sure the throttle link will fully open and close the carburetor throttle valve when actuated by the accelerator arm. Manually pull the accelerator arm (one end of the throttle link is hooked and it attaches to the accelerator arm) to the fully-open position and check that the throttle lever on the carburetor is also fully open. If adjustment is necessary, proceed as follows:

1. Locate and loosen the locknut on the throttle link.
2. Rotate the long link rod nut in order to adjust the link rod length, then retighten the locknut.
3. Recheck the adjustment by pushing the accelerator arm to the fully-closed side and making sure the throttle lever on the carburetor now moves to the fully-closed position.
4. With the lever fully closed, make sure the link rod end hook has some free-play on its mounting post otherwise the link must be readjusted or replaced.

CHECKING/ADJUSTING THE THROTTLE CABLE

◆ See Figure 183

Once the idle speed and throttle link are properly set, you should check to make sure the last link in the chain (the cable which connects the twist grip to the accelerator arm and linkages) operates properly. Operate the twist grip from idle to WOT and check to make sure the carburetor throttle lever has moved to the fully-open position. Checking for smooth cable operation. Visually inspect the cable for signs of damage or excessive wear and replace, if necessary. If the cable is replaced or does not fully open the carburetor throttle valve, adjust the throttle cable, as follows:

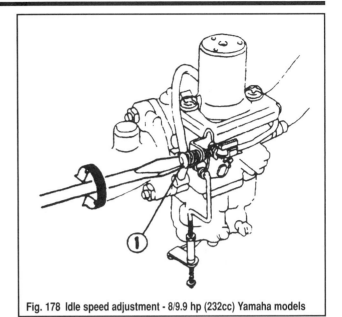

Fig. 178 Idle speed adjustment - 8/9.9 hp (232cc) Yamaha models

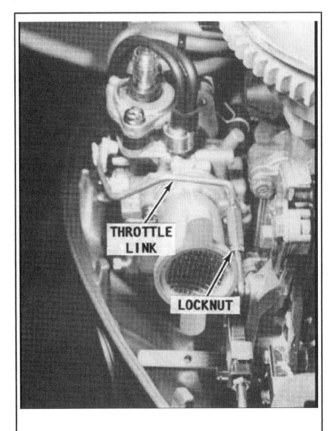

Fig. 179 Throttle link adjustment, start by loosening the locknut

1. The throttle cable connects to the powerhead with a linkage piece that is threaded onto the end of the cable and secured using a locknut. Start the adjustment by loosening the locknut and disconnecting the linkage from the powerhead.
2. Move the twist grip to the Wide-Open Throttle (WOT) position.
3. Pull the accelerator arm to the fully-open end of movement, then slowly turn the cable linkage piece inward or outward until it aligns with the pin on the accelerator arm which is marked with an **M**. Fit the cable over the pin, then push the end stopper forward to secure the pin.
4. Tighten the locknut and verify proper throttle operation.

2-58 MAINTENANCE & TUNE-UP

IGNITION TIMING

 MODERATE

◆ See Figure 184

Ignition timing on the 8/9.9 hp 4-stroke powerhead is adjusted (advanced/retarded) automatically by the ignition module to match rpm operating conditions. Because there are no adjustments for tuning, any problems or irregularities should be diagnosed and repaired as the result of defective mechanical or electrical components. No attempt should be made to perform adjustments in order to achieve the proper timing specifications, let the electronics do their job.

To check ignition timing, proceed as follows:

1. Supply a cooling water source. Although these motors used to be equipped with WOT timing marks, it appears that most are not anymore. Timing will be checked at idle speeds, however, to ensure proper idle speed it is still a good idea to keep the motor mounted in a test tank or the boat on a body of water.

✳✳ CAUTION

Never operate the powerhead above a fast idle with a flush attachment connected to the lower unit. Operating the powerhead at a high rpm with no load on the propeller shaft could cause the powerhead to runaway causing extensive damage to the unit.

2. In addition to the tachometer connected for Idle Speed adjustment, connect a timing light (with an inductive pickup) to the No. 1 cylinder (top) spark plug lead.

■ **Remember that the powerhead won't work unless the emergency tether is in place behind the kill switch knob.**

3. Find the timing marks on the flywheel and magneto base. If necessary clean them or apply a small dab of white paint (stationary corrective fluid, like "white-out"® works well) or white chalk to help make the mark(s) more visible.

■ **On all late-model versions of these engines there are normally only one set of timing marks visible on the magneto base. The timing marks on these motors consists of 2 lines spaced about 1/4 in. (6mm) apart. The set of lines should be considered the idle timing marks.**

4. Start and idle the powerhead until it reaches normal operating temperature.

5. Once the motor is warmed, aim the timing light at the flywheel checking to see if the flywheel mark is between the idle marks on the magneto base.

If the ignition timing is not operating within specification check the following potential causes:

- The pulse coil or CDI unit is faulty. Refer to information in the Ignition and Electrical System section for details on system and component testing.

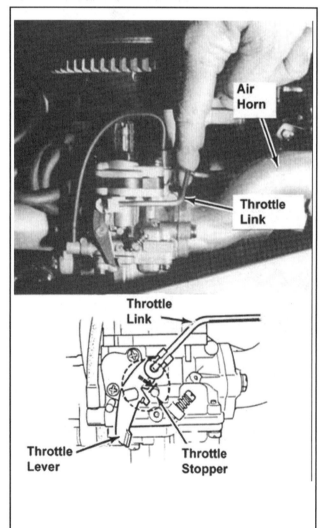

Fig. 180 Turn the adjuster until the throttle link holds the carb throttle lever fully open when the accelerator arm is pushed open...

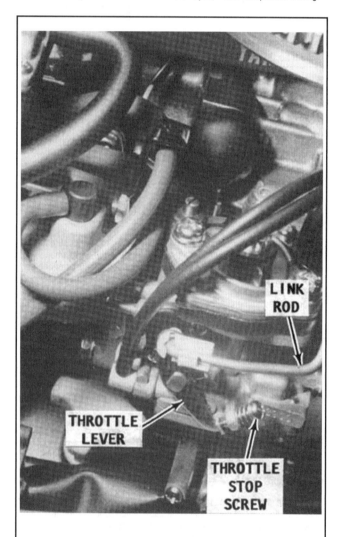

Fig. 181 ...and pulls the throttle lever closed when the accelerator arm is moved to the closed side

MAINTENANCE & TUNE-UP

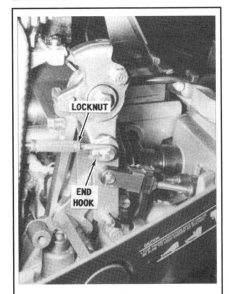

Fig. 182 With the lever closed, the rod end hook of the link must still have some play or the link must be readjusted or replaced

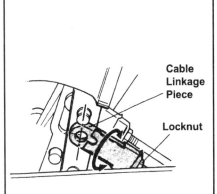

Fig. 183 Shifter adjustment - 8/9.9 hp (232cc) Yamaha models

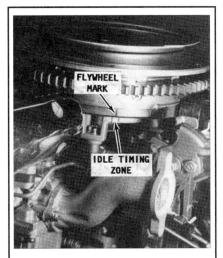

Fig. 184 Idle timing marks on a 8/9.9 hp (232cc) Yamaha models

- Valve clearance is out of spec.
- The camshaft timing is out due to a slipped or stretched timing belt.
- The timing and valve train (timing gears, camshaft lobes, valves, rocker arms or rocker shaft) have become worn beyond service.

8/9.9 Hp (232cc) Mercury/Mariner Models

Before making idle speed or mixture adjustments make sure the valve clearances have been properly checked and adjusted. For details, please refer to Valve Clearance, in this section.

TILLER HANDLE ADJUSTMENT

◆ See Figures 185 thru 188

1. Set the throttle grip to the idle position.
2. Check the tiller handle cables on the powerhead to make sure they are properly adjusted. There is a small pointer on the bell crank which, when the cables are adjusted correctly, will align with the appropriate mark or cutout on tab in the bell crank mounting bracket as follows:

- For side handle shift/single adjusting mark models, there is only 1 cutout between 11 o'clock and 12 o'clock when looking straight onto the bell crank bracket from the throttle link rod side. Make sure the pointer and the single cutout align.
- For side handle shift/triple adjusting mark platform, there are 3 cutouts on the tab, the pointer must align with the middle cutout which is between 11 o'clock and 12 o'clock when looking straight onto the bell crank bracket from the throttle link rod side. (We call it the middle cutout, but it is not centered, the left cutout is much further away from it than the right cutout). Make sure the pointer and the middle cutout align.
- For 8 hp models with tiller handle shift, there are 3 cutouts on the tab, HOWEVER the pointer must align with the front edge of the tab itself and NOT any of the cutouts.
- For 9.9 hp models with tiller handle shift, there are 3 cutouts on the tab, the pointer must align with the cutout at the rear of the bracket (further away from the other two cutouts and closest to the bracket mounting bolt). Make sure the pointer and this first (rear) cutout align.

3. If the mark and proper cutout (or edge of the bracket) do NOT align, loosen the jam nuts on either side of the cable adjuster, then turn the nuts as necessary to align the marks without any slack, but be careful NOT to over-tighten the cables. Once the cables are properly adjusted, retighten the jam nuts again the bracket.

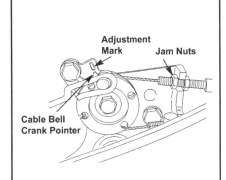

Fig. 185 Tiller handle adjustment - 8/9.9 Hp (232ccc) Mercury/Mariner Models with side handle shift/single adjusting mark

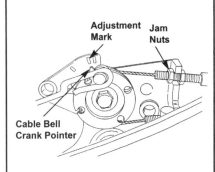

Fig. 186 Tiller handle adjustment - 8/9.9 Hp (232ccc) Mercury/Mariner Models with side handle shift/triple adjusting mark platform

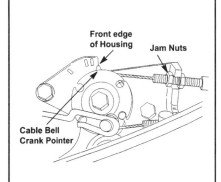

Fig. 187 Tiller handle adjustment - 8 Hp Mercury/Mariner Models with tiller handle shift

2-60 MAINTENANCE & TUNE-UP

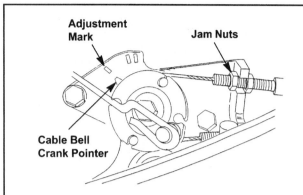

Fig. 188 Tiller handle adjustment - 9.9 Hp Mercury/Mariner Models with tiller handle shift

FULL THROTTLE LINK ROD ADJUSTMENT

◆ See Figures 189 and 190

1. For tiller models, place the control in the Wide Open Throttle (WOT) position.
2. For remote control models, disconnect the remote control cable from the throttle arm, than manually hold the arm in the WOT position.
3. Now check the throttle lever to make sure it is JUST touching the WOT stop. If necessary, loosen the throttle link rod jam nut and turn the adjustment nut until the lever is in the proper position. Once adjusted carefully tighten the jam nut to hold the adjustment nut in position.

IDLE MIXTURE (EXCEPT BODENSEE MODELS)

◆ See Figure 191

THE IDLE MIXTURE ADJUSTMENT IS NOT A PERIODIC MAINTENANCE SETTING. The screw should not require adjustment unless the carburetor has been rebuilt or replaced OR unless other mechanical operating parameters have changed drastically due to a significant amount of operating hours resulting in major engine wear OR a major engine repair. In other words, "if it ain't broke, don't fix it"

However, should an idle mixture adjustment become necessary, proceed as follows:

1. If not done during overhaul, turn the pilot mixture screw (located threaded vertically down into the top of the carburetor, near the mounting flange to the initial setting. This means you need to LIGHTLY seat the screw (do NOT tighten it in the bore as you will crush distort the end of the needle rendering it useless for adjustment), then back it out 2-4 turns.
2. Mount the engine in a test tank or provide a suitable source of cooling water.
3. Start the engine and allow it to run at idle in Neutral until it reaches normal operating temperature.
4. With the engine still running in Neutral turn the pilot screw inward (CLOCKWISE) VERY slowly until the engine JUST STARTS to loose rpm, then back it out 1/4 turn.
5. Adjust the screw further as necessary for best performance, HOWEVER do NOT adjust it any leaner (do not turn it inward) than necessary to maintain a smooth idle.
6. With the motor in a test tank or on a launched craft, place the outboard in gear and check performance between idle and WOT. If the engine hesitates upon acceleration, readjust the pilot screw.
7. Proceed with Idle Speed adjustment.

IDLE SPEED

◆ See Figures 192, 193 and 194

1. If not done already to adjust the Idle Mixture, mount the engine in a test tank or provide a suitable source of cooling water.
2. Connect a suitable tachometer to determine engine speed.
3. For tiller models, set the twist grip to the neutral/idle position.
4. For remote control models, disconnect the remote control cable from the throttle arm, then manually position the throttle lever to the idle position.
5. Locate the spring loaded idle speed screw (threaded horizontally into the throttle arm bracket). Back the idle speed screw off the throttle lever until the screw does NOT touch the lever.
6. Check the link rod. On tiller models there should be free play between the link rod and cam stud (on the round cable bevel). On remote control models the link rod itself should be centered in the throttle lever slot.
7. For an initial idle setting turn the idle speed screw inward (clockwise) until the screw JUST touches the throttle lever, THEN continue to rotate it inward one complete turn to just slightly open the throttle plate.
8. Start the engine and allow it to run at idle in neutral until it reaches normal operating temperature.
9. With the motor fully warmed and running in neutral, turn the idle speed screw as necessary to obtain an idle speed of 900-1000 rpm.
10. Place the motor in gear (the motor must be in a test tank or on a launched craft for this) and make sure idle speed drops to 800-900 rpm.
11. Adjust the idle speed screw further as necessary.
12. Shut the powerhead down and remove the tachometer.

IGNITION TIMING

The ignition timing is electronically controlled on these models. No information is provided by the manufacturer for even checking for proper operation of the electronic advance.

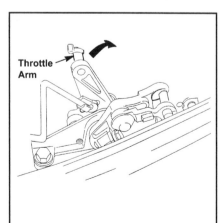

Fig. 189 On remote control models, disconnect the cable then hold the throttle arm in the WOT position

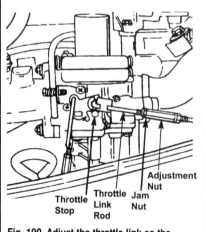

Fig. 190 Adjust the throttle link so the throttle lever JUST TOUCHES the WOT stop

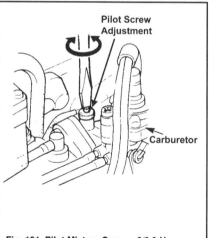

Fig. 191 Pilot Mixture Screw - 8/9.9 Hp (232ccc) Mercury/Mariner Models

MAINTENANCE & TUNE-UP 2-61

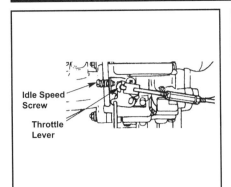

Fig. 192 Idle Speed Screw and Throttle Lever - 8/9.9 Hp (232ccc) Mercury/Mariner Models

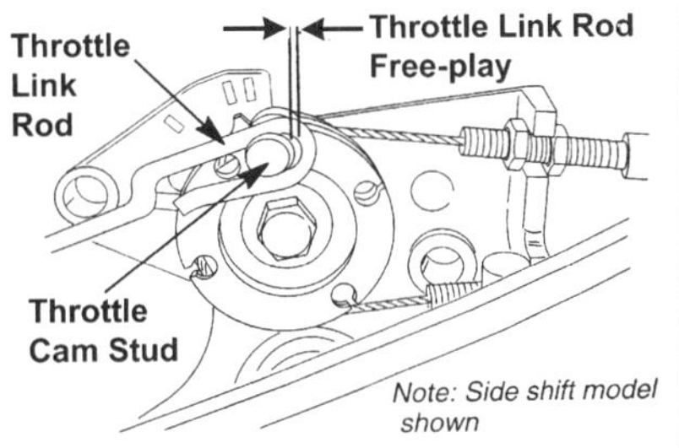

Fig. 193 Check the throttle link rod for a small amount of free-play - Tiller Models

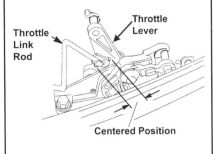

Fig. 194 Check to be sure the throttle link rod is centered in the throttle lever slot - Remote Control Models

15 Hp (323cc) Yamaha Models

CHECKING/ADJUSTING THE CONTROL LINKAGE

MODERATE

Tiller Control Models

◆ See Figures 195 and 196

■ Refer to the numbers in the accompanying illustration for component identification.

1. Visually inspect the throttle cables for damage or excessive wear and replace, if necessary (be sure to adjust after replacement). Even if the cables do not require replacement, inspect them for signs of excessive slack and adjust, if necessary as follows:
   a. Set the throttle lever to the Wide-Open Throttle (WOT) position.
   b. Make the stopper (2) on the bracket (1) contact the WOT stopper (4) on the throttle control cam (3), then loosen the 1st throttle cable (5) locknut and turn the adjusting nut (6) until the 1st cable is the appropriate length to hold the stoppers in contact. Tighten the locknut.
   c. Next, adjust the length of the 2nd throttle cable (7) by loosening the locknut and turning the adjusting nut (8). When finished, tighten the locknut for the 2nd throttle cable.
   d. Close the throttle and make sure the idle stopper (9) on the throttle cam (3) contacts the bracket (1) stopper (2). If necessary, readjust the cables.
   e. Set the throttle back to WOT position, and make sure the WOT stopper (4) contacts the bracket (1) stopper (2). If necessary, readjust the cables.
2. Once the throttle cables are properly checked or adjusted, visually inspect the throttle link rod for signs of loosening, pitting, bending or damage. If necessary, adjust the throttle link rod as follows:
   a. Loosen the throttle link rod lockscrew, which secures the rod to the throttle link (5).
   b. Set the throttle grip to the WOT position.
   c. With the free accelerator opening portion (3) of the throttle control cam (2) making contact with the throttle control lever (4), carefully push the throttle link (5) to the WOT position and holding everything in this position tighten the lockscrew.
   d. Close the throttle grip, then open it again to the WOT position and make sure the throttle link (5) moves to the fully-opened (WOT) position. If not, repeat the adjustment step.

Remote Control Models

◆ See Figures 197, 198 and 199

■ Refer to the numbers in the accompanying illustration for component identification.

1. Visually inspect the throttle cable for damage or excessive wear and replace, if necessary (be sure to adjust after replacement). Even if the cable does not require replacement, inspect it for signs of excessive slack and adjust, if necessary as follows:
   a. Set the shift lever to the forward position.
   b. Make the stopper (2) on the bracket contact the idle stopper (1) on the throttle control cam. Pull the inner portion of the remote control cable (3) to decrease the backlash, then install the remote control cable end (4) to the threaded part and turn the adjuster nut (5) to take up the slack and tighten it.
   c. Using the remote control lever, open the throttle to WOT position and then close it back to the idle position. Make sure the WOT stopper and the idle stopper on the throttle control cam (6) each contact the bracket stopper (7) in the appropriate positions. If necessary, readjust the cable.
2. Once the throttle cable is properly checked or adjusted, visually inspect the throttle link rod for signs of loosening, pitting, bending or damage. If necessary, adjust the throttle link rod as follows:
   a. Loosen the throttle link rod lockscrew (1), which secures the rod to the throttle link (5).
   b. With the throttle control lever opening portion (3) of the throttle control cam (2) making contact with the throttle control lever (4), carefully push the throttle link (5) to the WOT position and holding everything in this position tighten the lockscrew (1).
   c. Move the remote control to the idle and WOT positions. Make sure the throttle link (5) moves from the fully-closed (idle) to the fully-opened (WOT) positions. If not, repeat the adjustment step.
3. Once the throttle cable and link are properly adjusted, check and adjust the shift cable. First, visually inspect the shift cable for damage or excessive wear and replace, if necessary (be sure to adjust after replacement). Even if the cable does not require replacement, inspect it for signs of excessive slack and adjust, if necessary as follows:

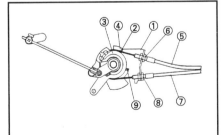

Fig. 195 Throttle cable adjustment - Tiller control 15 hp (323cc) Yamaha motors

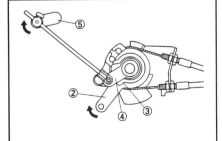

Fig. 196 Throttle link rod adjustment - Tiller control 15 hp (323cc) Yamaha motors

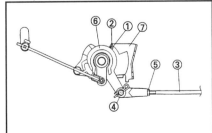

Fig. 197 Throttle cable adjustment - Remote control 15 hp (323cc) Yamaha motors

2-62 MAINTENANCE & TUNE-UP

a. Manually set the engine side of the shift link rod to the neutral position, then set the remote control lever to the neutral position.

b. Install the remote control cable end (3) to the threaded portion of the remote control cable (1) so that the remote control lever and shift link rod (2) are connected in the middle range of their free-play. Now, adjust the cable length and secure the remote control cable end using the adjust nut (4).

c. Move the shift control lever slowly back and forth from forward to reverse, checking that the engine lever properly engages (if any difficulty is encountered, try performing this with an assistant slowly rotating the propeller shaft and/or with the engine idling in a test tank or using the flush fitting). If the shifter does not properly engage, repeat this part of the adjustment procedure.

IDLE SPEED

◆ See Figures 200 and 201

1. Mount the engine in a test tank or move the boat to a body of water.
2. Remove the cowling and connect a tachometer to the powerhead.
3. Make sure the Control Linkage is properly adjusted.
4. For non-US models or any model whose carburetor contains an adjustable low-speed mixture (pilot) screw, adjust the pilot screw before starting the engine. Locate the pilot screw mounted horizontally in a bore on the top, side of the carburetor (just a little inboard on the carburetor body from the idle speed adjustment screw) and turn it inward until is just VERY lightly seats. Next, back out the screw the number of turns indicated in the Carburetor Set-Up Specifications chart, found in the Fuel System section.
5. Start the engine and allow it to warm to operating temperature.
6. Check engine speed at idle. The powerhead should idle at the rpm specified in the Tune-up Specifications chart.
7. Place the engine in gear and check engine trolling speed in the same manner.
8. If adjustment is necessary proceed as follows:
• For models equipped with an adjustable pilot screw, turn the pilot screw inward VERY slowly until the highest possible idle speed is obtained, then adjust the idle speed adjustment screw in the same manner as models NOT equipped with an adjustable pilot screw.
• For models not equipped with an adjustable pilot screw, rotate the idle speed adjustment screw (the horizontal screw mounted inline with the carburetor body and contacting the throttle lever, NOT the pilot screw) until the powerhead idles at the required rpm. Turning the screw **IN** will raise the idle speed, while turning the screw **OUT** will lower the idle speed.

CHECKING/ADJUSTING THE STARTER LOCKOUT

◆ See Figure 202

The hand-rewind start models of the 15 hp motor contain a starter lockout safety feature to prevent the motor from being started while in gear. The system should be checked at each tune-up to ensure that it is adjusted and functioning properly. Checking is a simple matter of placing the motor in gear and gently attempting to start the motor. If the starter will not rotate, the system is functioning and no further attention is required. However, if the motor rotates in gear, then adjust the cable as follows:

1. Begin with the engine not running and shifter positioned in **Neutral**.
2. Follow the starter lockout cable from the top of the manual starter housing down to an angled bracket mounted on the powerhead. Loosen the lockout cable locknut (positioned on top of the powerhead bracket).
3. Next, use the nut (located on the underside of the cable bracket) to change the length of the cable until the marking at the cable end aligns with the marking on the starter case.
4. Once positioned properly, retighten the locknut to hold the cable in this position.
5. Verify that the lockout is now working properly.

IGNITION TIMING

◆ See Figure 203

> ** WARNING
>
> **Remember, the motor MUST be operated in a test tank or with the boat launched in a body of water allowing it to run under load when running above idle. Failure to heed this warning will likely result in damage to the motor from overspeeding which will occur when the motor is raised toward WOT without a proper load on the propshaft.**

The ignition system on these models provides automatic ignition advance. Ignition timing is not adjustable. However, ignition timing should be checked periodically to ensure proper powerhead operation.

1. Check the idle speed and adjust as necessary.
2. In addition to the tachometer already installed for idle speed adjustment, connect a timing light to the spark plug lead.

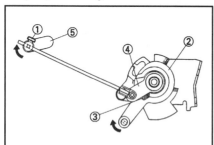

Fig. 198 Throttle link rod adjustment - Remote control 15 hp (323cc) Yamaha motors

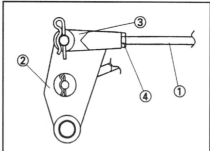

Fig. 199 Shift cable adjustment - Remote control 15 hp (323cc) Yamaha motors

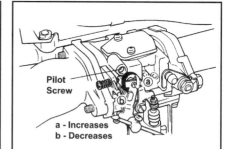

Fig. 200 Pilot screw adjustment - non-US 15 hp (323cc) Yamaha motors

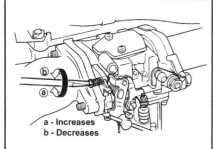

Fig. 201 Idle speed screw adjustment - 15 hp (323cc) Yamaha motors

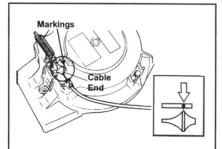

Fig. 202 Starter lockout adjustment - Hand-rewind start 15 hp (323cc) Yamaha motors

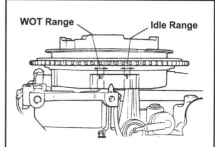

Fig. 203 Checking ignition Timing - 15 hp (323cc) Yamaha motors

MAINTENANCE & TUNE-UP

3. Start and run the engine, making sure it is fully warmed.

※※ WARNING

Remember, the motor MUST be operated in a test tank or with the boat launched in a body of water allowing it to run under load when running above idle. Failure to heed this warning will likely result in damage to the motor from overspeeding which will occur when the motor is raised toward WOT without a proper load on the propshaft.

4. Aim the timing light at the timing window, then run the motor at both idle and WOT to check the engine timing. The pointer should align with the appropriate markings on the flywheel for the noted engine operating speed. For timing specifications, please refer to the Tune-Up Specifications chart in this section.

5. Timing cannot be adjusted. If timing is incorrect, a fault has occurred in the CDI system.

9.9/15 Hp (323cc) Mercury/Mariner Models

LINKAGE ADJUSTMENTS

◆ See Figures 204 thru 208

Various adjustments are necessary to set up the throttle linkage, choke and accelerator lever. These adjustments would be required anytime a related component is replaced or the adjustment setting is in some way disturbed. However, none of these adjustments should be required as simple maintenance.

When these adjustments are required because of a repair, be sure to make/check all linkage adjustments before adjusting the Idle Speed.

1. For remote control models, check to make sure the end of the throttle linkage protrudes just past the lever boss (with the set-screw) the proper distance. Exactly 0.28 in. (7mm) of the linkage shaft must protrude past the boss. If necessary, loosen the set screw and reposition the linkage, then retighten the set screw to hold it in place.

2. For tiller control models open and close the throttle arm by hand, checking for free movement on the pulley. Make sure the throttle shaft is not sticking and that it returns to the idle position. Now, open the throttle at the grip to the WOT position, and check to make sure the throttle arm DOES NOT contact the throttle stop, but that there is no more than a 0.1 in. (2.54mm) gap between them. If necessary adjust the throttle link as follows:

a. Loosen the idle speed screw (counterclockwise) until it no longer touches the throttle shaft arm. Also, loosen the throttle linkage set-screw in the throttle arm boss.

b. Make sure the throttle link is free to move on the pulley boss (at the cable end) and the throttle boss/barrel (at the throttle arm end).

c. Slide the throttle link FORWARD until it comes to rest on the pulley boss. Hold the link in this position while you tighten the set-screw at the other end (the throttle arm boss/barrel) securely.

d. Now recheck the movement of the throttle arm and the gap at the WOT stop.

3. Make sure the manual choke cable fully opens the choke is OFF and fully closes the shutter when the choke is ON. If not, disconnect the shaft connector ball socket from the ball on the shutter lever. Then rotate the ball socket on the shaft threads until it will actuate the shutter lever properly. Carefully snap the socket back onto the ball and visually verify proper operation.

4. Lastly, check the accelerator pump actuator arm clearance. On the same side of the carburetor as the choke shutter linkage, but mounted a little lower and further back on the carburetor, is a threaded adjuster and stopper for the accelerator pump. There should be 0.010 in. (0.254mm) of clearance between the threaded adjuster and the top of the pump actuator itself (when the throttle is fully closed). If necessary, loosen the locknut and rotate the adjuster as necessary to place a slight drag on the proper sized feeler gauge when it is held between the adjuster and the actuator. Hold the adjuster from turning as you tighten the nut, then recheck the clearance.

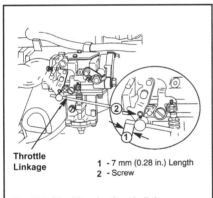

Fig. 204 Checking the throttle linkage - Remote Control 9.9/15 hp Models

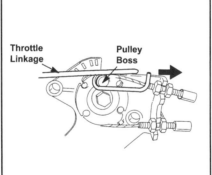

Fig. 205 When checking the throttle linkage on tiller models, make sure the link seats against the pulley boss

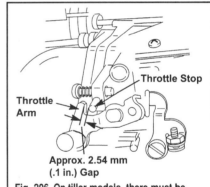

Fig. 206 On tiller models, there must be some clearance between the throttle arm and throttle stop

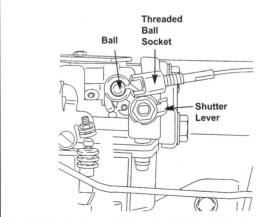

Fig. 207 Choke shutter adjustment - 9.9/15 hp (323ccc) Mercury/Mariner Models

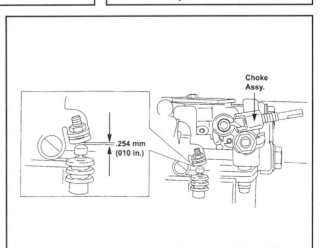

Fig. 208 Accelerator pump actuator adjustment - 9.9/15 hp (323ccc)

2-64 MAINTENANCE & TUNE-UP

IDLE SPEED

◆ See Figure 209

1. Mount the engine in a test tank or provide a suitable source of cooling water.
2. Connect a suitable tachometer to determine engine speed.
3. Start the engine and allow it to run at idle in forward gear until it reaches normal operating temperature.
4. Locate the spring loaded idle speed screw threaded horizontally into a bracket on the side of the carburetor in a position so that it's tip can hold the throttle arm open slightly.
5. Once the engine is fully warmed, turn the idle speed screw inward (clockwise) to increase speed or outward (counterclockwise) to decrease speed. Set the idle speed in neutral to 900-1000 rpm.
6. Either shut down the engine, or connect a timing light and proceed with Checking Ignition Timing.

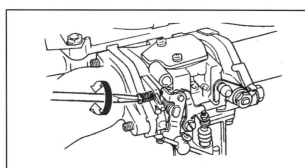

Fig. 209 Idle speed adjustment - 9.9/15 hp (323ccc) Mercury/Mariner Models

CHECKING IGNITION TIMING

◆ See Figure 210

■ The ignition timing on these models is controlled by the ECM unit and cannot be changed. However, a periodic timing check is still recommended to make sure all ignition components are functioning properly.

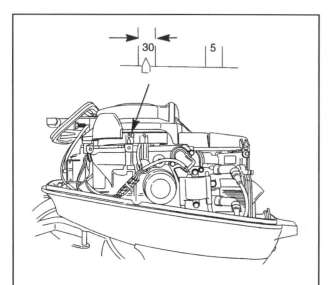

Fig. 210 Checking ignition timing (WOT shown) - 9.9/15 hp (323ccc) Mercury/Mariner Models

✴✴ CAUTION

Because the timing check includes running the motor at Wide Open Throttle (WOT) it CANNOT be done on a flush fitting or the motor will likely be badly damaged by running overspeed without a load. You'll have to either mount it in a test tank or launch the craft to which it is attached.

1. If not done already to adjust the Idle Speed, mount the engine in a test tank so that it can be run at WOT.
2. Connect a suitable tachometer to determine engine speed and a timing light to help observe the timing marks.
3. Start the engine and allow it to run at idle until it reaches normal operating temperature.
4. Point the timing light at the timing mark pointer on the side of the powerhead, just below the flywheel. The 5 degree BTDC mark on the flywheel should, more or less, align with the timing pointer.
5. Slowly increase the engine speed to WOT (or any speed 3000 rpm or higher) and visually check that the 30 degree BTDC mark on the flywheel is now roughly in alignment with the timing pointer.
6. If one or both of the timing marks are not in alignment as specified, EITHER the flywheel is mis-aligned OR there is a problem with the ignition system.
7. Shut the powerhead down, then remove the tachometer and timing light.

25 Hp Models

CHECKING/ADJUSTING THE LINKAGE

Throttle Link Rod (Yamaha Models Only)

◆ See Figure 211

Operate the throttle to the WOT position and visually check to ensure the carburetor throttle valve opens fully. If not adjust the throttle link rod, as follows:
1. Loosen the link rod lockscrew.
2. Turn the throttle control lever clockwise to the WOT position, when the lever contacts the stopper hold it in that position.
3. Now, turn the carburetor throttle lever to the WOT position and holding both levers in this position (another set of hands can be really helpful here), tighten the link rod lockscrew.
4. Operate the throttle control lever and visually make sure the carburetor throttle valve goes from idle to WOT.

Throttle & Shifter Cables (Mercury/Mariner Models Only)

◆ See Figures 212 and 213

No periodic adjustment should be necessary for the Shift and Throttle cables, however, should the cables be replaced they will require an initial set-up to place them in proper alignment with the linkage.
1. Make sure the throttle/shifter is in the idle/neutral position.
2. The shifter cable (the lower of the two cables) should be set so the distance between the cable barrel (on the threaded adjuster) and the cable end (where it attaches to the linkage) is the same as the distance between the cable barrel bracket and the linkage pin when both the gearcase and the shifter are in Neutral. If adjustment is necessary, free the cable and rotate the cable barrel/adjuster to set the proper distance.
3. Next move onto the throttle cable. Check the throttle cam, there is an alignment mark towards 12 o'clock on the rounded cam, it should be pointing almost straight up and should be centered with the roller on the carburetor throttle lever.
4. If adjustment is necessary, release the control cable latch (if not done already for shifter cable adjustment) by rotating clockwise about 1/4 turn, then adjust how the throttle cable (the upper of the two cables) sits in the barrel until the mark on the cam centers with the roller. Once the mark is properly aligned, rotate the latch counterclockwise to lock the cables back into position.
5. After adjustment, actuate the remote handle a few times from WOT back to idle and confirm the mark is still in alignment. Readjust, as necessary.

MAINTENANCE & TUNE-UP

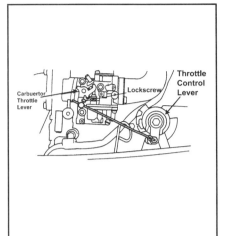

Fig. 211 Throttle link rod adjustment - 25 hp Yamaha motors

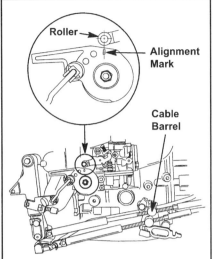

Fig. 212 Adjust the throttle cable so the mark on the cam aligns centers on the throttle roller - 25 hp Mercury/Mariner Models

Fig. 213 After adjustment, be sure to secure the cables with the latch - 25 hp Mercury/Mariner Models

IDLE SPEED

◆ See Figures 214 and 215

1. Mount the engine in a test tank or move the boat to a body of water.
2. Remove the cowling and connect a tachometer to the powerhead.
3. Start the engine and allow it to warm to operating temperature.
4. Check engine speed at idle. The powerhead should idle at 875-975 rpm for Yamaha motors or about 800-850 rpm for Mercury/Mariner models.
5. Place the engine in gear and check engine trolling speed in the same manner. Again the powerhead should idle at 775-875 rpm in gear for Yamaha motors or about 750-800 rpm in gear for Mercury/Mariner models.
6. If adjustment is necessary, rotate the idle adjustment screw (a spring-loaded screw threaded diagonally down against the throttle lever on the side of the carburetor) until the powerhead idles at the required rpm. Turning the screw INWARD will increase idle speed, while backing it OUTWARD will decrease speed.

CHECKING/ADJUSTING THE STARTER LOCKOUT (YAMAHA ONLY)

◆ See Figure 216

Manual start models of this motor contain a starter lockout safety feature to prevent the motor from being started while in gear. The system should be checked at each tune-up to ensure that it is adjusted and functioning properly. Checking is a simple matter of placing the motor in gear and gently attempting to start the motor. If the starter will not rotate, the system is functioning and no further attention is required. However, if the motor rotates in gear, then adjust the cable as follows:

1. Begin with the engine not running and shifter positioned in **Neutral**.
2. Loosen the screw securing the cable adjusting plate to the starter housing.
3. Reposition the adjusting plate until the mark on the cable connector (a) aligns with the mark on the flywheel cover (b), then retighten the adjusting plate screw.
4. Verify that the lockout is now working properly.

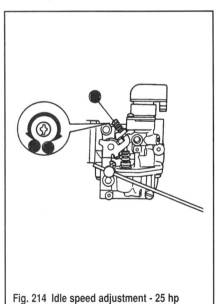

Fig. 214 Idle speed adjustment - 25 hp Yamaha motors

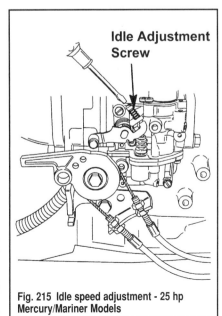

Fig. 215 Idle speed adjustment - 25 hp Mercury/Mariner Models

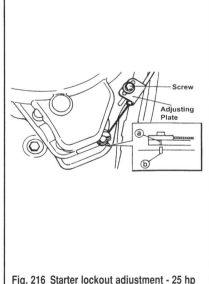

Fig. 216 Starter lockout adjustment - 25 hp Yamaha motors

2-66 MAINTENANCE & TUNE-UP

IGNITION TIMING

Both Yamaha and Mercury/Mariner provides no information on checking the timing for this model. Like most other 4-strokes, the ignition is controlled by the CDI unit/ECM unit and timing is not adjustable.

Carbureted 30/40 Hp 3-Cylinder Models

IDLE SPEED

◆ See Figure 217

1. Mount the engine in a test tank or move the boat to a body of water.
2. Remove the cowling and connect a tachometer to the powerhead.
3. Start the engine and allow it to warm to operating temperature.
4. Check engine speed at idle. The powerhead should idle at the rpm specified in the Tune-up Specifications charts.
5. Except for jet drives, place the engine in gear and check engine trolling speed in the same manner.
6. If adjustment is necessary, rotate the idle adjustment screw (NOT the pilot screw) until the powerhead idles at the required rpm. The idle adjustment screw is threaded diagonally downward into the linkage on the side of the bottom carburetor and contacts the throttle valve. Turning the screw INWARD will INCREASE idle speed, while turning it OUTWARD will DECREASE idle speed.

■ While making adjustments, turn the screw slowly, no more than 1/2-1 turn at a time, then rev the engine 2-3 times and allow idle to stabilize for about 15 seconds before adjusting it further.

CARBURETOR SYNCHRONIZATION

◆ See Figures 218 and 219

In order for an engine with multiple carburetors to operate smoothly all carburetors must open and close at the precise same time. Adjusting the carburetor linkage so that the throttle valves move in unison is a process referred to as Carburetor Synchronization. With some dual-carburetor motors this can be accomplished by eye, watching as the throttle valves move. But, for most motors with 3 or more carburetors a vacuum gauge set or a manometer (also known as a calibrated carburetor balance stick) allows for a much more exact setting (which is required for this motor).

The manufacturer wisely advises that it is not necessary to tamper with a carburetor that is running correctly. If the engine idles smoothly, do not bother adjusting the carburetor synchronization. However, if the carburetors have been removed, disassembled, replaced or the linkage otherwise disconnected or you are otherwise unable to obtain a smooth idle, take the time to synchronize the carburetors.

1. Check and adjust the idle speed. Leave the motor in the test tank (or leave the boat launched) and keep the tachometer connected to the powerhead.
2. Locate and remove the balance plugs from the side of the intake manifold and install a set of vacuum gauge adapters of the same thread threads as the plugs. Connect a set of vacuum gauges (or calibrated balance sticks) to the 3 adapters.
3. Mercury advises that you use a tubing clamp to pinch off each of the fuel system enrichener lines between the individual carburetors and the line leading to the enrichener valve itself.
4. Start the engine and allow it to re-warm to normal operating temperature
5. If necessary, turn the idle speed adjustment screw (which actuates the throttle valve of the bottom carburetor) until the engine speed is at the high end of the idle specification for Neutral operation or a little above. Temporarily set the idle to 900 rpm for Yamahas and 1000 rpm for Merc/Mariners.
6. With the engine idling at this test spec, measure and record the vacuum pressure indicated on the bottom carburetor. This is considered the base value, which will be used to calculate the vacuum generated by the settings of the other 2 carburetors.
7. For Yamaha powerheads, proceed as follows:
• Adjust Carb 2 (the middle) to a value of Carb 3's reading minus 0.2 in. Hg (5mmHg)
• Adjust Carb 1 (the top) to a value of Carb 3's reading minus 0.4 in. Hg (10mmHg)

■ So, on Yamaha powerheads, after adjustment the top carburetor should have the lowest vacuum reading and the vacuum of each carburetor as you move downward should increase by 0.2 in. hg (5mm hg) higher.

8. For Merc/Mariner powerheads, do your best to match the vacuum of the top 2 carburetors TO the base value of the bottom carburetor (Carb 3).
9. Use the throttle valve adjusting screws of the top and middle carburetors to adjust the vacuum gauge readings for each carburetor until the proper specifications are obtained.
10. Reset the Idle Speed, as detailed in this section.
11. Rev the engine a few times while watching the gauges, they should stay within 1.97 in. Hg (50mmHg) of each other. Otherwise, recheck your adjustment.
12. Shut the engine down, remove the gauges/adapters and reinstall the balance plugs to the intake manifold. On Mercury/Mariner models, remove the tubing clamps from the enrichener lines.

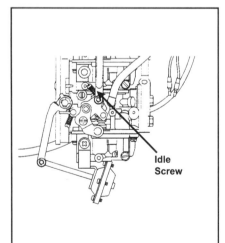

Fig. 217 Adjust idle speed using the screw on the linkage at the bottom carburetor - Carbureted 30/40 hp 3-cylinder motors

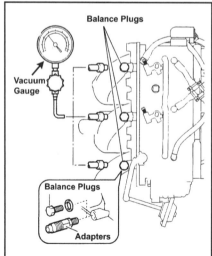

Fig. 218 Connect a set of vacuum gauges to adjust carburetor synchronization

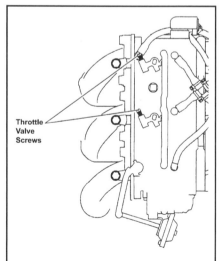

Fig. 219 Use the throttle valve screws to adjust the top and middle carbs

MAINTENANCE & TUNE-UP 2-67

DASH POT ADJUSTMENT (YAMAHA ONLY)

◆ See Figure 220

Improper dashpot operation will lead to engine stumbling or stalling upon quick deceleration. If the dash-pot remains undisturbed and the motor does not exhibit these symptoms leave it alone. Again, the motto is "if it ain't broke, don't fix it." However, if the dash-pot is replaced, the carburetors are removed/overhauled and the adjustment is disturbed or if symptoms lead you to suspect improper operation, adjust the dash-pot, as follows:

1. Check and adjust the idle speed. Leave the motor in the test tank (or leave the boat launched) and keep the tachometer connected to the powerhead.
2. Start the engine and allow it to re-warm to normal operating temperature
3. Hold the engine speed at about 3700-3800 rpm, then loosen the locknut and turn the dash-pot adjusting screw until the pot plunger is JUST touching the lever. Hold the screw in this position and retighten the locknut securely.
4. Check engine operation by running it at 3/4-WOT and then dropping it quickly to idle. The engine should not stumble or stall. Otherwise, inspect the dash-pot itself and/or recheck adjustment.
5. Shut the powerhead down.
6. Remove the tachometer.

CHECKING/ADJUSTING THE STARTER LOCKOUT (YAMAHA ONLY)

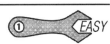

◆ See Figure 221

This model contains a starter lockout safety feature to prevent the motor from being started while in gear. The system should be checked at each tune-up to ensure that it is adjusted and functioning properly. Checking is a simple matter of placing the motor in gear and gently attempting to start the motor. If the starter will not rotate, the system is functioning and no further attention is required. However, if the motor rotates in gear, then adjust the cable as follows:

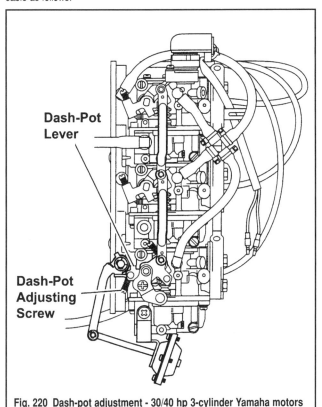

Fig. 220 Dash-pot adjustment - 30/40 hp 3-cylinder Yamaha motors

1. Begin with the engine not running and shifter positioned in **Neutral**.
2. Follow the cable from the top of the manual starter assembly, down to the point where it travels sort of vertically (well, diagonally at least) through a bracket protruding from the powerhead. The nut on top of the bracket is the locknut, the nut on the bottom is the adjustment nut.
3. Loosen the locknut (top), then turn the adjusting nut (bottom) until the point on the cable connector aligns with the mark on the manual starter housing.
4. Once positioned properly, retighten the locknut (top) to hold the cable in this position.
5. Verify that the lockout is now working properly.

THROTTLE CABLE ADJUSTMENT

Yamaha Models

◆ See Figure 222

■ **The engine Idle Speed and Carburetor Synchronization must be properly set before adjusting the throttle cable.**

Visually check the throttle valve movement in response to the throttle handle. When the handle is set to WOT, make sure the valves open fully and make sure, when the handle is closed to idle, the valves close. If the valves do not operate properly or smoothly, adjust, as follows:

1. Remove the retaining clip and disconnect the throttle cable joint from the powerhead linkage.
2. Set the throttle control lever or tiller handle grip to the fully closed position.
3. Loosen the locknut, then adjust the position of the throttle cable joint (by turning the joint inward or outward on the cable threads) until its mounting hole aligns with the set pin on the throttle control lever. Tighten the locknut to hold it in position on the cable.

※※ CAUTION

Make sure that after adjustment the cable joint is threaded over at LEAST 0.31 in. (8mm) of cable threads, otherwise there is a risk that the joint could separate in service causing a severe navigation/control hazard.

4. Position the cable joint over the control lever set pin and secure using the retaining clip.

Mercury/Mariner Models

◆ See Figure 223

No periodic adjustment should be necessary for the Shift and Throttle cables, however, should the cables be replaced they will require an initial set-up to place them in proper alignment with the linkage.

1. Make sure the throttle/shifter is in the idle/neutral position.
2. Check the throttle cam, in the idle/neutral position there should be a 1/16 in. (1.6mm) gap between the cam itself and an oval shaped boss immediately adjacent to the cam.
3. If adjustment is necessary, release the control cable latch (by rotating clockwise about 1/4 turn), then adjust how the throttle cable (the upper of the two cables) sits in the barrel until the proper gap is achieved. Once the proper gap is set, rotate the latch counterclockwise to lock the cables back into position.
4. After adjustment, actuate the remote handle a few times from WOT back to idle and confirm the proper gap is maintained. Readjust, as necessary.

SHIFT CABLE ADJUSTMENT

Yamaha Models

◆ See Figure 224

Visually check the shift linkage movement in response to the shift handle. If the shifter does not engage properly or smoothly, adjust, as follows:

1. Remove the retaining clip and disconnect the shift cable joint from the powerhead linkage.
2. Set the shift control lever to the **Neutral** position.

2-68 MAINTENANCE & TUNE-UP

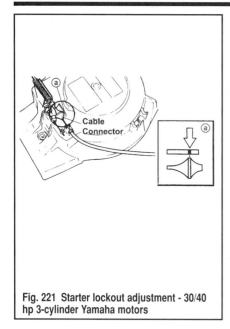

Fig. 221 Starter lockout adjustment - 30/40 hp 3-cylinder Yamaha motors

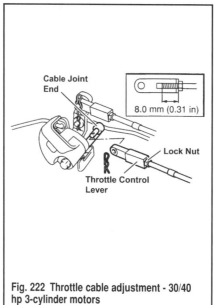

Fig. 222 Throttle cable adjustment - 30/40 hp 3-cylinder motors

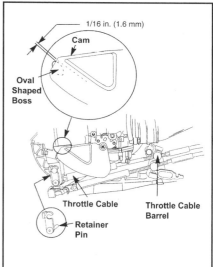

Fig. 223 Throttle cable adjustment for most Merc 3- and 4-cylinder motors (gap varies slightly by model)

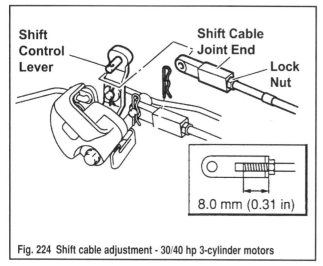

Fig. 224 Shift cable adjustment - 30/40 hp 3-cylinder motors

3. Loosen the locknut, then adjust the position of the shift cable joint (by turning the joint inward or outward on the cable threads) until its mounting hole aligns with the set pin on the shift control lever. Tighten the locknut to hold it in position on the cable.

✱✱ CAUTION

Make sure that after adjustment the cable joint is threaded over at LEAST 0.31 in. (8mm) of cable threads, otherwise there is a risk that the joint could separate in service causing a severe navigation/control hazard.

4. Position the cable joint over the control lever set pin and secure using the retaining clip.

Mercury/Mariner Models

No periodic adjustment should be necessary for the Shift and Throttle cables, however, should the cables be replaced they will require an initial set-up to place them in proper alignment with the linkage.

The shifter cable should be set so the distance between the cable barrel (on the threaded adjuster) and the cable end (where it attaches to the linkage) is the same as the distance between the cable barrel bracket and the linkage pin when both the gearcase and the shifter are in Neutral. If adjustment is necessary, free the cable and rotate the cable barrel/adjuster to set the proper distance.

IGNITION TIMING

Like most other 4-strokes, the ignition is controlled by the CDI unit/ECM unit and timing is not adjustable. Although Yamaha provides no information on checking the timing for this model, the Mercury/Mariner service information does specify that the flywheel is equipped with timing marks and the flywheel cover or manual starter housing is equipped with a timing pointer window. A timing light therefore can be used to verify proper operation of the ignition timing system (against the specifications listed in the Tune-Up Specifications charts). If so, make sure that you run the engine in gear, under load, with a suitable source of cooling water. If timing is out of specification, seek a mechanical or electrical reason (i.e. faulty component).

EFI 30/40 Hp 3-Cylinder and EFI 40/50/60 Hp 4-Cylinder Models

THROTTLE & SHIFT CABLE ADJUSTMENT

◆ See Figure 225

This is NOT a periodic adjustment and should only be necessary during initial rigging or after cable replacement.

1. Make sure the throttle/shifter is in the idle/neutral position.
2. The shifter cable (the lower of the two cables) should be set so the distance between the cable barrel (on the threaded adjuster) and the cable end (where it attaches to the linkage) is the same as the distance between the cable barrel bracket and the linkage pin when both the gearcase and the shifter are in Neutral. If adjustment is necessary, free the cable and rotate the cable barrel/adjuster to set the proper distance.
3. Next move onto the throttle cable. Check the throttle cam, there is a roller located on top of the cam. With the shifter/throttle in the idle position, the roller should be pointing almost straight up and should be centered with an alignment mark above it.
4. If adjustment is necessary, release the control cable latch (if not done already for shifter cable adjustment) by rotating clockwise about 1/4 turn, then adjust how the throttle cable (the upper of the two cables) sits in the barrel until the mark on the cam centers with the roller. Once the mark is properly aligned, rotate the latch counterclockwise to lock the cables back into position.
5. After adjustment, actuate the remote handle a few times from WOT back to idle and confirm the mark is still in alignment. Readjust, as necessary.

MAINTENANCE & TUNE-UP

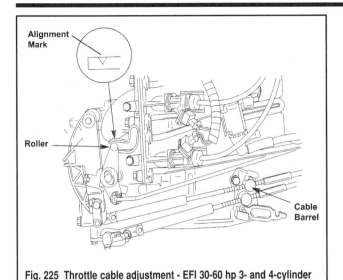

Fig. 225 Throttle cable adjustment - EFI 30-60 hp 3- and 4-cylinder models

THROTTLE LINK IDLE SETTING ADJUSTMENT

◆ See Figure 226

Like the Throttle and Shift Cables, adjustment should not be necessary on a periodic basis, however, should components be replaced or disturbed you should first check the alignment and, if necessary, perform the throttle link adjustment.

1. On remote control models, disconnect the throttle cable from the throttle link.

■ On tiller handle models the cables should remain connected as the tiller throttle twist grip should be used to advance throttle during the throttle link checking and adjustment procedure.

2. Lightly hold the throttle body arm against the throttle stop.
3. Slowly push the throttle lever (found on the side of the powerhead) forward until you JUST feel the throttle body arm start to move. At this point, check that the center of the throttle arm roller (the roller is at the top of the arm, near 11 o'clock) is in line with the throttle cam alignment mark (found in the linkage, just below the roller). The roller may be centered on the roller OR up to 1/8 inch (3.2mm) PAST the mark. If the roller is properly aligned, stop this procedure here.

4. If adjustment is necessary, remove the lower cowl for access.
5. Disconnect the throttle link rod from the throttle arm and throttle body arm.
6. If the throttle body arm engaged (the throttle shutter opened) BEFORE the center of the roller was in line with the alignment mark, SHORTEN the link rod by turning the socket CLOCKWISE.
7. If the throttle body arm engaged (the throttle shutter opened) AFTER the center of the roller was in line with the alignment mark, LENGTHEN the link rod by turning the socket COUNTERCLOCKWISE.
8. Snap the socket onto the throttle body arm and push the rod end into the throttle arm.
9. Recheck the mark alignment. Adjust as necessary, until the marks align properly.
10. On remote control models, reconnect the remote control throttle cable.

THROTTLE LINK WOT SETTING ADJUSTMENT

◆ See Figure 227

1. Check, and if necessary, adjust the Throttle Link Idle Setting, as necessary.
2. With the throttle cable(s) attached, use the remote control or tiller twist grip to fully open the throttle (place the powerhead at WOT, but engine NOT running). At WOT, the throttle stop should just LIGHTLY contact the adjoining surface.
3. To make sure the throttle shutter is fully open, back the throttle stop screw outward under there is a slight gap between the throttle stop screw and the adjoining surface (at the WOT position). Now, SLOWLY turn the throttle stop screw INWARD until the throttle stop lightly contacts the surface.

IGNITION TIMING

◆ See Figure 228

Like most other 4-strokes, the ignition is controlled by the ECM unit and timing is not adjustable. Mercury/Mariner service information does specify that the flywheel is equipped with timing marks and the flywheel cover is equipped with a timing pointer window. A timing light therefore can be used to verify proper operation of the ignition timing system (against the specifications listed in the Tune-Up Specifications charts). If so, make sure that you run the engine in gear, under load, with a suitable source of cooling water. This means in a test tank, on a dynamometer or on a boat secured to a trailer that has been backed into the water (with the transom straps in place!!!!). If timing is out of specification, seek a mechanical or electrical reason (i.e. faulty component).

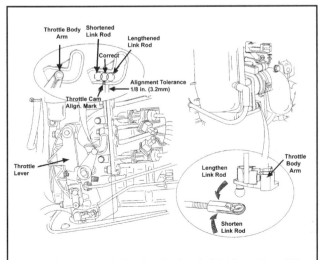

Fig. 226 Checking and adjusting the throttle link idle setting - EFI 30-60 hp 3- and 4-cylinder models

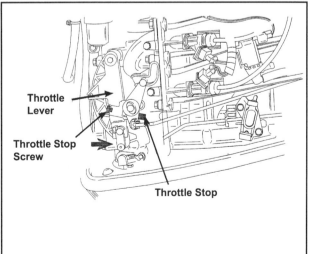

Fig. 227 Checking and adjusting the throttle link WOT setting - EFI 30-60 hp 3- and 4-cylinder models

2-70 MAINTENANCE & TUNE-UP

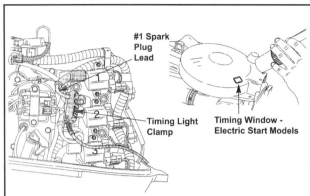

Fig. 228 Checking ignition timing - EFI 30-60 hp 3- and 4-cylinder models

40/45/50 Hp (935cc) and Carbureted 40/50/60 Hp (996cc) Models

Before making carburetor adjustments make sure the valve clearances have been properly checked and adjusted. For details, please refer to Valve Clearance (4-Stroke Models), in this section.

CHECKING/ADJUSTING THE THROTTLE & SHIFT CABLES

Yamaha

◆ See Figure 229

Operate the throttle and shifter checking for smooth, correct operation. Visually inspect the control cables for signs of damage or excessive wear and replace, if necessary. Check adjustment of both cables by setting the remote control lever (or shifter on tiller models) in **Neutral**. For tiller models, fully close the throttle grip. Now, visually inspect the markings on the shift and throttle linkage at the powerhead. The shift cable set pin should be aligned with the shift cable neutral mark at the center of the slider. Also, the alignment marks on the throttle cable cam should be aligned with the throttle cable neutral mark on the side, end of the shift cable slider bracket. In this position the throttle valves should be fully closed in the idle position. For more clarification on the alignment marks, refer to the accompanying illustration. If adjustment is necessary, proceed as follows:

1. Remove the retaining clip, then carefully pull the cable joint off the set pin.
2. Reposition the shift cable set pin slider and/or the throttle cam as necessary to properly align the neutral marks as detailed earlier and shown in the accompanying illustration. The throttle valves on the carburetor must be fully closed in the idle position.
3. Loosen the locknut on the cable (snugged against the cable joint to hold it in position), then rotate the cable joint inward or outward as necessary to align the cable joint with the repositioned set pin.

✱✱ CAUTION

To avoid a potential navigational/operational hazard, the cable joint must be installed at LEAST 0.31 in. (8mm) onto the end of the cable. If not, the cable joint could potentially separate during use, instantly removing control from the shifter or the throttle.

4. Once the adjustment is complete, tighten the locknut.
5. Install the cable joint back over the set pin and secure using the retaining clip.

Mercury/Mariner Models

◆ See Figure 223

No periodic adjustment should be necessary for the Shift and Throttle cables, however, should the cables be replaced they will require an initial set-up to place them in proper alignment with the linkage.

1. Make sure the throttle/shifter is in the idle/neutral position.
2. The shifter cable (the lower of the two cables) should be set so the distance between the cable barrel (on the threaded adjuster) and the cable end (where it attaches to the linkage) is the same as the distance between the cable barrel bracket and the linkage pin when both the gearcase and the shifter are in Neutral. If adjustment is necessary, free the cable and rotate the cable barrel/adjuster to set the proper distance.
3. Next move onto the throttle cable. Check the throttle cam, in the idle/neutral position there should be a 1/16 in. (1.6mm) gap for models through 2000 or a 1/8 in. (3.2mm) gap between the cam itself and an oval shaped boss immediately adjacent to the cam.
4. If adjustment is necessary, release the control cable latch (if not done already for shifter cable adjustment) by rotating clockwise about 1/4 turn, then adjust how the throttle cable (the upper of the two cables) sits in the barrel until the proper gap is achieved. Once the proper gap is set, rotate the latch counterclockwise to lock the cables back into position.
5. After adjustment, actuate the remote handle a few times from WOT back to idle and confirm the proper gap is maintained. Readjust, as necessary.

CARBURETOR ADJUSTMENTS

This section gives all of the adjustments that can be performed on the carburetors used for the 40/45/50 Hp and 40/50/60 Hp 4-stroke motors. However, most of these adjustments are not considered periodic maintenance and only are required after specific repairs or component replacements. To determine which adjustments are required, refer to Determining Necessary Adjustments.

Determining Necessary Adjustments

◆ See Figure 230

The only attention that should be periodically given to the carburetor settings are the idle speed, carburetor synchronization and dash-pot adjustments. Actually, with that said, it's really only idle speed that is checked and adjusted as a maintenance item and the other 2 adjustments should be performed only if problems occur. Carburetor Synchronization is only necessary on if a smooth and stable idle cannot be maintained. Dash-pot adjustments are performed if problems occur when you chop the throttles (upon deceleration).

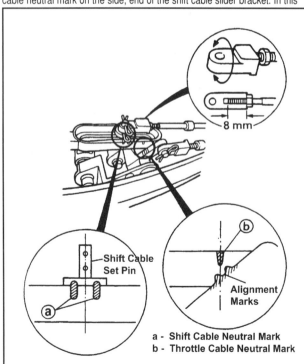

Fig. 229 Throttle and shift cable adjustment - 40/45/50 hp (935cc) and 40/50/60 hp (996cc) Yamaha motors

MAINTENANCE & TUNE-UP

■ The Idle Speed, Carburetor Synchronization, Dash-Pot and, when applicable, Pilot Screw adjustments are the only procedures in this section that apply to 2001 later models. It seems that engineers woke-up and realized it just didn't have to be that complicated. Therefore, the remaining adjustments (Choke Cam Angle, Fast Idle Screw, Unloader Screw and Choke Valve Closing Screw no longer apply).

The balance of the carburetor settings should remain undisturbed unless the carburetors or related components are removed, overhauled and/or replaced. If so, use the accompanying illustration to help determine which adjustments are required. Although all adjustments are not usually required, each of the on-board adjustments should only take place in the relative order in which it is listed in the illustration and in the accompanying text. Therefore, Idle Speed adjustment will always come before a Dash-Pot adjustment. However, there is one exception and that is that Idle Speed will have to be checked and possibly reset again after Carburetor Synchronization. The final four adjustments apply ONLY to models through 2000 (Choke Cam Angle, Fast Idle, Unloader and Choke Valve Closing) and are only performed if the No. 4 carburetor (bottom) is separated for replacement or overhaul or if one of the screws controlling that adjustment comes loose. The final four adjustments are listed last because they are rarely necessary, however, when needed they are performed with the carburetor removed from the powerhead so they are performed before the dynamic or on-board adjustments. The same goes for Pilot Screw adjustment, that is, perform it ONLY when it is truly necessary.

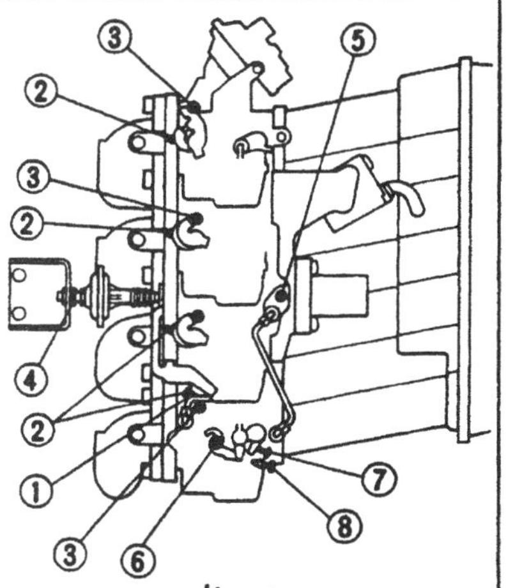

Item	Part	Periodic inspection and adjustment	Complete carburetor unit removed	Carburetor overhauled	Parts replaced/removed	#4 carburetor replaced	Electrothermal ram replaced	Choke control bracket replaced/overhauled	Static adjustment	Dynamic adjustment
Idle speed adjustment	1	●	●	●	●	●	–	–	—	On board
Carburetor synchronization	2	●	–	○	●	●	–	–	Removed	On board
	1									
Pilot screw adjustment *1	3	–	–	○	●	–	–	–	Removed	On board
Dash-pot adjustment	4	●	●	○	●	●	–	–	—	On board
Choke cam angle adjustment	5	–	–	○	●	●	●	●	Removed	—
Fast idle adjustment	6	–	–	○	●	●	–	●	Removed	—
Unloader adjustment	7	–	–	○	●	●	–	●	Removed	—
Choke valve closing adjustment	8	–	–	○	●	●	–	●	Removed	—

NOTE:

●: Service required.
○: Service not required. Do not loosen screws.
– : No need for adjustment.
Removed: Complete carburetor unit should be removed.
On board: Adjustment should be carried out with the engine running.

*1 : Applicable for all Yamaha motors exc. F54A and the 60J/60 hp motors
F45A pilot screw is factory adjusted and do not recalibrate.

Fig. 230 Use this chart to help determine what adjustments are necessary for 40/45/50 hp (935cc) and 40/50/60 hp (996cc) motors. BE SURE NOT to loosen or remove the screws numbered 5, 6, 7 and 8. If the No. 4 screw becomes loose or requires replacement, perform the adjustments starting with Part 5, Choke Cam Angle (NOTE carb linkage shown is used models through 2000 models, 2001 and later use a much simplified set-up)

2-72 MAINTENANCE & TUNE-UP

Idle Speed

◆ See Figures 231 and 232

To ensure proper adjustment the engine must be fully warmed to normal operating temperature before proceeding with idle speed adjustment. The idle adjustment screw mounted to the linkage on the lower carburetor is the only part which needs to be touched for this adjustment.

1. Remove the engine cowling and connect a shop tachometer to the motor.
2. Provide a suitable source of cooling water to the motor (tank, flush fitting or launch the boat), then start and warm the motor to normal operating temperature.
3. Observe engine idle speed with the engine fully warmed, running in **Neutral** and compare that to what is listed in the Tune-Up Specification Chart. If adjustment is necessary turn the idle speed screw very slowly, either clockwise to increase idle speed or counterclockwise to decrease idle speed.
4. Increase engine speed above idle (not too far if running on a flush fitting) 2 or 3 times, then allow the motor to return to idle for at least 15 seconds. Double-check that the motor has stabilized and is smoothly idling in the proper operating range.
5. If the engine is idling smoothly and within range, it will not be necessary to synchronize the carburetors.
6. If the engine does not idle smoothly, proceed with Carburetor Synchronization.

CARBURETOR SYNCHRONIZATION

Carburetor synchronization ensures that the throttle valves (butterflies) on each carburetor open and close at precisely the same time. If one throttle valve precedes the others or lags behind, the cylinders will not receive equal amounts of air/fuel mixture per combustion cycle, causing one or more cylinders to run sluggish or lag behind slightly in power. This will lead to rough running, hesitation or stumbling.

A static adjustment is specifically provided for most models (except the 2001 and later Mercury/Mariner 50/60 hp motors, for which the given procedures may also help preset the carbs, but we cannot guarantee it). When applicable a static adjustment should be performed if the complete carburetor assembly is removed from the motor. The dynamic adjustment fine-tunes the synchronization and is usually sufficient to correct unstable idle if the carburetor assembly has not been removed for service.

Although no special tools are required for the static adjustment, a vacuum gauge or monometer (carburetor vacuum synchronization tool also known as a set of carb sticks) is required for dynamic adjustment.

Preliminary Static Adjustment

◆ See Figure 233

This adjustment begins with the complete carburetor assembly removed from the powerhead. Two slightly different procedures are given by each of the manufacturers. You might find one to be easier than the other, and both SHOULD apply to all models.

1. To follow Yamaha's recommendations, proceed as follows:
2. Disconnect the throttle link rods, but do not remove anything else from the carburetors.
3. Adjust the throttle valve by first loosening the throttle stop screws for carburetors 1 (top), 2 (second one down) and 3 (third one down) along with the idle stop screw (idle adjustment screw) for the bottom carburetor. Loosen each of the screws until the lever comes free of the screw.
4. Invert the carburetor assembly (turn it upside down) in order to locate the bypass hole, then tighten the No. 4 idle stop screw (idle adjustment screw) until 1/2 of the bypass hole is visible.
5. Tighten the throttle stop screws starting with the 3rd carburetor down, next the 2nd carburetor from the top and finally the top carburetor in the same manner as the No. 4 idle stop screw.
6. Reconnect the link rods
7. To follow Mercury/Mariner's recommendations, proceed as follows:
8. Disconnect the throttle link rods, then disconnect the Auto Choke "Z" link (the roughly "Z" shaped link found at the lower right side of the carburetor assemblies when looking at them mounted on the side of the powerhead).

■ **The Auto Choke "Z" link is blacked out in the accompanying illustration.**

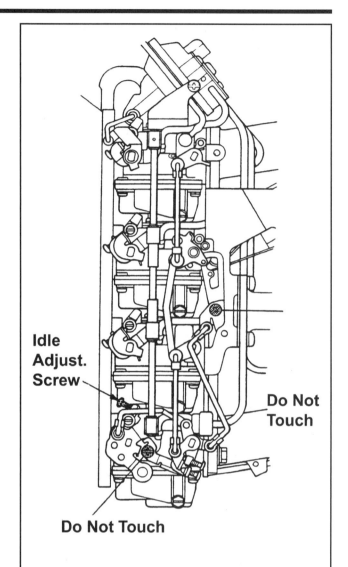

Fig. 231 Idle speed is altered using the adjustment screw, do NOT touch the other noted screws - Yamaha models through 2000 and all Merc/Mariners shown

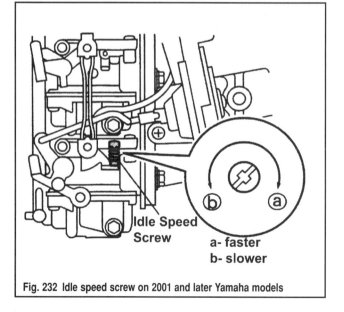

Fig. 232 Idle speed screw on 2001 and later Yamaha models

MAINTENANCE & TUNE-UP

9. Loosen the Idle Speed Adjustment screw at the No. 4 (bottom) carburetor (threaded diagonally downward into the carburetor linkage) until the throttle shutter for the carburetor covers HALF of the bypass hole.
10. Install the throttle link.
11. Working your way down from the top carb, adjust each of the throttle stop screws (threaded diagonally upward to each carburetor linkage) for the carburetors 1 (top), 2 (second one down) and 3 (third one down). Turn each of the throttle stop screws just enough to make sure the throttle shutter cover for each carburetor is obstructing 1/2 of the bypass hole.
12. Install the Auto Choke "Z" link.

Final Dynamic Adjustment
◆ See Figures 234, 235 and 236

This adjustment takes place with the carburetor assembly installed and the engine fully warmed to normal operating temperature. An initial idle speed adjustment may be required in order to get the motor run correctly during the adjustment. A slight difference in the adjustment occurs between models through 2000 and on 2001 or later models. For ALL models the vacuum adjustment for carburetors is a function of the vacuum measured at the No. 4 cylinder (bottom) carburetor. For models through 2000, each step up the powerhead from the No. 4 carburetor should be adjusted 0.4 in. Hg. (10mmHg) lower than the No. 4 setting (meaning the No. 3 is 10mmHg lower, No. 2 is 20mmHg lower and No. 1 is 30mmHg lower). However, starting for 2001 models engineers decided that is was better for the motor to be balanced by setting all carburetors (1-3) to the SAME vacuum reading as the No. 4 carburetor reading.

1. If not done already, install the carburetor assembly to the powerhead, then adjust the idle speed as detailed in this section.
2. With the motor fully warmed and running at idle, remove the M6 vacuum port screw from the intake manifold. If you're using multiple vacuum gauges or a calibrated carburetor balance stick (a tool that normally contains as many as 4 separate tubes that measure vacuum) remove all 4 screws so all of the vacuum indicators can be installed.

■ Keep in mind during this procedure that actual vacuum readings will vary based upon altitude, weather and engine condition. What is important during this procedure is not the actual reading so much as how much the reading varies from cylinder-to-cylinder.

3. For 2001 and later Mercury/Mariner models, Mercury advises that you use a tubing clamp to pinch off each of the fuel system enrichener lines between the individual carburetors and the line leading to the enrichener valve itself.
4. With the engine running slightly above idle, at least 1000 rpm (temporarily readjust the idle speed, if necessary), record the vacuum measurement for each carburetor. For 2001 and later models the carbs are synchronized by bringing carbs No. 1-3 to match the reading you get on

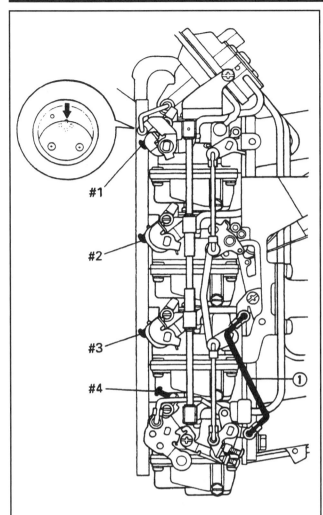

Fig. 233 Static carburetor synchronization involves loosening the throttle stop/idle stop screws on all the carburetors, then tightening them one at a time, starting at the bottom and working your way up (models through 2000 shown, 2001 and later similar, but screws facing different directions)

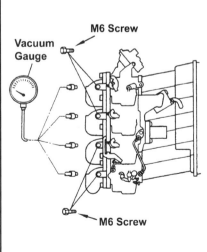

Fig. 234 Dynamic carburetor synchronization involves connecting vacuum gauges to the intake manifold vacuum ports...

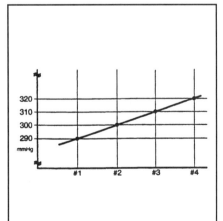

Fig. 235 ...and measuring for 10mmHg (0.4 in. Hg) differences in vacuum between the each cylinder with the No. 1 having the lowest vacuum reading and the No. 4 having the highest (on models through 2000)

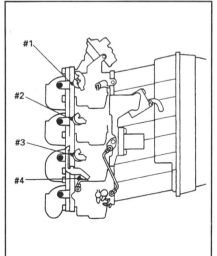

Fig. 236 Synchronization is achieved using the throttle stop/idle stop screws

2-74 MAINTENANCE & TUNE-UP

No. 4. However, for models through 2000 the carburetors are synchronized based on a formula which utilizes the No. 4 carburetor vacuum reading. The carburetors are considered synchronized if the readings are as follows:
- No. 4 is the base line reading (can equal almost anything, but the other carb readings must be the same reading minus a specified amount).
- No. 3 must equal the base line minus 10mmHg (0.4 in. Hg).
- No. 2 must equal the base line minus 20mmHg (0.8 in. Hg).
- No. 1 must equal the base line minus 30mmHg (1.2 in. Hg).

■ If adjustments are necessary, each time you turn an idle/throttle stop screw remember to rev the engine 2 or 3 times, then allow the engine to stabilize back at idle for about 15 seconds before taking a vacuum reading or adjusting the screw further.

5. If one or more of the cylinders are out of the specified variance (through 2000) or do not equal the reading (2001 and later) from the No. 4 reading, you'll have to adjust them all.

6. For models through 2001, start adjustments by determining the average variance by adding the highest and lowest readings and dividing that figure by 2, then adjust the idle adjustment screw until the reading on No. 4 is the average. Let's say, simply for sake of argument that the AVERAGE was 320 mmHg (12.6 in. Hg), then adjust the idle stop and throttle stop screws as follows:
- Adjust No. 4 to the average (320mmHg/12.6 in. Hg).
- Adjust No. 3 to the average (320mmHg/12.6 in. Hg) minus 10mmHg (0.4 in. Hg) or 310mmHg (12.2 in. Hg) in this example.
- Adjust No. 2 to the average (320mmHg/12.6 in. Hg) minus 20mmHg (0.8 in. Hg) or 300mmHg (11.8 in. Hg) in this example.
- Adjust No. 1 to the average (320mmHg/12.6 in. Hg) minus 30mmHg (1.2 in. Hg) or 290mmHg (11.4 in. Hg) in this example.

■ In this way, when adjustment is completed, the No. 1 carburetor will have the lowest vacuum reading, and each carburetor down from it will contain 10mmHg (0.4 in. Hg) more, per carburetor until you get to No. 4 which will have the highest vacuum reading.

7. For 2001 and later models, the engineers simplified things for us by allowing all carbs to simply equal the vacuum reading on the No. 4. This means the first adjustment you should make is to slowly attempt to bring the other carbs into agreement. However, if you have a problem achieving this on one or more carbs, you can try adjusting the No. 4 up or down slightly to make it easier.

8. Once the throttle stop screws are properly adjusted to achieve the proper vacuum variance, readjust the idle speed one LAST time back to the specification listed in the Tune-Up Specification Chart.

9. Rev the engine a few times while watching the gauges, they should stay within 1.97 in. Hg (50mmHg) of each other. Otherwise, recheck your adjustment.

10. Shut the engine down, remove the gauges/adapters and reinstall the balance plugs to the intake manifold.

Pilot Screw

◆ See Figure 237

First off, the carburetor pilot screw affects the air/fuel mixture for the cylinder fed by that carburetor. There may have been a time when this adjustment was necessary on a periodic basis, but it has long passed. A combination of improved engine/carburetor design and tougher EPA regulations has rendered this adjustment a thing of the past, UNLESS the carburetor has been overhauled and a stable idle cannot be obtained, even after carburetor synchronization.

Furthermore, the screws on some Yamaha models (specifically the 45 hp and 60J/60 hp) are adjusted and sealed at the factory and the manufacturer warns that they should not be tampered with. EPA regulations require that the screws be covered on all models sold in the US and that they are to be RECOVERED after adjustment.

Like carburetor synchronization, the pilot screw adjustment takes place in 2 phases. The Preliminary Static Pilot Screw Adjustment is the initial seating and backing out of the screw according to specifications in the Carburetor Set-Up Specifications chart. This is normally done during carburetor overhaul to give you the proper starting point for the Dynamic Adjustment. Once the carburetors are installed a Final Dynamic Pilot Screw Adjustment procedure dials in the mixture to the exact proper amount in order to compensate for differences in engine tolerances from build and wear variations.

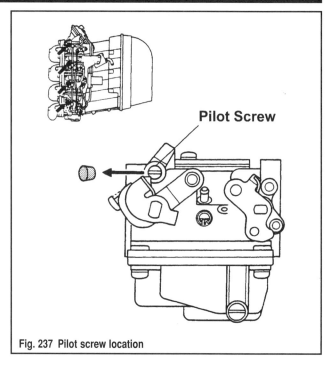

Fig. 237 Pilot screw location

Preliminary Static Pilot Screw Adjustment
◆ See Figure 237

1. Install the pilot screw (if removed), threading it very slowly and gently until it just lightly seats.

✷✷ WARNING

Tightening a pilot screw once it contacts the seat will most likely destroy the screw. As the screw tip is forced against the seat it will deform leading to an inability to properly adjust the air fuel mixture. Always be extra gentle with pilot screws.

2. Locate the Carburetor Set-Up Specifications Charts from the Fuel System section and back the pilot screw out slowly the specified number of turns.

3. Proceed with Final Dynamic Pilot Screw Adjustment.

Final Dynamic Pilot Screw Adjustment
◆ See Figure 237

■ Again, the manufacturers give various methods of making this adjustment. In theory, any of these procedures should work as well on the other models of this engine family, but we'd first suggest trying the one specifically labeled for your motor.

1. Connect a shop tachometer to the motor and provide a suitable source of cooling water (flush, tank or launch the boat).
2. Start and warm the engine to normal operating temperature.
3. For Mercury/Mariner 45 hp motors, proceed as follows:
   a. Remove the 4 plugs from the exhaust manifold (similar to the balance plugs used on the intake) and install an exhaust manifold hose adapter (Merc # 91-809905A1) to each hole.
   b. Connect a CO exhaust analyzer to each of the adaptors. If you do not have multiple sniffers (analyzer pickups) then each adapter that is NOT being monitored plugged. In other words, if you only have one analyzer pickup, you only need one adaptor and you should move it from port-to-port.
   c. Restart the motor and, with it running at about 925-975 rpm at normal operating temperatures, check the CO content at each exhaust port. It should be in the 4-6% range.
   d. Adjust each cylinder as necessary to keep it's reading in the 4-6% range. Turn the Pilot screw for that cylinder's carburetor CLOCKWISE to lean out the mixture (reduce the percentage) or COUNTERCLOCKWISE to enrichen the mixture (increase the percentage).

MAINTENANCE & TUNE-UP

■ Make sure the motor continues to run at about 950 rpm, and continue to adjust the Throttle stop/Idle RPM screw as necessary to keep the engine running at the proper speed.

  e. Changes to any one cylinder can affect the others, so if changes are made to ANY cylinder, all other cylinders (even previously checked ones) must be rechecked.
  f. Once finished, shut the powerhead down, then remove the analyzer and adaptor(s). Reinstall the exhaust manifold plugs.
  g. Recheck Carburetor Synchronization after any adjustments.
 4. For Mercury/Mariner 40/50 hp motors, proceed as follows:
  a. With the engine warm and idling, turn the No. 4 carburetor (bottom) mixture screw 1/8 of a turn in either direction (first one way, then the other way), while listening for a change in engine speed.
  b. If the engine speed drops while turning it in either direction, LEAVE THE SCREW ALONE, in the original position.
  c. If the engine speed increases in one direction, KEEP turning it in that direction UNTIL you get to the point of highest idle speed.
  d. Adjust the screws for the remaining carburetors in the same manner working your way down from the top No. 1, then No. 2 and finally No. 3.

■ Each time the motor is adjusted rev it 2-3 times, then give it about 15 seconds for the idle to stabilize.

 5. For all other motors, proceed as follows:
  a. With the engine idling begin the adjustment with the No. 4 carburetor (bottom) pilot screw. Slowly turn the screw inward until the idle speed drops about 50 rpm, then back the screw out again 5/8 turn in order to achieve the best engine speed.

■ As with all engine running adjustments, each time you make an adjustment rev the engine 2 or 3 times, then allow the engine to stabilize back at idle for about 15 seconds before adjusting the screw further.

  b. Repeat this adjustment at the No. 3, No. 2 and finally the No. 1 carburetor pilot screws respectively.
 6. Perform the Idle Speed Adjustment, as necessary.

Dash-pot

Normally Dash-pot adjustment is only required in response to certain performance symptoms or after specific repairs. If the carburetors have been removed, the No. 4 carburetor itself has been replaced or dash-pot itself is removed/replaced then the adjustment should be performed after the idle speed, carb synchronization and pilot screws are adjusted. Also if the following symptoms occur after these adjustments are performed:
• If the engine stalls or stumbles upon deceleration (chopping the throttle).
• If it is hard to shift to **Neutral** or **Reverse** upon deceleration.
Adjustment takes place with the carburetors installed and the engine running.

Models Through 2000
◆ See Figure 238

Because of engine operating speeds, this adjustment should take place with the engine operating in a test tank or on a launched craft.

■ For models approximately through 1998 (and possibly some later models, however the manufacturer is not specific), in order to properly adjust the dash-pot you'll have to start, run and fully warm the motor, then STOP IT, and RESTART again and perform the adjustment within 3 minutes, otherwise computer timing advance will be activated preventing proper adjustment.

 1. Provide a suitable source of cooling water to the motor (tank, flush fitting or launch the boat), then start and warm the motor to normal operating temperature.
 2. Once the engine is fully warmed, shut it **OFF**. Restart the motor again and finish the adjustment before 3 minutes have elapsed.
 3. Pull the retard plunger out fully.
 4. Loosen the locknut. Rotate the dash-pot until the ram is touching the lever and engine speed is 1650-1750 rpm, then retighten the locknut.

2001 and Later Yamaha Models
◆ See Figure 239

Because of engine operating speeds, this adjustment should take place with the engine operating in a test tank or on a launched craft.

 1. Attach a shop tachometer to the motor, then start and warm the engine for at least 5 minutes, making sure it reaches normal operating temperature.
 2. Check the engine speed and watch for operating of the dash-pot. Slowly open the throttle cam by hand, then check for and note the speed when the acceleration pump stopper comes into contact with the dash-pot, as shown in the accompanying illustration.
 3. The contact point should occur at a minimum of 2700 rpm for models up to and including 50 hp or 2000-2200 rpm for 60J/60 hp models. If not, manually bring the throttle cam forward to the point where the pump stopper comes into contact with the dash-pot and rotate the adjusting screw (located on the other end of the accelerator pump linkage from the dash-pot) until the minimum engine speed is reached.

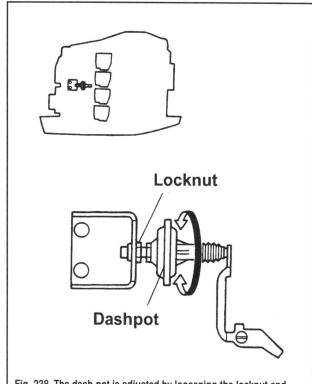

Fig. 238 The dash-pot is adjusted by loosening the locknut and turning the assembly - models through 2000

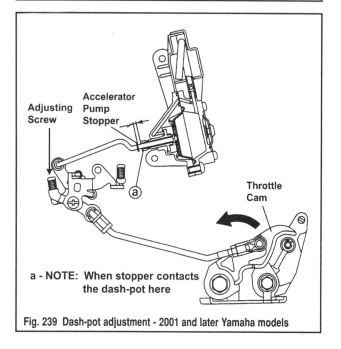

Fig. 239 Dash-pot adjustment - 2001 and later Yamaha models

2-76 MAINTENANCE & TUNE-UP

4. Open and close the throttle cam and couple of times, checking and noting the operational engine speed of the dash-pot again, to make sure it is in specification. Adjust again, as necessary.

2001 and Later Mercury/Mariner

Mercury/Mariner changed to recommending a static adjustment on these motors.

1. Open the throttle cam by hand until the dashpot linkage is fully open and the plunger is fully extended.
2. Using a feeler gauge, measure the clearance between the end of the dashpot plunger and the stop on the adjacent bracket. The clearance must be 0.050-0.060 in. (1.3-1.5mm).
3. If the clearance is out of spec, use a 12mm wrench to loosen the locknut, then use a 10mm wrench to slowly turn the dashpot inward or outward until the correct clearance is obtained.
4. Once the gap is correct, hold the dashpot steady with the 10mm wrench and tighten the jam nut (locknut) against it to hold the dashpot in place using the 12mm wrench.
5. Double check the gap after tightening the jam nut and, readjust, if necessary.

Choke Cam Angle

◆ See Figure 240

■ This adjustment only applies to models through 2000.

The Choke Cam Angle is normally adjusted with the carburetor assembly removed from the powerhead and only after the No. 4 carburetor has been removed/replaced or the Choke Cam Angle screw has loosened in service.

The choke cam angle will vary slightly based on ambient temperature. Adjustment is made by measuring the gap between the choke cam and bracket as shown in the accompanying illustration, then comparing the measurement to the accompanying graph of appropriate gaps at temperature. The gap measurements are in mm for ease of adjustment.

If adjustment is necessary, loosen the choke cam angle adjustment screw, adjust the gap and hold in position while retightening the adjustment screw. Verify that the gap did not change while you were tightening the screw.

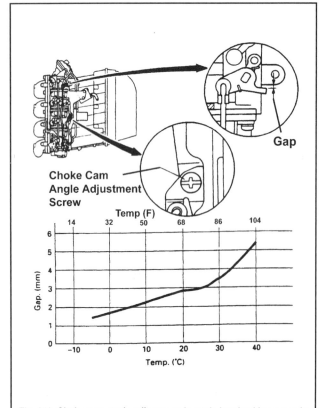

Fig. 240 Choke cam angle adjustment is made by checking a gap in relation to the ambient temperature

Fast Idle

1995-98 Yamaha Models and 1995-00 Mercury/Mariner Models

◆ See Figure 241

The Fast Idle is normally adjusted with the carburetor assembly removed from the powerhead and only after the No. 4 carburetor has been removed/replaced or the Fast Idle screw has loosened in service. Fast idle adjustment is required if the fast idle exceeds 1550 rpm or if the engine stalls during a cold start. The fast idle specification is 1250-1550 rpm for this motor.

1. Disconnect the link rods.
2. Loosen the fast idle lock screw.
3. Loosen the idle stop screw until it just comes free of the lever, then retighten the screw until the bypass hole is half visible.
4. Starting with the bypass hole half visible, tighten the idle stop screw an additional number of turns depending upon the ambient temperature. Refer to the accompanying illustration to determine precisely how many additional turns to give the idle stop screw.
5. Reconnect the link rods.
6. Hold the fast idle lever manually against the roller, then tighten the fast idle screw.

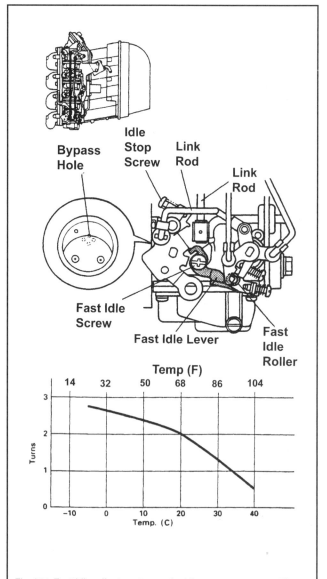

Fig. 241 Fast idle adjustment uses the idle stop screw to position the throttle lever. The idle stop screw is tightened a number of turns (which varies depending upon ambient temperature) from a fixed starting point - models through 1998

MAINTENANCE & TUNE-UP 2-77

1999-2000 Yamaha Models
◆ See Figure 242

The Fast Idle is normally adjusted with the carburetor assembly removed from the powerhead and only after the No. 4 carburetor has been removed/replaced or the Fast Idle screw has loosened in service. Fast idle adjustment is required if the fast idle exceeds 1550 rpm or if the engine stalls during a cold start. The fast idle specification is 1250-1550 rpm for this motor.

1. Disconnect the G/W lead for the electro-thermal ram from the wiring harness.
2. Loosen the fast idle lever lock-screw.
3. Loosen the idle adjust screw about 2 turns.
4. Start the engine and turn the idle adjust screw back inward until the specified fast idle speed is obtained. Shut the engine down.
5. Hold the fast idle lever against the roller and tighten the fast idle lever lock-screw.
6. Now, turn the idle adjust screw back outward a specified number of turns depending upon the ambient temperature. Refer to the accompanying illustration to determine precisely how many additional turns outward to give the idle adjust screw.
7. Reconnect the G/W lead for the electro-thermal ram to the wiring harness.
8. Restart the engine and allow it to fully re-warm, then check and adjust the Idle Speed, as detailed earlier in this section.

Unloader

◆ See Figure 243

■ This adjustment only applies to models through 2000.

The Unloader really only requires adjustment when the unloader screw has been loosened during service (or has worked loose during operation). This adjustment usually only takes place with the carburetor assembly removed from the powerhead and only after the No. 4 carburetor has been removed/replaced or the Unloader screw has loosened in service.

To adjust the unloader:

1. Loosen the unloader adjustment screw and fully close the choke valve.
2. At this point depress the No. 4 carburetor choke lever and make sure the No. 1 choke lever is also touching its stopper.
3. Turn the unloader adjustment screw until the tip of the screw just touches the lever.

Choke Valve Closing Gap

◆ See Figure 244

■ This adjustment only applies to models through 2000.

The Choke Valve Closing Gap is normally adjusted with the carburetor assembly removed from the powerhead and only after the No. 4 carburetor has been removed/replaced or the Choke Valve Closing Gap screw has loosened in service.

The Choke Valve Closing Gap is another measurement that varies with ambient temperature. Adjustment is performed by first measuring the gap between the screw and valve as noted in the accompanying illustration, then determining the proper gap for the current ambient temperature as follows:

1. Push the choke link down to close the choke valve.
2. Measure the gap at the end of the choke valve adjustment screw.
3. Tighten or loosen the screw as necessary to achieve the proper gap according to the accompanying graph.

IGNITION TIMING

Yamaha Models

Ignition timing on the 40/45/50 hp 4-stroke powerhead is adjusted (advanced/retarded) automatically by the ignition module to match rpm operating conditions. Because there are no adjustments for timing, any problems or irregularities should be diagnosed and repaired as the result of defective mechanical or electrical components. No attempt should be made to perform adjustments in order to achieve the proper timing specifications, let the electronics do their job, but give them the tools that they need.

On 2001-04 models a TDC timing mark and a timing pointer is provided in order to check ignition system operation. This is done using a timing light and an operating the engine at idle speed. Since the Idle Timing specification for these models is TDC give or take 1.5 degrees, the TDC timing mark should roughly align with the timing pointer at proper idle speed. If timing is

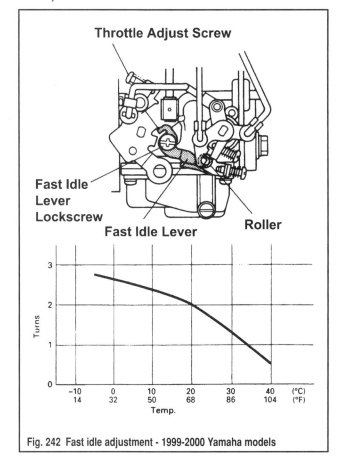

Fig. 242 Fast idle adjustment - 1999-2000 Yamaha models

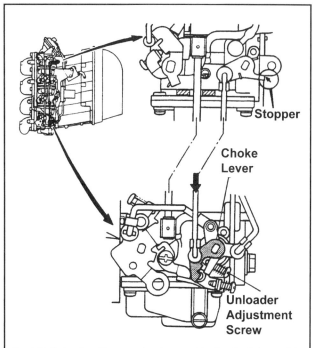

Fig. 243 Unloader adjustment involves positioning the linkage with the adjustment screw just touching the lever when the choke valve is fully closed on all carbs

2-78 MAINTENANCE & TUNE-UP

out of specification, first double check the timing pointer positioning using a dial-gauge to determine when the piston is really physically at TDC (the marks must align or the pointer is off), then check/diagnose the ignition system for problems.

Mercury/Mariner Models

Like most other 4-strokes, the ignition is controlled by the ECM unit and timing is not adjustable. Mercury/Mariner service information does specify that the flywheel is equipped with timing marks and the flywheel cover is equipped with a timing pointer window. A timing light therefore can be used to verify proper operation of the ignition timing system (against the specifications listed in the Tune-Up Specifications charts). If so, make sure that you run the engine in gear, under load, with a suitable source of cooling water. This means in a test tank, on a dynamometer or on a boat secured to a trailer that has been backed into the water (with the transom straps in place!!!!). If timing is out of specification, seek a mechanical or electrical reason (i.e. faulty component).

75/80/90/100 Hp Models

CARBURETOR ADJUSTMENTS

 MODERATE

The carburetor idle speed should be checked periodically to make sure that the motor is operating smoothly and within specification. In addition, the carburetor synchronization should be checked to make sure that the vacuum generated by each of the carburetors remains within the specified 1.97 in. Hg (50mmHg) throughout throttle operating range. Perform these two checks and adjustments at least on an annual basis and/or with each tune-up. However, adjustments should not be made to these settings unless the engine is operating out of specification (according to the checks) or idle is unstable.

Idle Speed (and Pilot Screws, Where Applicable)

◆ See Figure 245

To ensure proper adjustment the engine must be fully warmed to normal operating temperature.

1. Remove the engine cowling and connect a shop tachometer to the motor.
2. Provide a suitable source of cooling water and load to the motor (either mount it to a test tank or launch the boat), then start and warm the motor to normal operating temperature.
3. Observe engine idle speed with the engine fully warmed, running in **Neutral** and compare that to what is listed in the Tune-Up Specification Chart. Then, except for jet drives, place the engine in gear and check the trolling speed.
4. Shut the powerhead down. If the motor idled smoothly, no adjustment is needed to the pilot screws, however the idle adjust screw (on the linkage for the bottom carburetor) MAY be used to set the idle speed. However, since you should check Carburetor Synchronization, skip to that procedure and reset the idle speed itself after Synchronization is checked/adjusted.

■ **On most EPA regulated motors the pilot screw is plugged and therefore is not supposed to be adjustable.**

5. If the motor did not idle smoothly, locate the pilot screw for each carburetor (mounted horizontally into the side of the carburetor body). Thread each pilot screw very slowly and gently until it just lightly seats.

✱✱ WARNING

Tightening a pilot screw once it contacts the seat will most likely destroy the screw. As the screw tip is forced against the seat it will deform leading to an inability to properly adjust the air fuel mixture. Always be extra gentle with pilot screws.

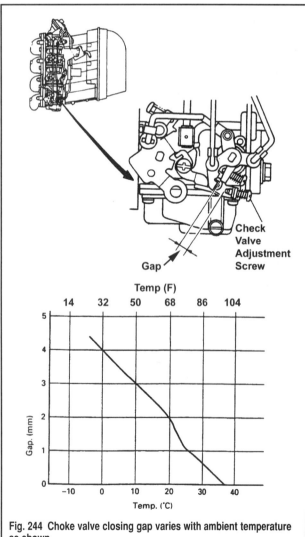

Fig. 244 Choke valve closing gap varies with ambient temperature as shown

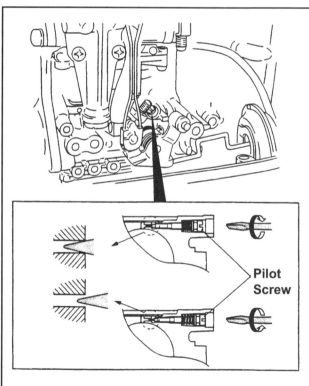

Fig. 245 Pilot screw adjustment - 75/80/90/100 hp motors

MAINTENANCE & TUNE-UP 2-79

6. Locate the Carburetor Set-Up Specifications Charts from the Fuel System section and back each pilot screw out slowly the specified number of turns.

7. With the tachometer still connected and the engine still mounted in the tank or boat still launched, restart the engine.

8. Allow the motor to fully re-warm to normal operating temperature.

9. If engine idle is unstable, turn each of the pilot screws inward until the engine is idling about 40 rpm below the desired specification (listed in the Tune-Up Specifications chart). Then rotate the screws back outward about 3/4 of a turn.

10. Turn the idle adjust screw on the No. 4 carburetor to bring engine idle speed within the desired specification. The idle adjust screw is threaded diagonally downward and contacts the throttle valve linkage. Turning the screw INWARD will INCREASE idle speed, while turning the screw OUTWARD will DECREASE idle speed.

■ As with all engine running adjustments, each time you make an adjustment rev the engine 2 or 3 times, then allow the engine to stabilize back at idle for about 15 seconds before adjusting the screw further.

11. Refer to the Carburetor Synchronization procedure to at least check the vacuum variation between the carbs. If the idle speed is correct and the variation is within specification, no further carburetor adjustment is necessary.

CARBURETOR SYNCHRONIZATION

◆ See Figures 246, 247 and 248

Carburetor synchronization ensures that the throttle valves (butterflies) on each carburetor open and close at precisely the same time. If one throttle valve precedes the others or lags behind, the cylinders will not receive equal amounts of air/fuel mixture per combustion cycle, causing one or more cylinders to run sluggish or lag behind slightly in power. This will lead to rough running, hesitation or stumbling.

A vacuum gauge or carburetor vacuum synchronization tool (such as a set of calibrated carb balance sticks) is required for this adjustment.

An initial idle speed check/adjustment is required in order to get the motor run correctly during carburetor synchronization. Adjustment is a relatively straightforward procedure in which you measure and record the vacuum created by the No. 4 (bottom) carburetor, then adjust the throttle valve screws of each other carburetor to match it.

1. If not done already, install the carburetor assembly to the powerhead, then adjust the idle speed as detailed in this section.

2. With the motor fully warmed and running at idle, remove the vacuum port screws from the intake manifold. If you're using multiple vacuum gauges or a carburetor balance stick (a tool that normally contains as many as 4 separate tubes that measure vacuum) remove all 4 screws so all of the vacuum indicators can be installed.

3. Rev the engine a few times while watching the gauges, they should stay within 1.97 in. Hg (50mmHg) of each other. If they do, you're done with this procedure, but if not, continue in order to properly resynchronize the carburetors.

■ Keep in mind during this procedure that actual vacuum readings will vary based upon altitude, weather and engine condition. What is important during this procedure is not the actual reading so much as how much the reading varies from cylinder-to-cylinder.

4. With the engine running slightly above idle, at least 850-950 rpm for Yamahas or 1000 rpm for Merc/Mariners (temporarily readjust the idle speed, if necessary), record the vacuum measurement for each carburetor. The carbs are synchronized by bringing carbs No. 1-3 to match the reading you get on No. 4. Adjust one carburetor at a time, by turning the throttle valve positioning screws slowly, no more than 1/2-1 turn at a time.

■ If adjustments are necessary, each time you turn an idle adjust/throttle valve screw remember to rev the engine 2 or 3 times, then allow the engine to stabilize back at idle for about 15 seconds before taking a vacuum reading or adjusting the screw further.

5. Once the throttle valve positioning screws are properly adjusted, check/readjust the idle speed one LAST time to the specification listed in the Tune-Up Specification Chart.

6. Rev the engine a few times while watching the gauges, they should stay within 1.97 in. Hg (50mmHg) of each other. Otherwise, recheck your adjustment.

7. Shut the engine down, remove the gauges/adapters and reinstall the balance plugs to the intake manifold.

CHECKING/ADJUSTING CARBURETOR PICKUP TIMING

Yamaha Models

◆ See Figures 249 and 250

1. To check the carburetor pickup timing, turn the throttle control lever to the idle position, then check that the accelerator cam idle mark aligns with the throttle roller. If not, continue with the procedure to adjust the pickup timing.

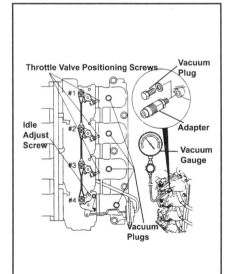

Fig. 246 Carburetor synchronization - 75/80/90/100 hp motors

Fig. 247 The carbs are synchronized using the vacuum ports and throttle positioning screws

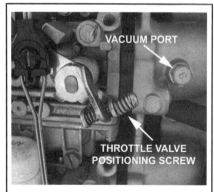

Fig. 248 View of a throttle valve positioning screw from a Mercury model

2-80 MAINTENANCE & TUNE-UP

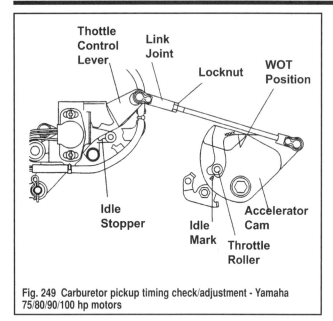

Fig. 249 Carburetor pickup timing check/adjustment - Yamaha 75/80/90/100 hp motors

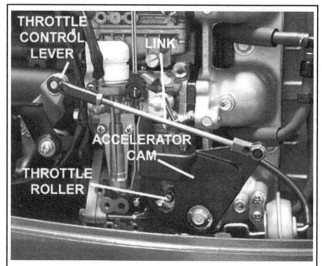

Fig. 250 Carb pickup timing is controlled by the link connecting the throttle control lever to the accelerator cam - Yamaha models shown

2. Loosen the locknut for the link joint, then disconnect the joint from the throttle control lever.

3. Move the throttle control lever until it contacts the idle stopper. Visually check the carburetor throttle valves to make sure they are fully closed.

4. Align the accelerator cam idle mark with the center of the throttle roller, then adjust the length of the link joint (threading inward or outward, as necessary) until the joint can be reinstalled over the ball on the throttle control lever with the marks still aligned. Tighten the locknut.

5. Move the throttle control lever, checking that the throttle valve opens and closes smoothly. When the lever is moved to the WOT position, make sure the throttle roller aligns with the WOT position on the accelerator cam.

6. Set the throttle to the idle position and recheck that the idle mark aligns with the roller.

■ Make sure there is no interference between the throttle control lever and the throttle position sensor lead.

Mercury/Mariner Models

◆ See Figures 251, 252 and 253

1. Check to make sure the Throttle Position Sensor (TPS) is rotated FULLY CLOCKWISE in the mounting brackets (the screws must be up against the end of the slots).

2. Manually turn the throttle control lever to the idle/throttle fully retarded position (with the throttle arm against the throttle stop).

3. In this position check that the pointer on the lower left side of the throttle cam (at about the 9 o'clock position when facing the cam) is aligned with the CENTER of the throttle lever shaft (where the carburetor synchronization linkage attaches to the shaft at the bottom carburetor).

4. If adjustment is necessary, carefully unsnap the ball connector from the throttle cam, then thread the connector inward or outward on the shaft as necessary to properly position the throttle cam when it is reattached. Turning the connector CLOCKWISE will move the pointer DOWNWARD, while turning the connector COUNTERCLOCKWISE will move the pointer UPWARD.

5. Snap the ball connector back onto the throttle cam and double-check alignment of the pointer and throttle shaft.

SHIFT CABLE ADJUSTMENT

Yamaha Models

◆ See Figure 254

Visually check the shift linkage movement in response to the shift handle. If the shifter does not engage properly or smoothly, adjust, as follows:

1. Set the shift control lever to the **Neutral** position. Visually check the powerhead linkage set pin to see if it aligns with the **Neutral** mark on the slider bracket. If it does and the shifter is not working properly, suspect a problem with the linkage itself. If the set pin is not aligned with the mark, adjust it as follows:

2. Manually align the set pin with the **Neutral** mark on the slider bracket.

3. Remove the retaining clip and disconnect the shift cable joint from the powerhead linkage.

■ If it is difficult to remove the cable joint from the set pin because of a lack of vertical free-play in the cable, follow the cable back a little ways to a securing bracket that is bolted down overtop of the cable. Remove the bolt and bracket to create some vertical freedom of movement.

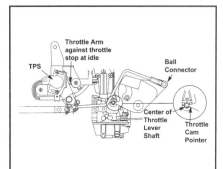

Fig. 251 Carburetor pickup timing check/adjustment - Mercury/Mariner 75/90 Hp Models

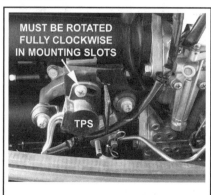

Fig. 252 Make sure the TPS is rotated as far clockwise as the mounting slots will allow

Fig. 253 If adjustment is necessary, unsnap the ball connector and thread it inward/outward on the link

MAINTENANCE & TUNE-UP 2-81

4. Move the shifter control lever to the **Neutral** position.
5. Loosen the locknut, then adjust the position of the shift cable joint (by turning the joint inward or outward on the cable threads) until its mounting hole aligns with the linkage set pin at the center of the bracket. Tighten the locknut to hold it in position on the cable.

✳✳ CAUTION

Make sure that after adjustment the cable joint is threaded over at LEAST 0.31 in. (8mm) of cable threads, otherwise there is a risk that the joint could separate in service causing a severe navigation/control hazard.

6. Position the cable joint over the control lever set pin and secure using the retaining clip.
7. If the cable retaining bracket was removed for play, reposition it and secure using the retaining bolt.
8. Operate the control lever to confirm proper operation and smooth adjustment.

Mercury/Mariner Models

◆ See Figure 255

No periodic adjustment should be necessary for the Shift and Throttle cables, however, should the cables be replaced they will require an initial set-up to place them in proper alignment with the linkage.

The shifter cable should be set so the distance between the cable barrel (on the threaded adjuster) and the cable end (where it attaches to the linkage) is the same as the distance between the cable barrel bracket and the linkage pin when both the gearcase and the shifter are in Neutral. If adjustment is necessary, free the cable and rotate the cable end to set the proper distance.

THROTTLE CABLE ADJUSTMENT

Yamaha Models

◆ See Figure 256

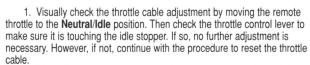

1. Visually check the throttle cable adjustment by moving the remote throttle to the **Neutral/Idle** position. Then check the throttle control lever to make sure it is touching the idle stopper. If so, no further adjustment is necessary. However, if not, continue with the procedure to reset the throttle cable.
2. Manually move the throttle control lever into contact with the idle stopper.
3. Loosen the locknut on the throttle cable joint, then remove the retaining clip and pull it from the set pin.
4. Make sure the remote throttle lever is in the **Neutral/Idle** position, then turn the cable joint inward or outward on the cable threads until it aligns

with the set pin while both the remote throttle and the powerhead throttle lever are in the idle position. Tighten the locknut to hold the cable joint in this position.

✳✳ CAUTION

Make sure that after adjustment the cable joint is threaded over at LEAST 0.31 in. (8mm) of cable threads, otherwise there is a risk that the joint could separate in service causing a severe navigation/control hazard.

5. Install the cable joint back over the set pin and secure using the retaining clip.
6. Open and close the throttle a couple of times using the remote while visually checking to be sure the throttle valves on the carburetor open and close smoothly and to the full extent of their travel. If necessary, repeat the adjustment procedure.

Mercury/Mariner Models

◆ See Figure 257

No periodic adjustment should be necessary for the Shift and Throttle cables, however, should the cables be replaced they will require an initial set-up to place them in proper alignment with the linkage.

The throttle cable should be set so the distance between the cable barrel (on the threaded adjuster) and the cable end (where it attaches to the linkage) is sufficient so that with the remote in Neutral the cable will gently hold the throttle arm on the powerhead against the idle stop. If adjustment is necessary, free the cable and rotate the cable barrel/adjuster to set the proper distance.

IGNITION TIMING

Like most other 4-strokes, the ignition is controlled by the CDI unit/ECM unit and timing is not adjustable. Although Yamaha provides no information on checking the timing for this model, the Mercury/Mariner service information does specify that the flywheel is equipped with timing marks and the flywheel cover or manual starter housing is equipped with a timing pointer window. And, our Yamaha teardown motors also had a timing window in the flywheel covers and marks on the flywheels, so we are relatively certain you'll be able to verify proper ignition system operation through timing checks on all models.

A timing light can be used to verify proper operation of the ignition timing system (against the specifications listed in the Tune-Up Specifications charts). If so, make sure that you run the engine in gear, under load, with a suitable source of cooling water (test tank, dynamometer, or secured to a trailer that is backed into the water). If timing is out of specification, seek a mechanical or electrical reason (i.e. faulty component).

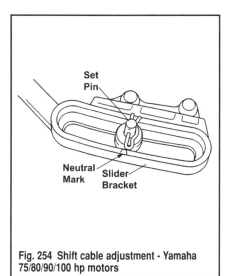

Fig. 254 Shift cable adjustment - Yamaha 75/80/90/100 hp motors

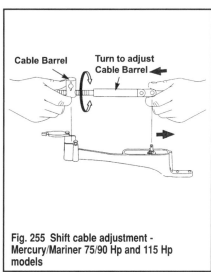

Fig. 255 Shift cable adjustment - Mercury/Mariner 75/90 Hp and 115 Hp models

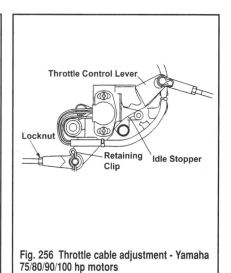

Fig. 256 Throttle cable adjustment - Yamaha 75/80/90/100 hp motors

2-82 MAINTENANCE & TUNE-UP

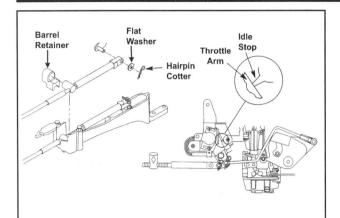

Fig. 257 Throttle cable adjustment - Mercury/Mariner 75/90 Hp Models shown (115 Hp Models similar, but the TPS is located elsewhere)

115 Hp Models

CHECKING/ADJUSTING THROTTLE BODY PICKUP TIMING

 MODERATE

◆ See Figures 258 thru 262

■ Although the shape of the throttle control lever varies slightly between Yamaha and Mercury/Mariner models, the function and adjustment is almost identical.

Because the throttle control lever and throttle cam can be seen through the runners of the air intake silencer assembly on Yamaha motors, you can usually CHECK pickup timing without having to remove the silencer. However, access is just a little too tight to make it worth trying to adjust it with the silencer installed, so if the pickup timing needs adjustment, unbolt and remove the silencer assembly. On Mercury/Mariner models, the intake runners obscure the control lever and the portion of the cam that contains the roller, therefore you must remove the intake silencer even for inspection.

Fig. 258 On Yamahas, the throttle control linkage is visible between the air intake tubes...

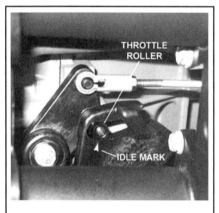

Fig. 259 ...and the throttle body pickup timing is correct if the roller aligns with the mark on the throttle cam

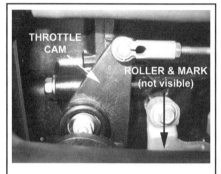

Fig. 260 On Mercs however, the mark and roller are obscured by the intake runners

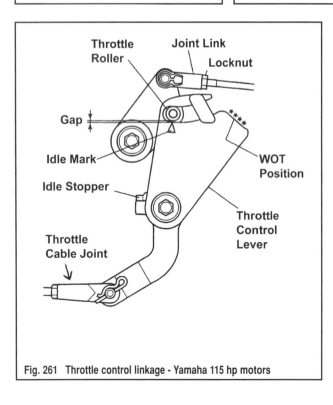

Fig. 261 Throttle control linkage - Yamaha 115 hp motors

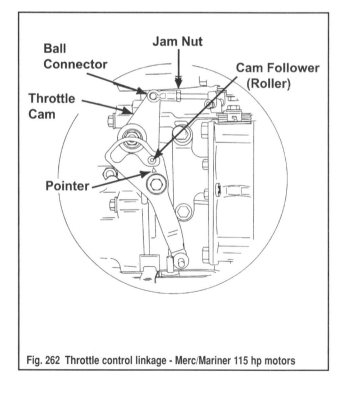

Fig. 262 Throttle control linkage - Merc/Mariner 115 hp motors

MAINTENANCE & TUNE-UP

1. On Mercury/Mariner models, remove unbolt and remove the air intake silencer assembly so that you can see the throttle control lever alignment mark and the throttle cam roller.
2. Check the throttle body pickup timing by setting the remote to the idle position and/or manually rotating the throttle control lever on the powerhead to the idle position. Look at the throttle linkage between the air intake tubes and verify that the idle mark on the throttle cam aligns with the center of the throttle roller. Also, use a feeler gauge to verify that there is a 0.02 in. (0.5mm) gap between the throttle roller and the throttle control lever. If so, no adjustment is necessary. If not, continue with the procedure to set the proper throttle body pickup timing.

■ Mercury also advises that while checking alignment of the mark you visually verify that the idle speed arm on the other end of the throttle linkage is against the speed screw. If you are unsure of the screw location, follow the link from the ball connector at the top of the throttle cam to the right (while looking at the powerhead) to the point where the link connects to the idle speed arm. The idle speed screw is the spring loaded screw threaded horizontally inward toward the powerhead and the arm. The screw is normally just underneath a small plastic cover.

3. If not done already, remove unbolt and remove the air intake silencer assembly for clearance.
4. Loosen the joint locknut and separate the throttle joint link rod from the linkage.
5. Manually set the throttle control lever on the powerhead in the idle position, making sure it contacts the idle stopper. Visually inspect the throttle body to make sure the throttle valve is fully closed.
6. Align the throttle control lever idle mark with the center of the throttle roller.
7. Adjust the length of the throttle joint link rod (by turning the joint inward or outward on the link threads) until there is a 0.02 in. (0.5mm) gap between the throttle roller and the throttle control lever.
8. Move the throttle control lever to check that the valve opens and closes smoothly. Check that the throttle roller aligns with the accelerator cam WOT position, then return the throttle to the idle position and double-check alignment with the idle mark.

SHIFT CABLE ADJUSTMENT

Yamaha Models

◆ See Figures 263 and 264

Visually check the shift linkage movement in response to the shift handle. If the shifter does not engage properly or smoothly, adjust, as follows:
1. Set the shift control lever to the **Neutral** position. Visually check the powerhead linkage set pin to see if it aligns with the **Neutral** mark on the slider bracket. If it does and the shifter is not working properly, suspect a problem with the linkage itself. If the set pin is not aligned with the mark, adjust it as follows:

2. Manually align the set pin with the **Neutral** mark on the slider bracket.
3. Remove the retaining clip and disconnect the shift cable joint from the powerhead linkage.

■ If it is difficult to remove the cable joint from the set pin because of a lack of vertical free-play in the cable, follow the cable back a little ways to a securing bracket that is bolted down overtop of the cable. Remove the bolt and bracket to create some vertical freedom of movement.

4. Move the shifter control lever to the **Neutral** position.
5. Loosen the locknut, then adjust the position of the shift cable joint (by turning the joint inward or outward on the cable threads) until its mounting hole aligns with the linkage set pin at the center of the bracket. Tighten the locknut to hold it in position on the cable.

✱✱ CAUTION

Make sure that after adjustment the cable joint is threaded over at LEAST 0.31 in. (8mm) of cable threads, otherwise there is a risk that the joint could separate in service causing a severe navigation/control hazard.

6. Position the cable joint over the control lever set pin and secure using the retaining clip.
7. If the cable retaining bracket was removed for play, reposition it and secure using the retaining bolt.
8. Operate the control lever to confirm proper operation and smooth adjustment.

Mercury/Mariner Models

◆ See Figure 255

No periodic adjustment should be necessary for the Shift and Throttle cables, however, should the cables be replaced they will require an initial set-up to place them in proper alignment with the linkage.

The shifter cable should be set so the distance between the cable barrel (on the threaded adjuster) and the cable end (where it attaches to the linkage) is the same as the distance between the cable barrel bracket and the linkage pin when both the gearcase and the shifter are in Neutral. If adjustment is necessary, free the cable and rotate the cable end to set the proper distance.

THROTTLE CABLE ADJUSTMENT

Yamaha Models

◆ See Figure 261

1. Visually check the throttle cable adjustment by moving the remote throttle to the **Neutral/Idle** position. Then check the throttle control lever to

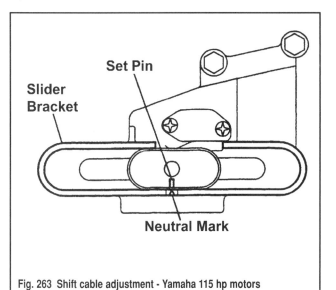

Fig. 263 Shift cable adjustment - Yamaha 115 hp motors

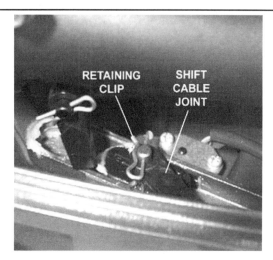

Fig. 264 The shift cable is adjusted by disconnecting it from the set pin and repositioning the cable joint

2-84 MAINTENANCE & TUNE-UP

make sure it is touching the idle stopper. If so, no further adjustment is necessary. However, if not, continue with the procedure to reset the throttle cable.

2. Manually move the throttle control lever into contact with the idle stopper.

3. Loosen the locknut on the throttle cable joint, then remove the retaining clip and pull it from the set pin.

4. Make sure the remote throttle lever is in the **Neutral/Idle** position, then turn the cable joint inward or outward on the cable threads until it aligns with the set pin while both the remote throttle and the powerhead throttle lever are in the idle position. Tighten the locknut to hold the joint in this position.

✱✱ CAUTION

Make sure that after adjustment the cable joint is threaded over at LEAST 0.31 in. (8mm) of cable threads, otherwise there is a risk that the joint could separate in service causing a severe navigation/control hazard.

5. Install the cable joint back over the set pin and secure using the retaining clip.

6. Open and close the throttle a couple of times using the remote while visually checking to be sure the throttle valves open and close smoothly and to the full extent of their travel. If necessary, repeat the adjustment procedure.

Mercury/Mariner Models

◆ See Figure 257

No periodic adjustment should be necessary for the Shift and Throttle cables, however, should the cables be replaced they will require an initial set-up to place them in proper alignment with the linkage.

The throttle cable should be set so the distance between the cable barrel (on the threaded adjuster) and the cable end (where it attaches to the linkage) is sufficient so that with the remote in Neutral the cable will gently hold the throttle arm on the powerhead against the idle stop. If adjustment is necessary, free the cable and rotate the cable barrel/adjuster to set the proper distance.

THROTTLE VALVE SYNCHRONIZATION

◆ See Figures 265 and 266

If the throttle is operating properly there is NO reason to perform this adjustment. Un-necessary tampering usually just leads to performance problems.

Throttle valve synchronization on a fuel injected motor is very similar to carburetor synchronization on carbureted motors. In both cases, throttle valve screws are adjusted in a manner that ensures each of the throttle valves (in the throttle body or carburetor, as applicable) open and close at the same instants. The most precise way to measure this is using a vacuum gauge on the intake manifold to measure the distance the valve is opened at any point. On these motors the vacuum for cylinder numbers 3 and 4 (bottom 2 cylinders) should be adjusted within 1.18 in. Hg (30mmHg) of the pressure found on cylinder number 1 and 2 (top 2 cylinders).

1. Remove the engine cowling and connect a shop tachometer to the motor.

2. Remove the air intake silencer assembly from the motor and the throttle bodies for greater access.

3. Remove the vacuum port screws from the intake manifold. Connect multiple vacuum gauges or a carburetor balance stick (a tool that normally contains as many as 4 separate tubes that measure vacuum). You'll need adapters that thread into the intake manifold vacuum ports.

✱✱ CAUTION

If you are using a liquid-filled set of carburetor balance sticks, use great care to prevent revving the motor too much or too suddenly as this motor makes sufficient vacuum to sometimes pull the liquid from the tubes.

4. Provide a suitable source of cooling water to the motor (mount it to a test tank, attach a flush fitting or launch the boat), then start and warm the motor to normal operating temperature.

5. Observe engine idle speed with the engine fully warmed, running in **Neutral** and compare that to what is listed in the Tune-Up Specification Chart.

6. Now, with the engine running at idle, note the vacuum readings on all 4 cylinders. Turn the synchronizing screw (found at the center of the throttle body assembly, between the air intakes for the top 2 and bottom 2 cylinders) in order to adjust the vacuum reading on the bottom 2 cylinders to something within 1.18 in. Hg (30mmHg) of the reading on the top 2 cylinders. Tightening the screw will increase the vacuum reading on the lower cylinders, while loosening the screw will decrease the vacuum on the lower cylinders.

■ **On most models a Torx anti-tamper T-25 bit (with a concave opening in the center) is needed to turn the synchronization screw.**

7. Once the adjustment is complete, remove the vacuum gauges and tachometer, then reinstall the vacuum port screws and the air intake silencer assembly.

■ **If you need to check/adjust the throttle position sensor, leave the air intake silencer assembly removed for access and leave the tachometer in place to measure engine rpm.**

THROTTLE POSITION SENSOR ADJUSTMENT

Maintenance Adjustments

◆ See Figures 267 and 268

■ **Mercury states that the TPS and throttle shaft are set at the factory and that NO periodic adjustments are necessary. Yamaha suggests that at least a check of the TPS output voltage should be performed annually, but we suspect that adjustment should not normally be necessary.**

With each tune-up, or at least annually, Yamaha recommends that should check the Throttle Position Sensor (TPS) output voltage with the throttle valves fully closed. If the reading is out of specification, the physical positioning of the sensor can be adjusted in order to bring the reading into spec.

However, the sensor does not generate a voltage, so much as the variable resistor within the sensor will change a reference signal from the Engine Control Module (ECM). Therefore, the circuit must be intact in order to test the sensor signal. You've got 2 options. One option is to attempt to back-probe the connector (insert the probes through the rear of the connector while it is still attached to the sensor). However this method risks damaging the connector or the wiring insulation (and could lead to problems with the circuit later). The better method is to disconnect the wiring harness and use 3 jumper wires (one for each terminal) to reconnect the harness. The jumpers must not contact each other (or you risk damage to the ECM from a short), however, some point on the jumper must be exposed so that you can probe the completed circuit using a DVOM. Yamaha makes 3-pin test harness for this application (#90890-06793). Mercury also sells a 3-pin test harness for this application (#91-881827).

1. Make sure the throttle valves are properly adjusted (check the Throttle Valve Synchronization) before proceeding.

2. If it is not still removed from checking the Throttle Valve Synchronization, remove the air intake assembly for access.

3. If not already installed, connect a tachometer and provide a source of cooling water to the motor.

4. Disconnect the throttle link from the throttle body.

5. Disconnect the wiring from the TPS and connect a test harness or jumper wires to complete the circuit.

6. Connect a Digital Volt Ohmmeter (DVOM) set to read DC volts to the Pink and Black wires of the jumper/test harness.

7. Turn the ignition switch **ON**, then measure the output voltage with the throttle valves fully closed. Voltage should read 0.718-0.746 volts.

8. Start the motor and recheck the voltage reading.

MAINTENANCE & TUNE-UP 2-85

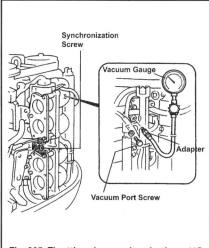

Fig. 265 Throttle valve synchronization - 115 hp motors

Fig. 266 There is a cutout on most motors (like this Merc) to access the throttle valve synchronization screw through the plastic cover

Fig. 267 Throttle position sensor and throttle link - 115 hp motors

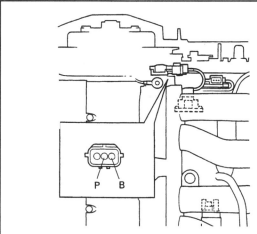

Fig. 268 TPS sensor adjustment involves measuring sensor output at the harness

9. If the sensor voltage is out of range, loosen the 2 mounting screws (using a Torx T-20 anti-tamper bit w/ a concave opening in the center for most motors) and reposition the sensor to obtain the correct voltage readings. Once the sensor is properly repositioned, tighten the screws again to hold it in position and shut down the powerhead.
10. Reconnect the throttle link to the throttle body.
11. Remove the tachometer.

Adjustment When Replacing or Overhauling the Throttle Body

◆ See Figures 261, 267 and 268 and 269

If the throttle body has been overhauled or replaced, a more involved sensor adjustment should take place to ensure proper operation.
1. Install the intake manifold assembly, but keep the air intake silencer disconnected for access.
2. Disconnect the wiring from the TPS and connect a test harness or jumper wires to complete the circuit.
3. Connect a Digital Volt Ohmmeter (DVOM) set to read DC volts to the Pink and Black wires of the jumper/test harness.
4. Turn the ignition switch **ON**
5. Loosen the synchronization screw and fully OPEN the throttle valve for the lower 2 throttle bodies (Cylinder #3 and #4).
6. Loosen the idle adjustment screw and fully CLOSE the throttle valve for the upper 2 throttle bodies (Cylinder #1 and #2).

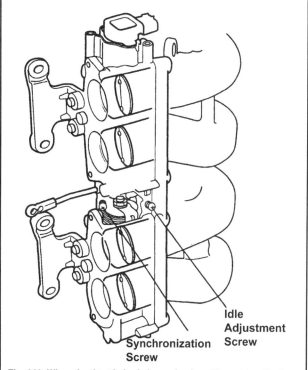

Fig. 269 When the throttle body is serviced you'll need to adjust the TPS sensor using the idle adjustment and synchronization screws

7. Adjust the sensor positioning until the DVOM shows that output voltage is 0.69-0.71 volts. Operate the throttle valve several times and record the value.
8. Tighten the synchronization screw SLOWLY until the recorded value in the previous step is changed. Then SLOWLY tighten the idle adjustment screw until there is a 0.028-0.036 volt increase over the previously recorded value.
9. Operate the throttle valve a couple of times, then allow the throttle to close, now check the position of the throttle roller and the idle mark to make sure they are properly aligned.
10. Install the air intake silencer and start the powerhead.
11. Measure the output voltage with the throttle valves fully closed, but the engine idling. Voltage should read 0.718-0.746 volts.
12. Check the Throttle Valve Synchronization, as detailed in this section. If necessary, recheck the TPS adjustment after following the synchronization procedure.

MAINTENANCE & TUNE-UP

IGNITION TIMING

Like most other 4-strokes, the ignition is controlled by the CDI unit/ECM unit and timing is not adjustable. Although Yamaha provides no information on checking the timing for this model, the Mercury/Mariner service information does specify that the flywheel is equipped with timing marks and the flywheel cover or manual starter housing is equipped with a timing pointer window. And, our Yamaha teardown motors also had a timing window in the flywheel covers and marks on the flywheels, so we are relatively certain you'll be able to verify proper ignition system operation through timing checks on all models.

A timing light can be used to verify proper operation of the ignition timing system (against the specifications listed in the Tune-Up Specifications charts). If so, make sure that you run the engine in gear, under load, with a suitable source of cooling water (test tank, dynamometer, or secured to a trailer that is backed into the water). If timing is out of specification, seek a mechanical or electrical reason (i.e. faulty component).

150 Hp Yamaha Models

THROTTLE VALVE SYNCHRONIZATION

◆ See Figures 270 and 271

If the throttle is operating properly there is NO reason to perform this adjustment. Un-necessary tampering usually just leads to performance problems.

Throttle valve synchronization on a fuel injected motor is very similar to carburetor synchronization on carbureted motors. In both cases, throttle valve screws are adjusted in a manner that ensures each of the throttle valves (in the throttle body or carburetor, as applicable) open and close at the same instants. The most precise way to measure this is using a vacuum gauge on the intake manifold to measure the distance the valve is opened at any point. On these motors the vacuum for cylinder numbers 3 and 4 (bottom 2 cylinders) should be adjusted within 1.18 in. Hg (30mmHg), also sometime expressed as 0.58 psi (4 kPa), of the pressure found on cylinder number 1 and 2 (top 2 cylinders) while the engine is running at idle speed.

1. Remove the engine cowling and connect a shop tachometer to the motor.
2. Remove the vacuum port caps from the intake manifold (there is one cap for each cylinder, found in a vertical line on the manifold, just to the right (when looking at the outboard) from the fuel cooler assembly).
3. Connect multiple vacuum gauges or a carburetor balance stick (a tool that normally contains as many as 4 separate tubes that measure vacuum) to the intake manifold vacuum ports.

✴✴ CAUTION

If you are using a liquid-filled set of carburetor balance sticks, use great care to prevent revving the motor too much or too suddenly as this motor may make sufficient vacuum to sometimes pull the liquid from the tubes.

4. Provide a suitable source of cooling water to the motor (mount it to a test tank, attach a flush fitting or launch the boat), then start and warm the motor to normal operating temperature.
5. Observe engine idle speed with the engine fully warmed, running in **Neutral** and compare that to what is listed in the Tune-Up Specification Chart.
6. Now, with the engine running at an idle of about 650-750 rpm, note the vacuum readings on all 4 cylinders. Turn the synchronizing screw (found at the center of the throttle body assembly, between the air intakes for the top 2 and bottom 2 cylinders) in order to adjust the vacuum reading on the bottom 2 cylinders to something within 0.58 psi (4 kPa)/1.18 in. Hg (30mmHg) of the reading on the top 2 cylinders. To say it another way the reading from the bottom 2 cylinders which is furthest away from the highest or lowest reading on the top 2 cylinders should be adjusted so the LARGEST difference is 0.58 psi (4 kPa) or less. Tightening the screw will increase the vacuum reading on the lower cylinders, while loosening the screw will decrease the vacuum on the lower cylinders.

■ A Torx anti-tamper bit (usually a T-25 bit with a concave opening in the center) is normally needed to turn the synchronization screw.

7. Once the adjustment is complete, remove the vacuum gauges and tachometer, then reinstall the vacuum port screws and the air intake silencer assembly.

■ If you need to check/adjust the throttle position sensor, leave the tachometer in place to measure engine rpm.

THROTTLE POSITION SENSOR ADJUSTMENT

Maintenance Adjustments

◆ See Figures 272 and 273

■ On other models, Yamaha suggests that a check of the TPS output voltage should be performed annually, but it is not mentioned in the maintenance specifications for the 150 hp motors. We suspect that even if you decide to check it annually, adjustment will not normally be necessary.

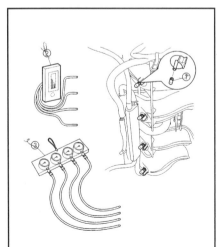

Fig. 270 Connect a set of vacuum gauges or sync tool to the intake manifold ports - Yamaha 150 Hp Models

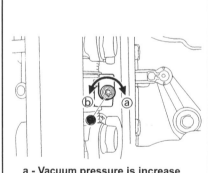

a - Vacuum pressure is increase
b - Vacuum pressure is decrease

Fig. 271 Throttle valve synchronization screw - Yamaha 150 Hp Models

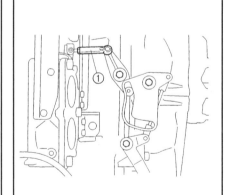

Fig. 272 Before adjusting the TPS you must disconnect the throttle joint link rod so it doesn't hold the throttle valves open - Yamaha 150 Hp Models

MAINTENANCE & TUNE-UP

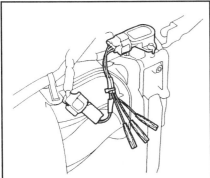

Fig. 273 Use a set of jumper wires which will allow you to take voltage readings without breaking the circuit - Yamaha 150 Hp Models

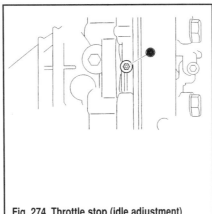

Fig. 274 Throttle stop (idle adjustment) screw - Yamaha 150 Hp Models

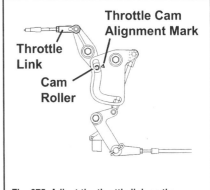

Fig. 275 Adjust the throttle link so the throttle cam roller is centered on the cam alignment mark - Yamaha 150 Hp Models

Should the throttle valve synchronization be performed, it is probably a good idea to at least check the Throttle Position Sensor (TPS) output voltage. This is normally accomplished by checking with the throttle valves fully closed. If the reading is out of specification, the physical positioning of the sensor can be adjusted in order to bring the reading into spec.

However, the sensor does not generate a voltage, so much as the variable resistor within the sensor will change a reference signal from the Engine Control Module (ECM). Therefore, the circuit must be intact in order to test the sensor signal. You've got 2 options. One option is to attempt to back-probe the connector (insert the probes through the rear of the connector while it is still attached to the sensor). However this method risks damaging the connector or the wiring insulation (and could lead to problems with the circuit later). The better method is to disconnect the wiring harness and use 3 jumper wires (one for each terminal) to reconnect the harness. The jumpers must not contact each other (or you risk damage to the ECM from a short), however, some point on the jumper must be exposed so that you can probe the completed circuit using a DVOM. Yamaha makes 3-pin test harness for this application (#YB-06793).

1. Make sure the throttle valves are properly adjusted (check the Throttle Valve Synchronization) before proceeding.
2. Remove the air intake assembly for access.
3. If not already installed, connect a tachometer and provide a source of cooling water to the motor.
4. Disconnect the throttle joint link by carefully snapping the connector free from the ball stud on the upper throttle lever.
5. Visually inspect the throttle valves in the throttle body bores to make sure they are fully closed.
6. Disconnect the wiring from the TPS (found at the top of the intake manifold above the throttle valves) and connect a test harness or jumper wires to complete the circuit.

■ Some OE sources say to test the circuit on 150 hp motors between the Pink and Orange wires of the test harness. We cannot be certain whether or not this is a typo or the sensor wires and test harness wires are not consistent with each other. Note that other sources say that the Orange and Black wires in the engine harness carry the 5 volt reference signal (and ground) for the TPS circuit.

7. Connect a Digital Volt Ohmmeter (DVOM) set to read DC volts to the Pink and Black wires of the jumper/test harness.
8. Turn the ignition switch **ON**, then measure the output voltage with the throttle valves fully closed. Voltage should read 0.68-0.72 volts.
9. Start the motor and recheck the voltage reading.
10. If the sensor voltage is out of range, loosen the 2 mounting screws (these may be of a Torx anti-tamper bit w/ a concave opening in the center design, usually a T-20 in size) and reposition the sensor to obtain the correct voltage readings. Once the sensor is properly repositioned, tighten the screws again to hold it in position and shut down the powerhead.
11. Reconnect the throttle link.
12. Remove the tachometer.

Adjustment When Replacing or Overhauling the Throttle Body

◆ See Figures 272 thru 275

If the throttle body has been overhauled or replaced, a more involved sensor adjustment should take place to ensure proper operation.

1. Install the intake manifold assembly, but keep the air intake silencer disconnected for access and keep the throttle link joint disconnected so the throttle valves remain fully closed.
2. Disconnect the wiring from the TPS and connect a test harness or jumper wires to complete the circuit.
3. Connect a Digital Volt Ohmmeter (DVOM) set to read DC volts to the Pink and Black wires of the jumper/test harness.
4. Turn the ignition switch **ON**
5. Loosen the throttle synchronization screw.
6. Loosen the idle adjustment (throttle stop) screw and fully CLOSE the throttle valve for the upper 2 throttle bodies (Cylinder #1 and #2).
7. Turn the ignition switch ON, but do not start the motor.
8. Adjust the sensor positioning until the DVOM shows that output voltage is 0.68-0.72 volts. Operate the throttle valve several times and record the value.
9. Tighten the synchronization screw SLOWLY and stop when the TPS voltage JUST begins to change.
10. Tighten the throttle stop screw until it JUST touches the throttle body lever.
11. Operate the throttle valve a couple of times, making sure the TP sensor output voltage dos not change, then tighten the TPS mounting screws securely.
12. Connect the throttle joint link rod. If necessary, adjust the length of the rod until the throttle cam roller is aligned with the alignment mark on the throttle cam.
13. Install the air intake silencer and start the powerhead.
14. Measure the output voltage with the throttle valves fully closed, but the engine idling. Voltage should read 0.68-0.72 volts.
15. Check the Throttle Valve Synchronization, as detailed in this section. If necessary, recheck the TPS adjustment after following the synchronization procedure.

THROTTLE CABLE AND LINK ADJUSTMENT

Method I (Without A Feeler Gauge)

◆ See Figures 276 thru 280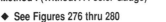

■ Before beginning this adjustment, make sure the Throttle Valve Synchronization is correct. If necessary, refer to the procedure earlier in this section to check and adjust Throttle Valve Synchronization.

1. If not done already for access, remove the air intake silencer assembly.
2. Loosen the locknut and remove the clip retaining the throttle cable joint to the bottom of the throttle lever/cam/link assembly. Disconnect the cable joint from the assembly.

2-88 MAINTENANCE & TUNE-UP

3. Now move to the top of the throttle cam/link assembly and loosen the locknut for the throttle link. Carefully unsnap the throttle link and the link joint each from the ball studs on the upper throttle lever and throttle cam (refer to the accompanying illustrations to help clarify).

4. By hand, turn the throttle lever and the throttle cam both fully counterclockwise (so the throttle lever roller is at the absolute top of the slot in the throttle cam).

5. Make sure the throttle valves (in the bores of the throttle bodies) are fully closed, then adjust the rod joint on the throttle link so the hole in the joint aligns with the ball stud at the top of the throttle lever.

6. Gently snap the joint for the throttle link onto the upper throttle lever and tighten the joint locknut.

7. Connect the link rod (from the lower throttle lever) back over the ball stud on the backside of the throttle cam.

8. Contact the lower throttle lever to the fully closed stopper on the cylinder block, then check to make sure there is clearance between the throttle cam roller (on the throttle lever) and the slot in the throttle cam. If there is no clearance, re-loosen the throttle link locknut, and turn the joint on the throttle link ONE TURN and recheck. Tighten the locknut again if it is adjusted.

9. Manually open and close the lower throttle lever, making sure that the valves open fully and close fully. When the throttle cam is rotated fully CLOCKWISE and the throttle roller on the upper lever has moved to the bottom end of the cam slot (anywhere in the straightened portion of the cam slot), the throttle valves must be FULLY opened.

10. Close the throttle, allowing the cam roller to move to the top end of the cam slot again. Now check that the roller is centered with the alignment mark on the throttle cam AND the lower throttle lever is in contact with the fully closed stopper. With the cam in this position adjust the cable end joint as necessary to connect it to the lower throttle lever.

■ The throttle cable joint must be screwed onto the cable a MINIMUM of 0.31 in. (8.0mm).

11. Reconnect the cable end joint and secure with the retaining clip. Also, be sure to tighten the locknut.

12. Check the throttle cable for smooth operation and, repeat the adjustment as necessary.

Method II (With A Feeler Gauge)

◆ See Figures 276 thru 280

■ Before beginning this adjustment, make sure the Throttle Valve Synchronization is correct. If necessary, refer to the procedure earlier in this section to check and adjust Throttle Valve Synchronization.

1. If not done already for access, remove the air intake silencer assembly.

2. Loosen the locknut and remove the clip retaining the throttle cable joint to the bottom of the throttle cam/link assembly. Disconnect the cable joint from the assembly.

3. By hand, move the LOWER throttle lever so that it contacts the stopper on the block. Make sure there is approximately 0.02 in. (0.5mm) of clearance between the roller on the upper throttle lever and the throttle cam. If not, disconnect the link rod from the ball stud at the top of the UPPER throttle lever and loosen the locknut, then adjust the link rod, as necessary to obtain the required clearance, then reconnect the link rod and tighten the locknut.

4. Next, operate the lower throttle lever so the throttle valves are fully opened, then check the position the throttle roller in the throttle cam. It must have traveled into the straightened portion of the cam slot at the other end of the cam from the fully closed alignment in the previous step.

5. Close the throttle again, allowing the cam roller to move to the top end of the cam slot. Now check that the roller is centered with the alignment mark on the throttle cam AND the lower throttle lever is in contact with the fully closed stopper. With the cam in this position adjust the cable end joint as necessary to connect it to the lower throttle lever.

■ The throttle cable joint must be screwed onto the cable a MINIMUM of 0.31 in. (8.0mm).

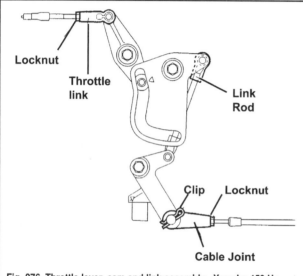

Fig. 276 Throttle lever, cam and link assembly - Yamaha 150 Hp Models

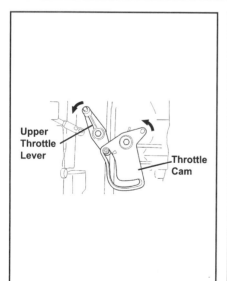

Fig. 277 With the links disconnected, rotate both the upper throttle lever and throttle cam fully counterclockwise

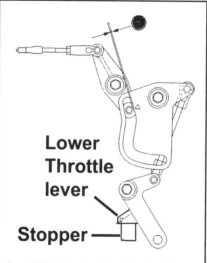

Fig. 278 With the throttles fully closed (lower throttle lever against the stopper) there should be some clearance between the throttle roller and throttle cam

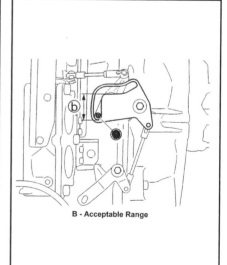

Fig. 279 With the throttle fully open the roller will be somewhere in the range shown

MAINTENANCE & TUNE-UP 2-89

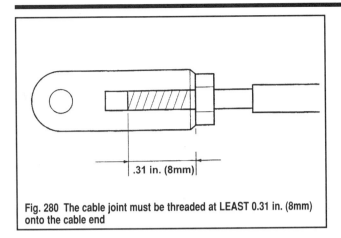

Fig. 280 The cable joint must be threaded at LEAST 0.31 in. (8mm) onto the cable end

6. Reconnect the cable end joint and secure with the retaining clip. Also, be sure to tighten the locknut.
7. Check the throttle cable for smooth operation and, repeat the adjustment as necessary.

SHIFT CABLE ADJUSTMENT

♦ See Figures 281 and 282

Visually check the shift linkage movement in response to the shift handle. If the shifter does not engage properly or smoothly, adjust, as follows:
1. Set the shift control lever to the **Neutral** position. Visually check the slider in the powerhead linkage to see if the pin aligns with the **Neutral** mark (diagonal line) on the shift position switch plate. Also, check the center of the linkage set pin to make sure it aligns with the mark on the bottom cowling. If it does and the shifter is not working properly, suspect a problem with the linkage itself. If the pins and alignment marks are not aligned, adjust as follows:
2. Again, start with the shift lever in the **Neutral** position.
3. Loosen the locknut and remove the clip both securing the shift cable end joint to the slider assembly. Disconnect the cable joint.
4. Manually align the set pin with the **Neutral** mark (diagonal line) on the shift position switch plate.
5. Next, manually align the center of the set pin on the shift lever with the alignment mark on the bottom of the cowl.
6. Adjust the position of the shift cable joint (by turning the joint inward or outward on the cable threads) until its mounting hole aligns with the linkage set pin at the center of the slider bracket. Tighten the locknut to hold it in position on the cable.

✱✱ CAUTION

Make sure that after adjustment the cable joint is threaded over at LEAST 0.31 in. (8mm) of cable threads, otherwise there is a risk that the joint could separate in service causing a severe navigation/control hazard.

7. Position the cable joint over the control lever set pin and secure using the retaining clip.
8. Operate the control lever to confirm proper operation and smooth adjustment.

IGNITION TIMING

♦ See Figures 283 and 284

Like most other 4-strokes, the ignition is controlled by the CDI unit/ECM unit and timing is not adjustable. The only details Yamaha provides concerning checking of the ignition timing is to make sure the motor is fully warmed by running it (using a suitable source of cooling water) for AT LEAST 5 minutes. Also, keep in mind that proper timing on this model is approximately TDC at idle speed. No check or specification is provided for

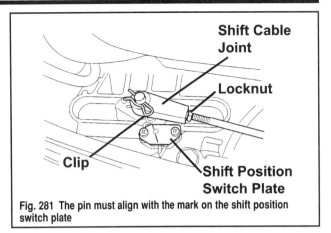

Fig. 281 The pin must align with the mark on the shift position switch plate

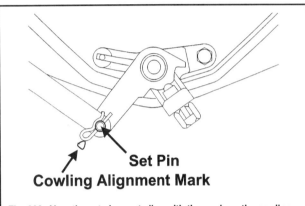

Fig. 282 Also, the set pin must align with the mark on the cowling

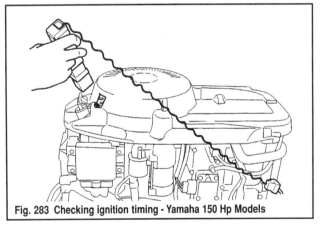

Fig. 283 Checking ignition timing - Yamaha 150 Hp Models

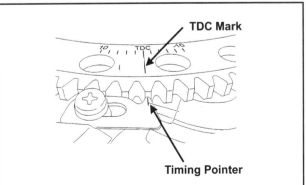

Fig. 284 Timing should be equal to TDC at idle speed - Yamaha 150 Hp Models

2-90 MAINTENANCE & TUNE-UP

WOT, though it is probably safe to assume that timing would ADVANCE somewhat with throttle position.

A timing light can be used to verify proper operation of the ignition timing system. Make sure that if you check timing at a speed other than idle, you also make sure that you run the engine in gear, under load, with a suitable source of cooling water (test tank, dynamometer, or secured to a trailer that is backed into the water). If timing is out of specification, seek a mechanical or electrical reason (i.e. faulty component).

200/225 Hp Models

THROTTLE VALVE SYNCHRONIZATION AND TPS ADJUSTMENT

◆ See Figures 285 thru 290

If the throttle is operating properly there is NO reason to perform this adjustment. Un-necessary tampering usually just leads to performance problems. However, if unsteady idle or performance problems lead you to suspect poor synchronization you can follow this procedure to ensure the linkage is properly set. As a quick check, you can attach a set of vacuum gauges to the vacuum ports and take measurements for each cylinder at idle. Cylinder on a single bank should not vary by more than 0.79 in. Hg (20mmHg) and the difference between the average ratings on the port and starboard banks should not exceed 1.57 in. Hg (40mmHg).

Throttle valve synchronization on fuel injected motor is very similar to carburetor synchronization on carbureted motors. In both cases, throttle valve screws are adjusted in a manner that ensures each of the throttle valves (in the throttle body or carburetor, as applicable) open and close at the same instants. The most precise way to measure this is using a vacuum gauge on the intake manifold to measure the distance the valve is opened at any point.

As part of a throttle valve synchronization you should also check that the Throttle Position Sensor (TPS) is putting out the proper signals to the Engine Control Module (ECM). In this way you are certain that the ECM knows when the throttle valves are fully closed and when (including how much) they are opened. This is checked by measuring the TPS output voltage with the throttle valves closed. If the reading is out of specification, the physical positioning of the sensor can be adjusted in order to bring the reading into spec.

However, the sensor does not generate a voltage, so much as the variable resistor within the sensor will change a reference signal from the Engine Control Module (ECM). Therefore, the circuit must be intact in order to test the sensor signal. You've got 2 options. One option is to attempt to back-probe the connector (insert the probes through the rear of the connector while it is still attached to the sensor). However this method risks damaging the connector or the wiring insulation (and could lead to problems with the circuit later). The better method is to disconnect the wiring harness and use 3 jumper wires (one for each terminal) to reconnect the harness. The jumpers must not contact each other (or you risk damage to the ECM from a short), however, some point on the jumper must be exposed so that you can probe the completed circuit using a DVOM. Yamaha makes a 3-pin test harness for this application (#YB-06793). Mercury also makes a 3-pin test harness for this application (#91-881827).

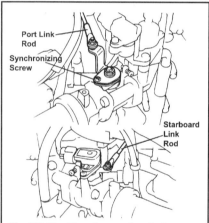

Fig. 285 Throttle valve synchronization on 200/225 hp motors begins with loosening the bank screw and disconnecting the link rods...

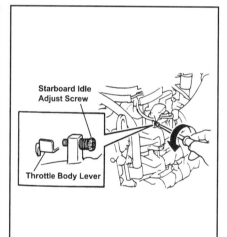

Fig. 286 ... then loosen the starboard idle adjust screw...

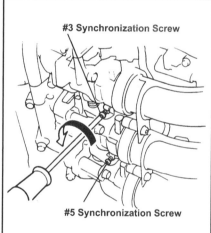

Fig. 287 ... and starboard synchronizing screws (and repeat for port side)

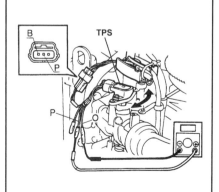

Fig. 288 TPS output voltage is used to measure starboard bank throttle valve movement

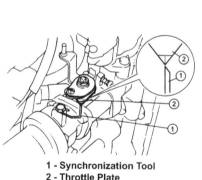

Fig. 289 Port bank movement is gauged using a wire synchronization tool

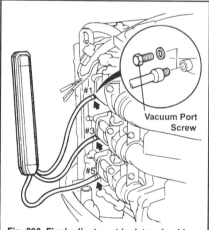

Fig. 290 Final adjustment is determined by measuring vacuum for each cylinder

MAINTENANCE & TUNE-UP 2-91

1. Remove the engine cowling and connect a shop tachometer to the motor (though it is not absolutely needed for synchronization, you'll need it to adjust the Idle Speed afterward).
2. Provide a suitable source of cooling water to the motor (mount it to a test tank, attach a flush fitting or launch the boat).
3. Loosen the bank synchronizing screw.
4. Disconnect the port and starboard link rods.
5. Loosen the starboard idle adjust screw (top) until it separates from the throttle body lever, then loosen the synchronizing screw for cylinder No. 3 (middle on the starboard side) in order to pen the throttle valve for that cylinder.
6. Loosen the starboard idle adjust screw (top) until it separates from the throttle body lever, then loosen the synchronizing screw for cylinder No. 5 (bottom on the starboard side) in order to open the throttle valve for that cylinder.
7. Loosen the port idle adjust screw until it separates from the throttle body lever, then loosen the synchronizing screw for cylinder No. 4 (middle on the port side) in order to open the throttle valves for those cylinders.
8. Loosen the port idle adjust screw until it separates from the throttle body lever, then loosen the synchronizing screw for cylinder No. 6 (bottom on the port side) in order to open the throttle valve for that cylinder.
9. Disconnect the wiring from the TPS and connect a test harness or jumper wires to complete the circuit.

✱✱ WARNING

After testing, make sure the sensor wiring is safely tucked in the same position, to prevent possible damage from moving components.

10. Connect a Digital Volt Ohmmeter (DVOM) set to read DC volts to the Pink and Black wires of the jumper/test harness.
11. Turn the ignition switch **ON**, then measure the output voltage with the throttle valves fully closed. The meter should show 645-665 mil-volts.
12. If the sensor voltage is out of range, loosen the 2 mounting screws and reposition the sensor (turn it slightly one direction or the other) to obtain the correct voltage readings. Once the sensor is properly repositioned, tighten the screws again to hold it in position.
13. Slowly tighten the synchronization screw for cylinder No. 3 (middle, starboard) while watching the sensor output on the DVOM. Stop tightening the screw as soon as output voltage starts to change.
14. Slowly tighten the synchronization screw for cylinder No. 5 (bottom, starboard) while watching the sensor output on the DVOM. Stop tightening the screw as soon as output voltage starts to change.
15. Now, slowly tighten the starboard idle adjust screw until the output voltage is 40 mil-volts higher than the initial adjustment (before tightening the screws for cylinder No. 3 and 5). At this point, the sensor voltage must be 685-705 mil-volts.
16. Attach a synchronization tool to the cylinder No. 2 (top, port) throttle body. The synchronization tool is essentially a wire pointer with a loop on one end so that it can be secured under a bolt head. The pointer is positioned in line with a pointer on the throttle plate and is used to make plate movement more obvious. You can easily form a tool out of a length of mechanic's wire.
17. Slowly tighten the synchronization screw for cylinder No. 4 (middle, port) while watching the throttle plate and synchronization tool. STOP tightening the screw as soon as the throttle plate starts to move.
18. Slowly tighten the synchronization screw for cylinder No. 6 (bottom, port) while watching the throttle plate and synchronization tool. STOP tightening the screw as soon as the throttle plate starts to move.
19. Tighten the port idle adjust screw (top) until it contacts the throttle body lever, then give it one additional turn inward.
20. Start and warm the motor to normal operating temperature. Make sure the motor is run for AT LEAST 5 minutes.
21. Remove the vacuum port screws from the intake manifold. Connect multiple vacuum gauges or a calibrated carburetor balance stick (a tool that normally contains as many as 4 separate tubes that measure vacuum).

■ If you've got 6 gauges, or 2 balance sticks, that's great, go ahead and hook them all up to the intake using adapters. However if you've only got 3 gauges or one balance stick, then check one bank at a time, starting with the starboard side (cylinders No. 1, 3 and 5).

22. Measure and record the vacuum readings of the starboard cylinders No. 1, 3 and 5 (top to bottom). There should be no more than a 0.79 in. Hg (20mmHg) difference in readings between each cylinder. If there is, you'll have to use the DVOM and go back to the point in the procedure where you adjusted the No. 3 (and then No. 5) synchronization screws.

23. Measure the vacuum of the port cylinders No. 2, 4 and 6 (top to bottom). Again, there should be no more than a 0.79 in. Hg (20mmHg) difference in readings between each cylinder. If there is, you'll have to use the DVOM and go back to the point in the procedure where you adjusted the No. 3 (and then No. 5) synchronization screws.

■ Yes, the manufacturer recommends that you go back and readjust the starboard side, even if only the port side is out of adjustment. We think that is odd too, and you MAY get acceptable results if you simply go back to the point when you adjust the No. 4 and 6 synchronization screws.

24. Now, calculate the average reading for each bank port and starboard (add the three readings together from the cylinders for each bank and divide by three). Compare the average of the port cylinders to the average of the starboard cylinders. The difference in average readings should be no more than 1.57 in. Hg (40mmHg) or further adjustment is necessary.
25. Recheck the TPS output voltage when the engine is running at idle speed. If necessary, adjust the throttle position sensor again to make sure it remains in the 685-705 mil-volt range (695-705 is ideal) with the motor idling at 700 rpm or slightly faster.
26. Turn the ignition switch **OFF**, then remove the DVOM and test harness. Reconnect the TPS wiring.
27. Reconnect the port and starboard link rods.
28. Tighten the bank synchronizing screw, making sure that you DO NOT tilt the plate.
29. Check to make sure there is no clearance between the idle adjust screws (on both port and starboard sides) and the throttle body levers that they are supposed to contact. If there is any clearance loosen and retighten the bank synchronization screw.
30. Restart the engine and recheck the average vacuum differences between the port and starboard cylinder banks.
31. When you are finished, shut the powerhead down and go have a beer, because you deserve it!
32. Keep the source of cooling water handy for Idle Speed adjustment, which can be made after the Throttle Link & Cable Adjustment is finished.

THROTTLE LINK & CABLE

Yamaha Models

Checking Link and Cable Operation
◆ See Figure 291

At least annually, be sure to check the feel and operation of the throttle cable, making sure it is smooth and free of sticking or binding. With the remote control set in the **Neutral/Idle** position, visually check the throttle control cam to make sure it is fully closed. The mark on the throttle cam should be positioned between the two **Neutral/Idle** alignment marks on the throttle control lever, as shown in the accompanying illustration. If so, no adjustment is necessary. However, if the marks do not align, adjust the length of the throttle cable and the position of the control lever using the throttle link.

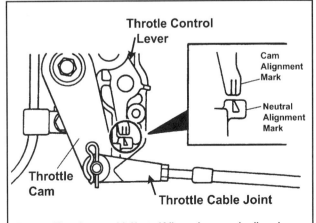

Fig. 291 Throttle cam with Neutral/Idle marks properly aligned on Yamaha models

2-92 MAINTENANCE & TUNE-UP

There are two ways to adjust the throttle cable and link adjustment. One method uses a stop bolt (a generic metric, 10mm bolt with a certain thread diameter 6mm and thread pitch 1mm) to hold the throttle control lever in position, keeping the marks aligned. The other method (if a bolt is not handy) simply requires that hold the throttle cam/lever aligned by hand. The stop bolt method is probably marginally easier.

■ The Throttle Valve Synchronization & TPS Adjustment procedure must be followed (or you must be sure that they are in spec) before adjusting the Throttle Link & Cable. Otherwise adjustment may be incorrect.

Adjustment

◆ See Figure 291 thru 294

1. Loosen the locknut on the throttle cable joint, then remove the retaining clip and pull it from the set pin.
2. Loosen the locknut, then disconnect the link rod from the magneto control lever (just below the flywheel).
3. If you're using stop bolt, proceed as follows:
   a. Obtain a 10mm bolt with 1mm thread pitch and 6mm thread diameter to use as a stop bolt. Thread the bolt into the bore provided next to the throttle linkage, just above the alignment marks adjacent to the throttle control lever.
   b. Push the link ball end of the magneto control lever toward the flywheel in order to eliminate any free-play, then adjust the link rod so it aligns with the magneto control lever ball stud. Reconnect the link rod and tighten the locknut.
   c. Remove the stop bolt and double-check that the alignment mark on the throttle control cam is between the marks on the throttle control lever. If not, repeat the adjustment procedure up to this point.
4. If you don't have a 10mm bolt handy, proceed as follows:
   a. Adjust the link rod so that the throttle cam alignment mark is centered between the marks on the throttle control lever.
   b. Keeping the link in this position, push the magneto control lever toward the flywheel to eliminate any free-play and install the link rod over the lever. Tighten the locknut.
   c. Double-check that the alignment mark on the throttle control cam is between the marks on the throttle control lever. If not, repeat the adjustment procedure up to this point.
5. Manually position the throttle cam against the idle stopper and check for a minimum clearance of 0.04 in. (1.0mm) between the cam and roller. Hold the throttle cam in this position, then adjust the throttle cable joint inward or outward on the cable threads until it aligns with the set pin. Tighten the locknut to hold the joint in this position.

✱✱ CAUTION

Make sure that after adjustment the cable joint is threaded over at LEAST 0.31 in. (8mm) of cable threads, otherwise there is a risk that the joint could separate in service causing a severe navigation/control hazard.

6. Install the cable joint back over the set pin and secure using the retaining clip.
7. Open and close the throttle a couple of times using the remote while visually checking to be sure the throttle valves open and close smoothly and to the full extent of their travel. If necessary, repeat the adjustment procedure.

Mercury/Mariner Models

◆ See Figures 295, 296 and 297

At least annually, be sure to check the feel and operation of the throttle cable, making sure it is smooth and free of sticking or binding. With the remote control set in the **Neutral/Idle** position, visually check the throttle control cam to make sure it is fully closed. The mark on the throttle cam should be aligned with the center of the roller, as shown in the accompanying illustration. If so, no adjustment is necessary. However, if the marks do not align, adjust the length of the throttle cable and the position of the control lever using the throttle link.

■ The Throttle Valve Synchronization & TPS Adjustment procedure must be followed (or you must be sure that they are in spec) before adjusting the Throttle Link & Cable. Otherwise adjustment may be incorrect.

1. Loosen the locknut, then disconnect the link rod from the magneto control lever (just below the flywheel).
2. Push the magneto control lever toward the flywheel to eliminate any freeplay.
3. With the remote control lever in the Idle/Neutral position, check the throttle cam lever to see if the alignment mark is centered with the roller. If not, follow the throttle cable back to the barrel retaining mechanism, release the cable and barrel from the mechanism. Now adjust the barrel until the alignment mark on the cam lever is centered on the roller, then secure the cable/barrel in the mechanism.
4. With the marks properly aligned, position the magneto control lever so that there is a minimum clearance of 0.04 in. (1.0mm) between the cam and roller. Hold the magneto control lever in this position, then loosen the locknut on the link rod and adjust the link rod to align with the ball joint on the magneto control rod. Tighten the locknut to hold the joint in this position.
5. Gently snap the link rod over the magneto control lever and confirm that the minimum amount of clearance has been maintained.
6. Open and close the throttle a couple of times using the remote while visually checking to be sure the throttle valves open and close smoothly and to the full extent of their travel. Double-check the roller alignment. If necessary, repeat the adjustment procedure.

SHIFT CABLE ADJUSTMENT

Yamaha Models

◆ See Figure 298

Visually check the shift linkage movement in response to the shift handle. If the shifter does not engage properly or smoothly, adjust, as follows:

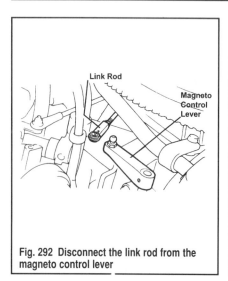

Fig. 292 Disconnect the link rod from the magneto control lever

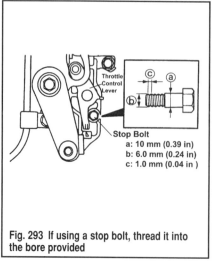

Fig. 293 If using a stop bolt, thread it into the bore provided

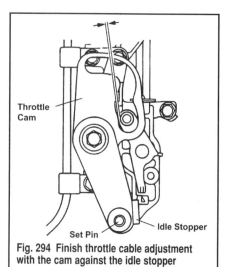

Fig. 294 Finish throttle cable adjustment with the cam against the idle stopper

MAINTENANCE & TUNE-UP 2-93

1. Set the remote control lever to the **Neutral** position. Visually check the powerhead linkage set pin to see if it aligns with the **Neutral** marks at the center of the slider bracket. If it does and the shifter is not working properly, suspect a problem with the linkage itself. If the set pin is not aligned with the mark, adjust it as follows:
2. Remove the retaining clip and disconnect the shift cable joint from the powerhead linkage.
3. Manually align the set pin with the **Neutral** mark on the slider bracket.
4. Make sure the remote control lever to the **Neutral** position.
5. Loosen the locknut, then adjust the position of the shift cable joint (by turning the joint inward or outward on the cable threads) until its mounting hole aligns with the linkage set pin at the center of the bracket. Tighten the locknut to hold it in position on the cable.

✶✶ CAUTION

Make sure that after adjustment the cable joint is threaded over at LEAST 0.31 in. (8mm) of cable threads, otherwise there is a risk that the joint could separate in service causing a severe navigation/control hazard.

6. Position the cable joint over the control lever set pin and secure using the retaining clip.
7. Operate the control lever to confirm proper operation and smooth adjustment.

Mercury/Mariner Models

◆ See Figure 299

No periodic adjustment should be necessary for the Shift cable, however, should the cable be replaced they will require an initial set-up to place them in proper alignment with the linkage.

The shifter cable should be set so the shift pin (and alignment mark at its base) is aligned with the center mark (or marks) on the shift slider bracket when both the gearcase and the remote control shifter are in Neutral. If adjustment is necessary, free the cable and adjust the cable barrel to properly position the slider/shift pin.

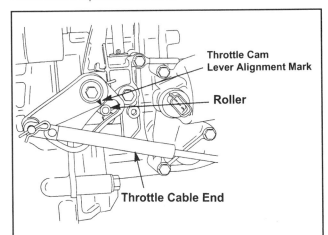

Fig. 295 The throttle link and cable is properly adjusted when the cam lever mark centers on the roller - Mercury/Mariner models

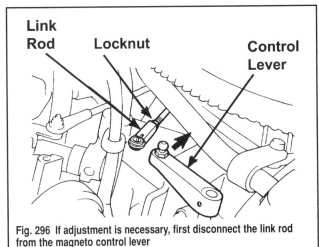

Fig. 296 If adjustment is necessary, first disconnect the link rod from the magneto control lever

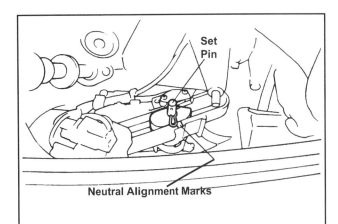

Fig. 298 The shift cable is properly adjusted when the set pin aligns with the Neutral marks (and the remote is also in Neutral)

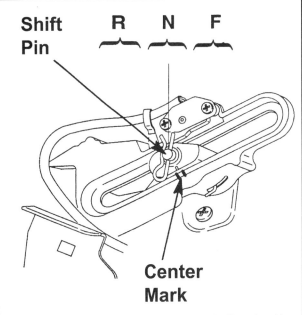

Fig. 299 Shift pin-to-slider bracket alignment in the Neutral position

Fig. 297 The link itself is properly adjusted when there is a minimum 0.040 in. (1mm) gap between the aligned mark and roller has

2-94 MAINTENANCE & TUNE-UP

IDLE SPEED

◆ See Figures 291, 292, 295 and 300

1. Be sure to check/set the Throttle Valve Synchronization and TPS Sensor Adjustment, before proceeding.
2. Mount the engine in a test tank or move the boat to a body of water.
3. Remove the cowling and the ignition coil cover, then connect a tachometer to the powerhead.
4. Start the engine and allow it to warm to operating temperature (meaning run it for AT LEAST 5 minutes).
5. Check engine speed at idle. The powerhead should idle at the rpm specified in the Tune-up Specifications chart.
6. Place the engine in gear and check engine trolling speed in the same manner.
7. If adjustment is necessary, shut the powerhead down.
8. Disconnect the link rod from the magneto control lever.
9. Loosen the PORT idle adjust screw (on top of the throttle body assembly) until it separates from the throttle body lever.
10. Restart and run the engine, making sure it is fully warmed.
11. Turn the STARBOARD idle adjust screw (found just below the Throttle Position Sensor) to obtain the desired idle speed. Turning the screw INWARD will INCREASE idle speed, while turning the screw OUTWARD will DECREASE idle speed.
12. Once the engine is idling in the proper range, slowly turn the PORT idle adjust screw back inward until it JUST touches the throttle body lever.
13. If an adjustment was made you'll have to recheck and/or adjust the Throttle Link & Cable, as detailed earlier in this section.

IGNITION TIMING

Ignition timing on these powerheads is adjusted (advanced/retarded) automatically by the Engine Control Module to match rpm operating conditions. Because there are no adjustments for timing, any problems or irregularities should be diagnosed and repaired as the result of defective

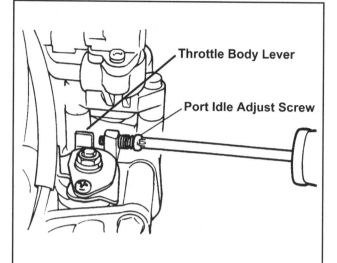

Fig. 300 The Port (shown) and Starboard idle adjust screws control the throttle valve linkage (and therefore the idle speed) - 200/225 hp motors

mechanical or electrical components. No attempt should be made to perform adjustments in order to achieve the proper timing specifications, let the electronics do their job, but give them the tools that they need.

However, most motors are equipped with a timing pointer and timing marks on the flywheel. If desired, you can use a timing light to confirm that the ECM is properly controlling timing. To do this, make sure the motor is either mounted in a test tank or the boat is launched as the motor must operate under load. Always check the ignition timing with the motor fully warmed to normal operating temperature. For details on timing specifications, please refer to the Tune-Up Specifications chart in this section. If the timing is out of specification, double-check your procedures and test conditions, then suspect a fault with the ignition/fuel injection/engine management system.

VALVE CLEARANCE

VALVE LASH

◆ See Figure 301

Valve lash is the amount of freeplay present in the valve train. On many of these motors it is measured as the clearance between the rocker arm and the valve stem, though it is measured between the camshaft and lifter or tappet itself on some larger motors). This distance is critical, as too little clearance can hold the valve open or keep it from fully contacting the valve seat, preventing it from cooling by transmitting heat to the cylinder head through contact. This would lead to a burnt valve, requiring overhaul and replacement. Conversely, too much valve clearance might keep the valve from opening fully, preventing the engine from making maximum horsepower. Of course poor driveability is arguably better than burning a valve. There's an old mechanic's saying that applies here, "a tappy valve, is a happy valve." That's not to say you necessarily WANT your valves a little too loose and tapping, but whenever you're in doubt, leave a valve a **little** too loose rather than a little too tight.

Adjusting valve lash is a part of normal maintenance to keep the valve clearance within a specified range accounting for the normal wear and tear of valve train components. Valve clearance is always measured when the rocker or tappet for the valve being measured is fully released, meaning that it is in a position where it contacts the base of the camshaft and not any part of the raised camshaft lobe. For this reason it is necessary to find when the piston for the valve being measured is at TDC of the compression stroke. Remember that TDC of the compression stroke is the point during the 4-stroke engine cycle when the piston travels upward while both valves close in order to seal the combustion chamber and compress the air/fuel mixture.

Although timing marks can normally be used to determine TDC, always double-check the timing marks by watching the valves for the cylinder on which you are working as the timing mark on the flywheel or camshaft

approaches the mark on the engine. During a normal cycle, the exhaust valve and then the intake valve will close as the piston begins its travel upward on the compression stroke. If instead the exhaust valve opens the motor is on the exhaust stroke and is one full turn of the crankshaft away from the start of the compression stroke. Both valves must be closed, and remain closed as the piston comes to the top of the cylinder and the timing marks align (if you're looking at the No. 1 piston). If so, the piston is on TDC and the valves for that cylinder may be adjusted.

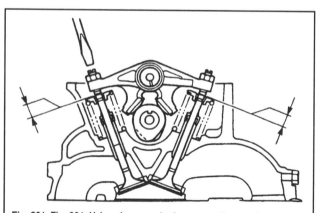

Fig. 301 Fig. 301 Valve clearance is the gap or distance between the top of the valve stem and the bottom of the rocker arm adjuster (or the bottom of the rocker arm and the base of the camshaft lobe, depending upon the model) when the valve in question is fully closed

MAINTENANCE & TUNE-UP

ADJUSTMENT

2.5 Hp Yamaha Models

◆ See Figure 302

A set of flat feeler gauges is necessary to check valve clearance on these motors.

Valve specifications are for an overnight cold engine. It is best to check and/or adjust the valves with the powerhead at approximately 68° F (20° C).

1. Remove the engine shut-off cord from the engine shut-off switch on the tiller handle.
2. Loosen the 2 bolts and remove the carrying handle from the engine.
3. Remove the bolts securing the fuel cock lever, then remove the lever.
4. Loosen the 4 bolts (to threaded upward from underneath and 2 threaded toward the powerhead on the inside), then separate the bottom cowling from the powerhead (on the valve cover side of the motor).
5. Disconnect the spark plug wiring and ground the lead for safety.
6. Remove the spark plug to relieve engine compression.
7. Remove the 2 fuel tank bolts and reposition the tank, as necessary for better access.

■ You've got to love Yamaha motors. When necessary most models have both torque sequences and torque values for important components embossed right on the part casting. When a single number appears next to each of a component's bolts, loosen the bolts using multiple passes in the reverse of the sequence and install/tighten using multiple passes following the sequence. Keep an eye out for torque values, as they are often embossed right on the component as well.

8. Remove the cylinder head/valve cover for access to the valve train. When removing the valve cover, slowly loosen all the bolts around the perimeter of the cover, using at least 2 passes of a crossing sequence to make sure it cannot warp and become damaged. For more details, please refer to the Powerhead overhaul section.
9. Carefully pull the valve cover from the cylinder head. Inspect the gasket for damage.
10. Using the starter rope, slowly and carefully rotate the flywheel (CLOCKWISE, in the normal direction of rotation) until the piston is at TDC of the compression stroke. When rotating the crankshaft, watch the valves for the cylinder and make sure they close to verify that you are at TDC.
11. Measure the clearance of the intake and exhaust valves for the cylinder at TDC. Insert feeler gauges of various sizes between the rocker arm and the valve stem for both valves being checked at this point. The size gauge that passes between the arm and stem with a slight drag indicates the valve clearance. Compare the clearance measured with the specifications of 0.0031-0.0047 in. (0.08-0.12mm) for both the intake and exhaust valve.
12. If adjustment is necessary, proceed as follows:
    a. Loosen the rocker arm locknut using a proper sized wrench, then turn the rocker arm adjuster until the clearance is correct. Turning the adjuster INWARD will DECREASE clearance, while turning the adjuster OUTWARD will INCREASE clearance.
    b. Hold the adjuster from turning and tighten the locknut to 7.4 ft. lbs. (10 Nm).
    c. Recheck the clearance to ensure the adjuster didn't move when you tightened the locknut.
13. Although it may not be absolutely necessary we highly recommend installing a new valve/cylinder head cover gasket/O-ring to ensure a proper seal. You don't want an oil leak do you?
14. Install the valve/cylinder head cover and tighten the bolts using a crossing pattern or following the embossed torque sequence using at least 2 passes. First tighten the bolts to 3.7 ft. lbs. (5.0 Nm), and then to 8.9 ft. lbs. (12 Nm).
15. Reposition and secure the fuel tank.
16. Install the spark plug and reconnect the lead.
17. Install and secure the bottom cowling.
18. Install the fuel cock lever and the carrying handle.
19. Provide a water source, then start the engine and check for oil leaks at the rocker arm cover mating surfaces.

4 Hp Yamaha Models

◆ See Figures 303 and 304

A set of flat feeler gauges is necessary to check valve clearance on these motors.

Valve specifications are for an overnight cold engine. It is best to check and/or adjust the valves with the powerhead at approximately 68° F (20° C).

1. Remove the engine top cover (cowling) for access.
2. Remove the carrying handle bolt, protector and handle assembly from the engine.
3. Loosen the retainers and then remove the bottom cowling from the powerhead for access.
4. Unbolt and remove the manual starter housing from the top of the powerhead.
5. Disconnect the spark plug wiring and ground the lead for safety.
6. Remove the spark plug to relieve engine compression.

■ You've got to love Yamaha motors. When necessary most models have both torque sequences and torque values for important components embossed right on the part casting. When a single number appears next to each of a component's bolts, loosen the bolts using multiple passes in the reverse of the sequence and install/tighten using multiple passes following the sequence. Keep an eye out for torque values, as they are often embossed right on the component as well.

7. Remove the cylinder head/valve cover for access to the valve train. When removing the valve cover, slowly loosen all the bolts around the perimeter of the cover, using at least 2 passes of a crossing sequence to make sure it cannot warp and become damaged. For more details, please refer to the Powerhead overhaul section.
8. Carefully pull the valve cover from the cylinder head. Inspect the gasket for damage.
9. Slowly and carefully rotate the flywheel (CLOCKWISE, in the normal direction of rotation) until the piston is at TDC of the compression stroke. You can do this by aligning the inner edge of the left projection on the igniter (when viewed from the top) with the rotor edge of the flywheel. When rotating the crankshaft, watch the valves for the cylinder and make sure they close to verify that you are at TDC.
10. Measure the clearance of the intake and exhaust valves for the cylinder at TDC. Insert feeler gauges of various sizes between the rocker arm and the valve stem for both valves being checked at this point. The size

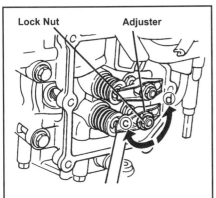

Fig. 302 Valve adjustment - 2.5 hp Yamaha Motors

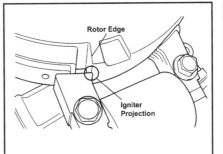

Fig. 303 The piston should be at TDC when the projection on the igniter aligns with the rotor edge of the flywheel - 4 hp Yamaha Motors

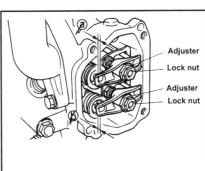

Fig. 304 Valve adjustment - 4 hp Yamaha Motors

2-96 MAINTENANCE & TUNE-UP

gauge that passes between the arm and stem with a slight drag indicates the valve clearance. Compare the clearance measured with the specifications of 0.0032-0.0048 in. (0.08-0.12mm) for both the intake and exhaust valve.

11. If adjustment is necessary, proceed as follows:

  a. Loosen the rocker arm locknut using a proper sized wrench and using another, larger wrench to hold the adjuster in place (otherwise the locknut will not loosen). Then turn the rocker arm adjuster until the clearance is correct. Turning the adjuster INWARD will DECREASE clearance, while turning the adjuster OUTWARD will INCREASE clearance.

  b. Hold the adjuster from turning and tighten the locknut to 7.2 ft. lbs. (10 Nm).

  c. Recheck the clearance to ensure the adjuster didn't move when you tightened the locknut.

12. Although it may not be absolutely necessary we highly recommend installing a new valve/cylinder head cover gasket/O-ring to ensure a proper seal. You don't want an oil leak do you?

13. Install the valve/cylinder head cover and tighten the bolts using a crossing pattern or following the embossed torque sequence using at least 2 passes. Tighten the bolts to 8 ft. lbs. (11 Nm).

14. Install the spark plug and reconnect the lead.

15. Install the manual starter assembly.

16. Install and secure the bottom cowling.

17. Install and secure the carrying handle/protector.

18. Provide a water source, then start the engine and check for oil leaks at the rocker arm cover mating surfaces.

19. Shut the powerhead down and install the engine top cover.

4/5/6 Hp Mercury/Mariner Models

◆ See Figures 305 and 306

A set of flat feeler gauges is necessary to check valve clearance on these motors.

Valve specifications are for an overnight cold engine. It is best to check and/or adjust the valves with the powerhead at approximately 68° F (20° C).

1. Disconnect the spark plug wiring and ground the lead for safety.
2. Remove the spark plug to relieve engine compression.
3. Remove the 4 cylinder head/valve cover for access to the valve train. When removing the valve cover, slowly loosen all the bolts around the perimeter of the cover, using at least 2 passes of a crossing sequence to make sure it cannot warp and become damaged.
4. Carefully pull the valve cover from the cylinder head. Inspect the gasket for damage.
5. Remove the fuel pump.
6. Using the starter rope, slowly and carefully rotate the flywheel (CLOCKWISE, in the normal direction of rotation) until the piston is at TDC of the compression stroke. When rotating the crankshaft, watch the valves for the cylinder and make sure they close to verify that you are at TDC.

■ There are multiple ways to know you are a TDC on these motors. Look through the fuel pump hole and watch for raised mark to be visible on the camshaft. Also, make sure the notch marked on the flywheel aligns with the TDC mark on the block casting (it's normally the larger mark to the far left of the other 5 and 25 degree BTDC marks when facing the powerhead). And, of course, both the intake and exhaust valves will be closed.

7. Measure the clearance of the intake (top) and exhaust (bottom) valves for the cylinder at TDC. Insert feeler gauges of various sizes between the rocker arm and the valve stem for both valves being checked at this point. The size gauge that passes between the arm and stem with a slight drag indicates the valve clearance. Compare the clearance measured with the specifications of 0.002-0.005 (0.06-0.14) for the intake valve and 0.004-0.007 (0.11-0.19) for the exhaust valve.

8. If adjustment is necessary, proceed as follows:

  a. Hold the pivot while loosening the rocker arm locknut using a proper sized wrench, then turn the rocker arm pivot (adjuster) until the clearance is correct. Turning the adjuster INWARD will DECREASE clearance, while turning the adjuster OUTWARD will INCREASE clearance.

  b. Hold the adjuster from turning and tighten the locknut to 90 inch lbs./7.4 ft. lbs. (10 Nm).

  c. Recheck the clearance to ensure the adjuster didn't move when you tightened the locknut.

9. Although it may not be absolutely necessary we highly recommend installing a new valve/cylinder head cover gasket to ensure a proper seal. You don't want an oil leak do you?

10. Install the valve/cylinder head cover and tighten the bolts to 70 inch lbs. (8 Nm) using at least 2 passes of a crossing pattern.

11. Reposition and secure the fuel pump. Tighten the 2 retaining screws securely.

12. Install the spark plug and reconnect the lead.

13. Provide a water source, then start the engine and check for oil leaks at the rocker arm cover mating surfaces.

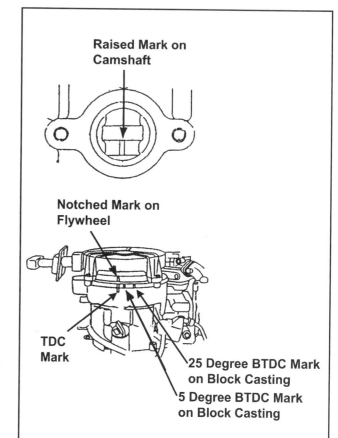

Fig. 305 The piston should be a TDC when the raised portion of the camshaft is visible through the fuel pump mounting bore and the flywheel notch aligns with the TDC cast mark - 4/5/6 Hp Mercury/Mariner Models

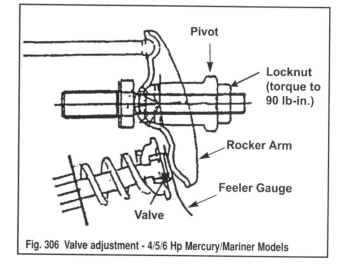

Fig. 306 Valve adjustment - 4/5/6 Hp Mercury/Mariner Models

MAINTENANCE & TUNE-UP 2-97

6/8 Hp Yamaha Models

◆ See Figures 307, 308 and 309

A set of flat feeler gauges is necessary to check valve clearance on these motors.

Valve specifications are for an overnight cold engine. It is best to check and/or adjust the valves with the powerhead at approximately 68° F (20° C).

On these motors, the valve clearance is measured between the tappet portion of the rocker arm and the base of the camshaft lobe, while adjustment occurs with a threaded adjuster placed on the opposite end of the rocker arm. Piston positioning can be determined using the embossed marks on the camshaft pulley. An embossed **1** represents No. 1 (top) cylinder TDC, while an embossed **2** represents No. 2 (bottom) cylinder TDC. In both cases, the cylinder is at TDC when the appropriate mark is aligned with the mark on the thermostat housing (the mark on the pulley is facing the crankshaft).

1. Remove the engine top cover (cowling) for access.
2. Unbolt and remove the manual starter housing from the top of the powerhead.
3. Remove the dust cover from the top of the powerhead, over the camshaft pulley.
4. Remove the fuel pump assembly. For details, please refer to the Fuel System section.
5. Tag and disconnect the spark plug wiring, then ground the leads for safety.
6. Remove the spark plugs to relieve engine compression.

■ You've got to love Yamaha motors. When necessary most models have both torque sequences and torque values for important components embossed right on the part casting. When a single number appears next to each of a component's bolts, loosen the bolts using multiple passes in the reverse of the sequence and install/tighten using multiple passes following the sequence. Keep an eye out for torque values, as they are often embossed right on the component as well.

7. Remove the cylinder head/valve cover for access to the valve train. When removing the valve cover, slowly loosen all the bolts around the perimeter of the cover, using at least 2 passes of a crossing sequence to make sure it cannot warp and become damaged. For more details, please refer to the Powerhead overhaul section.
8. Carefully pull the valve cover from the cylinder head. Inspect the gasket for damage.
9. Rotate the flywheel (CLOCKWISE, in the normal direction of rotation) until the embossed **1** on the camshaft gear aligns with the mark on the thermostat (and is therefore, facing the crankshaft). When rotating the crankshaft, watch the valves for the No. 1 (top) cylinder and see that they close to verify that you are at No. 1 cylinder TDC.

10. Measure the clearance of the intake and exhaust valves for the cylinder at TDC. Insert feeler gauges of various sizes between the rocker arm and base of the camshaft lobe for both valves being checked at this point. The size gauge that passes between the arm and stem with a slight drag indicates the valve clearance. Compare the clearance measured with the specifications of 0.006-0.008 in. (0.15-0.20mm) for the intake valve or 0.008-0.010 in. (0.20-0.25mm) for the exhaust valve. On this motor the intake is the top valve for each cylinder so it goes Intake, Exhaust, Intake, Exhaust, from top-to-bottom.
11. If adjustment is necessary, proceed as follows:
    a. Loosen the locknut using a proper sized wrench, then turn the adjuster screw inward or outward until the clearance is correct.
    b. Hold the screw from turning and tighten the locknut to 5.9 ft. lbs. (8 Nm).
    c. Recheck the clearance to ensure the adjuster didn't move when you tightened the locknut.
12. Rotate the flywheel clockwise one full revolution or 360 degrees in order to turn the camshaft pulley one half of a revolution or 180 degrees. The embossed No. 2 on the camshaft sprocket should now face the thermostat mark (will face toward the crankshaft) and the No. 2 valves should both be closed verifying that the No. 2 cylinder is at TDC. Adjust the No. 2 cylinder valves in the same manner as the No. 1 valves.
13. Although it may not be absolutely necessary we highly recommend installing a new valve/cylinder head cover gasket/O-ring to ensure a proper seal. You don't want an oil leak do you?
14. Install the valve/cylinder head cover and tighten the bolts securely using a crossing pattern or following the embossed torque sequence using at least 2 passes.
15. Install the fuel pump, as detailed in the Fuel System section.
16. Install the spark plugs and reconnect the leads, as tagged during removal.
17. Install the dust cover over the camshaft pulley.
18. Install the manual starter assembly.
19. Provide a water source, then start the engine and check for oil leaks at the rocker arm cover mating surfaces.
20. Shut the powerhead down and install the engine top cover.

8/9.9 Hp (232cc) Models

◆ See Figures 310 thru 314

A set of flat feeler gauges and the Yamaha Valve Adjusting Tool (#YM-08035 or Mercury #91-809498A1) are necessary to check valve clearance on these motors. The Adjusting Tool is used to rotate the valve lash adjuster on each rocker arm.

Valve specifications are for an overnight cold engine. It is best to check and/or adjust the valves with the powerhead at approximately 68° F (20° C).

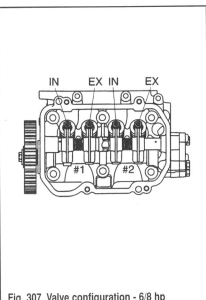

Fig. 307 Valve configuration - 6/8 hp Yamaha Motors

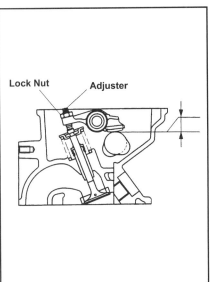

Fig. 308 Valve adjustment - 6/8 hp Yamaha Motors

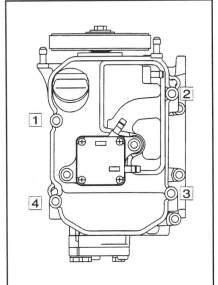

Fig. 309 Valve cover torque sequence - 6/8 hp Yamaha Motors

2-98 MAINTENANCE & TUNE-UP

1. Remove the flywheel cover. On Yamahas this is accomplished by removing the screw toward the center of the cover and pulling the hinge pin out of the cover end, then lifting the cover from the powerhead.

2. For Yamahas, from underneath the cowling (just underneath the cylinder head) remove the bolt(s) and the cowling clamp lever.

3. For safety, tag and disconnect the spark plug leads, then ground them to the powerhead.

4. Remove the spark plugs to relieve engine compression.

5. Tag and disconnect the hoses from the cylinder head/valve cover.

■ You've got to love Yamaha motors. When necessary most models have both torque sequences and torque values for important components embossed right on the part casting (The same is usually true for the Yamaha powerheads which are painted black and sold to Mercury). When a single number appears next to each of a component's bolts, loosen the bolts using multiple passes in the reverse of the sequence and install/tighten using multiple passes following the sequence. Keep an eye out for torque values, as they are often embossed right on the component as well.

6. Remove the cylinder head/valve cover for access to the valve train. When removing the valve cover, slowly loosen all the bolts around the perimeter of the cover, using multiple passes of a crossing or more often on these particular powerheads a rotating sequence to make sure it cannot warp and become damaged. For more details, please refer to the Powerhead overhaul section.

7. Carefully pull the cylinder head/valve cover from the cylinder head. Inspect the gasket for damage.

8. Rotate the flywheel (CLOCKWISE, in the normal direction of rotation) until the embossed **1** on the camshaft gear aligns with the mark on the powerhead AND with the bolt hole for the plastic flywheel cowling or cylinder head cast mark (as applicable). When rotating the crankshaft, watch the valves for the No. 1 (top) cylinder and see that they close to verify that you are at No. 1 cylinder TDC.

9. Measure the clearance of the intake and exhaust valves for the cylinder at TDC. Insert feeler gauges of various sizes between the rocker arm and the valve stem for both valves being checked at this point. The size gauge that passes between the arm and stem with a slight drag indicates the valve clearance. Compare the clearance measured with the specifications of 0.006-0.008 in. (0.15-0.20mm) for the intake valve or 0.008-0.010 in. (0.20-0.25mm) for the exhaust valve. To determine which valve is an intake and which is an exhaust, observe the position of the valves in relation to the other components on the powerhead. The intake valves are adjacent to the ports for the intake manifold attached to the powerhead and the exhaust valves are adjacent to the exhaust ports. For the 8/9.9 hp motor the exhaust valves are on the port side while the intake valves are on the starboard side.

10. If adjustment is necessary, proceed as follows:

   a. Loosen the locknut using a proper sized wrench (normally 9mm), then turn the adjuster screw (using #YM-08035 or #91-809498A1) until the clearance is correct.

Fig. 310 Remove the cylinder head cover...

Fig. 311 ...then turn the flywheel clockwise until the motor is at No. 1 TDC

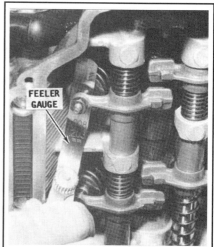
Fig. 312 Use a feeler gauge to check each of the valves

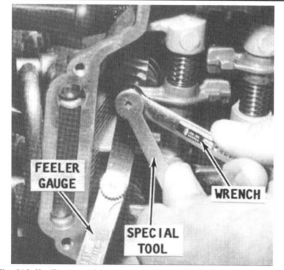
Fig. 313 If adjustment is necessary, loosen the locknut and rotate the adjuster

Fig. 314 During installation, be sure to tighten the cover bolts using the proper torque sequence (and to specification if one is molded on the cover)

MAINTENANCE & TUNE-UP

b. Hold the screw from turning and tighten the locknut to 70 inch lbs./5.8 ft. lbs. (8 Nm).

c. Recheck the clearance to ensure the adjuster didn't move when you tightened the locknut.

11. Rotate the flywheel clockwise one full revolution or 360 degrees in order to turn the camshaft pulley one half of a revolution or 180 degrees. The embossed No. 2 on the camshaft sprocket should now face the mark on the powerhead and the No. 2 valves should both be closed indicating that the No. 2 cylinder is at TDC. Adjust the No. 2 cylinder valves in the same manner as the No. 1 valves.

12. Apply a light coating of engine oil to the valve stem and adjustment bolts.

13. Although it may not be absolutely necessary we highly recommend installing a new valve/cylinder head cover gasket/O-ring to ensure a proper seal. You don't want an oil leak do you?

14. Install the valve/cylinder head cover and tighten the bolts using a crossing pattern or following the embossed torque sequence (usually a clockwise sequence starting at the upper left bolt on these models).

15. Reconnect the hoses to the cylinder head/valve cover, as tagged during removal.

16. If applicable, install the clamp lever and secure using the bolt(s).

17. Install the flywheel cover. On Yamahas and as equipped, use the bolt and hinge pin to secure it.

18. Install the spark plugs and leads, as tagged during removal.

19. Provide a water source, then start the engine and check for oil leaks at the rocker arm cover mating surfaces.

20. Shut down the powerhead and install the upper engine cover.

9.9/15 Hp (323cc) and 25 Hp 4-Stroke Models

◆ See Figures 315 and 316

A set of flat feeler gauges is necessary to check valve clearance on these motors.

Valve specifications are for an overnight cold engine. It is best to check and/or adjust the valves with the powerhead at approximately 68° F (20° C).

1. Remove the engine top cover (cowling) for access.
2. Remove the flywheel cover or manual starter cover, as applicable.
3. For 9.9/15 hp motors, remove the camshaft sprocket dust cover.
4. For safety, tag and disconnect the spark plug leads, then ground them to the powerhead.
5. Remove the spark plugs to relieve engine compression.
6. Tag and disconnect the fuel hoses from the pump at the cylinder head/valve cover.

■ **You've got to love Yamaha powerheads. When necessary most models have both torque sequences and torque values for important components embossed right on the part casting. When a single number appears next to each of a component's bolts, loosen the bolts using multiple passes in the reverse of the sequence and install/tighten using multiple passes following the sequence. Keep an eye out for torque values, as they are often embossed right on the component as well.**

7. Remove the cylinder head/valve cover for access to the valve train. When removing the valve cover, slowly loosen all the bolts around the perimeter of the cover, using multiple passes of a crossing sequence or the reverse of the embossed torque sequence to make sure it cannot warp and become damaged. For more details, please refer to the Powerhead overhaul section.

8. Carefully pull the cylinder head/valve cover from the cylinder head. Inspect the gasket for damage.

9. Rotate the flywheel (CLOCKWISE, in the normal direction of rotation) until the embossed **1** or the triangular mark on the camshaft gear aligns with the mark on the powerhead. When rotating the crankshaft, watch the valves for the No. 1 (top) cylinder and see that they close to verify that you are at No. 1 cylinder TDC.

10. Measure the clearance of the intake and exhaust valves for the cylinder at TDC. Insert feeler gauges of various sizes between the rocker arm and the valve stem for both valves being checked at this point. The size gauge that passes between the arm and stem with a slight drag indicates the valve clearance. Compare the clearance measured with the specifications of 0.006-0.010 in. (0.15-0.25mm) for the intake valve or 0.008-0.012 in. (0.20-0.30mm) for the exhaust valve on 9.9/15 hp motors and 0.010-0.014 in. (0.25-0.35mm) for 25 hp motors.

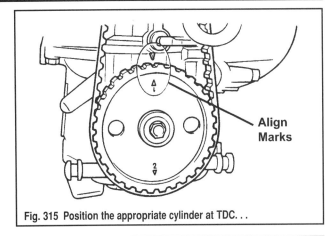

Fig. 315 Position the appropriate cylinder at TDC...

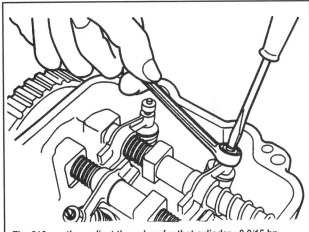

Fig. 316 ...then adjust the valves for that cylinder - 9.9/15 hp (323cc) Motors Shown

■ **To determine which valve is an intake and which is an exhaust, observe the position of the valves in relation to the other components on the powerhead. The intake valves are adjacent to the ports for the intake manifold attached to the powerhead and the exhaust valves are adjacent to the exhaust ports. For both the 9.9/15 hp and 25 hp motors the exhaust valves are on the port side while the intake valves are on the starboard side.**

11. If adjustment is necessary, proceed as follows:

a. Loosen the locknut using a proper sized wrench (while holding the adjuster screw).

b. Turn the adjusting screw inward or outward, as necessary to adjust clearance.

c. Hold the adjuster screw from turning and tighten the locknut to 124 inch lbs./10 ft. lbs. (14 Nm).

d. Recheck the clearance to ensure the adjuster didn't move when you tightened the locknut.

12. Rotate the flywheel clockwise one full revolution or 360 degrees in order to turn the camshaft pulley one half of a revolution or 180 degrees. The embossed No. 2 and/or the triangular mark (9.9/15 hp motors) or circular mark (25 hp motors) on the camshaft sprocket should now face the mark on the powerhead. More importantly, the No. 2 valves should both be closed indicating the No. 2 cylinder is at TDC. Adjust the No. 2 cylinder valves in the same manner as the No. 1 valves.

13. Although it may not be absolutely necessary we highly recommend installing a new valve/cylinder head cover gasket/O-ring to ensure a proper seal. You don't want an oil leak do you?

14. Install the valve/cylinder head cover and tighten the bolts using a crossing pattern or following the embossed torque sequence. Tighten the bolts in 2 passes to 71 inch lbs./5.8 ft. lbs. (8 Nm) for 9.9/15 hp motors AND for Mercury/Mariner 25 hp models, or to 4.4 ft. lbs. (6.0 Nm) on the first pass and 8.7 ft. lbs. (11.8 Nm) on the final pass for Yamaha 25 hp motors.

15. Reconnect the fuel hoses as tagged during removal.

16. Install the spark plugs and leads, as tagged during removal.

MAINTENANCE & TUNE-UP

17. For 9.9/15 hp motors, install the camshaft sprocket dust cover.
18. Install the flywheel cover or manual starter assembly, as applicable.
19. Provide a water source, then start the engine and check for oil leaks at the rocker arm cover mating surfaces.
20. Shut down the powerhead and install the upper engine cover.

30/40 Hp 3-Cylinder Models

◆ See Figures 317 thru 320

A set of flat feeler gauges is necessary to check valve clearance on these motors.

Valve specifications are for an overnight cold engine. It is best to check and/or adjust the valves with the powerhead at approximately 68° F (20° C).

1. Remove the engine top cover (cowling) for access.
2. On EFI motors, properly relieve the EFI fuel system pressure since the high-pressure line will need to be removed from the fuel distribution manifold later in this procedure. For details, please refer to the Fuel System section.
3. Remove the flywheel cover or manual starter cover, as applicable.
4. For safety, tag and disconnect the spark plug leads, then ground them to the powerhead.
5. Remove the spark plugs to relieve engine compression.
6. For carbureted motors, tag and disconnect the fuel hoses and blowby hose from the cylinder head/valve cover.
7. For EFI motors, proceed as follows:
   a. Tag and disconnect the breather hose, vent hose, fuel pump input/output hoses AND the fuel cooler input/output hoses.
   b. If necessary, unbolt and remove the fuel pump from the valve cover.
   c. With the high-pressure fuel system relieved of internal pressure, depress the locking tab and carefully disconnect the high-pressure fuel line from the fuel distribution manifold assembly.
   d. Remove the 3 coil plate fasteners, then swing the coil plate assembly to the side for access.
   e. Loosen the nut securing the water separating fuel filter assembly to the mounting bracket, then carefully reposition the assembly out of the way.

■ You've got to love Yamaha powerheads. When necessary most models have both torque sequences and torque values for important components embossed right on the part casting. When a single number appears next to each of a component's bolts, loosen the bolts using multiple passes in the reverse of the sequence and install/tighten using multiple passes following the sequence. Keep an eye out for torque values, as they are often embossed right on the component as well.

8. Remove the cylinder head/valve cover for access to the valve train. When removing the valve cover, slowly loosen all the bolts around the perimeter of the cover, using multiple passes of a crossing sequence (or the reverse of the embossed torque sequence, if present) to make sure it cannot warp and become damaged. For more details, please refer to the Powerhead overhaul section.
9. Carefully pull the cylinder head/valve cover from the cylinder head. Inspect the gasket/seal for damage.
10. Rotate the flywheel (CLOCKWISE, in the normal direction of rotation) until the embossed **1** and/or triangular mark on the camshaft gear aligns with the mark on the powerhead. When rotating the crankshaft, watch the valves for the No. 1 (top) cylinder and see that they close to verify that you are at No. 1 cylinder TDC.

11. Measure the clearance of the intake and exhaust valves for the cylinder at TDC. Insert feeler gauges of various sizes between the rocker arm and the valve stem for both valves being checked at this point. The size gauge that passes between the arm and stem with a slight drag indicates the valve clearance. Compare the clearance measured with the specifications of 0.006-0.010 in. (0.15-0.25mm) for the intake valve or 0.010-0.014 in. (0.25-0.35mm) for the exhaust valve.

■ To determine which valve is an intake and which is an exhaust, observe the position of the valves in relation to the other components on the powerhead. The intake valves are adjacent to the ports for the intake manifold attached to the powerhead and the exhaust valves are adjacent to the exhaust ports. For the 30/40 hp motors the exhaust valves are on the port side while the intake valves are on the starboard side.

12. If adjustment is necessary, proceed as follows:
    a. Loosen the locknut using a proper sized wrench (while holding the adjuster screw using a flathead screwdriver).
    b. Turn the adjusting screw inward or outward, as necessary to adjust clearance.
    c. Hold the adjuster screw from turning and tighten the locknut to 124 inch lbs./10 ft. lbs. (14 Nm).
    d. Recheck the clearance to ensure the adjuster didn't move when you tightened the locknut.

■ OK, here is a dilemma. Mercury and Yamaha manuals differ on this next step. You see Yamaha claims that when the camshaft rotates to the next timing mark, you'll bring No. 3 up to TDC (and then No. 2 on the last mark). However, Mercury claims that the firing order is 1-2-3 and that rotating the camshaft 120 degrees from No. 1 TDC will bring up No. 2 TDC. (and then No. 3 on the last mark). It is unlikely that the 2 manufacturers decided to use different camshafts AND crankshafts to achieve this, so one is probably an error. But the truth is, all you have to do is observe the intake and exhaust valves as you rotate the crankshaft/camshaft and adjust whichever cylinder's valves are BOTH closed (the exhaust will close first, then the intake) as the piston approaches TDC (and the timing marks align).

13. Slowly Rotate the flywheel clockwise in order to turn the camshaft pulley 1/3 a turn (120 degrees). The next embossed mark (usually a triangle) on the camshaft sprocket should now face the mark on the powerhead and the No. 3 valves or No. 2 valves should both be closed indicating the No. 3 (bottom) or No. 2 (middle) cylinder is at TDC. Adjust these valves in the same manner as the No. 1 valves.
14. Slowly Rotate the flywheel clockwise in order to turn the camshaft pulley an additional 1/3 turn (120 degrees). The final embossed mark on the camshaft sprocket should now face the mark on the powerhead and the final set of valves (either No. 2 or No. 3 depending upon the results of the previous step) should both be closed indicating the cylinder that is at TDC. Adjust the final cylinder valves in the same manner as the No. 1 valves.
15. Although it may not be absolutely necessary we highly recommend installing a new valve/cylinder head cover gasket/O-ring to ensure a proper seal. You don't want an oil leak do you?
16. Apply a light coating of marine grade grease to the valve/cylinder head cover seal.
17. Install the valve/cylinder head cover and tighten the bolts securely using a crossing pattern or following the embossed torque sequence. Tighten the bolts to 70-75 inch lbs. (8.0-8.5 Nm).

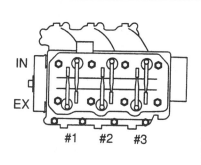

Fig. 317 Valve configuration - 30/40 Hp 3-Cylinder Motors

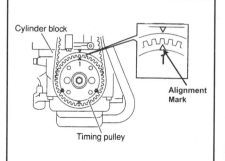

Fig. 318 Align the marks to TDC for the cylinder you are adjusting. . .

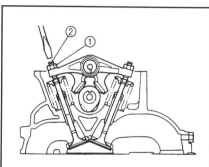

Fig. 319 . . .then check/adjust the valves for that cylinder

MAINTENANCE & TUNE-UP 2-101

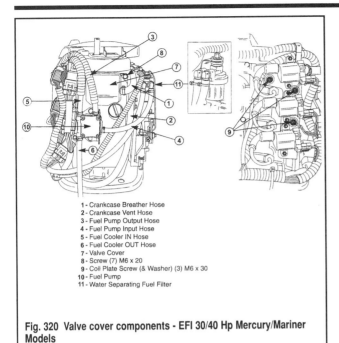

1 - Crankcase Breather Hose
2 - Crankcase Vent Hose
3 - Fuel Pump Output Hose
4 - Fuel Pump Input Hose
5 - Fuel Cooler IN Hose
6 - Fuel Cooler OUT Hose
7 - Valve Cover
8 - Screw (7) M6 x 20
9 - Coil Plate Screw (& Washer) (3) M6 x 30
10 - Fuel Pump
11 - Water Separating Fuel Filter

Fig. 320 Valve cover components - EFI 30/40 Hp Mercury/Mariner Models

18. For EFI motors, proceed as follows:
  a. Reposition the water separating fuel filter to the bracket and secure with the retaining nut.
  b. Reposition the coil plate and secure it using the screws and washers. Tighten the screws to 75 inch lbs. (8.5 Nm).
  c. If removed, install the fuel pump.
  d. As tagged, reconnect the fuel cooler inlet/outlet hoses, the fuel pump inlet/outlet hoses, as well as the crankcase breather and vent hoses. Secure all hoses with sta-straps (essentially wire ties).
  e. Reconnect the high-pressure fuel line to the fuel distribution manifold, making sure the locking tab engages.
19. For carbureted motors, reconnect the fuel and blowby hoses as tagged during removal.
20. Install the spark plugs and leads, as tagged during removal.
21. Install the flywheel cover or manual starter assembly, as applicable.
22. For EFI motors, properly pressurize the high-pressure fuel system and inspect thoroughly for leaks.
23. Provide a water source, then start the engine and check for oil leaks at the rocker arm cover mating surfaces.
24. Shut down the powerhead and install the upper engine cover.

40/45/50 Hp (935cc) and 40/50/60 Hp (996cc) Models

◆ See Figures 321 thru 324

A set of flat feeler gauges is necessary to check valve clearance on these motors.

Valve specifications are for an overnight cold engine. It is best to check and/or adjust the valves with the powerhead at approximately 68° F (20° C).

The valves for this motor are adjusted with the camshaft in 2 positions, half of the valves are adjusted with the engine at No. 1 TDC and the other half may be adjusted with the engine at No. 4 TDC.

■ On MOST models the camshaft pulley is thankfully embossed with one TDC mark per cylinder, so finding TDC on the cylinder for which you are working is relatively easy. With the flywheel cover removed, rotate the crankshaft in the normal direction of rotation until the embossed number on the camshaft pulley for the cylinder you want at TDC aligns with the mark on the bottom of the cylinder head split line. At TDC of the compression stroke for any given cylinder, the base of the camshaft lobe will be touching the rocker arms (the raised portion of the lobe will face away from the rockers). However, on some models there are only 2 marks, one for the No. 1 cylinder and the other, 180 degrees out from there, for the No. 4 cylinder. Details in the procedure will help you with either situation.

1. On EFI motors, properly relieve the EFI fuel system pressure since the high-pressure line will need to be removed from the fuel distribution manifold later in this procedure. For details, please refer to the Fuel System section.
2. Disconnect the negative battery cable for safety.
3. Tag and disconnect the spark plug wires, then remove the spark plugs in order to ease engine compression. For details, please refer to the Spark Plug procedure in this section.

✱✱ WARNING

Ground the spark plug leads to prevent damage if the engine is cranked for any reason while they are still disconnected.

4. Remove the flywheel cover.

■ On some carbureted models, the Dash-Pot is mounted to the cylinder head cover and should be removed along with the cover. DO NOT remove the Dash-Pot from the cover.

5. For carbureted models, proceed as follows:
  a. Tag and disconnect the fuel hoses from the fuel pump. Use a rag to catch any escaping fuel and make sure no-one squeezes the primer bulb while they are disconnected. It's not a bad idea to cap or plug the hoses to prevent the possibility of any fuel line contamination, especially if you might not reconnect everything right away.

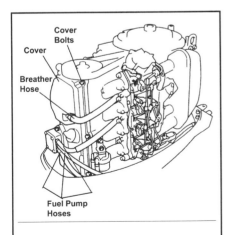

Fig. 321 To remove the rocker arm cover tag and disconnect the hoses, then remove the bolts

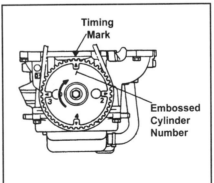

Fig. 322 Rotate the flywheel so the No. 1 or No. 4 cylinder mark on the camshaft pulley aligns with the timing mark on the powerhead

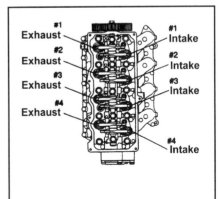

Fig. 323 Intake and exhaust valve identification - 40/45/50 Hp (935cc) and 40/50/60 Hp (996cc) Motors

2-102 MAINTENANCE & TUNE-UP

■ In some later publications Yamaha recommends removing the pump from the cylinder head cover instead of just disconnect the hoses (same is true for some earlier Mercury publications). However, it doesn't appear that there is a mechanical change on which this change in procedure is based. If that is the case, there should be no harm in leaving the pump attached (which saves you a couple of steps). However, if you're going to be completely cautious you can always remove the pump just to be sure. It's up to you.

   b. Disconnect the breather hose from the cylinder head cover.
   c. For Mercury/Mariner 40/50/60 hp carbureted (996cc) models, remove the throttle lever assembly from the starboard side of the cylinder head.

6. For EFI motors, proceed as follows:
   a. Tag and disconnect the breather hose, vent hose, fuel pump input/output hoses AND the fuel cooler input/output hoses.
   b. If necessary, unbolt and remove the fuel pump from the valve cover.
   c. With the high-pressure fuel system relieved of internal pressure, depress the locking tab and carefully disconnect the high-pressure fuel line from the fuel distribution manifold assembly.
   d. Remove the 3 coil plate fasteners, then swing the coil plate assembly to the side for access.
   e. Loosen the nut securing the water separating fuel filter assembly to the mounting bracket, then carefully reposition the assembly out of the way.

■ You've got to love Yamaha powerheads. When necessary most models have both torque sequences and torque values for important components embossed right on the part casting. When a single number appears next to each of a component's bolts, loosen the bolts using multiple passes in the reverse of the sequence and install/tighten using multiple passes following the sequence. Keep an eye out for torque values, as they are often embossed right on the component as well.

7. Remove the cylinder head/valve cover for access to the valve train. When removing the valve cover, slowly loosen all the bolts around the perimeter of the cover, using multiple passes of a crossing sequence (in the reverse of the embossed torque sequence, if present) to make sure it cannot warp and become damaged. For more details, please refer to the Powerhead overhaul section.

8. Carefully pull the cylinder head/valve cover from the cylinder head. Inspect the gasket for damage.

9. Rotate the flywheel (CLOCKWISE, in the normal direction of rotation) until the embossed **1** and/or the triangular mark on the camshaft gear aligns with the triangular mark at the cylinder head-to-block split line. When rotating the crankshaft, watch the valves for the No. 1 (top) cylinder. They should close to verify that you are at No. 1 cylinder TDC.

■ With the flywheel in the No. 1 TDC position, the raised portion of the camshaft lobes should face away from and not be in contact with the rockers. If the position is correct, both the valves for the No. 1 cylinder will be closed.

10. Measure the clearance of the intake and exhaust valves for the No. 1 cylinder, also measure the No. 2 cylinder intake valve and the No. 3 cylinder exhaust valve. Insert feeler gauges of various sizes between the rocker arm and the valve stem for each valve. The size that passes between the arm and stem with a slight drag indicates the valve clearance. Compare the clearance measured with the specifications of 0.006-0.010 in. (0.15-0.25mm) for the intake valve or 0.010-0.014 in. (0.25-0.35mm) for the exhaust valve.

■ To determine which valve is an intake and which is an exhaust, observe the position of the valves in relation to the other components on the powerhead. The intake valves are adjacent to the ports for the intake manifold attached to the powerhead and the exhaust valves are adjacent to the exhaust ports. For these motors (like with most Yamaha/Mercury 4-strokes) the exhaust valves are on the port side while the intake valves are on the starboard side.

11. If lash is out of specification on one or more valves, adjust it as follows:
   a. Loosen the locknut, then turn the adjusting screw using a suitable flat-head screwdriver until the clearance is correct.
   b. Hold the screw to keep it from turning while tightening the locknut to 120 inch lbs./10 ft. lbs. (13.5 Nm).
   c. Recheck the valve clearance to make sure the adjuster screw wasn't turned while tightening the locknut.

12. Rotate the flywheel clockwise one full revolution (360 degrees) so the No. 4 TDC mark (sometimes this mark is just a dot 180 degrees away from the No. 1 TDC mark on the sprocket) aligns with the mark on the powerhead (and the No. 1 cylinder is now on its exhaust stroke). The camshaft rotates at 1/2 the rate of the crankshaft/flywheel, so rotating the flywheel as directed will turn the camshaft sprocket only 180 degrees. This places the No. 1 TDC mark exactly 1/2 a turn away from the previous location. At this point, both of the valves for the No. 4 (bottom) cylinder should be closed.

13. Measure the clearance of the No. 4 cylinder intake and exhaust valves, the No 2 exhaust valve and the No. 3 intake valve. Insert feeler gauges of various sizes between the rocker arm and the valve stem for each valve. The size that passes between the arm and stem with a slight drag indicates the valve clearance. Compare the clearance measured with the specifications of 0.006-0.010 in. (0.15-0.25mm) for the intake valve or 0.010-0.014 in. (0.25-0.35mm) for the exhaust valve.

14. Install the cylinder head/valve cover using a new O-ring that has been lubricated with a light coating of an all purpose Marine Grease, then tighten the bolts until snug using a crossing pattern (or following the torque sequence if it is embossed on the cover). No torque specification was provided in the Yamaha literature, however there may be one embossed on the cover as well, so check for that. Mercury service literature gives a torque value of 70 inch lbs. (8 Nm). Either way, make sure the bolts are secure, but not so tight as to risk breaking them or damaging the cover.

15. For EFI motors, proceed as follows:
   a. Reposition the water separating fuel filter to the bracket and secure with the retaining nut.
   b. Reposition the coil plate and secure it using the screws and washers. Tighten the screws securely.
   c. If removed, install the fuel pump.
   d. As tagged, reconnect the fuel cooler inlet/outlet hoses, the fuel pump inlet/outlet hoses, as well as the crankcase breather and vent hoses. Secure all hoses with sta-straps (essentially wire ties).
   e. Reconnect the high-pressure fuel line to the fuel distribution manifold, making sure the locking tab engages.

16. For carbureted motors, proceed as follows:
   a. For Mercury/Mariner 40/50/60 hp carbureted (996cc) models, install the throttle lever assembly to the starboard side of the cylinder head. Double-check that the linkage does not require adjustment. For details, please refer to Timing and Synchronization in this section.
   b. Reconnect the breather hose to the rocker cover.
   c. Reconnect the fuel pump hoses, as tagged during removal (or reinstall the fuel pump assembly, if it was removed from the cover).

17. Install the flywheel cover.
18. Install the spark plugs and connect the leads as tagged during removal.
19. Connect the negative battery cable.
20. For EFI motors, properly pressurize the high-pressure fuel system and inspect thoroughly for leaks.
21. Connect a flushing device and start the engine to check for oil leaks at the rocker arm cover mating surfaces. Also, take a good look at the fuel line fittings to make sure you didn't cause any leaks which could lead to an explosive situation.
22. Install the engine top case.

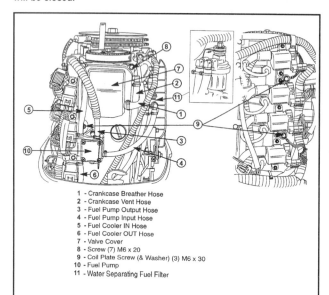

1 - Crankcase Breather Hose
2 - Crankcase Vent Hose
3 - Fuel Pump Output Hose
4 - Fuel Pump Input Hose
5 - Fuel Cooler IN Hose
6 - Fuel Cooler OUT Hose
7 - Valve Cover
8 - Screw (7) M6 x 20
9 - Coil Plate Screw (& Washer) (3) M6 x 30
10 - Fuel Pump
11 - Water Separating Fuel Filter

Fig. 324 Valve cover components - EFI 40/50/60 Hp Mercury/Mariner Models

MAINTENANCE & TUNE-UP 2-103

75/80/90/100 Hp and 115 Hp Models

◆ See Figures 325 thru 333  DIFFICULT

A set of flat feeler gauges is necessary to check valve clearance on these motors.

Valve specifications are for an overnight cold engine. It is best to check and/or adjust the valves with the powerhead at approximately 68° F (20° C).

Like with most 4-stroke motors, clearance measurement is a simple matter of aligning the crankshaft in a certain position and inserting a feeler gauge under the base of the crankshaft lobe on correct valves. On these motors, the valves are checked with the camshaft in 2 positions, half of the valves are checked with the engine at No. 1 TDC and the other half may be adjusted with the engine at No. 4 TDC.

Valve clearance on these motors is controlled by the thickness of the shim installed in the lifter that connects the camshaft lobe to the valve stem. If too great a clearance exists, you'll have to remove the old shim and install a thicker one. If too little clearance exists, remove the old shim and install a thinner one. Never use 2 shims in the same lifter, as the top could become dislodged and damage the camshaft and/or lifter.

Neither Yamaha nor Mercury mention any provision for changing shims with the camshaft still installed. So, although it is not uncommon in many 4-stroke motor designs to utilize a valve compressing tool to depress the lifter and slide the shim out without removing the camshaft, we cannot recommend the use of such a tool here. Either Yamaha could not leave sufficient surface area or tabs on the top of the lifters for such a tool to work, or they are concerned about possible scoring/damage to the crankshaft. Therefore, if one or more valves requires adjustment the timing belt and camshaft(s) will have to be removed for access.

Furthermore, Yamaha does not provide any information regarding the significance of the timing marks on the camshaft pulleys. It appears both from our various test engines AND from information published by Mercury that the No. 1 cylinder will be at TDC when the marks on both pulleys are aligned with and directly facing each other. When true, the No. 4 cylinder should then be at TDC when the marks are facing directly AWAY from each other.

However, there is no 100% guarantee that all models will be similar, so your best bet is simply to watch the camshafts as you slowly rotate the flywheel. The No. 1 cylinder will be at TDC when the exhaust AND intake valves are both closed. You will be able to tell because the top lobe on each camshaft will be facing away from the tappet.

You can also use a pencil inserted and held into the cylinder through the spark plug hole to watch piston travel. When the pencil is pushed outward to the furthest point and both the valves for that cylinder are closed, the cylinder is at TDC. Once you have found TDC for the No. 1 cylinder, TDC for the No. 4 cylinder is exactly one turn of the flywheel away. Again, you can watch the valves to be certain of your positioning.

The bottom line of valve adjustment is that it is REALLY not important that the motor be in a particular position except that in order to properly measure valve clearance the valve must be in the completely released position. So on this motor it means you can measure ANY valve, as long as the camshaft lobe is facing AWAY from the valve lifter/shim assembly. Positioning the motor at No. 1 and No 4. TDC is just a convenient way to do this for all the valves.

1. Disconnect the negative battery cable for safety.
2. Loosen the retaining bolts, then carefully pull the flywheel cover from the rubber grommets and/or retaining pins, as applicable.
3. Loosen and remove the Phillips head (Yamaha) or Hex head (Mercury) screws (there are usually 5 of them) securing the spark plug cover to the powerhead, then remove the cover for access.
4. Tag and disconnect the spark plug wires, then remove the spark plugs in order to ease engine compression. For details, please refer to the Spark Plug procedure in this section.

✳✳ WARNING

Ground the spark plug leads to prevent damage if the engine is cranked for any reason while they are still disconnected. Better yet, disconnect the negative battery cable and hide the keys to the ignition so no-one will accidentally turn it on.

Fig. 325 Loosen the retaining bolts...

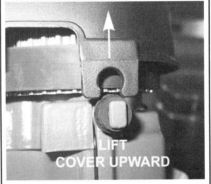

Fig. 326 ...and pull the flywheel cover upward off the rubber grommets...

Fig. 327 ...then gently pull it off the retaining pin(s)

Fig. 328 Loosen the screws and remove the spark plug cover...

Fig. 329 ...and tag/disconnect hoses from the valve cover

Fig. 330 When looking for TDC, remember the camshaft sprockets make 1/2 of at turn for each revolution of the flywheel

2-104 MAINTENANCE & TUNE-UP

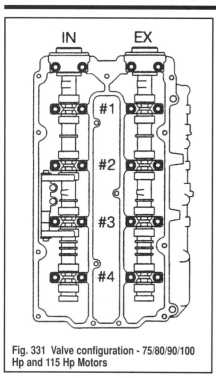

Fig. 331 Valve configuration - 75/80/90/100 Hp and 115 Hp Motors

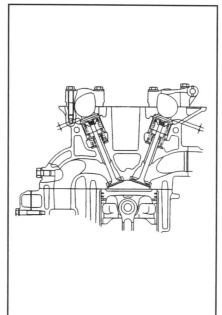

Fig. 332 Measuring valve clearance on a shim/tappet valve train

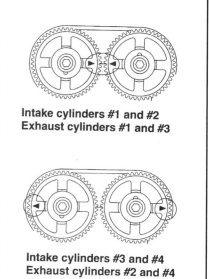

Fig. 333 Camshaft sprocket timing marks, relative to engine position - Mercury/Mariner shown (Yamaha SHOULD be the same)

5. Tag and disconnect the fuel hoses from the fuel pump (EFI or some Mercury Carb Models) or fuel pumps (most Yamaha carb models) mounted to the valve cover. Use a rag to catch any escaping fuel and make sure no-one squeezes the primer bulb while they are disconnected. It's not a bad idea to cap or plug the hoses to prevent the possibility of any fuel line contamination, especially if you might not reconnect everything right away.

■ In some later publications Yamaha recommends removing the pump from the cylinder head cover instead of just disconnect the hoses. However, it doesn't appear that there is a mechanical change on which this change in procedure is based (and Mercury still seems to recommend leaving the pump or pumps in place). So, there should be no harm in leaving the pump(s) attached (which saves you a couple of steps). However, if you're going to be completely cautious remove the pump on 2001 and later Yamaha models so that you're completely in compliance with Yamaha's recommendation for your motor. It's up to you.

6. For EFI motors, tag and disconnect any ventilation and vacuum hoses which will interfere with cylinder head cover removal.
7. Remove the cylinder head/valve cover for access to the valve train. When removing the valve cover, slowly loosen all 14 bolts around the perimeter (and through the spark plug valley) of the cover, using multiple passes of a crossing sequence to make sure it cannot warp and become damaged. For more details, please refer to the Powerhead overhaul section.
8. Carefully pull the cylinder head/valve cover from the cylinder head. Inspect the gasket for damage.
9. Rotate the flywheel (CLOCKWISE, in the normal direction of rotation) until No. 1 cylinder is at TDC (measure the height of the piston using a dial gauge or a small tool such as a pencil and watch the valves for the No. 1/top cylinder). When rotating the crankshaft, once the valves for the No. 1 cylinder both close, you're at or close to TDC. At No. 1 TDC both marks on the camshaft sprockets should normally be pointing directly inward, facing towards each other.

■ With the flywheel in the No. 1 TDC position, the raised portion of the camshaft lobes should face away from and not be in contact with the valve lifters/shims.

10. Measure and note the clearance of the intake and exhaust valves for the No. 1 cylinder, also measure the No. 2 cylinder intake valve and the No. 3 cylinder exhaust valve. Insert feeler gauges of various sizes between the base of the camshaft lobe and the valve lifter/shim for each valve. The size that passes between the camshaft and shim with a slight drag indicates the valve clearance. Compare the clearance measured with the specifications of 0.007-0.009 in. (0.17-0.23mm) for the intake valve or 0.012-0.015 in. (0.31-0.37mm) for the exhaust valve. Noting the exact clearance measurement is important if the valve is out of spec, since you'll need that measurement later, when deciding what size replacement shim to install.

■ To determine which valve is an intake and which is an exhaust, observe the position of the valves in relation to the other components on the powerhead. The intake valves are adjacent to the ports for the intake manifold attached to the powerhead (on the same side as the carbs or the fuel injectors). The exhaust valves are adjacent to the exhaust ports. For these motors (unlike most smaller Yamaha/Mercury 4-strokes) the intake valves are on the port side while the exhaust valves are on the starboard side.

11. Rotate the flywheel clockwise one full revolution (360 degrees) so the No. 4 cylinder is at TDC (meaning the valves for No. 4 now close and the piston is at the top of its travel). The camshaft rotates at 1/2 the rate of the crankshaft/flywheel, so rotating the flywheel as directed will turn the camshaft sprockets only 180 degrees (placing the two marks in the opposite position from where they were at No. 1 TDC, which should mean they are pointing straight out to the port and starboard sides of the motor.
12. Measure and note the clearance of the No. 4 cylinder intake and exhaust valves, the No 2 exhaust valve and the No. 3 intake valve. Insert feeler gauges of various sizes between the base of the camshaft lobe and the valve lifter/shim for each valve. The size that passes between the camshaft and shim with a slight drag indicates the valve clearance. Compare the clearance measured with the specifications of 0.007-0.009 in. (0.17-0.23mm) for the intake valve or 0.012-0.015 in. (0.31-0.37mm) for the exhaust valve.
13. If lash is within spec, you're in luck, no adjustment is necessary so you can button her back up. However, if the lash is out of spec for one or more valves, adjust it as follows:
   a. Loosen the timing belt tensioner, then remove the timing belt and camshaft sprocket for the camshaft whose valve(s) require(s) adjustment. If valves from BOTH camshafts require adjustment, remove both sprockets. For details, please refer to the Powerhead section.
   b. Remove the camshaft bearings caps, using multiple passes of a spiraling sequence for safety, then remove the Intake and/or Exhaust camshaft from the cylinder head, as necessary. Again, for details, please refer to the Powerhead section.
   c. Insert a thin screwdriver into the notch of the valve lifter whose shim must be replaced. Carefully pry outward to free the shim from the lifter.
   d. Measure the thickness of the shim using a micrometer. Then, use the clearance measurement taken earlier to determine how much larger or smaller a replacement shim is necessary.

MAINTENANCE & TUNE-UP 2-105

■ Take the clearance measurement made earlier, let's say it was larger than the spec. If so, subtract the spec from the measurement, the result tells you how much THICKER a shim you will need than what is already installed in order to reduce the clearance measurement to proper specification. Similarly, if the clearance measurement was smaller than the spec, subtract IT from the spec in order to get how much THINNER a shim you'll need than the one which is currently in use.

  e. Measure the thickness of the replacement shims and select one that will bring the clearance back into spec. Lubricate the replacement shim with engine assembly lube or some form of molybdenum disulfide grease and install it into the lifter.

  f. Repeat the shim removal, measurement and replacement shim installation procedure for each valve whose clearance was out of specification. It is best not to reuse shims removed from one lifter in a different lifter.

■ Obviously, you only need to remove shims from valves whose clearance was out of specification.

  g. Once finished, reinstall the camshaft(s), bearing caps, camshaft sprocket(s), timing belt and tensioner following the procedures in the Powerhead section.

  h. Double-check the clearance measurements on valves whose shims were replaced. If any valves are still out of specification (and they really shouldn't be unless you made a mistake somewhere), you'll have to remove the camshafts and try again.

14. Install the cylinder head/valve cover using a new rubber gasket that has been lubricated with a light coating of an all purpose Marine Grease, first tighten the bolts until just snug using a crossing pattern, then tighten the bolts to 70 inch lbs./5.8 ft. lbs. (8 Nm) using a torque wrench.

15. On EFI motors, reconnect the vacuum and ventilation hoses as tagged during removal.

16. Reconnect the fuel pump hoses to the fuel pump or pumps, again as tagged during removal.

17. Install the spark plugs and connect the leads as tagged during removal.

18. Install the spark plug cap cover and secure using the Phillips head (Yamaha) or Hex head (Mercury) screws (there are usually 5 of them).

19. Install the flywheel cover over the rubber grommets and/or retaining pins, as applicable, then secure using the retaining bolts.

20. Connect the negative battery cable, then connect a flushing device and start the engine to check for oil leaks at the rocker arm cover mating surfaces. Also, take a good look at the fuel line fittings to make sure you didn't cause any leaks which could lead to an explosive situation.

21. Shut down the powerhead and install the engine top case.

150 Hp Yamaha Models

◆ See Figures 332, 334 and 335

A set of flat feeler gauges is necessary to check valve clearance on these motors.

Valve specifications are for an overnight cold engine. It is best to check and/or adjust the valves with the powerhead at approximately 68° F (20° C).

Like with most 4-stroke motors, clearance measurement is a simple matter of aligning the crankshaft in a certain position and inserting a feeler gauge under the base of the crankshaft lobe on correct valves. On these motors, the valves are checked with the camshaft in 2 positions, half of the valves are checked with the engine at No. 1 TDC and the other half may be adjusted with the engine at No. 4 TDC.

Valve clearance on these motors is controlled by the thickness of the shim installed in the lifter that connects the camshaft lobe to the valve stem. If too great a clearance exists, you'll have to remove the old shim and install a thicker one. If too little clearance exists, remove the old shim and install a thinner one. Never use 2 shims in the same lifter, as the top could become dislodged and damage the camshaft and/or lifter.

Yamaha does not mention any provision for changing shims with the camshaft still installed. So, although it is not uncommon in many 4-stroke motor designs to utilize a valve compressing tool to depress the lifter and slide the shim out without removing the camshaft, we cannot recommend the use of such a tool here. Either Yamaha could not leave sufficient surface area or tabs on the top of the lifters for such a tool to work, or they are concerned about possible scoring/damage to the crankshaft. Therefore, if one or more valves requires adjustment the timing belt and camshaft(s) will have to be removed for access.

Furthermore, as with some of our smaller 4-cylinder test engines, the No. 1 cylinder will be at TDC when the marks on both pulleys are aligned with and directly facing each other. Also, the No. 4 cylinder should then be at TDC when the marks are facing directly AWAY from each other.

1. Disconnect the negative battery cable for safety.
2. Loosen the retaining bolts, then carefully pull the flywheel cover from the rubber grommets and/or retaining pins, as applicable.
3. Tag and disconnect the hoses from the fuel pumps. Use a rag to catch any escaping fuel and make sure no-one squeezes the primer bulb while they are disconnected. It's not a bad idea to cap or plug the hoses to prevent the possibility of any fuel line contamination, especially if you might not reconnect everything right away.
4. Loosen and remove the screws securing the spark plug cover to the powerhead, then remove the cover for access.
5. Tag and disconnect the spark plug wires, then remove the spark plugs in order to ease engine compression. For details, please refer to the Spark Plug procedure in this section.

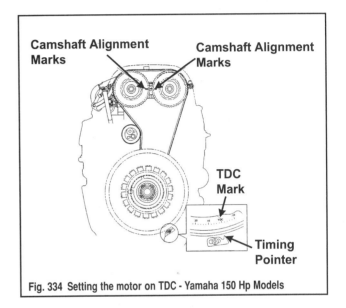

Fig. 334 Setting the motor on TDC - Yamaha 150 Hp Models

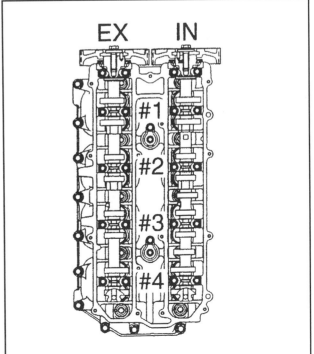

Fig. 335 Valve arrangement - Yamaha 150 Hp Models

2-106 MAINTENANCE & TUNE-UP

※※ WARNING

Ground the spark plug leads to prevent damage if the engine is cranked for any reason while they are still disconnected. Better yet, disconnect the negative battery cable and hide the keys to the ignition so no-one will accidentally turn it on.

6. Tag and disconnect the 2 blow-by hoses at the top of the valve cover.
7. Remove the cylinder head/valve cover for access to the valve train. When removing the valve cover, slowly loosen all 15 bolts around the perimeter (and through the spark plug valley) of the cover, using multiple passes of a crossing sequence to make sure it cannot warp and become damaged. For more details, please refer to the Powerhead overhaul section.
8. Carefully pull the cylinder head/valve cover from the cylinder head. Inspect the gasket for damage.
9. Rotate the flywheel (CLOCKWISE, in the normal direction of rotation) until No. 1 cylinder is at TDC (meaning the flywheel TDC mark aligns with the pointer and the marks on the 2 camshaft sprockets are facing each other). When rotating the crankshaft, once the valves for the No. 1 cylinder both close, you're at or close to TDC. At No. 1 TDC both marks on the camshaft sprockets should normally be pointing directly inward, facing towards each other.

■ With the flywheel in the No. 1 TDC position, the raised portion of the camshaft lobes should face away from and not be in contact with the valve lifters/shims.

10. Measure and note the clearance of the intake and exhaust valves for the No. 1 cylinder, also measure the No. 2 cylinder intake valve and the No. 3 cylinder exhaust valve. Insert feeler gauges of various sizes between the base of the camshaft lobe and the valve lifter/shim for each valve. The size that passes between the camshaft and shim with a slight drag indicates the valve clearance. Compare the clearance measured with the specifications of 0.007-0.009 in. (0.17-0.23mm) for the intake valve or 0.012-0.015 in. (0.31-0.37mm) for the exhaust valve. Noting the exact clearance measurement is important if the valve is out of spec, since you'll need that measurement later, when deciding what size replacement shim to install.

■ To determine which valve is an intake and which is an exhaust, observe the position of the valves in relation to the other components on the powerhead. The intake valves are adjacent to the ports for the intake manifold attached to the powerhead (on the same side as the carbs or the fuel injectors). The exhaust valves are adjacent to the exhaust ports. For the 150 hp motor (unlike most other Yamaha 4-cylinder engines) the intake valves are on the starboard side while the exhaust valves are on the port side.

11. Rotate the flywheel clockwise one full revolution (360 degrees) so the No. 4 cylinder is at TDC (meaning the valves for No. 4 now close and the piston is at the top of its travel). The camshaft rotates at 1/2 the rate of the crankshaft/flywheel, so rotating the flywheel as directed will turn the camshaft sprockets only 180 degrees (placing the two marks in the opposite position from where they were at No. 1 TDC, which should mean they are pointing straight out to the port and starboard sides of the motor.
12. Measure and note the clearance of the No. 4 cylinder intake and exhaust valves, the No 2 exhaust valve and the No. 3 intake valve. Insert feeler gauges of various sizes between the base of the camshaft lobe and the valve lifter/shim for each valve. The size that passes between the camshaft and shim with a slight drag indicates the valve clearance. Compare the clearance measured with the specifications of 0.007-0.009 in. (0.17-0.23mm) for the intake valve or 0.012-0.015 in. (0.31-0.37mm) for the exhaust valve.
13. If lash is within spec, you're in luck, no adjustment is necessary so you can button her back up. However, if the lash is out of spec for one or more valves, adjust it as follows:

a. Loosen the timing belt tensioner, then remove the timing belt and camshaft sprocket for the camshaft whose valve(s) require(s) adjustment. If valves from BOTH camshafts require adjustment, remove both sprockets. For details, please refer to the Powerhead section.

b. Remove the camshaft bearings caps, using multiple passes of a spiraling sequence for safety, then remove the Intake and/or Exhaust camshaft from the cylinder head, as necessary. Again, for details, please refer to the Powerhead section.

c. A small oiling hole is supplied in the top of the valve shim which is also used to free the shim from the lifter. Insert the nozzle of an air gun and use a small blast of low-pressure compressed air to push the shim from the lifter.

d. Measure the thickness of the shim using a micrometer. Then, use the clearance measurement taken earlier to determine how much larger or smaller a replacement shim is necessary.

■ Take the clearance measurement made earlier, let's say it was larger than the spec. If so, subtract the spec from the measurement, the result tells you how much THICKER a shim you will need than what is already installed in order to reduce the clearance measurement to proper specification. Similarly, if the clearance measurement was smaller than the spec, subtract IT from the spec in order to get how much THINNER a shim you'll need than the one which is currently in use.

e. Measure the thickness of the replacement shims and select one that will bring the clearance back into spec. Lubricate the replacement shim with engine assembly lube or some form of molybdenum disulfide grease and install it into the lifter.

f. Repeat the shim removal, measurement and replacement shim installation procedure for each valve whose clearance was out of specification. It is best not to reuse shims removed from one lifter in a different lifter.

■ Obviously, you only need to remove shims from valves whose clearance was out of specification.

g. Once finished, reinstall the camshaft(s), bearing caps, camshaft sprocket(s), timing belt and tensioner following the procedures in the Powerhead section.

h. Double-check the clearance measurements on valves whose shims were replaced. If any valves are still out of specification (and they really shouldn't be unless you made a mistake somewhere), you'll have to remove the camshafts and try again.

14. Install the cylinder head/valve cover using a new rubber gasket. Apply a dab of sealant (such as Yamaha #S1280B) to the 4 upper corners of the gasket. Position the cover carefully, then tighten the bolts until just snug using a crossing pattern and finally tighten the bolts to 70 inch lbs./5.8 ft. lbs. (8 Nm) using a torque wrench.
15. Install the spark plugs and connect the leads as tagged during removal.
16. Reconnect the 2 breather hoses to the top of the valve cover as tagged during removal.
17. Install the spark plug cover and secure using the retaining screws.
18. Reconnect the fuel pump hoses to the fuel pump or pumps, again as tagged during removal.
19. Install the flywheel cover over the rubber grommets and/or retaining pins, as applicable, then secure using the retaining bolts.
20. Connect the negative battery cable, then connect a flushing device and start the engine to check for oil leaks at the rocker arm cover mating surfaces. Also, take a good look at the fuel line fittings to make sure you didn't cause any leaks which could lead to an explosive situation.
21. Shut down the powerhead and install the engine top case.

200/225 Hp Models

◆ See Figures 332 and 336 thru 340

A set of flat feeler gauges is necessary to check valve clearance on these motors.

Valve specifications are for an overnight cold engine. It is best to check and/or adjust the valves with the powerhead at approximately 68° F (20° C).

Like with most 4-stroke motors, clearance measurement is a simple matter of aligning the crankshaft in a certain position and inserting a feeler gauge under the base of the crankshaft lobe on the correct valves. On these motors, the valves are checked with the camshaft in 3 positions, some of the valves are checked with the engine at No. 1 TDC, some with the motor at No. 2 TDC and the rest with the motor at No. 3 TDC.

Fortunately Yamaha/Mercury provides 3 separate TDC marks on the flywheel, for just this purpose, so finding TDC for each of the required cylinders should not be a problem. However, it is a good idea to simply watch the camshafts as you slowly rotate the flywheel. The No. 1 cylinder will be at TDC when the exhaust AND intake valves both close and the No. 1 TDC mark on the flywheel approaches the timing pointer are both closed. You will be able to tell that a valve is closed because the camshaft lobe will face away from the tappet for that valve.

MAINTENANCE & TUNE-UP

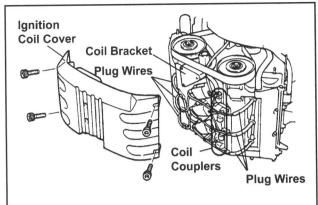

Fig. 336 Remove these components to measure the valve clearance - 200/225 hp Motors

The bottom line of valve adjustment is that it is REALLY not important that the motor be in a particular position except that in order to properly measure valve clearance the valve must be in the completely released position. So on this motor it means you can measure ANY valve, as long as the camshaft lobe is facing AWAY from the valve lifter/shim assembly. Positioning the motor at No. 1, then No. 2 and finally No 3. TDC is just a convenient way to do this for all the valves.

Valve clearance on these motors is controlled by the thickness of the shim installed in the lifter that connects the camshaft lobe to the valve stem. If too great a clearance exists, you'll have to remove the old shim and install a thicker one. If too little clearance exists, remove the old shim and install a thinner one. Never use 2 shims in the same lifter, as the top could become dislodged and damage the camshaft and/or lifter.

Neither Yamaha nor Mercury mentions any provision for changing shims with the camshaft still installed. So, although it is not uncommon in many 4-stroke motor designs to utilize a valve compressing tool to depress the lifter and slide the shim out without removing the camshaft, we cannot recommend the use of such a tool here. Either Yamaha could not leave sufficient surface area or tabs on the top of the lifters for such a tool to work, or they are concerned about possible scoring/damage to the crankshaft. Therefore, if one or more valves requires adjustment the timing belt and camshaft(s) will have to be removed for access.

1. Disconnect the negative battery cable for safety.
2. Loosen the retaining bolts, then carefully pull the flywheel cover from the rubber grommets and/or retaining pins, as applicable.
3. Loosen and remove the bolts (usually 4) securing the ignition coil cover to the powerhead, then remove the cover for access.
4. Disconnect the ignition coil couplers, then tag and disconnect the spark plug wires.
5. Remove the ignition coil bracket, then remove the spark plugs in order to ease engine compression. For details, please refer to the Spark Plug procedure in this section.

✶✶ WARNING

Ground the spark plug leads to prevent damage if the engine is cranked for any reason while they are still disconnected. Better yet, disconnect the negative battery cable and hide the keys to the ignition so no-one will accidentally turn it on.

6. Remove the cylinder head/valve covers for access to the valve trains. When removing the valve covers, slowly loosen all 14 bolts (for each cover) around the perimeter (and through the spark plug valley) of the cover, using multiple passes of a counterclockwise spiraling sequence that starts at the top outer bolt and works inward (the reverse of the torque sequence) to make sure it cannot warp and become damaged. For more details, please refer to the Powerhead overhaul section.
7. Carefully pull the cylinder head/valve covers from the cylinder heads. Discard the gaskets, as they should be replaced once the cover has been removed.
8. Rotate the flywheel (CLOCKWISE, in the normal direction of rotation) until No. 1 cylinder is at TDC (the **1TDC** mark on the flywheel will align with the timing pointer, but also the marks on each of the camshaft timing belt drive sprockets should also align with the marks on the powerhead). Of

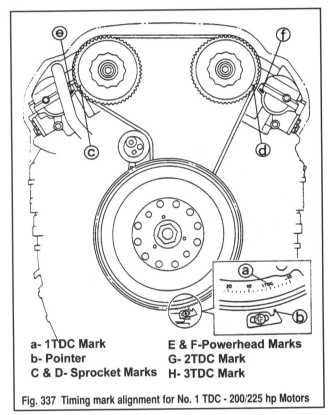

a- 1TDC Mark
b- Pointer
C & D- Sprocket Marks
E & F- Powerhead Marks
G- 2TDC Mark
H- 3TDC Mark

Fig. 337 Timing mark alignment for No. 1 TDC - 200/225 hp Motors

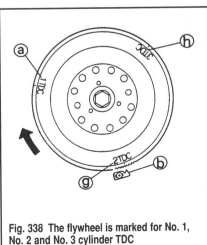

Fig. 338 The flywheel is marked for No. 1, No. 2 and No. 3 cylinder TDC

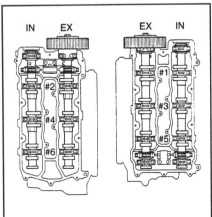

Fig. 339 Cylinder and valve configuration - 200/225 hp Motors

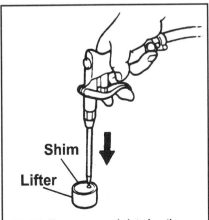

Fig. 340 Use compressed air to free the shim

2-108 MAINTENANCE & TUNE-UP

course, the way to know for sure is to watch the No. 1 cylinder valves as you rotate the flywheel, they should both close as the **1TDC** mark approaches the pointer.

■ **With the flywheel in the No. 1 TDC position, the raised portion of the camshaft lobes should face away from and not be in contact with the valve lifters/shims.**

9. Measure and note the clearance of the intake and exhaust valves for the No. 1 and No. 4 cylinders in this position. Insert feeler gauges of various sizes between the base of the camshaft lobe and the valve lifter/shim for each valve. The size that passes between the camshaft and shim with a slight drag indicates the valve clearance. Compare the clearance measured with the specifications of 0.007-0.009 in. (0.17-0.23mm) for the intake valve or 0.012-0.015 in. (0.31-0.37mm) for the exhaust valve. Noting the exact clearance measurement is important if the valve is out of spec, since you'll need that measurement later, when deciding what size replacement shim to install.

■ **To determine which valve is an intake and which is an exhaust, observe the position of the valves in relation to the other components on the powerhead. The intake valves are adjacent to the ports for the intake manifold attached to the powerhead while the exhaust valves are adjacent to the exhaust ports. For these motors the intake valves are one the outside of the V, while the exhaust valves are on the inside of the V.**

10. Rotate the flywheel 1/3 turn (120 degrees) so the No. 2 cylinder is at TDC (meaning the **2TDC** on the flywheel is now aligned with the timing pointer). Again, the No. 2 cylinder valves should close as the mark approaches the pointer.

11. Now measure and note the clearance of the No. 2 and No. 5 cylinder intake and exhaust valves. Insert feeler gauges of various sizes between the base of the camshaft lobe and the valve lifter/shim for each valve. The size that passes between the camshaft and shim with a slight drag indicates the valve clearance. Compare the clearance measured with the specifications of 0.007-0.009 in. (0.17-0.23mm) for the intake valve or 0.012-0.015 in. (0.31-0.37mm) for the exhaust valve.

12. Rotate the flywheel another 1/3 turn (120 degrees) so the No. 3 cylinder is at TDC (meaning the **3TDC** on the flywheel is now aligned with the timing pointer). Again, the No. 3 cylinder valves should close as the mark approaches the pointer.

13. Now measure and note the clearance of the No. 3 and No. 6 cylinder intake and exhaust valves. Insert feeler gauges of various sizes between the base of the camshaft lobe and the valve lifter/shim for each valve. The size that passes between the camshaft and shim with a slight drag indicates the valve clearance. Compare the clearance measured with the specifications of 0.007-0.009 in. (0.17-0.23mm) for the intake valve or 0.012-0.015 in. (0.31-0.37mm) for the exhaust valve.

14. If lash is within spec, you're in luck (BIG luck!), as no adjustment is necessary so you can button her back up. However, if the lash is out of spec for one or more valves, adjust it as follows:

 a. Turn the flywheel slowly CLOCKWISE until the **1TDC** mark is aligned with the pointer, and the marks on the camshaft timing belt sprockets are also aligned with their marks on the powerhead.

 b. Remove the timing belt and camshaft sprocket for the camshaft (or bank) whose valve(s) require(s) adjustment. If valves from BOTH banks require adjustment, remove both sprockets. For details, please refer to the Powerhead section.

 c. Remove the camshafts for whichever valves require adjustment. Again, for details, please refer to the Powerhead section.

 d. A small oiling hole is supplied in the top of the valve shim which is also used to free the shim from the lifter. Insert the nozzle of an air gun and use a small blast of low-pressure compressed air to push the shim from the lifter.

 e. Measure the thickness of the shim using a micrometer. Then, use the clearance measurement taken earlier to determine how much larger or smaller a replacement shim is necessary.

■ **Take the clearance measurement made earlier, let's say it was larger than the spec. If so, subtract the spec from the measurement, the result tells you how much THICKER a shim you will need than what is already installed in order to reduce the clearance measurement to proper specification. Similarly, if the clearance measurement was smaller than the spec, subtract IT from the spec in order to get how much THINNER a shim you'll need than the one which is currently in use.**

 f. Measure the thickness of the replacement shims and select one that will bring the clearance back into spec. Lubricate the replacement shim with engine assembly lube or some form of molybdenum disulfide grease and install it into the lifter.

 g. Repeat the shim removal, measurement and replacement shim installation procedure for each valve whose clearance was out of specification. It is best not to reuse shims removed from one lifter in a different lifter.

■ **Obviously, you only need to remove shims from valves whose clearance was out of specification.**

 h. Once finished, reinstall the camshaft(s), camshaft sprocket(s) and timing belt following the procedures in the Powerhead section.

 i. Double-check the clearance measurements on valves whose shims were replaced. If any valves are still out of specification, you'll have to remove the camshafts and try again.

15. Install the cylinder head/valve covers using a new gaskets, however, first apply a light coating of Yamabond No. 4, RTV 587 or an equivalent Silicone Sealer to the 4 top corners of each gasket to ensure a proper seal. Tighten the cover bolts, first until just snug using a crossing pattern, then tighten the bolts to 70 inch lbs./5.8 ft. lbs. (8 Nm) using a torque wrench and repeat, rightening each bolt to the same spec (without loosening them).

16. Install the spark plugs and the ignition coil bracket.

17. Connect the spark plug leads as tagged during removal and the ignition coil couplers.

18. Install the ignition coil cover and secure using the bolts (usually 4).

19. Install the flywheel cover over the rubber grommets and/or retaining pins, as applicable, then secure using the retaining bolts.

20. Connect the negative battery cable, then connect a flushing device and start the engine to check for oil leaks at the rocker arm cover mating surfaces. Also, take a good look at the fuel line fittings to make sure you didn't cause any leaks which could lead to an explosive situation.

21. Shut down the powerhead and install the engine top case.

STORAGE (WHAT TO DO BEFORE AND AFTER)

Winterization

◆ See Figure 341

Taking extra time to store the boat and motor properly at the end of each season or before any extended period of storage will greatly increase the chances of satisfactory service at the next season. Remember, that next to hard use on the water, the time spent in storage can be the greatest enemy of an outboard motor. Ideally, outboards should be used regularly. If weather in your area allows it, don't store the motor, enjoy it. Use it, at least on a monthly basis. It's best to enjoy and service the boat's steering and shifting mechanism several times each month. If a small amount of time is spent in such maintenance, the reward will be satisfactory performance, increased longevity and greatly reduced maintenance expenses.

But, in many cases, weather or other factors will interfere with time for enjoying a boat and motor. If you must place them in storage, take time to properly winterize the boat and outboard. This will be your best shot at making time stand still for them.

For many years there was a widespread belief simply shutting off the fuel at the tank and then running the powerhead until it stops constituted prepping the motor for storage. Right? Well, WRONG!

First, it is not possible to remove all fuel in the carburetor or fuel injection system by operating the powerhead until it stops. Considerable fuel will remain trapped in the carburetor float chamber (or vapor separator tank of fuel injected motors) and other passages, especially in the lines leading to carburetors or injectors. The only guaranteed method of removing all fuel from a carbureted motor is to take the physically drain the carburetors from the float bowls. And, though you should drain the fuel from the vapor separator tank on most fuel injected motors, you still will not be able to remove all of it from the sealed high-pressure lines.

Depending upon the length of storage you can also use fuel stabilizer as opposed to draining the fuel system, but if the motor is going to be stored for

MAINTENANCE & TUNE-UP 2-109

more than a couple of months at a time, draining the system is really the better option.

■ Up here in the northeastern U.S., we always start adding fuel stabilizer to the fuel tank with every fuel fill up starting sometime in September. That helps to make sure that we'll be at least partially protected if the weather takes a sudden turn and we haven't had a chance to complete winterization yet.

Proper storage involves adequate protection of the unit from physical damage, rust, corrosion and dirt. The following steps provide an adequate maintenance program for storing the unit at the end of a season.

■ If your outboard requires one or more repairs, PERFORM THEM NOW or during the off-season. Don't wait until the sun is shining, the weather is great and you want to be back on the water. That's not the time to realize you have to pull of the gearcase and replace the seals. Don't put a motor that requires a repair into storage unless you plan on making the repair during the off-season. It's too easy to let it get away from you and it will cost you in down time next season.

PREPPING FOR STORAGE

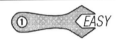

Where to Store Your Boat and Motor

Ok, a well lit, locked, heated garage and work area is the best place to store you precious boat and motor, right? Well, we're probably not the only ones who wish we had access to a place like that, but if you're like most of us, we place our boat and motor wherever we can.

Of course, no matter what storage limitations are placed by where you live or how much space you have available, there are ways to maximize the storage site.

If possible, select an area that is dry. Covered is great, even if it is under a carport or sturdy portable structure designed for off-season storage. Many people utilize canvas and metal frame structures for such purposes. If you've got room in a garage or shed, that's even better. If you've got a heated garage, God bless you, when can we come over? If you do have a garage or shed that's not heated, an insulated area will help minimize the more extreme temperature variations and an attached garage is usually better than a detached for this reason. Just take extra care to make sure you've properly inspected the fuel system before leaving your boat in an attached garage for any amount of time.

If a storage area contains large windows, mask them to keep sunlight off the boat and motor otherwise, use a high-quality, canvas cover over the boat, motor and if possible, the trailer too. A breathable cover is best to avoid the possible build-up of mold or mildew, but a heavy duty, non-breathable cover will work too. If using a non-breathable cover, place wooden blocks or length's of 2 x 4 under various reinforced spots in the cover to hold it up off the boat's surface. This should provide enough room for air to circulate under the cover, allowing for moisture to evaporate and escape.

Whenever possible, avoid storing your boat in industrial buildings or parks areas where corrosive emissions may be present. The same goes for storing your boat too close to large bodies of saltwater. Hey, on the other hand, if you live in the Florida Keys, we're jealous again, just enjoy it and service the boat often to prevent corrosion from causing damage.

Finally, when picking a place to store your motor, consider the risk or damage from fire, vandalism or even theft. Check with your insurance agent regarding coverage while the boat and motor is stored.

Storage Checklist (Preparing the Boat and Motor)

◆ See Figures 342 thru 345

The amount of time spent and number of steps followed in the storage procedure will vary with factors such as the length of planed storage time, the conditions under which boat and motor are to be stored and your personal decisions regarding storage.

But, even considering the variables, plans can change, so be careful if you decide to perform only the minimal amount of preparation. A boat and motor that has been thoroughly prepared for storage can remain so with minimum adverse affects for as short or long a time as is reasonably necessary. The same cannot be said for a boat or motor on which important winterization steps were skipped.

■ If possible it is best to store your motor vertically on the boat or on a suitable engine stand. If you can avoid it, do NOT lay a 4-stroke motor down for any length of time, as engine oil will seep past the rings causing extreme smoking upon startup. At best, burning that oil will promote spark plug and combustion chamber fouling, at worst it could cause a partial hydro-lock condition that could even mechanically damage the powerhead.

Although Yamaha and Mercury recommend storing all of their 4-stroke outboards in an upright position, both manufacturers say that you CAN store their smaller motors lying on their sides, but only one side and only under certain circumstances. Yamaha shows that all of their portable motors may be stored with the tiller handle side facing downward and, larger motors (even the non-portables) up to 60 hp CAN be stored with the Port side downward. Mercury only shows their portable units stored on their sides, and like the Yamahas, recommends that if this is done that the Tiller handle side (the Port side) be faced downward.

✱✱ CAUTION

On all 4-strokes, IF the motor is to be stored on the side, make sure it is first left in a vertical position long enough to ensure all water has been completely drained otherwise water could enter the motor through the exhaust ports possibly causing a corrosion problem. Obviously water left in the exhaust system of a motor which will be exposed to freezing conditions would be an even GREATER concern.

NO motor should be placed on its side until after ALL water has drained, otherwise water may enter a cylinder through an exhaust port causing corrosion (or worse, may become trapped in a passage and freeze causing cracks in the powerhead or gearcase!). Check your owner's manual for more details if you need to store a motor on its side. However if you do store it this way, be sure to return the motor vertically a few days before intended service and check the combustion chambers for oil before cranking the motor.

A word on engine fogging. We do it, we think you should too, but these days recommendations vary on how to perform it. Yamaha seems ok with spraying fogging oil either into the carbs/throttle bodies of running motors. Mercury seems that way too, at least on carbureted motors. On EFI motors they give a bizarre warning that the use of their Quicksilver Storage Seal may cause a buildup inside injectors. The only problem is, there SHOULD be

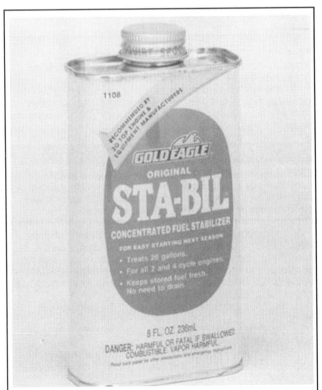

Fig. 341 Add fuel stabilizer to the system anytime it will be stored without complete draining

2-110 MAINTENANCE & TUNE-UP

Fig. 342 Be sure to fog the motor through the spark plug ports...

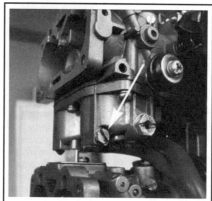

Fig. 343 ...and drain the carb float bowls before storage

Fig. 344 Multiple carburetors will mean multiple float bowls to drain

no way for the Storage Seal to get INTO the injectors, unless you put it in the fuel system and don't spray it down the bore of the throttle body. In Mercury service manuals they simply recommend manually adding a small amount of engine oil to each cylinder, through the spark plug opening. If that's the way you want to go FINE, but we'd probably elect to at least use a fogging oil spray which will be more evenly distributed throughout the cylinder.

1. Thoroughly wash the boat motor and hull. Be sure to remove all traces of dirt, debris or marine life. Check the water stream fitting, water inlet(s) and, on jet models, the impeller grate for debris. If equipped, inspect the speedometer opening at the leading edge of the gearcase or any other gearcase drains for debris (clean debris with low-pressure compressed air or a piece of thin wire).

■ As the season winds down and you are approaching your last outing, start treating the fuel system then to make sure it thoroughly mixes with all the fuel in the tank. By the last outing of the season you should already have a protected fuel system.

2. Change the engine crankcase oil and, if applicable, service the oil filter. Refill the crankcase and gearcase with fresh oil (for details, refer to the Oil and Filter Change procedures in this section). The motor should be run after the oil is changed to distribute the fresh/clean oil throughout the powerhead. Of course, the motor should be run for flushing and fogging anyway.

✱✱ WARNING

Besides treating the fuel system to prevent evaporation or clogging from deposits left behind, coating all bearing surfaces in the motor with FRESH, clean oil is the most important step you can take to protect the engine from damage during storage. NEVER leave the engine filled with used oil that likely contains moisture, acids and other damaging byproducts of combustion that will damage engine bearings over time.

3. Stabilize the engine's fuel supply using a high quality fuel stabilizer (of course the manufacturers recommend using Yamaha Fuel Conditioner and Stabilizer or Quicksilver Gasoline Stabilizer respectively) and take this opportunity to thoroughly flush the engine cooling system at the same time as follows:

   a. Add an appropriate amount of fuel stabilizer to the fuel tank (for most stabilizers it is normally about one ounce for each gallon of untreated fuel) and top off to minimize the formation of moisture through condensation in the fuel tank.

   b. Attach a flushing attachment as a cooling water/flushing source. For details, please refer to the information on Flushing the Cooling System, in this section.

   c. Start and run the engine at fast idle approximately 10-15 minutes. This will ensure the entire fuel supply system contains the appropriate storage mixtures.

   d. Just prior to stopping the motor fog the engine using Yamaha Stor-Rite Engine Fogging Oil, Mercury's Quicksilver Storage Seal, or an equivalent fogging spray. Spray the oil alternately into each of the carburetor

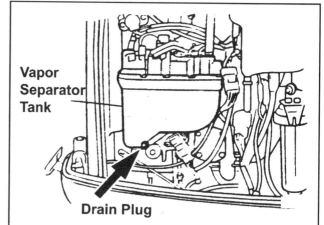

Fig. 345 HPDI and some EFI motors have a drain plug on the vapor separator tank, just like carburetor float bowls, they should be drained

or throttle body (EFI) throats (you'll likely have to remove an air intake silencer/flame arrestor for access). When properly fogged the motor will smoke excessively, will stumble and will almost stall.

   e. Stop the engine and remove the flushing source, keeping the outboard perfectly vertical. Allow the cooling system to drain completely, especially if the outboard might be exposed to freezing temperatures during storage.

✱✱ WARNING

NEVER keep the outboard tilted when storing in below-freezing temperatures as water could remain trapped in the cooling system. Any water left in cooling passages might freeze and could cause severe engine damage by cracking the powerhead or gearcase.

4. Drain and refill the engine gearcase while the oil is still warm (for details, refer to the Gearcase Oil procedures in this section). Take the opportunity to inspect for problems now, as storage time should allow you the opportunity to replace damaged or defective seals. More importantly, remove the old, contaminated gear oil now and place the motor into storage with fresh oil to help prevent internal corrosion.

5. Finish fogging the motor manually through the spark plug ports as follows:

   a. Tag and disconnect the spark plug leads, then remove the spark plugs as described under Spark Plugs.

   b. Spray a generous amount of fogging oil into the spark plug ports. Yamaha recommends a 5-10 second long spray of their Yamaha Stor-Rite Engine Fogging Oil for each cylinder.

MAINTENANCE & TUNE-UP 2-111

■ On most some Yamaha/Mercury motors you can disable the ignition system by leaving the safety lanyard disconnected but still use the starter motor to turn the motor. This is handy for things like compression tests or distributing fogging oil. To be certain use a spark plug gap tester on one lead and crank the motor using the keyswitch. If no spark is present, you're good to go.

c. Turn the flywheel slowly by hand (clockwise, in the normal direction of rotation) to distribute the fogging oil evenly across the cylinder walls. On electric start models, the starter can be used to crank the motor over in a few short bursts, but make sure the spark plugs leads remain disconnected and grounded to the powerhead (away from the spark plug ports) to prevent accidental combustion. Also, if the fuel system is not already drained and/or the EFI system is not disabled, you run the risk of fuel spray washing away the very same oil you're trying to leave on the inside of the motor. If necessary, re-spray into each cylinder when that cylinder's piston reaches the bottom of its travel. Reinstall and tighten the spark plugs, but leave the leads disconnected (and GROUNDED) to prevent further attempts at starting until the motor is ready for re-commissioning.

■ On motors equipped with a rope start handle, the rope can be used to turn the motor slowly and carefully using the rope starter. For other models, turn the flywheel by hand or using a suitable tool, but be sure to ALWAYS turn the engine in the normal direction of rotation (normally clockwise on these motors). Also, keep in mind that the flywheel on most Yamaha/Mercury outboards (even electric start models) is notched to accept an emergency starter rope. You can always use a knotted rope inserted into the notch and wound around the flywheel to help turn it.

6. On carbureted motors, if the motor is to be stored for any length of time more than one off-season you really MUST drain the carburetor float bowls. Honestly, it is a pretty easy task and we'd recommend doing that for all motors, even if they are only going to be stored for a few months. To drain the float bowls locate the drain screw on the bottom of each bowl, place a small container under the bowl and remove the screw. Repeat for the remaining float bowls on multiple carburetor motors.

7. For models equipped with portable fuel tanks, disconnect and relocate them to a safe, well-ventilated, storage area, away from the motor. Drain any fuel lines that remain attached to the tank. It's a tough call whether or not to drain a portable tank. Plastic tanks, drain 'em and burn the fuel in something else. Metal tanks, well, draining them will expose them to moisture and possible corrosion, while topping them off will help prevent this, so it probably makes more sense to top them off with treated fuel.

8. For boats with permanently installed fuel tanks, there's a huge debate going whether or not it is better to drain your fuel tanks completely during storage or to top them off. The first option allows you fill the tank with fresh fuel when the boat is removed from storage. But depending on temperature swings during storage you could amass a significant amount of water in the tank from condensation. You'd need to drain this water before re-commissioning. The later option is the easiest, especially for shorter term storage (one winter) and it prevents the formation of condensation, as there just isn't room for much air/moisture in the top of the filled tank.

9. Remove the battery or batteries from the boat and store in a cool dry place. If possible, place the battery on a smart charger or Battery Tender®, otherwise, trickle charge the battery once a month to maintain proper charge.

✳✳ WARNING

Remember that the electrolyte in a discharged battery has a much lower freezing point and is more likely to freeze (cracking/destroying the battery case) when stored for long periods in areas exposed to freezing temperatures. Although keeping the battery charged offers one level or protection against freezing; the other is to store the battery in a heated or protected storage area.

10. For models equipped with a boat mounted fuel filter or filter/water canister, clean or replace the boat mounted fuel filter at this time. If the fuel system was treated, the engine mounted fuel filters should be left intact, so the sealed system remains filled with treated fuel during the storage period.

11. Perform a complete lubrication service following the procedures in this section.

12. Except for Jet Drive models, remove the propeller and check thoroughly for damage. Clean the propeller shaft and apply a protective coating of grease.

13. On Jet models, thoroughly inspect the impeller and check the impeller clearance. Refer to the procedures in this section.

14. Check the motor for loose, broken or missing fasteners. Tighten fasteners and, again, use the storage time to make any necessary repairs.

15. Inspect and repair all electrical wiring and connections at this time. Make sure nothing was damaged during the season's use. Repair any loose connectors or any wires with broken, cracked or otherwise damaged insulation.

16. Clean all components under the engine cover and apply a corrosion preventative spray.

17. Too many people forget the boat and trailer, don't be one of them.
a. Coat the boat and outside painted surfaces of the motor with a fresh coating of wax, then cover it with a breathable cover
b. If possible place the trailer on stands or blocks so the wheels are supported off the ground.
c. Check the air pressure in the trailer tires. If it hasn't been done in a while, remove the wheels to clean and repack the wheel bearings.

18. Sleep well, since you know that your baby will be ready for you come next season.

Re-Commissioning

REMOVAL FROM STORAGE

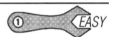

The amount of service required when recommissioning the boat and motor after storage depends on the length of non-use, the thoroughness of the storage procedures and the storage conditions.

At minimum, a thorough spring or pre-season tune-up and a full lubrication service is essential to getting the most out of your engine. If the engine has been properly winterized, it is usually no problem to get it in top running condition again in the springtime. If the engine has just been put in the garage and forgotten for the winter, then it is doubly important to perform a complete tune-up before putting the engine back into service. If you have ever been stranded on the water because your engine has died and you had to suffer the embarrassment of having to be towed back to the marina you know how it can be a miserable experience. Now is the time to prevent that from occurring.

■ Although you should normally replace your spark plugs at the beginning of each season, we like to start and run the engine (for the spring compression check) using the old spark plugs. Why? Well, on that first start-up you're going to be burning off a lot of fogging oil, why expose the new plugs to it? Besides you have to remove the plugs in order to perform the tune-up compression check anyway so you can install the new ones at that time.

Take the opportunity to perform any annual maintenance procedures that were not conducted immediately prior to placing the motor into storage. If the motor was stored for more than one off-season, pay special attention to inspection procedures, especially those regarding hoses and fittings. Check the engine gear oil for excessive moisture contamination. The same goes for engine crankcase oil. If necessary, change the gearcase or engine oil to be certain no bad or contaminated fluids are used.

■ Although not absolutely necessary, it is a good idea to ensure optimum cooling system operation by replacing the water pump impeller at this time. Impeller replacement was once an annual ritual, we now hear that most impellers seem to do well for 2-3 seasons, and some more. You'll have to make your own risk assessment here, but keep an eye on your cooling indicator stream get to know how strong the spray looks (and how warm it feels) to help you make your decision.

Other items that require attention include:
1. Install the battery (or batteries) if so equipped.

2-112 MAINTENANCE & TUNE-UP

2. Inspect all wiring and electrical connections. Rodents have a knack for feasting on wiring harness insulation over the winter. If any signs of rodent life are found, check the wiring carefully for damage, do not start the motor until damaged wiring has been fixed or replaced.

3. If not done when placing the motor into storage clean and/or replace the fuel filters at this time.

■ Portable fuel tanks should be emptied and cleaned using solvent. Take the opportunity to thoroughly inspect the condition of the tank. For more details on fuel tanks, please refer to the Fuel System section.

4. If the fuel tank was emptied, or if it must be emptied because the fuel is stale fill the tank with fresh fuel. Keep in mind that even fuel that was treated with stabilizer will eventually become stale, especially if the tank is stored for more than one off-season. Pump the primer bulb and check for fuel leakage or flooding at the carburetor.

5. Attach a flush device or place the outboard in a test tank and start the engine. Run the engine at idle speed and warm it to normal operating temperature. Check for proper operation of the cooling, electrical and warning systems.

✱✱ CAUTION

Before putting the boat in the water, take time to verify the drain plug is installed. Countless number of spring boating excursions have had a very sad beginning because the boat was eased into the water only to have the boat begin to fill with it.

CLEARING A SUBMRGED MOTOR

 DIFFICULT

Unfortunately, because an outboard is mounted on the exposed transom of a boat, and many of the outboards covered here are portable units that are mounted and removed on a regular basis, an outboard can fall overboard. Ok, it's relatively rare, but it happens often enough to warrant some coverage here. The best way to deal with such a situation is to prevent it, by keeping a watchful eye on the engine mounting hardware (bolts and/or clamps). But, should it occur, here's how to salvage, service and enjoy the motor again.

In order to prevent severe damage, be sure to recover an engine that is dropped overboard or otherwise completely submerged as soon as possible. It is really best to recover it immediately. But, keep in mind that once a submerged motor is recovered exposure to the atmosphere will allow corrosion to begin etching highly polished bearing surfaces of the crankshaft, connecting rods and bearings. For this reason, not only do you have to recover it right away, but you should service it right away too. Make sure the motor is serviced within about 3 hours of initial submersion.

OK, maybe now you're saying "3 hours, it will take me that long to get it to a shop or to my own garage." Well, if the engine cannot be serviced immediately (or sufficiently serviced so it can be started), re-submerge it in a tank of fresh water to minimize exposure to the atmosphere and slow the corrosion process. Even if you do this, do not delay any more than absolutely necessary, service the engine as soon as possible. This is especially important if the engine was submerged in salt, brackish or polluted water as even submersion in fresh water will not preserve the engine indefinitely. Service the engine, at the **MOST** within a few days of protective submersion.

After the engine is recovered, vigorously wash all debris from the engine using pressurized freshwater.

■ If the engine was submerged while still running, there is a good chance of internal damage (such as a bent connecting rod). Under these circumstances, don't start the motor, follow the beginning of this procedure to try turning it over slowly by hand, feeling for mechanical problems. If necessary, refer to Powerhead Overhaul for complete disassembly and repair instructions.

✱✱ WARNING

NEVER try to start a recovered motor until at least the first few steps (the ones dealing with draining the motor and checking to see it if is hydro-locked or damaged) are performed. Keep in mind that attempting to start a hydro-locked motor could cause major damage to the powerhead, including bending or breaking a connecting rod.

If the motor was submerged for any length of time it should be thoroughly disassembled and cleaned. Of course, this depends on whether water intruded into the motor or not. To determine this check the crankcase and gearcase oils for signs of contamination.

The extent of cleaning and disassembly that must take place depends also on the type of water in which the engine was submerged. Engines totally submerged, for even a short length of time, in salt, brackish or polluted water will require more thorough servicing than ones submerged in fresh water for the same length of time. But, as the total length of submerged time or time before service increases, even engines submerged in fresh water will require more attention. Complete powerhead disassembly and inspection is required when sand, silt or other gritty material is found inside the engine cover.

Many engine components suffer the corrosive effects of submersion in salt, brackish or polluted water. The symptoms may not occur for some time after the event. Salt crystals will form in areas of the engine and promote significant corrosion.

Electrical components should be dried and cleaned or replaced, as necessary. If the motor was submerged in salt water, the wire harness and connections are usually affected in a shorter amount of time. Since it is difficult (or nearly impossible) to remove the salt crystals from the wiring connectors, it is best to replace the wire harness and clean all electrical component connections. The starter motor, relays and switches on the engine usually fail if not thoroughly cleaned or replaced.

To ensure a through cleaning and inspection:

1. Remove the engine cover and wash all material from the engine using pressurized freshwater. If sand, silt or gritty material is present inside the engine cover, completely disassemble and inspect the powerhead.

2. Tag (except on single cylinder motors) and disconnect the spark plugs leads. Be sure to grasp the spark plug cap and not the wire, then twist the cap while pulling upward to free it from the plug. Remove the spark plugs. For more details, refer to the Spark Plug procedure in this section.

3. Disconnect the fuel supply line from the engine, then drain and clean all fuel lines. Depending on the circumstances surrounding the submersion, inspect the fuel tank for contamination and drain, if necessary.

✱✱ WARNING

When attempting to turn the flywheel for the first time after the submersion, be sure to turn it SLOWLY, feeling for sticking or binding that could indicate internal damage from hydro-lock. This is a concern, especially if the engine was cranked before the spark plug(s) were removed to drain water or if the engine was submerged while still running.

4. Support the engine horizontally with the spark plug port(s) facing downward, allowing water, if present, to drain. Force any remaining the water out by slowly rotating the flywheel by hand about 20 times or until there are no signs of water. If there signs of water present, spray some fogging oil into the spark plug ports before turning the flywheel. This will help dislodge moisture and lubricate the cylinder walls.

5. On carbureted models, drain the carburetor(s). The best method to thoroughly drain/clean the carburetor is to remove and disassemble it, but the float bowl drain screws are pretty accessible on most Yamahas and Mercury/Mariners and this is not absolutely necessary. It's not a bad idea to spray some fogging oil into the carburetors to make sure moisture is displaced. The oil will burn once the motor is started anyway. For more details on carburetor service refer to the Carburetor procedures under Fuel System.

MAINTENANCE & TUNE-UP

6. Support the engine in the normal upright position. Check the engine gearcase oil for contamination. Refer to the procedures for Gearcase Oil in this section. The gearcase is sealed and, if the seals are in good condition, should have survived the submersion without contamination. But, if contamination is found, look for possible leaks in the seals, then drain the gearcase and make the necessary repairs before refilling it. For more details, refer to the section on Gearcases.

7. Drain the crankcase engine oil and change the filter. Refer to the procedures in this section. If contaminated oil drains from the crankcase, flush the crankcase using a quart or two of fresh four-stroke engine oil (by pouring it into the motor as normal, but allowing it to drain as well) before refilling the crankcase.

8. Remove all external electrical components for disassembly and cleaning. Spray all connectors with electrical contact cleaner, then apply a small amount of dielectric grease prior to reconnection to help prevent corrosion. For electric start models, remove, disassemble and clean the starter components. For details on the electrical system components, refer to the Ignition and Electrical section.

9. Reassemble the motor and mount the engine or place it in a test tank. Start and run the engine for 1/2 hour. If the engine won't start, remove the spark plugs again and check for signs of moisture on the tips. If necessary, use compressed air to clean moisture from the electrodes or replace the plugs.

10. Stop the engine, then recheck the gearcase oil (and engine crankcase oil on 4-strokes).

11. Perform all other lubrication services.

12. Try not to let it get away from you (or anyone else) again!

SPECIFICATIONS

General Engine Specifications - Yamaha 4-Stroke Motors

Model (Hp)	No. of Cyl	Engine Type	Year	Displacement cu. in. (cc)	Bore and Stroke in. (mm)	Compression Ratio	Appx Weight lb. (kg) *
2.5	1	4-stroke	2003-04	4.4 (72)	2.13 x 1.24 (54.0 x 31.5)	9.0:1	37-38 (17.0-17.5)
4	1	4-stroke	1999-04	6.7 (112)	2.32 x 1.61 (59 x 41)	8.4:1	49-51 (22-23)
6	2	IL 4-stroke	2001-04	12 (197)	2.20 x 1.60 (56 x 40)	8.79:1	84-87 (38-40)
8	2	IL 4-stroke	2001-04	12 (197)	2.20 x 1.60 (56 x 40)	8.79:1	84-102 (38-47)
8	2	IL 4-stroke	1997-04	14 (232)	2.32 x 1.67 (59 x 42)	9.3:1	91-99 (42-45)
9.9	2	IL 4-stroke	1995-04	14 (232)	2.32 x 1.67 (59 x 42)	9.3:1	91-101 (42-46)
15	2	IL 4-stroke	1998-04	20 (323)	2.32 x 2.32 (59 x 59)	9.19:1	99-121 (45-55)
25	2	Standard	1998-04	30 (498)	2.56 x 2.95 (65 x 70)	9.87:1	136-152 (62-72)
	2	High Thrust	2001-04	30 (498)	2.56 x 2.95 (65 x 70)	9.87:1	185-199 (84-90)
30	3	IL 4-stroke	2001-04	46 (747)	2.55 x 2.95 (65 x 75)	9.87:1	189-199 (86-90)
40J	3	IL 4-stroke	2002-04	46 (747)	2.55 x 2.95 (65 x 75)	9.87:1	201 (91)
40	3	IL 4-stroke	2000-04	46 (747)	2.55 x 2.95 (65 x 75)	9.87:1	182-220 (83-100)
40	4	IL 4-stroke	1999-00	58 (935)	2.48 x 2.95 (63 x 75)	9.3:1	230-247 (104-112)
45	4	IL 4-stroke	1996-00	58 (935)	2.48 x 2.95 (63 x 75)	9.3:1	230-247 (104-112)
50J	4	IL 4-stroke	2002	58 (935)	2.48 x 2.95 (63 x 75)	9.3:1	254 (116)
50	4	IL 4-stroke	1995-04	58 (935)	2.48 x 2.95 (63 x 75)	9.3:1	233-247 (106-112)
60J	4	IL 4-stroke	2003-04	61 (996)	2.56 x 2.95 (65 x 75)	9.5:1	264 (120)
60	4	Standard	2002-04	61 (996)	2.56 x 2.95 (65 x 75)	9.5:1	244-251 (111-114)
	4	High thrust	2003-04	61 (996)	2.56 x 2.95 (65 x 75)	9.5:1	255-265 (116-120)
75	4	IL 4-stroke	2003-04	97 (1596)	3.11 x 3.20 (79.0 x 81.4)	8.9:1	370 (168)
80	4	IL 4-stroke	1999-02	97 (1596)	3.11 x 3.20 (79.0 x 81.4)	9.6:1	358 (162)
90J	4	IL 4-stroke	2003-04	97 (1596)	3.11 x 3.20 (79.0 x 81.4)	9.6:1	391 (178)
90	4	IL 4-stroke	2003-04	97 (1596)	3.11 x 3.20 (79.0 x 81.4)	8.9:1	370 (168)
100	4	IL 4-stroke	1999-02	97 (1596)	3.11 x 3.20 (79.0 x 81.4)	9.6:1	366-373 (166-169)
115J	4	IL 4-stroke	2003-04	106 (1741)	3.11 x 3.55 (79.0 x 88.8)	9.7:1	415 (189)
115	4	IL 4-stroke	2000-04	106 (1741)	3.11 x 3.55 (79.0 x 88.8)	9.7:1	402-433 (183-197)
150	4	IL 4-stroke	2004	163 (2670)	3.70 x 3.79 (94.0 x 96.2)	9.0:1	467-476 (212-216)
200	6	60 V 4-stroke	2002-04	205 (3352)	3.70 x 3.17 (94.0 x 80.5)	9.9:1	583-593 (265-269)
225	6	60 V 4-stroke	2002-04	205 (3352)	3.70 x 3.17 (94.0 x 80.5)	9.9:1	583-613 (265-278)

IL - inline motor V - "V" configuration motor

Because manufacturers differ with what specifications they choose to provide, please refer to the Mercury/Mariner charts for additional specifications which MAY apply to the Yamaha model in question as well

* NOTE: Measurements are approximate, minus fluids and propeller for most models.

2-114 MAINTENANCE & TUNE-UP

General Engine Specifications — Mercury/Mariner 4-Stroke Motors

Model (HP)	Engine Type	Year	Displacement cu.in. (cc)	Bore and Stroke in. (mm)	Appx Weight lb. (kg) *	Air/Fuel Induction System	Ignition System	Starting System	Cooling System	Battery ① Minimum CCA
4	1-cyl	1999-04	7.5 (123)	2.32 x 1.77 (59 x 45)	55 (25)	2-Valve/P-rod	CDI	Manual	WC w/ Tstat	-
5	1-cyl	1999-04	7.5 (123)	2.32 x 1.77 (59 x 45)	55 (25)	2-Valve/P-rod	CDI	Manual	WC w/ Tstat	-
6	1-cyl	2000-04	7.5 (123)	2.32 x 1.77 (59 x 45)	55 (25)	2-Valve/P-rod	CDI	Manual	WC w/ Tstat	-
8	2-cyl	1997-98	14 (232)	2.32 x 1.67 (59 x 42)	91-101 (42-46)	2-Valve/SOHC	CDI w/ Elec Spark Advance	Manual or Electric	WC w/ Tstat	350
9.9	2-cyl	1995-98	14 (232)	2.32 x 1.67 (59 x 42)	91-101 (42-46)	2-Valve/SOHC	CDI w/ Elec Spark Advance	Manual or Electric	WC w/ Tstat	350
9.9	2-cyl	1999-04	20 (323)	2.32 x 2.32 (59 x 59)	111-133 (50-60)	2-Valve/SOHC	CDI w/ Elec Spark Advance	Man or Elec (turnkey)	WC w/ Tstat	350
15	2-cyl	1999-04	20 (323)	2.32 x 2.32 (59 x 59)	111-128 (50-58)	2-Valve/SOHC	CDI w/ Elec Spark Advance	Man or Elec (turnkey)	WC w/ Tstat	350
25	2-cyl	1997-04	30 (498)	2.56 x 2.95 (65 x 75)	175-178 (80-81)	2-Valve/SOHC	CDI w/ Elec Spark Advance	Man or Elec (turnkey)	WC w/ Tstat	350
30 Carb	3-cyl	1999-01	46 (747)	2.56 x 2.95 (65 x 75)	204 (93)	2-Valve/SOHC	CDI w/ Elec Spark Advance	Man or Elec (turnkey)	WC w/ Tstat	350
	3-cyl	2002-03	46 (747)	2.56 x 2.95 (65 x 75)	204 (93)	2-Valve/SOHC	CDI w/ Elec Spark Advance	Manual	WC w/ Tstat	350
30 EFI	3-cyl	2003-04	46 (747)	2.56 x 2.95 (65 x 75)	204 (93)	2-Valve/SOHC	Digital Inductive (ECM555)	Electric (turnkey)	WC w/ Tstat	350
40 Carb	3-cyl	2002-04	46 (747)	2.56 x 2.95 (65 x 75)	216 (98)	2-Valve/SOHC	Digital Inductive (ECM555)	Electric (turnkey)	WC w/ Tstat	350
	3-cyl	1999-01	46 (747)	2.56 x 2.95 (65 x 75)	204 (93)	2-Valve/SOHC	CDI w/ Elec Spark Advance	Man or Elec (turnkey)	WC w/ Tstat	350
	3-cyl	2002-03	46 (747)	2.56 x 2.95 (65 x 75)	204 (93)	2-Valve/SOHC	CDI w/ Elec Spark Advance	Manual	WC w/ Tstat	350
	3-cyl	2003-04	46 (747)	2.56 x 2.95 (65 x 75)	204 (93)	2-Valve/SOHC	Digital Inductive (ECM555)	Electric (turnkey)	WC w/ Tstat	350
40 EFI	3-cyl	2002-04	46 (747)	2.56 x 2.95 (65 x 75)	216 (98)	2-Valve/SOHC	Digital Inductive (ECM555)	Electric (turnkey)	WC w/ Tstat	350
40	4-cyl	1997-00	58 (935)	2.48 x 2.95 (63 x 75)	238-247 (108-112)	2-Valve/SOHC	CDI w/ Elec Spark Advance	Electric	WC w/ Tstat	350
45	4-cyl	1995-00	58 (935)	2.48 x 2.95 (63 x 75)	238-247 (108-112)	2-Valve/SOHC	CDI w/ Elec Spark Advance	Electric	WC w/ Tstat	350
50	4-cyl	1995-00	58 (935)	2.48 x 2.95 (63 x 75)	238-247 (108-112)	2-Valve/SOHC	CDI w/ Elec Spark Advance	Electric	WC w/ Tstat	350
40 Carb	4-cyl	2001	61 (996)	2.56 x 2.95 (65 x 75)	252 (114)	2-Valve/SOHC	CDI w/ Elec Spark Advance	Electric	WC w/ Tstat	350
40 EFI	4-cyl	2002-04	61 (996)	2.56 x 2.95 (65 x 75)	240-264 (109-120)	2-Valve/SOHC	Digital Inductive (ECM555)	Electric (turnkey)	WC w/ Tstat	350
50 Carb	4-cyl	2001	61 (996)	2.56 x 2.95 (65 x 75)	224-252 (102-114)	2-Valve/SOHC	CDI w/ Elec Spark Advance	Electric	WC w/ Tstat	350
50 EFI	4-cyl	2002-04	61 (996)	2.56 x 2.95 (65 x 75)	248-264 (112-120)	2-Valve/SOHC	Digital Inductive (ECM555)	Electric (turnkey)	WC w/ Tstat	350
60 Carb	4-cyl	2001	61 (996)	2.56 x 2.95 (65 x 75)	236-252 (107-114)	2-Valve/SOHC	CDI w/ Elec Spark Advance	Electric	WC w/ Tstat	350
60 EFI	4-cyl	2002-04	61 (996)	2.56 x 2.95 (65 x 75)	248-264 (112-120)	2-Valve/SOHC	Digital Inductive (ECM555)	Electric (turnkey)	WC w/ Tstat	350
75	4-cyl	2000-04	97 (1596)	3.11 x 3.20 (79 x 81)	386 (175)	4-Valve/DOHC	CDI w/ Elec Spark Advance	Electric (turnkey)	WC w/ Tstat	350
90	4-cyl	2000-04	97 (1596)	3.11 x 3.20 (79 x 81)	386 (175)	4-Valve/DOHC	CDI w/ Elec Spark Advance	Electric (turnkey)	WC w/ Tstat	350
115 EFI	4-cyl	2001-04	106 (1741)	3.110x3.495 (79 x 89)	386 (175)	4-Valve/DOHC	CDI w/ Elec Spark Advance	Electric (turnkey)	WC w/ Tstat	350
225 EFI	V-6 (60°)	2003-04	205 (3352)	3.70x3.17 (94.0 x 80.5)	583 (265)	4-Valve/DOHC	Inductive Electronic	Electric (turnkey)	WC w/ Tstat & PC	350

CDI - Capacitor Discharge Ignition DOHC - Dual Overhead Camshafts PC - Pressure Controlled SOHC - Single Overhead Camshaft Tstat - Thermostat WC - Water Cooled

Because manufacturers differ with what specifications they choose to provide, please refer to the Yamaha charts for additional specifications which MAY apply as well

* NOTE: Measurements are approximate, minus fluids and propeller for most models. Generally speaking (especially on larger models), carbureted versions are lightest and EFI or BIGFOOT models are heaviest, but there are some exceptions. See owner's manual

① Minimum recommended Cold Cranking Amp (CCA) ratings for electric start models when motor/cables are new. Replacements must meet or exceed specification

General Engine System Specifications - Yamaha 4-Stroke Motors

Model (Hp)	No. of Cyl	Engine Type	Year	Displace cu. in. (cc)	Ignition System	Starting System	Fuel System	Battery ① CCA (AH)	Reserve
2.5	1	4-stroke	2003-04	4.4 (72)	TCI	Manual	Carb	n/a	n/a
4	1	4-stroke	1999-04	6.7 (112)	TCI	Manual	Carb	n/a	n/a
6	2	IL 4-stroke	2001-04	12 (197)	CDI	Manual	Carb	380 (70)	124 minutes
8	2	IL 4-stroke	2001-04	12 (197)	CDI	Manual	Carb	380 (70)	124 minutes
8	2	IL 4-stroke	1997-04	14 (232)	CDI	Manual and/or Elec	Carb	380 (70)	124 minutes
9.9	2	IL 4-stroke	1995-04	14 (232)	CDI	Manual and/or Elec	Carb	380 (70)	124 minutes
15	2	IL 4-stroke	1998-04	20 (323)	CDI	Manual and/or Elec	Carb	380 (70)	124 minutes
25	2	IL 4-stroke	1998-04	30 (498)	CDI Micro	Manual and/or Elec	Carb	380 (70)	124 minutes
30	3	IL 4-stroke	2001-04	46 (747)	CDI Micro	Manual and Elec	Carb	380 (70)	124 minutes
40J	3	IL 4-stroke	2002-04	46 (747)	CDI Micro	Manual and/or Elec	Carb	380 (70)	124 minutes
40	3	IL 4-stroke	2000-04	46 (747)	CDI Micro	Manual and Elec	Carb	380 (70)	124 minutes
40	4	IL 4-stroke	1999-00	58 (935)	CDI Micro	Electric	Carb	380 (70)	124 minutes
45	4	IL 4-stroke	1996-00	58 (935)	CDI Micro	Electric	Carb	380 (70)	124 minutes
50J	4	IL 4-stroke	2002	58 (935)	CDI Micro	Electric	Carb	380 (70)	124 minutes
50	4	IL 4-stroke	1995-04	58 (935)	CDI Micro	Electric	Carb	380 (70)	124 minutes
60J	4	IL 4-stroke	2003-04	61 (996)	CDI Micro	Electric	Carb	380 (70)	124 minutes
60	4	IL 4-stroke	2002-04	61 (996)	CDI Micro	Electric	Carb	380 (70)	124 minutes
75	4	IL 4-stroke	2003-04	97 (1596)	CDI Micro	Electric	Carb	380 (70)	124 minutes
80	4	IL 4-stroke	1999-02	97 (1596)	CDI Micro	Electric	Carb	380 (70)	124 minutes
90J	4	IL 4-stroke	2003-04	97 (1596)	CDI Micro	Electric	Carb	380 (70)	124 minutes
90	4	IL 4-stroke	2003-04	97 (1596)	CDI Micro	Electric	Carb	380 (70)	124 minutes
100	4	IL 4-stroke	1999-02	97 (1596)	CDI Micro	Electric	Carb	380 (70)	124 minutes
115J	4	IL 4-stroke	2003-04	106 (1741)	TCI Micro	Electric	EFI	512 (100)	182 minutes
115	4	IL 4-stroke	2000-04	106 (1741)	TCI Micro	Electric	EFI	512 (100)	182 minutes
150	4	IL 4-stroke	2004	163 (2670)	TCI Micro	Electric	EFI	512 (100)	182 minutes
200	6	60 V 4-stroke	2002-04	205 (3352)	TCI Micro	Electric	EFI	512 (100)	182 minutes
225	6	60 V 4-stroke	2002-04	205 (3352)	TCI Micro	Electric	EFI	512 (100)	182 minutes

AH - Amp Hours   CDI - Capacitor Discharge Ignition

Because manufacturers differ with what specifications they choose to provide, please refer to the Mercury/Mariner charts for additional specificitions which MAY apply to the Yamaha model in question as well

① Minimum recommended cca (mca) ratings for electric start models when motor/cables are new. Replacements must meet or exceed specification

MAINTENANCE & TUNE-UP

Maintenance Interval Chart

Component	Each Use or As Needed	Initial 10-Hour Check (Break-In)	Initial 50-Hour Check (3 months)	Every 6mths/100hrs ①	Off Season 12mths/200hrs ①
Anode(s)		I	I	I	I
Balancer (150 hp motors)					R (2000 hrs/10 yrs)
Battery condition and connections (if equipped)*	I (condition/connections), T			I	I
Battery charge / fluid level (if equipped)*	I (at least monthly)			I	I
Boat hull*	I			I	I
Bolts and nuts (all accessible fasteners)*		I, T		I, T	I
Carburetor (clean / inspect - if equipped)		C, I	C, I	C, I	C, I
Carburetor (adjust - if equipped) ②		A		A (inline motors)	A (V motors)
Case finish and condition (inspect & wash/wax)	C, salt/brackish/polluted water	I	I	I, C	I, C
Choke solenoid (if equipped)		A			A
Cylinder head (as applicable) ③		Check, T		Check, T	
Decarbon Pistons ⑧				P	P
Driveshaft splines (Mercury 8/9.9 hp & up) ⑨					L
Electricical wiring and connectors*		I	I	I	I
Emergency stop switch, clip &/or lanyard*	I	I		I	I
Engine mounting bolts		I		I	I
Engine crankcase oil	I		R		R	R
Engine crankcase oil filter (8/9.9 hp & up) ④		C or R (as applicable)		C/R (as applicable)	C/R (as applicable)
Flush cooling system	salt/brackish/polluted water		P	P	P
Fuel filter (clean or replace, as applicable)		P	P	P	P
Fuel hose and system components*	I	I	I (EFI)	I	I
Fuel tank (integral or portable)	I				C (portable), I
Fuel water separator (if equipped)	I (200/225)			I	I
Gear oil	I (for signs of leakage)	R	I	R	R
Idle speed	I (look/listen for changes)			A	A
Ignition timing (200/225 Ins. only at 1st 10hr)		I		I	I
Impeller clearance/intake grate (jet models)	I (visually inspect impeller)	I		I	I
Jet drive bearing lubrication ⑤	L, fill vent hose after each day	L		L	L
Lubrication points	I	⑥		⑥	⑥
Oil pump check (200/225)					I (400 hrs/2 yrs)
Power trim and tilt (if equipped)	I (check fluid level monthly)	I		I	I
Propeller	I	I	I	I	I
Propeller shaft and cotter pin/nut (or shear pin)	I	I		I, L / T	I, L / T
Remote control*	I	I		I, A	I, A
Spark plugs	R (as needed)	I	I	I	I
Start in gear lockout system (if applicable)	I	Check, A			Check, A
Steering cable*	L (as needed)	L		L	L
TCI air gap (Yamaha 2.5 & 4 hp)		I, A		I, A	
Thermostat and Pressure Control Valve		I			I
Throttle (tiller models)		Check, A			Check, A
Throttle position sensor (EFI)					I, A
Throttle valve switch (EFI)					I, A
Tune-up	A (as needed)			I (annually)	Pre-season tune-up
Timing belt (6/8-225 hp)				I	I (R 1000 hrs/5 yrs)
Timing chain (200/225 hp)				I	(R 1000 hrs/5 yrs)
Timing chain tensioner (200/225 hp) ⑩					I, R (1000 hrs/5 yrs)
Valve clearance (up to 60 hp)		I		I	
Valve clearance (75 hp & up, except 150)					I, A (400 hrs/2 yrs)
Valve clearance (150 hp)					I, A (500 hrs/2.5 yrs)
Water pump impeller					varies ⑦
Water pump intake grate and indicator	I				

A-Adjust
C-Clean
I-Inspect and Clean, Adjust, Lubricate or Replace, as necessary
L-Lubricate
R-Replace
T-Tighten
P - Perform

* Denotes possible safety item (although, all maintenance inspections/service can be considered safety related when it means not being stranded on the water should a component fail.)

① Many items are listed for both every 100 hours and off season. Since many boaters use their crafts less than 100 hours a year, these items should at least be performed annually. If you find yourself right around 100 hours per season, try to time the service so it occurs immediately prior to placing the motor in storage

② Recommendations vary regarding the need to adjust the carburetor (as opposed to only checking how it is operating). The most recent advice we've received from Yamaha/Merc is not to tamper with a carb that is running correctly

③ DO NOT retighten the cylinder head or crankcase bolts on the 200/225 4-stroke motors

④ For models that use replaceable filter elements the replacement interval may be extended up to 200 hours, but only during a single season

⑤ Lubricate the jet drive bearing at the end of EVERY day's use and every 10 hours. Also, every 50 hours replace the grease using the same lubrication fitting, but adding enough to purge the old grease

⑥ Varies with use, generally every 30 days when used in salt, brackish or polluted water and every 60 days when used in fresh water (refer to Lubrication Chart for more details)

⑦ Water pump impeller inspection recommendations vary from inspect every 100 hours (most Yamahas), to replace every 100 hours (Mercury 4/5/6), replace every 300 hours (Mercury 8/9.9-225) or replace every 500 hours/30 months (Yamaha 200/225 hp 4-st)

⑧ Mercury manuals for 25 hp and larger motors (except the EFI 30/40 3-cyl) recommend decarboning of the pistons using Power Tune Engine Cleaner annually or every 100 hours

⑨ Mercury manuals list lubrication of the driveshaft splines every year, but we recommend waiting for impeller replacement and doing both at the same time

⑩ Mercury manuals say to check the Timing Chain tensioner every 2 years/400 hours but that sounds excessive

MAINTENANCE & TUNE-UP

Mercury/Mariner Lubrication Chart

Component	Recommended Lubricant	Minimum Frequency ① Fresh Water	Minimum Frequency ① Salt, Polluted or Brackish Water
Clamp screws	Quicksilver 2-4-C w/ Teflon or Special Lube 101	every 100 hours or 60 days	every 50 hours or 30 days
Co-pilot shaft (tiller models)	Quicksilver 2-4-C w/ Teflon or Special Lube 101		
Cowling clamp lever/pivot	Quicksilver 2-4-C w/ Teflon or Special Lube 101	every 100 hours or 60 days	every 50 hours or 30 days
Engine oil	SAE 10W-30 (all temps) / 25W-40 (if above 40F/4C) API SE or greater (SF, SG or SH)	Check every outing / Replace every 100 hours	Check every outing / Replace every 100 hours
Jet drive bearing lubrication	Quicksilver Anti-Corrosion Grease	every 10 hours	every 5 hours
Jet drive bearing grease replacement	Quicksilver Anti-Corrosion Grease	every 50 hours	every 25 hours
Lower unit	Quicksilver Gear Lube - Premium Blend	every 100 hours or 60 days	every 50 hours or 30 days
Manual recoil starter (6 hp or larger motors)	Quicksilver 2-4-C w/ Teflon	Upon assembly only	Upon assembly only
Propeller shaft	Quicksilver 2-4-C w/ Teflon or Anti-Corrosion Grease	every 100 hours or 60 days	every 50 hours or 30 days
Shift cables &/or pivots	Quicksilver 2-4-C w/ Teflon or Special Lube 101	every 100 hours or 60 days	every 50 hours or 30 days
Shift handle detent 4/5/6 hp (requires disassembly)	Quicksilver 2-4-C w/ Teflon or Special Lube 101	every 100 hours or 60 days	every 50 hours or 30 days
Steeringt cables &/or pivots	Quicksilver 2-4-C w/ Teflon or Special Lube 101	every 100 hours or 60 days	every 50 hours or 30 days
Steering link rod pivot points	Light weight oil	every 100 hours or 60 days	every 50 hours or 30 days
Steering friction adjustment shaft (tiller 9.9/15 models)	Quicksilver 2-4-C w/ Teflon or Special Lube 101	every 100 hours or 60 days	every 50 hours or 30 days
Swivel bracket (zerk)	Quicksilver 2-4-C w/ Teflon or Special Lube 101	every 100 hours or 60 days	every 50 hours or 30 days
Throttle linkage	Quicksilver 2-4-C w/ Teflon or Special Lube 101	every 100 hours or 60 days	every 50 hours or 30 days
Tiller handle bushing 4/5/6 hp (requires disassembly)	Quicksilver 2-4-C w/ Teflon or Special Lube 101	every 100 hours or 60 days	every 50 hours or 30 days
Tilt lock pins & track (as applicable)	Quicksilver 2-4-C w/ Teflon or Special Lube 101	every 100 hours or 60 days	every 50 hours or 30 days
Tilt tube (w/ zerk fittings)	Light weight oil	every 100 hours or 60 days	every 50 hours or 30 days
Tilt tube (w/out zerk fittings)		every 100 hours or 60 days	every 50 hours or 30 days

NOTE: not all components used by all models. Refer to the section on Lubrication and/or your owner's manual for more details

① Lubrication points should be checked weekly or with each use, whichever is LESS frequent. Based upon individual motor/use needs frequency of actual lubrication should occur at recommended intervals or more often during season. Perform all lubrication procedures immediately prior to extended motor storage

Yamaha Lubrication Chart

Component	Recommended Lubricant	Minimum Frequency ① Fresh Water	Minimum Frequency ① Salt, Polluted or Brackish Water
Choke lever (2.5-4 hp only)	Yamaha all-purpose marine grease	every 100 hours or 60 days	every 50 hours or 30 days
Clamp screws	Yamaha all-purpose marine grease	every 100 hours or 60 days	every 50 hours or 30 days
Cowling clamp lever journal	Yamaha all-purpose marine grease	every 100 hours or 60 days	every 50 hours or 30 days
Engine oil	SAE 10W-30 / 10W-40 API SE or greater (SF, SG)	Check every outing / Replace every 100 hours	Check every outing / Replace every 100 hours
Jet drive bearing lubrication	Yamaha all-purpose marine grease	every 10 hours	every 5 hours
Jet drive bearing grease replacement	Yamaha all-purpose marine grease	every 50 hours	every 25 hours
Lower unit	Yamaha GEAR CASE LUBE (hypoid gear oil - SAE 90)	every 100 hours or 60 days	every 50 hours or 30 days
Manual recoil starter (2.5-4 hp motors)	Yamaha all-purpose marine grease	every 100 hours or 60 days	every 50 hours or 30 days
Manual recoil starter (6 hp or larger motors)	Yamaha all-purpose marine grease	Upon assembly only	Upon assembly only
Propeller shaft	Yamaha all-purpose marine grease	every 100 hours or 60 days	every 50 hours or 30 days
Shift lever journal / housing	Yamaha all-purpose marine grease	every 100 hours or 60 days	every 50 hours or 30 days
Shift mechanism	Yamaha all-purpose marine grease	every 100 hours or 60 days	every 50 hours or 30 days
Steering pivot shaft	Yamaha all-purpose marine grease	every 100 hours or 60 days	every 50 hours or 30 days
Swivel bracket	Yamaha all-purpose marine grease	every 100 hours or 60 days	every 50 hours or 30 days
Throttle control lever	Yamaha all-purpose marine grease	every 100 hours or 60 days	every 50 hours or 30 days
Throttle grip housing	Yamaha all-purpose marine grease	every 100 hours or 60 days	every 50 hours or 30 days
Throttle linkage	Yamaha all-purpose marine grease	every 100 hours or 60 days	every 50 hours or 30 days
Throttle link journal	Yamaha all-purpose marine grease	every 100 hours or 60 days	every 50 hours or 30 days
Tilt mechanism	Yamaha all-purpose marine grease	every 100 hours or 60 days	every 50 hours or 30 days

NOTE: not all components used by all models. Refer to the section on Lubrication and/or your owner's manual for more details

① Lubrication points should be checked weekly or with each use, whichever is LESS frequent. Based upon individual motor/use needs frequency of actual lubrication should occur at recommended intervals or more often during season. Perform all lubrication procedures immediately prior to extended motor storage

2-118 MAINTENANCE & TUNE-UP

Capacities - Yamaha 4-Stroke Motors

Model (Hp)	No. of Cyl	Engine Type	Year	Displace cu. in. (cc)	Gear Oil Oz (mL/cc①)	Crankcase Oil qt (L)
2.5	1	4-stroke	2003-04	4.4 (72)	2.5 (75)	0.37 (0.35)
4	1	4-stroke	1999-04	6.7 (112)	3.4 (100)	0.53 (0.50)
6	2	IL 4-stroke	2001-04	12 (197)	-	0.85 (0.80)
8	2	IL 4-stroke	2001-04	12 (197)	5.1 (150) ②	0.85 (0.80)
8	2	IL 4-stroke	1997-04	14 (232)	-	1.06 (1.0)
9.9	2	IL 4-stroke	1995-04	14 (232)	6.25 (185) ②	1.06 (1.0)
15	2	IL 4-stroke	1998-04	20 (323)	8.45 (250)	1.3 (1.2)
25	2	Standard	1998-04	30 (498)	10.8 (320)	1.8 (1.7)
25	2	High Thrust	2001-04	30 (498)	14.5 (430)	2.01 (1.9)
30	3	IL 4-stroke	2001-04	46 (747)	14.5 (430)	2.33 (2.2)
40J	3	IL 4-stroke	2002-04	46 (747)	-	2.33 (2.2)
40	3	IL 4-stroke	2000-04	46 (747)	14.5 (430)	2.33 (2.2)
40	4	IL 4-stroke	1999-00	58 (935)	14.5 (430)	2.33 (2.2)
45	4	IL 4-stroke	1996-00	58 (935)	14.5 (430)	2.33 (2.2)
50J	4	IL 4-stroke	2002	58 (935)	-	2.33 (2.2)
50	4	Standard	1995-04	58 (935)	14.5 (430)	2.33 (2.2)
50	4	High thrust	1995-04	58 (935)	20.6 (610)	2.33 (2.2)
60J	4	IL 4-stroke	2003-04	61 (996)	-	2.33 (2.2)
60	4	Standard	2002-04	61 (996)	14.5 (430)	2.33 (2.2)
60	4	High thrust	2003-04	61 (996)	22.8 (670)	2.33 (2.2)
75	4	IL 4-stroke	2003-04	97 (1596)	22.8 (670)	4.97 (4.7)
80	4	IL 4-stroke	1999-02	97 (1596)	22.8 (670)	4.97 (4.7)
90J	4	IL 4-stroke	2003-04	97 (1596)	-	4.97 (4.7)
90	4	IL 4-stroke	2003-04	97 (1596)	22.8 (670)	4.97 (4.7)
100	4	IL 4-stroke	1999-02	97 (1596)	22.8 (670)	4.97 (4.7)
115J	4	IL 4-stroke	2003-04	106 (1741)	-	4.97 (4.7)
115	4	IL 4-stroke	2000-04	106 (1741)	25.7 (760) ③	4.97 (4.7)
150	4	IL 4-stroke	2004	163 (2670)	33.1 (980) ④	5.7 (5.4)
200	6	60 V 4-stroke	2002-04	205 (3352)	38.9 (1150) ⑤	6.3 (6.0)
225	6	60 V 4-stroke	2002-04	205 (3352)	38.9 (1150) ⑤	6.3 (6.0)

n/a - not applicable

NOTE - all capacities are approximate. Always add lubricants gradually, checking the level often
Because manufacturers differ with what specifications they choose to provide, please refer to the Mercury/Mariner charts for additional specifications which MAY apply to the Yamaha model in question as well
① Yamaha provides this specification as CCs which is equal to milliliters (the measurement found on most bottles of gear lube)
② Specification is for standard F models, the capacity on T (high-thrust) models is 10.9 oz (320mL)
③ Specification is for all except LF/FL models (counter-rotating units) for which the spec is 24.2 oz (715mL)
④ Specification is for all except LF models (counter-rotating units) for which the spec is 29.4 oz (870mL)

Capacities - Mercury/Mariner 4-Stroke Motors

Model (Hp)	No. of Cyl	Engine Type	Year	Displace cu. in. (cc)	Gear Oil Oz (mL/cc①)	Crankcase Oil qt (L)
4	1	Single 4-st	1999-04	7.5 (123)	6.5 (195)	0.47 (0.45)
5	1	Single 4-st	1999-04	7.5 (123)	6.5 (195)	0.47 (0.45)
6	1	Single 4-st	2000-04	7.5 (123)	6.5 (195)	0.47 (0.45)
8	2	IL 4-stroke	1997-98	14 (232)	6.8 (200)	1.06 (1.0)
9.9	2	IL 4-stroke	1995-98	14 (232)	6.8 (200)	1.06 (1.0)
9.9	2	Standard / BF	1999-04	20 (323)	6.8 (200) / 7.8 (230)	1.3 (1.2)
15	2	Standard / BF	1999-04	20 (323)	6.8 (200) / 7.8 (230)	1.3 (1.2)
25	2	IL 4-stroke	1997-04	30 (498)	14.9 (440)	3.0 (2.8)
30 Carb	3	Standard / BF	1999-04	46 (747)	14.9 (440) / 22.5 (655)	3.0 (2.8)
30 EFI	3	Standard / BF	2002-04	46 (747)	14.9 (440) / 24 (710)	3.0 (2.8)
40 Carb	3	Standard / BF	1999-04	46 (747)	14.9 (440) / 22.5 (655)	3.0 (2.8)
40 EFI	3	Standard / BF	2002-04	46 (747)	14.9 (440) / 24 (710)	3.0 (2.8)
40	4	Standard	1997-00	58 (935)	14.9 (440)	3.0 (2.8)
45	4	Standard	1995-00	58 (935)	14.9 (440)	3.0 (2.8)
50	4	Standard / BF	1995-00	58 (935)	14.9 (440) / 22.5 (655)	3.0 (2.8)
40 Carb	4	Standard / BF	2001	61 (996)	11.5 (340) / 24 (710)	3.0 (2.8)
40 EFI	4	Standard / BF	2002-04	61 (996)	11.5 (340) / 24 (710)	3.0 (2.8)
50 Carb	4	Standard / BF	2001	61 (996)	11.5 (340) / 24 (710)	3.0 (2.8)
50 EFI	4	Standard / BF	2002-04	61 (996)	11.5 (340) / 24 (710)	3.0 (2.8)
60 Carb	4	Standard / BF	2001	61 (996)	11.5 (340) / 24 (710)	3.0 (2.8)
60 EFI	4	Standard / BF	2002-04	61 (996)	11.5 (340) / 24 (710)	3.0 (2.8)
75	4	IL 4-stroke	2000-04	97 (1596)	24 (710)	4.97 (4.7)
90	4	IL 4-stroke	2000-04	97 (1596)	24 (710)	4.97 (4.7)
115 EFI	4	IL 4-stroke	2001-04	106 (1741)	24 (710)	4.97 (4.7)
225 EFI	6	60 V 4-stroke	2003-04	205 (3352)	38.9 (1150) ②	6.3 (6.0)

BF - BigFoot
n/a - not applicable
Because manufacturers differ with what specifications they choose to provide, please refer to the Yamaha charts for additional specifications which MAY apply to the Mercury/Mariner model in question as well
NOTE - all capacities are approximate. Always add lubricants gradually, checking the level often
① Some manufacturers provide this specification as CCs which is equal to milliliters (the measurement found on most bottles of gear lube)
② Specification is for all except counter-rotating units for which the spec is 33.8 oz (1000mL)

MAINTENANCE & TUNE-UP

TUNE-UP SPECIFICATIONS CHART - Yamaha 4-Stroke Motors

Model (Hp)	No. of Cyl	Engine Type	Year	Displace cu. in. (cc)	Spark Plug NGK	Spark Plug Champion	Gap Inch(mm)	Ignition Timing Degrees Idle	Ignition Timing Degrees WOT	Idle Speed RPM Neutral	Idle Speed RPM Trolling	Tach Pole Setting	Full Throttle RPM	Test Propeller Kent Moore Part Number	Test Propeller New Engine RPM
2.5	1	4-stroke	2003-04	4.4 (72)	BR6HS	L86C	0.024-0.028 (0.6-0.7)	TCI cont.	30	1800-2000	-	-	5250-5750	-	-
4	1	4-stroke	1999-04	6.7 (112)	B7HS	L82C	0.024-0.028 (0.6-0.7)	TCI cont.	28-32	1450-1550	1300-1400	-	4000-5000	YB-1630	5250-5450
6	2	IL 4-stroke	2001-04	12 (197)	BR6HS-10	RL86C	0.035-0.039 (0.9-1.0)	2-8	27-33	1000-1100	-	6	4500-5500	YB-1625	4450-4650
8	2	IL 4-stroke	2001-04	12 (197)	BR6HS-10	RL86C	0.035-0.039 (0.9-1.0)	2-8	27-33	1000-1100	-	6	5000-6000	YB-1625	5000-5200
8	2	IL 4-stroke	1997-04	14 (232)	CR5HS	-	0.024-0.028 (0.6-0.7)	2-8	32-38	900-1000	800-900	6	4500-5000	YB-1619	3950-4150
9.9	2	IL 4-stroke	1995-04	14 (232)	CR6HS	-	0.024-0.028 (0.6-0.7)	2-8	32-38	①	①	6	②	③	③
15	2	IL 4-stroke	1998-04	20 (323)	DPR6EA-9	-	0.031-0.035 (0.8-0.9)	2-8	27-33	900-1000	800-900	6	4500-5500	YB-1619	5000-5400
25	2	IL 4-stroke	1998	30 (498)	DPR6EA-9	RL86C	0.031-0.035 (0.8-0.9)	9-11	27-33	875-975	775-875	12	5000-6000	YB-1621	4200-4400
25	2	IL 4-stroke	1999-04	30 (498)	DPR6EA-9	RL86C	0.031-0.035 (0.8-0.9)	4.5-10.5	27-33	875-975	775-875	12	5000-6000	YB-1621	4200-4400
30	3	IL 4-stroke	2001-04	46 (747)	DPR6EA-9	-	0.031-0.035 (0.8-0.9)	3-7	26-30	800-900	700-800	12	5000-6000	-	-
40J	3	IL 4-stroke	2002-04	46 (747)	DPR6EA-9	-	0.031-0.035 (0.8-0.9)	3-7	26-30	800-900	-	12	5000-6000	-	-
40	3	IL 4-stroke	2000-04	46 (747)	DPR6EA-9	-	0.031-0.035 (0.8-0.9)	3-7	26-30	800-900	700-800	12	5000-6000	-	-
40	4	IL 4-stroke	1999-00	58 (935)	DPR6EA-9	-	0.031-0.035 (0.8-0.9)	2-8	32-38	700-800	-	12	5000-6000	YB-1611	5250-5450
45	4	IL 4-stroke	1996	58 (935)	DPR6EA-9	-	0.031-0.035 (0.8-0.9)	2-8	22-28	900-1000	-	12	5000-6000	YB-1611	5250-5450
45	4	IL 4-stroke	1997-00	58 (935)	DPR6EA-9	-	0.031-0.035 (0.8-0.9)	2-8	22-28	700-800	-	12	5000-6000	YB-1611	5250-5450
50J	4	IL 4-stroke	2002	58 (935)	DPR6EA-9	-	0.031-0.035 (0.8-0.9)	2-8	32-38	700-800	-	12	5000-6000	-	-
50	4	Standard	1995	58 (935)	DPR6EA-9	-	0.031-0.035 (0.8-0.9)	3.5-6.5	27-33	700-800	-	12	5000-6000	YB-1611	5250-5450
50	4	High thrust	1995	58 (935)	DPR6EA-9	-	0.031-0.035 (0.8-0.9)	3.5-6.5	27-33	700-800	-	12	5000-6000	YB-1620	-
50	4	Standard	1996-00	58 (935)	DPR6EA-9	-	0.031-0.035 (0.8-0.9)	2-8	32-38	700-800	-	12	5000-6000	YB-1611	5250-5450
50	4	High thrust	1996-00	58 (935)	DPR6EA-9	-	0.031-0.035 (0.8-0.9)	2-8	32-38	900-1000	-	12	5000-6000	YB-1620	-
50	4	Standard	2001-04	58 (935)	DPR6EA-9	-	0.031-0.035 (0.8-0.9)	-1.5-1.5	32-38	900-1000	-	12	5000-6000	YB-1611	5250-5450
50	4	High thrust	2001-04	58 (935)	DPR6EA-9	-	0.031-0.035 (0.8-0.9)	-1.5-1.5	32-38	900-1000	-	12	5000-6000	-	-
60J	4	IL 4-stroke	2003-04	61 (996)	DPR5EA-9	-	0.031-0.035 (0.8-0.9)	-1-1	22-28	800-900	-	12	5000-6000	-	-
60	4	IL 4-stroke	2002-04	61 (996)	DPR5EA-9	-	0.031-0.035 (0.8-0.9)	-1-1	22-28	800-900	-	12	5000-6000	YB-1611	5700-5900
75	4	IL 4-stroke	2003-04	97 (1596)	LFR5A-11	-	0.043 (1.1) ④	3-7 ATDC	19-25	800-900	700-800	12	5000-6000	YB-1624	5900-6000
80	4	IL 4-stroke	1999-02	97 (1596)	LFR5A-11	-	0.043 (1.1) ④	3-7 ATDC	19-25	800-900	700-800	12	5000-6000	YB-1624	5900-6000
90J	4	IL 4-stroke	2003-04	97 (1596)	LFR5A-11	-	0.043 (1.1) ④	3-7 ATDC	19-25	800-900	700-800	12	5000-6000	-	-
90	4	IL 4-stroke	1999-02	97 (1596)	LFR5A-11	-	0.043 (1.1) ④	3-7 ATDC	19-25	800-900	700-800	12	5000-6000	YB-1624	5900-6000
100	4	IL 4-stroke	1999-02	97 (1596)	LFR5A-11	-	0.043 (1.1) ④	3-7 ATDC	19-25	800-900	700-800	12	5000-6000	YB-1624	5900-6000
115J	4	IL 4-stroke	2003-04	106 (1741)	LFR6A-11	-	0.043 (1.1) ④	3-5 ATDC	25-27	700-800	-	12	5000-6000	-	-
115	4	IL 4-stroke	2000-04	106 (1741)	LFR6A-11	-	0.043 (1.1) ④	3-5 ATDC	25-27	700-800	650-750	12	5000-6000	YB-1624	5600-5800
150	4	IL 4-stroke	2004	163 (2670)	LFR5A-11	-	0.039-0.043 (1.0-1.1)	TDC	-	650-750	650-750	12	5000-6000	-	-
200	6	60 V 4-st	2002-04	205 (3352)	LFR5A-11	-	0.035-0.039 (0.9-1.0)	-1-1	20-22	650-750	600-700	12	5000-6000	-	-
225	6	60 V 4-st	2002-04	205 (3352)	LFR5A-11	-	0.035-0.039 (0.9-1.0)	-1-1	20-22	650-750	600-700	12	5000-6000	-	-

Because manufacturers differ with what specifications they choose to provide, please refer to the Mercury/Mariner charts for additional specifications which MAY also apply

NOTE: It is recommended for optimum performance and engine life to adjust the wide open throttle engine speed to the top end of the recommended range

① Specifications vary with model, the standard F9.9 is 900-1000 idle or 800-900 trolling, while high-thrust 9.9 models, the FT9.9 or T9.9 are all 1100-1200 idle or 950-1050 trolling
② Specifications vary with model, the standard F9.9 is 4500-5500, while high thrust 9.9 models, the FT9.9 or T9.9 are all 4000-5000 rpm
③ Specifications vary with model, the standard F9.9 is YB-1619 and 3950-4150, while high thrust 9.9 models, the FT9.9 or T9.9 are all YB-1627 and 4000-4200
④ Specifications conflict in various Yamaha sources. The service guide and tune-up guides list the spec given in the chart, but the Yamaha website shows a range of 0.35-0.39 in. (0.9-1.0mm)

MAINTENANCE & TUNE-UP

TUNE-UP SPECIFICATIONS CHART - Mercury/Mariner 4-Stroke Motors

Model (Hp)	No. of Cyl	Engine Type	Year	Displace cu. in. (cc)	Spark Plug NGK	Spark Plug Champion	Gap Inch(mm)	Ignition Timing Degrees Idle	Ignition Timing Degrees WOT	Idle Speed RPM Neutral	Idle Speed RPM Trolling	Full Throttle RPM
4	1	Single 4-st	1999-00	7.5 (123)	DCPR6E	-	0.035 (0.9)	24-26	24-26	1250-1350	1050-1150	4500-5000
4	1	Single 4-st	2000-04	7.5 (123)	DCPR6E	-	0.035 (0.9)	24-26	24-26	1250-1350	1050-1150	5000-6000
5	1	Single 4-st	1999-00	7.5 (123)	DCPR6E	-	0.035 (0.9)	24-26	24-26	1250-1350	1050-1150	4500-5000
5	1	Single 4-st	2000-04	7.5 (123)	DCPR6E	-	0.035 (0.9)	24-26	24-26	1250-1350	1050-1150	5000-6000
6	1	Single 4-st	2000-05	7.5 (123)	DCPR6E	-	0.035 (0.9)	24-26	24-26	1250-1350	1050-1150	5000-6000
8	2	IL 4-stroke	1997-98	14 (232)	CR6HS	-	0.024-0.028 (0.6-0.7)	4-6	34-36	900-1000	800-900	5000-5500
9.9	2	IL 4-stroke	1995-98	14 (232)	CR6HS	-	0.024-0.028 (0.6-0.7)	4-6	34-36	900-1000	800-900	①
9.9	2	IL 4-stroke	1999-04	20 (323)	DPR6EA-9	-	0.031-0.035 (0.8-0.9)	5	30	900-1000	800-900	②
15	2	IL 4-stroke	1999-04	20 (323)	DPR6EA-9	-	0.031-0.035 (0.8-0.9)	5	30	900-1000	800-900	②
25	2	IL 4-stroke	1997-04	30 (498)	DPR6EA-9	-	0.031-0.035 (0.8-0.9)	5 ③	30	800-850	750-800	5500-6000
30 Carb	3	IL 4-stroke	1999-04	46 (747)	DPR6EA-9	-	0.031-0.035 (0.8-0.9)	10	28	875-925	775-825	5500-6000
30 EFI	3	IL 4-stroke	2002-04	46 (747)	-	RA8HC	0.040 (1.0)	ECM	28	750-800	750-800	5500-6000
40 Carb	3	IL 4-stroke	1999-04	46 (747)	DPR6EA-9	-	0.031-0.035 (0.8-0.9)	10	28	875-925	775-825	5500-6000
40 EFI	3	IL 4-stroke	2002-04	46 (747)	-	RA8HC	0.040 (1.0)	ECM	28	750-800	750-800	5500-6000
40	4	IL 4-stroke	1997-00	58 (935)	DPR6EA-9	-	0.031-0.035 (0.8-0.9)	5	25	800-850	700-750	5000-6000
45	4	IL 4-stroke	1995-00	58 (935)	DPR6EA-9	-	0.031-0.035 (0.8-0.9)	5	25	925-975	825-875	5500-6000
50	4	IL 4-stroke	1995-00	58 (935)	DPR6EA-9	-	0.031-0.035 (0.8-0.9)	5	35	800-850	700-750	5500-6000
40 Carb	4	IL 4-stroke	2001	61 (996)	DPR6EA-9	-	0.031-0.035 (0.8-0.9)	5	28	825-875	700-750	5000-6000
40 EFI	4	IL 4-stroke	2002-04	61 (996)	-	RA8HC	0.040 (1.0)	ECM	28	700-750	700-750	5500-6000
50 Carb	4	IL 4-stroke	2001	61 (996)	DPR6EA-9	-	0.031-0.035 (0.8-0.9)	5	28	825-875	700-750	5000-6000
50 EFI	4	IL 4-stroke	2002-04	61 (996)	-	RA8HC	0.040 (1.0)	ECM	28	700-750	700-750	5500-6000
60 Carb	4	IL 4-stroke	2001	61 (996)	DPR6EA-9	-	0.031-0.035 (0.8-0.9)	5	28	825-875	700-750	5000-6000
60 EFI	4	IL 4-stroke	2002-04	61 (996)	-	RA8HC	0.040 (1.0)	ECM	28	700-750	700-750	5500-6000
75	4	IL 4-stroke	2000-04	97 (1596)	LFR5A-11	-	0.043 (1.1) ④	4-6 ATDC	18-20	825-875	775-800	4500-5500
90	4	IL 4-stroke	2000-04	97 (1596)	LFR5A-11	-	0.043 (1.1) ④	4-6 ATDC	18-20	825-875	775-800	5000-6000
115 EFI	4	IL 4-stroke	2001-04	106 (1741)	LFR6A-11	-	0.039-0.043 (1.0-1.1)	4 ATDC	20	700-800	650-750	5000-6000
225 EFI	6	60 V 4-stroke	2003-04	205 (3352)	LFR5A-11	-	0.039-0.043 (1.0-1.1)	0	24	650-750	600-700	5000-6000

Because manufacturers differ with what specifications they choose to provide, please refer to the Yamaha charts for additional specifications which MAY also apply

NOTE: It is recommended for optimum performance and engine life to adjust the wide open throttle engine speed to the top end of the recommended range

① Specifications are 4500-5500 rpm except for "BODENSEE" models which are 5000-5500 rpm

② Specifications are 4500-5500 rpm except for "BODENSEE" models which are 5200-5700 rpm

③ Specifications are for all models EXCEPT those with ECM PN 855311 to 856057 which are 10 degrees BTDC @ idle

④ Specifications for the Yamaha version of this motor conflict in various Yamaha sources. The Merc service guide and Yamaha tune-up guides list the spec given in the chart, but the Yamaha website shows a range of 0.35-0.39 in. (0.9-1.0mm)

Valve Clearance Specifications - Mercury/Mariner Four-Stroke Engines

Model (Hp)	No. of Cyl	Engine Type	Year	Displace cu. in. (cc)	Intake in. (mm)	Exhaust in. (mm)
4	1	Single 4-st	1999-04	7.5 (123)	0.002-0.005 (0.06-0.14)	0.004-0.007 (0.11-0.19)
5	1	Single 4-st	1999-04	7.5 (123)	0.002-0.005 (0.06-0.14)	0.004-0.007 (0.11-0.19)
6	1	Single 4-st	2000-04	7.5 (123)	0.002-0.005 (0.06-0.14)	0.004-0.007 (0.11-0.19)
8	2	IL 4-stroke	1997-98	14 (232)	0.006-0.008 (0.15-0.20)	0.008-0.010 (0.20-0.25)
9.9	2	IL 4-stroke	1995-98	14 (232)	0.006-0.008 (0.15-0.20)	0.008-0.010 (0.20-0.25)
9.9	2	IL 4-stroke	1999-04	20 (323)	0.006-0.010 (0.15-0.25)	0.008-0.012 (0.20-0.30)
15	2	IL 4-stroke	1999-04	20 (323)	0.006-0.010 (0.15-0.25)	0.008-0.012 (0.20-0.30)
25	2	IL 4-stroke	1997-04	30 (498)	0.006-0.010 (0.15-0.25)	0.010-0.014 (0.25-0.35)
30 Carb	3-cyl	IL 4-stroke	1999-04	46 (747)	0.006-0.010 (0.15-0.25)	0.010-0.014 (0.25-0.35)
30 EFI	3-cyl	IL 4-stroke	2002-04	46 (747)	0.006-0.010 (0.15-0.25)	0.010-0.014 (0.25-0.35)
40 Carb	3-cyl	IL 4-stroke	1999-04	46 (747)	0.006-0.010 (0.15-0.25)	0.010-0.014 (0.25-0.35)
40 EFI	3-cyl	IL 4-stroke	2002-04	46 (747)	0.006-0.010 (0.15-0.25)	0.010-0.014 (0.25-0.35)
40	4	IL 4-stroke	1997-00	58 (935)	0.006-0.010 (0.15-0.25)	0.010-0.014 (0.25-0.35)
45	4	IL 4-stroke	1995-00	58 (935)	0.006-0.010 (0.15-0.25)	0.010-0.014 (0.25-0.35)
50	4	IL 4-stroke	1995-00	58 (935)	0.006-0.010 (0.15-0.25)	0.010-0.014 (0.25-0.35)
40 Carb	4	IL 4-stroke	2001	61 (996)	0.006-0.010 (0.15-0.25)	0.010-0.014 (0.25-0.35)
40 EFI	4	IL 4-stroke	2002-04	61 (996)	0.006-0.010 (0.15-0.25)	0.010-0.014 (0.25-0.35)
50 Carb	4	IL 4-stroke	2001	61 (996)	0.006-0.010 (0.15-0.25)	0.010-0.014 (0.25-0.35)
50 EFI	4	IL 4-stroke	2002-04	61 (996)	0.006-0.010 (0.15-0.25)	0.010-0.014 (0.25-0.35)
60 Carb	4	IL 4-stroke	2001	61 (996)	0.006-0.010 (0.15-0.25)	0.010-0.014 (0.25-0.35)
60 EFI	4	IL 4-stroke	2002-04	61 (996)	0.006-0.010 (0.15-0.25)	0.010-0.014 (0.25-0.35)
75	4	IL 4-stroke	2000-04	97 (1596)	0.007-0.009 (0.17-0.23)	0.012-0.015 (0.31-0.37)
90	4	IL 4-stroke	2000-04	97 (1596)	0.007-0.009 (0.17-0.23)	0.012-0.015 (0.31-0.37)
115 EFI	4	IL 4-stroke	2001-04	106 (1741)	0.007-0.009 (0.17-0.23)	0.012-0.015 (0.31-0.37)
225 EFI	6	60 V 4-stroke	2003-04	205 (3352)	0.007-0.009 (0.17-0.23)	0.012-0.015 (0.31-0.37)

Valve Clearance Specifications - Yamaha Four-Stroke Engines

Model (Hp)	No. of Cyl	Engine Type	Year	Displace cu. in. (cc)	Intake in. (mm)	Exhaust in. (mm)
2.5	1	4-stroke	2003-04	4.4 (72)	0.0031-0.0047 (0.08-0.12)	0.0031-0.0047 (0.08-0.12)
4	1	4-stroke	1999-04	6.7 (112)	0.0032-0.0048 (0.08-0.12)	0.0032-0.0048 (0.08-0.12)
6	2	IL 4-stroke	2001-04	12 (197)	0.006-0.008 (0.15-0.20)	0.008-0.010 (0.20-0.25)
8	2	IL 4-stroke	2001-04	12 (197)	0.006-0.008 (0.15-0.20)	0.008-0.010 (0.20-0.25)
8	2	IL 4-stroke	1997-04	14 (232)	0.006-0.008 (0.15-0.20)	0.008-0.010 (0.20-0.25)
9.9	2	IL 4-stroke	1995-04	14 (232)	0.006-0.008 (0.15-0.20)	0.008-0.010 (0.20-0.25)
15	2	IL 4-stroke	1998-04	20 (323)	0.006-0.010 (0.15-0.25)	0.008-0.012 (0.20-0.30)
25	2	IL 4-stroke	1998-04	30 (498)	0.006-0.010 (0.15-0.25)	0.010-0.014 (0.25-0.35)
30	3	IL 4-stroke	2001-04	46 (747)	0.006-0.010 (0.15-0.25)	0.010-0.014 (0.25-0.35)
40J	3	IL 4-stroke	2002-04	46 (747)	0.006-0.010 (0.15-0.25)	0.010-0.014 (0.25-0.35)
40	3	IL 4-stroke	2000-04	46 (747)	0.006-0.010 (0.15-0.25)	0.010-0.014 (0.25-0.35)
40	4	IL 4-stroke	1999-00	58 (935)	0.006-0.010 (0.15-0.25)	0.010-0.014 (0.25-0.35)
45	4	IL 4-stroke	1996-00	58 (935)	0.006-0.010 (0.15-0.25)	0.010-0.014 (0.25-0.35)
50J	4	IL 4-stroke	2002	58 (935)	0.006-0.010 (0.15-0.25)	0.010-0.014 (0.25-0.35)
50	4	IL 4-stroke	1995-04	58 (935)	0.006-0.010 (0.15-0.25)	0.010-0.014 (0.25-0.35)
60J	4	IL 4-stroke	2003-04	61 (996)	0.006-0.010 (0.15-0.25)	0.010-0.014 (0.25-0.35)
60	4	IL 4-stroke	2002-04	61 (996)	0.006-0.010 (0.15-0.25)	0.010-0.014 (0.25-0.35)
75	4	IL 4-stroke	2003-04	97 (1596)	0.007-0.009 (0.17-0.23)	0.012-0.015 (0.31-0.37)
80	4	IL 4-stroke	1999-02	97 (1596)	0.007-0.009 (0.17-0.23)	0.012-0.015 (0.31-0.37)
90J	4	IL 4-stroke	2003-04	97 (1596)	0.007-0.009 (0.17-0.23)	0.012-0.015 (0.31-0.37)
90	4	IL 4-stroke	2003-04	97 (1596)	0.007-0.009 (0.17-0.23)	0.012-0.015 (0.31-0.37)
100	4	IL 4-stroke	1999-02	97 (1596)	0.007-0.009 (0.17-0.23)	0.012-0.015 (0.31-0.37)
115J	4	IL 4-stroke	2003-04	106 (1741)	0.007-0.009 (0.17-0.23)	0.012-0.015 (0.31-0.37)
115	4	IL 4-stroke	2000-04	106 (1741)	0.007-0.009 (0.17-0.23)	0.012-0.015 (0.31-0.37)
150	4	IL 4-stroke	2004	163 (2670)	0.007-0.009 (0.17-0.23)	0.012-0.015 (0.31-0.37)
200	6	60 V 4-st	2002-03	205 (3352)	0.007-0.009 (0.17-0.23)	0.012-0.015 (0.31-0.37)
225	6	60 V 4-st	2002-03	205 (3352)	0.007-0.009 (0.17-0.23)	0.012-0.015 (0.31-0.37)

2-122 MAINTENANCE & TUNE-UP

SPARK PLUG DIAGNOSIS

Tracking Arc
High voltage arcs between a fouling deposit on the insulator tip and spark plug shell. This ignites the fuel/air mixture at some point along the insulator tip, retarding the ignition timing which causes a power and fuel loss.

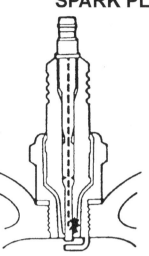

Wide Gap
Spark plug electrodes are worn so that the high voltage charge cannot arc across the electrodes. Improper gapping of electrodes on new or "cleaned" spark plugs could cause a similar condition. Fuel remains unburned and a power loss results.

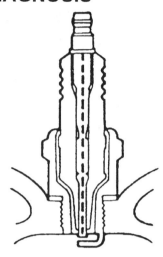

Flashover
A damaged spark plug boot, along with dirt and moisture, could permit the high voltage charge to short over the insulator to the spark plug shell or the engine. A buttress insulator design helps prevent high voltage flashover.

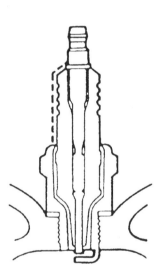

Fouled Spark Plug
Deposits that have formed on the insulator tip may become conductive and provide a "shunt" path to the shell. This prevents the high voltage from arcing between the electrodes. A power and fuel loss is the result.

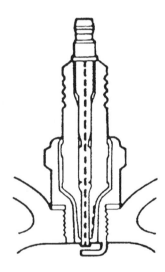

Bridged Electrodes
Fouling deposits between the electrodes "ground out" the high voltage needed to fire the spark plug. The arc between the electrodes does not occur and the fuel air mixture is not ignited. This causes a power loss and exhausting of raw fuel.

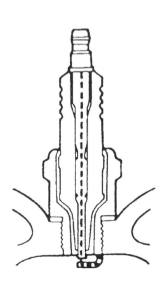

Cracked Insulator
A crack in the spark plug insulator could cause the high voltage charge to "ground out." Here, the spark does not jump the electrode gap and the fuel air mixture is not ignited. This causes a power loss and raw fuel is exhausted.

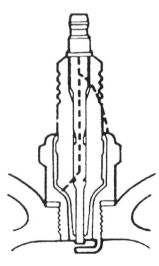

CARBURETED FUEL SYSTEM	3-11	
DESCRIPTION AND OPERATION	3-11	
BASIC FUNCTIONS	3-13	
CARBURETOR CIRCUITS	3-13	
GENERAL INFORMATION	3-11	
TROUBLESHOOTING	3-14	
COMBUSTION RELATED PISTON FAILURES	3-16	
COMMON PROBLEMS	3-15	
LOGICAL	3-15	
CARBURETOR SERVICE	3-17	
CARBURETOR IDENTIFICATION	3-17	
2.5 HP YAMAHA	3-17	
CLEANING & INSPECTION	3-18	
OVERHAUL	3-18	
REMOVAL & INSTALLATION	3-17	
4 HP YAMAHA	3-18	
CLEANING & INSPECTION	3-21	
OVERHAUL	3-20	
REMOVAL & INSTALLATION	3-18	
4/5/6 HP MERCURY/MARINER	3-21	
CLEANING & INSPECTION	3-23	
OVERHAUL	3-21	
REMOVAL & INSTALLATION	3-21	
6/8 HP YAMAHA	3-23	
CLEANING & INSPECTION	3-24	
OVERHAUL	3-24	
REMOVAL & INSTALLATION	3-23	
8/9.9 HP (232cc)	3-24	
CLEANING & INSPECTION	3-28	
OVERHAUL	3-26	
REMOVAL & INSTALLATION	3-25	
9.9/15 HP (323cc)	3-28	
CLEANING & INSPECTION	3-30	
OVERHAUL	3-29	
REMOVAL & INSTALLATION	3-28	
25 HP MODELS	3-30	
CLEANING & INSPECTION	3-33	
OVERHAUL	3-31	
REMOVAL & INSTALLATION	3-31	
30/40 HP MODELS	3-33	
OVERHAUL	3-35	
REMOVAL & INSTALLATION	3-33	
40/45/50 HP (935cc) AND 40/50/60 HP (996cc)	3-35	
CLEANING & INSPECTION	3-44	
OVERHAUL	3-40	
REMOVAL & INSTALLATION	3-35	
75/80/90/100 HP	3-44	
CLEANING & INSPECTION	3-49	
OVERHAUL	3-48	
REMOVAL & INSTALLATION	3-44	
THROTTLE POSITION SENSOR (TPS) ADJUSTMENT	3-47	
ELECTRONIC FUEL INJECTION	3-50	
COMPONENT LOCATIONS	3-86	
CONTROLLED COMBUSTION SYSTEM	3-52	
CRANKSHAFT POSITION SENSOR (CPS)/PULSER COIL	3-80	
ELECTRIC LOW-PRESSURE FUEL PUMP (V6 ONLY)	3-68	
REMOVAL & INSTALLATION	3-69	
TESTING	3-68	
ENGINE CONTROL MODULE (ECM/CDI)	3-77	
REMOVAL & INSTALLATION	3-77	
FUEL DISTRIBUTION MANIFOLD (MERCURY/MARINER)	3-76	
REMOVAL & INSTALLATION	3-76	
FUEL FLOW SCHEMATICS	3-95	
FUEL INJECTION BASICS	3-50	
FUEL INJECTORS	3-72	
REMOVAL & INSTALLATION	3-73	
60 HP AND SMALLER MOTORS (MERCURY/MARINER)	3-73	
115 HP AND LARGER MOTORS (YAMAHA)	3-74	
TESTING	3-73	
FUEL PRESSURE REGULATOR	3-76	
REMOVAL & INSTALLATION	3-77	
TESTING	3-76	
IDLE SPEED/AIR CONTROL (ISC/IAC) VALVE	3-84	
REMOVAL & INSTALLATION	3-84	
MERCURY/MARINER EFI	3-51	
MERCURY/MARINER TROUBLESHOOTING CHARTS	3-55	
LOW-PRESSURE LIFT PUMP PRESSURE TESTING	3-56	
LOW-PRESSURE LIFT PUMP VACUUM TESTING	3-55	
MERCURY EFI SYMPTOM TROUBLESHOOTING	3-57	
WARNING HORN/GUARDIAN SYSTEM FAULTS	3-57	
NEUTRAL OR SHIFT POSITION SWITCHES	3-84	
PRESSURE SENSORS (IAP/MAP)	3-83	
REMOVAL & INSTALLATION	3-84	
TESTING	3-83	
SELF DIAGNOSTIC SYSTEM	3-53	
READING TROUBLE CODES	3-53	
SENSOR AND CIRCUIT RESISTANCE/OUTPUT TESTS	3-57	
TESTING ELECTRONIC COMPONENTS	3-57	
TEMPERATURE SENSORS	3-80	
REMOVAL & INSTALLATION	3-83	
TESTING	3-81	
THROTTLE BODY AND INTAKE ASSEMBLY	3-61	
REMOVAL & INSTALLATION	3-61	
30-60 HP MERCURY/MARINER	3-61	
115 HP	3-63	
150 HP	3-65	
200/225 HP	3-66	
THROTTLE POSITION SENSOR (TPS)	3-78	
REMOVAL & INSTALLATION	3-80	
TESTING	3-78	
TROUBLESHOOTING	3-52	
VAPOR SEPARATOR TANK (VST) AND HIGH-PRESSURE FUEL PUMP	3-69	
TESTING	3-69	
REMOVAL & INSTALLATION	3-69	
OVERHAUL	3-72	
YAMAHA EFI	3-50	
FUEL AND COMBUSTION BASICS	**3-2**	
COMBUSTION	3-5	
ABNORMAL	3-5	
FACTORS AFFECTING	3-5	
FUEL	3-2	
ALCOHOL-BLENDED FUELS	3-3	
CHECKING FOR STALE/CONTAMINATED FUEL	3-3	
HIGH ALTITUDE OPERATION	3-3	
OCTANE RATING	3-2	
THE BOTTOM LINE WITH FUELS	3-3	
VAPOR PRESSURE	3-3	
FUEL SYSTEM PRESSURIZATION	3-4	
PRESSURIZING (CHECKING FOR LEAKS)	3-5	
RELIEVING PRESSURE (EFI)	3-4	
FUEL SYSTEM SERVICE CAUTIONS	3-2	
FUEL PUMP (LOW-PRESSURE) SERVICE	**3-97**	
FUEL PUMPS	3-97	
GENERAL INFORMATION	3-97	
TESTING	3-98	
PLUNGER ACTUATED FUEL PUMPS	3-100	
DESCRIPTION & OPERATION	3-100	
OVERHAUL	3-101	
REMOVAL & INSTALLATION	3-100	
FUEL TANK AND LINES	**3-6**	
FUEL LINES AND FITTINGS	3-8	
SERVICE	3-10	
TESTING	3-8	
FUEL TANK	3-6	
SERVICE	3-7	
SPECIFICATIONS	**3-105**	
MERCURY/MARINER	3-106	
YAMAHA	3-105	

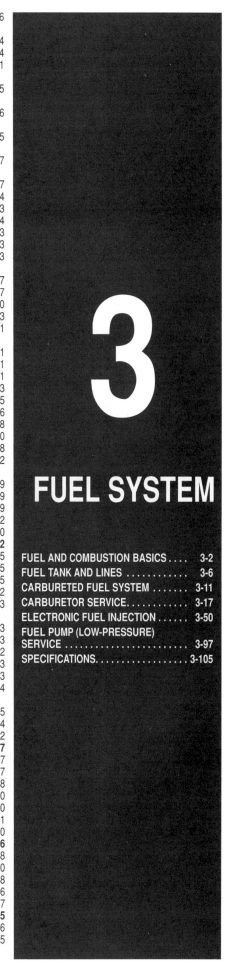

3

FUEL SYSTEM

FUEL AND COMBUSTION BASICS	3-2
FUEL TANK AND LINES	3-6
CARBURETED FUEL SYSTEM	3-11
CARBURETOR SERVICE	3-17
ELECTRONIC FUEL INJECTION	3-50
FUEL PUMP (LOW-PRESSURE) SERVICE	3-97
SPECIFICATIONS	3-105

3-2 FUEL SYSTEM

FUEL AND COMBUSTION BASICS

※※ CAUTION

If equipped, disconnect the negative battery cable ANYTIME work is performed on the engine, especially when working on the fuel system. This will help prevent the possibility of sparks during service (from accidentally grounding a hot lead or powered component). Sparks could ignite vapors or exposed fuel. Disconnecting the cable on electric start motors will also help prevent the possibility fuel spillage if an attempt is made to crank the engine while the fuel system is open.

※※ CAUTION

Fuel leaking from a loose, damaged, or incorrectly installed hose or fitting may cause a fire or an explosion. ALWAYS pressurize the fuel system and run the motor while inspecting for leaks after servicing any component of the fuel system.

To tune, troubleshoot or repair the fuel system of an outboard motor you must first understand the carburetion or fuel injection, and ignition principles of engine operation.

If you have any doubts concerning your understanding of engine operation, it would be best to study

The Basic Operating Principles of an engine as detailed under Troubleshooting in Section 1, before tackling any work on the fuel system.

The fuel systems used on engines covered by this guide range from single carburetors to multiple carburetors or various forms of electronic fuel injection systems. Carbureted motors utilize various means of enriching fuel mixture for cold starts, including a manual choke, manual primer or electric primer solenoid.

Fuel System Service Cautions

There is no way around it. Working with gasoline can provide for many different safety hazards and requires that extra caution is used during all steps of service. To protect yourself and others, you must take all necessary precautions against igniting the fuel or vapors (which will cause a fire at best or an explosion at worst).

※※ CAUTION

Take extreme care when working with the fuel system. NEVER smoke (it's bad for you anyhow, but smoking during fuel system service could kill you much faster!) or allow flames or sparks in the work area. Flames or sparks can ignite fuel, especially vapors, resulting in a fire at best or an explosion at worst.

For starters, disconnect the negative battery cable EVERY time a fuel system hose or fitting is going to be disconnected. It takes only one moment of forgetfulness for someone to crank the motor, possibly causing a dangerous spray of fuel from the opening. This is especially true on EFI systems that contain an electric fuel pump, where simply turning the key on (without cranking) will energize the pump and result in high-pressure fuel spray in certain fuel system lines/fittings.

Gasoline contains harmful additives and is quickly absorbed by exposed skin. As an additional precaution, always wear gloves and some form of eye protection (regular glasses help, but only safety glasses can really protect your eyes).

■ Throughout service, pay attention to ensure that all components, hoses and fittings are installed them in the correct location and orientation to prevent the possibility of leakage. Matchmark components before they are removed as necessary.

Because of the dangerous conditions that result from working with gasoline and fuel vapors

always take extra care and be sure to follow these guidelines for safety:
- Keep a Coast Guard-approved fire extinguisher handy when working.
- Allow the engine to cool completely before opening a fuel fitting. Don't allow gasoline to drip on a hot engine.
- The first thing you must do after removing the engine cover is to check for the presence of gasoline fumes. If strong fumes are present, look for leaking or damage hoses, fittings or other fuel system components and repair.
- Do not repair the motor or any fuel system component near any sources of ignition; including sparks, open flames, or anyone smoking.
- Clean up spilled gasoline right away using clean rags. Keep all fuel soaked rags in a metal container until they can be properly disposed of or cleaned. NEVER leave solvent, gasoline or oil soaked rags in the hull.
- Don't use electric powered tools in the hull or near the boat during fuel system service or after service, until the system is pressurized and checked for leaks.
- Fuel leaking from a loose, damaged or incorrectly installed hose or fitting may cause a fire or an explosion. ALWAYS pressurize the fuel system and run the motor while inspecting for leaks after servicing any component of the fuel system.

Fuel

◆ See Figure 1

Fuel recommendations have become more complex as the chemistry of modern gasoline changes. The major driving force behind the many of the changes in gasoline chemistry was the search for additives to replace lead as an octane booster and lubricant. These additives are governed by the types of emissions they produce in the combustion process. Also, the replacement additives do not always provide the same level of combustion stability, making a fuel's octane rating less meaningful.

In the 1960's and 1970's, leaded fuel was common. The lead served two functions. First, it served as an octane booster (combustion stabilizer) and second it served as a valve seat lubricant.

For decades now, all lead has been removed from the refining process. This means that the benefit of lead as an octane booster has been eliminated. Several substitute octane boosters have been introduced in the place of lead. While many are adequate in automobile engines, most do not perform nearly as well as lead did, even though the octane rating of the fuel is the same.

OCTANE RATING

◆ See Figure 1

A fuel's octane rating is a measurement of how stable the fuel is when heat is introduced. Octane rating is a major consideration when deciding whether a fuel is suitable for a particular application. For example, in an engine, we want the fuel to ignite when the spark plug fires and not before, even under high pressure and temperatures. Once the fuel is ignited, it must burn slowly and smoothly, even though heat and pressure are building up while the burn occurs. The unburned fuel should be ignited by the traveling flame front, not by some other source of ignition, such as carbon deposits or the heat from the expanding gasses. A fuel's octane rating is known as a

Fig. 1 Damaged piston, possibly caused by; using too-low an octane fuel or by insufficient oil

FUEL SYSTEM

measurement of the fuel's anti-knock properties (ability to burn without exploding). Essentially, the octane rating is a measure of a fuel's stability.

Usually a fuel with a higher octane rating can be subjected to a more severe combustion environment before spontaneous or abnormal combustion occurs. To understand how two gasoline samples can be different, even though they have the same octane rating, we need to know how octane rating is determined.

The American Society of Testing and Materials (ASTM) has developed a universal method of determining the octane rating of a fuel sample. The octane rating you see on the pump at a gasoline station is known as the pump octane number. Look at the small print on the pump. The rating has a formula. The rating is determined by the R+M/2 method. This number is the average of the research octane reading and the motor octane rating.

• The Research Octane Rating is a measure of a fuel's anti-knock properties under a light load or part throttle conditions. During this test, combustion heat is easily dissipated.

• The Motor Octane Rating is a measure of a fuel's anti-knock properties under a heavy load or full throttle conditions, when heat buildup is at maximum.

VAPOR PRESSURE

Fuel vapor pressure is a measure of how easily a fuel sample evaporates. Many additives used in gasoline contain aromatics. Aromatics are light hydrocarbons distilled off the top of a crude oil sample. They are effective at increasing the research octane of a fuel sample but can cause vapor lock (bubbles in the fuel line) on a very hot day. If you have an inconsistent running engine and you suspect vapor lock, use a piece of clear fuel line to look for bubbles, indicating that the fuel is vaporizing.

One negative side effect of aromatics is that they create additional combustion products such as carbon and varnish. If your engine requires high octane fuel to prevent detonation, de-carbon the engine more frequently with an internal engine cleaner to prevent ring sticking due to excessive varnish buildup.

ALCOHOL-BLENDED FUELS

When the Environmental Protection Agency mandated a phase-out of the leaded fuels in January of 1986, fuel suppliers needed an additive to improve the octane rating of their fuels. Although there are multiple methods currently employed, the addition of alcohol to gasoline seems to be favored because of its favorable results and low cost. Two types of alcohol are used in fuel today as octane boosters, methanol (wood alcohol) or ethanol (grain alcohol).

When used as a fuel additive, alcohol tends to raise the research octane of the fuel, so these additives will have limited benefit in an outboard motor. There are, however, some special considerations due to the effects of alcohol in fuel.

• Since alcohol contains oxygen, it replaces gasoline without oxygen content and tends to cause the air/fuel mixture to become leaner.

• On older outboards, the leaching affect of alcohol will, in time, cause fuel lines and plastic components to become brittle to the point of cracking. Unless replaced, these cracked lines could leak fuel, increasing the potential for hazardous situations. However, later model outboards (such as the models covered here) utilize fuel lines which are much more resistant to alcohol in the fuel mixture.

■ **Modern outboard fuel lines and plastic fuel system components have been specially formulated to resist alcohol leaching effects.**

• When alcohol blended fuels become contaminated with water, the water combines with the alcohol then settles to the bottom of the tank. This leaves the gasoline on a top layer.

THE BOTTOM LINE WITH FUELS

If we could buy fuel of the correct octane rating, free of alcohol and aromatics, this would be our first choice.

Yamaha and Mercury continue to recommend unleaded fuel. This is almost a redundant recommendation due to the near universal unavailability of any other type fuel.

According to the fuel recommendations that come with your outboard, there is no engine in the product line that requires more than 89 octane. Most Yamaha and Mercury/Mariner engines (including EFI models) need only 86 octane or less. An 89 octane rating generally means middle grade unleaded. Premium unleaded is more stable under severe conditions, but also produces more combustion products. Therefore, when using premium unleaded, more frequent de-carboning is necessary.

Regardless of the fuel octane rating you choose, try to stay with a name brand fuel. You never know for sure what kinds of additives or how much is in off brand fuel.

HIGH ALTITUDE OPERATION

At elevated altitudes there is less oxygen in the atmosphere than at sea level. Less oxygen means lower combustion efficiency and less power output. Power output is reduced three percent for every thousand feet above sea level. At ten thousand feet, power is reduced 30 percent from that available at sea level.

Re-jetting carburetors for high altitude does not restore this lost power. It simply corrects the air-fuel ratio for the reduced air density, and makes the most of the remaining available power. If you re-jet an engine, you are locked into the higher elevation. You cannot operate at sea level until you re-jet for sea level. Understand that going below the elevation jetted for your motor will damage the engine with overly lean fuel:air ratios. As a general rule, jet for the lowest elevation anticipated. Spark plug insulator tip color is the best guide for high altitude jetting.

If you are in an area of known poor fuel quality, you may want to use fuel additives. Today's additives are mostly alcohol and aromatics, and their effectiveness may be limited. It is difficult to find additives without ethanol, methanol, or aromatics. If you use octane boosters frequent de-carboning may be necessary. If possible, the best policy is to use name brand pump fuel with no additional additives except Yamaha fuel conditioner and Ring-Free(tm) or Quicksilver Fuel Stabilizer.

CHECKING FOR STALE/CONTAMINATED FUEL

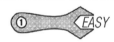

◆ See Figures 2 and 3

Outboard motors often sit weeks at a time making them the perfect candidate for fuel problems. Gasoline has a short life, as combustibles begin evaporating almost immediately. Even when stored properly, fuel starts to deteriorate within a few months, leaving behind a stale fuel mixture that can cause hard-starting, poor engine performance and even lead to possible engine damage.

Further more, as gasoline evaporates it leaves behind gum deposits that can clog filters, lines and small passages. Although the sealed high-pressure fuel system of an EFI motor is less susceptible to fuel evaporation, the low-pressure fuel systems of all engines can suffer the affects. However carburetors, due to their tiny passages and naturally vented designs are the most susceptible components on non-EFI motors.

As mentioned under Alcohol-Blended fuels, modern fuels contain alcohol, which is hydroscopic (meaning it absorbs water). And, over time, fuel stored in a partially filled tank or a tank that is vented to the atmosphere will absorb water. The water/alcohol settles to the bottom of the tank, promoting rust (in metal tanks) and leaving a non-combustible mixture at the bottom of a tank that could leave a boater stranded.

One of the first steps to fuel system troubleshooting is to make sure the fuel source is not at fault for engine performance problems. Check the fuel if the engine will not start and there is no ignition problem.

Stale or contaminated fuels will often exhibit an unusual or even unpleasant unusual odor.

■ **The best method of disposing stale fuel is through a local waste pickup service, automotive repair facility or marine dealership. But, this can be a hassle. If fuel is not too stale or too badly contaminated, it may be mixed with greater amounts of fresh fuel and used to power lawn/yard equipment or (as long as we're not talking about disposing pre-mix here) even an automobile (if greatly diluted so as to prevent misfiring, unstable idle or damage to the automotive engine). But we feel that it is much less of a risk to have a lawn mower stop running because of the fuel problem than it is to have your boat motor quit or refuse to start.**

Yamaha and Mercury/Mariner carburetors are normally equipped with a float bowl drain screw that can be used to drain fuel from the carburetor for storage or for inspection. The vapor separator tank on EFI models is also usually equipped with a drain screw, not unlike the float bowl screw on carburetor motors. The drain screws on these motors are usually pretty easy to access, but on a few motors it may be easier to drain a fuel sample from

3-4 FUEL SYSTEM

the hoses leading to or from the low-pressure fuel filter or fuel pump. Removal and installation instructions for the fuel filters are provided in the Maintenance Section, while fuel pump procedures are found in this section. To check for stale or contaminated fuel:

1. Disconnect the negative battery cable for safety. Secure it or place tape over the end so that it cannot accidentally contact the terminal and complete the circuit.

✳✳ CAUTION

Throughout this procedure, clean up any spilled fuel to prevent a fire hazard.

Fig. 2 Yamaha and Mercury/Mariner float bowls are usually equipped with drain screws which make draining the float bowls or taking a fuel sample a snap!

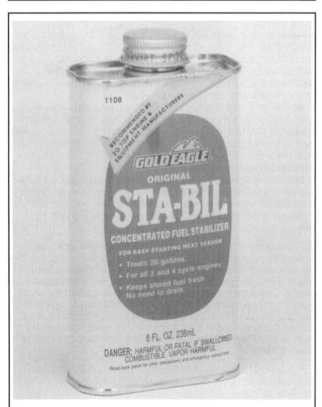

Fig. 3 Commercial additives, such as Sta-bil, may be used to help prevent "souring"

2. For carbureted motors, remove the float bowl drain screw (and orifice plug, if equipped), then allow a small amount of fuel to drain into a glass container.

3. For EFI motors, remove the float bowl drain screw from the bottom of the vapor separator tank or disconnect the fuel supply hose from the pump or low-pressure fuel filter (as desired).

4. If there is no fuel present in the float bowl, vapor tank or fuel line (as applicable) squeeze the fuel primer bulb to force fuel through the lines and to the opening. If necessary, you can always disconnect the fuel line between the low-pressure filter and pump or, on models with serviceable filter elements simply unscrew the housing to get a sample.

■ **If a sample cannot be obtained from the float bowls/separator tank, fuel filter or pump supply hose, there is a problem with the fuel tank-to-motor fuel circuit. Check the tank, primer bulb, fuel hose, fuel pump, fitting or carburetor float inlet needle (as necessary).**

5. Check the appearance and odor of the fuel. An unusual smell, signs of visible debris or a cloudy appearance (or even the obvious presence of water) points to a fuel that should be replaced.

6. If contaminated fuel is found, drain the fuel system and dispose of the fuel in a responsible manner, then clean the entire fuel system.

■ **If debris is found in the fuel system, clean and/or replace all fuel filters.**

7. When finished, reconnect the negative battery cable, then properly pressurize the fuel system and check for leaks.

Fuel System Pressurization

When it comes to safety and outboards, the condition of the fuel system is of the utmost importance. The system must be checked for signs of damage or leakage with every use and checked, especially carefully when portions of the system have been opened for service.

The best method to check the fuel system is to visually inspect the lines, hoses and fittings once the system has been properly pressurized.

All engines, carbureted or fuel injected utilize a low-pressure fuel circuit which delivers fuel from a portable or boat mounted tank to the powerhead via fuel hoses and the action of a diaphragm-displacement fuel pump.

EFI engines utilize a 2-stage fuel system, the low-pressure circuit (fuel tank-to-pump and pump-to-vapor separator), as well as a high-pressure circuit (vapor separator-to-fuel injectors).

The high-pressure circuit of all fuel injected motors works in a similar fashion using a submerged electric fuel pump located within the vapor separator tank to achieve fuel pressure somewhere in the 30-44 psi (207-304 kPa) range depending upon the year and model. Most smaller models (60 hp and smaller) are equipped with a fuel test port on the top of the separator tank, but there are exceptions (such as all 115 hp and larger motors) where the port is mounted to the top of the fuel rail assembly.

RELIEVING FUEL SYSTEM PRESSURE (EFI MOTORS)

◆ See Figure 4

As its name implies, the high-pressure circuit of EFI motors contains fuel under pressure that, if given the chance, will spray from a damaged/loose hose or fitting. When servicing components of the high-pressure system, the fuel pressure must first be relieved in a safe and controlled manner to help avoid the potential explosive and dangerous conditions that would result from simply opening a fitting and allowing fuel to spray uncontrolled into the work area.

Relieving fuel system pressure from the high-pressure fuel circuit is a relatively straight forward procedure on Yamaha and Mercury/Mariner motors, however Yamaha service literature varies slightly on a couple of models when it comes to recommendations.

All fuel injected motors are equipped with a high-pressure test port (a threaded Schrader type valve) to which you can connect a fuel pressure gauge. This port is usually found under a rubber cap on the top of the vapor separator tank for 60 hp and smaller motors or it is mounted to the fuel rail assembly for 115 and larger motors (on top of the rail for all but the 150 hp motor where it is found midway down the rail).

The BEST method of relieving fuel system pressure is to cover this fitting with a rag and then attach a fuel pressure gauge which is equipped with a

FUEL SYSTEM 3-5

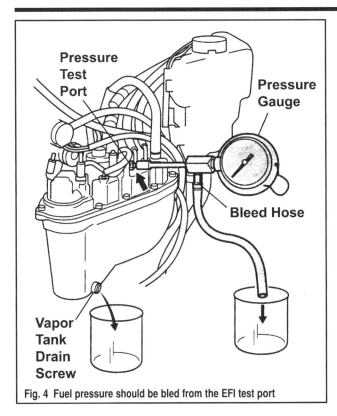

Fig. 4 Fuel pressure should be bled from the EFI test port

pressure bleed valve/hose. Direct the bleed hose into a suitable container, then slowly open the bleed valve, draining (spraying) fuel from the hose into the fuel storage container. Once the pressure has dropped, remove the drain screw and drain the remaining fuel from the vapor separator tank. At this point you will be able to disconnect any of the high-pressure fuel lines/fittings without worrying about spray, however fuel will still be present in lines and so you'll want to keep a rag handy.

Now in a few service publications Yamaha shows fuel pressure relief WITHOUT the use of a fuel gauge, by placing a rag over the fitting, then depressing the Schrader valve fitting using the tip of a screwdriver. We can't honestly recommend this method, because it is both more dangerous and messy than using a pressure gauge. However, if precautions are taken against fuel spray, it could in theory be a viable alternative to the use of a pressure gauge. But we emphasize that you'll need protection (rubber gloves, safety glasses), a thick shop rag, and a well-ventilated area, completely free of potential sources of ignition (sparks or flame).

✳✳ WARNING

Even after following these steps, some residual pressure (and definitely, some residual fuel) will be present in the fuel system. Always cover lines and fittings with a shop rag when disconnecting them to protect yourself from fuel spray.

PRESSURIZING THE FUEL SYSTEM (CHECKING FOR LEAKS)

Carbureted Motors

Carbureted engines are equipped with a low-pressure fuel system, making pressure release before service a relative non-issue. However, if the primer bulb is firm when you go to disconnect a fuel line, there will be some fuel leakage once the hose is disconnected, so always have a rag or handy during service. But, even a low-pressure fuel system should be checked following repairs to make sure that no leaks are present. Only by checking a fuel system under normal operating pressures can you be sure of the system's integrity.

✳✳ CAUTION

Fuel leaking from a loose, damaged or incorrectly installed hose or fitting may cause a fire or an explosion. ALWAYS pressurize the fuel system and run the motor while inspecting for leaks after servicing any component of the fuel system.

Most carbureted engines (except some integral tank models with gravity feed) utilize a fuel primer bulb mounted inline between the fuel tank and engine. On models so equipped, the bulb can be used to pressurize that portion of the fuel system. Squeeze the bulb until it and the fuel lines feel firm with gasoline. At this point check all fittings between the tank and motor for signs of leakage and correct, as necessary.

Once fuel reaches the engine it is the job of the fuel pump(s) to distribute it to the carburetor(s). No matter what system you are inspecting, start and run the motor with the engine top case removed, then check each of the system hoses, fittings and gasket-sealed components to be sure there is no leakage after service.

Fuel Injected Motors

◆ See Figure 4

The low-pressure circuit of an EFI motor is pressurized and checked using the primer bulb, in the same manner as the low-pressure circuit of a carbureted motor. For more details, refer to the section on Carbureted Motors.

However, the vapor separator tank contains an electric high-pressure pump which is used on the high-pressure circuit to build system pressure. Normally this pump is energized for a few seconds when the ignition switch is turned on (even before the engine is started). If so, you can usually hear the pump run for a few seconds when the switch is turned on. For these models, turn the ignition switch on, wait a few seconds, and then turn the switch off again. Repeat a few times until system pressure starts to build (you can monitor this using a pressure gauge attached to the fuel pressure test port).

■ The BEST way to make sure no leaks occur during service is to visually inspect the motor while it is running at idle (that way all fuel circuits are pressurized and operating as normal).

✳✳ CAUTION

As usual, use extreme caution when working around an operating engine. Keep hands, fingers, hair, loose clothing, etc as far away as possible from moving components. Severe injury or death could occur from getting caught in rotating components.

Combustion

A two cycle engine has a power stroke every revolution of the crankshaft. A four cycle engine has a power stroke every other revolution of the crankshaft. Therefore, the two cycle engine has twice as many power strokes for any given RPM. If the displacement of the two types of engines is identical, then the two cycle engine has to dissipate twice as much heat as the four cycle engine. In such a high heat environment, the fuel must be very stable to avoid detonation. If any parameters affecting combustion change suddenly (the engine runs lean for example), uncontrolled heat buildup occurs very rapidly in a two cycle engine.

ABNORMAL COMBUSTION

There are two types of abnormal combustion:
• Pre-ignition - Occurs when the air-fuel mixture is ignited by some other incandescent source other than the correctly timed spark from the spark plug.
• Detonation - Occurs when excessive heat and or pressure ignites the air/fuel mixture rather than the spark plug. The burn becomes explosive.

FACTORS AFFECTING COMBUSTION

The combustion process is affected by several interrelated factors. This means that when one factor is changed, the other factors also must be changed to maintain the same controlled burn and level of combustion stability.

3-6 FUEL SYSTEM

Compression

Determines the level of heat buildup in the cylinder when the air-fuel mixture is compressed. As compression increases, so does the potential for heat buildup. Over time, the build-up of carbon deposits can increase compression (by decreasing the available size of the combustion chamber). Similarly, machining the cylinder head (to correct for warpage) can increase compression. Increases of compression often bring with them the need to increase fuel octane to prevent abnormal combustion.

Ignition Timing

Determines when the ignited gasses will start to expand in relation to the motion of the piston. If the gasses begin to expand too soon, such as they would during pre-ignition or in an overly advanced ignition timing, the motion of the piston opposes the expansion of the gasses, resulting in extremely high combustion chamber pressures and heat.

As ignition timing is retarded, the burn occurs later in relation to piston position. This means that the piston has less distance to travel under power to the bottom of the cylinder, resulting in less usable power.

Fuel Mixture

Determines how efficient the burn will be. A rich mixture burns cooler and slower than a lean one. However, if the mixture is too lean, it can't become explosive. The slower the burn, the cooler the combustion chamber, because pressure buildup is gradual.

Fuel Quality (Octane Rating)

Determines how much heat is necessary to ignite the mixture. Once the burn is in progress, heat is on the rise. The unburned poor quality fuel is ignited all at once by the rising heat instead of burning gradually as a flame front of the burn passing by. This action results in detonation (pinging).

Other Factors

In general, anything that can cause abnormal heat buildup can be enough to push an engine over the edge to abnormal combustion, if any of the four basic factors previously discussed are already near the danger point, for example, excessive carbon buildup raises the compression and retains heat as glowing embers.

FUEL TANK AND LINES

✳✳ CAUTION

If equipped, disconnect the negative battery cable ANYTIME work is performed on the engine, especially when working on the fuel system. This will help prevent the possibility of sparks during service (from accidentally grounding a hot lead or powered component). Sparks could ignite vapors or exposed fuel. Disconnecting the cable on electric start motors will also help prevent the possibility fuel spillage if an attempt is made to crank the engine while the fuel system is open.

✳✳ CAUTION

Fuel leaking from a loose, damaged or incorrectly installed hose or fitting may cause a fire or an explosion. ALWAYS pressurize the fuel system and run the motor while inspecting for leaks after servicing any component of the fuel system.

If a problem is suspected in the fuel supply, tank and/or lines, by far the easiest test to eliminate these components as possible culprits is to substitute a known good fuel supply. This is known as running a motor on a test tank (as opposed to running a motor IN a test tank, which is an entirely different concept). If possible, borrow a portable tank, fill it with fresh gasoline and connect it to the motor.

■ When using a test fuel tank, make sure the inside diameter of the fuel hose and fuel fittings is at least equal to the size of the hose used on the usual supply line (otherwise you could cause fuel starvation problems or other symptoms which will make troubleshooting nearly impossible.

Fuel Tank

◆ See Figures 5 and 6

There are 3 different types of fuel tanks that might be used along with these motors. The very smallest motors covered by this guide may be equipped with an integral fuel tank mounted to the powerhead. But, most motors, even some of the smaller ones, may be rigged using either a portable fuel tank or a boat mounted tank. In both cases, a tank that is not mounted to the engine itself is commonly called a remote tank.

■ Although many dealers rig boats using Yamaha or Mercury fuel tanks, there are many other tank manufacturers and tank designs may vary greatly. Your outboard might be equipped with a tank from the engine manufacturer, the boat manufacturer or even another tank manufacturer. Although components used, as well as the techniques for cleaning and repairing tanks are similar for almost all fuel tanks be sure to use caution and common sense. Since design vary greatly we cannot provide step-by-step service procedures for the fuel tanks you might encounter. Instead, we've provided information about tanks and how you can decide when service is necessary. Refer to a reputable marine repair shop or marine dealership when parts are needed for aftermarket fuel tanks.

Whether or not your boat is equipped with a boat mounted, built-in tank depends mostly on the boat builder and partially on the initial engine installer. Boat mounted tanks can be hard to access (sometimes even a little hard to find if parts of the deck must be removed. When dealing with boat mounted tanks, look for access panels (as most manufacturers are smart or kind enough to install them for tough to reach tanks). At the very least, all manufacturers must provide access to fuel line fittings and, usually, the fuel level sender assembly.

No matter what type of tank is used, all must be equipped with a vent (either a manual vent or an automatic one-way check valve) which allows air in (but should prevent vapors from escaping). An inoperable vent (one that is blocked in some fashion) would allow the formation of a vacuum that could prevent the fuel pump from drawing fuel from the tank. A blocked vent could cause fuel starvation problems. Whenever filling the tank, check to make sure air does not rush into the tank each time the cap is loosened (which could be an early warning sign of a blocked vent).

If fuel delivery problems are encountered, first try running the motor with the fuel tank cap removed to ensure that no vacuum lock will occur in the tank or lines due to vent problems. If the motor runs properly with the cap removed but stalls, hesitates or misses with the cap installed, you know the problem is with the tank vent system.

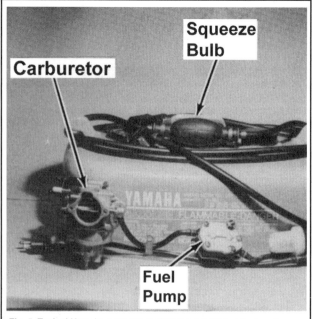

Fig. 5 Typical Yamaha portable fuel tank (and related fuel system components)

FUEL SYSTEM 3-7

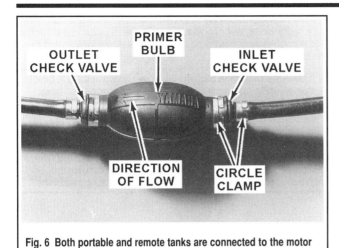

Fig. 6 Both portable and remote tanks are connected to the motor using a fuel line equipped with a primer bulb

SERVICE

Integral Fuel Tanks

 MODERATE

◆ See Figure 7

On small Yamaha powerheads equipped with an integral fuel tank above the powerhead, the filter in the tank may become plugged preventing proper fuel flow. We say Yamaha, because the small Mercury/Mariners do not use an tank mounted filter, opting to use a filter inline between the tank and carburetor.

Either way, to check fuel flow, disconnect the fuel line at the carburetor and fuel should flow from the line if the shut-off valve is open. If fuel is not present, the usual cause is a likely a plugged filter.

On Yamaha motors with an filter mounted inside the tank it can be very difficult to determine these filter are plugged simply by a visual inspection. The general appearance that the filter is satisfactory may be a false indication. Therefore, if the filter is suspected, the best remedy is replacement. The cost is very modest and this one area is thus eliminated as a problem source.

Would you believe, many times a lack of fuel at the carburetor is caused because the vent on the fuel tank was not opened or has become clogged/stuck? As mentioned under Fuel Tank in this section, it is important to make sure the vent is open at all times during engine operation to prevent formation of a vacuum in the tank that could prevent fuel from reaching the fuel pump or carburetor.

To remove the tank, proceed as follows:

1. Remove the upper and lower engine covers (as applicable). For more information, please refer to Engine Covers (Top Cover and Cowling) in the Engine Maintenance section.

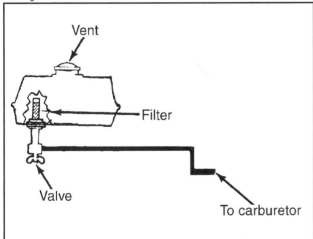

Fig. 7 Typical integral fuel tank and fuel supply circuit

2. If equipped, make sure the fuel tank valve is shut off.

■ **For models not equipped with a fuel tank mounted shut off valve, have a shop rag, large funnel and gas can handy to drain the fuel into when the hose is removed from the carburetor.**

3. For Mercury/Mariner models, disconnect the fuel supply line from the fuel tank shut-off valve (petcock).
4. For Yamaha motors trace the fuel line from the tank to the carburetor, then disconnect it from the tank or the carburetor fitting, whichever is easier and provides more access. For models without a shut off valve, immediately direct the hose into the funnel and gas can to prevent or minimize fuel spillage.
5. Remove the screws, washers and rubber mounts or spacers fastening the tank to the powerhead, then make sure no other wires, components or brackets will interfere with tank removal.
6. Carefully lift the fuel tank from the powerhead.
7. Check the tank, filler cap and gasket for wear or damage. Replace any components as necessary to correct any leakage (if found.)
8. On Yamaha models with a tank-mounted filter element, remove the filter and clean or replace as necessary. For more information refer to Fuel Filters in the Engine Maintenance section.
9. If cleaning is necessary, flush the tank with a small amount of solvent or gasoline, then drain and dispose of the flammable liquid properly.
10. If equipped, check the tank cushions for excessive deterioration, wear or damage and replace, as necessary.
11. When installing the tank, be sure all fuel lines are connected properly and that the retaining screws are securely bolted in place.
12. Refill the tank and pressure test the system by opening the fuel valve, then starting and running the engine.

Portable Fuel Tanks

Modern fuel tanks are vented to prevent vapor-lock of the fuel supply system, but are normally vented by a one-way valve to prevent pollution through the evaporation of vapors. A squeeze bulb is used to prime the system until the powerhead is operating. Once the engine starts, the fuel pump, mounted on the powerhead pull fuel from the tank and feeds the carburetor(s) or fuel vapor separator tank (EFI). The pickup unit in the tank is usually sold as a complete unit, but without the gauge and float.

To disassemble and inspect or replace tank components, proceed as follows:

1. For safety, remove the filler cap and drain the tank into a suitable container.
2. Disconnect the fuel supply line from the tank fitting.
3. Fuel pick-up units are usually 1 of 2 possible types:
    a. Threaded fittings are mounted to the tops of tanks using a Teflon sealant or tape to seal the threads. They may contain right or left-hand threads, depending upon the tank manufacturer so care should be used when removing or installing the fitting. On many plastic tanks the pick-up itself cannot be removed and the tank must be replaced if damaged.
    b. Bolted fittings are used, predominantly on metallic portable tanks. To replace the pickup unit, first remove the screws (normally 4) securing the unit in the tank. Next, lift the pickup unit up out of the tank. On models with an integrated fuel gauge on the pickup, there are normally screws securing the gauge to the bottom of the pickup unit.

■ **If the pickup unit is not being replaced, clean and check the screen for damage. It is possible to bend a new piece of screen material around the pickup and solder it in place without purchasing a complete new unit.**

4. If equipped with a level gauge assembly, check for smooth, and non-binding movement of the float arm and replace if binding is found. Check the float itself for physical damage or saturation and replace, if found.
5. Check the fuel tank for dirt or moisture contamination. If any is found use a small amount of gasoline or solvent to clean the tank. Pour the solvent in and slosh it around to loosen and wash away deposits, then pour out the solvent and recheck. Allow the tank to air dry, or help it along with the use of an air hose from a compressor.

✳✳ CAUTION

Use extreme care when working with solvents or fuel. Remember that both are even more dangerous when their vapors are concentrated in a small area. No source of ignition from flames to sparks can be allowed in the workplace for even an instant.

FUEL SYSTEM

To Install:

6. For bolted fuel fittings:

   a. Attach the fuel gauge to the new pickup unit and secure it in place with the screws.

   b. Clean the old gasket material from fuel tank and, if being used, the old pickup unit. Position a new gasket/seal, then work the float arm down through the fuel tank opening, and at the same time the fuel pickup tube into the tank. It will probably be necessary to exert a little force on the float arm in order to feed it all into the hole. The fuel pickup arm should spring into place once it is through the hole.

   c. Secure the pickup and float unit in place with the attaching screws.

7. For threaded fuel fittings, carefully clean the threads on the fitting and the tank, removing any traces of old sealant or tape. Apply a light coating of Teflon plumbers tape or sealant to the threads, then carefully thread the fitting into position until lightly seated.

8. If removed, connect the fuel line to the tank, then pressurize the fuel system and check for leaks.

Boat Mounted Fuel Tanks

The other type of remote fuel tank sometimes used on these models (usually only on the larger models) is a boat mounted built-in tank. Depending on the boat manufacturer, built-in tanks may vary greatly in actual shape/design and access. All should be of a one-way vented valve type to prevent a vacuum lock, but capped to prevent evaporation design.

Most boat manufacturers are kind enough to incorporate some means of access to the tank should fuel lines, fuel pickup or floats require servicing. But, the means of access will vary greatly from boat-to-boat. Some might contain simple access panels, while others might require the removal of one or more minor or even major components for access. If you encounter difficulty, seek the advice of a local dealer for that boat builder. The dealer or his/her techs should be able to set you in the right direction.

✱✱ CAUTION

Observe all fuel system cautions, especially when working in recessed portions of a hull. Fuel vapors tend to gather in enclosed areas causing an even more dangerous possibility of explosion.

Fuel Lines and Fittings

◆ See Figure 8

In order for an engine to run properly it must receive an uninterrupted and unrestricted flow of fuel. This cannot occur if improper fuel lines are used or if any of the lines/fittings are damaged. Too small a fuel line could cause hesitation or missing at higher engine rpm. Worn or damaged lines or fittings could cause similar problems (also including stalling, poor/rough idle) as air might be drawn into the system instead of fuel. Similarly, a clogged fuel line, fuel filter or dirty fuel pickup or vacuum lock (from a clogged tank vent as mentioned under Fuel Tank) could cause these symptoms by starving the motor for fuel.

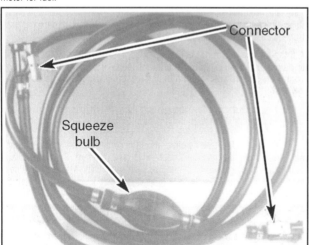

Fig. 8 Problems with the fuel lines or fittings can manifest themselves in engine performance problems, especially at WOT

If fuel delivery problems are suspected, check the tank first to make sure it is properly vented, then turn your attention to the fuel lines. First check the lines and valves for obvious signs of leakage, then check for collapsed hoses that could cause restrictions.

■ **If there is a restriction between the primer bulb and the fuel tank, vacuum from the fuel pump may cause the primer bulb to collapse. Watch for this sign when troubleshooting fuel delivery problems.**

✱✱ CAUTION

Only use the proper fuel lines containing suitable Coast Guard ratings on a boat. Failure to do so may cause an extremely dangerous condition should fuel lines fail during adverse operating conditions.

TESTING

A lack of adequate fuel supply will cause the powerhead to run lean or loose rpm. If a fuel tank other than the one supplied or recommended by Yamaha is being used you should make sure the fuel cap has an adequate air vent. Also, verify the size of the fuel line from the tank to be sure it is of sufficient size to accommodate the powerhead demands.

An adequate size line would be anywhere from 5/16-3/8 in. (7.94-9.52mm) Inner Diameter (ID). Check the fuel strainer on the end of the pickup in the fuel tank to be sure it is not tool small or clogged. Check the fuel pickup tube to make sure it is large enough and is not kinked or clogged. Be sure to check the inline filter to make sure sufficient quantities of fuel can pass through it.

If necessary, perform a Fuel Line Quick Check to see if the problem is fuel line or delivery system related.

Fuel Line Quick Check

◆ See Figure 8

Stalling, hesitation, rough idle, misses at high rpm are all possible results of problems with the fuel lines. A quick visual check of the lines for leaks, kinked or collapsed lengths or other obvious damage may uncover the problem. If no obvious cause is found, the problem may be due to a restriction in the line or a problem with the fuel pump.

If a fuel delivery problem due to a restriction or lack of proper fuel flow is suspected, operate the engine while attempting to duplicate the miss or hesitation. While the condition is present, squeeze the primer bulb rapidly to manually pump fuel from the tank to (and through) the fuel pump to the carburetors (or the vapor separator tank on fuel injected motors). If the engine then runs properly while under these conditions, suspect a problem with a clogged restricted fuel line, a clogged fuel filter or a problem with the fuel pump.

Checking Fuel Flow at Motor

◆ See Figures 8 thru 12

To perform a more thorough check of the fuel lines and isolate or eliminate the possibility of a restriction, proceed as follows:

1. For safety, disconnect the spark plug leads, then ground each of them to the powerhead to prevent sparks and to protect the ignition system.

2. Disconnect the fuel line from the engine (quick-connector on many models). Place a suitable container over the end of the fuel line to catch the fuel discharged. If equipped with a quick-connector, insert a small screwdriver into the end of the line to hold the valve open.

3. Squeeze the primer bulb and observe if there is satisfactory fuel flow from the line. If there is no fuel discharged from the line, the check valve in the squeeze bulb may be defective, or there may be a break or obstruction in the fuel line.

4. If there is a good fuel flow, reconnect the tank-to-motor fuel supply line and disconnect the fuel line from the carburetor(s) or fuel injection vapor separator tank (as applicable), directing that line into a suitable container. Crank the powerhead. If the fuel pump is operating properly, a healthy stream of fuel should pulse out of the line. If sufficient fuel does not pulse from the line, compare flow at either side of the inline fuel filter (if equipped) or check the fuel pump.

FUEL SYSTEM 3-9

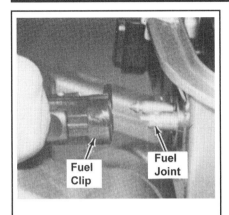

Fig. 9 Disconnect the fuel supply from the engine quick-connect to check fuel flow

Fig. 10 The fuel line can also be disconnected from the fuel pump. In either case, squeeze the primer bulb and watch for unrestricted fuel flow

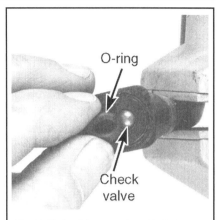

Fig. 11 Fuel quick-connector with O-ring and check valve visible (you'll have to depress the check valve in order to see fuel flow with the connector removed from the engine

  5. Continue cranking the powerhead and catching the fuel for about 15 pulses to determine if the amount of fuel decreases with each pulse or maintains a constant amount. A decrease in the discharge indicates a restriction in the line. If the fuel line is plugged, the fuel stream may stop. If there is fuel in the fuel tank but no fuel flows out the fuel line while the powerhead is being cranked, the problem may be in one of several areas:
  6. Plugged fuel line from the fuel pump to the carburetor(s) or vapor separator tank (EFI).
  7. Defective O-ring in fuel line connector into the fuel tank.
  8. Defective O-ring in fuel line connector into the engine.
  9. Defective fuel pump.
  10. The line from the fuel tank to the fuel pump may be plugged; the line may be leaking air; or the squeeze bulb may be defective.
  11. Defective fuel tank.
  12. If the engine does not start even though there is adequate fuel flow from the fuel line, the float inlet needle valve and the seat may be gummed together and prevent adequate fuel flow into the float bowl or fuel injection vapor separator tank.

Checking the Primer Bulb

◆ See Figuree 8 and 13

The way most outboards are rigged, fuel will evaporate from the system during periods of non-use. Also, anytime quick-connect fittings on portable tanks are removed, there is a chance that small amounts of fuel will escape and some air will make it into the fuel lines. For this reason, outboards are normally rigged with some method of priming the fuel system through a hand-operated pump (primer bulb).

When squeezed, the bulb forces fuel from inside the bulb, through the one-way check valve toward the motor filling the carburetor float bowl(s) or fuel injection vapor separator tank with the fuel necessary to start the motor. When the bulb is released, the one-way check valve on the opposite end (tank side of the bulb) opens under vacuum to draw fuel from the tank and refill the bulb.

When using the bulb, squeeze it gently as repetitive or forceful pumping may flood the carburetor(s) or potentially overfill the vapor separator tank on fuel injected motors. The bulb is operating normally if a few squeezes will cause it to become firm, meaning the float bowl/tank is full, and the float valve is closed. If the bulb collapses and does not regain its shape, the bulb must be replaced.

For the bulb to operate properly, both check valves must operate properly and the fuel lines from the check valves back to the tank or forward to the motor must be in good condition (properly sealed). To check the bulb and check valves use hand operated vacuum/pressure pump (available from most marine or automotive parts stores).
  1. Remove the fuel hose from the tank and the motor, then remove the clamps for the quick-connect valves at the ends of the hose.

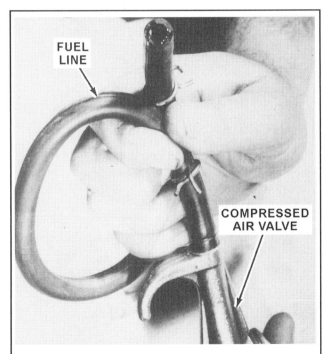

Fig. 12 Many times, restrictions such as foreign matter may be cleared from fuel lines using compressed air. But be careful to make sure the hose is pointed away from yourself or others

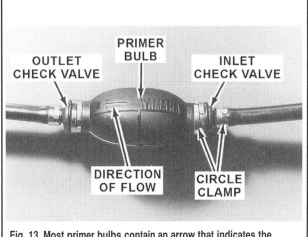

Fig. 13 Most primer bulbs contain an arrow that indicates the direction of fuel flow (points toward the motor)

3-10 FUEL SYSTEM

■ Many quick-connect valves are secured to the fuel supply hose using disposable plastic ties that must be cut and discarded for removal. If equipped, spring-type or threaded metal clamps may be reused, but be sure they are in good condition first. Do not over tighten threaded clamps and crack the valve or cut the hose.

2. Carefully remove the fuel line connector (fitting or quick-connect valve, as applicable) from the motor side of the fuel line, then place the end of the line into the filler opening of the fuel tank. Gently pump the primer bulb to empty the hose into the fuel tank.

■ Be careful when removing the quick-connect valve from the fuel line as fuel will likely still be present in the hose and will escape (drain or splash) if the valve is jerked from the line. Also, make sure the primer bulb is empty of fuel before proceeding.

3. Next, remove the fuel line connector (fitting or quick-connect valve) from the tank side of the fuel line, draining any residual fuel into the tank. Boat mounted fuel tanks often have a water separating fuel filter installed inline between the fuel tank and the motor, on these systems disconnect the tank/primer bulb line from the separator.

■ For proper orientation during testing or installation, the primer bulb is marked with an arrow that faces the engine side check valve.

4. Securely connect the pressure pump to the hose on the tank side of the primer bulb. Using the pump, slowly apply pressure while listening for air escaping from the end of the hose that connects to the motor. If air escapes, both one-way check valves on the tank side and motor side of the prime bulb are opening.
5. If air escapes prior to the motor end of the hose, hold the bulb, check valve and hose connections under water (in a small bucket or tank). Apply additional air pressure using the pump and watch for escaping bubbles to determine what component or fitting is at fault. Repair the fitting or replace the defective hose/bulb component.
6. If no air escapes, attempt to draw a vacuum from the tank side of the primer bulb. The pump should draw and hold a vacuum without collapsing the primer bulb, indicating that the tank side check valve remained closed.
7. Securely connect the pressure pump to the hose on the motor side of the primer bulb. Using the pump, slowly apply pressure while listening for air escaping from the end of the hose that connects to the tank. This time, the check valve on the tank side of the primer bulb should remain closed, preventing air from escaping or from pressurizing the bulb. If the bulb pressurizes, the motor side check valve is allowing pressure back into the bulb, but the tank side valve is operating properly.
8. Replace the bulb and/or check valves if they operate improperly.

SERVICE

◆ See Figures 13, 14 and 15

■ When replacing fuel lines, make sure the inside diameter of the fuel hose and fitting is at least the same diameter as the hose it is replacing. Also, be certain to use only marine fuel line the meets or exceeds United States Coast Guard (USCG) A1 or B1 guidelines.

Whenever work is performed on the fuel system, check all hoses for wear or damage. Replace hoses that are soft and spongy or ones that are hard and brittle. Fuel hoses should be smooth and free of surface cracks, and they should definitely not have split ends (there's a bad hair joke in there, but we won't sink that low). Do not cut the split ends of a hose and attempt to reuse it, whatever caused the split (most likely time and deterioration) will cause the new end to follow soon. Fuel hoses are safety items, don't scrimp on them, instead, replace them when necessary. If one hose is too old, check the rest, as they are likely also in need of replacement.

When replacing fuel lines only use replacement hoses or other marine fuel supply lines that meet United States Coast Guard (USCG) requirements A1 or B1 for marine applications. All lines must be of the same inner diameter as the original to prevent leakage and maintain the proper seal that is necessary for fuel system operation.

■ Using a smaller fuel hose than originally specified/equipped could cause fuel starvation problems leading to misfiring, hesitation, rough idling and possibly even engine damage.

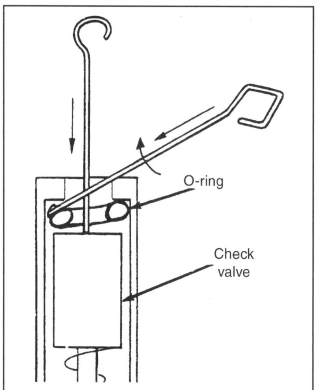

Fig. 14 Use two picks, punches or other small tool to replace quick-connect O-rings. One to push the valve and the other work the O-ring free

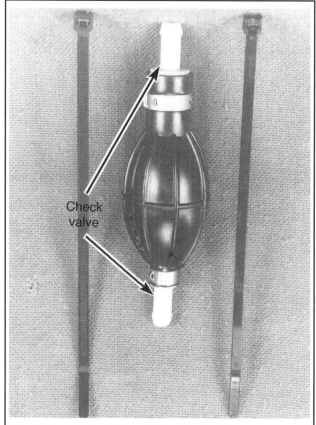

Fig. 15 A squeeze bulb kit usually includes the bulb, 2 check valves, and two tie straps (or other replacement clamps)

FUEL SYSTEM 3-11

The USCG ratings for fuel supply lines have to do with whether or not the lines have been testing regarding length of time it might take for them to succumb to flame (burn through) in an emergency situation. A line is "A" rated if it passes specific requirements regarding burn-through times, while "B" rated lines are not tested in this fashion. The A1 and B1 lines (normally recommended on these applications) are capable of containing liquid fuel at all times. The A2 and B2 rated lines are designed to contain fuel vapor, but not liquid.

> ** CAUTION
>
> **To help prevent the possibility of significant personal injury or death, do not substitute "B" rated lines when "A" rated lines are required. Similarly, DO NOT use "A2" or "B2" lines when "A1" or "A2" lines are specified.**

Various styles of fuel line clamps may be found on these motors. Many applications will simply secure lines with plastic wire ties or special plastic locking clamps. Although some of the plastic locking clamps may be released and reconnected, it is usually a good idea to replace them. Obviously wire ties are cut for removal, which requires that they be replaced.

Some applications use metal spring-type clamps, that contain tabs which are squeezed allowing the clamp to slid up the hose and over the end of the fitting so the hose can be pulled from the fitting. Threaded metal clamps are nice since they are very secure and can be reused, but do not over tighten threaded clamps as they will start to cut into the hose and they can even damage some fittings underneath the hose. Metal clamps should be replaced anytime they've lost tension (spring type clamps), are corroded, bent or otherwise damaged.

■ The best way to ensure proper fuel fitting connection is to use the same size and style clamp that was originally installed (unless of course the "original" clamp never worked correctly, but in those cases, someone probably replaced it with the wrong type before you ever saw it).

To avoid leaks, replace all displaced or disturbed gaskets, O-rings or seals whenever a fuel system component is removed.

CARBURETED FUEL SYSTEM

> ** CAUTION
>
> **If equipped, disconnect the negative battery cable ANYTIME work is performed on the engine, especially when working on the fuel system. This will help prevent the possibility of sparks during service (from accidentally grounding a hot lead or powered component). Sparks could ignite vapors or exposed fuel. Disconnecting the cable on electric start motors will also help prevent the possibility fuel spillage if an attempt is made to crank the engine while the fuel system is open.**

> ** CAUTION
>
> **Fuel leaking from a loose, damaged or incorrectly installed hose or fitting may cause a fire or an explosion. ALWAYS pressurize the fuel system and run the motor while inspecting for leaks after servicing any component of the fuel system.**

Carbureted motors covered by this repair guide are equipped with anything from one single barrel carburetor to 4 single-barrel carbs (one carburetor per cylinder). Although on initial inspection some of the larger carbureted motors may look somewhat complicated, they're actually very basic, especially when compared with other modern fuel systems (such as an automotive or marine fuel injection system).

The entire system essentially consists of a fuel tank, a fuel supply line, and a mechanical fuel pump assembly mounted to the powerhead all designed to feed the carburetor with the fuel necessary to power the motor. Cold starting is enhanced by the use of a choke plate or an enrichment circuit.

For information on fuels, tanks and lines please refer to the sections on Fuel System Basics and Fuel Tanks and Lines.

The most important fuel system maintenance that a boat owner can perform is to stabilize fuel supplies before allowing the system to sit idle for any length of time more than a few weeks (or better yet, to completely drain the carburetors. The next most important item is to provide the system with

On many installations, the fuel line is provided with quick-disconnect fittings at the tank and at the powerhead. If there is reason to believe the problem is at the quick-disconnects, the hose ends can be replaced as an assembly, or new O-rings may be installed. A supply of new O-rings should be carried on board for use in isolated areas where a marine store is not available (like dockside, or worse, should you need one while on the water). For a small additional expense, the entire fuel line can be replaced and eliminate this entire area as a problem source for many future seasons. (If the fuel line is replaced, keep the old one around as a spare, just in case).

If an O-ring must be replaced, use two small punches, picks or similar tools, one to push down the check valve of the connector and the other to work the O-ring out of the hole. Apply just a drop of oil into the hole of the connector. Apply a thin coating of oil to the surface of the O-ring. Pinch the O-ring together and work it into the hole while simultaneously using a punch to depress the check valve inside the connector.

The primer squeeze bulb can be replaced in a short time. A squeeze bulb assembly kit, complete with the check valves installed, may be obtained from your dealer or marine parts retailer. The replacement kit will usually also include two tie straps (or some other type of clamp) to secure the bulb properly in the line.

An arrow is clearly visible on the squeeze bulb to indicate the direction of fuel flow. The squeeze bulb must be installed correctly in the line because the check valves in each end of the bulb will allow fuel to flow in only one direction. Therefore, if the squeeze bulb should be installed backwards, in a moment of haste to get the job done, fuel will not reach the carburetor(s).

To replace the bulb, first undo the clamps on the hose at each end of the bulb. Next, pull the hose out of the check valves at each end of the bulb. New clamps are included with a new squeeze bulb.

If the fuel line has been exposed to considerable sunlight, it may have become hardened, causing difficulty in working it over the check valve. To remedy this situation, simply immerse the ends of the hose in boiling water for a few minutes to soften the rubber. The hose will then slip onto the check valve without further problems. After the lines on both sides have been installed, snap the clamps in place to secure the line. Check a second time to be sure the arrow is pointing in the fuel flow direction, towards the powerhead.

fresh gasoline if the system has stood idle for any length of time, especially if it was without fuel system stabilizer during that time.

If a sudden increase in gas consumption is noticed, or if the engine does not perform properly, a carburetor overhaul, including cleaning or replacement of the fuel pump may be required.

Description and Operation

◆ See Figures 16 and 17

GENERAL INFORMATION

The Role of a Carburetor

◆ See Figures 16 and 17

The carburetor is merely a metering device for mixing fuel and air in the proper proportions for efficient engine operation. At idle speed, an outboard engine requires a mixture of about 8 parts air to 1 part fuel. At high speed or under heavy duty service, the mixture may change to as much as 12 parts air to 1 part fuel.

Carburetors are wonderful devices that succeed in creating precise air/fuel mixture ratios based on tiny passages, needle jets or orifices and the variable vacuum that occurs as engine rpm and operating conditions vary.

Because of the tiny passages and small moving parts in a carburetor (and the need for them to work precisely to achieve exact air/fuel mixture ratios) it is important that the fuel system integrity is maintained. Introduction of water (that might lead to corrosion), debris (that could clog passages) or even the presence of unstabilized fuel that could evaporate over time can cause big problems for a carburetor. Keep in mind that when fuel evaporates it leaves behind a gummy deposit that can clog those tiny passages, preventing the carburetor (and therefore preventing the engine) from operating properly.

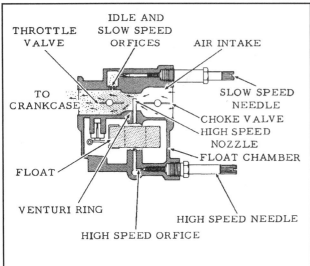

Fig. 16 Fuel flow through the venturi, showing principle and related parts controlling intake and outflow (carburetor with manual choke circuit shown)

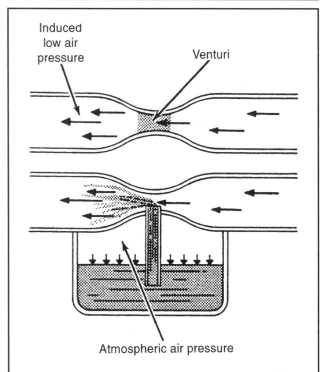

Fig. 17 Air flow principle of a modern carburetor, demonstrates how the low pressure induced behind the venturi draws fuel through the high speed nozzle

Float Systems

◆ See Figure 16

Ever lift the tank lid off the back of your toilet. Pretty simple stuff once you realize what's going on in there. A supply line keeps the tank full until a valve opens allowing all or some of the liquid in the tank to be drawn out through a passage. The dropping level in the tank causes a float to change position, and, as it lowers in the tank it opens a valve allowing more pressurized liquid back into the tank to raise levels again. OK, we were talking about a toilet right, well yes and no, we're also talking about the float bowl on a carburetor. The carburetor uses a more precise level, uses vacuum to draw out fuel from the bowl through a metered passage (instead of gravity) and, most importantly, store gasoline instead of water, but otherwise, they basically work in the same way.

A small chamber in the carburetor serves as a fuel reservoir. A float valve admits fuel into the reservoir to replace the fuel consumed by the engine. If the carburetor has more than one reservoir, the fuel level in each reservoir (chamber) is controlled by identical float systems.

Fuel level in each chamber is extremely critical and must be maintained accurately. Accuracy is obtained through proper adjustment of the float. This adjustment will provide a balanced metering of fuel to each cylinder at all speeds. Improper levels will lead to engine operating problems. Too high a level can promote rich running and spark plug fouling, while excessively low float bowl fuel levels can cause lean conditions, possibly leading to engine damage.

Following the fuel through its course, from the fuel tank to the combustion chamber of the cylinder, will provide an appreciation of exactly what is taking place. In order to start the engine, the fuel must be moved from the tank to the carburetor by a squeeze bulb installed in the fuel line. This action is necessary because the fuel pump does not have sufficient pressure to draw fuel from the tank during cranking before the engine starts.

The fuel for some small horsepower units is gravity fed from a tank mounted at the rear of the powerhead. Even with the gravity feed method, a small fuel pump may be an integral part of the carburetor for some manufacturers.

After the engine starts, the fuel passes through the pump to the carburetor. All systems have some type of filter installed somewhere in the line between the tank and the carburetor. Over the years many carburetors have utilized a filter as an integral part of the carburetor.

At the carburetor, the fuel passes through the inlet passage to the needle and seat, and then into the float chamber (reservoir). A float in the chamber rides up and down on the surface of the fuel. After fuel enters the chamber and the level rises to a predetermined point, a tang on the float closes the inlet needle and the flow entering the chamber is cut off. When fuel leaves the chamber as the engine operates, the fuel level drops and the float tang allows the inlet needle to move off its seat and fuel once again enters the chamber. In this manner, a constant reservoir of fuel is maintained in the chamber to satisfy the demands of the engine at all speeds.

A fuel chamber vent hole is located near the top of the carburetor body to permit atmospheric pressure to act against the fuel in each chamber. This pressure assures an adequate fuel supply to the various operating systems of the powerhead.

Air/Fuel Mixture

◆ See Figure 17

A suction effect is every OTHER time the piston moves downward in a 4-stroke motor (each time the piston moves downward on the intake stroke, with the intake valve open). This suction draws air through the throat of the carburetor. A restriction in the throat, called a venturi, controls air velocity and has the effect of reducing air pressure at this point.

The difference in air pressures at the throat and in the fuel chamber, causes the fuel to be pushed out of metering jets extending down into the fuel chamber. When the fuel leaves the jets, it mixes with the air passing through the venturi. This fuel/air mixture should then be in the proper proportion for burning in the cylinders for maximum engine performance.

In order to obtain the proper air/fuel mixture for all engine speeds, high- and low-speed orifices or needle valves are installed. On most modern powerheads the high-speed needle valve has been replaced with a fixed high-speed orifice (to more discourage tampering and to help maintain proper emissions under load). There is no adjustment with the orifice type. The needle valves are used to compensate for changing atmospheric conditions. The low-speed needles, on the other hand, are still provided so that air/fuel mixture can be precisely adjusted for idle conditions other than what occurs at atmospheric sea-level. Although the low-speed needle should not normally require periodic adjustment, it can be adjusted to compensate for high-altitude (river/lake) operation or to adjust for component wear within the fuel system.

A throttle valve controls the flow of air/fuel mixture drawn into the combustion chambers. A cold powerhead requires a richer fuel mixture to start and during the brief period it is warming to normal operating temperature. A choke valve is often placed ahead of the metering jets and venturi. As this valve begins to close, the volume of air intake is reduced, thus enriching the mixture entering the cylinders.

When this choke valve is fully closed, a very rich fuel mixture is drawn into the cylinders.

FUEL SYSTEM 3-13

■ **Some carburetors do not use a choke valve, but instead use an enriching circuit which mechanically provides more fuel to the carburetor or intake during cold start-up.**

The throat of the carburetor is usually referred to as the barrel. Carburetors with single, double, or four barrels have individual metering jets, needle valves, throttle and, if applicable, choke plates for each barrel. Single and two barrel carburetors are fed by a single float and chamber.

CARBURETOR CIRCUITS

The following section illustrates the circuit functions and locations of a typical marine carburetor.

Starting Circuit

◆ See Figure 18

The choke plate is closed, creating a partial vacuum in the venturi. As the piston rises, negative pressure in the crankcase draws the rich air-fuel mixture from the float bowl into the venturi and on into the engine.

■ **Some of these motors use an enrichener circuit instead of a choke plate. An enrichener allows the full amount of air to pass through the carburetor during cold-start up, but it fattens the air/fuel mixture by providing additional fuel through a dedicated (usually electronically controlled) passage.**

Low Speed Circuit

◆ See Figure 19

Zero-one-eighth throttle, when the pressure in the crankcase is lowered, the air-fuel mixture is discharged into the venturi through the pilot outlet because the throttle plate is closed. No other outlets are exposed to low venturi pressure. The fuel is metered by the pilot jet. The air is metered by the pilot air jet. The combined air-fuel mixture is regulated by the pilot air screw.

Mid-Range Circuit

◆ See Figure 20

One-eighth-three-eighths throttle, as the throttle plate continues to open, the air-fuel mixture is discharged into the venturi through the bypass holes. As the throttle plate uncovers more bypass holes, increased fuel flow results because of the low pressure in the venturi. Depending on the model, there could be two, three or four bypass holes.

High Speed Circuit

◆ See Figure 21

Three-eighths-wide-open throttle, as the throttle plate moves toward wide open, we have maximum air flow and very low pressure. The fuel is metered through the main jet, and is drawn into the main discharge nozzle. Air is metered by the main air jet and enters the discharge nozzle, where it combines with fuel. The mixture atomizes, enters the venturi, and is drawn into the engine.

BASIC FUNCTIONS

◆ See Figure 22

The carburetor systems on inline engines require careful cleaning and adjustment if problems occur. These carburetors may seem a little complicated but they are not too complex to understand. All carburetors operate on the same principles.

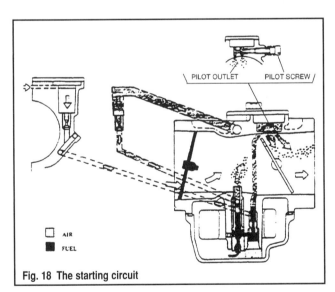
Fig. 18 The starting circuit

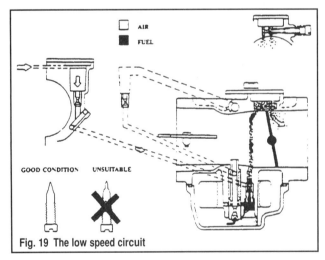
Fig. 19 The low speed circuit

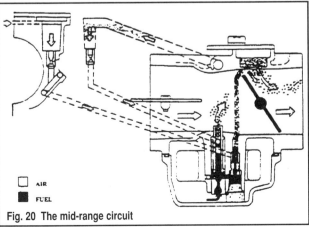
Fig. 20 The mid-range circuit

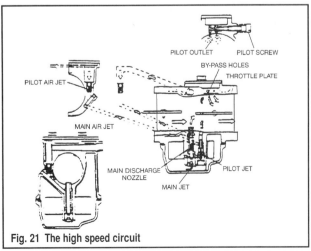
Fig. 21 The high speed circuit

3-14 FUEL SYSTEM

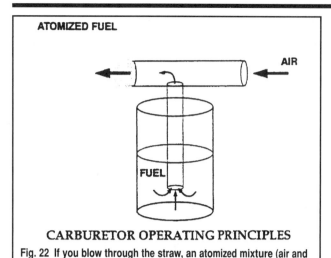

Fig. 22 If you blow through the straw, an atomized mixture (air and fuel droplets) comes out

Traditional carburetor theory often involves a number of laws and principles. To troubleshoot carburetors learn the basic principles, watch how the carburetor comes apart, trace the circuits, see what they do and make sure they are clean. These are the basic steps for troubleshooting and successful repair.

The diagram illustrates several carburetor basics. If you blow through the horizontal straw an atomized mixture (air and fuel droplets) comes out. When you blow through the straw a pressure drop is created in the straw column inserted in the liquid. In a carburetor this is mostly air and a little fuel. The actual ratio of air to fuel differs with engine conditions but is usually from 15 parts air to one part fuel at optimum cruise to as little as 7 parts air to one part fuel at full choke.

If the top of the container is covered and sealed around the straw what will happen? No flow. This is typical of a clogged carburetor bowl vent.

If the base of the straw is clogged or restricted what will happen? No flow or low flow. This represents a clogged main jet.

If the liquid in the glass is lowered and you blow through the straw with the same force what will happen? Not as much fuel will flow. A lean condition occurs.

If the fuel level is raised and you blow again at the same velocity what happens? The result is a richer mixture.

Many Yamaha and Mercury/Mariner carburetors control air flow semi-independently of RPM. This is done with a throttle plate. The throttle plate works in conjunction with other systems or circuits to deliver correct mixtures within certain RPM bands. The idle circuit pilot outlet controls from 0-1/8 throttle. The series of small holes in the carburetor throat called transition holes control the 1/8-3/8 throttle range. At wide open throttle the main jet handles most of the fuel metering chores, but the low and mid-range circuits continue to supply part of the fuel.

Enrichment is necessary to start a cold engine. Fuel and air mix does not want to vaporize in a cold engine. In order to get a little fuel to vaporize, a lot of fuel is dumped into the engine. On some engines a choke plate is used for cold starts. This plate restricts air entering the engine and increases the fuel to air ratio.

The Prime Start enrichment system used on many late-model engines is controlled by a heated wax pellet. This pellet is heated by current from the stator. Temperature is monitored and enrichment is automatic. Other engines use choke plates, or electric solenoid enrichment systems.

Troubleshooting the Carbureted Fuel System

◆ See Figure 23

Troubleshooting fuel systems requires the same techniques used in other areas. A thorough, systematic approach to troubleshooting will pay big rewards. Build your troubleshooting checklist, with the most likely offenders at the top. Use your experience to adjust your list for local conditions. At one time or another nearly everyone has been tempted to jump into the carburetor based on nothing more than a vague hunch. Pause a moment and review the facts when this urge occurs.

In order to accurately troubleshoot a carburetor or fuel system problem, you must first verify that the problem is fuel related. Many symptoms can have several different possible causes. Be sure to eliminate mechanical and electrical systems as the potential fault. Carburetion is a common cause of most engine problems, but there are many other possibilities.

One of the toughest tasks with a fuel system is the actual troubleshooting. Several tools are at your disposal for making this process very simple. A timing light works well for observing carburetor spray patterns. Look for the proper amount of fuel and for proper atomization in the two fuel outlet areas (main nozzle and bypass holes). The strobe effect of the lights helps you see in detail the fuel being drawn through the throat of the carburetor. On multiple carburetor engines, always attach the timing light to the cylinder you are observing so the strobe doesn't change the appearance of the patterns. If you need to compare two cylinders, change the timing light hookup each time you observe a different cylinder.

Pressure testing fuel pump output can determine whether sufficient fuel is being supplied for the fuel spray and if the fuel pump diaphragms are functioning correctly. A pressure gauge placed between the fuel pumps and the carburetors will test the entire fuel delivery system. Normally a fuel system problem will show up at high speed where the fuel demand is the greatest. A common symptom of a fuel pump output problem is surging at wide open throttle, while still operating normally at slower speeds. To check the fuel pump output, install the pressure gauge and accelerate the engine to wide open throttle. Observe the pressure gauge needle. It should always swing up to some value above 2 psi (14 kPa), usually something in the 5-6 psi (34-41 kPa) area and remain steady. This reading would indicate a system that is functioning properly.

If the needle gradually swings down toward zero, fuel demand is greater than the fuel system can supply. This reading isolates the problem to the fuel delivery system (fuel tank or line). To confirm this, an auxiliary tank should be installed and the engine retested. Be aware that a bad anti-siphon valve on a built-in tank can create enough restriction to cause a lean condition and serious engine damage.

If the needle movement becomes erratic, suspect a ruptured diaphragm in the fuel pump.

A quick way to check for a ruptured fuel pump diaphragm is, while the engine is at idle speed, to squeeze the primer bulb and hold steady firm pressure on it. If the diaphragm is ruptured, this will cause a rough running condition because of the extra fuel passing through the diaphragm into the crankcase. After performing this test you should check the spark plugs. If the spark plugs are OK, but the fuel pumps are still suspected, you should

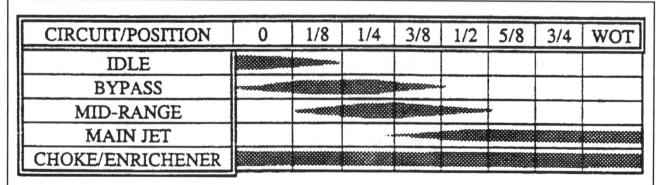

Fig. 23 Illustration of throttle position and carburetor operating circuits

FUEL SYSTEM 3-15

remove the fuel pumps and completely disassemble them. Rebuild or replace the pumps as needed.

To check the boat's fuel system for a restriction, install a vacuum gauge in the line before the fuel pump. Run the engine under load at wide open throttle to get a reading. Vacuum should read no more than 4.5 in. Hg (15.2 kPa) for most engines.

To check for air entering the fuel system, install a clear fuel hose between the fuel screen and fuel pump. If air is in the line, check all fittings back to the boat's fuel tank.

Spark plug tip appearance is a good indication of combustion efficiency. The tip should be a light tan. A white insulator or small beads on the insulator indicate too much heat. A dark or oil fouled insulator indicates incomplete combustion. To properly read spark plug tip appearance, run the engine at the rpm you are testing for about 15 second and then immediately turn the engine OFF without changing the throttle position.

Reading spark plug tip appearance is also the proper way to test jet verifications in high altitude.

The accompanying illustration explains the relationship between throttle position and carburetion circuits.

LOGICAL TROUBLESHOOTING

The following paragraphs provide an orderly sequence of tests to pinpoint problems in the fuel system.

1. Gather as much information as you can.
2. Duplicate the condition. Take the boat out and verify the complaint.
3. If the problem cannot be duplicated, you cannot fix it. This could be a product operation problem.
4. Once the problem has been duplicated, you can begin troubleshooting. Give the entire unit a careful visual inspection. You can tell a lot about the engine from the care and condition of the entire rig. What's the condition of the propeller and the lower unit? Remove the hood and look for any visible signs of failure. Are there any signs of head gasket leakage? Is the engine paint discolored from high temperature or are there any holes or cracks in the engine block? Perform a compression (or better yet, if you have the equipment, a leak down test). While cranking the engine during the compression test, listen for any abnormal sounds. If the engine passes these simple tests we can assume that the mechanical condition of the engine is good. All other engine mechanical inspection would be too time consuming at this point.
5. Your next step is to isolate the fuel system into two sub-systems. Separate the fuel delivery components from the carburetors. To do this, substitute a known good fuel supply for the regular boat's fuel supply. Use a 6 gallon portable tank and fuel line. Connect the portable fuel supply directly to the engine fuel pump, bypassing the boat fuel delivery system. Now test the engine. If the problem is no longer present, you know where to look. If the problem is still present, further troubleshooting is required.
6. When testing the engine, observe the throttle position when the problem occurs. This will help you pinpoint the circuit that is malfunctioning. Carburetor troubleshooting and repair is very demanding. You must pay close attention to the location, position and sometimes the numbering on each part removed. The ability to identify a circuit by the operating RPM it affects is important. Often your best troubleshooting tool is a can of cleaner. This can be used to trace those mystery circuits and find that last speck of dirt. Be careful and wear safety glasses when using this method.

■ **The last step of fuel system troubleshooting is to adjust or rebuild and then adjust the carburetor. We say it is the last step, because it is the most involved repair procedures on the fuel system and should only be performed after all other possible causes of fuel system trouble have been eliminated.**

COMMON PROBLEMS

Fuel Delivery

Many times fuel system troubles are caused by a plugged fuel filter, a defective fuel pump, or by a leak in the line from the fuel tank to the fuel pump. Aged fuel left in the carburetor and the formation of varnish could cause the needle to stick in its seat and prevent fuel flow into the bowl. A defective choke may also cause problems. Would you believe, a majority of starting troubles, which are traced to the fuel system, are the result of an empty fuel tank or aged fuel.

If fuel delivery problems are suspected, refer to the testing procedures in Fuel Tank and Lines to make sure the tank vent is working properly and that there are not leaks or restrictions that would prevent fuel from getting to the pump and/or carburetor(s).

A blocked low-pressure fuel filter causes hard starting, stalling, misfire or poor performance. Typically the engine malfunction worsens with increased engine speed. This filter prevents contaminants from reaching the low-pressure fuel pump. Refer to the Fuel Filter in the section on Maintenance and Tune-Up for more details on checking, cleaning or replacing fuel filters.

Sour Fuel

◆ See Figure 24

Under average conditions (temperate climates), fuel will begin to break down in about four months. A gummy substance forms in the bottom of the fuel tank and in other areas. The filter screen between the tank and the carburetor and small passages in the carburetor will become clogged. The gasoline will begin to give off an odor similar to rotten eggs. Such a condition can cause the owner much frustration, time in cleaning components, and the expense of replacement or overhaul parts for the carburetor.

Even with the high price of fuel, removing gasoline that has been standing unused over a long period of time is still the easiest and least expensive preventative maintenance possible. In most cases, this old gas can be used without harmful effects in an automobile using regular gasoline.

A gasoline preservative additive will usually keep the fuel fresh for up to twelve months (or longer).

Refer to the information on Fuel System Basics in this section, specifically the procedure under Fuel entitled Checking For Stale/Contaminated Fuel will provide information on how to determine if stale fuel is present in the system. If draining the system of contaminated fuel and refilling it with fresh fuel does not make a difference in the problem, look for restrictions or other problems with the fuel delivery system. If stale fuel was left in the tank/system for a long period of time and evaporation occurred, there is a good chance that the carburetor is gummed (tiny passages are clogged by deposits left behind when the fuel evaporated). If no fuel delivery problems are found, the carburetor(s) should be removed for disassembly and cleaning.

■ **Although there are some commercially available fuel system cleaning products that are either added to the fuel mixture or sprayed into the carburetor throttle bores, the truth is that although they can provide some measure of improvement, there is no substitute for a thorough disassembly and cleaning. The more fuel which was allowed to evaporate, the more gum or varnish may have been left behind and the more likely that only a disassembly will be able to restore proper performance.**

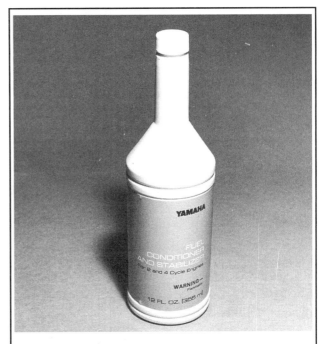

Fig. 24 The use of an approved fuel additive will prevent fuel from souring for up to twelve months

FUEL SYSTEM

Choke Problems

◆ See Figure 25

When the engine is hot, the fuel system can cause starting problems. After a hot engine is shut down, the temperature inside the fuel bowl may rise to 200 degrees F (94 degrees C) and cause the fuel to actually boil. All carburetors are vented to allow this pressure to escape to the atmosphere. However, some of the fuel may percolate over the high-speed nozzle.

If the choke should stick in the open position, the engine will be hard to start. If the choke should stick in the closed position, the engine will flood, making it very difficult to start.

In order for this raw fuel to vaporize enough to burn, considerable air must be added to lean out the mixture. Therefore, the best remedy for a flooded motor is to remove the spark plugs, ground the leads, crank the powerhead through about ten revolutions, clean the plugs, reinstall the plugs, and start the engine.

■ Another common solution to a flooded motor is to open the throttle all the way while cranking (using the additional air to help clear/burn the excess fuel in the combustion chambers), but the problem with marine engines is that this puts the gearcase in FORWARD which will prevent the starter from working on most models.

If the needle valve and seat assembly is leaking, an excessive amount of fuel may enter the intake tract. After the powerhead is shut down, the pressure left in the fuel line will force fuel past the leaking needle valve. This extra fuel will raise the level in the fuel bowl and cause fuel to overflow into the intake.

A continuous overflow of fuel into the intake tract may be due to a sticking inlet needle or to a defective float, which would cause an extra high level of fuel in the bowl and overflow.

Rough Engine Idle

■ As with all troubleshooting procedures, start with the easiest items to check/fix and work towards the more complicated ones.

If an engine does not idle smoothly, the most reasonable approach to the problem is to perform a tune-up to eliminate such areas as:
- Faulty spark plugs
- Timing and synchronization out of adjustment

Other problems that can prevent an engine from running smoothly include:
- An air leak in the intake manifold
- Uneven compression between the cylinders
- Problems with the valve train

Of course any problem in the carburetor affecting the air/fuel mixture will also prevent the engine from operating smoothly at idle speed. These problems usually include:
- Too high a fuel level in the bowl
- A heavy float
- Leaking needle valve and seat
- Defective automatic choke
- Improper adjustments for idle mixture or idle speed

"Sour" fuel (fuel left in a tank without a preservative additive) will cause an engine to run rough and idle with great difficulty.

Excessive Fuel Consumption

◆ See Figures 26 and 27

Excessive fuel consumption can result from one of four conditions, or some combination of the four.
1. Inefficient engine operation.
2. Damaged condition of the hull, outdrive or propeller, including excessive marine growth.
3. Poor boating habits of the operator.
4. Leaking or out of tune carburetor.

If the fuel consumption suddenly increases over what could be considered normal, then the cause can probably be attributed to the engine or boat and not the operator (unless he/she just drastically changed the manner in which the boat is operated).

Marine growth on the hull can have a very marked effect on boat performance. This is why sail boats always try to have a haul-out as close to race time as possible. While you are checking the bottom take note of the propeller condition. A bent blade or other damage will definitely cause poor boat performance.

If the hull and propeller are in good shape, then check the fuel system for possible leaks. Check the line between the fuel pump and the carburetor while the engine is running and the line between the fuel tank and the pump when the engine is not running. A leak between the tank and the pump many times will not appear when the engine is operating, because the suction created by the pump drawing fuel will not allow the fuel to leak. Once the engine is turned off and the suction no longer exists, fuel may begin to leak.

If a minor tune-up has been performed and the spark plugs and engine timing/synchronization are properly adjusted, then the problem most likely is in the carburetor, indicating an overhaul is in order. Check for leaks at the needle valve and seat. Use extra care when making any adjustments affecting the fuel consumption, such as the float level or automatic choke.

Engine Surge

If the engine operates as if the load on the boat is being constantly increased and decreased, even though an attempt is being made to hold a constant engine speed, the problem can most likely be attributed to the fuel pump (or a restriction between the tank and powerhead). Refer to Fuel Tank and Lines in this section for information on checking the lines for restrictions and checking fuel flow. Also, refer to Fuel Pump Service for more information on fuel pump testing, operation and repair.

COMBUSTION RELATED PISTON FAILURES

When an engine has a piston failure due to abnormal combustion, fixing the mechanical portion of the engine is the easiest part. The hard part is determining what caused the problem, in order to prevent a repeat failure. Think back to the four basic areas that affect combustion to find the cause of the failure.

Since you probably removed the cylinder head. Inspect the failed piston, look for excessive deposit buildup that could raise compression, or retain heat in the combustion chamber. Statically check the wide open throttle timing. Be sure that the timing is not over advanced. It is a good idea to seal these adjustments with paint to detect tampering.

If everything else looks good, the final possibility is poor quality fuel.

Fig. 25 Fouled spark plug, possibly caused by over-choking or a malfunctioning enrichment circuit

Fig. 26 Marine growth on the lower unit will create "drag" and seriously hamper boat performance

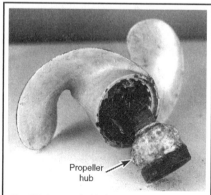

Fig. 27 A corroded hub on a small engine propeller. Hub and propeller damage will also cause poor performance

FUEL SYSTEM 3-17

CARBURETOR SERVICE

Carburetor Identification

◆ See Figures 28 and 29

A carburetor identification number is normally stamped somewhere on the housing (usually on a flange). Be sure to take down this number and have it handy when purchasing parts or overhaul kits. In some cases with Yamaha models adjustment specifications will vary dependant upon this number, we've cited those instances in the Carburetor Set-Up Specifications chart at the end of this section.

Fig. 28 Many carburetors (like this Yamaha model) have an identification number embossed the carburetor flange

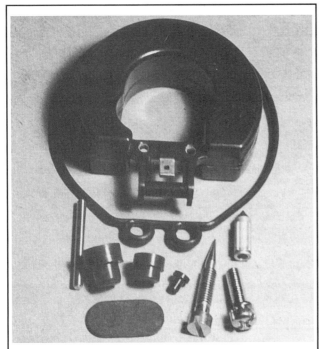

Fig. 29 Carburetor repair kits, available at your local service dealer, contain the necessary components to perform a carburetor overhaul. In most cases, an illustration showing the parts contained in the package is included in the cleaning and inspection portion for each carburetor.

2.5 Hp Yamaha Models

REMOVAL & INSTALLATION

◆ See Figures 30 and 31

1. Tag and disconnect the hoses from the air intake silencer and carburetor assembly. You'll have to loosen the clamps and/or cut the wire ties, depending on how the hoses are secured.
2. Loosen the set screw and disconnect the throttle cable from the throttle lever.
3. Support the carburetor assembly, then remove the 2 bolts and collars from the face of the air intake silencer which secure the silencer and carburetor to the powerhead.
4. Carefully remove the air intake silencer and position aside. Locate and remove the O-ring from the silencer or carburetor, then discard (as a new one should be used during assembly).
5. Pull the carburetor outward from the powerhead and upward slightly to free it from the fuel cock.
6. Remove the triple gasket assembly from the carburetor and/or powerhead, as applicable. Note the positioning and orientation of each gasket piece for assembly purposes, but new gaskets must be used during installation.
7. Clean and inspect and/or overhaul the carburetor, as applicable.

To Install:

8. Install a new O-ring for the air intake silencer.
9. Position 3 new gaskets by aligning the projections on the corners of the gaskets (and double-checking that the bolts holes are properly aligned as a result), then install the carburetor assembly and gaskets to the powerhead.

■ Make sure the carburetor fuel cock properly seats into the lever actuator on the powerhead as the carburetor is installed.

10. Install the air intake silencer using the 2 retaining bolts and collars, making sure the bolts are properly inserted through the triple gasket assembly and into the powerhead. Tighten the bolts securely.
11. Reconnect the hoses as tagged during removal and secure using the clamps or new wire ties.
12. Loosely install the throttle cable, the final cable adjustment is made once the Idle Speed is properly set.
13. Refer to the Timing and Synchronization adjustments in the Maintenance and Tune-Up section to make sure the Idle Speed is properly adjusted.

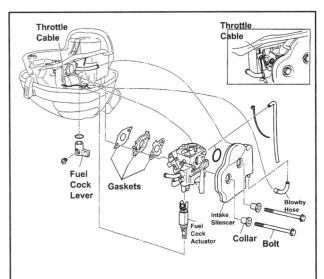

Fig. 30 Exploded view of the carburetor mounting - 2.5 Hp Yamaha Models

3-18 FUEL SYSTEM

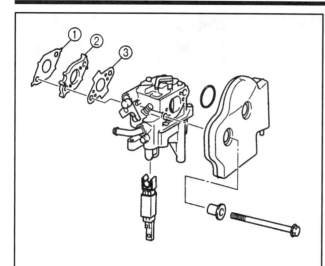

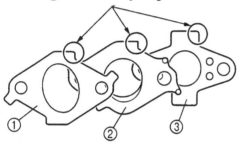

Align these projections

Fig. 31 Install the gaskets to the powerhead in the order shown, while aligning the projections

OVERHAUL

◆ See Figure 32

Good shop practice dictates purchasing a carburetor repair kit and using new parts any time the carburetor is disassembled. Make an attempt to keep the work area organized and to cover parts after they have been cleaned. This practice will prevent foreign matter from entering passageways or adhering to critical parts.

1. Remove the drain screw from the bottom of the carburetor float bowl and carefully drain the bowl of fuel.
2. Remove the carburetor from the powerhead, as detailed earlier in this section.
3. Loosen and remove the 3 top cover screws, then remove the top cover and gasket from the carburetor body. Discard the old gasket and carefully remove all traces from the mating surfaces.
4. Remove the set screw from the fuel cock boss on the lower side of the carburetor, then carefully lower the fuel cock from the underside of the boss. Remove and discard the old O-ring from the tip of the fuel cock.
5. Invert the carburetor body (with the float bowl facing upward), then loosen the 2 float bowl retaining screws. Remove the float bowl from the carburetor body, then remove and discard the old bowl-to-body gasket.
6. Remove the screw, then grasp the end of the float hinge pin and carefully pull it free of the carburetor body and float hinge.
7. Lift gently upward to remove the float and needle valve from the carburetor body.
8. Remove the bushing, then slowly unthread the pilot jet using a flathead screwdriver and counting the number of turns necessary to remove it from the carburetor body. Note the number of turns for assembly purposes.
9. Using a flathead screwdriver, carefully unthread the main jet, then remove main nozzle from the carburetor body.
10. Clean and inspect all components as detailed in this section. Replace any damaged, worn or defective components. Discard all O-rings or gaskets.

To Assemble:

11. Install the pilot screw the same number of turns as noted during disassembly, then install the bushing.
12. Install the main nozzle, then gently screw the main jet into position.
13. Connect the needle valve to the float, then lower the float and valve into position. Insert the hinge pin through the float and carburetor body, then install and tighten the screw.
14. With the carburetor still inverted, so the float is sitting gently on the needle valve (which is resting on the seat) measure the float height from the carburetor body-to-float bowl mating surface up to the top (actually the bottom, but it is on top now) of the float. Refer to the Carburetor Set-Up Specifications chart in this section for proper float drop/height specs. If necessary, gently bend the float hinge to achieve the proper measurement.

✱✱ WARNING

When measuring the float height DO NOT place any pressure downward on the needle valve or you could damage it (and/or you may make an incorrect adjustment).

15. Double-check that the float and valve move smoothly without sticking or binding.
16. Install the float bowl to the carburetor body using a new gasket, then gently secure using the 2 screws. If not done already, install the drain screw to the bowl.
17. Invert the carb, then install the carburetor cover to the top of the body using a new gasket and the 3 retaining screws. Tighten the screws securely.
18. Apply a light coating of marine grade grease to the shaft of the fuel cock and to the set screw. Install the fuel cock and a NEW O-ring to the underside of the fuel cock boss on the side of the carb, then install and tighten the set screw.
19. Install the carburetor and adjust for proper operation.

CLEANING & INSPECTION

◆ See Figures 32, 33 and 34

✱✱ CAUTION

Never dip rubber or plastic parts in carburetor cleaner. These parts should be cleaned only in solvent, and then blown dry with compressed air.

Place all metal parts in a screen-type tray and dip them in carburetor cleaner until they appear completely clean, then blow them dry with compressed air.

Blow out all passages in the castings with low-pressure compressed air. Check all parts and passages to be sure they are not clogged or contain any deposits. Never use a piece of wire or any type of pointed instrument to clean drilled passages or calibrated holes in a carburetor.

Move the throttle shaft back and forth to check for wear. If the shaft appears to be too loose, replace the complete carburetor body because individual replacement parts are not available.

Inspect the main body, air horn, and venturi cluster gasket surfaces for cracks and burrs which might cause a leak.

Check the float for deterioration. If any part of the float is damaged, the unit must be replaced.

Inspect the tapered section of the float needle and replace if it has developed a groove or is no longer evenly tapered.

As previously mentioned, most of the parts which should be replaced during a carburetor overhaul are included in overhaul kits available from your local marine dealer. Replace all components included in the kit for durability.

4 Hp Yamaha Models

REMOVAL & INSTALLATION

◆ See Figure 35

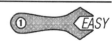

1. Tag and disconnect the choke and throttle cables from the carburetor.
2. Tag and disconnect the 2 breather hoses from the air intake silencer.
3. Disconnect the fuel hose from the carburetor assembly.

FUEL SYSTEM 3-19

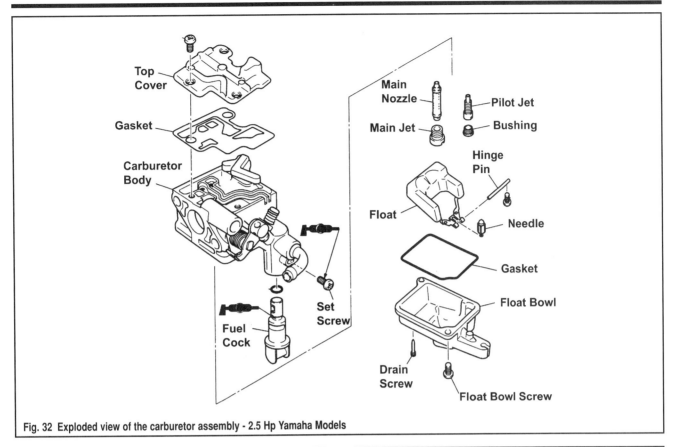

Fig. 32 Exploded view of the carburetor assembly - 2.5 Hp Yamaha Models

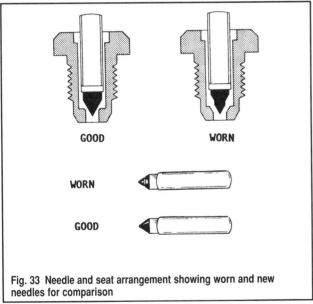

Fig. 33 Needle and seat arrangement showing worn and new needles for comparison

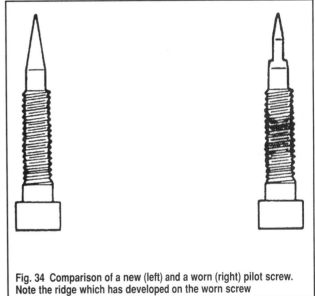

Fig. 34 Comparison of a new (left) and a worn (right) pilot screw. Note the ridge which has developed on the worn screw

4. Support the carburetor assembly. Remove the 2 grommets and the 2 bolts from the face of the air intake silencer which secure the silencer and carburetor to the powerhead.
5. Carefully remove the air intake silencer and carburetor assembly from the powerhead.
6. Separate the air silencer from the carburetor, then remove and discard the O-ring seal for the intake silencer.
7. The carburetor is mounted to the powerhead by a spacer with a gasket on either side. The spacer may have come off with the carburetor or may have been left behind on the powerhead. In either scenario, remove the spacer, then remove and discard the 2 gaskets.
8. Clean and inspect and/or overhaul the carburetor, as applicable.

To Install:
9. Position a NEW O-ring, then connect the air intake silencer to the carburetor.
10. Insert the 2 carburetor/intake silencer retaining bolts through the silencer and carburetor assembly. Install a NEW gasket to the rear of the carburetor over the 2 bolts, then install the spacer and the second NEW gasket.
11. Position the silencer, carburetor, spacer and gaskets assembly to the powerhead and thread the 2 retaining bolts. Tighten the bolts to 5.8 ft. lbs. (8 Nm) and install the grommets.
12. Reconnect the fuel hose to the carburetor and secure using the retaining clamp.

3-20 FUEL SYSTEM

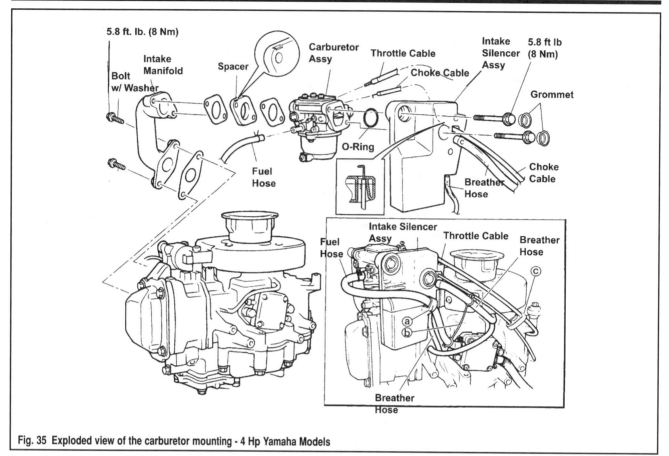

Fig. 35 Exploded view of the carburetor mounting - 4 Hp Yamaha Models

13. Connect the throttle and choke cables to the carburetor.
14. Refer to the Timing and Synchronization adjustments in the Maintenance and Tune-Up section to make sure the Idle Speed and Throttle Cable are both properly adjusted.

OVERHAUL

◆ See Figures 36 and 37

Good shop practice dictates purchasing a carburetor repair kit and using new parts any time the carburetor is disassembled. Make an attempt to keep the work area organized and to cover parts after they have been cleaned. This practice will prevent foreign matter from entering passageways or adhering to critical parts.

1. Remove the carburetor from the powerhead, as detailed earlier in this section.
2. If equipped, remove the float bowl drain plug and drain the carburetor of fuel.
3. Loosen and remove the 3 top cover screws, then remove the top cover plate and gasket from the carburetor body. Remove the plug from under the cover plate.
4. Remove the pilot screw from the side of the carburetor body.
5. Invert the carburetor body (with the float bowl facing upward), then loosen the float bowl retaining screws. Remove the float bowl from the carburetor body, then remove and discard the old bowl O-ring.
6. Grasp the end of the float hinge pin and carefully pull it free of the carburetor body and float hinge, then lift gently upward to remove the float and needle valve from the carburetor body.
7. Remove the plug, main nozzle, main jet and pilot jet from the carburetor body, keeping all parts separated and identified for installation or replacement purposes.
8. Clean and inspect all components as detailed in this section. Replace any damaged, worn or defective components. Discard all O-rings or gaskets.

To Assemble:

9. Install the pilot jet, main jet, main nozzle and plug to their positions in the carburetor body, as noted during removal.

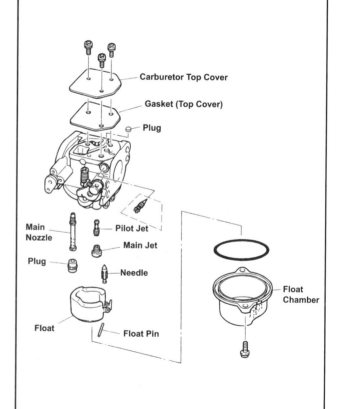

Fig. 36 Exploded view of the carburetor assembly - 4 Hp Yamaha Models

FUEL SYSTEM 3-21

10. Hook the needle valve to the float, then lower the float and valve into position. Insert the hinge pin through the float and carburetor body starting from the throttle lever side of the carburetor and gently tap it into position toward the fuel inlet pipe side of the carburetor.

11. With the carburetor still inverted, so the float is sitting gently on the needle valve (which is resting on the seat) gently tilt the carburetor down at a 45 degree angle from vertical (with the float hinge on top), then measure the float height from the carburetor body-to-float bowl mating surface up to the top (actually the bottom, but it is on top now even if pointed downward at an angle) of the float. Refer to the Carburetor Set-Up Specifications chart in this section for proper float drop/height specs. If necessary, gently bend the float hinge to achieve the proper measurement.

※※ WARNING

When measuring the float height DO NOT place any pressure downward on the needle valve or you could damage it (and/or you may make an incorrect adjustment). Tilting the carburetor 45 degrees downward from vertical (with the hinge toward the top) will make sure the needle valve is closed, but has NO LOAD applied.

12. Double-check that the float and valve move smoothly without sticking or binding.
13. Install the float bowl to the carburetor body using a new O-ring, then gently secure using the retaining screws. If not done already, install the drain screw to the bowl.
14. Invert the carb, then install the plug and the carburetor top cover plate to the top of the body. Tighten the 3 screws securely.
15. Install the pilot screw until is JUST gently seats, then back it out about 1 1/2 turns as an initial low speed setting.
16. Install the carburetor and adjust for proper operation.

CLEANING & INSPECTION

◆ See Figures 33, 34 and 36

※※ CAUTION

Never dip rubber or plastic parts in carburetor cleaner. These parts should be cleaned only in solvent, and then blown dry with compressed air.

Place all metal parts in a screen-type tray and dip them in carburetor cleaner until they appear completely clean, then blow them dry with compressed air.

Blow out all passages in the castings with low-pressure compressed air. Check all parts and passages to be sure they are not clogged or contain any deposits. Never use a piece of wire or any type of pointed instrument to clean drilled passages or calibrated holes in a carburetor.

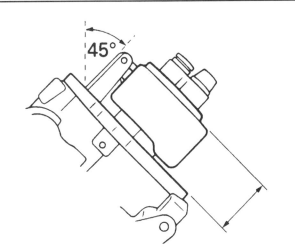

Fig. 37 Tilt the carburetor 45 degrees downward from vertical when measuring float height

Move the throttle shaft back and forth to check for wear. If the shaft appears to be too loose, replace the complete carburetor body because individual replacement parts are not available.

Inspect the main body, air horn, and venturi cluster gasket surfaces for cracks and burrs which might cause a leak.

Check the float for deterioration. If any part of the float is damaged, the unit must be replaced.

Inspect the tapered section of the float needle and pilot screw for wear or damage. Replace either if they have developed a groove or are no longer evenly tapered.

As previously mentioned, most of the parts which should be replaced during a carburetor overhaul are included in overhaul kits available from your local marine dealer. Replace all components included in the kit for durability.

4/5/6 Hp Mercury/Mariner Models

REMOVAL & INSTALLATION

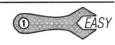

◆ See Figure 38

1. Make sure the fuel supply valve (if equipped) is turned off, then disconnect the fuel supply from the side of the carburetor. It is normally secured with a spring type clamp.
2. Remove the hand rewind starter assembly from the top of the motor.
3. Disconnect the breather hose from the top of the carburetor assembly. Like the fuel hose, it is normally secured by a spring type clamp.
4. Support the flame arrestor and carburetor assembly, then remove the 2 bolts holding the assembly to the intake manifold.
5. Disconnect the choke cable and throttle linkage.
6. Inspect the gasket assembly. At the manifold this model uses an insulator gasket, sealed with a gasket on either side. Although it is not always necessary, it is a good idea to use new gaskets during installation to ensure a proper seal.
7. Clean and inspect and/or overhaul the carburetor, as applicable.

To Install:

8. Support the carburetor assembly while reconnecting the choke cable and throttle linkage.
9. Install the carburetor to the intake manifold using the triple gasket setup (insulator gasket with a gasket on each side).
10. Install and tighten the carburetor mounting bolts to 70 inch lbs. (8 Nm).
11. Reconnect the fuel supply and breather hoses, making sure the spring clamps are placed inboard of the nipples to hold the lines in place.
12. Install the hand rewind starter.
13. Refer to the Timing and Synchronization adjustments in the Maintenance and Tune-Up section to make sure the Idle Speed and Throttle Cable are both properly adjusted.

OVERHAUL

◆ See Figures 39 and 40

Good shop practice dictates purchasing a carburetor repair kit and using new parts any time the carburetor is disassembled. Make an attempt to keep the work area organized and to cover parts after they have been cleaned. This practice will prevent foreign matter from entering passageways or adhering to critical parts.

1. Remove the carburetor from the powerhead, as detailed earlier in this section.
2. Remove the float bowl drain screw and drain as much fuel from the carburetor as you can.
3. Loosen the 2 screws securing the float bowl to the bottom of the carburetor, then separate the bowl from the carburetor body. Remove and discard the gasket.
4. Observe the location of the float hinge pin in relation to the fuel inlet nipple on the side of the carburetor. To remove the float, carefully push the hinge pin out TOWARD the fuel inlet nipple. Once the pin is removed carefully life the float and needle valve assembly from the carburetor.

3-22 FUEL SYSTEM

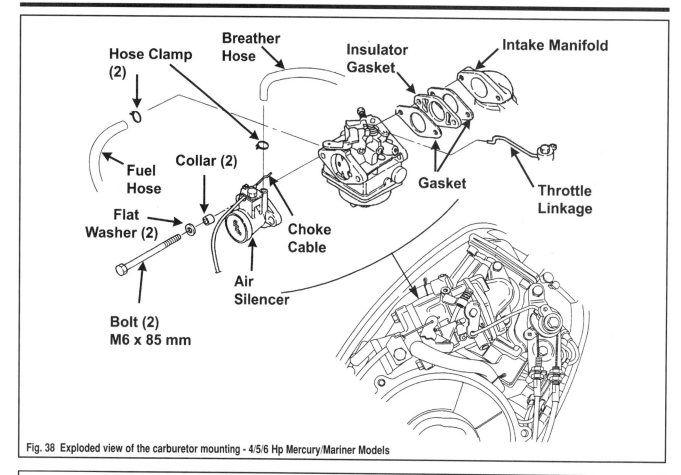

Fig. 38 Exploded view of the carburetor mounting - 4/5/6 Hp Mercury/Mariner Models

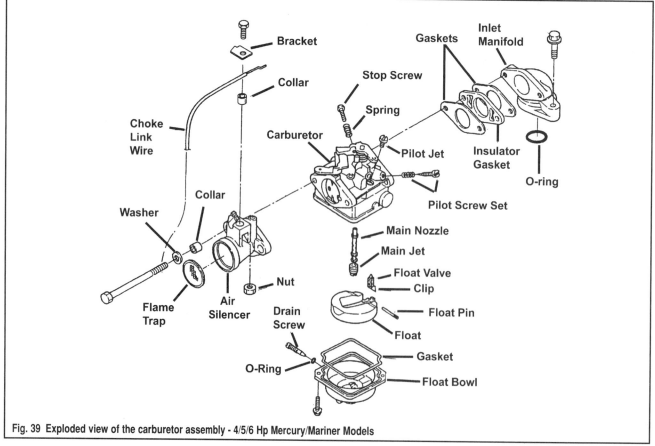

Fig. 39 Exploded view of the carburetor assembly - 4/5/6 Hp Mercury/Mariner Models

FUEL SYSTEM 3-23

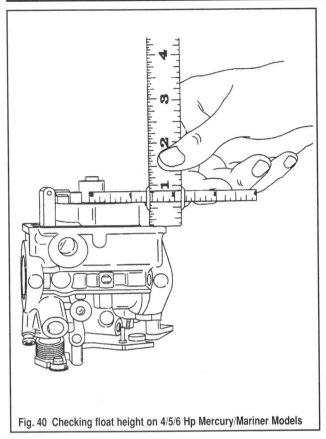

Fig. 40 Checking float height on 4/5/6 Hp Mercury/Mariner Models

■ The hinge pin may be knurled slightly on one end. When applicable, this end should face the fuel inlet nipple.

5. Unthread the main jet, then remove the main nozzle. The nozzle SHOULD drop out freely, however, IF NECESSARY, carefully push on the nozzle from the top using a screwdriver (tipped with a piece of electrical tape to protect the carburetor body/nozzle).

■ Before removing the pilot screw, slowly and gently thread the screw into the carburetor bore until it JUST LIGHTLY seats, counting the number of turns inward you've threaded the screw. If desired, the number of turns counted can be used as an alternate setting to the Initial Pilot Screw setting during assembly.

6. If necessary remove the idle screw and spring, the idle jet, and/or the pilot screw and spring from the carburetor body. Keep in mind that some US models may not utilize an adjustable pilot screw.

7. Clean and inspect all components as detailed in this section. Replace any damaged, worn or defective components. Discard all O-rings or gaskets.

To Assemble:

8. Install the float needle valve onto the metal tab on the float, then carefully position the assembly in the carburetor body.

9. Install the float hinge pin to secure the float, inserting the smooth end from the fuel nipple side of the carburetor body.

10. Use a sliding carburetor scale or precision measuring instrument to check the float height with the carburetor body inverted (float facing upward). The top of the float should be 0.35-0.39 in. (9-10mm) above the gasket surface of the carburetor body.

11. Install the float bowl using a new gasket and tighten the screws securely.

12. If removed install the idle screw and spring, the pilot jet and/or the pilot screw and spring to the carburetor body. For the pilot screw, carefully thread the screw into the carburetor bore until it JUST LIGHTLY seats, then either back the screw out to the initial setting as per the Carburetor Specifications chart and/or back it out the same number of turns as noted during removal to duplicate the adjusted setting to which it was set before overhaul.

13. Install the carburetor and adjust for proper operation.

CLEANING & INSPECTION

 MODERATE

◆ See Figures 33, 34 and 39

※※ CAUTION

Never dip rubber or plastic parts in carburetor cleaner. These parts should be cleaned only in solvent, and then blown dry with compressed air.

Place all metal parts in a screen-type tray and dip them in carburetor cleaner until they appear completely clean, then blow them dry with compressed air.

Blow out all passages in the castings with low-pressure compressed air. Check all parts and passages to be sure they are not clogged or contain any deposits. Never use a piece of wire or any type of pointed instrument to clean drilled passages or calibrated holes in a carburetor.

Move the throttle shaft back and forth to check for wear. If the shaft appears to be too loose, replace the complete carburetor body because individual replacement parts are normally not available.

Inspect the main body gasket surfaces for cracks and burrs which might cause a leak.

Check the float for deterioration. If any part of the float is damaged, the unit must be replaced.

Inspect the tapered section of the float needle and pilot screw for wear or damage. Replace either if they have developed a groove or are no longer evenly tapered.

As previously mentioned, most of the parts which should be replaced during a carburetor overhaul are included in overhaul kits available from your local marine dealer. Replace all components included in the kit for durability.

6/8 Hp Yamahas Models

REMOVAL & INSTALLATION

 EASY

◆ See Figure 41

1. Disconnect the throttle axle link and the choke rod from the carburetor.

2. Disconnect the breather hose from the back side of the air intake silencer assembly.

3. Depress the ears of the spring-clamp and slide it back on the hose until it is clear of the carburetor nipple, then disconnect the fuel hose from the carburetor assembly.

4. Support the carburetor assembly. Remove the 2 bolts from the face of the air intake silencer which secure the silencer and carburetor to the powerhead.

5. Carefully remove the air intake silencer and carburetor assembly from the powerhead.

6. Separate the air silencer from the carburetor, then remove and discard the O-ring seal for the intake silencer.

7. The carburetor is mounted to the powerhead by an insulator with a gasket on either side. The insulator may have come off with the carburetor or may have been left behind on the powerhead. In either scenario, remove it, then remove and discard the 2 gaskets.

8. Clean and inspect and/or overhaul the carburetor, as applicable.

9. If necessary, tag and disconnect the water pilot and water hoses, then loosen the intake manifold retaining bolts (using the reverse of the torque sequence, working from the outer bolts and moving inward in a counterclockwise spiraling patter that starts at the top bolt) and remove the intake manifold from the powerhead. Remove and discard the old intake manifold gasket.

To Install:

10. If removed, install the intake manifold using a new gasket, then tighten using the torque sequence (a spiraling, clockwise pattern that works from the inner toward the outer bolts). Tighten the bolts using at least 2 passes of the sequence, first to 4.4 ft. lbs. (6 Nm) and then to 8.8 ft. lbs. (12 Nm). Reconnect the water pilot and water hoses, as tagged during removal.

111. Position a NEW O-ring, then connect the air intake silencer to the carburetor.

3-24 FUEL SYSTEM

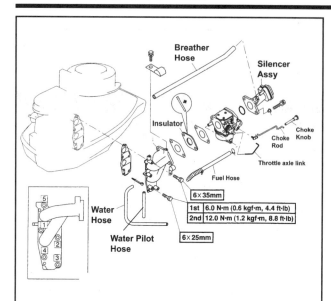

Fig. 41 Exploded view of the carburetor mounting - 6/8 Hp Yamahas Models

12. Insert the 2 carburetor/intake silencer retaining bolts through the silencer and carburetor assembly. Install a NEW gasket to the rear of the carburetor over the 2 bolts, then install the insulator and the second NEW gasket.
13. Position the silencer, carburetor, spacer and gaskets assembly to the powerhead, then thread the 2 retaining bolts and tighten them securely.
14. Reconnect the fuel hose to the carburetor and secure using the retaining clamp.
15. Connect the choke rod and throttle link to the carburetor.
16. Refer to the Timing and Synchronization adjustments in the Maintenance and Tune-Up section to make sure the Throttle Axle Link and Idle Speed are both properly adjusted.

OVERHAUL

◆ See Figure 37, 42 and 43

Good shop practice dictates purchasing a carburetor repair kit and using new parts any time the carburetor is disassembled. Make an attempt to keep the work area organized and to cover parts after they have been cleaned. This practice will prevent foreign matter from entering passageways or adhering to critical parts.

1. Remove the carburetor from the powerhead, as detailed earlier in this section.
2. Drain the carburetor of fuel by removing the float bowl drain plug and gasket. Discard the plug gasket and replace with a new one during assembly to prevent leaks.
3. Loosen and remove the 3 top cover screws, then remove the top cover plate and gasket from the carburetor body.
4. If necessary, remove the idle adjust (throttle stop) screw and spring from the boss on the side of the carburetor body.
5. Invert the carburetor body (with the float bowl facing upward), then loosen the float bowl retaining screws. Remove the float bowl from the carburetor body, then remove and discard the old bowl O-ring.
6. Using a small punch gently tap the end of the float hinge pin in the OPPOSITE direction of the molded arrow on the carburetor body to free it from the carb and float. Lift gently upward to remove the float and needle valve from the carburetor body.
7. Remove the main jet, pilot jet, plug and main nozzle from the carburetor body, keeping all parts separated and identified for installation or replacement purposes.
8. Clean and inspect all components as detailed in this section. Replace any damaged, worn or defective components. Discard all O-rings or gaskets.

To Assemble:

9. Install the main nozzle, plug, pilot jet and main jet, to their positions in the carburetor body, as noted during removal.

10. Connect the needle valve to the float, then lower the float and valve into position. Insert the hinge pin through the float and carburetor body in the direction of the arrow molded on the carburetor body, then gently tap it into position.
11. With the carburetor still inverted, so the float is sitting gently on the needle valve (which is resting on the seat) gently tilt the carburetor down at a 45 degree angle from vertical (with the float hinge on top), then measure the float height from the carburetor body-to-float bowl mating surface up to the top (actually the bottom, but it is on top now even if pointed downward at an angle) of the float. Refer to the Carburetor Set-Up Specifications chart in this section for proper float drop/height specs. If necessary, gently bend the float hinge to achieve the proper measurement.

※※ WARNING

When measuring the float height DO NOT place any pressure downward on the needle valve or you could damage it (and/or you may make an incorrect adjustment). Tilting the carburetor 45 degrees downward from vertical (with the hinge toward the top) will make sure the needle valve is closed, but has NO LOAD applied.

12. Double-check that the float and valve move smoothly without sticking or binding.
13. Install the float bowl to the carburetor body using a NEW O-ring, then gently secure using the retaining screws. If not done already, install the drain screw to the bowl using a NEW gasket.
14. Invert the carb, then the carburetor top cover plate to the top of the body. Tighten the 3 screws securely.
15. If removed, install the idle adjust (throttle stop) screw and spring until it just seats on the lever. You'll have to adjust the Idle Speed once the carburetor is installed.
16. Install the carburetor and adjust for proper operation.

CLEANING & INSPECTION

◆ See Figures 33, 34 and 42

※※ CAUTION

Never dip rubber or plastic parts in carburetor cleaner. These parts should be cleaned only in solvent, and then blown dry with compressed air.

Place all metal parts in a screen-type tray and dip them in carburetor cleaner until they appear completely clean, then blow them dry with compressed air.
Blow out all passages in the castings with low-pressure compressed air. Check all parts and passages to be sure they are not clogged or contain any deposits. Never use a piece of wire or any type of pointed instrument to clean drilled passages or calibrated holes in a carburetor.
Move the throttle shaft back and forth to check for wear. If the shaft appears to be too loose, replace the complete carburetor body because individual replacement parts are not available.
Inspect the main body, air horn, and venturi cluster gasket surfaces for cracks and burrs which might cause a leak.
Check the float for deterioration. If any part of the float is damaged, the unit must be replaced.
Inspect the tapered section of the float needle and pilot screw for wear or damage. Replace either if they have developed a groove or are no longer evenly tapered.
As previously mentioned, most of the parts which should be replaced during a carburetor overhaul are included in overhaul kits available from your local marine dealer. Replace all components included in the kit for durability.

8/9.9 Hp (232cc) Models

All 8/9.9 hp 4-stroke motors utilize a single barrel, side draft, single float carburetor. Starting sometime around 1992 or 1993 Yamaha replaced their un-necessarily complicated Bi-metallic Vacuum Switching [BVS] valves cold starting system with a single electro-thermal valve.
On all models covered by this guide, an electro-thermal valve is mounted to the top of the carburetor assembly. This valve receives power from a dedicated power coil once the motor is started. The current allows an internal element in the valve to gradually heat up (a process that is timed to take

FUEL SYSTEM 3-25

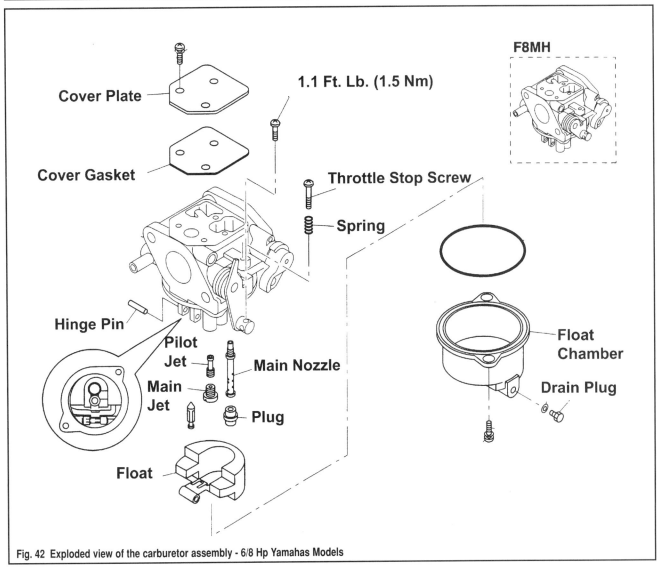

Fig. 42 Exploded view of the carburetor assembly - 6/8 Hp Yamahas Models

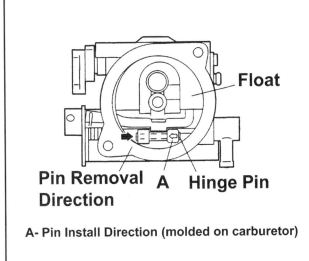

Fig. 43 Remove or install the hinge pin in the directions shown (installation is in the same direction as the arrow molded into the carb body)

about the same amount of time it should take a cold motor to warm to normal operating temperature). This temperature change causes the valve piston to move and hold a new position, cutting the enrichment circuit as the engine warms.

Although Yamaha and Mercury do not say specifically for this model how the enrichment circuit works, the same type of device on the 25 hp 4-stroke motors works by opening an additional fuel passage when the ram is retracted. As the heater element in the electro-thermal ram heats up the ram will begin to slowly extend, slowly sealing off the enrichment passage. It is likely that the valve works in the same manner on these motors.

■ Good shop practice dictates purchasing a carburetor repair kit and using new parts (especially O-rings and gaskets) any time the carburetor is disassembled. Make an attempt to keep the work area organized and to cover parts after they have been cleaned. This practice will prevent foreign matter from entering passageways or adhering to critical parts.

REMOVAL & INSTALLATION

◆ See Figure 44

1. Disconnect the throttle link from the carburetor.
2. Loosen the clamp (it's usually a spring-type clamp) and disconnect the fuel hose from the carburetor assembly.

3-26 FUEL SYSTEM

3. Tag and disconnect the electro-thermal valve wiring from the engine harness.
4. Loosen and remove the 2 bolts securing the air intake funnel to the carburetor, then remove the air intake and gasket.
5. Support the carburetor assembly, then remove the 2 nuts securing the carburetor to the engine. Remove the carburetor from the motor.
6. The carburetor is mounted to the powerhead by an insulator with a gasket on either side. The insulator may have come off with the carburetor or may have been left behind on the powerhead. In either scenario, remove it, then remove and discard the 2 gaskets.
7. Clean and inspect and/or overhaul the carburetor, as applicable.

To Install:
8. Position the insulator along with 2 new gaskets over the carburetor mounting studs, then position the carburetor to the powerhead and tighten the nuts securely.
9. Reconnect the electro-thermal valve wiring to the engine harness.
10. Reconnect the fuel hose to the carburetor and secure using the retaining clamp.
11. Connect the throttle link to the carburetor.
12. Install the air intake funnel and gasket to the carburetor, then secure using the retaining bolts.
13. Refer to the Timing and Synchronization adjustments in the Maintenance and Tune-Up section to make sure the Idle Speed and Throttle Link are both properly adjusted.

OVERHAUL

◆ See Figure 44 and 45 thru 49

Good shop practice dictates purchasing a carburetor repair kit and using new parts any time the carburetor is disassembled. Make an attempt to keep the work area organized and to cover parts after they have been cleaned. This practice will prevent foreign matter from entering passageways or adhering to critical parts.

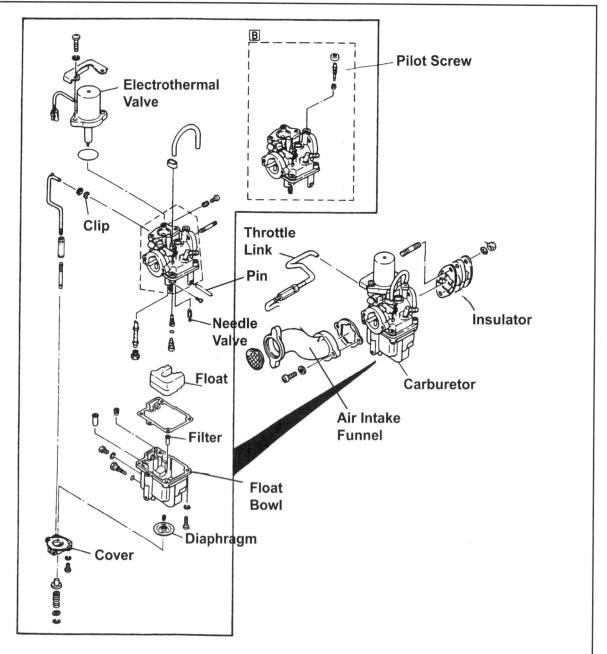

Fig. 44 Exploded view of the carburetor and carburetor mounting - 8/9.9 Hp (232cc) Models

FUEL SYSTEM 3-27

1. Remove the carburetor from the powerhead, as detailed earlier in this section.
2. If not done already, drain the carburetor of fuel by removing the float bowl drain plug and gasket.
3. Remove the 2 screws and bracket securing the electro-thermal valve to the top of the carburetor. Carefully pull the valve and O-ring up and off the carburetor.
4. Remove the clip from the diaphragm link on the side of the carburetor.
5. Except for 1998 and later US or Swiss models (and Mercury Bodensee models), remove the plug, then loosen and remove the pilot screw (and spring) from the vertical bore in the top of the carburetor assembly (immediately adjacent to the intake manifold mating surface).
6. Invert the carburetor body (with the float bowl facing upward), then loosen the 4 float bowl retaining screws. Remove the float bowl from the carburetor body, then remove and discard the old bowl gasket.
7. Using a small punch gently tap the end of the float hinge pin to free it from the carb and float. Lift gently upward to remove the float and needle valve from the carburetor body.
8. Remove the screw and main jet, the seal cap, O-ring and pilot jet, and the main nozzle from the carburetor body. Remove the filter and check valve from the float bowl. Keep all parts separated and identified for installation or replacement purposes.
9. Remove the 2 cover screws, then carefully remove the cover, diaphragm and spring from the bottom of the float bowl.
10. Clean and inspect all components as detailed in this section. Replace any damaged, worn or defective components. Discard all O-rings or gaskets.

To Assemble:
11. Position the spring and diaphragm to the bottom of the float bowl and secure using the cover and 2 retaining screws.
12. Install the main nozzle, check valve, pilot jet (with O-ring and seal cap) and main jet with screw, to their positions in the carburetor body, as noted during removal. Keep in mind the following:
- When installing the main nozzle be certain to face the cutout on the end of the nozzle TOWARD the intake manifold (away from the air intake funnel).
- When installing the main jet, align the slit in the jet with the screw hole in the carburetor, then install the screw.
13. Connect the needle valve to the float, then lower the float and valve into position. Insert the hinge pin through the float and carburetor body using a pair of needle-nose pliers.
14. With the carburetor still inverted and perfectly level, so the float is sitting gently on the needle valve (which is resting on the seat) measure the float height from the carburetor body-to-float bowl mating surface up to the top (actually the bottom, but it is on top now) of the float. Refer to the Carburetor Set-Up Specifications chart in this section however for all models of the 8/9.9 hp motor the float height should be about 1 in. (25.5mm). If necessary, gently bend the float hinge to achieve the proper measurement.

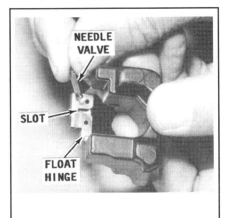

Fig. 45 Connect the needle valve to the float...

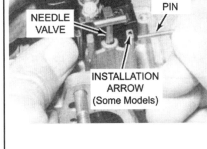

Fig. 46 ...then install the float/valve using the hinge pin

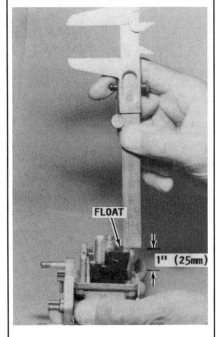

Fig. 47 Checking the float height

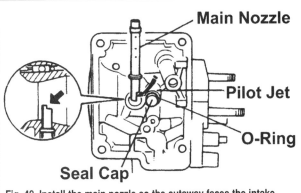

Fig. 48 Install the main nozzle so the cutaway faces the intake manifold

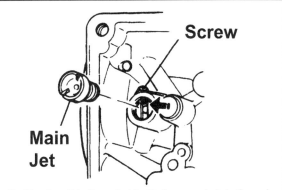

Fig. 49 Align the slit in the main jet with the screw hole in the carb body

FUEL SYSTEM

✴✴ WARNING

When measuring the float height DO NOT place any pressure downward on the needle valve or you could damage it (and/or you may make an incorrect adjustment).

15. Double-check that the float and valve move smoothly without sticking or binding.
16. Install the check valve and the filter to the float bowl.
17. Install the float bowl to the carburetor body using a NEW gasket, then gently secure using the retaining screws. If not done already, install the drain screw to the bowl.
18. Invert the carb, then reconnect the diaphragm link using the retaining clip and washer.
19. Install the electro-thermal valve and O-ring (it really should be replaced using a NEW O-ring) to the top of the carburetor. Secure the valve using the bracket and retaining screws.
20. If applicable, carefully install the pilot screw and spring, by threading the screw inward until it is JUST LIGHTLY seated, then backing it out the specified number of turns as listed in the Carburetor Set-Up Specifications chart (the spec for Yamahas is usually about 3 turns for standard models or 3 1/2 turns for high thrust models. The spec for adjustable Mercury/Mariner units is normally 2-4 turns).
21. Install the carburetor and adjust for proper operation.

CLEANING & INSPECTION

◆ See Figures 33, 34, 44, 50 and 51

✴✴ CAUTION

Never dip rubber or plastic parts in carburetor cleaner. These parts should be cleaned only in solvent, and then blown dry with compressed air.

Place all metal parts in a screen-type tray and dip them in carburetor cleaner until they appear completely clean, then blow them dry with compressed air.

Blow out all passages in the castings with low-pressure compressed air. Check all parts and passages to be sure they are not clogged or contain any deposits. Never use a piece of wire or any type of pointed instrument to clean drilled passages or calibrated holes in a carburetor.

Move the throttle shaft back and forth to check for wear. If the shaft appears to be too loose, replace the complete carburetor body because individual replacement parts are not available.

Inspect the main body, air horn, and venturi cluster gasket surfaces for cracks and burrs which might cause a leak.

Check the float for deterioration. If any part of the float is damaged, the unit must be replaced.

Inspect the tapered section of the float needle and pilot screw (if applicable) for wear or damage. Replace either if they have developed a groove or are no longer evenly tapered.

Check the main jet, pilot jet, check valve and main nozzle for signs of dirt or contamination. If they cannot be cleaned, they should be replaced. Again, NEVER clean these components using a wire or any pointer instrument, you'll change the calibration.

Check the diaphragm for signs of deterioration or damage and replace. Of course, like all diaphragms, they weaken over time so if there is any doubt, just replace it to be save.

Check the filter in the float bowl for signs of contamination. Like the nozzles, if it cannot be cleaned, it must be replaced. And like the diaphragm, it's an inexpensive part and you're here now, so go ahead.

■ **As previously mentioned, most of the parts which should be replaced during a carburetor overhaul are included in overhaul kits available from your local marine dealer. Replace all components included in the kit for durability.**

Inspect the electro-thermal valve, as follows:
• Using an ohmmeter check the valve element resistance. Connect the ohmmeter across the two valve leads, it should read about 4.8-7.2 ohms resistance at an ambient temperature of about 68°F (20°C).

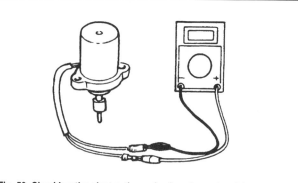

Fig. 50 Checking the electro-thermal valve element resistance

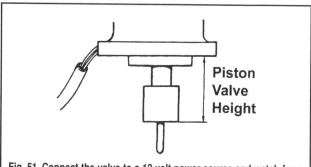

Fig. 51 Connect the valve to a 12 volt power source and watch for a change in piston height

• Check the distance that the piston is protruding from the valve. Apply a 12 volt power source to the 2 valve leads and leave it connected for several minutes, then recheck the valve piston height. If the valve is working properly the piston height should have changed as the internal valve element was gradually heated.

Always replace any and all worn parts.

9.9/15 Hp (323cc) Models

REMOVAL & INSTALLATION

◆ See Figure 52

1. Disconnect the breather hose from the air intake silencer.
2. For Mercury/Mariner models, matchmark, then remove the throttle linkage from the carburetor barrel.

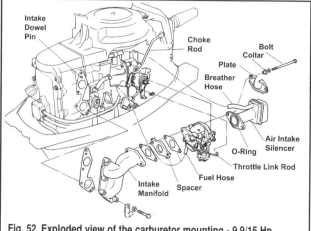

Fig. 52 Exploded view of the carburetor mounting - 9.9/15 Hp (323cc) Models

FUEL SYSTEM 3-29

3. Pull the choke knob outward, then disconnect the choke rod by carefully snapping the linkage off the side of the carburetor.

4. For Yamaha models, disconnect the throttle link rod.

5. Loosen the clamp (it's usually a spring-type clamp) and disconnect the fuel hose from the carburetor assembly.

6. Support the carburetor assembly, then loosen the 2 bolts and collars securing the plate, air intake silencer, carburetor and spacer to the intake manifold. Loosen the bolts just to the point where they pull free of the intake manifold, but keep them inserted to hold the carburetor and other components in place, then carefully remove the components as an assembly from the powerhead.

7. Pull components off from the rear of the assembly. First remove the spacer and gaskets (there should be one on either side). Of course, the spacer is sometimes left behind on the powerhead as it may stick to the gasket.

8. Separate the carburetor and the air intake silencer. Keep track of the air intake silencer O-ring.

9. Clean and inspect and/or overhaul the carburetor, as applicable.

10. If necessary, loosen the intake manifold retaining bolts (working from the outer bolts and moving inward in a counterclockwise spiraling patter that starts at the top bolt) and remove the intake manifold from the powerhead. Remove and discard the old intake manifold gasket.

To Install:

11. If removed, install the intake manifold using a new gasket and the dowel pin in the powerhead to align them. Tighten the bolts using a spiraling, clockwise pattern that works from the inner toward the outer bolts). Tighten the bolts using at least 2 passes of the sequence, to 5.8 ft. lbs. (8 Nm).

12. Make sure the O-ring is in good condition (or replace if it is cut or worn), then connect the intake air silencer to the carburetor, insert the 2 bolts (with collars) through the plate, silencer and carburetor to help hold the components aligned together.

13. From the back side of the carburetor, install a new inner gasket, the spacer and a new outer gasket, then align the assembly with the bolt holes on the intake manifold. Install the assembly to the intake and tighten the bolts securely.

14. Reconnect the fuel hose to the carburetor and secure using the retaining clamp.

15. Connect the throttle link rod and the choke rod. On Mercury/Mariner models, use the matchmark to help set the throttle link to the proper initial position.

16. Reconnect the breather hose to the air intake silencer.

17. Refer to the Timing and Synchronization adjustments in the Maintenance and Tune-Up section to make sure the Throttle Link Rod and Idle Speed are both properly adjusted.

OVERHAUL

◆ See Figures 53 and 54

Good shop practice dictates purchasing a carburetor repair kit and using new parts any time the carburetor is disassembled. Make an attempt to keep the work area organized and to cover parts after they have been cleaned. This practice will prevent foreign matter from entering passageways or adhering to critical parts.

1. Remove the carburetor from the powerhead, as detailed earlier in this section.

2. If equipped (most U.S. model are set at the factory and not adjustable), remove the pilot screw from the bore on the top, side of the carburetor body.

3. Remove the 2 screws securing the top cover to the carburetor, then remove the cover and discard the old gasket.

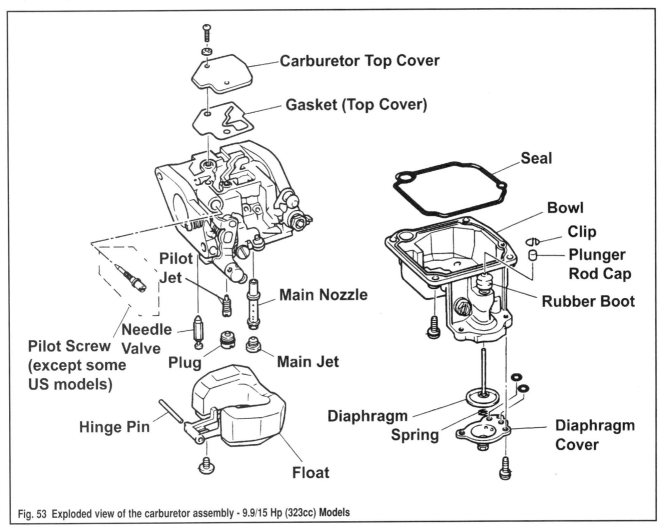

Fig. 53 Exploded view of the carburetor assembly - 9.9/15 Hp (323cc) Models

3-30 FUEL SYSTEM

4. Invert the carburetor body (with the float bowl facing upward), then loosen the 4 float bowl retaining screws. Remove the float bowl from the carburetor body, then remove and discard the old bowl seal.

5. Remove the cap and clip from the top of the accelerator pump plunger. If the float bowl is to be cleaned using solvent (or if necessary for replacement) remove the rubber boot from the side of the float bowl.

6. Remove the 3 screws securing the accelerator pump diaphragm cover to the bottom of the float bowl, then remove the cover, spring and diaphragm.

7. Loosen and remove the screw, then use a small punch to gently tap the end of the float hinge pin until it can be pulled free from the carb and float. Lift gently upward to remove the float and needle valve from the carburetor body.

8. Remove the main jet and main nozzle, then remove the plug and pilot jet from the carburetor body, keeping all parts separated and identified for installation or replacement purposes.

9. Clean and inspect all components as detailed in this section. Replace any damaged, worn or defective components. Discard all O-rings or gaskets.

To Assemble:

10. Install the pilot jet and plug, then the main nozzle and main jet each to the carburetor body in their original positions, as noted during removal.

11. If removed, install the rubber boot back into the carburetor float bowl. Install the diaphragm, spring and cover to the bottom of the float bowl, then secure using the 3 retaining screws.

12. Install the clip and plunger rod cap.

13. Connect the needle valve to the float, then lower the float and valve into position. Insert the hinge pin through the float and carburetor body using a pair of needle-nose pliers, then install the screw to secure the pin.

14. With the carburetor still inverted and perfectly level, so the float is sitting gently on the needle valve (which is resting on the seat) measure the float height from the carburetor body-to-float bowl mating surface up to the center of the float (directly opposite the hinge pin). Refer to the Carburetor Set-Up Specifications chart in this section for the exact range, however the float height should be about 0.4 in. (10mm) for Yamaha models and 0.61 in. (15.5mm) for Mercury/Mariner models. If necessary, gently bend the float hinge to achieve the proper measurement.

✳✳ WARNING

When measuring the float height DO NOT place any pressure downward on the needle valve or you could damage it (and/or you may make an incorrect adjustment).

15. Double-check that the float and valve move smoothly without sticking or binding.

16. Install the float bowl to the carburetor body using a NEW seal, then gently secure using the retaining screws.

17. Install the top cover to the top of the carburetor assembly using a new gasket and secure with the 2 retaining screws.

18. If applicable, carefully install the pilot screw by threading the screw inward until it is JUST LIGHTLY seated, then backing it out the specified number of turns as listed in the Carburetor Set-Up Specifications chart (the spec is about 1 1/4 turns).

19. Install the carburetor and adjust for proper operation.

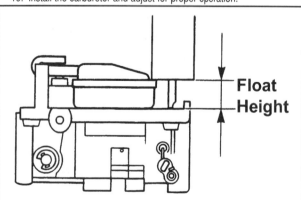

Fig. 54 Measure the float height from the gasket surface to the center of the float, as shown

CLEANING & INSPECTION

◆ See Figures 33, 34 and 53

✳✳ CAUTION

Never dip rubber or plastic parts in carburetor cleaner. These parts should be cleaned only in solvent, and then blown dry with compressed air.

Place all metal parts in a screen-type tray and dip them in carburetor cleaner until they appear completely clean, then blow them dry with compressed air.

Blow out all passages in the castings with low-pressure compressed air. Check all parts and passages to be sure they are not clogged or contain any deposits. Never use a piece of wire or any type of pointed instrument to clean drilled passages or calibrated holes in a carburetor.

Move the throttle shaft back and forth to check for wear. If the shaft appears to be too loose, replace the complete carburetor body because individual replacement parts are not available.

Inspect the main body, air horn, and venturi cluster gasket surfaces for cracks and burrs which might cause a leak.

Check the float for deterioration. If any part of the float is damaged, the unit must be replaced.

Inspect the tapered section of the float needle and pilot screw (if applicable) for wear or damage. Replace either if they have developed a groove or are no longer evenly tapered.

Check the main jet, pilot jet and main nozzle for signs of dirt or contamination. If they cannot be cleaned, they should be replaced. Again, NEVER clean these components using a wire or any pointer instrument, you'll change the calibration.

Check the diaphragm for signs of deterioration or damage and replace. Of course, like all diaphragms, they weaken over time so if there is any doubt, just replace it to be save. Make sure the spring has not deformed or lost its strength.

Check the rubber boot which protects the accelerator pump rod for deterioration or damage and replace, if necessary.

■ As previously mentioned, most of the parts which should be replaced during a carburetor overhaul are included in overhaul kits available from your local marine dealer. Replace all components included in the kit for durability.

Always replace any and all worn parts.

25 Hp

◆ See Figure 55

All 25 hp 4-stroke Yamahas utilize a single barrel, side draft, single float carburetor. An electro-thermal valve is mounted to the top of the carburetor assembly to provide additional fuel for cold start enrichment. When cold, the valve contains a piston which is retracted into the valve unblocking an additional fuel circuit in the carburetor (from the float bowl to the throttle bore). However, once the motor is started the power from the stator/charge coil is applied to a heater element in the valve in order to gradually build heat in the valve. This process is timed to take about the same amount of time it should take a cold motor to warm to normal operating temperature. The temperature change causes the valve piston to extend, gradually cutting the enrichment circuit as the engine warms.

■ Good shop practice dictates purchasing a carburetor repair kit and using new parts (especially O-rings and gaskets) any time the carburetor is disassembled. Make an attempt to keep the work area organized and to cover parts after they have been cleaned. This practice will prevent foreign matter from entering passageways or adhering to critical parts.

FUEL SYSTEM 3-31

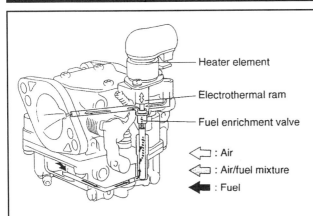

Fig. 55 An electro-thermal ram controls movement of fuel through an enrichment circuit for cold-starting

REMOVAL & INSTALLATION

◆ See Figure 56

1. Disconnect the breather hose from the side of the air intake silencer.
2. Locate the plastic wire tie that is wrapped around the air intake silencer (just behind where the breather hose was connected). The wire tie secures a hose to the side of the air silencer. Cut the wire tie so the carburetor and silencer assembly can be removed.
3. Disconnect the throttle link rod from the linkage (it is usually easier to leave it attached to the carburetor itself).
4. Loosen the clamp (it's usually a spring-type clamp) and disconnect the fuel hose from the carburetor assembly.
5. Tag and disconnect the electro-thermal valve wiring from the engine harness.
6. Loosen and remove the flange bolts which are threaded through the intake manifold and into the intake manifold bracket/carburetor assembly. Remove the carburetor from the intake manifold.
7. If further disassembly is required (such as for carburetor service), place the assembly on a clean work surface and remove the 2 bolts securing the air intake silencer to the carburetor.
8. Remove the air intake silencer and O-ring from the carburetor.
9. Separate the carburetor from the spacer and the intake manifold bracket. Remove the O-rings from the carburetor, spacer, bracket and intake flow separator. All O-rings really should be discarded and replaced, however they can USUALLY be reused if they are in good enough condition (as long as they are not worn or cut).
10. Clean and inspect and/or overhaul the carburetor, as applicable.

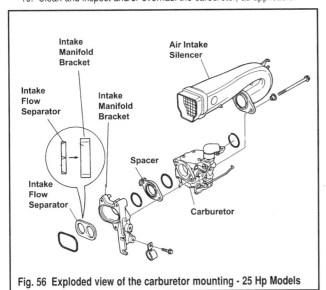

Fig. 56 Exploded view of the carburetor mounting - 25 Hp Models

To Install:

11. Install the intake flow separator into the intake manifold bracket and install the O-ring.
12. Position the spacer to the intake manifold bracket along with the O-rings (one for the bracket and one for the carburetor).
13. Position the O-ring for the air intake silencer, then install the carburetor to the spacer/bracket and the silencer to the carburetor. Secure the whole assembly using the 2 bolts and collars.
14. Install the carburetor assembly to the intake manifold and secure using the manifold-to-carburetor bracket retaining bolts.
15. Reconnect the electro-thermal valve wiring to the engine harness.
16. Reconnect the fuel hose to the carburetor and secure using the retaining clamp.
17. Connect the throttle link rod.
18. Refer to the Timing and Synchronization adjustments in the Maintenance and Tune-Up section to make sure the Throttle Link Rod and Idle Speed are both properly adjusted.

OVERHAUL

◆ See Figures 57, 58 and 59

This carburetor is found both on the 25 Hp and 30/40 Hp Models, the only difference however is that the twin uses 1 carburetor, while the 3-cylinder motor uses 3 carburetors. For the 3-cylinder carb assembly, the electro-thermal valve is installed on only the top carburetor.

Good shop practice dictates purchasing a carburetor repair kit and using new parts any time the carburetor is disassembled. Make an attempt to keep the work area organized and to cover parts after they have been cleaned. This practice will prevent foreign matter from entering passageways or adhering to critical parts.

1. Remove the carburetor from the powerhead and separate it from the intake manifold bracket or mounting plates, as detailed in this section.
2. If necessary on 25 hp Yamaha motors, loosen the set screw and remove the throttle link rod.
3. Remove the screw and the electro-thermal ram retainer (top carb only for 3-cylinder models), then carefully pull the electro-thermal valve and O-ring up and off the carburetor.
4. Loosen and remove the screws securing the top cover to the carburetor, then remove the top cover and discard the gasket.
5. For 25 hp motors, loosen and remove the pilot outlet jet and O-ring from the bore in the top of the carburetor body.

■ Although no details are provided, some Mercury/Mariner service literature shows additional components like nozzles and what appears to be a pilot screw for models with US serial No. OT409000 and up, as well as Belgium serial No. OP153500 and up. Unfortunately, these components are not mentioned in their procedures, nor are they even labeled in exploded views. Before disturbing any additional screws on these models, such as a pilot screw, turn the screw inward until lightly seated while counting the number of turns. In this way you can begin adjustment with the screw installed and backed out the same number of turns to the current position.

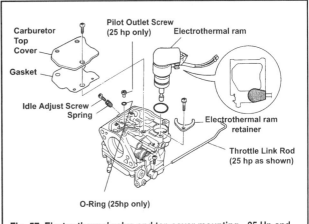

Fig. 57 Electro-thermal valve and top cover mounting - 25 Hp and 30/40 Hp Models

3-32 FUEL SYSTEM

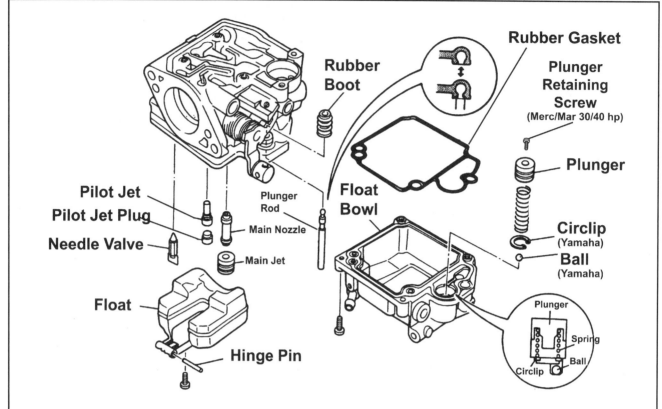

Fig. 58 Exploded view of the carburetor body and float assembly - 25 Hp and 30/40 Hp Models (Note some late-model Mercury/Mariners utilize additional needle/jet components)

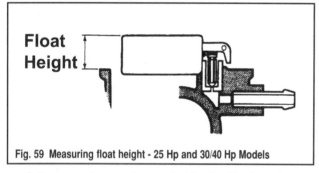

Fig. 59 Measuring float height - 25 Hp and 30/40 Hp Models

6. If necessary, loosen and remove the idle adjust (throttle stop) screw and spring.

7. Invert the carburetor body (with the float bowl facing upward), then loosen the float bowl retaining screws. Remove the float bowl from the carburetor body, then remove and discard the old bowl rubber gasket.

8. Remove the accelerator pump plunger rod. If the carburetor body is going to be immersed in cleaner or if the boot shows any signs of damage/deterioration, remove the plunger rod rubber boot from the carburetor body.

9. Remove the plunger, spring, circlip and check ball from the bore in the float bowl.

10. Remove the screw, then gently remove the float hinge pin to free it from the carb and float. Lift gently upward to remove the float and needle valve from the carburetor body.

11. Remove the main jet and nozzle, then remove the pilot jet plug and pilot jet from the carburetor body, keeping all parts separated and identified for installation or replacement purposes.

12. Clean and inspect all components as detailed in this section. Replace any damaged, worn or defective components. Discard all O-rings or gaskets.

To Assemble:

13. Install the pilot jet and pilot jet plug, then install the main nozzle and main jet, each to the carburetor body in the positions noted during removal.

14. Connect the needle valve to the float, then lower the float and valve into position. Insert the hinge pin through the float and carburetor body using a pair of needle-nose pliers and secure using the screw.

15. With the carburetor still inverted and perfectly level, so the float is sitting gently on the needle valve (which is resting on the seat) measure the float height from the carburetor body-to-float bowl mating surface up to the top (actually the bottom, but it is on top now) of the float. Refer to the Carburetor Set-Up Specifications chart in this section for the allowable range, but the height should be about 0.55 in. (14mm). If necessary, gently bend the float hinge to achieve the proper measurement.

✱✱ WARNING

When measuring the float height DO NOT place any pressure downward on the needle valve or you could damage it (and/or you may make an incorrect adjustment).

16. Double-check that the float and valve move smoothly without sticking or binding.

17. Assemble the accelerator pump components into the bore in the float bowl. Insert the check ball, circlip, spring and plunger.

18. If removed, install the rubber boot to the carburetor body.

19. Position the accelerator pump plunger rod.

20. Install the float bowl to the carburetor body using a NEW rubber gasket, making sure the plunger rod is in position, then gently secure using the retaining screws.

21. If removed, install the idle adjust (throttle stop) screw and spring.

22. For 25 hp motors, install the pilot outlet jet and O-ring.

23. Install the carburetor top cover using a new gasket, then secure using the retaining screws.

24. Install the electro-thermal valve and O-ring (it really should be replaced using a NEW O-ring) to the top of the carburetor (top carb only of the 3 carb assembly on 3-cylinder motors). Secure the valve using the retainer and screw.

25. If removed on 25 hp Yamaha motors, install the throttle link rod and loosely secure using the set screw (it will have to be adjusted once the carburetor is installed).

26. Install the carburetor and adjust for proper operation.

CLEANING & INSPECTION

◆ See Figures 33, 57 and 58

** CAUTION

Never dip rubber or plastic parts in carburetor cleaner. These parts should be cleaned only in solvent, and then blown dry with compressed air.

Place all metal parts in a screen-type tray and dip them in carburetor cleaner until they appear completely clean, then blow them dry with compressed air.

Blow out all passages in the castings with low-pressure compressed air. Check all parts and passages to be sure they are not clogged or contain any deposits. Never use a piece of wire or any type of pointed instrument to clean drilled passages or calibrated holes in a carburetor.

Move the throttle shaft back and forth to check for wear. If the shaft appears to be too loose, replace the complete carburetor body because individual replacement parts are not available.

Inspect the main body, air horn, and venturi cluster gasket surfaces for cracks and burrs which might cause a leak.

Check the float for deterioration. If any part of the float is damaged, the unit must be replaced.

Inspect the tapered section of the float needle for wear or damage. Replace if it has developed a groove or is no longer evenly tapered.

Check the main jet, pilot jet, check valve and main nozzle for signs of dirt or contamination. If they cannot be cleaned, they should be replaced. Again, NEVER clean these components using a wire or any pointer instrument, you'll change the calibration.

Check the accelerator pump plunger, rod and bore in the float bowl for signs of wear or damage and replace.

■ As previously mentioned, most of the parts which should be replaced during a carburetor overhaul are included in overhaul kits available from your local marine dealer. Replace all components included in the kit for durability.

Inspect the electro-thermal valve, as follows:
* For Mercury/Mariner models, use an ohmmeter to check the valve element resistance. Connect the ohmmeter across the two valve leads, it should read about 15-25 ohms resistance at an ambient temperature of about 68°F (20°C).

■ The ohmmeter test MAY well be valid for Yamaha models, unfortunately Yamaha does NOT provide a specification.

* Check the distance that the piston is protruding from the valve. Apply a 12 volt power source to the 2 valve leads and leave it connected for several minutes, then recheck the valve piston height. If the valve is working properly the heater element should have caused the height to increase sufficiently to block the enrichment passage if the valve was installed on top of the carburetor.

Always replace any and all worn parts.

30/40 Hp Models

◆ See Figure 55

The 30/40 Hp Models utilize a carburetor assembly consisting of 3 single barrel, side draft, single float carburetors. An electro-thermal valve is mounted to the top of the top carburetor assembly to provide additional fuel for cold start enrichment. When cold, the valve contains a piston which is retracted into the valve unblocking an additional fuel circuit in the carburetor (from the float bowl to the throttle bore). However, once the motor is started the power from the stator/charge coil is applied to a heater element in the valve in order to gradually build heat in the valve. This process is timed to take about the same amount of time it should take a cold motor to warm to normal operating temperature. The temperature change causes the valve piston to extend, gradually cutting the enrichment circuit as the engine warms.

■ Good shop practice dictates purchasing a carburetor repair kit and using new parts (especially O-rings and gaskets) any time the carburetor is disassembled. Make an attempt to keep the work area organized and to cover parts after they have been cleaned. This practice will prevent foreign matter from entering passageways or adhering to critical parts.

REMOVAL & INSTALLATION

◆ See Figures 60 and 61

On these motors the 3 carburetors are sandwiched between 2 mounting plates and the entire assembly is bolted to the intake manifold. Therefore, all 3 carburetors are removed or installed as an assembly, then they may be separated for individual service or replacement.

■ The fuel and breather hoses on these motors are normally secured using spring-type clamps. These are easily removed by gently squeezing the clamp ears, then sliding the clamp back up the hose until it is past the raised portion of the nipple on the fitting. Before reinstalling the spring-type clamps, make sure they have not deformed or weakened or the hose may leak at the fitting.

1. Disconnect the breather hose at the valve cover.
2. Tag and disconnect the 3 carburetor fuel hoses at the fuel pump outlet (also at the valve cover).
3. Trace the wiring for the electro-thermal valve and disconnect the bullets from the bullet connectors.
4. For Yamaha models, remove the 2 bolts, collars and 1 clamp from the side of the powerhead (at the air intake silencer).
5. For Mercury/Mariner models, remove the 3 bolts from the side of the powerhead (at the air intake silencer). Two of them are centrally located, in between the air intake runners, but the third is at the top right of the assembly, when facing the powerhead from the side.
6. Disconnect the powerhead throttle link from the bottom carburetor.
7. Loosen and remove the 6 flange bolts which are threaded through the intake manifold and insulator into the inner carburetor assembly mounting plate. Support the carburetor assembly as the last bolt is removed, then carefully remove the carburetor unit from the intake manifold alignment pins.
8. If further disassembly is required (such as for carburetor service), place the assembly on a clean work surface and remove the 6 bolts securing the air intake silencer to the carburetor, then remove the silencer.
9. At the other end of the carburetor unit, remove the 2 plastic holders from the insulator, then remove the insulator and the 6 O-rings (3 on the intake manifold side and 3 on the carburetor mounting plate side). Discard all 6 O-rings as they should be replaced once they are removed.
10. Tag and disconnect the hoses from carburetor unit. Keep in mind that there should be a total of 8 hoses connected to the assembly and there is NO WAY you're going to remember where each one guys (and Yamaha doesn't provide any diagrams for that either), so TAG EM now or don't complain later.
11. Remove the 3 cotter pins from the throttle link rod and the dash pot link, then remove the washers (keep track of them for installation purposes).

■ The best way to keep track of the cotter pins and washers is to place them back on their studs after the respective links are disconnected in the following 2 steps.

12. Disconnect the throttle link rod from the carburetors.
13. Remove the 2 retaining bolts and remove the dash-pot.
14. Back on the air intake silencer side of the assembly, remove the 6 bolts securing the carburetors between the 2 mounting plates. Carefully separate the carbs and mounting plate. It is a good idea to scribe the position (top, middle, bottom or cylinder number 1, 2, 3 from the top down) on each carburetor float bowl to make sure it is installed in the proper position after overhaul or service. Also, pay attention to plate orientation to ensure proper assembly.
15. Clean and inspect and/or overhaul the carburetor, as applicable.

To Install:

16. Place the inner carburetor mounting plate (the one that goes on the intake manifold side) on a clean work surface with the intake manifold side

3-34 FUEL SYSTEM

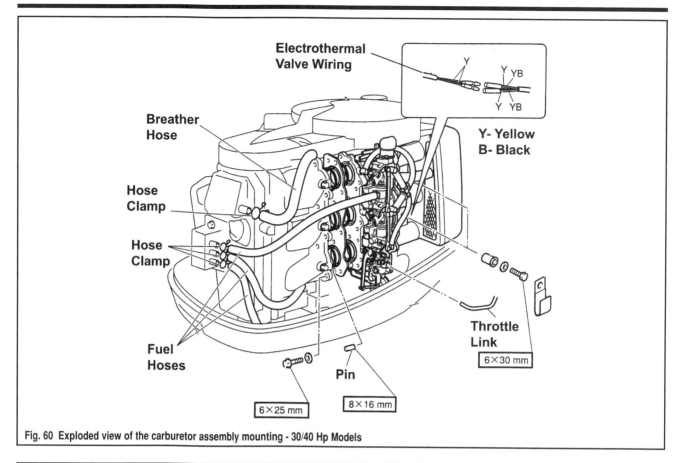

Fig. 60 Exploded view of the carburetor assembly mounting - 30/40 Hp Models

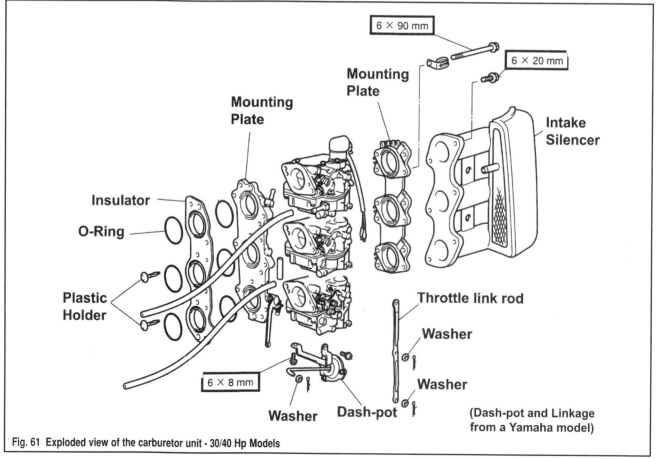

Fig. 61 Exploded view of the carburetor unit - 30/40 Hp Models

FUEL SYSTEM

facing downward. Place each of the carburetors in their proper position on the inner mounting plate, then install the outer mounting plate over the back of the carbs. Fasten the carbs to the mounting plates by threading and securely tightening the 6 mounting bolts.

17. Install the dash-pot and tighten the 2 retaining screws.
18. Install the throttle link rod.
19. Install the 3 washers and cotter pins to the throttle link and dash-pot link.
20. Reconnect the 8 hoses to the carburetor unit, as tagged during removal.

■ WHAT, you didn't tag the hoses (or somebody else removed them before you got stuck with the job). Well, Yamaha doesn't publish a diagram of these hoses, but take a good look at them, fuel hoses are run from the pump to the carburetor fuel inlets. The necessarily tight spaces under the engine cowling require that hoses are JUST the right length to get the job done. You shouldn't have much slack on any one hose once they are all connected.

21. Position 6 new O-rings, then install the insulator to the inner carb mounting plate, secure using the 2 plastic holders.
22. Install the air intake silencer to the outer carb mounting plate and secure using the 6 mounting bolts.
23. Make sure the 3 insulator-to-intake manifold O-rings are still in position on the insulator (you placed them there 2 steps ago), then carefully align the carburetor unit to the intake manifold using the 2 manifold pins. Install the 6 intake manifold-to-carburetor mounting plate retaining bolts and tighten securely.
24. Reconnect the throttle link to the bottom carburetor.
25. For Mercury/Mariner models, install and secure the 3 intake retaining bolts to the side of the powerhead.
26. For Yamaha models, install the clamp along with the 2 bolts and collars to the side of the powerhead.
27. Reconnect the electro-thermal valve wiring bullet connectors.
28. Reconnect the 3 fuel hoses to the fuel pump and secure using the retaining clamps.
29. Reconnect the breather hose to the valve cover and secure using the retaining clamp.
30. Refer to the Timing and Synchronization adjustments in the Maintenance and Tune-Up section to make sure the Idle Speed and Dash-Pot are properly adjusted.

OVERHAUL

 DIFFICULT

The carburetors used on the 30/40 hp motors are almost identical to the single carburetor used on the twin cylinder, 25 hp motors. Please refer to the Overhaul and Cleaning & Inspection procedures for the 25 Hp Models found earlier in this section.

40/45/50 Hp (935cc) and 40/50/60 Hp (996cc) Models

These 4-cylinder, 4-stroke motors are equipped with 4 carburetors, one per cylinder. The carburetors are sandwiched between mounting plates and secured to the intake manifold assembly. Therefore, all 4 carburetors are removed or installed as an assembly, then they may be separated for individual service or replacement. A significant change to the carburetor unit was made between the 2000 and 2001 model years, so we've broken them out into separate procedures where necessary.

REMOVAL & INSTALLATION

 MODERATE

Models Through 2000

◆ See Figures 62 and 63

■ Refer to the numbers in the accompanying procedure to help identify components listed in the text.

1. Disconnect the negative battery for safety.
2. Using a pair of pliers compress the tabs on the spring clip (1), then disconnect the blow-by hose (2) from the valve cover.

3. Release the spring clips (3) on the fuel hoses (4) in the same manner, then disconnect the fuel hoses from the fuel pump. Have a rag handy to catch any escaping fuel. Also, be sure to cap the open lines/fittings to prevent system contamination.
4. Tag and disconnect the solenoid (5) blue, black and electro-thermal ram (6) green, black lead bullet connectors.
5. Remove the 2 mounting bolts (7) and collars (8) threaded into the side of the powerhead at the air intake end of the carburetors.
6. Support the carburetor assembly (13), then loosen and remove the 8 carburetor mounting bolts (9) holding the insulator (10) and carburetor assembly to the intake manifold (threaded from the intake manifold side of the assembly). Once the bolts are removed, carefully lower the carburetor assembly and the insulator from the motor.
7. Check the 2 dowel pins (11) between the insulator and the manifold.
8. It is really best to remove and discard the 8 O-rings (8) sealing the manifold to the insulator and the insulator to the carburetors if they have been in service for any length of time. Regardless, check each O-ring for cuts or tears and replace if damaged.

To Install:

■ Some initial adjustments may be easier to perform (or must be performed) with the carburetor assembly removed from the powerhead. Before installation, review the Timing and Synchronization procedures in the Maintenance and Tune-Up section to determine which steps should be followed before proceeding.

9. Make sure the carburetor, insulator and intake manifold mounting surfaces are clean and free of damage.
10. Install the O-rings between the carburetor/insulator and the insulator/intake manifold.
11. Install the insulator to the intake manifold using the dowel pins to hold it in position. Make sure the O-rings are not disturbed as the insulator is positioned.
12. Install the carburetor assembly to the insulator and intake manifold. Install the 8 retaining bolts and washers from the intake manifold side, then tighten the bolts to 5.9 ft. lbs. (8 Nm).
13. Install the 2 collars and secure using the 2 bolts to the air intake side of the carburetor assembly.
14. Connect the green and black electro-thermal ram leads, then connect the blue and black solenoid leads.
15. Reconnect the fuel hoses to the fuel pump and secure using the spring clips.

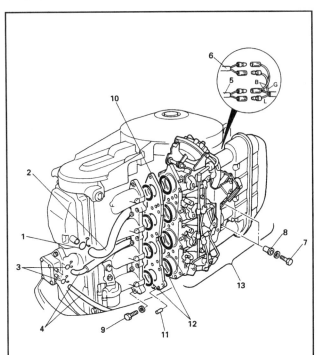

Fig. 62 Carburetor removal and installation for 40/45/50 hp (935cc) motors through 2000; follow the numbers in sequence for removal and reverse them for installation

3-36 FUEL SYSTEM

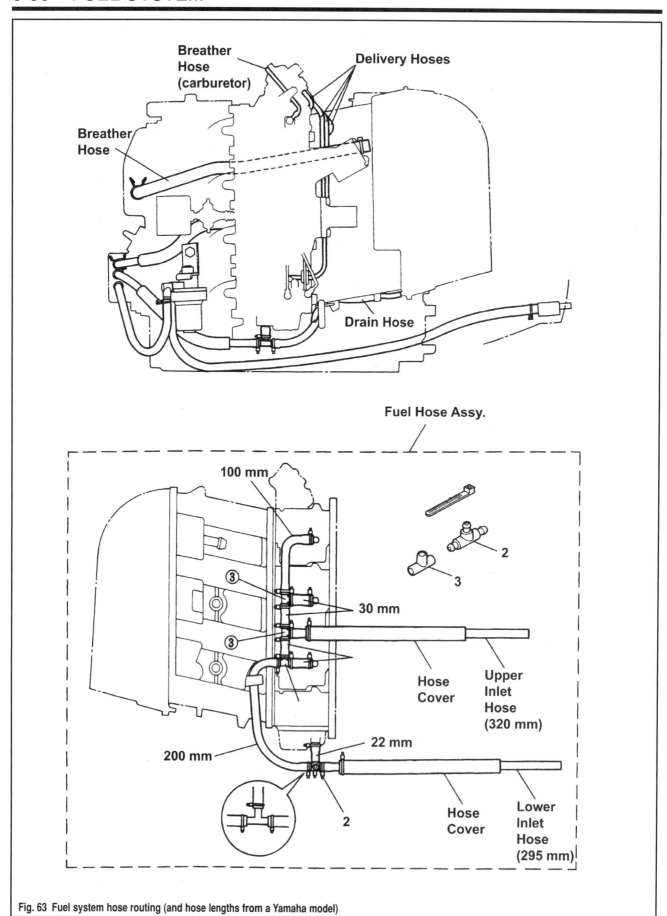

Fig. 63 Fuel system hose routing (and hose lengths from a Yamaha model)

FUEL SYSTEM 3-37

16. Reconnect the blow-by hose to the valve cover and secure using the spring clip.

17. Using the primer bulb pressurize the fuel system and check carefully at the fuel lines and float bowls for any sign of leakage.

18. Complete any additional Timing and Synchronization procedures as detailed in the Engine and Tune-Up section.

2001 and Later Models

◆ See Figures 64 thru 67

■ The fuel and breather hoses on these motors are normally secured using spring-type clamps. These are easily removed by gently squeezing the clamp ears, then sliding the clamp back up the hose until it is past the raised portion of the nipple on the fitting. Before reinstalling the spring-type clamps, make sure they have not deformed or weakened or the hose may leak at the fitting.

1. Disconnect the breather hose at the nipple jutting out from underneath the flywheel cover.
2. Tag and disconnect the 2 carburetor fuel hoses at the fuel pump outlet (at the valve cover).
3. Locate the wiring for the Prime Start valve and disconnect the coupler.
4. Remove the 2 bolts and collars securing the air intake silencer to the side of the powerhead.
5. Disconnect the powerhead throttle link from the bottom carburetor.
6. Loosen and remove the 8 flange bolts which are threaded through the intake manifold and insulator into the inner carburetor unit mounting plate. Support the carburetor assembly as the last bolt is removed, then carefully remove the carburetor unit from the intake manifold alignment pins. Remove and discard the 4 O-rings which seal the insulator to the intake manifold.
7. Remove the 2 couplers securing the insulator to the carburetor inner mounting plate. Remove and discard the 4 O-rings which seal the insulator to the inner mounting plate.
8. If further disassembly is required (such as for carburetor service), place the assembly on a clean work surface and remove the link rod from the carburetor throttle links.
9. Tag and disconnect the accelerator pump hoses.
10. Disconnect the accelerator pump link rod from the throttle linkage for the lower carburetor.
11. Loosen the 3 screws and remove the accelerator pump from the carburetor unit.
12. If necessary, cut the wire tie and disconnect the breather hose from the air intake silencer.
13. Cut the wires ties, as necessary, then tag and disconnect fuel lines between the individual carburetors. Refer to the illustration for more details.

■ There are 8 mounting bolts threaded through the air intake silencer, outer mounting plate and carburetors to the inner mounting plate. When removing the bolts it is probably best to have the assembly positioned with the inner mounting plate face down.

14. Remove the 8 bolts from the air intake silencer flange, then remove the silencer, followed by the carburetor outer mounting plate. Each of the carburetors are no free for removal.

■ Be sure to number each of the carbs as they are removed to ensure installation in the correct positions.

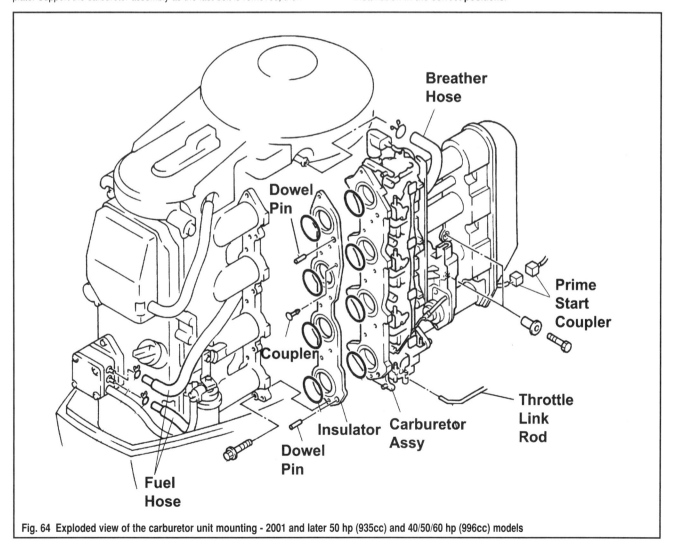

Fig. 64 Exploded view of the carburetor unit mounting - 2001 and later 50 hp (935cc) and 40/50/60 hp (996cc) models

3-38 FUEL SYSTEM

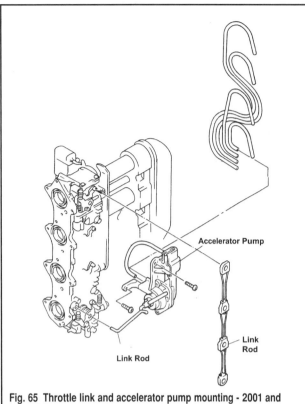

Fig. 65 Throttle link and accelerator pump mounting - 2001 and later 50 hp

15. Remove and discard the 4 O-rings which seal the air intake silencer to the outer mounting plate, the 4 O-rings which seal the outer mounting plate to the carburetors, and the 4 O-rings which seal the carburetors to the inner mounting plate. All 12 of these O-rings (along with the other 8 which were removed earlier from the insulator) must be replaced to ensure proper operation, especially at idle.
16. Clean and inspect and/or overhaul the carburetor, as applicable.

To Install:

■ O-rings seal the connection between each air intake component and the mating surface of the next component. If the carburetor unit has been completely disassembled that means you need 20 replacement O-rings to ensure there is no vacuum loss which would lead, at the very least, to problems obtaining a stable idle. Before beginning, place each of the replacement O-rings in position and take care during the procedure to make sure none are dislodged or pinched.

17. Place the carburetor inner carburetor mounting plate (the one that goes on the intake manifold side) on a clean work surface with the intake manifold side facing downward. Place each of the carburetors in their proper position on the inner mounting plate, then install the outer mounting plate over the back of the carbs.

■ Depending on how easy it was to disconnect the fuel hoses, you may want to hold off on installing the air intake silencer and securing the carburetors to the mounting plates until after the fuel hoses are reconnected.

18. Next, position the air intake silencer on top of the outer mounting plate and carefully thread the 8 retaining bolts through the assembly. Tighten the bolts securely.
19. Reconnect the fuel hoses between the individual carburetors, as tagged during removal.
20. If removed, secure the breather hose to the air intake silencer assembly using a new plastic wire tie.

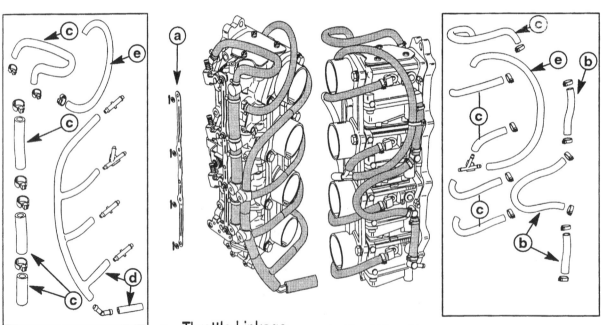

a - Throttle Linkage
b - Fuel Hoses
c - Enrichener/Balance Hoses
d - Bowl Vent Hoses
e - Enrichener Vent Hose – to "Y" of Bowl Vent Hose

Fig. 66 Fuel line routing and throttle link components - 2001 and later 40/50/60 hp (996cc) Mercury/Mariner models

FUEL SYSTEM 3-39

21. Position the carburetor unit on the work surface with the powerhead side facing downward. Install the accelerator pump to the assembly and secure using the 3 retaining bolts. Reconnect the accelerator pump link rod to the throttle linkage of the lower carburetor.
22. Install the carburetor throttle link rod to each of the carburetor throttle valves.
23. Make sure the new O-rings are still in position, then install the insulator to the inner carb mounting plate, secure using the 2 couplers.
24. Make sure the 4 insulator-to-intake manifold O-rings are still in position on the insulator (you placed them there at the beginning of the installation, correct), then carefully align the carburetor unit to the intake manifold using the 2 manifold pins. Install the 8 intake manifold-to-carburetor mounting plate retaining bolts and tighten securely.
25. Install the 2 bolts and collars for the air intake silencer to the side of the powerhead.
26. Reconnect the throttle link to the bottom carburetor.
27. Reconnect the Prime Start wiring coupler.
28. Reconnect the 2 fuel hoses from the carburetors to the fuel pump and secure using the retaining clamps.
29. Reconnect the breather hose to nipple at the top of the powerhead and secure using the retaining clamp.
30. Refer to the Timing and Synchronization adjustments in the Maintenance and Tune-Up section to make sure the carburetors are properly adjusted.

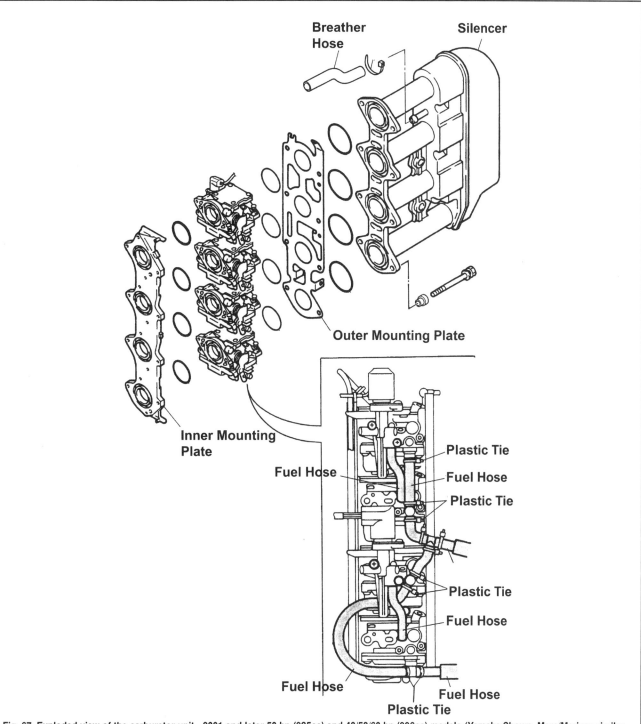

Fig. 67 Exploded view of the carburetor unit - 2001 and later 50 hp (935cc) and 40/50/60 hp (996cc) models (Yamaha Shown, Merc/Mariner similar with some fuel line differences)

3-40 FUEL SYSTEM

OVERHAUL

Models Through 2000

Linkage, Electro-Thermal Plunger and Solenoid Coil
◆ See Figure 68

■ Refer to the numbers in the accompanying procedure to help identify components listed in the text.

1. Carefully remove the upper (1) and lower (2) choke links from the carburetor assembly.
2. Remove the auto choke link (3) from the assembly.
3. Remove the unloader link (4) from the No. 4 cylinder (lower) carburetor.
4. Disconnect the throttle link (5).
5. Inspect the 8 link retainers (6), there are normally 5 black, 2 brown and 1 blue. Replace any that are damaged or no longer serviceable.
6. Remove the joint plate (7).
7. Disconnect the spring (8).
8. Loosen the 2 mounting screws (9), then carefully remove the choke solenoid (10).
9. Loosen the 2 mounting screws (11), then carefully remove the electro-thermal ram (12).
10. If further disassembly or overhaul is required, remove the Accelerator Pump, Dash-Pot and Silencer as detailed in this section.

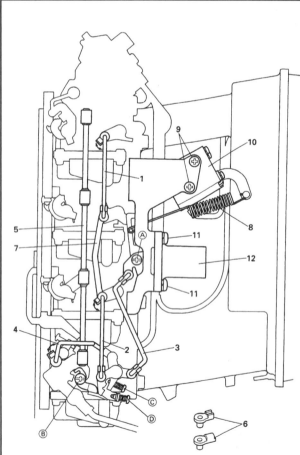

Do not loosen screws marked A, B, C, & D

Fig. 68 Linkage, electro-thermal plunger and Solenoid removal and installation for 40/45/50 hp (935cc) motors through 2000, follow the numbers in sequence for removal and reverse them for installation (Yamaha shown, Merc/Mariner similar with some small linkage differences)

To Assemble:

11. Position the electro-thermal ram and secure using the 2 mounting screws.
12. Position the choke solenoid and secure using the 2 mounting screw.
13. Connect the spring.
14. Install the join plate.
15. Connect the throttle link.
16. Connect the unloader link.
17. Connect the auto choke link, followed by the upper and lower choke links.
18. Install the carburetor and perform the necessary Timing and Synchronization procedures as detailed in the Engine and Tune-Up section.

Accelerator Pump, Dash-Pot and Silencer
◆ See Figure 69

■ Refer to the numbers in the accompanying procedure to help identify components listed in the text.

1. If necessary for complete disassembly/overhaul, remove the Linkage, Electro-Thermal Plunger and Solenoid Coil from the carburetor assembly, as detailed earlier in this section.
2. Remove the 3 screws from the side of the accelerator pump (1), then disconnect the pump linkage retainer (2). Remove the accelerator pump from the top of the carburetor assembly.
3. Remove the 2 bolts (4) securing the dash-pot mounting bracket, then remove the dash-pot (5) from the carburetor assembly.
4. Remove the 8 bolts (6) and washers that secure the air intake silencer (8) to the carburetor assembly through the collars (7). Remove the air intake silencer. Inspect the collars and the silencer O-rings (9) for damage and replace, as necessary.
5. If not done already, remove the plate (10) from the intake side of the carburetor. Inspect the O-rings (11) that go between the plate and carburetor assembly. Remove and replace if they are damaged or worn.
6. Remove the retaining plate (12) from the air intake silencer side of the carburetor assembly. Inspect the O-rings (13) hat go between the plate and carburetor assembly. Remove and replace if they are damaged or worn.

■ It is really best to remove and discard all of the O-rings used in this assembly if they have been in service for any length of time. Regardless, check each O-ring for cuts or tears and replace if damaged.

7. If the carburetors are to be separated for further disassembly and overhaul mark them to ensure they are properly positioned during assembly.

To Assemble:

8. Make sure the carburetors, if they were separated are placed in the same positions as marked prior to disassembly.
9. Install the retaining plate and O-rings to the air intake silencer side of the carburetor.
10. Install the plate and O-rings to the intake manifold side of the carburetor.
11. Install the air intake silencer assembly and O-rings to the retaining plate side of the carburetor, then secure using the 8 bolts, washers and collars. Tighten the bolts securely, but do not over-tighten and damage the housing or collars.
12. Install the dash-pot and secure loosely secure using the 2 retaining screws (Refer to Timing and Synchronization in the Maintenance and Tune-Up section as it will require adjustment).
13. Position the accelerator pump, installing the retainer and securing using the 3 screws.
14. If necessary, install the Linkage, Plunger and Solenoid Coil, as necessary and as detailed earlier in this section.

Overhaul of the Individual Carburetor Assemblies
◆ See Figures 70, 71 and 72

■ Refer to the numbers in the accompanying procedure to help identify components listed in the text.

If not done already, mark each of the carburetors to ensure installation in the same positions. Follow the appropriate steps for each of the carburetors. Keep the parts from each carburetor separate during the process. Do not mix and match parts from one carburetor assembly to another.

FUEL SYSTEM 3-41

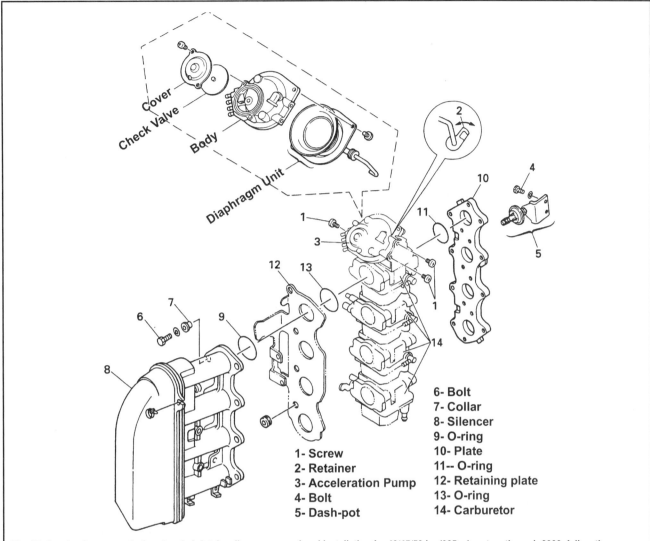

Fig. 69 Accelerator pump, dash-pot and air intake silencer removal and installation for 40/45/50 hp (935cc) motors through 2000, follow the numbers in sequence for removal and reverse them for installation

All rubber or plastic components and gaskets should be replaced to ensure proper operation. Pilot screws and springs may be reused if they are in good condition, however if replacements are provided in the overhaul kit there is really no good reason to reuse the old components.

1. Remove the Linkage, Electro-Thermal Plunger and Solenoid Coil from the carburetor assembly, as detailed earlier in this section.
2. Remove the Accelerator Pump, Dash-Pot and Air Intake Silencer from the carburetor assembly, as detailed earlier in this section.
3. Remove the drain bolt (1) and aluminum gasket (2) from the float bowl. If there is any fuel remaining in the bowl allow it to drain into a suitable approved container.
4. Remove the 2 screws (3) that secure the cover plate (4) and packing (5) from the top of the carburetor.
5. Except for 45 hp motors, remove the rubber plug (6) from the upper side of the carburetor, then carefully unthread the pilot screw (7) from the carburetor body. Remove the O-ring (8) and the spring (9) from the screw. Check the pilot screw and spring for wear or damage and replace, if necessary.
6. Remove the throttle stop screw (10) and spring (11).

■ On the No. 4 cylinder (lower) carburetor the throttle stop screw (10) is the idle adjustment screw.

7. From the No. 4 cylinder (lower) carburetor:
   a. Remove the choke closing adjustment screw (12) and spring (13).
   b. Remove the unloader screw (14) and spring (15).
   c. Remove the fast idle adjustment screw (16) and fast idle shoe plate (17).

8. Remove the 4 screws (18) securing the float bowl to the carburetor body.
9. Carefully separate the float bowl (19) from the carb body, then remove and discard the gasket (20) from the mating surfaces.
10. Remove the float pin retaining screw (21), then carefully withdraw the float pin (22) by hand or using a pair of needle-nose pliers.
11. Remove the float (23) and needle valve (24).
12. Remove the plug (25) which covers the main nozzle (26), then carefully remove the nozzle from the carburetor body.
13. Remove the main jet (27), followed by the pilot jet (28) from the carburetor body.
14. Clean and inspect all components as detailed under Cleaning & Inspection in this section.

To Assemble:
15. Install the pilot jet, followed by the main jet.
16. Install the main nozzle, O-ring and plug.
17. Install the needle valve and float (carefully placing the valve into the valve seat). Next, carefully insert the float pin through the slot in the carburetor. Secure the float pin using the float pin retaining screw. Check for smoother float movement.

✽✽ WARNING

To prevent possible damage to the float needle valve and valve seat, NEVER force the valve or float in towards the carburetor body.

3-42 FUEL SYSTEM

18. Using a straight edge, measure from the lip of the carburetor body/float bowl mating surface to a point half-way up the float on the opposite side of the needle valve. This point is level with the top of the needle valve and in this position the float should be just resting on the valve and NOT compressing the valve into the needle. This dimension should be 0.37-0.41 in. (9.5-10.5mm). No adjustment is possible, if this measurement is out of specification the needle valve and/or float must be replaced.

19. Install the float bowl using a new gasket, then secure using the 4 bowl screws.

20. If working on the No. 4 cylinder (lower) carburetor:
   a. Install the fast idle shoe plate, followed by the fast idle screw.
   b. Install the unloader screw and spring.
   c. Install the choke closing adjustment screw and spring.

21. Install the throttle stop or idle adjustment screw and spring, as applicable.

22. If equipped (meaning, normally except for 45 hp motors), install the pilot screw and spring using a new O-ring. Carefully and gently thread the screw into position until it JUST LIGHTLY seats, then back it out 1 3/4-2 3/4 turns for proper adjustment on Yamaha models or 2 1/2 turns for Mercury/Mariner models. Install the rubber plug over top of the pilot screw to seal it.

23. Install the packing and cover plate to the top of the carburetor body and secure using the 2 screws.

24. Install the drain bolt using a new aluminum gasket.

25. Install the Accelerator Pump, Dash-Pot and Air Intake Silencer from the carburetor assembly, as detailed earlier in this section.

26. Install the Linkage, Electro-Thermal Plunger and Solenoid Coil from the carburetor assembly, as detailed earlier in this section.

27. Install the carburetor assembly, as detailed earlier in this section.

2001 and Later Models

◆ See Figures 71, 72 and 73

Good shop practice dictates purchasing a carburetor repair kit and using new parts any time the carburetor is disassembled. Make an attempt to keep the work area organized and to cover parts after they have been cleaned. This practice will prevent foreign matter from entering passageways or adhering to critical parts.

1. Remove the carburetor unit from the powerhead and separate it from the mounting plates, as detailed earlier in this section.

2. If not already done, remove the drain plug and O-ring from the bottom side of the float bowl and drain the fuel from the carburetor.

3. For the top carb only, if necessary, remove the screw and the Prime Start bracket, then carefully pull the Prime Start valve and O-ring up and off the carburetor. If necessary, remove the 2 prime start valve body bolts and separate the valve body from the carburetor. Remove and discard the irregular-shaped valve body gasket.

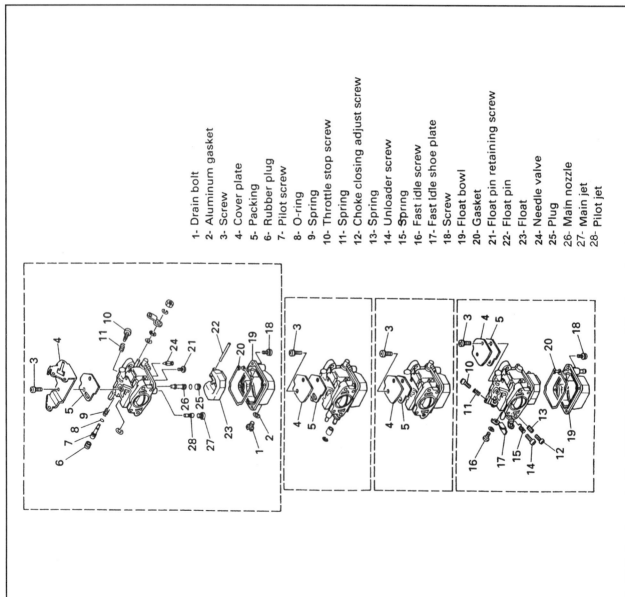

Fig. 70 Carburetor overhaul for 40/45/50 hp (935cc) motors, follow the numbers in sequence for disassembly and reverse them for assembly

1- Drain bolt
2- Aluminum gasket
3- Screw
4- Cover plate
5- Packing
6- Rubber plug
7- Pilot screw
8- O-ring
9- Spring
10- Throttle stop screw
11- Spring
12- Choke closing adjust screw
13- Spring
14- Unloader screw
15- Spring
16- Fast idle screw
17- Fast idle shoe plate
18- Screw
19- Float bowl
20- Gasket
21- Float pin retaining screw
22- Float pin
23- Float
24- Needle valve
25- Plug
26- Main nozzle
27- Main jet
28- Pilot jet

FUEL SYSTEM 3-43

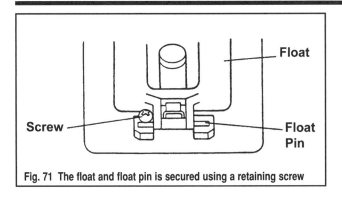

Fig. 71 The float and float pin is secured using a retaining screw

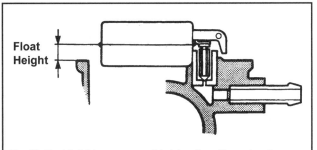

Fig. 72 Float height measurement is taken from the carburetor body (to float bowl mating surface) to the a point half-way up the float which is level with the top of the needle valve

4. Loosen and remove the 4 screws securing the top cover to the carburetor, then remove the top cover and discard the seal.

5. Invert the carburetor body (with the float bowl facing upward), then loosen the 4 float bowl retaining screws. Remove the float bowl from the carburetor body, then remove and discard the old bowl gasket.

6. Remove the screw, then gently remove the float hinge pin to free it from the carb and float. Lift gently upward to remove the float and needle valve from the carburetor body.

7. Remove the fuel metering components (main jet, pilot jet, plug, O-ring and main nozzle) each from the carburetor body, keeping all parts separated and identified for installation or replacement purposes.

8. Clean and inspect all components as detailed in this section. Replace any damaged, worn or defective components. Discard all O-rings or gaskets.

To Assemble:

9. Install fuel metering components to the carburetor body (main nozzle, plug with O-ring, pilot jet and main jet) in the positions noted during removal.

10. Connect the needle valve to the float, then lower the float and valve into position. Insert the hinge pin through the float and carburetor body using a pair of needle-nose pliers and secure using the screw.

11. With the carburetor still inverted and perfectly level, so the float is sitting gently on the needle valve (which is resting on the seat) measure the float height from the carburetor body-to-float bowl mating surface up to the top (actually the bottom, but it is on top now) of the float. Refer to the Carburetor Set-Up Specifications chart in this section for the allowable range. If necessary, gently bend the float hinge to achieve the proper measurement.

✱✱ WARNING

When measuring the float height DO NOT place any pressure downward on the needle valve or you could damage it (and/or you may make an incorrect adjustment).

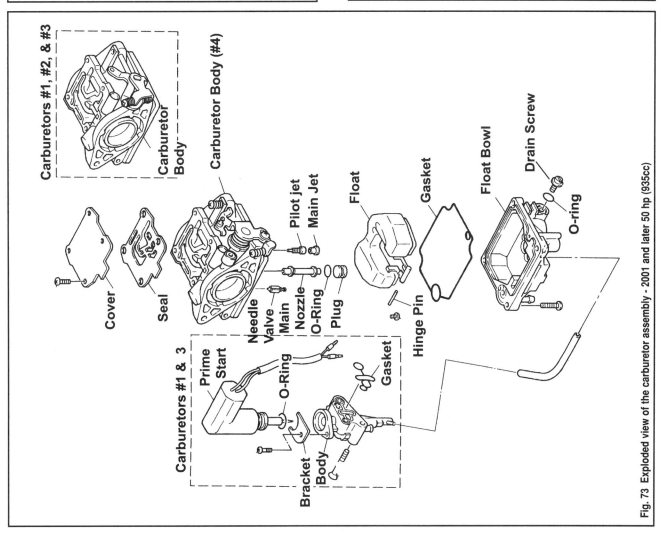

Fig. 73 Exploded view of the carburetor assembly - 2001 and later 50 hp (935cc)

3-44 FUEL SYSTEM

12. Double-check that the float and valve move smoothly without sticking or binding.
13. Install the float bowl to the carburetor body using a NEW gasket, then gently secure using the retaining screws. If not done already, install the drain screw using a NEW O-ring.
14. Install the carburetor top cover using a new seal, then secure using the retaining screws.
15. If working on the top carb, install the enrichment valve body using a new gasket and secure using the 2 retaining bolts. Install the Prime Start valve using a NEW O-ring, then secure using the bracket and retaining screw.
16. Assemble the carburetor unit, then install and adjust for proper operation.

CLEANING & INSPECTION

◆ See Figures 33, 34, 70, 73 and 74

✱✱ CAUTION

Never dip rubber or plastic parts in carburetor cleaner. These parts should be cleaned only in solvent, and then blown dry with compressed air. Actually, rubber and plastic parts should, whenever possible, be replaced.

Place all metal parts in a screen type tray and dip them in carburetor cleaner until they appear completely clean, then blow them dry with compressed air.

Blow out all passages in the castings with low-pressure compressed air. Check all parts and passages to be sure they are not clogged or contain any deposits. Never use a piece of wire or any type of pointed instrument to clean drilled passages or calibrated holes in a carburetor.

Move the throttle and choke shafts back and forth to check for wear. If the shaft appears to be too loose, replace the complete mixing chamber because individual replacement parts are not available.

Inspect the mixing chamber, and fuel bowl gasket surfaces for cracks and burrs which might cause a leak.

Check the floats for deterioration. Check the needle valve tip contacting surface and replace the needle valve if this surface has a groove worn in it.

Inspect the tapered section of the pilot screw and replace the screw if it has developed a groove.

Most of the parts which should be replaced during a carburetor overhaul are included in an overhaul kit available from your local marine dealer. Usually, one of these kits will contain a matched fuel inlet needle and seat. It is usually a good idea to replace this combination each time the carburetor is disassembled as a precaution against leakage.

Inspect the electro-thermal valve, as follows:
• For Mercury/Mariner models, use an ohmmeter to check the valve element resistance. Connect the ohmmeter across the two valve leads, it should read about 15-25 ohms resistance at an ambient temperature of about 68°F (20°C).

■ **The ohmmeter test MAY well be valid for Yamaha models, unfortunately Yamaha does NOT provide a specification.**

• Check the distance that the piston protrudes from the valve. Apply a 12 volt power source to the 2 valve leads (positive to the blue lead and negative to the black lead). Leave it connected for at least 5 minutes, then recheck the valve piston height. If the valve is working properly the heater element should have caused the height to increase. For Yamaha models, the manufacturer ads that the increase must be to a minimum height of 0.97 in. (24.6mm).

Always replace any and all worn parts.

75/80/90/100 Hp Models

REMOVAL & INSTALLATION

◆ See Figures 75 thru 84

The 75/80/90J/90/100 hp models are equipped with one carburetor per cylinder, mounted between spacers to an intake manifold. Although a single carburetor can be removed individually from the motor for replacement or service, the procedure given here is to remove all 4 carburetors, as a unit, along with the intake manifold assembly. Removing the intake manifold is the only way to remove all 4 carburetors as a unit and simplifies access to the remaining components of the carburetor and air intake silencer assembly.

1. For Yamaha models, remove the control linkage, as follows:
   a. Disconnect the throttle control shaft from either the accelerator cam and/or the throttle control lever.
   b. Remove the bolt from the center of the accelerator cam, then remove the cam itself.
   c. Remove the 2 screws securing the Throttle Position Sensor (TPS). Either lay the sensor in the cowling with the wiring still connected, or follow the wire back and disconnect it from the coupler, then remove the sensor completely.
   d. Remove the 2 screws from the TPS mounting bracket, then remove the bracket from the powerhead.
   e. Remove the 2 bolts securing the throttle control lever bracket to the side of the powerhead. Pull the bracket and throttle control lever free of the motor. If further disassembly of the bracket, lever and spacer is necessary, remove the spring pin and spacer from the throttle control lever, then remove the bolt from the back side of the bracket and separate the lever.
   f. There is a plastic wire tie used to hold some wires and/or hoses in position at the top of the air intake silencer. Locate and cut the plastic wire tie, then reposition the wires/hoses as necessary to prevent interference with removal of the silencer and carbs.

2. For Mercury/Mariner models, remove the control linkage, as follows:
   a. Locate the TPS connector, about halfway down the air intake runners, just about the TPS and linkage mounting bracket. Carefully insert a small flat-bladed screwdriver between the connector itself and the mount, then gently push downward with the screwdriver while pulling up on the connector to release it from the mount. Disengage the TPS wiring connector and position aside.
   b. Trace the TPS wiring down from the connector, just to the right of the TPS itself is a wire tie securing the wire to the powerhead, note the positioning and carefully cut that wire tie.
   c. Disconnect the throttle link (control shaft) from either the accelerator cam and/or the throttle control lever.
   d. Remove the bolt from the center of the accelerator cam, then remove the cam assembly. Keep track of the bushing (spacer), wave washer, flat washers and spring, noting their locations and positioning for installation purposes.

■ **In the next step you unbolt the entire throttle control lever and TPS mounting bracket as an assembly. With this method there is no need to disturb the TPS itself, leaving it secured in position to the bracket.**

   e. Now at the other end of the throttle link, remove the 2 bolts securing the throttle control lever and TPS mounting bracket assembly, then remove the entire control lever bracket assembly from the motor.
   f. A little above where the TPS wiring connector was mounted locate the auto enrichener wiring harness. It is positioned horizontally just above the air intake runners. Carefully disconnect the auto enrichener wiring.

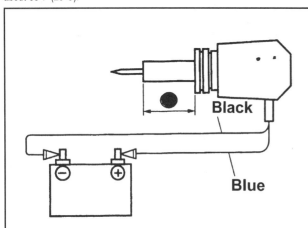

Fig. 74 Checking the prime start valve on 2001 and later 50 hp (935cc) and 40/50/60 hp (996cc) models

FUEL SYSTEM 3-45

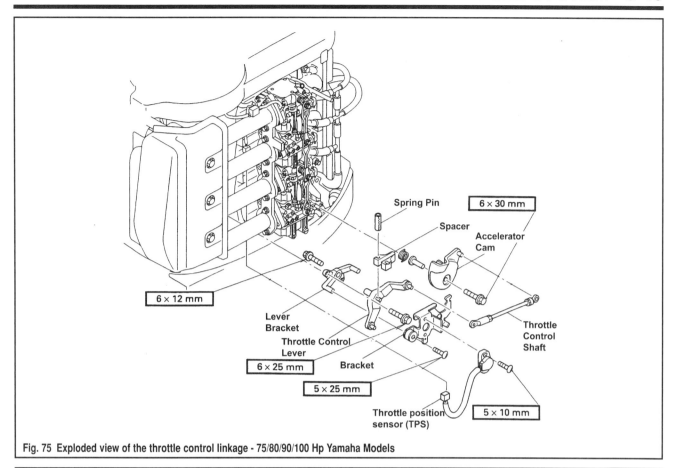

Fig. 75 Exploded view of the throttle control linkage - 75/80/90/100 Hp Yamaha Models

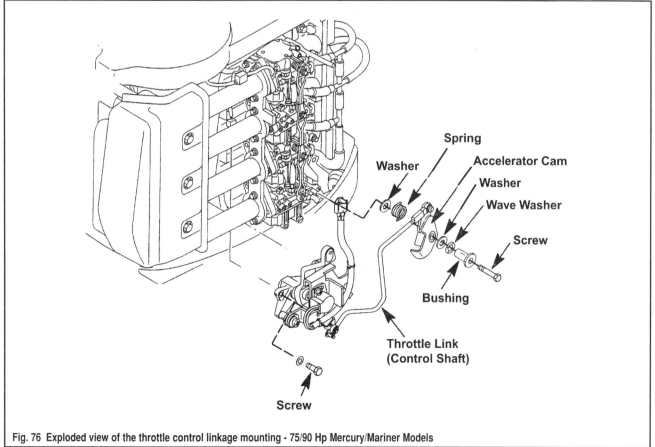

Fig. 76 Exploded view of the throttle control linkage mounting - 75/90 Hp Mercury/Mariner Models

3-46 FUEL SYSTEM

3. Locate and disconnect the breather hose at the top back side of the air intake silencer.

4. Tag and disconnect the fuel supply hose(s) for the carburetor assembly.

■ It may be easier to disconnect the fuel supply hose(s) from the fuel supply source (fuel pump/pumps) and leave them attached to the carburetor unit until it is disassembled.

5. Remove the 3 bolts from the side of the air intake silencer, securing it to the powerhead.

■ If the carburetors are going to be separated and replaced or serviced, it is probably easier to remove the air intake silencer assembly from the carbs at this point. If so, follow the next step to unbolt the silencer from the carbs. However, if everything is being removed for access to something else, skip the next step and go right to unbolting the intake manifold from the side of the powerhead.

6. If necessary, remove the air intake silencer assembly from the carburetors. Remove the 8 bolts from the air intake silencer flange, then remove the silencer from the carburetors. There are 4 O-rings which seal the silencer to the carburetor spacers. Locate the O-rings and inspect them (as they can be reused if they are not worn, cut or otherwise damaged).

7. Remove the 5 bolts from the intake manifold (using a counterclockwise spiraling pattern that starts at the top outer bolt and works toward the center bolts). Support the carburetor/intake manifold assembly as the last bolt is removed, then remove the assembly from the powerhead. Remove and discard the old intake manifold gasket.

8. You're done, unless further disassembly is necessary for carburetor or intake service/replacement. If it is necessary to remove one or more of the carburetors, remove the link rod from the carburetor throttle levers and continue the procedure.

9. Loosen the 2 screws securing the accelerator pump to the bottom of the intake manifold, then remove the pump from the manifold, either disconnect the pump pipe assembly from the pump itself, or tag and disconnect them pipes from the carburetors (this must be done if you're removing the carburetors anyway).

■ Accelerator pump pipe positioning on the carburetors is pretty straight forward, but don't assume anything, tag the pipes before disconnecting them from the carburetor.

10. For each carburetor that is to be removed, loosen the 2 mounting bolts then remove the outer spacer, carburetor and inner spacer. There is an O-ring sealing each spacer to the carb and the inner spacer to the intake manifold (16 O-rings in all, not including those that sealed the intake silencer to the outer spacer). All O-rings should be inspected and replaced, if worn or damaged, however the 8 O-rings downstream of the carburetor should ALWAYS be replaced to ensure proper operation and idle quality.

■ Be sure to number each of the carbs as they are removed to ensure installation in the correct positions.

11. Clean and inspect and/or overhaul the carburetor, as applicable.

To Install:

■ O-rings seal the connection between each air intake component and the mating surface of the next component. If the carburetor unit has been completely disassembled that means you need to install properly install a total of 16 O-rings during this procedure. Furthermore, 8 of those O-rings are positioned between the carburetor and intake manifold, making them especially important to engine vacuum and proper carburetor balance/operation. Make sure that you ALWAYS replace at least those 8 O-rings (if not the full 16). Before beginning, place each of the O-rings in position and take care during the procedure to make sure none are dislodged or pinched.

12. Install each carburetor to the intake manifold using the inner and outer spacers, O-rings and 2 retaining bolts. Tighten the bolts securely for each carb. Be sure to install each carb in the correct position, as noted during removal.

■ Depending on how easy it was to disconnect the fuel hoses, you may want to reconnect one or more of the fuel hoses/pipes before securing each of the carburetors to the intake manifold.

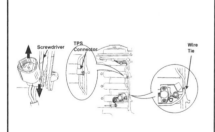

Fig. 77 Releasing the TPS connector from the mounting bracket - 75/90 Hp Mercury/Mariner Models

Fig. 78 On Merc models, cut this wire tie to free the TPS wiring

Fig. 79 Major components of the intake system (Yamaha shown)

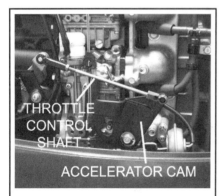

Fig. 80 Disconnect the throttle control link...

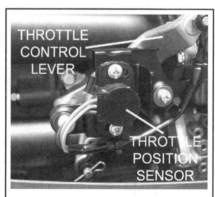

Fig. 81 ...then on Yamaha's unbolt and remove the TPS

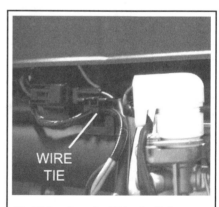

Fig. 82 Locate and cut this wire tie for Yamahas

FUEL SYSTEM 3-47

13. Reconnect the accelerator pump pipes, then position the pump to the intake manifold and secure using the 2 retaining screws.

14. Install the link rod to the carburetor throttle control levers.

15. Install the carburetor and intake manifold assembly to the powerhead using a new intake manifold gasket. Finger-tighten the 5 manifold retaining bolts, then tighten the bolts securely using one or more passes of a clockwise spiraling pattern that starts at the center bolt and works outward.

16. Make sure the 4 O-rings are in position, then install the air intake silencer to the carburetor assembly and secure using the 8 retaining bolts.

17. Reconnect the fuel hoses between the individual carburetors, as tagged during removal.

18. Connect the breather hose to the top, back side of the air intake silencer.

19. Reposition the wires/hoses to the air intake silencer, as noted during removal, and secure using a new plastic wire tie.

20. If the TPS was unbolted from it's mounting bracket (instead of the mounting bracket being unbolted and removed completely) or if the linkage was replaced/changed in some way, properly check/adjust the TPS output voltage, as detailed in this section.

21. Refer to the Timing and Synchronization adjustments in the Maintenance and Tune-Up section to make sure the carburetors and control linkage are properly adjusted.

THROTTLE POSITION SENSOR (TPS) ADJUSTMENT

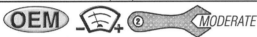

The Throttle Position Sensor (TPS) is a rotary potentiometer, meaning it sends a variable signal to the CDI/Ignition control unit based on physical throttle (and therefore sensor) positioning. The TPS sensor output is used

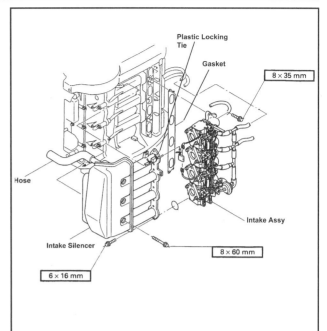

Fig. 83 Exploded view of the carburetor and intake manifold mounting - Yamaha shown (Merc/Mariner almost identical)

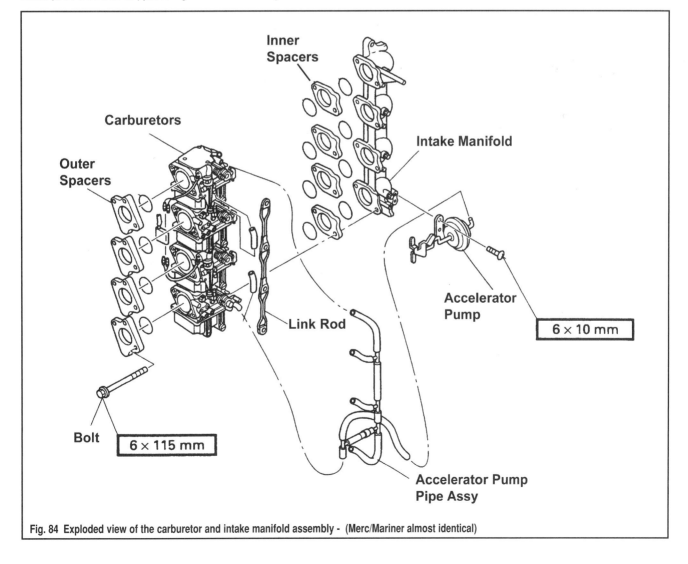

Fig. 84 Exploded view of the carburetor and intake manifold assembly - (Merc/Mariner almost identical)

3-48 FUEL SYSTEM

by the ignition system to fine-tune ignition timing functions. If the sensor is removed or if problems are suspected with the ignition system, you should check/adjust the TPS sensor output voltage to match proper base operating conditions.

The sensor itself receives a reference voltage (usually about 5 volts) from the CDI unit, and as the throttle lever is rotated, the CDI unit then receives a return voltage signal through a separate wire. This signal will change with the actual throttle position. As the throttle shaft opens the voltage increases, as the shaft closes voltage decreases. A third wire is used to complete the ground circuit back to the CDI unit.

Since the sensor does not generate a voltage, but instead acts upon a reference voltage from the CDI unit all testing must occur with the circuit complete. Therefore, you've got 2 options. One option is to attempt to backprobe the connector (insert the probes through the rear of the connector while it is still attached to the sensor). However this method risks damaging the connector or the wiring insulation (and could lead to problems with the circuit later). The better method is to disconnect the wiring harness and use 3 jumper wires (one for each terminal) to reconnect the harness. The jumpers must not contact each other (or you risk damage to the CDI unit from a short), however, some point on the jumper must be exposed so that you can probe the completed circuit using a DVOM. Yamaha and Mercury each sell a 3-pin test harness for this application.

Sensor checking and adjustment takes 2 parts, first checking the input signal from the CDI unit to make sure the circuit is functioning properly, then checking the output signal from the sensor itself. The later test tells you if the sensor is adjusted and functioning properly. If the later signal is not correct first try loosening the mounting screws and adjusting the sensor, but if the signal cannot be brought within specification, the sensor must be replaced.

On these motors, the TPS is mounted to the lower side of the powerhead, on the control linkage immediately adjacent to the carburetors.

1. Disconnect the wiring from the TPS (located about halfway up the powerhead, along the intake air silencer-to-carburetor mating surface) and connect a test harness or jumper wires to complete the circuit.

■ **Follow the harness from where it exits the sensor, up to the wiring harness connector, noting the routing for installation purposes. After testing, make sure the sensor wiring is safely tucked in the same position, to prevent possible damage from moving components.**

2. Connect a Digital Volt Ohmmeter (DVOM) set to read DC volts to the terminals for the Red and Orange wires in order to check the CDI reference voltage. Turn the ignition switch **ON** without starting the motor, the meter must show about 4.75-5.25 volts otherwise the test connections, the circuit back to the CDI unit or the CDI unit itself is faulty.

3. Move the DVOM test leads to the Pink and Orange wires in order to check the sensor output voltage. Start the engine (using a suitable source of cooling water) and allow it to warm to normal operating temperature. Once the motor is fully warmed, run at idle speed, then check the meter, it should read 0.68-0.72 volts for Yamaha models and 0.68-0.82 for Mercury/Mariner models. Slowly open and close the throttle valves while watching the meter, the voltage must gradually increase and decrease, free of any sharp voltage spikes/dips.

■ **Some meters with an auto range setting may not show the idle voltage or the gradual change properly. A scope or analog meter is best to really see this, so don't suspect a voltage spike/drop without thorough testing. However, when a clear spike/drop is shown, the sensor should be replaced to ensure proper function.**

4. For Yamaha models, if the sensor reads out of specification at idle speed, MAKE SURE the throttle control lever is in contact with the idle (fully-closed) stopper, then loosen the TPS mounting screws and carefully adjust the position of the sensor until the DVOM reads 0.68-0.72 volts. Once the sensor is set within specification, carefully tighten the mounting screws and double-check the meter to make sure the TPS did not move while securing the screws.

5. For Mercury/Mariner models the factory service information does NOT tell you to rotate the sensor in order to bring it into the proper voltage range of operation. We cannot figure out why. All Merc says is to set the sensor rotated fully clockwise on the slots provided for adjustment and to make sure the throttle control lever is contacting the fully-closed stopper in this position.

6. After testing turn the ignition keyswitch **OFF**, disconnect the test harness and reconnect the TPS wiring.

OVERHAUL

◆ See Figures 85, 86 and 87

Good shop practice dictates purchasing a carburetor repair kit and using new parts any time the carburetor is disassembled. Make an attempt to keep the work area organized and to cover parts after they have been cleaned. This practice will prevent foreign matter from entering passageways or adhering to critical parts.

1. Remove the carburetor unit from the powerhead and separate it from the mounting plates, as detailed earlier in this section.
2. If not already done, remove the drain plug from the bottom side of the float bowl and drain the fuel from the carburetor.
3. For the No. 1 and No. 3 (top and second from bottom) carbs, remove the screw and the electro-thermal enrichment valve bracket, then carefully pull the valve and O-ring up and off the carburetor. If necessary, remove the 2 enrichment valve body bolts and separate the valve body from the carburetor. Remove and discard the 4 O-rings from the back of the valve body.
4. Loosen and remove the 3 screws securing the top cover to the carburetor, then remove the top cover and discard the gasket.
5. From the side of the carburetor body, remove the plug, then remove the pilot screw, O-ring and spring.
6. Invert the carburetor body (with the float bowl facing upward), then loosen the 4 float bowl retaining screws. Remove the float bowl from the carburetor body, then remove and discard the old bowl rubber gasket.
7. Remove the screw, then gently remove the float hinge pin to free it from the carb and float. Lift gently upward to remove the float and needle valve from the carburetor body.
8. Remove the needle valve seat and O-ring.
9. Remove the fuel metering components (main jet and main nozzle, then the pilot jet plug and pilot jet) each from the carburetor body, keeping all parts separated and identified for installation or replacement purposes.
10. Clean and inspect all components as detailed in this section. Replace any damaged, worn or defective components. Discard all O-rings or gaskets.

To Assemble:

11. Install fuel metering components to the carburetor body (pilot jet and plug, then the main nozzle and main jet) in the positions noted during removal.
12. Install the needle valve seat using a new O-ring.
13. Connect the needle valve to the float, then lower the float and valve into position. Insert the hinge pin through the float and carburetor body using a pair of needle-nose pliers and secure using the screw.
14. With the carburetor still inverted and perfectly level, so the float is sitting gently on the needle valve (which is resting on the seat) measure the float height from the carburetor body-to-float bowl mating surface (with now gasket) to the top of the needle valve (actually the bottom, but it is on top now). The float height should be about 0.55 in. (14mm) on these models, though Mercury specifies an acceptable range of 0.51-0.59 in. (13-15mm). If necessary, gently bend the needle valve tab to achieve the proper measurement.

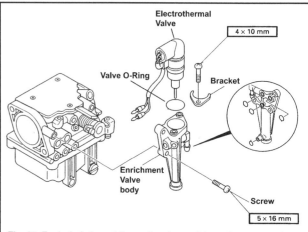

Fig. 85 Exploded view of the carburetor enrichment components - 75/80/90/100 models

FUEL SYSTEM 3-49

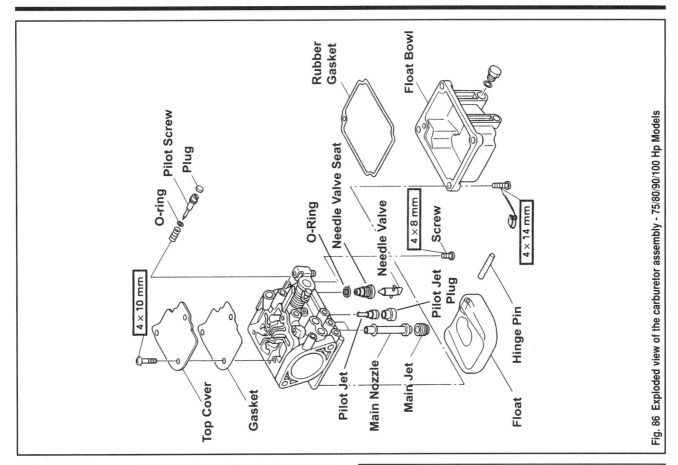

Fig. 86 Exploded view of the carburetor assembly - 75/80/90/100 Hp Models

※※ WARNING

When measuring the float height DO NOT place any pressure downward on the needle valve or you could damage it (and/or you may make an incorrect adjustment).

15. Double-check that the float and valve move smoothly without sticking or binding.
16. Install the float bowl to the carburetor body using a NEW rubber gasket, then gently secure using the retaining screws. If not done already, install the drain screw.
17. If applicable, install the pilot screw using a new spring and O-ring. Carefully and gently thread the screw into position until it JUST LIGHTLY seats, then back it out either 2-3 turns for 75/80 hp models or 1 1/2-2 1/2 turns for 90/100 hp motors as initial adjustment. Install the plug over top of the pilot screw to seal it.
18. Install the carburetor top cover using a new gasket, then secure using the retaining screws.
19. If working on the No. 1 or No. 3 carb, install the enrichment valve body using 4 NEW O-rings and secure using the 2 retaining bolts. Install the electro-thermal enrichment valve using a NEW O-ring, then secure using the bracket and retaining screw.
20. Assemble the carburetor unit, then install and adjust for proper operation.

CLEANING & INSPECTION

◆ See Figures 33, 34, 85 and 86

※※ CAUTION

Never dip rubber or plastic parts in carburetor cleaner. These parts should be cleaned only in solvent, and then blown dry with compressed air. Actually, rubber and plastic parts should, whenever possible, be replaced.

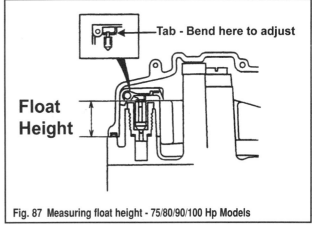

Fig. 87 Measuring float height - 75/80/90/100 Hp Models

Place all metal parts in a screen type tray and dip them in carburetor cleaner until they appear completely clean, then blow them dry with compressed air.

Blow out all passages in the castings with low-pressure compressed air. Check all parts and passages to be sure they are not clogged or contain any deposits. Never use a piece of wire or any type of pointed instrument to clean drilled passages or calibrated holes in a carburetor.

Move the throttle and choke shafts back and forth to check for wear. If the shaft appears to be too loose, replace the complete mixing chamber because individual replacement parts are not available.

Inspect the mixing chamber, and fuel bowl gasket surfaces for cracks and burrs which might cause a leak.

Check the floats for deterioration. Check the needle valve tip contacting surface and replace the needle valve if this surface has a groove worn in it.

Inspect the tapered section of the pilot screw and replace the screw if it has developed a groove.

3-50 FUEL SYSTEM

Most of the parts which should be replaced during a carburetor overhaul are included in an overhaul kit available from your local marine dealer. Usually, one of these kits will contain a matched fuel inlet needle and seat. It is usually a good idea to replace this combination each time the carburetor is disassembled as a precaution against leakage.

■ **In service literature for the 75/80/90/100 hp models, Yamaha does not describe the electro-thermal valve operation or provide any specs for testing. However, on all other Yamaha 4-strokes equipped with a similar electro-thermal valve, the valve operates by slowly changing position as an internal heater element warms in response to a connection to 12 volts. Also, the Mercury service literature confirms this. Therefore, the following test should tell you if the valve is operating or not.**

Inspect the electro-thermal valve, as follows:
• For Mercury/Mariner models, use an ohmmeter to check the valve element resistance. Connect the ohmmeter across the two valve leads, it should read about 15-25 ohms resistance at an ambient temperature of about 68°F (20°C).

■ **The ohmmeter test MAY well be valid for Yamaha models, unfortunately Yamaha does NOT provide a specification.**

• Check the distance that the piston protrudes from the valve. Apply a 12 volt power source to the 2 valve leads (blue and black wires). Leave it connected for at least 5 minutes, then recheck the valve piston height. If the valve is working properly the heater element should have caused the height to increase sufficiently to block the enrichment passage if the valve was installed on top of the carburetor.
Always replace any and all worn parts.

ELECTRONIC FUEL INJECTION SYSTEM

Beginning in 2000 Yamaha introduced the first of their EFI equipped 4-stroke outboards, followed shortly by Mercury/Mariner versions of the same motors in 2001. For 2002 Mercury/Mariner added their own EFI system to smaller 3- and 4-cylinder powerheads from 40-60 hp as the technology continues to expand to a number of motors in both manufacturer's lines of outboards.

The systems used on these motors are essentially all versions of an manifold-injection system with minor differences from model-to-model. In all cases, these systems improve low speed smoothness, reduce maintenance, improve durability, and increase fuel efficiency while lowering harmful engine emissions.

For all motors an engine control computer known as the Engine or Electronic Control Module (ECM) is capable of adjusting fuel delivery and ignition timing over a wide range to match engine operation. Unlike most carburetor-equipped engines, fuel-injected engines will automatically adjust for changes in altitude, air temperature, and barometric pressure. This precise control over engine operation results in crisp throttle response and maximum efficiency over a broad range of conditions.

■ **For some reason Yamaha SOMETIMES calls the ECM a CDI unit, however, since this unit does more than just control ignition, we think that is a little misleading and will refer to it as the ECM (the more commonly used term for fuel and ignition control units/models) throughout this repair guide. However, knowing that Yamaha MIGHT call it that on some motors could be handy when dealing with a parts counterperson.**

Fuel Injection Basics

◆ See Figures 88, 89 and 90

Fuel injection is not a new invention. Even as early as the 1950s, various automobile manufacturers experimented with mechanical-type injection systems. There was even a vacuum tube equipped control unit offered for one system! This might have been the first "electronic fuel injection system." Early problems with fuel injection revolved around the control components. The electronics were not very smart or reliable. These systems have steadily improved since. Today's fuel injection technology, responding to the need for better economy and emission control, has become amazingly reliable and efficient. Computerized engine management, the brain of fuel injection, continues to get more reliable and more precise.

Components needed for a basic computer-controlled system are as follows:
• A computer-controlled engine manager, which is the Electronic Control Module (ECM), with a set of internal maps to follow (you know, if this, then do that type of information for the fuel and ignition systems).
• A set of input devices to inform the ECM of engine performance parameters.
• A set of output devices. Each device is controlled by the ECM. These devices modify fuel delivery and timing. Changes to fuel injection and timing are based on input information matched to the map programs.

This list gets a little more complicated when you start to look at specific components. Some fuel injection systems may have twenty or more input devices. On many systems, output control can extend beyond fuel and timing. The Yamaha and Mercury/Mariner Fuel Injection Systems provide more than just the basic functions, but they are still straight forward in layout.

There are generally somewhere between seven and ten input devices and generally between eight and ten output controls (depending upon the system, year and model). The accompanying diagrams show the typical input and output devices of the various fuel injection systems.

There are several fuel injection delivery technologies in wide use around the world and the manufacturers have chosen one major form of them for these motors. A brief discussion of the various types (even those not used by Yamaha and Mercury/Mariner) would be helpful in understanding these systems.

Throttle body injection is relatively inexpensive and was used widely in early automotive systems. This is usually a low pressure system running at 15 PSI or less. Often an engine with a single carburetor was selected for throttle body injection. The carburetor was recast to hold a single injector and the original manifold was retained. Throttle body injection is not as precise or efficient as port injection.

Multi-port fuel injection is defined as an engine that uses one or more electrically activated solenoid injectors (or physically actuated injectors in the case of General Motors Central Multi-Port Injection used in some 1990's era vehicles) for each cylinder. Multi-port injection generally operates at higher pressures than throttle body systems. The Yamaha Mercury/Mariner EFI systems operate with a system pressure of somewhere in the 30-44 psi (207-304 kPa) range depending upon the year and model. All of these systems use a type of port injection where the fuel is injected into the air intake track behind the throttle bodies just in front of the intake valves.

Another type of Multi-port fuel injection used predominantly on 2-stroke marine engines (but also found on automotive and marine diesels) is Direct Injection, which is called so because fuel is introduced from the injector DIRECTLY into the combustion chamber (as opposed to somewhere on the intake tract).

No matter whether the fuel is injected to the manifolds, behind valves, or directly into the combustion chamber, the real advantage of a fuel injection system is the precise control afforded the ECM by the use of high response electronic components. This gives electronic fuel injectors are great advantage over mechanical injectors when it comes to emissions and operating efficiency. Port injectors can be triggered two ways. One system uses simultaneous injection. All injectors are triggered at once. The fuel "hangs around" until the pressure drop in the cylinder pulls the fuel into the combustion chamber. This type of system looses some of the advantage that is possible with an electronic fuel injector. The second type is more precise and follows the firing order or SEQUENCE of the engine. Each cylinder gets a squirt of fuel precisely when needed. If you haven't already guessed, this second type of fuel injection is known as Sequential Multi-port Fuel Injection.

Yamaha EFI 4-Stroke Injection

◆ See Figures 89 and 90

■ **This system is used on all Yamaha EFI motors as well as the Mercury/Mariner 115 hp 4-cylinder and on V6 models as well.**

The Yamaha EFI 4-stroke system is a tuned port, sequential method of fuel injection. The injectors are mounted a common fuel rail and the intake manifold, and are found in the induction track between the throttle air valve(s) and the cylinder head intake valves.

Like most marine fuel injection systems made during this timeframe, the 4-stroke EFI system does not contain an Oxygen sensor such as the one

FUEL SYSTEM 3-51

used by some Yamaha 2-stroke systems, therefore all engine fuel mapping decisions are based upon a combination of factors including engine speed from the pulser coil (which provides crankshaft position information), throttle opening from the Throttle Position Sensor (TPS) and intake air from the Intake Air Temperature (IAT) and Intake Air Pressure (IAP) sensors. In this way the ECM keeps track of just how much air is entering the engine (between the TPS and the temperature/pressure of the air) and can adjust air/fuel mixture ratios to suit operating conditions. The lack of Oxygen sensor makes this form of injection open-loop only (meaning the ECM receives no feedback on the combustion process).

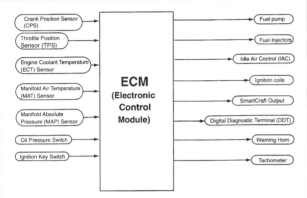

Fig. 88 Electronic Fuel Injection (EFI) schematic of inputs and outputs - 40-60 hp Mercury/Mariner motors

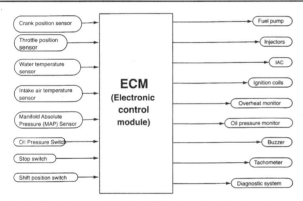

Fig. 89 Electronic Fuel Injection (EFI) schematic of inputs and outputs - 115 hp Mercury/Mariner motors

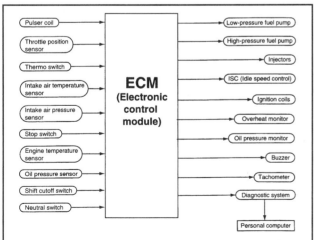

Fig. 90 Electronic Fuel Injection (EFI) schematic of inputs and outputs - V6 motors (and Yamaha 115-150 hp 4-cylinder motors, except they do not use an electric lift pump)

In fuel injection terms this type of system is known as a speed-density injection system. This is because the basic engine fuel mapping decisions are made by the ECM based on a comparison between the engine rpm (speed) and the air pressure (air density). The basic, pre-programmed, fuel mapping is then modified based upon input from the remaining sensors. Specifically, engine and air temperatures are taken into account.

Like most fuel injection systems, the Yamaha EFI 4-stroke system incorporates a return-to-port capability or limp mode. If there is a major sensor failure that prevents the ECM from processing one or more signals, the engine is run on a minimum performance map. In most instances the ECM will substitute a fixed value for the missing sensor signal. The only input that will shut the engine down completely is battery voltage. If the battery is disconnected or the battery voltage falls below a certain point, the fuel pump quits pumping so the engine stops.

Major components of this system include the ECM and Sensors along with the low- and high-pressure fuel delivery circuits. The electronic control sensors include:
- Throttle Position Sensor (TPS)
- Thermo Switch
- Intake Air Temperature (IAT) Sensor
- Intake Air Pressure (IAP) Sensor
- Water Temperature Sensor (WTS)
- Oil Pressure Sensor
- Neutral Switch
- Pulser Coil (instead of a CPS)
- Idle Speed Control Valve (ok, it's not a sensor, but it's an important piece of the system)

The low-pressure fuel circuit is similar to the fuel delivery circuit found on carbureted motors. A fuel line with primer bulb is used to connect the boat mounted or portable fuel tank to a powerhead mounted low-pressure fuel pump. The 4-cylinder EFI motors use a mechanical diaphragm-displacement low-pressure lift pump (or a pair of pumps for 150 hp motors), while the V6 4-strokes utilize an electric low-pressure lift pump. The low-pressure circuit is used to deliver fuel from the fuel tank to a powerhead mounted vapor separator tank which contains a submerged high-pressure electric fuel pump. The high-pressure pump then draws fuel from the vapor tank and forces it through high-pressure fuel lines to the fuel injectors for each cylinder. Excess fuel passes the injectors and goes through the fuel pressure regulator. From the regulator the fuel passes through a water-cooled fuel cooler in order to prevent excessive vaporization and then continues back into the vapor separator tank to be pumped back to the injectors again.

Mercury/Mariner EFI 4-Stroke Injection

◆ See Figure 88

■ All V6 models and 115 hp inline 4-cylinder models, including Mercury/Mariners, are equipped with the Yamaha EFI 4-Stroke Injection system.

For 30-60 hp inline 4-cylinder motors, Mercury/Mariner opted to install their own fuel injection system as Yamaha had not yet marketed those powerheads with an EFI system of their own.

The Mercury/Mariner EFI 4-stroke system is a tuned multi-port, method of fuel injection which is very similar to the Yamaha system differing mostly in the fuel hose routing/flow of the high pressure circuit (specifically the use of a fuel distribution manifold assembly and non re-circulating fuel at the injectors). Like the Yamaha system, the injectors are mounted to the intake manifold, however they are NOT mounted to a fuel rail assembly as they are connected by fuel lines to the fuel distribution manifold from which fuel makes a one way only trip to the injectors.

Also like the Yamaha system the injectors are found between the throttle air valve (throttle body) and the cylinder head intake valves. Like most marine fuel injection systems made during this timeframe, the 4-stroke EFI system does not contain an Oxygen sensor, therefore all engine fuel mapping decisions are based upon a combination of factors including engine speed from the pulser coil (which provides crankshaft position information and is referred to in Mercury service literature as the Crankshaft Position Sensor or CPS), throttle opening from the Throttle Position Sensor (TPS) and intake air from the Manifold Air Temperature (MAT) and Manifold Air Pressure (MAP) sensors. In this way the ECM keeps track of just how much air is entering the engine (between the TPS and the temperature/pressure of the air) and can adjust air/fuel mixture ratios to suit operating conditions. The lack of Oxygen sensor makes this form of injection open-loop only (meaning the ECM receives no feedback on the combustion process).

3-52 FUEL SYSTEM

In fuel injection terms this type of system is known as a speed-density injection system. This is because the basic engine fuel mapping decisions are made by the ECM based on a comparison between the engine rpm (speed) and the air pressure (air density). The basic, pre-programmed, fuel mapping is then modified based upon input from the remaining sensors. Specifically, engine and air temperatures are taken into account.

Like most fuel injection systems, the Mercury/Mariner EFI 4-stroke system incorporates a return-to-port capability or limp mode. If there is a major sensor failure that prevents the ECM from processing one or more signals, the engine is run on a minimum performance map. In most instances the ECM will substitute a fixed value for the missing sensor signal. The only input that will shut the engine down completely is battery voltage. If the battery is disconnected or the battery voltage falls below a certain point, the fuel pump quits pumping so the engine stops.

Major components of this system include the ECM and Sensors along with the low- and high-pressure fuel delivery circuits. The electronic control sensors include:
- Throttle Position Sensor (TPS)
- Crankshaft Position Sensor (CPS, also known as a Pulser Coil)
- Engine Coolant Temperature (ECT) Sensor
- Manifold Air Temperature (MAT) Sensor
- Manifold Air Pressure (MAP) Sensor
- Oil Pressure Switch
- Ignition Key Switch
- Idle Air Control Valve (ok, it's not a sensor, but it's an important piece of the system)

The low-pressure fuel circuit is similar to the fuel delivery circuit found on carbureted motors. A fuel line with primer bulb is used to connect the boat mounted or portable fuel tank to a powerhead mounted low-pressure fuel pump. A mechanical diaphragm-displacement low-pressure lift pump controls the low-pressure circuit and is used to deliver fuel from the fuel tank through an inline water-separating fuel filter through the pump and finally to a powerhead mounted vapor separator tank.

Inside the vapor separator tank is a submerged high-pressure electric fuel pump. The high-pressure circuit is where the Mercury system differs most from the Yamaha. The submerged high-pressure pump draws fuel from the vapor tank and forces it through a fuel cooler assembly, through the fuel distribution manifold to the injectors. Excess fuel pressure is bled off directly from the fuel cooler through a pressure regulator and back into the vapor tank. In this way fuel is not constantly pumped past the fuel injectors and recirculated back to the tank, but instead no fuel makes it to the injectors without actually waiting under pressure to be expelled by the injectors.

Controlled Combustion System

Various versions of a Controlled Combustion System (CCS) are employed by Yamaha fuel injected motors to reduce emissions, smooth idle and help improve shifting or low speed stability for trolling, but it is primarily only found on the V6 motors equipped with the Yamaha fuel injection system (meaning it includes the V6 Mercury/Mariners).

The CCS mapping in the ECM is designed to cut one or more cylinders under certain circumstances. This reduces noise, increases slow speed stability and improves fuel economy.

It is an inventive solution to the problems associated with running a large displacement engine at neutral or trolling speed. A large displacement engine is not working hard enough at low speeds to maintain peak combustion efficiency. It is also difficult to meter small amounts of fuel and air precisely through large diameter throttle valves and intakes. These problems can result in excessive fuel consumption and roughness when trolling for extended periods. The CCS effectively shrinks engine size by limiting fuel delivery to 4 or 5 cylinders, depending on the fuel system and engine RPM.

On V6 models of the 4-stroke EFI motors a version of this system is actuated whenever the motor is operating below 2000 rpm and the shift cut switch is activated. When engine speed below 850 rpm the ECM will misfire cylinders #1 or #4, and when engine speed is between 850-2000 rpm the ECM will misfire cylinders #1 and #2 or #4 and #5.

■ **Keep the CCS system in mind when attempting to diagnose potential fuel and ignition problems, as the system may make it appear that you have one or more dead cylinders when it may not necessarily be true under normal operating conditions. Always verify a suspected dead cylinder by running the motor in gear (under load) to eliminate the CCS as a potential cause.**

Troubleshooting Electronic Fuel Injection

On carbureted outboards fuel is metered through needles and valves that react to changes in engine vacuum as the amount of air drawn into the motor increases or decreases. The amount of air drawn into carbureted motors is controlled through throttle plates that effectively increase or decrease the size of the carburetor throat (as they are rotated open or closed).

In contrast, fuel injected engines use a computer control module to regulate the amount of fuel introduced to the motor. The Electronic Control Module (ECM) monitors input from various engine sensors in order to receive precise data on items like engine position (where each piston is on its 4-stroke cycle), engine speed, engine and/or air temperatures, ambient air or manifold pressure and throttle position. Analyzing the data from these sensors tells the engine exactly how much air is drawn into the motor at any given moment and allows the ECM to determine how much fuel is required.

The ECM will energize (open) the fuel injectors for the precise length of time required to spray the amount of fuel needed for engine operating conditions. In actuality, the injectors are not just activated and held-open as much as they are pulsed, opened and closed rapidly for the correct total amount of time necessary to spray the desired amount of fuel. This electronically controlled, precisely metered fuel spray or "fuel injection" is the heart of a modern fuel injection system and the main difference between a fuel injected and carburetor motor.

Troubleshooting a fuel injected motor contains similarities to carbureted motors. Mechanically, the powerhead of a 4-stroke fuel injected motor operates in the same way as that of its carbureted counterpart. There still must be good engine compression and mechanical timing for either engine to operate properly. Wear or physical damage will have virtually the same affect upon either motor. Furthermore, the low-pressure fuel system that supplies fuel to the reservoir in the vapor separator tank operates in the same manner as the fuel circuit that supplies gasoline to the carburetor float bowl.

The major difference in troubleshooting engine performance on EFI motors is the presence of the ECM and electronic engine controls. The complex interrelation of the sensors used to monitor engine operation and the ECM used to control both the fuel injection and ignition systems makes logical troubleshooting all that much more important.

Before beginning troubleshooting on an EFI motor, make sure the basics are all true. Make sure the engine mechanically has good compression (refer to the Compression Check procedure that is a part of a regular Tune-Up). Make sure the fuel is not stale. Check for leaks or restrictions in the Lines and Fittings of the low pressure fuel circuit, as directed in this section under Fuel Tank and Lines. EFI systems cannot operate properly unless the circuits are complete and a sufficient voltage is available from the battery and charging systems. A quick-check of the battery state or charge and alternator output with the engine running will help determine if these conditions are adversely affecting EFI operation.

■ **Loose/corroded connections or problems with the wiring harness cause a large percentage of the problems with EFI systems. Before getting too far into engine diagnostics, check each connector to make sure they are clean and tight. Visually inspect the wiring harness for visible breaks in the insulation, burn spots or other obvious damage.**

In order to help find electronic problems with the EFI system, the ECM contains a self-diagnostic system that constantly monitors and compares each of the signals from the various sensors. Should a value received by the ECM from one or more sensors fall outside certain pre-determined ranges the ECM will determine there is a problem with that sensor's circuit. Basically the ECM compares signals received from different sensors to each other and to real world possible values and makes a decision if it thinks one must be lying. For instance, if the ECM receives a crankshaft position sensor signal that shows the engine is rotating at a high rpm, but also receives a signal from the TPS that says the throttle is fully closed, it knows one is wrong. Similarly, if it receives a ridiculous signal, say the intake air temperature suddenly provides a signal above 338°F/170°C, the ECM will know there is something wrong with that signal. Depending on the severity of the fault or faults, the engine will continue to run, substituting fixed values for the sensors that are considered out of range. Under these circumstances, engine performance and economy may become drastically reduced.

■ **Check for the presence of diagnostic codes as described in this section BEFORE disconnecting the battery. If codes are not present, yet problems persist, use the symptom charts to help determine what further components or systems to check.**

FUEL SYSTEM 3-53

When a fault is present the ECM will display a diagnostic code when the proper diagnostic flash adapter for Yamaha EFI systems or the Quicksilver Digital Diagnostical Terminal (DDT) for Mercury/Mariner EFI systems is attached to the motor.

For 115 hp and larger motors, the flash adapter contains a flash indicator (essentially a small light) which will illuminate in specific patterns to communicate diagnostic codes to the technician. The codes can then be used to help determine what components and circuits should be checked for trouble.

For 30-60 hp motors, the Quicksilver DDT is a full function scan tool which must be used to interact with the ECM. Unfortunately it is NOT inexpensive and at the time of writing we know of no substitutes for the proprietary scan too. The good news is that component testing is still possible without the tool AND the Mercury/Mariner Warning Horn and Guardian system may give some clues as to which components to check first, depending upon the fault. For more details, please refer to the section on Mercury/Mariner EFI Troubleshooting Charts, later in this section.

For all systems, remember that a fault code doesn't automatically mean that a component (such as a sensor) is bad, it means that the signal received from the sensor circuit is missing or out of range. This can be caused by loose or corroded connections, problems with the wiring harness, problems with mechanical components (that are actually causing this condition to be true), or a faulty sensor.

Once components or circuits that require testing have been identified, use the testing procedures found in this section (or other sections, as applicable) and the wiring diagrams to test components and circuits until the fault has been determined.

Keep in mind that although a haphazard approach might find the cause of problems, only a systematic approach will prevent wasted time and the possibility of unnecessary component replacement. In some cases, installing an electronic component into a faulty circuit that damaged or destroyed the previous component, will instantly destroy the replacement. For various reasons, including this possibility, most parts suppliers do not accept returns on electrical components.

■ If a problem is speed related or intermittent, we HIGHLY suggest that you check for connector and/or contact related problems. Make sure all connectors are clean and tight. Check the connectors for loose wires, loose pins and signs of corrosion. Use dielectric grease to prevent future corrosion and manually secure connectors as necessary to ensure they will not be the problem.

At least the manufacturer states that on Mercury fuel injection systems, any sensor can be unplugged with the motor running without harming the ECM (which can be used as a diagnostic tool to see if there is a change in operation when unplugged, no change may mean the sensor or sensor circuit is the potential culprit). Keep in mind that unplugging the CPS will cause the motor to shut down.

Self Diagnostic System

READING TROUBLE CODES

115 Hp and Larger Motors

◆ See Figures 91 thru 96

■ Certain electrical equipment such as stereos and communication radios can interfere with the electronic fuel injection system. To be certain there is no interference, shut these devices off when troubleshooting. If a check engine light illuminates immediately after installing or re-rigging an existing accessory, reroute the accessory wiring to prevent interference.

When the electronic engine control system detects a problem with one of its circuits, the ECM will display diagnostic trouble codes when the appropriate diagnostic flash adapter is connected to the wiring harness. Flash adapter and connection locations vary slightly by model.

As a result of most faults, the ECM will ignore the circuit signal and enter a fail-safe mode designed to keeps the boat and motor from becoming stranded. During fail-safe operation the ECM will provide a fixed substitute value for the faulty circuit. During fail-safe operation the engine will run, but usually with reduced performance (power and economy).

Once a malfunction ceases (has been fixed or the circuit signal returns to something within the anticipated normal range), engine operation will return to normal and the diagnostic flash indicator will cease to display the trouble code for that circuit.

Proceed as follows to display stored codes:
1. Start with the engine and ignition switch turned OFF
2. Using the accompanying illustrations and information, locate the necessary wiring harness connector and install the diagnostic flash indicator:

• For Yamaha 115 hp motors, connect the diagnostic flash socket (#6Y5-83536-10) along with (#6Y5-83517-00) to the dedicated diagnostic harness connectors on the blue/white and red/black wires, as detailed in the accompanying illustration.

• For Mercury/Mariner 115 hp motors through 2001 connect the diagnostic flash socket, known as the Malfunction Indicator Lamp or MIL to Mercury dealers (#99056A1), to the dedicated diagnostic harness connectors on the blue/white and red/black wires, as detailed in the accompanying illustration.

• For Yamaha 150 hp motors connect the diagnostic flash adapter (#YB-06795) to the dedicated diagnostic harness connectors and the blue/white wires, as detailed in the accompanying illustration.

• For 2002 and later Mercury/Mariner 115 hp motors connect the Malfunction Indicator Lamp (MIL) (#91-884793A1) - which is essentially (#99056A1 along with an additional 3-pin terminal connector) to the dedicated diagnostic harness connectors. To do so, remove the rubber plug from the bullet connector, then connect the free pin bullet from the lamp wire. Next, remove the test plug cap from the 3-pin harness mounted just below the relays on the side of the powerhead. Connect the 3-pin harness from the lamp assembly to the 3-pin test harness connector on the powerhead. Refer to the accompanying illustration for details, as necessary.

• For Yamaha 200/225 hp EFI motors connect the diagnostic flash adapter (#YB-06794) along with the indicator (#YB-06765) to the dedicated diagnostic harness connectors on the blue/white and the red/black, white/black and black wires at the front base of the powerhead, as detailed in the accompanying illustration.

• For Mercury/Mariner 200/225 hp EFI motors connect the diagnostic flash adapter (#91-884793A1) along with the indicator (#91-99056A1) to the dedicated diagnostic harness connectors on the blue/white and the red/black, white/black and black wires at the front base of the powerhead, as detailed in the accompanying illustration.

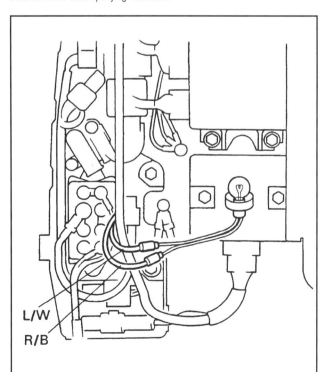

Fig. 91 Connecting the diagnostic flash indicator - 115 hp EFI motors (Mercury/Mariner through 2001 and all Yamaha models)

3-54 FUEL SYSTEM

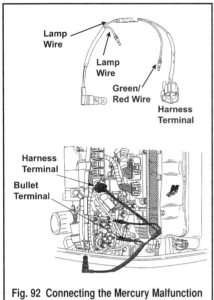

Fig. 92 Connecting the Mercury Malfunction Indicator Lamp (MIL) for 2002 and later 115 hp EFI motors

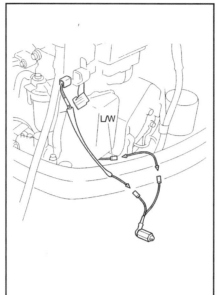

Fig. 93 Connecting the diagnostic flash indicator - 150 hp EFI motors

Fig. 94 Connecting the diagnostic flash indicator (1) and adaptor (2) - 200/225 hp EFI motors

3. Connect a suitable source of cooling water to the motor, then start the engine and allow it idle, observe the flash pattern on the diagnostic indicator to determine if there are any trouble codes.

4. Interpret codes by counting the flashes. Only one code will be displayed at a time. The first digit of the 2-digit code will display first, with one, two, three or possibly even four flashes (the light will remain on and off for the same amount of time, about 0.33 seconds, when flashing a code digit). Count the number of times on and you've got the first digit.

■ These motors use a single digit code 1 to indicate normal operation. When present the single 0.3 second flash will be separated from the next single flash by a long 5 second pause.

5. There will be a long pause between digits of the 2-digit code (the light will remain off for about 1.65 seconds), then the short on/off flashes will indicate the second digit of the 2-digit code. This second flash may consist of any number from one to nine. A single code will continue to flash until the problem is corrected, then the next higher code (if present) will flash until corrected. This pattern will repeat until all faults are corrected.

■ If more than one is present, codes will flash in the order of priority (lowest number to highest number) as listed in the accompanying code chart. Only a single code will be displayed until it is corrected, then the next higher code will flash.

6. Compare the codes with the accompanying code table to determine the defective circuit. Remember that a code does NOT necessarily mean a given component is at fault, it means that the ECM sees signal that is out of the normal, predetermined operating range. The problem may also lie with the wiring, another system or another component that would make a circuit read out of specification.

■ Make sure all connections for that circuit and related components of the electrical system are clean and tight before troubleshooting the circuit. Bad wiring or connections can cause out of range signals and set trouble codes.

7. Test the sensor or system components as described in this section. When troubleshooting, always start with the easiest checks/fixes and work toward the more complicated.

8. Once all of the problems have been corrected, connect a source of cooling water to the flushing system, then start and run the engine to verify proper operation. If the problems are gone, the diagnostic indicator either will remain off or will flash a code 1 (depending upon the model).

9. Shut the ignition switch **off** and remove the diagnostic connector.

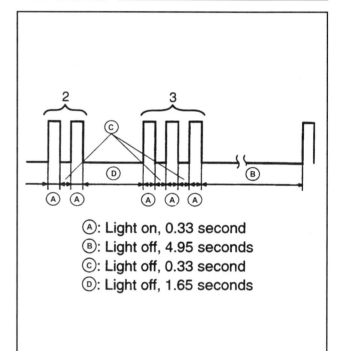

Ⓐ: Light on, 0.33 second
Ⓑ: Light off, 4.95 seconds
Ⓒ: Light off, 0.33 second
Ⓓ: Light off, 1.65 seconds

Fig. 95 Flash pattern for code 23

60 Hp and Smaller Motors

As stated earlier a full function scan tool designed to interact with the Mercury/Mariner ECM must be used on all models 60 hp or smaller to check for faults. There is no fault code list per se, as the Quicksilver DDT spells out the fault for the technician. The good news is that component testing is still possible without the tool AND the Mercury/Mariner Warning Horn and Guardian system may give some clues as to which components to check first, depending upon the fault. For more details, please refer to the section on Mercury/Mariner EFI Troubleshooting Charts, later in this section.

FUEL SYSTEM 3-55

YAMAHA EFI SYSTEM TROUBLE CODES [1]

Trouble Code	Fault
1	Normal
13	Incorrect trigger (pulser/crankshaft) coil input signal
15	Incorrect Engine Cooling Water Temperature Sensor (ECT/WTS) signal
18	Incorrect Throttle Position Sensor (TPS) signal
19	Low battery input voltage
23	Incorrect Intake Air Temperature (IAT) sensor signal
28	Incorrect shift position switch signal
29	Intake/Mainfold Air Pressure (IAP/MAP) sensor signal out of normal range
31-44	These codes are Micro-computer processing information and not necessarily faults
33	Ignition timing being slightly corrected - during a cold start
37	Intake passage air leakage / Incorrect Idle Speed Control (ISC) signal
39	Incorrect oil pressure sensor signal
44	Engine stop switch (and/or safety lanyard) control activated
45	Incorrect shift cut switch signal
46	Incorrect thermoswitch signal

NOTE: not all codes used by all model. Also remember a code means a problem with a CIRCUIT and not necessarily a component

[1] The Yamaha EFI system is used on all 115 hp and larger models, including Mercury/Mariner motors

Fig. 96 Trouble Codes For Yamaha EFI Systems (Yamaha & Merc/Mariner Models 115 Hp And Larger)

Mercury/Mariner EFI Troubleshooting Charts

LOW-PRESSURE LIFT PUMP VACUUM TESTING

◆ See Figures 97 and 98

The low-pressure lift pump is needed to transfer fuel from the portable or built-in fuel tank on the boat to the Vapor Separator Tank (VST) on the side of the powerhead. Unless the low-pressure pump keeps the VST filled to a proper level, the submerged high-pressure pump cannot provide sufficient fuel pressure for proper EFI operation.

The pump itself is designed to lift fuel (vertically) about 60 in. (1524mm), but ONLY if there are no other restrictions. The pump is specified to operate with a fuel supply hose that has at least a minimum diameter of 5/16 in. (7.9mm). As restrictions are added to the system (such as filters, valves, etc) the amount of total fuel pump lift available from the pump decreases.

Though safety regulations require boat builders and tank manufacturers to equip each vessel with an anti-siphon valve to prevent the tank from self-emptying should a problem occur with the rigging, these valves can be more a source of trouble.

Many valves are prone to clogging with debris, may be too small for the rigged motor and/or may contain too heavy an internal spring. All of which would result in restricting the fuel flow to a point where normal EFI operation could be hindered.

Some common symptoms of a restricted fuel supply include
- High speed surging
- Loss of fuel pump pressure
- Loss of power
- Motor cuts out or hesitates upon acceleration
- Motor runs rough
- Motor quits and cannot be started
- Motor will not start
- Pre-ignition/detonation (piston dome erosion)
- Vapor lock

The quickest way to eliminate the valve from suspicion is to temporarily rig an alternate fuel supply (test tank). If the valve is found to be faulty you'll have to either repair or replace it. One possibility is to install a solenoid-operated fuel shut-off valve.

If however the test tank does not eliminate the problem, the next step is to check the low-pressure fuel pump by conducting a vacuum (lift) test as follows:

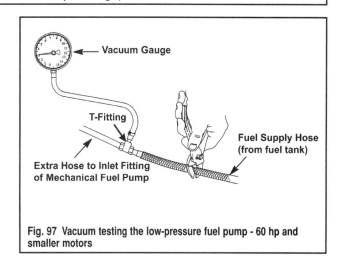

Fig. 97 Vacuum testing the low-pressure fuel pump - 60 hp and smaller motors

■ For this test the motor will be run at 1000 rpm and below, so it CAN be done on a flushing fitting, but a test tank or backing the trailer into the water may still be preferred in case other tests (at speed) are necessary.

You'll need a short length of fuel hose, a T-fitting and a vacuum gauge to conduct this test.

1. Provide a suitable source of cooling water to the motor.
2. Disconnect the fuel hose from the INLET fitting of the low-pressure fuel pump, then connect the short length of extra fuel hose to the pump fitting.
3. Install the T-fitting between the length of extra fuel hose and fuel line from the tank (which was removed from the pump).
4. Connect a vacuum gauge to the T-fitting.
5. Start and run the engine at 1000 rpm while checking the vacuum reading on the gauge. If should be about 1-2 in. Hg. (25-50mmHg).
6. Use a pair of pliers (with a rag over the jaws to help protect the fuel line) or a line restrictor to carefully pinch off the fuel supply hose between the T-fitting and the gauge. This will isolate the low-pressure pump from the tank and should allow the vacuum gauge to display the maximum vacuum the pump is capable of generating. If the pump is in good condition it will show 4 in. Hg. (101.6mm Hg).
7. If the vacuum reading is not within specifications, refer to the accompanying Low-Pressure Lift Pump Vacuum Troubleshooting Chart.

3-56 FUEL SYSTEM

Vacuum (Lift) Troubleshooting		
Condition	**Cause**	**Correction**
Fuel system vacuum (lift) above specification 1 - 2 in. Hg (25-50mm Hg) @1000 RPM	Restricted anti-siphon valve	Refer to "Checking for Restricted Fuel Flow caused by Anti-Siphon Valves" preceding
	Plugged fuel tank pick-up screen	Clean/replace fuel pick-up screen
	Pinched/collapsed fuel hose	Inspect/replace fuel hose(s)
	Dirty/plugged water separating fuel filter	Clean/replace water separating fuel filter
	Restriction in fuel line thru-hull fitting	Clean/replace fitting
	Restriction in fuel tank switching valve	Clean/replace valve
	Restriction within primer bulb	Rebuild/replace primer bulb
Fuel system vacuum (lift) below specifications 1 - 2 in. Hg (25-50mm Hg) @1000 RPM	Low fuel level in fuel tank	Fill tank with fuel
	Hole/cut in pick-up tube of fuel tank	Replace fuel pick-up tube
	Loose fuel line connection	Check/tighten all connections
	Hole/cut is fuel line	Inspect/replace fuel hose(s)
	Loose fuel pump screws	Torque screws to specification
	Fuel pump gasket(s) worn or leaking	Rebuild/replace fuel pump
	Fuel pump check valves/seals leaking	Rebuild/replace fuel pump
	Leaky fuel pump diaphragm	Rebuild/replace fuel pump
	Worn/broken fuel pump springs	Rebuild/replace fuel pump
	Leaky fuel pump seals	Rebuild/replace fuel pump
	Fuel filter bowl loose	Tighten fuel filter bowl
	Fuel filter gasket cut/worn	Replace gasket
	Fuel vaporization	Check for plugged fuel pump water cooling circuit

Fig. 98 Low-Pressure Fuel (Lift) Pump Vacuum Troubleshooting - 60 Hp And Smaller EFI Motors

LOW-PRESSURE LIFT PUMP PRESSURE TESTING

MODERATE

◆ See Figures 99 and 100

If the lift pump appears to be drawing sufficient vacuum, the next step is to verify delivery pressure from the mechanical pump to the Vapor Separator Tank (VST) as follows:

■ **For this test the motor will be run at 1000 rpm and below, so it CAN be done on a flushing fitting, but a test tank or backing the trailer into the water may still be preferred in case other tests (at speed) are necessary.**

You'll need a short length (about 4 in. /10 cm) of CLEAR fuel hose, a T-fitting and a fuel pressure gauge to conduct this test.
 1. Provide a suitable source of cooling water to the motor.
 2. Disconnect the fuel hose from the OUTLET fitting of the low-pressure fuel pump, then connect the short length of clear fuel hose to the pump fitting.
 3. Install the T-fitting between the length of extra fuel hose and the pump outlet fitting line which is still connected to the VST.
 4. Connect a vacuum gauge to the T-fitting.
 5. Start and run the engine at 1000 rpm while checking the pressure reading on the gauge. If should be about 2 psi (13.7 kPa).
 6. Use a pair of pliers (with a rag over the jaws to help protect the fuel line) or a line restrictor to carefully pinch off the fuel supply hose between the T-fitting and the VST. This will isolate the low-pressure pump from the VST and should allow the pressure gauge to display the maximum pressure that

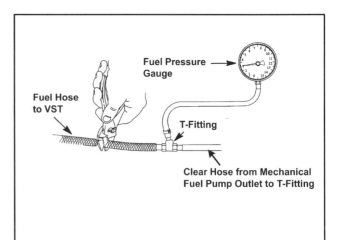

Fig. 99 Pressure testing the low-pressure fuel pump - 60 hp and smaller motors

the pump is capable of generating. If the pump is in good condition it will show at least 3 psi (20.7 kPa).
 7. If the pressure reading is not within specifications, refer to the accompanying Low-Pressure Lift Pump Pressure Testing Chart.

FUEL SYSTEM 3-57

Fuel Pressure Troubleshooting

Condition	Cause	Correction
Fuel system pressure below specification	Restricted anti-siphon valve	Refer to "Checking for Restricted Fuel Flow caused by Anti-Siphon Valves" preceding
	Low fuel level in fuel tank *	Fill tank with fuel
	Plugged fuel tank pick-up screen	Clean/replace fuel pick-up screen
	Hole/cut in pick-up tube of fuel tank *	Replace fuel pick-up tube
	Loose fuel line connection *	Check/tighten all connections
	Hole/cut in fuel line *	Inspect/replace fuel hose(s)
	Fuel line primer bulb check valves not opening	Replace fuel line primer bulb
	Fuel hose/line internal diameter too small	Use 5/16 in. (8mm) fuel hose
	Restriction in fuel line thru-hull fitting	Clean/replace fitting
	Restriction in fuel tank switching valve	Clean/replace valve
	Restriction within primer bulb	Rebuild/replace primer bulb
	Pinched/collapsed fuel hose	Inspect/replace fuel hose(s)
	Dirty/plugged water separating fuel filter	Clean/replace water separating fuel filter
	Fuel filter bowl loose *	Tighten fuel filter bowl
	Fuel filter gasket cut/worn *	Replace gasket
	Loose fuel pump screws *	Torque screws to specification
	Fuel pump gasket(s) worn or leaking *	Rebuild/replace fuel pump
	Fuel pump check valves/seals leaking	Rebuild/replace fuel pump
	Leaky fuel pump diaphragm *	Rebuild/replace fuel pump
	Worn/broken fuel pump springs	Rebuild/replace fuel pump
	Leaky fuel pump seals	Rebuild/replace fuel pump
	Fuel vaporization	Check for plugged fuel pump water cooling circuit

NOTE: * Air bubbles may also be visible as fuel passes through the clear fuel (test) hose installed between the mechanical fuel pump outlet fitting and the VST.

Fig. 100 Low-Pressure Fuel (Lift) Pump Pressure Testing - 60 Hp And Smaller EFI Motors

WARNING HORN/GUARDIAN SYSTEM FAULTS

◆ See Figure 101

The Guardian Protection System monitors critical engine functions and will reduce engine power under certain circumstances to try and keep the engine running while allowing minimal damage to critical systems. Unfortunately, depending upon the fault the Guardian system cannot ensure the engine will not be damaged. Mercury would rather allow the engine to keep running in order to get you to port than to shut it down in a potentially life-threatening situation.

As stated earlier, without the Quicksilver DDT you have no way of determining the actual faults which the ECM is detecting, HOWEVER, when there is an activation of the Guardian system, the number and frequency of warning sounds can be used as a hint as to the nature of the trouble.

For instance, six beeps every time the key is turned on or during operation means that the ECM has detected a problem with one of the sensor circuits that is important to performance, but not to safety of the motor. Under this circumstance the engine can still be operated, but should be serviced soon. In comparison, a continuous warning tone indicates that the ECM has detected a potentially engine damaging fault, such as low oil pressure or an overheating signal and should be shut down as soon as is safely possible.

Refer to the accompanying Warning Horn/Guardian System Operation chart for additional information on deciphering warning horn signals.

MERCURY EFI SYMPTOM TROUBLESHOOTING

◆ See Figures 102 thru 105

The accompanying symptom troubleshooting charts can be used to help narrow potential EFI system problems down, especially when you do not have access to a suitable marine scan tool, such as the Quicksilver DDT.

■ **ALWAYS, ALWAYS, ALWAYS, remember to check the wiring harness and connectors for corrosion or loose connections before assuming a component is faulty!**

Sensor and Circuit Resistance/Output Tests

TESTING ELECTRONIC COMPONENTS

Before conducting resistance or voltage checks on a component/circuit, make sure all electrical connections in the system are clean and tight. Visually check the wiring for obvious breaks, defects or other problems.

Keep in mind that although a haphazard approach might find the cause of problems, only a systematic approach will prevent wasted time and the

FUEL SYSTEM

IMPORTANT: In all instances check wiring harness integrity (especially ground connections) in boat and on engine.

Condition	Cause/Fault	Warning Mode	Check
Engine cranks but will not start	Lanyard stop switch is in the "OFF" position	None	Set lanyard stop switch to "RUN"
	Weak battery or bad starter motor. Battery voltage drops below 8 volts while cranking (ECM cuts out below 6 volts) (Fuel pump requires 8 volts).	3 Beeps every 4 minutes for low battery voltage.	Check condition of battery/starter solenoid terminals and cables. Charge/replace battery. Inspect condition of starter motor.
	Blown Fuse	None	Replace fuse. Inspect engine wiring harness and electrical components. Fuse #2 - Fuel Injectors/IAC/Fuel Pump Fuse #3 - Main Power Relay/Accessory Fuse #4 - Ignition Coils
	Main Power Relay	Intermittent Beeps	Listen for relay to "click" when key switch is turned to "ON" Between pin #22 (YEL/PUR) of port ECM connector and (RED/BLU) wire of fuse #3 (fuse removed) - or - Between pin #85 and pin #86 of relay 300 – 350 ohms
	Crank Position Sensor (CPS)	None Note: No rpm reading at tachometer	Between pin #5 (RED) and pin #6 (WHT) of starboard ECM connector. - or - Between pin #1 (RED) and pin #2 (WHT) of CPS connector.
	Electric fuel pump	Intermittent Beeps	Listen for pump. Fuel pump should run 2 seconds after key switch is turned to "RUN" position. 32 - 41 ohms Between pin #19 (BLK/BLU) and pin # 23 (RED/BLU) - or - Between pins of fuel pump connector.
	Flywheel misaligned	None	Remove flywheel and inspect flywheel key/key way
	Engine Coolant Temperature (ECT) sensor	3 Beeps every 4 minutes	See ECT sensor resistance Advancing the remote control fast idle feature or advancing the tiller handle throttle grip half way may assist starting.

Fig. 102 EFI Symptom Troubleshooting Chart No. 1 - 60 Hp And Smaller Mercury EFI Motors

Warning Horn/Guardian System Operation

Sound	Condition	Description
One Beep on key up	Normal	System Test
Six Beeps on key up, or during a running failure.	Failure detected with MAP, MAT*, TPS, or Flash Check Sum (ECM).	Engine should run well however, service will be required.
Three Beeps every 4 Minutes.	Failure detected with: • Battery Voltage * • EST * - Open detected at key up. Short detected with engine running • Fuel Injector - Detected while cranking/running * • Coolant Sensor * • IAC **	Engine will start hard, run rough and/or stall. Utilizing the neutral fast idle feature may assist starting. Service is required.
Intermittent Beeps	Failure detected with: • Fuel Pump - May start momentarily ** • Main Power Relay - No start ** • ECM Reference Voltage to MAP/TPS - Starts but stalls under load	Engine may or may not start. If engine starts it easily stalls. Service is required.
Continuous	Engine Overheat	Engine Guardian System is activated. Power limit will vary with level of overheat. Stop engine and check water intake for obstruction. Advancing throttle above idle may provide additional cooling.
	Low Oil Pressure	Guardian System is activated. Engine power is limited to 10% of maximum. Stop engine and check oil level. Add oil if necessary.
	Battery Voltage Less Than 10v or More Than 16v	Engine Guardian System is activated. Engine power is limited to 75% of maximum.
	Coolant Sensor Failure	Engine Guardian System is activated. Engine power is limited to 50% of maximum. Engine overheat protection is compromised.
	Engine Speed Limiter	Exceeding 6200 rpm cuts spark/injection on cylinders #2 and #3 (4 cylinder only) to reduce engine speed. Exceeding 6350 rpm cuts spark/injection on all cylinders to reduce engine speed.

* Horn Beeps once on key up, plus failure code.

** Sticky Fault requires key off to reset.

Fig. 101 Warning Horn/Guarding System Operation - 60 Hp And Smaller EFI Motors

FUEL SYSTEM

IMPORTANT: In all instances check wiring harness integrity (especially ground connections) in boat and on engine.

Condition	Cause/Fault	Warning Mode	Check
Engine cranks, starts and stalls	Remote control to engine wiring harness connection is poor	None	Clean and inspect male and female connections.
	Air in fuel system/lines	None	Crank and start engine several times.
	Manifold Absolute Pressure (MAP) sensor	6 Beeps at key up or failure	See MAP sensor resistance
	Throttle Position Sensor (TPS)	6 Beeps at key up or failure	Typical TPI range with DDT: Idle 0.39-1.0 volts, WOT 3.66-4.80 volts.
	Idle Air Control (IAC)	3 Beeps every 4 minutes	20 - 24 ohms Between pin #20 (WHT/ORG) and pin #23 (RED/BLU) of starboard ECM connector. - or - Between pin A and pin B of IAC.
	ECM reference voltage to MAP/TPS	Intermittent Beeps	5 volts Between PUR/YEL pin of MAP sensor wiring harness connector and engine ground (key switch to "RUN").
	Fuel pressure at VST fitting	None	See fuel pressure test
	Flywheel misaligned	None	Remove flywheel and inspect flywheel key and key way
Engine Idles Fast after warm-up (900-1100rpm)	Engine Coolant Temperature (ECT) sensor	3 Beeps every 4 minutes	See ECT sensor resistance
Poor at idle or WOT running quality	Fuel injector	3 Beeps every 4 minutes	10.0 - 13.5 ohms Between fuel injector pin #1 and pin #2. - or - Between (removed) fuse #2 (RED/BLU) wire and port ECM connector: Pin #17 (PNK/BRN) Fuel Injector #1 Pin #2 (PNK/RED) Fuel Injector #2 Pin #1 (PNK/ORG) Fuel Injector #3 Pin #18 (PNK/YEL) Fuel Injector #4(4 cyl)
	Ignition coil (EST) ***	None	See ignition coil resistance
	Fuel pressure at VST fitting	None	See fuel pressure test
	Fuel filter plugged	None	Replace fuel filter
	Improper spark plugs	None	Use recommended resistive spark plugs
	Loose grounds	None	Check all ground connections.
	Flywheel timing tooth pattern	None	Check tooth pattern for partially missing or damaged teeth
	Fouled spark plug(s)	None	Replace spark plug(s).

Fig. 103 EFI Symptom Troubleshooting Chart No. 2 - 60 Hp And Smaller Mercury EFI Motors

IMPORTANT: In all instances check wiring harness integrity (especially ground connections) in boat and on engine.

Condition	Cause/Fault	Warning Mode	Check
Poor idle quality	Crank Position Sensor (CPS)	None	300 - 350 ohms Between pin #5 (RED) and pin #6 (WHT) of starboard ECM connector. - or - Between pin #1 (RED) and pin #2 (WHT) of CPS connector.
	Manifold Absolute Pressure (MAP) sensor	6 Beeps at key up or failure	See MAP sensor resistance chart
	Throttle Position Sensor (TPS)	6 Beeps at key up or failure	Typical TPI range with DDT: Idle 0.39-1.0 volts, WOT 3.66-4.80 volts.
	Engine Coolant Temperature (ECT) sensor	3 Beeps every 4 minutes	See ECT sensor resistance chart
	Manifold Air Temperature (MAT) sensor	6 Beeps at key up or failure	See MAT sensor resistance
	Fuel injector	3 Beeps every 4 minutes	10.0 - 13.5 ohms Between fuel injector pin #1 and pin #2. - or - Between (removed) fuse #2 (RED/BLU) wire and port ECM connector: Pin #17 (PNK/BRN) Fuel Injector #1 Pin #2 (PNK/RED) Fuel Injector #2 Pin #1 (PNK/ORG) Fuel Injector #3 Pin #18 (PNK/YEL) Fuel Injector #4(4 cyl)
	Ignition coil (EST) ***	3 Beeps every 4 minutes	See ignition coil resistance
	Idle Air Control (IAC)	3 Beeps every 4 minutes	20 - 24 ohms Between pin #20 (WHT/ORG) and pin #23 (RED/BLU) of starboard ECM connector. - or - Between pin A and pin B of IAC.
	Fuel pressure at VST fitting.	None	See fuel pressure test
	Loose grounds	None	Check all ground connections.
	Fouled spark plug(s)	None	Replace spark plug(s).

*** The ECM will only monitor the EST connection to the ignition coil, use resistance tests and/or spark gap test to confirm an ignition coil failure.

Fig. 104 EFI Symptom Troubleshooting Chart No. 3 - 60 Hp And Smaller Mercury EFI Motors

3-60 FUEL SYSTEM

IMPORTANT: In all instances check wiring harness integrity (especially ground connections) in boat and on engine.

Condition	Cause/Fault	Warning Mode	Check
Engine runs rich	Fuel pressure regulator	None	42 - 44 psi (290 – 303 kPa) at VST fitting
	Engine Coolant Temperature (ECT) Sensor	3 Beeps every 4 minutes	See ECT sensor resistance chart
	Thermostat stuck open	None	Remove and inspect thermostat.
Speed Reduction Engine RPM Limited to 2000	Low oil pressure or grounded oil pressure switch lead	GUARDIAN Continuous Horn Above 10% Power Setting	Check engine oil level and add oil as needed. Remove oil pressure switch and install oil pressure gauge, (warm engine) oil pressure should be: Above 2.9 psi (20.0 kPa) at idle 30-40 psi. (207-278 kPa) at 3000 rpm. See Oil Pressure Switch test - Section 4B "Cylinder Block/Crankcase". Check for short between pin #7 (BLU) of starboard ECM connector and open connector of oil pressure switch.
Speed Reduction Engine RPM Limited	Engine Overheat	GUARDIAN Continuous	Engine Guardian System is activated. Power limit will vary with level of overheat. Stop engine and check water intake for obstruction. Advancing throttle above idle may provide additional cooling.
	Battery Voltage Less Than 10v or More Than 16v	GUARDIAN Continuous Horn Above 75% Power Setting	Engine Guardian System is activated. Engine power is limited to 75% of maximum.
	Engine Coolant Temperature (ECT) Sensor Failure	GUARDIAN Continuous Horn Above 50% Power Setting	Engine Guardian System is activated. Engine power is limited to 50% of maximum. Engine overheat protection is compromised

Fig. 105 EFI Symptom Troubleshooting Chart No. 4 - 60 Hp And Smaller Mercury EFI Motors

possibility of unnecessary component replacement. In some cases, installing an electronic component into a faulty circuit that damaged or destroyed the previous component, will instantly destroy the replacement. For various reasons, including this possibility, most parts suppliers do not accept returns on electrical components.

Make sure you perform each test procedure and especially, make all test connections as described. If a component tests out of range (faulty), but the reading is close to the service limit, bring the component to your marine dealer so they can verify your result before purchasing the replacement. In some cases, your dealer may be willing to check a sensor against the reading on a new one to verify whether or not your sensor is actually faulty.

Although resistance checks are relatively easy to conduct, their results can be difficult to interpret. Remember that resistance in any electrical component or circuit will vary with temperature. Also, ohmmeters will vary with quality, meaning that readings can vary on the same component between different meters. The specifications provided in this service are based upon the use of a high-quality digital multi-meter applied to a component that is currently at about 68°F (20°).

■ **More information on electrical test equipment is can be found in the General Information section, while additional information on basic electrical theory and electrical troubleshooting can be found in the Ignition and Electrical System section.**

As a general rule, the resistance of a circuit or component will increase (rise) as temperature increases. Although this is true for most resistor and windings, some manufacturers use negative temperature coefficient sensors for certain functions. A negative temperature coefficient sensor reacts in an opposite manner, meaning that its resistance decreases (lowers) as temperature increases (rises) and vice-versa.

Most components of the EFI systems are checked using resistance or voltages checks. This means that a voltmeter or ohmmeter is connected to the component or circuit under the proper circumstances.

For resistance checks, no voltage must be applied to the circuit (except that as provided by the meter to perform the check). Typically, the circuit or component wiring is isolated, by disconnecting the wiring from the component itself or somewhere else in the circuit, then applying the meter probes across 2 terminals that connect through the component. When conducting resistance checks, you **must** isolate the ECM from the meter (unless specifically directed otherwise) in order to prevent the possibility of damage to the control module. By identifying the proper circuit you can conduct tests of the entire component circuit right from the harness connectors that attach to the ECM. Otherwise, use the wiring diagrams and the test procedures to determine other points to disconnect the wiring and test the circuit/component.

We recommend you work in one of two directions. Either test the component first, then work your way back through the component wiring toward the ECM, or disconnect the harness connectors from the ECM, testing the entire circuit first, then working your way toward the component. A bad test reading at the ECM harness connector must be verified by making sure it is not the result of a bad wiring harness or connection along the way to the component. Likewise, a good reading at a component (when there is a bad reading on the circuit as noted by the presence of a trouble code or by a reading at the ECM harness) must be verified by checking the circuit wiring for trouble.

When checking the circuit wiring, isolate the harness by disconnecting it from both the component and the ECM, then using an ohmmeter to check resistance from one connector to the next on each wire. Wiggle the wire and connectors while conducting the test to see if reading fluctuate. Although no specifications are provided for wiring resistance, all sensor/switch wiring for these motors should have very little resistance. Also, be sure to check wiring for shorts to ground (by checking resistance between one terminal and a good engine ground), and for shorts to power (by checking a terminal using a voltmeter and connecting the other probe to a good engine ground or the negative battery cable).

✱✱ WARNING

Be VERY careful when checking circuits for electronic engine control components. Always isolate the ECM by disconnecting the wiring from it before using an ohmmeter on the wires. Keep in mind that making the wrong connections across the terminals of an ECM can fry the unit, requiring an expensive replacement.

When conducting voltage tests, the circuit must be complete in order to receive a reading. In most cases, this means the wiring must be left connected and the meter probes should be carefully inserted through the back of the connector (called back-probing) in order to get a reading without opening and disrupting circuit operation. However, back-probing should NOT be done on waterproof connectors as you will compromise the seal and jumper wires should instead by used to complete the circuit. The best method for manufacturing a jumper wire is to obtain two identical size/type connectors as the portion of the harness to which you will be inserting the jumper wires. Wire these jumper connectors together along with a second wire connected to EACH terminal. Each second wire from a terminal should route off to a female bullet connector which can be safely probed without the fear of accidental shorting by connecting 2 or more wires together.

Throttle Body and Intake Assembly

Like the throttle bore of a carburetor, the throttle body assembly controls engine speed by mechanically controlling the amount of air allowed to enter the engine. Whereas a carburetor meters both fuel and air, the throttle body assembly on multi-port fuel injected motors is only used to meter incoming air. As a matter of fact, a typical throttle body assembly shares the same basic design and function (to control engine speed through the control of incoming air) as the air throttle components of a carburetor.

The larger motors (115 hp and up) are equipped with a throttle body assembly which uses an individual throttle bore and throttle valve for each cylinder. On smaller motors (60 hp and down), a single throttle valve at the center of a single bore throttle body is used to meter the air entering all of the intake runners. However, in all cases, the throttle body assembly is found at the front of the intake manifold, behind the air intake silencer.

As stated earlier, the larger motors (115 hp and up) use multiple throttle body assemblies. The V6 4-stroke uses 2 stacks of 3 throttle valves/bores, one stack for either engine bank. On this motor each throttle valve is mounted in an individual throttle body. The 115 and 150 hp EFI motors utilizes 2 dual throat (for a total of 4 throats) throttle body housings. On these motors, each throttle body unit contains 2 throttle valves/bores.

In all cases, the manufacturer recommends removing the throttle body and intake as an assembly (and in the case of the 115 hp motor, removing the throttle body and fuel vapor separator as an assembly). It may be possible to remove one or more throttle bodies independently of the intake (or to remove the vapor separator independently of the throttle body assembly, except on the 115 hp, 4-stroke) if the necessary steps of the appropriate procedure is followed. However, if you choose to do this, use caution and proceed slowly, making sure that you will not damage either component this way. Details on removing the vapor separator from the assembly can be found later in this section under Vapor Separator Tank and High-Pressure Fuel Pump.

On smaller motors (60 hp and down) with the Mercury EFI system, the fuel distribution manifold is normally removed along with the intake manifold and throttle body assembly.

REMOVAL & INSTALLATION

30-60 Hp Mercury/Mariner EFI Motors

◆ See Figures 106, 107 and 108

On these motors, there are either 3 or 4 individual induction tracts (one per cylinder) molded into the intake manifold assembly, however ALL are fed by a single throttle body. Each induction tract has an injector which is fed by a line from the fuel distribution manifold.

The fuel vapor separator tank is mounted to the end of the powerhead, right next to the starter motor. So unlike the larger EFI counterparts, the vapor separator tank does NOT need to be disturbed in order to service the intake manifold and throttle body assembly.

1. Properly release the pressure from the high-pressure fuel system by using the pressure test port on the top of the vapor separator tank. For more details, please refer to Fuel System Pressurization in this section.
2. Disconnect the negative battery cable for safety.
3. Remove the flywheel cover for access.
4. Tag and disengage the wiring connector from each fuel injector. In order to safely release the connector, push IN on the connector retaining clip while then gently pulling the harness connector. It is NOT necessary to remove the retaining clip itself from the connector.

■ Note the routing of the injector wiring harness as a gently sweeping curve, passing just underneath the fuel distribution manifold, before repositioning it.

5. Tag and disconnect the wiring from the Idle Air Control (IAC) valve connector.

■ Have a small shop rag handy when disconnecting the high-pressure fuel line. Even though you've already released the system pressure, fuel will almost always still be present in the line.

6. Tag and disconnect the high-pressure fuel line (fuel delivery line) from the top side of the fuel distribution manifold (at about 10 o'clock when looking at the manifold). The fuel line is easily disconnected by pushing inward on the locking tab when gently pulling back on the line.

■ To prevent the possibility of warping or damage, loosen the intake manifold flange bolts using one or two passes in the reverse of the torque sequence. Since the torque sequence itself is a clockwise spiraling pattern that starts towards the center and works outward, you should loosen the bolts in a counterclockwise pattern that starts at the top bolt and works its way inward. If you are unsure, refer to the accompanying illustrations.

7. Loosen and remove the 7 (3-cylinder) or 9 (4-cylinder) bolts securing the intake manifold assembly to the powerhead. There are either 6 (3-cylinder) or 8 (4-cylinder) bolts securing the manifold flange itself (toward the end of the powerhead, near the fuel filter assembly). Keep track of the bolt locations, as the 2nd and 3rd bolts from the bottom are of a different length than the other flange bolts (usually longer). Also, the single bolt found at the complete other end of the manifold assembly is also a different length.
8. Once all of the 7 or 9 bolts (respectively) are removed, carefully lift up on the intake manifold assembly while you disconnect the throttle link rod from the ball socket (located toward the bottom of the throttle body assembly).
9. Reposition the intake manifold as necessary for additional access.
10. Tag and disconnect the ECM harness connectors. Then remove the ECM mounting screws (3) and washers, then remove the ECM from the mounting plate.
11. Tag and disconnect the wiring from the Manifold Air Pressure (MAP), Manifold Air Temperature (MAT) sensors and the Throttle Position Sensor (TPS), all at or near the throttle body.
12. Cut the wire tie, then disconnect the crankcase breather hose from the air intake silencer (sound attenuator).
13. Double-check to make sure all wire and hose connections are free and out of the way, then carefully remove the intake manifold and throttle body assembly from the powerhead. Keep track of the 4 intake manifold O-rings/gaskets, as they MAY be reused if they are in good condition.
14. If necessary, remove the throttle body mounting screws, then separate the throttle body assembly from the intake manifold. Keep track of the throttle body mounting O-ring for installation purposes. The O-ring may be reused IF it is in good condition.
15. If further disassembly is necessary, refer to the Fuel Injectors and or the Fuel Distribution Manifold procedures in this section.

To Install:
16. If removed, install the Fuel Injectors and or the Fuel Distribution Manifold, as detailed in those procedures.

■ The throttle body O-rings MAY be reused, as long as they are not damaged, worn or cut. However, they're cheap and replacement is inexpensive insurance. If you're in doubt as to their condition, install new O-rings.

17. Lubricate the throttle body O-ring with a light coating of engine oil or marine grade grease, then install the throttle body assembly to the intake manifold. Tighten the retaining bolts to 31 inch lbs. (3.5 Nm).

■ If access to the wiring and hoses is easier with the intake manifold unbolted, you MAY wish to just loosely place the manifold in position, then come back and secure it with the bolts after the ECM and throttle body electrical connectors have been attached. However, there is usually sufficient room, so we'll advise going ahead and bolting the intake in at this point.

18. Put a light coating of Loctite® 242 or an equivalent threadlocking material to the threads of the intake manifold retaining bolts.

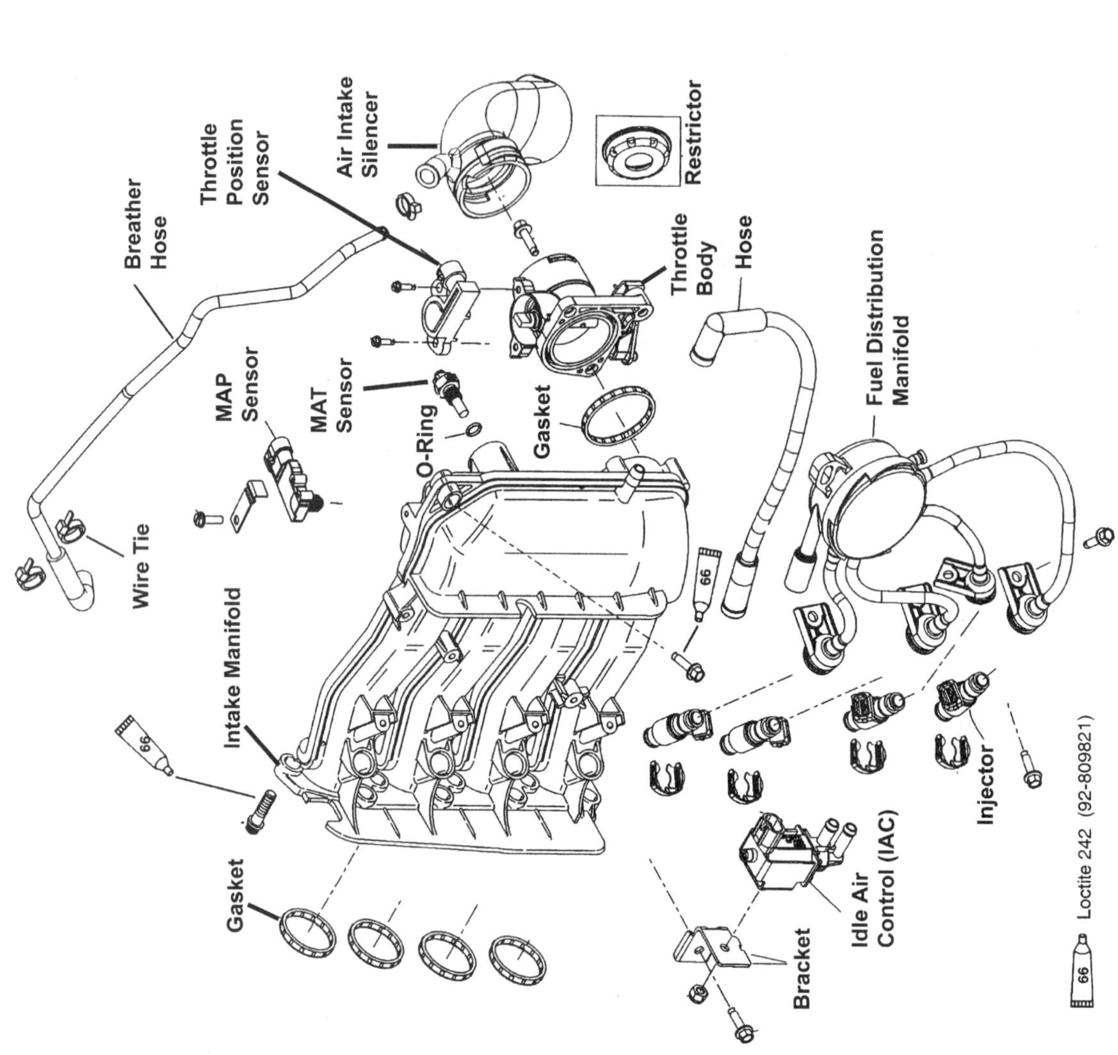

3-62 FUEL SYSTEM

Fig. 106 Exploded view of the intake manifold and throttle body assembly - 30-60 hp Mercury EFI motors (note 3-cyl models utilize one less injector, intake runner, gasket etc)

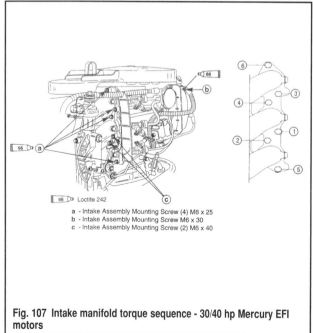

Fig. 107 Intake manifold torque sequence - 30/40 hp Mercury EFI motors

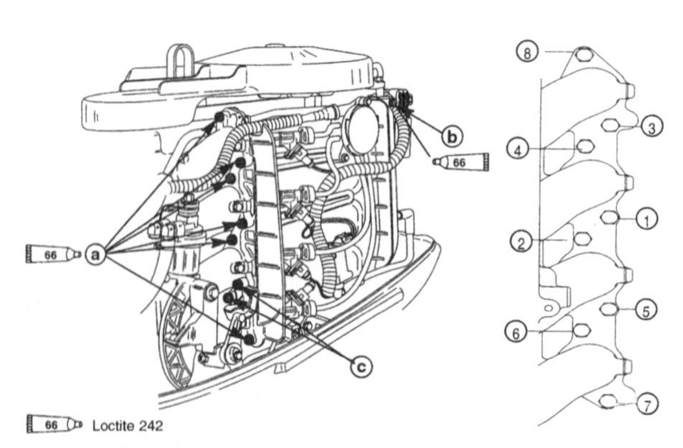

a - Intake Assembly Mounting Screw (4) M6 x 25
b - Intake Assembly Mounting Screw M6 x 30
c - Intake Assembly Mounting Screw (2) M6 x 40

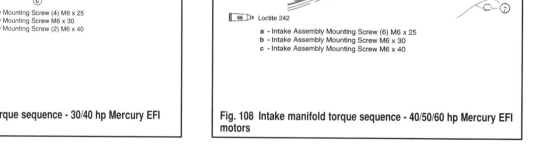

Fig. 108 Intake manifold torque sequence - 40/50/60 hp Mercury EFI motors

a - Intake Assembly Mounting Screw (6) M6 x 25
b - Intake Assembly Mounting Screw M6 x 30
c - Intake Assembly Mounting Screw M6 x 40

19. With the O-rings/gaskets in place, position the intake manifold and throttle body assembly to the powerhead.

20. By hand, slowly thread all 7 or 9 of the bolts in place, then tighten the bolts to 75 inch lbs. (8.5 Nm) using at least 2 passes of the intake manifold torque sequence (a clockwise spiraling sequence that starts towards the center of the flange bolts and works its way outward). Remember that the 2 longest bolts go in the 2nd and 3rd spots from the bottom on the flange (and that the other odd sized bolt is not in the torque sequence as it is mounted at the other end of the manifold and NOT in the flange, so tighten that bolt last).

21. Reconnect the throttle link rod to the ball socket on the throttle body. Be sure to double-check Throttle Link Rod Adjustment, as detailed under Timing and Synchronization section.

22. Reconnect the wiring for the MAP, MAT and TPS, as tagged during removal. If NOT tagged, refer to the accompanying illustrations and the Wiring Diagrams in the Ignition and Electrical System section for help finding the right connectors.

23. If removed, install the MAP sensor retaining clip and tighten the screw securely.

24. Reconnect the crankcase breather hose to the air intake silencer and secure using a wire tie.

25. Install the ECM to the mounting plate and tighten the 3 retaining screws to 54 inch lbs. (5.1 Nm). Reconnect the ECM harness wiring, as tagged during removal.

26. Reconnect the high-pressure fuel line to the fuel distribution manifold. Push on the connector until it snaps into place, then pull back gently to make sure the connector is locked.

27. Reroute the injector wiring harness as noted during removal, the reconnect each of the injector harness connectors as tagged. Make sure the retainers lock into place.

28. Reconnect the negative battery cable and properly pressurize the fuel system in order to check for leaks. Start and run the motor with the top cover removed and observe the fuel lines/fittings for any signs of weepage. Double-check all lines and fittings after the first outing with the motor.

29. Install the flywheel cover.

115 Hp EFI Motors

◆ See Figures 109 thru 113

On these models, there are 4 individual induction tracts molded into 2 throttle body assemblies mounted in a single bank. There is one tract for each cylinder. Each induction tract has an air throttle valve and an injector. The fuel vapor separator tank is mounted to the inside (powerhead side) of the intake manifold assembly, so the throttle body and intake must be removed for access. Like their 2-stroke Yamaha brethren (EFI OX66 motors), removal of the throttle body, intake and vapor separator as an assembly allows removal of the fuel injectors and high-pressure fuel lines without depressurizing the system. However, if further disassembly of the unit is required, fuel pressure must be released. This can be done before or after the assembly is removed from the powerhead, though we give it at the beginning of the procedure because if you're going to do it, there is no good reason to delay. Why work with pressurized fuel lines if you don't have to?

1. Disconnect the negative battery cable for safety.
2. Remove the flywheel cover for access.
3. Cut the plastic wire tie from the vapor separator vent hose where it connects to the back of the air intake silencer assembly. Tag and disconnect the separator vent hose and the cylinder head cover vent hose from the intake silencer.
4. Tag and disconnect the wiring from the Intake Air Temperature (IAT) sensor at the base of the air intake silencer.
5. Remove the 2 long bolts and collars threaded through the side of the air intake silencer and into the side of the powerhead.
6. Remove the 8 bolts from the silencer flange that secure the silencer to the throttle bodies.
7. Remove the air intake silencer assembly from the throttle bodies, then remove the 2 double O-rings which secure the silencer to the throttle bodies.
8. If necessary, loosen the 2 retaining bolts and remove the bracket from the side of the powerhead.
9. If the fuel injectors, rail, lines or vapor separator tank are going to be removed from the throttle body and intake assembly, properly relieve the fuel system pressure, as detailed under Relieving Fuel System Pressure earlier in this section.

10. Disconnect the throttle link rod (located at the center of the throttle body assembly) from the powerhead linkage.

11. Tag and disconnect the wiring for the Throttle Position Sensor (TPS), Idle Speed Control (ISC) valve, Atmospheric Pressure/Manifold Air Pressure Sensor (APS/MAP), fuel injector harness and the high-pressure fuel pump.

12. Tag and disconnect the hoses for the vapor separator tank. Because access is too tight at the vapor separator itself, trace the hoses from the separator to where they connect elsewhere on the powerhead. Cut the plastic wire ties and/or loosen the clamps, then carefully pull each of the hoses (fuel inlet, water hoses etc) from their retaining nipples.

13. Loosen and remove the 5 bolts retaining the intake manifold assembly to the powerhead. Keep track of the bolt locations, as the lower bolt is longer than the other 4. Carefully remove the entire throttle body, vapor separator and intake manifold assembly from the powerhead, then remove and discard the intake manifold gasket.

14. If necessary, remove the 2 bolts and the bracket from each of the throttle bodies, then remove the 6 bolts securing the throttle bodies themselves. Remove each of the throttle bodies and the 4 O-rings from the powerhead.

■ **Take care not to loose the spring on the idle speed (throttle stop) adjustment screw and the throttle joint link rod.**

15. If further disassembly is necessary, refer to the Fuel Rail and Injectors and the Vapor Separator Tank and High-Pressure Fuel Pump procedures in this section.

To Install:

16. If removed, install the Vapor Separator Tank and/or the Fuel Rail and Injectors to the Throttle Body Assembly, as detailed in those procedures.

■ **The throttle body O-rings MAY be reused, as long as they are not damaged, worn or cut. However, they're cheap and replacement is inexpensive insurance. If you're in doubt as to their condition, install new O-rings.**

17. If removed, install the throttle bodies to the intake manifold using the 4 O-rings and secure using the 6 retaining bolts. If removed, install the throttle link rod and/or the spring on the idle speed adjustment screw. Install the 2 brackets and 4 bolts to the throttle body assemblies.

18. Position the throttle body, vapor separator and intake manifold assembly to the powerhead using a new intake manifold gasket, then secure using the 5 retaining bolts.

■ **Remember, the long intake manifold bolt is installed through the collar in the lowest bolt hole.**

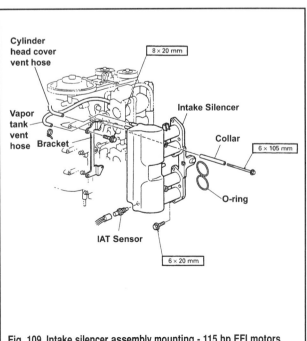

Fig. 109 Intake silencer assembly mounting - 115 hp EFI motors

3-64 FUEL SYSTEM

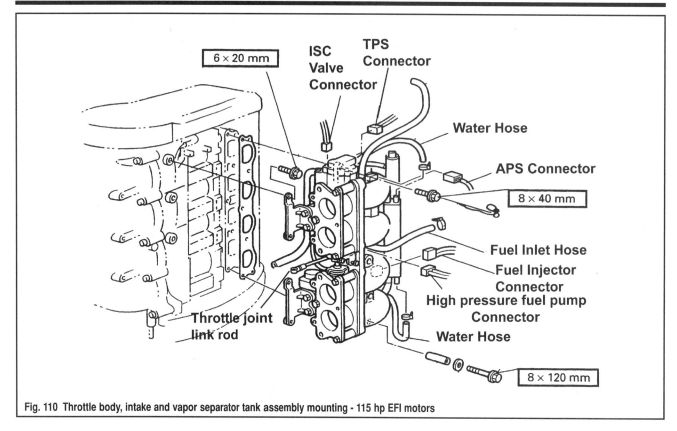

Fig. 110 Throttle body, intake and vapor separator tank assembly mounting - 115 hp EFI motors

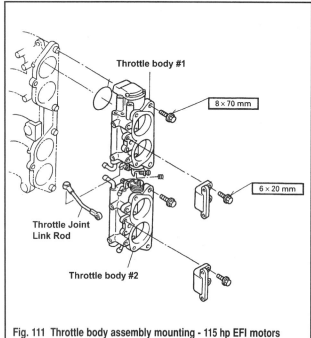

Fig. 111 Throttle body assembly mounting - 115 hp EFI motors

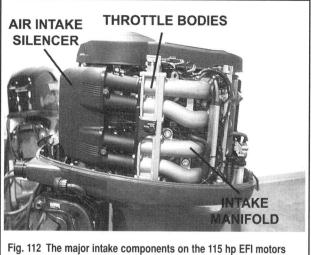

Fig. 112 The major intake components on the 115 hp EFI motors (Yamaha shown, Merc almost identical)

19. Reconnect the fuel and water hoses disconnected for removal. Secure using the clamps and/or using new plastic wire ties, as applicable.

20. Reconnect the wiring for the high-pressure fuel pump, fuel injector harness, APS/MAP, ISC valve and TPS, as tagged during removal. If NOT tagged, refer to the accompanying illustrations and the Wiring Diagrams in the Ignition and Electrical System section for help finding the right connectors.

21. Reconnect the throttle join link rod to the powerhead linkage.

22. If removed, install the air intake silencer bracket to the powerhead and secure using the 2 retaining bolts.

23. Install the air intake silencer assembly along with the 2 double O-rings. Secure the silencer using the 6 flange bolts and the 2 long bolts/collars which thread into the bracket on the side of the powerhead.

24. Reconnect the IAT sensor wiring connector.

25. Reconnect the vapor separator and cylinder head cover vent hoses to the fittings on the back of the air intake silencer. Secure the vapor separator tank hose using a new plastic wire ties.

26. Reconnect the negative battery cable and properly pressurize the fuel system in order to check for leaks. Start and run the motor with the top cover removed and observe the fuel lines/fittings for any signs of weepage. Double-check all lines and fittings after the first outing with the motor.

27. Install the flywheel cover.

FUEL SYSTEM 3-65

Fig. 113 Tag and disconnect the sensor wiring, such as from the TPS

150 Hp EFI Motors

◆ See Figure 114

On these models, there are 4 individual induction tracts molded into 2 throttle body assemblies mounted in a single bank. There is one tract for each cylinder. Each induction tract has an air throttle valve and an injector.

Unlike their slightly smaller cousins (the 115 hp motors) the fuel vapor separator tank is mounted to the powerhead (instead of the to the inside of the manifold), which means that you can remove the throttle bodies and/or intake manifold assembly independently of the vapor separator tank (which is not really true on the 115 hp motors).

On these models you may be able to remove either an individual throttle body or BOTH throttle bodies as an assembly with or without removing the intake manifold. However, if you are removing both the manifold and the throttle bodies in order to service the powerhead itself, it is easiest to leave them connected. Follow only the steps necessary to service the components necessary for your repair.

1. Disconnect the negative battery cable for safety.
2. Properly relieve the fuel system pressure at the test port on the lower portion of the fuel rail assembly. For details, please refer to the procedure found earlier in this section.
3. If necessary, remove the flywheel cover for additional access.
4. Remove the air intake silencer assembly.
5. Tag and disconnect the fuel line from the vapor separator tank to the fuel rail. There is normally a tab-released, spring-type clamp on this line, but other clamps may also be installed.
6. Tag and disconnect the wiring from the Throttle Position Sensor (TPS), on top of the top throttle body.
7. If removing the intake manifold as well, tag and disconnect the wiring from the Idle Speed Control (ISC) unit and Intake Air Pressure (IAP) sensor on top of the manifold. Also, tag and disconnect the wiring from the fuel injectors.

■ The lead for the electric fuel pump inside the vapor separator tank may be in the way, if so, tag and disconnect that lead as well.

8. Carefully unsnap the throttle joint link rod free of the ball connector at either end (don't disturb the locknut or alter the length of the link).
9. If you're removing the throttle bodies as a unit from the intake, loosen and remove the bolts threaded from the air intake silencer side of the throttle bodies, through them into the intake manifold. However, leave the bolts threaded perpendicular to them through the bracket assembly which secures them to each other. Separate each throttle body from the intake manifold, while removing and discarding the figure 8 shaped O-ring for each assembly.

■ If you're only removing one throttle body for some reason, loosen and remove the 2 screws securing the bracket on the side of the throttle body.

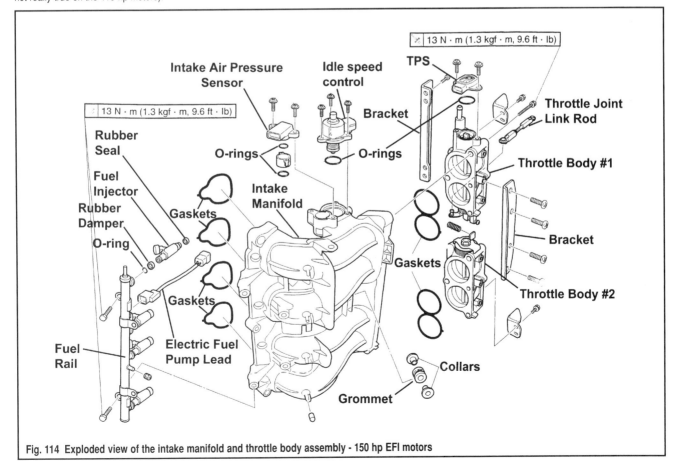

Fig. 114 Exploded view of the intake manifold and throttle body assembly - 150 hp EFI motors

3-66 FUEL SYSTEM

10. Gradually loosen and remove the bolts retaining the intake manifold assembly to the powerhead. Carefully remove the intake manifold assembly, then remove and discard the 4 intake manifold gaskets.
11. If further disassembly is necessary, refer to the Fuel Rail and Injectors in this section.

To Install:

12. If removed, install the Fuel Rail and Injectors to the Throttle Body Assembly, as detailed elsewhere in this section.
13. If removed, install the throttle bodies to the intake manifold using the 2 figure 8 shaped O-rings and secure using the 6 retaining bolts.
14. If removed, install the bracket to the side of the throttle bodies and secure using the 2 screws per throttle body.
15. Position the throttle body and intake manifold assembly to the powerhead using 4 new intake manifold gaskets, then secure using the retaining bolts.
16. Reconnect the fuel hose to the top of the injector fuel rail. Secure using the clamp.
17. Reconnect the wiring for the IAP sensor, ISC unit, injectors and/or TPS, as applicable. If removed, reconnect the electric fuel pump lead.
18. Reconnect the throttle joint link rod.
19. Install the air intake silencer assembly.
20. Reconnect the negative battery cable and properly pressurize the fuel system in order to check for leaks. Start and run the motor with the top cover removed and observe the fuel lines/fittings for any signs of weepage. Double-check all lines and fittings after the first outing with the motor.
21. If removed for additional clearance, install the flywheel cover.

■ If the TPS sensor was disturbed or if one or both of the throttle body assemblies were replaced, be sure to check the TPS adjust and throttle body synchronization.

200/225 Hp EFI Motors

◆ See Figures 115 thru 118

On these models, there are 6 individual induction tracts molded into 6 throttle body assemblies mounted in 2 banks (one per cylinder bank). There is one tract for each cylinder. Each induction tract has an air throttle valve and an injector. Unlike some other motors equipped with that Yamaha EFI systems, the distribution of the air/fuel intake components on either side of the V6 4-stroke make removal and installation of the components as a unit impractical. Instead, you will have to tag and disconnect a myriad of hoses in order to remove the throttle bodies and manifold from each bank.

Positioning of fuel and air intake components makes removal of a single throttle body impractical as well. The best way to proceed is to remove the fuel rail, throttle body and intake manifold assembly, then separate components as necessary. Most of the fuel rails and lines may also be serviced separately, without removal of the throttle body and intake manifold assemblies.

The low-pressure fuel lines are normally secured by spring-loaded clamps, which are easily removed by squeezing the tabs and sliding the metal clamps up the hoses. The high-pressure fuel lines however are secured by metallic clamps that are crimped in place. These clamps must be cut for removal (use a pair of cutters to cut the crimp, positioned perpendicular to the crimp itself) and must be replaced with suitable crimp-type clamps during installation. However, MOST of these clamps are attached to fuel joints. The high-pressure fuel joints at the top and bottom of each of the fuel rails are BOLTED into position and sealed with an O-ring. It is usually much more convenient to simply unbolt the fuel joint, leaving the crimp clamps undisturbed.

Use the accompanying illustrations to help locate the various hoses and fittings, as well as determine which ones must be removed depending upon the amount of disassembly you desire.

1. Disconnect the negative battery cable for safety.
2. Properly relieve the fuel system pressure, as detailed under Relieving Fuel System Pressure earlier in this section.
3. Remove the flywheel cover for access.
4. Tag and disconnect the vent hose from the port and/or starboard air intake silencer.
5. If removing the starboard remove the bolts and disconnect the battery cable.
6. If removing the port air intake silencer, tag and disconnect the wiring from the Intake Air/Manifold Air Temperature (IAT/MAT) sensor at the base of the air intake silencer.

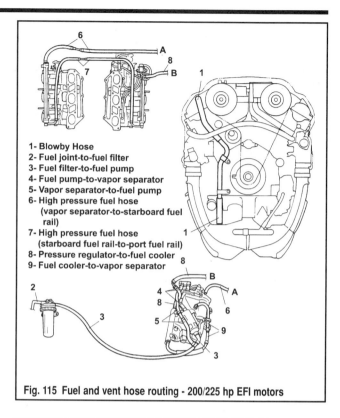

1- Blowby Hose
2- Fuel joint-to-fuel filter
3- Fuel filter-to-fuel pump
4- Fuel pump-to-vapor separator
5- Vapor separator-to-fuel pump
6- High pressure fuel hose (vapor separator-to-starboard fuel rail)
7- High pressure fuel hose (starboard fuel rail-to-port fuel rail)
8- Pressure regulator-to-fuel cooler
9- Fuel cooler-to-vapor separator

Fig. 115 Fuel and vent hose routing - 200/225 hp EFI motors

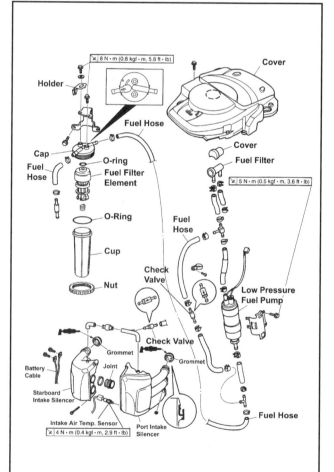

Fig. 116 Intake silencer, fuel filter and low-pressure fuel pump mounting - 200/225 hp EFI motors

FUEL SYSTEM 3-67

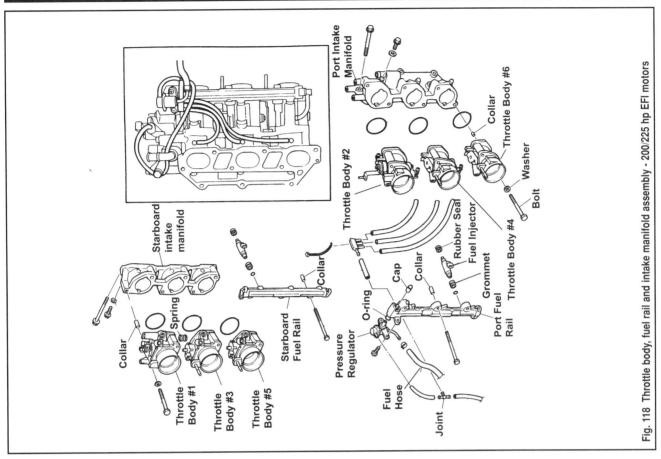

Fig. 118 Throttle body, fuel rail and intake manifold assembly - 200/225 hp EFI motors

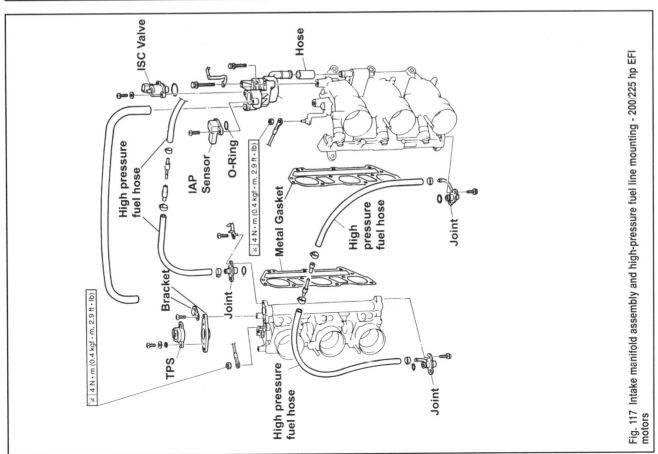

Fig. 117 Intake manifold assembly and high-pressure fuel line mounting - 200/225 hp EFI motors

3-68 FUEL SYSTEM

7. Remove the bolts and nuts securing the port and/or starboard air intake silencer assembly, then carefully remove the silencer(s) while disconnecting the joints and grommets.

8. If necessary, tag and disconnect any of the low-pressure fuel lines which are in the way.

9. If removing the port assembly, tag and disconnect the wiring for the Intake Air/Manifold Air Pressure (IAP/MAP) sensor and the Idle Speed Control (ISC) valve.

10. If removing the starboard assembly, tag and disconnect the wiring for the Throttle Position Sensor (TPS).

11. Remove the nut and disconnect the throttle link from the top throttle body.

12. Tag and disconnect the high-pressure fuel hoses at each end of the fuel rail on the bank that is being removed. The easiest way to disconnect the hoses is to remove the 2 bolts securing each fuel joint to the rail, then remove the joint and hose as an assembly (that way you don't have to mess with the crimped clamps). Remove and discard the O-ring from each joint.

13. Remove the bolts securing the throttle body, fuel rail, intake manifold assemblies to the powerhead. Remove each assembly, then remove and discard the metal gasket.

14. If further disassembly is necessary, remove the bolts and collars securing the fuel rails to the assembly, then carefully pull the rails off the injectors, or the rails and injectors off the assembly. It is probably easier to leave the injectors in place, pulling the rails free, but without an extra set of hands the injectors usually make your mind up for you.

■ The rubber seal which connects the injector to the throttle body and intake manifold assembly may be reused, however the same is not true for the grommet and O-ring which seals the other end of each injector to the fuel rail. For this reason, if you can remove the fuel rail with the injectors attached, you'll save yourself having to replace the grommets and O-rings. Otherwise, if the injector pulls free of the rail, you'll need to replace both the grommet and O-ring for that injector.

15. If necessary unbolt and remove each of the throttle bodies from the intake manifolds, then remove and discard the seal for each.

To Install:

16. If removed, install each of the throttle bodies to the intake manifold using a new seal for each. Secure each throttle body using the retaining bolts.

17. If removed, install the fuel injectors and fuel rails to the throttle body and intake manifold assembly using the rubber seals. Remember the rubber grommets and O-rings on the rail side of the injector must be replaced if removed, but don't ignore the rubber seals on the other end either, replace them if they show any signs of wear or damage.

18. Reconnect the high-pressure fuel lines and joints to the fuel rails using a new O-ring for each joint. Tighten the joint bolts securely.

19. Reconnect the wiring for the IAP/MAP and ISC (on the port side) and/or the TPS wiring (on the starboard side), as necessary.

20. Reconnect the throttle links to the top of each throttle body stack and secure each using the retaining nut.

21. Reconnect and low-pressure fuel lines which were disconnected or repositioned for access.

22. Apply a light coating of a suitable marine grade grease to the air intake silencer grommets, then install the port and/or starboard silencers, making sure the grommets and joints seat properly. Secure the silencer(s) using the bolts and nuts.

23. Reconnect the vent hose to the port and/or starboard air intake silencer housing(s).

24. Reconnect the negative battery cable and properly pressurize the fuel system in order to check for leaks. Start and run the motor with the top cover removed and observe the fuel lines/fittings for any signs of weepage. Double-check all lines and fittings after the first outing with the motor.

25. Install the flywheel cover.

Electric Low-Pressure Fuel Pump (V6 only)

◆ See Figure 119

Whereas most Yamaha and Mercury/Mariner motors use a mechanical, diaphragm-displacement low-pressure fuel pump, the 4-stroke V6 motor utilizes an electric pump to serve the same purpose. In all cases, the low-pressure fuel pump draws a fuel supply from the tank and feeds a float bowl. The differences between EFI and carbureted motors comes in what happens in the float bowl, since it is mounted in a vapor separator tank with a high-

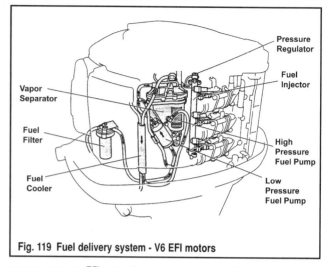

Fig. 119 Fuel delivery system - V6 EFI motors

pressure pump on EFI motors instead of under the throttle bore of a carburetor. Also, the electric pump used on V6 motors operates in a slightly different fashion than the mechanical pumps.

On most motors the mechanical lift pump will supply fuel to the float bowl as long as the engine is operating (meaning the 4-stroke engine's pump is being actuated by the valve train). However, the needle inlet valve in the float bowl can shut fuel flow off against this low-pressure whenever fuel rises to a level determined by float adjustment.

The electric low-pressure pump incorporates one change to that type of fuel delivery. When the engine is operated at speeds above 1200 rpm or when the engine is first started it will operate constantly, providing a steady stream of fuel to the float bowl in the vapor separator tank. However, when the engine is operated at low speeds (below 1200 rpm) the fuel pump will cycle **ON** for 10 seconds, then **OFF** for 20 seconds to prevent overfilling the float bowl (or putting undue stress on the float needle valve).

TESTING

◆ See Figure 120

The problem most often seen with most fuel pumps is fuel starvation, hesitation or missing due to inadequate fuel pressure/delivery. In extreme cases, this might lead to a no start condition as all but total failure of the pump prevents fuel from reaching and filling the vapor separator tank). More likely, pump mechanical failures are not total, and the motor will start and run fine at idle, only to miss, hesitate or stall at speed when pump performance falls short of the greater demand for fuel at high rpm. Of course, because this motor utilizes electric fuel pumps, electrical failures can be sudden and complete. So the first and most important test is to listen/feel for fuel pump operation.

Locate the brown starter relay lead (at the junction block immediately next to the low-pressure fuel filter/strainer assembly and disconnect it to make sure the engine will not start. Then turn the engine start switch **ON** and listen for the fuel pumps to run. The low-pressure pump is mounted to the powerhead right next to the vapor separator tank and you should be able to hear them both run when the keyswitch is first turned **ON**. The low-pressure pump will run for 10 seconds and the high-pressure pump will run for about 5 seconds. This is done to prime the fuel system immediately before start-up.

■ You can use a short length of hose, or a solid object such as a screwdriver held against the pump to help isolate the sound (remember, sound is transmitted best through a solid). You can also touch the pump or vapor tank housing to feel for a very slight vibration.

If the one or both of the pumps do not run, check for power at the pump terminals when the keyswitch is activated. If necessary, troubleshoot the pump circuit.

Before replacing a suspect fuel pump, be absolutely certain the problem is the pump and NOT with fuel tank, lines or filter. A plugged tank vent could create vacuum in the tank that will overpower the pump's ability to create vacuum and draw fuel through the lines. An obstructed line or fuel filter could

FUEL SYSTEM 3-69

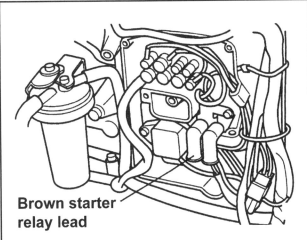

Fig. 120 Disconnect the brown starter relay lead to keep the engine from starting when checking the fuel pump circuits - V6 EFI motors

also keep fuel from reaching the pump. Any of these conditions could partially restrict fuel flow, allowing the pump to deliver fuel, but at a lower pressure/rate. Before replacing the fuel pump, refer to the testing procedures found under Fuel Lines and Fitting to ensure there are no problems with the tank, lines or filter.

If inadequate fuel delivery is suspected and no problems are found with the tank, lines or filters, a conduct a quick-check to see how the pump affects performance. Use the primer bulb to supplement fuel pump. This is done by operating the motor under load and otherwise under normal operating conditions to recreate the problem. Once the motor begins to hesitate, stumble or stall, pump the primer bulb quickly and repeatedly while listening for motor response. Pumping the bulb by hand like this will force fuel through the lines to the vapor separator tank, regardless of the fuel pump's ability to draw and deliver fuel. If the engine performance problem goes away while pumping the bulb, and returns when you stop, there is a good chance you've isolated the low-pressure fuel pump as the culprit.

✳✳ WARNING

Never run a motor without cooling water. Use a test tank, a flush/test device or launch the craft. Also, never run a motor at speed without load, so for tests running over idle speed, make sure the motor is either in a test tank with a test wheel or on a launched craft with the normal propeller installed.

REMOVAL & INSTALLATION

◆ See Figure 116

1. Properly relieve the fuel system pressure and disconnect the negative battery cable for safety.
2. Locate the low-pressure fuel pump mounted to the side of the powerhead, at the vapor separator tank assembly. Disconnect the pump connector from the wiring harness.
3. Loosen the spring-type fuel hose clamps, then carefully pull the fuel inlet and outlet lines from the pump nipples. Keep a rag or small suitable container handy to catch any fuel in the lines that will escape as the lines pull free.
4. Loosen the 2 bolts securing the pump clamp, then remove the clamp and pump from the powerhead.

   To Install:

5. Position the pump to the powerhead with the wiring harness on top, then secure it using the clamp and 2 retaining bolts.
6. Reconnect the fuel lines to the fuel pump nipples and secure using the clamps. Gently squeeze the clamp ears and slide the clamp over the hose until it is past the raised portion of the pump nipple, but still shy of the hose end.
7. Reconnect the pump wiring.
8. Reconnect the negative battery cable and properly pressurize the fuel system in order to check for leaks. Start and run the motor with the top cover removed and observe the fuel lines/fittings for any signs of weepage. Double-check all lines and fittings after the first outing with the motor.

Vapor Separator Tank (VST) and High-Pressure Fuel Pump

The first major difference between most EFI fuel systems and a carbureted system (as far as fuel delivery is concerned) comes at the fuel Vapor Separator Tank (VST). The vapor separator is mounted on the powerhead, often on or near the throttle body assembly (though this doesn't hold true for the 60 hp and smaller motors equipped with the Mercury EFI system.)

The separator tank functions as a bizarre cross between a very large float bowl and a very tiny gas tank. It receives fuel from the low-pressure pump via a float and needle valve assembly (in the same manner as a carburetor's float bowl). The level is maintained within the vapor separator tank so that it serves as a reservoir for the high-pressure electric fuel pump mounted to the separator cover.

■ **The float and needle valve are serviced in the same manner as a carburetor's float bowl. They can be accessed once the vapor separator cover/fuel pump assembly is removed.**

The high-pressure fuel system works by keeping the injectors constantly supplied with more fuel than the maximum demand requires. A pressure regulator controls this over supply by diverting excess high pressure fuel back to the separator.

■ **On all models, pressurizing the fuel in the injector fuel rail or fuel distribution manifold retards gum formation. Often, EFI engines will crank after a year of storage. But that is not to encourage improper storage. Remember that the fuel injectors contain tiny passages and are very sensitive to any gum formation, so don't risk it, always store the outboard properly for periods of non-use.**

To help prevent fuel problems like vapor lock or hot soak, the separator is normally vented to the air intake silencer so vapors can be drawn into the throttle body and burned when the engine is running.

The tank cover contains a replaceable high-pressure fuel pump assembly. The pump itself also contains a replaceable filter screen mounted at the fuel pickup. Some of the larger EFI motors (specifically the 115 hp and V6 models) utilize a remote, fuel rail mounted pressure regulator assembly.

A fuel reservoir drain screw is normally found on the bottom of the tank and can be used to drain the tank of fuel.

TESTING

Refer to the Maintenance and Tune-Up section for High-Pressure Fuel Pump Inspection and pressure checking procedures.

✳✳ CAUTION

The high pressure fuel system, as it name might imply, is capable of spraying fuel under extreme pressure. This means that the fuel will spray free under high pressure if a fitting is opened without first relieving pressure (which makes for good fuel atomization and a highly combustible condition). It will also spray fuel if the pump is actuated for any reason while a fitting is disconnected. These could lead to extremely dangerous work conditions. Do not allow ANY source of ignition (sparks, flames, etc) anywhere near the work area when servicing the fuel system.

REMOVAL & INSTALLATION

◆ See Figures 121 thru 125

On some Yamaha EFI system equipped models (specifically the 115 hp motors from both Yamaha and Merc/Mariner) the vapor separator tank assembly can be removed along with the throttle body and intake assembly. This allows you to remove the entire fuel injection unit (tank, throttle body, intake, fuel rails and injectors) without depressurizing the system.

However, on most models (such as the 150 hp and V6 models and all

3-70 FUEL SYSTEM

Mercury/Mariner EFI system models such as the 60 hp and smaller motors) the fact that the injectors and throttle bodies are installed separately from the VST (in 2 separate banks on V motors) makes this impractical.

The decision whether to remove the VST and intake as an assembly or separately on 115 hp motors should come down to what is the reason for removal. If you need to service or replace the vapor separator, fuel lines or related components, then the big advantage of removing them as an assembly may be lost. However, access to the vapor separator tank and related components may be much easier with the entire unit removed from the powerhead, so it may save you some time/trouble in the long run.

Also, keep in mind that when servicing a 115 hp motor the vapor tank assembly is mounted to the back (powerhead) side of the intake manifold and simply CANNOT be serviced without first removing the throttle body and intake manifold assembly. So, it's up to you, but if you decide to remove the vapor tank along with the Throttle Body and Intake assembly, refer to that procedure first and then to the appropriate steps of this procedure to separate the vapor tank from the fuel injection unit.

1. Properly relieve the fuel system pressure, then disconnect the negative battery cable for safety.
2. For 60 hp and smaller motors, remove the lower engine cowl for better access.
3. If not done already (such as to remove the throttle body and intake assembly), tag and disconnect the wiring from the vapor tank assembly. All models have at least an electric fuel pump wiring harness running to the vapor tank cover.
4. Tag and disconnect the fuel lines from the top of the vapor tank cover. The number of lines, locations, clamping methods all vary from model-to-model. Keep in mind that fittings with crimp style clamps are removed by cutting the crimp itself, perpendicular to the crimp. Also keep in mind that clamps should be replaced using the same style as what was originally installed (meaning you'll need a special crimping tool for crimp style clamps).

■ On certain EFI motors it is possible to remove the vapor separator tank and fuel rail/injectors as an assembly. Either without depressurizing the system and/or without cutting the crimp style clamps usually found on the fuel rail lines. Of course, you'll still need to depressurize the system if you are going to open us the vapor separator for service. For more details about Fuel Rail and Injector service, refer to the procedures later in this section.

5. Loosen and remove the bolts (usually 3) securing the vapor separator tank to the intake manifold (115 hp motors) or to the powerhead/mounting bracket (all other models).

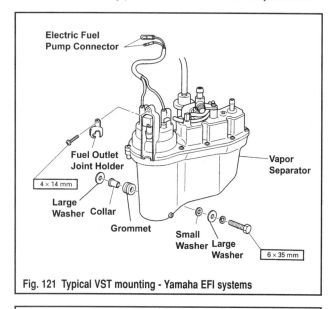

Fig. 121 Typical VST mounting - Yamaha EFI systems

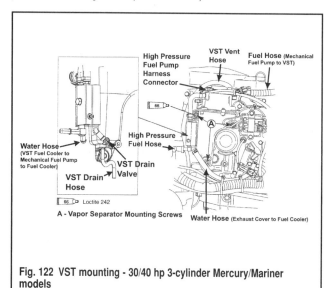

Fig. 122 VST mounting - 30/40 hp 3-cylinder Mercury/Mariner models

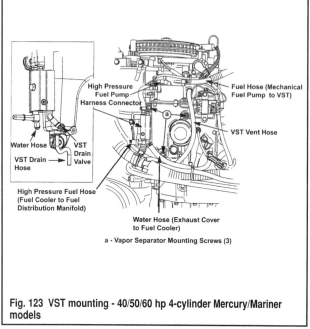

Fig. 123 VST mounting - 40/50/60 hp 4-cylinder Mercury/Mariner models

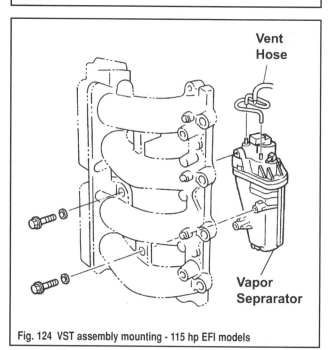

Fig. 124 VST assembly mounting - 115 hp EFI models

FUEL SYSTEM 3-71

■ On V6 models the vapor tank retaining bolts are threaded through the back of a bracket into the rear of the tank assembly. To access these bolts you must usually first remove the 4 bolts securing the bracket and relocate the tank, fuel cooler and bracket assembly.

6. If the high-pressure fuel pump, pump pickup filter screen or float and needle valve assembly require service, please refer to the Overhaul procedure in this section.
7. If the fuel pressure regulator requires service, refer to Fuel Pressure Regulator, in this section.

To Install:
8. If separated for overhaul, install the cover to the vapor separator tank assembly using a new O-ring, as detailed under Overhaul
9. For 30/40 hp motors Mercury recommends coating the threads of the VST retaining bolts with Loctite® 242 or an equivalent thread-locking compound. Frankly, we can't see how this would be a bad idea for the mounting bolts on any of the motors.
10. Install the vapor tank to the throttle body assembly, intake manifold, powerhead or bracket, as applicable and secure using the fasteners. Mercury/Mariner is the only one who provides a torque spec for the bolts of 45 inch lbs. (5 Nm) on 60 hp and smaller motors and 70 inch lbs. (8 Nm) for 115 hp and larger motors.
11. Reconnect the fuel lines to the vapor tank cover as tagged during removal. If you run into a snag (a tag fell off or something wasn't marked) refer to the art accompanying the Throttle Body and Intake procedures to see if the hose is labeled in the artwork. Hose locations vary slightly from model-to-model. In most cases the hoses are JUST the right length to reach their fittings, so if two hoses appear kinked or bowed with extra length (or not enough length) suspect that you might have them switched. Secure the hoses using new crimp type clamps or wire ties, when applicable or using the spring-type clamps, when used.

✳✳ CAUTION
Don't install a different style fuel hose clamp than was originally used in that location, or you're risking a leak and a severe fire/explosion hazard.

12. Unless you've removed the entire Throttle Body and Intake assembly, reconnect the harness wiring to the vapor tank components such as the high-pressure fuel pump.
13. If removed, install the Throttle Body and Intake assembly, as detailed earlier in this section.
14. For 60 hp and smaller motors, install the lower engine cowling.
15. Reconnect the negative battery cable and properly pressurize the fuel system in order to check for leaks. Start and run the motor with the top cover removed and observe the fuel lines/fittings for any signs of weepage. Double-check all lines and fittings after the first outing with the motor.

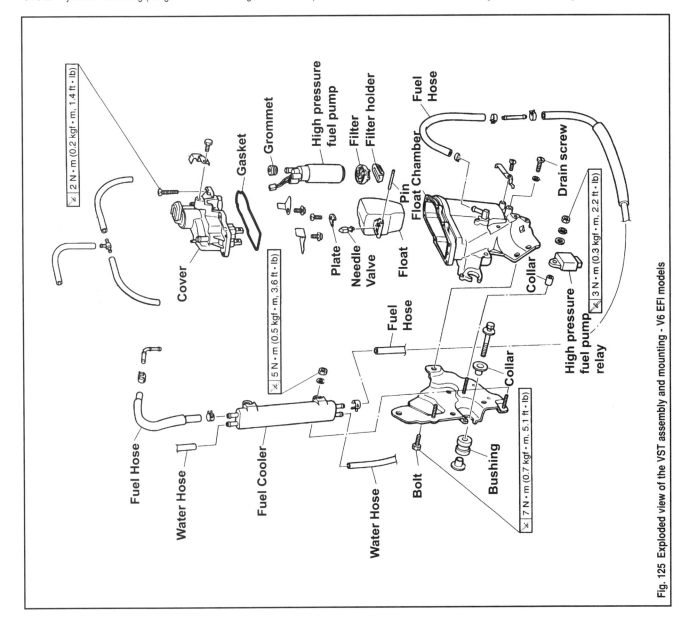

Fig. 125 Exploded view of the VST assembly and mounting - V6 EFI models

3-72 FUEL SYSTEM

OVERHAUL

◆ See Figure 125, 126 and 127

1. Remove the vapor separator assembly from the powerhead. For details, please refer to the procedure earlier in this section.
2. If not done already, drain the VST of any remaining fuel as follows:
   a. For 60 hp and smaller motors there is a drain valve mounted horizontally in a boss at the bottom of the assembly. Loosen the screw and fuel will flow from the drain hose on the bottom of the tank.
   b. For most 115 hp and larger motors, remove the drain screw and gasket or O-ring from the bottom, side of the vapor tank, then drain the fuel into a suitable container. We say most because in some sources Yamaha indicates that there may be a drain hose or fitting for a drain hose on the 150 hp models, so check for that first and don't completely remove the screw, first back it out a couple of turns to check for fuel flow from a drain hose/fitting.
3. Remove the 5 (60 hp and smaller) or 7 (115 hp and larger) screws from the vapor separator cover flange, then carefully lift the cover separating it from the float bowl. Remove and discard the old O-ring seal from the cover-to-bowl mating surface.

■ Although the manufacturer does not always specify that the bowl O-ring cannot be reused (Yamaha does on some models, like the 150 hp motor), we recommend replacing it whenever the assembly is overhauled for safety (to prevent possible leaks that might occur is a worn or deformed seal is reused).

4. Remove the filter holder and then the filter from the bottom of the electric fuel pump.

■ Check the filter element for contamination or damage. Clean or replace the filter, as necessary.

5. If necessary, remove the fuel pump and/or float assembly, as follows:
   a. To remove the float valve assembly, carefully slide the hinge pin free of the hinge and cover, then lower the float and needle valve from the cover.
   b. On 115 hp motors, remove the screw and retainer that hold the needle valve seat collar in position. Remove the collar and O-ring.
   c. To remove the fuel pump, carefully pull it down and out of the vapor separator cover. Remove the grommet and disengage the wiring connector.
   d. On 60 hp and smaller motors, there is a pump seat and gasket mounted in the tube on the underside of the VST cover.
6. Carefully clean all metallic parts in carburetor cleaner or a suitable solvent. DO NOT immerse rubber or electrical parts in solvent. Replace any worn, damaged or questionable components. Although Yamaha does not make specific recommendations regarding O-ring replacement, we recommend that ALL O-rings be replaced to ensure proper and reliable operation.

To Assemble:
7. If removed, install the fuel pump and/or float assembly.
   a. Engage the fuel pump wiring connector, then insert the fuel pump and grommet into the vapor cover.
   b. On 60 hp and smaller motors, if removed, install the pump seat and gasket into the tube on the underside of the VST cover.
   c. If applicable, install the needle valve seat collar and O-ring, then secure using the retainer and screw.
   d. Install the float and needle valve assembly, then secure using the hinge pin.
8. Install the filter and holder to the bottom of the electric fuel pump.
9. Install the cover to the float bowl making sure the cover O-ring seal is not dislodged or pinched during installation. Install the 5 (60 hp and smaller) or 7 (115 hp and larger) cover screws and tighten them securely. Mercury/Mariner gives a torque specification on 60 hp and smaller motors of 32 inch lbs. (3.5 Nm).
10. Either tighten the drain valve and/or install the drain screw using a new gasket or O-ring (as applicable).
11. Install the fuel vapor separator assembly, as detailed in this section.

Fuel Injectors

◆ See Figure 128

A fuel injector is a small, solenoid valve that is designed to open against spring pressure when power is applied to the circuit. An internal spring snaps the valve closed the instant that power is removed from the circuit.

Fuel injectors are supplied with a constant supply of high pressure fuel. Because the pressure is held constant, the amount of fuel that sprays through the injector is a function of time (the less each injector is actuated, the less fuel is delivered, the more each injector is actuated, the more fuel is delivered).

Each cylinder is equipped with an individual electronic fuel injector to deliver metered amounts of fuel, matching engine operating conditions. The exact fuel metering made possible by the fuel injection system is responsible for the EFI engine's ability to maximize both engine performance and fuel economy.

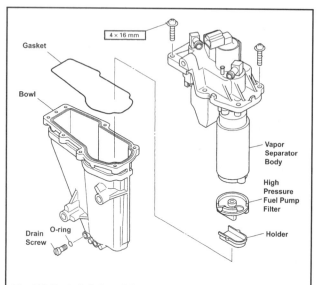

Fig. 126 Exploded view of the vapor separator assembly used on 115 hp and larger motors (60 hp and smaller motors similar)

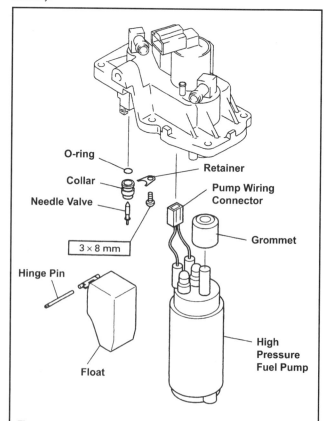

Fig. 127 Exploded view of the high-pressure pump and vapor tank float assembly - 115 hp models shown (60 hp and smaller motors similar)

FUEL SYSTEM 3-73

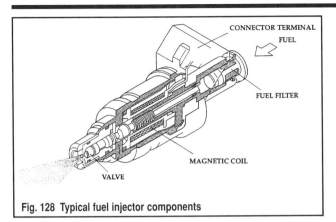

Fig. 128 Typical fuel injector components

The most important fuel injection system maintenance is a combination of periodic filter changes and the use of fuel stabilizer if the motor is stored for any amount of time (more than a few weeks). This is true because the passages inside a fuel injector are very small, and are easily clogged by dirt or debris in the fuel system.

On these smaller of these motors (60 hp and down) the fuel injectors are secured to the intake manifold by a retaining clip and a fuel injector cap (which is part of the fuel line from the fuel distribution manifold) that is bolted in place. A special tool is available from Mercury (#91-883877A1) which can be used to pry free the injector caps from the injectors.

On the larger of these motors (115 hp and up) the fuel injectors are secured between the fuel rail and a throttle body/intake manifold assembly.

In all cases, they deliver fuel to the cylinder head intake valves.

TESTING

Injector Operational Test

The fastest way to check for inoperable injectors is to listen or feel for solenoid operation. If available, use a mechanic's stethoscope, but because solid matter transmits sound and vibration, a long screwdriver can also be used to amplify and/or feel each injector.

1. Provide the motor with a source of cooling water.
2. Start and run the engine at idle.
3. Position the stethoscope or screwdriver against the body of each fuel injector.

■ **If using a screwdriver to amplify the sounds of the injector, place your ear near the handle or hold the driver lightly while feeling for the light tapping of the solenoid valve.**

4. If an injector is operating properly you will hear or feel a slight clicking from it. This tells you that the valve is opening and closing.
5. If there is a noticeably different noise or no clicking is felt at all from an individual injector, perform the Injector Resistance Test and check the wiring between the injector and ECM. If resistance is within specification, check the harness for opens or shorts. Before replacing an ECM, substitute a known good injector.

■ **For test purposes, fuel injectors can be switched from cylinder-to-cylinder to see if the problem follows the injector or remains behind. If the problem remains behind, look to the harness and signal for trouble. If the problem follows the injector, the problem IS the injector.**

Injector Resistance Test

◆ See Figure 129

Another quick-check of a fuel injector is made using an ohmmeter to measure the resistance of the winding inside the injector itself. It is important to remember that a correct reading does not mean the injector is operating. Mechanical damage or clogs within the injector could prevent it from opening or closing properly which would lead to engine performance problems.

■ **The injector does not need to be removed from the engine to check its resistance.**

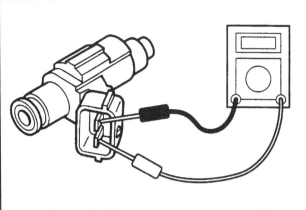

Fig. 129 Checking fuel injector resistance - Injector from 150 hp and V6 models shown, others similar

6. In order to protect the test equipment, disconnect the negative battery cable.
7. Disengage the wire harness connector from the injector.
8. Set the DVOM to the resistance scale, then apply the meter probes across the 2 terminals on the top of the injector as interpret the results as follows:

   a. For 115 hp and larger models (all of which are equipped with the Yamaha EFI system), unfortunately Yamaha isn't consistent with the specifications they provide from motor-to-motor. The only specs we could find were for the injectors on these models was for the 150 hp and V6 models, which should be 14-15 ohms at an ambient temperature of about 68 degrees F (20 degrees C). For other motors with the Yamaha EFI system, take readings on all the injectors and compare them to one another to decide if one seems greatly out of line with the rest. Replace the injector if resistance seems WAY out of line.

   b. For 60 hp and smaller models (all of which are equipped with the Mercury/Mariner EFI system) the injectors should be 10-13.5 ohms at an ambient temperature of about 68 degrees F (20 degrees C).

9. If the injector is inoperable, but resistance is within specification, check the circuit and signal to help determine if it is a circuit fault or a mechanical fault within the injector itself.
10. When finished, reconnect the wiring harness to the injector.

REMOVAL & INSTALLATION

60 Hp and Smaller Motors (Mercury/Mariner EFI Systems)

◆ See Figure 130

The smaller motors are equipped with a Mercury EFI system. On models so equipped the fuel injectors are secured to the intake manifold by a retaining clip and a fuel injector cap (which is part of the fuel line from the fuel distribution manifold) that is bolted in place. A special tool is available from Mercury (#91-883877A1) which can be used to pry the injector caps free from the injectors.

1. Properly release the fuel system pressure, as detailed earlier in this section.
2. Disconnect the negative battery cable for safety.
3. If more than one fuel injector is being removed tag the wiring and the fuel lines from the distribution manifold to ensure proper installation.
4. Carefully remove the injector retaining clip(s).
5. Remove the injector cap retaining bolt(s).
6. Disengage the injector wiring connector(s). To do so, push inward on the connector retaining clip and pull back on the connector housing.
7. Using the special service tool or a small pickle fork shaped prytool, CAREFULLY pry the injector cap(s) free from the injector(s).

■ **Have a rag handy because fuel will still be present in the manifold and fuel lines even after the pressure has been released.**

8. Grasp the injector and pull it straight back, out of the manifold.

3-74 FUEL SYSTEM

To Install:

9. Install each injector cap by pushing it straight onto the back of the injector until it bottoms.

10. Lubricate the injector O-rings using a light coating of clean engine oil, then install each injector which was removed by pushing it straight into the intake manifold.

11. Secure each injector and cap assembly using the cap retaining screw. Tighten the retaining screw to 31 inch lbs. (3.5 Nm).

12. Install the injector cap retaining clips so that the locking teeth align with the locking rub on the fuel injector. For details, please refer to the accompanying illustration.

13. Reconnect each injector harness connector, pushing the connector housing onto the injector until the retaining clip locks.

14. Reconnect the negative battery cable, then properly pressurize the fuel system and check for leaks.

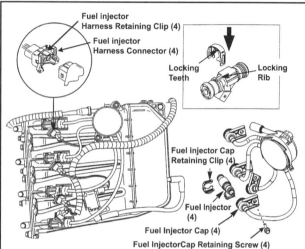

Fig. 130 Fuel injector mounting for Mercury/Mariner EFI systems (4-cyl shown, 3-cyl identical, with one less injector of course)

115 hp and Larger Motors (Yamaha EFI Systems)

◆ See Figures 114, 117, 118 and 131 thru 135

OK, you've got some decisions to make. For some larger EFI motors (specifically the 115 hp models), Yamaha recommends removing the entire Throttle Body and Intake assembly, along with the fuel rail, injectors and vapor separator tank, THEN disassembling the unit. In some cases, this makes more sense than others (as access to fuel line fittings or certain bolts can be difficult, but NOT impossible with the assembly installed).

The truth is that removing the whole assembly will require that you replace a couple more gaskets or seals and give you the opportunity to visually inspect more components. However, it is not absolutely necessary on most models.

If you decide to remove the Throttle Body and Intake assembly, please refer to the appropriate procedure found earlier in this section for details, then follow the appropriate steps of this procedure to disassemble the fuel rail and injectors. In all cases you're going to have to depressurize the fuel system since you'll be removing injectors from the fuel rail assembly.

■ One factor which may influence your decision on how to proceed is how the fuel lines are secured to the rail(s). In some cases there are crimp-type clamps on both ends of the fuel lines, so unless there is sufficient free-play to move the fuel rail enough to free the injector without disconnecting the lines, you HAVE to either disconnect the lines OR remove/reposition the vapor separator tank. Luckily, most of the fittings on these motors are attached to fuel joints that unbolt from the fuel rail. So you've got a couple of options, including leaving the vapor tank in place and removing fuel lines via the fittings or by unbolting the joints. The latter option requires the replacement of an O-ring instead of (when used) a crimp-style clamp). Obviously any fitting with a reusable spring-type clamp is easily disconnected and reconnected using the same clamp. And, lastly, don't read us wrong, the crimp type clamps are easy to remove and not that hard to install, however you will need a special crimping tool to perform the job, so we're just trying to give you other options.

1. Depressurize the fuel system and disconnect the negative battery cable for safety.

2. If necessary for access, remove the flywheel cover.

3. For all except the 115 and 150 hp motors remove the air intake silencer for access.

■ Take a good look at the fuel lines and mounting points for the fuel rail. Decide how you'd like to proceed. You can remove the entire Throttle Body and Intake for additional access. On all motors but the 115 hp model you can just remove and/or reposition the Vapor Separator Tank. OR, you can unbolt the fuel rail with the lines still attached and see if there is sufficient slack to reposition it enough to remove the fuel injectors (it has to move outward from the powerhead a little more than 1 in. (25mm), give or take.

4. There is usually at least one plastic wire tie holding wiring or fuel/vapor lines to the fuel rail assembly. Note the positioning of the wire tie(s) and components secured by the tie(s), then cut the tie and reposition the wires or lines out of the way.

5. Tag and disconnect the wiring from the fuel injector connectors. This usually involves depressing the lock-tab while gently pulling on the connector.

6. Remove the fuel rail retaining bolts (usually 3 bolts on these motors, but not all since the rail on the 150 hp motor is held in position by 2 bolts. Also keep in mind that the fuel cooler on 115 hp motor will also have 2 bolts that must also be removed. These same fuel cooler bolts may or may not be threaded through tabs on the 150 hp fuel rails, depending on the model).

7. If there is sufficient free-play, gently pull back on the fuel rail assembly to free the injectors from the powerhead, then reposition the rail for access to the injectors.

■ If one or more of the fuel lines are binding, holding the fuel rail in place or keeping you from repositioning it for access, you'll have to disconnect that (or those) line(s).

8. If necessary to reposition or completely remove the fuel rail disconnect one or more of the fuel lines from it. You can undo the clamps, or you can unbolt the joint from the rail. Unfortunately on the 115 hp model, the top joint is the pressure regular and the bolt faces inward toward the powerhead while the bolts on the bottom joint face downward, if you can reposition the rail slightly there should be access for a slim line wrench or a small 1/4 driver set.

9. Place a matchmark on the fuel rail immediately adjacent to each injector electrical connector to ensure the injectors are installed facing the right direction. The fuel injectors are gently pressed into the fuel rail with a rubber damper and an O-ring to seal them. Grasp each injector you wish to remove and pull it from the fuel rail.

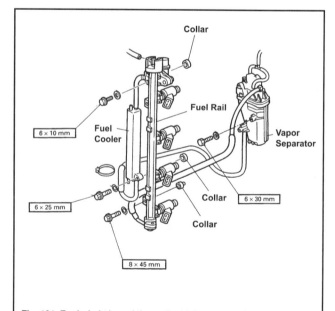

Fig. 131 Exploded view of the entire high-pressure fuel system - 115 hp EFI motors

FUEL SYSTEM 3-75

■ Yamaha and Merc do not specify that you MUST replace the injector seal, damper or O-ring for most models (they DO specify that you must on 150 hp units), however it is never a bad idea to ensure proper sealing and operation. If you opt NOT to replace any of these components, check them carefully for signs of wear, distortion, damage or just plain degradation from age.

To Install:

10. If the fuel injectors were removed from the fuel rail assembly each with the rubber damper and small O-ring on the fuel rail side and the thick rubber seal on the powerhead side. Gently push each injector into the fuel rail with the wiring connected facing outward toward the matchmark made before removal.

11. Position the fuel rail assembly to the powerhead and reconnect the fuel lines. Be sure to use NEW O-rings on motors where one or more fuel joints were unbolted from the rail. Tighten fuel joint bolts securely.

12. Push gently inward toward the powerhead to seat the fuel rail, then thread and finger-tighten the retaining bolts. Make sure that all fuel lines and wires are properly routed/positioned, then tighten the fuel rail retaining bolts securely.

13. Reconnect the fuel injector wiring as tagged during removal.
14. Reposition any wires or hoses and secure using a new wire tie, as noted during removal.
15. If removed, install the air intake silencer and/or flywheel cover.
16. Reconnect the negative battery cable and properly pressurize the fuel system in order to check for leaks. Start and run the motor with the top cover removed and observe the fuel lines/fittings for any signs of weepage. Double-check all lines and fittings after the first outing with the motor.

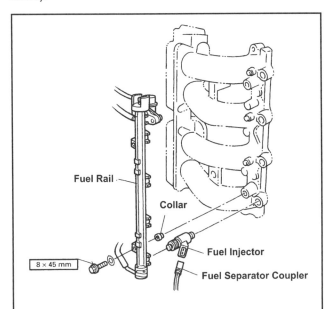

Fig. 132 Exploded view of the fuel rail and injector mounting - 115 hp EFI motors (others similar)

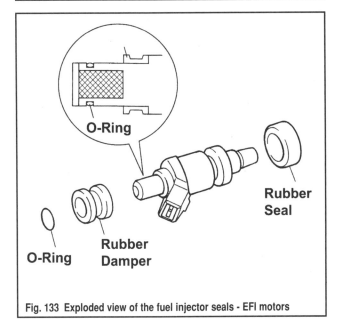

Fig. 133 Exploded view of the fuel injector seals - EFI motors

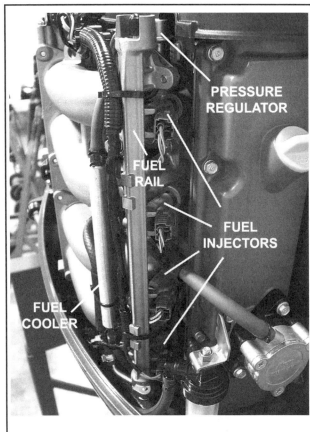

Fig. 134 Fuel rail and injectors - 115 hp EFI motors shown

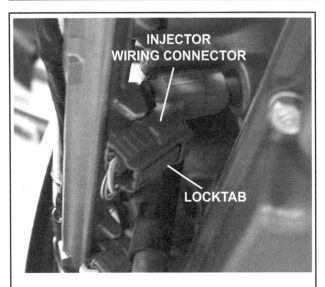

Fig. 135 On most motors, depress the lock-tab while gently pulling on the injector connector

3-76 FUEL SYSTEM

Fuel Distribution Manifold (Mercury/Mariner Systems Only)

REMOVAL & INSTALLATION

◆ See Figures 130 and 136

The smaller 4-stroke motors are equipped with a Mercury EFI system. On models so equipped the fuel injectors are fed by a fuel distribution manifold that is bolted to the upper right corner of the intake manifold assembly (when facing the side of the powerhead). The fuel distribution manifold contains one fuel line per injector. The lines have built-in caps that are secured to the secured to the injector by a retaining clip, as well as an interference fit, and the caps are also secured to the manifold by a retaining screw. A special tool is available from Mercury (#91-883877A1) which can be used to pry the injector caps free from the injectors.

1. Properly release the fuel system pressure, as detailed earlier in this section.
2. Disconnect the negative battery cable for safety.
3. Tag or note the routing of each of the fuel distribution manifold fuel lines.
4. Remove the injector cap retaining bolt(s).
5. Carefully remove the injector retaining clip(s).
6. Using the special service tool or a small pickle fork shaped prytool, CAREFULLY pry the injector cap(s) free from the injector(s).

■ Have a rag handy because fuel will still be present in the manifold and fuel lines even after the pressure has been released.

7. Remove the bolts (normally 2) securing the fuel distribution manifold to the intake manifold assembly, then carefully remove the manifold.
8. Disconnect the high-pressure fuel line from the fuel distribution manifold by gently pressing downward on the locking tab while pulling the line/manifold free of each other.

To Install:

9. Position the fuel distribution manifold to the intake and secure using the retaining screws. Tighten the screws to 31 inch lbs. (3.5 Nm).
10. Route the fuel lines as tagged/noted during removal, then install each injector cap by pushing it straight onto the back of the injector until it bottoms.
11. Secure each injector and cap assembly using the cap retaining screw. Tighten the retaining screw to 31 inch lbs. (3.5 Nm).
12. Install the injector cap retaining clips so that the locking teeth align with the locking rub on the fuel injector. For details, please refer to the accompanying illustration.
13. Reconnect the high-pressure fuel line to the distribution manifold. Carefully push inward until the locking tab snaps into place. Tug back gently on the connection to make sure it is secured.
14. Reconnect the negative battery cable, then properly pressurize the fuel system and check for leaks.

Fuel Pressure Regulator

◆ See Figure 137

All EFI fuel systems utilize an atmospheric pressure or intake manifold vacuum balanced pressure regulator whose job it is to control fuel pressure in the high-pressure circuit. The smaller (60 hp and down) motors along with the 150 hp units usually utilize a regulator which is mounted to the Vapor Separator Tank (VST), while the remaining larger motors (115 hp and V6 models) usually utilize a remote, fuel rail mounted pressure regulator assembly. Regardless, the function is essentially identical.

The job of the electric fuel pump is to deliver fuel at a maximum pressure which meets or exceeds the demands of the motor. The job of the pressure regulator is to vary the opening on the return line to restrict flow (and thereby maintain pressure at a constant level). Essentially the pressure regulator works like a thumb placed over the end of an open garden hose, reducing the diameter of the hose and thereby raising pressure of the water stream exiting from the hose.

The design and mounting of the pressure regulator is basically the same, regardless of the mounting location. And service is a relatively simple matter of depressurizing the system, disconnecting the hoses and removing the regulator.

TESTING

◆ See Figure 137

In typical fashion, the manufacturers do NOT provide specific testing parameters for all EFI pressure regulators. On all motors there is a Pressure Check that can be made the high-pressure system to determine if it is operating within normal parameters. For starters, refer to the Fuel System Checks under Maintenance and Tune-Up for details on how to perform a

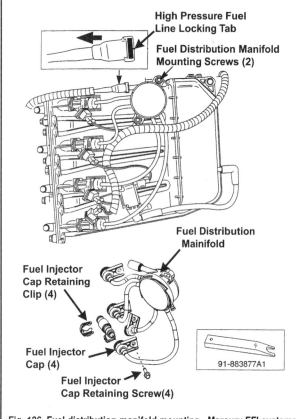

Fig. 136 Fuel distribution manifold mounting - Mercury EFI systems (4-cyl shown, 3-cyl identical except one less fuel line, of course)

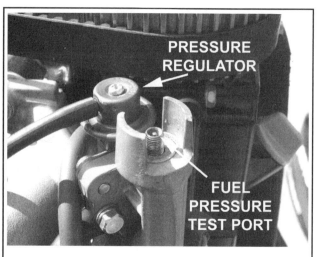

Fig. 137 Typical remote mounted, vacuum balanced pressure regulator - 115 hp motor shown

FUEL SYSTEM 3-77

pressure check. If fuel pressure is ABOVE the specification, check for restrictions in the fuel lines and, if none are found, the pressure regulator is likely the culprit. Similarly, if pressure is below specification you've got 2 possible culprits, the fuel pump or the pressure regulator. The quick check is to squeeze a fuel line and see if you can bring pressure up to specification by restricting fuel flow. If this works, the pressure regulator is the likely culprit, however if even after manually restricting a fuel line pressure remains below specification it is time to look at the fuel pump as the possible cause.

On 150 hp models Yamaha recommends that you set up the fuel Pressure Check AND connect a hand-held vacuum pump/gauge to the fuel pressure regulator (simply disconnect the vacuum line from the regulator and connect the vacuum pump in its place). With the engine running observe the fuel system operating pressure and gradually apply vacuum to the pressure regulator. Apply vacuum (Yamaha doesn't specify how much for this model, but on 2-stroke EFI motors they recommend 7.4025 in Hg./25 kPa of pressure). Fuel pressure should decrease as the vacuum supplied to the pressure regulator is increased. If not, the pressure regulator is faulty. We cannot see why this same operational test would not apply to the other 4-stroke EFI motors which are equipped with the Yamaha EFI system and a vacuum balanced pressure regulator (one that is attached to an intake manifold vacuum line), however use caution in interpreting results, as Yamaha leaves it out of the remaining 4-stroke factory technical information.

REMOVAL & INSTALLATION

◆ See Figures 138, 139 and 140

1. Properly relieve fuel system pressure and disconnect the negative battery cable for safety.
2. For 115 hp and larger motors, tag and disconnect the vacuum hose from the pressure regulator.
3. Also for 115 hp and larger motors, tag and disconnect the fuel hose from the pressure regulator. In most cases this fitting uses a spring-type clamp which is easily removed by holding the clamp ears to expand it while sliding it back up the hose. These types of clamps are reusable, as long as they have not deformed or lost their spring tension.
4. Loosen the 2 retaining bolts, then carefully pull the pressure sensor directly up from the top of the VST (60 hp and smaller motors, as well as 150 hp models) or out from the fuel rail (115 hp and V6 motors).
5. Remove the O-ring seal from the pressure regulator, normally found under the mounting flange. The VST mounted regulators for Mercury/Mariner models usually have 2 O-rings, a small and a large ring. The VST mounted regulator on 150 hp Yamahas has only the one small O-ring.

To Install:

6. Install the pressure regulator using a NEW O-ring (or new O-rings, as applicable) on under the mounting flange.
7. For 60 hp and smaller motors Mercury recommends coating the threads of the retaining bolts using Loctite® 242 or an equivalent threadlocking compound. We can't see the harm in doing this for all motors, whether or not it was specifically recommend.
8. Secure the regulator assembly using the 2 retaining bolts.

9. On 115 hp and larger motors, reconnect the fuel hose and secure using the retaining clamp.
10. On 115 hp and larger motors, connect the vacuum hose to the pressure regulator.
11. Reconnect the negative battery cable and properly pressurize the fuel system in order to check for leaks. Start and run the motor with the top cover removed and observe the fuel lines/fittings for any signs of weepage. Double-check all lines and fittings after the first outing with the motor.

Engine Control Module (ECM/CDI)

The Engine Control Module (ECM) controls all functions of the EFI and ignition systems. Probably as a holdover from the carbureted 2-stroke motors, the earliest of the Yamaha fuel injection systems refer to this as the CDI Unit. However, since it also controls the fuel injection system we think this is a little misleading and will refer to it as the ECM throughout this guide.

■ **Mercury always seems to refer to this as an ECM on their 4-stroke motors.**

Problems with the ECM are rare, but when they occur can cause a no-start, stumbling, misfire, hesitation, incorrect engine timing, rough idle or incorrect speed limiting through improper control of the ignition and/or fuel injection systems.

Unfortunately, solid state components like the ECM cannot be directly tested in many ways. One exception comes with checking the Capacitor Discharge Ignition (CDI) peak output voltages actuated by the ignition control circuits of the ECM on Yamaha's and some Mercs with Yamaha EFI systems. Since this test is the same for carbureted and fuel injected motors, we cover it under CDI Unit, Testing in the Ignition and Electrical System sections. However, remember that the ECM on EFI motors is what actually performs the function of the CDI unit on carb engines.

In most cases, ECM testing involves a process of elimination, testing all other possible causes of a symptom. Condemn the ECM only if all other components that could cause a problem have been eliminated. Remember that many of the circuits used by the ECM for information or for direct control of the motor are sensitive to changes in resistance. Simple problems such as loose, dirty or corroded connectors, even pinched wires or interference caused by marine radios or other electronic accessories can cause symptoms making an otherwise good ECM seem bad.

REMOVAL & INSTALLATION

◆ See Figures 141 thru 145

The module is normally mounted to the near end (front end) of the powerhead. It is easy to access, sometimes mounted under a small protective cover and sometimes just mounted out in the open.

Once accessed ECM removal and installation is a relatively simple matter of disconnecting the wiring and removing any fasteners which secure it to the mounting bracket.

1. Disconnect the negative battery cable for safety.

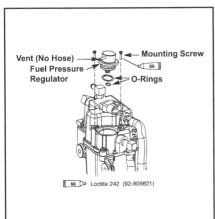

Fig. 138 Exploded view of the fuel pressure regulator mounting - 30-60 hp EFI motors

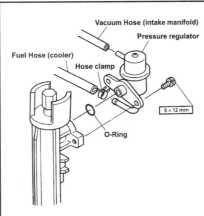

Fig. 139 Exploded view of the fuel pressure regulator mounting - 115 hp EFI motors

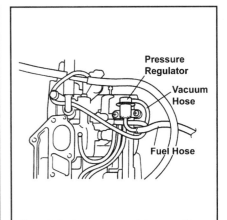

Fig. 140 Fuel pressure regulator mounting - V6 motors

3-78 FUEL SYSTEM

Fig. 141 The ECM is mounted to the back of the powerhead on some models - 115 hp model shown

2. For V6 motors, Mercury advises removing the air intake silencer for access.

3. Locate the ECM on the powerhead assembly. It is at the opposite end of the powerhead from the VST on 60 hp and smaller motors. It is just port side of the air intake silencer on 115 hp motors. The ECM is mounted on the upper, front, port side of the powerhead for 150 hp motors, just a little in front of the starter motor in the top part of the electrical junction box. It is to the starboard side of the starter motor on V6 engines. If it is under a cover and carefully remove the cover.

■ In the case of the 115 hp and most 150 hp motors there is no cover blocking access to the ECM itself, however the harness connectors are obscured by the electrical junction box cover just to the port side of the ECM (115 hp models) or just below the ECM (150 hp models). There is normally a locktab on the cover (at the top of that cover for 115 hp models) which should be gently depressed while pulling outward on the cover to free it from the junction box.

4. Tag and disconnect the wiring harness connectors from the rear (60 hp and smaller), bottom (150 hp and V6 models) or side (115 hp) of the ECM.

5. Loosen and remove the 3 (hp and smaller) or 4 (115 hp and larger) bolts securing the ECM to the powerhead. Take note of the following, depending upon the model:

a. For 60 hp and smaller motors there rear of the ECM faces outward and the mounting bracket is actually further out from the powerhead than the module itself. There are 2 bolts at the top and one at the bottom corner. Pay special attention to the bottom bolt, as it secures a J-clip for the wiring harness which must be positioned properly during installation.

b. On 115 hp motors the bolts thread through an isolating collar and grommet assembly directly into the powerhead. Keep track of the collars and grommets, noting orientation for installation purposes.

c. On 115 hp motors the bolts are threaded into the electrical junction box itself, which helps perform the protective/isolating function for the ECM.

d. On V6 motors the bolts are threaded into an isolating bracket mounted on the powerhead.

6. Carefully remove the ECM from the powerhead, checking for any ground straps that might be attached and, if so, trace and unbolt them.

7. Installation is essentially the reverse of the removal procedure. Make sure the retaining bolts are snug, but don't over tighten and distort any of the fasteners or mounts.

Throttle Position Sensor (TPS)

The Throttle Position Sensor (TPS) is a rotary potentiometer, meaning it sends a variable signal to the ECM based on physical throttle (and therefore sensor) positioning.

The sensor itself receives a reference voltage (usually about 5 volts) from the ECM, and as the throttle lever is rotated, the ECM receives a return voltage signal through a separate wire. This signal will change with throttle position. As the throttle shaft opens the voltage increases, as the shaft closes voltage decreases. A third wire is used to complete the ground circuit back to the ECM.

Sensor location varies slightly from model-to-model, but on all motors, the TPS sensor is located on the top or top side of the throttle body assembly, in a position where it can be directly, mechanically actuated by a throttle valve.

The sensor is mounted in a fixed position on Mercury EFI systems (60 hp and smaller) or in an adjustable position on Yamaha EFI systems (115 hp and larger). When applicable, adjustment, which is covered under Timing and Synchronization in Maintenance and Tune-Up involves checking the output voltage (in a similar manner to the Testing covered here) while repositioning the switch.

TESTING

◆ See Figures 146 and 147

Checking the TPS output should be part of each tune-up (or should at least be performed annually) on the Yamaha EFI systems (115 hp and larger motors). If sensor output is slightly out of spec, it can usually be unbolted and adjusted slightly to bring it within specifications, however, if it cannot be adjusted the sensor must be replaced (unless there is a problem with the ECM reference voltage).

Mercury does not provide any information on TPS testing for their 60 hp and smaller motors (equipped with the Merc EFI system). Partially this is because the TPS is mounted in a fixed position for those motors. But also, it is because they expect the technician to use the Quicksilver DDT scan tool to check the sensor. They DO provide a specs for TPS circuit voltage of about 0.39-1.0 volts at idle and 3.66-4.80 volts at WOT, so in theory if you construct a jumper harness you could check to verify this voltage.

For 115 hp and larger motors testing the sensor is virtually identical to the adjustment procedure, with the possible exception that sensor input voltage

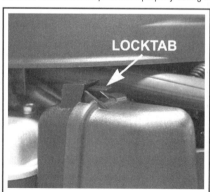

Fig. 142 This cover is secured with a locktab...

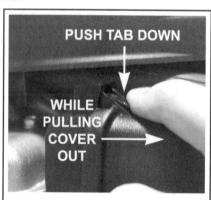

Fig. 143 ...push down on the tab while pulling outward...

Fig. 144 ...to remove the electrical junction box cover

FUEL SYSTEM 3-79

Fig. 145 Disconnect the wiring and loosen the bolts to remove the ECM

is checked on some models and on others the sensor output voltage is checked at idle (150 hp motors) or both Idle and WOT (V6 models). Even if it is not specified to do so for a given model, we suggest that if you suspect problems with the TPS circuit, you should slowly open and close the throttle while watching the TPS output voltage to make sure there are no sudden spikes or drops which could lead to intermittent performance problems.

Since the sensor does not generate a voltage, but instead acts upon a reference voltage from the ECM, all testing must occur with the circuit complete. Therefore, you've got 2 options. One option is to attempt to back-probe the connector (insert the probes through the rear of the connector while it is still attached to the sensor). However this method risks damaging the connector or the wiring insulation (and could lead to problems with the circuit later). The better method is to disconnect the wiring harness and use 3 jumper wires (one for each terminal) to reconnect the harness. The jumpers must not contact each other (or you risk damage to the ECM from a short), however, some point on the jumper must be exposed so that you can probe the completed circuit using a DVOM. Yamaha makes a 3-pin test harness for this application (#90890-06793) for 115 hp motors, and (#YB-06793) for 150 hp and V6 models. Mercury also sells a 3-pin test harness for this application (#91-881827) for both the 115 hp and V6 motors.

1. If necessary, remove the flywheel cover for additional access.
2. If necessary, remove the air intake assembly for access.
3. Disconnect the wiring from the TPS and connect a test harness or jumper wires to complete the circuit. Disconnect the harness directly at the switch itself.

■ For 150 hp motors, Yamaha recommends making sure that the throttle valves in the throttle body bores are closed completely. One way to ensure this is to unsnap the throttle rod link joint from the ball connector at one end.

4. Connect a Digital Volt Ohmmeter (DVOM) set to read DC volts to the appropriate terminals and turn the ignition switch **ON** without starting the motor to check for the following results depending upon motor and throttle setting:
• For 60 hp and smaller motors, the TPS circuit should show 0.39-1.0 volts with the throttle at idle and 3.66-4.80 volts with the throttle wide open.
• 115 hp motors, voltage output across the Pink and Black wires should be 0.718-0.746 volts with the key on and engine not running but the throttle closed (in the idle position). On this model, start the engine (using a source of cooling water to prevent damage to the impeller and powerhead) and recheck the setting while running at idle, it must remain within specification.

■ Some OE sources say to test the circuit on 150 hp motors between the Pink and Orange wires of the test harness. We cannot be certain whether or not this is a typo or the sensor wires and test harness wires are not consistent with each other. Note that other sources say that the Orange and Black wires in the engine harness carry the 5 volt reference signal (and ground) for the TPS circuit.

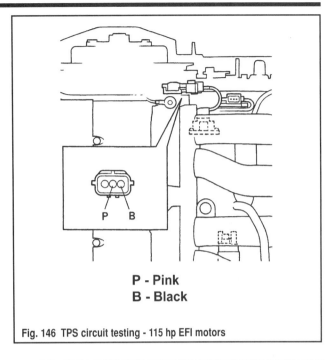

P - Pink
B - Black

Fig. 146 TPS circuit testing - 115 hp EFI motors

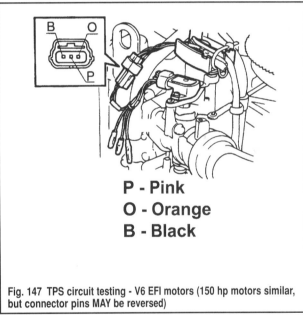

P - Pink
O - Orange
B - Black

Fig. 147 TPS circuit testing - V6 EFI motors (150 hp motors similar, but connector pins MAY be reversed)

• 150 hp motors, voltage output across the Pink and Black wires should be 0.68-0.72 volts with the key on and engine not running. On this model, start the engine (using a source of cooling water to prevent damage to the impeller and powerhead) and recheck the setting while running at idle, it must remain within specification.
• For V6 motors, voltage output across the Pink and Black wires of the harness should be about 695-705 mV (mill volts) at idle setting (throttle closed). Next, check the sensor input voltage the Orange and Black terminals, it should be 5 volts.

5. If the sensor voltage (not reference voltage) is out of range, refer to the appropriate adjustment procedure in the Timing and Synchronization section. If the sensor cannot be adjusted and no problems are found with the ECM, wiring or reference voltage, the sensor must be replaced.
6. If the reference voltage is out of specification, check the wiring harness thoroughly for problems, make sure the adapter/tests connections are good, then suspect the ECM only after all other options have been verified good.
7. After testing turn the ignition keyswitch **OFF**, disconnect the test harness and reconnect the TPS wiring.

3-80 FUEL SYSTEM

REMOVAL & INSTALLATION

◆ See Figures 148, 149 and 150

Removal and installation of the TPS is relatively straightforward. On a few motors the sensor may be obscured by a cover or a bracket, so if necessary, remove the cover until you can clearly see the sensor body, wiring and retaining screws. Besides a dedicated sensor cover, depending upon positioning the air intake silencer and/or the flywheel cover may be in the way.

Once you have clear access to the sensor all you need to do is disconnect the wiring and remove the bolts/screws (there are usually 2) that secure the sensor to the bracket or throttle body. If an adjustable sensor (115 hp and larger motors) is not being replaced, you might want to scribe a quick matchmark before removal between the sensor and mounting point. This will provide a handy reference point to start adjustment after reinstallation. Upon installation of an adjustable sensor you should leave the retaining screws just slightly loosened so that you can pivot the sensor for proper adjustment. Details on TPS adjustment can be found under the appropriate Timing and Synchronization procedure in the Maintenance and Tune-Up section.

Crankshaft Position Sensor (CPS)/Pulser Coil

The ECM needs to know exactly where each piston is in its 4-stroke cycle in order to accomplish certain ignition timing and fuel injection functions. In addition, voltage must be generated by the flywheel magnets and powerhead mounted coils in order to power the ignition system.

The smaller EFI motors (60 hp and down) utilize a single Crankshaft Position Sensor mounted immediately adjacent to the flywheel. The larger EFI motors (115 hp and up) use only the Pulser Coils (one winding for every pair of cylinders) incorporated into the stator to perform the same function. In this way, the ignition systems are operated in much the same manner as the carbureted Yamaha and Mercury Capacitor Discharge Ignition (CDI) systems also covered in this guide. For this reason, all testing and service information on the CPS and the Pulser Coils can be found in the Ignition and Electrical System section.

Temperature Sensors (Air and Water/Engine)

◆ See Figures 151 thru 155

Yamaha sometimes refers to the Water Temperature Sensor (WTS) as the Engine Temperature Sensor (ETS) or engine cooling water temperature sensor. Mercury normally refers to the same sensor as an Engine Coolant Temperature (ECT) sensor (unless the motor is equipped with a Yamaha fuel injection system, like the 115 hp and larger models, then they call it a WTS or ETS just like Yamaha). We just wanted you to know in case you say WTS and the parts guy gives you a funny looks and says, "do you mean ETS or ECT." They are the same sensor and perform the same function, but the names vary in different Yamaha and Mercury publications.

Similarly, though Yamaha usually calls the air temperature sensor an Intake Air Temperature (IAT) sensor, Mercury normally refers to it as a Manifold Air Temperature (MAT) sensor. But what is in a name, a rose by any other name would smell as sweet (our apologies to the great bard) and a temperature sensor by any other name, still varies resistance with the temperature of whatever it is touching.

Signals from the air temperature (IAT/MAT) and water temperature (WTS/ETS/ECT) sensor are used by the EFI system to manage engine operation. As their names suggest, the sensors are used to monitor air and engine/water temperatures. The air temperature sensor is used by the ECM to help determine air/fuel ratios. The water temperature sensor provides essential information during cold starts and warm-up.

■ In addition to these sensors, a thermo-switch is used on the Yamaha engine cooling overheat warning system (as well as on the V6 Mercury/Mariner models). It differs from a sensor in that it is a simple ON/OFF switch. Whereas a sensor gives a varying, but constant signal, a thermo-switch either has continuity or not, and therefore returns a signal only a certain temperatures. Because the basic design of the thermo-switch is the same for all motors (carbureted and fuel injected) it is covered in the Lubrication and Cooling section of this guide.

Temperature sensors for modern fuel injection systems are normally thermistors, meaning that they are variable resistors or electrical components that change their resistance value with changes in temperature. From the test data provided by the manufacturer it appears that both Yamaha and Mercury sensors are usually Negative Temperature Coefficient (NTC) thermistors. Whereas the resistance of most thermistors (and most electrical circuits) increases with temperature increases (or lowers as the temperature goes down), an NTC sensor operates in an opposite manner. The resistance of an NTC thermistor goes down as temperature rises (or goes up when temperature goes down).

Sensor locations vary slightly by engine and are as follows:
- 60 hp and smaller motors. The MAT is mounted to the backside of the intake manifold, just above the throttle body assembly (unfortunately this can make access to the sensor itself difficult/impossible when the manifold is installed, though the harness can be reached for testing). The ECT is threaded into the exhaust cover, immediately adjacent to the oil filter (access is MUCH easier).
- For 115 hp EFI motors, the IAT is mounted at the front of the powerhead, on the bottom, front of the air intake silencer assembly. The WTS is mounted toward the middle of the starboard side of the powerhead (just aft of the electric starter motor and a little above/behind the oil filter).
- For 150 hp EFI motors, the IAT is mounted toward the port side on the front of the powerhead (just below the ECM in the electrical junction box, and directly across from the front of the air intake silencer assembly). The ETS is mounted on the top of the powerhead, toward the starboard side, between the timing belt tensioner and the Throttle Position Sensor (TPS).
- For V6 EFI motors, the IAT is mounted at the front of the powerhead, toward the inside, bottom of the port half of the air intake silencer assembly. The WTS is mounted to the centrally to the top of the powerhead, a little port of the stator, almost in line (perhaps a little forward) of the throttle body assembly).

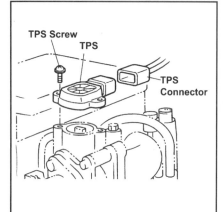

Fig. 148 TPS sensor mounting - 115 and 150 hp EFI motors (V6 similar)

Fig. 149 Some sensors are mounted under a protective cover...

Fig. 150 ...but they're always mounted on the throttle body assembly

FUEL SYSTEM 3-81

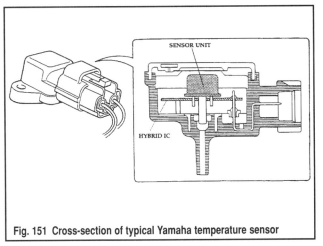

Fig. 151 Cross-section of typical Yamaha temperature sensor

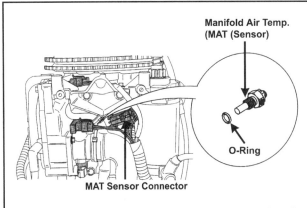

Fig. 152 Manifold Air Temperature (MAT) sensor - 60 hp and smaller motors

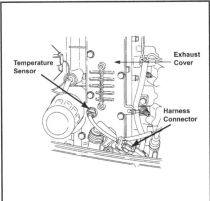

Fig. 153 Engine Coolant Temperature (ECT) sensor - 60 hp and smaller motors

Fig. 154 Typical Yamaha IAT sensor - 115 hp shown

Fig. 155 Typical Yamaha WTS - 115 hp shown

TESTING

♦ See Figures 151 thru 155

Temperature sensors are among the easiest components of the EFI system to check for proper operation. That is because the operation of an NTC thermistor is basically straightforward. In general terms, raise the temperature of the sensor and resistance should go down. Lower the temperature of the sensor and resistance should go up. The only real concern during testing is to make clean test connections with the probe and to use accurate (high quality) testing devices including a DVOM and a relatively accurate thermometer or thermosensor.

A quick check of the circuit and/or sensor can be made by disconnecting the sensor wiring and checking resistance (comparing specifications to the ambient temperature of the motor and sensor at the time of the test). Keep in mind that this test can be misleading as it could mask a sensor that reads incorrectly at other temperatures. Of course, a cold engine can be warmed and checked again in this manner.

More detailed testing involves removing the sensor and suspending it in a container of liquid (Yamaha and Mercury both recommend water), then slowly heating the liquid while watching sensor resistance changes on a DVOM. This method allows you to check for problems in the sensor as it heats across its entire operating range.

■ Testing would be easier if Yamaha decided to provide the same specs for all sensors. However, although they do provide a certain amount of resistance specs for most motors, they DO NOT supply them for ALL motors. For some unbeknownst reason, the Yamaha literature gives NO data on the 115 hp 4-cylinder, 4-stroke motors, but provides some test specs for 150 hp models. In contrast, Mercury was good enough to provide complete specifications for sensor resistance on models equipped with their system (meaning 60 hp and smaller models), as well as charts for the ECT on 115 hp motors and test specifications for the ECT on V6 motors.

Since the sensor specifications available vary, check the Test Specifications for your motor, before determining which test(s) you will follow.

Sensor Test Specifications

♦ See Figures 156 and 157

■ For models equipped with the Mercury EFI system (60 hp and smaller motors) refer to the accompanying chart.

It appears that the calibration of the sensors used on different motors with the Yamaha EFI system may be the same, however, we cannot be sure of this since Yamaha and Mercury do not provide a uniform set of specs for most of the EFI motors. For this reason, use the following specifications to determine which tests you can conduct with the motor on which you are working.

Intake Air Temperature (IAT) sensor testing specifications:
• **60 hp and smaller motors**, please refer to the accompanying illustration for details, however sensor resistance should vary between 0-1,000 ohms at 270 degrees F and more than 32,000 ohms at about 30 degrees F. At an ambient temperature of around 60 degrees F the sensor should read about 14,500 ohms.
• **115 hp Motors**: SORRY, but we could not find any published specs from Yamaha or Mercury for this sensor on this motor. However, either the Quick Check or the Comprehensive Test should tell you something about this sensor. First of all, resistance MUST fluctuate with temperature or it is bad, right? Also, it is still likely an NTC sensor. Lastly, keep in mind that you should be able to talk a cooperative parts counterperson into allowing you to put a DVOM across a replacement sensor (at least at ambient (counter) temperatures.
• **150 hp motors** sensor resistance should be about 2,200-2,700 ohms at 68 degrees F (20 degrees C). Readings should obviously vary with changes in temperature, but no further specifications were available from Yamaha.

3-82 FUEL SYSTEM

- **V6 EFI Motors**: should read about 5,400-6,600 ohms at 32°F (0°C) and 290-390 ohms at 80°F (176°C). The Quick Test and/or Comprehensive Test may be made across the sensor terminals.

Water Temperature Sensor (WTS) testing specifications:

- **60 hp and smaller motors**: please refer to the accompanying illustration for details, however sensor resistance should vary between 0-1,000 ohms at 270 degrees F and more than 32,000 ohms at about 30 degrees F. At an ambient temperature of around 60 degrees F the sensor should read about 14,500 ohms.
- **115 hp EFI Motors**: please refer to the accompanying illustration for details, however sensor resistance should vary should read about 4,620 ohms at 41°F (5°C), 2,440 ohms at 68°F (20°C) or 190 ohms at 212°F (100°C). The Quick Test and/or Comprehensive Test may be made across the Black/Yellow and Black wires of the sensor pigtail.
- **150 hp motors**: sensor resistance should be about 54,200-69,000 ohms at 68 degrees F (20 degrees C), and 3,120-3,480 ohms at 212 degrees F (100 degrees C). Readings should obviously vary with changes in temperature, but no further specifications were available from Yamaha. However, keep in mind that we can deduce from these specifications that the sensor IS indeed of a Negative Temperature Coefficient (NTC) design, meaning that resistance should continue to go DOWN as temperatures go UP (and vice versa).
- **V6 EFI Motors**: should read about 54,000-69,000 ohms at 68°F (20°C) and 3,120-3,480 ohms at 212°F (100°C). The Quick Test and/or Comprehensive Test may be made across the Black/Yellow and Black wires of the sensor pigtail.

Quick Test

 MODERATE

When resistance specifications are available, a quick check of a temperature sensor can be made using a DVOM set to the resistance scale and applied across the sensor terminals. The DVOM can be connected to directly to the sensor, or to the sensor pigtail, as the wiring varies by model. Use a thermometer or a thermo-sensor to determine ambient engine/sensor temperature before checking resistance.

Even if the sensor tests ok cold, the sensor might read incorrectly hot (or anywhere in between). If trouble is suspected, reconnect the circuit, then start and run the engine to normal operating temperature. After the engine is fully warmed, shut the engine **OFF** and recheck the sensor hot. If the sensor checks within specification hot, it is still possible that another temperature point in between cold and fully-warmed specifications could be causing a problem, but not likely. The sensor can be removed and checked using the Comprehensive Test in this section or other causes for the symptoms can be checked. If the sensor was checked directly and looks good, but there are still problems with the circuit, be sure to check the wiring harness between the ECM and the sensor for continuity. Excessive resistance due to loose connections or damage in the wiring harness can cause the sensor signals to read out of range. Remember, never take resistance readings on the ECM harness without first disconnecting the harness from the ECM.

Comprehensive Test

  MODERATE

It is important to EFI operation that the temperature sensors provides accurate signals across the entire operating range and not just when fully hot or fully cold. For this reason, when resistance specifications are available, it is best to test the sensor by watching resistance constantly as the sensor is heated from a cold temperature to the upper end of the engine's operating range. The most accurate way to do this is to suspend the sensor in a container of water, connect a DVOM and slowly heat the liquid while watching resistance on the meter.

To perform this check, you will need a high quality (accurate) DVOM, a thermometer (or thermo-sensor, some multi-meters are available with thermo-sensor adapters), a length of wire, a metal or laboratory grade glass container and a heat source (such as a hot plate or camp stove). A DVOM with alligator clip style probes will make this test a lot easier, otherwise alligator clip adapters can be used, but check before testing to make sure they do not add significant additional resistance to the circuit. This check is performed by connecting the two alligator clips together and checking for a very low or 0 resistance reading. If readings are higher than 0, record the value to subtract from the sensor resistance readings that are taken with the clips in order to compensate for the use of the alligator clips.

1. Remove the temperature switch as detailed in this section.
2. Suspend the sensor and the thermometer or thermo-sensor probe in a container of cool water or four-stroke engine oil.

■ **To ensure accurate readings make sure the temperature sensor and the thermometer are suspended in the liquid and are not touching the bottom or sides of the container (as the temperature of the container may vary somewhat from the liquid contained within and sensor or thermometer held in suspension).**

3. Set the DVOM to the resistance scale, then attach the probes to the sensor or sensor pigtail terminals, as applicable.
4. Allow the temperature of the sensor and thermometer to stabilize, then note the temperature and the resistance reading. If a resistance specification is provided for low-temperatures, you may wish to add ice to the water in order to cool it down and start the test well below ambient temperatures.
5. Use the hot plate or camp stove to slowly raise the temperature. Watch the display on the DVOM closely (in case there are any sudden dips or spikes in the reading which could indicate a problem). Continue to note resistance readings as the temperature rises to 68°F (20°C) and 212°F (100°C). Of course by 212 degrees, your water should be boiling, so it is time to stop the test before you make a mess or get scalded. Up to that point, the meter should show a steady decrease in resistance that is proportional to the rate at which the liquid is heated. Extreme peaks or valleys in the sensor signal should be rechecked to see if they are results of sudden temperature increases or a possible problem with the sensor.
6. Compare the readings to the Sensor Test Specifications.

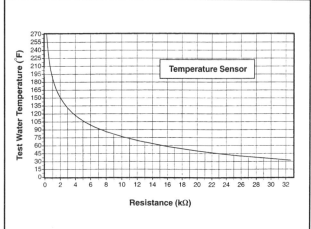

Fig. 156 Temperature sensor resistance specifications (ECT/MAT) for Mercury/Mariner EFI systems

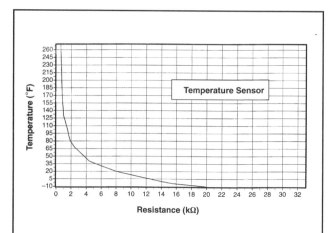

Fig. 157 Temperature sensor resistance specifications (ECT) for 115 Hp motors

FUEL SYSTEM 3-83

REMOVAL & INSTALLATION

 MODERATE

◆ See Figures 151 thru 155

There are basically two ways that temperature sensors are normally mounted on these motors. Some sensors are secured to the powerhead or a mounting bracket via a retention plate and mounting screws/bolts (usually 2, one on either side of the sensor). Others contain a hex on the sensor itself and are threaded into position. The later (threaded into place) is the method always used on the Mercury EFI system (60 hp and smaller motors).

Removal and installation is therefore pretty straightforward, as long as access is not restricted. For most of the sensors mounted on top of the powerhead the flywheel cover must be removed for access and that is usually sufficient. In some rare instances, another cover or an additional component must be removed for better access. This includes some air intake silencer mounted IAT sensors to which access is difficult or impossible with the silencer installed.

■ BAD NEWS, the Manifold Air Temperature sensor used on 60 hp and smaller motors is mounted to the backside of the intake manifold, just above the TBI unit. The manufacturer recommends removing the Intake Manifold and Throttle Body assembly for access. You MAY be able to access it on some models without doing this, and it's worth a try. But if you cannot get a socket on the sensor, you'll have to follow the Intake Manifold removal procedure located earlier in this section.

Once you have clear access to the sensor all you need to do is disconnect the wiring and remove the bolts/screws (there are usually 2) that secure the sensor to the bracket or unthread the sensor, as applicable.

Some sensors are equipped with an O-ring or gasket. If applicable, be sure to keep track of it for installation purposes. Also, replace any gasket or O-ring that shows signs of damage or deterioration.

Pressure Sensors (IAP/MAP)

◆ See Figure 158

All fuel injection systems are equipped with some form of an air pressure sensor. The Yamaha EFI system 4-stroke motors call it an Intake Air Pressure (IAP) sensor, while the Mercury EFI system (60 hp and smaller) call it a Manifold Air Pressure (MAP) sensor.

Although the mounting point differs slightly from motor-to-motor and may have an affect on exactly what the sensor is measuring (and for what the ECM is calibrated) the basic function performed by these sensors is the same. The sensor works like a barometer and reports barometric air pressure. This sensor allows the computer to compensate for barometric pressure changes at sea level and the normal reduction in atmospheric pressure found at high altitudes.

Fig. 158 Typical IAP/MAP sensor - 115 hp EFI shown

During normal EFI operation, a large portion of fuel mapping decisions (injector on-time strategy) is made based upon signals from the pressure sensor.

You should also note that these motors utilize oil pressure sensors to monitor internal engine lubrication functions (however for more information on Oil Pressure sensors, please refer to the Lubrication and Cooling section as it is not a direct component of the fuel injection system).

Sensor locations vary slightly by engine and are as follows:
- For 60 hp and smaller motors the MAP sensor is located on top of a boss in the intake manifold, toward the top rear of the manifold, near the overhang of the flywheel. Unfortunately, like the MAT sensor, Mercury recommends removing the Intake Manifold and Throttle Body assembly for access.
- For 115 hp EFI motors, the IAP sensor is mounted to the top of the powerhead, on the port side, directly above the intake manifold runners (a little behind the TPS and Idle Speed Control Valve).
- For 150 hp EFI motors, the IAP sensor is mounted to the top of the powerhead, on the starboard side, directly above the intake manifold runners (a little behind the TPS and Idle Speed Control Valve).
- For V6 EFI motors, the IAP sensor is mounted to the top of the port throttle body assembly, adjacent to (and just in front of) the Idle Speed Control Valve.

TESTING

 DIFFICULT

◆ See Figure 158 and 159

■ Unfortunately, we could not locate any testing/operational specifications for the IAP sensor used on Yamaha EFI motors. Mercury was kind enough to provide specs for the sensor used on Mercury EFI systems (60 hp and smaller motors), those specs are included here.

Pressure sensors operate using a reference voltage from the ECM, as a result, there is no way to test them unless the circuit is complete and operating. For the IAP/MAP sensor this means with the ignition keyswitch **ON** and the engine not running.

Again, although Yamaha does not provide any information specifically for their 4-stroke motors, we found that nearly all of the EFI and HPDI 2-stroke motors seem to use the same IAP sensor. On those 2-strokes Yamaha provides for testing across the Pink and Black wires (using a 3-wire adapter harness or jumpers which allow the circuit to remain complete). For most Yamaha 2-strokes the sensor output should read 3.2-4.6 volts with the engine off, or 2.8-3.2 volts with the engine at idle. We cannot guarantee that these specs apply to the 115 hp and larger 4-strokes, but it is a good place to start.

For 60 hp and smaller motors (Mercury EFI systems), testing occurs across the various pins of the 3-pin connector attached to the MAP sensor. Using a DVOM set to read resistance, check the following pins/connections.
- Black/Orange (Pin A)-to-Yellow (Pin B), the meter should show 95,000-105,000 ohms resistance.
- Black/Orange (Pin A)-to-Purple/Yellow (Pin C), the meter should show 3,900-4,300 ohms resistance.
- Yellow (Pin B)-to-Purple/Yellow (Pin C), the meter should show 95,000-105,000 ohms resistance.

■ Use the accompanying illustration and/or the associated harness wire colors to help determine pin locations. With the sensor positioned as it sits on the manifold (with the sensor protrusion facing downward and the electrical connector locktab facing upward) the pins are C, B and A from RIGHT to LEFT when looking in the connector at the pins.

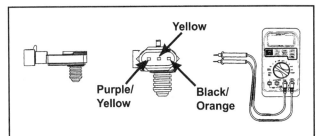

Fig. 159 Testing the MAP sensor on Mercury EFI systems (60 hp and smaller motors)

3-84 FUEL SYSTEM

REMOVAL & INSTALLATION

◆ See Figure 158

Removal and installation of the IAP/MAP sensor is fairly straightforward on most Yamaha motors. Locate the sensor and determine if any covers or other components need to be removed. The sensors are mounted on top of the throttle body assembly and, although they may be visible, access is usually improved with the flywheel cover removed. Removal requires disconnecting the wiring and removing the mounting screws.

■ BAD NEWS, the Manifold Air Pressure sensor used on 60 hp and smaller motors is mounted to a protrusion on the backside of the intake manifold, near the MAT sensor and the TBI unit. The manufacturer recommends removing the Intake Manifold and Throttle Body assembly for access. You MAY be able to access it on some models without doing this, and it's worth a try. But if you cannot get a socket on the sensor, you'll have to follow the Intake Manifold removal procedure located earlier in this section. Once you've accessed the sensor removal on these models is as simple as removing the retaining bolt and clip, then disconnecting the wire and gently pulling the sensor free. Upon installation it is normally a good idea to place a small amount of lubricant on the portion of the sensor that seats in the manifold.

Idle Speed/Air Control (ISC/IAC) Valve

◆ See Figure 160

Motors equipped with the Yamaha EFI system are equipped with an Idle Speed Control (ISC) valve, mounted to the top of the throttle body assembly. Motors equipped with the Mercury EFI system utilize a similar component called the Idle Air Control (IAC) valve mounted to the Intake Manifold and Throttle Body assembly.

As both names suggest, the purpose of this valve is to give the ECM direct control over engine idling speed. Since engine speed is controlled by the amount of air allowed past the throttle plates into the motor, these valves generally work in one of two ways. Either the valve is designed to mechanically actuate the throttle plates, opening or closing them slightly as desired by the ECM, or they are designed to provide an idle air bypass, allowing air to move around the throttle plate and enter the throttle bodies and air intake stream from another point. The later type of valve generally works by moving a stepper-motor pintle inward or outward to obstruct or open an idle air passage.

Fig. 160 Typical Yamaha ISC valve - The valve and IAP sensor are mounted to a valve body on top of the throttle body assembly (port side of the motor)

Yamaha does not tell us which design is used on their EFI motors, however we suspect from the air plumbing attached to the valve body it is probably the later.

Mercury tells us that their valve is an electrically operated, spring-loaded, solenoid valve which controls the amount of air allowed to bypass the closed throttle shutter. It is designed to hold the bypass open in an ECM regulated duty cycle of 0-100% open or to remain closed (via spring pressure) when not actuated.

As stated earlier, on Yamaha EFI systems the ISC valve is mounted to the top of the throttle body assembly. To be more specific, the valve is installed on a valve body that is then installed to the throttle body assembly on the port side of the motor. The IAP sensor is also installed to the same valve body.

For Mercury EFI systems the IAC valve is found on the lower portion of the intake manifold and throttle body assembly.

■ Although Yamaha doesn't give ISC test values, Mercury states that on 60 hp and smaller motors the IAC valve coil winding should show 24-30 ohms resistance.

REMOVAL & INSTALLATION

◆ See Figures 161 thru 165

Servicing the ISC/IAC valve is pretty simple on most motors, it's a matter of tagging and disconnecting the wiring and hoses and unbolting the assembly. That's it. The differences come in the location and the number of hoses.

For Mercury EFI systems (60 hp and smaller motors) there is normally one wiring connector and one hose. Though it may be easier on these models to unbolt the assembly first and THEN disconnect the hose.

For Yamaha EFI systems (used on all 115 hp and larger motors), you've got a couple of options when it comes to servicing the valve. Removal of the flywheel cover will usually give you a little better access to the valve, valve body or even the IAP sensor. The ISC valve itself is fastened to the top of the valve body on these models using 3 retaining screws. If desired you can disconnect the wiring and then just unbolt and remove the valve leaving the valve body and related hoses behind. When the valve is removed from the valve body you should carefully inspect the O-ring which MAY be reused as long as it is not damaged or worn.

On some Yamaha EFI systems, if desired, the whole valve body assembly (along with the valve and IAP sensor) can also be removed from the powerhead. It is a simple matter of tagging and disconnecting the wiring, then doing the same for any hoses (there are 4 air/vacuum hoses on the 115 hp EFI motor, but only one large hose and one small hose on the V6 EFI engine). It appears that the valve body assembly is secured using 3 bolts on 115 hp motor and by 2 bolts on V6 engines, but we cannot be sure from the diagrams as there is also a long bracket bolt threaded into the body on V6 engines which might go all the way through the valve body.

■ On 150 hp motors, Yamaha changed the design so the valve body is incorporated into the top of the intake manifold, so separate removal of the body or ISC and IAP sensor is no longer possible (unless you're taking the whole intake off).

If there are troubles with the ISC/IAC valve, remember that they could be mechanical as well as electrical, so it is usually a good idea to remove the whole valve body so you can inspect the passageways for dirt, corrosion or contamination and clean, as necessary. In addition, the valve body used on the 115 hp motor is equipped with a small filter element (to which the intake manifold vacuum hose is connected), the element should inspected and replaced if trouble is suspected with the system.

Neutral or Shift Position Switches

◆ See Figures 166

Most fuel injected motors are equipped with one or more neutral or shift switches. For Yamaha EFI systems, these switches are typically used to tell the ECM whether or not the motor is in gear/neutral. Different fuel and ignition mapping decisions may be made depending upon these conditions.

When equipped, these switches are normally of the simple ON/OFF one switch position design. Again, at least on Yamaha EFI systems, this USUALLY means that there should be NO CONTINUITY across the switch

FUEL SYSTEM 3-85

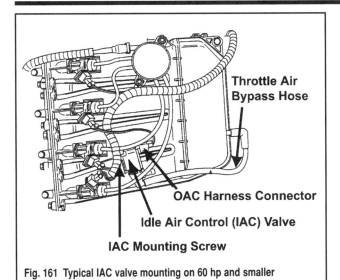

Fig. 161 Typical IAC valve mounting on 60 hp and smaller Mercury/Mariner motors

Fig. 162 Exploded view of the ISC valve body mounting - 115 hp EFI motors

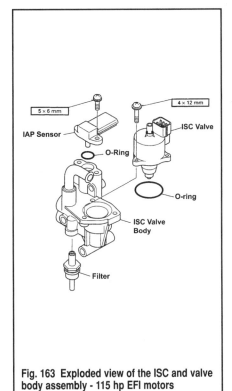

Fig. 163 Exploded view of the ISC and valve body assembly - 115 hp EFI motors

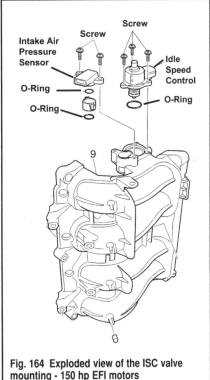

Fig. 164 Exploded view of the ISC valve mounting - 150 hp EFI motors

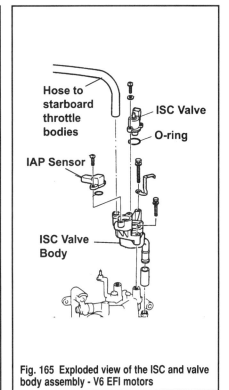

Fig. 165 Exploded view of the ISC and valve body assembly - V6 EFI motors

terminals when the plunger is released (extended). Conversely, when the switch is depressed (plunger is down against the switch body) there should be continuity across the switch contacts.

Using a DVOM set to read resistance you can easily check the function of these switches, it is not important that they match this description perfectly, but it is important that there is little or no resistance in one position and infinite resistance in the opposite position.

■ Mercury doesn't mention the Neutral Safety Switch in the EFI sections of their service Information and we cannot see from the wiring diagrams any wire where the ECM might be connected to that portion of the starter circuit, so we suspect that the ECM does not know or care when the gearcase is in Neutral for 60 hp and smaller models, but we cannot be certain. It does appear that the switch used on these systems operates in the same basic fashion as the one on Yamaha EFI systems, meaning it is a normally open switch which closes once the shifter is in neutral and the plunger is pressed inward.

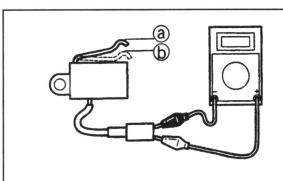

Fig. 166 Testing a typical lever type Yamaha Neutral or Shift Position switch

3-86 FUEL SYSTEM

Fuel Injection and Electronic Component Locations

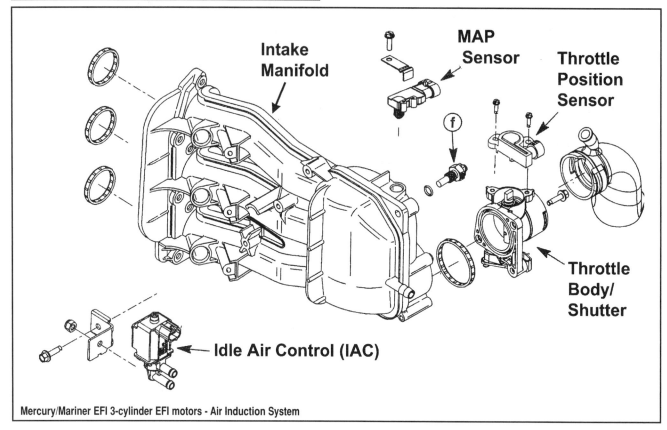

Mercury/Mariner EFI 3-cylinder EFI motors - Air Induction System

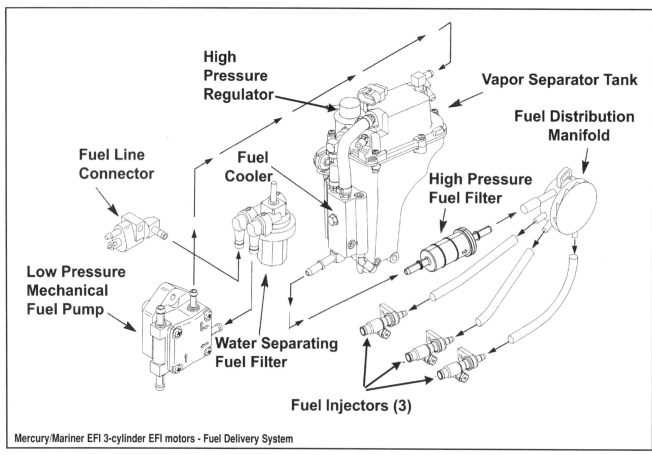

Mercury/Mariner EFI 3-cylinder EFI motors - Fuel Delivery System

FUEL SYSTEM 3-87

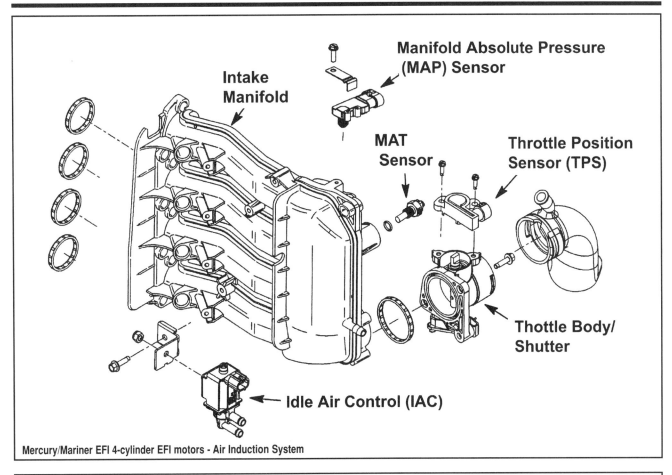

Mercury/Mariner EFI 4-cylinder EFI motors - Air Induction System

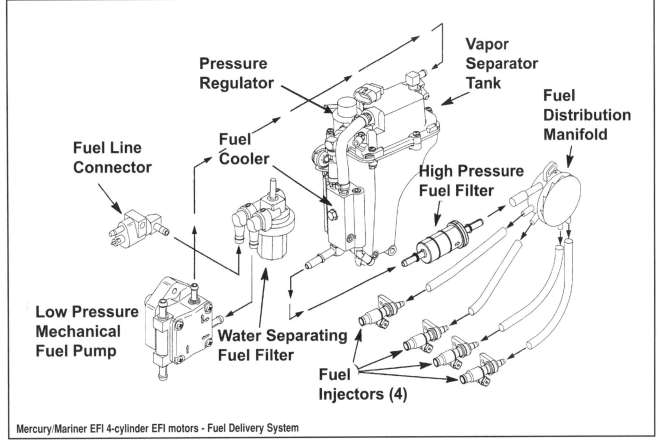

Mercury/Mariner EFI 4-cylinder EFI motors - Fuel Delivery System

3-88 FUEL SYSTEM

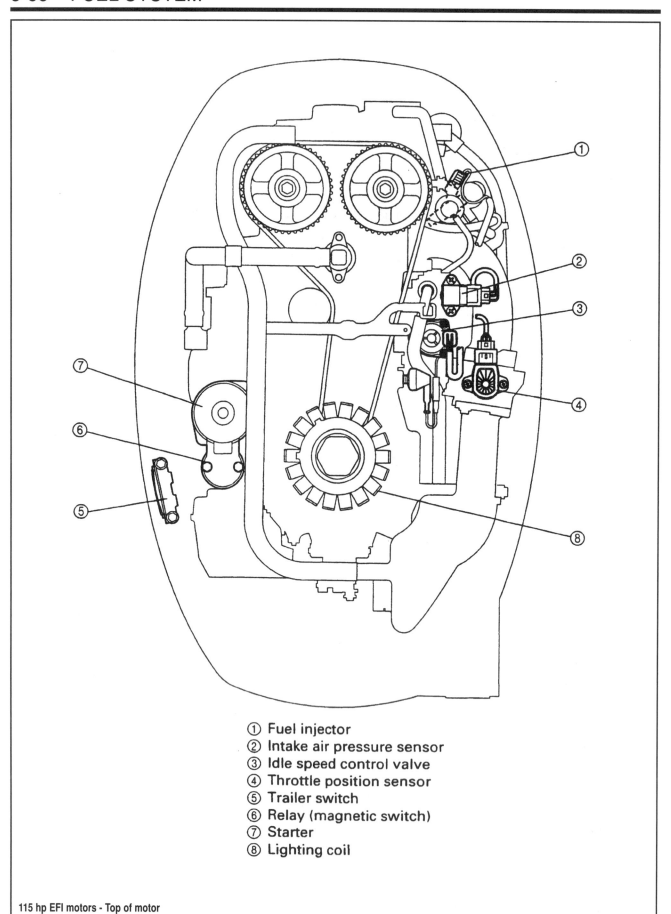

① Fuel injector
② Intake air pressure sensor
③ Idle speed control valve
④ Throttle position sensor
⑤ Trailer switch
⑥ Relay (magnetic switch)
⑦ Starter
⑧ Lighting coil

115 hp EFI motors - Top of motor

FUEL SYSTEM 3-89

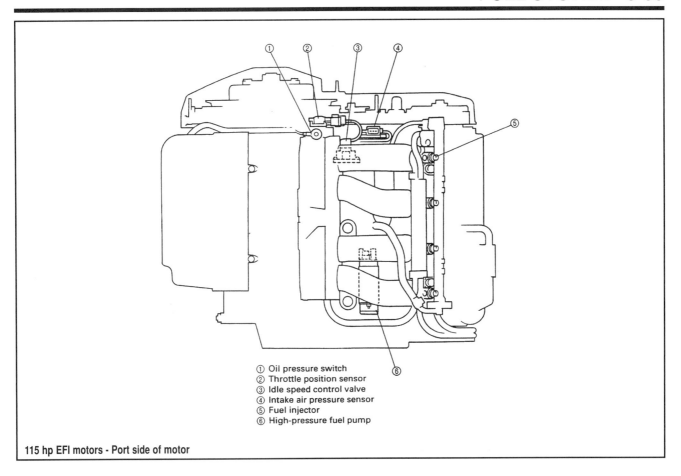

① Oil pressure switch
② Throttle position sensor
③ Idle speed control valve
④ Intake air pressure sensor
⑤ Fuel injector
⑥ High-pressure fuel pump

115 hp EFI motors - Port side of motor

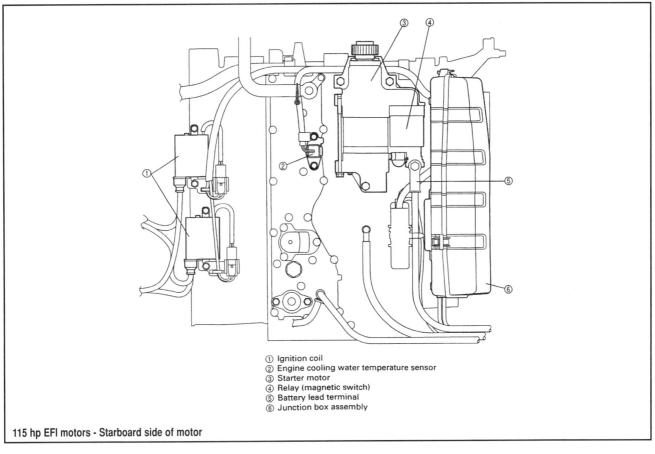

① Ignition coil
② Engine cooling water temperature sensor
③ Starter motor
④ Relay (magnetic switch)
⑤ Battery lead terminal
⑥ Junction box assembly

115 hp EFI motors - Starboard side of motor

3-90 FUEL SYSTEM

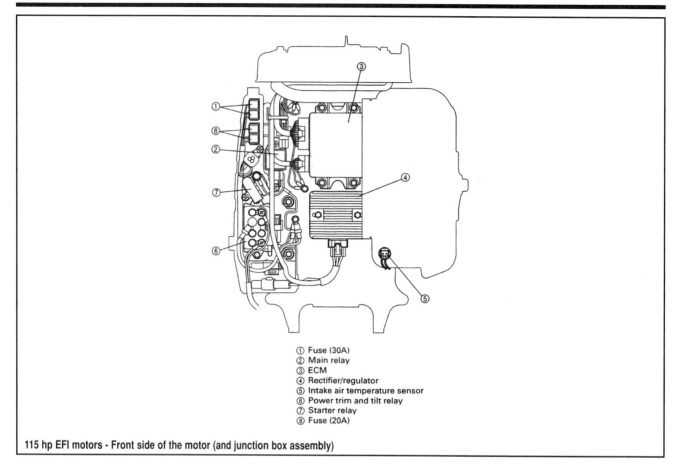

① Fuse (30A)
② Main relay
③ ECM
④ Rectifier/regulator
⑤ Intake air temperature sensor
⑥ Power trim and tilt relay
⑦ Starter relay
⑧ Fuse (20A)

115 hp EFI motors - Front side of the motor (and junction box assembly)

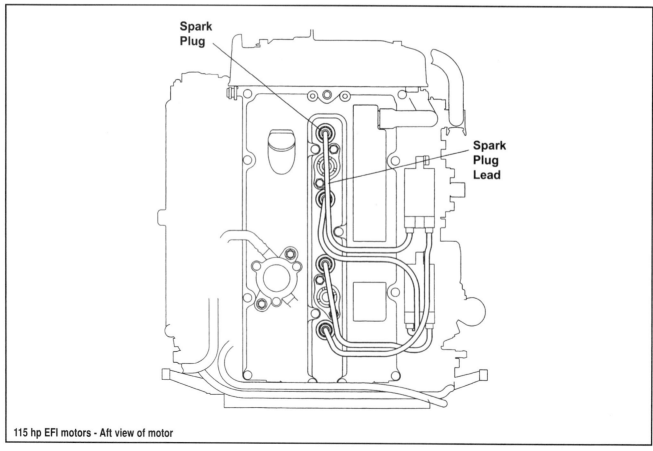

115 hp EFI motors - Aft view of motor

FUEL SYSTEM 3-91

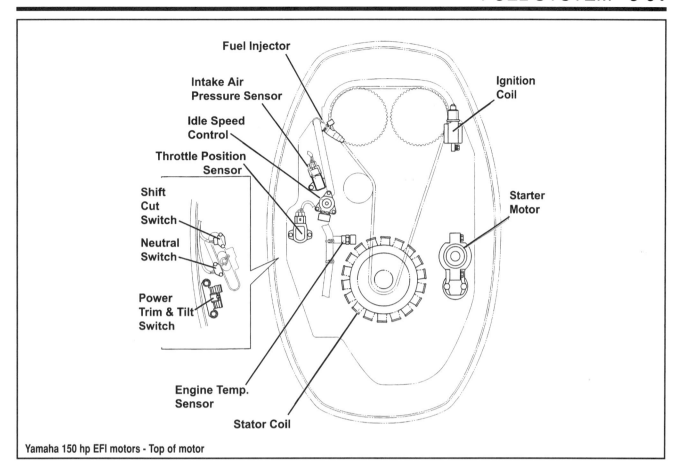

Yamaha 150 hp EFI motors - Top of motor

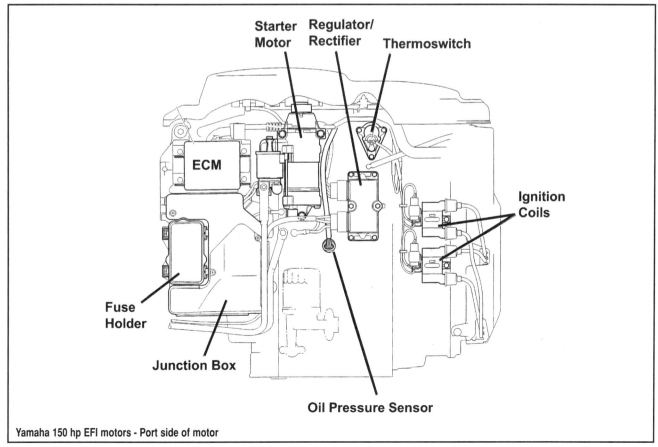

Yamaha 150 hp EFI motors - Port side of motor

3-92 FUEL SYSTEM

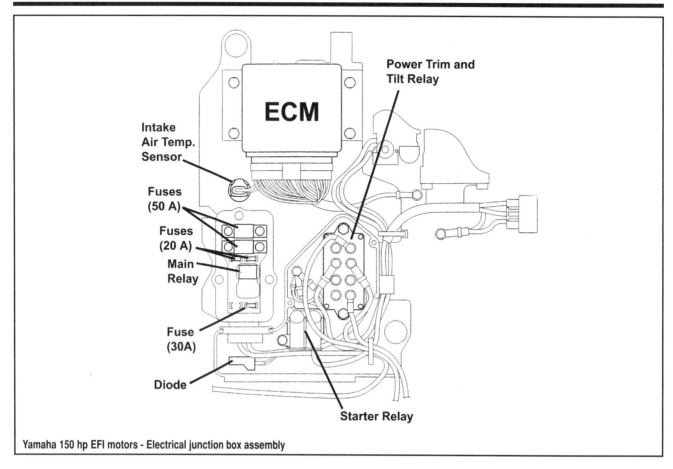

Yamaha 150 hp EFI motors - Electrical junction box assembly

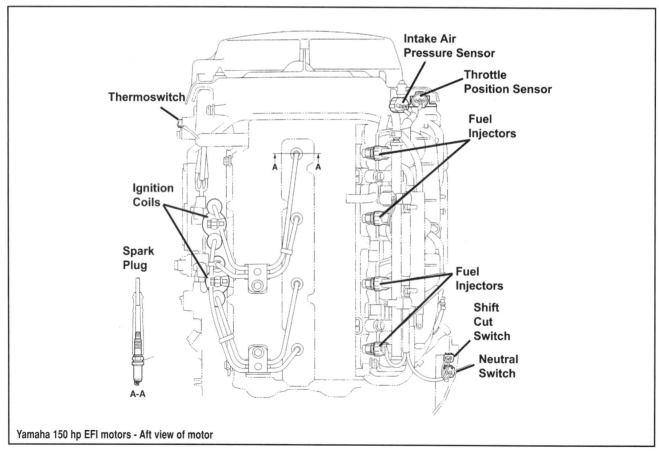

Yamaha 150 hp EFI motors - Aft view of motor

FUEL SYSTEM 3-93

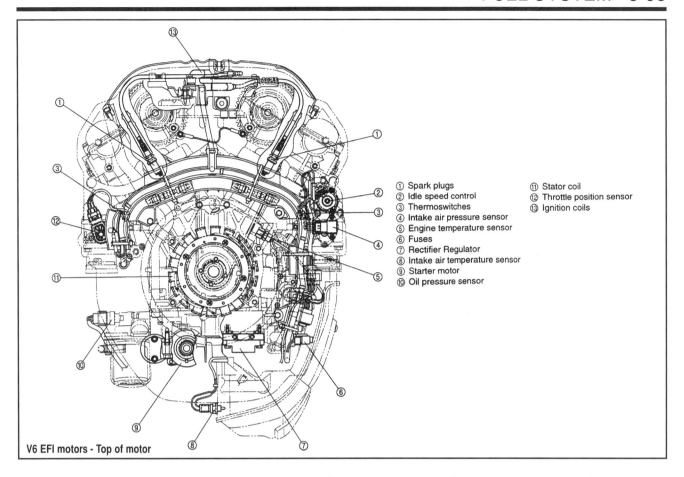

① Spark plugs
② Idle speed control
③ Thermoswitches
④ Intake air pressure sensor
⑤ Engine temperature sensor
⑥ Fuses
⑦ Rectifier Regulator
⑧ Intake air temperature sensor
⑨ Starter motor
⑩ Oil pressure sensor
⑪ Stator coil
⑫ Throttle position sensor
⑬ Ignition coils

V6 EFI motors - Top of motor

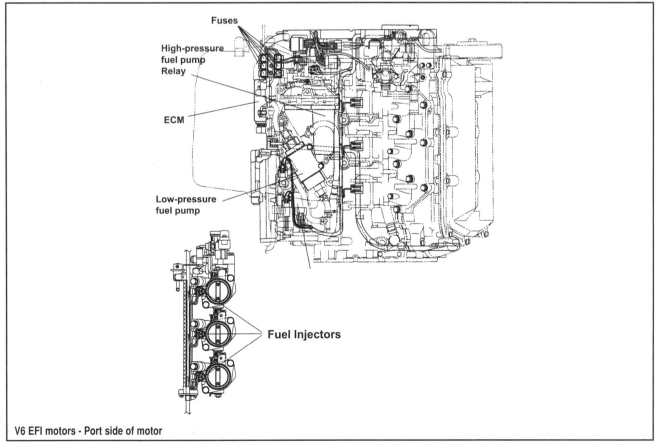

V6 EFI motors - Port side of motor

3-94 FUEL SYSTEM

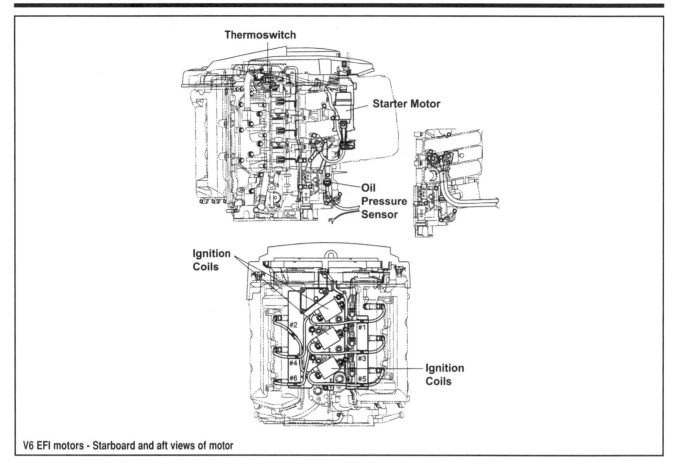

V6 EFI motors - Starboard and aft views of motor

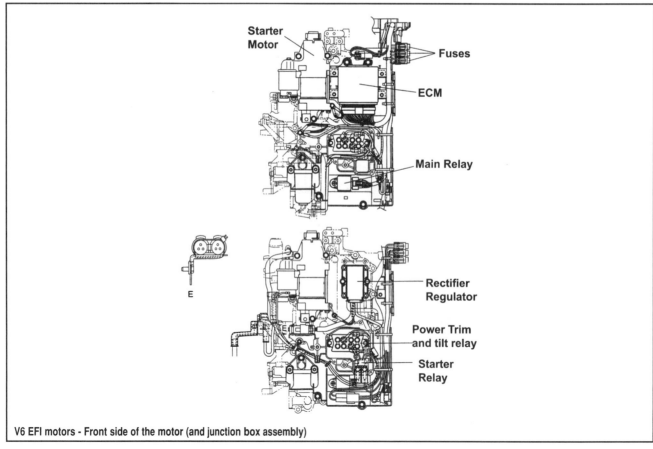

V6 EFI motors - Front side of the motor (and junction box assembly)

FUEL SYSTEM

EFI Fuel Flow Schematics

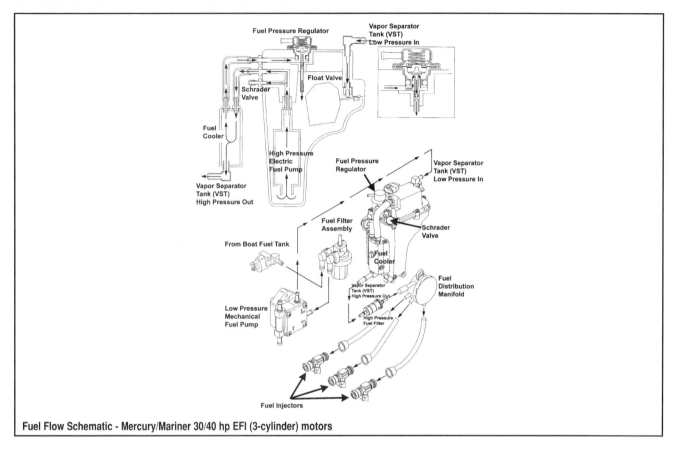

Fuel Flow Schematic - Mercury/Mariner 30/40 hp EFI (3-cylinder) motors

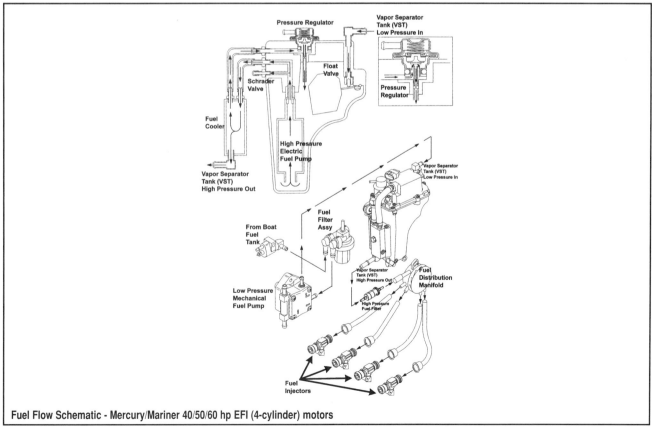

Fuel Flow Schematic - Mercury/Mariner 40/50/60 hp EFI (4-cylinder) motors

3-96 FUEL SYSTEM

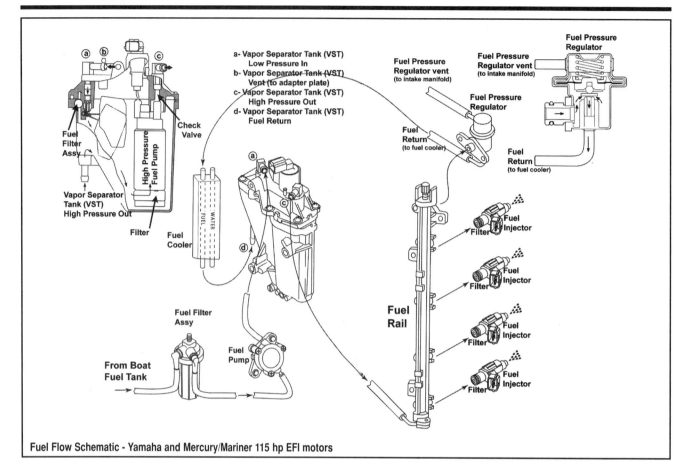

Fuel Flow Schematic - Yamaha and Mercury/Mariner 115 hp EFI motors

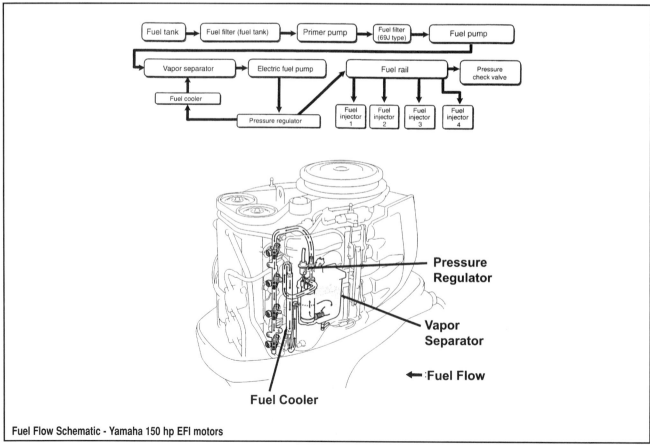

Fuel Flow Schematic - Yamaha 150 hp EFI motors

FUEL SYSTEM 3-97

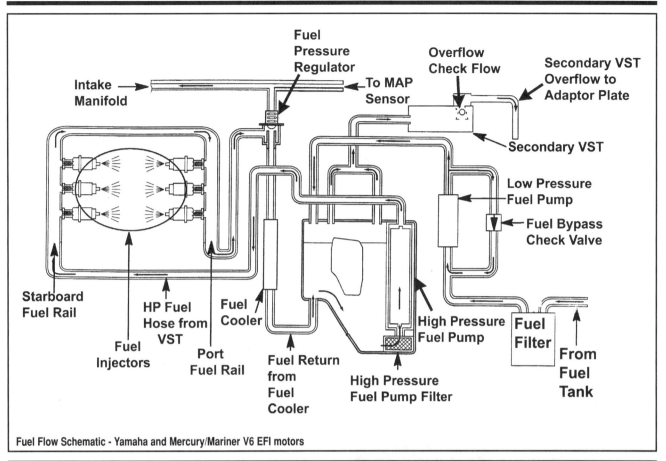

Fuel Flow Schematic - Yamaha and Mercury/Mariner V6 EFI motors

FUEL PUMP (LOW-PRESSURE) SERVICE

This section deals with the low-pressure, mechanical, diaphragm-displacement fuel pumps found on all carbureted motors and which act as a lift pump for most fuel-injected motors. The electric lift pump used on V6 EFI motors and the electric high-pressure pump used on all EFI motors are both covered in the section on Electronic Fuel Injection Systems.

Fuel Pumps

GENERAL INFORMATION

◆ See Figures 168 and 169

A fuel pump is a basic mechanical device used to pull fuel from the tank and feed it to the float bowl(s) in the carburetor(s) or the fuel vapor separator assembly on EFI motors.

On most 4-stroke motors the pump(s) is(are) mounted to the valve cover in a position so that a rocker arm can actuate a plunger in the pump itself. A major difference between the design of 2-stroke and 4-stroke motors is the fact that the vacuum and compression created by the cylinder, STAYS in the cylinder (and intake tract) of the 4-stroke motor. (Stated conversely, the crankcase is NOT part of the intake tract on 4-stroke motors and therefore does not have the same degree of alternating vacuum/pressure of a 2-stroke motors). For this reason, the pump on 4-stroke motors creates its own vacuum/pressure through the motion of the plunger activated diaphragm. The pumps work in the same basic manner, except again that they are mechanically actuated through a camshaft operated, spring loaded plunger.

The diaphragm-displacement fuel pump used on both types (2- and 4-stroke motors) is a reliable method to move fuel but can have several problems. The diaphragm and valves are moving parts subject to wear. The flexibility of the diaphragm material can go away over time, reducing or stopping flow. Rust or dirt can hang a valve open and reduce or stop fuel flow. And, obviously, if a diaphragm develops one or more holes it will loose the ability to generate sufficient amounts of vacuum/pressure to overcome the check valves.

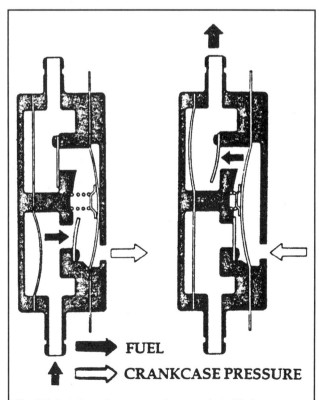

Fig. 168 4-stroke motors use a rocker arm actuated fuel pump mounted to the cylinder head/valve cover

3-98 FUEL SYSTEM

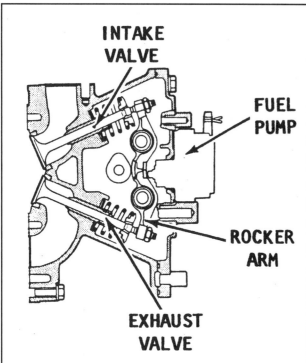

Fig. 169 A fuel pump is a basic mechanical device that pulls fuel from the tank. (the one shown is a 2-stroke pump which works on crankcase pressure, where the ones on 4-strokes create their own vacuum/pressure using a plunger activated diaphragm)

✴✴ CAUTION

The hose fittings on some pumps can be particularly fragile and can break if you try twisting or pulling off the hoses. It is usually safest to slowly pry off the hoses using a small screwdriver or prytool.

TESTING

The problem most often seen with fuel pumps is fuel starvation, hesitation or missing due to inadequate fuel pressure/delivery. In extreme cases, this might lead to a no start condition, but that is pretty rare as the primer bulb should at least allow the operator to fill the float bowl or vapor separator tank). More likely, pump failures are not total, and the motor will start and run fine at idle, only to miss, hesitate or stall at speed when pump performance falls short of the greater demand for fuel at high rpm.

Before replacing a suspect fuel pump, be absolutely certain the problem is the pump and NOT with fuel tank, lines or filter. A plugged tank vent could create vacuum in the tank that will overpower the pump's ability to create vacuum and draw fuel through the lines. An obstructed line or fuel filter could also keep fuel from reaching the pump. Any of these conditions could partially restrict fuel flow, allowing the pump to deliver fuel, but at a lower pressure/rate. A pump delivery or pressure test under these circumstances would give a low reading that might be mistaken for a faulty pump. Before testing the fuel pump, refer to the testing procedures found under Fuel Lines and Fitting to ensure there are no problems with the tank, lines or filter.

A quick check of fuel pump operation is to gently squeeze the primer bulb with the motor running. If a seemingly rough or lean running condition (especially at speed) goes away when the bulb is squeezed, the fuel pump is suspect.

If inadequate fuel delivery is suspected and no problems are found with the tank, lines or filters, a conduct a quick-check to see how the pump affects performance. Use the primer bulb to supplement fuel pump. This is done by operating the motor under load and otherwise under normal operating conditions to recreate the problem. Once the motor begins to hesitate, stumble or stall, pump the primer bulb quickly and repeatedly while listening for motor response. Pumping the bulb by hand like this will force fuel through the lines to the vapor separator tank, regardless of the fuel pump's ability to draw and deliver fuel. If the engine performance problem goes away while pumping the bulb, and returns when you stop, there is a good chance you've isolated the low pressure fuel pump as the culprit. Depending upon the model you may be able to perform a pressure or vacuum check (Yamaha and Mercury tend to recommend the later) or you'll have to disassemble the pump to physically inspect the check valves and diaphragms (your only option on carburetor integrated pumps, but it's not that difficult and can be done on ALL pumps).

✴✴ WARNING

Never run a motor without cooling water. Use a test tank, a flush/test device or launch the craft. Also, never run a motor at speed without load, so for tests running over idle speed, make sure the motor is either in a test tank with a test wheel or on a launched craft with the normal propeller installed.

Vacuum Checking the Delivery System

◆ See Figure 170

Fuel system vacuum testing is an excellent way to pinpoint air leaks, restricted fuel lines and fittings or other fuel supply related performance problems.

When a fuel starvation problem is suspected such as engine hesitation or engine stopping, perform the following fuel system test to see if you should check the fuel tank, filter(s) and lines or check the pump itself:

■ **For comprehensive pressure and vacuum testing on the low-pressure fuel pump for 60 hp and smaller Mercury/Mariner EFI motors, please refer to the Electronic Fuel Injection section.**

1. Connect the piece of clear fuel hose to a side barb of a "T" fitting.
2. Connect one end of a long piece of fuel hose to the vacuum gauge and the other end to the center barb of the "T" fitting.

■ **Use a long enough piece of fuel hose so the vacuum gauge may be read at the helm.**

3. Remove the existing fuel hose from the fuel tank side of the fuel pump (that usually means between the fuel filter and the fuel pump), and connect the remaining barb of the "T" fitting to the fuel hose.
4. Connect the short piece of clear fuel hose to the fuel check valve leading from the fuel filter. If a check valve does not exist, connect the clear fuel hose directly to the fuel filter.
5. Check the vacuum gauge reading after running the engine long enough to stabilize at full power.

■ **The vacuum is to not exceed 4.5 in. Hg (15.2 kPa) for these motors. On the bright side, if the vacuum DOES exceed this figures, the problem is more than likely NOT the pump!**

6. An anti-siphon valve (required if the fuel system drops below the top of the fuel tank) will generally cause a 1.5 to 2.5 in. Hg (8.4 kPa) increase in vacuum.
7. If high vacuum is noted, move the T-fitting to the fuel filter inlet and retest.

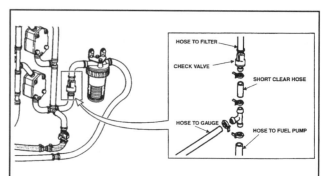

Fig. 170 Connecting a vacuum gauge inline in preparation for isolating fuel system problems

FUEL SYSTEM 3-99

8. Continue to the fuel filter inlet and along the remaining fuel system until a large drop/change in vacuum locates the problem.

9. A good clean water separator fuel filter will increase vacuum about 0.5 in. Hg (1.7 kPa).

10. Small internal passages inside a fuel selector valve, fuel tank pickup, or fuel line fittings may cause excessive fuel restriction and high vacuum.

11. Unstable and slowly rising vacuum readings, especially with a full tank of fuel, usually indicates a restricted vent line.

■ Bubbles in the clear fuel line section indicate an air leak, making for an inaccurate vacuum test. Check all fittings for tightened clamps and a tight fuel filter.

■ Vacuum gauges are not calibrated and some may read as much as 2 in. Hg (6.8 kPa) lower than the actual vacuum. It is recommended to perform a fuel system test while no problems exist to determine vacuum gauge accuracy.

Checking Pump Pressure

■ For comprehensive pressure and vacuum testing on the low-pressure fuel pump for 60 hp and smaller Mercury/Mariner EFI motors, please refer to the Electronic Fuel Injection section.

A basic low-pressure fuel pump pressure check is included in the Fuel System Checks found under the Maintenance and Tune-Up section of this guide For most 4-stroke motors Yamaha publishes various pressure or, more likely, fuel delivery specifications. Mercury tends to publish more pressure than delivery specifications. Either way, when available they are contained in the Engine Specifications charts located in the Powerhead section.

Check the appropriate chart to see what specification is available. If a pressure specification is provided (they are often pump MAX delivery pressure and NOT a required minimum pressure) you can use it along with the Pressure Check procedure to decide if further testing is required.

You'll notice one problem with most Yamaha fuel delivery specs, they're unrealistic for testing purposes. When provided, Yamaha often tends to give you the amount (in Gallons/Liters) that the pump should deliver over a predetermined amount of time, specifically an HOUR. Chances are you're not going to run the engine for an hour (on another fuel source, such as a gravity feed) just to measure the amount of fuel displaced by the pump in that amount of time. But, you've got an option, you can divide the specification by 60 and determine what portion of a gallon will be delivered in one minute. That's much more realistic. You'll also likely wind up with a number which should be converted into Oz (L may still work) to match whatever measuring device you've got at your disposal (there are 128 ounces in a Gallon, so multiply the GPH by 128 and divide by 60 to get ounces per minute). For example, specifications for some motors are 4.75 gallons (18.0L)/hour for 9.9 hp motors or more commonly 18.5 gallons (70L)/hour for most 3-cylinder and larger 4-stroke motors. This translates into about 10.1 fl. Oz. (300ml)/minute for 9.9 hp motors and 39.5 fl. Oz. (1.2L) for most larger motors.

You can check fuel pump delivery by measuring the amount of fuel that is expelled from a disconnected fuel pump outlet hose while the motor runs at speed for one minute. In order to safely conduct this test the motor must either be in a test tank or on a launched craft as engine speed should be maintained toward the high end of mid-range (about 3000 rpm, but check the Engine Specifications charts for exact requirements). Before starting the test you'll need to run the engine and make sure the carburetor float bowl(s) are full (since they won't continue to receive fuel once the test is started). Carefully disconnect the fuel pump outlet line from the carburetor(s) and direct it into an approved container. Start and run the engine for one minute, then shut down the powerhead and measure the amount of fuel collected using a graduated cylinder or beaker.

■ The fuel consumption rate of many multi-cylinder 4-strokes makes the test a little more complicated, since the motor may consume all of the fuel in the float bowls before one minute expires. If this is the case you've got 2 choices. For one, if the motor only runs for another 30 seconds, multiply the amount of fuel measured by 2 and compare it to the specification provide. Your other option is to connect a separate fuel tank line directly to the carburetors and gently squeeze the primer bulb on this auxiliary tank a couple of times as the test is run to continue feeding the carburetor float bowls during the test.

Fuel Pump Diaphragm and Check Valve Testing

◆ See Figures 171 and 172

For almost all non-carburetor integrated fuel pumps you can use a simple hand-held vacuum/pressure pump and gauge to perform basic tests of the fuel pump diaphragm and check valve conditions.

Sometimes we have a hard time getting a grip on why a manufacturer chooses to do some of the things they do. When it comes to Yamaha service information, the format of what is or what is not available for a motor sometimes make no sense. In this case we bring this up because sometimes it might just be who at Yamaha wrote the given service publication or update, because they specifically give this vacuum/pressure check for all V4 and V6 2-strokes, but only a few of the inline 2-strokes (such as the 48 hp twin), even when there appear to be no differences in the designs of the pumps themselves.

The same seems to go for 4-stroke motors, where Yamaha only mentions this check on the 9.9 hp twin, the 40/45/50J/50 and 60J/60 hp 4-cylinder motors as well as on the 115 and 150 hp EFI 4-cylinder motors, despite the fact that the fuel pumps are of identical design on virtually all 15-100 hp 4-stroke motors?

To make matters worse, Mercury doesn't seem to mention this test on any of their motors, except the 115 hp (which is a Yamaha powerhead, complete with Yamaha EFI system).

For this reason, we feel that this test is applicable to all Yamaha, non-carburetor integrated diaphragm-displacement fuel pumps (and potentially applicable to Mercury/Mariner pumps as well). However, if the motor on which you are working is not listed above, we recommend that you DO NOT condemn it based on these test results until after you've also disassembled it (the only way that Yamaha and Mercury recommends testing it in that case). However, conversely, if the pump DOES pass this test, we feel it is VERY likely that the pump is not your problem.

Attach a hand-held vacuum/pressure pump (like the Mity Vac®) to the fuel pump inlet fitting (the fitting to which the filter/tank line connects). Using the pump apply 7 psi (50 kPa) of pressure while manually restricting the outlet fitting (or fittings, there are 2 on some pumps) using your finger. If the diaphragm is in good condition it will hold the pressure for at least 10 seconds.

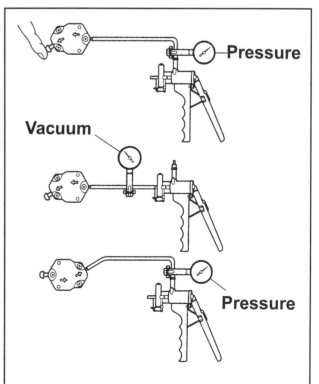

Fig. 171 Vacuum/pressure checking a typical Yamaha fuel pump (the shape of the pump body will vary)

3-100 FUEL SYSTEM

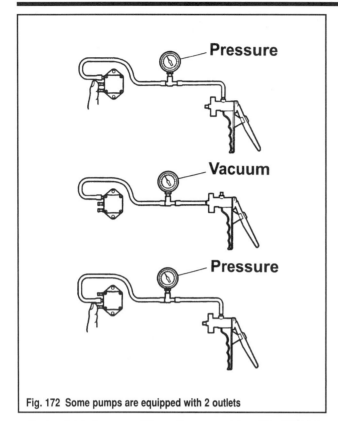

Fig. 172 Some pumps are equipped with 2 outlets

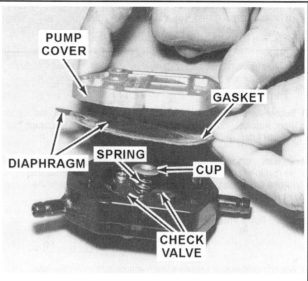

Fig. 173 Most pumps contain replaceable diaphragms and check valves

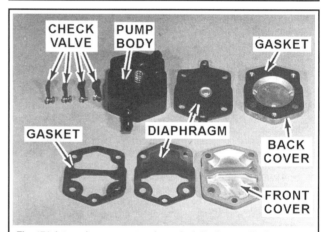

Fig. 174 Internal components of a typical diaphragm-displacement pump (vacuum actuated 2-stroke pump shown, 4-stroke very similar but plunger actuated)

Next, switch to the vacuum fitting on the pump and draw 4.3 psi (30 kPa) of negative pressure (vacuum) on the fuel pump inlet fitting. This checks if the one-way check valve in the pump remains closed. It should hold vacuum for at least 10 seconds.

Now, move the pressure pump to the fuel pump outlet fitting. This time cover the inlet fitting (and other outlet fitting, if applicable) with your finger and apply 7 psi (50 kPa) of pressure. Again, the pressure must hold for at least 10 seconds.

■ If pressure does not hold, verify that it is not leaking past your finger (on the opposite fitting or fittings) when applicable or from a pump/hose test connection. If leakage is occurring in the diaphragm, the fuel pump should be overhauled.

Visually Inspecting the Pump Components

◆ See Figures 173 and 174

The only way that Yamaha recommends to inspect MOST of their diaphragm-displacement fuel pumps is through disassembly and visual inspection. Mercury usually provides pressure specs, but also recognizes that the sometimes the best method is simply to disassemble and check the components out for yourself.

To perform a visual inspection remove and/or disassemble the pump according to the procedures found either in this section.

Wash all metal parts thoroughly in solvent, and then blow them dry with compressed air. Use care when using compressed air on the check valves. Do not hold the nozzle too close because the check valve can be damaged from an excessive blast of air.

Inspect each part for wear and damage. Visually check the pump body/cover assembly for signs of cracks or other damage. Verify that the valve seats provide a flat contact area for the valve. Tighten all check valve connections firmly as they are replaced.

Check the diaphragms for pin holes by holding it up to the light. If pin holes are detected or if the diaphragm is not pliable, it MUST be replaced.

■ If you've come this far and are uncertain about pump condition, replace the diaphragms and check valves, you've already got to replace any gaskets or O-rings which were removed. Once the pump is rebuilt, you can remove it from your list of potential worries for quite some time.

Plunger Actuated Fuel Pumps

DESCRIPTION & OPERATION

Although the basic function of the fuel pump used on 4-stroke motors is similar to the pumps used on 2-strokes, the method of pump actuation is vastly different. On 4-stroke motors the fuel pump is located on the cylinder head cover and is operated by one of the rocker arms. As the rocker arm moves, a plunger inside the pump is alternately depressed and released.

A diaphragm is attached to the plunger which creates alternate vacuum and pressure to operate the suction and discharge check valves. This configuration ensures efficient delivery of fuel to the carburetor(s) or EFI vapor separator tank at all engine speeds and operating conditions.

REMOVAL & INSTALLATION

◆ See Figures 175 and 176

Fuel pump service is very straightforward on these models. Although the shape of the pump may vary slightly from model-to-model (or year-to-year), they are all assembled units consisting of a front (or outer) cover, pump body and, in some cases, a separate rear cover, along with internal diaphragms, check valves and a plunger assembly.

FUEL SYSTEM 3-101

Fig. 175 Fuel pumps on 4-stroke motors mount to the valve cover so they can be actuated by a rocker arm

Fig. 176 These pumps are fastened to the motor by 2 bolts through the rear cover

The pumps on all Yamaha and Mercury 4-stroke motors are secured to the valve cover by 2 bolts threaded through the a flange on the PUMP BODY or REAR COVER. Don't confuse these bolts with the 3 or 4 screws (normally Philips head) that are threaded through the front cover and hold the pump assembly together.

■ The pump used on the 8/9.9 hp (232cc) motor differs VERY slightly in 2 ways. First, there are 2 Philips head screws (not bolts) holding it to the valve cover. And second, the 4 cover screws are threaded from the REAR cover (powerhead side) out toward the outer cover.

12. For safety, disconnect the negative battery cable, if equipped, and/or remove and ground the spark plug lead(s). Also for safety, on motors with portable tanks, disengage the quick-connect fitting from the motor.

■ These actions will help prevent the possibility of sparks that could potentially ignite fuel vapors, help prevent accidental starting of the motor while you are working on it, and lastly, help prevent raw fuel from spraying through an open fitting (should someone squeeze or step on the primer bulb).

✳✳ CAUTION

The hose fittings on some pumps can be particularly fragile and can break if you try twisting or pulling off the hoses. It is usually safest to slowly pry off the hoses using a small screwdriver or prytool.

13. Tag and disconnect the inlet and outlet hoses from the fuel pump assembly. Many pumps used by Yamaha and Mercury are already labeled **IN** or **OUT** to help with tagging (or at least have arrows on the fittings showing a direction of fuel flow). These models normally use spring-type clamps or plastic wire ties on the fuel pump hose fittings. The clamps are removed by gently squeezing using a pair of pliers and then sliding the clamps back up the hose, past the point into which the fitting protrudes. Obviously, the wire ties must be carefully cut for removal, making sure you don't nick and damage the fuel hose itself.

■ A small amount of fuel will normally still be present in the lines. Have a rag handy to catch any escaping fuel. Also, be sure to cover the lines (either cap them with golf-tees or cover them with plastic bags) to help keep contaminants out of the fuel lines.

14. Remove the 2 bolts (screws on the 8/9.9 hp motor) securing the fuel pump (via the rear pump cover) to the valve cover. Pull the pump carefully out from the cover.
15. Remove and discard the O-ring from the back (valve cover) side of the fuel pump assembly. Yes, in theory the O-ring could be reused if it is not deformed or damaged, but why take the risk of an oil leak.
16. Refer to Overhaul in this section for details on disassembling and assembling the fuel pump assembly.

To Install:

17. Install a NEW O-ring over the piston side of the fuel pump (over the protrusion that extends into the valve cover when installed).
18. Carefully position the fuel pump to the valve cover, inserting the piston into the cover where it will be actuated by the rocker arm.

■ If the rocker arm is in a position that places pressure on the fuel pump piston is may be difficult to properly seat the pump and align the bolt holes. In this case, slowly rotate the flywheel in the normal direction of rotation (CLOCKWISE) by hand until the rocker arm moves to the point of travel furthest away from the piston.

19. Thread the fuel pump retaining bolts through the pump and into the valve cover. Make sure the pump is properly aligned and seated, then tighten the bolts securely.
20. Reconnect the fuel pump inlet and outlet lines, as tagged during removal. Luckily the pump cover on these motors is often (but not always) embossed with word or arrows showing fuel direction. But when equipped, the line from the tank and fuel filter should go to the arrow pointing inward toward the pump, while the fuel line for the carburetor(s) or vapor separator tank should attach to the fitting whose arrow is pointing outward, away from the pump.
21. Secure the hoses using the spring clamps or using new plastic wire ties (as applicable). Be sure to replace any clamp that has lost its tension.
22. Reconnect the fuel inlet quick-connect, the spark plug wires and/or the negative battery cable, as applicable.
23. Properly pressurize the fuel system using the primer bulb and verify that there are no leaks. It is a good idea to actually run the motor with the top case removed and observe that there are no leaks under normal operating conditions as well, just to be sure.

OVERHAUL

◆ See Figures 177 thru 187

Fuel pump overhaul is relatively straightforward on these models. Just be sure that the work area is clean and free of dust/dirt or other contaminants that could get into the fuel pump. Take your time, paying attention to the direction components are facing, especially check valves. All components which are reused must be returned to their original position. Also, take your time to make sure the diaphragm is properly aligned and seated so it does not become pinched or damaged during assembly.

Once the pump is disassembled, wash all metal components with solvent and then blow them dry using low-level compressed air. Use of excessive

3-102 FUEL SYSTEM

pressure on the diaphragm or check valves could destroy them, so use care. Check each component for wear or damage. Verify that valve seats provide a flat contact area for the valve. Visually inspect the diaphragm for pin holes by holding it up to the light. The diaphragm MUST be replaced if it has stiffened or if it contains any pin holes/tears.

To ensure proper fuel pump operation always use new gaskets. The fuel pumps are assembled dry, meaning that no sealant should be used of any type on the gaskets, however you may wish to wet the diaphragm with a small amount of gasoline.

Although the shape of the pump varies slightly from model-to-model (or year-to-year) all are overhauled in essentially the same manner.

1. Remove the fuel pump assembly from the powerhead, as detailed in this section.

2. Check the Engine Specification charts in the Powerhead section to determine if there is a pump plunger stroke specification. If so, measure the distance the plunger travels before disassembly to determine if it is in spec.

3. Since the components (front cover, rear cover and body) on some pumps can align in different ways, scribe a matchmark across all 3 (or 2 in some cases) pieces to make sure the covers and body are properly aligned during assembly.

4. Loosen and remove the 3 or 4 screws (most pumps use 4 screws, except the round-body late-model pump found on some larger 4-cylinder motors) that secure the pump body and cover in position. All pumps (except the 8/9.9 hp/232cc) thread these screws from the front (outer) cover toward the rear (inner) cover. The 4 screws on the 8/9.9 hp motor are threaded from the rear cover toward the front cover. Hold the housing together as the last of the bolts is removed, as spring/diaphragm pressure may try to force them apart.

5. Remove the front cover (the cover which does not face the powerhead) from the pump body, then remove the gasket/packing and/or the outer diaphragm, as equipped.

6. Carefully pull the pump body away from the rear cover (the cover with the plunger, which attaches to the powerhead).

■ Be sure to note check valve placement and orientation for installation purposes. Most Yamaha and Mercury/Mariner models use check valves which are screwed into position, however a few Mercury/Mariner models (specifically all of the 30-60 hp EFI models, as well as SOME of the 30/40 hp 3-cylinder carburetor models) instead use a push fit check valve assembly.

7. Except for late-model (round-bodied) pumps and 8/9.9 hp pumps and for 30-60 hp Mercury/Mariners with push fit check valves, loosen the 2 screws (and nuts on many pumps) that secure the check valves to the pump body (note that one screw and valve is located on either side of the pump body and the nut that secures it on the other side).

8. For 30-60 hp Mercury/Mariners with push fit check valves, ONLY IF IT IS NECESSARY FOR REPLACMENT use a pair of pliers to grasp and pull the check valves free of the pump body.

9. For 8/9.9 hp (232cc) motors, remove the Philips head screw securing the plate and check valve to the pump cover. Note the plate and valve positioning for installation purposes.

■ On most models the plunger assembly uses a set pin which must be withdrawn in order to remove the diaphragm and plunger components. On most pumps, you must turn the diaphragm 90 degrees in order to pivot the piston and plunger retaining pin to a point where the pin can be removed. Although Yamaha doesn't always detail a method for this, Mercury normally tells you to slowly rotate the plunger until the pin and slots in the side of the plunger/housing/cover align. At that point carefully compress the plunger and allow the pin to slide out or gently pull it free. Upon assembly you'll have to do the reverse to secure the plunger.

10. Working on the rear cover, remove the retaining pin from the diaphragm shaft, then carefully remove the plunger and plunger spring (taller, thinner spring). Pull the diaphragm out of the rear cover and remove the diaphragm spring (which has a larger diameter than the plunger spring).

11. If equipped (on some non-US For 8/9.9 hp models), loosen the 2 screws securing the fuel inlet fitting assembly and seal to the pump cover. Remove the assembly and discard the seal.

12. Thoroughly clean and inspect all components. Replace any that are worn or damaged.

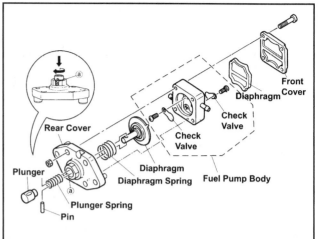

Fig. 177 Exploded view of the fuel pump assembly used on 4 hp Yamaha motors

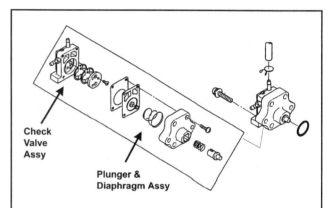

Fig. 178 Exploded view of the fuel pump assembly used on 4/5/6 hp Mercury/Mariner motors

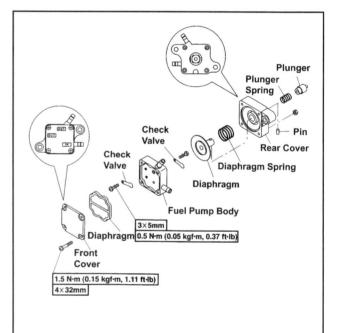

Fig. 179 Exploded view of the fuel pump assembly used on 6/8 hp Yamaha motors

FUEL SYSTEM 3-103

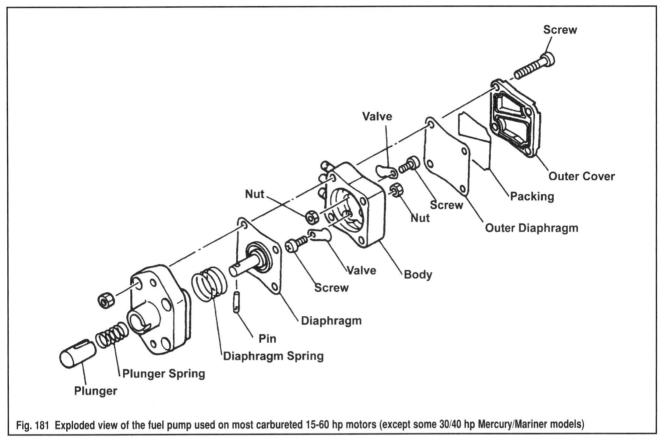

Fig. 181 Exploded view of the fuel pump used on most carbureted 15-60 hp motors (except some 30/40 hp Mercury/Mariner models)

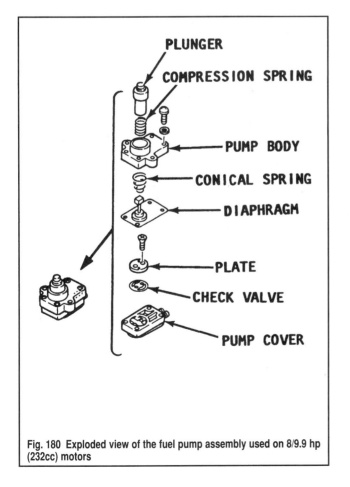

Fig. 180 Exploded view of the fuel pump assembly used on 8/9.9 hp (232cc) motors

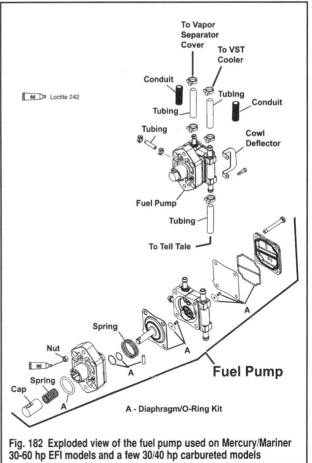

Fig. 182 Exploded view of the fuel pump used on Mercury/Mariner 30-60 hp EFI models and a few 30/40 hp carbureted models

3-104 FUEL SYSTEM

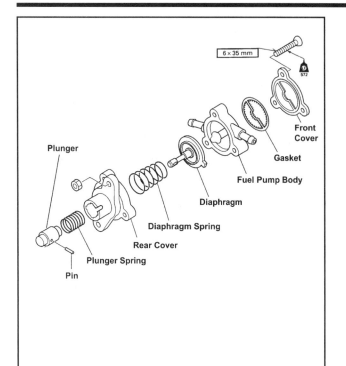

Fig. 183 ...but on the 8/9.9 hp (232cc) motors, the cover screws are usually

To Assemble:

13. Position the diaphragm spring (larger diameter) and carefully compress it into the rear cover while positioning the diaphragm and inserting the diaphragm shaft through the pump body.

■ On some pumps the diaphragm spring is conical (tapered) meaning one end is larger than the other. On these pumps the spring will only seat properly when installed facing the correct direction.

14. Install the plunger and plunger spring, then insert the retaining pin through the diaphragm shaft to hold the assembly in place. On most models you'll then have to rotate the diaphragm 90 degrees to secure the pin.
15. For 8/9.9 hp (232cc) motor pumps, install the check valve, plate and screws to the pump cover. Be sure to align the recess in the check valve and plate with the projection of the pump cover.
16. For 30-60 hp Mercury/Mariners with push fit check valves, Mercury recommends a home-made check valve installation tool for most of their models. This tool can be made from a metal rod or wooden dowel which is 0.375 in. (9.5mm) in diameter, by drilling a 5/64 in. (2mm) hole in the center of one end, then inserting a peg of the same size into the hole until only 0.302-0.322 in. (7.7-8.2mm) of the peg protrudes from the end of the tool. The peg will insert through the center of the check valve. If the check valves were removed on one of these models, use the aforementioned installation tool to carefully push the check valve assembly into the carburetor body.
17. For all except the 8/9.9 hp (232cc) motor or the late-model (round-bodies) pumps and 30-60 hp Mercury/Mariner models with press-fit valves, install the check valves to the pump body in the positions noted during removal. Tighten the retaining screw and nut to secure each valve.
18. Coat the threads of the 3 or 4 cover screws with Loctite® 572 or an equivalent thread sealant and place aside for use in the next steps.
19. If applicable, place the fuel pump body over the rear cover, then hold the assembly as you install the outer diaphragm and/or gasket/packing, as applicable.
20. Install the front (outer) cover in position over the packing and outer diaphragm, then install the cover bolts (and nuts, if applicable). Tighten the fasteners securely.
21. Check the pump plunger for smooth movement.
22. Install the pump, as detailed in this section.

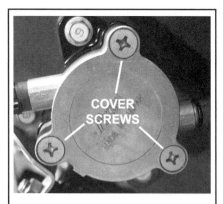

Fig. 184 On most pumps, the cover screws are threaded from the front...

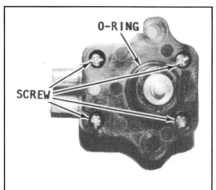

Fig. 185 ...but on the 8/9.9 hp (232cc) motors, the cover screws are usually threaded from behind

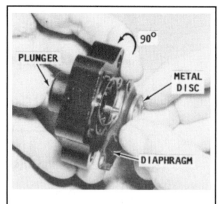

Fig. 186 On most pumps, the diaphragm is removed by rotating 90 degrees

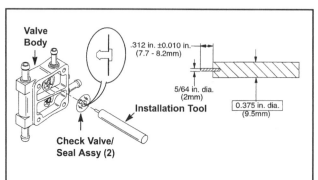

Fig. 187 Press fit check valve installation with fabricated tool for 30-60 hp Mercury/Mariner motors

FUEL SYSTEM

SPECIFICATIONS

Carburetor Set-Up Specifications - Yamaha 4-Stroke Motors

Model (Hp)	No. of Cyl	Engine Type	Year	Displace cu. in. (cc)	Pilot Screw Initial Low Speed Setting	Float Height/Drop Setting In. (mm)	Valve Seat Size In. (mm)
2.5	1	4-stroke	2003-04	4.4 (72)	1 1/2 - 1 3/4	0.41-0.45 (10.5-11.5)	-
4	1	4-stroke	1999-04	6.7 (112)	1 1/2 - 1 3/4 ①	0.71-0.75 (18.0-19.0)	-
6	2	IL 4-stroke	2001-04	12 (197)	not adjustable	0.75-0.79 (19.0-20.0)	-
8	2	IL 4-stroke	2001-04	12 (197)	not adjustable	0.75-0.79 (19.0-20.0)	-
8	2	IL 4-stroke	1997	14 (232)	2 - 4	0.96-1.04 (24.5-26.5)	0.05 (1.2)
	2	IL 4-stroke	1998-04	14 (232)	2 1/2 - 4 1/2	0.96-1.04 (24.5-26.5)	0.05 (1.2)
9.9	2	IL 4-stroke	1995-97	14 (232)	2 - 4 ②	0.96-1.04 (24.5-26.5)	0.05 (1.2)
	2	Standard	1998-04	14 (232)	2 - 4 ②	0.96-1.04 (24.5-26.5)	0.05 (1.2)
	2	High thrust	1998-04	14 (232)	2 1/2 - 4 1/2 ②	0.96-1.04 (24.5-26.5)	0.05 (1.2)
15	2	IL 4-stroke	1998-04	20 (323)	3/4 - 1 3/4 ①	0.38-0.42 (9.5-10.5)	-
25	2	IL 4-stroke	1998-04	30 (498)	not adjustable	0.53-0.57 (13.5-14.5)	-
30	3	IL 4-stroke	2001-04	46 (747)	not adjustable	0.53-0.57 (13.5-14.5)	-
40J	3	IL 4-stroke	2002-04	46 (747)	not adjustable	0.53-0.57 (13.5-14.5)	-
40	3	IL 4-stroke	2000-04	46 (747)	not adjustable	0.53-0.57 (13.5-14.5)	-
40	4	IL 4-stroke	1999-00	58 (935)	1 3/4 - 2 3/4	0.37-0.41 (9.5-10.5)	-
45	4	IL 4-stroke	1996-00	58 (935)	not adjustable	0.37-0.41 (9.5-10.5)	-
50J	4	IL 4-stroke	2002	58 (935)	1 3/4 - 2 3/4 ①	0.37-0.41 (9.5-10.5)	-
50	4	IL 4-stroke	1995-04	58 (935)	1 3/4 - 2 3/4 ①	0.37-0.41 (9.5-10.5)	-
60J	4	IL 4-stroke	2003-04	61 (996)	not adjustable	0.20 (5.2)	-
60	4	IL 4-stroke	2002-04	61 (996)	not adjustable	0.20 (5.2)	-
75	4	IL 4-stroke	2003-04	97 (1596)	2 - 3 ①	0.55 (14)	-
80	4	IL 4-stroke	1999-02	97 (1596)	2 - 3 ①	0.55 (14)	-
90J	4	IL 4-stroke	2003-04	97 (1596)	1 1/2 - 2 1/2 ①	0.55 (14)	-
90	4	IL 4-stroke	2003-04	97 (1596)	1 1/2 - 2 1/2 ①	0.55 (14)	-
100	4	IL 4-stroke	1999-02	97 (1596)	1 1/2 - 2 1/2 ①	0.55 (14)	-

Because manufacturers differ with what specifications they choose to provide, please refer to the Mercury/Mariner charts for additional specificitions which MAY apply to the Yamaha model in question as well

Initial low speed setting turn(s): back (counterclockwise) from a *lightly* seated position

① Specification is for adjustable carburetors normally sold worldwide. Some EPA regulated models sold in the US (including most sold after 1998) may be sealed and may not allow adjustment

② Specification is for adjustable carburetors normally sold worldwide. Some EPA regulated models sold in the US (including most sold after 1998) and most Swiss models may be sealed and may not allow adjustment

Carburetor Set-Up Specifications - Mercury/Mariner 4-Stroke Motors

Model (Hp)	No. of Cyl	Engine Type	Year	Displace cu. in. (cc)	Pilot Screw Initial Low Speed Setting	Float Height/Drop Setting In. (mm)
4	1	Single 4-st	1999-00	7.5 (123)	2 1/2 - 3 1/2	0.35-0.39 (9-10)
	1	U.S.	2001-04	7.5 (123)	Not Adjustable	0.35-0.39 (9-10)
	1	World	2001-04	7.5 (123)	1 5/8 - 2 5/8	0.35-0.39 (9-10)
5	1	Single 4-st	1999-00	7.5 (123)	2 1/2 - 3 1/2	0.35-0.39 (9-10)
	1	U.S.	2001-04	7.5 (123)	Not Adjustable	0.35-0.39 (9-10)
	1	World	2001-04	7.5 (123)	1 - 2	0.35-0.39 (9-10)
6	1	Single 4-st	2000	7.5 (123)	2 1/2 - 3 1/2	0.35-0.39 (9-10)
	1	U.S.	2001-04	7.5 (123)	Not Adjustable	0.35-0.39 (9-10)
	1	World	2001-04	7.5 (123)	2 1/4 - 3 1/4	0.35-0.39 (9-10)
8	2	IL 4-stroke	1997-98	14 (232)	Not Adjustable	0.96-1.04 (24.5-26.5)
9.9	2	IL 4-stroke	1995-98	14 (232)	2 - 4 ②	0.96-1.04 (24.5-26.5)
9.9	2	IL 4-stroke	1999-04	20 (323)	not adjustable	0.57-0.65 (14.5-16.5)
15	2	IL 4-stroke	1999-04	20 (323)	not adjustable	0.57-0.65 (14.5-16.5)
25	2	IL 4-stroke	1997-04	30 (498)	not adjustable	0.47-0.63 (12-16)
30 Carb	3-cyl	IL 4-stroke	1999-04	46 (747)	not adjustable	0.47-0.63 (12-16)
40 Carb	3-cyl	IL 4-stroke	1999-04	46 (747)	not adjustable	0.47-0.63 (12-16)
40	4	IL 4-stroke	1997	58 (935)	2 1/2	0.37-0.41 (9.5-10.5)
	4	IL 4-stroke	1998-00	58 (935)	not adjustable	0.37-0.41 (9.5-10.5)
45	4	IL 4-stroke	1995-00	58 (935)	2 1/2	0.37-0.41 (9.5-10.5)
50	4	IL 4-stroke	1995-97	58 (935)	2 1/2	0.37-0.41 (9.5-10.5)
	4	IL 4-stroke	1998-00	58 (935)	not adjustable	0.37-0.41 (9.5-10.5)
40 Carb	4	IL 4-stroke	2001	61 (996)	1 turn out	0.47-0.63 (12-16)
50 Carb	4	IL 4-stroke	2001	61 (996)	1 turn out	0.47-0.63 (12-16)
60 Carb	4	IL 4-stroke	2001	61 (996)	1 turn out	0.47-0.63 (12-16)
75	4	IL 4-stroke	2000-04	97 (1596)	2 - 3 ①	0.51-0.59 (13-15)
90	4	IL 4-stroke	2000-04	97 (1596)	1 1/2 - 2 1/2 ①	0.51-0.59 (13-15)

Because manufacturers differ with what specifications they choose to provide, please refer to the Yamaha charts for additional specificitions which MAY apply to the Mercury/Mariner model in question as well

Note: Initial low speed setting turn(s): back (counterclockwise) from a *lightly* seated position

① Specification is for adjustable carburetors normally sold worldwide. Some EPA regulated models sold in the US may be sealed and may not allow adjustment

② Specification is for all except "BODENSEE" models which are not adjustable

4

IGNITION AND ELECTRICAL SYSTEMS

ELECTRICAL SYSTEMS	4-2
IGNITION SYSTEMS	4-8
CHARGING SYSTEM	4-35
CRANKING SYSTEM	4-39
ELECTRICAL SWITCH/SOLENOID SERVICE	4-51
SPECIFICATIONS	4-54
WIRING DIAGRAMS	4-62

CHARGING SYSTEM ... 4-35
- BATTERY ... 4-37
  - BATTERY CABLES ... 4-39
  - BATTERY CHARGERS ... 4-39
  - CONSTRUCTION ... 4-38
  - MARINE BATTERIES ... 4-37
  - LOCATION ... 4-38
  - RATINGS ... 4-38
  - SAFETY PRECAUTIONS ... 4-38
- CHARGING CIRCUIT ... 4-35
- CHARGING SYSTEMS ... 4-35
  - DESCRIPTION & OPERATION ... 4-35
  - TROUBLESHOOTING ... 4-36

CRANKING SYSTEM ... 4-39
- STARTING CIRCUITS ... 4-39
  - DESCRIPTION AND OPERATION ... 4-39
  - FAULTY SYMPTOMS ... 4-40
  - MAINTENANCE ... 4-40
- STARTER MOTOR ... 4-43
  - DESCRIPTION ... 4-43
  - EXPLODED VIEWS ... 4-45
  - OVERHAUL ... 4-49
  - REMOVAL & INSTALLATION ... 4-43
- STARTER MOTOR CIRCUIT ... 4-40
  - TROUBLESHOOTING ... 4-40
- STARTER MOTOR RELAY/SOLENOID ... 4-42
  - DESCRIPTION & OPERATION ... 4-42
  - REMOVAL & INSTALLATION ... 4-43
  - TESTING ... 4-42

ELECTRICAL SWITCH/SOLENOID SERVICE ... 4-51
- KILL SWITCH ... 4-51
- MERCURY MAIN KEYSWITCH ... 4-51
- NEUTRAL SAFETY SWITCH ... 4-52
- START BUTTON ... 4-52
- YAMAHA MAIN KEYSWITCH ... 4-51
- WARNING BUZZER/HORN ... 4-53

IGNITION SYSTEMS ... 4-8
- CDI/TCI/DIGITAL INDUCTIVE IGNITION SYSTEMS ... 4-8
  - DESCRIPTION & OPERATION ... 4-8
  - TROUBLESHOOTING ... 4-12
  - COMPONENT TESTING ... 4-12
- FLYWHEEL AND STATOR (STATOR/CHARGE/LIGHTING/PULSER COILS) ... 4-21
  - CLEANING & INSPECTION ... 4-27
  - REMOVAL & INSTALLATION ... 4-21
- CDI/TCI UNIT AND IGNITION COILS ... 4-28
  - REMOVAL & INSTALLATION ... 4-28
- YAMAHA MICROCOMPUTER IGNITION SYSTEMS (YMIS) ... 4-31
  - TROUBLESHOOTING ... 4-31
- CRANKSHAFT POSITION SENSOR (CPS) ... 4-32
  - DESCRIPTION & OPERATION ... 4-32
  - GAP ADJUSTMENT & SERVICE ... 4-32
- THERMO-SENSOR & THERMO-SWITCH ... 4-32
  - CHECKING RESISTANCE ... 4-34
  - DESCRIPTION & OPERATION ... 4-33
  - OPERATIONAL CHECK ... 4-33

UNDERSTANDING AND TROUBLESHOOTING ELECTRICAL SYSTEMS ... 4-2
- BASIC ELECTRICAL THEORY ... 4-2
  - HOW DOES ELECTRICITY WORK: THE WATER ANALOGY ... 4-2
  - OHM'S LAW ... 4-2
- ELECTRICAL COMPONENTS ... 4-2
- ELECTRICAL SYSTEM PRECAUTIONS ... 4-8
- ELECTRICAL TESTING ... 4-6
  - OPEN CIRCUITS ... 4-6
  - RESISTANCE ... 4-6
  - SHORT CIRCUITS ... 4-7
  - VOLTAGE 4-6
  - VOLTAGE DROP ... 4-6
- TEST EQUIPMENT ... 4-4
  - JUMPER WIRES ... 4-4
  - MULTI-METERS ... 4-5
  - TEST LIGHTS ... 4-5
- TROUBLESHOOTING ... 4-6
- WIRE AND CONNECTOR REPAIR ... 4-7

SPECIFICATIONS ... 4-54
- CHARGE & PRIMARY COIL OUTPUT - MERCURY/MARINER INLINE ENGINES ... 4-58
- CHARGING SYSTEM - MERCURY/MARINER ... 4-61
- CHARGING SYSTEM - YAMAHA ... 4-59
- IGNITION SYSTEM - 1-CYL ... 4-55
- IGNITION SYSTEM - MERCURY/MARINER INLINE ... 4-57
- IGNITION SYSTEM - YAMAHA INLINE ... 4-54
- IGNITION CONTROL SYSTEM - YAMAHA INLINE ... 4-56
- IGNITION SYSTEM - V-ENGINES ... 4-56

WIRING DIAGRAMS ... 4-62
- INDEX ... 4-62
- MERCURY/MARINER
  - 4/5/6 HP ... 4-63
  - 8/9.9 HP (232cc) ... 4-65
  - 9.9/15 HP (323cc) ... 4-72
  - 25 HP ... 4-78
  - 30/40 HP ... 4-86
  - 40/45/50 HP (935cc) ... 4-92
  - 40/50/60 HP (996cc) ... 4-97
  - 75/90 HP ... 4-102
  - 2001 115 HP EFI ... 4-106
  - 225 HP EFI ... 4-112
- YAMAHA
  - 2.5 HP AND 4 HP ... 4-63
  - 6/8 HP ... 4-64
  - 8/9.9 HP (232cc) ... 4-66
  - 9.9/15 HP (323cc) ... 4-76
  - 25 HP ... 4-82
  - 30/40J/40 HP ... 4-91
  - 45/50/50J HP (935cc) ... 4-93
  - 60J/60 HP (996cc) 4-102
  - 75/80/90J/90/100 HP ... 4-105
  - 115J/115 HP EFI ... 4-109
  - 150 HP EFI ... 4-110
  - 200/225 HP EFI ... 4-111
  - 200/225 HP EFI ... 4-114

4-2 IGNITION AND ELECTRICAL SYSTEMS

UNDERSTANDING AND TROUBLESHOOTING ELECTRICAL SYSTEMS

Basic Electrical Theory

◆ See Figure 1

For any 12-volt, negative ground, electrical system to operate, the electricity must travel in a complete circuit. This simply means that current (power) from the positive terminal (+) of the battery must eventually return to the negative terminal (-) of the battery. Along the way, this current will travel through wires, fuses, switches and components. If, for any reason, the flow of current through the circuit is interrupted, the component fed by that circuit would cease to function properly.

Perhaps the easiest way to visualize a circuit is to think of connecting a light bulb (with two wires attached to it) to the battery - one wire attached to the negative (-) terminal of the battery and the other wire to the positive (+) terminal. With the two wires touching the battery terminals, the circuit would be complete and the light bulb would illuminate. Electricity would follow a path from the battery to the bulb and back to the battery. It's easy to see that with wires of sufficient length, our light bulb could be mounted nearly anywhere on the boat. Further, one wire could be fitted with a switch inline so that the light could be turned on and off without having to physically remove the wire(s) from the battery.

The normal marine circuit differs from this simple example in two ways. First, instead of having a return wire from each bulb to the battery, the current travels through a single ground wire that handles all the grounds for a specific circuit. Secondly, most marine circuits contain multiple components that receive power from a single circuit. This lessens the overall amount of wire needed to power components.

HOW DOES ELECTRICITY WORK: THE WATER ANALOGY

Electricity is the flow of electrons - the sub-atomic particles that constitute the outer shell of an atom. Electrons spin in an orbit around the center core of an atom. The center core is comprised of protons (positive charge) and neutrons (neutral charge). Electrons have a negative charge and balance out the positive charge of the protons. When an outside force causes the number of electrons to unbalance the charge of the protons, the electrons will split off the atom and look for another atom to balance out. If this imbalance is kept up, electrons will continue to move and an electrical flow will exist.

Many people find electrical theory easier to understand when using an analogy with water. In a comparison with water flowing through a pipe, the electrons would be the water and the wire is the pipe.

The flow of electricity can be measured much like the flow of water through a pipe. The unit of measurement used is amperes, frequently abbreviated as amps (a). You can compare amperage to the volume of water flowing through a pipe (for water that would mean a measurement of mass usually measured in units delivered over a set amount of time such as gallons or liters per minute). When connected to a circuit, an ammeter will measure the actual amount of current flowing through the circuit. When relatively few electrons flow through a circuit, the amperage is low. When many electrons flow, the amperage is high.

Water pressure is measured in units such as pounds per square inch (psi). The electrical pressure is measured in units called volts (v). When a voltmeter is connected to a circuit, it is measuring the electrical pressure.

The actual flow of electricity depends not only on voltage and amperage, but also on the resistance of the circuit. The higher the resistance, the higher the force necessary to push the current through the circuit. The standard unit for measuring resistance is an ohm (Ω). Resistance in a circuit varies depending on the amount and type of components used in the circuit. The main factors that determine resistance are:

• Material - some materials have more resistance than others. Those with high resistance are said to be insulators. Rubber materials (or rubber-like plastics) are some of the most common insulators used, as they have a very high resistance to electricity. Very low resistance materials are said to be conductors. Copper wire is among the best conductors. Silver is actually a superior conductor to copper and is used in some relay contacts, but its high cost prohibits its use as common wiring. Most marine wiring is made of copper.

• Size - the larger the wire size being used, the less resistance the wire will have (just as a large diameter pipe will allow small amounts of water to just trickle through). This is why components that use large amounts of electricity usually have large wires supplying current to them.

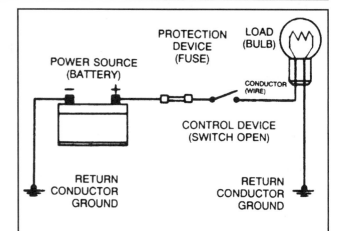

Fig. 1 This example illustrates a simple circuit. When the switch is closed, power from the positive (+) battery terminal flows through the fuse and the switch, and then to the light bulb. The electricity illuminates the bulb and the circuit is completed through the ground wire back to the negative (-) battery terminal.

• Length - for a given thickness of wire, the longer the wire, the greater the resistance. The shorter the wire, the less the resistance. When determining the proper wire for a circuit, both size and length must be considered to design a circuit that can handle the current needs of the component.

• Temperature - with many materials, the higher the temperature, the greater the resistance (positive temperature coefficient). Some materials exhibit the opposite trait of lower resistance with higher temperatures (these are said to have a negative temperature coefficient). These principles are used in many engine control sensors (especially those found on microcomputer controlled ignition and fuel injection systems).

OHM'S LAW

There is a direct relationship between current, voltage and resistance. The relationship between current, voltage and resistance can be summed up by a statement known as Ohm's law.

Voltage (E) is equal to amperage (I) times resistance (R): $E = I \times R$

Other forms of the formula are $R = E/I$ and $I = E/R$

In each of these formulas, E is the voltage in volts, I is the current in amps and R is the resistance in ohms. The basic point to remember is that if the voltage of a circuit remains the same, as the resistance of that circuit goes up, the amount of current that flows in the circuit will go down.

The amount of work that electricity can perform is expressed as power. The unit of power is the watt (w). The relationship between power, voltage and current is expressed as:

Power (W) is equal to amperage (I) times voltage (E): $W = I \times E$

This is only true for direct current (DC) circuits; the alternating current formula is a tad different, but since the electrical circuits in most vessels are DC type, we need not get into AC circuit theory.

Electrical Components

POWER SOURCE

◆ See Figure 2

Typically, power is supplied to a vessel by two devices: The battery and the stator (or battery charge coil). The stator supplies electrical current anytime the engine is running in order to recharge the battery and in order to operate electrical devices of the vessel. The battery supplies electrical power during starting or during periods when the current demand of the vessel's electrical system exceeds stator output capacity (which includes times when the motor is shut off and stator output is zero).

IGNITION AND ELECTRICAL SYSTEMS 4-3

The Battery

In most modern vessels, the battery is a lead/acid electrochemical device consisting of six 2-volt subsections (cells) connected in series, so that the unit is capable of producing approximately 12 volts of electrical pressure. Each subsection consists of a series of positive and negative plates held a short distance apart in a solution of sulfuric acid and water.

The two types of plates in each battery cell are of dissimilar metals. This sets up a chemical reaction, and it is this reaction which produces current flow from the battery when its positive and negative terminals are connected to an electrical load. Power removed from the battery in use is replaced by current from the stator and restores the battery to its original chemical state.

The Stator

Alternators and generators are devices that consist of coils of wires wound together making big electromagnets. The coil is normally referred to as a stator or battery charge coil. Either, one group of coils spins within another set (or a set of permanently charged magnets, usually attached to the flywheel, are spun around a set of coils) and the interaction of the magnetic fields generates an electrical current. This current is then drawn off the coils and fed into the vessel's electrical system.

■ **Some vessels utilize a generator instead of an alternator. Although the terms are often misused and interchanged, the main difference is that an alternator supplies alternating current that is changed to direct current for use on the vessel, while a generator produces direct current. Alternators tend to be more efficient and that is why they are used on almost all modern engines.**

GROUND

Two types of grounds are used in marine electric circuits. Direct ground components are grounded to the electrically conductive metal through their mounting points. All other components use some sort of ground wire that leads back to the battery. The electrical current runs through the ground wire and returns to the battery through the ground or negative (-) cable; if you look, you'll see that the battery ground cable connects between the battery and a heavy gauge ground wire.

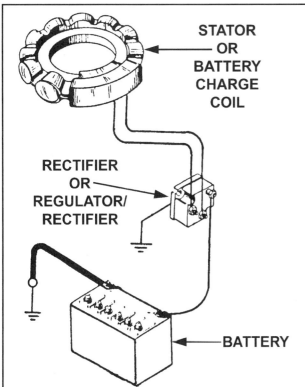

Fig. 2 Functional diagram of a typical charging circuit showing the relationship between the stator (battery charge coil), rectifier (or regulator/rectifier) and battery

■ **A large percentage of electrical problems can be traced to bad grounds.**

If you refer back to the basic explanation of a circuit, you'll see that the ground portion of the circuit is just as important as the power feed. The wires delivering power to a component can have perfectly good, clean connections, but the circuit would fail to operate if there was a damaged ground connection. Since many components ground through their mounting or through wires that are connected to an engine surface, contamination from dirt or corrosion can raise resistance in a circuit to a point where it cannot operate.

PROTECTIVE DEVICES

Problems can occur in the electrical system that will cause large surges of current to pass through the electrical system of your vessel. These problems can be the fault of the charging circuit, but more likely would be a problem with the operating electrical components that causes an excessively high load. An unusually high load can occur in a circuit from problems such as a seized electric motor (like a damaged starter) or the excessive resistance caused by a bad ground (from loose or damaged wires or connections). A short to ground that bypasses the load and allows the battery to quickly discharge through a wire can also cause current surges.

If this surge of current were to reach the load in the circuit, the surge could burn it out or severely damage it. It can also overload the wiring, causing the harness to get hot and melt the insulation. To prevent this, fuses, circuit breakers and/or fusible links are connected into the supply wires of the electrical system. These items are nothing more than a built-in weak spot in the system. When an abnormal amount of current flows through the system, these protective devices work as follows to protect the circuit:

- Fuse - when an excessive electrical current passes through a fuse, the fuse blows (the conductor melts) and opens the circuit, preventing current flow.
- Circuit Breaker - a circuit breaker is basically a self-repairing fuse. It will open the circuit in the same fashion as a fuse, but when the surge subsides, the circuit breaker can be reset and does not need replacement. Most circuit breakers on marine engine applications are self-resetting, but some that operate accessories (such as on larger vessels with a circuit breaker panel) must be reset manually (just like the circuit breaker panels in most homes).
- Fusible Link - a fusible link (fuse link or main link) is a short length of special, high temperature insulated wire that acts as a fuse. When an excessive electrical current passes through a fusible link, the thin gauge wire inside the link melts, creating an intentional open to protect the circuit. To repair the circuit, the link must be replaced. Some newer type fusible links are housed in plug-in modules, which are simply replaced like a fuse, while older type fusible links must be cut and spliced if they melt. Since this link is very early in the electrical path, it's the first place to look if nothing on the vessel works, yet the battery seems to be charged and is otherwise properly connected.

✳✳ CAUTION

Always replace fuses, circuit breakers and fusible links with identically rated components. Under no circumstances should a component of higher or lower amperage rating be substituted. A lower rated component will disable the circuit sooner than necessary (possibly during normal operation), while a higher rated component can allow dangerous amounts of current that could damage the circuit or component (or even melt insulation causing sparks or a fire).

SWITCHES & RELAYS

◆ See Figure 3

Switches are used in electrical circuits to control the passage of current. The most common use is to open and close circuits between the battery and the various electric devices in the system. Switches are rated according to the amount of amperage they can handle. If a sufficient amperage rated switch is not used in a circuit, the switch could overload and cause damage.

Some electrical components that require a large amount of current to operate use a special switch called a relay. Since these circuits carry a large amount of current, the thickness of the wire in the circuit is also greater. If this large wire were connected from the load to the control switch, the switch would have to carry the high amperage load and the space needed for wiring in the vessel would be twice as big to accommodate the increased size of the wiring harness. A relay is used to prevent these problems.

4-4 IGNITION AND ELECTRICAL SYSTEMS

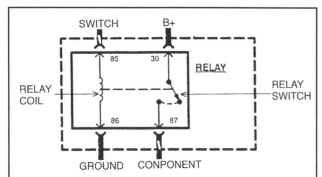

Fig. 3 Relays are composed of a coil and a switch. These two components are linked together so that when one is operated it actuates the other. The large wires in the circuit are connected from the battery to one side of the relay switch (B+) and from the opposite side of the relay switch to the load (component). Smaller wires are connected from the relay coil to the control switch for the circuit and from the opposite side of the relay coil to ground

Think of relays as essentially "remote controlled switches." They allow a smaller current to throw the switch that operates higher amperages devices. Relays are composed of a coil and a set of contacts. When current is passed through the coil, a magnetic field is formed that causes the contacts to move together, closing the circuit. Most relays are normally open, preventing current from passing through the main circuit until power is applied to the coil. But, relays can take various electrical forms depending on the job for which they are intended. Some common circuits that may use relays are horns, lights, starters, electric fuel pumps and other potentially high draw circuits.

LOAD

Every electrical circuit must include a load (something to use the electricity coming from the source). Without this load, the battery would attempt to deliver its entire power supply from one pole to another. This is called a short circuit. All this electricity would take a short cut to ground and cause a great amount of damage to other components in the circuit (including the battery) by developing a tremendous amount of heat. This condition could develop sufficient heat to melt the insulation on all the surrounding wires and reduce a multiple wire cable to a lump of plastic and copper. A short can allow sparks that could ignite fuel vapors or other combustible materials in the vessel, causing an extremely hazardous condition.

WIRING & HARNESSES

The average vessel contains miles of wiring, with hundreds of individual connections. To protect the many wires from damage and to keep them from becoming a confusing tangle, they are organized into bundles, enclosed in plastic or taped together and called wiring harnesses. Different harnesses serve different parts of the vessel. Individual wires are color coded to help trace them through a harness where sections are hidden from view.

Marine wiring or circuit conductors can be either a single strand wire, multi-strand wire or printed circuitry. Single strand wire has a solid metal core and is usually used inside such components as stator coil windings, motors, relays and other devices. Multi-strand wire has a core made of many small strands of wire twisted together into a single conductor. Most of the wiring in a marine electrical system is made up of multi-strand wire, either as a single conductor or grouped together in a harness. All wiring is color coded on the insulator, either as a solid color or as a colored wire with an identification stripe. A printed circuit is a thin film of copper or other conductor that is printed on an insulator backing. Occasionally, a printed circuit is sandwiched between two sheets of plastic for more protection and flexibility. A complete printed circuit, consisting of conductors, insulating material and connectors is called a printed circuit board. Printed circuitry is used in place of individual wires or harnesses in places where space is limited, such as behind 1-piece instrument clusters.

Since marine electrical systems are very sensitive to changes in resistance, the selection of properly sized wires is critical when systems are repaired. A loose or corroded connection or a replacement wire that is too small for the circuit will add extra resistance and an additional voltage drop to the circuit.

The wire gauge number is an expression of the cross-section area of the conductor. Vessels from countries that use the metric system will typically describe the wire size as its cross-sectional area in square millimeters. In this method, the larger the wire, the greater the number. Another common system for expressing wire size is the American Wire Gauge (AWG) system. As gauge number increases, area decreases and the wire becomes smaller. Using the AWG system, an 18 gauge wire is smaller than a 4 gauge wire. A wire with a higher gauge number will carry less current than a wire with a lower gauge number. Gauge wire size refers to the size of the strands of the conductor, not the size of the complete wire with insulator. It is possible, therefore, to have two wires of the same gauge with different diameters because one may have thicker insulation than the other.

It is essential to understand how a circuit works before trying to figure out why it doesn't. An electrical schematic shows the electrical current paths when a circuit is operating properly. Schematics break the entire electrical system down into individual circuits. In most schematics no attempt is made to represent wiring and components as they physically appear on the vessel; switches and other components are shown as simply as possible. But, this is **not** always the case on Yamaha schematics and some of the wiring diagrams provided here. So, when using a Yamaha schematic if the component in question is represented by something more than a small square or rectangle with a label, it is likely a intended to be a representation of the actual component's shape. On most schematics, the face views of harness connectors show the cavity or terminal locations in all multi-pin connectors to help locate test points.

Test Equipment DVOM

Pinpointing the exact cause of trouble in an electrical circuit is usually accomplished by the use of special test equipment, but the equipment does not always have to be expensive. The following sections describe different types of commonly used test equipment and briefly explains how to use them in diagnosis. In addition to the information covered below, be sure to read and understand the tool manufacturer's instruction manual (provided with most tools) before attempting any test procedures.

JUMPER WIRES

◆ See Figure 4

✱✱ CAUTION

Never use jumper wires made from a thinner gauge wire than the circuit being tested. If the jumper wire is of too small a gauge, it may overheat and possibly melt. Never use jumpers to bypass high resistance loads in a circuit. Bypassing resistances, in effect, creates a short circuit. This may, in turn, cause damage and fire. Jumper wires should only be used to bypass lengths of wire or to simulate switches.

Jumper wires are simple, yet extremely valuable, pieces of test equipment. They are basically test wires that are used to bypass sections of a circuit. Although jumper wires can be purchased, they are usually fabricated from lengths of standard marine wire and whatever type of connector (alligator clip, spade connector or pin connector) that is required for the particular application being tested. In cramped, hard-to-reach areas, it is advisable to have insulated boots over the jumper wire terminals in order to prevent accidental grounding. It is also advisable to include a standard marine fuse in any jumper wire. This is commonly referred to as a fused jumper. By inserting an in-line fuse holder between a set of test leads, a fused jumper wire is created for bypassing open circuits. Use a 5-amp fuse to provide protection against voltage spikes.

Jumper wires are used primarily to locate open electrical circuits, on either the ground (-) side of the circuit or on the power (+) side. If an electrical component fails to operate, connect the jumper wire between the component and a good ground. If the component operates only with the jumper installed, the ground circuit is open. If the ground circuit is good, but the component does not operate, the circuit between the power feed and component may be open. By moving the jumper wire successively back from the component toward the power source, you can isolate the area of the circuit where the open is located. When the component stops functioning, or the power is cut off, the open is in the segment of wire between the jumper and the point previously tested.

You can sometimes connect the jumper wire directly from the battery to the hot terminal of the component, but first make sure the component uses a full 12 volts in operation. Some electrical components, such as sensors, are designed to operate on smaller voltages like 4 or 5 volts, and running 12 volts directly to these components can damage or destroy them.

TEST LIGHTS

◆ See Figure 5

The test light is used to check circuits and components while electrical current is flowing through them. It is used for voltage and ground tests. To use a 12-volt test light, connect the ground clip to a good ground and probe connectors the pick where you are wondering if voltage is present. The test light will illuminate when voltage is detected. This does not necessarily mean that 12 volts (or any particular amount of voltage) is present; it only means that some voltage is present. It is advisable before using the test light to touch its ground clip and probe across the battery posts or terminals to make sure the light is operating properly and to note how brightly the light glows when 12 volts is present.

✱✱ WARNING

Do not use a test light to probe electronic ignition, spark plug or coil wires, as the circuit is much, much higher than 12 volts. Also, never use a pick-type test light to probe wiring on electronically controlled systems unless specifically instructed to do so. Whenever possible, avoid piercing insulation with the test light pick, as you are inviting shorts or corrosion and excessive resistance. But, any wire insulation that is pierced by necessity, must be sealed with silicone and taped after testing.

Like the jumper wire, the 12-volt test light is used to isolate opens in circuits. But, whereas the jumper wire is used to bypass the open to operate the load, the 12-volt test light is used to locate the presence or lack of voltage in a circuit. If the test light illuminates, there is power up to that point in the circuit; if the test light does not illuminate, there is an open circuit (no power). Move the test light in successive steps back toward the power source until the light in the handle illuminates. The open is between the probe and the point that was previously probed.

The self-powered test light is similar in design to the 12-volt test light, but contains a 1.5 volt penlight battery in the handle. It is most often used in place of a multi-meter to check for open or short circuits when power is isolated from the circuit (thereby performing a continuity test).

The battery in a self-powered test light does not provide much current. A weak battery may not provide enough power to illuminate the test light even when a complete circuit is made (especially if there is high resistance in the circuit). Always make sure that the test battery is strong. To check the battery, briefly touch the ground clip to the probe; if the light glows brightly, the battery is strong enough for testing.

■ A self-powered test light should not be used on any electronically controlled system or component. Even the small amount of electricity transmitted by the test light is enough to damage many electronic components.

MULTI-METERS

◆ See Figure 6

Multi-meters are extremely useful for troubleshooting electrical problems. They can be purchased in both analog or digital form and have a price range to suit nearly any budget. A multi-meter is a voltmeter, ammeter and ohmmeter (along with other features) combined into one instrument. It is often used when testing solid state circuits because of its high input impedance (usually 10 mega ohms or more). A high-quality digital multi-meter or Digital Volt Ohm Meter (DVOM) helps to ensure the most accurate test results and, although not absolutely necessary for electronic components such as computer controlled ignition systems and charging systems, is highly recommended. A brief description of the main test functions of a multi-meter follows:

• Voltmeter - the voltmeter is used to measure voltage at any point in a circuit or to measure the voltage drop across any part of a circuit. Voltmeters usually have various scales and a selector switch to allow metering and display of different voltage ranges. The voltmeter has a positive and a negative lead. To avoid damage to the meter, connect the negative lead to the negative (-) side of the circuit (to ground or nearest the ground side of the circuit) and connect the positive lead to the positive (+) side of the circuit (to the power source or the nearest power source). This is mostly a concern on analog meters, as DVOMs are not normally adversely affected (as they are usually designed to take readings even with reverse polarity and display accordingly). Note that the negative voltmeter lead will always be black and that the positive voltmeter will always be some color other than black (usually red).

• Ohmmeter - the ohmmeter is designed to read resistance (measured in ohms) in a circuit or component. Most ohmmeters will have a selector switch which permits the measurement of different ranges of resistance (usually the selector switch allows the multiplication of the meter reading by 10, 100, 1,000 and 10,000). Most modern ohmmeters (especially DVOMs) are auto-ranging which means the meter itself will determine which scale to use. Since ohmmeters are powered by an internal battery, the ohmmeter can be used like a self-powered test light. When the ohmmeter is connected, current from the ohmmeter flows through the circuit or component being tested. Since the ohmmeter's internal resistance and voltage are known values, the amount of current flow through the meter depends on the resistance of the circuit or component being tested. The ohmmeter can also be used to perform a continuity test for suspected open circuits. When using the meter for continuity checks, do not be concerned with the actual resistance readings. Zero resistance, or any ohm reading, indicates continuity in the circuit. Infinite resistance indicates an opening in the circuit. A high resistance reading where there should be none indicates a problem in the circuit. Checks for short circuits are made in the same manner as checks for open circuits, except that the circuit must be isolated from both power and normal ground. Infinite resistance indicates no continuity, while zero resistance indicates a dead short.

✱✱ WARNING

Never use an ohmmeter to check the resistance of a component or wire while there is voltage applied to the circuit. Voltage in the circuit can damage or destroy the meter.

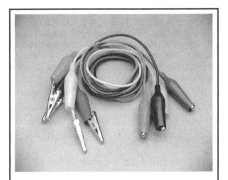

Fig. 4 Jumper wires are simple, but valuable pieces of test equipment

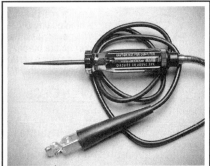

Fig. 5 A 12-volt test light is used to detect the presence of voltage in a circuit

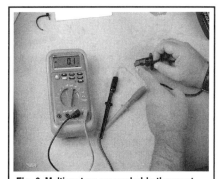

Fig. 6 Multi-meters are probably the most versatile and handy tools for diagnosing faulty electrical components or circuits

4-6 IGNITION AND ELECTRICAL SYSTEMS

• **Ammeter** - an ammeter measures the amount of current flowing through a circuit in units called amperes or amps. At normal operating voltage, most circuits have a characteristic amount of amperes, called current draw that can be measured using an ammeter. By referring to a specified current draw rating, and then measuring the amperes and comparing the two values, you can determine what is happening within the circuit to aid in diagnosis. An open circuit, for example, will not allow any current to flow, so the ammeter reading will be zero. A damaged component or circuit will have an increased current draw, so the reading will be high. The ammeter is always connected in series with the tested circuit. All of the current that normally flows through the circuit must also flow through the ammeter; if there is any other path for the current to follow, the ammeter reading will not be accurate. The ammeter itself has very little resistance to current flow and, therefore, will not affect the circuit, but it will measure current draw only when the circuit is closed and electricity is flowing. Excessive current draw can blow fuses and drain the battery, while a reduced current draw can cause motors to run slowly, lights to dim and other components to not operate properly.

Troubleshooting Electrical Systems

When diagnosing a specific problem, organized troubleshooting is a must. The complexity of a modern marine vessel demands that you approach any problem in a logical, organized manner. There are certain troubleshooting techniques, however, which are standard:

• **Establish when the problem occurs**. Does the problem appear only under certain conditions? Were there any noises, odors or other unusual symptoms? Isolate the problem area. To do this, make some simple tests and observations, and then eliminate the systems that are working properly. Check for obvious problems, such as broken wires and loose or dirty connections. Always check the obvious before assuming something complicated is the cause.

• **Test for problems systematically to determine the cause once the problem area is isolated**. Are all the components functioning properly? Is there power going to electrical switches and motors? Performing careful, systematic checks will often turn up most causes on the first inspection, without wasting time checking components that have little or no relationship to the problem.

• **Test all repairs after the work is done to make sure that the problem is fixed**. Some causes can be traced to more than one component, so a careful verification of repair work is important in order to pick up additional malfunctions that may cause a problem to reappear or a different problem to arise. A blown fuse, for example, is a simple problem that may require more than another fuse to repair. If you don't look for a problem that caused a fuse to blow, a shorted wire (for example) may go undetected and cause the new fuse to blow right away (if the short is still present) or during subsequent operation (as soon as the short returns if it is intermittent).

Experience shows that most problems tend to be the result of a fairly simple and obvious cause, such as loose or corroded connectors, bad grounds or damaged wire insulation that causes a short. This makes careful visual inspection of components during testing essential to quick and accurate troubleshooting.

Electrical Testing

VOLTAGE

♦ See Figure 7

This test determines the voltage available from the battery and should be the first step in any electrical troubleshooting procedure after visual inspection. Many electrical problems, especially on electronically controlled systems, can be caused by a low state of charge in the battery. Many circuits cannot function correctly if the battery voltage drops below normal operating levels.

Loose or corroded battery cable terminals can cause poor contact that will prevent proper charging and full battery current flow.

1. Set the voltmeter selector switch to the 20V position.
2. Connect the meter negative lead to the battery's negative (-) post or terminal and the positive lead to the battery's positive (+) post or terminal.
3. Turn the ignition switch **ON** to provide a small load.
4. A well charged battery should register over 12 volts. If the meter reads below 11.5 volts, the battery power may be insufficient to operate the electrical system properly. Check and charge or replace the battery as detailed under Engine Maintenance before further tests are conducted on the electrical system.

VOLTAGE DROP

♦ See Figure 8

When current flows through a load, the voltage beyond the load drops. This voltage drop is due to the resistance created by the load and also by small resistances created by corrosion at the connectors (or by damaged insulation on the wires). Since all voltage drops are cumulative, the maximum allowable voltage drop under load is critical, especially if there is more than one load in the circuit.

1. Set the voltmeter selector switch to the 20 volts position.
2. Connect the multi-meter negative lead to a good ground.
3. Operate the circuit and check the voltage prior to the first component (load).
4. There should be little or no voltage drop in the circuit prior to the first component. If a voltage drop exists, the wire or connectors in the circuit are suspect.
5. While operating the first component in the circuit, probe the ground side of the component with the positive meter lead and observe the voltage readings. A small voltage drop should be noticed. This voltage drop is caused by the resistance of the component.
6. Repeat the test for each component (load) down the circuit.
7. If an excessively large voltage drop is noticed, the preceding component, wire or connector is suspect.

RESISTANCE

♦ See Figure 9

✱✱ WARNING

Never use an ohmmeter with power applied to the circuit. The ohmmeter is designed to operate on its own power supply. The normal 12-volt electrical system voltage will damage or destroy many meters!

1. Isolate the circuit from the vessel's power source.
2. Ensure that the ignition key is **OFF** when disconnecting any components or the battery.
3. Where necessary, also isolate at least one side of the circuit to be checked, in order to avoid reading parallel resistances. Parallel circuit resistances will always give a lower reading than the actual resistance of either of the branches.
4. Connect the meter leads to both sides of the circuit (wire or component) and read the actual measured ohms on the meter scale. Make sure the selector switch is set to the proper ohm scale for the circuit being tested, to avoid misreading the ohmmeter test value.

■ The resistance reading of most electrical components will vary with temperature. Unless otherwise noted, specifications given are for testing under ambient conditions of 68°F (20°C). If the component is tested at higher or lower temperatures, expect the readings to vary slightly. When testing engine control sensors or coil windings with smaller resistance specifications (less than 1000 ohms) it is best to use a high quality DVOM and be especially careful of your test results. Whenever possible, double-check your results against a known good part before purchasing the replacement. If necessary, bring the old part to the marine parts dealer and have them compare the readings to prevent possibly replacing a good component.

OPEN CIRCUITS

♦ See Figure 10

This test already assumes the existence of an open in the circuit and it is used to help locate position of the open.

1. Isolate the circuit from power and ground.
2. Connect the self-powered test light or ohmmeter ground clip to the ground side of the circuit and probe sections of the circuit sequentially.

IGNITION AND ELECTRICAL SYSTEMS 4-7

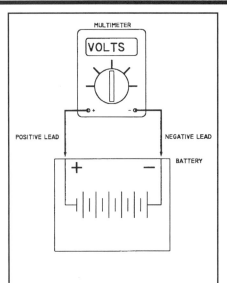

Fig. 7 A voltage check determines the amount of battery voltage available and, as such, should be the first step in any troubleshooting procedure

Fig. 8 Voltage drops are due to resistance in the circuit, from the load or from problems with the wiring

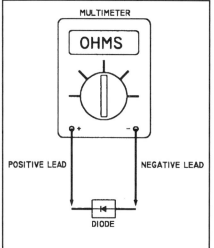

Fig. 9 Resistance tests must be conducted on portions of the circuit, isolated from battery power

3. If the light is out or there is infinite resistance, the open is between the probe and the circuit ground.
4. If the light is on or the meter shows continuity, the open is between the probe and the end of the circuit toward the power source.

SHORT CIRCUITS

◆ See Figure 11

■ **Never use a self-powered test light to perform checks for opens or shorts when power is applied to the circuit under test. The test light can be damaged by outside power.**

1. Isolate the circuit from power and ground.
2. Connect the self-powered test light or ohmmeter ground clip to a good ground and probe any easy-to-reach point in the circuit.
3. If the light comes on or there is continuity, there is a short somewhere in the circuit.

4. To isolate the short, probe a test point at either end of the isolated circuit (the light should be on or the meter should indicate continuity).
5. Leave the test light probe engaged and sequentially open connectors or switches, remove parts, etc. until the light goes out or continuity is broken.
6. When the light goes out, the short is between the last two circuit components that were opened.

Wire And Connector Repair

Almost anyone can replace damaged wires, as long as the proper tools and parts are available. Wire and terminals are available to fit almost any need. Even the specialized weatherproof, molded and hard shell connectors used by many marine manufacturers.

Be sure the ends of all the wires are fitted with the proper terminal hardware and connectors. Wrapping a wire around a stud is not a permanent solution and will only cause trouble later. Replace wires one at a time to avoid confusion. Always route wires in the same manner of the manufacturer.

When replacing connections, make absolutely certain that the connectors are certified for marine use. Automotive wire connectors may not meet United States Coast Guard (USCG) specifications.

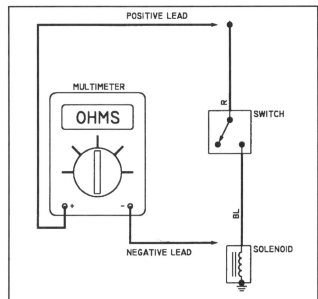

Fig. 10 The easiest way to illustrate an open circuit is to picture a circuit in which the switch is turned OFF (creating an opening in the circuit) that prevents power from reaching the load

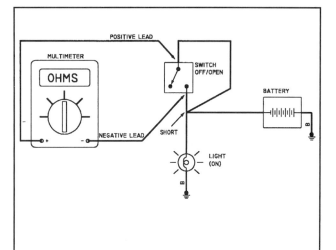

Fig. 11 In this illustration a load (the light) is powered when it should not be (since the switch should be creating an open condition), but a short to power (battery) is powering the circuit. Shorts like this can be caused by chaffed wires with worn or broken insulation

4-8 IGNITION AND ELECTRICAL SYSTEMS

■ If connector repair is necessary, only attempt it if you have the proper tools. Weatherproof and hard shell connectors may require special tools to release the pins inside the connector. Attempting to repair these connectors with conventional hand tools will damage them.

Electrical System Precautions

• Wear safety glasses when working on or near the battery.
• Don't wear a watch with a metal band when servicing the battery or starter. Serious burns can result if the band completes the circuit between the positive battery terminal (or a hot wire) and ground.
• Be absolutely sure of the polarity of a booster battery before making connections. Remember that even momentary connection of a booster battery with the polarity reversed will damage charging system diodes. Connect the cables positive-to-positive, and negative (of the good battery)-to-a good ground on the engine (away from the battery to prevent the possibility of an explosion if hydrogen vapors are present from the electrolyte in the discharged battery). Connect positive cables first (starting with the discharged battery), and then make the last connection to ground on the body of the booster vessel so that arcing cannot ignite hydrogen gas that may have accumulated near the battery. • Disconnect both vessel battery cables before attempting to charge a battery.
• Never ground the alternator or generator output or battery terminal. Be cautious when using metal tools around a battery to avoid creating a short circuit between the terminals.
• When installing a battery, make sure that the positive and negative cables are not reversed.
• Always disconnect the battery (negative cable first) when charging.
• Never smoke or expose an open flame around the battery. Hydrogen gas is released from battery electrolyte during use and accumulates near the battery. Hydrogen gas is **highly** explosive.

IGNITION SYSTEMS

◆ See Figure 12

The less an outboard engine is operated, the more care it needs. Allowing an outboard engine to remain idle will do more harm than if it is used regularly. To maintain the engine in top shape and always ready for efficient operation at any time, the engine ideally should be operated every 3 to 4 weeks throughout the year.

The carburetion and ignition principles of 4-stroke engine operation must be understood in order to perform a proper tune-up on an outboard motor. If you have any doubts concerning your understanding of engine operation, it would be best to study the operation theory before tackling any work on the ignition system.

CDI/TCI/Digital Inductive Ignition Systems - General Information

DESCRIPTION & OPERATION

◆ See Figures 13 and 14

All Yamaha and Mercury/Mariner 4-stroke motors are equipped with some form of a Capacitor Discharge Ignition (CDI) or Transistor Controlled Ignition (TCI) electronic ignition system. Some models are equipped with a micro-computer controlled version of the system (called CDI or TCI Micro by Yamaha or Digital Inductive/ECM555 by Mercury/Mariner) which means the CDI/TCI unit or ECM (as the unit is known on most fuel injected motors) utilizes additional input (sensor input such as a crank position sensor, temperature sensor or even oil pressure sensor on some motors) to help make spark timing decisions, however, the basic function of the various Yamaha and Mercury/Mariner ignition systems are all similar.

The first thing these ignitions have in common is that a flywheel/magneto assembly is used to generate power/signals over pulser (also known as a trigger coil or CPS) and/or charge (or stator) coils. In all cases, the power and signals are fed to a control unit which is used to "trigger" the ignition coils and spark plugs.

Although the basic function of the components is the same across most systems, their actual jobs may be performed by different physical components or their functions may be combined into a single unit on some motors.

Generally speaking, charge coils are used to generate system power, however, their function may be performed by the stator coil in some cases (either through dedicated charge coil windings in the stator coil or possibly even by the lighting coil windings when separate charge coil windings are not provided, meaning the battery/charging circuit). Similarly, pulser coils are normally used to generate signals triggering the CDI/TCI unit to fire the ignition coils, however on some single cylinder models no pulser coil is needed, as the plugs are fired each time the charge coil generates power.

The CDI/TCI unit takes in power from the charge coil (or stator) and signals from the pulser coil(s) and then uses the power and signals to control function of the ignition coils. On the smallest of the 4-stroke Yamahas the TCI unit is mounted underneath the flywheel which contains both and internal charge coil AND an ignition coil. While fuel injected motors normally combine the function of the CDI/TCI unit into the Electronic Control Module (ECM) which also controls the fuel injection system.

Fig. 12 The ignition system must be properly adjusted and synchronized for optimum powerhead performance

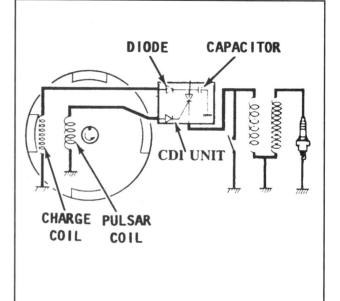

Fig. 13 Functional diagram of a typical CDI ignition system

IGNITION AND ELECTRICAL SYSTEMS 4-9

To understand any more about any one motor's ignition system, refer to the system descriptions below. If you have any doubt as to the components utilized by a given motor, please refer to the Wiring Diagrams and the Ignition System Component testing charts in this section as well.

■ Throughout this section we'll refer to the CDI unit when talking about the CDI unit, TCI unit or ECM, whichever is responsible for the ignition control module. Don't be confused if you're researching a TCI motor and we're calling it CDI, that's just to keep from having to constantly write (and for you to constantly read "CDI/TCI unit or ECM").

1-Cylinder Yamaha TCI Ignition System

The TCI system found on single-cylinder Yamaha motors is among the simplest form of an ignition system and is composed of the following elements:
- Flywheel mounted magnet (magneto)
- Igniter (TCI) box (with integral ignition charge coil)
- Engine stop switch
- Spark plug

The engine stop switch, though not necessary for basic TCI operation, must be included in the circuit if only to stop TCI operation.

To understand basic TCI operation, it is important to understand the basic theory of induction. Induction theory states that if we move a magnet (magnetic field) past a coil of wire (or the coil by the magnet), AC current will be generated in the coil.

The amount of current produced depends on several factors:
- How fast the magnet moves past the coil
- The size of the magnet (strength)
- How close the magnet is to the coil
- Number of turns of wire and the size of the windings

These models utilize a TCI unit mounted under the flywheel which incorporates the functions of the charge/pulser coils, TCI unit and ignition coil all inside a single housing.

Like most ignition systems, the ignition coil (inside the TCI unit in this case) is a step-up transformer. It turns the relatively low voltage entering the primary windings into high voltage at the secondary windings. This occurs due to a phenomena known as induction.

As the flywheel rotates, the primary coil (inside the TCI box), generates a voltage and an electric current. This electric current opens the first transistor and, as a result, an electric current flows to the primary coil.

As the flywheel continues to rotate, the voltage generated by the primary coil increases. When the voltage reaches the level necessary for the operation of a second transistor an electric current flows to that second transistor (opening it). Ignition timing occurs in this instant. The electric current at then flowing through the primary coil through the first transistor is cut off and the secondary coil generates a high-voltage through electric induction, which fires the spark plug.

Once the complete cycle has occurred, the spinning flywheel magnet immediately starts the process over again.

The engine stop switch on these models is normally connected on the wire in between the TCI box primary ignition circuit and ground. When the main switch or stop switch is turned to the OFF position, the switch is closed. This closed switch short-circuits the primary ignition current to ground rather than allowing it to induce a high-voltage signal in the secondary winding, therefore there is no spark and the engine stops.

- 3-5 hp, 2-stroke motors: utilize a low- and high-speed pulser coils in addition to the balance of the standard CDI ignition components.

1-Cylinder Mercury/Mariner Ignition System

◆ See Figures 13 and 14

The CDI system found on single-cylinder Mercury/Mariner motors is basically a 1 cylinder version of the system found on most Mercury and Yamaha motors composed of the following elements:
- Flywheel magneto
- Charge (exciter) coil
- Pulser (trigger) coil
- CDI Unit
- Ignition coil
- Spark plug

Other components such as an engine stop switch or warning lamp are included in the circuit, though, are not necessary for basic CDI operation.

To understand basic CDI operation, it is important to understand the basic theory of induction. Induction theory states that if we move a magnet (magnetic field) past a coil of wire (or the coil by the magnet), AC current will be generated in the coil.

The amount of current produced depends on several factors:
- How fast the magnet moves past the coil
- The size of the magnet (strength)
- How close the magnet is to the coil
- Number of turns of wire and the size of the windings

The current produced in the charge coil goes to the CDI box where it is converted to DC current by a diode. This DC current is stored in the capacitor located inside the box. As the charge coil produces current, the capacitor stores it.

At a specific time in the magneto's revolution, the magnets go past the pulser coil (which is basically a smaller version of a charge coil, so it has less current output). The current from the pulser also goes into the CDI box. This current signals the CDI box when to fire the capacitor (the pulser may be called a trigger coil for the obvious reason that it pulls the "trigger" on the circuit). The current from the capacitor flows out to the ignition coil and spark plug. The pulser acts much like the points in older ignitions systems.

When the pulser signal reaches the CDI box, all the electricity stored in the capacitor is released at once. This current flows through the ignition coil's primary windings.

The ignition coil is a step-up transformer. It turns the relatively low voltage entering the primary windings into high voltage at the secondary windings. This occurs due to a phenomena known as induction.

The high voltage generated in the secondary windings leaves the ignition coil and goes to the spark plug. The spark in turn ignites the air-fuel charge in the combustion chamber.

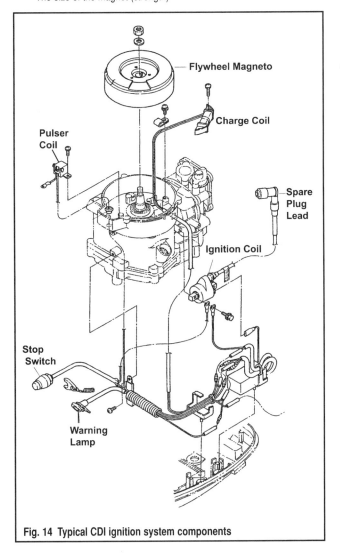

Fig. 14 Typical CDI ignition system components

4-10 IGNITION AND ELECTRICAL SYSTEMS

Once the complete cycle has occurred, the spinning magneto immediately starts the process over again.

Main switches, engine stop switches, and the like are usually connected on the wire in between the CDI box and the ignition coil. When the main switch or stop switch is turned to the **OFF** position, the switch is closed. This closed switch short-circuits the pulser (trigger) coil current to ground rather than sending it through the CDI box, there is no spark and the engine stops.

On these models the charge coil and pulser coil are mounted independently under the flywheel. The separate ignition coil and CDI unit are each mounted to the powerhead itself.

2-Cylinder Ignition Systems

◆ See Figures 13 and 14

Yamaha produces 4 different 2-cylinder 4-stroke outboards engine families ranging from smaller 6 hp units to slightly larger 25 hp motors. Mercury/Mariner sells versions of 3 families ranging from the 8/9.9 (232cc) to that same 25 hp motor. These outboards are almost always equipped with one, dual-lead ignition coil.

In addition to the dual-lead ignition coil, the system typically uses one pulser coil (called a CPS on Mercury/Mariner models), one charge coil (or stator coil on 8/9.9 hp/232cc and 25 hp motors) and a CDI box (known as an ECM on some models, including the Mercury/Mariner 25 hp motor).

When equipped with only one pulser coil, both spark plugs are fired at the same time. Although both cylinders spark at the same time, only one cylinder is actually producing power. The crankshaft is a 180 degree type which means that as piston number one is at top dead center, piston number two is at bottom dead center. The piston at TDC is compressing a fuel charge that the spark then ignites.

At the same moment, the piston at BDC isn't compressing a fuel charge. Actually, there are still exhaust gases going out. The spark in this cylinder has no effect on power production. This combination of engine and ignition design is called waste spark system (which is something as a misnomer, because no efficiency is actually lost by the spark in the non-compressed cylinder).

After the crankshaft has rotated another 180 degrees, the two pistons have reversed position. The spark fires again to ignite the fuel charge compressed in cylinder number two and sparks to no effect in cylinder one.

On a traditional ignition coil, the current leaves the coil, goes to the spark plug, then through the cylinder head to ground.

On a dual-lead coil, the current usually leaves the coil, goes through one of the spark plugs, travels through the cylinder head to the second spark plug, and returns to the coil itself. This way one coil can fire two cylinders.

This type of system requires about 30% more voltage than a standard system to fire the second spark plug. This is because more energy is required for the spark to jump from the bridge of the spark plug to the center electrode.

■ **If this system has weak spark, the first sign is the second spark plug will not having a hot enough spark. This plug will foul even though the other plug is working fine.**

A quick check to tell if the plug fouling on one cylinder is due to weak ignition is to switch the ignition coil leads. If the plug fouling goes to the other cylinder, weak ignition components are the problem.

3-Cylinder Ignition Systems

◆ See Figures 13 and 14

The 30/40 hp 3-cylinder motors sold by both Yamaha and Mercury/Mariner all use an expanded version of the CDI systems described earlier for 1 and 2-cylinder models.

The system consists of a charge coil (incorporated into a stator coil), a pulser coil (or called on Mercury/Models), CDI unit (known as an ECM on Mercury/Mariner models) and 3 separate ignition coils.

In addition to the usual ignition components (CDI unit, ignition coils and spark plugs) the system also uses input from the engine temperature sensor and oil pressure switch for fine tuning ignition control.

Components such as the ignition coils themselves will likely vary between the carbureted and EFI versions of the Mercury/Mariner powerheads in this range, the later of which is normally equipped with a version of the system referred to as Digital Inductive (ECM555) ignition.

Inline 4-Cylinder Ignition System

◆ See Figures 13 and 14

The ignition system used on the 40-60 Hp 4-cylinder models utilizes one pulser coil, one stator mounted charge coil winding, a CDI unit (usually an ECM on these models) and either 2 dual-lead ignition coils (all Mercury/Mariner models 75 hp and larger or all Yamaha motors) or 4 single-lead ignition coils (40-60 hp Mercury/Mariner models). The system is something of a cross between the ignition used on single cylinder motors and the one used on twins.

All ignition and timing functions are controlled by the CDI based on input from the pulser coil and voltage from the charge coil. On models with dual-lead ignition coils, the spark plugs are fired in pairs (one cylinder at TDC of the compression stroke while the other is on TDC of the exhaust stroke) using a traditional "waste-spark" configuration. Typically, one dual-lead ignition coil is used for the #1 and #4 cylinders, while the other is used to fire the #2 and #3 cylinders.

The ignition system used on the 75 hp and larger motors is very similar to the 40-60 hp motors, with the following differences. The 75-150 hp 4-strokes utilize 2 pulser coils. It appears from Yamaha test data that each sends the trigger signal for an individual coil (so one fires cylinders #1 and #4, while the other fires cylinders #2 and #3). In addition to thermo-switch and oil pressure switch input, microcomputer control on these models also uses input from the Throttle Position Sensor (TPS). Lastly, on the 115 hp and 150 hp EFI models, the CDI unit functions are integrated into the fuel injection ECM itself.

V6 Ignition System

◆ See Figures 13 and 14

EFI 4-stroke V6 motors utilize 3 separate sets of pulser coil windings, all located in the stator itself, to trigger the ignition. In addition, power for the ignition system comes from the lighting coil windings of the stator. The ECM controlled system actuates 3 dual lead ignition coils which are fired using a waste-spark strategy. Microcomputer control utilizes most of the fuel injection sensor (IAT/MAT, WTS/ECT, IAP/MAP, TPS) along with other inputs such as the oil pressure sensor, thermo-switch, shift cut switch and neutral switch.

Charge Circuit

Power For The Ignition

◆ See Figures 15, 16 and 17

Two basic circuits are used with the CDI/TCI system; the first of these is the charge circuit. The charge circuit typically consists of the flywheel magnet, a charge coil (or lighting coil on some models), a diode (usually contained within the CDI unit), and a capacitor (also within the CDI unit). As the flywheel magnet passes by the charge coil, a voltage is induced in the coil windings. As the flywheel continues to rotate and the coil is no longer influenced by the magnet, the magnetic field collapses. Therefore, an alternating current is produced at the charge coil. This AC current is changed to DC by a diode inside the CDI unit.

■ **A diode is a solid state unit which permits current to flow in one direction but prevents flow in the opposite direction. A diode may also be known as a rectifier.**

The current then passes to a capacitor, also located inside the CDI unit, where it is stored.

POWER FOR ACCESSORIES

◆ See Figure 18

The battery charging system consists of the flywheel magnets, a lighting coil (the outboard equivalent of an alternator), a rectifier, voltage regulator (on some motors), and the battery.

The lighting coil, so named because it may be used to power the lights on the boat when used together with a rectifier (or regulator/rectifier), allows the powerhead to generate additional electrical current to charge the battery.

As the flywheel magnets rotate, voltage is induced in the lighting coil (either a separate coil located next to the pulser coils, or the lighting coil windings may be found in the stator coil). This alternating current passes through a series of diodes and emerges as DC current. Therefore, it may be stored in the battery. A lighting coil may be identified by its clean laminated copper windings (all other coils are normally wrapped in tape and their

IGNITION AND ELECTRICAL SYSTEMS 4-11

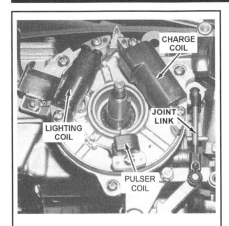

Fig. 15 Example of a typical stator plate with 1 pulser, 1 charge and 1 lighting coil. The stator plate is rotated through the joint link to advance the timing

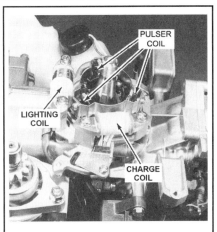

Fig. 16 Stator plate from a typical multi-cylinder motor (this one with 3 pulser coils, 1 charge and one lighting coil)

Fig. 17 The CDI unit utilizes output from the charge coil to power the ignition coils

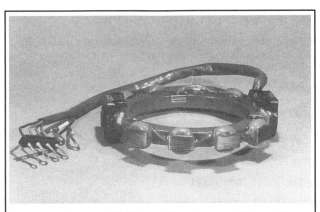

Fig. 18 The charge and lighting coil (stator) removed from a V6 powerhead for bench testing

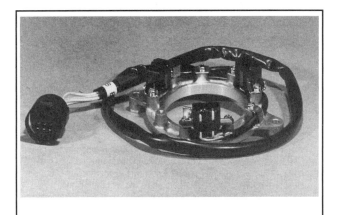

Fig. 19 A typical Yamaha 3-pulser coil assembly

windings are not visible) or by the wire colors. On most Yamaha and Mercury/Mariner 4-stroke motors the light coil wiring is normally Green, Yellow or White (or some combination). For details on a specific motor, refer to the Charging System Testing charts and/or the Wiring Diagrams found in this section.

The rectifier converts the Alternating Current (AC) voltage into DC voltage, which may then be stored in the battery. The rectifier is a sealed unit and generally contains four diodes. If one of the diodes is defective, the entire unit must be replaced.

When equipped, the voltage regulator stabilizes the power output of the coils and extends the life of the light bulbs and battery by preventing power surges or overcharging.

Pulser Circuit

◆ See Figure 19

The second circuit used in CDI/TCI systems is the pulser circuit. The pulser circuit usually has its own flywheel magnet (or magnets), one or more pulser coils, a diode, and a thyristor. Those last 2 components (thyristor and diode) are almost always internal components to the CDI unit. In case you were wondering, a thyristor is a solid state electronic switching device which permits voltage to flow only after it is triggered by another voltage source.

At the point in time when the ignition timing marks align, an alternating current is induced in the pulser coil, in the same manner as previously described for the charge coil. This current is then passed to a second diode located in the CDI unit where it becomes DC current and flows on to the thyristor. This voltage triggers the thyristor to permit the voltage stored in the capacitor to be discharged. The capacitor voltage passes through the thyristor and on to the primary windings of the ignition coil.

In this manner, a spark at the plug may be accurately timed by the timing marks on the flywheel relative to the magnets in the flywheel and to provide as many as 100 sparks per second for a powerhead operating at 6000 rpm.

Different ignition strategies on Yamaha and Mercury/Mariner motors may utilize as many as 6 pulser coils (up to one per cylinder) on 2-stroke V6 motors or 2 pulser coils on a single cylinder motor (on which the composite wave form is used for ignition control). Differences are spelled out as clearly as we can in earlier sections on Ignition by motor type (and in the Ignition System Component testing charts and Wiring Diagrams, in this section).

Ignition Coils

◆ See Figure 20

Yamaha and Mercury/Mariner ignition coils, usually one per cylinder or one per pair of cylinders, boost the DC voltage instantly to over 20,000 volts (and potentially as much as 50,000 volts) for spark. This completes the primary side of the ignition circuit.

Once the voltage is discharged from the ignition coil, the secondary circuit begins and only stretches from the ignition coil to the spark plugs via extremely large high tension leads. At the spark plug end, the voltage arcs in the form of a spark, across from the center electrode to the outer electrode, and then to ground via the spark plug threads. This completes the ignition circuit.

4-12 IGNITION AND ELECTRICAL SYSTEMS

Fig. 20 Many Yamaha and Mercury/Mariner motors utilize one ignition coil for each cylinder

Ignition Timing

At the point in time when the ignition timing marks align, an alternating current is induced in the pulser coil, in the same manner as previously described for the charge coil. This current is then passed to a second diode located in the CDI unit where it becomes DC current and flows on to the thyristor. This voltage triggers the thyristor to permit the voltage stored in the capacitor to be discharged. The capacitor voltage passes through the thyristor and on to the primary windings of the ignition coil.

In this manner, a spark at the plug may be accurately timed by the timing marks on the flywheel relative to the magnets in the flywheel and to provide as many as 100 sparks per second for a powerhead operating at 6000 rpm.

Almost all 4-stroke units however are equipped with an automatic advance type CDI/TCI system and have no moving or sliding parts. Ignition advance is accomplished electronically - by the electric waves emitted by the pulser coils. These coils increase their wave output proportionately to engine speed increase, thus advancing the timing.

Units equipped with a mechanical advance type CDI system use a link rod between the carburetor and the ignition base plate assembly. At the time the throttle is opened, the ignition base plate assembly is rotated by means of the link rod, thus advancing the timing.

Ignition and Charging System Troubleshooting

Accurate troubleshooting of an ignition system is viewed by many as mission impossible. But this really does not have to be the case. Having a systematic approach to troubleshooting is the key. Your first priority is to understand how the system works.

Get yourself mentally prepared for ignition troubleshooting before you touch the motor. Make notes about how the motor behaves. Is it totally dead? Misfiring? If it is dead, what were the circumstances that led to its demise? Did it run poorly before it died? If it is running, but not well, is this a permanent condition? Only when cold? Only when warm? Under load? Ask yourself a lot of questions. The more information you can gather the faster the repair can be made.

It is easy to forget to ask yourself if the root cause of the failure is truly the ignition? Could it be something else? Look at the plug(s) and check the spark. Once you have done the basics, the hands-on troubleshooting begins.

The key to successful troubleshooting is a systematic approach. Do not skip around. On new or unfamiliar equipment try writing up a check list on a 3 x 5 note card. This makes a handy reference for future troubleshooting.

Use a spark tester to observe spark quality. Does it jump the gap? Is it bright and blue? Orange spark usually means trouble in the ignition system.

Do not forget to check the ignition controlling components such as the main or stop switches. These can be disconnected from the system. These switches can cause failures that checks of the individual ignition components may not reveal.

Motors with microcomputer controlled ignition systems utilize other sensor inputs to help determine spark timing. A thermoswitch or oil pressure sensor giving an incorrect reading may prevent spark timing advance. Don't forget these and other ignition control sensors.

Always attempt to proceed with the troubleshooting in an orderly manner. The shot in the dark approach will only result in wasted time, incorrect diagnosis, replacement of unnecessary parts, and frustration.

COMPONENT TESTING

Resistance Testing Precautions

When performing an electrical resistance test on coil windings, a rectifier unit, or on a device where the resistance varies with electrical polarity or applied voltage the following points should be noted:

• There are no standard specifications for the design of either analog or digital test meters.
• When performing resistance testing, some test meter manufacturers will apply positive battery power to the Red test lead, while others apply the positive voltage to the Black lead.
• Some meters (especially digital meters) will apply a very low voltage to the test leads that may not properly activate some devices (e.g., CDI units).
• Be aware there are ohmmeters that have reverse polarity. If your test results all differ from the chart, swap + and - leads and retest.

■ **Digital meter resistance values are not reliable when testing CDI units, rectifier units, or any device containing semiconductors, transistors, and diodes.**

• Although resistance test specifications can be helpful, they are also very misleading in ignition circuits/components. A component may test within specification during a static resistance check, yet still fail during normal operating conditions. Also, remember that readings will vary both with temperature and by meter.

Spark Plugs

◆ See Figures 21, 22 and 23

1. Check the plug wires to be sure they are properly connected. Check the entire length of the wires from the plugs to the coil. If the wires are to be removed from the spark plug, always use a pulling and twisting motion as a precaution against damaging the connection.
2. Attempt to remove the spark plug by hand. This is a rough test to determine if the plug is tightened properly. The attempt to loosen the plug by hand should fail. The plug should be tight and require the proper socket size tool. Remove the spark plug and evaluate its condition. Reinstall the spark plug and tighten with a torque wrench to the proper specification.
3. Use a spark tester and check for spark. If a spark tester is not available (and for Pete's sake they're cheap and can be found in almost all auto parts stores) hold the plug wire about 1/4 in. (6.4mm) from the engine (leave the plug in the port for safety) or better yet, use a spare or old spark plug in the spark wire and hold the plug (using rubber handled pliers) against the powerhead.
4. Carefully operate the starter and check for spark. A strong spark over a wide gap must be observed when testing in this manner, because under compression a strong spark is necessary in order to ignite the air-fuel mixture in the cylinder. This means it is possible to think a strong spark is present, when in reality the spark will be too weak when the plug is installed. If there is no spark, or if the spark is weak, the trouble is most likely under the flywheel in the magneto.

Compression

◆ See Figure 24

1. Before spending too much time and money attempting to trace a problem to the ignition system, a compression check of the cylinder should be made. If the cylinder does not have adequate compression,

IGNITION AND ELECTRICAL SYSTEMS 4-13

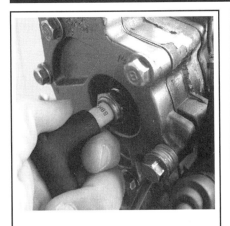
Fig. 21 Disconnect the spark plug wire...

Fig. 22 ...and check spark using a tester

Fig. 23 Remove and inspect the spark plug

troubleshooting and attempted service of the ignition or fuel system will fail to give the desired results of satisfactory engine performance.

For details, please refer to Compression Testing in the Maintenance and Tune-Up section.

Pulser Coils

◆ See Figures 25 and 26

There are basically 2 overall methods used in testing pulser coils, a static resistance check and a dynamic cranking or running output voltage check. Specifications vary by year and model so significantly that it precludes trying to list them in this procedure. In addition, the method of conducting the static check will vary with the specifications that are provided. Generally speaking cranking or running tests are more accurate and more likely to show intermittent problems. Dynamic checks are done under different circumstances, with load (circuit complete) and without load (circuit incomplete), again, depending upon the specifications provided by the manufacturer. Please refer to the Ignition System Component Testing specification charts in this section for details.

■ If you're in doubt about any test connections verify them using the information in the Ignition Component Testing specifications charts AND the Wiring Diagrams, found in this section.

The basic test for a pulser coil is continuity. This measures the actual resistance from one end of the pulser coil to the other. When available, the correct specification for each coil can be found in the Ignition Testing Specifications charts.

1. Adjust the meter to read resistance (ohms).
2. Connect the tester across the pulser leads (as noted in the testing charts and/or the wiring diagrams) and note the reading.
3. Compare the coil reading to the specifications.
4. If the reading is well above or below the correct value it should be checked dynamically (if output specifications are available) and/or be replaced.

■ Remember that temperature has an affect on resistance. Most resistance specifications are given assuming a temperature of 68°F (20°C).

There are two types of pulser coils. The first type has one coil lead connected directly to ground through a Black wire or a grounded bolt hole. The other end of the coil has a color-coded wire.

The second type of pulser coil (with a separate ground lead) is not grounded at the powerhead. Instead, both leads go to the CDI unit. Sometimes both coil leads are often White with a colored tracer, but other times the ground lead will be Black or Black with a tracer.

When checking either type pulser coil for continuity, connect the ohmmeter to the color wires listed in the Ignition Testing Specifications chart.

A second check must be made on pulser coils that are not grounded on one end. This second check is called a short-to-ground check and is designed to make sure that this type of coil not only has correct resistance (continuity) but also is not shorted to ground (and leaking current to ground).

5. Connect one tester lead to a coil lead.
6. Touch the other tester lead to the engine ground or the mounting point of the pulser.

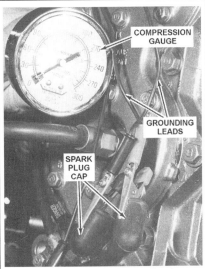
Fig. 24 Preparing to check powerhead cranking compression. Note the jumper leads used to ground the cylinders

Fig. 25 Testing a typical Yamaha or Mercury/Mariner pulser coil

Fig. 26 Coil testing may include cranking or running voltage and/or resistance tests (this shows a typical resistance test)

4-14 IGNITION AND ELECTRICAL SYSTEMS

7. If the pulser is good, the reading on the ohmmeter should be infinity.
8. Any ohm reading other than infinity (O.L on some digital meters) indicates a bad coil. Repair or replace the coil.

■ **This second check is just as critical as the standard continuity check.**

The final, and most important check, is a dynamic cranking or running output voltage check. If specifications are provided, proceed as follows:

9. Set the meter to read volts in the proper scale (as determined by the specification range in the chart).
10. Connect the meter to the pulser coil lead(s) under the appropriate conditions. For with load specifications the circuit must be complete meaning you'll have to either backprobe the wiring harness or (if this cannot be done without damaging the connectors) use jumper wires between the disconnected ends of the harness. For without load specifications you can just disconnect the harness and probe the coil sides.
11. If the coil fails to produce sufficient voltage for specifications, double-check all connections and wire colors. Once you are certain the coil is out of spec it must be replaced.

Charge Coils

◆ See Figures 27, 28 and 29

The charge coil checks are essentially the same as those for the pulser coil. There are normally 2 methods used in testing charge coils, a static resistance check and a dynamic cranking or running output voltage check. Specifications vary by year and model so significantly that it precludes trying to list them in this procedure. In addition, the method of conducting the static check will vary with the specifications that are provided. Generally speaking cranking or running tests are more accurate and more likely to show intermittent problems. Dynamic checks are done under different circumstances, with load (circuit complete) and without load (circuit incomplete), again, depending upon the specifications provided by the manufacturer. Please refer to the Ignition System Component Testing specification charts in this section for details.

There are two types of charge coils. One type has a charge coil lead connected to a powerhead ground. The other type charge coil, which represents the VAST majority of these motors, has both leads go directly to the CDI box. Both charge coil types allow for a continuity test. This checks the coil's internal resistance. On charge coils that do not have a direct powerhead grounded lead, the short-to-ground test must also be done

■ **If you're in doubt about any test connections verify them using the information in the Ignition Component Testing specifications charts AND the Wiring Diagrams, found in this section.**

The basic test for a charge coil is continuity. This measures the actual resistance from one end of the charge coil to the other. The correct specification for each coil can be found in the Ignition System Component Testing specifications chart.

1. Adjust the meter to read resistance (ohms).
2. Connect the tester across the charge coil leads and note the reading.
3. Compare the coil reading to the specifications. If the reading is well above or below the correct value it should be checked dynamically (if output specifications are available) and/or be replaced.

■ **Remember that temperature has an affect on resistance. Most resistance specifications are given assuming a temperature of 68°F (20°C).**

There are generally two types of charge coils. One type has one coil lead connected directly to a powerhead ground through a Black wire or a grounded bolt hole. The other end of the coil has a color-coded wire. These type coils often have one lead listed as Black and the other is often a White wire with a colored tracer.

The second type of charge coil is not grounded directly to the powerhead, but through the circuitry in the CDI unit. Therefore, both leads go to the CDI box. Both coil leads are color coded, often including Blue or White with a colored tracer.

When checking either type charge coil for continuity, connect the ohmmeter to the color wires listed in the specifications.

A second check must be made on charge coils that are not grounded directly to the powerhead on one end. This second check is called a short-to-ground check and is designed to make sure that this type of coil not only has correct resistance (continuity) but also is not shorted to ground (and leaking current to ground).

4. Connect one tester lead to a coil lead.
5. Touch the other tester lead to the engine ground or the mounting point of the coil.
6. If the coil is good, the reading on the ohmmeter should be infinity. Any ohm reading other than infinity (O.L on some digital meters) indicates a bad coil.
7. Repair or replace the coil.

■ **This second check is just as critical as the standard continuity check.**

The final, and most important check, is a dynamic cranking or running output voltage check. If specifications are provided, proceed as follows:

8. Set the meter to read volts in the proper scale (as determined by the specification range in the chart).
9. Connect the meter to the charge coil lead(s) under the appropriate conditions. For with load specifications the circuit must be complete meaning you'll have to either back-probe the wiring harness or (if this cannot be done without damaging the connectors) use jumper wires between the disconnected ends of the harness. For without load specifications you can just disconnect the harness and probe the coil sides.
10. If the coil fails to produce sufficient voltage for specifications, double-check all connections and wire colors. Once you are certain the coil is out of spec it must be replaced.

Fig. 27 Locate the charge coil leads...

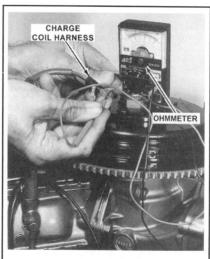

Fig. 28 ...then conduct static (ohm)...

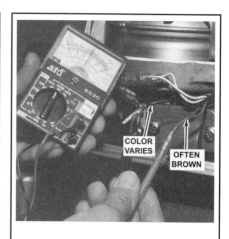

Fig. 29 ...and/or dynamic (voltage) tests

IGNITION AND ELECTRICAL SYSTEMS 4-15

Lighting Coil

◆ See Figures 30 and 31

The lighting coil windings really aren't usually part of the ignition system (at least they aren't when separate charge coil windings are provided to power the ignition), however it operates in the same fashion as the charge and pulser coils, and it is replaced in the same fashion, so it makes sense to cover it here.

On many newer Yamaha and Mercury/Mariner motors the charge coil winding are incorporated into the stator (along with the lighting coil), and on some there are no distinct charge coil windings. On these late-model motors, the entire stator assembly must be replaced if a problem is isolated to the ignition charge circuit.

The lighting coil checks are the same as those for the pulser and charge coils. There are normally 2 methods used in testing lighting coils, a static resistance check and a dynamic cranking or running output voltage check. Specifications vary by year and model so significantly that it precludes trying to list them in this procedure. In addition, the method of conducting the static check will vary with the specifications that are provided. Generally speaking cranking or running tests are more accurate and more likely to show intermittent problems. Dynamic checks are done under different circumstances, with load (circuit complete) and without load (circuit incomplete), again, depending upon the specifications provided by the manufacturer.

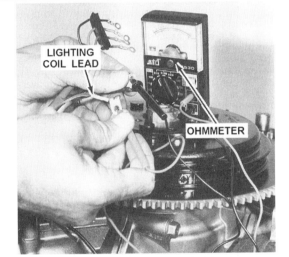

Fig. 30 Lighting coils are also tested statically (ohms)...

Fig. 31 ...or dynamically (voltage) - most models use Green and/or Yellow leads

■ **If you're in doubt about any test connections verify them using the information in the Charging System Testing specifications charts AND the Wiring Diagrams, found in this section.**

The basic test for a lighting coil is continuity. This measures the actual resistance from one end of the lighting coil to the other. The correct specification for each coil can be found in the Charging System Testing specifications chart.

1. Disconnect the two wires (often, but not always, Green) between the stator and the rectifier at the rectifier. Connect the ohmmeter leads to the wires and measure the resistance.
2. Compare the coil reading to the specifications. If the reading is well above or below the correct value it should be checked dynamically (if output specifications are available) and/or be replaced.

■ **Remember that temperature has an affect on resistance. Most resistance specifications are given assuming a temperature of 68°F (20°C).**

3. If the resistance is not within specification, the battery will not hold a charge and the boat accessories which depend on this coil for power may not function properly.

A second check must be made on lighting coils. This second check is called a short-to-ground check and is designed to make sure that this type of coil not only has correct resistance (continuity) but also is not shorted to ground (and leaking current to ground).

4. Connect one tester lead to a coil lead.
5. Touch the other tester lead to the engine ground or the mounting point of the pulser.
6. If the pulser is good, the reading on the ohmmeter should be infinity.
7. Any ohm reading other than infinity (O.L on some digital meters) indicates a bad coil. Repair or replace the coil.

■ **This second check is just as critical as the standard continuity check.**

■ **Unless specifically directed by the test charts, never attempt to verify the charging circuit by operating the powerhead with the battery disconnected. In most cases, such action would force current (normally directed to charge the battery), back through the rectifier and damage the diodes in the rectifier.**

The final, and most important check, is a dynamic cranking or running output voltage check. If specifications are provided, proceed as follows:

8. Set the meter to read volts in the proper scale (as determined by the specification range in the chart).
9. Connect the meter to the charge coil lead(s) under the appropriate conditions. Unless otherwise directed, protect the rectifier by taking test readings with the circuit complete. You'll have to either back-probe the wiring harness or (if this cannot be done without damaging the connectors) use jumper wires between the disconnected ends of the harness.
10. If the coil fails to produce sufficient voltage for specifications, double-check all connections and wire colors. Once you are certain the coil is out of spec it must be replaced.

Ignition Coils

◆ See Figures 32, 33 and 34

Although the best test for an ignition coil is on a dynamic ignition coil tester, resistance checks can also be performed to help determine condition. As with other coils, remember that a static test within specification cannot absolutely rule out a problem with the ignition coil under load in dynamic conditions.

There are two circuits in an ignition coil, the primary winding circuit and the secondary winding circuit. Whenever possible, both need to be checked.

The tester connection procedure for a continuity check will depend on how the coil is constructed. Generally, the primary circuit is the small gauge wire or wires, while the secondary circuit contains the high tension or plug lead.

When there are two primary wires running to the ignition coil, the primary circuit test is performed across both of those wire terminals (for the ignition coil). When there is only one primary wire, the circuit is checked between that wire and ground.

4-16 IGNITION AND ELECTRICAL SYSTEMS

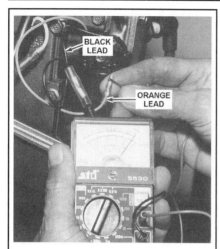

Fig. 32 Test the primary coil circuit either across the 2 small coil leads or directly at the coil terminals

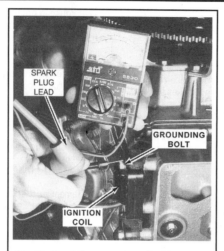

Fig. 33 Check the secondary circuit across the single spark plug lead and primary circuit ground (wire or grounding point)...

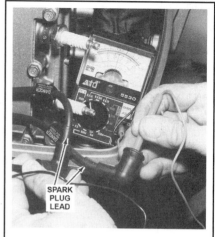

Fig. 34 ... or across the 2 spark plug leads (as applicable)

When an ignition coil is designed to fire only one spark plug, the secondary circuit is normally either through the spark plug lead (without the resistor cap, when equipped) and ground or ground wire terminal (depending upon the primary circuit wiring). If, however, the ignition coil contains 2 spark plug leads, the secondary circuit is checked across both leads (again, without the resistor caps, if equipped).

Some ignition coils have the primary and/or secondary circuits grounded on one end. On these type coils, only the continuity check is done. On ignition coils that are not grounded on one end (which have 2 wires going back to the CDI unit) a short-to-ground test must also be done (a check from the primary side of the circuit to ground in order to make sure there is no continuity). Regardless of the coil type, compare the resistance with the Ignition System Component Testing specification charts found in this section.

■ When checking the secondary side, remove the spark plug caps to make the measurement (most models utilize resistor caps). When equipped with resistor caps, in some cases the cap is bad, not the coil. Bad resistor caps can be the cause of high-speed misfire. Unscrew the cap and check the resistance (usually about 5 killi-ohms, give or take a kill-ohm). Leaving the cap on during measurement could condemn an otherwise in spec ignition coil.

The other method used to test ignition coils is with a Dynamic Ignition Coil Tester. Since the output side of the ignition coil has very high voltage, a regular voltmeter can not be used. While resistance reading can be valuable, the best tool for checking dynamic coil performance is a dynamic ignition coil tester.

1. Connect the coil to the tester according to the manufacturer's instructions.
2. Set the spark gap according to the specifications.
3. Operate the coil for about 5 minutes.
4. If the spark jumps the gap with the correct spark color, the coil is probably good.

■ If you're in doubt about any test connections verify them using the information in the Ignition Component Testing specifications charts AND the Wiring Diagrams, found in this section.

CDI Unit

◆ See Figure 35

There are potentially 2 methods of testing the CDI/TCI unit, and one of them is active and second passive. Probably the most common and one of the most accurate methods of testing the unit is to check the output voltage. For most Yamaha motors specifications are available for CDI/TCI unit output tests, while Mercury/Mariner calls them Primary Ignition Coil Circuit testing. In either case, this dynamic check is the preferred method of CDI unit testing, as it is more likely that faults will appear under load conditions. The test connections and specifications are provided in the Ignition System Component Testing and the Charge & Primary Coil Output Testing specifications charts in this section.

The 2nd method of CDI/TCI unit testing is probably the most important. It is a process of elimination where the remaining ignition system components are first tested and proven good. If there are no problems found anywhere else in the system components or wiring, you are usually safe replacing the CDI/TCI unit. Just be certain that you haven't overlooked another possible component or wire when condemning the unit.

■ In the past Yamaha published resistance specifications for testing CDI units. However, no resistance testing specifications are available for ANY of the modern 4-stroke Yamahas, nor for any unit (even 2-strokes) introduced (or revised) after 1997 or 1998 (ish).

Yamaha Rectifiers or Regulator/Rectifiers

◆ See Figures 36 and 37

When an outboard is equipped with a lighting coil, unless it is an AC lighting model (which is designed to power AC accessories), it must also have a rectifier to convert the voltage to DC and/or a regulator to control the battery charging voltage. Over the years Yamaha has changed what tests it provides for the rectifier or regulator/rectifier. As they have moved away from resistance tests, for questions of accuracy they've also begun to favor output voltage checks for regulator/rectifiers.

For all models, except the 8/9.9 hp (232cc) motors, output specifications for the regulator/rectifier are available and are listed in the Charging System Testing specifications chart in this section. This specification is the BEST way to troubleshoot the regulator/rectifier on Yamaha Models.

■ Remember that a regulator/rectifier does NOT create voltage, it only controls voltage. Whenever the Lighting Coil output should always be checked before suspecting that a regulator/rectifier is the culprit for undercharging a battery. However, since the regulator/rectifier's primary job is to prevent OVERcharging, the lighting coil is not usually the first suspect under those conditions.

For the 8/9.9 hp (232cc) motors all Yamaha provides are some resistance/diode checking charts. If trouble is suspected on these motors you can use these charts to help determine the condition of the rectifier/regulator. However, keep in mind that if the system is undercharging you should first check Lighting Coil output to make sure there is sufficient voltage available to the system.

■ The accompanying charts for 8/9.9 hp motors outline resistance or voltage drop testing procedures for the rectifier. The unit may remain installed on the powerhead, or it may be removed for testing. In either case, the testing procedures are identical.

1. For all motors except the 8/9.9 hp (232cc) models, use the specifications given in the Lighting Coil chart to check cranking and/or running output for the Lighting Coil and the Rectifier.
2. For standard thrust 8/9.9 hp (232cc) models, select the appropriate scale on either an analog (needle-type) or digital ohmmeter (depending upon

IGNITION AND ELECTRICAL SYSTEMS 4-17

the information in the test chart). Make contact with the Red meter lead to the leads called out as (+) and the Black meter lead to the leads called out as (-). Proceed slowly and carefully in the order given.

3. For high-thrust 9.9 hp (232cc) models, select the diode inspection mode of a digital voltmeter. Make contact with the Red meter lead to the leads called out as (+) and the Black meter lead to the leads called out as (-). Proceed slowly and carefully in the order given checking voltage drops as noted in the accompanying chart.

■ **The manufacturer notes for some models as specified in the charts that a digital tester will not perform properly or will perform differently. Again, refer to the charts for details.**

4. If voltage output or resistance (depending upon the model) is not as specified, the rectifier is faulty and should be replaced.

Mercury/Mariner Rectifiers or Regulator/Rectifiers

When an outboard is equipped with a lighting coil, unless it is an AC lighting model (which is designed to power AC accessories), it must also have a rectifier to convert the voltage to DC and/or a regulator to control the battery charging voltage.

Whereas Yamaha prefers voltage output checks, Mercury/Mariner still prefers to rely on resistance and/or diode checks for the rectifiers and regulator/rectifiers on their motors.

In the following sub-section we've provided either step-by-step procedures and/or charts which include the testing procedures, sorted by models/charging systems.

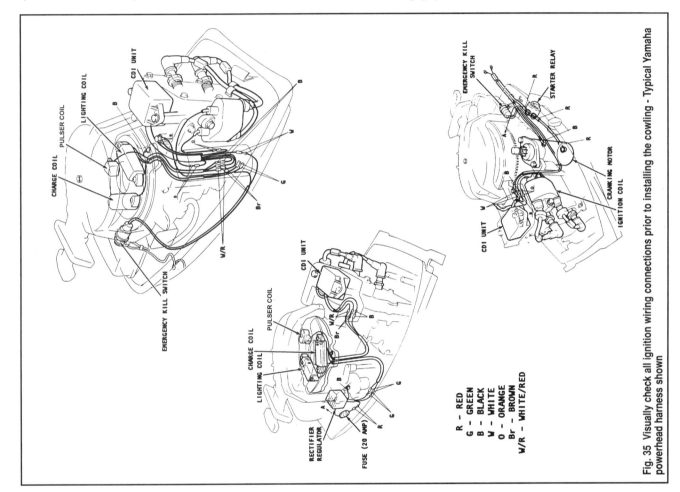

Fig. 35 Visually check all ignition wiring connections prior to installing the cowling - Typical Yamaha powerhead harness shown

Fig. 36 Regulator/rectifier resistance testing - standard thrust 8/9.9 hp (232cc) models only

Tester ⊕ ⊖ Tester	Green	Green/White	Red	Black
Green		∞	O	∞
Green/White	∞		O	∞
Red	∞	∞		∞
Black	O	O	O	

O : Continuity.
∞ : Discontinuity.

Fig. 37 Regulator/rectifier diode voltage drop testing - high-thrust 9.9 hp (232cc) models only

Tester ⊕ ⊖ Tester	① Black	② Green/White	③ Green	④ White/Green	⑤ Red
① Black		0.50 ~ 0.70	*	0.75 ~ 1.05	*
② Green/White	0.30 ~ 0.50		*	0.30 ~ 0.50	*
③ Green	0.30 ~ 0.50	0.75 ~ 1.05		0.95 ~ 1.25	*
④ White/Green	*	*	*		*
⑤ Red	0.65 ~ 0.85	0.30 ~ 0.50	0.30 ~ 0.50	0.65 ~ 0.85	

* Measured value is not affected by tester leads connection.

4-18 IGNITION AND ELECTRICAL SYSTEMS

■ The accompanying procedures outline resistance/diode testing procedures for the rectifier. The unit may remain installed on the powerhead, or it may be removed for testing. In either case, the testing procedures are identical.

1. Select the appropriate scale on either an analog (needle-type) or digital ohmmeter (depending upon the information in the test chart). Make contact with the Red meter lead to the leads called out as (+) and the Black meter lead to the leads called out as (-). Proceed slowly and carefully in the order given.
2. If resistance is not as specified, the rectifier is faulty and should be replaced.

■ Remember that a regulator/rectifier does NOT create voltage, it only controls voltage. Whenever the Lighting Coil output should always be checked before suspecting that a regulator/rectifier is the culprit for undercharging a battery. However, since the regulator/rectifier's primary job is to prevent OVERcharging, the lighting coil is not usually the first suspect under those conditions.

4/5/6 Hp Mercury/Mariner Models

For the Mercury/Mariner 4/5/6 hp motors the rectifier test consists of a simple diode check. Unplug the rectifier wiring, then use an ohmmeter across the Red and Yellow wire terminals. Reverse the leads and check again. If the diodes are operating properly there should only be continuity in one direction and not both. Replace the rectifier on these models if the reading is the same in both directions.

8/9.9 Hp (232cc) Models

◆ See Figure 38

For these models Mercury recommends checking the rectifier/regulator using a digital voltmeter with a "diode inspection" setting. When checking, keep in mind that due to manufacturing differences the internal polarity of some ohmmeters may vary resulting in test readings which would be completely opposite of the accompanying chart. If so, reverse the leads and recheck.

■ A slight variance from the readings does not indicate a defective component.

9.9/15 Hp (323cc) Models

◆ See Figures 39, 40 and 41

For these models Mercury recommends checking the rectifier/regulator using a digital voltmeter with a "diode inspection" setting. When checking, keep in mind that due to manufacturing differences the internal polarity of some ohmmeters may vary resulting in test readings which would be completely opposite of the accompanying chart. If so, reverse the leads and recheck.

Two separate charts are provided, one for models with a 6-amp charging system and one for models with a 10-amp charging system. Use the accompanying illustration and rectifier/regulator shape to help visually identify the proper system.

■ A slight variance from the readings does not indicate a defective component.

Rectifier/Regulator Diode Test Chart

	Red (+) Meter Lead To:				
Black (-) Meter Lead To:	Black	Green/White	Green	White/Green	Red
Black		0.50 to 1.80	*	0.75 to 2.00	*
Green/White	0.30 to 0.80		*	0.30 to 0.80	*
Green	0.30 to 0.80	0.75 to 2.00		0.95 to 2.00	*
White/Green	*	*	*		*
Red	0.65 to 0.95	0.30 to 0.80	0.30 to 0.80	0.65 to 1.07	

*Measured value is not affected by tester lead connection.

Fig. 38 Regulator/rectifier diode voltage drop testing - high-thrust 9.9 hp (232cc) models only

25-60 Hp Models, Except 1995-2000 40/45/50 Hp (935cc) Motors

◆ See Figure 42

These motors are all equipped with the same voltage rectifier/regulator, Part No. 854514 or 854514-1. For this component, Mercury provides specific resistance testing of the various internal circuits. Although the component itself may be physically left attached to the powerhead for testing, the wires should be disconnected from the powerhead harness, and realistically, it doesn't take much more effort to just unbolt it and perform the test on the bench.

Because results will vary, two separate sets of tests and results have been provided. The first for use with an Analog (needle) ohmmeter and the second for a DVOM with a diode testing setting. For the later, keep in mind that due to manufacturing differences the internal polarity of some ohmmeters may vary resulting in test readings which would be completely opposite of the accompanying chart. If so, reverse the leads and recheck.

Testing With an Analog Meter

1. To test the Diodes:
    a. Set the ohmmeter to read resistance on the R x 10 scale.
    b. Connect the positive (red) meter probe to the RED wire on the regulator.
    c. Connect the negative (black) meter probe to the one of the YELLOW wires on the regulator.
    d. The meter should show 100-400 ohms.
    e. Leave the positive probe connected, then move the negative (black) meter probe to the OTHER YELLOW wire on the regulator.
    f. The meter should again show 100-400 ohms.
    g. Next, set the ohmmeter to read resistance on the R x 1000 (1k) scale.
    h. Connect the negative (black) meter probe to the RED wire on the regulator.
    i. Connect the positive (red) meter probe to one of the YELLOW wires on the regulator. Note the reading, then move the positive (red) meter probe to the OTHER YELLOW wire on the regulator.
    j. The meter should show 20,000 ohms or more (including infinite) for one of the readings and should show infinite resistance on the other.
2. To test the Silicon Controlled Rectifier (SCR) Circuits:
    a. If not already done for the last of the Diode Tests, set the ohmmeter to read resistance on the R x 1000 (1k) scale.
    b. Connect the positive (red) meter probe to the regulator case.
    c. Connect the negative (black) meter probe to the one of the YELLOW wires on the regulator.
    d. The meter should show 8,000-15,000 ohms (8-15k).

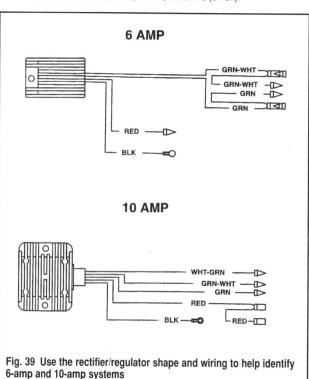

Fig. 39 Use the rectifier/regulator shape and wiring to help identify 6-amp and 10-amp systems

IGNITION AND ELECTRICAL SYSTEMS 4-19

e. Connect the negative (black) meter probe to the OTHER YELLOW wire on the regulator.

f. The meter should again show 8,000-15,000 ohms (8-15k).

3. To test the Tachometer Circuit:

a. If not already done for the SCR Tests, set the ohmmeter to read resistance on the R x 1000 (1k) scale.

b. Connect the positive (red) meter probe to the GREY wire on the regulator.

c. Connect the negative (black) meter probe to the to the regulator case.

d. The meter should show 10,000-50,000 ohms (10-50k).

4. If the rectifier/regulator is not performing properly and fails one or more of these tests, it should be replaced.

Testing With a Digital Meter

5. To test the Diodes:

a. Set the DVOM to the diode test (normally represented as an arrow facing to the right intersecting a perpendicular line).

b. Connect the negative (black) meter probe to the RED wire on the regulator.

c. Connect the positive (red) meter probe to the one of the YELLOW wires on the regulator.

d. The meter should show a reading of 0.4-0.8 (the setting should display voltage drop).

e. Leave the negative probe connected, then move the positive (red) meter probe to the OTHER YELLOW wire on the regulator.

f. The meter should again show a reading of 0.4-0.8.

g. Next, connect the positive (red) meter probe to the RED wire on the regulator.

h. Connect the negative (black) meter probe to one of the YELLOW wires on the regulator. Note the reading, then move the positive (red) meter probe to the OTHER YELLOW wire on the regulator.

i. The meter should show infinite resistance, OUCH or OL (Out of Limit) depending upon the meter.

6. To test the Silicon Controlled Rectifier (SCR) Circuits:

a. If not already done for the last of the Diode Tests, set the ohmmeter to the diode test setting (normally represented as an arrow facing to the right intersecting a perpendicular line).

b. Connect the negative (black) meter probe to the regulator case.

c. Connect the positive (red) meter probe to the one of the YELLOW wires on the regulator.

d. The meter should show 1.5V, infinity, OUCH or OL (Out of Limit), depending upon the meter.

e. Connect the red (positive) meter probe to the OTHER YELLOW wire on the regulator.

f. The meter should again show 1.5V, infinity, OUCH or OL (Out of Limit), depending upon the meter.

7. To test the Tachometer Circuit:

a. Set the ohmmeter to read resistance on the R x 1000 (1k) scale.

b. Connect the positive (red) meter probe to the GREY wire on the regulator.

c. Connect the negative (black) meter probe to the to the regulator case.

d. The meter should show 10,000-50,000 ohms (10-50k).

8. If the rectifier/regulator is not performing properly and fails one or more of these tests, it should be replaced.

1995-2000 40/45/50 Hp (935cc) Models
◆ See Figures 43 and 44

For these models Mercury recommends checking the rectifier/regulator using a digital voltmeter with a "diode inspection" setting. When checking, keep in mind that due to manufacturing differences the internal polarity of some ohmmeters may vary resulting in test readings which would be completely opposite of the accompanying chart. If so, reverse the leads and recheck.

Use the accompanying schematic to help identify test points.

■ **A slight variance from the readings does not indicate a defective component.**

6 AMP RECTIFIER/REGULATOR DIODE TEST – DIGITAL METER (→|− SCALE)

Black (−) Meter Lead To:	Red (+) Meter Lead To:			
	Black	Green/White	Green Red	Red Red
Green/White	0.30 to 0.80	–	OUCH or ∞	OUCH
Green	0.30 to 0.80	OUCH	–	OUCH
Red	0.65 to 2.0	0.30 to 0.80	0.30 to 0.80	–

OUCH = OL = Full Meter Deflection

Fig. 40 Regulator/rectifier diode testing - 9.9/15 hp (323cc) models with 6-amp charging system

10 AMP RECTIFIER/REGULATOR DIODE TEST - DIGITAL METER (→|− SCALE)

Black (−) Meter Lead To:	Red (+) Meter Lead To:				
	Black	Green/White	Green	White/Green	Red
Black	–	1.0 to OUCH	1.0 to OUCH	1.2 to OUCH	OUCH
Green/White	0.30 to 0.80	–	1.2 to OUCH	0.30 to 0.80	OUCH
Green	0.30 to 0.80	1.2 to OUCH	–	1.5 to OUCH	OUCH
White/Green	OUCH	OUCH	OUCH	–	OUCH
Red	0.65 to 1.2	0.30 to 0.80	0.30 to 0.80	0.65 to 1.3	–

OUCH = OL = Full Meter Deflection

Fig. 41 Regulator/rectifier diode testing - 9.9/15 hp (323cc) models with 10-amp charging system

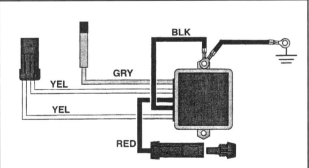

Fig. 42 Schematic for the regulator/rectifier found on all 25-60 hp Mercury/Mariner motors (except the 1995-00 40/45/50 hp/935cc models)

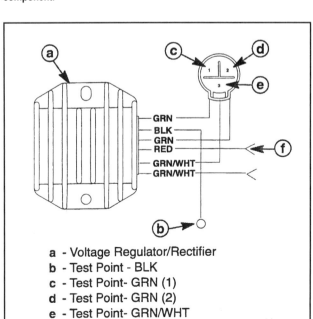

a - Voltage Regulator/Rectifier
b - Test Point - BLK
c - Test Point- GRN (1)
d - Test Point- GRN (2)
e - Test Point- GRN/WHT
f - Test Point- RED

Fig. 43 Regulator/rectifier schematic for 1995-00 40/45/50 hp (935cc) models

4-20 IGNITION AND ELECTRICAL SYSTEMS

Rectifier/Regulator Diode Test Chart-Digital Meter (→⊢Scale)

Black (-) Meter Lead To:	Red (+) Meter Lead To:					
		BLK	GRN (1)	GRN (2)	GRN/WHT	RED
	BLK	-	OUCH	OUCH	OUCH	OUCH
	GRN (1)	0.3 to 0.8	-	OUCH	OUCH	OUCH
	GRN (2)	0.3 to 0.8	OUCH	-	OUCH	OUCH
	GRN/WHT	0.3 to 0.8	OUCH	OUCH	-	OUCH
	RED	0.8 to 1.3	0.3 to 0.8	0.3 to 0.8	0.3 to 0.8	-

NOTE: OUCH=OL=Full Meter Deflection

NOTE: All rectifier/regulator leads **MUST BE** disconnected to obtain accurate diode test readings.

Fig. 44 Regulator/rectifier diode testing - 1995-00 40/45/50 hp/935cc models

75/90 and 115 Hp Models
◆ See Figures 45 and 46

These motors are equipped with the same voltage rectifier/regulator. For this component, Mercury provides specific resistance testing of the various internal circuits. Although the component itself may be physically left attached to the powerhead for testing, the wires should be disconnected from the powerhead harness, and realistically, it doesn't take much more effort to just unbolt it and perform the test on the bench.

Because results will vary, two separate sets of tests and results have been provided. The first for use with an Analog (needle) ohmmeter and the second for a DVOM with a diode testing setting. For the later, keep in mind that due to manufacturing differences the internal polarity of some ohmmeters may vary resulting in test readings which would be completely opposite of the accompanying chart. If so, reverse the leads and recheck.

■ Although this test can be performed directly at the regulator/rectifier harness, Mercury recommends using test harness 91-881824 for 115 hp motors when the component remains unplugged, but physically mounted on the powerhead.

Testing With an Analog Meter
1. To test the Diodes:
  a. Set the ohmmeter to read resistance on the R x 10 scale.
  b. Connect the positive (red) meter probe to the RED wire on the regulator.
  c. Connect the negative (black) meter probe in turn to each of the GREEN and, on 75/90 hp motors, GREEN/WHITE wires on the regulator.
  d. The meter should show 100-300 ohms across each connection.
  e. Next, set the ohmmeter to read resistance on the R x 1000 (1k) scale.
  f. Connect the negative (black) meter probe to the RED wire on the regulator.
  g. Connect the positive (red) meter probe in turn to each of the GREEN and, on 75/90 hp motors, GREEN/WHITE wires on the regulator.
  h. The meter should show infinite resistance.
2. To test the ground side of the Bridge Diode:
  a. If not already done for the last of the previous Diode Tests, set the ohmmeter to read resistance on the R x 1000 (1k) scale.
  b. Connect the positive (red) meter probe to the BLACK ground wire on the regulator.
  c. Connect the negative (black) meter probe in turn to each of the GREEN and, on 75/90 hp motors, GREEN/WHITE wires on the regulator.
  d. The meter should show infinite resistance.
3. To further test the Diodes:
  a. Set the ohmmeter to read resistance on the R x 10 scale.
  b. Connect the negative (black) meter probe to the to the BLACK ground wire for the regulator.
  c. Connect the positive (red) meter probe in turn to each of the GREEN and, on 75/90 hp motors, GREEN/WHITE wires on the regulator.
  d. The meter should show 100-300 ohms across each connection.
4. For 75/90 hp motors, to test the Tachometer Circuit:
  a. Set the ohmmeter to read resistance on the R x 1000 (1k) scale.
  b. Connect the ohmmeter across the GREEN/WHITE wires.
  c. The meter should show continuity (little or NO resistance).
5. If the rectifier/regulator is not performing properly and fails one or more of these tests, it should be replaced.

Testing With a Digital Meter
6. To test the Diodes:
  a. Set the DVOM to the diode test (normally represented as an arrow facing to the right intersecting a perpendicular line).
  b. Connect the positive (red) meter probe to the RED wire on the regulator.
  c. Connect the negative (black) meter probe in turn to each of the GREEN and, on 75/90 hp motors, GREEN/WHITE wires on the regulator.
  d. The meter should show infinite resistance, OUCH or OL (Out of Limit) depending upon the meter.
  e. Connect the negative (black) meter probe to the RED wire on the regulator.
  f. Connect the positive (red) meter probe in turn to each of the GREEN and, on 75/90 hp motors, GREEN/WHITE wires on the regulator.
  g. The meter should show a reading of 0.4-0.85.
7. To test the ground side of the Bridge Diode:
  a. If not done already to test the other Diode Circuits, set the DVOM to the diode test (normally represented as an arrow facing to the right intersecting a perpendicular line).
  b. Connect the positive (red) meter probe to the BLACK ground wire on the regulator.
  c. Connect the negative (black) meter probe in turn to each of the GREEN and, on 75/90 hp motors, GREEN/WHITE wires on the regulator.
  d. The meter should show a reading of 0.4-0.85.
8. To further test the Diodes:
  a. If not done already to test the other Diode Circuits or the Bridge Diode, set the DVOM to the diode test.
  b. Connect the negative (black) meter probe to the BLACK ground wire for the regulator.
  c. Connect the positive (red) meter probe in turn to each of the 3 GREEN and, wires on the regulator.

■ Here's where we run into a snag. Mercury advises connecting the positive meter probe in the last step to each of the 3 Green regulator wires. However, on the 75/90 hp motors, schematics show only 2 Green leads and 2 Green/White leads. No further information is currently available.

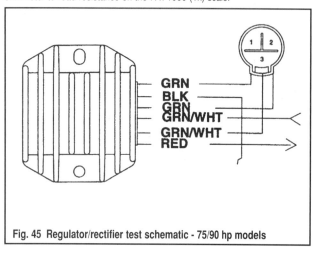

Fig. 45 Regulator/rectifier test schematic - 75/90 hp models

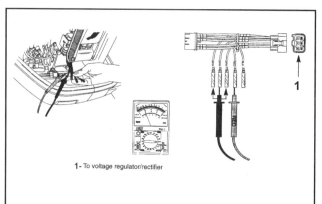

Fig. 46 Regulator/rectifier test schematic (showing the first of the analog meter diode tests) using the Mercury test harness for a 115 hp model

IGNITION AND ELECTRICAL SYSTEMS   4-21

d. The meter should show no continuity, either infinity, OUCH or OL (Out of Limit), depending upon the meter.

9. To test the Tachometer Circuit:
10. For 75/90 hp motors, to test the Tachometer Circuit:
   a. Connect the ohmmeter across the GREEN/WHITE wires.
   b. The meter should show continuity (little or NO resistance).
11. If the rectifier/regulator is not performing properly and fails one or more of these tests, it should be replaced.

225 Hp Models

◆ See Figure 47

On these models, which are VERY, VERY, VERY closely related to their Yamaha counterparts, Mercury has opted for Yamaha style testing. There are no resistance values, instead, measurements should be taken of unloaded (circuit open) output with the engine running from 1500-3500 rpm.

In order to easily access the necessary output wires of the rectifier/regulator circuit, Mercury recommends using test harness 91-888862 (or a suitable home-made jumper harness).

To test voltage output proceed as follows:

1. Using a source of cooling water, start and run the engine to normal operating temperature, then shut the engine down.
2. Remove the air intake silencer for access.
3. Unplug the regulator wiring harness, then connect the MALE end of the test harness to the side of connector whose wires run back to the regulator. DO NOT connect the female side of the test harness to the other side of the powerhead harness connector.
4. Install the air intake silencer for safety.
5. Connect a voltmeter set to read DC volts to the test harness. Connect the positive (red) meter probe each in turn to the RED test harness and to the BLACK harness wires. Both times, have the negative (black) meter probe connected to a good engine ground.
6. Restart and run the engine to 1500 rpm, noting the meter reading, then slowly increasing engine speed to 3500 rpm, again noting the meter reading.
7. In both instances, with the engine running from 1500-3500 rpm, the meter must show AT LEAST 10.5 volts across the Red test harness wire-to-ground, and the Black test harness wire-to-ground.

■ Although Mercury for some reason doesn't give maximum output, Yamaha lists it as about 13 volts for this model with the circuit open. We would also suggest that with the circuit completed the regulator should limit voltage to ABOUT 15 volts or less, anything more we would be concerned might cook your battery.

Flywheel and Stator (Stator/Charge/Lighting/Pulser Coils)

The following service section lists the procedures required to pull the flywheel and remove the stator or stator plate in order to service the ignition system components (pulser, charge, lighting coils, as applicable). Removal and installation of the stator or stator plate itself is often necessary in order to gain access to the wiring harness retainer underneath the stator plate. Cleaning and Inspecting procedures in addition to proper assembling and installation steps are also included.

REMOVAL & INSTALLATION

1 & 2-Cylinder Powerheads

◆ See Figures 48 thru 55

■ On 2.5 and 4 hp Yamaha motors there are no separate pulser or charge coils, as their windings are incorporated into the TCI unit itself. The flywheel usually does need to be removed to service the TCI unit on 2.5 hp motors, but usually does NOT need to be removed in order to service the unit on 4 hp motors. However, both motors are included in this procedure, since flywheel removal may be necessary for other repairs.

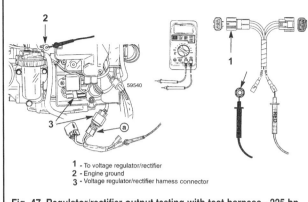

Fig. 47 Regulator/rectifier output testing with test harness - 225 hp models

Once the flywheel is removed, use the wire colors from the Wiring Diagrams or from the Ignition System Component testing charts in order to help identify the individual coils.

■ The Pulser coil used on many Mercury/Mariner motors is mounted along the perimeter of the flywheel, NOT UNDERNEATH it, so the flywheel does not usually need to be removed for access on those models.

1. Remove the cowling from the powerhead.
2. Disconnect and ground the spark plug lead(s) for safety. On twins, be sure to tag the spark plug wires to ensure proper installation.
3. Remove the mounting hardware securing the hand rewind starter or flywheel cover to the powerhead. Remove the hand starter or flywheel cover for access to the flywheel.

■ Do not remove the handle from the hand starter rope because the rope would immediately rewind inside the starter. Such action would require considerable time and effort to correct.

4. A few motors, such as the 25 hp Yamaha, may be equipped with a flywheel mounted cover and/or hand starter pulley which must be removed for access to the flywheel nut or the holes used to keep the flywheel from turning when loosening the nut (and/or used to thread the puller into the top of the flywheel). If applicable, scribe a matchmark between the cover/pulley and flywheel, then carefully loosen the screws and remove the pulley/cover for access.

■ On 4/5/6 hp Mercury/Mariner models the easiest way to hold the flywheel from turning when loosening the nut is to use a strap wrench on the flywheel starter pulley (and this might be true with other models as well). If so, loosen the flywheel retaining nut BEFORE removing the starter pulley.

5. Obtain a flywheel holder or large strap wrench (but check the torque value on the flywheel nut first, refer to the installation steps, as a strap wrench isn't going to work on any but the smallest motors covered here). On these motors a flywheel holder is normally used by inserting 2 indexing pins on the tool arms through the holes in the flywheel or flywheel cover. This type of tool can usually be fabricated from a couple of pieces of heavy stock and some tempered bolts.
6. If not done already (before removing the starter pulley, if so equipped), secure the flywheel using the holder or strap wrench in order to keep it from turning (remember, you don't want to turn the flywheel counterclockwise as you could damage the vanes of the water pump impeller). Hold the flywheel from rotating and at the same time remove the flywheel nut.
7. Obtain a threaded flywheel/steering wheel puller tool. Ensure the puller will apply force at the bolt holes in the flywheel and not from around the perimeter of the flywheel.

✳✳ WARNING

Never attempt to use a jawed puller (which would put stress on the outside edge of the flywheel).

IGNITION AND ELECTRICAL SYSTEMS

8. Install the puller onto the flywheel, take a strain on the puller with the proper size wrench. Now, continue to tighten on the tool and at the same time, shock the crankshaft with a gentle to moderate tap using a hammer or mallet on the end of the tool. This shock will assist in breaking the flywheel loose from the crankshaft.

9. Lift the flywheel free of the crankshaft. Remove and save the Woodruff key from the recess in the crankshaft.

■ On most motors the stator coil or the separate charge/lighting coil(s) can now be unbolted and removed without further component disassembly.

10. Tag and disconnect the stator or individual coil harness at the connector fittings (many use bullet connectors).

■ Pay close attention to how the wiring harness is routed for installation purposes. It is not uncommon for someone to install the wiring incorrectly only to have the flywheel rub through the insulation, shorting the wiring and destroying the coil.

11. Unbolt and remove the stator or individual coils.

To Install:

12. Apply a light coating of threadlock to the threads of the stator or individual coil (charge/lighting etc) retaining bolts and place aside for use in the next step.

13. Install the individual coil(s) or the stator assembly to the powerhead, then install the retaining bolts and tighten securely.

14. Route the stator wire harness as noted during removal, then reconnect them as tagged (though even without tagging MOST wires connect color-to-like color, however, if the harness was ever repaired this may no longer be true).

15. Place a tiny dab of thick lubricant on the curved surface of the Woodruff key to hold it in place while the flywheel is being installed. Press the Woodruff key into place in the crankshaft recess. Wipe away any excess lubricant to prevent the flywheel from walking during powerhead operation.

16. Check the flywheel magnets to ensure they are free of any metal particles. Double check the taper in the flywheel hub and the taper on the crankshaft to verify they are clean and contain no oil.

17. Slide the flywheel down over the crankshaft with the keyway in the flywheel aligned with the Woodruff key in place on the crankshaft. Rotate the flywheel carefully clockwise to be sure it does not contact any part of the stator plate or wiring.

18. Slide the washer onto the crankshaft, and then on some models the manufacturer recommends applying a thin coating of clean engine oil to the threads of the flywheel nut. This is the case for the smaller Yamaha motors, 8.8/9.9 hp (232cc) and SMALLER, as well as for the 8/9.9 hp (232cc) and 25 hp Mercury/Mariner modes.

■ On some models it might be necessary to install the starter pulley before installing the flywheel in order to give you to which you can secure the strap wrench.

19. Thread the flywheel nut onto the crankshaft. Secure the flywheel from turning using the holding tool, then tighten the flywheel nut to the following torque value for the models listed:

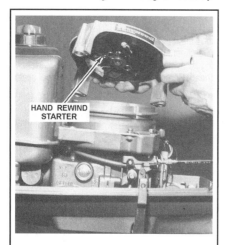

Fig. 48 Remove the hand rewind starter

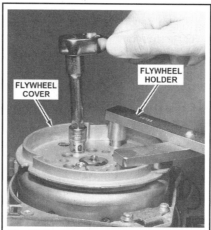

Fig. 49 If equipped, loosen the flywheel cover bolts. . .

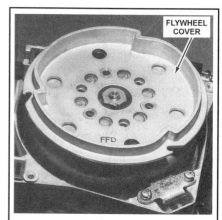

Fig. 50 . . .so you can remove the cover for access to the flywheel nut

Fig. 51 Loosen and remove the flywheel nut. . .

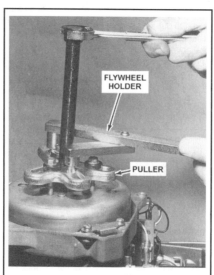

Fig. 52 . . .then install a threaded puller to free the flywheel

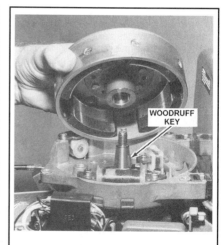

Fig. 53 Remove the flywheel and woodruff key

IGNITION AND ELECTRICAL SYSTEMS 4-23

- 2.5 and 4 hp Yamaha motors - 32 ft. lbs. (44 Nm)
- 4/5/6 hp Mercury/Mariner motors - 40 ft. lbs. (54 Nm)
- 6/8 Yamaha and all 8/9.9 hp (232cc) motors - 72 ft. lbs. (100 Nm)
- 9.9/15 hp (323cc) motors - 80 ft. lbs. (110 Nm)
- 25 hp motors - 116 ft. lbs. (157 Nm)

20. If equipped (and if not done already for flywheel nut installation), position the pulley/cover, aligning the matchmarks made earlier, then thread the retaining bolts and tighten them securely.
21. Install the hand rewind starter or flywheel cover, as applicable.
22. Reconnect the spark plug lead(s).
23. Install the cowling to the powerhead.

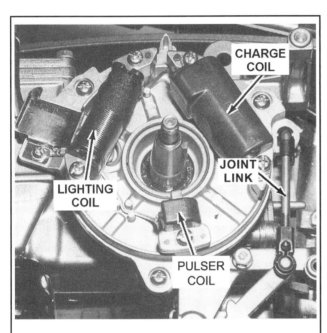

Fig. 54 Once the flywheel is removed you can unbolt the individual coils (typical Yamaha shown) or the one piece stator

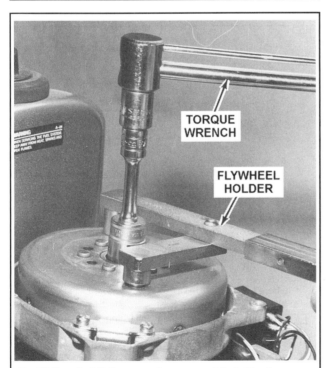

Fig. 55 Upon installation, use a torque wrench to tighten the flywheel nut

3- and 4-Cylinder Powerheads

◆ See Figures 56 thru 67

All 3- and 4-cylinder models utilize a 1-piece stator. For the smaller 30-60 hp motors, the pulser coil (also known as the trigger coil or CPS) is mounted on the top of the powerhead adjacent to (and not underneath the stator) so it may be easily removed for replacement without even disturbing the stator (and often without disturbing the flywheel). However, on the larger 75-150 hp motors, the pulser coil windings are integrated into the stator coil base, so the stator must obviously be unbolted to remove it (though it is also usually possible to unbolt the base and stator, removing them as an assembly).

1. Disconnect the negative battery cable for safety.
2. Loosen the bolts (usually 2) securing the flywheel cover to the top of the powerhead, then carefully pull it free of the mounting. If there are bolts on only one end of the cover there are usually one or more locating pins or grommets on the opposite end, gently pull the cover so the pins come free.
3. A few of the smallest models may be equipped with a removable starter pulley. If so, matchmark the pulley to the flywheel, then loosen the bolts and remove it from the flywheel for access to use a flywheel holding tool and puller. A prybar can usually be used across bolt heads to keep the flywheel from turning when loosening the pulley/cover bolts.

■ The flywheel nut is tightened to 116-199 ft. lbs. (157-270 Nm) and therefore requires considerable force to loosen. Yamaha calls for using flywheel tool (#YB-06139) as does Mercury (#91-83163M) to hold the flywheel from turning. In both cases, the tool is basically a heavy bar stock with locating pins that fit into two or three holes on the flywheel. The tool contains an adjustable, hinged bar that can be pivoted so a pin can be inserted into another hole in the flywheel without interfering with the flywheel nut. A substitute can be made from heavy bar stock and bolts which will fit into those holes (but make sure they're not deeper than the flywheel itself in case the holes are not blind, you don't want to catch on anything under the flywheel). Use the illustrations showing the tools as a reference).

4. Use a flywheel holding tool to keep the flywheel from turning while you use a large socket and breaker bar to loosen the flywheel nut. DO NOT use an impact gun on this nut (as you risk damage to the flywheel magnets). Remove the nut and washer from the center of the flywheel.
5. Install a threaded puller, a steering-wheel type puller may be used for this purpose, to the flywheel, making sure to keep it parallel with the flywheel. Slowly tighten the puller screw to free the flywheel from the crankshaft.
6. Remove the puller and the flywheel, keeping track of the crankshaft woodruff key. Keep in mind that the strength of the magnets may make it a little more difficult to lift the flywheel, just watch your fingers to keep them from getting pinched.
7. As necessary, tag and disconnect the stator and/or pulser coil wiring connectors.

■ Be sure to note the stator and pulser coil wire routing for installation purposes. It is critical that they are positioned in a fashion that will NOT interfere with moving components, such as the flywheel or the wiring and possibly the coils will be destroyed.

8. To remove ONLY the pulser coil on models through 60 hp, loosen the 2 screws securing it to the stator base, then remove the coil.

■ Although most pulser coils on these motors are mounted in a fixed (non-adjustable) position a few (including some of the mid-range motors, like the 935cc 4-cylinders) have an adjustable gap. For models with an adjustable gap you might want to matchmark the coil to the powerhead to ease installation (even if the coil is being replaced, you can always transfer the mark).

9. To remove the stator coil, loosen the 3 bolts or screws holding it to the stator base.

■ Remember, the stator base IS the pulser coil assembly on 75 hp and larger motors.

10. To remove the stator base (with or without the coils attached) on motors so equipped, loosen and remove the 4 screws and washers, then carefully lift it from the powerhead. On models through 60 hp and on some of the larger models (such as the 150 hp Yamaha) keep track of the 2 collars

4-24 IGNITION AND ELECTRICAL SYSTEMS

which are normally placed underneath 2 of the stator base screws. Many of the smaller Mercury/Mariners don't have a separate removable base, but may have a load ring placed under the stator itself.

11. Clean and inspect the flywheel, coils and stator base, as detailed in this section.

To Install:

The following procedures pickup the work after the flywheel and stator assembly have been cleaned, inspected, serviced, and assembled.

■ For most models through 60 hp and some of the larger ones (as equipped, such as the 150 hp motor), make sure the collars used on two of the stator base retaining screws are in position under the stator base legs before installation.

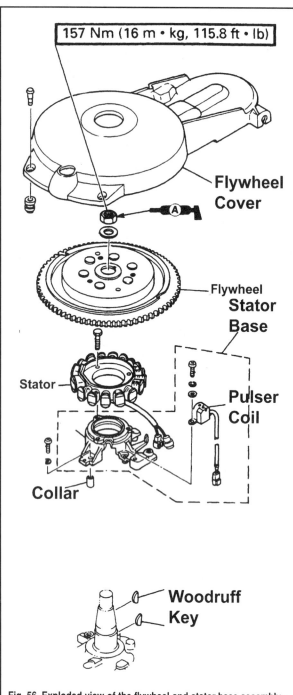

Fig. 56 Exploded view of the flywheel and stator base assembly - 40-60 hp Yamaha motors and 4-cylinder Mercury/Mariner motors through 2000

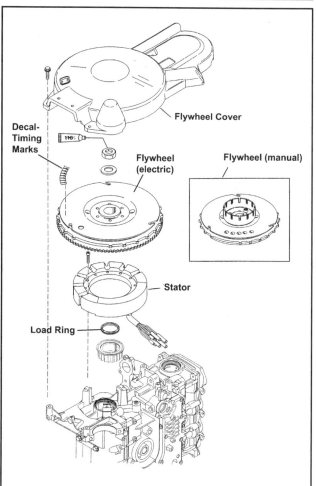

Fig. 57 Exploded view of the flywheel and stator base assembly - 30-60 hp Mercury/Mariner (except 4-cylinder models through 2000)

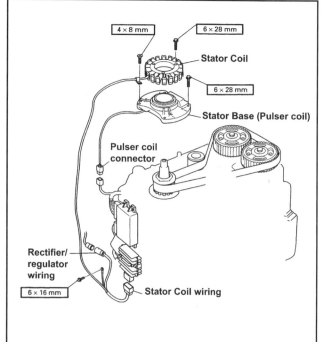

Fig. 58 Exploded view of the stator base assembly - 75-115 hp motors (150 hp models very similar)

IGNITION AND ELECTRICAL SYSTEMS 4-25

12. If applicable, position the stator base assembly to the powerhead. Install and tighten the 4 stators base retaining screws.

■ Mercury recommends applying a light coating of a threadlocking compound, such as Loctite® 222 to the threads of the stator mounting screws. We cannot see why this would be a problem on any of the motors and recommend that you do it on Yamaha's as well.

13. If removed, install the stator coil and secure using the 3 retaining screws. For models so equipped, be sure to position the stator coil load ring underneath the stator during installation.

14. If removed on models through 60 hp, install the adjacent pulser coil and secure using the 2 retaining screws.

■ On models with an adjustable sensor you'll have to install the flywheel before tightening the sensor screws so that you can check and set the gap.

15. Route the wiring as noted during removal, then reconnect the stator and/or pulser coil connectors.

Fig. 59 Loosen the flywheel cover bolts...

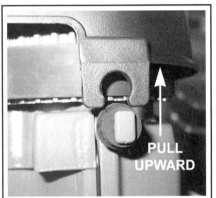

Fig. 60 ...then pull the cover up off any grommets...

Fig. 61 ...or out, off any retaining pins

Fig. 62 Coils are mounted to a base under the flywheel

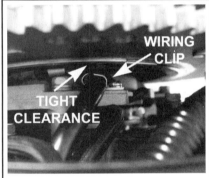

Fig. 63 Note the tight clearance for the wiring

Fig. 64 Loosen the nut to remove the flywheel

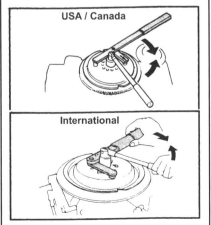

Fig. 65 Use a holder to keep the flywheel from turning - Yamaha tools illustrated, Mercury similar

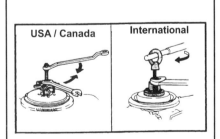

Fig. 66 Loosen the flywheel using a threaded puller - Yamaha shown, Merc similar

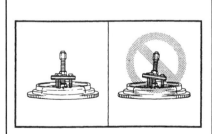

Fig. 67 Make sure the puller remains parallel to the flywheel - Yamaha shown, Merc similar

4-26 IGNITION AND ELECTRICAL SYSTEMS

✳✳ WARNING

It is critical that wiring is positioned in a fashion that will NOT interfere with moving components, such as the flywheel or the wiring and possibly the coils will be destroyed.

16. Make sure the crankshaft woodruff key is in position, then align and install the flywheel carefully over the crankshaft.

17. Apply a very light coating of clean engine oil to the threads of the flywheel retaining nut and the contacting surface of the washer. Install the washer and nut.

18. Using a suitable flywheel holding tool carefully tighten the flywheel retaining nut to 116 ft. lbs. (157 Nm) for 30-60 hp motors, to 140 ft. lbs. (190 Nm) for 75-115 hp motors or to 199 ft. lbs. (270 Nm) for 150 hp motors.

19. If equipped, install the starter pulley and secure the retaining bolts.

20. For Mercury/Mariner 40/45/50 hp (935cc) motors if the CPS (Pulser Coil) was disturbed, properly adjust the air gap between the sensor and the flywheel to 0.030 in. (0.77mm) using a feeler gauge, then tighten the retaining screws to hold it in that position.

21. Position the flywheel cover using the grommets/retaining pin(s), then secure using the retaining bolts. Tighten the bolts gently, but securely.

22. Reconnect the negative battery cable.

V6 Powerheads

◆ See Figures 68 thru 72

Modern Yamaha and Mercury/Mariner V6 motors are 1-piece stator coil that performs the functions of the charge and lighting coils. The stator is mounted to a stator base that contains an integral pulser coil assembly. This stator base is mounted in a fixed position on these motors, meaning that all timing functions are performed by the CDI unit or ECM.

Like most outboards, service to the flywheel and coil windings is pretty straightforward. The flywheel must be removed for access to the coil windings and care must be taken to observe and match wire routing during installation. Other than that, it is a relatively easy takes, if you have a method of holding the flywheel steady.

Yamaha and Mercury recommend a universal flywheel holding tool that works on most of their outboards (#YB-06139 for Yamaha or #91-83163M for Mercury). The shape and function of both tools are virtually identical. In either case the tool is basically heavy bar stock with locating pins that fit into two or three holes on the flywheel. The tool contains an adjustable, hinged bar that can be pivoted so a pin can be inserted into another hole in the flywheel without interfering with the flywheel nut. A substitute can be made from heavy bar stock and bolts which will fit into those holes (but make sure they're not deeper than the flywheel itself in case the holes are not blind, you don't want to catch on anything under the flywheel).

1. Disconnect the negative battery cable for safety.

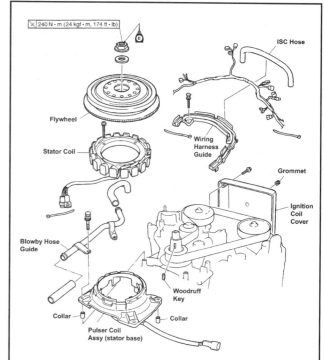

Fig. 68 Exploded view of the stator and pulser coil assembly used on V6 models

2. Loosen the bolts securing the flywheel cover to the top of the powerhead, then carefully pull it free of the mounting. Most covers contain locating pins or grommets from which the cover must be gently pulled.

3. Insert a flywheel holding tool into the two holes provided in the flywheel and hold it steady while the flywheel nut is loosened with the correct size socket. Remove the flywheel nut and washer.

■ There is considerable force on the flywheel nut. It is safest to position the holder tool and the breaker bar so that you are pulling each in toward the other. Brace yourself and apply force, but avoid the use of an impact gun which could potentially damage the flywheel or crankshaft.

4. Install a threaded puller, a steering-wheel type puller may be used for this purpose, to the flywheel, making sure to keep it parallel with the flywheel. Slowly tighten the puller screw to free the flywheel from the crankshaft. If necessary, lightly tap the end of the puller screw with a mallet

Fig. 69 Remove the plastic flywheel cover for access

Fig. 70 Loosen the flywheel nut...

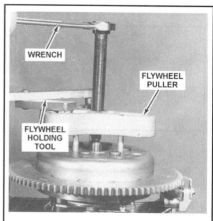

Fig. 71 ...then install a puller to free the flywheel

IGNITION AND ELECTRICAL SYSTEMS 4-27

as this gentle shock will assist in breaking the flywheel loose from the crankshaft taper.

✱✱ WARNING

A violent strike to the center bolt may cause damage to the crankshaft oil seals. Therefore, use only a gentle to moderate tap with the hammer. Also, never attempt to use a jawed puller, which would potentially damage the flywheel by applying force to the outside edges.

5. Remove the puller and the flywheel, keeping track of the crankshaft woodruff key. Keep in mind that the strength of the magnets may make it a little more difficult to lift the flywheel, just watch your fingers to keep them from getting pinched.

6. Trace the pulser coil assembly and/or stator wiring. As necessary, tag and disconnect the stator and/or pulser coil wiring connectors (from the CDI unit, regulator/rectifier etc).

7. The way Yamaha formats their information they make you think it might be necessary to tag and disconnect a number of components for access. Mercury however, does NOT list this as necessary on nearly identical models, but we'll list the information here, JUST in case you run into a snag. ONLY IF NECESSARY for clearance, tag and disconnect the following:
- Ignition coil connectors.
- Idle Speed/Air Control (ISC/IAC) connector.
- Intake Air/Manifold Air Pressure (IAP/MAP) sensor connector.
- Throttle Position Sensor (TPS) connector.
- Thermo-switch leads.
- Idle speed control hose.
- High-pressure fuel hoses.
- Neutral switch connector.
- Shift cut switch connector.
- Starboard fuel injector connectors.

■ **Be sure to note the stator and pulser coil wire routing for installation purposes. It is critical that they are positioned in a fashion that will NOT interfere with moving components, such as the flywheel or the wiring and possibly the coils will be destroyed.**

8. Matchmark the relative positions of the pulser coil assembly (stator base) and/or stator coil to the powerhead before removal. Marking their positions will help make sure wires are routed in the same positions.

9. Remove the bolts (usually 4) securing the stator assembly to the powerhead. Carefully life the stator coil assembly off the powerhead.

10. If necessary, remove the stator base/pulser coil assembly retainers (usually 4). On these models (with fixed position pulser assemblies) the 4 bolts/screws fasten the assembly directly down to the powerhead. Lift off the pulser assembly

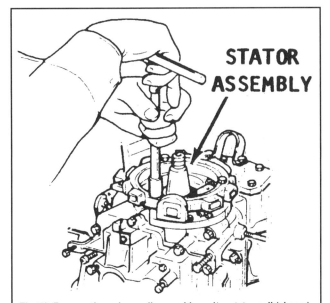

Fig. 72 Remove the pulser coil assembly and/or stator coil (shown) from the powerhead

To Install:
The following procedures pickup the work after the flywheel and stator assembly have been cleaned, inspected, serviced, and assembled.

■ **When a pulser or stator coil is replaced, transfer the matchmarks made on the old coil during removal to the replacement in order to ensure proper positioning.**

11. If removed, install the stator base/pulser coil assembly. It is a simple matter of positioning the assembly (With the wiring facing as noted during removal), then installing and tightening the retaining bolts. Secure the pulser assembly using the 4 retaining tabs and screws.

12. Position the stator coil over the crankshaft, aligning the matchmarks made earlier. The 4 bolt holes are sometimes unevenly spaced which may help with alignment. Install the retainers and tighten to secure the stator coil assembly.

13. Carefully thread the wiring harnesses through the grommets or retainers, as noted during removal and connect them as tagged. If any questions occur, refer to the Wiring Diagrams in this section for clarification.

✱✱ WARNING

Remember the wiring must be positioned so there will be NO contact with moving components, otherwise the wiring and possibly the components which it connects may be damaged.

14. If any sensors were disconnected for clearance, reconnect the wiring as tagged during removal.

15. Install the Woodruff key into the crankshaft. A tiny dab of grease will help hold the key in place while the flywheel is installed.

16. Lower the flywheel over the crankshaft with the Woodruff key indexing into the slot in the flywheel.

✱✱ CAUTION

Use care as you lower the flywheel into place not to get you fingers pinched as the magnets apply some pull to the plate.

17. Apply a very light coating of clean engine oil to the threads and washer mating surface of the flywheel nut, then position the washer and thread the nut onto the end of the flywheel.

18. Keep the flywheel from rotating using a flywheel holder tool and tighten the flywheel nut to 177 ft. lbs. (240 Nm).

19. Slowly rotate the flywheel by hand feeling for binding and visually checking for interference with wiring or any other powerhead mounting components.

20. Install the flywheel cover and secure using the retaining bolts.

21. Reconnect the negative battery cable.

CLEANING & INSPECTION

◆ See Figure 73

Inspect the flywheel for cracks or other damage, especially around the inside of the center hub. Check to be sure metal parts have not become attached to the magnets. Verify each magnet has good magnetism by using a screwdriver or other suitable tool.

Thoroughly clean the inside taper of the flywheel and the taper on the crankshaft to prevent the flywheel from walking on the crankshaft during operation.

Check the top seal around the crankshaft to be sure no oil has been leaking onto the stator plate. If there is any evidence the seal has been leaking, it must be replaced.

Test the stator assembly to verify it is not loose. Attempt to lift each side of the plate. There should be little or no evidence of movement.

Inspect the stator plate oil seal and the O-ring on the underside of the plate.

Some models have a retainer, a retainer ring, and a friction plate located under the stator. The retainer ring is a guard around the retainer, subject to cracking and wear. Inspect the condition of this guard and replace if it is damaged.

4-28 IGNITION AND ELECTRICAL SYSTEMS

Fig. 73 Always check the flywheel carefully to be sure particles of metal have not stuck to the magnets

CDI/TCI Unit and Ignition Coils

The ignition coil(s) and CDI/TCI unit have to be one of the easiest components to locate on the outboard (ok, with the exception of the flywheel, propeller and spark plugs, but that's still pretty easy). To find the ignition coil for a given cylinder, simply follow the spark plug lead back from the plug directly to the coil. To locate the CDI unit, following the wiring back from the coil to the CDI. Simple, right?

Well it is, for the most part. Generally speaking, the CDI unit is normally mounted either on the side of the powerhead (inline motors) or on the rear of the powerhead, between the cylinder head banks (V6 motors) which makes sense considering the job of the CDI unit is to power/trigger the ignition coils and the job of the ignition coils is to fire the spark plugs, so keeping everything in the same general area is a good idea. However, there are some notable exceptions.

The 2.5 and 4 hp Yamaha motors use a vastly simplified TCI ignition system where the TCI module itself is located on TOP of the powerhead, adjacent to or under the flywheel. Why? Well, easy, it is because the TCI unit on these models integrates the functions of the pulser/charge coils, TCI module AND the ignition coil, all in one component. So we deal with them in a separate removal and installation procedure.

Another exception is most EFI motors, since the CDI/TCI unit on these motors is incorporated into the ECM itself. Furthermore on many fuel injected motors the ECM is mounted under one or more covers at the rear of the powerhead. We've provided detailed procedures for accessing the ECM on EFI models, it can be found in the Electronic Fuel Injection portion of the Fuel System section.

As for EFI models, other than the differences for accessing and removing the ECM, the ignition coils are pretty much serviced in the same manner as the carbureted counterparts to those motors listed here. For more details on ignition coil service, continue with the procedures in this section.

Perhaps the most overlooked portions of ignition coil service has to do with the coil mounting and wiring (which are related on some motors). The ignition coil wiring is divided into 2 parts, the power/trigger primary circuit from the CDI (the small wires) and the secondary high-tension circuit (the large wires, the spark plug leads). Some coils have a ground wire in the primary circuit, while others are grounded to the powerhead (this is where the wiring and mounting are one and the same). On these models, corrosion can cause problems with the circuit as it prevents proper circuit grounding. For all motors, make sure the coil mounting points are clean and free of corrosion and make sure that ALL wiring contains insulation that is not broken, brittle or otherwise damaged.

REMOVAL & INSTALLATION

2.5 and 4 Hp Yamaha Motors

◆ See Figures 74, 75 and 76

On these models the TCI unit is located on top of the powerhead under (2.5 hp) or immediately adjacent to (4 hp motors) the flywheel. The ignition coil is integrated into the TCI unit.

1. For 2.5 hp motors, remove the fuel tank from the top of the powerhead.
2. Remove the hand-rewind starter assembly from the top of the powerhead.
3. For 2.5 hp motors, remove the Flywheel, as detailed in the Ignition and Electrical System section.

■ As always with outboard servicing, take close note of the routing for all wires before they are disturbed. They must be positioned so that they will not come into contact with and be damaged by moving parts (like the flywheel).

■ For illustrations, please refer to TCI Unit Air Gap adjustment in the Maintenance and Tune-Up section.

4. Disconnect the spark plug lead.
5. Disengage the bullet connector.
6. Locate the ground lead(s), then remove the retaining bolt to free the lead(s).
7. Loosen the 2 retaining bolts, then remove the TCI unit from the powerhead.

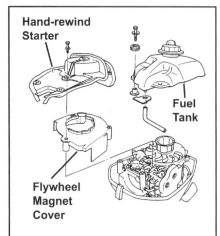

Fig. 74 To access the TCI unit you'll have to remove the flywheel (2.5 hp motor shown)

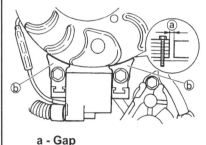

a - Gap
b - TCI Unit Projections

Fig. 75 Upon installation you'll need to adjust the TCI unit gap (the distance between the 2 projections on the TCI unit and the flywheel)

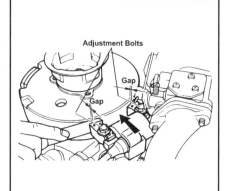

Fig. 76 Slide the projections gently against a feeler gauge held against the flywheel, then tighten the TCI unit retaining bolts

IGNITION AND ELECTRICAL SYSTEMS 4-29

To Install:

8. Position the TCI unit to the powerhead and loosely install the retaining bolts (since you'll have to adjust the TCI Unit Air Gap for proper operation).

9. Route the wiring as noted during removal to prevent contact with moving components such as the flywheel.

10. Make sure the mounting point for the ground lead(s) and the ring terminal(s) are clean and free of corrosion. Position the ground lead(s) and secure using the retaining bolt. Place a small dab of dielectric or marine grade grease over the terminals and bolt to help prevent corrosion.

11. Engage the bullet connector, then connect the spark plug lead.

12. There are 2 projections coming off the TCI unit, facing the flywheel (the flywheel ends of the projections are curved slightly to match the curve of the flywheel). The TCI unit air gap is the distance between these projections and the flywheel itself. Use an appropriate sized feeler gauge to check the gap as follows:
   - For 2.5 hp motors the gap should be 0.016-0.024 in. (0.4-0.6mm).
   - For 4 hp motors the gap should be 0.012-0.024 in. (0.3-0.6mm).

■ When checking the gap, a feeler gauge larger than the specification should not fit between the tips of the TCI unit and the flywheel. A gauge within specification should fit, but with a slight drag. A smaller gauge should fit without any interference whatsoever.

13. If adjustment is necessary, loosen the 2 mounting bolts (one is located over each of the TCI unit projections) and gently reposition by sliding each projection toward or away from the flywheel while holding the feeler gauge in position. Tighten the mounting bolts and recheck the gap to ensure the projections did not move while tightening the fasteners.

14. Install the flywheel, hand-rewind starter and fuel tank, as applicable.

All 1-4 Cylinder Inline Motors (Except 2.5 and 4 Hp Yamahas)

◆ See Figures 77 thru 84

■ The ECM on EFI motors performs the functions of the CDI unit. For details on removal and installation, refer to the Engine Control Module (ECM/CDI) procedure in the Fuel System section. However, the steps of this procedure regarding the ignition coils are applicable to the EFI motors and may be used.

On most 1-4 cylinder inline motors (except the 2.5 and 4 hp Yamahas) the ignition coils and CDI unit are mounted either to the side or end of the powerhead. On most 9.9/15 hp (323cc) motors, the CDI unit is mounted at the front of the powerhead, either on the crankcase or on the electric starter motor bracket (when equipped). However, keep in mind that if you have any trouble finding the components, the rules of thumb for locating the ignition coils and CDI unit by tracing the wiring back from the spark plugs still apply.

■ The CDI unit and coil(s) may be removed independently of each other, but in most cases, labeling the wires before proceeding is critical to ensuring proper operation once the CDI unit and/or coil(s) are reinstalled.

1. If equipped, disconnect the negative battery cable for safety.
2. To remove the CDI unit, proceed as follows:
   a. Start by locating the CDI unit itself. If necessary, follow the wiring back from the ignition coil(s) to the unit. On some models the unit is mounted under a cover, if so remove the retaining screw(s) and remove the cover for

Fig. 77 Trace the wiring and locate the CDI unit...

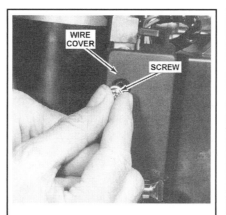

Fig. 78 ...then, if equipped, remove the cover

Fig. 79 Tag and disconnect the wiring, including grounding bolts...

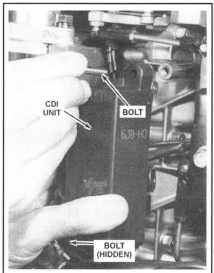

Fig. 80 ...then remove the CDI unit

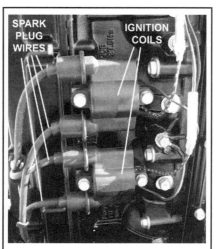

Fig. 81 Ignition coils are both easy to find and easy to remove

Fig. 82 Typical single-lead ignition coil

4-30 IGNITION AND ELECTRICAL SYSTEMS

access. If the CDI unit is mounted on top of the powerhead you may need to remove the flywheel cover or hand-rewind starter for better access.

b. Tag and disconnect the CDI unit wiring. Take close note of how each wire is routed to make sure it can be returned to the original position. On some smaller powerheads this is not a big deal as there are only a couple of wires, but as you get to the larger 3- and 4-cylinder motors there may be as many as almost a dozen wiring connections so it is a good idea to take some pictures with a digital camera (or old Polaroid).

■ At least one wire connection normally goes to a ground strap that is bolted to the powerhead.

c. For multi-cylinder motors, check for white plastic sleeves on the wires leading from the CDI unit to each ignition coil. When present, these sleeves are numbered 1, 2, 3, etc. Disconnect each wire at the quick disconnect fitting. However, if there are no sleeves (or markings directly on the wires themselves), then mark each wire to identify to which coil the lead must be reconnected. If these wires are not connected properly, the powerhead cylinders will not fire in the correct sequence. The powerhead may operate, but very poorly.

■ On many models the ignition coil wires from the CDI unit are all the same color, however they may not be interchangeable. This means that there may be no other way to determine which lead is for which cylinder other than the sleeves or tracing the leads. If the leads are disconnected with no sleeves, and if the leads are not identified by markings or pieces of tape, then a trial and error method must be used to find the correct firing order for the powerhead. Save yourself the potential hassle now and make sure all wires are tagged.

3. To remove an ignition coil, proceed as follows:
a. On multi-cylinder motors, although the coils are usually arranged in a common-sense stack (indicating which coil fires which cylinder), don't assume anything, tag the coils and wires now.

■ On some models, such as the 75-150 hp motors, you must first remove the spark plug cover for access to the ends of the plug wires.

b. Twist off the spark plug lead(s) from the plug(s).
c. Remove the bolts securing the ignition coil to the powerhead (there are usually 1 or 2 bolts securing a coil).
d. Repeat for the remaining coil(s).

■ One of the mounting bolts of some coils will also secure a Black grounding lead with an eye connector to the powerhead. This lead may come from the coil itself or may come from a connector plug in the lower cowling.

To Install:

4. If removed, install each ignition coil to the powerhead using the attaching bolts (usually 2). Make sure the grounding point (lead and/or coil mounting surfaces) are clean and free of corrosion. If applicable, make sure the ground lead is positioned when installing the coil. Install and securely tighten the coil mounting bolts, then reconnect the spark plug lead(s), as tagged during removal on multi-cylinder motors.

5. If the CDI unit was removed, proceed as follows:
a. Position the CDI unit to the powerhead and secure using the fasteners.
b. Carefully route the wiring as noted and tagged during removal, reconnect the ignition coil, charge coil, pulser coil and ground leads, as applicable. Make sure any ring terminal and mounting point for a ground lead is clean and free of corrosion.
c. If applicable, connect the CDI timing linkage.
d. If equipped, install and secure the CDI unit cover.
e. If removed for access, install the hand-rewind starter or flywheel cover.

V6 Powerheads

◆ See Figures 85 and 86

On V6 motors there are 3 dual-lead ignition coils (one per pair of cylinders), which are mounted to a central coil mounting bracket in the valley between the cylinder heads.

As stated earlier, for all fuel injected motors the functions of the CDI are performed by the Electronic Control Module, so details on removal and installation of those modules are covered under Engine Control Module (ECM/CDI) in the Fuel System section. However, ignition coil service, once covers are removed for access to them, is virtually identical and you can use the appropriate portions of the procedures found here for ignition coils on these EFI motors.

■ The ECM unit and coils may be removed independently of each other, but labeling the wires before proceeding is critical to ensuring proper operation once the ECM unit and/or coils are reinstalled. However, the ignition coils can easily be removed as an assembly by disconnecting their spark plug leads and unbolting the brackets on which they are mounted.

1. If equipped, disconnect the negative battery cable for safety.
2. Loosen the bolts securing the ignition coil cover (there are usually 4 bolts/screws found at the corners of the cover), then carefully pull the cover

Fig. 83 Typical dual-lead ignition coils

Fig. 84 Upon installation, make sure all wires are properly secured using any provided retainers and wire loops or wire ties, as applicable

IGNITION AND ELECTRICAL SYSTEMS 4-31

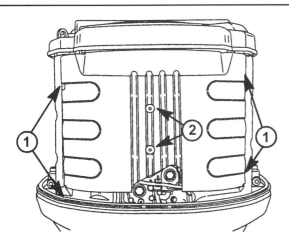

1 - Ignition coil cover mounting screws
2 - Grommets

Fig. 85 Ignition coil cover for V6 motors - Mercury/Mariner shown (Yamaha similar)

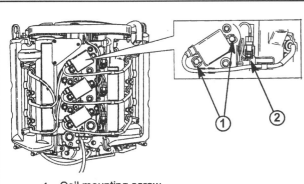

1 - Coil mounting screw.
2 - Coil primary wire harness connector

Fig. 86 Ignition coil mounting on V6 motors

from the mounting grommets (there are at least 2 positioned toward the center of the cover on most models).

3. To remove an ignition coil, tag and disconnect the wiring from the coil to the ECM. Also, take and disconnect the spark plug wiring for the coil.
4. Loosen and remove the coil mounting bolts (normally 2 per coil), watching for ground leads as ring terminals may be secured by a coil mounting bolt on some models.
5. Remove the ignition coil from the bracket.

To Install:
6. If removed, install each ignition coil to the bracket, routing the wiring as noted during removal. If there was a ground terminal connected to a mounting bolt, make sure that both the ring terminal and the mounting surface are clean and free of corrosion. Tighten the mounting bolt securely.
7. Make sure the ignition coil wiring is connected as noted during removal.
8. Install the ignition coil cover.
9. Reconnect the negative battery cable.

YAMAHA MICROCOMPUTER IGNITION SYSTEMS (YMIS)

At the end of the day it is tough to put a single label on the Yamaha Microcomputer Ignition Systems. We could say that an outboard equipped with YMIS utilizes some form of electronic control over the spark timing curve of the CDI/TCI system, based on input from one or more sensors. And though that is a true statement, it also applies to some systems that Yamaha does not label as CDI Micro. We're a little puzzled, by this, but don't let that confuse you. If the system Wiring Diagram shows that a thermo switch finds its way back to the CDI unit, then regardless of the system label, it may be taken into consideration when controlling spark timing and the relevant information here applies to that system as well.

For the most part, all 25 hp and larger 4-strokes are equipped either with a CDI Micro ignition (carbureted motors) or TCI Micro ignition (EFI motors). For 4-stroke motors the CDI/TCI unit takes into consideration the thermoswitch (or WTS) and oil pressure sensor for spark mapping. On EFI models the ECM (which performs the function of the CDI/TCI unit) may also use input from other fuel injection system sensors.

The various forms of the YMIS allows the ignition control module to react to various inputs in order to improve engine performance.

Thermo switches (carb) or the WTS (fuel injection) allow the engine to limit engine rpm and timing advancement both during engine warm-up and anytime there is an overheat warning signal. In both cases, this allows the ignition module to protect the powerhead or at least limit it from potential operational damage.

In a similar way to the temperature sensor/switches, the oil level or oil pressure sensors also allow the ignition module to protect the powerhead. In either case, when conditions occur that would trigger a warning signal (low 2-stroke oil level or low 4-stroke oil pressure), the module can limit engine rpm and timing advancement, hopefully protecting the powerhead, while still allowing the operator to limp the boat/motor back to port.

■ Mercury/Mariner motors also use computer control on nearly ALL of the 4-stroke motors. The differences come in that they refer to it as the Guardian Protection system, as opposed to renaming the ignition system itself.

TROUBLESHOOTING

When working on these systems, keep in mind that most engine problems are NOT caused by the computer. If ignition timing changes are noticed at the same time as an engine variation, the computer is probably just reacting to a carburetion/fuel injection or basic ignition system problem, not a computer problem. Always check the fuel system fully before troubleshooting the computer.

■ On EFI engines the ignition and fuel systems are integrated, before proceeding, refer to the section on Electronic Fuel Injection for tips on troubleshooting the entire system, including the use of the EFI Self-Diagnostic system.

An intermittent problem can cause severe changes in ignition timing. For most systems, when the computer is bypassed, the ignition timing will be retarded and engine performance will be reduced. This will result in reduced fuel flow and any existing carburetion problem may not occur at this reduced fuel flow.

The first objective of troubleshooting the YMIS system is to determine if the problem is ignition timing related. A weak spark or no spark is usually caused by the basic ignition components (stator, pulser, ignition coil, CDI, or wiring), not the computer system.

The Yamaha high voltage spark has different characteristics than the spark produced by some other manufacturer's engines. When looking for a spark, especially when in direct sunlight, the Yamaha spark may be hardly visible. If using an aftermarket spark checker, it is possible for electrical radio interference from this checker to interfere with proper computer operation and cause an erratic spark. If you are unsure of your test results, repeat the test in the shade using a resistor spark plug.

If the computer system is suspected, experience has shown that the problem is usually not the computer unit itself. Most computer problems are caused by either a bad sensor, poor connections or grounds or basic engine problems.

Nothing can be more frustrating than an intermittent failure of a component. The resistance tests provided under certain components are procedures to test parts in a laboratory situation mostly at room temperature in ideal working conditions. Unfortunately, this does not provide 100% reliable test results. The sensor, switch, wire harness, and the connector, are all subjected to vibration and temperature extremes during actual operation of the powerhead. If the component could be tested in these operating

4-32 IGNITION AND ELECTRICAL SYSTEMS

conditions it might register readings which are borderline with the specifications. If the heat or the vibration becomes excessive, the part may temporarily fail. Perhaps even if tested as outlined in the various sections of this guide, the conditions may not be severe enough to induce the part to fail. The best advice is to attempt to duplicate the conditions under which the part failed before troubleshooting can be successful.

Other than the CDI output test, no test exists, either in resistance values or operational performance for the microcomputer. If all the sensor signals are processed correctly, all the sensor signals are processed correctly, all engine control circuits function, and the timing is correct then the microcomputer is alive and well.

If the following tests and conditions have been performed or met:
All tests performed on the sensors.
All test performed on the switches.
Operational checks on the system have been conducted and are satisfactory.
All connectors have been checked and rechecked.
The battery is ample in size and power.
Then it is time to consider the computer as a possible fault.

Additional Troubleshooting Tips for Carbureted Motors

The carbureted motors which are equipped with a CDI Micro system utilize input from the oil pressure sensor and thermo switch. In both cases, signals can be used to limit engine rpm and spark timing. Keep this fact in mind when troubleshooting the system. Although the warning system SHOULD be activated when either sensor/switch is closed by operating conditions, it is conceivable that the CDI module could receive a grounded signal (causing it to limit rpm and timing) while a part of the warning circuit is also malfunctioning, preventing the warning horn from sounding. Both switches/sensors (along with their wiring) are easy enough to check and eliminate as potential problems if this is suspected.

Crankshaft Position Sensor (CPS)

DESCRIPTION & OPERATION

◆ See Figure 87

Modern ignition and fuel systems utilize an Engine Control Module (ECM) computer to automatically adjust items like ignition timing (and fuel delivery on EFI systems). In order to affectively accomplish this task, the computer must know at all times what physical position the crankshaft is in. This is accomplished through the use of a Crankshaft Position Sensor (CPS), which is also referred to in charts and procedures as a Trigger or Pulser Coil.

The reason for the nomenclature is that the sensor is in fact a coil that generates voltage using the flywheel magneto in the same basic manner as both the charge and lighting coils, but with a slightly different purpose. Whereas the voltage generated by the charge and lighting coils go to power the ignition and charging systems respectively, the voltage generated by the CPS is really used as a signal, variances in the voltage (caused by variances in the magnetic field) tell the computer precisely where the crankshaft is out of its 360 degrees of rotation.

Basic service, including testing and replacement is covered earlier in this section under Component Testing, Pulser Coil and under Flywheel and Stator (Stator/Charge/Lighting/Pulser Coils), respectively. However, for some Mercury/Mariner models there is one additional step which is necessary. Although most models mount the CPS in a fixed position, the 8/9.9 hp (232cc), and 40/45/50 hp (935cc) powerheads mount the coil using slightly elliptical bolt holes in order to allow for an air gap adjustment between the sensor and the flywheel.

The adjustment should not be needed on a periodic basis, but will become necessary anytime the CPS is repositioned or removed.

GAP ADJUSTMENT & SERVICE

MODERATE

◆ See Figure 87

For most motors the CPS (a moniker really only used by Mecury/Mariner for these motors, Yamaha calling it either a Trigger Coil or Pulser Coil) is mounted in a fixed position. Meaning that the gap between it and the flywheel magneto is not adjustable.

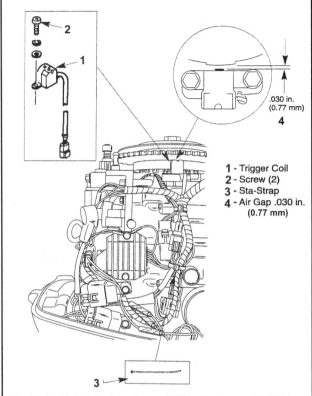

Fig. 87 Crankshaft Position Sensor (CPS) adjustment - 40/45/50 hp (935cc) Mercury/Mariner motors shown (8/9.9 hp model similar)

1 - Trigger Coil
2 - Screw (2)
3 - Sta-Strap
4 - Air Gap .030 in. (0.77 mm)

However, on the 8/9.9 hp (232cc) and 40/45/50 hp (935cc) Mercury/Mariner motors, the sensor bracket allows for a slight forward and aft adjustment in relation to the flywheel. For these motors, check the sensor gap by trying to insert a 0.030 in. (0.77mm) feeler gauge between the sensor and the flywheel. When measuring a gap with a feeler gauge, remember that the gap is equal to the size gauge that will fit through the gap with a slight drag. The next larger size gauge should not fit, while the next smaller gauge should fit without touching or dragging.

If adjustment is necessary, loosen the 2 CPS mounting screws and gently slide the sensor against a 0.030 in. (0.77mm) feeler gauge inserted between the flywheel and sensor. Tighten the mounting screws and double-check the gap using the gauge set.

Thermo-sensor & Thermo-switch

Ok, let's try to explain this best we can. Most Yamaha outboards are equipped with either a thermo-switch (and on/off switch which is activated by temperature) or a thermo-sensor (a variable resistor whose values change with changes in temperatures).

For Yamaha the breakdown goes as follows:
• The 25-40 hp (2 and 3-cylinder) Yamaha motors are equipped with only a single thermo-sensor. The sensor functions with the YMIS system. If the motor is also equipped with an overheat warning system, the signals are provided by this sensor as well.
• The 40-60 hp (4-cylinder) Yamaha motors are equipped with only a single thermo-switch which is used both for the overheat warning system and the YMIS.
• The 75-115 hp (4-cylinder) Yamaha motors are equipped with a thermo-sensor (known as a WTS on fuel injected models), however specifications for the motor also mention overheat warning system on/off temperatures (even though no switch is shown in the wiring diagrams). This either means the switch listing is an oversight, OR the warning system uses data from the sensor.
• The Yamaha 150 hp fuel injected motors are equipped with both a thermo-sensor (known as a WTS) for the fuel and ignition system as well as a thermo-switch. The switch is located in a bore in the thermostat housing (which in turn is mounted to the top of the exhaust cover), while the sensor is mounted to the top of the powerhead, near the TPS and IAP sensor.

IGNITION AND ELECTRICAL SYSTEMS 4-33

- The Yamaha V6 fuel injected motors are equipped with both a thermo-sensor (known as a WTS) for the fuel and ignition system as well as a pair of thermo-switches (one at the top of each cylinder head) for use with the warning system.

When it comes to Mercury/Mariner models, all 25 hp and larger motors are equipped with a single thermo-sensor (called an Engine Coolant Temperature, or ECT sensor most of the time), except for the 40/45/50 hp (935cc) models, which use a single thermo-switch. V6 models are also equipped with a pair of thermo-switches, one per cylinder head, used by the engine overheat monitoring system.

All Mercury/Mariner models equipped with a thermo-sensor utilize the sensor as part of the overheat protection system. When the ECM detects a temperature above a specified range, it will limit engine RPM.

This section deals with the thermo-sensors found on CARBURETED motors only, as the WTS/ECT sensor used on fuel injected motors is covered in the Fuel System section. Furthermore, since the thermo-switch is really more a component of the cooling and warning systems, they are covered (for all motors, carbureted and fuel injected) in the Lubrication and Cooling section.

DESCRIPTION & OPERATION

◆ See Figures 88, 89 and 90

On certain engines (identified earlier) a thermo-sensor is located somewhere on the powerhead for use with either the overheat warning system and/or the ignition system. This sensor is a thermistor, which is a variable resistor. A thermistor's resistance changes with temperature. There are generally 2 types of thermistors; a negative temperature coefficient thermistor and a positive temperature coefficient thermistor. The difference comes in how each type of thermistor responds to temperature changes. The resistance of a negative temperature coefficient sensor will change in the opposite direction of temperature changes (i.e. if the temperature goes UP, the resistance goes DOWN and vice versa). The resistance of a positive temperature coefficient sensor will change WITH the direction of temperature changes (meaning that as temperature goes UP, resistance goes UP and vice versa). Most temperature sensors used by both Yamaha and Mercury/Mariner are of the negative coefficient design.

This information can be useful to the ignition system for one of two reasons. For starters, it can alert the control module as to whether or not the engine has reached normal operating temperatures. A YMIS module or Mercury/Mariner ECM may have the programming to limit rpm and spark advance until the motor reaches normal temperatures. Perhaps more importantly, the sensor can also be used to limit rpm and spark advance if an overheat signal is received, helping to slow and hopefully allow the powerhead to cool (preventing damage).

There is a distinct difference between a thermo-sensor (covered here) and a thermo-switch (covered in the Lubrication and Cooling section). A thermo-sensor was identified as a thermistor - an electronic device capable of detecting changes in temperature and transmitting these changes as voltage signals to the microcomputer. However, a thermo-switch is simply a bimetal type switch which has 2 settings ON and OFF and nothing in between. It cannot tell a control module at what temperature the engine is operating, it can only say whether or not it has reached a specified point which is considered overheating.

To locate the thermo-sensor on carbureted models so equipped, proceed as follows:
- On 25 hp Yamaha motors, the sensor is found toward the top of the powerhead's port side, directly below the pulser coil. It's a little above and about halfway between the starter relay and the regulator/rectifier.
- On 30-40 hp Yamaha motors, the sensor is found almost directly in the middle of the powerhead's port side, below and inline with the pulser coil and immediately below the regulator/rectifier. Its position puts it about halfway between the starter motor and the ignition coils.
- On 25-60 hp Mercury/Mariner motors, the sensor is threaded into the side of the exhaust cover on the side of the powerhead.
- On all 75-100 hp motors, the sensor is found toward the bottom or middle (depending upon the year and model) of the powerhead's starboard side. It is just a little aft of the starter motor.

OPERATIONAL THERMOSENSOR CHECK

 MODERATE

◆ See Figures 88 thru 91

■ This section deals with the thermo-sensors found on CARBURETED motors only, as the WTS/ECT sensor used on fuel injected motors is covered in the Fuel System section. Furthermore, since the thermo-switch is really more a component of the cooling and warning systems, they are covered (for all motors, carbureted and fuel injected) in the Lubrication and Cooling section.

1. Mount the engine in a test tank, or on a boat in a body of water.
2. Connect a timing light and tachometer to the powerhead.
3. Start the engine and allow it to warm to operating temperature.
4. Aim the timing light at the timing pointer and ensure the initial timing is at specification.
5. Disconnect the thermo-sensor wire harness at the quick disconnect fitting and use a small jumper cable and connect the two leads of the harness (motor side of the harness not sensor side) together. The timing should now retard and the engine rpm decrease.

■ Since the thermo-sensor is of a Negative Temperature Coefficient (NTC) design, the resistance in the circuit will go DOWN, as the temperature goes UP. By using a small jumper cable to complete the circuit without the sensor you fool the ECM into thinking that the temperature has gone through the roof, and it will retard timing/limit engine rpm to prevent potential damage to the powerhead it THINKS is overheating. This technique would in theory work on EFI motors as well, but we don't recommend it there in case it would also set a trouble code which might require a scan tool to reset on some models.

6. Remove the thermo-sensor from the motor. Reconnect the leads to the sensor and hold the sensing end of the sensor against an ice cube. The timing should now advance and the engine rpm increase.
7. To replace the sensor, simply unscrew or unbolt it from the powerhead.

Fig. 88 The thermo-sensor is usually threaded...

Fig. 89 ...or bolted...into position on the powerhead

Fig. 90 In contrast, a Yamaha thermo-switch is usually press fit into a bore at the top of a cylinder head

4-34 IGNITION AND ELECTRICAL SYSTEMS

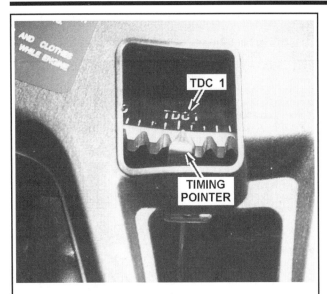

Fig. 91 Mark the timing locations (to make them easier to read) on the flywheel prior to starting the thermo-sensor test

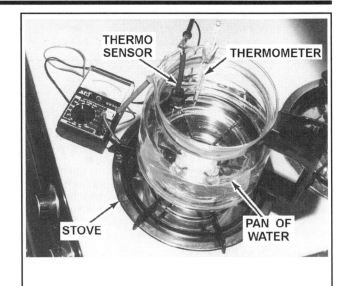

Fig. 92 Thermo-sensor resistance should decrease smoothly as the temperature rises and increase smoothly as it falls

CHECKING THERMOSENSOR RESISTANCE

◆ See Figures 88, 89 and 92 thru 94

■ This section deals with the thermo-sensors found on CARBURETED motors only, as the WTS/ECT sensor used on fuel injected motors is covered in the Fuel System section. Furthermore, since the thermo-switch is really more a component of the cooling and warning systems, they are covered (for all motors, carbureted and fuel injected) in the Lubrication and Cooling section.

Resistance tests can be performed while the sensor is installed in the powerhead, but you won't have the best control over the temperatures so for most accurate testing, remove the sensor from the block.

1. Disconnect the wire harness connector, at the quick disconnect fitting and remove the sensor from the powerhead.
2. The resistance of the sensor is monitored at different temperatures while heating the sensing end of the sensor in a body of water.
3. Place the container of water, at room temperature, on a stove. Secure the thermometer in the water in such a manner to prevent the bulb from contacting the sides or bottom of the pan (so you'll be assured to get the water temperature and not the temperature of the container)
4. Immerse the sensing ends of the sensor into the water up to the shoulder (suspend the sensor so it too is not touching the side or bottom of the container so it will become the same temperature as the water and thermometer).
5. Check resistance across the sensor terminals and compare the readings for the appropriate water temperature. The specifications vary somewhat by engine family so please refer to the appropriate chart. For Yamaha motors, see the Ignition Control System Component Testing - 4-Stroke Inline Engines, found in this section. For Mercury/Mariner models, please refer to one of the accompanying illustrations.
6. Resistance should rise and fall evenly, but in the opposite direction of the temperature. Remember, the thermo-sensor is a thermistor which characteristically varies resistance (increases or decreases voltage in a circuit) with a change in temperature. Slowly heat or chill the water and watch to make sure the sensor shows values within specification over the anticipated operating range.
7. If resistance is not within specification or does not fluctuate evenly, the sensor is faulty.
8. To replace the sensor, simply unscrew or unbolt it from the powerhead (unless of course, you already did to test the sensor).

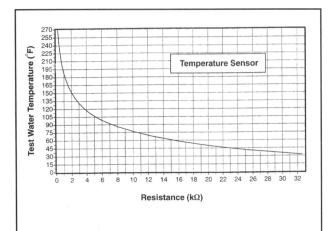

Fig. 93 Engine Coolant Temperature (ECT) sensor resistance specifications for 25-60 hp Mercury/Mariner motors

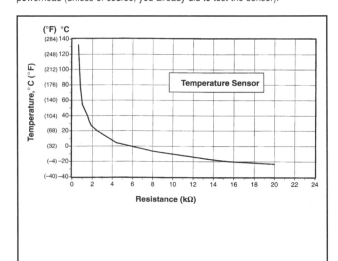

Fig. 94 Engine Coolant Temperature (ECT) sensor resistance specifications for 75-90 hp Mercury/Mariner motors

IGNITION AND ELECTRICAL SYSTEMS

CHARGING SYSTEM

Charging Circuit

◆ See Figure 95

For many years, single-phase, full-wave charging systems were the dominant design. Most manufacturers used single-phase systems because they were simple and reliable. The drawback to these systems is low output. A typical single-phase lighting coil system is capable of only 10 to 15 amps. On many larger rigs, the electrical demand is more than 15 amps. New electronic and electrical devices are arriving on the market every day, so the demand for higher amperage output continues to rise.

In response to higher electrical demands, multi-phase (usually 3-phase) systems were introduced on most larger Yamaha outboards by the early 1990s and since they've found their way onto many of the smaller motors, especially the 4-strokes. These systems produce 25 to 45 amps. At the heart of a three-phase charging system is an interconnected three coil winding. By using three coils instead of one, output is more than doubled. The multi-phase rectifier/regulator works in a similar way to the single-phase system. The difference is the addition of more diodes.

■ On many Yamahas and nearly all Mercury/Mariners the lighting coils are part of a 1-piece stator unit which is also an integral part of the ignition system. Even on motors that utilize separate charge and lighting coils, access to the coils themselves and testing procedures are similar enough that all are covered in the Ignition System section.

Charging Systems

DESCRIPTION & OPERATION

The single-phase charging system found on the smallest of motors covered here provides basic battery maintenance. Single-phase, full wave systems like these are found on a variety of products. Many outboard engines, water vehicles, motorcycles, golf carts and snowmobiles use similar systems.

This charging system produces electricity by moving a magnet past a fixed coil. Alternating current is produced by this method. Since a battery cannot be charged by AC (alternating current), the AC current produced by the lighting coil is rectified or changed into DC (direct current) to charge the battery.

To control the charging rate an additional device called a regulator is used. When the battery voltage reaches approximately 14.6 volts the regulator sends the excess current to ground. This prevents the battery from overcharging and boiling away the electrolyte.

The charging system consists of the following components:
- A flywheel containing magnets
- The lighting coil or stator coil
- The battery, fuse assembly and wiring
- A regulator/rectifier

However, most 4-strokes including all EFI and V6 motors are equipped with a three-phase charging system. The charging system operates in essentially the same manner as the single-phase system. The differences in the system come in output and testing, as it uses 3 coils wired to each other in a Y-configuration (and 3 stator coil wires for the charging system circuit all of which must be tested for shorts to ground and for proper output).

■ Some 3-phase systems may also have a battery isolator integrated into the charging system circuit for the use of multiple batteries.

Servicing charging systems is not difficult if you follow a few basic rules. Always start by verifying the problem. If the complaint is that the battery will not stay charged do not automatically assume that the charging system is at fault. Something as simple as an accessory that draws current with the key off will convince anyone they have a bad charging system. Another culprit is the battery. Remember to clean and service your battery regularly. Battery abuse is the number one charging system problem.

The regulator/rectifier is the brains of the charging system. This assembly controls current flow in the charging system. If battery voltage is below about 14.5-14.6 volts the regulator sends the available current from the rectifier to the battery. If the battery is fully charged (is at or above 14.5-14.6 volts) the regulator diverts most of the current from the rectifier back to the lighting coil through ground.

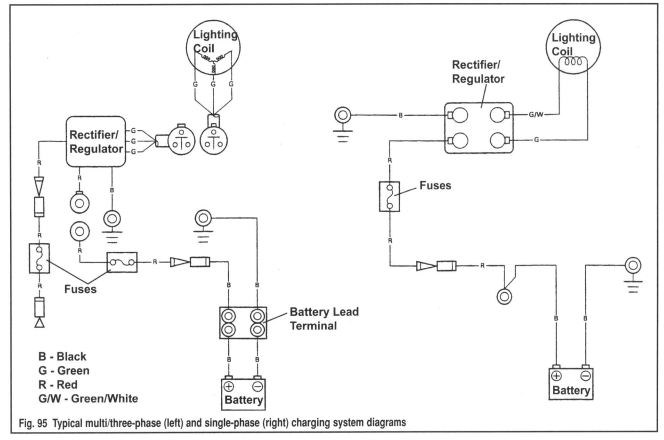

Fig. 95 Typical multi/three-phase (left) and single-phase (right) charging system diagrams

4-36 IGNITION AND ELECTRICAL SYSTEMS

Do not expect the regulator/rectifier to send current to a fully charged battery. You may find that you must pull down the battery voltage below 12.5 volts to test charging system output. Running the power trim and tilt will reduce the battery voltage. Even a pair of 12 volt sealed beam lamps hooked to the battery will reduce the battery voltage quickly.

In the charging system the regulator/rectifier is both the hardest and easiest item to troubleshoot. This is normally accomplished by checking around it. Check the battery and charge or replace it as needed. Check the AC voltage output of the lighting coil. If AC voltage is low check the charge coil for proper resistance and insulation to ground and repair/replace, as necessary. If these check OK measure the resistance of the Black wire from the rectifier/regulator to ground and for proper voltage output on the Red lead coming from the rectifier/regulator going to the battery. If all the above check within specification replace the rectifier/regulator and verify the repair by performing a charge rate test. This same check around method is used on other components like the CDI unit or ECM.

TROUBLESHOOTING

◆ See Figure 96

There are several conditions you can find when troubleshooting a charging system:
- **Normal** - In a normal system the charge rate is correct. There could be a problem elsewhere, such as an electrical short, too many accessories or a bad battery. Whatever the problem, there is no need to work on the charging system. It is working correctly!
- **Overcharging** - The charging system continues to supply current to the battery even after the battery is fully charged. Usually this is caused by a bad regulator portion of the rectifier. However, in some cases it could also potentially be due to the wiring the regulator uses to determine battery voltage.
- **Undercharging** - The charging system is not producing current at all or is producing it at a reduced rate. In this circumstance the question is whether or not the lighting coils are producing what they should and, if they ARE, the fault is in the wiring or the regulator.
- **Battery drain during non-use** - Isolate the battery draw to the motor or to the boat accessories by removing loads one at a time. Although it is rare, one of the diodes in the rectifier can short and slowly drain the battery.

Keep the above possibilities in mind when troubleshooting. By simplifying your troubleshooting approach you may get to the solution more quickly.

The first step in testing a charging system is to confirm there is a charging problem.

■ **The absolute number 1 charging system problem is a faulty battery!**

The real problem may be too many accessories or a draw with the key off. The charging system may really be OK, but if the draw on the system exceeds the charging output, the charging system will appear faulty.

There are two methods for checking three-phase charging systems, the DC amp check and the DC voltage check. The DC amp check method is usually preferred, but both methods are useful.

DC Amperage Check

1. Eliminate the battery as the problem by using a known good one. Now check for any amperage drain on the battery with an ammeter. Connect the ammeter between the positive battery cable and the positive post on the battery. Take your first reading with the key switch **OFF**. Your reading should be at or near zero (does your boat have a clock or radio with a memory?). Then take a reading with the key switch **ON**. This reading will vary according to how many accessories are connected to key-on power. Record the amperage reading.

2. Determine actual DC amperage output from the alternator. To determine alternator output, subtract the key-on/engine-off amperage draw from the maximum alternator output. Start this process by disconnecting the isolator lead from the accessory battery and protect it from grounding out. Next, connect a DC ammeter (0-40 amps) in series with the rectifier/regulator output lead (usually found at the starter solenoid).

■ **The motor should be in the water for this test to provide adequate cool water supply and to prevent a possible engine overspeed condition.**

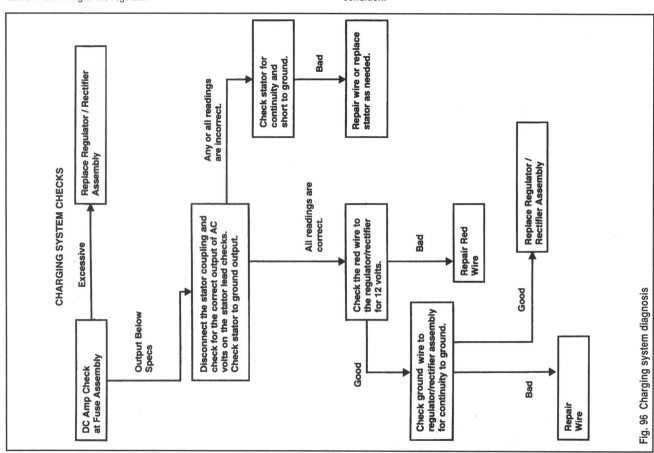

Fig. 96 Charging system diagnosis

IGNITION AND ELECTRICAL SYSTEMS

3. Now, start and rev the motor to 3000 rpm. Observe the alternator output.

■ **When making the DC amperage output check, do not use a fully charged battery! A fully charged battery will not receive full system output. It may only show half the expected amperage output. Reduce the battery state of charge by operating the trim/tilt motor through several cycles. This reduces the charge on the battery and permits higher alternator output.**

Voltage Check

1. Connect a suitable source of cooling water or lower the outboard into a test tank or into the water.
2. The voltage check begins just like the DC amperage check. Begin the voltage check by establishing the condition of the battery and by checking for any key-off amperage draw.
3. Verify that you have 12 volts through the fuse and Red wire. Reading voltage through the fuse establishes continuity and fuse condition.
4. Check the system carefully for good grounds at all the ground points for the Black wires. Remember that there are several splices within the harness that can corrode and create high resistance.

■ **The voltage check is either AC or DC depending upon where in the system you check. Remember, it is the job of the regulator/rectifier to change AC voltage into DC voltage through the diodes. Therefore, you are checking AC voltage when you check the circuit BEFORE the regulator (such as straight from the lighting/stator coil) and you are checking DC voltage when you check AFTER the regulator.**

5. At this point move to the lighting coil wires and disengage the connector(s). Crank the engine and check the AC voltage output among the wires (across different combinations of wires when dealing with a multi-phase system). Check each leg to ground. The AC reading should be nearly the same from each leg to ground. If not, shut off the engine, switch the meter to ohms and look for a coil leg that is open or shorted to ground.
6. Start and run the engine and continue to watch the meter. The voltage should be above battery voltage at idle and rise with engine rpm.

■ **Refer to the Charging System Testing specifications charts in this section for details on lighting coil voltage output under various conditions.**

7. The rectifier/regulator is purposely left for last. In reality, these checks aren't usually necessary. By the process of elimination you can decide if the rectifier is bad. If the stator, battery and fuse are all good, and if the Red and Black wires are OK, the only thing left is the rectifier/regulator assembly. Since you can't disassemble the unit to fix it, replace it.

■ **For more details about checking Lighting Coil output or testing the Regulator/Rectifier please refer to the Component Testing procedures found in under Ignition and Charging System Troubleshooting earlier in this section.**

Charging System Checks

Excessive charging

There is really only one cause for this type of failure, the regulator is not working (is not controlling voltage supplied to the battery), either because it is not getting a proper battery voltage signal or because it is broken. If there are no wiring problems, since there is no repair of regulator/rectifier, replace it.

Undercharging

If there is an undercharge condition after running the DC amperage check at the fuse assembly, then disconnect the stator coupling from the harness and perform AC voltage checks between the stator leads. Check between two stator leads at a time on 3-phase systems, but be sure to check across all possible combinations (3 separate checks). On 3-phase systems all three readings should be equal, within a volt or two.

■ **Refer to the Charging System Testing specifications charts in this section for details on lighting coil voltage output under various conditions.**

Stator/lighting coil shorts to ground can be checked by doing a voltage test between one stator lead and ground, engine running. There should be roughly half the normal stator voltage check reading.

If the readings are all within specification, the stator is working correctly. Proceed to the Red wire and Black wire checks.

If any or all readings are below normal, turn the engine **OFF** and check the stator windings using an ohmmeter. An isolated continuity check and a short to ground check should be done. If the stator is bad, replace it since it can't be repaired.

Battery

The battery is one of the most important (and vulnerable) components of the electrical system. In addition to providing electrical power to start the engine, it also provides power for operation of the running lights, radio, and electrical accessories.

Because of its job and the potential consequences of a failure (especially in an emergency), the best advice is to purchase a well-known brand, with a sufficient warranty period, from a reputable dealer.

The usual warranty covers a pro-rated replacement policy, which means the purchaser would be entitled to a consideration for the time left on the warranty period if the battery should prove defective before its time.

Do not consider a battery of less than the rating provided in the General Engine System Specifications chart from the Maintenance and Tune-Up section. For most Yamaha motors (except the smallest and largest, including all fuel injected motors) this means a battery of 70- amp/hour or 100-minute reserve capacity. For Mercury/Mariner motors, most literature specifies a battery of at least 350 Cold Cranking Amps (CCAs), but also adds that a larger capacity unit, one of at least 775 CCAs be used in conditions under 32 degrees F (0 degrees C). But it never hurts to buy a battery of larger capacity. Especially as the boat and motor's electrical system ages, cables crack internally and resistance increases.

MARINE BATTERIES

◆ See Figure 97

Because marine batteries are required to perform under much more rigorous conditions than automotive batteries, they are constructed differently than those used in automobiles or trucks. Therefore, a marine battery should always be the No. 1 unit for the boat and other types of batteries used only in an emergency.

Marine batteries have a much heavier exterior case to withstand the violent pounding and shocks imposed on it as the boat moves through rough water and in extremely tight turns. The plates are thicker and each plate is securely anchored within the battery case to ensure extended life. The caps are spill proof to prevent acid from spilling into the bilges when the boat heels to one side in a tight turn, or is moving through rough water. Because of these features, the marine battery will recover from a low charge condition and give satisfactory service over a much longer period of time than any type intended for automotive use.

Fig. 97 A fully charged battery, filled to the proper level with electrolyte, is the heart of the ignition and electrical systems. Engine cranking and efficient performance of electrical items depend on a full rated battery

4-38 IGNITION AND ELECTRICAL SYSTEMS

✲✲ WARNING

Avoid the use an automotive type batteries with an outboard unit. It is not built for the pounding or deep cycling to which this use will subject it and it may be quickly damaged.

BATTERY CONSTRUCTION

◆ See Figure 98

A battery consists of a number of positive and negative plates immersed in a solution of diluted sulfuric acid. The plates contain dissimilar active materials and are kept apart by separators. The plates are grouped into elements. Plate straps on top of each element connect all of the positive plates and all of the negative plates into groups.

The battery is divided into cells holding a number of the elements apart from the others. The entire arrangement is contained within a hard plastic case. The top is a one-piece cover and contains the filler caps for each cell. The terminal posts protrude through the top where the battery connections for the boat are made. Each of the cells is connected to its neighbor in a positive-to-negative manner with a heavy strap called the cell connector.

BATTERY RATINGS

◆ See Figure 99

Typically, at least three different methods are used to measure and indicate battery electrical capacity:
- Amp/hour rating
- Cold cranking performance
- Reserve capacity

The amp/hour rating of a battery refers to the battery's ability to provide a set amount of amps for a given amount of time under test conditions at a constant temperature. Therefore, if the battery is capable of supplying 4 amps of current for 20 consecutive hours, the battery is rated as an 80 amp/hour battery. The amp/hour rating is useful for some service operations, such as slow charging or battery testing.

Cold cranking performance is measured by cooling a fully charged battery to 0°F (-17°C) and then testing it for 30 seconds to determine the maximum current flow. In this manner the cold cranking amp rating is the number of amps available to be drawn from the battery before the voltage drops below 7.2 volts.

The illustration depicts the amount of power in watts available from a battery at different temperatures and the amount of power in watts required of the engine at the same temperature. It becomes quite obvious - the colder the climate, the more necessary for the battery to be fully charged.

Reserve capacity of a battery is considered the length of time, in minutes, at 80°F (27°C), a 25 amp current can be maintained before the voltage drops below 10.5 volts. This test is intended to provide an approximation of how long the engine, including electrical accessories, could operate satisfactorily if the stator assembly or lighting coil did not produce sufficient current. A typical rating is 100 minutes.

■ Mercury often mentions a 4th type of battery output classification, Marine Cranking Amps (MCA). The MCA numbers are higher for a given application than CCAs. Mercury's minimum recommendation for these motors is 465 MCAs, but raises it to 1000 MCAs for operation in ambient temperatures below 32 degrees F (0 degrees C).

If possible, the new battery should always have a power rating equal to or higher than the original unit.

BATTERY LOCATION

Every battery installed in a boat must be secured in a well protected, ventilated area. If the battery area lacks adequate ventilation, hydrogen gas, which is given off during charging, is very explosive. This is especially true if the gas is concentrated and confined.

Remember, the boat is going to take a pounding at some point and it is critical that the battery is secured so that wire terminals are not subjected to stresses which could break the cables or battery (and could provide for other dangerous conditions, including the potential for shorts, sparks and even an explosion of that hydrogen gas we just mentioned). There are many types of battery tray mounting assemblies available and you will normally find the perfect one for your application on the shelves of boat supply stores. Make sure the tie down bracket or strap is snug, but not over-tightened to the point of distorting or cracking the case.

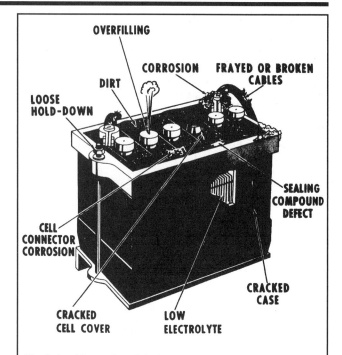

Fig. A visual inspection of the battery should be made each time the boat is used. Such a quick check may reveal a potential problem in its early stages. A dead battery in a busy waterway or far from assistance could have serious consequences

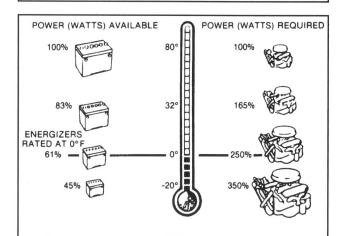

Fig. 99 Comparison of battery efficiency and engine demands at various temperatures

BATTERY AND CHARGING SAFETY PRECAUTIONS

◆ See Figure 100

Always follow these safety precautions when charging or handling a battery:
- Wear eye protection when working around batteries. Batteries contain corrosive acid and produce explosive gas as a byproduct of their operation. Acid on the skin should be neutralized with a solution of baking soda and water made into a paste. In case acid contacts the eyes, flush with clear water and seek medical attention immediately.
- Avoid flame or sparks that could ignite the hydrogen gas produced by the battery and cause an explosion. Connection and disconnection of cables to battery terminals is one of the most common potential causes of sparks. Use extreme caution with jumper cables and make sure all switches are **OFF** before making any other connections.
- Always turn a battery charger **OFF**, before connecting or disconnecting the leads. When connecting the leads, connect the positive lead first, then the negative lead, to avoid sparks.

IGNITION AND ELECTRICAL SYSTEMS

Fig. 100 Explosive hydrogen gas is released from the batteries in a discharged state. This one exploded when something ignited the gas. Explosions can be caused by a spark from the battery terminals or jumper cables

- When lifting a battery, use a battery carrier or lift at opposite corners of the base.
- Ensure there is good ventilation in a room where the battery is being charged.
- Do not attempt to charge or load-test a maintenance-free battery when the charge indicator dot is indicating insufficient electrolyte.
- Disconnect the negative battery cable if the battery is to remain in the boat during the charging process.
- Be sure the ignition switch is **OFF** before connecting or turning the charger **ON**. Sudden power surges can destroy electronic components.
- Use proper adapters to connect charger leads to batteries with non-conventional terminals.

BATTERY CHARGERS

Before using any battery charger, consult the manufacturer's instructions for its use. Battery chargers are electrical devices that change Alternating Current (AC) to a lower voltage of Direct Current (DC) that can be used to charge a marine battery. There are two types of battery chargers - manual and automatic.

A manual battery charger must be physically disconnected when the battery has come to a full charge. If not, the battery can be overcharged, and possibly fail. Excess charging current at the end of the charging cycle will heat the electrolyte, resulting in loss of water and active material, substantially reducing battery life.

■ As a rule, on manual chargers, when the ammeter on the charger registers half the rated amperage of the charger, the battery is fully charged. This can vary, and it is recommended to use a hydrometer to accurately measure state of charge.

Automatic battery chargers have an important advantage - they can be left connected (for instance, overnight) without the possibility of overcharging the battery. Automatic chargers are equipped with a sensing device to allow the battery charge to taper off to near zero as the battery becomes fully charged. When charging a low or completely discharged battery, the meter will read close to full rated output. If only partially discharged, the initial reading may be less than full rated output, as the charger responds to the condition of the battery. As the battery continues to charge, the sensing device monitors the state of charge and reduces the charging rate. As the rate of charge tapers to zero amps, the charger will continue to supply a few milliamps of current - just enough to maintain a charged condition.

✳✳ WARNING

Even with automatic storage chargers, don't assume the charger is working properly. When a battery is connected for any time longer than a day, check it frequently (at least daily for the first week and weekly thereafter). Feel the battery case for excessive heat, listen and look for gassing, check the electrolyte levels.

REPLACING BATTERY CABLES

Battery cables don't go bad very often, but like anything else, they can wear out. If the cables on your boat are cracked, frayed or broken, they should be replaced.

When working on any electrical component, it is always a good idea to disconnect the negative (-) battery cable. This will prevent potential damage to many sensitive electrical components.

Always replace the battery cables with one of the same length, or you will increase resistance and possibly cause hard starting. Smear the battery posts with a light film of dielectric grease, or a battery terminal protectant spray once you've installed the new cables. If you replace the cables one at a time, you won't mix them up.

■ Any time you disconnect the battery cables, it is recommended that you disconnect the negative (-) battery cable first. This will prevent you from accidentally grounding the positive (+) terminal when disconnecting it, thereby preventing damage to the electrical system.

Before you disconnect the cable(s), first turn the ignition to the **OFF** position. This will prevent a draw on the battery which could cause arcing. When the battery cable(s) are reconnected (negative cable last), be sure to check all electrical accessories are all working correctly.

CRANKING (ELECTRIC STARTER) SYSTEM

Starting Circuits

DESCRIPTION AND OPERATION

In the early days, all outboard engines were started by simply pulling on a rope wound around the flywheel. As time passed (and outboards became bigger and bigger) owners became more reluctant to use muscle power (or less capable, some of these motors are HUGE), so it was necessary to replace the rope starter with some form of power cranking system. Today, many small engines are still started by pulling on a rope, but most have a powered cranking motor installed.

The system utilized to replace the rope method was an electric cranking motor coupled with a mechanical gear mesh between the cranking motor and the powerhead flywheel, similar to the method used to crank an automobile engine.

As the name implies, the sole purpose of the cranking motor circuit is to control operation of the starter motor to turn or "crank" the powerhead until the engine is operating. The circuit includes a solenoid or magnetic switch to connect or disconnect the motor from the battery. The operator controls the switch with a key switch.

A neutral safety switch is normally installed into the circuit to permit operation of the cranking motor only if the shift control lever is in neutral. This switch is a safety device to prevent accidental engine start when the engine is in gear.

The cranking motor is a series wound electric motor which draws a heavy current from the battery. It is designed to be used only for short periods of time to crank the engine for starting. To prevent overheating the motor, cranking should not be continued for more than 30-seconds without allowing the motor to cool for at least three minutes. Actually, this time can be spent in making preliminary checks to determine why the engine fails to start (such as checking the safety lanyard, the fuel supply, the fuel line primer bulb, etc).

Power is transmitted from the cranking motor to the powerhead flywheel through a Bendix drive. This drive has a pinion gear mounted on screw threads. When the motor is operated, the pinion gear moves upward and meshes with the teeth on the flywheel ring gear.

When the powerhead starts, the pinion gear is driven faster than the shaft, and as a result, it screws out of mesh with the flywheel. A rubber cushion is built into the Bendix drive to absorb the shock when the pinion meshes with the flywheel ring gear. The parts of the drive must be properly assembled for efficient operation. If the drive is removed for cleaning or

4-40 IGNITION AND ELECTRICAL SYSTEMS

overhaul, take care to assemble the parts as shown in the accompanying illustrations in this section. If the screw shaft assembly is reversed, it will strike the splines and the rubber cushion will not absorb the shock.

The sound of the motor during cranking is a good indication of whether the cranking motor is operating properly or not. Naturally, temperature conditions will affect the speed at which the cranking motor is able to crank the engine. The speed when cranking a cold engine will be much slower than when the same starter and battery are used to crank a warm engine. An experienced operator will learn to recognize the favorable sounds of the powerhead cranking under various conditions.

MAINTENANCE

The cranking motor does not require periodic maintenance or lubrication. If the motor fails to perform properly, the checks outlined in the previous paragraph should be performed. The frequency of starts governs how often the motor should be removed and reconditioned. Although Mercury doesn't recommend a set overhaul timeframe, Yamaha recommends removal and reconditioning (or replacement) every 1000 hours. Naturally, the motor will have to be replaced if corrective actions do not restore the motor to satisfactory operation.

FAULTY SYMPTOMS

If the cranking motor spins, but fails to crank the engine, the cause is usually a corroded or gummy Bendix drive. The drive should be removed, cleaned, and given an inspection.

If the cranking motor cranks the engine too slowly, the following are possible causes and the corrective actions that may be taken:
- Battery charge is low. Charge the battery to full capacity.
- High resistance connections at the battery, solenoid, or motor. Clean and tighten all connections. (A loose or bad ground/battery negative cable is often the culprit here).
- Undersize battery cables (creating too much resistance). Replace cables with sufficient size.
- Battery cables too long (creating too much resistance). Relocate the battery to shorten the run to the solenoid.

Starter Motor Circuit

◆ See Figures 101, 102 and 103

Before wasting too much time troubleshooting the cranking motor circuit, the following checks should be made. Many times, the problem will be corrected.
- Battery fully charged.
- Shift control lever in neutral (most models use a Neutral Safety Switch).
- Are there any blown fuses?
- All electrical connections clean and tight.
- Wiring in good condition, insulation not worn or frayed.

Two more areas may cause the powerhead to crank slowly even though the cranking motor circuit is in excellent condition: a tight or frozen powerhead and water in the lower unit. The following troubleshooting procedures are presented in a logical sequence, with the most common and easily corrected areas listed first in each problem area. The connection number refers to the numbered positions in the accompanying illustrations.

TROUBLESHOOTING

◆ See Figure 104

The starter circuit may seem complicated at first, but it is actually pretty simple when it comes down to it. Power from the positive side of the battery is normally connected to one side of the Starter Relay/Solenoid (No. 7 in the accompanying diagram). A ground cable from the battery either connects to a common ground on the powerhead or to the Starter Motor itself. When the Starter Relay/Solenoid is activated (remember, a relay is essentially a remote controlled switch), it closes an internal switch connecting power from the positive battery cable to a cable (No. 7 to No. 1) which runs to the Starter Motor itself (No. 6).

The Starter Relay/Solenoid is activated when power is applied to it (No. 3) from a Starter Button or Ignition Switch (No. 5 and 4). In many cases the circuit contains a normally open Neutral Safety Switch which closes only when the switch plunger is depressed by the shift linkage (which should only occur when the linkage is in the neutral position).

Obviously since this circuit will vary slightly depending upon the motor and rigging you should also refer to the Wiring Diagram for the particular model on which you are working to verify components and wire colors. Many of the Mercury/Mariner and some of the Yamaha models ground the starter through the mounting assembly instead of through a separate cable and many of the wire colors will vary from the accompanying diagram.

✶✶ WARNING

Although we've given some resistance tests to illustrate how a component can be checked on or off the motor, DON'T perform resistance checks with the wires still connected. The meter could be damaged if you attempt to take resistance readings on a live circuit. If the component is still installed, use the voltage checks.

1. Cranking Motor Rotates Slowly

■ In this case you KNOW that the activation circuit is working. You don't know how well the power circuit is doing (i.e. is there too much resistance somewhere in the circuit) and you don't know if the problem is the starter motor itself or a mechanical problem with the powerhead.

    a. Battery charge is low. Charge the battery to full capacity.
    b. Electrical connections corroded or loose. Clean and tighten.
    c. Defective cranking motor. Perform an amp draw test. Lay an amp draw-gauge on the cable leading to the cranking motor. Turn the key on and

Fig. 101 Typical Yamaha starter motor and starter mounted solenoid

Fig. 102 Many late-model motors place relays and fuses in an electrical junction box

Fig. 103 The inside of the box is usually labeled to help with component identification

IGNITION AND ELECTRICAL SYSTEMS

attempt to crank the engine. If the gauge indicates an excessive amperage draw, the cranking motor must be replaced or rebuilt.

2. Cranking Motor Fails To Crank Powerhead

■ In this case, you don't know whether the problem is the activation portion of the circuit or the starter motor itself? Pick a spot in the circuit and trace the power back from the starter motor and solenoid toward the battery.

a. Disconnect the cranking motor lead from the solenoid (No. 1) to prevent the powerhead from starting during the testing process.

■ This lead is to remain disconnected from the solenoid during tests No. 2—6.

b. Use a voltmeter to check for approximate battery voltage at the output (starter) side of the starter relay/solenoid (No. 1) when the starter switch or button is turned to the start position. If there is sufficient voltage, the problem lies in the cable or starter (or ground circuit between the starter and the powerhead/battery). If there is no voltage, check the battery side of the solenoid (No. 7), just to verify that the solenoid is getting battery power to the starter circuit.

c. Check the solenoid side of the ground circuit by disconnecting the Black ground wire from the No. 2 terminal. Connect a voltmeter between the No. 2 terminal and a common engine ground. Turn the key switch to the start position. Observe the voltmeter reading. If there is the slightest amount of reading, check the Black ground wire connection or check for an open circuit.

3. Test Cranking Motor Relay/Solenoid

■ The solenoid can easily be tested while it is on or off the motor. The basis for the test is to apply 12 volts across the activation circuit (No. 3 and No. 2 in the illustration) and then to check and see if there is continuity across the starter motor power supply circuit (No. 7 and No. 1). If there is continuity without power applied to the activation circuit the switch is stuck closed and faulty, or alternately, if there is no continuity with power applied the switch is stuck open and faulty.

a. Check to see if the activation circuit (neutral safety switch and/or ignition switch/button) is supplying voltage to the starter solenoid. Connect a voltmeter between the engine common ground and the No. 3 terminal. Turn the ignition key switch to the start position. Observe the voltmeter reading. If the meter gives a significant voltage reading (something more than at least 0.3 volt), the solenoid is defective and must be replaced (this is true since you've already verified the ground and battery voltage to power the circuit). If however there is little or no voltage, trace the circuit back toward the ignition start button/switch (through the neutral start switch, if equipped).

■ Some motors are equipped with both a starter circuit relay AND a starter mounted solenoid. For details, please refer to the Wiring Diagrams section.

4. Test Neutral Start Switch

■ Like the relay/solenoid, the neutral start (also known as a neutral safety) switch can be checked either on or off the motor. When equipped, most Yamahas (except the EFI motors) and most Mercury/Mariners use a switch with a plunger which is normally open, but closes the switch contacts when the plunger is depressed. The Yamaha EFI motors normally use a switch with a lever, but they are activated in the same way, closing normally open switch contacts when the lever is depressed. The most simple test you can perform on the switch is to check for resistance across the switch terminals when the plunger (or lever) is depressed vs. when it is released. If the switch closes (meter shows little/no resistance) when it is depressed and opens (meter shows no continuity, meaning infinite resistance) when released, the switch is operating properly.

a. Connect a voltmeter between the common engine ground and the No. 4. Turn the ignition key switch to the start position. Observe the voltmeter. If there is any indication of a reading at No 4, but NOT at No. 3, the neutral start switch is open or the lead is open between the No. 3 and No. 4. Repair the switch or lead, as applicable. If there was no power at No. 4, it is time to check the switch/button and power supply to the switch.

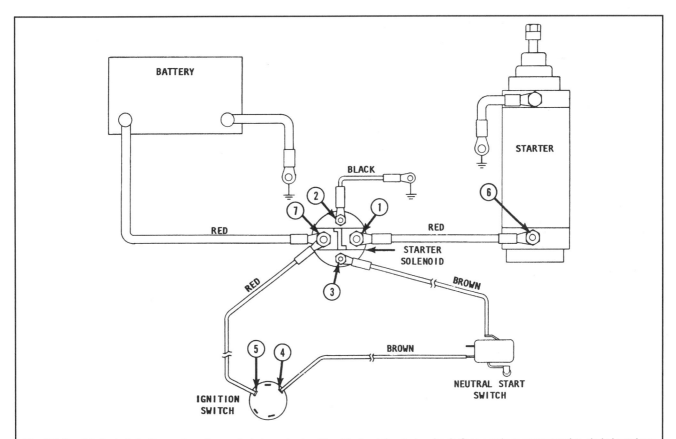

Fig. 104 Use this typical starting system diagram to help understand/troubleshoot the starter circuit. Step numbers correspond to circled numbers in the diagram (note that wiring colors may vary)

4-42 IGNITION AND ELECTRICAL SYSTEMS

5. Test for power to the Ignition Switch/Button

a. Connect a voltmeter between the common engine ground and No. 5.

b. The voltmeter should indicate approximate battery voltage (about 12-volts). If the meter needle flickers (fails to hold steady), check the circuit between No. 5 and common engine ground. If meter fails to indicate voltage, replace the positive battery cable.

c. If power is good to the button/switch, but not on the other (No 4, neutral switch) side, the button or switch is suspect.

6. Test Large Red Power Supply Cables

a. Connect the Red cable to the cranking motor solenoid.

b. Connect the voltmeter between the engine common ground and No. 6.

c. Turn the ignition key switch to the start position, or depress the start button.

d. Observe the voltmeter. If there is no reading, check the Red cable for a poor connection or an open circuit. If there is any indication of a reading, and the cranking motor does not rotate, make sure the starter motor ground is good, otherwise the cranking motor must serviced or replaced.

Starter Motor Relay/Solenoid

DESCRIPTION & OPERATION

◆ See Figure 105

The starter cranking motor relay is actually a remote controlled switch located in the wiring between the battery and the powerhead. Although it can be tested, it cannot be repaired, therefore, if troubleshooting indicates the switch to be faulty, it must be replaced.

Before beginning any work on the relay, disconnect the positive (+) and negative (-) leads from the battery terminal for safety. Keep in mind that the positive lead, where it connects to the solenoid, should always be hot and it is too easy to accidentally ground the tool you are using to the powerhead. This not only will give you quite a shock (we've actually seen it all but weld a wrench in place), but the resultant sparks could create a dangerous condition.

✹✹ WARNING

Disconnecting the battery leads is most important because the cranking motor relay lead will be disconnected and allowed to hang free. The other end of this lead is connected to the battery. If the leads are not disconnected from the battery and the free relay end should happen to come in contact with any metal part on the powerhead, sparks would fly and the end of the lead would be burned.

As detailed in the Starter Motor Circuit section, the starter relay/solenoid has the all important function of responding to the activation circuit in order to physically close the switch contacts that will apply battery power to the starter. The starter motor power circuit contains heavy gauge leads capable of large amp loads in order to power the motor. The activation circuit contains smaller gauge leads, since it only need carry sufficient amperage to activate the relay itself. The activation circuit applies battery power to the solenoid through a neutral safety switch (when equipped) and a starter button or switch.

TESTING

◆ See Figures 105, 106 and 107

✹✹ WARNING

The following tests must be conducted with the relay wiring disconnected (to prevent accidental starting and to prevent potential damage to the meter).

■ The wire colors provided in this procedure are true for most years and models, but if yours differs, check the Wiring Diagrams in this section. In some cases wires may not agree with either the procedure or the diagram as all or part of a harness may have been replaced previously on older outboards, or on motors rigged with an electrical starter after the factory.

The test is relatively simple. Since the relay is simply a remote controlled electric switch, first you check that the remote control (activation) circuit has continuity and the main switch (starter power) circuit does NOT. Then you apply 12 volts to that remote control circuit and make sure the main switch contacts close only when power is applied.

1. Obtain an ohmmeter or DVOM set to read resistance. Check for continuity across the terminals for the small relay leads (The remote control/activation circuit). In most cases these are black and brown leads (black for ground and brown from the neutral or ignition switch). The meter should indicate continuity. If the meter registers no continuity, the relay is defective and must be replaced. No service or adjustment is possible.

2. Connect one test lead of the ohmmeter to each of the large relay terminals. There should be NO continuity.

3. Now, use a 12-volt battery connected to the remote control circuit of the relay in order to activate the relay while continuing to check for continuity

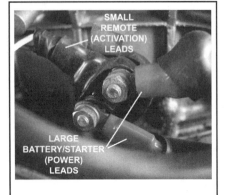

Fig. 105 Typical starter relay (the large terminal leads are exposed as the boots are pulled back)

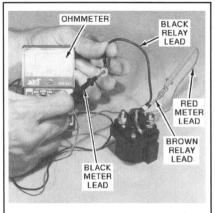

Fig. 106 Check for continuity across the small relay lead terminals (then check for continuity across the large relay lead terminals). With no voltage applied, there should only be continuity across the small terminals

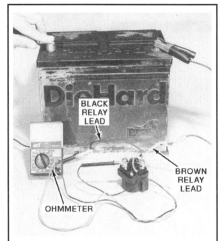

Fig. 107 Connect a 12-volt battery to the small relay lead terminals and recheck the large lead terminals for continuity. If the relay is good the contacts for the large terminals will close and there will be continuity as long as 12 volts are applied to the small terminals

IGNITION AND ELECTRICAL SYSTEMS   4-43

across the large relay terminals. Connect the positive (+) lead from a fully charged 12-volt battery to the small terminal Brown lead and momentarily make contact with the ground lead from the battery to the small terminal Black lead. If a loud click sound is heard, and the ohmmeter indicates continuity, the solenoid is in serviceable condition. If, however, a click sound is not heard, and/or the ohmmeter does not indicate continuity, the solenoid is defective and must be replaced with a suitable marine type solenoid.

REMOVAL & INSTALLATION

◆ See Figure 105

■ **The wire colors provided in this procedure are true for most years and models, but if yours differs, check the Wiring Diagrams in this section. In some cases wires may not agree with either the procedure or the diagram as all or part of a harness may have been replaced previously on older outboards, or on motors rigged with an electrical starter after the factory.**

Most Yamaha and some Mercury/Mariner starter relays are press-fit into a mounting bracket. The rest are bolted into position. When they are bolted in position, use care not to over-tighten the fasteners and crack the case. The same goes for the terminal nuts, make sure they are snug but do not over-tighten them.

On 75 hp and larger models the solenoid function is performed by 2 related, but separate components, one a remote mounted relay and the other a magnetic switch mounted piggy-back on the starter motor assembly itself. It is usually easier to remove the starter completely from the powerhead before attempting to remove the later from the starter itself.

  1. Push down the boots from the two large leads (they are usually Red) at the relay. Take time tag and/or note each of the leads and to which terminal they are connected, then remove each of the leads from the solenoid. Disconnect the smaller leads (usually Brown and Black).

  2. Slide the relay from its mounting bracket. In some cases there will be a grounding bolt on the bracket behind the relay to release the Black relay grounding wire.

To Install:

  3. Slide the relay back onto its mounting bracket. If applicable, secure the Black lead behind the relay by installing the grounding bolt on the bracket.

  4. Connect the smaller relay leads (usually Brown and Black) as tagged during removal.

  5. Connect the two large leads (usually Red) to the front of the cranking motor relay, as tagged during removal. One is from the terminal on the cranking motor, while the other is hot, coming from the positive battery terminal. This second lead also usually has a small Red lead attached to the relay terminal.

  6. Install the elbow boots onto the relay leads.

  7. If the work is complete, then reconnect the battery terminals. If further work is to be carried out on the cranking system, then leave the battery cables disconnected until the work is complete.

Starter Motor

DESCRIPTION

One basic type of cranking motor is used on all powerheads covered here. However, the housing/mounting and exterior appearance of the motor itself will vary slightly from model-to-model. Therefore, the accompanying illustrations may differ slightly from the unit being serviced, but the procedures and maintenance instructions are valid.

Marine cranking motors are very similar in construction and operation to the units used in the automotive industry.

Yamaha and Mercury/Mariner marine cranking motors typically use the inertia type drive assembly. This type assembly is mounted on an armature shaft with external spiral splines which mate with the internal splines of the drive assembly.

As with most starter motors, the housing is not designed to provide airflow which would be necessary for continuous or repeated operation. Therefore, never operate a cranking motor for more than 30 seconds without allowing it to cool for at least three minutes. Continuous operation without the cooling period can cause serious damage to or destroy the cranking motor.

REMOVAL & INSTALLATION

2-Cylinder Powerheads

◆ See Figures 108, 109 and 110

  1. Disconnect the battery cables for safety. It is a good idea to place a plastic bag over the cables or secure them away from the battery to prevent them from accidentally contacting the terminals again during the procedure.

  2. Remove the cowling from the powerhead.

  3. Some models also contain a hand-rewind starter or flywheel cover that might interfere with access to the starter bolts. If necessary, remove the hand-rewind starter or flywheel cover for access.

  4. Disconnect the starter motor wiring. For Yamahas, unless the motor is grounded through the mounting there should be 2 large leads, one for power and the other for ground. On most Mercury/Mariner models the unit IS grounded through the mounting and there is only one large lead, furthermore it is sometimes secured by one of the mounting bolts.

■ **On some Yamaha models, you cannot reach both leads while the starter is still installed. If necessary, you can disconnect the lead on the other end (at the starter relay or the powerhead) or you can wait until the starter is unbolted and repositioned to access the lead(s) at the starter.**

  5. Remove the mounting bolts and washers from the starter motor housing. Most starters on these size Yamaha engines are secured by 2 bolts at the top of the motor (threaded downward on 8/9.9 (232cc) models, or

Fig. 108 The starters on most some 2-cylinder powerheads are secured using 2 mounting bolts threaded downward from the top

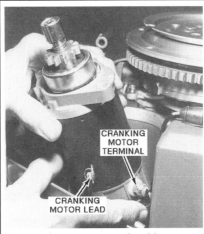

Fig. 109 On some motors, the wiring may not be accessible until the motor is repositioned slightly

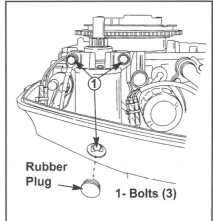

Fig. 110 On 9.9/15 hp (323cc) Mercury/Mariners, there is a rubber access plug to access the lower starter mounting bolt

4-44 IGNITION AND ELECTRICAL SYSTEMS

threaded inward toward the powerhead on 15 hp and larger 4-stroke motors). Also, note that some Yamahas and most Mercury/Mariner models have an additional bolt threaded from the bottom of the motor sideways into the powerhead, so make sure you've got all the bolts before you try to remove the motor.

■ **On some 9.9/15 hp (323cc) models (including all Mercury/Mariners of this size) there is a third starter mounting bolt threaded sideways into the powerhead at the bottom of the starter. However the Mercury thought to include a small rubber plug in the side of the lower engine cowling that can be removed for access.**

6. Remove the cranking motor from the powerhead. If not done already, disconnect the large Red lead from the motor.

To Install:

7. Move the starter motor close to position on the powerhead. If access is not possible once the motor is bolted to powerhead, connect the Red lead to the cranking motor terminal before positioning it to the mounting bracket.

8. Hold the motor in position as you thread the 2 (or 3) retaining bolts and washers. Tighten the mounting bolts securely.

9. Connect the starter motor wiring (or alternately the wiring from the motor to the powerhead and/or relay, as applicable). Tighten the terminals securely, but be careful not to over-tighten and strip the bolts or crack the housing.

10. Mount the outboard unit in a test tank, on the boat in a body of water, or connect a flush attachment and hose to the lower unit.

✲✲ CAUTION

Water must circulate through the lower unit to the engine any time the engine is run to prevent damage to the water pump in the lower unit. Just five seconds without water will damage the water pump.

Never, again, never operate the engine at high speed with a flush device attached. The engine, operating at high speed with such a device attached, would runaway from lack of a load on the propeller, causing extensive damage.

11. Crank the powerhead with the starter motor and start the unit. Shut the powerhead down and restart it several times to check operation of the cranking motor.

3-Cylinder and 4-Cylinder Powerheads

◆ See Figures 111 thru 114

1. Disconnect the battery cables for safety. It is a good idea to place a plastic bag over the cables or secure them away from the battery to prevent them from accidentally contacting the terminals again during the procedure.

2. Remove the cowling from the powerhead.

3. Locate the starter motor, on some 3-and 4-cylinder motors you can remove the flywheel cover for better access, but it is usually not necessary. When equipped, the flywheel cover is usually retained by one or more bolts and rubber mounting grommets/pins.

4. On 75-115 hp motors (and 150 hp motors, if necessary), remove the electrical cover from the side of the powerhead for access to the starter solenoid wiring. On these models a magnetic switch is attached to the starter itself and some of the wiring may be located behind the cover.

5. Disconnect the hot lead and ground lead (if equipped, since some of these motors ground the starter circuit through the starter mounting). Typically, the hot lead is connected toward the bottom of the starter and the ground lead toward the top, but this is not always the case (such as the larger 4-strokes which attach the ground to the bottom), so watch the wire colors and tag them if necessary to ensure proper installation. The smaller Mercury/Mariner motors are usually grounded through the mounting bracket and only contain one lead on the starter itself.

6. Remove the bolts threaded through the starter and into the powerhead or mounting bracket. The mounting bolt patterns vary greatly from model-to-model. Most motors are secured by 2 or 3 mounting bolts as follows:

• On 30/40 hp 3-cylinder motors and the Mercury/Mariner 40-60 hp 4-cylinder motors (except SOME of the Mercury/Mariner 935cc models which may be equipped with the same set-up as Yamaha 40-60 hp motors) there are 2 bolts at the top of the starter motor and one at the bottom. All 3 bolts are threaded sideways into the mounting bracket. Upon installation tighten these bolts to 21 ft. lbs. (29 Nm) for all except Mercury/Mariner 935cc motors with this configuration, whose bolts should be tightened to 16.5 ft. lbs. (22.3 Nm).

• On 40-60 hp Yamaha motors, as well as some 40/45/50 hp (935cc) Mercury/Mariner models, there are 2 bolts threaded downward into the top of the starter motor and, usually, a third bolt threaded sideways into the powerhead. The bolts should be tightened to about one 22 ft. lbs. (30 Nm) upon installation.

• On 75-150 hp motors there are 2 bolts at the top of the starter motor and one at the bottom. All 3 bolts are threaded sideways into the mounting bracket. Although Yamaha does not give a torque specification for most of these models, Mercury states that the bolts should be tightened to 13 ft. lbs./156 inch lbs. (18 Nm) for the 75-115 hp motors. Yamaha does provide a specification for the 150 hp units, at the starter retaining bolts on those models should be tightened to 21 ft. lbs. (29 Nm).

7. Once all the bolts are removed, carefully lift the motor free of the bracket.

8. Installation is essentially the reverse of the removal procedure. Tighten the bolts to specification. Visually inspect the condition of the cables and the terminals before installation. Replace any cables whose insulation has become cracked, brittle or otherwise damaged. Carefully clean all terminals and ground mounting points of any signs of corrosion. This is especially true for starter motors whose circuit grounds through the mount itself. Also, use care when reconnecting the cables. Make sure the cables are routed as noted during removal and are not pinched or damaged in anyway.

9. Once you are finished, mount the outboard unit in a test tank, on the boat in a body of water, or connect a flush attachment and hose to the lower unit.

Fig. 111 Some starters are secured by 2 bolts threaded downward at the top of the housing...

Fig. 112 ...but most use 2 bolts at the top and 1 or 2 at the bottom

Fig. 113 On 75-115 hp models (and some 150 hp models) the wiring is usually partially obscured by the junction box cover

IGNITION AND ELECTRICAL SYSTEMS 4-45

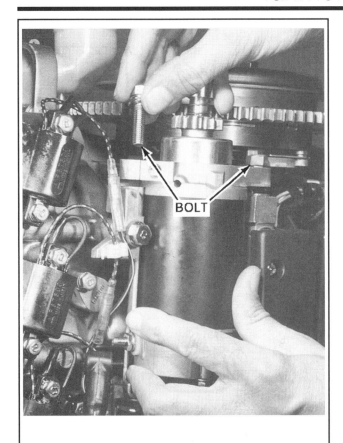

Fig. 114 Once the bolts are removed, lift off the starter

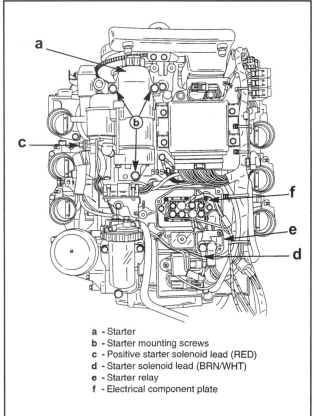

a - Starter
b - Starter mounting screws
c - Positive starter solenoid lead (RED)
d - Starter solenoid lead (BRN/WHT)
e - Starter relay
f - Electrical component plate

Fig. 115 V6 starter motor mounting - Mercury/Mariner shown (Yamaha very similar)

✳✳ CAUTION

Water must circulate through the lower unit to the powerhead anytime the powerhead is operating to prevent damage to the water pump in the lower unit. Just five seconds without water will damage the water pump impeller.

Never, again, never operate the engine at high speed with a flush device attached. The engine, operating at high speed with such a device attached, would runaway from lack of a load on the propeller, causing extensive damage.

10. Crank the powerhead with the cranking motor and start the unit. Shut the powerhead down and restart it several times to check operation of the cranking motor.

V6 Powerheads

◆ See Figure 115

1. Disconnect the battery cables for safety. It is a good idea to place a plastic bag over the cables or secure them away from the battery to prevent them from accidentally contacting the terminals again during the procedure.
2. Remove the cowling from the powerhead.
3. Remove the electrical component cover from the powerhead for access to the wiring and components.
4. Tag and disconnect the wiring for the starter and attached solenoid. On most models this means a large RED lead attached to the bottom of the solenoid and the BROWN/WHITE lead which runs down to the starter relay.
5. Remove the bolts threaded through the starter and into the powerhead or mounting bracket. There are 2 bolts at the top of the starter motor and one at the bottom. All 3 bolts are threaded sideways into the mounting bracket. The bolts should be tightened to 21 ft. lbs. (29 Nm) upon installation.

6. Once all the bolts are removed, carefully lift the motor free of the bracket.
7. Installation is essentially the reverse of the removal procedure. Tighten the bolts to specification. Visually inspect the condition of the cables and the terminals before installation. Replace any cables whose insulation has become cracked, brittle or otherwise damaged. Carefully clean all terminals and ground mounting points of any signs of corrosion. This is especially true for starter motors whose circuit grounds through the mount itself. Also, use care when reconnecting the cables. Make sure the cables are routed as noted during removal and are not pinched or damaged in anyway.
8. Once you are finished, mount the outboard unit in a test tank, on the boat in a body of water, or connect a flush attachment and hose to the lower unit.

✳✳ CAUTION

Water must circulate through the lower unit to the powerhead anytime the powerhead is operating to prevent damage to the water pump in the lower unit. Just five seconds without water will damage the water pump impeller.

Never, again, never operate the engine at high speed with a flush device attached. The engine, operating at high speed with such a device attached, would runaway from lack of a load on the propeller, causing extensive damage.

9. Crank the powerhead with the cranking motor and start the unit. Shut the powerhead down and restart it several times to check operation of the cranking motor.

STARTER MOTOR EXPLODED VIEWS

◆ See Figures 116 thru 121

4-46 IGNITION AND ELECTRICAL SYSTEMS

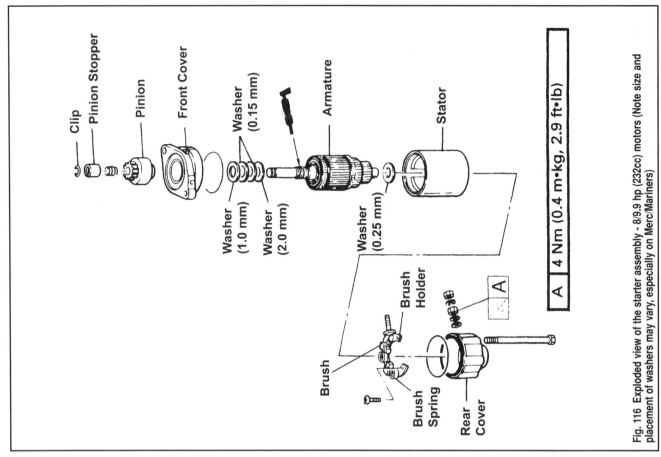

Fig. 116 Exploded view of the starter assembly - 8/9.9 hp (232cc) motors (Note size and placement of washers may vary, especially on Merc/Mariners)

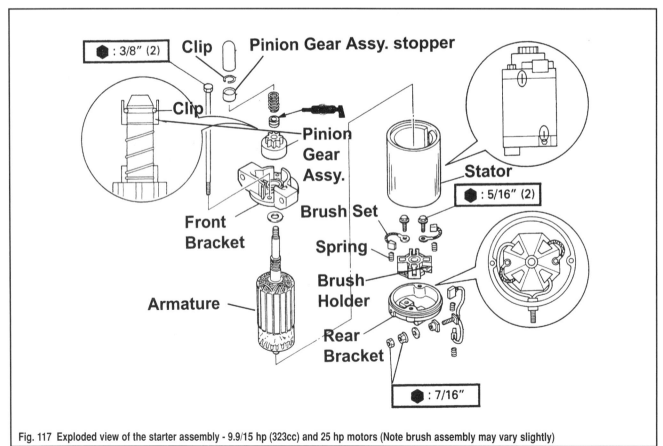

Fig. 117 Exploded view of the starter assembly - 9.9/15 hp (323cc) and 25 hp motors (Note brush assembly may vary slightly)

IGNITION AND ELECTRICAL SYSTEMS

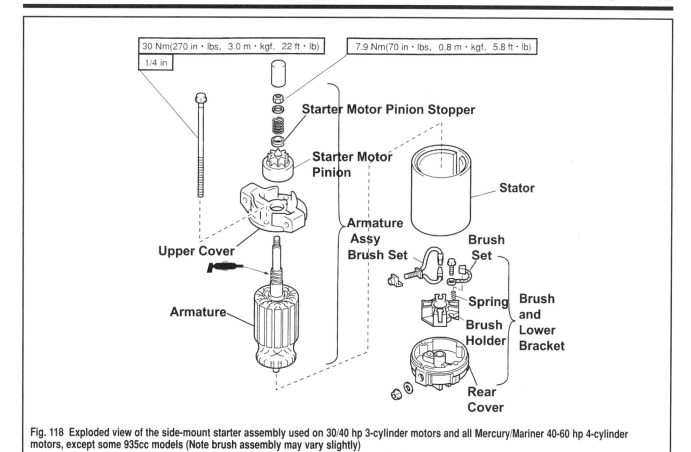

Fig. 118 Exploded view of the side-mount starter assembly used on 30/40 hp 3-cylinder motors and all Mercury/Mariner 40-60 hp 4-cylinder motors, except some 935cc models (Note brush assembly may vary slightly)

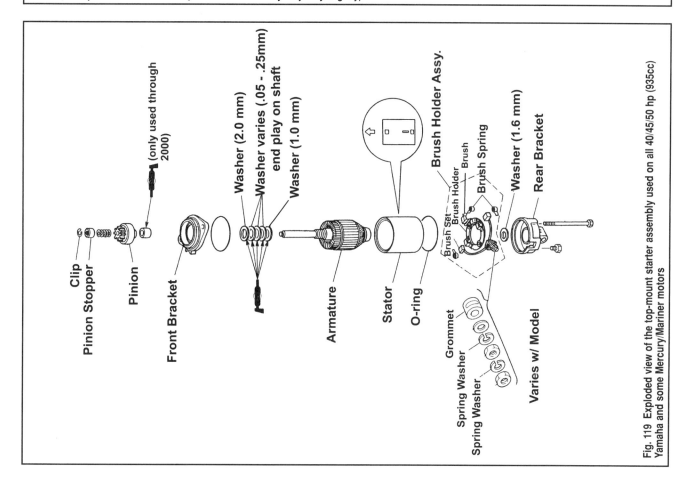

Fig. 119 Exploded view of the top-mount starter assembly used on all 40/45/50 hp (935cc) Yamaha and some Mercury/Mariner motors

4-48 IGNITION AND ELECTRICAL SYSTEMS

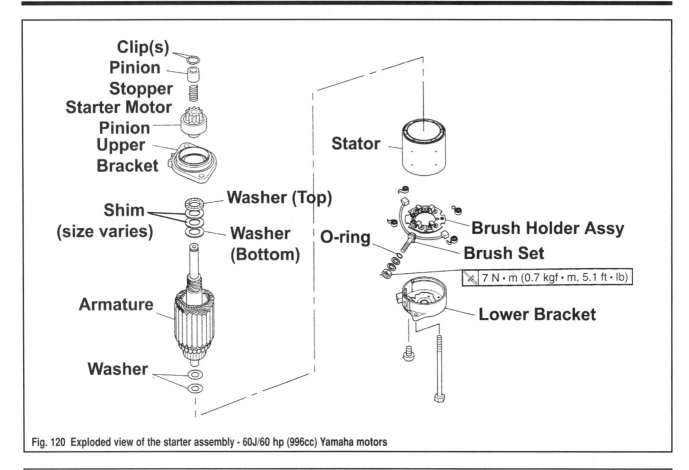

Fig. 120 Exploded view of the starter assembly - 60J/60 hp (996cc) Yamaha motors

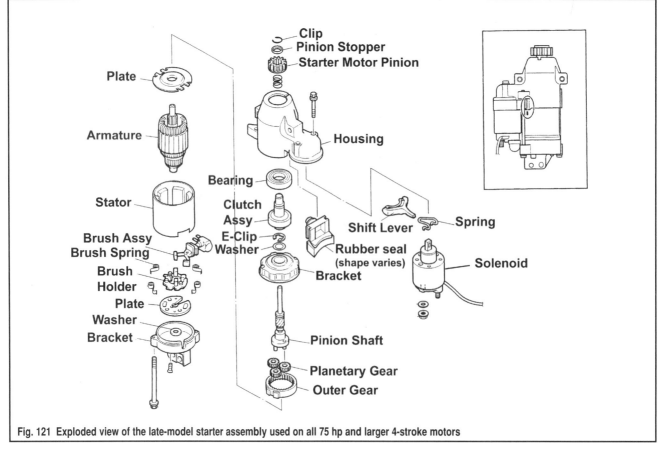

Fig. 121 Exploded view of the late-model starter assembly used on all 75 hp and larger 4-stroke motors

IGNITION AND ELECTRICAL SYSTEMS

OVERHAUL

◆ See Figures 116 thru 121 and 122 thru 131

Although the wide availability of rebuilt starters on the marine parts market today dissuades most technicians and many DIYers from overhauling the starter, rebuild kits area normally available. If you decide to undertake this, keep close track of all starter components as you disassemble it. Digital cameras are relatively cheap these days, so if you've got access to one, work slowly taking pics as you go. They can be a great reference when it comes time for assembly. Be sure to keep track of all washers and shims so that they are placed in the same positions upon assembly.

■ Yamaha and Mercury/Mariner reuses essentially the same starter assembly on multiple powerheads. Besides small differences in the top or bottom housing, differences will occur with the thickness of shims used between armature washers from year-to-year or model-to-model. For this reason, it is critical that you keep track of all washers and shims for installation in the same positions.

Before disassembling the starter motor, scribe matchmarks between the upper/lower (front/rear) covers and the starter housing (also sometimes known as the stator housing). This will ensure easy alignment during assembly.

Keep in mind that the magnets which make up a large part of the starter motor are quite strong. Once the through bolts are removed it can be difficult to separate the field frame/housing and the armature shaft. It is even more important to keep this in mind when assembling the starter as it is easy to badly pinch a finger during that part of the process. So take your time and watch the piggys.

Once disassembled, carefully clean all metal components with a mild solvent and either blow dry with compressed air or allow them to air dry. Thoroughly inspect all components as follows:
• Inspect the pinion gear for wear or damage to the teeth and replace, as necessary.
• Check the clutch for freedom of movement in one direction and stiff movement in the other (generally it is clockwise-free, counterclockwise-stiff).
• A dirty commutator may be cleaned using #600 grit sandpaper, but should be thoroughly cleaned afterwards using solvent and compressed air.
• Be sure to inspect the housing and end caps for signs of wear or damage such as cracks.
• If you have access to a set of precision V-blocks and a dial-gauge, check the commutator (armature shaft) run-out. Generally no more than 0.0020 in. (0.05mm) of run-out is considered serviceable.
• Carefully exam the mica undercut (depth of the grooves) on the end of the commutator. Build specifications vary greatly, however service limits are generally 0.01-0.03 in. (0.2-0.8mm). If the undercut depth is less than this spec, remove some metal from between the commutator segments using a hacksaw blade, then remove all metal and mica particles using compressed air (and safety goggles!).
• Inspect the brush lengths (height of each brush from top to bottom, if the point where the wire attaches is considered the side). Again, build specifications vary greatly, however service limits are generally 0.35-0.49 in. (9.0-12.5mm) on most motors. Some, but not all, of the smaller motors use proportionally smaller brushes (as small as 0.18-0.25 in./4.5-6.5mm), so if you are unsure as to whether or not the brushes are serviceable, check with a local parts supplier to see if you can measure the height of a replacement set (or just spring for the replacements, since you've come this far).

■ Mercury/Mariner lists the following brush limits for starters on their models. A minimum of 0.35-0.49 in. (9.0-12.5mm) for 8/9.9 hp (232cc) motors, 0.47 in. (12mm) for top-mount 935cc motors, 0.25 in. (6.4mm) for all other motors through 60 hp and 0.374 in. (9.5mm) for all 75 hp and larger motors.

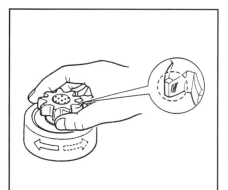

Fig. 122 Check pinion movement and inspect the teeth for wear

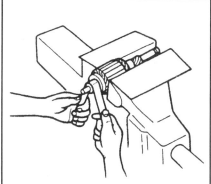

Fig. 123 If necessary, clean the commutator with sandpaper

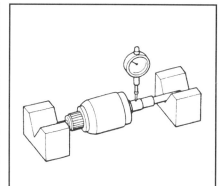

Fig. 124 Use a dial-gauge to check armature shaft run-out

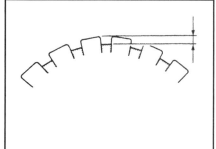

Fig. 125 Check the mica undercut depth

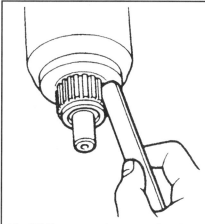

Fig. 126 If necessary, increase undercut depth using a hacksaw blade or file

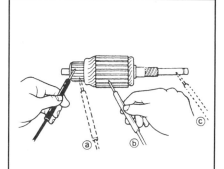

Fig. 127 Use a ohmmeter to check for continuity between the commutator segments (a), the segments and the armature core (b) and lastly, between the segments and armature shaft (c)

4-50 IGNITION AND ELECTRICAL SYSTEMS

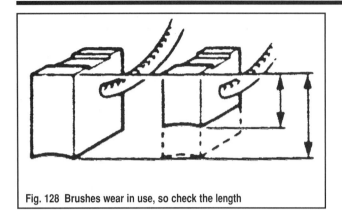

Fig. 128 Brushes wear in use, so check the length

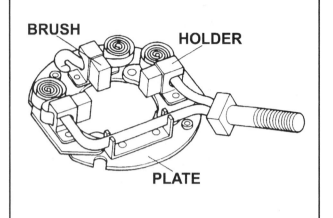

Fig. 129 Using an ohmmeter check brush continuity - typical brush plate assembly shown

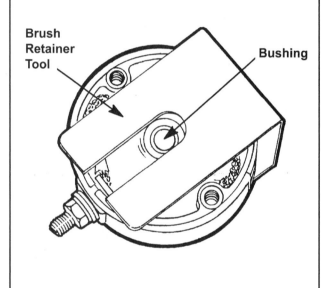

Fig. 131 When applicable, use the brush holder to keep the brushes in position as the end cap is installed, then carefully slide the tool out at the last minute

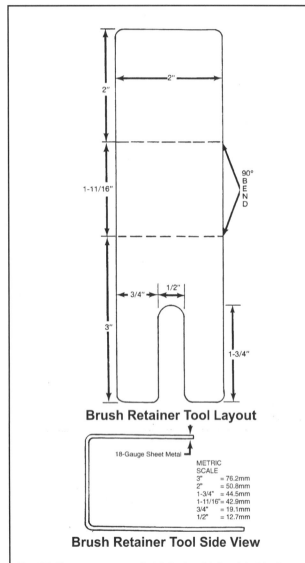

Fig. 130 Mercury recommends fabricating this brush holder for assembly of the starter on their 9.9/15 hp (323cc) to 60 hp motors (except for top-mount 935cc models)

- Use an ohmmeter to inspect armature continuity. First check for continuity between each of the segments on the commutator, there should be continuity. Next, check between the commutator segments and the armature core (the thick main shaft of the armature), there should be NO continuity. Finally check between the commutator segments and the armature shaft (the long, thin shaft at the other end of the armature from the commutator segments), again there should be NO continuity. Replace the armature if it fails this armature continuity test.
- Use an ohmmeter to inspect brush set continuity. Use the accompanying illustration of a typical brush set to help identify components. Check that there is continuity between brush No. 1 and No. 2. Next make sure there is NO continuity between those brushes and brush No. 3. Also thee should be no continuity between each brush holder (No. 4) and the brush assembly holder or brush plate (No. 5) as the holder is used to isolate the brushes from the plate.

Replace any components which do not appear to be serviceable.

IGNITION AND ELECTRICAL SYSTEMS

ELECTRICAL SWITCH/SOLENOID SERVICE

This short section provides testing procedures for other electrical parts installed on the powerhead. If a unit fails the testing, the faulty part must be replaced. In most cases, removal and installation is through attaching hardware.

Mercury Main Keyswitch

For models equipped with the Mercury Commander 2000 Key Switch, please refer to the Wiring Diagrams in the Remote Controls section for a diagram and continuity test chart. Using a DVOM set to read resistance or show continuity, check across each of the given combinations of wires for each of the relevant key positions. Replace the switch if continuity is not as shown in the diagram.

Yamaha Main Keyswitch

TESTING

◆ See Figures 132 and 133

The Yamaha main keyswitch is normally located inside the control box or it is rigged to the dash of the boat. When installed in the control box, the box must be normally be opened to gain access to the main switch leads.

■ **The wire colors on this test are only applicable to Yamaha harnesses and keyswitches. Depending upon boat rigging, the craft on which you are working could deviate from these colors, at least on the switch side of the harness.**

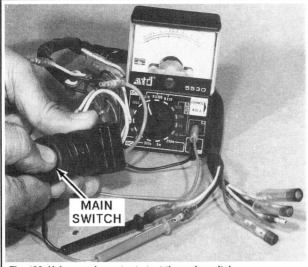

Fig. 132 Using an ohmmeter to test the main switch

Switch position	Lead color				
	White (W)	Black (B)	Red (R)	Pink or Yellow (P or Y)	Brown (Br)
OFF	O—	—O			
ON			O—	—O	
START			O—	—O—	—O

Fig. 133 Yamaha keyswitch testing, the diagram shows continuity between leads in various switch positions (NOTE, some switches use Yellow, instead of a Pink lead)

Disconnect the five leads from the main switch at their quick disconnect fittings: the White, Black, Red, Pink or Yellow (depending upon the switch/model), and Brown leads. Obtain an ohmmeter or a DVOM set to read resistance. Check to be sure the switch is in the off position. Make contact with the Red meter lead to the White switch lead, and the Black meter lead to the Black switch lead. The meter should indicate continuity. Keep both meter leads in place. Rotate the switch to the on position, and then to the start position. If the meter indicates continuity in either or both switch positions, the switch is defective and must be replaced.

■ **Likewise, check for any continuity between the White or Black leads and each of the three additional wire leads. There should be no continuity with the Red, Yellow (or Pink) and/or Brown leads and the White or Black leads with the switch in ANY position.**

Move the Red meter lead to make contact with the Red switch lead, and the Black meter lead to make contact with the Yellow (or Pink) switch lead, with the switch OFF there should be no continuity. Now rotate the switch to the ON position. The meter should indicate continuity. Turn the switch back to the off position and then on to the start position. If the meter fails to indicate continuity in either the ON or START position, the switch is defective and must be replaced.

Keep the Red meter lead in contact with the Red switch lead, but move the Black meter lead to the Brown switch lead. There should be NO continuity in any switch position except START.

Keep the Black meter lead on the Brown switch lead and move the Red meter lead to the Yellow (or Pink) switch lead. Again, there should be no continuity in any position except START.

■ **If the meter fails to indicates continuity when specified OR shows continuity in other positions/lead combinations, the switch is defective and must be replaced.**

✱✱ CAUTION

A faulty main switch could cause the powerhead to start accidentally, while being serviced, and cause personal injury.

Once testing has been satisfactorily completed connect the five leads, color to color, at their quick connect fittings. Tuck the leads neatly inside the control box, away from moving parts, and replace the outer cover.

Kill Switch

TESTING

◆ See Figure 134

The Kill switch is normally located on the front panel of any powerhead not equipped with a control box. For powerheads equipped with a control box, the kill switch is often mounted on the forward side of the box. The Kill switch must have the emergency tether in place before testing.

■ **Depending upon how the boat is rigged, the kill switch may be attached to the boat dash itself.**

Trace the Kill switch button harness containing two wires from the switch to their nearest quick disconnect fitting. The colors may vary for different models, but it shouldn't make a difference, because it is only a 2 wire switch and your tests will be with one meter lead attached to each wire. Generally the wires are White, Black or White/Black. For more details, refer to the Wiring Diagrams, in this section.

Disengage the two leads, then connect an ohmmeter or DVOM set to read resistance across the disconnected leads.

Verify the emergency tether is in place behind the Kill switch button. Check the meter, with the button released and the tether installed there should be NO continuity (continuity would shut down the powerhead if the switch was connected and the powerhead was operating).

Now, depress the kill button. The meter should indicate continuity.

Release the button (making sure the tether is still in place). The meter should now indicate NO continuity.

4-52 IGNITION AND ELECTRICAL SYSTEMS

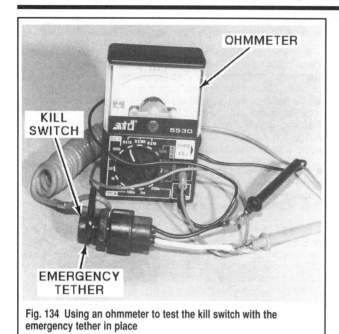

Fig. 134 Using an ohmmeter to test the kill switch with the emergency tether in place

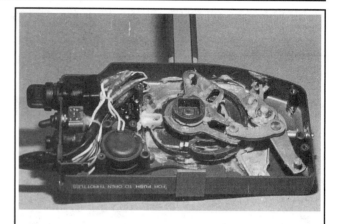

Fig. 135 The most difficult task involving testing of the remote control box components is stuffing the electrical leads back into the box in an orderly manner

Remove the emergency tether, and the meter should again indicate continuity.

All tests must be successful. If the switch fails any one test, the switch is defective and must be replaced. The switch is a one piece sealed unit and cannot be serviced.

Start Button

TESTING

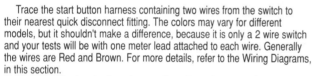

◆ See Figures 135 and 136

Trace the start button harness containing two wires from the switch to their nearest quick disconnect fitting. The colors may vary for different models, but it shouldn't make a difference, because it is only a 2 wire switch and your tests will be with one meter lead attached to each wire. Generally the wires are Red and Brown. For more details, refer to the Wiring Diagrams, in this section.

Disengage the two leads and connect an ohmmeter across the disconnected leads. Depress the start button. The meter should register continuity. Release the button and the meter should now register NO continuity. Both tests must be successful. If the tests are not successful, the start button must be replaced. The start button is a one piece sealed unit and cannot be serviced.

Neutral Safety Switch

TESTING

◆ See Figures 137, 138 and 139

The neutral safety switch on some manual start models is replaced by the no-start-in-gear protection system located at the hand rewind starter. For other models the switch is either located on the powerhead or in the control box. In either case, it is positioned where it can be physically activated by movement of the shift linkage.

On most models so equipped, the switch is a normally open switch (should exhibit no continuity when the plunger or lever is released). Therefore, it is positioned so that the shift linkage will push inward on the plunger (or downward on the lever for some models) in any position EXCEPT neutral.

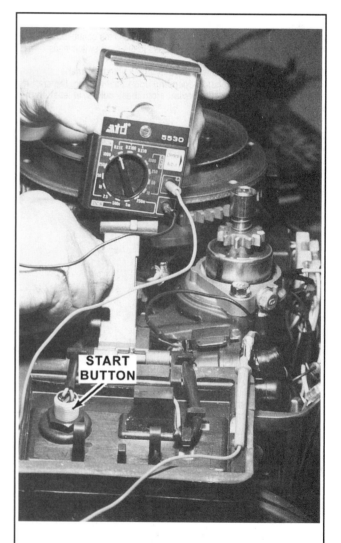

Fig. 136 Using an ohmmeter to test the a typical start button

IGNITION AND ELECTRICAL SYSTEMS 4-53

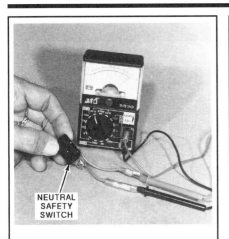

Fig. 137 Using an ohmmeter to test the neutral safety switch

Fig. 138 When powerhead mounted, the neutral safety switch is sometimes hidden, but is usually located on the shift linkage just inside the lower cowling pan (typical Mercury/Mariner shown)

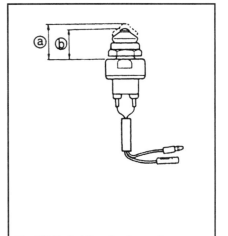

Fig. 139 Typical Yamaha plunger type neutral safety switch showing the two switch positions

Yamaha switches of the plunger design normally have a plunger free-length (measured from the switch face) of 0.77-0.81 in. (19.5-20.5mm). Once the plunger is pushed inward the switch will show continuity across both contacts and plunger length should be reduced to about 0.73-0.77 in. (18.5-19.5mm)

To test the switch, trace the neutral safety switch leads from the switch to their nearest quick disconnect fitting. Though the colors may vary, both of these leads are often Brown. Of course, it shouldn't make a difference, because it is only a 2 wire switch and your tests will be with one meter lead attached to each wire. For more details, refer to the Wiring Diagrams, in this section.

Disengage the two leads, then connect an ohmmeter or a DVOM set to read resistance across the two disconnected leads. When the shift lever is in the neutral position (the plunger or lever is free), the meter should indicate continuity.

When the lower unit is shifted into either forward or reverse gear (the plunger or lever is depressed), the meter should indicate no continuity.

The switch must pass all three tests to verify the safety aspect of the switch is functioning properly. If the switch fails any one or more of the tests, the switch must be adjusted (if the physical positioning of the switch allows this) or replaced (more likely on most Yamaha and Mercury/Mariner models).

Remember this is a safety switch. A faulty switch may allow the powerhead to be started with the lower unit in gear - an extremely dangerous situation for the boat, crew, and passengers.

Warning Buzzer/Horn

TESTING

◆ See Figure 140

The buzzer or horn is a warning device to indicate low oil pressure, an over rev. condition, or overheating of the powerhead (depending upon the model). The buzzer is usually located inside the control box (for remote units).

Remove the control box cover and identify the two leads from the buzzer, one is normally Yellow and the other is normally Pink. Of course, it shouldn't make a difference, because it is only a 2 wire component and your tests will be with one meter lead attached to each wire. For more details, refer to the Wiring Diagrams, in this section.

Disconnect the two buzzer leads at their quick disconnect fittings, and ease the buzzer out from between the four posts which anchor it in place.

Obtain a 12-volt battery. Momentarily connect the buzzer leads to the battery (generally the Yellow buzzer lead would connect to the negative battery terminal, while the Pink buzzer lead would connect to the positive terminal). As soon as the leads are connected, the buzzer should sound. If the buzzer is silent, or the sound emitted does not capture the helmsperson's attention immediately, the buzzer should be replaced. Service or adjustment is not possible. If the sound is satisfactory and immediate, install the buzzer between the four posts and connect the two leads matching color to color. Tuck the leads to prevent them from making contact with any moving parts inside the control box. Replace the cover.

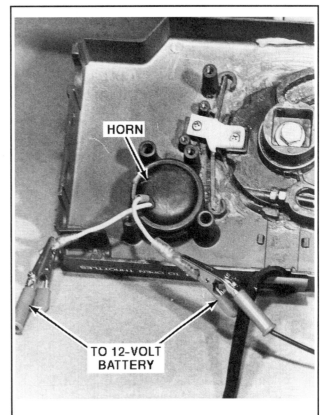

Fig. 140 The warning buzzer/horn is tested by applying 12 volts and listening for activation

4-54 IGNITION AND ELECTRICAL SYSTEMS

SPECIFICATIONS

IGNITION SYSTEM COMPONENT TESTING - YAMAHA 4-STROKE INLINE ENGINES

HP	No. of Cyl	Engine Type	Model Year	Displace cu. in. (cc)	Pulser Coils Wire leads (+)	Pulser Coils Wire leads (-)	Pulser Coils Resistance Ohms ①	Pulser Coils Peak Voltage With Load	Pulser Coils Peak Voltage No Load	Pulser Coils Peak Voltage V @ RPM	Charge Coil Wire leads (+)	Charge Coil Wire leads (-)	Charge Coil Resistance Ohms ①	Charge Coil Peak Voltage With Load	Charge Coil Peak Voltage No Load	Charge Coil Peak Voltage V @ RPM	CDI Output Wire leads (+)	CDI Output Wire leads (-)	CDI Output Peak Voltage With Load	CDI Output Peak Voltage No Load	CDI Output Peak Voltage V @ RPM	Ignition Coil Wire leads (+)	Ignition Coil Wire leads (-)	Ignition Coil Resistance Primary Ohms ①	Ignition Coil Resistance Secondary Ohms ①
6	2	IL 4-stroke	2001-04	12 (197)	W/R	B	②	3.5	7.0	7 @ 1500 ③	BR	Blue	②	154	112	185 @ 1500	O	B	140	53.2	168	O	B	0.3-0.6	6240-9360
8	2	IL 4-stroke	2001-04	12 (197)	W/R	B	②	3.5	7.0	7 @ 1500 ③	BR	Blue	②	154	112	185 @ 1500	O	B	140	53.2	168	O	B	0.3-0.6	6240-9360
8	2	IL 4-stroke	1997	14 (232)	W/R	B	165-252	5	4.0	8 @ 1500	BR	Blue	280-420	160	95	220 @ 1500	O	B	150	-	210 @ 1500	O	B	-	3845-4715
8	2	IL 4-stroke	1998-04	14 (232)	W/R	B	165-252	2.5	3.5	7.5@1500 ④	BR	Blue	280-420	90	95	205@1500 ④	O	B	85	85	195@1500 ④	O	B	-	3485-4715
9.9	2	IL 4-stroke	1995	14 (232)	W/R	B	168-252	2.5	4	7.5 @ 1500	BR	Blue	280-420	90	95	205 @ 1500	O	B	85	-	195 @ 1500	O	B	0.08-0.11	3485-4715
9.9	2	IL 4-stroke	⑫	14 (232)	W/R	B	⑬	8	8	10 @ 1500	BR	Blue	⑮	160	160	220	O	B	150	-	210 @ 1500	O	B	0.08-0.11	3485-4715
9.9	2	IL 4-stroke	⑫	14 (232)	W/R	B	⑭	5	5	7 @ 1500	BR	Blue	②	160	160	230	O	B	150	-	210 @ 1500	O	B	0.08-0.11	3485-4715
Standard	2	IL 4-stroke	1997	14 (232)	W/R	B	165-252	5	4.0	8 @ 1500	BR	Blue	280-420	160	95	220 @ 1500	O	B	150	-	210 @ 1500	O	B	-	3845-4715
High thrust	2	IL 4-stroke	1997	14 (232)	W/R	B	165-252	7	4.0	10 @ 1500	BR	Blue	280-420	160	95	230 @ 1500	O	B	150	-	210 @ 1500	O	B	-	3845-4715
Standard	2	IL 4-stroke	1998-04	14 (232)	W/R	B	165-252	2.5	3.5	7.5 @1500④	BR	Blue	280-420	90	95	205@1500④	O	B	85	85	195@1500④	O	B	-	3845-4715
High thrust	2	IL 4-stroke	1998-04	14 (232)	W/R	B	165-252	2.5	4.0	7.5@1500④	BR	Blue	280-420	150	100	205@1500④	O	B	90	95	205@1500④	O	B	-	3845-4715
15	2	IL 4-stroke	1998-04	20 (323)	W/G	B	234-348	3.5	4.0	11@1500⑦	BR	Blue	272-408	135	130	180@1500⑦	O	B	115	120	160@1500⑦	O	B	-	3928-5892
25	2	Standard	1998-04	30 (498)	R	W	260-390	90 ⑧	90 ⑧	210 @1500⑧	G/W	W/G	660-710	240 ⑧	210 ⑧	210@1500⑧	O	B/W	200 ⑧	180 ⑧	190@1500⑧	O	B	0.08-0.70	3500-4700 ⑨
25	2	High thrust	2001-03	30 (498)	R	W	260-390	6	6	17.5@1500⑨	G/W	W/G	660-710	193	189	194@1500⑨	O	B/W	101	152	123@1500⑨	O	B	-	3280-4920
30	3	IL 4-stroke	2001-04	46 (747)	R	W	240-360	6	6	18@1500⑩	G/W	W/G	528-792	193	193	194@1500⑩	O	B/W	100	151	123@1500⑩	O	B	0.17-0.25	2700-3700
40J	3	IL 4-stroke	2002-04	46 (747)	R	W	240-360	6	6	18@1500⑩	G/W	W/G	528-792	193	193	194@1500⑩	O	B/W	100	151	123@1500⑩	O	B	0.17-0.25	2700-3700
40	3	IL 4-stroke	2000-04	46 (747)	R	W	240-360	6	6	18@1500⑩	G/W	W/G	528-792	193	193	194@1500⑩	O	B/W	100	151	123@1500⑩	O	B	0.17-0.25	2700-3700
40	4	IL 4-stroke	1999-04	58 (935)	W/R	W/B	340-510	7	-	14@1500 ⑪	Br	Blue	272-408	140	-	150@1500⑪	⑫	B	105	-	100@1500⑫	O	B	0.078-0.106	3280-4944
45	4	IL 4-stroke	1996	58 (935)	W/R	W/B	326-524	7	-	⑮	Br	Blue	272-408	140	140	150 @ 1500	②	B	105	-	⑫	O	B	0.08-0.11	3280-4920
45	4	IL 4-stroke	1997-00	58 (935)	W/R	W/B	340-510	7	-	14@1500 ⑪	Br	Blue	272-408	140	-	150@1500⑪	②	B	105	-	100@1500⑪	O	B	0.078-0.106	3280-4944
50J	4	IL 4-stroke	2002	58 (935)	W/R	W/B	396-594	3.5	6.3	7.4@1500⑨	Br	Blue	272-408	137	144	169@1500⑨	②	B	150	126	151@1500⑨	O	B	0.078-0.106	3500-4700
50	4	IL 4-stroke	1995-96	58 (935)	W/R	W/B	326-524	7	-	⑲	Br	Blue	272-408	140	140	150 @ 1500	②	B	105	-	⑳	O	B	0.08-0.11	3280-4920
50	4	IL 4-stroke	1997-00	58 (935)	W/R	W/B	340-510	7	-	14@1500 ⑪	Br	Blue	272-408	140	-	150@1500⑪	②	B	105	-	100@1500⑪	O	B	0.078-0.106	3280-4944
50	4	IL 4-stroke	2001-04	58 (935)	W/R	W/B	396-594	3.5	6.3	7.4@1500⑨	Br	Blue	272-408	137	144	169@1500⑨	②	B	150	126	151@1500⑨	O	B	0.078-0.106	3500-4700
60J	4	IL 4-stroke	2003-04	61 (996)	W/R	W/B	396-596	3.5	6.3	7.4@1500⑨	Br	Blue	272-408	137	144	165@1500⑨	②	B	141	126	150@1500⑨	O	B	0.078-0.106	3500-4700
60	4	IL 4-stroke	2002-04	61 (996)	W/R	W/B	396-596	3.5	6.3	7.4@1500⑨	Br	Blue	272-408	137	144	165@1500⑨	②	B	141	126	150@1500⑨	O	B	0.078-0.106	3500-4700
75	4	IL 4-stroke	2003-04	97 (1596)	②	B	396-596	2.7	4.5	15@1500⑥	-	-	Lighting Coil	Charging Sys	in	Charging Sys	②	B	107	107	110@1500⑥	②	B	0.078-0.106	3280-4920
80	4	IL 4-stroke	1999-02	97 (1596)	②	B	396-596	2.7	4.5	15@1500⑥	-	-	Lighting Coil	Charging Sys	in	Charging Sys	②	B	107	107	110@1500⑥	B/W	B	0.078-0.106	3280-4920
90J	4	IL 4-stroke	2003-04	97 (1596)	②	B	396-596	2.7	4.5	15@1500⑥	-	-	Lighting Coil	Charging Sys	in	Charging Sys	②	B	107	107	110@1500⑥	B/W	B	0.078-0.106	3280-4920
90	4	IL 4-stroke	2003-04	97 (1596)	②	B	396-596	2.7	4.5	15@1500⑥	-	-	Lighting Coil	Charging Sys	in	Charging Sys	②	B	107	107	110@1500⑥	B/W	B	0.078-0.106	3280-4920
100	4	IL 4-stroke	1999-02	97 (1596)	②	B	396-596	2.7	4.5	15@1500⑥	-	-	Lighting Coil	Charging Sys	in	Charging Sys	②	B	107	107	110@1500⑥	B/W	B	0.078-0.106	3280-4920
115J	4	IL 4-stroke	2003-04	106 (1741)	②	B	-	3.0	3.5	26@1500⑰	-	-	Lighting Coil	Charging Sys	in	Charging Sys	②	B	122	5.0	242@1500⑰	-	-	-	-
115	4	IL 4-stroke	2000-04	106 (1741)	②	B	-	3.0	3.5	26@1500⑰	-	-	Lighting Coil	Charging Sys	in	Charging Sys	②	B	122	5.0	242@1500⑰	-	-	-	-
150	4	IL 4-stroke	2004	163 (2670)	②	B	459-561	3.6	3.5	23.9@1500㉓	-	-	Lighting Coil	Charging Sys	in	Charging Sys	㉓	B	260	-	260@1500㉓	R	B/W	1.53-2.07	12,500-16,910

IGNITION AND ELECTRICAL SYSTEMS 4-55

Because manufacturers differ with what specifications they choose to provide, please refer to the Mercury/Mariner charts for additional specifications which MAY apply to the Yamaha model in question as well as to the Yamaha model in question as well

NOTE: Unless stated otherwise, With Load tests are made with circuit connected as normal, while No Load tests are conducted with the component disconnected from the wiring harness

① Unless noted otherwise all resistance specifications are at an ambient temperature 68 degrees F (20 degrees C). Keep in mind that all resistance readings will vary with temperature and from meter-to-meter

② Specifications conflict; service literature calls for pulser coil resistance of 240-360 ohms and charge coil resistance of 232-348 ohms. However, the Yamaha tune-up guide calls for pulser coil resistance of 92-112 ohms and charge coil of 81-99 ohms

③ Specification is also a minimum of 10.5 volts with the engine running at 3500 rpm

④ Specification is 13 volts with the engine running at 3500 rpm for standard models and 12 volts @ 3500 rpm for high thrust models

⑤ Specification is 180 volts with the engine running at 3500 rpm for standard models and 210 volts @ 3500 rpm for high thrust models

⑥ Specification is 170 volts with the engine running at 3500 rpm for standard models and 195 volts @ 3500 rpm for high thrust models

⑦ Specifications for running at 3500 rpm are 23 volts for the pulser coil, 180 volts for the charge coil and 165 volts for the CDI unit

⑧ Specifications conflict; service literature lists specs above for standard models (not high thrust), however, the factory tune-up guide shows values listed in the chart as high thrust apply to the standard models also

⑨ Specifications for running at 3500 rpm are 26 volts for the pulser coil, 194 volts for the charge coil and 161 volts for the CDI unit

⑩ Specifications for running at 3500 rpm are 35 volts for the pulser coil, 194 volts for the charge coil and 161 volts for the CDI unit

⑪ Specifications for running at 3500 rpm are 20 volts for the pulser coil, 135 volts for the charge coil and 100 volts for the CDI unit

⑫ Wire colors are B/W for Cyl #'s 2 & 3, B/O for Cyl #'s 1 & 4

⑬ Specifications for running at 3500 rpm are 11.2 volts for the pulser coil, 129 volts for the charge coil and 116 volts for the CDI unit

⑭ Specifications for running at 3500 rpm are 11.2 volts for the pulser coil, 125 volts for the charge coil and 112 volts for the CDI unit

⑮ Wire colors are normally W/B for Cyl #'s 2 & 3, W/R for Cyl #'s 1 & 4

⑯ Specifications for running at 3500 rpm are 18 volts for the pulser coils and 111 volts for the CDI unit

⑰ Specifications for running at 3500 rpm are 44 volts for the pulser coils and 245 volts for the CDI unit

⑱ Wire colors are B/R for Cyl #'s 1 & 2, B/W for Cyl #'s 3 & 4

⑲ Specifications on 1996 models conflict in some OE sources. The tune-up guide list the same specifications as listed above for 1992-95 models. However, the specifications given in this row are provided in the service manual for all except T and A models

⑳ Specifications on 1996 models conflict in some OE sources. The tune-up guide list the same specifications as listed above for 1992-95 models. However, the specifications given in this row are provided in the service manual for 1996 T and A models

㉑ Specifications conflict in various factory sources, the tune up guide sasy 14 volts @ 1500 rpm and 20 @ 3500 rpm, while the service manual only lists a max of about 7.7 volts @ 1500 rpm

㉒ Wire colors are B/W for Cyl #'s 2 & 3, B/O for Cyl #'s 1 & 4

㉓ Specifications conflict in various factory sources, the tune up guide says 110 volts @ 1500 rpm and 100 @ 3500 rpm, while the service manual lists a max of about 134 volts @ 1500 rpm

㉔ Specifications for running at 3500 rpm are 49.7 volts for the pulser coils and 270 volts for the CDI unit/ECM

㉕ Specifications conflict in various factory sources, one portion of the manual says B/O-B and B/W-B, but the wiring diagram shows B/O-R/Y and B/W-R/Y

IGNITION SYSTEM COMPONENT TESTING - SINGLE CYLINDER MODELS

HP	No. of Cyl	Engine Type	Model Year	Displace cu. in. (cc)	Trigger Coil Wire leads (+)	Trigger Coil Wire leads (−)	Trigger Coil Resistance Ohms ①	Charge Coil Wire leads (+)	Charge Coil Wire leads (−)	Charge Coil Resistance Ohms ①	Ignition Coil / TCI Unit Primary Wire Leads (+)	Ignition Coil / TCI Unit Primary Wire Leads (−)	Ignition Coil / TCI Unit Primary Resistance Ohms ①	Ignition Coil / TCI Unit Secondary Resistance Ohms ①
2.5	1	Yamaha	2003-04	4.4 (72)	-	-	-	-	-	-	W	B	0.56-0.84	11600-17400
4	1	Yamaha	1999-04	6.7 (112)	-	-	-	-	-	-	W ②	B ②	0.56-0.84	11600-17400
4	1	Merc/Mar	1999-04	7.5 (123)	R/W	B	149-243	W	B/R	95-134	③	③	0.02-0.38	3000-4000 ④
5	1	Merc/Mar	1999-04	7.5 (123)	R/W	B	149-243	W	B/R	95-134	③	③	0.02-0.38	3000-4000 ④
6	1	Merc/Mar	2000-04	7.5 (123)	R/W	B	149-243	W	B/R	95-134	③	③	0.02-0.38	3000-4000 ④

NOTE: Unless stated otherwise, With Load tests are made with circuit connected as normal, while No Load tests are conducted with the component disconnected from the wiring harness

① Unless noted otherwise all resistance specifications are at an ambient temperature 68 degrees F (20 degrees C). Keep in mind that all resistance readings will vary with temperature and from meter-to-meter

② Peak TCI unit cranking output at 300 rpm is 126 volts

③ Test primary circuit directly across coil terminals

④ Specification is without Spark Plug Boot. Plug Boot is 3500-5200 ohms.

4-56 IGNITION AND ELECTRICAL SYSTEMS

IGNITION CONTROL SYSTEM COMPONENT TESTING - YAMAHA 4-STROKE INLINE ENGINES

HP	No. of Cyl	Engine Type	Model Year	Displace cu. in. (cc)	Leads (+)	Leads (-)	Resistance Ohms [1]	Power Coil Peak Voltage - Closed Circuit Crank	V @ RPM	V @ RPM	Power Coil Peak Voltage - Open Circuit Crank	V @ RPM	V @ RPM	Engine Temperature Sensor Resitance @ temp F (C) Ohms	Ohms	Ohms	Oil Pressure Switch Activate @ PSI (kPa)	Pressure Control rpm
6	2	IL 4-stroke	2001-04	12 (197)	-	-	-	-	-	-	-	-	-	-	-	-	2.13 (14.7)	1850-2150
8	2	IL 4-stroke	2001-04	12 (197)	-	-	-	-	-	-	-	-	-	-	-	-	2.13 (14.7)	1850-2150
8	2	IL 4-stroke	1997-04	14 (232)	-	-	-	-	-	-	-	-	-	-	-	-	7-9 (49-69)	-
9.9	2	IL 4-stroke	1995-04	14 (232)	-	-	-	-	-	-	-	-	-	-	-	-	7-9 (49-69)	-
15	2	IL 4-stroke	1998-04	20 (323)	-	-	-	-	-	-	-	-	-	-	-	-	2.13 (14.7)	-
25	2	IL 4-stroke	1998-04	30 (498)	-	-	-	-	-	-	-	-	-	2500 @ 41 (5)	1250 @ 68 (20)	176 @ 158 (70)	2.25 (15.5)	-
30	3	IL 4-stroke	2001-04	46 (747)	Y/B	Y/B	6.5-7.2	7	25 @ 1500	67 @ 3500	15	25 @ 1500	67 @ 3500	25000 @ 41 (5)	12000 @ 68 (20)	2000 @ 158 (70)	2.25 (15.5)	-
40J	3	IL 4-stroke	2002-04	46 (747)	Y/B	Y/B	6.5-7.2	7	25 @ 1500	67 @ 3500	15	25 @ 1500	67 @ 3500	25000 @ 41 (5)	12000 @ 68 (20)	2000 @ 158 (70)	2.25 (15.5)	-
40	3	IL 4-stroke	2000-04	46 (747)	Y/B	Y/B	6.5-7.2	7	25 @ 1500	67 @ 3500	15	25 @ 1500	67 @ 3500	25000 @ 41 (5)	12000 @ 68 (20)	2000 @ 158 (70)	2.25 (15.5)	-
40	4	IL 4-stroke	1999-00	58 (935)	-	-	-	-	-	-	-	-	-	See	Cooling	System	1.4-2.8 (10-20)	-
45	4	IL 4-stroke	1996-00	58 (935)	-	-	-	-	-	-	-	-	-	See	Cooling	System	1.4-2.8 (10-20)	-
50J	4	IL 4-stroke	2002	58 (935)	-	-	-	-	-	-	-	-	-	See	Cooling	System	7.11 (50)	-
50	4	IL 4-stroke	1995-00	58 (935)	-	-	-	-	-	-	-	-	-	See	Cooling	System	1.4-2.8 (10-20)	-
50	4	IL 4-stroke	2001-04	58 (935)	-	-	-	-	-	-	-	-	-	See	Cooling	System	7.11 (50)	-
60J	4	IL 4-stroke	2003-04	61 (996)	-	-	-	-	-	-	-	-	-	See	Cooling	System	4.2-8.4 (29-59)	-
60	4	IL 4-stroke	2002-04	61 (996)	-	-	-	-	-	-	-	-	-	See	Cooling	System	4.2-8.4 (29-59)	-
75	4	IL 4-stroke	2003-04	97 (1596)	-	-	-	-	-	-	-	-	-	5790 @ 32 (0)	2450 @ 68 (20)	1500 @ 104 (40)	2.13 (15)	-
80	4	IL 4-stroke	1999-02	97 (1596)	-	-	-	-	-	-	-	-	-	5790 @ 32 (0)	2450 @ 68 (20)	1500 @ 104 (40)	2.13 (15)	-
90J	4	IL 4-stroke	2003-04	97 (1596)	-	-	-	-	-	-	-	-	-	5790 @ 32 (0)	2450 @ 68 (20)	1500 @ 104 (40)	2.13 (15)	-
90	4	IL 4-stroke	2003-04	97 (1596)	-	-	-	-	-	-	-	-	-	5790 @ 32 (0)	2450 @ 68 (20)	1500 @ 104 (40)	2.13 (15)	-
100	4	IL 4-stroke	1999-02	97 (1596)	-	-	-	-	-	-	-	-	-	5790 @ 32 (0)	2450 @ 68 (20)	1500 @ 104 (40)	2.13 (15)	-
115J	4	IL 4-stroke	2003-04	106 (1741)	-	-	-	-	-	-	-	-	-	4620 @ 41 (5)	2440 @ 68 (20)	190 @ 212 (100)	19-25 (130-170)	-
115	4	IL 4-stroke	2000-04	106 (1741)	-	-	-	-	-	-	-	-	-	4620 @ 41 (5)	2440 @ 68 (20)	190 @ 212 (100)	19-25 (130-170)	-
150	4	IL 4-stroke	2004	163 (2670)	-	-	-	-	-	-	-	-	-	54,200-69,000 @ 68 (20)		3120-3480@212(100)	Sensor	-

NOTE: Unless stated otherwise, With Load tests are made with circuit connected as normal, while No Load tests are conducted with the component disconnected from the wiring harness

[1] Unless noted otherwise all resistance specifications are at an ambient temperature 68 degrees F. (20 degrees C). Keep in mind that all resistance readings will vary with temperature and from meter-to-meter

IGNITION SYSTEM COMPONENT TESTING - V ENGINES

HP	No. of Cyl	Engine Type	Model Year	Displace cu. in. (cc)	Pulser Coils Wire leads (+)	Wire leads (-)	Resistance Ohms [1]	With Load	No Load	Peak Voltage V @ RPM	V @ RPM	Charge Coil (Stator)	CDI Output Wire leads (+)	Wire leads (-)	With Load	No Load	Peak Voltage V @ RPM	V @ RPM	Ignition Coil Wire leads (+)	Wire leads (-)	Resistance Primary Ohms [1]	Secondary Ohms [1]
200	6	Yamaha	2002-04	205 (3352)	[1]	B	459-561	5.3	5.3	20 @ 1500	43 @ 3500	[2]	[3]	B	252	-	260 @ 1500	260 @ 3500	[4]	R/Y	1.5-1.9	[5]
225	6	Yamaha	2002-04	205 (3352)	[1]	B	459-561	5.3	5.3	20 @ 1500	43 @ 3500	[2]	[3]	B	252	-	260 @ 1500	260 @ 3500	[4]	R/Y	1.5-1.9	[5]
225	6	Merc/Mar	2003-04	205 (3352)	[1]	B	459-561	5.3	5.3	20 @ 1500	43 @ 3500	[2]	[3]	B	160	-	260 @ 1500	260 @ 3500	[4]	R/Y	1.5-2.5	[5]

NOTE: Unless stated otherwise, With Load tests are made with circuit connected as normal, while No Load tests are conducted with the component disconnected from the wiring harness

[1] There are 3 separate pulser coil windings, the leads are W/R, W/B and W/G

[2] The Charge Coil is part of the Stator Assembly and it's functions are performed by the Lighting Coil. For more details, please refer to Lighting Coil in the Charging System section

[3] The ignition is ECM controlled and output should be checked across 3 pairs of wires the B and each of the following: B/O, B/Y and B/W (some Merc diagrams depict the B/O as B/R, either way, it is going to the ignition coil).

[4] The ignition coil primary resistance is checked between R/Y and each of the following: B/O, B/Y and B/W (some Merc diagrams depict the B/O as B/R, either way, it is going to the ignition coil).

[5] The ignition coil secondary resistance is checked from secondary wire lead-to-secondary wire lead and should be 19,600-35,400 ohms

IGNITION AND ELECTRICAL SYSTEMS

IGNITION SYSTEM COMPONENT TESTING - MERCURY/MARINER 4-STROKE INLINE ENGINES

HP	No. of Cyl	Engine Type	Model Year	Displace cu. in. (cc)	Trigger (Pulser/Crankshaft Position) Coils							Charge Coil							Ignition Coil			
					Wire leads		Resistance Ohms ①	Peak Voltage ②			Wire leads		Resistance Ohms ①	Peak Voltage ②			Wire leads		Resistance Primary Ohms ①	Secondary Ohms ①
					(+)	(-)		With Load	No Load	V @ RPM	(+)	(-)		With Load	No Load	V @ RPM	(+)	(-)		
8	2	IL 4-stroke	1997-98	14 (232)	W/R	B	168-252	2.5	4	7 @ 1500	BR	Blue	280-420	90	95	185 @ 1500	O	B	0.08-0.22	13,000-16,000 ③
9.9	2	IL 4-stroke	1995-98	14 (232)	W/R	B	168-252	2.5	4	7 @ 1500	BR	Blue	280-420	90	95	185 @ 1500	O	B	0.08-0.22	13,000-16,000 ③
9.9	2	IL 4-stroke	1999-04	20 (323)	W/G	B	230-350	3.5	4.0	11@1500④	BR	Blue	270-410	135	130	180@1500④	O	B	0.16-0.24	3900-5900 ③
15	2	IL 4-stroke	1999-04	20 (323)	W/G	B	230-350	3.5	4.0	11@1500④	BR	Blue	270-410	135	130	180@1500④	O	B	0.16-0.24	3900-5900 ③
25	2	IL 4-stroke	1997-04	30 (498)	R	W	300-350	6	6	17.5@1500⑤	G/W	W/G	660-710	193	189	194@1500⑤	O	B	0.08-0.70	3500-4700 ③
30	3	Carb	1999-04	46 (747)	R	W	300-350	6	6	18@1500⑥	G/W	W/G	660-710	193	193	194@1500⑥	B/W	B	0.08-0.70⑦	3500-4700 ⑦
	3	EFI	2002-04	46 (747)	R	W	300-350	-	-	-	-	-	-	-	-	-	⑧	⑧	⑧	3000-7000 ⑧
40	3	Carb	1999-04	46 (747)	R	W	300-350	6	6	18@1500⑥	G/W	W/G	660-710	193	193	194@1500⑥	B/W	B	0.08-0.70⑦	3500-4700 ⑦
	3	EFI	2002-04	46 (747)	R	W	300-350	-	-	-	-	-	-	-	-	-	⑧	⑧	⑧	3000-7000 ⑧
40	4	IL 4-stroke	1997-00	58 (935)	W/R	W/B	396-594	3.5	6.3	7.4@1500⑨	Br	Blue	272-408	137	144	169@1500⑨	O	B	0.1-0.7	3500-4700 ③
45	4	IL 4-stroke	1995-00	58 (935)	W/R	W/B	396-594	3.5	6.3	7.4@1500⑨	Br	Blue	272-408	137	144	169@1500⑨	O	B	0.1-0.7	3500-4700 ③
50	4	IL 4-stroke	1995-00	58 (935)	W/R	W/B	396-594	3.5	6.3	7.4@1500⑨	Br	Blue	272-408	137	144	169@1500⑨	O	B	0.1-0.7	3500-4700 ③
40	4	Carb	2001	61 (996)	R	W	300-350	6	6	18@1500⑥	G/W	W/G	660-710	193	193	194@1500⑥	B/W	B	0.08-0.70	3500-4700 ③
	4	EFI	2002-04	61 (996)	R	W	300-350	-	-	-	-	-	-	-	-	-	⑧	⑧	⑧	3000-7000 ⑧
50	4	Carb	2001	61 (996)	R	W	300-350	6	6	18@1500⑥	G/W	W/G	660-710	193	193	194@1500⑥	B/W	B	0.08-0.70	3500-4700 ③
	4	EFI	2002-04	61 (996)	R	W	300-350	-	-	-	-	-	-	-	-	-	⑧	⑧	⑧	3000-7000 ⑧
60	4	Carb	2001	61 (996)	R	W	300-350	6	6	18@1500⑥	G/W	W/G	660-710	193	193	194@1500⑥	B/W	B	0.08-0.70	3500-4700 ③
	4	EFI	2002-04	61 (996)	R	W	300-350	-	-	-	-	-	-	-	-	-	⑧	⑧	⑧	3000-7000 ⑧
75	4	IL 4-stroke	2000-04	97 (1596)	⑩	B	445-545	-	-	⑪	W	W	0.32-0.48	-	-	⑫	B/W	B	0.078-0.106	3500-4700 ⑯
90	4	IL 4-stroke	2000-04	97 (1596)	⑩	B	445-545	-	-	⑪	W	W	0.32-0.48	-	-	⑫	B/W	B	0.078-0.106	3500-4700 ⑯
115	4	EFI	2001-04	106 (1741)	⑩	B	-	-	-	⑬	W	W	0.20-0.80	-	-	⑭	B/W	R/Y	1.8-2.6	⑮

Because manufacturers differ with what specifications they choose to provide, please refer to the Yamaha charts for additional specificiations which MAY apply to the Mercury/Mariner model in question as well

NOTE: Unless stated otherwise, With Load tests are made with circuit connected as normal, while No Load tests are conducted with the component disconnected from the wiring harness

① Unless noted otherwise all resistance specifications are at an ambient temperature 68 degrees F (20 degrees C). Keep in mind that all resistance readings will vary with temperature and from meter-to-meter

② Specs not always provided by Mercury. These are from the Yamaha for the same models, use as reference (additional specs available for some models in the Mercury/Mariner Charge Coil Output chart)

③ Specification is across high tension lead with the appox 5,000 ohm resistor cap REMOVED for test

④ Specifications for running at 3500 rpm are 23 volts for the pulser coil, 180 volts for the charge coil and 165 volts for the CDI unit

⑤ Yamaha specifications for running at 3500 rpm are 26 volts for the pulser coil, 194 volts for the charge coil and 161 volts for the CDI unit

⑥ Yamaha specifications for running at 3500 rpm are 35 volts for the pulser coil, 194 volts for the charge coil and 161 volts for the CDI unit

⑦ Specifications in the Merc manuals conflict. Specs in chart are found in reference charts, while table in the Ignition chapter lists Primary resistance as 0.18-0.24 and secondary resistance as 2720-3680 ohms

⑧ Specifications and wire colors vary with the test - Internal Shielding (Pin A & Mounting Braket) 0-10,000 ohms, Electronic Spark Trigger (Pin B & Pin C) 8,500-12,000 ohms, Secondary test spec listed is between Pin A & Coil Tower), High Tension Lead/Boot 600-1,100 ohms

⑨ Yamaha specifications for running at 3500 rpm are 11.2 volts for the pulser coil, 129 volts for the charge coil and 116 volts for the CDI unit

⑩ Specs on wire colors disagree. For all Yamaha and 115 Mercs sources say normally W/B for Cyl #'s 2 & 3, W/R for Cyl #'s 1 & 4, except the Merc manual for the 75/90 switches those wires/cylinders

⑪ Mercury specifications are 2.8-3.4 volts at 400 rpm (cranking), 6.5-7.8 volts at 1500 rpm, and 10.5-12.0 volts at 3500 rpm

⑫ Mercury specifications are 6-9 volts at 400 rpm (cranking), 12-18 volts at 1500 rpm, and 14-35 volts at 3500 rpm, ALSO Merc gives CDI output voltages over the B/O to B wires (Cyl #'s 1 & 4) and B/W to B wires (Cyl #'s 2 & 3) should be 165-190 V cranking, or 175-200 running 1500-3500 rpm

⑬ Mercury specifications are 3.0-6.3 volts at 400 rpm (cranking), 9-16 volts at 750 rpm (idle), 18-28 volts at 1500 rpm, and 35-55 volts at 3500 rpm

⑭ Mercury specifications are 10-18 volts at 400 rpm (cranking), 16-24 volts at 750-3500 rpm, ALSO Merc gives ECM/CDI output voltages over the appropriate test harness B/W wire to ground for each pair of cylinders should be 180-320 V @ 400 rpm (cranking), 180-235 V @ 759 rpm (idle), 230-290 @ 1500 rpm or 280-340 @ 3500 rpm

⑮ Resistance is tested from end of high tension lead (spark plug book)-to-end of high tension lead. Specs however differ by cylinder pairs. 18,970-35,230 for Cyl #'s 1 & 4 or 18,550-34,450 for Cyl #'s 2 & 3

⑯ Resistance is tested from spark plug lead tower-to-tower on the coil itself, with the high-tension wires removed

4-58 IGNITION AND ELECTRICAL SYSTEMS

CHARGE & PRIMARY COIL OUTPUT TESTING - MERCURY/MARINER 4-STROKE INLINE ENGINES

HP	No. of Cyl	Engine Type	Model Year	Displace cu. in. (cc)	Charge Coil Wire leads (+)	Charge Coil Wire leads (-)	Charge Coil Peak Voltage V @ RPM	Primary Ignition Coil Wire leads (+)	Primary Ignition Coil Wire leads (-)	Primary Ignition Coil Peak Voltage V @ RPM
4	1	Merc/Mar	1999-04	7.5 (123)	B/R	Ground	150-325 @ 300-2000	Ground	B/Y	120-300 @ 300-2000
5	1	Merc/Mar	1999-04	7.5 (123)	B/R	Ground	150-325 @ 300-2000	Ground	B/Y	120-300 @ 300-2000
6	1	Merc/Mar	2000-04	7.5 (123)	B/R	Ground	150-325 @ 300-2000	Ground	B/Y	120-300 @ 300-2000
8	2	IL 4-stroke	1997-98	14 (232)	Br	Ground	210-340 @ 300-3000	B	O	200-340 @ 300-3000
9.9	2	IL 4-stroke	1995-98	14 (232)	Br	Ground	210-340 @ 300-3000	B	O	200-340 @ 300-3000
9.9	2	IL 4-stroke	1999-04	20 (323)	Br	Ground	230-280 @ 300-3000	B	O	200-280 @ 300-3000
	2	IL 4-stroke	1999-04	20 (323)	Blue	Ground	20-45 @ 300-3000	-	-	-
15	2	IL 4-stroke	1999-04	20 (323)	Br	Ground	230-280 @ 300-3000	B	O	200-280 @ 300-3000
	2	IL 4-stroke	1999-04	20 (323)	Blue	Ground	20-45 @ 300-3000	-	-	-
25	2	IL 4-stroke	1997-04	30 (498)	G/W	Ground	250-330 @ 300-3000	B	O	250-320 @ 300-3000
	2	IL 4-stroke	1997-04	30 (498)	W/G	Ground	250-330 @ 300-3000	-	-	-
30	3	Carb	1999-04	46 (747)	G/W	Ground	250-330 @ 300-3000	B	O	250-320 @ 300-3000
	3	Carb	1999-04	46 (747)	W/G	Ground	250-330 @ 300-3000	-	-	-
	3	EFI	2002-04	46 (747)	-	-	-	-	-	-
40	3	Carb	1999-04	46 (747)	G/W	Ground	250-330 @ 300-3000	B	O	250-320 @ 300-3000
	3	Carb	1999-04	46 (747)	W/G	Ground	250-330 @ 300-3000	-	-	-
	3	EFI	2002-04	46 (747)	-	-	-	-	-	-
40	4	IL 4-stroke	1997-00	58 (935)	Br	Ground	160-280 @ 300-3000	B	O	160-280 @ 300-3000
45	4	IL 4-stroke	1995-00	58 (935)	Br	Ground	160-280 @ 300-3000	B	O	160-280 @ 300-3000
50	4	IL 4-stroke	1995-00	58 (935)	Br	Ground	160-280 @ 300-3000	B	O	160-280 @ 300-3000
40	4	Carb	2001	61 (996)	G/W	Ground	250-330 @ 300-3000	B	O	250-320 @ 300-3000
	4	Carb	2001	61 (996)	W/G	Ground	250-330 @ 300-3000	-	-	-
	4	EFI	2002-04	61 (996)	-	-	-	-	-	-
50	4	Carb	2001	61 (996)	G/W	Ground	250-330 @ 300-3000	B	O	250-320 @ 300-3000
	4	Carb	2001	61 (996)	W/G	Ground	250-330 @ 300-3000	-	-	-
	4	EFI	2002-04	61 (996)	-	-	-	-	-	-
60	4	Carb	2001	61 (996)	G/W	Ground	250-330 @ 300-3000	B	O	250-320 @ 300-3000
	4	Carb	2001	61 (996)	W/G	Ground	250-330 @ 300-3000	-	-	-
	4	EFI	2002-04	61 (996)	-	-	-	-	-	-
75	4	IL 4-stroke	2000-04	97 (1596)	W/B	W	165-190 V Cranking	W/B	W	170-200 @ 1500-3500
	4	IL 4-stroke	2000-04	97 (1596)	W/G	W	165-190 V Cranking	W/G	W	170-200 @ 1500-3500
90	4	IL 4-stroke	2000-04	97 (1596)	W/G	W	165-190 V Cranking	W/G	W	170-200 @ 1500-3500
	4	IL 4-stroke	2000-04	97 (1596)	W/G	W	165-190 V Cranking	W/G	W	170-200 @ 1500-3500
115	4	EFI	2001-04	106 (1741)	B/W	Ground	180-320 V Cranking	B/W	Ground	180-235 @ idle
	4	EFI	2001-04	106 (1741)	B/W	Ground	230-290 @ 1500	B/W	Ground	280-340 @ 3500

Because manufacturers differ with what specifications they choose to provide, please refer to the Yamaha charts for additional specificitions which MAY apply to the Mercury/Mariner model in question as well

① Unless readings will vary with temperature and from meter-to-meter

IGNITION AND ELECTRICAL SYSTEMS

CHARGING SYSTEM TESTING - Yamaha 4-Stroke Motors

Model (Hp)	No. of Cyl	Engine Type	Year	Disp cu. in. (cc)	Wire Colors	Lighting Coil Resistance Ohms ①	Voltage @ RPM Minimum	Voltage @ RPM Peak
6	2	1 lighting coil	2001-04	12 (197)	G - G	0.24-0.36	3.5 cranking	14 @ 1500/33 @ 3500
	2	Dual light coil	2001-04	12 (197)	G - G	0.61-0.91	8.4 cranking	33 @ 1500/77 @ 3500
	2	Rectifier output	2001-04	12 (197)	R - B	-	7.7 cranking (open)	27 @ 1500-3500
8	2	1 lighting coil	2001-04	12 (197)	G - G	0.24-0.36	3.5 cranking	14 @ 1500/33 @ 3500
	2	Dual light coil	2001-04	12 (197)	G - G	0.61-0.91	8.4 cranking	33 @ 1500/77 @ 3500
	2	Rectifier output	2001-04	12 (197)	R - B	-	7.7 cranking (open)	27 @ 1500-3500
8	2	IL 4-stroke	1997	14 (232)	G - G	-	8 cranking	11 @ 1500
	2	IL 4-stroke	1998-04	14 (232)	G - G	-	7.5-8 cranking	30 @ 1500/65 @ 3500
9.9	2	Standard	1995	14 (232)	G - G	0.73-1.09	-	-
	2	High thrust	1996	14 (232)	G-G	0.91-1.37	-	-
	2	High thrust	1996	14 (232)	G-G	0.96-1.44	-	-
	2	Standard	1996-97	14 (232)	G - G	-	8 cranking	11 @ 1500
	2	High thrust	1996-97	14 (232)	G - G	-	8 cranking	10 @ 1500
	2	Standard	1998-04	14 (232)	G - G	-	7.5-8 cranking	30 @ 1500/60 @ 3500
	2	High thrust	1998-04	14 (232)	G - G	-	9 cranking	35 @ 1500/75 @ 3500
	2	High thrust	1998-04	14 (232)	W/G - B	-	8 cranking	30 @ 1500/65 @ 3500
15	2	10 Amp	1998-04	20 (323)	G - G/W	0.24-0.36	6.0-6.5 cranking	21 @ 1500/46 @ 3500
	2	6 Amp	1998-04	20 (323)	G - G	0.48-0.72	6.0-6.5 cranking	21 @ 1500/46 @ 3500
	2	Rectifier output	1998-04	20 (323)	R - B	-	6.0 cranking	20 @ 1500/45 @ 3500
25	2	Standard	1998-04	30 (498)	Y - Y	0.22-0.24	10 cranking	25 @ 1500/65 @ 3500
	2	High thrust	2001-04	30 (498)	Y - Y	0.22-0.24	8 cranking	26 @ 1500/60 @ 3500
		Rectifier/standard	1998-04	30 (498)	R - B	-	9 cranking ②	35@1500/15@3500 ②
	2	Rectifier/H thrust	2001-04	30 (498)	R - B	-	9 cranking ③	12.5 @ 1500-3500
30	3	6 amp	2001-04	46 (747)	Y - Y	0.9-1.1	8 cranking	25@1500/59@3500 ④
	3	15 amp	2001-04	46 (747)	Y - Y	0.25-0.27	8 cranking	25@1500/59@3500 ④
	3	Rectifier output	2001-04	46 (747)	-	-	8 cranking	12.5 @ 1500-3500
40J	3	6 amp	2001-04	46 (747)	Y - Y	0.9-1.1	8 cranking	25@1500/59@3500 ④
	3	15 amp	2001-04	46 (747)	Y - Y	0.25-0.27	8 cranking	25@1500/59@3500 ④
	3	Rectifier output	2002-04	46 (747)	-	-	8 cranking	12.5 @ 1500-3500
40	3	6 amp	2001-04	46 (747)	Y - Y	0.9-1.1	8 cranking	25@1500/59@3500 ④
	3	15 amp	2001-04	46 (747)	Y - Y	0.25-0.27	8 cranking	25@1500/59@3500 ④
	3	Rectifier output	2000-04	46 (747)	-	-	8 cranking	12.5 @ 1500-3500
40	4	IL 4-stroke	1999-00	58 (935)	G - G	1.2-1.8	10 cranking	16 @ 1500/19 @ 3500
	4	Rectifier output	1999-00	58 (935)	R - B	-	9.5 cranking closed	15 @ 1500/15 @ 3500
45	4	IL 4-stroke	1996-97	58 (935)	G - G	1.2-1.8	-	about 8.9 @ 1500
	4	Rectifier output	1996-97	58 (935)	R - B	-	-	15 @ 1500/15 @ 3500
	4	IL 4-stroke	1998-00	58 (935)	G - G	1.2-1.8	10 cranking	16 @ 1500/19 @ 3500
	4	Rectifier output	1998-00	58 (935)	R - B	-	9.5 cranking closed	15 @ 1500/15 @ 3500
50J	4	IL 4-stroke	2002	58 (935)	G - G	1.2-1.8	11.9 cranking ⑤	42@1500/127@3500 ⑤
	4	Rectifier output	2002	58 (935)	R - B	-	-	18.9 @ 1500/19.5 @ 3500
50	4	IL 4-stroke	1996-97	58 (935)	G - G	1.2-1.8	-	about 8.9 @ 1500
	4	Rectifier output	1996-97	58 (935)	R - B	-	-	15 @ 1500/15 @ 3500
	4	IL 4-stroke	1998-00	58 (935)	G - G	1.2-1.8	10 cranking	16 @ 1500/19 @ 3500
	4	Rectifier output	1998-00	58 (935)	R - B	-	9.5 cranking closed	15 @ 1500/15 @ 3500
	4	IL 4-stroke	2001-04	58 (935)	G - G	1.2-1.8	11.9 cranking ⑤	42@1500/127@3500 ⑤
	4	Rectifier output	2001-04	58 (935)	R - B	-	-	18.9 @ 1500/19.5 @ 3500

4-60 IGNITION AND ELECTRICAL SYSTEMS

CHARGING SYSTEM TESTING - Yamaha 4-Stroke Motors

Model (Hp)	No. of Cyl	Engine Type	Year	Disp cu. in. (cc)	Wire Colors	Lighting Coil Resistance Ohms ①	Voltage @ RPM Minimum	Voltage @ RPM Peak
60J	4	IL 4-stroke	2003-04	61 (996)	G - G	1.2-1.8	14 cranking ⑤	38@1500/86@3500 ⑤
	4	Rectifier output	2003-04	61 (996)	R - B	-	-	22@1500/27@3500 ⑤
60	4	Standard	2002-04	61 (996)	G - G	1.2-1.8	14 cranking ⑤	38@1500/86@3500 ⑤
	4	High thrust	2003-04	61 (996)	G - G	1.2-1.8	14 cranking ⑤	45@1500/90@3500 ⑤
	4	Rectifier output	2002-04	61 (996)	R - B	-	-	22@1500/27@3500 ⑥
75	4	IL 4-stroke	2003-04	97 (1596)	W - W	0.32-0.48	7.1 cranking open	15 @ 1500/18 @ 3500
	4	Rectifier output	2003-04	97 (1596)	R - B	-	-	14 @ 1500/18 @ 3500
80	4	IL 4-stroke	1999-02	97 (1596)	W - W	0.32-0.48	7.1 cranking open	15 @ 1500/18 @ 3500
	4	Rectifier output	1999-02	97 (1596)	R - B	-	-	14 @ 1500/18 @ 3500
90J	4	IL 4-stroke	2003-04	97 (1596)	W - W	0.32-0.48	7.1 cranking open	15 @ 1500/18 @ 3500
	4	Rectifier output	2003-04	97 (1596)	R - B	-	-	14 @ 1500/18 @ 3500
90	4	IL 4-stroke	2003-04	97 (1596)	W - W	0.32-0.48	7.1 cranking open	15 @ 1500/18 @ 3500
	4	Rectifier output	2003-04	97 (1596)	R - B	-	-	14 @ 1500/18 @ 3500
100	4	IL 4-stroke	1999-02	97 (1596)	W - W	0.32-0.48	7.1 cranking open	15 @ 1500/18 @ 3500
	4	Rectifier output	1999-02	97 (1596)	R - B	-	-	14 @ 1500/18 @ 3500
115J	4	IL 4-stroke	2003-04	106 (1741)	W - W	-	7.4 cranking open⑦	37@1500/89@3500 ⑤
	4	Rectifier output	2003-04	106 (1741)	R - B	-	-	12.5 @ 1500/13 @ 3500
115	4	IL 4-stroke	2000-04	106 (1741)	W - W	-	7.4 cranking open⑦	37@1500/89@3500 ⑤
	4	Rectifier output	2000-04	106 (1741)	R - B	-	-	12.5 @ 1500/13 @ 3500
150	4	IL 4-stroke	2004	163 (2670)	G - G	0.20-0.30	12 cranking ⑤	50@1500/110@3500 ⑤
	4	Rectifier output	2004	163 (2670)	R - B	-	-	13 @ 1500-3500 ⑤
200	6	60 V 4-stroke	2002-04	205 (3352)	G - G	0.24-0.41	10 cranking ⑤	42@1500/93@3500 ⑤
	6	60 V 4-stroke	2002-04	205 (3352)	G/W - G/W	0.21-0.30	9 cranking ⑤	34@1500/78@3500 ⑤
	6	Rectifier output	2002-04	205 (3352)	R-B, R/Y-B	-	-	13 @ 1500-3500 ⑤
225	6	60 V 4-stroke	2002-04	205 (3352)	G - G	0.24-0.41	10 cranking ⑤	42@1500/93@3500 ⑤
	6	60 V 4-stroke	2002-04	205 (3352)	G/W - G/W	0.21-0.30	9 cranking ⑤	34@1500/78@3500 ⑤
	6	Rectifier output	2002-04	205 (3352)	R-B, R/Y-B	-	-	13 @ 1500-3500 ⑤

Because manufacturers differ with what specifications they choose to provide, please refer to the Mercury/Mariner charts for additional specificitions which MAY apply to the Yamaha model in question as well

① Unless noted otherwise all resistance specifications are at an ambient temperature 68 degrees F (20 degrees C). Keep in mind that all resistance readings will vary with temperature and from meter-to-meter

② Specifications are for open circuit, with circuit complete all readings should be 12 volts

③ Specifications are for open circuit, with circuit complete the reading should be 12.5 volts

④ Specifications are for open circuit, with circuit complete both readings should be 12.5 volts

⑤ All specifications are for open circuit only

⑥ Specifications are for open circuit only on standard models, specs for open circuit testing on high thrust models are 20 volts @ 1500 and 25 volts @ 3500

IGNITION AND ELECTRICAL SYSTEMS 4-61

CHARGING SYSTEM TESTING - Mercury/Mariner 4-Stroke Motors

Model (Hp)	No. of Cyl	Engine Type	Year	Disp cu. in. (cc)	Wire Colors	Resistance Ohms ①	Lighting Coil Amperage Output @ RPM Minimum	Max (At/Near Peak)
4	1	Single 4-st	1999-04	7.5 (123)	Y/R - Y/R	0.31-0.47	-	-
5	1	Single 4-st	1999-04	7.5 (123)	Y/R - Y/R	0.31-047	-	-
6	1	Single 4-st	2000-04	7.5 (123)	Y/R - Y/R	0.31-047	-	-
8	2	6-amp	1997-98	14 (232)	G - G	0.73-1.09	1 amp @ idle	5.91 amps @ 4000
	2	6-amp	1997-98	14 (232)	W/G - B	0.96-1.44	1 amp @ idle	5.91 amps @ 4000
	2	10-amp	1997-98	14 (232)	G - G	0.91-1.37	2.4 amps @ idle	8.3 amps @ 4000
	2	10-amp	1997-98	14 (232)	W/G - B	0.96-1.44	2.4 amps @ idle	8.3 amps @ 4000
9.9	2	6-amp	1995-98	14 (232)	G - G	0.73-1.09	1 amp @ idle	5.91 amps @ 4000
	2	6-amp	1995-98	14 (232)	W/G - B	0.96-1.44	1 amp @ idle	5.91 amps @ 4000
	2	10-amp	1995-98	14 (232)	G - G	0.91-1.37	2.4 amps @ idle	8.3 amps @ 4000
	2	10-amp	1995-98	14 (232)	W/G - B	0.96-1.44	2.4 amps @ idle	8.3 amps @ 4000
9.9	2	6-amp	1999-04	20 (323)	G - G	0.48-0.72	2 amps @ idle	5.5 amps @ 5000
	2	10-amp	1999-04	20 (323)	G - G/W	0.24-0.36	1 amp @ idle	9.1 amps @ 5000
15	2	6-amp	1999-04	20 (323)	G - G	0.48-0.72	2 amps @ idle	5.5 amps @ 5000
	2	10-amp	1999-04	20 (323)	G - G/W	0.24-0.36	1 amp @ idle	9.1 amps @ 5000
25	2	6-amp	1997-05	30 (498)	Y - Y	0.9-1.1	1-2 amps @ idle	6.2 amps @ 5000
	2	6-amp	1997-05	30 (498)	Y/B - Y/B	6.7-7.1	-	-
	2	15-amp	1997-05	30 (498)	Y - Y	0.22-0.24	1-2 amps @ idle	15 amps @ 4000
	2	15-amp	1997-05	30 (498)	Y/B - Y/B	6.7-7.1	-	-
30 Carb	3	6-amp	1999-04	46 (747)	Y - Y	0.9-1.1	1-2 amps @ idle	6.2 amps @ 5000
	3	6-amp	1999-04	46 (747)	Y/B - Y/B	6.7-7.1	-	-
	3	15-amp	1999-04	46 (747)	Y - Y	0.22-0.24	1-2 amps @ idle	15 amps @ 4000
	3	15-amp	1999-04	46 (747)	Y/B - Y/B	6.7-7.1	-	-
30 EFI	3	IL 4-stroke	2002-04	46 (747)	Y - Y	0.20-0.30	1-2 amps @ idle	19 amps @ 4000
40 Carb	3	6-amp	1999-04	46 (747)	Y - Y	0.9-1.1	1-2 amps @ idle	6.2 amps @ 5000
	3	6-amp	1999-04	46 (747)	Y/B - Y/B	6.7-7.1	-	-
	3	15-amp	1999-04	46 (747)	Y - Y	0.22-0.24	1-2 amps @ idle	15 amps @ 4000
	3	15-amp	1999-04	46 (747)	Y/B - Y/B	6.7-7.1	-	-
40 EFI	3	IL 4-stroke	2002-04	46 (747)	Y - Y	0.20-0.30	1-2 amps @ idle	19 amps @ 4000
40	4	IL 4-stroke	1997-00	58 (935)	G - G	1.2-3.2	4.2 amps @ idle	8.6 amps @ 5000
45	4	IL 4-stroke	1995-00	58 (935)	G - G	1.2-3.2	4.2 amps @ idle	8.6 amps @ 5000
50	4	IL 4-stroke	1995-00	58 (935)	G - G	1.2-3.2	4.2 amps @ idle	8.6 amps @ 5000
40 Carb	4	6-amp	2001	61 (996)	Y - Y	0.9-1.1	1-2 amps @ idle	6.2 amps @ 5000
	4	6-amp	2001	61 (996)	Y/B - Y/B	6.7-7.1	-	-
	4	15-amp	2001	61 (996)	Y - Y	0.22-0.24	1-2 amps @ idle	15 amps @ 4000
	4	15-amp	2001	61 (996)	Y/B - Y/B	6.7-7.1	-	-
40 EFI	4	IL 4-stroke	2002-04	61 (996)	Y - Y	0.20-0.30	1-2 amps @ idle	19 amps @ 4000
50 Carb	4	IL 4-stroke	2001	61 (996)	Y - Y	0.9-1.1	1-2 amps @ idle	6.2 amps @ 5000
	4	6-amp	2001	61 (996)	Y/B - Y/B	6.7-7.1	-	-
	4	15-amp	2001	61 (996)	Y - Y	0.22-0.24	1-2 amps @ idle	15 amps @ 4000
	4	15-amp	2001	61 (996)	Y/B - Y/B	6.7-7.1	-	-
50 EFI	4	IL 4-stroke	2002-04	61 (996)	Y - Y	0.20-0.30	1-2 amps @ idle	19 amps @ 4000

4-62 IGNITION AND ELECTRICAL SYSTEMS

CHARGING SYSTEM TESTING - Mercury/Mariner 4-Stroke Motors

Model (Hp)	No. of Cyl	Engine Type	Year	Disp cu. in. (cc)	Wire Colors	Resistance Ohms ①	Lighting Coil Amperage Output @ RPM Minimum	Max (At/Near Peak)
60 Carb	4	IL 4-stroke	2001	61 (996)	Y - Y	0.9-1.1	1-2 amps @ idle	6.2 amps @ 5000
	4	6-amp	2001	61 (996)	Y/B - Y/B	6.7-7.1	-	-
	4	15-amp	2001	61 (996)	Y - Y	0.22-0.24	1-2 amps @ idle	15 amps @ 4000
	4	15-amp	2001	61 (996)	Y/B - Y/B	6.7-7.1	-	-
60 EFI	4	IL 4-stroke	2002-04	61 (996)	Y - Y	0.20-0.30	1-2 amps @ idle	19 amps @ 4000
75	4	IL 4-stroke	2000-04	97 (1596)	W - W	0.32-0.48	6-9 @ 400 (cranking)	12-18 @ 1500/14-25 @ 3500
90	4	IL 4-stroke	2000-04	97 (1596)	W - W	0.32-0.48	6-9 @ 400 (cranking)	12-18 @ 1500/14-25 @ 3500
115 EFI	4	IL 4-stroke	2001-04	106 (1741)	W - W	0.20-0.80	11 amps @ idle	22 amps @ 6000
225	6	60 V 4-stroke	2003-04	205 (3352)	G - G	0.10-0.60	8 volts cranking ②	42V@1500/93V@3500 ②
	6	60 V 4-stroke	2003-04	205 (3352)	G/W - G/W	0.10-0.50	7 volts cranking ②	34V@1500/78V@3500 ②

Because manufacturers differ with what specifications they choose to provide, please refer to the Yamaha charts for additional specifications which MAY apply to the Mercury/Mariner model in question as well

① Unless noted otherwise all resistance specifications are at an ambient temperature 68 degrees F (20 degrees C). Keep in mind that all resistance readings will vary with temperature and from meter-to-meter

② All specifications are for Voltage reading on an open circuit only. Amperage tests for this model should show 18-28 amps @ idle and 40-41 amps by 3000 rpm

WIRING DIAGRAM INDEX

The following wiring diagrams represent the most common models of outboard engines covered in this manual. Models with accessory options may not be depicted here.

Mercury/Mariner

4/5/6 hp models	4-63
8/9.9 hp (232cc) models	4-65
9.9/15 hp (323cc) models	4-72
25 hp models	4-78
30/40 hp models	4-86
40/45/50 hp (935cc) models	4-92
40/50/50 hp (996cc) models	4-97
75/90 hp models	4-102
115 hp models	4-106
225 hp models	4-112

Yamaha

2.5 hp and 4 hp models	4-63
6/8 hp models	4-64
8/9.9 hp (232cc) models	4-66
9.9/15 hp (323cc) models	4-76
25 hp models	4-82
30/40J/40 hp models	4-91
45/50 hp (935cc) models	4-93
40/45/50J/50 hp (935cc) models	4-94
60J/60 hp models	4-102
75/80/90J/90/100 hp models	4-105
115J/115 hp models	4-109
150 hp models	4-110
200/225 hp EFI models	4-111
200/225 hp models	4-114

IGNITION AND ELECTRICAL SYSTEMS 4-63

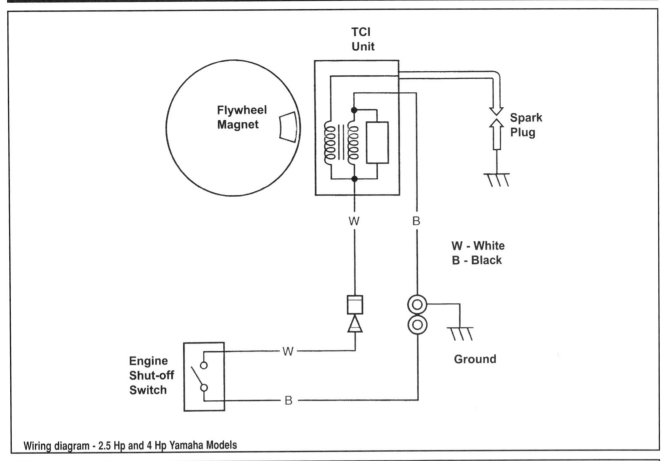

Wiring diagram - 2.5 Hp and 4 Hp Yamaha Models

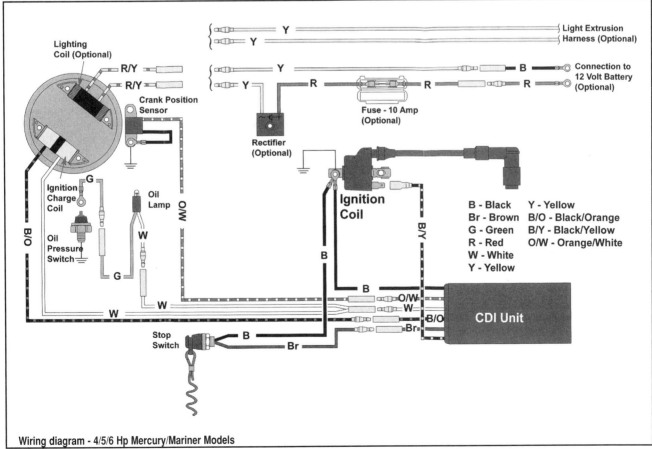

Wiring diagram - 4/5/6 Hp Mercury/Mariner Models

4-64 IGNITION AND ELECTRICAL SYSTEMS

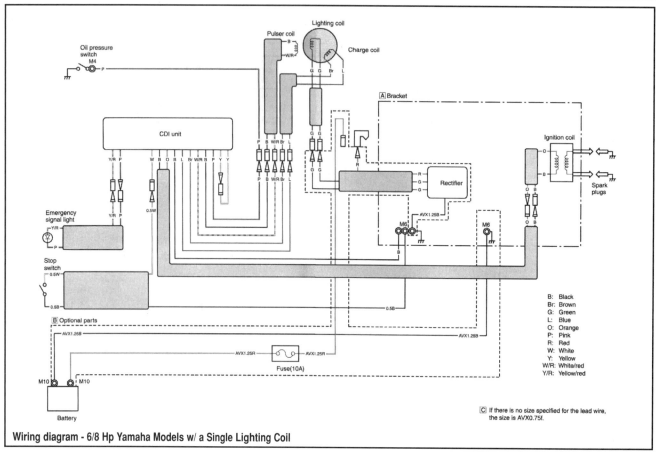

Wiring diagram - 6/8 Hp Yamaha Models w/ a Single Lighting Coil

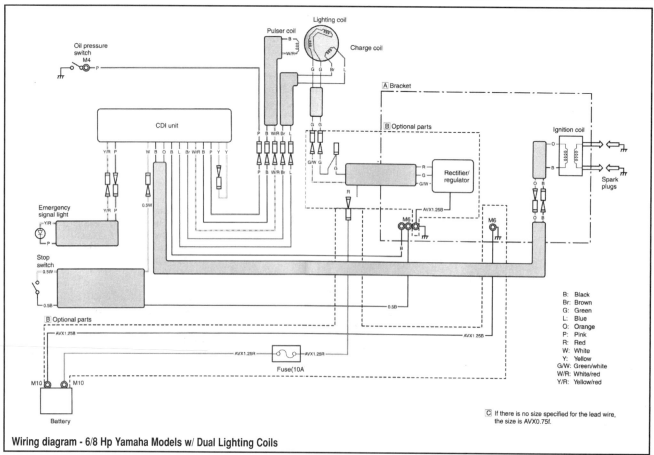

Wiring diagram - 6/8 Hp Yamaha Models w/ Dual Lighting Coils

IGNITION AND ELECTRICAL SYSTEMS

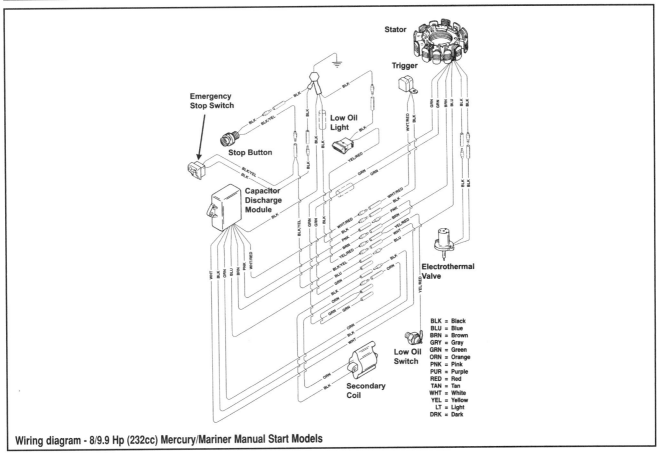

Wiring diagram - 8/9.9 Hp (232cc) Mercury/Mariner Manual Start Models

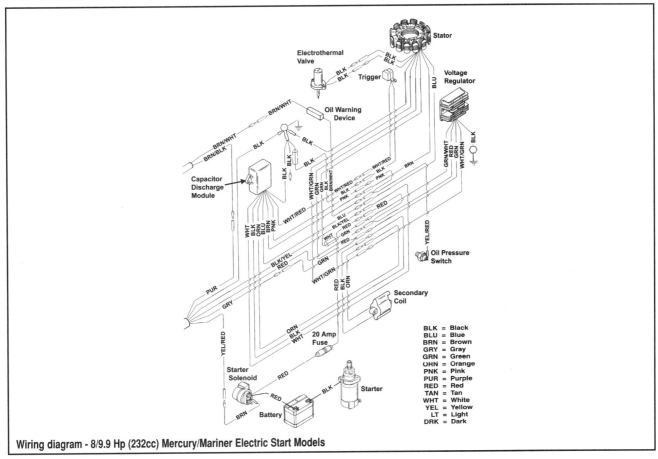

Wiring diagram - 8/9.9 Hp (232cc) Mercury/Mariner Electric Start Models

4-66 IGNITION AND ELECTRICAL SYSTEMS

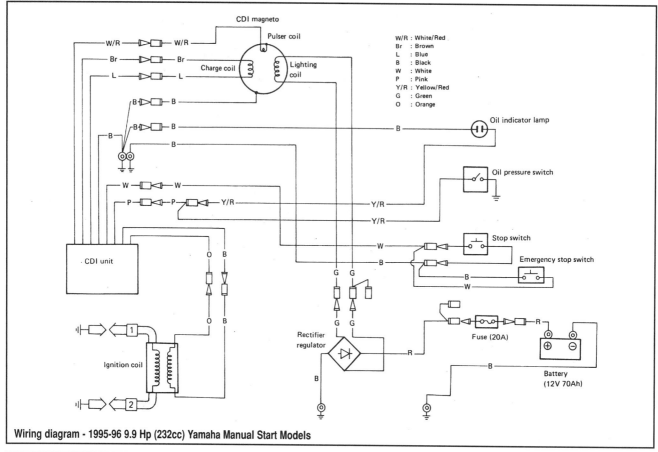

Wiring diagram - 1995-96 9.9 Hp (232cc) Yamaha Manual Start Models

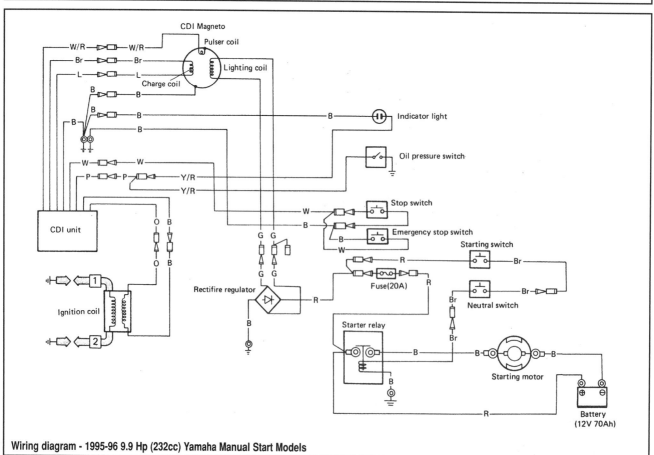

Wiring diagram - 1995-96 9.9 Hp (232cc) Yamaha Manual Start Models

IGNITION AND ELECTRICAL SYSTEMS 4-67

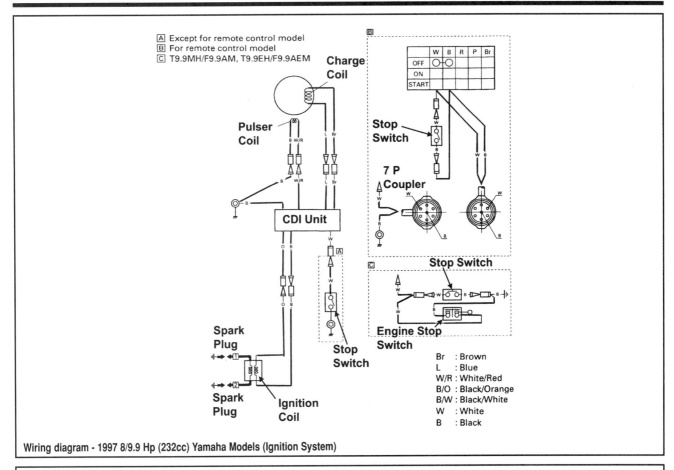

Wiring diagram - 1997 8/9.9 Hp (232cc) Yamaha Models (Ignition System)

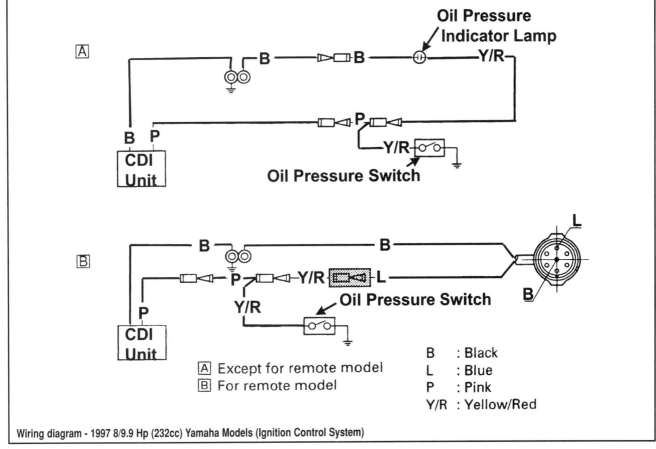

Wiring diagram - 1997 8/9.9 Hp (232cc) Yamaha Models (Ignition Control System)

4-68 IGNITION AND ELECTRICAL SYSTEMS

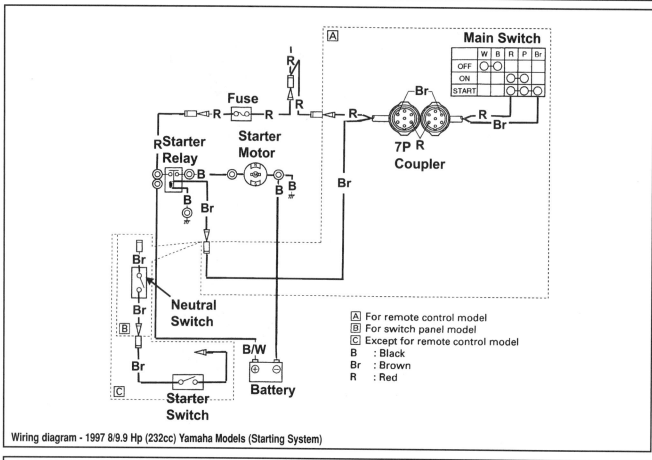

Wiring diagram - 1997 8/9.9 Hp (232cc) Yamaha Models (Starting System)

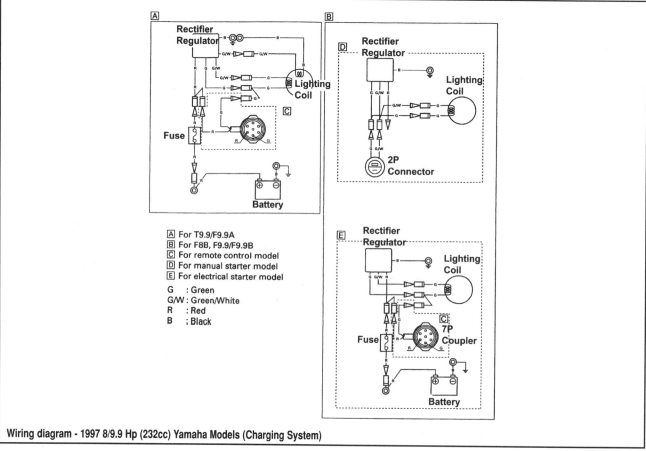

Wiring diagram - 1997 8/9.9 Hp (232cc) Yamaha Models (Charging System)

IGNITION AND ELECTRICAL SYSTEMS 4-69

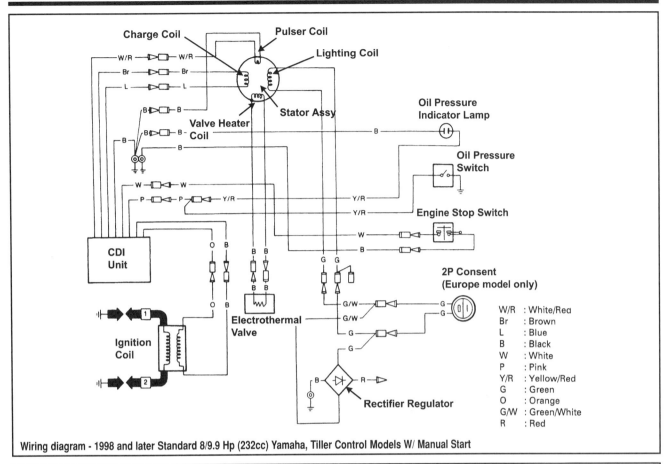

Wiring diagram - 1998 and later Standard 8/9.9 Hp (232cc) Yamaha, Tiller Control Models W/ Manual Start

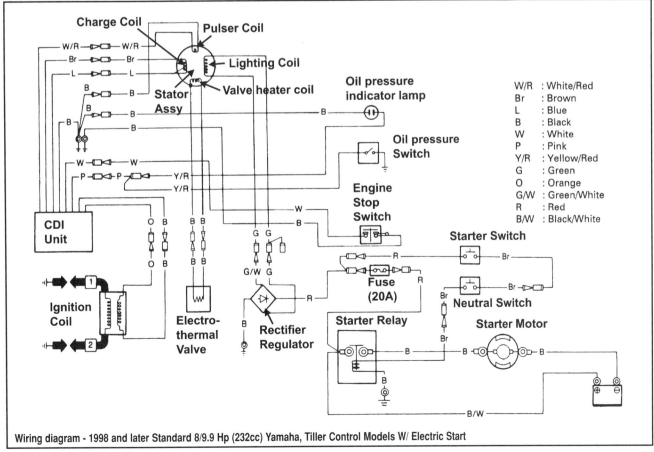

Wiring diagram - 1998 and later Standard 8/9.9 Hp (232cc) Yamaha, Tiller Control Models W/ Electric Start

4-70 IGNITION AND ELECTRICAL SYSTEMS

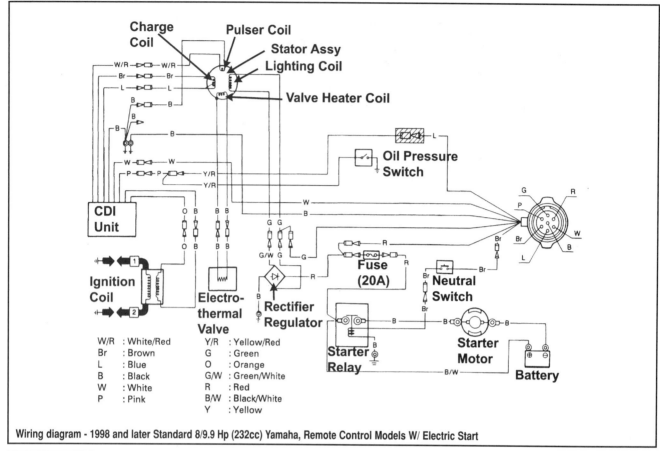

Wiring diagram - 1998 and later Standard 8/9.9 Hp (232cc) Yamaha, Remote Control Models W/ Electric Start

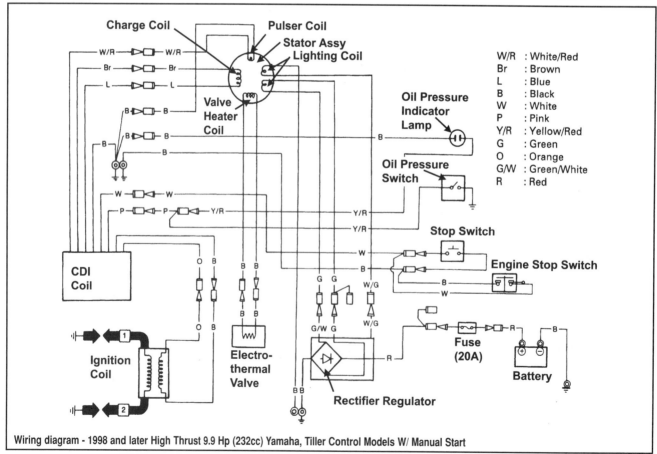

Wiring diagram - 1998 and later High Thrust 9.9 Hp (232cc) Yamaha, Tiller Control Models W/ Manual Start

IGNITION AND ELECTRICAL SYSTEMS 4-71

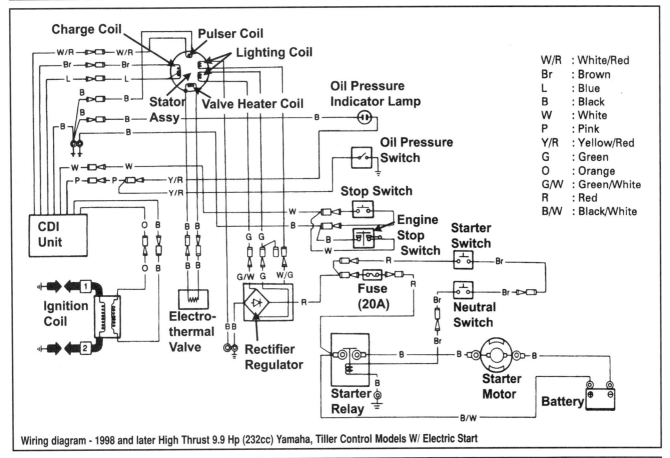

Wiring diagram - 1998 and later High Thrust 9.9 Hp (232cc) Yamaha, Tiller Control Models W/ Electric Start

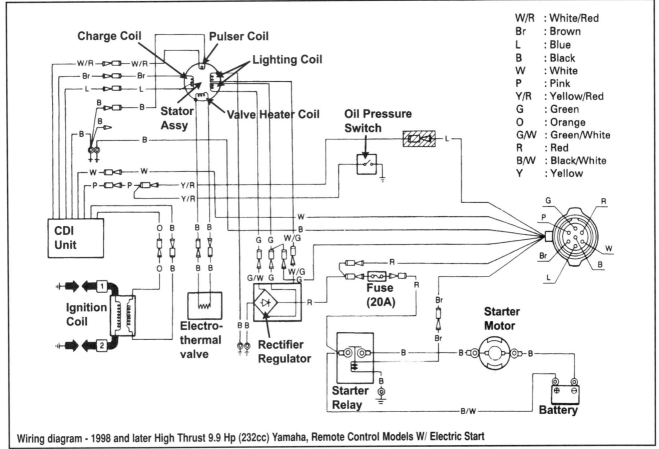

Wiring diagram - 1998 and later High Thrust 9.9 Hp (232cc) Yamaha, Remote Control Models W/ Electric Start

4-72 IGNITION AND ELECTRICAL SYSTEMS

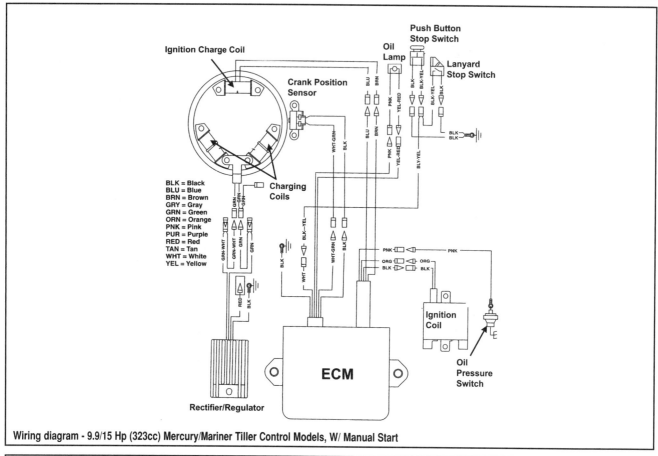

Wiring diagram - 9.9/15 Hp (323cc) Mercury/Mariner Tiller Control Models, W/ Manual Start

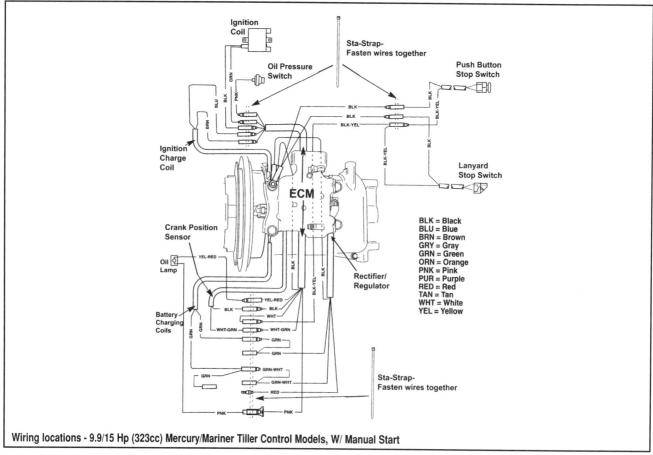

Wiring locations - 9.9/15 Hp (323cc) Mercury/Mariner Tiller Control Models, W/ Manual Start

IGNITION AND ELECTRICAL SYSTEMS

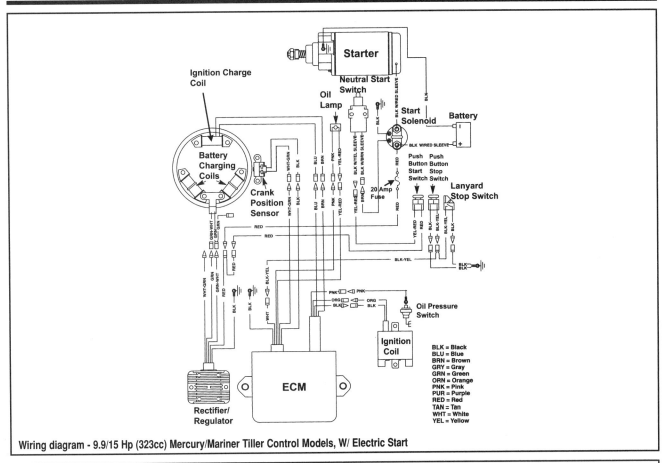

Wiring diagram - 9.9/15 Hp (323cc) Mercury/Mariner Tiller Control Models, W/ Electric Start

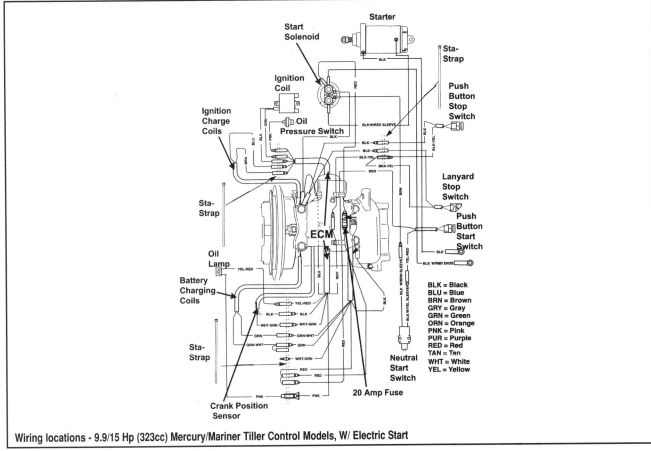

Wiring locations - 9.9/15 Hp (323cc) Mercury/Mariner Tiller Control Models, W/ Electric Start

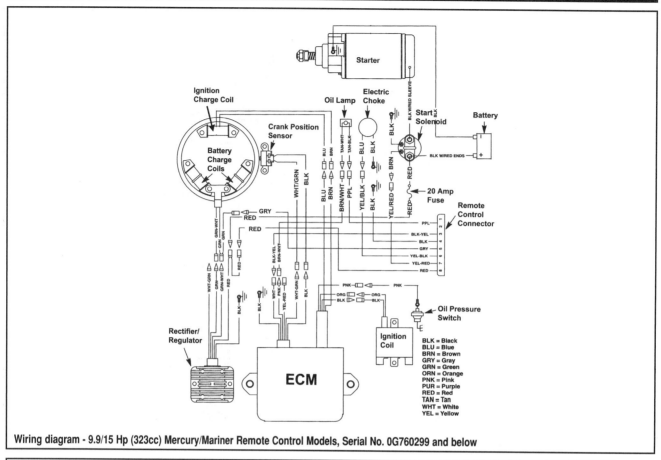
Wiring diagram - 9.9/15 Hp (323cc) Mercury/Mariner Remote Control Models, Serial No. 0G760299 and below

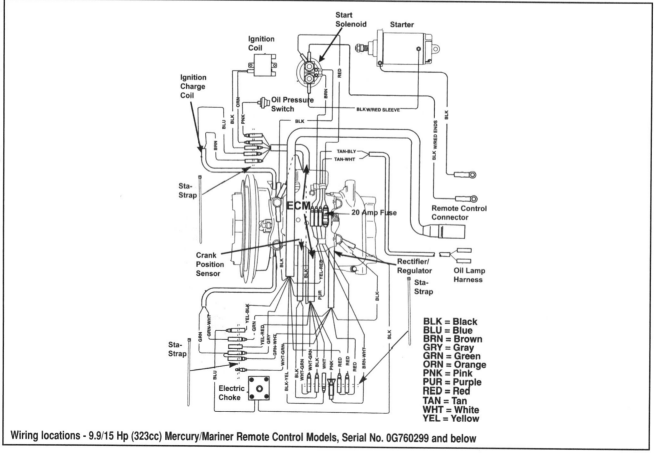
Wiring locations - 9.9/15 Hp (323cc) Mercury/Mariner Remote Control Models, Serial No. 0G760299 and below

IGNITION AND ELECTRICAL SYSTEMS 4-75

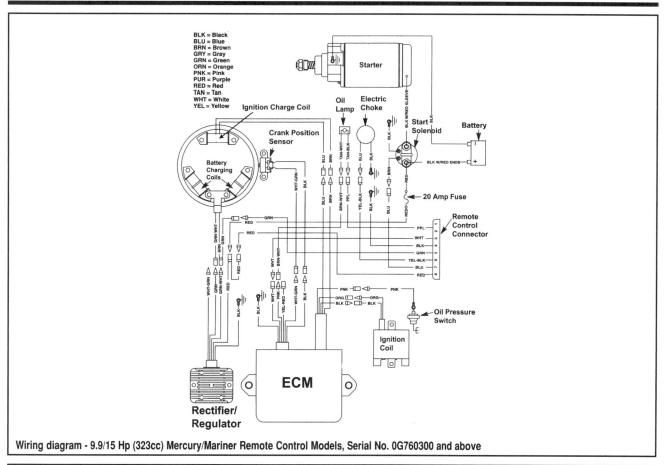

Wiring diagram - 9.9/15 Hp (323cc) Mercury/Mariner Remote Control Models, Serial No. 0G760300 and above

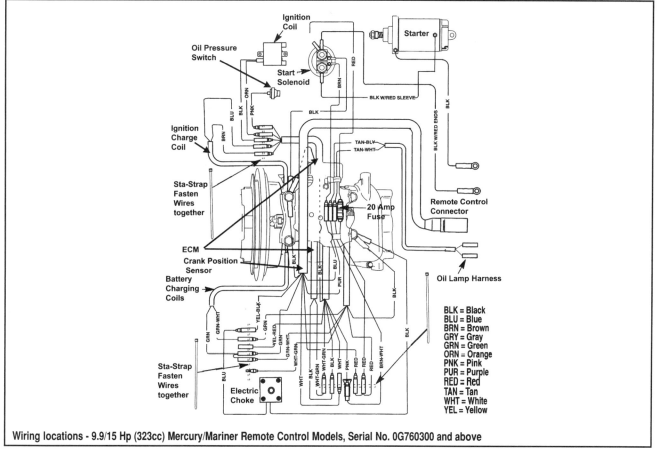

Wiring locations - 9.9/15 Hp (323cc) Mercury/Mariner Remote Control Models, Serial No. 0G760300 and above

4-76 IGNITION AND ELECTRICAL SYSTEMS

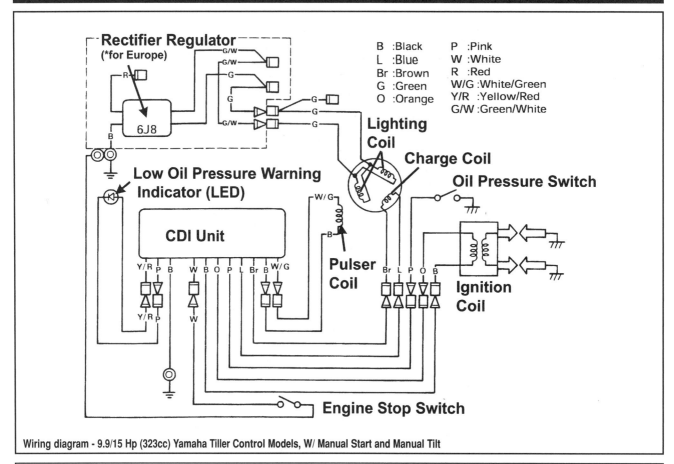

Wiring diagram - 9.9/15 Hp (323cc) Yamaha Tiller Control Models, W/ Manual Start and Manual Tilt

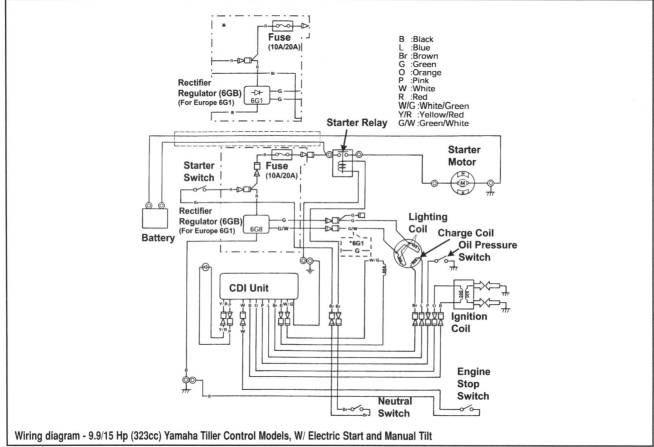

Wiring diagram - 9.9/15 Hp (323cc) Yamaha Tiller Control Models, W/ Electric Start and Manual Tilt

IGNITION AND ELECTRICAL SYSTEMS

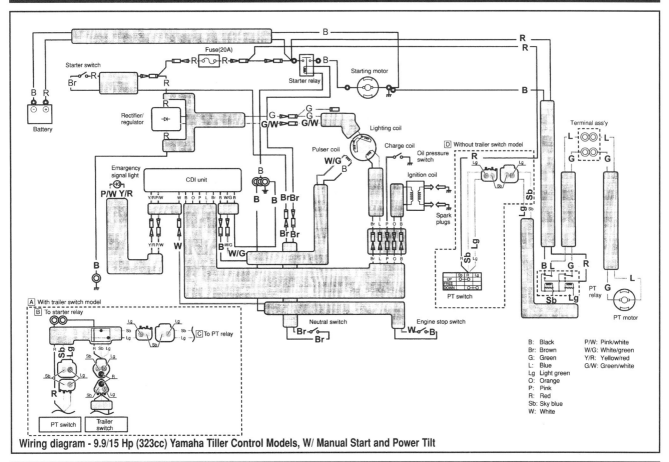

Wiring diagram - 9.9/15 Hp (323cc) Yamaha Tiller Control Models, W/ Manual Start and Power Tilt

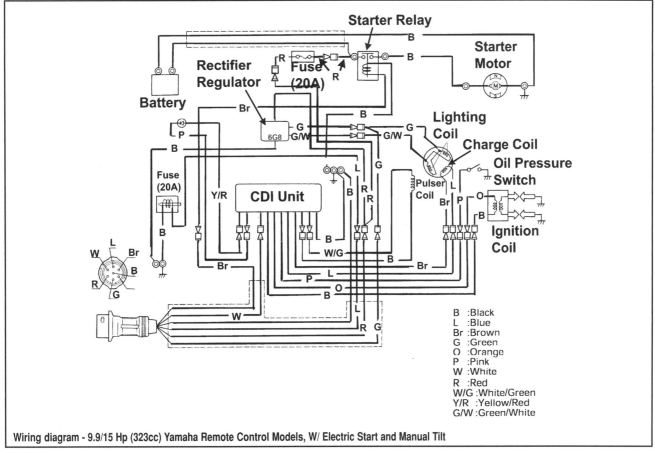

Wiring diagram - 9.9/15 Hp (323cc) Yamaha Remote Control Models, W/ Electric Start and Manual Tilt

4-78 IGNITION AND ELECTRICAL SYSTEMS

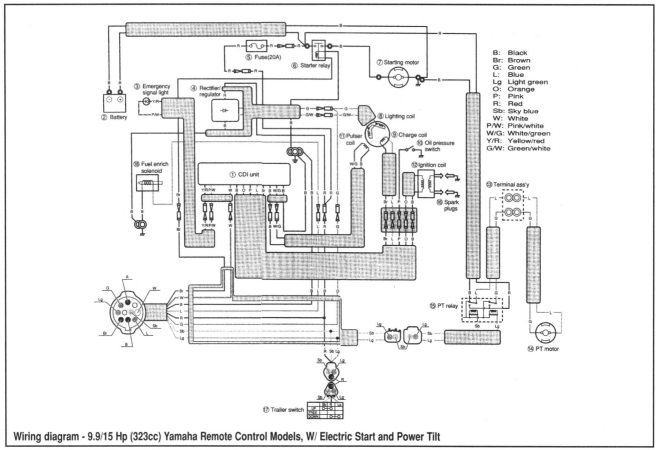

Wiring diagram - 9.9/15 Hp (323cc) Yamaha Remote Control Models, W/ Electric Start and Power Tilt

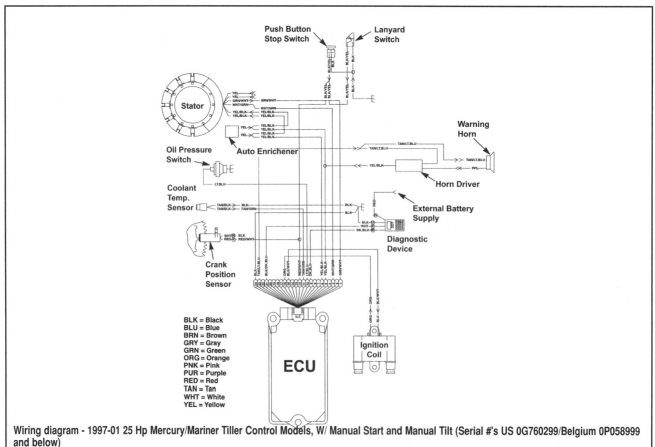

Wiring diagram - 1997-01 25 Hp Mercury/Mariner Tiller Control Models, W/ Manual Start and Manual Tilt (Serial #'s US 0G760299/Belgium 0P058999 and below)

IGNITION AND ELECTRICAL SYSTEMS

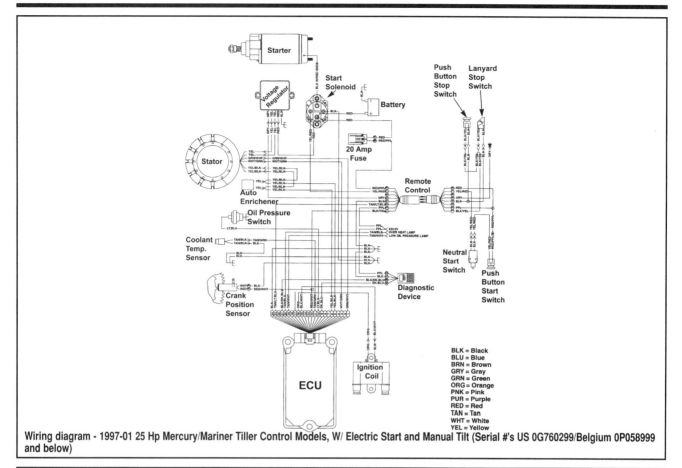

Wiring diagram - 1997-01 25 Hp Mercury/Mariner Tiller Control Models, W/ Electric Start and Manual Tilt (Serial #'s US 0G760299/Belgium 0P058999 and below)

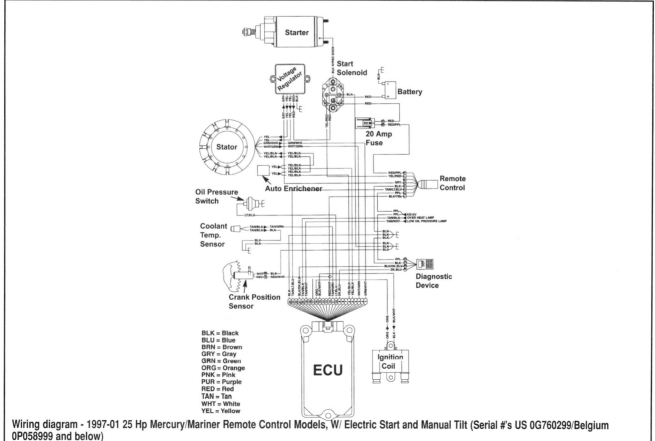

Wiring diagram - 1997-01 25 Hp Mercury/Mariner Remote Control Models, W/ Electric Start and Manual Tilt (Serial #'s US 0G760299/Belgium 0P058999 and below)

4-80 IGNITION AND ELECTRICAL SYSTEMS

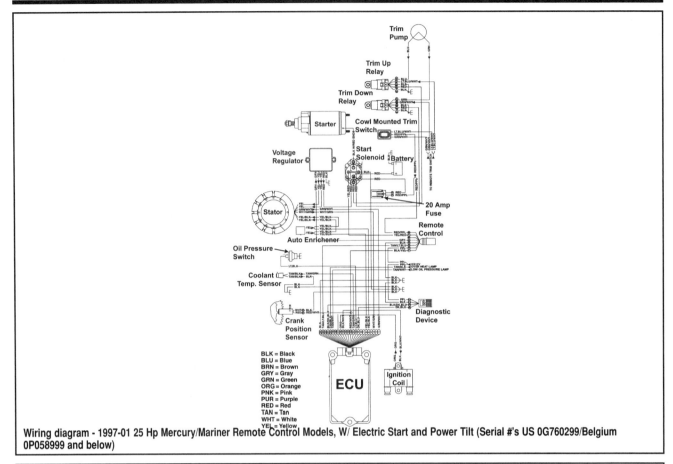

Wiring diagram - 1997-01 25 Hp Mercury/Mariner Remote Control Models, W/ Electric Start and Power Tilt (Serial #'s US 0G760299/Belgium 0P058999 and below)

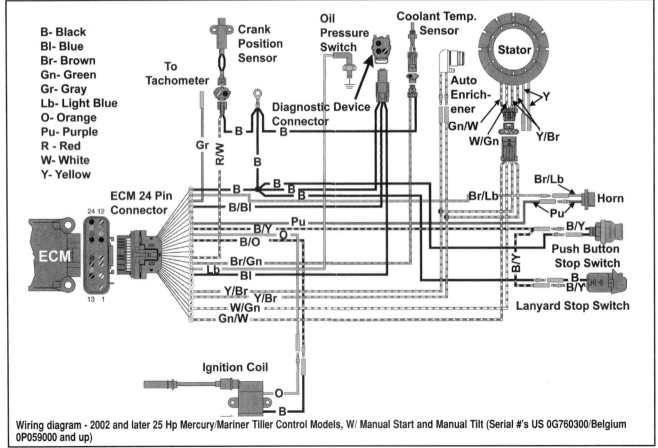

Wiring diagram - 2002 and later 25 Hp Mercury/Mariner Tiller Control Models, W/ Manual Start and Manual Tilt (Serial #'s US 0G760300/Belgium 0P059000 and up)

IGNITION AND ELECTRICAL SYSTEMS 4-81

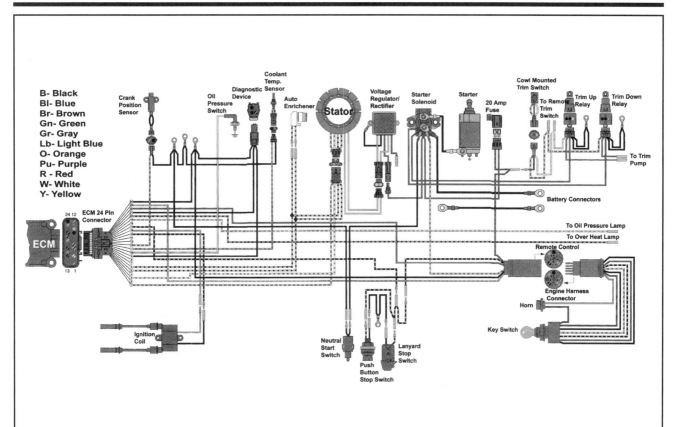

Wiring diagram - 2002 and later 25 Hp Mercury/Mariner Tiller Control Models, W/ Electric Start and Power Tilt (Serial #'s US 0G760300/Belgium 0P059000 and up)

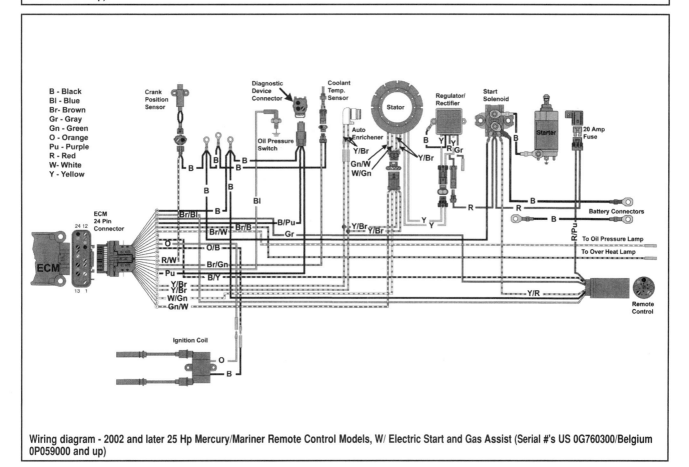

Wiring diagram - 2002 and later 25 Hp Mercury/Mariner Remote Control Models, W/ Electric Start and Gas Assist (Serial #'s US 0G760300/Belgium 0P059000 and up)

4-82 IGNITION AND ELECTRICAL SYSTEMS

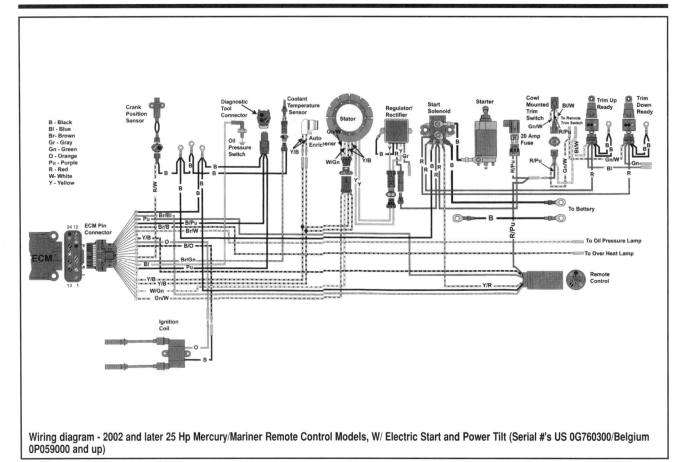

Wiring diagram - 2002 and later 25 Hp Mercury/Mariner Remote Control Models, W/ Electric Start and Power Tilt (Serial #'s US 0G760300/Belgium 0P059000 and up)

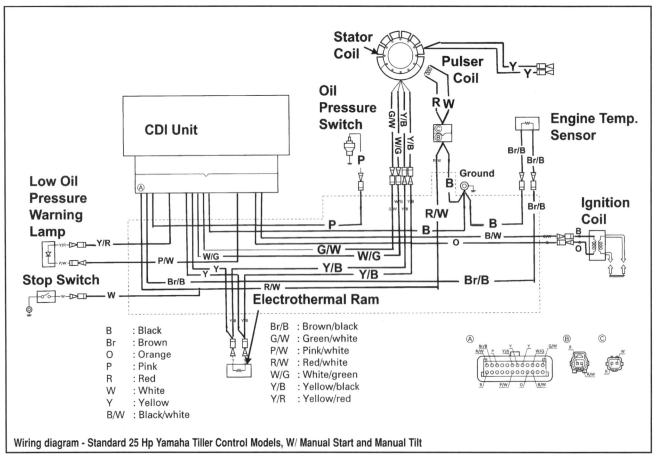

Wiring diagram - Standard 25 Hp Yamaha Tiller Control Models, W/ Manual Start and Manual Tilt

IGNITION AND ELECTRICAL SYSTEMS 4-83

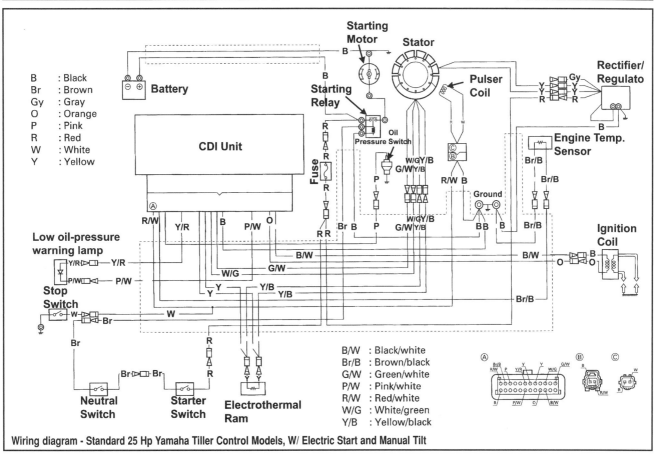

Wiring diagram - Standard 25 Hp Yamaha Tiller Control Models, W/ Electric Start and Manual Tilt

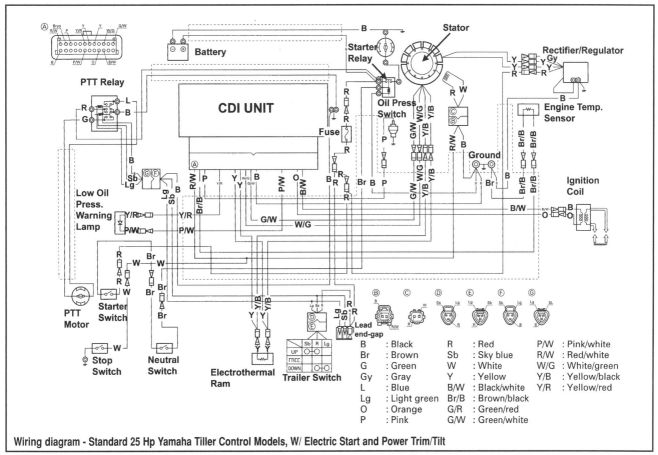

Wiring diagram - Standard 25 Hp Yamaha Tiller Control Models, W/ Electric Start and Power Trim/Tilt

4-84 IGNITION AND ELECTRICAL SYSTEMS

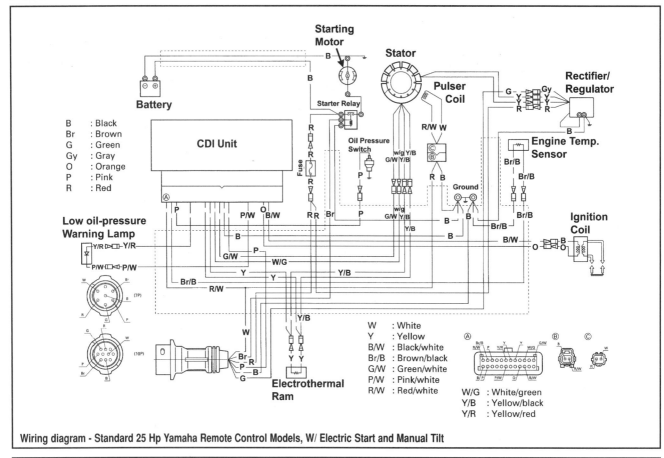

Wiring diagram - Standard 25 Hp Yamaha Remote Control Models, W/ Electric Start and Manual Tilt

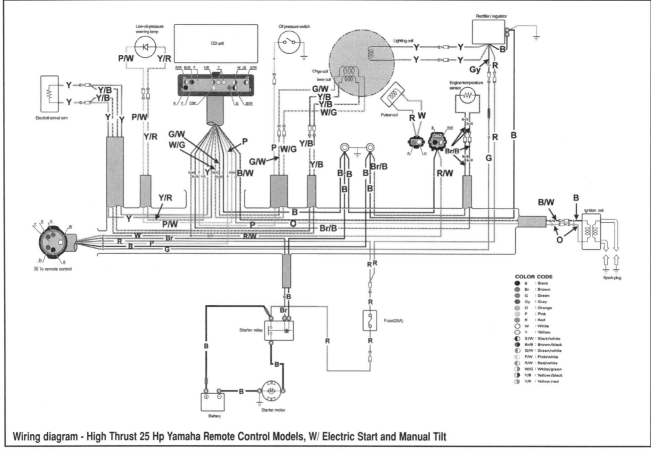

Wiring diagram - High Thrust 25 Hp Yamaha Remote Control Models, W/ Electric Start and Manual Tilt

IGNITION AND ELECTRICAL SYSTEMS 4-85

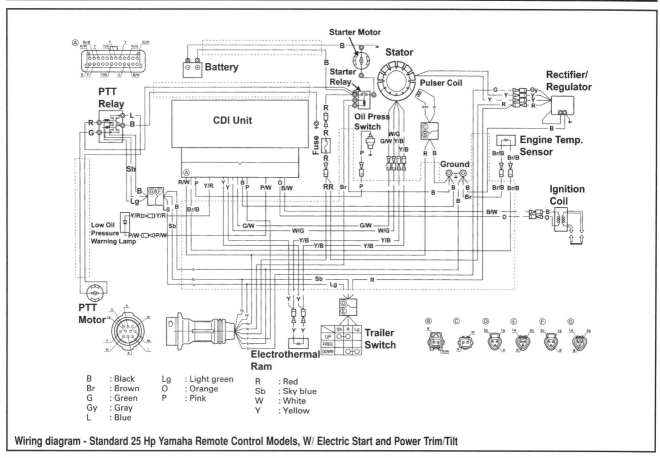

Wiring diagram - Standard 25 Hp Yamaha Remote Control Models, W/ Electric Start and Power Trim/Tilt

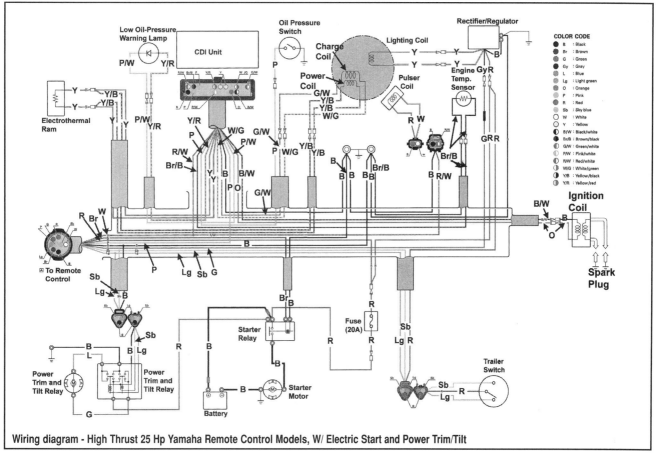

Wiring diagram - High Thrust 25 Hp Yamaha Remote Control Models, W/ Electric Start and Power Trim/Tilt

4-86 IGNITION AND ELECTRICAL SYSTEMS

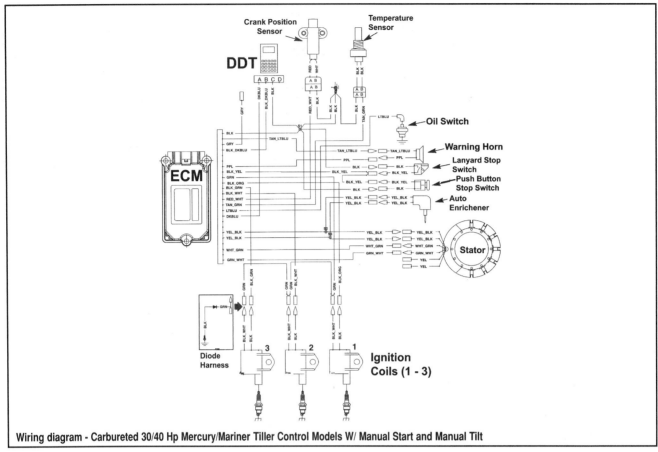

Wiring diagram - Carbureted 30/40 Hp Mercury/Mariner Tiller Control Models W/ Manual Start and Manual Tilt

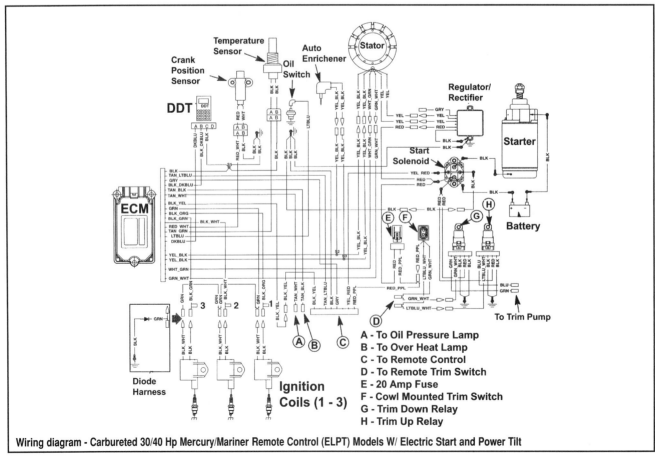

Wiring diagram - Carbureted 30/40 Hp Mercury/Mariner Remote Control (ELPT) Models W/ Electric Start and Power Tilt

IGNITION AND ELECTRICAL SYSTEMS

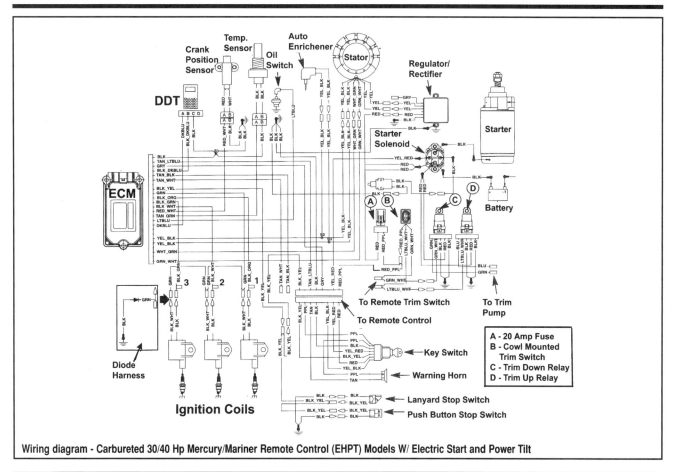

Wiring diagram - Carbureted 30/40 Hp Mercury/Mariner Remote Control (EHPT) Models W/ Electric Start and Power Tilt

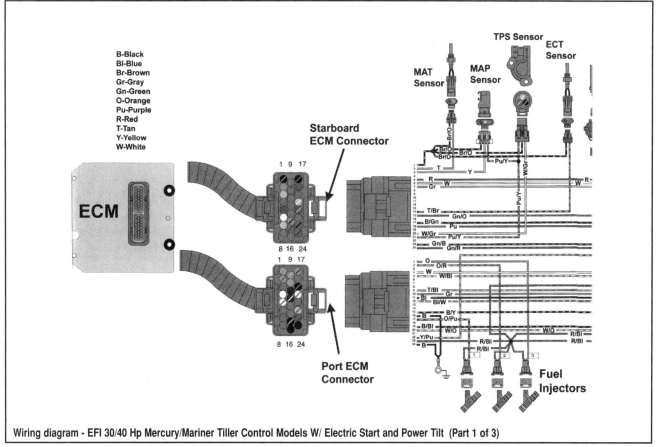

Wiring diagram - EFI 30/40 Hp Mercury/Mariner Tiller Control Models W/ Electric Start and Power Tilt (Part 1 of 3)

4-88 IGNITION AND ELECTRICAL SYSTEMS

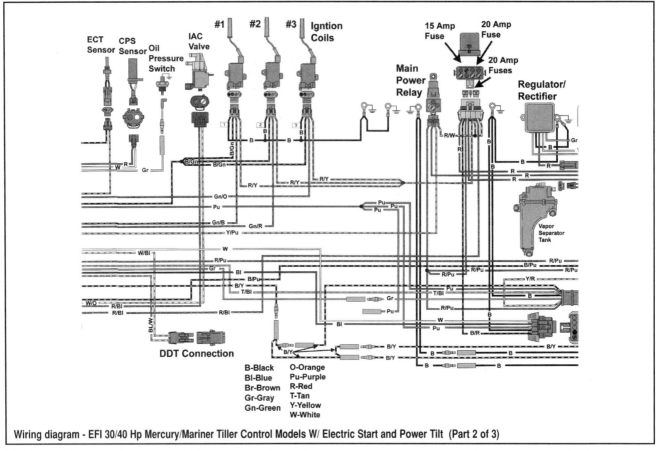

Wiring diagram - EFI 30/40 Hp Mercury/Mariner Tiller Control Models W/ Electric Start and Power Tilt (Part 2 of 3)

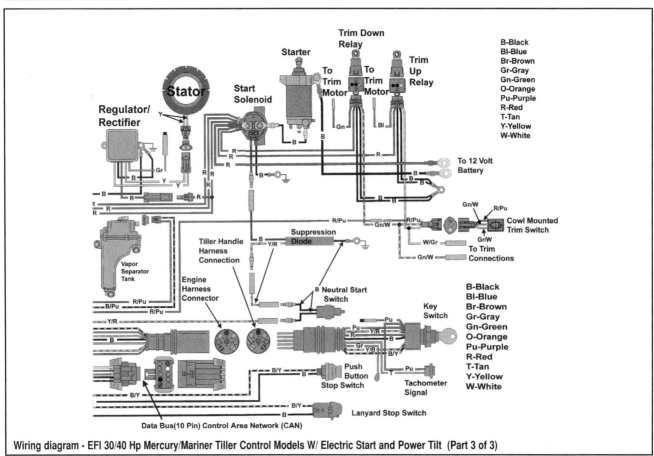

Wiring diagram - EFI 30/40 Hp Mercury/Mariner Tiller Control Models W/ Electric Start and Power Tilt (Part 3 of 3)

IGNITION AND ELECTRICAL SYSTEMS 4-89

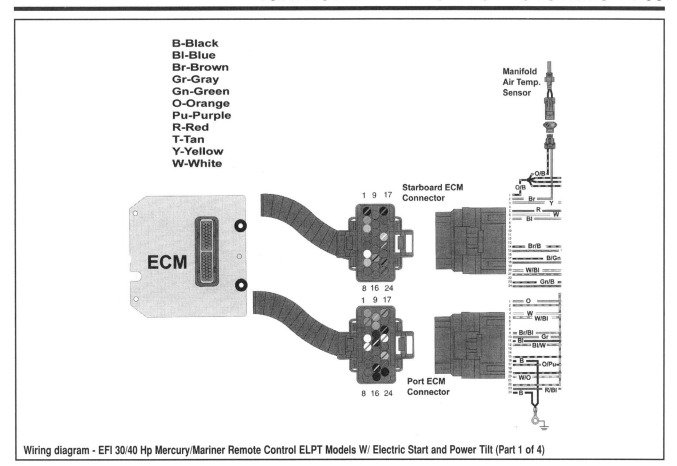

Wiring diagram - EFI 30/40 Hp Mercury/Mariner Remote Control ELPT Models W/ Electric Start and Power Tilt (Part 1 of 4)

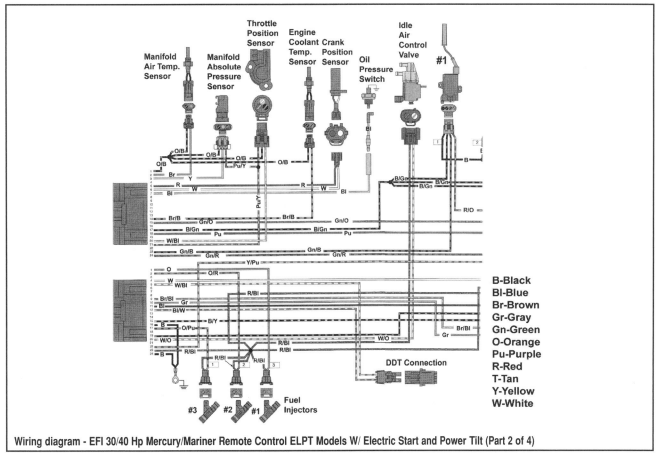

Wiring diagram - EFI 30/40 Hp Mercury/Mariner Remote Control ELPT Models W/ Electric Start and Power Tilt (Part 2 of 4)

4-90 IGNITION AND ELECTRICAL SYSTEMS

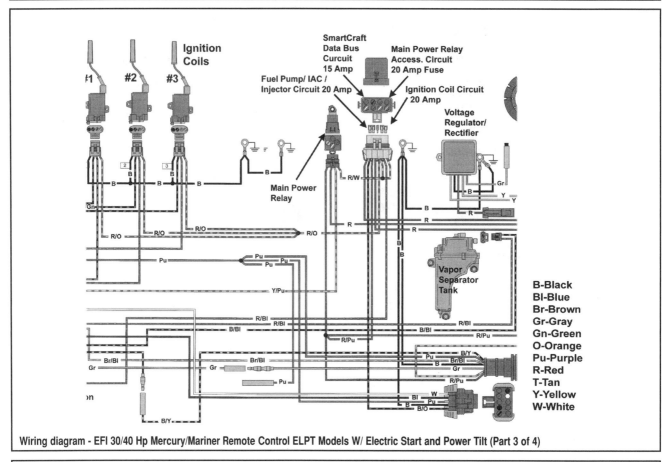

Wiring diagram - EFI 30/40 Hp Mercury/Mariner Remote Control ELPT Models W/ Electric Start and Power Tilt (Part 3 of 4)

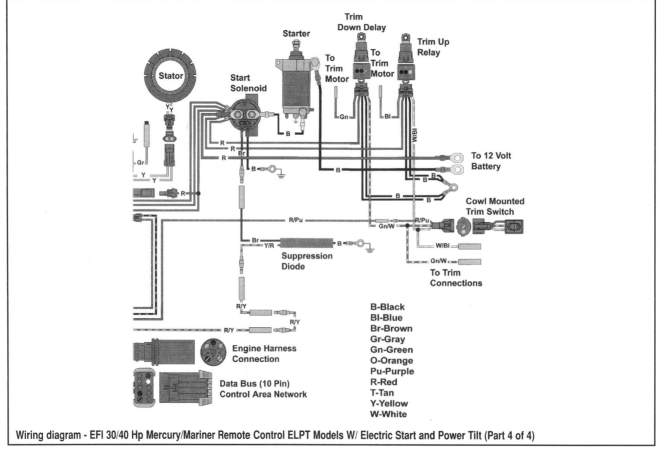

Wiring diagram - EFI 30/40 Hp Mercury/Mariner Remote Control ELPT Models W/ Electric Start and Power Tilt (Part 4 of 4)

IGNITION AND ELECTRICAL SYSTEMS   4-91

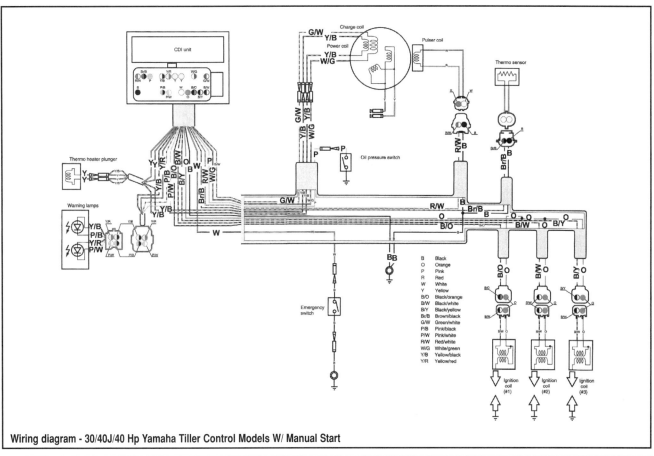

Wiring diagram - 30/40J/40 Hp Yamaha Tiller Control Models W/ Manual Start

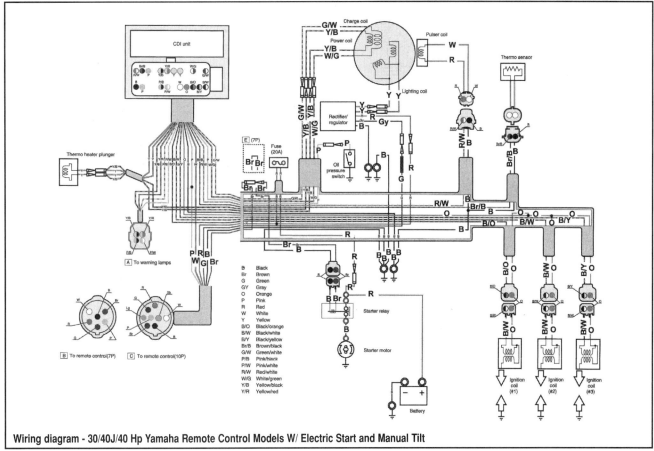

Wiring diagram - 30/40J/40 Hp Yamaha Remote Control Models W/ Electric Start and Manual Tilt

4-92 IGNITION AND ELECTRICAL SYSTEMS

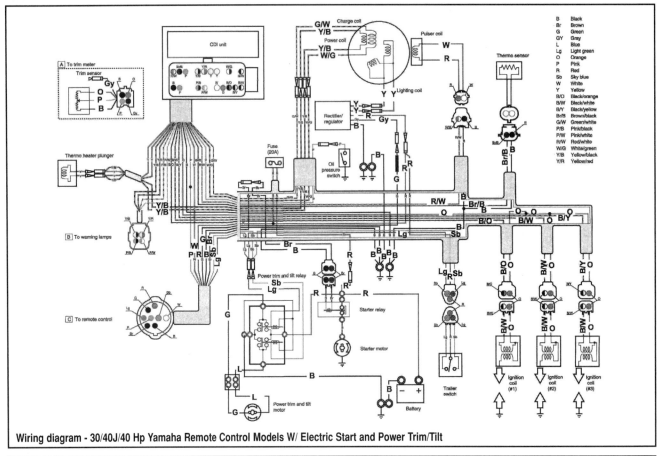

Wiring diagram - 30/40J/40 Hp Yamaha Remote Control Models W/ Electric Start and Power Trim/Tilt

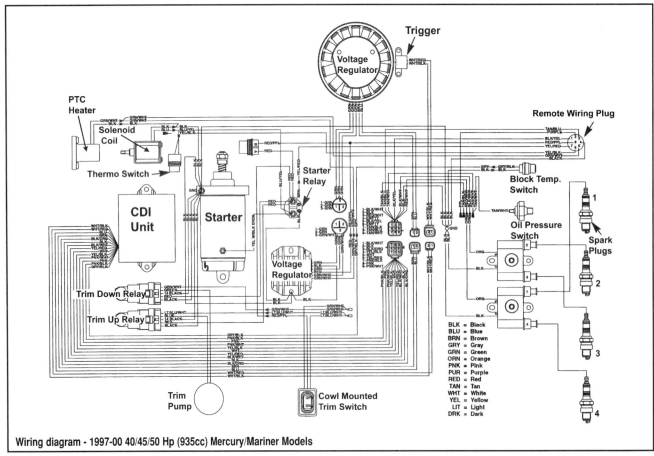

Wiring diagram - 1997-00 40/45/50 Hp (935cc) Mercury/Mariner Models

IGNITION AND ELECTRICAL SYSTEMS 4-93

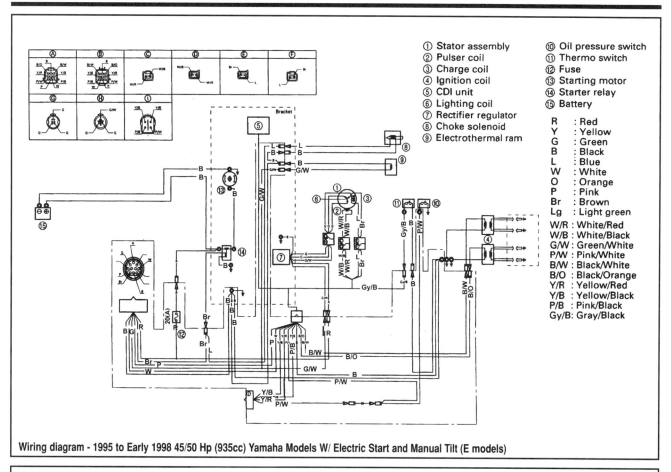

Wiring diagram - 1995 to Early 1998 45/50 Hp (935cc) Yamaha Models W/ Electric Start and Manual Tilt (E models)

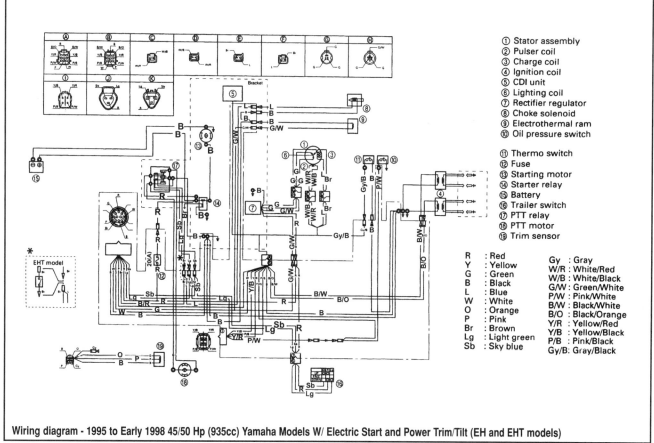

Wiring diagram - 1995 to Early 1998 45/50 Hp (935cc) Yamaha Models W/ Electric Start and Power Trim/Tilt (EH and EHT models)

4-94 IGNITION AND ELECTRICAL SYSTEMS

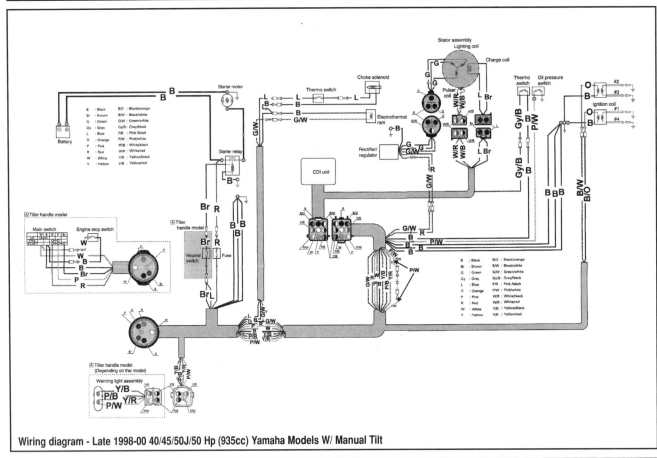

Wiring diagram - Late 1998-00 40/45/50J/50 Hp (935cc) Yamaha Models W/ Manual Tilt

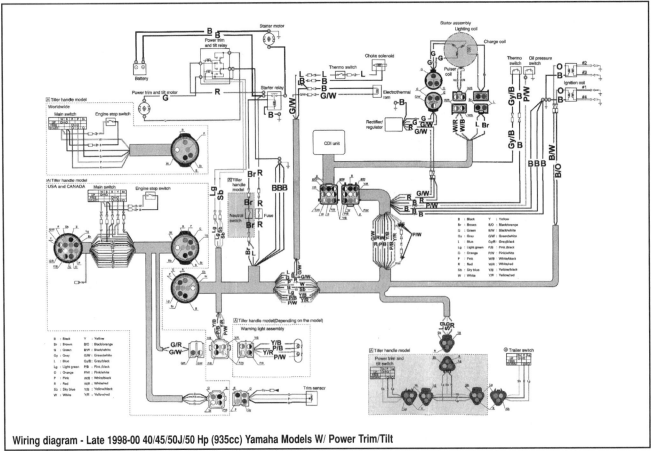

Wiring diagram - Late 1998-00 40/45/50J/50 Hp (935cc) Yamaha Models W/ Power Trim/Tilt

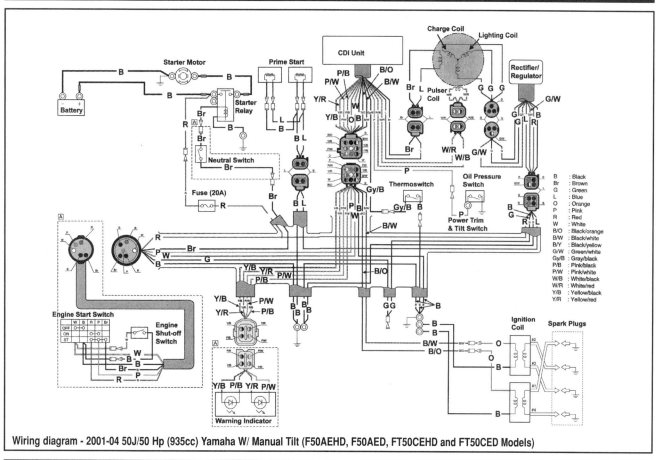

Wiring diagram - 2001-04 50J/50 Hp (935cc) Yamaha W/ Manual Tilt (F50AEHD, F50AED, FT50CEHD and FT50CED Models)

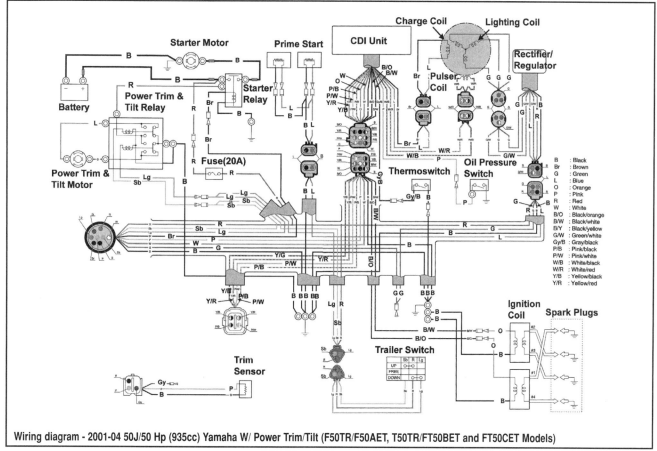

Wiring diagram - 2001-04 50J/50 Hp (935cc) Yamaha W/ Power Trim/Tilt (F50TR/F50AET, T50TR/FT50BET and FT50CET Models)

4-96 IGNITION AND ELECTRICAL SYSTEMS

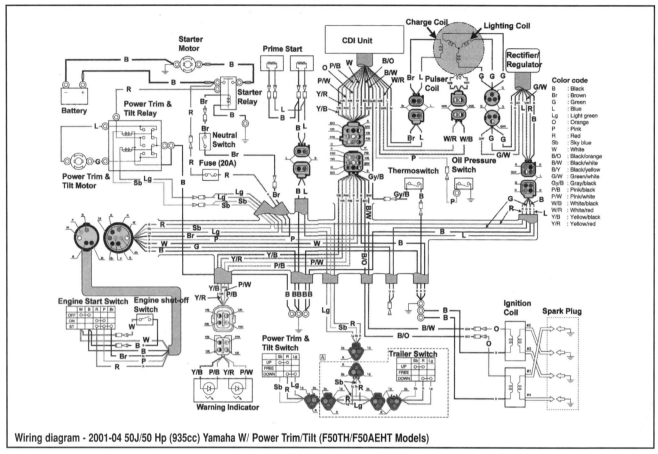

Wiring diagram - 2001-04 50J/50 Hp (935cc) Yamaha W/ Power Trim/Tilt (F50TH/F50AEHT Models)

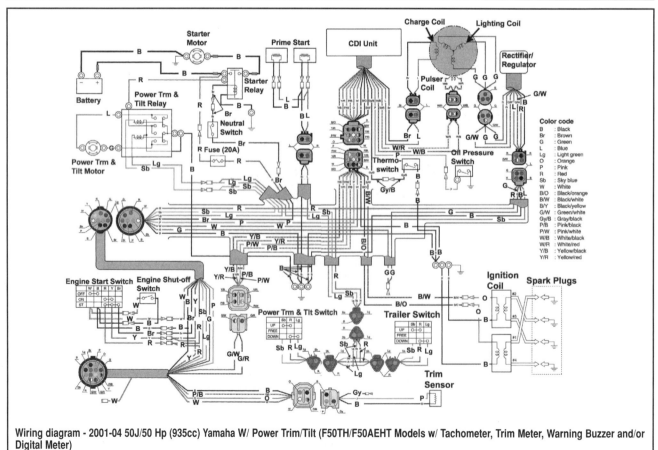

Wiring diagram - 2001-04 50J/50 Hp (935cc) Yamaha W/ Power Trim/Tilt (F50TH/F50AEHT Models w/ Tachometer, Trim Meter, Warning Buzzer and/or Digital Meter)

IGNITION AND ELECTRICAL SYSTEMS

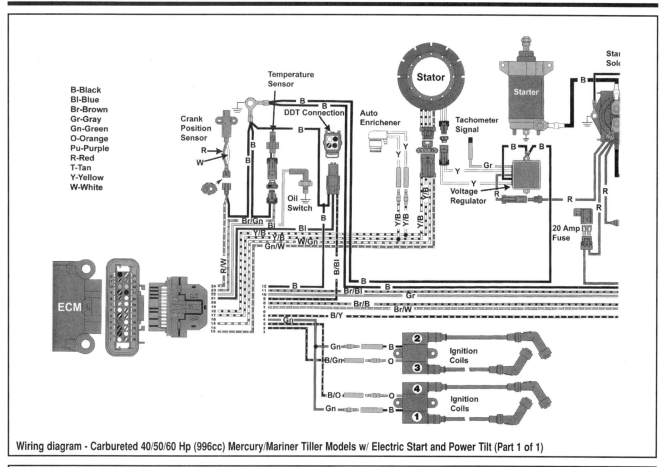

Wiring diagram - Carbureted 40/50/60 Hp (996cc) Mercury/Mariner Tiller Models w/ Electric Start and Power Tilt (Part 1 of 1)

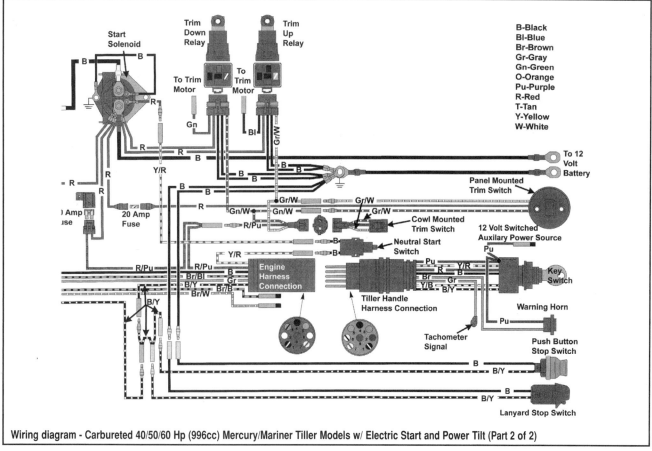

Wiring diagram - Carbureted 40/50/60 Hp (996cc) Mercury/Mariner Tiller Models w/ Electric Start and Power Tilt (Part 2 of 2)

4-98 IGNITION AND ELECTRICAL SYSTEMS

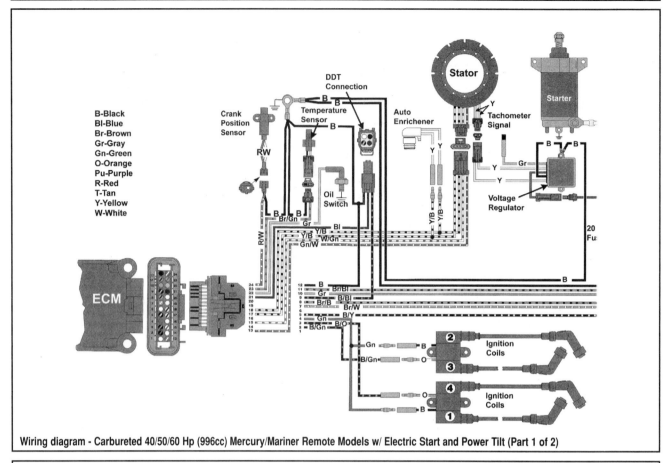

Wiring diagram - Carbureted 40/50/60 Hp (996cc) Mercury/Mariner Remote Models w/ Electric Start and Power Tilt (Part 1 of 2)

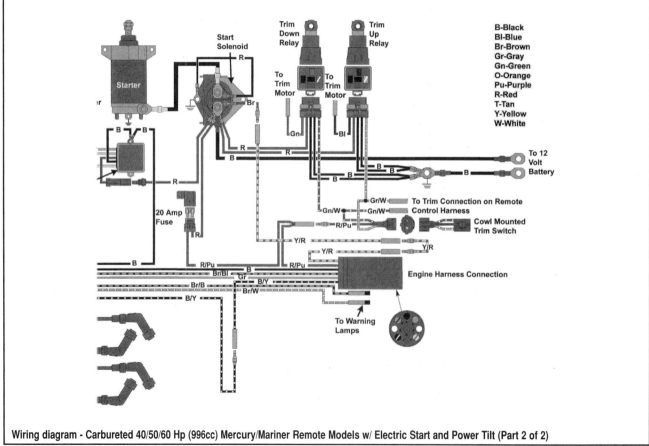

Wiring diagram - Carbureted 40/50/60 Hp (996cc) Mercury/Mariner Remote Models w/ Electric Start and Power Tilt (Part 2 of 2)

IGNITION AND ELECTRICAL SYSTEMS

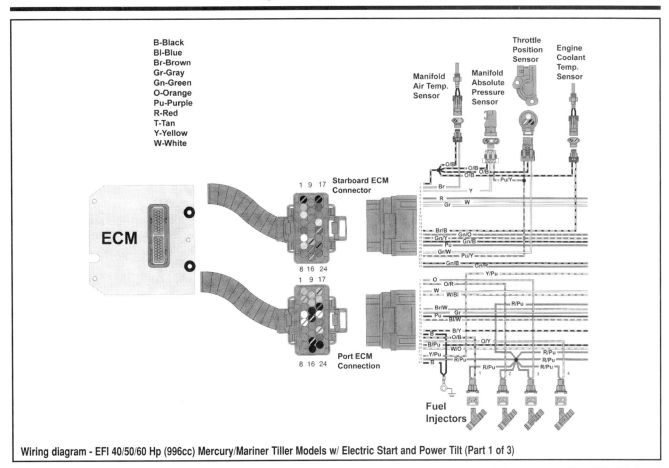

Wiring diagram - EFI 40/50/60 Hp (996cc) Mercury/Mariner Tiller Models w/ Electric Start and Power Tilt (Part 1 of 3)

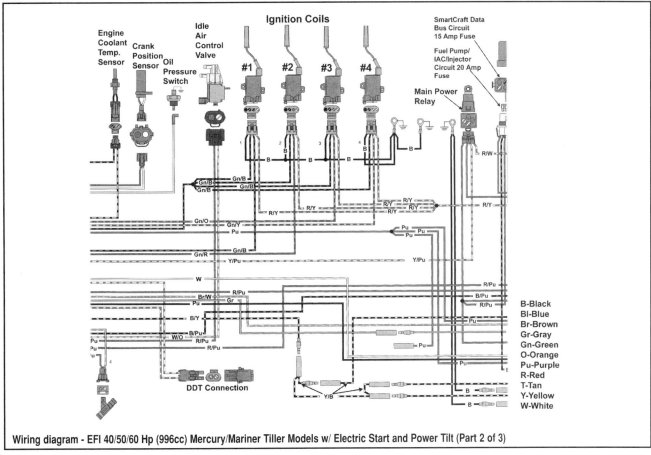

Wiring diagram - EFI 40/50/60 Hp (996cc) Mercury/Mariner Tiller Models w/ Electric Start and Power Tilt (Part 2 of 3)

4-100 IGNITION AND ELECTRICAL SYSTEMS

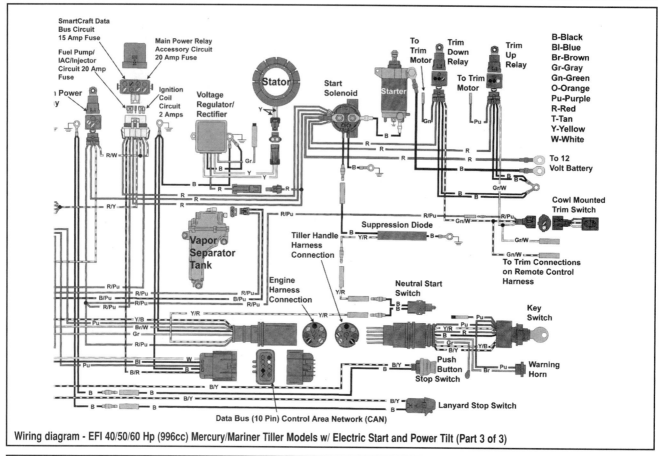

Wiring diagram - EFI 40/50/60 Hp (996cc) Mercury/Mariner Tiller Models w/ Electric Start and Power Tilt (Part 3 of 3)

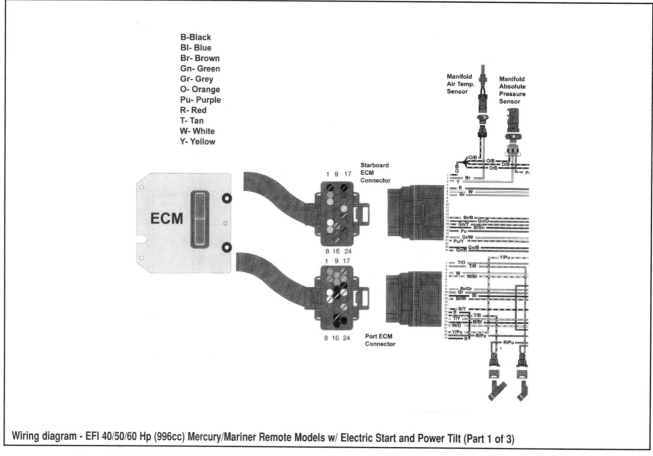

Wiring diagram - EFI 40/50/60 Hp (996cc) Mercury/Mariner Remote Models w/ Electric Start and Power Tilt (Part 1 of 3)

IGNITION AND ELECTRICAL SYSTEMS 4-101

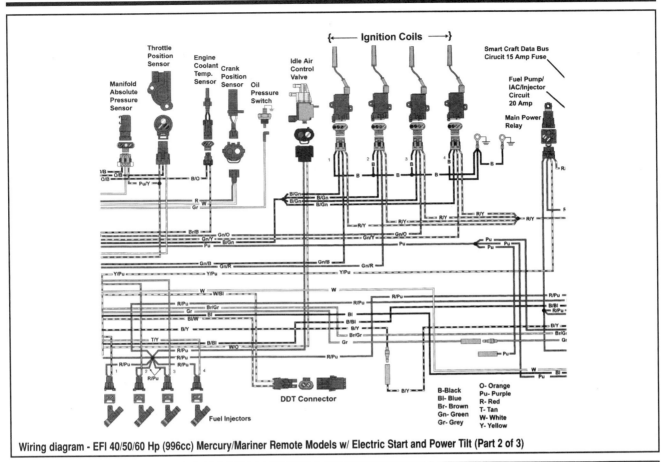

Wiring diagram - EFI 40/50/60 Hp (996cc) Mercury/Mariner Remote Models w/ Electric Start and Power Tilt (Part 2 of 3)

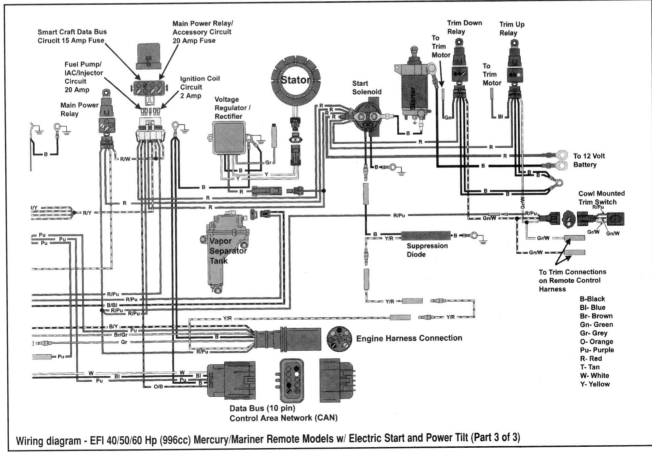

Wiring diagram - EFI 40/50/60 Hp (996cc) Mercury/Mariner Remote Models w/ Electric Start and Power Tilt (Part 3 of 3)

4-102 IGNITION AND ELECTRICAL SYSTEMS

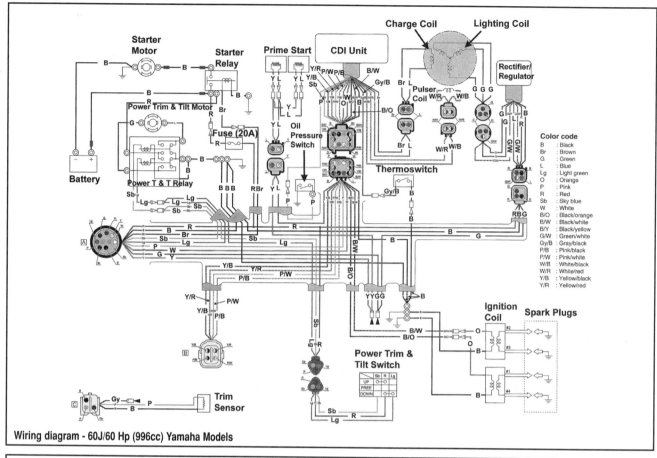

Wiring diagram - 60J/60 Hp (996cc) Yamaha Models

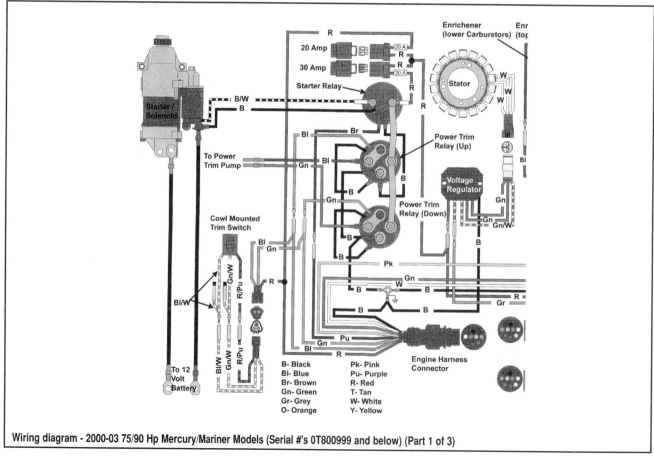

Wiring diagram - 2000-03 75/90 Hp Mercury/Mariner Models (Serial #'s 0T800999 and below) (Part 1 of 3)

IGNITION AND ELECTRICAL SYSTEMS

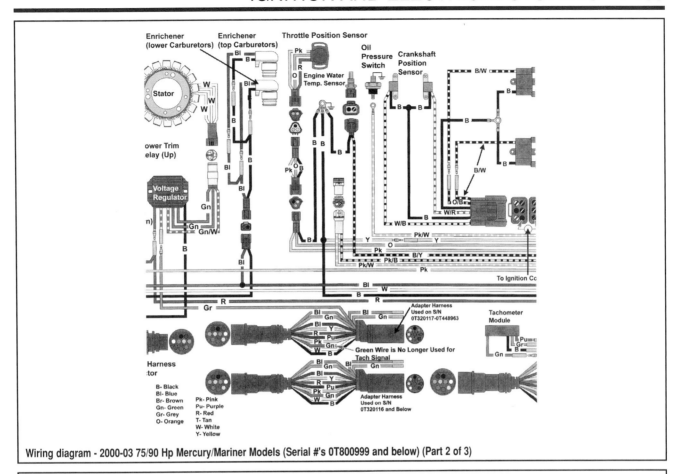

Wiring diagram - 2000-03 75/90 Hp Mercury/Mariner Models (Serial #'s 0T800999 and below) (Part 2 of 3)

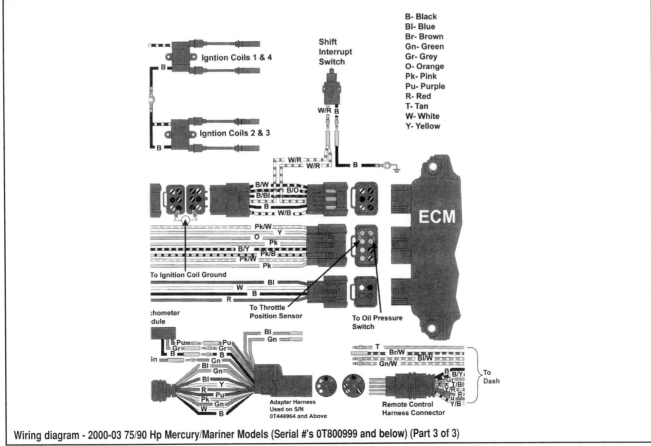

Wiring diagram - 2000-03 75/90 Hp Mercury/Mariner Models (Serial #'s 0T800999 and below) (Part 3 of 3)

4-104　IGNITION AND ELECTRICAL SYSTEMS

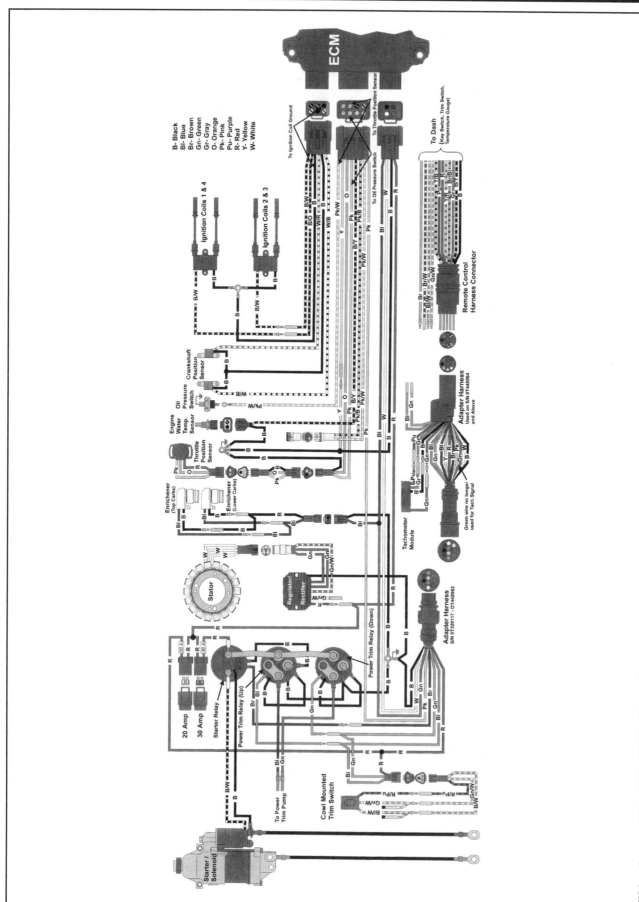

Wiring diagram - 2004 and later 75/90 Hp Mercury/Mariner Models (Serial #'s 0T801000 and up)

IGNITION AND ELECTRICAL SYSTEMS 4-105

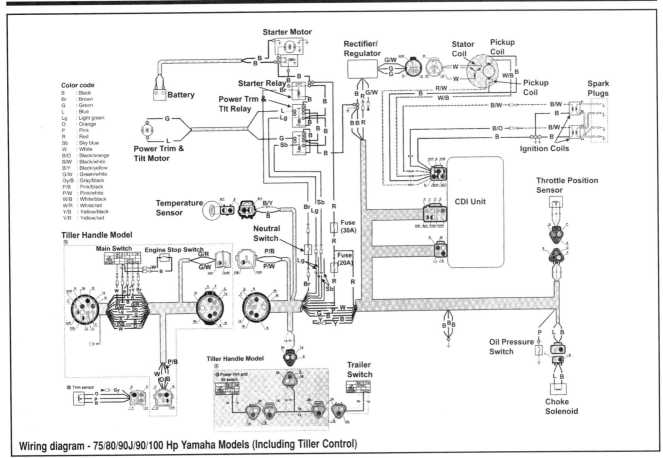

Wiring diagram - 75/80/90J/90/100 Hp Yamaha Models (Including Tiller Control)

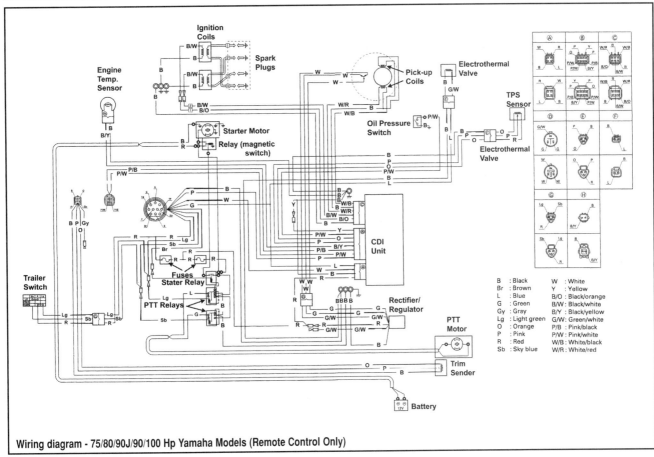

Wiring diagram - 75/80/90J/90/100 Hp Yamaha Models (Remote Control Only)

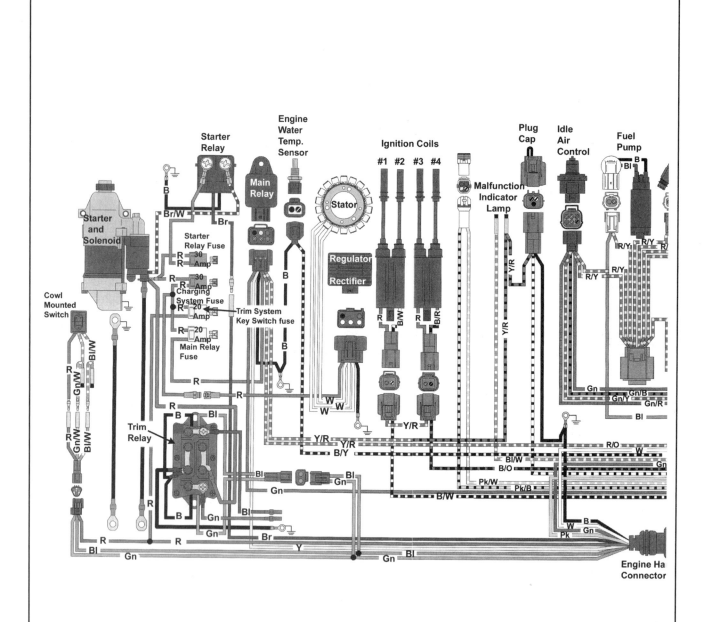

Wiring diagram - 2001 115 Hp EFI Mercury/Mariner Models (Part 1 of 2)

IGNITION AND ELECTRICAL SYSTEMS

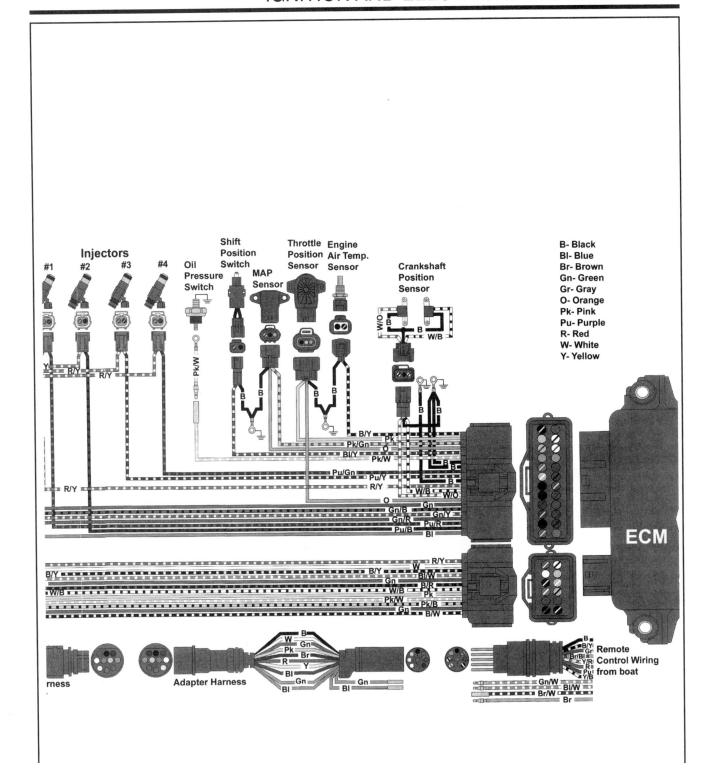

Wiring diagram - 2001 115 Hp EFI Mercury/Mariner Models (Part 2 of 2)

4-108 IGNITION AND ELECTRICAL SYSTEMS

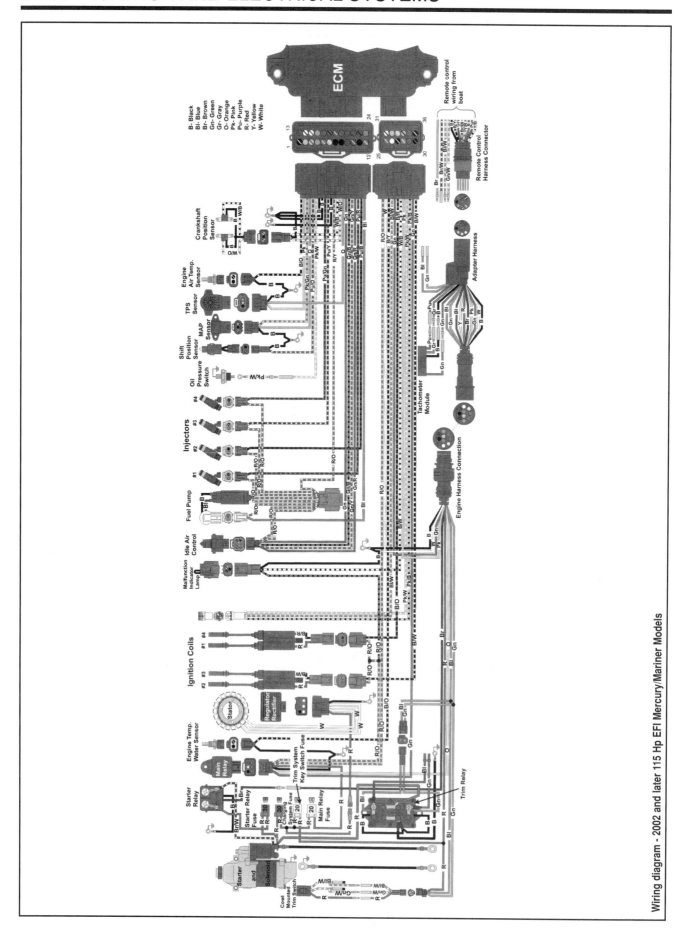

Wiring diagram - 2002 and later 115 Hp EFI Mercury/Mariner Models

IGNITION AND ELECTRICAL SYSTEMS 4-109

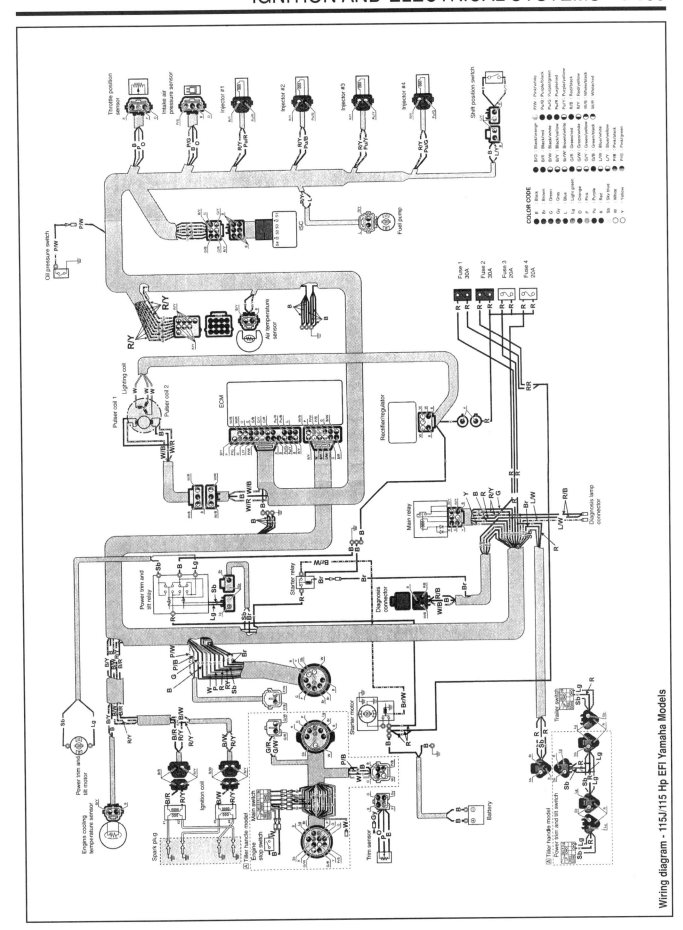

Wiring diagram - 115J/115 Hp EFI Yamaha Models

4-110 IGNITION AND ELECTRICAL SYSTEMS

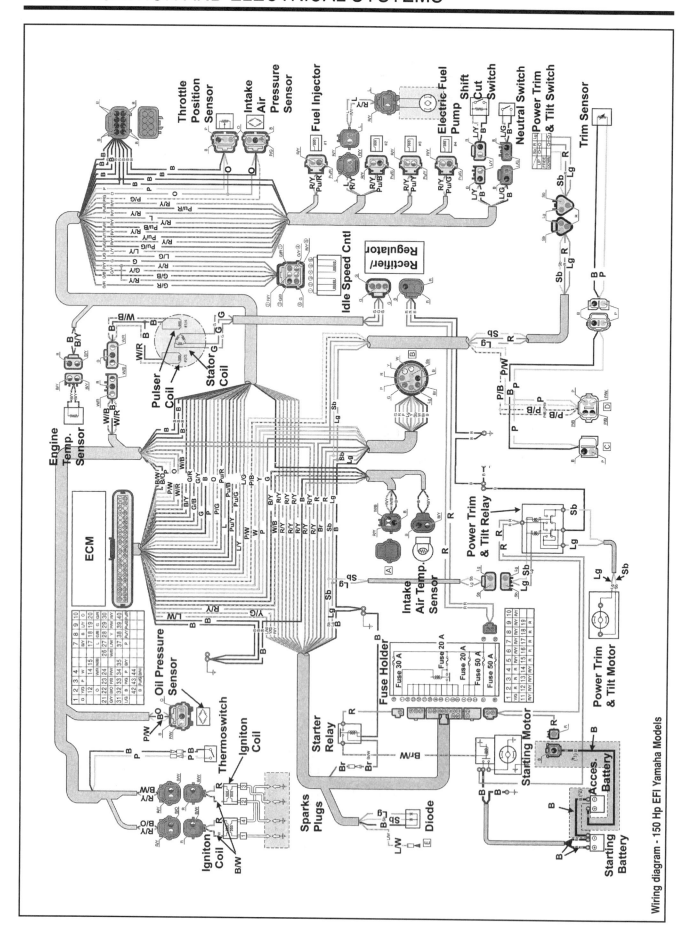

Wiring diagram - 150 Hp EFI Yamaha Models

IGNITION AND ELECTRICAL SYSTEMS 4-111

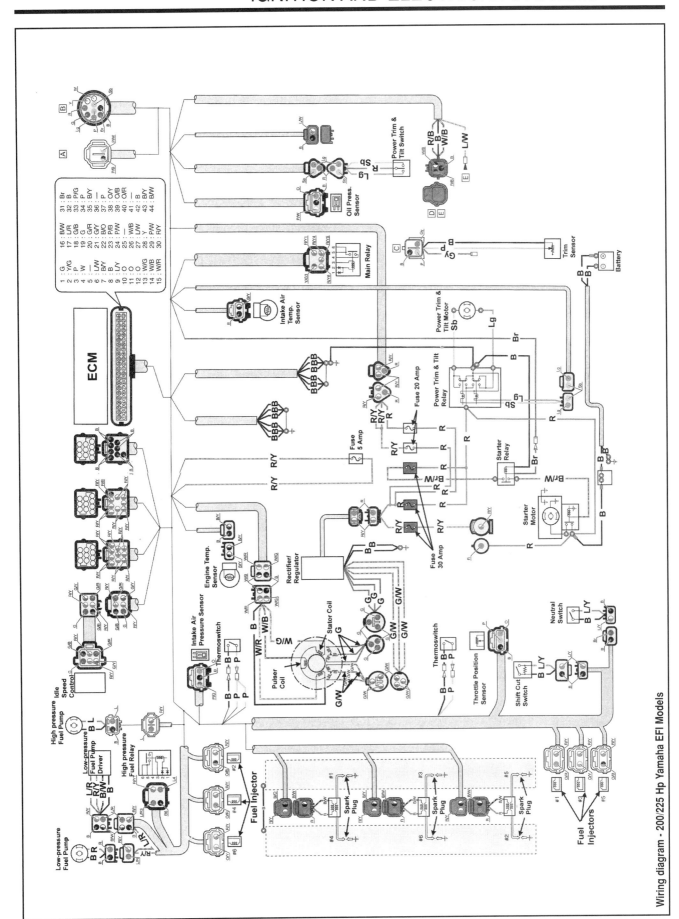

Wiring diagram - 200/225 Hp Yamaha EFI Models

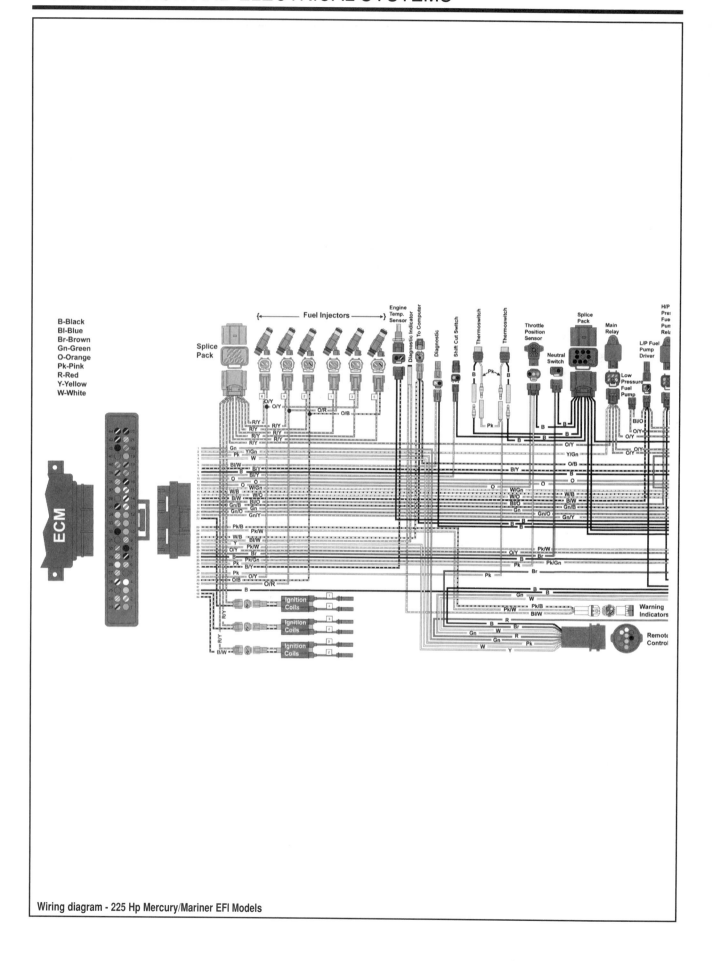

Wiring diagram - 225 Hp Mercury/Mariner EFI Models

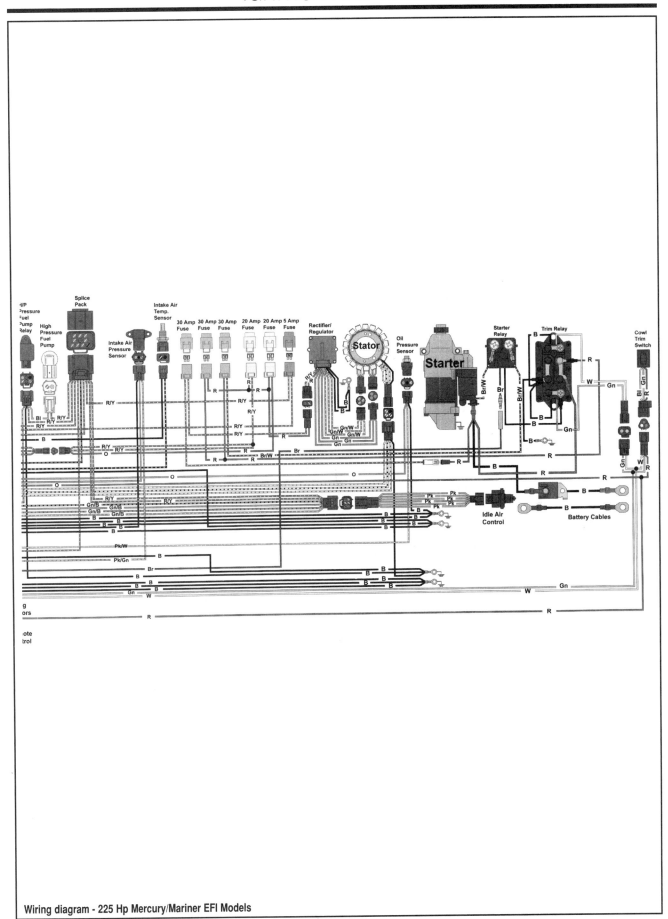

Wiring diagram - 225 Hp Mercury/Mariner EFI Models

4-114 IGNITION AND ELECTRICAL SYSTEMS

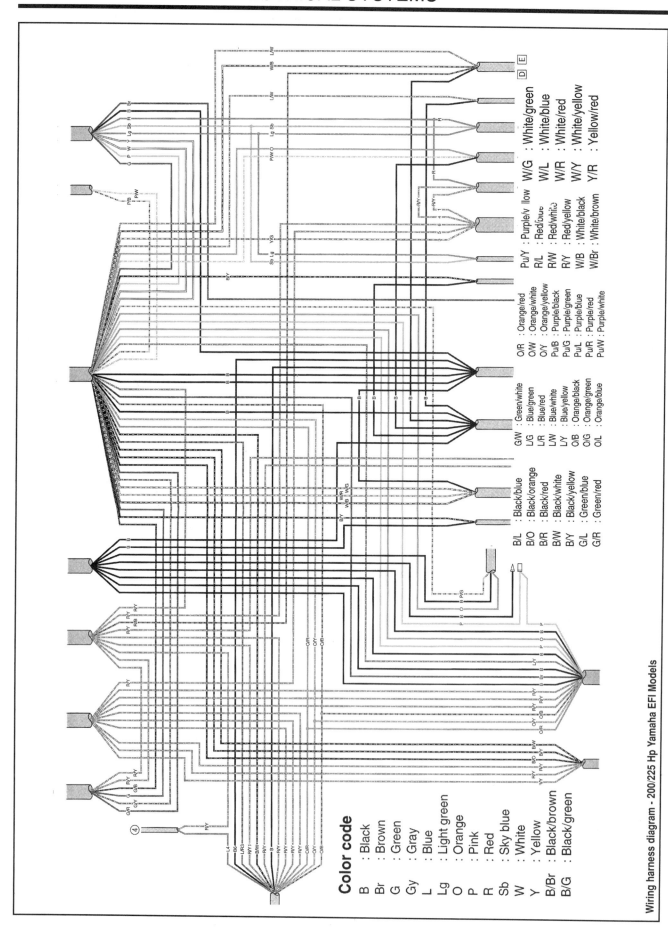

Wiring harness diagram - 200/225 Hp Yamaha EFI Models

5

LUBRICATION AND COOLING

LUBRICATION SYSTEM 5-2
COOLING SYSTEM 5-10
SPECIFICATIONS 5-31

COOLING SYSTEM **5-10**
  DESCRIPTION AND OPERATION.. 5-11
  THERMOSTAT... 5-27
    REMOVAL & INSTALLATION.. 5-27
  THERMO-SWITCH.. 5-28
    CHECKING CONTINUITY... 5-30
    DESCRIPTION & OPERATION... 5-28
    OPERATIONAL CHECK... 5-30
  TROUBLESHOOTING... 5-12
    TESTING COOLING SYSTEM EFFICIENCY...................................... 5-13
    TESTING THE THERMOSTAT.. 5-14
  WATER PUMP.. 5-14
    EXPLODED VIEWS.. 5-19
    INSPECTION & OVERHAUL... 5-17
    REMOVAL & INSTALLATION.. 5-15
LUBRICATION SYSTEM **5-2**
  DESCRIPTION AND OPERATION.. 5-2
  OIL PRESSURE.. 5-2
    TESTING... 5-2
  OIL PRESSURE SWITCH/SENSOR... 5-3
    DESCRIPTION & OPERATION.. 5-3
    REMOVAL & INSTALLATION... 5-4
    TESTING.. 5-3
      EXCEPT 150 HP AND V6 MOTORS... 5-3
      150 HP AND V6 MOTORS.. 5-4
  OIL PUMP.. 5-5
    REMOVAL, OVERHAUL & INSTALLATION....................................... 5-5
      1-CYL 4/5/6 HP MERCURY/MARINER....................................... 5-5
      2-4 CYL, 6-60 HP... 5-6
      75-225 HP.. 5-9
SPECIFICATIONS.. **5-31**
  MERCURY/MARINER... 5-32
  YAMAHA.. 5-31

5-2 LUBRICATION AND COOLING

LUBRICATION SYSTEM

Description and Operation

◆ See Figures 1 and 2

The smallest of the Yamaha 4-stroke motors (2.5 hp and 4 hp models) use a lubrication system common to small engine applications (such as the type found in most smaller 4-stroke lawnmowers). It is called a splash lubrication system. Essentially a small paddle-wheel is driven off a gear on the bottom of the camshaft. The wheel itself is partially immersed in the crankcase oil so as the wheel rotates it literally throws or splashes the oil about the crankcase, onto the camshaft, crankshaft and cylinder wall beneath the piston. This is a purely mechanical system which is normally only checked when the powerhead is disassembled for overhaul. For more details, please refer to the Powerhead section.

All Mercury/Mariners and all 6 hp or larger Yamaha motors use a pressurized (forced) lubrication system, which is resembles the system used on most large 4-stroke engines like the one in your car or truck. Its primary components consist of an oil sump located under the powerhead, an oil pickup tube submerged in the sump, an oil pump, an oil filter (8/9.9 hp/232cc and larger models) and various oil passage ways in the engine.

Oil is drawn from the sump by the pump and routed through passages to the oil filter (on 8/9.9 hp/232cc and larger models), where dirt and metal particles are removed. To prevent system over-pressurization in the event of a clogged filter or passage most models contain either a separate relief valve or one may be incorporated into the pump. The relief valve diverts excess pressure back to the sump. For details, please refer to the Engine Specifications charts in the Powerhead section.

After the pump (and filter in most cases) the oil is then delivered under pressure to various parts of the engine via oil galleys machined into the engine. Oil from the oil pump flows through drilled passages in the camshaft(s) to the rockers and valves. The oil also travels through one or more galleries in the cylinder block. When equipped, pressurized oil will travel through a passage to the powerhead mounted oil pressure switch in order to open the switch contacts (which stops the warning system from activating whenever sufficient oil pressure exists in the passage, which is hopefully, anytime the engine is running).

The oil pump itself is a Trochoid design and consists of an inner rotor, outer rotor and pump body. Normally, the inner rotor is driven by the camshaft or crankshaft (depending upon the model) while the outer rotor is free in the pump body and is driven by the inner rotor. As the rotors spin, the volume of the oil between them changes and provides the force to push oil out into the oil gallery.

Pump mounting varies slightly by powerhead design. The Mercury/Mariner 4/5/6 hp motors utilize an oil pump which is mounted in the oil pan assembly, but is still driven directly off the end of the camshaft. All Mercury/Mariner and Yamaha 2 and 3 cylinder motors, along with the 4-cylinder motors through 60 hp utilize a camshaft driven pump which is mounted to the bottom of the cylinder head. The 75-115 hp 4-cylinder motors utilize a crankshaft mounted pump which is mounted to the top of the sump in the engine cowling/bracket assembly, directly under the powerhead. The 150 hp and V6 motors also utilize a crankshaft driven pump, located in the same basic position as the 75-115 hp motors, however the pump itself is fastened to the underside of the powerhead instead of to the engine bracket.

Oil Pressure

TESTING

◆ See Figure 3

■ The oil pressure sensor on most models uses 1/8 in. standard pipe thread and require a pressure gauge or adapter with that thread.

If there is any doubt as to whether or not a 4-stroke engine is receiving sufficient oil, the oil pressure should be checked using a suitable oil pressure gauge. However Yamaha is all over the board whether or not they publish standard specifications as to what pressure should be present for many of their models. They do publish specs for most of the 4-cylinder and some V6 models and they are included in the Engine Specifications charts from the Powerhead section.

Mercury/Mariner is a little more consistent, publishing specifications for nearly ALL of their motors, with the exception of the 8/9.9 hp (232cc) models. However, even if a specification is not provided for the model on which you are working, you can make some educated guesses based on related figures. For one, the 2-6 cylinder Mercury/Mariner and Yamaha powerheads, especially the oil pumps, are nearly identical. So there is a good chance that the oil pressure specs from Mercury/Mariner models will apply, at least as a preliminary check, for the associated Yamaha model.

Also, all motors with a pressurized lubrication system also utilize an oil pressure indicator/warning system that will sound a buzzer and/or illuminates an oil warning light until the pressure reaches a certain level. For instance, in the case of the Yamaha 8/9.9 hp (232cc) motor the oil pressure switch is calibrated to close switch contacts (stopping the warning buzzer) once oil pressure rises above 7.1-9.9 psi (49-69kPa). The same system utilizes a pressure relief valve that opens to prevent over-pressurization at about 56-64 psi (388-450kPa), depending upon the year and model. With those numbers we can deduce that oil pressure should normally be above 10 psi (69kPa) and below about 56 psi. (388kPa).

From experience we'd add the oil pressure will vary with engine temperature, engine speed and oil viscosity, so don't be alarmed to see pressures all over this wide range. You should however be alarmed if pressures, even above idle speed are hovering too close to the 10 psi (69kPa) mark, so as to possibly activate the warning system. If increasing oil viscosity and engine speed do not rectify low oil pressures, the oil pump and internal engine components should be checked for excessive clearance.

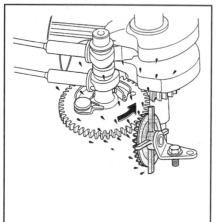

Fig. 1 Splash lubrication is used on Yamaha 2.5 and 4 hp motors

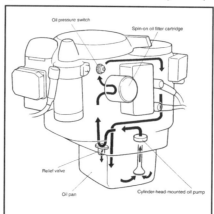

Fig. 2 Pressurized lubrication is used on 6 hp and larger Yamahas, as well as ALL Mercury/Mariner motors

Fig. 3 An oil pressure gauge can be connected to the powerhead by threading it into mounting point for the oil pressure switch

LUBRICATION AND COOLING 5-3

■ A gauge with a range of 0-100 psi (0-690kPa) will be sufficient for all motors covered here. However, a gauge with a more limited operating range, one that is the oil pump pressure specification (if provided) or the relief pressure valve blow-off spec plus 10% (if operating pressure is not provided) should be sufficient.

To properly test the oil pressure the engine must be run both at idle and at or near Wide Open Throttle (WOT), therefore a flush fitting is not suitable for this test. Either mount the engine in a test tank (and use a suitable test wheel) or launch the craft using an assistant to navigate while you check oil pressures.

Locate the oil pressure switch, usually found on the side of the powerhead (use the Wiring Diagrams from the Ignition and Electrical System section to help identify the sensor if there are any doubts). For V6 motors, the sensor is obscured by the air intake silencer assembly, which must be temporarily removed for access.

1. With the engine cold, locate and remove the oil pressure switch. Use tape to cover the pressure switch wiring terminal (to prevent it from grounding), then tie the wiring back out of the way of any moving components.
2. Following the tool manufacturer's instructions, connect a mechanical oil pressure gauge the switch port.
3. Also following the tool manufacturer's instructions, connect a suitable shop tachometer.
4. Start the engine while watching the oil pressure gauge. Within 10 seconds of start-up, sufficient pressure must be established in order to warrant continuing to run and test the engine. Cold oil pressure is normally toward the top end of the engine's oil pressure capability and typically, pressure will drop slightly as the engine warms. Either way, sufficient pressure must be available or the engine cannot be safely run/tested.

✱✱ WARNING

Watch the gauge to make sure enough oil pressure is being generated to warrant continuing the test. If this test is a result of an activation of the warning circuit, DO NOT allow the engine to idle until it is warm if there is little or no oil pressure. If there is little or no pressure and the oil level is sufficient, oil pump and related components (pickup tube/inlet screen) must be removed for examination to determine the cause of the problem. Before performing extensive tear-down work, make sure there are no clogged passages (use low pressure compressed air to blow out the oil pressure passage).

5. While the engine idles, check the gauge and fitting for leaks. If present, shut the engine OFF and correct the oil leak before proceeding.
6. Allow the engine to idle until it reaches normal operating temperature while continuing to watch the oil gauge. It may take as much as 10 minutes to fully warm the motor.
7. Once the engine is fully warmed, record the oil pressure throughout the engine operating range from idle to WOT.

■ Low oil pressure readings may be the result of low oil level, diluted oil, incorrect type or grade of oil, a faulty oil pump or damage/wear to the crankshaft and bearings (excessive clearance). High oil pressure readings typically indicate a damaged or clogged oil pressure relief valve, blockage at the powerhead mating surface or in cylinder oil passages. As usual with troubleshooting, inspect the simple things first before suspecting and inspecting the complicated items.

8. Once the oil pressure has been recorded at the appropriate throttle setting(s), slowly return the engine to idle, then shut the engine OFF.
9. Remove the oil pressure gauge and adapter, as well as the tachometer.
10. Carefully thread the switch or port into the opening by hand, then tighten until snug (but do not over-tighten and damage the switch or powerhead threads). Then reconnect the switch wiring.

Oil Pressure Switch/Sensor

DESCRIPTION & OPERATION

◆ See Figures 4 and 5

All Mercury/Mariner motors, as well as all 6 hp and larger Yamaha motors utilize an oil pressure sensor as part of the pressure lubrication system. The sensor is threaded into a powerhead oil passage. For all except the 150 hp and V6 motors, it is actually not a variable sensor per se, but a simple on/off switch. The switch contacts are normally CLOSED meaning they are completing the warning circuit unless oil pressure is applied to the switch to open the contacts. However, on 150 hp and V6 motors the sensor IS a variable resistor whose value will change in response to changes in pressure.

TESTING

Except 150 Hp and V6 Motors

◆ See Figures 4 and 5

Since the oil pressure sensor usually found on these motors contains normally closed switch contacts, testing is a relatively simple matter of determining if the switch contacts are closed and if they open with the application of system oil pressure.

Fig. 4 Most Yamaha oil pressure sensors are positioned under a small rubber protective cover and threaded into the side of the powerhead

Fig. 5 On some of the larger motors, like this 90 hp Mercury, it's located just below the timing belt

5-4 LUBRICATION AND COOLING

■ Because one or both of the manufacturers does not give a testing specification for the switch it COULD be possible (though unlikely) that the switch on some motors is designed to be normally OPEN, closing only with the application of pressure. So if specifications are not available, the most important test is to determine whether or not switch status changes with the application of pressure.

There are multiple ways to test the switch. But in all cases, start by locating and disconnecting the wiring harness (it's usually a single wire terminal as the switch is used to provide or deny ground to the warning circuit). Use an ohmmeter or DVOM set to read resistance to measure between the switch body (or powerhead) and the switch terminal. There should be resistance without the engine running.

Next you should check to see when (and if) the switch position changes (the contacts open). To accomplish this, you can remove it from the motor and use a hand pressure pump to apply the specified amount of pressure while watching a DVOM to see when the switch contacts open. Compare the results to the specifications. For Yamaha motors, please refer to the Ignition Control System Component Testing chart in the Ignition and Electrical System section for details on switch activation pressures (the amount of oil pressure it should normally take to open the switch contacts).

■ For some reason we could not always locate the switch activation points for some motors. But keep in mind it should be at the bottom of the oil pump operating range.

For Mercury/Mariner models if the switch activation pressure is not available from the manufacturer you can use the specification from the associated Yamaha model as a starting point. When available, the switch activation pressures on Merc/Mariner models are as follows:
• 4/5/6 hp motors: there should be continuity below 2.8-4.2 psi (19.5-29.5 kPa).
• 9.9/15 hp (323cc) and 25 hp motors: there should be continuity below 2.5 psi (17 kPa).
• 40-60 hp 4-cylinder and EFI 30/40 hp 3-cylinder motors: there should be continuity below 2.9 psi (20 kPa).
• 75/90 hp motors through 2000 (Serial #'s 0T178499 and below): the switch should be stamped with a ".15" and there should be continuity below 2.2 psi (15 kPa).
• 75/90 hp motors 2001 and later (Serial #'s 0T178500 and up): the switch should be stamped with a "1.5" and there should be continuity below 21.78 psi (150 kPa).
• 115 hp motors: there should be continuity below 21.78 psi (150 kPa).

If however a hand pressure pump is not available AND if the pressure lubrication system is supplying the proper amount of pressure (this can be checked using an oil pressure gauge threaded into the port for the pressure sensor) you can operate the motor and use the DVOM to verify that the switch contacts open shortly after start-up.

If the oil pressure at the switch port is known to be good, start and run the engine (using a flush fitting or test tank to prevent damage to the powerhead) and watch the ohmmeter. Once the engine is started the meter which WAS showing continuity should now show infinite resistance (no continuity) showing that the switch contacts have opened in response to the system pressure.

The switch must be replaced if it is not functioning properly in response to system or test pump pressure.

150 Hp and V6 Motors

◆ See Figure 6

The pressure sensor used on 150 hp and V6 motors is NOT simply an on/off switch, but instead is of a variable resistor design (similar in operation to the temperatures sensors used by the EFI system, however different in that it changes internal resistance in response to pressure instead of temperature).

Unfortunately neither manufacturer publishes any resistance specifications, they do however publish voltage output specifications (in relation to pressure). The problem is that since the voltage output is also dependent upon voltage INPUT (a 5-volt reference signal from the ECM), the circuit must be complete to test this sensor.

There are 3 basic methods of taking a voltage reading from a complete circuit. The first and least desirable is to try and back-probe the connectors (stick the meter probes in through the back of the wiring connectors, while they are still connected). The best method is to disconnect the circuit

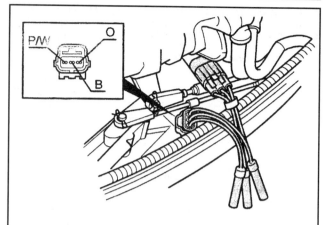

Fig. 6 The best method to check the oil pressure sensor is to use a test harness, Yamaha V6 harness shown (150 hp and Mercury V6 similar, but colors may vary)

temporarily and reconnect it, either with the appropriate test harness (which will contain test points) or using a home-made jumper harness. If you choose the last harness, just be careful to be sure the various jumpers do not touch each other and short the circuit, as this could lead to ECM damage.

■ Yamaha makes a 3-pin test harness for this application (#YB-06769). A 3-pin test harness (#91-888860) is also available from Mercury.

To test the sensor and circuit, proceed as follows:
1. On V6 motors remove the air intake silencer assembly for access.

■ The oil pressure sensor is located on the port side of 150 hp motors, just below the starter motor, therefore the air intake assembly is not in the way.

2. If using a test or jumper harness carefully disengage the sensor wiring, then connect the test/jumper harness.
3. Provide a suitable source of cooling water for the motor, then start and allow the motor to idle for 5 minutes (until fully warmed).
4. Set the DVOM to read DC volts. Place black voltmeter probe on the black ground wire for the circuit, then carefully check each of the other 2 wires. One of them should be a 5-volt reference signal and the other should be the sensor output signal (which is about 2.8 volts @ an idle speed 700 rpm for 150 hp motors or about 3.8 volts @ an idle speed 700 rpm for V6 motors). The signals should be as follows:
• Yamaha models: Pink/White wire is the reference signal from the ECM, while the Orange wire is the output signal from the sensor.
• Mercury models: Usually Green/White wire is the reference signal from the ECM, while the Green wire is the output signal from the sensor.
5. If not done already, measure the Oil Pressure, as detailed earlier in this section (to be sure it is within specification).
6. Replace the sensor if oil pressure and reference voltage is within specification but sensor output voltage is incorrect.

REMOVAL & INSTALLATION

◆ See Figures 4 and 5

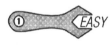

Switch (or sensor, on 150 hp and V6 models) removal and installation is a simple matter of locating the switch, disconnecting the wiring and carefully unthreading it from the port. The switch is normally located high on the side of the powerhead (except for models like the Mercury/Mariner 4/5/6 hp motors where it is threaded into the side of the oil pan on the bottom of the powerhead).

On some models, like the V6 motors, the sensor is obscured by a cover or component. On the V6 motors it is necessary to remove the air intake silencer assembly for access.

Use the switch wire harness coloring (as shown in the Wiring Diagrams of the Ignition and Electrical section) for help if you are unsure as to switch identification.

LUBRICATION AND COOLING 5-5

✳✳ WARNING

When installing the switch be careful not to overtighten and damage the switch or powerhead oil passage.

■ Typically sealant is not used on the switch threads as it could insulate the switch from the powerhead, preventing proper warning system function (the switch must ground the circuit to the powerhead for it to work).

Oil Pump

Although the basic design of the oil pump is the same for all Yamaha and Mercury/Mariner motors with pressurized lubrication systems, the location and service procedures differ slightly between 3 groups.

For the smallest Mercury/Mariner motors, the 4/5/6 hp single cylinder models, the pump is mounted inside the oil pan assembly which is bolted to the bottom of the powerhead. On all 2-4 cylinder, 6-60 hp motors, the oil pump is mounted to the bottom of the cylinder head where it is driven directly off the end of the camshaft. For 75 hp and larger motors, the pump is mounted under the powerhead, in a position where it can be driven by the crankshaft.

REMOVAL, OVERHAUL & INSTALLATION

 DIFFICULT

1-Cylinder 4/5/6 Hp Mercury/Mariner Motors

◆ See Figures 7, 8 and 9

The positioning of the oil pump on these models means that the powerhead itself must be removed from the cowling/engine bracket/lower unit assembly for access. The pump is located inside the oil pan assembly which is bolted to the under side of the powerhead. Luckily, all pump components can be accessed from the underside of the pan, so it should not be necessary to separate the pan from the motor.

1. Disconnect the negative battery cable for safety.
2. For access, remove the powerhead from the cowling/engine bracket/lower unit assembly, as detailed in the Powerhead section.

■ If only the oil pump is being worked on, you need to position the powerhead so the bottom end is facing upward. That can be tough, and will usually require removing one or two components from the top of the powerhead, or positioning the powerhead on wooden blocks in such a way as to prevent damage to components on the top of the powerhead.

3. Remove the 2 bolts securing the oil pump cover to the underside of the oil pan assembly, then remove the cover.
4. Remove and discard the O-ring seal from the pan.
5. Remove the inner pump components (outer rotor, inner rotor and pin).

■ There is a small M6 x 16 bolt and a retaining clip/cover on a boss to the side of the oil pan (right next to the oil pressure switch). This bolt and cover retain the components of the oil pressure blow off valve (Seat, O-ring, Spring and Plunger). It is usually a good idea to at least remove and clean/inspect them at this time. Although it is probably a better idea to just go head and replace the O-ring, spring and plunger).

6. Clean the rotors, housing and cover.
7. Inspect the rotors, housing and cover for signs of obvious damage, contamination or excessive wear. Each of the pump components, including matched rotor sets should be available from your parts dealer. Replace worn or damaged components or the pump assembly, as necessary and desired.
8. Using a sliding caliper with depth gauge, check both the oil pump body (in the oil pan) depth and inside diameter. If they are out of specification the oil pan assembly will have to be replaced. For more details, please refer to the Powerhead section.
9. Check the height of the oil pump outer rotor. If the height is below the wear limit, the oil pump internals will have to be replaced.
10. Temporarily reinstall the rotors and pin to the housing in order to check clearances.
11. Using a feeler gauge gently inserted between the outer rotor and the pump housing, carefully measure and record the outer rotor-to-pump housing (side) clearance.

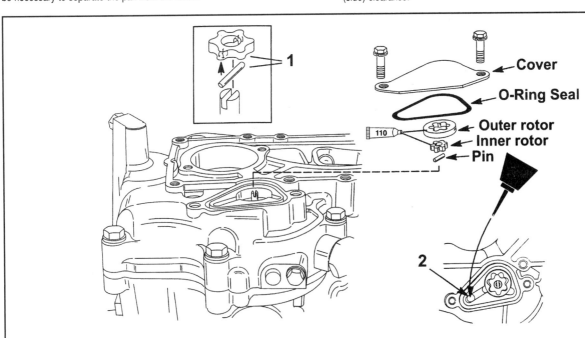

110 ▢ 4-Stroke Outboard Oil

1- Match the Pin with the Pocket in the Inner Rotor
2- Pour Approximately 1 fl. oz. (30 ml) of Engine Oil into the Oil Pump Inlet side

Fig. 7 Exploded view of the oil pump assembly showing installation - 4/5/6 hp Mercury/Mariner motors

5-6 LUBRICATION AND COOLING

■ Remember when using a feeler gauge set, the gauge of the proper clearance measurement will always fit with a very slight drag (do NOT force a gauge into place). The next size smaller gauge will normally have a loose fit, allowing some movement perpendicular to the angle of insertion. The next size larger gauge will normally NOT fit.

12. Using a feeler gauge gently inserted between the outer rotor and the inner rotor, carefully measure and record the outer rotor-to-inner rotor clearance.

13. Using a feeler gauge gently inserted between the top of the outer rotor and a precision straight edge placed across the housing's cover mating surface, carefully measure and record the outer rotor-to-pump cover clearance. Repeat this measurement moving the straightedge to couple different positions.

14. Compare all measurements to the specs provided in the Engine Specifications charts found in the Powerhead section. In some cases, just the rotor set can be replaced, reusing the oil pan/pump housing. Replace the entire oil pan/pump assembly if the housing is damaged or out of specification.

To install:

15. Apply a light coating of clean 4-stroke engine oil to the inner and outer rotors and the drive pin.

16. Position the pin in the driveshaft, then install the inner and outer rotor, respectively.

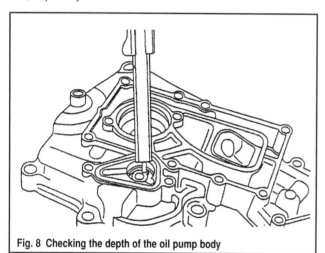

Fig. 8 Checking the depth of the oil pump body

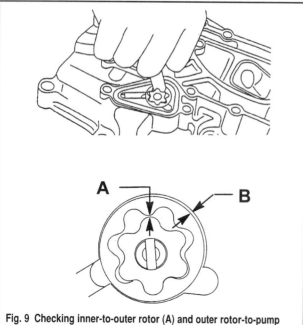

Fig. 9 Checking inner-to-outer rotor (A) and outer rotor-to-pump housing (B) clearance

■ If the pressure relief valve assembly was removed, coat the plunger with engine oil, then install the assembly (using a new O-ring) and secure using the cover and retaining bolt.

17. When looking down at the pump components you'll see a long passage leading away from the pump, that is exposed since the cover is removed. Prime the pump by pouring about 1 oz. (30ml) of fresh engine oil into the passage.

18. Position the new pump cover O-ring, then install the cover and tighten the retaining bolts to 70 inch lbs. (8 Nm).

19. Install the Powerhead to the motor, as detailed in the Powerhead section.

20. Connect the negative battery cable and install an oil pressure gauge, then perform a pressure test to ensure the pump is working properly.

2-4 Cylinder, 6-60 Hp Motors

◆ See Figures 10 thru 19

Unfortunately, clearance is tight (to say the least) and oil pump inspection or service normally requires first removing the cylinder head from the motor for access. On some smaller powerheads, you may want to double-check to see if there is sufficient clearance. Also, depending upon the powerhead removal procedure, you can also opt to remove the whole assembled powerhead (usually complete with fuel and electrical components if the work is done carefully) in order to gain sufficient clearance to remove the oil pump from the cylinder head, while the head remains attached to the cylinder block. This gives you the ability to remove the oil pump without disturbing the flywheel, timing belt, valve train and cylinder head.

1. Disconnect the negative battery cable for safety.
2. For access, either remove the cylinder from the motor or the complete, assembled powerhead from the cowling/engine bracket, as detailed in the Powerhead section.
3. Remove the bolts (usually 3 on 6-15 hp motors or 4 on 25-60 hp motors) securing the pump assembly to the bottom end of the cylinder head. This will free the assembly from the cylinder head.
4. Pull the pump assembly straight out from the bottom of the cylinder head to disconnect the housing from the head and the pump shaft from the end of the camshaft.
5. Remove and discard the gasket and/or O-ring seals from pump housing. The seals vary slightly by model as follows:
   - 6/8 hp motors usually are equipped with a single gasket.
   - 8/9.9 hp (232cc) motors are equipped with a large O-ring for the pump housing, 2 small O-rings (one for the suction and one for the discharge ports) and usually an additional O-ring for the rocker arm shaft.
   - 9.9/15 hp (323cc) motors are either equipped with a single gasket or a gasket AND an O-ring.
   - 25-60 hp motors are normally equipped with a large O-ring for the pump housing and 2 small O-rings (one for the suction and one for the discharge ports).

■ If the pump is being replaced, skip to the appropriate steps of the installation procedure. However, if necessary, the pump may be disassembled for inspection or overhaul. Although it is normally shop practice to matchmark the pump cover to the pump body, the irregular shape of the pumps ands cover used on these motors normally precludes the necessity for this. However, it is still a good idea to matchmark the rotors to the housing before removal to ensure ease of proper assembly.

6. Remove the bolts (usually 2) securing the oil pump cover to the housing, then carefully lift cover from the housing. Remove and discard the packing or O-ring which seals the cover to the housing.

■ The inner and/or outer rotors on some models are equipped with a small punch mark to assist in installation. When equipped, the marks must face a certain direction (the pump housing or pump cover) so note their positioning during removal.

7. Matchmark the inner and outer rotors to each other and the housing ensure proper assembly. Then remove the inner pump components, rotors, driveshaft and, if equipped, shaft pin.
8. Clean the rotors, housing, cover and shaft assembly using a suitable solvent.
9. Inspect the rotors, housing, cover and oil pump shaft for signs of obvious damage, contamination or excessive wear. Each of the pump components, including matched rotor sets should be available from your parts dealer. Replace worn or damaged components or the pump assembly, as necessary and desired.

LUBRICATION AND COOLING 5-7

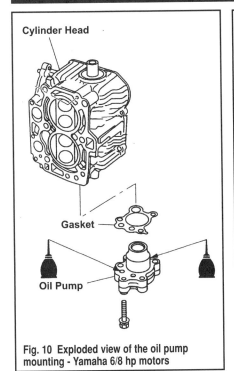

Fig. 10 Exploded view of the oil pump mounting - Yamaha 6/8 hp motors

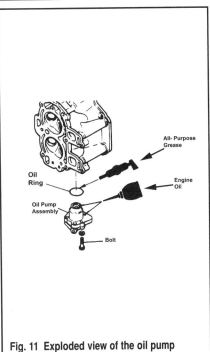

Fig. 11 Exploded view of the oil pump mounting - 8/9.9 hp (232cc) motors

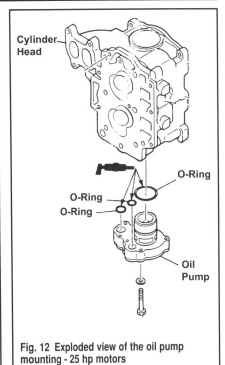

Fig. 12 Exploded view of the oil pump mounting - 25 hp motors

10. Temporarily reinstall the shaft and rotors to the housing in order to check clearances.

11. Using a feeler gauge gently inserted between the outer rotor and the pump housing, carefully measure and record the outer rotor-to-pump housing (side) clearance.

■ Remember when using a feeler gauge set, the gauge of the proper clearance measurement will always fit with a very slight drag (do NOT force a gauge into place). The next size smaller gauge will normally have a loose fit, allowing some movement perpendicular to the angle of insertion. The next size larger gauge will normally NOT fit.

12. Using a feeler gauge gently inserted between the outer rotor and the inner rotor, carefully measure and record the outer rotor-to-inner rotor clearance.

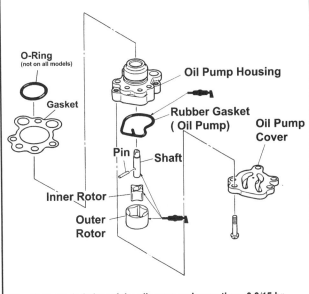

Fig. 13 Exploded view of the oil pump and mounting - 9.9/15 hp (323cc) motors (note pump internals are typical of most Yamaha and Mercury/Mariner pumps, though housing and seal shapes will vary)

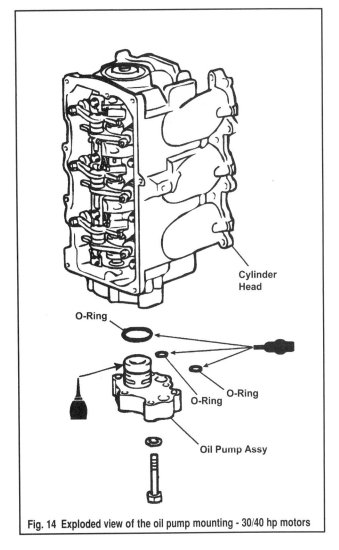

Fig. 14 Exploded view of the oil pump mounting - 30/40 hp motors

5-8 LUBRICATION AND COOLING

13. Using a feeler gauge gently inserted between the top of the outer rotor and a precision straight edge placed across the housing's cover mating surface, carefully measure and record the outer rotor-to-pump cover clearance. Repeat this measurement moving the straightedge to couple different positions.

14. Compare all measurements to the specs provided in the Engine Specifications charts found in the Powerhead section. In some cases, just the rotor set can be replaced, reusing the pump housing. Replace the entire pump assembly if the housing is damaged or out of specification.

To Install:

15. Apply a light coating of clean 4-stroke engine oil or an all-purpose (marine grade) grease to the inner and outer rotors and the oil pump shaft.

16. Install the pump shaft (and pin, if used) along with the inner and outer rotors, aligning any matchmarks made during removal.

■ If the inner rotor contains a punch mark from the manufacturer it must normally face toward the oil pump housing. Conversely, with the outer rotor contains a punch mark from the manufacturer it must normally face toward the oil pump cover.

17. Coat the NEW pump cover O-ring using a suitable marine grease, then position it and the cover over the pump body. Secure the cover to the pump housing using the retaining bolts. Tighten the bolts snugly.

18. Apply a light coating of an All-Purpose marine grease to the NEW pump housing O-rings, then install them to the pump housing and oil passage recesses.

19. For 6/8 hp motors, although Yamaha does not mention the use of any O-rings, they do mention a suction and discharge port on the housing. Yamaha recommends adding 1cc of oil to each port before installation.

20. If equipped with a gasket, install it dry with no oil, grease or sealant.

21. Install the pump assembly to the cylinder head making sure not to pinch or damage the O-rings. As the pump is installed, be sure to align the recess in the oil pump shaft with the camshaft projection.

22. Install the retaining bolts (usually 3 or 4 depending upon the model) securing the pump housing assembly to the cylinder head. Tighten the mounting bolts securely.

23. Install the cylinder head to the motor or the powerhead to the outboard bracket, as applicable and as detailed in the Powerhead section.

24. Connect the negative battery cable and install an oil pressure gauge, then perform a pressure test to ensure the pump is working properly.

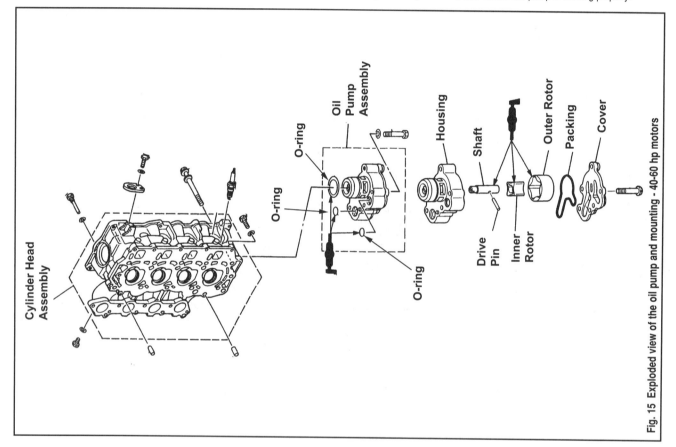

Fig. 15 Exploded view of the oil pump and mounting - 40-60 hp motors

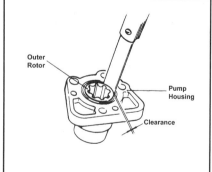

Fig. 16 Measuring outer rotor-to-pump housing (side) clearance

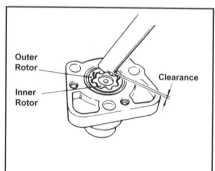

Fig. 17 Measuring outer rotor-to-inner rotor clearance

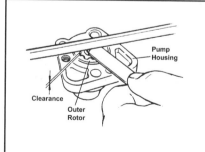

Fig. 18 Measuring outer rotor-to-pump cover (top) clearance

LUBRICATION AND COOLING 5-9

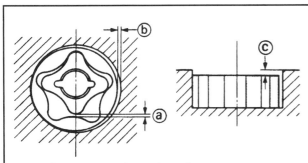

a - Inner-to-outer rotor clearance
b - Outer rotor-to-housing (side) clearance
c - Outer rotor-to-cover clearance

Fig. 19 Oil pump clearance measurements

75-225 Hp Motors

◆ See Figures 20, 21 and 22

The positioning of the oil pump on these models means that the powerhead itself must be removed from the cowling/engine bracket/lower unit assembly for access. On the 75-115 hp motors, the pump is bolted to the top of the engine bracket, while on the 150 hp and V6 models the pump is bolted to the underside powerhead.

1. Disconnect the negative battery cable for safety.
2. For access, remove the powerhead from the cowling/engine bracket/lower unit assembly, as detailed in the Powerhead section.

■ On 150 hp and V6 motors you need to support the powerhead in such a fashion that you can comfortably access the oil pump bolted to the bottom of the assembly. The powerhead is HEAVY, so make sure it is completely secure before working on it. It is best to use an engine stand or place it on a sturdy workbench (rather than keeping it on the hoist), but it must be positioned so there is still access to the pump.

✱✱ CAUTION

If you decide to work on the 150 hp or V6 motor while it is still on a hoist, be sure to block the host in the raised position to ensure it cannot lower or release suddenly potentially causing serious injury or even death.

3. On 75-115 hp models, remove the pump as follows:
   a. Unbolt and remove the lower cowling for additional access.
   b. Remove the 6 bolts securing the oil pump to the top of the upper case assembly. This will free the assembly from the upper case (engine bracket housing), pull it carefully upward and off the outboard.
   c. Remove and discard the 2 O-rings from the bottom of the pump or the top of the case.

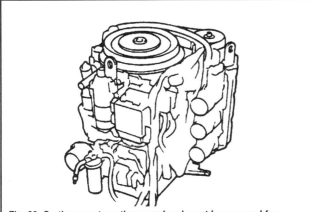

Fig. 20 On these motors, the powerhead must be removed for access (200/225 hp shown)

   d. Double-check to make sure the 2 dowel pins are in position (in the top of the case or bottom of the pump) and are not cracked or damaged.
4. For 150 hp and V6 motors, remove the pump as follows:
   a. Loosen the 4 bolts securing the pump assembly to the underside of the powerhead. This will free the assembly from the motor.
   b. CAREFULLY, pull the assembly straight downward and off the bottom of the crankshaft. Be careful not to cock the crankshaft, potentially scoring it which could lead to oil leaks.
   c. Remove and discard the 2 small O-rings from the ports on either side of the pump.
5. Remove and discard the crankshaft oil seal (75-115 hp motors), the either 1 or 2 oil seals (150 Hp) or the 2 seals for V6 motors. We say 1 or 2 seals for the 150 hp motors because only SOME of the factory illustrations show both the inner and outer seals on top of the oil pump housing and it is possible (although we think unlikely) that the outer seal is not utilized on this application.

■ If the pump is being replaced, skip to the appropriate steps of the installation procedure. However, if necessary, the pump may be disassembled for inspection or overhaul. Although it is normally shop practice to Matchmark the pump cover to the pump body, the irregular shape of the pump and cover used on these motors normally precludes the necessity for this. However, it is still a good idea to Matchmark the rotors to the housing before removal to ensure proper assembly.

■ It is not clear if overhaul components are available from Yamaha or Mercury/Mariner for 75-115 hp motors and no specification for clearances or internal pump tolerances are available for any of these motors. If you decide to try and overhaul the pump, FIRST check with a parts supplier to find if seals, rotors, etc are available, then proceed as follows:

6. Remove the bolts (usually 2) securing the oil pump cover to the housing, then carefully lift cover from the housing. Remove and discard the packing or O-ring which seals the cover to the housing.
7. Matchmark the inner and outer rotors to each other and the housing ensure proper assembly. Then remove the inner pump components, rotors, driveshaft and, if equipped, shaft pin.
8. On 150 hp and V6 motors there are 3 additional seals mounted in the pump housing, underneath the rotors. Carefully remove and discard each of these seals (noting the direction the seal lips face for installation purposes).
9. Clean the rotors, housing, cover and shaft assembly using a suitable solvent.

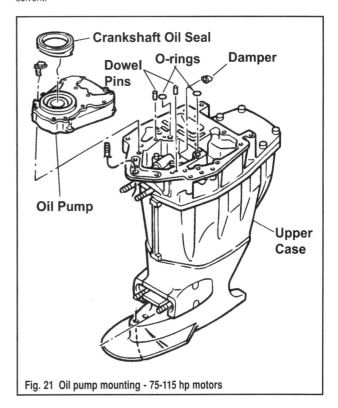

Fig. 21 Oil pump mounting - 75-115 hp motors

5-10 LUBRICATION AND COOLING

10. Inspect the rotors, housing, cover and oil pump shaft for signs of obvious damage, contamination or excessive wear. Each of the pump components, including matched rotor sets should be available from your parts dealer. Replace worn or damaged components or the pump assembly, as necessary and desired.

To install:

11. On 150 hp and V6 motors, install the 3 new seals which are normally mounted in the pump housing, underneath the inner and outer rotors. Coat the lips of the bottom 2 seals using marine grade grease and the lips of the top most of the 3 seals with clean 4-stroke engine oil. Use a suitable driver or smooth socket/length of pipe whose outer diameter matches the structural diameter of the seal to gently tap them squarely in position in the cover facing as noted during removal or as shown in the cutaway portions of the accompanying illustration.

12. Apply a light coating of engine oil to the inner and outer rotors and to the oil pump cover O-ring/packing.

13. Install the inner and outer rotors, aligning any matchmarks made during removal.

14. Carefully position the NEW O-ring and the cover over the pump body. Secure the cover to the pump housing using the retaining bolts. Tighten the bolts snugly.

15. For 150 hp and V6 motors, proceed as follows:
   a. Apply a light coating of engine oil to the 1 or 2 new crankshaft seals (as applicable) and carefully install them to the pump cover using a suitable driver or smoother socket/length of pipe (just like the seals installed inside the pump housing). Make sure the seals are not cocked during installation.
   b. Apply a light coating of marine grade grease to the 2 new small O-rings and install them into the ports on either side of the pump housing.
   c. Carefully install the oil pump assembly to the bottom of the powerhead, sliding it directly upward over the end of the crankshaft (taking care not to score the shaft or damage the oil seals). Install the bolts which secure the pump to the powerhead and tighten securely.

16. For 75-115 hp motors, install the pump assembly as follows:
   a. Apply a light coating of marine grade grease to 2 new O-rings, then position them in the bores on the top of the outboard's upper case.
   b. Make sure the dowel pins are in position in the upper case, then install the oil pump assembly over the dowel pins and secure it to the upper case by tightening the bolts until snug.
   c. Apply a light coating of clean 4-stroke engine oil to a NEW crankshaft seal, then install the seal into the top of the pump. Use a suitable driver or smooth-edged socket/length of pipe which contacts the structural portion of the seal housing to carefully tap the seal into position.

17. Install the Powerhead to the motor, as detailed in the Powerhead section.

18. Connect the negative battery cable and install an oil pressure gauge, then perform a pressure test to ensure the pump is working properly.

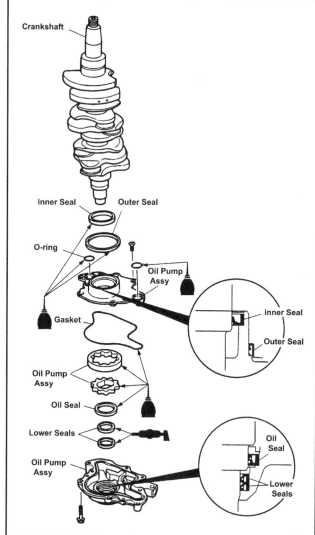

Fig. 22 Exploded view of the oil pump assembly and crankshaft - 200\225 hp motors shown (150 hp motors almost identical, though they may not use the outer seal)

COOLING SYSTEM

◆ See Figure 23

All Yamaha and Mercury/Mariner outboard engines are equipped with a raw water cooling system, meaning that sea, lake or river water is drawn through a water intake in the gearcase lower unit and pumped through the powerhead by a water pump impeller. The exact mounting and location of the pump/impeller varies only slightly from model-to-model, but on all of these motors it is mounted to the lower unit along the gearcase-to-intermediate section split line.

For many boaters, annual replacement of the water pump impeller is considered cheap insurance for a trouble-free boating season. This is probably a bit too conservative for most people, but after a number of trouble-free seasons, an impeller doesn't owe you anything and you should consider taking the time and a little bit of money necessary to replace it. Remember that should an impeller fail you'll be stranded. Not to mention that a worn impeller will simply supply less cooling water than required by specification, allowing the powerhead to run hot placing unnecessary stress on components and best or risking overheating the powerhead at worst. And, to make matters even worse, should an impeller disintegrate the broken bits will likely travel up through the system, potentially forming blocks/clogs in cooling passages.

■ While impeller replacement is no longer usually an annual task, most people we talk to still recommend the impeller be replaced every 2nd or 3rd season to ensure trouble-free operation.

All of these motors are also equipped with a thermostat that restricts the amount of cooling water allowed into the powerhead until the powerhead reaches normal operating temperature. The purpose of the thermostat is to increase engine performance and reduce emissions by making sure the engine warms as quickly as possible to operating temperature and remains there during use under all conditions.

Running a motor without a thermostat may prevent it from fully warming, not only increasing emissions and reducing fuel economy, but it will likely lead to carbon fouling, stumbling and poor performance in general. It can even damage the motor, especially if the motor is then run under load (such as full-throttle operation) without allowing it to thoroughly warm). A restricted thermostat can promote engine overheating. The good news is that should you be caught on the water with a restricted thermostat, you should be able to easily remove it and get back to shore, just make sure you replace it before the next outing.

The water intake grate and cooling passages throughout the powerhead and gearcase comprise the balance of the cooling system. Both components

LUBRICATION AND COOLING 5-11

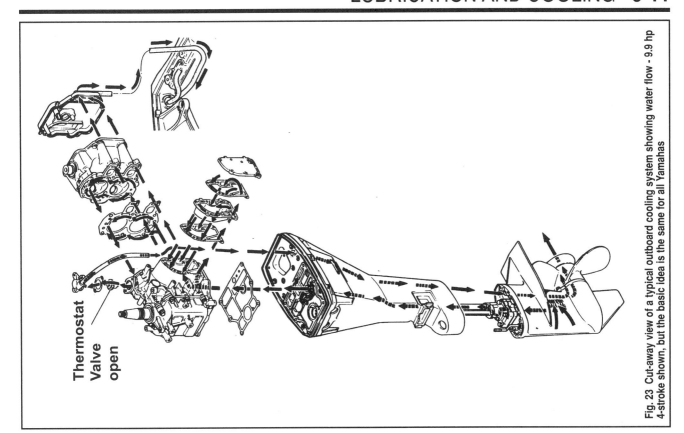

Fig. 23 Cut-away view of a typical outboard cooling system showing water flow - 9.9 hp 4-stroke shown, but the basic idea is the same for all Yamahas

require the most simple, but most frequent maintenance to ensure proper cooling system operation. The water intake grate should be inspected before and after each outing to make sure it is not clogged or damaged. A damaged grate could allow debris into the motor that could clog passages or damage the water pump impeller (both conditions could lead to overheating the powerhead).

Cooling passages have the tendency to become clogged gradually over time by debris and corrosion. The best way to prevent this is to flush the cooling system **after each use** regardless of where you boat (salt or freshwater). But obviously, this form of maintenance is even more important on vessels used in salt, brackish or polluted waters that will promote internal corrosion of the cooling passages.

■ Most V6 (along with a few of the inline) motors are equipped with a spring-loaded pressure relief valve which gives the cooling water an additional escape path to prevent system over-pressurization during warm-up or if there is a clog somewhere in the system.

Description and Operation

◆ See Figures 24 thru 27

The water pump uses an impeller driven by the driveshaft, sealing between an offset housing and lower plate to create a flexing of the impeller blades. The rubber impeller inside the pump maintains an equal volume of water flow at most operating speeds.

At low speeds the pump acts like a full displacement pump with the longer impeller blades following the contour of the pump housing. As pump speed increases, and because of resistance to the flow of water, the impeller vanes bend back away from the pump housing and the pump acts like a centrifugal pump. If the impeller blades are short, they remain in contact throughout the full RPM range, supplying full pressure.

✳✳ WARNING

The outboard should never be run without water, not even for a moment. As the dry impeller tips come in contact with the pump housing or insert, the impeller will be damaged. In most cases, damage will occur to the impeller in seconds.

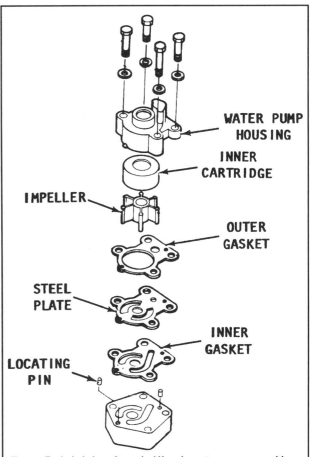

Fig. 24 Exploded view of a typical Yamaha water pump assembly (Merc/Mariner similar)

5-12 LUBRICATION AND COOLING

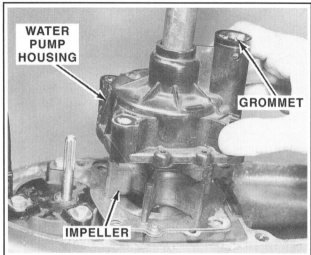

Fig. 25 The water pump housing is mounted to the top of the gearcase lower unit, with the impeller just underneath the housing

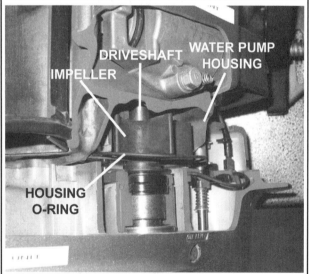

Fig. 26 This cutaway of a Yamaha outboard shows the water pump housing, impeller, O-ring and the driveshaft which turns the impeller

Fig. 27 The water intake grate and cooling passages should be checked and cleaned with each use

On most powerheads, if the powerhead overheats, a warning circuit is triggered by a temperature switch to signal the operator of an overheat condition. This should happen before major damage can occur. Reasons for overheating can be as simple as a plastic bag over the water inlet, or as serious as a leaking head gasket.

Whenever the powerhead is started and the cooling system begins pumping water through the powerhead, a water indicator stream will appear from a cooling system indicator in the engine cover. The water stream fitting commonly becomes blocked with debris (especially when lazy operators fail to flush the system after each use, yes we said LAZY, does this mean YOU?) and ceases flowing. This leads one to suspect a cooling system malfunction. Clean the opening in the fitting using a stiff piece of wire before testing or inspecting other cooling system components.

Whenever water is pumped through the powerhead it absorbs and removes excessive heat. This means that anytime a motor begins to overheat, there is not enough (or no) water flowing to the powerhead. This can happen for various reasons, including a damaged or worn impeller, clogged intake or water passages or a stuck closed/restricted thermostat. A sometimes overlooked cause of overheating is the inability of the linings of the cooling passage to conduct heat. Over time, large amounts of corrosion deposits will form, especially on engines that have not received sufficient maintenance. Corrosion deposits can insulate the powerhead passages from the raw water flowing through them.

Troubleshooting the Cooling System

◆ See Figures 24 thru 27

■ When troubleshooting the cooling system, especially for overheat conditions, the motor should be run in a test tank or on a launched vessel (to simulate normal running conditions.). Running the motor on a flushing device may provide both a higher volume of water than the system would deliver (and often times with water that is much colder than would be found in your local lake or bay).

A water-cooled powerhead has a lot of potential problems to consider when talking about overheating. The most overlooked tends to be the simplest, clogged cooling passages or water intake grate. Although a visual inspection of the intake grate will go a long way, the cooling passage condition can really only be checked by operating the motor or disassembling it to observe the passages.

Damaged or worn cooling system components tend to cause most other problems. And since there are relatively few components, they are easy to discuss. The most obvious is a thermostat that is damaged or corroded, which will often cause the motor to run hot or cold (depending on the position in which the thermostat is stuck, closed or open). The other most obvious component is the water pump impeller is really the heart of the cooling system and it is easy to check, easy that is once it is accessed.

Periodic inspection and replacement of the water pump impeller is a mainstay for many mariners. There are those that wouldn't really consider launching their vessel at the beginning of a season without performing this check. If the water pump is removed for inspection, check the impeller and housing for wear, grooves or scoring that might prevent proper sealing. Check for grooves in the driveshaft where the seal rides. Any damage in these areas may cause air or exhaust gases to be drawn into the pump, putting bubbles into the water. In this case, air does not aid in cooling. When inspecting the pump, consider the following:

• Is the pump inlet clear and clean of foreign material or marine growth? Check that the inlet screen is totally open. How about the impeller?

• Try and separate the impeller hub from the rubber. If it shows signs of loosening or cracking away from the hub, replace the impeller.

• Has the impeller taken a set, and are the blade tips worn down or do they look burned? Are the side sealing rings on the impeller worn away? If so, replace the impeller.

Remember that the life of the powerhead depends on this pump, so don't reuse any parts that look damaged. Are any parts of the impeller missing! If so, they must be found. Broken pieces will migrate up the water tube into the water jacket passages and cause a restriction that could block a water passage. It can be expensive or time consuming to locate the broken pieces in the water passages, but they must be found, or major damage could occur.

■ Impellers are made of highly durable materials these days and although some people still replace them annually, it is much more common to get 2 or 3 seasons out of an impeller before replacing it. And, even after 2-3 seasons, impellers are more often replaced out of preventive maintenance than for necessary repairs.

LUBRICATION AND COOLING 5-13

The best insurance against breaking the impeller is to replace it at the beginning of each boating season (or after a couple of seasons, are you sensing a pattern here?), and to NEVER run it out of the water. If installing a metallic body pump housing, coat all screws with non-hardening sealing compound to retard galvanic corrosion. A water tube normally carries the water from the pump to the powerhead. Grommets seal the water tube to the water pump and exhaust housing at each end of the tube, and can deteriorate. Also, the water tube(s) should be checked for holes through the side of the tube, for restrictions, dents, or kinks.

Overheating at high RPM, but not under light load, may indicate a leaking head gasket. If a head gasket it leaking, water can go into the cylinder, or hot exhaust gases may go into the water jacket, creating exhaust bubbles and excessive heat. Remember that aluminum heads have a tendency to warp, and need to be checked (if not re-surfaced) each time they are removed. If necessary, they can be resurfaced by using emery paper and a surface block moving in a figure-eight motion. If this is done, you should also inspect the cylinders and pistons for damage. Other areas to consider are the exhaust cover gaskets and plate. Look for corrosion pin holes. This is rare, but if the outboard has been operated in salt water over the years, there may just be a problem.

If the outboard is mounted too high on the transom, air may be drawn into the water inlet or sufficient water may not be available at the water inlet. When underway the outboard anti-ventilation plate should be running at or near the bottom of the boat and parallel to the surface of the water. This will allow undisturbed water to come to the lower unit, and the water pick-up should be able to draw sufficient water for proper cooling.

Whenever the outboard has been run in polluted, brackish or saltwater, the cooling system should be flushed. Follow the instructions provided under Flushing the Cooling System in the Engine Maintenance section for more details. But in most cases, the outboard should be flushed for at least five minutes. This will wash the salt from the castings and reduce internal corrosion. If the outboard is small and there is no flushing tool that will fit, run the outboard in a tank, drum, or large, sturdy bucket.

There is no need to run in gear during the flushing operation. After the flushing job is done, rinse the external parts of the outboard off to remove the salt spray.

When service work is done on the water pump or lower unit, all the bolts that attach the lower unit to the exhaust housing, and bolts that hold the water pump housing (unless otherwise specified), should be coated with non-hardening gasket sealing compound to guard against corrosion. If this is not done, the bolts may become seized by galvanic corrosion and may become extremely difficult to remove the next time service work is performed.

Last but not least, check to be sure that the overheat warning system is working properly. On most motors, grounding the wire at the sending unit will cause the horn to sound and/or a light should turn on.

TESTING COOLING SYSTEM EFFICIENCY

◆ See Figure 28

If trouble is suspected, cooling system efficiency can be checked by running the motor in a test tank or on a launched boat (while an assistant navigates) and monitoring cylinder head temperatures. There are 2 common methods available to monitor cylinder head temperature, the use of a heat sensitive marker or an electronic pyrometer.

The Stevens Instrument company markets a product known as the Markal Thermomelt Stik®. This is a physical marker that can be purchased to check different heat ranges. The marker is designed to leave a chalky mark behind on a part of the motor that will remain solid and chalky until it is warmed to a specific temperature, at which point the mark will melt appearing liquid and glossy. When using a Thermomelt Stik or equivalent indicator, markers of 2 different heat specifications are necessary for this test. One, the low temperature range marker, should be rated just equal to or very slightly less than the range of the Thermostat Opening rating. The second, the high temperature range marker should be just equal to or slightly less than the Overheat Thermoswitch On rating.

■ **To determine normal operating temperatures for you motor, please refer to the Cooling System Specifications chart in this section. Generally speaking, engines should run at or slightly above the Thermostat Opening Temperature or anywhere in the range below the Overheat Sensor Off temperature.**

Generally speaking this means you'll want either a 125° (52°C) marker or a 158° (70°C) marker to determine if the motor is reaching normal operating temperature and a higher marker, at least a 163°F (73°C) marker to check for overheating.

Alternately, an electronic pyrometer may be used. Many DVOMs are available with thermo-sensor adapters that can be touched to the cylinder head in order to get a reading. Also, some instrument companies are now producing relatively inexpensive infra-red pyrometers (such as the Raytek® MiniTemp®) of a point-and-shoot design. These units are simply pointed toward the cylinder head while holding down the trigger and the electronic display will give cylinder head temperature. For ease of use and relative accuracy of information, it is hard to beat these infra-red pyrometers. Be sure to follow the tool manufacturer's instructions closely when using any pyrometer to ensure accurate readings.

■ **The infra-red pyrometers are EXTREMELY useful when attempting to find a cooling system blockage. You can, in a matter of seconds, take temperature readings of all different parts/locations on the outboard. Once you've used a pyrometer, you won't go back to the markers.**

To test the cooling system efficiency, obtain either a Thermomelt Stik (or equivalent temperature indicating marker) or a pyrometer and proceed as follows:

1. If available, install a shop tachometer to gauge engine speed during the test.
2. Make sure the proper propeller or test wheel is installed on the motor.
3. Place the motor in a test tank or on a launched craft.

■ **In order to ensure proper readings, water temperature must be approximately 60-80°F (18-24°C).**

4. Start and run the engine at idle until it warms, then run it and about 3000 rpm for **at least** five minutes.
5. Reduce engine speed to a low idle and as proceed as follows depending on the test equipment:

• If using Thermomelt Stiks, make 2 marks on the cylinder head (at the thermostat housing), one with the low-range marker and one with the high-range marker. Continue to operate the motor at idle. The low-range mark must turn liquid and glossy or the engine is being overcooled (if equipped, check the thermostat for a stuck open condition). The high-range mark must remain chalky, or the motor is overheating (if equipped, check the thermostat for a stuck closed condition and then check the cooling system passages and the water pump impeller).

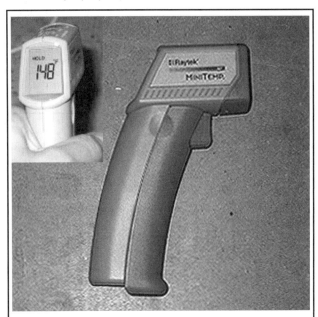

Fig. 28 Bar far the easiest way to check cylinder head temperature is with a hand-held pyrometer like the MiniTemp® from Raytek® pictured herea

5-14 LUBRICATION AND COOLING

■ **For most models, temperature readings should be taken on the cylinder head or powerhead (at or near the thermostat housing). For details on thermostat opening temperature specifications please refer to the Cooling System Specifications chart in this section.**

• If using a pyrometer, take temperature readings on the cylinder head and powerhead. The meter should indicate temperatures between the Thermostat Opening Temp and the Overheat Sensor Off specification (when available). On smaller motors which do not use a thermostat and/or an overheat sensor temperatures should usually be in approximately 125-155°F (53-67°C) range, otherwise it is likely that the engine is being over/under cooled. If equipped, check the thermostat first for either condition and then suspect the cooling water passages and/or the impeller.

6. Increase engine speed to Wide Open Throttle (WOT) rpm and continue to watch the markers or the reading on the pyrometer. The engine must not overheat at this speed either or the system components must be examined further.

■ **When checking the engine at speed expect temperatures to vary slightly from the idle test. Some models will run slightly hotter and some slightly cooler due to the differences in volume of water delivered by the cooling system when compared with engine load. This is normal. In all cases (when equipped), temperatures should not exceed the Overheat Warning Sensor On temperature.**

TESTING THE THERMOSTAT

◆ See Figures 29 and 30

All 2.5 hp and larger motors are equipped with a thermostat that restricts the amount of cooling water allowed into the powerhead until the powerhead reaches normal operating temperature. The purpose of the thermostat is to prevent the water from starting to cool the powerhead until the powerhead has warmed to normal operating temperature. In doing this the thermostat will increase engine performance and reduce emissions.

However, this means that the thermostat is vitally important to proper cooling system operation. A thermostat can fail by seizing in either the open or closed positions, or it can, due to wear or deterioration, open or close at the wrong time. All failures would potentially affect engine operation.

A thermostat that is stuck open or will not fully close, may prevent a powerhead from ever fully warming, this could lead to carbon fouling, stumbling, hesitation and all around poor performance. Although these symptoms could occur at any speed, they are more likely to affect most motors at idle when high water flow through the open orifice will allow for more cooling than the lower production of heat in the powerhead requires.

A thermostat that is stuck closed will usually reveal itself right away as the engine will not only come up to temperature quickly, but the temperature warning circuit may be triggered shortly thereafter. However, a thermostat that is stuck partially closed may be harder to notice. Cooling water may reach the powerhead and keep it within a normal operating temperature range at various engine rpm, but allow heat to build up at other rpm. Generally speaking, engines suffering from this type of thermostat failure will show symptoms at part or full throttle, but problems can occur at idle as well. Symptoms, besides overheating, may include hesitation, stumbling, increased noise and smoke from the motor and, general, poor performance.

Testing a thermostat is a relatively easy proposition. Simply remove the thermostat from the powerhead and suspend it in a container of water, then heat the water watching for the thermostat element to move (open) and noting at what temperature it accomplishes this. If you suspect a faulty or inoperable thermostat and cannot seem to verify proper opening/closing temperatures, it may be a good idea (especially since you've already gone through the trouble of removing the thermostat) to simply replace it (it's a relatively low cost part, that performs an important function). Doing so should remove it from suspicion for at least a couple of seasons.

1. Locate and remove the thermostat from the powerhead, as detailed in this section.

2. Suspend the thermostat and a thermometer in a container of water. For most accurate test results, it is best to hang the thermostat and a thermometer using lengths of string so that they are not touching the bottoms or sides of the container (this ensures that both components remain at the same temperature as the water and not the container).

3. Slowly heat the water while observing the thermostat for movement. The moment you observe movement, check the thermometer and note the temperature. If the water begins to boil (reaches about 212°F/100°C at normal atmospheric pressure) and NO movement has occurred, discontinue the test and throw the piece of junk thermostat away (if you are SURE there was no movement).

4. Remove the source of heat and allow the water to cool (you can speed this up a little by adding some cool water to the container, but if you're using a glass container, don't add too much or you'll risk breaking the container). Observe the thermostat again for movement as the water cools. When movement occurs, check the thermometer and record the temperature.

5. Compare the opening temperature with the specifications provided in the Cooling System Specifications charts in this section.

6. Specifications for closing temperatures are not specifically provided by the manufacturer, but typically a thermostat must close at a temperature near, but below the temperature for the opening specification. A slight modulation (repeated opening and closing) of the thermostat will normally occur at borderline temperatures during engine operation keeping the powerhead in the proper operating range.

7. Replace the thermostat if it does not operate as described, or if you are unsure of the test results and would like to eliminate the thermostat as a possible problem. Refer to the removal and installation procedure for Thermostat in this section for more details.

Water Pump

On all of these Yamaha and Mercury/Mariner motors, the water pump is attached to the inside of the gearcase at the gearcase-to-intermediate housing split-line. The pump itself usually consists of a composite material impeller attached to a portion of the driveshaft that runs through a pump housing and/or cover. Usually there is a replaceable insert installed into the cover which is replaced along with the impeller as a matched set whenever the pump requires service.

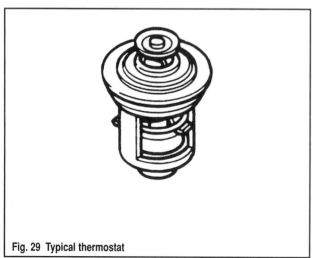

Fig. 29 Typical thermostat

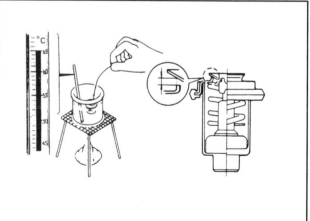

Fig. 30 Test a thermostat by heating it and observing at what temperatures it opens/closes

LUBRICATION AND COOLING 5-15

REMOVAL & INSTALLATION

◆ See Figures 24, 25 and 26, and 31 thru 36

■ For more specific art showing the water pump and related components, please refer to the Lower Unit section.

Since the water pump is mounted at the top of the lower unit (at the case split line) the entire lower unit must be removed from the outboard for access. This is not as difficult as it may sound, but it varies by model so for details, please refer to the Lower Unit section.

1. Remove the lower unit from the intermediate housing for access to the water pump housing.

■ If the water tube came out with the lower unit, remove the tube from the pump housing grommet (so the grommet can be removed and replaced).

2. Loosen and remove the bolts securing the water pump housing to the lower unit, then slide or work the cover up off the driveshaft.

■ Depending on whether or not the impeller pulls off the driveshaft with the housing, either remove it from the housing using a pair of needle-nose pliers or gently pry it upward from the shaft using a prybar.

3. If the impeller did not come off with the cover, work the impeller off the driveshaft.
4. Remove the impeller drive pin or drive key from the driveshaft.

■ Although people have been known to reuse some of the serviceable water pump components, it just doesn't make sense. If you've come this far to fix a water pump problem (or for other service requirements) it is a good idea to replace the impeller, insert, grommet(s) and all O-rings or gaskets to ensure proper water pump operation.

5. Disassemble and service the water pump housing by removing the insert (cartridge), grommet(s), gasket(s) and/or O-ring(s), as applicable depending upon the model being serviced as follows:

• 2.5 Hp Yamaha Models - Above the impeller there is a large O-ring under the cartridge in the water pump housing. Below the impeller there is the outer plate cartridge and a lower plate (through which the shift rod is also installed). There is a shift rod O-ring between the outer plate cartridge and the lower plate. It is usually a good idea to at least remove the outer plate to replace the O-ring, but if the lower plate is removed you'll have to apply a coat of Yamabond No. 4 sealant to both sides of a new lower plate gasket to ensure a proper seal with the lower unit (and you'll also want to make sure the 2 dowel pins are in position). During installation apply a light coat of marine grade grease to the O-ring(s), cartridge insert, impeller and the impeller mating surface of the outer cartridge plate. During installation, align the projection on the top of the cartridge with the hole in the water pump housing.

• 4 Hp Yamaha Models - On these models there is a water tube grommet mounted to the top of the pump, an insert cartridge in the pump housing and a lower cartridge installed with dowel pins (and the water pump housing bolts) to a lower plate. The lower plate is sealed to the gearcase with a gasket, but removal is not necessary and it should not be disturbed by water pump removal. During installation apply a light coating of marine grease to the water tube grommet, insert cartridge, the impeller and the impeller mating surface of the lower cartridge. Also, although the water pump housing is installed to the lower cartridge using a gasket, no mention is made of sealant. If the old gasket was dry sealed, use the same method during installation.

• 4/5/6 Hp Mercury/Mariner Models - On these models there is a water tube grommet on top of the pump housing, as well as a pump pickup tube and pickup tube grommet mounted horizontally to the rear of the housing. The impeller is contained in a replaceable insert under the housing. The inside of the housing, the impeller drive pin and the water tube/pickup grommets should be coated lightly with Quicksilver 2-4-C with Teflon or another suitable marine lubricant during assembly. The impeller rides on a replaceable guide plate. The plate itself does not need to be

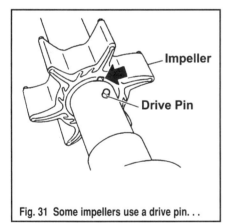

Fig. 31 Some impellers use a drive pin...

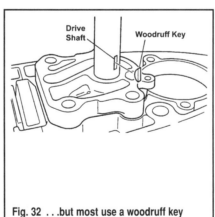

Fig. 32 ...but most use a woodruff key

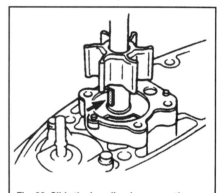

Fig. 33 Slide the impeller down over the pin/key

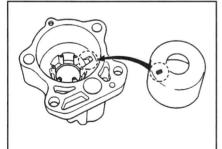

Fig. 34 If applicable, install the insert indexing any tabs, as shown

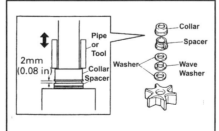

Fig. 35 On 115 hp and larger Yamahas as well as all V6 motors, install the spacer and collar as shown

Fig. 36 Slide the pump housing down over the impeller while turning the driveshaft clockwise

5-16 LUBRICATION AND COOLING

removed/disturbed unless replacement is necessary. A gasket is installed both above and below the guide plate. Lastly, the water tube grommet normally contains 2 small protrusions which must be indexed with holes in the water pump housing during installation.

- **6/8 Hp Yamaha Models** - On these models there is a water tube grommet mounted to the top of the pump, a housing O-ring, an insert cartridge in the pump housing and a lower cartridge (also called a cartridge outer plate) installed with dowel pins to a lower plate on the gearcase. During installation apply a light coating of marine grease to the water tube grommet, housing O-ring, insert cartridge, the impeller and the impeller mating surface of the lower cartridge. Also, keep in mind that the cartridge is indexed to the water pump housing. During replacement, make sure the protrusion on top of the cartridge is inserted in the hole in the water pump housing.
- **8/9.9 Hp (232cc) Yamaha Models** - On these models a water tube grommet is located under a small plate bolted to the top of the housing. The impeller and cartridge insert are mounted between the water pump housing and an outer plate. The outer plate uses a gasket on each side (top and bottom). During installation, be sure to apply a light coating of marine grade grease to the water tube grommet, insert, impeller and the impeller side (top) of the top outer plate gasket.
- **8/9.9 Hp (232cc) Mercury/Mariner Models** - On these models there is a water tube guide and seal position on top of the pump housing. No mention is made of a replaceable insert, which indicates the pump housing IS the impeller/pump mating surface. The impeller itself is mounted over a drive key in the driveshaft, with a washer positioned above and below which helps to lock the key assembly in place. During installation the manufacturer recommends that you coat the outer diameter of the seal on top of the housing with Quicksilver Special Lubricant 101. The inside of the housing and the impeller drive pin should be coated lightly with Quicksilver 2-4-C with Teflon or another suitable marine lubricant during assembly. The impeller rides on a face (guide) plate that is secured to the top of the bearing housing by the water pump retaining bolts. The face plate only uses a gasket on the underside and does not need to be removed/disturbed unless replacement is necessary. Lastly, during assembly, the manufacturer recommends the use of Loctite® 271 or an equivalent threadlocking material on the threads of the housing bolts, making sure to tighten them using a criss-crossing pattern.
- **9.9/15 Hp (323cc) Yamaha Models** - On these models there is a water tube grommet mounted to the top of the pump, a housing O-ring and a lower cartridge (also called a cartridge outer plate) installed with dowel pins and 2 retaining bolts to a lower plate on the gearcase. The 4-stroke 15 hp motors definitely use an insert. During installation apply a light coating of marine grease to the water tube grommet, housing O-ring, insert cartridge (if used) and the impeller.
- **9.9/15 Hp (323cc) Mercury/Mariner Models** - On these models there are two slightly different water pump assemblies, depending upon the model. Standard thrust models use the same gearcase and water pump as the 8/9.9 hp (232cc) Mercury/Mariner motors covered earlier. However for the Bigfoot (high-thrust) models, there is a water tube grommet and a driveshaft seal positioned in the top of the pump housing. A replaceable insert is installed into the water pump housing and the assembly is sealed to the face plate using an O-ring. The impeller is mounted over a drive key in the driveshaft, with a nylon washer positioned above and below. During installation the manufacturer recommends that you coat the inside of the housing with Quicksilver 2-4-C with Teflon or another suitable marine lubricant during assembly. The impeller rides on a face (guide) plate that is secured to the top of the bearing housing by the water pump retaining bolts. The face plate only uses a gasket on the underside and does not need to be removed/disturbed unless replacement is necessary. Lastly, during assembly, the manufacturer recommends the use of Loctite® 271 or an equivalent threadlocking material on the threads of the housing bolts.
- **25 Hp Yamaha Models** - On standard versions of these motors the water pump housing is usually found just under an extension plate and depending upon the model (short or long shaft) there may be a long or short water tube. Either tube uses O-rings on each end. As usual a water seal grommet for the water tube is found on top of the housing. In contrast, the high-thrust versions of these motors utilize a plastic collar and rubber spacer on top of the pump. Under the housing for all types of motors there is an insert cartridge, O-ring, Impeller and lower cartridge plate. On high-thrust versions of this motor the lower plate contains 2 dowel pins and is sealed to the gearcase using a gasket (which should be replaced if the plate is disturbed during this procedure). The lower plate gasket is normally installed dry (but if it was sealed from the factory, we'd recommend using gasket making sealant on it). For all models, during installation be sure to coat the grommet, O-ring(s) and impeller with marine grade grease.
- **25-115 hp Mercury/Mariner Models** - These models may be equipped with one of 2 major designs, generally distinguished by Bigfoot (High-Thrust) vs. Non-Bigfoot (Standard-thrust) models. Though the terminology of Bigfoot is not attached to the larger of these motors 75 hp and above and all of the gearcases for those models can be designated as roughly the same design as the Bigfoot models of the smaller motors. Generally speaking all models are equipped with a water tube seal on top of the pump housing. No insert is shown in any of the procedures or exploded view for these motors, meaning that the impeller working surface is part of the pump housing itself. Also for all models, the impeller is mounted over a drive key in the driveshaft, however only the standard thrust models are equipped with a nylon washer positioned above and below. During installation the manufacturer recommends that you coat the inside of the housing with Quicksilver 2-4-C with Teflon or another suitable marine lubricant during assembly. For standard thrust models only (models WITH the nylon washers above and below the impeller) the manufacturer also recommends that you use Loctite® 405 when installing the water tube seal. For all models the impeller rides on a base (face/guide) plate that is secured to the top of the bearing housing by the water pump retaining bolts. The face plate contains a gasket on both sides (one to seal with the water pump housing and the other to seal with the driveshaft bearing housing). Inspect the base plate on all models to make sure it is not excessively worn or scored. The water pump housing itself on these models should be inspected for wear and replaced if the thickness of the steel at the discharge slots is 0.060 in. (1.5mm) or less and/or if the grooves (other than the impeller sealing bead groove) in the cover roof are more than 0.030 in. (0.75mm) deep. Lastly, during assembly, the manufacturer recommends the use of Loctite® 271 or an equivalent threadlocking material on the threads of the housing bolts.
- **30/40J/40 Hp Yamaha Models** - These models appear to use the same water pump assembly as the high-thrust 25 hp Yamaha motors. On top of the pump is a plastic collar and rubber spacer (also called a seal cover and seal rubber grommet on some models). Under the housing there is an insert cartridge, O-ring, impeller and lower cartridge plate. The lower plate contains 2 dowel pins and is sealed to the gearcase using a gasket (which should be replaced if the plate is disturbed during this procedure). The lower plate gasket is normally installed dry (but if it was sealed from the factory, we'd recommend using gasket making sealant on it). During installation be sure to coat the water tube attaching point, O-ring and impeller with marine grade grease.
- **40/45/50J/50 Hp and 60J/60 Hp Yamaha Models** - Except for the high-thrust 60 hp motors, these models appear to use the same water pump assembly as the high-thrust 25 hp 4-stroke motors. On top of the pump is a plastic collar and rubber spacer. Under the housing there is an insert cartridge, O-ring, impeller and lower cartridge plate. The lower plate contains 2 dowel pins and is sealed to the gearcase using a gasket (which should be replaced if the plate is disturbed during this procedure). The lower plate gasket is normally installed dry (but if it was sealed from the factory, we'd recommend using gasket making sealant on it). During installation be sure to coat the water tube attaching point, insert, O-ring and impeller with marine grade grease. The water pump on the high-thrust 60 hp motor is the same as used on many of Yamaha's 3-cylinder 2-stroke motors. These models are usually, but not always equipped with a replaceable water tube grommet on top of the water pump housing. The housing itself contains an insert (which is 2-piece on a few models) and the impeller. The housing and impeller are sealed to an outer plate cartridge using a gasket. The outer plate is then secured to either a lower water pump housing or oil seal housing using a second gasket (held in position during assembly by at least one dowel pin). The lower water pump housing/oil seal housing itself is ALSO sealed to the lower gearcase using a gasket. You do not need to remove and replace the gaskets on the lower housing or outer plate cartridge, as long as they are undisturbed, however once removed, the gaskets must be replaced. During installation be sure to coat the insert, impeller, impeller mating surface of the outer plate and water tube mounting points with marine grade grease.
- **75/80/90J/90/100 Hp Yamaha Models** - On top of the pump is a grommet (which is secured in place by 2 self-tapping screws) and a rubber seal. When reinstalling self-tapping screws, position them and turn them COUNTERCLOCKWISE (as if loosening them) very slowly until you just feel them fall into place, then turn them slowly clockwise in the old threads). The housing itself contains an insert and the impeller. The housing and impeller are sealed to a cartridge plate using a gasket (the plate also contains 2 dowel pins). The plate is then secured to an oil seal housing using a second gasket. The lower oil seal housing itself is ALSO sealed to the lower gearcase using both an O-ring and a gasket. You do not need to remove and replace the gaskets on the underside of the lower housing or cartridge plate, as long as they are undisturbed, however once removed, the gaskets must

be replaced. During installation be sure to coat the rubber seal, insert, impeller, impeller mating surface of the outer plate and, if removed, the oil seal O-ring with marine grade grease.

• 115 hp and 150 Hp Yamaha Models as well as all (Yamaha/Mercury/Mariner) V6 Models - The water pumps used on all 115 hp and larger Yamaha models, as well as the V6 Mercury/Mariners are of the same basic design. There is a grommet and spacer mounted to the top of the pump housing. Inside the housing there is a small O-ring, an impeller insert and then a large O-ring. The impeller is held in position on the driveshaft by a combination of components listed (from the top down) as follows: A collar, spacer, washer, wave washer and second washer. When installed properly the collar is a tight fit on the spacer but leaving a 0.08 in. (2mm) gap as shown in the accompanying illustration. Underneath the impeller on these models is a lower impeller plate that is sealed to the gearcase using a gasket and at least one (sometimes 2) dowel pin(s). During installation, apply a light coating of marine grade grease (Yamaha grease is obviously recommended by Yamaha while Mercury recommends Quicksilver 2-4-C with Teflon) to the impeller, insert, and O-rings, then slide the impeller carefully down over the driveshaft and into position over the impeller woodruff drive key. Next install the washer, wave washer, washer assembly followed by the spacer and finally the collar. Use a smooth length of pipe (slipped over the driveshaft, but small enough to evenly contact the diameter of the collar) to carefully tap the collar down over the spacer until there is only a 0.08 in. (2.0mm) gap between the collar and spacer. As usual, when replacing the insert, index the tab or tabs on the top of the insert with the hole(s) in the water pump housing and take care not to damage the small O-ring which is should already be installed in the housing at that point.

6. Thoroughly inspect the impeller, cover, insert (if applicable) and impeller plate for signs of damage or wear and replace, as necessary. For more details, please refer to Inspection & Overhaul, in this section.

■ **Regardless of condition, it is always a good idea to replace the impeller (as well as the insert and the O-rings) to ensure proper cooling system operation.**

To install:

7. Replace all O-rings and/or gaskets according to the specific model instructions earlier in this procedure. Coat all appropriate components with marine grade grease.

8. If the lower housing and/or lower cartridge plate was removed, install it now using a new gasket or O-ring, as applicable. Most Yamaha and Mercury/Mariner motors use a gasket, installed dry, however if the old gasket was coated with sealant, take this as a sign that you can/should use sealant this time.

9. Apply a dab of marine grade grease to the impeller drive pin or woodruff key (this will help hold it in position) and insert it into the driveshaft.

10. Carefully slide the impeller down the driveshaft and over the pin or woodruff key.

11. If not done already, assemble the water pump housing. Install any O-rings or grommets and, if removed/equipped, the cartridge insert. Most Yamaha motors and a few of the larger Mercury/Mariners use one or more raised tabs on the top of the insert which must be indexed with a corresponding hole in the water pump housing to ensure proper positioning.

12. On 115 hp and larger Yamahas as well as the V6 Merc/Mariner motors, be sure to install the washers, spacer and collar over top of the impeller as detailed earlier in this procedure.

13. If a gasket is used to seal the housing (as opposed to an O-ring), place the gasket on the lower cartridge insert at this time.

■ **On some motors (except the 115 hp and larger Yamahas and all V6 models, which use the washers, spacer and collar assembly) it may be possible to pre-install the impeller into the housing, but then you've got the issue of locating it properly over the drive pin/woodruff key as the housing is slid into position. Still, if you're having a problem positioning the housing down over the impeller this might be a suitable alternative.**

14. Carefully slide the water pump housing down over the driveshaft until it is JUST above the impeller. Then, while slowly rotating the driveshaft CLOCKWISE when viewed from above, slowly slide the housing down over top of the impeller. The driveshaft must be turned so the blades are inserted into the housing facing the normal direction of rotation.

※※ WARNING

Once the impeller is installed into the water pump NEVER turn the driveshaft counterclockwise, as the impeller vanes will likely become bent back in the wrong direction and could be damaged either by this motion or by the sudden movement bending them back to the proper direction when the outboard is cranked or started. Should the driveshaft be rotated the wrong direction, don't panic, you don't need to disassemble everything again, just make sure you rotate the driveshaft (or the whole outboard is assembled) SLOWLY by hand in the normally direction of rotation to properly reset the impeller vanes before the motor is started.

15. Once the impeller is started correctly into the housing, continue to turn the driveshaft slowly CLOCKWISE and seat the pump to the gearcase.

16. Apply a light coating of Loctite® 572 (Yamaha's recommendation) or Loctite®271 (Mercury/Mariner's usual recommendation) or suitable threadlocking material to the threads of the water pump housing retaining bolts. The threadlock will not only keep the bolts from loosening in service, but will also help prevent the possibility of galvanic corrosion locking them in place when it comes time to service the pump again.

17. Install the water retaining bolts and tighten them securely to fasten the pump to the gearcase.

18. Install the Lower Unit, as detailed in the Lower Unit section. Be careful to properly align the shift linkage, water tube and/or driveshaft as the gearcase is raised into position.

INSPECTION & OVERHAUL

◆ See Figures 37 and 38

■ **Replace the impeller, gaskets and any O-rings/seal whenever the water pump is removed for inspection or service. There is no reason to use questionable parts. Keep in mind that damage to the powerhead caused by an overheating condition (and the subsequent trouble that can occur from becoming stranded on the water) quickly overtakes the expense of a water pump service kit.**

Let's face it, you've gone through the trouble of removing the water pump for a reason. Either, you've already had cooling system problems and you're looking to fix it, or you are looking to perform some preventive maintenance. Although the truth is that you can just remove and inspect the impeller, replacing only the impeller (or even reusing the impeller if it looks to be in good shape). But WHY would you? The cost of a water pump rebuild kit is very little when compared with the even the time involved to get this far. If you've misjudged a component, or an O-ring (which by the way, never reseal quite the same way the second time), then you'll be taking this lower unit off again in the very near future to replace these parts. And, at best this will be because the warning system activated or you noticed a weak coolant indicator streams or, at worst, it will be cause you're dealing with the results of an overheated powerhead.

In short, if there is one way to protect you and your engine, it is to replace the impeller and insert (if used) along with all O-ring seals, grommets and gaskets, anytime the pump housing is removed. If not, take time to thoroughly clean and inspect the old impeller, housing and related components before assembly and installation. New components should be checked against the old. Seek explanations for differences with your parts supplier (but keep in mind that some rebuild kits may contain upgrades or modifications.)

■ **Although Mercury/Mariner occasionally mentions reinstalling a used water pump impeller, they almost NEVER fail to add that they DON'T RECOMMEND IT.**

1. Remove and disassemble the water pump assembly as detailed in this section.

2. If used, carefully remove all traces of gasket or sealant material from components. If some material is stubborn, use a suitable solvent in the next steps to help clean material. Avoid scraping whenever possible, especially on plastic components whose gasket surfaces are easily scored and damaged.

5-18 LUBRICATION AND COOLING

3. Clean all metallic components using a mild solvent (such as Simple Green® cut with water or mineral spirits), then dry using compressed air (or allow them to air dry).

4. Clean plastic components using isopropyl alcohol.

■ Skip the next step, we really mean it, don't INSPECT the impeller, REPLACE IT. Ok, we've been there before, if you absolutely don't want to replace the impeller. Let's say it's only been used one season or so and you're here for another reason, but you're just being thorough and checking the pump, then perform the next step.

5. Check the impeller for missing, brittle or burned blades. Inspect the impeller side surfaces and blade tips for cracks, tears, excessive wear or a glazed (or melted) appearance. Replace the impeller if these defects are found. Next, squeeze the vanes toward the hub and release them. The vanes should spring back to the extended position. Replace the impeller if the vanes are set in a curled position and do not spring back when released.

6. The water pump impeller should move smoothly up or downward on the driveshaft. If not, it could become wedged up against the housing or down against the impeller plate, causing undue wear to the top or bottom of the blades and hub. Check the impeller on the driveshaft and, if necessary, clean the driveshaft contact surface (inside the impeller hub) using emery cloth.

7. Inspect the water pump body or the lining inside the water pump body, as applicable, for burned, worn or damaged surfaces. Replace the impeller lining, if equipped, or the water pump body if any defects are noted.

8. Visually check the water pump housing (and insert, if equipped), along with the impeller wear plate for signs of overheating including warpage (especially on the plate) or melted plastic. Some wear is expected on the impeller plate and, if equipped, on the housing insert, but deep grooves (with edges) are signs of excessive wear requiring component replacement.

■ A groove is considered deep or edged if it catches a fingernail.

9. Check the water tube grommets and seals for a burned appearance or for cracked or brittle surfaces. Replace the grommets and seals if any of these defects are noted.

Fig. 38 A typical water pump/impeller replacement kit - it's worth the $$

After 60 seconds at 1500 rpm.

After 90 seconds at 1500 rpm.

After 30 seconds at 2000 rpm.

After 45 seconds at 2000 rpm.

After 60 seconds at 2000 rpm

Fig. 37 The impeller can be damaged by as little as 30-90 seconds of engine operation without a suitable source of cooling water

LUBRICATION AND COOLING 5-19

WATER PUMP EXPLODED VIEWS

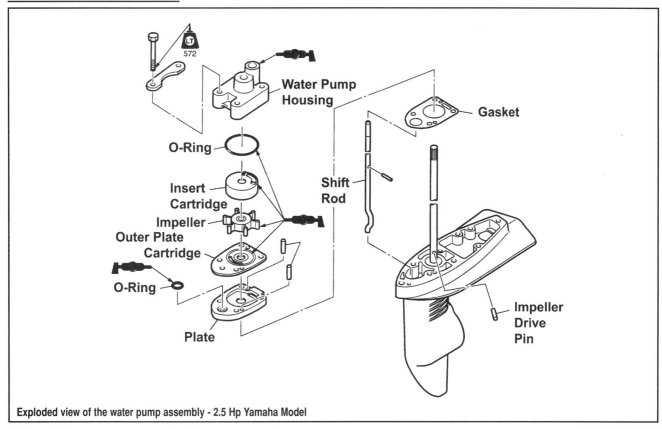

Exploded view of the water pump assembly - 2.5 Hp Yamaha Model

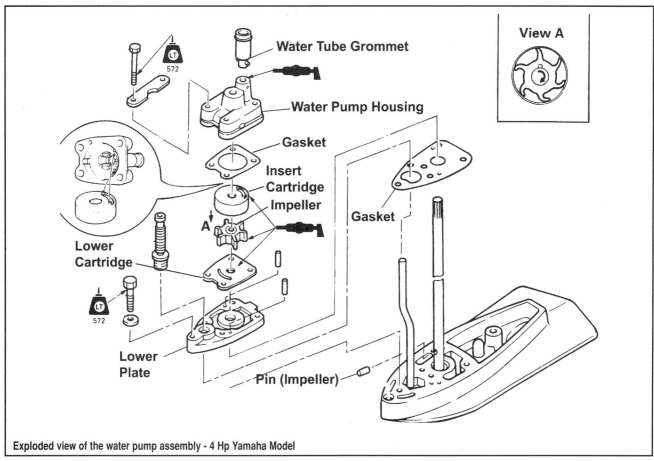

Exploded view of the water pump assembly - 4 Hp Yamaha Model

5-20 LUBRICATION AND COOLING

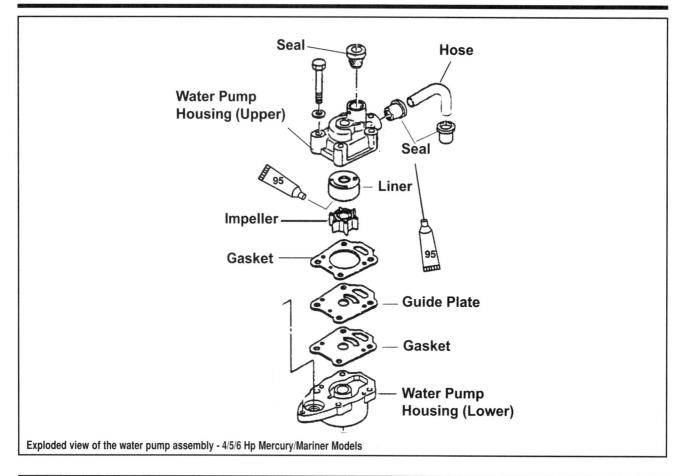

Exploded view of the water pump assembly - 4/5/6 Hp Mercury/Mariner Models

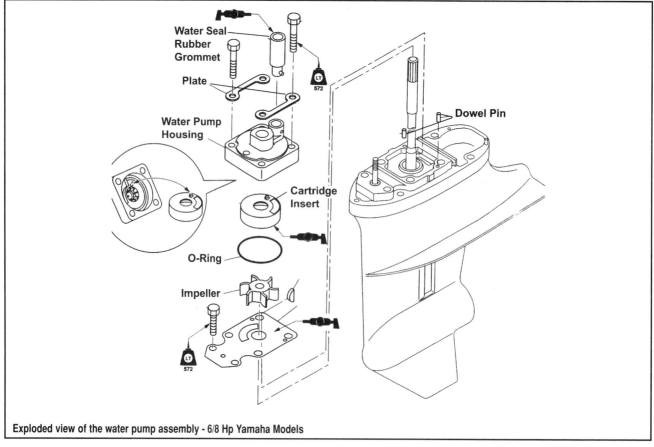

Exploded view of the water pump assembly - 6/8 Hp Yamaha Models

LUBRICATION AND COOLING 5-21

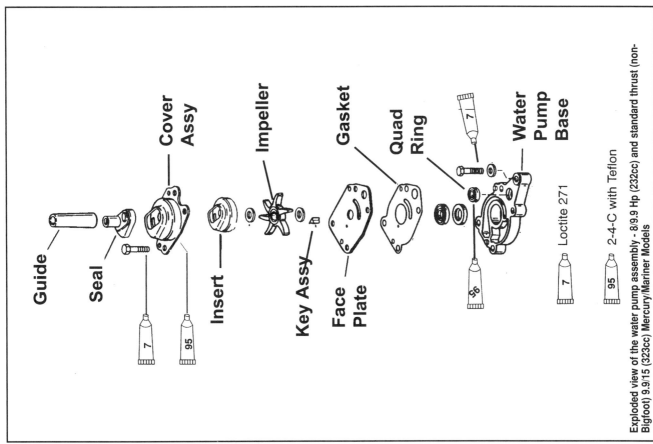

Exploded view of the water pump assembly - 8/9.9 Hp (232cc) and standard thrust (non-Bigfoot) 9.9/15 (323cc) Mercury/Mariner Models

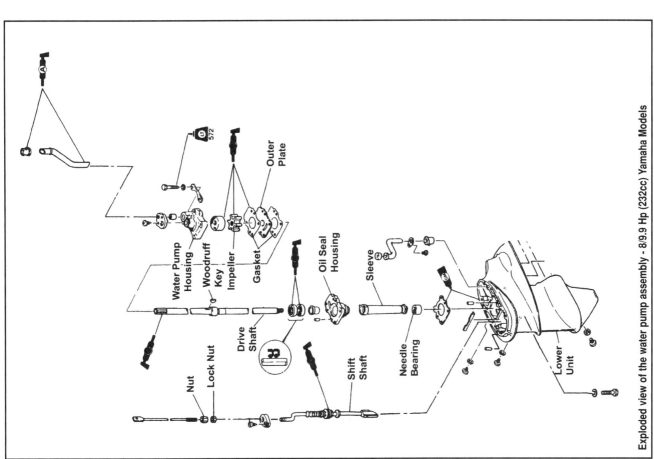

Exploded view of the water pump assembly - 8/9.9 Hp (232cc) Yamaha Models

5-22 LUBRICATION AND COOLING

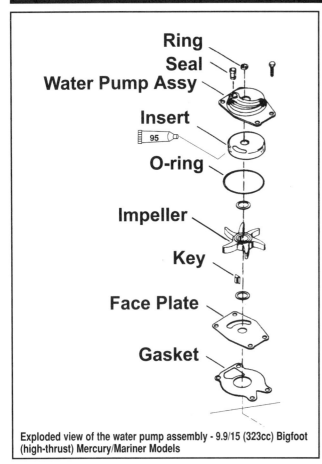

Exploded view of the water pump assembly - 9.9/15 (323cc) Bigfoot (high-thrust) Mercury/Mariner Models

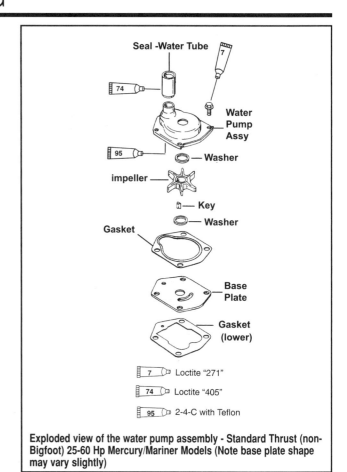

Exploded view of the water pump assembly - Standard Thrust (non-Bigfoot) 25-60 Hp Mercury/Mariner Models (Note base plate shape may vary slightly)

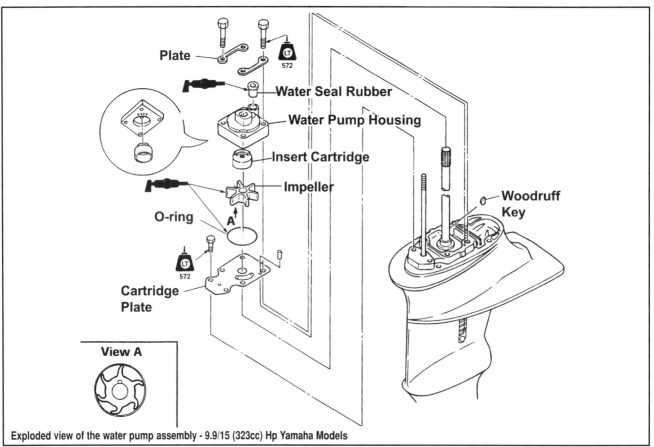

Exploded view of the water pump assembly - 9.9/15 (323cc) Hp Yamaha Models

LUBRICATION AND COOLING 5-23

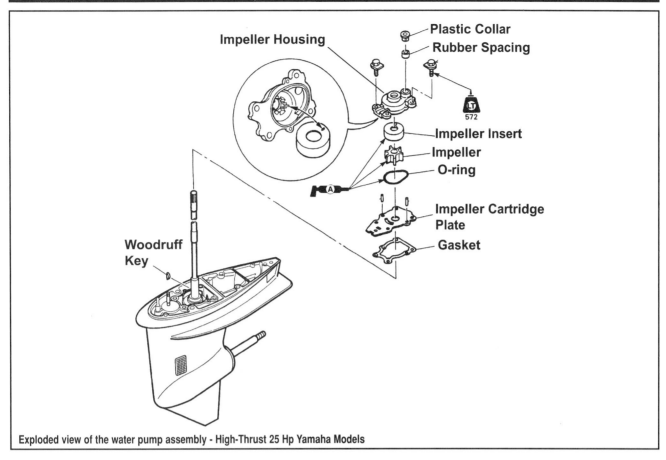

Exploded view of the water pump assembly - High-Thrust 25 Hp Yamaha Models

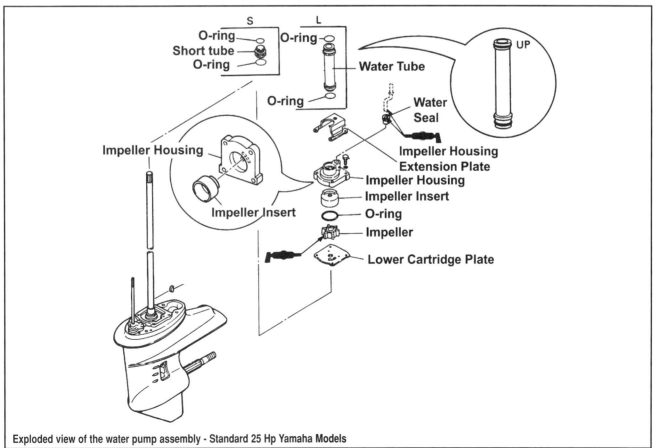

Exploded view of the water pump assembly - Standard 25 Hp Yamaha Models

5-24 LUBRICATION AND COOLING

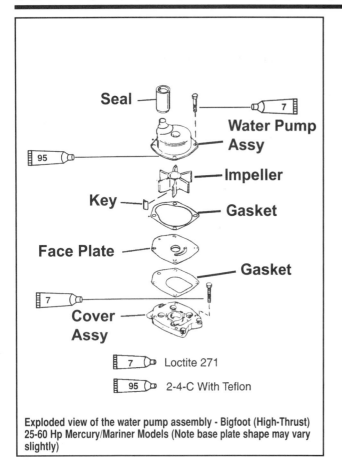

Exploded view of the water pump assembly - Bigfoot (High-Thrust) 25-60 Hp Mercury/Mariner Models (Note base plate shape may vary slightly)

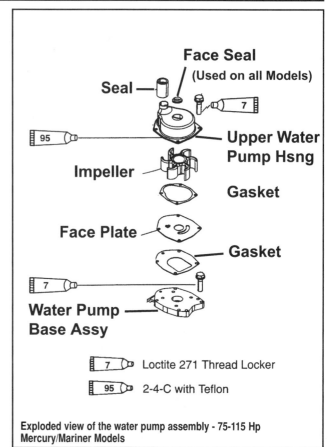

Exploded view of the water pump assembly - 75-115 Hp Mercury/Mariner Models

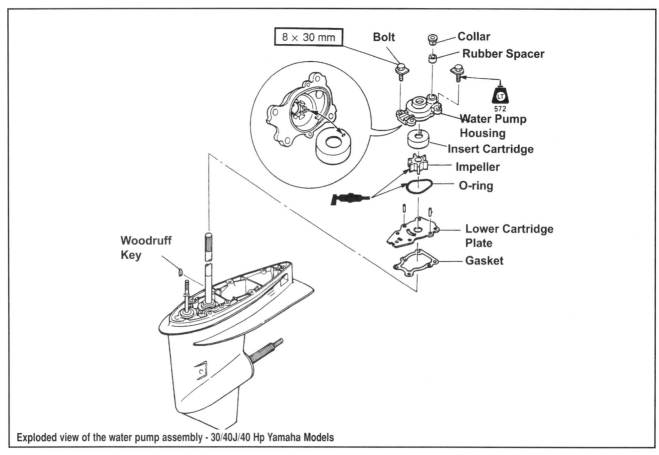

Exploded view of the water pump assembly - 30/40J/40 Hp Yamaha Models

LUBRICATION AND COOLING 5-25

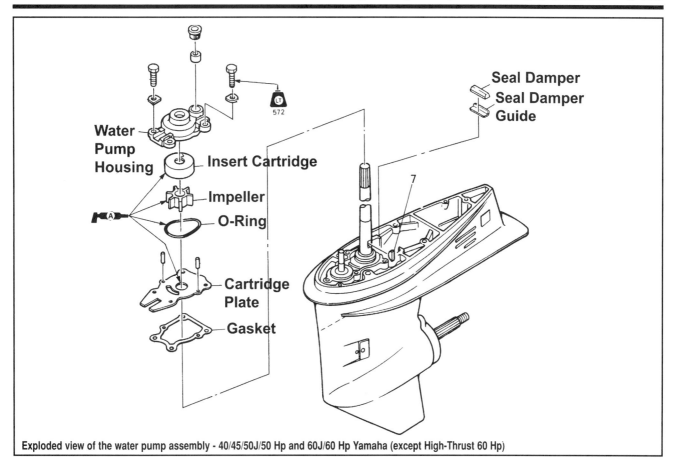

Exploded view of the water pump assembly - 40/45/50J/50 Hp and 60J/60 Hp Yamaha (except High-Thrust 60 Hp)

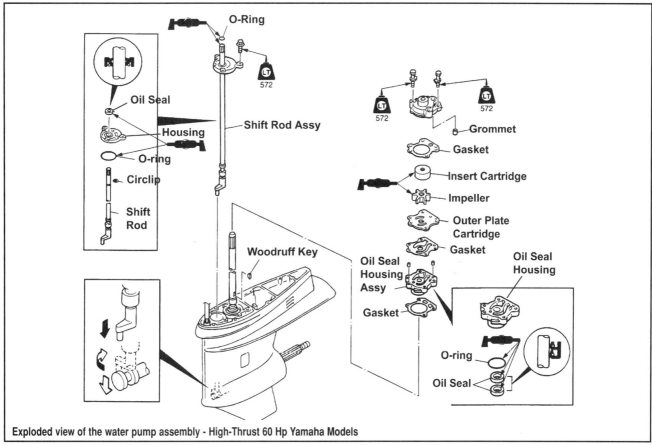

Exploded view of the water pump assembly - High-Thrust 60 Hp Yamaha Models

5-26 LUBRICATION AND COOLING

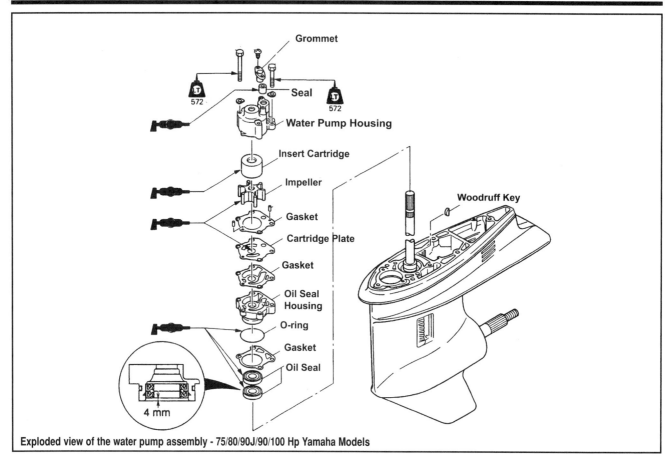

Exploded view of the water pump assembly - 75/80/90J/90/100 Hp Yamaha Models

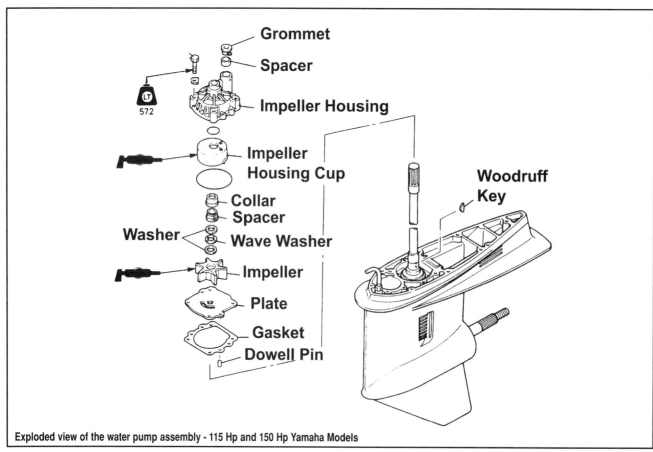

Exploded view of the water pump assembly - 115 Hp and 150 Hp Yamaha Models

LUBRICATION AND COOLING 5-27

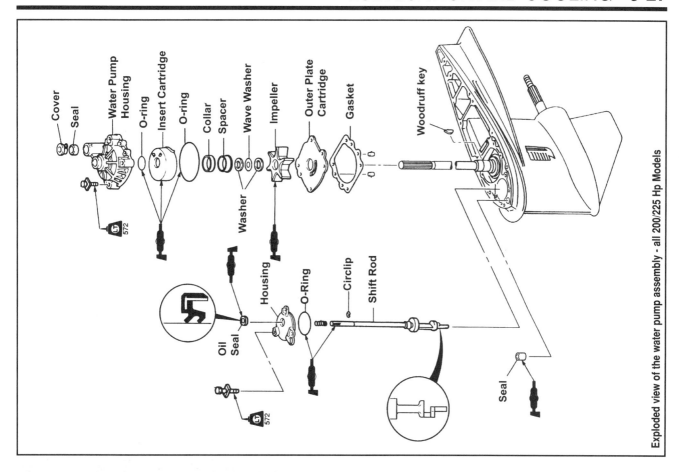

Exploded view of the water pump assembly - all 200/225 Hp Models

Thermostat

All of these motors are equipped with a thermostat that restricts the amount of cooling water allowed into the powerhead until the powerhead reaches normal operating temperature. The purpose of the thermostat is to prevent cooling water from reaching the powerhead until the powerhead has warmed sufficiently. In doing this the thermostat will increase engine performance and reduce emissions.

On all models the thermostat components are mounted somewhere on the powerhead, in a cooling passage, under an access cover that is sealed using a gasket or an O-ring. On these motors the thermostats are normally mounted in one of 3 basic areas of the powerhead, the top of the cylinder head, the top of the block or the side of the block, on the exhaust cover.

REMOVAL & INSTALLATION

◆ See Figures 39, 40 and 41

■ **For more specific art showing the thermostat and cover mounting, please refer to the Powerhead section.**

On all models, the thermostat assembly is mounted under a cover on the powerhead. The size, shape and location of this cover, including the number of components and seals found underneath varies slightly by model. However, even with this said the service procedures are virtually identical for all motors.

Fig. 39 On most models the thermostats are mounted at the top of the cylinder head (pictured), exhaust cover or powerhead

Fig. 40 Most thermostat covers are secured using 2 or 3 bolts - 75-115 hp motors shown

Fig. 41 The corrosion found in this housing shows you why a thermostat could fail

5-28 LUBRICATION AND COOLING

■ On some motors, the cover can be installed facing different directions. For these models, matchmark the cover to the mating surface or otherwise make a note of cover orientation to ensure installation facing the proper direction. This is especially important for covers to which hoses attach (that have been removed).

1. Disconnect the negative battery cable (if equipped) and/or remove the spark plug wire(s) from the plug(s) and ground them on the powerhead for safety.
2. Remove the engine top cover for access.

■ On some motors (including most smaller models under 15 hp and most larger models, 75-115 hp and V6 motors, a hose is attached to the thermostat housing cover. On these motors, either the hose can either be disconnected or, in some cases if there is sufficient play in the hose, it can be left attached while the cover is removed and pushed aside for access to the thermostat. If necessary or desired, cut the wire tie or loosen the hose clamp, then carefully push the hose from the cover fitting.

3. Locate the thermostat housing on the powerhead and remove any interfering components (flywheel or side cover), then remove the cover and thermostat as follows:
   - 2.5 Hp Yamaha Model - The thermostat is located under a small cover on the top of the powerhead, near the oil fill. There is a small hose attached to the irregular-shaped cover which is secured by 3 bolts and a gasket. NOTE that an anode is screwed to the underside of the cover and should be inspected or replaced whenever the cover is removed.
   - 4 Hp Yamaha Model - The thermostat is located under a small cover on the top side of the powerhead, near the oil level plug. There is a small hose attached to the irregular-shaped cover which is secured by 3 bolts and a gasket. NOTE that an anode is screwed to the underside of the cover and should be inspected or replaced whenever the cover is removed.
   - 4/5/6 Hp Mercury/Mariner Models - The thermostat is located under a small, irregular-shaped cover on the top of the cylinder head. The cover is secured by 2 bolts and a gasket.
   - 6/8 Hp Yamahas and all 8/9.9 Hp (232cc) Models - The thermostat is located under a small, irregular-shaped cover on the top of the powerhead, between the drive and driven sprockets for the timing belt. The cover is secured by either 2 or 3 bolts and a gasket (it also has a hose attached to one end).
   - All 9.9/15 Hp (323cc), 25 Hp Twin and 30/40 Hp 3-Cylinder Models - The thermostat is located under a small, irregular-shaped cover bolted to the top of the exhaust cover on side of the powerhead, just slightly above the oil filter. The cover is secured by 2 bolts and a gasket.
   - All 40/45/50 Hp (935cc) and 40/50/60 Hp (996cc) 4-Cylinder Models - The thermostat is located under a small, irregular-shaped cover bolted to the top of the exhaust cover on the side of the powerhead, just a little above the oil filter. The cover is secured by 2 bolts and a gasket. On some models, including all 2001 and later Yamahas motors, the cover is obscured by an electrical component cover which must first be removed.
   - All 75/80/90J/90/100 Hp and 115J/115 Hp Models - The thermostat is located under a small cover directly on top of the cylinder head. The cover has one hose attached to it and can usually be installed in either direction, so be sure to matchmark it to the cylinder head before removal. Otherwise the cover is secured by 2 bolts and is easy to remove.
   - 150 Hp Yamaha Models - The thermostat is located under a small, irregular-shaped cover bolted to the top of the exhaust cover on the side of the powerhead, a ways above the oil filter. The cover is secured by 2 bolts and a gasket.
   - All 200/225 Hp V6 Models - A thermostat is mounted under a cover on top of each side of the block, directly behind the cylinder head mating surface. There is a thermo-switch mounted into the top of each cover and a hose coming off the bottom. Each cover is also secured to the powerhead by 2 bolts and a gasket. There is normally enough play in the hoses to reposition the covers for access without disconnecting them, however, it is usually necessary to remove the thermo-switch or disconnect the wiring.

4. For 150 hp motors, if you are going to remove the cover completely from the motor (as opposed to simply repositioning it for access) you'll have to either unplug the wiring for the thermo-switch and/or remove the switch itself. If you decide to remove the switch it is secured by a holder (a small clip) that is bolted into position.
5. Loosen and remove the bolts (as noted in the previous step) securing the cover, then carefully pull the cover from the powerhead. If necessary, tap around the outside of the cover using a rubber or plastic mallet to help loosen the seal.
6. Check if the seal or gasket was removed with the cover. When a composite gasket and/or sealant was used, make sure all traces of gasket and sealant material are removed from the cover and the powerhead mounting surface. For installation purposes, take note of the direction in which the thermostat is facing before removal.

■ Most Yamaha and Mercury/Mariner motors do NOT use sealant on the thermostat housing gasket. Also, the thermostats on these motors is normally installed with the flange side facing the cover.

7. Visually inspect the thermostat for obvious damage including corrosion, cracks/breaks or severe discoloration from overheating. Make sure any springs have not lost tension. If necessary, refer to the Testing the Thermostat in this section for details concerning using heat to test thermostat function.

To install:
8. Install each of the thermostat components in the reverse of the removal procedure. Replace any gaskets, seals and/or O-rings. Pay close attention to the direction each component is installed.

■ You've got to love Yamaha. When a powerhead torque value is important or a torque sequence is necessary, it is normally molded on the powerhead component itself. Check the thermostat housing cover for numbers near the bolt holes and, if present, follow that order during installation and tightening. Also, check the cover itself for an embossed torque specification and, if you've got a torque wrench available, follow it as well.

9. Install the thermostat housing cover using a new gasket or seal. Tighten the bolts until snug.
10. If removed, connect the hose to the thermostat cover fitting and secure using the clamp or a new wire tie.
11. For 150 hp motors, if removed either reinstall and/or reconnect the wiring for the thermo-switch.
12. Connect the negative battery cable and/or spark plug lead(s), then verify proper cooling system operation.

Thermo-switch

DESCRIPTION & OPERATION

◆ See Figures 42, 43 and 44

All 20 hp or larger Yamaha and Mercury/Mariner outboards are equipped with either a thermo-switch (and on/off switch which is activated by temperature) and/or a thermo-sensor (a variable resistor whose values change with changes in temperatures). The vast majority of Mercury/Mariner models utilize a thermo-sensor, which is covered in the Electronic Fuel Injection system section (for EFI models) or in the Ignition and Electrical System section (for carbureted models). However, the 40/45/50 hp (935cc) Mercury/Mariners utilize a Yamaha style thermo-switch instead of the ECT. Also, V6 Mercury/Mariner models utilize both the ECT for the EFI system AND a pair of thermo-switches (one per cylinder head) for the overheat warning system.

As for Yamaha models, the breakdown goes as follows:
- The 25-40 hp (2 and 3-cylinder) Yamaha motors are equipped with only a single thermo-sensor. The sensor functions both with the YMIS system.
- The 40-60 hp (4-cylinder) Yamaha motors are equipped with only a single thermo-switch which is used both for the overheat warning system and the YMIS.
- The 75-115 hp and larger (4-cylinder) Yamaha motors are equipped with a thermo-sensor (known as a WTS on fuel injected models), however specifications for the motor also mention overheat warning system on/off temperatures (even though no switch is shown in the wiring diagrams). This either means the switch listing is an oversight, OR the warning system uses data from the sensor.
- The Yamaha 150 hp fuel injected motors are equipped with both a thermo-sensor (known as a WTS) for the fuel and ignition system as well as a thermo-switch. The switch is located in a bore in the thermostat housing (which in turn is mounted to the top of the exhaust cover), while the sensor is mounted to the top of the powerhead, near the TPS and IAP sensor.
- All V6 fuel injected motors are equipped with both a thermo-sensor (known as a WTS) for the fuel and ignition system as well as a pair of thermo-switches (usually there is one at the top of each cylinder head) for use with the warning system.

LUBRICATION AND COOLING 5-29

This section deals with the thermo-switches only (on/off temperature activated switches). For information on the variable resistance temperature sensors such as the WTS/ECT used on fuel injected motors, please refer to the Fuel System section. For more information on the variable resistance temperature sensors used on some Yamaha and many Mercury/Mariner CARBURETED motors, please refer to the Ignition and Electrical System section.

A thermo-switch is a bimetallic type on/off switch. Typically a thermo-switch will consist of two different type metals attached together in a small strip. Because each metal has a different coefficient of expansion - each will expand a different amount within a given temperature range, and the strip will bend rather than extend. Depending upon whether the switch is designed to be normally closed or normally open in a given temperature range will determine how the metal switch and contacts are arranged.

Yamaha and Mercury/Mariner motors USUALLY (but not always) use a normally open switch assembly that provides no continuity across the switch contacts until a given temperature is reached. The internals of the switch are designed to bend in such a manner that the switch contacts will only close at or above a certain temperature. In this way, only once the switch physically reaches a given temperature will the switch contacts close, activating the warning circuit.

■ **Specifications for the 40/45/50 hp (935cc) Mercury/Mariners show that is uses a normally CLOSED switch that opens above a certain temperature. Although this MIGHT be true, we're a little skeptical.**

The thermo-switch therefore only sends an ON or OFF signal, not a variable voltage as in the thermo-sensor signal. On V6 powerheads, two thermo-switches are installed - one on each of the cylinder banks. A thermo-switch is required on each cylinder bank, because it is possible for the temperature of one bank to vary considerably from the other bank. The cause for such a difference may be due to blockage of a cooling passage or blockage of an oil hose leading to one cylinder. Either one of these conditions will contribute to increased friction, raising the temperature in or around a particular cylinder. The failure of a mechanical part within a cylinder, such as a broken ring, will cause added friction and increased temperature in the area.

If the powerhead temperature exceeds a specified level, the thermo-switch will close a warning circuit activating a buzzer. Generally, the buzzer will continue to sound until the temperature drops to an acceptable level.

The thermo-switches used by these motors normally has two calibrations. Each has an overheat warning **ON** temperature setting (the temperature at which the switch contacts will close), but they also have an overheat warning **OFF** temperature (which is usually significantly lower than the **ON** temperature). The switches are designed not to close until a certain point has been reached, but once reached, they remain closed until the powerhead (switch) has cooled well below the activation point. For details on Yamaha motors, please refer to the Cooling System Specifications charts in this section.

■ **On some models a separate thermo-sensor is used to signal the CDI unit or control module to limit engine RPM in order to protect the powerhead.**

For Mercury/Mariner models the specifications are as follows:
• Specifications for the 40/45/50 hp (935cc) Mercury/Mariners show that is uses a normally CLOSED switch that opens above a certain temperature. Although this MIGHT be true, we're a little skeptical (if you find it to be the opposite, there is no problem, as long as the warning system seems to function properly. Just remember that if it is true, testing will yield the opposite results of the other thermo-switches we've described here). Specifications claim that the switch will show CONTINUITY (closed circuit) at temperatures below 104 degrees F (40 degrees C) and NO CONTINUITY (open circuit) at or above 131 degrees F (50 degrees C). It should reset somewhere in the middle.
• For V6 Mercury/Mariners the thermo-switches should have NO CONTINUITY (open circuit) at temperatures below 183 degrees F (84 degrees C) and CONTINUITY (closed circuit) above that temperature. Once the circuit is closed it will reset somewhere between 154-179 degrees F (68-82 degrees C).

Fig. 42 Most Yamahas (and a few of the Mercury/Mariners) utilize a thermo-switch mounted to the cylinder head/head cover

Fig. 43 To remove a thermo-switch, grasp and gently pull it from the mounting bore

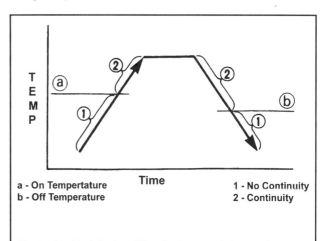

a - On Temperature
b - Off Temperature
1 - No Continuity
2 - Continuity

Fig. 44 Graphical display of Yamaha thermo-switch operation showing switch position (on/off, i.e. continuity/discontinuity) in relation to temperature and time

5-30 LUBRICATION AND COOLING

OPERATIONAL THERMOSWITCH CHECK

◆ See Figures 42, 43 and 44

 MODERATE

■ This section deals with the thermo-switches only (on/off temperature activated switches). For information on the variable resistance temperature sensors such as the WTS/ECT used on fuel injected motors, please refer to the Fuel System section. For more information on the variable resistance temperature sensors used on some Yamaha and many Mercury/Mariner CARBURETED motors, please refer to the Ignition and Electrical System section.

1. Identify the leads (usually Pink and Black, but check the Wiring Diagrams in the Ignition and Electrical section for the motor on which you are working to be sure) from the thermo-switch.
2. Disconnect the two leads at the quick disconnect fittings. This won't have an affect on most Yamaha and Mercury/Mariner motors because the circuit is broken by the switch during normal operation.

■ On the 40/45/50 hp (935cc) Mercury/Mariners if you believe the specs you'll want to leave the wires connected for the first part of the test and disconnect them later when we say to touch them together on other models, that is because NORMAL operation on 935cc Mercs should occur with the circuit complete and ABNORMAL operation should occur with it broken (the opposite of the rest of the motors covered here).

3. Start and run the motor (because you'll be running at above idle, it is really best to use a test tank or launch the craft to prevent the possibility of engine overspeed). Increase engine speed to just above 2500 rpm.
4. Touch the two leads from the switch together (or separate them on 40/45/50 hp (935cc) Mercury/Mariners). The warning buzzer should sound continuously and, on most models, the powerhead rpm should drop to 2000 rpm.
5. Reconnect the wiring to the switch leads, back the throttle down to idle rpm and the buzzer should be silent.
6. On V-configuration motors, repeat this test for the second thermo-switch installed in the other cylinder bank.
7. If the thermo-switch does not operate as specified, either the switch or its circuit are faulty.

CHECKING THERMOSWITCH CONTINUITY

◆ See Figures 42 thru 45

DIFFICULT

■ This section deals with the thermo-switches only (on/off temperature activated switches). For information on the variable resistance temperature sensors such as the WTS/ECT used on fuel injected motors, please refer to the Fuel System section. For more information on the variable resistance temperature sensors used on some Yamaha and many Mercury/Mariner CARBURETED motors, please refer to the Ignition and Electrical System section.

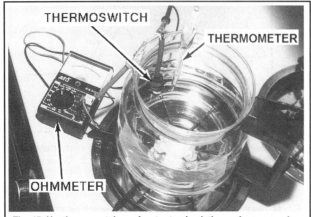

Fig. 45 Heating a container of water to check the performance of the thermo-switch

Locate and remove the thermo-switch from the powerhead. The thermo-switch resembles a condenser, usually with Pink and Black leads (but check the Wiring Diagrams from the Ignition and Electrical System section for the motor on which you are working). The thermo-switch can normally be removed from the powerhead easily, by pulling it out of its recess. However, on a few models, such as on the 40/45/50 hp (935cc) Mercury/Mariners, the thermo-switch may be secured by a cover that is bolted in place. So, if present, make sure you remove any covers before attempting to full the switch free. The thermo-switch is a very fragile sensor and should never be dropped or handled in a rough manner.

Remove the thermo-switch from the block and disconnect the two leads at their quick disconnect fittings. Since the switch should have no continuity (for most models, except the 40/45/50 hp Mercury/Mariners) until it is heated past a point which could damage the powerhead, there are no resistance tests to be performed on this switch while it is installed.

■ Ok, that last statement was too much of a generalization. If the warning system is activated and remains activated on a cold motor, you can check the thermo-switch for continuity across the contacts. If there is continuity even with a cold motor, then the switch is bad, get rid of it! That's is, UNLESS it is a 40/45/50 hp (935cc) Mercury/Mariners for which the specifications say there SHOULD be continuity when switch temperature is below 104 degrees F (40 degrees C).

Testing is a simple matter of watching for continuity across the switch contacts (using an ohmmeter or DVOM set to read resistance across the switch wiring) as the switch is slowly heated to activation temperature. The best method for this is to suspend the switch in water which is slowly heated to temperature while you watch the meter. Obtain a thermometer, an ohmmeter, and a suitable pan in which to boil water.

Place the container of water, at room temperature, on a stove. Secure the thermometer in the water in such a manner to prevent the thermometer bulb from contacting the bottom or sides of the pan (this ensures the thermometer will read the temperature of the water and NOT the pan, which could be considerably higher).

Immerse the sensing end of the thermo-switch into the water up to the shoulder and suspend it, also NOT touching the pan for the same reason as the thermometer.

Make contact with the Red meter lead to the Pink thermo-switch lead, and the Black meter lead to the Black thermo-switch lead. The meter should indicate no continuity, as long as the water temperature is below the Overheat Thermo-switch On temperature listed in the Yamaha Cooling System Specifications charts (or listed earlier on Mercury/Mariners) in this section. Again, in the case of the 40/45/50 hp (935cc) Mercury/Mariners the switch should actually read continuous while the temperature is below specification and then OPEN (read no continuity) once switch temperatures reach 131 degrees F (50 degrees C) or above.

Slowly heat the water, while watching the thermometer and meter. When the temperature rises above the activation point listed in the Yamaha chart (or listed earlier in this section for Mercury/Mariners), the meter should indicate continuity (except on 40/45/50 hp (935cc) Mercury/Mariners, we don't really have to explain it here again do we?).

■ A slight variation is allowable above or below the stated temperatures listed. But keep in mind that too great a variation could either allow damage to the powerhead by operating it under conditions that are too hot or could allow activation of the warning system when the powerhead is not in danger.

Remove the pan from heat and allow the sensor to slowly cool (you can add a little cool water or, if using a metal pan, immerse the pan itself in a second pan of cooler water to help speed the cooling process). Watch the meter and the thermometer as the water cools. The switch contacts must open (showing no continuity) at or near the switch Off temperature listed in the Yamaha charts or listed earlier in this section for Mercury/Mariners. Again, the switches for 40/45/50 hp (935cc) Mercury/Mariners should act the opposite and CLOSE again (showing continuity) below the specified point listed earlier.

If the meter reading is acceptable, the thermo-switch is functioning correctly. If the meter readings are not acceptable, the thermo-switch can be easily replaced with a new unit.

LUBRICATION AND COOLING 5-31

COOLING SYSTEM SPECIFICATIONS - Yamaha 4-Stroke Motors

Model (Hp)	No. of Cyl	Engine Type	Year	Displace cu. in. (cc)	Thermostat Open Temp. Degrees F(C)	Overheat Thermoswitch Degrees F(C) On	Overheat Thermoswitch Degrees F(C) Off
2.5	1	4-stroke	2003-04	4.4 (72)	118-126 (48-52)	-	-
4	1	4-stroke	1999-04	6.7 (112)	136-144 (58-62)	-	-
6	2	IL 4-stroke	2001-04	12 (197)	136-144 (58-62)	-	-
8	2	IL 4-stroke	2001-04	12 (197)	136-144 (58-62)	-	-
8	2	IL 4-stroke	1997-04	14 (232)	136-144 (58-62)	-	-
9.9	2	IL 4-stroke	1995-04	14 (232)	136-144 (58-62)	-	-
15	2	IL 4-stroke	1998-04	20 (323)	136-144 (58-62)	-	-
25	2	IL 4-stroke	1998-04	30 (498)	140-158 (60-70)	See Ignition ①	See Ignition ①
30	3	IL 4-stroke	2001-04	46 (747)	140-158 (60-70)	See Ignition ①	See Ignition ①
40J	3	IL 4-stroke	2002-04	46 (747)	140-158 (60-70)	See Ignition ①	See Ignition ①
40	3	IL 4-stroke	2000-04	46 (747)	140-158 (60-70)	See Ignition ①	See Ignition ①
40	4	IL 4-stroke	1999-00	58 (935)	136-144 (58-62)	169-183 (76-84)	145-170 (63-77)
45	4	IL 4-stroke	1996-00	58 (935)	140-158 (60-70)	169-183 (76-84)	145-170 (63-77)
50J	4	IL 4-stroke	2002	58 (935)	140-158 (60-70)	169-183 (76-84)	145-170 (63-77)
50	4	IL 4-stroke	1995-04	58 (935)	140-158 (60-70)	169-183 (76-84)	145-170 (63-77)
60J	4	IL 4-stroke	2003-04	61 (996)	140-158 (60-70)	169-183 (76-84)	145-170 (63-77)
60	4	IL 4-stroke	2002-04	61 (996)	140-158 (60-70)	169-183 (76-84)	145-170 (63-77)
75	4	IL 4-stroke	2003-04	97 (1596)	140-158 (60-70)	140 (60) ①	118 (48) ①
80	4	IL 4-stroke	1999-02	97 (1596)	140-158 (60-70)	140 (60) ①	118 (48) ①
90J	4	IL 4-stroke	2003-04	97 (1596)	140-158 (60-70)	140 (60) ①	118 (48) ①
90	4	IL 4-stroke	2003-04	97 (1596)	140-158 (60-70)	140 (60) ①	118 (48) ①
100	4	IL 4-stroke	1999-02	97 (1596)	140-158 (60-70)	140 (60) ①	118 (48) ①
115J	4	IL 4-stroke	2003-04	106 (1741)	122-140 (50-60)	140 (60) ①	118 (48) ①
115	4	IL 4-stroke	2000-04	106 (1741)	122-140 (50-60)	140 (60) ①	118 (48) ①
150	4	IL 4-stroke	2004	163 (2670)	136-158 (58-70)	183-194 (84-90)	154-179 (68-82)
200	6	60 V 4-st	2002-03	205 (3352)	140-158 (60-70)	183-194 (84-90)	154-179 (68-82)
225	6	60 V 4-st	2002-03	205 (3352)	140-158 (60-70)	183-194 (84-90)	154-179 (68-82)

Because manufacturers differ with what specifications they choose to provide, please refer to the Mercury/Mariner charts for additional specificitions which MAY apply to the Yamaha model in question as well

① These models are equipped with a thermosensor which supplies the ECM with engine temperatures. Based on these signals, the ECM activates or deactivates the warning signals at the approximate temperatures listed.

LUBRICATION AND COOLING

COOLING SYSTEM SPECIFICATIONS - Mercury/Mariner 4-Stroke Motors

Model (Hp)	No. of Cyl	Engine Type	Year	Displace cu. in. (cc)	Thermostat Open Temp. Degrees F(C)	Thermostat Full Open Temp Degrees F(C)	Valve Lift Minimum in. (mm)
4	1	Single 4-st	1999-04	7.5 (123)	122-129 (50-54)	145-153 (63-67)	0.12 (3.0)
5	1	Single 4-st	1999-04	7.5 (123)	122-129 (50-54)	145-153 (63-67)	0.12 (3.0)
6	1	Single 4-st	2000-04	7.5 (123)	122-129 (50-54)	145-153 (63-67)	0.12 (3.0)
8	2	IL 4-stroke	1997-98	14 (232)	136-144(58-62)②	144 (62) ②	-
9.9	2	IL 4-stroke	1995-98	14 (232)	136-144(58-62)②	144 (62) ②	-
9.9	2	IL 4-stroke	1999-04	20 (323)	136-144 (58-62)	158 (70)	0.12 (3.0)
15	2	IL 4-stroke	1999-04	20 (323)	136-144 (58-62)	158 (70)	0.12 (3.0)
25	2	IL 4-stroke	1998-04	30 (498)	136-144 (58-62)	158 (70)	0.12 (3.0)
30 Carb	3	IL 4-stroke	1999-04	46 (747)	136-144 (58-62)	158 (70)	0.12 (3.0)
30 EFI	3	IL 4-stroke	2002-04	46 (747)	118-123 (48-51)	145 (63)	-
40 Carb	3	IL 4-stroke	1999-04	46 (747)	136-144 (58-62)	158 (70)	0.12 (3.0)
40 EFI	3	IL 4-stroke	2002-04	46 (747)	118-123 (48-51)	145 (63)	-
40	4	IL 4-stroke	1997-00	58 (935)	136-144 (58-62)	158 (70)	0.12 (3.0)
45	4	IL 4-stroke	1995-00	58 (935)	136-144 (58-62)	158 (70)	0.12 (3.0)
50	4	IL 4-stroke	1995-00	58 (935)	136-144 (58-62)	158 (70)	0.12 (3.0)
40 Carb	4	IL 4-stroke	2001	61 (996)	118-123 (48-51)	145 (63)	0.12 (3.0)
40 EFI	4	IL 4-stroke	2002-04	61 (996)	118-123 (48-51)	145 (63)	0.12 (3.0)
50 Carb	4	IL 4-stroke	2001	61 (996)	118-123 (48-51)	145 (63)	0.12 (3.0)
50 EFI	4	IL 4-stroke	2002-04	61 (996)	118-123 (48-51)	145 (63)	0.12 (3.0)
60 Carb	4	IL 4-stroke	2001	61 (996)	118-123 (48-51)	145 (63)	0.12 (3.0)
60 EFI	4	IL 4-stroke	2002-04	61 (996)	118-123 (48-51)	145 (63)	0.12 (3.0)
75	4	IL 4-stroke	2000-04	97 (1596)	140 (60)	158 (70)	0.12 (3.0)
90	4	IL 4-stroke	2000-04	97 (1596)	140 (60)	158 (70)	0.12 (3.0)
115 EFI	4	IL 4-stroke	2001-04	106 (1741)	122 (50)	140 (60)	0.17 (4.3)
225 EFI	6	60 V 4-stroke	2003-04	205 (3352)	140 (60)	158 (70)	0.17 (4.3)

Because manufacturers differ with what specifications they choose to provide, please refer to the Yamaha charts for additional specificitions which MAY apply to the Mercury/Mariner model in question as well

① These models are equipped with a thermosensor which supplies the ECM with engine temperatures. Based on these signals, the ECM activates or deactivates the warning signals at the approximate temperatures listed.

② Mecury provides no specifications for this model, so specs are from the Yamaha manuals for the equivalent model

6
POWERHEAD

POWERHEAD MECHANICAL	6-2
POWERHEAD REFINISHING	6-121
POWERHEAD BREAK-IN	6-136
SPECIFICATIONS	6-137

POWERHEAD BREAK-IN ... 6-136
- BREAK-IN PROCEDURES ... 6-136

POWERHEAD MECHANICAL ... 6-2
- GENERAL INFORMATION ... 6-2
  - CLEANLINESS ... 6-3
  - IMPORTANT POINTS FOR POWERHEAD OVERHAUL ... 6-2
  - POWERHEAD COMPONENTS ... 6-3
- 2.5 HP AND 4 HP YAMAHA ... 6-4
  - POWERHEAD REMOVAL & INSTALLATION ... 6-3
  - CLEANING & INSPECTION ... 6-12
  - CYLINDER BLOCK OVERHAUL ... 6-9
  - CYLINDER HEAD OVERHAUL ... 6-8
  - CYLINDER HEAD REMOVAL & INSTALLATION ... 6-5
  - POWERHEAD REMOVAL & INSTALLATION ... 6-3
- 4/5/6 HP MERCURY/MARINER ... 6-12
  - CLEANING & INSPECTION ... 6-18
  - CYLINDER BLOCK OVERHAUL ... 6-15
  - CYLINDER HEAD OVERHAUL ... 6-14
  - CYLINDER HEAD REMOVAL & INSTALLATION ... 6-13
  - POWERHEAD REMOVAL & INSTALLATION ... 6-12
- 6/8 HP YAMAHA AND ALL 8/9.9 HP (232cc) ... 6-18
  - CLEANING & INSPECTING ... 6-36
  - CYLINDER BLOCK OVERHAUL ... 6-30
  - CYLINDER HEAD OVERHAUL ... 6-26
  - CYLINDER HEAD REMOVAL & INSTALLATION ... 6-24
  - POWERHEAD REMOVAL & INSTALLATION ... 6-18
  - TIMING BELT/SPROCKET REMOVAL & INSTALLATION ... 6-21
- 9.9/15 HP (323cc) ... 6-36
  - CLEANING & INSPECTING ... 6-45
  - CYLINDER BLOCK OVERHAUL ... 6-42
  - CYLINDER HEAD OVERHAUL ... 6-41
  - CYLINDER HEAD REMOVAL & INSTALLATION ... 6-40
  - POWERHEAD REMOVAL & INSTALLATION ... 6-36
  - TIMING BELT/SPROCKET REMOVAL & INSTALLATION ... 6-38
- 25 HP ... 6-45
  - CLEANING & INSPECTING ... 6-56
  - CYLINDER BLOCK OVERHAUL ... 6-51
  - CYLINDER HEAD OVERHAUL ... 6-50
  - CYLINDER HEAD REMOVAL & INSTALLATION ... 6-49
  - POWERHEAD REMOVAL & INSTALLATION ... 6-45
  - TIMING BELT/SPROCKET REMOVAL & INSTALLATION ... 6-47
- 30/40 HP 3-CYLINDER ... 6-56
  - CLEANING & INSPECTING ... 6-66
  - CYLINDER BLOCK OVERHAUL ... 6-63
  - CYLINDER HEAD OVERHAUL ... 6-61
  - CYLINDER HEAD REMOVAL & INSTALLATION ... 6-59
  - POWERHEAD REMOVAL & INSTALLATION ... 6-56
  - TIMING BELT/SPROCKET REMOVAL & INSTALLATION ... 6-58
- FOUR-CYLINDER 40/45/50 HP (935cc) AND 30/40/60 HP (966cc) ... 6-66
  - CYLINDER BLOCK OVERHAUL ... 6-78
  - CYLINDER HEAD OVERHAUL ... 6-76
  - CYLINDER HEAD REMOVAL & INSTALLATION ... 6-74
  - POWERHEAD REMOVAL & INSTALLATION ... 6-66
    - MERCURY/MARINER MODELS ... 6-66
    - YAMAHA MODELS ... 6-67
  - STRIPPING THE POWERHEAD FOR REBUILD ... 6-71
    - MERCURY/MARINER MODELS ... 6-71
    - YAMAHA MODELS - ELECTRICAL UNIT REMOVAL & INSTALLATION ... 6-71
  - TIMING BELT/SPROCKET REMOVAL & INSTALLATION ... 6-69
- 75/80/90/100 HP AND 115 HP POWERHEADS ... 6-81
  - CLEANING & INSPECTING ... 6-97
  - CYLINDER BLOCK OVERHAUL ... 6-92
  - CYLINDER HEAD REMOVAL & INSTALLATION ... 6-88
  - POWERHEAD REMOVAL & INSTALLATION ... 6-81
    - MERCURY/MARINER MODELS ... 6-81
    - YAMAHA MODELS ... 6-83
  - TIMING BELT/SPROCKET REMOVAL & INSTALLATION ... 6-86
  - VALVE TRAIN OVERHAUL ... 6-92
- 150 HP YAMAHA ... 6-97
  - CLEANING & INSPECTING ... 6-107
  - CYLINDER BLOCK OVERHAUL ... 6-103
  - CYLINDER HEAD REMOVAL & INSTALLATION ... 6-100
  - POWERHEAD REMOVAL & INSTALLATION ... 6-97
  - TIMING BELT/SPROCKET REMOVAL & INSTALLATION ... 6-98
  - VALVE TRAIN OVERHAUL ... 6-102
- 200/225 HP V6 ... 6-107
  - CLEANING & INSPECTING ... 6-121
  - CYLINDER BLOCK OVERHAUL ... 6-116
  - CYLINDER HEAD REMOVAL & INSTALLATION ... 6-112
  - POWERHEAD REMOVAL & INSTALLATION ... 6-107
  - TIMING BELT/SPROCKET REMOVAL & INSTALLATION ... 6-109
  - VALVE TRAIN OVERHAUL ... 6-115

POWERHEAD REFINISHING ... 6-121
- CONNECTING RODS ... 6-125
  - INSPECTION ... 6-125
- CRANKSHAFT ... 6-123
  - INSPECTION ... 6-123
  - MEASURING BEARING CLEARANCE USING PLASTIGAGE ... 6-124
- CYLINDER BLOCK ... 6-128
  - BLOCK & CYLINDER HEAD WARPAGE ... 6-130
  - HONING CYLINDER WALLS ... 6-129
  - INSPECTION ... 6-128
- CYLINDER HEAD COMPONENTS (CAMSHAFT/VALVE TRAIN) ... 6-132
  - CHECKING CAMSHAFT ... 6-132
  - CHECKING ROCKER ARMS & SHAFTS ... 6-133
  - CHECKING VALVE ASSEMBLIES ... 6-134
  - CHECKING VALVE LEAKAGE ... 6-132
- EXHAUST COVER ... 6-123
  - INSPECTION ... 6-123
- GENERAL INFORMATION ... 6-121
  - CLEANING ... 6-121
- PISTONS ... 6-126
  - INSPECTION ... 6-126
  - OVERSIZE PISTONS & RINGS ... 6-128
  - PISTON RING SIDE CLEARANCE ... 6-128
  - RING END-GAP CLEARANCE ... 6-127

SPECIFICATIONS ... 6-137
- MERCURY/MARINER
  - 4/5/6 HP (123cc) 1-CYL ... 6-139
  - 8/9.9 HP (232cc) 2-CYL ... 6-142
  - 9.9/15 HP (323cc) 2-CYL ... 6-144
  - 25 HP (498cc) 2-CYL ... 6-146
  - 30/40 HP (747cc) 3-CYL ... 6-148
  - 40/45/50 HP (935cc) 4-CYL ... 6-150
  - 40/50/60 HP (996cc) 4-CYL ... 6-152
  - 75/90 HP (1596cc) 4-CYL ... 6-155
  - 115 HP (1741cc) 4-CYL ... 6-157
  - 225 HP (3352cc) V6 ... 6-161
- YAMAHA ... 6-137
  - 2.5 HP (72cc) 1-CYL ... 6-137
  - 4 HP (112cc) 1-CYL ... 6-138
  - 6/8 HP (197cc) 2-CYL ... 6-140
  - 8/9.9 HP (232cc) 2-CYL ... 6-141
  - 15 HP (323cc) 2-CYL ... 6-143
  - 25 HP (498cc) 2-CYL ... 6-145
  - 30/40J/40 HP (747cc) 3-CYL ... 6-147
  - 40/45/50J/50 HP (935cc) 4-CYL ... 6-149
  - 60J/60 HP (996cc) 4-CYL ... 6-151
  - 75/80/90J/90/100 HP (1596cc) 4-CYL ... 6-153
  - 115J/115 HP (1741cc) 4-CYL ... 6-156
  - 150 HP (2670cc) 4-CYL ... 6-158
  - 200/225 HP (3352cc) V6 ... 6-159

POWERHEAD MECHANICAL

General Information

You can compare the major components of an outboard with the engine and drivetrain of your car or truck. In doing so, the powerhead is the equivalent of the engine and the gearcase is the equivalent of your drivetrain (the transmission/transaxle). The powerhead is the assembly that produces the power necessary to move the vehicle, while the gearcase is the assembly that transmits that power via gears, shafts and a propeller (instead of tires).

Speaking in this manner, the powerhead is the "engine" or "motor" portion of your outboard. It is an assembly of long-life components that are protected through proper maintenance. Lubrication, the use of high-quality oils and frequent oil inspection/changes are the most important ways to preserve powerhead condition. Similarly, proper tune-ups that help maintain proper air/fuel mixture ratios and prevent pinging, knocking or other potentially damaging operating conditions are the next best way to preserve your motor. But, even given the best of conditions, components in a motor begin wearing the first time the motor is started and will continue to do so over the life of the powerhead.

Eventually, all powerheads will require some repair. The particular broken or worn component, plus the age and overall condition of the motor may help dictate whether a small repair or major overhaul is warranted. As much as you can generalize about mechanical work, the complexity of the job will vary with 2 major factors:
- The age of the motor (the older OR less well maintained the motor is) the more difficult the repair
- The larger and more complex the motor, the more difficult the repair.

Again, these are generalizations and, working carefully, a skilled do-it-yourself boater can disassemble and repair a 225 hp, fuel injected V6 powerhead, as well as a seasoned professional. But both DIYers and professionals must know their limits. These days, many professionals will leave portions of machine work (from cylinder block and piston disassembly, cleaning and inspection to honing and assembly up to a machinist). This is not because they are not capable of the task, but because that's what a machinist does day in and day out. A machinist is naturally going to be more experienced with the procedures.

If a complete powerhead overhaul is necessary on your outboard, we recommend that you find a local machine shop that has both an excellent reputation and that specializes in marine work. This is just as important and handy a resource to the professional as a DIYer. If possible, consult with the machine shop before disassembly to make sure you follow procedures or mark components, as they would desire. Some machine shops would prefer to perform the disassembly themselves. In these cases, you can usually remove the powerhead from the gearcase and deliver the entire unit to the shop for disassembly, inspection, machining and assembly.

■ **There are a number of major outboard powerhead re-manufacturers throughout the U.S. and you may be able to quite cost effectively purchase a replacement powerhead for installation making it an attractive option. At least it is an option you should consider before undertaking a major overhaul.**

If you decide to perform the entire overhaul yourself, proceed slowly, taking care to follow instructions closely. Consider using a digital camera (if available) to help document component mounting during the removal and disassembly procedures. This can be especially helpful if the overhaul or rebuild is going to take place over an extended amount of time. If this is your first overhaul, don't even THINK about trying to get it done in one weekend, YOU WON'T. It is better to proceed slowly, asking for help when necessary from your trusted parts counterman or a tech with experience on these motors.

Keep in mind that anytime pistons, rings and bearings have been replaced, the powerhead must be broken-in again, as if it were a brand-new motor. Once a major overhaul is completed, refer to the section on Powerhead Break-In for details on how to ensure the rings set properly without damage or scoring to the new cylinder wall or the piston surfaces. Careful break-in of a properly overhauled motor will ensure many years of service for the trusty powerhead.

IMPORTANT POINTS FOR POWERHEAD OVERHAUL

Repair Procedures

Service and repair procedures will vary slightly between individual models, but the basic instructions are quite similar. Special tools may be called out in certain instances. These tools may be purchased from the local marine dealer.

✹✹ CAUTION

When removing the powerhead, always secure the gearcase in a suitable holding fixture to prevent injury or damage if the powerhead releases from the gearcase suddenly (which often occurs on outboards that have been in service for some time or that are extensively corroded).

On most of the single cylinder motors, and a FEW of the twins, the powerhead is actually light enough to be lifted by hand. BUT, it is really a better idea to use an engine hoist when lifting the powerhead assembly. This not only makes sure that the powerhead is secured at all times, but helps prevent injuries, which could occur if the powerhead releases suddenly. Also, keep in mind the components such as the driveshaft could be damaged if the powerhead is removed or installed at an angle other than perfectly perpendicular to the gearcase. It is a lot easier to align the powerhead when it is supported, than when you are holding the powerhead and trying to raise it from or lower it into position.

✹✹ WARNING

If powerhead removal is difficult, first check to make sure there are no missed fasteners between the gearcase (midsection/exhaust housing/intermediate housing, as applicable) and the powerhead itself. If none are found, apply suitable penetrating oil, like WD-40® or our new favorite, PB Blaster® to the mating surfaces. Give a few minutes (or a few hours) for the oil to work, then carefully pry the powerhead free while lifting it from the gearcase. Be careful not to damage the mating surfaces by using any sharp-tipped prybars or other by prying on thin/weak gearcase or powerhead bosses/surfaces.

When working on a powerhead, use either a very sturdy workbench or an engine stand to hold the assembly. Keep in mind that larger Hp motors utilize powerheads that can weigh several hundred pounds.

Although removal and installation is relatively straightforward on most models, the overhaul procedures can be quite involved. Whatever portion you decide to tackle, always, always, ALWAYS take good notes and tag as many parts/hoses/connections as you can during the removal process. As they are removed, arrange components along the work surface in the same orientation to each other as they are when installed.

Torque Sequence and Values

◆ See Figures 1 and 2

You've just GOT to love Yamaha (and remember, even if you're working on a Mercury/Mariner, the powerheads for ALL of these motors were manufactured by YAMAHA). We've worked on a lot of engines from a lot of manufacturers, but Yamaha has to be one of the best for designing serviceable motors. Not only are most of the components laid out logically for service, but Yamaha has gone as far as giving the technician some of the necessary information RIGHT ON the motor itself. Most major powerhead components (usually including Cylinder head covers, cylinder heads, exhaust covers, intake manifolds, thermostat housings, etc) have a torque sequence molded into the cover itself. Take a close look at some of these components and you see small numbers next to each of the retaining bolts. When present, this is a torque sequence that should be followed using multiple passes.

A torque sequence means that when installing that component you tighten bolt No. 1 a couple of turns, then No. 2, then 3, and so on. Once the bolts start to snug, you can precede one of two ways. If you don't have a torque value, you tighten the bolts securely, a 1/4 turn or so at a time still following that sequence until you feel they are securely fastened. However, if you have do have a torque value, you tighten the bolts just to the point when they start to snug, then set a torque wrench to the proper value and continue to tighten

POWERHEAD 6-3

each bolt in 1-2 passes of that sequence until the proper amount of torque is placed on each of the bolts.

To take a moment and praise Yamaha even a step further, they've been kind enough to also mold the torque value on these same components. Now, this is where our praise ends, because you'll have to keep a calculator handy. Someone didn't tell the Yamaha engineers that, at least in the U.S., our torque wrenches say Foot Pounds (ft. lbs.) and usually the common metric conversion of Newton Meters (Nm). So they must have gone with the measurement they work with the most, Kilogram Meter (KGM). Luckily, it's a simple enough conversion (simple enough with a calculator). To convert a KGM value to ft. lbs. simply multiply the KGM value by 7.233058 (of course 7.2 is probably accurate enough for most torque wrenches). To convert a KGM spec to Nm, simply multiply by 9.80665 (again, 9.8 should be sufficient). For example, the exhaust cover pictured in the accompanying illustration shows a torque value spec of 0.8 KGM, which is 5.7864464 ft. lbs. (5-6 ft. lbs. should be sufficient for a typical torque wrench). The same 0.8 KGM converts to 7.84532 Nm (or about 7-8 Nm).

■ **Mercury normally advises 3 passes of a torque sequence (using an increasing torque value for each pass) in order to reach a given torque specification.**

Fig. 1 Most torque sequences are molded onto the component - and many of them follow a spiraling pattern such as this cylinder head cover

Fig. 2 Yamaha was also kind enough to mold the torque value into most powerhead components

POWERHEAD COMPONENTS

Service procedures for the carburetors, fuel pumps, starter, and other powerhead components are given in their respective Sections of this technical guide.

CLEANLINESS

◆ See Figure 3

Make a determined effort to keep parts and the work area as clean as possible. Parts must be cleaned and thoroughly inspected before they are assembled, installed, or adjusted. Use proper lubricants, or their equivalent, whenever they are recommended.

Assembly Sealants and Lubricants

Over the years Yamaha seems to vary or change their recommendations for fastener and component installation, not only from model-to-model but from year-to-year on even the same model. Some items, such as a head gasket, NEVER receive sealant, but others, such as the powerhead-to-intermediate housing gasket is sometimes installed using a gasket making sealant applied to both sides and other times installed dry. Similarly, some bolts are installed dry, some are to be coated with engine oil and others with Loctite. Some seals should be lubricated with clean engine oil and others with marine grade grease. And lastly, internal engine components are usually coated with engine oil, but many bearing surfaces may also be coated with a molybdenum disulfide engine assembly lube.

■ **Mercury seems to be a little more consistent with oiling or threadlocking bolts and with the use of sealants or greases.**

We honestly don't believe that coating a seal with grease instead of oil or vice versa is going to cause any problem. Nor do we believe that Loctite on bolt threads instead of oil is ever an issue. When it comes to engine oil vs. engine assembly lube, the assembly lube may have some advantages and can almost always be substituted for oil on bearing surfaces. Pay attention to the instructions and accompanying illustrations. When an illustration calls for engine oil, there is a small symbol of an oil can. When the application calls for grease the illustration will show a small grease gun, however the same grease gun (sometimes with a letter M in the middle) is meant to mean an engine assembly lube when directed towards an internal engine component other than a seal.

Fig. 3 The exterior and interior of the powerhead must be kept clean, well lubricated and properly tuned for the owner to receive maximum enjoyment

6-4 POWERHEAD

2.5 Hp and 4 Hp Yamaha Powerheads

POWERHEAD REMOVAL & INSTALLATION

◆ See Figures 4, 5 and 6

These powerheads are small enough that you CAN remove them from the intermediate housing without removing most fuel and electrical components. However, Yamaha specifically directs their technicians to remove those components, though it seems more likely that the do this because removal will be necessary for whatever type of service is conducted on the powerhead after removal.

We advise that you consider why the powerhead is being removed. If you are planning on disassembling the powerhead for inspection or overhaul, then you'll have to remove the fuel and electrical components anyway. If this is the case, removing them before powerhead removal is a good idea, if only to protect them from potential damage when lifting and moving the powerhead itself. However, the choice remains yours whether or not you follow all of the initial steps for stripping the powerhead of these components.

1. Remove the engine top cowling.
2. If desired, remove the hand-rewind starter assembly and flywheel from the powerhead.
3. For 2.5 hp models, remove the lower cowling as follows:
   a. Loosen the throttle cable stop screw on the carburetor throttle valve, then remove the throttle cable from the carburetor.
   b. . Loosen the bolts and remove the outboard carrying handle.
   c. Unbolt and remove the fuel cock from the carburetor fuel cock lever. Keep track of the lever O-ring and replace if warn or damaged.
   d. Locate and remove the fasteners securing the front and rear halves of the bottom cowling. There are usually 4 bolts threaded upward from underneath the cowling and 2 bolts at the top flange threaded sideways from one cowling half to the other. Once the fasteners are all removed, carefully separate the halves.
4. For 4 hp models, remove the lower cowling as follows:
   a. Disconnect the choke wire from the carburetor.
   b. Tag and disconnect the engine stop switch wiring.
   c. Remove the 4 bolts and washers threaded sideways through the rear carrying handle/protector into the rear cowling half.
   d. Loosen and remove the 3 bolts and washers threaded upward into the bottom of the rear cowling half, then remove the carrying handle (protector) and rear cowling half.
   e. Loosen the screw and remove the shift lever.
   f. Remove the 4 bolts and washers threaded sideways through the front carrying handle into the front cowling half.
   g. Remove the 3 bolts and washers threaded upward from underneath the bottom of the front cowling half.
   h. Remove the front carrying handle.
   i. Loosen and remove the small bottom center screw on the front cowling half.
   j. Remove the fuel cock assembly
   k. Loosen the bolt and washer threaded sideways through the front cowling half for the fuel joint assembly. Remove the fuel joint.
   l. Remove the front cowling half.
5. If desired, remove the electrical components from the powerhead.
6. If desired, remove the fuel components from the powerhead.
7. On 4 hp motors, disconnect the throttle cable at the carburetor, then disconnect the start-in-gear protection cable from the bracket on the powerhead.
8. From the underside of the powerhead-to-intermediate housing mounting flange, loosen and remove the 6 (2.5 hp models) or 7 (4 hp models) bolts securing the powerhead.
9. Tap the side of the powerhead with a soft head mallet to shock the gasket seal, and then lift the powerhead from the intermediate housing.

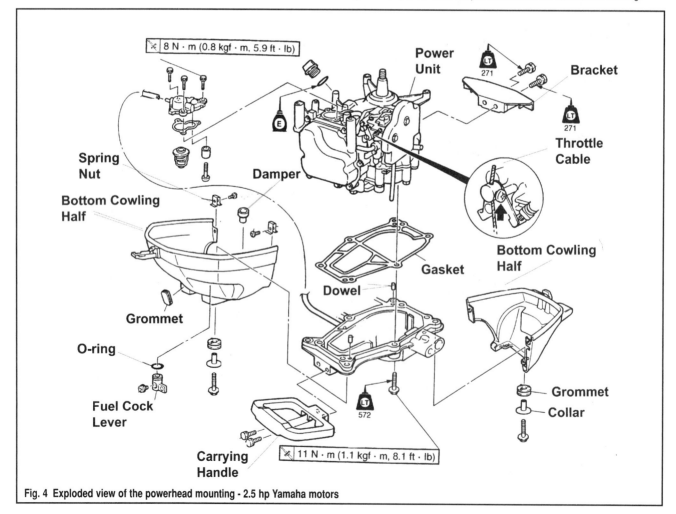

Fig. 4 Exploded view of the powerhead mounting - 2.5 hp Yamaha motors

POWERHEAD 6-5

10. Remove all traces of gasket and sealant from the mating surfaces.
11. There is usually a locating dowel pin in the mating surface, remove and take care not to drop or lose the 2 dowel pins located at opposite corners of the mating surface.

To Install:

12. Make sure the dowel pins are in position at the opposite corners of the intermediate housing.
13. Place the new gasket in position over the 2 dowel pins, this should make sure the gasket does not cock during installation and does not obscure any water passages.
14. Apply a light coating of marine grade grease to the driveshaft-to-crankshaft splines
15. Carefully install the powerhead onto the intermediate housing, using the dowel pins to align the two surfaces. If necessary, slowly rotate the flywheel/crankshaft (clockwise) to align the crankshaft and driveshaft splines. Alternately the propeller shaft may be rotated (if the gearcase is in gear), but again, only in the normal direction of rotation to prevent potential damage to the water pump impeller.
16. Once the powerhead is properly seated (the crankshaft is properly splined to the driveshaft) it is time to install the retaining bolts. Apply a light coating of Loctite® 572 to the threads of the bolts that secure the powerhead to the intermediate housing. Tighten the bolts starting with the center bolts then working outward to 8.1 ft. lbs. (11 Nm) for 2.5 hp motors or to 5.8 ft. lbs. (8 Nm) for 4 hp motors.
17. If removed, install the fuel components to the powerhead.
18. If removed, install the electrical components to the powerhead.
19. Install the lower cowling halves, carrying handle(s) and related components in the reverse of the sub-steps given earlier in this procedure for removal.
20. If removed, install the flywheel, followed by the hand-rewind starter assembly.
21. Reconnect the choke and/or throttle cables, as applicable. Refer to the Timing and Synchronization adjustments in the Engine and Maintenance section for adjustment procedures.

■ **Be sure to run the motor without the top cowling installed in order to check for potential fuel or oil leaks. Remedy any leaks before proceeding.**

22. Install the top cowling to the powerhead.

■ **If the powerhead was rebuilt or replaced with a remanufactured unit, don't forget to follow the proper Break-In procedures.**

CYLINDER HEAD REMOVAL & INSTALLATION

◆ See Figures 7 thru 10

It is not entirely clear whether or not the cylinder head can be removed with the powerhead assembly still installed, but it looks likely on the 4 hp motors. On the 2.5 hp motor, it is probably not possible, as there are 2 bolts threaded upward through the bottom of a projection in the crankcase and into the bottom end of the cylinder head. Even if you can get to the bolts heads, it is unlikely that you could back them out sufficiently to remove the head

1. Of course, it is true on both motors that in order to get to that point that you could remove the cylinder head you've likely got to do some of the work required for powerhead removal anyway. So the question really comes down to the lower cowling assembly. Once that is out of the way, you're pretty much home free. Of course, if work conditions would improve by placing everything on the bench while working, just go ahead and remove the powerhead.
Remember, when loosening the retainers on manifolds, covers and other major components, always try to follow the reverse of the torque sequence indicated in the assembly procedure or molded on the component itself.
When applicable tighten all bolts in using the proper torque sequence and to the proper specification. When possible we've included it in the procedure, but on most Yamahas both the sequence and value is molded into the casting on critical components. You'll notice that MOST (but not all) Yamaha torque sequences are clockwise spirals starting somewhere near the center of the component.
2. If not done already, strip the necessary electrical components from the powerhead.
3. Remove the spark plug from the cylinder head.
4. On 2.5 hp motors, if necessary tag and disconnect the blow-by hose from the cylinder head.
5. Remove the 6 (2.5 hp) or 4 (4 hp) bolts from the cylinder head cover using the reverse of the torque sequence (either molded in the cover or, for 2.5 hp motors included in the accompanying illustrations). On 4 hp motors use a criss-cross sequence.
6. Remove the cylinder head cover, then remove and discard the gasket.
7. On 2.5 hp motors, there are two bolts threaded upward from underneath the powerhead through a projection on the crankcase and into the bottom of the cylinder head. Loosen and remove these 2 bolts.
8. Again, using multiple passes in the reverse of the torque sequence (this time it is a criss-cross pattern for 2 hp motors or use the reverse of the molded/illustrated torque sequence on 4 hp motors), loosen the 4 (2.5 hp) or 5 (4 hp) cylinder head bolts.

■ **On 4 hp motors, keep close track of the bolts and their locations as there are 3 different sizes. Bolts 1, 3 and 4 of the torque sequence are the longest (M8 x 60mm) bolts, while the bolt for position 2 is slightly shorter (M8 x 40mm) and the bolt for position number 5 is the smallest, though not shortest (M6 x 45mm).**

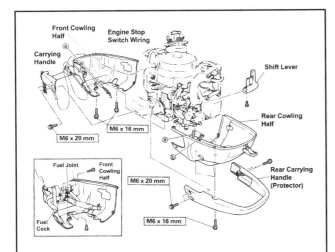

Fig. 5 Exploded view of the lower cowling assembly - 4 hp Yamaha motors

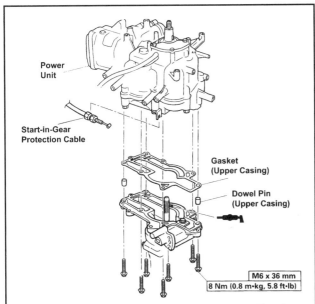

Fig. 6 Exploded view of the powerhead mounting - 4 hp Yamaha motors

6-6 POWERHEAD

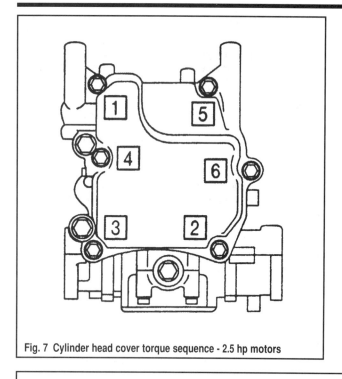

Fig. 7 Cylinder head cover torque sequence - 2.5 hp motors

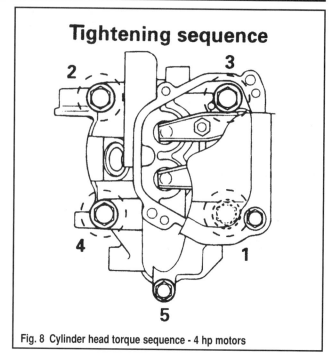

Fig. 8 Cylinder head torque sequence - 4 hp motors

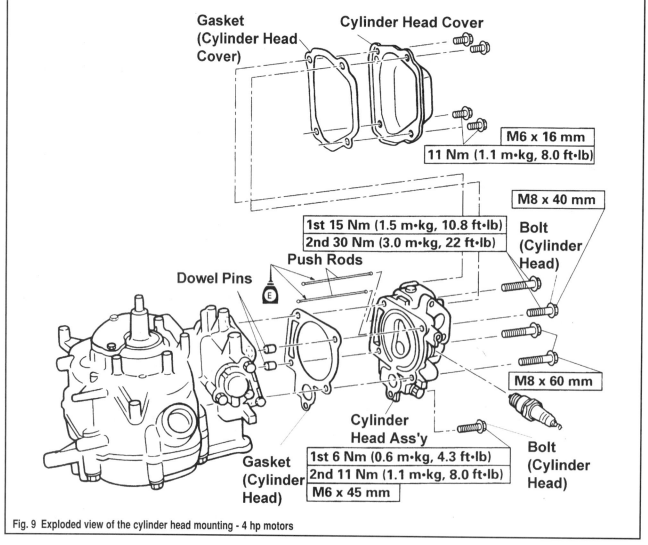

Fig. 9 Exploded view of the cylinder head mounting - 4 hp motors

POWERHEAD 6-7

9. Carefully break the gasket seal and remove the cylinder head from the cylinder block. On these motors there is normally one or more pry tabs positioned at the lower corner(s) of the cylinder head. Use a small prytool on the tabs to carefully loosen the cylinder head breaking the gasket seal.

✳✳ WARNING

These motors use pushrods to actuate the cylinder head mounted rocker arms via a crankcase mounted camshaft. Therefore be sure to pull the cylinder head STRAIGHT off the crankcase until the pushrods are clear to prevent possible damage to the rocker arm assemblies.

10. Remove the pushrods, noting both their positioning and orientation since they must be returned to their original locations if they are to be reused.
11. Remove and discard the old cylinder head gasket. Carefully remove all traces of the gasket material from the mating surfaces.
12. Check the condition of the cylinder head dowel pins (there are normally 2 at opposite corners of the cylinder head) and replace, if necessary.
13. Refer to Cylinder Head Overhaul in this section for details on removal and installation of the valve train components. Also, be sure to check the information on gasket mating surfaces to make sure the cylinder head is not warped.

To Install:

14. Make sure the cylinder head dowel pins are in position on the block, then position a new cylinder head gasket over the pins.
15. If the cylinder head was not overhauled, loosen the rocker arm pivot nuts so you can install the pushrods AFTER the cylinder head is bolted in position.
16. Carefully place the cylinder head over the gasket and onto the block.
17. On 2.5 hp motors, apply a light coating of clean engine oil to the threads of the 4 main cylinder head bolts (M8 x 60mm).
18. Thread the cylinder head bolts (paying attention to bolt positioning on 4 hp motors). The dowel pins should have held the gasket in position, but make a quick visual inspection to be certain it did not dislodge before you torque the head.
19. On 2.5 hp motors thread and finger-tighten the 2 small cylinder head bolts from underneath the powerhead up into the bottom end of the cylinder head.
20. Tighten the cylinder head bolts using at least 2 passes of the molded torque sequence (or if not present using either a criss-cross pattern for 2.5 hp motors or the sequence provided in an accompanying illustration for 4 hp motors). Tighten the bolts first to 11 ft. lbs. (15 Nm) and then to 22 ft. lbs. (30 Nm) for all except the small No. 5 bolt on 4 hp motors. For the smallest bolt (No.5 in the torque sequence) on 4 hp motors, tighten the bolt first to 4.3 ft. lbs. (6 Nm) and then to 8.0 ft. lbs. (11 Nm).

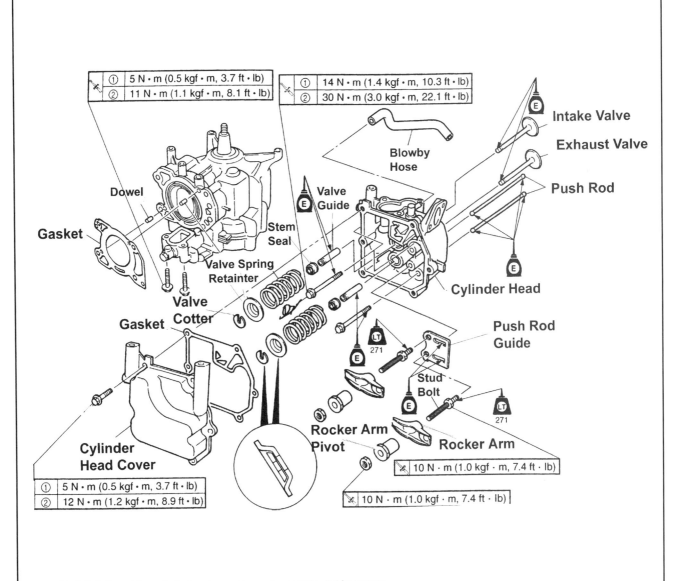

Fig. 10 Exploded view of the cylinder head mounting and assembly - 2.5 hp motors

6-8 POWERHEAD

21. On 2.5 hp motors tighten the 2 small cylinder head bolts (threaded from underneath the powerhead up into the bottom end of the cylinder head) alternately and evenly first 3.7 ft. lbs. (5 Nm) and then to 8.0 ft. lbs. (11 Nm).
22. Apply a light coating of clean engine oil to the pushrods and pushrod guides.
23. Reposition the rocker arms as necessary to insert the pushrods. If you are reusing the original pushrods be certain to install them to their original position, with the same end facing inward toward the pushrod holder in the crankcase. Then reposition and loosely secure the rocker arms.
24. Rotate the crankshaft in the normal direction of rotation (clockwise when viewed from above) through a couple of full revolutions, then adjust the Valve Clearance, as detailed in the Maintenance and Tune-Up section.
25. Install the cylinder head cover using a new gasket. If the valves have been properly adjusted install and tighten the cylinder head cover bolts either using a criss-cross sequence (4 hp motors) or using the molded/illustrated torque sequence (2.5 hp motors). Tighten the bolts in 2 steps, first to 3.7 ft. lbs. (5 Nm) and then to 8.0 ft. lbs. (11 Nm).
26. If removed for some reason during overhaul, install the thermostat assembly.
27. If removed, install the breather hose.
28. Install the spark plug to the cylinder head.

■ If the powerhead was not removed, install any remaining electrical components which may have been removed for access at this time.

29. If removed, install the powerhead assembly.

CYLINDER HEAD OVERHAUL

◆ See Figures 10 and 11

1. Remove the Cylinder Head from the powerhead assembly, as detailed in this section.
2. Loosen and remove the locknut securing each rocker arm assembly, then remove the adjuster and rocker arm. Unthread the pivot. Be sure to reassemble all parts to keep track of not only their positioning, but from what assembly they were removed. Once the second rocker arm assembly and pivot is removed, remove the pushrod guide plate.

■ Valve train components such as rocker arms, adjusters, pivots, valves and valve seats/cotters (keepers) should not be mixed and matched from one assembly to another. This means that if they are to be reused, they must all be labeled prior to removal or VERY carefully sorted during service to prevent installation to the wrong positions.

3. To remove the each valve, proceed as follows:

✳✳ CAUTION

Wear eye protection when working with valve springs and cotters. They have a nasty habit of flying all over the place at the most unexpected times.

   a. Using a clamp-type valve spring compressor, carefully place just enough pressure on the valve spring so the seat moves down sufficiently to expose the cotter (keeper).
   b. Remove the pair of valve cotter, then slowly ease pressure on the valve spring until the compressor can be removed.
   c. Remove the valve spring seat and valve spring from the shaft of the valve.
   d. Remove the valve from the underside (combustion chamber side) of the cylinder head.
   e. If necessary, carefully pry the old valve stem seal from its position on the cylinder head.
   f. Repeat this procedure for the remaining valve.
4. Refer to Powerhead Refinishing, later in this section, for details on cleaning and inspecting cylinder head components.

To Install:

5. If removed, install each of the valve and spring assemblies as follows:
   a. If removed, install new valve spring seals to the cylinder head.

■ I know we said IF removed when it comes to the valve spring seals. But by now, you should know our mantra. If you've come this far, you might as well replace wear components such as these.

   b. Apply a light coat of clean engine oil to the valve stems and springs.
   c. Position the valve into the cylinder head (into the original guide and valve seat unless the valve or valve seat was replaced).
   d. Position the spring over the valve and against the cylinder head, then place the seat over the spring and use the valve spring compressor tool to press downward on the seat so you can install the cotter.

■ Use a thin screwdriver with a dab of marine grease to hold an install the valve cotters (keepers).

   e. Insert the valve cotter, then slowly back off the valve spring compressor, making sure the cotter is fully in position. Remove the valve spring compressor and tap lightly on the valve spring seat using a plastic or rubber mallet to make sure the cotter is securely set.

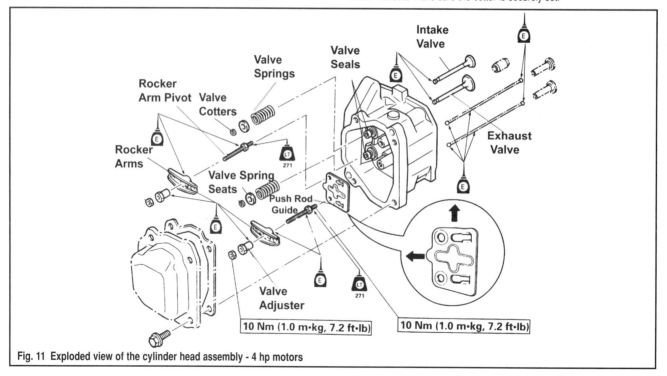

Fig. 11 Exploded view of the cylinder head assembly - 4 hp motors

POWERHEAD

f. Repeat for the remaining valve.
6. Install each of the rocker arm assemblies as follows:
   a. Apply a light coating of Loctite® 271 or an equivalent threadlocking compound to the lower (cylinder head) end of the rocker arm pivot threads.
   b. Apply a light coat of engine oil to the pushrod contact surfaces, then position the pushrod guide plate to the cylinder head and secure using one of the rocker arm pivots.
   c. Install the rocker arm pivot and tighten to 7.2 ft. lbs. (10 Nm).
   d. Loosely install the rocker arm, adjuster and locknut.

■ **The rocker arms must be only loosely installed at this point since they must be repositioned during installation to insert the pushrods.**

   e. Repeat for the remaining rocker arm pivot, arm, adjuster and locknut assembly.
7. Install the Cylinder Head, to the powerhead assembly as detailed earlier in this section.

CYLINDER BLOCK OVERHAUL

◆ See Figures 12 thru 20

Remember, when loosening the retainers on manifolds, covers and other major components, always try to follow the reverse of the torque sequence indicated in the assembly procedure or molded on the component itself.

When applicable tighten all bolts in using the proper torque sequence and to the proper specification. When possible we've included it in the procedure, but on many Yamahas both the sequence and value is molded into the casting on critical components.

1. Drain the engine oil from the powerhead, as detailed in the Maintenance and Tune-Up section.
2. Remove the Powerhead, as detailed in this section.

■ **Be sure to strip the powerhead of fuel and electrical components.**

3. Remove the Cylinder Head, as detailed in this section.
4. For 4 hp motors, remove the Thermostat assembly from the cylinder block, as detailed in the Lubrication and Cooling section. Also on these models, loosen and remove the breather cover (retained by 3 bolts and a non-reusable gasket) from the top of the crankcase. A screw underneath the cover retains a removable breather valve plate.
5. Unbolt and remove the oil seal housing from the bottom of the crankcase. If the oil seals are to be replace (and they really should be since you're here) remove them from the housing noting their positioning. All models use an O-ring to seal the housing and 3 oil seals (with the largest seal positioned closest to the crankshaft). On 2.5 hp motors all 3 seals are installed with their lips facing upward toward the crankshaft. On 4 hp motors only the large seal lips face the crankshaft, while the lips of the 2 smaller seals are installed with their lips facing downward toward the gearcase.
6. Remove the crankcase cover retaining bolts using multiple passes in the reverse of the torque sequence.
7. Carefully break the gasket seal and remove the crankcase from the cylinder block. On most models there is one or more prytabs located at the crankcase cover-to-cylinder block mating surface. If so, insert a small prybar and carefully lever the components apart.
8. Remove the camshaft and valve lifters from the cylinder block.

■ **Components within the crankcase must be installed in the same positions and facing the same directions from which they were removed. Always matchmark components and place alignment marks on them to ensure proper orientation. To prevent confusion, always place marks on the flywheel (top) side of components.**

9. Matchmark the big end cap to the connecting rod, then remove the 2 connecting rod cap bolts. Remove the cap and place it aside for installation back on the original connecting rod as soon as the crankshaft has been removed.

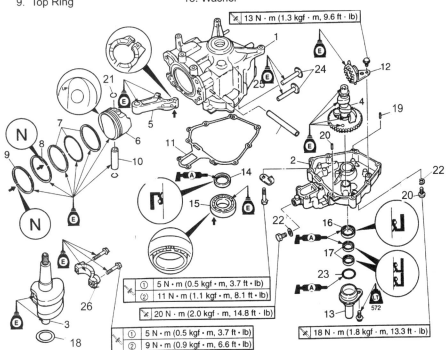

1. Cylinder Block
2. Crankcase Cover
3. Crankshaft
4. Camshaft
5. Connecting Rod Assembly
6. Piston
7. Oil Ring Set
8. Second Ring
9. Top Ring
10. Piston Pin
11. Gasket
12. Oil Splasher Gear
13. Oil Seal Housing
14. Oil Seal
15. Ball Bearing
16. Oil Seal
17. Oil Seal
18. Washer
19. Dowel
20. Drain Bolt
21. Piston Pin Clip
22. Metal Gasket
23. O-Ring
24. Valve Lifter
25. Hose
26. End Cap

Fig. 12 Exploded view of the powerhead assembly - 2.5 hp Yamaha motors

6-10 POWERHEAD

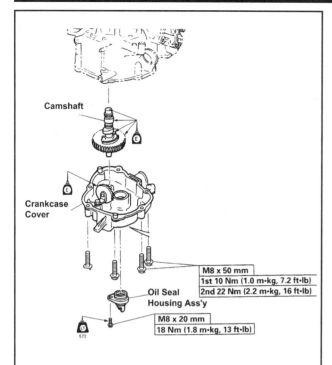

Fig. 13 Exploded view of the crankcase cover and camshaft - 4 hp Yamaha motors

10. Unless an excessive ridge is formed at the top of the cylinder bore (which you can feel for using a fingernail), carefully push the piston and connecting rod up and out the top of the cylinder block. If a ridge has formed the concern is that a ring may become hung up on it and break, however you may still be able to remove the piston if you can carefully compress the rings and work the piston out of the bore. Your other option is to position the piston at the top of its travel and turn the crankshaft so as to achieve maximum clearance, then attempt to remove the crankshaft first so that you can pull the piston out of the bottom of the cylinder block. If you choose to try this, take care not to score or damage the polished crankshaft surfaces.

■ These pistons and connecting rods are normally already marked from the factory as to which side faces upward, identify the marks and perhaps, remark them to ensure they are found easily during assembly.

11. Remove the plate washer, then carefully remove the crankshaft from the block. Place it on a protective surface.

✳✳ WARNING

Handle the crankshaft carefully. Remember it is an expensive and vital piece to the motor. The polished surfaces should be covered to prevent scratches or other damage and the whole shaft should be protected against impact or other damage which could destroy it.

12. Once removed, immediately reinstall the connecting rod end cap to it piston while aligning the matchmark made during removal to ensure proper orientation. This will make sure the orientation is not lost if the marks are dulled or removed by cleaning and inspection procedures.
13. Carefully pry the crankshaft oil seal from the top of the cylinder block. Discard the seal.
14. If necessary, disassemble the piston, as follows:
    a. Carefully pry the clips free from the piston pin bore on either side of the piston assembly.

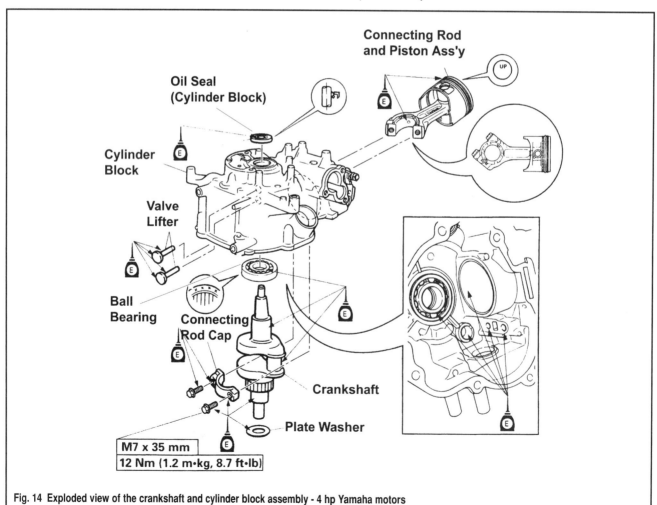

Fig. 14 Exploded view of the crankshaft and cylinder block assembly - 4 hp Yamaha motors

POWERHEAD 6-11

b. Slide the piston pin free of the bore (specifications tell us that this is NOT an interference fit, so it should slide free with ease).

c. Using a piston ring spreading tool, carefully remove each of the rings (top and second compression rings, followed by the oil ring/rings) from the piston.

■ Some models have stamp marks on one side of the piston ring to denote which side must face upward. Check the rings as they are removed for the presence of such marks and make a note for installation purposes.

d. Keep all components together, sorted by positioning and marked for orientation for reuse or replacement.

15. A ball bearing assembly is pressed into the top of the cylinder block for the crankshaft. The bearing should NOT be removed unless it is going to be replaced (as removal normally places sufficient stress on the bearing cage to damage it beyond use). If replacement is desired, use a slide-hammer and jawed puller to remove the bearing from the inside top of the cylinder block.

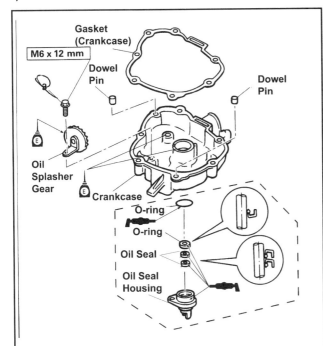

Fig. 15 Exploded view of the oil splasher gear and oil seal housing - 4 hp Yamaha motors

16. Loosen the retaining bolt and remove the oil splasher gear from the crankcase cover. Be sure to thoroughly inspect the teeth/vanes of the splasher gear to make sure it is in serviceable condition, or the gear must be replaced to ensure proper internal lubrication.

17. Check the placement and condition of the 2 cylinder block dowel pins (they're normally found at opposite corners of the block-to-cover mating surface and generally installed in the cover itself to hold the gasket in position during installation). They should be replaced if they are damaged.

18. Refer to Powerhead Refinishing, later in this section, for details on cleaning and inspecting cylinder block components. There are probably more components and specifications to measure on Yamaha 4-stroke motors than similar output/sized 2-stroke counterparts. So take your time and make sure you've made measurements for all components listed in the Engine Specifications chart for this model.

To Assemble:

19. Make sure the mating surfaces of the crankcase and cylinder block are completely clean and free of debris, gasket material or damage. The dowel pins should already be installed in the crankcase cover, but if not, obtain new pins and install them in the 2 positions at opposite sides of the crankcase cover.

20. Apply a light coat of engine oil to the oil splasher gear, then install it to the crankcase cover.

21. If removed, apply a light coating of engine oil to the replacement crankshaft ball bearing, then use a suitable sized driver to carefully tap it into position in the top of the cylinder block.

■ Before assembling the piston, be sure to check the Connecting Rod Big-End Oil Clearance by assembling the connecting rod to the crankshaft using Plastigage® or an equivalent gauging compound. For more details, please refer to the Powerhead Refinishing section.

22. Apply a light coating of marine grade grease to the lips of a new upper crankshaft seal, then position the seal to the top of the cylinder block with the lips facing downward toward the crankshaft. Using a suitably sized driver, carefully tap the seal into position on the block.

23. If disassembled, prepare the piston assembly for installation as follows:

a. Using a ring expander, carefully install each of the rings with their gaps as noted in the accompanying illustration. While looking at the piston dome with the top (flywheel end) facing 12 o'clock, the bottom oil ring should be positioned with the gap facing between 4 and 5 o'clock. The middle oil ring (scrubber) is positioned with the gap facing between 7 and 8 o'clock. The upper oil ring should be positioned with the gap facing between 10 and 11 o'clock. The lower (second) compression ring should be positioned with its gap between 1 and 2 o'clock. The upper (top) compression ring should be positioned with its gap facing between 7 and 8 o'clock, the same as the middle oil ring.

■ Some models may use a single oil control ring. If so, install the gap 180° from the gap of the lower (second) compression ring.

b. Align the mark made on the connecting rod (facing up/flywheel) with the mark on the piston (also facing up/flywheel), then install the connecting

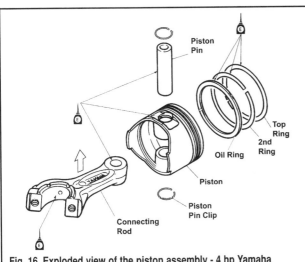

Fig. 16 Exploded view of the piston assembly - 4 hp Yamaha motors

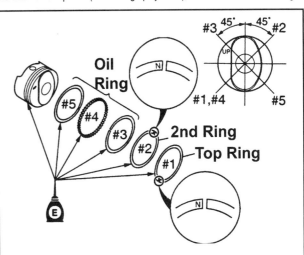

Fig. 17 Piston ring positioning - 2.5 hp motors shown, 4 hp similar (though oil control ring/s may vary)

6-12 POWERHEAD

rod to the piston using the piston pin. On some models the connecting rod is marked with **YAMAHA** on the side that faces upward. Similarly, some pistons are stamped with the word **UP** on the dome at the portion that should be facing upward toward the flywheel.

    c. Install the piston pin clips in either side of the pin bore to secure the piston pin in place.

✴✴ WARNING

Use extreme care to prevent the connecting rod from contacting and scoring the cylinder wall during piston installation. It is a good idea to wrap the ends in a shop cloth or something else to protect the cylinder.

  24. Apply a light coating of engine oil to the crankshaft, especially on the drive gear and bearing mating surfaces, then carefully insert the crankshaft into the cylinder block. Oil and install the washer over the bottom end of the crankshaft.

  25. Remove the connecting rod end cap and bolts so the piston can be installed, positioning the connecting rod over the crankshaft

  26. Install the piston (with the correct mark/side facing UP toward the flywheel) to the cylinder block using a ring compressor. Apply a very light coating of clean engine oil to the inside of the compressor, to the cylinder bore and the connecting rod mating surface, then place the compressor around the piston. Insert the piston through the cylinder head side of the bore and use a wooden hammer handle to gently tap the piston down through the compressor and into the cylinder bore.

✴✴ WARNING

Be VERY careful when installing the piston. Get the lower portion of the skirt below the ring compressor and insert it into the cylinder bore to make sure everything is aligned. Then, GENTLY tap the piston through the compressor, stopping if it seems to hang up even slightly to make sure that a ring is not caught on the deck of the cylinder block.

  27. Apply a light coating of engine oil to both the connecting rod end cap bearing surface as well as the threads of the cap bolts. Install the end cap and tighten the connecting rod retaining bolts alternately and evenly first to 3.7 ft. lbs. (5 Nm), and then either to 6.6 ft. lbs. (9 Nm) for 2.5 hp motors or to 8.7 ft. lbs. (12 Nm) for 4 hp motors.

  28. Oil and install the valve lifters to the bores in the cylinder block. If the lifters are being reused, be sure to install them to the same locations from which they were removed.

  29. Apply a light coating of engine oil to the camshaft, then install the shaft to the cylinder block aligning the stamped timing mark on the crankshaft gear with the mark on the meshing camshaft gear.

  30. Apply a light coating of marine grade grease to 3 NEW O-seals and a new O-ring, then install them to the oil seal housing. On 2.5 hp motors, all seals are positioned with the seal lips facing upward towards the crankshaft. On 4 hp motors the 2 smaller seals (positioned deeper in the housing) should be faced with their lips downward toward the crankcase and only the larger seal (installed on top of the smaller seals) should be faced with its lips upward toward the crankshaft.

  31. Apply a light coating of Loctite® 572 or an equivalent threadlocking compound to the threads of the oil seal housing bolt. Install the oil seal housing to the crankcase cover and tighten the retaining bolt securely.

  32. Position a new gasket to the crankcase cover and make sure it is aligned over the dowel pins.

  33. Carefully install the crankcase cover to the cylinder block making sure the oil splasher gear teeth mesh with the teeth of the camshaft gear. Also, be sure the gasket does not become dislodged as the cover and block come into alignment.

  34. For 2.5 hp motors, apply a light coating of engine oil to the threads of the crankcase cover bolts.

  35. Install the crankcase cover bolts and tighten them using multiple passes of the torque sequence. For 2.5 hp motors tighten the bolts first to 3.7 ft. lbs. (5 Nm) and then to 8.1 ft. lbs. (11 Nm). For 4 hp motors tighten the bolts first to 7.2 ft. lbs. (10 Nm) and then to 16 ft. lbs. (22 Nm).

  36. Refill the powerhead with clean engine oil.

  37. Install the Cylinder Head, as detailed in this section.

  38. Install the Powerhead, as detailed in this section.

  39. Once the powerhead is fully assembled, if cylinder block components were replaced, especially the piston and/or rings, be sure to operate the engine as directed for new component break-in. For details, please refer to Powerhead Break-In, in this section.

CLEANING & INSPECTION

Cleaning and inspecting the components is virtually the same for most 2- and 4-stroke outboards and varies mostly by specifications (which are listed in the Engine Specifications charts) or by component type. A section detailing the proper procedures, sorted mostly by component, can be found under Powerhead Refinishing.

4/5/6 Hp Mercury/Mariner Powerheads

POWERHEAD REMOVAL & INSTALLATION

◆ See Figure 21

These powerheads are small enough that you CAN remove them from the intermediate housing without removing most fuel and electrical components. However, you might want to consider doing this to protect the components, depending upon the reason you're removing the powerhead in the first place.

If you are planning on disassembling the powerhead for inspection or overhaul, then you'll have to remove the fuel and electrical components anyway. If this is the case, removing them before powerhead removal is a good idea, if only to protect them from potential damage when lifting and moving the powerhead itself. However, the choice remains yours whether or not you follow all of the initial steps for stripping the powerhead of these components.

  1. Remove the engine top cowling.

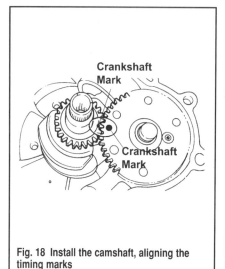

Fig. 18 Install the camshaft, aligning the timing marks

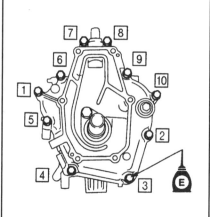

Fig. 19 Crankcase cover bolt torque sequence - 2.5 hp motors

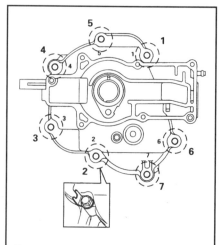

Fig. 20 Crankcase cover bolt torque sequence - 4 hp motors

POWERHEAD 6-13

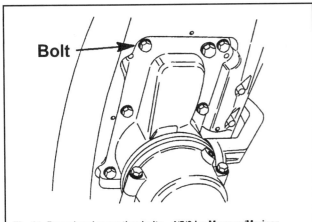

Fig. 21 Powerhead mounting bolts - 4/5/6 hp Mercury/Mariner motors

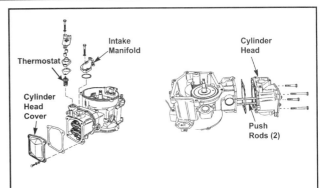

Fig. 22 Exploded view of the cylinder head mounting - 4/5/6 hp Mercury/Mariner motors

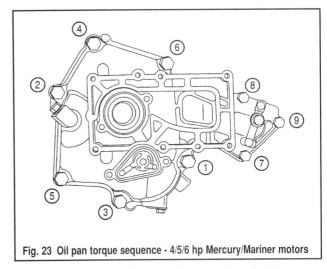

Fig. 23 Oil pan torque sequence - 4/5/6 hp Mercury/Mariner motors

2. Drain the engine oil. For details, please refer to the Maintenance and Tune-Up section.
3. Disconnect the starter lock rod.
4. Disconnect the choke wire.
5. Disconnect the throttle cables.
6. Tag and disconnect the wiring for the stop switch, oil warning light and oil pressure sensor. If necessary, refer to the Wiring Diagrams to help identify the proper wire colors.
7. Tag and disconnect the ground wires at the ignition coil and the excite coil wiring.
8. Disconnect the fuel inlet hose. Have a rag handy in case there is still some fuel in the line.
9. From underneath the cowling loosen and remove the 6 M8 x 40mm retaining bolts that are threaded upward securing the powerhead to the intermediate housing.
10. Carefully lift the powerhead upward and free of the intermediate housing.
11. Remove and discard the old powerhead mounting gasket.

To Install:
12. Apply a light coating of marine grade grease to the sides (NOT TOP) of the driveshaft splines.
13. Place a new gasket in position on the intermediate housing.
14. Carefully install the powerhead onto the intermediate housing, while carefully aligning the two surfaces (and making sure the gasket does not move). If necessary, slowly rotate the flywheel/crankshaft (clockwise) to align the crankshaft and driveshaft splines. Alternately the propeller shaft may be rotated (if the gearcase is in gear), but only in the normal direction of rotation to prevent potential damage to the water pump impeller.
15. Install the powerhead retaining bolts and tighten to 70 inch lbs. (8 Nm).
16. Reconnect the fuel line.
17. Reconnect the wiring as tagged during removal.
18. Reconnect the throttle cables, choke wire and starter lock rod. Refer to the Timing and Synchronization adjustments in the Engine and Maintenance section for adjustment procedures.

■ **Be sure to run the motor without the top cowling installed in order to check for potential fuel or oil leaks. Remedy any leaks before proceeding.**

19. Install the top cowling to the powerhead.

■ **If the powerhead was rebuilt or replaced with a remanufactured unit, don't forget to follow the proper Break-In procedures.**

CYLINDER HEAD REMOVAL & INSTALLATION

◆ See Figures 22 and 23

Unfortunately, the cylinder head is bolted to the oil pan assembly on the bottom of the powerhead, so it is not possible to remove it with the powerhead installed.

Of course, the good news is that in order to get to the point where you could remove the cylinder head, you've got to do most of the work required for powerhead removal anyway (and it is not that extensive a job on this motor).

Remember, when loosening the retainers on manifolds, covers and other major components, always try to follow the reverse of the torque sequence (if provided) or by using a crossing pattern (if not provided). When applicable tighten all bolts in using the proper torque sequence and to the proper specification (or again, using a crossing pattern).

1. Remove the Powerhead assembly, as detailed earlier in this section.
2. If not done already, strip the necessary electrical components from the powerhead.
3. Remove the spark plug from the cylinder head.
4. If necessary, remove the thermostat assembly from the top of the cylinder head.
5. Remove the carburetor and intake manifold from the top of the cylinder head.
6. Invert the powerhead so the oil pan is facing upward. Locate the 3 oil pan-to-cylinder head bolts (numbers 7, 8 and 9 in the oil pan torque sequence). Loosen and remove those 3 oil pan bolts, then invert the assembly again to the top of the powerhead is facing upward.
7. Remove the 4 bolts from the cylinder head cover using a crossing sequence.
8. Remove the cylinder head cover, then remove and discard the gasket.
9. Using multiple passes of a crossing sequence loosen and remove the 4 cylinder head bolts.

■ **Keep close track of the bolts and their locations as there are 2 different size (length, M8 x 90mm and M8 x 60mm) cylinder head bolts on this model.**

10. Carefully break the gasket seal and remove the cylinder head from the cylinder block.

6-14 POWERHEAD

※※ WARNING

These motors use pushrods to actuate the cylinder head mounted rocker arms via a crankcase mounted camshaft. Therefore be sure to pull the cylinder head STRAIGHT off the crankcase until the pushrods are clear to prevent possible damage to the rocker arm assemblies.

11. Remove the pushrods, noting both their positioning and orientation since they must be returned to their original locations if they are to be reused.
12. Remove and discard the old cylinder head gasket. Carefully remove all traces of the gasket material from the mating surfaces.
13. Check the condition of the 2 cylinder head dowel pins and replace, if necessary.
14. Refer to Cylinder Head Overhaul in this section for details on removal and installation of the valve train components. Also, be sure to check the information on gasket mating surfaces to make sure the cylinder head is not warped.

To Install:
15. Make sure the 2 cylinder head dowel pins are in position on the block.
16. Position a new cylinder head gasket over the dowel pins.

■ You've got a choice here. We'd normally recommend installing the pushrods AFTER cylinder head installation, just to simplify matters (as detailed in the next step). However, it is a SMALL cylinder head and there are only 2 pushrods, so you may well be able to position the rods and install the head over them. It's up to you.

17. If the cylinder head was not overhauled, loosen the rocker arm pivot nuts so you can install the pushrods AFTER the cylinder head is bolted in position.
18. Carefully place the cylinder head over the gasket and onto the block.
19. Apply a light coating of clean engine oil to the threads of the 4 main cylinder head bolts NOT the oil pan bolts.
20. Thread the cylinder head bolts (paying attention to bolt positioning). Make a quick visual inspection to be certain the gasket did not dislodge before you torque the head.
21. Tighten the cylinder head bolts to 18 ft. lbs. (24.5 Nm).
22. Invert the powerhead assembly and install the 3 oil pan-to-cylinder head bolts. Tighten them to 80 inch lbs. (9 Nm).
23. If not installed already, apply a light coating of clean engine oil to the pushrods and pushrod guides. Reposition the rocker arms as necessary to insert the pushrods. If you are reusing the original pushrods be certain to install them to their original position, with the same end facing inward toward the pushrod holder in the crankcase. Then reposition and loosely secure the rocker arms.
24. Rotate the crankshaft in the normal direction of rotation (clockwise when viewed from above) through a couple of full revolutions, then adjust the Valve Clearance, as detailed in the Maintenance and Tune-Up section.
25. Install the cylinder head cover using a new gasket. If the valves have been properly adjusted, install and tighten the cylinder head cover bolts using a criss-cross sequence. Tighten the bolts to 70 inch lbs. (8 Nm).
26. If removed for some reason during overhaul, install the thermostat assembly.
27. If removed, install the carburetor and intake manifold assembly.
28. Install the spark plug to the cylinder head.

■ If the powerhead was not removed, install any remaining electrical components which may have been removed for access at this time.

29. Install the powerhead assembly.

CYLINDER HEAD OVERHAUL

◆ See Figures 24 and 25

1. Remove the Cylinder Head from the powerhead assembly, as detailed in this section.
2. Loosen and remove the locknut securing each rocker arm assembly, then remove the pivot and rocker arm. Unthread the pivot bolts (stud) in order to remove the guide plate. Be sure to reassemble all parts to keep track of not only their positioning, but from what assembly they were removed. Once the second rocker arm assembly and pivot bolt is removed, remove the pushrod guide plate.

■ Valve train components such as rocker arms, pivots, valves and valve seats/cotters (keepers) should not be mixed and matched from one assembly to another. This means that if they are to be reused, they must all be labeled prior to removal or VERY carefully sorted during service to prevent installation to the wrong positions.

3. To remove the each valve, proceed as follows:

※※ CAUTION

Wear eye protection when working with valve springs and cotters. They have a nasty habit of flying all over the place at the most unexpected times.

a. Place a rag under the back of the cylinder head to catch the valves and help prevent them from becoming damaged should they fall from the head.
b. Placer a box-end wrench (just large enough to pass over the keepers, but small enough to apply pressure to the valve spring retainer) over the keepers and valve spring retainers for one of the valves.
c. Using the box-end wrench, carefully place just enough pressure on the valve spring/retainer so the retainer moves down sufficiently to expose the cotters (keepers).

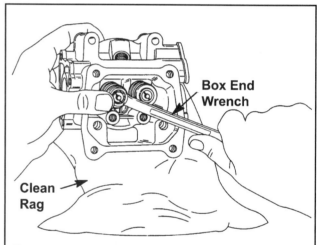

Fig. 24 Use a box-end wrench to depress the spring when removing or installing keepers

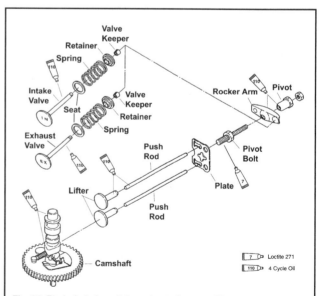

Fig. 25 Exploded view of the valve train assembly - 4/5/6 hp Mercury/Mariner motors

POWERHEAD 6-15

d. Remove the pair of valve cotters, then slowly ease pressure on the valve spring until the wrench can be removed.

e. Remove the valve spring retainer, valve spring and valve spring seat from the shaft of the valve.

f. Remove the valve from the underside (combustion chamber side) of the cylinder head.

g. Repeat this procedure for the remaining valve.

4. Refer to Powerhead Refinishing, later in this section, for details on cleaning and inspecting cylinder head components.

To Install:

5. If removed, install each of the valve and spring assemblies as follows:

a. Apply a light coat of clean engine oil to the valve stems, springs and retainers.

b. Position the valve into the cylinder head (into the original guide and valve seat unless the valve or valve seat was replaced).

c. Position the valve spring seat and then spring over the valve and against the cylinder head, then place the spring retainer on top of the assembly.

d. Use the box-end wrench to press downward on the retainer so you can install the cotters.

■ Use a thin screwdriver with a dab of marine grease to hold an install the valve cotters (keepers).

e. Insert the valve cotters, then slowly ease off the valve spring making sure the cotter is fully in position. Remove the valve spring compressor and tap lightly on the valve spring seat using a plastic or rubber mallet to make sure the cotter is securely set.

f. Repeat for the remaining valve.

6. Install each of the rocker arm assemblies as follows:

a. Apply a light coating of Loctite® 271 or an equivalent threadlocking compound to the threads on the lower (cylinder head) end of the rocker arm pivot bolt (stud).

b. Apply a light coat of engine oil to the pushrod contact surfaces (as well as the rocker arm and pivot contact surfaces), then position the pushrod guide plate to the cylinder head and secure using one of the rocker arm pivot studs.

c. Install the other rocker arm pivot stud and tighten them both to 18 ft. lbs. (24.5 Nm).

d. Loosely install the rocker arms, pivots and locknuts.

■ The rocker arms must be only loosely installed at this point since they must be repositioned during installation to insert the pushrods.

7. Install the Cylinder Head, to the powerhead assembly as detailed earlier in this section.

CYLINDER BLOCK OVERHAUL

◆ See Figure 23 and 26 thru 31

Throughout assembly procedures, components should be coated with fresh, clean engine oil. When a torque specification is provided, Mercury advises that a torque wrench must be used in a minimum of 3 stepped passes in order to reach the specification.

1. Remove the Powerhead, as detailed in this section (be sure to drain the engine oil).

2. Strip the powerhead for overhaul by removing the following components:
- Manual (hand rewind) starter assembly
- Flywheel and Ignition Coils (exciter, pulser and ignition coils)
- Fuel pump
- Carburetor (or carburetor and intake manifold, as desired)

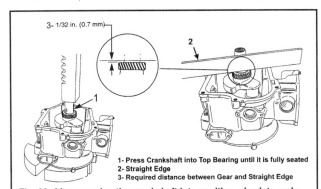

Fig. 26 After pressing the crankshaft into position, check to make sure it and the bearing are fully seated

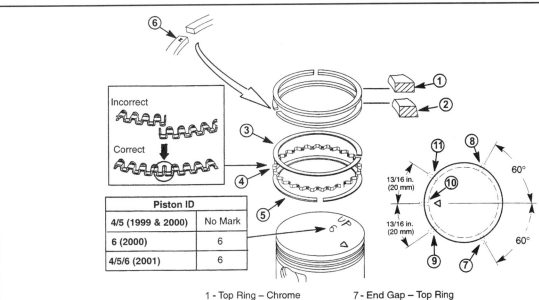

Fig. 27 Piston ring installation and gap positioning

6-16 POWERHEAD

3. Place the powerhead upside-down (flywheel side down) on the work surface.

4. Loosen the 2 bolts securing the oil seal housing to the bottom of the oil pan. Carefully break the gasket seal and remove the housing. If necessary use a rubber mallet to gently tap on the housing and loosen the gasket seal. Remove and discard the old gasket.

5. Using a jawed puller remove the two oil seals and spacer from the top of the oil seal housing. For installation purposes, take note of which direction the seal lips are facing before removal.

6. Using multiple passes in the reverse of the torque sequence, carefully loosen and remove the oil pan retaining bolts (they are threaded into the underside of the cylinder block AND cylinder head).

7. Gently and carefully pry at the provided tab areas on the oil pan to help break the gasket seal and free it from the cylinder head and block. Remove the oil pan, then remove and discard the pan-to-block and pan-to-head gaskets.

8. If necessary, disassemble the oil pan components. Take your time noting the location of each component for installation purposes.

9. If necessary, disassemble the oil pump. For details, please refer to the Lubrication and Cooling section.

10. Note the direction the seal lips are facing, then remove the lower seal from the crankshaft bore in the bottom of the oil pan.

11. Grasp and carefully withdraw the camshaft from the bottom of the cylinder block. Be sure to remove and keep track of both lifters (since if they are reused, they must be installed in the same positions, facing the same direction).

12. Remove the Cylinder Head, as detailed in this section.

13. Remove the piston assembly from the crankshaft, as follows:

   a. Remove the wrist pin retaining clips from either side of the piston pin bore. The easiest way to do this is to spin the clip in the bore so the gap in the clip aligns with the notch in the side of the piston, then CAREFULLY pry the clip free.

   b. Carefully remove the piston pin, freeing the piston from the connecting rod.

   c. Matchmark the end cap to the connecting rod, then if possible, unbolt and remove the bearing cap from the crankshaft end of the connecting rod. Remove the crankshaft cap and bearing.

   d. Unless an excessive ridge is formed at the top of the cylinder bore (which you can feel for using a fingernail), carefully push the piston and connecting rod up and out the top of the cylinder block. If a ridge has formed the concern is that a ring may become hung up on it and break, however you may still be able to remove the piston if you can carefully compress the rings and work the piston out of the bore. Your other option is to position the piston at the top of its travel and turn the crankshaft so as to achieve maximum clearance, then attempt to remove the crankshaft first so that you can pull the piston out of the bottom of the cylinder block. If you choose to try this, take care not to score or damage the polished crankshaft surfaces.

■ Once removed, immediately reinstall the connecting rod end cap to the piston while aligning the matchmark made during removal to ensure proper orientation. This will make sure the orientation is not lost if the marks are dulled or removed by cleaning and inspection procedures.

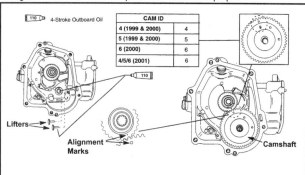

Fig. 28 Camshaft installation and timing mark alignment

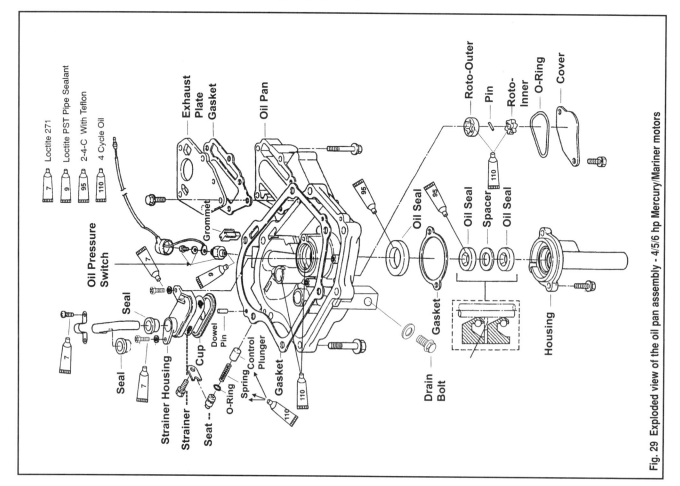

Fig. 29 Exploded view of the oil pan assembly - 4/5/6 hp Mercury/Mariner motors

POWERHEAD 6-17

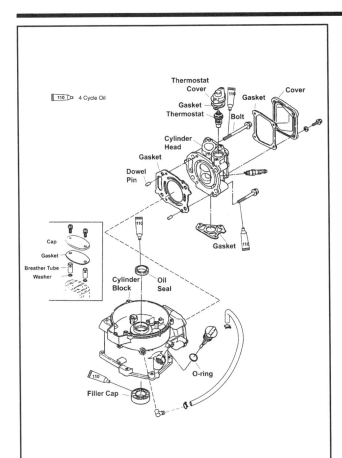

Fig. 30 Exploded view of the cylinder head and block mounting - 4/5/6 hp Mercury/Mariner motors

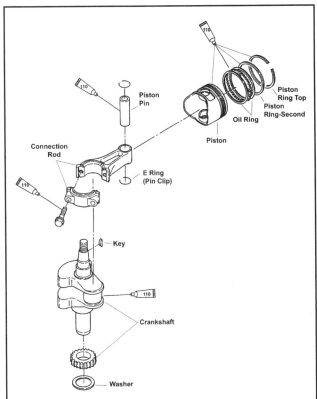

Fig. 31 Exploded view of the crankshaft and piston assembly - 4/5/6 hp Mercury/Mariner motors

e. Remove the connecting rod.

■ These pistons and connecting rods are normally already marked from the factory for orientation. A stamping on the connecting rod 3H6 must face downward toward the oil pan, while a stamped triangular mark or arrow on the piston dome should face the camshaft. Some pistons will also include the stamped word UP to show which side faces the flywheel. Regardless, it is a good idea to make marks which will face upward on both components to ensure proper orientation during assembly.

14. Using an arbor press, carefully press the crankshaft out of the cylinder block (pushing it downward from the flywheel side and out the bottom of the block). Don't loose the large washer on the bottom of the crankshaft assembly. Remove the crankshaft from the block. Place it on a protective surface.

✴✴ WARNING

Handle the crankshaft carefully. Remember it is an expensive and vital piece to the motor. The polished surfaces should be covered to prevent scratches or other damage and the whole shaft should be protected against impact or other damage which could destroy it.

15. Carefully pry the crankshaft oil seal from the top of the cylinder block. Discard the seal.
16. Only if replacement is necessary, use a jawed puller with a puller bridge for support to remove the upper ball bearing from the case. The bearing is removed from the oil pan side of the block.
17. Remove the 2 bolts securing the breather cap in the top of the block, then remove the cap. Remove and discard the cap gasket. Remove the 2 breather tubes and 2 wave washers.
18. If necessary, disassemble the piston, as follows:
    a. Using a piston ring spreading tool, carefully remove each of the rings (top and second compression rings, followed by the oil ring/rings) from the piston.

■ Some models have stamp marks on one side of the piston ring to denote which side must face upward. Check the rings as they are removed for the presence of such marks and make a note for installation purposes.

    b. Keep all components together, sorted by positioning and marked for orientation for reuse or replacement.
19. Refer to Powerhead Refinishing, later in this section, for details on cleaning and inspecting cylinder block components. Take your time and make sure you've made measurements for all components listed in the Engine Specifications chart for this model.

To Assemble:
20. Make sure the mating surfaces of the crankcase and cylinder block are completely clean and free of debris, gasket material or damage.
21. Using a suitable driver, install a new upper oil seal into the top of the cylinder block with the seal lips facing as noted during removal (which should mean lips facing downward toward the block). Drive the seal into position slowly until it bottoms in the bore.
22. If removed install the breather assembly. First install the 2 wave washers, followed by the 2 breather tubes. Install the breather cap using a new gasket and tighten the screws to 70 inch lbs. (8 Nm).
23. If removed, apply a light coating of engine oil to the replacement crankshaft ball bearing, then use a suitable sized driver to carefully press or tap it into position in the top of the cylinder block. The bearing is installed from the underside (oil pan side) of the block. Make sure the driver presses on the OUTER bearing race or the bearing could be damaged. Tap the bearing into position until it bottoms out in the bore.

■ Before assembling the piston and/or installing the crankshaft or connecting rod, be sure to check the Connecting Rod Big-End Oil Clearance by assembling the connecting rod to the crankshaft using Plastigage® or an equivalent gauging compound. For more details, please refer to the Powerhead Refinishing section.

24. Apply a light coating of engine oil to the crankshaft, especially on the drive gear and bearing mating surfaces, then carefully insert the crankshaft into the cylinder block. Press the crankshaft fully into the top bearing.
25. Check to make sure the crankshaft and top bearing are both fully seated. To do this place a precision straight edge across the case so that it

6-18 POWERHEAD

passes over the crankshaft timing (crankshaft-to-camshaft drive gear). The gap between the straight edge and the gear should be about 1/32 in./0.031 in. (0.7mm) and it should remain constant from one side of the gear to the other.

26. Oil and install the washer over the bottom end of the crankshaft.
27. If disassembled, prepare the piston assembly for installation as follows:

   a. If you made your own alignment marks during assembly (to orient the connecting rod and piston facing upward), align those marks at this time.

   b. Make sure the stamped mark on the connecting rod 3H6 faces downward and the stamped mark on the piston dome will face the camshaft.

   c. Oil and insert the connecting rod into the piston, then carefully install the piston pin.

   d. Install the piston pin clips in either side of the pin bore to secure the piston pin in place. Once each clip is inserted, be sure to spin the clip in the bore so the gap DOES NOT align with the notch in the side of the piston. A solid portion of the clip must bridge the notch.

   e. Using a ring expander, carefully install each of the rings with their gaps as noted in the accompanying illustration. Keep in mind that the illustration is with the camshaft mark to your left (at 9 o'clock) which should place the UP mark or flywheel side of the piston at the bottom (6 o'clock).

■ Most rings are stamped with a mark which should face upward. When present the mark 1N means it is the top compression ring, 2N means the second compression ring. In both cases a stamped 50 means oversize. Similarly, a piston whose dome is stamped with 0.5 is an oversize piston.

   f. Start with the oil ring assembly. The lower side rail of the oil ring should be positioned with the gap facing between 10 and 11 o'clock. The oil ring expander should be placed with the gap facing directly toward 9 o'clock (and the camshaft mark on the dome). The top side rail of the oil ring should be positioned with the gap facing between 7 and 8 o'clock. In this position the gaps of the side rails are each about 13/16 in. (20mm) from the expander gap.

   g. Install the second (lower of the 2 compression rings, which is the tapered ring in this case) with the gap facing at approximately 1 o'clock (60 degrees upward from 3 o'clock).

   h. Install the first (top of the compression rings, which is the squared ring in this case) with the gap facing at approximately 5 o'clock (60 degrees downward from 3 o'clock).

   i. Once installed, make sure all rings are free to move in their grooves, then double-check the ring gap positions as detailed earlier.

✸✸ WARNING

Use extreme care to prevent the connecting rod from contacting and scoring the cylinder wall during piston installation. It is a good idea to wrap the ends in a shop cloth or something else to protect the cylinder.

28. If installed for storage purposes, remove the connecting rod end cap and bolts so the piston can be installed, positioning the connecting rod over the crankshaft
29. Install the piston (with the correct mark/side facing UP toward the flywheel and the stamping 3H6 on the connecting rod facing the oil pan) to the cylinder block using a ring compressor. Apply a very light coating of clean engine oil to the inside of the compressor, to the cylinder bore and the connecting rod mating surface, then place the compressor around the piston. Insert the piston through the cylinder head side of the bore and use a wooden hammer handle to gently tap the piston down through the compressor and into the cylinder bore.

✸✸ WARNING

Be VERY careful when installing the piston. Get the lower portion of the skirt below the ring compressor and insert it into the cylinder bore to make sure everything is aligned. Then, GENTLY tap the piston through the compressor, stopping if it seems to hang up even slightly to make sure that a ring is not caught on the deck of the cylinder block.

30. Apply a light coating of engine oil to both the connecting rod end cap bearing surface as well as the threads of the cap bolts. Install the end cap and tighten the connecting rod retaining bolts alternately and evenly first to 53 inch lbs. (6 Nm), and then to 106 inch lbs. (12 Nm).
31. Oil and install the valve lifters to the bores in the cylinder block. If the lifters are being reused, be sure to install them to the same locations from which they were removed.
32. Apply a light coating of engine oil to the camshaft, then install the shaft to the cylinder block aligning the stamped timing mark on the crankshaft gear with the mark on the meshing camshaft gear.
33. Install the Cylinder Head, as detailed in this section.
34. If disassembled, install the components installed in the oil pan (the oil pump should be installed AFTER the pan is installed so that you can insert the drive pin into the end of the camshaft). Keep in mind the following points when servicing the oil pan:

• If the crankshaft oil seal was removed, install a new seal from the bottom side (outside or gearcase side of the pan) using a suitable driver. Position the seal lips as noted during removal (this is normally facing upward toward the crankshaft) and carefully drive the seal into position until it seats in the bore.

• Apply a light coating of clean engine oil to all bores, seals and to the components of the pressure relief valve such as the cup and plunger.

• Clean the threads and bores for all internal retaining screws, then coat the threads with Loctite® 271 or an equivalent threadlocking compound.

• If the oil pressure switch was removed, apply a light coating of Loctite Pipe Sealant W/ Teflon-567 or an equivalent pipe sealant, then install and carefully tighten the switch.

35. Make sure the lower crankshaft washer is still in position, then place a new gaskets on the oil pan-to-block and pan-to-cylinder head mating surfaces.
36. Lower the pan into position over the gaskets and carefully thread the bolts (remember the smaller bolts are used on the cylinder head). Tighten the bolts using multiple passes of the torque sequence until the small bolts are tightened to 80 inch lbs. (9 Nm) and the large bolts are tightened to 18.5 ft. lbs. (25 Nm).
37. Install the oil pump assembly, as detailed in the Lubrication and Cooling System section.
38. Install the 2 new oils seals to the oil seal housing (along with the spacer). The seals are normally positioned with the lips facing downward (toward the gearcase). Install the first seal with a suitable driver and gently tap it into position until it is seated in the bore, then install the spacer and, finally, install the second seal in the same manner as the first. Stop tapping when the second seal is seated against the spacer.
39. Install the oil seal housing to the bottom of the oil pan and tighten the 2 retaining bolts to 70 inch lbs. (8 Nm).
40. You can install the remaining components now or after the powerhead is installed, the call is yours. When ready, install the following components:

• Carburetor (or carburetor and intake manifold, as desired)
• Fuel pump
• Flywheel and Ignition Coils (exciter, pulser and ignition coils)
• Manual (hand rewind) starter assembly

41. Refill the powerhead with clean engine oil.
42. Install the Powerhead, as detailed in this section.
43. Once the powerhead is fully assembled, if cylinder block components were replaced, especially the piston and/or rings, be sure to operate the engine as directed for new component break-in. For details, please refer to Powerhead Break-In, in this section.

CLEANING & INSPECTION

Cleaning and inspecting the components is virtually the same for most 4-stroke outboards and varies mostly by specifications (which are listed in the Engine Specifications charts) or by component type. A section detailing the proper procedures, sorted mostly by component, can be found under Powerhead Refinishing.

6/8 Hp Yamaha and All 8/9.9 Hp (232cc) Powerheads

POWERHEAD REMOVAL & INSTALLATION

◆ See Figures 32 thru 39

MODERATE

Although these powerheads are still small enough that you can usually manage to lift them manually from the intermediate housing (and with most fuel and electrical components still installed) they're starting to move into the larger and more bulky assemblies which make this more difficult.

As usual, we advise that you consider why the powerhead is being removed. If you are planning on disassembling the powerhead for inspection

POWERHEAD 6-19

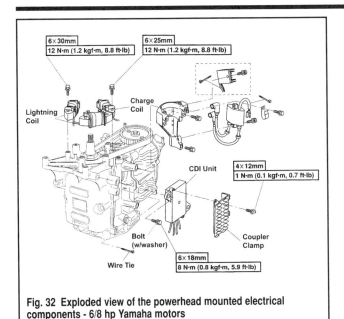

Fig. 32 Exploded view of the powerhead mounted electrical components - 6/8 hp Yamaha motors

or overhaul, then you'll have to remove the fuel and electrical components anyway. If this is the case, removing them before powerhead removal is a good idea, if only to protect them from potential damage when lifting and moving the powerhead itself. However, the choice remains yours whether or not you follow all of the initial steps for stripping the powerhead of these components. In this procedure we'll do our best to lay out the bare minimum steps necessary to remove the entire powerhead from the intermediate housing.

1. Remove the engine top cover for access to the powerhead.
2. Tag and disconnect the necessary hoses either from the powerhead or from the intermediate housing, as necessary. On all models it is usually easiest to disconnect the fuel line at the carburetor/powerhead. On 8/9.9 hp (232cc) motors, you'll also normally have to disconnect the blow-by hose and 3 water hoses. On 6/8 hp Yamaha motors it is usually sufficient to just to disconnect the cooling system indicator hose.
3. Remove the Hand-Rewind Starter assembly or timing belt/flywheel cover from the powerhead. On 8/9.9 hp (232cc) motors you must usually remove the starter grip. For more details, refer to the Hand-Rewind Starter section. On these same models, disconnect the neutral starter-lockout cable.
4. For Yamaha motors, tag and disconnect the shift and/or throttle cable(s), as applicable. On 6/8 hp Yamaha motors there are 2 throttle cables and a separate shift cable.
5. For Mercury/Mariner motors disconnect the throttle linkage either by unsnapping the throttle link from the remote slide (remote control models) or by un-clipping the throttle link age from the cable pulley stud (tiller handle models).

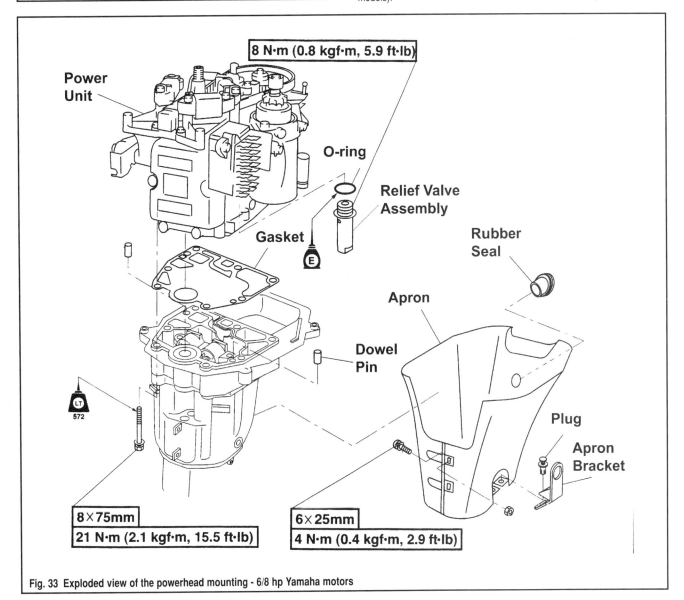

Fig. 33 Exploded view of the powerhead mounting - 6/8 hp Yamaha motors

6-20 POWERHEAD

6. Drain the Engine Oil, as detailed in the Maintenance and Tune-Up section, then remove the oil dipstick.

7. Tag and disconnect the necessary electrical connections. This usually means the wiring for the engine stop switch and emergency signal light (3 connectors and a ring terminal for 6/8 hp Yamaha motors). On 8/9.9 hp (232cc) motors with electric start this should also include the battery cable leads.

8. If desired, strip the powerhead of the fuel system components (specifically the carburetor), as detailed in the Fuel System section.

9. If desired, remove the flywheel and strip the powerhead of the remaining electrical system components.

10. For access, remove the lower engine cowling/apron, as follows:
• For 6/8 hp Yamaha motors, loosen the 2 screws threaded sideways between the 2 halves of the lower apron into retaining nuts. Also remove the 2 plugs and apron bracket from the lower side of the assembly. Remove the apron halves for access.
• For 8/9.9 hp (232cc) motors, remove the screw and apron cover from the lower cowling on either side of the motor.

11. Gradually loosen and remove the powerhead retaining bolts starting at the outer fasteners and working inward. Although all models use 6 main powerhead mounting bolts, some models, such as the 6/8 hp Yamaha motors utilize two additional fasteners which must also be removed (bolt No.'s 8 and 9 in the torque sequence).

■ If the unit is several years old, or if it has been operated in salt water, or has not had proper maintenance, or shelter, or any number of other factors, then separating the powerhead from the intermediate housing may not be a simple task. An air hammer may be required on the bolts to shake the corrosion loose; heat may have to be applied to the casting to expand it slightly, or other devices employed in order to remove the powerhead. One very serious condition would be the driveshaft "frozen" with the lower end of the crankshaft. In this case a circular plug type hole must be drilled and a torch used to cut the driveshaft.

12. Carefully lift and remove the powerhead assembly from the intermediate housing and place it on a suitable work surface. If necessary, a piece of wood may be inserted between the powerhead and the intermediate housing as a means of using leverage to force them apart.

■ Take care not to lose the two dowel pins. The pins may come away with the powerhead or they may stay in the intermediate housing. Be ESPECIALLY careful not to drop them into the lower unit.

13. If necessary on 6/8 hp Yamaha motors, unthread and remove the relief valve assembly from the lower portion of the powerhead. If removed, discard the O-ring.

To Install:

14. If removed on 6/8 hp Yamaha motors, apply a light coating of engine oil to the NEW relief valve assembly O-ring, then install the valve and tighten it to 6 ft. lbs. (8 Nm).

15. Make sure the dowel pins are in position at the opposite sides of the intermediate housing.

16. On 8/9.9 hp (232cc) Yamaha motors, apply a light coating of gasket making sealant to both sides of a new powerhead mounting gasket. The Mercury/Mariner manuals do not mention the use of sealant and, if none was present when you removed the powerhead, then we'd say you should go ahead and install the gasket dry on those models.

■ No sealing agent is required on the intermediate housing gasket for 6/8 hp Yamaha motors.

17. Place the new gasket in position over the 2 dowel pins, this should make sure the gasket does not cock during installation and does not obscure any water passages, but keep a close eye on the gasket as the powerhead is positioned anyway.

18. Apply a light coating of marine grade grease to the driveshaft-to-crankshaft splines

19. On 8/9.9 hp (232cc) motors there is an O-ring and 2 seals on the intermediate housing. Make sure new seals and O-rings are coated with marine grade grease and installed in position, as shown in the accompanying line drawing.

20. Carefully lower the powerhead onto the intermediate housing using the dowel pins to align the two surfaces. The propeller (if in gear) or the flywheel/crankshaft (if not) may have to be rotated slightly (only in the normal direction of rotation, CLOCKWISE on these models) to permit the crankshaft to index with the driveshaft and allow the powerhead to seat properly.

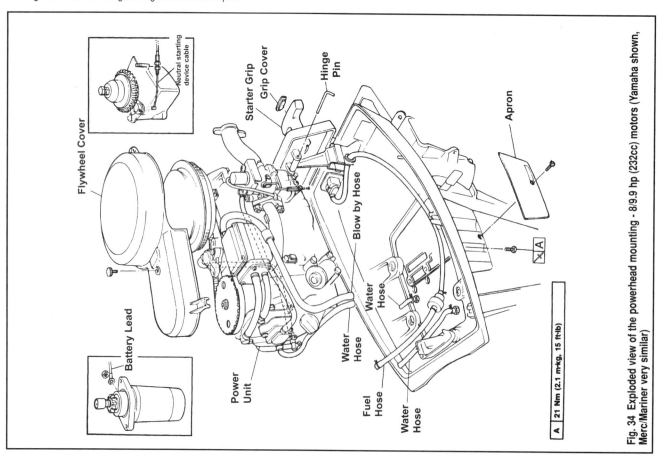

Fig. 34 Exploded view of the powerhead mounting - 8/9.9 hp (232cc) motors (Yamaha shown, Merc/Mariner very similar)

A | 21 Nm (2.1 m•kg, 15 ft•lb)

POWERHEAD 6-21

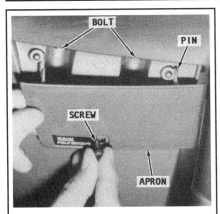

Fig. 35 Remove the aprons for access to the powerhead bolts - 8/9.9 hp (232cc) motors shown

Fig. 36 Once the bolts are removed, carefully lift the powerhead from the cowling

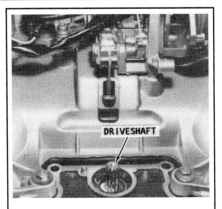

Fig. 37 Apply a light coating of grease to the driveshaft splines before powerhead installation

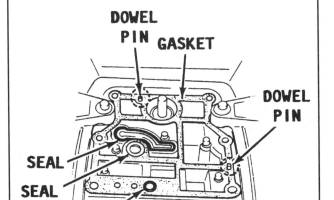

Fig. 38 On 8/9.9 hp (232cc) motors make sure the dowel pins, seals, O-ring and gasket are all in their proper positions

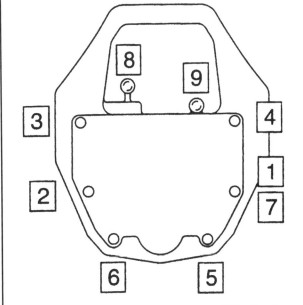

Fig. 39 Powerhead mounting torque sequence - 6/8 hp Yamaha motors

21. Apply a light coating of Loctite® 572 or equivalent threadlocking compound to the threads of the powerhead attaching bolts.

■ Mercury does not mention using threadlock on the powerhead mounting bolts, but we don't see the harm and actually would go so far as to recommend it anyway.

22. Start the powerhead attaching bolts, with washers. Tighten the bolts alternately and evenly to a torque value of 15 ft. lbs. (21 Nm). On 6/8 hp Yamaha motors use multiple passes of the torque sequence shown in the accompanying illustration. For 8/9.9 hp (232cc) motors, use a spiraling pattern that starts at the inside fasteners and works outward.
23. Install the lower cowling/apron assembly to either side of the outboard.
24. If removed, install the electrical components and the flywheel.
25. If removed, install the carburetor and fuel system components.
26. Reconnect the electrical wiring as tagged during removal. On 6/8 hp Yamaha motors this usually means wiring for the engine stop switch and emergency signal light (3 connectors and a ring terminal). On 8/9.9 hp (232cc) motors with electric start this should also include the battery cable leads.
27. Refill the motors with engine oil and install the dipstick.
28. Reconnect the shift and/or throttle cable(s) or linkage, as applicable and as tagged during removal.
29. Install the Hand-Rewind Starter assembly or timing belt/flywheel cover to the powerhead. On 8/9.9 hp (232cc) motors install the starter grip and reconnect the neutral starter-lockout cable.
30. Reconnect the breather, water and/or fuel hoses, as tagged during removal.
31. Refer to the Timing and Synchronization adjustments in the Engine and Maintenance section for adjustment procedures.

■ Be sure to run the motor without the top cowling installed in order to check for potential fuel or oil leaks. Remedy any leaks before proceeding.

32. Install the top cowling to the powerhead.

■ If the powerhead was rebuilt or replaced with a remanufactured unit, don't forget to follow the proper Break-In procedures.

TIMING BELT/SPROCKET REMOVAL & INSTALLATION

◆ See Figures 40 thru 49

This procedure can be performed with the powerhead assembly removed or installed on the motor, but if the crankshaft sprocket must be removed with the powerhead installed you'll have to lock the driveshaft and propeller shaft from spinning in order to loosen the crankshaft nut. Luckily the crankshaft sprocket nut is not tightened to a very high torque on these models, so it is usually possible to hold the crankshaft from turning either

6-22 POWERHEAD

using a strap wrench or possibly by locking the propeller shaft using a block of wood between an old propeller and the anti-cavitation plate (don't use a good propeller for fear or damage to the hub). If the powerhead is removed from the outboard, the best method for locking the crankshaft is to use an old driveshaft inserted into the splines on the bottom of the crankshaft.

If the timing belt is to be reused, be sure to mark the direction of rotation on the old belt before removal. Also, if either the cylinder head or cylinder block is not to be disassembled further, be sure to properly align the timing marks before removal. This will significantly ease installation.

■ Camshaft sprocket removal requires the use of a holder tool which inserts into the 2 holes in the sprocket itself. The same tool that is used on the flywheel can normally be used on the camshaft sprocket.

1. If the powerhead is still installed and on models so equipped, disconnect the negative battery cable for safety.
2. Remove the Flywheel and Stator Plate, as detailed in the Ignition and Electrical Systems section.
3. If either the cylinder head or cylinder block is not being overhauled, rotate crankshaft sprocket in the normal direction of rotation (clockwise) in order to align the timing marks on the crankshaft and camshaft pulleys with the marks on the powerhead.

4. Check the condition of the timing belt prior to removing it. On 8/9.9 hp (232cc) motors, if the belt can be deflected more than 0.39 in. (10mm) at the midway point, the belt has stretched and **must** be replaced.

■ Remember, if the timing belt is to be reused, be sure to mark it showing the normal direction of rotation to ensure installation in the proper position.

5. Slip the timing belt free of the camshaft pulley **first,** and remove it from the crankshaft pulley.

If necessary to strip the powerhead for rebuild or to replace one or more damaged sprockets, remove the camshaft (driven) or crankshaft (drive) sprocket from the motor as detailed in the following steps:

6. To remove the camshaft pulley, obtain a suitable flywheel holder tool (Yamaha #YB-6139 or Mercury #91-83163M). Place the two arms of the tool over the two holes of the camshaft pulley. Hold the pulley steady with the tool and at the same time loosen, then remove the camshaft sprocket bolt and washer.
7. Lift the camshaft pulley straight up and free of the camshaft. Remove and save the Woodruff key from the recess in the camshaft.
8. To remove the crankshaft pulley tap the upturned ears of the large tabbed washer down flat against the crankshaft pulley. Remove the large

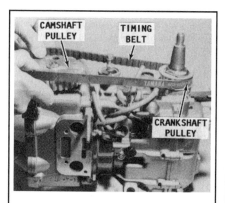

Fig. 40 Remove the timing belt - if it is to be reused, label the direction of rotation

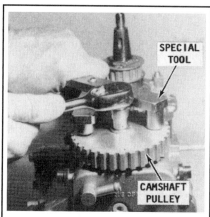

Fig. 41 Using a special tool to hold the camshaft, loosen the pulley nut...

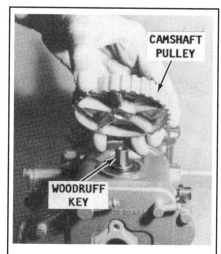

Fig. 42 ... then lift the pulley off the camshaft woodruff key

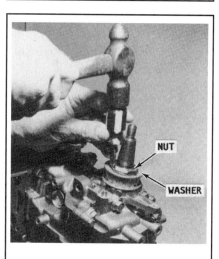

Fig. 43 If applicable, Use a small chisel to un-stake the ears of the crankshaft pulley tabbed washer...

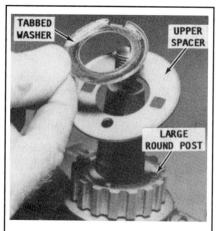

Fig. 44 ... remove the nut, washer and upper spacer from the crankshaft pulley - 8/9.9 hp (232cc) motors shown

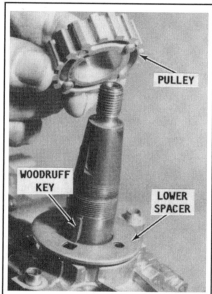

Fig. 45 Next, lift the pulley from the crankshaft woodruff key - again, 8/9.9 hp (232cc) motors shown

crankshaft nut. It will be necessary to hold and prevent the crankshaft from rotating while the nut is loosened and removed. This may be accomplished employing one of two methods:

- If the powerhead is removed from the outboard, the BEST method is to use a holding fixture from Yamaha (Mercury doesn't list one) or, if an old lower driveshaft is available, temporarily index the splines on the upper end of the driveshaft with the splines on the lower end of the crankshaft. Hold the old lower driveshaft in a vise or with a pair of vice grip type pliers.
- If the powerhead is installed first try a strap wrench on the crankshaft pulley, however if necessary, a block of wood may usually be used to keep the propeller (use only an old propeller since you risk hub damage) from turning.

9. After the crankshaft nut has been removed, lift off the large tabbed washer and then the upper spacer/plate washer. Note how a tab was inserted into one hole of the upper spacer/plate washer. Also, note that one some models holes in the upper spacer/plate washer are a indexed over round posts on the crankshaft pulley.

10. Slide the pulley straight up and free of the crankshaft. Remove and save the Woodruff key from the recess in the crankshaft. If equipped, remove the lower spacer. When equipped, note that there are usually two round posts on the underside of the pulley which are indexed into the two round holes of the lower spacer.

To Install:

11. If equipped with a separate lower belt spacer, slide it down the crankshaft with the convex (domed) side of the spacer facing **upward** toward the threaded end of the crankshaft. To clarify, this means the lip of the spacer is extended facing downward toward the powerhead).

12. Apply just a dab of lubricant to the Woodruff key, and then install the key into the lower cutout in the crankshaft. The lubricant will help hold the key in place.

13. Install the crankshaft sprocket over the retaining plate and woodruff key. If the crankshaft was not disturbed the sprocket should be positioned so the mark on the drive gear or cutout on the upper spacer/plate washer (once installed) will face the timing mark on the powerhead (and face directly toward the camshaft).

■ **When installing the sprocket on 8/9.9 hp (232cc) motors, make sure the stamped word TOP is facing UPWARD. Also, be sure to align two round posts on the bottom of the pulley indexed into the two round holes of the lower spacer.**

14. Install the belt upper spacer/plate washer over the crankshaft sprocket with the beveled edge facing upward, away from the powerhead. The timing marks mentioned in the last step must be aligned, or the crankshaft will need to be rotated once the nut is installed and tightened.

■ **Remember, if applicable position the upper spacer over the pulley with the two large round posts on the pulley indexed in the two square holes in the spacer and the one small round post on the pulley indexed with one of the round holes in the spacer.**

15. Slide the tabbed washer over the crankshaft with the tab facing **downward** to index with the other round hole in the spacer.

16. For 6/8 hp Yamaha motors, apply a light coat of clean engine oil to the threads of the crankshaft sprocket nut.

17. Thread the large nut onto the end of the crankshaft. Hold the crankshaft from turning and then tighten the nut to a value of 26.5 ft. lbs. (36 Nm) for 6/8 hp Yamaha motors or to 17 ft. lbs. (23 Nm) for 8/9.9 hp (232cc) motors. The crankshaft must be held firm while the nut is being tightened. Use the same device which was used during disassembly - an old section of driveshaft or with a strap wrench.

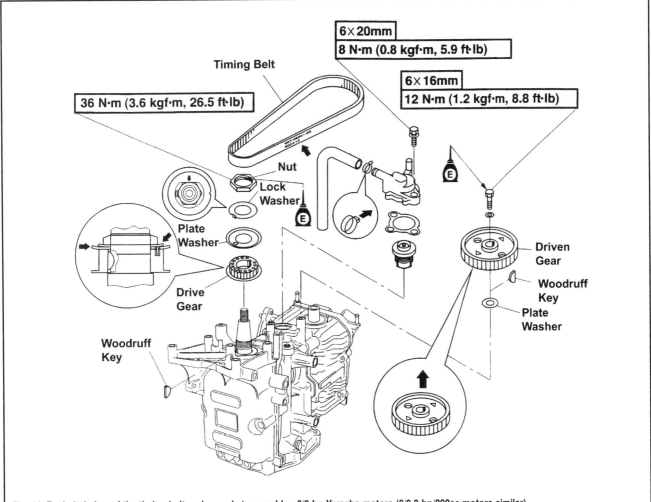

Fig. 46 Exploded view of the timing belt and sprocket assembly - 6/8 hp Yamaha motors (8/9.9 hp/232cc motors similar)

6-24 POWERHEAD

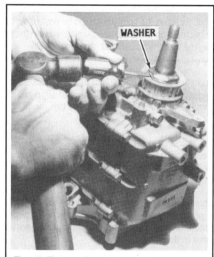

Fig. 47 Tighten the pulley nut, then carefully bend the washer ear to lock the nut in place

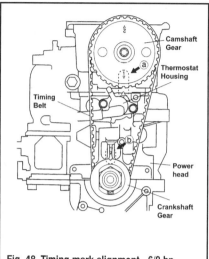

Fig. 48 Timing mark alignment - 6/8 hp Yamaha motors

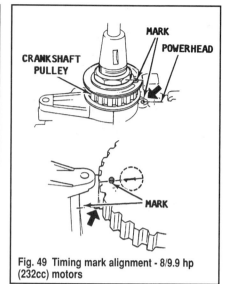

Fig. 49 Timing mark alignment - 8/9.9 hp (232cc) motors

18. After the nut has been tightened to the required torque value, tap the side of the washer up against the nut to lock the nut in place.
19. Make sure the mark on the crankshaft pulley assembly is aligned with the mark on the powerhead, facing the camshaft. If necessary, slowly rotate the crankshaft in the normal direction of rotation (clockwise) until the timing marks align. If the camshaft was disturbed, stop turning if you feel any interference.
20. Place the washer over the camshaft. Apply just a dab of lubricant onto the Woodruff key, and then insert the key into the camshaft slot. The lubricant should hold the key in place. Lower the camshaft pulley onto the camshaft and align the keyway in the pulley over the Woodruff key.
21. For 6/8 hp Yamaha motors, apply a light coating of engine oil to the threads of the camshaft sprocket retaining bolt.
22. Hold the camshaft sprocket from turning using the flywheel holder tool, then install and tighten the sprocket retaining bolt to 8.8 ft. lbs. (12 Nm) for 6/8 hp Yamaha motors or to 113 inch lbs./9.4 ft. lbs. (13 Nm) for 8/9.9 hp (232cc) motors.
23. If the camshaft was not disturbed, the No. 1 timing mark on the pulley should align with the triangle or mark on the powerhead, facing the crankshaft. If not, SLOWLY turn the crankshaft in the normal direction of rotation (also clockwise when viewed from above) until the timing marks are appropriately aligned.
24. Place the timing belt, with the word **YAMAHA** and/or the part number facing the right way up, over the crankshaft pulley, without moving the pulley. After the belt is in place stretch it over the camshaft pulley without moving the pulley.

■ Remember, if the belt is being reused, be sure to face it in the same direction as normal rotation (clockwise).

25. Turn the crankshaft clockwise until it completes 2 full revolutions, then recheck the timing marks.
26. Install the Flywheel and Stator Plate assembly, as detailed in the Ignition and Electrical Systems section.
27. If removed, install the powerhead assembly.
28. If applicable, reconnect the negative battery cable.

CYLINDER HEAD REMOVAL & INSTALLATION

◆ See Figures 50 thru 59

It is not entirely clear whether or not the cylinder head can be removed with the powerhead assembly still installed, but it looks likely (and Mercury literature lists separate procedures for it). Of course, in order to get to that point you've got to do so much of the work, it might be wise to just go ahead and remove the powerhead anyway (especially if work conditions would improve by placing everything on the bench while working).

Remember, when loosening the retainers on manifolds, covers and other major components, always try to follow the reverse of the torque sequence indicated in the assembly procedure or molded on the component itself.

When applicable tighten all bolts in using the proper torque sequence and to the proper specification. When possible we've included it in the procedure, but on many Yamaha powerheads both the sequence and value is molded into the casting on critical components. You'll notice that most Yamaha torque sequences are clockwise spirals starting somewhere near the center of the component.

■ Because of the high temperatures and pressures developed, the sealing surfaces of the cylinder head and the block are the most prone to water leaks. No sealing agent is recommended because it is almost impossible to apply an even coat of sealer. An even coat would be essential to ensure an air/water tight seal.

✳✳ WARNING

Never, never, use automotive type head gasket sealer. The chemicals in the sealer will cause electrolytic action and eat the aluminum faster than you can get to the bank for money to buy a new cylinder block.

Some head gaskets are supplied with a tacky coating on both surfaces applied at the time of manufacture. This tacky substance will provide an even coating all around. Therefore, no further sealing agent is required.

However, if a slight water leak should be noticed following completed assembly work and powerhead start up, **do not** attempt to stop the leak by tightening the head bolts beyond the recommended torque value. Such action will only aggravate the problem and most likely distort the head.

Furthermore, tightening the bolts, which are case hardened aluminum, may force the bolt beyond its elastic limit and cause the bolt to **fracture.** This would be **Bad news, very bad news** indeed. A fractured bolt must usually be drilled out and the hole retapped to accommodate an oversize bolt, etc. Avoid such a situation.

Probable causes and remedies of a new head gasket leaking are:
    a. Sealing surfaces not thoroughly cleaned of old gasket material. Disassemble and remove **all** traces of old gasket.
    b. Damage to the machined surface of the head or the block. The remedy for this damage is the same as for the next case "c".
    c. Permanently distorted head or block. Spray a light **even** coat of any type metallic spray paint on both sides of a new head gasket. Use only metallic paint - any color will do. Regular spray paint does not have the particle content required to provide the extra sealing properties this procedure requires.

Assemble the block and head with the gasket while the paint is till **tacky.** Install the head bolts and tighten in the recommended sequence and to the proper torque value and **no** more!

Allow the paint to set for at least 24 hours before starting the powerhead.

POWERHEAD 6-25

Consider this procedure as a temporary "band aid" type solution until a new head may be purchased or other permanent measures can be performed.

Under normal circumstances, if procedures have been followed to the letter, the head gasket will not leak. **ok, END SEALING SURFACE LECTURE**

1. Remove the Flywheel and Stator Plate, as detailed in the Ignition and Electrical Systems section.
2. Remove the Timing Belt, as detailed in this section. If the cylinder head is to be overhauled follow the necessary steps to remove the camshaft sprocket.
3. If not done already, strip the necessary electrical components from the powerhead.
4. Remove the spark plugs from the cylinder head.
5. Using multiple passes of a crossing pattern, remove the bolts (usually 4) securing the cylinder head cover to the cylinder head. Remove the cover, then remove and discard the seal.
6. Locate and remove the nine (6/8 hp Yamaha motors) or eight (8/9.9 hp (232cc) motors) cylinder head bolts from inside the head and from the head flange. Be sure to slowly loosen the bolts using the reverse of the torque sequence (essentially this results in a counterclockwise spiraling pattern that starts at the outer bolts and ends at the most central bolts).

■ **The bolts used for the cylinder heads on these motors are of 3 different lengths. Take note of the lengths NOW and tag or arrange them (you can stick them through holes punched in a shoe box to keep track of them) to ensure they'll be returned to their proper positions.**

7. Lift the head up and free of the block. If the head will not release from the block easily, insert a small prytool between the tabs usually provided for this purpose and CAREFULLY pry the two surfaces apart. **Never** pry at gasket sealing surfaces. Such action would very likely damage the sealing surface of an aluminum powerhead. If no tabs are provided, use a rubber mallet to gently tap around the perimeter of the surface to help break the gasket seal.

■ **Take care not to lose the two dowel pins at the mating surfaces. These pins may remain either in the cylinder head or in the block.**

8. Remove and discard the old cylinder head gasket.
9. If necessary for further cylinder head disassembly or for oil pump service, loosen the fasteners and remove the pump from the head. This will be necessary if you plan on removing the camshaft and valve train, but it is a good idea to remove and inspect the pump since you've come this far regardless. For details on removal, installation, inspection and overhaul of the oil pump assembly, please refer to Oil Pump in the portion of the Lubrication and Cooling section.
10. Refer to Cylinder Head Overhaul in this section for details on removal and installation of the valve train components. Also, be sure to check the information on gasket mating surfaces to make sure the cylinder head is not warped.

To Install:

11. If removed, refer to the Lubrication and Cooling section for details on Oil Pump assembly and installation.
12. Position a NEW head gasket over the two dowel pins. On 8/9.9 hp (232cc) motors, install the O-ring through the opening in the gasket, at the bottom center of the cylinder head.

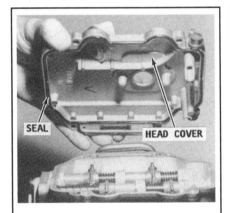

Fig. 50 Remove the valve cover...

Fig. 51 ...then remove the cylinder head bolts and...

Fig. 52 ...remove the cylinder head from the cylinder block

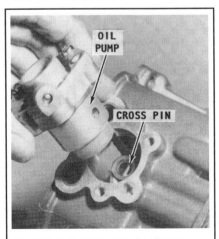

Fig. 53 Unbolt and remove the oil pump from the cylinder head

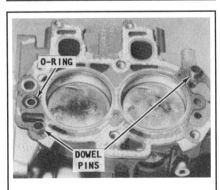

Fig. 54 Position a NEW head gasket over the dowel pins

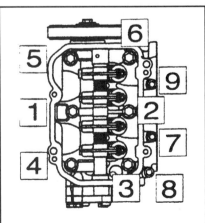

Fig. 55 Cylinder head torque sequence - 6/8 hp Yamaha motors

6-26 POWERHEAD

13. Install the cylinder head onto the block with the holes in the head indexed over the two dowel pins. Apply a thin coating of clean engine oil to the threads of the eight or nine attaching bolts (as applicable).
14. For 6/8 hp Yamaha motors thread the head attaching bolts according to the accompanying illustrations and the following tables.
    - Numbers 1, 4, and 5 - are usually 6x70mm bolts
    - Numbers 2, 3, and 6 - are usually the Long 8mm bolts
    - Numbers 7, 8 and 9 - two of these are Short 8mm bolts, while ONE is a Long bolt. It appears from Yamaha's information that number 8 is probably the long bolt, but we can't be sure, so compare the depth of the threads in the powerhead and/or the height of the bolt bosses on the cylinder head before installation and you'll be fine!
15. for 8/9.9 hp (232cc) motors thread the head attaching bolts according to the accompanying illustrations and the following tables.
    - Numbers 1, 4, and 5 - Long 8mm bolts
    - Numbers 2, 3, and 6 - Short 8mm bolts
    - Numbers 7 and 8 - 6mm bolts

■ The bolt locations for 8/9.9 (232cc) motors was taken from a tear-down model, so be sure to confirm it against your own notes before proceeding. If you're unsure, compare the size and depth of the bolt threads on the powerhead and the height of the bolt bosses on the cylinder head to confirm bolt locations.

16. Tighten the head bolts in the sequence shown in the accompanying illustration to the following torque values.
    First Sequence:
    - 8mm bolts - 11 ft. lbs. (15 Nm)
    - 6mm bolts - 2.2 ft. lbs. (3 Nm) for all Yamaha motors or 53 inch lbs. (6 Nm) for Mercury/Mariner motors
    Second Sequence
    - 8mm bolts - 22 ft. lbs. (30 Nm)
    - 6mm bolts - 106 inch lbs./8.8 ft. lbs. (12 Nm) for 6/8 hp Yamaha motors and 8/9.9 hp Mercury/Mariner models or 5.8 ft. lbs. (8 Nm) for 8/9.9 hp Yamaha motors

■ The reason for the large difference in torque values between the two bolt sizes is based on the fact the 8mm bolts are case hardened, grade 11, while the 6mm bolts are not. We don't have a good explanation as to why Yamaha and Mercury seem to disagree on the final torque for the smaller bolts on the same size motor.

17. Install a new sealing ring around the inside of the cylinder head cover.

■ If the valve train components were disturbed in any way you are going to have to adjust the valves once the timing belt is installed, so either leave the cylinder head cover off or just finger-tighten the bolts.

18. Install the cylinder head cover and new O-ring. If the valves were not disturbed, install and tighten the 4 cylinder head cover bolts securely either using the molded torque sequence (or if not present using a clockwise spiraling pattern that begins at the top left bolt, near the oil filler cap and works around to the bottom left bolt).

■ Make sure the new O-ring remains seated in the cover groove and does not become pinched during installation or there will be oil leaks!

19. Install any electrical components which were removed to the cylinder head and/or powerhead.
20. Install the Timing Belt and, if removed, Camshaft Sprocket, as detailed in this section.
21. If the valves were disturbed, adjust them at this time, as detailed under Valve Lash in the Maintenance and Tune-Up section, then install the cylinder head cover.
22. Install the spark plugs to the cylinder head.
23. Install the Flywheel and Stator Plate, as detailed in the Ignition and Electrical Systems section.
24. If removed, install the powerhead assembly.

CYLINDER HEAD OVERHAUL

◆ See Figures 60 thru 70

1. Remove the Cylinder Head from the powerhead assembly, as detailed in this section.
2. Remove the Oil Pump from the cylinder head, as detailed in the Lubrication and Cooling section.

■ For 8/9.9 hp (232cc) motors you'll need to obtain valve adjustment tool (#YM8035 from Yamaha or #91-809498A-1 from Mercury) in order to turn the valve adjusters. On these models, hold the 10mm nut with a box-end wrench, and at the same time loosen the square center locknut.

3. For 8/9.9 hp (232cc) motors remove the oil separator and intake manifold assemblies. Be sure to remove and discard all gaskets from the

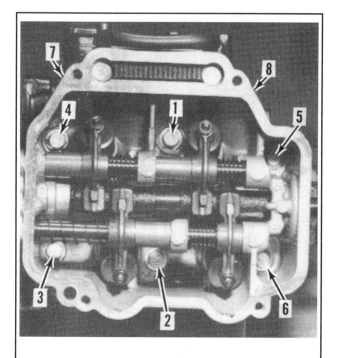

Fig. 56 Cylinder head torque sequence - 8/9.9 hp (232cc) motors

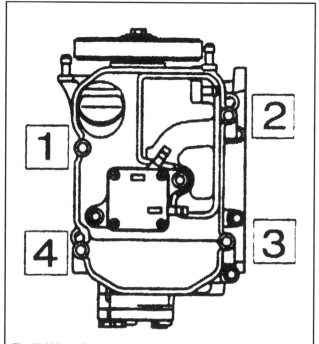

Fig. 57 Valve/cylinder head cover torque sequence - 6/8 hp Yamaha motors shown (8/9.9 hp motors similar)

POWERHEAD 6-27

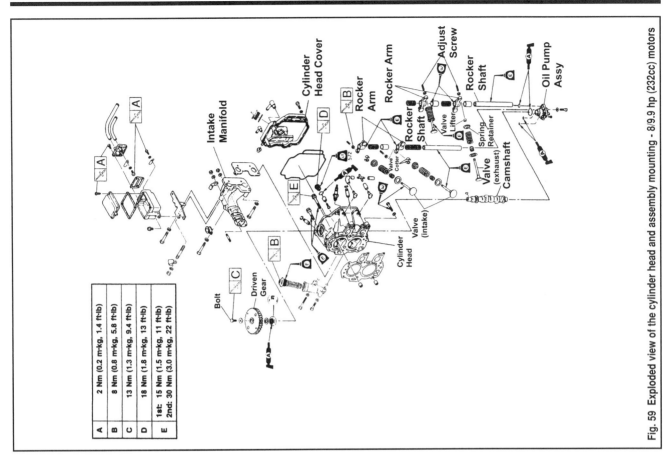

Fig. 59 Exploded view of the cylinder head and assembly mounting - 8/9.9 hp (232cc) motors

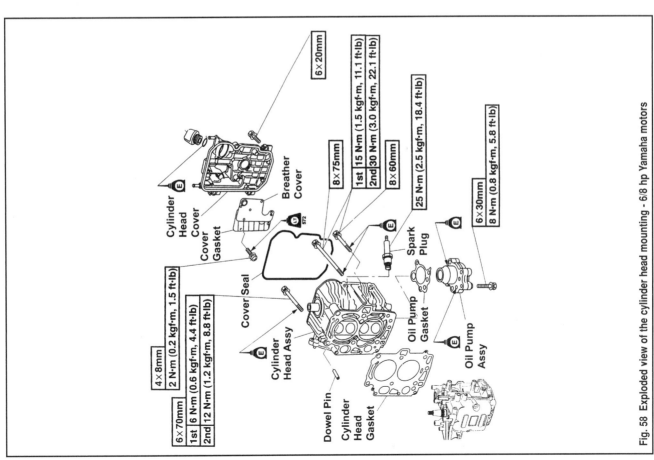

Fig. 58 Exploded view of the cylinder head mounting - 6/8 hp Yamaha motors

6-28 POWERHEAD

mating surfaces. If necessary, the oil separator can be disassembled to clean, inspect and/or replace the reed valve. Again, internal gaskets must all be replaced upon assembly and installation.

4. Using a suitable adjustment tool and a wrench, back off each valve lash adjuster until the rocker arm, when placed on the high point of the camshaft, will clear the valve stem.

■ The 6/8 hp Yamaha motors utilize one rocker shaft for all 4 valves, while the 8/9.9 hp (232cc) motors utilize 2 shafts, one for the intake and the other for the exhaust valves.

5. Slide the rocker shaft out and free of the cylinder head. As the shaft is withdrawn carefully note the arrangement of the rockers, springs, and the spacers. If difficulty is experienced in withdrawing the shaft, thread a bolt (usually 10mm) into the upper end of the shaft and use a bolt to gently pull the rocker shaft free. On 8/9.9 hp (232cc) motors, repeat for the remaining rocker arm shaft.

■ When reused, ALL valve train components must be installed in their original locations. For this reason be sure to tag and/or sort all rocker arms, springs, spacers, valves, valve springs, keepers etc and keep them separated during service.

6. Guide the camshaft carefully out of the head. **Take care** not to damage the camshaft lobes. If necessary, remove the oil seal with a slide hammer and expanding jaw attachment puller.

7. To remove the each valve, proceed as follows:
  a. Using a clamp-type valve spring compressor, carefully place just enough pressure on the valve spring so the retainer moves down sufficiently to expose the keepers.
  b. Remove the single valve keeper (used on 6/8 hp Yamaha motors and sometimes also called a lock) or the pair of valve keepers (used on 8/9.9 hp motors), then slowly ease pressure on the valve spring until the compressor can be removed.
  c. Remove the valve spring retainer, valve spring and valve spring seat from the shaft of the valve.
  d. Remove the valve from the underside (combustion chamber side) of the cylinder head.
  e. If necessary, carefully pry the old valve stem seal from its position on the cylinder head.
  f. Repeat this procedure for each of the remaining valves.

8. Some cylinder heads, such as the 8/9.9 hp (232cc) motors are equipped with a cylinder head mounted anode which looks like an additional bolt attached to the head itself. If equipped, remove the anode and gasket from the cylinder head.

9. Refer to Powerhead Refinishing, later in this section, for details on cleaning and inspecting cylinder head components.

To Assemble:

10. If applicable, install a new anode and washer securely into the head.
11. If removed, install each of the valve and spring assemblies as follows:
  a. If removed, install new valve spring seals to the cylinder head.

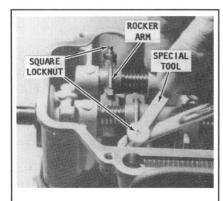

Fig. 60 Loosen the locknut on the valve last adjusters - 8/9.9 hp (232cc) motors shown

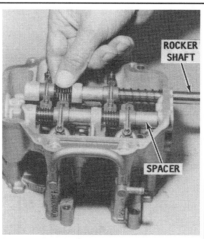

Fig. 61 Carefully remove the rocker shaft(s) and components...

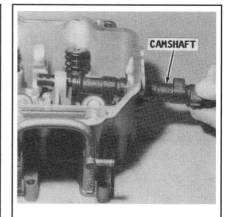

Fig. 62 ...then carefully slide out the camshaft

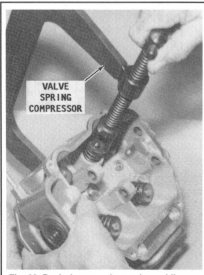

Fig. 63 Push down on the retainer while you remove the keeper(s)

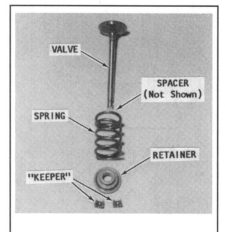

Fig. 64 Exploded view of a typical valve assembly

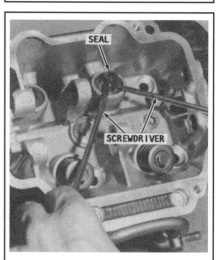

Fig. 65 Carefully pry to remove the valve stem seals

POWERHEAD 6-29

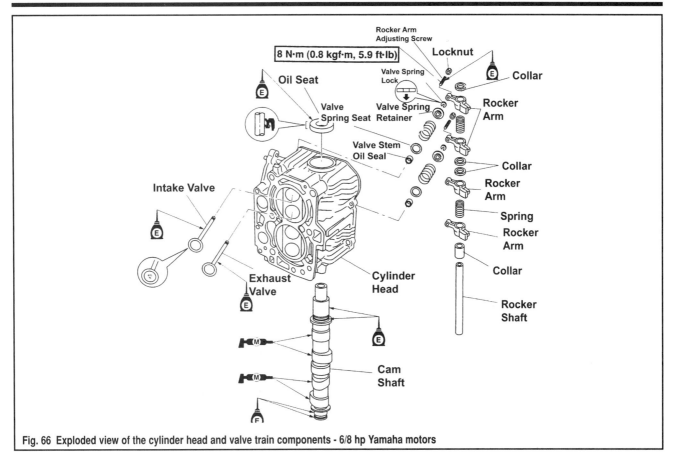

Fig. 66 Exploded view of the cylinder head and valve train components - 6/8 hp Yamaha motors

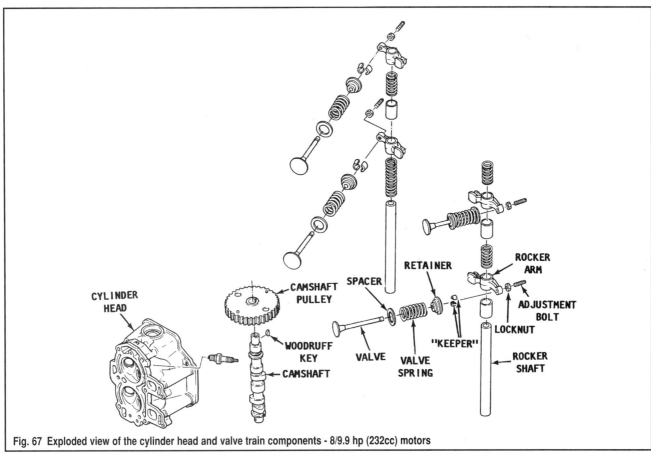

Fig. 67 Exploded view of the cylinder head and valve train components - 8/9.9 hp (232cc) motors

6-30 POWERHEAD

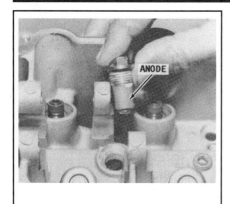

Fig. 68 If applicable, remove the cylinder head anode for inspection/replacement

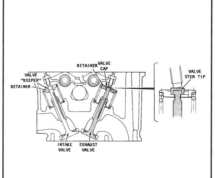

Fig. 69 Cross-sectional view of the valve train assembly showing keeper-type valves (left) and cap-type valves (right)

Fig. 70 Install new oil seals using a suitable driver and soft mallet

　b. Apply a light coating of engine oil to the stem, then position the valve into the cylinder head (into the original guide and valve seat unless the valve or valve seat was replaced).
　c. Position the valve spring seat down onto the valve shaft and cylinder head, then oil and install the spring.
　d. Place the retainer over the spring, then use the valve spring compressor tool to press downward on the retainer and spring so you can install the keeper(s) or lock.
　e. Insert the valve keepers/lock, then slowly back off the valve spring compressor, making sure the keeper(s) are fully in position. Then remove the valve spring compressor.
　f. Repeat for each of the remaining valves.
　12. If the camshaft oil seal was removed during disassembly, coat the lips of a new seal either with a good grade of clean engine oil (6/8 hp Yamaha motors) or a light coating of marine grade grease (8/9.9 hp motors). Install the seal into the head using a suitable driver. On all motors the seal should be installed with the lips facing downward into the cylinder head. On most applications this means the seal manufacturer's marks or numbers are facing outward.

■ Do not attempt to rotate the camshaft at this time. At this time the camshaft is supported at just one end. The camshaft should ONLY be rotated after the oil pump has been installed and the shaft is then supported at both ends.

　13. Apply a light coating of engine oil to the front and rear bearing surfaces of the camshaft, then apply a light coating of a molybdenum disulfide engine assembly lube to the camshaft lobes and inner camshaft journals. Next, **carefully** insert the "plain" end (without the cross pin) of the camshaft into the head from the oil pump end of the head. Continue inserting the camshaft until other end of the shaft is seated in the camshaft oil seal against the head at the other end.

■ The ends of the rocker arm shafts are threaded on some models, to ease removal. When so equipped the threaded ends should always be left toward the oil pump end of the cylinder head.

　14. Insert the rocker arm shaft into the cylinder head. For 8/9.9 hp (232cc) motors, start by installing the exhaust rocker arm shaft into the port side of the head. For these motors, slide the shaft through a spacer, exhaust rocker arm, small compression spring, through a boss in the head, another spacer, the other exhaust rocker arm, another small compression spring, and finally into a boss in the head at the opposite end. The 6/8 hp Yamaha motors utilize only one rocker arm shaft so slide the shaft through the spacer, exhaust rocker arm, spring, intake rocker arm, collar, boss, collar, exhaust rocker arm, spring, intake rocker arm collar and finally the remaining boss.

■ Remember to lightly coat all the rocker arm and shaft components with clean engine oil as they are installed.

　15. For 8/9.9 hp (232cc) motors, insert the intake rocker arm shaft into the starboard side of the head. Slide it through a long compression spring, an intake rocker arm, a spacer, through the boss in the head, a small compression spring, another intake rocker arm, and finally into the boss in the head at the opposite end.
　16. Install the Oil Pump assembly to the cylinder head, as detailed in the Lubrication and Cooling section.
　17. For 8/9.9 hp (232cc) motors install the oil separator and intake manifold assemblies using new gaskets.
　18. Install the Cylinder Head as detailed in this section.
　19. Once the powerhead is fully assembled, if valve train components were replaced, be sure to operate the engine as directed for new component break-in. For details, please refer to Powerhead Break-In, in this section.

CYLINDER BLOCK OVERHAUL

◆ See Figures 71 thru 91

　Remember, when loosening the retainers on manifolds, covers and other major components, always try to follow the reverse of the torque sequence indicated in the assembly procedure or molded on the component itself.
　When applicable tighten all bolts in using the proper torque sequence and to the proper specification. When possible we've included it in the procedure, but on many Yamahas powerheads both the sequence and value is molded into the casting on critical components. You'll notice that most Yamaha torque sequences are clockwise spirals starting somewhere near the center of the component.
　1. Remove the Powerhead Assembly, as detailed in this section.
　2. Remove the Timing Belt/Sprocket Assembly, as detailed in this section.
　3. Remove the Cylinder Head Assembly, as detailed in this section.
　4. On some models the oil pressure sensor is threaded into the cylinder block. Carefully loosen and remove the sensor from the block to prevent damage to the sensor and to make sure the oil passages can be thoroughly cleaned.
　5. Remove the Thermostat assembly from the cylinder block to prevent damage and so the cooling passages can be thoroughly cleaned. You CAN reuse the old thermostat upon installation, but you're rebuilding or repairing the BLOCK right, so we'd just break-down and replace it.

■ The exhaust cover should always be removed during a powerhead overhaul. Many times water in the powerhead is caused by a leaking exhaust cover gasket or plate.

　6. Remove the exhaust cover bolts using the reverse of the molded torque sequence (or if not present using a counterclockwise spiraling pattern that begins at the top cover bolt and works inward).

POWERHEAD 6-31

7. Remove the exhaust cover, then remove and discard the old gasket. On 8/9.9 hp (232cc) motors there is also an inner cover and second gasket. If the cover is stuck to the powerhead, insert a small pry-tool between the tabs provided for this purpose and pry the two surfaces apart. **Never** pry at gasket sealing surfaces. Such action would very likely damage the sealing surface of an aluminum powerhead.

8. Remove the crankcase retaining bolts using the reverse of the molded torque sequence/sequences (as applicable). If no sequence is molded on the crankcase, the procedure varies very slightly by model as follows:

• For 6/8 hp Yamaha motors there are only 6 crankcase retaining bolts, loosen them using multiple passes of a counterclockwise, spiraling sequence that starts at the upper right bolts and works inward.

• For 8/9.9 hp (232cc) motors, use multiple passes of a counterclockwise spiraling sequence that starts at the upper right bolt of the 6 smaller (M6) bolts and then continues at the lower left bolt of the larger (M8) bolts.

■ Remember that on 8/9.9 hp (232cc) motors there are two different size bolts are used so keep them separated for identified for installation purposes. Bolts identified as 1, 2, 3, and 4 by an embossed number on the crankcase (the torque sequence) are 8mm bolts. These are found at the top and bottom center of the crankcase. Bolts identified as 5, 6, 7, 8, 9, and 10 are 6mm and are found on the outer flange of the crankcase (3 to a side). An accompanying illustration clearly shows these embossed numbers to identify the bolts.

9. After all bolts have been removed, separate the crankcase from the powerhead. **Take care** not to lose the four (6/8 hp Yamaha models) or two (8/9.9 hp models) dowel pins. These pins may remain with powerhead or

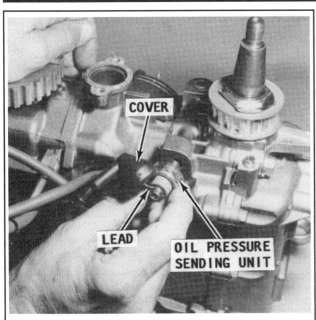

Fig. 71 Strip the cylinder block of components such as the thermostat and/or oil pressure sending unit (pictured)

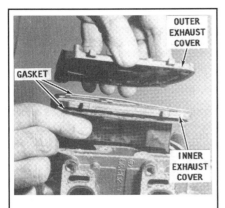

Fig. 72 Unbolt and remove the exhaust cover(s)

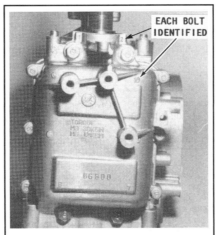

Fig. 73 Loosen the crankcase bolts in the reverse of the embossed torque pattern

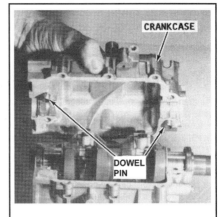

Fig. 74 Then separate the crankcase from the cylinder block - 8/9.9 hp (232cc) motor shown

Fig. 75 Remove the bearings inserts (note the locating tangs for installation)

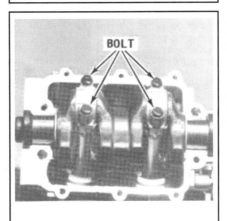

Fig. 76 Matchmark and remove the connecting rod caps

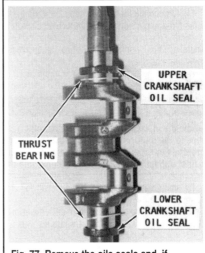

Fig. 77 Remove the oils seals and, if applicable thrust washers

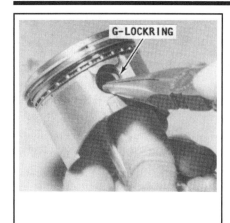

Fig. 78 Remove the lockrings from the piston bores...

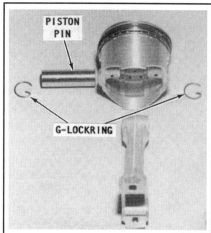

Fig. 79 ...then remove the pin and connecting rod from the piston

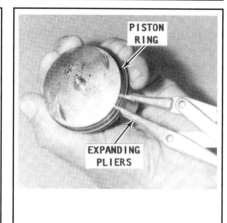

Fig. 80 Use a ring expander to remove the old piston rings

come away with the crankcase. If difficulty is encountered in releasing the crankcase from the powerhead, insert a small pry-tool between the tabs provided for this purpose and pry the two surfaces apart. **Never** pry at gasket sealing surfaces. Such action would very likely damage the sealing surface of an aluminum powerhead.

■ **If the bearing caps are fit for further service, keep them with the crankcase, resting on the spot from which they were removed.**

10. Slide the metal bearing caps out of the top and bottom crankcase halves. Note how each cap has an indexing tang. This tang **must** be aligned with the slot in the crankcase.

※※ **WARNING**

The crankshaft bearing caps absolutely MUST be installed back in the same position from which they were removed. These bearing caps are matched sets - half in each crankcase half. They must be installed with their matching halves on the same crankshaft bearing surfaces.

■ Most parts inside the cylinder block are matched sets and should not be moved from one cylinder to another. This includes the pistons as well as each of the connecting rod/end cap combination. For this reason, prior to disassembly it is essential to matchmark each of the connecting rods to their end caps (this ensures that not only will each cap be installed on the proper rod, but facing the correct direction). Also, be sure to label each piston as to what cylinder it belongs and place an arrow or some indicator facing upward to ensure they are installed in the same direction.

11. Matchmark the big end caps to the connecting rods (or identify the marks already placed there by the manufacturer), then remove the 4 connecting rod cap bolts. Remove the caps and place them aside for reinstallation back on their original connecting rods as soon as the crankshaft has been removed.
12. Carefully lift the crankshaft from the block. Place it on a protective surface.

※※ **WARNING**

Handle the crankshaft carefully. Remember it is an expensive and vital piece to the motor. The polished surfaces should be covered to prevent scratches or other damage and the whole shaft should be protected against impact or other damage which could destroy it.

13. Immediately reinstall the connecting rod end caps to their respective pistons.
14. Remove the top and bottom oil seals from the crankshaft and, if applicable the thrust washer(s). Discard the seals.

■ **These pistons and connecting rods are normally already marked from the factory as to what side faces up, identify the marks and perhaps re-mark them to ensure they are found easily during assembly.**

15. Label and remove each of the pistons from the bores. Number the pistons and place or identify the mark on each piston and connecting rod facing up (facing the flywheel) for installation purposes. If necessary, disassemble each of the pistons, as follows:

   a. Carefully pry the clips free from the piston pin bore on either side of the piston assembly.
   b. Slide the piston pin free of the bore (specifications tell us that this is NOT an interference fit, so it should slide free with ease).
   c. Using a piston ring spreading tool, carefully remove each of the rings (top and second compression rings, followed by the oil ring/rings) from the piston. Good shop practice dictates the rings be replaced during a powerhead overhaul. However, if the rings are to be used again, expand them **only** enough to clear the piston and the grooves because used rings are brittle and break very easily.

■ The manufacturer does not recommend using the oil ring assembly a second time. The oil ring consists of a top rail, an expander section, and a lower rail. If the oil ring is to be used a second time, the ring MUST be installed with the upper rail and the lower rail in the same position from which they were removed. If the oil ring is installed in the reverse position, oil consumption will be increased.

Therefore, when removing the oil ring assembly, lay out the rings one by one in the order and direction from which they were removed - regardless if they are to be used again or not. This action will act as a guide to match new rails of the new rings.

   d. Keep all components from a single piston together for reuse or replacement, but don't mix and match components from other pistons.

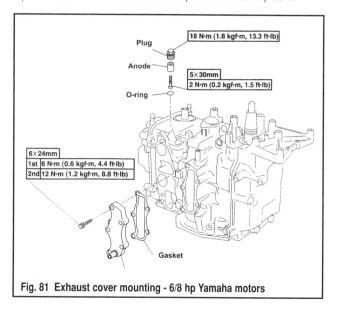

Fig. 81 Exhaust cover mounting - 6/8 hp Yamaha motors

POWERHEAD 6-33

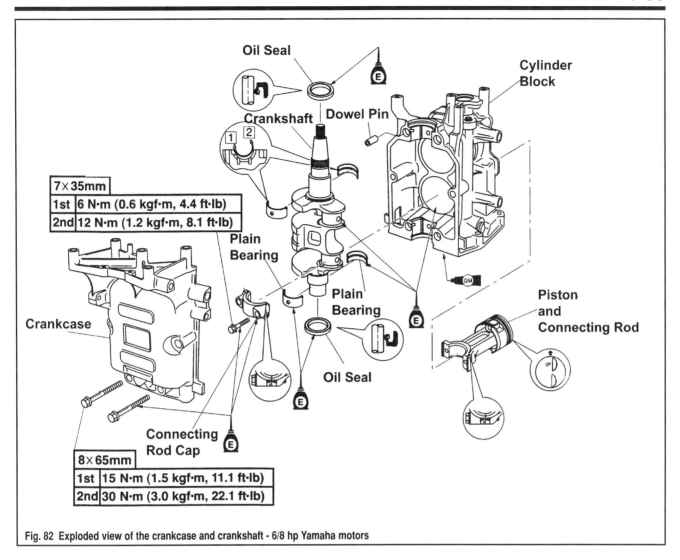

Fig. 82 Exploded view of the crankcase and crankshaft - 6/8 hp Yamaha motors

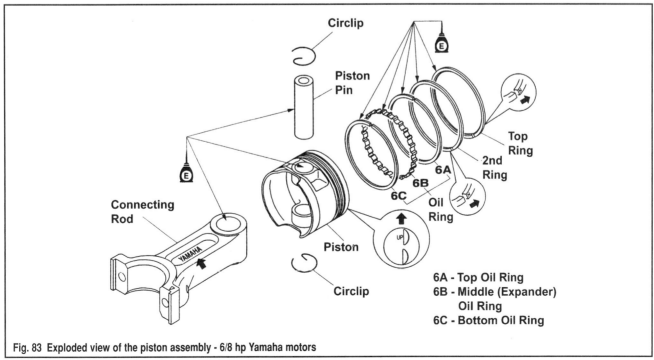

Fig. 83 Exploded view of the piston assembly - 6/8 hp Yamaha motors

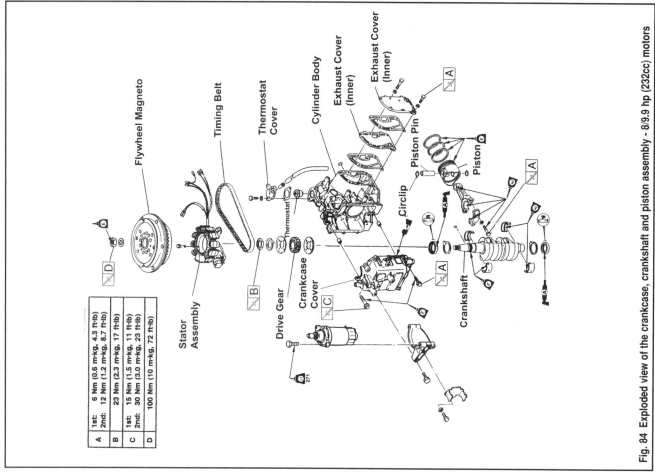

Fig. 84 Exploded view of the crankcase, crankshaft and piston assembly - 8/9.9 hp (232cc) motors

16. Refer to Powerhead Refinishing, later in this section, for details on cleaning and inspecting cylinder block components. There are probably more components and specifications to measure these 4-stroke motors than on any 2-stroke Yamahas of similar size. Take your time and make sure you've made measurements for all components listed in the Engine Specifications chart for this model.

To Assemble:

17. Make sure the mating surfaces of the crankcase and cylinder block are completely clean and free of sealant or damage. The dowel pins should already be installed in the block, but if not, obtain new pins and install them in the positions at opposite ends of the block.

■ **If the old main bearings are to be installed for further service, each must be installed in the same location from which it was removed.**

18. Place the main bearing inserts into the crankcase and cylinder block. There are normally 2 things to look out for when doing this. Most bearing inserts will be equipped with locating tabs, make sure they are seated in the appropriate grooves. Also, inserts with oil holes should be aligned with an oil passage. If a new set of bearings has been purchased for installation, match the colors on the side of the new bearings with the letters embossed on the outside of the crankcase.
- Letter **A** matches the Blue bearing.
- Letter **B** matches the Black bearing.
- Letter **C** matches the Brown bearing.

19. If they were disassembled, prepare each piston assembly for installation as follows:

■ **Apply a light coat of clean engine oil to each piston component as or right after it is installed.**

a. Using a ring expander, carefully install each of the rings with their gaps as noted here and as shown in the accompanying illustrations. All descriptions and illustrations are depicted while looking at the piston dome with the top (flywheel end) facing 12 o'clock.

• For 6/8 hp Yamaha motors, the bottom oil ring should be positioned with the gap facing between 4 and 5 o'clock. The middle oil ring (expander) should be positioned with the gap facing between 7 and 8 o'clock. The upper oil ring should be positioned with the gap facing between 10 and 11 o'clock. The lower (second) compression ring should be positioned with its gap at between 1 and 2 o'clock. The upper (top) compression ring should be positioned with its gap facing between 7 and 8 o'clock (the same as the oil expander ring).

• For 8/9.9 hp (232cc) motors, the oil ring assembly contains an expander ring with 2 side rail rings one above and one below the expander). The side rail ring gaps should be positioned 180 degrees apart, one at about 1 o'clock and the other at about 7 o'clock. Meanwhile the oil expander ring (middle oil ring) should be positioned at about 4 o'clock. The lower (second) compression ring should be positioned with its gap at 10 o'clock. The upper (top) compression ring should be positioned with its gap facing 2 o'clock.

b. Align the mark on the connecting rod (facing up/flywheel) with the mark on the piston (also facing up/flywheel), install the connecting rod to the piston using the piston pin.

■ **On some models there is an embossed YAMAHA on one side of the connecting rod, this side should be positioned facing upward toward the flywheel.**

c. Install the piston pin clips in either side of the pin bore to secure the piston pin in place.

d. Repeat for the remaining piston.

※※ **WARNING**

Use extreme care to prevent the connecting rods from contacting and scoring the cylinder walls during piston installation. It is a good idea to wrap the ends in a shop cloth or something else to protect the cylinders.

20. Install each of the pistons (with the correct mark/side facing UP toward the flywheel) to the cylinder block using a ring compressor. Apply a

POWERHEAD 6-35

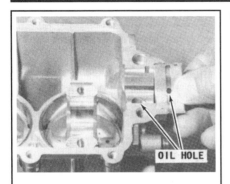

Fig. 85 Install the bearing inserts to the block, making sure to align the oil passages

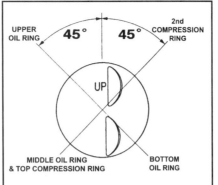

Fig. 86 Ring gap positioning - 6/8 hp Yamaha motors

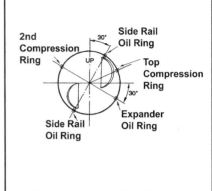

Fig. 87 Ring gap positioning - 8/9.9 hp (232cc) motors

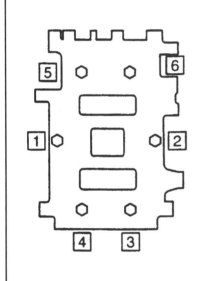

Fig. 88 Crankcase torque sequence - 6/8 hp Yamaha motors

Fig. 89 Crankcase torque sequence - 8/9.9 hp (232cc) motors

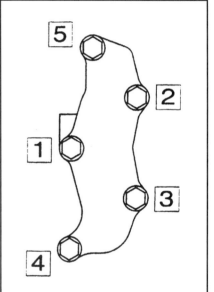

Fig. 90 Exhaust cover torque sequence - 6/8 hp Yamaha motors

very light coating of clean engine oil to the inside of the compressor and to the cylinder bore, then place the compressor around the piston. Insert the piston through the cylinder head side of the bore and use a wooden hammer handle to gently tap the piston down through the compressor and into the cylinder bore.

✳✳ WARNING

Be VERY careful when installing the piston. Get the lower portion of the skirt below the ring compressor and insert it into the cylinder bore to make sure everything is aligned. Then, GENTLY tap the piston through the compressor, stopping if it seems to hang up even slightly to make sure that a ring is not caught on the deck of the cylinder block.

21. Remove the connecting rod end caps and position them so the crankshaft can be installed. Place the connecting rod bearing inserts into position on the connecting rod big end and cap.

■ Here's where procedures differ slightly depending upon whether you're installing bearings of a known clearance or measuring clearance. If you've already determined that bearing clearances are fine, apply alight coat of engine oil to each of the connecting rod, main bearing and crankshaft journal surfaces. If however, you need to measure and determine bearing clearance, Temporarily INSTALL THESE COMPONENTS DRY for measurement purposes. For more details, please refer to Measuring Bearing Clearance Using Plastigage®, in the Powerhead Refinishing section.

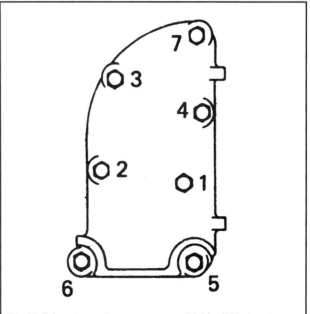

Fig. 91 Exhaust cover torque sequence - 8/9.9 hp (232cc) motors

6-36 POWERHEAD

22. If applicable, slide a thrust bearing onto each end of the crankshaft. The oil grooves **must** face toward the center (toward the throws) of the crankshaft.

■ NEVER attempt to install the upper or lower crankshaft oil seals with the crankcase already installed and tightened otherwise the seal will likely be damaged.

23. Coat the lips of new upper and lower oil seals with a clean engine oil, then slide the oil seals onto the crankshaft with their lips facing toward the center (toward the shaft throws). Seat the seals as best you can at this time. To ensure the seals are not cocked you can install the crankcase and FINGER-TIGTHEN the bolts then lightly knock the seals using a suitably sized driver and rubber mallet. Once this is done, remove the crankcase again and continue the cylinder block assembly.
24. Bring up the connecting rod ends around the crankshaft throws. If you've already measured bearing clearances, apply a light coat of engine oil or engine assembly lube to the connecting rod, cap and crankshaft bearing surfaces.
25. Align the matchmarks and reinstall the connecting rod bearing caps. Install the bolts alternately and evenly first to 53 inch lbs./4.4 ft. lbs. (6 Nm) and then to 106 inch lbs./8.1 ft. lbs. (12 Nm).
26. Confirm that the dowel pins are still in position, then apply a thin bead of Gasket Maker sealant (such as Loctite 514 Master Gasket or equivalent) to the cylinder block side of the crankcase half mating surfaces.
27. Apply a light coating of engine oil to threads of the crankcase mating bolts.
28. Carefully install the crankcase cover over the cylinder block using the dowel pins to align the two surfaces. Install and finger-tighten the retaining bolts (remember that the 8/9.9 uses 2 different size bolts, depending upon location, No. 1-4 of the torque sequence are the larger M8 bolts).
29. Tighten the crankcase retaining bolts using multiple passes of the clockwise, spiraling torque sequence which starts at the middle and works outward on 6/8 hp Yamaha motors, but starts with the larger bolts on 8/9.9 hp (232cc) motors and then works outward from the middle of the smaller bolts. Tighten all bolts on 6/8 hp Yamaha motors and the large (M8 bolts) on 8/9.9 hp (232cc) motors first to 11 ft. lbs. (15 Nm) and then to 22 ft. lbs. (30 Nm). Tighten the small (M6 bolts) on 8/9.9 hp (232cc) motors first to 53 inch lbs./4.3 ft. lbs. (6 Nm) and then to 106 inch lbs./8.7 ft. lbs. (12 Nm).
30. Install the exhaust cover(s) and gasket(s). Tighten the bolts using multiple passes of a clockwise spiraling pattern that starts at the center bolts and works outward. Tighten the bolts first to 53 inch lbs./4.3 ft. lbs. (6 Nm) and then to 106 inch lbs./8.7 ft. lbs. (12 Nm).
31. Install a new oil filter to the powerhead.
32. Install the Cylinder Head Assembly, as detailed in this section.
33. Install the Timing Belt/Sprocket Assembly, as detailed in this section.
34. Install the Powerhead Assembly, as detailed in this section.
35. Once the powerhead is fully assembled, if cylinder block components were replaced, especially the pistons and/or rings, be sure to operate the engine as directed for new component break-in. For details, please refer to Powerhead Break-In, in this section.

CLEANING & INSPECTING

Cleaning and inspecting internal engine components is virtually the same for all Yamaha and Mercury/Mariner outboards with the exception of a few components that are unique to certain 4-stroke motors. The main variance between different motors comes in specifications (which are listed in the Engine Specifications charts) or by component type. A section detailing the proper procedures, sorted mostly by component, can be found under Powerhead Refinishing.

9.9/15 Hp (323cc) Powerheads

POWERHEAD REMOVAL & INSTALLATION

◆ See Figures 92 and 93

Although these powerheads are still small enough that you can usually manage to lift them manually from the intermediate housing (and with most fuel and electrical components still installed) they're starting to move into the larger and more bulky assemblies which make this more difficult.

As usual, we advise that you consider why the powerhead is being removed. If you are planning on disassembling the powerhead for inspection or overhaul, then you'll have to remove the fuel and electrical components anyway. If this is the case, removing them before powerhead removal is a good idea, if only to protect them from potential damage when lifting and moving the powerhead itself. However, in this procedure we've pretty much only included the steps which are necessary in order to remove the assembled powerhead. Additional steps should be taken if you desire to strip the powerhead of all fuel and electrical components before proceeding. If so, refer to components in the Fuel System and Ignition and Electrical System sections for details.

1. Remove the engine top cover for access to the powerhead.
2. On electric start models, tag and disconnect the battery cables on the starter motor side of the powerhead.
3. Remove the Hand-Rewind Starter Assembly for access.
4. On Yamaha models (and Merc/Mariners, if applicable), remove the bolt securing the fitting plate to the front of the powerhead. Remove the plate.
5. For most Yamaha models through 2003, tag and disconnect the throttle cables from the push/pull linkage on the powerhead. Disconnect the shift link rod from the linkage, just below the push/pull assembly. Starting on some late 2003 models there is a throttle cable, shift cable and shift link rod which must be disconnected.
6. For late 2003 Yamaha models so equipped, disconnect the power tilt motor lead and 2 relay leads (at the relay, just in front of the starter motor).
7. For Mercury/Mariner models disconnect the throttle linkage (all models), shift lever (remote models) and choke linkage (tiller models).
8. For tiller control models, disconnect the engine stop switch lead in the junction box at the front of the powerhead.
9. For most electric start models, disconnect the neutral switch lead. For all electric start models, disconnect the fuse lead. On or both of these leads should be in the junction box at the front of the powerhead.
10. If applicable (for most except for late 2003 or later Yamaha models), tag and disconnect the crankcase breather hose from the top, exhaust cover side of the powerhead (between the crankshaft and camshaft pulleys).
11. Tag and disconnect the fuel filter-to-pump hose.
12. If applicable (for most except for late 2003 or later Yamaha models), disconnect the cooling system indicator hose from the fitting on the lower rear side of the powerhead, a little below and behind the oil filter.
13. For late 2003 or later Yamaha models, if equipped, disconnect the trailer switch lead at the lower rear of the cowling, a little below the spark plugs. Then tag and disconnect both the cooling system indicator hose and the water flushing hose, which are located at the side of the powerhead, just a little above and behind the trailer switch.
14. Remove the engine oil dipstick.
15. If applicable, remove the driveshaft housing covers. On Mercury/Mariner models they are secured by 4 bolts and 4 nuts.
16. From underneath the cowling, gradually loosen and remove the 6 powerhead retaining bolts starting at the outer fasteners and working inward.

■ If the unit is several years old, or if it has been operated in salt water, or has not had proper maintenance, or shelter, or any number of other factors, then separating the powerhead from the intermediate housing may not be a simple task. An air hammer may be required on the bolts to shake the corrosion loose; heat may have to be applied to the casting to expand it slightly, or other devices employed in order to remove the powerhead.

17. Carefully lift and remove the powerhead assembly from the intermediate housing and place it on a suitable work surface. Before putting it down on a bench, be sure to remove the oil strainer assembly. If necessary, a piece of wood may be inserted between the powerhead and the intermediate housing as a means of using leverage to force them apart.

■ Take care not to lose the two dowel pins. The pins may come away with the powerhead or they may stay in the intermediate housing. Be ESPECIALLY careful not to drop them into the lower unit.

18. Remove and discard the old powerhead gasket.

■ Mercury/Mariner literature shows an oil deflection plate that is mounted between the powerhead and the intermediate/driveshaft housing. There are 2 powerhead mounting gaskets utilized with this set-up, one above and one below the oil deflection plate. Both gaskets must be replaced EVERYTIME the powerhead is removed.

POWERHEAD 6-37

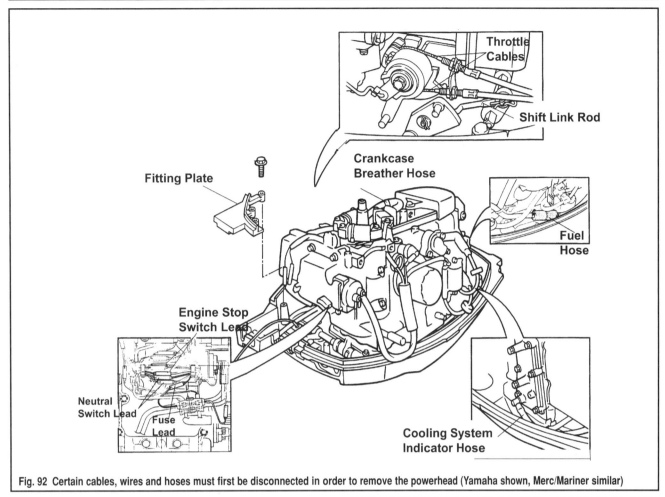

Fig. 92 Certain cables, wires and hoses must first be disconnected in order to remove the powerhead (Yamaha shown, Merc/Mariner similar)

To Install:

19. Make sure the dowel pins are in position at the opposite sides of the intermediate housing.

20. For Mercury/Mariner models position the first of the 2 NEW powerhead gaskets on the intermediate housing (over the dowels) followed by the oil deflection plate.

21. Place the NEW gasket (second of the 2 gaskets on Merc/Mariners) in position over the 2 dowel pins, this should make sure the gasket does not cock during installation and does not obscure any water passages, but keep a close eye on the gasket as the powerhead is positioned anyway.

22. Apply a light coating of marine grade grease to the driveshaft-to-crankshaft splines

23. Install the oil strainer assembly to the underside of the powerhead.

24. Carefully lower the powerhead onto the intermediate housing making sure not to cock or damage the oil strainer assembly and using the dowel pins to align the two surfaces. The propeller (if in gear) or the flywheel/crankshaft (if not) may have to be rotated slightly (only in the normal direction of rotation, CLOCKWISE on these models) to permit the crankshaft to index with the driveshaft and allow the powerhead to seat properly.

25. Apply a light coating of Loctite® 572 or equivalent threadlocking compound to the threads of the powerhead attaching bolts.

■ **Mercury does not mention using threadlock on the powerhead mounting bolts, but we don't see the harm and actually would go so far as to recommend it anyway.**

26. Start the powerhead attaching bolts, with washers.

27. For Yamaha models, tighten the bolts alternately and evenly to a torque value of 15 ft. lbs. (21 Nm) using a spiraling pattern that starts at the inside fasteners and works outward.

28. For Mercury/Mariner models, tighten the bolts alternately and evenly using at least 2 passes of a criss-crossing torque sequence. First tighten the bolts to a value of 11 ft. lbs. (15 Nm) and finally to a value of 22 ft. lbs. (30 Nm).

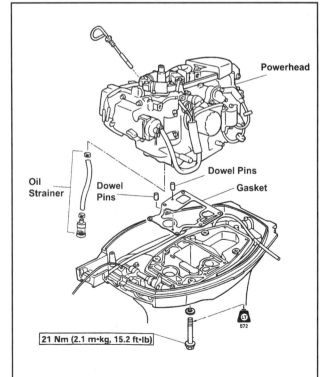

Fig. 93 Exploded view of the powerhead mounting - 9.9/15 Hp (323cc) motors

6-38 POWERHEAD

29. If applicable, install the driveshaft/intermediate housing side covers/shrouds.
30. Install the oil level dipstick. If the outboard was drained, refill the crankcase engine oil at this time.
31. Reconnect the cooling system indicator hose and, if applicable, the water flushing hose.
32. On late 2003 or later models, or as equipped, reconnect the trailer switch wiring.
33. Reconnect the fuel line from the filter to the pump.
34. If applicable, reconnect the crankcase breather hose to the top, exhaust cover side of the powerhead.
35. Reconnect the fuse, neutral switch and/or engine stop switch leads (as equipped) at the junction block at the front of the powerhead.
36. If equipped on late 2003 or later models, reconnect the power tilt relay leads and the motor lead.
37. Reconnect the shift and throttle linkage and/or cables, as equipped.
38. If equipped, install the fitting plate and retaining bolt.
39. Install the Hand-Rewind Starter Assembly.
40. On electric start models, reconnect the battery cables to the starter motor side of the powerhead.
41. Refer to the Timing and Synchronization adjustments in the Engine and Maintenance section for cable/linkage adjustment procedures.

■ **Be sure to run the motor without the top cowling installed in order to check for potential fuel or oil leaks. Remedy any leaks before proceeding.**

42. Install the top cowling to the powerhead.

■ **If the powerhead was rebuilt or replaced with a remanufactured unit, don't forget to follow the proper Break-In procedures.**

TIMING BELT/SPROCKET REMOVAL & INSTALLATION

◆ See Figures 94 thru 98

This procedure can be performed with the powerhead assembly removed or installed on the motor, but if the crankshaft sprocket must be removed with the powerhead installed you'll have to lock the driveshaft and propeller shaft from spinning in order to loosen the crankshaft nut. Luckily the crankshaft sprocket nut is not tightened to a very high torque on these models (29 ft. lbs./40 Nm), so it is usually possible to hold the crankshaft from turning either using a strap wrench or possibly by locking the propeller shaft using a block of wood between an old propeller and the anti-cavitation plate (don't use a good propeller for fear or damage to the hub). If the powerhead is removed from the outboard, the best method for locking the crankshaft is to use an old driveshaft inserted into the splines on the bottom of the crankshaft or to use the Yamaha special shaft holder (#90890-06069).

If the timing belt is to be reused, be sure to mark the direction of rotation on the old belt before removal. Also, if either the cylinder head or cylinder block is not to be disassembled further, be sure to properly align the timing marks before removal. This will significantly ease installation.

■ **Camshaft sprocket removal requires the use of a holder tool which inserts into the 2 holes in the sprocket itself. The same tool that is used on the flywheel can normally be used on the camshaft sprocket.**

1. If the powerhead is still installed and on electric start models, disconnect the negative battery cable for safety.
2. Remove the Flywheel and Stator Plate, as detailed in the Ignition and Electrical Systems section.
3. If either the cylinder head or cylinder block is not being overhauled, rotate crankshaft sprocket in the normal direction of rotation (clockwise) in order to align the timing marks on the crankshaft and camshaft pulleys with the marks on the powerhead.

■ **The camshaft sprocket has a No. 1 stamp mark which must align with the triangular mark on the powerhead (at the cylinder head split line) when the motor is at No. 1 TDC. Similarly, when the crankshaft is turned so the woodruff key hole (and usually a small cutout on the pulley and/or washer) aligns with a triangular mark on the powerhead, the motor is set to No. 1 piston TDC.**

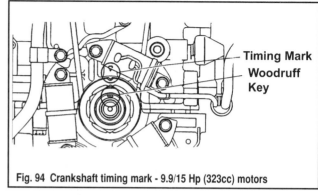

Fig. 94 Crankshaft timing mark - 9.9/15 Hp (323cc) motors

4. Check the condition of the timing belt prior to removing it. If the belt can be deflected more than 0.40 in. (10mm) at the midway point, the belt has stretched and **must** be replaced.

■ **Remember, if the timing belt is to be reused, be sure to mark it showing the normal direction of rotation to ensure installation in the proper position.**

5. Slip the timing belt free of the camshaft pulley **first,** and remove it from the crankshaft pulley. If the fit is too tight or you have trouble slipping the belt from the pulley, move to the next step and unbolt the camshaft (driven) pulley.

If necessary to strip the powerhead for rebuild or to replace one or more damaged sprockets, remove the camshaft (driven) or crankshaft (drive) sprocket from the motor as detailed in the following steps:

6. To remove the camshaft pulley, obtain a suitable flywheel holder tool (Yamaha #YB-06139 or Mercury #91-83163M). Place the two arms of the tool over the two holes of the camshaft pulley. Hold the pulley steady with the tool and at the same time loosen, then remove the camshaft sprocket bolt and washer.
7. Lift the camshaft pulley straight up and free of the camshaft. Remove and save the Woodruff key from the recess in the camshaft.
8. To remove the crankshaft pulley loosen and remove the large crankshaft nut using a DEEP 36mm socket or box wrench. It will be necessary to hold and prevent the crankshaft from rotating while the nut is loosened and removed. This may be accomplished employing one of two methods:
• If the powerhead is removed from the outboard, the BEST method is to use a holding fixture from Yamaha or, if an old lower driveshaft is available, temporarily index the splines on the upper end of the driveshaft with the splines on the lower end of the crankshaft. Hold the old lower driveshaft in a vise or with a pair of vice grip type pliers.
• If the powerhead is installed first try a strap wrench on the crankshaft pulley, however if necessary, a block of wood may usually be used to keep the propeller (use only an old propeller since you risk hub damage) from turning.
9. After the crankshaft nut has been removed, lift off the large washer/belt guide.
10. Slide the pulley straight up, off of the crankshaft and woodruff key. Remove and save the Woodruff key from the recess in the crankshaft.

To Install:

11. Apply just a dab of lubricant to the Woodruff key, and then install the key into the lower cutout in the crankshaft. The lubricant will help hold the key in place.
12. Install the crankshaft sprocket over the retaining plate and woodruff key. If the crankshaft was not disturbed the sprocket should be positioned so the small cutout (if equipped) on the drive gear or washer/belt guide (once installed) will face the timing mark on the powerhead (and face directly toward the camshaft). In all cases, the woodruff key itself will be facing the timing mark when the No. 1 piston is at TDC.
13. Install the belt guide/washer over the crankshaft sprocket with the beveled edge facing upward, away from the powerhead.
14. Thread the large nut onto the end of the crankshaft. Hold the crankshaft from turning and then tighten the nut to 29 ft. lbs. (40 Nm). The crankshaft must be held firm while the nut is being tightened. Use the same device which was used during disassembly - an old section of driveshaft or with a strap wrench.

POWERHEAD 6-39

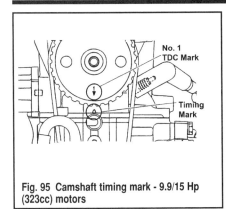

Fig. 95 Camshaft timing mark - 9.9/15 Hp (323cc) motors

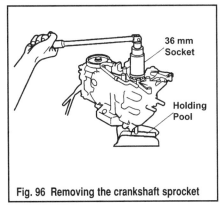

Fig. 96 Removing the crankshaft sprocket

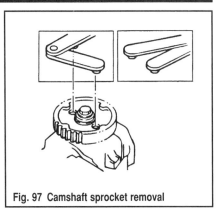

Fig. 97 Camshaft sprocket removal

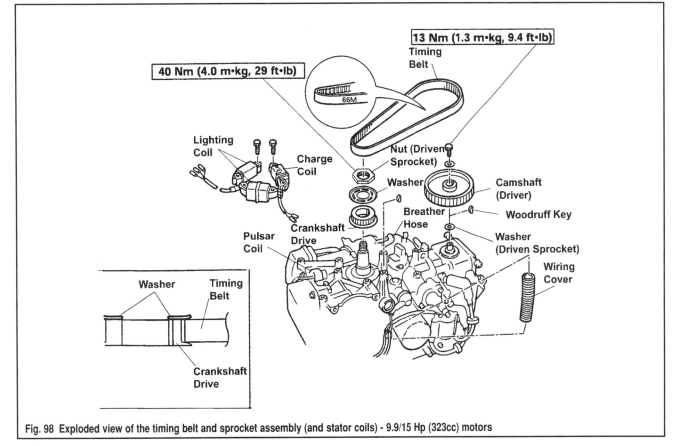

Fig. 98 Exploded view of the timing belt and sprocket assembly (and stator coils) - 9.9/15 Hp (323cc) motors

15. Make sure the cutout mark on the crankshaft pulley or washer or the woodruff keyway is aligned with the mark on the powerhead, facing the camshaft. If necessary, slowly rotate the crankshaft in the normal direction of rotation (clockwise) until the timing marks align. If the camshaft was disturbed, stop turning if you feel any interference.

16. Place the lower washer over the camshaft. Apply just a dab of lubricant onto the Woodruff key, and then insert the key into the camshaft slot. The lubricant should hold the key in place. Slide the camshaft pulley down onto the camshaft and align the keyway in the pulley over the Woodruff key.

17. Hold the camshaft sprocket from turning using the flywheel holder tool, then install and tighten the sprocket retaining bolt to 124 inch lbs./9.4 ft. lbs. (13 Nm).

18. If the camshaft was not disturbed, the No. 1 timing mark on the pulley should align with the triangle or mark on the powerhead, facing the crankshaft. If not, SLOWLY turn the crankshaft in the normal direction of rotation (also clockwise when viewed from above) until the timing marks are appropriately aligned.

■ Remember, if the belt is being reused, be sure to face it in the same direction as normal rotation (clockwise).

19. Place the timing belt, with the imprinted part number facing the right way up, over the crankshaft pulley, without disturbing timing mark alignment. After the belt is in place stretch it over the camshaft pulley without moving the pulley.

■ If belt installation is too difficult, remove the camshaft sprocket bolt and reposition the sprocket. With the sprocket properly aligned and the belt fully installed over it, slowly push downward on the belt and sprocket to slide it over the camshaft and woodruff key. Then hold the camshaft sprocket steady while you install and tighten the bolt.

20. Turn the crankshaft clockwise until it completes 2 full revolutions, then recheck the timing marks.
21. Install the Flywheel and Stator Plate assembly, as detailed in the Ignition and Electrical Systems section.
22. If removed, install the powerhead assembly.
23. If applicable, reconnect the negative battery cable.

CYLINDER HEAD REMOVAL & INSTALLATION

◆ See Figure 99

Although it is not absolutely clear in Yamaha literature, it looks likely that the cylinder head can be removed on these models without removing the powerhead assembly. Mercury gives a separate procedure just for cylinder head removal an installation which helps to confirm this, as long as the lower cowling doesn't interfere on Yamaha models.

Remember, when loosening the retainers on the cylinder head, cover or any other major components, always try to follow the reverse of the torque sequence indicated in the assembly procedure or molded on the component itself.

When applicable tighten all bolts in using the proper torque sequence and to the proper specification.

■ Because of the high temperatures and pressures developed, the sealing surfaces of the cylinder head and the block are the most prone to water leaks. No sealing agent is recommended because it is almost impossible to apply an even coat of sealer. An even coat would be essential to ensure an air/water tight seal.

※※ WARNING

Never, never, use automotive type head gasket sealer. The chemicals in the sealer will cause electrolytic action and eat the aluminum faster than you can get to the bank for money to buy a new cylinder block.

1. Remove the Flywheel and Stator Plate, as detailed in the Ignition and Electrical Systems section.
2. Remove the Timing Belt, as detailed in this section. If the cylinder head is to be overhauled follow the necessary steps to remove the camshaft sprocket.
3. Remove the spark plugs from the cylinder head. Remove any other electrical components which may still be attached to the cylinder head (or at least disconnect the wiring).
4. Using multiple passes of a crossing pattern, remove the 4 bolts securing the cylinder head cover to the head. Remove the cover, then remove and discard the seal.
5. Locate and remove the 3 short cylinder head bolts from the head flange. Loosen the bolts alternately and evenly.
6. Locate and remove the 6 long cylinder head bolts from inside the head. Be sure to slowly loosen the bolts using the reverse of the torque sequence. This essentially results in a counterclockwise spiraling pattern that starts at the top right outer bolt when looking at the head and ends at the most central bolts.

※※ WARNING

NEVER pry at gasket sealing surfaces. Such action would very likely damage the sealing surface of an aluminum powerhead.

7. Lift the head up and free of the block. Although Yamaha normally casts pry-tabs on heads, we could not locate any on this model. If the head will not release from the block looks for pry-tabs. If no tabs are provided, use a rubber mallet to gently tap around the perimeter of the surface to help break the gasket seal.

■ Take care not to lose the two dowel pins at the mating surfaces. These pins may remain either in the cylinder head or in the block.

8. Remove and discard the old cylinder head gasket.
9. If necessary for further cylinder head disassembly or for oil pump service, loosen the fasteners and remove the pump from the head. This will be necessary if you plan on removing the camshaft and valve train, but it is a good idea to remove and inspect the pump since you've come this far regardless. For details on removal, installation, inspection and overhaul of the oil pump assembly, please refer to Oil Pump in the portion of the Lubrication and Cooling section.
10. Refer to Cylinder Head Overhaul in this section for details on removal and installation of the valve train components. Also, be sure to check the information on gasket mating surfaces to make sure the cylinder head is not warped.

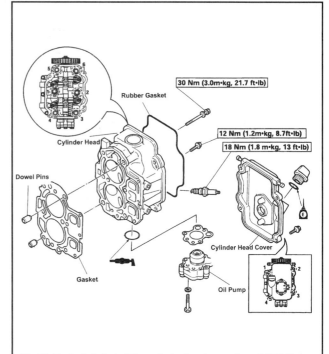

Fig. 99 Exploded view of the cylinder head mounting - 9.9/15 Hp (323cc) motors

To Install:

11. If removed, refer to the Lubrication and Cooling section for details on Oil Pump assembly and installation.
12. Position a NEW head gasket over the two dowel pins.
13. Install the cylinder head onto the block with the holes in the head indexed over the two dowel pins. Apply a thin coating of clean engine oil to the threads of the eight or nine attaching bolts (as applicable).
14. Thread and finger-tighten the 6 long cylinder head bolts which are positioned inside the head, then thread and finger-tighten the 3 short cylinder head flange bolts.
15. Tighten the 6 long cylinder head bolts using multiple passes of a clockwise, spiraling torque sequence that starts on the left, center bolt and works towards the outer bolts. Tighten the long bolts first to 11 ft. lbs. (15 Nm) and then to 22 ft. lbs. (30 Nm).
16. After the long bolts are tightened, tighten the short head flange bolts alternately and evenly first to 53 inch lbs./4.5 ft. lbs. (6 Nm) and then to 104 inch lbs./8.7 ft. lbs. (12 Nm).

■ If the valve train components were disturbed in any way you are going to have to adjust the valves once the timing belt is installed, so either leave the cylinder head cover off or just finger-tighten the bolts.

17. Install the cylinder head cover and new O-ring. If the valves were not disturbed, install and tighten the 4 cylinder head cover bolts securely either using the molded torque sequence (or if not present using a clockwise spiraling pattern that begins at the top left bolt, near the oil filler cap and works around to the bottom left bolt).

■ Make sure the new O-ring remains seated in the cover groove and does not become pinched during installation or there will be oil leaks!

18. Install the Timing Belt and, if removed, Camshaft Sprocket, as detailed in this section.
19. If the valves were disturbed, adjust them at this time, as detailed under Valve Lash in the Maintenance and Tune-Up section, then install the cylinder head cover.
20. Install the spark plugs to the cylinder head.
21. Install the Flywheel and Stator Plate, as detailed in the Ignition and Electrical Systems section.
22. If removed, install the powerhead assembly.

POWERHEAD

CYLINDER HEAD OVERHAUL

◆ See Figure 100

1. Remove the Cylinder Head from the powerhead assembly, as detailed in this section.
2. Remove the Oil Pump from the cylinder head, as detailed in the Lubrication and Cooling section.
3. Loosen the locknuts and back off each valve lash adjuster until the rocker arm, when placed on the high point of the camshaft, will clear the valve stem.

■ When reused, ALL valve train components must be installed in their original locations. For this reason be sure to tag and/or sort all rocker arms, springs, spacers, valves, valve springs, keepers etc and keep them separated during service.

4. Slide the rocker shaft out and free of the cylinder head. As the shaft is withdrawn **carefully** note the arrangement of the rockers, springs, and the spacers. Repeat for the remaining rocker arm shaft.
5. Guide the camshaft carefully out of the head. **Take care** not to damage the camshaft lobes. If necessary, remove the oil seal with a slide hammer and expanding jaw attachment puller.
6. To remove the each valve, proceed as follows:
   a. Using a clamp-type valve spring compressor, carefully place just enough pressure on the valve spring so the upper spring seat (retainer) moves down sufficiently to expose the keepers.
   b. Remove the pair of valve keepers, then slowly ease pressure on the valve spring until the compressor can be removed.
   c. Remove the upper seat (retainer), valve spring and lower spring seat from the shaft of the valve.
   d. Remove the valve from the underside (combustion chamber side) of the cylinder head.
   e. If necessary, carefully pry the old valve stem seal from its position on the cylinder head.
   f. Repeat this procedure for each of the remaining valves.
7. Refer to Powerhead Refinishing, later in this section, for details on cleaning and inspecting cylinder head components.

To Assemble:

8. If removed, install each of the valve and spring assemblies as follows:
   a. If removed, install new valve spring seals to the cylinder head.
   b. Apply a light coating of engine oil to the stem, then position the valve into the cylinder head (into the original guide and valve seat unless the valve or valve seat was replaced).
   c. Position the lower spring seat down onto the valve shaft and cylinder head, then oil and install the spring.
   d. Place the upper spring seat (retainer) over the spring, then use the valve spring compressor tool to press downward on the seat and spring so you can install the keepers.
   e. Insert the valve keepers, then slowly back off the valve spring compressor, making sure the keepers are fully in position. Then remove the valve spring compressor.
   f. Repeat for each of the remaining valves.
9. If the camshaft oil seal was removed during disassembly, coat the lips of a new seal with a light coating of clean engine oil. Install the seal into the head using a suitable driver. The seal should be installed with the lips facing downward into the cylinder head. On most applications this means the seal manufacturer's marks or numbers are facing outward.

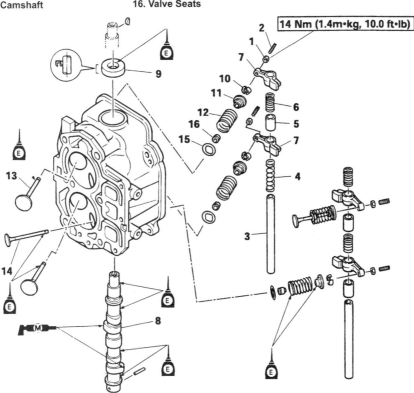

1. Locknuts
2. Valve Adjusting Screws
3. Rocker Arm Shaft
4. Large Spring
5. Collar
6. Small Spring
7. Rocker Arms
8. Camshaft
9. Oil Seal
10. Valve Keepers
11. Upper Spring Seat (Retainer)
12. Valve Springs
13. Intake Valves
14. Exhaust Valves
15. Lower Spring Seats
16. Valve Seats

14 Nm (1.4m·kg, 10.0 ft·lb)

Fig. 100 Exploded view of the cylinder head assembly - 9.9/15 Hp (323cc) motors

6-42 POWERHEAD

■ Do not attempt to rotate the camshaft at this time. At this time the camshaft is supported at just one end. The camshaft should ONLY be rotated after the oil pump has been installed and the shaft is then supported at both ends.

10. Apply a light coating of engine oil to the camshaft bearing journals, then apply a light coating of a molybdenum disulfide engine assembly lube to the camshaft lobes. Next, **carefully** insert the sprocket end (with the keyway) of the camshaft into the head from the oil pump end of the head. Continue inserting the camshaft until end of the shaft is seated in the camshaft oil seal against the head at the other end.

■ The ends of the rocker arm shafts are threaded on some models, to ease removal. When so equipped the threaded ends should always be left toward the oil pump end of the cylinder head.

11. Insert each of the rocker arm shafts into the cylinder head. When installing the intake valve rocker shaft be sure to position the long spring, No. 2 intake rocker, collar, short spring and No. 1 intake rocker. When installing the exhaust valve rocker shaft start by installing the lower collar, then the No. 2 exhaust rocker, short spring, collar, No. 1 exhaust rocker and finally the other short spring.

■ Remember to lightly coat all the rocker arm and shaft components with clean engine oil as they are installed.

12. Install the Oil Pump assembly to the cylinder head, as detailed in the Lubrication and Cooling section.
13. Install the Cylinder Head as detailed in this section.
14. Once the powerhead is fully assembled, if valve train components were replaced, be sure to operate the engine as directed for new component break-in. For details, please refer to Powerhead Break-In, in this section.

CYLINDER BLOCK OVERHAUL

◆ See Figures 101 thru 107

Remember, when loosening the retainers on manifolds, covers and other major components, always try to follow the reverse of the torque sequence indicated in the assembly procedure or molded on the component itself.

When applicable tighten all bolts in using the proper torque sequence and to the proper specification. When possible we've included it in the procedure, but on many Yamahas both the sequence and value is molded into the casting on critical components. You'll notice that most Yamaha torque sequences are clockwise spirals starting somewhere near the center of the component.

1. Remove the Powerhead Assembly, as detailed in this section.
2. Remove the Timing Belt/Sprocket Assembly, as detailed in this section.
3. Remove the Cylinder Head Assembly, as detailed in this section.

■ The exhaust cover should always be removed during a powerhead overhaul. Many times water in the powerhead is caused by a leaking exhaust cover gasket or plate.

4. Although not absolutely necessary, we recommend you strip the block of the following components for inspection and passage cleaning both of which will help ensure a thorough rebuild:
   a. Remove the oil filter, and filter plug.
   b. Gradually loosen and remove the exhaust cover retaining bolts using the reverse of the torque pattern (roughly a counterclockwise spiraling pattern that starts at the upper most bolts and works towards the center). The thermostat assembly is mounted under a housing to the top of the exhaust cover (the top 2 exhaust cover bolts also retain the thermostat housing).
   c. Remove the thermostat assembly from the exhaust cover. For details, please refer to the Lubrication and Cooling section. Remove and discard the old thermostat gasket.
   d. Remove the exhaust cover and gasket from the side of the powerhead. Remove and discard the old exhaust cover gasket.
   e. Remove the 3 bolts securing the breather cover to the top of the cylinder block. Remove the breather cover, then remove and discard the cover gasket.
   f. On the opposite side of the powerhead from the oil filter, remove the bolt and plate from the top of the tear-dropped shaped anode cover. Then remove the bolt, cover, rubber packing and anode from the cylinder block.

5. Remove the crankcase retaining bolts using multiple passes in the reverse of the torque sequence (either molded on the crankcase or as illustrated later in this procedure). Essentially the crankcase has 6 M6 bolts along the flanges and 4 M8 bolts (2 each at the top and bottom centers of the crankcase). For each pass of the sequence start at the upper right M6 bolt and move in a counterclockwise spiral towards the center, then continue that counterclockwise spiral starting with the bottom left M8 bolt, ending at the top left M8 bolt.

■ Remember that since there two different size bolts are used keep them separated for identified for installation purposes. Bolts identified as 1, 2, 3, and 4 in the torque sequence are 8mm bolts. These are found at the top and bottom center of the crankcase. Bolts identified as 5, 6, 7, 8, 9, and 10 are 6mm and are found on the outer flange of the crankcase (3 to a side).

6. After all bolts have been removed, separate the crankcase from the powerhead. **Take care** not to lose the two dowel pins. These pins may remain with cylinder block or come away with the crankcase cover.

✱✱ WARNING

If difficulty is encountered, use a rubber mallet to tap around the perimeter of the crankcase cover. Some Yamahas have molded pry tabs to help break the gasket seal, however NEVER pry at gasket sealing surfaces. Such action would very likely damage the sealing surface of an aluminum powerhead.

■ If the bearing caps are fit for further service, keep them with the crankcase, resting on the spot from which they were removed.

7. Slide the metal bearing caps out of the top and bottom crankcase halves.

✱✱ WARNING

The crankshaft bearing caps absolutely MUST be installed back in the same position from which they were removed. These bearing caps are matched sets - half in each crankcase half. They must be installed with their matching halves on the same crankshaft bearing surfaces.

■ Most parts inside the cylinder block are matched sets and should not be moved from one cylinder to another. This includes the pistons as well as each of the connecting rod/end cap combination. For this reason, prior to disassembly it is essential to matchmark each of the connecting rods to their end caps (this ensures that not only will each cap be installed on the proper rod, but facing the correct direction).

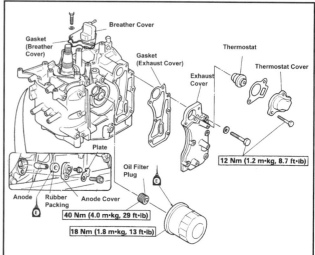

Fig. 101 Exploded view of the thermostat, exhaust cover, oil filter and breather mounting - 9.9/15 Hp (323cc) motors

POWERHEAD 6-43

8. Matchmark the big end caps to the connecting rods (or identify the marks already placed there by the manufacturer), then remove the 4 connecting rod cap bolts. Remove the caps and place them aside for reinstallation back on their original connecting rods as soon as the crankshaft has been removed.

■ When loosening or tightening the connecting rod cap bolts, alternate between each bolt every 1/2-1 turn.

9. Carefully lift the crankshaft from the block. Place it on a protective surface.

✶✶ WARNING

Handle the crankshaft carefully. Remember it is an expensive and vital piece to the motor. The polished surfaces should be covered to prevent scratches or other damage and the whole shaft should be protected against impact or other damage which could destroy it.

10. Immediately reinstall the connecting rod end caps to their respective pistons (to prevent loss, mixing them up or potential damage to the bearing surfaces).
11. Remove the top and bottom oil seals from the crankshaft. Discard the seals.

■ These pistons and connecting rods are normally already marked from the factory as to what side faces up, identify the marks (UP on the piston dome and Y on the con rod) and perhaps, remark them to ensure they are found easily during assembly. Also, each piston and connecting rod must be returned to the original bore and crankshaft journal, so be sure to place a number on each to ensure proper positioning.

12. Label and remove each of the pistons from the bores. Number the pistons and place or identify the mark on each piston and connecting rod which faces up (faces the flywheel) for installation purposes. If necessary, disassemble each of the pistons, as follows:

   a. Carefully pry the clips free from the piston pin bore on either side of the piston assembly.
   b. Slide the piston pin free of the bore (specifications tell us that this is NOT an interference fit, so it should slide free with ease).
   c. Using a piston ring spreading tool, carefully remove each of the rings (top and second compression rings, followed by the oil rings) from the piston. Good shop practice dictates the rings be replaced during a powerhead overhaul. However, if the rings are to be used again, expand them **only** enough to clear the piston and the grooves because used rings are brittle and break very easily.
   d. Keep all components from a single piston together for reuse or replacement, but don't mix and match components from other pistons.

13. Refer to Powerhead Refinishing, later in this section, for details on cleaning and inspecting cylinder block components. There are probably more components and specifications to measure these 4-stroke motors than on any 2-stroke Yamahas of similar size. Take your time and make sure you've made measurements for all components listed in the Engine Specifications chart for this model.

To Assemble:

14. Make sure the mating surfaces of the crankcase and cylinder block are completely clean and free of sealant or damage. The dowel pins should already be installed in the block or crankcase cover, but if not, obtain new pins and install them in the positions at opposite ends of the block.

■ If the old main bearings are to be installed for further service, each must be installed in the same location from which it was removed.

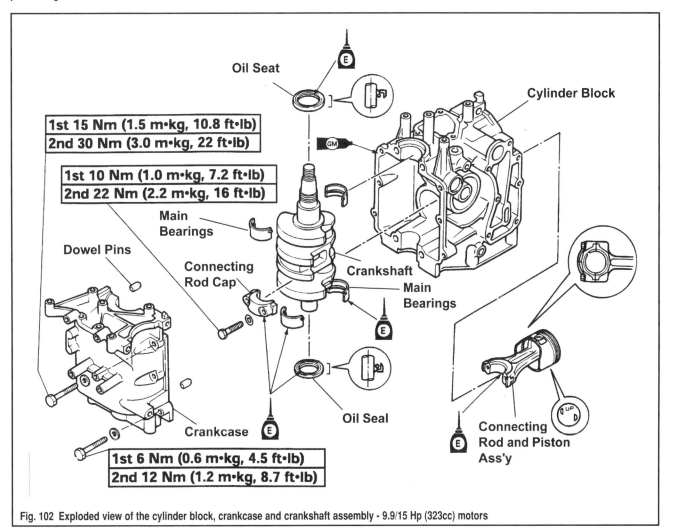

Fig. 102 Exploded view of the cylinder block, crankcase and crankshaft assembly - 9.9/15 Hp (323cc) motors

6-44 POWERHEAD

15. Place the main bearing inserts into the crankcase and cylinder block. There are normally 2 things to look out for when doing this. Most bearing inserts will be equipped with locating tabs, make sure they are seated in the appropriate grooves. Also, inserts with oil holes should be aligned with an oil passage. If a new set of bearings has been purchased for installation, match the colors on the side of the new bearings with the letters embossed on the outside of the crankcase.
- Letter **A** matches the Blue bearing.
- Letter **B** matches the Black bearing.
- Letter **C** matches the Brown bearing.

16. If they were disassembled, prepare each piston assembly for installation as follows:

■ Apply a light coat of clean engine oil to each piston component as or right after it is installed.

   a. Using a ring expander, carefully install each of the rings with their gaps as noted here and as shown in the accompanying illustration. The description and illustration are depicted while looking at the piston dome with the top (flywheel end) facing 12 o'clock.

■ UNFORTUNATELY Mercury and Yamaha sometimes have a different philosophy on exactly how to set up the ring gaps prior to piston installation. This does go to reinforce a popular opinion we've often heard from machinists, that since the rings are going to move around in service, it really shouldn't matter. But, we've gone through the trouble of giving you each manufacturer's preferred set-up method.

   b. The oil ring is an assembly which consists of bottom and top side rails with an expander ring in the middle. Yamaha recommends that middle oil ring (expander) should be positioned with the gap facing between 1 and 2 o'clock, while the top and bottom side rail gaps should be 90 degrees on either side of it, the upper between 10 and 11 o'clock with the lower between 4 and 5 o'clock. Mercury recommends that the, gap on the expander ring should face directly to 6 o' clock, while the top and bottom side rail gaps should each be 90 degrees from it, the top at 9 o'clock and the bottom at 3 o'clock.

   c. The lower (second) compression ring should be positioned with its gap at 3 o'clock (as recommended by Yamaha) or between 1 and 2 o'clock (as recommended by Mercury).

   d. The upper (top) compression ring should be positioned with its gap at 9 o'clock (as recommended by Yamaha) or between 10 and 11 o'clock (as recommended by Mercury).

   e. Align the **Y** mark on the connecting rod (facing up/flywheel) with the **UP** mark on the piston (also facing up/flywheel), install the connecting rod to the piston using the piston pin.

   f. Install the piston pin clips in either side of the pin bore to secure the piston pin in place.

   g. Repeat for the remaining piston.

✳✳ WARNING

Use extreme care to prevent the connecting rods from contacting and scoring the cylinder walls during piston installation. It is a good idea to wrap the ends in a shop cloth or something else to protect the cylinders.

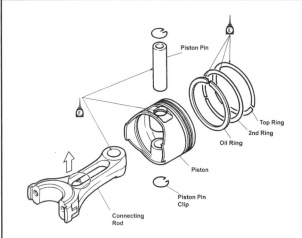

Fig. 103 Exploded view of the piston assembly - 9.9/15 Hp (323cc) motors

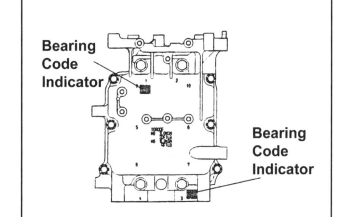

Fig. 104 Select replacement bearings using the code indicators, the letter corresponds to a color marking on the side of the bearing insert (and a bearing size)

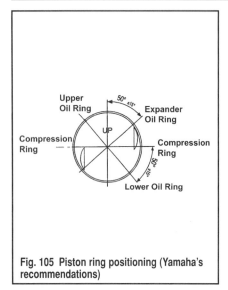

Fig. 105 Piston ring positioning (Yamaha's recommendations)

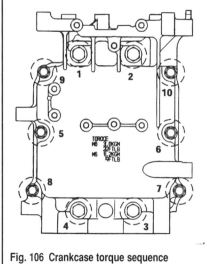

Fig. 106 Crankcase torque sequence

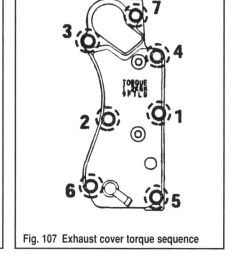

Fig. 107 Exhaust cover torque sequence

POWERHEAD 6-45

17. Install each of the pistons (with the correct mark/side facing UP toward the flywheel) to the cylinder block using a ring compressor. Apply a very light coating of clean engine oil to the inside of the compressor and to the cylinder bore, then place the compressor around the piston. Insert the piston through the cylinder head side of the bore and use a wooden hammer handle to gently tap the piston down through the compressor and into the cylinder bore.

✱✱ WARNING

Be VERY careful when installing the piston. Get the lower portion of the skirt below the ring compressor and insert it into the cylinder bore to make sure everything is aligned. Then, GENTLY tap the piston through the compressor, stopping if it seems to hang up even slightly to make sure that a ring is not caught on the deck of the cylinder block.

18. Remove the connecting rod end caps and position them so the crankshaft can be installed. Place the connecting rod bearing inserts into position on the connecting rod big end and cap.

■ Here's where procedures differ slightly depending upon whether you're installing bearings of a known clearance or measuring clearance. If you've already determined that bearing clearances are fine, apply alight coat of engine oil or engine assembly lube to each of the connecting rod, main bearing and crankshaft journal surfaces. If however, you need to measure and determine bearing clearance, Temporarily INSTALL THESE COMPONENTS DRY for measurement purposes. For more details, please refer to Measuring Bearing Clearance Using Plastigage®, in the Powerhead Refinishing section.

■ NEVER attempt to install the upper or lower crankshaft oil seals with the crankcase already installed and tightened otherwise the seal will likely be damaged.

19. Coat the lips of new upper and lower oil seals with a clean engine oil, then slide the oil seals onto the crankshaft with their lips facing toward the center (toward the shaft throws). Seat the seals as best you can at this time. To ensure the seals are not cocked you can install the crankcase and FINGER-TIGTHEN the bolts then lightly knock the seals using a suitably sized driver and rubber mallet. Once this is done, remove the crankcase again and continue the cylinder block assembly.

20. Bring up the connecting rod ends around the crankshaft throws. If you've already measured bearing clearances, apply a light coat of engine oil or engine assembly lube to the connecting rod, cap and crankshaft bearing surfaces.

21. Align the matchmarks and reinstall the connecting rod bearing caps. Install the bolts alternately and evenly first to 7.2 ft. lbs. (10 Nm) and then to 16 ft. lbs. (22 Nm).

22. Confirm that the dowel pins are still in position, then apply a thin bead of Gasket Maker sealant (such as Loctite 514 Master Gasket or equivalent) to the cylinder block side of the crankcase half mating surfaces.

23. Carefully install the crankcase cover over the cylinder block using the dowel pins to align the two surfaces. Install and finger-tighten the retaining bolts (remember that there are 2 different size bolts, depending upon location, No. 1-4 of the torque sequence are the larger M8 bolts).

24. Tighten the crankcase retaining bolts using multiple passes of the clockwise, spiraling torque sequence which starts with the larger bolts on the top and bottom of the crankcase and then works outward from the middle of the smaller bolts. Tighten the large (M8 bolts) on first to 11 ft. lbs. (15 Nm) and then to 22 ft. lbs. (30 Nm). Tighten the small (M6 bolts) first to 53 inch lbs./4.5 ft. lbs. (6 Nm) and then to 104 inch lbs./8.7 ft. lbs. (12 Nm).

25. If removed, install the exhaust cover, thermostat assembly and gaskets (one for the exhaust cover and one for the thermostat housing). Tighten the bolts using multiple passes of a clockwise spiraling pattern that starts at the center bolts and works outward. For Yamaha models, tighten the bolts first to 4.3 ft. lbs. (6 Nm) and then to 8.7 ft. lbs. (12 Nm). Mercury recommends that you tighten the bolts in multiple passes to 71 inch lbs. (8 Nm), but keep in mind, these are all powerheads manufactured by Yamaha, so either method and torque should work.

26. If removed, install the breather assembly to the top of the crankcase using a new gasket. Tighten the 3 retaining bolts securely.

■ There is no excuse not to at least CHECK the anode while you're here, but we'd recommend that you replace it for good measure.

27. If removed, insert a new anode into the bore in the cylinder block. Apply a light coating of engine oil to the rubber packing, then install the anode cover and secure using the retaining screw. Finally, apply a light coating of Loctite® 572 or equivalent threadlocking material to the threads of the plate screw to keep in from loosening in service, then install the plate and plate screw over the cover.

28. If removed, install the oil filter plug and tighten, then install a new oil filter.

29. Install the Cylinder Head Assembly, as detailed in this section.
30. Install the Timing Belt/Sprocket Assembly, as detailed in this section.
31. Install the Powerhead Assembly, as detailed in this section.
32. Once the powerhead is fully assembled, if cylinder block components were replaced, especially the pistons and/or rings, be sure to operate the engine as directed for new component break-in. For details, please refer to Powerhead Break-In, in this section.

CLEANING & INSPECTING

Cleaning and inspecting internal engine components is virtually the same for all Yamaha and Mercury/Mariner outboards with the exception of a few components that are unique to certain 4-stroke motors. The main variance between different motors comes in specifications (which are listed in the Engine Specifications charts) or by component type. A section detailing the proper procedures, sorted mostly by component, can be found under Powerhead Refinishing.

25 Hp Powerheads

POWERHEAD REMOVAL & INSTALLATION

◆ See Figures 108 thru 111

Although these powerheads are still small enough that you can usually manage to lift them manually from the intermediate housing (and with most fuel and electrical components still installed) they're starting to move into the larger and more bulky assemblies which make this more difficult.

As usual, we advise that you consider why the powerhead is being removed. If you are planning on disassembling the powerhead for inspection or overhaul, then you'll have to remove the fuel and electrical components anyway. If this is the case, removing them before powerhead removal is a good idea, if only to protect them from potential damage when lifting and moving the powerhead itself. However, in this procedure we've pretty much only included the steps which are necessary in order to remove the assembled powerhead. Additional steps should be taken if you desire to strip the powerhead of all fuel and electrical components before proceeding. If so, refer to components in the Fuel System and Ignition and Electrical System sections for details.

1. Remove the engine top cover for access to the powerhead.
2. On electric start models, disconnect the battery cables from the battery.
3. For Yamaha Models, proceed as follows:
  a. Tag and disconnect the throttle cables.
  b. For tiller control models, tag and disconnect the engine stop switch leads.
  c. Tag and disconnect the wiring for the low-oil-pressure warning lamp.
  d. Remove the bolt securing the fitting plate to the front of the powerhead. Remove the plate.
  e. If equipped, disconnect the Blue and Green leads for the power trim/tilt motor from the junction panel on the starboard side of the motor.
  f. Carefully cut the plastic wire tie, then disconnect the oil pan breather hose from the top of the powerhead.
  g. Tag and disconnect the fuel filter-to-pump hose from the pump itself.
  h. Disconnect the cooling system indicator hose (at the rear, port side of the powerhead).
  i. Disconnect the trailer switch coupler (found at the rear, port side of the cowling, near the trailer switch).
  j. At the front starboard side of the powerhead, remove the retaining clip and disconnect the shift link rod. Also in the same general area, if equipped, remove the retaining clip and disconnect the start-in-gear protection cable.
  k. For electric start models, at the front of the powerhead tag and disconnect the starter switch and neutral switch leads.

6-46 POWERHEAD

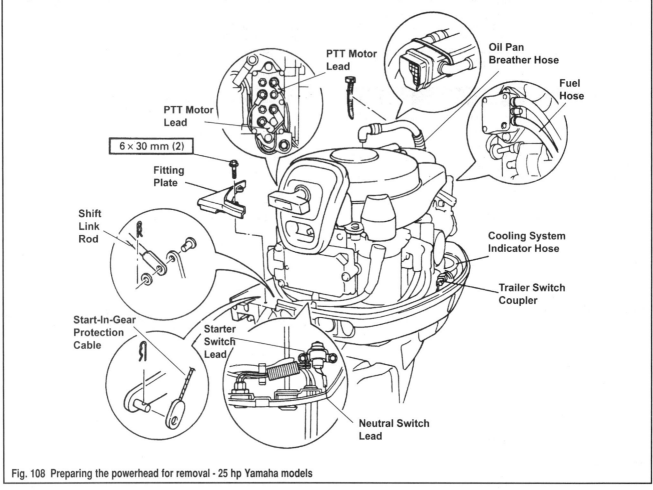

Fig. 108 Preparing the powerhead for removal - 25 hp Yamaha models

l. Remove the 4 bolts threaded upward securing the apron at the rear of the powerhead to the underside of the cowling. Remove the apron for access to the powerhead mounting bolts.

4. For Mercury/Mariner models, proceed as follows:
  a. If equipped with Power Trim/Tilt (PTT), disconnect the wiring.
  b. Remove the 7 screws securing the bottom cowling, then separate the cowling from the intermediate housing for access.
  c. Locate the vent hose attached at the base of the powerhead, between the spark plugs and the dipstick, then carefully cut the sta-strap (wire tie) and disconnect the hose.
  d. Cut the Sta-strap and disconnect the fuel inlet hose from the fuel pump (check the hose routing or the marking on the pump to be sure, however the inlet hose is usually the lower hose on these models).
  e. Locate the vertical shift shaft linkage under the front center of the powerhead. Disengage the shift shaft by first pushing the retainer up and then sliding it over toward the starboard side.

5. Remove the engine oil dipstick.
6. From underneath the cowling, gradually loosen and remove the 8 powerhead retaining bolts starting at the outer fasteners and working inward.

■ If the unit is several years old, or if it has been operated in salt water, or has not had proper maintenance, or shelter, or any number of other factors, then separating the powerhead from the intermediate housing may not be a simple task. An air hammer may be required on the bolts to shake the corrosion loose; heat may have to be applied to the casting to expand it slightly, or other devices employed in order to remove the powerhead.

7. Carefully lift and remove the powerhead assembly from the intermediate housing and place it on a suitable worksurface. Keep in mind that the powerhead weighs a couple of pounds, so an assistant may be advisable. If necessary, a piece of wood may be inserted between the powerhead and the intermediate housing as a means of using leverage to force them apart.

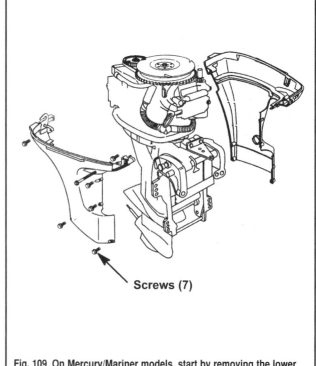

Fig. 109 On Mercury/Mariner models, start by removing the lower cowling

POWERHEAD 6-47

■ If equipped, take care not to lose the dowel pins (usually 2). The pins may come away with the powerhead or they may stay in the intermediate housing. Be ESPECIALLY careful not to drop them into the lower unit.

8. Remove and discard the old powerhead gasket.

To Install:

9. If equipped, make sure the dowel pins are in position (usually at the opposite sides of the intermediate housing.)

10. Place the new gasket in position (over the dowel pins, if utilized to make sure the gasket does not cock during installation and does not obscure any water passages). Keep a close eye on the gasket as the powerhead is positioned.

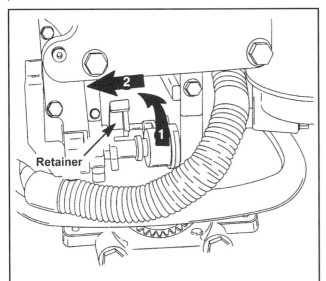

Fig. 110 To release the vertical shift shaft on Mercury/Mariner models push the retainer up, then slide it over

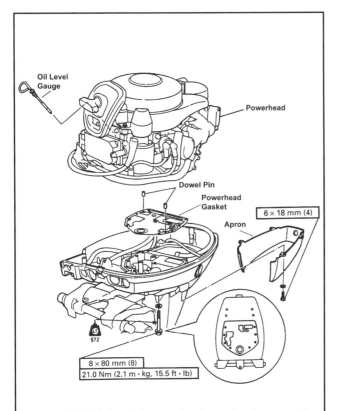

Fig. 111 Exploded view of the powerhead mounting - 25 hp models

11. Apply a light coating of marine grade grease to the driveshaft-to-crankshaft splines

12. Carefully lower the powerhead onto the intermediate housing using the dowel pins to align the two surfaces. The propeller (if in gear) or the flywheel/crankshaft (if not) may have to be rotated slightly (only in the normal direction of rotation, CLOCKWISE on these models) to permit the crankshaft to index with the driveshaft and allow the powerhead to seat properly.

13. Apply a light coating of Loctite® 572 or equivalent threadlocking compound to the threads of the powerhead attaching bolts.

■ Mercury does not mention using threadlock on the powerhead mounting bolts, but we don't see the harm and actually would go so far as to recommend it anyway.

14. Start the powerhead attaching bolts, with washers. Tighten the bolts alternately and evenly to a torque value of 15.5 ft. lbs. (21 Nm) for Yamaha models or to 20 ft. lbs. (38 Nm) for Mercury/Mariner models.

15. For Mercury/Mariner models, proceed as follows:

 a. Secure the vertical shift shaft with the latch (slide the retainer toward the port side, then rotate it downward to lock it in position).

 b. Reconnect the inlet fuel hose to the fuel pump and the vent hose to the fitting, securing each with new wire ties.

 c. Install the bottom cowling assembly and secure using the 7 screws.

 d. If equipped, reconnect the PTT wiring.

16. For Yamaha models, proceed as follows:

 a. Install the apron to the rear, lower side of the cowling and secure using the 4 retaining bolts. Be sure to tighten the fasteners securely, but do not over-tighten and crack the apron.

 b. Reconnect the shift link rod and, if equipped, the start-in-gear protection cable and secure each using the retaining clip.

 c. On electric start models, reconnect the neutral switch and starter switch wiring.

 d. Reconnect the trailer tilt switch coupler.

 e. Reconnect the cooling system indicator hose.

 f. Reconnect the fuel hose from the filter to the fuel pump.

 g. Reconnect the oil pan breather hose and secure using a new plastic wire tie.

 h. If equipped with power trim/tilt, reconnect the Blue and Green motor leads.

 i. Install the fitting plate and retaining bolt.

 j. Reconnect the wiring for the low oil pressure warning lamp.

 k. On tiller control models, reconnect the wiring for the engine stop switch.

 l. Reconnect the throttle cables.

17. Install the oil level dipstick. If the outboard was drained, refill the crankcase engine oil at this time.

18. On electric start models, reconnect the battery cables to the starter motor side of the powerhead.

19. Refer to the Timing and Synchronization adjustments in the Engine and Maintenance section for cable/linkage adjustment procedures.

■ Be sure to run the motor without the top cowling installed in order to check for potential fuel or oil leaks. Remedy any leaks before proceeding.

20. Install the top cowling to the powerhead.

■ If the powerhead was rebuilt or replaced with a remanufactured unit, don't forget to follow the proper Break-In procedures.

TIMING BELT/SPROCKET REMOVAL & INSTALLATION

♦ See Figures 112 and 113

This procedure can be performed with the powerhead assembly removed or installed on the motor. Unlike most of the Yamaha produced 4-stroke powerheads, the crankshaft sprocket on the 25 hp model does not utilize its own retaining nut, therefore no special tool is necessary to hold the driveshaft from turning to remove the crankshaft sprocket. Like most other Yamahas and Mercury/Mariners, the camshaft sprocket can accept the same tool used to hold the flywheel (or a sturdy strap wrench).

If the timing belt is to be reused, be sure to mark the direction of rotation on the old belt before removal. Also, if either the cylinder head or cylinder

6-48 POWERHEAD

block is not to be disassembled further, be sure to properly align the timing marks before removal. This will significantly ease installation.

■ Camshaft sprocket removal requires the use of a holder tool which inserts into the 2 holes in the sprocket itself. The same tool that is used on the flywheel can normally be used on the camshaft sprocket.

1. If the powerhead is still installed and on electric start models, disconnect the negative battery cable for safety.
2. Remove the Hand-Rewind Starter Assembly for access.
3. Remove the Flywheel and Stator Plate, as detailed in the Ignition and Electrical Systems section.
4. If either the cylinder head or cylinder block is not being overhauled, rotate crankshaft sprocket in the normal direction of rotation (clockwise) in order to align the timing marks on the crankshaft and camshaft pulleys with the marks on the powerhead.

■ The camshaft sprocket has a triangular stamp mark which must align with the triangular mark on the powerhead (at the cylinder head split line) when the motor is at No. 1 TDC. Similarly, when the crankshaft is turned so the woodruff key hole aligns with a triangular mark on the powerhead, the motor is set to No. 1 piston TDC.

5. Check the condition of the timing belt prior to removing it. If the belt appears loose or has excessive play at the midway point, the belt has stretched and **must** be replaced.

■ Remember, if the timing belt is to be reused, be sure to mark it showing the normal direction of rotation to ensure installation in the proper position.

6. Slip the timing belt free of the camshaft pulley **first,** and remove it from the crankshaft pulley. If the fit is too tight or you have trouble slipping the belt from the pulley, move to the next step and unbolt the camshaft (driven) pulley.

If necessary to strip the powerhead for rebuild or to replace one or more damaged sprockets, remove the camshaft (driven) or crankshaft (drive) sprocket from the motor as detailed in the following steps:

7. To remove the camshaft pulley, obtain a suitable flywheel holder tool (Yamaha #YB-06139 or Mercury #91-83163M). Place the two arms of the tool over the two holes of the camshaft pulley. Hold the pulley steady with the tool and at the same time loosen, then remove the camshaft sprocket bolt and washer.
8. Lift the camshaft pulley straight up and free of the camshaft. Remove and save the dowel pin from the recess in the camshaft.
9. To remove the crankshaft pulley simply pull is straight upward off the crankshaft and woodruff key. Remove and save the Woodruff key from the recess in the crankshaft.

To Install:

10. Apply just a dab of lubricant to the Woodruff key, and then install the key into the lower cutout in the crankshaft. The lubricant will help hold the key in place.
11. Install the crankshaft sprocket over the retaining plate and woodruff key. If the crankshaft was not disturbed the sprocket should be positioned so sprocket keyway is facing directly toward the camshaft sprocket. In this position the No. 1 piston is at TDC.
12. Make sure the woodruff keyway is aligned with the mark on the powerhead, facing the camshaft. If necessary, slowly rotate the crankshaft in the normal direction of rotation (clockwise) until the timing marks align. If the camshaft was disturbed, stop turning if you feel any interference.
13. Apply a light coating of grease to the threads of the camshaft retaining bolt and position aside for use in the next step or two.
14. Insert the dowel pin into the bore in the side, top of the camshaft. Position the camshaft pulley down onto the camshaft aligning it with the dowel pin.

■ Because there is no adjustment for the belt, installation can sometimes be difficult with both pulleys already installed, you may want to install the belt before threading the retaining bolt. If difficulty is encountered, remove the pulley from the camshaft and insert it into the belt, then carefully place tension on the belt as you install the pulley and belt assembly.

15. Hold the camshaft sprocket from turning using the flywheel holder tool, then install and tighten the sprocket retaining bolt to 28 ft. lbs. (38 Nm).

16. If the camshaft was not disturbed, the triangular mark on the pulley should align with the triangular mark on the powerhead, facing the crankshaft. If not, SLOWLY turn the camshaft in the normal direction of rotation (also clockwise when viewed from above) until the timing marks are appropriately aligned.

■ Remember, if the belt is being reused, be sure to face it in the same direction as normal rotation (clockwise).

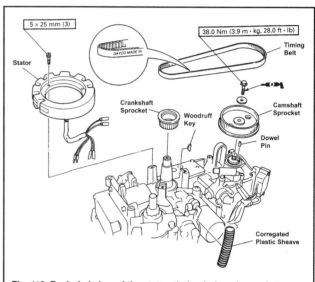

Fig. 112 Exploded view of the stator, timing belt and sprocket assembly

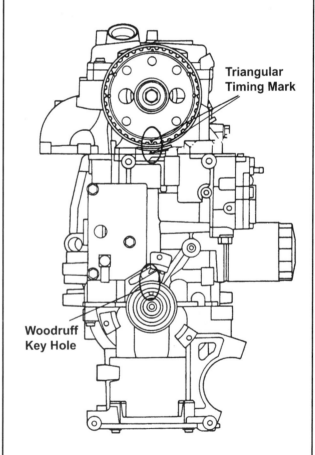

Fig. 113 Aligning the timing marks at No. 1 TDC - 25 hp motors

POWERHEAD 6-49

17. Place the timing belt, with the imprinted part number facing the right way up, over the crankshaft pulley, without disturbing timing mark alignment. After the belt is in place stretch it over the camshaft pulley without moving the pulley.

■ If belt installation is too difficult, remove the camshaft sprocket bolt and reposition the sprocket. With the sprocket properly aligned and the belt fully installed over it, slowly push downward on the belt and sprocket to slide it over the camshaft and dowel pin. Then hold the camshaft sprocket steady while you install and tighten the bolt.

18. Turn the crankshaft clockwise until it completes 2 full revolutions, then recheck the timing marks.
19. Install the Flywheel and Stator Plate assembly, as detailed in the Ignition and Electrical Systems section.
20. Install the Hand-Rewind Starter Assembly.
21. If removed, install the powerhead assembly.
22. If applicable, reconnect the negative battery cable.

CYLINDER HEAD REMOVAL & INSTALLATION

◆ See Figures 114 and 115

Although it is not absolutely clear in Yamaha literature, it looks likely that the cylinder head can be removed on these models without removing the powerhead assembly. It is definitely true on Mercury/Mariner models, though access requires removal of the lower cowling shrouds.

Remember, when loosening the retainers on the cylinder head, cover or any other major components, always try to follow the reverse of the torque sequence indicated in the assembly procedure or molded on the component itself.

When applicable tighten all bolts in using the proper torque sequence and to the proper specification.

■ Because of the high temperatures and pressures developed, the sealing surfaces of the cylinder head and the block are the most prone to water leaks. No sealing agent is recommended because it is almost impossible to apply an even coat of sealer. An even coat would be essential to ensure an air/water tight seal.

✱✱ WARNING

Never, never, use automotive type head gasket sealer. The chemicals in the sealer will cause electrolytic action and eat the aluminum faster than you can get to the bank for money to buy a new cylinder block.

1. For Mercury/Mariner models, loosen the 7 screws securing the lower cowling shroud to the intermediate housing, then remove the shroud assembly for access.
2. Remove the Flywheel and Stator Plate, as detailed in the Ignition and Electrical Systems section.
3. Remove the Timing Belt, as detailed in this section. If the cylinder head is to be overhauled follow the necessary steps to remove the camshaft sprocket.
4. Remove any other electrical components which may still be attached to the cylinder head (or at least disconnect the wiring). Remove the spark plugs from the cylinder head. If equipped, remove and/or disconnect the wiring and reposition the regulator/rectifier from the cylinder head.
5. If desired, remove the fuel pump. It is sometimes easier to reinstall the cylinder head cover if the pump is removed (or if the camshaft is turned so that the pump plunger is not fully depressed when you are trying to reinstall the pump or, if the pump was not separated, the cylinder head cover).
6. Using multiple passes of a star pattern, remove the 5 bolts securing the cylinder head cover to the head. Remove the cover and rubber seal. Neither Yamaha nor Mercury says that the seal cannot be reused, so we'll leave this to your judgment, however, remember this keeps oil from mucking up your cowling. Inspect the seal thoroughly for wear or damage and replace it if you are uncertain of its condition.
7. Locate and remove the 3 short cylinder head bolts from the head flange. Loosen the bolts alternately and evenly.
8. Locate and remove the 6 long cylinder head bolts from inside the head. Be sure to slowly loosen the bolts using the reverse of the torque sequence. This essentially results in a clockwise spiraling pattern that starts and ends at the center bolts (since they're the first AND last bolts in the sequence).

✱✱ WARNING

NEVER pry at gasket sealing surfaces. Such action would very likely damage the sealing surface of an aluminum powerhead.

9. Lift the head up and free of the block. Although Yamaha normally casts pry-tabs on heads, we could not locate any on this model. If the head will not release from the block looks for pry-tabs. If no tabs are provided, use a rubber mallet to gently tap around the perimeter of the surface to help break the gasket seal.

■ Take care not to lose the two dowel pins at the mating surfaces. These pins may remain either in the cylinder head or in the block.

10. Remove and discard the old cylinder head gasket.
11. If necessary for further cylinder head disassembly or for oil pump service, loosen the fasteners and remove the pump from the head. This will be necessary if you plan on removing the camshaft and valve train, but it is a good idea to remove and inspect the pump since you've come this far regardless. For details on removal, installation, inspection and overhaul of the oil pump assembly, please refer to Oil Pump in the portion of the Lubrication and Cooling section.

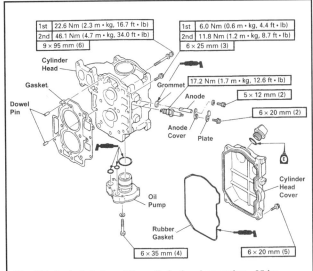

Fig. 114 Exploded view of the cylinder head mounting - 25 hp motors

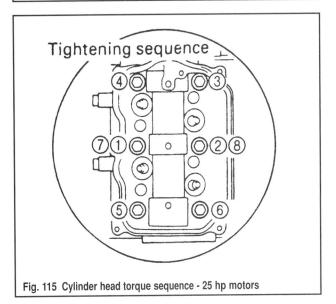

Fig. 115 Cylinder head torque sequence - 25 hp motors

12. A little outboard from the spark plug ports there are 2 cylinder head anode installed into bores in the side of the head. Remove the retaining bolt and plate from each anode cover, then remove the bolt securing each anode cover, then remove the cover, anode and grommet from each bore.

13. Refer to Cylinder Head Overhaul in this section for details on removal and installation of the valve train components. Also, be sure to check the information on gasket mating surfaces to make sure the cylinder head is not warped.

To Install:

14. Apply a light coating of marine grade grease the anode grommets, then install a grommet, anode and anode cover to each bore in the cylinder head. Secure each cover using the retaining bolt, followed by the plate and plate bolt.

15. If removed, refer to the Lubrication and Cooling section for details on Oil Pump assembly and installation.

16. Position a NEW head gasket over the two dowel pins.

17. Install the cylinder head onto the block with the holes in the head indexed over the two dowel pins.

18. Thread and finger-tighten the 6 long cylinder head bolts which are positioned inside the head, then thread and finger-tighten the 3 short cylinder head flange bolts.

19. Tighten the 6 long cylinder head bolts using multiple passes of a mostly counterclockwise, spiraling torque sequence that starts on the left, center bolt and works towards the outer bolts (then ends with the left center and right center bolts again, in that order, at least Yamaha recommends tightening the 2 center bolts again, even if Mercury does not specify to). Tighten the long bolts first to 16.7 ft. lbs. (22.6 Nm) and then to 34 ft. lbs. (46.1 Nm).

20. After the long bolts are tightened, tighten the short head flange bolts alternately and evenly first to 53 inch lbs./4.4 ft. lbs. (6 Nm) and then to 106 inch lbs./8.7 ft. lbs. (11.8 Nm).

■ **If the valve train components were disturbed in any way you are going to have to adjust the valves once the timing belt is installed, so either leave the cylinder head cover off or just finger-tighten the bolts.**

21. If removed, install the O-ring seal into the cover groove, then apply a light coating of marine grade grease to the seal.

22. Install the cylinder head cover and O-ring. If the valves were not disturbed, install and tighten the 4 cylinder head cover bolts securely either using the molded torque sequence (or if not present using a star-shaped crossing pattern that begins).

■ **Make sure the new O-ring remains seated in the cover groove and does not become pinched during installation or there will be oil leaks!**

23. Install the Timing Belt and, if removed, Camshaft Sprocket, as detailed in this section.

24. If the valves were disturbed, adjust them at this time, as detailed under Valve Lash in the Maintenance and Tune-Up section, then install the cylinder head cover.

25. Install the spark plugs to the cylinder head. Install any electrical components, such as the regulator/rectifier to the cylinder head, as applicable.

26. Install the Flywheel and Stator Plate, as detailed in the Ignition and Electrical Systems section.

27. On Mercury/Mariner models, install the lower cowling shrouds.

28. If removed, install the powerhead assembly.

CYLINDER HEAD OVERHAUL

◆ See Figure 116

1. Remove the Cylinder Head from the powerhead assembly, as detailed in this section.

2. Remove the Oil Pump from the cylinder head, as detailed in the Lubrication and Cooling section.

■ **When reused, ALL valve train components must be installed in their original locations. For this reason be sure to tag and/or sort all shaft retainers, rocker arms, valves, valve springs, keepers etc and keep them separated during service.**

3. Loosen the 3 rocker arm shaft retainer bolts, then remove each of the retainers. Note the center retainer is of a different shape than the outer 2. Keep the retainers sorted for installation in their original locations.

4. Remove the rocker shaft and arm assembly from the cylinder head. If necessary, mark and remove each of the rockers from the shaft itself.

5. To remove the each valve, proceed as follows:

a. Using a clamp-type valve spring compressor, carefully place just enough pressure on the valve spring so the upper spring seat (spring retainer) moves down sufficiently to expose the keepers.

b. Remove the pair of valve keepers, then slowly ease pressure on the valve spring until the compressor can be removed.

c. Remove the upper seat (retainer), valve spring and lower spring seat from the shaft of the valve.

d. Remove the valve from the underside (combustion chamber side) of the cylinder head.

e. If necessary, carefully pry the old valve stem seal from its position on the cylinder head.

f. Repeat this procedure for each of the remaining valves.

6. If necessary, remove the bolt and gasket from the top of the cylinder head.

7. Carefully slide the camshaft from the cylinder head, out through the bottom (oil pump) end of the head.

8. If necessary, remove the oil seal from the top of the cylinder head.

9. Refer to Powerhead Refinishing, later in this section, for details on cleaning and inspecting cylinder head components.

To Assemble:

10. If the camshaft oil seal was removed during disassembly, coat the lips of a new seal with a light coating of clean engine oil. Install the seal into the head using a suitable driver. The seal should be installed with the lips facing downward into the cylinder head. On most applications this means the seal manufacturer's marks or numbers are facing outward.

11. Apply a light coating of engine oil to the camshaft bearing journals, then apply a light coating of a molybdenum disulfide engine assembly lube to the camshaft lobes. Next, **carefully** insert the sprocket end (with the dowel pin hole) of the camshaft into the head from the oil pump end of the head. Continue inserting the camshaft until end of the shaft is seated in the camshaft oil seal against the head at the other end.

■ **Do not attempt to rotate the camshaft at this time. At this time the camshaft is supported at just one end. The camshaft should ONLY be rotated after the oil pump has been installed and the shaft is then supported at both ends.**

12. Install the bolt and gasket to the top of the cylinder head.

13. If removed, install each of the valve and spring assemblies as follows:

a. If removed, install new valve spring seals to the cylinder head.

b. Apply a light coating of engine oil to the stem, then position the valve into the cylinder head (into the original guide and valve seat unless the valve or valve seat was replaced).

c. Position the lower spring seat down onto the valve shaft and cylinder head, then oil and install the spring.

d. Place the upper spring seat (retainer) over the spring, then use the valve spring compressor tool to press downward on the seat and spring so you can install the keepers.

e. Insert the valve keepers, then slowly back off the valve spring compressor, making sure the keepers are fully in position. Then remove the valve spring compressor.

f. Repeat for each of the remaining valves.

■ **Remember to lightly coat all the rocker arm and shaft components with clean engine oil as they are installed.**

14. If disassembled, apply a light coat of clean engine oil or engine assembly lube to the rocker arms and shaft, then install the rocker arm assemblies to the rocker shaft, in the SAME positions from which they were removed.

15. Carefully lower the rocker arm and shaft assembly to the cylinder head. If the lash adjusters were not loosened, the preload on one or more of the valve springs may prevent you from properly seating the shaft. Determine which adjusters must be loosened (note if the valves were serviced or replaced, you should just go ahead and back off all of them, besides, they will all require checking and adjustment regardless).

16. Apply a light coating of engine oil to the threads of the shaft retaining bolts and to the shaft retainer clips themselves.

POWERHEAD 6-51

17. Install the rocker shaft retainers and secure using the retaining bolts. Since the center retainer is symmetrically shaped there is normally an embossed arrow that should be positioned facing the top/flywheel end of the cylinder head). Tighten the each of the shaft retainer bolts to 13 ft. lbs. (18 Nm).
18. Install the Oil Pump assembly to the cylinder head, as detailed in the Lubrication and Cooling section.
19. Install the Cylinder Head as detailed in this section.
20. Once the powerhead is fully assembled, if valve train components were replaced, be sure to operate the engine as directed for new component break-in. For details, please refer to Powerhead Break-In, in this section.

CYLINDER BLOCK OVERHAUL

◆ See Figures 117 thru 124

This motor is unique among most Yamaha and Mercury/Mariner outboards in that it is a 2-cylinder motor that contains 3 pistons! We know, you're asking, "how is this possible." But the answer is simple, the motor contains 2 pistons that participate in the Otto cycle of engine operation and a 3rd balance piston whose sole purpose (as suggested by the name) is to

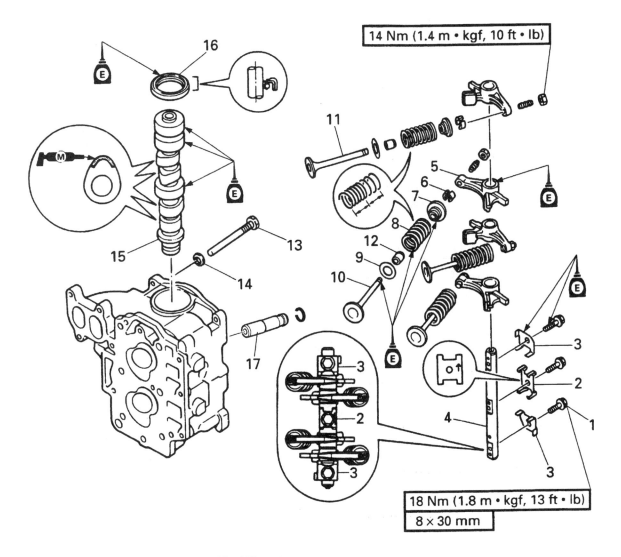

1. Bolt
2. Rocker Arm Retainer
3. Rocker Arm Retainer
4. Rocker Arm Shaft
5. Rocker Arm
6. Valve Keepers
7. Spring Retainer
8. Valve Spring
9. Valve Spring Seat
10. Intake Valve
11. Exhaust Valve
12. Valve Stem Seal
13. Bolt
14. Gasket
15. Camshaft
16. Oil Seal
17. Valve Guide

Fig. 116 Exploded view of the cylinder head assembly - 25 hp motors

6-52 POWERHEAD

smooth out crankshaft rotational forces and prevent engine vibration. The balance piston is attached to the top of the crankshaft and faces in the opposite direction from the other 2 pistons. The piston itself must be removed in order to separate the crankcase from the cylinder block. Besides, complete crankcase overhaul includes inspection of the balance piston and bore, so the piston would have to be removed anyway.

Remember, when loosening the retainers on manifolds, covers and other major components, always try to follow the reverse of the torque sequence indicated in the assembly procedure or molded on the component itself.

When applicable tighten all bolts in using the proper torque sequence and to the proper specification. When possible we've included it in the procedure, but on many Yamahas both the sequence and value is molded into the casting on critical components. You'll notice that most Yamaha torque sequences are generally clockwise spirals starting somewhere near the center of the component.

1. Remove the Powerhead Assembly, as detailed in this section.
2. Remove the Timing Belt/Sprocket Assembly, as detailed in this section.
3. Remove the Cylinder Head Assembly, as detailed in this section.
4. Remove any electrical components which were not stripped during the previous steps.

■ The exhaust cover should always be removed during a powerhead overhaul. Many times water in the powerhead is caused by a leaking exhaust cover gasket or plate.

5. Although not absolutely necessary, we recommend you strip the block of the following components for inspection and passage cleaning both of which will help ensure a thorough rebuild:
   a. If not already done, remove and discard the old oil filter.
   b. Gradually loosen and remove the exhaust cover retaining bolts using the reverse of the torque pattern (very roughly a counterclockwise spiraling pattern that starts with the two center side bolts, then continues with the upper most bolts and works back towards the center bolts again). The thermostat assembly is mounted under a housing to the top of the exhaust cover (the top 2 exhaust cover bolts also retain the thermostat housing).
   c. Remove the thermostat assembly from the exhaust cover. For details, please refer to the Lubrication and Cooling section. Remove and discard the old thermostat gasket.
   d. Remove the exhaust cover and gasket from the side of the powerhead. Remove and discard the old exhaust cover gasket.

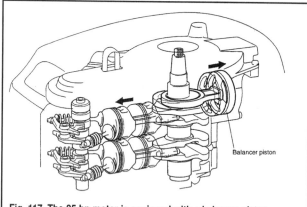

Fig. 117 The 25 hp motor is equipped with a balancer piston

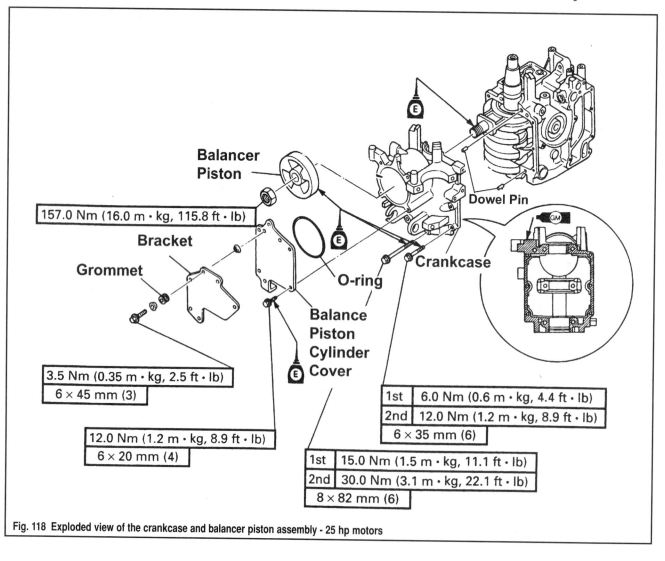

Fig. 118 Exploded view of the crankcase and balancer piston assembly - 25 hp motors

e. Remove the 3 bolts securing the breather cover to the top of the cylinder block. Remove the breather cover, then remove and discard the cover gasket. Remove the filter breather element and clean using a suitable solvent. Inspect for damage and replace, as necessary.

f. On the top of the block, in the side of the boss just behind the exhaust cover and thermostat there is a plug and gasket. The plug is reusable, the gasket it not. Remove the plug for passageway inspection and cleaning.

6. Remove the 3 bolts, washers and grommets from the bracket mounted to the balancer piston cylinder cover, then remove the bracket.

7. Loosen and remove the 4 bolts securing the balancer piston cylinder cover. Remove the balancer piston cylinder cover, then remove the old O-ring. Yamaha does not say that the O-ring cannot be reused, so we'll leave this to your judgment. Inspect the seal thoroughly for wear or damage and replace it if you are uncertain of its condition.

8. Loosen and remove the large nut and the center of the balancer piston, then remove the piston.

9. Remove the crankcase retaining bolts using multiple passes in the reverse of the torque sequence (either molded on the crankcase or as illustrated later in this procedure. Essentially the crankcase has 6 long bolts at the center of the crankcase and 6 short bolts along the sides of the flange. The torque sequence begins and ends at the center two bolts. The reverse of the sequence, would start with the 2 center bolts, then work in 2 counterclockwise spirals inward, starting at first the top right flange bolt and then the top right center bolt.

■ Remember that since there two different size bolts are used keep them separated for identified for installation purposes. Long bolts to the center, short bolts to the flange.

10. After all bolts have been removed, separate the crankcase from the powerhead. **Take care** not to lose the two dowel pins. These pins may remain with cylinder block or come away with the crankcase cover.

✱✱ WARNING

If difficulty is encountered, use a rubber mallet to tap around the perimeter of the crankcase cover. Some Yamahas have molded pry tabs to help break the gasket seal, however NEVER pry at gasket sealing surfaces. Such action would very likely damage the sealing surface of an aluminum powerhead.

■ Most parts inside the cylinder block are matched sets and should not be moved from one cylinder to another. This includes the pistons as well as each of the connecting rod/end cap combinations. For this reason, prior to disassembly it is essential to matchmark each of the connecting rods to their end caps (this ensures that not only will each cap be installed on the proper rod, but facing the correct direction).

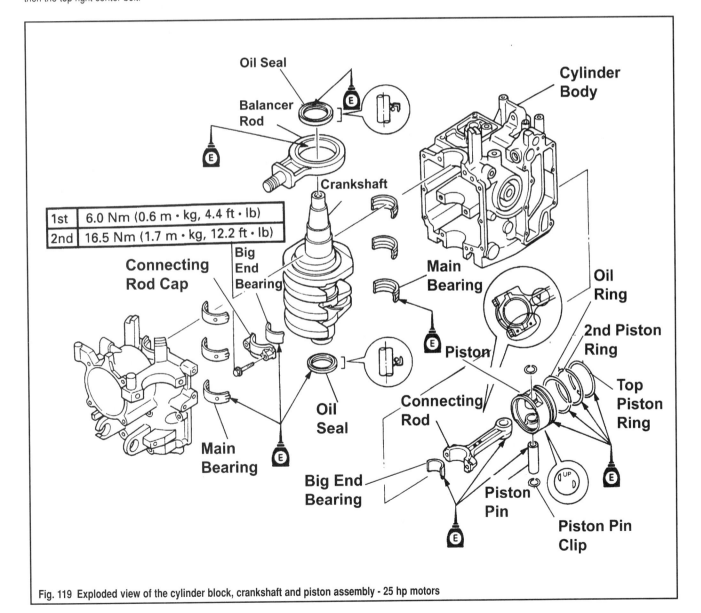

Fig. 119 Exploded view of the cylinder block, crankshaft and piston assembly - 25 hp motors

6-54 POWERHEAD

11. Matchmark the big end caps to the connecting rods (or identify the marks already placed there by the manufacturer), then remove the 4 connecting rod cap bolts. Remove the caps and their bearing inserts, then place them aside for reinstallation back on their original connecting rods as soon as the crankshaft has been removed.

■ **When loosening or tightening the connecting rod cap bolts, alternate between each bolt every 1/2-1 turn.**

12. Carefully lift the crankshaft from the block. Place it on a protective surface.

✱✱ WARNING

Handle the crankshaft carefully. Remember it is an expensive and vital piece to the motor. The polished surfaces should be covered to prevent scratches or other damage and the whole shaft should be protected against impact or other damage which could destroy it.

13. Immediately reinstall the connecting rod end caps and bearing inserts to their respective pistons (to prevent loss, mixing them up or potential damage to the bearing surfaces).
14. Remove the top and bottom crankshaft oil seals. Discard the seals.

■ **If the main bearing inserts are fit for further service, keep them with the crankcase, resting on the spot from which they were removed.**

15. Slide the 3 sets of metal bearing inserts out of the top and bottom crankcase halves.

✱✱ WARNING

The crankshaft bearings absolutely MUST be installed back in the same position from which they were removed. These bearings are matched sets - half in each crankcase half (or half in each connecting rod/cap half). If they are to be reused, they must be installed with their matching halves on the same crankshaft bearing surfaces.

■ **These pistons and connecting rods are normally already marked from the factory as to what side faces up, identify the marks (UP on the piston dome and Y on the con rod) and perhaps, remark them to ensure they are found easily during assembly. Also, each piston and connecting rod must be returned to the original bore and crankshaft journal, so be sure to place a number on each to ensure proper positioning.**

16. Remove the balancer rod from the crankshaft. Place a small mark on the rod itself, on the side that faces upward.
17. Label and remove each of the pistons from the bores. Number the pistons and place marks on the piston/con rod which face up or identify the mark already on each piston and connecting rod which faces up (faces the flywheel) for installation purposes. If necessary, disassemble each of the pistons, as follows:
    a. Carefully pry the clips free from the piston pin bore on either side of the piston assembly.
    b. Slide the piston pin free of the bore (specifications tell us that this is NOT an interference fit, so it should slide free with ease).
    c. Using a piston ring spreading tool, carefully remove each of the rings (top and second compression rings, followed by the oil rings) from the piston. Good shop practice dictates the rings be replaced during a powerhead overhaul. However, if the rings are to be used again, expand them **only** enough to clear the piston and the grooves because used rings are brittle and break very easily.
    d. Keep all components from a single piston together for reuse or replacement, but don't mix and match components from other pistons.
18. Refer to Powerhead Refinishing, later in this section, for details on cleaning and inspecting cylinder block components. There are probably more components and specifications to measure these 4-stroke motors than on any 2-stroke outboards of similar size. Take your time and make sure you've made measurements for all components listed in the Engine Specifications chart for this model.

To Assemble:

19. Make sure the mating surfaces of the crankcase and cylinder block are completely clean and free of sealant or damage. The dowel pins should already be installed in the block or crankcase cover, but if not, obtain new pins and install them in the positions at opposite ends of the block.

■ **If the old main bearings are to be installed for further service, each must be installed in the same location from which it was removed.**

20. Place the main bearing inserts into the crankcase and cylinder block. There are normally 2 things to look out for when doing this. Most bearing inserts will be equipped with locating tabs, make sure they are seated in the appropriate grooves. Also, inserts with oil holes should be aligned with an oil passage. If a new set of bearings has been purchased for installation, match the colors on the side of the new bearings with the letters embossed on the bottom side of the crankcase. The marks are stacked on the crankcase, to correspond to the 3 main bearing positions. The top being No. 1, the middle being No. 2 and the bottom being No. 3.
    - Letter **A** matches the Blue bearing.
    - Letter **B** matches the Black bearing.
    - Letter **C** matches the Brown bearing.
21. If they were disassembled, prepare each piston assembly for installation as follows:

■ **Apply a light coat of clean engine oil to each piston component as or right after it is installed.**

a. Using a ring expander, carefully install each of the rings with their gaps as noted here and as shown in the accompanying illustration. The description and illustration are depicted while looking at the piston dome with the top (flywheel end) facing 12 o'clock.

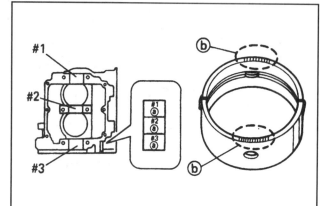

Fig. 120 The crankcase is stamped and bearing inserts are marked with color codes to help select crankshaft bearings

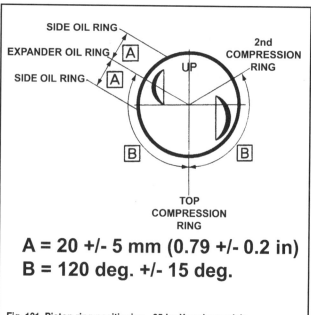

Fig. 121 Piston ring positioning - 25 hp Yamaha models

POWERHEAD 6-55

■ UNFORTUNATELY Mercury and Yamaha sometimes have a different philosophy on exactly how to set up the ring gaps prior to piston installation. This does go to reinforce a popular opinion we've often heard from machinists, that since the rings are going to move around in service, it really shouldn't matter. But, we've gone through the trouble of giving you each manufacturer's preferred set-up method.

   b. The oil ring is an assembly which consists of bottom and top side rails with an expander ring in the middle. Yamaha recommends that the gap on the expander ring should face between 10 and 11 o' clock, while the top and bottom side rail gaps should be just on either side of it, one at between 11 and 12 o'clock with the other between 9 and 10 o'clock. Mercury recommends that the gap on the expander ring should face directly to 6 o' clock, while the top and bottom side rail gaps should each be 90 degrees from it, the top at 9 o'clock and the bottom at 3 o'clock.

   c. The lower (second) compression ring should be positioned with its gap at about 2 o'clock (as recommended by Yamaha) or between 1 and 2 o'clock (as recommended by Mercury).

   d. The upper (top) compression ring should be positioned with its gap at 6 o'clock (as recommended by Yamaha) or between 10 and 11 o'clock (as recommended by Mercury).

   e. Align the **Y** mark on the connecting rod (facing up/flywheel) with the **UP** mark on the piston (also facing up/flywheel), install the connecting rod to the piston using the piston pin.

   f. Install the piston pin clips in either side of the pin bore to secure the piston pin in place.

   g. Repeat for the remaining piston.

✱✱ WARNING

Use extreme care to prevent the connecting rods from contacting and scoring the cylinder walls during piston installation. It is a good idea to wrap the ends in a shop cloth or something else to protect the cylinders.

  22. Install each of the pistons (with the correct mark/side facing UP toward the flywheel) to the cylinder block using a ring compressor. Apply a very light coating of clean engine oil to the inside of the compressor and to the cylinder bore, then place the compressor around the piston. Insert the piston through the cylinder head side of the bore and use a wooden hammer handle to gently tap the piston down through the compressor and into the cylinder bore.

✱✱ WARNING

Be VERY careful when installing the piston. Get the lower portion of the skirt below the ring compressor and insert it into the cylinder bore to make sure everything is aligned. Then, GENTLY tap the piston through the compressor, stopping if it seems to hang up even slightly to make sure that a ring is not caught on the deck of the cylinder block.

  23. Remove the connecting rod end caps and position them so the crankshaft can be installed. Place the connecting rod bearing inserts into position on the connecting rod big end and cap. Keep in mind that the connecting rods and bearing inserts are also marked with stampings and color codes to help assist in bearing selection. The codes and colors correspond to the other crankshaft codes for this motor (A: blue, B: black, C: brown).

■ Here's where procedures differ slightly depending upon whether you're installing bearings of a known clearance or measuring clearance. If you've already determined that bearing clearances are fine, apply a light coat of engine oil or engine assembly lube to each of the connecting rod, main bearing and crankshaft journal surfaces. If however, you need to measure and determine bearing clearance, Temporarily INSTALL THESE COMPONENTS DRY for measurement purposes. For more details, please refer to Measuring Bearing Clearance Using Plastigage®, in the Powerhead Refinishing section.

  24. Apply a light coating of clean engine oil or engine assembly lube to the bearing surface of the balancer piston rod, then install it over the crankshaft, with the mark made during assembly facing the flywheel end of the shaft.

  25. Carefully position the crankshaft to the cylinder block, over top the main bearing inserts.

  26. Bring up the connecting rod ends around the crankshaft throws. If you've already measured bearing clearances, apply a light coat of engine oil or engine assembly lube to the connecting rod, cap and crankshaft bearing surfaces.

  27. Align the matchmarks and reinstall the connecting rod bearing caps. Install the bolts alternately and evenly first to 53 inch lbs./4.4 ft. lbs. (6 Nm) and then to 150 inch lbs./12.2 ft. lbs. (16.5 Nm).

■ **NEVER attempt to install the upper or lower crankshaft oil seals with the crankcase already installed and tightened otherwise the seal will likely be damaged.**

  28. Coat the lips of new upper and lower oil seals with a clean engine oil, then slide the oil seals onto the crankshaft with their lips facing toward the center (toward the shaft throws). Seat the seals as best you can at this time. To ensure the seals are not cocked you can install the crankcase and FINGER-TIGTHEN the bolts then lightly knock the seals using a suitably sized driver and rubber mallet. Once this is done, remove the crankcase again and continue the cylinder block assembly.

  29. Confirm that the dowel pins are still in position, then apply a thin bead of Gasket Maker sealant (such as Loctite 514 Master Gasket or equivalent) to the crankcase side of the crankcase-to-cylinder block mating surfaces.

  30. Apply a light coating of engine oil to the threads of the crankcase retaining bolt threads.

  31. Carefully install the crankcase cover over the cylinder block using the dowel pins to align the two surfaces. Install and finger-tighten the retaining bolts (remember that there are 2 different size bolts, long bolts go to the center, while short bolts go to the flange.

  32. Tighten the crankcase retaining bolts using multiple passes of the clockwise, spiraling torque sequence which starts at the center of the long bolts and works outward, then starts again at the center of the short bolts and also works outwards. Finish the torque sequence each time by again tightening the center two long bolts. Tighten the long center bolts first to 11 ft. lbs. (15 Nm) and then to 22 ft. lbs. (30 Nm). Tighten the small flange bolts first to 53 inch lbs./4.4 ft. lbs. (6 Nm) and then to 106 inch lbs./8.9 ft. lbs. (12 Nm).

  33. Apply a light coating of engine oil to the balancer piston and the threads of the balancer rod, then carefully install the piston and secure using the retaining nut. Tighten the nut to 115.8 ft. lbs. (157 Nm).

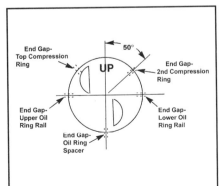

Fig. 122 Piston ring positioning - Most Mercury/Mariner models

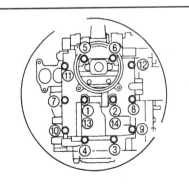

Fig. 123 Crankcase torque sequence - 25 hp motors

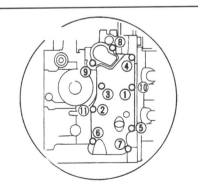

Fig. 124 Exhaust cover torque sequence - 25 hp motors

6-56 POWERHEAD

34. Install the balancer piston cylinder cover and O-ring. Apply a light coating of engine oil to the threads of the 4 bolts, then install and tighten to 106 inch lbs./8.9 ft. lbs. (12 Nm).
35. Install the bracket to the cylinder cover using the 3 sets of bolts, washers and grommets. Tighten the bolts to 2.5 ft. lbs. (3.5 Nm).
36. If removed from the top of the powerhead, install the breather filter, then install the breather cover using a new gasket.
37. If removed from the bore at the top of the powerhead, just behind the top of the exhaust cover mating surface, install the plug using a new gasket.
38. If removed, install the exhaust cover, thermostat assembly and gaskets (one for the exhaust cover and one for the thermostat housing). Tighten the bolts using multiple passes of a roughly clockwise spiraling pattern that starts at the center bolts and works outward. Tighten the bolts first to 53 inch lbs./4.4 ft. lbs. (6 Nm) and then to 106 inch lbs./8.9 ft. lbs. (12 Nm).
39. If removed, install a new oil filter.
40. Install the Cylinder Head Assembly, as detailed in this section.
41. Install the Timing Belt/Sprocket Assembly, as detailed in this section.
42. Install the Powerhead Assembly, as detailed in this section.
43. Install any electrical components which were not installed during the previous steps.
44. Once the powerhead is fully assembled, if cylinder block components were replaced, especially the pistons and/or rings, be sure to operate the engine as directed for new component break-in. For details, please refer to Powerhead Break-In, in this section.

CLEANING & INSPECTING

Cleaning and inspecting internal engine components is virtually the same for all Yamaha and Mercury/Mariner outboards with the exception of a few components that are unique to certain 4-stroke motors. The main variance between different motors comes in specifications (which are listed in the Engine Specifications charts) or by component type. A section detailing the proper procedures, sorted mostly by component, can be found under Powerhead Refinishing.

30/40 Hp 3-Cylinder Powerheads

POWERHEAD REMOVAL & INSTALLATION

◆ See Figures 109 and 110, and 125, 126 and 127

Starting with the 3-cylinder powerheads the Yamaha and Mercury/Mariner 4-stroke motors are becoming large enough that stripping the powerhead before removal (for both the weight savings and prevention of damage to the powerhead mounted components of the fuel and electrical systems) is advisable (though still not necessary IF you are using a lifting device). Conversely, although a lifting device is not absolutely necessary, one is still advisable to make for easy removal and to help when positioning the powerhead over the mating surface and crankshaft splines (especially if you decide NOT to strip the powerhead).

Consider why the powerhead is being removed. If you are planning on disassembling the powerhead for inspection or overhaul, then you'll have to remove the fuel and electrical components anyway. If this is the case, removing them before powerhead removal is a good idea, if only to protect them from potential damage when lifting and moving the powerhead itself. As usual, we've pretty much only included the steps in this procedure which are necessary in order to remove the assembled powerhead. Additional steps should be taken if you desire to strip the powerhead of all fuel and electrical components before proceeding. If so, refer to components in the Fuel System and Ignition and Electrical System sections for details.

1. Remove the engine top cover for access to the powerhead.
2. On electric start models, disconnect the battery cable for safety, then disconnect the cables at the powerhead. For most models this means at or near the starter/starter solenoid on the port side of the motor.
3. For Yamaha models, proceed as follows:
   a. For power trim/tilt models, disconnect the blue and green leads from the ring terminals just aft of the oil filter (also on the port side of the motor).
   b. At the front of the starboard side of the motor remove the retaining clips, then disconnect the shift and throttle cables.
   c. As applicable, at the front of the cowling disconnect the main harness connector or the engine stop switch bullet connectors.

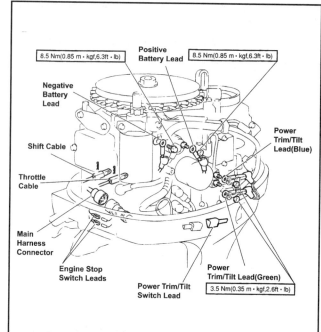

Fig. 125 Harness and cable connections - 30/40 hp Yamaha 3-cylinder motors (Merc similar)

   d. If equipped with power trim/tilt, disengage the switch lead found in the lower cowling along the port side, just below the oil filter.
   e. At the aft end of the starboard side, loosen the bolt and remove the plate that holds the fuel line/filter in position, then loosen the spring clamp and disconnect the fuel hose from the filter.
   f. Carefully cut the plastic wire ties, then disconnect the 2 waters hoses from the fitting on the lower port side of the powerhead (a little below and aft of the oil filter).
   g. Loosen the bolts and the front and rear of the cowling apron (they are threaded vertically at one end and horizontally at the other), then carefully remove the apron from the underside of the cowling for access to the powerhead mounting bolts.
4. For Mercury/Mariner models, proceed as follows:
   a. If equipped with Power Trim/Tilt (PTT), disconnect the wiring.
   b. Remove the 7 screws securing the bottom cowling, then separate the cowling from the intermediate housing for access.
   c. Drain the engine oil from the crankcase.
   d. For remote control EFI models, remove the control cable anchor bracket screws (on the lower starboard side of the powerhead, toward the front). Then remove the screw at the pivot point on top of the throttle/shift lever (also on the starboard side of the powerhead, but toward the stern of the motor). With the bolt removed, swing the entire shift linkage to the side.
   e. For all models, locate the vertical shift shaft linkage under the front center of the powerhead. Disengage the shift shaft by first pushing the retainer up and then sliding it over toward the starboard side.
   f. For Carbureted Tiller models, remove the throttle cable and shift rod, then disconnect the tiller handle wiring.
   g. For EFI Tiller models, remove the 2 screws securing the irregular shaped shift guide bracket.
   h. Locate the crankcase vent hose attached at the base of the powerhead, between the spark plugs and the dipstick, then carefully cut the sta-strap (wire tie) and disconnect the hose.
   i. For carbureted models, cut the sta-strap and disconnect the fuel inlet hose from the fuel pump (check the hose routing or the marking on the pump to be sure, however the inlet hose is usually the bottom hose on these models).
   j. For EFI models, disconnect the fuel supply hose from the water separating fuel filter.
5. Remove the engine oil dipstick.
6. From underneath the cowling, gradually loosen and remove the powerhead mounting bolts (there are usually 8 on this model) starting at the outer fasteners and working inward.

POWERHEAD 6-57

■ If the unit is several years old, or if it has been operated in salt water, or has not had proper maintenance, or shelter, or any number of other factors, then separating the powerhead from the intermediate housing may not be a simple task. An air hammer may be required on the bolts to shake the corrosion loose; heat may have to be applied to the casting to expand it slightly, or other devices employed in order to remove the powerhead.

7. Carefully lift and remove the powerhead assembly from the intermediate housing and place it on a suitable worksurface.

■ Take care not to lose the dowel pins (usually 2). The pins may come away with the powerhead or they may stay in the intermediate housing. Be ESPECIALLY careful not to drop them into the lower unit.

8. Remove and discard the old powerhead gasket.

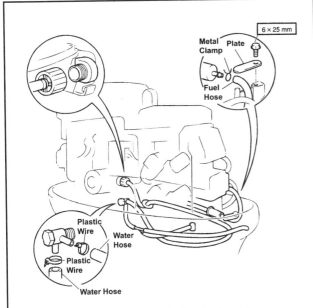

Fig. 126 Hose connections - 30/40 hp Yamaha 3-cylinder motors (Merc similar)

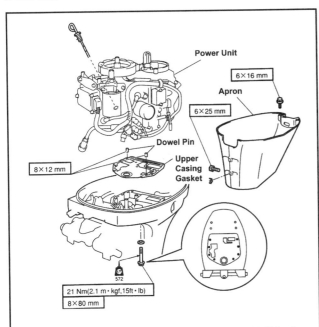

Fig. 127 Exploded view of the powerhead mounting - 30/40 hp 3-cylinder motors (Merc similar)

To Install:

9. Make sure the dowel pins (usually 2) are in position at the opposite sides of the intermediate housing.
10. Place the new gasket in position over the dowel pin(s), this should make sure the gasket does not cock during installation and does not obscure any water passages, but keep a close eye on the gasket as the powerhead is positioned anyway.
11. Apply a light coating of marine grade grease to the driveshaft-to-crankshaft splines
12. Carefully lower the powerhead onto the intermediate housing using the dowel pins to align the two surfaces. The propeller (if in gear) or the flywheel/crankshaft (if not) may have to be rotated slightly (only in the normal direction of rotation, CLOCKWISE on these models) to permit the crankshaft to index with the driveshaft and allow the powerhead to seat properly.
13. Apply a light coating of Loctite® 572 or equivalent threadlocking compound to the threads of the powerhead attaching bolts.

■ Mercury does not mention using threadlock on the powerhead mounting bolts, but we don't see the harm and actually would go so far as to recommend it anyway.

14. Start the powerhead attaching bolts, with washers.
15. For Yamaha models, tighten the bolts alternately and evenly to a torque value of 15 ft. lbs. (21 Nm).
16. For Mercury/Mariner models, tighten the bolts using multiple passes of a side-to-side crossing pattern that starts at the center and works outward. Tighten the bolts first to 28 ft. lbs. (38 Nm) for carbureted motors or FIRST to 27.5 ft. lbs. (37 Nm) and THEN to 34 ft. lbs. (46 Nm) for EFI motors.
17. For Mercury/Mariner models, proceed as follows:
   a. Reconnect the fuel hose and the vent hose to the appropriate fittings, securing each with new wire ties.
   b. For EFI Tiller models, secure the shift guide bracket using the 2 mounting screws.
   c. For Carbureted Tiller models, reconnect the tiller handle wiring, the shift rod and the throttle cable.
   d. For all models, secure the vertical shift shaft with the latch (slide the retainer toward the port side, then rotate it downward to lock it in position).
   e. For remote control EFI models, swing the throttle/shift lever back into position, then secure using the bolt at the top of the lever. Install and tighten, the control cable anchor bracket screws.
   f. Install the bottom cowling assembly and secure using the 7 screws.
   g. If equipped, reconnect the PTT wiring.
18. For Yamaha models, proceed as follows:
   a. Install the apron to the underside the cowling and secure using the retaining bolts. Be sure to tighten the fasteners securely, but do not overtighten and crack the apron.
   b. Reconnect the water hoses to the fitting on the lower, port side of the motor. Secure them using new wire ties.
   c. Reconnect the fuel hose to the filter and secure using the spring clip (make sure the clip hasn't lost tension or become damaged). Install the plate and retaining bolt to hold the fuel line/filter in position.
   d. If equipped with power trim/tilt, reconnect the switch lead found in the lower cowling along the port side, just below the oil filter.
   e. Reconnect either the bullet connectors for the engine stop switch or the main engine harness connector (located at the front of the cowling).
   f. Connect the throttle and shift cables, then secure with the retaining clips.
   g. If equipped, reconnect the green and blue power trim/tilt leads to the terminals on the port side of the powerhead.
19. Install the oil level dipstick. If the outboard was drained, refill the crankcase engine oil at this time.
20. Reconnect the positive and finally, negative battery cables to their connections on (normally to the starter/solenoid on the port side of the powerhead).
21. Refer to the Timing and Synchronization adjustments in the Engine and Maintenance section for cable/linkage adjustment procedures.

■ Be sure to run the motor without the top cowling installed in order to check for potential fuel or oil leaks. Remedy any leaks before proceeding.

22. Install the top cowling to the powerhead.

■ If the powerhead was rebuilt or replaced with a remanufactured unit, don't forget to follow the proper Break-In procedures.

6-58 POWERHEAD

TIMING BELT/SPROCKET REMOVAL & INSTALLATION

◆ See Figures 128 and 129

This procedure can be performed with the powerhead assembly removed or installed on the motor. Unlike most of the Yamaha produced 4-stroke powerheads, the crankshaft sprocket on the 30/40 hp 3-cylinder model does not utilize its own retaining nut, therefore no special tool is necessary to hold the driveshaft from turning to remove the crankshaft sprocket. Like most other Yamaha and Mercury/Mariners, the camshaft sprocket can accept the same tool used to hold the flywheel (or a sturdy strap wrench).

If the timing belt is to be reused, be sure to mark the direction of rotation on the old belt before removal. Also, if either the cylinder head or cylinder block is not to be disassembled further, be sure to properly align the timing marks before removal. This will significantly ease installation.

■ **Camshaft sprocket removal requires the use of a holder tool which inserts into the 2 holes in the sprocket itself. The same tool that is used on the flywheel can normally be used on the camshaft sprocket.**

1. If the powerhead is still installed and on electric start models, disconnect the negative battery cable for safety.
2. Remove the flywheel cover/Hand-Rewind Starter Assembly for access.
3. Remove the Flywheel and Stator Plate, as detailed in the Ignition and Electrical Systems section.
4. Remove the belt guide plate from the top of the crankshaft pulley. On most models holes in the plate index with protrusions on the top of the crankshaft pulley. Note orientation for installation purposes.
5. If either the cylinder head or cylinder block is not being overhauled, rotate crankshaft sprocket in the normal direction of rotation (clockwise) in order to align the timing marks on the crankshaft and camshaft pulleys with the marks on the powerhead.

■ **The camshaft sprocket has a triangular stamp mark which must align with the triangular mark on the cylinder head when the motor is at No. 1 TDC. Similarly, the crankshaft is positioned with the No. 1 cylinder at TDC when the mark/hole on the crankshaft sprocket is aligned with the engine centerline, facing directly toward the camshaft sprocket.**

6. Check the condition of the timing belt prior to removing it. If the belt appears loose or has excessive play at the midway point, the belt has stretched and **must** be replaced.

■ **Remember, if the timing belt is to be reused, be sure to mark it showing the normal direction of rotation to ensure installation in the proper position.**

7. Slip the timing belt free of the camshaft pulley **first,** and remove it from the crankshaft pulley. If the fit is too tight or you have trouble slipping the belt from the pulley, move to the next step and unbolt the camshaft (driven) pulley.

If necessary to strip the powerhead for rebuild or to replace one or more damaged sprockets, remove the camshaft (driven) or crankshaft (drive) sprocket from the motor as detailed in the following steps:

8. To remove the camshaft pulley, obtain a suitable flywheel holder tool (Yamaha #YB-06139 or Mercury #91-83163M). Place the two arms of the tool over the two holes of the camshaft pulley. Hold the pulley steady with the tool and at the same time loosen, then remove the camshaft sprocket bolt and washer.
9. Lift the camshaft pulley straight up and free of the camshaft. Remove and save the dowel pin from the recess in the camshaft.
10. To remove the crankshaft pulley simply pull is straight upward off the crankshaft and woodruff key. Remove and save the Woodruff key from the recess in the crankshaft.

To Install:

11. Apply just a dab of lubricant to the Woodruff key, and then install the key into the lower cutout in the crankshaft. The lubricant will help hold the key in place.

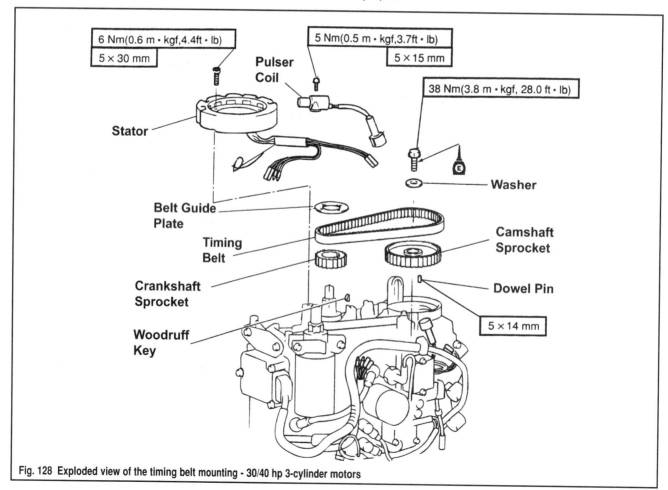

Fig. 128 Exploded view of the timing belt mounting - 30/40 hp 3-cylinder motors

POWERHEAD

12. Install the crankshaft sprocket over the retaining plate and woodruff key. If the crankshaft was not disturbed the sprocket should be positioned so the whole/mark is facing directly toward the camshaft sprocket (along the centerline of the engine). In this position the No. 1 piston is at TDC.
13. Make sure the mark/hole on the crankshaft pulley is facing the camshaft. If necessary, slowly rotate the crankshaft in the normal direction of rotation (clockwise) until the timing marks align. If the camshaft was disturbed, stop turning if you feel any interference.
14. Apply a light coating of clean engine oil to the threads of the camshaft retaining bolt and position aside for use in the next step or two.
15. Insert the dowel pin into the bore in the side, top of the camshaft. Position the camshaft pulley down onto the camshaft aligning it with the dowel pin.

■ Because there is no adjustment for the belt, installation can sometimes be difficult with both pulleys already installed, you may want to install the belt before threading the retaining bolt. If difficulty is encountered, remove the pulley from the camshaft and insert it into the belt, then carefully place tension on the belt as you install the pulley and belt assembly.

16. Hold the camshaft sprocket from turning using the flywheel holder tool, then install and tighten the sprocket retaining bolt to 28 ft. lbs. (38 Nm).
17. If the camshaft was not disturbed, the triangular mark on the pulley should align with the triangular mark on the cylinder head, facing the crankshaft. If not, SLOWLY turn the camshaft in the normal direction of rotation (also clockwise when viewed from above) until the timing marks are appropriately aligned.

■ Remember, if the belt is being reused, be sure to face it in the same direction as normal rotation (clockwise).

18. Place the timing belt over the crankshaft pulley, without disturbing timing mark alignment. After the belt is in place stretch it over the camshaft pulley without moving the pulley.

■ If belt installation is too difficult, remove the camshaft sprocket bolt and reposition the sprocket. With the sprocket properly aligned and the belt fully installed over it, slowly push downward on the belt and sprocket to slide it over the camshaft and dowel pin. Then hold the camshaft sprocket steady while you install and tighten the bolt.

19. Turn the crankshaft clockwise until it completes 2 full revolutions, then recheck the timing marks.
20. Install the belt guide plate to the top of the crankshaft sprocket.
21. Install the Flywheel and Stator Plate assembly, as detailed in the Ignition and Electrical Systems section.
22. Install the flywheel cover/Hand-Rewind Starter Assembly.
23. If removed, install the powerhead assembly.
24. If applicable, reconnect the negative battery cable.

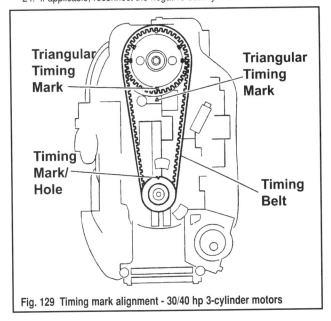

Fig. 129 Timing mark alignment - 30/40 hp 3-cylinder motors

CYLINDER HEAD REMOVAL & INSTALLATION

◆ See Figures 130, 131 and 132

Although it is not absolutely clear in Yamaha literature, it looks likely that the cylinder head can be removed on these models without removing the powerhead assembly. And Mercury certainly thinks so from their literature, though there are some differences in the lower cowlings, it's a good sign for Yamaha models.

Remember, when loosening the retainers on the cylinder head, cover or any other major components, always try to follow the reverse of the torque sequence indicated in the assembly procedure or molded on the component itself.

When applicable tighten all bolts in using the proper torque sequence and to the proper specification.

■ Because of the high temperatures and pressures developed, the sealing surfaces of the cylinder head and the block are the most prone to water leaks. No sealing agent is recommended because it is almost impossible to apply an even coat of sealer. An even coat would be essential to ensure an air/water tight seal.

※※ **WARNING**

Never, never, use automotive type head gasket sealer. The chemicals in the sealer will cause electrolytic action and eat the aluminum faster than you can get to the bank for money to buy a new cylinder block.

1. Remove the Flywheel and Stator Plate, as detailed in the Ignition and Electrical Systems section.
2. Remove the Timing Belt, as detailed in this section. If the cylinder head is to be overhauled follow the necessary steps to remove the camshaft sprocket.
3. Remove any other electrical components which may still be attached to the cylinder head (or at least disconnect the wiring). Remove the spark plugs from the cylinder head.
4. Remove either the Carburetor Assembly or the Throttle Body and Intake Assembly, as equipped. For details, please refer to the Fuel System section.
5. Tag and disconnect the fuel and breather hoses attached to the cylinder head cover.
6. Using multiple passes in the reverse of the torque sequence (an irregular spiraling pattern that starts at the outer and ends at the inner bolts), remove the 7 bolts securing the cylinder head cover to the head. Remove and discard the cover packing (rubber seal).
7. Using multiple passes in the reverse of the torque sequence, loosen and remove the cylinder head bolts. There are 4 short flange bolts to the left of the head and 8 longer bolts found under the valve cover. During each pass, loosen the flange bolts first, starting at the ends and working inward, then do the same for the longer bolts inside the head.

※※ **WARNING**

NEVER pry at gasket sealing surfaces. Such action would very likely damage the sealing surface of an aluminum powerhead.

8. Lift the head up and free of the block. Although Yamaha normally casts pry-tabs on heads, we could not locate any on this model. If the head will not release from the block looks for pry-tabs. If no tabs are provided, use a rubber mallet to gently tap around the perimeter of the surface to help break the gasket seal.

■ Take care not to lose the two dowel pins at the mating surfaces. These pins may remain either in the cylinder head or in the block.

9. Remove and discard the old cylinder head gasket.
10. If necessary for further cylinder head disassembly or for oil pump service, loosen the fasteners and remove the pump from the head. This will be necessary if you plan on removing the camshaft and valve train, but it is a good idea to remove and inspect the pump since you've come this far regardless. For details on removal, installation, inspection and overhaul of the oil pump assembly, please refer to Oil Pump in the portion of the Lubrication and Cooling section.

6-60 POWERHEAD

11. Refer to Cylinder Head Overhaul in this section for details on removal and installation of the valve train components. Also, be sure to check the information on gasket mating surfaces to make sure the cylinder head is not warped.

To Install:

12. If removed, refer to the Lubrication and Cooling section for details on Oil Pump assembly and installation.
13. Position a NEW head gasket over the two dowel pins.
14. Install the cylinder head onto the block with the holes in the head indexed over the two dowel pins.
15. Thread and finger-tighten the 8 long cylinder head bolts which are positioned inside the head along with the 4 short cylinder head flange bolts.
16. Tighten all 12 bolts using multiple passes of the torque sequence (which is essentially a clockwise spiraling pattern that starts at the inner long bolts and works towards the outer long bolts, then proceeds to the flange bolts, starting at the inner 2 bolts and ending at the outer 2 bolts). Tighten the long bolts first to 17 ft. lbs. (22.6 Nm) and then to 34 ft. lbs. (46.1 Nm). Tighten the short flange bolts first to 53 inch lbs./4.4 ft. lbs. (6 Nm) and then to 106 inch lbs./8.7 ft. lbs. (11.8 Nm).

■ If the valve train components were disturbed in any way you are going to have to adjust the valves once the timing belt is installed, so either leave the cylinder head cover off or just finger-tighten the bolts.

17. Apply a light coating of marine grade grease (Yamaha's recommendation) or a light coating of clean engine oil (Mercury's recommendation) to a new valve cover seal (packing) and position it in the cover groove.
18. Install the cylinder head cover and seal/packing. If the valves were not disturbed, install and tighten the 7 cylinder head cover bolts securely either using the molded or illustrated torque sequence (a roughly spiraling pattern that starts counterclockwise at the upper, inner bolt and continues to the other 2 inner bolts, then jumps to the upper left bolt and continues clockwise for the remaining three bolts). Tighten the bolts to 70 inch lbs. (8 Nm).

■ Make sure the new seal remains seated in the cover groove and does not become pinched during installation or there will be oil leaks!

19. Install either the Carburetor Assembly or the Throttle Body and Intake Assembly, as equipped. For details, please refer to the Fuel System section.
20. Reconnect the fuel and breather hoses to the cylinder head cover as tagged during removal.
21. Install the Timing Belt and, if removed, Camshaft Sprocket, as detailed in this section.
22. If the valves were disturbed, adjust them at this time, as detailed under Valve Lash in the Maintenance and Tune-Up section, then install the cylinder head cover using the steps outlined earlier.
23. Install the spark plugs to the cylinder head. Install any electrical components which were stripped from the powerhead during service.
24. Install the Flywheel and Stator Plate, as detailed in the Ignition and Electrical Systems section.
25. If removed, install the powerhead assembly.

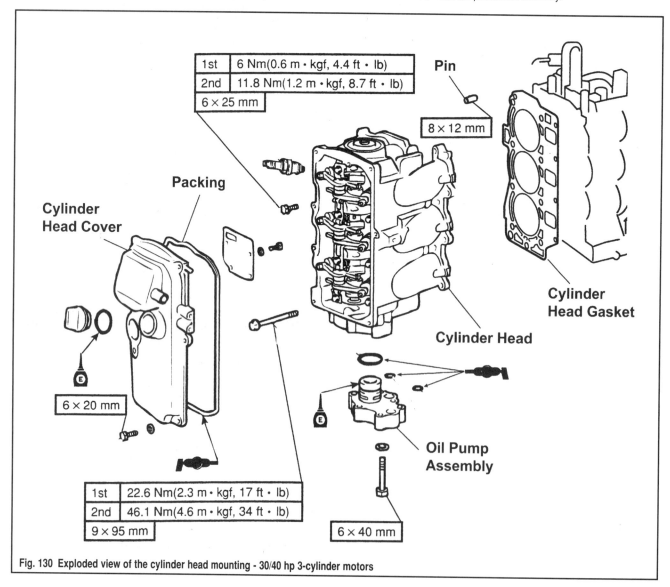

Fig. 130 Exploded view of the cylinder head mounting - 30/40 hp 3-cylinder motors

POWERHEAD 6-61

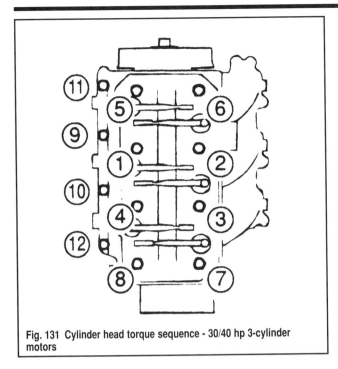

Fig. 131 Cylinder head torque sequence - 30/40 hp 3-cylinder motors

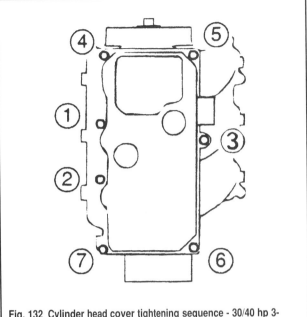

Fig. 132 Cylinder head cover tightening sequence - 30/40 hp 3-cylinder motors

CYLINDER HEAD OVERHAUL

◆ See Figure 133

■ The camshafts used in some versions of the 30/40 hp motors (specifically, manual start models) are equipped with decompression actuators on the camshaft assembly. This doesn't affect overhaul procedures, except that it adds an additional component to be checked for damage or wear. If there are problems with the actuators (if they do not move freely or are otherwise worn/damaged), the camshaft must be replaced.

1. Remove the Cylinder Head from the powerhead assembly, as detailed in this section.
2. Remove the Oil Pump from the cylinder head, as detailed in the Lubrication and Cooling section.

■ When reused, ALL valve train components must be installed in their original locations. For this reason be sure to tag and/or sort all shaft/rocker arm retainers, rocker arms, valves, valve springs, keepers etc and keep them separated during service.

3. Loosen the 4 rocker arm/shaft retainer and tensioner bolts, then remove each of the retainers/tensioners. Note that the retainers are of slightly different shapes and sizes, so mark/note their positioning before removal to ensure they can be installed in their original locations.
4. Remove the rocker shaft and arm assembly from the cylinder head. If necessary, mark and remove each of the rockers from the shaft itself.
5. If necessary, remove the bolt and gasket from the top of the cylinder head.
6. Carefully slide the camshaft from the cylinder head, out through the bottom (oil pump) end of the head.
7. If necessary, remove the oil seal from the top of the cylinder head.
8. To remove the each valve, proceed as follows:
    a. Using a clamp-type valve spring compressor, carefully place just enough pressure on the valve spring so the upper spring seat (spring retainer) moves down sufficiently to expose the keepers.
    b. Remove the pair of valve keepers, then slowly ease pressure on the valve spring until the compressor can be removed.
    c. Remove the upper seat (retainer), valve spring and lower spring seat from the shaft of the valve.

■ Some factory sources conflict on whether or not all the 30/40 Mercury/Mariner models utilize a separate lower spring seat, or if the seat is integral with the cylinder head and/or the valve stem seal. DO NOT remove valve stem seals unless you plan on replacing them.

    d. Remove the valve from the underside (combustion chamber side) of the cylinder head.
    e. If necessary, carefully pry the old valve stem seal from its position on the cylinder head.
    f. Repeat this procedure for each of the remaining valves.
9. Refer to Powerhead Refinishing, later in this section, for details on cleaning and inspecting cylinder head components.

To Assemble:
10. If removed, install each of the valve and spring assemblies as follows:
    a. If removed, install new valve spring seals to the cylinder head.
    b. Apply a light coating of engine oil to the stem, then position the valve into the cylinder head (into the original guide and valve seat unless the valve or valve seat was replaced).
    c. Position the lower spring seat down onto the valve shaft and cylinder head, then oil and install the spring.
    d. Place the upper spring seat (retainer) over the spring, then use the valve spring compressor tool to press downward on the seat and spring so you can install the keepers.
    e. Insert the valve keepers, then slowly back off the valve spring compressor, making sure the keepers are fully in position. Then remove the valve spring compressor.
    f. Repeat for each of the remaining valves.
11. If the camshaft oil seal was removed during disassembly, coat the lips of a new seal with a light coating of clean engine oil. Install the seal into the head using a suitable driver. The seal should be installed with the lips facing downward into the cylinder head.
12. Apply a light coating of engine oil to the camshaft bearing journals, then apply a light coating of a molybdenum disulfide engine assembly lube to the camshaft lobes. Next, **carefully** insert the sprocket end (with the dowel pin hole) of the camshaft into the head from the oil pump end of the head. Continue inserting the camshaft until end of the shaft is seated in the camshaft oil seal against the head at the other end.

■ Do not attempt to rotate the camshaft at this time. At this time the camshaft is supported at just one end. The camshaft should ONLY be rotated after the oil pump has been installed and the shaft is then supported at both ends.

6-62 POWERHEAD

13. Install the bolt and gasket to the top of the cylinder head.

■ **Remember to lightly coat all the rocker arm and shaft components with clean engine oil as they are installed.**

14. If disassembled, apply a light coat of clean engine oil or engine assembly lube to the rocker arms and shaft, then install the rocker arm assemblies to the rocker shaft, in the SAME positions from which they were removed.

15. Carefully lower the rocker arm and shaft assembly to the cylinder head. If the lash adjusters were not loosened, the preload on one or more of the valve springs may prevent you from properly seating the shaft. Determine which adjusters must be loosened (note if the valves were serviced or replaced, you should just go ahead and back off all of them, besides, they will all require checking and adjustment regardless).

16. Apply a light coating of engine oil to the threads of the shaft retaining bolts and to the shaft retainer/tensioner clips themselves.

17. Install the rocker shaft retainers and tensioners to their original positions as noted/marked during disassembly and secure using the retaining bolts. Tighten the each of the shaft retainer bolts to 160 inch lbs./13 ft. lbs. (18 Nm).

18. Install the Oil Pump assembly to the cylinder head, as detailed in the Lubrication and Cooling section.

19. Install the Cylinder Head as detailed in this section.

20. Once the powerhead is fully assembled, if valve train components were replaced, be sure to operate the engine as directed for new component break-in. For details, please refer to Powerhead Break-In, in this section.

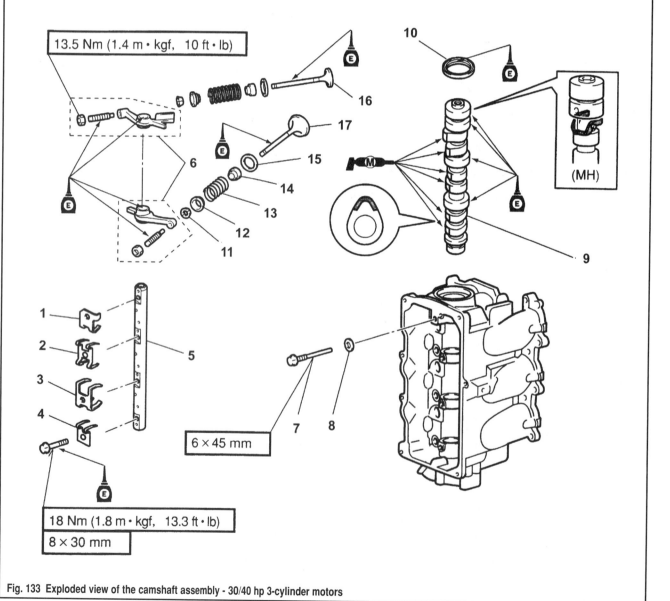

1. Rocker Arm Retainer
2. Tensioner
3. Rocker Arm Retainer
4. Tensioner
5. Rocker Shaft
6. Rocker Arm Assembly
7. Retaining Bolt
8. Gasket
9. Camshaft
10. Oil Seal
11. Valve Keepers
12. Spring Retainer
13. Valve Spring
14. Valve Stem Seal
15. Lower Spring Seat
16. Intake Valve
17. Exhaust Valve

Fig. 133 Exploded view of the camshaft assembly - 30/40 hp 3-cylinder motors

POWERHEAD

CYLINDER BLOCK OVERHAUL

◆ See Figures 122 and 134 thru 140

Remember, when loosening the retainers on manifolds, covers and other major components, always try to follow the reverse of the torque sequence indicated in the assembly procedure or molded on the component itself.

When applicable tighten all bolts in using the proper torque sequence and to the proper specification. When possible we've included it in the procedure, but on many Yamahas both the sequence and value is molded into the casting on critical components. You'll notice that most Yamaha torque sequences are generally clockwise spirals starting somewhere near the center of the component.

1. Remove the Powerhead Assembly, as detailed in this section.
2. Remove the Timing Belt/Sprocket Assembly, as detailed in this section.
3. Remove the Cylinder Head Assembly, as detailed in this section.
4. Remove any electrical components which were not stripped during the previous steps.

■ The exhaust cover should always be removed during a powerhead overhaul. Many times water in the powerhead is caused by a leaking exhaust cover gasket or plate.

5. Although not absolutely necessary, we recommend you strip the block of the following components for inspection and passage cleaning both of which will help ensure a thorough rebuild:
   a. If not already done, remove and discard the old oil filter.
   b. Gradually loosen and remove the exhaust cover retaining bolts using the reverse of the torque pattern (very roughly a counterclockwise spiraling pattern that starts with the upper right bolt and moves toward the center bolts). The thermostat assembly is mounted under a housing to the top of the exhaust cover (2 of the exhaust cover bolts also retain the thermostat housing).
   c. Remove the thermostat assembly from the exhaust cover. For details, please refer to the Lubrication and Cooling section. Remove and discard the old thermostat gasket.

■ The engine temperature switch is installed in the cover itself. If you did not remove it (as per Step 4, stripping the powerhead of electrical components), be sure to unplug the wiring before removing the exhaust cover.

   d. Remove the exhaust cover and gasket from the side of the powerhead. Remove and discard the old exhaust cover gasket.
6. Remove the crankcase retaining bolts using multiple passes in the reverse of the torque sequence (either molded on the crankcase or as illustrated later in this procedure. Essentially the crankcase has 8 long bolts at the center of the crankcase and 8 short bolts along the sides of the flange. Loosen the bolts using a clockwise spiraling pattern that begins with the lower left flange bolt, continues through the rest of the flange bolts, then moves to the lower left of the longer center bolts and finally ends at the short bolts at the center of the case.

■ Remember that since there two different size bolts are used keep them separated for identified for installation purposes. Long bolts to the center, short bolts to the flange.

7. After all bolts have been removed, separate the crankcase from the powerhead. **Take care** not to lose the two dowel pins. These pins may remain with cylinder block or come away with the crankcase cover.

※※ WARNING

If difficulty is encountered, use a rubber mallet to tap around the perimeter of the crankcase cover. Some Yamahas have molded pry tabs to help break the gasket seal, however NEVER pry at gasket sealing surfaces. Such action would very likely damage the sealing surface of an aluminum powerhead.

■ Most parts inside the cylinder block are matched sets and should not be moved from one cylinder to another. This includes the pistons as well as each of the connecting rod/end cap combinations. For this reason, prior to disassembly it is essential to matchmark each of the connecting rods to their end caps (this ensures that not only will each cap be installed on the proper rod, but facing the correct direction).

8. Matchmark the big end caps to the connecting rods (or identify the marks already placed there by the manufacturer), then remove the 6 connecting rod cap bolts. Remove the caps and their bearing inserts, then place them aside for reinstallation back on their original connecting rods as soon as the crankshaft has been removed.

■ When loosening or tightening the connecting rod cap bolts, alternate between each bolt every 1/2-1 turn.

9. Carefully lift the crankshaft from the block. Place it on a protective surface.

※※ WARNING

Handle the crankshaft carefully. Remember it is an expensive and vital piece to the motor. The polished surfaces should be covered to prevent scratches or other damage and the whole shaft should be protected against impact or other damage which could destroy it.

10. Immediately reinstall the connecting rod end caps and bearing inserts to their respective pistons (to prevent loss, mixing them up or potential damage to the bearing surfaces).

■ If the main bearing inserts are fit for further service, keep them with the crankcase, resting on the spot from which they were removed.

11. Slide the 4 sets of metal bearing inserts out of the top and bottom crankcase halves.

※※ WARNING

The crankshaft bearings absolutely MUST be installed back in the same position from which they were removed. These bearings are matched sets - half in each crankcase half (or half in each connecting rod/cap half). If they are to be reused, they must be installed with their matching halves on the same crankshaft bearing surfaces.

12. Remove the top and bottom crankshaft oil seals. Discard the seals.

■ These pistons and connecting rods may or may not be marked from the factory as to what side faces up, identify the marks or add marks of your own to be used during assembly. Also, each piston and connecting rod must be returned to the original bore and crankshaft journal, so be sure to place a number on each to ensure proper positioning.

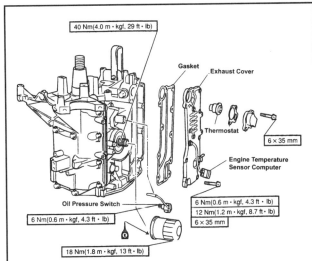

Fig. 134 Exhaust cover, thermostat and oil filter mounting - 30/40 hp 3-cylinder motors

6-64 POWERHEAD

13. Label and remove each of the pistons from the bores. Number the pistons and place marks on the piston/con rod to identify which side faces up (faces the flywheel) for installation purposes. If necessary, disassemble each of the pistons, as follows:

   a. Carefully pry the clips free from the piston pin bore on either side of the piston assembly.

   b. Slide the piston pin free of the bore (specifications tell us that this is NOT an interference fit, so it should slide free with ease).

   c. Using a piston ring spreading tool, carefully remove each of the rings (top and second compression rings, followed by the oil rings) from the piston. Good shop practice dictates the rings be replaced during a powerhead overhaul. However, if the rings are to be used again, expand them **only** enough to clear the piston and the grooves because used rings are brittle and break very easily.

   d. Keep all components from a single piston together for reuse or replacement, but don't mix and match components from other pistons.

14. Refer to Powerhead Refinishing, later in this section, for details on cleaning and inspecting cylinder block components. There are probably more components and specifications to measure these 4-stroke motors than on any 2-stroke Yamahas of similar size. Take your time and make sure you've made measurements for all components listed in the Engine Specifications chart for this model.

To Assemble:

15. Make sure the mating surfaces of the crankcase and cylinder block are completely clean and free of sealant or damage. The dowel pins should already be installed in the block or crankcase cover, but if not, obtain new pins and install them in the positions at opposite ends of the block.

■ If the old main bearings are to be installed for further service, each must be installed in the same location from which it was removed.

16. Place the main bearing inserts into the crankcase and cylinder block. There are normally 2 things to look out for when doing this. Most bearing inserts will be equipped with locating tabs, make sure they are seated in the appropriate grooves. Also, inserts with oil holes should be aligned with an oil passage. If a new set of bearings has been purchased for installation, match the colors on the side of the new bearings with the letters embossed on the bottom side of the crankcase. The marks are stacked on the crankcase, to correspond to the 3 main bearing positions. The top being No. 1, the top middle being No. 2, bottom middle being No. 3 and bottom bearing being No 4.

- Letter **A** matches the Blue bearing.
- Letter **B** matches the Black bearing.
- Letter **C** matches the Brown bearing.

17. If they were disassembled, prepare each piston assembly for installation as follows:

■ Apply a light coat of clean engine oil to each piston component as or right after it is installed.

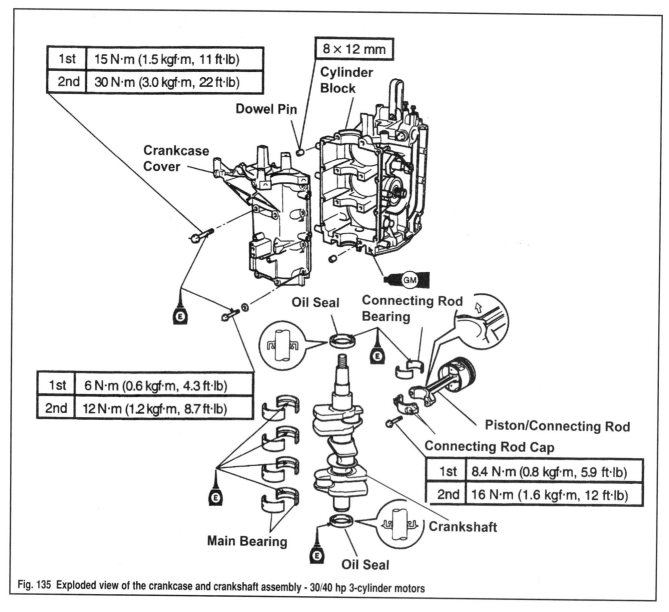

Fig. 135 Exploded view of the crankcase and crankshaft assembly - 30/40 hp 3-cylinder motors

POWERHEAD

a. Using a ring expander, carefully install each of the rings with their gaps as noted here and as shown in the accompanying illustration. The description and illustration are depicted while looking at the piston dome with the top (flywheel end) facing 12 o'clock.

■ UNFORTUNATELY Mercury and Yamaha sometimes have a different philosophy on exactly how to set up the ring gaps prior to piston installation. This does go to reinforce a popular opinion we've often heard from machinists, that since the rings are going to move around in service, it really shouldn't matter. But, we've gone through the trouble of giving you each manufacturer's preferred set-up method.

b. The bottom oil rail ring should be positioned with its gap between 4 and 5 o'clock (as recommended by Yamaha) or at 3 o'clock (as recommended by Mercury).

c. The middle (expander) oil ring should be positioned with its gap between 1 and 2 o'clock (as recommended by Yamaha) or at 6 o'clock (as recommended by Mercury).

d. The top oil rail ring should be positioned with its gap between 10 and 11 o'clock (as recommended by Yamaha) or at 9 o'clock (as recommended by Mercury).

e. The lower (second) compression ring should be positioned with its gap at 3 o'clock (as recommended by Yamaha) or between 1 and 2 o'clock (as recommended by Mercury).

f. The upper (top) compression ring should be positioned with its gap at 9 o'clock (as recommended by Yamaha) or between 10 and 11 o'clock (as recommended by Mercury).

g. Align the flywheel/up mark made on the connecting rod (facing up/flywheel) with the mark on the piston (also facing up/flywheel), install the connecting rod to the piston using the piston pin.

h. Install the piston pin clips in either side of the pin bore to secure the piston pin in place.

i. Repeat for the remaining piston.

✳✳ WARNING

Use extreme care to prevent the connecting rods from contacting and scoring the cylinder walls during piston installation. It is a good idea to wrap the ends in a shop cloth or something else to protect the cylinders.

18. Install each of the pistons (with the correct mark/side facing UP toward the flywheel) to the cylinder block using a ring compressor. Apply a very light coating of clean engine oil to the inside of the compressor and to the cylinder bore, then place the compressor around the piston. Insert the piston through the cylinder head side of the bore and use a wooden hammer handle to gently tap the piston down through the compressor and into the cylinder bore.

✳✳ WARNING

Be VERY careful when installing the piston. Get the lower portion of the skirt below the ring compressor and insert it into the cylinder bore to make sure everything is aligned. Then, GENTLY tap the piston through the compressor, stopping if it seems to hang up even slightly to make sure that a ring is not caught on the deck of the cylinder block.

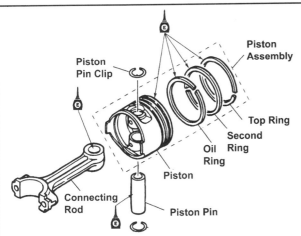

Fig. 136 Exploded view of the piston assembly - 30/40 hp 3-cylinder motors

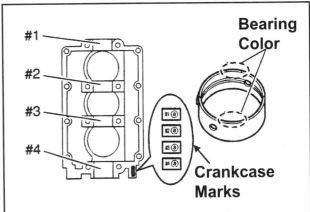

Fig. 137 Crankcase and main bearing insert selection codes/markings

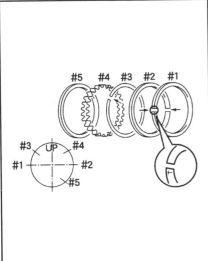

Fig. 138 Piston ring gap positioning - 30/40 hp 3-cylinder motors

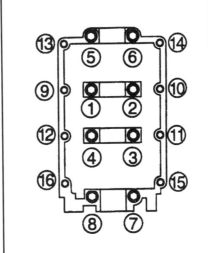

Fig. 139 Crankcase torque sequence - 30/40 hp 3-cylinder motors

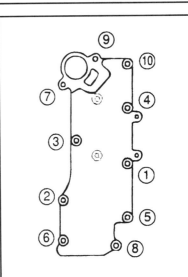

Fig. 140 Exhaust cover torque sequence - 30/40 hp 3-cylinder motors

19. Remove the connecting rod end caps and position them so the crankshaft can be installed. Place the connecting rod bearing inserts into position on the connecting rod big end and cap. Keep in mind that the connecting rods and bearing inserts are also marked with stampings and color codes to help assist in bearing selection. The codes and colors correspond to the other crankshaft codes for this motor (A: blue, B: black. C: brown).

■ Here's where procedures differ slightly depending upon whether you're installing bearings of a known clearance or measuring clearance. If you've already determined that bearing clearances are fine, apply a light coat of engine oil or engine assembly lube to each of the connecting rod, main bearing and crankshaft journal surfaces. If however, you need to measure and determine bearing clearance, Temporarily INSTALL THESE COMPONENTS DRY for measurement purposes. For more details, please refer to Measuring Bearing Clearance Using Plastigage®, in the Powerhead Refinishing section.

■ NEVER attempt to install the upper or lower crankshaft oil seals with the crankcase already installed and tightened otherwise the seal will likely be damaged.

20. Coat the lips of new upper and lower oil seals with a clean engine oil, then slide the oil seals onto the crankshaft with their lips facing toward the center (toward the shaft throws). Seat the seals as best you can at this time. To ensure the seals are not cocked you can install the crankcase and FINGER-TIGTHEN the bolts then lightly knock the seals using a suitably sized driver and rubber mallet. Once this is done, remove the crankcase again and continue the cylinder block assembly.
21. If not done in the previous step to seat the seals, carefully position the crankshaft to the cylinder block, over top the main bearing inserts.
22. Bring up the connecting rod ends around the crankshaft throws. If you've already measured bearing clearances, apply a light coat of engine oil or engine assembly lube to the connecting rod, cap and crankshaft bearing surfaces.
23. Align the matchmarks and reinstall the connecting rod bearing caps. Install the bolts alternately and evenly first to 53 inch lbs./5.9 ft. lbs. (8 Nm) and then to 150 inch lbs/12 ft. lbs. (16 Nm).

■ The first literature published by Yamaha called for connecting rod bolts to be tightened first to the torque spec of 5.9 ft. lbs. (8 Nm) and then tightened an additional 40°. It is not clear why later editions now say a specific torque specification instead of the original angle spec. If you have an angle gauge and would prefer to use that spec, you should be fine. However, if you don't have an angle gauge handy, go ahead and use the 12 ft. lbs. (16 Nm) spec. You'll also notice that we included the torque specification of 150 inch lbs. which REALLY equals 17 Nm, however, it is the spec that Mercury provided so we've included it anyway. In any case, we suspect most torque wrenches are not so precisely adjusted as to matter significantly. The 2 things you are most concerned with is that the bolts are not over-tightened and thereby weakened or broken and that the bolts are tightened uniformly.

24. Confirm that the dowel pins are still in position, then apply a thin bead of Gasket Maker sealant (such as Loctite 514 Master Gasket or equivalent) to the cylinder block side of the crankcase-to-cylinder block mating surfaces.
25. Apply a light coating of engine oil to the threads of the crankcase retaining bolt threads.
26. Carefully install the crankcase cover over the cylinder block using the dowel pins to align the two surfaces. Install and finger-tighten the retaining bolts (remember that there are 2 different size bolts, long bolts go to the center, while short bolts go to the flange).
27. Tighten the crankcase retaining bolts using multiple passes of the clockwise, spiraling torque sequence which starts at the upper, left center of the central long bolts and works outward, then starts again at the upper, left center of the short flange bolts and also works outwards. Tighten the long center bolts first to 11 ft. lbs. (15 Nm) and then to 22 ft. lbs. (30 Nm). Tighten the small flange bolts first to 53 inch lbs./4.3 ft. lbs. (6 Nm) and then to 106 inch lbs./8.7 ft. lbs. (12 Nm).
28. If removed, install the exhaust cover, thermostat assembly and gaskets (one for the exhaust cover and one for the thermostat housing). Tighten the bolts using multiple passes of a roughly clockwise spiraling pattern that starts at the center bolts and works outward. Tighten the bolts first to 53 inch lbs./4.3 ft. lbs. (6 Nm) and then to 106 inch lbs./8.7 ft. lbs. (12 Nm).

29. If removed, install a new oil filter.
30. Install the Cylinder Head Assembly, as detailed in this section.
31. Install the Timing Belt/Sprocket Assembly, as detailed in this section.
32. Install the Powerhead Assembly, as detailed in this section.
33. Install any electrical components which were not installed during the previous steps.
34. Once the powerhead is fully assembled, if cylinder block components were replaced, especially the pistons and/or rings, be sure to operate the engine as directed for new component break-in. For details, please refer to Powerhead Break-In, in this section.

CLEANING & INSPECTING

Cleaning and inspecting internal engine components is virtually the same for all Yamaha and Mercury/Mariner outboards with the exception of a few components that are unique to certain 4-stroke motors. The main variance between different motors comes in specifications (which are listed in the Engine Specifications charts) or by component type. A section detailing the proper procedures, sorted mostly by component, can be found under Powerhead Refinishing.

Four-Cylinder 40/45/50 Hp (935cc) and 30/40/60 Hp (966cc) Powerheads

POWERHEAD REMOVAL & INSTALLATION

Mercury/Mariner Models

◆ See Figures 109, 110 and 141, 142 and 143

The powerhead may be removed or installed as an almost fully assembled unit. The components mentioned here are the bare minimums that must be removed before separating the powerhead assembly from the rest of the motor. However, if overhaul or replacement of the cylinder head and/or block is desired, you may wish to remove additional components before removing the powerhead assembly, not only to lighten the load, but to help prevent potential damage to the external fuel and electrical system components mounted to the powerhead. The choice is yours, so if necessary, refer to other component procedures in this section and in the Fuel System and/or Ignition and Electrical Systems sections before proceeding.

1. Disconnect the negative battery cable at the battery itself for safety.
2. For electric start models, disconnect the battery cables at the engine.
3. If equipped with Power Trim/Tilt, locate and disconnect the wiring at the on the powerhead. For 40/45/50 hp (935cc) models the wiring is found just a little above the latch assembly for the vertical shift-shaft at the front of the motor. On some models you may have to cut a wire tie in order to reposition the wiring sufficiently for access.
4. Remove the lower engine shrouds. For most models this means removing 7 screws (6 short and 1 long), then separating the halves of the housing.
5. For 40/45/50 hp (935cc) models, proceed as follows:
   a. Locate the vertical shift shaft linkage under the front center of the powerhead. Disengage the shift shaft by first pushing the retainer up and then sliding it over toward the starboard side.
   b. Disconnect the vent hose from the fitting at the lower front of the motor (toward the base of the starboard side).
6. Drain the engine oil.
7. For 40/50/60 hp (996cc) models, proceed as follows:
   a. For remote control models, remove the control cable anchor bracket screws (on the lower starboard side of the powerhead, toward the front). Then remove the screw at the pivot point on top of the throttle/shift lever (also on the starboard side of the powerhead, but toward the stern of the motor). With the bolt removed, swing the entire shift linkage to the side.
   b. For tiller models, remove the 2 screws securing the irregular shaped shift guide bracket (at the front, starboard side base of the powerhead).
   c. Locate the crankcase vent hose attached to the top right of the valve cover, then carefully cut the Sta-strap (wire tie) at the fitting (on either end, the choice is yours, though we usually find the lower end of the hose easier in the long run) and disconnect the hose.

POWERHEAD 6-67

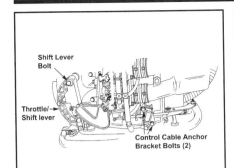

Fig. 141 Disconnecting the throttle/shift linkage - Remote Control 40/50/60 hp (966cc) Mercury/Mariner motors

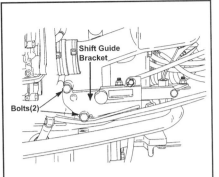

Fig. 142 Disconnecting the shift guide bracket - Tiller Control 40/50/60 hp (966cc) Mercury/Mariner motors

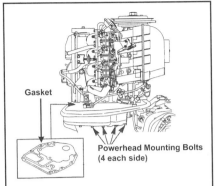

Fig. 143 Powerhead mounting bolts and gasket - 40/45/50 hp (935cc) and 40/50/60 hp (966cc) Mercury/Mariner motors

d. Cut the Sta-strap and disconnect the fuel inlet hose from the fuel pump (check the hose routing or the marking on the pump to be sure, however the inlet hose is usually the bottom hose on these models).

✷✷ CAUTION

The hose fittings on the pump can be particularly fragile and can break if you try twisting or pulling off the hoses. It is usually safest to slowly pry off the hoses using a small screwdriver or pry-tool.

8. Remove the bolts (4 on each side) threaded upward into the powerhead from the intermediate housing. Some illustrations make it look like there are more than 4 bolts in that portion of the cowling, but they point to the first 4 in a line starting from the front of the motor moving back, so if there are more, only loosen the first 4 on each side.

9. Attach a suitable lifting device and carefully lift the powerhead assembly from the intermediate housing. In order to prevent potential damage to the crankshaft and driveshaft be certain to lift the powerhead STRAIGHT up and off the intermediate housing.

10. Remove and discard the upper casing gasket from the powerhead-to-intermediate housing mating surface. Carefully clean the mating surfaces of all traces of old gasket.

11. Check for placement of any dowel pins, if used. Inspect the pins for damage and replace, if necessary.

12. Position the powerhead on a suitable workbench for further disassembly or overhaul, as necessary. Refer to the additional procedures in this section as needed.

To Install:

13. With any dowel pins utilized in place, position a NEW upper casing gasket.

14. Using a suitable lifting device, carefully lift the powerhead and lower it slowly, STRAIGHT onto the driveshaft and intermediate housing. If necessary, rotate the flywheel slightly in a clockwise direction in order to mate the crankshaft to the driveshaft.

■ Although Mercury does not mention it, we like the thought of coating the powerhead retaining bolts with a medium strength threadlocking compound, both for security AND to act as a barrier against corrosion.

15. Install the 8 as applicable powerhead retaining bolts and tighten to 28 ft. lbs. (38 Nm) for 40/45/50 hp (935cc) motors or to 22 ft. lbs. (48 Nm) for 40/50/60 hp (966cc) motors.

16. For 40/50/60 hp (996cc) models, proceed as follows:
   a. Reconnect the fuel inlet hose to the fuel pump and secure using a new wire tie (or with the clamp, as applicable).
   b. For tiller models, secure the shift guide bracket to the front, starboard side, base of the powerhead using the 2 retaining screws.
   c. For remote control models, swing the throttle/shift lever back into position and secure using the bolt at the top, pivot point. Tighten the bolt to 100 inch lbs. (11 Nm). Then install the cable anchor bracket screws and tighten to 75 inch lbs. (8.5 Nm).

17. Reconnect the crankcase vent hose and secure using a new wire tie.
18. Install the lower engine shrouds.
19. For 40/45/50 hp (935cc) models reconnect secure the vertical shift shaft with the latch (slide the retainer toward the port side, then rotate it downward to lock it in position).
20. If equipped, reconnect the PTT wiring.
21. Refill the engine crankcase.
22. For electric start models, reconnect the battery cables at the powerhead, then reconnect battery cable negative battery cable to the battery itself.
23. If the powerhead was overhauled (replacing components such as the valve train, main bearings, crankshaft, pistons and/or rings) be sure to subject the motor to a complete new engine break-in period to ensure long, trouble free life.

Yamaha Models

◆ See Figure 144

The powerhead may be removed or installed as an almost fully assembled unit. The components mentioned here are the bare minimums that must be removed before separating the powerhead assembly from the rest of the motor. However, if overhaul or replacement of the cylinder head and/or block is desired, you may wish to remove additional components before removing the powerhead assembly, not only to lighten the load, but to help prevent potential damage to the external fuel and electrical system components mounted to the powerhead. The choice is yours, so if necessary, refer to other component procedures in this section and in the Fuel System and/or Ignition and Electrical Systems sections before proceeding.

Throughout the procedure we will refer to the accompanying diagrams by numbering the referenced components. Note that the wiring harness and electrical components will vary slightly by model (tiller or remote,

1. Disconnect the negative battery cable at the battery itself for safety.
2. If applicable, disconnect the main switch lead coupler (1) usually found just below the electric cranking starter motor.
3. For ET, EHD and EHT models (as applicable) disconnect the extension lead/warning lamp coupler (2) and trailer switch lead coupler (3) both located toward the front of the powerhead, below the regulator.
4. For models equipped with Power Trim/Tilt (PTT), such as the ET and EHT models, remove the 2 nuts (4) illustrated securing the 2 PTT motor leads (5) to the relay. For 1999 and later models, disengage the 2 bullet connectors for the PTT relay.
5. For all models, remove the bolt (6) securing the battery ground cable (7) to the upper end of the starter motor. Next remove the bolt (8) securing the positive battery cable (9) to the starter solenoid.
6. For 1999 and later models, remove the bolt securing the ground cable to the front of the powerhead, just a little below the voltage regulator.
7. On some models, including all 1999 and alter, you should remove the PTT relay from the front of the powerhead, then remove the shift rod and shift rod bolts.
8. At the front, lower starboard side of the motor, tag and disconnect the shift and throttle cables. The cable bracket shape varies slightly by model/year, however the cables are removed in essentially the same manner. Remove the shift cable clip (10). Next remove the usually 2 bolts (11) and the bracket (12), followed by the throttle cable clip (13).

6-68 POWERHEAD

9. Slide the fuel hose clip (14) back over the raised portion of the fuel filter nipple, then disconnect the fuel inlet line (15) from the filter assembly.
10. Disconnect the pilot water (cooling system indicator) hose (16). On models equipped with a flushing device, disconnect the flushing hose.
11. If applicable, disconnect the throttle link rod.
12. Remove the engine oil dipstick (17).
13. Remove the 4 screws (18) securing the apron (19) around the underside of the powerhead/engine cowling. Usually 2 of screws are threaded horizontally at the front of cowling and are secured using nuts and washers, while the other 2 are threaded vertically downward into the top of the apron from the top side of the lower cowling. Keep track of all nuts and washers for installation purposes. Remove the apron from the motor for access to the powerhead mounting bolts.

14. Remove the bolts (20) threaded upward into the powerhead from the intermediate housing. On some early models there are only 4 bolts (2 per side), however most models (including all 1999 and later) are equipped with 8 powerhead mounting bolts.
15. Attach a suitable lifting device and carefully lift the powerhead assembly (21) from the intermediate housing. In order to prevent potential damage to the crankshaft and driveshaft be certain to lift the powerhead STRAIGHT up and off the intermediate housing.
16. Remove and discard the upper casing gasket (22) from the powerhead-to-intermediate housing mating surface. Carefully clean the mating surfaces of all traces of old gasket.
17. Check for placement of the dowel pins (23). Inspect the pins for damage and replace, if necessary.

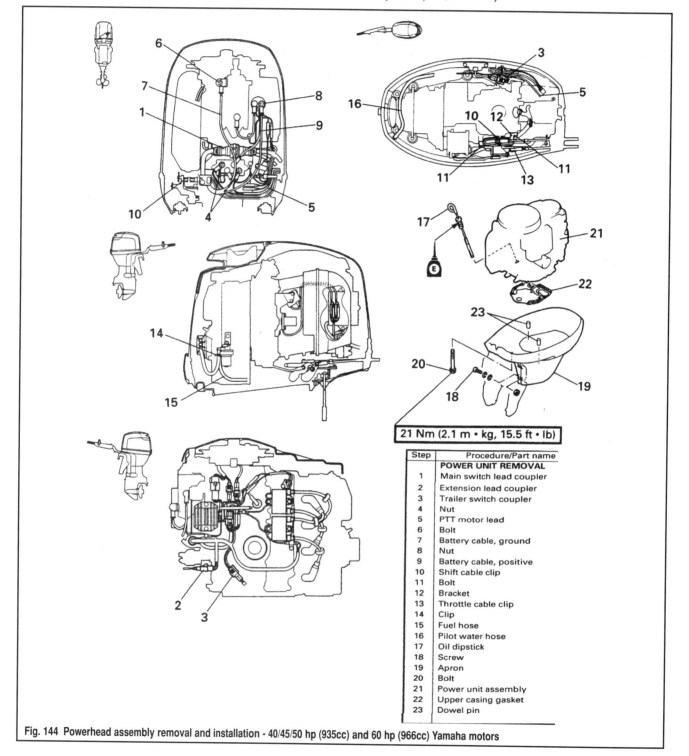

Step	Procedure/Part name
	POWER UNIT REMOVAL
1	Main switch lead coupler
2	Extension lead coupler
3	Trailer switch coupler
4	Nut
5	PTT motor lead
6	Bolt
7	Battery cable, ground
8	Nut
9	Battery cable, positive
10	Shift cable clip
11	Bolt
12	Bracket
13	Throttle cable clip
14	Clip
15	Fuel hose
16	Pilot water hose
17	Oil dipstick
18	Screw
19	Apron
20	Bolt
21	Power unit assembly
22	Upper casing gasket
23	Dowel pin

Fig. 144 Powerhead assembly removal and installation - 40/45/50 hp (935cc) and 60 hp (966cc) Yamaha motors

POWERHEAD 6-69

18. Position the powerhead on a suitable workbench for further disassembly or overhaul, as necessary. Refer to the additional procedures in this section as needed.

To Install:

19. With the dowel pins (23) in place, position a NEW upper casing gasket (22).
20. Using a suitable lifting device, carefully lift the powerhead and lower it slowly, STRAIGHT onto the driveshaft and intermediate housing. If necessary, rotate the flywheel slightly in a clockwise direction in order to mate the crankshaft to the driveshaft.
21. Apply a light coating of Loctite® 572 to the threads of the powerhead mounting bolts.
22. Install the 4 or 8 (as applicable) powerhead retaining bolts (20) and tighten to 15.5 ft. lbs. (21 Nm).
23. Install the apron (19) to the engine cowling and secure using the 4 screws (along with the nuts and washers on the horizontal screws at the front of the apron). Tighten the screws securely, but do not overtighten and damage the cowling or apron.
24. Install the engine oil dipstick.
25. If applicable, reconnect the throttle link rod.
26. Reconnect the pilot water hose and, if equipped, the flushing hose.
27. Reconnect the fuel inlet line to the fuel filter and secure using the clip.
28. Reconnect the throttle and shift cables. For most models you should first reconnect the throttle cable clip (13), then install the bracket (12) and tighten the retaining bolts (usually 2), then install the shift cable clip (10).
29. If removed, secure the shift rod using the retaining bolts, then install the PTT relay to the front of the powerhead.
30. For 1999 and later models, install the ground cable to the front of the powerhead, below the voltage regulator, and secure using the retaining bolt.
31. Reconnect the positive battery cable (9) to the starter solenoid, followed by the negative battery (ground) cable (7) to the upper end of the starter.
32. If equipped with PTT, connect the 2 PTT motors leads to the relay terminals, tighten the 2 nuts (4). If removed, reconnect the PTT relay bullet connectors.
33. As applicable, reconnect the trailer switch coupler (3) and/or the extension lead/warning lamp coupler (2).
34. If applicable, connect the main switch lead coupler.
35. Reconnect the negative battery cable to the battery itself.
36. If the powerhead was overhauled (replacing components such as the valve train, main bearings, crankshaft, pistons and/or rings) be sure to subject the motor to a complete new engine break-in period to ensure long, trouble free life.

TIMING BELT/SPROCKET REMOVAL & INSTALLATION

◆ See Figures 145 thru 148

This procedure can be performed with the powerhead assembly removed or installed on the motor, but if the crankshaft sprocket must be removed with the powerhead installed (on Yamaha models or on Mercury/Mariner models that utilize a crankshaft nut) you'll have to lock the driveshaft and propeller shaft from spinning in order to loosen the crankshaft nut. And since

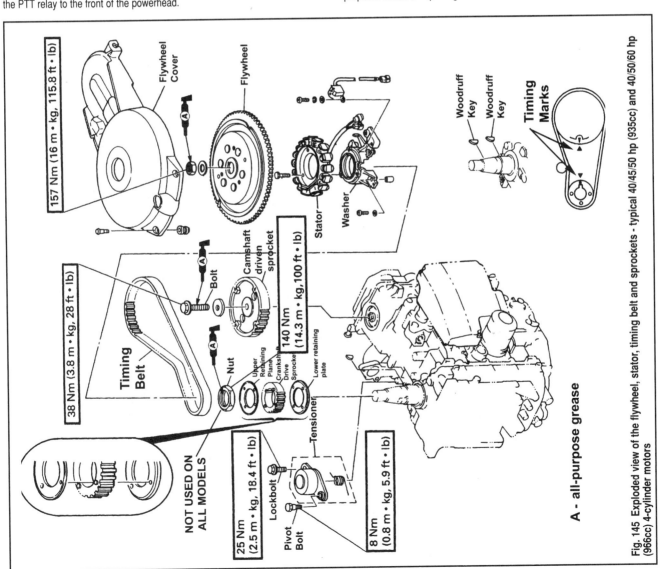

Fig. 145 Exploded view of the flywheel, stator, timing belt and sprockets - typical 40/45/50 hp (935cc) and 40/50/60 hp (966cc) 4-cylinder motors

6-70 POWERHEAD

we're talking about a significant amount of torque, we'd be worried about the potential for damage to the shafts/splines.

■ Some Mercury/Mariner models use a crankshaft drive sprocket assembly which is secured under the flywheel by a load ring and spacer. The good news is that, when equipped, there is no separate nut so removal and installation is easy. The bad news is that the load ring should be replaced each time it is removed (not a bad trade off).

If the timing belt is to be reused, be sure to mark the direction of rotation on the old belt before removal. Also, if either the cylinder head or cylinder block is not to be disassembled further, be sure to properly align the timing marks before removal. This will significantly ease installation.

■ Camshaft sprocket removal requires the use of a holder tool which inserts into the 2 holes in the sprocket itself. The same tool that is used on the flywheel can normally be used on the camshaft sprocket. However, on models with a crankshaft sprocket nut, removal requires the use of a crankshaft holding tool which is inserted into the driveshaft end of the crankshaft splines. This tool can be fabricated from a discarded driveshaft (if you can find one from a gearcase rebuilder or marine bone yard). If the powerhead is still installed you can usually place it in gear and lock the propshaft from spinning, but you do risk some damage to the prop as we're talking about 100 ft. lbs. (140 Nm) of torque. Also, crankshaft sprocket removal will require the use of a LARGE 42mm socket which is at least 76mm deep (that load ring and spacer used on some Mercury/Mariner models is sounding better and better, eh?).

1. If the powerhead is still installed, disconnect the negative battery cable for safety.
2. Remove the Flywheel and Stator Plate, as detailed in the Ignition and Electrical Systems section.
3. If either the cylinder head or cylinder block is not being overhauled, rotate crankshaft sprocket in the normal direction of rotation (clockwise) in order to align the timing marks on the crankshaft and camshaft pulleys with the marks on the powerhead.
4. Remove the 8 x 20mm lockbolt from the camshaft side of the timing belt tensioner, allowing the tensioner to pivot and loosen the timing belt.

■ Remember, if the timing belt is to be reused, be sure to mark it showing the normal direction of rotation to ensure installation in the proper position.

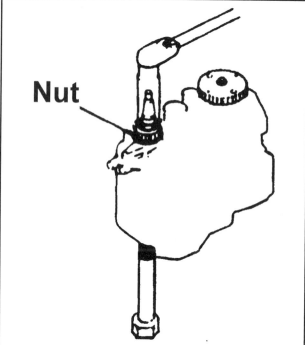

Fig. 146 When equipped with a crankshaft sprocket nut, if the powerhead is removed, the crankshaft must be locked from the driveshaft spline end in order to remove or install the nut

5. Remove the timing belt from the sprockets.
6. If necessary for replacement or to strip the powerhead for overhaul, remove the pivot bolt (usually a 6mm bolt) from the crankshaft side of the timing belt tensioner, then remove the tensioner from the powerhead.
7. Using a flywheel holder tool inserted into the 2 holes in the camshaft sprocket, lock the sprocket from turning and remove the retaining bolt. Then remove the washer and camshaft sprocket.

■ A Yamaha special tool (#YW-06355 for models through 1999 or #YW-06562 for 2000 and later models) can be inserted into the driveshaft end of the crankshaft and held in order to keep the crankshaft from turning when removing or installing the crankshaft sprocket nut. A 42mm DEEP socket (at least 76mm deep!) is necessary for the crankshaft nut itself. Mercury doesn't mention a special tool, but then again, many of Mercury/Mariner models (though not all) are equipped with a load ring instead of a crankshaft nut.0

8. If equipped with a crankshaft nut, lock the crankshaft from spinning using the propeller shaft (if the powerhead is installed) or the crankshaft-to-driveshaft splines (if removed), then loosen and remove the crankshaft nut.
9. Remove the upper retaining plate, load ring with spacer (some Mercury/Mariner models), crankshaft sprocket, crankshaft woodruff key and lower retaining plate from the top of the crankshaft.

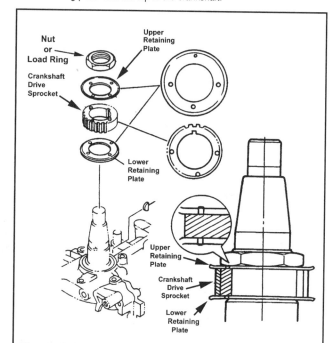

Fig. 147 Component damage and camshaft timing problems can result from a failure to properly align the sprocket with the holes in the timing belt upper and lower retaining plates

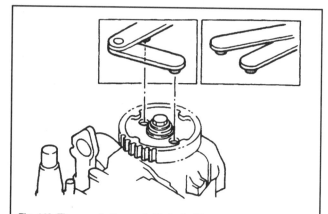

Fig. 148 The camshaft sprocket is locked into position using the flywheel holder tool

POWERHEAD 6-71

To Install:

10. Install the sprocket to the camshaft. If the camshaft position was not disturbed, make sure the number 1 on the sprocket aligns with the triangular mark on the cylinder head.
11. Apply a light coating of marine grade grease to the threads of the camshaft sprocket retaining bolt, then install the bolt and washer and tighten to 28 ft. lbs. (38 Nm).
12. Position the timing belt lower retaining plate over the crankshaft with the beveled edge turned facing downward toward the powerhead.
13. Using a small dab of grease to retain it, position the crankshaft sprocket woodruff key in the crankshaft.
14. Install the crankshaft sprocket over the retaining plate and woodruff key. Be sure to align the projections on the side of the sprocket with the holes in the retaining plate. Also, be sure to align the timing mark hole on the sprocket with the triangle on the cylinder block.
15. Install the belt upper retaining plate over the crankshaft sprocket with the beveled edge facing upward, away from the powerhead. Be sure to align the projections in the side of the sprocket with the holes in the retaining plate.

■ Failure to align the sprocket projections with the holes in the retaining plates can cause misalignment which will not only potentially damage the components but also cause difficulty in properly timing the camshaft.

16. If equipped, install the crankshaft sprocket retaining nut and tighten to 100 ft. lbs. (140 Nm).
17. For Mercury/Mariner models that use a load ring instead of a crankshaft nut, position the NEW load ring and then spacer on top of the crankshaft sprocket.
18. If removed, install the timing belt tensioner using the pivot bolt and tighten the bolt to 5.9 ft. lbs. (8 Nm).
19. Install the timing belt, making sure the timing marks are still aligned. Remember, if the belt is being reused, be sure to face it in the same direction as normal rotation (clockwise).
20. Turn the crankshaft clockwise until it completes 2 full revolutions, then recheck the timing marks.
21. Install the timing belt tensioner lockbolt and tighten to 18.4 ft. lbs. (25 Nm).
22. Install the Flywheel and Stator Plate assembly, as detailed in the Ignition and Electrical Systems section.
23. If removed, install the powerhead assembly.

STRIPPING THE POWERHEAD FOR REBUILD

Mercury/Mariner Models

◆ See Figure 149

Stripping the powerhead for rebuild is essentially a compilation of procedures covered elsewhere. When necessary refer to the appropriate procedures, mostly found in the Fuel System and/or Ignition and Electrical System sections.

1. Remove the Powerhead from the outboard and place it on a workbench or engine stand.
2. Remove the Flywheel and Ignition Coils.
3. Remove the Timing Belt and, as necessary, Sprockets.
4. Remove the Air Intake Silencer Assembly.
5. For carbureted models, remove the entire Carburetor Assembly.
6. For EFI models, remove the Intake Assembly, then remove the Vapor Separator assembly.
7. For 40/45/50 hp (935cc) motors, proceed as follows:
   a. Remove the control cable anchor bracket.
   b. Remove the shift and throttle lever.
   c. Remove the cylinder head.
   d. Remove the 4 bolts securing the starter bracket.
   e. Cut the Sta-straps (wire ties) and lift the starter bracket along with the electrical components and wiring harness from the engine. Tagging and disconnecting wiring, as necessary.
8. For 40/45/50 hp (935cc) motors, proceed as follows:
   a. Remove the 3 ECM bracket bolts.
   b. Remove the main harness ground wire bolts (near the bottom of the starter).

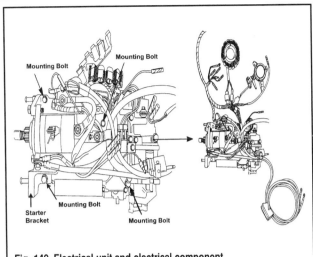

Fig. 149 Electrical unit and electrical component removal/installation - 40/45/50 hp (935cc) Mercury/Mariner motors

   c. Remove the starter solenoid mounting bolts.
   d. Remove the oil pressure switch, Engine Cooling Temperature (ECT) sensor and Crankshaft Position Sensor (CPS).
   e. Remove the regulator/rectifier mounting screws.
   f. Remove the ignition coil mounting bolts.
   g. Remove the electrical components as an assembly, tagging and disconnecting wiring and cutting sta-straps (wire ties), as necessary.

To Install:

9. Reinstall the electrical component assemblies, reconnecting the wiring as tagged and securing wiring with new wire ties. Secure the components using the mounting bolts loosened or removed during removal.
10. For EFI models, install the Vapor Separator assembly, then install the Intake Assembly.
11. For carbureted models, install the entire Carburetor Assembly.
12. Install the Air Intake Silencer Assembly.
13. Install the Timing Belt and, as necessary, Sprockets.
14. Install the Flywheel and Ignition Coils.
15. Install the Powerhead assembly as detailed earlier in this section.

Yamaha Models - Electrical Unit Removal & Installation

The electrical components on the powerhead must be removed for overhaul. A few of those components (including the starter motor, starter relay, regulator/rectifier, CDI unit and possibly even the PTT relay on some models) are attached to a common electrical unit bracket which can be removed as an assembly (if desired) and disassembled as necessary. The components on this bracket can normally all be separated from the bracket assembly while the bracket is still installed on the powerhead.

Component and, more likely, wiring connector locations will vary slightly by model and more so, by year. Not all components are used by all models.

Models Through 1998

◆ See Figures 150 and 151

To strip the powerhead of electrical components (other than the stator assembly which is removed before overhaul with the flywheel and timing belt/sprockets), proceed as follows:

Throughout the procedure we will refer to the accompanying diagrams by numbering the referenced components. Refer to the diagrams for component location and identification.

■ It is a REALLY good idea to tag all electrical connections before they are removed. Trust us, you THINK you'll remember, but then you'll stand there scratching your head saying I SHOULDA' TAGGED those wires.

1. If the powerhead is still installed, disconnect the negative battery cable for safety.
2. Disconnect the pulser coil coupler (1), charge coil coupler (2) and lighting coil coupler (3).
3. Disconnect the CDI unit coupler (4), followed by the 2 regulator/rectifier electrical connectors (5).

6-72 POWERHEAD

4. Disconnect the 2 thermo switch wiring harnesses (6), then just below that wiring, disconnect the lead from the oil pressure switch (7).
5. Disengage the 2 ignition coil connectors (8).
6. On ET and EHT models, disengage the 2 Power Trim Tilt (PTT) connectors (9).
7. Disengage the fuse connector (10), followed on the EHT models by the 2 neutral safety switch connectors.
8. Disengage the 2 choke solenoid connectors (12) and the 2 electro-thermal ram connectors (13).
9. Disconnect the 4 wiring harness grounds (14).
10. Remove the nut (15) above the starter solenoid and the 3 bolts (16) securing the electrical unit assembly.
11. Remove the electrical unit assembly (17).
12. If necessary, remove the 4 bolts (18) securing the ignition coils (19), then remove the 2 coils from the powerhead.
13. Remove the bolt (20) and the retainer (21), then remove the overheat switch (22).
14. Remove the oil pressure switch (23).
15. If necessary, remove the some or all of the components from the electrical unit, as follows:
    a. Remove the 2 bolts threaded from the top of the starter assembly, then remove the motor from the bracket.
    b. Remove the 2 bolts securing the CDI unit, then remove the CDI unit.
    c. Remove the 2 bolts securing the regulator/rectifier, then remove the regulator from the bracket.
    d. Remove the bolt, bracket and socket mounting the starter solenoid relay to the electrical bracket assembly. Remove the starter relay.
    e. Remove the 2 bolts securing the PTT relay, then remove the relay from the bracket.
    f. Installation of these components to the bracket is essentially the reverse of removal. Tighten each of the fasteners securely, however torque specifications are provided for the CDI unit bolts which should be tightened to 5.2 ft. lbs. (7 Nm). The 2 bolts at the top of the starter motor should be tightened to 21.8 ft. lbs. (29.5 Nm).

To Install:

16. If disassembled, secure the components to the electrical bracket assembly.

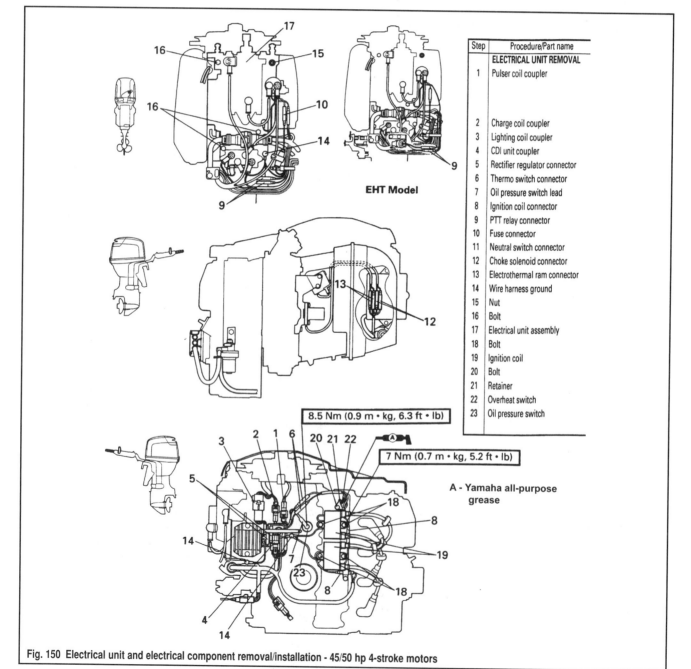

Fig. 150 Electrical unit and electrical component removal/installation - 45/50 hp 4-stroke motors

POWERHEAD 6-73

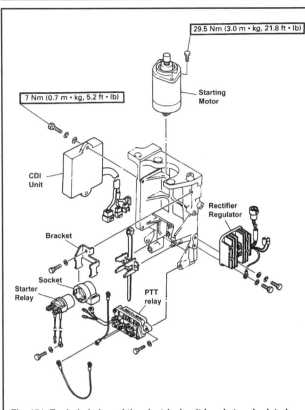

Fig. 151 Exploded view of the electrical unit bracket and related components - 45/50 hp 4-stroke motors

17. If removed, install the oil pressure switch.
18. Install the overheat switch and secure using the retainer and bolt.
19. If removed, install the ignition coils and tighten the retaining bolts to 5.2 ft. lbs. (7 Nm).
20. Position the assembled electrical unit bracket to the powerhead and secure using the 3 bolts and one nut. Tighten the fasteners securely.
21. Connect the 4 wiring harness grounds (14).
22. Engage the 2 harness connectors for the electrothermal ram and the 2 for the choke solenoid.
23. On EHT models, reconnect the 2 neutral switch wiring connectors.
24. Engage the fuse connector.
25. On ET and EHT models, engage the 2 PTT relay connectors.
26. Engage the 2 ignition coil connectors.
27. Reconnect the wiring to the oil pressure switch.
28. Engage the 2 thermo switch connectors.
29. Engage the 2 regulator/rectifier connectors.
30. Engage the CDI, lighting coil, charge coil and pulser coil couplers.
31. If the powerhead is installed, reconnect the negative battery cable.

1999 and Later Models

◆ See Figures 152, 153 and 154

Use the accompanying illustrations for guidance to strip the powerhead of electrical components (other than the stator assembly which is removed before overhaul with the flywheel and timing belt/sprockets). The illustrations can also be useful when installing components after overhaul or when replacing a powerhead with a short block. For more details on the major electrical components (such as the Starter, Starter Solenoid, Voltage Regulator, etc), please refer to the Ignition and Electrical Systems section.

■ ALWAYS tag wire connections before removal, for both ease of installation and to prevent potentially damaging mix-ups. You can use a Sharpie® or similar permanent marker on the harness or connector or apply tape/tags to both sides of each connection.

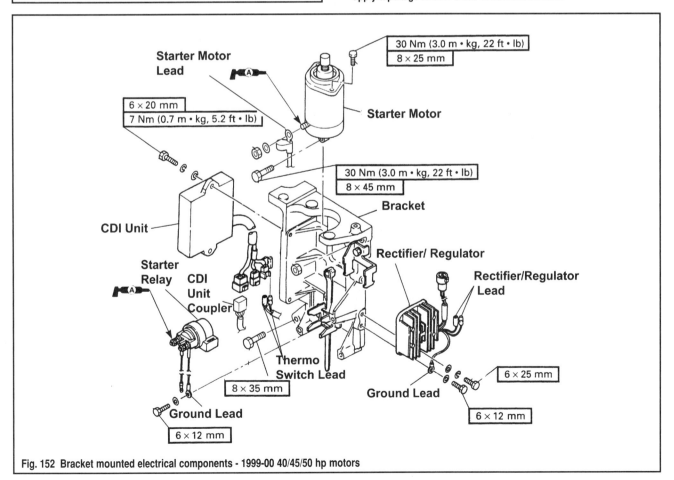

Fig. 152 Bracket mounted electrical components - 1999-00 40/45/50 hp motors

6-74 POWERHEAD

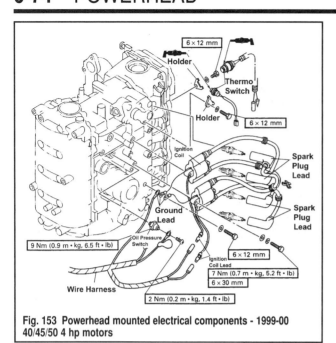

Fig. 153 Powerhead mounted electrical components - 1999-00 40/45/50 4 hp motors

CYLINDER HEAD REMOVAL & INSTALLATION

◆ See Figures 155, 156 and 157

As usual, it is not entirely clear from Yamaha literature whether or not the cylinder head can be removed with the powerhead assembly still installed, but it looks likely. And, as usual, Mercury has been kind enough to point out that it is possible on their motors. We cannot be 100% sure if that is because of differences in cowling, but we suspect that you can remove the Yamaha cylinder heads just as easily.

Of course, in order to get to that point you've got to do so much of the work, it might be wise to just go ahead and remove the powerhead anyway (especially if work conditions would improve by placing everything on the bench while working).

Remember, when loosening the retainers on manifolds, covers and other major components, always try to follow the reverse of the torque sequence indicated in the assembly procedure or molded on the component itself.

When applicable tighten all bolts in using the proper torque sequence and to the proper specification. When possible we've included it in the procedure, but on most Yamahas manufactured powerheads both the sequence and value is molded into the casting on critical components. You'll notice that most Yamaha torque sequences are clockwise spirals starting somewhere near the center of the component.

1. Remove the Flywheel and Stator Plate, as detailed in the Ignition and Electrical Systems section.

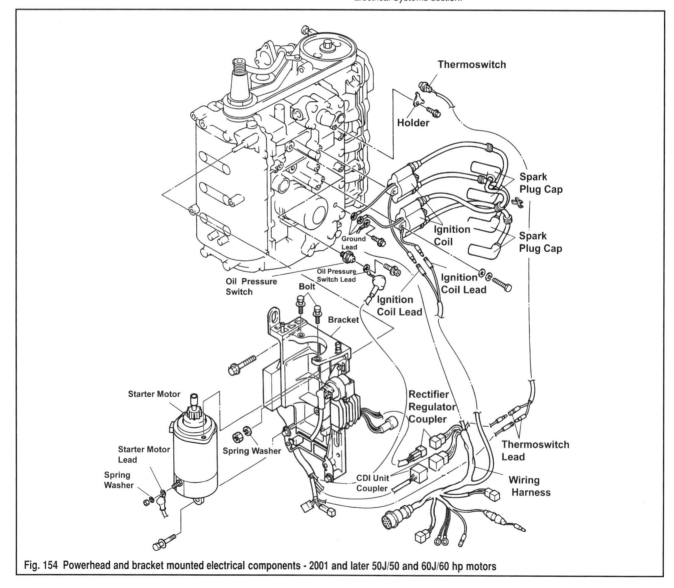

Fig. 154 Powerhead and bracket mounted electrical components - 2001 and later 50J/50 and 60J/60 hp motors

POWERHEAD 6-75

2. Remove the Timing Belt, as detailed in this section. If the cylinder head is to be overhauled follow the necessary steps to remove the camshaft sprocket.

■ Before removing the Timing Belt, align the timing marks and note the camshaft dowel position when the marks are aligned. In this way you'll be able to set the camshaft up properly before cylinder head installation.

3. Strip the necessary electrical components from the powerhead. If the cylinder block is also being serviced, remove the Electrical Unit bracket, as detailed in this section.
4. Remove the spark plugs from the cylinder head.
5. If necessary on models through 1998, remove the 2 bolts securing the dash pot bracket, then remove the dash pot and bracket assembly from the valve cover. Do not disturb the dash pot itself if you wish to preserve the adjustment.
6. Remove the 7 bolts from the cylinder head cover using the reverse of the molded or illustrated torque sequence (this results in a counterclockwise spiraling pattern that begins at the bottom left of the cover and works inward).

■ There are usually a number of components which must be removed in order to access the cylinder head cover, including but not limited to the fuel/breather hoses, the fuel pump and/or ignition coil bracket assemblies. For more details, please refer to the Valve Adjustment procedures in the Maintenance and Tune-Up section.

7. Remove the cylinder head cover, then remove and discard the O-ring seal.
8. If the carburetors or EFI Throttle body assembly are still installed, loosen and remove the intake manifold-to-carb or TBI assembly retaining bolts using the reverse of the torque sequence (since the torque sequence is normally a back and forth pattern starting at the center bolts and working outward, the reverse would start at the ends and work inward).
9. Remove the 5 short (6 x 25mm) bolts from the base flange of the cylinder head using the reverse of the molded torque sequence (or if not present using a back and forth pattern, working from the bottom and top bolts, inward).
10. Remove the long (9 x 95mm) cylinder head bolts from underneath the cylinder head cover also using the reverse of the molded torque sequence (or if not present using a clockwise spiraling pattern that begins at the bottom right bolts and works inward). There are normally 8 head bolts on Yamaha models through 1998 and 10 head bolts on all Mercury/Marine models and on 1999 and later Yamaha models.
11. Carefully break the gasket seal and remove the cylinder head from the cylinder block.
12. Remove and discard the old cylinder head gasket. Carefully remove all traces of the gasket material from the mating surfaces.
13. Check the condition of the cylinder head dowel pins and replace, if necessary.
14. If necessary for further cylinder head disassembly or for oil pump service, refer to Oil Pump in the portion of the Lubrication and Cooling section for details on removal, installation, inspection and overhaul of the oil pump assembly.

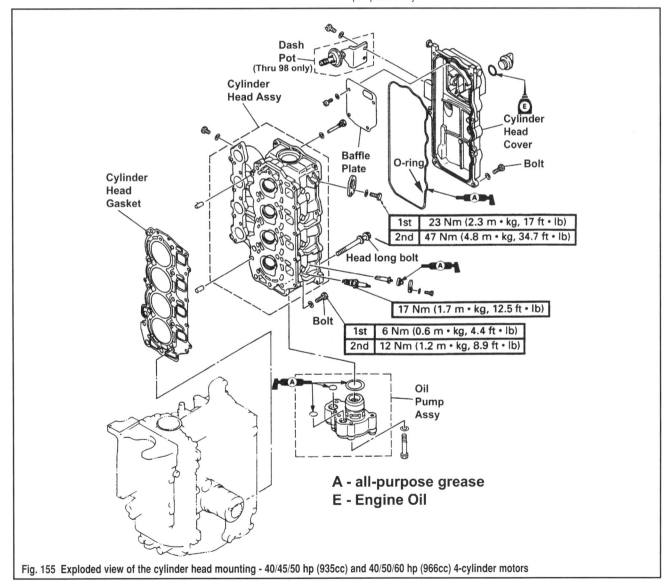

Fig. 155 Exploded view of the cylinder head mounting - 40/45/50 hp (935cc) and 40/50/60 hp (966cc) 4-cylinder motors

6-76 POWERHEAD

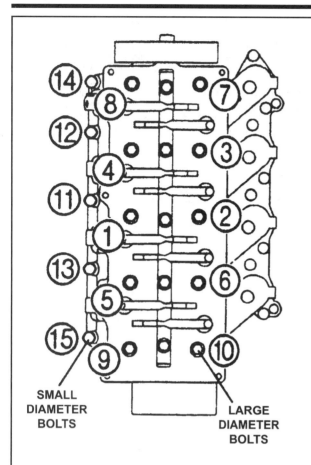

SMALL DIAMETER BOLTS

LARGE DIAMETER BOLTS

Fig. 156 Cylinder head torque sequence - All Mercury/Mariners, as well as 1999 and later Yamaha models shown (earlier Yamahas similar, with fewer larger head bolts)

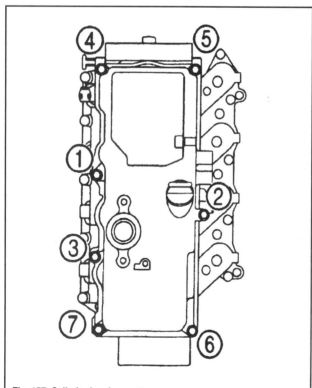

Fig. 157 Cylinder head cover torque sequence

15. Refer to Cylinder Head Overhaul in this section for details on removal and installation of the valve train components. Also, be sure to check the information on gasket mating surfaces to make sure the cylinder head is not warped.

To Install:

16. If removed, refer to the Lubrication and Cooling section for details on Oil Pump assembly and installation.

17. For 40/50/60 hp (966cc) Mariner models or all 2001 and later Yamaha models, apply a light coating of clean engine oil to the threads of the long cylinder head bolts.

18. Position a new cylinder head gasket on the cylinder block, then carefully position the cylinder head over the gasket and onto the block. Thread the long and short cylinder head bolts making sure the gasket remained perfectly in position.

■ Although the two torque sequences are ESSENTIALLY the same, Mercury advises you to completely torque the large diameter/long cylinder head bolts first, THEN to go back and torque the flange and intake manifold bolts in their own separate torque sequences.

19. Tighten the long and short cylinder head bolts using at least 2 passes of the molded torque sequence or the sequence in the accompanying illustration. The sequence is essentially a counterclockwise spiraling pattern that starts at the left, center of the long bolts and works outward through the long bolts and continues with the center of the short flange bolts, then goes back and forth working toward the top and bottom of the flange bolts. Tighten the long bolts first to 17 ft. lbs. (23 Nm) and then to 34.7 ft. lbs. (47 Nm). Tighten the short flange bolts first to 53 inch lbs./4.4 ft. lbs. (6 Nm) and then to 106 inch lbs./8.9 ft. lbs. (12 Nm). Lastly, if the carburetor or TBI assembly is in position, torque the carb or TBI flange bolts to 75 inch lbs. (8.5 Nm) on 40/45/50 hp (935cc) models or to 70 inch lbs. (8 Nm) on 40/50/60 hp (966cc) models.

■ If the valve train components were disturbed in any way you are going to have to adjust the valves once the timing belt is installed, so either leave the cylinder head cover off or just finger-tighten the bolts.

20. Apply a light coating of marine grade grease to a NEW cylinder head cover O-ring. Install the cover and new O-ring. If the valves were not disturbed, install and tighten the 7 cylinder head cover bolts securely either using the molded torque sequence (or if not present using a clockwise spiraling pattern that begins at the center, left cover bolt and works outward).

■ Make sure the new O-ring remains seated in the cover groove and does not become pinched during installation or there will be oil leaks!

21. For models so equipped through 1998, install the dash pot and bracket, then secure using the 2 bracket bolts.
22. Install any electrical components which were removed to the powerhead.
23. Install the Timing Belt and, if removed, Camshaft Sprocket, as detailed in this section.
24. If the valves were disturbed, adjust them at this time, as detailed under Valve Lash in the Maintenance and Tune-Up section, then install the cylinder head cover.
25. Install the spark plugs to the cylinder head.
26. Install the Flywheel and Stator Plate, as detailed in the Ignition and Electrical Systems section.
27. If removed, install the powerhead assembly.

CYLINDER HEAD OVERHAUL

◆ See Figures 158 and 159

■ Like some other Mercury/Mariner models, the literature is not clear whether or not all of these motor utilize separately replaceable lower valve spring seats or if the seats are also a part of the valve stem seals. Either way, just don't remove the SEALS unless you are planning on replacing them.

1. Remove the Cylinder Head from the powerhead assembly, as detailed in this section.
2. Remove the Oil Pump from the cylinder head, as detailed in the Lubrication and Cooling section.

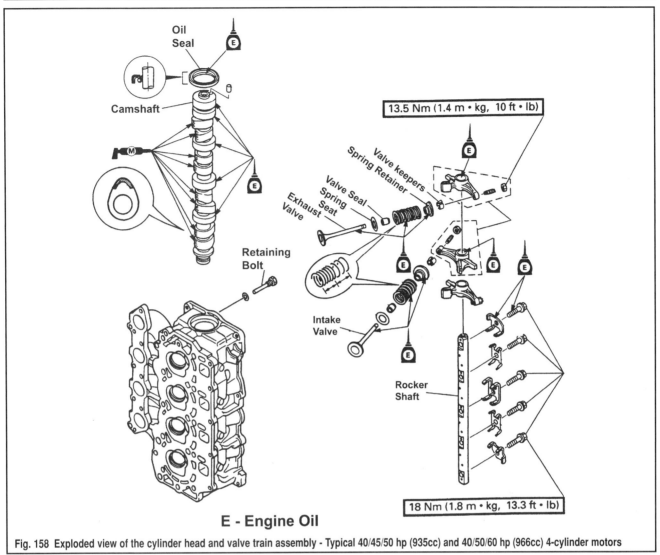

Fig. 158 Exploded view of the cylinder head and valve train assembly - Typical 40/45/50 hp (935cc) and 40/50/60 hp (966cc) 4-cylinder motors

3. Remove the 5 bolts securing the rocker shaft and rocker arm retainers/tensioners. The clips at either end of the shaft and at center are known as retainers, while the other two (on either side of the center retainer) are normally called tensioners. If they are reused the retainers and tensioners must not only be installed in the same positions, but also facing with the same orientation. In some cases the tensioners may have arrows stamped on their surface, which must be faced toward the flywheel end of the cylinder head.

■ On some 60J/60 hp motors each of the tensioners ALSO has a stopper guide installed over top of the tensioner (using the same bolt). When equipped, the stopper guides are also directional, and also marked as to which side should face the flywheel end of the head.

4. Remove the rocker arm shaft along with the 8 rocker arm assemblies. However, whenever possible, keep the rockers and retainers assembled to the shaft in order to prevent accidentally mixing parts during cleaning and inspection. When working on the assembly during cleaning either tag all parts as they are removed or keep them arranged carefully to prevent mix-up during assembly.

■ Valve train components such as rocker arms, retainers, tensioners, valves and keepers should not be mixed and matched from one assembly to another. This means that if they are to be reused, they must all be labeled prior to removal or VERY carefully sorted during service to prevent installation to the wrong positions.

5. Remove the retainer bolt and washer found in the valve train valley toward the top end of the cylinder head, then carefully withdraw the camshaft from the head (out through the oil pump drive opening).

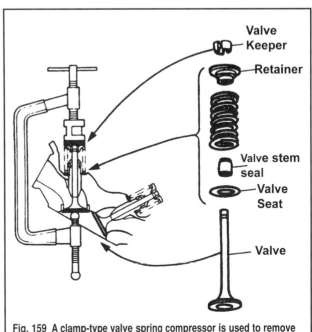

Fig. 159 A clamp-type valve spring compressor is used to remove the valves assemblies from the cylinder head

6-78 POWERHEAD

6. Remove and discard the oil seal from the top end of the camshaft. Check the dowel and replace, if necessary.

7. To remove the each valve, proceed as follows:

a. Using a clamp-type valve spring compressor, carefully place just enough pressure on the valve spring so the retainer moves down sufficiently to expose the keepers.

b. Remove the pair of valve keepers, then slowly ease pressure on the valve spring until the compressor can be removed.

c. Remove the valve spring retainer, valve spring and valve spring seat from the shaft of the valve.

d. Remove the valve from the underside (combustion chamber side) of the cylinder head.

e. If necessary, carefully pry the old valve stem seal from its position on the cylinder head.

f. Repeat this procedure for each of the remaining valves.

8. Refer to Powerhead Refinishing, later in this section, for details on cleaning and inspecting cylinder head components.

To Install:

9. If removed, install each of the valve and spring assemblies as follows:

a. If removed, install new valve spring seals to the cylinder head.

b. Position the valve into the cylinder head (into the original guide and valve seat unless the valve or valve seat was replaced).

c. Position the valve spring seat down onto the valve shaft and cylinder head, then oil and install the spring with the fine pitch side of the spring facing down toward the seat.

d. Place the retainer over the spring, then use the valve spring compressor tool to press downward on the retainer and spring so you can install the keepers.

e. Insert the valve keepers, then slowly back off the valve spring compressor, making sure the keepers are fully in position. Then remove the valve spring compressor.

f. Repeat for each of the remaining valves.

10. Make sure the camshaft dowel pin is installed, then apply a light coating of engine oil to the lips of a new camshaft oil seal and position it over the end of the camshaft. Apply a light coating of engine oil to the journals (not lobes) of the camshaft.

11. Apply a light coating of an engine assembly lube which contains Molybdenum Disulfide to the camshaft lobes.

12. Carefully slide the camshaft up into position in the cylinder head, making sure the oil seal remains seated properly, then install the retainer bolt and washer.

13. If removed from the rocker arm shaft, apply a light coating of engine oil to the inside rocker arm bores and to the threads of the rocker arm shaft retainer bolts. Slide the rocker arms back into their original positions (if they are being reused). Install the rocker arm and shaft assembly to the cylinder head, placing each of the retainers and tensioners (along with the stopper guides, if equipped) in place and finger-tightening the bolts.

14. Tighten the 5 rocker arm shaft retainer bolts to 160 inch lbs/13.3 ft. lbs. (18 Nm).

15. Install the Oil Pump assembly to the cylinder head, as detailed in the Lubrication and Cooling section.

16. Install the Cylinder Head as detailed in this section.

17. Once the powerhead is fully assembled, if valve train components were replaced, be sure to operate the engine as directed for new component break-in. For details, please refer to Powerhead Break-In, in this section.

CYLINDER BLOCK OVERHAUL

◆ See Figures 122 and 160 thru 167

Remember, when loosening the retainers on manifolds, covers and other major components, always try to follow the reverse of the torque sequence indicated in the assembly procedure or molded on the component itself.

When applicable tighten all bolts in using the proper torque sequence and to the proper specification. When possible we've included it in the procedure, but on most Yamaha manufactured powerheads both the sequence and value is molded into the casting on critical components. You'll notice that most Yamaha torque sequences are clockwise spirals starting somewhere near the center of the component.

1. Remove the Powerhead assembly, as detailed in this section.

2. Remove the Timing Belt/Sprocket assembly, as detailed in this section.

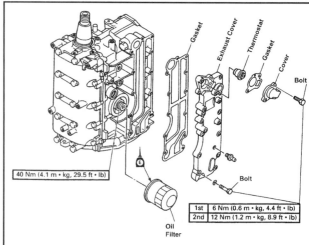

Fig. 160 Exploded view of the exhaust cover mounting - 40/45/50 hp (935cc) and 40/50/60 hp (966cc) 4-cylinder motors

3. Strip the powerhead for disassembly, as detailed earlier in this section.

4. Remove the Cylinder Head assembly, as detailed in this section.

5. Remove and discard the old oil filter from the cylinder block.

6. Gradually loosen and remove the exhaust cover retaining bolts using the reverse of the torque pattern (roughly a counterclockwise spiraling pattern that starts with the lower right bolt and moves toward the center bolts). The thermostat assembly is mounted under a housing to the top of the exhaust cover (2 of the exhaust cover bolts also retain the thermostat housing).

7. Remove the thermostat assembly from the exhaust cover. For details, please refer to the Lubrication and Cooling section. Remove and discard the old thermostat gasket.

■ You CAN reuse the old thermostat, but you're rebuilding or repairing the BLOCK right, so just break-down and replace it.

8. Remove the exhaust cover and gasket from the side of the powerhead. Remove and discard the old exhaust cover gasket.

■ The exhaust cover on some models (including all 2001 and later Yamaha models) contains a anode mounted under a small cap towards the base of the exhaust cover. Loosen the screw and clip mounted to the cap, then remove the bolt and cap so that you can withdraw the anode and grommet. During installation, be sure to coat the grommet lightly with marine grade grease and the threads of the cap retaining bolt with Loctite® 572 or an equivalent threadlocking compound.

9. Remove the crankcase flange and center bearing retaining bolts using the reverse of the molded torque sequence. This results in a counterclockwise spiraling sequence that starts each time with the flange bolts (at the upper right bolt and works inward) and then moves each time to the center bearing bolts (again, starting at the upper right bolt and working inward).

10. Carefully break the gasket seal and remove the crankcase from the cylinder block.

■ Most parts inside the cylinder block are matched sets and should not be moved from one cylinder to another. This includes the pistons as well as each of the connecting rod/end cap combination. For this reason, prior to disassembly it is essential to matchmark each of the connecting rods to their end caps (this ensures that not only will each cap be installed on the proper rod, but facing the correct direction). Also, be sure to label each piston and connecting rod as to what cylinder it belongs and identify the factory marks and/or place an arrow or some indicator facing upward to ensure they are installed in the same direction.

11. Matchmark the big end caps to the connecting rods, then remove the 8 connecting rod cap bolts (working on one cap at a time, alternating between each of the cap bolts after each turn). Remove the caps and place them aside for reinstallation back on their original connecting rods as soon as the crankshaft has been removed.

POWERHEAD 6-79

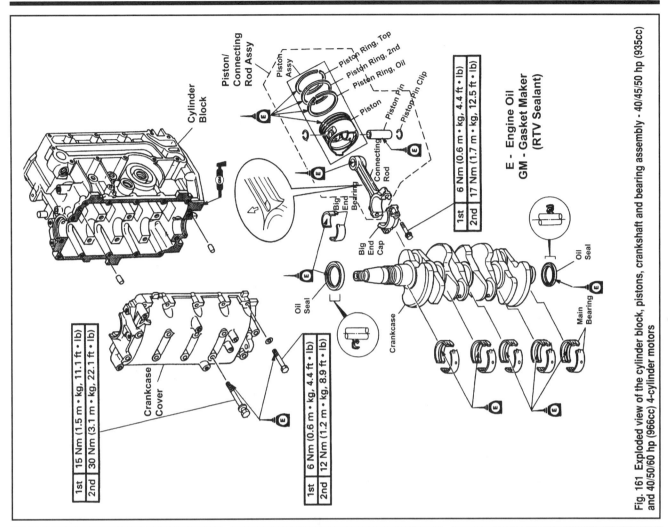

Fig. 161 Exploded view of the cylinder block, pistons, crankshaft and bearing assembly - 40/45/50 hp (935cc) and 40/50/60 hp (966cc) 4-cylinder motors

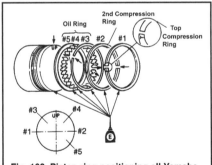

Fig. 162 Piston ring positioning all Yamaha models, as well as the 40/45/50 hp (935cc) Mercury/Mariner models

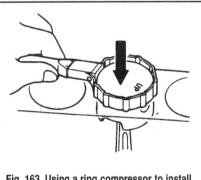

Fig. 163 Using a ring compressor to install the 4-stroke piston

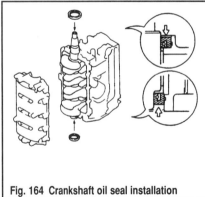

Fig. 164 Crankshaft oil seal installation

12. Carefully lift the crankshaft from the block. Place it on a protective surface.

✱✱ WARNING

Handle the crankshaft carefully. Remember it is a heavy, expensive and vital piece to the motor. The polished surfaces should be covered to prevent scratches or other damage and the whole shaft should be protected against impact or other damage which could destroy it.

13. Immediately reinstall the connecting rod end caps and their bearing inserts to their respective pistons.
14. Remove the top and bottom oil seals from the crankshaft. Discard the seals.

■ These pistons and connecting rods are normally already marked from the factory as to what side faces up, identify the marks and perhaps, remark them to ensure they are found easily during assembly.

15. Label and remove each of the pistons from the bores. Number the pistons and place a mark on each piston and connecting rod facing up (facing the flywheel) for installation purposes. If necessary, disassemble each of the pistons, as follows:
   a. Carefully pry the clips free from the piston pin bore on either side of the piston assembly.
   b. Slide the piston pin free of the bore (specifications tell us that this is NOT an interference fit, so it should slide free with ease).
   c. Using a piston ring spreading tool, carefully remove each of the rings (top and second compression rings, followed by the oil ring assembly) from the piston.

6-80 POWERHEAD

d. Keep all components from a single piston together for reuse or replacement, but don't mix and match components from other pistons.

16. Remove the crankshaft main (big end) bearings from the crankcase and cylinder block. Again, note their positions and orientation if they are going to be reused.

17. Check the placement and condition of the 2 cylinder block dowel pins. They should be replaced if they are damaged.

18. Refer to Powerhead Refinishing, later in this section, for details on cleaning and inspecting cylinder block components. There are probably more components and specifications to measure on these 4-stroke motors than on any smaller Yamaha or Mercury/Mariner engine and most larger 2-strokes. Take your time and make sure you've made measurements for all components listed in the Engine Specifications chart for this model.

To Install:

19. Make sure the mating surfaces of the crankcase and cylinder block are completely clean and free of sealant or damage. The dowel pins should already be installed in the block, but if not, obtain new pins and install them in the 2 positions at opposite ends on the oil filter housing side of the block.

20. Place the main bearing inserts into the crankcase and cylinder block. There are normally 2 things to look out for when doing this. Most bearing inserts will be equipped with locating tabs, make sure they are seated in the appropriate grooves. Also, inserts with oil holes should be aligned with an oil passage.

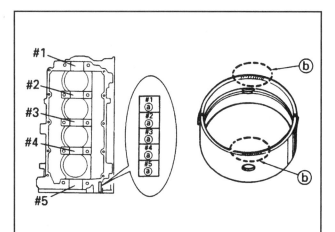

Fig. 165 The cylinder block and main bearings are color and/or alpha coded to indicate crankcase diameter (a) and bearing thickness

■ UNFORTUNATELY Mercury and Yamaha sometimes have a different philosophy on exactly how to set up the ring gaps prior to piston installation. This does go to reinforce a popular opinion we've often heard from machinists, that since the rings are going to move around in service, it really shouldn't matter. But, we've gone through the trouble of giving you each manufacturer's preferred set-up method. To make matters more confusing Mercury accepted Yamaha's method for the 40/45/50 hp (935cc) models but went back to the more common Mercury recommendations for the 40/50/60 hp (966cc) models.

21. If they were disassembled, prepare each piston assembly for installation as follows:

a. Using a ring expander, carefully install each of the rings with their gaps as noted here and as shown in the accompanying illustration. The description and illustration are depicted while looking at the piston dome with the top (flywheel end) facing 12 o'clock.

b. The bottom oil ring should be positioned with the gap facing between 4 and 5 o'clock on Yamaha models or Mercury/Mariner 935cc models or it should be positioned at 3 o'clock for Mercury/Mariner 966cc models.

c. The middle oil ring (expander ring) should be positioned with the gap facing between 1 and 2 o'clock on Yamaha models or Mercury/Mariner 935cc models or it should be positioned directly to 6 o'clock for Mercury/Mariner 966cc models.

d. The upper oil ring should be positioned with the gap facing between 10 and 11 o'clock (there may be a locating pin for this gap on some pistons) on Yamaha models or Mercury/Mariner 935cc models or it should be positioned at 9 o'clock for Mercury/Mariner 966cc models.

e. The lower (second) compression ring should be positioned with its gap at 3 o'clock on Yamaha models or Mercury/Mariner 935cc models or it should be positioned between 1 and 2 o'clock for Mercury/Mariner 966cc models.

f. The upper (top) compression ring should be positioned with its gap facing 9 o'clock on Yamaha models or Mercury/Mariner 935cc models or it should be positioned between 10 and 11 o'clock for Mercury/Mariner 966cc models.

g. Align the mark on the connecting rod (facing up/flywheel) with the mark on the piston (also facing up/flywheel), install the connecting rod to the piston using the piston pin.

h. Install the piston pin clips in either side of the pin bore to secure the piston pin in place.

i. Repeat for the remaining pistons.

✱✱ WARNING

Use extreme care to prevent the connecting rods from contacting and scoring the cylinder walls during piston installation. It is a good idea to wrap the ends in a shop cloth or something else to protect the cylinders.

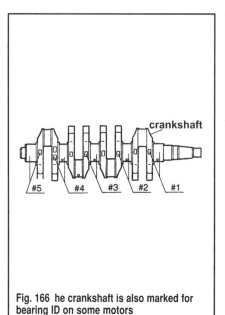

Fig. 166 The crankshaft is also marked for bearing ID on some motors

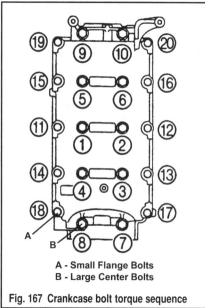

A - Small Flange Bolts
B - Large Center Bolts

Fig. 167 Crankcase bolt torque sequence

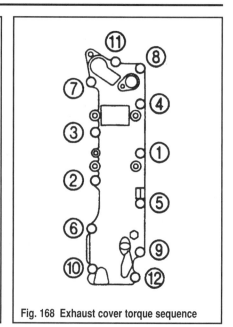

Fig. 168 Exhaust cover torque sequence

POWERHEAD 6-81

22. Install each of the pistons (with the correct mark/side facing UP toward the flywheel) to the cylinder block using a ring compressor. Apply a very light coating of clean engine oil to the inside of the compressor and to the cylinder bore, then place the compressor around the piston. Insert the piston through the cylinder head side of the bore and use a wooden hammer handle to gently tap the piston down through the compressor and into the cylinder bore.

✱✱ WARNING

Be VERY careful when installing the piston. Get the lower portion of the skirt below the ring compressor and insert it into the cylinder bore to make sure everything is aligned. Then, GENTLY tap the piston through the compressor, stopping if it seems to hang up even slightly to make sure that a ring is not caught on the deck of the cylinder block.

23. Remove the connecting rod end caps and position them so the crankshaft can be installed. Place the connecting rod bearing inserts into position on the connecting rod big end and cap.

■ Here's where procedures differ slightly depending upon whether you're installing bearings of a known clearance or measuring clearance. If you've already determined that bearing clearances are fine, apply alight coat of engine oil to each of the connecting rod, main bearing and crankshaft journal surfaces. If however, you need to measure and determine bearing clearance, Temporarily INSTALL THESE COMPONENTS DRY for measurement purposes.

24. Position the crankshaft to the cylinder block, and carefully pull each of the connecting rods into position.

■ NEVER install the upper or lower crankshaft oil seals with the crankcase installed and tightened otherwise the seal will likely be damaged.

25. Install a new upper and lower oil seal to crankshaft. Seat the seals as best you can at this time. To ensure the seals are not cocked you can install the crankcase and FINGER-TIGTHEN the bolts then lightly knock the seals using a suitably sized driver and rubber mallet. Once this is done, remove the crankcase again and continue the cylinder block assembly.

26. Install each of the match-marked connecting rod caps and tighten the bolts alternating back and forth between the 2 bolts on each cap. Tighten the bolts first to 53 inch lbs./4.4 ft. lbs. (6 Nm) and then to 150 inch lbs./12.5 ft. lbs. (17 Nm).

27. If you've already measured bearing clearances, apply a light coating of Gasket Maker (such as Loctite 514 Master Gasket, RTV sealant or equivalent) to the entire perimeter of the cylinder block-to-crankcase mating surface.

28. With the main bearing inserts in place, carefully lower the crankcase onto the cylinder block, then JUST thread the long center bolts and shorter flange bolts.

29. Tighten the long center bolts and short flange bolts using 2 or more passes of the molded torque sequence (or if not present, using a clockwise, spiraling pattern that starts at the left, center of the long bolts and works toward the ends of the crankcase, then continues with the left center of the short flange bolts and again works towards the ends of the crankcase). During the torque sequence, tighten the long bolts first to 11.1 ft. lbs. (15 Nm) and then to 22.1 ft. lbs. (30 Nm) while you tighten the short bolts first to 53 inch lbs./4.4 ft. lbs. (6 Nm) and then to 106 inch lbs./8.9 ft. lbs. (12 Nm).

30. If removed, install the exhaust cover, thermostat assembly and gaskets (one for the exhaust cover and one for the thermostat housing). Tighten the bolts using multiple passes of a clockwise spiraling pattern that starts at the right center bolt and works outward. Tighten the bolts first to 53 inch lbs./4.4 ft. lbs. (6 Nm) and then to 106 inch lbs./8.9 ft. lbs. (12 Nm).

31. Install a new oil filter to the powerhead.
32. Install the Cylinder Head assembly, as detailed in this section.
33. Install the components stripped from the powerhead for rebuilding purposes. For details, refer to the procedure found earlier in this section.
34. Install the Timing Belt/Sprocket assembly, as detailed in this section.
35. Install the Powerhead assembly, as detailed in this section.
36. Once the powerhead is fully assembled, if cylinder block components were replaced, especially the pistons and/or rings, be sure to operate the engine as directed for new component break-in. For details, please refer to Powerhead Break-In, in this section.

75/80/90/100 Hp and 115 Hp Powerheads

POWERHEAD REMOVAL & INSTALLATION

Mercury/Mariner Models

◆ See Figures 169 thru 173

The powerhead may be removed or installed as an almost fully assembled unit. The components mentioned here are the bare minimums that must be removed before separating the powerhead assembly from the rest of the motor. However, if overhaul or replacement of the cylinder head and/or block is desired, you may wish to remove additional components before removing the powerhead assembly, not only to lighten the load, but to help prevent potential damage to the external fuel and electrical system components mounted to the powerhead. The choice is yours, so if necessary, refer to other component procedures in this section and in the Fuel System and/or Ignition and Electrical Systems sections before proceeding.

1. Disconnect the negative battery cable at the battery itself for safety.
2. Remove the top engine cowl.
3. Loosen the 2 bolts securing the flywheel cover, then carefully free the cover from the grommets.
4. On the starboard side, near the trailering Power Trim/Tilt (PTT) switch (just below the oil filter), locate and disconnect the wiring bullet connectors for the cowl PTT switch.
5. On 115 hp motors, remove the 2 halves of the rubber cowl grommet from the top of the engine shrouds.
6. At the front, port side of the engine shroud (lower engine cowling) loosen the 2 screws securing the grommet cover to the cowling, then remove the cover and round wiring/cable grommet.
7. Remove the 8 screws securing the 2 halves of the lower engine cowling together, then remove separate the halves and remove the cowling.
8. Remove the engine oil dipstick. Depending upon the amount of overhaul and service condition of the motor, it is probably a good idea to drain the engine oil at this time.
9. For carbureted motors proceed as follows:
   a. On the starboard side of the motor toward the rear (at the base of the exhaust cover, just a little behind and below the anode cover), locate the cooling hose run vertically downward and wire tied to a fitting at the base of the powerhead. This is the coolant hose for the driveshaft bushing, carefully disconnect the hose from the lower fitting.

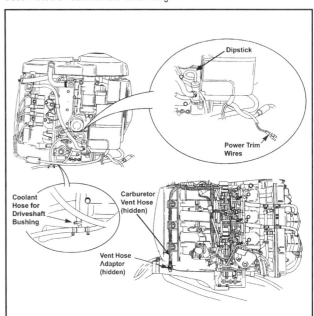

Fig. 169 Hose and wiring connections for powerhead removal - 75/90 Hp (carbureted) models

POWERHEAD

b. On the port side of the motor toward the front, locate the carburetor vent hose wire tied to the top of the air intake silencer/intake runners. The hose turns downward behind the silencer and runs all the way down to a fitting. Carefully disconnect the hose from the fitting.

10. For EFI motors, proceed as follows:
   a. Locate the vapor separator tank vent hose. It can be found at the base of the air intake silencer/intake runner assembly at the front port side of the powerhead. Carefully disconnect the vent hose.
   b. Locate the shift switch harness connector, on the port side, rear of the motor at the base of the powerhead, just below the fuel cooler. Tag and disconnect the shift switch harness.

11. At the base of the starboard, front, side of the powerhead, locate the PTT pump wires (normally Blue and Green). Tag and disconnect the pump wires.

12. Tag and disconnect the incoming fuel line. For EFI motors it is at the front of the powerhead on the fuel filter fitting.

13. Remove the 6 nuts and 2 bolts (along with their washers) threaded upward into the powerhead (or onto powerhead studs) from the intermediate housing.

14. Attach a suitable lifting device to the flywheel assembly (via the 3 puller holes) and carefully lift the powerhead assembly from the intermediate housing. In order to prevent potential damage to the crankshaft and driveshaft be certain to lift the powerhead STRAIGHT up and off the intermediate housing.

15. Remove and discard the upper casing gasket from the powerhead-to-intermediate housing mating surface. Carefully clean the mating surfaces of all traces of old gasket.

16. Check for placement of the 2 dowel pins. They may have remained with the intermediate housing or have come away with the powerhead itself. Inspect the pins for damage and replace, if necessary.

17. Position the powerhead on a suitable workbench or engine stand for further disassembly or overhaul, as necessary. Refer to the additional procedures in this section as needed.

To Install:

18. With the 2 dowel pins in place, position a NEW gasket to the intermediate housing.

19. Apply a light coat of marine grade grease to the driveshaft-to-crankshaft mating splines.

20. Using a suitable lifting device, carefully lift the powerhead and lower it slowly, STRAIGHT onto the driveshaft and intermediate housing. If necessary, rotate the flywheel slightly in a clockwise direction in order to mate the crankshaft to the driveshaft.

21. Although Mercury doesn't specify it, we think it is a good idea to apply a light coating of Loctite® 572 to the threads of the powerhead mounting bolts and studs.

22. Install the 6 nuts and 2 bolts securing the powerhead. Tighten the nuts to 40 ft. lbs. (54.5 Nm) and the bolts to 20 ft. lbs. (27 Nm) using at least 2 passes of the proper torque sequence (as shown in the accompanying illustration).

23. Reconnect the fuel supply hose.

24. For EFI motors, reconnect the vapor separator tank vent hose and the shift switch harness connector.

25. For Carb motors, reconnect the carburetor vent hose and the driveshaft bushing coolant hose.

26. Install the engine oil dipstick and, if drained, refill the engine with fresh engine oil.

27. Reconnect the PTT pump wires.

28. Install the lower engine cowling assembly and secure using the 8 screws. Install the round wiring/cable harness and secure the grommet cover.

29. For 115 hp motors, install the 2 halves of the lower cowling rubber grommet.

30. Reconnect the PTT trailering switch harness bullet connectors.

31. Install the flywheel cover to the grommets, then secure the 2 retaining bolts.

32. Reconnect the negative battery cable to the battery itself.

33. Be sure to check all Timing and Synchronization adjustments as detailed in the Engine and Maintenance section.

34. If the powerhead was overhauled (replacing components such as the valve train, main bearings, crankshaft, pistons and/or rings) be sure to subject the motor to a complete new engine break-in period to ensure long, trouble free life.

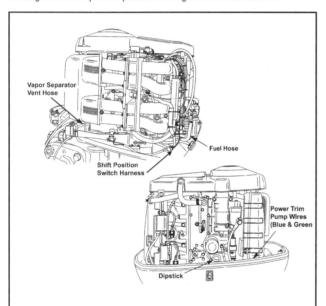

Fig. 170 Hose and wiring connections for powerhead removal - 115 Hp (EFI) models

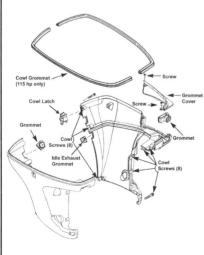

Fig. 171 Exploded view of the lower cowling assembly - 75/90 Hp and 115 Hp models

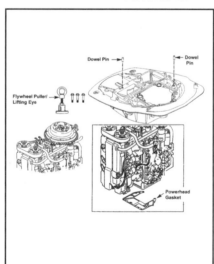

Fig. 172 Powerhead mounting - 75/90 Hp and 115 Hp models

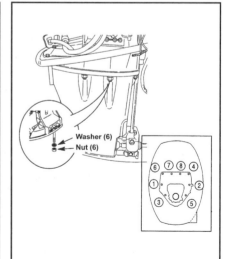

Fig. 173 Powerhead mounting torque sequence - 75/90 Hp and 115 Hp models

POWERHEAD 6-83

Yamaha Models

◆ See Figures 174 thru 189

The powerhead may be removed or installed as an almost fully assembled unit. The components mentioned here are the bare minimums that must be removed before separating the powerhead assembly from the rest of the motor. However, if overhaul or replacement of the cylinder head and/or block is desired, you may wish to remove additional components before removing the powerhead assembly, not only to lighten the load, but to help prevent potential damage to the external fuel and electrical system components mounted to the powerhead. The choice is yours, so if necessary, refer to other component procedures in this section and in the Fuel System and/or Ignition and Electrical Systems sections before proceeding.

1. Disconnect the negative battery cable at the battery itself for safety.
2. Disconnect the throttle and shift cables as follows:
  a. Loosen the 2 bolts securing the grommet retaining plate to the front corner of the engine cowling.
  b. Loosen the bolt and remove the hose locating plate.
  c. Remove the 2 clips securing the cable ends to the linkage studs.
  d. Remove the bolt securing the cable clamp, then reposition the cables along with the grommet off the lower cowling and out of the way.
3. Remove the cover from the electrical junction box for access to wiring.
4. For carbureted motors, disconnect the necessary hoses and wires from the powerhead, as follows:

  a. Loosen the 3 clamp bolts for the fuel line, then disconnect the fuel line and reposition it for access.
  b. If necessary, remove the fuel pumps from the powerhead.
  c. Loosen the 2 bolts and remove or reposition the clamp plates which secure the flushing water hose.
  d. Loosen the bolt and remove the metal clamp from the side of the powerhead, just above the oil level dipstick.
  e. Tag and disengage the following wiring connectors: remote control wire harness coupler, tachometer coupler, trim meter coupler and the Power Trim/Tilt (PTT) coupler.
  f. Remove the bolt and disconnect the negative battery lead from the starter motor, then remove the nut and disconnect the positive battery lead from the starter solenoid (relay).
  g. Remove the 2 nuts and washers securing the PTT motor leads (one is blue and one is green) to the relays.
5. For EFI motors, disconnect the necessary hoses and wires from the powerhead, as follows:
  a. Tag and disconnect the large remote control wire harness coupler at the front of the powerhead, then tag and disconnect the smaller warning lamp coupler found just behind the remote harness.
  b. Remove the bolt and disconnect the negative battery lead from the starter motor, then remove the nut with washers and disconnect the positive battery lead from the starter solenoid (relay).
  c. Just on the other side of the bracket from the main harness and starter wiring (in the electrical junction box itself) remove the 2 bolts securing the Power Trim/Tilt (PTT) sky blue and light green leads to the PTT relay assembly. Tag and disconnect the leads.
  d. Locate the pilot water hoses at the top and bottom of the fuel rail cooling tube. Squeeze the ears of the spring-loaded clamps and slide each of the clamps back onto the pilot water hoses, then disconnect the hoses from the cooling tube.
6. Remove the engine oil dipstick. Depending upon the amount of overhaul and service condition of the motor, it is probably a good idea to drain the engine oil at this time.
7. Carefully cut the plastic wire tie that secures the pilot water (cooling system indicator) hose and the flushing water hose to the fittings on the side of the powerhead. Tag and disconnect the 2 hoses.
8. Locate the 2 apron bolts and nuts threaded horizontally at the front of the powerhead. Loosen and remove the bolts.
9. Next, locate and remove the 5 bolts which area threaded downward through the lower engine cowling into the top of the apron. There are 2 on the port side of the motor and 3 on the starboard side of the motor. Once the bolts are removed, carefully remove the apron for access to the powerhead mounting bolts.
10. Remove the 8 bolts threaded upward into the powerhead from the intermediate housing. Keep track of the bolts as there are 2 different sizes. There are 3 long bolts on each side and 2 short bolts at the center, rear of the powerhead.
11. Attach a suitable lifting device and carefully lift the powerhead assembly from the intermediate housing. In order to prevent potential damage to the crankshaft and driveshaft be certain to lift the powerhead STRAIGHT up and off the intermediate housing.
12. Remove and discard the upper casing gasket from the powerhead-to-intermediate housing mating surface. Carefully clean the mating surfaces of all traces of old gasket.

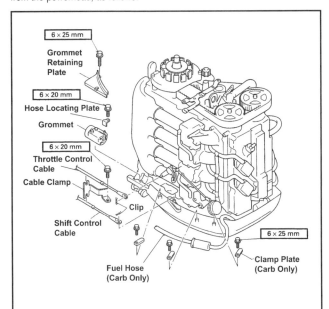

Fig. 174 Exploded view of the control cable mounting - 75/80/90/100 Hp and 115 Hp models

Fig. 175 Remove the electrical box cover

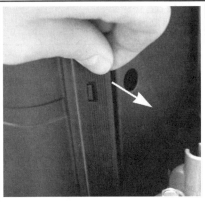

Fig. 176 On this carb model, pull out on the flange...

Fig. 177 ...or press in on the edge, to free the respective tabs

6-84 POWERHEAD

Fig. 178 The cover has a handy diagram (and fuse tool)

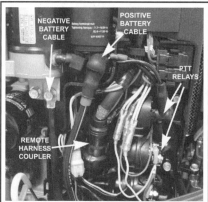
Fig. 179 Tag and disconnect the necessary wiring (carb shown)

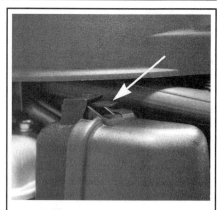
Fig. 180 The cover on this EFI motor uses locktabs...

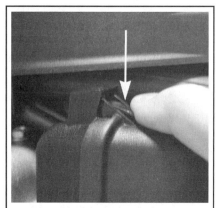
Fig. 181 ...which must be pressed to release

Fig. 182 The cover on EFI motors is also labeled

Fig. 183 Tag and disengage connectors, like for the remote harness

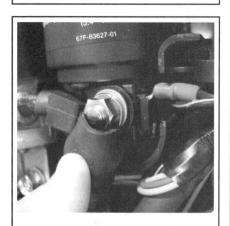
Fig. 184 Disconnect the battery leads

Fig. 185 Disconnect the PTT motor leads (EFI shown)

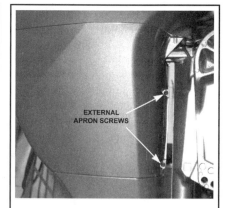
Fig. 186 Remove the apron for access

13. Check for placement of the 2 dowel pins. They may have remained with the intermediate housing or have come away with the powerhead itself. Inspect the pins for damage and replace, if necessary.

14. Position the powerhead on a suitable workbench or engine stand for further disassembly or overhaul, as necessary. Refer to the additional procedures in this section as needed.

To Install:

15. With the 2 dowel pins in place, position a NEW gasket to the intermediate housing.

16. Apply a light coat of marine grade grease to the driveshaft-to-crankshaft mating splines.

17. Using a suitable lifting device, carefully lift the powerhead and lower it slowly, STRAIGHT onto the driveshaft and intermediate housing. If necessary, rotate the flywheel slightly in a clockwise direction in order to mate the crankshaft to the driveshaft.

18. Apply a light coating of Loctite® 572 to the threads of the powerhead mounting bolts. Although the manufacturer only specifies that the threadlocking compound is necessary on the long bolts, it is not a bad idea to use it on the short powerhead bolts in order to reduce the possibility of corrosion locking them in place in the future.

19. Install the 8 powerhead retaining bolts and tighten to 31 ft. lbs. (42 Nm).

20. Install the apron to the engine cowling and secure using the 5 vertically threaded bolts and then with the 2 horizontal bolts (along with the nuts and washers on the horizontal bolts at the front of the apron). Tighten the screws securely, but do not over-tighten and damage the cowling or apron.

21. Install the engine oil dipstick and, if drained, refill the engine with fresh engine oil.

POWERHEAD 6-85

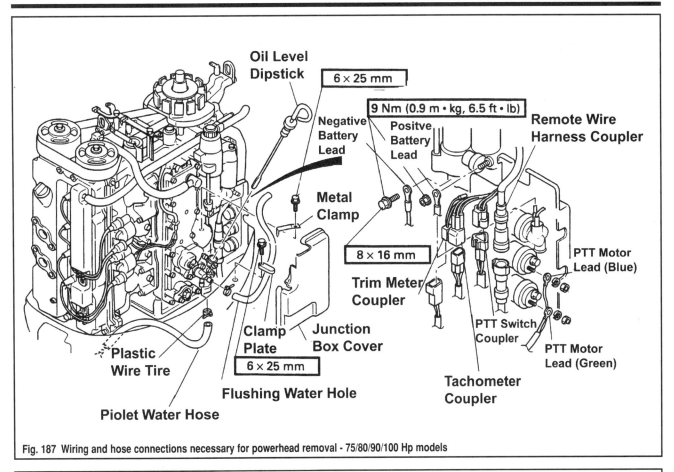

Fig. 187 Wiring and hose connections necessary for powerhead removal - 75/80/90/100 Hp models

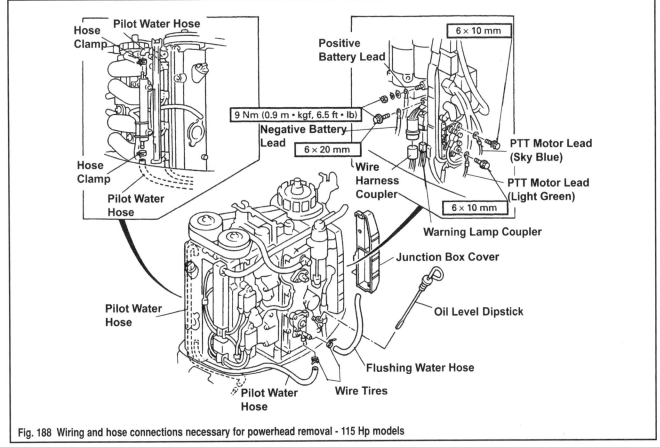

Fig. 188 Wiring and hose connections necessary for powerhead removal - 115 Hp models

6-86 POWERHEAD

22. Reconnect the pilot water (cooling system indicator) hose and the flushing hose. Secure the hoses using new wire ties.
23. For EFI motors, reconnect the necessary hoses and wires to the powerhead, as follows:
    a. Install the pilot water hoses to the top and bottom of the fuel rail cooling tube and position the spring clamps to secure the hoses.
    b. Reconnect the sky blue and light green leads to the PTT relay terminals, as noted during removal. Secure the leads using the bolts.
    c. Reconnect the positive battery lead to the starter solenoid (relay) and secure using the nut with washers. Reconnect the negative battery cable to the starter motor and secure using the retaining bolt.
    d. Reconnect the small warning lamp coupler and the larger remote control wire harness coupler.
24. For carbureted motors, disconnect the necessary hoses and wires from the powerhead, as follows:
    a. Connect the blue and green PTT motor leads to the relays (as noted during removal) and secure using the 2 nuts.
    b. Reconnect the positive battery lead to the starter solenoid (relay) and secure using the nut. Reconnect the negative battery cable to the starter motor and secure using the retaining bolt.
    c. Engage the following wiring connectors, as tagged during removal: remote control wire harness coupler, tachometer coupler, trim meter coupler and the Power Trim/Tilt (PTT) coupler.
    d. Install the metal clamp and bolt to the side of the powerhead, just above the oil level dipstick.
    e. Install the 2 clamp plates for the flushing water hose, then secure using the retaining bolts.
    f. If removed, install the fuel pumps to the powerhead.
    g. Reconnect the fuel line, then secure using the 3 clamps and retaining bolts
25. Install the electrical junction box cover.
26. Reconnect the throttle and shift cables as follows:
    a. Reposition the grommet and cables to the cowling along with the cable clamp (NOT the grommet retainer plate). Install the bolt to secure the cable clamp.
    b. Connect the cables to the appropriate linkage studs and secure using the 2 clips (keep in mind that if adjustment is necessary you may wish to hold off on the clamps).
    c. Install the hose locating plate and secure using the bolt.
    d. Install the grommet retaining plate and secure using the 2 bolts.
27. Reconnect the negative battery cable to the battery itself.
28. Be sure to check all Timing and Synchronization adjustments as detailed in the Engine and Maintenance section.
29. If the powerhead was overhauled (replacing components such as the valve train, main bearings, crankshaft, pistons and/or rings) be sure to subject the motor to a complete new engine break-in period to ensure long, trouble free life.

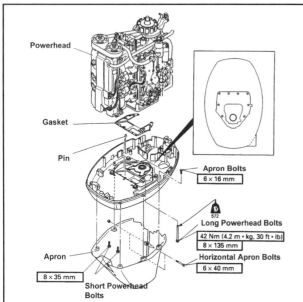

Fig. 189 Exploded view of the powerhead mounting - 75/80/90/100 Hp and 115 Hp models

TIMING BELT/SPROCKET REMOVAL & INSTALLATION

◆ See Figures 190 thru 197

This procedure can be performed with the powerhead assembly removed or installed on the motor, but if the crankshaft sprocket must be removed with the powerhead installed you'll have to lock the driveshaft and propeller shaft from spinning in order to loosen the crankshaft nut. And since we're talking about a significant amount of torque, we'd be worried about the potential for damage to the shafts/splines.

✱✱ CAUTION

The high torque of the crankshaft nut REALLY requires the powerhead to be removed, IF the crankshaft sprocket must be replaced. We just can't in good conscience recommend risking potential damage to the propshaft.

If the timing belt is to be reused, be sure to mark the direction of rotation on the old belt before removal. Also, if either the cylinder head or cylinder block is not to be disassembled further, be sure to properly align the timing marks before removal. This will significantly ease installation.

■ Camshaft sprocket removal requires the use of a holder tool which inserts into the 2 holes in the sprocket itself. The same tool that is used on the flywheel can normally be used on the camshaft sprocket. Crankshaft sprocket removal requires the use of a crankshaft holding tool which is inserted into the driveshaft end of the crankshaft splines. This tool can be fabricated from a discarded driveshaft (if you can find one from a gearcase rebuilder or marine bone yard). If the powerhead is still installed you can usually place it in gear and lock the propshaft from spinning, but you do risk some damage to the prop as we're talking about 116 ft. lbs. (160 Nm) of torque for carbureted Yamaha motors or 195 ft. lbs (265 Nm) for EFI Yamaha motors and all Mercury/Mariner models (Carb or EFI)! Also, crankshaft sprocket removal will require the use of a LARGE 46mm socket which is at least 76mm deep.

1. If the powerhead is still installed, disconnect the negative battery cable for safety.
2. Remove the Flywheel and Stator Plate, as detailed in the Ignition and Electrical Systems section.
3. If either the cylinder head or cylinder block is not being overhauled, rotate crankshaft sprocket in the normal direction of rotation (clockwise) in order to align the timing marks on the camshaft pulleys at the center (facing toward each other) and the marks on the crankshaft pulley and pickup coil rotor with the mark on the powerhead (facing toward the camshaft sprockets, aligned with the centerline of the powerhead).
4. See the hose which attaches to a fitting on top of the powerhead, it's in the way and will prevent timing belt removal. Carefully cut the wire tie and disconnect the hose from the fitting on top of the powerhead. Reposition the hose out of the way, but if necessary you can disconnect the host at the other end and remove it completely.

■ Although most models use wire ties to secure this hose at both ends, some models (including some EFI units) may be equipped with a spring-type clamp at one or both ends of the hose.

5. Remove the 2 bolts securing the flywheel cover mounting bracket to the center top of the powerhead, then remove the bracket.
6. Remove the bolt from the center of the timing belt tensioner, then unhook the spring and remove the tensioner itself.

■ Remember, if the timing belt is to be reused, be sure to mark it showing the normal direction of rotation to ensure installation in the proper position. Most OE timing belts they are normally installed with the labeled part numbers facing upward (so that you can read the part number) when installed.

7. Remove the timing belt from the sprockets.
8. If necessary, remove one or both of the camshaft sprockets using a flywheel holder tool inserted into 2 of the holes in the sprocket, then holding the sprocket from turning while removing the retaining bolt. Remove the large washer, driven sprocket and dowel pin from the camshaft.

POWERHEAD

■ A special tool (Yamaha #YB-06552 or Mercury #91-804770A1 for 75/90 Hp models or Mercury #91-804776A1 for 115 Hp models) can be inserted into the driveshaft end of the crankshaft and held in order to keep the crankshaft from turning when removing or installing the crankshaft sprocket nut. In all cases, a 46mm DEEP socket (at least 76mm deep!) is necessary for the crankshaft nut itself.

9. Lock the crankshaft from spinning using the propeller shaft (if the powerhead is installed and you ABSOLUTELY are sure you're willing to risk it) or the crankshaft-to-driveshaft splines (if removed, and that's how we recommend you proceed), then loosen and remove the crankshaft nut. Remove the crankshaft sprocket and the pick-up coil rotor from the top of the crankshaft.

10. Remove the woodruff key from the crankshaft and retain for use during installation.

To Install:

11. Apply a small dab of marine grade grease to the woodruff key (in order to help hold it in place) and position it into the slot in the crankshaft.

12. With the timing mark facing upward, slide the pick-up coil rotor over the crankshaft. Align the slot in the rotor with the keyway as the rotor is seated on the powerhead.

■ If the crankshaft is positioned properly with the No. 1 cylinder at TDC, the crankshaft key and the timing mark on the rotor will align with the timing mark on the powerhead (facing toward the camshafts). If necessary, rotate the crankshaft SLOWLY clockwise (when viewed from above) in order to properly align the keyway and timing mark.

13. Install the crankshaft sprocket over the pick-coil rotor and woodruff key. Again, the timing mark on the sprocket should align with the timing marks/keyway.

14. Apply a light coating of engine oil to the threads of the crankshaft sprocket retaining nut. Install the crankshaft sprocket retaining nut and tighten to 116 ft. lbs. (160 Nm) for carbureted Yamaha motors or to 192 ft. lbs. (265 Nm) for EFI Yamaha motors and ALL Mercury/Mariner motors (Carb or EFI).

15. If removed, install each of the sprockets to the camshaft over the locating dowel pin in each shaft. If the camshaft positions were not disturbed, make sure the number triangular marks on the sprockets are aligning with each other at a point closest together in their travel.

16. Apply a light coating of engine oil to the threads of the camshaft sprocket retaining bolts. Install a bolt and washer to each camshaft sprocket, then using the flywheel tool to hold each sprocket from turning tighten the bolt to 44 ft. lbs. (60 Nm).

17. Install the timing belt, making sure the timing marks are still aligned. Remember, if the belt is being reused, be sure to face it in the same direction as normal rotation (clockwise). New belts from Yamaha should be installed with the part numbers facing so they are legible when the belt is in place, and though Mercury doesn't mention it, that's probably wouldn't hurt with a Mercury supplied belt as well.

18. Install the timing belt tensioner by just lightly threading the bolt, connecting the spring and then finally tightening tensioner bolt to 29 ft. lbs. (40 Nm).

19. Turn the crankshaft clockwise until it completes 2 full revolutions, then recheck the timing marks.

Fig. 190 Remove the flywheel cover for access

Fig. 191 These marks are not aligned, they should face each other

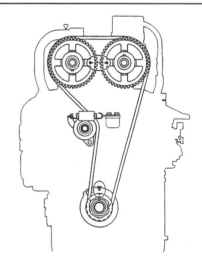

Fig. 192 Align the timing marks as shown before belt removal

Fig. 193 Cut the wire tie and disconnect the hose...

Fig. 194 ...then remove the flywheel cover bracket

Fig. 195 Remove the bolt, undo the spring and lift the tensioner off the pivot

6-88 POWERHEAD

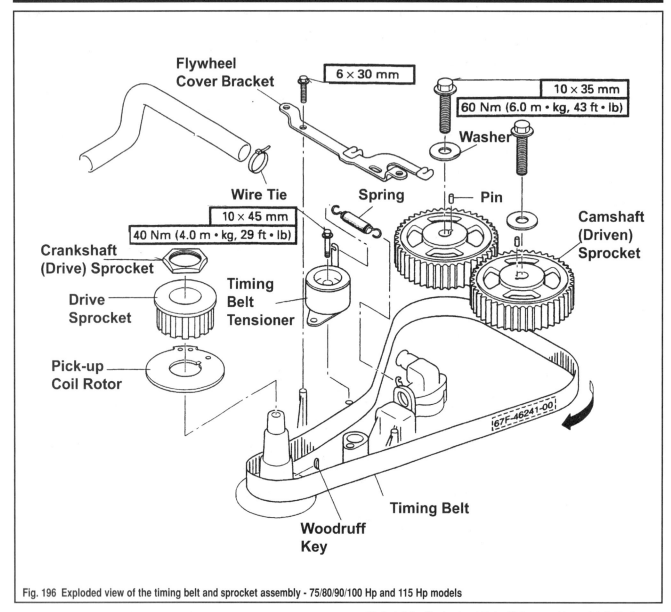

Fig. 196 Exploded view of the timing belt and sprocket assembly - 75/80/90/100 Hp and 115 Hp models

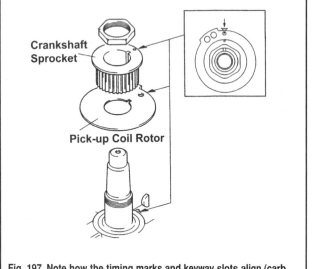

Fig. 197 Note how the timing marks and keyway slots align (carb shown, EFI similar, but rotor is slightly different)

20. Install the Flywheel and Stator Plate assembly, as detailed in the Ignition and Electrical Systems section.
21. If removed, install the powerhead assembly.

CYLINDER HEAD REMOVAL & INSTALLATION

◆ See Figures 198 thru 204

■ Unique to the large Yamahas and Mercury/Mariner motors, the cylinder head bolts require a torque angle meter for tightening to proper specifications. Although the use of an inexpensive gauge is the best way to proceed, since the specification on these motors is for another 90 degrees, it IS possible simply to paint a mark on the socket and eyeball 1/4 additional turn, but don't be cheap, get the gauge.

We cover a lot of ground in this procedure. Although it is a cylinder head assembly removal procedure, there are many instances where you may need to follow only parts of the procedure to service other components, such as the camshafts and valve train.

First off, it should NOT be necessary to remove the powerhead from the intermediate housing in order to service the cylinder head. Secondly, not only is it NOT necessary to remove the cylinder head in order to service the camshafts, it is actually NECESSARY to remove the camshafts in order to

POWERHEAD 6-89

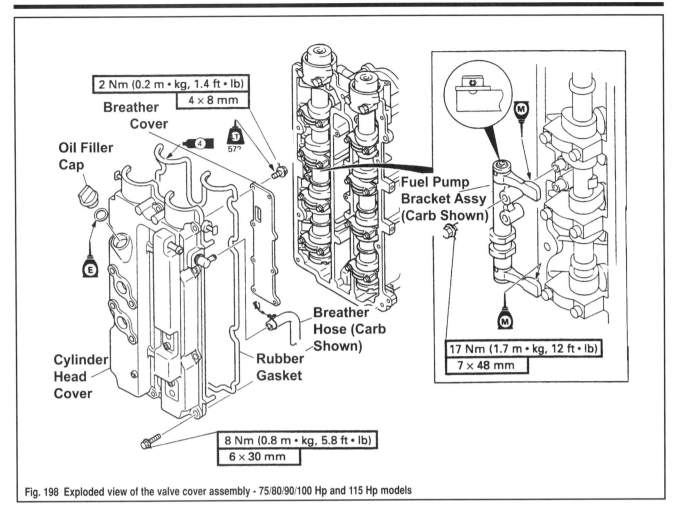

Fig. 198 Exploded view of the valve cover assembly - 75/80/90/100 Hp and 115 Hp models

remove the cylinder head (since they're in the way of some cylinder head mounting bolts). Lastly, it is not clear if the timing belt and sprockets MUST be removed in order to remove the cylinder head cover on all models, but it does make it easier and is probably advisable to ensure proper cover installation and sealing.

■ Only perform the steps which are appropriate for your reason for working on the cylinder head. Therefore, if you are removing the valve cover for valve adjustment or for valve train inspection, there is no need to follow steps like removing the flywheel.

Remember, when loosening the retainers on manifolds, covers and other major components, always try to follow the reverse of the torque sequence indicated in the assembly procedure or molded on the component itself.

When applicable tighten all bolts in using the proper torque sequence and to the proper specification. When possible we've included it in the procedure, but on most Yamaha manufactured powerheads both the sequence and value is molded into the casting on critical components. You'll also notice that most Yamaha torque sequences are clockwise spirals starting somewhere near the center of the component.

■ Most cylinder head bolts for these models require a Torx® T55 socket driver.

1. Disconnect the negative battery cable for safety.
2. Remove the Flywheel and Stator Plate, as detailed in the Ignition and Electrical Systems section.
3. Remove the Timing Belt and the camshaft sprockets, as detailed in this section.

■ Before removing the Timing Belt, align the timing marks and note the camshaft timing mark positions when the marks are aligned. In this way you'll be able to set the camshafts up properly before cylinder head installation.

4. Remove the spark plug cover, then tag and disconnect the plug wires. If removing the cylinder head completely, remove the spark plugs as well. You may also wish to remove the plugs in order to relieve engine compression for tasks like Valve Adjustment.
5. If removing the cylinder head assembly, proceed as follows:
   a. Strip the necessary electrical components from the powerhead.
   b. Remove the Throttle Body and Intake Assembly, as detailed in the Fuel System section.
6. If applicable, loosen the retaining clamp and disconnect the breather hose from the cylinder head cover.
7. Remove the 14 bolts from the cylinder head cover using a counterclockwise spiraling pattern that begins at top right of the cover and works inward.

■ Neither Yamaha nor Mercury states that the cover gasket cannot be reused, however you'll have to use common sense here, especially considering the fact that sealant may have been used the last time it was installed and you may not be able to clean it properly to ensure a good seal. Remember that a leak here can be messy, as that gasket keeps a significant amount of oil (used to lubricate the valve train) from leaking into your cowling. Take a close inspection of the gasket and replace it if you see any signs of distortion, cracking, deterioration or damage.

8. Remove the cylinder head cover, along with the gasket.

■ If you are removing the cylinder head cover for access to the valve train (for valve lash adjustment) refer to the Valve Adjustment procedure under General Information and Maintenance for your next steps.

9. If necessary, loosen the 2 bolts securing the fuel pump bracket assembly (the fuel pump actuator assembly) to the cylinder head and remove the assembly.

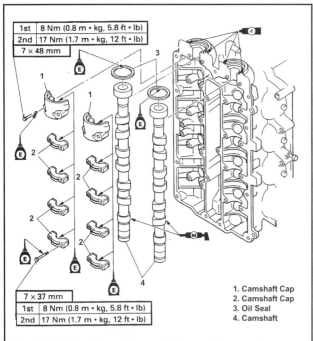

Fig. 199 Exploded view of the camshaft assembly mounting - 75/80/90/100 Hp and 115 Hp models

■ If you are not planning on camshaft or camshaft bearing replacement it is a good idea to measure valve clearances before proceeding, since they can be easily adjusted during installation by replacing the offending shims.

10. To remove the camshafts, loosen the 20 camshaft retaining bolts (10 per shaft) using the reverse of the illustrated torque sequence. Keep track of the bolts because there are two different sizes. The lower 8 bolts on each camshaft are shorter than the top 2.

■ As camshaft caps are removed, be sure to note their positioning for installation in the same locations and facing the same directions. Stamped numbers are already provided on the tops of the caps to accomplish this, identify the numbers as they are removed and compare them to the illustrations that are provided.

11. Remove the oil seal at the top of each camshaft. The same advise we gave for the valve cover gasket goes for the camshaft seals.
12. Remove each of the camshafts.
13. There are 15 cylinder head bolts on these motors, 10 long (10 x 145mm, 1.5mm thread pitch) bolts in a double-row down the head, under the camshafts and 5 short (8 x 55mm) flange bolts on the right side flange. Remove the bolts using multiple passes in the reverse of the molded or illustrated torque sequence. This results in a pattern that starts at the outer flange bolts (bottom bolt) and works inward alternately from top and bottom, then continues with the long bolts starting at the top right and working in a counterclockwise spiral toward the inner bolts).
14. Carefully break the gasket seal and remove the cylinder head from the cylinder block.
15. Remove and discard the old cylinder head gasket. Carefully remove all traces of the gasket material from the mating surfaces.

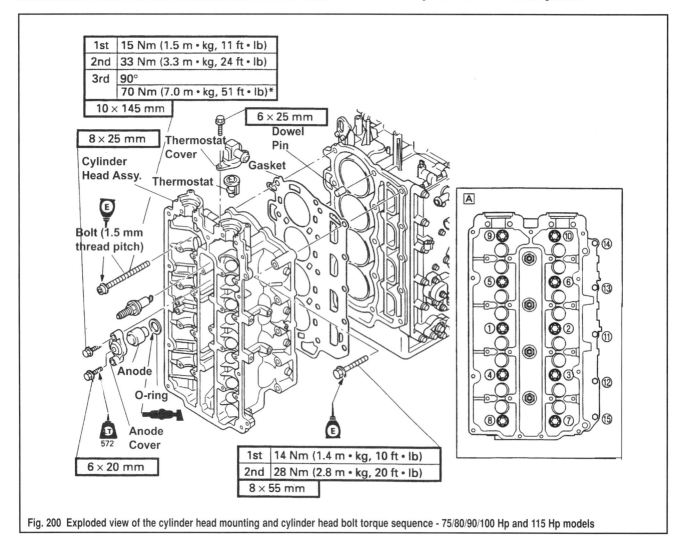

Fig. 200 Exploded view of the cylinder head mounting and cylinder head bolt torque sequence - 75/80/90/100 Hp and 115 Hp models

POWERHEAD 6-91

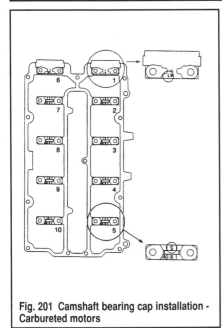

Fig. 201 Camshaft bearing cap installation - Carbureted motors

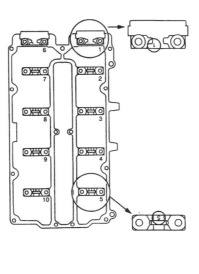

Fig. 202 Camshaft bearing cap installation - EFI motors

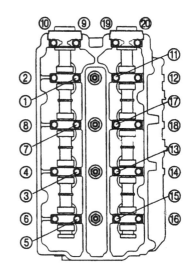

Fig. 203 Camshaft bearing cap bolt torque sequence - Carbureted motors and Mercury/Mariner EFI motors

16. Check the condition of the 2 cylinder head dowel pins and replace, if necessary.

17. If necessary for further cylinder head disassembly remove the thermostat assembly from the top of the head and/or the anode cover(s), anode(s) and O-ring(s) from the center of the head.

18. Refer to Valve Train Overhaul in this section for details on removal and installation of the remaining valve train components. Also, be sure to check the information on gasket mating surfaces to make sure the cylinder head is not warped.

To Install:

19. If removed, install the anode assembly or assemblies to the center of the cylinder head. For each assembly, apply a light coating of marine grade grease (Yamaha) or 2-4-C with Teflon (Mercury/Mariner) to the O-ring and Loctite® 572 (Yamaha) or Loctite 271 (Mercury/Mariner) or equivalent threadlocking compound to the threads of the anode cover retaining bolt. Install the O-ring, anode and cover, then tighten the retaining bolts securely.

20. If removed, install the thermostat assembly to the top of the cylinder head. For details, please refer to the Lubrication and Cooling section.

■ If crankshaft positioning in the cylinder block was disturbed, slowly turn the crankshaft in the normal direction of rotation (clockwise when viewed from above) until the No. 1 cylinder is at TDC.

21. Apply a light coating of clean engine oil to the threads of the cylinder head bolts.

22. Position a new cylinder head gasket on the cylinder block over the 2 dowel pins, then carefully position the cylinder head over the gasket and onto the block. Thread the long and short cylinder head bolts making sure the gasket remained perfectly in position.

23. Tighten the long and short cylinder head bolts using multiple passes of the molded or illustrated torque sequence. The sequence is essentially a clockwise spiraling pattern that starts at the left, center of the long bolts and works outward through the long bolts and then continues with the center of the short flange bolts, then goes back and forth working toward the bottom and top of the flange bolts. Tighten the long bolts first to 132 inch lbs./11 ft. lbs. (15 Nm), next to 24 ft. lbs. (33 Nm) for 75-100 hp motors or 22 ft. lbs. (30 Nm) for 115 hp motors, and finally for all motors in the third pass to an additional 90° using a torque angle meter (both Yamaha and Mercury give a torque value for the third tightening of 51 ft. lbs./70 Nm, but warn it is to be used only as a reference). Tighten the short flange bolts first to 120 inch lbs./10 ft. lbs. (14 Nm) and then to 20 ft. lbs. (28 Nm).

24. Apply a light coating of a molybdenum disulfide engine assembly lube to the journals and lobes of the camshafts and bearing surfaces of the cylinder head and caps.

25. Position the camshafts to the cylinder head, position them so the No. 1 cylinder would be at TDC (valves closed, timing marks aligned if the sprockets were to be installed).

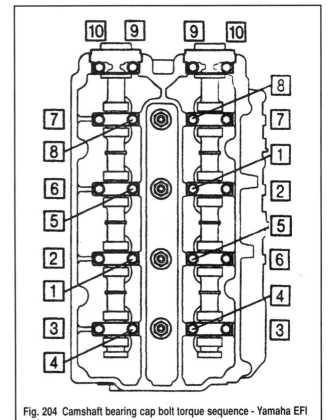

Fig. 204 Camshaft bearing cap bolt torque sequence - Yamaha EFI motors

■ Alternately, Mercury advises that you turn the crankshaft 90 degrees FROM TDC (in either direction) before camshaft installation, as this position will protect the valves. Also, Mercury notes that the camshaft with the tang and pink identifying mark should go on the exhaust side of the cylinder head.

26. Apply a light coating of engine oil to the camshaft oil seals and position the seals. The manufacturers do not say they MUST be replaced, but we recommend it.

27. Apply a light coating of engine oil to the 20 camshaft cap retaining

6-92 POWERHEAD

bolts (don't mix them up, but if you do the 4 long ones go on the 2 top bearing caps).

28. Apply a light coating of Yamabond No.4, RTV 587 Silicone Sealant, or equivalent sealant/adhesive to the mating surface for the oil seal housing/top bearing cap for each camshaft.

29. Position the bearing caps as noted during removal and as stamped. This typically means the stamped numbers are positioned upside down and the caps are installed numbers 1-5 top-to-bottom on the right side (when looking at the camshafts) and numbers 6-10 on the left side. Install and finger-tighten the bolts.

30. Tighten the cylinder head cap bolts using multiple passes of the appropriate, illustrated torque sequence. Tighten the bolts first to 70 inch lbs./5.8 ft. lbs. (8 Nm) and then to 150 inch lbs./12 ft. lbs. (17 Nm).

■ If the valve train components were replaced or if you failed to take measurements before the camshafts were removed, take the time now to check and adjust the valve clearances.

31. If removed, apply a light coating of molybdenum disulfide to the wear surfaces on the arms of the fuel pump bracket assembly, then install the assembly and tighten the 2 retaining bolts to 70 inch lbs./12 ft. lbs. (17 Nm).

32. Apply a light coating of Yamabond No.4, RTV 587 Silicone Sealant or equivalent sealant/adhesive to the valve cover gasket, then carefully install the cover and gasket to the cylinder head. Tighten the cylinder head cover bolts to 70 inch lbs./5.8 ft. lbs. (8 Nm) using multiple passes of a clockwise spiraling pattern that begins at the center bolts and works outward.

33. If applicable, connect the breather hose to the cylinder head cover and secure using the retaining clamp.

34. Install the spark plugs to the cylinder head, then install the spark plug cover.

35. If the cylinder head was removed, proceed as follows:
  a. If removed, install the Throttle Body and Intake Assembly, as detailed in the Fuel System section.
  b. Install the electrical components which were stripped from the powerhead and/or cylinder head.

36. Install the Timing Belt/Sprocket Assembly, as detailed in this section.
37. Install the Flywheel and Stator Plate, as detailed in the Ignition and Electrical Systems section.
38. Reconnect the negative battery cable.

VALVE TRAIN OVERHAUL

◆ See Figure 205

■ If you only need to check/replace the valve shims or lifters and not the valves themselves there is no need to remove the cylinder head.

1. Remove the Cylinder Head from the powerhead assembly, as detailed in this section.

■ Valve train components such as valve lifters, retainers, keepers, springs, seats and valves should not be mixed and matched from one assembly to another. This means that if they are to be reused, they must all be labeled prior to removal or VERY carefully sorted during service to prevent installation to the wrong positions.

To remove the each valve assembly, proceed as follows:
2. Remove the valve shim and lifter from the bore in the cylinder head to expose the valve step and keepers.
3. Using a clamp-type valve spring compressor, carefully place just enough pressure on the valve spring so the retainer moves down sufficiently to expose the keepers.
4. Remove the pair of valve keepers, then slowly ease pressure on the valve spring until the compressor can be removed.
5. Remove the valve spring retainer, valve spring and valve spring seat from the shaft of the valve.
6. Remove the valve from the underside (combustion chamber side) of the cylinder head.
7. If necessary, carefully pry the old valve stem seal from its position on the cylinder head.
8. Repeat this procedure for each of the remaining valves.
9. Refer to Powerhead Refinishing, later in this section, for details on cleaning and inspecting cylinder head components.

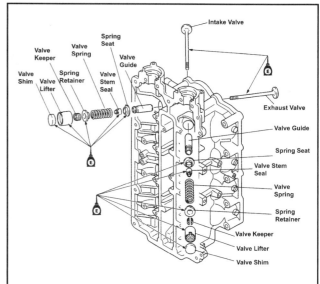

Fig. 205 Exploded view of the valve train assembly - 75/80/90/100 Hp and 115 Hp models

To Install:

■ Apply a light coating of clean engine oil to each of the valve train components, except the valve spring and the valve keepers as they are installed. You can use a dab of marine grade grease on the keepers in order to help keep them in position during installation. One popular method is to actually use the grease to temporarily adhere the keepers to the installation tool itself, using the retainer to free it from the tool and place it in position.

10. If removed, install new valve spring seals to the cylinder head.
11. Position the valve into the cylinder head (into the original guide and valve seat unless the valve or valve seat was replaced).
12. Position the valve spring seat down onto the valve shaft and cylinder head, then oil and install the spring with the fine pitch side of the spring facing down toward the seat.
13. Place the retainer over the spring, then use the valve spring compressor tool to press downward on the retainer and spring so you can install the keepers.
14. Insert the valve keepers, then slowly back off the valve spring compressor, making sure the keepers are fully in position. Then remove the valve spring compressor.
15. Install the valve lifter and shim.
16. Repeat for each of the remaining valves.
17. Install the Cylinder Head as detailed in this section.
18. Once the powerhead is fully assembled, if valve train components were replaced, be sure to operate the engine as directed for new component break-in. For details, please refer to Powerhead Break-In, in this section.

CYLINDER BLOCK OVERHAUL

◆ See Figures 206 thru 215

■ Unique to the large 4-stroke Yamahas and Mercury/Mariners, the crankcase bearing and connecting rod bolts require a torque angle meter for tightening to proper specifications. Although some people will visually site the spec on when it is 90 degrees, the use of an inexpensive gauge is really necessary on the crankcase bearings for all motors, as well as the connecting rod cap bolts for 115 hp motors (since the spec is 60 degrees and that is hard to eyeball).

Remember, when loosening the retainers on manifolds, covers and other major components, always try to follow the reverse of the torque sequence indicated in the assembly procedure or molded on the component itself.

When applicable tighten all bolts in using the proper torque sequence and to the proper specification. When possible we've included it in the procedure,

POWERHEAD

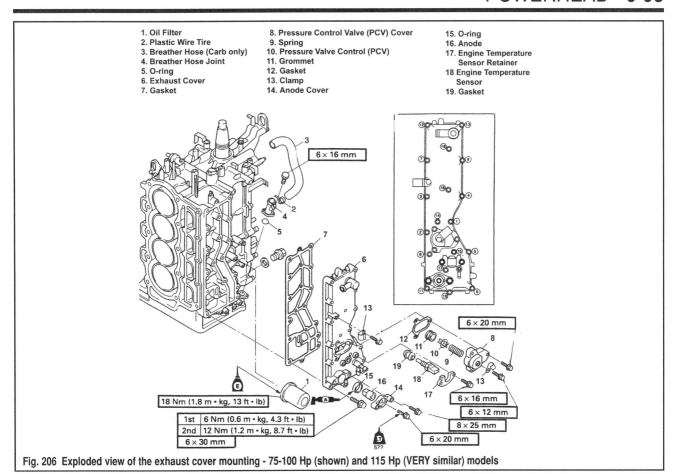

Fig. 206 Exploded view of the exhaust cover mounting - 75-100 Hp (shown) and 115 Hp (VERY similar) models

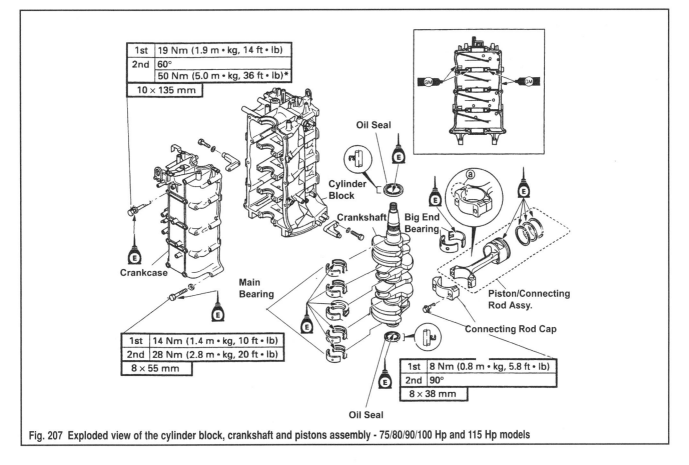

Fig. 207 Exploded view of the cylinder block, crankshaft and pistons assembly - 75/80/90/100 Hp and 115 Hp models

6-94 POWERHEAD

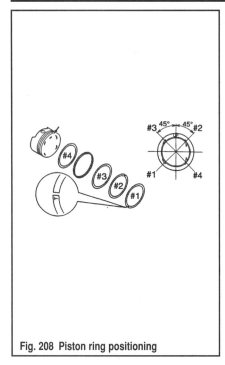

Fig. 208 Piston ring positioning

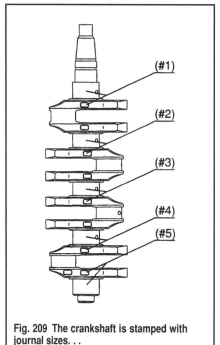

Fig. 209 The crankshaft is stamped with journal sizes...

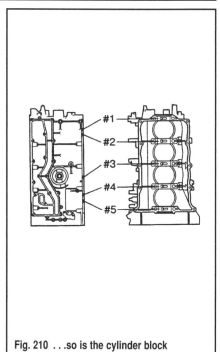

Fig. 210 ...so is the cylinder block

but on most Yamaha manufactured powerheads both the sequence and value is molded into the casting on critical components. You'll notice that most Yamaha torque sequences are clockwise spirals starting somewhere near the center of the component.

■ Most connecting rod cap screws require a Torx E10 socket for 75-100 hp motors or a 5/16 in. 12-point socket for 115 hp motors.

1. Remove the Powerhead assembly, as detailed in this section.
2. Remove the Timing Belt/Sprocket assembly, as detailed in this section.
3. Strip all electrical components from the powerhead, then remove the Cylinder Head assembly, as detailed in this section.
4. Remove and discard the old oil filter from the cylinder block.
5. Gradually loosen and remove the exhaust cover retaining bolts using the reverse of the torque pattern (roughly a back and forth pattern starting with the top and bottom bolts along the center of the cover, working toward the middle bolts along the center, then a counterclockwise spiraling pattern of the bolts along the flange that starts with the top right bolt and moves toward the center of the flange).
6. Remove the exhaust cover and gasket from the side of the powerhead. Remove and discard the old exhaust cover gasket.

■ There are multiple components mounted to the exhaust cover on these motors, the and anode at the lower left corner of the cover, a pressure control valve assembly above the anode and an temperature sensor either between the two (on carb motors) or up toward the top right of the cover (on EFI models). These components should ALL be removed from the cover for inspection and/or replacement. See the accompanying illustration for more details.

7. Remove the crankcase flange and center bearing retaining bolts using the reverse of the molded torque sequence. Start each sequence with the 2 center bearing bolts, then use a counterclockwise spiraling sequence that begins with the flange bolts (at the upper right bolt and works inward) and then moves each time to the center bearing bolts (again, starting at the upper right bolt and working inward).

■ Once removed keep the long 1.5mm thread pitch center bolts separated from the shorter flange bolts.

8. Carefully break the seal and separate the crankcase from the cylinder block.

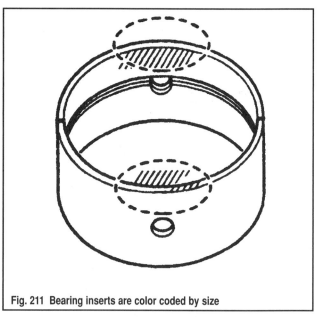

Fig. 211 Bearing inserts are color coded by size

■ Most parts inside the cylinder block are matched sets and should not be moved from one cylinder to another. This includes the bearings and pistons as well as each of the connecting rod/end cap combination. For this reason, prior to disassembly it is essential to matchmark each of the connecting rods to their end caps (this ensures that not only will each cap be installed on the proper rod, but facing the correct direction). Also, be sure to label each piston and connecting rod as to what cylinder it belongs and identify the factory marks and/or place an arrow or some indicator facing upward to ensure they are installed in the same direction. Yamaha often marks pistons with UP on the dome and connecting rods with the letter Y or word YAMAHA and/or alignment dots at the con rod-to-cap mating surface, all of which should face the flywheel.

9. Matchmark the big end caps to the connecting rods, then remove the 8 connecting rod cap bolts (working on one cap at a time, alternating between each of the cap bolts after each turn). Remove the caps and place them aside for reinstallation back on their original connecting rods as soon as the crankshaft has been removed.

POWERHEAD 6-95

10. Carefully lift the crankshaft from the block. Place it on a protective surface.

※※ WARNING

Handle the crankshaft carefully. Remember it is a heavy, expensive and vital piece to the motor. The polished surfaces should be covered to prevent scratches or other damage and the whole shaft should be protected against impact or other damage which could destroy it.

11. Immediately reinstall the connecting rod end caps and their bearing inserts to their respective pistons.
12. Remove the top and bottom oil seals from the crankshaft. Discard the seals.

■ These pistons and connecting rods are normally already marked from the factory as to what side faces up, identify the marks and perhaps, remark them to ensure they are found easily during assembly.

13. Label and remove each of the pistons from the bores. Number the pistons and place a mark (or identify the factory marks) on each piston and connecting rod facing up (facing the flywheel) for installation purposes. If necessary, disassemble each of the pistons, as follows:
    a. If equipped, carefully pry the clips free from the piston pin bore on either side of the piston assembly.
    b. Slide the piston pin free of the bore (specifications tell us that this should not be an interference fit, but the Yamaha service publications do not give any procedures, so we cannot be certain. If necessary, bring the assembly to a machine shop to determine if it is an interference fit and/or if the piston pin should even be removed).
    c. Using a piston ring spreading tool, carefully remove each of the rings (top and second compression rings, followed by the oil ring assembly) from the piston.
    d. Keep all components from a single piston together for reuse or replacement, but don't mix and match components from other pistons.
14. Remove the crankshaft main (big end) bearings from the crankcase and cylinder block. Again, note their positions and orientation if they are going to be reused.

■ On these models different color code bearings may be installed on the same crankshaft journal (one in the crankcase and a different one in the block) in order to achieve the proper clearance measurement. Pay close attention to bearing positions and codes.

15. Refer to Powerhead Refinishing, later in this section, for details on cleaning and inspecting cylinder block components. There are probably more components and specifications to measure on these 4-stroke motors than on any smaller Yamaha or Mercury/Mariner engine and even most larger 2-strokes. Take your time and make sure you've made measurements for all components listed in the Engine Specifications chart for this model.

Connecting rod bearing selection table (20 °C (68 °F))

Measurement value using a Plastigauge® (mm)	Upper bearing	Lower bearing
0.020 ~ 0.022	Yellow	Yellow
0.023 ~ 0.029	Yellow	Green*
0.030 ~ 0.036	Green	Green
0.037 ~ 0.044	Green	Blue*
0.045 ~ 0.052	Blue	Blue
0.053 ~ 0.058	Blue	Red*

CAUTION:
The (*) mark indicates that the color of the upper and lower bearings are different.

Fig. 212 Connecting rod bearing selection chart - Carbureted motors

To Install:

16. Make sure the mating surfaces of the crankcase and cylinder block are completely clean and free of sealant or damage. The dowel pins should already be installed in the block, but if not, obtain new pins and install them in the 2 positions at opposite ends on the oil filter housing side of the block.
17. Place the main bearing inserts into the crankcase and cylinder block. There are normally 2 things to look out for when doing this. Most bearing inserts will be equipped with locating tabs, make sure they are seated in the appropriate grooves. Also, inserts with oil holes should be aligned with an oil passage.
18. If they were disassembled, prepare each piston assembly for installation as follows:
    a. Using a ring expander, carefully install each of the rings with their gaps as noted in the accompanying illustration. While looking at the piston dome with the top (flywheel end) facing 12 o'clock, the bottom oil ring should be positioned with the gap facing between 4 and 5 o'clock. There is no specified placement for the gap (if any is present) on the middle oil ring (expander ring). The upper oil ring should be positioned with the gap facing between 10 and 11 o'clock. The lower (second) compression ring should be positioned with its gap at between 1 and 2 o'clock. The upper (top) compression ring should be positioned with its gap facing between 7 and 8 o'clock.

■ Some rings, like the top compression ring, will be equipped with a stamp mark (T in the case of the top compression ring) which should be faced upward toward the top of the piston dome.

    b. Align the mark(s) on the connecting rod (facing up/flywheel) with the mark on the piston (also facing up/flywheel), install the connecting rod to the piston using the piston pin.
    c. If applicable, install the piston pin clips in either side of the pin bore to secure the piston pin in place.
    d. Repeat for the remaining pistons.

Connecting rod bearing selection table (20 °C (68 °F))

Measurement value using a Plastigauge® (mm)	Upper bearing	Lower bearing
0.025 - 0.031	Yellow	Yellow
0.032 - 0.039	Yellow	Green*
0.040 - 0.046	Green	Green
0.047 - 0.052	Green	Blue*
0.053 - 0.058	Blue	Blue
0.059 - 0.063	Blue	Red*

Fig. 213 Connecting rod bearing selection chart - EFI motors

Crankshaft bearing selection table (20 °C (68 °F))

Cylinder block journal diameters – crankshaft journal diameters (mm)	Bearing (cylinder side)/ thrust bearing	Bearing (crankcase side)
6.023 - 6.026	Green	Yellow*
6.027 - 6.034	Blue	Green*
6.035 - 6.042	Blue	Blue
6.043 - 6.049	Red	Blue*

Fig. 214 Crankshaft main bearing selection chart

6-96 POWERHEAD

※※ WARNING

Use extreme care to prevent the connecting rods from contacting and scoring the cylinder walls during piston installation. It is a good idea to wrap the ends in a shop cloth or something else to protect the cylinders.

19. Install each of the pistons (with the correct mark/side facing UP toward the flywheel) to the cylinder block using a ring compressor. Apply a very light coating of clean engine oil to the inside of the compressor and to the cylinder bore, then place the compressor around the piston. Insert the piston through the cylinder head side of the bore and use a wooden hammer handle to gently tap the piston down through the compressor and into the cylinder bore.

※※ WARNING

Be VERY careful when installing the piston. Get the lower portion of the skirt below the ring compressor and insert it into the cylinder bore to make sure everything is aligned. Then, GENTLY tap the piston through the compressor, stopping if it seems to hang up even slightly to make sure that a ring is not caught on the deck of the cylinder block.

20. Remove the connecting rod end caps and position them so the crankshaft can be installed. Place the connecting rod bearing inserts into position on the connecting rod big end and cap.

■ Here's where procedures differ slightly depending upon whether you're installing bearings of a known clearance or measuring clearance. If you've already determined that bearing clearances are fine, apply alight coat of engine oil to each of the connecting rod, main bearing and crankshaft journal surfaces. If however, you need to measure and determine bearing clearance, Temporarily INSTALL THESE COMPONENTS DRY for measurement purposes.

A few notes on connecting rod and crankshaft main bearing selection for these models. Use your Plastigage® measurements and the accompanying tables to help select replacement connecting rod bearings.

When it comes to the crankshaft and cylinder block, you can reach a good starting point for bearing selection based on the stamped codes on the shaft itself. For each crankshaft journal take the base number 47.900 and ADD the number stamped on the journal divided by 1000 (move the decimal point 3 spots to the left). So if the number stamped was 91 move the decimal points and get 0.091, then add that to 47.900 and get 47.991. Easy right? Same goes for the cylinder block, except start with the base number 54.000.

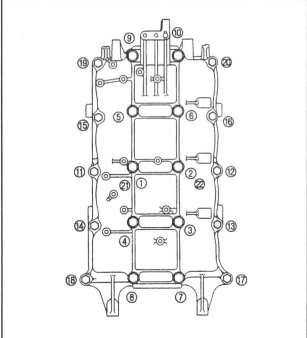

Fig. 215 Crankcase torque sequence

Also, note the following measurements will require replacement of the crankshaft and connecting rods:
• If the measured difference between the cylinder block journal diameter and crankshaft journal diameters is above 6.058mm.
• If the measurement of the connecting rod bearing clearance on carbureted motors is more than 0.058mm.
• If the measurement of the connecting rod bearing clearance on EFI motors is more than 0.071mm.

21. Position the crankshaft to the cylinder block, and carefully pull each of the connecting rods into position.

■ NEVER install the upper or lower crankshaft oil seals with the crankcase installed and tightened otherwise the seal will likely be damaged.

22. Oil and install a new upper and lower oil seal to crankshaft. Both seals should be positioned so the seal lips are facing inward toward the crankshaft. Seat the seals as best you can at this time. To ensure the seals are not cocked you can install the crankcase and FINGER-TIGTHEN the bolts then lightly knock the seals using a suitably sized driver and rubber mallet. Once this is done, remove the crankcase again and continue the cylinder block assembly.

23. Install each of the match-marked connecting rod caps and tighten the bolts alternating back and forth between the 2 bolts on each cap. For 75-100 hp motors, tighten the bolts first to 70 inch lbs./5.8 ft. lbs. (8 Nm) and then, using a torque angle gauge tighten the bolts an additional 90 degrees each. For 115 hp motors, tighten the bolts first to 132 inch lbs./11 ft. lbs. (15 Nm) and then, using a torque angle gauge tighten the bolts an additional 60 degrees each.

24. If you've already measured bearing clearances, apply a light coating of Gasket Maker (such as Loctite 514 Master Gasket, RTV sealant or equivalent) to the entire perimeter of the cylinder block-to-crankcase mating surface (that includes the bearing holder mating surface along the center of the crankcase/block). A small paint roller is really handy for this application.

25. Apply a light coating of engine oil to the threads of the long and short crankcase bolts.

26. With the main bearing inserts in place, carefully lower the crankcase onto the cylinder block, then JUST thread the long center bolts and shorter flange bolts.

27. Tighten the long center bolts and short flange bolts using 2 or more passes of the molded torque sequence (or if not present, a clockwise, spiraling pattern that starts at the left, center of the long bolts and works toward the ends of the crankcase, then continues with the left center of the short flange bolts and again works towards the ends of the crankcase then ends with re-tightening the 2 center flange bolts). During the torque sequence, tighten the long bolts first to 168 inch lbs./14. ft. lbs. (19 Nm) and an additional 60 degrees (both Yamaha and Mercury give a reference of 36 ft. lbs./50 Nm for the angle torque, but they note it is for reference only, use an angle meter). During the sequence tighten the short bolts first to 120 inch lbs./10 ft. lbs. (14 Nm) and then to 20 ft. lbs. (28 Nm).

28. Install the exhaust cover using a new gasket. Tighten the bolts using multiple passes of a clockwise spiraling pattern that starts with the flange bolts (at the right bolt, 3rd from the bottom) and works outward, then continues with the bolts down the center of the cover (again working from the center, alternating bottom-to-top). Tighten the bolts first to 53 inch lbs./4.3 ft. lbs. (6 Nm) and then to 106 inch lbs./8.7 ft. lbs. (12 Nm).

■ If removed, install the pressure relief valve, temperature sensor and/or anode to the exhaust cover. In the case of the anode, be sure to apply a light coat of marine grade grease (Yamaha) or Quicksilver 2-4-C with Teflon (Mercury/Mariner) to the O-ring and a light coat of Loctite®572 (Yamaha), Loctite® 271 (Mercury/Mariner) or an equivalent threadlocking compound to the threads of the 2 cover retaining bolts.

29. Install a new oil filter to the powerhead.
30. Install the Cylinder Head assembly (along with the camshafts and any electrical components stripped from the powerhead) as detailed in this section.
31. Install the Timing Belt/Sprocket assembly, as detailed in this section.
32. Install the Powerhead assembly, as detailed in this section.
33. Once the powerhead is fully assembled, if cylinder block components were replaced, especially the pistons and/or rings, be sure to operate the engine as directed for new component break-in. For details, please refer to Powerhead Break-In, in this section.

POWERHEAD 6-97

CLEANING & INSPECTING

Cleaning and inspecting internal engine components is virtually the same for all Yamaha and Mercury/Mariner outboards with the exception of a few components that are unique to certain 4-stroke motors. The main variance between different motors comes in specifications (which are listed in the Engine Specifications charts) or by component type. A section detailing the proper procedures, sorted mostly by component, can be found under Powerhead Refinishing.

150 Hp Yamaha Powerheads

POWERHEAD REMOVAL & INSTALLATION

◆ See Figures 216, 217 and 218

The powerhead may be removed or installed as an almost fully assembled unit. The components mentioned here are the bare minimums that must be removed before separating the powerhead assembly from the rest of the motor. However, if overhaul or replacement of the cylinder head and/or block is desired, you may wish to remove additional components before removing the powerhead assembly, not only to lighten the load, but to help prevent potential damage to the external fuel and electrical system components mounted to the powerhead. The choice is yours, so if necessary, refer to other component procedures in this section and in the Fuel System and/or Ignition and Electrical Systems sections before proceeding.

■ **Yamaha recommends breaking the flywheel nut loose (just slightly) BEFORE removing the powerhead. This is probably because it will be easier than trying it in a bench. However, if you're using a sturdy enough engine stand, it probably doesn't matter.**

1. Disconnect the negative battery cable at the battery itself for safety.
2. Remove the top engine cowl.
3. Remove the flywheel cover.

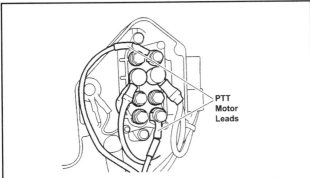

Fig. 216 Power Trim/Tilt (PTT) leads in the electrical junction box - 150 Hp models

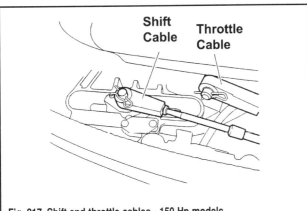

Fig. 217 Shift and throttle cables - 150 Hp models

4. If desired, use a flywheel holder tool (like #YB-06139) to break the flywheel nut loose before proceeding.
5. Tag and disconnect the battery leads at the starter and just below the starter on the powerhead.
6. Disconnect the throttle and shift cables by removing the retaining clops and pulling the cables free of the linkage pins.
7. Remove the cover from the electrical junction box for access to wiring.
8. Tag and disengage the PTT motor leads from the relays mounted in the junction box. The relays are found in an assembly with 8 large terminals, the motor leads are at the top right and bottom right of those terminals, but use the Wiring Diagrams to check the lead colors if you are uncertain.
9. Disconnect the fuel inlet hose from the fuel filter on the powerhead.
10. Remove the engine oil dipstick, then unbolt and remove the dipstick guide. Depending upon the amount of overhaul and service condition of the motor, it is probably a good idea to drain the engine oil at this time. There is an O-ring that seals the dipstick guide tube, locate and discard the O-ring (using a NEW O-ring for assembly).
11. Carefully cut the plastic wire ties that secure the pilot water (cooling system indicator) hose and the flushing water hose to the fittings on the powerhead. Tag and disconnect the hoses.
12. Tag and disconnect the wiring for the PTT switch coupler, the shift cut switch coupler and the neutral switch coupler.
13. Unbolt and remove the lower engine apron for access to the powerhead mounting bolts. The apron is normally a flexible, 1-piece assembly that is secured at the front of the motor assembly using 2 screws and 2 nuts threaded horizontally at the front of the powerhead, plus 4 or 5 bolts threaded downward through the lower engine cowling into the apron. First loosen and remove the 2 screws and nuts at the front of the motor.
14. Next, locate and remove the 4-5 bolts which area threaded downward through the lower engine cowling into the top of the apron. There are normally 2 or 3 on the port side of the motor and 2 on the starboard side of the motor (located toward the rear of the assembly). Once the bolts are removed, carefully remove the apron for access to the powerhead mounting bolts.
15. Remove the 15 bolts threaded upward into the powerhead from the intermediate housing. Keep track of the bolts as there are 3 different sizes. There should be 4 long bolts on each side, 4 thick, short bolts at the center front of the powerhead, and 3 thinner, short bolts at the rear of the powerhead.
16. Attach a suitable lifting device and carefully lift the powerhead assembly from the intermediate housing. In order to prevent potential damage to the crankshaft and driveshaft be certain to lift the powerhead STRAIGHT up and off the intermediate housing.
17. Remove and discard the upper casing gasket from the powerhead-to-intermediate housing mating surface. Carefully clean the mating surfaces of all traces of old gasket.
18. Check for placement of the 2 dowel pins. They may have remained with the intermediate housing or have come away with the powerhead itself. Inspect the pins for damage and replace, if necessary.

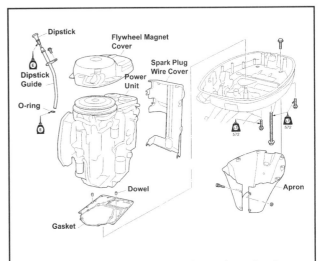

Fig. 218 Exploded view of the powerhead mounting and engine apron - 150 Hp models

6-98 POWERHEAD

19. Position the powerhead on a suitable workbench or engine stand for further disassembly or overhaul, as necessary. Refer to the additional procedures in this section as needed.

To Install:

20. With the 2 dowel pins in place, position a NEW gasket to the intermediate housing.
21. Apply a light coat of marine grade grease to the driveshaft-to-crankshaft mating splines.
22. Using a suitable lifting device, carefully lift the powerhead and lower it slowly, STRAIGHT onto the driveshaft and intermediate housing. If necessary, rotate the flywheel slightly in a clockwise direction in order to mate the crankshaft to the driveshaft.
23. Apply a light coating of Loctite® 572 to the threads of the powerhead mounting bolts. Although the manufacturer only specifies that the threadlocking compound is necessary on the long bolts, it is not a bad idea to use it on the short powerhead bolts in order to reduce the possibility of corrosion locking them in place in the future.
24. Install the 15 powerhead retaining bolts. Tighten the thick bolts (M10 both short and long) at the sides and front of the powerhead to 31 ft. lbs. (42 Nm).
25. Install the apron to the engine cowling and secure using the 4-5 vertically threaded bolts and then with the 2 horizontal screws (along with the nuts and washers on the horizontal bolts at the front of the apron). Tighten the vertically mounted bolts to 5.9 ft. lbs. (8 Nm) and the horizontally mounted screws/nuts to 3.0 ft. lbs. (4 Nm).
26. Reconnect the PTT switch coupler, shift cut switch coupler and neutral switch coupler.
27. Reconnect and secure the cooling water pilot hose and flushing hose. These hoses are normally secured using new wire ties.
28. Apply a light coat of engine oil to a new O-ring and to the dipstick guide tube, then install the guide tube to the powerhead and secure using the retaining bolt. Install the engine oil dipstick and, if drained, refill the engine with fresh engine oil.
29. Reconnect the fuel line to the filter assembly on the powerhead.
30. Inside the electrical junction box, reconnect the PTT motor leads to the relay terminals from which they were removed (top right and bottom right of the relay assembly). Install the electrical junction box cover.
31. Reconnect the throttle and shift cables to the linkage pins and secure using the retaining clips.
32. Reconnect the battery leads to the starter and to the powerhead ground point right below the starter assembly.
33. If loosened or removed, install and/or tighten the flywheel retaining nut to 199 ft. lbs. (270 Nm).
34. Reconnect the negative battery cable to the battery itself.
35. Be sure to check all Timing and Synchronization adjustments as detailed in the Engine and Maintenance section.
36. If the powerhead was overhauled (replacing components such as the valve train, main bearings, crankshaft, pistons and/or rings) be sure to subject the motor to a complete new engine break-in period to ensure long, trouble free life.

TIMING BELT/SPROCKET REMOVAL & INSTALLATION

◆ See Figures 219 thru 222

This procedure can be performed with the powerhead assembly removed or installed on the motor. And, unlike the smaller 4-cylinder Yamahas, if the crankshaft sprocket need to be removed for any reason, it is not a big deal. This is because the crankshaft sprocket is secured using 4 low-torque bolts instead of one high torque nut.

If the timing belt is to be reused, be sure to mark the direction of rotation on the old belt before removal. Also, if either the cylinder head or cylinder block is not to be disassembled further, be sure to properly align the timing marks before removal. This will significantly ease installation.

■ **Camshaft sprocket removal requires the removal of the valve cover so that you can hold the camshaft itself from turning using a wrench.**

1. If the powerhead is still installed, disconnect the negative battery cable for safety.
2. Remove the Flywheel and Stator Plate, as detailed in the Ignition and Electrical Systems section.
3. If either the cylinder head or cylinder block is not being overhauled, rotate crankshaft sprocket in the normal direction of rotation (clockwise) in order to align the timing marks on the camshaft pulleys at the center (facing toward each other) and the marks on the crankshaft pulley and plate with the mark on the powerhead (facing toward the camshaft sprockets, aligned with the centerline of the powerhead).
4. If removing the camshaft sprocket(s) and/or the cylinder head itself, remove the valve cover, as follows:
   a. If not done already, remove the spark plug cover.
   b. Tag and disconnect the spark plug wires.
   c. Tag and disconnect the hoses for the fuel pumps.
   d. Disconnect the breather hose from the top center of the valve cover.
   e. Loosen and remove the 15 valve cover retaining bolts using multiple passes either in the reverse of the embossed torque sequence (if equipped) or using a criss-cross or spiraling pattern that starts at the edges and works inward.
   f. Carefully remove the valve cover, then remove and DISCARD the valve cover gasket. Clean the top of the cylinder head and valve cover where sealant was applied to the original gasket (on the top corners of the gasket, near the camshaft oil seals).
5. Insert a hex key into the top of the timing belt tensioner and use the key to turn the tensioner CLOCKWISE against spring pressure, relieving belt tension. Insert a 0.2 in. (5mm) pin into the hole next to the hex in the tensioner in order to hold the tensioner steady in this position.

■ **Remember, if the timing belt is to be reused, be sure to mark it showing the normal direction of rotation to ensure installation in the proper position. Most OE timing belts they are normally installed with the labeled part numbers facing upward (so that you can read the part number) when installed.**

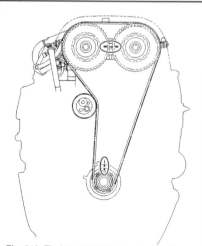

Fig. 219 Timing mark alignment - 150 Hp models

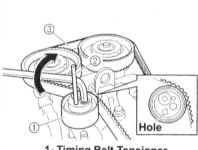

1- Timing Belt Tensioner
2- Pin
3- Timing Belt

Fig. 220 Use a hex wrench to rotate the tensioner (taking pressure off the belt), then put a 5mm pin in the tensioner to hold that position

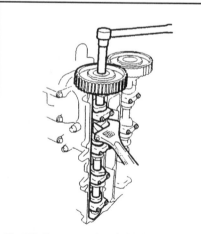

Fig. 221 Use a wrench to hold the camshaft from turning when removing/installing the drive sprocket

6. Remove the timing belt from the sprockets.
7. If necessary, unbolt and remove the tensioner from the powerhead (but don't remove the pin).
8. If necessary, remove one or both of the camshaft sprockets using a wrench on the flat provided toward the top of the camshaft to hold the shaft (and sprocket) from turning while you loosen the sprocket retaining bolt.
9. If necessary, loosen each of the crankshaft sprocket bolts (they are relatively low torque and shouldn't be a problem, but if necessary, place a prybar across the heads of 2 bolts to keep the crankshaft from rotating backwards while breaking each bolt loose. Once the bolts are all loosened, remove the crankshaft sprocket and the lower plate from the top of the crankshaft. Remove the woodruff key from the crankshaft and retain for use during installation.

To Install:

10. If removed, apply a small dab of marine grade grease to the woodruff key (in order to help hold it in place) and position it into the slot in the crankshaft. With the triangular timing mark facing upward, slide the crankshaft pulley base plate over the crankshaft. Align the slot in the plate with the keyway as the plate is seated on the powerhead. Install the crankshaft sprocket over the plate and woodruff key. Again, the timing mark on the sprocket should align with the timing marks/keyway. Apply a light coating of engine oil to the threads of the retaining bolts, then install and tighten the bolts to 5.2 ft. lbs. (7 Nm).
11. If removed, install each of the sprockets to the camshaft over the locating dowel pin in each shaft. If the camshaft positions were not disturbed, make sure the timing marks on the sprockets are aligning with each other at a point closest together in their travel. Apply a light coating of engine oil to the threads of the camshaft sprocket retaining bolts. Install a bolt and washer to each camshaft sprocket, then using a wrench to hold each shaft from turning tighten the bolt to 44 ft. lbs. (60 Nm).
12. If removed, install the timing belt tensioner to the powerhead and tighten the retaining bolt to 28.8 ft. lbs. (39 Nm).
13. Install the timing belt, making sure the timing marks are still aligned. Remember, if the belt is being reused, be sure to face it in the same direction as normal rotation (clockwise). New belts from Yamaha should be installed with the part numbers facing so they are legible when the belt is in place. Position the belt over the sprockets, starting with the crankshaft sprocket and working your way around the camshaft sprocket and the tensioner.
14. Use a hex to relieve the spring pressure on the pin, then remove the pin and SLOWLY allow the tensioner to apply tension to the timing belt. Remove the hex key.
15. Turn the crankshaft clockwise until it completes 2 full revolutions, then recheck the timing marks.
16. If the valve cover was removed for access to the camshafts, install it using a NEW gasket as follows:
    a. Apply a light coating of adhesive/sealant to the 4 top corners of the valve cover gasket, then place the gasket in position.
    b. Install the valve cover to the cylinder head, making sure not to disturb the gasket. If the fuel pumps were not removed, camshaft positioning MAY make the valve cover more difficult to install, but since the timing belt is reinstalled you can rotate the crankshaft CLOCKWISE as necessary to reposition the cams making it easier.
    c. Install the retaining bolts and tighten using at least 2 passes of an embossed (or if there is no embossed a spiraling or criss-crossing pattern that starts at the center and works outward). Tighten the bolts to 5.9 ft. lbs. (8 Nm) and then REPEAT to the same torque.
    d. Reconnect the fuel hoses, breather hose and spark plug wires, all as tagged during removal.
    e. Install the spark plug cover.
17. Install the Flywheel and Stator Plate assembly, as detailed in the Ignition and Electrical Systems section.

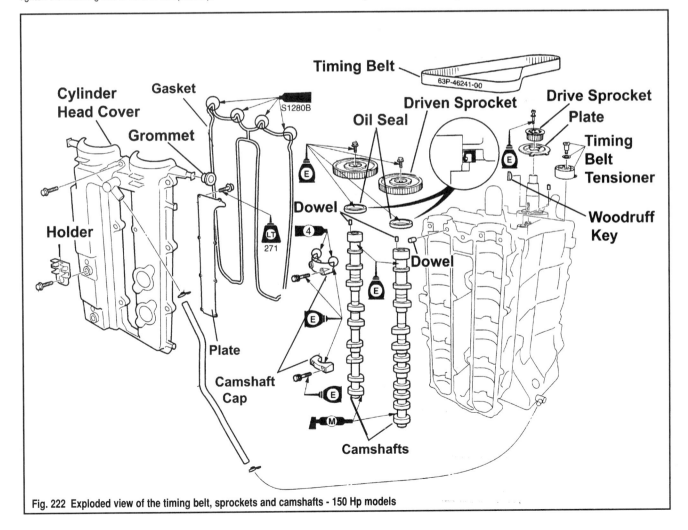

Fig. 222 Exploded view of the timing belt, sprockets and camshafts - 150 Hp models

6-100 POWERHEAD

CYLINDER HEAD REMOVAL & INSTALLATION

♦ See Figures 222 thru 227

■ Unique to the large Yamaha motors, the cylinder head bolts require a torque angle meter for tightening to proper specifications. Although the use of an inexpensive gauge is the best way to proceed, since the specification on these motors is for another 90 degrees, it IS possible simply to paint a mark on the socket and eyeball 1/4 additional turn, but don't be cheap, get the gauge.

We cover a lot of ground in this procedure. Although it is a cylinder head assembly removal procedure, there are many instances where you may need to follow only parts of the procedure to service other components, such as the camshafts and valve train.

First off, it should NOT be necessary to remove the powerhead from the intermediate housing in order to service the cylinder head. Secondly, not only is it NOT necessary to remove the cylinder head in order to service the camshafts, it is actually NECESSARY to remove the camshafts in order to remove the cylinder head (since they're in the way of some cylinder head mounting bolts).

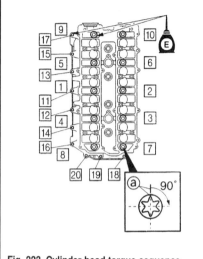

Fig. 223 Cylinder head torque sequence - 150 Hp models

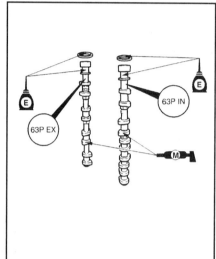

Fig. 224 Camshafts are stamped for identification - 150 Hp models

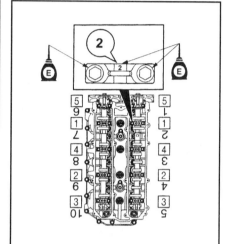

Fig. 225 Camshaft bearing cap torque sequence (in boxes) and stamp markings (upside-down) - 150 Hp models

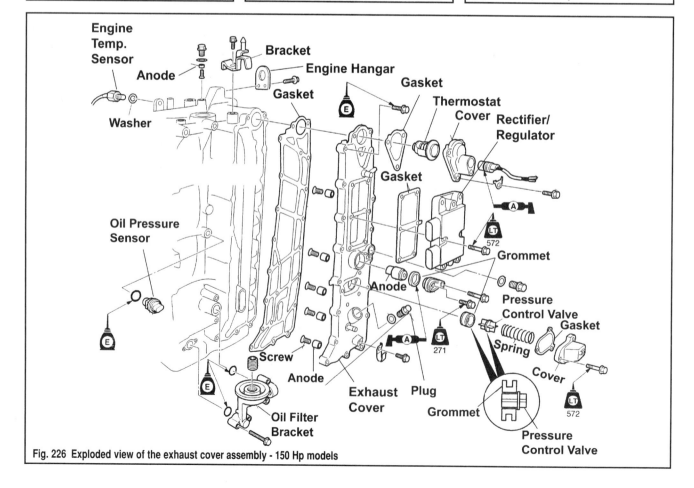

Fig. 226 Exploded view of the exhaust cover assembly - 150 Hp models

POWERHEAD 6-101

■ Only perform the steps which are appropriate for your reason for working on the cylinder head. Therefore, if you are removing the valve cover for valve adjustment or for valve train inspection, there is no need to follow steps like removing the flywheel.

Remember, when loosening the retainers on manifolds, covers and other major components, always try to follow the reverse of the torque sequence indicated in the assembly procedure or molded on the component itself.

When applicable tighten all bolts in using the proper torque sequence and to the proper specification. When possible we've included it in the procedure, but on most Yamaha manufactured powerheads both the sequence and value is molded into the casting on critical components. You'll also notice that most Yamaha torque sequences are clockwise spirals starting somewhere near the center of the component.

1. Disconnect the negative battery cable for safety.
2. Remove the Flywheel and Stator Plate, as detailed in the Ignition and Electrical Systems section.

■ If you are not planning on camshaft or camshaft bearing replacement it is a good idea to measure valve clearances before proceeding, since they can be easily adjusted during installation by replacing the offending shims.

3. Remove the Timing Belt and, if necessary, the camshaft sprockets, as detailed in this section.

■ Before removing the Timing Belt, align the timing marks and note the camshaft timing mark positions when the marks are aligned. In this way you'll be able to set the camshafts up properly before cylinder head installation.

4. If removing the cylinder head assembly, proceed as follows:
   a. Strip the necessary electrical components from the powerhead (specifically from the cylinder head and exhaust cover).
   b. Remove the Throttle Body and Intake Assembly, as detailed in the Fuel System section.

5. To remove the camshafts, loosen the 20 camshaft retaining bolts (10 per shaft) using the reverse of the illustrated torque sequence. Keep track of the bolts because there are two different sizes. The lower 8 bolts on each camshaft are shorter than the top 2.

■ As camshaft caps are removed, be sure to note their positioning for installation in the same locations and facing the same directions. Stamped numbers are already provided on the tops of the caps to accomplish this, identify the numbers as they are removed and compare them to numbers listed upside-down in the torque sequence (the numbers are listed that way, as it is the same way they should be oriented on the caps, when the caps are in place).

6. Remove and discard the oil seal at the top of each camshaft.
7. Remove each of the camshafts.
8. There are 10 cylinder head bolts on these motors, 10 long (10 x 120mm) bolts in a double-row down the head, under the camshafts and 10 short (8 x 55mm) flange bolts on the left side and bottom flanges. Remove the bolts using multiple passes in the reverse of the molded or illustrated torque sequence. This results in a pattern that starts at the outer flange bolts (bottom bolt) and works inward alternately from top and bottom, then continues with the long bolts starting at the top right and working in a counterclockwise spiral toward the inner bolts).
9. Carefully break the gasket seal and remove the cylinder head from the cylinder block.
10. Remove and discard the old cylinder head gasket. Carefully remove all traces of the gasket material from the mating surfaces.
11. Check the condition of the 2 cylinder head dowel pins and replace, if necessary.
12. If necessary for further cylinder head disassembly remove the thermostat assembly from the top, or anode cover or pressure valve assembly from the bottom of the exhaust cover. Also, note there are normally two anode covers, anodes and O-ring assemblies toward the center of the head.

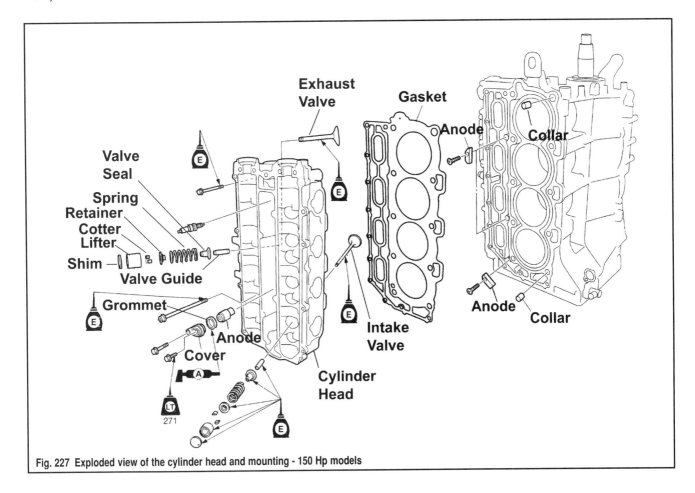

Fig. 227 Exploded view of the cylinder head and mounting - 150 Hp models

6-102 POWERHEAD

13. Refer to Valve Train Overhaul in this section for details on removal and installation of the remaining valve train components. Also, be sure to check the information on gasket mating surfaces to make sure the cylinder head is not warped.

14. If necessary, remove the exhaust cover from the cylinder head and discard the gasket. There are 7 anodes which are secured to the underside of the exhaust cover by 7 bolts.

To Install:

15. If removed, install the exhaust cover to the cylinder head using a new gasket. Apply a light coat of clean engine oil to the threads of the cover retaining bolts, then install and tighten the bolts using at least 2 passes of a sequence that starts at the center bolts and works outwards. Tighten the bolts first to 4.4 ft. lbs. (6 Nm) and then to 8.9 ft. lbs. (12 Nm).

16. If removed, install the anode assemblies to the center of the cylinder head and/or the bottom of the exhaust cover. For each assembly, apply a light coating of marine grade grease to the O-ring and Loctite® 572 (equivalent threadlocking compound) to the threads of the anode cover retaining bolt. Install the O-ring, anode and cover, then tighten the retaining bolts securely.

17. If removed, install the thermostat assembly to the top of the exhaust cover. For details, please refer to the Lubrication and Cooling section.

■ **If crankshaft positioning in the cylinder block was disturbed, slowly turn the crankshaft in the normal direction of rotation (clockwise when viewed from above) until the No. 1 cylinder is at TDC.**

18. Apply a light coating of clean engine oil to the threads of the cylinder head bolts.

19. Position a new cylinder head gasket on the cylinder block over the 2 dowel pins, then carefully position the cylinder head over the gasket and onto the block. Thread the long and short cylinder head bolts making sure the gasket remained perfectly in position.

20. Tighten the long and short cylinder head bolts using multiple passes of the molded or illustrated torque sequence. The sequence is essentially a clockwise spiraling pattern that starts at the left, center of the long bolts and works outward through the long bolts and then continues with the center of the short flange bolts, then goes back and forth working toward the bottom and top of the flange bolts, ending from RIGHT to LEFT with the 3 flange bolts on the bottom. Tighten the long bolts first to 14 ft. lbs. (19 Nm), next to 27.3 ft. lbs. (37 Nm), and finally to an additional 90° using a torque angle meter. Meanwhile, tighten the short flange bolts first to 120 inch lbs./10 ft. lbs. (14 Nm) and then to 21 ft. lbs. (28 Nm).

21. Apply a light coating of a molybdenum disulfide engine assembly lube to the journals and lobes of the camshafts and bearing surfaces of the cylinder head and caps.

22. Install the camshafts to the cylinder head using NEW oil seals. Position them so the No. 1 cylinder would be at TDC (valves closed, timing marks aligned if the sprockets were to be installed). There are stamp marks towards the top of the camshafts (shortly below the oil seals). Be sure the shaft marked **63P IN** is installed on the INTAKE side of the head, while the shaft marked **63P EX** is installed on the EXHAUST side of the head.

■ **It APPEARS that the dowels in the camshafts are normally aligned with the timing marks, but double-check by holding each of the drive sprockets in position when checking camshaft positioning.**

23. Apply a light coating of engine oil to the 20 camshaft cap retaining bolts (don't mix them up, but if you do the 4 long ones go on the 2 top bearing caps).

24. Apply a light coating of Yamabond No.4 or equivalent sealant/adhesive to the mating surface for the oil seal housing/top bearing cap for each camshaft.

25. Position the bearing caps as noted during removal and as stamped. This typically means the stamped numbers are positioned upside down and the caps are installed numbers 1-5 top-to-bottom on the right side (when looking at the camshafts) and numbers 6-10 on the left side. Install and finger-tighten the bolts.

26. Tighten the cylinder head cap bolts using multiple passes of the appropriate, illustrated torque sequence. Tighten the bolts first to 70 inch lbs./5.9 ft. lbs. (8 Nm) and then to 150 inch lbs./12.5 ft. lbs. (17 Nm).

■ **If the valve train components were replaced or if you failed to take measurements before the camshafts were removed, install the Timing Belt and take the to check and adjust the valve clearances.**

27. Install the Timing Belt and Sprocket (the latter part, IF removed), as detailed in this section.

28. If the cylinder head was removed, proceed as follows:
   a. If removed, install the Throttle Body and Intake Assembly, as detailed in the Fuel System section.
   b. Install the electrical components which were stripped from the powerhead and/or cylinder head.

29. Install the Flywheel and Stator Plate, as detailed in the Ignition and Electrical Systems section.

30. Reconnect the negative battery cable.

VALVE TRAIN OVERHAUL

◆ See Figure 227

■ **If you only need to check/replace the valve shims or lifters and not the valves themselves there is no need to remove the cylinder head.**

1. Remove the Cylinder Head from the powerhead assembly, as detailed in this section.

■ **Valve train components such as valve lifters, retainers, keepers, springs and valves should not be mixed and matched from one assembly to another. This means that if they are to be reused, they must all be labeled prior to removal or VERY carefully sorted during service to prevent installation to the wrong positions.**

To remove the each valve assembly, proceed as follows:
2. Remove the valve shim and lifter from the bore in the cylinder head to expose the valve step and keepers.
3. Using a clamp-type valve spring compressor, carefully place just enough pressure on the valve spring so the retainer moves down sufficiently to expose the keepers.
4. Remove the pair of valve keepers, then slowly ease pressure on the valve spring until the compressor can be removed.
5. Remove the valve spring retainer and valve spring from the shaft of the valve.
6. Remove the valve from the underside (combustion chamber side) of the cylinder head.
7. If necessary, carefully pry the old valve stem seal from its position on the cylinder head.
8. Repeat this procedure for each of the remaining valves.
9. Refer to Powerhead Refinishing, later in this section, for details on cleaning and inspecting cylinder head components.

To Install:

■ **Apply a light coating of clean engine oil to each of the valve train components, except the valve spring and the valve keepers as they are installed. You can use a dab of marine grade grease on the keepers in order to help keep them in position during installation. One popular method is to actually use the grease to temporarily adhere the keepers to the installation tool itself, using the retainer to free it from the tool and place it in position.**

10. If removed, install new valve spring seals to the cylinder head.
11. Position the valve into the cylinder head (into the original guide and valve seat unless the valve or valve seat was replaced).
12. Position the valve spring seat down onto the valve shaft and cylinder head, then oil and install the spring with the fine pitch side of the spring facing down toward the seat.
13. Place the retainer over the spring, then use the valve spring compressor tool to press downward on the retainer and spring so you can install the keepers.
14. Insert the valve keepers, then slowly back off the valve spring compressor, making sure the keepers are fully in position. Then remove the valve spring compressor.
15. Install the valve lifter and shim.
16. Repeat for each of the remaining valves.
17. Install the Cylinder Head as detailed in this section.
18. Once the powerhead is fully assembled, if valve train components were replaced, be sure to operate the engine as directed for new component break-in. For details, please refer to Powerhead Break-In, in this section.

POWERHEAD 6-103

CYLINDER BLOCK OVERHAUL

◆ See Figures 228 thru 240

■ Unique to the large Yamahas and Mercury/Mariners, the crankcase, main bearing and connecting rod bolts require a torque angle meter for tightening to proper specifications. Although the use of an inexpensive gauge is the best way to proceed, since the specification on these motors is for another 90 degrees, it IS possible simply to paint a mark on the socket and eyeball 1/4 additional turn, but don't be cheap, get the gauge.

Remember, when loosening the retainers on manifolds, covers and other major components, always try to follow the reverse of the torque sequence indicated in the assembly procedure or molded on the component itself.

When applicable tighten all bolts in using the proper torque sequence and to the proper specification. When possible we've included it in the procedure, but on most Yamaha manufactured powerheads both the sequence and value is molded into the casting on critical components. You'll notice that most Yamaha torque sequences are clockwise spirals starting somewhere near the center of the component.

■ Yamaha states that you CAN reuse the main bearing cap and the crankcase bolts up to 3 times each. HOWEVER, they specifically state that you CANNOT reuse the connecting rod cap bolts, so be sure to obtain new ones before installation.

1. Remove the Powerhead assembly, as detailed in this section.
2. Remove the Timing Belt/Sprocket assembly, as detailed in this section.
3. Strip all electrical components from the powerhead, then remove the Cylinder Head assembly, as detailed in this section.
4. Remove and discard the old oil filter from the cylinder block.
5. Remove the Oil Pump assembly, as detailed in the Lubrication and Cooling System section.

6. Remove the crankshaft balancer assembly bolts from the belly of the block using the reverse of the molded torque sequence, which should result in a counterclockwise spiraling sequence that begins with the top and bottom bolts, then works its way inward. Carefully separate the balancer assembly.

■ Once removed, check the teeth on the balancer shaft for cracks or excessive wear and replace the assembly, as necessary.

7. Remove the crankcase flange bolts using the reverse of the molded torque sequence, which should result in a counterclockwise spiraling sequence that begins with the top right corner and works it's way inward (HOWEVER, the sequence skips the top 2 center and bottom 2 center bolts, returning to them at the end of the spiral).

■ The top 2 center and bottom 2 center crankcase flange bolts are actually the top and lower MAIN BEARING cap bolts. Also, DO NOT remove the 2 plugs from the lower center of the crankcase.

8. Carefully break the seal and separate the crankcase from the cylinder block.

■ Most parts inside the cylinder block are matched sets and should not be moved from one cylinder to another. This includes the bearings and pistons as well as each of the connecting rod/end cap combination. For this reason, prior to disassembly it is essential to matchmark each of the connecting rods to their end caps (this ensures that not only will each cap be installed on the proper rod, but facing the correct direction). Also, be sure to label each piston and connecting rod as to what cylinder it belongs and identify the factory marks and/or place an arrow or some indicator facing upward to ensure they are installed in the same direction. Yamaha often marks pistons with UP on the dome and connecting rods with the letter Y or word YAMAHA and/or alignment dots at the con rod-to-cap mating surface, all of which should face the flywheel.

9. Matchmark the big end caps to the connecting rods, then remove the 8 connecting rod cap bolts (working on one cap at a time, alternating between each of the cap bolts after each turn). Remove the caps and place them aside for reinstallation back on their original connecting rods as soon as the crankshaft has been removed.

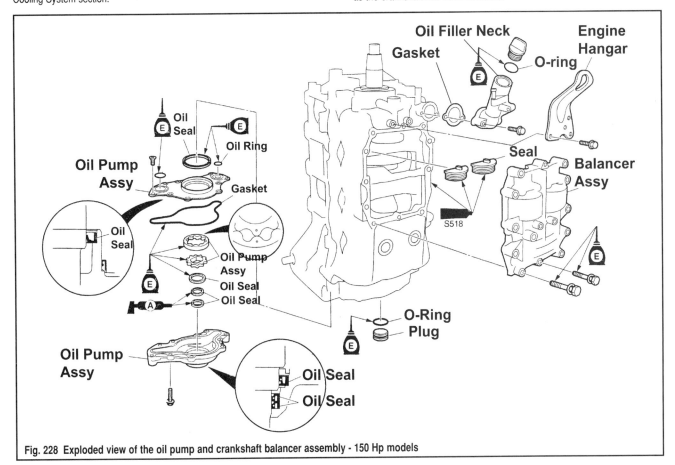

Fig. 228 Exploded view of the oil pump and crankshaft balancer assembly - 150 Hp models

6-104 POWERHEAD

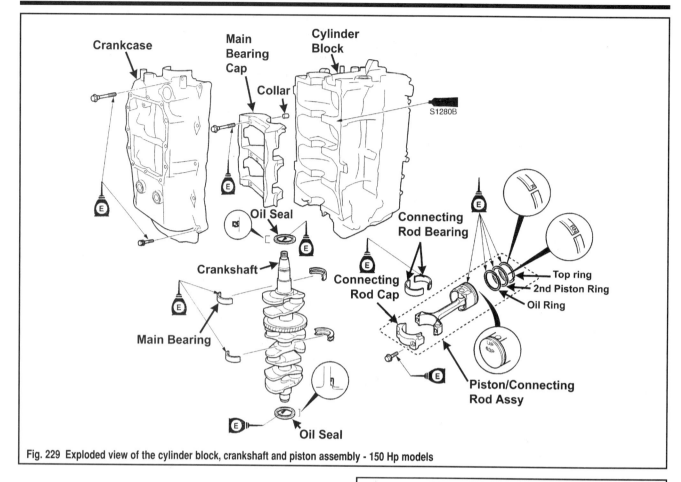

Fig. 229 Exploded view of the cylinder block, crankshaft and piston assembly - 150 Hp models

10. Remove the main bearing cap assembly retaining bolts using the reverse of the torque sequence, which should result in a counterclockwise spiraling sequence that begins with the top right corner and works it's way inward.

11. Carefully lift the crankshaft from the block. Place it on a protective surface.

✱✱ WARNING

Handle the crankshaft carefully. Remember it is a heavy, expensive and vital piece to the motor. The polished surfaces should be covered to prevent scratches or other damage and the whole shaft should be protected against impact or other damage which could destroy it.

12. Immediately reinstall the connecting rod end caps and their bearing inserts to their respective pistons.

13. Remove the top and bottom oil seals from the crankshaft. Discard the seals.

■ These pistons and connecting rods are probably already marked from the factory as to what side faces up, identify the marks and perhaps, remark them to ensure they are found easily during assembly.

14. Label and remove each of the pistons from the bores. Number the pistons and place a mark (or identify the factory marks) on each piston and connecting rod facing up (facing the flywheel) for installation purposes. If necessary, disassemble each of the pistons, as follows:

■ There is no indication in the Yamaha service literature that the connecting rods can be removed from the pistons, no indication that is, EXCEPT a single piston pin bore diameter spec. However, no exploded view or procedure even shows retaining clips or a removable piston pin (even though that's pretty much how every other Yamaha 4-stroke does it). If in doubt, check with a Yamaha parts supplier to see what replacement parts are available for these pistons.

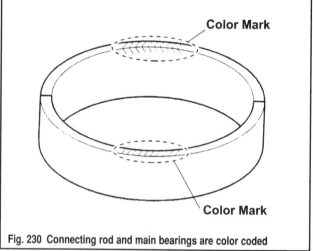

Fig. 230 Connecting rod and main bearings are color coded

a. If equipped, carefully pry the clips free from the piston pin bore on either side of the piston assembly. Slide the piston pin free of the bore (experience tells us that this should not be an interference fit, but the Yamaha service publications do not give any procedures, so we cannot be certain. If necessary, bring the assembly to a machine shop to determine if it is an interference fit and/or if the piston pin should even be removed).

b. Using a piston ring spreading tool, carefully remove each of the rings (top and second compression rings, followed by the oil ring assembly) from the piston.

c. Keep all components from a single piston together for reuse or replacement, but don't mix and match components from other pistons.

15. Remove the crankshaft main (big end) bearings from the crankcase and cylinder block. Again, note their positions and orientation if they are going to be reused.

POWERHEAD 6-105

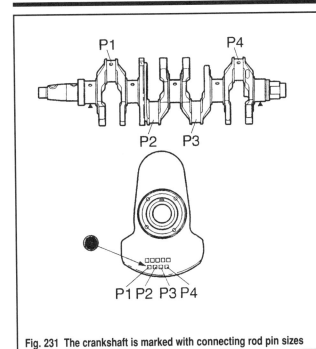

Fig. 231 The crankshaft is marked with connecting rod pin sizes

■ On these models different color code bearings may be installed on the same crankshaft journal (one in the crankcase and a different one in the block) in order to achieve the proper clearance measurement. Pay close attention to bearing positions and codes. Also, keep in mind that numerical codes are placed on the block which can be used to help select replacement bearings.

16. Refer to Powerhead Refinishing, later in this section, for details on cleaning and inspecting cylinder block components. There are probably more components and specifications to measure on these 4-stroke motors than on any smaller Yamaha engine and even most larger 2-strokes. Take your time and make sure you've made measurements for all components listed in the Engine Specifications chart for this model.

To Install:

17. Make sure the mating surfaces of the crankcase, bearing cap, balancer assembly and cylinder block are completely clean and free of sealant or damage.

18. Place the main bearing inserts into the main bearing cap and cylinder block. There are normally 2 things to look out for when doing this. Most bearing inserts will be equipped with locating tabs, make sure they are seated in the appropriate grooves. Also, inserts with oil holes should be aligned with an oil passage.

19. If they were disassembled, prepare each piston assembly for installation as follows:

   a. Using a ring expander, carefully install each of the rings with their gaps as noted in the accompanying illustration. While looking at the piston dome with the top (flywheel end) facing 12 o'clock, the bottom oil ring should be positioned with the gap facing between 4 and 5 o'clock. The middle oil ring (expander ring) should be positioned with the gap facing between 7 and 8 o'clock. The upper oil ring should be positioned with the gap facing between 10 and 11 o'clock. The lower (second) compression ring should be positioned with its gap at between 1 and 2 o'clock. And lastly, the upper (top) compression ring should be positioned with its gap facing between 7 and 8 o'clock (like the middle/expander oil ring).

■ Some rings, like the compression rings, will be equipped with a stamp mark (R in the case of the top compression ring, or RN for the second ring) which should be faced upward toward the top of the piston dome.

   b. If they could be (and were) separated, align the mark(s) on the connecting rod (facing up/flywheel) with the mark on the piston (also facing up/flywheel), install the connecting rod to the piston using the piston pin. If applicable, install the piston pin clips in either side of the pin bore to secure the piston pin in place.

   c. Repeat for the remaining pistons.

✳✳ WARNING

Use extreme care to prevent the connecting rods from contacting and scoring the cylinder walls during piston installation. It is a good idea to wrap the ends in a shop cloth or something else to protect the cylinders.

20. Install each of the pistons (with the correct mark/side facing UP toward the flywheel) to the cylinder block using a ring compressor. Apply a very light coating of clean engine oil to the inside of the compressor and to the cylinder bore, then place the compressor around the piston. Insert the piston through the cylinder head side of the bore and use a wooden hammer handle to gently tap the piston down through the compressor and into the cylinder bore.

✳✳ WARNING

Be VERY careful when installing the piston. Get the lower portion of the skirt below the ring compressor and insert it into the cylinder bore to make sure everything is aligned. Then, GENTLY tap the piston through the compressor, stopping if it seems to hang up even slightly to make sure that a ring is not caught on the deck of the cylinder block.

21. Remove the connecting rod end caps and position them so the crankshaft can be installed. Place the connecting rod bearing inserts into position on the connecting rod big end and cap.

■ Here's where procedures differ slightly depending upon whether you're installing bearings of a known clearance or measuring clearance. If you've already determined that bearing clearances are fine, apply a light coat of engine oil to each of the connecting rod, main bearing and crankshaft journal surfaces. If however, you need to measure and determine bearing clearance, Temporarily INSTALL THESE COMPONENTS DRY for measurement purposes.

A few notes on connecting rod and crankshaft main bearing selection for these models. Use your Plastigage® measurements and the accompanying tables to help select replacement connecting rod bearings.

When it comes to the main bearings use the measurements taken with Plastigage® OR use the marks on the crankshaft and block along with the accompanying bearing selection table to pick a suitable set of bearings. Of course, even after using the table, you should still use a gauging material to measure and confirm that clearances are within specification.

When it comes to the connecting rod bearings, use the measurements taken with Plastigage® OR use a measurement of the big end inner diameter along with the accompanying bearing selection table and the mark on the crankshaft to select a suitable set of bearings. Like with the main bearings, even after using the table, you should still use a gauging material to measure and confirm that clearances are within specification.

22. If not done already, it is normally easier to oil and install new oil seals onto the shaft BEFORE it is positioned in the block.

■ NEVER install the upper or lower crankshaft oil seals with the crankcase installed and tightened otherwise the seal will likely be damaged.

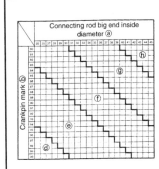

Fig. 232 Use the crankshaft marking and the measured rod big end diameter to help select the proper bearings

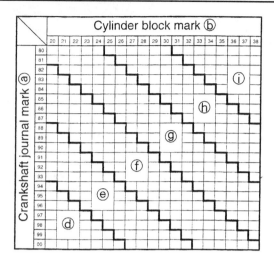

Fig. 234 Again, use the markings to help select main bearings

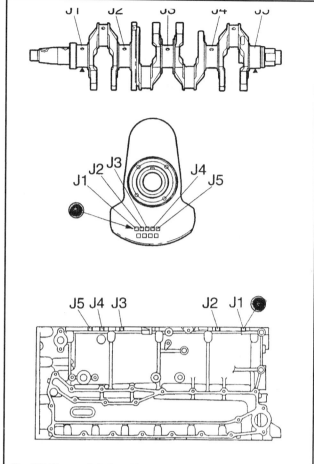

Fig. 233 The crankshaft and block are marked for main bearing sizes

23. Position the crankshaft to the cylinder block, and carefully pull each of the connecting rods into position. If not done already, it is normally easier to oil and install new oil seals onto the shaft BEFORE it is positioned in the block.

24. Install the main bearing cap assembly. Apply a light coating of engine oil to the threads of the main bearing cap bolts, then tighten them using multiple passes of the clockwise spiraling torque sequence that starts at the left, inner bolt and works it's way outward. Tighten the bolts first to 22 ft. lbs. (30 Nm) and finally an additional 90 degrees using a torque angle meter.

25. Install each of the match-marked connecting rod caps. Apply a light coating of engine oil to the threads of the connecting rod bolts, then tighten them alternating back and forth between the 2 bolts on each cap in at least 3 stages. Tighten the bolts first to 17 ft. lbs. (23 Nm), next to 32 ft. lbs. (43 Nm) and then, using a torque angle gauge tighten the bolts an additional 90 degrees each.

■ Since the connecting rod bolts should not be reused, we suggest skipping the 90 degree angle torquing IF you are only temporarily assembling the connecting rod caps in order to check bearing clearances using a gauging material, but be sure to follow that step during final assembly.

26. If you've already measured bearing clearances, apply a light coating of a Gasket Maker (such as Yamaha S518, or Loctite 514 Master Gasket or equivalent) to the entire outer perimeter of the cylinder block-to-crankcase mating surface. A small paint roller is really handy for this application. DO NOT get any sealer on the top and bottom main bearings, apply engine oil to them instead.

27. Apply a light coating of engine oil to the threads of the crankcase flange (and the 2 upper and 2 lower main bearing cap) bolts.

28. With the upper and lower main bearing inserts in place, carefully lower the crankcase onto the cylinder block, then JUST thread the center bolts and flange bolts.

29. Tighten the bolts using multiple passes of the molded torque sequence, or if not present, a clockwise, spiraling pattern that starts at the lower right center bolts (larger/main bearing cap bolts) and continues through the rest of the 4 large bearing cap bolts, then continues with the smaller flange bolts (working from the center outward to the corners). During the torque sequence, tighten the larger M10 bolts first to 22. ft. lbs. (30 Nm) and an additional 90 degrees using an angle meter. During the sequence tighten the smaller M8 bolts first to 120 inch lbs./10 ft. lbs. (14 Nm) and then to 19 ft. lbs. (26 Nm).

30. Align the crankshaft sprocket keyway with the cylinder block-to-crankcase mating surface.

31. Next align the marks on the balancer shafts with the alignment marks on the balancer assembly.

32. Apply a light coating of a Gasket Maker (such as Yamaha S518, or Loctite 514 Master Gasket or equivalent) to the outer perimeter of the balancer assembly-to-crankcase mating surface. A small paint roller is handy for this application also.

33. Apply a light coating of engine oil to the balancer assembly retaining bolts.

34. With the marks aligned as noted earlier, install the balancer assembly to the crankcase and then tighten the bolts using multiple passes of the molded torque sequence, or if not present, a clockwise, spiraling pattern that starts at the left, center bolts and works it's way outward, ending with the bolts at the top and bottom center of the assembly. During the torque sequence, tighten the larger M8 bolts first to 13 ft. lbs. (18 Nm) and then to 23 ft. lbs. (31 Nm). During the sequence tighten the smaller M6 bolts first to 5 ft. lbs. (7 Nm) and then to 10 ft. lbs. (13 Nm).

35. After the balancer assembly is installed and tightened apply a light coating of the same sealant used to seal the assembly around the periphery of the top seals, then install the seals into the assembly.

36. Install the Oil Pump assembly, as detailed in the Lubrication and Cooling System section.

37. Install a new oil filter to the powerhead.

38. Install the Cylinder Head assembly (along with the camshafts and any electrical components stripped from the powerhead) as detailed in this section.

39. Install the Timing Belt/Sprocket assembly, as detailed in this section.

POWERHEAD 6-107

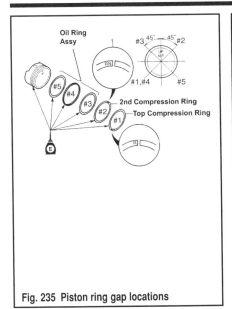
Fig. 235 Piston ring gap locations

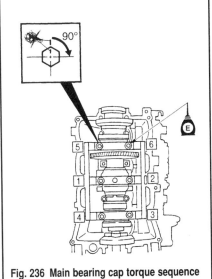
Fig. 236 Main bearing cap torque sequence

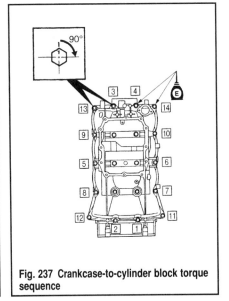
Fig. 237 Crankcase-to-cylinder block torque sequence

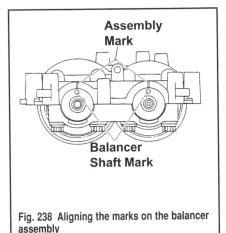
Fig. 238 Aligning the marks on the balancer assembly

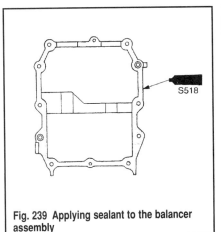
Fig. 239 Applying sealant to the balancer assembly

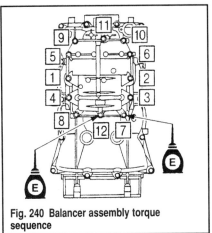
Fig. 240 Balancer assembly torque sequence

40. Install the Powerhead assembly, as detailed in this section.
41. Once the powerhead is fully assembled, if cylinder block components were replaced, especially the pistons and/or rings, be sure to operate the engine as directed for new component break-in. For details, please refer to Powerhead Break-In, in this section.

CLEANING & INSPECTING

Cleaning and inspecting internal engine components is virtually the same for all Yamaha outboards with the exception of a few components that are unique to certain 4-stroke motors. The main variance between different motors comes in specifications (which are listed in the Engine Specifications charts) or by component type. A section detailing the proper procedures, sorted mostly by component, can be found under Powerhead Refinishing.

200/225 Hp V6 Powerheads

POWERHEAD REMOVAL & INSTALLATION

◆ See Figures 241 thru 244

The powerhead may be removed or installed as an almost fully assembled unit. The components mentioned here are the bare minimums that must be removed before separating the powerhead assembly from the rest of the motor. However, if overhaul or replacement of the cylinder head and/or block is desired, you may wish to remove additional components before removing the powerhead assembly, not only to lighten the load, but to help prevent potential damage to the external fuel and electrical system components mounted to the powerhead. The choice is yours, so if necessary, refer to other component procedures in this section and in the Fuel System and/or Ignition and Electrical Systems sections before proceeding.

1. Disconnect the negative battery cable at the battery itself for safety.
2. Remove the flywheel cover for access.
3. Remove the air intake silencer assembly. For details, refer to the air silencer portion of the Throttle Body and Intake Assembly procedure found in the Fuel System section.
4. At the front, starboard side of the motor loosen the bolt securing the retaining plate to the lower cowling, then remove the retaining plate. Next loosen the 3 bolts securing the various cable and wiring grommet holders.
5. Tag and disconnect the throttle and shift cables by removing the 2 clips securing the cable ends to the linkage studs, then carefully disconnect the cables from the linkage.
6. Remove the engine oil dipstick. Depending upon the amount of overhaul and service condition of the motor, it is probably a good idea to drain the engine oil at this time.
7. Carefully cut the plastic wire tie securing the fuel supply line to the fuel filter at the lower front (starboard corner) of the powerhead. Disconnect the fuel line from the filter nipple.
8. Loosen the bolts securing the small electric component cover found at the front of the motor, just to the right of the fuel filter. Remove the cover, then tag and disconnect the 2 PTT motor leads.
9. If equipped with a flushing device, locate the T-fitting at the lower front, port side of the powerhead, then carefully cut the wire tie and disconnect the hose.

6-108 POWERHEAD

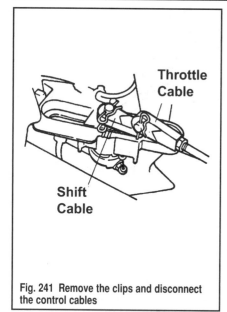

Fig. 241 Remove the clips and disconnect the control cables

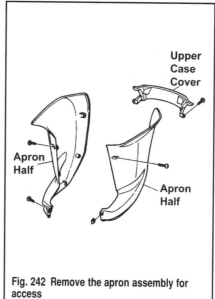

Fig. 242 Remove the apron assembly for access

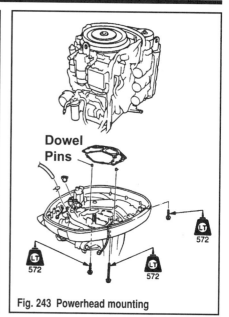

Fig. 243 Powerhead mounting

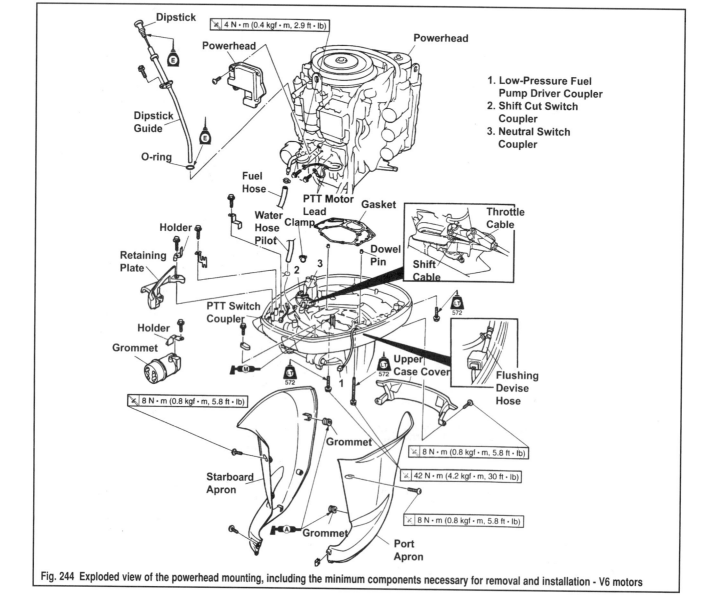

Fig. 244 Exploded view of the powerhead mounting, including the minimum components necessary for removal and installation - V6 motors

10. Also at the front port side of the motor, locate, tag and then disconnect the wiring coupler for the low-pressure fuel pump driver.

11. At the front starboard side of the motor, near the linkage and fuel filter, locate the PTT switch coupler, and a little behind it both the neutral switch coupler and shift cut switch coupler. Tag and disengage each of these connectors.

12. Finally, at the lower, front, starboard side of the motor tag and disconnect the pilot water (cooling system indicator) hose.

13. Locate and remove the 8 bolts securing the upper case cover and starboard/port apron halves. There are usually 2 bolts on the upper case cover (one at each lower corner) which is found at the rear of the apron assembly. Then there are usually 6 bolts (Yamaha) or 12 screws (Mercury/Mariner) threaded sideways through the port and starboard halves of the apron. Remove the upper case cover first, then remove the apron halves.

14. Remove the 13 bolts threaded upward into the powerhead from the intermediate housing. Keep track of the bolts as there are 2 different sizes. Yamaha tells us that there are 7 short bolt sand 6 long bolts, however Mercury literature suggests the count is 6 long and 7 short (we're not sure if the motors sold by each manufacturer are actually different, or if there is a typo for one).

15. Attach a suitable lifting device and carefully lift the powerhead assembly from the intermediate housing. In order to prevent potential damage to the crankshaft and driveshaft be certain to lift the powerhead STRAIGHT up and off the intermediate housing.

16. Remove and discard the upper casing gasket from the powerhead-to-intermediate housing mating surface. Carefully clean the mating surfaces of all traces of old gasket.

17. Check for placement of the 2 dowel pins. They may have remained with the intermediate housing or have come away with the powerhead itself. Inspect the pins for damage and replace, if necessary.

18. Position the powerhead on a suitable workbench or engine stand for further disassembly or overhaul, as necessary. Refer to the additional procedures in this section as needed.

To Install:

19. With the 2 dowel pins in place, position a NEW gasket to the intermediate housing.

20. Apply a light coat of an engine assembly lube which contains molybdenum disulfide to the driveshaft-to-crankshaft mating splines.

21. Using a suitable lifting device, carefully lift the powerhead and lower it slowly, STRAIGHT onto the driveshaft and intermediate housing. If necessary, rotate the flywheel slightly in a clockwise direction in order to mate the crankshaft to the driveshaft.

■ Mercury does not mention using threadlock on the powerhead mounting bolts, but we don't see the harm and actually would go so far as to recommend it anyway.

22. Apply a light coating of Loctite® 572 to the threads of the powerhead mounting bolts.

23. Install the 13 powerhead retaining bolts and tighten to 31 ft. lbs. (42 Nm).

24. Install the apron halves and upper case cover, then tighten the bolts to 70 inch lbs./5.8 ft. lbs. (8 Nm).

25. Install the engine oil dipstick and, if drained, refill the engine with fresh engine oil.

26. Reconnect the pilot water (cooling system indicator) hose at the lower, front, starboard side of the motor.

27. Reconnect the PTT switch, neutral switch and shift cut switch couplers, as tagged during removal.

28. Reconnect the wiring for the low-pressure fuel pump driver (at the front port side of the motor).

29. If equipped with a flushing device, reconnect the flushing hose and secure using a new wire tie.

30. Reconnect the 2 PTT motor leads, as tagged during removal, then install the small electric cover.

31. Reconnect the fuel supply line to the fuel filter, position the line over the nipple, then secure using a new plastic wire tie.

32. Reconnect the throttle and shift cables and secure using the 2 retaining clips.

33. Install the various cable and wiring grommet holders, then secure using the 3 retaining bolts.

34. Install the retaining plate to the front, starboard side of the lower cowling and secure using the retaining bolts.

35. Install the air intake silencer assembly.

36. Remove the flywheel cover for access.

37. Reconnect the negative battery cable to the battery itself.

38. Be sure to check all Timing and Synchronization adjustments as detailed in the Engine and Maintenance section.

39. If the powerhead was overhauled (replacing components such as the valve train, main bearings, crankshaft, pistons and/or rings) be sure to subject the motor to a complete new engine break-in period to ensure long, trouble free life.

TIMING BELT/SPROCKET REMOVAL & INSTALLATION

◆ See Figures 245 thru 250

The V6 Yamahas and Mercury/Mariners probably have the most complicated valve train of any Yamaha produced powerhead. A crankshaft driven timing belt rotates two camshaft sprockets (the inner or exhaust camshaft from each bank). In turn, each camshaft bank contains a timing chain and tensioner assembly that is used to drive the outer (intake) camshaft off the inner (exhaust) shaft. Service of the camshafts and timing chain can be found under the Cylinder Head Removal & Installation procedure found later in this section.

This procedure can be performed with the powerhead assembly removed or installed on the motor, regardless of whether or not the crankshaft or camshaft sprockets must be removed. In the case of the camshaft sprockets, a hex is provided on each camshaft so that you can use a wrench to hold it from turning. The crankshaft sprocket is secured by 4 relatively low torque bolts so that you can use a pry-tool across 2 of the bolt heads and/or use a strap wrench on the sprocket to keep it from rotating while you are loosening sprocket bolts.

If the timing belt is to be reused, be sure to mark the direction of rotation on the old belt before removal. Also, if either the cylinder head or cylinder block is not to be disassembled further, be sure to properly align the timing marks before removal. This will significantly ease installation.

■ In order to hold the timing belt tensioner in a tension relieved position during removal or installation of the timing belt you'll need a small, metal pin of approximately 0.2 in. (5mm) in diameter. The pin is inserted into the top of the tensioner once it is moved to a certain position.

1. If the powerhead is still installed, disconnect the negative battery cable for safety.

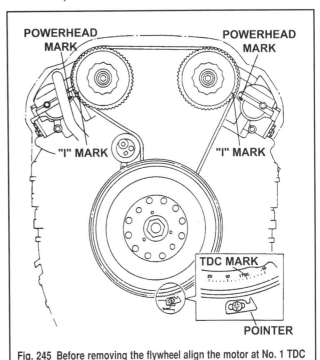

Fig. 245 Before removing the flywheel align the motor at No. 1 TDC

6-110 POWERHEAD

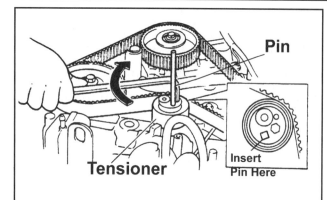

Fig. 246 Use a hex wrench to pivot the tensioner pulley, then lock it in position using a pin

2. If either the cylinder head or cylinder block is not being overhauled, rotate flywheel in the normal direction of rotation (clockwise) until the TDC mark is aligned with the timing pointer. For the motor to be at No. 1 TDC, the single line mark on each camshaft pulley should be facing the marks on the cylinder head (this places the mark on each camshaft pulley facing away from the opposite bank's pulley). If not, rotate the flywheel one full additional turn until all marks are aligned as noted.

■ When you remove the flywheel, double-check the timing mark on the crankshaft sprocket itself. The triangular mark on the crankshaft belt plate (and sometimes on the sprocket itself) should align with the triangular mark on the block (located on the tensioner side of the sprocket).

3. Remove the Flywheel and Stator Plate, as detailed in the Ignition and Electrical Systems section.

✶✶ WARNING

To prevent damage, never rotate the crankshaft sprocket counterclockwise.

■ The timing belt tensioner is mounted to the top of the powerhead using a single bolt inserted through the top of an idler pulley. The mounting hole is offset from the center of the pulley in order to allow internal spring pressure to rotate the pulley into contact with the timing belt where it will remain in contact and placing the spring strength tension on the belt. Since the mounting hole is offset, rotating the pulley results in an elliptical arc which increases or decreases tension on the belt.

4. Insert a hexagon wrench into the top of the timing belt tensioner pulley, then rotate the hex wrench slowly CLOCKWISE to take tension off the timing belt (placing the spring tension on the pulley itself). Holding the pulley in this position, insert a 0.2 in. (5mm) diameter pin through the other opening in the top of the pulley in order to lock it in this position.

5. With the belt tension relieved slide the timing belt off one of the camshaft sprockets, then remove it from the crankshaft and remaining camshaft sprockets. Remove the belt from the motor. If the belt alone is being replaced, skip to the appropriate steps of the installation procedure.

✶✶ WARNING

NEVER pull the pin from the tensioner when it is locked in position without first using a hex wrench to keep it from moving. And, even then, once the pin is removed, control the pulley movement to make sure the tensioner assembly is not damaged by a sudden release of pressure.

6. If further service is required to the sprockets or powerhead, don't risk potential damage to the tensioner which could occur if the pin was suddenly removed. Use the hex wrench to hold the pulley in position, then remove the pin and slowly allow the pulley to rotate with the spring tension. If necessary, once the tension has been released, loosen the tensioner retaining bolt, then remove the bolt, washer and tensioner from the top of the powerhead.

7. If it is necessary to remove the drive (crankshaft) sprocket, loosen the 4 retaining bolts, then remove the sprocket and plate from the powerhead. As stated earlier, either use a strap wrench to keep the sprocket from turning, or use a prybar across 2 bolt heads as you just break loose the other two bolts, then switch the prybar until all 4 bolts are loosened sufficiently that they can be unthreaded. Lift the crankshaft and drive plate from the crankshaft. Remove and retain the crankshaft woodruff key.

8. If necessary to remove one or both of the driven (camshaft) sprockets, remove the cylinder head cover for the sprocket in question. Use a wrench to hold the camshaft from turning while you loosen and remove the camshaft sprocket retaining bolt. Lift the sprocket from the end of the camshaft and the dowel pin. Check the condition of the dowel drive pin and replace, if necessary.

■ Yamaha recommends that the valve cover gaskets NOT be reused. Remove and discard the gaskets once the valve cover is removed.

To Install:

9. If removed, install the dowel pin to the top of the camshaft.
10. To install a camshaft sprocket, align the sprocket with the dowel pin and carefully seat it to the end of the camshaft. Apply a light coating of clean engine oil to the sprocket retaining bolt, then install and tighten to 43 ft. lbs. (60 Nm) while holding the camshaft from turning using a wrench.

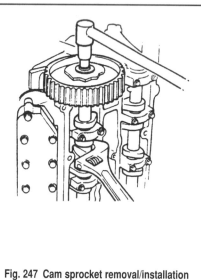

Fig. 247 Cam sprocket removal/installation

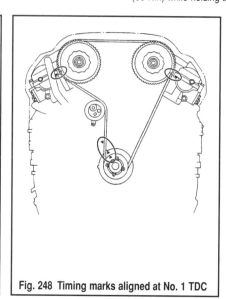

Fig. 248 Timing marks aligned at No. 1 TDC

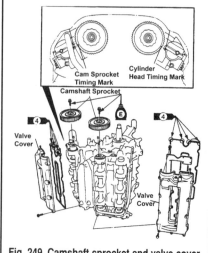

Fig. 249 Camshaft sprocket and valve cover installation

POWERHEAD 6-111

11. Install a NEW gasket to each valve cover and apply a light coating of Yamabond No. 4 (Yamaha), RTV 587 Silicone Sealer (Mercury) or an equivalent adhesive to the 4 top corners of the valve cover gasket, then install the valve cover and tighten the retaining bolts to 70 inch. lbs./5.8 ft. lbs. (8 Nm), then again to 70 inch. lbs./5.8 ft. lbs. (8 Nm) using two passes of either the molded torque sequence or using a clockwise spiraling pattern that starts at the center of the valve cover and works outward.

12. Make sure each camshaft sprocket timing mark is aligned with the mark on the cylinder head (facing away from the opposite bank's camshaft). If necessary, carefully and slowly rotate the camshaft and sprocket until the marks align. Be sure to turn the camshaft CLOCKWISE while feeling for any interference or binding. If any is felt, stop turning until the cause can be determined.

13. If the crankshaft sprocket and drive plate was removed, install them now. Use a dab of marine grade grease to hold the woodruff key in place, then align and install the plate followed by the sprocket over top of the crankshaft. Apply a light coat of engine oil to the threads of the 4 sprocket and plate retaining bolts, then install and tighten each of the bolts to 62 inch. lbs./5.1 ft. lbs. (7 Nm).

■ If the crankshaft is positioned properly with the No. 1 cylinder at TDC, the crankshaft sprocket drive plate (and possibly sprocket) triangular mark(s) will face the triangular timing mark on the powerhead (towards the tensioner). If necessary, rotate the crankshaft SLOWLY clockwise (when viewed from above) in order to properly align the keyway and timing mark.

14. If removed, install the timing belt tensioner using the bolt and washer. Tighten the bolt to 28 ft. lbs. (39 Nm). Using the hexagon wrench to once again turn the tensioner clockwise to a position where belt tension would be relieved, then install the pin to hold the tensioner.

15. Install the timing belt, starting from the drive (crankshaft) sprocket end and making sure the timing marks are still aligned. Remember, if the belt is being reused, be sure to face it in the same direction as normal rotation (clockwise). New belts from Yamaha or Mercury should be installed with the part numbers facing so they are legible when the belt is in place.

** WARNING

Again, NEVER pull the pin from the tensioner when it is locked in position without first using a hex wrench to keep it from moving. And, even then, once the pin is removed, control the pulley movement to make sure the tensioner assembly is not damaged by a sudden release of pressure.

16. Use the hex wrench to hold the pulley in position, then remove the pin and slowly allow the spring tension to push the pulley toward the belt, tensioning the belt as the pulley sets into position.

17. Turn the crankshaft clockwise until it completes 2 full revolutions, then recheck the timing marks.

18. Install the Flywheel and Stator Plate assembly, as detailed in the Ignition and Electrical Systems section.

19. If removed, install the powerhead assembly.

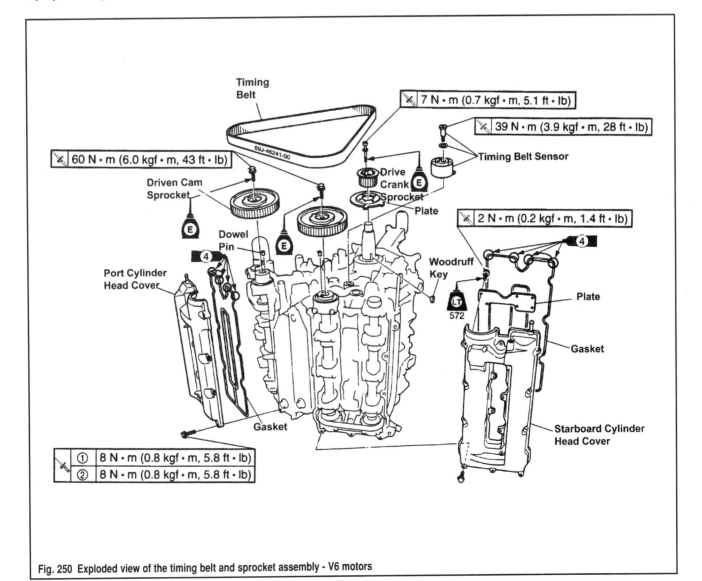

Fig. 250 Exploded view of the timing belt and sprocket assembly - V6 motors

6-112 POWERHEAD

CYLINDER HEAD REMOVAL & INSTALLATION

◆ See Figures 251 thru 257

■ Unique to the large Yamaha and Mercury/Mariner motors, the cylinder head bolts require a torque angle meter for tightening to proper specifications. Although the use of an inexpensive gauge is the best way to proceed, since the specification on these motors is for another 90 degrees, it IS possible simply to paint a mark on the socket and eyeball 1/4 additional turn, but don't be cheap, get the gauge.

We cover a lot of ground in this procedure. Although it is a cylinder head assembly removal procedure, there are many instances where you may need to follow only parts of the procedure to service other components, such as the camshafts and valve train.

First off, it should NOT be necessary to remove the powerhead from the intermediate housing in order to service the cylinder head. Secondly, not only is it NOT necessary to remove the cylinder head in order to service the camshafts, it is actually NECESSARY to remove the camshafts in order to remove the cylinder head (since they're in the way of most cylinder head mounting bolts).

■ Only perform the steps which are appropriate for your reason for working on the cylinder head. Therefore, if you are removing the valve cover simply to check the valve adjustment or for valve train inspection, there is no need to follow steps like removing the flywheel. Of course, if you learn that you need to actually adjust the valves or replace a lifter, then you're going to have to remove the flywheel in order to remove the camshafts.

Remember, when loosening the retainers on manifolds, covers and other major components, always try to follow the reverse of the torque sequence indicated in the assembly procedure or, in some cases, molded on the component itself.

When applicable tighten all bolts in using the proper torque sequence and to the proper specification. When possible we've included it in the procedure, but on most Yamaha produced powerheads both the sequence and value is molded into the casting on critical components. You'll notice that most Yamaha torque sequences are clockwise spirals starting somewhere near the center of the component.

■ If you are not planning on planning on replacing the camshafts or valve train components, check the Valve Clearance as detailed in the Maintenance and Tune-Up section before beginning work. This way you'll be able to decide before hand if any valve lifter shims must be replaced.

1. Disconnect the negative battery cable for safety.

■ Before removing the Flywheel, align the timing marks and note the camshaft timing mark positions when the marks are aligned. In this way you'll be able to set the camshafts up properly before cylinder head installation.

2. Remove the Flywheel and Stator Plate, as detailed in the Ignition and Electrical Systems section.
3. Remove the Timing Belt and the camshaft sprockets, as detailed in this section.
4. To completely remove the cylinder head, first remove the Throttle Body and Intake Assembly. For details, please refer to the procedure found in the Fuel System section.
5. If you are removing the cylinder head (and not just checking/adjusting valves), remove the exhaust covers. To prevent possible warpage or damage, gradually loosen the retaining bolts for each exhaust cover using multiple passes in the reverse of the torque sequence (which essentially results in a spiraling pattern that starts at the outer bolts and works inward). Once the bolts are removed, carefully break the gasket seals, then remove the inner and outer cover, along with their gaskets from the cylinder head.

■ The camshaft caps are normally marked from the factory for identification and orientation, before removal either identify these marks or make some of your own to ensure they will be reinstalled in the same positions and facing the same direction.

6. For each cylinder head, first loosen and remove the timing chain tensioner found at one end of the head, between the two camshaft chain gears, then loosen and remove the camshaft retaining caps in the reverse of the installation sequence. Loosen the bolts on each cap alternately and evenly until they can be turned without a tool.
7. Support the camshafts as the last caps are removed, then remove the two camshafts and timing chain from the cylinder head as an assembly. Repeat for the other bank, as necessary.
8. Disconnect the ground lead from the inner (valley) side top corner of each cylinder head.
9. Remove and discard the oil seal at the top of each inner (exhaust) camshaft.

■ The cylinder head bolts (at least the large M10s) normally require a larger Torx® bit driver.

10. Loosen the cylinder head bolts using multiple passes in the reverse of the torque sequence (which essentially results in a spiraling pattern that starts at the outer bolts and works inward).

■ There are 3 different length/size bolts used on the cylinder heads for these models. Be sure to keep track of all bolt positions for installation purposes. The longest/largest bolts are the top 8 bolts, while 2 shortest/smallest and one medium sized bolt are installed at the bottom of each head.

11. Carefully break the gasket seal and remove the cylinder head from the cylinder block.
12. Remove and discard the old cylinder head gasket. Carefully remove all traces of the gasket material from the mating surfaces.
13. Check the condition of the 2 cylinder head dowel pins (collars) for each head and replace, if necessary.
14. If necessary for further cylinder head disassembly remove the 2 anode covers, 2 anodes and 2 grommets from the center of the each head.

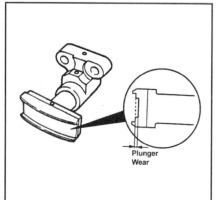

Fig. 251 Check the timing chain tensioner for wear

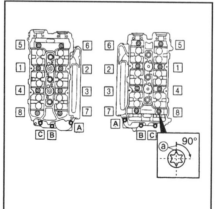

Fig. 252 Cylinder head torque sequence

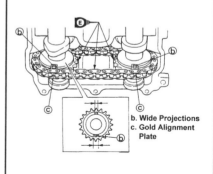

Fig. 253 Align the wide projections with the gold alignment plates

POWERHEAD 6-113

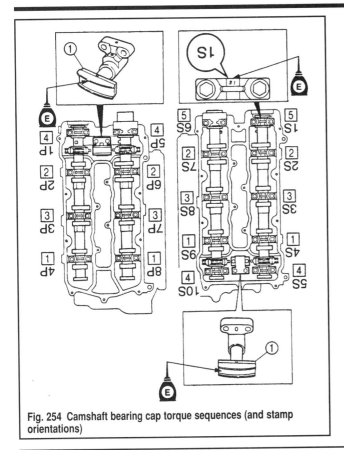

Fig. 254 Camshaft bearing cap torque sequences (and stamp orientations)

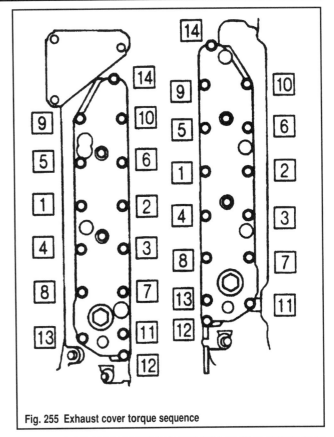

Fig. 255 Exhaust cover torque sequence

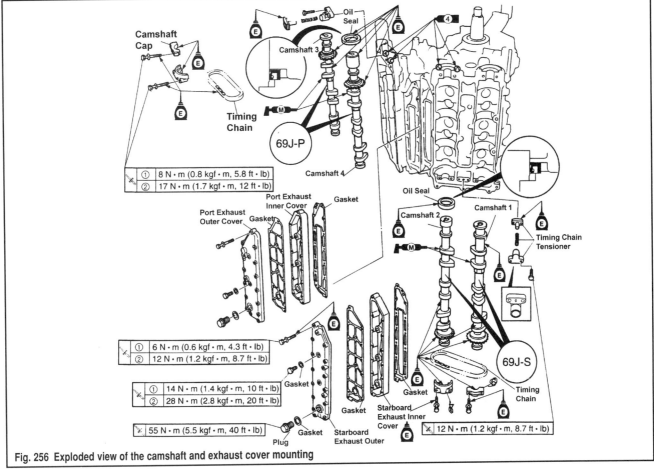

Fig. 256 Exploded view of the camshaft and exhaust cover mounting

6-114 POWERHEAD

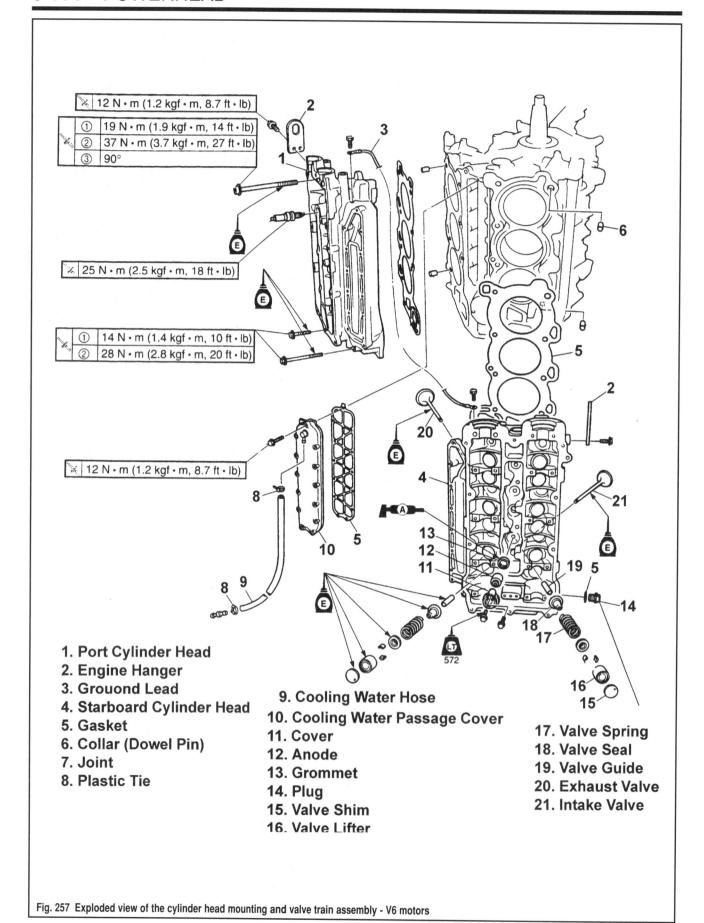

Fig. 257 Exploded view of the cylinder head mounting and valve train assembly - V6 motors

POWERHEAD 6-115

15. Refer to Valve Train Overhaul in this section for details on removal and installation of the remaining valve train components. Also, be sure to check the information on gasket mating surfaces to make sure the cylinder head is not warped.
16. The manufacturers both recommend checking the tensioner's plunger extended length after every 400 hours of operation. Replace the tensioner if the extended length has been reduced by 0.04 in. (1mm) or more from standard by wear.

To Install:

17. If removed, install the anode assembly or assemblies to the center of the cylinder head. For each assembly, apply a light coating of marine grade grease to the grommet (Yamaha All-Purpose or Quicksilver 2-4-C with Teflon) and Loctite® 572 (Yamaha), Loctite® 242 (Mercury) or equivalent threadlocking compound to the threads of the anode cover retaining bolt. Install the grommet, anode and cover, then tighten the retaining bolts securely.

■ **If cylinder block positioning was disturbed, slowly turn the crankshaft in the normal direction of rotation (clockwise when viewed from above) until the No. 1 cylinder is at TDC.**

18. Apply a light coating of clean engine oil to the threads of the cylinder head bolts.
19. Position a new cylinder head gasket on the cylinder block over the 2 dowel pins, then carefully position the cylinder head over the gasket and onto the block. Thread the various long and short cylinder head bolts in their proper places, making sure the gasket remained perfectly in position.
20. Finger-tighten all bolts, then tighten the bolts using multiple passes of the 2 torque sequences. First you'll tighten the 8 large (M10) bolts using at least 3 passes of steps 1-8 of the sequence, then you'll tighten the 1 medium and 2 smaller bolts using at least 2 passes of steps A-C of the sequence.
21. Tighten the 8 large bolts first to 168 inch lbs./14 ft. lbs. (19 Nm) and next to 27 ft. lbs. (37 Nm). Lastly, tighten the large bolts an additional 90 degrees each using a torque angle meter.
22. Tighten the 1 medium and 2 small bolts first to 120 inch lbs./10 ft. lbs. (14 Nm) and next to 21 ft. lbs. (28 Nm).
23. Connect the ground lead to each cylinder head and secure using its retaining bolt.

■ **Both the camshafts and the bearing caps should have stamp marks from the factory to help identify their positioning. The center of each camshaft should be stamped 69J- either P for Port or S for Starboard. The bearing caps are numbered 1-8 P or S. The caps are installed with the numbers facing upside down. They are positioned with numbers 1-4P or 1-5S top-to-bottom (1 on top and 4 or 5 respectively on the bottom) for the outer (intake) camshaft. The caps for the inner camshaft are numbered 5-8P or 6-10S (again with 5 or 6 on top and 8 or 10 on the bottom).**

24. Apply a light coating of engine oil to the new oil seal, camshaft bearing journals (NOT lobes), camshaft bearing caps, the threads of the bearing cap bolts, the contact surface of he timing chain tensioner, the timing chain and the camshaft chain gears. Apply a light coating of an engine assembly lube that contains molybdenum disulfide to the camshaft lobes.
25. Assemble two camshafts and a timing chain by aligning the wide projections along the camshaft chain gears with the gold alignment plates on the chain, then install the camshafts along with a new oil seal to the cylinder head.
26. Place the camshaft caps in position on the cylinder head and thread a couple of the cap bolts to hold them in position. Install the balance of the caps and finger thread all of the bolts.
27. Using two passes of the illustrated torque sequence, alternately and evenly tighten each pair of bearing cap bolts (meaning work on one cap at a time, in the illustrated sequence, but when tightening the 2 bolts on that cap, go back and forth each turn between the 2 bolts). Tighten the bolts first to 70 inch lbs./5.8 ft. lbs. (8 Nm) and then to 150 inch lbs./12 ft. lbs. (17 Nm).

■ **If the valve train components were replaced or if you failed to take measurements before the camshafts were removed, take the time now to check and adjust the valve clearances.**

28. Install the timing chain tensioner to each cylinder head and tighten the retaining bolts to 105 inch lbs./8.7 ft. lbs. (12 Nm).
29. If the exhaust cover assemblies were removed, install the inner and outer cover to each cylinder head using NEW gaskets. Apply a light coat of engine oil to the exhaust cover bolts, then install and tighten the cover bolts using at least two passes of the torque sequence. Tighten the bolts first to 53 inch lbs./4.3 ft. lbs. (6 Nm) and then to 105 inch lbs./8.7 ft. lbs. (12 Nm).
30. If removed, install the Throttle Body and Intake Assembly, as detailed in the Fuel System section.
31. Install the Timing Belt/Sprocket Assembly, as detailed in this section.
32. Remove the Flywheel and Stator Plate, as detailed in the Ignition and Electrical Systems section.
33. Reconnect the negative battery cable.

VALVE TRAIN OVERHAUL

◆ See Figure 257, 258 and 259

■ **If you only need to check/replace the valve shims or lifters and not the valves themselves there is no need to remove the cylinder head.**

1. Remove the Cylinder Head from the powerhead assembly, as detailed in this section.

■ **Valve train components such as valve lifters, retainers, keepers, springs and valves should not be mixed and matched from one assembly to another. This means that if they are to be reused, they must all be labeled prior to removal or VERY carefully sorted during service to prevent installation to the wrong positions.**

To remove the each valve assembly, proceed as follows:
2. Remove the valve shim and lifter from the bore in the cylinder head to expose the valve step and keepers.

■ **The top of the valve shim normally has a small oil hole. One easy way to separate the shim from the lifter is to blow a small amount of compressed air into the oil hole, the air pressure will push the shim upward from beneath.**

3. Using a clamp-type valve spring compressor, carefully place just enough pressure on the valve spring so the retainer moves down sufficiently to expose the keepers.
4. Remove the pair of valve keepers, then slowly ease pressure on the valve spring until the compressor can be removed.
5. Remove the valve spring retainer and valve spring from the shaft of the valve.
6. Remove the valve from the underside (combustion chamber side) of the cylinder head.

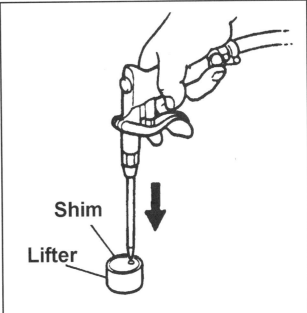

Fig. 258 You may use compressed air to free the shim from the lifter

6-116 POWERHEAD

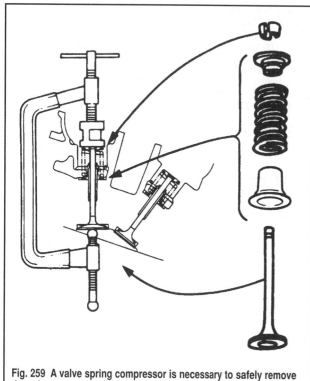

Fig. 259 A valve spring compressor is necessary to safely remove the valves

7. If necessary, carefully pry the old valve stem seal from its position on the cylinder head.

8. Repeat this procedure for each of the remaining valves.

9. Refer to Powerhead Refinishing, later in this section, for details on cleaning and inspecting cylinder head components.

To Install:

■ Apply a light coating of clean engine oil to each of the valve train components, except the valve spring and the valve keepers as they are installed. You can use a dab of marine grade grease on the keepers in order to help keep them in position during installation. One popular method is to actually use the grease to temporarily adhere the keepers to the installation tool itself, using the retainer to free it from the tool and place it in position.

10. If removed, install new valve spring seals to the cylinder head.

11. Position the valve into the cylinder head (into the original guide and valve seat unless the valve or valve seat was replaced).

12. Oil and install the spring over the valve stem.

13. Place the retainer over the spring, then use the valve spring compressor tool to press downward on the retainer and spring so you can install the keepers.

14. Insert the valve keepers, then slowly back off the valve spring compressor, making sure the keepers are fully in position. Then remove the valve spring compressor.

15. Install the valve lifter and shim.

16. Repeat for each of the remaining valves.

17. Install the Cylinder Head as detailed in this section.

18. Once the powerhead is fully assembled, if valve train components were replaced, be sure to operate the engine as directed for new component break-in. For details, please refer to Powerhead Break-In, in this section.

CYLINDER BLOCK OVERHAUL

 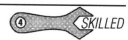

◆ See Figures 260 thru 270

■ Unique to the large Yamaha and Mercury/Mariner motors, the crankcase bearing and connecting rod bolts require a torque angle meter for tightening to proper specifications. Although the use of an inexpensive gauge is the best way to proceed, since the specification on these motors is for another 90 degrees, it IS possible simply to paint a mark on the socket and eyeball 1/4 additional turn, but don't be cheap, get the gauge.

Remember, when loosening the retainers on manifolds, covers and other major components, always try to follow the reverse of the torque sequence indicated in the assembly procedure or molded on the component itself.

When applicable tighten all bolts in using the proper torque sequence and to the proper specification. When possible we've included it in the procedure, but on many Yamaha manufactured powerheads both the sequence and value is molded into the casting on critical components. You'll notice that most Yamaha torque sequences are clockwise spirals starting somewhere near the center of the component.

1. Remove the Powerhead assembly, as detailed in this section.
2. Remove the Timing Belt/Sprocket assembly, as detailed in this section.
3. Strip all electrical components from the powerhead, then remove the Cylinder Head assembly, as detailed in this section.
4. Remove the Oil Pump from the bottom of the powerhead assembly. Service the pump, O-rings and seals, as detailed in the Lubrication and Cooling section.
5. If not done already, remove and discard the old oil filter from the cylinder block.
6. Locate the cooling water passage cover which is found in the center of the cylinder block V, between the cylinder head mating surfaces. A cooling water hose attaches to a fitting on the upper end of this cover. If the hose is not already disconnected at one end, cut a plastic wire tie and disconnect the hose from one of the hose joints.
7. Gradually loosen and remove the 13 cover bolts using the reverse of the torque pattern (this is roughly a counterclockwise spiraling pattern that starts at the lower left bolt and works it's way inward).
8. Remove the cooling water passage cover from the cylinder block. Remove and discard the old cover gasket.
9. Gradually loosen and remove the 17 crankcase cover flange bolts using the reverse of the torque pattern (this is roughly a counterclockwise spiraling pattern that starts at the top, right bolt and works it's way inward).

■ There is NO need to remove the timing pointer from the cover!

10. Carefully break the seal and separate the crankcase cover from the cylinder block, then remove and discard the gasket. Remove the 6 nuts and separate the baffle plate from the studs on the bottom of the crankcase.

■ Most parts inside the cylinder block are matched sets and should not be moved from one cylinder to another. This includes the bearings and pistons as well as each of the connecting rod/end cap combination. For this reason, prior to disassembly it is essential to number and matchmark each of the connecting rods to their end caps (this ensures that not only will each cap be installed on the proper rod, but facing the correct direction). Also, be sure to label each piston and connecting rod as to what cylinder it belongs and identify the factory marks and/or place an arrow or some indicator facing upward to ensure they are installed in the same direction. Yamaha often marks pistons with UP on the dome and connecting rods with the letter Y or word YAMAHA and/or alignment dots at the con rod-to-cap mating surface, all of which should face the flywheel.

POWERHEAD 6-117

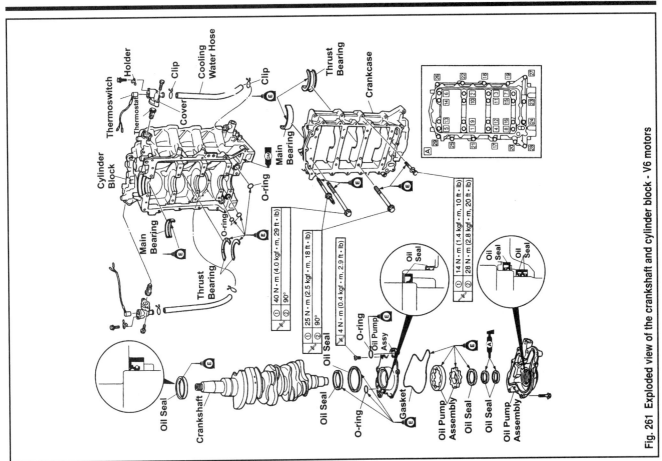

Fig. 261 Exploded view of the crankshaft and cylinder block - V6 motors

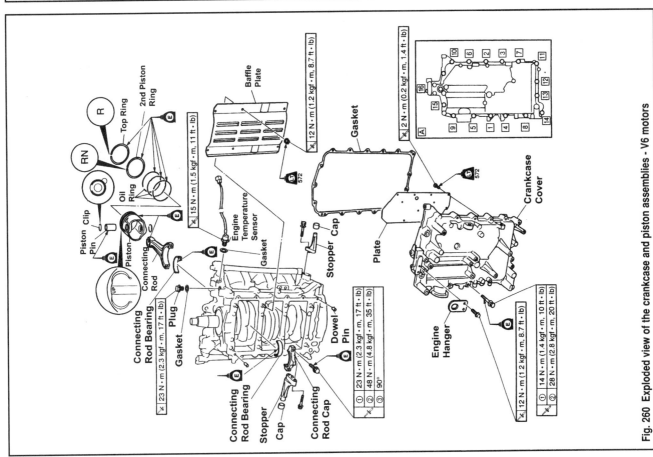

Fig. 260 Exploded view of the crankcase and piston assemblies - V6 motors

6-118 POWERHEAD

■ As soon as the pistons and/or crankshaft is removed, reinstall the connecting rod end caps and bearing halves to the con rods to ensure there is no mix-up.

11. Matchmark the big end caps to the connecting rods, then remove the 12 connecting rod cap bolts (working on one cap at a time, alternating between each of the cap bolts after each turn). Remove the caps and place them aside for reinstallation back on their original connecting rods as soon as the crankshaft has been removed. DISCARD the old connecting rod cap bolts as they are too distorted by the angle torque process and cannot be reused.

■ Yamaha recommends that the pistons be removed out the top of the cylinder bore before the crankshaft is unbolted and removed. However, in order to do this, you'll need to make sure there is no ridge at the top of the bore on which a piston ring may catch. If necessary, you either use a ridge reamer before the piston is removed OR you can position all the pistons at the top of their travel and VERY carefully remove the crankshaft so the pistons can be removed from the bottom. However, remember, that if a ridge is present, you'll still have to remove it with a ridge reamer before piston installation.

12. Gradually loosen and remove the crankcase bearing/flange bolts using the reverse of the torque sequence. Keep close track of the bolts since there are 4 different types, all of which must be installed in their original positions. This includes 13 short M8 flange bolts, 6 stud bolts, 2 M8 bearing bolts and 8 M10 bearing bolts. Remove the crankcase from the block.
13. Carefully lift the crankshaft from the block. Place it on a protective surface.

✲✲ WARNING

Handle the crankshaft carefully. Remember it is a heavy, expensive and vital piece to the motor. The polished surfaces should be covered to prevent scratches or other damage and the whole shaft should be protected against impact or other damage which could destroy it.

14. If not done already, immediately reinstall the connecting rod end caps and their bearing inserts to their respective pistons.
15. Remove the top oil seal from the crankshaft and discard.
16. If you didn't remove the pistons from the top of the bores earlier, remove them from the bottom now.
17. If necessary, disassemble each of the pistons, as follows:
    a. Carefully pry the clips free from the piston pin bore on either side of the piston assembly. Discard the clips as they are normally distorted during removal and even the slightest distortion makes them unsuitable for further service.
    b. Slide the piston pin free of the bore (specifications tell us that this should not be an interference fit, however if the pin does not readily slide free use a rubber mallet and soft faced driver to carefully tap it free of the piston).
    c. Using a piston ring spreading tool, carefully remove each of the rings (top and second compression rings, followed by the oil ring assembly) from the piston.
    d. Keep all components from a single piston together for reuse or replacement, but don't mix and match components from other pistons.
18. Remove the crankshaft main (big end) bearings from the crankcase and cylinder block. Again, note their positions and orientation if they are going to be reused.

■ The bearings on these models are coded by color (connecting rod) or bearing number (main/thrust bearing). Codes are stamped on the outside of one crankshaft throw. Four codes on top, left-to-right correspond to the 4 main journals, top-to-bottom. Then 6 codes below the aforementioned four correspond to the 6 connecting rod journals, also top-to-bottom. For the main journals a second set of stamps are found along the crankcase mating flange of the cylinder block. The combination of the code on the crankshaft and the code on the cylinder block are cross-referenced with a provided table in order to help select the proper number code replacement bearing.

19. Refer to Powerhead Refinishing, later in this section, for details on cleaning and inspecting cylinder block components. This is easily the most complicated of the Yamaha motors when it comes to number of internal components that must be checked during a rebuild. Take your time and make sure you've made measurements for all components listed in the Engine Specifications chart for this model.

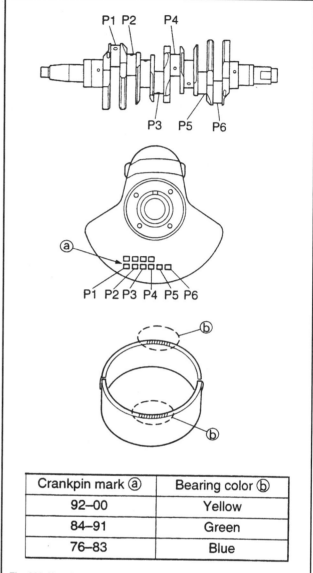

Crankpin mark ⓐ	Bearing color ⓑ
92–00	Yellow
84–91	Green
76–83	Blue

Fig. 262 Use the crankshaft marks and bearing colors to determine the starting point for replacement connecting rod bearings

To Install:

20. Make sure the mating surfaces of the crankcase and cylinder block are completely clean and free of sealant or damage.
21. Remove and discard the 3 O-rings from the bottom of the cylinder block-to-crankcase mating surface. Apply a light coat of engine oil to the replacement O-rings and position them in the block.
22. If they were disassembled, prepare each piston assembly for installation as follows:

■ For some unbeknown reason Mercury recommends switching the gap positioning of the upper and lower compression rings with each other (as well as the upper and lower oil rings) as compared with the information from Yamaha that is described below and illustrated. As we've said elsewhere in this section, we believe this is "much ado about nothing" and there are many machinists and engine assemblers who agree, however you've now been warned and can make up your own minds.

   a. Using a ring expander, carefully install each of the rings with their gaps as noted in the accompanying illustration. While looking at the piston dome with the top (flywheel end) facing 12 o'clock, the bottom oil ring should be positioned with the gap facing between 4 and 5 o'clock. The middle oil ring (expander ring) should be positioned with the gap facing between 7 and 8 o'clock.. The upper oil ring should be positioned with the gap facing

POWERHEAD 6-119

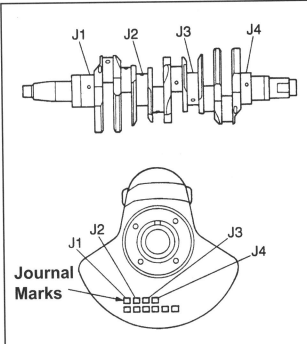

Fig. 263 The crankshaft is marked to help in selecting replacement main bearings

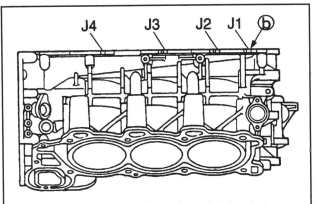

Fig. 264 The cylinder block is also marked to help in selecting replacement main bearings

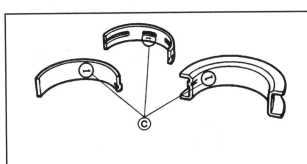

between 10 and 11 o'clock. The lower (second) compression ring should be positioned with its gap at between 1 and 2 o'clock. The upper (top) compression ring should be positioned with its gap facing between 7 and 8 o'clock.

■ The compression rings should be stamped with an identifying mark which should be faced upward toward the top of the piston dome. On the 2nd compression ring the stamped mark is usually RN, while the mark on the top compression ring is usually R.

    b. Align the mark(s) on the connecting rod (facing up/flywheel) with the mark on the piston (also facing up/flywheel), install the connecting rod to the piston using the piston pin.
    c. Install new piston pin clips in either side of the pin bore to secure the piston pin in place. Once the clip is installed make sure the gap in the clip ends DOES NOT align with the piston pin slot (the small elliptical cutout).
    d. Repeat for the remaining pistons.

■ We don't really like Yamaha's recommendations here, but we'll pass them along. Yamaha wants you to install and tighten the crankshaft THEN install the pistons to their bores. Our only problem with this is we find it is too easy to accidentally score the piston wall or damage the crankshaft while installing the pistons in this manner (with the crank already in place). The choice is yours. If you decide to install the pistons now, before the crank, then take great care when installing the crank. However, if you follow this procedure (and Yamaha's recommendations) take great care when installing the pistons.

    23. Place the main bearing inserts into the crankcase and cylinder block. There are normally 2 things to look out for when doing this. Most bearing inserts will be equipped with locating tabs, make sure they are seated in the appropriate grooves. Also, inserts with oil holes should be aligned with an oil passage.

■ Here's where procedures differ slightly depending upon whether you're installing bearings of a known clearance or measuring clearance. If you've already determined that bearing clearances are fine, apply alight coat of engine oil to each of the connecting rod, main bearing and crankshaft journal surfaces. If however, you need to measure and determine bearing clearance, Temporarily INSTALL THESE COMPONENTS DRY for measurement purposes.

    24. Position the crankshaft along with a new upper oil seal (positioned so the seal lips are facing inward, toward the crankshaft) and thrust bearings into position in the cylinder block.

Fig. 265 Cross reference the cylinder block and crankshaft markings on this chart to determine the proper replacement bearing number (but remember you still need to check clearances)

    25. If you've already measured bearing clearances, apply a light coating of Gasket Maker (such as Loctite 514 Master Gasket, RTV sealant or equivalent) to the entire outer perimeter of the cylinder block-to-crankcase mating surface.

✱✱ WARNING

DO NOT get any of the gasket maker in the journal bearings!

    26. Apply a light coating of engine oil to the threads of all 4 types of crankcase bolts.

6-120 POWERHEAD

■ Yamaha notes that bolts 1-16 in the crankcase torque pattern may be reused up to 5 times, but NO MORE than that.

27. With the main bearing inserts in place, carefully lower the crankcase onto the cylinder block, then JUST thread the 4 types of bolts into their respective positions.

28. Tighten bolts 1-16 of the torque sequence using at least two passes of the sequence. During this portion of the crankcase sequence tighten bolts 1-8 (M8 bolts) first to 18 ft. lbs. (25 Nm) and bolts 9-16 (M10 bolts) first to 29 ft. lbs. (40 Nm). Next tighten bolts 1-16 and additional 90 degrees each using a torque angle meter.

29. Once bolts 1-16 are fully tightened, tighten bolts 17-29 (the flange bolts) using at least two passes of the sequence, first to 120 inch lbs./10 ft. lbs. (14 Nm) and then to 20 ft. lbs. (28 Nm).

※※ **WARNING**

Use extreme care to prevent the connecting rods from contacting and scoring the cylinder walls during piston installation. It is a good idea to wrap the ends in a shop cloth or something else to protect the cylinders AND the crankshaft.

30. Install each of the pistons (with the correct mark/side facing UP toward the flywheel) to the cylinder block using a ring compressor. Apply a very light coating of clean engine oil to the inside of the compressor, to the cylinder bore and to the sides of the pistons, then place the compressor around the piston. Insert the piston through the cylinder head side of the bore and use a wooden hammer handle to gently tap the piston down through the compressor and into the cylinder bore. Proceed slowly, checking the position of the connecting rod end with respect to the crankshaft as you are going.

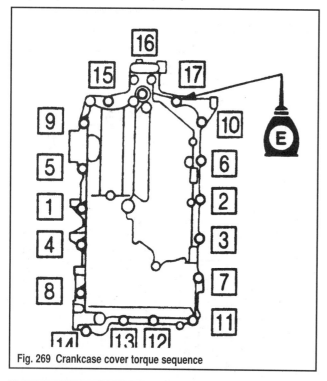
Fig. 269 Crankcase cover torque sequence

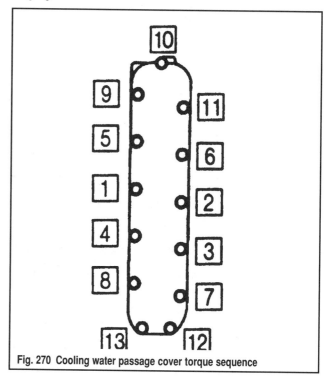
Fig. 270 Cooling water passage cover torque sequence

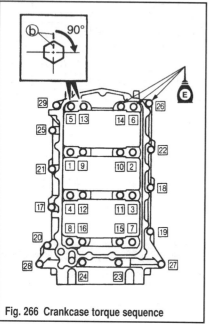
Fig. 266 Crankcase torque sequence

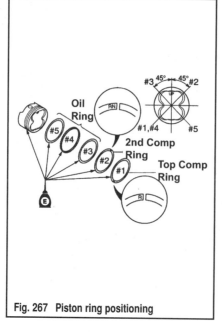
Fig. 267 Piston ring positioning

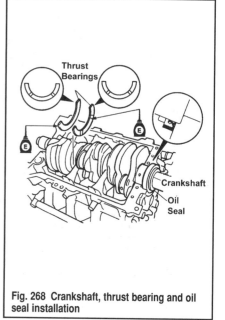
Fig. 268 Crankshaft, thrust bearing and oil seal installation

POWERHEAD 6-121

> ※※ **WARNING**
>
> Be VERY careful when installing the piston. Get the lower portion of the skirt below the ring compressor and insert it into the cylinder bore to make sure everything is aligned. Then, GENTLY tap the piston through the compressor, stopping if it seems to hang up even slightly to make sure that a ring is not caught on the deck of the cylinder block.

31. Remove the connecting rod end caps and position them so the crankshaft can be installed. Place the connecting rod bearing inserts into position on the connecting rod big end and cap.

■ It's probably smarter to install the connecting rod cap and bearings to the crankshaft for each piston as it is being installed before moving on to the next piston. In this way, should you rotate the crankshaft you won't risk damaging the shaft, rod or cylinder wall if the connecting rod makes unwanted contact with either.

32. Install each of the match-marked connecting rod caps using NEW connecting rod retaining bolts, alternating back and forth between the 2 bolts on each cap. Tighten the bolts first to 17 ft. lbs. (23 Nm), next to 35 ft. lbs. (48 Nm) and then, using a torque angle gauge tighten the bolts an additional 90 degrees each.
33. Install the baffle plate to the studs on the bottom of the crankcase, then tighten the retaining nuts to 106 inch lbs./8.7 ft. lbs. (12 Nm).
34. Apply a light coating of engine oil to the 17 crankcase retaining bolts, then install the crankcase cover using a new gasket. Tighten the cover bolts using at least 2 passes of the clockwise spiraling torque sequence that starts at the left center bolt and works outward. Tighten the bolts first to 120 inch lbs./10 ft. lbs. (14 Nm) and then to 20 ft. lbs. (28 Nm).
35. Properly install the Oil Pump assembly to the bottom of the powerhead as detailed in the Lubrication and Cooling section.
36. Install the cooling water passage cover using a new gasket. Tighten the bolts to 106 inch lbs./8.7 ft. lbs. (12 Nm), using multiple passes of a clockwise spiraling pattern that starts with the left center bolt and works outward.
37. Refill the engine with clean engine oil through the passage in the oil filter bracket, then install a NEW oil filter.
38. Install the Cylinder Head assembly (along with the camshafts and any electrical components stripped from the powerhead) as detailed in this section.
39. Install the Timing Belt/Sprocket assembly, as detailed in this section.
40. Install the Powerhead assembly, as detailed in this section.
41. Once the powerhead is fully assembled, if cylinder block components were replaced, especially the pistons and/or rings, be sure to operate the engine as directed for new component break-in. For details, please refer to Powerhead Break-In, in this section.

CLEANING & INSPECTING

Cleaning and inspecting internal engine components is virtually the same for all Yamaha and Mercury/Mariner outboards with the exception of a few components that are unique to certain 4-stroke motors. The main variance between different motors comes in specifications (which are listed in the Engine Specifications charts) or by component type. A section detailing the proper procedures, sorted mostly by component, can be found under Powerhead Refinishing.

POWERHEAD REFINISHING

General Information

The success of the overhaul work is largely dependent on how well the cleaning and inspecting tasks are completed. If some parts are not thoroughly cleaned, or if an unsatisfactory unit is allowed to be returned to service through negligent inspection, the time and expense involved in the work will not be justified with peak powerhead performance and long operating life. Therefore, the procedures in the following sections should be followed closely and the work performed with patience and attention to detail.

■ When measuring components, remember that specifications may vary with temperature. Unless otherwise noted, specifications are for components at an ambient room temperature of 68°F (20°C).

CLEANING

All powerhead components must be clean and free of gasket material, oil and carbon deposits before they are inspected. Take your time when cleaning components. Before using solvent to clean something, make sure the chemical is compatible with the material of which the component is constructed. Also, check each component for matchmarks or ID marks before cleaning (as some solvents may remove even permanent marker). If necessary, re-ID the component as soon it has been cleaned and dried (before moving onto the next component). Whenever you match-marked 2 components in order to ensure exact alignment during installation, if possible, fasten them together during cleaning (in case the marks come off). In this way, they can be re-marked after the cleaning process is through.

When it comes to inspection, this section will provide information on how to check various components. But, keep in mind that not all components will be found on all motors. If in doubt whether a component is used (or should be checked) refer to the Disassembly, Assembly or Overhaul procedures (as applicable). Also, be sure to check the Specifications Chart for your motor. If a component is not listed in the specification chart, it is either not used, or, does not have a specific tolerance for which it must be inspected. Whether or not a tolerance is provided, all components must be clean and free of obvious defects (deep cracks, scoring, excessive carbon deposits that cannot be removed, warpage, etc). If in doubt whether or not a component is serviceable, seek someone with more Yamaha or Mercury/Mariner engine rebuilding experience than yourself.

■ The 4-stroke motor, by design, tends to be much more complicated than a 2-stroke motor, containing a greater number of precision machined parts that require inspection. This is especially true in the cylinder head and related components (camshaft and valve train) most of which are not found on 2-strokes.

Generally speaking, the larger the hp model, the more involved will be the cleaning and inspection process. This is not because of size, as much as it has to do with the larger motors being designed to produce more hp and sell for much more money, so they are usually designed with more components/systems.

When inspection involves precision measurements, not only must the components be clean, but they must be measured roughly at room temperature, using the appropriate measurement equipment to ensure accurate results.

It is very important to keep in mind that all wear components (pistons, rings, shafts, springs, valves, bearings, etc) **must** be reinstalled in their original locations whenever they are being reused. Wear patterns form on all contact surfaces during use. Mismatching wear patterns will accelerate wear, while matching wear patterns helps ensure a durable and reliable repair.

Cleaning

◆ See Figures 271 thru 276

> ※※ **WARNING**
>
> Avoid removing excessive amounts of metal when removing carbon deposits.

1. Use a blunt-tip scraper or dulled chisel to loosen carbon deposits from various components of the combustion chamber and ports/valves. Work slowly and carefully to prevent damaging or excessively scoring the surfaces. Then use a Scotchbrite® pad and mild solvent to remove most/all of the remaining deposits. Remove deposits from the following components, as applicable:
   • Remove carbon from the combustion chambers in the cylinder head. Use great care on these 4-stroke motors, since the valve seating surface will be exposed after disassembly. DO NOT score or damage the valve seating surfaces. Another option is to protect the valve seating surface by leaving the valves in place while you are removing the carbon deposits, but the same care must be taken to prevent deep scores on the bottoms of the valves.
   • Remove carbon deposits or corrosion from the exhaust cover(s).

6-122 POWERHEAD

■ Deposits on pistons can be removed while they are still installed in the bores. This is handy when the cylinder head is removed for service without completely disassembling the crankcase. If this is to be accomplished, position the piston to be cleaned at TDC and cover the other piston bore(s) using rags and plastic. Thoroughly clean all debris using solvent and compressed air (WHILE WEARING SAFETY GLASSES) before moving to the next piston.

• Remove carbon deposits from the top of the piston(s). When working on the piston domes, use a light touch to prevent scratching, or worse, gouging the piston.

✳✳ **WARNING**

Wire brushes are not recommended for cleaning piston domes since particles of steel could become lodged into the piston surface. If this occurs, they could glow hot when the piston is returned to service, causing pre-ignition or detonation that could damage the piston and combustion chamber.

• Remove carbon from the ring grooves either using a ring groove cleaner, or, better yet, using a broken piece of the piston ring with an angle ground on the end. When using a ring to clean the piston grooves, use the ring actually removed from the groove (if it is being replaced) or one from the same groove on another piston.

✳✳ **WARNING**

When cleaning piston ring grooves, use the same caution as with the pistons. Do NOT remove excessive amount of material or the piston will be damaged beyond use. Some ring groove cleaning tools are heavy duty and will easily remove too much material, so use them with care. Believe it or not some manufacturer recommend the filed broken ring method.

2. Inspect all water passages for corrosion deposits, debris or blockage. Remove debris and clean corrosion as needed and accessible. Use low-pressure compressed air to blow out all water passages.

Fig. 271 Although a blunt chisel is preferred, a wire brush can be used WITH CARE to clean most carbon deposits

Fig. 272 Pistons CAN be cleaned while still installed, but again use care, and leave NO metallic deposits behind

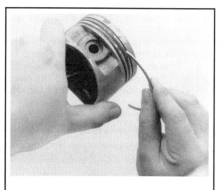

Fig. 273 The preferred method for cleaning piston grooves is to use the filed end of a broken ring...

Fig. 274 ... but a ring groove cleaning tool can be used, with care to prevent damaging the piston

Fig. 275 The cylinder walls must show an obvious cross-hatching (tiny grooves, criss-crossed in a pattern around the bore)...

Fig. 276 ... otherwise a cylinder hone must be used to break the smooth glazed surface into a cross-hatch pattern

POWERHEAD 6-123

3. Clean and degrease all regularly oiled surfaces (including the crankshaft, pistons and connecting rods) using solvent or degreaser. Use low-pressure compressed air to remove all build-up from shaft and rod oil holes.

4. Remove all traces of gasket or sealant using a spray gasket remover or equivalent gasket removing solvent. Whenever possible, avoid the use of gasket scrapers to help avoid the possibility of scoring and damaging the gasket mating surfaces.

■ Honing the cylinder walls can wait until after measurements are taken to determine if boring for oversized pistons/rings will be necessary. If honing is pushed off until after inspection, be sure to follow the remaining steps in order to finish cleaning the block, then repeat the appropriate steps again, after honing.

5. Check the cylinder walls for glazing (a smooth, glassy appearance) and, if found, hone the cylinder walls using a medium grit cylinder hone. Use a slow rpm while raising and lowering the hone through the cylinder in order to cross-hatch the cylinder walls for maximum oil retention.

※※ WARNING

Use the cylinder hone slowly, carefully and as little as possible to avoid the possibility of removing too much material from the cylinder walls. If the bores are over-honed, it could cause the pistons to be below specification for the new cylinder measurement, or worse, cause the cylinder bores to be overspec for what pistons are available.

6. Wash the entire cylinder block, crankcase and head using warm, soapy water to remove all traces of contaminants. Use low-pressure compressed air to blow dry all passageways.

7. Apply a light coating of clean engine oil to all machined surfaces that are not about to be measured right away. When you return to the task of measuring components that have been oiled, use a solvent covered rag to wipe away the oil before measurements are taken.

8. Cover all components using a plastic sheet to keep dust, dirt or debris from contaminating the cleaned and especially the oiled surfaces.

Exhaust Cover

INSPECTION

◆ See Figures 277 and 278

The exhaust covers are one of the most neglected items on any outboard powerhead. Seldom are they checked and serviced. Many times a powerhead may be overhauled and returned to service without the exhaust covers ever having been removed.

One reason the exhaust covers are not removed is because the attaching bolts usually become corroded in place. This means they are very difficult to remove, but the work should be done. Heat applied to the bolt head and around the exhaust cover will help in removal. However, some bolts may still be broken. If the bolt is broken it must be drilled out and the hole tapped with new threads.

The exhaust covers are installed over the exhaust ports to allow the exhaust to leave the powerhead and be transferred to the exhaust housing. If the cover was the only item over the exhaust ports, they would become so hot from the exhaust gases they might cause a fire or a person could be severely burned if they came in contact with the cover.

Therefore, depending upon the inner plumbing of the powerhead, MOST are equipped with an inner plate to help dissipate the exhaust heat. Two gaskets are normally installed - one on either side of the inner plate. Water is channeled to circulate between the exhaust cover and the inner plate. This circulating water cools the exhaust cover and prevents it from becoming a hazard.

A thorough cleaning of the inner plate behind the exhaust covers should be performed during a major powerhead overhaul. If the integrity of the exhaust cover assembly is in doubt, replace the inner plate.

Fig. 277 The exhaust cover must be removed...

Fig. 278 ... and inspected for warpage or clogged passages in order to ensure proper powerhead operation

Crankshaft

INSPECTION

◆ See Figures 279 and 280

Clean the crankshaft with solvent and wipe the journals dry with a lint free cloth. Inspect the main journals and connecting rod journals for cracks, scratches, grooves, or scores. Inspect the crankshaft oil seal surface for nicks, sharp edges or burrs which might damage the oil seal during installation or might cause premature seal wear. Always handle the crankshaft carefully to avoid damaging the highly finished journal surfaces. Blow out all oil passages with compressed air. The oil passageway leads from the rod to the main bearing journal. Take care not to blow dirt into the main bearing journal bore.

Inspect the internal splines at one end and threads at the other end for signs of abnormal wear. Check the crankshaft for run-out by supporting it on two V-blocks at the main bearing surfaces.

Install a dial indicator gauge above the main bearing journals. Rotate the crankshaft and measure the run-out (or the out-of-round) and the taper at both ends and in the two center journals (center journal on all two-cylinder models).

6-124 POWERHEAD

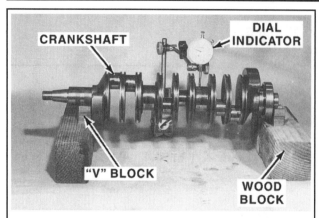

Fig. 279 Crankshaft set up with V-blocks and a dial indicator to measure run-out

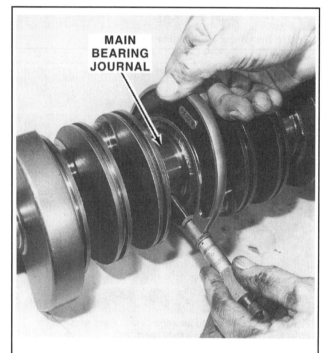

Fig. 280 You can also use a micrometer to check run-out and taper on the main bearing journals (and connecting rod bearing journals depending upon what specs are available)

If V-blocks or a dial indicator are not available, a micrometer may be used to measure the diameter of the journal. Make a second measurement at right angles to the first. Check the difference between the first and second measurement for out-of-round condition. If the journals are tapered, ridged, or out-of-round by more than the specification allows, the journals should be reground, or the crankshaft replaced.

Any out-of-round or taper shortens bearing life. Good shop practice dictates new main bearings be installed with a new or reground crankshaft.

■ Some models utilize pressed together crankshaft assemblies. On these models, normally the connecting rods would only be pressed from the crankshaft if either the crankshaft and/or the connecting rods were to be replaced. Therefore, the connecting rod axial play is checked at the piston end to determine the amount of wear at the crankshaft end of the connecting rod.

Specifications on some models are provided for different crankshaft widths. These measurements are made to determine if excessive wear has caused the need to replace the crankshaft and/or connecting rods. When specifications are provided they vary greatly by year and model. Also, the measurements may be around a single crankpin, the distance between to crankpins or the distance around all crankpins. Be sure to use a precision instrument when making these measurements. Refer to the Engine Specifications chart to see which measurements apply.

■ If the measurement is not within specification, check the Connecting Rod Big End Side Clearance and Small-End Free-play measurements of the connecting rod(s) involved. This measurement will give an indication if the counterweight is walking on the crank pin or if the clearance is due to worn parts. Normally the connecting rod would wear before the edge of the counterweight because the rod is manufactured from a much softer material.

Inspect the crankshaft oil seal surfaces to be sure they are not grooved, pitted, or scratched. Replace the crankshaft if it is severely damaged or worn. Check all crankshaft bearing surfaces for rust, water marks, chatter marks, uneven wear, or overheating. Clean the crankshaft surfaces with 320-grit wet/dry sandpaper dampened with solvent. Never spin-dry a crankshaft ball bearing with compressed air.

Clean the crankshaft and, if equipped, crankshaft ball bearing with solvent. Dry the parts, but not the ball bearing, with compressed air. Check the crankshaft surfaces a second time. Replace the crankshaft if the surfaces cannot be cleaned properly for satisfactory service. If the crankshaft is to be installed for service, lubricate the surfaces with light oil. Do not lubricate the crankshaft ball bearing at this time.

On models so equipped, after the crankshaft has been cleaned, grasp the outer race of the crankshaft ball bearing installed on the lower end of the crankshaft, and attempt to work the race back and forth. There should not be excessive play. A very slight amount of side play is acceptable because there is only about 0.001 inch (0.025mm) clearance in the bearing.

Lubricate the ball bearing with light oil. Check the action of the bearing by rotating the outer bearing race. The bearing should have a smooth action and no rust stains. If the ball bearing sounds or feels rough or catches, the bearing should be removed and discarded.

For models equipped with bearing inserts, use a strip of Plastigage® or some equivalent compressible gauging material across the center of the crankshaft journal (lengthwise from flywheel-to-driveshaft end of the journal) and bolt the crankcase/bearing retainers in position, then remove again to exam the gauging material. For more details, please refer to Measuring Bearing Clearance Using Plastigage® later in this section.

MEASURING BEARING CLEARANCE USING PLASTIGAGE®

◆ See Figures 281, 282 and 283

Most of the larger Yamaha and Mercury/Mariner 4-strokes use automotive style bearing inserts for the crankshaft and connecting rods. Besides a visual inspection for obvious defects or flaws, the most important check that can be made to bearing inserts is to measure their installed oil clearance (the distance between the bearing and crankshaft journal) to determine if they are still fit for service. Even when selecting replacement bearings using color or number codes provided on the motor components, this check must be made to ensure the assembled motor will be within design tolerances.

To check bearing oil clearances you need a way of measuring a tiny gap (thousands of an inch or hundredths of a millimeter) between otherwise inaccessible components. The basic method is to apply a strip of a gauging material such as Plastigage® to the crankshaft journal and then to bolt the crankcase/main bearing retainer and/or connecting rod cap together. When the components are bolted together the gauging material is compressed (squished) between the bearing and journal. The key is that the material is designed to expand at a uniform rate so that you can then remove the bearing and measure the width of the squished gauging material. The width of the material will correspond to a specific clearance measurement (a width-to-clearance conversion scale is provided with the gauging material).

Using Plastigage® is a simple, inexpensive, and fairly reliable method to measure main bearing and connecting rod bearing clearance. However, Plastigage® is soluble in oil, so clean the crankshaft journal thoroughly of all traces of oil, and then turn the crankshaft until the oil hole is down away from the cap to prevent getting any oil on the Plastigage®. To measure bearing clearances, proceed as follows:

1. Obtain a package of Plastigage® from the local auto or marine parts outlet.

2. Cut a one inch (2.5cm) section of the bead for each journal/bearing combination that is being measured.

POWERHEAD 6-125

3. Place each one inch section on the crankshaft journal surface at the main bearings and on each connecting rod bearing surface.

■ DO NOT rotate the crankshaft until the measurements have been completed. Such action will smear the Plastigage® bead and the measurement will be useless.

4. Install the connecting rod caps along with the bearing inserts and tighten to specification (for details, refer to the Powerhead, Overhaul procedures found earlier in this section). **Take your time** to be sure each is installed back into the same position from which it was removed according to the marks made during disassembling.

✳✳ WARNING

For motors whose connecting rod caps are angle torqued and should not be reused, you can either install the old bolts (JUST for the test) or the new bolts, but in either case, tighten the bolts to specification BUT refrain from the angle torquing step to prevent the possibility of breaking the old bolts or stretching the new bolts beyond use.

5. Install the main bearings and tighten the crankcase/bearing retainers to specification (again, for details, please refer to the Powerhead, Overhaul procedures for the particular model on which you are working).

6. Remove the crankcase/bearing retainer bolts, the crankcase, the connecting rod bolts, and the connecting rod caps to gain access to the Plastigage®.

7. Measure the flattened Plastigage® at its widest point using the scale printed on the Plastigage® envelope. The number within the graduation which most closely corresponds to the width of the Plastigage® indicates the bearing clearance in thousands of an inch (or hundredths of a millimeter, depending upon the scale). A metric scale in mm is normally given on the back of the package. The widest point of the Plastigage® is the minimum clearance and the narrowest point is the maximum clearance. The difference between the readings indicates the taper of the journals. Compare the Plastigage® measurements with the specifications found in the Engine Specifications chart in this section.

8. If the clearance is not within specifications, the bearings or the crankshaft, or the connecting rods **must** be replaced as required. Obviously, you'd start with new bearing inserts to see if one can be obtained to return that particular journal to specified clearances.

9. Once you are finished checking and selecting bearings, clean all traces of the Plastigage® from the bearing surfaces, and then follow necessary steps again (oiling the bearing surfaces and skipping the Plastigage®) to make a final installation of the connecting rod caps and the crankcase. Check to be **sure** the crankshaft rotates freely. If any binding is discovered, the thrust washer tabs may be misaligned or one or more of the connecting rod caps may be installed backwards.

Connecting Rods

INSPECTION

◆ See Figures 284 thru 288

■ Specifications available for the connecting rods vary by model, so check the Engine Specifications charts to determine which steps are applicable to the motor on which you are working.

Each connecting rod should be cleaned and, once cleaned, should be thoroughly inspected to make sure it is not damaged, bent or excessively worn. Obvious defects, such as visible warpage or cracks are signs of unserviceability. On many motors, the connecting rod small-end (piston end) Free-play should be measured using a dial gauge, while connecting rod big-end (crankshaft end) must be checked for Side Clearance using a feeler gauge. On some models measurements may be provided for small end diameter (which will correspond to a measurement slightly larger than piston pin diameter). To thoroughly inspect the connecting rods, proceed as follows:

1. Clean the inside diameter of the piston pin end of the connecting rod with crocus cloth. Clean the connecting rod only enough to remove marks. Do not continue, once the marks have disappeared.

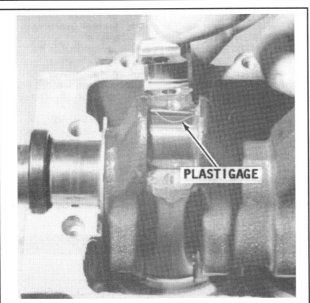

Fig. 281 To check bearing clearances, apply a strip of gauge material to each journal and bolt the bearings/retainers in place...

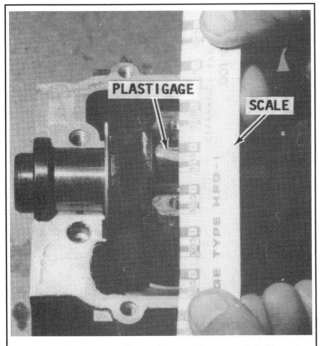

Fig. 282 ...then remove the crankcase and compare it to the scale

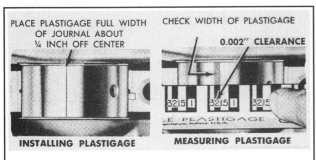

Fig. 283 Once the crankcase and/or rod caps are properly tightened and subsequently removed again compare the now flattened gauging material to the scale that came with the package

2. Assemble the piston end of the connecting rod with the loose needle bearings, piston pin, retainers, and C-lockrings. Insert the piston pin and check for vertical play. The piston pin should have no noticeable vertical play.

3. If the pin is loose or there is vertical play check for and replace the worn part/s.

4. Inspect the piston pin and matching rod end for signs of heat discoloration. Overheating is identified as a bluish bearing surface color and is caused by inadequate lubrication or by operating the powerhead at excessive high rpm.

5. On all bolt together connecting rods, check the bearing surface of the rod and rod cap for signs of chatter marks. This condition is identified by a rough bearing surface resembling a tiny washboard. The condition is caused by a combination of low-speed low-load operation in cold water. The condition is aggravated by inadequate lubrication and improper fuel.

Under these conditions, the crankshaft journal is hammered by the connecting rod. As ignition occurs in the cylinder, the piston pushes the connecting rod with tremendous force. This force is transferred to the connecting rod journal. Since there is little or no load on the crankshaft, it bounces away from the connecting rod. The crankshaft then remains immobile for a split second, until the piston travel causes the connecting rod to catch up to the waiting crankshaft journal, then hammers it. In some instances, the connecting rod crank pin bore becomes highly polished.

While the powerhead is running, a whir and/or chirp sound may be heard when the powerhead is accelerated rapidly - say from idle speed to about 1500 rpm, then quickly returned to idle. If chatter marks are discovered, the crankshaft and the connecting rods should be replaced.

6. Inspect the bearing surface of the rod and rod cap for signs of uneven wear and possible overheating. Overheating is identified as a bluish bearing surface color and is caused by inadequate lubrication or by operating the powerhead at excessive high rpm.

7. Use a feeler gauge to check Big-End Side Clearance on the crankshaft/connecting rod assembly by inserting a feeler gauge between the connecting rod and the counterweight of the crankshaft.

If the Side-Clearance measurement is not within specification, measure the crank width (if specifications are provided). Such a measurement may give an indication if the counterweight is walking on the crank pin or if the clearance is due to worn parts. The connecting rod would wear before the edge of the counterweight because the rod is manufactured from a much softer material.

Keep in mind that even if only one connecting rod shows excessive Freeplay or Side-Clearance, good shop practice dictates the entire set of rods be replaced. Rods are sold in sets.

8. Some models utilize replaceable automotive style connecting rod cap with plain bearing inserts. On these models there is normally an Oil Clearance provided in the engine specification chart. Be sure to measure this clearance using Plastigage® or an equivalent gauging material as detailed earlier in this section under Crankshaft, Measuring Bearing Clearances using Plastigage®.

Pistons

INSPECTION

◆ See Figures 289 thru 294

Inspect each piston for evidence of scoring, cracks, metal damage, cracked piston pin boss, or worn pin boss. Be especially critical during inspection if the outboard unit has been submerged. If the piston pin is bent, the piston and pin must be replaced as a set for two reasons. First, a bent pin will damage the boss when it is removed. Secondly, a piston pin is not sold as a separate item.

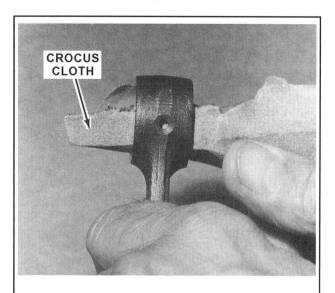

Fig. 284 Cleaning the piston end of a connecting rod with crocus cloth

Fig. 285 Checking the piston end of a connecting rod for vertical free-play using a piston pin

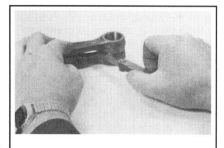

Fig. 286 Testing two rods for warpage at the piston end using a feeler gauge

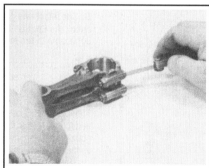

Fig. 287 Testing two rods for warpage at the crankshaft end using a feeler gauge

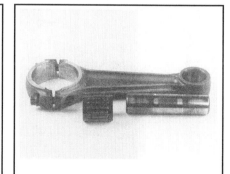

Fig. 288 Visually inspect the connecting rod for signs of damage

POWERHEAD 6-127

Check the piston ring grooves for wear, burns distortion, or loose locating pins. During an overhaul, the rings should be replaced to ensure lasting repair and proper powerhead performance after the work is completed. Clean the piston dome, ring grooves, and the piston skirt. Clean carbon deposits from the ring grooves using the recessed end of a broken piston ring.

Never use a rectangular ring to clean the groove for a tapered ring, or use a tapered ring to clean the groove for a rectangular ring.

Unless you are VERY cautious, avoid the use an automotive type ring groove cleaner, because such a tool may loosen the piston ring locating pins.

Clean carbon deposits from the top of the piston using a soft wire brush, carbon removal solution or by sand blasting. If a wire brush is used, Take care not to burr or round machined edges. Clean the piston skirt with crocus cloth.

Install the piston pin through the first boss only. Check for vertical free-play. There should be no vertical free-play. The presence of play is an indication the piston boss is worn. The piston is manufactured from a softer material than the piston pin. Therefore, the piston boss will wear more quickly than the pin.

Excessive piston skirt wear cannot be visually detected. Therefore, good shop practice dictates the piston skirt diameter be measured with a micrometer.

Piston skirt diameter should be measured at right angles to the piston pin axis at a point above the bottom edge of the piston. Refer to the Engine Specifications chart for Measuring Point specifications.

RING END-GAP CLEARANCE

◆ See Figures 295 and 296

Before the piston rings are installed onto the piston, the ring end-gap clearance for each ring must be determined. The purpose of the piston rings is to prevent the blow-by of gases in the combustion chamber. This cannot be achieved unless the correct oil film thickness is left on the cylinder wall.

This thin coating of oil acts as a seal between the cylinder wall and the face of the piston ring. An excessive end-gap will allow blow-by and the cylinder will lose compression. An inadequate end-gap will scrape too much oil from the cylinder wall and limit lubrication. Lack of adequate lubrication will cause excessive heat and wear.

Ideally the ring end-gap measurement should be taken after the cylinder bore has been measured for wear and taper and after any corrective work, such as boring or honing, has been completed.

If the ring end-gap is measured with a taper to the cylinder wall, the diameter at the lower limit of ring travel will be smaller than the diameter at the top of the cylinder.

If the ring is fitted to the upper part of a cylinder with a taper, the ring end-gap will not be great enough at the lower limit of ring travel. Such a condition could result in a broken ring and/or damage to the cylinder wall and/or damage to the piston and/or damage to the cylinder head.

If the cylinder is to be only honed, not bored, or if only cleaned, not honed, the ring end-gap should be measured at the lower limit of ring travel.

Fig. 289 This piston seized at high rpm (the connecting rod ripped the lower part of the piston from the dome)

Fig. 290 This pitting is a result of foreign matter in the cylinder

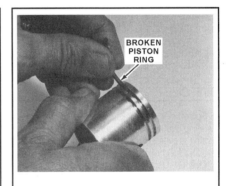

Fig. 291 Cleaning the ring groove with a broken ring

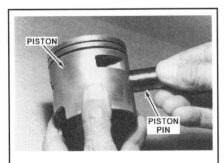

Fig. 292 There should be no play on the piston pin

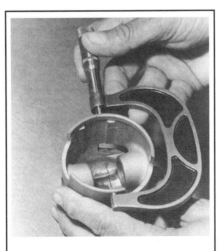

Fig. 293 Use a micrometer to check the piston skirt diameter

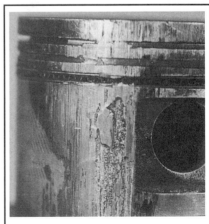

Fig. 294 The rings on this piston became stuck due to lack of lubrication

6-128 POWERHEAD

Some manufacturers give the precise depth the ring should be inserted into the cylinder and assume the cylinder walls are parallel with no taper.

Rings may be inserted from the top or bottom (depending upon the measurement point). In most cases the piston can be inserted and used to square the ring up in the bore. Once the ring is in the proper position, measure the ring end-gap with a feeler gauge.

If the end-gap is greater than the amount listed in the Engine Specification chart, replace the entire ring set.

If the end-gap is less than the amount listed, carefully file the ends of the ring - just a little at a time - until the correct end-gap is obtained.

■ If equipped, inspect the piston ring locating pins (these are much more common on 2-stroke motors than on the 4-strokes covered here) to be sure they are tight. When used, there is only one locating pin in each ring groove. If the locating pins are loose, the piston must be replaced.

PISTON RING SIDE CLEARANCE

◆ See Figure 297

After the rings are installed on the piston, the clearance in the grooves needs to be checked with a feeler gauge. Check the clearance between the ring and the upper land and compare your measurement with the specifications. Ring wear forms a step at the inner portion of the upper land. If the piston grooves have worn to the extent to cause high steps on the upper land, the step will interfere with the operation of new rings and the ring clearance will be too much. Therefore, if steps exist in any of the upper lands, the piston should be replaced.

On a plain ring this clearance may be measured either above or below the ring. On a Keystone ring - one with a taper at the top - the clearance must be measured below the ring.

OVERSIZE PISTONS & RINGS

Scored cylinder blocks can be saved for further service by boring and installing oversize pistons and piston rings. However, in most cases, if the scoring is over 0.0075 inch (0.13mm) deep, the block cannot be effectively re-bored for continued use.

Oversize pistons and rings are not available for all powerheads. Check with the parts department at your local Yamaha or Mercury/Mariner dealer for the model being serviced.

If oversize pistons are not available, the local marine shop may have the facilities to knurl the piston, making it larger.

A honed cylinder block is just surface finished. It is not parallel and therefore if an oversized or knurled piston is installed in this bore, the piston will soon seize-at the lower narrow end of the bore.

Cylinder Block

INSPECTION

◆ See Figures 298 thru 301

Inspect the cylinder block and cylinder bores for cracks or other damage. Remove carbon with a fine wire brush on a shaft attached to an electric drill or use a carbon remover solution.

✱✱ WARNING

If the cylinder block is to be submerged in a carbon removal solution, components like the crankcase bleed system must be removed from the block to prevent damage to hoses and check valves.

Use an inside micrometer or telescopic gauge and micrometer to check the cylinders for wear. Check the bore for out-of-round and/or oversize bore. If the bore is tapered, out-of-round or worn more than the wear limit specified by the manufacturer, the cylinders should be rebored - provided oversize pistons and rings are available.

Check with a local Yamaha or Mercury/Mariner dealer prior to boring. If oversize pistons and matching rings are not available, the block must be replaced.

■ Oversize piston weight is approximately the same as a standard size piston. Therefore, it is not necessary to rebore all cylinders in a block just because one cylinder requires boring.

Cylinder sleeves are an integral part of the die cast cylinder block and cannot be replaced. In other words, the cylinder cannot be sleeved.

Four inside cylinder bore measurements must be taken for each cylinder to determine an out-of-round condition, the maximum taper, and the maximum bore diameter.

In the accompanying illustration, measurements D1 and D2 are diameters measured at about 0.8 inch (20mm) from the top of the cylinder at right angles to each other. Measurements D3 and D4 are diameters measured at about 2.4 inch (60mm) from the top of the cylinder at right angles to each other.

Out-of-Round

◆ See Figure 301

Measure the cylinder diameter at D1 and D2. The manufacturer requires the difference between the two measurements should be less than 0.002 inch (0.050mm) for most models. Refer to the Engine Specifications chart.

Fig. 295 Using a feeler gauge to check ring end-gap

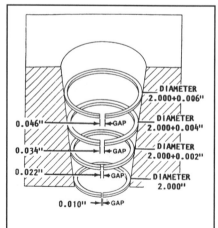

Fig. 296 Cylinder wall taper drastically effects ring end-gap

Fig. 297 Using a feeler gauge to check piston ring side clearance

POWERHEAD 6-129

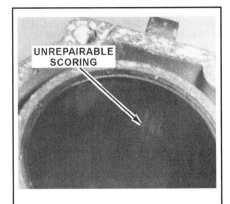

Fig. 298 This cylinder was scored beyond repair when a piston ring broke and worked its way into the combustion chamber

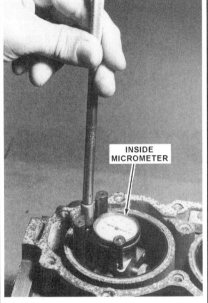

Fig. 299 Check the cylinder bore for taper using an inside micrometer. Measure near the near the top, in the middle and a near the bottom

Fig. 300 Using a drill mounted wire brush to clean carbon deposits from the cylinder head

Maximum Taper

◆ See Figure 301

Measure the cylinder diameter at D1, D2, D3, and D4. Take the largest of the D1 or D2 measurements and subtract the smallest measurement at D3 or D4. The result is the cylinder taper - it should be less than 0.003 inch (0.08mm) for most models. Refer to the Engine Specifications chart.

Bore Wear Limit

◆ See Figure 301

The maximum cylinder diameter D1, D2, D3, and D4 must not exceed the bore wear limits (when indicated in the Engine Specifications chart) before the bore is re-bored for the first time. These limits are only imposed on original parts because the sealing ability of the rings would be lost, resulting in power loss, increased powerhead noise, unnecessary vibration, piston slap, and excessive oil consumption.

The limits above the standard bore are usually 0.003 to 0.005 inch (0.08mm to 0.127mm) on all powerheads except V4 and V6, or 0.007 inch (0.1mm) on V4 and V6 powerheads. For details, refer to the Engine Specifications chart.

Piston Clearance

◆ See Figure 301

Piston clearance is the difference between a maximum piston diameter and a minimum cylinder bore diameter. If this clearance is excessive, the powerhead will develop the same symptoms as for excessive cylinder bore wear - loss of ring sealing ability, loss of power, increased powerhead noise, unnecessary vibration, and excessive oil consumption.

Maximum piston diameter was described earlier in this section. Minimum cylinder bore diameter is usually determined by measurement D3 or D4 also described earlier in this section.

If the piston clearance exceeds the limits in the specifications chart, either the piston or the cylinder block must be replaced.

Calculate the piston clearance by subtracting the maximum piston skirt diameter from the maximum cylinder bore measurement and compare the results for the model being serviced.

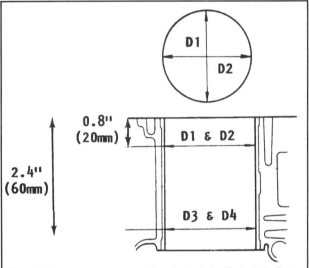

Fig. 301 Top view and cross section of a typical cylinder to indicate where measurements should be made for wear, taper and out of round

HONING CYLINDER WALLS

◆ See Figures 302 and 303

Hone the cylinder walls lightly to seat the new piston rings, as outlined in this section. If the cylinders have been scored, but are not out-of-round or the bore is rough, clean the surface of the cylinder with a cylinder hone as described in the following procedures.

■ **If overheating has occurred, check and resurface the spark plug end of the cylinder block, if necessary. This can be accomplished with 240-grit sandpaper and a small flat block of wood.**

To ensure satisfactory powerhead performance and long life following the overhaul work, the honing work should be performed with patience, skill, and in the following sequence:

1. Follow the hone manufacturer's recommendations for use of the hone and for cleaning and lubricating during the honing operation. A Christmas tree hone may also be used.

Pump a continuous flow of honing oil into the work area. If pumping is not practical, use an oil can. Apply the oil generously and frequently on both the stones and work surface.

2. Begin the stroking at the smallest diameter. Maintain a firm stone pressure against the cylinder wall to assure fast stock removal and accurate results.

3. Expand the stones as necessary to compensate for stock removal and stone wear. The best cross hatch pattern is obtained using a stroke rate of 30 complete cycles per minute. Again, use the honing oil generously.

4. Hone the cylinder walls only enough to deglaze the walls.

5. After the honing operation has been completed, clean the cylinder bores with hot water and detergent. Scrub the walls with a stiff bristle brush and rinse thoroughly with hot water. The cylinders must be thoroughly cleaned to prevent any abrasive material from remaining in the cylinder bore. Such material will cause rapid wear of new piston rings, the cylinder bore, and the bearings.

6. After cleaning, swab the bores several times with engine oil and a clean cloth, and then wipe them dry with a clean cloth. Never use kerosene or gasoline to clean the cylinders.

7. Clean the remainder of the cylinder block to remove any excess material spread during the honing operation.

■ If oversize pistons are not available, the local marine shop may have the facilities to knurl the piston, making it larger. If installing oversize or knurled pistons, it should be remembered that a honed cylinder block is just surface finished. It is not parallel and therefore if an oversized or knurled piston is installed in this bore, the piston will soon seize at the lower narrow end of the bore. When installing oversize or knurled pistons, the block must be bored oversize.

BLOCK & CYLINDER HEAD WARPAGE

◆ See Figures 304 thru 308

First, check to be sure all old gasket material has been removed from the contact surfaces of the block and the cylinder head. Clean both surfaces down to shiny metal, to ensure a true measurement.

Next, place a straightedge across the gasket surface. Check under the straightedge with a suitably sized feeler gauge (see the Engine Specifications chart). The warpage limit is normally 0.004 inch (0.1mm) on Yamaha and Mercury/Mariner motors. Move the straightedge to at least eight different locations. If the feeler gauge can pass under the straightedge - anywhere contact with the other is made - the surface will have to be resurfaced.

The block or the cylinder head may be resurfaced by placing the warped surface on 400-600 grit wet sandpaper, with the sandpaper resting on a flat machined surface. If a machined surface is not available a large piece of glass or mirror may be used. Do not attempt to use a workbench or similar surface for this task. A workbench is never perfectly flat and the block or cylinder head will pickup the imperfections of the surface and the warpage will be made worse.

Sand - work - the warped surface on the wet sandpaper using large figure 8 motions. Rotate the block, or head, through 180° (turn it end for end) and spend an equal amount of time in each position to avoid removing too much material from one side.

Fig. 302 Refinishing the cylinder wall using a ball hone. Always keep the home moving and constantly flood the cylinder with honing oil

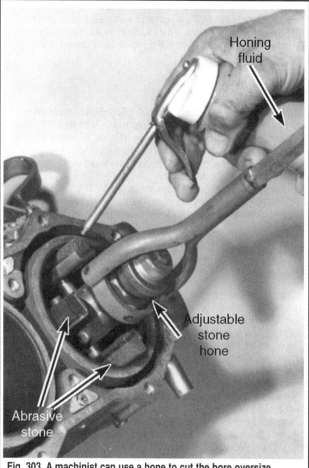

Fig. 303 A machinist can use a hone to cut the bore oversize

POWERHEAD 6-131

If a suitable flat surface is not available, the next best method is to wrap 400-600 grit wet sandpaper around a large file. Draw the file as evenly as possible in one sweep across the surface. Do not file in one place. Draw the file in many, many directions to get as even a finish as possible.

As the work moves along, check the progress with the straightedge and feeler gauge. Once the 0.004 inch (0.1mm) feeler gauge will no longer slide under the straightedge, consider the work completed.

If the warpage cannot be reduced using one of the described methods, the block or cylinder head should be replaced (or resurfaced by a machine shop, if possible).

Assembling a Warped Cylinder Head or Block

If the warpage cannot be reduced and it is not possible to obtain new items - and the warped part must be assembled for further use - there is a strong possibility of a water leak at the head gasket. In an effort to prevent a water leak, follow the instructions outlined in the following paragraphs.

■ **Because of the high temperatures and pressures developed, the sealing surfaces of the cylinder head and the block are the most prone to water leaks. No sealing agent is recommended because it is almost impossible to apply an even coat of sealer. An even coat would be essential to ensure an air/water tight seal.**

Never, never, use automotive type head gasket sealer. The chemicals in the sealer will cause electrolytic action and eat the aluminum faster than you can get to the bank for money to buy a new cylinder block.

Some head gaskets are supplied with a tacky coating on both surfaces applied at the time of manufacture. This tacky substance will provide an even coating all around. Therefore, no further sealing agent is required.

However, if a slight water leak should be noticed following completed assembly work and powerhead start up, do not attempt to stop the leak by tightening the head bolts beyond the recommended torque value. Such action will only aggravate the problem and most likely distort the head.

Furthermore, tightening the bolts, which are case hardened aluminum, may force the bolt beyond its elastic limit and cause the bolt to fracture. bad news, very bad news indeed. A fractured bolt must usually be drilled out and the hole re-tapped to accommodate an oversize bolt, etc. Avoid such a situation.

Probable causes and remedies of a new head gasket leaking are:

a. Sealing surfaces not thoroughly cleaned of old gasket material. Disassemble and remove all traces of old gasket.

b. Damage to the machined surface of the head or the block. The remedy for this damage is the same as for the next case.

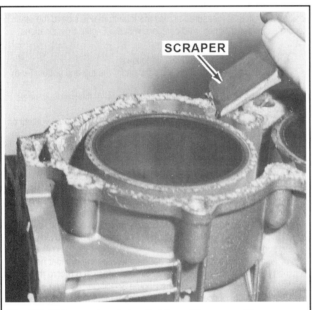
Fig. 304 All traces of gasket material must be removed to ensure accuracy

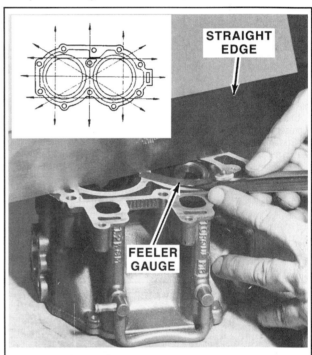
Fig. 305 Use a straightedge and feeler gauge to check the cylinder head-to-powerhead mating surfaces

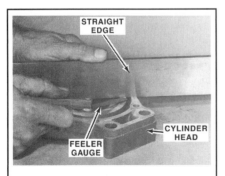
Fig. 306 Use a straightedge and feeler gauge to check for cylinder head warpage

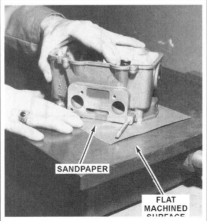
Fig. 307 Use only a flat machined surface and sandpaper to correct warpage on a cylinder head

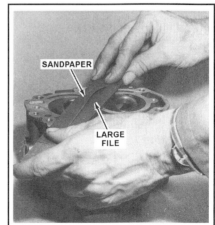
Fig. 308 When a flat machined surface is not available, a large flat file wrapped with sandpaper may be used

6-132 POWERHEAD

c. Permanently distorted head or block. Spray a light even coat of any type metallic spray paint on both sides of a new head gasket. Use only metallic paint - any color will do. Regular spray paint does not have the particle content required to provide the extra sealing properties this procedure requires.

Assemble the block and head with the gasket while the paint is till tacky. Install the head bolts and tighten in the recommended sequence and to the proper torque value and no more!

Allow the paint to set for at least 24 hours before starting the powerhead.

Consider this procedure as a temporary band aid type solution until a new head may be purchased or other permanent measures can be performed.

Under normal circumstances, if procedures have been followed to the letter, the head gasket will not leak.

Cylinder Head Components (Camshaft/Valve Train)

■ Before going any further, if you haven't already, check the Engine Specifications chart for the motor on which you are working. Keep it handy when reviewing the inspection process, both to determine which checks are applicable to your motor and to provide the required specification.

CHECKING FOR VALVE LEAKAGE

The cylinder heads on 4-stroke motors should be checked for valve leakage before disassembly for overhaul.

Check the cylinder head for valve leakage. If leakage or an obvious defect is present, disassemble the cylinder head (as detailed in this section under the Powerhead or Cylinder Head Overhaul procedures) and inspect the valve seating surfaces.

1. Place the cylinder head on a suitable work surface. Make sure the intake manifold ports are located above the valves.
2. Pour solvent or kerosene into the intake and exhaust ports, and watch the valves. There should be little or no leakage from a closed valve.
3. If significant leakage is present, check the valve seats, valve faces and springs for signs of defects, wear or damage.
4. Pour the solvent or kerosene out of the cylinder head before proceeding.

CHECKING THE CAMSHAFT

◆ See Figures 309 thru 314

Most 4-stroke motors utilize a camshaft that is timed to the crankshaft via a timing belt (or a set of cams timed by BOTH a belt and then to each other with a chain, on V6 motors). Although it is true that the smallest 4-strokes covered here utilize camshafts that are directly gear driven off the crankshaft instead of with a timing belt.

In all cases, the camshaft is used to open and close the intake/exhaust valves at the appropriate time allowing the engine to draw air/fuel mixtures into the combustion chamber, seal the chamber when producing power, and open the chamber afterwards to release the unburned remains of the air/fuel charge.

The height of in the intake and exhaust lobes are probably the most important specification on a camshaft, as they determine the maximum possible amount the shaft can open the valves. A shaft with worn lobes will not fully open the affected valves, leading to performance problems (by preventing the combustion chamber from breathing properly, preventing it from drawing in sufficient air/fuel charges and/or from completely venting the exhaust gasses).

In addition to camshaft lobe specs, many more camshaft specifications are available for some larger 4-stroke motors than are available for the smaller, carbureted 4-stroke Yamahas and Mercury/Mariners.

On all engines, the general condition of the shaft itself must be checked. Look for scored or worn areas, signs of discoloration from heat or excessive wear possibly caused by a lack of oiling. Replace the camshaft is obvious defects are found or if any surfaces measure out of specification.

As with all inspection procedures, make sure the surfaces are clean and dry.

1. Using a micrometer, measure the height and width of each camshaft lobe. When the lobe (raised portion of the camshaft) is positioned upward, the height is the vertical measurement from base to lobe tip. In the same position, the width is measured across the lobe from one side to the other. Record each measurement and compare with the Engine Specifications chart.

2. Check the camshaft run-out, as follows:
   a. Support the camshaft on V-blocks under the top and bottom (end) journals.
   b. Position a dial indicator with the tip on one of the middle journals and zero the dial.
   c. Rotate the camshaft slowly and record the maximum movement on the dial.
   d. Compare the reading to the Engine Specifications chart. Replace the camshaft if readings are greater than specified.

3. If specifications are available, check the bearing surface dimensions, as follows:
   a. Using a micrometer, measure the thickness of the camshaft bearing journals (at the point where they contact the cylinder head surface). Compare the measurements with the Engine Specifications chart. If the journal thickness is within specification, the camshaft is good, check the bearing surfaces in the cylinder head.
   b. Using a micrometer or bore gauge, measure the cylinder head camshaft bore (journal holder) at each journal surface and record the results. Compare the measurements with the Engine Specifications chart. If the measurement is out of specification, the cylinder head must be replaced.

■ As with all precision measurements, double-check your work before condemning a part (especially a big one like the cylinder head or the camshaft.

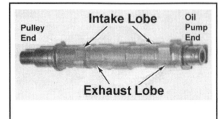

Fig. 309 Identify the intake and exhaust lobes of the camshaft by comparing their positions with the ports on the cylinder head - 9.9 hp shown

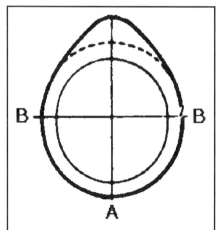

Fig. 310 Cut-away view of a camshaft lobe showing height (A) versus base width (B)

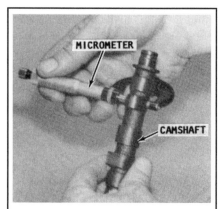

Fig. 311 Measure camshaft lobe width (dimension B, across the base of the lobe) using a micrometer as shown

POWERHEAD 6-133

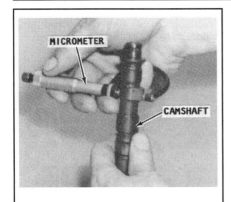

Fig. 312 Measure camshaft lobe height (dimension A, from top-to-bottom of the lobe) using a micrometer as shown

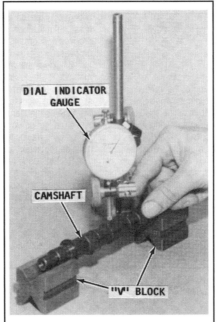

Fig. 313 Use a dial indicator and pair of V-blocks to check camshaft run-out

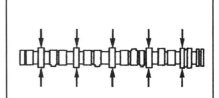

Fig. 314 If the specification is provided, measure the bearing journals (shown) and cylinder head journal holders (bores) to see if they are in spec

CHECKING THE ROCKER ARMS AND SHAFTS

◆ See Figures 315 and 316

The valve train of some Yamaha and Mercury/Mariner motors consists of a camshaft that actuates the valves through shaft mounted, adjustable rocker arms. As with all inspection procedures, make sure the surfaces are clean and dry. Refer to the Engine Specifications chart for the motor on which you are working and the appropriate inspection steps in the following procedure:

1. Visually check for signs of obviously damaged (cracked, chipped), discolored or excessively worn surfaces. The contact surface with the cam should be smooth. Use a flat file to finish the surface, if necessary. Any material filed away will probably affect the valve lash. The other end of the rocker arm (with the bolt and locknut can be adjusted).

■ During inspection, expect to find some wear on the rocker arm face. But, there should be no pitting. If wear is bad enough to show pitting on the face, replace the rocker arm to ensure component durability.

2. On some of the smaller motors (like the 8/9.9 hp, 232cc models) the rocker arms are free to walk along the cam lobe between the spacers. This walking can quickly wear a groove in a lobe with a high point on the contact surface. Most of the larger motors utilize rocker arm retainers, but they still allow enough movement that some grooving can occur. Inspect the shaft closely to determine if it is still serviceable.

3. Check the shaft for obvious warpage. Replace damaged or excessively worn parts.

4. Check the condition of the center adjustment bolt threads and the internal threads of the rocker arm and locknut. Replace any defective components.

5. Check any oiling holes in the rocker or shaft for clogging and clean, as necessary.

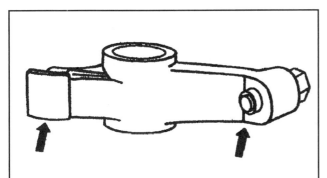

Fig. 315 Check the rocker arm faces for pitting and replace, as necessary

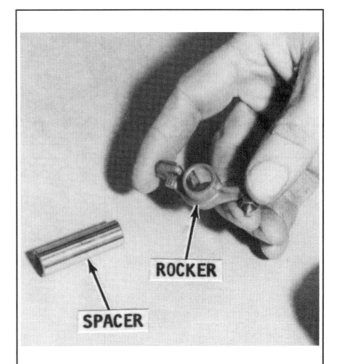

Fig. 316 Examine each rocker arm carefully as defects in the rocker arm will show in quick camshaft wear/damage - 9.9 hp rocker and spacer shown

6-134 POWERHEAD

CHECKING THE VALVE ASSEMBLIES

◆ See Figures 317 thru 329

Probably the most important key to cylinder head overhaul is the measurement and resurfacing or replacement (if necessary) of the valves, seats and guides. The tools and experience necessary to properly perform this phase of powerhead overhaul are USUALLY limited to machine shops. A talented mechanic can successfully perform these measurements, and with access to the right tools, repair a damaged cylinder head. But, these days, most wont bother, instead referring the work to a reputable marine machine shop, since the time and effort is not worth the risk balance.

If a machine shop is not available or desired, take your time during this procedure. Make ALL listed checks and, when specs are available, measurements.

Carbon should be removed from the valves and combustion chambers, as detailed under Cleaning in this section. The valve springs, keepers/caps, seats, retainers etc should all be cleaned using a suitable solvent and blown dry with low-pressure compressed air.

1. On all motors, visually check the valve faces and seats for signs of pitting, cracked, corroded or damaged surfaces. Then check the valve stems for signs of obvious wear or damage. Inspect the face and head for pits, grooves or scores. Check the stems for wear and check the ends for grooves.

2. Check the valve stems for signs of warpage. There are 2 acceptable methods for this, either use a flat piece of glass (such as a table top without and edge) to roll the valve stems (with the faces hanging over the edge to allow the full stem to contact the glass surface). You should be able to feel warpage as the stem is rolled. The better method is to use a pair of V-blocks and a dial-gauge (in a fashion similar to checking the camshaft or crankshaft for run-out). Valves with more than the specified maximum allowed run-out must be replaced.

3. If not done already, CAREFULLY remove carbon deposits from the valve guides. Using a small bore gauge, measure the inside diameter of each guide and record the measurement. Compare to the Engine Specifications chart. If guide replacement is necessary, refer it to a reputable machine shop.

4. Using a micrometer, measure the valve stem outer diameter at various points along the shaft. Record each measurement and compare to specifications. Then use these measurements to determine the stem-to-

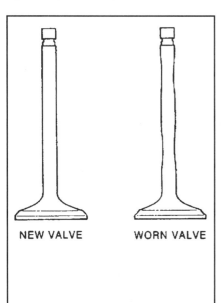

Fig. 317 Visually inspect the valves for signs of wear or damage

Fig. 318 Rolling the stem across a flat surface quickly checks valve stem run-out

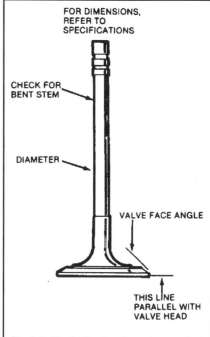

Fig. 319 Check all valve dimensions against spec

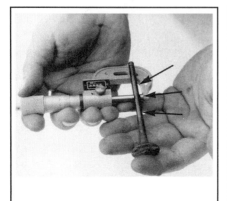

Fig. 320 Measure the stem thickness at various points

Fig. 321 Lower a split ball micrometer (or small bore gauge) into the valve guide, then turn until it just makes contact with the guide

Fig. 322 ... then withdraw the bore guide and measure outside diameter using a micrometer

POWERHEAD 6-135

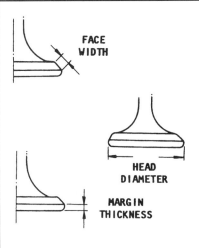

Fig. 323 Common valve terminology - Measurements are usually made with a calipers and/or micrometers

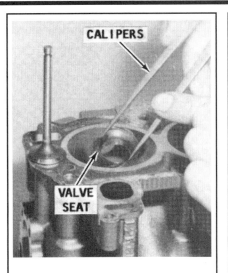

Fig. 324 Measuring the valve seat diameter using a pair of calipers

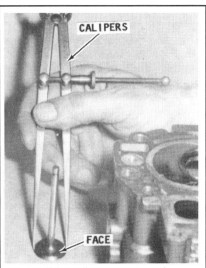

Fig. 325 Using calipers to measure the valve face diameter

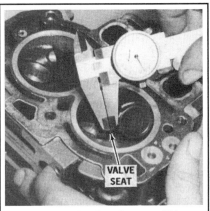

Fig. 326 Measuring the valve seat width using a pair of calipers

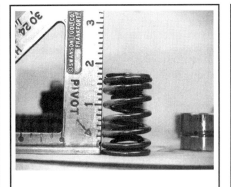

Fig. 327 Check the valve spring for distortion using a square

Fig. 328 On EFI motors, make sure the free length is not below the minimum spec

guide oil clearance (the amount of gap that exists between the stem and the guide for oil). In this case, oil clearance is equal to the size of the guide bore, minus the size of the stem. Subtract each stem measurement from the appropriate bore size and compare to the service limit for the guide-to-stem clearance specifications. Again, if guide replacement is necessary, refer it to a reputable machine shop.

5. Measure each valve seat and face width and record the readings. If it is out of spec, have the valve seat surfaced at a machine shop.

6. If specs are given, measure each valve head thickness and record the readings. Replace any valve with a thickness below spec.

7. If specs are given, use a sliding caliper to measure free length (completely uncompressed length) of each valve spring. Record the results and compare to the Engine Specifications chart. Replace any spring that is less than the minimum spec.

8. Using a spring tension measurement device, check the tension and the specified height. Replace the valve spring is tension is less than specified for the given height.

9. Position each valve spring on a flat surface, then slide a square up next to each to measure the deviation from squareness (the maximum distance the top of the spring is from the square when the bottom is butt up against the square). Replace any spring that is out of spec.

10. For some motors a specification is given for spring installed height (valve closed). In order to measure this you'll have to reassemble the cylinder head, but keep in mind that the spring pressure test at the same height should give you some idea how the spring will make out during this test, but there is also a question about the valve stem itself which can only be answered by installing the valve.

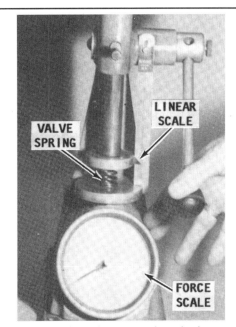

Fig. 329 Using a valve spring tester to determine force required to compress a spring to a specified length

6-136 POWERHEAD

POWERHEAD BREAK-IN

Break-in Procedures

Anytime a new or rebuilt powerhead is installed (this includes a powerhead whose wear components such as pistons and rings, main bearings or cylinder head components have been replaced), the motor must undergo proper break-in.

By following break-in procedures largely consisting of specific engine operating limitations during the first 10 hours of operation, you will help you will help ensure a long and trouble-free life. Failure to follow these recommendations may allow components to seat improperly, causing accelerated wear and premature powerhead failure.

Special attention is required to the engine oil during initial break-in, expect increased oil consumption during that period. Don't be alarmed, but be sure to check the engine oil level frequently to prevent the possibility of powerhead damage from operation with a low oil level. Oil consumption should lessen as the piston rings seat to the cylinder walls.

Especially during break-in, pay close attention to all pre and post operation checks. This goes double when checking for fuel, oil or water leaks. At each start-up and frequently during operation, check for presence of the cooling indicator stream.

At the completion of break-in, double-check the tightness of all exposed engine fasteners.

Even though Yamaha and Mercury technically end the break-in period of a motor after the first 10 hours of operation, it is a good idea to continue to pay closer attention to the motor until the completion of the entire first 20 hours of engine operation (and completion of a 20 hour service). During break-in, one of the most important things you can do is to **vary** the engine speed. This allows parts to wear in under conditions throughout the powerband, not just at idle or mid-throttle.

■ During break-in, check your hourmeter or a watch frequently and be sure to change the engine speed at least every 15 minutes (that means between every 2-3 tenths on the hourmeter).

Be sure to **always** allow the engine to reach operating temperature before setting the throttle anywhere above idle. This means you should always start and run the motor for at least 5 minutes before advancing the throttle.

✳✳ WARNING

NEVER run the engine out of the water, unless a flush fitting is used to provide a source of cooling. Remember that the water pump can be destroyed in less than a minute just from a lack of water. The powerhead will suffer damage in very little time as well, but even if it is not overheated out of water, reduced cooling from a damaged water pump impeller could destroy it later. Don't risk it.

■ Although a flush fitting can be used to run an engine, it CANNOT be used to break-in a motor, as you cannot run an engine much above idle on a flush fitting without risking damage from a runaway powerhead (the motor running overspeed).

✳✳ WARNING

Do not operate the engine at full throttle except for very short periods, until after 10 hours of operation.

To properly break-in a powerhead, proceed as follows:
1. Each time you start the motor, IMMEDIATELY check to be sure the water pump is operating. If the water pump is operating, a fine stream will be discharged from the exhaust relief hole at the rear of the driveshaft housing. Also, each time you start the powerhead, allow it to idle until it reaches normal operating temperature (at least 5 minutes) before increasing speed above idle.

■ If the cooling water indicator stream is missing, shut the powerhead down and figure out why before proceeding.

2. The first time you start the motor, allow it to idle for **10** minutes. Make sure the motor runs at the slowest possible speed (a fast idle in neutral is best).
3. For the remainder of the first hour operate the motor in gear at various speeds, but DO NOT exceed 1/2 throttle. The one exception is if you have an easily planning boat, accelerate at full throttle (or close to it) JUST long enough to get the boat up onto plane, then immediately reduce speed below 3000 rpm. Vary engine speeds below 1/2 throttle, but try to keep the boat on plane (as that lessons the amount of load on the motor).

■ DO NOT exceed 2000 rpm for the first hour or 3000 rpm for the second hour.

4. For the second hour of operation accelerate at full throttle onto plane, then reduce speed to 3/4 throttle. Continue to vary the engine speed and operate the motor under different load conditions. At least a couple of times during the hour, run the engine AT FULL THROTTLE for ONE MINUTE, then allow at least 10 minutes of operation at 3/4 throttle or less to allow the motor to cool. Yamaha recommends running the motor at FULL THROTTLE for ONE MINUTE, every 10 minutes, so 5-6 times during this second hour of operation.

■ Build good habits now. It is NEVER good for the motor to operate at full throttle for any length of time then suddenly drop off to idle. You should always decelerate the motor gradually, allowing the pistons to cool as the speed is reduced.

5. For the remainder of the first 10 hours (for hours 3-10) continue the ritual of warming the motor and varying engine speed. Full throttle operation should NOT exceed 5 minutes allowing sufficient time to slowly cool the pistons and operate under other conditions before the next full throttle run.
6. After the 10th hour of operation, run the engine as normal.

■ While the engine is operating during the initial period, check the fuel, exhaust, and water systems for leaks. Tilt the motor up out of the water at the end of each day to closely inspect the drive unit and look for signs of damage or leakage (of course, if you're trailering, this should be done to protect the skeg before pulling it out on the trailer anyway, right?).

POWERHEAD

SPECIFICATIONS

ENGINE SPECIFICATIONS - YAMAHA 2.5 HP (72cc) 4-STROKE ENGINES

Component	U.S. (in.)①	Metric (mm)①
Camshaft		
Lobe Height (top of lobe to bottom of journal)		
Intake and Exhaust	1.0290-1.0330	26.139-26.239
Lobe Width (side-to-side / non-raised portion of lobe)		
Intake and Exhaust	0.8642-0.8681	21.950-22.050
Journal Diameter	0.5892-0.5902	14.965-14.990
Journal Oil Clearance	0.0004-0.0021	0.010-0.053
Runout - Max	0.0012	0.03
Compression		
Minimum Compression	102 psi	700 kPa
Cylinder Bore		
Standard Bore Diameter	2.1260-2.1266	54.000-54.015
Out-of-round Service Limit	0.002	0.05
Taper Service Limit	0.003	0.08
Cylinder Head		
Gasket Surface Warpage Limit	0.004	0.1
Crankshaft		
Connecting Rod Small-End Inside Diameter	0.4727-0.4732	12.006-12.020
Connecting Rod Big-End Inside Diameter	0.9449-0.9455	24.000-24.015
Connecting Rod Big-End Side Clearance	0.008-0.024	0.2-0.6
Connecting Rod Big-End (Crankpin) Oil Clearance	0.0006-0.0018	0.016-0.046
Crankshaft Journal Diameter	0.8654-0.8659	21.980-21.993
Crankpin Diameter	0.9437-0.9443	23.969-23.984
Crankpin Width	0.827-0.831	21.0-21.1
Runout - Max	0.0004	0.01
Piston		
Standard Diameter	2.1240-2.1246	53.950-53.965
Measuring Point (above bottom of skirt)	0.0	0.0
Piston-to-Cylinder Clearance	0.0014-0.0026	0.035-0.065
Piston Pin Boss Bore	0.4728-0.4731	12.009-12.017
Piston Pin Outer Diameter	0.4723-0.4724	11.996-12.000
Piston Rings		
Width (measured from outer-to-inner edge of ring)		
Top	0.0768-0.0846	1.95-2.15
Middle	0.0906-0.0984	2.30-2.50
Bottom (oil control)	0.0827-0.0945	2.10-2.40
Thickness (measured vertically top-to-bottom of ring)		
Top	0.0382-0.0390	0.97-0.99
Middle	0.0461-0.0468	1.17-1.19
Bottom (oil control)	0.0736-0.0768	1.87-1.95
End Gap		
Top	0.0059-0.0118	0.15-0.30
Middle	0.0118-0.0177	0.30-0.45
Bottom (oil control)	0.0079-0.0276	0.20-0.70

ENGINE SPECIFICATIONS - YAMAHA 2.5 HP (72cc) 4-STROKE ENGINES

Component	U.S. (in.)①	Metric (mm)①
Ring Side Clearance		
Top	0.0016-0.0031	0.04-0.08
Middle	0.0008-0.0024	0.02-0.06
Bottom (oil control)	0.0024-0.0063	0.06-0.16
Valves		
Valve Dimensions		
Head Diameter (IN / EX)	0.941-0.949 / 0.862-0.870	23.9-24.1 / 21.9-22.1
Face Width (IN and EX)	0.0724-0.0890	1.84-2.26
Seat Contact Width (IN and EX)	0.024-0.031	0.6-0.8
Margin Thickness (IN / EX)	0.028 / 0.040	0.7 / 1.0
Stem Thickness (IN)	0.2156-0.2161	5.475-5.490
Stem Thickness (EX)	0.2150-0.2156	5.460-5.475
Valve Guide Inner Diameter (IN and EX)	0.2165-0.2170	5.500-5.512
Stem-to-guide Clearance (IN)	0.0004-0.0015	0.010-0.037
Stem-to-guide Clearance (EX)	0.0010-0.0020	0.025-0.052
Valve Stem Runout	0.0012	0.03
Valve Springs		
Free Length	1.378	35.0
Out-of-Square (Spring Warpage/Tilt)	0.05	1.2

① Unless otherwise noted

POWERHEAD

ENGINE SPECIFICATIONS - YAMAHA 4 HP (112cc) 4-STROKE ENGINES

Component	U.S. (in.)①	Metric (mm)①
Camshaft		
Lobe Height (top of lobe to bottom of journal)		
Intake	1.0526-1.0565	26.736-26.836
Exhaust	1.0442-1.0481	26.532-26.623
Lobe Width (side-to-side / non-raised portion of lobe)		
Intake and Exhaust	0.8642-0.8681	21.950-22.050
Journal Diameter	0.5880-0.5899	14.934-14.984
Runout - Max	0.001	0.03
Compression		
Minimum Compression	93.87 psi	660 kPa
Cylinder Bore		
Standard Bore Diameter	2.323-2.324	59.00-59.02
Wear Limit	2.33	59.10
Out-of-round Service Limit	0.002	0.05
Taper Service Limit	0.003	0.08
Cylinder Head		
Gasket Surface Warpage Limit	0.004	0.1
Crankshaft		
Connecting Rod Big-End Oil Clearance	0.5120-0.5126	13.006-13.020
Connecting Rod Small-End Inside Diameter	0.0006-0.0018	0.016-0.046
Crankshaft Width	2.535-2.539	64.40-64.50
Crankshaft Big-End Side Clearance	0.008-0.024	0.20-0.60
Crankpin Diameter (Wear Limit)	1.022	25.95
Runout - Max	0.0012	0.03
Fuel Pump		
Discharge Rate	2.11 Gph @ 5000 rpm	8Lph @ 5000 rpm
Pressure	7.1 psi	50 kPa
Plunger Stroke	0.07-0.09	1.8-2.2
Piston		
Standard Diameter	2.3209-2.3215	58.950-58.965
Measuring Point (above bottom of skirt)	0.40	10.0
Piston-to-Cylinder Clearance	0.0014-0.0026	0.035-0.065
Piston Pin Boss Bore	0.5122-0.5125	13.009-13.017
Piston Pin Outer Diameter	0.5116-0.5118	12.995-13.000
Oversize		
1st	2.333	59.25
2nd	2.343	59.50
Piston Rings		
Width (measured from outer-to-inner edge of ring)		
Top and Middle	0.10	2.6
Bottom (oil control)	0.10	2.5
Thickness (measured vertically top-to-bottom of ring)		
Top and Middle	0.06	1.5
Bottom (oil control)	0.09	2.4

ENGINE SPECIFICATIONS - YAMAHA 4 HP (112cc) 4-STROKE ENGINES

Component	U.S. (in.)①	Metric (mm)①
End Gap		
Top - Standard / Limit	0.004-0.008 / 0.016	0.10-0.20 / 0.40
Middle - Standard / Limit	0.010-0.016 / 0.024	0.25-0.40 / 0.60
Bottom (oil control) - Standard / Limit	0.008-0.028 / 0.04	0.20-0.70 / 0.90
Ring Side Clearance		
Top	0.002-0.003	0.04-0.08
Middle	0.001-0.002	0.02-0.06
Pushrods		
Runout - Max	0.02	0.50
Valves		
Valve Dimensions		
Head Diameter (IN / EX)	0.94-0.95 / 0.86-0.87	23.9-24.1 / 21.9-22.1
Face Width (IN and EX)	0.072-0.089	1.84-2.26
Seat Contact Width (IN and EX)	0.02-0.03	0.6-0.8
Margin Thickness (IN / EX)	0.028 / 0.039	0.7 / 1.0
Stem Thickness (IN)	0.2156-0.2161	5.475-5.490
Stem Thickness (EX)	0.2150-0.2156	5.460-5.475
Valve Guide Inner Diameter (IN and EX)	0.2165-0.2170	5.500-5.512
Stem-to-guide Clearance (IN)	0.0004-0.0015	0.010-0.037
Stem-to-guide Clearance (EX)	0.0010-0.0020	0.025-0.052
Valve Stem Runout	0.0012	0.03
Valve Lifters		
Outer Diameter	0.3136-0.3142	7.965-7.980
Valve Springs		
Free Length (Standard / Limit)	1.38 / 1.34	35.0 / 34.0
Pressure @ Length	15.87 lbs. @ 0.85 in.	7.2 kg @ 21.6mm
Out-of-Square (Spring Warpage/Tilt)	0.05	1.2
① Unless otherwise noted		

POWERHEAD 6-139

ENGINE SPECIFICATIONS - MERCURY/MARINER 4/5/6 HP (123cc) 4-STROKE ENGINES

Component	U.S. (in.) ①	Metric (mm) ①
Camshaft		
Lobe Height (top of lobe to bottom of journal)		
Intake and Exhaust		
4 hp (1999-2000)	0.993	25.24
5 hp (1999-2000)	1.047	26.59
6 hp (2000)	1.115	28.33
4/5/6 hp (2001 and later)+A69	1.115	28.33
Journal Diameter	0.5515	13.98
Oil Pan Bearing Inner Diameter	0.55	14.01
Camshaft-to-Oil Pan Bearing Clearance	0.0008-0.0020	0.02-0.05
Compression		
Min/Max (with Decompression)	36-50 psi	248-345 kPa
Min/Max (without Decompression - exhaust rocker off)	100-130 psi	690-896 kPa
Cylinder Bore		
Standard Bore Diameter	2.323	59.00
Oversize 0.020 in. (0.5mm)	2.343	59.50
Out-of-round Service Limit	0.003	0.076
Taper Service Limit	0.003	0.076
Cylinder Head		
Gasket Surface Warpage Limit	0.0012	0.03
Crankshaft		
Connecting Rod Big-End Oil Clearance	0.002-0.003	0.053-0.079
Connecting Rod Big-End-to-Crank Pin Side Clearance	0.008-0.016	0.200-0.400
Connecting Rod Small-End Inside Diameter	0.6303	16.01
Crankshaft Journal Diameter (in oil pan bearing)	0.983-0.982	24.98-24.96
Oil Pan Bearing Inner Diameter	0.985	25.01
Crankshaft-to-Oil Pan Bearing Clearance	0.0006-0.0015	0.015-0.040
Crankpin Journal Diameter	1.179-1.177	29.94-29.91
Crankshaft Runout - Max	0.002	0.05
Fuel Pump		
Pressure	2.5-5.0 psi	17-35 kPa
Diaphragm Stroke	0.059	1.5
Plunger Stroke	0.059	1.5
Oil Pump		
Oil Pressure (engine warm)		
Min @ 1300 rpm	4 psi	27.6 kPa
Min @ 5000 rpm	21 psi	144.8 kPa
Relief Valve Operating Pressure	31-40 psi	216-275 kPa
Outer Rotor Height Spec/Limit	0.236 / 0.234	5.99 / 5.96
Depth of Pump Body Spec/Limit	0.236 / 0.238	5.99 / 6.06
Inner Diameter of Pump Body Spec/Limit	0.909 / 0.910	23.09 / 23.13
Inner-to-Outer Rotor Clearance Spec/Limit	0.006 / 0.008	0.15 / 0.20
Outer Rotor-to-Pump Housing Clearance (side of rotor)	0.005-0.010	0.12-0.25
Outer Rotor-to-Pump Cover Clearance (flat top of rotor)	0.0008-0.0040	0.02-0.10

ENGINE SPECIFICATIONS - MERCURY/MARINER 4/5/6 HP (123cc) 4-STROKE ENGINES

Component	U.S. (in.) ①	Metric (mm) ①
Piston		
Standard Diameter (at skirt)	2.321	58.960
Oversize 0.020 in. (0.5mm)	2.341	59.460
Piston-to-Cylinder Clearance Standard (Max)	0.001-0.002 (0.006)	0.020-0.055 (0.150)
Piston Pin Boss Bore	0.6300	16.002
Piston Pin Outer Diameter	0.6299	16.000
Piston Pin-to-Bore Clearance	0.0001-0.0005	0.002-0.012
Piston Rings		
End Gap		
Top	0.006-0.014	0.15-0.35
Middle	0.012-0.020	0.30-0.50
Bottom (oil control)	0.008-0.016	0.20-0.40
Ring Side Clearance		
Top	0.0015-0.0030	0.04-0.08
Middle	0.0012-0.0030	0.03-0.07
Bottom (oil control)	0.0004-0.0070	0.01-0.18
Valves		
Valve Dimensions		
Head Diameter (IN / EX)	0.980-0.988 / 0.941-0.949	24.9-25.1 / 23.9-24.1
Face Width (IN and EX)	0.102	2.6
Seat Contact Width (IN and EX)	0.031	0.8
Margin Thickness (IN / EX)	0.028 / 0.047	0.7 / 1.2
Stem Thickness (IN / EX)	0.216 / 0.214	5.48 / 5.44
Valve Guide Inner Diameter (IN and EX)	0.2165	5.5
Stem-to-guide Clearance (IN)	0.0008-0.0017	0.020-0.044
Stem-to-guide Clearance (EX)	0.0018-0.0028	0.045-0.072
Valve Stem Runout	0.0006	0.016
Valve Springs		
Free Length / Out-of-Square Limit (Tilt)		
4 hp (1999)	1.260 / 0.044	32.0 / 1.12
4/5/6 hp (2000 and later)	1.378 / 0.044	35.0 / 1.12
Compressed Pressure (Installed)		
Closed Height 0.965 in. (24.4mm)		
4/5 hp (1999-2000)	17 lbs.	7.7 kg
6 hp (2000)	24 lbs.	10.8kg
4/5/6 hp (2001 and later)	24 lbs.	10.8kg
Open Height 0.709 in. (27.4mm)		
4/5 hp (1999-2000)	31 lbs.	14 kg
6 hp (2000)	39.4 lbs.	17.9 kg
4/5/6 hp (2001 and later)	39.4 lbs.	17.9 kg
Direction of Winding	Right Hand	Right Hand

① Unless otherwise noted

ENGINE SPECIFICATIONS - YAMAHA 6/8 HP (197cc) 4-STROKE ENGINES

Component	U.S. (in.) [1]	Metric (mm) [1]
Camshaft		
Lobe Height (top of lobe to bottom of journal) [2]		
Intake and Exhaust	0.9232-0.9271	23.45-23.55
Lobe Width (side-to-side / non-raised portion of lobe) [2]		
Intake and Exhaust	0.7854-0.7893	19.95-20.05
Journal Diameter		
Cylinder Head Journal	0.7077-0.7083	17.975-17.991
Oil Pump Housing Journal	0.6289-0.6296	15.975-15.991
Runout - Max	0.001	0.03
Compression		
Minimum Compression	135.1 psi	950 kPa
Cylinder Bore		
Standard Bore Diameter	2.2047-2.2053	56.000-56.015
Out-of-round Service Limit	0.002	0.05
Taper Service Limit	0.003	0.08
Cylinder Head		
Gasket Surface Warpage Limit	0.004	0.1
Crankshaft		
Connecting Rod Big-End Inside Diameter	1.0642-1.0646	27.030-27.042
Connecting Rod Big-End Oil Clearance	0.0012-0.0017	0.030-0.042
Connecting Rod Big-End-to-Crank Pin Side Clearance	0.002-0.009	0.05-0.22
Connecting Rod Small-End Inside Diameter	0.5518-0.5523	14.015-14.029
Crankshaft Journal Diameter	1.1810-1.1815	29.997-30.009
Crankcase Journal Inside Diameter	1.3000-1.3008	33.016-33.040
Crankshaft Journal Oil Clearance (w/ Bearing Installed)	0.0004-0.0015	0.011-0.039
Crankpin Journal Diameter	1.0629-1.0633	26.997-27.009
Crankshaft Runout - Max	0.0012	0.03
Crankshaft Bearing Mark/Color		
A: Blue	1.3005-1.3008	33.032-33.040
B: Black	1.30016-1.3005	33.024-33.032
C: Brown	1.2998-1.30016	33.016-33.024
Fuel Pump		
Discharge Rate (Min)	3.17 Gph @ 2750 rpm	12Lph @ 2750 rpm
Pressure (Max)	17.1 psi @ 2750 rpm	117.6 kPa @ 2750 rpm
Diaphragm Stroke	0.095-0.189	2.4-4.8
Plunger Stroke	0.139-0.259	3.525-6.575
Oil Pump		
Discharge Rate (Min)	0.925 gph @ 3000 rpm	3.5Lph @ 3000 rpm
Inner-to-Outer Rotor Clearance	0.002-0.006	0.04-0.14
Outer Rotor-to-Pump Housing Clearance (side of rotor)	0.004-0.006	0.10-0.15
Outer Rotor-to-Pump Cover Clearance (flat top of rotor)	0.001-0.004	0.03-0.09
Piston		
Standard Diameter	2.2028-2.2033	55.950-55.965
Measuring Point (above bottom of skirt)	0.06	1.5
Piston-to-Cylinder Clearance	0.0014-0.0026	0.035-0.065

ENGINE SPECIFICATIONS - YAMAHA 6/8 HP (197cc) 4-STROKE ENGINES

Component	U.S. (in.) [1]	Metric (mm) [1]
Piston Pin Boss Bore	0.5513-0.5518	14.004-14.015
Piston Pin Outer Diameter	0.5510-0.5512	13.996-14.000
Oversize		
1st	2.2126-2.2132	56.200-56.215
2nd	2.2224-2.2230	56.450-56.465
Piston Rings		
Width (measured from outer-to-inner edge of ring)		
Top	0.08	2.05
Middle	0.10	2.50
Bottom (oil control)	0.10	2.50
Thickness (measured vertically top-to-bottom of ring)		
Top	0.05	1.20
Middle	0.06	1.50
Bottom (oil control)	0.10	2.45
End Gap		
Top	0.006-0.012	0.15-0.30
Middle	0.012-0.018	0.30-0.45
Bottom (oil control)	0.008-0.028	0.20-0.70
Ring Side Clearance		
Top	0.0016-0.0031	0.04-0.08
Middle	0.0120-0.0028	0.03-0.07
Rocker Arms and Shaft		
Rocker Arm Inner Diameter	0.5118-0.5125	13.000-13.018
Rocker Arm Shaft Outside Diameter	0.5095-0.5099	12.941-12.951
Valves		
Valve Dimensions		
Head Diameter (IN / EX)	0.94-0.95 / 0.86-0.87	23.9-24.1 / 21.9-22.1
Face Width (IN and EX)	0.072-0.089	1.84-2.26
Seat Contact Width (IN and EX)	0.023-0.031	0.6-0.8
Margin Thickness (IN / EX)	0.027 / 0.039	0.7 / 1.0
Stem Thickness (IN)	0.2156-0.2161	5.475-5.490
Stem Thickness (EX)	0.2150-0.2156	5.460-5.475
Valve Guide Inner Diameter (IN and EX)	0.2165-0.2170	5.500-5.512
Stem-to-guide Clearance (IN)	0.0004-0.0015	0.010-0.037
Stem-to-guide Clearance (EX)	0.0010-0.0020	0.025-0.052
Valve Stem Runout	0.0012	0.03
Valve Springs		
Free Length (Standard / Limit)	1.09 / 1.09	27.6 / 27.6
Out-of-Square (Spring Warpage/Tilt)	0.04	1.0

[1] Unless otherwise noted
[2] Uspecifications are our best guess considering an obvious error in the Yamaha factory information. Yamaha lists both the height and width of the camshaft lobes to be the same (which would make it round and not a lobe). Double-check this spec with your local parts supplier before condemning a used camshaft

POWERHEAD 6-141

ENGINE SPECIFICATIONS - YAMAHA 8/9.9 HP (232cc) 4-STROKE ENGINES

Component	U.S. (in.) [1]	Metric (mm) [1]
Camshaft		
Lobe Height (top of lobe to bottom of journal)		
Intake	0.966-0.970	24.541-24.641
Exhaust	0.968-0.972	24.578-24.678
Lobe Width (side-to-side / non-raised portion of lobe)		
Intake	0.793-0.797	20.137-20.237
Exhaust	0.794-0.798	20.178-20.278
Belt Deflection - Max	0.4	10
Runout - Max	0.004	0.1
Cylinder Bore		
Standard Bore Diameter	2.323-2.324	59.00-59.02
Limit	2.326	59.10
Out-of-round Service Limit	0.002	0.05
Taper Service Limit	0.003	0.08
Cylinder Head		
Gasket Surface Warpage Limit	0.004	0.1
Crankshaft [2]		
Crankshaft Width (Journal to Journal, outside of each)	4.87-4.88	123.7-123.9
Connecting Rod Big-End Oil Clearance	0.0008-0.0018	0.021-0.045
Main Bearing Clearance		
1992-97	0.0002-0.0017	0.005-0.043
1998-03	0.0000-0.0011	0.000-0.027
Runout - Max	0.0008	0.02
Crankshaft Bearing Mark/Color		
A: Blue	1.3005-1.3008	33.032-33.040
B: Black	1.30016-1.3005	33.024-33.032
C: Brown	1.2998-1.30016	33.016-33.024
Fuel Pump		
Discharge Rate (Min)	4.75 Gph @ 2750 rpm	18Lph @ 2750 rpm
Diaphragm Stroke	0.0945	2.4
Plunger Stroke	0.2283	5.8
Oil Pump		
Inner-to-Outer Rotor Clearance	0.0008-0.0059	0.02-0.15
Outer Rotor-to-Pump Housing Clearance (side of rotor)	0.0024-0.0043	0.06-0.11
Outer Rotor-to-Pump Cover Clearance (flat top of rotor)	0.0008-0.0028	0.02-0.07
Relief Valve Operating Pressure	56-64 psi	388-450 kPa
Piston		
Standard Size	2.3209-2.3215	58.950-58.965
Measuring Point (above bottom of skirt)	0.40	10.0
Clearance	0.0014-0.0026	0.035-0.065
Oversize		
1st	2.333	59.25
2nd	2.343	59.50

ENGINE SPECIFICATIONS - YAMAHA 8/9.9 HP (232cc) 4-STROKE ENGINES

Component	U.S. (in.) [1]	Metric (mm) [1]
Piston Rings		
Width (measured from outer-to-inner edge of ring)		
Top	0.0906	2.3
Middle	0.0945	2.4
Bottom (oil control)	0.1000	2.5
Thickness (measured vertically top-to-bottom of ring)		
Top and Middle	0.0591	1.5
Bottom (oil control)	0.09	2.4
End Gap		
Top and Middle	0.0060-0.0120	0.15-0.30
Bottom (oil control)	0.0080-0.0276	0.20-0.70
Ring Side Clearance		
Top	0.0016-0.0031	0.04-0.08
Middle	0.0012-0.0028	0.03-0.07
Valves		
Valve Dimensions		
Head Diameter (IN / EX)	1.020-1.028 / 0.862-0.870	25.9-26.1 / 21.9-22.1
Face Width (IN and EX)	0.078-0.122	1.980-3.110
Seat Width (IN and EX)	0.024-0.031	0.6-0.8
Margin Thickness (IN and EX)	0.020-0.035	0.5-0.9
Stem Thickness (IN)	0.2156-0.2161	5.475-5.490
Stem Thickness (EX)	0.2150-0.2156	5.460-5.475
Valve Guide Inner Diameter (IN and EX)	0.2165-0.2170	5.500-5.512
Stem-to-guide Clearance (IN)	0.0004-0.0015	0.010-0.037
Stem-to-guide Clearance (EX)	0.0010-0.0020	0.025-0.052
Valve Stem Runout	0.0006	0.016
Valve Springs		
Installed Height (Valve Closed)	0.96	24.4
Pressure @ Installed Height	19.8-22.0 lbs.	90-100 N
Out-of-Square (Spring Warpage/Tilt)	0.043	1.1
Direction of Winding	Left Hand	Left Hand

Because manufacturers differ with what specifications they choose to provide, please refer to the Mercury/Mariner charts for additional specifications (or another mechanical point of view) which MAY apply as well

[1] Unless otherwise noted
[2] Crankcase mark bearing color codes - A - Blue, B - Black and C - Brown

POWERHEAD

ENGINE SPECIFICATIONS - MERCURY/MARINER 8/9.9 HP (232cc) 4-STROKE ENGINES

Component	U.S. (in.) [1]	Metric (mm) [1]
Piston		
Standard Size (at skirt)	2.3209-2.3216	58.950-58.965
Clearance	0.0014-0.0026	0.035-0.065
Oversize		
1st	2.333	59.25
2nd	2.343	59.50
Piston Rings		
End Gap		
Top and Middle (limit)	0.0060-0.0120 (0.0200)	0.15-0.30 (0.50)
Bottom "oil control" (limit)	0.0080-0.0276 (0.0200)	0.20-0.70 (0.50)
Ring Side Clearance		
Top (limit)	0.0016-0.0031 (0.0200)	0.04-0.08 (0.50)
Middle (limit)	0.0012-0.0028 (0.0200)	0.03-0.07 (0.50)
Valves		
Valve Dimensions		
Head Diameter (IN / EX)	1.020-1.028 / 0.862-0.870	25.9-26.1 / 21.9-22.1
Face Width (IN and EX)	0.079-0.124	2.00-3.14
Seat Width (IN and EX)	0.024-0.031	0.6-0.8
Margin Thickness (IN and EX)	0.020-0.035	0.5-0.9
Stem Thickness (IN)	0.2156-0.2161	5.475-5.490
Stem Thickness (EX)	0.2150-0.2156	5.460-5.475
Valve Guide Inner Diameter (IN and EX)	0.2165-0.2170	5.500-5.512
Stem-to-guide Clearance (IN)	0.0004-0.0015	0.010-0.037
Stem-to-guide Clearance (EX)	0.0010-0.0020	0.025-0.052
Valve Stem Runout	0.0006	0.016
Valve Springs		
Installed Height (Valve Closed)	0.96	24.4
Pressure @ Installed Height	19.8-22.0 lbs.	9.0-10 kg
Out-of-Square (Spring Warpage/Tilt)	0.043	1.1
Direction of Winding	Left Hand	Left Hand

Because manufacturers differ with what specifications they choose to provide, please refer to the Yamaha charts for additional specifications (or another mechanical point of view) which MAY apply as well
[1] Unless otherwise noted
[2] Crankcase mark bearing color codes - A - Blue, B - Black and C - Brown

ENGINE SPECIFICATIONS - MERCURY/MARINER 8/9.9 HP (232cc) 4-STROKE ENGINES

Component	U.S. (in.) [1]	Metric (mm) [1]
Camshaft		
Lobe Height (top of lobe to bottom of journal)		
Intake	0.966-0.970	24.541-24.641
Exhaust	0.968-0.972	24.578-24.678
Lobe Width (side-to-side / non-raised portion of lobe)		
Intake	0.793-0.797	20.137-20.237
Exhaust	0.794-0.798	20.178-20.278
Belt Deflection - Max	0.4	10
Runout - Max	0.004	0.1
Cylinder Bore		
Standard Bore Diameter	2.323-2.324	59.00-59.02
Limit	2.326	59.10
Out-of-round Service Limit	0.003	0.08
Taper Service Limit	0.003	0.08
Cylinder Head		
Gasket Surface Warpage Limit	0.004	0.1
Crankshaft [2]		
Crankshaft Width (Journal to Journal, outside of each)	4.87-4.88	123.7-123.9
Connecting Rod Big-End Oil Clearance	0.0008-0.0018	0.021-0.045
Main Bearing Clearance	0.0002-0.0017	0.005-0.043
Crankshaft Main Bearing Journal Diameters	1.1809-1.1811	29.997-30.003
Connectinf Rod Journal Diameter	1.1019-1.1021	27.997-28.003
Runout - Max	0.0008	0.02
Crankshaft Bearing Mark/Color/Bearing Size		
A: Blue	0.0594-0.0595	1.508-1.512
B: Black	0.0592-0.0594	1.504-1.508
C: Brown	0.0591-0.0592	1.500-1.504
Crankshaft Bearing Mark/Color/Housing Size		
A: Blue	1.3005-1.3008	33.032-33.040
B: Black	1.30016-1.3005	33.024-33.032
C: Brown	1.2998-1.30016	33.016-33.024
Fuel Pump		
Pump Operating Pressure	3-6 psi	20-41 kPa
Diaphragm Stroke	0.0945	2.4
Plunger Stroke	0.2283	5.8
Oil Pump		
Inner-to-Outer Rotor Clearance	0.0008-0.0059	0.02-0.15
Outer Rotor-to-Pump Housing Clearance (side of rotor)	0.0024-0.0043	0.06-0.11
Outer Rotor-to-Pump Cover Clearance (flat top of rotor)	0.0008-0.0028	0.02-0.07
Relief Valve Operating Pressure	55-64 psi	388-450 kPa

POWERHEAD

ENGINE SPECIFICATIONS - YAMAHA 15 HP (323cc) 4-STROKE ENGINES

Component	U.S. (in.) ①	Metric (mm) ①
Camshaft		
Lobe Height (top of lobe to bottom of journal)		
Intake ③	0.9407-0.9447	23.895-23.995
Intake ④	1.0865-1.0904	27.596-27.696
Exhaust ③	0.9416-0.9456	23.917-24.017
Exhaust ④	1.0872-1.0912	27.616-27.716
Lobe Width (side-to-side / non-raised portion of lobe)		
Intake and Exhaust ③	0.7854-0.7894	19.950-20.050
Intake and Exhaust ④	0.9429-0.9468	23.950-24.050
Camshaft Journal Diameter		
Large Journal ③	0.7077-0.7088	17.975-17.991
Small Journal ③	0.6289-0.6293	15.973-15.984
Cylinder Head Journal Diameter		
All ③	0.7087-0.7094	18.000-18.018
Large Center Journal ④	1.3754-1.3762	34.935-34.955
Small Driven Gear Journal ④	0.7076-0.7080	17.973-17.984
Oil Pump Housing Journal Diameter		
③	0.6299-0.6307	16.000-16.019
④	0.6289-0.6293	15.973-15.984
Timing Belt Deflection	0.0-0.4	0-10
Runout - Max	0.001	0.03
Compression		
Minimum Compression - ③ / ④	173.5 / 139.4 psi	1196 / 961 kPa
Cylinder Bore		
Standard Bore Diameter	2.3228-2.3234	59.000-59.015
Out-of-round Service Limit	0.002	0.05
Taper Service Limit	0.003	0.08
Cylinder Head		
Gasket Surface Warpage Limit	0.004	0.1
Crankshaft ②		
Crankshaft Width (Journal to Journal, outside of each)	4.99-5.00	126.70-126.90
Radial Clearance ③	0.002	0.05
Crankshaft Big-End Side Clearance	0.002-0.009	0.05-0.22
Crankshaft Journal Clearance	0.0005-0.0018	0.012-0.045
Runout - Max	0.0012	0.03
Crankshaft Bearing Mark/Color		
A: Blue	1.4974-1.4976	38.033-38.040
B: Black	1.4970-1.4973	38.025-38.032
C: Brown	1.4967-1.4970	38.016-38.024
Fuel Pump		
Discharge Rate	6.6 Gph @ 3000 rpm	25Lph @ 3000 rpm
Pressure	17.07 psi	117.6 kPa
Diaphragm Stroke	0.094-0.189	2.4-4.8
Plunger Stroke	0.139-0.259	3.525-6.575
Oil Pump		
Inner-to-Outer Rotor Clearance	0.002-0.006	0.04-0.14
Outer Rotor-to-Pump Housing Clearance (side of rotor)	0.004-0.006	0.10-0.15
Outer Rotor-to-Pump Cover Clearance (flat top of rotor)	0.001-0.004	0.03-0.09
Relief Valve Operating Pressure	55-64 psi	388-450 kPa

ENGINE SPECIFICATIONS - YAMAHA 15 HP (323cc) 4-STROKE ENGINES

Component	U.S. (in.) ①	Metric (mm) ①
Piston		
Standard Size	2.3206-2.3215	58.950-58.965
Measuring Point (above bottom of skirt)	0.20	5
Clearance	0.0014-0.0026	0.035-0.065
Piston Pin Boss Bore Inner Diameter	0.5513-0.5518	14.004-14.015
Piston Pin Outside Diameter	0.5510-0.5512	13.996-14.000
Oversize		
1st	2.333	59.25
2nd	2.343	59.50
Piston Rings		
Width (measured from outer-to-inner edge of ring)		
Top	0.090	2.30
Middle	0.100	2.60
Bottom (oil control)	0.100	2.45
Thickness (measured vertically top-to-bottom of ring)		
Top	0.050	1.20
Middle	0.060	1.50
Bottom (oil control)	0.090	2.40
End Gap		
Top - Standard / Limit	0.0060-0.0120 / 0.0200	0.15-0.30 / 0.50
Middle - Standard / Limit	0.0120-0.0200 / 0.0280	0.30-0.50 / 0.70
Bottom (oil control) - Standard / Limit	0.0080-0.0276 / 0.0400	0.20-0.70 / 0.90
Ring Side Clearance		
Top	0.0005-0.0013	0.013-0.035
Middle	0.0010-0.0020	0.020-0.040
Rocker Arms and Shaft		
Rocker Arm Inner Diameter	0.5118-0.5125	13.000-13.018
Rocker Arm Shaft Outside Diameter	0.5095-0.5099	12.941-12.951
Valves		
Valve Dimensions		
Head Diameter (IN / EX)	1.10-1.11 / 0.86-0.87	27.9-28.1 / 21.9-22.1
Face Angle	90.5-91.5 degrees	90.5-91.5 degrees
Face Width (IN and EX)	0.079-0.122	2.0-3.1
Seat Width (IN and EX)	0.02-0.03	0.6-0.8
Margin Thickness (IN and EX)	0.020-0.035	0.5-0.9
Stem Thickness (IN)	0.2156-0.2161	5.475-5.490
Stem Thickness (EX)	0.2150-0.2156	5.460-5.475
Valve Guide Inner Diameter (IN and EX)	0.2165-0.2170	5.500-5.512
Stem-to-guide Clearance (IN)	0.0004-0.0015	0.010-0.037
Stem-to-guide Clearance (EX)	0.0010-0.0020	0.025-0.052
Valve Stem Runout	0.0006	0.016
Valve Springs		
Free Length (Standard / Limit)	1.35 / 1.29	34.4 / 32.7
Pressure @ Set Length	24.2 lbs @ 1.00	11 kg @ 25.4
Out-of-Square (Spring Warpage/Tilt)	0.06	1.5

Because manufacturers differ with what specifications they choose to provide, please refer to the Mercury/Mariner charts for additional specifications (or another mechanical point of view) which MAY apply as well

① Unless otherwise noted
② Crankcase mark bearing color codes - A - Blue, B - Black and C - Brown
③ Specification applies to all 1998-02 models and 03 F15AMH/F15MHW, F15AEH/F15EHW, and F15AE
④ Specification applies to all 2003 F15AEP/F15PR and F15AEHP/F15PH

6-144 POWERHEAD

ENGINE SPECIFICATIONS - MERCURY/MARINER 9.9/15 HP (323cc) 4-STROKE ENGINES

Component	U.S. (in.) ①	Metric (mm) ①
Camshaft		
Lobe Height (top of lobe to bottom of journal)		
Intake and Exhaust ③	0.941-0.945	23.90-24.00
Intake and Exhaust ④	1.089-1.090	27.66-27.69
Lobe Width (side-to-side / non-raised portion of lobe)		
Intake and Exhaust ③	0.785-0.789	19.95-20.05
Intake and Exhaust ④	0.943-0.946	23.96-24.02
Camshaft Journal Diameter		
Large (Lower/Oil Pump End) Journal	0.707-0.708	17.97-17.99
Small (Upper/Cylinder Head) Journal	0.628-0.629	15.97-15.98
Cylinder Head Camshaft Bore Diameter	0.7087-0.7094	18.000-18.018
Oil Pump Housing Journal Dimensions		
Pump Journal Height From Housing ③/④	0.937 / 0.842	23.8 / 21.39
Pump Journal Outer Diameter ③/④	1.260 / 1.378	32.00 / 34.98
Pump Journal Inner Diameter	0.6299-0.6307	16.000-16.019
Timing Belt Deflection	0.0-0.4	0-10
Runout - Max	0.001	0.03
Compression		
Max (cold motor/no compression release)	185-190 psi	1276-1310 kPa
Manual models with compression	40-60 psi	276-414 kPa
Cylinder Bore		
Standard Bore Diameter	2.3228-2.3236	59.00-59.02
Oversize Bore Diameter (0.010 in. / 0.25mm)	2.3327-2.3335	59.25-59.27
Oversize Bore Diameter (0.020 in. / 0.50mm)	2.3425-2.3433	59.50-59.52
Out-of-round Service Limit	0.003	0.08
Taper Service Limit	0.003	0.08
Cylinder Head		
Gasket Surface Warpage Limit	0.004	0.1
Crankshaft ②		
Connecting Rod Big-End Oil Clearance	0.021-0.045	0.0008-0.0018
Connecting Rod Small-End Inner Diameter	0.5518-0.5523	14.015-14.029
Crankshaft Main Journal Clearance	0.0004-0.0015	0.011-0.039
Runout - Max	0.0008	0.02
Fuel Pump		
Pump Operating Pressure	3-6 psi	21-41 kPa
Diaphragm Stroke	0.14-0.27	3.52-6.58
Oil Pump		
Pump Operating Pressure (warm engine)	30-40 psi @ 3000 rpm	207-279 kPa @ 3000
Inner-to-Outer Rotor Clearance	0.0008-0.0059	0.02-0.15
Outer Rotor-to-Pump Housing Clearance (side of rotor)	0.0024-0.0043	0.06-0.11
Outer Rotor-to-Pump Cover Clearance (flat top of rotor)	0.0008-0.0030	0.02-0.07
Relief Valve Operating Pressure	55-64 psi	388-450 kPa

ENGINE SPECIFICATIONS - MERCURY/MARINER 9.9/15 HP (323cc) 4-STROKE ENGINES

Component	U.S. (in.) ①	Metric (mm) ①
Piston		
Measuring Point (above bottom of skirt)	0.20	5
Standard Size (measured at skirt)	2.3209-2.3214	58.950-58.965
Oversize Piston Diameter (0.010 in. / 0.25mm)	2.3307-2.3313	59.200-59.215
Oversize Bore Diameter (0.020 in. / 0.50mm)	2.3406-2.3411	59.450-59.465
Clearance	0.0014-0.0026	0.035-0.065
Piston Pin Outside Diameter	0.5510-0.5512	13.996-14.000
Piston Rings		
End Gap		
Top - Standard / Limit	0.006-0.012	0.15-0.30
Middle - Standard / Limit	0.012-0.020	0.30-0.50
Bottom (oil control) - Standard / Limit	0.008-0.028	0.20-0.70
Ring Side Clearance		
Top	0.0016-0.0032	0.04-0.08
Middle	0.0012-0.0028	0.03-0.07
Rocker Arms and Shaft		
Rocker Arm Inner Diameter	0.5118-0.5125	13.000-13.018
Rocker Arm Shaft Outside Diameter	0.5095-0.5099	12.941-12.951
Valves		
Valve Dimensions		
Face Angle	90.5-91.5 degrees	90.5-91.5 degrees
Face Width (IN and EX)	0.079-0.124	2.00-3.14
Seat Width (IN and EX)	0.024-0.031	0.6-0.8
Margin Thickness (IN and EX)	0.020-0.035	0.5-0.9
Stem Thickness (IN)	0.2156-0.2161	5.475-5.490
Stem Thickness (EX)	0.2150-0.2156	5.460-5.475
Valve Guide Inner Diameter (IN and EX)	0.2165-0.2170	5.500-5.512
Stem-to-guide Clearance (IN)	0.0004-0.0015	0.010-0.037
Stem-to-guide Clearance (EX)	0.0010-0.0020	0.025-0.052
Valve Stem Runout	0.0006	0.016
Valve Springs		
Free Length (Standard)	1.354	34.4
Pressure @ Installed Height	23.1-25.4 lbs	10.5-11.5 kg
Out-of-Square (Spring Warpage/Tilt)	0.043	1.1
Direction of Winding	Right Hand	Right Hand

Because manufacturers differ with what specifications they choose to provide, please refer to the Yamaha charts for additional specifications (or another mechanical point of view) which MAY apply as well

① Unless otherwise noted
② Crankcase mark bearing color codes: A - Blue, B - Black and C - Brown
③ Specification applies to all Manual start models through 1999 and Electric start models through 04
④ Specification applies to all 2001 and later Manual start models

POWERHEAD 6-145

ENGINE SPECIFICATIONS - YAMAHA 25 HP (498cc) 4-STROKE ENGINES

Component		U.S. (in.)①	Metric (mm)①
Camshaft			
Lobe Height (top of lobe to bottom of journal)			
Intake		1.2159-1.2198	30.884-30.984
Exhaust	1998	1.2130-1.2170	30.820-30.920
	1999-03	1.2159-1.2198	30.884-30.984
Lobe Width (side-to-side / non-raised portion of lobe)			
Intake and Exhaust		1.0220-1.0250	25.950-26.050
Camshaft Journal Diameter		1.4537-1.4545	36.925-36.945
Cylinder Head Journal Diameter		1.4567-1.4577	37.000-37.025
Runout - Max		0.0039	0.1
Compression			
Minimum Compression		171-181 psi	1,180-1,250 kPa
Cylinder Bore			
Standard Bore Diameter		2.5590-2.5596	65.000-65.015
Out-of-round Service Limit		0.003	0.08
Taper Service Limit		0.003	0.08
Cylinder Head			
Gasket Surface Warpage Limit		0.004	0.1
Crankshaft			
Crankshaft Outer Diameter		1.692-1.693	42.984-43.000
Main Journal Clearance		0.0005-0.0017	0.012-0.044
Crankshaft Big-End Side Clearance		0.002-0.009	0.05-0.22
Crankshaft Journal Clearance		0.0008-0.0020	0.020-0.052
Connecting Rod Small-End Inner Diameter		0.6293-0.6298	15.985-15.998
Connecting Rod Big-End Oil Clearance		0.0008-0.0020	0.020-0.052
Runout - Max		0.0012	0.03
Crankshaft Bearing Mark/Color			
A: Blue		1.4973-1.4976	38.032-38.040
B: Black		1.4970-1.4973	38.024-38.032
C: Brown		1.4967-1.4970	38.016-38.024
Crankshaft Balancer			
Balancer Piston Diameter		3.7360	94.893
Crankcase Balancer Cylinder - Inner Diameter		3.7409	95.018
Connecting Rod Inner Diameter		2.6791	68.049
Fuel Pump			
Discharge Rate		18.5 Gph @ 6000 rpm	70Lph @ 6000 rpm
Pressure		3-6 psi	20-40 kPa
Diaphragm Stroke		0.14-0.20	3.50-5.10
Plunger Stroke		0.23-0.36	5.85-9.05
Oil Pump			
Inner-to-Outer Rotor Clearance (Max)		0.005	0.12
Outer Rotor-to-Pump Housing Clearance (side of rotor)		0.001-0.006	0.03-0.15
Outer Rotor-to-Pump Cover Clearance (flat top of rotor)		0.001-0.003	0.03-0.08
Shaft-to-housing clearance		0.0002-0.0013	0.006-0.034
Relief Valve Operating Pressure ②		0.54-0.63 psi	3.82-4.42 kPa

ENGINE SPECIFICATIONS - YAMAHA 25 HP (498cc) 4-STROKE ENGINES

Component		U.S. (in.)①	Metric (mm)①
Piston			
Standard Size		2.5570-2.5573	64.950-64.965
Oversize Piston Assembled at Factory		2.5768-2.5774	65.450-65.465
Measuring Point (above bottom of skirt)		0.08	2
Clearance		0.0014-0.0026	0.035-0.065
Piston Pin Boss Bore Inner Diameter		0.6289-0.6293	15.974-15.985
Piston Pin Outside Diameter		0.6285-0.6287	15.965-15.970
Oversize 1st		+ 0.0098	+ 0.25
2nd		+ 0.0198	+ 0.50
Piston Rings			
Width (measured from outer-to-inner edge of ring)			
Top		0.094	2.40
Middle		0.106	2.70
Bottom (oil control)		0.108	2.75
Thickness (measured vertically top-to-bottom of ring)			
Top		0.047	1.20
Middle		0.059	1.50
Bottom (oil control)		0.095	2.42
End Gap			
Top - Standard / Limit		0.006-0.012 / 0.020	0.15-0.30 / 0.50
Middle - Standard / Limit		0.012-0.020 / 0.028	0.30-0.50 / 0.70
Bottom (oil control)		0.008-0.028	0.20-0.70
Ring Side Clearance			
Top 1998		0.0008-0.0024	0.02-0.06
1999-03		0.0020-0.0031	0.04-0.08
Middle 1998		0.0008-0.0024	0.02-0.06
1999-03		0.0012-0.0028	0.03-0.07
Rocker Arms and Shaft			
Rocker Arm Inner Diameter		0.6299-0.6306	16.000-16.018
Rocker Arm Shaft Outside Diameter		0.6288-0.6296	15.971-15.991
Valves			
Valve Dimensions			
Head Diameter (IN / EX)		1.255-1.263 / 1.020-1.027	31.9-32.1 / 25.9-26.1
Face Angle		180, 90, 60 degrees	180, 90, 60 degrees
Face Width (IN / EX)		0.091-0.106 / 0.098-0.114	2.3-2.7 / 2.5-2.9
Seat Width (IN and EX)		0.035-0.043	0.9-1.1
Margin Thickness (IN / EX)		0.024-0.039 / 0.026-0.043	0.6-1.0 / 0.7-1.1
Stem Thickness (IN)		0.2155-0.2161	5.475-5.490
Stem Thickness (EX)		0.2150-0.2155	5.460-5.475
Valve Guide Inner Diameter (IN and EX)		0.2165-0.2170	5.500-5.512
Stem-to-guide Clearance (IN)		0.0004-0.0015	0.010-0.037
Stem-to-guide Clearance (EX)		0.0010-0.0020	0.025-0.052
Valve Stem Runout		0.0006	0.016
Valve Springs			
Free Length (Standard / Limit)		1.491-1.569 / 1.491	37.85-39.85 / 37.85
Out-of-Square (Spring Warpage/Tilt)		0.07	1.7

Because manufacturers differ with what specifications they choose to provide, please refer to the Mercury/Mariner charts for additional specifications (or another mechanical point of view) which MAY apply as well

① Unless otherwise noted
② Spec provided by Yamaha, but it looks suspicious to us. Most other Yamaha 4-strokes are 55-64 psi

6-146 POWERHEAD

ENGINE SPECIFICATIONS - MERCURY/MARINER 25 HP (498cc) 4-STROKE ENGINES

Component	U.S. (in.) ①	Metric (mm) ①
Camshaft		
Lobe Height (top of lobe to bottom of journal) Intake and Exhaust	1.216-1.220	30.89-30.99
Lobe Width (side-to-side / non-raised portion of lobe) Intake and Exhaust	1.022-1.025	25.95-26.05
Camshaft Journal Diameter	1.4541-1.4549	36.935-36.955
Cylinder Head Journal Diameter	1.4567-1.4577	37.000-37.025
Runout - Max	0.0039	0.1
Compression		
Max (cold motor/electric start only)	180-210 psi	1,241-1,448 kPa
Cylinder Bore		
Standard Bore Diameter	2.5591-2.5597	65.000-65.015
Oversize Bore Diameter (0.010 in. / 0.25mm)	2.5689	65.25
Oversize Bore Diameter (0.020 in. / 0.50mm)	2.5787	65.50
Out-of-round Service Limit	0.003	0.08
Taper Service Limit	0.003	0.08
Cylinder Head		
Gasket Surface Warpage Limit	0.004	0.1
Crankshaft ②		
Crankshaft Main Journal Clearance	0.0005-0.0017	0.012-0.044
Connecting Rod Small-End Inner Diameter	0.6293-0.6298	15.985-15.998
Connecting Rod Big-End Oil Clearance	0.0008-0.0020	0.020-0.052
Runout - Max	0.0012	0.03
Crankshaft Balancer		
Balancer Piston Minimum Diameter	3.735	94.883
Crankcase Cover		
Bore Maximum Diameter	3.802	95.045
Fuel Pump		
Pressure	3-6 psi	21-41 kPa
Plunger Stroke	0.23-0.38	5.85-9.65
Oil Pump		
Pump Operating Pressure (warm engine)	30-40 psi @ 3000 rpm	207-279 kPa @ 3000
Inner-to-Outer Rotor Clearance (Max)	0.005	0.12
Outer Rotor-to-Pump Housing Clearance (side of rotor)	0.004-0.006	0.09-0.15
Outer Rotor-to-Pump Cover Clearance (flat top of rotor)	0.001-0.003	0.03-0.08
Piston		
Standard Size	2.5570-2.5578	64.950-64.965
Oversize Piston Diameter (0.010 in. / 0.25mm)	2.5669-2.5675	65.200-65.215
Oversize Bore Diameter (0.020 in. / 0.50mm)	2.5768-2.5774	65.450-65.465
Measuring Point (above bottom of skirt)	0.02	5
Clearance	0.0014-0.0026	0.035-0.065
Piston Pin Outside Diameter	0.6285-0.6287	15.965-15.970

ENGINE SPECIFICATIONS - MERCURY/MARINER 25 HP (498cc) 4-STROKE ENGINES

Component	U.S. (in.) ①	Metric (mm) ①
Piston Rings		
End Gap		
Top - Standard	0.006-0.012	0.15-0.30
Middle - Standard	0.012-0.020	0.30-0.50
Bottom (oil control)	0.008-0.028	0.20-0.70
Ring Side Clearance		
Top and Middle	0.0008-0.0024	0.02-0.06
Rocker Arms and Shaft		
Rocker Arm Inner Diameter	0.6299-0.6306	16.000-16.018
Rocker Arm Shaft Outside Diameter	0.6288-0.6296	15.971-15.991
Valves		
Valve Dimensions		
Head Diameter (IN / EX)	1.256-1.264 / 1.020-1.028	31.9-32.1 / 25.9-26.1
Face Width (IN and EX)	0.079-0.124	2.00-3.14
Seat Width (IN and EX)	0.035-0.043	0.9-1.1
Margin Thickness (IN and EX)	0.020-0.035	0.5-0.9
Stem Thickness (IN)	0.2156-0.2161	5.475-5.490
Stem Thickness (EX)	0.2150-0.2156	5.460-5.475
Valve Guide Inner Diameter (IN and EX)	0.2165-0.2170	5.500-5.512
Stem-to-guide Clearance (IN)	0.0004-0.0015	0.010-0.037
Stem-to-guide Clearance (EX)	0.0010-0.0020	0.025-0.052
Valve Stem Runout	0.0006	0.016
Valve Springs		
Free Length	1.491-1.569	37.85-39.85
Pressure @ Installed Height	19.8-22.0 lbs	9.0-10.0 kg
Out-of-Square (Spring Warpage/Tilt)		
Mercury gives two conflicting limits A:	0.043	1.1
B:	0.060	1.7
Direction of Winding	Left Hand	Left Hand

Because manufacturers differ with what specifications they choose to provide, please refer to the Yamaha charts for additional specifications (or another mechanical point of view) which MAY apply as well
① Unless otherwise noted
② Crankcase mark bearing color codes - A - Blue, B - Black and C - Brown

POWERHEAD 6-147

ENGINE SPECIFICATIONS - YAMAHA 30/40/J40 HP (747cc) 4-STROKE ENGINES

Component	U.S. (in.) ①	Metric (mm) ①
Camshaft		
Lobe Height (top of lobe to bottom of journal) Intake and Exhaust	1.2139-1.2179	30.834-30.934
Lobe Width (side-to-side / non-raised portion of lobe) Intake and Exhaust	1.0200-1.0280	25.900-26.100
Camshaft Journal Diameter		
Journal # 1	1.4537-1.4545	36.925-36.945
Journal #'s 2, 3 and 4	1.4541-1.4549	36.935-36.955
Camshaft Journal Oil Clearance		
Journal # 1	0.0031-0.0039	0.08-0.10
Journal #'s 2, 3 and 4	0.0028-0.0035	0.07-0.09
Runout - Max	0.0039	0.1
Compression		
Minimum Compression	175.45 psi	1,210 kPa
Cylinder Bore		
Standard Bore Diameter	2.5590-2.5596	65.000-65.015
Out-of-round Service Limit	0.003	0.08
Taper Service Limit	0.003	0.08
Cylinder Head		
Gasket Surface Warpage Limit	0.004	0.1
Crankshaft		
Crankshaft Journal Diameter	1.6923-1.6929	42.984-43.000
Crankpin Diameter	1.2986-1.2992	32.984-33.000
Crankcase Main Journal Inside Diameter		
A	1.8117-1.8120	46.017-46.024
B	1.8114-1.8117	46.009-46.017
C	1.8110-1.8113	46.000-46.008
Crankcase Main Journal Oil Clearance	0.0004-0.0017	0.012-0.044
Runout - Max	0.001	0.03
Crankshaft Bearing Mark/Color		
A: Blue	1.8116-1.8120	46.016-46.024
B: Black	1.8113-1.8116	46.008-46.016
C: Brown	1.8110-1.8113	46.000-46.008
Connecting Rods		
Small-End Inner Diameter	0.6293-0.6298	15.985-15.998
Big-End Inner Diameter	1.4173-1.4183	36.000-36.024
Connecting Rod Big-End Oil Clearance	0.0008-0.0020	0.020-0.052
Crankshaft Big-End Side Clearance	0.0059-0.0087	0.15-0.22
Crankshaft Bearing Mark/Color		
A: Blue	1.4180-1.4183	36.016-36.024
B: Black	1.4176-1.4180	36.008-36.016
C: Brown	1.4173-1.4176	36.000-36.008
Fuel Pump		
Discharge Rate	18.5 Gph @ 3000 rpm	70 Lph @ 3000 rpm
Pressure (Minimum)	3-6 psi @ 3000 rpm	20-40 kPa @ 3000 rpm
Plunger Stroke	0.23-0.36	5.85-9.05

ENGINE SPECIFICATIONS - YAMAHA 30/40/J40 HP (747cc) 4-STROKE ENGINES

Component	U.S. (in.) ①	Metric (mm) ①
Oil Pump		
Inner-to-Outer Rotor Clearance (Max)	0.005	0.12
Outer Rotor-to-Pump Housing Clearance (side of rotor)	0.001-0.006	0.03-0.15
Outer Rotor-to-Pump Cover Clearance (flat top of rotor)	0.001-0.003	0.03-0.08
Piston		
Standard Size	2.5570-2.5573	64.950-64.965
Measuring Point (above bottom of skirt)	0.08	2
Clearance	0.0014-0.0026	0.035-0.065
Piston Pin Boss Bore Inner Diameter	0.6289-0.6293	15.974-15.985
Piston Pin Outside Diameter	0.6285-0.6287	15.965-15.970
Oversize 1st	+ 0.0098	+ 0.25
2nd	+ 0.0198	+ 0.50
Piston Rings		
Width (measured from outer-to-inner edge of ring)		
Top	0.094	2.40
Middle	0.106	2.70
Bottom (oil control)	0.108	2.75
Thickness (measured vertically top-to-bottom of ring)		
Top	0.047	1.20
Middle	0.059	1.50
Bottom (oil control)	0.098	2.50
End Gap		
Top - Standard	0.006-0.012	0.15-0.30
Middle - Standard	0.012-0.020	0.30-0.50
Bottom (oil control)	0.008-0.028	0.20-0.70
Ring Side Clearance		
Top and Middle	0.0004-0.0012	0.01-0.03
Oil Control	0.0016-0.0071	0.04-0.18
Rocker Arms and Shaft		
Rocker Arm Inner Diameter	0.6299-0.6306	16.000-16.018
Rocker Arm Shaft Outside Diameter	0.6288-0.6296	15.971-15.991
Valves		
Valve Dimensions		
Head Diameter (IN / EX)	1.255-1.263 / 1.020-1.027	31.9-32.1 / 25.9-26.1
Face Width (IN / EX)	0.091-0.106 / 0.098-0.114	2.3-2.7 / 2.5-2.9
Seat Width (IN and EX)	0.035-0.043	0.9-1.1
Margin Thickness (IN / EX)	0.024-0.039 / 0.026-0.043	0.6-1.0 / 0.7-1.1
Stem Thickness (IN)	0.2155-0.2161	5.475-5.490
Stem Thickness (EX)	0.2150-0.2155	5.460-5.475
Valve Guide Inner Diameter (IN and EX)	0.2165-0.2170	5.500-5.512
Stem-to-guide Clearance (IN)	0.0004-0.0015	0.010-0.037
Stem-to-guide Clearance (EX)	0.0010-0.0020	0.025-0.052
Valve Stem Runout	0.0006	0.016
Valve Springs		
Free Length (Standard / Limit)	1.491-1.569 / 1.491	37.85-39.85 / 37.85
Out-of-Square (Spring Warpage/Tilt)	0.07	1.7

Because manufacturers differ with what specifications they choose to provide, please refer to the Mercury/Mariner charts for additional specifications (or another mechanical point of view) which MAY apply as well

① Unless otherwise noted

6-148 POWERHEAD

ENGINE SPECIFICATIONS - MERCURY/MARINER 30/40 HP (747cc) 4-STROKE ENGINES

Component		U.S. (in.)①	Metric (mm)①
Camshaft			
Lobe Height (top of lobe to bottom of journal)			
Intake and Exhaust		1.216-1.220	30.89-30.99
Lobe Width (side-to-side / non-raised portion of lobe)		1.214-1.222	30.83-31.03
Intake and Exhaust			
Camshaft Journal Diameter	Carb	1.022-1.025	25.95-26.05
	EFI	1.020-1.028	25.90-26.10
Camshaft Journal Oil Clearance		1.4541-1.4549	36.935-36.955
Runout - Max		1.4567-1.4577	37.000-37.025
		0.0039	0.1
Compression			
Minimum (EFI motors only)		138 psi	950 kPa
Max (cold motor/electric start only)		180-210 psi	1,241-1,448 kPa
Cylinder Bore			
Standard Bore Diameter		2.5591	65.00
Oversize Bore Diameter (0.010 in. / 0.25mm)		2.5689	65.25
Oversize Bore Diameter (0.020 in. / 0.50mm)		2.5787	65.50
Out-of-round Service Limit		0.003	0.08
Taper Service Limit		0.003	0.08
Cylinder Head			
Gasket Surface Warpage Limit		0.004	0.1
Crankshaft②			
Crankshaft Main Journal Clearance		0.0005-0.0017	0.012-0.044
Connecting Rod Small-End Inner Diameter		0.6293-0.6298	15.985-15.998
Connecting Rod Big-End Oil Clearance		0.0008-0.0020	0.020-0.052
Runout - Max		0.0018	0.046
Fuel Pump (Mechanical Pump on all motors)			
Operating Pressure		3-6 psi	20-40 kPa
Plunger Stroke (Carb)		0.23-0.38	5.85-9.65
Oil Pump			
Pump Operating Pressure (warm engine)		30-40 psi @ 3000 rpm	207-279 kPa @ 3000
Inner-to-Outer Rotor Clearance (Max)		0.005	0.12
Outer Rotor-to-Pump Housing Clearance (side of rotor)			
	Carb	0.0035-0.0060	0.09-0.15
	EFI	0.0045-0.0090	0.11-0.23
Outer Rotor-to-Pump Cover Clearance (flat top of rotor)			
	Carb	0.0010-0.0030	0.03-0.08
	EFI	0.0015-0.0030	0.04-0.08

ENGINE SPECIFICATIONS - MERCURY/MARINER 30/40 HP (747cc) 4-STROKE ENGINES

Component	U.S. (in.)①	Metric (mm)①
Piston		
Standard Size	2.5570-2.5578	64.950-64.965
Oversize Piston Diameter (0.010 in. / 0.25mm)	2.5669-2.5675	65.200-65.215
Oversize Bore Diameter (0.020 in. / 0.50mm)	2.5768-2.5774	65.450-65.465
Measuring Point (above bottom of skirt)	0.02	5
Clearance	0.0014-0.0026	0.035-0.065
Piston Pin Outside Diameter	0.6285-0.6287	15.965-15.970
Piston Rings		
End Gap		
Top - Standard	0.006-0.012	0.15-0.30
Middle - Standard	0.012-0.020	0.30-0.50
Bottom (oil control)	0.008-0.028	0.20-0.70
Ring Side Clearance		
Top and Middle	0.0008-0.0024	0.02-0.06
Rocker Arms and Shaft		
Rocker Arm Inner Diameter	0.6299-0.6306	16.000-16.018
Rocker Arm Shaft Outside Diameter	0.6288-0.6296	15.971-15.991
Valves		
Valve Dimensions		
Head Diameter (IN / EX)	1.256-1.264 / 1.020-1.028	31.9-32.1 / 25.9-26.1
Face Width (IN and EX)	0.079-0.124	2.00-3.14
Seat Width (IN and EX)	0.035-0.043	0.9-1.1
Margin Thickness (IN and EX)	0.020-0.035	0.5-0.9
Stem Thickness (IN)	0.2156-0.2161	5.475-5.490
Stem Thickness (EX)	0.2150-0.2156	5.460-5.475
Valve Guide Inner Diameter (IN and EX)	0.2165-0.2170	5.500-5.512
Stem-to-guide Clearance (IN)	0.0004-0.0015	0.010-0.037
Stem-to-guide Clearance (EX)	0.0010-0.0020	0.025-0.052
Valve Stem Runout	0.0006	0.016
Valve Springs		
Free Length	1.491-1.569	37.85-39.85
Pressure @ Installed Height	19.8-22.0 lbs	9.0-10.0 kg
Out-of-Square (Spring Warpage/Tilt)		
Mercury gives two conflicting limits A:	0.043	1.1
B:	0.060	1.7
Direction of Winding	Left Hand	Left Hand

Because manufacturers differ with what specifications they choose to provide, please refer to the Yamaha charts for additional specifications (or another mechanical point of view) which MAY apply as well

① Unless otherwise noted
② Crankcase mark bearing color codes - A - Blue, B - Black and C - Brown

POWERHEAD 6-149

ENGINE SPECIFICATIONS - YAMAHA 40/45/50J/50 HP (935cc) 4-STROKE ENGINES

Component			U.S. (in.) ①	Metric (mm) ①
Camshaft				
Lobe Height (top of lobe to bottom of journal)				
Intake			1.2161-1.2200	30.888-30.988
Exhaust			1.2135-1.2175	30.824-30.924
Lobe Width (side-to-side / non-raised portion of lobe)				
Intake and Exhaust			1.0217-1.0256	25.95-26.05
Camshaft Journal Outer Diameter				
Cylinder # 1	1995-00		1.4537-1.4545	36.925-36.945
	2001-03		1.4539-1.4543	36.930-36.950
Cylinder #s 2-4	1995-00		1.4541-1.4549	36.935-36.955
	2001-03		1.4543-1.4547	36.940-36.950
Cylinder Head Cam Journal Inner Diameter			1.4567-1.4577	37.000-37.025
Cylinder Head Cam Journal Oil Clearance				
1998-00 ③	Cylinder # 1		0.0018-0.0035	0.045-0.090
	Cylinder #s 2-4		0.0022-0.0039	0.055-0.100
2001-03 ③	Cylinder # 1		0.0022-0.0039	0.055-0.100
	Cylinder #s 2-4		0.0018-0.0035	0.045-0.090
Runout - Max	1995-00		0.0040	0.10
	2001-03		0.0016	0.04
Compression				
Minimum Compression	1995-98		184 psi	1,270 kPa
	1999-00		128 psi	900 kPa
	2001-03		122 psi	840 kPa
Cylinder Bore				
Standard Bore Diameter			2.4803-2.4809	63.000-63.015
Out-of-round	Service Limit		0.003	0.08
Taper	Service Limit		0.003	0.08
Cylinder Head				
Gasket Surface Warpage Limit			0.004	0.1
Crankshaft				
Crankshaft Journal Outer Diameter			1.6923-1.6929	42.894-43.000
Crankpin Journal Outer Diameter (1998-01)			1.2986-1.2992	32.984-33.000
Crankcase Main Journal Inner Diameter (1998-01)			1.8110-1.8120	46.000-46.024
Crankcase Main Bearing Journal Clearance	1995-00		0.0005-0.0017	0.012-0.044
	2001-03		0.0005-0.0014	0.012-0.036
Crankshaft Journal Bearing Thickness (Mark/Color)				
1995-00		A: Blue	0.0590-0.0591	1.498-1.502
		B: Black	0.0588-0.0590	1.494-1.498
		C: Brown	0.0587-0.0588	1.490-1.494

ENGINE SPECIFICATIONS - YAMAHA 40/45/50J/50 HP (935cc) 4-STROKE ENGINES

Component			U.S. (in.) ①	Metric (mm) ①
2001-03		A: Yellow	0.0591-0.0593	1.502-1.506
		B: Red	0.0590-0.0591	1.498-1.502
		C: Pink	0.0588-0.0590	1.494-1.498
		D: Green	0.0587-0.0588	1.490-1.494
Runout - Max			0.0012	0.03
Connecting Rods				
Connecting Rod Small-End Inside Diameter				
	1995-98		0.6293-0.6298	15.985-15.998
	1999-00		0.6293-0.6294	15.985-15.988
	2001-03		0.6293-0.6298	15.985-15.998
Connecting Rod Big-End Inside Diameter (1998-03)			1.4173-1.4183	36.000-36.024
Crank Pin (Big-End) Journal Clearance				
	1995-00		0.0008-0.0020	0.020-0.052
	2001-03		0.0006-0.0015	0.016-0.040
Crank Pin (Big-End) Side Clearance				
	1995-00		0.0020-0.0087	0.050-0.220
	2001-03		no spec provided	no spec provided
Conn Rod Big-End Bearing Thickness (Mark/Color)				
1995-00		A: Blue	0.0588-0.0590	1.494-1.498
		B: Black	0.0587-0.0588	1.490-1.494
		C: Brown	0.0585-0.0587	1.486-1.490
2001-03		A: Yellow	0.0591-0.0592	1.500-1.504
		B: Red	0.0589-0.0591	1.496-1.500
		C: Pink	0.0587-0.0589	1.492-1.496
		D: Green	0.0586-0.0587	1.488-1.492
Conn Rod Small-End Oil Clearance (1997-98 only)			0.0008-0.0020	0.020-0.052
Fuel Pump				
Discharge Rate			18.5 Gph	70Lph
Pressure (Max)			7 psi	49 kPa
Plunger Stroke			0.23-0.36	5.85-9.05
Oil Pump				
Inner-to-Outer Rotor Clearance				
	1995-98		about 0.004	about 0.12
	1999-00		0-0.004	0-0.12
	2001-03		0.0004-0.0039	0.01-0.10
Outer Rotor-to-Pump Housing Clearance (side of rotor)				
	1995-98		0.001-0.006	0.03-0.05
	1999-03		0.004-0.006	0.09-0.15
Outer Rotor-to-Pump Cover Clearance (flat top of rotor)			0.001-0.003	0.03-0.08
Operating Parameters				
1995-98 (Relief Valve Operating Pressure)			56-64 psi	382-442 kPa
1999-00 (Discharge @ 68 degrees F/20 C)			5.68 Gpm	21.5Lpm
2001-03 (Pressure @ 131 degree F/55 C)			14 psi @ 900 rpm	100 kPa @ 900 rpm
Piston				
Standard Size			2.4783-2.4789	62.950-62.965
Measuring Point (above bottom of skirt)			0.2	5.0
Clearance			0.0014-0.0026	0.035-0.065
Piston Pin Outer Diameter			0.6285-0.6287	15.965-15.970
Pin Boss Inside Diameter			0.6289-0.6293	15.974-15.985

6-150 POWERHEAD

ENGINE SPECIFICATIONS - YAMAHA 40/45/50/J50 HP (935cc) 4-STROKE ENGINES

Component		U.S. (in.) ①	Metric (mm) ①
Oversize			
1st		+ 0.0098	+ 0.25
2nd		+ 0.0198	+ 0.50
Piston Rings			
Width (measured from outer-to-inner edge of ring)			
Top		0.094-0.095	2.39-2.41
Middle		0.098-0.099	2.49-2.51
Bottom (oil control)		about 0.108	about 2.75
Thickness (measured vertically top-to-bottom of ring)			
Top		0.046-0.047	1.17-1.19
Middle		0.058-0.059	1.47-1.49
Bottom (oil control)		0.092-0.097	2.34-2.46
End Gap			
Top			
1995-98 (Build/Wear Limit)		0.006-0.012 / 0.020	0.15-0.30 / 0.50
1999-03		0.006-0.012	0.15-0.30
Middle			
1995-98 (Build/Wear Limit)		0.012-0.020 / 0.027	0.30-0.50 / 0.70
1999-03		0.012-0.020	0.30-0.50
Bottom (oil control)		0.008-0.028	0.20-0.70
Ring Side Clearance			
Top		0.002-0.003	0.04-0.08
Middle		0.001-0.003	0.03-0.07
Bottom (oil control) - 1999-03 only		0.002-0.008	0.05-0.19
Valves			
Valve Dimensions			
Head Diameter (IN / EX)		1.177-1.185 / 1.020-1.027	29.9-30.1 / 25.9-26.1
Face Width (IN / EX)			
	1995-98	0.091-0.106 / 0.099-0.114	2.3-2.7 / 2.5-2.9
	1999-00	0.128-0.150 / 0.139-0.161	3.25-3.82 / 3.54-4.10
	2001-03	0.072-0.117 / 0.078-0.122	1.84-2.97 / 1.98-3.11
Seat Width (IN and EX or IN / EX)			
	1995-98	0.036-0.043	0.9-1.1
	1999-00	0.041-0.054 / 0.026-0.046	1.03-1.38 / 0.67-1.17
	2001-03	0.035-0.043	0.9-1.1
Margin Thickness (IN / EX)		0.024-0.039 / 0.028-0.043	0.6-1.0 / 0.7-1.1
Stem Thickness (IN)		0.2156-0.2161	5.475-5.490
Stem Thickness (EX)		0.2150-0.2156	5.460-5.475
Valve Guide Inner Diameter (IN and EX)		0.2165-0.2170	5.500-5.512
Stem-to-guide Clearance (IN)		0.0004-0.0015	0.010-0.037
Stem-to-guide Clearance (EX)		0.0010-0.0020	0.025-0.052
Valve Stem Runout			
	1995-00	0.0006	0.016
	2001-03	0.0012	0.030
Valve Springs			
Free Length (Build Spec / Limit)		1.569 / 1.491	39.85 / 37.85
Out-of-Square (Spring Warpage/Tilt)		0.07	1.7

Because manufacturers differ with what specifications they choose to provide, please refer to the Mercury/Mariner charts for additional specifications (or another measurement point of view) which MAY apply as well
① Unless otherwise noted
② Crankcase mark bearing color codes vary with year

ENGINE SPECIFICATIONS - MERCURY/MARINER 40/45/50 HP (935cc) 4-STROKE ENGINES

Component	U.S. (in.) ①	Metric (mm) ①
Camshaft		
Lobe Height (top of lobe to bottom of journal)		
Intake	1.216-1.220	30.89-30.99
Exhaust	1.213-1.217	30.82-30.92
Lobe Width (side-to-side / non-raised portion of lobe)		
Intake and Exhaust	1.022-1.025	25.95-26.05
Camshaft Journal Outer Diameter		
Large Journal on End	1.4541-1.4549	36.935-36.955
Small Journals	1.4537-1.4545	36.925-36.945
Cylinder Head Cam Journal Inner Diameter	1.4567-1.4577	37.000-37.025
Runout - Max	0.0039	0.1
Compression		
Max (cold motor/WOT)	170-190 psi	1,172-1,310 kPa
Cylinder Bore		
Standard Bore Diameter	2.4803-2.4809	63.000-63.015
Oversize 0.020 in. (0.50mm)	2.5003-2.5009	63.500-63.515
Out-of-round		
Service Limit	0.003	0.08
Taper		
Service Limit	0.003	0.08
Cylinder Head		
Gasket Surface Warpage Limit	0.004	0.1
Crankshaft ②		
Crankshaft Main Journal Clearance	0.0005-0.0017	0.012-0.044
Connecting Rod Small-End Inner Diameter	0.6293-0.6298	15.985-15.998
Connecting Rod Big-End Oil Clearance	0.0008-0.0020	0.020-0.052
Runout - Max	0.0012	0.03
Fuel Pump		
Operating Pressure	3-6 psi	21-41 kPa
Diaphragm Stroke	0.14-0.20	3.50-5.10
Plunger Stroke	0.23-0.38	5.85-9.65
Oil Pump		
Pump Operating Pressure (warm engine)	30-40 psi @ 3000 rpm	207-279 kPa @ 3000
Inner-to-Outer Rotor Clearance (Max)	0.005	0.12
Outer Rotor-to-Pump Housing Clearance (side of rotor)	0.001-0.006	0.03-0.15
Outer Rotor-to-Pump Cover Clearance (flat top of rotor)	0.001-0.003	0.03-0.08
Piston		
Standard Size	2.4783-2.4789	62.950-62.965
Oversize 0.020 in. (0.50mm)	2.4983-2.4989	63.450-63.465
Measuring Point (above bottom of skirt)	0.2	5.0
Clearance	0.0014-0.0026	0.035-0.065
Piston Pin Outer Diameter	0.6285-0.6287	15.965-15.970

POWERHEAD 6-151

ENGINE SPECIFICATIONS - YAMAHA 60J/60 HP (996cc) 4-STROKE ENGINES

Component	U.S. (in.)①	Metric (mm)①
Camshaft		
Lobe Height (top of lobe to bottom of journal)		
Intake	1.2161-1.2200	30.89-30.99
Exhaust	1.2135-1.2175	30.82-30.92
Lobe Width (side-to-side / non-raised portion of lobe)		
Intake and Exhaust	1.0217-1.0256	25.95-26.05
Camshaft Journal Diameter		
Journal #1	1.4539-1.4543	36.93-36.94
Journal #'s 2, 3 and 4	1.4543-1.4547	36.94-36.95
Camshaft Journal Oil Clearance		
Journal #1	0.0023-0.0039	0.06-0.10
Journal #'s 2, 3 and 4	0.0020-0.0035	0.05-0.09
Cylinder Head Journal Inside Diameter	1.4567-1.4577	37.000-37.025
Runout - Max	0.0016	0.04
Compression		
Minimum Compression		
F60 (Standard)	125 psi	880 kPa
T60 (High Thrust)	134 psi	940 kPa
Cylinder Bore		
Standard Bore Diameter	2.5591-2.5594	65.00-65.01
Out-of-round Service Limit	0.0004	0.01
Taper Service Limit	0.0031	0.08
Cylinder Head		
Gasket Surface Warpage Limit	0.004	0.1
Crankshaft		
Crankshaft Journal Diameter	1.6923-1.6929	42.984-43.000
Crankpin Diameter	1.2986-1.2992	32.984-33.000
Crankcase Main Journal Inside Diameter	1.8110-1.8120	46.000-46.024
Crankcase Main Journal Bearing Thickness		
Yellow	0.0591-0.0593	1.502-1.506
Red	0.0590-0.0591	1.498-1.502
Pink	0.0588-0.0590	1.494-1.498
Green	0.0587-0.0588	1.490-1.494
Crankcase Main Journal Oil Clearance	0.0005-0.0014	0.012-0.036
Runout - Max	0.0016	0.04
Connecting Rods		
Small-End Inner Diameter	0.6293-0.6298	15.985-15.998
Big-End Inner Diameter	1.4173-1.4183	36.000-36.024
Crankpin (Connecting Rod Big-End) Oil Clearance	0.0006-0.0015	0.016-0.040
Crankshaft Bearing Mark/Color		
Yellow	0.0591-0.0592	1.500-1.504
Red	0.0589-0.0591	1.496-1.500
Pink	0.0587-0.0589	1.492-1.496
Green	0.0586-0.0587	1.488-1.492
Fuel Pump		
Discharge Rate	18.5 Gph @ 6000 rpm	70Lph @ 6000 rpm
Pressure	7 psi	49 kPa
Plunger Stroke	0.23-0.36	5.85-9.05

ENGINE SPECIFICATIONS - MERCURY/MARINER 40/45/50 HP (935cc) 4-STROKE ENGINES

Component	U.S. (in.)①	Metric (mm)①
Piston Rings		
End Gap		
Top - Standard	0.006-0.012	0.15-0.30
Middle - Standard	0.012-0.020	0.30-0.50
Bottom (oil control)	0.008-0.028	0.20-0.70
Ring Side Clearance		
Top	0.002-0.003	0.04-0.08
Middle	0.001-0.003	0.03-0.08
Valves		
Valve Dimensions		
Head Diameter (IN / EX)	1.177-1.185 / 1.020-1.027	29.9-30.1 / 25.9-26.1
Face Width (IN and EX)	0.079-0.124	2.00-3.14
Seat Width (IN and EX)	0.035-0.043	0.9-1.1
Margin Thickness (IN and EX)	0.020-0.035	0.5-0.9
Stem Thickness (IN)	0.2156-0.2161	5.475-5.490
Stem Thickness (EX)	0.2150-0.2156	5.460-5.475
Valve Guide Inner Diameter (IN and EX)	0.2165-0.2170	5.500-5.512
Stem-to-guide Clearance (IN)	0.0004-0.0015	0.010-0.037
Stem-to-guide Clearance (EX)	0.0010-0.0020	0.025-0.052
Valve Stem Runout	0.0006	0.016
Valve Springs		
Free Length	1.491-1.569	37.85-39.85
Pressure @ Installed Height	19.8-22.0 lbs	9.0-10.0 kg
Out-of-Square (Spring Warpage/Tilt)	0.043	1.1
Mercury gives two conflicting limits A:	0.060	1.7
B:		
Direction of Winding	Left Hand	Left Hand

Because manufacturers differ with what specifications they choose to provide, please refer to the Yamaha charts for additional specifications (or another mechanical point of view) which MAY apply as well
① Unless otherwise noted
② Crankcase mark bearing color codes - A - Blue, B - Black and C - Brown

POWERHEAD

ENGINE SPECIFICATIONS - MERCURY/MARINER 40/50/60 HP (996cc) 4-STROKE ENGINES

Component	U.S. (in.) [1]	Metric (mm) [1]
Camshaft		
Lobe Height (top of lobe to bottom of journal)		
Intake and Exhaust	1.214-1.223	30.83-31.03
Lobe Width (side-to-side / non-raised portion of lobe)		
Intake and Exhaust	1.020-1.028	25.90-26.10
Camshaft Journal Diameter	1.4541-1.4549	36.935-36.955
Cylinder Head Journal Inside Diameter	1.4567-1.4577	37.000-37.025
Runout - Max	0.0039	0.1
Compression		
Minimum	138 psi	950 kPa
Max (cold motor/electric start only)	180-210 psi	1,241-1,448 kPa
Cylinder Bore		
Standard Bore Diameter	2.5591	65.00
Oversize 0.010 in. (0.25mm)	2.5689	65.25
Oversize 0.020 in. (0.50mm)	2.5787	65.50
Out-of-round Service Limit	0.003	0.08
Taper Service Limit	0.003	0.08
Cylinder Head		
Gasket Surface Warpage Limit	0.004	0.1
Crankshaft [2]		
Crankshaft Main Journal Clearance	0.0005-0.0017	0.012-0.044
Connecting Rod Small-End Inner Diameter	0.6293-0.6298	15.985-15.998
Connecting Rod Big-End Oil Clearance	0.0008-0.0020	0.020-0.052
Runout - Max	0.0018	0.046
Fuel Pump		
Operating Pressure	3-6 psi	21-41 kPa
Plunger Stroke (Carb)	0.23-0.38	5.85-9.65
Oil Pump		
Delivery Pressure @ 3000 rpm (Engine Warm)	30-40 psi	207-278 kPa
Inner-to-Outer Rotor Clearance (Max)	0.005	0.12
Outer Rotor-to-Pump Housing Clearance (side of rotor)	0.0045-0.0090	0.11-0.23
Outer Rotor-to-Pump Cover Clearance (flat top of rotor)	0.0015-0.0030	0.04-0.08
Piston		
Standard Size	2.5570-2.5578	64.950-64.965
Oversize 0.010 in. (0.25mm)	2.5669-2.5675	65.200-65.215
Oversize 0.020 in. (0.50mm)	2.5768-2.5774	65.450-65.465
Measuring Point (above bottom of skirt)	0.2	5
Clearance	0.0014-0.0026	0.035-0.065

ENGINE SPECIFICATIONS - YAMAHA 60J/60 HP (996cc) 4-STROKE ENGINES

Component	U.S. (in.) [1]	Metric (mm) [1]
Oil Pump		
Delivery Pressure @ 900 rpm	about 16 psi	about 880 kPa
Inner-to-Outer Rotor Clearance (Max)	0.0047	0.12
Outer Rotor-to-Pump Housing Clearance (side of rotor)	0.0035-0.0059	0.09-0.15
Outer Rotor-to-Pump Cover Clearance (flat top of rotor)	0.0012-0.0031	0.03-0.08
Piston		
Standard Size	2.5570-2.5574	64.95-64.96
Measuring Point (above bottom of skirt)	0.2	5
Clearance	0.0014-0.0026	0.035-0.065
Piston Pin Boss Bore Inner Diameter	0.6289-0.6293	15.974-15.985
Piston Pin Outside Diameter	0.6285-0.6287	15.965-15.970
Oversize Piston Diameter		
1st	2.5669-2.5673	65.20-65.21
2nd	2.5768-2.5771	65.45-65.46
Oversize Piston Sizes		
1st	+ 0.0098	+ 0.25
2nd	+ 0.0198	+ 0.50
Piston Rings		
Width (measured from outer-to-inner edge of ring)		
Top	0.0906-0.0984	2.30-2.50
Middle	0.1024-0.1102	2.60-2.70
Bottom (oil control)	0.1083	2.75
Thickness (measured vertically top-to-bottom of ring)		
Top	0.0461-0.0468	1.17-1.19
Middle	0.0578-0.0586	1.47-1.49
Bottom (oil control)	0.0929-0.0976	2.36-2.48
End Gap		
Top - Standard	0.0060-0.0118	0.15-0.30
Middle - Standard	0.0118-0.0196	0.30-0.50
Bottom (oil control)	0.0079-0.0275	0.20-0.70
Ring Side Clearance		
Top and Middle	0.0008-0.0023	0.02-0.06
Oil Control	0.0016-0.0071	0.04-0.18
Rocker Arms and Shaft		
Rocker Arm Inner Diameter	0.6299-0.6303	16.00-16.01
Rocker Arm Shaft Outside Diameter	0.6291-0.6295	15.98-15.99
Valves		
Valve Dimensions		
Head Diameter (IN / EX)	1.2560-1.2637 / 1.0472-1.0551	31.9-32.1 / 26.6-26.8
Face Width (IN / EX)	0.0780-0.0945 / 0.0850-0.1098	1.98-2.40 / 2.16-2.79
Seat Width (IN and EX)	0.0512-0.0590	1.3-1.5
Margin Thickness (IN / EX)	0.0315-0.0472 / 0.0394-0.0551	0.8-1.2 / 1.0-1.4
Stem Thickness (IN)	0.2157-0.2161	5.48-5.49
Stem Thickness (EX)	0.2150-0.2153	5.46-5.47
Valve Guide Inner Diameter (IN and EX)	0.2165-0.2169	5.50-5.51
Stem-to-guide Clearance (IN)	0.0004-0.0012	0.01-0.03
Stem-to-guide Clearance (EX)	0.0012-0.0020	0.03-0.05
Valve Stem Runout (IN / EX)	0.0020 / 0.0012	0.05 / 0.03
Valve Springs		
Free Length (Standard / Limit)	1.5689 / 1.4901	39.85 / 37.85
Out-of-Square (Spring Warpage/Tilt)	0.067	1.7

Because manufacturers differ with what specifications they choose to provide, please refer to the Mercury/Mariner charts for additional specifications (or another mechanical point of view) which MAY apply as well

[1] Unless otherwise noted

POWERHEAD

ENGINE SPECIFICATIONS - MERCURY/MARINER 40/50/60 HP (996cc) 4-STROKE ENGINES

Component	U.S. (in.) ①	Metric (mm) ①
Piston Rings		
End Gap		
Top - Standard	0.006-0.012	0.15-0.30
Middle - Standard	0.012-0.020	0.30-0.50
Bottom (oil control)	0.008-0.028	0.20-0.70
Ring Side Clearance		
Top and Middle	0.0008-0.0024	0.02-0.06
Rocker Arms and Shaft		
Rocker Arm Inner Diameter	0.6299-0.6306	16.000-16.018
Rocker Arm Shaft Outside Diameter	0.6288-0.6296	15.971-15.991
Valves		
Valve Dimensions		
Head Diameter (IN / EX)	1.256-1.264 / 1.020-1.028	31.9-32.1 / 25.9-26.1
Face Width (IN and EX)	0.079-0.124	2.00-3.14
Seat Width (IN and EX)	0.035-0.043	0.9-1.1
Margin Thickness (IN and EX)	0.020-0.035	0.5-0.9
Stem Thickness (IN)	0.2156-0.2161	5.475-5.490
Stem Thickness (EX)	0.2150-0.2156	5.460-5.475
Valve Guide Inner Diameter (IN and EX)	0.2165-0.2170	5.500-5.512
Stem-to-guide Clearance (IN)	0.0004-0.0015	0.010-0.037
Stem-to-guide Clearance (EX)	0.0010-0.0020	0.025-0.052
Valve Stem Runout	0.0006	0.016
Valve Springs		
Free Length	1.491-1.569	37.85-39.85
Pressure @ Installed Height	19.8-22.0 lbs	9.0-10.0 kg
Out-of-Square (Spring Warpage/Tilt)		
Mercury gives two conflicting limits A:	0.043	1.1
B:	0.060	1.7
Direction of Winding	Left Hand	Left Hand

Because manufacturers differ with what specifications they choose to provide, please refer to the Yamaha charts for additional specifications (or another mechanical point of view) which MAY apply as well.

① Unless otherwise noted
② Crankcase mark bearing color codes - A - Blue, B - Black and C - Brown

ENGINE SPECIFICATIONS - YAMAHA 75/80/90J/90/100 HP (1596cc) 4-STROKE ENGINES

Component	U.S. (in.) ①	Metric (mm) ①
Camshaft		
Lobe Height (top of lobe to bottom of journal)		
Intake	1.526-1.529	38.76-38.84
Exhaust	1.514-1.517	38.46-38.54
Lobe Width (side-to-side / non-raised portion of lobe)		
Intake and Exhaust	1.178-1.184	29.92-30.08
Valve Lift (IN / EX)	0.273 / 0.259	6.94 / 6.58
Camshaft Journal Diameter	0.9827-0.9835	24.96-24.98
Camshaft Journal Oil Clearance	0.0008-0.0024	0.020-0.061
Cylinder Head Journal Inside Diameter	0.9843-0.9850	25.000-25.021
Runout - Max	0.0016	0.04
Compression		
Minimum Compression	138 psi	950 kPa
Cylinder Bore		
Standard Bore Diameter	3.110-3.111	79.00-79.02
Out-of-round Service Limit	0.003	0.08
Taper Service Limit	0.003	0.08
Cylinder Head		
Gasket Surface Warpage Limit	0.004	0.1
Crankshaft		
Crankshaft Journal Diameter (Build Spec)	1.8892-1.8898	47.985-48.000
Crankshaft Journal Diameter (Min)	1.8887	47.972
Crankpin Diameter (Build Spec)	1.7316-1.7323	43.982-44.000
Crankpin Diameter (Min)	1.7311	43.971
Crankcase Main Journal Inside Diameter	2.1269-2.1276	54.023-54.042
Crankcase Main Journal Oil Clearance	0.0009-0.0017	0.024-0.044
Lower Crankcase Main Journal Bearing Thickness		
Yellow	0.1185-0.1188	3.010-3.017
Green	0.1188-0.1191	3.017-3.024
Blue	0.1191-0.1193	3.024-3.031
Red	0.1193-0.1196	3.031-3.038
Upper Crankcase Main Journal Bearing Thickness		
Green	0.1178-0.1181	2.992-2.999
Blue	0.1181-0.1183	2.999-3.006
Red	0.1183-0.1186	3.006-30.13
No. 3 Main Journal Bearing Thickness		
Green	0.1178-0.1181	2.992-2.999
Blue	0.1181-0.1183	2.999-3.006
Red	0.1183-0.1186	3.006-3.013
Runout - Max	0.001	0.03

6-154 POWERHEAD

ENGINE SPECIFICATIONS - YAMAHA 75/80/90J/90/100 HP (1596cc) 4-STROKE ENGINES

Component	U.S. (in.) ①	Metric (mm) ①
Connecting Rods		
Small-End Inner Diameter	0.7073-0.7081	17.965-17.985
Big-End Inner Diameter	1.8514-1.8518	47.025-47.035
Crankpin (Connecting Rod Big-End) Oil Clearance	0.0009-0.0014	0.023-0.035
Crankshaft Bearing Mark/Color		
Yellow	0.0590-0.0593	1.499-1.506
Green	0.0593-0.0596	1.506-1.513
Blue	0.0596-0.0598	1.513-1.520
Red	0.0598-0.0601	1.520-1.527
Fuel Pump		
Discharge Rate	17.16 Gph @ 3000 rpm	65Lph @ 3000 rpm
Pressure (Max)	7 psi	49 kPa
Plunger Stroke	0.23-0.36	5.85-9.05
Oil Pump		
Discharge at 212 degrees F (100C) w/ 10W-30	1.56 Gpm	5.9Lpm
Pressure	16.78 psi	118 kPa
Relief Valve Opening Pressure	71 psi	490 kPa
Piston		
Standard Size	3.1074-3.1082	78.928-78.949
Measuring Point (above bottom of skirt)	0.51	13
Clearance	0.0028-0.0031	0.070-0.080
Piston Pin Boss Bore Inner Diameter	0.7090-0.7093	18.008-18.015
Piston Pin Outside Diameter	0.7085-0.7087	17.997-18.000
Oversize Piston Diameter	3.120	79.25
Piston Rings		
Width (measured from outer-to-inner edge of ring)		
Top	0.114-0.115	2.89-2.91
Middle	0.118-0.126	3.00-3.20
Bottom (oil control)	0.094	2.40
Thickness (measured vertically top-to-bottom of ring)		
Top	0.046-0.047	1.17-1.19
Middle	0.058-0.059	1.47-1.49
Bottom (oil control)	0.094-0.098	2.38-2.48
End Gap		
Top - Standard	0.006-0.012	0.15-0.30
Middle - Standard	0.028-0.035	0.70-0.90
Bottom (oil control)	0.008-0.028	0.20-0.70
Ring Side Clearance		
Top	0.001-0.003	0.02-0.08
Middle	0.001-0.003	0.03-0.07
Oil Control	0.001-0.006	0.03-0.15
Valve Lifters		
Cylinder Head Lifter Bore Inner Diameter	1.1020-1.1030	28.000-28.021
Valve Lifter Outside Diameter	1.1010-1.1016	27.965-27.980
Valve Lifter-to-Cylinder Head Bore Clearance	0.0008-0.0022	0.020-0.056
Valve Shims (in 0.0010 in. / 0.025mm increments)	0.0787-0.1299	2.000-3.300

ENGINE SPECIFICATIONS - YAMAHA 75/80/90J/90/100 HP (1596cc) 4-STROKE ENGINES

Component	U.S. (in.) ①	Metric (mm) ①
Valves		
Valve Dimensions		
Head Diameter (IN / EX)	1.142-1.150 / 0.945-0.953	29.0-29.2 / 24.0-24.2
Face Angle (IN)	120, 91, 110 degrees	120, 91, 110 degrees
Face Angle (EX)	140, 91, 110 degrees	140, 91, 110 degrees
Face Width (IN / EX)	0.079-0.096 / 0.090-0.107	2.00-2.43 / 2.28-2.71
Seat Width (IN and EX)	0.014-0.022	0.35-0.55
Margin Thickness (IN / EX)	0.018-0.026 / 0.026-0.033	0.45-0.65 / 0.65-0.85
Stem Thickness (IN)	0.2352-0.2358	5.975-5.990
Stem Thickness (EX)	0.2346-0.2352	5.960-5.975
Valve Guide Inner Diameter (IN and EX)	0.2364-0.2369	6.005-6.018
Stem-to-guide Clearance (IN)	0.0006-0.0017	0.015-0.043
Stem-to-guide Clearance (EX)	0.0012-0.0023	0.030-0.058
Valve Stem Runout	0.001	0.03
Valve Springs		
Free Length (Standard / Limit)	2.094 / 2.057	53.20 / 52.25
Out-of-Square (Spring Warpage/Tilt)	0.10	2.6

Because manufacturers differ with what specifications they choose to provide, please refer to the Mercury/Mariner charts for additional specifications (or another mechanical point of view) which MAY apply as well
① Unless otherwise noted

POWERHEAD 6-155

ENGINE SPECIFICATIONS - MERCURY/MARINER 75/90 HP (1596cc) 4-STROKE ENGINES

Component	U.S. (in.) ①	Metric (mm) ①
Camshaft		
Lobe Height (top of lobe to bottom of journal)		
Intake	1.465-1.472	37.22-37.38
Exhaust	1.453-1.459	36.90-37.06
Lobe Width (side-to-side / non-raised portion of lobe)		
Intake and Exhaust	1.178-1.184	29.92-30.08
Valve Lift (IN / EX)	0.273 / 0.259	6.94 / 6.58
Camshaft Journal Diameter	0.9827-0.9835	24.96-24.98
Camshaft Journal Oil Clearance	0.0008-0.0024	0.020-0.061
Cylinder Head Journal Inside Diameter	0.9843-0.9850	25.000-25.021
Runout - Max	0.0039	0.10
Compression		
Minimum Compression	138 psi	950 kPa
Cylinder Bore		
Standard Bore Diameter	3.110-3.111	79.00-79.02
Oversize 0.010 in. (0.25mm)	3.120-3.121	79.25-79.27
Out-of-round Service Limit	0.003	0.08
Taper Service Limit	0.003	0.08
Cylinder Head		
Gasket Surface Warpage Limit	0.004	0.1
Crankshaft		
Crankshaft Journal Diameter (Build Spec)	1.8892-1.8898	47.985-48.000
Crankshaft Journal Diameter (Min)	1.8887	47.972
Crankpin Diameter (Build Spec)	1.7316-1.7323	43.982-44.000
Crankpin Diameter (Min)	1.7311	43.971
Crankcase Main Journal Inside Diameter	2.1269-2.1276	54.023-54.042
Crankcase Main Journal Oil Clearance	0.0009-0.0017	0.024-0.044
Lower Crankcase Main Journal Bearing Thickness		
Yellow	0.1185-0.1188	3.010-3.017
Green	0.1188-0.1191	3.017-3.024
Blue	0.1191-0.1193	3.024-3.031
Red	0.1193-0.1196	3.031-3.038
Upper Crankcase Main Journal Bearing Thickness		
Green	0.1178-0.1181	2.992-2.999
Blue	0.1181-0.1183	2.999-3.006
Red	0.1183-0.1186	3.006-30.13
No. 3 Main Journal Bearing Thickness		
Green	0.1178-0.1181	2.992-2.999
Blue	0.1181-0.1183	2.999-3.006
Red	0.1183-0.1186	3.006-3.013
Runout - Max	0.001	0.03

ENGINE SPECIFICATIONS - MERCURY/MARINER 75/90 HP (1596cc) 4-STROKE ENGINES

Component	U.S. (in.) ①	Metric (mm) ①
Connecting Rods		
Small-End Inner Diameter	0.7073-0.7081	17.965-17.985
Big-End Inner Diameter	1.8514-1.8518	47.025-47.035
Crankpin (Connecting Rod Big-End) Oil Clearance	0.0009-0.0014	0.023-0.035
Crankshaft Bearing Mark/Color		
Yellow	0.0590-0.0593	1.499-1.506
Green	0.0593-0.0596	1.506-1.513
Blue	0.0596-0.0598	1.513-1.520
Red	0.0598-0.0601	1.520-1.527
Fuel Pump		
Discharge Rate	17.16 Gph @ 3000 rpm	65Lph @ 3000 rpm
Pressure (Max)	7 psi	49 kPa
Plunger Stroke	0.23-0.36	5.85-9.05
Oil Pump		
Discharge at 212 degrees F (100C) w/ 10W-30	1.56 Gph @ 1000 rpm	5.9Lph @ 1000 rpm
Pressure Warm Engine @ Idle (850 rpm)	30-60 psi	207-414 kPa
Relief Valve Opening Pressure	71 psi	490 kPa
Piston		
Standard Size	3.10734-3.1082	78.928-78.949
Oversize 0.010 in. (0.25mm)	3.1174-3.1182	79.178-79.199
Measuring Point (above bottom of skirt)	0.51	13
Clearance	0.0028-0.0031	0.070-0.080
Piston Pin Boss Bore Inner Diameter	0.7090-0.7093	18.008-18.015
Piston Pin Outside Diameter	0.7083-0.7087	17.997-18.000
Piston Rings		
Width (measured from outer-to-inner edge of ring)		
Top	0.114-0.115	2.89-2.91
Middle	0.118-0.126	3.00-3.20
Bottom (oil control)	0.094	2.40
Thickness (measured vertically top-to-bottom of ring)		
Top	0.046-0.047	1.17-1.19
Middle	0.058-0.059	1.47-1.49
Bottom (oil control)	0.094-0.098	2.38-2.48
End Gap		
Top - Standard	0.006-0.012	0.15-0.30
Middle - Standard	0.028-0.035	0.70-0.90
Bottom (oil control)	0.008-0.028	0.20-0.70
Ring Side Clearance		
Top	0.001-0.003	0.02-0.08
Middle	0.001-0.003	0.03-0.07
Oil Control	0.001-0.006	0.03-0.15

ENGINE SPECIFICATIONS - MERCURY/MARINER 75/90 HP (1596cc) 4-STROKE ENGINES

Component	U.S. (in.)①	Metric (mm)①
Valve Lifters		
Cylinder Head Lifter Bore Inner Diameter	1.1020-1.1030	28.000-28.021
Valve Lifter Outside Diameter	1.1010-1.1016	27.965-27.980
Valve Lifter-to-Cylinder Head Bore Clearance	0.0008-0.0022	0.020-0.056
Valve Shims (in 0.025mm increments)	0.0787-0.1299	2.000-3.300
Valves		
Valve Dimensions		
Head Diameter (IN / EX)	1.142-1.150 / 0.945-0.953	29.0-29.2 / 24.0-24.2
Face Angle (IN)	120, 91, 110 degrees	120, 91, 110 degrees
Face Angle (EX)	140, 91, 110 degrees	140, 91, 110 degrees
Face Width (IN / EX)	0.079-0.096 / 0.090-0.107	2.00-2.43 / 2.28-2.71
Seat Width (IN and EX)	0.014-0.022	0.35-0.55
Margin Thickness (IN / EX)	0.018-0.026 / 0.026-0.033	0.45-0.65 / 0.65-0.85
Stem Thickness (IN)	0.2352-0.2358	5.975-5.990
Stem Thickness (EX)	0.2346-0.2352	5.960-5.975
Valve Guide Inner Diameter (IN and EX)	0.2364-0.2369	6.005-6.018
Stem-to-guide Clearance (IN)	0.0006-0.0017	0.015-0.043
Stem-to-guide Clearance (EX)	0.0012-0.0023	0.030-0.058
Valve Stem Runout	0.001	0.03
Valve Springs		
Free Length (Standard / Limit)	2.094 / 2.057	53.20 / 52.25
Out-of-Square (Spring Warpage/Tilt)	0.10	2.6

Because manufacturers differ with what specifications they choose to provide, please refer to the Yamaha charts for additional specifications (or another mechanical point of view) which MAY apply as well
① Unless otherwise noted

ENGINE SPECIFICATIONS - YAMAHA 115J/115 (1741cc) 4-STROKE ENGINES

Component	U.S. (in.)①	Metric (mm)①
Camshaft		
Lobe Height (top of lobe to bottom of journal)		
Intake	1.465-1.472	37.22-37.38
Exhaust	1.453-1.459	36.90-37.06
Lobe Width (side-to-side / non-raised portion of lobe)		
Intake and Exhaust	1.178-1.184	29.92-30.08
Valve Lift (IN / EX)	0.287 / 0.275	7.30 / 6.98
Camshaft Journal Diameter	0.9827-0.9835	24.96-24.98
Camshaft Journal Oil Clearance	0.0008-0.0024	0.020-0.061
Cylinder Head Journal Inside Diameter	0.9843-0.9850	25.000-25.021
Runout - Max	0.004	0.1
Compression		
Minimum Compression	135 psi	950 kPa
Cylinder Bore		
Standard Bore Diameter	3.110-3.111	79.00-79.02
Out-of-round Service Limit	0.003	0.08
Taper Service Limit	0.003	0.08
Cylinder Head		
Gasket Surface Warpage Limit	0.004	0.1
Crankshaft		
Crankshaft Journal Diameter (Build Spec)	1.8891-1.8898	47.984-48.000
Crankshaft Journal Diameter (Min)	1.8887	47.972
Crankpin Diameter	1.6535-1.6528	42.000-41.982
Crankcase Main Journal Inside Diameter	2.1269-2.1276	54.023-54.042
Crankcase Main Journal Oil Clearance	0.0009-0.0017	0.024-0.044
Lower Crankcase Main Journal Bearing Thickness		
Yellow	0.1185-0.1188	3.010-3.017
Green	0.1188-0.1191	3.017-3.024
Blue	0.1191-0.1193	3.024-3.031
Red	0.1193-0.1196	3.031-3.038
Upper Crankcase Main Journal Bearing Thickness		
Green	0.1178-0.1181	2.992-2.999
Blue	0.1181-0.1183	2.999-3.006
Red	0.1183-0.1186	3.006-3.013
No. 3 Main Journal Bearing Thickness		
Green	0.1178-0.1181	2.992-2.999
Blue	0.1181-0.1183	2.999-3.006
Red	0.1183-0.1186	3.006-3.013
Runout - Max	0.001	0.03
Connecting Rods		
Small-End Inner Diameter	0.7073-0.7081	17.965-17.985
Big-End Inner Diameter	1.7726-1.7734	45.025-45.045
Crankpin (Connecting Rod Big-End) Oil Clearance	0.0010-0.0012	0.025-0.031
Crankshaft Bearing Mark/Color		
Yellow	0.0591-0.0594	1.502-1.508
Green	0.0594-0.0596	1.508-1.514
Blue	0.0596-0.0598	1.514-1.520
Red	0.0598-0.0601	1.520-1.526

POWERHEAD

ENGINE SPECIFICATIONS - YAMAHA 115J/115 (1741cc) 4-STROKE ENGINES

Component	U.S. (in.) ①	Metric (mm) ①
Oil Pump		
Discharge at 212 degrees F (100C) w/ 10W-30	1.56 Gpm	5.9Lpm
Pressure	16.78 psi	118 kPa
Relief Valve Opening Pressure	69.69 psi	490 kPa
Piston		
Standard Size	3.1074-3.1082	78.928-78.949
Measuring Point (above bottom of skirt)	0.51	13
Clearance (Build Spec / Limit)	0.0028-0.0031 / 0.0081	0.070-0.080 / 0.206
Piston Pin Outside Diameter	0.7085-0.7087	17.997-18.000
Oversize Piston Diameter	3.120	79.25
Piston Rings		
Width (measured from outer-to-inner edge of ring)		
Top	0.114-0.115	2.89-2.91
Middle	0.118-0.126	3.00-3.20
Bottom (oil control)	0.094	2.40
Thickness (measured vertically top-to-bottom of ring)		
Top	0.046-0.047	1.17-1.19
Middle	0.058-0.059	1.47-1.49
Bottom (oil control)	0.094-0.098	2.38-2.48
End Gap		
Top - Standard	0.006-0.012	0.15-0.30
Middle - Standard	0.028-0.035	0.70-0.90
Bottom (oil control)	0.008-0.028	0.20-0.70
Ring Side Clearance		
Top	0.001-0.003	0.02-0.08
Middle	0.001-0.003	0.03-0.07
Oil Control	0.001-0.006	0.03-0.15
Valve Lifters		
Cylinder Head Lifter Bore Inner Diameter	1.1020-1.1030	28.000-28.021
Valve Lifter Outside Diameter	1.1010-1.1016	27.965-27.980
Valve Lifter-to-Cylinder Head Bore Clearance	0.0008-0.0022	0.020-0.056
Valve Shims (in 0.0010 in. / 0.025mm increments)	0.0787-0.1299	2.000-3.300
Valves		
Valve Dimensions		
Head Diameter (IN / EX)	1.142-1.150 / 0.945-0.953	29.0-29.2 / 24.0-24.2
Face Angle (IN)	91, 120, 160 degrees	91, 120, 160 degrees
Face Angle (EX)	90, 140 degrees	90, 140 degrees
Face Width (IN / EX)	0.078-0.094 / 0.089-0.106	1.98-2.40 / 2.26-2.69
Seat Width (IN / EX)	0.062-0.076 / 0.071-0.080	1.58-1.94 / 1.80-2.02
Margin Thickness (IN / EX)	0.031-0.047 / 0.039-0.055	0.8-1.2 / 1.0-1.4
Stem Thickness (IN)	0.2352-0.2358	5.975-5.990
Stem Thickness (EX)	0.2346-0.2352	5.960-5.975
Valve Guide Inner Diameter (IN and EX)	0.2364-0.2369	6.005-6.018
Stem-to-guide Clearance (IN)	0.0006-0.0017	0.015-0.043
Stem-to-guide Clearance (EX)	0.0012-0.0023	0.030-0.058
Valve Stem Runout	0.001	0.03
Valve Springs		
Free Length (Standard / Limit)	2.094 / 2.057	53.20 / 52.25
Out-of-Square (Spring Warpage/Tilt)	0.10	2.6

Because manufacturers differ with what specifications they choose to provide, please refer to the Mercury/Mariner charts for additional specifications (or another mechanical point of view) which MAY apply as well

① Unless otherwise noted

ENGINE SPECIFICATIONS - MERCURY/MARINER 115J/115 (1741cc) 4-STROKE ENGINES

Component	U.S. (in.) ①	Metric (mm) ①
Camshaft		
Lobe Height (top of lobe to bottom of journal)		
Intake	1.465-1.472	37.22-37.38
Exhaust	1.453-1.459	36.90-37.06
Lobe Width (side-to-side / non-raised portion of lobe)		
Intake and Exhaust	1.178-1.184	29.92-30.08
Valve Lift (IN / EX)	0.287 / 0.275	7.30 / 6.98
Camshaft Journal Diameter	0.9827-0.9835	24.96-24.98
Camshaft Journal Oil Clearance	0.0008-0.0024	0.020-0.061
Cylinder Head Journal Inside Diameter	0.9843-0.9850	25.000-25.021
Runout - Max	0.004	0.1
Compression		
Minimum Compression	138 psi	950 kPa
Cylinder Bore		
Standard Bore Diameter	3.110-3.111	79.00-79.02
Oversize 0.010 in. (0.25mm)	3.120-3.121	79.25-79.27
Out-of-round Service Limit	0.003	0.08
Taper Service Limit	0.003	0.08
Cylinder Head		
Gasket Surface Warpage Limit	0.004	0.1
Crankshaft		
Crankshaft Journal Diameter (Build Spec)	1.8891-1.8898	47.984-48.000
Crankshaft Journal Diameter (Min)	1.8887	47.972
Crankpin Diameter (Build Spec)	1.6535-1.6528	42.000-41.982
Crankpin Diameter (Min)	1.7311	43.971
Crankcase Main Journal Inside Diameter	2.1269-2.1276	54.023-54.042
Crankcase Main Journal Oil Clearance	0.0009-0.0017	0.024-0.044
Lower Crankcase Main Journal Bearing Thickness		
Yellow	0.1185-0.1188	3.010-3.017
Green	0.1188-0.1191	3.017-3.024
Blue	0.1191-0.1193	3.024-3.031
Red	0.1193-0.1196	3.031-3.038
Upper Crankcase Main Journal Bearing Thickness		
Green	0.1178-0.1181	2.992-2.999
Blue	0.1181-0.1183	2.999-3.006
Red	0.1183-0.1186	3.006-3.013
No. 3 Main Journal Bearing Thickness		
Green	0.1178-0.1181	2.992-2.999
Blue	0.1181-0.1183	2.999-3.006
Red	0.1183-0.1186	3.006-3.013
Runout - Max	0.001	0.03
Connecting Rods		
Small-End Inner Diameter	0.7073-0.7081	17.965-17.985
Big-End Inner Diameter	1.7726-1.7734	45.025-45.045
Crankpin (Connecting Rod Big-End) Oil Clearance	0.0010-0.0012	0.025-0.031
Crankshaft Bearing Mark/Color		
Yellow	0.0591-0.0594	1.502-1.508
Green	0.0594-0.0596	1.508-1.514
Blue	0.0596-0.0598	1.514-1.520
Red	0.0598-0.0601	1.520-1.526

POWERHEAD

ENGINE SPECIFICATIONS - MERCURY/MARINER 115J/115 (1741cc) 4-STROKE ENGINES

Component	U.S. (in.) ①	Metric (mm) ①
Oil Pump		
Discharge at 212 degrees F (100C) w/ 10W-30	1.56 Gph @ 1000 rpm	5.9Lph @ 1000 rpm
Pressure Warm Engine @ Idle (750 rpm)	50.75 psi	350 kPa
Relief Valve Opening Pressure	71 psi	490 kPa
Piston		
Standard Size	3.10734-3.1082	78.928-78.949
Oversize 0.010 in. (0.25mm)	3.1174-3.1182	79.178-79.199
Measuring Point (above bottom of skirt)	0.51	13
Clearance (Build Spec)	0.0028-0.0031	0.070-0.080
Piston Pin Boss Bore Inner Diameter	0.7090-0.7093	18.008-18.015
Piston Pin Outside Diameter	0.7085-0.7087	17.997-18.000
Piston Rings		
Width (measured from outer-to-inner edge of ring)		
Top	0.114-0.115	2.89-2.91
Middle	0.118-0.126	3.00-3.20
Bottom (oil control)	0.094	2.40
Thickness (measured vertically top-to-bottom of ring)		
Top	0.046-0.047	1.17-1.19
Middle	0.058-0.059	1.47-1.49
Bottom (oil control)	0.094-0.098	2.38-2.48
End Gap		
Top - Standard	0.006-0.012	0.15-0.30
Middle - Standard	0.028-0.035	0.70-0.90
Bottom (oil control)	0.008-0.028	0.20-0.70
Ring Side Clearance		
Top and Middle	0.001-0.003	0.03-0.08
Oil Control	0.001-0.006	0.03-0.15
Valve Lifters		
Cylinder Head Lifter Bore Inner Diameter	1.1020-1.1030	28.000-28.021
Valve Lifter Outside Diameter	1.1010-1.1016	27.965-27.980
Valve Lifter-to-Cylinder Head Bore Clearance	0.0008-0.0022	0.020-0.056
Valve Shims (in 0.025mm increments)	0.0787-0.1299	2.000-3.300
Valves		
Valve Dimensions		
Head Diameter (IN / EX)	1.142-1.150 / 0.945-0.953	29.0-29.2 / 24.0-24.2
Face Width (IN / EX)	0.078-0.094 / 0.089-0.106	1.98-2.40 / 2.26-2.69
Seat Width (IN / EX)	0.062-0.076 / 0.071-0.080	1.58-1.94 / 1.80-2.02
Margin Thickness (IN / EX)	0.031-0.047 / 0.039-0.055	0.8-1.2 / 1.0-1.4
Stem Thickness (IN)	0.2352-0.2358	5.975-5.990
Stem Thickness (EX)	0.2346-0.2352	5.960-5.975
Valve Guide Inner Diameter (IN and EX)	0.2364-0.2369	6.005-6.018
Stem-to-guide Clearance (IN)	0.0006-0.0017	0.015-0.043
Stem-to-guide Clearance (EX)	0.0012-0.0023	0.030-0.058
Valve Stem Runout	0.001	0.03
Valve Springs		
Free Length (Standard / Limit)	2.094 / 2.057	53.20 / 52.25
Out-of-Square (Spring Warpage/Tilt)	0.10	2.6

Because manufacturers differ with what specifications they choose to provide, please refer to the Yamaha charts for additional specifications (or another mechanical point of view) which MAY apply as well

① Unless otherwise noted

ENGINE SPECIFICATIONS - YAMAHA 150 (2670cc) 4-STROKE ENGINES

Component	U.S. (in.) ①	Metric (mm) ①
Camshaft		
Lobe Height (top of lobe to bottom of journal)		
Intake	1.7835-1.7874	45.30-45.40
Exhaust	1.7461-1.7500	44.35-44.45
Lobe Width (side-to-side / non-raised portion of lobe)		
Intake and Exhaust	1.4154-1.4193	35.95-36.05
Camshaft Cap Inner Diameter	0.9843-0.9851	25.00-25.02
Camshaft Journal Diameter	0.9827-0.9835	24.96-24.98
Camshaft Journal Oil Clearance	0.0008-0.0024	0.020-0.060
Cylinder Head Journal Inside Diameter	0.9827-0.9835	24.96-24.98
Runout - Max	0.0012	0.03
Compression		
Minimum Compression	128 psi	880 kPa
Cylinder Bore		
Standard Bore Diameter	3.7008-3.7014	94.000-94.017
Out-of-round Service Limit	0.0020	0.05
Taper Service Limit	0.0031	0.08
Cylinder Head		
Gasket Surface Warpage Limit	0.004	0.1
Crankshaft		
Crankshaft Journal Diameter	2.0465-2.0472	51.980-52.000
Crankpin Diameter	1.9677-1.9685	49.980-50.000
Crankpin Width	0.8661-0.8701	22.00-22.10
Crankcase Main Journal Oil Clearance	0.0008-0.0020	0.021-0.050
Lower Crankcase Main Journal Bearing Thickness		
1 (Green)	0.0987-0.0988	2.506-2.509
2 (Red)	0.0989-0.0990	2.512-2.515
3 (Yellow)	0.0991-0.0993	2.518-2.521
4 (Blue+Green)	0.0994-0.0995	2.524-2.527
Upper Crankcase Main Journal Bearing Thickness		
1 (Green)	0.0987-0.0988	2.506-2.509
2 (Red)	0.0989-0.0990	2.512-2.515
3 (Yellow)	0.0991-0.0993	2.518-2.521
No. 3 Main Journal Bearing Thickness		
1 (Green)	0.0986-0.0988	2.504-2.509
2 (Red)	0.0988-0.0990	2.510-2.515
3 (Yellow)	0.0991-0.0993	2.516-2.521
Runout - Max	0.0012	0.03
Connecting Rods		
Big-End Inner Diameter	2.0876-2.0884	53.025-53.045
Crankpin (Connecting Rod Big-End) Oil Clearance	0.0011-0.0020	0.027-0.052
Crankshaft Bearing Mark/Color		
Green	0.0589-0.0591	1.496-1.502
Blue	0.0593-0.0595	1.505-1.511
Red	0.0596-0.0598	1.514-1.520

ENGINE SPECIFICATIONS - YAMAHA 150 (2670cc) 4-STROKE ENGINES

Component	U.S. (in.)①	Metric (mm)①
Oil Pump		
Discharge at 207-217 degrees F (97-103C) w/ 10W-30	2.113 Gpm @ 700 rpm	8.0Lpm @ 700 rpm
Pressure	19.1-23.5 psi	132-162 kPa
Pressure @ idle speed	65.3 psi	450 kPa
Relief Valve Opening Pressure	56.84-71.05 psi	392-490 kPa
Piston		
Standard Size	3.6979-3.6982	93.928-93.934
Measuring Point (above bottom of skirt)	0.20	5.0
Clearance	0.0030-0.0031	0.075-0.080
Piston Pin Bore Inner Diameter	0.8269-0.8274	21.004-21.015
Piston Rings		
Width (measured from outer-to-inner edge of ring)		
Top	0.110-0.118	2.80-3.00
Middle	0.146-0.154	3.70-3.90
Bottom (oil control)	0.091-0.106	2.30-2.70
Thickness (measured vertically top-to-bottom of ring)		
Top and Middle	0.0461-0.0469	1.17-1.19
Bottom (oil control)	0.0945-0.0972	2.40-2.47
End Gap		
Top - Standard	0.0059-0.0118	0.15-0.30
Middle - Standard	0.0118-0.0177	0.30-0.45
Bottom (oil control)	0.0059-0.0236	0.15-0.60
Ring Side Clearance		
Top	0.0016-0.0031	0.04-0.08
Middle	0.0012-0.0028	0.03-0.07
Oil Control	0.0016-0.0051	0.04-0.13
Valve Lifters		
Valve Lifter Outside Diameter	1.2985-1.2990	32.982-32.997
Valve Lifter-to-Cylinder Head Bore Clearance	0.0008-0.0022	0.020-0.055
Valve Shims (in 0.0008 in. / 0.020mm increments)	0.09-0.12	2.3-2.9
Valves		
Valve Dimensions		
Head Diameter (IN)	1.3720-1.3839	34.85-35.15
Head Diameter (IN)	1.1752-1.1870	29.85-30.15
Face Width (IN / EX)	0.0831 / 0.0957	2.11 / 2.43
Seat Width (IN / EX)	0.043-0.055 / 0.055-0.067	1.1-1.4 / 1.4-1.7
Margin Thickness (IN / EX)	0.028 / 0.039	0.7 / 1.0
Stem Thickness (IN)	0.2156-0.2162	5.477-5.492
Stem Thickness (EX)	0.2151-0.2157	5.464-5.479
Valve Guide Inner Diameter (IN and EX)	0.2167-0.2174	5.504-5.522
Stem-to-guide Clearance (IN and EX)	0.0010-0.0023	0.025-0.058
Valve Stem Runout	0.0004	0.01
Valve Springs		
Free Length (Standard / Limit)	1.740 / 1.677	44.20 / 42.60
Out-of-Square (Spring Warpage/Tilt)	0.06	1.5

① Unless otherwise noted

ENGINE SPECIFICATIONS - YAMAHA 200/225 (3352cc) 4-STROKE ENGINES

Component	U.S. (in.)①	Metric (mm)①
Camshaft		
Lobe Height (top of lobe to bottom of journal)		
Intake	1.7835-1.7874	45.30-45.40
Exhaust	1.7854-1.7894	45.35-45.45
Lobe Width (side-to-side / non-raised portion of lobe)		
Intake and Exhaust	1.4154-1.4193	35.95-36.05
Camshaft Journal Diameter	0.9827-0.9835	24.96-24.98
Camshaft Journal Oil Clearance	0.0008-0.0024	0.020-0.061
Cylinder Head Journal Inside Diameter	0.9843-0.9850	25.00-25.02
Runout - Max	0.004	0.1
Compression		
Minimum Compression	128 psi	880 kPa
Cylinder Bore		
Standard Bore Diameter	3.7008-3.7016	94.00-94.02
Out-of-round Service Limit	0.0004	0.01
Taper Service Limit	0.0020	0.05
Cylinder Head		
Gasket Surface Warpage Limit	0.004	0.1
Crankshaft		
Crankshaft Journal Diameter	2.4791-2.4800	62.968-62.992
Crankpin Diameter	1.9676-1.9685	49.976-50.000
Crankpin Width	0.8465-0.8484	21.50-21.55
Crankcase Main Journal Oil Clearance	0.0010-0.0020	0.025-0.050
Lower Crankcase Main Journal Bearing Thickness		
1	0.0981-0.0984	2.494-2.500
2	0.0983-0.0986	2.498-2.504
3	0.0985-0.0987	2.502-2.508
Upper Crankcase Main Journal Bearing Thickness		
1	0.0981-0.0984	2.494-2.500
2	0.0983-0.0986	2.498-2.504
3	0.0985-0.0987	2.502-2.508
No. 3 Main Journal Bearing Thickness		
1	0.0980-0.0984	2.492-2.500
2	0.0982-0.0986	2.496-2.504
3	0.0984-0.0987	2.500-2.508
Runout - Max	0.0012	0.03
Connecting Rods		
Small-End Inner Diameter	0.827	21.00
Big-End Inner Diameter	2.087	53.00
Crankpin (Connecting Rod Big-End) Oil Clearance	0.0014-0.0028	0.035-0.071
Crankshaft Bearing Mark/Color		
Yellow	0.0587-0.0588	1.492-1.496
Green	0.0588-0.0591	1.496-1.500
Blue	0.0591-0.0592	1.500-1.504

① Unless otherwise noted

ENGINE SPECIFICATIONS - YAMAHA 200/225 (3352cc) 4-STROKE ENGINES

Component	U.S. (in.) ①	Metric (mm) ①
Oil Pump		
Discharge at 212 degrees F (100C) w/ 10W-40	2.32 Gpm @ 700 rpm	8.8Lpm @ 700 rpm
Pressure	19.62 psi	138 kPa
Relief Valve Opening Pressure	75.22-92.00 psi	529-647 kPa
Piston		
Standard Size	3.6977-3.6985	93.921-93.941
Measuring Point (above bottom of skirt)	0.20	5.0
Clearance	0.0030-0.0031	0.075-0.080
Piston Pin Outside Diameter	0.827	21
Piston Pin Bore Inner Diameter	0.8276-0.8280	21.02-21.03
Piston Rings		
Width (measured from outer-to-inner edge of ring)		
Top	0.110-0.118	2.80-3.00
Middle	0.142-0.150	3.60-3.80
Bottom (oil control)	0.091-0.106	2.30-2.70
Thickness (measured vertically top-to-bottom of ring)		
Top and Middle	0.0461-0.0469	1.17-1.19
Bottom (oil control)	0.0945-0.0972	2.40-2.47
End Gap		
Top - Standard	0.0059-0.0118	0.15-0.30
Middle - Standard	0.0118-0.0177	0.30-0.45
Bottom (oil control)	0.0059-0.0236	0.15-0.60
Ring Side Clearance		
Top	0.0016-0.0031	0.04-0.08
Middle	0.0012-0.0028	0.03-0.07
Oil Control	0.0016-0.0051	0.04-0.13
Valve Lifters		
Valve Lifter Outside Diameter	1.2984-1.2992	32.98-33.00
Valve Lifter-to-Cylinder Head Bore Clearance	0.0008-0.0020	0.020-0.050
Valve Shims (in 0.0008 in. / 0.020mm increments)	0.0913-0.1165	2.320-2.960
Valves		
Valve Dimensions		
Head Diameter (IN)	1.3720-1.3839	34.85-35.15
Head Diameter (IN)	1.1752-1.1870	29.85-30.15
Face Width (IN / EX)	0.0831 / 0.0957	2.11 / 2.43
Seat Width (IN / EX)	0.043-0.055 / 0.055-0.067	1.1-1.4 / 1.4-1.7
Margin Thickness (IN / EX)	0.028 / 0.039	0.7 / 1.0
Stem Thickness (IN)	0.2156-0.2162	5.477-5.492
Stem Thickness (EX)	0.2151-0.2157	5.464-5.479
Valve Guide Inner Diameter (IN and EX)	0.2169-0.2173	5.51-5.52
Stem-to-guide Clearance (IN and EX)	0.0004-0.0008	0.01-0.02
Valve Stem Runout	0.0004	0.01
Valve Springs		
Free Length (Standard / Limit)	1.740 / 1.677	44.20 / 42.60
Out-of-Square (Spring Warpage/Tilt)	0.06	1.5

Because manufacturers differ with what specifications they choose to provide, please refer to the Mercury/Mariner charts for additional specifications (or another mechanical point of view) which MAY apply as well

① Unless otherwise noted

ENGINE SPECIFICATIONS - MERCURY/MARINER 225 (3352cc) 4-STROKE ENGINES

Component	U.S. (in.) ①	Metric (mm) ①
Camshaft		
Lobe Height (top of lobe to bottom of journal)		
Intake	1.7835-1.7874	45.30-45.40
Exhaust	1.7854-1.7894	45.35-45.45
Lobe Width (side-to-side / non-raised portion of lobe)		
Intake and Exhaust	1.4154-1.4193	35.95-36.05
Camshaft Journal Diameter	0.9827-0.9835	24.96-24.98
Camshaft Journal Oil Clearance	0.0008-0.0023	0.020-0.060
Cylinder Head Journal Inside Diameter	0.9843-0.9850	25.00-25.02
Runout - Max	0.004	0.1
Compression		
Minimum Compression	128 psi	880 kPa
Cylinder Bore		
Standard Bore Diameter	3.7008-3.7016	94.00-94.02
Out-of-round		
Service Limit	0.0004	0.01
Taper		
Service Limit	0.0020	0.05
Cylinder Head		
Gasket Surface Warpage Limit	0.004	0.1
Crankshaft		
Crankshaft Journal Diameter	2.4791-2.4800	62.968-62.992
Crankpin Diameter	1.9676-1.9685	49.976-50.000
Crankpin Width	0.8465-0.8484	21.50-21.55
Crankcase Main Journal Oil Clearance	0.0010-0.0020	0.025-0.050
Lower Crankcase Main Journal Bearing Thickness		
1	0.0981-0.0984	2.494-2.500
2	0.0983-0.0986	2.498-2.504
3	0.0985-0.0987	2.502-2.508
Upper Crankcase Main Journal Bearing Thickness		
1	0.0981-0.0984	2.494-2.500
2	0.0983-0.0986	2.498-2.504
3	0.0985-0.0987	2.502-2.508
No. 3 Main Journal Bearing Thickness		
1	0.0980-0.0984	2.492-2.500
2	0.0982-0.0986	2.496-2.504
3	0.0984-0.0987	2.500-2.508
Runout - Max	0.0012	0.03
Connecting Rods		
Small-End Inner Diameter	0.827	21.00
Big-End Inner Diameter	2.087	53.00
Crankpin (Connecting Rod Big-End) Oil Clearance	0.0014-0.0028	0.035-0.071
Crankshaft Bearing Mark/Color		
Yellow	0.0587-0.0588	1.492-1.496
Green	0.0588-0.0591	1.496-1.500
Blue	0.0591-0.0592	1.500-1.504
Oil Pump		
Discharge at 207-217 degrees F (97-103C) w/ 10W-40	2.32 Gpm @ 1000 rpm	8.8 Lpm @ 1000 rpm
Pressure	19.62 psi	138 kPa
Relief Valve Opening Pressure	76.87-94.27 psi	530-650 kPa

ENGINE SPECIFICATIONS - MERCURY/MARINER 225 (3352cc) 4-STROKE ENGINES

Component	U.S. (in.) ①	Metric (mm) ①
Piston		
Standard Size	3.6977-3.6985	93.921-93.941
Measuring Point (above bottom of skirt)	0.20	5.0
Clearance	0.0030-0.0031	0.075-0.080
Piston Pin Outside Diameter	0.827	21
Piston Pin Bore Inner Diameter	0.8276-0.8280	21.02-21.03
Piston Rings		
Width (measured from outer-to-inner edge of ring)		
Top	0.110-0.118	2.80-3.00
Middle	0.142-0.150	3.60-3.80
Bottom (oil control)	0.091-0.106	2.30-2.70
Thickness (measured vertically top-to-bottom of ring)		
Top and Middle	0.0461-0.0469	1.17-1.19
Bottom (oil control)	0.0945-0.0972	2.40-2.47
End Gap		
Top - Standard	0.0059-0.0118	0.15-0.30
Middle - Standard	0.0118-0.0177	0.30-0.45
Bottom (oil control)	0.0059-0.0236	0.15-0.60
Ring Side Clearance		
Top	0.0016-0.0031	0.04-0.08
Middle	0.0012-0.0028	0.03-0.07
Oil Control	0.0016-0.0051	0.04-0.13
Valve Lifters		
NOTE Merc gives 2 sets of seemingly conflicting specs when it comes to Valve Lifters as follows:		
Valve Lifter Outside Diameter	1.2984-1.2992	32.98-33.00
Valve Lifter Bore Outside Diameter	1.2984-1.2992	32.98-33.00
Valve Lifter-to-Cylinder Head Clearance	0.0008-0.0020	0.020-0.050
Valve Lifter-to-Lifter Bore Clearance	0.0004-0.0008	0.010-0.020
Valve Shims (in 0.020mm increments)	0.0913-0.1165	2.320-2.960
Valves		
Valve Dimensions		
Head Diameter (IN)	1.3720-1.3839	34.85-35.15
Head Diameter (IN)	1.1752-1.1870	29.85-30.15
Face Width (IN / EX)	0.0831 / 0.0957	2.11 / 2.43
Seat Width (IN / EX)	0.043-0.055 / 0.055-0.067	1.1-1.4 / 1.4-1.7
Margin Thickness (IN / EX)	0.028 / 0.039	0.7 / 1.0
Stem Thickness (IN)	0.2156-0.2162	5.477-5.492
Stem Thickness (EX)	0.2151-0.2157	5.464-5.479
Valve Guide Inner Diameter (IN and EX)	0.2169-0.2173	5.51-5.52
Stem-to-guide Clearance (IN and EX)	0.0004-0.0008	0.01-0.02
Valve Stem Runout	0.0004	0.01
Valve Springs		
Free Length (Standard / Limit)	1.740 / 1.677	44.20 / 42.60
Out-of-Square (Spring Warpage/Tilt)	0.06	1.5

Because manufacturers differ with what specifications they choose to provide, please refer to the Yamaha charts for additional specifications (or another mechanical point of view) which MAY apply as well

① Unless otherwise noted

GEARCASE SERVICE - MERCURY/MARINER	7-59
GEARCASE REPAIR AND OVERHAUL TIPS	7-59
GEARCASE - 4/5/6 HP (1-CYL)	7-60
ASSEMBLY	7-62
CLEANING & INSPECTION	7-62
OVERHAUL	7-61
REMOVAL & INSTALLATION	7-60
SHIMMING	7-64
GEARCASE - 8-15 HP (2-CYL)	7-64
ASSEMBLY	7-66
CLEANING & INSPECTION	7-66
OVERHAUL	7-65
REMOVAL & INSTALLATION	7-64
SHIMMING	7-70
GEARCASE - 25 HP (BIGFOOT), 30/40 HP (NON-BIGFOOT) AND 40/45/50 HP (NON-BIGFOOT, 935cc)	7-70
ASSEMBLY	7-73
CLEANING & INSPECTION	7-72
OVERHAUL	7-71
REMOVAL & INSTALLATION	7-70
SHIMMING	7-74
GEARCASE - 30-115 HP (BIGFOOT), 40/50/60 HP (996cc) AND 75-115 HP (NON-BIGFOOT)	7-74
ASSEMBLY	7-79
CLEANING & INSPECTION	7-78
OVERHAUL	7-76
REMOVAL & INSTALLATION	7-75
SHIMMING	7-84
GEARCASE - V6 MODELS	7-87
GEARCASE SERVICE - YAMAHA	**7-5**
GEARCASE REPAIR AND OVERHAUL TIPS	7-5
GEARCASE - 2.5 AND 4 HP (1-CYL)	7-5
ASSEMBLY	7-12
CLEANING & INSPECTING	7-8
DISASSEMBLY	7-6
REMOVAL & INSTALLATION	7-5
GEARCASE - 6-40 HP (2- AND 3-CYL)	7-13
ASSEMBLY	7-21
CLEANING & INSPECTING	7-18
DISASSEMBLY	7-15
GEARCASE SHIMMING	7-27
REMOVAL & INSTALLATION	7-13
GEARCASE - 40-225 HP (4- AND 6-CYL) AND V6 MERCURY/MARINER	7-35
ASSEMBLY	7-43
CLEANING & INSPECTING	7-40
DISASSEMBLY	7-36
GEARCASE SHIMMING	7-48
REMOVAL & INSTALLATION	7-35
JET DRIVE	**7-87**
JET DRIVE ADJUSTMENTS	7-96
GATE LINKAGE ADJUSTMENTS	7-96
SHIFT ARM/PIVOT LINKAGE (EARLY-MODELS)	7-97
CAM AND ROLLER LINKAGE (LATE-MODELS)	7-99
JET DRIVE ASSEMBLY	7-87
ASSEMBLY	7-93
CLEANING & INSPECTING	7-90
DESCRIPTION & OPERATION	7-87
MODEL IDENTIFICATION (YAMAHA ONLY)	7-88
REMOVAL & DISASSEMBLY	7-88
TRIM ADJUSTMENT	7-100
LOWER UNIT	**7-2**
EXHAUST GASES	7-4
GENERAL INFORMATION	7-2
SHIFTING PRINCIPLES	7-2
COUNTER-ROTATING UNIT	7-3
STANDARD ROTATING UNIT	7-2
TROUBLESHOOTING	7-4

7

LOWER UNIT

LOWER UNIT	7-2
GEARCASE SERVICE - YAMAHA	7-5
GEARCASE SERVICE - MERCURY/MARINER	7-59
JET DRIVE	7-87

LOWER UNIT

General Information

Generally speaking, the lower unit is considered to be the part of the outboard below the exhaust housing. The unit contains the propeller shaft, the driven and pinion gears, the driveshaft from the powerhead and the water pump. Torque is transferred from the powerhead's crankshaft to the gearcase by a driveshaft. A pinion gear at the bottom of the driveshaft meshes with a drive gear in the gearcase to change the vertical power flow into a horizontal flow through the propeller shaft. The power head driveshaft rotates clockwise continuously when the engine is running, but propeller rotation is controlled by the gear train shifting mechanism.

The lower units normally found on all but a few of the smallest motors (specifically some 2-strokes not covered here) are equipped with shifting capabilities. The forward and, if equipped, reverse gears together with the clutch, shift assembly and related linkage are all housed within the lower unit.

On outboards with a reverse gear, a sliding clutch engages the appropriate gear in the gearcase when the shift mechanism is placed in forward or reverse respectively. This creates a direct coupling that transfers the power flow from the pinion to the propeller shaft. A similar clutch mechanism is used on a handful of small hp Yamaha motors that contain both a forward and neutral shift position, but no reverse gear.

Generally lower units are available in a 1 or 2 major types, depending upon the size of the motor. Most of the mid-range units are available with standard or high-thrust (Yamaha)/Bigfoot (Mercury/Mariner) models. The high-thrust or Bigfoot units are simply beefier units with higher gear ratios, designed for more push and less speed. The larger motors are generally available in both conventional and counter-rotating models.

On counter-rotating models (designated with an "L" in front of the horsepower model designation on Yamahas), the propeller rotates in the opposite direction than on regular models and is used when dual engines are mounted in the boat to equalize the propeller's directional churning force on the water. This allows the operator to maintain a true course instead of being pulled off toward one side while underway.

Another type of lower unit available on some mid-range units is a jet drive propulsion system. Water is drawn in from the forward edge of the lower unit and forced out under pressure to propel the boat forward.

In all cases, the lower unit is easily removed without removing the entire outboard from the boat.

Shifting Principles

STANDARD ROTATING UNIT

Non-Reverse Type
◆ See Figure 1

There are generally 2 types of non-reversing type lower units normally used on the smallest of outboards. Usually in the 2-4 hp range. The first is a direct drive unit - the pinion gear on the lower end of the driveshaft is in constant mesh with the forward gear (none of these are found on Yamaha or Mercury/Mariner 4-stroke motors). The second (the type used on 2.5 hp Yamaha units) is a Forward-Neutral shifting unit, in which shift assembly allows the drive gear on the propshaft to move in and out of contact with the pinion gear on the lower end of the driveshaft.

In both cases, reverse action of the propeller is accomplished by the operator swinging the engine with the tiller handle 180° and holding it in this position while the boat moves sternward. When the operator is ready to move forward again, they simply swing the tiller handle back to the normal forward position.

Reverse Type
◆ See Figures 2, 3 and 4

The most common lower unit design (whether it be standard, high-thrust/Bigfoot, or counter-rotating) is equipped with a clutch dog permitting operation in neutral, forward and reverse. Although some of the smaller units use a step shifter mechanism, most mid-range and larger units use a rotating shifter cam. On these this mid-range and larger units unit when the outboard is shifted from neutral to reverse, the shift rod is rotated. This action moves

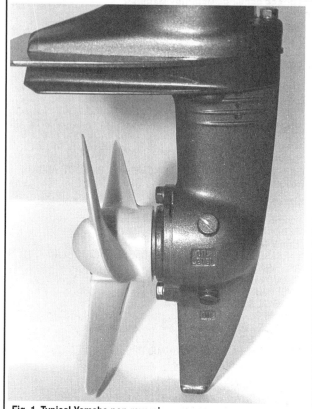

Fig. 1 Typical Yamaha non-reversing gearcase

the plunger and clutch dog toward the reverse gear. When the unit is shifted from reverse to neutral and then to forward gear, the rotation of the shift rod is reversed and the clutch dog is moved toward the forward gear.

The shift mechanism may vary slightly on some models but typically consists of a cam on the shift rod, a shifter (plunger), a shift slide and a compression spring. Historically on Yamahas the shift slide is a hollow tube, closed at one end. The closed end of this tube is necked and the head or knob of the tube fits into the shifter. The shifter has a vertical hole all the way through it. The shift rod passes through this hole in the shifter. The cam in the shift rod engages the shifter at the side notch, so any rotating movement in the shift rod will translate to a back and forth movement in the shifter and the shift slide. The Mercury design is not all that different and operates in the same general fashion as the Yamaha units.

In the neutral position, the cam is centered in the shifter.

When the lower unit is shifted into forward gear, the shift cam rotates (usually in a counterclockwise direction) and pulls on the shifter and shift slide. Because the shifter and shift slide are connected, they are moved out of the propeller shaft - moving the clutch dog toward the forward gear. The shift into forward gear is complete.

When the lower unit is shifted into reverse gear, the unit first moves into the neutral position. The shift cam usually rotates in a clockwise direction. The cam pushes the shifter and the shift slide in toward the propeller shaft. The clutch dog moves toward the reverse gear and the shift movement to reverse gear is complete.

The pinion gear on the lower end of the driveshaft is in constant mesh with both the forward and reverse gear. These three gears constantly rotate anytime the powerhead is operating.

A sliding clutch dog is mounted on the propeller shaft. A shifting motion at the control box will translate into a back and forth motion at the clutch dog via a series of shift mechanism components. When the clutch dog is moved forward, it engages with the forward gear. Because the clutch is secured to the propeller shaft with a pin, the shaft rotates at the same speed as the clutch. The propeller is thereby rotated to move the boat forward.

When the clutch dog is moved aft, the clutch engages only the reverse gear. The propeller shaft and the propeller are thus moved in the opposite direction to move the boat sternward.

LOWER UNIT

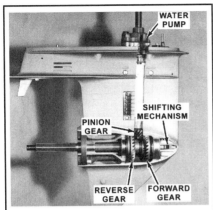

Fig. 2 Cutaway view of a lower unit with major parts, including the shift mechanism and water pump, identified. Note how the forward, reverse and pinion gears are all bevel cut

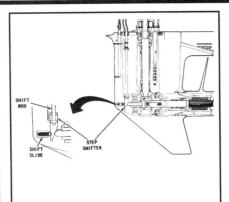

Fig. 3 Cross section of a typical lower unit showing a step shifter mechanism

Fig. 4 Cut away view of a standard rotation lower unit with major parts identified

When the clutch dog is in the neutral position, neither the forward nor reverse gear is engaged with the clutch and the propeller shaft does not rotate.

From this explanation, an understanding of wear characteristics can be appreciated. The pinion gear and the clutch dog receive the most wear, followed by the forward gear, with the reverse gear receiving the least wear. All three gears, the forward, reverse and pinion, are spiral bevel type gears.

A mixture of ball bearings, tapered roller bearings and caged or loose needle bearings is used in each unit. The type bearing used is normally indicated in the procedures.

COUNTER-ROTATING UNIT

◆ See Figures 5 thru 9

■ Counter-rotational shifting is usually accomplished without modification to the shift cable at the shift box. In fact, the normal setup is essential for correct shifting. The only real special equipment the counter-rotating unit requires is the installation of a left-hand propeller.

The same shifting mechanisms used on standard units may also be found on their counter-rotating counterparts. There are a few minor differences between most standard and counter-rotating units. The biggest differences are in the shifting mechanism (in the gearcase itself). The counter-rotating shift mechanism is a mirror image of the standard shift mechanism. Mirror image shifting mechanisms produce counter-rotation of the propeller shaft.

Part of this difference is that normally the forward and reverse gears are reversed in position when compared to standard units (well, there's no actual difference in the gears, but the gears that perform the functions are reversed): what would be the forward gear on a standard unit becomes the reverse gear on a counter-rotating unit and what would normally be the reverse gear on a standard unit, becomes the forward gear on a counter-rotating unit. The pinion gear remains the same and driveshaft rotation remains the same as on a standard lower unit.

On a standard Yamaha lower units, the cam on the shift shaft is normally located on the starboard side of the shifter. Therefore, when the rod is rotated counterclockwise, the clutch dog is pulled forward and the forward gear is engaged.

On a typical counter-rotating Yamaha lower unit, the cam on the shift rod is normally located on the port side of the shifter. Therefore, when the rod is rotated counterclockwise, the clutch dog is pushed back and the gear in the aft end of the housing (which normally is the reverse gear) is engaged. In this manner, the rotation of the propeller shaft is reversed. The same logic applies to the selection of reverse gear.

Fig. 5 Cutaway view of a typical counter-rotating lower unit (Yamaha shown) with the clutch dog in the neutral position

Fig. 6 Cutaway view of typical counter-rotating lower unit with the clutch dog in the forward position

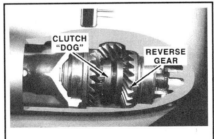

Fig. 7 Cutaway view of typical counter-rotating lower unit with the clutch dog in the reverse position

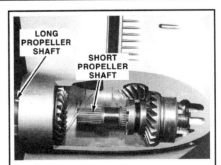

Fig. 8 On older Yamaha motors the counter-rotating lower unit used a two piece propeller shaft. The longer section attached to the propeller, the shorter section was internal

7-4 LOWER UNIT

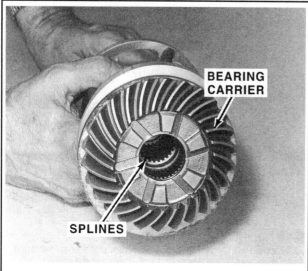

Fig. 9 Bearing carrier of an older Yamaha counter-rotating lower unit with the internal splines for the short section of the propeller shaft clearly visible

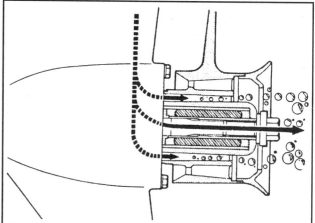

Fig. 10 Cross section of a typical lower unit showing route of the exhaust gases with the unit in forward (top) and reverse (bottom) gears

Exhaust Gases

◆ See Figure 10

At low powerhead speed, the exhaust gases from the powerhead escape from an idle hole in the intermediate housing on many motors. As powerhead rpm increases to normal cruising rpm or high-speed rpm, these gases are forced down through the intermediate housing and lower unit, then out with cycled water through the propeller.

Water for cooling is pumped to the powerhead by the water pump and is expelled with the exhaust gases. The water pump impeller is installed on the driveshaft. Therefore, the water output of the pump is directly proportional to powerhead rpm.

Troubleshooting

◆ See Figures 11 and 12

Troubleshooting (other than disassembly and inspection) must be done before the unit is removed from the powerhead to permit isolating the problem to one area. Always attempt to proceed with troubleshooting in an orderly manner. The shot-in the-dark approach will only result in wasted time, incorrect diagnosis, frustration and unnecessary replacement of parts.

The following steps are presented in a logical sequence with the most prevalent, easiest and less costly items to be checked listed first.

1. Check the propeller and the rubber hub. See if the hub is shredded or loosen and slipping. If the propeller has been subjected to many strikes against underwater objects, it could slip on its hub. If the hub appears to be damaged, replace it with a new hub. Replacement of the hub must be done by a propeller rebuilding shop equipped with the proper tools and experience for such work.

2. Verify the ignition switch is **OFF**, to prevent possible personal injury, should the engine start. Shift the unit into reverse gear and at the same time have an assistant turn the propeller shaft to ensure the clutch is fully engaged.

3. If the shift handle is hard to move, the trouble may be in the lower unit shift rod or in the cable itself, requiring an adjustment or a repair, or in the shift box.

4. Disconnect the remote control cable at the engine and then remove the remote control shift cable. Operate the shift lever. If shifting is still hard, the problem is in the shift cable or control box.

5. If the shifting feels normal with the remote control cable disconnected, the problem must be in the lower unit. To verify the problem is in the lower unit, have an assistant turn the propeller and at the same time move the shift cable back and forth. Determine if the clutch engages properly.

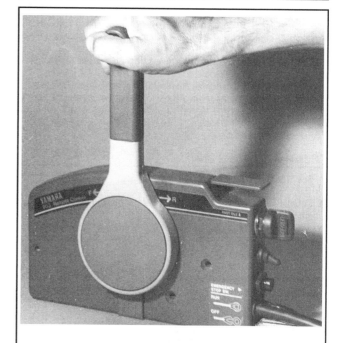

Fig. 11 Test the shift mechanism to see that the shifter fully engages. . .

Fig. 12 . . . if it does not check cable adjustment, or try shifting the mechanism by hand

LOWER UNIT 7-5

GEARCASE SERVICE - YAMAHA

Gearcase Repair and Overhaul Tips

All Yamaha gearcases use the same basic inner design in that pinion gear is splined to the bottom of a vertical driveshaft and secured by a retaining nut. The driveshaft uses the pinion gear to transfer the rotational power from the crankshaft to the propeller shaft. Both the driveshaft and propeller shaft ride in replaceable bushing (on the smaller gearcases) or bearings and a bath of gear oil. Gear oil is kept inside the case itself (and water is kept out) by propeller shaft and driveshaft housing seals at either end (bottom rear and top center) of the gearcase.

The size and type of the shafts, bearings, gears and seals will vary slightly from model-to-model. Also, all gearcases contain some form of Neutral or Neutral and Reverse shifting capabilities. Generally speaking one of 2 major designs are employed, a vertical push-pull shifter retaining by nuts on threaded rods (or a bolted coupler assembly) is used by most smaller gearcases, while a horizontally ratcheting shifter which connected through splines to the shift lever in the intermediate housing is used on most larger outboards.

We've divided coverage from removal and installation to complete overhaul procedures into 3 major categories. First a procedure for 2.5 and 4 hp (1 cylinder) motors covers the smallest Yamaha gearcases (most of which do NOT contain a Reverse gear). Second, a procedure for 6-40 hp (2- and 3-cylinder) models covers most small and intermediate sized gearcases that do contain Reverse gears AND that also utilize push-pull shifter assemblies. Lastly we've provided a procedure which applies to all 4-cylinder and larger (mid-range and high-output) motors. This last category includes the larger and more complicated gearcases that contain Reverse gears and splined, ratcheting shifter assemblies.

We've included diagrams and photographs from as many gearcases on which we could realistically get our hands, however keep in mind that subtle differences will occur from model-to-model or year-to-year. To ensure a successful overhaul a few general notes should be followed throughout the overhaul procedure, as follows:

- As with all repair or overhaul procedures, make sure the work area is clean and free of excessive dirt or moisture.
- Take your time when removing and disassembling components taking a moment with each to either matchmark its positioning or note its orientation in relation to other components. This goes ESPECIALLY for the placement of any shims which might be used. Absent specific instructions for checking and re-shimming a gearcase (which occurs when bearings, gears and/or shafts are replaced on some gearcases) all shims must be returned to their original positions during assembly.
- Seal placement and orientation will vary on some models. Before removing any seal be sure to first note the direction in which the seal lips are facing and be sure to install the replacement seal (and you ALWAYS replace seals when they are removed) facing the same direction.
- Unless otherwise directed seal lips are packed with a light coating of Yamaha Marine Grease or an equivalent water resistant grease. When used, O-rings are also coated with marine grease during installation.
- Needle bearings which are pressed in place are generally destroyed by the removal process and should be replaced when separated from a gear, shaft or bore. Ball Bearings can usually be freed without damage, but use common sense. Once the bearing is free, clean it thoroughly, then rotate the inner/outer cage feeling for rough spots. Replace any bearing that is not in top shape.
- The correct size driver must be used for bearing or seal installation. Generally speaking a driver must be smooth (especially for seals, to prevent it from cutting the sealing surface) and should be sized to spread as much of the force over the seal or bearing surface as possible. When a bearing is installed ON a shaft, the driver MUST contact the inner race (never push on only the outer race in this case or the bearing will be damaged). When a bearing is installed IN a bore, the driver MUST contact the outer race (in this circumstance, do NOT push only on the inner race or the bearing will be damaged).
- Driveshaft bearings or bushings are often pressed straight into position from above and/or pulled into position from below. Although Yamaha often has a special tool available to help draw a bushing or bearing into position in the lower portion of the gearcase, all you REALLY need to accomplish this is a sufficiently long threaded bolt with a nut and a washer that is the size of the driver you'd want to contact that bushing or bearing. You'll also need a nut and a plate to place on top of the gearcase. Place the threaded rod through one nut and plate and then down into the gearcase through the bushing/bearing (so the plate is resting against the top of the gearcase, the upper nut is resting against the plate and the threads of the rod are just protruding down into the pinion mounting area of the gearcase). Next install the washer and lower nut over the bottom threads of the rod to support the bearings or busing. Slowly draw the bearing or bushing into position in the gearcase by holding the threaded rod from turning while at the same time turning the upper nut to draw the entire bolt, bushing/bearing, washer and nut assembly upward until the bushing or bearing seats. Then loosen the nut on the bottom of the threaded rod in the gearcase and remove the threaded rod, plate, nut and washer.

In all cases the water pump, driveshaft oil seal or propeller shaft oil seal may be replaced without complete gearcase disassembly.

Although Yamaha calls for various special tools for gearcase overhaul MOST of them come down to the appropriate sized driver (see the bullet about drivers earlier in this section), OR a universal puller or a slide hammer generally with inside jaws. Whenever possible, we prefer the use of a puller over a slide hammer, just because the force is applied more linearly, however there are simply times when the impact of a slide hammer makes the job easier, so use common sense when determining which tool you'd prefer to use for bearing, race or potentially even shaft removal.

Lastly, we've tried to provide the best procedures possible for gearcase overhaul, and in the most logical order. However, in most cases procedures for some of the earlier procedures can be performed alone, without additional procedures. For instance water pump or propeller shaft bearing carrier/seal housing procedures can normally be performed completely independently of each other. However, since the driveshaft oil seal housing is found UNDER the water pump, you'd obviously have to remove the pump first for access.

Gearcase - 2.5 and 4 Hp (1-Cylinder) Models

REMOVAL & INSTALLATION

# MODERATE

◆ See Figures 13 thru 18

Gearcase removal or installation is a relatively straightforward procedure on most models. Generally speaking it involves disconnecting the shift linkage and unbolting the gearcase from the intermediate housing, then carefully lowering the gearcase (along with the shift rod, driveshaft and water tube) straight down and off the motor. Because the water pump housing is mounted to the top of the gearcase, between the case and intermediate housing, removal is necessary for impeller inspection and/or replacement.

Because the gearcase is a sealed unit when it comes to gear oil, the fluid only needs to be drained if the case itself is being opened for further service (some form of seal or internal parts replacement). However, if this is one of the reasons the case is being removed, it is usually easier to go ahead and drain the gearcase oil before the assembly is removed.

On these models the gearcase is bolted to the intermediate housing with a handful (usually 2-3) bolts, mostly threaded upward from the underside of the anti-cavitation plate. The shift linkage is accessed by removing a cover or covers from the side of the intermediate housing.

1. If the gearcase is to be overhauled or resealed, drain the gearcase fluid as detailed in this Maintenance and Tune-Up section. After the lubricant has drained, temporarily reinstall both the drain and oil level screws.

■ **As the lubricant drains, catch some with your fingers from time to time and rub it between your thumb and finger to determine if any metal particles are present. If any significant amount of metal is detected in the lubricant, the unit must be completely disassembled, inspected and the damaged parts replaced. Check the color of the lubricant as it drains. A whitish or creamy color indicates the presence of water in the lubricant. Check the drain pan for signs of water separation from the lubricant. The presence of any water in the gear lubricant is bad news. The unit must be completely disassembled, inspected and the cause of the problem determined and corrected.**

2. Remove the propeller for better access to the gearcase retaining bolt(s) threaded upward from underneath the anti-cavitation plate. For details, please refer to the Maintenance and Tune-Up section.

3. Remove the round side cover(s) from the intermediate housing for access to the shifter bolt. Loosen and remove the shifter linkage bolt.

7-6 LOWER UNIT

4. Loosen and remove the bolts securing the gearcase to the intermediate housing. Most models use about 2-3 bolts, all threaded upward from under the anti-cavitation plate. You may have to remove the bolt securing the anode as well since (even though usually not), because it sometimes threads all the way through to the intermediate housing. Of course, there is no harm in removing and inspecting the anode at this time anyway.

5. Separate the lower unit from the intermediate housing by pulling carefully STRAIGHT downward. When the two units are separated watch for and save any dowel pins (there are usually 2) used to help secure the gearcase. Normally the water tube will come out of the grommet and remain with the intermediate housing, while the driveshaft will remain with the lower unit.

To Install:

6. Apply very a light coating of Yamaha Marine Grease or an equivalent water resistant lubricant to the splines on the driveshaft.

7. Begin to bring the intermediate housing and the lower gear housing together. As the two units come closer, rotate the propeller shaft slightly to index the upper end of the lower driveshaft tube with the upper rectangular driveshaft. At the same time, feed the water tube into the water tube seal (and, if equipped, align the shifter shaft with the upper half of the shaft located in the intermediate housing).

8. Push the lower gear housing and the intermediate housing together. The dowel pin(s) will index with a matching hole as the two pieces are compressed together.

9. Apply Loctite® 572 or an equivalent threadlocking compound to the threads of the bolts used to secure the lower unit to the intermediate housing. A threadlocking compound can also be used on the anode retaining bolt. Install and tighten the bolts securely to 5.8 ft. lbs. (8 Nm).

10. Install and tighten the shifter linkage bolt, then install the side cover(s) to the intermediate housing.

11. Apply Yamaha All Purpose Grease, or equivalent anti-seize compound to the propeller shaft.

■ The compound will prevent the propeller from freezing to the shaft and permit the propeller to be removed, without difficulty, the next time removal is required.

12. Install the propeller, as detailed in the Maintenance and Tune-Up section.

13. If drained for service, properly refill the gearcase with lubricant, as detailed in the Maintenance and Tune-Up section.

DISASSEMBLY

■ For gearcase exploded views, please refer to Cleaning & Inspection in this section.

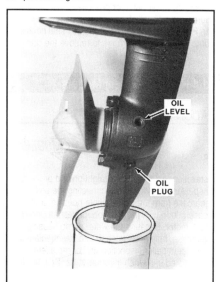

Fig. 13 If the gearcase is being repaired, start by draining the gearcase oil

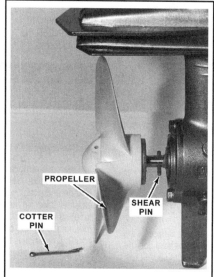

Fig. 14 Remove the propeller for access

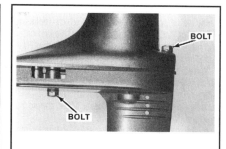

Fig. 15 Remove the gearcase bolts (2 hp motors shown)

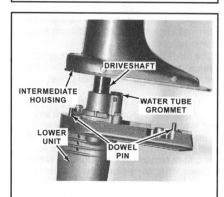

Fig. 16 Carefully lower the gearcase straight downward off the housing

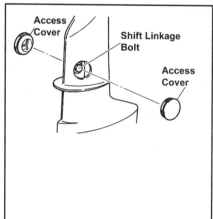

Fig. 17 Typical shift linkage access - 2.5-4 Hp Models

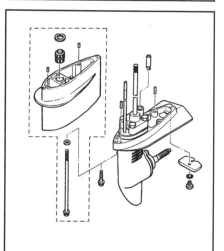

Fig. 18 Exploded view of a typical gearcase mounting - 2.5-4 Hp Models

LOWER UNIT 7-7

1. Remove the Propeller from the gearcase, as detailed under Maintenance and Tune-Up.
2. Drain the gearcase oil and remove the gearcase from the outboard as detailed earlier in this section under Gearcase (Lower Unit).
3. If necessary to access the driveshaft or upper oil seal assembly, remove the Water Pump, as detailed in the Lubrication and Cooling system section.
4. Remove the two bolts securing the bearing/seal carrier cap.
5. Either rotate the cap 90° so the flanges protrude past the side of the gearcase and then gently tap the bearing carrier cap using a rubber mallet to break it free from the lower unit OR, some models contain a small slit at the lower portion of the cap-to-gearcase sealing surface where you can use a small pry tool to carefully pry the cap free.

■ **The propeller shaft MAY come free with the bearing/seal cap. If it does, no big deal, just carefully separate it from the cap for attention later. On 4 hp motors, keep track of the reverse gear and 2 thrust washers if they are separated from the shaft.**

6. Remove the cap from the lower unit (on some models the entire propeller shaft assembly will come out with the cap, this is fine). Remove and discard the bearing/seal carrier O-ring. For the record, the O-ring CAN be reused on some models, however it is a cheap part and it is not worth the risk of a water leak ruining the gearcase.

■ **Perform the next step only if the seal has been damaged and is no longer fit for service. Removal of the seal destroys its sealing qualities and it cannot be installed a second time. Therefore, be absolutely sure a new seal is available before removing the old seal in the next step.**

7. If necessary, disassemble the bearing/seal carrier. On most models the seal(s) can be removed using either a small pry tool (or hooked seal removal tool) or by using a slide hammer with expanding jaw attachment. If equipped with a bushing, it is usually removed with a driver, however bearings must be removed either using a puller or the same slide-hammer with expanding jaw attachment which can be used to remove the seals. Components and access varies as follows:
- For 2.5 hp motors, dual seals are installed between the propeller and a propeller shaft bushing. Note the seal lips normally both face toward the propeller. The seals themselves may be pried free from the propeller side of the cap. The seals must be removed in order to service the bushing. Turn the cap over so the propeller side is facing downward (it is easiest if the cap is supported on some blocks by the cap mounting ears). Insert a driver from the gearcase side of the cap and carefully drive the bushing out through the opening in the propeller side of the cap.
- For 4 hp motors, the dual seals and the bearing assembly are normally removed or installed from the same (gearcase) side of the cap. Therefore it is necessary to replace the bearing if the seals must be serviced. The preferred method for bearing removal is an internal jawed puller and a standard puller bridge installed against the gearcase side of the cap or a slide hammer with internal expanding jaws. The same assembly can be used to remove the dual seals assembly (after noting that the both seal lips are normally positioned facing outward, toward the propeller).

■ **As noted earlier, on 4 hp motors be sure to keep track of the Reverse gear and the 2 thrust washers when the propeller shaft is disassembled.**

8. If the propeller shaft did not come out with the bearing/seal cap, remove it from the gearcase at this time. Carefully disassemble the propeller shaft and shifter assembly. On most models this involves removing the shift plunger from the end of the shaft, then using a small flat-bladed tool (such as a small screwdriver) inserted from the side of the shaft to carefully compress the spring while you push the dog clutch out of the side of the shaft's opening.

■ **Take careful note as to the dog clutch positioning on some models, such as the 4 hp, there is an orientation mark on the clutch which must be installed in the same position. In the case of the 4 hp motor a mark on one side of the clutch must be installed so that it faces upward when the propeller shaft is installed.**

9. At the bottom end of the driveshaft (inside the gearcase) carefully pry the circlip free of the groove in the shaft using a small pry tool. This clip holds the pinion gear onto the driveshaft. The clip may not come free on the first try, but have patience and it will come free. With one hand, remove driveshaft up and out of the lower unit housing and at the same time, with the other hand, catch the pinion gear, washer and any shim material from behind the gear. The shim material is critical to obtaining the correct backlash during installation. Using the old shim material will save considerable time over the alternative of starting with no shim material or by guessing on the shim amount.

■ **Remember removal of an oil seal destroys its sealing qualities and it cannot be installed a second time. Therefore, remove the oil seal(s) only if it is unfit for further service and be absolutely sure a new seal is available before removing the old seal.**

10. Inspect the condition of the upper oil seals for the gearcase. If replacement is required, carefully pry them free or use a slide hammer and internal jaw adapter to remove the seal. Remove them oil seal or seals from the top of the gearcase housing. As usual, note the seal orientation. The 2.5 hp motors use dual lipped seals so orientation is not as big a deal, but they are not completely symmetrical and have an outer casing lip that faces downward toward the gearcase. The 4 hp motors utilize a dual seal assembly, both seals are normally positioned with the lips facing upward toward the top of the gearcase.
11. Remove the Forward gear from the gearcase.
12. On most of these gearcases the driveshaft rides in 2 or 3 bushings which may be removed for replacement purposes only (since the action of removing them will make them unfit for further service). Lower bushings can normally be driven out of the gearcase using a long rod/driver inserted from the top at a slight angle. Tap a lower bushing down and into the gearcase where it can be fished out and removed. Upper bushings may be pulled upward and out using a small internal jawed puller and slide-hammer or using a threaded rod assembled with a suitably sized washer, plate and a couple of nuts (as described earlier under Overhaul Tips, it's essentially the same setup that can be used to install lower bushings). The 2.5 hp actually uses 3 bushings; a lower, upper and middle bushing, however most other models only use a lower and upper.
13. If necessary for replacement remove the Forward Propeller Shaft Bearing assembly from the nose of the gearcase.

Forward Propeller Shaft Bearing

■ **Removal of the forward propeller shaft bearing in the next step will almost always destroy the bearing. Therefore, remove the bearing only if it is no longer fit for service and be absolutely sure a replacement bearing is available prior to removal.**

Rotate the forward propeller shaft bearing. If any roughness is felt, the bearing must be removed and a new one installed.

1. Removal of the bearing is accomplished using a slide hammer with jaw expander attachment (#YB-6096), however if one if not available there is an alternative that we've found usually works.

■ **The following 3 steps are to be performed only if the forward propeller shaft bearing must be removed and a suitable bearing puller is not available. The procedure will change the temperature between the bearing retainer and the housing substantially, hopefully about 80°F (50°C). This change will contract one metal - the bearing retainer and expand the other metal - the housing, giving perhaps as much as 0.003 inch (0.08mm) clearance to allow the bearing to fall free. Read the complete steps before commencing the work because three things are necessary: a freezer, refrigerator, or ice chest; some ice cubes or crushed ice; and a container large enough to immerse about 1-1/2 in. (3.8cm) of the forward part of the lower gear housing in boiling water.**

2. After all parts, including all seals, grommets, etc., have been removed from the lower gear housing, place the gear housing in a freezer, preferably overnight. If a freezer is not available try an electric refrigerator or ice chest. The next morning, obtain a container of suitable size to hold about 1-1/2 in. (3.8cm) of the forward part of the lower gear housing. Fill the container with water and bring to a rapid boil. While the water is coming to a boil, place a folded towel on a flat surface for padding.
3. After the water is boiling, remove the lower gear housing from its cold storage area. Fill the propeller shaft cavity with ice cubes or crushed ice. Hold the lower gear housing by the trim tab end and the lower end of the housing. Now, immerse the lower unit in the boiling water for about 20 or 30 seconds.

7-8 LOWER UNIT

4. Quickly remove the housing from the boiling water; dump the ice; and with the open end of the housing facing downward, slam the housing onto the padded surface. Presto, the bearing should fall out.

■ If the bearing fails to come free, try the complete procedure a second time. Failure on the second attempt will require the use of a bearing puller.

CLEANING & INSPECTING

◆ See Figures 19 thru 25

✳✳ WARNING

If an old impeller is installed be sure the impeller is installed in the same manner from which it was removed - the blades will rotate in the same direction. Never turn the impeller over thinking it will extend its life. On the contrary, the blades would crack and break after just a short time of operation.

Clean all water pump parts with solvent and then dry them with compressed air. Inspect the water pump cover and base for cracks and distortion. If possible, always install a new water pump impeller while the lower unit is disassembled. A new impeller will ensure extended satisfactory service and give peace of mind to the owner. If the old impeller must be returned to service, never install it in reverse to the original direction of rotation. Installation in reverse will cause premature impeller failure.

Inspect the ends of the impeller blades for cracks, tears and wear. Check for a glazed or melted appearance, caused from operating without sufficient water. If any question exists, as previously stated, install a new impeller if at all possible.

Inspect the bearing surface of the propeller shaft. Check the shaft surface for pitting, scoring, grooving, imbedded particles, uneven wear and discoloration.

Check the straightness of the propeller shaft with a set of V-blocks. Rotate the propeller on the blocks

Good shop practice dictates installation of new O-rings and oil seals regardless of their appearance.

Clean the pinion gear and the propeller shaft with solvent. Dry the cleaned parts with compressed air.

Check the pinion gear and the drive gear for damage or abnormal wear.

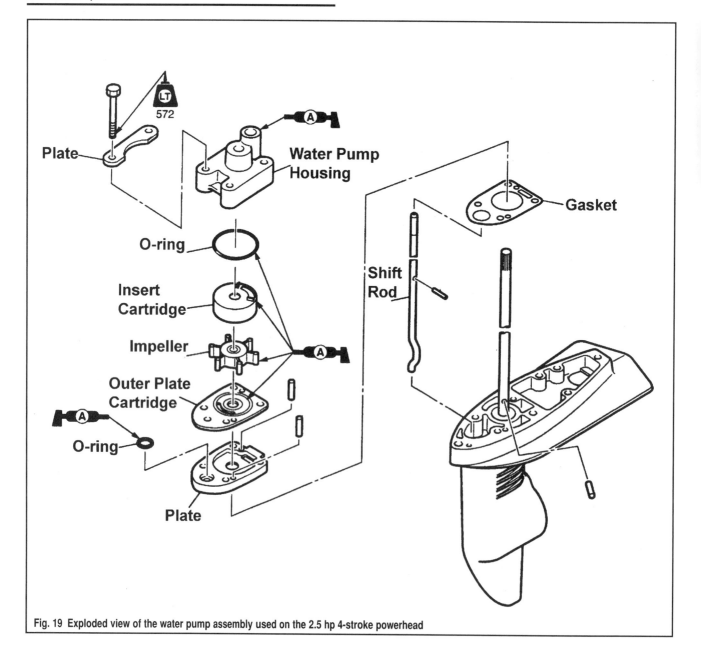

Fig. 19 Exploded view of the water pump assembly used on the 2.5 hp 4-stroke powerhead

LOWER UNIT

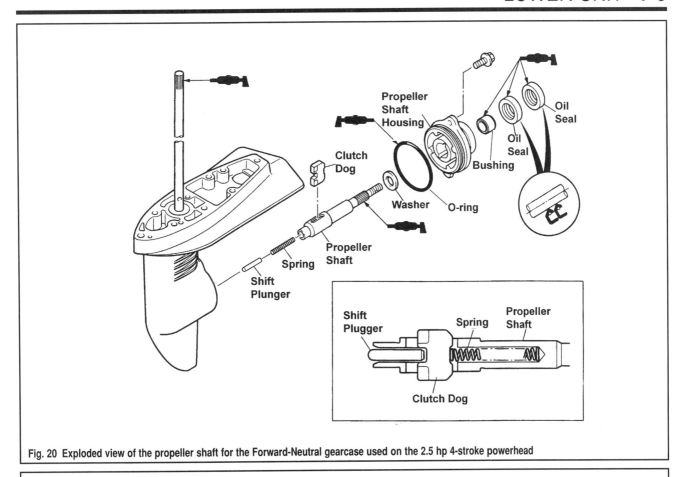

Fig. 20 Exploded view of the propeller shaft for the Forward-Neutral gearcase used on the 2.5 hp 4-stroke powerhead

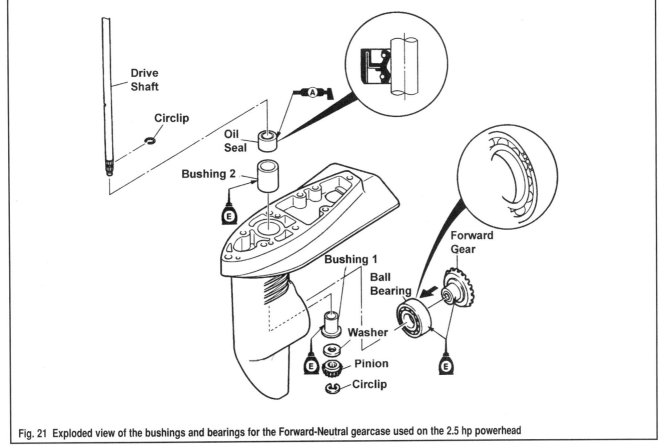

Fig. 21 Exploded view of the bushings and bearings for the Forward-Neutral gearcase used on the 2.5 hp powerhead

7-10 LOWER UNIT

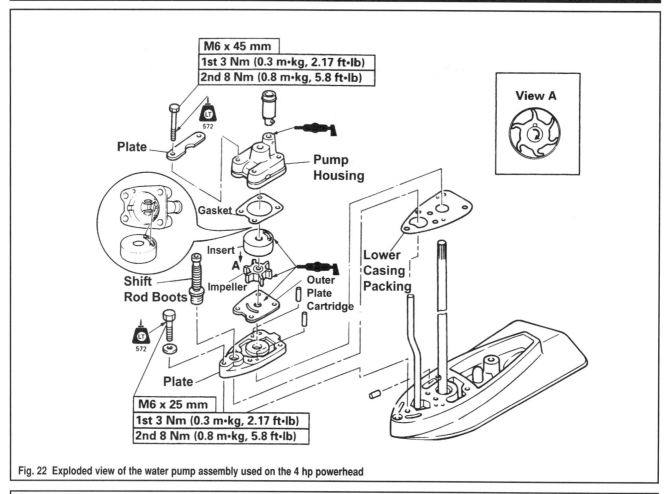

Fig. 22 Exploded view of the water pump assembly used on the 4 hp powerhead

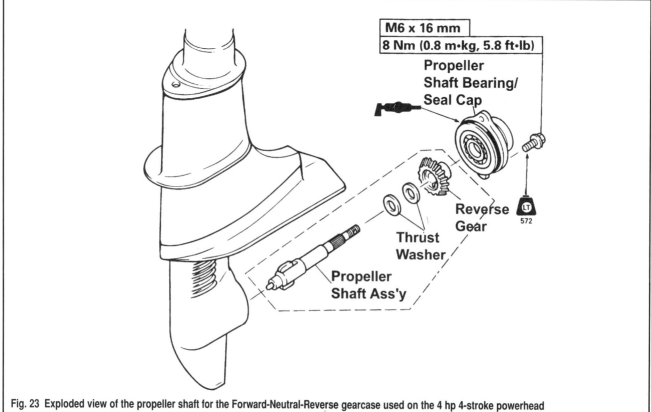

Fig. 23 Exploded view of the propeller shaft for the Forward-Neutral-Reverse gearcase used on the 4 hp 4-stroke powerhead

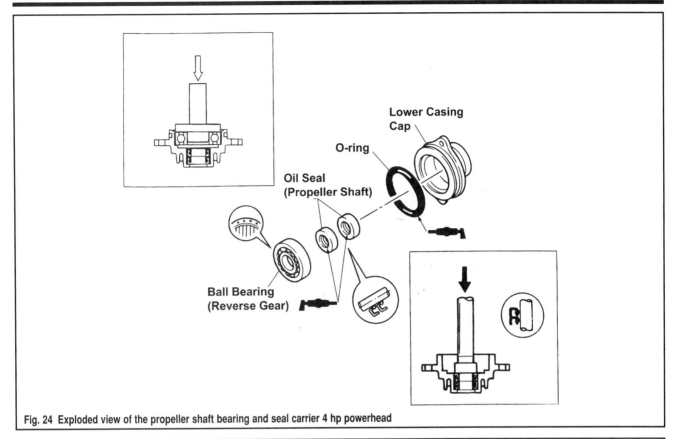

Fig. 24 Exploded view of the propeller shaft bearing and seal carrier 4 hp powerhead

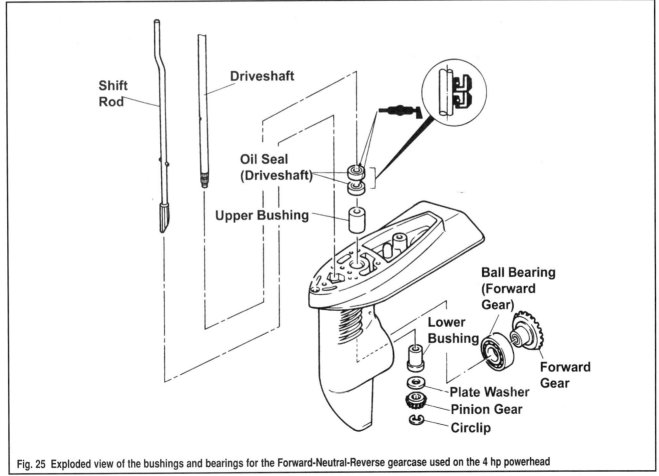

Fig. 25 Exploded view of the bushings and bearings for the Forward-Neutral-Reverse gearcase used on the 4 hp powerhead

7-12 LOWER UNIT

ASSEMBLY

◆ See Figures 19 thru 25

The following procedures are intended to guide you step-by-step through reinstallation and setup of the major gearcase components. If one or more of the components was not removed you can skip that section and proceed to the next following component.

Forward Propeller Shaft Bearing

This first section applies only if the forward propeller shaft bearing was removed from the housing.

1. Place the propeller shaft forward bearing squarely into the housing with the side embossed with the bearing size facing outward (back toward the gear and propeller shaft). Drive the bearing into the housing until it is fully seated using a suitable driver that contacts the appropriate portion of the bearing cage (the outer cage on this bearing should be in contact with the gearcase, so the driver should apply force to that portion of the bearing cage).

2. If the bearing is an unusually tight fit you may use the boiling water trick to help ease installation as follows:

   a. Place a new forward propeller shaft bearing in a freezer, refrigerator, or ice chest, preferably overnight. The next morning, boil water in a container of sufficient size to allow about 1-1/2 in. (3mm) of the forward part of the lower gear housing to be immersed. While the water is coming to a boil, place a folded towel on a flat surface for padding. Immerse the forward part of the gear housing in the boiling water for about a minute.

   b. Quickly remove the lower gear housing from the boiling water and place it on the padded surface with the open end facing upward. At the same time, have an assistant bring the bearing from the cold storage area. Continue working rapidly. Carefully place the bearing squarely into the housing as far as possible, again with the embossed side facing outward. Push the bearing into place. Obtain a blunt punch or piece of tubing to bear on the complete circumference of the retainer. Carefully tap the bearing retainer all the way into its forward position - until it bottoms-out. Tap evenly around the outer perimeter of the retainer, shifting from one side to the other to ensure the bearing is going squarely into place. The bearing must be properly installed to receive the forward end of the propeller shaft.

Driveshaft Bushings

■ **The following steps are to be performed if any of the driveshaft bushings were removed.**

Lower bushings are installed into a gearcase by drawing the bushing upward into position from underneath. This is normally accomplished using a long threaded bushing installer (#YB-06029). However, it is possible to fabricate this tool using a long threaded bolt, a suitably sized washer, plate and a couple of nuts (as described earlier under Overhaul Tips).

1. Apply a light coating of gearcase or engine oil to the outside of the bushing to help ease installation.

2. To install a lower bushing:

   a. Place a nut and a plate toward the top of the long threaded bushing installer, then position the installer into the gearcase from the top until the threads protrude into the lower cavity of the lower unit.

   b. Now, if not already down, lower the plate so it sits on the top of the gearcase and thread the nut down until it is snug against the plate.

   c. Slide the first bushing over the threads at the lower end of the driveshaft and then install the bushing retainer (or suitably sized washer and nut) onto the threads.

   d. Generally these tools work by holding the rod from turning with one wrench, while you turn the nut against the plate with another wrench. This will slowly draw the entire threaded rod along with the bushing retainer (or nut and washer and the bottom) along with the bushing upward into the gearcase. Continue turning the nut until the bushing is drawn up into the lower unit and seats properly.

3. To install an upper bushing, start the bushing into the opening on the top of the gearcase by hand, then use a suitable driver to gently push or tap the bushing squarely into position. Make sure the bushing either fully seats against the stop in the gearcase. On 2.5 hp motors, the upper driveshaft bushing should be installed to a depth of 0.03-0.04 in. (0.9-1.0mm) below the bottom of the mounting deck for the upper oil seal.

Upper Oil Seal

Once removed, oil seals MUST be replaced, like many bearings, the very action of removing the seal will make it unfit for further use.

1. Obtain a suitable driver or seal installer. If one is not available a socket or smooth piece of pipe of the right size will suffice. The correct size driver, socket etc is one that will place the force on the outside of the seal (the portion of the seal that contacts the gearcase).

2. Apply a light coating of marine grade grease to the outer diameter of the oil seal casing.

■ **When two seals are used in this application it is usually easier and safer to install one seal at a time. Seat the first one with the driver, then install and seat the second seal against the first one.**

3. Place the oil seal over the end of the installer with the lips facing as noted during removal. The 2.5 hp motors use dual lipped seals so orientation is not as big a deal, but they are not completely symmetrical and have an outer casing lip that faces downward toward the gearcase. The 4 hp motors utilize a dual seal assembly; both seals are normally positioned with the lips facing upward toward the top of the gearcase.

4. Lower the seal installer and the seal squarely into the seal recess. Tap the end of the handle with a hammer until the seal is fully seated or the seal is installed to the appropriate depth below the top of the gearcase (if applicable), as follows:

   • For 2.5 hp motors, make sure the seal is installed to a depth of 0.06-0.08 in. (1.5-2.0mm) below the surface of the gearcase deck.

   • For 4 hp motors, the first seal should be mounted to a depth of 0.3 in. (8mm) below the gearcase deck, then the second seal should be installed on top of the first and carefully driven to a depth of 0.04 in. (1.0mm) below the surface of the gearcase deck.

5. After installation, pack the seal with Yamaha all-purpose marine grease or equivalent water resistant lubricant.

Propeller Shaft and Forward Gear

■ **The forward gear is installed separately from the propeller shaft, which inserts through the gear and bearing and engages using a clutch dog/shifter assembly. The propeller shaft is installed later, AFTER the driveshaft and pinion gear.**

1. If the propeller shaft was disassembled, take a moment to prepare it for installation. Install and compress the spring, then install the clutch dog. On some models there is a marking on the clutch which must be positioned so it faces upward when the propeller shaft is installed. Once the clutch dog is in position, install the shift plunger in the end of the shaft.

2. Install the forward gear into the lower unit housing. Push the propeller shaft and/or gear into the forward propeller shaft bearing as far as possible.

Driveshaft (and Pinion Gear)

1. Lower the driveshaft into the lower unit until the splines on the lower end of the driveshaft protrude into the lower unit cavity.

2. Slide the thrust washer and any shim material saved during disassembly onto the lower end of the driveshaft. Slide the pinion gear up onto the end of the driveshaft. The splines of the pinion gear will index with the splines of the driveshaft and the gear teeth will mesh with the teeth of the forward gear. Rotating the pinion gear slightly will permit the splines to index and the gears to mesh. Now, comes the hard part. Snap the circlip into the groove on the end of the driveshaft to secure the pinion gear in place. If the first attempt is not successful, try again. Take a break, have a cup of coffee, tea, (a beer, but JUST ONE) whatever, then give it another go. With patience, the task can be accomplished.

Propeller Shaft Bearing and/or Seal Carrier

At this point the gearcase is all but ready for final assembly, except that you still need to actually install the propeller shaft assembly and you need to prepare the bearing/seal carrier for installation.

■ **Replacement oil seals should be coated and packed with a suitable marine grade grease. Replacement bearings or bushings should be coated lightly with clean engine or gear oil.**

LOWER UNIT

1. Start by installing any oil seal or bearing that was removed from the bearing/seal cap. Use a suitable driver to install a replacement bearing, bushing and/or seal(s), depending upon the model as follows:
   - For 2.5 hp motors, dual seals are installed between the propeller and a propeller shaft bushing. Install the bushing from the gearcase side of the cap (even though it was pressed out through the other side). Make sure the slit in the bushing is facing about 90 degrees from the small semi-circular cutouts that are found on 2 sides of the bore (180 degrees apart from each other). Once the bushing is in position, invert the housing and install each of the 2 oil seals. The seal lips normally both face toward the propeller. Once installed the outer of the 2 seals must be at a depth of 0.04-0.06 in. (1.0-1.5mm) below the edge of the bushing/seal cap.
   - For 4 hp motors, the dual seals and the bearing assembly are installed from the same (gearcase) side of the cap. Install the two oil seals (each individually) with their lips facing outward, toward the propeller until they are both seated. Next, oil and install the bearing, with the embossed markings on the bearing facing upward toward the driver (which would be toward the Reverse gear and gearcase when the cap is installed).

2. Apply a light coating of marine grade grease to a NEW O-ring, then install the O-ring to the groove in the bearing/bushing and seal cap.

3. On models with shifting capability you can either pre-position the propeller shaft in the gearcase THEN install the cap OR, we usually find it is easier to insert the shaft through the cap, then install the shaft and cap as an assembly. Remember that these models will normally utilize one or more thrust washers on the shaft AND, in the case of the 4 hp motor, will also utilize a Reverse gear that needs to be installed between the shaft and the cap. Assemble the cap and shaft.

4. Install the bearing carrier cap (and propeller shaft on shifting models) to the lower unit, then carefully seat the cap making sure it is flush with the gearcase evenly all the way around.

5. Apply Loctite, or equivalent, to the threads of the bearing carrier attaching bolts. Install and tighten the bolts securely alternating back and forth until they are tight. Yamaha does not publish a torque spec for these bolts on any of the motors in this procedure except the 4 hp, on which the bolts should be tightened alternately and evenly to 5.8 ft. lbs. (8 Nm).

■ **No backlash checking or adjustment procedures are provided by Yamaha for the motors covered here.**

6. Before installing the gearcase, check the action of the shifter to make sure it engages and disengages smoothly without hitches. Slowly rotate the driveshaft CLOCKWISE when viewed from above.

7. If possible, pressure test the gearcase. Yamaha does not provides specs for these motors (however they do for a similar gearcase used on 3 hp 2-strokes, so the theory behind the test should work for all models). With the oil fill/drain screw installed and tightened, attach a threaded pressure tester to the oil level hole, then use a hand pump to apply 14.22 psi (100 kPa) of pressure to the gearcase. Watch that the pressure remains steady for at least 10 seconds. If pressure falls before that point one or more of the seals require further attention. You can immerse the gearcase in water and look for bubbles to determine where the leak is occurring. Be sure of gearcase integrity before installation.

8. If not done already, install the Water Pump to the top of the gearcase as detailed in the Lubrication and Cooling section.

9. Install the Gearcase to the intermediate housing, as detailed in this section.

10. Install the Propeller and properly refill the Gearcase Lubricant, both as detailed in the Maintenance and Tune-Up section.

Gearcase - 6-40 Hp (2- and 3-Cylinder) Models

REMOVAL & INSTALLATION

◆ See Figures 26 thru 33

Gearcase removal or installation is a relatively straightforward procedure on most models. Generally speaking on these models it involves disconnecting the shift linkage and unbolting the gearcase from the intermediate housing, then carefully lowering the gearcase (along with the shift rod, driveshaft and water tube) straight down and off the motor. Because the water pump housing is mounted to the top of the gearcase, between the case and intermediate housing, removal is necessary for impeller inspection and/or replacement.

Because the gearcase is a sealed unit when it comes to gear oil, the fluid only needs to be drained if the case itself is being opened for further service (some form of seal or internal parts replacement). However, if this is one of the reasons the case is being removed, it is usually easier to go ahead and drain the gearcase oil before the assembly is removed.

On these models the gearcase is bolted to the intermediate housing with 4-5 bolts threaded upward from the underside of the anti-cavitation plate. Generally, if equipped with a trim tab, the tab must be removed in order to access one of the bolts. The shift linkage on these models is normally connected by a set of nuts (one connecting nut and one locknut) which attach to the threaded ends of the upper and lower shift rods. Shift linkage is found at the front of the gearcase and is normally accessible somewhere on the front of the outboard, above the anti-cavitation plate (though occasionally under an intermediate housing front cover or under the engine cowling).

1. If the gearcase is to be overhauled or resealed, drain the gearcase fluid as detailed in this Maintenance and Tune-Up section. After the lubricant has drained, temporarily reinstall both the drain and oil level screws.

■ **As the lubricant drains, catch some with your fingers from time to time and rub it between your thumb and finger to determine if any metal particles are present. If any significant amount of metal is detected in the lubricant, the unit must be completely disassembled, inspected and the damaged parts replaced. Check the color of the lubricant as it drains. A whitish or creamy color indicates the presence of water in the lubricant. Check the drain pan for signs of water separation from the lubricant. The presence of any water in the gear lubricant is bad news. The unit must be completely disassembled, inspected and the cause of the problem determined and corrected.**

2. Remove the propeller for better access to the gearcase retaining bolt(s) threaded upward from underneath the anti-cavitation plate. For details, please refer to the Maintenance and Tune-Up section.

3. For 6/8 hp motors remove the bolt securing the anode to the underside of the anti-cavitation plate, then remove the anode. Also on these motors, remove the bolt and nut securing the water inlet covers to the side of the gearcase housing, then remove the covers.

4. Locate and disconnect the shifter linkage at the shift rod union (at the front of the intermediate housing, just above the gearcase, possibly under a small cover on some models). Most models utilize threaded shift rods with a locknut and an adjusting nut (it's easy to tell the difference, the locknut is the small one). On these models mark the position of the adjuster nut on the threads, loosen the locknut, and then unthread the adjuster nut so the 2 rods are disconnected.

■ **On some of the larger models so equipped the trim tab obscures access to one of the gearcase retaining bolts.**

5. For most 25 hp and larger motors, matchmark the position of the trim tab to the gearcase, as an aid to installing it back in its original location, then loosen the retaining bolt and remove the trim tab from the underside of the anti-cavitation plate.

6. Some models, may be equipped with a speedometer pilot water hose at the top front of the gearcase. If equipped, carefully cut the wire tie and gently pull the two ends of the tube apart, at their connector.

7. Remove the 4 bolts threaded upward (2 on each side) from the gearcase to the intermediate housing. On models with a trim tab that was removed for access, remove the 5th bolt from the center of the gearcase, under the trim tab mounting area.

8. Carefully separate the lower unit from the intermediate housing by pulling STRAIGHT downward from the intermediate housing. Watch for and save the dowel pins (usually 2) when the two units are separated. The water tube will normally come out of the grommet and remain with the intermediate housing. The driveshaft will remain with the lower unit. The lower portion of the shift linkage should remain with the lower housing.

To Install:

9. Apply a dab of Yamaha Marine Grease, or equivalent water resistant lubricant to the driveshaft splines (the sides of the driveshaft's upper end, NOT the top surface itself).

■ **An excessive amount of lubricant on top of the driveshaft will be trapped in the clearance between the top of the driveshaft and bottom of the crankshaft. This trapped lubricant may not allow the driveshaft to fully engage with the crankshaft.**

7-14 LOWER UNIT

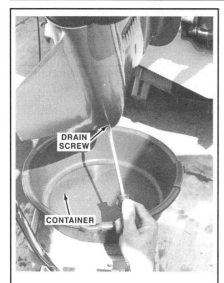
Fig. 26 If necessary, drain the gearcase before removal

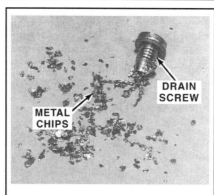
Fig. 27 Metal shards in the oil indicates a need for overhaul

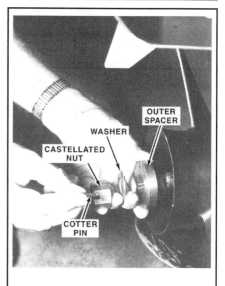
Fig. 28 Remove the propeller for access...

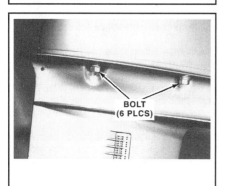
Fig. 29 ...then remove the gearcase retaining bolts

Fig. 30 Larger models have a bolt under the trim tab

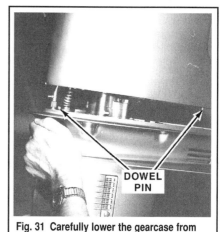
Fig. 31 Carefully lower the gearcase from the intermediate housing

10. Apply some of the same lubricant to the end of the water tube in the intermediate housing, and to the dowel pin(s) on the mating surface of the lower unit.

11. Set the gearcase in gear so that you can rotate the propeller shaft slightly to assist with crankshaft-to-driveshaft spline alignment. Yamaha recommends that both the shifter and the gearcase be placed in **Reverse** for ease of installation.

12. Begin to bring the intermediate housing and lower gear housing together.

■ The next step takes time and patience. If it's your first time, success will probably not be achieved on the first attempt. Two items must mate at the same time before the lower unit can be seated against the intermediate housing.

• The top of the driveshaft on the lower unit indexes with the lower end of the crankshaft.
• The water tube in the intermediate housing slides into the grommet on the water pump housing.

13. As the two units come closer, rotate the propeller shaft slightly to index the splines on the upper end of the lower driveshaft with the crankshaft splines. Be sure to turn the propeller shaft only in the correct direction of rotation for the gear in which the gearcase is set. (This means that the propeller shaft is turned clockwise if the shifter is in forward or counterclockwise if the shifter is in reverse. This is done to make sure the water pump impeller is not damaged). At the same time, feed the water tube into the water tube grommet and make sure the upper and lower shift rods are positioned so they can be reconnected.

14. Push the lower unit housing and the intermediate housing together. The dowel pin(s) should index as the surfaces mate.

■ If all items appear to mate properly, but the lower unit seems locked in position about 4 in. (10cm) away from the intermediate housing and it is not possible, with ease, to bring the two housings closer, the driveshaft has missed the cylindrical lower oil seal housing leading to the crankshaft. Move the lower unit out of the way. Shine a flashlight up into the intermediate housing and find the oil seal housing. Now, on the next attempt, find the edges of the oil seal housing with the driveshaft before trying to mate anything else - the water tube and the top of the shift rod. If the driveshaft can be made to enter the oil seal housing, the driveshaft can then be easily indexed with the crankshaft.

15. Install a finger-tighten 2 of the gearcase retainers just to hold the assembly in place while you turn your attention to the shifter assembly.

16. Reconnect the shift linkage and secure using the nut, adjuster nut and locknut or connector and retaining bolt as applicable. On some models there are specifications for adjuster nut installation as follows:

• On 6/8 hp motors, first thread the locknut down onto the shift cam in the gearcase as far as it will go. Push down on the shift cam (making sure the dog clutch is fully meshed with reverse gear) and double check the shifter is set in the Reverse position. Screw the adjuster nut down onto the gearcase shift cam until it covers about 0.31-0.35 in (8-9mm) of the gearcase shift cam. In this position it should align with the mark made during disassembly. Tighten the locknut to hold it in position.

• On 8/9.9 hp motors, the locknut is threaded onto the shift cam in the gearcase, run it down the threads to make room for adjuster nut installed,

LOWER UNIT 7-15

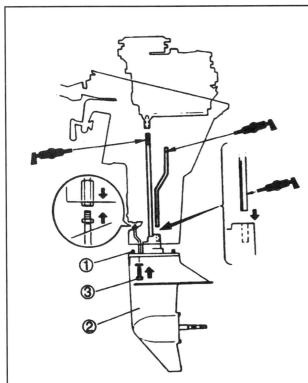

Fig. 32 Typical gearcase mounting, showing most common shift linkage

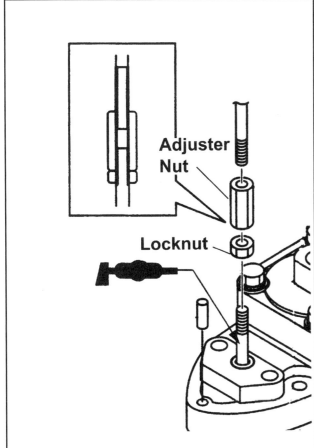

Fig. 33 Threaded shift linkage like this may have the lock nut on the top or the bottom, depending upon the model

then with both the shifter and gearcase in Reverse, thread the adjuster nut onto the gearcase shift cam AT LEAST 5 turns (checking the mark made during removal for alignment), then tighten the locknut.

- On 15 hp or larger motors, no specification is given for the distance the adjuster nut is threaded onto each rod. Make sure both the gearcase and shifter are in Reverse, then install the adjuster nut until you reach the alignment mark made during removal. If there is no mark, try to make sure there is about an equal amount of thread from each rod in the adjuster nut. Tighten the locknut to hold the adjuster in position once you are happy with adjustment.

17. Apply a light coating of Loctite® 572 or an equivalent threadlocking compound to the threads of the gearcase retaining bolts. Remember, one bolt passes through the area covered by the trim tab. Install and finger-tighten the bolts as you apply the compound, then withdraw the 2 bolts you'd previously installed to hold the gearcase in place so that you can apply the compound to those threads as well. Once all retainers are installed either tighten them securely. A torque specification is not available for most 4-15 hp motors, however the retaining bolts on all larger models should be tightened to 29 ft. lbs. (39 Nm).

18. Apply a light coating of Loctite® 572 or an equivalent threadlocking compound to the threads of the trim tab retaining bolt, then install the trim tab, aligning the matchmarks made earlier and tighten the bolt securely.

19. On models equipped with a speedometer pickup tube, slide the two ends of the pilot tube together at the push-on fitting. Secure the tube using a new wire tie.

20. For 6/8 hp motors, if removed, install the water inlet covers to the side of the gearcase and the anode to the underside of the anti-cavitation plate.

21. Operate the shift lever through all gears. The shifting should be smooth and the propeller should rotate in the proper direction when the flywheel is rotated by hand in a clockwise direction. Naturally the propeller should not rotate when the unit is in neutral.

22. Apply Yamaha All Purpose Grease, or equivalent anti-seize compound to the propeller shaft.

■ **The compound will prevent the propeller from freezing to the shaft and permit the propeller to be removed, without difficulty, the next time removal is required.**

23. Install the propeller, as detailed in the Maintenance and Tune-Up section.

24. If drained for service, properly refill the gearcase with lubricant, as detailed in the Maintenance and Tune-Up section.

DISASSEMBLY

■ **For gearcase exploded views, please refer to Cleaning & Inspection in this section.**

1. Remove the Propeller from the gearcase, as detailed under Maintenance and Tune-Up.
2. Drain the gearcase oil and remove the gearcase from the outboard as detailed earlier in this section under Gearcase (Lower Unit).
3. If necessary to access the driveshaft or upper oil seal assembly, remove the Water Pump assembly from the top of the gearcase, as detailed in the Lubrication and Cooling section.

■ **On all motors the impeller and water pump housing mounts to a water pump plate on top of the gearcase and driveshaft oil seal housing.**

4. Remove the water pump plate from the top of the gearcase. For most models the water pump housing bolts were holding the plate in position, but for some gearcases (such as the ones used on 6/8 hp 4-strokes) the plate is secured with bolts of its own. If so loosen and remove the bolts, then carefully lift the plate from the top of the gearcase.
5. If the water pump plate was sealed to the oil seal housing using a gasket, remove and discard the gasket at this time. Generally speaking only the gearcases for the following motors used gaskets under the water pump plate: the 8/9.9 hp, T25 hp and the 30/40 hp motors.

7-16 LOWER UNIT

■ The driveshaft oil seal housing can be removed from MOST (but not all) models without removing the driveshaft itself from the gearcase. This is because on most motors the upper portion of the driveshaft is a smooth shaft of constant outer diameter, so the oil seal housing can be slid upward and off the top of the driveshaft. Unfortunately the driveshafts used on some gearcases (such as the ones typically found on 8/9.9 hp and 15 hp 4-strokes) are of a larger outer-diameter than the housing, requiring that the housing be removed from the bottom of the shaft. Therefore, on some models it will be necessary to remove the propeller shaft and pinion gear, but on most, if there is no problem with the propeller shaft bearing or seal, you CAN skip the Oil Seal Housing procedure.

Propeller Shaft Bearing/Seal Carrier Cap

The rear of the propeller shaft rides in a bearing/seal carrier cap which mounts in the lower rear of the gearcase. As the name suggest, the purpose of the carrier is to house oil seals as well as one or more propeller shaft bearings.

Seal or bearing replacement is a relatively straightforward task that can occur with the gearcase still installed on the outboard (if it is first drained of oil or tilted fully upward). Seal/bearing carrier removal is required for most gearcase services (other than water pump and driveshaft oil seal on most models).

The bearing/seal carrier is secured to the gearcase by one of two methods. On most of the models covered in this procedure the carrier is bolted to the rear of the gearcase using 2 flange bolts.

1. Loosen and remove the 2 flange bolts which are securing the bearing/seal carrier. Either rotate the cap 90° so the flanges protrude past the side of the gearcase and then gently tap the bearing carrier with a rubber mallet to break it free from the lower unit OR, some models contain a small slit at the lower portion of the cap-to-gearcase sealing surface where you can use a small pry-tool to carefully pry the cap free.

■ The propeller shaft MAY come free with the bearing/seal cap. If it does, no big deal, just carefully separate it from the cap for attention later.

2. Remove the bearing seal/carrier from the lower unit (on some models the entire propeller shaft assembly will come out with the cap, this is fine). Remove and discard the bearing/seal carrier O-ring(s). For the record, the O-ring(s) CAN be reused on some models, however it is a cheap part and it is not worth the risk of a water leak ruining the gearcase.

■ On models whose bearing carrier uses the ring nut for retention, take care not to lose the small key fitted into the side of the bearing carrier.

3. On stubborn carriers, you can use a universal type puller, such as Yamaha's (#YB-06234) and a set of J-bolts which hook to the carrier in order to free the assembly from the gearcase. Hook the ends of the J-bolts into the bearing carrier ribs across from each other. Insert the threaded ends through the puller and then install the washers and nuts to take up the slack. Rotate the center threaded shaft clockwise, pressing against the propeller shaft to separate the bearing carrier from the lower unit. Remove the tools and withdraw the bearing carrier along with the reverse gear assembly.

■ If bearing service is not required, it may not be necessary to remove the reverse gear assembly at all, as long as the oil seals are accessible from the opposite end of the housing. It depends on the extent of the rebuild or reseal you are performing.

4. If possible, remove the reverse gear from the bearing carrier. Watch for and save any shim material from the back side of the reverse gear. The shim material is critical in obtaining the correct backlash during assembling. Using the old shim material will save considerable time, especially starting with no shim material. If the gear won't come free, proceed with the next step.

5. On most models the reverse ball bearing assembly is pressed into the bearing carrier and on many the ball bearing is pressed onto the back of the reverse gear. If this is the case, the gear and bearing assembly must be separated from the carrier (or from each other) either using a slide hammer with jaw expander attachment or using a bridged puller and a bearing separator. Use a slide hammer with jaw expander attachment and pull the bearing or the bearing and reverse gear from the bearing carrier. When using the tool check to be sure the jaws are hooked onto the inner race. If the bearing and gear come out as an assembly you'll need to use a bearing separator and a shop press to separate the two.

6. Inspect the condition of the two seals in the bearing carrier. If the seals appear to be damaged and replacement is required, note the direction the seal lips are facing and then remove them from the housing. In almost all cases these gearcases are equipped with dual oil seals whose lips are both faced in the same direction, towards the propeller. Seal removal can be accomplished by various means. On most models you can use the same slide hammer and jaw attachments which were used on the gear or bearing assembly to remove the seals. On some models you can drive the seals out from the opposite side of the carrier. And, on a couple of models a hooked seal remover or small prytool may be used to free the seal. Whatever method you use, just be sure not to score or damage the inside diameter of the carrier (the sealing surface).

7. On all models except 6/8 hp models, a caged needle bearing set is pressed into the bearing carrier. To remove the bearing set use a suitable driver to push the bearing set free of the bearing carrier. In most, but not all cases, the bearing should be driven in the direction of the oil seals. Inspect both sides of the bearing to determine the proper direction.

8. On the 6/8 hp models, a solid bushing is installed in the carrier, between the bearing set and the oil seals. Using a suitable driver, carefully push the bushing free of the bearing carrier. The direction the bushing is to be driven out does not matter, unless the bushing has a shoulder. In this case, drive the bushing from the flush side.

Driveshaft

The driveshaft is used to transmit the clockwise (when viewed from the top) rotational motion of the crankshaft through the pinion gear to the propeller shaft (either through the Forward or Reverse gears). The driveshaft is splined at both ends, the top splines mating with the lower end of the crankshaft and the bottom end mating with the pinion gear. The pinion gear itself is held in place either by a circlip (6/8 hp motors) or by a retaining nut (8/9.9 hp and larger motors).

Exact positioning of the pinion gear on the shaft (so that it meshes properly with the Forward and Reverse gears) is often adjusted through the use of different thickness shim packs. Although adjustments may also take place at the gears themselves with additional shim pack (or instead of a shim pack used on the pinion gear). Also, keep in mind that the exact placement of the shim pack in relation to other potential gear components such as a plate washer and/or thrust bearing will vary, so pay CLOSE attention to the order of removal of components mounted above the pinion gear. In other words, on some models there will only be a gear, on others a gear and a shim pack. On still others there will be a shim pack above or below a thrust bearing. And, on a few there may even be a shim, thrust washer and plate washer (usually in that order from bottom-to-top).

Before the driveshaft can be removed, the pinion gear at the lower end of the shaft must be removed.

■ Driveshaft removal is necessary for access gearcase mounted components like driveshaft bearings or bushings and/or the Forward gear/bearing. However, on some models (where the oil seal housing must be removed from the bottom of the driveshaft), shaft removal may also be necessary.

1. On 6/8 hp models, pry the circlip free from the end of the driveshaft with a thin pry-tool. This clip holds the pinion gear onto the lower end of the driveshaft. The clip may not come free on the first try, but have patience and it will come free. With one hand, lift the driveshaft up and out of the lower unit housing and at the same time, with the other hand, catch the pinion gear and SAVE any shim material from behind the gear. The shim material is critical to obtaining the correct backlash during installation. Using the old shim material will save considerable time, especially starting with no shim material.

■ In most cases, when working with tools, a nut is rotated to remove or install it to a particular bolt, shaft, etc. In the next two steps, the reverse is usually required because there is very little-to-no room to move a wrench inside the lower unit cavity. Normally, the nut on the lower end of the driveshaft is held steady and the shaft is rotated until the nut is free.

2. Obtain a pinion nut wrench (Yamaha #YB-06078). This wrench is used to prevent the nut from rotating while the driveshaft is rotated. It's a LONG handled wrench of the same size as the pinion nut. You may be able

LOWER UNIT 7-17

to use a wrench or a breaker bar and socket already in your box. Also, you'll need the correct driveshaft holder (it's essentially a nut with internal splines that fits on top of the driveshaft so that you can hold or turn it using a wrench or large socket). On 15 hp and smaller motors (with a nut retained pinion gear), obtain a driveshaft tool (#YB-06228). On 25 hp and larger models, obtain a driveshaft tool (#YB-06079).

3. Now, hold the pinion nut steady with the tool and at the same time install the proper driveshaft tool, for the model being serviced, on top of the driveshaft. With both tools in place, one holding the nut and the other on the driveshaft, rotate the driveshaft counterclockwise to break the nut free.

■ **Save any shim material from behind the pinion gear. The shim material is critical to obtaining the correct backlash during installation. Using the old shim material will save considerable time, especially starting with no shim material.**

4. Remove the pinion nut and gently pull up on the driveshaft and at the same time rotate the driveshaft. The pinion gear, followed by any shim material and washer or thrust bearing will come free from the lower end of the driveshaft. As we noted earlier, pay CLOSE attention to the order in which these components come free from the driveshaft (not easy to do unless you hold them in position of pull them down one at a time).

5. Pull the driveshaft up out of the lower unit housing. Keep the parts from the lower end of the driveshaft in order, as an aid during assembling.

■ **On T25 hp and 30/40 hp motors the small round driveshaft oil seal housing should come out if the gearcase along with the driveshaft as the tapered bearing which is pressed onto the shaft will pull it free (or the driveshaft won't come out). Set the oil seal housing aside for seal inspection and/or service later.**

6. On some larger models there is a tapered roller bearing in the center of the driveshaft. If it is unfit for further service, press the bearing free using the proper size mandrel. Take care not to bend or distort the driveshaft because of its length.

Driveshaft Oil Seal Housing

At the very top of the gearcase is a small housing that contains the oil seals. On some models that housing may also contain the driveshaft's upper bearing or bushing assembly. The size a shape of the housing varies GREATLY from model-to-model and may be tiny as a small round housing just a little bit larger than the seal or as large as a plate which stretches from the driveshaft to the shift shaft (and also containing a bushing or boot for the shifter assembly). Servicing the oil seal and bearing or bushing is essentially accomplished in the same manner as one the propeller bearing/seal carrier.

On motors where the oil seal housing may be slid over the top of the driveshaft, you do not NEED to remove the pinion gear and driveshaft if you are ONLY going to service the seals. However, it is a little more difficult to protect the seals during installation, so if you've got the time, it's never a bad idea to pull the driveshaft. Your choice. If you decide NOT to pull the driveshaft, take great care when freeing the housing to pull straight upward and, more importantly, when installing to keep the housing squared to the driveshaft so as not to damage the seals (or bearings/bushing on some models).

1. Most oil seal housings (except the round one used on T25 hp and 30/40 hp motors) are secured by the water pump housing and/or plate bolts. However, a few are also secured by one or two additional bolts. This is usually true housings that also contain a passage for the shifter shaft and the additional bolt(s) is(are) normally found at the front of the oil seal housing, near the shifter. Remove any bolt or bolts that are still threaded through the housing assembly.

2. Carefully lift upward to remove the seal housing from the top of the gearcase. Remove and discard the housing gasket.

■ **All of these models use dual oil seals which are positioned with their lips facing upwards (away from the gearcase). Double-check this to make sure there is no variance on your gearcase before removal. Also, remember that removal of the seals from the housing will destroy their sealing qualities and they cannot be installed a second time. Therefore, remove a seal only if it is unfit for further service and be absolutely sure a new seal is available before removing the old seal.**

3. Seal removal can be accomplished by various means. On most models you can use a hooked seal remover or small pry-tool in order to free the each seal. However, if access is difficult you can always use a slide hammer and expanding jaw attachment to remove the seals. Sometimes you can drive them out from the other side of the housing, but use care not to damage the housing or the bushing (if used and if it is not being replaced). Whatever method you use, just be sure not to score or damage the inside diameter of the carrier (the sealing surface).

4. On 6-15 hp models there is also a solid bushing in the housing. If the bushing must be replaced use the appropriate sized driver to carefully push it from the oil seal housing.

■ **The direction the bushing is to be driven out does not matter, UNLESS the bushing has a shoulder. In this case, drive the bushing from the flush side.**

Driveshaft Bushings/Bearings

The driveshaft for all models rides in multiple bushings or bearings. For many models the upper bushing or bearing is installed in the oil seal housing. However, for some models the upper bushing (some smaller models) or upper bearing (25 hp and larger) is mounted directly in the gearcase. Although for 25-40 hp motors, the upper bearing is actually pressed onto the driveshaft and not into the gearcase. On the T25 hp and 30/40 hp motors there is a bearing race in the gearcase which can be removed for service.

All of these gearcases use a lower driveshaft bushing (6/8 hp motors) or a lower shaft bearing (8/9.9 hp and larger motors) which is pressed directly into the gearcase. In addition, most gearcases have some sort of driveshaft sleeve mounted between the upper and lower bearing assemblies. The sleeves are easily removed (by hand) once the upper bearing or bushing is out of the way (this includes models where the bearing or bushing is installed in the oil seal housing).

Upper bearings or bushings (or even races on a motors) are removed from the top of the gearcase using expanding jaw adapters attached to either a slide-hammer or a bridged puller, while the lower bearing or bushing is normally removed by pushing it into the propeller shaft cavity from above using a suitable driver.

If the driveshaft bushing, upper or lower, on some models, or the needle bearing on other models, is unfit for service, perform the following steps, depending on the model being serviced.

■ **Although the ball bearings used on the upper end of some driveshafts/gearcases can usually be removed and reinstalled without damage, all needle bearings and bushings should be left undisturbed unless they are damaged and in need of replacement. The act of removing most bearings (at least needle bearings, but including ball bearings that are mishandled) and bushings normally makes them unfit for further service.**

■ **A few models, such as the 25 hp and larger motors, place the pinion gear shim directly beneath the upper bearing. Watch for and save any shim material from behind the bearing. The shim material is critical to obtaining the correct backlash during installation. Using the old shim material will save considerable time, especially starting with no shim material.**

1. On models where the upper driveshaft bushing or bearing (or removable race in the case of the T25 hp and 30/40 hp motors) is mounted directly to the gearcase, use a slide hammer with expanding jaw attachment (or a bridged puller) to remove the bearing/bushing from the top of the gearcase. Check the bottom of the bearing and the top of the mounting in the gearcase for shim material and, if found, retain for installation purposes.

2. Grasp and remove the driveshaft sleeve from the top of the gearcase. Although some are completely round with no identifying marks or variance in shape, many are equipped with a tab, flat or marking which must be faced forward during installation. Check the sleeve as it is removed and, if applicable, note the direction of installation for use later.

3. If removal is necessary, use a suitable long-handled driver to push the lower bushing or needle bearing downward (into the propeller shaft cavity).

Forward Gear

All models contain a Forward gear that rotates in a dedicated forward bearing. All 6 hp and larger motors are normally equipped with a tapered roller bearing that is pressed onto the Forward gear. The assembly can be

7-18 LOWER UNIT

removed or installed by hand in a race which is pressed into the nose cone of the gearcase. Shimming material is placed between the bearing race and the mounting surface in the nose cone on these models.

1. Reach into the gearcase, then remove the Forward gear and bearing assembly.

■ **If a two piece bearing is to be replaced, the bearing and the race must be replaced as a matched set. Remove the bearing only if it is unfit for further service.**

2. If the Forward bearing is not longer fit for service, proceed as follows:
   a. Position a bearing separator between the forward gear and the tapered roller bearing. Using a hydraulic press, separate the gear from the bearing.
   b. Obtain a slide hammer with a jaw attachment. Use the tool to pull the bearing race from the lower unit housing. Be sure to hold the slide hammer at right angles to the driveshaft while working the slide hammer. Watch for and save any shim material found in front of the forward gear bearing race (deeper into the gearcase nose cone). The shim material is critical to obtaining the correct backlash during installation. Using the old shim material will save considerable time, especially starting with no shim material.

Clutch Dog

The clutch dog and shifting assembly may be removed from the propeller shaft for cleaning, inspection and replacement, as necessary. The clutch dog itself is splined to the shaft to allow for movement back and forth to engage/disengage the Forward and Reverse gears. Movement along the shaft is limited by a cross-pin. A shift plunger and spring are used to maintain contact with the linkage found in the gearcase nose cone, in front of the Forward gear bearing.

If propeller shaft shifter component disassembly is necessary, proceed as follows:

1. Insert an awl under the end loop of the cross pin ring and carefully pry the ring free of the clutch dog.

2. Use a long pointed punch to press out the cross pin, then remove the plunger and spring from the end of the propeller shaft.

3. Slide the clutch dog from the shaft. Observe how the clutch dog was installed. It must be installed in the same direction. Most clutch dogs are embossed with a marking (such as **F** to denote which edge faces front, toward the Forward gear).

CLEANING & INSPECTING

◆ See Figures 34 thru 47

Good shop practice requires installation of new O-rings and oil seals regardless of their appearance.

Clean all water pump parts with solvent and then dry them with compressed air. Inspect the water pump housing and oil seal housing for cracks and distortion, possibly caused from overheating. Inspect the inner and outer plates and water pump cartridge for grooves and/or rough surfaces. If possible, always install a new water pump impeller while the lower unit is disassembled. A new impeller will ensure extended satisfactory service and give peace of mind to the owner. If the old impeller must be returned to service, never install it in reverse to the original direction of rotation. Installation in reverse will cause premature impeller failure.

If installation of a new impeller is not possible, check the seal surfaces. All must be in good condition to ensure proper pump operation. Check the upper, lower and ends of the impeller vanes for grooves, cracking and wear. Check to be sure the indexing notch of the impeller hub is intact and will not allow the impeller to slip.

Clean around the Woodruff key or impeller pin. Clean all bearings with solvent, dry them with compressed air and inspect them carefully. Be sure there is no water in the air line. Hold the bearing to keep it from turning and direct the air stream through the bearing. After the bearings are clean and dry, lubricate them with oil. Do not lubricate tapered bearing cups until after they have been inspected.

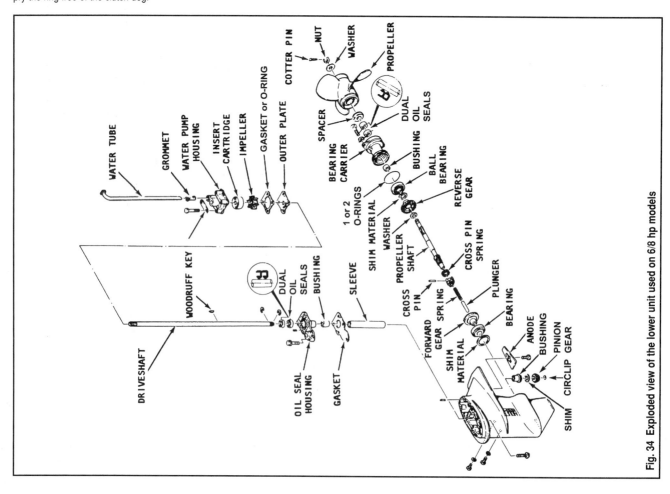

Fig. 34 Exploded view of the lower unit used on 6/8 hp models

LOWER UNIT 7-19

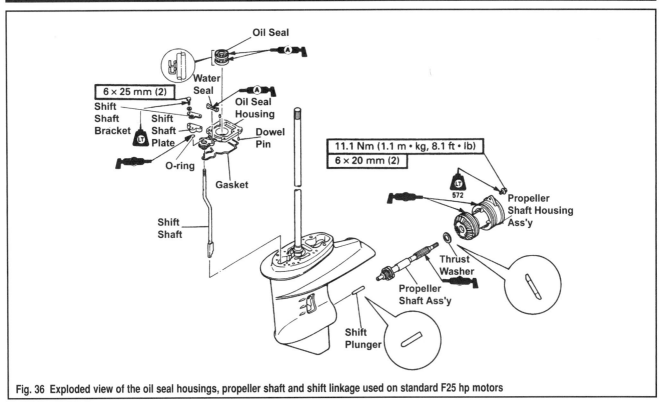

Fig. 36 Exploded view of the oil seal housings, propeller shaft and shift linkage used on standard F25 hp motors

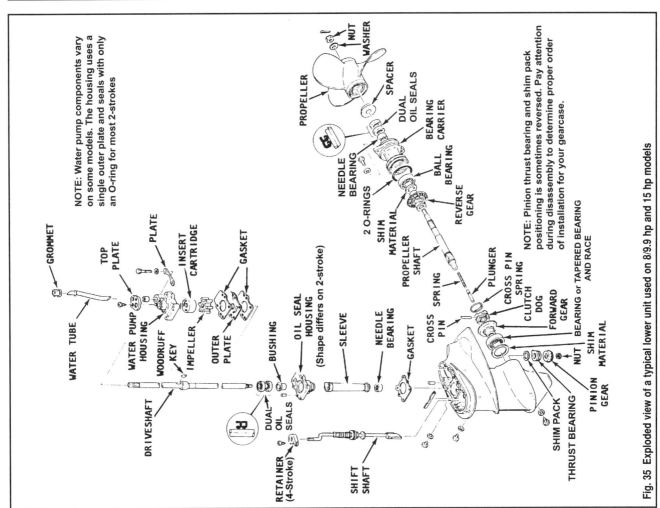

Fig. 35 Exploded view of a typical lower unit used on 8/9.9 hp and 15 hp models

7-20 LOWER UNIT

✸✸ WARNING

Never spin a bearing with compressed air. Such action is highly dangerous and may cause the bearing to score from lack of lubrication.

Inspect all ball bearings for roughness, scratches and bearing race side wear. Hold the outer race and work the inner bearing race in-and-out, to check for side wear.

Determine the condition of tapered bearing rollers and inner bearing race, by inspecting the bearing cup for pitting, scoring, grooves, uneven wear, imbedded particles and discoloration caused from overheating. Always replace tapered roller bearings and their race as a set.

Clean the forward gear with solvent and then dry it with compressed air. Inspect the gear teeth for wear. Under normal conditions the gear will show signs of wear but it will be smooth and even.

Clean the bearing carrier or cap with solvent and then dry it with compressed air. Check the gear teeth of the reverse gear for wear. The wear should be smooth and even.

Check the clutch dogs to be sure they are not rounded-off, or chipped. Such damage is usually the result of poor operator habits and is caused by shifting too slowly or shifting while the engine is operating at high rpm. Such damage might also be caused by improper shift rod adjustments.

Rotate the reverse gear and check for catches and roughness. Check the bearing for side wear of the bearing races.

Inspect the roller bearing surface of the propeller shaft. Check the shaft surface for pitting, scoring, grooving, embedded particles, uneven wear and discoloration caused from overheating.

Clean the driveshaft with solvent and then dry it with compressed air. Inspect the bearing for roughness, scratches, or side wear. If the bearing shows signs of such damage, it should be replaced. If the bearing is satisfactory for further service coat it with oil.

Inspect the driveshaft splines for excessive wear. Check the oil seal surfaces above and below the water pump drive pin or Woodruff key area for grooves. Replace the shaft if grooves are discovered.

Inspect the driveshaft bearing surface above the pinion gear splines for pitting, grooves, scoring, uneven wear, embedded metal particles and discoloration caused by overheating.

Inspect the propeller shaft oil seal surface to be sure it is not pitted, grooved, or scratched. Inspect the roller bearing contact surface on the propeller shaft for pitting, grooves, scoring, uneven wear, embedded metal particles and discoloration caused from overheating.

Inspect the propeller shaft splines for wear and corrosion damage. Check the propeller shaft for straightness.

As equipped, inspect the following parts for wear, corrosion, or other signs of damage:
- Shift shaft boot
- Shift shaft retainer
- Shift cam
- All bearing bores for loose fitting bearings.
- Gear housing for impact damage.
- Threads for cross-threading and corrosion damage.
- If used, check the circlip to be sure it is not bent or stretched. If the clip is deformed, it must be replaced.
- Check the pinion nut corners for wear or damage. This nut is a special locknut. Therefore, do not attempt to replace it with a standard nut. Obtain the correct nut from an authorized Yamaha dealer.

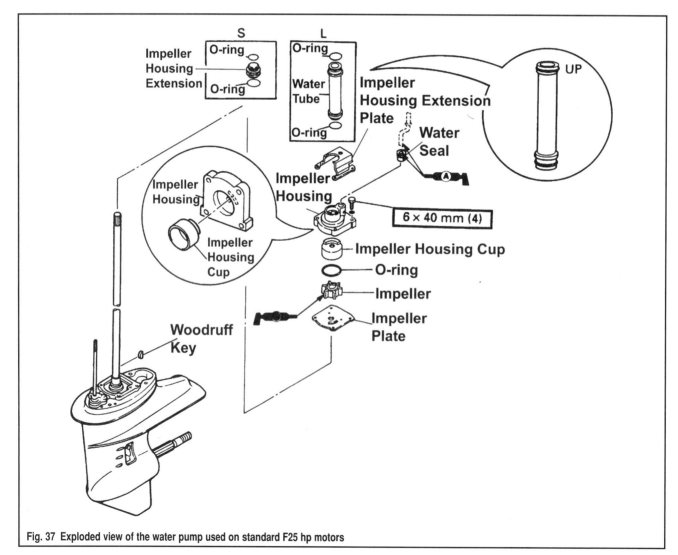

Fig. 37 Exploded view of the water pump used on standard F25 hp motors

LOWER UNIT

ASSEMBLY

◆ See Figures 34 thru 47

The following procedures are intended to guide you step-by-step through reinstallation and setup of the major gearcase components. If one or more of the components was not removed you can skip that section and proceed to the next following component.

Gearcase shimming procedures vary slightly from model-to-model and they depend upon the components that are being replaced. However, generally speaking:

• NO SHIMMING is required if the original case and original inner parts (gears, bearings and shafts) are installed.

• NUMERIC CALCULATION RESHIMMING should be performed on 25 hp and larger gearcases when the original inner parts (gears, bearings and shafts) are installed in a NEW case. Many original and replacement cases are marked with various gear measurements. Typically an embossed F (forward gear), R (reverse gear) and/or P (pinion) gear measurements. When these marks are present the differences between the marks on the new case vs. the marks on the original case can be used to determine new shim sizes BEFORE assembly (potentially saving you from having to disassemble/reassemble the gearcase a second time).

• SPECIAL TOOL MEASUREMENT RESHIMMING is performed prior to assembly on certain models and should be used when a shaft, bearing/bushing or gear is replaced. In this case, special tools are used to measure pre-assembled or partially assembled components and determine if a change in shim sizes is necessary. A special tool procedure is available for the gearcase found on the 8/9.9 hp and 15 hp motors. A complete pinion gear shimming tool set is used on gearcases whose driveshaft is equipped with a pressed on tapered roller bearing. In addition, one or more shimming gauges are available for all 25 hp and larger motors.

• BACKLASH MEASUREMENT RESHIMMING should be performed and adjusted when one or more of the original inner components (gears, bearings and shafts) are installed. Backlash measurement is normally

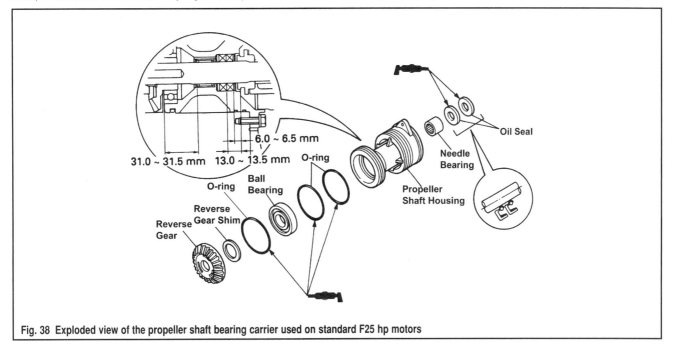

Fig. 38 Exploded view of the propeller shaft bearing carrier used on standard F25 hp motors

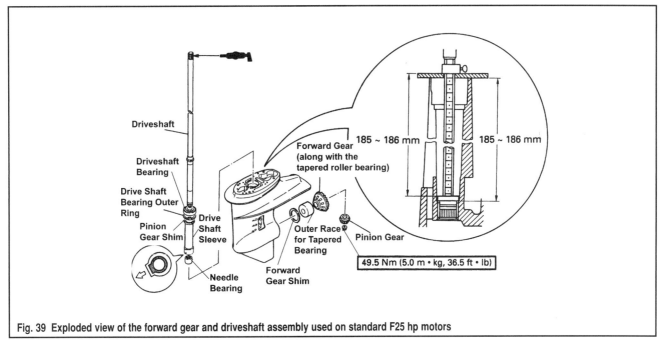

Fig. 39 Exploded view of the forward gear and driveshaft assembly used on standard F25 hp motors

7-22 LOWER UNIT

performed on an almost fully assembled gearcase using a dial gauge to read the amount of play in the driveshaft. This procedure can check the placement of all 3 gears through this measurement. Though adjustment is much more hit and miss than the other methods, since you need to then re-disassemble the gearcase, swap shims, re-assemble and re-check, then repeat until you get it right. Fortunately, formulas are available to use the measured backlash in helping to pick shim sizes and small shim changes are all that is normally necessary. It is rare that you would have to disassemble the gearcase more than one additional time.

■ **If the case, gear(s), bearing(s) and/or shaft(s) have been replaced, refer to the Gearcase Shimming procedures later in this section BEFORE starting reassembly to determine if there are any changes to shims that can be made before your start assembly so that you can potentially avoid having to disassemble the gearcase a second time to change shims after checking backlash.**

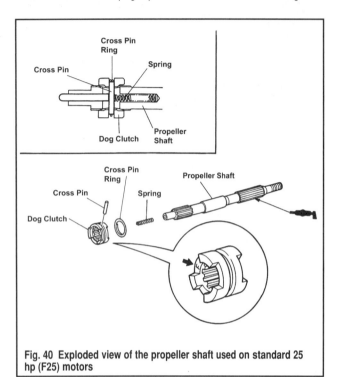

Fig. 40 Exploded view of the propeller shaft used on standard 25 hp (F25) motors

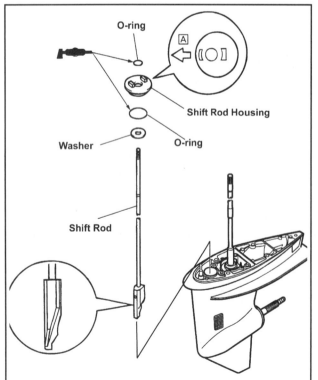

Fig. 42 Exploded view of the shifter assembly used on 30/40 hp and High-Thrust 25 hp (T25) motors

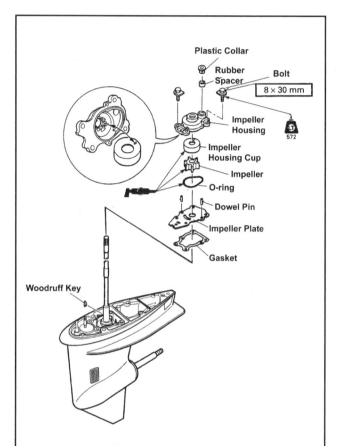

Fig. 41 Exploded view of the water pump assembly used on 30/40 hp and High-Thrust 25 hp (T25) motors

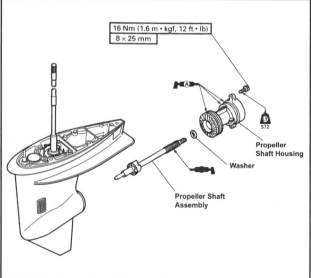

Fig. 43 Exploded view of the propeller shaft and bearing carrier mounting used on 30/40 hp and High-Thrust 25 hp (T25) motors

Throughout installation you will need to make a couple of choices. If you have simply replaced O-rings and seals (not bearings, shafts or gears) then there is probably little chance that the gear lash will require adjustment. If so you can lubricate all components and apply sealant or threadlocking compounds as called for in the procedures. However, if you have replaced components such as the bearings, shafts or gears you may wish to check gear contact patterns using a Machine Dye or suitable powder, and all components should then be temporarily installed dry. Also, when components such as these are replaced you'll need to check and possibly adjust gear lash, though reusing shims which were removed during disassembly can often give you a good starting point and MAY prevent you from having to partially disassemble the gearcase again (but only if you've chosen to lubricate components during installation). As we said, the choice is yours and can only be made based upon the circumstances of your overhaul or reseal and the condition of the unit before tear-down.

Unless otherwise noted, upon final installation be sure to lubricate all bearings with gear oil, shafts with oil or a marine grade grease and seals with oil, packing their lips with a suitable marine grade grease.

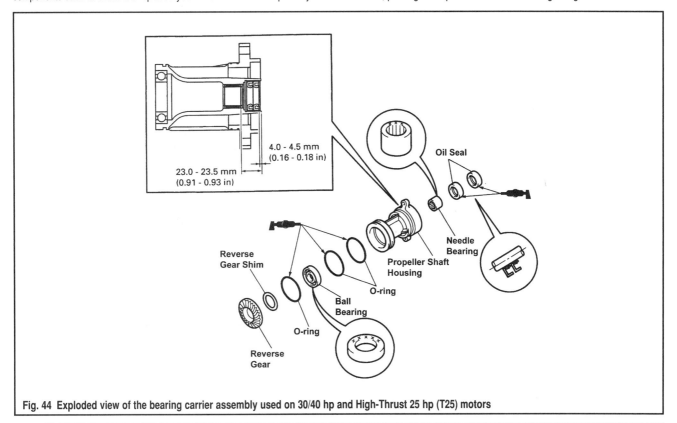

Fig. 44 Exploded view of the bearing carrier assembly used on 30/40 hp and High-Thrust 25 hp (T25) motors

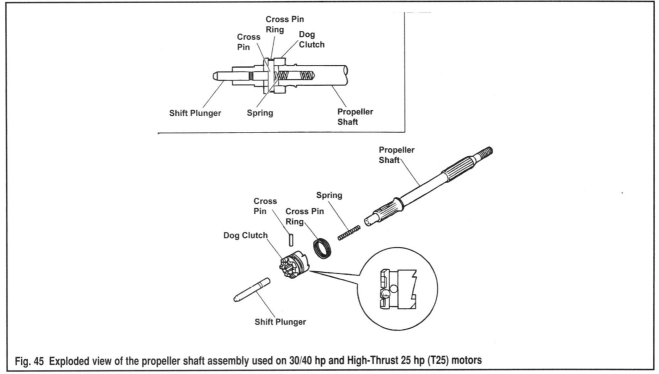

Fig. 45 Exploded view of the propeller shaft assembly used on 30/40 hp and High-Thrust 25 hp (T25) motors

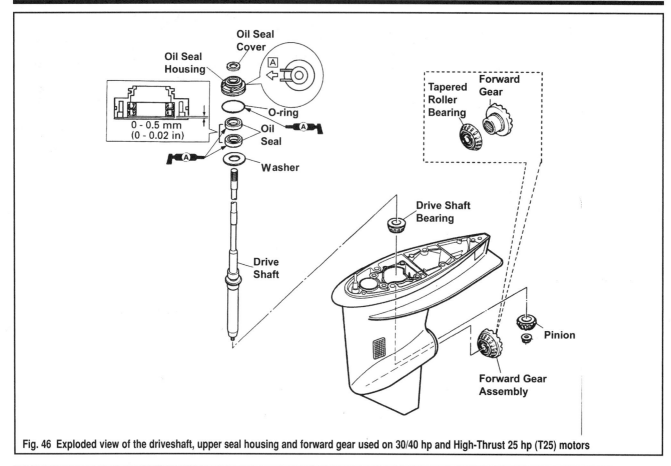

Fig. 46 Exploded view of the driveshaft, upper seal housing and forward gear used on 30/40 hp and High-Thrust 25 hp (T25) motors

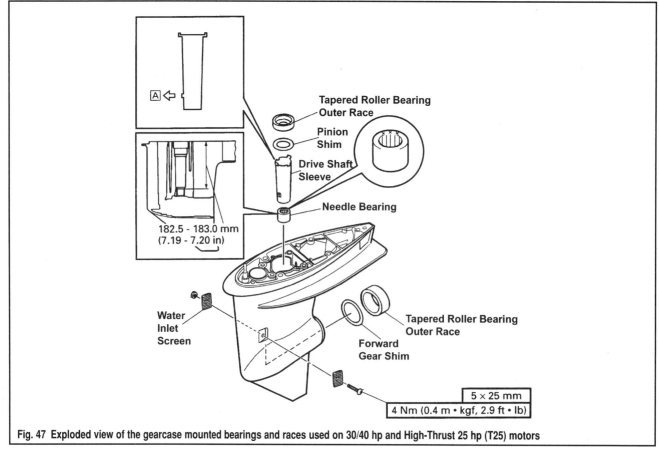

Fig. 47 Exploded view of the gearcase mounted bearings and races used on 30/40 hp and High-Thrust 25 hp (T25) motors

LOWER UNIT 7-25

Clutch Dog

1. Most clutch dogs are embossed with an **F** on the face or side (at one end) of the dog itself. During installation on the propeller shaft, the marking on the clutch dog must face the forward gear (toward the front of the gearcase).
2. Slide the spring down into the propeller shaft. Insert a narrow screwdriver into the slot in the shaft and compress the spring until approximately 1/2 in. (12mm) is obtained between the top of the slot and the screwdriver.
3. Hold the spring compressed and at the same time, slide the clutch dog over the splines of the propeller shaft with the hole in the dog aligned with the slot in the shaft. Again, the embossed letter **F** on the dog must face toward the forward gear.
4. Insert the cross pin into the clutch dog and through the space held open by the screwdriver. Center the pin and then remove the screwdriver allowing the spring to pop back into place.
5. Fit the cross pin ring into the groove around the clutch dog to retain the cross pin in place.
6. Insert the flat end of the plunger into the propeller shaft, with the rounded end protruding to permit the plunger to slide along the cam of the shift rod.

Forward Gear Bearing

As noted earlier, all models contain a Forward gear that rotates in a dedicated forward bearing. All 6 hp and larger motors are normally equipped with a tapered roller bearing that is pressed onto the Forward gear. The assembly can be installed by hand in a race which is pressed into the nose cone of the gearcase. Shimming material is placed between the bearing race and the mounting surface in the nose cone on these models.

The next step applies only if the forward gear bearing was removed.

1. If the Forward bearing and/or race was removed, proceed as follows:
    a. Insert the shim material saved during disassembly into the lower unit bearing race cavity. Depending upon what components were changed, the shim material should give the same amount of backlash between the pinion gear and the forward gear as before disassembling (or at least put you in the right ballpark).
    b. Insert the bearing race squarely into the gearcase nose cone, then use a suitable driver to gently tap the race into the housing until it is fully seated.

■ **Remember the bearing and race are a set here. Although you CAN replace a damaged race and keep the original bearing, there is a good chance that a damaged race will also have a damaged bearing and both should be replaced. Of course, if the bearing is replaced the race MUST also be replaced.**

    c. Position the forward gear tapered bearing over the forward gear. Use a suitable mandrel and press the bearing flush against the shoulder of the gear. Always press on the inner race, never on the cage or the rollers.
2. If you wish to visually check gear contact patterns, obtain a suitable substance which can be used to indicate a wear pattern on the forward and pinion gears as they mesh. Machine dye may be used and if this material is not available, Desenex® Foot Powder (obtainable at the local Drug Store/Pharmacy), or equivalent may be substituted. Desenex® is a white powder available in an aerosol container. Before assembling either gear, apply a light film of the dye, Desenex®, or equivalent, to the driven side of the gear. After the gears are assembled and rotated several times, they will be disassembled and the wear pattern can be examined. The substance will be removed from the gears prior to final assembly.
3. If this is the final assembly, be sure to lubricate the bearing with fresh gear oil.
4. Position the forward gear assembly into the forward bearing race.

Driveshaft Bushings/Bearings

As we stated earlier, the driveshaft for all models rides in multiple bushings or bearings. For many models the upper bushing or bearing is installed in the oil seal housing. However, for some models the upper bearing (25 hp and larger motors) is mounted directly in the gearcase. Although actually on the 25-40 hp 4-strokes the upper bearing is pressed onto the driveshaft and not into the gearcase. On the T25 hp and 30/40 hp motors there is a bearing race in the gearcase which can be removed for service.

Also, all of these gearcases use a lower driveshaft bushing (6/8 hp motors) or bearing (8/9.9 hp and larger motors) which is pressed directly into the gearcase. In addition, most gearcases have some sort of driveshaft sleeve mounted between the upper and lower bearing assemblies. The sleeves are easily removed (by hand) once the upper bearing or bushing is out of the way (this includes models where the bearing or bushing is installed in the oil seal housing).

The lower bearing or bushing is usually (but not always) installed from above using a driver and a long driver handle to position it to the proper depth. In some cases however, the lower bearing/bushing must be installed from below, using a threaded rod and a driver to pull it upward into the bore from the bottom (from inside the gearcase propeller shaft cavity).

■ **Yamaha often has a special tool available to help draw a bushing or bearing upward and into position in the lower portion of the gearcase, all you REALLY need to accomplish this is a sufficiently long threaded bolt with a nut and a washer that is the size of the driver you'd want to contact that bushing or bearing. You'll also need a nut and a plate to place on top of the gearcase. Place the threaded rod through one nut and plate and then down into the gearcase through the bushing/bearing (so the plate is resting against the top of the gearcase, the upper nut is resting against the plate and the threads of the rod are just protruding down into the pinion mounting area of the gearcase.). Next install the washer and lower nut over the bottom threads of the rod to support the bearings or busing. Slowly draw the bearing or bushing into position in the gearcase by holding the threaded rod from turning while at the same time turning the upper nut to draw the entire bolt, bushing/bearing, washer and nut assembly upward until the bushing or bearing seats. Then loosen the nut on the bottom of the threaded rod in the gearcase and remove the threaded rod, plate, nut and washer.**

Upper bearings or bushings (or even races on a few 4-strokes) are always installed from the top of the gearcase simply using a suitable driver to seat them.

■ **When bearings are installed from above using a driver, the only potential issue is that if you are not using the specific Yamaha driver for that gearcase would be the depth to which the bushing/bearing is installed. On some models there are shoulders against which the bushing or bearing seats. If not, we will provide a dimension to measure installed depth, ensuring a proper installation.**

If the driveshaft bushing, upper or lower, on some models, or the needle bearing on other models, was removed because they were unfit for service, perform the following steps, depending on the model being serviced.

• The 6/8 hp motors have two bushing for the driveshaft. One (the upper) is located in the Oil Seal Housing, and details on its installation can be found under that procedure. The other (the lower) is positioned just above the pinion gear. The lower bushing is installed from underneath by pulling it upward using a threaded rod. The bushing driver is slightly larger than the bushing and will stop at the correct depth in the gearcase as it contacts a slightly tapered flange. Once the lower bushing is installed, slide the driveshaft sleeve into position from above.

• The driveshafts on 8/9.9 hp and 15 hp motors use a lower needle bearing assembly that is driven down into the gearcase from above using a suitable driver and long driver handle. The bearing installation depth is determined by the use of a special tool, or by measuring the distance the top edge of the bearing sits below the gearcase deck. The proper installation depth varies slightly by model. For all standard thrust 8/9.9 (F8/F9.9) motors, the depth should be 7.17-7.19 in. (182.2-182.7mm). For all high-thrust 9.9 hp (T9.9) motors, the depth should be 7.86-7.88 in. (199.6-200.1mm). For all 15 hp motors the depth should be 6.80-6.82 in. (172.7-173.2mm). A driveshaft sleeve is then installed over top of the bearing. Standard versions of these models should also have a replaceable upper bushing that is mounted in the oil seal housing.

• For 25 hp and larger motors the driveshaft rides on a lower needle bearing that is pressed into the gearcase, as well as an upper tapered roller bearing that is pressed directly onto the driveshaft itself. A bearing race and shim pack (for the pinion gear) is installed under the race. In addition, a driveshaft sleeve is installed in the gearcase below the tapered bearing race and above the needle bearing. To install the needle bearing place it squarely into the gearcase bore with any embossed markings facing upward (toward the driver). Use a long-handled driver to carefully tap the bearing downward into the gearcase until the top edge of the bearing (embossed edge) reaches the specified depth below the gearcase deck. For standard thrust 25 hp (F25) motors, drive the bearing to a depth of 7.28-7.32 in. (185-186mm). For all 30/40 hp motors, as well as all high-thrust 25 hp (T25) motors, install the bearing to a depth of 7.19-7.20 in. (182.5-183.0mm). Next install the driveshaft sleeve with the tab or cutout facing forward, as noted during disassembly. Position the pinion gear shim pack and the tapered roller bearing race. Gently tap the bearing race into the gearcase. Finally, if the tapered roller bearing was removed from the driveshaft, using a suitable bearing separator plate to carefully press the bearing (pressing on the inside race) into position on the driveshaft.

Driveshaft Oil Seal Housing

All of these motors utilize some form of oil seal housing which mounts to the top of the gearcase. On some motors this housing may also include an upper driveshaft bushing or bearing assembly.

1. On 20 hp or larger motors that contain a needle bearing in the oil seal housing, use a suitable driver to install the bearing from the top of the housing. Make sure the driver contacts the outer bearing cage to make sure the bearing is not damaged and install the bearing to a depth of 1.22-1.24 in. (31.0-31.5mm) measured from the top of the housing to the top of the bearing cage.

2. On 6-15 hp motors equipped with a bushing in the oil seal housing, use a driver to carefully install the bushing to the housing. Most, but not all, bushings are installed from the oil seal (top) side of the housing. Most housings do not give an installation depth for the bushing, which probably indicates that the bushing is installed against a shoulder or stop in the housing. On all other motors either carefully seat the bushing against the stop, or drive the bushing down into the lower portion of the housing, making sure it does not extend beyond the lower edge of the bore.

■ **Remember to apply a light coating of gear oil to the oil seal lips and cases before installation. Then after installation, pack the seal lips with a small amount of marine grade grease.**

3. Oil seal installation varies only slightly by model. For all gearcases covered in this procedure a dual oil seal assembly is installed with BOTH sets of lips facing upward, toward the water pump, away from the gearcase. All oil seals are installed from the top of the housing, except the small round housing used on 30/40 hp and high-thrust 25 hp (T25) motors, on which the seals are installed from beneath the housing. If no dimension is provided, drive the seals into the housing until they are at least flush with or slightly below the top deck of the housing mating surface. For the following models, drive the seals slowly downward until the TOP edge of the top most seal is the specified distance below the housing deck:
- On 6/8 hp motors, install the seals to a depth of 0.24-0.26 in. (6.0-6.5mm).
- On 15 hp motors, install the seals to a depth of 0.28-0.31 in. (7.0-8.0mm).
- On standard thrust 25 hp (F25) motors, install the first (bottom) of the seals to a depth of 0.49-0.51 in. (12.5-13.0mm) and the second (top) of the seals to a depth of 0.22-0.24 in. (5.5-6.0mm).
- On 30/40 hp motors and high-thrust 25 hp (T25) motors, install the seals to a depth of 0.00-0.02 in. (0.0-0.5mm) measured from the BOTTOM of the housing to the bottom of the LOWER seal on these models.

■ **If you are going to measure gear backlash you may wish to leave the oil seal housing off the motor until after lash is measured (and potentially the gearcase is partially re-disassembled for adjustment). However, on models where the upper driveshaft bearing or bushing is installed in the oil seal housing, it is a good idea to at least temporarily install the housing to make sure there is no additional lateral play on the driveshaft. It really shouldn't affect your lash readings, but it does not hurt to be sure.**

4. For all except 30/40 hp motors and high-thrust 25 hp (T25) motors, if you are ready to install the oil seal housing position it to the top of the gearcase. Some models use dowels and/or a gasket, and if applicable, be sure they are already in position.

5. For 30/40 hp motors and high-thrust 25 hp (T25) motors, grease and install the O-ring to the oil seal housing, then slide the washer and housing down into position on the driveshaft (taking care not to damage the new seals) and place the assembly aside for installation.

Driveshaft (and Pinion Gear)

The driveshaft can be installed to the gearcase with or without the upper seal housing in place. In either case you need to take great care when sliding the splines through the seals to make sure the seals are not damaged.

For all except 6/8 hp motors the driveshaft pinion is secured by a retaining nut. In order to install and tighten this nut you'll need to obtain a pinion nut wrench (#YB-06078). This wrench is used to prevent the nut from rotating while the driveshaft is rotated. It's a LONG handled wrench of the same size as the pinion nut. You may be able to use a wrench or a breaker bar and socket already in your box. Also, you'll need the correct driveshaft holder (it's essentially a nut with internal splines that fits on top of the driveshaft so that you can turn the driveshaft using a torque wrench and large socket in order to tighten the pinion nut). On 15 hp and smaller motors (with a nut-retained pinion gear), obtain a driveshaft tool (#YB-06228). On 25 hp and larger models, obtain a driveshaft tool (#YB-06079).

1. Lower the driveshaft down through the sleeve in the upper end of the lower unit. If the upper seal housing is already on the gearcase, take extra care not to damage the seals with the splines on the lower end of the shaft.

2. If you are looking to visually check gear mesh patterns, coat the pinion gear with a fine spray of Desenex®, as instructed in the opening paragraphs for Assembly and again under Forward Gear Bearing. Handle the gear carefully to prevent disturbing the powder.

■ **When installing the pinion gear onto the driveshaft, it may be necessary to rotate the driveshaft slightly to allow the splines on the driveshaft to index with the internal splines of the pinion gear. The teeth of the pinion gear will index with the teeth of the forward gear.**

3. Hold the driveshaft in place with one hand and at the same time assemble the parts onto the lower end of the driveshaft with the other hand in the same order as they were removed. Hopefully they were kept in order on the work surface, but since the order varies slightly by model, here's how to assemble the pinion gear and shim pack and/or thrust washer/bearing:
- On 6/8 hp motors, install the shim washer followed by the pinion gear and the circlip.

■ **On 6/8 hp models, slipping the circlip into the groove on the end of the driveshaft is not an easy task, but have patience and success will be the reward.**

- On 8/9.9 hp motors, install the shim followed by the thrust bearing, pinion gear and retaining nut. Tighten the nut to 19 ft. lbs. (26 Nm).
- On 15 hp motors, install the plate washer, followed by the thrust bearing, shim, pinion gear and retaining nut. Tighten the nut to 18 ft. lbs. (25 Nm).
- On 25 hp and larger motors (all motors whose gearcase uses a tapered roller bearing that is pressed to the driveshaft) the pinion gear shim was already installed underneath the tapered bearing race. For these motors, install the pinion gear, followed by the retaining nut (there are no shims, washers or other bearings). Tighten the retaining nut to 36 ft. lbs. (50 Nm) for all 25 hp models or to 54 ft. lbs. (75 Nm) for all 30/40 hp models.

■ **Where used, the same amount of shim material which was removed during disassembly should be reused. However, if the gearcase itself or a shaft, bearing or gear was replaced, the Gearcase Shimming procedure(s) must be followed to ensure proper lash adjustment.**

How well the pinion gear can mesh with the forward or reverse gear is partially a factor of how far the pinion gear protrudes into the gearcase. Before proceeding with gearcase assembly, take a moment to visually inspect the pinion gear depth.

4. Push downward slightly on the top of the driveshaft (pre-loading it inward toward the gearcase). At the same time, shine a flashlight into the gearcase and look at the pinion gear-to-forward gear contact areas. Look at the pinion gear tooth engagement with the forward gear teeth to be sure contact is made the full length of the tooth. If the pinion gear depth is not correct, the amount of shim material behind the pinion gear must be adjusted. Refer to the Gearcase Shimming procedures for more information Pinion Gear selection.

■ **On 8/9.9 and 15 hp motors there is a pinion gear shimming tool which can be inserted into the gearcase and used, along with a feeler gauge, to check the exact pinion gear depth setting. Obviously if this is necessary, this must be done BEFORE the propeller shaft and bearing seal/carrier cap is installed.**

LOWER UNIT 7-27

Propeller Shaft Bearing/Seal Carrier Cap

With the exception of potential Gearcase Shimming procedures which may need to be performed if certain parts were replaced, the Bearing/Seal Carrier (Cap) and propeller shaft installation is the last major piece of the gearcase overhaul.

If the bearing/seal carrier cap was disassembled to replace a bearing, bushing and/or oil seal it must be properly prepared for installation. This means you'll need to use suitably sized drivers to install replacement seals, bearing or bushing, depending upon what was removed. As usual, remember the driver must contact the part of the seal, bearing or bushing that is in contact with the housing. On most installations that means the outside of the seal, bearing or bushing, HOWEVER on some models the reverse gear bearing is installed ONTO the gear and not INTO the housing. When a gear bearing is pressed onto the outer diameter of a shaft on the gear, the driver needs to contact the INNER race instead.

■ **Remember to oil the contact surfaces of bearings, bushings or seals before installation. Be sure to pack the seal lips with a suitable marine grade grease once they are installed.**

Keep in mind that seals, bushing or bearing installation varies slightly from model-to-model. Although all seals used on these gearcases are of the dual variety and install with BOTH of their seal lips facing the rear of the outboard (the propeller), they do not all install from the same side of the Bearing/Seal Carrier Cap. Also, although some parts will install against a shoulder inside the carrier, others must be driven to a specified depth. If disassembled, prepare the bearing/seal carrier cap and the propeller shaft components for installation, as follows:

• For 6/8 hp motors, start by installing the solid bushing to the carrier from the oil seal (propeller side). Gently tap or press the bushing into place, then install each oil seal individually (again from the propeller side), checking the installation depth on each seal as you go. The first (inner) seal installed should go to a depth of 0.39-0.41 in. (10.0-10.5mm) measured from the propeller THRUST BEARING surface (NOT the absolute top of the carrier housing) to the top of the seal. The second (outer) seal installed should go to a depth of 0.12-0.14 in. (3.0-3.5mm) measured the same way. Lastly, turn the carrier over and press the reverse gear ball bearing into position, making sure the embossed markings on the bearing face outward (toward the driver).

• For 8/9.9 hp and 15 hp motors start by installing the needle bearing to the carrier. For all except high-thrust 9.9 models install the needle bearing from the oil seal (propeller) side. On high-thrust 9.9 models, install the needle bearing from the reverse gear (gearcase) side. Either way, position the bearing so the embossed side faces toward the propeller side and gently tap or press it into place (until it is flush with the mounting surface for the reverse gear ball bearing). Next install the oil seals until the top seal is at the proper depth when measured from the propeller THRUST BEARING surface (NOT the absolute top of the carrier housing) to the top of the seal. The seals are properly installed when the second (outer) installed is at a depth of 0.12-0.14 in. (3.0-3.5mm) for all except high-thrust 9.9 hp models on which the seals should be installed to a depth of 0.18-0.20 in. (4.5-5.0mm). Lastly, turn the carrier over and press the reverse gear ball bearing into position, making sure the embossed markings on the bearing face the gearcase (outward the driver).

• On standard thrust 25 hp (F25) models, start by installing the needle bearing to the carrier from the gearcase (reverse gear) side of the carrier. Make sure any embossed markings on the bearing are facing outward toward the gearcase side (driver). Press until the top of the bearing reaches a depth of 1.22-1.24 in. (31.0-31.5mm) from the gearcase end of the housing. Next invert the carrier and install each oil seal individually (from the propeller side), checking the installation depth on each seal as you go. The first (inner) seal installed should go to a depth of 0.51-0.53 in. (13.0-13.5mm) measured from the top of the seal bore (NOT the absolute top of the carrier housing) to the top of the seal. The second (outer) seal installed should go to a depth of 0.24-0.26 in. (6.0-6.5mm) measured the same way. The reverse gear bearing on these motors is pressed directly onto a bearing, which is in turn pressed into the housing. Yamaha recommends assembling the gear, shim and ball bearings, then installing the whole assembly to the bearing carrier. Make sure that you face the embossed side of the bearing toward the reverse gear (gearcase) side of the housing.

• On 30/40 hp models and high-thrust 25 hp (T25) models, start by installing the needle bearing to the carrier from the propeller (oil seal) side of the carrier. Make sure any embossed markings on the bearing are facing outward toward the propeller side (driver). For world production models (Except USA and Canada) press until the top of the bearing reaches a depth just a little less than 0.91-0.93 in. (23.0-23.5mm) from the propeller end of the housing. We say just a little less because Yamaha provides the spec from the TOP of the depth plate (#YB-90890-06603) that is placed on top of the carrier while installing the seal. On USA and Canada models Yamaha recommends you install the bearing using their driver rod and bearing/oil seal attachment (#YB-06071 and #YB-06111 respectively). If this tool is used drive the bearing into the carrier until the flange of the bearing/oil seal attachment is 0.12-0.14 in. (3.0-3.5mm) from the propeller end of the carrier. If their diagrams are accurate, this places the bearing approximately centered in the bore between the oil seal mounting area and the tapered bore which leads to the reverse gear bearing mounting surface. Next, install the oil seals until the top (outer) of the two seals is 0.16-0.18 in (4.0-4.5mm) from the top of the carrier housing. The reverse gear bearing on these motors is pressed directly onto a bearing, which is in turn pressed into the housing. Yamaha recommends assembling the gear, shim and ball bearings, then installing the whole assembly to the bearing carrier. Make sure that you face the embossed side of the bearing toward the reverse gear (gearcase) side of the housing.

1. The bearing carrier itself is sealed to the gearcase bore using one or two O-rings (depending upon the model). If this is the final assembly (and not a temporary assembly before Gearcase Shimming or to check gear contact patterns), for each O-ring used, install a NEW O-ring to ensure proper seal, then apply a light coating of marine grade grease.

2. If this assembly is to check gear contact patterns, coat the teeth of the reverse gear with a fine spray of Desenex®, as mentioned earlier. Handle the gear carefully to prevent disturbing the powder.

3. If applicable, place the washer over the reverse gear, then slide the threaded end of the assembled propeller shaft into the bearing carrier.

4. Install the propeller shaft and bearing carrier into the gearcase making sure the shaft inserts through the forward gear and into contact with the shifter. If equipped, align the keyway in the lower unit housing with the keyway in the bearing carrier. Insert the key into both grooves and then push the bearing carrier into place in the lower unit housing.

5. If this is the final installation (again, not temporary for shimming or checking contact patterns), apply a light coating of Loctite® 572 or an equivalent threadlocking compound to the bearing/seal carrier cap retaining screws.

6. Install and tighten the bearing/seal carrier cap retaining bolts or ring nut and tighten to specification as follows:

• For all 6-15 hp motors tighten the bolts to 5.8 ft. lbs. (8 Nm).

• For standard thrust 25 hp (F25) motors, tighten the bolts to 8.1 ft. lbs. (11.1 Nm).

• For 30/40 hp 4-stroke motors and high-thrust 25 hp (T25) motors, tighten the bolts to 12 ft. lbs. (16 Nm).

7. If components other than seals were replaced (such as the case itself, gears, bearings and/or shafts) proceed to the appropriate section of the Gearcase Shimming procedures to check and/or adjust gear lash.

8. If this is final assembly, install the Water Pump, as detailed in the Lubrication and Cooling section.

9. Refill the gearcase, install the propeller and install the lower unit assembly (in whatever order you'd prefer).

GEARCASE SHIMMING

■ **Since Yamaha shims (and gearcase markings) are all in mm we have kept most of the procedures in metric measurements only.**

In order for the gearcase to function properly the pinion gear teeth must fully engage with the teeth of the forward or reverse gear (depending on shifter position). Because build tolerances for the gearcase, gears, shafts and bearings/bushings will vary slightly the manufacturer has positioned shims strategically at various points in the gearcase to give you the ability to make up for the differences in these tolerances. In this way, a gear that is not fully engaging can be repositioned slightly by the use of a thicker shim. Similarly, a gear that does not fully disengage can be repositioned slightly further away by using a thinner shim.

Gearcase shimming procedures vary slightly from model-to-model and they depend upon the components that are being replaced. However, generally speaking:

• NO SHIMMING is required if the original case and original inner parts (gears, bearings and shafts) are installed.

7-28 LOWER UNIT

- NUMERIC CALCULATION RESHIMMING should be performed on 25 hp and larger gearcases when the original inner parts (gears, bearings and shafts) are installed in a NEW case. Many original and replacement cases are marked with various gear measurements. Typically an embossed F (forward gear), R (reverse gear) and/or P (pinion) gear measurements. When these marks are present the differences between the marks on the new case vs. the marks on the original case can be used to determine new shim sizes BEFORE assembly (potentially saving you from having to disassemble/reassemble the gearcase a second time).
- SPECIAL TOOL MEASUREMENT RESHIMMING is performed prior to assembly on certain models and should be used when a shaft, bearing/bushing or gear is replaced. In this case, special tools are used to measure pre-assembled or partially assembled components and determine if a change in shim sizes is necessary. A special tool procedure is available for the gearcase found on the 8/9.9 hp and 15 hp motors. A complete pinion gear shimming tool set is used on gearcases whose driveshaft is equipped with a pressed on tapered roller bearing. In addition, one or more shimming gauges are available for all 25 hp and larger motors.
- BACKLASH MEASUREMENT RESHIMMING should be performed and adjusted when one or more of the original inner components (gears, bearings and shafts) are installed. Backlash measurement is normally performed on an almost fully assembled gearcase using a dial gauge to read the amount of play in the driveshaft. This procedure can check the placement of all 3 gears through this measurement. Though adjustment is much more hit and miss than the other methods, since you need to then re-disassemble the gearcase, swap shims, re-assemble and re-check, then repeat until you get it right. Fortunately, formulas are available to use the measured backlash in helping to pick shim sizes and small shim changes are all that is normally necessary. It is rare that you would have to disassemble the gearcase more than one additional time.

Follow the appropriate Shimming or Backlash measurement procedure, depending upon your needs as outlined above.

Numeric Calculation Re-shimming

The gearcase for all 25 hp and larger models SHOULD be marked with measurements representing the case's deviation from standard. Typically an embossed F (forward gear), R (reverse gear) and/or P (pinion) and a + or - with a number. The codes can be found on the underside of the anti-cavitation plate, usually underneath the trim tab (so if you look under there and can't find them, remove the trim tab and check again).

The stamped number should be interpreted as hundredths of a mm. This means that if there is a marking such as F +5, it means that the case requires a shim that it 5/100mm (0.05mm) larger than standard. The + or - accompanying the number tells you whether you should add (+) or take away (-) the number from the standard measurement. If there is NO number after the F, R or P, assume the value is 0 and the case meets standard tolerance. If the number is there, but illegible, you'll have to work from 0 and make adjustments for that shim by checking lash.

Now, when moving all of the internal components to a new case, compare the numbers on the 2 cases and adjust the shims accordingly. If the numbers are all the same (lets say they both are F +5, R-1 and P+8), you've got it easy, use the original shims and make no changes. But, if the new case is, lets say an F+3, then subtract the new value from the old value to determine how much of a change you need to make to the forward gear shim. In this example, 5 - 3 = 2. This means the NEW gearcase needs a shim that is 0.02mm SMALLER than the forward gear shim which was originally used. Measure the old shim and obtain a new shim which is about 0.02mm smaller for reassembly.

Once you've selected the proper shims, assemble the gearcase as detailed earlier in this section and, before the water pump is installed, check the gearcase Backlash as detailed later in this section. As long as the backlash is within spec, no further adjustment is necessary. However, if backlash is out of specification you'll need to further adjust one or more of the shims to correct the problem.

Special Tool Measurement Re-shimming

On some models one ore more special tools are available to assist in determining shim sizes before installation is complete. A pinion height gauge is available for the gearcase found on the 8/9.9 hp and 15 hp motors. A complete pinion gear shimming tool set is used on gearcases whose driveshaft is equipped with a pressed on tapered roller bearing. In addition, one or more shimming gauges are available for all 25 hp and larger motors.

8/9.9 and 15 Hp motors
◆ See Figures 48 thru 52

On the US and Canadian versions of these gearcases, a pinion height gauge is available (#YB-34232 for standard thrust models or #YB-6299 for high-thrust models), in order to help determine pinion shim size. Then, forward and reverse gear shim selection is made by checking backlash and adjusting the shims according to backlash readings.

On world production models (those made for sale/use outside the US and Canada), pinion gear shim selection is determined using a formula (and a sliding caliper to measure the thickness of the current shim and bearing). Following pinion gear shim selection on world production models, additional measurement are taken using a flat shimming plate (#YB-90890), which is 10mm thick is used to take measurements (again using the sliding caliper) to help determine forward and reverse shim thickness. Of course backlash readings should still be taken to verify shim selection on the world production models.

1. On US and Canadian models, make sure the gearcase is assembled to the point where the driveshaft and pinion are in place. Insert the appropriate pinion height gauge, then use a feeler gauge between the pinion gear and the gauge itself to check clearance. Clearance should be 1.15-1.25mm on all standard thrust models, or 0.45-0.55mm on high-thrust models. If the measurement is out of specification, remove the pinion gear and measure the shim material that was installed. Once you've got the old shim thickness, and the clearance measurement, proceed as follows to select the ideal new shim thickness which would put you toward the center of the range:

- For standard thrust models, if the clearance was LESS than 1.15mm, subtract the measurement from 1.20mm. The result is the amount you need to DECREASE the pinion gear shim. For example, if the measurement was 1.10mm, subtract it from 1.20mm to get 0.10mm. The new shim should be about 0.10mm SMALLER than the shim used during the measurement.
- For standard thrust models, if the clearance was MORE than 1.25mm, subtract 1.20mm from the measurement. The result is the amount you need to INCREASE the pinion gear shim. For example, if the measurement was 1.35mm, subtract 1.20mm from it to get 0.15mm. The new shim should be about 0.15mm LARGER than the shim used during the measurement.
- For high-thrust models, if the clearance was LESS than 0.45mm, subtract the measurement from 0.50mm. The result is the amount you need to DECREASE the pinion gear shim. For example, if the measurement was 0.30mm, subtract it from 0.50mm to get 0.20mm. The new shim should be about 0.20mm SMALLER than the shim used during the measurement.
- For high-thrust models, if the clearance was MORE than 0.55mm, subtract 0.50mm from the measurement. The result is the amount you need to INCREASE the pinion gear shim. For example, if the measurement was 0.65mm, subtract 0.50mm from it to get 0.15mm. The new shim should be about 0.15mm LARGER than the shim used during the measurement.

2. For World production models (except USA and Canada), proceed as follows:

a. Pinion shim selection is made by measuring the combined thickness of the thrust bearing and plate washer, then subtracting that measurement from a standard spec of 6.05mm for standard models or 5.99mm for high-thrust models. In other words, the shim material, plate washer and thrust bearing together must be a combined total thickness equal to the standard spec.

b. Now measure the thickness of the forward gear bearing (at the very outer race of the bearing cage). Take the measurement at 2-3 spots around the bearing cage, then average the measurements. Take the measurement and subtract it from the required bearing and shim thickness spec of 16.60mm for standard thrust models and 16.50mm for high-thrust models. The result is the ideal forward gear shim size, select one that is as close as possible.

c. To determine the reverse gear shim size, first place the shimming plate (or other 10mm thick piece of flat stock) across the back of the assembled propeller shaft bearing carrier and gear assembly. Measure the distance from the top of the plate to the back (gearcase NOT propeller) side of the carrier mounting flange (through which the carrier retaining bolts are threaded). Subtract this measurement from the required carrier mount-to-gear total thickness of 80.57mm for standard thrust models or 81.00mm for standard thrust models. The result is the ideal reverse gear shim size that would place the reverse gear at a suitable distance in front of the mounting flange when the assembly is installed in the gearcase.

3. Once you've selected the proper shims, assemble the gearcase as detailed earlier in this section and, before the water pump is installed, check the gearcase Backlash as detailed later in this section. As long as the

LOWER UNIT 7-29

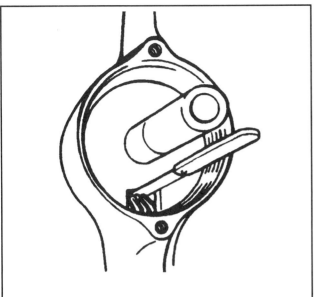

Fig. 48 Pinion gear height measurement on US and Canadian high-thrust models

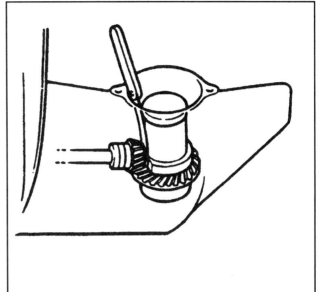

Fig. 49 Pinion gear height measurement on US and Canadian standard thrust models

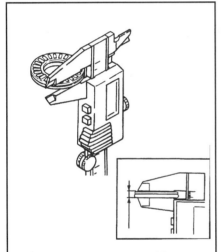

Fig. 50 On world production models measure the thrust bearing and plate washer thickness

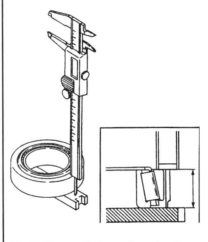

Fig. 51 Measure the forward gear bearing thickness to determine forward gear shim size

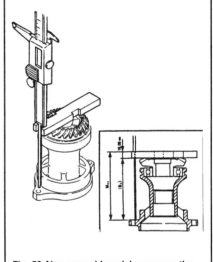

Fig. 52 Also on world models measure the assembled bearing carrier-to-gear thickness

backlash is within spec, no further adjustment is necessary. However, if backlash is out of specification you'll need to further adjust one or more of the shims to correct the problem.

Shim Selection on US and Canadian 25 Hp and Larger Motors With A Tapered Roller Driveshaft Bearing

◆ See Figures 53, 54, and 56 thru 58

On US and Canadian models that use a tapered roller bearing pressed onto the driveshaft (this includes all 25 hp and larger motors), a special pinion gauge block, adaptor plate and base is used to measure clearance. This is basically a method of pre-assembling the driveshaft, bearing, pinion gear and selected shims outside the gearcase in order to precisely check the assembled distance between the bearing and the pinion. The components of the gauge set used vary slightly from model-to-model and are detailed in the procedure.

On these models both the forward and reverse gear bearing shim thicknesses are determined using a shimming gauge, the bearing and a calculation that includes the stamped gearcase deviation from standard.

1. First determine the pinion gear shim thickness, as follows:
    a. Check the stamped code on the underside of the anti-ventilation plate (it is stamped in the trim tab mounting area, so if you haven't already, matchmark and remove the tab). The code **P** followed by a **+** or **-** and a number is the gearcase's pinion bearing mounting deviation from the standard spec, in hundredths of a mm. This means that if there is a marking such as P +5, it means that the case requires a shim that it 5/100mm (0.05mm) larger than standard. The **+** or **-** accompanying the number tells you whether you should add (**+**) or take away (**-**) the number from the standard measurement. If there is NO number after the **P**, assume the value is 0 and the case meets standard tolerance. If the number is there, but illegible, you'll have to work from 0 and make adjustments for that shim by checking lash.
    b. Determine the measured pinion clearance specification for installation in the gearcase in question. This is done by adding the stamped variant value to the standard specification of 0.05mm for standard thrust 25 hp (F25) motors or to the standard spec of 0.03mm for 30/40 hp motors or high-thrust 25 hp (T25) motors. The result is your target measured clearance between the pinion gear and the gauge block once the driveshaft assembly is installed in the next step. Adjustment takes place by using a thicker or thinner shim until the measured clearance equals this measurement.
    c. Assemble the driveshaft, bearing, shim and pinion gear in the gauge base (#YB-34432-11) along with the adapter plate (#YB-34432-10) and the appropriate gauge block. For the standard thrust 25 hp (F25) models use gauge block (#YB-34432-16). For 30/40 hp motors and high-thrust 25 hp

7-30 LOWER UNIT

(T25) motors, use gauge block (#YB-34432-8). On all except the standard thrust 25 hp (F25) motors you'll also need a clamp (#YB-34432-17).

    d. Once assembled, use a feeler gauge to check the gap between the top of the pinion gear and the bottom of the gauge block. Change shims as necessary until this measured gap equals the value calculated earlier in this procedure.

  2. Next, for standard thrust 25 hp (F25) motors, use the appropriate gauge to measure the forward gear bearing and the assembled reverse gear/bearing assembly and determine the necessary shim thicknesses (also using the case deviation from standard markings) as follows:

    a. Place the forward gear bearing a flat work surface (without the shim), then position the Yamaha shimming gauge (#YB-06344) over top of it. Using a feeler gauge, measure the gap between the tab protruding from the underside of the gauge and the bearing. This measurement will be used along with the standard spec and gearcase deviation from spec marking to determine the size of the shim which should be used.

    b. To determine the size of the necessary forward gear shim ADD the standard spec to the measurement AND the gearcase deviation. The standard spec is 1.0mm for 25 (F25) motors. For example, if you were working on a gearcase marking of P+5 AND measured a clearance of 0.15mm you would add 0.15mm (measured clearance), plus 1.0mm (standard spec) and 0.05 (gearcase deviation) to come up with a shim size of 1.20mm.

    c. Place the assembled propeller shaft bearing carrier and reverse gear/bearing assembly with the propeller side facing downward on a suitable work surface. Place the appropriate shimming gauge over the top of the reverse gear so that the gauge legs contact the housing. Use a Yamaha shimming gauge (#YB-39799), which has long legs that contact the backside (gearcase side) of the mounting flange at the other end of the bearing carrier.

    d. Hold the shimming gauge firmly against the housing while using a feeler gauge to measure the gap between the protrusion on the underside of the shimming gauge and the gear itself. For all motors the desired gap should be equal to the standard spec of 1.0mm MINUS the gearcase deviation stamp. Now remember, if the spec is a R-5 you would actually ADD 0.05mm to the standard spec because subtracting a negative is the same as adding the positive, remember? Adjust the shim size(s) until the specified feeler gauge is a proper fit.

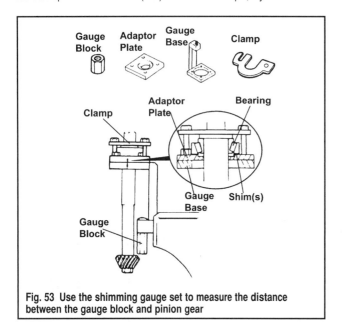

Fig. 53 Use the shimming gauge set to measure the distance between the gauge block and pinion gear

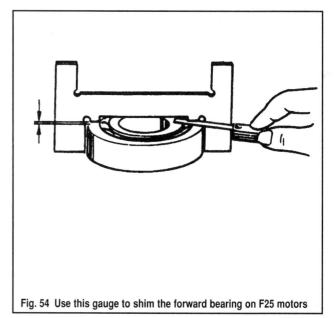

Fig. 54 Use this gauge to shim the forward bearing on F25 motors

Fig. 56 A different gauge is used for the reverse gear on the F25

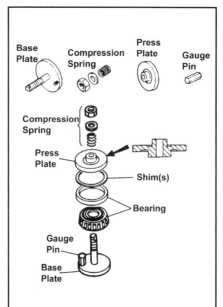

Fig. 57 This gauge set is used to measure the forward gear shim for 30/40 hp and T25 motors...

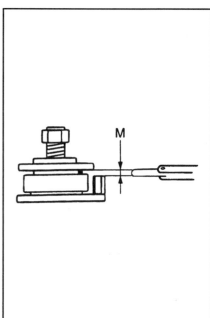

Fig. 58 ...measure clearance between the gauge pin and press plate

3. For models other than the standard thrust 25 hp (F25) motor, use the appropriate gauge to measure the forward gear bearing and determine the necessary shim thickness (also using the case deviation from standard markings) as follows:

   a. Assemble the forward gear bearing and original shim(s) in the Yamaha gauging set consisting of the base plate (#YB-34446-1), compression spring (#YB-34446-3) and press plate (#YB-34446-4). You'll also need to use either gauge pin (#YB-34446-7) for 30/40 hp motors or gauge pin (#YB-34446-15) for high-thrust 25 hp (T25) motors.

■ **If the original shims are not available, start measurement using a 0.50mm shim.**

   b. Once assembled, use a feeler gauge to check the gap between the press plate and the gauge pin. The goal is for this gap to be equal to the standard spec PLUS the gearcase deviation (or as close as possible to this). The standard spec for these gearcases is 0.75mm for high-thrust 25 hp (T25) motors or 0.06mm for 30/40 hp motors.
   c. Add or remove shims as necessary to reach the appropriate gap in the shimming tool.
   d. On these models Reverse gear shim selection is performed after assembly by checking backlash and making adjustments accordingly.

4. Once you've selected the proper shims, assemble the gearcase as detailed earlier in this section and, before the water pump is installed, check the gearcase Backlash as detailed later in this section. As long as the backlash is within spec, no further adjustment is necessary. However, if backlash is out of specification you'll need to further adjust one or more of the shims to correct the problem.

Shim Selection on World Production 25 Hp and Larger Motors With A Tapered Roller Driveshaft Bearing

◆ See Figures 51, 59 and 60

On World Production models that use a tapered roller bearing pressed onto the driveshaft (this includes all 25 hp and larger motors), a different special pinion gauge set (consisting only of a driveshaft holder and height gauge) than used by US and Canadian models. But the concept is the same in that using the gauge set is a method of pre-assembling the driveshaft, bearing, pinion gear and selected shims outside the gearcase in order to precisely check the assembled distance between the bearing and the pinion.

On these models both the forward and reverse gear bearing shim thicknesses are determined calculated based on measurements taken of the bearing or assembled gear/bearing housing assembly and the stamped gearcase deviation from standard.

1. To select the pinion gear bearing shim, proceed as follows:
   a. Check the stamped code on the underside of the anti-ventilation plate (it is stamped in the trim tab mounting area, so if you haven't already, matchmark and remove the tab). The code **P** followed by a **+** or **-** and a number is the gearcase's pinion bearing mounting deviation from the standard spec, in hundredths of a mm. This means that if there is a marking such as P +5, it means that the case requires a shim that it 5/100mm (0.05mm) larger than standard. The **+** or **-** accompanying the number tells you whether you should add (**+**) or take away (**-**) the number from the standard measurement. If there is NO number after the **P**, assume the value is 0 and the case meets standard tolerance. If the number is there, but illegible, you'll have to work from 0 and make adjustments for that shim by checking lash.

   b. Assemble the driveshaft, bearing and pinion gear (without any shim) in the Yamaha driveshaft holder (#90890-06517) along with the pinion height gauge (#90890-06702).

   c. Once assembled, use a sliding caliper to check the gap between the top of the pinion gear and the bottom of the height gauge. Rotate the driveshaft in the bearing and repeat this measurement at 2-3 spots around the pinion, then average the results.

   d. Now, determine the thickness of the shim which is necessary by using the following formula and standard spec. It varies slightly from model-to-model but in all cases you are subtracting a series of numbers, including the stamped gearcase deviation spec, the standard for the given gearcase and the measurement taken in the previous step. Remember that if the stamped deviation spec was a negative number that when you subtract a negative you are actually adding the absolute value (adding the positive of that number). Calculate the necessary pinion gear shim thickness as follows:

• For standard thrust 25 hp (F25) motors, the shim thickness is equal to the measurement, MINUS the standard spec for this gearcase of 27mm, MINUS the stamped gearcase deviation. For example, if the measurement was 28.2mm and the stamping was P+5 (plus 0.05mm) the shim thickness would be equal to 28.2mm, minus 27mm, minus 0.05, which comes to 1.15mm. In this case there is no 1.15mm shim, so you would use the closest, which is 1.2mm.

• For high-thrust 25 hp (T25) motors, the shim thickness is equal to the standard spec for this gearcase of 12mm, MINUS the measurement, MINUS the stamped gearcase deviation.

• For 30/40 hp motors, the shim thickness is equal to the measurement, MINUS the standard spec for this gearcase of 11.3mm, MINUS the stamped gearcase deviation.

■ **On these gearcases, Yamaha recommends that you use the smallest or nearest small shim available during calculations and adjustments.**

2. Next, determine the thickness of the forward gear bearing shim by measuring the bearing itself (at the very outer race of the bearing cage). Take the measurement at 2-3 spots around the bearing cage, then average the measurements. Use the measurement in the appropriate equation for your gearcase (along with the standard spec for that model and the stamped

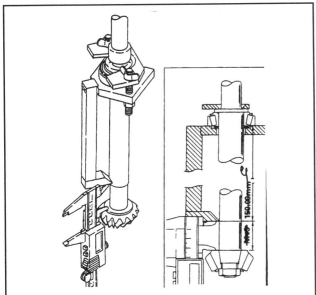

Fig. 59 Use the pinion height gauge to for the pinion shimming measurement

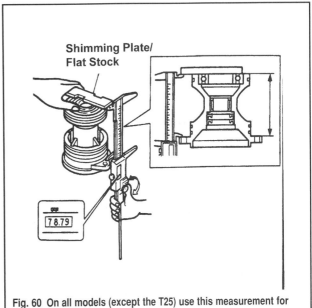

Fig. 60 On all models (except the T25) use this measurement for reverse gear shimming

7-32 LOWER UNIT

gearcase deviation, although in this case you are adding the deviation so a negative deviation does reduce the amount of shimming material needed).

Calculate the necessary pinion gear shim thickness as follows:

- For standard thrust 25 hp (F25) motors, the shim thickness is equal to the standard specification of 17.5mm, PLUS the stamped gearcase deviation, MINUS the measurement. For example, if the measurement was 16.25mm and the stamping was F+5 (plus 0.04mm) the shim thickness would be equal to 17.5mm, PLUS 0.05, minus 16.25, which comes to 1.30mm.
- For 30/40 hp motors and high-thrust 25 hp (T25) motors, the shim thickness is equal to the standard spec for this gearcase of 22.75mm, PLUS the stamped gearcase deviation, MINUS the measurement.

■ **For all high-thrust 25 hp (T25) motors the reverse gear shim is determined by checking and adjusting backlash. If possible, reuse the original shim or, if not available, start with a 0.50mm shim.**

3. Finally, for all except the high-thrust 25 hp (T25) motors, use a measurement and a calculation to determine the thickness of the reverse gear bearing shim as follows:

- For standard thrust 25 hp (F25) motors, use place a shimming plate or piece of perfectly flat bar stock over top of the reverse gear in the assembled reverse gear/bearing and propeller shaft bearing carrier assembly. Use a sliding caliper to measure the distance from the BOTTOM of the shimming plate to the inside (gearcase side) of the bearing carrier flange (at the mounting bolt ears). Calculate the required shim thickness by the following equation, standard spec of 80mm, PLUS the gearcase deviation, MINUS the measurement. For example, if the measurement was 78.79mm and the stamping was R+5, add 80mm and 0.05mm, then subtract 78.79mm to get 1.26mm and use the 1.20mm shim (as this is smallest of the shims closest in measurement as there are 1.20 and 1.30mm).
- For 30/40 hp motors, measure the distance from the gear to the bearing carrier flange in the same manner as the F25, however and use the same equation, except that the standard spec if different. For these motors take the standard spec of 98mm and add the stamped gearcase deviation, then subtract the measurement.

4. Once you've selected the proper shims, assemble the gearcase as detailed earlier in this section and, before the water pump is installed, check the gearcase Backlash as detailed later in this section. As long as the backlash is within spec, no further adjustment is necessary. However, if backlash is out of specification you'll need to further adjust one or more of the shims to correct the problem.

Backlash

◆ See Figures 61, 62 and 63

Backlash (also known as lash or play) is the acceptable clearance between two meshing gears, in order to take into account possible errors in machining, deformation due to load, expansion due to heat generated in the lower unit and center-to-center distance tolerances. A no backlash condition is unacceptable, as such a condition would mean the gears are locked together or are too tight against each other which would cause phenomenal wear and generate excessive heat from the resulting friction.

Excessive backlash which cannot be corrected with shim material adjustment indicates worn gears. Such worn gears must be replaced. Excessive backlash is usually accompanied by a loud whine when the lower unit is operating in neutral gear.

The backlash measurements are taken before the water pump is installed. If the amount of backlash needs to be adjusted, the lower unit must be partially disassembled to change the amount of shim material behind one or more of the gears. Of course, it usually does not make sense to change shims on BOTH the pinion AND either the forward or reverse gears at the same time, since the changes will have an aggregate affect on each other's readings and you're just making it harder on yourself.

As a general rule, if the lower unit was merely disassembled, cleaned and then assembled with only a new water pump impeller, new gaskets, seals and O-rings, there is no reason to believe the backlash would have changed. Therefore, it is safe to say this procedure may be skipped.

However, if any one or more of the following components were replaced, the gear backlash should be checked for possible shim adjustment (even if shimming calculations were performed):

- New lower unit housing - check forward and reverse gear shim material.
- New forward gear bearing - check forward gear backlash.
- New pinion gear - check pinion gear depth and forward/reverse gear backlash.
- New forward gear - check forward gear backlash.
- New reverse gear - check reverse gear backlash.
- New bearing carrier - check reverse gear backlash.
- New thrust washer on 25 hp and larger models - check reverse gear backlash.

■ **The lower unit backlash is measured with the unit inverted (upside down) for all 4-15 hp models or in the upright position for most 25 hp and larger models.**

To perform this check you'll need a dial gauge on a magnetic or threaded base, as well as a gear lash indicator. The lash indicator is essentially a hose clamp with a straight lever attached at a 90 degree angle to the clamp body. The clamp is attached to the driveshaft in a position that any movement in the shaft will result in movement at the lever. The dial indicator is then installed to the gearcase itself in a position contacting the lever to read the amount of movement. The only thing that stops you from fabricating the indicator yourself if the fact that the amount of lever movement (when compared to shaft rotation) increases dramatically as you move outward away from the clamp further down the lever. And Yamaha simply directs you to use the "mark" provided on the indicator when measuring lash. For all motors you can use the Yamaha Indicator (#YB-06265).

1. Position the gearcase in a suitable workstand either upright or inverted, depending upon the model. The lower unit backlash is measured with the unit inverted (upside down) for all 4-15 hp models or in the upright position for most 25 hp and larger models.

■ **Some people prefer to set up the backlash indicator and dial gauge while the gearcase is still upright (just because it is easier than working on it while it is upside down). If so, perform the next couple of steps first, then invert the gearcase on 4-15 hp motors.**

2. Make sure the gearcase is in **Neutral** by slowly spinning the driveshaft and checking for a lack of motion at the propeller shaft.

■ **We're not really sure why Yamaha chooses to preload gears by pushing on the prop shaft when checking Forward gear backlash and pulling on the prop shaft when checking Reverser gear backlash. Some other manufacturers (and on some other Yamahas) the preferred method is simply to shift the gearcase into the gear whose lash is being checked. This is always an alternative if you can't seem to get the prop shaft and gear loaded sufficiently to check lash.**

3. Since you normally start by checking Forward gear backlash, install a bearing carrier puller and J bolts onto the ribs of the carrier. With the bearing carrier attaching bolts or the bearing carrier locknut secured, tighten the puller just enough to push on the propeller shaft preloading it.

■ **The bearing carrier puller is used to both keep the propeller shaft from turning and to preload the forward gear, thus freeing up one of your hands. However, you can simply push inward and hold the shaft (or have an assistant hold the shaft) while taking readings.**

4. Obtain and install a backlash indicator gauge (#YB6265) onto the driveshaft. If the hose clamp for the indicator is too large to fit the propeller shaft snugly, insert some kind of packing into the clamp to allow it to be tightened. The clamp must be secure to permit no movement of the indicator on the shaft.

■ **Yamaha recommends using the backlash adjusting plate (#YB-07003) to mount the dial indicator to the case. Although it is not absolutely necessary, it does provide a level, fixed position to which the dial gauge can be mounted. If you do NOT use this plate, make sure the gauge is secure and will not move at all during lash measurement.**

5. Install the gauge to the gearcase. Position the dial gauge against the mark on the indicator tool.

6. If you are not using a puller to hold the propeller shaft from turning, grab a hold of the propeller shaft in order to keep it from turning, then twist the driveshaft in one direction (either clockwise or counterclockwise, it doesn't matter at this time) and zero the dial gauge.

LOWER UNIT 7-33

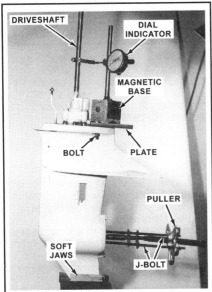

Fig. 61 Yamaha gearcase set-up for Forward gear lash measurement

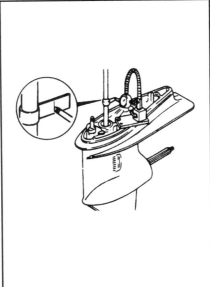

Fig. 62 Position the dial gauge so it contacts the mark on the lash indicator...

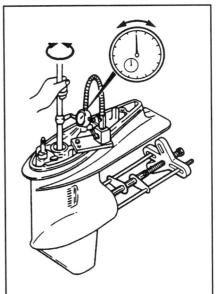

Fig. 63 ...turn the driveshaft one way, zero the gauge, then rotate to the opposite stop

■ See the specifications listed below before attempting to read gear backlash. On some models other special conditions are necessary, such as pulling or pushing on the driveshaft as it is rotated or installing the propeller without the spacer to pre-load the reverse gear when checking reverse gear lash.

7. Now, continue to hold the propeller shaft to prevent the shaft from rotating. With the other hand gently rock the driveshaft back and forth. In this way, you'll move the driveshaft back and forth between gear contact points with the forward or reverse gear (which can't move because you're holding the propeller shaft). Note the maximum deflection of the dial gauge needle, this is the amount of gear lash present in the gearcase for the current shims in use and compare them to the specifications, as follows:

• On 6/8 hp motors, with the gearcase inverted and while pulling downward slightly on the driveshaft, forward gear backlash should be about 0.88-1.26mm (0.0346-0.0496 in.). The center bolt on the puller which is used to preload the forward gear should be tightened to 1.5 ft. lbs. (2 Nm). If forward gear lash is LESS than 0.88mm, determine the amount of shim to remove by subtracting the measurement from 1.07mm and dividing the result by 2.52mm. If the amount of lash is MORE than 1.26mm, determine the amount of shim to add by subtracting 1.07mm from the measurement and dividing the result by 2.52mm. To check reverse gear backlash, remove the puller and install the propeller without the propeller nut side spacer, then tighten the nut to 1.5 ft. lbs. (2 Nm) to preload the reverse gear, then take the readings in the same manner, while pulling downward on the driveshaft. Reverse gear backlash should be about 1.01-1.38mm (0.0398-0.0543 in.). If lash is LESS than 1.01mm, determine the amount of shim to remove by subtracting the measurement from 1.20mm and dividing the result by 2.52mm. If the amount of lash is MORE than 1.38mm, determine the amount of shim to add by subtracting 1.20mm from the measurement and dividing the result by 2.52mm.

■ When making gear lash adjustments. Remember, too much lash means there is too much space between the gear in question and the pinion, therefore ADDING shim material will DECREASE backlash. Too little lash means there is not a large enough gap, therefore REMOVING shim material will INCREASE backlash.

• On standard thrust 8/9.9 hp motors, with the gearcase inverted and while pulling downward slightly on the driveshaft, forward gear backlash should be about 0.23-0.70mm (0.0091-0.0276 in.). The center bolt on the puller which is used to preload the forward gear or the propeller nut which is used to preload the reverse gear should be tightened by hand just until the gear restricts movement sufficiently to measure lash. If forward gear lash is LESS than 0.23mm, determine the amount of shim to remove by subtracting the measurement from 0.47mm and dividing the result by 2.30mm. If the amount of lash is MORE than 0.70mm, determine the amount of shim to add by subtracting 0.47mm from the measurement and dividing the result by 2.30mm. To check reverse gear backlash, remove the puller and install the propeller without the front side spacer, tighten the nut lightly by hand to preload the reverse gear, then take the readings in the same manner, while pulling downward on the driveshaft. Reverse gear backlash should be about 0.82-1.16mm (0.0323-0.0457 in.). If lash is LESS than 0.82mm, determine the amount of shim to remove by subtracting the measurement from 0.99mm and dividing the result by 2.30mm. If the amount of lash is MORE than 1.16mm, determine the amount of shim to add by subtracting 0.99mm from the measurement and dividing the result by 2.30mm.

• On high-thrust 9.9 hp motors, with the gearcase inverted and while pulling downward slightly on the driveshaft, forward gear backlash should be about 0.26-0.77mm (0.0102-0.0303 in.). The center bolt on the puller which is used to preload the forward gear or the propeller nut which is used to preload the reverse gear should be tightened by hand just until the gear restricts movement sufficiently to measure lash. If forward gear lash is LESS than 0.26mm, determine the amount of shim to remove by subtracting the measurement from 0.51mm and dividing the result by 2.50mm. If the amount of lash is MORE than 0.77mm, determine the amount of shim to add by subtracting 0.51mm from the measurement and dividing the result by 2.50mm. To check reverse gear backlash, remove the puller and install the propeller without the front side spacer, tighten the nut lightly by hand to preload the reverse gear, then take the readings in the same manner, while pulling downward on the driveshaft. Reverse gear backlash should be about 0.51-1.02mm (0.0201-0.0402 in.). If lash is LESS than 0.51mm, determine the amount of shim to remove by subtracting the measurement from 0.77mm and dividing the result by 2.50mm. If the amount of lash is MORE than 1.02mm, determine the amount of shim to add by subtracting 0.77mm from the measurement and dividing the result by 2.30mm.

• On 15 hp motors, with the gearcase inverted and while pulling downward slightly on the driveshaft, forward gear backlash should be about 0.19-0.86mm (0.0070-0.0340 in.). The center bolt on the puller which is used to preload the forward gear should be tightened to 3.6 ft. lbs. (5 Nm). If forward gear lash is LESS than 0.19mm, determine the amount of shim to remove by subtracting the measurement from 0.53mm and dividing the result by 2.30mm. If the amount of lash is MORE than 0.86mm, determine the amount of shim to add by subtracting 0.53mm from the measurement and dividing the result by 2.30mm. To check reverse gear backlash, remove the puller and install the propeller without the front side spacer, then tighten the nut to 3.6 ft. lbs. (5 Nm) to preload the reverse gear, then take the readings in the same manner, while pulling downward on the driveshaft. Reverse gear backlash should be about 0.95-1.65mm (0.0370-0.0640 in.). If lash is LESS than 0.95mm, determine the amount of shim to remove by subtracting the measurement from 1.30mm and dividing the result by 2.30mm. If the amount of lash is MORE than 1.65mm, determine the amount of shim to add by subtracting 1.30mm from the measurement and dividing the result by 2.30mm.

7-34 LOWER UNIT

- On standard thrust 25 hp (F25) motors, all measurements are taken with the gearcase upright. The center bolt on the puller which is used to preload the forward gear should be tightened to 3.6 ft. lbs. (5 Nm) so the gear restricts movement sufficiently to measure lash. The forward gear backlash should be about 0.31-0.72mm (0.0120-0.0280 in.). If lash is LESS than 0.31mm, determine the amount of shim to remove by subtracting the measurement from 0.51mm and then multiplying the result by 0.49mm. If the amount of lash is MORE than 0.72mm, determine the amount of shim to add by subtracting 0.51mm from the measurement and then multiplying the result by 0.49mm. To check reverse gear backlash, remove the puller and set the shift shaft to the **Reverse** position, then install the propeller without the spacer/collar and tighten the nut to 3.6 ft. lbs. (5 Nm) to preload the reverse gear. Take the readings in the same manner as forward gear. Reverse gear backlash should be about 0.93-1.65mm (0.0370-0.0650 in.). If lash is LESS than 0.93mm, determine the amount of shim to remove by subtracting the measurement from 1.29mm and then multiplying the result by 0.49mm. If the amount of lash is MORE than 1.65mm, determine the amount of shim to add by subtracting 1.29mm from the measurement and then multiplying the result by 0.49mm.

- On high-thrust 25 hp (T25) motors, all measurements are taken with the gearcase upright. The center bolt on the puller which is used to preload the forward gear should be tightened to 3.6 ft. lbs. (5 Nm) so the gear restricts movement sufficiently to measure lash. The forward gear backlash should be about 0.10-0.30mm (0.0040-0.0120 in.). If lash is LESS than 0.10mm, determine the amount of shim to remove by subtracting the measurement from 0.20mm and then multiplying the result by 0.44mm. If the amount of lash is MORE than 0.30mm, determine the amount of shim to add by subtracting 0.20mm from the measurement and then multiplying the result by 0.44mm. To check reverse gear backlash, remove the puller, then install the propeller, washer (facing backwards), spacer, washer and then tighten the nut to 3.6 ft. lbs. (5 Nm) to preload the reverse gear. Take the readings in the same manner as forward gear. Reverse gear backlash should be about 0.40-0.60mm (0.0160-0.0240 in.). If lash is LESS than 0.40mm, determine the amount of shim to remove by subtracting the measurement from 0.50mm and then multiplying the result by 0.44mm. If the amount of lash is MORE than 0.60mm, determine the amount of shim to add by subtracting 0.50mm from the measurement and then multiplying the result by 0.44mm.

- On 30/40 hp motors, all measurements are taken with the gearcase upright. The center bolt on the puller which is used to preload the forward gear should be tightened to 3.6 ft. lbs. (5 Nm) so the gear restricts movement sufficiently to measure lash. The forward gear backlash should be about 0.10-0.30mm (0.0040-0.0120 in.). If lash is LESS than 0.19mm, determine the amount of shim to remove by subtracting the measurement from 0.38mm and then multiplying the result by 0.53mm. If the amount of lash is MORE than 0.56mm, determine the amount of shim to add by subtracting 0.38mm from the measurement and then multiplying the result by 0.53mm. To check reverse gear backlash, remove the puller and install the propeller without the collar/spacer that normally goes between the propeller and gearcase, then tighten the nut to 3.6 ft. lbs. (5 Nm) to preload the reverse gear. Take the readings in the same manner as forward gear. Reverse gear backlash should be about 0.40-0.60mm (0.0160-0.0240 in.). If lash is LESS than 0.75mm, determine the amount of shim to remove by subtracting the measurement from 0.94mm and then multiplying the result by 0.53mm. If the amount of lash is MORE than 1.13mm, determine the amount of shim to add by subtracting 0.94mm from the measurement and then multiplying the result by 0.53mm.

■ If the lash is too low on BOTH the forward and reverse gears, the pinion height may be too low. If the lash is too high on BOTH gears, the pinion may be too high in the gearcase. Check the installation of the bearings/bushings and the shim materials or spacers used.

12. If adjustment is necessary, partially disassemble the gearcase as necessary to access the forward or reverse gear shims, then ad or take away shim material, as necessary to bring backlash with spec. However, if the backlash specification cannot be reached by adding and removing shim material, the gears may have to be replaced.

13. If possible, pressure test the gearcase. With the oil fill/drain screw installed and tightened, attach a threaded pressure tester to the oil level hole, then use a hand pump to apply 14.22 psi (100 kPa) of pressure to the gearcase. Watch that the pressure remains steady for at least 10 seconds. If pressure falls before that point one or more of the seals require further attention. You can immerse the gearcase in water and look for bubbles to determine where the leak is occurring. Be sure of gearcase integrity before installation.

✱✱ WARNING
Don't over-pressurize the gearcase as this could CAUSE leaks.

14. Install the Water Pump to the top of the gearcase as detailed in the Lubrication and Cooling section.
15. Install the Gearcase to the intermediate housing, as detailed in this section.
16. Install the Propeller and properly refill the Gearcase Lubricant, both as detailed in the Maintenance and Tune-Up section.

Gear Mesh Pattern

◆ See Figure 64

As noted earlier, you can use a machinist dye or other powder to check the gear mesh pattern. This is useful when determining if there is too much wear on a gear set to return them to service (especially if a dial gauge set and lash indicator is not available to check gear backlash).

The basic method used to check a gear mesh pattern is to coat the gears with a dye or powder, then install the gears, carefully rotate the shafts/gears and carefully disassemble the case again to examine the gears.

All models should be held with the lower unit in the upright (normal) position.

Grasp the driveshaft and pull upward. At the same time, rotate the propeller shaft counterclockwise through about six or eight complete revolutions. This action will establish a wear pattern on the gears with the dye/Desenex® powder.

1. Disassemble the unit and compare the pattern made on the gear teeth with the accompanying illustrations. The pattern should almost be oval on the drive side and be positioned about halfway up the gear teeth.

If the pattern appears to be satisfactory, clean the dye or powder from the gear teeth and assemble the unit one final time.

If the pattern does not appear to be satisfactory, add or remove shim material, as required. Adding or removing shim material will move the gear pattern towards or away from the center of the teeth.

After the gear mesh pattern is determined to be satisfactory, assemble the bearing carrier one final time.

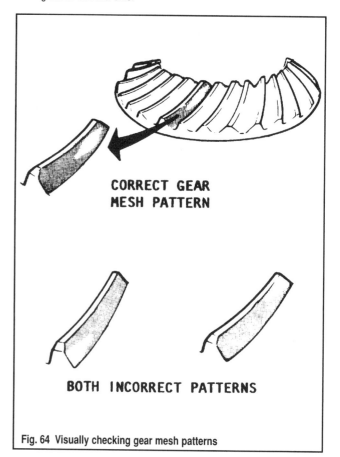

Fig. 64 Visually checking gear mesh patterns

LOWER UNIT

Gearcase - 40-225 Hp (4- and 6-Cylinder) Models and V6 Mercury/Mariner Models

REMOVAL & INSTALLATION

 MODERATE

◆ See Figures 26, 27, 28 and 65 thru 68

Gearcase removal or installation is a relatively straightforward procedure on most models. Generally speaking on all but the smallest of the Yamaha outboards it involves disconnecting the shift linkage and unbolting the gearcase from the intermediate housing, then carefully lowering the gearcase (along with the shift rod and/or driveshaft and water tube) straight down and off the motor. Because the water pump housing is mounted to the top of the gearcase, between the case and intermediate housing, removal is necessary for impeller inspection and/or replacement.

Because the gearcase is a sealed unit when it comes to gear oil, the fluid only needs to be drained if the case itself is being opened for further service (some form of seal or internal parts replacement). However, if this is one of the reasons the case is being removed, it is usually easier to go ahead and drain the gearcase oil before the assembly is removed.

On these models the gearcase is bolted to the intermediate housing with anywhere from 5 to 8 bolts and/or nuts threaded upward from the underside of the anti-cavitation plate. Generally, if equipped with a trim tab, the tab must be removed in order to access one of the bolts. The shift linkage on these models works by rotation and not by lifting or pressing downward as on the smaller Yamaha gearcases. As a result the top of the gearcase shift linkage is actually splined for connection with the linkage found in the intermediate housing and can therefore simply be pulled apart when the gearcase is unbolted. A few of these models may also equipped with a speedometer pickup/water tube at the front of the gearcase which must be disconnected before gearcase removal.

■ The V6 Mercury/Mariner models are ALSO equipped with a Yamaha built gearcase. Mercury sells Yamaha special tools (though usually under Mercury's own part numbers) for use on this gearcase. Whenever possible, we've listed both tool numbers, so they can be acquired from either source.

1. If the gearcase is to be overhauled or resealed, drain the gearcase fluid as detailed in this Maintenance and Tune-Up section. After the lubricant has drained, temporarily reinstall both the drain and oil level screws.

■ As the lubricant drains, catch some with your fingers from time to time and rub it between your thumb and finger to determine if any metal particles are present. If any significant amount of metal is detected in the lubricant, the unit must be completely disassembled, inspected and the damaged parts replaced. Check the color of the lubricant as it drains. A whitish or creamy color indicates the presence of water in the lubricant. Check the drain pan for signs of water separation from the lubricant. The presence of any water in the gear lubricant is bad news. The unit must be completely disassembled, inspected and the cause of the problem determined and corrected.

2. A few models utilize a speedometer pilot water hose at the top front of the gearcase. If equipped, carefully cut the wire tie and gently pull the hose off the fitting.
3. Remove the propeller for better access to the gearcase retaining bolt(s) threaded upward from underneath the anti-cavitation plate. For details, please refer to the Maintenance and Tune-Up section.

■ The trim tab obscures access to one of the gearcase retaining bolts.

4. Matchmark the position of the trim tab to the gearcase, as an aid to installing it back in its original location.

■ Generally speaking, on smaller motors (most in the 40-60 hp range) the trim tab is retained by a bolt threaded upward from underneath the anti-cavitation plate. However on most larger motors the tab is installed by a bolt threaded downward through the intermediate housing flange (it can be found by removing a small rubber cap/grommet).

5. For larger outboards, remove the plastic cap above the trim tab for access to the trim tab retaining bolts.
6. For all motors, use the correct size socket (and a long extension when threaded from above the tab) to remove the bolt and the trim tab.
7. Remove the bolts (or nuts) threaded upward (there are usually 2 or 3 on each side) securing the gearcase to the intermediate housing. Also, remove the bolt from the center of the gearcase which was exposed by removing the trim tab.

■ On models with studs and nuts instead of bolts retaining the gearcase there is normally a intermediate housing extension piece which mounts between the housing and the gearcase. This extension is normally free to come off once the gearcase is removed, so take care that it does not drop and become damaged.

8. Carefully separate the lower unit from the intermediate housing by pulling STRAIGHT downward from the intermediate housing. Watch for and save the dowel pins (usually 2) when the two units are separated. The water

Fig. 65 The trim tab hides one of the gearcase bolts

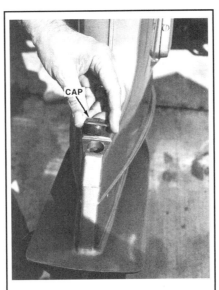

Fig. 66 On most models, access the tab bolt from the top

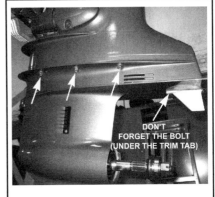

Fig. 67 Remove the bolts and carefully lower the gearcase

LOWER UNIT

tube will normally come out of the grommet and remain with the intermediate housing. The driveshaft will remain with the lower unit. The lower portion of the shift linkage should remain with the lower housing.

■ V6 models there is a normally a rubber exhaust seal mounted to the top of the gearcase or bottom of the intermediate housing. On the larger V6 models there may also be a seal plate between the seal and water pump housing. Keep track of these components for installation purposes. Also, take a good look at the rubber seal before deciding that it is fit for further service.

To Install:

9. If removed on and V6 models install the plate (if equipped) and the rubber exhaust seal. Place a light coating of marine grade grease on the rubber seal before installation.
10. Apply a dab of lubricant such as a Molybdenum lubricant or Yamaha Marine Grease, or equivalent water resistant lubricant to the driveshaft splines (the sides of the driveshaft's upper end, NOT the top surface itself).

■ An excessive amount of lubricant on top of the driveshaft will be trapped in the clearance between the top of the driveshaft and bottom of the crankshaft. This trapped lubricant may not allow the driveshaft to fully engage with the crankshaft.

11. Apply some of the same lubricant to the end of the water tube in the intermediate housing, and to the dowel pin(s) on the mating surface of the lower unit.
12. Set the gearcase in gear so that you can rotate the propeller shaft slightly to assist with crankshaft-to-driveshaft spline alignment. Be sure to place both the gearcase and the shifter in the same gear for ease of installation.
13. Begin to bring the intermediate housing and lower gear housing together.

■ The next step takes time and patience. If it's your first time, success will probably not be achieved on the first attempt. Three items must mate at the same time before the lower unit can be seated against the intermediate housing.

- The top of the driveshaft on the lower unit indexes with the lower end of the crankshaft.
- The water tube in the intermediate housing slides into the grommet on the water pump housing.
- The top splines of the lower shift rod in the lower unit slide into the internal splines of the upper shift rod in the intermediate housing.

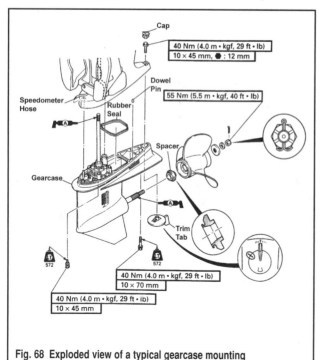

Fig. 68 Exploded view of a typical gearcase mounting

14. As the two units come closer, rotate the propeller shaft back and forth EVER-SO-SLIGHTLY to index the upper end of the lower driveshaft tube with the crankshaft. At the same time, feed the water tube into the water tube grommet and feed the lower shift rod into the upper shift rod.
15. Push the lower unit housing and the intermediate housing together. The dowel pin(s) should index ensuring the gearcase is properly aligned.

■ If all items appear to mate properly, but the lower unit seems locked in position about 4 inches (10cm) away from the intermediate housing and it is not possible, with ease, to bring the two housings closer, the driveshaft has missed the cylindrical lower oil seal leading to the crankshaft. Move the lower unit out of the way. Shine a flashlight up into the intermediate housing and find the oil seal housing. Now, on the next attempt, find the edges of the oil seal housing with the driveshaft before trying to mate anything else - the water tube and the top of the shift rod. If the driveshaft can be made to enter the oil seal housing, the driveshaft can then be easily indexed with the crankshaft.

16. Apply a light coating of Loctite® 572, or an equivalent threadlocking compound to the threads of the fasteners used to secure the lower unit to the intermediate housing. Remember, one bolt passes through the area covered by the trim tab. Install and tighten the bolts to a torque value of 26-29 ft. lbs. (37-40 Nm) for all except the 150 hp and larger motors on which the retainers should be tightened to 35 ft. lbs. (48 Nm).

■ Although Yamaha recommends threadlocking compound on most of their gearcase retainers, they occasionally do not specifically call for it. We prefer to ALWAYS use it in these applications because it offers to benefits. Besides helping to prevent the fasteners from loosening in service threadlocking compound also insulates the fastener threads from the threads of the housing or stud, helping to prevent corrosion that could seize the fastener in place. In this way it also acts as something of an anti-seize that also holds the bolt in place, a win, win situation if you ask us.

17. Apply a light coating of Loctite® 572 or an equivalent threadlocking compound to the threads of the trim tab retaining bolt, then install the trim tab, aligning the matchmarks made earlier and tighten the bolt securely. On larger models where the bolt is threaded from above the tab, install the rubber grommet into the intermediate housing over the bolt.
18. If equipped reconnect the speedometer pilot hose and secure using a new wire tie.
19. Operate the shift lever through all gears. The shifting should be smooth and the propeller should rotate in the proper direction when the flywheel is rotated by hand in a clockwise direction. Naturally the propeller should not rotate when the unit is in neutral.
20. Apply Yamaha All Purpose Grease, or equivalent anti-seize compound to the propeller shaft.

■ The compound will prevent the propeller from freezing to the shaft and permit the propeller to be removed, without difficulty, the next time removal is required.

21. Install the propeller, as detailed in the Maintenance and Tune-Up section.
22. If drained for service, properly refill the gearcase with lubricant, as detailed in the Maintenance and Tune-Up section.

DISASSEMBLY

■ For gearcase exploded views, please refer to Cleaning & Inspection in this section.

The overhaul procedure the gearcases used on the largest of Yamaha motors (4- and 6-cylinder units) is essentially the same. Differences between the units have been called out in the procedures as necessary.

Counter-rotating models are, for the most part, the same as the normal rotation models of the same HP. There are 2 exceptions. First off, some counter-rotation models (mostly in the 115 hp range) utilize a 2-piece propeller shaft assembly. And, secondly, whereas the Reverse gear on normal rotation models is mounted in the bearing carrier (at the rear of the gearcase) and the Forward gear is mounted in the nose cone (at the front of the gearcase), the positions of these two gears are SWITCHED on counter-

LOWER UNIT 7-37

rotation models. Throughout this procedure, we will call the gears by their names as if they were mounted in a normal rotation model, so unless we specify otherwise, when working on a counter-rotation model, remember that what we are calling the Reverse gear is really the Forward gear and vice versa.

1. Remove the Propeller from the gearcase, as detailed under Maintenance and Tune-Up.
2. Drain the gearcase oil and remove the gearcase from the outboard as detailed earlier in this section under Gearcase (Lower Unit).
3. If necessary to access the driveshaft or upper oil seal assembly, remove the Water Pump assembly from the top of the gearcase, as detailed in the Lubrication and Cooling section.

■ **On all motors the impeller and water pump housing mounts to a water pump plate on top of the gearcase and driveshaft oil seal housing.**

4. Remove the water pump plate from the top of the gearcase (on most models the bolts which secured the water pump housing also secured the plate). Remove and discard the plate gasket.

■ **The driveshaft oil seal housing can be removed from these models without removing the driveshaft itself from the gearcase. This is because the upper portion of the driveshaft is a smooth shaft of constant outer diameter, so the oil seal housing can be slid upward and off the top of the driveshaft. Therefore, if there is no problem with the propeller shaft bearing or seal, you CAN skip the Oil Seal Housing procedure.**

Shift Rod

The shifter assembly on these models rotates back and forth (as opposed to the smaller Yamaha gearcases on which it lifts up and down). On some models a straight 1-piece shifter shaft is splined at both ends, the bottom of which splines to a removable shift cam inside the housing which can be removed independently of the propeller shaft. However, on some models (including the high-thrust 50/60 hp and 75-115 hp motors, as well some V6 models) utilize a shaft that is only splined on the top end. For these models the lower end of the shaft contains a pivot pin that is inserted through or around a shift joint that attaches to the propeller shaft. On models that utilize a pivot pin, the shifter assembly must be removed from the gearcase before the propeller shaft can be safely withdrawn.

■ **On some models, like the 150 hp motors, although the lower end of the shift rod is not splined, it contains an offset protrusion that inserts through a slot in the propeller shaft shifter assembly, which negates the need to rotate it for removal or installation.**

Either way, the shifter should be removed for inspection and O-ring/seal replacement.

1. Check to ensure the lower unit is in neutral gear. On most models the shift rod cannot (or should not) be removed if the unit is in forward or reverse. If the shifter is in gear, rotate it clockwise or counterclockwise, as necessary to disengage the gear. A special tool with splines that are matched to the top of the shift shaft is available to assist in rotation the shaft. If an alternate tool is used, take care not to damage the shaft or splines.
2. Loosen and remove the shift rod housing retaining bolts (there are normally either 2 or 3 bolts securing the housing to the gearcase). Carefully lift straight up to disengage the lower end of the shift rod from the shift cam or shift rod joint (as applicable) on most motors. On high-thrust 50/60 hp (T50/T60) motors, remove the shift rod by lifting VERY slightly and rotating the shift rod 90° counterclockwise, then pulling it the rest of the way from the gearcase. On 75-100 hp motors, lift up slightly on the shift rod, then rotate the shift rod 90° clockwise so that the pivot pin can pass through a hole in the lower portion of the case, then gently pull the shift rod the rest of the way from the gearcase.
3. Remove the shift rod and housing, then remove and discard the O-ring(s).

■ **As with most oil seals, do NOT remove the oil seal from the shift rod housing unless you are planning to replace it.**

4. Separate the shift rod from the shift rod housing so that you can remove the oil seal. Note the positioning of the seal lips (they are normally faced downward toward the gearcase), then carefully remove and discard the old seal.

Driveshaft Oil Seal Housing

At the very top of the gearcase is a small housing that contains the oil seals. On some models the housing also contains the driveshaft's upper needle bearing. The size a shape of the housing varies slightly from model-to-model, but there are generally 3 designs used on these gearcases.

The smallest of these motors (typically the 40-60 hp, except the high-thrust 50/60 hp [T50/T60]) are equipped with a small round housing just a little bit larger than the seal. The housing is not secured by any bolts, but sits in the gearcase, just above a bearing that is pressed onto the center of the driveshaft. Dual oil seals are mounted into the bottom of this housing, usually with the lips of both seals facing upward toward the water pump. The housing itself is sealed to the crankcase using an O-ring.

■ **We say usually in the previous paragraph because some Yamaha factory sources conflict. In some service manuals the seals are pictured at one point with the both lips facing upwards toward the water pump, but in another illustration with both lips facing downward toward the gearcase. For this reason it is CRITICAL on these motors to pay attention to and note seal lip direction BEFORE removing the seals from the housing! However, on almost all other Yamaha gearcases, the seal lips do face UPWARD, so in a pinch, use that as a guide.**

Mid-range motors (including the high-thrust 50/60 hp [T50/T60] and all 75-100 hp motors) utilize an irregularly-shaped housing which matches the shape of the water pump plate. This housing is normally secured to the gearcase by the water pump assembly bolts. Dual oil seals are also mounted into the bottom of this housing, both with the lips facing upward toward the water pump. This housing is sealed to the top of the gearcase using a gasket and, in most (but not all) cases, is also sealed to the driveshaft bore using an O-ring.

The largest of these motors (including all 115 hp and larger models) utilize a small oil seal and bearing housing that is square on top and round on the bottom. For these motors dual oil seals are mounted into the top of the housing with both their sets of lips facing upward toward the water pump, while a needle bearing is pressed into the housing from below. The housing is secured into the top of the gearcase normally using 4 bolts and is sealed to the gearcase using an O-ring.

With the exception of how the housing is secured and the side from which seals are installed, all housings are basically serviced in the same general manner. To remove and service the housing, proceed as follows:

On these motors the oil seal housing may be slid over the top of the driveshaft, therefore you do not NEED to remove the pinion gear and driveshaft if you are ONLY going to service the seals. However, it is a little more difficult to protect the seals during installation, so if you've got the time, it's never a bad idea to pull the driveshaft. It's your choice. If you decide NOT to pull the driveshaft, take great care when freeing the housing to pull straight upward and, more importantly, when installing to keep the housing squared to the driveshaft so as not to damage the seals (and bearings on some models).

1. For 115 hp and larger motors, remove the four bolts securing the oil seal housing.
2. Carefully lift upward to remove the seal housing from the top of the gearcase. If removal is difficult, use 2 small pry-tools to gently pry up on both sides of the oil seal housing. Lift the housing up and free of the driveshaft. If the housing is stubborn and cannot be removed you can always disassemble the lower portion of the gearcase and remove the driveshaft (this can be especially helpful on the smallest of motors covered here as the housing sits on top of a bearing and shoulder on the driveshaft and MUST come free of the gearcase if the driveshaft is removed).

■ **The small round oil housing used on the smallest of motors in this section sometimes contains an oil seal cover (it looks something like a small washer on top of the housing). If it comes loose on models so equipped, be sure to retain it for installation.**

3. The housing on most motors is sealed with an O-ring. If so, remove and discard the old O-ring. On the mid-range motors listed earlier, the housing is also sealed with a gasket, remove and discard the old housing gasket.

■ **On MOST of these models use dual oil seals which are positioned with their lips facing upwards (away from the gearcase). Double-check this to make sure there is no variance on your gearcase before removal. Also, remember that removal of the seals from the housing will destroy their sealing qualities and they cannot be installed a second time. Therefore, remove a seal only if it is unfit for further service and be absolutely sure a new seal is available before removing the old seal. The same goes for the needle bearing used on the largest of these motors.**

7-38 LOWER UNIT

4. Seal removal can be accomplished by various means. On most models you can use a hooked seal remover or small pry-tool in order to free the each seal. However, if access is difficult you can always use a slide hammer and expanding jaw attachment to remove the seals. Sometimes you can drive them out from the other side of the housing, but use care not to damage the housing or the bearing (on the larger models if it is not being replaced). Whatever method you use, just be sure not to score or damage the inside diameter of the carrier (the sealing surface).

5. On 115 hp and larger models, if the driveshaft needle bearing must be replaced, position the housing with the seal installation surface facing upwards, then use a suitable driver to carefully tap or push the needle bearing out the bottom of the housing. Also on these models there is normally shim material, a plate washer and thrust bearing positioned on top of the driveshaft (right below where the housing mounts to the gearcase). If possible, remove the shims, washer and thrust bearing by sliding them up/off the shaft. Keep the shim material for reuse or reference during installation.

Propeller Shaft Bearing Carrier

The rear of the propeller shaft rides in a bearing and seal carrier which mounts in the lower rear of the gearcase. As the name suggest, the purpose of the carrier is to house oil seals as well as propeller shaft bearings (usually 2, a large diameter ball bearing at the front of the housing and a smaller diameter needle bearing towards the center or rear of the housing).

■ **Some models, such as the 150 hp counter-rotating unit use a tapered roller bearing at the front of the propeller shaft bearing carrier, instead of a ball bearing.**

Seal or bearing replacement is a relatively straightforward task that can usually occur with the gearcase still installed on the outboard (if it is first drained of oil and/or tilted fully upward) and as long as the carrier can be carefully slide back, off the propeller shaft. Bearing carrier removal is required for most gearcase services (other than water pump and driveshaft oil seal on most models).

The bearing/seal carrier is secured to the gearcase by one of two methods. On about half of the models covered in this procedure (mostly the smallest motors, but also on the largest v6 engines) the carrier is bolted to the rear of the gearcase using 2 flange bolts. However on most of the mid-range to large engine engines (other than the V6 motors) the carrier is instead secured by a large ring nut and tabbed washer. On these models a special tool is required to remove the ring nut. The tool looks like a very deep socket (since it fits over the propeller shaft) that contains multiple tabs which contact and transmit rotational force against the ring nut. Since the nut is tightened to a specification of somewhere between 76-105 ft. lbs. (105-145 Nm) depending upon the model in question, we cannot recommend the use of another spanner to safely loosen the nut. For all models use #YB-34447.

■ **If you're working on a counter-rotation model, remember the actual gear positions are opposite of a normal rotation model. On counter-rotation models, the Forward gear is in the bearing carrier and the Reverse gear is in the gearcase nose cone.**

1. For models equipped with a bolt retained bearing carrier, proceed as follows:
   a. For V6 motors, loosen and remove the 2 bolts securing the retaining ring over the carrier. Remove the retaining ring for access to the carrier flange mounting bolts.
   b. Loosen and remove the 2 flange bolts which secure the bearing carrier.

2. For models equipped with a lock-ring retained bearing carrier, proceed as follows:
   a. Straighten the tabs on the lock-washer which are bent over the ring nut.
   b. There is normally the word **OFF** embossed on the ring nut along with an arrow indicating the correct direction to loosen the nut. Using a suitable ring nut spanner wrench, rotate the ring in the direction indicated (also normally counterclockwise) until the nut is free.
   c. Remove the tabbed lock-washer.

■ **Lock-ring mounted bearing carriers normally utilize a tiny key to help keep the carrier from turning inside the gearcase. The key, is usually installed into the underside of the carrier, be sure to locate and retain it as the carrier is removed.**

3. Attempt to remove the bearing carrier. If the carrier will not come out easily, it will have to be pulled. On most models Yamaha recommends or illustrates the use of a universal type puller, however on some V6 models they show the use of a slide hammer, the choice is really yours as both should work on all of these gearcases. But proceed slowly and carefully to make sure you do not damage anything.

■ **The propeller shaft (or rear half of the shaft on some counter-rotating models) MAY come free with the bearing carrier. If it does, no big deal, just carefully separate it from the carrier for attention later.**

4. Watch for and save any shim material or spacer/thrust washer from the back side of the reverse gear (forward gear on counter-rotation units). When utilized, the shim material is critical in obtaining the correct backlash during assembling. Using the old shim material will save considerable time, especially starting with no shim material.

5. Remove and discard the O-ring(s) around the bearing carrier. There may be as few as 1 or as many as 3 O-rings depending upon the model.

■ **The reverse gear (forward on counter-rotation units) and ball bearing assembly (on most models) is pressed into the bearing carrier. In addition, the ball bearing is pressed onto the back of the reverse gear. Usually all of these components must be separated using a slide hammer and/or bearing separator. However, the tapered roller bearing used on the forward gear of some counter-rotating units (such as the 150 hp motor) is normally pressed on the gear, but the assembly can be freely removed from the carrier without tools.**

6. The reverse gear (forward on counter-rotation units) and ball bearing assembly is pressed into the bearing carrier on most models. In addition the ball bearing itself (or tapered roller bearing on some models) was pressed onto the gear. Removal usually involves first pulling the gear from the bearing and housing, and then pulling the bearing from the housing. However, the bearing has been known to come out with the gear, depending upon the method used for removal. It's your choice as you CAN use either a slide hammer with internal expanding jaw attachment or you can use a bridged puller with our without a bearing separator. If the gear and bearing come out as an assembly, use a bearing separator to free the bearing from the gear. If they do NOT come out as an assembly, use the puller or slide-hammer to then extract the bearing from the carrier.

■ **On the V6 counter-rotation models the propeller shaft is part of the carrier, retained by a tapered roller bearing, race, tabbed washer and ring nut. On these models, once the forward gear is removed, turn the carrier around and install it back to the gearcase backwards (with the forward gear mounting area pointing outward). Use a ring nut wrench (like Yamaha #YB-06578 or Mercury #91-888880T) to then loosen the ring nut. The propeller shaft, shims, bearing and race can then be removed using a shop press.**

7. If the oils seals AND needle bearing are all being replaced, the easiest way to remove them is to drive them out the propeller side of the housing all together by tapping with a suitable driver from the gearcase side of the housing. However, DO NOT remove the needle bearing unless it is being replaced, as removal normally makes it unfit for further service. In the later case, if only the oil seals must be replaced, use a hooked seal removal tool, small pry bar or at worst, an internal expanding jawed puller attachment and a puller or slide-hammer to remove the seals.

Propeller Shaft and Clutch Dog

The job of the propeller shaft is straight forward enough, change the direction of the clockwise, vertical rotating driveshaft to clockwise (or counterclockwise) horizontal rotation in order to drive the propeller. At all times the pinion gear on the end of the driveshaft is in contact with and rotating 2 gears (a forward and reverse), but because of their positioning the gears are rotating in different directions. A clutch dog assembly which is splined to the propeller shaft (at its center) is positioned in the middle of the 2 gears.

A shift mechanism is attached to the clutch dog with a cross pin. This mechanism extends forward from the propeller shaft, through the gear and bearing assembly in the gearcase nose cone. The shift rod mechanism works by rotating forward or backward, transmitting this force through the shifter in the propeller shaft to the clutch dog, moving it toward one or the other gear. When the clutch dog comes in contact with one of the gears it locks into place and begins to spin with that gear, transmitting the force of the gear to the propeller shaft through the clutch dog-to-shaft splines.

LOWER UNIT 7-39

Propeller shaft disassembly involves removal of the clutch dog and internal shift components, which vary slightly from model-to-model. Most smaller gearcases (generally found on 60 hp and smaller motors) contain a simple spring, plunger and slider mechanism. However there are some exceptions, including the high-thrust 50/60 hp (T50/T60) motors, which uses a shift mechanism more similar to the larger motors.

Conversely, motors larger than 60 hp (again, there are a few exceptions) utilize a slightly more complicated shift mechanism that includes a shift rod joint, joint slider, spring (sometimes with spring nuts and/or washers on either end) and anywhere from 2-6 check balls. At minimum 2 check balls are installed at the front end of the joint slider, though many models also 2 balls of the same size at the rear of the joint slider. Still other models use 2 larger check balls positioned in the propeller shaft behind the joint slider and on either side of the spring.

Work slowly and carefully, making sure no check balls, springs, spring nuts, etc are lost when disassembling the shift mechanism. For more details, refer to the exploded views found in the Cleaning and Inspection section.

1. If the shaft did not come out with the bearing carrier, pull the propeller shaft, clutch and shifter assembly straight back and free of the lower unit.

■ **If you're working on a counter-rotation model, remember the actual gear positions are opposite of a normal rotation model. On counter-rotation models, the Forward gear is in the bearing carrier and the Reverse gear is in the gearcase nose cone.**

2. Insert an awl under the end loop of the cross pin ring and carefully pry the ring free of the clutch dog.
3. Use a long pointed punch to press out the cross pin.
4. Slide the clutch dog from the shaft.

■ **Observe how the clutch dog was installed. On some models it must be installed in the same direction from which it was removed. On these models the clutch dogs are embossed with a marking (such as F to denote which edge faces front, toward the Forward gear). On other models you might notice that both shoulders of the dog are an equal width and the dog is not stamped. On those models the dog may be reinstalled either way, usually with the least worn side facing the forward gear. Remember this fact, as an aid during assembling.**

5. Carefully disassemble the balance of the shifting components, by withdrawing them from the propeller shaft. On some models (those equipped with a separate shim cam that splines onto the bottom of the shift rod), this might only be a shift plunger, slider and spring, but on others (those with shift pins on the bottom of the shift rods) there will be other components, including a shifter (pin receiver), separate slider, 2-6 shifter check balls, spring and possible a few spring nuts and/or washers. For models equipped with 2 or more shifter check balls, slowly withdraw the shift slider, recovering the balls as they JUST pull free of the propeller shaft. Remember there may be as many as 4 mounted 2 on each end of the shifter, plus there may be 2 balls with the spring (one on either side of the spring).

■ **As the work proceeds in the disassembly of the shift slide, the check balls used in various shifters are often of different sizes (but grouped in pairs of similar sizes). Take care to store these balls separately to avoid confusion during assembly.**

Driveshaft and Bearings

The driveshaft is used to transmit the clockwise (when viewed from the top) rotational motion of the crankshaft through the pinion gear to the propeller shaft (either through the Forward or Reverse gears). The driveshaft is splined at both ends, the top splines mating with the lower end of the crankshaft and the bottom end mating with the pinion gear.

Exact positioning of the pinion gear on the shaft (so that it meshes properly with the Forward and Reverse gears) is adjusted through the use of different thickness shim packs. There are essentially 2 configurations for shims on these models. For many of the larger gearcases which contain a needle bearing for the driveshaft in the driveshaft oil housing, there is normally a shim, followed by a thrust washer and thrust bearing all mounted toward the top of the driveshaft, just underneath the housing. They may have been removed earlier during Driveshaft Oil Seal Housing service.

Most other gearcases found on these models use a tapered roller bearing that is pressed directly on the driveshaft. For these models a bearing race and shim pack may be found in the gearcase, just above the driveshaft sleeve.

Regardless of the upper bearing configuration, all models use a needle bearing assembly at the lower end of the driveshaft. Many early models and some late-models used a bearing with loose needles, however the majority of Yamahas these days use only a one-piece needle bearing assembly.

■ **Driveshaft removal is necessary for access gearcase mounted components like driveshaft bearings or bushings and/or the Forward (Reverse on counter-rotating units) gear/bearing (the gear and bearing mounting in the case nose-cone).**

Before the driveshaft can be freed from the gearcase, the pinion gear at the lower end of the shaft must first be removed. On all models a driveshaft holder tool (essentially a large nut with slides over the driveshaft's splines in order to provide a means of holding or turning the driveshaft) must be used. This tool varies slightly by model as follows:
  • For 40/45/50 hp models through 2000 use Yamaha #YB-06079.
  • For 2001 and later 50 hp and 60 hp motors use Yamaha #YB-06049 on US and Canadian models or #90890-06158 for World Production models.
  • For 75-115 hp motors, use Yamaha #YB-06151.
  • For 150 hp and V6 models, use Yamaha #YB-06201, or Mercury #91-888889 (may be stamped 06520)

■ **In most cases, when working with tools, a nut is rotated to remove or install it to a particular bolt, shaft, etc. In the next step, the reverse is required because there is really is too little room to move a wrench inside the lower unit cavity. The nut on the lower end of the driveshaft is held steady and the shaft is rotated (using the special tool) until the nut is free.**

1. Place a large, long handled wrench or breaker bar with socket (usually 22mm on these motors) over the pinion nut. Slide the driveshaft holder tool over the splines on the top of the driveshaft and place a large wrench or socket over the tool. Now, hold the pinion nut steady with the tool, rotate the driveshaft counterclockwise to break the nut free.
2. Remove the pinion nut.
3. Gently pull up on the driveshaft and at the same time rotate the driveshaft. The pinion gear will come free from the lower end of the driveshaft.
4. Pull the driveshaft up out of the lower unit housing.
5. If the oil seal housing and the components beneath it were not removed the housing at this point will still be on the driveshaft and can now be easily removed. Go back and perform the steps which were omitted.
6. On models with a tapered roller bearing pressed onto the driveshaft, use a suitable internal expanding jawed puller attachment on either a bridged puller or a slide-hammer to remove the bearing race from the gearcase. Keep the shim pack found under the race for installation purposes.
7. Lift out the driveshaft sleeve for access to the lower needle bearing.

■ **For all inline 4-cylinder motors (except the 115 hp and 150 hp models), you'll need to install the replacement needle bearing at the same depth below the top of the gearcase as the current bearing. Because specifications may vary slightly on gearcases or in case you don't have all the necessary special bearing depth tools for Yamaha, take a measurement of the installed height of the needle bearing set (from top of the gearcase to the top of the bearing) before removal for reference during installation.**

8. The needle bearing set (with loose needle on some models) is pressed into the driveshaft housing. Use a suitable driver and long driver handle carefully push the needle bearing assembly down and into the gearcase propeller shaft cavity.

Nose Gear/Bearing

All gearcases use one gear mounted to the propeller shaft bearing carrier assembly and one gear which turns the opposite direction mounted just forward of the driveshaft in the gearcase nose cone.

On standard rotation models the Reverse gear is found at the propeller shaft bearing carrier, while the Forward gear is mounted in the nose cone.

On counter-rotation models the Forward gear is found in the propeller shaft bearing carrier, while the Reverse gear is mounted in the nose cone.

■ **Gear and bearing designs used in the nose cone vary slightly between standard and counter-rotation models**

7-40 LOWER UNIT

Forward Gear (Standard Rotation Units)

On most standard rotation units a tapered roller bearing is pressed onto the back of the Forward gear (found in the nose cone). The assembly is inserted into a race that is pressed into the gearcase (on top of the shim material necessary for proper Forward gear placement). If the tapered roller bearing is pressed off the gear, it should be replaced. And, anytime the bearing is replaced, the race must also be replaced.

■ **Most of the larger standard-rotation gearcases ALSO use one or two needle bearings (or a bushing) pressed inside the Forward gear. Like most needle bearings, they must be replaced if removed.**

1. Lift out the Forward gear and tapered roller bearing assembly from the gearcase mounted race. Inspect the gear and bearing assembly to determine if further disassembly is necessary for component replacement.
2. If necessary, use an expanding jaw attachment with a suitable slide hammer or puller to remove the bearing race from the lower unit housing. Be sure to hold the tool at right angles when pulling the race.
3. Watch for and save any shim material found behind the forward gear bearing race. The shim material is critical to obtaining the correct backlash during installation. Using the old shim material will save considerable time, especially starting with no shim material.

■ **Remove the bearing only if it is unfit for further service.**

4. If necessary to replace the bearing, position a bearing separator between the Forward gear and the tapered roller bearing. Using a hydraulic press, separate the gear from the bearing.
5. On models equipped with one or more needle bearings pressed into the gear, if replacement is necessary, use an expanding jaw attachment on a puller or slide hammer to remove the bearing(s) from the gear.
6. On models equipped with a bushing pressed into the gear, if replacement is necessary, use a suitable driver to push it out from the rear of the gear.

Reverse Gear (Counter-Rotation Units)

■ **If you're working on a counter-rotation model, remember the actual gear positions are opposite of a normal rotation model. On counter-rotation models, the Forward gear is in the bearing carrier and the Reverse gear is in the gearcase nose cone.**

On most counter-rotation units instead of the tapered roller bearing (found on other models) the Reverse gear in the nose cone uses a needle bearing and retainer or roller bearing that is pressed into the gearcase itself. Reverse gear shim material is found mounted underneath the bearing and retainer assembly (similar to the material found under the tapered bearing race on standard rotation models).

In addition, the Reverse gear is normally equipped with a thrust bearing with a large washer and/or spacer on the shaft of the gear (between the gear and bearing assembly). On models that use a roller bearing pressed into the gearcase (as opposed to a needle bearing and retainer) a roller bearing inner race is pressed onto the back of the Reverse.

These gearcases ALSO use one or two needle bearings pressed inside the Forward gear. Like most needle bearings, they must be replaced if removed.

1. Lift out the reverse gear, then lift out the thrust bearing. Take care not to confuse this thrust bearing with the one previously removed (with the propeller shaft assembly). If these thrust bearings are to be reused, they must be installed in their original locations, because each develops a unique wear pattern. If a bearing is installed in a different location, the wear on the individual tiny roller bearings would be greatly accelerated and lead to premature failure of the bearing.
2. Lift out the large thrust washer and/or spacer, as equipped.
3. Obtain a slide hammer (or puller) with expanding jaw attachment to pull the needle bearing and retainer or roller bearing (as applicable) from the lower unit housing. Be sure to hold the tool right angles to the driveshaft while working.
4. Watch for and save any shim material found behind the reverse gear bearing. The shim material is critical to obtaining the correct backlash during installation. Using the old shim material will save considerable time, especially starting with no shim material.
5. If necessary due to bearing replacement on roller bearing models, position a bearing separator between the Reverse gear and the inner bearing race (on the back of the gear). Using a hydraulic press, carefully separate the gear from the race.

6. If necessary due to bearing replacement on needle bearing models, use a suitable driver and a hydraulic press (or hammer) to push (or tap) the needle bearing out of the retainer.
7. On models equipped with one or more needle bearings pressed into the gear, if replacement is necessary, use an expanding jaw attachment on a puller or slide hammer to remove the bearing(s) from the gear.

■ **On some models the needle bearings can be carefully driven out from the tooth side of the year using a punch or chisel.**

CLEANING & INSPECTING

Gearcase Exploded Views

◆ See Figures 69 thru 72

Because of the shear number of gearcases available on these models it is impossible to include exploded view of each and every minor variation. The vast majority of gearcases used on these models follow the same basic design with small variances in the type or placement of shims, washers and bearings. Small differences also may be found on some shifter mechanisms. We've included a couple of the most representative exploded views for these gearcases. But, during disassembly, keep comparing what you're pulling out of the gearcase with what is pictured here and note any differences for assembly purposes.

Checking Gearcase Components

■ **If you're working on a counter-rotation model, remember the actual gear positions are opposite of a normal rotation model. On counter-rotation models, the Forward gear is in the bearing carrier and the Reverse gear is in the gearcase nose cone.**

Good shop practice requires installation of new O-rings and oil seals regardless of their appearance.

Clean all water pump parts with solvent and then dry them with compressed air. Inspect the water pump housing and oil seal housing for cracks and distortion, possibly caused from overheating. Inspect the plate and water pump cartridge for grooves and/or rough surfaces. If possible, always install a new water pump impeller while the lower unit is disassembled. A new impeller will ensure extended satisfactory service and give peace of mind to the owner. If the old impeller must be returned to service, never install it in reverse to the original direction of rotation. Installation in reverse will cause premature impeller failure.

If installation of a new impeller is not possible, check the seal surfaces. All must be in good condition to ensure proper pump operation. Check the upper, lower and ends of the impeller vanes for grooves, cracking and wear. Check to be sure the indexing notch of the impeller hub is intact and will not allow the impeller to slip.

Clean around the Woodruff key. Clean all bearings with solvent, dry them with compressed air and inspect them carefully. Be sure there is no water in the air line. Hold the bearing to keep it from turning and direct the air stream through the bearing. After the bearings are clean and dry, lubricate them with oil. Do not lubricate tapered bearing cups until after they have been inspected.

✸✸ WARNING

Never spin a bearing with compressed air. Such action is highly dangerous and may cause the bearing to score from lack of lubrication.

Inspect all ball bearings for roughness, scratches and bearing race side wear. Hold the outer race and work the inner bearing race in-and-out, to check for side wear.

Determine the condition of tapered bearing rollers and inner bearing race, by inspecting the bearing cup for pitting, scoring, grooves, uneven wear, imbedded particles and discoloration caused from overheating. Always replace tapered roller bearings and their race as a set.

Clean the forward gear with solvent and then dry it with compressed air. Inspect the gear teeth for wear. Under normal conditions the gear will show signs of wear but it will be smooth and even.

Clean the bearing carrier with solvent and then dry it with compressed air. Check the gear teeth of the reverse gear for wear. The wear should be smooth and even.

LOWER UNIT 7-41

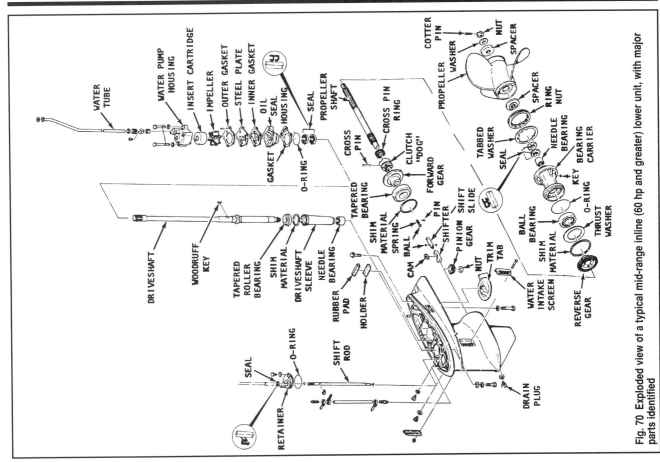

Fig. 70 Exploded view of a typical mid-range inline (60 hp and greater) lower unit, with major parts identified

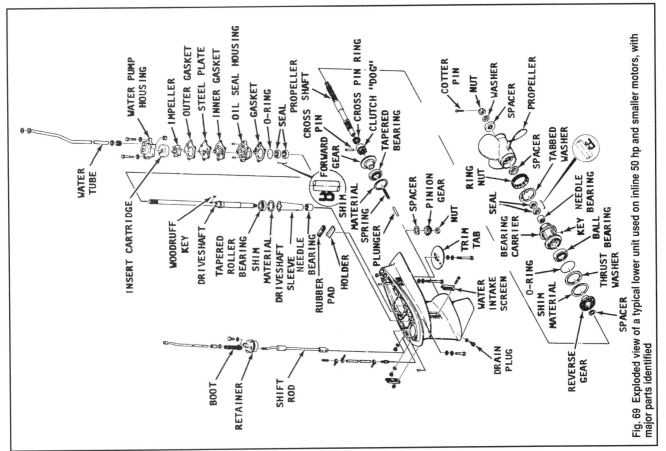

Fig. 69 Exploded view of a typical lower unit used on inline 50 hp and smaller motors, with major parts identified

7-42 LOWER UNIT

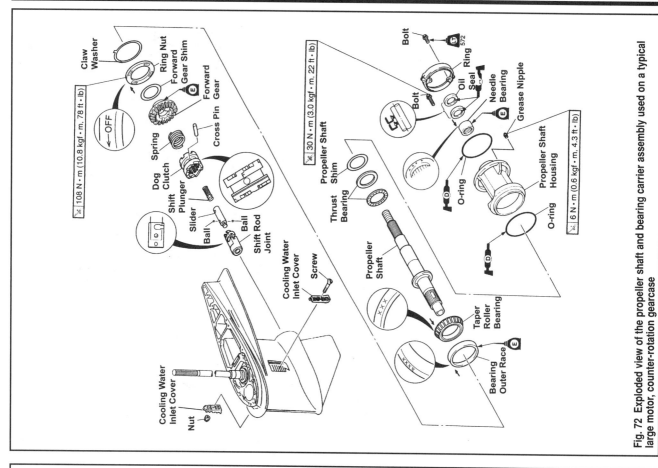

Fig. 72 Exploded view of the propeller shaft and bearing carrier assembly used on a typical large motor, counter-rotation gearcase

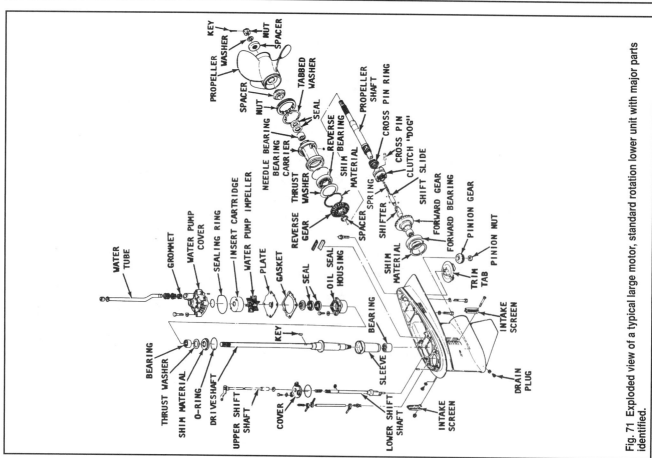

Fig. 71 Exploded view of a typical large motor, standard rotation lower unit with major parts identified.

LOWER UNIT 7-43

Check the clutch dogs to be sure they are not rounded-off, or chipped. Such damage is usually the result of poor operator habits and is caused by shifting too slowly or shifting while the engine is operating at high rpm. Such damage might also be caused by improper shift rod adjustments.

Rotate the reverse gear and check for catches and roughness. Check the bearing for side wear of the bearing races.

Inspect the bearing surfaces of the propeller shaft. Check the shaft surface for pitting, scoring, grooving, embedded particles, uneven wear and discoloration caused from overheating.

Clean the driveshaft with solvent and then dry it with compressed air.

Inspect the driveshaft splines for excessive wear. Check the oil seal surfaces above and below the water pump drive pin or Woodruff key area for grooves. Replace the shaft if grooves are discovered.

Inspect the driveshaft bearing surface above the pinion gear splines for pitting, grooves, scoring, uneven wear, embedded metal particles and discoloration caused by overheating.

Inspect the propeller shaft oil seal surface to be sure it is not pitted, grooved, or scratched. Inspect the bearing contact surface on the propeller shaft for pitting, grooves, scoring, uneven wear, embedded metal particles and discoloration caused from overheating.

Inspect the propeller shaft splines for wear and corrosion damage. Check the propeller shaft for and driveshaft for run-out.

Place each shaft on V blocks and measure the run-out with a dial indicator gauge at a point midway between the V-blocks. Typically the maximum acceptable run-out for a propeller shaft is around 0.0008 in. (0.02mm), while the maximum acceptable run-out for the driveshaft is about 0.02 in. (0.5mm). However, keep in mind that Yamaha does not publish specs for most of these gearcases, so you may have to use a judgment call.

As equipped, inspect the following parts for wear, corrosion, or other signs of damage:
- Shift shaft assembly
- Shift cam
- All bearing bores for loose fitting bearings.
- Gear housing for impact damage.
- Threads for cross-threading and corrosion damage.
- Check the pinion nut corners for wear or damage. This nut is a normally a special locknut. Therefore, do not attempt to replace it with a standard nut. Obtain the correct nut from an authorized Yamaha dealer.

If the lower unit case is to be repainted. Mask off the threads engaging the cover nut. If this nut is installed against painted threads, a false torque value will be obtained upon tightening the nut and it is possible the nut could back off with continued use.

ASSEMBLY

◆ See Figures 69 thru 72

The following procedures are intended to guide you step-by-step through reinstallation and setup of the major gearcase components. If one or more of the components was not removed you can skip that section and proceed to the next following component.

Gearcase shimming procedures vary slightly from model-to-model and they depend upon the components that are being replaced. However, generally speaking:
- NO SHIMMING is required if the original case and original inner parts (gears, bearings and shafts) are installed.
- NUMERIC CALCULATION RE-SHIMMING should be performed on all gearcases when the original inner parts (gears, bearings and shafts) are installed in a NEW case. Most original and replacement cases are marked with various gear measurements. Typically an embossed **F** (forward gear), **R** (reverse gear) and/or **P** (pinion) gear measurements. When these marks are present the differences between the marks on the new case vs. the marks on the original case can be used to determine new shim sizes BEFORE assembly (potentially saving you from having to disassemble/reassemble the gearcase a second time).
- SPECIAL TOOL MEASUREMENT RE-SHIMMING is performed prior to assembly on all models and should be used when a specific shaft, bearing/bushing or gear is replaced. In this case, special tools are used to measure pre-assembled or partially assembled components and determine if a change in shim sizes is necessary. If the special shimming tools are not available, you can often reuse shim sizes from the previous assembly to at least get you in the ball-park and then use the Backlash Measurement Re-shimming procedures to dial it in.
- BACKLASH MEASUREMENT RE-SHIMMING should be performed and adjusted when one or more of the original inner components (gears, bearings and shafts) are installed. Backlash measurement is normally performed on an almost fully assembled gearcase using a dial gauge to read the amount of play in the driveshaft. This procedure can check the placement of all 3 gears through this measurement. Though adjustment is much more hit and miss than the other methods, since you need to then re-disassemble the gearcase, swap shims, re-assemble and re-check, then repeat until you get it right. Fortunately, formulas are available to use the measured backlash in helping to pick shim sizes and small shim changes are all that is normally necessary. It is rare that you would have to disassemble the gearcase more than one additional time.

■ **If the case, gear(s), bearing(s) and/or shaft(s) have been replaced, refer to the Gearcase Shimming procedures later in this section BEFORE starting reassembly to determine if there are any changes to shims that can be made before your start assembly so that you can potentially avoid having to disassemble the gearcase a second time to change shims after checking backlash.**

Throughout installation you will need to make a couple of choices. If you have simply replaced O-rings and seals (not bearings, shafts or gears) then there is probably little chance that the gear lash will require adjustment. If so you can lubricate all components and apply sealant or threadlocking compounds as called for in the procedures. However, if you have replaced components such as the bearings, shafts or gears you may wish to check gear contact patterns using a Machine Dye or suitable powder, and all components should then be temporarily installed dry. Also, when components such as these are replaced you'll need to check and possibly adjust gear lash, though reusing shims which were removed during disassembly can often give you a good starting point and MAY prevent you from having to partially disassemble the gearcase again (but only if you've chosen to lubricate components during installation). As we said, the choice is yours and can only be made based upon the circumstances of your overhaul or reseal and the condition of the unit before tear-down.

Unless otherwise noted, upon final installation be sure to lubricate all bearings with gear oil, shafts with oil or a marine grade grease and seals with oil, packing their lips with a suitable marine grade grease.

Assembly generally means preparing the gearcase (by installing any bearings or races which were pressed or pulled out of the case itself) and the various sub-assemblies such as the propeller shaft, bearing carrier, oil seal housing etc. Then assembly means carefully bringing these components back together in the proper order inside the gearcase. The order in which the sub-assemblies are readied for installation doesn't matter, as long as they are not needed in the gearcase at a certain point. Typically speaking, first you would install bearings or races into the gearcase itself, then prepare all of the sub-assemblies. The nose cone gear must be installed into the gearcase BEFORE the driveshaft (since the pinion gear which holds the driveshaft in position will also hold the nose cone gear in position. On dual-propeller models this nose-cone gear is actually mounted to the inner propeller shaft, so the inner propeller shaft must be installed BEFORE the driveshaft. Also, the water pump assembly, generally the last component installed, is left OFF the gearcase until after backlash has been check and adjusted.

Nose Gear/Bearing

All gearcases use one gear mounted to the propeller shaft bearing carrier assembly and one gear which turns the opposite direction mounted just forward of the driveshaft in the gearcase nose cone.

■ **Gear and bearing designs used in the nose cone vary slightly between standard and counter-rotation models**

On standard rotation models the Reverse gear is found at the propeller shaft bearing carrier, while the Forward gear is mounted in the nose cone.

On counter-rotation models the Forward gear is found in the propeller shaft bearing carrier, while the Reverse gear is mounted in the nose cone.

■ **On some smaller gearcases covered in this section the shifter cam is a small, tab-like component that splines onto the end of the shift shaft and sits otherwise freely in the nose cone. On some of these models installation of the shift shaft to the cam can be very difficult once the nose cone gear is retained by the driveshaft and pinion nut, so you may wish to assemble and install the shift-shaft at this time or sometime before the driveshaft is installed.**

7-44 LOWER UNIT

Reverse Gear (Counter-Rotation Units)

■ If you're working on a counter-rotation model, remember the actual gear positions are opposite of a normal rotation model. On counter-rotation models, the Forward gear is in the bearing carrier and the Reverse gear is in the gearcase nose cone.

On most counter-rotation units instead of the tapered roller bearing (found on other models) the Reverse gear located in the nose cone uses a needle bearing and retainer or roller bearing that is pressed into the gearcase itself. Reverse gear shim material is found mounted underneath the bearing and retainer assembly (similar to the material found under the tapered bearing race on standard rotation models).

In addition, the Reverse gear is normally equipped with a thrust bearing that utilizes a large washer and/or spacer on the shaft of the gear (between the gear and bearing assembly). On models that use a roller bearing pressed into the gearcase (as opposed to a needle bearing and retainer) a roller bearing inner race is pressed onto the back of the Reverse.

These gearcases ALSO use one or two needle bearings pressed inside the Forward gear. Like most needle bearings, they must be replaced if removed.

1. On models equipped with one or more needle bearings pressed into the gear, install the new bearing(s) using a suitable driver. The bearings are normally installed from the rear of the gear (not the toothed side) and normally with any embossed markings on the bearing facing back towards the driver. Install the bearings to the specified depth in the gear depending upon the model, as follows:
   - On 115 hp motors, install the bearing to a depth of 0.098-0.138 in. (2.5-3.5mm) below the rear edge of the gear.
   - On 150 hp motors, install the first bearing to a depth of 0.825-0.844 in. (20.95-21.45mm) and the second bearing 0.175-0.195 in. (4.45-4.95mm) below the rear edge of the gear.
   - On 3.3L V6 motors, install the first bearing to a depth of 0.817-0.836 in. (20.75-21.25mm) and the second bearing 0.167-0.187 in. (4.25-4.75mm) below the rear edge of the gear.

2. On roller bearing models (these cases are not USUALLY found on these 4-stroke models, but you never know so we mention it just in case), if the bearing was replaced, use a hydraulic press to carefully push the replacement bearing inner race onto the back of the gear. Be sure to position the bearing with any markings embossed on the race facing back towards the driver.

3. On needle bearing models, if the gearcase retainer mounted bearing is being replaced, use a hydraulic press to install the bearing into the retainer. Position the bearing so any embossed markings are facing away from the retainer (toward the gear). Install the bearing to the specified depth in the retainer depending upon the model, as follows:
   - On 115 hp motors, install the bearing to a depth of 0.030-0.049 in. (0.75-1.25mm) below the edge of the retainer.
   - On 3.3L V6 motors, install the bearing so that it is flush, give or take 0.010 in. (0.25mm) meaning a depth of between -0.010 and +0.010 in. (-0.25 and +0.25mm) from the edge of the retainer.

■ On 150 hp needle bearing models there is no mention of a retainer, install the outer bearing housing assembly and shim material to the gearcase housing using a suitable driver. Be sure to position any markings on the housing facing forward (into the nose of the gearcase, away from the gear).

4. Position the roller bearing, needle bearing or needle bearing and retainer to the gearcase, along with the shims that were removed during disassembly OR with the proper calculated shims if the gearcase has been replaced (for details please refer to Gearcase Shimming, later in this section).

■ Roller bearings should be positioned into the nose cone with any embossed markings facing back toward the gear.

5. Use a suitable long handled driver to gently tap the bearing or bearing and retaining assembly into position until they are seated in the gearcase.

■ If the driveshaft is ready for installation (the bearings were not removed or are already replaced), proceed with the next step in anticipation of driveshaft installation. However, if you still need to install the lower driveshaft bearing, you'll probably want to hold off on positioning the gear and wait until you ARE ready to install the driveshaft.

6. Position the reverse gear in the nose cone bearing/race along with the thrust washer and plane washer/spacer (the order of insertion to the case it washer/spacer, thrust washer and finally gear).

■ If you wish to visually check gear contact patterns, obtain a suitable substance which can be used to indicate a wear pattern on the forward and pinion gears as they mesh. Machine dye may be used and if this material is not available, Desenex® Foot Powder (obtainable at the local Drug Store/Pharmacy), or equivalent may be substituted. Desenex® is a white powder available in an aerosol container. Before assembling the gears, apply a light film of the dye, Desenex®, or equivalent, to the driven side of the each. After the gears are assembled and rotated several times, they will be disassembled and the wear pattern can be examined. The substance will be removed from the gears prior to final assembly.

Forward Gear (Standard Rotation Units)

On most standard rotation units a tapered roller bearing is pressed onto the back of the Forward gear (found in the nose cone). The assembly is inserted into a race that is pressed into the gearcase (on top of the shim material necessary for proper Forward gear placement). If the tapered roller bearing is pressed off the gear, it should be replaced. And, anytime the bearing is replaced, the race must also be replaced.

■ Most of the larger standard-rotation gearcases ALSO use one or two needle bearings (or possibly even a bushing) pressed inside the Forward gear. Like most needle bearings, they must be replaced if removed.

1. On models equipped with one or more needle bearings (or a bushing) pressed into the gear, install the new bearing(s) or bushing using a suitable driver. The bearings are normally installed from the rear of the gear (not the toothed side) and normally with any embossed markings on the bearing facing back towards the driver. Install the bearing(s) to the specified depth in the gear depending upon the model, as follows:
   - On all 115 hp motors, install the bearing(s) so the bearing (or top/outer bearing if there are two) is at a depth of 0.098-0.138 in. (2.5-3.5mm) below the rear edge of the gear.
   - On 150 hp motors, install the first bearing to a depth of 0.825-0.844 in. (20.95-21.45mm) and the second bearing 0.175-0.195 in. (4.45-4.95mm) below the rear edge of the gear.

2. If the tapered roller bearing was removed from the outside of the Forward gear use a hydraulic press to carefully push the replacement bearing into position on the back of the gear.

3. Position the forward gear bearing race to the gearcase, along with the shims that were removed during disassembly OR the proper calculated shims if the gearcase has been replaced (for details please refer to Gearcase Shimming, later in this section).

4. Use a suitable long handled driver to gently tap the bearing race into position until it is seated in the gearcase.

■ If the driveshaft is ready for installation (the bearings were not removed or are already replaced), proceed with the next step in anticipation of driveshaft installation. However, if you still need to install the lower driveshaft bearing, you'll probably want to hold off on positioning the gear and wait until you ARE ready to install the driveshaft.

5. Position the forward gear and bearing assembly into the bearing race in the nose cone.

■ If you wish to visually check gear contact patterns, obtain a suitable substance which can be used to indicate a wear pattern on the forward and pinion gears as they mesh. Machine dye may be used and if this material is not available, Desenex® Foot Powder (obtainable at the local Drug Store/Pharmacy), or equivalent may be substituted. Desenex® is a white powder available in an aerosol container. Before assembling the gears, apply a light film of the dye, Desenex®, or equivalent, to the driven side of the each. After the gears are assembled and rotated several times, they will be disassembled and the wear pattern can be examined. The substance will be removed from the gears prior to final assembly.

Driveshaft Bearings

As noted earlier during disassembly, the driveshaft is used to transmit the clockwise (when viewed from the top) rotational motion of the crankshaft through the pinion gear to the propeller shaft (either through the Forward or Reverse gears). The driveshaft is splined at both ends, the top splines mating with the lower end of the crankshaft and the bottom end mating with the pinion gear.

Exact positioning of the pinion gear on the shaft (so that it meshes properly with the Forward and Reverse gears) is adjusted through the use of different thickness shim packs. There are essentially 2 configurations for shims on these models. For many of the larger gearcases which contain a needle bearing for the driveshaft in the driveshaft oil housing, there is normally a shim, followed by a thrust washer and thrust bearing all mounted toward the top of the driveshaft, just underneath the housing. They may have been removed earlier during Driveshaft Oil Seal Housing service.

Most other gearcases found on these models use a tapered roller bearing that is pressed directly on the driveshaft. For these models a bearing race and shim pack may be found in the gearcase, just above the driveshaft sleeve.

Regardless of the upper bearing configuration, all models use a needle bearing assembly at the lower end of the driveshaft. Many early models and some late-models used a bearing with loose needles, however the majority of Yamahas these days use only a one-piece needle bearing assembly. But don't let that fool you, as there are still a couple of large gearcases that use loose needle bearings.

Additionally the lower needle bearing is installed in one of two ways, depending upon the model. For all smaller gearcases (used on inline 4-cylinder motors, except the 115 hp and 150 hp motors) the bearing is driven in from the top of the gearcase to a specified depth. However, for larger gearcases (used on the 115 hp, 150 hp and V6 models) the bearing must be pulled up into position. either used a special tool or a long threaded bolt with appropriately sized washers and nuts.

On the larger gearcases, although Yamaha often has a special tool available to help draw the bearing into position in the lower portion of the gearcase, all you REALLY need to accomplish this is a sufficiently long threaded bolt with a nut and a washer that is the size of the driver you'd want to contact that bushing or bearing. You'll also need a nut and a plate to place on top of the gearcase. Place the threaded rod through one nut and plate and then down into the gearcase through the bushing/bearing (so the plate is resting against the top of the gearcase, the upper nut is resting against the plate and the threads of the rod are just protruding down into the pinion mounting area of the gearcase.). Next install the washer and lower nut over the bottom threads of the rod to support the bearings or busing. Slowly draw the bearing or bushing into position in the gearcase by holding the threaded rod from turning while at the same time turning the upper nut to draw the entire bolt, bushing/bearing, washer and nut assembly upward until the bushing or bearing seats. Then loosen the nut on the bottom of the threaded rod in the gearcase and remove the threaded rod, plate, nut and washer.

1. For models whose gearcase contains loose needle bearings, apply a light coating of marine grade grease to the bearing housing. The loose needles are normally installed after the housing is pulled into position, HOWEVER, if you have something which can be used as a bushing (a piece of plastic pipe or something that is just a little smaller in outer diameter than the driveshaft, but still big enough to allow the bearing installation rod to go through the center) the needles can be positioned now.

2. For large gearcases (as noted earlier) where the needle bearing (or bearing housing) is installed from below, assemble an installation tool and use the tool to carefully draw the bearing up into the gearcase until it is seated. On some models there may be a part number embossed on one end of the bearing, make sure the embossed surface is positioned facing downward, toward the gear. Once the bearing (or bearing housing) is pulled into position, loosen the retainer on the bottom and remove the tool. If not done earlier on loose needle bearing models, apply a light coating of grease to the inside of the bearing carrier and install the needles. Keep an eye on the needles as the driveshaft is installed later to make sure none are dislodged.

■ **On smaller gearcases, always compare the specification to the depth measured before disassembly to make sure there are no gearcase design variances from published spec.**

3. For smaller gearcases (as noted earlier) use a suitable driver and long driver handle to carefully tap the bearing in from the top. Since the bearing must be installed to a specific depth, you should probably insert the driver and handle into the bore, then measure and mark on the driver a spot JUST above the spec, so you'll know to measure and double-check the installed bearing height right before you finish tapping the bearing into position. Install the bearing to the suitable depth as follows:

• On 40-60 hp motors, the bearing is normally positioned with the embossed numbers facing upward and set to a depth (from the top of the deck) of 5.22-5.24 in. (132.5-133.0mm) for standard thrust US and Canadian models, 7.19-7.21 in. (182.5-183.0mm) for standard thrust World production models, to 7.34-7.43 in. (186.4-188.6mm) for 50 hp (T50) high-thrust models or to 7.39-7.43 in. (187.6-188.6mm) for 60 hp (T60) high-thrust models.

• On 75-100 hp motors, the bearing is normally positioned with the embossed numbers facing upward and set to a depth (from the top of the deck) of 7.385-7.425 in. (187.6-188.6mm).

4. Slide the driveshaft sleeve into position from the upper end of the lower unit. Some sleeves contain a notch or tab at the front which should be faced toward the front of the gearcase.

5. On models with a tapered roller bearing pressed onto the driveshaft, position the shim pack (either which was removed during disassembly or determined by Numerical Calculation or Special Tool Measurement shimming, as detailed under Gearcase Shimming) followed by the bearing race into the top of the gearcase. Using a suitable driver, carefully seat the race.

6. If the tapered bearing was removed from the driveshaft, on models so equipped, use a suitable driver to press the replacement bearing onto the shaft. A "suitable" driver in this case will have an outer-diameter JUST larger than the driveshaft so that it ONLY contacts the inner bearing race (and does NOT contact the rollers or outer race). The driver must also obviously fit over the driveshaft (meaning be long enough).

7. As mentioned earlier, if you want to visually check gear mesh, obtain a suitable substance which can be used to indicate a wear pattern on the forward and pinion gears as they mesh. Machine dye may be used and if this material is not available, Desenex® Foot Powder (obtainable at the local Drug Store/Pharmacy), or equivalent may be substituted. Desenex® is a white powder available in an aerosol container. Before assembling either gear, apply a light film of the dye, Desenex®, or equivalent, to the driven side of the gear. After the gears are assembled and rotated several times, they will be disassembled and the wear pattern can be examined. The substance will be removed from the gears prior to final assembly.

■ **If this is final assembly, already with the needed shims, be sure to oil all gears and bearings.**

8. Position the forward gear (reverse gear on counter-rotating models or forward gear along with the inner prop shaft on dual-propeller models) and tapered bearing assembly into the gearcase.

■ **On counter-rotating lower units, remember to insert the spacer or washer and thrust bearing to rest up against the reverse gear bearing race.**

You're now ready to install the driveshaft and pinion gear.

Pinion Gear and Driveshaft

1. Lower the driveshaft down through the sleeve in the upper end of the lower unit. On models with loose needle bearings it is usually a good idea to apply a fresh coat of grease to the accessible portion of the needles to help hold them in place as you slide the driveshaft down into position. Also on models with loose needles, work slowly making sure no needle becomes dislodged and/or damaged.

■ **If you're planning on visually checking gear mesh, coat the pinion gear with a fine spray of Desenex®. Handle the gear carefully to prevent disturbing the powder.**

2. Insert the pinion gear into the gearcase and raise the pinion gear to allow the driveshaft to pass fully through the gear. It may be necessary to rotate the driveshaft slightly to allow the splines on the driveshaft to index with the internal splines of the pinion gear. The teeth of the pinion gear will index with the teeth of the forward gear (reverse gear on counter-rotating units).

3. Once the pinion gear is in place, start the threads of the pinion gear nut. Tighten the nut as much as possible by rotating the driveshaft with one hand and holding the nut with the other hand.

7-46 LOWER UNIT

4. Tighten the pinion gear nut to specification either using a wrench or pinion nut holder to keep the nut steady with a torque wrench on the driveshaft holder OR by turning the driveshaft with a wrench and holding the pinion nut with the torque wrench and socket. Tighten the nuts to spec as follows:

- On all 40-50 hp motors (except high-thrust 50 hp T50 model), tighten the pinion nut to 54 ft. lbs. (75 Nm).
- On all 60-90 hp motors and 75-100 hp motors, along with high-thrust 50/60 hp (T50/T60) motors, tighten the pinion nut to 69 ft lbs. (95 Nm).
- On 115 hp and 150 hp motors, tighten the pinion nut to 67 ft. lbs. (93 Nm).
- On 3.3L V6 motors, tighten the pinion nut to 103 ft. lbs. (142 Nm).

✱✱ WARNING

The pinion nut torque is CRITICAL for the well-being of the gearcase. Should the nut come loose in service, the gear would fall striking the forward and reverse gears (wreaking havoc on the gears and teeth, possibly damaging other parts of the case and components as well).

5. For models where the upper driveshaft bearing is located in the oil seal housing (as opposed to tapered roller bearing models where the bearing was pressed onto the shaft), position the thrust bearing, thrust washer and shim material (either that which was removed or that which was calculated or measured during Gearcase Shimming, as detailed later in this section) over the driveshaft and against the driveshaft sleeve.

Driveshaft Oil Seal Housing

As noted during disassembly there are basically 3 types of oil seal housings used on these motors. However, all 3 types have a few things in common. For starters, dual oil seals (both of whose seal lips are NORMALLY positioned facing upwards, towards the water pump) are used to seal the driveshaft. Additionally, an O-ring is used to help seal the housing to the gearcase. The rest of the design features vary, some use a gasket in addition to the O-ring, others do not. Some contain a needle bearing pressed into the housing, others do not.

In all cases, seals, the gasket and/or O-ring should be replaced to ensure your gearcase components are protected. Additionally, care must be given to make sure the new seals are not damaged by the driveshaft splines when the housing is slid into position. Otherwise, it is a relatively straight forward reseal and installation.

1. On 115 hp and larger motors, if the driveshaft needle bearing was removed, use a suitable installer to drive the replacement bearing in from the underside of the housing. Install the bearing to the proper depth below the bottom edge of the oil seal housing. The bearing should be installed to a depth of 0.226-0.246 in. (5.75-6.25mm) for all 4-cylinder motors or to a depth of 0.167-0.187 in. (4.25-4.75mm) for V6 motors.

■ **Remember to pack the seal lips with grease upon installation.**

2. If the oil seals were removed, place the housing on the worksurface with the seal bore facing upward (this means with the housing upright on the largest motors such as the 115 hp and larger models, but the housing inverted on smaller motors, where the gearcase uses a driveshaft with a pressed on tapered roller bearing). Install the replacement oil seals using a suitable driver, making sure their lips are facing upward (toward the water pump, unless a variance was noted during disassembly) and the seals are installed to the proper depth as follows depending upon the type of oil seal housing used by this gearcase:

- On the smallest of these motors (typically the 40-60 hp motors, except the high-thrust 50/60 hp) which are equipped with a small round housing just a little bit larger than the seal (not retained by bolts) and a driveshaft with a tapered roller bearing pressed onto the shaft, the oil seals are installed from UNDERNEATH the housing. As stated earlier during disassembly some Yamaha sources conflict as to which direction the seal lips should be facing, but we think it should normally be upward toward the water pump. On these models, drive the seals into position until the outer of the two (bottom in this case, since they are installed from underneath the housing) is 0.00-0.02 in. (0.0-0.5mm) below the edge of the housing.
- On mid-range motors (including the high-thrust 50/60 hp and all 75-100 hp motors) which utilize an irregularly-shaped housing that matches the shape of the water pump plate, both seals are installed from underneath the housing, with their lips facing upward. On MOST of these gearcases Yamaha does NOT give a dimension for seal installation depth, meaning that the seals should be installed under they are seated or flush. However, we did find 1 exploded view for the 75-100 hp models that showed the outer (lower) seal at a dimension of 0.2 in. (4mm) below the surface (recessed into the bore from the lower edge of the housing).
- On the largest of these motors (including all 115 hp, 150 hp and V6 models) a small oil seal and bearing housing mounts to the top of the gearcase. On these the housings are designed so the seals are installed from the TOP of the housing (since the bearing would be in the way from the bottom). As usual, both seal lips should be positioned so they are facing upward. On all models they should be installed so the top most seal is 0.010-0.030 in. (0.25-0.75mm) below the top edge of the housing.

3. Install a new O-ring around the oil seal housing and coat the O-ring with marine grade grease.

■ **If you are going to check Gear Backlash on mid-range models where the seal housing follows the shape of the water pump, it may be easier to leave the housing off until after checking lash.**

4. If equipped, position a new oil seal housing gasket.
5. CAREFULLY lower the oil seal housing down over the driveshaft making sure not to damage the seals on the driveshaft splines. Keep the housing squared to the driveshaft and gearcase, then tap it lightly to seat it properly in the lower unit.
6. On all but the smallest of gearcases noted earlier, check to be sure the housing bolt holes align. On housings that use their own retaining bolts (as opposed to housings which are retained by water pump retaining bolts), install and tighten the bolts securely.

Propeller Shaft and Clutch Dog

The job of the propeller shaft is straight forward enough, change the direction of the clockwise, vertical rotating driveshaft to clockwise (or counterclockwise) horizontal rotation in order to drive the propeller. At all times the pinion gear on the end of the driveshaft is in contact with and rotating 2 gears (a forward and reverse), but because of their positioning the gears are rotating in different directions.

A clutch dog assembly which is splined to the propeller shaft is positioned in the middle of the 2 gears. A shift mechanism is attached to the clutch dog with a cross pin. This mechanism extends forward from the propeller shaft, through the gear and bearing assembly in the gearcase nose cone.

The shift rod mechanism works by rotating forward or backward, transmitting this force through the shifter in the propeller shaft to the clutch dog, moving it toward one or the other gear. When the clutch dog comes in contact with one of the gears it locks into place and begins to spin with that gear, transmitting the force of the gear to the propeller shaft through the clutch dog-to-shaft splines.

Propeller shaft assembly involves installation of the clutch dog and internal shift components, which vary slightly from model-to-model. Most smaller gearcases (generally found on 60 hp and smaller motors) contain a simple spring, plunger and slider mechanism. However there are some exceptions, including the high-thrust 50/60 hp (T50/T60) motors, which uses a shift mechanism more similar to the larger motors.

Conversely, motors larger than 60 hp (again, there are a few exceptions) utilize a slightly more complicated shift mechanism that includes a shift rod joint, joint slider, spring (sometimes with spring nuts and/or washers on either end) and anywhere from 2-6 check balls. At minimum 2 check balls are installed at the front end of the joint slider, though many models also 2 balls of the same size at the rear of the joint slider. Still other models use 2 larger check balls positioned in the propeller shaft behind the joint slider and on either side of the spring.

Work slowly and carefully, making sure no check balls, springs, spring nuts, etc are lost when assembling the shift mechanism. For more details, refer to the exploded views found in the Cleaning and Inspection section.

1. Install the compression spring into the shift slide or end of the prop shaft (as applicable). If the compression spring is surrounded by large check balls, washers or spacers, be sure to position them properly as it is inserted.
2. For models equipped with 2 or more side mounted check balls, apply just a dab of grease to the two smallest balls to help keep them in place and then insert them into the holes nearest the open end of the shift slide.
3. Insert the shift slide into the propeller shaft with the large hole in the slide aligned with the slot in the propeller shaft.
4. For models which use check balls on the shift slide, continue to insert the shift slide into the propeller shaft until both small balls enter the shaft - then stop. If the shifter uses 2 more check balls at the opposite end of the shaft apply a dab of grease to those balls (they may be the same size as the

LOWER UNIT 7-47

small balls or they may be medium sized, depending upon the model) to help keep them in place and then insert them into the holes nearest the necked end of shift slide. Continue to insert the shift slide into the propeller shaft until both medium balls are about to enter the shaft - then stop.

5. For models with a separate shifter joint, hook the shifter onto the knob on the end of the shift slide. Push in the shift slide until seated. Again, on models with check balls work slowly feeling for the balls to click slightly into place in the grooves inside the shaft. Stop the action when a click is heard, indicating the two small balls have moved into the neutral groove in the propeller shaft. There is only one groove, so wiggle the slide back and forth until the ball is definitely seated in the groove.

6. Slide the clutch dog onto the propeller shaft, with the hole in the dog aligned with the two holes (hopefully already aligned), of the shaft and shift slide. Check the clutch dog for markings, often there is a stamped mark (usually a **F** on the clutch dog which should be positioned facing forward, toward the shifter).

7. Insert the cross pin into the hole and push it through until the pin ends are flush with the clutch dog groove surface.

8. Wrap the cross pin retaining ring around the groove to retain the cross pin. The ring may be wrapped in either direction. Check to be sure the ring turns do not overlap one another.

9. On some motors it is easier to guide the assembled propeller shaft into the lower unit at this time, on others it is easier to wait and install it as an assembly with the propeller shaft bearing carrier (a lot depends on the gearcase and whether the shaft came out with the carrier). On dual-propeller models you should be reading this step THEN going back to perform driveshaft installation.

Shift Rod

The shifter assembly on these models rotates back and forth (as opposed to the smaller Yamaha gearcases on which it lifts up and down). On some models a straight 1-piece shifter shaft is splined at both ends, the bottom of which splines to a removable shift cam inside the housing which can be removed independently of the propeller shaft. However, most models (including the high-thrust 50/60 hp and 75-115 hp motors, as well as most V6 models) utilize a shaft that is only splined on the top end. For these models the lower end of the shaft contains a pivot pin that is inserted through or around a shift joint that attaches to the propeller shaft.

The best time for shifter assembly installation will vary with the design of the lower shift mechanism. For models where the shift rod slides into a splined cam installation occurs either right before the propeller shaft is installed (if the cam sits freely in the housing, so you can insert your hand or a tool to position the cam while the shift rod splines are mated to it) or right AFTER the propeller shaft is installed (on models where the cam rides in a small housing on the end of the prop shaft shifter assembly. On models that utilize a pivot pin, the shifter assembly must be installed AFTER the propeller shaft is in place (because the rod/pin must be lowered into or around the shift slider, again depending upon the design).

■ **For most of the gearcases covered here that means shift rod installation CAN occur at this point, but there are exceptions where the shift rod should already be installed or cannot be installed until later, after the bearing carrier. If necessary come back to this procedure after the propeller shaft and bearing carrier are in position.**

1. If removed, install a new oil seal with the lips facing the same direction as noted during removal (or as shown in the exploded views under Cleaning & Inspection, earlier in this section).

2. Install new housing O-ring(s) and apply a light coating of marine grade grease.

3. Lower the shift rod down into the lower unit and through the hole in the shifter. For splined units, make sure the splines on the bottom of the rod engage with the splines on the shift cam. On pin-type shifters, the cam on the shift rod must normally face to starboard (for standard lower units) or to port (for counter-rotating lower units) and the cam or pin on the end of the rod will insert directly into the hole in the shift slider. But there are some exceptions, for instance:
   - On high-thrust 50/60 hp (T50/T60) motors, install the shift rod with the cam facing directly forward by lowering the rod until it is ALMOST in position, then turning it 90° counterclockwise and insert it the rest of the way, pushing the cam over the shifter joint.
   - On 75-100 hp motors, insert the shifter into the gearcase with the cam facing toward port, then partway through the gearcase it will catch on a hole in the bore. At this point turn the shift rod 90° clockwise so that the pivot pin can pass through the hole and, as soon as it does, turn it 90° clockwise so it will fit over the shifter joint, then lower it into contact with the shifter.

4. Once the shift rod is properly engaged on the lower end, install and tighten the bolts that secure the housing.

Propeller Shaft Bearing Carrier

■ **If you're working on a counter-rotation model, remember the actual gear positions are opposite of a normal rotation model. On counter-rotation models, the Forward gear is in the bearing carrier and the Reverse gear is in the gearcase nose cone.**

With the exception of potential Gearcase Shimming procedures which may need to be performed if certain parts were replaced, the bearing carrier and propeller shaft installation is the last major piece of the gearcase overhaul.

If the bearing carrier cap was disassembled to replace a bearing, bushing and/or oil seal it must be properly prepared for installation. This means you'll need to use suitably sized drivers to install replacement seals, bearing or bushing, depending upon what was removed. As usual, remember the driver must contact the part of the seal, bearing or bushing that is in contact with the housing. On most installations that means the outside of the seal, bearing or bushing, HOWEVER on some installations a gear bearing is installed ONTO the gear and not INTO the housing. When a gear bearing is pressed onto the outer diameter of a shaft on the gear, the driver needs to contact the INNER race instead.

■ **Remember to oil the contact surfaces of bearings, bushings or seals before installation. Be sure to pack the seal lips with a suitable marine grade grease once they are installed.**

Keep in mind that seals, bushing or bearing installation varies slightly from model-to-model. Although all seals used on these gearcases are of the dual variety they are all installed with BOTH of their seal lips facing the rear of the outboard (the propeller).

All models (even including the counter-rotation units) are equipped with a needle bearing and dual seal assembly that installs from the propeller side of the carrier. However, that's where the similarities end, as the counter models utilize slightly different components on the other end of the housing. If disassembled, prepare the bearing/seal carrier cap and the propeller shaft components for installation, as follows:

■ **Unless otherwise noted, install all bearings with any embossed markings facing back toward the propeller.**

- For standard thrust 40-60 hp motors, start by installing the needle bearing through the propeller side of the housing until seated. On US and Canadian models, if you have access to the Yamaha bearing attachment (#YB-06111) you would drive the bearing into position until the flange on the bearing tool is 0.12-0.14 in. (3.0-3.5mm) from the contact surface of the carrier. On world production models the top of the bearing should be roughly 0.91-0.93 in. (23.0-23.5mm) below the surface of the housing (but this dimension includes the thickness of the Yamaha bearing depth plate). Install the 2 seals over top of the needle bearing so that the top of the 2 seals is 0.16-0.18 in. (4.0-4.5mm) below the end of the carrier. Position the gear shim on the reverse gear, then use a suitable driver to carefully tap the ball bearing into position on the gear. Place the bearing carrier over the assembled gear and use a shop press to install the gear and bearing assembly to the carrier.
- For high-thrust 50/60 hp (T50/T60) motors, start by installing the needle bearing through the propeller side of the housing until seated. On US and Canadian models through 2000, if you have access to the Yamaha bearing attachment (#YB-06152) you would drive the bearing into position until the flange on the bearing tool is just in contact with the surface of the carrier. For world production models, as well as 2001 and later US and Canadian models, the top of the bearing should be roughly 0.98-1.00 in. (25.0-25.5mm) below the surface of the housing. Install the 2 seals over top of the needle bearing so that the top of the outer seal is approximately 0.18-0.22 in. (4.5-5.5mm) below the end of the housing. Position the thrust washer on the reverse gear, then use a suitable driver to carefully tap the ball bearing into position on the gear. Place the bearing carrier over the assembled gear and use a shop press to install the gear and bearing assembly to the carrier.

7-48 LOWER UNIT

- For 75-100 hp 4-cylinder motors, start by installing the needle bearing through the propeller side of the housing until the top of the bearing is about 1.004-0.984 in. (25.00-25.50mm) below the surface of the housing. Install the 2 seals over top of the needle bearing so that the top of the outer seal is approximately 0.177-0.216 in. (4.5-5.5mm) below the end of the housing. Position the thrust washer on the reverse gear, then use a suitable driver to carefully tap the ball bearing into position on the gear. Place the bearing carrier over the assembled gear and use a shop press to install the gear and bearing assembly to the carrier.
- For standard rotation models of the 115 hp and 150 hp motors, start by installing the needle bearing through the propeller side of the housing until the top of the bearing is about 0.974-0.994 in. (24.75-25.25mm) below the surface of the housing. Install the 2 seals over top of the needle bearing so that the top of the outer seal is approximately 0.187-0.207 in. (4.75-5.25mm) below the end of the housing. Position the thrust washer on the reverse gear, then use a suitable driver to carefully tap the ball bearing into position on the gear. Place the bearing carrier over the assembled gear and use a shop press to install the gear and bearing assembly to the carrier.
- For counter-rotation models of the 115 hp 4-cylinder motor, start by installing the first of the 2 carrier needle bearings through the gearcase end of the housing until the top of the bearing is about 1.348-1.368 in. (34.25-34.75mm) below the surface of the housing (gearcase end). Then, invert the housing so the propeller end is facing upward and install the other needle bearing so the top of the bearing is about 0.974-0.994 in. (24.75-25.25mm) below the propeller end of the housing. Install the 2 seals over top of the needle bearing so that the top of the outer seal is approximately 0.187-0.207 in. (4.75-5.25mm) below the end of the housing. Position the thrust washer on the forward gear, then use a suitable driver to carefully tap the ball bearing into position on the gear. Place the bearing carrier over the assembled gear and use a shop press to install the gear and bearing assembly to the carrier. Finally, prepare and install the forward gear assembly. Insert the needle bearing into the back of the gear and press until it is seated with the top of the bearing about 0.40-0.42 in. (10.25-10.75mm) below the rear surface of the gear. Install the rear propeller shaft, tapered roller bearing and outer race to the gear end of the carrier housing using a driver, then use a shop press to install the forward gear and thrust washer to the assembly.
- For counter-rotation models of the 150 hp motor, start by installing the needle bearing through the propeller side of the housing until the top of the bearing is about 0.974-0.994 in. (24.75-25.25mm) below the surface of the housing. Install the 2 seals over top of the needle bearing so that the top of the outer seal is approximately 0.187-0.207 in. (4.75-5.25mm) below the end of the housing. Next, prepare and install the forward and propeller shaft assembly. From the propeller end of the shaft slide the thrust bearing along with the washer and/or shim(s) over the shaft (in that order). Insert the propeller shaft to the bearing carrier. Position the assembly (carrier and shaft) into a press, then carefully install a NEW taper roller bearing and outer bearing race into the housing using the press and a driver. Using a suitable driver, install the thrust washer, forward gear and clutch dog onto the assembly (making sure the dog clutch is positioned with the "F" mark facing toward the forward gear. Finish installing the shifter components as detailed earlier in this section under Clutch Dog assembly.
- For standard rotation models of the 3.3L V6 motors, start by installing the needle bearing through the propeller side of the housing until the top of the bearing is about 0.986-1.006 in. (25.05-25.55mm) below the surface of the housing. Install the 2 seals over top of the needle bearing so that the top of the outer seal is approximately 0.187-0.207 in. (4.75-5.25mm) below the end of the housing. Use a suitable driver to carefully tap the ball bearing into position in the bearing carrier, then position the reverse gear shim on the gear and use a driver or press to install the gear to the bearing and carrier assembly.
- For counter-rotation models of the 3.3L V6 motors, start by installing the needle bearing into the propeller end of the carrier so the top of the bearing is about 0.986-1.006 in. (25.05-25.55mm) below the propeller end of the housing. Install the 2 seals over top of the needle bearing so that the top of the outer seal is approximately 0.187-0.207 in. (4.75-5.25mm) below the end of the housing. Next, prepare and install the forward gear assembly. From the propeller end of the shaft slide the thrust bearing, then washer and finally the propeller shaft shim (if used and of proper size) over the shaft. Insert the propeller shaft to the bearing carrier. Using a shop press, install the tapered roller bearing and outer race to the carrier and shaft assembly. In order to hold the carrier steady for ring nut installation, temporarily install the bearing carrier in the reverse of the normal position (with the gearcase side facing outward) into the bottom of the gearcase. Install the claw washer and ring nut, then tighten the nut to 80 ft. lbs. (108 Nm) using a ring nut wrench

(Yamaha #YB-06578 or Mercury #91-888880T). Remove the carrier from the gearcase, then use a shop press and driver to install the forward gear shim, the gear and the clutch dog (making sure the embossed V on the clutch dog is facing the gear) onto the front of the propeller shaft. Finally, install the Clutch Dog assembly and shifter joint (as detailed earlier in this section).

1. If it is time for final assembly (not if you are just assembling to check shimming and backlash) install a new O-ring into each groove provided in the bearing carrier (these carriers may use 1-3 O-rings, depending upon the model. Apply a light coating of marine grade grease to the O-ring(s).
2. If this assembly is to check gear contact patterns, coat the teeth of the reverse gear with a fine spray of Desenex®, as mentioned earlier. Handle the gear carefully to prevent disturbing the powder.
3. Install the bearing carrier or carrier and propeller shaft assembly (depending upon model and method of installation) into the lower unit. If applicable, align the keyway in the lower unit housing with the keyway in the bearing carrier. Insert the key into both grooves and then push the bearing carrier into place in the lower unit housing.

■ On some assemblies, such as the dual-propeller models, the bearing carrier is equipped with a stamping that should be faced upward. Of course, since many carriers use a keyway, it is easy to figure out correct installation and orientation.

4. If this is the final installation (again, not temporary for shimming or checking contact patterns), apply a light coating of Loctite® 572 or an equivalent threadlocking compound to the bearing carrier retaining screws or a light coating of marine grade grease to the threads of the lock-ring (as equipped).

■ Many of the bearing carriers on the larger models (though not all) are retained by a ring nut and lock-washer. On these models install a tabbed washer against the bearing carrier and then install the ring nut with the embossed marks facing outward, away from the bearing carrier.

5. Install the bearing carrier retaining bolts or ring nut and tighten to specification as follows:
- For standard thrust 40-60 hp motors tighten the retaining bolts to 11 ft. lbs. (16 Nm).
- For 75-115 hp and high-thrust 50/60 hp (T50/T60) motors, tighten the ring nut to 76 ft. lbs. (105 Nm).
- For 150 hp motors, tighten the ring nut to 105 ft. lbs. (142 Nm).
- For 3.3L V6 motors tighten the carrier retaining bolts to 22 ft. lbs. (30 Nm), then, install the ring cap and tighten cap bolts securely.
6. If components other than seals were replaced (such as the case itself, gears, bearings and/or shafts) proceed to the appropriate section of the Gearcase Shimming procedures to check and/or adjust gear lash.
7. If this is final assembly, install the Water Pump, as detailed in the Lubrication and Cooling section.
8. Refill the gearcase, install the propeller and install the lower unit assembly (in whatever order you'd prefer).

GEARCASE SHIMMING

◆ See Figure 73

■ Since Yamaha shims (and gearcase markings) are all in mm we have kept most of the procedures in metric measurements only.

In order for the gearcase to function properly the pinion gear teeth must fully engage with the teeth of the forward or reverse gear (depending on shifter position). Because build tolerances for the gearcase, gears, shafts and bearings/bushings will vary slightly the manufacturer has positioned shims strategically at various points in the gearcase to give you the ability to make up for the differences in these tolerances. In this way, a gear that is not fully engaging can be repositioned slightly by the use of a thicker shim. Similarly, a gear that does not fully disengage can be repositioned slightly further away by using a thinner shim.

Gearcase shimming procedures vary slightly from model-to-model and they depend upon the components that are being replaced. However, generally speaking:

- NO SHIMMING is required if the original case and original inner parts (gears, bearings and shafts) are installed after cleaning, inspection and/or seal replacement.

LOWER UNIT 7-49

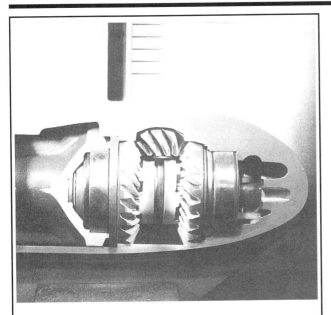

Fig. 73 Gearcase shimming is all about getting the forward and reverse gears to mesh properly with the pinion gear

- NUMERIC CALCULATION RE-SHIMMING should be performed on all gearcases when the original inner parts (gears, bearings and shafts) are installed in a NEW case. Most original and replacement cases are marked with various gear measurements. Typically an embossed **F** (forward gear), **R** (reverse gear) and/or **P** (pinion) gear measurements. When these marks are present the differences between the marks on the new case vs. the marks on the original case can be used to determine new shim sizes BEFORE assembly (potentially saving you from having to disassemble/reassemble the gearcase a second time).
- SPECIAL TOOL MEASUREMENT RE-SHIMMING is performed prior to assembly on all models and should be used when a specific shaft, bearing/bushing or gear is replaced. In this case, special tools are used to measure pre-assembled or partially assembled components and determine if a change in shim sizes is necessary. If the special shimming tools are not available, you can often reuse shim sizes from the previous assembly to at least get you in the ball-park and then use the Backlash Measurement Re-shimming procedures to dial it in.
- BACKLASH MEASUREMENT RE-SHIMMING should be performed and adjusted when one or more of the original inner components (gears, bearings and shafts) are installed. Backlash measurement is normally performed on an almost fully assembled gearcase using a dial gauge to read the amount of play in the driveshaft. This procedure can check the placement of all 3 gears through this measurement. Though adjustment is much more hit and miss than the other methods, since you need to then re-disassemble the gearcase, swap shims, re-assemble and re-check, then repeat until you get it right. Fortunately, small shim changes are all that is normally necessary and it is rare that you would have to disassemble the gearcase more than one additional time.

Follow the appropriate Shimming or Backlash measurement procedure, depending upon your needs as outlined above.

Numeric Calculation Re-shimming

The gearcase for all of these models SHOULD be marked with measurements representing the case's deviation from standard. Typically an embossed **F** (forward gear), **R** (reverse gear) and/or **P** (pinion) and a **+** or **-** with a number. The codes can be found on the underside of the anti-cavitation plate, usually underneath the trim tab (so if you look under there and can't find them, remove the trim tab and check again).

The stamped number should be interpreted as hundredths of a mm. This means that if there is a marking such as **F +5**, it means that the case requires a shim that it 5/100mm (0.05mm) larger than standard. The **+** or **-** accompanying the number tells you whether you should add (**+**) or take away (**-**) the number from the standard measurement. If there is NO number after the **F, R** or **P**, assume the value is 0 and the case meets standard tolerance. If the number is there, but illegible, you'll have to work from 0 and make adjustments for that shim by checking lash.

Now, when moving all of the internal components to a new case, compare the numbers on the 2 cases and adjust the shims accordingly. If the numbers are all the same (lets say they both are F +5, R-1 and P+8), you've got it easy, use the original shims and make no changes. But, if the new case is, lets say an **F+3**, then subtract the new value from the old value to determine how much of a change you need to make to the forward gear shim. In this example, 5-3=2. This means the NEW gearcase needs a shim that is 0.02mm SMALLER than the forward gear shim which was originally used. Measure the old shim and obtain a new shim which is about 0.02mm smaller for reassembly.

Once you've selected the proper shims, assemble the gearcase as detailed earlier in this section and, before the water pump is installed, check the gearcase Backlash as detailed later in this section. As long as the backlash is within spec, no further adjustment is necessary. However, if backlash is out of specification you'll need to further adjust one or more of the shims to correct the problem.

Special Tool Measurement Re-shimming

On all models one or more special tools (pinion height gauges, bearing press plates/gauges and/or shimming gauges) are available to assist in determining shim sizes before installation is complete. All shim changes must still be verified through backlash measurement, but when available, the use of shimming gauges may help select the right shim the first time out, saving you un-necessary disassembly/reassembly of the gearcase during overall.

Tool sets will vary first and foremost by market. The tools used for USA and Canadian models are slightly different than those that are available for World Production models. Also, tools will vary slightly from model-to-model.

The one constant about re-shimming procedures is that you will pre-assemble some portion of the shaft and/or bearings or gears and then measure a dimension of the assembly using the special tool. The measurement is then applied to a standard formula (which often includes the gearcase deviation from standard measurements) and the result is the desired shim size (or amount of change to the shim which was used during the measurements).

Pinion Gear Shim Selection

◆ See Figures 74 thru 80

For all models, the quickest way to achieve proper gear mesh and pinion gear height is the use of Yamaha's pinion height gauge set. In most cases, different tools, formulas and specifications are used for US/Canadian models than are used for World Production models, however the actual measurement procedure is the similar for all. This is basically a method of pre-assembling the driveshaft, bearing (tapered roller on some or thrust bearing on others depending on how the gearcase is constructed), pinion gear and selected shims outside the gearcase in order to precisely check the assembled distance between the bearing and the pinion. The components of the gauge set used vary slightly from model-to-model and are detailed in the procedure.

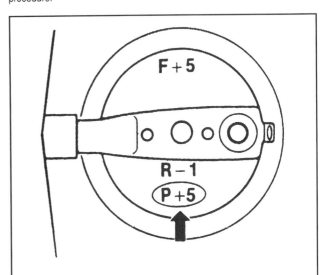

Fig. 74 All pinion shimming formulas use the stamped gearcase 'P' deviation

7-50 LOWER UNIT

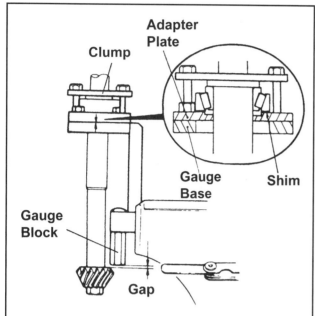

Fig. 75 Checking the assembled driveshaft on US & Canadian models that utilizes a bearing pressed onto the driveshaft

■ Unless otherwise noted, all US and Canadian models use components of Yamahas #YB-34432 gauge set. Parts of this set will be represented as -## (such as -9 or -10) for short-hand, but the full number of each part would be YB-34432-## (meaning YB-34432-9 or YB-34432-10).

■ The pinion gear shim selection procedure varies slightly from model-to-model. The variances are essentially 2 different ways around a square (2 equivalent processes that approach the problem from the opposite directions). However, because of the limited specifications provided by Yamaha you must adhere to the method for the particular model in question.

Determine the pinion gear shim thickness, as follows:
1. Check the stamped code on the underside of the anti-ventilation plate (it is stamped in the trim tab mounting area, so if you haven't already, matchmark and remove the tab). The code **P** followed by a **+** or **-** and a number is the gearcase's pinion bearing mounting deviation from the standard spec, in hundredths of a mm. This means that if there is a marking such as **P +5**, it means that the case requires a shim that it 5/100mm (0.05mm) larger than standard. The **+** or **-** accompanying the number tells you whether you should add (**+**) or take away (**-**) the number from the standard measurement. If there is NO number after the **P**, assume the value is 0 and the case meets standard tolerance. If the number is there, but illegible, you'll have to work from 0 and make adjustments for that shim by checking lash.

■ All measurements should be taken at no less than 3 points around the bearing and the measurements should be averaged.

2. For US and Canadian Models (along with ALL 150 hp and larger motors, as no world production specs/tools were available), assemble the appropriate portions of the gauge set (#YB-34432) along with the driveshaft, bearing, shim, pinion and pinion gear nut (tighten the nut to the spec listed in the Pinion Gear and Driveshaft procedure), then take the measurement using a feeler gauge between the top of the gear and the bottom of the gauge block and calculate the shim changes depending upon the model as follows:
• For 40-60 hp motors, assemble the gauge block **-9** (except for 1999 and later standard thrust models which use block **-8**), adapter plate **-10**, gauge base **-11** and clamp **-17**. If the original shim is not available start measurement with the largest shim (0.50mm). The measured gap must be 0.30mm (for standard thrust models) or 0.20mm (for high-thrust models), PLUS the **P** stamped deviation. If not add or subtract shim material to achieve the specified gap.
• For 75-100 hp motors, assemble the gauge block **-9**, adapter plate **-10**, gauge base **-11-A** and clamp **-17-A**. If the original shim is not available start measurement with the largest shim (0.50mm). The measured gap must be 0.20mm PLUS the **P** stamped gearcase deviation. If not add or subtract shim material to achieve the specified gap.
• For 115 hp motors, assemble the gauge block **-6** and gauge base **-11** along with the driveshaft, upper oil seal/bearing housing, thrust washer, thrust bearing and shim. If the original shim is not available start measurement with the largest shim (0.50mm). The measured gap must be 1.00mm PLUS the **P** stamped gearcase deviation. If not add or subtract shim material to achieve the specified gap.
• For 150 hp motors, assemble the gauge block '-7' and gauge base '-11A' along with the driveshaft, upper oil seal/bearing housing, thrust washer, thrust bearing and shim. If the original shim is not available start measurement with the largest shim (0.50mm). The measured gap must be 1.00mm PLUS the **P** stamped gearcase deviation. If not add or subtract shim material to achieve the specified gap.
• For 3.3L V6 motors, assemble the driveshaft (along with the driveshaft oil seal/bearing housing, thrust bearing and shims, pinion gear and nut) in Yamaha a pinion height gauge (#YB-06441) and tighten the pinion nut to spec, then use a feeler gauge to check the gap between the pinion gear and the gauge. If the original shim is not available start measurement with the largest shim (0.50mm). The measured gap must be 1.00mm, PLUS the **P** stamped gearcase deviation. If not add or subtract shim material to achieve the specified gap.

3. For World Production Models, assemble the appropriate portions of the listed gauge set along with the driveshaft, bearing, shim, pinion and pinion nut (tighten the nut to the spec listed in the Pinion Gear and Driveshaft procedure), then take the measurement with a feeler gauge and calculate the shim changes depending upon the model as follows:

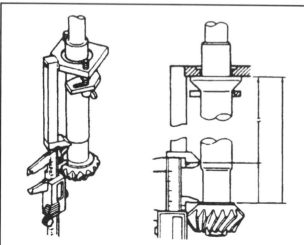

Fig. 76 On most worldwide units you measure from the pinion to the height gauge...

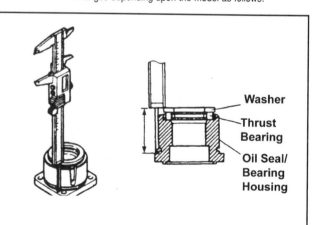

Fig. 77 ...but on models where the bearing mounts in a housing, you also need to check the distance from the mounting flange to the thrust washer

LOWER UNIT 7-51

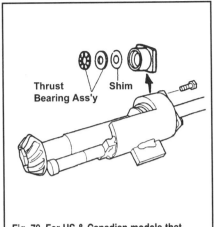

Fig. 78 For US & Canadian models that utilize a bearing housing, a different pinion height gauge is available...

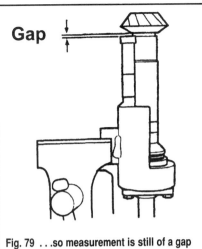

Fig. 79 ...so measurement is still of a gap using a feeler gauge

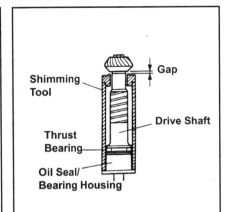

Fig. 80 The largest gearcases utilize a different height gauge, but it works the same way as the gauges on other US and Canadian models

• For 40-60 hp motors, assemble the driveshaft components along with height gauge (#90890-06702), then use a sliding caliper to measure the distance from the top of the pinion gear to the underside of the height gauge. To select the proper shim on standard thrust models through 1998, take the measurement and SUBTRACT 11.3mm, then ADD the **P** stamped gearcase deviation. To select the proper shim on 1999 and later standard thrust models, take the measurement and SUBTRACT 11.3mm, then SUBTRACT the **P** stamped gearcase deviation (remember that if the standard deviation is a negative number, that subtracting a negative is the same as adding the positive or absolute value of that number). To select the proper shim on high-thrust models, take the measurement and SUBTRACT 31.5mm, then SUBTRACT the **P** stamped gearcase deviation (again remembering to subtract a negative number is really to add the absolute value). If the resulting value is zero, no shim change is necessary otherwise add or subtract the amount shim equal to the value.

• For 75-100 hp motors, assemble the driveshaft components along with a height gauge (#90890-06702), then use a sliding caliper to measure the distance from the top of the pinion gear to the underside of the height gauge. To select the proper shim, take the measurement, then SUBTRACT either 31.5mm or 3.15mm (we think it is the first measurement) and SUBTRACT the **P** stamped gearcase deviation (remember that if the standard deviation is a negative number, that subtracting a negative is the same as adding the positive or absolute value of that number). We say either because although Yamaha says both in the factory service manual and the larger specification (which you'll note is any easy typo from the smaller spec) is the same measurement as the 2-strokes of similar size that essentially use the same gearcase. We suspect the larger measurement is probably the best starting point (especially since we can't see why Yamaha would want you to use a sliding caliper for such a small measurement as the smaller possible spec). If the measurement is out of spec add or subtract shim material to achieve the specified gap.

• For 115 hp motors, you're going to take 2 measurements. For the first, invert the driveshaft upper oil seal/bearing housing, then place the thrust bearing and washer in that order on top of the inverted housing (in their normal positions relative to the housing). Now, use a sliding caliper to measure the distance from the top of the thrust washer to the flange on the oil seal/bearing housing (the portion of the flange that would contact the gearcase when right-side-up and installed). Now for the second measurement, assemble the driveshaft components along with a height gauge (#90890-06702). After the wing nuts contact the fixing plate tighten them an additional 1/4 turn. Next, use a sliding caliper to measure the distance from the top of the pinion gear to the underside of the height gauge. To select the proper shim, take the specification 62.5mm, then ADD the **P** stamped gearcase deviation. Now from that sum, SUBTRACT both measurements to determine proper shim thickness.

4. Once you've selected the proper shims, assemble the gearcase as detailed earlier in this section and, before the water pump is installed, check the gearcase Backlash as detailed later in this section. As long as the backlash is within spec, no further adjustment is necessary. However, if backlash is out of specification you'll need to further adjust one or more of the shims to correct the problem.

Nose Gear Shim Selection

◆ See Figures 81 thru 86

For all models, the quickest way to achieve proper nose gear (Forward gear on standard units or Reverse on counter-rotating units) backlash is the use of Yamaha's gear height gauge set. Different tools, formulas and specifications are used for US/Canadian models than are used for World Production models, however the actual measurement procedure is basically the same in that you are checking the actual manufactured height of the bearing.

On most standard rotation US and Canadian models this means placing the bearing and shim in a special height gauge (#YB-34446) along with various components such as a press plate, compression spring and gauge pin, then checking the gap between the plate and pin.

On most counter-rotation US and Canadian models this means using a sliding caliper to measure the height of the bearing either with the shim and/or thrust bearing assembly, as noted in this procedure by model.

On World Production models this means placing the bearing on a flat work surface and measuring the height of the bearing from one end of the bearing cage to the other edge of a roller (not to the other end of the cage, unless otherwise noted).

■ Unless otherwise noted, all standard rotation US and Canadian models use components of Yamaha's #YB-34446 gauge set. Parts of this set will be represented as -## (such as -1 or -7) for short-hand, but the full number of each part would be YB-34446-## (meaning YB-34446-1 or YB-34446-7).

■ The nose gear shim selection procedure varies slightly from model-to-model. The variances are essentially 2 different ways around a square (2 equivalent processes that approach the problem from the opposite directions). However, because of the limited specifications provided by Yamaha you must adhere to the method for the particular model in question.

Determine the nose gear shim thickness, as follows:

1. Check the stamped code on the underside of the anti-ventilation plate (it is stamped in the trim tab mounting area, so if you haven't already, matchmark and remove the tab). The code **F** followed by a + or - and a number is the gearcase's front or nose cone bearing mounting deviation from the standard spec, in hundredths of a mm (this turns out to be the Forward gear/bearing assembly on standard rotation motors and the Reverse gear/bearing assembly on counter-rotation motors).

A stamping such as **F +5**, means that the case requires a shim that it 5/100mm (0.05mm) larger than standard. The + or - accompanying the number tells you whether you should add (+) or take away (-) the number from the standard measurement. If there is NO number after the **F**, assume the value is 0 and the case meets standard tolerance. If the number is there, but illegible, you'll have to work from 0 and make adjustments for that shim by checking lash.

7-52 LOWER UNIT

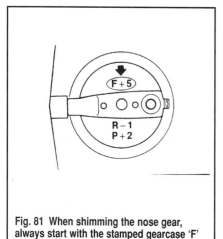

Fig. 81 When shimming the nose gear, always start with the stamped gearcase 'F' deviation

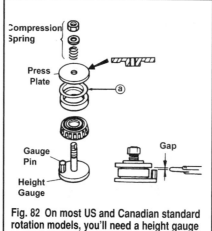

Fig. 82 On most US and Canadian standard rotation models, you'll need a height gauge set...

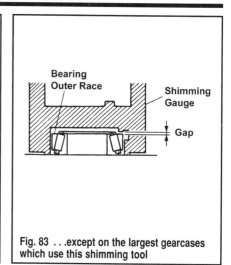

Fig. 83 ...except on the largest gearcases which use this shimming tool

■ All measurements should be taken at no less than 3 points around the bearing and the measurements should be averaged.

2. For US and Canadian Models, assemble the appropriate portions of the gauge set (#YB-34446) (at least on standard rotation units), along with the bearing and gear (see the instructions below for counter-rotation units). Install the bearing and shim into the gauge set (tightening the nut 4 FULL turns after it contacts the spring, unless otherwise noted). Measure the gap between the press plate and the gauge pin using a feeler gauge, then calculate the shim changes depending upon the model as follows:

• For 40-60 hp motors, assemble the bearing and shim along with height gauge '-1', compression spring '-3', press plate '-4' for standard thrust models or plate '-5' for high-thrust models, and the gauge pin '-7' (for all except 2001 and later standard thrust models which use pin '-15'). If the original shim is not available, start using the largest 0.50mm shim. The gap must be a required spec (which varies by year and model), PLUS the stamped **F** deviation. The required spec is 0.06mm for standard thrust models through 2000, 0.95mm for 2001 and later standard thrust models, 1.50mm for high-thrust models through 2000, or 1.70mm for 2001 and later high-thrust models. If not add or subtract shim material to achieve the specified gap.

• For 75-100 hp motors, assemble the bearing and shim along with height gauge/base plate '-1', compression spring '-3', press plate '-5', and the gauge pin '-7'. If the original shim is not available, start using the largest 0.50mm shim. The gap should be either 0.69mm or 1.69mm, PLUS the stamped **F** deviation. We say either because Yamaha says both in the same factory service manual, under specs they list 0.69, but under the sample measurement calculations they give 1.69, which is the same measurement as the 2-strokes of similar size and which essentially use the same gearcase. We suspect the larger measurement is probably the best starting point. If the measurement is out of spec add or subtract shim material to achieve the specified gap.

• For standard-rotation 115 hp motors, assemble the bearing and shim along with height gauge/base plate '-1', compression spring '-3', press plate '-5', and the gauge pin '-7'. If the original shim is not available, start using the largest 0.50mm shim. The gap should be 1.80mm, PLUS the stamped **F** deviation. If the measurement is out of spec add or subtract shim material to achieve the specified gap.

• For counter-rotation 115 hp motors, make a stack out of the reverse gear washer, thrust bearing, roller bearing and shim, then use a micrometer or sliding caliper to measure the thickness of the stack. The measured value must be equal to 25.30mm, PLUS the stamped **F** deviation (yes, they don't know during manufacture and measurement that the gearcase will be used on a counter-rotating unit so the F stamp is still for the nose gear). If the measurement is out of spec add or subtract shim material to achieve the specified gap.

• For standard-rotation 150 hp motors, assemble the bearing and shim along with height gauge/base plate '-1', compression spring '-3', press plate '-4', and the gauge pin '-8'. If the original shim is not available, start using the largest 0.50mm shim. The gap should be 1.60mm, PLUS the stamped **F** deviation. If the measurement is out of spec add or subtract shim material to achieve the specified gap.

• For counter-rotation 150 hp motors, make a stack out of the reverse gear washer, thrust bearing, roller bearing and shim, then use a micrometer or sliding caliper to measure the thickness of the stack. The measured value must be equal to 29.10mm, PLUS the stamped **F** deviation (yes, they don't know during manufacture and measurement that the gearcase will be used on a counter-rotating unit so the F stamp is still for the nose gear, think of the stamp as the FRONT gear/bearing stamp instead of "forward" as it normally is on the standard rotation units). If the measurement is out of spec add or subtract shim material to achieve the specified gap.

• For standard rotation 3.3L V6 motors, place the forward gear bearing and shim down on a perfectly flat work surface with the tapered portion of the bearing (wider portion of the roller) downward. Place the Yamaha

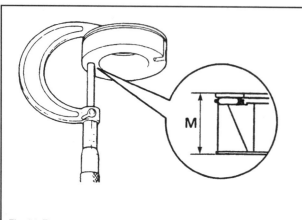

Fig. 84 For most counter-rotation units you make and measure a stack of the bearing and shim (US & Canada)...

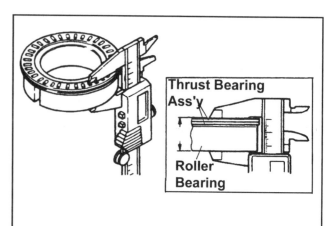

Fig. 85 ...or the bearing and thrust bearing assembly (worldwide production or ALL 150 hp motors)

shimming gauge (#YB-06439) over top of the bearing and use a feeler gauge to check the gap between the edge of the bearing and the tab on the underside of the gauge. The gap must be equal to 0.50mm, MINUS the **F** stamped gearcase deviation (remember that if the deviation is a negative number that subtracting a negative is the same as adding the absolute value). If not add or subtract shims as necessary to achieve the specified gap.

• For counter-rotating 3.3L motors, make a stack out of the reverse gear washer, thrust bearing, roller bearing and shim, then use a micrometer or sliding caliper to measure the thickness of the stack. The measured value must be equal to 30.60mm, PLUS the stamped **F** deviation (yes, they don't know during manufacture and measurement that the gearcase will be used on a counter-rotating unit so the **F** stamp is still for the nose gear). If the measurement is out of spec add or subtract shim material to achieve the specified gap.

3. For World Production Models (other than 150 hp and larger motors, as no separate specs were provided for them, so you'll have to use US and Canadian specs/procedures), place the bearing down on a completely flat work surface, then measure the distance from the top of the roller (not the bearing cage) to the portion of the bearing that is in contact with the work surface. Calculate the shim changes depending upon the model as follows:

• For 40-60 hp motors, the forward gear shim should be equal to the 22.75mm (for standard thrust models) or 24.50mm (for high-thrust models), PLUS the **F** stamped gearcase deviation, MINUS the bearing measurement.

• For 75-100 hp motors, the forward gear shim should be equal to the 24.50mm, PLUS the **F** stamped gearcase deviation, MINUS the bearing measurement.

• For standard rotation 115 hp motors, the forward gear shim should be equal to the 24.60mm, PLUS the **F** stamped gearcase deviation, MINUS the bearing measurement.

• For counter-rotating 115 hp motors, make a stack out of the reverse gear ball/roller bearing, thrust bearing and washer, then measure the thickness of the stack using a micrometer or sliding caliper. The proper reverse gear shim thickness is then calculated by taking the specification of 25.3mm, PLUS the **F** stamped gearcase deviation (the F stamping on the gearcase indicates the nose cone bearing/gear mounting deviation since MOST gearcases are used on standard rotation units), MINUS the stack measurement.

4. Once you've selected the proper shims, assemble the gearcase as detailed earlier in this section and, before the water pump is installed, check the gearcase Backlash as detailed later in this section. As long as the backlash is within spec, no further adjustment is necessary. However, if backlash is out of specification you'll need to further adjust one or more of the shims to correct the problem.

Bearing Carrier Gear Shim Selection

◆ See Figures 87 thru 92

For all models, the quickest way to achieve proper bearing carrier gear (Reverse gear on standard units or Forward on counter-rotating units) backlash is the use of Yamaha's gear height gauge. Different tools, formulas and specifications are used for US/Canadian models than are used for some World Production models, however the actual measurement procedure is basically the same in that you are checking the actual assembled height of the gear on the carrier.

On most US and Canadian models this means placing the gauge set over the gear in contact with the carrier and using a feeler gauge between the tab on the underside of the gauge and the gear teeth.

On most World Production models this usually means placing a flat-stock shimming plate over top of the gear and then measuring the distance from the TOP of the shimming plate to the top edge (thrust plate) of the carrier. Although you could substitute another piece of flat-stock, it would have to be the exact same thickness as the shimming plate.

■ **The bearing carrier gear shim selection procedure varies slightly from model-to-model. The variances are essentially 2 different ways around a square (2 equivalent processes that approach the problem from the opposite directions). However, because of the limited specifications provided by Yamaha you must adhere to the method for the particular model in question.**

Determine the bearing carrier gear shim thickness, as follows:
1. Check the stamped code on the underside of the anti-ventilation plate (it is stamped in the trim tab mounting area, so if you haven't already, matchmark and remove the tab). The code **R** followed by a + or - and a number is the gearcase's rear (located in the prop shaft carrier) bearing mounting deviation from the standard spec, in hundredths of a mm. This means that if there is a marking such as **R +5**, it means that the case requires a shim that it 5/100mm (0.05mm) larger than standard. The + or - accompanying the number tells you whether you should add (+) or take away (-) the number from the standard measurement. If there is NO number after the **R**, assume the value is 0 and the case meets standard tolerance. If the number is there, but illegible, you'll have to work from 0 and make adjustments for that shim by checking lash.

■ **All measurements should be taken at no less than 3 points around the bearing and the measurements should be averaged.**

2. For US and Canadian Models, place the shimming gauge squarely over top of the gear with the legs in contact with the bearing carrier and press down firmly on the gauge as you insert a feeler gauge between the gear and the tab on the shimming gauge. Measure and calculate the shim changes depending upon the model as follows:

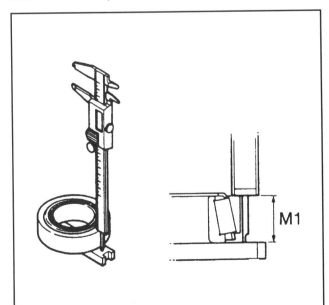

Fig. 86 On standard rotation worldwide units you normally measure bearing height from a particular side of the bearing case, to the opposite end (thin end) of the roller

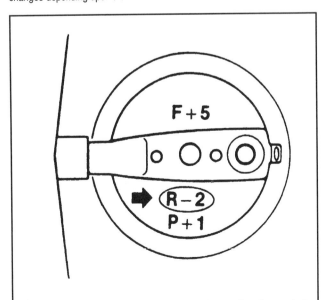

Fig. 87 When shimming the prop shaft bearing carrier, always start with the stamped gearcase 'R' deviation

7-54 LOWER UNIT

- For 40-60 hp motors, a shim measurement tool/procedure is only available for high-thrust models through 2000. All shim changes on other models should be made using Backlash measurements. If the original shim is not available, start adjustment using the largest of the shims (0.50mm). For high-thrust models through 2000, use a Yamaha shimming gauge (#YB-34468-5). The gap between the gear and the tab on the gauge should be 1.00mm MINUS the stamped **R** deviation (remember that if the standard deviation is a negative number, that subtracting a negative is the same as adding the positive or absolute value of that number). If not add or subtract shim material to achieve the specified gap.

- For 75-100 hp motors, these models do not use a reverse gear shim per-se, instead using a standard sized thrust washer between the reverse gear and bearing. Yamaha therefore states that "shim selection" is "unnecessary" for these models.

- For 115 hp motors, use the gauge (#YB-34468-2). The gap between the gear and the tab on the gauge should be 1.80mm (for the reverse gear on standard rotation units) or 1.10mm (for the forward gear on counter-rotation units), MINUS the stamped **R** deviation (remember that if the standard deviation is a negative number, that subtracting a negative is the same as adding the positive or absolute value of that number). If not add or subtract shim material to achieve the specified gap.

On counter-rotation units there is an additional shim found in the propeller shaft bearing carrier (between the carrier and the thrust bearing/thrust washer). To determine the proper shim size to use here, assemble the propeller shaft with tapered roller bearing, thrust bearing, washer, old shim and the bearing carrier, then place the assembly gearcase side down in a vise. Attach a dial gauge to the very end of the propeller shaft and check shaft free-play by pulling and pushing on the shaft. Free-play must be 0.25-0.35mm (0.010-0.014 in.) or it must be adjusted using the propeller shaft shim.

- For 150 hp motors, use a gauge (#YB-06338). The gap between the gear and the tab on the gauge should be either 1.60mm (standard rotation models) or 1.30mm (counter-rotation models), MINUS the stamped **R** deviation (remember that if the standard deviation is a negative number, that subtracting a negative is the same as adding the positive or absolute value of that number). If the gap does not equal this spec minus the deviation, add or subtract shim material to achieve the specified gap.

Furthermore, on counter-rotation units there is an additional shim or there are additional shims found in the propeller shaft bearing carrier (between the carrier and the thrust bearing and thrust washer). To determine the proper shim size to use here, assemble the propeller shaft with tapered roller bearing, thrust bearing, washer, old shim(s) and the bearing carrier, then place the assembly gearcase side down in a vise. Attach a dial gauge to the very end of the propeller shaft and check shaft free-play by pulling and pushing on the shaft. Free-play must be 0.25-0.35mm (0.010-0.014 in.) or it must be adjusted using the propeller shaft shim.

- For standard rotation 3.3L V6 motors, use the Yamaha shimming gauge (#YB-06439). The gauge fits over the partially assembled propeller shaft carrier (the bearing is installed but NOT the gear) to contact the mounting flange of the carrier and provide a gap at a tab over the bearing (at the other end from the mounting flange) which can then be measured with a feeler gauge. The gap must be equal to 0.50mm, MINUS the stamped **R** deviation (remember that if the standard deviation is a negative number, that subtracting a negative is the same as adding the positive or absolute value of that number). If the gap does not equal this spec minus the deviation, add or subtract shim material to achieve the specified gap.

- For counter-rotation 3.3L motors, you'll need to assemble the rear propeller shaft and bearing carrier assembly (without the gear) before measurements can be taken. For details, refer to the Assembly procedures found earlier in this section, but once the propeller shaft is installed to the bearing carrier along with the old shim material, thrust washer, thrust bearing, tapered roller bearing and outer race, the claw washer and ring nut should be installed and tightened to spec. First, use a Yamaha shimming gauge (#YB-06440-A) placed over the rear (gearcase side) of the housing and measure the gap between the gauge and the bearing carrier mounting flange. This gap should be equal to 2.50mm, PLUS the stamped **R** deviation (remember that even though this is the forward bearing for this gearcase, most cases are manufactured and measured assuming the reverse bearing is installed in the carrier, and so are stamped assuming this as well). Add or remove forward gear shims to achieve the proper gap.

Furthermore, on these counter-rotation units there is an additional shim found in the propeller shaft bearing carrier (between the carrier and the thrust bearing/thrust washer). To determine the proper shim size to use here, place the assembled carrier with the gearcase side down in a vise. Attach a dial gauge to the very end of the propeller shaft and check shaft free-play by pulling and pushing on the shaft. Free-play must be 0.25-0.35mm (0.010-0.014 in.) or it must be adjusted using the propeller shaft shim.

3. For World Production Models, use place the flat-stock shimming plate over the gear (assembled into the bearing carrier already) and measure the distance from the TOP of the shimming plate to the thrust plate on top of the carrier. Calculate the shim changes depending upon the model as follows:

- For 40-60 hp motors, a shim measurement tool/procedure is only available for high-thrust models through 2000. All shim changes on other models should be made using Backlash measurements. If the original shim is not available, start adjustment using the largest of the shims (0.50mm). For high-thrust models through 2000, shim changes are equal to the measurement, MINUS 28.00mm MINUS the **R** stamped gearcase deviation (remember that if the standard deviation is a negative number, that subtracting a negative is the same as adding the positive or absolute value of that number).

- For 75-100 hp motors, these models do not use a reverse gear shim per-se, instead using a standard sized thrust washer between the reverse gear and bearing. Yamaha therefore states that "shim selection" is "unnecessary" for these models.

- For 115 hp motors, shim changes are equal to the measurement, MINUS 27.4mm (for standard rotation units) or 28.1mm (for counter-rotation units), MINUS the **R** stamped gearcase deviation (remember that if the standard deviation is a negative number, that subtracting a negative is the same as adding the positive or absolute value of that number).

On counter-rotation units there is an additional shim found in the propeller shaft bearing carrier (between the carrier and the thrust bearing/thrust washer). To determine the proper shim size to use here, assemble the

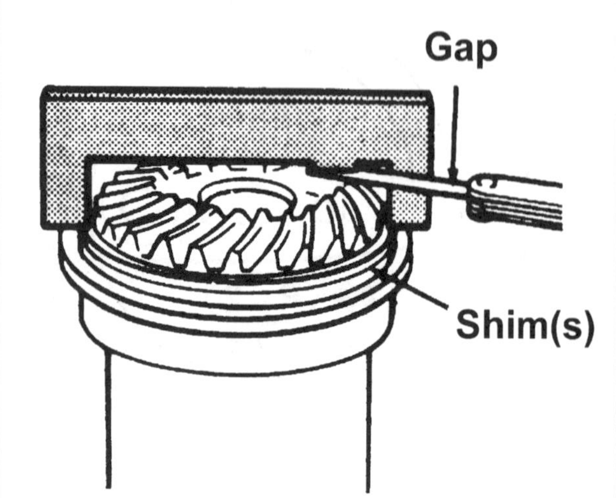

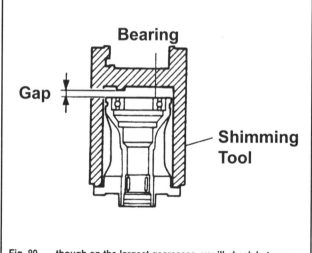

Fig. 88 On many US and Canadian models check the gap between a shim gauge and the gear assembled in the carrier...

Fig. 89 ...though on the largest gearcases, you'll check between the gauge and bearing (no gear)

LOWER UNIT

propeller shaft with tapered roller bearing, thrust bearing, washer (if used), old shim and the bearing carrier, then place the assembly gearcase side down in a vise. Attach a dial gauge to the very end of the propeller shaft and check shaft free-play by pulling and pushing on the shaft. Free-play must be 0.25-0.35mm (0.010-0.014 in.) or it must be adjusted using the propeller shaft shim.

4. Once you've selected the proper shims, assemble the gearcase as detailed earlier in this section and, before the water pump is installed, check the gearcase Backlash as detailed later in this section. As long as the backlash is within spec, no further adjustment is necessary. However, if backlash is out of specification you'll need to further adjust one or more of the shims to correct the problem.

Backlash

◆ See Figures 93 and 94

■ If you're working on a counter-rotating unit, remember the gear positions are reversed so you don't get too confused. The nose-cone gear, referred to as the Forward gear throughout this procedure, is actually the Reverse gear on counter-rotating gearcases. Similarly, the propeller shaft bearing carrier gear, referred to as the Reverse gear throughout this procedure, is actually the Forward gear.

Backlash (also known as lash or play) is the acceptable clearance between two meshing gears, in order to take into account possible errors in machining, deformation due to load, expansion due to heat generated in the lower unit and center-to-center distance tolerances. A no backlash condition is unacceptable, as such a condition would mean the gears are locked together or are too tight against each other which would cause phenomenal wear and generate excessive heat from the resulting friction.

Excessive backlash which cannot be corrected with shim material adjustment indicates worn gears. Such worn gears must be replaced. Excessive backlash is usually accompanied by a loud whine when the lower unit is operating in neutral gear.

The backlash measurements are taken before the water pump is installed. If the amount of backlash needs to be adjusted, the lower unit must be partially disassembled to change the amount of shim material behind one or more of the gears. Of course, it usually does not make sense to change shims on BOTH the pinion AND either the forward or reverse gears at the same time, since the changes will normally have an aggregate affect on each other's readings and you're just making it harder on yourself.

As a general rule, if the lower unit was merely disassembled, cleaned and then assembled with only a new water pump impeller, new gaskets, seals and O-rings, there is no reason to believe the backlash would have changed. Therefore, under those circumstances it should be safe to say this procedure may be skipped.

However, if any one or more of the following components were replaced, the gear backlash should be checked for possible shim adjustment (even if shimming calculations were performed):
- New lower unit housing - check forward and reverse gear backlash.
- New forward gear bearing - check forward gear backlash.
- New pinion gear - check forward/reverse gear backlash.
- New forward gear - check forward gear backlash.
- New reverse gear - check reverse gear backlash.
- New bearing carrier - check reverse gear backlash.
- New thrust washers - check the affected gear's backlash.

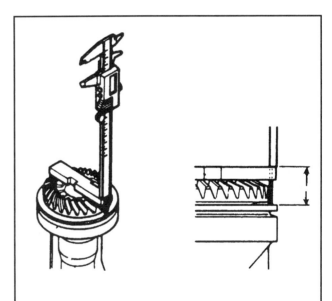
Fig. 91 For most worldwide units, you'll measure from the top of a shimming plate to the thrust plate on the carrier

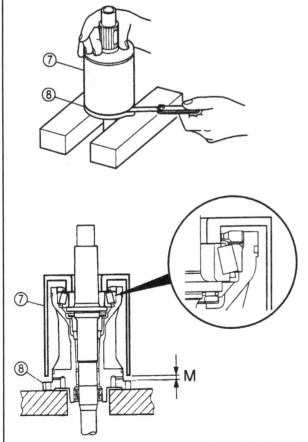
Fig. 90 On the largest counter-rotating units, the gap is checked between the gauge and carrier flange

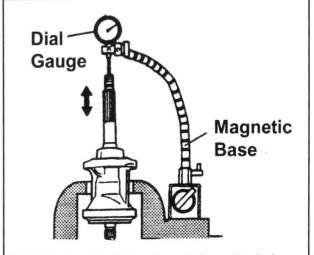
Fig. 92 On all counter-rotating units, you'll also need to check propeller shaft free-play to determine shaft-to-carrier shimming

7-56 LOWER UNIT

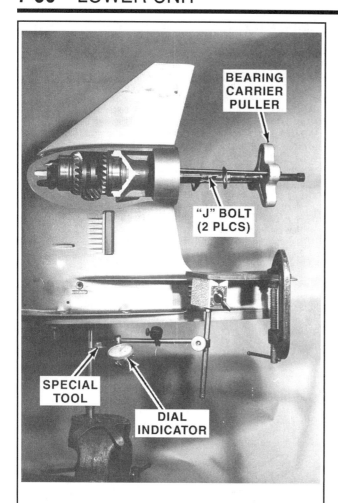

Fig. 93 Use a puller to preload the gear when checking forward gear lash

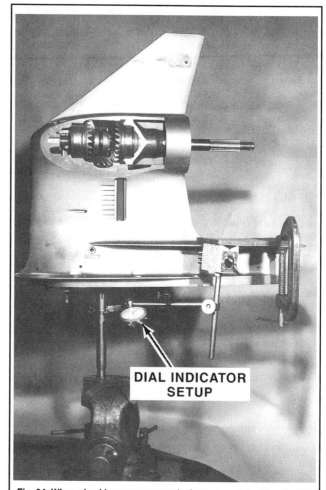

Fig. 94 When checking reverse gear lash most models either install the propeller minus a particular spacer and/or shift the gearcase into reverse to preload the gear

■ The lower unit backlash is generally measured with the unit right side up for all inline motors, except the 115 hp and 150 hp motors. On the 115 hp and larger motors, the backlash is measured with the gearcase inverted (upside down).

To perform this check you'll need a dial gauge on a magnetic or threaded base, as well as a gear lash indicator. The lash indicator is essentially a hose clamp with a straight lever attached at a 90 degree angle to the clamp body. The clamp is attached to the driveshaft in a position that any movement in the shaft will result in movement at the lever. The dial indicator is then installed to the gearcase itself in a position contacting the lever to read the amount of movement. The only thing that stops you from fabricating the indicator yourself if the fact that the amount of lever movement (when compared to shaft rotation) increases dramatically as you move outward away from the clamp further down the lever. And Yamaha simply directs you to use the "mark" provided on the indicator when measuring lash. For all motors you can use the Yamaha Indicator (#YB-06265).

In order to perform this check it is necessary to properly pre-load the Forward or Reverse gears. The method of achieving this is similar on all models. For the nose cone gear (Forward on most units) this is done by attaching a universal puller to the bearing carrier and tightening it lightly against the propeller shaft. For the propeller shaft bearing carrier gear this is normally achieved by installing the propeller in some particular fashion (reversed, without a specific spacer or with a spacer repositioned or facing some specific direction). Specifics will be given in the bullet lists for each model.

1. Position the gearcase in a suitable workstand either upright or inverted, depending upon the model. The lower unit backlash is measured with the unit inverted (upside down) for the 115 hp and larger models or upright position for all 100 hp and smaller motors.

■ Some people prefer to set up the backlash indicator and dial gauge while the gearcase is still upright (just because it is easier than working on it while it is upside down). If so, perform the next couple of steps first, then invert the gearcase.

2. Make sure the gearcase is in **Neutral** by slowly spinning the driveshaft and checking for a lack of motion at the propeller shaft. Although most gearcases are checked with the shifter in Neutral, there are some exceptions which are listed later in this procedure.

■ We're not really sure why Yamaha chooses to preload gears by pushing on the prop shaft when checking Forward gear backlash and pulling on the prop shaft when checking Reverser gear backlash. Some other manufacturers (and on some other Yamahas) the preferred method is simply to shift the gearcase into the gear whose lash is being checked. This is always an alternative if you can't seem to get the prop shaft and gear loaded sufficiently to check lash.

3. Since you normally start by checking Forward gear backlash, install a bearing carrier puller and J bolts onto the ribs of the carrier. With the bearing carrier attaching bolts or the bearing carrier locknut secured, tighten the puller just enough to push on the propeller shaft preloading it (if a specific torque is required on a given model, it will be listed in the bullet list later in this procedure with the backlash specifications).

■ The bearing carrier puller is used to both keep the propeller shaft from turning and to preload the forward gear, thus freeing up one of your hands. However, you can simply push inward and hold the shaft (or have an assistant hold the shaft) while taking readings.

LOWER UNIT 7-57

4. Obtain and install a backlash indicator gauge (#YB-06265) onto the driveshaft. If the hose clamp for the indicator is too large to fit the propeller shaft snugly, insert some kind of packing into the clamp to allow it to be tightened. The clamp must be secure to permit no movement of the indicator on the shaft.

5. Install the dial gauge to the gearcase and position the pointer against the mark on the indicator tool.

6. If you are not using a puller to hold the propeller shaft from turning, grab a hold of the propeller shaft in order to keep it from turning, then twist the driveshaft in one direction (either clockwise or counterclockwise, it doesn't matter at this time) and zero the dial gauge.

■ **See the specifications listed below before attempting to read gear backlash. On some models other special conditions are necessary, such as pulling or pushing on the driveshaft as it is rotated or installing the propeller without the spacer to pre-load the reverse gear when checking reverse gear lash.**

7. Now, continue to hold the propeller shaft to prevent the shaft from rotating. With the other hand gently rock the driveshaft back and forth. In this way, you'll move the driveshaft back and forth between gear contact points with the forward or reverse gear (which can't move because you're holding the propeller shaft). Note the maximum deflection of the dial gauge needle, this is the amount of gear lash present in the gearcase for the current shims in use and compare them to the specifications, as follows:

• For standard thrust 40-50 hp motors through 2000, check backlash with the gearcase upright and while pulling upward lightly on the driveshaft. The center bolt on the puller which is used to preload the forward gear should be tightened to 3.6 ft. lbs. (5 Nm) so the gear restricts movement sufficiently to measure lash. To check forward gear lash, set the shift shaft in the Forward position, and check lash. The forward gear backlash should be about 0.18-0.45mm (0.007-0.018 in.). If lash is LESS than 0.18mm, determine the amount of shim to remove by subtracting the measurement from 0.31mm and then multiplying the result by 1.85mm. If the amount of lash is MORE than 0.45mm, determine the amount of shim to add by subtracting 0.31mm from the measurement and then multiplying the result by 1.85mm. To check reverse gear backlash, remove the puller and install the propeller facing backwards and tighten the nut to 3.6 ft. lbs. (5 Nm) to preload the reverse gear. For some reason we can't quite figure out, Yamaha recommends also setting the shift shaft to Forward for this measurement (but it might be a typo and we suspect they meant to set the shifter in Reverse), if necessary try taking readings with the shaft in Forward and then Reverse to compare the two. Reverse gear backlash should be about 0.71-0.98mm (0.028-0.039 in.). If lash is LESS than 0.71mm, determine the amount of shim to remove by subtracting the measurement from 0.85mm and then multiplying the result by 1.85mm. If the amount of lash is MORE than 0.98mm, determine the amount of shim to add by subtracting 0.85mm from the measurement and then multiplying the result by 1.85mm.

■ **When making gear lash adjustments. Remember, too much lash means there is too much space between the gear in question and the pinion, therefore ADDING shim material will DECREASE backlash. Too little lash means there is not a large enough gap, therefore REMOVING shim material will INCREASE backlash.**

• For 2001 and later standard thrust 50-60 hp motors, check backlash with the gearcase upright and while pulling upward lightly on the driveshaft. The center bolt on the puller which is used to preload the forward gear should be tightened until the propeller shaft can not longer be turned. The forward gear backlash should be about 0.18-0.54mm (0.007-0.021 in.). If lash is LESS than 0.18mm, determine the amount of shim to remove by subtracting the measurement from 0.36mm and then multiplying the result by 0.56mm. If the amount of lash is MORE than 0.54mm, determine the amount of shim to add by subtracting 0.61mm from the measurement and then multiplying the result by 0.56mm. To check reverse gear backlash, remove the puller and install the propeller facing backwards, but REMOVE the spacer that is normally installed between the propeller and gearcase. Also, install the large diameter washer backwards (with the lips facing the propeller), followed by the smaller diameter washer also facing the same direction and finally the prop nut, then tighten the propeller nut until the shaft can no longer be turned in order to preload the reverse gear. Take the readings in the same manner as forward gear. Reverse gear backlash should be about 0.71-1.07mm (0.028-0.042 in.). If lash is LESS than 0.71mm, determine the amount of shim to remove by subtracting the measurement from 0.89mm and then multiplying the result by 0.56mm. If the amount of lash is MORE than 1.07mm, determine the amount of shim to add by subtracting 0.89mm from the measurement and then multiplying the result by 0.56mm.

• For high-thrust 50 hp (T50) motors through 2000, check backlash with the gearcase upright and while pulling upward lightly on the driveshaft. The center bolt on the puller which is used to preload the forward gear should be tightened to 3.6 ft. lbs. (5 Nm) so the gear restricts movement sufficiently to measure lash, but the shift shaft should also be set to Forward (placing the lower unit in gear). The forward gear backlash should be about 0.09-0.28mm (0.004-0.011 in.). If lash is LESS than 0.09mm, determine the amount of shim to remove by subtracting the measurement from 0.19mm and then dividing the result by 1.86mm. If the amount of lash is MORE than 0.28mm, determine the amount of shim to add by subtracting 0.19mm from the measurement and then dividing the result by 1.86mm. To check reverse gear backlash, remove the puller and install the propeller facing backwards without the spacer that normally goes between the gearcase and propeller, then tighten the nut to 3.6 ft. lbs. (5 Nm) to preload the reverse gear. Set the shift shaft in Reverse and then take the readings in otherwise the same manner as forward gear. Reverse gear backlash should be about 0.75-0.13mm (0.030-0.044 in.). If lash is LESS than 0.75mm, determine the amount of shim to remove by subtracting the measurement from 0.94mm and then dividing the result by 1.86mm. If the amount of lash is MORE than 1.13mm, determine the amount of shim to add by subtracting 0.94mm from the measurement and then dividing the result by 1.86mm.

• For 2001 and later high-thrust 50 hp (T50) motors, check backlash with the gearcase upright and while pulling upward lightly on the driveshaft. The center bolt on the puller which is used to preload the forward gear should be tightened until the propeller shaft can not longer be turned. The forward gear backlash should be about 0.12-0.45mm (0.005-0.018 in.). If lash is LESS than 0.12mm, determine the amount of shim to remove by subtracting the measurement from 0.29mm and then multiplying the result by 0.57mm. If the amount of lash is MORE than 0.45mm, determine the amount of shim to add by subtracting 0.29mm from the measurement and then multiplying the result by 0.57mm. To check reverse gear backlash, remove the puller and install the propeller facing backwards, but REMOVE the spacer that is normally installed between the propeller and gearcase. Also, install the large diameter washer with the lips facing the propeller nut, followed by the smaller diameter washer with the lips facing the propeller (the washer lips are positioned facing each other) and finally the prop nut, then tighten the propeller nut until the shaft can no longer be turned in order to preload the reverse gear. Take the readings in the same manner as forward gear. Reverse gear backlash should be about 0.71-1.11mm (0.028-0.044 in.). If lash is LESS than 0.71mm remove sufficient shim material to place it within the acceptable range. If the amount of lash is MORE than 1.11mm, add shims until lash is in the acceptable range.

• For high-thrust 60 hp (T60) motors, check backlash with the gearcase upright and while pulling upward lightly on the driveshaft. The center bolt on the puller which is used to preload the forward gear should be tightened until the propeller shaft can no longer be turned. The forward gear backlash should be about 0.09-0.62mm (0.004-0.025 in.). If lash is LESS than 0.09mm, determine the amount of shim to remove by subtracting the measurement from 0.36mm and then multiplying the result by 0.53mm. If the amount of lash is MORE than 0.62mm, determine the amount of shim to add by subtracting 0.36mm from the measurement and then multiplying the result by 0.53mm. To check reverse gear backlash, remove the puller and install the propeller facing backwards and tighten the nut to 3.6 ft. lbs. (5 Nm) to preload the reverse gear. Take the readings in the same manner as forward gear. Reverse gear backlash should be about 0.71-0.98mm (0.028-0.039 in.). If lash is LESS than 0.71mm, determine the amount of shim to remove by subtracting the measurement from 0.85mm and then multiplying the result by 0.56mm. If the amount of lash is MORE than 0.98mm, determine the amount of shim to add by subtracting 0.85mm from the measurement and then multiplying the result by 0.56mm. To check reverse gear backlash, remove the puller and install the propeller facing backwards, but REMOVE the spacer that is normally installed between the propeller and gearcase. Also, install the large diameter washer with the lips facing the propeller nut, followed by the smaller diameter washer with the lips facing the propeller (the washer lips are positioned facing each other) and finally the prop nut, then tighten the propeller nut until the shaft can no longer be turned in order to preload the reverse gear. Take the readings in the same manner as forward gear. Reverse gear backlash should be about 0.71-1.11mm (0.028-0.044 in.). If lash is LESS than 0.71mm remove sufficient shim material to place it within the acceptable range. If the amount of lash is MORE than 1.11mm, add shims until lash is in the acceptable range.

7-58 LOWER UNIT

• 75-100 hp motors, check backlash with the gearcase upright and while pulling upward lightly on the driveshaft. The center bolt on the puller which is used to preload the forward gear should be tightened to 3.6 ft. lbs. (5 Nm). The forward gear backlash should be about 0.13-0.46mm (0.005-0.018 in.). If lash is LESS than 0.13mm, determine the amount of shim to remove by subtracting the measurement from 0.30mm and then multiplying the result by 0.60mm. If the amount of lash is MORE than 0.46mm, determine the amount of shim to add by subtracting 0.30mm from the measurement and then multiplying the result by 0.60mm. There is no need to check reverse gear backlash on these models.

• For standard-rotation 115 hp motors, check backlash with the gearcase inverted (upside down) and with the backlash indicator attached to the 22.4mm (0.88 in.) diameter portion of the driveshaft. The center bolt on the puller which is used to preload the forward gear should be tightened to 7.2 ft. lbs. (10 Nm). The forward gear backlash should be about 0.10-0.31mm (0.008-0.012 in.). If lash is LESS than 0.20mm, determine the amount of shim to remove by subtracting the measurement from 0.26mm and then multiplying the result by 0.58mm. If the amount of lash is MORE than 0.31mm, determine the amount of shim to add by subtracting 0.26mm from the measurement and then multiplying the result by 0.58mm.

Reverse gear backlash is checked by removing the puller and then installing the propeller, but without the spacer that normally installs between the propeller and the gearcase. Tighten the propeller retaining nut to 7.2 ft. lbs. (10 Nm) in order to preload the reverse gear. Otherwise check reverse gear backlash in the same way as the forward gear, it should be about 0.50-0.73mm (0.020-0.029 in.). If lash is LESS than 0.50mm, determine the amount of shim to remove by subtracting the measurement from 0.62mm and then multiplying the result by 0.58mm. If the amount of lash is MORE than 0.73mm, determine the amount of shim to add by subtracting 0.62mm from the measurement and then multiplying the result by 0.58mm.

• For counter-rotation 115 hp motors, check backlash with the gearcase inverted (upside down) and with the backlash indicator attached to the 22.4mm (0.88 in.) diameter portion of the driveshaft. The center bolt on the puller which is used to preload the forward gear should be tightened to 7.2 ft. lbs. (10 Nm). The forward gear backlash should be about 0.15-0.30mm (0.006-0.012 in.). If lash is LESS than 0.15mm, determine the amount of shim to remove by subtracting the measurement from 0.23mm and then multiplying the result by 0.58mm. If the amount of lash is MORE than 0.30mm, determine the amount of shim to add by subtracting 0.23mm from the measurement and then multiplying the result by 0.58mm.

Reverse gear backlash is checked by removing the puller and then installing the propeller, but without the spacer that normally installs between the propeller and the gearcase. Tighten the propeller retaining nut to 7.2 ft. lbs. (10 Nm), then shift the gearcase into Reverse and SLOWLY turn the driveshaft clockwise until the clutch dog fully engages Reverse gear ensuring the gear is fully preloaded. Now turn the shift rod back into Neutral and turn the driveshaft about 30° counterclockwise. Finally turn the shift rod BACK into Reverse and slowly check backlash from one stop to the other, it should be about 0.50-0.70mm (0.020-0.028 in.). If lash is LESS than 0.50mm, determine the amount of shim to remove by subtracting the measurement from 0.60mm and then multiplying the result by 0.58mm. If the amount of lash is MORE than 0.70mm, determine the amount of shim to add by subtracting 0.60mm from the measurement and then multiplying the result by 0.58mm.

• For standard rotation 150 hp motors, check backlash with the gearcase inverted (upside down) and with the backlash indicator attached to the 22.4mm (0.88 in.) diameter portion of the driveshaft. The center bolt on the puller which is used to preload the forward gear should be tightened until the driveshaft can no longer be turned between gears (only pivoted back and forth slightly to check lash). The forward gear backlash should be about 0.14-0.46mm (0.006-0.018 in.). If lash is LESS than 0.14mm, determine the amount of shim to remove by subtracting the measurement from 0.30mm and then multiplying the result by 0.67mm. If the amount of lash is MORE than 0.46mm, determine the amount of shim to add by subtracting 0.30mm from the measurement and then multiplying the result by 0.67mm.

Reverse gear backlash is checked by removing the puller and then installing the propeller, but without the spacer that normally installs between the propeller and the gearcase (and without the large washer that normally goes between the nut and the rear spacer). Tighten the propeller retaining nut until the driveshaft is locked in the gear (in the same fashion as the puller when checking forward gearlash). Check reverse gear backlash in the same way as the forward gear, it should be about 0.32-0.67mm (0.013-0.026 in.). If lash is LESS than 0.32mm, determine the amount of shim to remove by subtracting the measurement from 0.50mm and then multiplying the result by 0.67mm. If the amount of lash is MORE than 0.67mm, determine the amount of shim to add by subtracting 0.50mm from the measurement and then multiplying the result by 0.67mm.

• For counter-rotation 150 hp motors, check backlash with the gearcase inverted (upside down) and with the backlash indicator attached to the 22.4mm (0.88 in.) diameter portion of the driveshaft. The center bolt on the puller which is used to preload the forward gear should be tightened until the driveshaft can no longer be turned between gears (only pivoted back and forth slightly to check lash). The forward gear backlash should be about 0.14-0.42mm (0.006-0.017 in.). If lash is LESS than 0.14mm, determine the amount of shim to remove by subtracting the measurement from 0.28mm and then multiplying the result by 0.67mm. If the amount of lash is MORE than 0.42mm, determine the amount of shim to add by subtracting 0.28mm from the measurement and then multiplying the result by 0.67mm.

Reverse gear backlash is checked by removing the puller and then installing the propeller, but without the spacer that normally installs between the propeller and the gearcase (and without the large washer that normally goes between the nut and the rear spacer). Tighten the propeller retaining nut until the driveshaft is locked in the gear (in the same fashion as the puller when checking forward gear lash). Check reverse gear backlash in the same way as the forward gear, it should be about 0.23-0.58mm (0.009-0.023 in.). If lash is LESS than 0.23mm, determine the amount of shim to remove by subtracting the measurement from 0.41mm and then multiplying the result by 0.67mm. If the amount of lash is MORE than 0.58mm, determine the amount of shim to add by subtracting 0.41mm from the measurement and then multiplying the result by 0.67mm.

• For standard rotation 3.3L V6 motors, check backlash with the gearcase inverted (upside down) and with the backlash indicator attached to the 22.4mm (0.88 in.) diameter portion of the driveshaft. The center bolt on the puller which is used to preload the forward gear should be tightened until it keeps the propeller shaft from turning, thus preloading the forward gear. The forward gear backlash should be about 0.21-0.44mm (0.008-0.017 in.). If lash is LESS than 0.21mm, determine the amount of shim to remove by subtracting the measurement from 0.33mm and then multiplying the result by 0.71mm. If the amount of lash is MORE than 0.44mm, determine the amount of shim to add by subtracting 0.33mm from the measurement and then multiplying the result by 0.71mm.

Reverse gear backlash is checked by removing the puller and then installing the propeller, but with the spacer that normally installs between the propeller and the gearcase moved to sit between the propeller and small washer normally used with the propeller nut (this installation leaves out the large thrust washer normally found between the small nut washer and the propeller, did you get all that?). Tighten the propeller while rotating the driveshaft until it holds the shaft from turning more than between gears. Check reverse gear backlash in the same way as the forward gear, it should be about 0.70-1.03mm (0.028-0.041 in.). If lash is LESS than 0.70mm, determine the amount of shim to remove by subtracting the measurement from 0.87mm and then multiplying the result by 0.71mm. If the amount of lash is MORE than 1.03mm, determine the amount of shim to add by subtracting 0.87mm from the measurement and then multiplying the result by 0.71mm.

• For counter-rotating 3.3L V6 motors, check backlash with the gearcase inverted (upside down) and with the backlash indicator attached to the 22.4mm (0.88 in.) diameter portion of the driveshaft. The center bolt on the puller which is used to preload the forward gear should be tightened until it keeps the propeller shaft from turning, thus preloading the forward gear. The forward gear backlash should be about 0.35-0.70mm (0.014-0.028 in.). If lash is LESS than 0.35mm, determine the amount of shim to remove by subtracting the measurement from 0.53mm and then multiplying the result by 0.71mm. If the amount of lash is MORE than 0.70mm, determine the amount of shim to add by subtracting 0.53mm from the measurement and then multiplying the result by 0.71mm.

Reverse gear backlash is checked by removing the puller and then shifting the gearcase into Reverse and SLOWLY turning the driveshaft clockwise until the clutch dog fully engages Reverse gear ensuring the gear is fully preloaded. Now turn the shift rod back into Neutral and turn the driveshaft about 30° counterclockwise. Finally turn the shift rod BACK into Reverse and slowly check backlash from one stop to the other, it should be about 0.70-1.03mm (0.028-0.041 in.). If lash is LESS than 0.70mm, determine the amount of shim to remove by subtracting the measurement from 0.87mm and then multiplying the result by 0.71mm. If lash is MORE than 1.03mm, determine the amount of shim to add by subtracting 0.87mm from the measurement and then multiplying the result by 0.71mm.

LOWER UNIT 7-59

■ If the lash is too low on BOTH the forward and reverse gears, the pinion height may be too low. If the lash is too high on BOTH gears, the pinion may be too high in the gearcase. Check the installation of the bearings/bushings and the shim materials or spacers used.

8. If adjustment is necessary, partially disassemble the gearcase as necessary to access the forward or reverse gear shims, then ad or take away shim material, as necessary to bring backlash with spec. However, if the backlash specification cannot be reached by adding and removing shim material, the gears may have to be replaced.

9. If possible, pressure test the gearcase. With the oil fill/drain screw installed and tightened, attach a threaded pressure tester to the oil level hole, then use a hand pump to apply 14.22 psi (100 kPa) of pressure to the gearcase. Watch that the pressure remains steady for at least 10 seconds. If pressure falls before that point one or more of the seals require further attention. You can immerse the gearcase in water and look for bubbles to determine where the leak is occurring. Be sure of gearcase integrity before installation.

※※ WARNING

Don't over-pressurize the gearcase as this could CAUSE leaks.

10. Install the Water Pump to the top of the gearcase as detailed in the Lubrication and Cooling section.
11. Install the Gearcase to the intermediate housing, as detailed in this section.
12. Install the Propeller and properly refill the Gearcase Lubricant, both as detailed in the Maintenance and Tune-Up section.

Gear Mesh Pattern

◆ See Figure 95

As noted earlier, you can use a machinist dye or other powder to check the gear mesh pattern. This is useful when determining if there is too much wear on a gear set to return them to service (especially if a dial gauge set and lash indicator is not available to check gear backlash).

The basic method used to check a gear mesh pattern is to coat the gears with a dye or powder, then install the gears, carefully rotate the shafts/gears and carefully disassemble the case again to examine the gears.

All models should be held with the lower unit in the upright (normal) position.

Grasp the driveshaft and pull upward. At the same time, rotate the propeller shaft counterclockwise through about six or eight complete

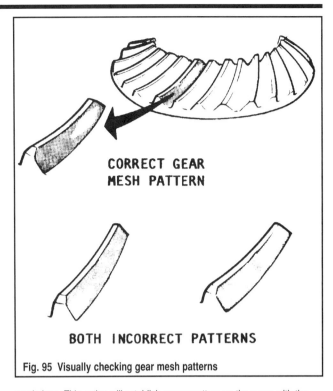

Fig. 95 Visually checking gear mesh patterns

revolutions. This action will establish a wear pattern on the gears with the dye/Desenex® powder.

1. Disassemble the unit and compare the pattern made on the gear teeth with the accompanying illustrations. The pattern should almost be oval on the drive side and be positioned about halfway up the gear teeth.

If the pattern appears to be satisfactory, clean the dye or powder from the gear teeth and assemble the unit one final time.

If the pattern does not appear to be satisfactory, add or remove shim material, as required. Adding or removing shim material will move the gear pattern towards or away from the center of the teeth.

After the gear mesh pattern is determined to be satisfactory, assemble the bearing carrier one final time.

GEARCASE SERVICE - MERCURY/MARINER

Gearcase Repair and Overhaul Tips

All Mercury/Mariner gearcases use the same basic inner design in that pinion gear is splined to the bottom of a vertical driveshaft and secured by a retaining nut. The driveshaft uses the pinion gear to transfer the rotational power from the crankshaft to the propeller shaft. Both the driveshaft and propeller shaft ride in replaceable bearings and a bath of gear oil. Gear oil is kept inside the case itself (and water is kept out) by propeller shaft and driveshaft housing seals at either end (bottom rear and top center) of the gearcase.

The size and type of the shafts, bearings, gears and seals will vary slightly from model-to-model. Also, all gearcases contain Neutral and Reverse shifting capabilities. Generally speaking one of 2 major designs are employed, a vertical push-pull shifter retaining by nuts on threaded rods (or a bolted coupler assembly) is used by mostly by smaller gearcases, but can be found on models up to 50 hp, while a horizontally ratcheting shifter which connected through splines to the shift lever in the intermediate housing is used on most larger outboards (but can be found on some smaller gearcases, mostly Bigfoot models, down to 30 hp).

We've divided coverage from removal and installation to complete overhaul procedures into a few major categories. First a procedure for 4/5/6 Hp (1-Cylinder) motors covers the smallest models. Second, a procedure for 8-15 Hp (2-Cylinder) models covers MOST, but not all of the 2-cylinder models (it does not cover the 25 hp motor which uses one of the mid-range style gearcases). A few of the smaller mid-range models are all covered together in the third procedure, including the 25 Hp (Bigfoot), 30/40 Hp (Non-Bigfoot) and 40/45/50 hp (Non-Bigfoot, 935cc) models. The balance of inline motors share a similar enough design that they are all covered in the fourth set of procedures, this includes the 30-115 Hp (Bigfoot), as well as 40/50/60 Hp (996cc) and 75-115 Hp (non-Bigfoot) models. Finally, V6 models, coming equipped from Mercury with a Yamaha gearcase are covered in a combined procedure for both Mercury/Mariner and Yamaha motors.

We've included diagrams and photographs from as many gearcases on which we could realistically get our hands, however keep in mind that subtle differences will occur from model-to-model or year-to-year. To ensure a successful overhaul a few general notes should be followed throughout the overhaul procedure, as follows:

• As with all repair or overhaul procedures, make sure the work area is clean and free of excessive dirt or moisture.

• Take your time when removing and disassembling components taking a moment with each to either matchmark its positioning or note its orientation in relation to other components. This goes ESPECIALLY for the placement of any shims which might be used (keep in mind that all models are NOT equipped with shims). Absent specific instructions for checking and re-shimming a gearcase (which occurs when bearings, gears and/or shafts are replaced on some gearcases) all shims must be returned to their original positions during assembly.

• Seal placement and orientation will vary on some models. Before removing any seal be sure to first note the direction in which the seal lips are facing and be sure to install the replacement seal (and you ALWAYS replace seals when they are removed) facing the same direction.

• Unless otherwise directed seal lips are packed with a light coating of Quicksilver 2-4-C W/ Teflon or an equivalent water resistant lubricant. When used, O-rings are also coated with 2-4-C or some form of marine grease during installation.

7-60 LOWER UNIT

- Needle bearings which are pressed in place are generally destroyed by the removal process and should be replaced when separated from a gear, shaft or bore. Ball Bearings can usually be freed without damage, but use common sense. Once the bearing is free, clean it thoroughly, then rotate the inner/outer cage feeling for rough spots. Replace any bearing that is not in top shape.
- The correct size driver must be used for bearing or seal installation. Generally speaking a driver must be smooth (especially for seals, to prevent it from cutting the sealing surface) and should be sized to spread as much of the force over the seal or bearing surface as possible. When a bearing is installed ON a shaft, the driver MUST contact the inner race (never push on only the outer race in this case or the bearing will be damaged). When a bearing is installed IN a bore, the driver MUST contact the outer race (in this circumstance, do NOT push only on the inner race or the bearing will be damaged).
- Driveshaft bearings or bushings are often pressed straight into position from above and/or pulled into position from below. Although Mercury often has a special tool available to help draw a bushing or bearing into position in the lower portion of the gearcase, all you REALLY need to accomplish this is a sufficiently long threaded bolt with a nut and a washer that is the size of the driver you'd want to contact that bushing or bearing. You'll also need a nut and a plate to place on top of the gearcase. Place the threaded rod through one nut and plate and then down into the gearcase through the bushing/bearing (so the plate is resting against the top of the gearcase, the upper nut is resting against the plate and the threads of the rod are just protruding down into the pinion mounting area of the gearcase.). Next install the washer and lower nut over the bottom threads of the rod to support the bearings or busing. Slowly draw the bearing or bushing into position in the gearcase by holding the threaded rod from turning while at the same time turning the upper nut to draw the entire bolt, bushing/bearing, washer and nut assembly upward until the bushing or bearing seats. Then loosen the nut on the bottom of the threaded rod in the gearcase and remove the threaded rod, plate, nut and washer.

In all cases the water pump, driveshaft oil seal or propeller shaft oil seal may be replaced without complete gearcase disassembly.

Although Mercury calls for various special tools for gearcase overhaul MOST of them come down to the appropriate sized driver (see the bullet about drivers earlier in this section), OR a universal puller or a slide hammer generally with inside jaws. Whenever possible, we prefer the use of a puller over a slide hammer, just because the force is applied more linearly, however there are simply times when the impact of a slide hammer makes the job easier, so use common sense when determining which tool you'd prefer to use for bearing, race or potentially even shaft removal.

Lastly, we've tried to provide the best procedures possible for gearcase overhaul, and in the most logical order. However, in most cases procedures for some of the earlier procedures can be performed alone, without additional procedures. For instance water pump or propeller shaft bearing carrier/seal housing procedures can normally be performed completely independently of each other. However, since the driveshaft oil seal housing is found UNDER the water pump, you'd obviously have to remove the pump first for access.

Gearcase - 4/5/6 Hp (1-Cylinder) Models

REMOVAL & INSTALLATION

◆ See Figures 96 thru 99

This section covers the 3/4/5 hp Mercury/Mariner single cylinder motors.

1. For safety, disconnect the high tension spark plug lead and remove the spark plug before working on the lower unit.
2. Shift the gearcase into Reverse.
3. If necessary for access or for overhaul purposes, remove the Propeller, as detailed in the Maintenance and Tune-Up section.
4. If necessary for gearcase overhaul purposes, position a suitable clean container under the lower unit. Remove the FILL screw on the bottom of the lower unit, and then the VENT screw. The vent screw must be removed to allow air to enter the lower unit behind the lubricant. Allow the gear lubricant to drain into the container.

As the lubricant drains, catch some with your fingers from time-to-time and rub it between your thumb and finger to determine if there are any metal particles present. Examine the fill plug. A small magnet imbedded in the end of the plug will pickup any metal particles. If metal is detected in the lubricant, the unit must be completely disassembled, inspected, and the damaged parts replaced.

Check the color of the lubricant as it drains. A whitish or creamy color indicates the presence of water in the lubricant. Check the container for signs of water separation from the lubricant. The presence of any water in the lubricant is bad news. The unit must be completely disassembled; inspected; the cause of the problem determined; and then corrected.

※※ WARNING

Check to be sure the spark plug has been removed; the RUN/STOP switch is in the STOP position and the battery has been disconnected to prevent any possibility of the powerhead starting during propeller removal.

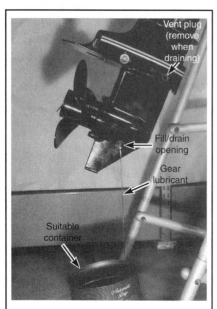

Fig. 96 If necessary for overhaul, drain the gearcase oil

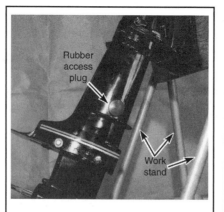

Fig. 97 Remove the rubber plug from the midsection...

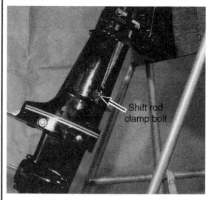

Fig. 98 ... for access to the shift linkage

LOWER UNIT

5. Rotate and swing the lower end unit in the full UP tilt position.
6. Remove the rubber access plug on the starboard side of the upper gear housing.
7. Loosen, but do not remove the bolt for the shift rod clamp. Loosen the bolt only enough to allow the shift shaft to slide free of the clamp when the lower unit is removed from the exhaust housing.
8. Remove the two attaching bolts and washers securing the lower unit to the exhaust housing. Pull down on the lower unit gear housing and separate the lower unit from the exhaust housing. Guide the shift rod and driveshaft out of the exhaust housing as the lower unit is lowered.

To Install:

9. If not already in that position, swing the exhaust housing outward until the tilt lock lever can be actuated, and then engage the tilt lock.
10. Both the lower unit and the shift lever must be in matching positions when assembling the lower unit. If necessary, move the shift lever to the REVERSE gear position. This will enable the upper shift shaft to align with the clamp on the lower shift shaft.
11. Rotate the propeller shaft while pushing down on the shift shaft to shift the lower unit. When in reverse, the propeller shaft will not rotate more than a few degrees in either direction.
12. Coat the driveshaft splines with a light coating of Quicksilver 2-4-C or a suitable multi-purpose lubricant. Take care not to use an excessive amount of lubricant on the driveshaft splines on any Mercury/Mariner Outboard unit.

■ An excessive amount of lubricant on top of the driveshaft to crankshaft splines will be trapped in the clearance space. This trapped lubricant will not allow the driveshaft to fully engage with the crankshaft.

13. Insert the driveshaft into the exhaust housing and at the same time align the water tube with the water pump cover outlet.
14. Feed the lower shift rod into the coupler on the upper shift rod. Maintain the lower unit mating surface parallel with the exhaust housing mating surface.
15. Push the lower unit toward the exhaust housing and at the same time rotate the flywheel to permit the crankshaft splines to index with the driveshaft splines.
16. Secure the lower unit in place with the two attaching bolts and washers. Tighten the bolts alternately to 70 inch lbs. (8Nm).
17. Tighten the shift shaft coupler bolt securely. Install the plug into the intermediate housing to cover the coupler bolt. Check the complete work for proper shifting.
18. If the gearcase was drained refill it with fresh gear oil.
19. If the propeller was removed, install the propeller.
20. Reconnect the spark plug wire.

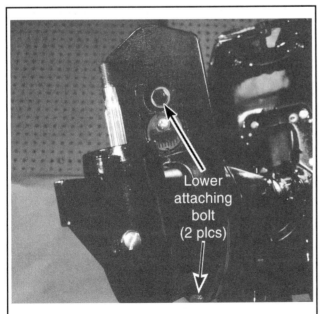

Fig. 99 Remove the gearcase attaching bolts

OVERHAUL

Bearing Carrier & Propeller Shaft

■ For gearcase exploded views, please refer to Cleaning & Inspection in this section.

1. Secure the lower unit in a vise equipped with soft jaws, in a holding fixture, or clamp it to something solid with a wood clamp. Begin disassembly of the bearing carrier by removing the two bolts securing the bearing carrier to the lower unit.
2. Grasp the propeller shaft and pull the shaft and bearing carrier assembly out of the lower unit.

■ If the bearing carrier is difficult to remove from the lower unit housing, the following alternate method will generally prove effective.

3. Clamp the propeller shaft in a vise equipped with soft jaws. Obtain a soft headed mallet and strike the lower unit just above the bearing carrier to jar the bearing carrier free.

■ Another method which can be used but must be done very carefully is to rotate the bearing carrier cap 90° (1/4 turn). Now, gently tap the bearing carrier cap with a soft headed mallet. Tap evenly on both sides of the cap to jar the bearing carrier free.

4. With the bearing carrier removed from the lower unit housing, pull the cam follower out the end of the propeller shaft.
5. Tap the end of the propeller shaft on the table and remove the small round guide from inside the propeller shaft.
6. Slide the bearing carrier off the end of the propeller shaft.
7. Slide the reverse gear and washer off the propeller shaft. If a small shim is on the back side of the reverse gear, be sure to save this shim.
8. Place the cam follower back into the end of the propeller shaft and rotate the propeller shaft in a vertical position.
9. Press down on the propeller shaft to release the spring tension against the sliding clutch. Using a 1/8 in. pin punch, push the cross pin out through the opposite side of the sliding clutch.
10. Remove the punch and withdraw the cross pin from the sliding clutch. When the cross pin is removed, the sliding clutch should fall free of the propeller shaft.
11. Remove the spring from inside the propeller shaft.

■ In most cases, the propeller shaft bearing will be damaged during the removal procedure. Examine the bearing carefully, perform the following work only if the propeller bearing is damaged and is no longer fit for further service.

12. Remove the O-ring from the bearing carrier and discard the O-ring.
13. Grasp the inner race of the propeller shaft ball bearing. Attempt to work the race back-and-forth. The attempt should result in very little, if any "play". A very slight amount of side "play" is acceptable, because there is only about 0.001 in. (0.025mm) clearance in the bearing.
14. Lubricate the ball bearing with light oil. Check the action of the bearing by rotating the outer bearing race. The bearing should have a smooth action and no rust stains. If the ball bearing sounds or feels rough or catches, the bearing should be removed and discarded.
15. Removal of the bearing or seal will distort it and therefore, it cannot be installed a second time. Be absolutely sure a new part is available before removing the bearing. Unfortunately, the bearing, good or bad, must be removed before the propeller shaft seal can be driven out, and of course they must be replaced with new ones.
16. Note the position of the bearing in relation to the carrier. Look for any embossed numbers or letters and observe how one shoulder of the bearing is flush with the carrier.
17. Remove the bearing using puller (#91-83165M) or a universal slide puller.

7-62 LOWER UNIT

Driveshaft & Water Pump Base

1. Remove the Water Pump assembly as detailed in the Lubrication and Cooling System section.
2. Remove the bolt securing the shift shaft assembly to the top of the driveshaft and water pump base. Carefully pull upward to remove the shift shaft and bushing. Remove the shaft O-ring from the bore in the base.
3. Grasp the driveshaft, then lift up and remove the driveshaft and water pump base as an assembly from the lower unit housing. Remove and discard the gasket.
4. Reach inside the lower unit bearing carrier cavity and remove the pinion gear.

■ Be alert for any shims behind the pinion gear. Save these shims, because they will be critical during the assembling work. Also be aware of any shims which may be installed on top of the driveshaft upper bearing.

5. Slide the water pump base off the driveshaft bearing assembly.
6. Remove the seals, O-rings and gasket materials from the water pump base assembly. This base is normally equipped with a dual-lipped seal that is removed or installed from underneath the housing. Use a internal jawed seal puller to remove it from the bore in the underside of the base.

■ If possible use the Mercury tool (#91-17351) for removal and installation of the driveshaft lower needle bearing. However, a suitable sized driver MAY be substituted.

7. If the driveshaft lower needle bearing must be removed, double-check the installed height of the current bearing before disturbing it (as the replacement is driven in from the top to a certain depth using the specified tool, OR if the special tool is not available, using a suitable driver and the measurement). If the bearing is to be replaced, use a driver to tap it out of its bore and into the gearcase. Reach in and remove the bearing.

Forward Gear and Bearing

1. Tilt the lower unit. The forward gear will fall into your hand. If the gear fails to fall free, strike the open end of the lower unit on a block of wood to jar the gear free.
2. Although Mercury claims this gearcase, when installed on these powerheads, is NOT adjustable when it comes to lash, watch for and save any shim material (if any is found behind the forward gear). If present, this shim material is critical to obtaining the correct backlash during installation. The shim material will be located between the forward gear and the bearing.
3. The forward gear bearing is a one piece ball bearing and will not - or should not - just fall out of the lower unit cavity. The bearing must be "pulled" from the lower unit only if it is unfit for further service.
4. Obtain a puller (#91-27780) or a slide hammer with a jaw expander attachment. Hook the jaws inside the inner bearing race and use the slide hammer to jar the race free.

CLEANING & INSPECTION

◆ See Figure 100

1. Clean all water pump parts with solvent, and then dry them using compressed air.
2. Inspect the water pump cover and base for cracks and distortion. If possible always install a new water pump impeller while the lower unit is disassembled. A new impeller will insure extended satisfactory service. If the old impeller must be used, never install it in reverse to the original direction of rotation. Reverse installation could lead to premature impeller failure and subsequent powerhead damage.
3. Inspect the bearing surface of the propeller shaft.
4. Check the shaft surface of propeller shaft for pitting, scoring, grooving, imbedded particles, uneven wear and discoloration.
5. Check the straightness of the propeller shaft with a set of machinist V-blocks.
6. Clean the pinion gear and the propeller shaft with solvent and dry the components with compressed air.
7. Check the pinion gear and drive gear for abnormal wear.

ASSEMBLY

Sliding Clutch Assembly

1. Slide the spring down into the propeller shaft. Insert a narrow screwdriver into the slot and compress the spring until approximately 1/4 in. (12mm) is obtained between the top of the slot and the screwdriver.
2. Hold the spring compressed, and at the same time, guide the sliding clutch over the splines of the propeller shaft with the hole in the clutch aligned with the hole in the propeller shaft. The clutch may be installed either way, preferably the side with the least amount of wear should face the forward gear.
3. Insert the cross pin into the sliding clutch and through the space held open by the screwdriver.
4. Center the pin and then remove the screwdriver allowing the spring to pop back into place.
5. Install the guide into the end of the propeller shaft, followed by the cam follower. As both ends of the cam follower are equally rounded, it may be installed either way.

Forward Gear and Bearing

1. Obtain the following special tools: Mandrel (#91-84532M) and Driver (#91-84529M). Install the ball bearing assembly with the numbered side facing toward the installation tool.

■ If shims were removed during disassembly, be sure to insert them between the housing and bearing assembly,

2. Thoroughly lubricate the bearing and gear with Mercury Super Duty Gear Lubricant, or equivalent. Insert any shim material saved during disassembly (if present), onto the forward gear. The shim material should give the same amount of backlash between the pinion gear and the forward gear as before disassembly.
3. Insert the lubricated forward gear and shim material into the lower unit with the teeth of the gear facing outward.

Driveshaft Needle Bearing

The following steps apply only if the driveshaft needle bearing was removed.

1. Obtain the needle bearing removal and installation tool (#91-17351), or a suitable driver (however, if not using the factory tool you'll need the measurement taken during disassembly of how deep the needle bearing sits in the gearcase).
2. Slide the new needle bearing onto the end of the tool with the embossed numbers facing up.
3. Slide the tool and the bearing into the top of the driveshaft cavity. Take care to ensure the bearing does not tilt. Drive the bearing into place. The bearing will seat itself at the correct location when the head of the drive rod seats against the guide bushing. (When not using the factory tool you'll have to stop periodically and check installed depth. One trick is to mark the driver where it will be at the appropriate depth so you'll know as your tapping.)

■ If the bearing position is not correct, both the upper and lower bearings will be reloaded by the water pump base and cause early bearing failure.

Water Pump Base Plate

1. Obtain a driver or equivalent size socket and the appropriate driver handle.
2. Place the water pump plate on the workbench with the bottom side facing up.
3. Coat the seal lip with a good grade of water resistant lubricant.
4. Install the oil seal using the driver with the oil seal (spring side), lip facing up. Tap the seal down into the pump base until the seal bottoms in the bore.

LOWER UNIT 7-63

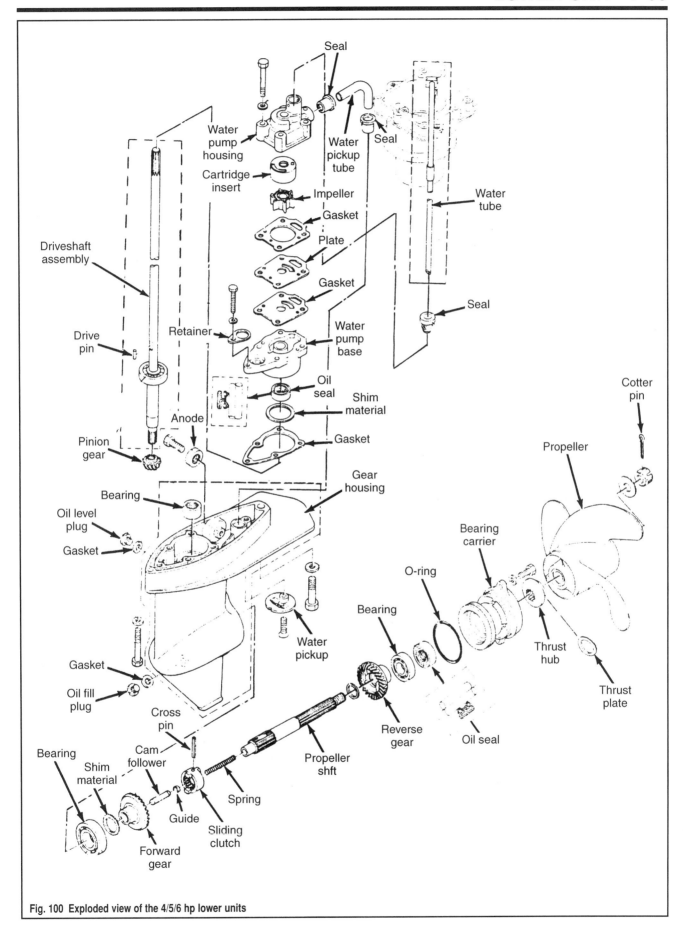

Fig. 100 Exploded view of the 4/5/6 hp lower units

7-64 LOWER UNIT

Driveshaft, Pinion Gear and Water Pump Base

1. Slide any shim material saved during disassembly, over the driveshaft and down onto the bearing. If present, the shim material should give the same amount of backlash between the pinion gear and the other two gears as before disassembling.

2. Lubricate the seal in the pump base lightly with a good grade of water resistant lubricant. Slide the water pump base down the driveshaft and over the upper bearing.

■ **Mercury normally directs technicians to assemble the shifter shaft components AFTER the driveshaft and water pump base assembly has been installed to the gearcase, but from our experience we've found you can position the shifter before the assembly is installed. The choice is really yours.**

3. Place a new O-ring into the pump base opening for the shift shaft. Slide a new shift shaft bushing down onto the shift shaft.
4. Rotate the tapered side of the shift cam toward the pinion shaft.
5. Place a new water pump base gasket on top of the lower unit. Be sure the pins and bolt holes are correctly aligned.
6. Insert the pinion gear into the lower unit. Now, with the other hand, lower the driveshaft, shift shaft and pump base down into the lower unit with the shift cam going in first. Slowly rotate the driveshaft until the driveshaft splines index with the splines of the pinion gear.
7. Install the Water Pump assembly as detailed in the Lubrication and Cooling System section.

Bearing Carrier and Propeller Shaft

1. Place the bearing carrier on the workbench with the propeller end facing down.
2. Obtain a driver/installation tool (#91-83174M). Apply a coating of good grade water resistant lubricant to the lip of the seal, and then install the seal with the flat side down.
3. Place the bearing, numbered side up, into the bearing carrier. Obtain a driver (#91-84536M) and tap the bearing down into the bearing carrier.
4. Install a new O-ring around the bearing carrier. Apply a coating of water resistant lubricant to the O-ring.

■ **Do not forget to install any shim material removed from behind the reverse gear during disassembling. This shim material, and in some cases a thrust washer, must be installed between the reverse gear and the ball bearing assembly in order to ensure proper mesh between the reverse gear and the pinion gear.**

5. Slide the washer (if equipped), and reverse gear onto the end of the propeller shaft with the teeth of the gear facing away from the bearing carrier.
6. Slide the bearing carrier over the propeller shaft.
7. Apply a liberal amount of water resistant lubricant on the cam follower and insert the follower into the end of the propeller shaft.
8. Check to be sure the cam follower is on the end of the propeller shaft. This is a loose piece and easily forgotten. Insert the propeller shaft and bearing carrier into the lower unit until the end of the shaft indexes into the forward gear.

9. Using a soft head mallet, lightly tap around the circumference of the carrier until it is fully seated in place. Check to be sure the two bolt holes are aligned with the lower unit holes.
10. Install the two washers and bolts. Tighten the bolts alternately to 70 inch lbs. (8Nm).

SHIMMING

■ **The manufacturer gives no specific instructions for setting up the backlash on these units. As a matter of fact, Mercury claims this gearcase, when installed on these powerheads, is NOT adjustable when it comes to lash. However, IF any shim material found during disassembly was placed in its original locations, the backlash should be acceptable. The manufacturer states: "The amount of play between the gears is not critical, but no play is unacceptable." Therefore, if after the assembly work is complete, the gears are indeed locked, the unit must be disassembled and shim material removed from behind the forward or reverse gears, using a "trial and error" method. If the lower unit is allowed to operate without "some" backlash, very heavy wear on the three gears will take place almost immediately.**

Gearcase - 8-15 Hp (2-Cylinder) Models

REMOVAL & INSTALLATION

 MODERATE

◆ See Figures 101, 102 and 103

1. For safety and to prevent accidental engine start, disconnect and ground the spark plug leads to the powerhead or for electric start models, tag and disconnect the battery cables.
2. If necessary for overhaul purposes, remove the Propeller, as detailed in the Maintenance and Tune-Up section.
3. If necessary for gearcase overhaul purposes, position a suitable clean container under the lower unit. Remove the FILL screw on the bottom of the lower unit, and then the VENT screw. The vent screw must be removed to allow air to enter the lower unit behind the lubricant. Allow the gear lubricant to drain into the container.

As the lubricant drains, catch some with your fingers from time-to-time and rub it between your thumb and finger to determine if there are any metal particles present. Examine the fill plug. A small magnet imbedded in the end of the plug will pickup any metal particles. If metal is detected in the lubricant, the unit must be completely disassembled, inspected, and the damaged parts replaced.

Check the color of the lubricant as it drains. A whitish or creamy color indicates the presence of water in the lubricant. Check the container for signs of water separation from the lubricant. The presence of any water in the lubricant is bad news. The unit must be completely disassembled; inspected; the cause of the problem determined; and then corrected.

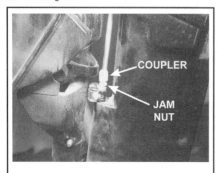

Fig. 101 Loosen the jam nut and coupler on the shift rod

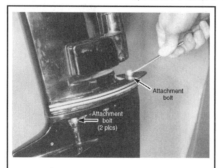

Fig. 102 Then unbolt the gearcase (non Bigfoot models shown)

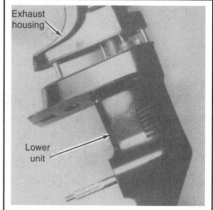

Fig. 103 Upon installation, guide the 2 shafts up into the housing as the lower unit is lifted into place

4. Rotate the outboard unit to the full UP position and engage the tilt lock pin.

5. Locate the shift shaft coupling on the exterior front of the motor, near the swivel housing (if the coupling is not readily visible, the motor MAY have a lower shroud which must first be removed for access).

6. Although the shift shaft coupling designs vary slightly, all should have a small jam nut located underneath a large coupler. Loosen the jam nut, then unthread the coupler to free the shift shaft.

■ **By measuring or marking the final coupler position on the threads you may save yourself some trouble when adjusting the shifter during installation.**

7. Remove the bolts and washers securing the lower unit to the exhaust housing. For non-Bigfoot models there should be 3 bolts, one threaded upward on each side and one threaded downward at the front of the gearcase. On Bigfoot models there should be 2 bolts threaded upward on either side of the gearcase.

8. Pull down on the lower unit gear housing and separate the lower unit from the exhaust housing.

9. Guide the shift rod and driveshaft out of the exhaust housing as the lower unit is removed.

To Install:

10. If not done already, swing the exhaust housing outward until the tilt lock lever can be actuated, and then engage the tilt lock.

11. Apply a liberal coating of 2-4-C Marine Lubricant With Teflon to the driveshaft splines (but NOT to the top of the shaft as it may prevent the shaft from fully seating in the crankshaft).

12. Position the driveshaft and shift shaft into the exhaust housing. Guide the two shafts up into the housing as the lower unit is lifted to meet the exhaust housing. Maintain the lower unit mating surface parallel with the exhaust housing mating surface.

13. As the lower unit approaches closer to the exhaust housing, align the water tube with the water pump tube guide.

14. If necessary rotate the flywheel slightly to permit the crankshaft splines to index with the driveshaft splines (alternately you COULD put the gearcase in gear and slowly turn the propeller shaft in the proper direction).

15. Secure the lower unit in place with the 3 or 4 attaching bolts. Tighten the bolts to 180 inch lbs./15 ft. lbs. (20 Nm) for non-Bigfoot models or to 40 ft. lbs. (54 Nm) for Bigfoot models.

16. Bring the two halves of the shift shaft together and thread the coupler back into position to join them. Either match the old marks/measurement or adjust the shifter so that the gearcase properly engages and releases Forward and Reverse gears.

17. If the gearcase was drained refill it with fresh gear oil.

18. If the propeller was removed, install the propeller.

19. Reconnect the spark plug wires and/or the battery cables.

OVERHAUL

Propeller Shaft and Bearing Carrier

■ **For gearcase exploded views, please refer to Cleaning & Inspection in this section.**

✱✱ CAUTION

The threads on the bearing carrier are LEFT HAND. Be sure to turn the bearing carrier tool in CLOCKWISE for removal and in the COUNTERCLOCKWISE for assembly.

1. For Bigfoot models, loosen and remove the 3 screws that secure the O-ring retainer plate and O-ring to the bearing carrier. Then, remove the plate and O-ring from the gear housing.

2. Obtain the appropriate bearing carrier tool (either #91-13664 for non-Bigfoot models or #91-93843—1 for Bigfoot models). Unscrew the bearing carrier by turning CLOCKWISE, then grasp the propeller shaft and pull to remove the bearing carrier and propeller shaft assembly from the lower unit gear housing.

■ **At this point the cam follower on the front (inner) end of the propeller shaft is free to slide out of the shaft. Recover the follower if it falls into the gearcase during removal.**

3. Remove and discard the O-ring on the bearing carrier.

4. Slide the reverse gear from the propeller shaft. Although Mercury claims most of these gearcases, when installed on these powerheads, are NOT adjustable when it comes to lash, watch for and save any shim material from the back side of the reverse gear. If present, the shim material is critical in obtaining the correct backlash during assembling.

5. On Bigfoot models, use a screwdriver or punch to carefully unwind the cross-pin retainer spring from the sliding clutch. To remove the cross-pin on thee models, insert the cam follower back into the end of the propeller shaft (with the flat end facing inward toward the cross-pin). Gently push the propeller shaft and cam follower against a flat surface then use a punch to push the cross-pin out of the sliding clutch.

6. For Non-Bigfoot models, use a punch to carefully tap the cross-pin out of the sliding clutch. Be sure to tap on the end of the cross-pin that is NOT grooved. Remove and discard the pin (which must be replaced during assembly).

7. When the cross pin is removed the sliding clutch should fall free of the propeller shaft. Examine the clutch on both ends for broken, chipped or rounded off teeth.

8. Check the cam follower for worn, flat or blunt end.

9. Remove the cam follower spring from inside the propeller shaft. Check the spring to be sure it is not broken or distorted.

✱✱ WARNING

Perform the following work only if the seal/s have been damaged and are no longer fit for service. Removal of the seal/s destroys their sealing qualities. Therefore, the seal/s cannot be installed a second time. Be absolutely sure new seal/s are available before removing the old ones.

10. Inspect the condition of the seals in the bearing carrier. If the seals appear to be damaged and replacement is required there are multiple methods available to remove them. In some cases, you can use a punch to drive the 2 seals from the rear (gearcase) side of the carrier, in other cases a seal remover or pry tool can be used to carefully extract them. If necessary a slide hammer and expanding jaw attachment will do the trick. However in all cases, it is usually easiest to proceed with the carrier carefully mounted in a soft-jawed vise for stability when removing the seal.

■ **Perform the following work only if the needle bearing is damaged and is no longer fit for further service. Removal of the bearing will distort it. Therefore, bearings with prior service cannot be installed a second time. Be absolutely sure a new part is available before removing the bearing. Unfortunately, the oil seals, good or bad, must be removed before the bearing can be driven out, and of course they must be replaced with new seals.**

11. Note the position of the bearing in relation to the carrier. Look for any embossed numbers or letters and which shoulder of the bearing is flush in the carrier. Remove the bearing using a punch and a hammer, or obtain a suitable sized socket and driver, and then drive the bearing from the carrier.

Driveshaft and Pinion Gear Removal

If not done already, remove the water pump assembly, as detailed in the Lubrication and Cooling section.

On non-Bigfoot models, pull upward on the driveshaft with one hand and with the other hand, reach into the lower unit cavity and grasp the pinion gear. As the driveshaft is removed the pinion gear and thrust washer will fall free in your hand. When removing the driveshaft watch for and save any shim material found on top of the pinion gear. Although Mercury claims most of these gearcases, when installed on these powerheads, are NOT adjustable when it comes to lash, if present, this shim material is critical to obtain the correct backlash during installation.

On Bigfoot models, clamp the driveshaft in a vise equipped with soft jaws and remove the bolt on the end of the driveshaft securing the pinion gear to the shaft. Pull the driveshaft out of the lower unit housing. Remove the pinion gear and tapered roller bearing.

7-66 LOWER UNIT

Forward Gear and Bearing

1. Tilt the lower unit. With the pinion removed, the forward gear and tapered roller bearing should fall into your hand. If the gear and bearing fail to fall free, strike the open end of the lower unit on a block of wood to jar the gear and bearing free.
2. Examine the tapered bearing for rust, pitted or discolored (bluish), rollers. If the bearing is worn or damaged than it must be replaced.
3. If the bearing must be replaced, use a Universal Puller Plate/Bearing Separator (such as #91-37241 or equivalent), and press the forward gear off the tapered roller bearing. Remove and save any shim material.
4. If the bearing is replaced or if the race mounted in the gearcase nose cone otherwise requires replacement use a suitable expanding (internal) jawed puller and either puller plate or a slide hammer to remove the race from the gearcase.

Driveshaft Seals, Bushings and/or Needle Bearings

The configuration of the Bigfoot and non-Bigfoot gearcases differ slightly when it comes to the configuration of the driveshaft seals, bushing (if applicable) and/or bearings.

Non-Bigfoot models utilize dual back-to-back seals mounted in the top of the water pump base plate. Under the plate an upper bushing is installed in the top of the gear housing with a lubrication sleeve underneath it. At the bottom of the housing, just beneath the sleeve and right above the pinion gear is a needle bearing. Be sure to measure the installed height of the needle bearing before removal to ensure proper installation.

Bigfoot models utilize dual back-to-back mounted upper seal assembly that is installed directly into the top of the gearcase housing. Underneath the seals you'll find a roller bearing and a sleeve. The sleeve should not be removed from the gearcase, and if you're not using the special tools, you should check the installed height of the bearing before removal.

■ **Perform the following work only if the seals, bushing and/or needle bearing(s) have been damaged and are no longer fit for service. Removal of the seals and bearings destroy the sealing qualities of the seal, and the bearing cages are distorted. Therefore, the seals and bearings cannot be installed a second time. Be absolutely sure a new seal or bearing is available before removing the old component.**

1. Examine the upper seal for deterioration, hardness and wear. Only one seal is visible for inspection, but if the seal appears to be damaged or excessively worn, both the seals should be replaced. The seals are installed back-to-back. The upper seal prevents water from entering the lower unit and the other seal prevents lubricant from escaping.
2. Remove the drive shaft seals, needle bearings and forward gear bearing race. Use a punch or seal removal tool to remove the seals from the water pump base plate on non-Bigfoot models. Use a Slide Hammer tool (#91-34569A1) and a universal slide hammer with a jaw expander attachment to remove the seals from the top of the gearcase on Bigfoot models.
3. Inspect the driveshaft bushing and bearing surface to determine the condition of the bushing and bearing. If worn or damaged, replace the shaft, bushing AND bearing. HOWEVER, IF either the bushing or bearing appears to be spinning in their bores, the gear housing itself must be replaced. If the bushing and/or bearing requires replacement on non-Bigfoot models, proceed as follows:
   a. Using an expanding rod (such as Snap-On® #CG 40-4) and collet (such as Snap-On #CG 40A6) and a suitable slide hammer, remove the upper bushing from the top of the gearcase.
   b. Check the top of the driveshaft bore for a burr that is sometimes formed when the upper bushing is installed. If present it may prevent the sleeve from being removed, so carefully remove the burr with a knife.
   c. Remove the lubrication sleeve for access to the lower bearing.
   d. Use a suitable driver to tap the bearing out into the bottom of the gearcase.
4. If the upper bearing requires replacement on Bigfoot models, proceed as follows:
   a. Using a suitable driver carefully tap the bearing THROUGH the sleeve and out into the bottom of the gearcase.
   b. To remove the lower bearing cup, use a threaded rod to draw the bearing cup puller (#91-44385) up into the cup from within the gearcase (by tightening the nut). Then, tap on the top of the threaded rod to drive the cup out into the bottom of the gearcase. Loosen and remove the threaded rod, then remove the cup and puller tool.

CLEANING & INSPECTION

◆ See Figures 104 thru 107

1. Clean all water pump parts with solvent, and then dry them using compressed air.
2. Inspect the water pump cover and base for cracks and distortion. If possible always install a new water pump impeller while the lower unit is disassembled. A new impeller will insure extended satisfactory service. If the old impeller must be used, never install it in reverse to the original direction of rotation. Reverse installation could lead to premature impeller failure and subsequent powerhead damage.
3. Inspect the bearing surface of the propeller shaft.
4. Check the shaft surface of propeller shaft for pitting, scoring, grooving, imbedded particles, uneven wear and discoloration.
5. Check the straightness of the propeller shaft with a set of machinist V-blocks.
6. Clean the pinion gear and the propeller shaft with solvent and dry the components with compressed air.
7. Check the pinion gear and drive gear for abnormal wear.

ASSEMBLY

Sliding Clutch

1. Apply a small amount of lower unit lubricant onto the spring and cam follower. Slide the spring down into the propeller shaft and either insert the cam follower with the flat end of the cam follower against the spring (Bigfoot models) or insert a small punch or screwdriver into the end of the shaft against the spring (non-Bigfoot models).
2. Slip the sliding clutch into position over the shaft. On Bigfoot models, the clutch must be installed with the longer end of the clutch towards the reverse gear. Guide the sliding clutch over the splines of the propeller shaft with the hole in the sliding clutch aligned with the slot in the propeller shaft.
3. Rotate the propeller shaft vertically and compress the spring by pushing the shaft and the cam follower (Bigfoot models) or punch/screwdriver (non Bigfoot models) against a solid object. For non-Bigfoot models it is easier to put the other end of the shaft or punch in a vise so you can pull the shaft back of the tool with the vise holding the tool still.
4. While holding the shaft against spring tension install the cross-pin as follows:
   a. For non-Bigfoot models, insert the NON-grooved end of a NEW pin in through the sliding clutch and shaft until it contacts the punch or screwdriver. Continue to push the pin into place as you slowly and carefully pull the shaft back from the vise, withdrawing the punch or screwdriver.
   b. For Bigfoot models, hold the spring compressed, and at the same time, insert the cross pin into the sliding clutch pushing the screwdriver or punch out the other side. Center the cross-pin in the sliding clutch. Install the cross-pin retainer spring into the recess on the sliding clutch. Pull out the cam follower.
5. Place a dab of multi-purpose lubricant onto both ends of the cam follower. Insert the cam follower with the flat end of the cam follower against the spring and the rounded end protruding. This position will allow smooth operation between the shift cam and the follower.
6. Slide the reverse gear and washer onto the propeller shaft. Do not forget to install any shim material removed from behind the reverse gear during disassembling. This shim material and, in some cases a thrust washer, must be installed between the reverse gear and the ball bearing assembly in order to ensure proper mesh between the reverse gear and the pinion gear.
7. Install a new O-ring on the bearing carrier and slip the bearing carrier onto the propeller shaft.

Forward Gear and Bearing

1. Insert any shim material saved during disassembly into the lower unit bearing race cavity. The shim material should give the same amount of backlash between the pinion gear and the forward gear as before disassembling. Apply a small amount of multipurpose lubricant to the bore for the forward bearing race.

LOWER UNIT 7-67

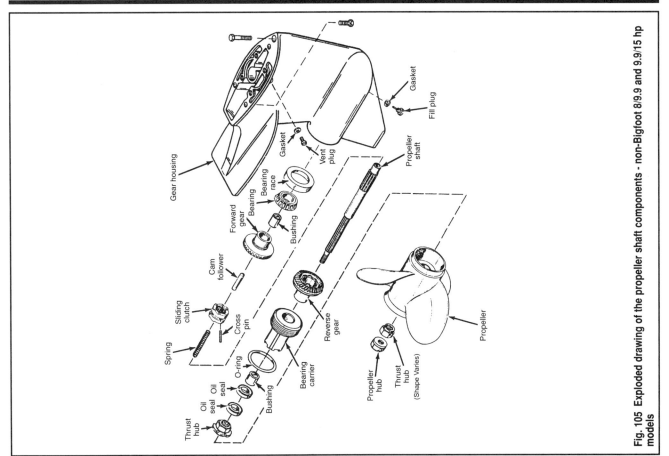

Fig. 105 Exploded drawing of the propeller shaft components - non-Bigfoot 8/9.9 and 9.9/15 hp models

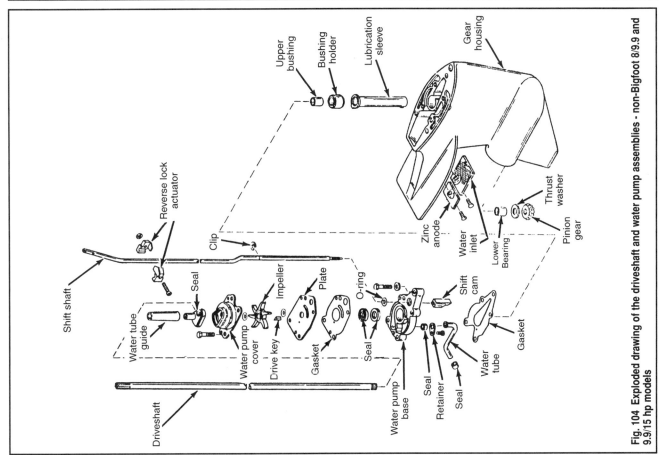

Fig. 104 Exploded drawing of the driveshaft and water pump assemblies - non-Bigfoot 8/9.9 and 9.9/15 hp models

7-68 LOWER UNIT

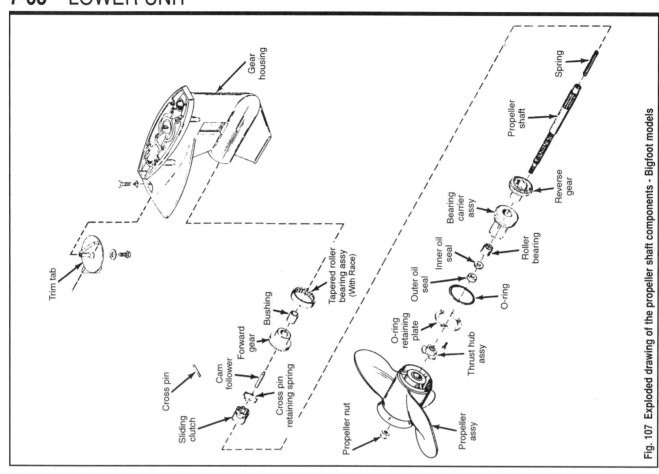

Fig. 107 Exploded drawing of the propeller shaft components - Bigfoot models

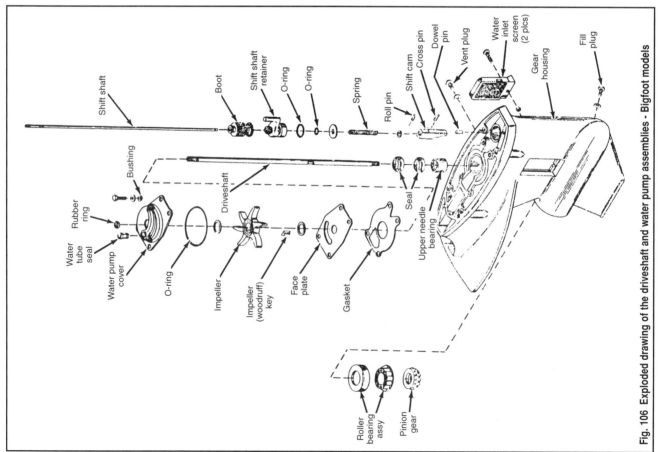

Fig. 106 Exploded drawing of the driveshaft and water pump assemblies - Bigfoot models

LOWER UNIT 7-69

2. Insert the bearing race squarely into the lower unit housing. Using a mandrel (#91-13658 on non-Bigfoot models or #91-36571 on Bigfoot models) and a suitable driver, press the bearing race down into the bore recess.

■ Mercury states that you can use the bearing carrier tool as a driver on non-Bigfoot models, or even the propeller shaft (if a suitable nut is threaded onto the end in order to take the tapping or pressing force) on Bigfoot models.

3. Position the forward gear tapered bearing over the forward gear. Use a suitable mandrel and press the bearing flush against the shoulder of the gear. Always press on the inner race, never on the cage or the rollers.
4. Thoroughly lubricate the tapered roller bearing and gear with Mercury Super Duty Gear Lubricant, or equivalent. Insert the lubricated forward gear and bearing into the lower unit with the teeth of the gear facing outward.

Driveshaft Needle Bearing

The lower bearing or bearing cup on these models is drawn up into position in the gearcase from underneath (near the pinion positioning). Although Mercury sells special tools for this, all you REALLY need to accomplish this is a sufficiently long threaded bolt with a nut and a washer that is the size of the driver you'd want to contact that bushing or bearing. You'll also need a nut and a plate to place on top of the gearcase. Place the threaded rod through one nut and plate and then down into the gearcase through the bushing/bearing (so the plate is resting against the top of the gearcase, the upper nut is resting against the plate and the threads of the rod are just protruding down into the pinion mounting area of the gearcase.). Next install the washer and lower nut over the bottom threads of the rod to support the bearings or bushing. Slowly draw the bearing or bushing into position in the gearcase by holding the threaded rod from turning while at the same time turning the upper nut to draw the entire bolt, bushing/bearing, washer and nut assembly upward until the bushing or bearing seats. Then loosen the nut on the bottom of the threaded rod in the gearcase and remove the threaded rod, plate, nut and washer.

■ One potential problem with using something other than the special tool to install the bearing or bearing cup is that IF there is no shoulder against which the bearing or cup is mounting you'll need to measure to ensure it is installed to the correct depth (height), we recommended that you take these measurements before removal.

1. For non-Bigfoot models, proceed as follows:
   a. The lower needle bearing has to be pulled up into the housing from the housing gear cavity. Therefore, a special tool (#91-824790A1) is the best way to do this. However, a threaded rod about the only way this bushing can be installed.
   b. Using a special tool (such as the ones mentioned earlier), position the plate on top of the lower unit. Lower the long threaded bushing installer into the lower unit from the top and the threads protrude into the lower unit cavity.
   c. Now, insert the bearing onto the mandrel and install this over the threaded rod in the gear cavity. Insert and hold the threaded nut holder onto the threaded rod. With a wrench, turn the rod nut clockwise and draw the bushing up into the lower housing.
   d. Remove the special tool and insert the lubrication sleeve down into the housing.
   e. Using the same installation tool install, but with the mandrel position reversed (on top), install the upper bushing down into the top of the gearcase. Alternately, a driver CAN be used to CAREFULLY tap the bushing into position.
2. For Bigfoot models, proceed as follows:
   a. Place the upper bearing over the bearing retainer bore with the lettered side facing upward, then using a suitable driver and mandrel carefully press or tap the bearing into the gearcase. Continue to press or tap the bearing into the housing sleeve until the bearing is JUST below the oil hole, which should place it a total depth of about 1 in. (25.4mm) below the top of the gearcase.
   b. Next, install the lower bearing cup into the driveshaft bore using either the special tool set from Mercury (#91-31229, #91-29310, #11-24156, #91-889853 and #12-29310) or a suitable set threaded rod and set of washers/nuts as described earlier. Using the tools, carefully draw the bearing cup into the same position as noted during removal.

■ Since the seals are mounted in the water pump base plate on non-Bigfoot models the seals are serviced later in this procedure. Obviously, since the seals are installed directly in the gearcase for Bigfoot models they needed to be removed for access to the bearings and must be replaced regardless of their original condition.

   c. Apply a light coating of Loctite® 274 to the outer diameter of 2 NEW driveshaft seals. Install the oil seals back-to-back using a suitable size socket and hammer. The upper seal prevents water from entering the lower unit and the lower seal prevents lubricant from escaping. Place the first oil seal into the bore with the lip of the seal facing down.
   d. Tap the seals down into position. Install the first seal until it is slightly past the bore opening. Place the second seal into the bore with the lip of the seal facing up. Now, tap both seals down into the bore until the top seal is 3/16 in. (4.7mm) below the bore.
   e. Wipe away any excess Loctite®.

Driveshaft and Pinion Gear

3. Make sure the forward gear is positioned in the gearcase.
4. Install the pinion gear assembly to the driveshaft as follows:
   a. For non-Bigfoot models, carefully insert the driveshaft into the gearcase, then lift the driveshaft slightly as you position the thrust washer (with the grooved side facing downward toward the pinion gear). Next, position the pinion gear over the splines. Rock or rotate the driveshaft slightly as necessary to align the splines.
   b. For Bigfoot models, insert the lower roller bearing into the cut, then position the pinion gear into the housing with the teeth meshed with the forward gear. While holding the gear in position carefully insert the driveshaft, rotating it slightly back and forth until the splines align. Apply a light coating of Loctite® 271 to the threads of the pinion nut, then install and tighten the nut to 180 inch lbs. (20 Nm).

Water Pump Plate

The manufacturer has produced a new style water pump base plate and gasket for the non-Bigfoot models. The older style plate and gasket could fail under certain conditions causing excessive water leakage. A newer style plate and silicon impregnated gasket has replaced the previous model which is no longer available. Replacements will always be the new style.

1. On the non-Bigfoot models, prepare the water pump base plate as follows:
   a. Obtain a mandrel (#91-13655 and the appropriate driver handle or suitable substitutes). Place the water pump plate on the workbench as it is to be installed in the lower unit. The seals are installed back-to-back. The upper seal prevents water from entering the lower unit and the lower seal prevents lubricant from escaping. Coat both seal lips with a good grade of water resistant lubricant and coat the outer diameter of both seals using Loctite® 271.
   b. Install the first oil seal using the driver with the oil seal lip facing down. Install the second oil seal with the oil seal lip facing up.
   c. Install a new seal onto the end of the water tube. Turn the base plate over and insert the water tube into the bottom of the base plate. Secure the water tube and seal with the retaining ring and Phillips head screw.
   d. Coat the surface of a new O-ring with a good grade of water resistant lubricant. Install the O-ring into the recess in the base plate.
   e. Slide the end of the shift shaft through the O-ring and pump base plate.
   f. Install the E-clip into the slotted ring on the shift shaft. This clip must be installed after the shift shaft has been installed through the water pump base plate.
   g. Screw the threaded shift cam onto the end of the shift shaft. The tip of the shaft should be visible through the hole in the shift cam.
   h. Place a new base plate gasket over the driveshaft and onto the lower unit housing. Be sure the small bypass hole is aligned with the lower unit housing.
   i. Slide the assembled shift shaft and base plate onto the lower unit with the cam end going in first. Be sure the cam tapered surface is pointing towards the forward gear. Lower the pump base down onto the lower unit housing guiding the water tube into the grommet at the rear of the housing.
   j. Apply Loctite®Grade "A" or equivalent to the threads of the bolt and secure the water pump base plate to the lower unit housing. Tighten the bolt to 40 inch lbs. (4.5 Nm).

7-70 LOWER UNIT

2. For Bigfoot models, proceed as follows:
   a. Assemble the shift shaft components. Be sure new inner and outer O-ring are installed. Lightly coat the O-rings with multipurpose lubricant to assist in assembly.
   b. Carefully examine the retainer boot, if there is any evidence of deterioration or excessive wear, replace the boot.
   c. Slide the assembled shift shaft into the lower unit with the cam end going in first. Be sure the tapered end of the cam is facing towards the propeller.
3. Install the Water Pump assembly as detailed in the Lubrication and Cooling System section.

Bearing Carrier

■ The bearing carrier used on these models utilize two seals installed back-to-back. One seal prevents lubricant in the lower unit from escaping and the other prevents water from entering. Be sure to install the seals correctly with the lips of the seals facing in the proper direction.

Prepare the bearing carrier for installation with the propeller shaft.
1. For non-Bigfoot models, proceed as follows:
   a. If the bushing was removed, position the carrier with the threaded side (gearcase side) facing upward. Apply a light coating of gear lube to the bushing, then press or carefully tap the bushing down into the housing using a suitable mandrel (like #91-824785A1).
   b. If the seals were removed, invert the carrier with the threaded (gearcase) side facing downward. Apply a light coating of Loctite® 271 or equivalent to the OUTER diameter of the 2 new seals. Install the first (inner) seal with the lip facing inward toward the bushing. Use the same mandrel as recommended for bushing installation to carefully push or tap the seal down into position. Next install the outer (fish line cutter) seal with the lip facing upward toward the mandrel.
   c. Apply a light coating of Special Lubricant 101 between the lips of the seals, to the threads of the bearing carrier and pilot diameter and to a NEW bearing carrier O-ring.
   d. Install the ring onto the bearing carrier.
2. For Bigfoot models, proceed as follows:
   a. If the bearing was removed, place the bearing carrier onto a hard surface with the threaded end down. Apply a light coating of oil to the bearing bore in the carrier. Insert the needle roller bearing into the bearing bore with the lettered end of the bearing facing up. Using a suitable size socket and driver, press or tap the needle bearing down into the bore until the lower edge of the bearing is 1.1406 in. (28.9mm) from the threaded end (currently the bottom) of the bearing carrier housing.
   b. Apply a light coating of Loctite® 271 or equivalent to the OUTER diameter of the 2 new seals. Position the inner seal into the seal bore with the lip of the seal facing down. Using a suitable size socket and driver, press or tap the seal down into the bore until the top of the seal is at a depth of 0.875 in. (22.2mm) from the end (currently the top) of the bearing carrier housing.
   c. Apply a light coating of 2-4-C with Teflon between the seals.
   d. Position the outer seal into the seal bore with the lip of the seal facing up. Using a suitable size socket and driver, press or tap the seal down into the bore until the top of the seal is at a depth of 0.875 in. (22.2mm) from the end (currently the top) of the bearing carrier housing.

   e. Apply a light coating of 2-4-C with Teflon to the lips of the seals. Also apply this lubricant to the bearing carrier threads, to the O-ring groove and to the pilot diameter.
3. Insert the propeller shaft with the reverse gear, through the bearing carrier. Check the cam follower on the end of the propeller shaft. This is a loose piece and is easily forgotten.
4. Using the same method as was used in the disassembly process, insert the bearing carrier into the lower unit.
5. Screw the bearing carrier into the lower unit housing COUNTERCLOCKWISE (since it has left-hand thread). Using the carrier tool referenced during disassembly, tighten the bearing carrier to 85 ft. lbs. (115 Nm) for non-Bigfoot models or to 80 ft. lbs. (109 Nm) for Bigfoot models.
6. On Bigfoot models, install a NEW O-ring and the O-ring retainer plate to the bearing carrier (making sure not to pinch the O-ring), then secure it with the 3 screws. Tighten retainer plate screws to 65 inch lbs. (7 Nm).

SHIMMING

■ The manufacturer gives no specific instructions for setting up the backlash on these units. As a matter of fact, Mercury claims most of these gearcases, when installed on these powerheads, are NOT adjustable when it comes to lash. However, if shim material was found during disassembly and was placed back in its original location, the backlash should be acceptable. The manufacturer simply states: "The amount of play between the gears is not critical, but NO play is unacceptable." Therefore, if after the assembly work is complete, the gears are "locked", the unit must be disassembled and shim material placed behind all three gears, using a "trial and error" method. If the lower unit is allowed to operate without "some" backlash, very heavy wear on the three gears will take place almost immediately.

Gearcase - 25 Hp (Bigfoot), 30/40 Hp (Non-Bigfoot) and 40/45/50 hp (Non-Bigfoot, 935cc) Models

REMOVAL & INSTALLATION

◆ See Figures 108, 109 and 110

1. For safety and to prevent accidental engine start, disconnect and ground the spark plug leads to the powerhead or for electric start models, tag and disconnect the battery cables.
2. If necessary for overhaul purposes, remove the Propeller, as detailed in the Maintenance and Tune-Up section.
3. If necessary for gearcase overhaul purposes, position a suitable clean container under the lower unit. Remove the FILL screw on the bottom of the lower unit, and then the VENT screw. The vent screw must be removed to allow air to enter the lower unit behind the lubricant. Allow the gear lubricant to drain into the container.

As the lubricant drains, catch some with your fingers from time-to-time and rub it between your thumb and finger to determine if there are any metal

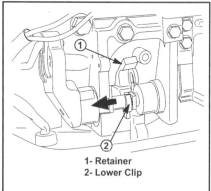

Fig. 108 Releasing the shift shaft coupling - 25 Hp (Bigfoot), 30/40 Hp (Non-Bigfoot) and 40/45/50 hp (Non-Bigfoot, 935cc) models

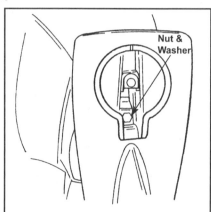

Fig. 109 Don't forget the nut and washer located under the trim tab

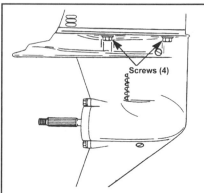

Fig. 110 In addition to the nut, these gearcases use 4 retaining bolts (2 on either side)

LOWER UNIT 7-71

particles present. Examine the fill plug. A small magnet imbedded in the end of the plug will pickup any metal particles. If metal is detected in the lubricant, the unit must be completely disassembled, inspected, and the damaged parts replaced.

Check the color of the lubricant as it drains. A whitish or creamy color indicates the presence of water in the lubricant. Check the container for signs of water separation from the lubricant. The presence of any water in the lubricant is bad news. The unit must be completely disassembled; inspected; the cause of the problem determined; and then corrected.

4. Rotate the outboard unit to the full UP position and engage the tilt lock pin.
5. Shift the gearcase into Neutral.
6. Locate the shift shaft coupling at the front of the motor, near the center at the base of the powerhead. Rotate the retaining clip upward then slide the clip toward the starboard side of the motor to release the shift shaft coupling.
7. Matchmark the trim tab to the bottom of the gearcase housing (to preserve alignment during installation), then remove the screw and washer securing the trim tab to the case. Remove the trim tab for access.
8. Remove the gearcase retaining nut and washer that is mounted high in the trim tab cavity.
9. Remove the 4 bolts and washers securing the lower unit to the exhaust housing.
10. Pull down on the lower unit gear housing and separate the lower unit from the exhaust housing.
11. Guide the shift rod and driveshaft out of the exhaust housing as the lower unit is removed.

To Install:

12. If not done already, swing the exhaust housing outward until the tilt lock lever can be actuated, and then engage the tilt lock.
13. Apply a liberal coating of 2-4-C Marine Lubricant With Teflon to the driveshaft splines (but NOT to the top of the shaft as it may prevent the shaft from fully seating in the crankshaft.
14. Position the driveshaft and shift shaft into the exhaust housing. Guide the two shafts up into the housing as the lower unit is lifted to meet the exhaust housing. Maintain the lower unit mating surface parallel with the exhaust housing mating surface.
15. As the lower unit approaches closer to the exhaust housing, align the water tube with the water pump tube guide.
16. If necessary rotate the flywheel slightly to permit the crankshaft splines to index with the driveshaft splines (alternately you COULD put the gearcase in gear and slowly turn the propeller shaft in the proper direction).
17. Secure the lower unit in place with the 4 attaching bolts. Tighten the bolts to 40 ft. lbs. (54 Nm).
18. Reconnect the shift shaft by sliding the retainer back toward the shift shaft, then rotating it 1/4 turn downward over the shaft coupling.
19. Install the gearcase retaining nut and washer located in the trim tab cavity. Tighten the nut to 40 ft. lbs. (54 Nm).
20. Align the matchmark made earlier, then install the trim tab and secure using the bolt and washer. Tighten the retaining bolt to 16 ft. lbs. (22 Nm).
21. If the gearcase was drained refill it with fresh gear oil.
22. If the propeller was removed, install the propeller.
23. Reconnect the spark plug wires and/or the battery cables.

OVERHAUL

Bearing Carrier and Propeller Shaft

■ For gearcase exploded views, please refer to Cleaning & Inspection in this section.

1. If not done already, clamp the skeg in a vice equipped with soft jaws or between two pieces of soft wood.
2. The bearing carrier on these gearcases are retained by 2 bolts. Sometimes the bolts are also secured by lock tabs. If so equipped, bend the lock tabs away from the bolts with a punch.
3. Loosen and remove the 2 bearing carrier retaining bolts.
4. Two methods are available to remove the bearing carrier from these lower units. The first and safest method is to use an internal jawed puller or slide hammer with internal jaws (such as #91-27780). The second and less desirable method is to tap the carrier so the bolt ears are no longer aligned at 12 and 6 o'clock and to use a mallet to actually drive the lower unit off the carrier. In both methods, heat carefully applied to the outside of the lower unit, will assist in removing the bearing carrier.
5. Remove the propeller shaft and bearing carrier gearcase. When pulling the propeller shaft free, be aware that the cam follower will be free to fall from the front of the shaft (and often does), so locate and keep track of it. Set the assembly aside for disassembly later.
6. Remove and discard the bearing carrier O-ring.
7. To replace bearings or seals in the bearing carrier, proceed as follows:
    a. Clamp the bearing carrier in a vise equipped with soft jaws. Use a slide hammer and internal jawed puller to remove the reverse gear. If the reverse gear ball bearing remained in the bearing carrier, remove the bearing using a slide hammer.

■ You've got 2 choices on seal removal. IF the needle bearing inside the carrier needs to be replaced you can drive it out the rear (propeller side) of the carrier and, in doing so, it will take the seals out with it. However, if ONLY seal replacement is necessary, use small internal jawed puller or seal remover (or if absolutely necessary a small pry-tool) to remove the seals from the propeller end of the carrier.

    b. If the needle bearing requires replacement, place the carrier so the oil seal end is facing downward, then insert a driver through the reverse gear side of the carrier. Drive the bearing, along with the oil seals out the back of the carrier.
    c. If ONLY the oil seals need to be removed for replacement, remove then by carefully pulling or prying them from the bore in the carrier. Take note of seal orientation, most models use two seals mounted back-to-back.
8. To disassemble the propeller shaft for cleaning, inspection and/or parts replacement, proceed as follows:
    a. Position the propeller shaft cam follower against a solid object. Insert a thin blade screwdriver or an awl under the first coil of the cross-pin retainer spring and rotate the propeller shaft to unwind the spring from the sliding clutch. Take care not to over-stretch the spring.
    b. Push against the cam follower and at the same time push the cross-pin out of the sliding clutch with a punch.
    c. Release the pressure on the cam follower and slide the clutch forward off the propeller shaft. Now, tip the propeller shaft and allow the cam follower and spring to slide out of the propeller shaft. Set the unit aside for cleaning and inspecting.

Driveshaft, Shift Shaft and Bearings

1. Remove the Water Pump assembly as detailed in the Lubrication and Cooling System section.
2. Obtain a Driveshaft Holding Tool (#91-83180M), which is essentially a large nut with internal splines that match the splines on top of the driveshaft. Use a box end wrench through the bearing carrier cavity on the pinion gear nut and the special tool on the end of the driveshaft with a breaker bar. Hold the pinion nut with the box end wrench and rotate the driveshaft counterclockwise until the nut is released from the driveshaft.
3. Rotate the driveshaft back and forth slightly while pulling upward on it to free the pinion gear and pinion bearing, while starting to withdraw the shaft itself from the gearcase.
4. Remove the pinion gear and the tapered roller bearing from the lower unit.
5. Withdraw the driveshaft from the lower unit. The ball bearing which is pressed on the shaft should come out along with the round water pump base/seal carrier. Lift the water pump base and seal carrier off the shaft and place it aside.
6. Remove the forward gear and bearing assembly from the race in the gearcase nose cone.
7. If necessary, pull upward and remove the shift shaft assembly from the gearcase.
8. If the driveshaft bearing MUST be replaced it must be pushed off the shaft (which will destroy the bearing, so don't do it UNLESS it must be replaced). Believe it or not, the recommended method from Mercury is to place the driveshaft in a vise with the jaws closed just tight enough to support the bearing (but not contact the shaft), then tap on the top of the driveshaft to separate the shaft and bearing (supporting the shaft from underneath so it doesn't fall and become damaged when the bearing gives way). Alternately, a Universal Puller Plate (#C-91-37241) can be positioned on a press to free the shaft from the bearing.

7-72 LOWER UNIT

9. If the pinion bearing requires replacement you'll have to push the lower bearing race out of the gearcase. Mercury has a special tool (#91-825200A1), which looks a little like a funny spring-loaded piece of exercise equipment designed to strengthen your grip when you squeeze. The tool is put in position through the lower portion of the gearcase, squeezed and inserted up into the race. A driver (#91-13779) is then inserted down from the top of the gearcase and into the middle of the spring loaded tool. Between the size of the driver and the spring loaded halves of the tool, the assembly is wedged into the race, then the driver can be tapped to push the race down into the bottom of the gearcase.

■ The water pump base seals are normally installed back-to-back, but double-check the old orientation before removal, just to be sure.

10. The round water pump base/seal carrier should be serviced to ensure proper sealing. Using a suitable driver, push or tap the seals out top of the base assembly from underneath. Also, remove and discard the old base O-ring.

Forward Gear and Bearing

1. After the pinion gear is removed, the forward gear and bearing can be lifted out of the lower unit. The tapered bearing race will remain within the lower unit.
2. If bearing and/or race replacement is necessary, remove the forward bearing race and shim material using a slide hammer and an internal jawed puller (such as #91-27780).

■ The forward gear does not have to be removed in order to perform an adequate job of cleaning and inspecting. Therefore, it is not necessary to remove the gear unless it is unfit for further service. And, keep in mind that IF the gear and bearing are separated, you'll have to replace the bearing.

3. If bearing replacement is necessary, position a universal puller plate (#91-37241) between the forward gear and the tapered bearing. Place the puller plate and gear on a press with the gear on the bottom.
4. Press the gear out of the bearing with a suitable mandrel. If the bearing cannot be removed, clamp the forward gear in a vise equipped with soft jaws. Use a punch and hammer to drive the roller bearing out of the forward gear.

■ Once the bearing has been removed it cannot be used a second time because the roller cage will be damaged during the removal operation.

CLEANING & INSPECTION

◆ See Figures 111 and 112

1. Clean all water pump parts with solvent, and then dry them using compressed air.
2. Inspect the water pump cover and base for cracks and distortion. If possible always install a new water pump impeller while the lower unit is disassembled. A new impeller will insure extended satisfactory service. If the old impeller must be used, never install it in reverse to the original direction of rotation. Reverse installation could lead to premature impeller failure and subsequent powerhead damage.
3. Inspect the bearing surface of the propeller shaft.
4. Check the shaft surface of propeller shaft for pitting, scoring, grooving, imbedded particles, uneven wear and discoloration.
5. Check the straightness of the propeller shaft with a set of machinist V-blocks.

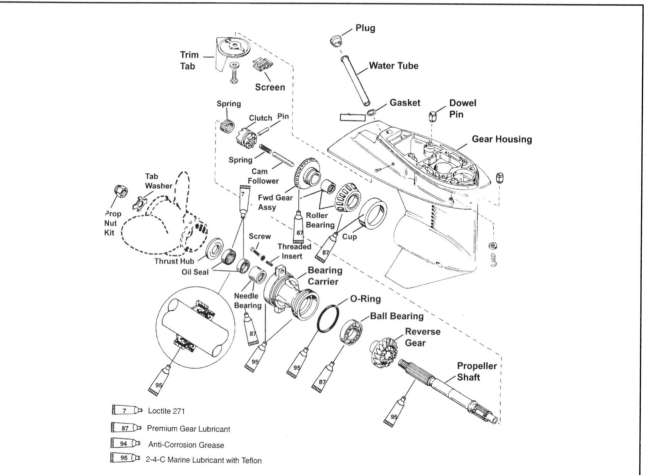

Fig. 111 Exploded drawing of the propeller shaft components - 25 Hp (Bigfoot), 30/40 Hp (Non-Bigfoot) and 40/45/50 hp (Non-Bigfoot, 935cc) models

LOWER UNIT 7-73

6. Clean the pinion gear and the propeller shaft with solvent and dry the components with compressed air.
7. Check the pinion gear and drive gear for abnormal wear.

ASSEMBLY

Forward Gear, Bearing and Race

1. If the bearings were removed from the forward gear, place the forward gear on a press with the gear teeth down.
2. Position the forward gear tapered bearing over the gear. Now, press the bearing onto the gear with a suitable mandrel until the bearing is firmly seated. MAKE SURE the mandrel pushes on the INNER bearing race. Check for clearance (gap) between the inner bearing race and the shoulder of the gear. There should be zero clearance.
3. Position the needle roller bearing over the center bore of the forward gear with the numbered side of the bearing facing up. Use a suitable mandrel and press the roller bearing into the gear until the bearing is seated against the shoulder. Make sure the mandrel pushes on the OUTER bearing race.

■ The bearing carrier is used as a pilot while installing the forward gear bearing cup in the next step. Therefore, the bearing carrier must have been assembled to include at least the propeller shaft roller bearing.

■ Earlier versions of these gearcases may be equipped with shims behind the forward bearing race. Mercury tells us that the shims should have been eliminated on these models, however, if you came across any during disassembly, be sure to reinstall them behind the new race during assembly.

4. Position the tapered bearing race squarely over the bearing bore in the front portion of the lower unit.
5. Obtain a Bearing Driver Mandrel (#91-36571). Place the tool over the tapered bearing race.
6. Insert the propeller shaft into the hole in the center of the bearing cup. Lower the bearing carrier assembly down over the propeller shaft, and then lower it into the lower unit. The bearing carrier will service as a pilot to ensure proper bearing race alignment.
7. Use a brass mallet (so as to protect the prop shaft) to drive the propeller shaft against the bearing driver cup until the tapered bearing race is seated.
8. Withdraw the propeller shaft and bearing carrier, then lift out the driver cup.
9. Position the forward gear assembly into the forward bearing race.

Driveshaft, Shift Shaft and Bearings

1. Apply a coating of 2-4-C Marine Lubricant With Teflon to the NEW O-rings, then install them on the shift shaft and carrier.

※※ WARNING

If the shift shaft is bottomed out in the gearcase the cross pin in the clutch dog will be bent by the cam follower when tightening the bearing carrier screws.

2. Position the carrier (with boot) on the shaft, then install the shaft and carrier assembly into the gear housing. Don't bottom the shaft out in the housing, pull upward on the shaft until the boot is NOT deformed.
3. Secure the boot using a new wire tie, then position the shift shaft so the ramp (tapered portion on the bottom) faces toward the propeller shaft. If necessary, look in through the propeller shaft opening as you twist the shift rod into the proper position.

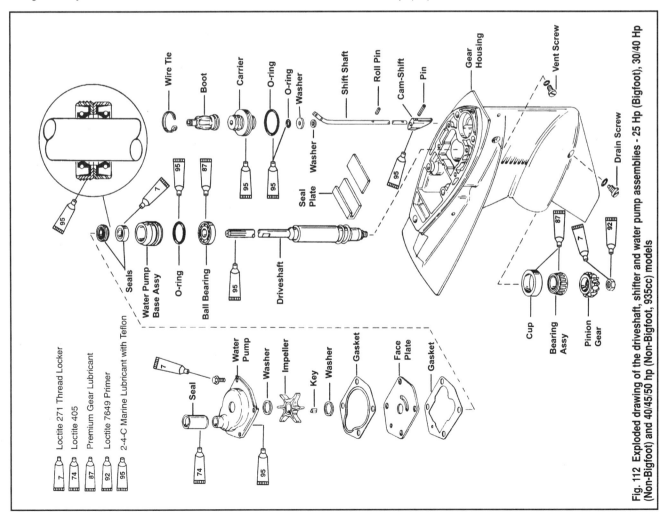

Fig. 112 Exploded drawing of the driveshaft, shifter and water pump assemblies - 25 Hp (Bigfoot), 30/40 Hp (Non-Bigfoot) and 40/45/50 hp (Non-Bigfoot, 935cc) models

7-74 LOWER UNIT

4. If the pinion bearing race was removed, you'll need to draw the replacement up into position in the gearcase using a long threaded rod (#91-31229), a support plate (#91-293100), some washers (#12-34961), a nut (#11-24156), a thrust bearing (#31-855560), a pilot (#91-825199) to keep everything centered) AND finally, a bearing cup sized mandrel (#91-825198) or equivalents. The right combination of a long threaded rod with nuts and washers SHOULD do the trick, but pick the sizes carefully. Position the race into the housing with the numbers facing UP and the tapered side facing DOWN, then assemble the tools and draw the face up into the gearcase until it is seated.

5. If the driveshaft upper roller bearing was removed, position the new bearing over the shaft and support the inner race with a bearing separator or, if necessary, a soft-jawed vise (just making sure the vice doesn't touch the shaft). Thread the old pinion nut 3/4 of the way onto the shaft in order to protect the splines, then use a press to carefully seat the bearing on the shaft.

6. Remove the driveshaft and bearing assembly from the press, then remove the pinion nut. Clean the threads of the pinion nut and the driveshaft with Loctite® 7649 Primer.

7. Carefully insert the driveshaft into the gearcase while holding the pinion gear and bearing in position, engage the splines of pinion gear to the driveshaft.

8. Apply a coating of Loctite® 271 or equivalent threadlocking compound to the threads of the pinion nut, then thread the nut onto the bottom of the driveshaft.

9. Using the splined driveshaft holding tool (#91-83180M) with a large enough socket and torque wrench on top and a pinion nut wrench on the bottom, tighten the pinion nut to 50 ft. lbs. (68 Nm).

10. If the water pump base seals were removed, apply a light coating of Loctite® 271 to the OUTER DIAMETER of the metal cased seal. Position it with the lips facing downward, then press it in through the top of the pump base until it bottoms. Next, with the lips of the neoprene outer diameter seal facing upward, press it into the case until it seats against the first seal.

11. Apply a light coating of 2-4-C Marine Lubricant With Teflon to the lips of both seals in the carrier and to a NEW water pump base O-ring, then install the O-ring to the groove in the carrier.

12. CAREFULLY position the carrier over the driveshaft and slide it down into position in the gearcase.

13. Install the Water Pump assembly as detailed in the Lubrication and Cooling System section.

Bearing Carrier Assembly

1. Position the propeller shaft needle roller bearing into the propeller end of the bearing carrier and apply a light coating of 2-4-C Marine Lubricant With Teflon to the outer diameter of the bearing.

2. Press the bearing into the bearing carrier with an Oil Seal and Bearing Driver (#91-817007) or a suitable mandrel.

■ **If driver is not available, press the bearing into the carrier to a depth of 0.82 in. (20mm) below the propeller shaft end of the carrier, then install the first seal to a depth of 0.44 in. (11mm) and the second seal to a depth of 0.04 in. (1mm) below the end of the carrier.**

3. Coat the outer diameter of both propeller shaft oil seals with Loctite® 271 or equivalent threadlocking compound.

4. Obtain a Oil Seal Driver (#91-817007) or an equivalent mandrel. Place one seal on the longer shoulder side of the driver tool with the lip of the seal away from the shoulder (facing inward toward the bearing and the carrier).

5. Press the seal into the bearing carrier until the seal driver bottoms against the bearing carrier. Place the second seal on the short shoulder side of the seal driver with the lip of the seal toward the shoulder.

6. Press the seal into the bearing carrier until the seal driver bottoms against the bearing carrier. Clean excess Loctite® from the seals.

7. Position the reverse gear on a press with the gear teeth facing down and apply a light coating of gear lube to the bearing mating surface. Place the ball bearing over the gear and press the bearing onto the gear with a suitable mandrel (1 1/4 in. socket).

8. Place the bearing carrier over the gear and bearing assembly. Press the bearing carrier onto the bearing using a suitable mandrel. If you position the gear facing upward in the press, you can use a 3/4 in. socket to press against the gear itself.

9. Place a new O-ring over the bearing carrier and position it between the bearing carrier and the thrust washer.

10. Coat the O-ring and oil seals with multipurpose lubricant, or equivalent.

Propeller Shaft Assembly

1. Insert the spring into the end of the propeller shaft.
2. Insert the flat end of the cam follower into the front end of the propeller shaft.
3. Slip the sliding clutch over the clutch splines with the cross-pin hole aligned with the cross-pin slot in the propeller shaft.
4. Position the cam follower against a solid object and push against the cam follower to compress the spring. Hold the propeller shaft in this position and at the same time, insert the cross-pin. Release pressure on the spring.

■ **If you have trouble compressing the follower spring sufficiently using the follower, remove it and substitute a 3/16 in. Allen wrench or similar sized tool.**

5. Install the cross-pin retainer spring over the sliding clutch. Take care not to over-stretch the spring.
6. Remove the cam follower and insert just a little Multipurpose Lubricant into the end of the propeller shaft to help hold the follower in position during assembly. Install the cam follower.

Propeller Shaft and Bearing Carrier Installation

1. Insert the propeller shaft into the center of the forward gear assembly, making sure the cam follower does not dislodge.
2. Slide the bearing carrier into the lower unit. Take care not to damage the propeller shaft oil seals.
3. Push the bearing carrier into the lower unit and at the same time slowly rotate the driveshaft to allow the pinion gear teeth to engage with the reverse gear teeth.
4. DISCARD THE ORIGINAL TAB WASHERS or the thin 0.063 in. (1.60mm) flat washers and original 25MM LONG SCREWS. Instead, install thicker flat washers and longer screws. Specifically obtain two 1.18 in. (30mm) long screws (10-855940-30) and two 0.090 in. (2.29mm) thick washers.
5. Apply a light coating of Loctite® 271 or equivalent threadlocking compound to the threads of the new screws. Install the screws and washers, then tighten the screws to 19. ft. lbs. (26 Nm).

SHIMMING

■ **The manufacturer gives no specific instructions for setting up the backlash on these units. As a matter of fact, Mercury claims these gearcases, when installed on these powerheads, are NOT adjustable when it comes to lash. However, if shim material was found during disassembly and was placed back in its original location, the backlash should be acceptable. The manufacturer simply states: "The amount of play between the gears is not critical, but NO play is unacceptable." Therefore, if after the assembly work is complete, the gears are "locked", the unit must be disassembled and shim material placed behind all three gears, using a "trial and error" method. If the lower unit is allowed to operate without "some" backlash, very heavy wear on the three gears will take place almost immediately.**

Gearcase - 30-115 Hp (Bigfoot), as well as 40/50/60 Hp (996cc) and 75-115 Hp (non-Bigfoot) Models

There are basically 3 similar design gearcases covered in this section. The first is the gearcase used on all 30-115 hp Bigfoot models. The second is found on the 40/50/60 hp (996cc) non-Bigfoot models. The third is found on 75-115 hp non-Bigfoot models. In all cases, there are far more similarities than differences.

Similarities between the 3 gearcase designs include:
• Water pump, base and seal assemblies (nearly identical on all but the 40/50/60 hp non-Bigfoots, which are still very similar)
• Rotating/ratcheting shift assemblies (again, nearly identical on all but the 40/50/60 hp non-Bigfoots, which are still very similar)

LOWER UNIT   7-75

- Propeller shaft bearing carrier (along with seals and bearings, again, nearly identical except the 40/50/60 hp non-Bigfoots, which are, yes you guessed it, still very similar).

Most differences between the gearcases come in the form of the driveshaft and bearing design and/or the propshaft shifter components. Other differences are minor, limited to a single bearing, shim or washer. All differences are clearly stated throughout the service and overhaul procedures.

■ **For gearcase exploded views, please refer to Cleaning & Inspection in this section.**

REMOVAL & INSTALLATION

◆ See Figures 113 and 114

Unlike most of the other Mercury gearcases, the rotating/ratcheting shift mechanism used on these models utilizes a splined shift shaft connection, making it un-necessary to locate and disconnect shift linkage before gearcase removal.

1. For safety and to prevent accidental engine start, disconnect and ground the spark plug leads to the powerhead or tag and disconnect the battery cables.
2. If necessary for overhaul purposes, remove the Propeller, as detailed in the Maintenance and Tune-Up section.
3. If necessary for gearcase overhaul purposes, position a suitable clean container under the lower unit. Remove the FILL screw on the bottom of the lower unit, and then the VENT screw. The vent screw must be removed to allow air to enter the lower unit behind the lubricant. Allow the gear lubricant to drain into the container.

As the lubricant drains, catch some with your fingers from time-to-time and rub it between your thumb and finger to determine if there are any metal particles present. Examine the fill plug. A small magnet imbedded in the end of the plug will pickup any metal particles. If metal is detected in the lubricant, the unit must be completely disassembled, inspected, and the damaged parts replaced.

Check the color of the lubricant as it drains. A whitish or creamy color indicates the presence of water in the lubricant. Check the container for signs of water separation from the lubricant. The presence of any water in the lubricant is bad news. The unit must be completely disassembled; inspected; the cause of the problem determined; and then corrected.

4. Rotate the outboard unit to the full UP position and engage the tilt lock pin.
5. Shift the gearcase into FORWARD.
6. All models utilize 4 gearcase fasteners toward the front and middle of the housing (2 on either side), in addition to a locknut and washer underneath the anti-cavitation plate. For some models (including the 40/50/60 hp 996cc motors) the locknut and washer obscured by the trim tab, if so scribe a line between the trim tab and the anti-cavitation plate. This mark will ensure the trim tab will be installed back at the original angle, then unbolt and remove the trim tab for access.

7. Remove the gearcase fastening locknut and washer from the rear of the anti-cavitation plate (either from directly in front of the trim tab or from the recess above the tab, as applicable).
8. Remove the 4 bolts (or nuts, depending upon the model) and washers securing the lower unit (2 on either side) to the exhaust housing.
9. Pull down on the lower unit gear housing and CAREFULLY separate the lower unit from the exhaust housing. On 75-115 hp motors use EXTREME caution when guiding the driveshaft downward and through the driveshaft bushing to avoid scoring the bushing surface.
10. Guide the shift rod and driveshaft out of the exhaust housing as the lower unit is removed.
11. Place the lower unit in a suitable holding fixture or work area.

To Install:

■ **Always fill the lower unit with lubricant and check for leaks before installing the unit to the driveshaft housing (could save you the trouble of removing it again right away).**

12. If a pressure tester is available, thread the adapter into the vent plug bore, then slowly pressurize the gearcase housing to 10-12 psi (69-83 kPa) pressure. Watch the gauge, as pressure must hold for at least 5 minutes. Slowly rotate the driveshaft and propshaft while watching the gauge. If the pressure drops, submerge the unit in water and recheck, watching for air bubbles to determine the pressure leak. Fix the leak before proceeding.
13. Take time to remove any old gasket material from the fill and vent recesses and from the screws.
14. Place the lower unit in an upright vertical position. Fill the lower unit with Super-Duty Lubricant, or equivalent, through the fill opening at the bottom of the unit. Never add lubricant to the lower unit without first removing the vent screw and having the unit in its normal operating position - vertical. Failure to remove the vent screw will result in air becoming trapped within the lower unit. Trapped air will not allow the proper amount of lubricant to be added.
15. Continue filling slowly until the lubricant begins to escape from the vent opening with no air bubbles visible.
16. Use a new gasket and install the vent screw. Slide a new gasket onto the fill screw. Remove the lubricant tube and quickly install the fill screw.
17. If no pressure tester was available, visibly check the lower unit for leaks. Recheck for leaks after the first time the unit is warmed to normal operating temperatures from use.
18. If not done already, swing the exhaust housing outward until the tilt lock lever can be actuated, and then engage the tilt lock.
19. Apply a coating of 2-4-C Marine Lubricant with Teflon to the inner diameter of the water tube seal, to the shift shaft coupler (and spacer on 40/50/60 non-Bigfoot models), and to the driveshaft splines

■ **DO NOT put lubricant on the ENDS of the shafts, as it could prevent the shaft splines from fully seating in the shift coupler and the crankshaft.**

20. Finish preparing the gearcase for installation, depending upon the model/gearcase, proceed as follows (not all gearcases require additional steps):

Fig. 113 These gearcases are secured by 4 fasteners on the side...

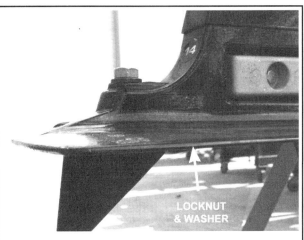

Fig. 114 ...and one locknut underneath the anti-cavitation plate

7-76 LOWER UNIT

• For Bigfoot models, apply a light bead of RTV 587 or equivalent silicone sealer to the top of the gearcase housing, to the seal area just behind the rear edge of the water pump housing.
• For 75-115 hp non-Bigfoot models, install the non-slotted end of the shift shaft bushing over the UPPER shift shaft (in the underside of the exhaust housing).
21. For all gearcases, apply a light coating of Loctite® 271 or equivalent threadlocking compound to the threads of the 4 gearcase fasteners (not on the locknut which goes under the anti-cavitation plate).

✱✱ WARNING

On 75-115 hp motors use EXTREME care while installing the driveshaft through the bushing to avoid scoring the surface.

22. Carefully bring the gearcase into alignment with the exhaust housing and slowly insert it STRAIGHT UP and into position, while aligning the driveshaft, shift shaft and water tube seal.

■ It may be necessary to move the shift block (located under the engine cowl) slightly to help align the upper shift shaft splines with the shaft coupler splines during gearcase installation. Likewise, it may be helpful to slowly turn the propeller in the normal direction of rotation to align the driveshaft-to-crankshaft splines (and on 75-115 hp motors the driveshaft-to-oil pump splines).

23. Thread the gearcase fasteners until they are all hand-tight, then alternately and evenly tighten the fasteners to 40 ft. lbs. (54 Nm).
24. If removed, align the matchmarks made earlier and install the trim tab.
25. If the propeller was removed, install the propeller.
26. Reconnect the spark plug wires and/or the battery cables.

OVERHAUL

Unlike most of the smaller Mercury gearcases, the lower unit found on these models contains various shims for adjusting pinion height and forward/reverse gear backlash. Always keep track of the shims during removal, since if the gearcase or the shafts/bearings are being reused the shims should be returned to the same positions. Also, if the gearcase and/or shafts/bearings are replaced, the shims can be sometimes be used as a good starting point for the shimming procedures.

■ For gearcase exploded views, please refer to Cleaning & Inspection in this section.

Bearing Carrier Removal & Disassembly

■ For 75-115 hp non-Bigfoot models, the shift shaft assembly splines into a shift cam that is installed on the front of the propeller shaft and for this reason the Bearing Carrier and Propeller Shaft assembly cannot be removed until AFTER the Shift Shaft is removed. For those models, please refer to Shift Shaft Removal later in this section and then continue with this Bearing Carrier procedure.

✱✱ WARNING

If you don't have access to the special Mercury mandrels designed to install bearings to the proper depth inside the bearing carrier, measure the installed height of all bearings before removal (except of course the reverse gear bearing on 40/50/60 hp non-Bigfoot models, since it is pressed directly onto the gear). Use the measured depths at installation time to ensure the bearings are positioned properly.

1. Remove the two bolts and washers (nuts and washers on some applications) securing the bearing carrier in the lower unit.

■ There are multiple ways to remove the bearing carrier on these models. The method recommended by Mercury (and therefore the preferred method) involves the use of a threaded puller with long internal jaws that will gently grab and pull the carrier off while pushing against the propshaft. We've listed a number of other possibilities, depending upon the tools at hand, however keep in mind that all of the alternatives involve some measure of risk to the gearcase and/or carrier assembly.

2. Free the bearing carrier assembly from the gearcase using one of the following methods:
   a. Install an internal jawed puller such as Long Puller Jaws (91-46086A1) and a suitable drive bolt. Position the propeller thrust hub to maintain an outward pressure on the puller jaws. Tighten the puller bolt to break the seal and free the bearing carrier.
   b. If the puller is having a problem freeing a frozen carrier, remove the puller bolt and connect a slide hammer (such as #91-34569A1 or equivalent) to the Puller Jaws (#91-46086A1). Use a few sharp quick blows with the slide hammer to hopefully break the gasket seal and free the bearing carrier.
   c. If no puller jaws are available, CAREFULLY use a soft head mallet and tap the ears of the bearing carrier to offset the carrier from the housing. Now, tap opposite "ears" alternately and evenly on the back side to remove the carrier from the lower unit housing.
   d. The final, and least desirable method is using a mallet to free a frozen carrier. Clamp the propeller shaft in a vise equipped with soft jaws in a horizontal position. Use a mallet and strike the lower unit with quick sharp blows midway between the anti-cavitation plate and the propeller shaft. This action will drive the lower unit off the bearing carrier. Take care not to drop the lower unit when the unit finally comes free of the carrier.

■ If the lower unit refuses to move, it may be necessary to carefully apply heat to the lower unit in the area of the carrier and at the same time attempt to move it off the carrier.

3. Once the seal is broken, remove the propeller shaft and bearing carrier from the gearcase. Separate the propeller shaft and bearing carrier assembly, placing each aside for disassembly.
4. For all but the 40/50/60 hp (996cc) non-Bigfoot models, the reverse gear should pull free of the carrier, so lift it from the carrier along with the thrust bearing and thrust washer. Replace any damaged components. The bearing on these models is pressed into the carrier. If the bearing does not roll freely or shows any sign of corrosion, clamp the carrier in a vise equipped with soft jaws. Obtain and use a Slide Hammer (#91-34569A1) or an equivalent slide hammer with internal jawed puller to free the bearing from the carrier.
6. For 40/50/60 hp (996cc) non-Bigfoot models, the reverse gear is pressed into the carrier. Check for damage. If the bearing or gear is damaged and requires replacement use a slide hammer with internal jaws or a suitable bearing puller such as (#91-27780). On these models, if the bearing requires replacement you'll need to use a press, universal bearing separator (like #91-37241) and suitable driver (like #91-37312) to separate the bearing from the gear.
7. Inspect the two seals and the condition of the bearing at the rear end of the carrier. If the seals have failed and have allowed water to enter the lower unit, the bearings are no longer fit for further service.
8. If only the seals are being replaced, use a small pry-tool or seal extractor to carefully remove both seals. On most models these seals are installed back-to-back but double-check seal orientation before removal.
9. If both the bearing and seals are being removed you can drive or press them all out together. Obtain a Bearing Removal and Installation Kit (#91-31229A-7) and a Driver Rod (#91-37323) and either Mandrel (#91-36569) for all except the 40/50/60 hp non-Bigfoot models, Mandrel (#91-37312) for the 40/50/60 hp non-Bigfoots or another suitable substitute mandrel. Insert the removal tools into the forward end of the carrier and press or drive the needle bearing and seals together out the rear end of the carrier.

Propeller Shaft Disassembly

To disassemble the propeller shaft for cleaning, inspection and/or parts replacement, proceed as follows:
1. Insert a thin blade screwdriver or an awl under the first coil of the cross-pin retainer spring and rotate the propeller shaft to unwind the spring from the sliding clutch. Take care not to over-stretch the spring.
2. For all except the 75-115 hp non-Bigfoot models, position the propeller shaft cam follower against a solid object. Push against the cam follower to depress the clutch spring and hold against the spring pressure during the next step.
3. Push the cross-pin out of the sliding clutch with a punch, then slide the clutch forward off the propeller shaft. For all, except the 75-115 hp non-Bigfoot models, pull the shaft slowly back away from the solid object to gently release spring pressure.
4. Now, tip the propeller shaft and allow the cam follower and spring components to slide out of the propeller shaft. The components vary slightly by gearcase model as follows:

LOWER UNIT

• For 40/50/60 hp (996cc) non-Bigfoot models, remove the cam follower, then remove the guide block and the spring.
• For 75-115 non-Bigfoot models, remove the slotted-cam follower, then remove the clutch actuator rod, spring and finally the washer.
• For 30-115 hp Bigfoot models, remove the cam follower, then remove the 3 metal balls, the guide block and finally, the spring.

5. For all models, check the propeller shaft for straightness using a dial-gauge and a pair of V-blocks. Set the shaft in the blocks so it is resting on the bearing surfaces, then set and zero the dial gauge to check run-out right behind (toward the gearcase side) of the propeller splines. Shaft run-out must not exceed 0.006 in. (0.152mm) for 40/50/60 hp non-Bigfoot models or 0.009 in. (0.228mm) for all other models.

6. Inspect the oil seal surface for grooving or pitting. Mercury only provides specs for the surface of the shaft on 40/50/60 hp non-Bigfoot models, on which they say grooving must not exceed 0.005 in. (0.12mm).

Driveshaft and Bearings

1. If not done already, remove the Water Pump assembly as detailed in the Lubrication and Cooling System section.

2. Remove the water pump base, as follows:
• For 40/50/60 hp non-Bigfoot models, use a small pry-tool to carefully pry under the ears or in the slot(s) provided on either side of the housing. Once free, lift the base carefully off the driveshaft, then remove and discard the base O-ring. Check the condition of the seal and plate on the gearcase and, if necessary, remove them for replacement.
• For all except the 40/50/60 hp non-Bigfoot models, loosen and remove the 6 water pump base retaining bolts, then CAREFULLY pry at the tabs provided (one at the front and one at the rear of the base) to free the base from the gearcase. Lift the base carefully from the driveshaft, then remove and discard the base gasket.

3. For installation purposes, note the directions in which the oil seals are facing (they are usually installed back-to-back) then carefully pry them from the water pump base.

4. Select a box end wrench the same size as the pinion nut. This tool will be used to prevent the nut from turning as the driveshaft is rotated. Obtain the proper driveshaft holding tool for the model gearcase being serviced as follows:
• For 30/40 hp (747cc) and 40/45/50 hp Bigfoot models use #91-56775 or #91-56775T.
• For all 40/50/60 hp (996cc) models use #91-87740A1 or #91-817070 (though the later MIGHT only work on Bigfoot models, we can't be 100% certain).
• For all 75-115 hp models use #91-804776A1.

5. Install the holding tool onto the end of the driveshaft. Now, with the box end wrench on the pinion gear nut, use an appropriate wrench and rotate the tool and driveshaft counterclockwise until the pinion gear nut is free.

6. Remove the nut and the pinion gear assembly. On all except the 40/50/60 hp non-Bigfoot models the pinion bearing should come free with the assembly. On 40/50/60 hp non-Bigfoot models the bearing is pressed up into the gearcase.

■ **The forward gear and bearing assembly is now free, remove it from the gearcase and set it aside for attention later.**

7. Obtain and position a Universal Puller Plate (#C-91-37241) between the pinion gear and the tapered roller bearing. Place the puller plate and gear, with the gear on the bottom, in an arbor press. Use a suitable mandrel and press the gear free of the bearing. The mandrel must contact the gear collar, but clear the bearing cage.

■ **Once the bearing has been removed, it cannot be used a second time. The roller cage will be distorted during the removal process.**

8. Pull the driveshaft up and out of the lower unit housing. For 40/50/60 hp non-Bigfoot models will require the use of an upper driveshaft bearing retainer tool (such as #91-43506 or equivalent) to loosen the round bearing retainer nut.

■ **When removing the driveshaft on 40/50/60 hp motors, watch for and retain the shims used to properly position the pinion gear, they are located under the upper driveshaft bearing. On all other models, the shims used to position the pinion gear are mounted above the bearing race.**

9. For 40/50/60 hp non-Bigfoot models, proceed as follows to service the driveshaft bearings:

a. The upper bearing is pressed onto the driveshaft. If replacement is necessary, use a universal puller plate to separate the bearing from the shaft.

b. The lower bearing is pressed into the gearcase from the top. If replacement is necessary, carefully drive the bearing down the rest of the way and out of the driveshaft bore into the propshaft bore of the gearcase using a driver (such as #91-817058A1). It is usually a good idea to measure the installed depth of the bearing before removal, to use as a reference during installation.

10. For all Bigfoot models, as well as 75-115 hp non-Bigfoot models, service the driveshaft and bearings as follows:

a. Inspect the condition of the wear sleeve at the lower end of the driveshaft. If the sleeve is worn or distorted it will allow water to enter the lower unit. If replacement is necessary, support the driveshaft in a universal bearing separator tool, resting over an open vice. Carefully, using a soft head mallet, tap the upper splined driveshaft end to force the wear sleeve up and free of the driveshaft.

b. Remove and discard the rubber sealing ring around the driveshaft.

c. If the upper driveshaft bearing requires replacement, use a suitable internal jawed puller assembly and puller bridge or support (such as #91-83165M or equivalent) to carefully pull the bearing from the gearcase bore. It is usually a good idea to measure the installed depth of the bearing before removal, to use as a reference during installation.

d. If the oil sleeve requires replacement use the same puller assembly that was used to remove the upper bearing to free it from the gearcase. Again, checking the installed height of the sleeve before removal is usually a good idea.

■ **The upper driveshaft bearing and sleeve do NOT need to be removed in order to service the pinion bearing race, but be careful not to damage them if they remain in position. ALSO, after removal, remember to retain any shims installed above the race.**

e. Lastly, if the pinion bearing and/or race requires replacement you'll have to push the lower bearing race out of the gearcase. Mercury has a number of special tools each of which look a little like a funny spring-loaded piece of exercise equipment designed to strengthen your grip when you squeeze. The tool is put in position through the lower portion of the gearcase, squeezed and inserted up into the race. A driver is then inserted down from the top of the gearcase and into the middle of the spring loaded tool. Between the size of the driver and the spring loaded halves of the tool, the assembly is wedged into the race, then the driver can be tapped to push the race down into the bottom of the gearcase. The actual tool number varies with the model gearcase, gear ratio and the number of teeth on the pinion gear.

f. For a gearcase ratio of 2.07:1 with a 14 tooth pinion gear part number 43-19672, use bearing race tool #91-13778T1 (no stamp).

g. For a gearcase ratio of 2.07:1 with a 14 tooth pinion gear part number 43-881259, use bearing race tool #91-889622A01 (stamped 91-889622).

h. For a gearcase ratio of 2.31:1 with a 13 tooth pinion gear, use bearing race tool #91-13778T1 (no stamp).

i. For a gearcase ratio of 2.33:1 with a 12 tooth pinion gear, use bearing race tool #91-889622A01 (stamped 91-889622).

■ **For all 75-115 non-Bigfoot models, use bearing race tool #91-889622A01 (stamped #91-889622).**

Forward Gear, Bearing, and Race

The forward gear and bearing assembly is inserted in a race in the nose cone and basically held in position by the pinion gear. A tapered bearing which rolls in the nose cone race is pressed onto the outside of the gear and a needle bearing is pressed inside the gear. The needle bearing supports the forward end of the propeller shaft.

As usual, removal of the bearings will normally render them unfit for service, so make sure they are shot and replacements are available before bearings are separated from the gear.

After the pinion gear and driveshaft have been removed, the forward gear assembly and associated bearing can be lifted out of the housing.

7-78 LOWER UNIT

■ The forward gear tapered roller bearing is pressed onto the short shaft of the gear assembly.

1. Most, but not all, forward gears utilize a small snapring at the bearing end (not toothed end) of the gear. The snapring is used to secure the internal needle bearing. If the bearing must be removed, first check for the presence of this snapring and remove it, if present.

2. If the needle bearing is no longer fit for further service, it can be removed by tapping, with a blunt punch and hammer, around the bearing cage from the aft end - the gear end - of the bearing. This action will destroy the bearing cage. Therefore, be sure a replacement bearing is on hand, before attempting to remove the defective bearing.

3. If the forward gear tapered roller bearing must be replaced, use a universal bearing separator and a press to push the bearing off the gear.

■ Remember that the bearing and race are a set, so if the bearing is replaced you'll also need to replace the nose cone race. A wear pattern will have been worn in the race by the old bearing. Therefore, if the race is not replaced, the new bearing will be quickly worn by the old worn race.

4. Remove the forward gear tapered roller bearing race using a slide hammer.

5. SAVE any shim material from behind the race after the race is removed. The same amount of shim material will probably be used during assembly.

Shift Shaft Removal

For 40/50/60 hp non-Bigfoot motors the shift shaft and seal assembly is inserted into the housing and is secured by a slight interference fit only, however on all other models the assembly is secured in a housing that is bolted to the top of the gearcase.

1. Remove the shift shaft coupler and spacer from the top of the shift shaft splines.

2. For 40/50/60 hp non-Bigfoot models, to remove the shaft, grasp it just below or toward the bottom of the splines using a pair of pliers (take care to prevent damaging the shaft/splines). Pull upward slightly to start to free the shaft and seal housing from the gearcase.

3. For all Bigfoot models, as well as 75 hp and larger non-Bigfoot motors, remove the 2 bolts securing the shaft housing, then use 2 small pry-tools to carefully pry the housing (and shaft) upward to just start to free the shaft assembly from the gearcase.

4. With one hand reaching into the gearcase to hold the shift cam in position and to help protect the orientation of the cam (on all except 75-115 hp non-Bigfoot models, since the shift cam is installed to the propeller shaft assembly on those models) pull upward to remove free the shaft and seal housing from the shift coupler and from the gearcase.

5. Separate the seal housing from the shaft. Remove the old clip from the shaft (located just below the seal housing).

6. Remove and discard the seal housing O-ring, then use a small pry-tool to carefully remove the seal from the top of the shaft (if damaged). Note the direction the seal is facing before removal.

7. For all except the 75-115 hp non-Bigfoot models, reach into the housing and pull out the shift cam. Attempt to preserve the orientation of the cam (which side is facing up and which side faces down) for reference come installation.

CLEANING & INSPECTION

◆ See Figures 115 thru 120

1. Clean all water pump parts with solvent, and then dry them with compressed air.

2. Inspect the water pump cover and base for cracks and distortion, possibly caused from overheating.

3. Inspect the face plate and water pump insert for grooves and/or rough surfaces. If possible, always install a new water pump impeller while the lower unit is disassembled. A new impeller will ensure extended satisfactory service and give "peace of mind" to the owner. If the old impeller must be returned to service, never install it in reverse to the original direction of rotation. Installation in reverse will cause premature impeller failure.

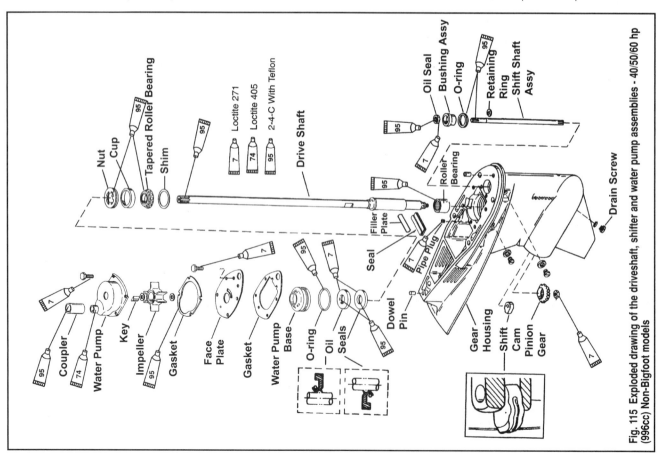

Fig. 115 Exploded drawing of the driveshaft, shifter and water pump assemblies - 40/50/60 hp (996cc) Non-Bigfoot models

LOWER UNIT

4. Inspect the impeller side seal surfaces and the ends of the impeller blades for cracks, tears, and wear. Check for a glazed or melted appearance, caused from operating without sufficient water. If any question exists, and as previously stated, install a new impeller if at all possible.

5. Clean all bearings with solvent, dry them with compressed air, and inspect them carefully. Be sure there is no water in the air line. Direct the air stream through the bearing. Never spin a bearing with compressed air. Such action is highly dangerous and may cause the bearing to score from lack of lubrication. After the bearings are clean and dry, lubricate them with Quicksilver Formula 50-D lubricant or equivalent. Do not lubricate tapered bearing cups until after they have been inspected.

6. Inspect all ball bearings for roughness, catches, and bearing race side wear. Hold the outer race, and work the inner bearing race in-and-out, to check for side wear.

7. Determine the condition of tapered bearing rollers and inner bearing race, by inspecting the bearing cup for pitting, scoring, grooves, uneven wear, imbedded particles, and discoloration caused from overheating. Always replace tapered roller bearings as a set.

8. Inspect the bearing surface of the shaft roller bearing support. Check the shaft surface for pitting, scoring, grooving, imbedded particles, uneven wear and discoloration caused from overheating. The shaft and bearing must be replaced as a set if either is unfit for continued service.

Inspect the sliding clutch of the propeller shaft. Check the reverse gear side clutch "dogs". If the "dogs" are rounded one of three causes may be to blame
   a. Improper shift cable adjustment.
   b. Running engine at too high an rpm while shifting.
   c. Shifting from neutral to reverse gear too quickly.

9. Inspect the cam follower and replace it and the shift cam if there is any evidence of pitting, scoring, or rough surfaces.

10. Inspect the propeller shaft roller bearing surfaces for pitting, rust marks, uneven wear, imbedded metal particles or signs of overheating caused by lack of adequate lubrication.

■ **Good shop practice requires installation of new O-rings and oil seals regardless of their appearance.**

11. Clean the bearing carrier, pinion gear, drive gear clutch spring, and the propeller shaft with solvent. Dry the cleaned parts with compressed air.

12. Check the pinion gear and the drive gear for abnormal wear. Apply a coating of light-weight oil to the roller bearing. Rotate the bearing and check for cracks or catches.

13. Inspect the propeller shaft oil seal surface to be sure it is not pitted, grooved, or scratched. Inspect the roller bearing contact surface on the propeller shaft for pitting, grooves, scoring, uneven wear, imbedded metal particles, and discoloration caused from overheating.

ASSEMBLY

✽✽ WARNING

Before beginning the installation work, count the number of teeth on the pinion gear and on the reverse gear. Not necessary to count the forward gear. Knowing the number of teeth on the pinion and reverse gear will permit obtaining the correct tool setup for the shimming procedure.

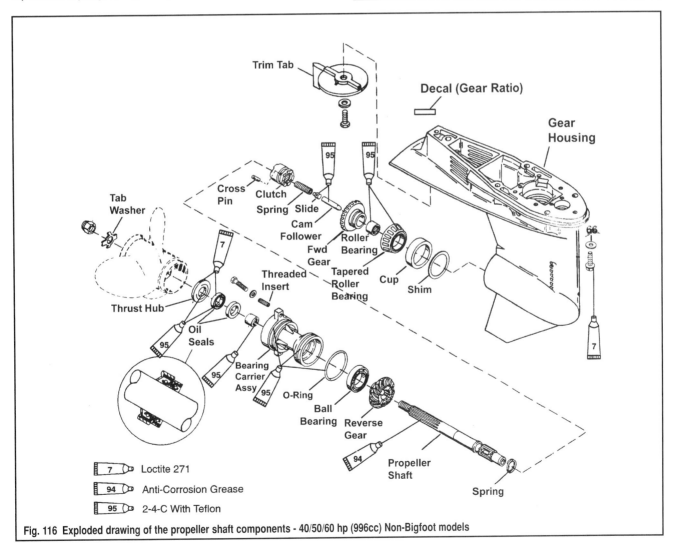

Fig. 116 Exploded drawing of the propeller shaft components - 40/50/60 hp (996cc) Non-Bigfoot models

7-80 LOWER UNIT

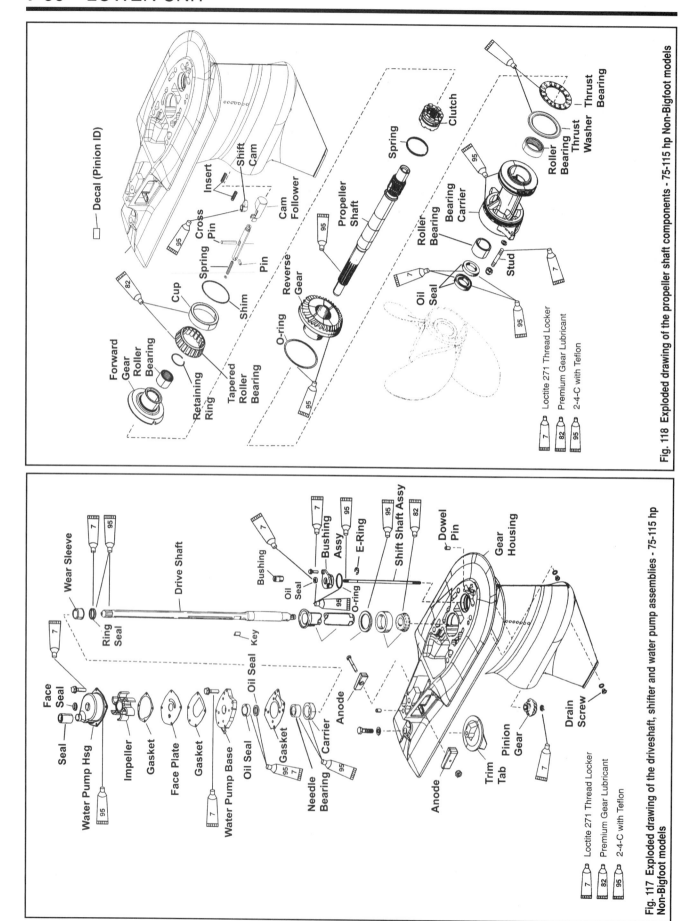

Fig. 118 Exploded drawing of the propeller shaft components - 75-115 hp Non-Bigfoot models

Fig. 117 Exploded drawing of the driveshaft, shifter and water pump assemblies - 75-115 hp Non-Bigfoot models

LOWER UNIT 7-81

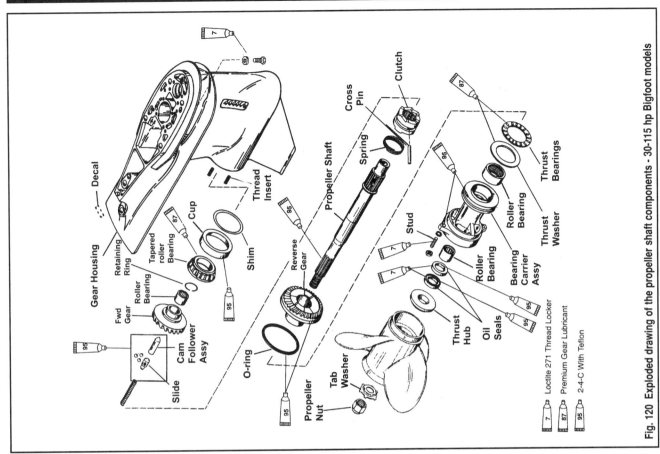

Fig. 120 Exploded drawing of the propeller shaft components - 30-115 hp Bigfoot models

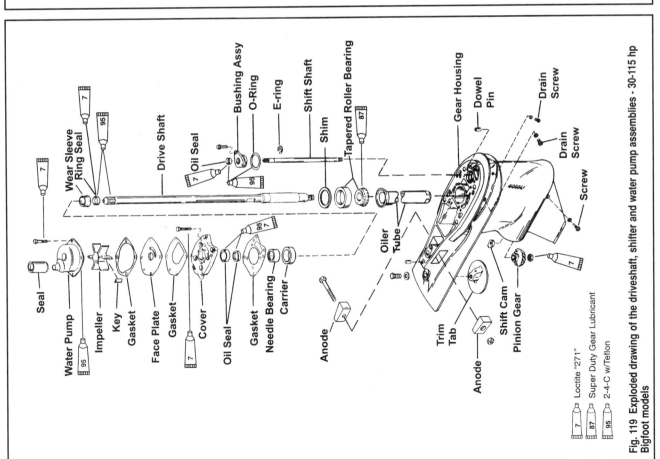

Fig. 119 Exploded drawing of the driveshaft, shifter and water pump assemblies - 30-115 hp Bigfoot models

LOWER UNIT

Shift Shaft Installation

■ For 75-115 hp non-Bigfoot models, the shift cam assembly is installed on the front of the propeller shaft and for this reason the Shift Shaft cannot be installed without FIRST installing the Bearing Carrier and Propeller Shaft assembly. For those models, continue with assembly and return to this procedure after the Bearing Carrier and Propeller Shaft has been installed.

1. For all except the 75-115 hp non-Bigfoot models, place the cam into the forward portion of the lower unit between the cast webbing while aligning the hole in the cam with the shift shaft pilot bore in the gearcase. The stamped (numbered) face of the shift cam must the same direction as noted during removal. If you couldn't tell the original orientation during disassembly (such as if the cam became dislodged before you could check), position the cam with the numbers facing downward on 30-60 hp models or upward on 75-115 hp models.
2. Install a new O-ring around the shift rod seal housing.
3. Apply a light coating of Loctite® 271 or equivalent to the outer diameter of a new shift shaft seal then position it into the top of the shift shaft seal retainer (usually with the lips facing upward, but orient it the same way as noted during removal). Tap the seal gently downward into position.
4. Apply a light coating of 2-4-C with Teflon or equivalent lubricant to the new O-ring and to the inner diameter of the new seal.
5. Install a new clip onto the shift shaft, then install the seal retainer CAREFULLY over the top of the splines.
6. Place the lower shift shaft, the short spline end, into the shift shaft cavity. Rotate the shift shaft to allow the shaft splines to index with the cam splines.
7. Tap the retainer gently and evenly into the shift shaft cavity.
8. For all except 40/50/60 hp (996cc) non-Bigfoot models, apply Loctite® 271 or equivalent threadlocking compound to the threads of both shaft retainer securing bolts. Install and tighten the bolts to 60 inch lbs. (7 Nm).
9. Install the coupler and nylon spacer to the top of the shaft.

Forward Gear, Bearing and Race

※※ WARNING

The bearing carrier is used as a pilot while installing the forward gear bearing race in the next step. Therefore, the bearing carrier must be temporarily assembled to include at least the propeller shaft roller bearing.

1. Place the same amount of shim material saved during disassembly into the lower unit. If the shim material was lost, or if a new lower unit is being used, begin with approximately 0.010 in. (0.25mm) material. Coat the forward bearing race bore with Quicksilver 2-4-C with Teflon or equivalent lubricant.
2. Position the tapered bearing race squarely over the bearing bore in the front portion of the lower unit. Obtain a Bearing Driver cup tool (#91-31106 for all except 40/50/60 hp non-Bigfoot models, or driver cup tool #91-817009 for the 40/50/60 hp non-Bigfoot models). Place the tool over the tapered bearing race.
3. Insert the propeller shaft into the hole in the center of the bearing race. Lower the bearing carrier assembly down over the propeller shaft, and then lower it into the lower unit. The bearing carrier will serve as a pilot to ensure proper bearing race alignment.
4. Use a soft-faced mallet (to protect the propeller shaft) and drive the propeller shaft against the bearing driver cup until the tapered bearing race is seated against the shim material. Withdraw the propeller shaft and bearing carrier, then lift out the driver cup.
5. If the forward gear bearings were removed, place the forward gear on a press with the gear teeth down. Apply a light coating of 2-4-C with Teflon or an equivalent lubricant to the bearing and cage surfaces.
6. Position the forward gear tapered bearing over the gear, then press the bearing onto the gear with a suitable mandrel until the bearing is firmly seated.

■ Press on the inner bearing race only. Pressing on the bearing cage will distort the bearing. Because the gear shoulder is longer than the bearing on most models be sure to use a tube type mandrel to be sure the bearing is fully seated.

7. Check for clearance (gap) between the inner bearing race and the shoulder of the gear. There should be zero clearance.
8. On Bigfoot models check the reverse gear side of the sliding clutch, it will have either 3 or 6 jaws. On models with a 3 jaw clutch, the needle bearing should be driven into the forward gear until it is 0.155 in. (3.94mm) below the surface of the gear. On 6 jaw models, drive the bearing in until flush with the surface.
9. Position the needle roller bearing over the center bore of the forward gear with the numbered side of the bearing facing up. Use a suitable mandrel and press the roller bearing into position either flush with the surface or slightly below (as noted earlier) or until seated on non-Bigfoot models.
10. If equipped, install a retaining ring into the groove on the inside of the forward gear to secure the needle bearing.
11. Insert the forward gear assembly into the forward gear bearing race.

Driveshaft and Bearings

1. If the pinion gear and/or upper bearings were removed on 40/50/60 hp non-Bigfoot install the replacements, as follows:
   a. Use a driver (such as #91-817058A1) or equivalent and a suitable mandrel to tap the new pinion bearing (which has been lubricated with 2-4-C and positioned with the numbered side facing upward) down into the gearcase from the top of the bore. Continue to gently tap the bearing until the top of the bearing is 7.05-7.07 in. (179.0-179.5mm) below the upper surface of the gearcase housing.
   b. Lubricate a new upper driveshaft bearing with 2-4-C, then use a suitable mandrel to push the bearing onto the driveshaft until it seats against the driveshaft's shoulder.
   c. Upon installation of the driveshaft and pinion gear, also be sure to install the race and shim or shims (saved from disassembly, or if lost start with a 0.15 in./0.361mm shim). Tighten the upper driveshaft bearing retainer using the special tool (#91-43506 or equivalent high-strength spanner) to 75 ft. lbs. (102 Nm).
2. For all Bigfoot models and for 75-115 hp non-Bigfoot models, install the driveshaft bearings as follows:
   a. If the Pinion bearing was removed from the gear, install a new bearing. Place the pinion gear on a press with the gear teeth down. Lubricate (using 2-4-C or equivalent) and position the pinion tapered bearing over the gear. Press the bearing onto the gear with a suitable mandrel until the bearing is firmly seated.

■ Press on the inner bearing race only. Pressing on the bearing cage will distort the bearing.

   b. Check for clearance (gap) between the pinion inner bearing race and the shoulder of the gear. There should be zero clearance.
   c. Install a new rubber sealing ring around the driveshaft, and then apply a light coating of Loctite® 271 or equivalent threadlocking compound around the ring.
   d. Obtain a Wear Sleeve Installation Tool (#91-14310A1), which is essentially a long, tubular driver which will fit over the driveshaft. Insert the new wear sleeve into the sleeve holder. Slide the bottom end of the driveshaft into the sleeve and holder. Slide the long collar portion of the installation tool over the top end of the driveshaft.
   e. Move the assembled tool and driveshaft to an arbor press. Position the base of the sleeve holder onto a suitable support which will allow the bottom end of the driveshaft to pass through. Press on the top surface of the long collar until the bottom surface of the collar seats against the top surface of the sleeve holder.
   f. Set the driveshaft aside for later installation.
   g. If the pinion gear bearing was replaced, use the Bearing Installation Tool Kit (#91-14309T01 or #T02), including the long threaded rod (#91-31229), nut (#11-24156), upper bore mandrel (#13781) and the appropriate bearing race mandrel, depending upon the gear ratio and number or pinion gear teeth.
   h. For a gearcase ratio of 2.07:1 with a 14 tooth pinion gear part number 43-19672, use bearing race tool #91-13780.
   i. For a gearcase ratio of 2.07:1 with a 14 tooth pinion gear part number 43-881259, use bearing race tool #91-889623.
   j. For a gearcase ratio of 2.31:1 with a 13 tooth pinion gear, use bearing race tool #91-13780.
   k. For a gearcase ratio of 2.33:1 with a 12 tooth pinion gear, use bearing race tool #91-889623.

LOWER UNIT 7-83

■ For all 75-115 non-Bigfoot models, use bearing race tool #91-889623.

Keep in mind that in the absence of these special tools you can usually substitute a long threaded rod, some nuts and suitably sized washers.

l. Using the threaded rod and drivers, set the pinion gear bearing race onto the shorter mandrel with the taper toward the mandrel. Place the same amount of shim material, removed during disassembly, on top of the race.

■ If the shim material was misplaced or not recorded, begin with material totaling 0.025 in. (0.635mm). Remember, shim material must be measured individually and the thickness of each added to arrive at the total thickness. Never use a micrometer to measure the total thickness of several pieces of shim material. Such a procedure results in a very inaccurate measurement and will lead to incorrect shimming of the lower unit.

m. Insert the threaded rod thru the shim material, the bearing race, and into the lower mandrel (or through the lower washer and secure using two nuts tightened against each other to keep them from loosening).

■ Lubricate the outer diameter of the bearing race using 2-4-C with Teflon or an equivalent lubricant.

n. At the top of the bore install the upper mandrel or a large washer, then secure the assembly with a nut. Tighten the nut to draw the threaded rob (and bearing race assembly) upward pulling the race into the bore until seated.

o. Apply a light coating of 2-4-C with Teflon or an equivalent lubricant to the inner and outer surfaces of the bearing sleeve.

p. Identify the end of the new roller bearing with the embossed numbers. Identify the end of the bearing sleeve with a slight taper. Place the tapered end down on the arbor press support plate.

q. Insert the new roller bearing down into the larger end of the bearing sleeve with the numbered end facing upward. Obtain a suitable mandrel and press the bearing in until it is flush with the sleeve.

r. If the oil sleeve was removed from the driveshaft bore, install the sleeve into the bore with the tab on the upper portion of the sleeve facing aft. This sleeve must be in place before the upper driveshaft bearing is installed.

s. To install the upper driveshaft bearing, obtain the same special tools used to install the pinion gear bearing race earlier in this procedure.

t. Thread the nut onto the rod until the nut is about 2/3 of the way up. Hold the shorter of the two mandrels inside the lower unit under the driveshaft bore.

u. Insert the threaded rod into the bore and thread the mandrel onto the rod until the mandrel is secure. Place the assembled roller bearing and sleeve over the threaded rod, into the cavity, and with the flush/numbered side facing upward.

v. Slide the other mandrel over the threaded rod with the shoulder side down to fit inside the sleeve.

x. Thread the nut on the rod down against the upper mandrel until all slack (clearance) is removed from the setup. With the proper size wrench on the top of the rod, prevent the rod from rotating, and at the same time rotate the nut clockwise down against the upper mandrel. As the nut is rotated the two mandrels will come closer together. This action will seat the bearing and sleeve assembly. When the shoulder of the upper mandrel makes contact with the upper face of the lower unit, the bearing/sleeve assembly is correctly positioned in the driveshaft bore.

y. Remove the special tools.

3. For all models, it's time to install the driveshaft itself. With one hand, lower the driveshaft into the top of the lower unit. With the other hand, hold the pinion gear up in the lower unit below the driveshaft cavity and with the teeth of the gear indexed (meshed) with the teeth of the forward gear.

4. Rotate and insert the driveshaft until the driveshaft splines align and engage with the splines of the pinion gear. Continue to insert the driveshaft into the pinion gear until the tapered bearing is against the bearing race.

■ Mercury sells new pinion nuts with a drylock patch on the threads. If you've got a new nut for installation don't install it until after backlash has been checked and possibly adjusted. Instead, use the old pinion nut for this step, and save the new nut for final assembly. If you don't have a new nut, no biggie, Mercury just advises the use of Loctite® 271, or equivalent threadlocking compound during final assembly.

5. Install the old pinion gear nut onto the driveshaft. Hold the pinion nut with a socket wrench. Pad the area where the socket wrench flex handle will contact the lower unit while the pinion nut is being tightened.

■ On some models there is a recessed portion of the pinion nut and, when present, it should face TOWARD the pinion gear.

6. Obtain the proper driveshaft holding tool for the model gearcase being serviced as follows:
• For 30/40 hp (747cc) and 40/45/50 hp Bigfoot models use #91-56775 or #91-56775T.
• For all 40/50/60 hp (996cc) models use #91-87740A1 or #91-817070 (though the later MIGHT only work on Bigfoot models, we can't be 100% certain).
• For all 75-115 hp models use #91-804776A1.

7. Place the special wrench over the upper end of the driveshaft. Use a torque wrench and socket to tighten the nut to 50 ft. lbs. (67 Nm) for 40/50/60 hp non-Bigfoot models or to 70 ft. lbs. (95 Nm) for all Bigfoot models and for 75-115 hp non-Bigfoot models.

※※ **WARNING**

After the pinion gear depth and the forward gear backlash have been set, the old nut must either be replaced with a new pinion gear nut OR the nut must be removed and the threads coated with Loctite® 271 or an equivalent threadlocking compound.

8. Now, if you've replaced the gearcase housing, or if you've replaced one or more of the bearings SKIP ahead to the Shimming procedures in order to check backlash and determine if some of the shims must be replaced. However, if only seals have been replaced, then you can proceed with the assembly (assuming you used a new nut or coated the old pinion nut with Loctite during assembly).

Propeller Shaft Assembly

1. Align the hole through the clutch "dog" with the slot in the propeller shaft, and then slide the "dog" onto the clutch with the "square" teeth facing the propeller end of the shaft (the reverse gear). The teeth with the "slanted" ramps (the ratcheting clutch teeth) face forward.

2. On 75-115 hp non-Bigfoot models, insert the washer into the propeller shaft.

3. Insert the spring into the end of the propeller shaft. If equipped, insert the guide block, stepped end, into the front end of the propeller shaft with the cross-pin hole aligned with the cross-pin hole in the sliding clutch.

■ On 75-115 hp non-Bigfoot models there is a small retaining pin which installs in the clutch actuator rod (the slotted rod to which the cam follower attaches). The flat on this retaining pin should face toward the spring.

4. For 30-115 hp Bigfoot models, install the 3 metal balls. Use a light coating of 2-4-C with Teflon or an equivalent lubricant to hold the balls in position.

5. For all models, apply a light coating of 2-4-C with Teflon to the cam follower, then insert the follower into the front end of the propeller shaft.

6. Position the cam follower against a solid object and push against the cam follower to compress the spring. (This is normally not necessary on the 75-115 hp non-Bigfoot models, but still might make aligning the cross-pin easier.)

7. Hold the propeller shaft in this position and at the same time, use a punch to align the guide block cross-pin opening with the sliding clutch cross-pin hole.

8. Remove the punch and insert the cross-pin.

9. Install the cross-pin ring over the sliding clutch. Take care not to over stretch the spring.

■ When handling the assembled propeller shaft, Take care to keep the forward end tilted slightly upward to keep the small balls or other loose parts in place inside the shaft.

Bearing Carrier Assembly and Installation

Be sure to lubricate the inner and outer diameters of the bearings, as well as the inner diameter only of the oil seals using 2-4-C with Teflon or an equivalent lubricant prior to installation. The outer diameter of the oil seals should be coated using Loctite® 271 or an equivalent threadlocking compound. After installing the seals, wipe away any excess Loctite before proceeding.

7-84 LOWER UNIT

1. Position the propeller shaft needle roller bearing into the aft (propeller) end of the bearing carrier with the numbered side toward the aft end (toward the driver). Press the needle bearing into the bearing carrier with a suitable mandrel. For 40/50/60 hp non-Bigfoot models, mandrel #91-817011 will give you the correct depth when the upper edge of the tool seats against the carrier. For all Bigfoot models, as well as 75-115 hp non-Bigfoot models, position mandrel #91-13945 underneath the carrier (at the reverse gear side) and then use mandrel #91-15755 to push the bearing down until it seats.

■ **If the Mercury tools are not available, position the bearing to the depth measured prior to removal.**

2. Obtain an Oil Seal Driver (#91-31108 for all except the 40/50/60 hp non-Bigfoot models which use #91-817007). Place one seal on the longer shoulder side of the driver tool with the lip of the seal away from the shoulder.
3. Press the seal into the bearing carrier until the seal driver bottoms against the bearing carrier for all except 40/50/60 hp non-Bigfoot models where the seal should be installed to the proper depth as noted.

■ **On 40/50/60 hp non-Bigfoot models if the Mercury mandrel is not available, make sure the first (inner) seal is installed to a depth of 0.82 in. (20mm) below the end of the bearing carrier and the second (outer) seal to a depth of 0.04 in. (1mm) below the end of the carrier.**

4. Place the second seal on the short shoulder side of the seal driver with the lip of the seal toward the shoulder. Press the seal into the bearing carrier until the seal driver bottoms against the bearing carrier for all except 40/50/60 hp non-Bigfoot models where the seal should be installed to the proper depth as noted. Clean excess Loctite from the seals.
5. Install the reverse gear and bearing as follows, depending upon the model:
   a. For 40/50/60 hp non-Bigfoot gearcases, use a suitable mandrel to press the roller bearing onto the back of the reverse gear until it bottoms. As usual, make sure the mandrel only presses on the portion of the bearing in contact with the gear (inner). Then use a mandrel to press the assembled gear and bearing into the forward (gearcase) end of the bearing housing until it bottoms out.
   b. For all Bigfoot models, as well as 75-115 hp non-Bigfoot models, position the larger propeller shaft roller bearing into the forward end of the bearing carrier with the numbered side toward the forward end. Press the roller bearing into the bearing carrier using a mandrel (#91-13945) or equivalent substitute. Slide the reverse gear and thrust bearing and thrust washer onto the propeller shaft (or conversely position them to the bearing housing, whichever you prefer).
6. Install a new O-ring around the carrier. Apply a coating of 2-4-C with Teflon or and equivalent Lubricant onto the installed O-ring.

■ **For all Bigfoot models and 75-115 hp non-Bigfoot models there is a trick to keep the propeller shaft, reverse gear and bearing all in position in the bearing carrier to install them as an assembly. Obtain a length of 1 1/4-1 1/2 in. (32-38mm) diameter pipe which is 6 in. (152mm) in length, then assembly the propeller shaft to the bearing carrier and fasten the pipe to the outer end (over the propeller splines) and secure using the propeller nut and tabbed washer. This will hold everything securely in position so you can install the bearing carrier and propeller shaft as an assembly.**

7. Insert the assembled propeller shaft (with reverse gear and thrust bearing if desired on Bigfoot and 75-115 hp non-Bigfoot models) into the lower unit or use the tip listed earlier so you can install the shaft and carrier as an assembly.
8. Rotate the bearing carrier, as necessary, to position the word "TOP" facing upward after installation (if the carrier is so stamped).
9. Following final installation, coat the threads of the attaching bolts with Loctite ® 271.

■ **Check the washers and bolts for the bearing carrier. For 40/50/60 hp non-Bigfoot motors, if you find 25mm long bolts and/or 0.063 in. (1.6mm) thick washers discard them and replace them with 30mm long bolts and 0.090 in. (2.29mm) thick washers. On other models you may use either the 0.060 in. (1.52mm) thick washers or 0.090 in. (2.29mm) thick washers, but the later (thicker washers are preferred).**

10. For 40/50/60 hp non-Bigfoot motors, install and tighten the bolts (or nuts on models equipped with studs) to a torque value of 225 inch lbs./18.8 ft. lbs. (19 Nm).

11. For all Bigfoot models, as well as 75-115 hp non-Bigfoot motors, install and tighten the bolts (or nuts on models equipped with studs) to 22 ft. lbs. (30 Nm) if the thicker washers are used or to 25 ft. lbs. (34 Nm) if the thinner washers are used.
12. For 75-115 hp non-Bigfoot models, if the preload has already been checked or a check is not necessary, install the Shift Shaft assembly as detailed earlier in this section.
13. If not done earlier during driveshaft installation (assuming no preload check needed to be performed) install the water pump base, followed by the water pump assembly, keeping the following points in mind:
   • The dual seals should be coated on the outer-diameter with Loctite® 271 or equivalent and installed back-to-back so the inner or lower seal faces the gearcase and the upper or outer seal faces the water pump.
   • A NEW O-ring should be used, and coated (along with the inner lips of the seals) with 2-4-C with Teflon or an equivalent lubricant.
   • For 40/50/60 hp non-Bigfoot motors the seals should be installed so the upper seal is 0.04 in. (1.02mm) from the top edge of the water pump base.
   • For all Bigfoot models, as well as 75-115 hp non-Bigfoot motors apply a light coating of the same threadlocking material used on the seals to the threads of the water pump base retaining bolts, then install the bolts and tighten to 60 inch lbs. (7 Nm).

SHIMMING

Backlash

◆ See Figure 121

For any gearcase to operate properly the pinion gear on the driveshaft must mesh properly with the forward and reverse gears. And, remember that these gears remain in contact with the pinion gear at all times, regardless of whether or not the sliding clutch has locked one of them to the propeller shaft. If these gears are not properly meshed, there is a greater risk of breaking teeth or, more commonly, rapid wear leading to gearcase failure.

On some of the smaller Mercury gearcases there is little or no adjustment for the positioning of these gears. However, the gearcases used on these medium to large sized Mercury motors are all equipped with shim material that is used to perform minute adjustments on the positioning of the bearings (and therefore the bearings and gears). Shim material can be found between the forward gear bearing and race assembly and the nose cone on all models. For 40/50/60 hp non-Bigfoot models pinion adjusting shim material is found at the upper driveshaft bearing. For all Bigfoot models as well as 75-115 hp non-Bigfoot models the pinion gear shim material is found directly above the pinion gear race (in the same general fashion as the nose cone gear).

If a gearcase is disassembled and reassembled replacing only the seals, then no shimming should be necessary. However, if shafts, gears/bearings and/or the gearcase housing itself is replaced it is necessary to check the Pinion Gear Depth and the Forward Gear Backlash to ensure proper gear positioning.

Pinion Gear Depth

◆ See Figures 122 thru 125

■ **This procedure starts with the gearcase mostly reassembled (following our procedures up to and including driveshaft reassembly, but with the propeller shaft, bearing carrier, water pump plate and water pump still removed).**

Obtain a Bearing Preload Tool (#91-14311A1 for 40/50/60 hp non-Bigfoot models or #A04 for all Bigfoot and 75-115 hp non-Bigfoot models). This tool consists of a spring, which seats against the lower unit and is used to simulate the upward driving force on the pinion gear bearing. Install the following, one by one, over the driveshaft.
   a. Plate (used on 40/50/60 hp non-Bigfoot models only)
   b. Adaptor, with the flat face down against the lower unit
   c. Thrust bearing
   d. Thrust washer
   e. Spring
   f. Special large bolt, with the nut on the bolt threaded all the way to the top of the threads.
   g. Sleeve (the proper sleeve varies with the model)

LOWER UNIT 7-85

■ Select the appropriate sleeve by model, as follows. For 40/50/60 hp non-Bigfoot models use the 3/4 in. (19mm) ID sleeve with 2 holes. For 30-60 hp Bigfoot models 7/8 in. (22mm) ID sleeve with 2 holes. For all 75-115 hp motors use the 5/8 in. (16mm) split sleeve.

1. Align the two holes in the sleeve (or the slit or slit and hole) with the Allen set screws in the special large bolt head. These set screws grip the driveshaft and prevent the tool from slipping on the driveshaft. Tighten the two set screws securely with the Allen key provided with the tool kit.
2. Measure the distance between the bottom surface of the bolt head and the top surface of the nut. Record this measurement.
3. Use a wrench to hold the bolt head steady and another wrench to tighten the nut down against the spring. Continue tightening the nut until the distance between the bottom surface of the bolt head and the top surface of the nut equals 1.00 in. (25.4mm) plus the distance recorded when the tool was installed.
4. The driveshaft has now been forced upwards placing an upward preload on the pinion gear bearing.
5. Obtain the correct Pinion Gear Locating Tool (either #91-817008A2 for 40/50/60 hp non-Bigfoot models or #91-12349A2 for Bigfoot models or for 75-115 hp non-Bigfoot models). For the later tool you'll need to assemble it by placing sliding the gauging block and the split collar onto the tool (with the gauging block numbers facing away from the split collar and toward the cross-hatched section of the tool shaft, on the opposite end of the tool from the snapring). Do NOT tighten the collar retaining screw at this time, however, use the 2 screws to secure the gauging block to the split collar.
6. Insert the tool into the installed forward gear assembly. For all except the 40/50/60 hp non-Bigfoot models, if positioned correctly, one of the flats on the gauging block should be positioned directly under the pinion gear teeth. On the 40/50/60 hp non-Bigfoot models the gauging surface is already part of the tool and it should be located under the near edge of the pinion gear (where the reverse gear would contact).

■ Notice the gauging block on the tool (#91-12349A2) has a series of flat sides, identified by the numbers 1 thru 8. The eight different numbered faces are needed for use with different combinations of gear ratios. Only two gear ratios are used on units covered here. At the beginning of this section, we asked you to count and record the number of teeth on the pinion and reverse gears, now it's time to use that data (yes it's true that there may be a sticker on the gearcase which lists the ratio, but if gears were replaced, well, you never know right).

7. For all Bigfoot models, and for 75-115 hp non-Bigfoot models, the gauging block accepts various gauging flats. After inserting it once and making sure the collar/block is under the near edge of the reverse gear, remove the tool and tighten the collar retaining bolt, THEN select and install the appropriate numbered gauge flat and locating disc depending upon the model (and more specifically the gear ratio or number of pinion teeth vs. number of reverse gear teeth, as follows):
• For 30-60 hp motors with 2.31:1 (13/30 gear teeth) or 2.33:1 (12/28 gear teeth) use flat No. 8 and Locating disc No. 3.
• For 75-90 hp motors with 2.33:1 (12/28 gear teeth) use flat No. 8 and Locating disc No. 3.
• For 75-115 hp motors with 2.07:1 (14/29 gear teeth) use flat No. 2 and Locating disc No. 3.

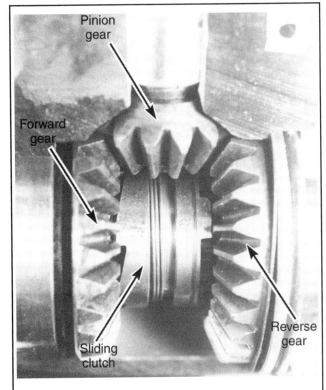

Fig. 121 Backlash is the amount of play allowed by the position of the forward and reverse gears in relation to the pinion gear

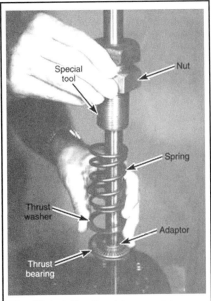

Fig. 122 Installing the pinion gear pre-loading tool

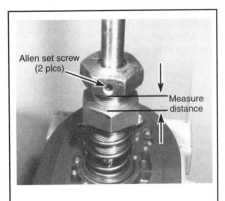

Fig. 123 Set the nut to the proper distance to set pre-load

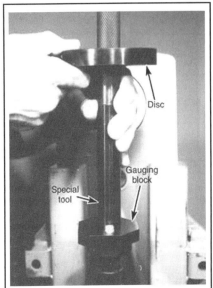

Fig. 124 Installing the proper flat/locating disc (all except 40/50/60 hp non-Bigfoot models)

7-86 LOWER UNIT

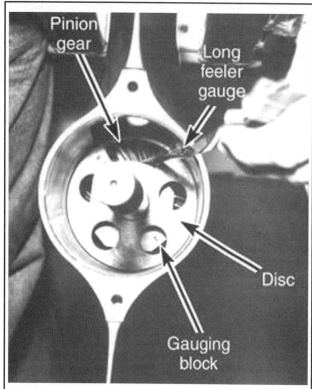

Fig. 125 Checking the gap between the pinion gear and the gauging surface

■ No other gear ratio combinations are possible for units covered in this manual. If another "odd-ball" gear ratio is encountered, somewhere, sometime, someone has purposely changed the gear ratio to suit their own requirements, possibly for operation of the outboard at different altitudes. Therefore, these factory specifications would no longer apply.

8. Let's assume everything is as it should be and the appropriate tool face to use is determined. Position this face (either No. 8 or No. 2 or the single face on the tool for 40/50/60 hp non-Bigfoot models) under the pinion gear teeth.

9. On all Bigfoot models and 75-115 hp non-Bigfoot models, install the locating disc over the tool shank with the access hole facing upward. Push the disc all the way into the lower unit until it rests against a machined shoulder of the bore. This disc supports the tool shank in the lower unit bore.

10. The pinion depth is the distance between the bottom of the pinion gear teeth and the flat surface of the gauging block.

11. Measure this distance by inserting a long feeler gauge into the access hole and between the pinion gear teeth and gauging block flat face.

12. For all units covered in this section, the correct clearance is 0.025 in. (0.64mm). If the clearance is correct, remove only the Pinion Gear Locating Tools from the inside of the gearcase. Leave the Pinion Gear Bearing Preload Tool in place on the driveshaft in order to check Forward Gear Backlash.

13. If the clearance is not correct, add or subtract shim material from behind the pinion gear bearing race or upper driveshaft bearing (as applicable) to lower or raise the gear.
• If the pinion gear depth was found to be greater than 0.025 in. (0.64mm), then that amount of shim material must be added to lower the gear.
• If the pinion gear depth was found to be less than 0.025 in. (0.64mm), then that amount of shim material must be removed to raise the gear.

■ The difference between 0.025 in. (0.64mm) and the actual clearance measured is the amount of shim material needed to correct the pinion gear depth.

Forward Gear Backlash

◆ See Figures 126 and 127

■ The propeller shaft is temporarily installed for this procedure.

Final installation of the shaft is usually performed later because an old pinion gear nut is normally installed on the driveshaft without the use of Loctite to determine the pinion gear depth (in case it needs to be disassembled again to make changes to the pinion gear bearing or forward gear bearing shims.

■ Remember the trick we mentioned in Bearing Carrier Assembly and Installation. For all Bigfoot models or 75-115 hp non-Bigfoot models, obtain a piece of PVC pipe, 6 in. (15.2cm) long by 1 1/4 -1 1/2 in. (3.2-3.8cm) diameter and a washer large enough to slide over the propeller shaft threads. Before installing the assembled bearing carrier into the lower unit, install the PVC pipe over the propeller end of the shaft, followed by the large washer and propeller nut to hold it all together. Hand tighten the nut. The pipe will exert a pressure on the bearing carrier, thrust washer and reverse gear and will allow each component to be held in alignment while the shaft is installed. Once the shaft is well seated inside the lower unit, the propeller nut, washer and PVC pipe is removed.

1. Insert the assembled propeller shaft and bearing carrier to the gearcase. Rotate the bearing carrier, if necessary, to position the word "TOP" facing upward after installation.

■ Check the washers and bolts for the bearing carrier. For 40/50/60 hp non-Bigfoot motors, if you find 25mm long bolts and/or 0.063 in. (1.6mm) thick washers discard them and replace them with 30mm long bolts and 0.090 in. (2.29mm) thick washers. On other models you may use either the 0.060 in. (1.52mm) thick washers or 0.090 in. (2.29mm) thick washers, but the later (thicker washers are preferred).

2. For 40/50/60 hp non-Bigfoot motors, install and tighten the bolts (or nuts on models equipped with studs) to 225 inch lbs./18.8 ft. lbs. (19 Nm).
3. For all Bigfoot models, as well as 75-115 hp non-Bigfoot motors, install and tighten the bolts (or nuts on models equipped with studs) to 22 ft. lbs. (30 Nm) if the thicker washers are used or to 25 ft. lbs. (34 Nm) if the thinner washers are used. On these models, remove the propeller nut, tabbed washer and length of PVC pipe at this point.
4. Install a bearing carrier puller with the arms of the puller on the carrier and the center bolt on the end of the propeller shaft. Tighten the puller center bolt to 45 inch lbs. (5 Nm). This action places a preload on the forward gear, pushing the forward gear into the forward gear bearing.
5. Rotate the driveshaft about five to ten full revolutions to properly seat the forward gear tapered roller bearing. Then recheck the torque value on the center puller bolt.
6. Attach the appropriate Backlash Dial Indicator Rod to the driveshaft and position a dial indicator (using a magnetic base or a threaded rod with nuts and plates to position it alongside the Indicator Rod) and THEN point the dial indicator to the appropriate number on the indicator rod tool, depending upon the model as follows:
• For 40/50/60 hp non-Bigfoot models use backlash indicator tool #91-19660-1 and point the tool to either the No. 4 mark or the 0.366 in. (9.3mm) mark on the indicator rod tool.

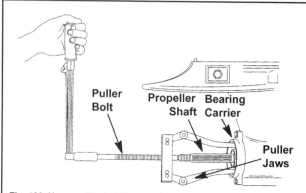

Fig. 126 Use a puller to fully seat the propeller shaft and forward gear. . .

LOWER UNIT

- For 30-60 hp Bigfoot models with 2.31:1 (13/30 gear teeth) or 2.33:1 (12/28 gear teeth) use backlash indicator tool #91-78473 and point the tool to the No. 4 mark on the indicator rod tool.
- For 75-90 hp models with 2.33:1 (12/28 gear teeth) use backlash indicator tool #91-78473 and point the tool to the No. 4 mark on the indicator rod tool.

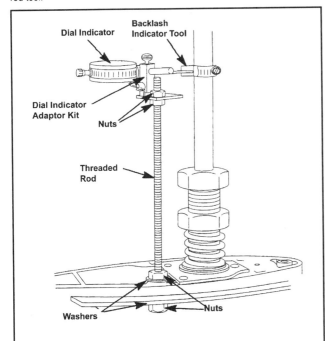

Fig. 127 ...then check forward gear backlash using the appropriate indicator tool

- For 75-115 hp motors with 2.07:1 (14/29 gear teeth) use backlash indicator tool #91-19660—1 and point the tool to the No. 1 mark on the indicator rod tool.

7. Rotate the driveshaft back-and-forth and observe movement of the dial indicator. Total movement is the forward gear backlash. Check the following listing proper amount of backlash permissible for the unit being serviced. If the backlash is too great, add shim material behind the forward gear bearing race. If the backlash is too small, remove shim material from behind the forward gear bearing race.

- For 40/50/60 hp non-Bigfoot models there should be 0.011-0.017 in. (0.28-0.43mm) of backlash.
- For 30/40 hp (747cc) and 40/45/50 hp (935cc) Bigfoot models with 2.31:1 (13/30 gear teeth) there should be 0.012-0.019 in. (0.30-0.48mm) of backlash.
- For 40/50/60 hp (966cc) and 75/90 hp models with 2.33:1 (12/28 gear teeth) there should be 0.013-0.019 in. (0.33-0.48mm) of backlash.
- For 75-115 hp models with 2.07:1 (14/29 gear teeth) there should be 0.013-0.019 in. (0.33-0.48mm) of backlash.

■ **For each 0.001 in. (0.025mm) of shim which is added or removed expect backlash to increase or decrease 0.00125 in. (0.032mm) for 40/50/60 hp non-Bigfoot models or to change 0.001 in. (0.025mm) for all Bigfoot and for 75-115 hp non-Bigfoot models.**

8. Remove all special tools. Remove the bearing carrier and propeller shaft.
9. Remove the pinion nut and either replace it with a new nut which has threadlocking compound already applied OR apply a light coating of Loctite® 271 or equivalent threadlocking compound to the threads. Proceed with final assembly from the pinion nut tightening through installation of the Bearing Carrier and Propeller Shaft assembly.

Gearcase - V6 Models

■ **The V6 Mercury/Mariner models are equipped with a Yamaha gearcase, please refer to that section for removal, installation and overhaul details.**

JET DRIVE

Jet Drive Assembly

DESCRIPTION & OPERATION

◆ See Figure 128

The jet drive unit is designed to permit boating in areas that would be impossible for a boat equipped with a conventional propeller drive system. The housing of the jet drive barely extends below the hull of the boat allowing passage in ankle deep water, white water rapids and over sand bars or in shoal water which would foul a propeller drive.

The jet drive provides reliable propulsion with a minimum of moving parts. Simply stated, water is drawn into the unit through an intake grille by an impeller driven by a driveshaft off the crankshaft of the powerhead. The water is immediately expelled under pressure through an outlet nozzle directed away from the stern of the boat.

As the speed of the boat increases and reaches planing speed, the jet drive discharges water freely into the air and only the intake grille makes contact with the water.

The jet drive is provided with a gate arrangement and linkage to permit the boat to be operated in reverse. When the gate is moved downward over the exhaust nozzle, the pressure stream is reversed by the gate (it bounces off or is turned around the thrust the opposite direction by the gate) and the boat moves sternward.

Conventional controls are used for powerhead speed, movement of the boat, shifting and power trim and tilt.

Fig. 128 Jet drive mounted to the intermediate housing of a 40 hp powerhead (a Yamaha 28J model)

7-88 LOWER UNIT

MODEL IDENTIFICATION & SERIAL NUMBERS (YAMAHA ONLY)

◆ See Figure 129

A model letter identification is stamped on the rear, port side of the jet drive housing. A serial number for the unit is stamped on the starboard side of the jet drive housing, as indicated in the accompanying illustration.

Since their introduction by Yamaha in 1984, jet drive units have been installed on various 3-cylinder and V4 and V6 units, however starting in 2002 all jet drives were moved from 2-stroke platforms to 4-stroke platforms, specifically inline 3 and 4-cylinder powerheads.

Typically four different size jet drives are used with Yamaha outboards in these model years: Model Z and Model Y, Model AA4 and AA6. These letters are embossed on the port side of the jet drive housing.

Model Z is normally used with the 28J and 35J (Model 40 hp and 50 hp motors respectively with jet drives). Model Y is usually used with the 65J (90 hp motors). Model AA4 is typically used on the 80J (115 hp V4), while the Model AA6 is typical used only on the largest of jet drive motors, the 105J and 140J (150 hp and 200 hp V6 units, respectively). On late-model motors (the 4-strokes), you will likely find the Model Z on the 40J, 50J and possibly 60J, the Model Y on the 60J and/or 90J and the Model AA4 on the 115J, but there could be exceptions.

For the most part, jet drive units are identical in design, function and operation. Differences lie in size and securing hardware, so just be sure to take the drive model number and the powerhead model number with you when ordering parts.

REMOVAL & DISASSEMBLY

◆ See Figures 130 thru 143

MODERATE

1. Unbolt the retainer securing the shift cable to the shift cable support bracket. There are normally 2 bolts.
2. Remove the locknut, bolt and washer securing the shift cable to the shift arm. Try not to disturb the length of the cable.
3. Remove the six bolts securing the intake grille to the jet casing.
4. Ease the intake grille from the jet drive housing.
5. Pry the tab or tabs of the tabbed washer away from the nut to allow the nut to be removed.
6. Loosen and then remove the nut.
7. Remove the tabbed washer and spacers. Make a careful count of the spacers behind the washer. If the unit is relatively new, there could be as many as eight spacers stacked together. If less than eight spacers are removed from behind the washer, the others will be found behind the jet impeller, which is removed in the following step. A total of eight spacers will normally be found.
8. Remove the jet impeller from the shaft. If the impeller is frozen to the shaft, obtain a block of wood and a hammer. Carefully tap the impeller in a clockwise direction to release the shear key.

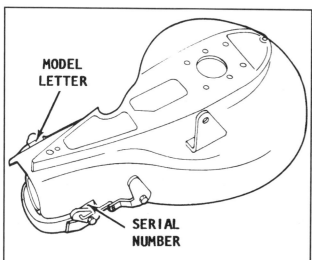

Fig. 129 The model letter designation and the serial numbers are embossed on the jet drive housing

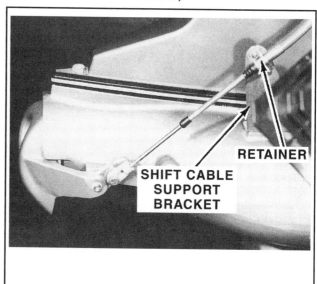

Fig. 130 Unbolt the shift cable from the support bracket...

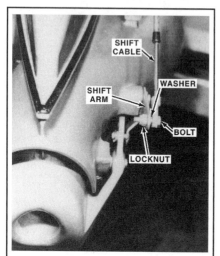

Fig. 131 ...then disconnect the cable from the shift arm

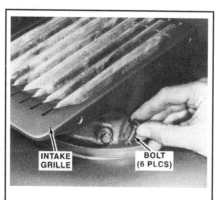

Fig. 132 Unbolt and remove the water intake grille for access to the impeller

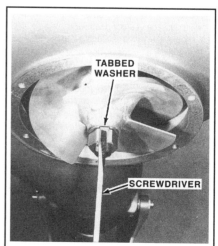

Fig. 133 To remove the impeller, bend back the locktab...

LOWER UNIT 7-89

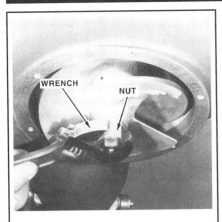

Fig. 134 ...then loosen and remove the nut

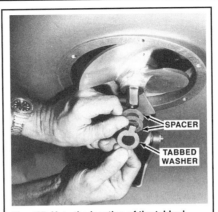

Fig. 135 Note the location of the tabbed washer and shims...

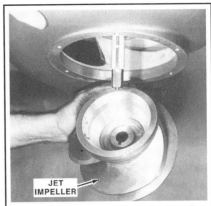

Fig. 136 ...then lower the impeller from the drive

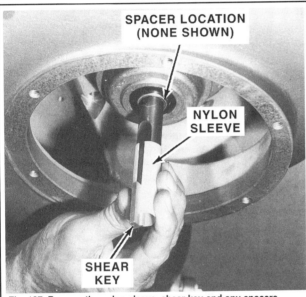

Fig. 137 Remove the nylon sleeve, shear key and any spacers which were above the impeller

9. Slide the nylon sleeve and shear key free of the driveshaft and any spacers found behind the impeller. Make a note of the number of spacers at both locations - behind the impeller and on top of the impeller, under the nut and tabbed washer.
10. Typically 1 external bolt and 4 internal bolts are used to secure the jet drive to the intermediate housing. The external bolt is located at the aft end of the anti-cavitation plate.
11. The four internal bolts are located inside the jet drive housing, as indicated in the accompanying illustration. Remove the five attaching bolts.

■ DO NOT mistake the 4 bearing housing bolts for the 4 internal jet drive assembly retaining bolts. The jet drive bolts are found outboard of the bearing housing.

12. Lower the jet drive from the intermediate housing. Remove the locating pin from the forward starboard side (or center forward, depending on the model being serviced) of the upper jet housing.

■ There area a number of locating pins (as many as 6 on some models) to be removed in the following steps. Make careful note of the size and location of each when they are removed, as an assist during assembling.

13. Remove the locating pin from the aft end of the housing. This pin and the one removed in the previous step should be of identical size.
14. Remove the four bolts and washers from the water pump housing.

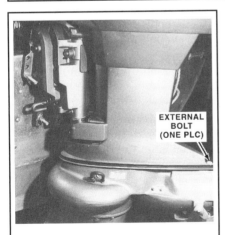

Fig. 138 To unbolt the drive remove any external bolts (usually 1)...

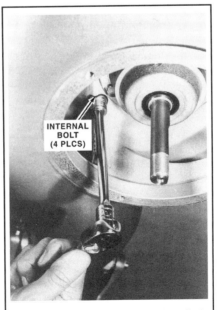

Fig. 139 ...and any internal bolts (usually 4 around the outside of the impeller housing)

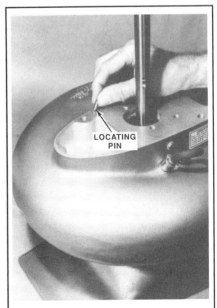

Fig. 140 When you lower the jet drive, check for the dowel (locating) pins...

7-90 LOWER UNIT

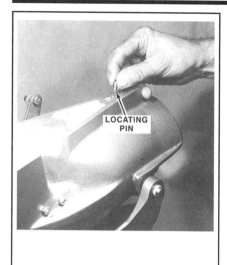

Fig. 141 ...there is usually one at each end

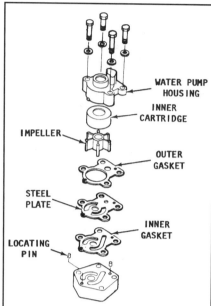

Fig. 142 Exploded view of a typical jet drive water pump assembly

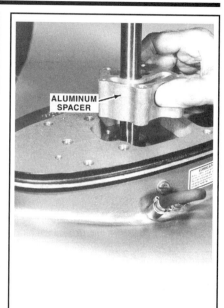

Fig. 143 If necessary, remove the pump shaped aluminum spacer

15. Pull the water pump housing, the inner cartridge and the water pump impeller, up and free of the driveshaft. Remove the Woodruff key from its recess in the driveshaft. Next, remove the outer gasket, the steel plate and the inner gasket.

16. Remove the two small locating pins and lift the aluminum spacer up and free of the driveshaft.

Remove the driveshaft and bearing assembly from the housing.

■ Model Z installed has two #10-24x5/8 in. screws and lock-washers securing the driveshaft assembly to the housing. Model Y has four 1/4-20x7/8 in. bolts and lock-washers securing the driveshaft assembly to the housing.

17. Remove the large thick adaptor plate from the intermediate housing. This plate is secured with seven bolts and lock-washers. Lower the adaptor plate from the intermediate housing and remove the two small locating pins, one on the forward port side and another from the last aft hole in the adaptor plate. Both pins should be of identical size.

CLEANING & INSPECTING

◆ See Figures 144 and 145

Wash all parts, except the driveshaft assembly, in solvent and blow them dry with compressed air. Rotate the bearing assembly on the driveshaft to inspect the bearings for rough spots, binding and signs of corrosion or damage.

Saturate a shop towel with solvent and wipe both extensions of the driveshaft.

Bearing Assembly

◆ See Figures 144, 145 and 146

Lightly wipe the exterior of the bearing assembly with the same shop towel. Do not allow solvent to enter the three lubricant passages of the bearing assembly. The best way to clean these passages is not with solvent - because any solvent remaining in the assembly after installation will continue to dissolve good useful lubricant and leave bearings and seals dry. This condition will cause bearings to fail through friction and seals to dry up and shrink - losing their sealing qualities.

The only way to clean and lubricate the bearing assembly is after installation to the jet drive - via the exterior lubrication fitting.

If the old lubricant emerging from the hose coupling is a dark, dirty, gray color, the seals have already broken down and water is attacking the

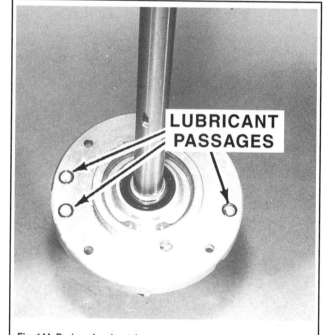

Fig. 144 During cleaning take care to prevent solvent from entering the lubrication passages

bearings. If such is the case, it is recommended the entire driveshaft bearing assembly be taken to the dealer for service of the bearings and seals.

Disassembling the Bearing Carrier

◆ See Figures 147 and 148

A slightly involved procedure must be followed to dismantle the bearing carrier including "torching" off the bearing housing. Naturally, excessive heat will destroy the seals and possibly damage the bearings. Therefore, this procedure should be avoided unless you are planning (or at least open to the possibility) of a complete overhaul to the Jet Drive. Otherwise, use great care or leave this part of the service work to someone who is more familiar with Jet Drives.

LOWER UNIT 7-91

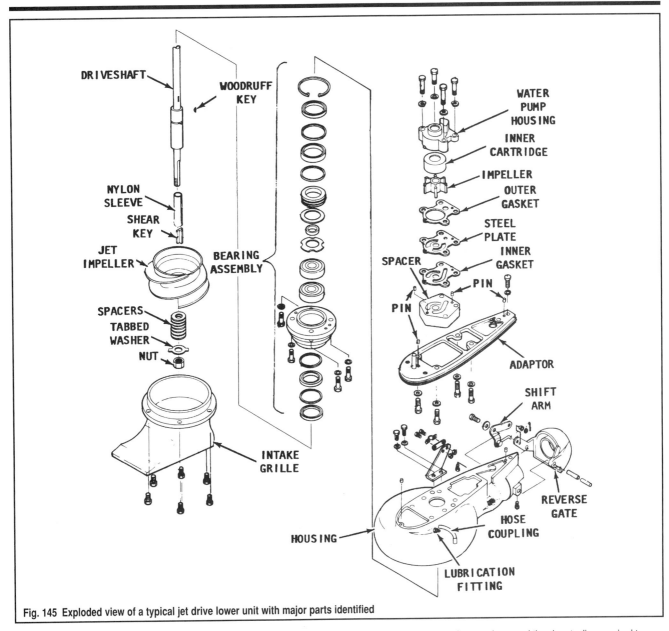

Fig. 145 Exploded view of a typical jet drive lower unit with major parts identified

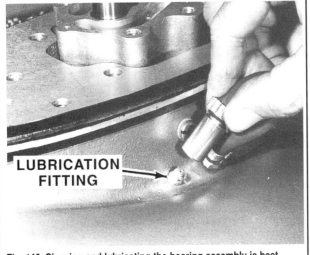

Fig. 146 Cleaning and lubricating the bearing assembly is best accomplished by completely replacing the old lubricant

However, if you have the experience and the shop tooling required to perform the torch heating and bearing press tasks, use the following procedures and illustrations to perform this work. Refer to the exploded view provided under the Cleaning & Inspecting heading in this section to help identify internal components, but remember that assemblies may vary slightly. A photographic exploded view is also provided in this section.

The following procedures pickup the work after the water pump assembly has been removed from the jet pump housing:

1. Remove the bolts (usually 4) securing the bearing carrier assembly in the jet pump housing. Lift out and discard the four small O-rings from the bolt holes on top of the bearing carrier flange.

✶✶ CAUTION

The retaining ring is under considerable force when installed. Use caution and wear eye protection when removing or installing this retaining ring.

2. Reach inside the top of the bearing carrier with a pair of snap ring pliers and remove the Truarc retaining ring.

3. Apply heat evenly all around the bearing carrier using a hand held propane torch or equivalent item. Apply only enough heat to the bearing carrier so it is very warm to the touch. Now, invert the assembly in a bearing press and support the carrier between two plates. Press on the impeller end

7-92 LOWER UNIT

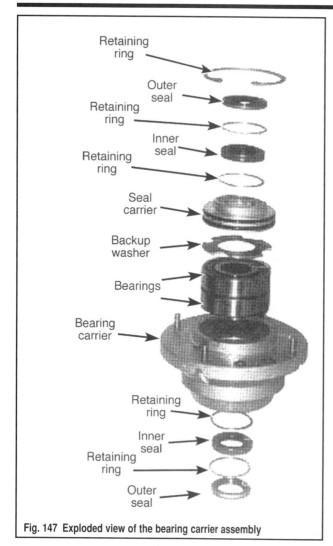

Fig. 147 Exploded view of the bearing carrier assembly

Fig. 148 After heating, use a shop press to separate the driveshaft from the carrier

of the driveshaft and drive the shaft and bearings out the bottom of the carrier. Have an assistant standing by to catch the shaft as it falls free of the bearing carrier, otherwise the splined end of the shaft could be damaged if it strikes a concrete floor.

■ **Use gloves or hand protection when handling the components such as the shaft, as should still be a bit hot.**

4. Remove the outer seal in the bearing carrier by placing the blunt end of a punch down through the bearing carrier onto the seal metal case. Tap the seal down and out the bottom of the bearing carrier. Next, grasp the tab end of the spiral retaining ring below the inner seal. Unwind the spiral retaining ring from the bearing carrier.

Again, place the blunt end of a punch down through the bearing carrier onto the seal metal case. Tap the seal down and out the bottom of the bearing carrier. Discard both seals and remove the remaining spiral retaining ring inside the bearing carrier. Thoroughly clean the bearing carrier in cleaning solvent and blow dry with compressed air.

Disassembling the Driveshaft

◆ See Figures 149 and 150

This procedure picks up where Disassembling the Bearing Carrier leaves off, at least, after the part where the driveshaft is removed from the carrier.
1. Slide the seal carrier and backup (thrust) washer off the end of the driveshaft.
2. Remove the two O-rings from the outer grooves and discard the O-rings.
3. Carefully, and closely, examine the seals inside the seal carrier. If there is any doubt as to the sealing action of the seals, replace the seal carrier with a new unit. Individual serviceable parts for the seal carrier are not available.
4. Inspect the driveshaft bearings carefully for roughness, dragging, or excessive clearance. If the bearings are defective or are removed from the driveshaft for any reason, they must be replaced with new bearings.

✶✶ WARNING

The driveshaft bearings need not be removed, unless they are unfit for further service. Once they are removed, they cannot be installed and used a second time.

5. To remove the bearings, first remove the Truarc snapring using safety glasses and a suitable pair of snapring pliers, then support the shaft in a shop press (with the impeller end up) using a bearing separator and press the bearings off the shaft. Discard both driveshaft bearings.
6. If necessary, remove and discard the retaining ring in the shaft itself.

Driveshaft and Associated Parts

Inspect the threads and splines on the driveshaft for wear, rounded edges, corrosion and damage.

Carefully check the driveshaft to verify the shaft is straight and true without any sign of damage.

Inspect the jet drive housing for nicks, dents, corrosion, or other signs of damage. Nicks may be removed with No. 120 and No. 180 emery cloth.

Reverse Gate

Inspect the gate and its pivot points. Check the swinging action to be sure it moves freely the entire distance of travel without binding.

Inspect the slats of the water intake grille for straightness. Straighten any bent slats, if possible. Use the utmost care when prying on any slat, as they tend to break if excessive force is applied. Replace the intake grille if a slat is lost, broken, or bent and cannot be repaired. The slats are spaced evenly and the distance between them is critical, to prevent large objects from passing through and becoming lodged between the jet impeller and the inside wall of the housing.

LOWER UNIT 7-93

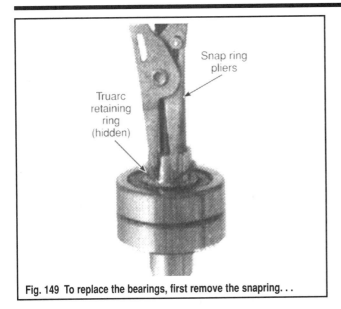

Fig. 149 To replace the bearings, first remove the snapring...

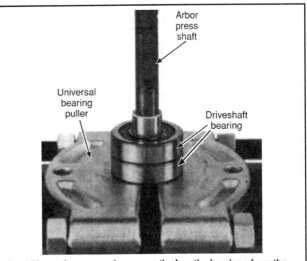

Fig. 150 ...then use a shop press the free the bearings from the shaft

Jet Impeller

◆ See Figures 151 and 152

The jet impeller is a precisely machined and dynamically balanced aluminum spiral. Observe the drilled recesses at exact locations to achieve this delicate balancing.

Excessive vibration of the jet drive may be attributed to an out-of-balance condition caused by the jet impeller being struck excessively by rocks, gravel or cavitation burn.

The term cavitation burn is a common expression used throughout the world among people working with pumps, impeller blades and forceful water movement.

Burns on the jet impeller blades are caused by cavitation air bubbles exploding with considerable force against the impeller blades. The edges of the blades may develop small dime size areas resembling a porous sponge, as the aluminum is actually eaten by the condition just described.

Excessive rounding of the jet impeller edges will reduce efficiency and performance. Therefore, the impeller should be inspected at regular intervals.

If rounding is detected, the impeller should be placed on a work bench and the edges restored to as sharp a condition as possible, using a file. Draw the file in only one direction. A back-and-forth motion will not produce a smooth edge. Take care not to nick the smooth surface of the jet impeller. Excessive nicking or pitting will create water turbulence and slow the flow of water through the pump.

Inspect the shear key. A slightly distorted key may be reused although some difficulty may be encountered in assembling the jet drive. A cracked shear key should be discarded and replaced with a new key.

Water Pump

Clean all water pump parts with solvent and then blow them dry with compressed air. Inspect the water pump housing for cracks and distortion, possibly caused from overheating. Inspect the steel plate, the thick aluminum spacer and the water pump cartridge for grooves and/or rough spots. If possible always install a new water pump impeller while the jet drive is disassembled. A new water pump impeller will ensure extended satisfactory service and give peace of mind to the owner. If the old water pump impeller must be returned to service, never install it in reverse of the original direction of rotation. Installation in reverse will cause premature impeller failure.

If installation of a new water pump impeller is not possible, check the sealing surfaces and be satisfied they are in good condition. Check the upper, lower and ends of the impeller vanes for grooves, cracking and wear. Check to be sure the indexing notch of the impeller hub is intact and will not allow the impeller to slip.

ASSEMBLY

◆ See Figure 153

1. Identify the two small locating pins used to index the large thick adaptor plate to the intermediate housing. Insert one pin into the last hole aft on the topside of the plate. Insert the other pin into the hole forward toward the port side.

2. Lift the plate into place against the intermediate housing with the locating pins indexing with the holes in the intermediate housing. Secure the plate with the five (or seven) bolts.

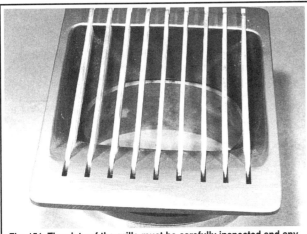

Fig. 151 The slats of the grille must be carefully inspected and any bent slats straightened for maximum jet drive performance

Fig. 152 The edges of the jet impeller should be kept as sharp as possible for maximum jet drive efficiency

7-94 LOWER UNIT

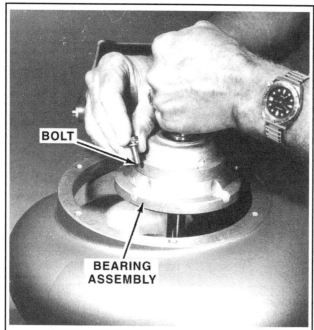

Fig. 153 If removed, install the bearing assembly into the jet drive housing

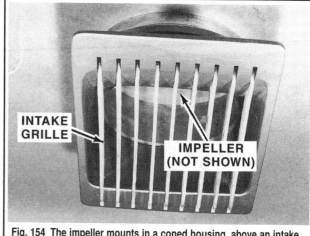

Fig. 154 The impeller mounts in a coned housing, above an intake grate

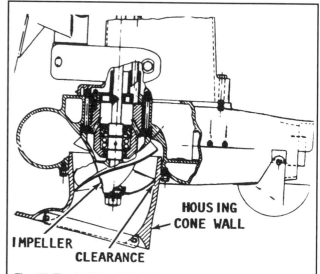

Fig. 155 The depth in which the impeller sits in the tapered housing determines impeller clearance, so adjustment is performed using shim

■ On the five bolt model, one of the five bolts is shorter than the other four. Install the short bolt in the most aft location.

3. Tighten the long bolts to 22 ft. lbs. (30 Nm). Tighten the short bolt to 11 ft. lbs. (15 Nm).
4. Place the driveshaft bearing assembly into the jet drive housing. Rotate the bearing assembly until all bolt holes align. There is only one correct position.

Model Z has two securing screws and lock-washer. Tighten these screws just good and snug by hand.

Model Y has four securing bolts and lock-washer. Tighten these four bolts to 5 ft. lbs. (7 Nm).

■ If installing a new jet impeller, place all eight spacers at the lower or nut end of the impeller, then check and, if necessary, adjust the impeller clearance by moving one or more of the shims above the impeller.

Shimming Jet Impeller

◆ See Figures 154 and 155

1. The clearance between the outer edge of the jet drive impeller and the water intake housing cone wall should be maintained at approximately 1/32 in. (0.79mm). This distance can be visually checked by shining a flashlight up through the intake grille and estimating the distance between the impeller and the casing cone, as indicated in the accompanying illustrations. But, it is not humanly possible to accurately measure this clearance by eye. Close observation between outings is fine to maintain a general idea of impeller condition, but, at least annually, the clearance must be measured using a set of long feeler gauges.

■ After continued use, the clearance will increase. The spacers previously removed are used to position the impeller along the driveshaft with a desired clearance of 1/32 in. (0.79mm) between the jet impeller and the housing wall.

2. A total of either eight (Model Z, Model AA4 and Model AA6) or nine (Model Y) spacers are normally used depending on the model being serviced. When new, all spacers are normally located at the tapered (or nut) end of the impeller. As the clearance increases, the spacers are transferred from the tapered (nut) end and placed at the wide (intermediate housing) end of the jet impeller.

■ This procedure is most easily accomplished while the jet drive is removed from the intermediate housing, but it can be done with the drive attached to the outboard.

3. Secure the driveshaft with the attaching hardware. Installation of the shear key and nylon sleeve is not vital to this procedure. Place the unit on a convenient work bench. Shine a flashlight through the intake grille into the housing cone and eyeball the clearance between the jet impeller and the cone wall, as indicated in the accompanying line drawing. Move spacers one-at-a-time from the tapered end to the wide end to obtain a satisfactory clearance. Dismantle the driveshaft and note the exact count of spacers at both ends of the bearing assembly. This count will be recalled later during assembly to properly install the jet impeller.

Water Pump Assembly

◆ See Figures 156 and 157

1. Place the aluminum spacer over the driveshaft with the two holes for the indexing pins facing upward. Fit the two locating pins into the holes of the spacer.

■ The manufacturer recommends no sealant be used on either side of the water pump gaskets.

LOWER UNIT 7-95

2. Slide the inner water pump gasket (the gasket with two curved openings) over the driveshaft. Position the gasket over the two locating pins. Slide the steel plate down over the driveshaft with the tangs on the plate facing downward and with the holes in the plate indexed over the two locating pins.

3. Check to be sure the tangs on the plate fit into the two curved openings of the gasket beneath the plate. Now, slide the outer gasket (the gasket with the large center hole) over the driveshaft. Position the gasket over the two locating pins.

4. Fit the Woodruff key into the driveshaft. Just a dab of grease on the key will help to hold the key in place. Slide the water pump impeller over the driveshaft with the rubber membrane on the top side and the keyway in the impeller indexed over the Woodruff key. Take care not to damage the membrane. Coat the impeller blades with Yamalube Grease or equivalent water resistant lubricant.

5. Install the insert cartridge, the inner plate and finally the water pump housing over the driveshaft. Rotate the insert cartridge counterclockwise over the impeller to tuck in the impeller vanes. Seat all parts over the two locating pins.

■ On some models, two different length bolts are used at this location. The Model Y uses special D shaped washers. The Model Z uses plain washers.

6. Tighten the four bolts to 11 ft. lbs. (15 Nm).

Jet Drive Installation

◆ See Figures 158 thru 169

1. Install one of the small locating pins into the aft end of the jet drive housing.
2. Install the other small locating pin into the forward starboard side (or center forward end, depending on the model being serviced).
3. Raise the jet drive unit up and align it with the intermediate housing, with the small pins indexed into matching holes in the adaptor plate. Install the internal bolts (normally 4).
4. Install the one external bolt at the aft end of the anti-cavitation plate. Tighten all bolts to 11 ft. lbs. (15 Nm).
5. Place the required number of spacers up against the bearing housing. Slide the nylon sleeve over the driveshaft and insert the shear key into the slot of the nylon sleeve with the key resting against the flattened portion of the driveshaft.

Fig. 156 If removed, install the water-pump shaped spacer to the jet drive

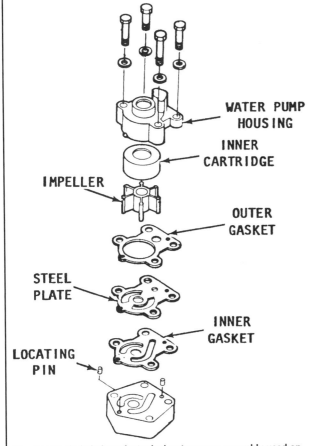

Fig. 157 Exploded view of a typical water pump assembly used on these jet drives

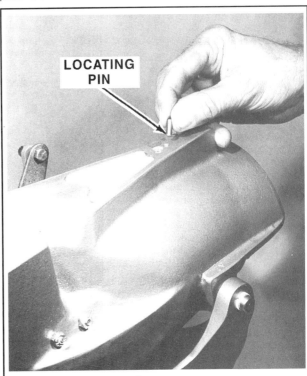

Fig. 158 Install the locating pins...

7-96 LOWER UNIT

6. Slide the jet impeller up onto the driveshaft, with the groove in the impeller collar indexing over the shear key.

7. Place the remaining spacers over the driveshaft. The number of spacers will be nine less the number used in previous step, for the Model Y and eight minus the number used during jet drive installation.

8. Tighten the nut to 17 ft. lbs. (23 Nm). If neither of the two tabs on the tabbed washer aligns with the sides of the nut, remove the nut and washer. Invert the tabbed washer. Turning the washer over will change the tabs by approximately 15°. Install and tighten the nut to the required torque value. The tabbed washer is designed to align with the nut in one of the two positions described.

9. Bend the tabs up against the nut to prevent the nut from backing off and becoming loose.

10. Install the intake grille onto the jet drive housing with the slots facing aft. Install and tighten the six securing bolts. Tighten 1/4 in. bolts to 5 ft. lbs. (7 Nm). Tighten 5/16 in. bolts to 11 ft. lbs. (15 Nm).

11. Slide the bolt through the end of the shift cable, washer and into the shift arm. Install the locknut onto the bolt and tighten the bolt securely.

12. Install the shift cable against the shift cable support bracket and secure it in place with the two bolts.

Jet Drive Adjustments

GATE LINKAGE ADJUSTMENTS

MODERATE

There are 2 potential reverse gate linkage types used on most Jet Drives. The first, predominantly found on early-model jet drives, uses a shift arm that rides on a pivot. The second, normally found on late-model jet drives, uses a roller that rides within a shift cam. Use the accompanying illustrations to determine the type of gate linkage, then follow the appropriate set of adjustment procedures.

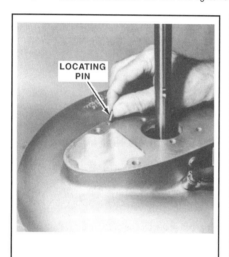

Fig. 159 ...at either end of the housing as noted during disassembly

Fig. 160 Install and tighten the 4 internal drive mounting bolts...

Fig. 161 ...along with any external bolts (there is usually just 1)

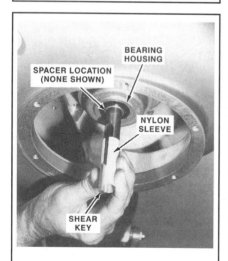

Fig. 162 Install any shims, the nylon sleeve and shear key...

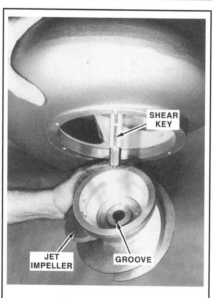

Fig. 163 ...then slide the impeller into position...

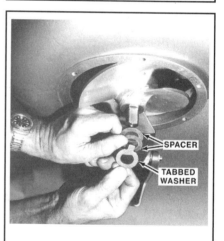

Fig. 164 ...followed by the shims and tabbed washer

LOWER UNIT 7-97

Shift Arm/Pivot Linkage (Early-Models)

Cable Alignment and Free-play
◆ See Figures 170 and 171

1. Move the shift lever downward into the forward position. The leaf spring should snap over on top of the lever to lock it in position.

2. Remove the locknut, washer and bolt from the threaded end of the shift cable. Push the reverse gate firmly against the rubber pad on the underside of the jet drive housing.

3. Check to be sure the link between the reverse gate and the shift arm is hooked into the LOWER hole on the gate.

4. Hold the shift arm up until the link rod and shift arm axis form an imaginary straight line, as indicated in the accompanying illustration. Adjust the length of the shift cable by rotating the threaded end, until the cable can be installed back onto the shift arm without disturbing the imaginary line. Pass the nut through the cable end, washer and shift arm. Install and tighten the locknut.

Neutral Stop Adjustment
◆ See Figures 172, 173 and 174

In the forward position, the reverse gate is neatly tucked underneath and clear of the exhaust jet stream.

In the reverse position, the gate swings up and blocks the jet stream deflecting the water in a forward direction under the jet housing to move the boat sternward.

In the neutral position, the gate assumes a happy medium - a balance between forward and reverse when the powerhead is operating at IDLE speed. Actually, the gate is deflecting some water to prevent the boat from moving forward, but not enough volume to move the boat sternward.

✲✲ WARNING

The gate must be properly adjusted for safety of boat and passengers. Improper adjustment could cause the gate to swing up to the reverse position while the boat is moving forward causing serious injury to boat or passengers.

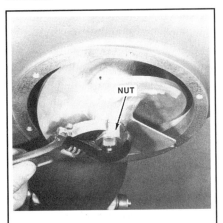

Fig. 165 Install and tighten the impeller retaining nut. . .

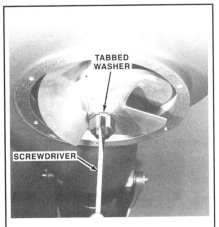

Fig. 166 . . .then bend the washer tab into place to secure the nut

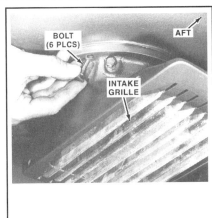

Fig. 167 Install the intake grille

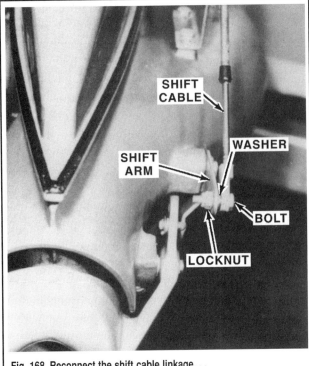

Fig. 168 Reconnect the shift cable linkage. . .

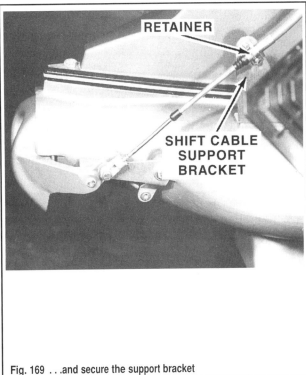

Fig. 169 . . .and secure the support bracket

7-98 LOWER UNIT

Fig. 170 When the shift lever moves down, the leaf spring should snap over and lock it in position for safety

1. Loosen, but do not remove the locknut on the neutral stop lever. Check to be sure the lever will slide up and down along the slot in the shift lever bracket.

■ The following procedure must be performed with the boat and jet drive in a body of water. Only with the boat in the water can a proper jet stream be applied against the gate for adjustment purposes.

✱✱ CAUTION

Water must circulate through the lower unit to the powerhead anytime the powerhead is operating to prevent damage to the water pump in the lower unit. Just five seconds without water will damage the water pump impeller.

2. Start the powerhead and allow it to operate only at IDLE speed. With the neutral stop lever in the down position, move the shift lever until the jet stream forces on the gate are balanced. Balanced means the water discharged is divided in both directions and the boat moves neither forward nor sternward. The gate is then in the neutral position with the powerhead at idle speed.
3. Move the neutral stop lever up against the shift lever until the stop lever barely makes contact with the shift lever. Tighten the locknut to maintain this new adjusted position. Shut down the powerhead.

■ The reverse gate may not swing to the full up position in reverse gear after the previous steps have been performed. Do not be concerned. This condition is acceptable, because water pressure in reverse will close the gate fully under normal operation.

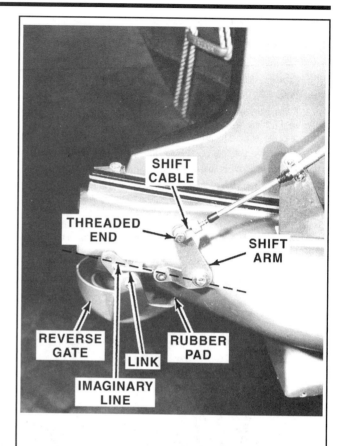

Fig. 171 Adjust the threaded end of the shift cable so that the reverse gate sits properly against the rubber pad as shown

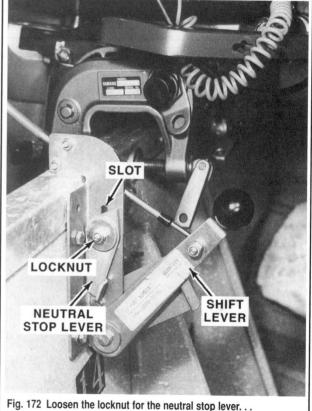

Fig. 172 Loosen the locknut for the neutral stop lever...

LOWER UNIT 7-99

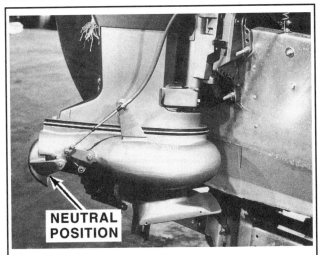

Fig. 173 ...and set the shift lever until the gate is in a balanced position...

Cam and Roller Linkage (Late-Models)

Gate Adjustment
◆ See Figures 175, 176 and 177

To adjust the reverse gate on the jet drive housing, proceed as follows:
1. Set the shifter handle to the **Neutral** detent.
2. Hold the reverse gate up and check to make sure there is about a 15/32 in. (12mm) gap between the top of the gate and the water flow passage.
3. If adjustment is necessary, have an assistant continue to hold the reverse gate up and check the clearance, while you loosen the cam screw and rotate the eccentric nut until the clearance is correct.
4. Tighten the cam screw, then recheck the adjustment.
5. Move the shifter handle to the **Forward** detent.
6. With the gate now in the **Forward** detent try to lift up on the reverse gate by hand (moving it back toward **Neutral**). The gate should NOT move. If necessary, adjust the eccentric nut again, so the **Neutral** gap is correct, but the gate will not move upward in the **Forward** position.

Remote Cable Adjustment
◆ See Figures 175 thru 179

Proper cable adjustment is necessary in order to keep water pressure (caused by boat movement) from moving the reverse gate upward blocking normal thrust.

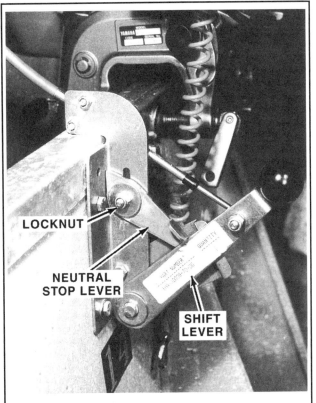

Fig. 174 ...then position the stop lever just barely against the shift lever and tighten the locknut

✴✴ CAUTION

Should be reverse gate shift suddenly and unexpectedly over the thrust nozzle, the boat will stop violently, shifting and possibly ejecting passengers and cargo. Proper adjustment is critical.

This procedure starts with the cable disconnected, for replacement, re-rigging or installation of a jet drive assembly.
1. Pull upward on the shift cam (1) until the roller is in the far end of the **Forward** range (2).
2. Shift the remote control to the **Forward** position. Temporarily, push the insert the cable guide onto the shift cam stud, then pull firmly on the

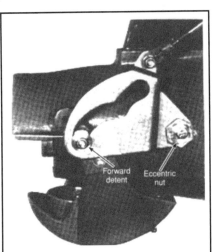

Fig. 175 Reverse gate in the FORWARD position with the cam roller at the lower end of the cutout in the bracket

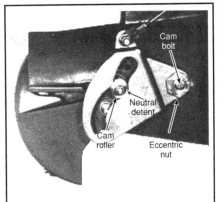

Fig. 176 Reverse gate in the NEUTRAL position, with the cam roller indexed into the neutral detent

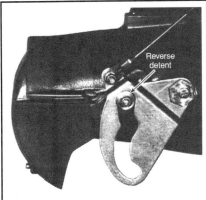

Fig. 177 Reverse gate in the REVERSE position, with the cam roller at the upper end of the cutout in the bracket

7-100 LOWER UNIT

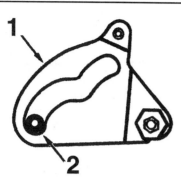

Fig. 178 When adjusting the remote cable/shift rod first move the cam (1) to the FORWARD position (2) . . .

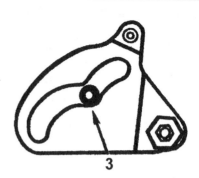

Fig. 179 . . . next, check that the cam snaps into the NEUTRAL position (3)

cable casing to remove all free-play. Next, adjust the cable trunnion until it aligns with the anchor bracket, then remove the cable from the cam stud.

3. Connect the cable trunnion to the anchor bracket, then turn 90° to lock the trunnion in position.
4. Push the cable guide back onto the cam stud, then install the washer and finger-tighten the locknut.
5. Shift the remote control to the **Neutral** position (3). The cam roller must snap into the **Neutral** detent when you pull upward on the reverse gate with moderate pressure. If the gate does not shift properly into **Neutral**, lengthen the cable slightly and recheck the adjustment.

■ Remember, it is CRITICAL that the reverse gate remains locked into the FORWARD position when the remote is shifted into FORWARD. You should NOT be able to move the gate out of this position by hand.

6. Tighten the cam locknut securely, then loosen it 1/8-1/4 of a turn in order to allow free movement of the cam.
7. Secure the cable loosely to the steering cable using a wire tie, but be sure NOT to restrict shift cable movement.

■ With the engine running in NEUTRAL and the warm-up lever raised, the engine will run rough due to exhaust restriction from the reverse gate. This is normal.

Tiller Shift Rod Adjustment
◆ See Figures 175 thru 179

Proper shift rod adjustment is necessary in order to keep water pressure (from boat movement) from moving the reverse gate upward blocking normal thrust.

✴✴ CAUTION

Should be reverse gate shift suddenly and unexpectedly over the thrust nozzle, the boat will stop violently, shifting and possibly ejecting passengers and cargo. Proper adjustment is critical.

This procedure starts with the shift rod (and connector pin) disconnected, for replacement, re-rigging or installation of a jet drive assembly.
1. Pull upward on the shift cam (1) until the roller is in the far end of the **Forward** range (2).
2. Move the shift handle to the **Forward** position, then thread the connector onto the shift rod until it is aligned with the hole in the shift cam.
3. Install the connector pin, then secure using the cotter pin and washer.
4. Move the shifter to the **Neutral** position (3). The cam roller must snap into the **Neutral** detent when you pull upward on the reverse gate with moderate pressure. If the gate does not shift properly into **Neutral**, remove the connector pin again and rotate the connector in order to lengthen the shift rod VERY slightly, then recheck the adjustment.

■ Remember, it is CRITICAL that the reverse gate remains locked into the FORWARD position when the handle is shifted into FORWARD. You should NOT be able to move the gate out of this position by hand.

5. Tighten the adjuster locknut securely.

■ With the engine running in NEUTRAL and the twist grip in the START position or higher, the engine will run rough due to exhaust restriction from the reverse gate. This is normal.

TRIM ADJUSTMENT

◆ See Figure 180

During operation, if the boat tends to pull to port or starboard, the flow fins may be adjusted to correct the condition. These fins are located at the top and bottom of the exhaust tube.
If the boat tends to pull to starboard, bend the trailing edge of each fin approximately 1/16 in. (1.5mm) toward the starboard side of the jet drive. Naturally, it follows logically that if the boat tends to pull to port, you should bend the fins toward the port side.

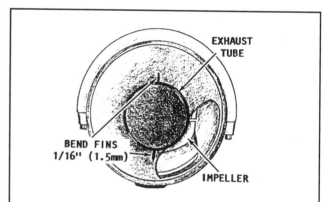

Fig. 180 Fins at the top and bottom of the exhaust tube allow for minor port/starboard trim thrust adjustment

8

TRIM AND TILT

YAMAHA HYDRO TILT LOCK SYSTEM	8-2
YAMAHA TRIM & TILT SYSTEMS	8-2
YAMAHA SINGLE TILT RAM POWER TRIM/TILT SYSTEMS	8-3
YAMAHA LARGE MOTOR POWER TRIM/TILT SYSTEM	8-15
MERCURY/MARINER TRIM & TILT SYSTEMS	8-28
MERCURY/MARINER GAS ASSIST TILT SYSTEM	8-31
MERCURY/MARINER SINGLE RAM INTEGRAL POWER TILT/TRIM	8-35

MERCURY/MARINER GAS ASSIST TILT SYSTEM ... 8-31
- DESCRIPTION & OPERATION ... 8-31
- GAS ASSIST DAMPER ... 8-31
  - OVERHAUL (30-60 HP ONLY) ... 8-32
  - REMOVAL & INSTALLATION ... 8-31
  - TESTING ... 8-31

MERCURY/MARINER SINGLE RAM INTEGRAL POWER TILT/TRIM ... 8-35
- DESCRIPTION & OPERATION ... 8-35
- PTT ASSEMBLY ... 8-42
  - HYDRAULIC SYSTEM BLEEDING ... 8-43
  - OVERHAUL (25-115 HP ONLY) ... 8-44
  - REMOVAL & INSTALLATION ... 8-42
- TROUBLESHOOTING THE POWER TRIM/TILT SYSTEM ... 8-35
  - COMPREHENSIVE TESTING ... 8-36
  - GETTING STARTED (FIRST THINGS TO CHECK) ... 8-35

MERCURY/MARINER TRIM & TILT SYSTEMS ... 8-28
- INTRODUCTION ... 8-28
- MECHANICAL TILT ... 8-28
  - SERVICING ... 8-29

YAMAHA HYDRO TILT LOCK SYSTEM ... 8-2
- DESCRIPTION & OPERATION ... 8-2
  - LOWERING OUTBOARD UNIT ... 8-2
  - RAISING OUTBOARD UNIT ... 8-2
  - UNDERWATER STRIKE ... 8-3
- SERVICING THE HYDRO TILT SYSTEM ... 8-3

YAMAHA LARGE MOTOR POWER TRIM/TILT SYSTEM ... 8-15
- DESCRIPTION & OPERATION ... 8-15
  - MANUAL OPERATION ... 8-16
  - SHOCK ABSORBER ACTION ... 8-16
  - TILT DOWN ... 8-16
  - TILT UP ... 8-15
  - TRIM DOWN ... 8-15
  - TRIM UP ... 8-15
- MAINTENANCE ... 8-16
  - CHECKING HYDRAULIC FLUID LEVEL ... 8-16
  - PURGING AIR FROM THE SYSTEM ... 8-16
- POWER TRIM/TILT SWITCH ... 8-20
  - TESTING ... 8-20
- PTT ASSEMBLY (DUAL TRIM RAM) ... 8-22
  - ASSEMBLY ... 8-27
  - CLEANING & INSPECTION ... 8-24
  - DISASSEMBLY ... 8-22
  - REMOVAL & INSTALLATION ... 8-22
- TRIM SENSOR ... 8-21
  - ADJUSTMENT ... 8-21
  - TESTING ... 8-21
- TRIM/TILT MOTOR ... 8-21
  - TESTING ... 8-21
- TRIM/TILT RELAY ... 8-19
  - TESTING ... 8-19
- TROUBLESHOOTING ... 8-17
  - HYDRAULIC TESTING ... 8-18
  - PRELIMINARY INSPECTION ... 8-17
  - SYMPTOM DIAGNOSIS ... 8-17

YAMAHA SINGLE TILT RAM POWER TRIM/TILT SYSTEMS ... 8-3
- DESCRIPTION & OPERATION ... 8-3
  - MANUAL OPERATION ... 8-4
  - SHOCK ABSORBER ACTION ... 8-5
  - TILT DOWN OPERATION ... 8-3
  - TILT UP OPERATION ... 8-3
- MAINTENANCE ... 8-5
  - CHECKING HYDRAULIC FLUID LEVEL ... 8-5
  - PURGING AIR FROM THE SYSTEM ... 8-5
- PTT ASSEMBLY ... 8-9
  - COMPONENT REPLACEMENT ... 8-12
  - REMOVAL & INSTALLATION ... 8-9
- TILT MOTOR ... 8-8
  - TESTING ... 8-8
- TILT RELAY ... 8-6
  - TESTING ... 8-6
- TILT SWITCH ... 8-7
  - TESTING ... 8-7
- TRIM SENSOR ... 8-9
  - TESTING ... 8-9
- TROUBLESHOOTING ... 8-5

YAMAHA TRIM & TILT SYSTEMS ... 8-2
- INTRODUCTION ... 8.22

8-2 TRIM AND TILT

YAMAHA TRIM & TILT SYSTEMS

Introduction

All outboard installations are equipped with some means of raising or lowering (pivoting), the complete unit for efficient operation under various load, boat design, water conditions, and for trailering to and from the water.

The correct trim angle ensures maximum performance and fuel economy as well as a more comfortable ride for the crew and passengers.

The most simple form of tilt is a mechanical tilt adjustment consisting of a series of holes in the transom mounting bracket through which an adjustment pin passes to secure the outboard unit at the desired angle.

Such a mechanical arrangement works quite well for the smaller units, but with larger (and heavier) outboard units some form of assist or power system is required. A simple hydraulic tilt assist system, known as the Yamaha Hydro Tilt system is used on larger motors that are NOT equipped with a power trim-tilt assembly. The main component of the Hydro Tilt system is a gas filled shock absorber that is used to both provide mechanical lift assistance as well as act to protect the outboard and transom from shock should the motor strike an underwater object.

Most 40 hp and larger motors (as well as a few smaller models such as high-thrust 15 hp 4-strokes and a few other 25 or 30 hp motors) use some form of a Power Trim/Tilt (PTT) system. The Yamaha power systems are hydraulically operated and electrically controlled from the helmsperson's position. There are basically 2 forms of the PTT system. The first uses a single trim/tilt rod to control all functions and provide shock absorption, while the second, found primarily on larger motors, uses a tilt rod for tilting and shock absorption, while using two dedicated trim cylinders for fine trim adjustment.

■ The single tilt rod PTT system is normally found on smaller motors, including all inline 4-strokes, except the 115 and 150 hp motors. The large motor PTT system (with single tilt rod and dual trim rods) is normally on all 115 hp and larger Yamaha 4-strokes (including the V6 Mercury/Mariner motors).

All trim and tilt systems are installed between the two large clamp brackets. On power systems, the trim/tilt relay is usually mounted in the upper cowling pan where it is fairly well protected from moisture.

All power trim/tilt systems contain a manual release valve to permit movement of the outboard unit in the remote event the trim/tilt system develops a malfunction, either hydraulic or electrical (or in case the battery or power source fails), preventing use of power.

This section covers three different types of trim/tilt units which may be installed on Yamaha outboards. Each system is described in a separate section. Troubleshooting, filling the system with hydraulic fluid and purging (bleeding) procedures are included, as applicable.

■ If you haven't figured it out yet, the 3.3L V6 4-stroke Mercury/Mariner motor IS a Yamaha, lock, stock and barrel. For this reason the trim & tilt system is covered under Large Motor Power Trim/Tilt System (Yamaha), earlier in this section. All other Mercury/Mariner motors utilize a Mercury trim & tilt system and therefore are covered in separate sections.

YAMAHA HYDRO TILT LOCK SYSTEM

Description & Operation

◆ See Figure 1

The Hydro Tilt Lock system is a mechanical assist lift and lock system for tilting medium-to-large sized outboards that are NOT equipped with a PTT system. It consists of a single shock absorber and tilt lever assembly.

■ The Hydro Tilt Lock system is normally found on medium-to-large sized commercial outboards or motors designed for high-reliability and low price (lacking extra features such as electric start) for reasons of simplicity and low expense of production cost.

The shock absorber contains a high pressure gas chamber located in the upper portion of the cylinder bore above the piston assembly. The piston contains a down relief valve and an absorber relief valve. Below the piston assembly, the lower cylinder bore contains an oil chamber. This lower chamber is connected to the upper chamber above the piston by a hydraulic line with a manual check valve. This check valve is located about half way down the hydraulic line.

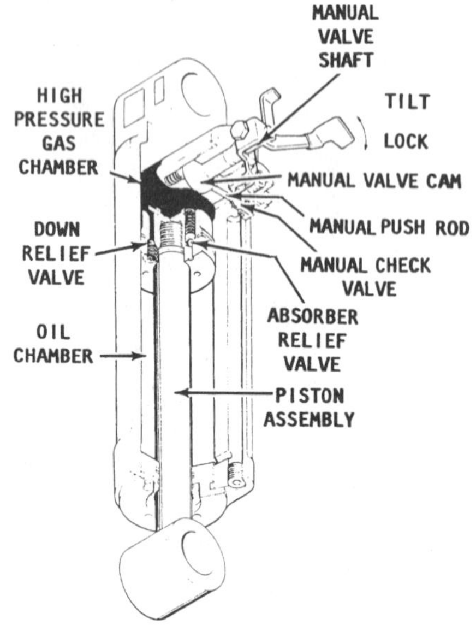

This manual check valve is activated by the tilt lever, when the lever is rotated from the lock (down) position to the tilt (up) position. The check valve cam rotates and pushes the manual check valve push rod against the check valve. This action opens the check valve and allows hydraulic fluid to flow from the lower chamber through the hydraulic line, past the open manual check valve and into the upper gas chamber.

RAISING OUTBOARD UNIT

When the outboard unit is tilted up, the volume below the piston decreases, and at the same time, the volume above the piston increases until the piston has reached the bottom of its stroke. In this position all fluid is contained above the piston.

The tilt lever is then rotated to the lock (down) position to engage with the clamp bracket. When the tilt lever is in the lock position, the manual valve push rod rests on a flat spot of the manual valve cam and releases pressure on the check valve. Releasing pressure on the check valve closes off the hydraulic line and the flow of hydraulic fluid. The outboard unit is now in the trailering position.

LOWERING OUTBOARD UNIT

To lower the outboard unit from the full up and locked position, the tilt lever is again rotated from the lock (down) position to the tilt (up) position. The manual check valve cam rotates and pushes the manual check valve push rod against the check valve and opens the valve.

■ When the manual check is open, the valve will allow hydraulic fluid to flow in only. One direction - from the lower chamber to the upper chamber.

As the outboard unit is tilted down, the piston moves up and compresses the fluid in the upper chamber. The fluid pressure overcomes the down relief valve spring and opens the relief valve. The valve in the open position permits hydraulic fluid to flow through the piston from the upper chamber to the lower chamber.

During normal cruising, the tilt lever is set in the lock (down) position. The manual check valve is closed to prevent the outboard unit from being tilted up by water pressure against the propeller when the unit is in reverse gear. When the unit is in forward gear, the outboard is held in position by the tilt pin through the swivel bracket.

Fig. 1 Cutaway view of a hydro tilt lock system with major parts identified

TRIM AND TILT 8-3

UNDERWATER STRIKE

In the event the outboard lower unit should strike an underwater object while the boat is underway, the piston would be forced down. The hydraulic fluid below the piston would be under pressure with no escape because the manual check valve is closed. The valve is closed because the tilt lever is in the lock (down) position.

To prevent rupture of the hydraulic line, a safety relief valve is incorporated in the piston. This relief valve permits fluid to pass through the piston from the lower chamber to the upper chamber through the absorber relief valve. After the outboard has passed the obstacle, the fluid returns to the lower chamber through the piston and the down relief valve, because the piston is pushed up.

Servicing the Hydro Tilt System

◆ See Figure 2

Service procedures for the Hydro Tilt Lock system are normally confined to periodic inspection for signs of extreme corrosion (especially on the shock absorber rod) and signs of oil leaks. Although small amounts of corrosion MAY be polished away in some cases, extreme corrosion and/or oil leaks will normally require that the entire unit be replaced.

Although it is possible to remove the unit from the steering and clamp bracket assembly on a few motors, for MOST motors it will require disassembly of the clamp bracket itself, a potentially time consuming task.

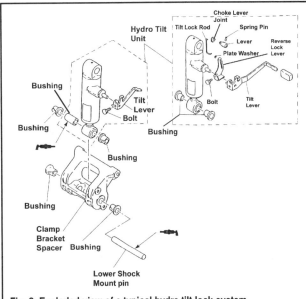

Fig. 2 Exploded view of a typical hydro tilt lock system

YAMAHA SINGLE TILT RAM POWER TRIM/TILT SYSTEMS

Description & Operation

◆ See Figures 3, 4 and 5

■ **The single tilt rod PTT system is normally found on smaller motors, including all inline 4-strokes, except the 115 hp and 150 hp motors.**

The PTT system found on most medium sized Yamaha outboards incorporates a single hydraulic cylinder and piston. Although on some older units this system was used only for tilt, on most modern Yamahas the system allows for some trim adjustment of the motor once lowered into operating position by the system. The system consists of an electric motor mounted on top of a gear driven hydraulic pump, a small fluid reservoir (which is an integral part of the pump) and a single hydraulic piston and cylinder used to move the outboard unit up or down, as required.

■ **The positioning of the tilt cylinder and the pump are reversed on some models. That is to say the pump may be found on either side (port or starboard) or the hydraulic ram.**

Unlike other power trim and tilt units, all hydraulic circuits are routed inside the unit.

■ **Three safety relief valves are incorporated into the hydraulic passageways as protection against excessive pressurization. Each of these valves has a different pressure release factor. The valves are not interchangeable. The up relief valve and the down relief valve are normally located, one on each side of the pump. The third, main relief valve, is normally found above the main valve assembly.**

Each valve is secured in place with an Allen head screw accessible from the exterior of the pump. The distance the Allen head screws are sunk into the pump housing is critical. Therefore, do not remove and examine the valves without good cause. If a valve is accidentally removed, refer to Troubleshooting, in this section.

TILT UP OPERATION

When the up portion of the tilt switch on the remote control handle is depressed, the electric motor rotates (usually in a clockwise direction). The drive gear, on the end of the motor shaft, indexed with the driven gear act as an oil pump. This action is very similar to the action in an automobile oil circulation pump.

The hydraulic fluid is forced through a series of valves into the lower chamber of the cylinder. The fluid fills the lower chamber and forces the piston upward and the outboard unit rises. As the piston continues to extend, oil in the upper chamber is routed back through the suction side of the pump until the tilt piston reaches the top of its stroke.

TILT DOWN OPERATION

When the down portion of the tilt switch on the remote control handle is depressed, the electric motor rotates in the opposite direction (therefore normally counterclockwise). The drive gear on the end of the motor shaft indexed with the driven gear are now rotating in the opposite direction. This action forces the fluid into the upper cylinder chamber under pressure causing the piston to retract, lowering the outboard. The fluid under pressure

Fig. 3 Typical Yamaha single tilt ram PTT unit

8-4 TRIM AND TILT

beneath the piston is routed back through a series of valves to the pump until the tilt piston reaches the bottom of its stroke and the outboard unit is in the full down position.

MANUAL OPERATION

◆ See Figures 6 and 7

※※ WARNING

If outboard unit is in the up position when the manual release valve is opened, the outboard will drop to the full down position rapidly. Therefore, ensure all persons stand clear.

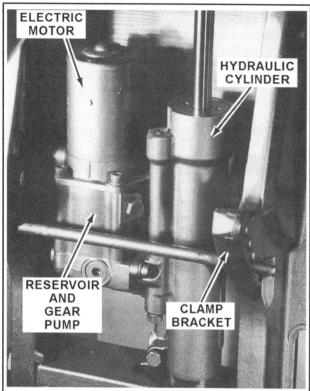

Fig. 4 PTT unit with major parts identified (note the position of the motor and cylinder is reversed on many late-model outboards)

Fig. 5 Mounting for the upper piston end of the power tilt unit

The outboard unit may be raised or lowered manually should the battery fail to provide sufficient current to operate the electric motor or should electric/hydraulic components of the PTT system fail for any other reason. A manual relief valve is provided to permit manual operation.

This manual relief valve is located on the lower end of the gear pump beneath the electric motor either facing aft (toward the motor/gearcase) or facing sideways (so that it is accessed through one of the clamp brackets). On most older Yamaha motors the valve faced aft and contained an Allen (hex key) head. On these models the valve usually contains left-hand threads, meaning it is OPENED or LOOSENED by turning CLOCKWISE. However, on MOST late-model Yamaha outboards (where the valve faces to one side, and is accessed through a clamp bracket), the valve is equipped with a slotted head and contains normal right-hand threads, meaning it is OPENED or LOOSENED by turning COUNTERCLOCKWISE.

Either way, opening the valve releases pressure in both the upper and lower cylinder chambers. With a complete loss of pressure, the piston may be moved up or down in the cylinder without resistance.

After the outboard unit has been moved to the desired position, the valve should be rotated in the opposite direction (counterclockwise for Allen head valves, clockwise for slotted valves) to close the valve and lock the outboard against movement.

Fig. 6 Fill plug and manual relief valve on an early-model PTT system

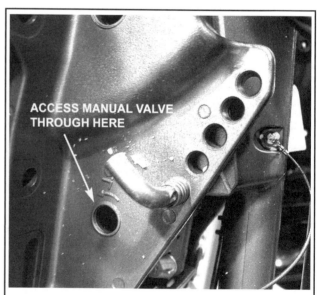

Fig. 7 On late-model PTT systems, the manual plug is accessed through the clamp bracket

TRIM AND TILT

SHOCK ABSORBER ACTION

The lower end of the tilt piston is capped with a free piston. This free piston normally moves up and down with the tilt piston. In the event the outboard lower unit should strike an underwater object while the boat is underway, the tilt piston would be suddenly and forcibly moved upward.

The free piston also moves upward but at a much slower rate than the tilt piston. The action of the tilt piston separating from the free piston causes two actions. First, the hydraulic fluid in the upper chamber above the piston is compressed and pressure builds in this area. Second, a vacuum is formed in the area between the tilt piston and the free piston.

This vacuum in the area between the two pistons sucks fluid from the upper chamber. The fluid fills the area slowly and the shock of the lower unit striking the object is absorbed. After the object has been passed, the weight of the outboard unit tends to retract the piston. The fluid between the tilt piston and the free piston is compressed and forced through check valves to the reservoir until the free piston reaches its original neutral position.

Maintenance

CHECKING HYDRAULIC FLUID LEVEL

◆ See Figures 8 and 9

■ If one of the hydraulic valves has been removed, or the system opened for any other reason, the system must be purged of any trapped air. To purge (bleed) the system, please refer to Purging Air From The System in this section.

The fluid in the power trim/tilt reservoir should be checked periodically to ensure it is full and is not contaminated. To check the fluid, tilt the motor upward to the full tilt position, then manually engage the tilt support for safety and to prevent damage.

In order to keep contaminants from entering the system when the plug is removed, clean the area around the fill plug (found on the reservoir, just below the motor on these models). Loosen and remove the plug using a suitable socket or wrench and make a visual inspection of the fluid. It should seem clear and not milky. The level is proper if, with the motor at full tilt, the level is even with the bottom of the filler plug hole.

■ Yamaha or Yamalube power trim and tilt fluid is a highly refined hydraulic fluid. This product has a high detergent content and additives to keep seals pliable. A high grade automatic transmission fluid, Dexron® II may also be used if the Yamalube fluid is not available.

PURGING AIR FROM THE SYSTEM

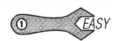

◆ See Figures 6 thru 9

Air must be purged (bled) from the system if any of the following conditions exist:
 a. Erratic motion is felt during operation of the system.
 b. Fluid has been contaminated with moisture or other foreign material.
 c. A component in the system has been replaced.
 d. The system has been opened for any reason.
 e. The outboard, when been left in the full up position, lowers slowly over a period of time.

■ When the reservoir is full to a normal level it should be even with the bottom of the plug threads when the outboard is in the fully upward, tilted position, meaning that when the outboard is lowered the level may be above the fill plug. One way to start refilling an empty or nearly empty system is to add fluid while the outboard is down, then open the manual release valve and manually lift the engine, drawing fluid into the system as the piston inside the tilt cylinder moves upward (as the ram is extended).

 1. Begin by making sure there is fluid inside the reservoir, even with the outboard in the normal running position (not tilted). Clean the area around

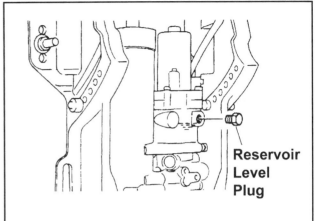

Fig. 8 With the outboard locked at full tilt, remove the reservoir plug. . .

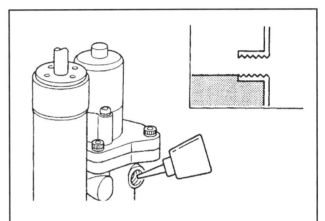

Fig. 9 . . .and make sure the fluid level is even with the bottom of the plug bore

the fill plug to prevent contaminants entering the system when the plug is loosened or removed. Slowly start to loosen the fill plug, if fluid starts to run (or spray) out, stop loosening the plug and retighten it until the outboard is tilted. However, if no fluid escapes, go ahead and remove the plug, then add fluid until it is level with the bottom of the screw threads. Lightly reinstall the reservoir plug.
 2. Open the manual release valve (either by rotating the Allen head of the valve clockwise, or the slotted head counterclockwise, as applicable), rotating it slowly in the proper direction until it stops.
 3. With the manual release valve open, slowly pull the outboard upward by hand, extending the tilt piston. As the tilt piston is extended, the piston in the cylinder will draw fluid into the cylinder, thus lowering the level in the chamber.
 4. Carefully release the outboard, allowing it to lower by its own weight, then once the outboard is lowered fully, retighten the manual release valve.
 5. Allow the PTT system fluid to settle for about 5 minutes.
 6. Push the PTT switch (either on the remote or the side of the motor) allowing the system to move the motor to the fully upward position.
 7. Push the tilt stop lever(s) to support the outboard.
 8. Again, allow the PTT system fluid to settle for about 5 minutes.
 9. Remove the reservoir cap and check the fluid level. Add fluid, as necessary until it is even with the bottom of the fill plug, then install and tighten the reservoir cap.
 10. Repeat as necessary until the fluid remains at the correct level.

Troubleshooting

◆ See Figure 10

If an operational problem develops and the determination is made the trouble is in the hydraulic system, only a few simple checks may be made. Valves, check valves, and relief valves, are all an integral part of the tilt unit.

8-6 TRIM AND TILT

Before assuming a serious fault in the system exists, make the checks listed below. Conduct an operational performance after each check has been completed for possible correction of the problem.

1. Check the quantity and quality of the hydraulic fluid in the reservoir. With the outboard unit in the full down position, the level of fluid should reach to the bottom of the fill plug threads. Add fluid, as required. If the fluid is murky drain all fluid from the system through the two drain plug openings. Fill the system with Yamalube Power Trim/Tilt Fluid or a good grade of automatic transmission fluid (Dexron or Type F). Purge the system of air.
2. Check the manual release valve to verify it is just snug (either counterclockwise on Allen head valves which are left-hand thread, or clockwise for slotted head valves).
3. Check the battery for a full charge.
4. Verify the quick disconnect plugs on the up/down relay are tight and making good contact. The relay is mounted in the bottom cowling pan, starboard side, next to the cranking motor relay.
5. Verify the wires are matched properly, color to color.
6. Test the up/down relay, the electric motor, and tilt button.

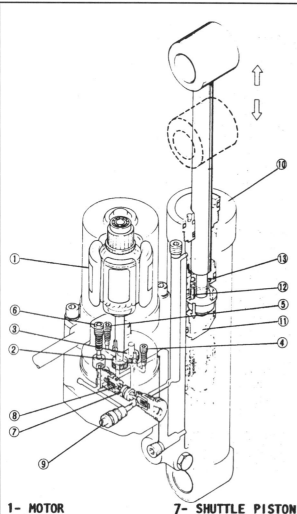

1- MOTOR	7- SHUTTLE PISTON
2- DRIVE GEAR	8- MAIN VALVE
3- DRIVEN GEAR	9- MANUAL VALVE
4- DOWN RELIEF VALVE	10- TILT CYLINDER
5- UP RELIEF VALVE #1	11- FREE PISTON
6- UP RELIEF VALVE #2	12- ABSORBER VALVE
13- CHECK VALVE	

Fig. 10 Cutaway drawing of a typical PTT unit with flow direction indicated and major parts identified

Tilt Relay

TESTING

◆ See Figures 11, 12 and 13

The PTT motor runs either clockwise or counterclockwise depending upon which direction you want the motor to move and also depending upon how the power is applied to the motor's terminals. If you apply positive current to one terminal and negative to the other, the motor will run in one direction. If you switch and apply positive and negative to the opposite terminals, the motor will run in the other direction. Control of this circuit is achieved through one or more PTT relays. The job of the relay, like all relays, is to close certain switch circuits depending upon the signals received from the tilt switch on either the engine or the remote.

There are generally 3 major designs of PTT relays that you might find on these motors. The first and most common relay used on motors in this size range is a rectangular piece that looks like it has 8 terminals and a 3 wire harness that comes off one end (though the harness may use a single connector or individual wire bullets/terminals). Four of the visible terminals (identified by position) and the 3 wire harness are used during testing to with an ohmmeter (and a 12-volt battery) to determine if the relay is operating properly. For testing purposes, the exposed terminal at one end of the relay (where the wiring harness attaches) will be referred to as Terminal 1. This terminal is also identified because it is diagonally adjacent to the relay Negative terminal (the black ground wire was connected to it before removal). At the far diagonal end of the relay from Terminal 1 is the position for Terminal 2. Again, Terminal 2 is easily identified because it is diagonally adjacent to the relay Positive terminal (the terminal to which the red power wire was attached before removal). For more details, please see the accompanying illustration.

A few models, including the 15 hp 4-stroke utilize a rectangular relay box that does not have exposed terminals, but which contains a 6 wire harness (usually one 2-wire connector and 4 individual ring terminals or bullets). On these relays wire colors are used to identify proper test connections, but testing is otherwise the same, using an ohmmeter and a 12-volt battery.

Lastly, some models (such as the 75-100 hp 4-strokes) may be equipped with 2 individual round-bodied relays (like the type once more commonly found on the large motor PTT system). Both wire color and terminal positions are used to test these relays in much the same fashion as the rectangular bodied units.

■ **On all relays, testing normally occurs with the relay completely removed from the powerhead, OR at least with all wiring disconnected (even if the relay is still physically bolted in position).**

1. If not done already, tag and disconnect the wiring from the PTT relay(s).
2. Obtain a 12-volt battery with jumper leads and an ohmmeter for continuity testing.
3. First check terminal-to-terminal continuity as detailed by relay type:
• For rectangular, exposed terminal type relays, use the ohmmeter to confirm there is continuity between the following pairs of wires: Terminal 1 and the Negative terminal, Terminal 2 and the Negative terminal, The Sky Blue wire and the Black wire and finally, the Light Green wire and the Black wire. Next use the ohmmeter to confirm there is NO continuity between Terminal 1 and the Positive terminal, as well as between Terminal 2 and the Positive terminal.
• For rectangular, non-exposed terminal type relays (such as those used on the 15 hp 4-stroke), perform the SAME tests as listed for the exposed terminal relays, EXCEPT do it by wire color. Use the ohmmeter to confirm there is continuity between the following pairs of wires: Sky Blue and Black, Light Green and Black, Blue and Black, and finally, Green and Black. Next use the ohmmeter to confirm there is NO continuity between Blue and Red, as well as between Green and Red.
• For models with 2 round-bodied relays (like the 75-100 hp 4-strokes), all testing occurs with a battery connected, so skip this and go on to the next step.
4. If the initial relay continuity tests pan out, you still need to check how the relay functions once power is applied to it. Remember, this is how a relay works, it throws an internal switch, changing a condition of non-continuity to continuity once power is applied to another portion of the relay. Connect a

TRIM AND TILT

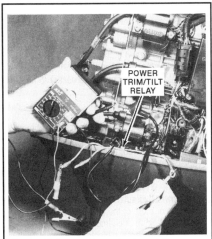

Fig. 11 Using an ohmmeter to test the power trim/tilt relay

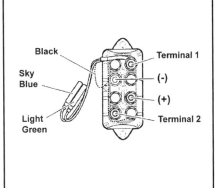

Fig. 12 Terminal identification on rectangular, exposed terminal PTT relays

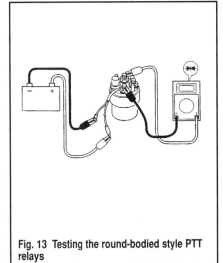

Fig. 13 Testing the round-bodied style PTT relays

12-volt battery to the proper terminals of the relay and use the ohmmeter to check for continuity changes. Again, terminal or wire connections vary slightly by relay type as follows:

• For rectangular, exposed terminal type relays, if there are 3 wires coming off the harness make the battery connections on the wires only (if not, make the Negative battery connection directly to the Negative terminal on the relay). First connect the positive battery lead to the Light Green wire and the Negative battery lead to the Black wire, then use the ohmmeter to check between Terminal 1 and the Positive relay terminal, there should now be continuity when there was none before. Move the Positive battery lead from the Light Green wire to the Sky Blue wire, then check for continuity between relay Terminal 2 and the Positive relay terminal. Again, there should now be continuity when there was none before.

• For rectangular, non-exposed terminal type relays (such as those used on the 15 hp 4-stroke), connect the positive battery lead to the Light Green wire and the Negative battery lead to the Black wire, then use the ohmmeter to check between the Green and Red wires, there should now be continuity when there was none before. Move the Positive battery lead from the Light Green wire to the Sky Blue wire, then check for continuity between the Blue and Red Wires. Again, there should now be continuity when there was none before.

• For models with 2 round-bodied relays (like the 75-100 hp 4-strokes), connect an ohmmeter across the relay terminals as shown in the accompanying illustration. Next, connect the negative lead from the battery to the Black relay lead and the positive lead from the battery to the Sky Blue or Light Green lead on each relay. The ohmmeter should ONLY show continuity with the battery connected as noted.

5. If the relay should fail to show continuity when noted or should continuity on the switched circuits not open after power is removed, the relay must be replaced.

Tilt Switch

TESTING

♦ See Figures 14 and 15

There is normally one power tilt switch located at the top of the remote control handle. The harness from the switch is routed down the handle to the base of the control box, and then into the box through a hole. The control box cover must be removed to gain access to the quick disconnect fittings and permit testing of the switch.

An auxiliary tilt switch is also normally found mounted on the exterior surface of the starboard side lower cowling pan. This switch is convenient for performing tests on the tilt unit, when the need to observe the unit in operation is required.

In either case, the tilt switch is a simple, 3 wire, 3 position switch and testing is relatively straightforward. There should be no continuity between any of the leads when the switch is released, however there should be continuity between 2 of the leads when the switch is pressed in the UP position, and there should be continuity between a different pair of leads when the switch is pressed in the DOWN position.

1. Disconnect the three leads for the switch: the Red, Light Green, and Sky Blue, leads either at their quick disconnect fittings or using the single switch connector (depending on the type of switch).

2. Connect the Red ohmmeter lead to the Red switch lead, and keep this connection for the following two resistance tests.

3. Connect the Black meter lead to the Sky Blue switch lead, then depress and hold the upper portion of the toggle switch (the UP button). The meter should indicate continuity. Release the switch to the neutral position. The meter should indicate no continuity. Depress the lower portion of the switch. The meter should still indicate no continuity.

4. With the Red meter lead still connected to the Red switch lead, move the Black meter lead to make contact with the Light Green switch lead. Depress the lower portion of the toggle switch. The meter should indicate continuity. Release the switch to the neutral position. The meter should indicate no continuity. Depress the upper portion of the switch. The meter should still indicate no continuity.

5. If the switch fails one or both resistance tests, replace the switch.

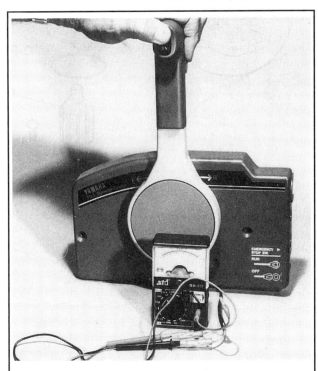

Fig. 14 Using an ohmmeter to test the power trim/tilt switch

8-8 TRIM AND TILT

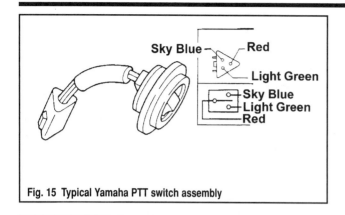

Fig. 15 Typical Yamaha PTT switch assembly

Tilt Motor

TESTING

♦ See Figure 16

To run the motor without a fully hydraulic load, set the manual release valve to the manual tilt position. This means make sure it is rotated approximately three full turns from the power tilt position, as evidenced by the embossed words and directional arrow on the housing (remember Allen head valves are normally left-hand thread and turn clockwise to loosen, while slotted head valves are normally regular, right-hand thread and turn counterclockwise to loosen).

1. Disconnect the 2 leads from the PTT motor at their quick disconnect fittings (the leads are normally Blue and Green on these Yamaha motors). Momentarily make contact with the two disconnected leads to the posts of a fully charged battery. Make the contact only as long as necessary to hear the electric motor rotating.

2. Reverse the leads on the battery posts and again listen for the sound of the motor rotating (when the leads are reversed the motor operates in the opposite direction). The motor should rotate with the leads making contact with the battery in either direction.

3. If the motor fails to operate in one or both directions, remove the electric motor for service or replacement as follows:

   a. First, place a suitable container under the unit to catch the hydraulic fluid as it drains.

   b. Next, disconnect the two lower hydraulic lines from the bottom of the housing (if equipped) or remove one or more of the valves, then remove the fill plug from the reservoir. Permit the fluid to drain into the container.

   c. After the fluid has drained, disconnect the electrical leads at the harness plug, and then remove the electric motor through the attaching hardware.

■ **This motor is very similar in construction and operation to a cranking motor. The arrangement of the brushes differs in order to allow the trim/tilt motor to operate in opposite directions, but otherwise, it is almost identical to the cranking motor.**

   d. Check armature commutator continuity, brush condition etc, in the same manner as the Yamaha electric starter motors. Repair or replace motor components, as necessary.

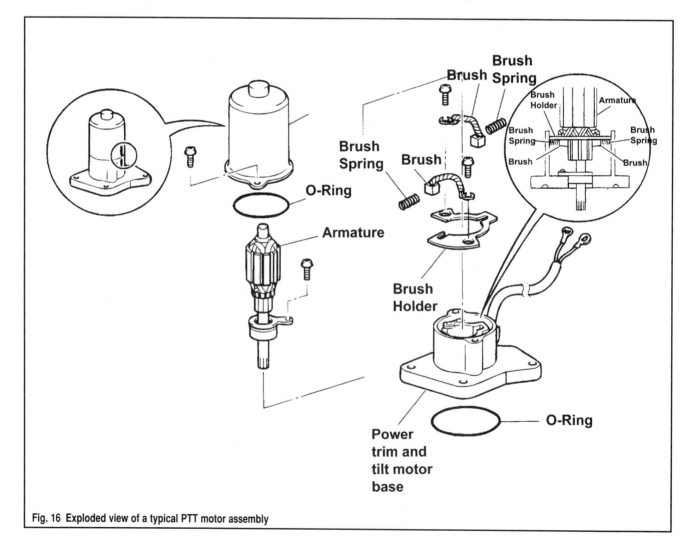

Fig. 16 Exploded view of a typical PTT motor assembly

TRIM AND TILT 8-9

Trim Sensor

TESTING

◆ See Figure 17

Most models are equipped with a trim sensor attached to the tilt bracket. The job of the sensor is to provide feedback to the operator concerning the degree of engine tilt. The sensor consists of one or two potentiometer circuits. A potentiometer is basically a device whose electrical resistance varies with mechanical movement. In this case, the sensor contains a small arm which moves with the up or down movement of the outboard.

Sensor testing is relatively straightforward and is conducted strictly with an ohmmeter. Start by disconnecting the sensor wiring and identifying the appropriate terminals for testing (this varies slightly by model). Then connect and ohmmeter and move the sensor arm through the full range of movement (either by hand with the sensor removed from the outboard or by tilting the outboard itself). Watch the meter for acceptable resistance within the specified range (again, this varies slightly by model) and make sure there are not sudden opens or shorts where continuity drops or spikes. All changes to resistance should be relatively linear.

■ **Remember that resistance specifications will vary slightly by meter and even more by temperature. All specs given are taken with a high quality Digital Ohmmeter, at an ambient/component temperature of approximately 68°F (20°C).**

When testing the trim sensor, look for values within the following ranges, based on model:
- For 40/45/50 hp 4-cylinder 4-strokes through 2000, first check resistance across the Pink and Black wires, it should vary from about 360-540 ohms through the range of motion. Then check resistance across the Orange and Black wires, it should vary from about 800-1200 ohms through the range of motion.
- For 2001 and later 50/50J/60/60J hp 4-strokes, check resistance across the Pink and Black wires, it should vary from about 10-278 ohms through the range of motion.
- For 30/40 hp 3-cylinder 4-strokes and 75-100 hp 4-cylinder 4-strokes, first check resistance across the Pink and Black wires, it should vary from about 582-873 ohms through the range of motion. Then check resistance across the Orange and Black wires, it should vary from about 800-1200 ohms through the range of motion.

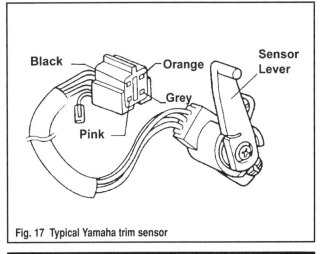

Fig. 17 Typical Yamaha trim sensor

PTT Assembly

REMOVAL & INSTALLATION

◆ See Figures 18 thru 22

1. Tilt the outboard to the fully upward position then secure in place using the tilt lever stop (trailering lock).
2. Tag and disconnect the wiring for the PTT motor. Also, note the mounting point and disconnect the ground lead(s) from the unit. There are often 2 ground leads mounted to the bottom of the assembly and, on some models the bolts retaining the ground leads also secure an anode to the bottom of the assembly.

■ **It will be necessary to cut one or more wire ties in order to reposition and free the necessary wiring. On some models there may be as many as 5 wire ties.**

3. On 15 hp motors, be sure to remove the bolt and washer threaded upward into the bottom of the PTT assembly.

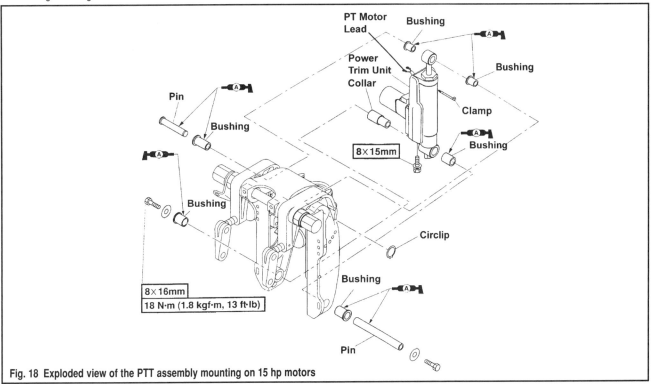

Fig. 18 Exploded view of the PTT assembly mounting on 15 hp motors

8-10 TRIM AND TILT

4. If equipped with a tilt rod assembly (such as most 40/50 hp models), note the location (through which holes it passes in the bracket), then remove the rod.

■ **Removal of some pins will require the removal of the bushings.**

5. Remove the pivot components securing the PTT assembly at the bottom and top of the unit. The components vary slightly by model as follows:

• For 15 hp motors, start at the top removing the circlip and pin securing the top of the tilt rod. Then move to the bottom. Loosen and remove the bolt and washer from each side of the lower clamp bracket. Support the PTT unit as you withdraw the lower pin, then remove the unit from the motor.

• For standard thrust 25 hp (F25) motors, start at the bottom, removing the bolt and washer from each side of the clamp bracket, then carefully remove the shaft (pin). Then move to the top to locate a single circlip on the starboard side of the top attaching point for the tilt ram, remove the clip, then withdraw the shaft pin to free the PTT unit (obviously you'll need to support the unit as you pull this pin).

• For 30/40 hp motors and high-thrust 25 hp (T25) motors, start at the bottom, removing the bolt and washer from each side of the clamp bracket, then carefully remove the shaft (pin). Then move to the top of the tilt ram where there are 2 circlips securing the shaft pin. Remove both (to prevent losing them), then support the unit as you carefully drive the pin free of the tilt ram.

• For 40/45/50 hp 4-cylinder motors through 2000, start at the bottom, removing the nut and plane washer from the port side of the clamp bracket, then carefully withdraw the stud bolt along with the large nut from the starboard side of the clamp bracket. Next, move to the top of the tilt ram and remove the 1 or 2 circlips securing the shaft pin. Support the unit as you carefully pull the pin free of the tilt ram.

• For 2001 and later 50J/50 and 60J/60 hp 4-cylinder motors, start at the top, where there are 2 circlips securing the shaft pin. Remove both (to prevent losing them), then drive the pin free of the tilt ram. Next, move to the bottom. Remove the bolt and washer from each side of the clamp bracket, then support the PTT unit while you carefully remove the pivot shaft.

• For 75-100 hp 4-cylinder motors, start at the top. Remove the circlip and withdraw the upper shock pin. Then move to the bottom and remove the 2 circlips securing the lower shock mount pin. Support the PTT unit as you carefully remove the mount pin.

6. If necessary remove the bushing(s) from the PTT unit and/or clamp bracket for installation purposes. Grease the replacement bushings to help ease installation.

7. Before installing the unit apply a light coating of marine grade grease to the inner diameter of each bushing.

8. Installation is essentially the reverse of the removal process. In most cases it really doesn't matter whether you remove or install the upper or lower shaft pin first, just make sure you properly secure the pins using the circlips or bolts, as applicable. For models that utilize bolts or nuts to secure the lower shaft mounting pin, apply a light coating of Loctite®242 or equivalent threadlock to the bolts before installation.

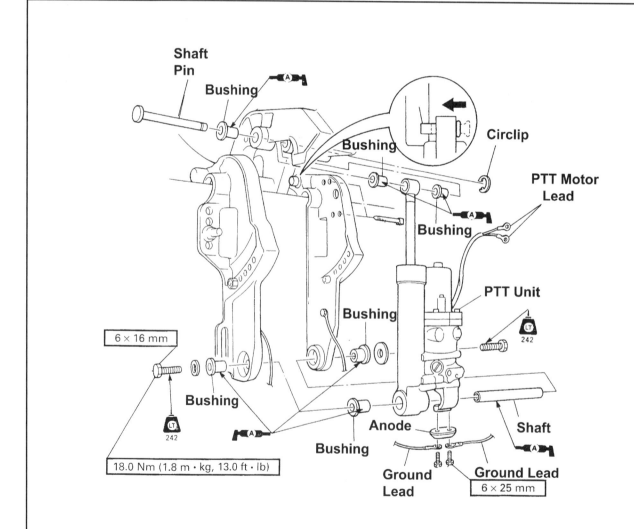

Fig. 19 Exploded view of the PTT assembly mounting on standard thrust 25 hp (F25) motors

TRIM AND TILT

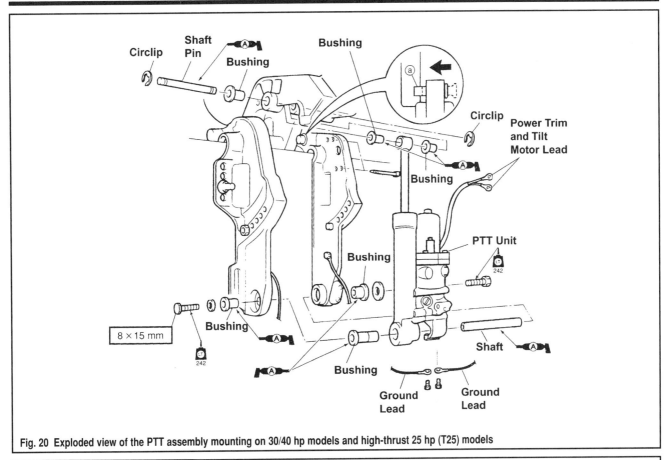

Fig. 20 Exploded view of the PTT assembly mounting on 30/40 hp models and high-thrust 25 hp (T25) models

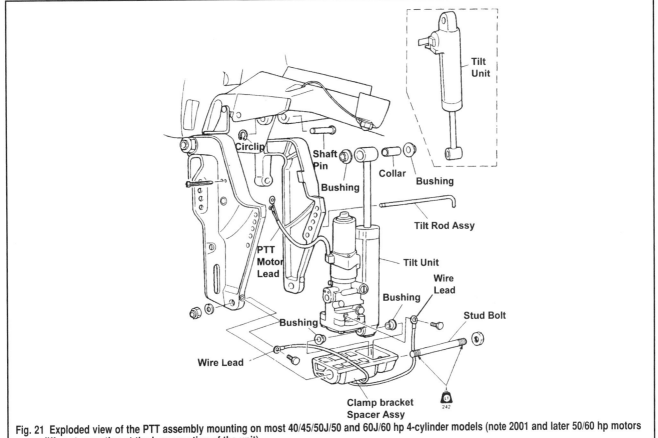

Fig. 21 Exploded view of the PTT assembly mounting on most 40/45/50J/50 and 60J/60 hp 4-cylinder models (note 2001 and later 50/60 hp motors use a different mounting at the lower portion of the unit)

8-12 TRIM AND TILT

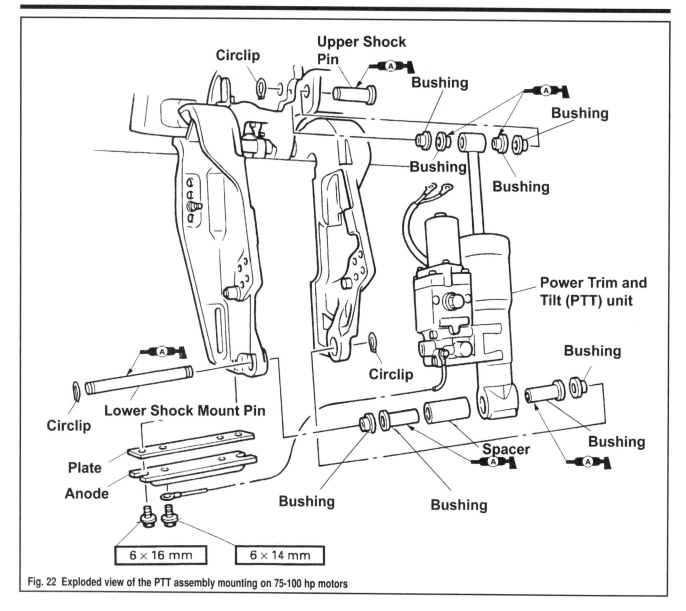

Fig. 22 Exploded view of the PTT assembly mounting on 75-100 hp motors

COMPONENT REPLACEMENT

◆ See Figures 23 thru 27

Service procedures for the power tilt system are confined to basic maintenance and some minor rebuilding. Rebuilding the unit involves removal of the end cap, removing the piston and replacing the O-rings.

A spanner wrench is required to remove the end cap. Even with the tool, removal of the end cap is not a simple task. The elements, especially if the unit has been used in a salt water atmosphere, will have their corrosive affect on the threads. The attempt with the special tool to break the end cap loose may very likely elongate the two holes provided for the tool. Once the holes are damaged, all hope of removing the end are lost. The only solution in such a case is to replace the unit.

If disassembly is required, work slowly keeping close track of the positioning for all components. Refer to the accompanying exploded views for help in identifying components and positions, but keep in mind that pump, cylinder and valve assemblies will vary greatly.

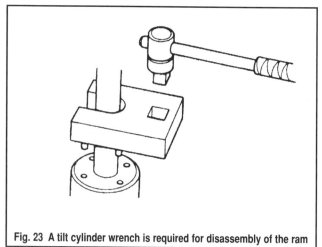

Fig. 23 A tilt cylinder wrench is required for disassembly of the ram

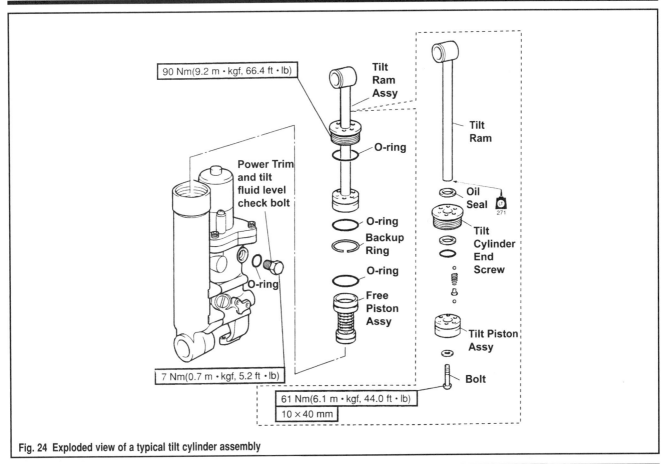

Fig. 24 Exploded view of a typical tilt cylinder assembly

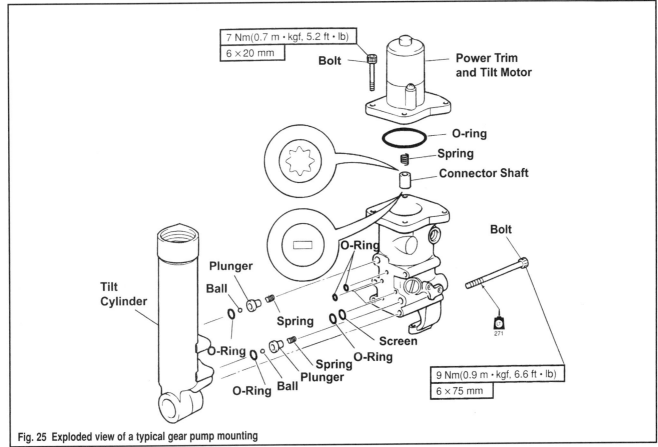

Fig. 25 Exploded view of a typical gear pump mounting

8-14 TRIM AND TILT

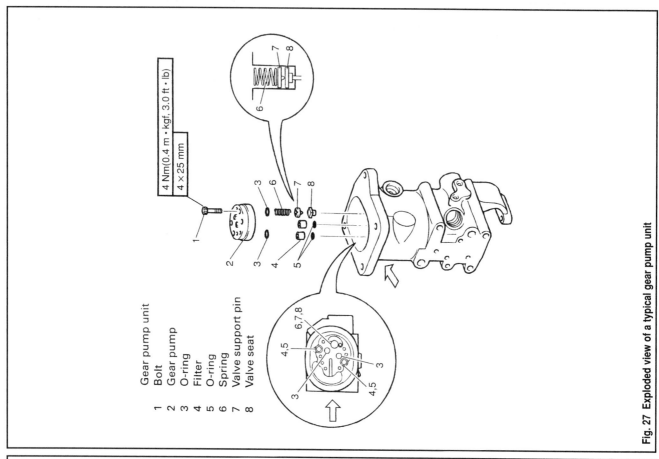

Fig. 27 Exploded view of a typical gear pump unit

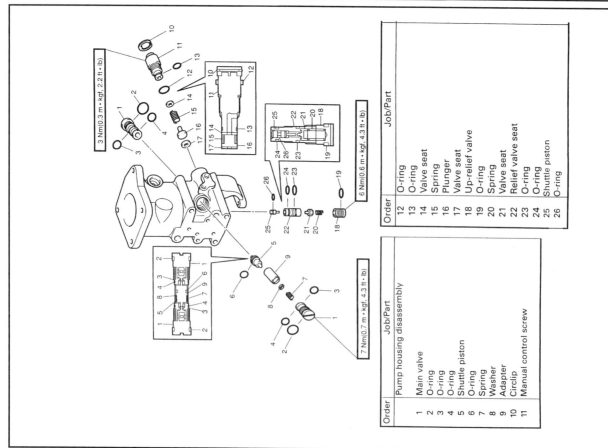

Fig. 26 Exploded view of a typical pump housing

TRIM AND TILT 8-15

YAMAHA LARGE MOTOR POWER TRIM/TILT SYSTEM

Description & Operation

◆ See Figures 28 and 29

■ The large motor PTT system (with single tilt rod and dual trim rods) is normally found on all 115 hp and larger Yamaha motors as well as the V6 Mercury/Mariner models).

The multi-cylinder power trim/tilt system consists of a housing with: an electric motor, gear driven hydraulic pump, hydraulic reservoir, two trim cylinders, and one tilt cylinder. Like the tilt cylinder on the single ram systems, the large tilt cylinder performs a double function as tilt ram and also as a shock absorber, should the lower unit strike an underwater object while the boat is underway.

The necessary valves, check valves, relief valves, and hydraulic passageways are incorporated internally and externally for efficient operation. A manual release valve is provided, on the side of the housing (usually to the port side of late-model motors), to permit the outboard unit to be raised or lowered should the battery fail to provide the necessary current to the electric motor or if a malfunction should occur in the hydraulic system.

The gear driven pump operates in much the same manner as an oil circulation pump installed on motor vehicles. However, the gears may revolve in either direction, depending on the desired direction of cylinder movement, up or down. One side of the pump is considered the suction side, and the other the pressure side, when the gears rotate in a given direction. These sides are reversed, the suction side becomes the pressure side and the pressure side becomes the suction side when gear movement is changed to the opposite direction.

Reversing of pump direction is achieved simply by swapping polarity on the pump wiring. When power is applied to one side, the pump will operate in a given direction. However, if you provide power to the other motor wire, the pump will operate in the opposite direction.

Depending on the model, one or two relays for the electric motor are located at the bottom cowling pan, where they are fairly well protected from moisture.

■ Yamaha engineers have been constantly working to improve the operational performance of the trim and tilt system installed on their outboard units. Therefore, many of the units will appear to be quite similar but internal hydraulic passages and check valves have been changed as well as the external routing of hydraulic lines.

The basic principles described apply to all multi-cylinder power trim/tilt units, but the specific location and number of components, may vary, and the routing of hydraulic lines may not be exactly as viewed on the unit being serviced. As such, it is critical that you note the placement of all internal check valves or seals if the unit is being disassembled for overhaul.

TRIM UP

◆ See Figure 30

When the up portion of the trim/tilt switch on the remote control handle (or on the cowling) is depressed, the up circuit, through the relay, is closed and the electric motor rotates (usually in a clockwise direction).

The pump sucks in fluid from the reservoir through a check valve and forces the fluid out the pressure side of the pump.

From the pump, the pressurized fluid passes through a series of valves to the lower chamber of both trim cylinders and the pistons are extended. The outboard unit rises. The fluid in the upper chamber of the pistons is routed back to the reservoir as the piston is extended. When the desired position for trim is obtained and the switch is released, the outboard is held stationary.

On most motors, a trim sender unit is installed on the port side clamp bracket. This unit sends a signal to an indicator on the control panel to advise the helmsperson of the relative position of the outboard unit.

If the switch is not released when the trim cylinders are fully extended, the tilt cylinder will continue to move the outboard unit upward until the fully tilted position is reached and there is no more physical travel left in the system.

TRIM DOWN

When the down portion of the trim/tilt switch on the remote control handle, or the auxiliary switch, is depressed, the down circuit, through the relay, is closed and the electric motor rotates in the opposite direction (therefore normally counterclockwise). The pressure side of the pump now becomes the suction side and the original suction side becomes the pressure side.

Fluid is forced through a series of check valves to the upper chamber of each trim cylinder and the pistons begin to retract, moving the outboard unit downward. Fluid from the lower chamber of each trim cylinder is routed back through the pump and a relief valve to the reservoir.

TILT UP

The first phase of the tilt up movement is the same as for the trim up function.

When the pistons of the trim cylinders are fully extended, fluid is forced through the system to the lower chamber of the tilt cylinder. Pressure increases in the lower chamber and the piston is extended, raising the outboard unit.

As fluid pressure in the upper chamber of the tilt cylinder increases, the fluid is routed through check valves back to the pump and the reservoir.

When the tilt piston is fully extended, fluid pressure in the lower chamber of the trim cylinders increases. This increase in pressure opens an up relief valve and the fluid is routed to the reservoir.

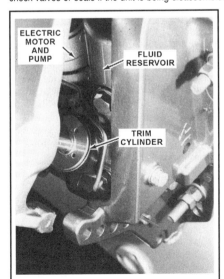

Fig. 28 Power trim/tilt system installed on a typical Yamaha

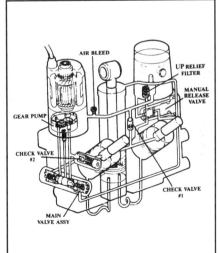

Fig. 29 Cutaway view of a typical Yamaha PTT system

Fig. 30 Close-up of the auxiliary PTT switch found on the cowling of most outboards

8-16 TRIM AND TILT

■ When the tilt piston becomes fully extended, the outboard is in the full up position. The sound of the electric motor and the pump will have a noticeable change. The switch on the remote control handle should be released immediately.

If the switch is not released, the motor will continue to rotate, the pump will continue to pump, but the up relief valve will open and the pressurized fluid will be routed to the reservoir.

If the boat is underway when the tilt cylinder is extended and powerhead rpm is increased beyond a very slow speed, the forward thrust of the propeller will increase the pressure on the tilt piston. This increase in pressure will cause the up relief valve to open and the outboard unit will begin a downward movement.

TILT DOWN

When the down portion of the trim/tilt switch on the remote control handle, or the auxiliary trim/tilt switch, is depressed, the down circuit is closed through the relay and the electric motor rotates in the opposite direction it would when the up switch is pressed (usually counterclockwise), as in the case of Trim Down.

The hydraulic pump sucks fluid from the reservoir. Fluid, under pressure is then routed to the upper chamber of the tilt cylinder and the piston begins to retract and the outboard unit moves downward.

Fluid in the lower chamber of the tilt cylinder is routed through the lower chamber of each trim cylinder and then back to the pump. When the outboard unit makes physical contact with the ends of the trim cylinders, the trim cylinders also retract until the outboard is in the full down position.

MANUAL OPERATION

◆ See Figure 7

※※ WARNING

If outboard unit is in the up position when the manual release valve is opened, the outboard will drop to the full down position rapidly. Therefore, ensure all persons stand clear.

The outboard unit may be raised or lowered manually should the battery fail to provide sufficient current to operate the electric motor or should electric/hydraulic components of the PTT system fail for any other reason. A manual relief valve is provided to permit manual operation.

This manual relief valve is located on the lower end of the gear pump beneath the electric motor either facing aft (toward the motor/gearcase) or facing sideways (so that it is accessed through one of the clamp brackets). On some older Yamaha motors the valve faced aft and often contained left-hand threads, meaning it is OPENED or LOOSENED by turning CLOCKWISE. However, on MOST late-model Yamaha outboards (where the valve faces to one side, and is accessed through a clamp bracket), the valve is equipped normal right-hand threads, meaning it is OPENED or LOOSENED by turning COUNTERCLOCKWISE.

Either way, opening the valve releases pressure in both the upper and lower cylinder chambers. With a complete loss of pressure, the piston may be moved up or down in the cylinder without resistance.

After the outboard unit has been moved to the desired position, the valve should be rotated in the opposite direction to close the valve and lock the outboard against movement.

SHOCK ABSORBER ACTION

The lower end of the tilt piston is capped with a free piston. This free piston normally moves up and down with the tilt piston - goes along for the ride.

In the event the outboard lower unit should strike an underwater object while the boat is underway, the tilt piston would be suddenly and forcibly extended, moved upward.

The free piston also moves upward but at a much slower rate than the tilt piston. The action of the tilt piston separating from the free piston causes two actions. First, the hydraulic fluid in the upper chamber above the piston is compressed and pressure builds in this area. Second, a vacuum is formed in the area between the tilt piston and the free piston.

This vacuum in the area between the two pistons sucks fluid from the upper chamber. The fluid fills the area slowly and the shock of the lower unit striking the object is absorbed. After the object has been passed the weight of the outboard unit tends to retract the piston. The fluid between the tilt piston and the free piston is compressed and forced through check valves to the reservoir until the free piston reaches its original neutral position.

Maintenance

CHECKING HYDRAULIC FLUID LEVEL

◆ See Figures 31 and 32

■ If one of the hydraulic valves has been removed, or the system opened for any other reason, the system must be purged of any trapped air. To purge (bleed) the system, please refer to Purging Air From The System in this section.

The fluid in the power trim/tilt reservoir should be checked periodically to ensure it is full and is not contaminated. To check the fluid, tilt the motor upward to the full tilt position, then manually engage the tilt support for safety and to prevent damage.

In order to keep contaminants from entering the system when the plug is removed, clean the area around the fill plug (found on the reservoir, just below the motor on these models). Loosen and remove the plug using a suitable socket or wrench and make a visual inspection of the fluid. It should seem clear and not milky. The level is proper if, with the motor at full tilt, the level is even with the bottom of the filler plug hole.

■ Yamaha or Yamalube power trim and tilt fluid is a highly refined hydraulic fluid. This product has a high detergent content and additives to keep seals pliable. A high grade automatic transmission fluid, Dexron® II may also be used if the Yamalube fluid is not available.

PURGING AIR FROM THE SYSTEM

◆ See Figure 33

Air must be purged (bled) from the system if any of the following conditions exist:
   a. Erratic motion is felt during operation of the system.
   b. Fluid has been contaminated with moisture or other foreign material.
   c. A component in the system has been replaced.
   d. The system has been opened for any reason.
   e. The outboard, when been left in the full up position, lowers slowly over a period of time.

■ When the reservoir is full to a normal level it should be even with the bottom of the plug threads when the outboard is in the fully upward, tilted position, meaning that when the outboard is lowered the level may be above the fill plug. One way to start refilling an empty or nearly empty system is to add fluid while the outboard is down, then open the manual release valve and manually lift the engine, drawing fluid into the system as the piston inside the tilt cylinder moves upward (as the ram is extended). Another method is to use is to refill the reservoir while the PTT unit is still completely removed from the outboard, and then use a battery with jumpers to cycle the motor and rams through their full motion of travel a couple of times, topping off the reservoir, as necessary.

1. Begin by making sure there is fluid inside the reservoir, even with the outboard in the normal running position (not tilted). Clean the area around the fill plug to prevent contaminants entering the system when the plug is loosened or removed. Slowly start to loosen the fill plug, if fluid starts to run (or spray) out, stop loosening the plug and retighten it until the outboard is tilted. However, if no fluid escapes, go ahead and remove the plug, then add fluid until it is level with the bottom of the screw threads. Lightly reinstall the reservoir plug.

2. Open the manual release valve (either by rotating the Allen head of the valve clockwise, or the slotted head counterclockwise, as applicable), rotating it slowly in the proper direction until it stops.

3. With the manual release valve open, slowly pull the outboard upward by hand, extending the tilt piston. As the tilt piston is extended, the piston in the cylinder will draw fluid into the cylinder, thus lowering the level in the chamber.

TRIM AND TILT 8-17

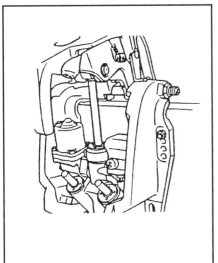

Fig. 31 To check fluid level, raise and lock the outboard to full tilt...

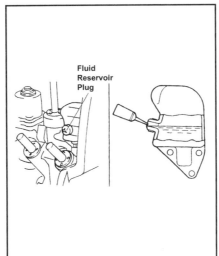

Fig. 32 ...then remove the reservoir plug and make sure the fluid level is even with the bottom of the bore

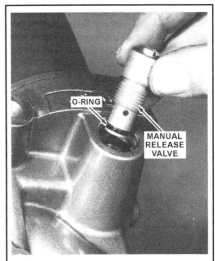

Fig. 33 A few manual relief valves used on early-model PTT system have left-hand threads

4. Carefully release the outboard, allowing it to lower by its own weight, then once the outboard is lowered fully, retighten the manual release valve.
5. Allow the PTT system fluid to settle for about 5 minutes.
6. Close the manual release valve.
7. Push the PTT switch (either on the remote or the side of the motor) allowing the system to move the motor to the fully upward position.
8. Push the tilt stop lever(s) to support the outboard.
9. Again, allow the PTT system fluid to settle for about 5 minutes.
10. Remove the reservoir cap and check the fluid level. Add fluid, as necessary until it is even with the bottom of the fill plug, then install and tighten the reservoir cap.
11. Repeat as necessary until the fluid remains at the correct level.

Troubleshooting

■ When moving through the listed troubleshooting procedures, always stop and check the system after each task. The problem may have been corrected, intentionally, or not.

PRELIMINARY INSPECTION

1. Check to be sure the manual release head is tightened snugly in the proper position.
2. Verify the hydraulic reservoir is filled with fluid. The level of fluid should reach to the lower edge of the fill opening. Replenish as required with Yamaha or Yamalube power trim and tilt fluid is a highly refined hydraulic fluid. This product has a high detergent content and additives to keep seals pliable. A high grade automatic transmission fluid, Dexron® II may also be used if the Yamalube fluid is not available.

✱✱ WARNING

The PTT system is pressurized! Unless instructed otherwise, do not remove fill screw unless outboard unit is raised to full up position. Tighten fill screw securely before lowering outboard.

3. Inspect the hydraulic system, lines and fittings, for leaks. If an external leak is discovered, correct the condition.
4. Purge air from the system, if there is any indication the hydraulic fluid contains air.
5. To check for air in the system, first check to be sure the manual release valve is snug in the power tilt position. Next, activate the up circuit and raise the outboard slightly with the trim cylinders. Now, exert a heavy, steady, downward force on the lower unit.
6. If the trim pistons retract into the trim cylinders more than about 1/8 in. (3.2mm) there is air in the system

7. Purge (bleed) air from the system.
8. If the system fails to hold the outboard unit in the full tilted (trailering) position, service the tilt cylinder.
9. If the system fails to hold the outboard unit in the desired trim position, service the trim cylinders.
10. If the hydraulic pump whines during operation, there is probably air in the system. Purge air from the system.
11. If the electric motor makes strange sounds or seems to be laboring, the electric motor may require service.

■ If a problem is encountered with the PTT system, it is important to determine, if possible, whether the malfunction is in the hydraulic system or in an electrical circuit.

SYMPTOM DIAGNOSIS

Symptoms of a Hydraulic Problem

The most common problem in the hydraulic system is failure of an O-ring to hold pressure.
- **Outboard Behaves Abnormally** - Ensure the battery is adequately charged; the fluid level in the reservoir reaches the lower edge of the fill opening, with the outboard fully raised; the hydraulic fluid is not contaminated (murky color); the system does not contain air; the unit does not physically bind somewhere due to an accident.
- **Outboard Fails to Trim Up or Down** - Ensure the unit has adequate fluid in the reservoir; manual release valve is tightened counterclockwise snugly to the power tilt position, approximately three full turns from the manual tilt position. Once cause may be hydraulic pump failure; O-rings in trim cylinders failing to hold pressure; trim cylinders damaged due to accident.
- **Outboard Trims/Tilts Up but Fails to Trim/Tilt Down** - Manual release valve is leaking; O-rings in trim cylinder or in the tilt cylinder failing to hold pressure; main valve has sticky or damaged shuttle piston; sticky or contaminated check valves; down relief valve has weak spring; damaged check ball or seat; or contamination is holding a valve open.
- **Outboard Trims/Tilts Down but Fails to Trim/Tilt Up** - Manual release valve is leaking; O-rings in trim cylinders or tilt cylinder failing to hold pressure; shuttle piston in main valve assembly is sticking; contaminated check valves; up relief valve has damaged seat or contamination holds the valve open.
- **Outboard Shudders When Shifted From One Gear to Another** - Hydraulic system contaminated with air or foreign matter; internal cylinder leaks; O-rings failing to hold pressure.
- **Outboard Fails to Hold Set Trim or Tilt Position** - O-rings in trim and/or tilt cylinder failing to hold pressure; check valves in tilt piston

8-18 TRIM AND TILT

contaminated requiring cleaning; external leak - fitting or part; manual release valve damaged; shuttle piston in main valve assembly sticking or contaminated check valve; up relief valve damaged or contaminated causing a slow leak.
- **Outboard Tilts Up When Unit In Reverse Gear** - Tilt piston has leaky absorber valve or metering valve; main valve assembly has sticky or damaged shuttle piston or leaky check valves; manual release valve is leaking; O-rings in tilt piston failing to hold pressure.
- **Outboard Makes Excessive Noise** - Hydraulic fluid level is low; fluid is contaminated with air.
- **Outboard Begins To Trail Out When Throttle Backed Off at High Speed** - Manual release valve not tightened snugly counterclockwise to power tilt position; O-rings in tilt cylinder failing to hold pressure.
- **Outboard Fails to Hold Trim Position When Unit Operating In Reverse** - Manual release valve not tightened snugly counterclockwise to power tilt position; O-rings in trim cylinders failing to hold pressure.
- **Outboard Moves With Jerky Motion** - System contaminated with air; internal leaks in cylinders.
- **Outboard Fails to Reach Full Down Position** - System contaminated with air or internal leaks in cylinders.

Symptoms of an Electrical Problem

If any of the following problems are encountered, troubleshoot the electrical system.
- Outboard trims up and down, but the electrical motor grinds.
- Outboard will not trim up or down.
- Outboard trims up, but will not trim down.
- Outboard trims down, but will not trim up.

HYDRAULIC TESTING

The health of the PTT unit's hydraulic system (check valves, pistons, O-rings, seals and motor) can be checked using a high-pressure hydraulic gauge (extremely high pressure on some systems, so pre-read the procedure) and a 12-volt battery with jumpers to run the motor. Although Yamaha does not specify as to whether or not the test MUST occur with the PTT unit removed from the outboard, all of their illustrations show it performed this way. In addition, the mounting point for gauges on most PTT units is the manual valve bore, which often cannot be accessed with the PTT unit installed. However, if access is possible, you'll still want to at least remove the outboard load from the unit as it is tested. To do this tilt and lock the outboard in the full tilt position, then disconnect the tilt ram from the outboard itself.

Inline 4-Cylinder Motors

◆ See Figures 34 and 35

For these motors you'll need an extremely high pressure Hydraulic Pressure Gauge (Yamaha #90890-06776, the Up Relief Valve #90890-06773 and the Down Relief Valve #90890-06774) or their equivalents. The up or down relief valve is installed into the manual release valve bore (depending upon the portion of the system being testing), then the gauge is attached to the valve.

When performing this test, attach a 12-volt battery to the PTT motor leads using jumpers. Attach the positive jumper to the Green motor lead and negative to the Blue motor lead in order to run the PTT motor downward. Reverse the connections (Green to negative and Blue to positive) in order to run the motor upward.

1. Run the motor to place the tilt and trim rams at the full UP position. Temporarily use the trailering bracket to secure the motor in position while installing the gauge.
2. Check and top off the hydraulic fluid, as necessary.
3. Install the Hydraulic Pressure Gauge to the Up Relief Valve, then tighten the gauge to 6.5 ft. lbs. (9 Nm).

■ In the next step, minimize the amount of fluid lost and the amount of air that might enter the system by having the gauge and relief valve attachment ready to install AS SOON AS the manual release valve is removed from its bore.

4. Remove the circlip securing the manual release valve in its bore, then loosen and remove the valve. Quickly install the Up Relief Valve and Pressure Gauge assembly, then tighten them to 2.2-2.9 ft. lbs. (3-4 Nm).

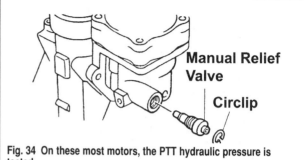

Fig. 34 On these most motors, the PTT hydraulic pressure is tested...

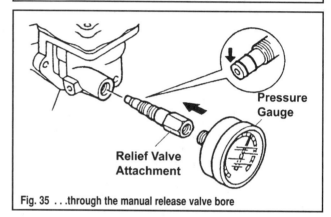

Fig. 35 ...through the manual release valve bore

5. Connect a 12-volt battery to the PTT motor using jumpers in order to run the trim and tilt rams to the fully downward position, then reverse the motor leads and run the rams to the fully extended (upward) position and observe hydraulic pressure, it should be 1421-1711 psi (9800-11,800 kPa / 9.8-11.8 mPa) for 115 hp motors or 1450-1740 psi (10,000-12,000 kPa / 10-12 mPa) for 150 hp motors.
6. Lower the PTT motor partially from the full up position, then carefully release pressure and remove the gauge assembly. Temporarily reinstall the manual release valve to minimize the amount of fluid loss. Meanwhile remove the Gauge from the Up Relief Valve attachment and connect it to the Down Relief Valve attachment, again, tighten the gauge to 6.5 ft. lbs. (9 Nm).
7. Remove the manual release valve from the bore again and quickly install the Down Relief Valve and Pressure Gauge assembly, then tighten them to 2.2-2.9 ft. lbs. (3-4 Nm).
8. Run the PTT motor to the full up position, then check and top off the fluid reservoir, as necessary. If too much fluid was lost, bleed the system of air by following the steps for Purging Air From The System, located earlier in this section.
9. Connect a 12-volt battery to the PTT motor using jumpers in order to run the trim and tilt rams to the fully downward position and observe hydraulic pressure, it should be 856-1276 psi (5900-8,800 kPa / 5.9-8.8 mPa) for 115 hp motors or 870-1305 psi (6,000-9,000 kPa / 6-9 mPa) for 150 hp motors.
10. Run the PTT assembly back to the fully extended position and support using the trailering bracket, then carefully release pressure and remove the gauge assembly. Quickly install the manual release valve and secure using the circlip.
11. Check the fluid level and bleed the system, as necessary.

3.3L V6 Motors

◆ See Figures 36 and 37

For these motors you'll need an extremely high pressure Hydraulic Pressure Gauge (Yamaha #YB-06580 or Mercury kit #91-52915A6). The gauge will install in place of one of the external hydraulic lines either at the motor connection (when testing downward pressure) or at the fitting on the base of the assembly (when testing upward pressure).

When performing this test, attach a 12-volt battery to the PTT motor leads using jumpers. Attach the positive jumper to the Green motor lead and negative to the Blue motor lead in order to run the PTT motor downward. Reverse the connections (Green to negative and Blue to positive) in order to run the motor upward.

TRIM AND TILT 8-19

1. Run the motor to place the tilt and trim rams at the full UP position. Temporarily use the trailering bracket to secure the outboard while installing the gauge.
2. Place a small rag or suitable container under the fittings to catch any spilled hydraulic fluid.
3. Loosen both of the hydraulic fittings for the line that runs from the bottom of the motor to the top of the tilt cylinder. Have the pressure gauge ready to install, then disconnect the line from the motor and install the pressure gauge in its place. Position the disconnected line end in a plastic bag or a small rag to catch any fluid which might leak out.
4. Connect a 12-volt battery to the PTT motor using jumpers in order to run the trim and tilt rams to the fully downward position and observe hydraulic pressure, it should be 972-1262 psi (6700-8700 kPa / 6.7-8.7 mPa).
5. Reverse the battery connections and run the PTT cylinders to the full up (extended) positions.
6. Remove the pressure gauge and reconnect the hydraulic line to the motor, then carefully retighten both hydraulic fittings to 133 inch. lbs./11 ft. lbs. (15 Nm).
7. Reconnect the battery jumpers to run the PTT cylinders to the full down (retracted) positions.
8. Loosen both of the hydraulic fittings for the line that runs from the bottom of the tilt cylinder to the bottom of the PTT assembly. Have the pressure gauge ready to install, then disconnect the line from the PTT assembly (not the bottom of the tilt cylinder) and install the pressure gauge in its place. Position the disconnected line end in a plastic bag or a small rag to catch any fluid which might leak out.
9. Connect a 12-volt battery to the PTT motor using jumpers in order to run the trim and tilt rams to the fully upward position and observe hydraulic pressure, it should be 1769-2059 psi (12,200-14,200 kPa / 12.2-14.2 mPa).

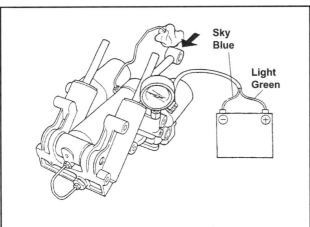

Fig. 36 Testing the PTT unit on a 3.3L motor for downward pressure

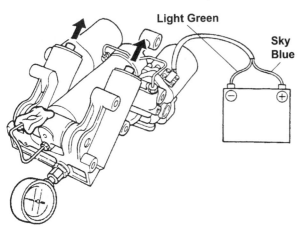

Fig. 37 Checking upward hydraulic pressure on a PTT unit for a 3.3L motor

10. After noting the gauge reading, reverse the battery connection and run the PTT rams back to the fully downward (retracted position).
11. Remove the pressure gauge and reconnect the hydraulic line to the PTT assembly, then carefully retighten both hydraulic fittings to 133 inch. lbs./11 ft. lbs. (15 Nm).
12. Run the PTT rams to the full upward position again, then remove the reservoir plug to check and possibly, top off the system, as necessary. If necessary, bleed the PTT system of air.

Trim/Tilt Relay

TESTING

◆ See Figures 38 and 39

The PTT motor runs either clockwise or counterclockwise depending upon which direction you want the motor to move and also depending upon how the power is applied to the motor's terminals. If you apply positive current to one terminal and negative to the other, the motor will run in one direction. If you switch and apply positive and negative to the opposite terminals, the motor will run in the other direction. Control of this circuit is achieved through one or more PTT relays. The job of the relay, like all relays, is to close certain switch circuits depending upon the signals received from the tilt switch on either the engine or the remote.

There are generally 2 major designs of PTT relays used by Yamaha motors, though you SHOULD only find the first on these motors. The first and most common relay used on late-model Yamahas is a rectangular piece that looks like it has 8 terminals and either a 2 or 3 wire harness that comes off one end (though the harness may use a single connector or individual wire bullets/terminals). Four of the visible terminals (identified by position) and the 2 or 3 wire harness are used during testing to with an ohmmeter (and a 12-volt battery) to determine if the relay is operating properly. For testing purposes, the terminal identification is slightly different on the 2 and 3 wire harness relays however, the 3-wire version should only be found on 2-strokes, so we'll only deal with the 2-wire version here.

The 2 wire harness rectangular relay is normally found on 4-stroke motors (both inline 4 cylinder and V6 models) as well as on HPDI V6 motors (including 3.1L and 3.3L models). On these relays the exposed terminal on the wire harness end of the relay is also on the same side as the wiring harness itself and is referred to in the testing procedure as Terminal 1. This terminal is also easily identified since it is RIGHT next to the Positive relay terminal (the terminal to which the red power wire was attached before removal). At the far end of the relay (not diagonally on this type of relay), on the same side of the relay as Terminal 1 is Terminal 2. The other way to identify Terminal 2 is that it is diagonally adjacent to the relay Negative terminal (the terminal to which the black ground wire was connected before removal). For more details, please see the accompanying illustration.

The second type of relay found on Yamahas is a round-bodied model. However, this type of relay should NOT be found on 4-stroke motors, so we do not cover it here.

■ On all relays, testing normally occurs with the relay completely removed from the powerhead, OR at least with all wiring disconnected (even if the relay is still physically bolted in position).

1. If not done already, tag and disconnect the wiring from the PTT relay(s).
2. Obtain a 12-volt battery with jumper leads and an ohmmeter for continuity testing.
3. First check terminal-to-terminal continuity. Use the ohmmeter to confirm there is continuity between the following pairs of wires: Terminal 1 and the Negative terminal, Terminal 2 and the Negative terminal, The Sky Blue wire and the Black wire and finally, the Light Green wire and the Black wire. Next use the ohmmeter to confirm there is NO continuity between Terminal 1 and the Positive terminal, as well as between Terminal 2 and the Positive terminal.
4. If the initial relay continuity tests pan out, you still need to check how the relay functions once power is applied to it. Remember, this is how a relay works, it throws an internal switch, changing a condition of non-continuity to continuity once power is applied to another portion of the relay. Connect a 12-volt battery to the proper terminals of the relay and use the ohmmeter to check for continuity changes. If there are 3 wires coming off the harness make the battery connections on the wires only (if not, make the Negative

8-20 TRIM AND TILT

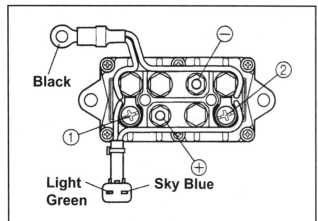

Fig. 38 2-wire PTT relay with Terminal IDs (Positive terminal may also be one to the right, negative may be one to the left, but in both cases they are usually bridged terminals)

battery connection directly to the Negative terminal on the relay or the black wire coming off the Negative terminal). First connect the positive battery lead to the Light Green wire and the Negative battery lead to the Black wire. Then use the ohmmeter to check between Terminal 1 and the Positive relay terminal, there should now be continuity when there was none before. Move the Positive battery lead from the Light Green wire to the Sky Blue wire, then check for continuity between relay Terminal 2 and the Positive relay terminal. Again, there should now be continuity when there was none before.

5. If the relay should fail to show continuity when noted or should continuity on the switched circuits not open after power is removed, the relay(s) must be replaced.

Power Trim/Tilt Switch

TESTING

◆ See Figure 40

There is normally one power tilt switch located at the top of the remote control handle. The harness from the switch is routed down the handle to the base of the control box, and then into the box through a hole. The control box cover must be removed to gain access to the quick disconnect fittings and permit testing of the switch.

An auxiliary tilt switch is also normally found mounted on the exterior surface of the starboard side lower cowling pan. This switch is convenient for performing tests on the tilt unit, when the need to observe the unit in operation is required.

In either case, the tilt switch is a simple, 3 wire, 3 position switch and testing is relatively straightforward. There should be no continuity between any of the leads when the switch is released, however there should be continuity between 2 of the leads when the switch is pressed in the UP position, and there should be continuity between a different pair of leads when the switch is pressed in the DOWN position.

1. Disconnect the three leads for the switch: the Red, Light Green, and Sky Blue, leads either at their quick disconnect fittings or using the single switch connector (depending on the type of switch).

2. Connect the Red ohmmeter lead to the Red switch lead, and keep this connection for the following two resistance tests.

3. Connect the Black meter lead to the Sky Blue switch lead, then depress and hold the upper portion of the toggle switch (the UP button). The meter should indicate continuity. Release the switch to the neutral position. The meter should indicate no continuity. Depress the lower portion of the switch. The meter should still indicate no continuity.

4. With the Red meter lead still connected to the Red switch lead, move the Black meter lead to make contact with the Light Green switch lead. Depress the lower portion of the toggle switch. The meter should indicate continuity. Release the switch to the neutral position. The meter should indicate no continuity. Depress the upper portion of the switch. The meter should still indicate no continuity.

5. If the switch fails one or both resistance tests, replace the switch.

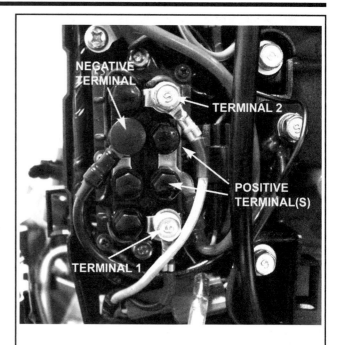

Fig. 39 The Red and Black wires help identify Positive and Negative terminals

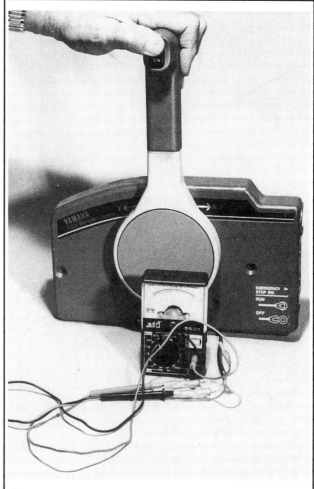

Fig. 40 Using an ohmmeter to test the power trim/tilt switch

TRIM AND TILT

Trim/Tilt Motor

TESTING

◆ See Figure 41

To run the motor without a fully hydraulic load, set the manual release valve to the manual tilt position. This means make sure it is rotated approximately three full turns from the power tilt position, as evidenced by the embossed words and directional arrow on the housing.

1. Disconnect the 2 leads from the PTT motor at their quick disconnect fittings (the leads are normally Blue and Green on these motors). Momentarily make contact with the two disconnected leads to the posts of a fully charged battery. Make the contact only as long as necessary to hear the electric motor rotating.

2. Reverse the leads on the battery posts and again listen for the sound of the motor rotating (when the leads are reversed the motor operates in the opposite direction). The motor should rotate with the leads making contact with the battery in either direction.

3. If the motor fails to operate in one or both directions, remove the electric motor for service or replacement as follows:

    a. First, place a suitable container under the unit to catch the hydraulic fluid as it drains.

    b. Next, disconnect the two lower hydraulic lines from the bottom of the housing (if equipped) or remove one or more of the valves, then remove the fill plug from the reservoir. Permit the fluid to drain into the container.

    c. After the fluid has drained, disconnect the electrical leads at the harness plug, and then remove the electric motor through the attaching hardware.

■ **This motor is very similar in construction and operation to a cranking motor. The arrangement of the brushes differs in order to allow the trim/tilt motor to operate in opposite directions, but otherwise, it is almost identical to the cranking motor.**

   d. Check armature commutator continuity, brush condition etc, in the same manner as the Yamaha electric starter motors. Repair or replace motor components, as necessary.

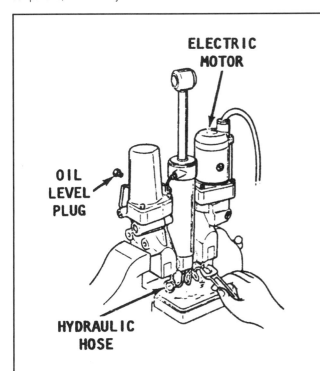

Fig. 41 Draining the hydraulic system prior to removing the electric motor for testing and/or servicing

Trim Sensor

ADJUSTMENT

◆ See Figure 42

The trim sensor is equipped with a lever that moves in relation to the position of the outboard. Since the sensor is a potentiometer (an electrical resistor whose value varies with the physical position of its lever) the exact positioning of the sensor on the outboard will can be set so the trim/tilt gauge displays properly. Adjustment will vary slightly from gauge-to-gauge, however all adjustment involves loosening the sensor mounting screws and rotating the housing slightly so the gauge reads properly.

Most trim tilt gauges are designed to only display the trim range of motor movement. This is the portion of movement which is controlled by the duel trim rods, not the tilt ram. In order to check adjustment compare the position of the motor at the top and bottom ends of the TRIM motion (not TILT) with the display on the gauge. If adjustment is necessary, loosen the screws and reposition the sensor slightly until the gauge properly reflects motor trim position.

■ **If during adjustment the gauge shows a sudden spike or jump in position, test the sensor for potential resistance problems.**

If the boat is equipped with Yamaha's digital meter, the following procedure can normally be used to ensure proper adjustment:

1. Raise the outboard to the full tilt position and support it in this position.
2. Loosen the 2 screws securing the trim sensor so it can move slightly.
3. Fully tilt the outboard to the full down position.
4. Set the main switch to the **ON** position to power the Yamaha digital meter.
5. With the outboard in the full down position, use a screwdriver and adjust the trim sensor so only one segment on the trim indicator on the digital meter goes on. Tighten the sensor screws while holding the outboard in this position.
6. Raise the outboard to the full tilt position, watching the display on the digital meter as the motor moves through the trim range. Make sure the trim sensor does not move from the original position. If necessary, tweak the adjustment to ensure proper display.

TESTING

◆ See Figure 42

Most models are equipped with a trim sensor attached to the tilt bracket. The job of the sensor is to provide feedback to the operator concerning the degree of engine tilt. The sensor consists of one or two potentiometer circuits. A potentiometer is basically a device whose electrical resistance varies with mechanical movement. In this case, the sensor contains a small arm which moves with the up or down movement of the outboard.

Sensor testing is relatively straightforward and is conducted strictly with an ohmmeter. Start by disconnecting the sensor wiring and identifying the

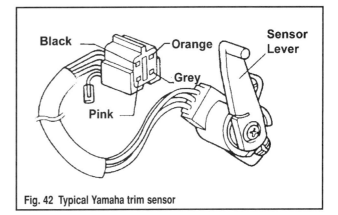

Fig. 42 Typical Yamaha trim sensor

appropriate terminals for testing (this varies slightly by model but it is usually the Pink and Black wires). Then connect an ohmmeter and move the sensor arm through the full range of movement (either by hand with the sensor removed from the outboard or by tilting the outboard itself). Watch the meter for acceptable resistance within the specified range (again, this varies slightly by model) and make sure there are not sudden opens or shorts where continuity drops or spikes. All changes to resistance should be relatively linear.

■ Remember that resistance specifications will vary slightly by meter and even more by temperature. All specs given are taken with a high quality Digital Ohmmeter, at an ambient/component temperature of approximately 68°F (20°C).

When testing the trim sensor, look for values within the following ranges, based on model:
• For all 115 hp motors, check resistance across the Pink and Black wires, it should vary from about 582-873 ohms through the range of motion. Next, if equipped check resistance across the Orange and Black wires, it should vary between 800-1200 ohms.
• For all 150 hp motors, check resistance across the Pink and Black wires, it should vary from about 9-11 ohm of resistance at one end of the range of motion to 238.8-378.8 ohms at the other end of the range.
• For all 3.3L V6 motors, check resistance across the Pink and Black wires, it should vary from about 9-11 ohms at one end of the sensor arm's travel to about 247.6-387.6 ohms at the other end of travel.

PTT Assembly (Dual Trim Ram)

REMOVAL & INSTALLATION

◆ See Figures 43 and 44

MODERATE

1. Tilt the outboard to the fully upward position then secure in place using the tilt lever stop (trailering lock).
2. Tag and disconnect the wiring for the PTT motor. Also, note the mounting point and disconnect the ground lead(s) from the unit. There are often 1 or 2 ground leads mounted to the bottom of the assembly and, on some models the bolts retaining the ground leads also help secure an anode to the bottom of the assembly.

■ It will be necessary to cut one or more wire ties in order to reposition and free the necessary wiring.

3. If equipped, remove the anode from the bottom of the PTT assembly.

■ Removal of some pins will require the removal of the bushings.

4. The PTT unit is secured by two pivot pin assemblies, one mounted on top and one of the bottom. The types of pins used vary greatly from model-to-model. Some are secured at one or both ends by a circlip, while others by a bolt or nut at either end. If there is only a bolt, circlip, nut etc at ONE end, it must be removed before the shaft can be withdrawn from the opposite end. If there are bolts, nuts, clips etc on both ends, then usually the pin can be withdrawn from either side. However, keep in mind that some models may actually use a captive nut on one end and obviously for those, the shaft must be withdrawn from the side of the captive nut. Start at one end (top or bottom of the PTT assembly usually doesn't matter) and remove the pivot shaft, then support the assembly as you remove the shaft from the other end.
5. If necessary remove the bushing(s) from the PTT unit and/or clamp bracket for installation purposes. Grease the replacement bushings to help ease installation.
6. Before installing the unit apply a light coating of marine grade grease to the inner diameter of each bushing.
7. Installation is essentially the reverse of the removal process. In most cases it really doesn't matter whether you remove or install the upper or lower shaft pin first, just make sure you properly secure the pins using the circlips or bolts, as applicable. For models that utilize bolts or nuts to secure the lower shaft mounting pin, apply a light coating of Loctite®241, Loctite®242, or an equivalent threadlock to the bolts before installation. Also, be sure to apply a light coat of marine grade grease to the friction points of the pins/bushing before installation.

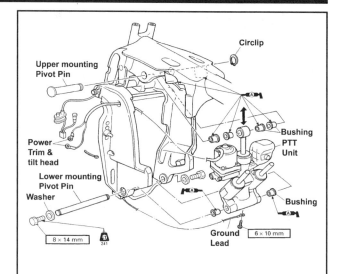

Fig. 43 Exploded view of a typical dual trim ram PTT unit mounting (normally found on 4-cylinder motors)

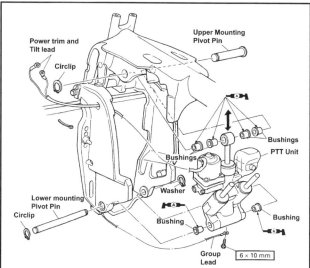

Fig. 44 Exploded view of another typical PTT unit (normally found on V6 motors, note the differences in pin mounting)

DISASSEMBLY

◆ See Figures 45 thru 50

The following procedures provide detailed instructions to service most accessible parts of the power trim/tilt system. Many of the instructions require the end cap of the tilt cylinder and/or the end caps of the trim cylinders to be removed. Potentially, this is not an easy task as salt or environmental exposure and corrosion often take their toll on motors over the years.

✱✱ WARNING

If the holes provided in the end cap for the special tool are damaged, the trim/tilt unit will probably have to be replaced.

If disassembly is required, work slowly keeping close track of the positioning for all components. Refer to the exploded views under Cleaning & Inspection for help in identifying components and positions, but keep in mind that pump, cylinder and valve assemblies will vary greatly.

TRIM AND TILT 8-23

■ Some of the accompanying illustrations were made with the trim/tilt unit on the work bench for photographic clarity. However, the work described, with the exception of the tilt cylinder removal, may be performed without removing the unit from the clamp bracket.

Do your BEST to keep parts identified as they are removed. Many parts may appear similar, but components should be installed into the same location from which they are removed.

1. If attempting to service the unit while installed on the outboard, power or manually, raise the outboard unit to the full up position and lock it in place.
2. Obtain a suitable container to receive the hydraulic fluid from the trim/tilt system.
3. If equipped, remove the two lines from the bottom of the trim/tilt housing in order to drain the hydraulic fluid. Remove the fill plug from the reservoir and allow the hydraulic fluid to drain into the container.

■ If there are no lines on the bottom of the PTT unit, fluid will drain through one or more of the valves when they are removed a little later in this procedure.

4. After the fluid has drained, replace the reservoir plug and loosely install the fluid line or valve body to assist in preventing contamination from entering the reservoir.
5. Obtain Yamaha special tool (#P/N YB-06548) or an equivalent spanner to unthread the tilt/trim rod caps. Observe the pattern of pins and the words etched on each side of the special tool. The side with the three pins and the word TRIM is used to remove/install the end cap on the trim cylinders. The side with the four pins and the word tilt is used for the end cap on the tilt cylinder.

■ On many late-model Yamahas, #YB-06175-1A (a taller, cylindrical looking tool) is used on the trim rods instead of the old style #YB-06548 which is then only used on the tilt cylinder. Mercury sells two tools, both of which look more like the Yamaha #YB-06548. Mercury provides #91-888867 for use on the tilt cylinder and #91-888869 for use on the trim cylinder.

6. Using the trim side of the special tool (or the dedicated trim tool, whichever you obtained) index the pins into the recess holes provided in the end cap. Index a socket wrench adaptor into the square hole in the tool and remove the end cap from the trim cylinder to be serviced. This is not an easy task, but under certain favorable conditions it can be accomplished.
7. Remove the end cap and, if equipped, the spring. Most late-models do not have the spring, but when present, the spring is partially attached to the end cap and will come free with the cap.
8. Withdraw the trim piston straight up and out of the cylinder. Repeat the last few steps in order to remove the other trim piston.

■ If the tilt piston must be removed from the cylinder for servicing, if not done already, the trim/tilt unit must be removed from the clamp bracket assembly. Clamp the unit in a vise equipped with soft jaws or pad the jaws with a couple pieces of wood. The vise will provide stability and access to almost all parts.

9. Using the tilt side of the same tool as for the trim cylinder end cap (if you used the universal tool, or use the dedicated tilt tool, as applicable), whichever you obtained), index the pins into the recess holes provided in the end cap. Remove the end cap in the same manner with a socket adaptor and wrench as was used for the end cap of the trim cylinders.

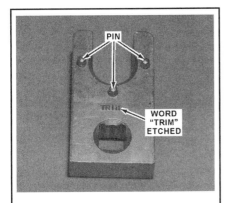

Fig. 45 A special spanner is used to remove the tilt or trim ram caps

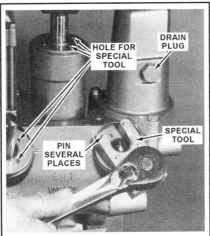

Fig. 46 Carefully loosen the caps using the spanner. . .

Fig. 47 . . .then unthread and remove the cap (with the spring IF equipped)

Fig. 48 Carefully withdraw the trim rod from the bore, keeping track of all components

Fig. 49 Removal of the tilt end cap and piston are similar

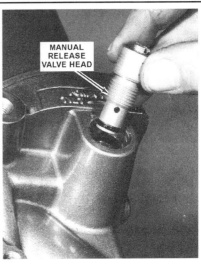

Fig. 50 Carefully unthread the valve bodies

8-24 TRIM AND TILT

10. Withdraw the tilt piston straight up and out of the cylinder.
11. Remove the circlip and then remove the manual release valve. The inner parts of the manual release valve, the ball, release rod, seat, spring, and pin, are all normally secured by the valve seat screw. Removal of this screw is extremely difficult. Without good cause, an attempt to remove the seat screw should not be made. Individual replacement parts are not available for the items behind the screw. Therefore, the only gain in removing them would be cleaning. In most cases replacement of the O-ring in the lower groove of the manual release head will solve a problem in this area.

■ In the majority of cases, service of the hydraulic items removed thus far will solve any rare problems encountered with the trim/tilt system.

12. The reservoir can be removed through the attaching hardware and cleaned if the system was considered contaminated with foreign material, which is highly unlikely, because the system is a closed system. The only route for entry of foreign material would be through the fill opening.
13. If necessary, remove the remaining valve bodies, seals and valve components. Take it slow, noting the positioning of each piece for installation purposes.

CLEANING & INSPECTION

◆ See Figures 51 thru 55

Keep the work area as clean as possible to prevent contamination through foreign material entering the system on the parts to be installed.

Clean all parts thoroughly with solvent and blow them dry with compressed air.

Carefully inspect the trim pistons and the tilt piston for any sign of damage. Scoring of the cylinders which would allow hydraulic pressure to escape past rings and seals will require parts or housing replacement.

Purchase, if available, new O-rings and discard the old items, but only after the replacement O-ring is verified as correct for the intended installation.

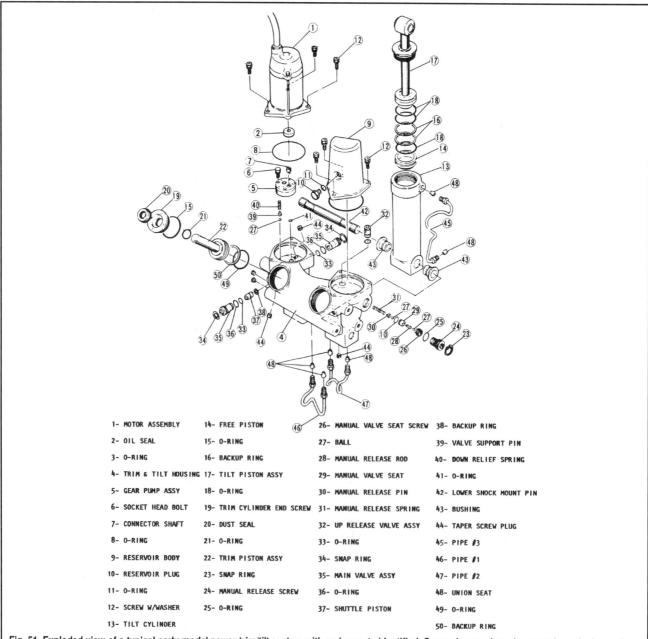

Fig. 51 Exploded view of a typical early-model power trim/tilt system with major parts identified. Some changes have been made to the internal routing of the hydraulic fluid and the number and location of internal check valves on various models

TRIM AND TILT 8-25

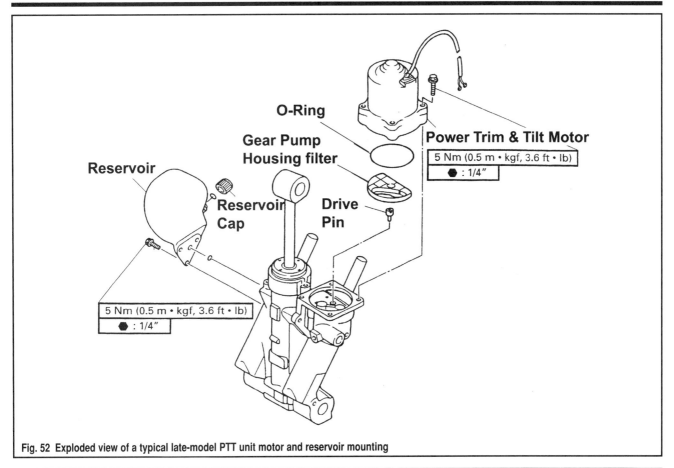

Fig. 52 Exploded view of a typical late-model PTT unit motor and reservoir mounting

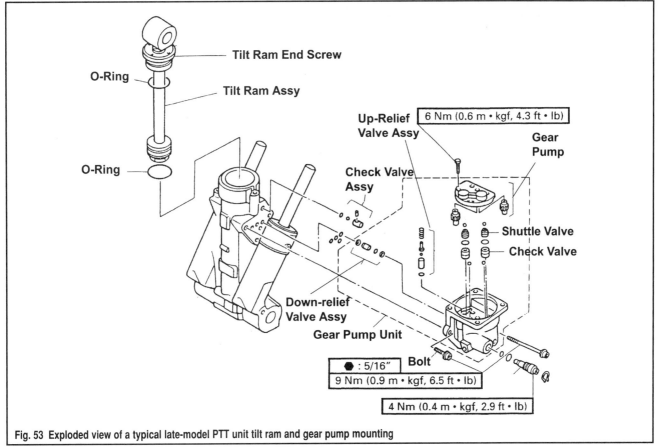

Fig. 53 Exploded view of a typical late-model PTT unit tilt ram and gear pump mounting

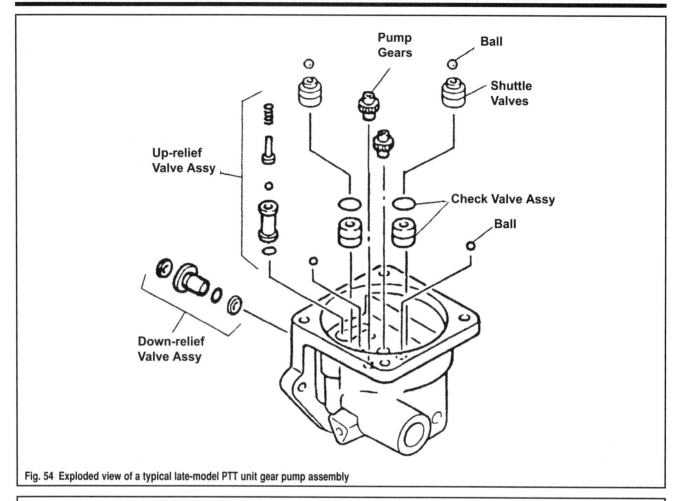

Fig. 54 Exploded view of a typical late-model PTT unit gear pump assembly

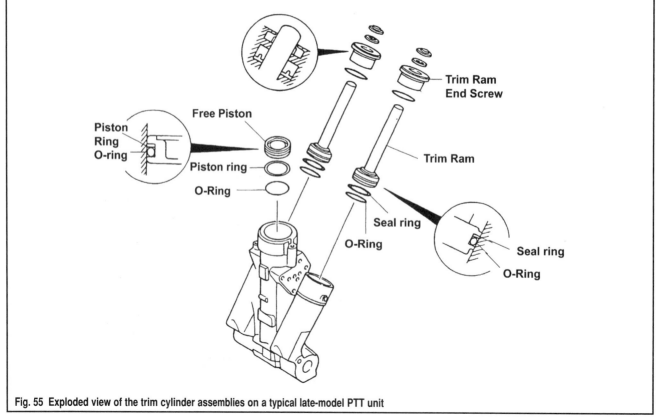

Fig. 55 Exploded view of the trim cylinder assemblies on a typical late-model PTT unit

TRIM AND TILT 8-27

ASSEMBLY

◆ See Figures 51 thru 62

Good shop practice dictates new O-rings be installed anytime the unit is disassembled and the rings are exposed.

1. Coat all new O-rings with Yamalube Power Trim and Tilt Fluid or a good grade of automatic transmission fluid, and then install the O-rings into the grooves in the valve bodies, manual release heads, trim/tilt rods, etc.

2. If removed, install all valve bodies and related components in the exact order and to the exact positions from which they were removed.

3. Insert and thread the head into the manual release valve opening. Remember some units are standard right hand threads and others have left hand threads. Tighten the head just snug because it will be rotated for power tilt and manual tilt operation.

4. Secure the head in the trim/tilt housing with a circlip or Tru-arc snapring, depending on the trim/tilt unit being serviced.

5. Check to be sure the back-up rings are in place in the groove of the tilt piston and the free piston, or install the rings if they were removed. Apply a coating of Yamalube Power Trim and Tilt Fluid to new O-rings, and then install the O-rings into the grooves of the tilt piston, free piston, and the tilt piston end cap.

■ Normally, with the piston rod facing up the O-ring of the free piston must be installed under the back-up ring. The O-ring of the tilt piston must be installed on top of the back-up ring.

6. Insert the tilt piston into the piston and thread the end cap into the housing.

7. Clamp the trim/tilt housing in a vise equipped with soft jaws or pad the jaws with a couple pieces of wood. Using the tilt side of Yamaha special tool YB-06175 with the pins indexed into the recess holes of the end cap, tighten the end cap securely.

8. Check to be sure the back-up ring is properly installed in the trim piston groove, or install the ring if it was removed. Coat a new O-ring with Yamalube Power Trim and Tilt Fluid, and then install the O-ring into the groove of the trim piston. With the trim piston rod facing up, the O-ring must normally be installed UNDER the back-up ring.

9. Insert the trim piston straight down into the cylinder. Push the piston as far down as possible. Repeat the past few steps for the other trim piston.

10. Coat a new O-ring with Yamalube Power Trim and Tilt Fluid, and then install the O-ring into the groove of the trim end cap. On early models so equipped, check to be sure the spring is properly seated. If the two seals in the end cap were removed, coat new seals with hydraulic fluid, and then insert them into the individual grooves of the end cap.

11. Slide the spring down over the trim piston rod, and then thread the end cap into the trim/tilt housing.

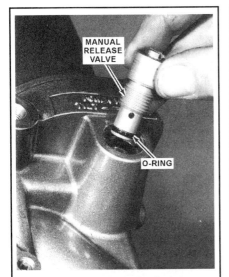

Fig. 56 Install the valve bodies in their original positions, using new O-rings

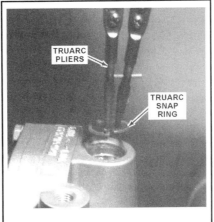

Fig. 57 Where used, carefully install the snap-rings

Fig. 58 Install the tilt cylinder and cap assembly

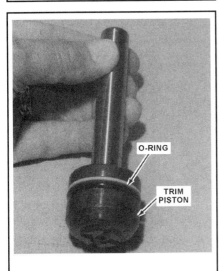

Fig. 59 Install new O-rings to the trim pistons. . .

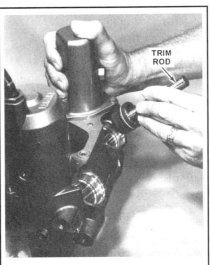

Fig. 60 . . .then carefully insert them into the bores

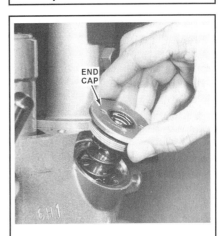

Fig. 61 Carefully thread the end cap (and spring, if equipped), by hand. . .

8-28 TRIM AND TILT

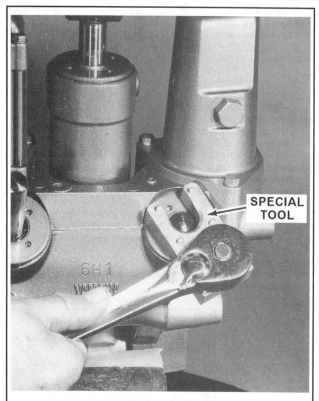

Fig. 62 ...then use the spanner to tighten the cap securely

12. Using the trim side of Yamaha special tool YB-06175 with the pins indexed into the recess holes of the end cap, tighten the end cap securely. Again, repeat the necessary steps for the other trim cylinder.

13. If the trim/tilt unit was not removed from the clamp brackets, fill the system with Yamalube Power Trim/Tilt Fluid or a good grade of automatic transmission fluid. Purge the system of air following the procedure found earlier in this section.

14. If the trim/tilt unit was removed from the clamp bracket, the system is filled and purged with the unit on the bench. Fill the system with Yamalube Power Trim/Tilt Fluid or a good grade of automatic transmission fluid and purge the system of air as follows:

  a. First pull the tilt cylinder to its fully extended position. Check the fluid level and add fluid as required to bring the level up to the bottom of the fill opening.

  b. Next, rotate the manual release valve to the power tilt position, approximately three full turns from the manual position.

  c. Now, obtain a 12 volt battery and connect the Blue lead from the trim/tilt unit to the positive lead from the battery. Now, momentarily make contact with the Green lead from the unit to the negative post of the battery. The trim cylinders should move to the fully extended position. Check the fluid level and add fluid, as required.

  d. Next, connect the Green lead from the unit to the positive post on the battery. Momentarily make contact with the Blue lead from the trim/tilt unit to the negative post of the battery. The trim cylinders and the tilt cylinder should retract to the full down position.

  e. Repeat the extension and retraction of the cylinders two or three times until the fluid level remains stable when checked with the tilt ram in the fully extended position.

15. To check for air in the system, extend the tilt cylinder, and then apply a downward heavy force manually to the end of the piston. The piston should feel solid. If the piston retracts more than about 1/8 in. (3.2mm), the system still contains air. Perform the purging sequence until the tilt cylinder is solid.

16. If removed, install the PTT assembly to the outboard.

17. Perform an operational check of the system.

TRIM & TILT SYSTEMS (MERCURY/MARINER)

Introduction

■ If you haven't figured it out yet, the 3.3L V6 4-stroke Mercury/Mariner motor IS a Yamaha, lock, stock and barrel. For this reason the trim & tilt system is covered under Large Motor Power Trim/Tilt System (Yamaha), earlier in this section. All other Mercury/Mariner motors utilize a Mercury trim & tilt system and therefore are covered in separate sections.

All outboard installations are equipped with some means of raising or lowering (pivoting), the complete unit for efficient operation under various load, boat design, water conditions, and for trailering to and from the water.

The correct trim angle ensures maximum performance and fuel economy as well as a more comfortable ride for the crew and passengers.

The most simple form of tilt is a mechanical tilt adjustment consisting of a series of holes in the transom mounting bracket through which an adjustment pin passes to secure the outboard unit at the desired angle.

Such a mechanical arrangement works quite well for the smallest of units, but the larger (and heavier) outboard become, the more they require some form of assist or power system. Pretty much all Mercury 4-strokes except the single cylinder models and the smallest of the twins, the 8/9.9 hp (232cc) motors, utilize either a manual gas assist or a power trim/tilt system. The 25-60 hp models may be equipped with either system, but the 9.9/15 hp (323cc) motors, as well as the 75 hp and larger motors should all be equipped with only Power Trim/Tilt (PTT).

The PTT system which may be found on most 9.9-115 hp motors is essentially the same basic single ram unit design. However, there are 3 major related versions of the system, one used by 9.9/15 hp (323cc) motors, one used by 25 hp motors, and one used by 30-115 hp motors.

Mechanical Tilt (Unassisted-Mercury/Mariner Models)

◆ See Figures 63 and 64

The system employed by the smallest 4-stroke Mercury/Mariners consists of a simple mechanical tilt pin arrangement.

On 4/5/6 hp single cylinder motors, as well as the smallest twins, the 8/9.9 hp (232cc) motors, a change in the tilt angle of the outboard unit is accomplished by inserting the tilt adjustment pin through one of a series of holes in the transom mounting bracket. These holes allow the operator to obtain the desired boat trim under various speeds and loading conditions.

The tilt angle of a lower unit is properly set when the anti-cavitation plate is approximately parallel with the bottom of the boat. The boat trim is corrected by stopping the boat, removing the adjustment pin, tilting the

TRIM AND TILT 8-29

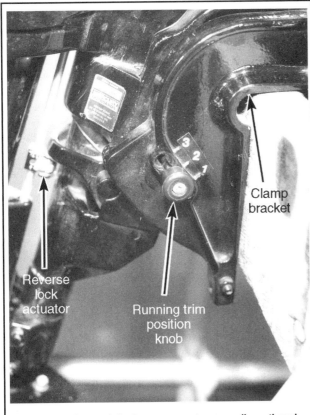

Fig. 63 Typical manual tilt pin arrangement on a smaller outboard unit

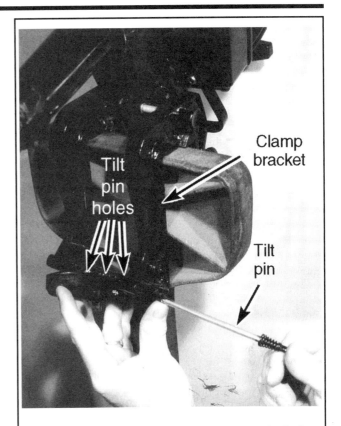

Fig. 64 With the outboard in the proper position, insert the tilt pin into the tilt pin hole

outboard upward or downward, as desired and then installing the pin through the new hole exposed in the transom mounting bracket.

To raise the bow of the boat, the outboard is raised one hole at-a-time until the operator is satisfied with the boat's performance. If the bow is to be lowered, the lower unit is lowered one hole at-a-time.

Performance will generally be improved if the bow is lowered during operation in rough water. The boat should never be operated with the lower unit set at an excessive raised position. Such a tilt angle will cause the boat to "porpoise", which is very dangerous in rough water. Under such conditions the helmsperson does not have complete control at all times.

Instead of making extreme changes in the lower unit angle, it is far better to shift passengers and/or the load to obtain proper performance.

In order to obtain maximum efficiency and safety from the boat and outboard unit, the tilt pin must be installed in the proper position. The wide range of boat designs with their various transom angles, requires a determination be made for each outboard installation.

Actually, the tilt pin is only required if the boat handles improperly in the full trimmed "in" position at WOT (wide open throttle). Usually this occurs when the transom "angle" is too large.

1. Move the outboard unit inward or outward until the anti-cavitation plate is parallel to the boat bottom. With the outboard in this position, notice the position of the swivel bracket in relation to the clamp bracket tilt pin holes. Now, install the tilt pin into the first full pin hole closer to the transom.

2. Lock the tilt pin in position by pushing in on the pin compressing the lock spring. Close the clevis on the end of the pin and release the tilt pin. The tilt pin is secured by the clevis on the end of the tilt pin.

Some earlier models utilize a cotter pin and washer on the end of the tilt pin. After installing the tilt pin through the desired clamp bracket hole, slide the washer over the end of the tilt pin and insert the cotter pin through the hole at the end of the tilt pin. Open the ends of the cotter pin to secure the tilt pin in place.

The angle of the lower unit is properly set when the boat is operating to give maximum performance, including comfort and safety.

SERVICING

◆ See Figures 65 and 66

Service procedures for the manual tilt system are confined to general lubrication and inspection. If individual components should wear or break, replacement of the defective components is necessary.

8-30 TRIM AND TILT

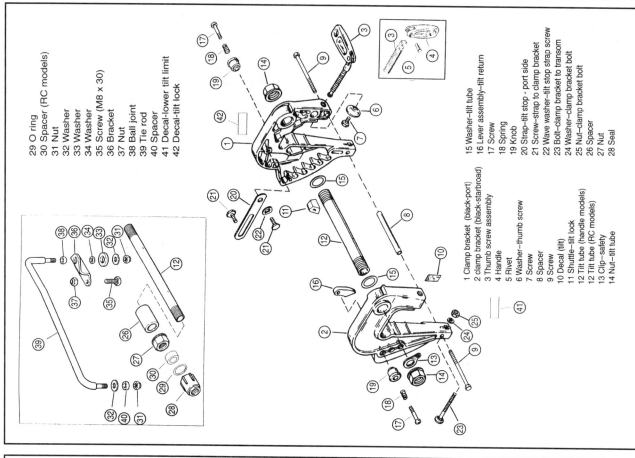

Fig. 66 Typical clamp bracket assembly - 8/9.9 hp (232cc) motors

1 Clamp bracket (black-port)
2 clamp bracket (black-starboard)
3 Thumb screw assembly
4 Handle
5 Rivet
6 Washer–thumb screw
7 Screw
8 Spacer
9 Screw
10 Decal (tilt)
11 Shuttle–tilt lock
12 Tilt tube (handle models)
13 Clip–safety
14 Nut–tilt tube
15 Washer–tilt tube
16 Lever assembly–tilt return
17 Screw
18 Spring
19 Knob
20 Strap–tilt stop - port side
21 Screw–strap to clamp bracket
22 Wave washer–tilt stop strap screw
23 Bolt–clamp bracket to transom
24 Washer–clamp bracket bolt
25 Nut–clamp bracket bolt
26 Spacer
27 Nut
28 Seal
29 O ring
30 Spacer (RC models)
31 Nut
32 Washer
33 Washer
34 Washer
35 Screw (M8 x 30)
36 Bracket
37 Nut
38 Ball joint
39 Tie rod
40 Spacer
41 Decal–lower tilt limit
42 Decal–tilt lock

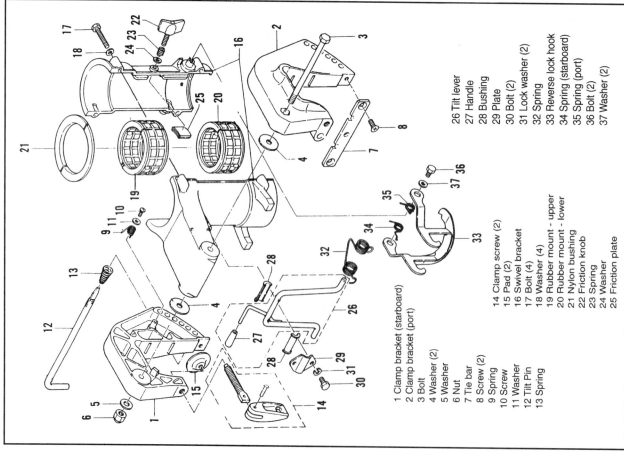

Fig. 65 Typical engine clamp bracket/swivel bracket components and lubrication points - 4/5/6 hp Mercury/Mariner motors

1 Clamp bracket (starboard)
2 Clamp bracket (port)
3 Bolt
4 Washer (2)
5 Washer
6 Nut
7 Tie bar
8 Screw (2)
9 Spring
10 Screw
11 Washer
12 Tilt Pin
13 Spring
14 Clamp screw (2)
15 Pad (2)
16 Swivel bracket
17 Bolt (4)
18 Washer (4)
19 Rubber mount - upper
20 Rubber mount - lower
21 Nylon bushing
22 Friction knob
23 Spring
24 Washer
25 Friction plate
26 Tilt lever
27 Handle
28 Bushing
29 Plate
30 Bolt (2)
31 Lock washer (2)
32 Spring
33 Reverse lock hook
34 Spring (starboard)
35 Spring (port)
36 Bolt (2)
37 Washer (2)

TRIM AND TILT 8-31

GAS ASSIST TILT SYSTEM (MERCURY/MARINER)

Description & Operation

The budget method of tilting a smaller outboard (usually found on tiller control models) is the manual gas assist tilt system. It can be found on Mercury/Mariner 4-stroke motors from 25-60 hp. The system found on the 25 hp models is a unique assembly to those motors but operates in BASICALLY the same fashion (even though the internals are different and it is a non-serviceable unit). The only service possible on the 25 hp models is removal and installation. The systems found on the 30-60 hp motors are mechanically identical to each other and can be disassembled for service.

The gas assist tilt system consists of the shock rod cylinder, valve block, nitrogen filled accumulator (on 30-60 hp models) and a manual release lever.

When the manual release lever is opened, the nitrogen filled accumulator assists the operator while tilting the outboard for shallow water operation or trailering. The entire system is located between the transom brackets.

If the lower unit strikes an underwater object, the force will apply a high pressure on the down side of the cylinder. A shock valve located in the piston will relieve this pressure to the up side of the piston and if necessary, to allow excess oil to return to the reservoir.

If the outboard has trouble staying in the selected tilt position, check for external fluid leaking from the accumulator or shock rod cylinder. Make repairs as necessary. Make sure that the manual release lever and control rod operate freely and make any adjustments if needed.

Gas Assist Damper

TESTING

◆ See Figure 67

For all models, if the unit does not appear to be functioning properly, start with a visual inspection. If debris or leaking is found the assembly must be replaced or, for 30-60 hp motors, the unit MAY also be disassembled for inspection or repairs

To check function of the gas assist cylinder proceed as follows:
1. Raise the outboard and secure using the tilt lock lever.
2. Check the adjustment on the manual release cam (the linkage just to the side of the cylinder and accumulator). The cam must open and close freely. If necessary, adjust the length of the linkage as necessary for the system to function properly.
3. For 30-60 hp motors, check a discharged accumulator as follows:
   a. Using a spring-loaded scale, tilt the outboard by applying pressure to the top of the engine cowling.
   b. It must take 30-50 ft. lbs. (47-68 Nm) of pulling force to tilt the outboard from a full DOWN position to a full UP position.
   c. If if requires more than 50 ft. lbs. (68 Nm) of torque, the nitrogen charge in the accumulator is discharged and the accumulator will need to be replaced.

REMOVAL & INSTALLATION

 MODERATE

◆ See Figures 68 and 69

1. Support the outboard in the up position using the tilt lock lever (30-60 hp) or the tilt stop pin (25 hp).
2. For 30-60 hp models, remove the link rod, then position a piece of wood under the transom bracket (instead of the tilt lock) for access to remove the upper tri-lobe pin.
3. Remove the upper tri-lobe pin (which is used to secure the upper pivot pin). On 25 hp motors, you can usually grab the pin with a set of needle-nose pliers and pull downward to free it. On 30-60 hp motors, it is usually best to use a suitable punch to remove (drive down) the upper dowel pin.

■ After removing the upper tri-lobe pin on 30-60 hp motors, reposition the tilt lock and remove the piece of wood.

4. Use a suitable punch to drive out the upper pivot pin freeing the top of the gas cylinder.
5. On 30-60 hp motors, proceed as follows:
   a. Remove the anode.
   b. Using a punch drive out the lower tri-lobe pin from the lower pivot pin (in the same manner as the upper tri-lobe pin was removed).
6. On 25 hp motors, proceed as follows:
   a. Disconnect the link rod from the cam lever.
   b. Remove the nuts and washers securing the lower pivot pin.

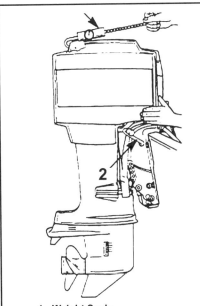

1 - Weight Scale
2 - Valve lever (open position)

Fig. 67 Using a scale to check the nitrogen charge in the accumulator - 30-60 hp motors

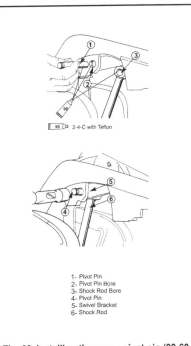

1- Pivot Pin
2- Pivot Pin Bore
3- Shock Rod Bore
4- Pivot Pin
5- Swivel Bracket
6- Shock Rod

Fig. 68 Installing the upper pivot pin (30-60 hp shown, 25 hp similar)

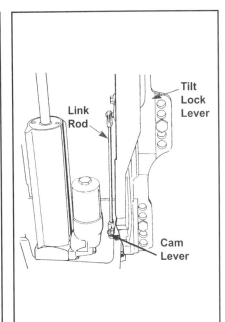

Fig. 69 Check and adjust the release linkage, as necessary (30-60 hp shown, 25 hp similar)

8-32 TRIM AND TILT

c. Remove the anode screw in order to free the ground strap.

7. Use a punch to carefully drive/tap out the lower pivot pin. On 25 hp motors, retain the anchor pin bushing from the clamp bracket and trim unit.

8. Tilt the shock absorber assembly (top first) out from the clamp bracket and remove the assembly.

■ **For 25 hp motors, stop here, no further disassembly is possible. If the unit is damaged, it must be replaced. For 30-60 hp motors, disassembly and component replacement is possible, refer to the Overhaul procedure later in this section.**

To install:

9. Apply a light coating of Quicksilver 2-4-C with Teflon, or an equivalent lubricant to the lower pivot pin hole and pivot pin surface (and on the 25 hp motors, to the bushings).

10. Start the lower pivot pin (and bushings on the 25 hp motors) into the pivot pin hole. At the same time, position the gas assist cylinder assembly as you guide the pivot pin fully into position.

■ **For 30-60 hp motors, use a small punch to gently drive the lower pivot pin into position until it is flush with the outer surface.**

11. For 25 hp motors, install the nuts and washers to anchor the lower pivot pin, then tighten them securely. Install the ground strap and anode screw. Install the tilt lock pin.

12. For 30-60 hp motors, use a small punch to drive the lower tri-lobe pin into position, securing the lower pivot pin.

13. Apply a light coating of Quicksilver 2-4-C with Teflon, or an equivalent marine lubricant to the surface of the upper pivot pin, pivot pin hole and the shock rod (gas tube) hole.

14. Rotate the assembly so the shock rod hole is aligned for installation of the upper pivot pin.

15. Using a soft mallet, carefully drive the upper pivot pin into the swivel bracket and through the shock rod until the pivot pin is flush with the swivel bracket.

16. Drive the upper tri-lobe (dowel) pin into its hole until seated.

17. Reconnect the release valve linkage.

18. For 30-60 hp motors, install the anode.

19. Check the manual release cam adjustment. The cam must open and close freely. Make adjustments to the link rod as necessary.

OVERHAUL (30-60 HP ONLY)

◆ See Figures 70 and 71

※※ WARNING

The tilt system is pressurized. The accumulator itself is sealed and contains a high-pressure nitrogen charge. Although the accumulator can be removed (ONLY when the tilt rod is in the full UP position), it cannot be disassembled further and must be replaced if faulty.

As with all rebuilds, be sure to remove and discard all O-rings during disassembly. Make sure the replacement O-rings are the proper size.

1. Remove the gas tilt assist assembly from the outboard and place it in a soft-jawed vise for access.

2. To remove the accumulator assembly, proceed as follows:
a. Make sure the tilt rod is in the FULL UP position.

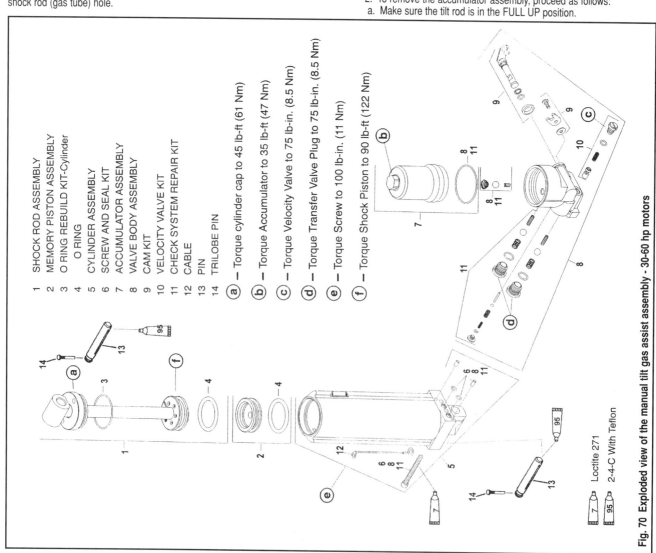

Fig. 70 Exploded view of the manual tilt gas assist assembly - 30-60 hp motors

b. Open the cam shaft valve (down position).

c. Position a small drain pan under the assembly, then locate the velocity valve (the slotted screw-head valve at the base of the assembly, right under the accumulator). Loosen the velocity valve JUST until it starts to drip fluid. Once the dripping stops loosen and remove the accumulator.

d. Check the plunger on the bottom of the accumulator. If it can be compressed into the accumulator by hand, the assembly is defective and must be replaced.

e. With the accumulator removed, remove the O-ring (it usually still in the bore into which the accumulator was threaded), then remove the conical spring, steel ball and plunger. All of these components should be available in a service kit and it is a good idea to replace them regardless of condition.

3. To remove and disassemble the shock rod (tilt rod), proceed as follows:

a. Using the tool (#91-74951) or an equivalent spanner with two adjustable pins that insert into the rod (the wrench has 1/4 in. x 5/16 in. long pegs) unscrew and remove the tilt rod end cap assembly.

b. Carefully lift the rod and end cap from the cylinder.

■ **The ONLY serviceable items on the shock rod assembly are the O-rings, wiper ring and the piston. If the shock requires ANY other repair, replace it as an assembly.**

c. Place the shock rod in a soft-jawed vise, then loosen and remove the 3 screws on the lower plate.

d. Carefully lift the plate from the rod, keeping CLOSE track of the 5 balls, 5 seats and 5 springs retained under the plate.

e. Remove and discard the O-ring from the lower plate.

f. With the shock rod still (or back) in the vise, use a heat gun to GENTLY apply heat to loosen the piston (still at the bottom of the rod).

g. While the piston is hot, use the spanner wrench you used on the cap loosen the piston, but do not remove it completely. Allow the piston to cool and remove it from the shock rod.

h. Inspect the check valve in the piston for debris and clean, as necessary. IF the check valve cannot be cleaned, replace the PISTON as an assembly.

i. Clean the shock and components with compressed air (but DON'T loose the springs, seats and balls).

j. Remove the inner O-ring from the piston bore.

k. Remove the cylinder end cap assembly from the shock rod and inspect. If the wiper (located in the cap) has failed to keep the rod cleaned, replace the wiper.

l. Place the end cap on a clean work surface, then use a small prytool to carefully remove the wiper from the top center of the cap. Remove the inner and outer O-rings and discard.

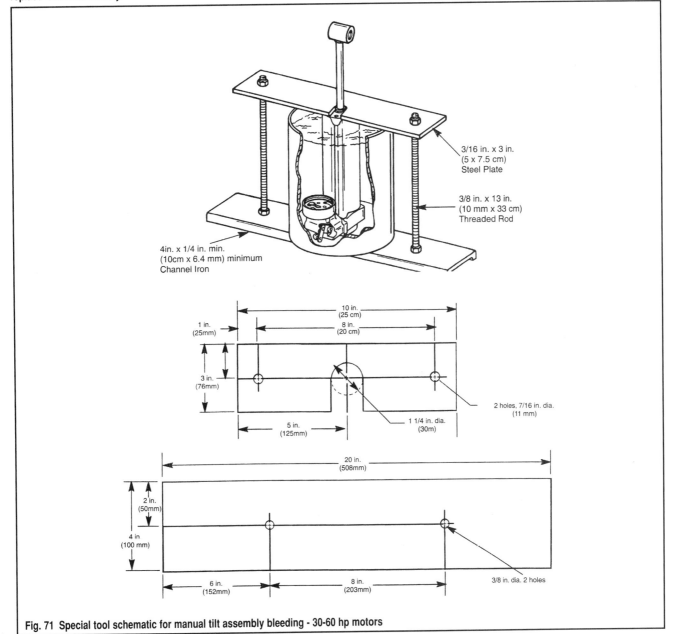

Fig. 71 Special tool schematic for manual tilt assembly bleeding - 30-60 hp motors

8-34 TRIM AND TILT

4. To remove/service the valve block, proceed as follows:
   a. Remove the 2 Allen head screws (threaded horizontally) from the base of the shock rod cylinder in order to separate the valve block. Carefully remove the valve block.
   b. Remove the 2 O-rings and 2 dowels.
5. If the memory piston requires service, you can remove it from the bore either using a pair of lock-ring pliers to grab it and pull it upward OR by using compressed air blown into the center O-ring hole. When using the second method, point the cylinder bore facing downward toward a shop rag or towel to both prevent injury AND to help prevent damage. Expect some fluid to be expelled at the same time. Remove and discard the O-ring from the memory piston.
6. To disassemble the valve block, proceed as follows:
   a. Remove the check retainer plug from the center side of the body. Behind the plug expect to find a small diameter spring, a larger diameter spring a steel ball and a small rod. Keep the pieces in order.
   b. Remove each of the hydraulic oil transfer valve plugs and their components. Behind each plug expect to find an O-ring, a spring, a large steel ball and a rod.
   c. On the opposite side of the valve block from the transfer valves, remove the velocity valve, O-ring, spring and spool (surge valve) assembly.
   d. Ninety degrees from either the transfer valves or velocity valve assembly, remove the screw, retainer clip and spacer retainer clip, along with the cam and shaft seal.
7. Clean and inspect all components which are to be reused using engine cleaner and low-pressure compressed air. DO NOT use cloth rags which may leave residue behind that could clog a valve.
8. Inspect all machined surfaces for burrs or scoring and repair or replace, as necessary.

To assemble:

9. Identify all of the O-rings for assembly purposes. Lubricate the O-rings using Quicksilver Power Trim and Steering Fluid (or, if not available, using automatic transmission fluid), that is all EXCEPT the cam shaft O-ring which should be lubricated with 2-4-C with Teflon or equivalent.
10. Assemble the valve block, as follows:
    a. Install the lubricated O-ring and back-up seal to the cam.
    b. Install the shaft seal in the valve block with the LIPS FACING OUTWARD.
    c. Install the cam shaft assembly into the valve block, then secure using the insulator, retainer plate and screw. Tighten the screw securely.
    d. Install the velocity valve (surge valve assembly) by positioning the spool, spring, lubricated O-ring and the velocity screw plug into the block. Tighten the plug to 75 inch lbs. (8.5 Nm).
    e. Install the check retainer assembly by positioning the rod (plunger), large diameter spring, ball, small diameter spring and finally the plug.
    f. Install each of the transfer valve assemblies by positioning the components in each bore. Install the rod (plunger), steel ball, spring and then screw plug with new, lubricated O-ring. Tighten each of the screw plugs to 75 inch lbs. (8.5 Nm).
    g. Install the new lubricated O-rings and dowel pins to the mating surface of the valve block itself, then position the block to the shock rod cylinder.
    h. Install the valve block retaining screws (Allen head screws) and tighten to 100 inch lbs. (11 Nm).
11. To assemble the shock (tilt) rod, proceed as follows:
    a. Install the lubricated O-rings to the end cap and to the piston.
    b. Install a new rod wiper to the center of the cap.
    c. Clamp the shock rod in a soft-jawed vise, then position the end cap onto the rod (with the O-ring facing the bottom end of the rod and the threads facing the top end of the rod).
    d. Apply a light coating of Loctite® 271 or equivalent thread-locking compound to the threads on the shock rod, then install the shock rod piston.

✳✳ CAUTION

To prevent damage to the shock rod piston, MAKE SURE the spanner wrench has 1/4 in. x 5/16 in. (6.4mm x 8mm) long pegs.

   e. Using the spanner, tighten the piston securely. If an adapter is available for a torque wrench, tighten the piston to 90 ft. lbs. (122 Nm).
   f. Install the 5 sets of balls, seats and springs to the shock rod piston, then position the plate to secure them. Tighten the 3 plate screws to 35 inch lbs. (4.0 Nm).
   g. Remove the shock assembly from the vise.

■ There are 2 possible methods for shock rod installation and fluid filling/bleeding. The first method involves a soft-jawed vise, while the second uses a specially fabricated tool assembly.

12. To install the shock rod with the accumulator and fill/bleed the fluid using only a soft-jawed vise, proceed as follows:
    a. Position the assembly in a soft-jawed vise.
    b. With the manifold cam lever in the closed (up position), slowly fill the cylinder and manifold to the top using Quicksilver Power Trim and Steering Fluid (or equivalent). Allow time for the bubbles to disperse.
    c. Install the lubricated O-ring to the memory piston, then use a pair of lock-ring pliers (like Snap-On® SRP2) position the memory piston on top of the cylinder. Now, open the cam lever (down position) and carefully push the memory piston down just below the cylinder threads and CLOSE the cam lever (up position).
    d. Fill the top of the cylinder again with fluid all the way to the top, then install the shock rod assembly on top of the memory piston. Again open the cam lever (down position) as you push the shock rod down to 1/8 inch below the cylinder threads, then again CLOSE the cam lever (up position).
    e. Fill the top of the shock rod assembly with fluid to the top of the cylinder. Open the cam lever (down position) and screw the cylinder cap down into position.
    f. Tighten the end cap securely using the spanner wrench. If a torquing adapter is available, tighten the end cap to 45 ft. lbs. (61 Nm).
    g. Close the cam lever (up position).
    h. Open and close the cam lever, watching for bubbles coming from the accumulator check ball hole. When the bubbles stop, fill the accumulator opening to the top with fluid.
    i. Grease the threads on the accumulator and opening using 2-4-C with Teflon or equivalent, then start the accumulator into the threads and OPEN the cam lever (down position). Tighten the accumulator securely or, if an adapter is available, tighten it to 35 ft. lbs. (47 Nm).

■ If the filling procedure was done correctly it should be difficult to turn the cylinder rod assembly by hand.

13. To install the shock rod with the accumulator and fill/bleed the fluid using the specially fabricated tool (as detailed in the accompanying illustration), proceed as follows:
    a. Install the lubricated O-ring to the memory piston and install the piston into the bore.
    b. Install the shock rod assembly. Tighten the end cap securely using the spanner wrench. If a torquing adapter is available, tighten the end cap to 45 ft. lbs. (61 Nm).
    c. With the shock rod in the full UP position and the manifold cam lever open (facing down), secure the tilt system to the retaining tool and container (a No. 10 can or 3 lbs. coffee can may be used to fabricate the heart of the container).
    d. Fill the container nearly full using Quicksilver Power Trim and Steering Fluid (or equivalent).

■ The fluid level MUST remain above the accumulator opening during the entire bleeding process.

   e. Bleed the unit by pushing the rod down slowly (taking 18-20 seconds PER stroke) until the rod is stopped at the base. Hold this position until all air bubbles exit the accumulator base, then pull upward slowly on the rod 3 in. (76mm) from the base again, wait for all air bubbles to exit the accumulator base.
   f. Continue to cycle the rod down and up in the same pattern, using short strokes, again 3 in. (76mm) from the base, allowing bubbles to dissipate each time. Repeat about 5-8 times.
   g. Allow the unit to stand for 5 minutes, then proceed to cycle the unit another 2-3 more times using short strokes. NO BUBBLES should appear from the port this time.
   h. With the oil level WELL above the accumulator port, SLOWLY pull the rod up to the FULL UP position.
   i. Install the accumulator making sure NO air bubbles enter the system.
   j. Tighten the accumulator snugly at this time.
   k. With the cam lever still open (facing down) remove the tilt assembly from the oil and secure in a soft-jawed vise. Using an adapter, tighten the accumulator to 35 ft. lbs. (47 Nm).
14. Install the gas assist tilt assembly as detailed earlier in this section.

TRIM AND TILT

SINGLE RAM INTEGRAL POWER TILT/TRIM (MERCURY/MARINER)

Description & Operation

Although a purely mechanical system is used on many smaller outboards, the larger (and therefore heavier) and outboard becomes, the more likely they are to utilize an electro-hydraulic Power Trim and Tilt (PTT) system. Therefore, most motors 9.9/15 hp (323cc) and larger utilize some form of a PTT system. Actually, all 232cc motors, as well as the 75 hp and larger motors should only be equipped with only Power Trim/Tilt (PTT), as well as most of the 25-60 hp models.

■ **The PTT system which may be found on most 9.9-115 hp motors is essentially the same basic single ram unit design. However, there are 3 major related versions of the system, one used by 9.9/15 hp (323cc) motors, one used by 25 hp motors, and one used by 30-115 hp motors. Separate procedures or steps are included along this division where necessary.**

Basically, this power trim/tilt systems consist of a housing with an electric motor, a gear driven hydraulic pump, hydraulic reservoir and one trim/tilt cylinder. The cylinder performs a double function as trim/tilt cylinders and also as shock absorbers, should the lower unit strike an underwater object while the boat is underway.

The necessary valves, check valves, relief valves and hydraulic passageways are incorporated internally and externally for efficient operation. A manual release valve is provided to permit the outboard unit to be raised or lowered should the battery fail to provide the necessary current to the electric motor or if a malfunction should occur in the hydraulic system.

The gear driven pump operates in much the same manner as an oil circulation pump installed on motor vehicles. The gears rotate in either direction, depending on the desired cylinder movement. One side of the pump is considered the suction side and the other the pressure side, when the gears rotate in a given direction. These sides are reversed, the suction side becomes the pressure side and the pressure side becomes the suction side when gear movement is changed to the opposite direction.

Depending on the model, up to two relays may be used for the electric motor. The relays are usually located at the bottom cowling pan, where they are fairly well protected from moisture.

■ **As a convenience, most models contain an auxiliary trim/tilt switch installed on the exterior cowling.**

When the up portion of the trim/tilt switch is depressed, the up circuit, through the relay, is closed and the electric motor rotates in a clockwise direction. Pressurized oil from the pump passes through a series of valves to the lower chamber of the trim cylinders, the pistons are extended and the outboard unit is raised. The fluid in the upper chamber of the pistons is routed back to the reservoir as the piston is extended. When the desired position for trim is obtained, the switch on the control handle is released and the outboard is held stationary.

If the trim cylinder pistons should become fully extended, such as in a tilt up situation, fluid pressure in the lower chamber of the trim cylinders increases. This increase in pressure opens an up relief valve and the fluid is routed to the reservoir. The sound of the electric motor and the pump will have a noticeable change.

When the down portion of the trim/tilt switch is depressed, the down circuit, through the relay, is closed and the electric motor rotates in a counterclockwise direction. The pressure side of the pump now becomes the suction side and the original suction side becomes the pressure side. Pressurized oil from the pump passes through a series of valves to the upper chamber of the trim cylinders, the pistons are retracted and the outboard unit is lowered. The fluid in the lower chamber of the pistons is routed back to the reservoir as the retracted is extended. When the desired position for trim is obtained, the switch on the control handle is released and the outboard is held stationary.

If the trim cylinder pistons should become fully retracted, such as in a tilt down situation, fluid pressure in the upper chamber of the trim cylinders increases. This increase in pressure opens an up relief valve and the fluid is routed to the reservoir. The sound of the electric motor and the pump will have a noticeable change.

In the event the outboard lower unit should strike an underwater object while the boat is underway, the tilt piston would be suddenly and forcibly extended, moved upward. For this reason, the lower end of the tilt piston is capped with a free piston. This free piston normally moves up and down with the tilt piston.

The free piston also moves upward but at a much slower rate than the tilt piston. The action of the tilt piston separating from the free piston causes two actions. First, the hydraulic fluid in the upper chamber above the piston is compressed and pressure builds in this area. Second, a vacuum is formed in the area between the tilt piston and the free piston.

This vacuum in the area between the two pistons sucks fluid from the upper chamber. The fluid fills the area slowly and the shock of the lower unit striking the object is absorbed. After the object has been passed the weight of the outboard unit tends to retract the piston. The fluid between the tilt piston and the free piston is compressed and forced through check valves to the reservoir until the free piston reaches its original neutral position.

A manual relief valve, located on the stern bracket, allows easy manual tilt of the outboard should electric power be lost. The valve opens when the screw is turned counterclockwise, allowing fluid to flow through the manual passage. When the relief valve screw is turned fully clockwise, the manual passage is closed and the outboard is locked in position (unless the hydraulic system changes the position of the trim/tilt ram).

A thermal valve is used to protect the trim/tilt motor and allow it to maintain a designated trim angle. Oil in the upper chamber is pressurized when force is applied to the outboard from the rear while cruising. Oil is directed through the right side check valve and activates the thermal valve to release oil pressure and lessen the strain on the motor and pump.

Troubleshooting the Power Trim/Tilt System

GETTING STARTED (FIRST THINGS TO CHECK)

Any time a problem develops in the power trim/tilt system the first step is to determine whether it is electrical or hydraulic in nature. After the determination is made, then the appropriate steps can be taken to remedy the problem.

The first step in troubleshooting is to make sure all the connectors are properly plugged in and that all the terminals and wires are free of corrosion. The simple act of disconnecting and connecting a terminal may sometimes loosen corrosion that preventing a proper electrical connection. Inspect each terminal carefully and coat each with dielectric grease to prevent corrosion.

The next step is to make sure the battery is fully charged and in good condition. While checking the battery, perform the same maintenance on the battery cables as you did on the electrical terminals. Disconnect the cables (negative side first), clean and coat them and then reinstall them. If the battery is past its useful life, replace it. If it only requires a charge, charge it.

Check the power trim/tilt fuse, as appropriate. Many systems will have a fuse to prevent large current draws from damaging the system. If this fuse is blown, the system will cease to function. This is a good indicator that you may have problems elsewhere in the electric system. Fuses don't blow without cause.

After inspecting the electrical side of the system, check the hydraulic fluid level and top it off as necessary. Remember to position the motor properly (full tilt up or down) to get an accurate measurement of fluid level. A slight decrease in the level of hydraulic fluid may cause the system to act sporadically.

Finally, make sure the manual release valve is in the proper position. A slightly open manual release valve may prevent the system from working properly and mimic other more serious problems.

Just remember to check the simple things first. If these simple tests do not diagnose the cause of the problem, then it is time to investigate more deeply. Perform the hydraulic pressure tests in this section to determine if the pump is making adequate pressure. Inspect the entire power trim/tilt electrical harness with a multimeter, checking for excessive resistance and proper voltage.

8-36 TRIM AND TILT

COMPREHENSIVE TESTING

9.9/15 Hp Motors - Symptom Diagnosis

For 9.9/15 hp motors, depending upon the symptom, check the following:

Trim motor runs, but trim system does not move up or down
- Low oil level
- Pump assembly faulty
- Broken motor/pump driveshaft

System does not trim down or does not trim up
- Low oil level
- Pump assembly faulty

System is Jerky up/down or operates only partially in either direction
- Low oil level
- Tilt ram piston O-ring leaking or cut

Thump noise when shifting
- Pump assembly faulty
- Tilt ram piston O-ring leaking or cut

Does not trim under load
- Low oil level
- Pump assembly faulty

Does not hold trim position under load
- Pump assembly faulty
- Tilt ram piston O-ring leaking or cut

Trails out when backing off from high speed
- Pump assembly faulty

Leaks down and does not hold trim
- Pump assembly faulty
- Tilt ram piston O-ring leaking or cut

Trim motor working hard and trims slow up and down
- Low oil level
- Pump assembly faulty

- Battery weak or discharged

Starts to trim up from full down position when IN trim button is pressed
- Pump assembly faulty

Trim position will not hold in reverse
- Pump assembly faulty
- Tilt ram piston O-ring leaking or cut

Trim motor does not run when trim button is depressed
- Weak or discharged battery
- Open circuit in trim wiring
- Wire harness corroded through
- Internal motor problem (brushes, shorter armature)
- Blown fuse(s)
- Trim switch failure
- Relay failure

Trim system trims opposite of buttons
- Wiring harness reversed in remote or panel control

cut this next 4-head for print and use in web only, it's just ONE freakin' model and we have given them some troubleshooting

9.9/15 Hp Motors - Wiring Diagram and Circuit Testing

◆ See Figures 72 thru 76

Use the accompanying wiring diagrams and troubleshooting charts to diagnose and repair electrical problems with the PTT system on 9.9/15 hp models.

Before beginning, check the following:
- Check for disconnected wires
- Make sure all connections are clean (of corrosion) and tight
- Make sure the plug-in connectors are fully engaged
- Make sure the battery is fully charged

The numbers listed in each of the troubleshooting charts correspond to the numbered locations on the wiring diagrams.

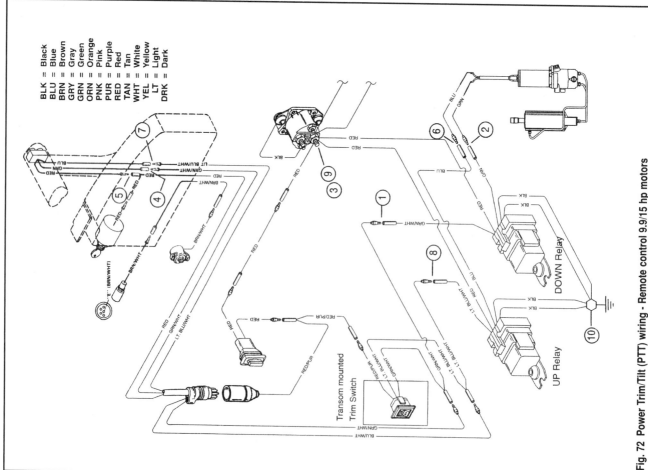

Fig. 72 Power Trim/Tilt (PTT) wiring - Remote control 9.9/15 hp motors

TRIM AND TILT 8-37

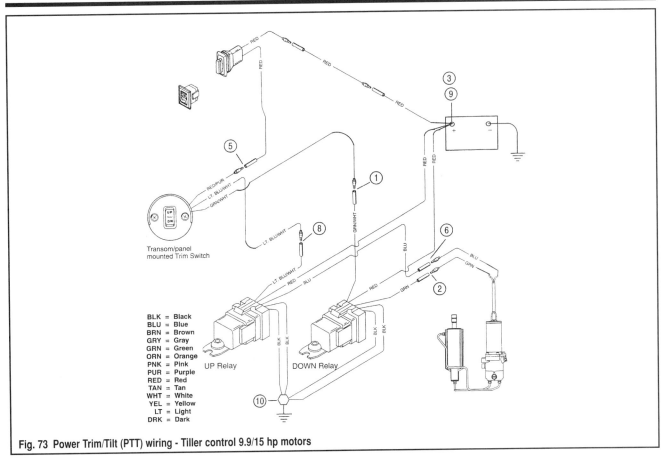

Fig. 73 Power Trim/Tilt (PTT) wiring - Tiller control 9.9/15 hp motors

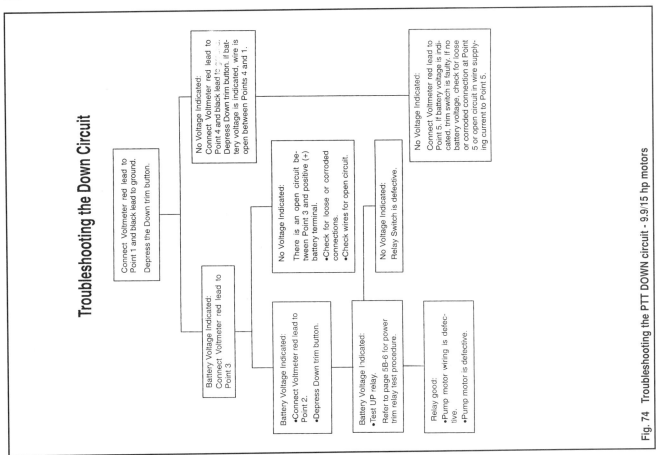

Fig. 74 Troubleshooting the PTT DOWN circuit - 9.9/15 hp motors

8-38 TRIM AND TILT

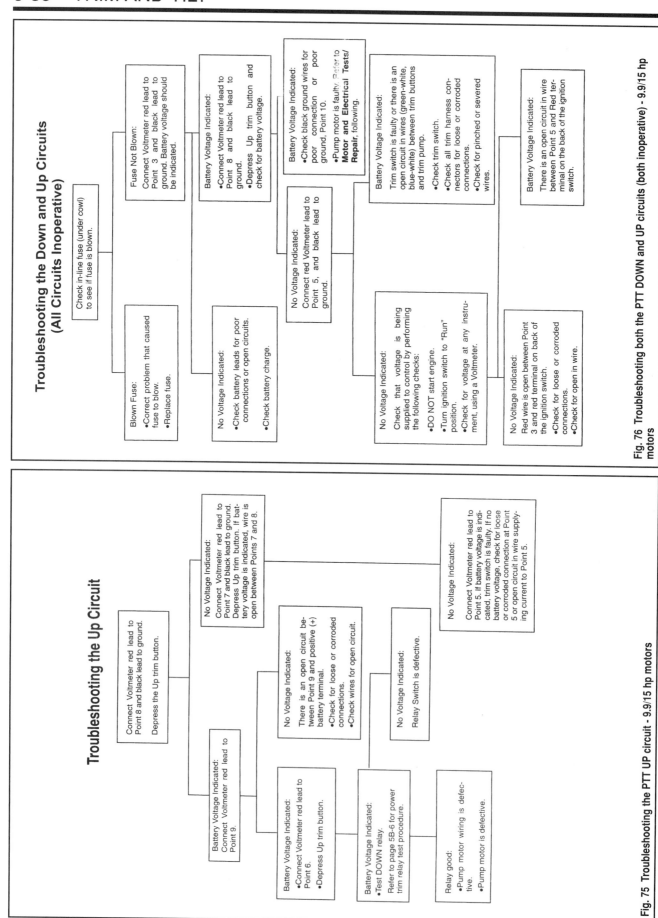

Fig. 76 Troubleshooting both the PTT DOWN and UP circuits (both inoperative) - 9.9/15 hp motors

Fig. 75 Troubleshooting the PTT UP circuit - 9.9/15 hp motors

TRIM AND TILT

25-115 Hp Motors - Symptom Diagnosis

◆ See Figures 77 and 78

Use the accompanying symptom charts to diagnose and repair problems with the PTT system used on 25-115 hp motors. Before starting, make be sure to visually inspect the system for leaks or damage. Also, make sure the oil level is correct.

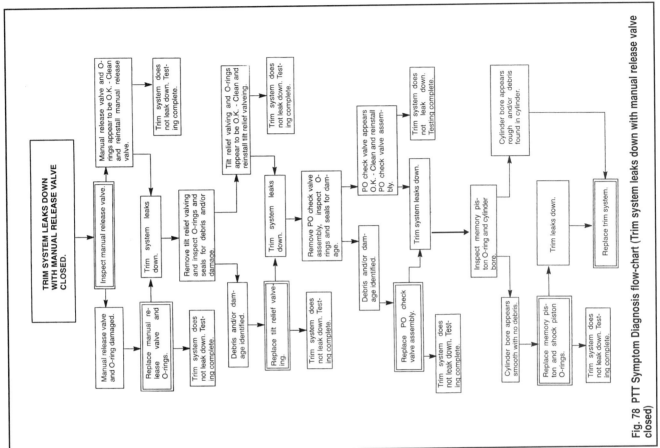

Fig. 78 PTT Symptom Diagnosis flow-chart (Trim system leaks down with manual release valve closed)

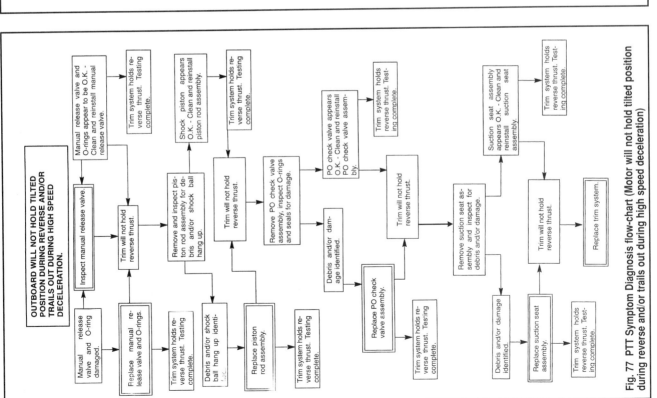

Fig. 77 PTT Symptom Diagnosis flow-chart (Motor will not hold tilted position during reverse and/or trails out during high speed deceleration)

8-40 TRIM AND TILT

25-115 Hp Motors - Wiring Diagram and Circuit Testing

◆ See Figures 79 thru 82

Use the accompanying wiring diagrams and troubleshooting charts to diagnose and repair electrical problems with the PTT systems on 25-115 hp motors.

Before starting any electrical troubleshooting, check the following:
- Check for disconnected wires
- Make sure all connections are clean (of corrosion) and tight
- Make sure the plug-in connectors are fully engaged
- Make sure the battery is fully charged

The numbers listed in the electrical troubleshooting chart corresponds to the numbered locations on the appropriate wiring diagrams.

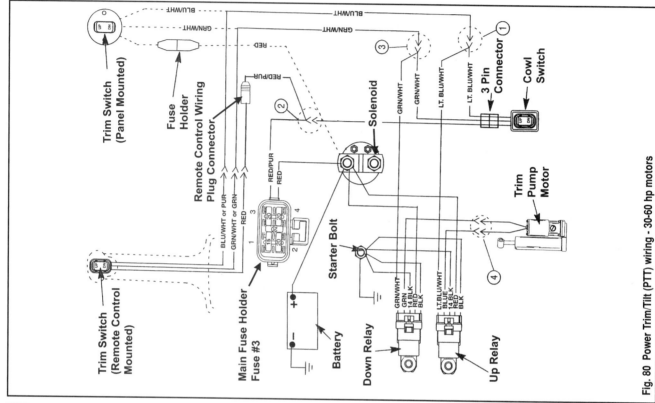

Fig. 80 Power Trim/Tilt (PTT) wiring - 30-60 hp motors

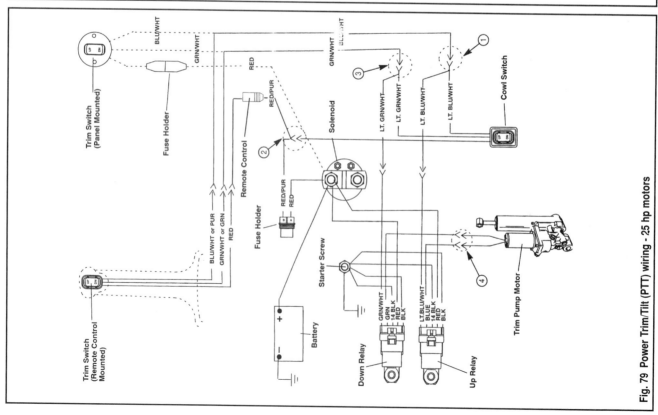

Fig. 79 Power Trim/Tilt (PTT) wiring - 25 hp motors

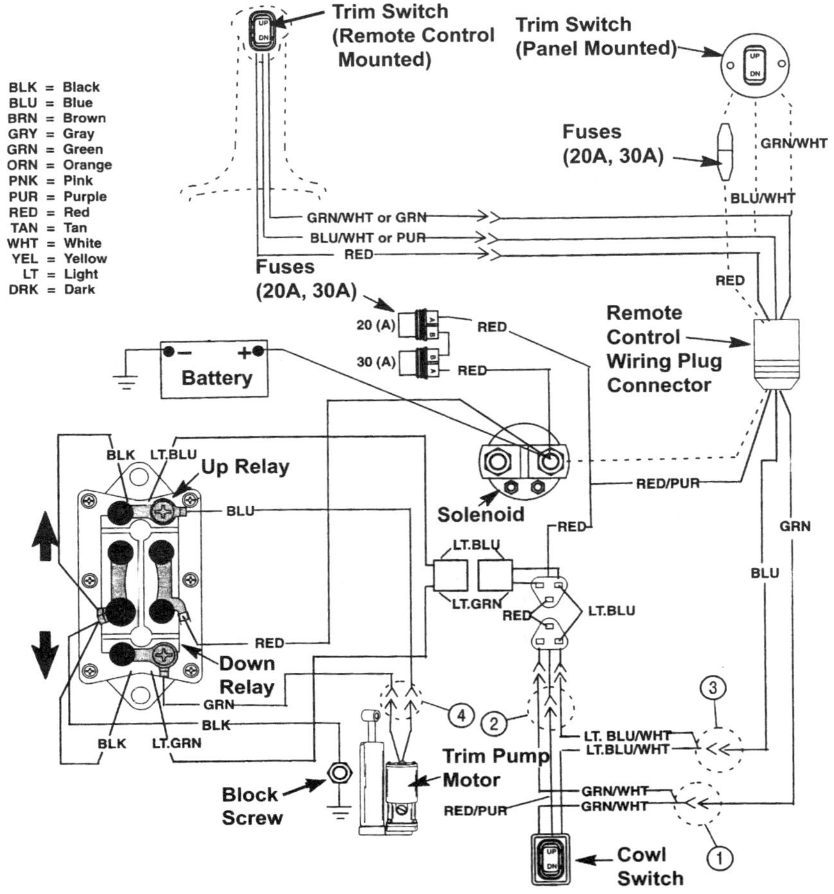

TRIM AND TILT

Problem	Possible Cause	Remedy
Trim Switch "UP" is inoperative, but the Cowl Switch "UP" does operate.	1. Open wire between Wire Connection (1) and Trim Switch. 2. Faulty Trim Switch.	1. Check for an open connection or cut wire. 2. Replace.
Cowl Switch "UP" is inoperative, but the Trim Switch "UP" does operate.	1. Open wire between Wire Connection (2) and Solenoid. 2. Faulty Cowl Switch.	1. Check for an open connection or cut wire. 2. Replace.
Trim Switch "UP" and Cowl Switch "UP" are both inoperative.	1. Open wire between Wire Connection (1) and the "Up" Relay. 2. Open BLK wire between ground and "UP" Relay. 3. Open RED wire between Solenoid and "UP" Relay. 4. Faulty "UP" Relay.	1. Check for an open connection. 2. Check for an open connection. 3. Check for an open connection. 4. Replace.
Trim Switch "DOWN" is inoperative, but the Cowl Switch "DOWN" does operate.	1. Open wire between Wire Connection (3) and Trim Switch. 2. Faulty Trim Switch.	1. Check for an open connection or cut wire. 2. Replace.
Cowl Switch "DOWN" is inoperative, but the Trim Switch "DOWN" does operate.	1. Open wire between Wire Connection (2) and Solenoid. 2. Faulty Cowl Switch.	1. Check for a open connection or cut wire. 2. Replace
Trim Switch "DOWN" and Cowl Switch "DOWN" are both inoperative.	1. Open wire between Wire Connection (3) and the "UP" Relay. 2. Open BLK wire between ground and "DOWN" Relay. 3. Open RED wire between Solenoid and "DOWN" Relay. 4. Faulty "DOWN" Relay	1. Check for an open connection. 2. Check for an open connection. 3. Check for an open connection. 4. Replace.
Trim Switch "UP" and "DOWN" are both inoperative, but the Cowl Switch does operate.	1. 20 AMP Fuse blown. 2. Faulty trim switch. 3. Wire is open between fuse holder and solenoid. 4. Wire is open between fuse holder and trim switch.	1. Replace fuse. Locate the cause of the blown fuse. Check electrical wiring for a shorted circuit. 2. Replace. 3. Check for an open connection or cut wire. 4. Check for a loose or corroded connection.
Trim Switch and Cowl Switch are both inoperative.	1. One of the Trim Pump Motor wires is open between the motor and the Relays. 2. Faulty trim pump motor.	1. Check wire connections (4) for loose or corroded condition. 2. If voltage is present at connections (4) when the appropriate trim button is pressed, then motor is faulty. Replace motor.
Trim system operates (motor runs) without pressing the switches.	1. The Trim or Cowl switch is shorted.	1. Replace.

Fig. 82 Electrical Troubleshooting PTT systems on 25-115 hp motors (use along with appropriate wiring diagram)

Fig. 81 Power Trim/Tilt (PTT) wiring - 75-115 hp motors

BLK = Black
BLU = Blue
BRN = Brown
GRY = Gray
GRN = Green
ORN = Orange
PNK = Pink
PUR = Purple
RED = Red
TAN = Tan
WHT = White
YEL = Yellow
LT = Light
DRK = Dark

8-42 TRIM AND TILT

PTT Assembly

REMOVAL & INSTALLATION

9.9/15 Hp Motors

◆ See Figures 83 and 84

■ On these models the pump and the trim cylinder are not mounted to a common bracket and are connected by external hydraulic lines. If an attempt it made to remove them as an assembly there is a good chance that the lines would be damaged by the flexing and force that would be applied to the connections. For this reason we recommend removing the pump and/or cylinder separately, as follows:

1. Raise the outboard unit to the full up position. If the hydraulic system is inoperative, first rotate the manual release valve counterclockwise about 3-4 complete turns and then manually lift the unit to the full up position.
2. Support the outboard in this position. The safest method is to use an engine hoist a support strap or chain attached to the lower unit.
3. To remove the tilt cylinder, proceed as follows:
   a. Remove the tilt stop strap retaining bolts and nuts, then remove the upper bracket.
   b. Remove the upper pivot pin from the bracket.
   c. Loosen, but DO NOT remove, the hydraulic lines where they connect to the trim/tilt cylinder, THEN disconnect the lines from the other end (from the PUMP).
   d. Remove the nuts securing the lower pivot pin.
   e. Use a punch to carefully drive/tap out the lower pivot pin.
   f. Remove the tilt cylinder from the outboard. If necessary remove the lines from the cylinder.
4. To remove the pump, proceed as follows:
   a. Tag and disconnect the trim motors wires from the solenoid harness.
   b. If not done already, loosen and disconnect the hydraulic lines from the pump.
   c. Support the pump and remove the 3 bolts securing the pump to the support bracket, then remove the pump.

■ No further overhaul, disassembly or service is possible on these models. If the pump or tilt cylinder is defective the component must be replaced.

To install:

5. If removed, install the pump, as follows:
   a. Position the pump to the bracket and loosely install the 3 bolts to support it, but do NOT tighten them at this time.
   b. If the tilt cylinder was not removed, reconnect the hydraulic lines to the pump, taking great care NOT to cross thread them. Finger-tighten the lines.

■ If the tilt cylinder is removed from the assembly, install that now, then come back to tighten the pump mounting bolts and hydraulic lines.

   c. Tighten the pump mounting bolts to 80 inch lbs. (9 Nm).
   d. Tighten the hydraulic lines at the pump.
   e. Reconnect the motor wires to the solenoid harness.
6. If removed, install the tilt cylinder, as follows:
   a. Loosely install, but DO NOT, tighten the hydraulic lines to the cylinder. Take care not to cross-thread them.
   b. Lubricate (using 2-4-C with Teflon or an equivalent lubricant), then insert the nylon bushing into the lower pivot bore.
   c. Insert the lower pivot pin into the support bracket, then install the spacer.
   d. Align the tilt cylinder with the lower pivot pin, then push the pin through the tilt cylinder to it is flush with the cylinder.
   e. Insert the remaining (lubricated) spacer and push the pivot pin through.
   f. Install the pivot nuts and tighten to 120 inch lbs. (13.6 Nm).
   g. Loosely install, but DO NOT, tighten the hydraulic lines to the pump. Take care not to cross-thread them.
   h. If the pump was removed, stop here, go back and tighten the pump retaining screws.
   i. Tighten all hydraulic lines.
   j. Install the upper pivot pin into the upper bracket.
   k. Guide the upper pivot through the trim cylinder rod eye and through the upper bracket.
   l. Install the upper bracket to the swivel bracket, then secure the bracket with the 2 shoulder bolts inserted through the tilt stop brackets.
   m. Install a washer and nylock nut on the shoulder bolt and tighten to 144 inch lbs. (16.3 Nm).
7. Properly refill and bleed the hydraulic system, as detailed in this section.
8. Check to be sure the battery has a full charge and then move the outboard unit through several cycles, full down to full up for trailering. Observe the system for satisfactory operation electrical and hydraulic. Check the system for possible leaks.

25-115 Hp Motors

◆ See Figures 85 and 86

1. Raise the outboard unit to the full up position. If the hydraulic system is inoperative, first rotate the manual release valve counterclockwise about 3-4 complete turns and then manually lift the unit to the full up position.
2. Set the tilt lock pin in place to support the outboard.
3. Tag and disconnect the PTT motor electrical wires (usually Blue and Green) and release any clip(s). Be sure to note the wire routing and the position of any clamp(s) for installation purposes.
4. Remove the upper tri-lobe pin (which is used to secure the upper pivot pin). You can usually grab the pin with a set of needle-nose pliers and pull downward to free it, however, on some models you may want to use a small punch to drive the pin downward and free of the pivot pin.

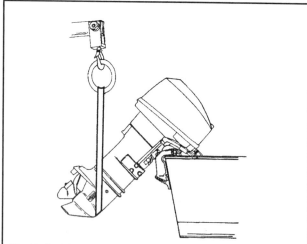

Fig. 83 On 9.9/15 hp motors, you must support the outboard during this procedure

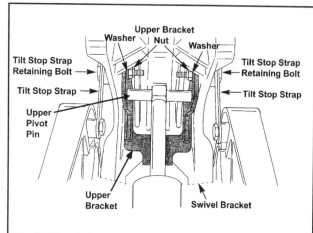

Fig. 84 Tilt cylinder upper mounting - 9.9/15 hp motors

TRIM AND TILT 8-43

5. Use a suitable punch to drive out the upper pivot pin freeing the top of the trim/tilt rod.
6. Remove the anode.
7. For 25 hp motors, remove the nuts and washers securing the lower pivot pin.
8. For 30-115 hp motors, using a punch drive out the lower tri-lobe pin from the lower pivot pin.
9. Use a punch to carefully drive/tap out the lower pivot pin. On 25 hp motors, retain the anchor pin bushing from the clamp bracket and trim unit.
10. Tilt the shock absorber assembly (top first) out from the clamp bracket and remove the assembly.

To install:

■ Be sure to lubricate the pivot pins and bores using 2-4-C with Teflon or an equivalent marine lubricant during assembly.

11. For 25 hp motors, lubricate and install the lower pivot pin bushings into the clamp brackets and trim unit.
12. Position the trim cylinder assembly between the clamp brackets. Tilt the bottom inward as you insert it into position.

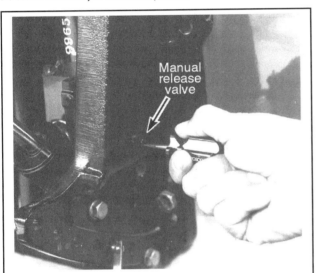

Fig. 85 If necessary, use the manual release valve so you can reposition the outboard

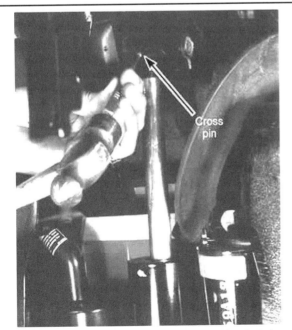

Fig. 86 Use a hammer and a soft punch to remove the pivot pin

■ On 75-115 hp motors be sure to route the pump wiring harness through the access hole in the starboard clamp bracket as the assembly is positioned.

13. Lubricate and install the lower pivot pin. On 30-115 hp motors you'll have to use a punch to drive the pin into the bracket until it is flush with the outside surface.
14. For 25 hp motors, install the nuts and washers to anchor the lower pivot pin, then tighten them to 18 ft. lbs. (24.5 Nm).
15. For 30-115 hp motors, use a small punch to drive the lower tri-lobe pin into position, securing the lower pivot pin.
16. Install the sacrificial anode. On some models one of the anode bolts is used to secure a ground strap, if applicable, be sure to position the strap between the anode and the bracket during installation.
17. Rotate the assembly so the shock rod hole is aligned for installation of the upper pivot pin.

■ Before installing the upper pivot pin on 75-115 hp motors, make sure the cross hole in the tilt rod eye is positioned facing downward toward the rear of the motor. If the trim ram is reversed (so the lower portion of the cross hole faces forward, the trim sender if equipped, will not operate.

18. Using a soft mallet, carefully drive the lubricated upper pivot pin into the swivel bracket and through the shock rod until the pivot pin is flush with the swivel bracket.
19. Drive the upper tri-lobe (dowel) pin into its hole until seated.
20. Route the pump wiring back into position and secure using the clamps, as noted during removal. Reconnect the wiring as tagged during removal.
21. Check the fluid level and top off or fill and bleed the system, as necessary. For details, please refer to Hydraulic Bleeding in this section.
22. Check to be sure the battery has a full charge and then move the outboard unit through several cycles, full down to full up for trailering. Observe the system for satisfactory operation electrical and hydraulic. Check the system for possible leaks.

HYDRAULIC SYSTEM BLEEDING

◆ See Figure 87

This trim/tilt system with one large cylinder has been designed with a "self bleeding" system. Fluid is pumped from one side of the cylinder to the opposite side as the piston extends or retracts. Therefore, any air trapped on one end of the cylinder, is pushed out back to the reservoir, when the piston bottoms out. Usually two or three cycles of the system will remove all entrapped air from the cylinder. However, if the components were replaced or the system was completely drained for some reason, it might take a LOT of cycling. Under those circumstances it might make more sense to open the manual release valve first and fill the system.

The following short section outlines the necessary steps to perform during the cycling sequence just mentioned. For most models (except the 9.9/15 hp motors where the system cannot be removed/installed as an assembly) the system may be bled on or off the motor.

Whenever possible use Quicksilver Power Trim and Steering Fluid to fill the system. The Mercury PTT fluid is a highly refined hydraulic fluid, containing a high detergent content and additives to keep seals pliable. A high grade automatic transmission fluid, Dexron® II, may also be used if the recommended fluid is not available.

1. Either install the system (all motors) or secure it in a soft-jawed vise (25-115 hp motors), whichever you prefer.
2. For 9.9/15 hp motors (or for all motors if the system was drained completely), open the manual release valve at least 3 turns.
3. Make sure the area around the fill plug is clean and free of any dirt or corrosion, then remove the fill plug and add fluid to the system until it is topped off, then reinstall the plug.
4. For 9.9/15 hp motors, cycle the system to fully raise and lower the unit several times, then check and top off the fluid as necessary.

■ For 25-115 hp motors the manufacturer does not specifically recommend initially bleeding an empty system by cycling the pump with the release valve still open, however it should speed up the process on those models as well.

8-44 TRIM AND TILT

5. Close the manual release valve.
6. Cycle the system to fully raise and lower the unit 3 times. If you're bleeding a unit from a 25-115 hp motor that is NOT installed on the outboard, manually connect the wires to a 12 volt power source (battery) as follows:
• To extend the trim ram (move upward) connect the BLUE lead to POSITIVE and the GREEN lead to NEGATIVE.
• To retract the trim ram (move downward) connect the GREEN lead to POSITIVE and the BLUE lead to NEGATIVE.
7. Fully extend the trim rod and recheck the fluid level. Top off the fluid, as necessary. If you needed to add a significant amount of fluid, repeat cycling the unit up and down another 3 times, then recheck the fluid level. Continue this pattern until the fluid remains at the proper level without topping it off again.

OVERHAUL (25-115 HP MOTORS ONLY)

◆ See Figures 88 thru 98

As with all rebuilds, be sure to remove and discard all O-rings during disassembly. Make sure the replacement O-rings are the proper size.
Removal of certain components requires special tools from Snap-On (#CG 41-11 and #CG 41-14) with the 5/16 in. end. HOWEVER, you can fabricate a suitable tool out of a length of 0.060 stainless steel rod. The final tool will have a 4 1/2 in. long shaft which leads to a 90° bend and a 3 1/4 in. long shaft that then ends in a 1/4 in. loop which faces back towards the original shaft.
1. Remove the PTT unit assembly from the outboard.
2. Remove the reservoir cap and the manual release valve. Tilt the unit slightly and drain the hydraulic fluid out the manual release valve bore.
3. Place it in a soft-jawed vise for access.
4. To remove the trim motor, proceed as follows:
    a. Loosen and remove the 4 Philips or Allen screws securing the trim motor to the assembly.
    b. Remove the motor and seal from the assembly. For 30-115 hp units the motor/reservoir comes off as one unit, then the coupler is also removed. The coupler contains valves and other components and may also be disassembled.
5. To disassemble the pump, coupler and manifold assemblies on 30-115 hp motors, proceed as follows:

■ To service the coupler, use the special tools from Snap-On (#CG 41-11 and #CG 41-14) with the 5/16 in. end to remove the spool (or fabricate one from a 0.060 stainless steel rod.). Also, be sure to inspect the poppet assembly for debris under the valve tip, near the rubber seat, and replace if any is found.

    a. Remove the plugs and components from each side of the housing. Behind each plug there should be (in this order) a spring, a check valve with a separate poppet and a valve seat. In addition, on one side, there is a spool under the valve seat. Keep all components sorted/arranged in order for assembly purposes.
    b. Remove the 3 screws that secure the pump to the coupler base, then remove the pump.
    c. Remove the filter and seal from under the pump.
    d. Remove the suction seat assembly from the bore in the base (seat, ring, ball and spring).
    e. Turn your attention to the manifold at the base of the cylinder. Remove the 2 Allen head screws (threaded horizontally) from the base of the cylinder in order to separate the manifold. Carefully remove the manifold.
    f. Remove tilt relief components from the side of the manifold, in the following order, the tilt limit spool, spool housing, poppet and springs.
6. To remove the pump and disassemble the manifold/check valve assemblies on 25 hp motors, proceed as follows:

■ On 25 hp models the pump itself IS NOT serviceable (although it can be replaced) make no attempt to disassemble it.

    a. Remove the 2 Allen head screws securing the oil pump to the base assembly. Remove the pump, O-ring and filter. Then remove the two springs and metal ball(s) from the housing.

■ Use the special tools from Snap-On (#CG 41-11 and #CG 41-14) with the 5/16 in. end (or fabricate one from a 0.060 stainless steel rod) to remove the pilot valve in the next step.

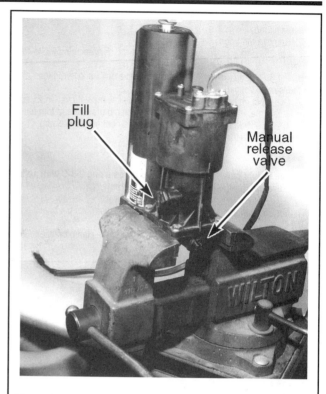

Fig. 87 Bench bleeding a PTT unit (the manual release valve may be opened and closed to expedite the process)

    b. Unscrew the plug from the side of the manifold/housing assembly (near the manual release valve) and remove the tilt relief valve components, in order. Behind the plug you'll find a shim (on most, but not all models), a spring, a poppet assembly, a pilot valve (with 2 O-rings) and an actuator pin.

■ Be sure to inspect each of the poppet valves for debris under the valve tip, near the rubber seat, and replace if any is found.

    c. At the base of the manifold assembly, facing downward, remove the plug and the suction seat assembly. Behind the plug you'll find a ball, the suction seat and a filter.
    d. On the front of rear of the manifold assembly remove the plugs for the pilot check valves. Behind each plug you'll find a spring and a poppet. Once they are removed, use the 0.060 wire or removal tool to push out the spool and one seat. Then from the opposite side, use a punch to push out the remaining seat.

■ The pilot check valve assembly components go through the manifold starting at either side, in this order, plug, spring, poppet, seat, spool, seat, poppet, spring and finally plug. In this way the assemblies are mirror images of each other with a spool in the middle.

7. To remove and disassemble the shock (trim/tilt) rod, proceed as follows:
    a. Using the special tool (#91-74951) or an equivalent spanner with two adjustable pins that insert into the rod (the wrench has 1/4 in. x 5/16 in. long pegs) unscrew and remove the tilt rod end cap assembly.
    b. Carefully lift the rod and end cap from the cylinder.

■ The ONLY serviceable items on the shock rod itself are the O-rings and wiper ring. If the shock rod requires ANY other repair, replace it as an assembly.

    c. Place the shock rod in a soft-jawed vise, then loosen and remove the 3 screws on the lower plate.
    d. Carefully lift the plate from the rod, keeping CLOSE track of the 5 balls, 5 seats and 5 springs retained under the plate.
    e. Remove and discard the O-ring from the lower plate.
    f. With the shock rod still (or back) in the vise, use a heat gun to GENTLY apply heat to loosen the piston (still at the bottom of the rod).

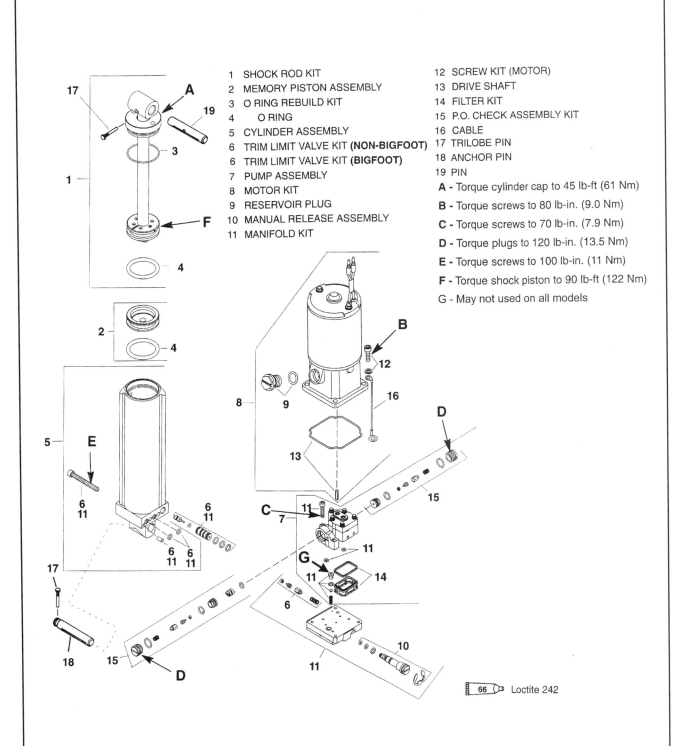

Fig. 88 Exploded view of the PTT assembly - 30-115 hp motors

8-46 TRIM AND TILT

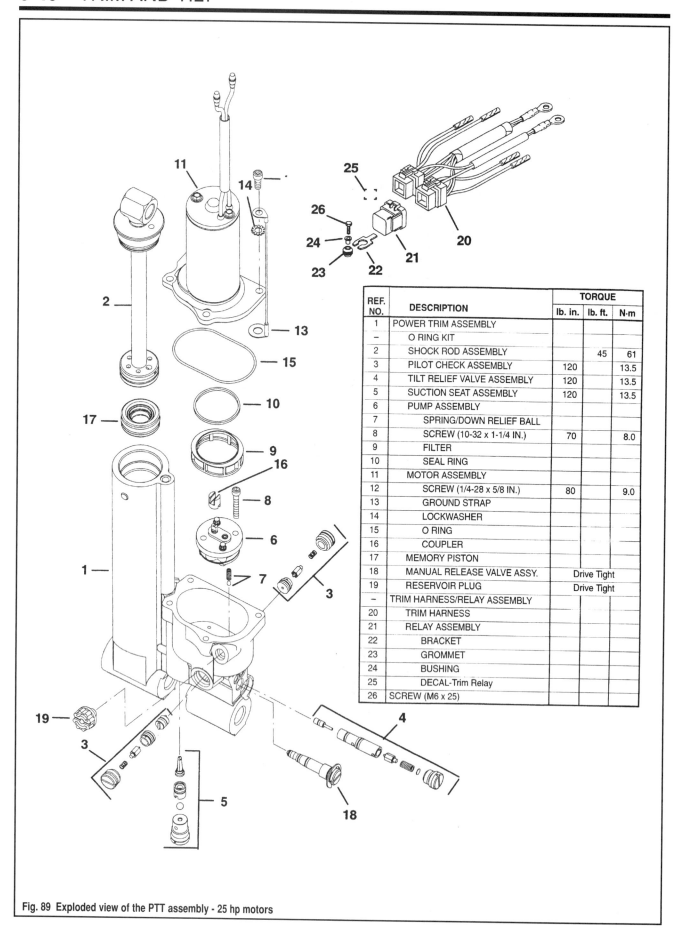

Fig. 89 Exploded view of the PTT assembly - 25 hp motors

REF. NO.	DESCRIPTION	TORQUE		
		lb. in.	lb. ft.	N·m
1	POWER TRIM ASSEMBLY			
–	O RING KIT			
2	SHOCK ROD ASSEMBLY		45	61
3	PILOT CHECK ASSEMBLY	120		13.5
4	TILT RELIEF VALVE ASSEMBLY	120		13.5
5	SUCTION SEAT ASSEMBLY	120		13.5
6	PUMP ASSEMBLY			
7	SPRING/DOWN RELIEF BALL			
8	SCREW (10-32 x 1-1/4 IN.)	70		8.0
9	FILTER			
10	SEAL RING			
11	MOTOR ASSEMBLY			
12	SCREW (1/4-28 x 5/8 IN.)	80		9.0
13	GROUND STRAP			
14	LOCKWASHER			
15	O RING			
16	COUPLER			
17	MEMORY PISTON			
18	MANUAL RELEASE VALVE ASSY.	Drive Tight		
19	RESERVOIR PLUG	Drive Tight		
–	TRIM HARNESS/RELAY ASSEMBLY			
20	TRIM HARNESS			
21	RELAY ASSEMBLY			
22	BRACKET			
23	GROMMET			
24	BUSHING			
25	DECAL-Trim Relay			
26	SCREW (M6 x 25)			

TRIM AND TILT 8-47

g. While the piston is hot, use the spanner wrench you used on the cap loosen the piston, but do not remove it completely. Allow the piston to cool and remove it from the shock rod.

h. Inspect the check valve in the piston for debris and clean, as necessary. IF the check valve cannot be cleaned, replace the PISTON as an assembly.

i. Clean the shock and components with compressed air (but DON'T loose the springs, seats and balls).

j. Remove the inner O-ring from the piston bore.

k. Remove the cylinder end cap assembly from the shock rod and inspect. If the wiper (located in the cap) has failed to keep the rod cleaned, replace the wiper.

l. Place the end cap on a clean work surface, then use a small pry tool to carefully remove the wiper from the top center of the cap. Remove the inner and outer O-rings and discard.

8. If the memory piston requires service, you can remove it from the bore either using a pair of lock-ring pliers to grab it and pull it upward OR by using compressed air blown into the center O-ring hole. When using the second method, point the cylinder bore facing downward toward a shop rag or towel to both prevent injury AND to help prevent damage. Expect some fluid to be expelled at the same time. Remove and discard the O-ring from the memory piston.

9. Clean and inspect all components which are to be reused using engine cleaner and low-pressure compressed air. DO NOT use cloth rags which may leave residue behind that could clog a valve.

10. Inspect all machined surfaces for burrs or scoring and repair or replace, as necessary.

To assemble:

11. Identify all of the O-rings for assembly purposes. Lubricate the O-rings using Quicksilver Power Trim and Steering Fluid (or, if not available, using automatic transmission fluid).

12. To assemble the shock (trim/tilt) rod, proceed as follows:

a. Install the lubricated O-rings to the end cap and to the piston.

b. Install a new rod wiper to the center of the cap.

c. Clamp the shock rod in a soft-jawed vise, then position the end cap onto the rod (with the O-ring facing the bottom end of the rod and the threads facing the top end of the rod).

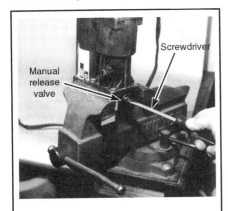

Fig. 90 Drain the hydraulic fluid through the release valve bore

Fig. 91 To remove the trim rod, loosen the cap with a spanner wrench...

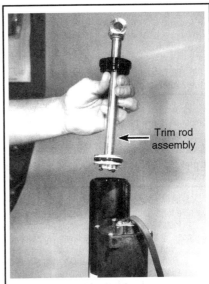

Fig. 92...then lift the trim rod from the cylinder

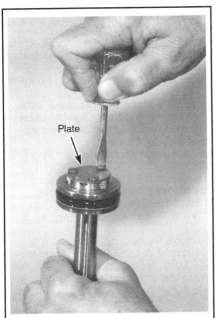

Fig. 93 Remove the plate, springs, seats and balls from the bottom of the piston

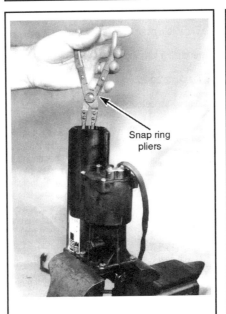

Fig. 94 Either grasp the memory piston with pliers...

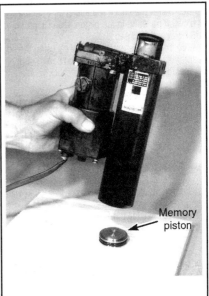

Fig. 95 Or invert it the unit and use compressed air to push it out

8-48 TRIM AND TILT

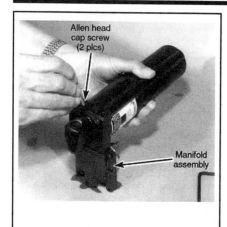

Fig. 96 To separate the manifold, remove the 2 Allen screws...

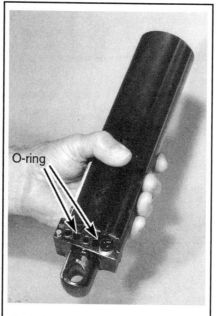

Fig. 97 Remove the old O-rings...

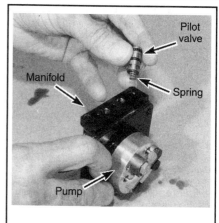

Fig. 98 ...and all the valve components from the manifold assembly

    d. Apply a light coating of Loctite® 271 or equivalent thread-locking compound to the threads on the shock rod, then install the shock rod piston.

✱✱ CAUTION

To prevent damage to the shock rod piston, MAKE SURE the spanner wrench has 1/4 in. x 5/16 in. (6.4mm x 8mm) long pegs.

    e. Using the spanner, tighten the piston securely. If an adapter is available for a torque wrench, tighten the piston to 90 ft. lbs. (122 Nm).
    f. Install the 5 sets of balls, seats and springs to the shock rod piston, then position the plate to secure them. Tighten the 3 plate screws to 35 inch lbs. (4.0 Nm).
    g. Remove the shock assembly from the vise.
  13. To install the shock (trim/tilt) rod to the cylinder, proceed as follows:
    a. Position the cylinder assembly in a soft-jawed vise.
    b. Install the lubricated O-ring to the memory piston and install the piston into the bore. Push the piston all the way to the bottom of the bore.
    c. Fill the cylinder to within 3 in. (76.2mm) from the top with PTT fluid.
    d. Install the shock rod assembly until the PTT fluid flows through the oil blow off passage in the piston, then fill the rest of the cylinder to a point JUST below the threads.
    e. Tighten the end cap securely using the spanner wrench. If a torquing adapter is available, tighten the end cap to 45 ft. lbs. (61 Nm).
  14. To assemble the manifold/check valves and install the pump on 25 hp motors, proceed as follows:
    a. Starting with the tilt relief valve assembly, install the new, lubricated O-rings onto the actuator pin and the pilot valve.
    b. Place the seat onto the pilot valve, then insert the assembly into the bore in the side of the manifold. Seat the valve using a 9/32 in. or 7mm socket in contact with the OUTER diameter of the pilot valve.
    c. Next, install the poppet, spring and shim (if equipped). Install the plug and tighten to 120 inch lbs. (13.5 Nm).
    d. Install the suction seat assembly to the bore which faces downward out the bottom of the manifold assembly. Start with installing the new, lubricated O-ring(s) for the seat components.
    e. Next, install the filter and suction seat using a 9/32 in. or 7mm socket in contact with the OUTER diameter of the suction seat.
    f. Install the ball and plug, tighten the plug to 120 inch lbs. (13.5 Nm).
    g. Now install the pilot check valve assembly, as usual with new, lubricated, O-rings. Install one of the seats into the manifold and push it into place using a 9/32 in. or 7mm socket in contact with the OUTER diameter of the seat. Once in position install the corresponding poppet, spring and plug.
    h. Move to the opposite side of the manifold assembly and position the spool and other seat, again using the socket to push the seat into place.
    i. Install the remaining poppet, spring and plug. Tighten both of the pilot check valve plugs to 120 inch lbs. (13.5 Nm).

    j. Finally, it is time to install the pump assembly. Position the relief ball and spring into the manifold.
    k. Make sure the O-rings are in position in the bottom of the pump, then place the square cut O-ring on the filter and position the filter over the pump.
    l. Install the pump to the manifold and secure using the 2 Allen head screws. Tighten the screws to 70 inch lbs. (8 Nm).
  15. To assemble the pump, coupler and manifold assemblies on 30-115 hp motors, proceed as follows:
    a. Starting with the trim limit assembly, install the new, lubricated O-rings on the spool housing and trim limit spool.
    b. Install the proper spring into the bore. If the spring is being replaced using a kit with multiple options the heavy spring is used on 75-115 hp motors, the medium spring is used on 40-60 hp Bigfoot models, and the light spring is used on standard thrust 30-60 hp models.
    c. After the spring, position the poppet, spool housing and trim limit spool.
    d. Now, lubricate the O-rings and install the manual release valve. IF the "E" clip was removed, be sure to install the clip over the valve before installing it into the manifold.
    e. Install the new lubricated O-rings and dowel pins to the base of the cylinder (at the mating surface for the manifold), then position the manifold to the cylinder.
    f. Install the manifold retaining screws (Allen head screws) and tighten to 100 inch lbs. (11 Nm).
    g. Now it's time to prepare the oil pump and coupler. Start by installing the spring, ball, lubricated O-ring and plastic seat to the top of the coupler.
    h. Make sure the O-rings are in position in the bottom of the pump, then install the filter and filter seal UNDER the pump.
    i. Position the pump and filter assembly to the coupler and tighten the 3 retaining screws to 70 inch lbs. (8 Nm).
    j. Lastly, install the valve assemblies into either side of the pump housing. Install the spool, followed by the seat (with lubricated O-ring), poppet, check valve, spring and plug (also with lubricated O-ring). Repeat on the other side starting with the seat. Tighten the check valve plugs to 120 inch lbs. (13.5 Nm).
  16. If removed, install the coupler into the top of the pump.
  17. Position a new seal, then install the motor, aligning the motor-to-pump coupler.
  18. Install the motor retaining bolts (making sure the ground wire is in position under one of the bolts. Tighten the motor retaining bolts to 80 inch lbs. (9 Nm).
  19. Install and Bleed the PTT assembly as detailed in both procedures located earlier in this section.

9 REMOTE CONTROLS

MERCURY/MARINER REMOTE CONTROLS 9-12
  SIDE MOUNT REMOTE CONTROL BOX 9-12
    ASSEMBLY 9-15
    CLEANING & INSPECTING 9-15
    DESCRIPTION & OPERATION 9-12
    DISASSEMBLY 9-14
    REMOVAL & INSTALLATION 9-12
    TROUBLESHOOTING 9-12
TILLER HANDLE 9-16
  DESCRIPTION & OPERATION 9-16
  TILLER HANDLE 9-16
    REMOVAL & INSTALLATION 9-16
  TROUBLESHOOTING 9-16
YAMAHA REMOTE CONTROLS 9-2
  CHOKE SWITCH 9-2
    TESTING 9-2
  CONTROL BOX ASSEMBLY 9-4
    ASSEMBLY & INSTALLATION 9-9
    CLEANING & INSPECTION 9-7
    REMOVAL & DISASSEMBLY 9-4
  KILL SWITCH 9-3
    TESTING 9-3
  NEUTRAL SAFETY SWITCH 9-3
    TESTING 9-3
  REMOTE CONTROL BOX 9-2
    DESCRIPTION & OPERATION 9-2
  START BUTTON 9-3
    TESTING 9-3
  WARNING BUZZER/HORN 9-2
    TESTING 9-2
WIRING DIAGRAMS - MERCURY/MARINER 9-21
  TYPICAL COMMANDER REMOTE (MANUAL START) 9-21
  TYPICAL COMMANDER REMOTE (ELECTRIC START) 9-22
  TYPICAL COMMANDER 2000 REMOTE (ELECTRIC START) 9-22
  VIEW OF A TYPICAL COMMANDER 3000 PANEL MOUNT CONTROL 9-23
  TYPICAL COMMANDER 3000 PANEL MOUNT CONTROL 9-23
  TYPICAL INSTRUMENT/STOP SWITCH - MID-RANGE MOTORS 9-24
  TYPICAL INSTRUMENT/STOP SWITCH - MID-RANGE MOTORS/DUAL-OUTBOARD
  INSTALLATION 9-25
  WIRING DIAGRAM AND CONTINUITY TEST CHART FOR
  COMMANDER 2000 KEYSWITCHES 9-26
  MECHANICAL PANEL CONTROL (MPC) 4000 (CARB. MID-RANGE OR LARGER MOTORS) 9-26
  TYPICAL SMARTCRAFT (CAN) INSTALLATION (EFI MID-RANGE 30-60 HP MOTORS) 9-27
  TYPICAL GAUGE INSTALLATION (75 HP AND LARGER MOTORS WITH OR
  W/O TACH CONVERTER) 9-28
WIRING DIAGRAMS - YAMAHA 9-20
  TYPICAL DIGITAL METER ASSEMBLY 9-20
  TYPICAL REMOTE CONTROL BOX 9-20
  TYPICAL REMOTE HARNESS ASSEMBLY 9-21

YAMAHA REMOTE CONTROLS 9-2
MERCURY/MARINER REMOTE
CONTROLS 9-12
TILLER HANDLE 9-16
WIRING DIAGRAMS - YAMAHA 9-20
WIRING DIAGRAMS -
MERCURY/MARINER 9-21

9-2 REMOTE CONTROLS

YAMAHA REMOTE CONTROLS

Remote Control Box

DESCRIPTION & OPERATION

◆ See Figures 1 and 2

■ The type of remote control mated to the outboard is determined by the boat manufacturer or the rigging dealer. Because there are such a range of remote control units that could be mated to Yamaha outboards, there is no way to cover them all in detail here. This section deals with a typical Yamaha remote control unit. Illustrations are of the most common example of this unit. Although the procedures apply to most Yamaha remotes and many aftermarket units, care and observation must be used when working on any unit as differences may exist in assemblies. When a component or mounting differs from the components covered here, use common sense, or contact the manufacturer of the unit for more information.

The remote control unit allows the helmsperson to control throttle operation and shift movements from a location other than where the outboard unit is mounted.

In most cases, the remote control box is mounted approximately halfway forward (midship) on the starboard side of the boat or on the starboard side of the center console (depending upon the style/type of boat).

The Yamaha control unit normally houses a key switch, a choke switch, a kill switch, a neutral safety switch, a warning horn, and the necessary wiring and cable hardware to connect the control box to the outboard unit.

A safety feature is incorporated in the unit. The control arm can be shifted out of the **NEUTRAL** position if and only if the neutral lever is squeezed into the control arm. This feature prevents the arm from being accidentally moved from Neutral into either forward or reverse gear. Unintentional movement of the shift lever could be dangerous, resulting in personal injury to the operator, passengers, or the boat.

Starting from the upright **NEUTRAL** position, when the control arm is moved forward to about 30° from the vertical position, the unit shifts into forward gear. At this point the throttle plate is fully closed. As the control arm is moved past the 30° position, forward and downward, the throttle will open, until the wide open position, approximately 90° from the vertical position, is reached.

To shift into reverse gear, the control arm is first returned to the full upright position (Neutral) momentarily. From the upright position, the control arm is moved aft about 30°, and the unit shifts in reverse gear. At this point, the throttle plate is fully closed. If the arm is moved further aft and downward, the throttle will be opened until the wide open position is reached at about 60°.

The remote control unit is equipped with a free acceleration lever. This lever can be moved up to open the throttle and down to close the throttle. This lever is utilized only during powerhead startup and when the control arm is in the full upright (Neutral) position. When the free acceleration lever is not in the full down, idling position, the control lever cannot be moved from the **NEUTRAL** position.

Choke Switch

TESTING

◆ See Figure 3
MODERATE

■ On MOST late model control boxes, the choke switch is integrated into the main switch for easier operation. On these, the choke circuit operates when the main starter switch is depressed as it is rotated to start.

Some early-model control boxes utilize a separate choke switch. On these models the choke switch is a spring loaded toggle type switch located in the forward side of the control box. The control box must be opened to gain access to the switch leads. To check the switch, proceed as follows:

1. Disconnect the leads (normally Blue and Yellow, but check the Wiring Diagrams to be sure) from the choke switch at the nearest quick disconnect fitting.
2. Select the 1000 ohm scale on the meter. Make contact with the meter leads, one to each of the disconnected leads.
3. In the normal **OFF** position the meter should indicate no continuity. Move the switch to the **ON** position. The meter should indicate continuity.
4. If the switch fails either of these two tests, it must be replaced. The switch cannot be serviced or adjusted.

Warning Buzzer/Horn

TESTING

EASY

The buzzer is a warning device to indicate low oil pressure on 4-stroke models, an over rev. condition, or overheating of the powerhead. The buzzer is normally located inside the control box on remote control powerheads.

1. Remove the control box cover and identify the two leads from the buzzer, one is normally Yellow and the other is normally Pink, but check the Wiring Diagrams to be sure.
2. Disconnect these two leads at their quick disconnect fittings, and ease the buzzer out from between the four posts which anchor it in place.
3. Obtain a 12-volt battery. Connect the Yellow buzzer lead to the negative battery terminal. Momentarily make contact with the Pink buzzer lead to the positive battery terminal.
4. As soon as the Pink buzzer lead makes contact with the positive battery terminal, the buzzer should sound. If the buzzer is silent, or the sound emitted does not capture the helmsperson's attention immediately, the buzzer should be replaced. Service or adjustment is not possible.
5. If the sound is satisfactory and immediate, install the buzzer between the four posts and connect the two leads matching color to color. Tuck the leads to prevent them from making contact with any moving parts inside the control box. Replace the cover.

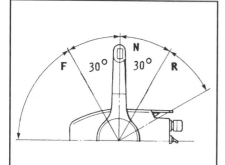

Fig. 1 Shift lever positioning for the remote control unit

Fig. 2 A typical Yamaha remote control box

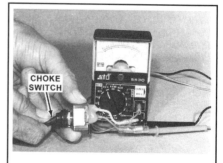

Fig. 3 Using an ohmmeter to test the choke switch

REMOTE CONTROLS 9-3

Neutral Safety Switch

TESTING

◆ See Figures 4 and 5

※※ CAUTION

Remember this is a safety switch. A faulty switch may allow the powerhead to be started with the lower unit in gear - an extremely dangerous situation for the boat and persons aboard. And just imagine what might happen if someone was in the water nearby and a kid turned the switch?

1. Trace the neutral safety switch leads from the switch to their nearest quick disconnect fitting. Both of these leads are usually Brown, but may vary for different models. Refer to the wiring diagram in the Ignition and Electrical System section for the proper color identification.

2. Disconnect the two leads and connect an ohmmeter across the two disconnected leads. When the shift lever is in the **NEUTRAL** position, the meter should register continuity.

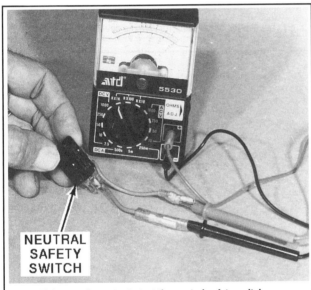

Fig. 4 Using an ohmmeter to test the neutral safety switch

Fig. 5 The neutral safety switch is sometimes hidden, but is usually located on the axis of the shift lever just inside the lower cowling pan

3. When the lower unit is shifted to either forward or reverse, the meter should register no continuity.

4. The switch must pass all three tests to indicate the safety switch is functioning properly. If the switch fails any one of the tests, the switch must be replaced.

Start Button

TESTING

◆ See Figures 6 and 7

1. Trace the start button harness containing two wires from the switch to their nearest quick disconnect fitting. The colors may vary for different models. Refer to the Wiring Diagrams in the Ignition and Electrical System section to confirm proper color identification.

2. Disconnect the two leads and connect an ohmmeter across the disconnected leads. Depress the start button. The meter should register continuity. Release the button and the meter should now register no continuity.

3. Both tests must be successful. If the tests are not successful, the start button must be replaced. The start button is a one piece sealed unit and cannot be serviced.

Kill Switch

TESTING

1. Trace the kill switch button harness containing two wires from the switch to their nearest quick disconnect fitting. The colors may vary for different models, but the hot wire is usually White/Black. Refer to the Wiring Diagrams in the Ignition and Electrical System section to confirm proper color identification.

Disconnect the two leads and connect an ohmmeter across the disconnected leads. Verify the emergency tether is in place behind the kill switch button. Select the 1000 ohm scale on the meter.

2. Depress the kill button. The meter should register continuity. Release the button. The meter should now register no continuity.

3. Both tests must be successful. If the switch fails either test, the switch is defective and must be replaced. The switch is a one piece sealed unit and cannot be serviced.

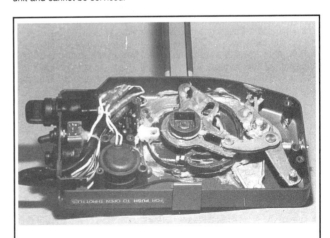

Fig. 6 The hardest part of control box service is stuffing the electrical leads back into the box so they are not damaged

9-4 REMOTE CONTROLS

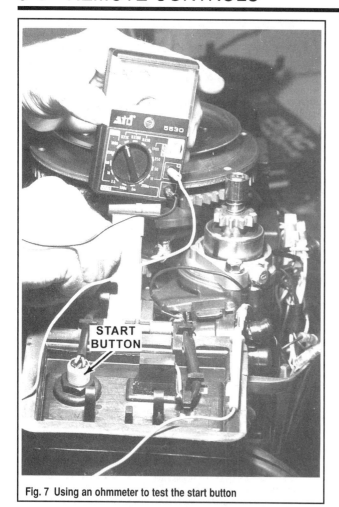

Fig. 7 Using an ohmmeter to test the start button

Fig. 8 Loosen the screws and remove the cover...

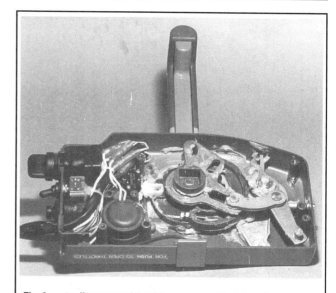

Fig. 9 ...to disconnect the cables or access the internal components

Control Box Assembly

REMOVAL & DISASSEMBLY

◆ See Figures 8 thru 18   ③ ◁DIFFICULT

1. Removal of the control box is accomplished by first disconnecting the throttle and shift cables at one end, and then disconnecting the main wiring harness at the quick disconnect fitting. Next, the mounting screws (usually 3) securing the control box to the boat are removed.

■ Disconnect the cables from the control box usually involves following the next 2 steps. So if you're going to remove the cable from the box end (as opposed to the powerhead end), then perform those steps before fully removing the box from the boat.

2. Move the control arm to the full upright (Neutral) position. Remove the Phillips head screws (usually 5) securing the upper and lower parts of the back plate.
3. Pry off the circlip retaining the throttle cable end to the throttle arm and the circlip retaining the shift cable end to the shift arm. Take care not to lose these two small circlips. Lift off the cable ends from the arms and be careful not to alter their length at the turnbuckles. Remove both cable ends from the box.

■ One of the most difficult tasks during assembling of the control unit is attempting to return all the wires back into their original positions.

Some wires are tucked neatly under switches, others are routed into neat bundles secured with plastic retainers, some are looped and double back, but believe-it-or-not, with patience and some good words, they all fit into one side of the box away from moving parts.

■ Therefore, it would be most advantageous to take a digital or Polaroid picture of the unit as an aid during assembling.

4. Remove the retaining nuts on the main key switch, the electric choke switch, and the kill switch (as equipped) on the side of the control box. Ease these three switches out of their grommets and holes. Disconnect the following leads (wire colors may vary slightly, refer to the Wiring Diagrams for more details) at their quick disconnect fittings:
   • The wire colors Pink and Yellow leads to the neutral safety switch (if mounted in the housing).
   • For PTT models, the Yellow, Green, and Black leads encased in a small harness leading to the up/down button on the upper control arm.
5. The entire harness, switches, and horn may now be lifted free of the control box, as shown in the accompanying illustration. The horn is normally not secured with hardware, but is simply retained between four bosses.
6. Lift out the throttle arm, consisting of three pieces hinged together. A plastic bushing will probably remain on the underneath side of the throttle arm. This bushing was indexed with the throttle friction bands.

REMOTE CONTROLS 9-5

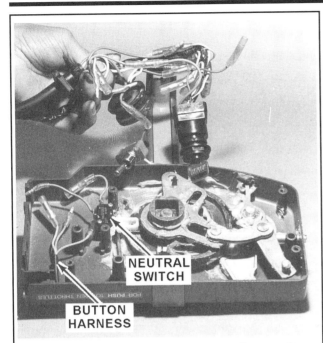

Fig. 10 Tag and disconnect the wiring, then carefully remove the harness

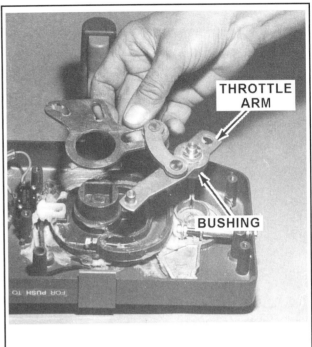

Fig. 11 Lift out the hinged throttle arm assembly (and bushing)

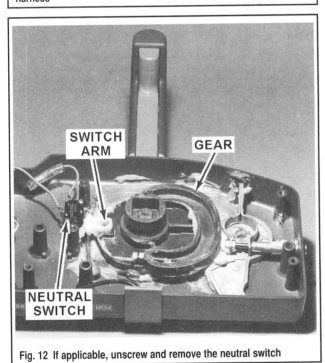

Fig. 12 If applicable, unscrew and remove the neutral switch

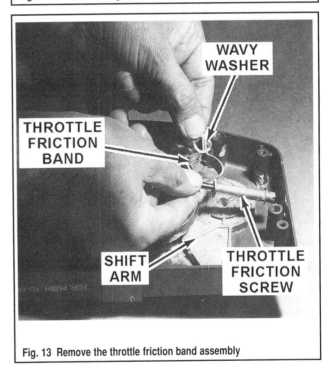

Fig. 13 Remove the throttle friction band assembly

    7. If mounted in the housing, remove the Phillips head screws (usually 2) securing the neutral switch, and then remove the switch. Remove the Phillips head screw and switch arm from the gear.

    8. Lift the two halves of the throttle friction band up and out of the control box. The throttle friction screw, with the circlip attached, will come away with the band.

    9. Lift out the small wavy washer from the center of the shift arm.

    10. Remove the Phillips screw securing the spring retainer, and then remove the spring retainer. Grasp one end of the leaf spring pack with a pair of needle nose pliers. Pull up on the pack and at the same time allow the pack to straighten to release tension on the springs. Lift out the detent roller from under the gear.

    11. Loosen the bolt in the center of the gear 1/2 turn only. After the center bolt has been loosened, lightly tap on the bolt to free it from the gear. Now, remove the bolt.

    12. Turn the control box over. Three things are now to be performed, almost simultaneously. Support the gear - now underneath - with one hand and at the same time unsnap the bottom harness cover from the front plate, as the control arm is lifted from the box. Remove the attaching screws, and then the **NEUTRAL** position plate.

    13. Support the gear and at the same time turn the control box over, and then lift the gear free of the box.

    14. Lift off the shift arm, a large flat washer, and two bushings from the front plate. This step concludes the disassembly of the inside of the front plate.

9-6 REMOTE CONTROLS

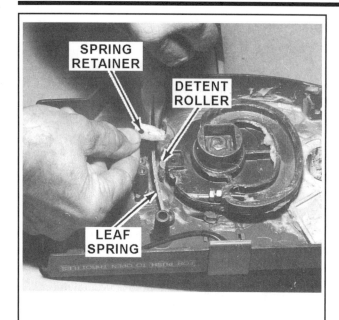

Fig. 14 Remove the spring retainer, release spring tension and free the detent roller

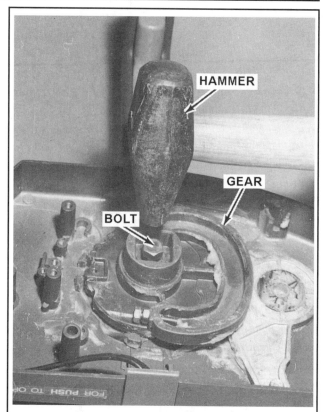

Fig. 15 After loosening it slightly, tap the gear center bolt to free the gear

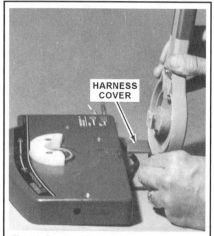

Fig. 16 Remove the control arm and neutral plate from the box, while supporting the gear underneath

Fig. 17 Turn the box over and remove the gear

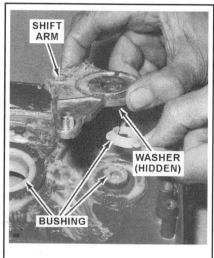

Fig. 18 Remove the shift arm and bushings

Acceleration Lever

◆ See Figures 19 thru 25

1. Place the upper back plate on the work bench with the lever facing down. Remove the Phillips screw and lift off the detent roller retainer.
2. Pry out the detent roller from the free acceleration disc.
3. Turn the back plate over and remove the Phillips screws (usually 2) securing the free acceleration lever to the disc. Lift off the lever and the wavy washer under the lever.
4. Turn the back plate over again, and then lift off the free acceleration disc and another wavy washer.

5. Remove the **NEUTRAL** position lever retainer and slide the lever from the arm. Remove the small spring between the top of the lever and the top of the arm.
6. Remove the small Phillips screws (usually 2) securing the handle to the control arm.
7. Pry up on the front disc and separate the disc from the control lever arm. Unthread the harness, to the up/down button, from the cavity of the arm.
8. Grasp the handle in one hand and the shaft of the control arm in the other. Gently slide them apart about 2 inches (5cm) or until the trim/tilt button with the harness attached can be removed from the handle.

REMOTE CONTROLS 9-7

Fig. 19 Remove the detent roller retainer...

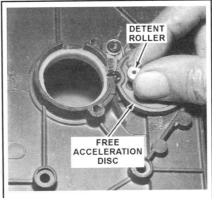

Fig. 20 ...then carefully pry the roller free of the acceleration disc

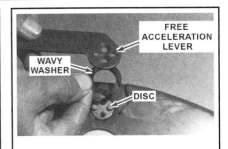

Fig. 21 Remove the screws and the free acceleration lever (with washer) from the disc

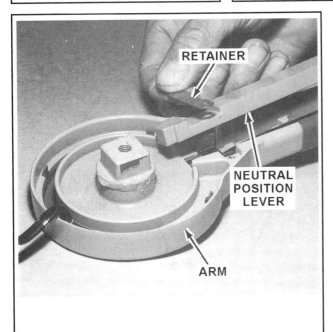

Fig. 22 Remove the retainer and slide the neutral position lever from the arm

CLEANING & INSPECTION

◆ See Figures 26 and 27

Clean all metal parts with solvent, and then blow them dry with compressed air.

Never allow nylon bushings, plastic washers, nylon retainers, wiring harness retainers, and the like, to remain submerged in solvent more than just a few moments. The solvent will cause these type parts to expand slightly. They are already considered a tight fit and even the slightest amount of expansion would make them very difficult to install. If force is used, the part is most likely to be distorted.

Inspect the control housing plastic case for cracks or other damage that would allow moisture to enter and cause problems with the mechanism.

Carefully check the teeth on the gear and shift arm for signs of wear. Inspect all ball bearings for nicks or grooves which would cause them to bind and fail to move freely.

Closely inspect the condition of all wires and their protective insulation. Look for exposed wires caused by the insulation rubbing on a moving part, cuts and nicks in the insulation and severe kinking which could cause internal breakage of the wires.

Inspect the bosses on both ends of the leaf spring. If either end shows signs of failure, the front plate must be replaced.

Inspect the edges of the cut-out in the **NEUTRAL** position plate. Replace this plate if the corners of the cut-out show any sign of rounding. If rounded, a slight pressure on the **NEUTRAL** position lever could throw the lower unit into gear. Check and double check all components of the **NEUTRAL** position

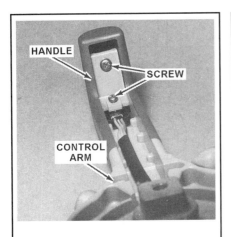

Fig. 23 Remove the screws securing the handle to the control arm

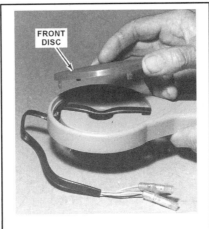

Fig. 24 Carefully pry the front disc free (for access to the PTT switch wiring)...

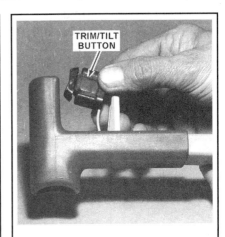

Fig. 25 ...then carefully free the switch from the handle

9-8 REMOTE CONTROLS

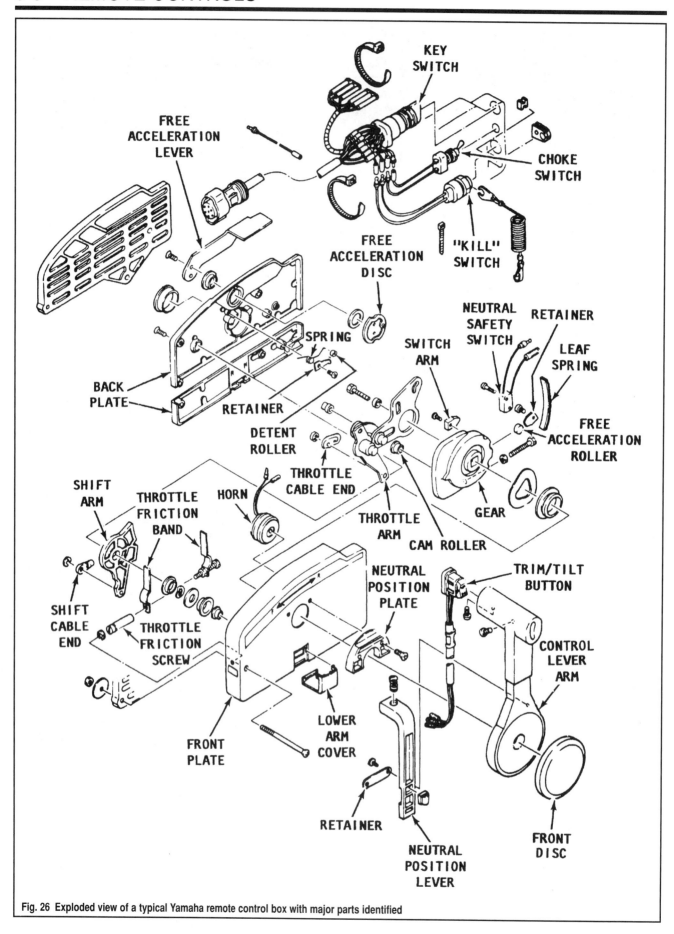

Fig. 26 Exploded view of a typical Yamaha remote control box with major parts identified

REMOTE CONTROLS 9-9

system from the spring at the top of the lever down to the extension at the bottom of the lever which indexes into the cut-out of the **NEUTRAL** position plate.

ASSEMBLY & INSTALLATION

 MODERATE

The following procedures provide complete detailed instructions to assemble all parts of the remote control unit. If certain areas were not disturbed during disassembling, simply bypass the steps involved and proceed with the work.

Again, remember this is based on a common Yamaha remote control unit, but not all remotes (Yamaha or not) will have the exact same construction and internal components.

Trim/Tilt Button and Harness

◆ See Figures 28 thru 31

1. Feed the harness wires for the power tilt/trim button down from the top in between the handle and the arm. Push the face of the button into the handle and observe the button from the operator's position. The word **UP** should be up and the **DN** should be down. Slide the control lever arm up into the handle.

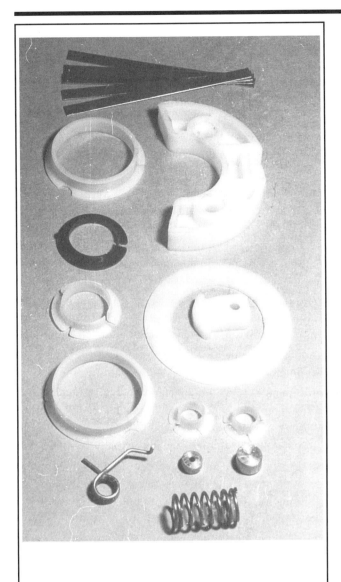

Fig. 27 Just some of the small metal and nylon parts from the interior of a remote control box

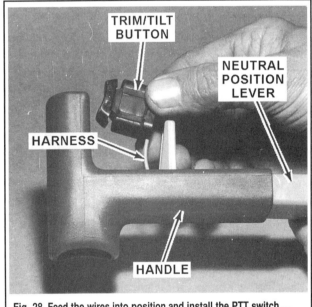

Fig. 28 Feed the wires into position and install the PTT switch...

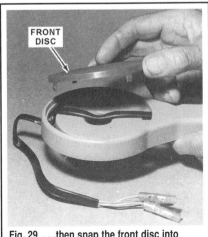

Fig. 29 ...then snap the front disc into place over the wires

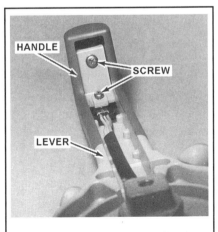

Fig. 30 Secure the arm to the handle using the screws...

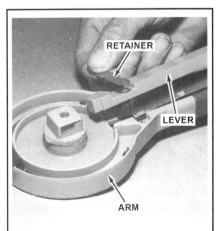

Fig. 31 ...then install the neutral position lever and retainer

9-10 REMOTE CONTROLS

2. Continue to feed the harness wires along the length of the arm - in and out of the cavity of the arm. Check to be sure the harness lies flat with no kinks. A kink in the wire will damage the harness over a period of time. Snap the front disc over the cavity.

3. Secure the arm to the handle with the Phillips head screws.

4. Hold the return spring in place in the recess at the top of the **NEUTRAL** position lever and at the same time slide the lever along the length of the arm up into the handle. Install the lever retainer across the lever and secure it to the arm with the small Phillips head screws.

Free Acceleration Lever

◆ See Figures 32 thru 46

1. Apply just a dab of Yamalube or an equivalent marine grade grease to the wavy washer and the free acceleration disc. Place the wavy washer and the free acceleration disc over the smaller hole in the inside of the top back plate. The single large post on the disc must face away from the notch cut in the plate for the free acceleration arm.

2. Hold the washer and disc (the Yamalube will help), in place - from falling away - and at the same time, turn the plate over on the work bench.

3. Place another wavy washer and the free acceleration arm over the installed disc.

■ The arm can only be installed one way because of the groove cut into the face of the upper back plate to accommodate the arm. If the free acceleration disc has been installed correctly in the previous step, the two beveled holes in the arm will align with the two threaded holes in the disc. Also, the two round holes in the arm will index over the two posts on the disc.

4. Install and tighten the two Phillips head screws. Turn the plate over again with the inside facing upward.

5. Install the upturned portion of the spring between the installed disc and the center boss, as shown. Hook the coiled part of the spring over the threaded post and at the same time direct the free end of the spring to anchor between the threaded and plain posts.

6. The detent roller must be positioned over the upturned end of the spring and be seated into the curved notch of the free acceleration disc. A small screwdriver inserted along the spring under the disc may aid in pushing the end into place to allow the roller to drop down over the spring end and against the disc.

7. Install the retainer over the detent roller. Secure the roller in place with the Phillips head screw threaded into the post with the spring around it.

8. Place the small bushing onto the shift arm boss on the inside back plate. Next, place the large bushing onto the same boss. The small and large bushings must be installed face-to-face. After both bushings are in position, slide the large flat washer onto the boss. Now, install the shift arm over the flat washer with the post for the shift cable facing upward - pointed toward the lower part of the case.

9. Lower the shift gear into the case with the post indexing into the notch in the shift gear. Support the shift gear inside the cover and turn the front plate over.

10. Place the **NEUTRAL** position plate onto the front plate and secure it in place with the two Phillips head screws. Position the control arm over the front plate with the **NEUTRAL** position lever indexed into the cutout of the **NEUTRAL** position plate.

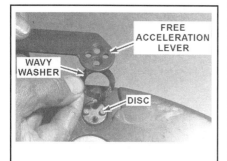

Fig. 32 Install the free acceleration lever to the disc using a new wavy washer

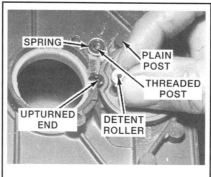

Fig. 33 Install the spring and detent roller...

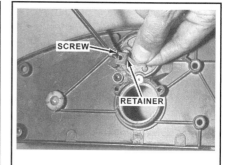

Fig. 34 ...then secure using the detent roller retainer

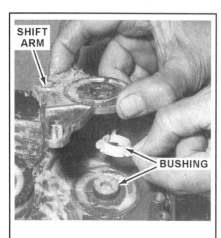

Fig. 35 Install the shift arm and bushings...

Fig. 36 ...then position the shift gear and support...

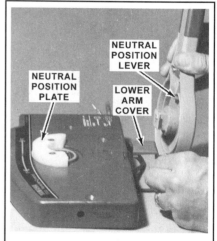

Fig. 37 ...while you install the neutral plate and the control arm

REMOTE CONTROLS 9-11

✷✷ WARNING

If the NEUTRAL position lever is not seated properly, a slight pressure on the handle could shift the lower unit into gear - forward or reverse. Such action could cause serious injury to crew, passengers, or the boat.

11. Guide the wire harness into the groove below the arm and snap on the lower arm cover to secure the harness in place.

12. Support the control arm and turn the front plate over. Install the washer and bolt into the square recess of the gear. Tighten the bolt to 5.8 ft. lbs. (8 Nm).

13. Slide the free acceleration roller into the center left notch in the shift gear. Grasp the pack of leaf springs with a pair of needle nose pliers. Insert one end of the pack into the lower boss. Bend the pack around the roller. Guide the other end into the upper boss. Install the small metal leaf spring retainer onto the threaded post to the left of the leaf spring. Secure the retainer in place with a Phillips head screw.

14. Install the throttle friction band over the shift arm. Move the two ends of the band into the boss on the front plate. Center the opening of the band squarely over the shift arm. Insert a wavy washer into the center of the band on the shift arm.

15. Install the neutral safety switch with the tab on the switch facing toward the shift gear. Secure the switch in place with the two Phillips head screws. Position the white plastic switch arm into the slot provided on the shift gear. Install and tighten the Phillips head screw to secure the arm to the gear.

16. Install the throttle arm. Try not to disturb the wavy washer in the center of the throttle friction band when the bushing on the underneath side of the throttle arm indexes with the band.

■ If you think some of the other assembling procedures were tricky on this unit, stand by! This next one will put you to the test.

17. Connect the neutral safety switch leads, usually Pink and Yellow (male) at the quick disconnect fittings - color-to-color. For models with PTT, connect the button harness leads Yellow (female), Green and Black at the quick disconnect fittings - color to color. Position the horn between the four tall posts next to the neutral safety switch. Install the main key switch, the electric choke switch, and the kill switch into their respective openings on the side of the remote control box. Arrange the mess of wires neatly.

■ Remember, some wires are tucked neatly under switches, others are routed into neat bundles secured with plastic retainers, and some are looped then doubled back. Believe-it-or-not, with patience and some good words, they will all fit into one side of the box. The wires must be clear of moving parts.

18. Secure the electrical switches to the side of the box with the retaining nuts. Slide the rubber boot over the choke switch. If possible, secure a tie-wrap around the bundle of wires to prevent loose wires from rubbing against and interfering with moving parts.

19. Thread the cable joint 0.3 in. (8mm) onto the shift cable. Hold the joint to prevent it from rotating and at the same time hook the cable end onto the shift arm post. Secure the cable joint with the restraining circlip.

20. Install the throttle cable in the same manner as the shift cable. Bring both halves of the back plate together with the front plate. Secure it all together with the five Phillips head screws.

21. Check operation of the remote control lever to and between each of the three positions - Neutral, forward, and reverse. The lever should move smoothly and with a definite action. The shift cable and the throttle cable should move in and out of the sheathing without any indication of binding. The movement can be checked at the powerhead end of the cable.

Control Box

1. If the cable ends were removed from the powerhead:
   a. Guide the cables along a selected path and secure them in place with the retainers. Do not bend any cable into a diameter smaller than 16 in. (40cm).

Fig. 38 Now support the control arm as you install and tighten the retaining bolt through the square recess in the gear

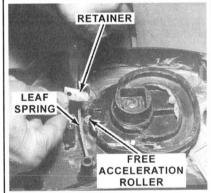

Fig. 39 Install the free acceleration roller, then position the leaf springs and secure the retainer

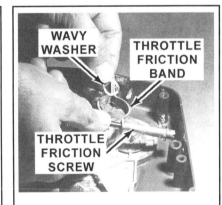

Fig. 40 Install the throttle friction assembly using a new wavy washer

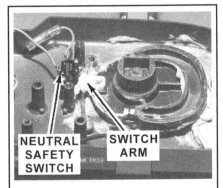

Fig. 41 If equipped, install and secure the neutral switch assembly

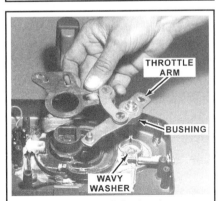

Fig. 42 Carefully install the throttle arm assembly and bushing to the throttle friction band (and wavy washer)

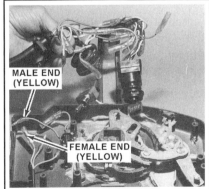

Fig. 43 Install the wiring harness as tagged during removal...

9-12 REMOTE CONTROLS

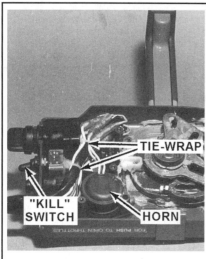

Fig. 44 ...then carefully position and secure the wiring as noted during removal

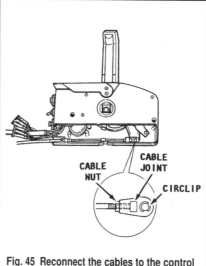

Fig. 45 Reconnect the cables to the control unit...

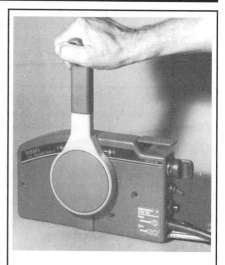

Fig. 46 ...then confirm movement and operation of the throttle/shifter lever

b. Connect the shift cable to the shift lever on the powerhead. Insert the outer throttle cable wire into the throttle cable bracket. Connect the cable joint onto the throttle control attachment.

c. Do not confuse the cables. If in doubt, operate the control lever on the control box. The cable moving first is the shift cable. Tilt the outboard unit from full up to the full down position, and from hard over starboard to hard over port, to verify the cables move smoothly without binding, bending, or buckling.

2. Install the control box to the boat and secure using the retainers.
3. Align the mark on the wire harness plug with the mark on the coupler plug, and then connect the plugs.
4. Install the safety tether to the back of the kill switch button.
5. Remember, the powerhead will not start without the tether in place.

MERCURY/MARINER REMOTE CONTROLS

Side Mount Remote Control Box (Mercury)

DESCRIPTION & OPERATION

■ The type of remote control mated to the outboard is determined by the boat manufacturer or the rigging dealer. Because there are such a range of remote control units that could be mated to Mercury/Mariner outboards, there is no way to cover them all in detail here. This section deals with a typical Mercury remote control unit. Illustrations are of the most common example of this unit. Although the procedures apply to most Mercury/Mariner remotes and many aftermarket units, care and observation must be used when working on any unit as differences may exist in assemblies. When a component or mounting differs from the components covered here, use common sense, or contact the manufacturer of the unit for more information.

The remote control unit allows the helmsperson to control throttle operation and shift movements from a location other than where the outboard unit is mounted.

In most cases, the remote control box is mounted approximately halfway forward (midship) on the starboard side of the boat or on the starboard side of the center console (depending upon the style/type of boat).

Several control box types are used, but all usually house a key switch, engine stop switch, choke switch, neutral safety switch, warning buzzer and the necessary wiring and cable hardware to connect the control box to the outboard unit.

TROUBLESHOOTING THE REMOTE CONTROLS

One of the things taken for granted on most boats is the engine controls and cables. Depending on how they are originally routed, they will either last the life of the vessel or can be easily damaged. These cables should be routinely maintained by careful inspection for kinks or other damage and lubrication with a marine grade grease.

If the cables do not operate properly, have a helper operate the controls at the helm while you observe the cable and linkage operation at the powerhead. Make sure nothing is binding, bent or kinked. Check the hardware that secures the cables to the boat and powerhead to make sure they are tight. Inspect the clevis and cotter pins in the ends of the cables and also give some attention to the cable release hardware.

Another area of concern is the neutral start switch that is sometimes located in the control box and sometimes on the powerhead. This switch may fall out of adjustment and prevent the engine from being started. It is easily inspected using a multimeter by performing a continuity check.

REMOVAL & INSTALLATION

◆ See Figures 47 thru 52

1. Turn the ignition key to the off position. For safety and to protect from accidental attempt to start if wires are crossed, disconnect the high tension leads from the spark plugs, with a twisting motion.
2. Disconnect the remote control wiring harness plug from the outboard trim/tilt motor and pump assembly.
3. If applicable, disconnect the tachometer wiring plug from the forward end of the control housing.
4. Remove the locknuts, flat washers, and bolts (usually 3) securing the control housing to the mounting panel. One normally is located next to the run button (the ignition safety stop switch), and a second is usually beneath the control handle on the lower portion of the plastic case. The third is located behind the control handle when the handle is in the neutral position. Shift the handle into forward or reverse position to remove the bolt, then shift it back into the neutral position for the following steps.
5. Pull the remote control housing away and free of the mounting panel. Remove the plastic cover from the back of the housing. Lift off the access cover from the housing. (Some "Commander" remote control units do not have an access cover.)
6. Remove the screws (usually 2) securing the cable retainer over the throttle cable, wiring harness, and shift cable. Unscrew the Phillips-head screws (usually 2) securing the back cover to the control module, and then lift off the cover.

REMOTE CONTROLS

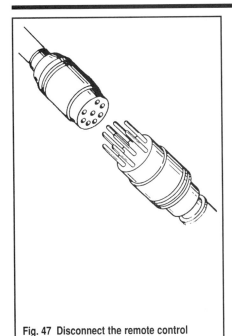

Fig. 47 Disconnect the remote control wiring harness

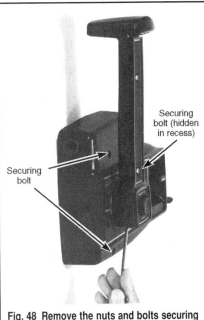

Fig. 48 Remove the nuts and bolts securing the control housing to the mounting panel

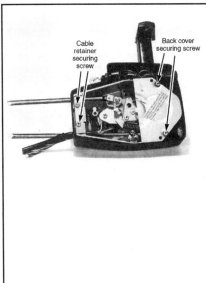

Fig. 49 You now have access to the back cover or the cable retainer

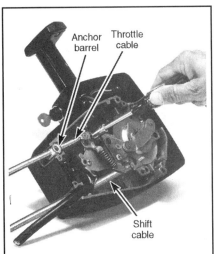

Fig. 50 If necessary, disconnect the cables from the remote

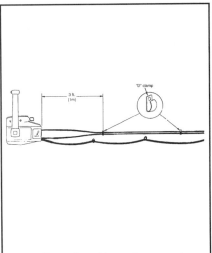

Fig. 51 Route the cables and secure using Sta-Straps or D-clamps

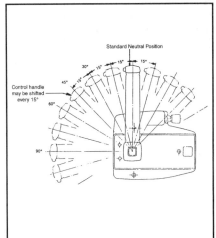

Fig. 52 The neutral position of the remote control handle may be changed to meet the owner's preference

7. To disconnect the throttle cable, proceed as follows:
  a. Loosen the cable retaining nut and raise the cable fastener enough to free the throttle cable from the pin. Lift the cable from the anchor barrel recess.
  b. Remove the grommet.
8. To disconnect the shift cable, proceed as follows:
  a. Shift the outboard unit into reverse gear by depressing the neutral lock bar on the control handle and moving the control handle into the reverse position.
  b. Loosen, but do not remove, the shift cable retainer nut (you'll usually need a 3/8 in. deep socket) as far as it will go without removing it. Raise the shift cable fastener enough to free the shift cable from the pin.

■ Do not attempt to shift into reverse while the cable fastener is loose. An attempt to shift may cause the cable fastener to strike the neutral safety micro-switch and cause it damage.

  c. Lift the wiring harness out of the cable anchor barrel recess and remove the shift cable from the control housing.

To install:
9. Position the control housing in place on the mounting panel and secure it with the long bolts, flat washers, and locknuts. Usually, one is located next to the run button (the ignition safety stop switch). The second is beneath the control handle on the power portion of the plastic case. The third bolt goes in behind the control handle when the handle is in the neutral position.
10. In order to install this bolt, shift the handle into the forward or reverse position, and then install the bolt. After the bolt is secure, shift the handle back to the neutral position for the next few steps.
11. The Commander shift box is now ready for installation.
12. Connect the tachometer wiring plug to the forward end of the control housing.

■ Clean the prongs of the connector with crocus cloth to ensure the best connection possible. Exercise care while cleaning to prevent bending the prongs.

13. Connect the remote control wiring harness plug from the outboard trim/tilt motor and pump assembly.
14. Install the high-tension leads to their respective spark plugs.
15. Route the wiring harness alongside the boat and fasten with the "Sta-Straps". Check to be sure the wiring will not be pinched or chafe on any moving part and will not come in contact with water in the bilge. Route the shift and throttle cables the best possible way to make large bends and as

9-14 REMOTE CONTROLS

few as possible. Secure the cables approximately every three feet (one meter).

16. The neutral position of the remote control handle may be changed to any one of a number of convenient angles to meet the owner's preference. The change is accomplished by shifting the handle one spline on the shaft at a time. Each spline equals 15° of arc, as shown.

DISASSEMBLY

 DIFFICULT

◆ See Figure 53

■ **For non-power trim/tilt units, it is not necessary to remove the cover of the control handle.**

1. Depress the neutral lock bar on the control handle and shift the control handle back to the neutral position. Remove the Phillips head screws (usually 2) which secure the cover to the handle, and then lift off the cover. The push button trim switch will come free with the cover, the toggle trim switch will stay in the handle body.
2. Unsnap and then remove the wire retainer. Carefully unplug the trim wires and straighten them out from the control panel hub for ease of removal later.
3. Back-off the set screw at the base of the control handle to allow the handle to be removed from the splined control shaft.
4. Grasp the "throttle only" button and pull it off the shaft.

✳✳ CAUTION

Take care not to damage the trim wires when removing the control handle, on power trim models.

5. Remove the control handle.
6. Lift the neutral lockring from the control housing.

■ **Take care to support the weight of the control housing to avoid placing any unnecessary stress on the control shaft during the following disassembling steps.**

7. Remove the Phillips-head screws (usually 3) securing the control module to the plastic case. Normally two are located on either side of the bearing plate and one is in the recess where the throttle cable enters the control housing.
8. Back-out the detent adjustment screw and the control handle friction screw until their heads are flush with the control module casing. This action will reduce the pre-load from the two springs on the detent ball for later removal.

■ **As this next step is performed, count the number of turns for each screw as they are backed-out and record the figure somewhere. This will be a tremendous aid during assembling.**

9. Remove the locknuts (usually 2) securing the neutral safety switch to the plate assembly and lift out the micro-switch from the recess in the assembly.
10. Remove the Phillips-head screw securing the retaining clip to the control module.
11. Support the module in your hand and tilt it until the shift gear spring, shift nylon pin (earlier models have a ball), shift gear pin, another ball the shift gear ball (inner), fall out from their recess. If the parts do not fall out into your hand, attach the control handle and ensure the unit is in the neutral position. The parts should come free when the handle is in the neutral position.

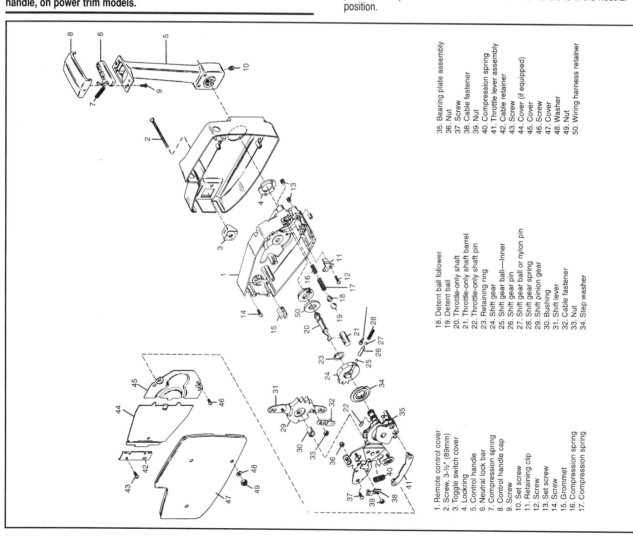

Fig. 53 Exploded view of a typical Mercury control box

1. Remote control cover
2. Screw, 3-½" (89mm)
3. Toggle switch cover
4. Lockring
5. Control handle
6. Neutral lock bar
7. Compression spring
8. Control handle cap
9. Screw
10. Set screw
11. Retaining clip
12. Screw
13. Set screw
14. Screw
15. Grommet
16. Compression spring
17. Compression spring
18. Detent ball follower
19. Detent ball
20. Throttle-only shaft
21. Throttle-only shaft barrel
22. Throttle-only shaft pin
23. Retaining ring
24. Shift gear
25. Shift gear ball—Inner
26. Shift gear pin
27. Shift gear ball or nylon pin
28. Shift gear spring
29. Shift pinion gear
30. Bushing
31. Shift lever
32. Cable fastener
33. Nut
34. Step washer
35. Bearing plate assembly
36. Nut
37. Screw
38. Cable fastener
39. Nut
40. Compression spring
41. Throttle lever assembly
42. Cable retainer
43. Screw
44. Cover (if equipped)
45. Cover
46. Screw
47. Cover
48. Washer
49. Nut
50. Wiring harness retainer

12. Arrange the parts from the control module recess. As the parts are removed and cleaned, keep them in order, ready for installation.
13. Remove the Phillips-head screws (usually 3) securing the bearing plate assembly to the control module housing.
14. Lift out the bearing plate assembly from the control module housing.
15. Uncoil the trim wires from the recess in the remote control module housing and lift them away with the trim harness bushing attached.
16. Remove the detent ball, the detent ball follower, and the two compression springs (located under the follower), from their recess in the control module housing.
17. If it is not part of the friction pad, remove the control handle friction sleeve from the recess in the control module housing.
18. Pull the throttle link assembly from the module. Remove the compression spring from the throttle lever. It is not necessary to remove this spring unless there is cause to replace it. At this point there is the least amount of tension on the compression spring. Therefore, now would be the time to replace it, if required.
19. Lift the shift pinion gear (with attached shift lever), off the pin on the bearing plate. The nylon bushing may come away with the shift lever and shift pinion gear as an assembly. Do not attempt to separate them. Both are replaced if one is worn.
20. On the non-power trim/tilt units only, remove the trim harness bushing and wiring harness retainer from the control shaft. (On non-power trim/tilt units these two items act as spacers.)
21. On all other units, remove the shift gear retaining ring from its groove with a pair of Circlip pliers.

■ If the Circlip slipped out of its groove, this would allow the shift gear to ride up on the shaft and cause damage to the small parts contained in its recess. The shift gear ball (inner), the shift gear pin, the shift gear ball (outer), or the nylon pin and particularly the shift gear spring must be inspected closely.

22. Lift the gear from the control shaft.
23. Remove the "throttle only" shaft pin and "throttle only" shaft from the control shaft.
24. Remove the step washer from the base of the bearing plate.

CLEANING & INSPECTING

◆ See Figure 53

1. Clean all metal parts with solvent, and then blow them dry with compressed air.

■ Never allow nylon bushings, plastic washers, nylon pins, wiring harness retainers, and the like, to remain submerged in solvent more than just a few moments. The solvent will cause these type parts to expand slightly. They are already considered a "tight fit" and even the slightest amount of expansion would make them very difficult to install. If force is used, the part is most likely to be distorted.

2. Inspect the control housing plastic case for cracks or other damage allowing moisture to enter and cause problems with the mechanism.
3. Carefully check the teeth on the shift gear and shift lever for signs of wear. Inspect all ball bearings for nicks or grooves which would cause them to bind and fail to move freely.
4. Closely inspect the condition of all wires and their protective insulation. Look for exposed wires caused by the insulation rubbing on a moving part, cuts and nicks in the insulation and severe kinking which could cause internal breakage of the wires.
5. Inspect the surface area above the groove in which the Circlip is positioned for signs of the Circlip rising out of the groove. Such action would occur if the clip had lost its "spring" or if it had worn away the top surface of the groove. If the Circlip slipped out of its groove, the shift gear would be able to ride up on the shaft and cause damage to the small parts contained in its recess.
6. The shift gear ball (inner), the shift gear pin, the shift gear ball (outer), or the nylon pin and particularly the shift gear spring must be inspected closely.
7. Inspect the "throttle only" shaft for wear along the ramp. In the early model units, this shaft was made of plastic. Later models have a shaft of stainless steel. Check for excessive wear or cracks on the ramp portion of the shaft. Also check the lower "stop" tab to be sure it has not broken away.

■ Good shop practice dictates a thin coat of multipurpose lubricant be applied to all moving parts as a precaution against the "enemy" moisture. Of course the lubricant will help to ensure continued satisfactory operation of the mechanism.

ASSEMBLY

◆ See Figure 53

■ The Commander control shift box, like others, has a number of small parts that must be assembled in only one order - the proper order. Therefore, the work should not be "rushed" or attempted if the person assembling the unit is "under pressure". Work slowly, exercise patience, read ahead before performing the task, and follow the steps closely.

1. Place the step washer over the control shaft and ensure the steps of the washer seat onto the base of the bearing plate.
2. Rotate the control shaft until the "throttle only" shaft pin hole is aligned centrally between the neutral detent notch and the control handle friction pad. Lower the "throttle only" shaft into the barrel of the control shaft. Secure the shaft in this position with the "throttle only" shaft pin.

■ When the pin is properly installed, it should protrude slightly in line with the plastic bushing.

3. Make an attempt to gently pull the "throttle only" shaft out of the control shaft. If the shaft and pin are properly installed, the attempt should fail.
4. Place the shift gear over the control shaft, and check to be sure the "throttle only" shaft pin clears the gear.
5. Install the retaining ring over the control shaft with a pair of Circlip pliers. Check to be sure the ring snaps into place within the groove.
6. On non-power tilt/trim units only, slide the wiring harness retainer and the trim harness bushing over the control shaft. The trim harness bushing is placed "stepped side" UP and the notched side toward the forward side of the control housing.
7. On power tilt/trim units only, insert the trim harness bushing into the recess of the remote control housing and carefully coil the wires, as shown in the illustration. Ensure the black line on the trim harness is positioned at the exact point shown for correct installation. The purpose of the coil is to allow slack in the wiring harness when the control handle is shifted through a full cycle. The bushing and wires move with the handle.
8. Position the bushing, shift pinion gear and shift lever onto the pin on the bearing plate, with the shift gear indexing with the shift pinion gear.
9. Install the two compression springs, the detent ball follower and the detent ball into their recess in the control module housing.
10. If the friction sleeve is not a part of the friction pad, then place the control handle friction sleeve into its recess in the control module housing.
11. Thread the detent adjustment screw and the control handle friction screw the exact number of turns as was turned out during disassembly. A fine adjustment may be necessary after the unit is completely assembled.
12. Place the compression spring (if removed), in position on the shift lever and shift pinion gear assembly against the bearing plate. Use a rubber band to secure the shift pinion gear to the bearing plate. Lower the complete bearing plate assembly into the control module housing.
13. Secure the bearing plate assembly to the control module housing with the three Phillips head screws, and then remove the rubber band.
14. Insert the gear shift ball (inner), into the recess of the shift gear and hole in the "throttle only" shaft barrel. Now, insert the shift gear pin into the recess with the rounded end of the pin away from the control shaft. Insert the nylon pin or shift gear ball (outer), into the same recess. Next, insert the shift gear spring.
15. Hold these small parts in place and at the same time secure them with the retaining clip and the Phillips head screw. On power trim/tilt units, this retaining clip also secures the trim wire to the control module.
16. Insert the neutral safety micro-switch into the recess of the plate assembly and secure it with the locknuts (usually 2).
17. Arrange the parts so that they are cleaned and ready for installation into the control module recess.
18. Secure the control module to the plastic control housing case with the Phillips head screws (usually 3). Normally, two are located on either side

of the bearing plate and the third in the recess where the throttle cable enters the control housing.

19. Temporarily install the control handle onto the control shaft. Shift the unit into forward detent only, not full forward, to align the holes for installation of the throttle link. After the holes are aligned, remove the handle. Install the throttle link.

20. Again, temporarily install the control handle onto the control shaft. This time shift the unit into the neutral position, and then remove the handle.

21. Place the neutral lock-ring over the control shaft, with the index mark directly beneath the small boot on the front face of the cover.

22. Install the control onto the splines of the control shaft, taking care not to cut, pinch, or damage the trim wires on the power trim/tilt unit.

※※ WARNING

When positioning the control handle, ensure the trim wire bushing is aligned with its locating pin against the corresponding slot in the control handle. On a Power Trim/Tilt unit: if this bushing is not installed correctly, it will not move with the control handle as it is designed to move - when shifted. This may pinch or cut the trim wires, causing serious problems. On a Non-Power Trim/Tilt unit, misplacement of this bushing (it is possible to install this bushing upside down) will not allow the control handle to seat properly against the lock-ring and housing. This situation will lead to the Allen screw at the base of the control handle to be incorrectly tightened to seat against the splines on the control shaft, instead of gripping the smooth portion of the shaft. Subsequently the control handle will feel "sloppy" and could cause the neutral lock to be ineffective.

※※ WARNING

If this handle is not seated properly, a slight pressure on the handle could throw the lower unit into gear, causing serious injury to crew, passengers, and the boat.

23. Push the "throttle only" button in place on the control shaft.
24. Ensure the control handle has seated properly, and then tighten the set screw at the base of the handle to 70 inch lbs. (7.9 Nm).

■ FAILURE to tighten the set screw to the required torque value, could allow the handle to disengage with a loss of throttle and shift control.

25. If the handle cover was removed during disassembly, slide the hooked end of the neutral lock rod into the slot in the neutral lock release. Route the trim wires in the control handle in their original locations. Connect them with the wires remaining in the handle and secure the connections with the wire retainer. Install the handle cover and tighten the Phillips head screws (usually 2).

26. Move the wiring harness clear of the barrel recess. Thread the shift cable anchor barrel to the end of the threads, away from the cable converter, and place it into the recess. Hook the pin on the end of the cable fastener through the outer hole in the shift lever.

27. Depress the neutral lock bar on the control handle and shift the handle into the reverse position. Take care to ensure the cable fastener will clear the neutral safety micro-switch. The access hole is now aligned with the locknut.

■ Check to be sure the pin on the cable fastener is all the way through the cable end and the shift lever. A pin partially engaging the cable and the shift lever may cause the cable fastener to bend when the nut is tightened.

28. Tighten the locknut (it normally requires a 3/8 in. deep socket) to 20-25 inch lbs. (2.26-2.82 Nm). Position the wiring harness over the installed shift cable.
29. Install the grommet into the throttle cable recess.
30. Thread the throttle cable anchor barrel to the end of the threads, away from the cable connector, and then place it into the recess over the grommet. Hook the pin on the end of the cable fastener through the outer hole in the shift lever.

■ Check to be sure the pin on the cable fastener is all the way through the cable end and the throttle lever. A pin partially engaging the cable and throttle lever may cause the cable fastener to bend when the nut is tightened.

31. Tighten the locknut to 20-25 inch lbs. (2.26-2.82 Nm).
32. Position the control module back cover in place and secure it with the Phillips-head screws (usually 2). Tighten the screws to 60 inch lbs. (6.78 Nm). Install the cable retainer plate over the two cables and secure it in place with the Phillip-head screws (also usually 2).
33. Place the plastic access cover over the control housing.

TILLER HANDLE

Description & Operation

Steering control for most outboards begins at the tiller handle and ends at the propeller. Tiller steering is the most simple form of small outboard control. All components are mounted directly to the engine and are easily serviceable.

Throttle control is performed via a throttle grip mounted to the tiller arm. As the grip is rotated a cable opens and closes the throttle lever on the engine. An adjustment thumbscrew is usually located near the throttle grip to allow adjustment of the turning resistance. In this way, the operator does not have to keep constant pressure on the grip to maintain engine speed.

An emergency engine stop switch is used on most outboards to prevent the engine from continuing to run without the operator in control. This switch is controlled by a small clip which keeps the switch open during normal engine operation. When the clip is removed, a spring inside the switch closes it and completes a ground connection to stop the engine. The clip is connected to a lanyard that is worn around the helmsman's wrist.

Some tiller systems utilize a throttle stopper system which limits throttle opening when the shift lever is in neutral and reverse. This prevents over revving the engine under no-load conditions and also limits the engine speed when in reverse.

Troubleshooting The Tiller Handle 2Wrench

If the tiller steering system seems loose, first check the engine for proper mounting. Ensure the engine is fastened to the transom securely. Next, check the tiller hinge point where it attaches to the engine and tighten the hinge pivot bolt as necessary.

Excessively tight steering that cannot be adjusted using the tension adjustment is usually due to a lack of lubrication. Once the swivel case bushings run dry, the steering shaft will get progressively tighter and eventually seize. This condition is generally caused by a lack of periodic maintenance.

Correct this condition by lubricating the swivel case bushings and working the outboard back and forth to spread the lubricant. However, this may only be a temporary fix. In severe cases, the swivel case bushings may need to be replaced.

Tiller Handle

REMOVAL & INSTALLATION

4/5/6 Hp Motors

◆ See Figure 54

■ Pay close attention to wire and cable routing during removal of the tiller handle. During installation be sure to position the wires and cables in the same position, re-securing them as necessary using wire ties.

1. Remove the J-clip screw which secures the throttle cables at the front of the powerhead just underneath the manual starter handle.
2. Follow the cables around the rear of the powerhead and remove the throttle drum retaining bolt (threaded straight downward where the cables connect to the throttle drum/bracket.
3. Loosen the throttle cable adjustment nuts, then remove the cables from the drum.

REMOTE CONTROLS 9-17

4. Remove the 2 nuts and lock-washers securing the tiller arm retaining plate.
5. Remove the cable ends from the throttle shaft pulley (at the underside pivot end of the arm), then remove the tiller handle from the lower engine cowling.
6. If necessary remove the bushing assembly from the cowling.

To install:
7. If removed, install the inner bushing into the bottom cowling. Lubricate the bushing with 2-4-C with Teflon, or an equivalent marine grade lubricant.
8. Route the throttle cables through the plate and inner bushing and out of the lower cowling.
9. If removed, install the nylon washer, spacer and outer bushing onto the tiller handle.
10. Install the tiller handle, connecting the cables to the drum in the handle (while keeping the cables in their respective locating pockets). Once the cables are seated, secure the tiller arm using the 2 nuts and lock-washers. Tighten the nuts to 70 inch lbs. (8 Nm).
11. Route the cables to the carburetor and secure using the J-clip right below the manual starter handle.
12. Slide the throttle cable ends into the throttle drum and secure the drum using the retaining bolt.
13. Place the end of the throttle cable jackets into the retaining bracket.
14. With the throttle arm of the carburetor against the idle speed screw, carefully remove the slack from the throttle cables using the adjustment nuts.

8-15 Hp Motors

◆ See Figure 55

■ Pay close attention to wire and cable routing during removal of the tiller handle. During installation be sure to position the wires and cables in the same position, re-securing them as necessary using wire ties.

1. Tag and disconnect the wiring for the tiller handle:
   a. For 8/9.9 hp motors this means the black and black/yellow stop button wires.
   b. For 9.9/15 hp motors this means the push start button and the stop wires. For details, please refer to the wiring diagrams.
2. Disconnect the throttle cables from the pulley.

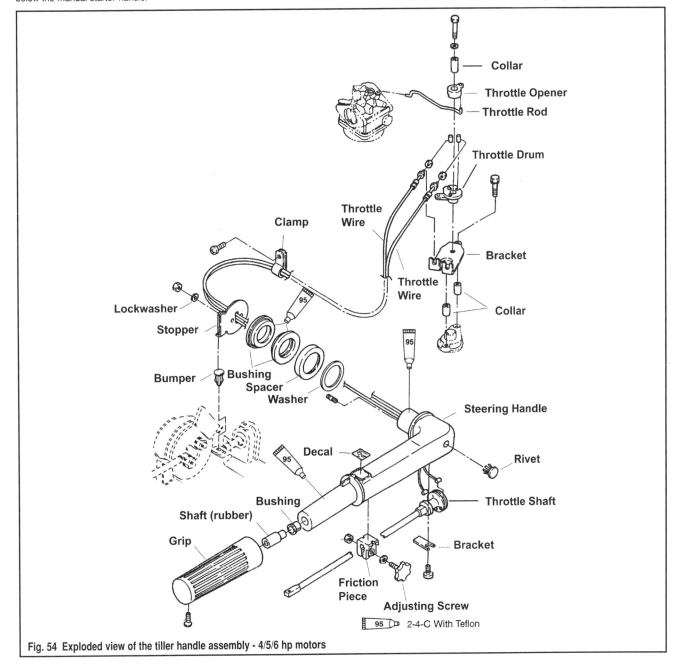

Fig. 54 Exploded view of the tiller handle assembly - 4/5/6 hp motors

9-18 REMOTE CONTROLS

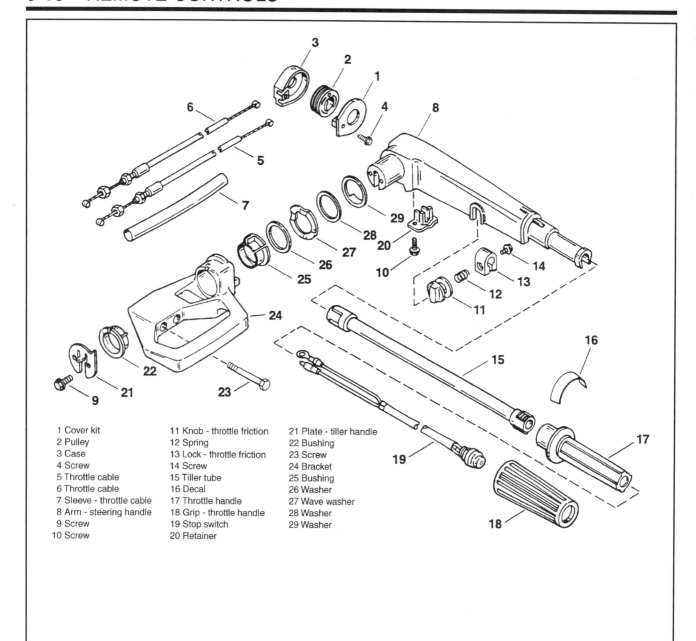

Fig. 55 Exploded view of a typical tiller handle assembly used on 8/9.9 hp (232cc) and 9.9/15 hp (323cc) motors (note, not all components used by all models)

1 Cover kit
2 Pulley
3 Case
4 Screw
5 Throttle cable
6 Throttle cable
7 Sleeve - throttle cable
8 Arm - steering handle
9 Screw
10 Screw
11 Knob - throttle friction
12 Spring
13 Lock - throttle friction
14 Screw
15 Tiller tube
16 Decal
17 Throttle handle
18 Grip - throttle handle
19 Stop switch
20 Retainer
21 Plate - tiller handle
22 Bushing
23 Screw
24 Bracket
25 Bushing
26 Washer
27 Wave washer
28 Washer
29 Washer

3. Loosen and remove the 2 bolts toward the cowling side of the pivot, securing the tiller arm to the bracket assembly, then remove the arm.
4. Installation is essentially the reverse of the removal procedure. During installation be sure to tighten the tiller arm retaining screws to 50 inch lbs. (5.6 Nm).

■ When installing the tiller arm on 8/9.9 hp motors, be sure to align the tabs of the inner and outer flanged bushings with the slots in the anchor bracket.

25-60 Hp Motors

◆ See Figure 56

■ Pay close attention to wire and cable routing during removal of the tiller handle. During installation be sure to position the wires and cables in the same position, re-securing them as necessary using wire ties.

1. If equipped, unbolt and remove the lower engine cowling (this is specifically recommended for all except 25 hp models).
2. Remove the shift handle by loosening and removing the bolt that is threaded through a bushing located in the base of the handle itself.
3. Loosen the jam nuts, then disconnect the throttle cables from the throttle lever on the side of the powerhead.
4. Tag and disconnect the tiller handle wiring.
5. At the front, center of the steering bracket, locate the 2 tiller handle retaining bolts. They are held in position by an oval shaped tabbed-washer. Carefully unbend the tabs, then loosen and remove the bolts.
6. Carefully pull the tiller handle assembly off the steering bracket.
7. Installation is essentially the reverse of removal. During installation, first inspect the tabbed washer to make sure it is not weak or damaged and, if necessary, replace the washer. After tightening the tiller handle retaining bolts to 35 ft. lbs. (47.5 Nm), bend the washer tabs up against the bolt heads to keep them from loosening in service. When reconnecting the wires, be sure to route them and secure them with wires ties as necessary.

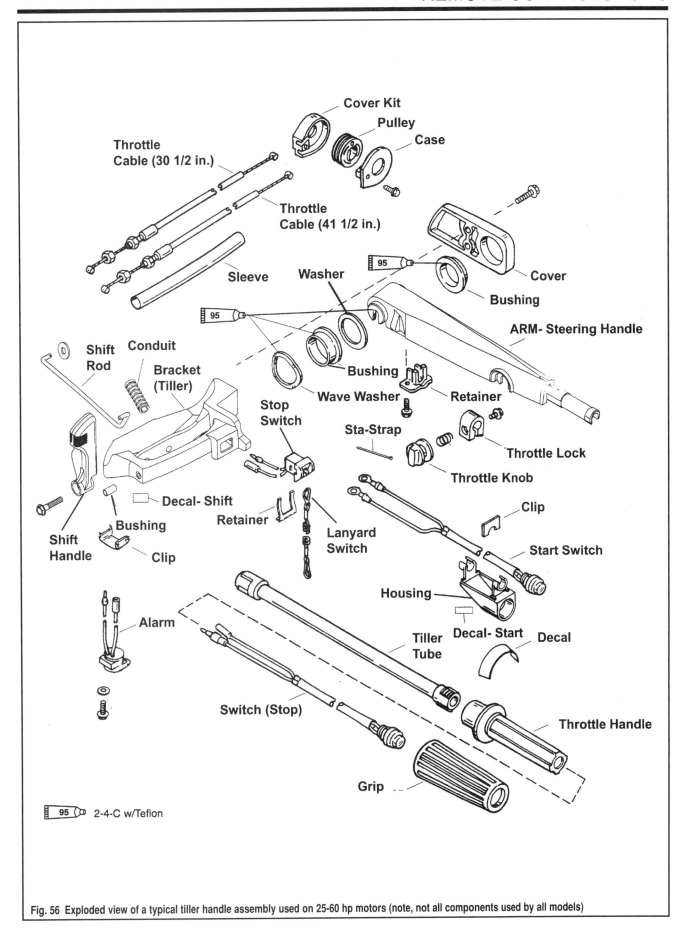

Fig. 56 Exploded view of a typical tiller handle assembly used on 25-60 hp motors (note, not all components used by all models)

9-20 REMOTE CONTROLS

REMOTE CONTROL/GAUGE WIRING

Yamaha Remote Control/Gauge Diagrams

◆ See Figures 57, 58 and 59

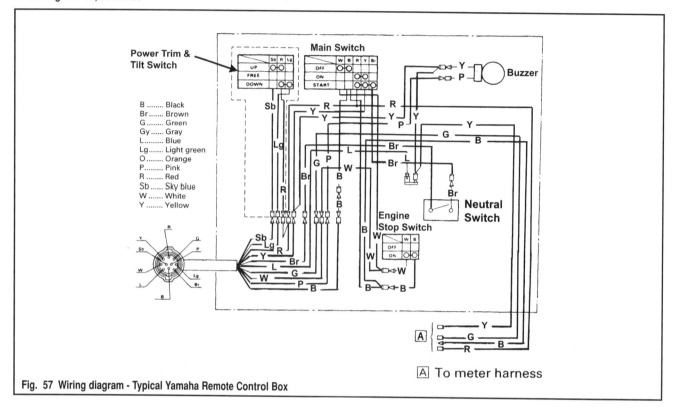

Fig. 57 Wiring diagram - Typical Yamaha Remote Control Box

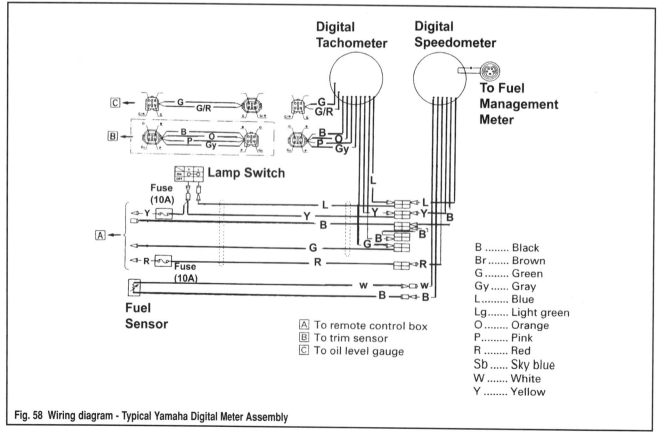

Fig. 58 Wiring diagram - Typical Yamaha Digital Meter Assembly

REMOTE CONTROLS 9-21

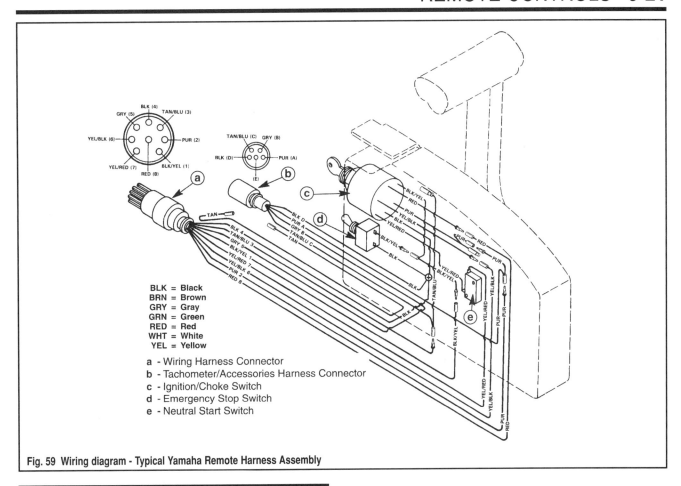

Fig. 59 Wiring diagram - Typical Yamaha Remote Harness Assembly

Mercury/Mariner Remote Control/Gauge Diagrams

◆ See Figures 60 thru 70

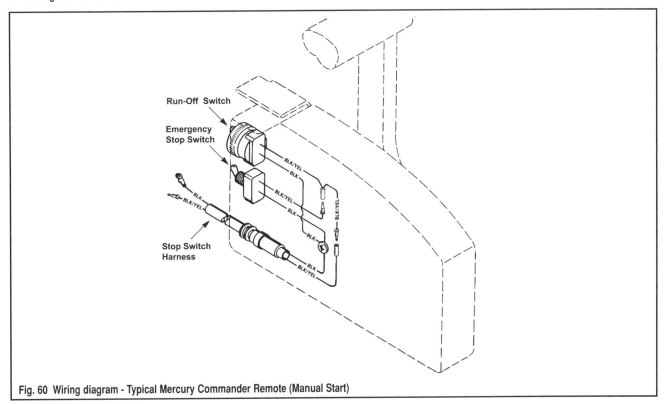

Fig. 60 Wiring diagram - Typical Mercury Commander Remote (Manual Start)

9-22 REMOTE CONTROLS

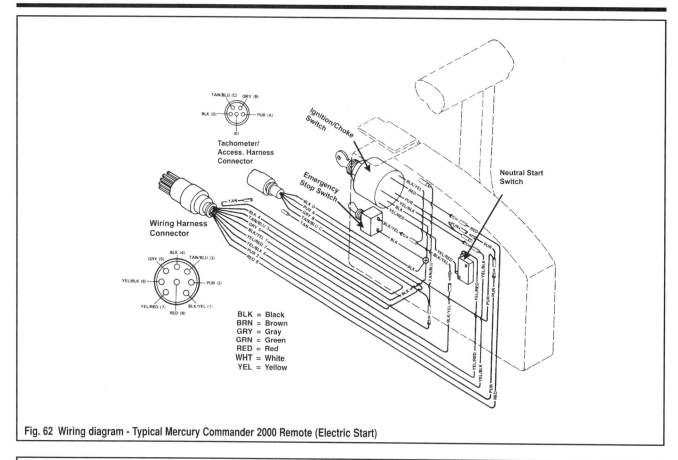

Fig. 62 Wiring diagram - Typical Mercury Commander 2000 Remote (Electric Start)

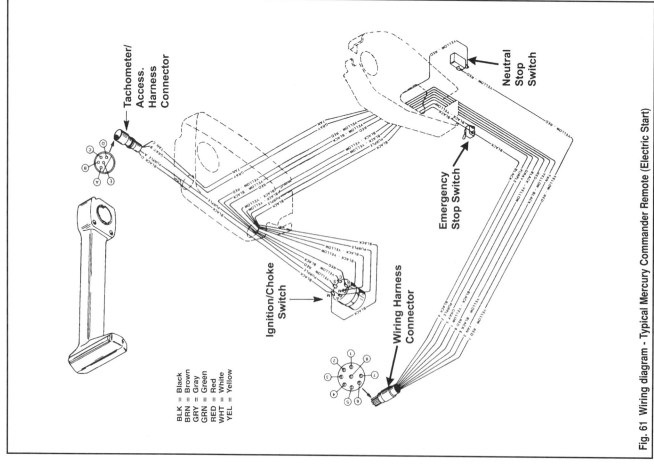

Fig. 61 Wiring diagram - Typical Mercury Commander Remote (Electric Start)

REMOTE CONTROLS 9-23

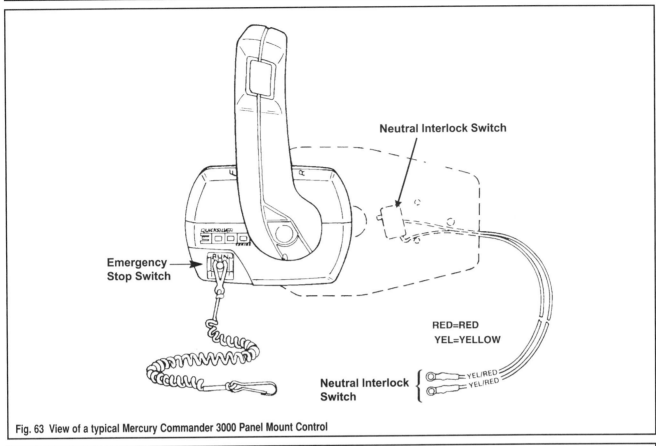

Fig. 63 View of a typical Mercury Commander 3000 Panel Mount Control

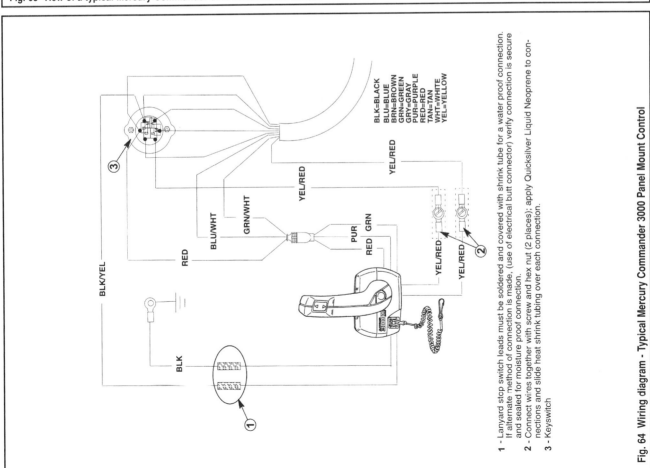

Fig. 64 Wiring diagram - Typical Mercury Commander 3000 Panel Mount Control

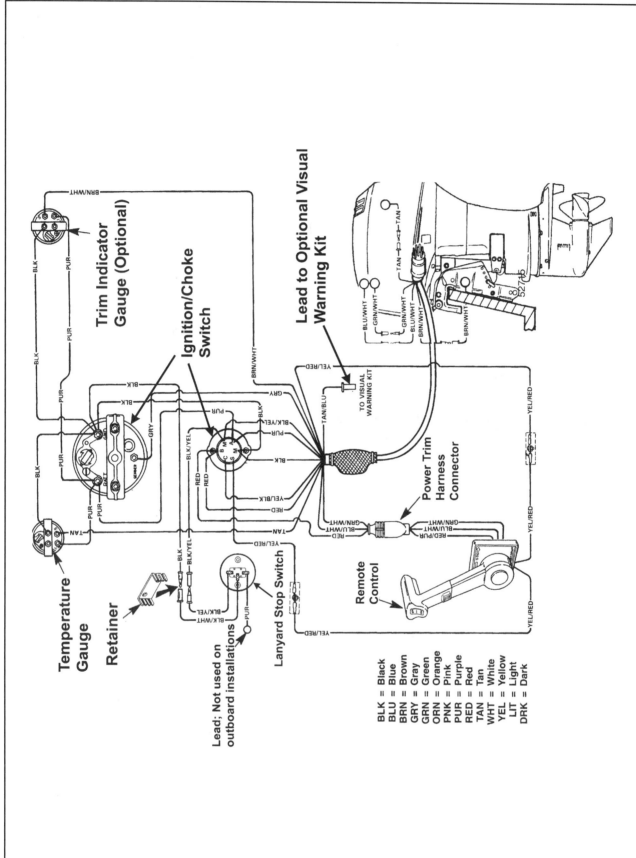

Fig. 65 Wiring diagram - Typical Instrument/Stop Switch Wiring for Mercury Mid-Range Motors

REMOTE CONTROLS 9-25

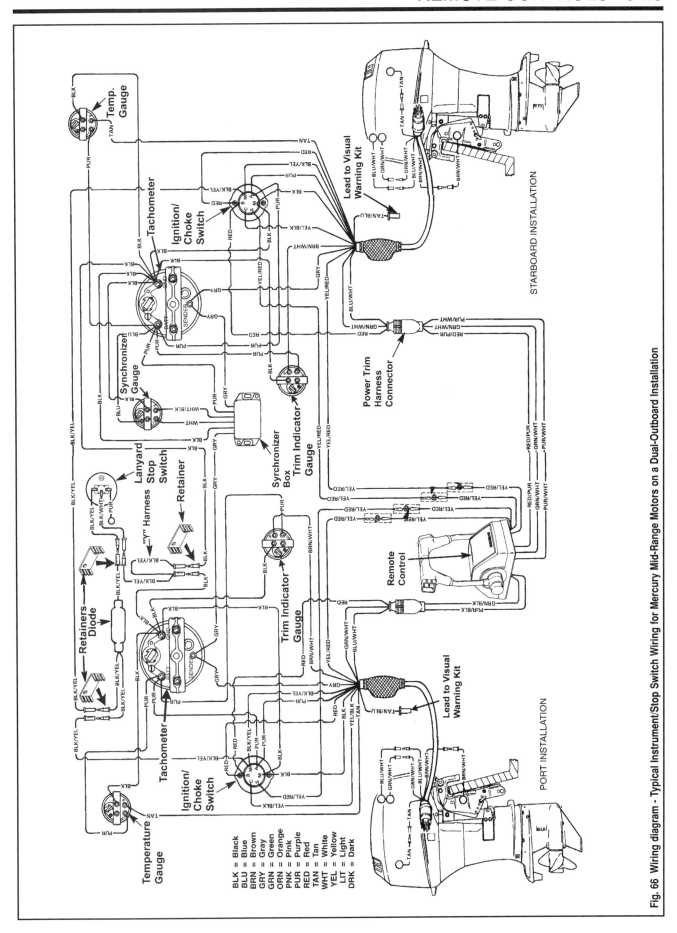

Fig. 66 Wiring diagram - Typical Instrument/Stop Switch Wiring for Mercury Mid-Range Motors on a Dual-Outboard Installation

9-26 REMOTE CONTROLS

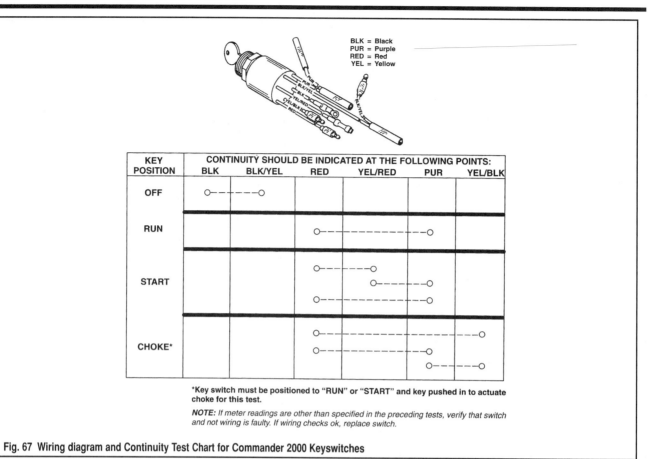

Fig. 67 Wiring diagram and Continuity Test Chart for Commander 2000 Keyswitches

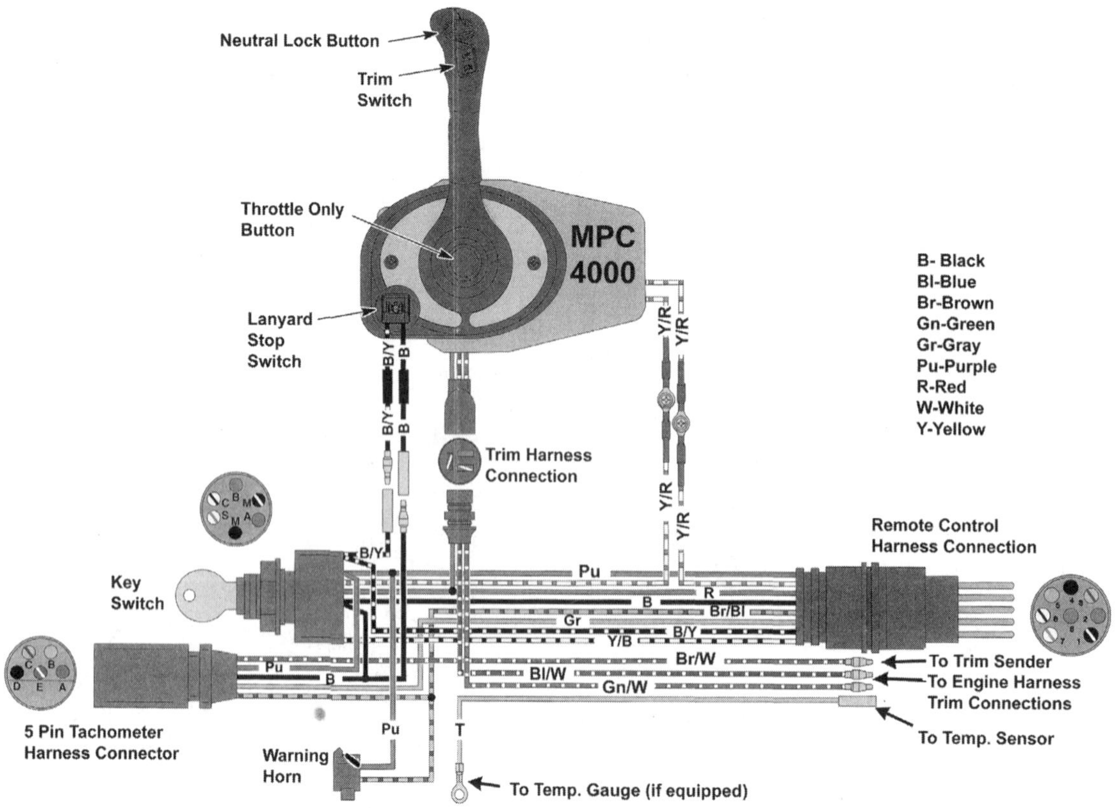

Fig. 68 Wiring diagram - Mechanical Panel Control (MPC) 4000 (normally found on carbureted Mercury Mid-Range or larger motors)

REMOTE CONTROLS 9-27

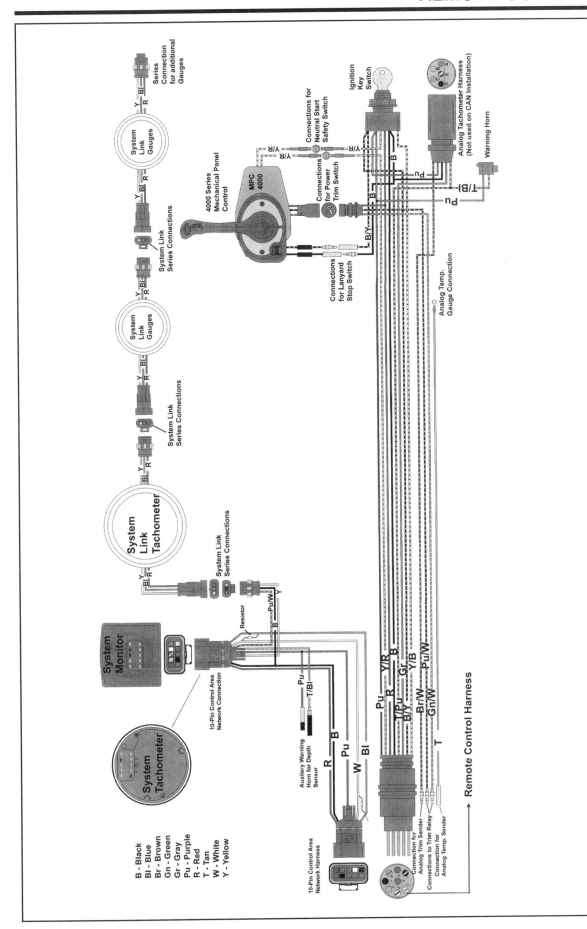

Fig. 69 Wiring diagram - Typical SmartCraft (CAN) Installation (Normally found on EFI Mercury Mid-Range 30-60 Hp motors)

9-28 REMOTE CONTROLS

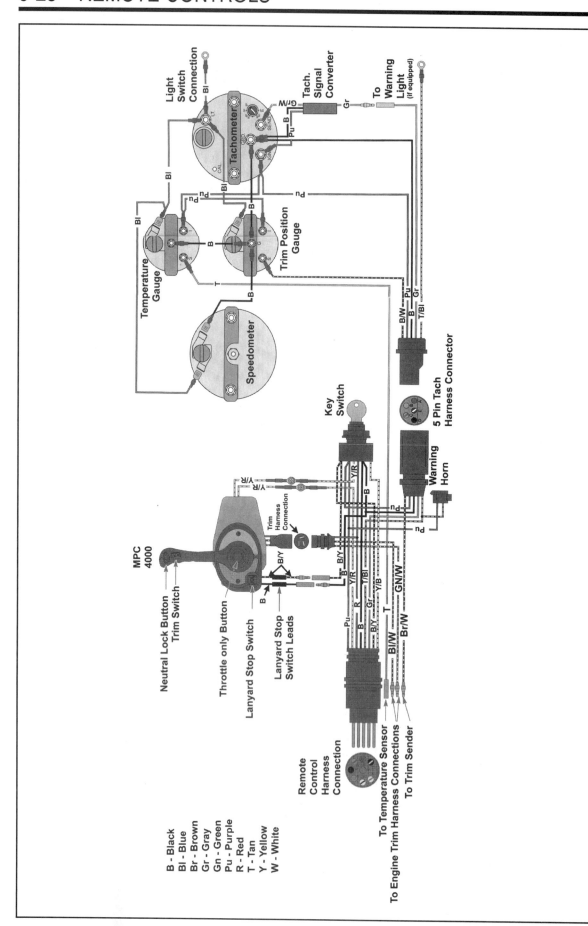

Fig. 70 Wiring diagram - Typical Gauge Installation (Normally found on Carb or EFI Mercury 75 Hp and larger motors with or without Tach Converter)

10

HAND REWIND STARTER

HAND REWIND STARTERS 10-2

HAND REWIND STARTERS **10-2**
  2.5 HP AND 4 HP YAMAHA 10-2
    ASSEMBLY & INSTALLATION 10-2
    CLEANING & INSPECTION 10-2
    REMOVAL & DISASSEMBLY 10-2
  4/5/6 HP MERCURY/MARINER 10-3
    ASSEMBLY & INSTALLATION 10-4
    CLEANING & INSPECTION 10-3
    REMOVAL & DISASSEMBLY 10-3
  6/8 HP YAMAHA MODELS 10-4
    ASSEMBLY & INSTALLATION 10-5
    CLEANING & INSPECTION 10-5
    REMOVAL & DISASSEMBLY 10-4
  8/9.9 HP (232CC) YAMAHA AND MERCURY/MARINER 10-6
    ASSEMBLING & INSTALLATION 10-8
      CHECKING THE NO START-IN-GEAR SYSTEM 10-11
      NEW SPRING 10-8
      OLD SPRING 10-8
    CLEANING & INSPECTION 10-6
    REMOVAL & DISASSEMBLY 10-6
  9.9/15 HP (323cc) YAMAHA AND MERCURY/MARINER 10-11
    ASSEMBLY & INSTALLATION 10-13
    CLEANING & INSPECTION 10-13
    REMOVAL & DISASSEMBLY 10-11
  25 HP YAMAHA AND 25-40 HP MERCURY/MARINER 10-13
    ASSEMBLY & INSTALLATION 10-15
    CLEANING & INSPECTION 10-15
    REMOVAL & DISASSEMBLY 10-13
  30/40J/40 HP YAMAHA 10-16
    ASSEMBLY & INSTALLATION 10-17
    CLEANING & INSPECTION 10-16
    REMOVAL & DISASSEMBLY 10-16

10-2 HAND REWIND STARTER

HAND REWIND STARTERS

2.5 Hp and 4 Hp Yamaha Models

REMOVAL & DISASSEMBLY

◆ See Figures 1 thru 4

1. Remove the engine cover/cowling assembly.
2. If equipped, disconnect start-in-gear protection cable from the starter assembly.
3. On 2.5 hp motors, free the choke wire from the carburetor. On 4 hp motors, free the choke wire from the side of the starter housing.
4. Remove the mounting bolts (usually 3) securing the legs at the corners of the hand rewind starter to the powerhead. On 4 hp motors, note the bolt locations as one of the 3 (the one placed furthest from the starter handle) is of a different length (is shorter).
5. Lift the hand rewind starter free of the powerhead.
6. Invert the starter assembly on a suitable work surface, then loosen and remove the drive plate retaining screw.
7. Remove the drive plate (along with the plate clip), then remove the drive pawls and pawl springs. Take care not to loose the pawl springs during removal.

※※ WARNING

If you go any further than this, WEAR heavy leather gloves as well as head/eye protection. Once the sheave drum is removed the sharp edged metal spiral spring can suddenly pop free and with great force. Use caution when handling the housing and spring.

8. Carefully lift the sheave drum from the assembly. In most cases the spiral spring will come out with the drum on these models. If you DO NOT wish to remove the spring from the drum, wrap it with some strong wire ties or a length of rope to secure it to the drum.

※※ CAUTION

The spring usually comes free of the drum with a bang and a sudden, uncontrollable expansion/lashing. Make sure there is nothing to damage and no-one around to be hurt by the spring.

9. To separate the spring from the drum, invert the drum (close to the floor) and tap it to loosen the spring. An alternate method is to place the drum on a work surface, covered by heavy rags and then pry to tap the spring free.

10. If it is necessary to remove/replace the rope, push the knot out from the sheave and from the handle where you can easily untie or cut it. If the spring is still in place, use caution (and some strong wire ties as noted earlier) to keep it from coming loose while the rope is being replaced.
11. Closely examine the remaining components in the housing (such as the rope guide collar, start-in-gear protection lever or spring/sheave plate, as equipped) to determine if any other components must be removed or replaced.

CLEANING & INSPECTION

◆ See Figures 1 and 2

Remove any trace of corrosion and wipe all metal parts with an oil dampened cloth.

Inspect the rope. Replace the rope if it appears to be weak or frayed. If the rope is frayed, check the holes through which the rope passes for rough edges or burrs. Remove the rough edges or burrs with a file and polish the surface until it is smooth. Inspect the starter spring end hooks. Replace the spring if it is weak, corroded or cracked. Inspect the inside surface of the sheave rewind recess for grooves or roughness. Grooves may cause erratic rewinding of the starter rope.

Coat the entire length of the used rewind spring (a new spring will be coated with lubricant from the package), with low-temperature lubricant.

ASSEMBLY & INSTALLATION

◆ See Figures 1 and 2

■ If the pull rope requires replacement be sure to use one which is the same length as the rope which was removed (or the total length of the segments if a rope broke). On 4 hp motors Yamaha supplies a replacement rope specification of 70.9 in. (180cm), however we could not locate a spec for the 2.5 hp motor.

The authors, the manufacturers, and almost anyone else who has handled the spring from this type rewind starter strongly recommend a pair of safety goggles or a face shield be worn while the spring is being installed. As the work progresses a tiger is being forced into a cage - in some cases up to 14 ft. (4.3 m) of spring steel wound into about 4 in. (10.2cm) circumference. If the spring is accidentally released, it will lash out with tremendous ferocity

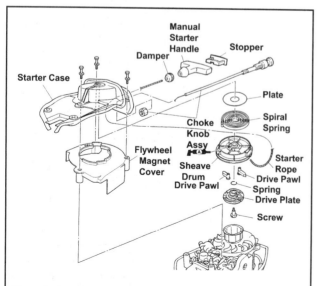

Fig. 1 Exploded view of the hand rewind starter used on 2.5 hp Yamaha motors

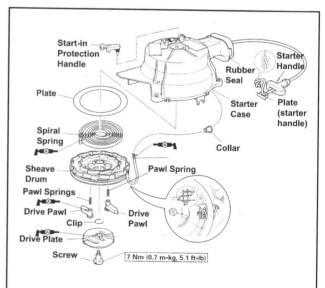

Fig. 2 Exploded view of the hand rewind starter used on 4 hp Yamaha motors

HAND REWIND STARTER 10-3

and very likely could cause personal injury to the installer or other persons nearby.

■ **Replacement spiral springs are normally supplied pre-wound and secured by wire loop(s). If using a replacement spring, position it on the drum sheave before removing the wire loop(s)**

1. If removed, pass a new rope through the rope guide, then insert the ends into the drum sheave and starter handle. Tie figure 8 knots in either end of the rope to secure them to the sheave and handle.

■ **When tying the figure 8 knots, be sure to leave about 0.20-0.39 in. (5-10mm) at the end of the starter rope beyond the knot in order to ensure it will not loosen.**

2. Wind the starter rope 2 times (for 2.5 hp motors) or 2 1/2 times (for 4 hp motors) around the sheave drum in a counterclockwise direction when viewed from above with the spring mounting side facing downward. Position the rope in the notch provided along the edge of the drum.
3. Install the spiral spring to the drum sheave. If reinstalling an unwound spring, carefully wind it by hand (wearing heavy leather gloves for protection). Bend the outer end of the spiral spring so that it hooks the cutout on the drum.
4. Carefully install the sheave drum and spring assembly to the starter housing so that the inner end of the spiral spring can be hooked onto the cutout in the housing.
5. Install the drive pawls and springs, then install the drive plate and clip. Securely tighten the drive plate retaining screw.
6. Using the starter rope positioned in the sheave notch, wind the spiral spring by turning the drum counterclockwise 3 turns (for 2.5 hp motors) or 2 1/2 turns (for 4 hp motors). Then holding tension on the pulley, carefully remove the rope from the notch and allow it to retract slowly into the housing.

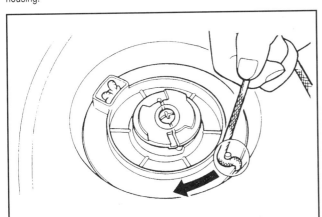

Fig. 3 Use the rope and sheave cutout to carefully unwind the spring

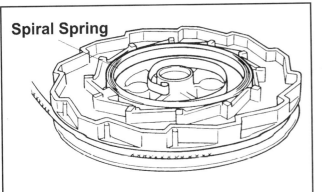

Fig. 4 The spiral spring usually comes out with the drum on these models

7. Pull the starter handle slowly out and hold as it retracts back a few times to verify that the drum turns smoothly and the spring is properly tensioned.
8. For 2.5 hp motors, if removed from the housing install the choke knob assembly.
9. Install the starter assembly to the top of the powerhead and secure using the 3 retaining bolts. Tighten the bolts securely.
10. For 2.5 hp motors, reconnect the choke knob assembly to the carburetor. For 4 hp motors, snap the choke cable into the guide on the side of the housing.
11. If equipped, connect the start-in-gear protection wire to the starter housing.
12. Slowly confirm the operation of the starter assembly.
13. Install the engine cover/cowling.

4/5/6 Hp Mercury/Mariner Models

REMOVAL & DISASSEMBLY

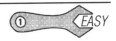

◆ See Figure 5

1. Remove the engine top cover assembly.
2. Disconnect start-in-gear interlink rod.
3. Remove the 3 mounting bolts securing the legs at the corners of the hand rewind starter to the powerhead.
4. Lift the hand rewind starter free of the powerhead.
5. Invert the starter assembly on a suitable work surface, then slowly pull the starter rope until it is fully unwound from the sheave. Hold the sheave in this position (you may be able to do this with thumb pressure alone) and at the same time lift the end of the rope from the knot recess in the sheave (opposite end from the handle). Untie or cut off the old knot so you can remove the rope from the sheave. Pull the starter rope from the sheave, then slowly allow the sheave to unwind, releasing the internal spring tension.
6. If necessary, pull the knot and retainer out of the starter handle, then untie or cut the knot at that end to pull the rope from the handle.
7. Back on the starter assembly itself, remove the bolt from the center of the sheave, then carefully lift the sheave from the housing. Take care not to loose the friction spring located under the sheave.
8. If necessary remove the return springs and ratchets.

※※ WARNING

Even though most of the spring tension is released, it's still a good idea to WEAR heavy leather gloves as well as head/eye protection. The spring can still suddenly pop free and it usually contains sharp edges. Use caution when handling the housing and spring.

9. Lift the spring housing and coiled spring out of the starter housing.
10. If necessary, remove the rope guide from the starter housing.
11. To disassemble the starter interlock components, remove the larger screw followed by the lock-washer, flat washer, spring, collar and cam. Then remove the smaller screw, spring, collar and starter lock.
12. If it is necessary to remove the rewind spring from the spring housing CAREFULLY unwind the coiled spring.
13. Closely examine all of the starter assembly components to determine if any other components must be removed or replaced. Refer to Cleaning & Inspection, in this section.

CLEANING & INSPECTION

◆ See Figure 5

Clean all components with solvent and dry with compressed air.
On metal components, be sure to remove any trace of corrosion and wipe all metal parts with an oil dampened cloth.
Inspect the rope. Replace the rope if it appears to be weak or frayed. If the rope is frayed, check the holes through which the rope passes for rough edges or burrs. Remove the rough edges or burrs with a file and polish the surface until it is smooth.

10-4 HAND REWIND STARTER

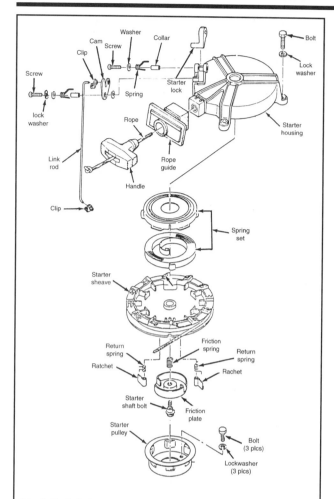

Fig. 5 Exploded view of the hand rewind starter used on 4/5/6 hp Mercury/Mariner motors

Replace the spring if it is weak, corroded, burred or cracked.
Check the interlock components, springs, ratchet pawls and friction plate for wear or damage and replace, as necessary.
Inspect the inside surface of the sheave rewind recess for grooves or roughness. Grooves may cause erratic rewinding of the starter rope.

ASSEMBLY & INSTALLATION

◆ See Figures 5 and 6

1. Apply a small amount of 2-4-C with Teflon to the center of the starter housing where the spring housing sheave install.
2. Apply a thin coat of non-metallic, low temperature grease to the spring cavity of the starter housing.
3. Place the OUTER hook of the starter spring into the slot of the housing, then slowly wind the spring CLOCKWISE into the spring cavity.
4. Install the spring/housing assembly into the starter housing, while positioning the inner spring hook onto the starter housing tab.
5. Apply a small amount of 2-4-C with Teflon in the starter housing (at the sheave friction surface) and install the sheave.
6. Install the ratchet pawls and pawl springs into the starter sheave.
7. Apply a light coating of Loctite® 271 or equivalent thread-locking compound to the sheave retaining bolt, then install the bolt and tighten to 70 inch lbs. (8 Nm).
8. If removed, install the rope guide.
9. If removed, install the interlock components, starting with those secured by the smaller headed (but longer) screw, the starter lock, collar, spring, flat washer and screw. Then install the cam, collar, spring, flat washer, lock-washer and finally larger head screw (HOWEVER, be sure to

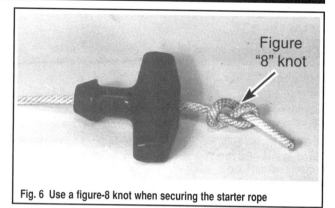

Fig. 6 Use a figure-8 knot when securing the starter rope

coat the threads of the larger head screw with Loctite® 271 or equivalent thread-locking compound before installation). Tighten the screws securely.
10. Tie a figure-8 knot in the end of the rope, then thread the rope through the retainer and manual starter handle.
11. Thread the rope through the rope guide, starter housing and starter sheave, then tie a figure-8 knot and pull the rope tight into the sheave.
12. Rotate the sheave 2 1/2 turns CLOCKWISE so the rope starts to coil into the sheave. Hold the rope and then rotate the sheave 3 turns COUNTERCLOCKWISE, then slowly allow the sheave to rewind, drawing in the remainder of the loose rope.
13. Pull the starter handle and check for smooth operation. The pawls must operate smoothly, the rope must pull freely and rewind consistently.
14. Install the manual starter assembly to the top of the powerhead and secure using the 3 retaining bolts. Tighten the bolts to 70 inch lbs. (8 Nm).
15. Reconnect the interlock link rod.
16. Install the top cowling.

6/8 Hp Yamaha Models

REMOVAL & DISASSEMBLY

◆ See Figures 7 and 8

1. Remove the engine top cover.
2. Disconnect start-in-gear protection cable at one end. For these models, it is sometimes easier to disconnect it from the shift linkage end and leave it attached to the starter assembly, however it is not difficult at the starter end either (just disconnect the cable end and the return spring).
3. Remove the 3 mounting bolts and washers securing the corners of the hand rewind starter to the powerhead.
4. Lift the hand rewind starter free of the powerhead. If necessary, remove the dust cover and grommets.
5. Invert the starter assembly on a suitable work surface
6. Turn the sheave drum until the cutout in the outer side of the drum faces the starter handle, then pull out the starter rope from the side of the housing (between the drum and handle). Hook the rope into the cutout in the side of the drum. Using the rope caught in the cutout, slowly wind the sheave drum clockwise until the spring tension is released.
7. Loosen and remove the drive plate retaining screw, then lift the drive plate and drive spring from the assembly.

■ When removing the drum, be sure to turn it upside down in order to keep the starter spring from popping up at you. If the spring is not being replaced, do not leave the drum out of the housing any longer than necessary OR use something (like a large heavy block of wood or strong wire ties) to help lock the spring in place.

8. Carefully pull the drum assembly from the starter housing. Which steps you follow from this point will depend upon which components require service or replacement. The spring, the rope or other starter components can each be individually replaced.
9. Remove the circlip from the spring side of the drum, then remove the drive pawl, return spring and spring from the pawl assembly.
10. Wearing heavy leather gloves to protect your hands, slowly and carefully remove the return spring from the start housing by unwinding it, one turn of the winding each time.

HAND REWIND STARTER 10-5

11. If rope replacement is necessary, carefully push the knots out of the drum and starter handle, then untie or cut the rope to free it at each end. A damper is installed on the rope at the handle end, be sure to keep track of the damper positioning for installation purposes.

12. Remove any other components from the housing necessary for replacement, such as the start-in-gear protection stopper, collar or return spring.

CLEANING & INSPECTION

◆ See Figures 7 and 8

Remove any trace of corrosion and wipe all metal parts with an oil dampened cloth.

Inspect the rope. Replace the rope if it appears to be weak or frayed. If the rope is frayed, check the holes through which the rope passes for rough edges or burrs. Remove the rough edges or burrs with a file and polish the surface until it is smooth. Inspect the starter spring end hooks. Replace the spring if it is weak, corroded or cracked. Inspect the inside surface of the sheave rewind recess for grooves or roughness. Grooves may cause erratic rewinding of the starter rope.

Coat the entire length of the used rewind spring (a new spring will be coated with lubricant from the package), with low-temperature lubricant.

ASSEMBLY & INSTALLATION

◆ See Figures 7 thru 11

■ If the pull rope requires replacement be sure to use one which is 59.1 in. (150cm) in length. When cutting a rope to the specified length, be sure to burn the ends so that it will not continue to unwind. Also, apply a light coating of marine grade grease to the new rope.

The authors, the manufacturers, and almost anyone else who has handled the spring from this type rewind starter strongly recommend a pair of safety goggles or a face shield be worn while the spring is being installed. As the work progresses a tiger is being forced into a cage - in some cases up to 14 ft. (4.3 m) of spring steel wound into about 4 in. (10.2cm) circumference. If the spring is accidentally released, it will lash out with tremendous ferocity and very likely could cause personal injury to the installer or other persons nearby.

■ Replacement spiral springs are normally supplied pre-wound and secured by wire loop(s). If using a replacement spring, position it on the drum sheave before removing the wire loop(s)

1. If removed, install the spring, return spring and drive pawl to the sheave drum and secure using the circlip.
2. If using a replacement spring, position it to the starter housing, then carefully remove the wire loop(s).
3. If using an unwound spring, carefully wind it into the starter housing, starting at the outside and working inward, one turn at a time.
4. If removed, pass a new rope through the rope guide, then insert the ends into the drum sheave and starter handle. Tie figure 8 knots in either end of the rope to secure them to the sheave and handle.

■ When tying the figure 8 knots, be sure to leave about 0.20-0.39 in. (5-10mm) at the end of the starter rope beyond the knot in order to ensure it will not loosen.

5. Position the starter rope in the cutout on the drum, then carefully install the drum assembly to the starter housing and over the spring. While installing the drum be sure to position the inner end of the spring over the retainer post on the drum.
6. If removed, install the drive spring to the drive plate.
7. Insert the tip of the drive spring through the coil portion of the return spring, then install the drive plate to the sheave drum.
8. Install the drive plate retaining screw and tighten to 4.5 ft. lbs. (6 Nm).
9. Using the starter rope in the drum cutout, slowly wind the starter spring by turning the sheave 5 turns COUNTERCLOCKWISE. Once pre-

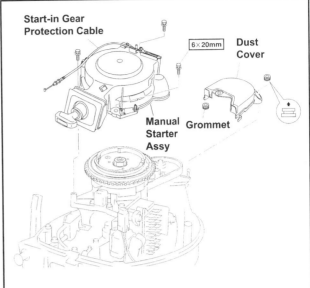

Fig. 7 Exploded view of the starter assembly mounting - 6/8 hp Yamaha motors

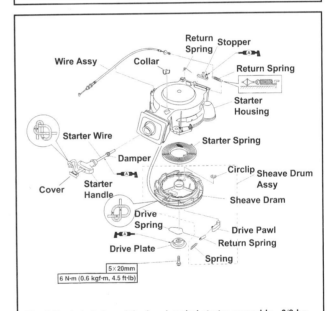

Fig. 8 Exploded view of the hand rewind starter assembly - 6/8 hp Yamaha motors

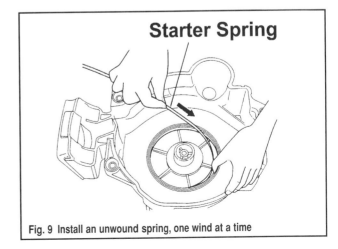

Fig. 9 Install an unwound spring, one wind at a time

10-6 HAND REWIND STARTER

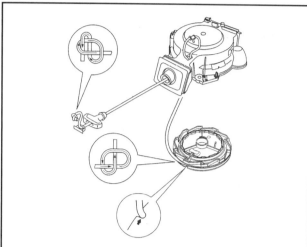

Fig. 10 Installing a replacement rope with figure 8 knots

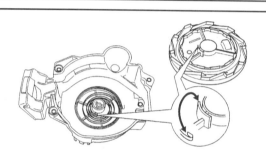

Fig. 11 When installing the drum, position it over the spring as shown

wound, remove the rope from the cutout and allow the spring to slowly wind it back into the housing.

10. Make sure the start-in-gear protection device is released, then check drum rotation and spring return using the starter rope and handle.

11. Install the starter assembly to the top of the powerhead (along with the dust cover and grommets, if removed). Secure the starter assembly using the 3 retaining bolts and washers. Tighten the bolts securely.

12. Reconnect the start-in-gear protection wire to the starter housing or the shift linkage (as applicable).

13. Slowly confirm the operation of the starter assembly.

14. Install the engine cover.

8/9.9 Hp (232cc) Yamaha and Mercury/Mariner Models

REMOVAL & DISASSEMBLY

♦ See Figures 12 thru 24

■ For exploded views, please refer to Cleaning & Inspection, later in this section.

1. Pry the small cover from the starter handle and pull out the knot. Untie the knot and **gently** allow the rope to be pulled into the starter housing. Once the rope has wound into the housing, most of the tension on the spring will be relaxed.

■ If the rope has broken, the rope rewound into the housing with considerable speed at the time of the break. This normally means damage to the spring or related housing components.

2. Loosen the two nuts and lift the cable from the mount on the starter body. Remove the cable end from the stopper arm. With the cable loose, take care not to lose the small plastic grommet which holds the cable in place away from the powerhead. This grommet is not visible in the accompanying illustration.

3. Use two 10mm wrenches to loosen and remove the top bolt and locknut. Remove the large washer.

4. If equipped, remove the plastic plug from the outer cowling. Removal of the plug will reveal an access hole for one of the starter mounting bolts. The other two bolts are hidden, but may be removed by reaching in with a socket extension following the path indicated on the accompanying illustration.

5. Lift the hand rewind starter free of the powerhead.

6. Place the rewind starter on a suitable work surface. Tap out the pinion pin, and then lift off the pinion gear assembly. The friction spring will come free with the pinion. Slide the bushing and washer up and free of the sheave shaft.

7. Remove the following parts in order: first, the nut, then the lock-washer, washer, stopper arm, and finally the arm spring.

8. Remove the cotter pin, then the washer, and finally the collar and stopper spring. After these parts have been removed, the stopper may then be removed from the bottom of the starter.

9. Remove the two bolts securing the guide plate to the starter body.

10. Remove the pulley bolt, lock-washer, washer, bushing, and then the starter pulley.

✳✳ WARNING

The rewind spring is a potential hazard. The spring is under tremendous tension when it is wound - a real tiger in a cage! If the spring should accidentally be released, severe personal injury could result from being struck by the spring with force. Therefore, the following steps MUST be performed with care to prevent personal injury to self and others in the area.

Do not attempt to remove the spring unless it is unfit for service and a new spring is to be installed.

11. Remove the three bolts and washers securing the starter cover to the starter body. **Very carefully** separate the cover from the sheave. If the sheave is pulled upward accidentally, the spring will be disturbed and could be released with extreme **force**; causing possible personal injury.

■ If the only service to be performed is to replace a broken rope, make every attempt NOT to disturb the sheave and rewind spring under the sheave. Let a sleeping tiger sleep!

✳✳ WARNING

For safety, wear a good pair of heavy gloves and safety glasses while performing the following tasks.

12. Place a screwdriver into the access hole in the sheave and hold down on the spring end while the sheave is removed. **Carefully** lift the sheave out of the housing **without** disturbing the rewind spring. Remove the rope from the sheave.

13. With one hand in a heavy glove, hold the starter body upside down over a bucket or similar container. Strike the base of the starter body a moderate rap with a mallet. The spring will be released from the starter body with force but should be confined inside the bucket. Remove the nylon washer from the starter body.

CLEANING & INSPECTION

♦ See Figures 25 and 26

Wash all parts except the rope and the handle in solvent, and then blow them dry with compressed air.

Remove any trace of corrosion and wipe all metal parts with an oil-dampened cloth.

Inspect the rope. Replace the rope if it appears to be weak or frayed. If the rope is frayed, check the holes through which the rope passes for rough edges or burrs. Remove the rough edges or burrs with a file and polish the surface until it is smooth. Inspect the starter spring end hooks. Replace the spring if it is weak, corroded or cracked. Inspect the inside surface of the sheave rewind recess for grooves or roughness. Grooves may cause erratic rewinding of the starter rope.

HAND REWIND STARTER 10-7

Fig. 12 Untie the knot from the rope in the starter handle

Fig. 13 Remove the cable from the starter arm

Fig. 14 Remove the top bolt and locknut

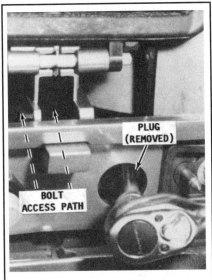
Fig. 15 Remove the starter mounting bolts (Yamaha shown)

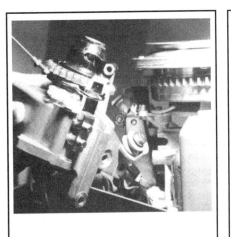
Fig. 16 Remove the starter from the powerhead

Fig. 17 Remove the pinion gear and bushing from the sheave shaft

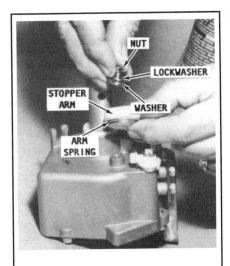
Fig. 18 Remove the nut, lockwasher, washer, stopper arm and spring

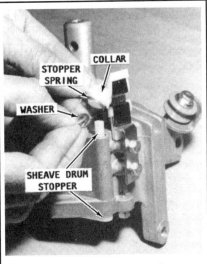
Fig. 19 Remove the cotter pin, washer, collar and stopper spring

Fig. 20 Unbolt the guide plate

HAND REWIND STARTER

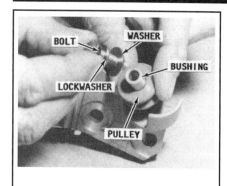

Fig. 21 Remove the pulley bolt, lock-washer, washer, bushing, and starter pulley

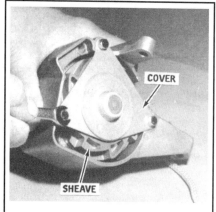

Fig. 22 Unbolt and remove the cover from the starter body

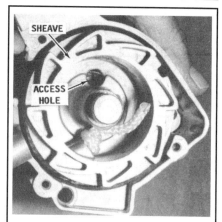

Fig. 23 Place a screwdriver in the access hole to hold down the spring end while the sheave is removed

Coat the entire length of the used rewind spring (a new spring will be coated with lubricant from the package), with low-temperature lubricant.

Coat the following parts with Yamaha All Purpose Lubricant, or equivalent: the arm spring, stopper spring, stopper shaft, pulley bushing, pinion bushing, and the sheave shaft.

ASSEMBLING & INSTALLATION

Old Spring

◆ See Figures 25 and 27

■ The authors, the manufacturers, and almost anyone else who has handled the spring from this type rewind starter STRONGLY recommend a pair of safety goggles or a face shield be worn while the spring is being installed. As the work progresses a tiger is being forced into a cage - in some cases as much as 14 ft. (4.3 m) of spring steel wound into about 4 in. (10.2cm) circumference. If the spring is accidentally released, it will lash out with tremendous ferocity and very likely could cause personal injury to the installer or other persons nearby.

Wear a good pair of gloves while winding and installing the spring. The spring will develop tension and the edges of the spring steel are extremely sharp. The gloves will prevent cuts to the hands and fingers.

Insert the nylon washer in the center recess of the starter housing. Place the outer loop of the used spring into the recess of the starter. Rotate the housing **counterclockwise** and at the same feed the spring one turn at a time into the recess. Push each turn outward toward the edge to keep the spring in place. Work slowly, carefully, and with **patience**. After the last winding is forced into place, tuck the free end with the loop, as shown.

Now follow the appropriate steps of the New Spring procedure to finish assembly and installation.

New Spring

◆ See Figures 25 and 28 thru 42

A new spring in the package will be tightly wound with a restraining wire hoop around the outside. This wire is **not** to be removed until the spring is in place in the housing, as instructed in the following step.

1. Insert the nylon washer into the center recess of the starter housing. **Carefully** place the new spring, with the wire hoop still in place, inside the starter housing. Press the outer loop of the spring into the recess and allow the outer turn of the spring to clear the restraining wire hoop and fall into place in the housing. Obtain the help of an assistant. Hold the spring in place with three wide blade screwdrivers spaced evenly around the spring. While holding the spring in place, work the wire hoop up and free of the spring.

2. Tie a figure 8 knot in one end of the rope. Thread the rope through the sheave. Wrap the rope around the sheave in a **clockwise** direction 5-1/2 turns. Place the rope in the notch in the sheave. Now, lower the sheave into the starter housing. Place the notch at the opening in the housing. Seat the

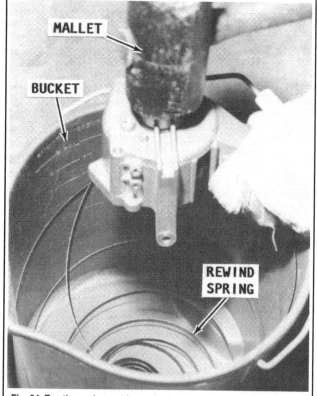

Fig. 24 Tap the spring out into a bucket for safety

sheave and at the same time hook the inner loop of the spring into the recess on the underside of the sheave. If necessary, insert a screwdriver into the access hole to guide the inner loop into place. Believe it or not, this is not a difficult task. Surprisingly, the loop will slip into place with ease.

3. Install the starter cover onto the sheave shaft with the longer plastic bushing on the outside. Secure the cover with the three bolts and lockwashers.

4. Place the stopper spring into place on the collar. Loop the hooked end of the spring around the collar, as shown.

5. Insert the stopper from the underneath side of the starter body. Install the collar and spring over the stopper. Now, wind the spring **counterclockwise** about 1/2 turn or so, to give the spring some tension, and then slide the straight end of the spring over the edge of the starter body. Check collar action to be sure the spring tension is in the proper direction. Place the washer over the stopper. Slide the cotter pin through the stopper, and then spread the ends to hold the assembly in place.

6. Slide the straight end of the arm spring into the hole in the stopper

HAND REWIND STARTER 10-9

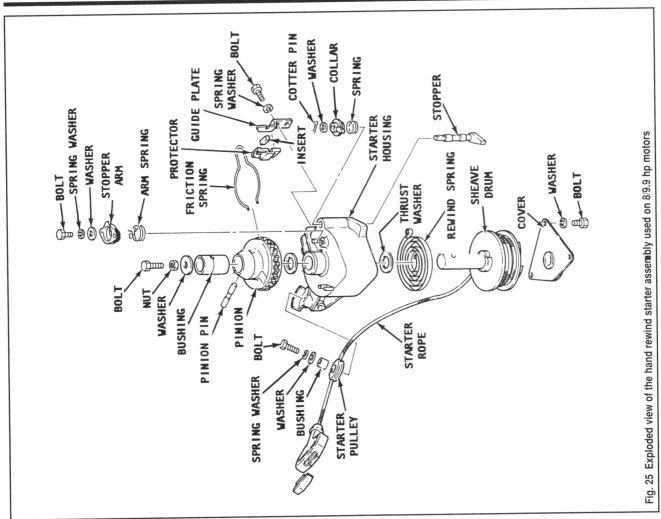

Fig. 25 Exploded view of the hand rewind starter assembly used on 8/9.9 hp motors

Fig. 26 Check the pinion gear for chipped or missing teeth. The friction spring should be held tightly against the pinion (if this spring is loose, it should be replaced)

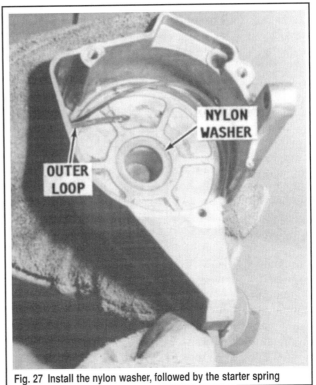

Fig. 27 Install the nylon washer, followed by the starter spring

10-10 HAND REWIND STARTER

arm.

7. Hold the spring and stopper arm together. Wind the spring about 1/2 turn or so to place a little tension on the spring. Now, lower the looped end of the arm spring to hook around the post on the starter body. Align the embossed arrow on the stopper arm with the arrow embossed on the collar. Install the washer and nut to secure the assembly in place. Check stopper arm action to be sure the spring tension is in the proper direction.

8. Install the guide plate with the two lock-washers and nuts. Place the bushing through the starter pulley and install it onto the starter body with the washer, lock-washer, and bolt. Pull the rope out until the starter sheave has rotated four full turns. This action places tension on the spring. Hold the sheave with the tension, and at the same time thread the rope between the pulley and the starter extension. Tie a temporary knot in the rope to hold tension on the sheave.

9. Slide the bushing and pinion over the sheave shaft. Index the friction spring over the guide plate. Align the holes. Tap the pinion pin through all the holes and center the pin.

10. Position the assembled rewind starter in place against the powerhead.

11. Install the three starter housing mounting bolts and washers. Tighten

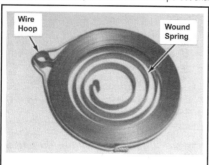

Fig. 28 A new spring will normally arrive restrained in a wire hoop (which is not removed until the spring is positioned inside the housing)

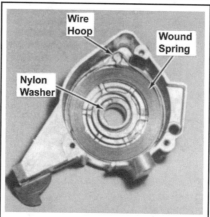

Fig. 29 When installing a new spring, insert the washer and pre-wound spring to the housing

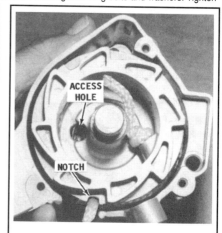

Fig. 30 Place the rope in the notch and carefully lower the sheave into the housing

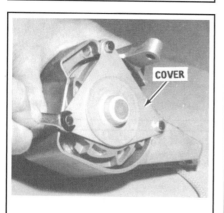

Fig. 31 Bolt the starter cover into place

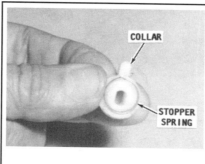

Fig. 32 Place the stopper spring into place on the collar as shown

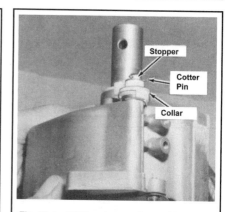

Fig. 33 Install the stopper along with the collar and cotter pin

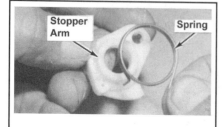

Fig. 34 Slide the straight end of the arm spring into the hole in the stopper arm

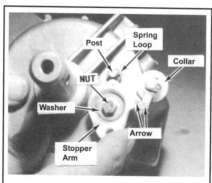

Fig. 35 Install the stopper arm and secure it using the nut and washer

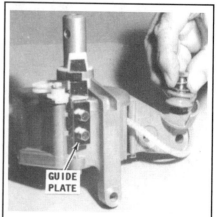

Fig. 36 After installing the guide plate, install the bushing and starter pulley with the washer, lock-washer and bolt

HAND REWIND STARTER

the bolts to 13 ft. lbs. (18 Nm).

12. Install the washer, locknut and bolt onto the sheave shaft. Hold the locknut with one wrench and tighten the bolt to 13 ft. lbs. (18 Nm) with another wrench.

13. Retain the other end of the cable in the grommet which holds the cable away from the powerhead. Slide the cable end into the slot in the stopper arm. Position the two nuts on both sides of the mount on the starter body. Tighten the two nuts to secure the cable in place.

14. Hold starter rope to maintain tension on the sheave and spring, and at the same time untie the temporary knot at the starter pulley. Continue holding the rope and thread it through the hole in the front panel. Pass the rope through the starter handle and tie a figure "8" knot in the end. Push the end into the starter handle and conceal the knot with the handle cover.

15. Check the position of the two embossed arrows. These arrows should be properly aligned. If they are not, adjust the length of the cable by shifting the two nuts on both sides of the mount to bring the arrows into alignment.

Checking the No Start-In-Gear System

1. Disconnect the spark plug leads from the spark plugs to prevent the powerhead from starting.

2. Shift the lower unit into **neutral** and attempt to **slowly** pull on the starter rope handle. The attempt should succeed. The starter pinion gear should rise and mesh with the teeth on the flywheel. The flywheel should rotate. Release tension on the rope and the rewind spring should pull the rope back into the starter housing with the rope handle tight against the front panel.

3. Shift the lower unit into **forward** gear. Attempt to pull on the starter handle to rotate the rewind starter and flywheel. The attempt should fail.

4. Shift the lower unit into **reverse** gear and again attempt to pull on the starter handle to rotate the rewind starter and flywheel. The attempt should fail.

9.9/15 Hp (323cc) Yamaha and Mercury/Mariner Models

REMOVAL & DISASSEMBLY

◆ See Figures 43, 44 and 45

1. Remove the engine top cover.
2. If equipped, disconnect start-in-gear protection cable at one end. It is usually easier to disconnect it from the starter housing.
3. For Yamaha models, disconnect the choke rod linkage from the starter housing, then follow the warning lamp wiring harness from the back of the rope guide down to the bullet connectors. Disengage the wiring harness connectors (Pink and Yellow/Red wires) for the warning lamp.
4. For Mercury/Mariner models, remove the 2 screws securing the handle rest, then rotate the rest forward for access.
5. Remove the 3 mounting bolts securing the corners of the hand rewind starter to the powerhead.
6. Lift the hand rewind starter free of the powerhead. If equipped, keep track of the 3 bolt/starter mounting collars. There is a separate camshaft sprocket cover which may pull free when removing the starter assembly.
7. Invert the starter assembly on a suitable work surface
8. Turn the sheave drum until the cutout in the outer side of the drum faces the starter handle, then pull out the starter rope from the side of the housing (between the drum and handle). Hook the rope into the cutout in the side of the drum. Using the rope caught in the cutout, slowly wind the sheave drum clockwise until the spring tension is released.
9. Loosen and remove the drive plate retaining screw, then lift the drive plate and drive spring from the assembly.

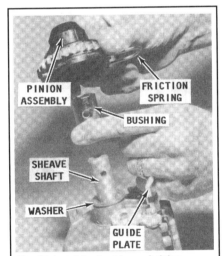

Fig. 37 Install the bushing and pinion assembly, aligning the holes to install the pinion pin

Fig. 38 Place the starter assembly in position on the powerhead...

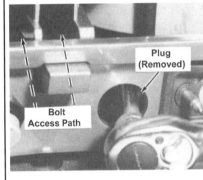

Fig. 39 ...then install and tighten the bolts (Yamaha with access hole shown)

Fig. 40 Install the washer, locknut and bolt to the top of the sheave shaft

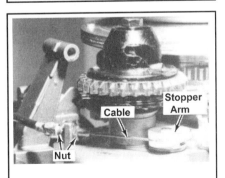

Fig. 41 Install the cable to the stopper arm

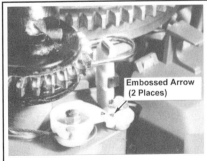

Fig. 42 Make sure the embossed arrows align or the adjust the cable length

10-12 HAND REWIND STARTER

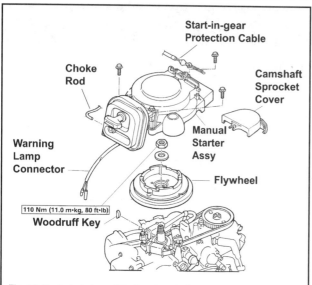

Fig. 43 Exploded view of the hand rewind starter mounting - 9.9/15 hp (323cc) motors

■ When removing the drum, be sure to turn it upside down in order to keep the starter spring from popping up at you. If the spring is not being replaced, do not leave the drum out of the housing any longer than necessary OR use something (like a large heavy block of wood or strong wire ties) to help lock the spring in place.

10. Carefully pull the drum assembly from the starter housing. Which steps you follow from this point will depend upon which components require service or replacement. The spring, the rope or other starter components can each be individually replaced.

11. Remove the circlip from the spring side of the drum, then remove the drive pawl, return spring and spring from the pawl assembly.

12. Wearing heavy leather gloves to protect your hands, slowly and carefully remove the return spring from the start housing by unwinding it, one turn of the winding each time.

13. If rope replacement is necessary, carefully push the knots out of the drum and starter handle, then untie or cut the rope to free it at each end. For some models, a damper is installed on the rope at the handle end, be sure to keep track of the damper positioning for installation purposes.

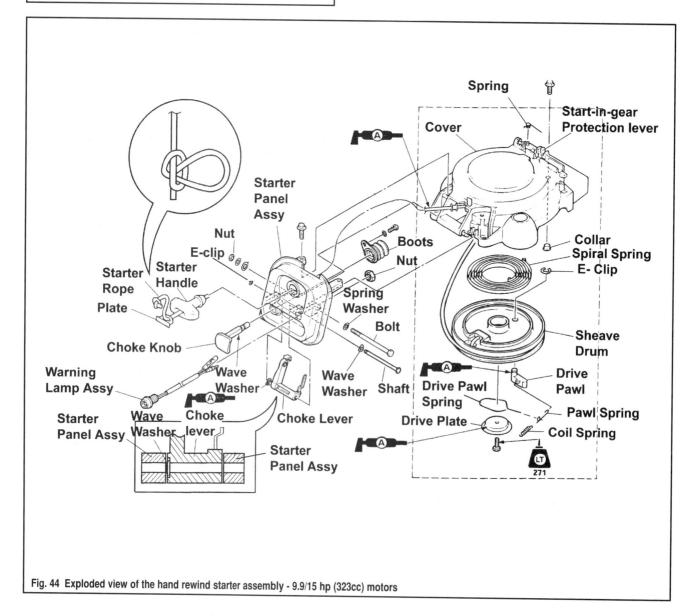

Fig. 44 Exploded view of the hand rewind starter assembly - 9.9/15 hp (323cc) motors

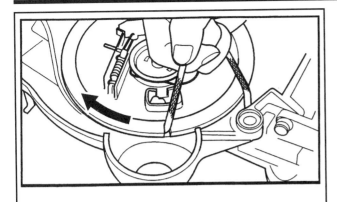

Fig. 45 Using the starter rope to wind/unwind the starter spring

CLEANING & INSPECTION

◆ See Figures 43 and 44

Remove any trace of corrosion and wipe all metal parts with an oil-dampened cloth.

Inspect the rope. Replace the rope if it appears to be weak or frayed. If the rope is frayed, check the holes through which the rope passes for rough edges or burrs. Remove the rough edges or burrs with a file and polish the surface until it is smooth. Inspect the starter spring end hooks. Replace the spring if it is weak, corroded or cracked. Inspect the inside surface of the sheave rewind recess for grooves or roughness. Grooves may cause erratic rewinding of the starter rope.

Coat the entire length of the used rewind spring (a new spring will be coated with lubricant from the package), with low-temperature lubricant.

ASSEMBLY & INSTALLATION

◆ See Figures 43, 44 and 45

■ If the pull rope requires replacement be sure to use one which measures the same length as the original. Unfortunately the Yamaha manuals list an error for these motors, showing a rope which is 150cm in length OR 70.9 in. You see the problem is we can't tell if they MEANT 180cm which would be 70.9 in or 150cm which would be 59.1 in. However, the Mercury/Mariner manuals say 180cm/71 in. so we're pretty sure it is the later. Either way, do your best bet is to measure the old rope, JUST to confirm. When cutting a rope to the specified length, be sure to burn the ends so that it will not continue to unwind. Also, apply a light coating of marine grade grease to the new rope.

The authors, the manufacturers, and almost anyone else who has handled the spring from this type rewind starter strongly recommend a pair of safety goggles or a face shield be worn while the spring is being installed. As the work progresses a tiger is being forced into a cage - in some cases up to 14 ft. (4.3 m) of spring steel wound into about 4 in. (10.2cm) circumference. If the spring is accidentally released, it will lash out with tremendous ferocity and very likely could cause personal injury to the installer or other persons nearby.

■ Replacement spiral springs are normally supplied pre-wound and secured by wire loop(s). If using a replacement spring, position it on the drum sheave before removing the wire loop(s)

1. If removed, install the spring, return spring and drive pawl to the sheave drum and secure using the circlip.
2. If using a replacement spring, position it to the starter housing, then carefully remove the wire loop(s).
3. If using an unwound spring, carefully wind it into the starter housing, starting at the outside and working inward, one turn at a time.

4. If removed, pass a new rope through the rope guide, then insert the ends into the drum sheave and starter handle. Tie figure "8" knots in either end of the rope to secure them to the sheave and handle.

■ When tying the figure "8" knots, be sure to leave about 0.20-0.39 in. (5-10mm) at the end of the starter rope beyond the knot in order to ensure it will not loosen.

5. Position the starter rope in the cutout on the drum, then carefully install the drum assembly to the starter housing and over the spring. While installing the drum be sure to position the inner end of the spring over the retainer post on the drum.
6. If removed, install the drive spring to the drive plate.
7. Insert the tip of the drive spring through the coil portion of the return spring, then install the drive plate to the sheave drum.
8. Install the drive plate retaining screw and tighten securely.
9. Using the starter rope in the drum cutout, slowly wind the starter spring by turning the sheave 3-4 turns COUNTERCLOCKWISE. Once pre-wound, remove the rope from the cutout and allow the spring to slowly wind it back into the housing.
10. Make sure the start-in-gear protection device is released, then check drum rotation and spring return using the starter rope and handle.
11. Install the starter assembly to the top of the powerhead (along with the camshaft sprocket cover if removed). Secure the starter assembly using the 3 retaining bolts. Tighten the bolts to 70 inch lbs. (8 Nm).
12. For Mercury/Mariner models, secure the handle rest using the 2 retaining screws and tighten to 30 inch lbs. (3.4 Nm).
13. For Yamaha models, reconnect the wiring for the warning lamp and attach the choke rod linkage.
14. Reconnect the start-in-gear protection wire to the starter housing or the shift linkage (as applicable).
15. Slowly confirm the operation of the starter assembly.
16. Install the engine cover.

25 Hp Yamaha and 25-40 Hp Mercury/Mariner Models

REMOVAL & DISASSEMBLY

◆ See Figures 46 and 47

The hand rewind starter used on 25 hp Yamahas and 25-40 hp Mercury/Mariners is a little unique to Yamaha and Mercury/Mariner outboards in that it uses a cartridge type starter spring instead of a completely free coiled spring. This makes service a little easier and a little safer (less chance of cutting yourself on the spring). However, keep in mind that a damaged spring may also mean a damaged housing. Though the cartridge is affective in keeping the spring contained, damage may allow it to escape. So, regardless of this cartridge still wear heavy leather gloves and face/eye protection when handling the spring!

1. Remove the engine top cover.
2. For Yamaha models, disconnect the 2 wire leads for the low-oil-pressure warning lamp mounted in the rope guide. Follow the leads down from the back on the lamp to the bullet connectors, then tag and disengage the leads.
3. Disconnect start-in-gear protection cable. For these models, it is usually easiest to disconnect the cable from the starter itself. To do this, remove the cotter pin in order to free the cable end, then remove the bolt and washer securing the cable guide to the starter housing.
4. Remove the mounting bolts (usually 4 on this model) securing the corners of the hand rewind starter/flywheel cover to the powerhead.
5. Lift the hand rewind starter free of the powerhead, and place it on a suitable work surface.
6. If necessary for rope replacement or for better access to the drum components, carefully pry the cap out of the starter handle, then push the rope out and untie or cut the knot. Remove the starter handle and damper. Also only if necessary, remove the 2 retaining bolts and separate the rope guide from the starter housing.

■ Once the rope knot is untied, the spring may want to pull the rope into the housing. This is no big deal if the sheave drum is being removed to service the spring or the rope. You can also use the opportunity to help relieve any remaining spring pressure by slowly turning the drum clockwise (the normal direction of rotation when the rope is being retracted by the spring).

10-14 HAND REWIND STARTER

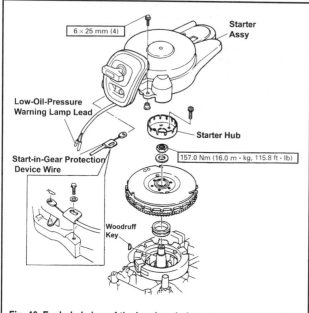

Fig. 46 Exploded view of the hand rewind starter mounting - 25 hp Yamaha motors shown (25-40 hp Mercury/Mariners almost identical)

7. If necessary, disassemble the lock cam assembly. Start by remove the 3 bolts that secure the lock cam retainer, then remove the retainer. Free the spring, then remove the lock cam. Finally remove the spring and lock cam base plate.

8. Turn the sheave drum until the cutout in the outer side of the drum faces the starter handle, then pull out the starter rope from the side of the housing (between the drum and handle). Hook the rope into the cutout in the side of the drum. Using the rope caught in the cutout, slowly wind the sheave drum clockwise until the spring tension is released.

9. Loosen the drive plate retaining screw, then remove the screw and drive plate.

10. Pull the drum assembly from the starter housing, then remove the drive plate spring. Which steps you follow from this point will depend upon which components require service or replacement. The spring, the rope or other starter components can each be individually replaced.

11. Wearing heavy leather gloves to protect your hands in case the spring should come free, carefully remove the cartridge spring followed by the spring washer from the start housing.

■ Note the orientation of the spring washer for installation purposes.

12. If rope replacement is necessary, carefully push the knot out of the drum, then untie or cut the rope to free it.

13. If necessary, remove one or both of the drive pawls from the sheave drum by carefully prying free the circlip, then removing the pawl and spring.

14. If necessary, remove the bolt and washer securing the collar and recoil starter roller to the top of the housing.

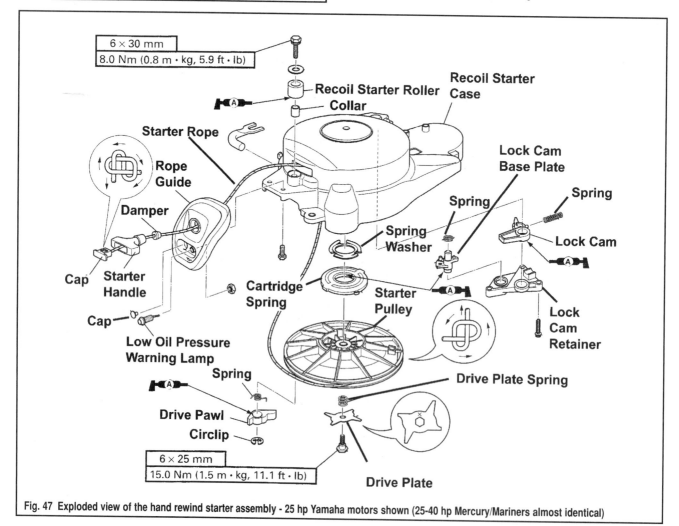

Fig. 47 Exploded view of the hand rewind starter assembly - 25 hp Yamaha motors shown (25-40 hp Mercury/Mariners almost identical)

HAND REWIND STARTER 10-15

CLEANING & INSPECTION

◆ See Figures 46 and 47

Remove any trace of corrosion and wipe all metal parts with an oil-dampened cloth.

Inspect the rope. Replace the rope if it appears to be weak or frayed. If the rope is frayed, check the holes through which the rope passes for rough edges or burrs. Remove the rough edges or burrs with a file and polish the surface until it is smooth.

Inspect the starter spring cartridge. Make sure there are no signs of damage or excessive wear. Be sure to replace the cartridge if the spring is weak or if the spring/housing is corroded or cracked.

Inspect the inside surface of the sheave rewind recess for grooves or roughness. Grooves may cause erratic rewinding of the starter rope.

Coat the inside diameter of the spring cartridge, the drive pawls, the lock cam and the recoil starter roller using a suitable marine grade grease during assembly.

ASSEMBLY & INSTALLATION

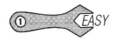

◆ See Figures 46 thru 50

■ If the pull rope requires replacement be sure to use one which is the same length as the original. Unfortunately we could not find a replacement specification from Yamaha, so you'll have to measure the old rope to be sure, reconstructing its original length if it was broken or cut. However, Mercury/Mariner lists a rope spec of 168 cm (66 in.) which should be correct for the Yamaha models as well, but we'd still double-check by measuring the old rope. When cutting a replacement rope to length, be sure to burn the ends so that it will not continue to unwind. Also, apply a light coating of marine grade grease to the new rope.

As noted earlier, this hand rewind starter uses a cartridge type starter spring instead of a completely free coiled spring. However there may be instances where the spring could still come free, so, regardless of this cartridge wear heavy leather gloves and face/eye protection when handling the spring!

1. If removed, apply a light coating of marine grade grease to the recoil starter roller, then install the roller and collar using the bolt and washer. Tighten the bolt to 5.9 ft. lbs. (8 Nm).

2. If removed, install one or both of the spring and drive pawl assemblies to the sheave drum using the circlip.

3. If removed, install the rope guide to the housing and secure using the 2 retaining bolts.

4. Apply a light coating of marine grade grease to the inner diameter of the spring cartridge. Install the spring washer to the starter housing with the projections facing downward (toward the housing), then position the spring cartridge making sure to index the inner end of the spring cartridge over the cutout in the housing.

■ Whereas Yamaha recommends installing the spring to the housing, then installing the drum over the spring. Mercury/Mariner instructs the opposite procedure, where you first install the spring cartridge and spring washer to the sheave, THEN install the sheave and spring assembly into the starter housing, making sure the looped end of the spring engages the V-notch in the hub of the starter housing. Either method should work, just take care when aligning and fitting the spring.

5. Install the drum assembly over the spring cartridge indexing the cutouts in the drum with the square tabs on the cartridge.

6. Install the drive plate spring and drive plate, then install and tighten the retaining bolt to 11.1 ft. lbs. (15 Nm).

7. If removed, tie a figure 8 knot in the end of a new starter rope, then wind the sheave drum counterclockwise until it stops. Slowly turn the pulley clockwise until the rope hole aligns with the recoil starter roller (if it takes less than 90° to turn from the stopped position to align the rope hole and roller, turn the pulley one more FULL turn clockwise before proceeding). Now hold the drum in this position as you pass a new rope through the rope guide and dampener. Finally feed the rope through the guide, dampener and handle, then tie a second figure 8 knot to secure it to the cap.

■ When tying the figure 8 knots, be sure to leave about 0.20-0.39 in. (5-10mm) at the end of the starter rope beyond the knot in order to ensure it will not loosen.

8. Make sure the start-in-gear protection device is released, then check drum rotation and spring return using the starter rope and handle.

9. Install the starter assembly to the top of the powerhead and secure using the retaining bolts (usually 4). Tighten the bolts to 60 inch lbs. (7 Nm).

10. Reconnect the start-in-gear protection wire to the starter housing. Secure the guide using the bolt and washer. Secure the cable end using a cotter pin.

11. Slowly confirm the operation of the starter assembly.

12. For Yamaha models, reconnect the wiring for the low-oil-pressure warning lamp.

13. Install the engine cover.

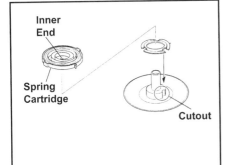

Fig. 48 On Yamahas the manufacturer recommends installing the spring cartridge with the inner end indexed to the cutaway in the housing...

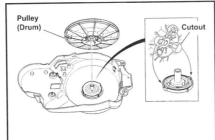

Fig. 49 ...then installing the pulley (drum) indexing the cutouts with the tabs on the spring cartridge

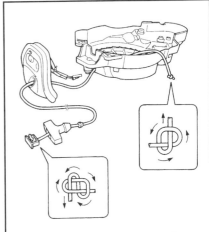

Fig. 50 Use figure 8 knots to secure the rope

10-16 HAND REWIND STARTER

30/40J/40 Hp Yamaha Models

REMOVAL & DISASSEMBLY

◆ See Figures 51 thru 54

1. Remove the engine top cover.
2. Carefully un-snap and remove the camshaft sprocket cover.
3. Disconnect start-in-gear protection cable and spring from the starter housing.
4. Remove the 3 mounting bolts securing the corners of the hand rewind starter to the powerhead.
5. Lift the hand rewind starter free of the powerhead. If necessary, remove the guide cam and spring.
6. Invert the starter assembly on a suitable work surface.
7. Manually release the starter lockout, then pull the starter rope out from the side of the housing (between the drum and handle). Hook the rope into the cutout in the side of the drum. Using the rope caught in the cutout, slowly wind the sheave drum clockwise until the spring tension is released.
8. Push the plate out of the starter handle then either untie or cut the rope to free it from the plate and handle.
9. Loosen and remove the drive plate retaining screw, then lift the drive plate and drive pawl spring from the assembly. Remove the coil spring and pawl spring from the drive pawl and sheave drum assembly.

■ When removing the drum, work slowly to keep the spiral spring from popping up at you. If the spring is not being replaced, do not leave the drum out of the housing any longer than necessary OR use something (like a large heavy block of wood or strong wire ties) to help lock the spring in place.

10. Carefully pull the drum assembly from the starter housing. Which steps you follow from this point will depend upon which components require service or replacement. The spring, the rope or other starter components can each be individually replaced.
11. Remove the circlip from the spring side of the drum, then remove the drive pawl from the sheave.
12. Wearing heavy leather gloves to protect your hands, slowly and carefully remove the return spring from the start housing by unwinding it, one turn of the winding each time.
13. If rope replacement is necessary, carefully push the knot out of the drum, then untie or cut the rope to free it.

CLEANING & INSPECTION

◆ See Figures 51 and 52

Remove any trace of corrosion and wipe all metal parts with an oil-dampened cloth.

Inspect the rope. Replace the rope if it appears to be weak or frayed. If the rope is frayed, check the holes through which the rope passes for rough edges or burrs. Remove the rough edges or burrs with a file and polish the surface until it is smooth. Inspect the starter spring end hooks. Replace the spring if it is weak, corroded or cracked. Inspect the inside surface of the sheave rewind recess for grooves or roughness. Grooves may cause erratic rewinding of the starter rope.

Coat the entire length of the used rewind spring (a new spring will be coated with lubricant from the package), with low-temperature lubricant.

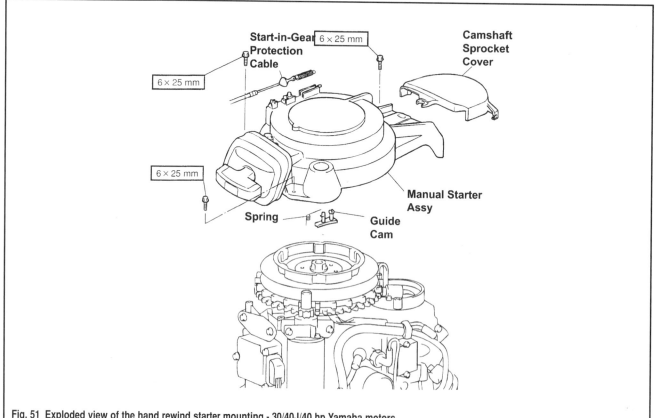

Fig. 51 Exploded view of the hand rewind starter mounting - 30/40J/40 hp Yamaha motors

HAND REWIND STARTER 10-17

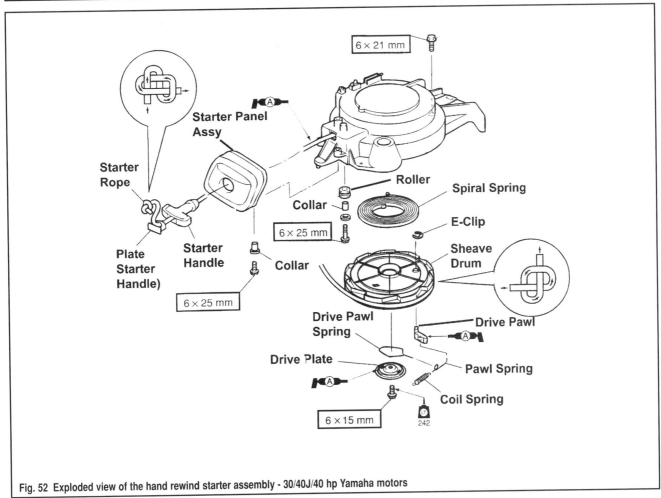

Fig. 52 Exploded view of the hand rewind starter assembly - 30/40J/40 hp Yamaha motors

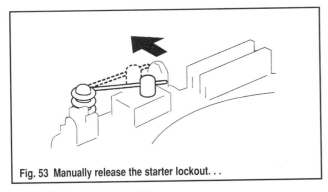

Fig. 53 Manually release the starter lockout...

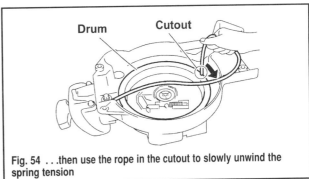

Fig. 54 ...then use the rope in the cutout to slowly unwind the spring tension

ASSEMBLY & INSTALLATION

◆ See Figures 51, 52 and 55

■ If the pull rope requires replacement be sure to use one which measures the same length as the original. We could not locate a spec for this motor, so your best bet is to measure to be sure. When cutting a rope to the measured length, be sure to burn the ends so that it will not continue to unwind. Also, apply a light coating of marine grade grease to the new rope.

The authors, the manufacturers, and almost anyone else who has handled the spring from this type rewind starter strongly recommend a pair of safety goggles or a face shield be worn while the spring is being installed. As the work progresses a tiger is being forced into a cage - in some cases up to 14 ft. (4.3 m) of spring steel wound into about 4 in. (10.2cm) circumference. If the spring is accidentally released, it will lash out with tremendous ferocity and very likely could cause personal injury to the installer or other persons nearby.

■ Replacement spiral springs are normally supplied pre-wound and secured by wire loop(s). If using a replacement spring, position it on the drum sheave before removing the wire loop(s)

10-18 HAND REWIND STARTER

1. If removed, install the drive pawl to the sheave drum and secure using the circlip.
2. If using a replacement spring, position it to the starter housing, then carefully remove the wire loop(s).
3. If using an unwound spring, carefully wind it into the starter housing, starting at the outside and working inward, one turn at a time.
4. If removed, pass a new rope through the rope guide, then insert the ends into the drum sheave and starter handle. Tie figure "8" knots in either end of the rope to secure them to the sheave and handle.

■ When tying the figure "8" knots, be sure to leave about 0.20-0.39 in. (5-10mm) at the end of the starter rope beyond the knot in order to ensure it will not loosen.

5. Position the starter rope in the cutout on the drum, then carefully install the drum assembly to the starter housing and over the spring. While installing the drum be sure to position the inner end of the spring over the retainer post on the drum.
6. If removed, mount the drive pawl spring to the drive plate.
7. Insert the drive pawl spring through the hole in the pawl spring, then attach the assembly with the drive plate to the sheave drum.
8. Apply a light coating of Loctite®242 or equivalent thread-locking compound to the threads of the drive plate retaining bolt, then install and tighten the bolt securely.
9. If not done already, connect the coil spring to the drive pawl assembly.
10. Using the starter rope in the drum cutout, slowly wind the starter spring by turning the sheave 3 turns COUNTERCLOCKWISE. Once pre-wound, remove the rope from the cutout and allow the spring to slowly wind it back into the housing.
11. Make sure the start-in-gear protection device is released, then check drum rotation and spring return using the starter rope and handle.
12. If removed, install the guide cam and spring.
13. Install the starter assembly to the top of the powerhead, then secure the starter assembly using the 3 retaining bolts.
14. Reconnect the start-in-gear protection wire to the starter housing.
15. Install the camshaft sprocket cover.
16. Slowly confirm the operation of the starter assembly.
17. Install the engine cover.

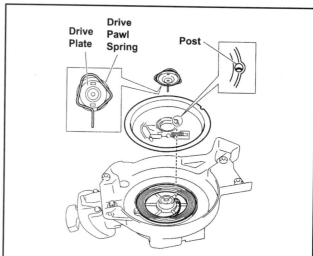

Fig. 55 When installing the drum, make sure the post engages the spiral spring

MASTER INDEX 10-19

ANODES (ZINCS)	2-29
INSPECTION	2-29
SERVICING	2-30
BOAT MAINTENANCE	**2-32**
BATTERIES	2-32
FIBERGLASS HULL	2-35
INTERIOR	2-35
CLEARING A SUBMERGED MOTOR	**2-112**
BATTERY	4-37
CABLES	4-39
CHARGERS	4-39
CONSTRUCTION	4-38
MARINE BATTERIES	4-37
LOCATION	4-38
PRECAUTIONS	4-38
RATINGS	4-38
BOATING EQUIPMENT (NOT REQUIRED BUT RECOMMENDED)	**1-10**
ANCHORS	1-10
BAILING DEVICES	1-10
COMPASS	1-11
FIRST AID KIT	1-10
OAR/PADDLE (SECOND MEANS OF PROPULSION)	1-10
TOOLS AND SPARE PARTS	1-12
VHF-FM RADIO	1-10
BOATING SAFETY	**1-4**
COURTESY MARINE EXAMINATIONS	1-10
REGULATIONS FOR YOUR BOAT	1-4
REQUIRED SAFETY EQUIPMENT	1-6
CARBURETED FUEL SYSTEM	**3-11**
DESCRIPTION AND OPERATION	3-11
TROUBLESHOOTING	3-14
CARBURETOR SERVICE	**3-17**
CARBURETOR IDENTIFICATION	3-17
2.5 HP YAMAHA	3-17
4 HP YAMAHA	3-18
4/5/6 HP MERCURY/MARINER	3-21
6/8 HP YAMAHA	3-23
8/9.9 HP (232cc)	3-24
9.9/15 HP (323cc)	3-28
25 HP MODELS	3-30
30/40 HP MODELS	3-33
40/45/50 HP (935cc)	3-35
40/50/60 HP (996cc)	3-35
75/80/90/100 HP	3-44
CDI/TCI UNIT AND IGNITION COILS	4-28
REMOVAL & INSTALLATION	4-28
CHARGING SYSTEM	**4-35**
BATTERY	4-37
CHARGING CIRCUIT	4-35
CHARGING SYSTEMS	4-35
CLEARING A SUBMERGED MOTOR	2-112
COMPRESSION TESTING	2-37
COMPRESSION CHECK	2-37
LOW COMPRESSION	2-38
CONNECTING RODS	6-125
COOLING SYSTEM	**5-10**
DESCRIPTION AND OPERATION	5-11
FLUSHING JET DRIVES	2-15
FLUSHING THE COOLING SYSTEM	2-13
THERMOSTAT	5-27
THERMO-SWITCH	5-28
TROUBLESHOOTING	5-12
WATER PUMP	5-14
CRANKING SYSTEM	**4-39**
STARTING CIRCUITS	4-39
STARTER MOTOR	4-43
STARTER MOTOR CIRCUIT	4-40
STARTER MOTOR RELAY/SOLENOID	4-42
CRANKSHAFT	6-123
CRANKSHAFT POSITION SENSOR (CPS)	4-32
DESCRIPTION & OPERATION	4-32
GAP ADJUSTMENT & SERVICE	4-32
CYLINDER BLOCK	6-128
CYLINDER HEAD	6-132
ELECTRICAL SYSTEM CHECKS	2-45
CHECKING THE BATTERY	2-45
CHECKING THE INTERNAL WIRING HARNESS	2-46
CHECKING THE STARTER	2-46
ELECTRICAL SWITCH/SOLENOID SERVICE	**4-51**
KILL SWITCH	4-51
MERCURY MAIN KEYSWITCH	4-51
NEUTRAL SAFETY SWITCH	4-52
START BUTTON	4-52
YAMAHA MAIN KEYSWITCH	4-51
WARNING BUZZER/HORN	4-53
ELECTRONIC FUEL INJECTION	**3-50**
COMPONENT LOCATIONS	3-86
CONTROLLED COMBUSTION SYSTEM	3-52
CRANKSHAFT POSITION SENSOR (CPS)/PULSER COIL	3-80
ELECTRIC LOW-PRESSURE FUEL PUMP (V6 ONLY)	3-68
ENGINE CONTROL MODULE (ECM/CDI)	3-77
FUEL DISTRIBUTION MANIFOLD (MERCURY/MARINER)	3-76
FUEL FLOW SCHEMATICS	3-95
FUEL INJECTION BASICS	3-50
FUEL INJECTORS	3-72
FUEL PRESSURE REGULATOR	3-76
IDLE SPEED/AIR CONTROL (ISC/IAC) VALVE	3-84
MERCURY/MARINER EFI	3-51
MERCURY/MARINER TROUBLESHOOTING CHARTS	3-55
NEUTRAL OR SHIFT POSITION SWITCHES	3-84
PRESSURE SENSORS (IAP/MAP)	3-83
SELF DIAGNOSTIC SYSTEM	3-53
SENSOR AND CIRCUIT RESISTANCE/OUTPUT TESTS	3-57
TEMPERATURE SENSORS	3-80
THROTTLE BODY AND INTAKE ASSEMBLY	3-61
THROTTLE POSITION SENSOR (TPS)	3-78
TROUBLESHOOTING	3-52
VAPOR SEPARATOR TANK (VST) AND HIGH-PRESSURE FUEL PUMP	3-69
YAMAHA EFI	3-50
ELECTRONIC IGNITION SYSTEMS	2-45
ENGINE COVERS	2-12
ENGINE IDENTIFICATION	2-2
ENGINE IDS	2-2
ENGINE MODEL & SERIAL NUMBERS	2-3
ENGINE MAINTENANCE	**2-12**
ANODES (ZINCS)	2-29
COOLING SYSTEM	2-13
ENGINE COVERS	2-12
ENGINE OIL AND FILTER (4-STROKE)	2-15
FUEL FILTER	2-19
JET DRIVE IMPELLER	2-28
PROPELLER	2-21
TIMING BELT	2-31
ENGINE OIL AND FILTER	2-15
CHECKING OIL LEVEL	2-15
OIL CHANGE & FILTER SERVICE	2-16
OIL RECOMMENDATIONS	2-15
EXHAUST COVER	6-123
FASTENERS, MEASUREMENTS AND CONVERSIONS	**1-27**
BOLTS, NUTS AND OTHER THREADED RETAINERS	1-27
TORQUE	1-27
STANDARD AND METRIC MEASUREMENTS	1-27
FLYWHEEL AND STATOR	4-21
CLEANING & INSPECTION	4-27
REMOVAL & INSTALLATION	4-21
FUEL	3-2
ALCOHOL-BLENDED FUELS	3-3
CHECKING FOR STALE/CONTAMINATED FUEL	3-3
HIGH ALTITUDE OPERATION	3-3
OCTANE RATING	3-2
THE BOTTOM LINE WITH FUELS	3-3
VAPOR PRESSURE	3-3
FUEL AND COMBUSTION BASICS	**3-2**
COMBUSTION	3-5
FUEL	3-2
FUEL SYSTEM PRESSURIZATION	3-4
FUEL SYSTEM SERVICE CAUTIONS	3-2
FUEL FILTER	2-19
FUEL INJECTORS	3-72

MASTER INDEX

REMOVAL & INSTALLATION 3-73
TESTING 3-73
FUEL LINES AND FITTINGS 3-8
FUEL PUMP (LOW-PRESSURE) SERVICE **3-97**
  FUEL PUMPS 3-97
  PLUNGER ACTUATED FUEL PUMPS 3-100
FUEL PRESSURE REGULATOR 3-76
  REMOVAL & INSTALLATION 3-77
  TESTING 3-76
FUEL SYSTEM CHECKS 2-46
FUEL SYSTEM PRESSURIZATION 3-4
  PRESSURIZING (CHECKING FOR LEAKS) 3-5
  RELIEVING PRESSURE (EFI) 3-4
FUEL TANK AND LINES **3-6**
  FUEL LINES AND FITTINGS 3-8
  FUEL TANK 3-6
GAS ASSIST TILT SYSTEM - MERCURY/MARINER **8-31**
  DESCRIPTION & OPERATION 8-31
  GAS ASSIST DAMPER 8-31
GEARCASE (LOWER UNIT) OIL 2-9
  CHECKING OIL LEVEL 2-9
  DRAINING AND FILLING 2-10
  OIL RECOMMENDATIONS 2-9
GEARCASE SERVICE - MERCURY/MARINER **7-59**
  REPAIR AND OVERHAUL TIPS 7-59
  4/5/6 HP (1-CYL) 7-60
  8-15 HP (2-CYL) 7-64
  25 HP (BIGFOOT) 7-70
  30/40 HP (NON-BIGFOOT) 7-70
  40/45/50 HP (NON-BIGFOOT, 935cc) 7-70
  30-115 HP (BIGFOOT) 7-74
  40/50/60 HP (996cc) 7-74
  75-115 HP (NON-BIGFOOT) 7-74
  V6 MODELS 7-87
GEARCASE SERVICE - YAMAHA **7-5**
  REPAIR AND OVERHAUL TIPS 7-5
  2.5 AND 4 HP (1-CYL) 7-5
  6-40 HP (2- AND 3-CYL) 7-13
  40-225 HP (4- AND 6-CYL) 7-35
GENERAL INFORMATION **2-2**
  BEFORE/AFTER EACH USE 2-4
  ENGINE IDENTIFICATION 2-2
  MAINTENANCE COVERAGE IN THIS MANUAL 2-2
  MAINTENANCE EQUALS SAFETY 2-2
  OUTBOARDS ON SAIL BOATS 2-2
HAND REWIND STARTERS **10-2**
  2.5 HP AND 4 HP YAMAHA 10-2
  4/5/6 HP MERCURY/MARINER 10-3
  6/8 HP YAMAHA 10-4
  8/9.9 HP (232cc) 10-6
  9.9/15 HP (323cc) 10-11
  25 HP YAMAHA 10-13
  25-40 HP MERCURY/MARINER 10-13
  30/40J/40 HP YAMAHA 10-16
HOW TO USE THIS MANUAL **1-2**
  AVOIDING THE MOST COMMON MISTAKES 1-4
  AVOIDING TROUBLE 1-2
  CAN YOU DO IT? 1-2
  DIRECTIONS AND LOCATIONS 1-3
  MAINTENANCE OR REPAIR? 1-2
  PROFESSIONAL HELP 1-3
  PURCHASING PARTS 1-3
  WHERE TO BEGIN 1-2
  WHY COMBINE YAMAHA AND MERCURY/MARINER? 1-2
HYDRO TILT LOCK SYSTEM - YAMAHA **8-2**
  DESCRIPTION & OPERATION 8-2
  SERVICING 8-3
IDLE SPEED ADJUSTMENT 2-48
IDLE SPEED/AIR CONTROL (ISC/IAC) VALVE 3-84
  REMOVAL & INSTALLATION 3-84
IGNITION SYSTEMS **4-8**
  CDI/TCI/DIGITAL INDUCTIVE IGNITION SYSTEMS 4-8
  CDI/TCI UNIT AND IGNITION COILS 4-28
  CRANKSHAFT POSITION SENSOR (CPS) 4-32

FLYWHEEL AND STATOR
  (STATOR/CHARGE/LIGHTING/PULSER COILS) 4-21
  THERMO-SENSOR & THERMO-SWITCH 4-32
  TROUBLESHOOTING 4-12
  YAMAHA MICROCOMPUTER IGNITION SYSTEMS (YMIS) 4-31
JET DRIVE **7-87**
  BEARING 2-10
  ADJUSTMENTS 7-96
  ASSEMBLY 7-87
  IMPELLER 2-28
JET DRIVE ADJUSTMENTS 7-96
JET DRIVE ASSEMBLY 7-87
  ASSEMBLY 7-93
  CLEANING & INSPECTING 7-90
  DESCRIPTION & OPERATION 7-87
  MODEL IDENTIFICATION 7-88
  REMOVAL & DISASSEMBLY 7-88
  TRIM ADJUSTMENT 7-100
JET DRIVE BEARING 2-10
  DAILY BEARING LUBRICATION 2-11
  GREASE REPLACEMENT 2-11
  RECOMMENDED LUBRICANT 2-10
JET DRIVE IMPELLER 2-28
LOWER UNIT **7-2**
  EXHAUST GASES 7-4
  GENERAL INFORMATION 7-2
  SHIFTING PRINCIPLES 7-2
  TROUBLESHOOTING 7-4
LUBRICANTS 2-6
LUBRICATION **2-6**
  GEARCASE (LOWER UNIT) OIL 2-9
  JET DRIVE BEARING 2-10
  LUBRICANTS 2-6
  LUBRICATING THE MOTOR 2-7
  LUBRICATION INSIDE THE BOAT 2-8
  POWER TRIM/TILT RESERVOIR 2-11
LUBRICATION SYSTEM **5-2**
  DESCRIPTION AND OPERATION 5-2
  OIL PRESSURE 5-2
  OIL PRESSURE SWITCH/SENSOR 5-3
  OIL PUMP 5-5
NEUTRAL OR SHIFT POSITION SWITCHES 3-84
OIL PRESSURE SWITCH/SENSOR 5-3
  DESCRIPTION & OPERATION 5-3
  REMOVAL & INSTALLATION 5-4
  TESTING 5-3
OIL PUMP 5-5
  REMOVAL, OVERHAUL & INSTALLATION 5-5
PISTONS 6-126
POWER TRIM/TILT RESERVOIR 2-11
  CHECKING FLUID LEVEL 2-11
  RECOMMENDED LUBRICANT 2-11
POWER TRIM/TILT SYSTEM - YAMAHA (LARGE MOTOR) ... **8-15**
  DESCRIPTION & OPERATION 8-15
  MAINTENANCE 8-16
  POWER TRIM/TILT SWITCH 8-20
  PTT ASSEMBLY (DUAL TRIM RAM) 8-22
  TRIM SENSOR 8-21
  TRIM/TILT MOTOR 8-21
  TRIM/TILT RELAY 8-19
  TROUBLESHOOTING 8-17
POWERHEAD BREAK-IN **6-136**
  BREAK-IN PROCEDURES 6-136
POWERHEAD MECHANICAL **6-2**
  GENERAL INFORMATION 6-2
  2.5 HP AND 4 HP YAMAHA 6-4
  4/5/6 HP MERCURY/MARINER 6-12
  6/8 HP YAMAHA AND ALL 8/9.9 HP (232cc) 6-18
  9.9/15 HP (323cc) 6-36
  25 HP 6-45
  30/40 HP 3-CYLINDER 6-56
  FOUR-CYLINDER 40/45/50 HP (935cc) AND 30/40/60 HP (966cc) . 6-66
  75/80/90/100 HP AND 115 HP POWERHEADS 6-81
  150 HP YAMAHA 6-97
  200/225 HP V6 6-107

MASTER INDEX 10-21

POWERHEAD REFINISHING 6-121
  CONNECTING RODS 6-125
  CRANKSHAFT 6-123
  CYLINDER BLOCK 6-128
  CYLINDER HEAD COMPONENTS (CAMSHAFT/VALVE TRAIN) ... 6-132
  EXHAUST COVER 6-123
  GENERAL INFORMATION 6-121
  PISTONS 6-126
PRESSURE SENSORS (IAP/MAP) 3-83
  REMOVAL & INSTALLATION 3-84
  TESTING 3-83
PROPELLER 2-21
  GENERAL INFORMATION 2-22
  INSPECTION 2-24
  REMOVAL & INSTALLATION 2-25
RE-COMMISSIONING 2-111
  REMOVAL FROM STORAGE 2-111
REMOTE CONTROLS 9-2
  MERCURY/MARINER 9-12
  YAMAHA 9-2
REMOTE CONTROLS - MERCURY/MARINER 9-12
  ASSEMBLY 9-15
  CLEANING & INSPECTING 9-15
  DESCRIPTION & OPERATION 9-12
  DISASSEMBLY 9-14
  REMOVAL & INSTALLATION 9-12
  TROUBLESHOOTING 9-12
  WIRING DIAGRAMS 9-21
REMOTE CONTROLS - YAMAHA 9-2
  CHOKE SWITCH 9-2
  CONTROL BOX ASSEMBLY 9-4
  KILL SWITCH 9-3
  NEUTRAL SAFETY SWITCH 9-3
  REMOTE CONTROL BOX 9-2
  START BUTTON 9-3
  WARNING BUZZER/HORN 9-2
  WIRING DIAGRAMS 9-20
SAFETY IN SERVICE 1-13
  DO'S 1-13
  DON'TS 1-13
SELF DIAGNOSTIC SYSTEM 3-53
SHOP EQUIPMENT 1-17
  CHEMICALS 1-17
  SAFETY TOOLS 1-17
SINGLE RAM INTEGRAL POWER TILT/TRIM -
MERCURY/MARINER 8-35
  DESCRIPTION & OPERATION 8-35
  PTT ASSEMBLY 8-42
  TROUBLESHOOTING 8-35
SINGLE TILT RAM POWER TRIM/TILT SYSTEM - YAMAHA 8-3
  DESCRIPTION & OPERATION 8-3
  MAINTENANCE 8-5
  PTT ASSEMBLY 8-9
  TILT RELAY 8-6
  TILT SWITCH 8-7
  TRIM SENSOR 8-9
  TROUBLESHOOTING 8-5
SPARK PLUG WIRES 2-43
SPARK PLUGS 2-39
SPECIFICATIONS
  BOLT TORQUE - METRIC 1-28
  BOLT TORQUE - U.S. STANDARD 1-28
  CONVERSION FACTORS 1-28
  CAPACITIES 2-118
  CARB - MERCURY/MARINER 3-106
  CARB - YAMAHA 3-105
  CHARGING SYSTEM - MERCURY/MARINER 4-61
  CHARGING SYSTEM - YAMAHA 4-59
  COOLING SYSTEM 5-31
  CONVERSION FACTORS 1-28
  GENERAL ENGINE 2-113
  IGNITION SYSTEM - 1-CYL 4-55
  IGNITION SYSTEM - MERCURY/MARINER INLINE 4-57
  IGNITION SYSTEM - YAMAHA INLINE 4-54
  IGNITION CONTROL SYSTEM - YAMAHA INLINE 4-56

  IGNITION SYSTEM - V-ENGINES 4-56
  LUBRICATION 2-117
  MAINTENANCE 2-116
  POWERHEAD 6-137
  TUNE-UP 2-119
  VALVE CLEARANCE - YAMAHA 2-121
  VALVE CLEARANCE - MERCURY/MARINER 2-121
STARTER MOTOR 4-43
  DESCRIPTION 4-43
  EXPLODED VIEWS 4-45
  OVERHAUL 4-49
  REMOVAL & INSTALLATION 4-43
STARTER RELAY/SOLENOID 4-42
  DESCRIPTION & OPERATION 4-42
  REMOVAL & INSTALLATION 4-43
  TESTING 4-42
STORAGE 2-108
  WINTERIZATION 2-108
  RE-COMMISSIONING 2-111
TCI UNIT AIR GAP 2-44
TEMPERATURE SENSORS 3-80
  REMOVAL & INSTALLATION 3-83
  TESTING 3-81
THERMOSTAT 5-27
  REMOVAL & INSTALLATION 5-27
THERMO-SWITCH 5-28
  CHECKING CONTINUITY 5-30
  DESCRIPTION & OPERATION 5-28
  OPERATIONAL CHECK 5-30
THERMO-SENSOR & THERMO-SWITCH 4-32
  CHECKING RESISTANCE 4-34
  DESCRIPTION & OPERATION 4-33
  OPERATIONAL CHECK 4-33
THROTTLE BODY AND INTAKE ASSEMBLY 3-61
  REMOVAL & INSTALLATION 3-61
THROTTLE POSITION SENSOR (TPS) 3-78
  REMOVAL & INSTALLATION 3-80
  TESTING 3-78
TILLER HANDLE 9-16
  DESCRIPTION & OPERATION 9-16
  REMOVAL & INSTALLATION 9-16
  TROUBLESHOOTING 9-16
TIMING AND SYNCHRONIZATION 2-49
  GENERAL INFORMATION 2-49
  PREPPING THE MOTOR 2-50
  SYNCHRONIZATION 2-50
  TIMING 2-49
  2.5 HP YAMAHA 2-50
  4 HP YAMAHA 2-51
  4/5/6 HP MERCURY/MARINER 2-52
  6/8 HP YAMAHA 2-54
  8/9.9 HP (232cc) YAMAHA 2-55
  8/9.9 HP (232cc) MERCURY/MARINER 2-59
  15 HP (323cc) YAMAHA 2-61
  9.9/15 HP (323cc) MERCURY/MARINER 2-63
  25 HP 2-64
  30/40 HP 3-CYL (CARB) 2-66
  30/40 HP 3-CYL (EFI) 2-68
  40/50/60 HP 4-CYL (EFI) 2-68
  40/45/50 HP (935cc) 2-70
  40/50/60 HP (996cc) - CARB 2-70
  75/80/90/100 HP 2-78
  115 HP MODELS 2-82
  150 HP YAMAHA 2-86
  200/225 HP MODELS 2-90
TIMING BELT 2-31
  INSPECTON 2-31
TOOLS 1-19
  HAND TOOLS 1-19
  OTHER COMMON TOOLS 1-22
  SPECIAL TOOLS 1-23
  ELECTRONIC TOOLS 1-23
  MEASURING TOOLS 1-25
TRIM/TILT SYSTEMS
  MERCURY/MARINER GAS ASSIST 8-31

MASTER INDEX

MERCURY/MARINER SINGLE RAM INTEGRAL POWER	8-35
YAMAHA HYDRO TILT LOCK	8-2
YAMAHA LARGE MOTOR POWER	8-15
YAMAHA SINGLE TILT RAM POWER	8-3
TRIM & TILT SYSTEMS - MERCURY/MARINER	**8-28**
INTRODUCTION	8-28
MECHANICAL TILT	8-28
TRIM & TILT SYSTEMS - YAMAHA	**8-2**
INTRODUCTION	8.2
TROUBLESHOOTING	**1-13**
BASIC OPERATING PRINCIPLES	1-13
CARBURETOR	3-14
COOLING SYSTEM	5-12
EFI	3-52
ELECTRICAL	4-6
IGNITION	4-12
MERCURY/MARINER EFI	3-55
STARTER	4-40
TILLER HANDLE	9-16
TUNE-UP	**2-36**
COMPRESSION TESTING	2-37
ELECTRICAL SYSTEM CHECKS	2-45
ELECTRONIC IGNITION SYSTEMS	2-45
FUEL SYSTEM CHECKS	2-46
IDLE SPEED ADJUSTMENT (MINOR ADJUSTMENTS)	2-48
INTRODUCTION	2-36
SPARK PLUG WIRES	2-43
SPARK PLUGS	2-39
TCI UNIT AIR GAP	2-44
TUNE-UP SEQUENCE	2-36
TUNE-UP SEQUENCE	2-36
UNDERSTANDING AND TROUBLESHOOTING ELECTRICAL SYSTEMS	**4-2**
BASIC ELECTRICAL THEORY	4-2
ELECTRICAL COMPONENTS	4-2
ELECTRICAL SYSTEM PRECAUTIONS	4-8
ELECTRICAL TESTING	4-6
TEST EQUIPMENT	4-4
TROUBLESHOOTING	4-6
WIRE AND CONNECTOR REPAIR	4-7
VALVE CLEARANCE	**2-94**
2.5 HP YAMAHA	2-95
4 HP YAMAHA	2-95
4/5/6 HP MERCURY/MARINER	2-96
6/8 HP YAMAHA	2-97
8/9.9 HP (232cc)	2-97
9.9/15 HP (323cc)	2-99
25 HP	2-99
30/40 HP 3-CYLINDER	2-100
40/45/50 HP (935cc)	2-101
40/50/60 HP (996cc)	2-101
75/80/90/100 HP	2-103
115 HP	2-103
150 HP YAMAHA	2-105
200/225 HP	2-106
WATER PUMP	**5-14**
EXPLODED VIEWS	5-19
INSPECTION & OVERHAUL	5-17
REMOVAL & INSTALLATION	5-15
WINTERIZATION	**2-108**
PREPPING FOR STORAGE	2-109
WIRING DIAGRAMS	**4-62**
MERCURY/MARINER	
4/5/6 HP	4-63
8/9.9 HP (232cc)	4-65
9.9/15 HP (323cc)	4-72
25 HP	4-78
30/40 HP	4-86
40/45/50 HP (935cc)	4-92
40/50/60 HP (996cc)	4-97
75/90 HP	4-102
2001 115 HP EFI	4-106
225 HP EFI	4-112
YAMAHA	
2.5 HP AND 4 HP	4-63
6/8 HP	4-64
8/9.9 HP (232cc)	4-66
9.9/15 HP (323cc)	4-76
25 HP	4-82
30/40J/40 HP	4-91
45/50/50J HP (935cc)	4-93
60J/60 HP (996cc)	4-102
75/80/90J/90/100 HP	4-105
115J/115 HP EFI	4-109
150 HP EFI	4-110
200/225 HP EFI	4-111
200/225 HP EFI	4-114